DIRECTORY OF RESEARCH GRANTS 2010

33rd Edition
Volume 2

Schoolhouse Partners LLC
Nashville, Indiana

Why 'Partners'? Partners work together as a team, with each person contributing different skills and expressing his or her individual interests and opinions to the unity and efficiency of the group in order to achieve common goals. Our aim is to make worthwhile resources available to you so that you and your institution or organization can and will achieve those goals.

Schoolhouse Partners, Post Office Box 2059, Nashville, IN 47448
www.schoolhousepartners.net
Published 1975, 33rd edition 2010

Printed and bound in the United States of America

ISBN 978-0-9841725-3-5
0-9841725-3-X

.able of Contents

Introduction

For over three decades the GRANTS Database and its print complements, including the annual Directory of Research Grants, have provided the research community with current, accurate information regarding funding for research-related programs and projects, scholarships, fellowships, conferences, and internships. To meet the increased information needs of professionals, the GRANTS Database, researched and edited by members of the Schoolhouse Partners editorial staff, has grown significantly. This 33rd edition of the annual Directory of Research Grants offers factual and concise descriptions of more than 5,800 research funding programs. With this new edition both experienced and novice grantseekers will be better able to target only the most applicable programs for their research needs.

Directory of Research Grants 2010

The *Directory of Research Grants 2010* features listings of programs that offer non-repayable research funding for projects in medicine, the physical and social sciences, the arts and humanities, education, community and economic development, and children and youth programs. Listings in the main section of the *Directory* contain annotations describing each program's focuses and goals, program requirements explaining eligibility, funding amounts, deadlines, *Catalog of Federal Domestic Assistance* program number (for U.S. government programs only), sponsor name and address, contact information, and the sponsor's Internet Web address. Grantseekers with access to the Internet can use the addresses to locate further information about the organizations and their application procedures. Internet addresses are also provided, when available, within listings. Some of the grant programs listed in the main section have geographic restrictions for applicants.

Indexes. The Subject Index of this *Directory* lists all program titles—with accession numbers—under their applicable subject terms. Other indexes follow, including the Sponsoring Organizations Index, which lists program sponsors alphabetically along with their programs; the Grants by Program Type Index, which lists 38 program categories, such as Basic Research, Fellowships, Travel Grants, etc., with the grants that fall within their scope; and the Geographic Index, which lists programs that have state, regional, or international focus. See How to Use This Directory on pages v-vii for sample index entries.

Using the *Directory* for Grantseeking

By using the *Directory of Research Grants*, grantseekers can match the needs of their particular programs with those sponsors offering funding in the researchers' area of interest. The information listed here is meant to eliminate the costs incurred by both grantseekers and grantmakers when inappropriate proposals are submitted for a sponsor's funding program. However, because the GRANTS Database is updated daily basis, with program listings continually added, deleted, and revised, grantseekers using this *Directory* may also search the GRANTS database online through *GrantSelect* (www.grantselect.com).

All new and revised information has been taken from (1) the sponsor's updated of previously published program statements included in earlier editions of GRANTS publications, (2) questionnaires sent to sponsors whose programs were not listed in previous editions, or (3) other materials published by the sponsor and furnished to Schoolhouse Partners. Updated information for U.S. government programs includes new and revised program information published in the *Federal Register*, the latest edition of the *Catalog of Federal Domestic Assistance*; the *NIH Guide*, published weekly by the National Institutes of Health; and the *NSF E-Bulletin*, a monthly publication of the National Science Foundation. Included in this edition are identifying document numbers from the NIH and NSF publications. Located at the ends of the program descriptions in certain entries, the numbers indicated the ongoing NIH program number (PA) or the request for applications number (RFA). For programs of the National Science Foundation, the *NSF Bulletin* number appears. This information will help users identify the programs when seeking additional information from program staff.

While Schoolhouse Partners has made every effort to verify that all information is both accurate and current within the confines of format and scope, the publisher does not assume and hereby disclaims any liability to any party for loss or damages caused by errors or omissions in this *Directory*, whether such errors or missions result from accident, negligence, or any other cause. Anyone having questions regarding the content, format, or any other aspect of the *Directory of Research Grants 2010*, GrantSelect, or the GRANTS Database should contact the Editors, GRANTS, Schoolhouse Partners or editor@schoolhousepartners.net.

How to Use This Directory

The *Directory of Research Grants 2010* is designed to allow the user quick and easy access to information regarding funding programs in a researcher's specific area of interest. This *Directory* is composed of a main section, Grant Programs, which lists grant programs in alphabetical order, and three indexes: the Subject Index, the Grants by Program Type Index, and the Geographic Index.

GRANT PROGRAMS

Each listing in this section consists of the following elements: an annotation describing each program's focus and goals, requirements explaining eligibility, funding amounts, application and renewal dates, the *Catalog of Federal Domestic Assistance* program number (for U.S. government programs only), sponsor information, contact information, and Internet address.

GRANT TITLE — **Paul G. Allen Family Foundation Grants** 4298 — ACCESSION NUMBER

The single foundation, created through the consolidation of Allen's six previous foundations (The Allen Foundation for the Arts, The Paul G. Allen Charitable Foundation, The Paul G. Allen Foundation for Medical Research, The Paul G. Allen Forest Protection Foundation, The Allen Foundation for Music, and The Paul G. Allen Virtual Education Foundation), will continue to focus on the Allen family's philanthropic interests in the areas of arts and culture, youth engagement, community development and social change, and scientific and technological innovation. The Arts and Culture Program fosters creativity and promotes critical thinking by helping strong arts organizations become sustainable and supporting projects that feature innovative and diverse artistic forms. — GRANT DESCRIPTION The Youth Engagement Program improves the way young people learn by supporting organizations that use innovative teaching strategies and provide opportunities for children to address issues relevant to their lives. The Community Development and Social Change Program promotes individual and community development by supporting initiatives and organizations that provide access to resources and opportunities. The Scientific and Technological Innovation Program advances promising scientific and technology research that has the potential to enhance understanding and stewardship of the world in which we live. Organizations may only receive one grant per year. Organizations must not have any delinquent final reports due to any of the Paul G. Allen Foundations for previous grants. Grantseekers are encouraged to apply through the online application process, where basic organizational and project information will be requested. Guidelines are available online.

REQUIREMENTS — *Requirements* 501(c)3 tax-exempt organizations, status from the Internal Revenue government entities, and IRS-recognized tribes are eligible. Eligible organizations must be located in, or serving populations of, the Pacific Northwest, which includes Alaska, Idaho, Montana, Oregon, and Washington.

RESTRICTIONS — *Restrictions* In general, the foundation will not consider requests for general fund drives, annual appeals, or federated campaigns; special events or sponsorships; direct grants, scholarships, or loans for the benefit of specific individuals; projects of organizations whose policies or practices discriminate on the basis of race, ethnic origin, sex, creed, or sexual orientation; contributions to sectarian or religious organizations whose principle activity is for the benefit of their own members or adherents; loans or debt retirement; projects that will benefit the students of a single school; general operating support for ongoing activities; or projects not aligned with the foundation's specified program areas. 509(a) private foundations are ineligible.

APPLICATION/ DUE DATE — *Date(s) Application Is Due* Mar 31; Sep 30.

Contact Grants Administrator, (206) 342-2030; fax: (206) 342-3030; email: info@pgafamilyfoundation.org — CONTACT

INTERNET ADDRESS — *Internet* http://www.pgafamilyfoundation.org

Sponsor Paul G. Allen Family Foundation — SPONSOR INFORMATION
505 Fifth Ave S, Ste 900
Seattle, WA 98104

SUBJECT INDEX

The most effective way to access specific funding programs is through the Subject Index. This index lists the subject terms with applicable grants program titles – and their accession numbers – alphabetically under each term. Terms were assigned to target the specific area of research designated in the description of each program. Cross-references are used to link subjects and assist the user in finding specific grant information.

Following are general guidelines that can make your search of this index more successful. First, check under the specific topic of interest rather than a more general term. For instance, if you are interested in chemical engineering, look under "Chemical Engineering" rather than "Engineering." Items indexed under "Engineering" indicate funding in broad areas of engineering.

Use general headings when you want grants covering broader areas or if you can't find a specific topic. For example, many grants list funding for humanities research, health programs, or science and technology. To find these grants use such headings as "Humanities," "Medical Programs," "Science," or "Technology." For additional grant information on more specific humanities research opportunities, such as in American History or Cultural Anthropology, also check under the topics "United States History" and "Anthropology, Cultural."

Many of the grant programs provide funding for research-related scholarships, faculty fellowships, dissertations, undergraduate education, conferences, or internships. If grant funds are designated for specific disciplines, you will find the items under the specific subject. Scholarships and fellowships are also listed under the terms "Native American Education," "African Americans (Student Support)," "Hispanic Education," "Minority Education," and "Women's Education."

Grants concerning study of a particular country are listed under the name of the country. Grants concerning the history, literature, art and language of a country are listed under the name of the country also, e.g., "Chinese Art" and "Chinese Language/Literature."

SUBJECT TERM —— **Education**

A.L. Mailman Family Foundation Grants, 5
AAAS Science & Technology Policy Fellowships - Health, Education & Human Services, 22
AARP Andrus Foundation Grants, 104
Abbott Laboratories Fund Grants, 125
ACIE Host University Edmund S. Muskie/Freedom Support Act Graduate Fellowships, 156
ACT Awards, 189
Akonadi Foundation Anti-Racism Grants, 351 —— **PROGRAM TITLE**
Akron Community Foundation Grants, 352
Albert and Margaret Alkek Foundation Grants, 370
Albuquerque Community Foundation Grants, 379
Alcoa Foundation Grants, 380
Alcon Foundation Grants Program, 382

GEOGRAPHIC INDEX

This index lists programs that have state, regional, or international geographic focus. The Geographic Index is arranged by state, followed by Canadian programs, then by international programs by country, and lists grant program titles and their corresponding accession numbers.

COUNTRY —— **United States**

Alabama —— **STATE**

3M Fndn Grants, 2
Alabama Humanities Fndn Grants Program, 368
Arkema Inc. Fndn Science Teachers Program, 705
CDC Injury Control Research Centers Grants, 1198
PROGRAM TITLE —— DOE Experimental Program to Stimulate Competitive Research (EPSCoR), 1653
Hill Crest Fndn Grants, 2293
Linn-Henley Charitable Trust Grants, 2777
NOAA Community-Based Restoration Program (CRP) Grants, 3843
Southern Company's Longleaf Pine Reforestation Fund, 4726

PROGRAM TYPE INDEX

This index is broken into 38 categories according to the type of program funded:

- Adult Basic Education
- Adult/Family Literacy Training
- Awards/Prizes
- Basic Research
- Building Construction and/or Renovation
- Capital Campaigns
- Centers: Research/Demonstration/Service
- Citizenship Instruction
- Community Development
- Consulting/Visiting Personnel
- Cultural Outreach
- Curriculum Development/Teacher Training
- Demonstration Grants
- Development (Institutional/Departmental)
- Dissertation/Thesis Research Support
- Educational Programs
- Environmental Programs
- Exhibitions, Collections, Performances, Video/Film Production
- Faculty/Professional Development
- Fellowships
- General Operating Support
- Graduate Assistantships
- International Exchange Programs
- International Grants
- Job Training/Adult Vocational Programs
- Matching/Challenge Funds
- Materials/Equipment Acquisition (Computers, Books, Tapes, etc.)
- Preservation/Restoration
- Publishing/Editing/Translating
- Religious Programs
- Scholarships
- Seed Grants
- Service Delivery Programs
- Symposia, Conferences, Workshops, Seminars
- Technical Assistance
- Training Programs/Internships
- Travel Grants
- Vocational Education

Government and Organization Acronyms

AAAAI American Academy of Allergy Asthma and Immunology
AAAS American Association for the Advancement of Science
AACAP American Academy of Child and Adolescent Psychiatry
AACN American Association of Critical Care Nurses
AACR American Association of Cancer Research
AAFCS American Association of Family and Consumer Sciences
AAFP American Academy of Family Physicians Foundation
AAFPRS American Academy of Facial Plastic and Reconstructive Surgery
AAR American Academy in Rome
AAS American Antiquarian Society
AASL American Association of School Libraries
AAUW American Association of University Women
ACC Asian Cultural Council
ACLS American Council of Learned Societies
ACM Association for Computing Machinery
ACS American Cancer Society
ADA American Diabetes Association
ADHF American Digestive Health Foundation
AF Arthritis Foundation
AFAR American Federation for Aging Research
AFOSR Air Force Office of Scientific Research
AFUD American Foundation for Urologic Disease
AFUW Australian Federation of University Women
AGS American Geriatrics Society
AHA American Heart Association
AHAF American Health Assistance Foundation
AFHMR Alberta Heritage Foundation for Medical Research
AHRQ Agency for Healthcare Research and Quality
AICR American Institute for Cancer Research
AIIS American Institute for Indian Studies
AJA American Jewish Archives
ALA American Library Association
ALISE Association for Library and Information Science Education
AMNH American Museum of Natural History
AMS American Musicological Society
ANL Argonne National Library
ANS American Numismatic Society
AOA American Osteopathic Association
APA American Psychological Association
APSA American Political Science Association
ARIT American Research Institute in Turkey
ARO Army Research Office
ASA American Statistical Association
ASCSA American School of Classical Studies at Athens
ASECS American Society for Eighteenth-Century Studies
ASF American-Scandinavian Foundation
ASHRAE American Society of Heating, Refrigerating, and Air Conditioning Engineers
ASME American Society of Mechanical Engineers
ASNS American Society for Nutritional Sciences
ASPRS American Society of Photogrammetry and Remote Sensing
AWU Associated Western Universities
AWWA American Water Works Association
BA British Academy
BWF Burroughs Wellcome Fund
CBIE Canadian Bureau for International Education
CCFF Canadian Cystic Fibrosis Foundation
CDC Centers for Disease Control and Prevention
CES Council for European Studies
CF The Commonwealth Fund
CFF Cystic Fibrosis Foundation
CFPC College of Family Physicians of Canada
CFUW Canadian Federation of University Women
CHEA Canadian Home Economics Association
CIES Council for International Exchange of Scholars
CIUS Canadian Institute of Ukrainian Studies
CLA Canadian Lung Association
CLF Canadian Liver Foundation
CRI Cancer Research Institute
DAAD Deutscher Akademische Austauschdienst (German Academic Exchange Service)
DHHS Department of Health and Human Services
DOA Department of Agriculture
DOC Department of Commerce
DOD Department of Defense
DOE Department of Energy
DOI Department of the Interior
DOJ Department of Justice
DOL Department of Labor
DOS Department of State
DOT Department of Transportation
EFA Epilepsy Foundation of America
EPA Environmental Protection Agency
ESF European Science Foundation
ETS Educational Testing Service
FCAR Formation de Chercheurs et L'Aide a la Recherche
FDA Food and Drug Administration
FIC Fogarty International Center
GAAC German American Academic Council

HHMI	Howard Hughes Medical Institute
HRSA	Health Resources and Services Administration
HUD	Department of Housing and Urban Development
IRA	International Reading Association
IREX	International Research and Exchanges Board
IUCP	Indiana University Center on Philanthropy
JDF	Juvenile Diabetes Foundation International
JMO	John M. Olin Foundation
JSPS	Japan Society for the Promotion of Science
KFC	Kidney Foundation of Canada
LSA	Leukemia Society of America
MHRC	Manitoba Health Research Council
MLA	Medical Library Association
MMA	Metropolitan Museum of Art
MSSC	Multiple Sclerosis Society of Canada
NAACP	National Association for the Advancement of Colored People
NARSAD	National Alliance for Research on Schizophrenia and Depression
NASM	National Air and Space Museum
NATO	North Atlantic Treaty Organization
NCCAM	National Center for Complementary and Alternative Medicine
NCI	National Cancer Institute
NCIC	National Cancer Institute of Canada
NCRR	National Center for Research Resources
NEH	National Endowment for the Humanities
NEI	National Eye Institute
NFID	National Foundation for Infectious Diseases
NFWF	National Fish and Wildlife Foundation
NHGRI	National Human Genome Research Institute
NHLBI	National Heart, Lung and Blood Institute
NIA	National Institute on Aging
NIAAA	National Institute on Alcohol Abuse and Alcoholism
NIAF	National Italian American Foundation
NIAID	National Institute of Allergy and Infectious Diseases
NIAMS	National Institute of Arthritis and Musculoskeletal Skin Diseases
NICHD	National Institute of Child Health and Human Development
NIDA	National Institute on Drug Abuse
NIDCD	National Institute on Deafness and Other Communication Disorders
NIDCR	National Institute of Dental and Craniofacial Research
NIDDK	National Institute of Diabetes, and Digestive and Kidney Diseases
NIDRR	National Institute on Disability and Rehabilitation Research
NIEHS	National Institute of Environmental Health Sciences
NIGMS	National Institute of General Medical Sciences
NIH	National Institutes of Health
NIJ	National Institute of Justice
NIMH	National Institute of Mental Health
NINDS	National Institute of Neurological Disorders and Stroke
NINR	National Institute of Nursing Research
NIOSH	National Institute for Occupational Safety and Health
NIST	National Institute of Standards and Technology
NKF	National Kidney Foundation
NL	Newberry Library
NLM	National Library of Medicine
NMF	National Medical Fellowships, Inc.
NMSS	National Multiple Sclerosis Society
NOAA	National Oceanic and Atmospheric Administration
NRC	National Research Council
NSERC	Natural Sciences and Engineering Research Council of Canada
NSF	National Science Foundation
OAH	Organization of American Historians
ONF	Oncology Nursing Foundation
ONR	Office of Naval Research
OREF	Orthopaedic Research and Education Foundation
ORISE	Oak Ridge Institute for Science and Education
OSI	Open Society Institute
PDF	Parkinson's Disease Foundation
PhRMA	Pharmaceutical Research and Manufacturers of American Foundation
PHSC	The Photographic Historical Society of Canada
RCPSC	Royal College of Physicians and Surgeons of Canada
RSC	Royal Society of Canada
RWJF	Robert Wood Johnson Foundation
SLA	Special Libraries Association
SSHRC	Social Sciences and Humanities Research Council of Canada
SSRC	Social Science Research Council
STTI	Sigma Theta Tau International
USHMM	United States Holocaust Memorial Museum Research Institute
USIA	United States Information Agency
USIP	United States Institute of Peace
WAWH	Western Association of Woman Historians

Kanner Fellowship in British Studies 3065
This three-month fellowship supports research at the Clark Library in any area pertaining to British history and culture. The fellowship is open to both postdoctoral and predoctoral scholars.
Amount of Grant: $6000 for three month tenure
Date(s) Application is Due: Feb 1
Contact: Fellowship Administrator, (310) 206-8552; fax: (310) 206-8577; email: clarklib@humnet.ucla.edu
Internet: http://www.humnet.ucla.edu/humnet/c1718cs/applic5.htm
Sponsor: UCLA Center for 17th- and 18th-Century Studies
405 Hilgard Avenue, 310 Royce Hall
Los Angeles, CA 90095-1404

Kansas Health Foundation Grants 3066
The foundation provides grants to Kansas organizations that share its mission to improve the quality of health in the state. Through solicited requests for proposals, the foundation proactively works as a partner with others in the state on programs that address the greatest health needs and achieve the most positive results. Grants are made in five primary categories, including primary care education, rural health, health promotion and disease prevention, public health, and health policy and research. At times, the foundation may respond to unsolicited proposals that are consistent with current programs. Types of support include fellowships, general operating budgets, special projects, and technical assistance. The foundation also awards recognition grants to honor Kansas grassroots organizations doing innovative work to improve the health of the people of the state.
Requirements: Kansas 501(c)3 nonprofit organizations are eligible.
Restrictions: The foundation does not support clinical or medical research; contributions to capital campaigns; organizations that practice discriminat on by race, color, creed, sex, age or national origin; operating deficits or retirement of debt; endowment programs; construction projects or real estate acquisitions; political projects of any kind; and vehicles, such as vans or buses, or emergency medical services equipment.
Geographic Focus: Kansas
Date(s) Application is Due: Mar 15; Sep 15
Contact: Grants Manager, (800) 373-7681 or (316) 262-7676 fax: (316) 262-2044; email: info@khf.org
Internet: http://www.kansashealth.org/grants/recognition_grants.jsp
Sponsor: Kansas Health Foundation
309 E Douglas
Wichita, KS 67202-3405

Kappa Alpha Theta Scholarships 3067
Each year approximately 100 one-year scholarships are awarded to provide financial assistance to members who are pursuing undergraduate and graduate study. Application materials are available on the Web site.
Requirements: Applicants must be female undergraduates and/or graduates of an accredited college or university, carrying a full course of study, citizens of the United States or Canada, and members in good standing.
Amount of Grant: $10,000-$25,000
Date(s) Application is Due: Feb 1
Contact: Cindy Thoennes, Program Coordinator, (888) 526-1870 ext 119 or (317) 876-1870; fax: (317) 876-1925; email: CThoennes@kappaalphatheta.org
Internet: http://www.kappaalphatheta.org/kat_v35/super_frames.cfm?area=thefou&body=thefou&extension=p1
Sponsor: Kappa Alpha Theta Foundation
8740 Founders Road
Indianapolis, IN 46268

Kappa Tau Alpha Frank Luther Mott KTA Research/Book Award 3068
The annual award recognizes the best book published in the field of journalism and mass communications research. Applicants should submit six copies of the published work for consideration to: Dr. Keith Sanders, KTA Executive Director, School of Journalism, 76 Gannett Hall, University of Missouri, Columbia MO 65211-1200.
Requirements: Entries must carry a 2005 copyright.
Restrictions: Edited volumes (i.e., collected works by several authors), textbooks and revised editions previously entered are not eligible.
Amount of Grant: $1000
Date(s) Application is Due: Dec 8
Contact: Dr. Keith Sanders, Executive Director, (573) 882-7685; fax: (573) 884-1720; email: umcjourkta@missouri.edu
Internet: http://www.missouri.edu/~ktahq/mott-kta.html
Sponsor: Kappa Tau Alpha
University of Missouri, School of Journalism, 3 Neff Hall
Columbia, MO 65211

Karst Research Fellowships and Grants 3069
The foundation annually awards graduate fellowships and grants for graduate research in any karst-related discipline anywhere in the world. Proposals involving surface or subsurface research are acceptable. The stipend is administered directly between the awardee and the foundation; the study itself is to be supervised at the graduate school of the awardee's choice by the thesis advisor. Application information is available upon request.
Requirements: Applicants must be graduate students at the master's or doctoral level and be in good standing with the graduate school.
Amount of Grant: $10,000 maximum
Date(s) Application is Due: Apr 15
Contact: Dr.Tom Poulson, Grants & Fellowships Chairman, (561) 630-3643; email: pnkambesis@juno.com
Internet: http://www.cave-research.org/grants/grants.html
Sponsor: Cave Research Foundation
345 Middlefield Road, MS 975
Menlo Park, CA 94025

Kate Neal Kinley Memorial Fellowship 3070
The fellowships support students of fine and applied arts who have received degrees in art, architecture, or music. Three major fellowships will be awarded, two each in art, architecture, or music and an additional one in music. The fellowships are to be used toward defraying the expenses of advanced study of the fine arts in America or abroad. Two or three additional fellowships of lesser amounts may also be granted. Contact the committee chair for an application form and instructions.
Requirements: The fellowship is open to graduates of the College of Fine and Applied Arts at the university and to graduates of similar national and international institutions of equal educational standing whose principal or major studies have been in the fields of architecture, art, or music. Preference is given to applicants under 25 years of age.
Date(s) Application is Due: Dec 3
Contact: Dr. Kathleen Conlin, Chair, (217) 333-1661; fax: (217) 244-8381; email: nacox.3@uiuc.edu
Internet: http://www.faa.uiuc.edu/about_faa/funding_opportunities.html
Sponsor: University of Illinois, College of Fine and Applied Arts
100 Architecture Building, 608 East Lorado Taft Drive
Champaign, IL 61820

Katrin H. Lamon Resident Scholar Program for Native Americans 3071
This resident scholar program for Native Americans awards one fellowship each year to a scholar who needs time to think and write. The major benefit to the recipient is the opportunity to work productively in a campus atmosphere that combines solitude with intellectual interaction with fellow scholars, visiting anthropologists, and school staff. The length of tenure is usually nine months, nonrenewable, beginning September 1 and ending at the end of May. Stipend, living quarters, and an office are provided. Applicant should send a proposal of no more than four pages in length, double-spaced; a professional vita; and three letters of recommendation received by November 15.
Requirements: Applications are welcome from Native American scholars working in any world area or on any topic within anthropology or from allied perspectives such as those of history, sociology, art, law, and philosophy. Postdoctoral scholars, retired scholars, or PhD candidates who have completed all requirements except for the dissertation are encouraged to apply. Preference will be given to those applicants whose fieldwork or basic research and analysis are complete and who need time to write up their results. Applicants should submit a brief statement of tribal affiliation.
Date(s) Application is Due: Nov 15
Contact: Resident Scholar Coordinator, (505) 982-2919; email: scholar@sarsf.org
Internet: http://www.sarweb.org/scholars/scholars.htm
Sponsor: School of American Research
P.O. Box 2188
Santa Fe, NM 87504-2188

Kavli Foundation Research Grants 3072
The Kavli Foundation supports research in the areas of astrophysics, nanoscience, neuroscience, and theoretical physics - three scientific frontiers that promise great discoveries in the 21st century and beyond. This mission is implemented through an international program of research institutes, professorships, symposia and workshops, as well as prizes in the fields of astrophysics, nanoscience, and neuroscience. Dedicated to advancing science for the benefit of humanity and promoting increased public understanding and support for scientists and their work, the Foundation also makes available materials on these sciences at no cost to educational institutions.
Contact: Grants Administrator; (805) 983-6000; fax: (805) 988-4800
Internet: http://www.kavlifoundation.org/areasofscience/
Sponsor: Kavli Foundation
1801 Solar Drive, Suite 250
Oxnard, CA 93031

Kavli Foundation Research Prizes 3073

The Kavli Prize includes three separate prizes, which are awarded for outstanding scientific research in the fields of astrophysics, neuroscience and nanoscience, and each consists of $1,000,000 in cash, a medal and a diploma. The Kavli Prize will be awarded in Oslo every two years. The Prize is intended not only as an accolade for the individual scientists concerned, but also as a recognition of research efforts that are of the utmost importance for the future of mankind. It aims to promote international cooperation between researchers and to increase awareness of the importance of research among the general public. The call for nominations will be open September 1 - December 15.
Samples: Maarten Schmidt, California Institute of Technology, and Donald Lynden-Bell, Cambridge University, UK, Astrophysics Prize Recipients (2008); Louis E. Brus, Columbia University, U.S., and Sumio Iijima, Meijo University in Japan, Nanoscience Prize Recipients (2008); Pasko Rakic, Yale University School of Medicine, Thomas Jessell, Columbia University, and Sten Grillner, Karolinska Institute in Sweden, Neuroscience Prize Recipients (2008).
Amount of Grant: $1,000,000 each
Contact: Prize Administrator; (805) 983-6000; fax: (805) 988-4800
Internet: http://www.kavlifoundation.org/kavli-prize
Sponsor: Kavli Foundation
1801 Solar Drive, Suite 250
Oxnard, CA 93031

Kennan Institute Regional Scholar Exchange Program Grants 3074

Regional six-month scholarships are available for individuals from the New Independent States to conduct research in the fields of the humanities or social sciences. Scholarships include monthly stipend, food allowance, lodging, and cost of one round-trip ticket for air travel from the home country to the District of Columbia plus financial support for excursionary travel within the United States.
Requirements: Eligible candidates include researchers and lecturers who hold at least kandidat degrees. Particular preference is given to junior faculty members and to individuals pursuing projects on contemporary issues.
Samples: Viktoria Antonova, Associate Professor, Department of History, Sociology, and Political Science, Pavolzhe's Academy for Civil Service (Saratov)—for research titled Public Service Reform and Training Public Servants in a Multi-Cultural Society: A Comparative Analysis of the United States, Canada, and Russia.; Elena Smilianskaia, Librarian, Department of Rare Books and Manuscripts, Moscow State U (Moscow)—for research titled Popular Christianity in Russia: New Sources and Comparative Approaches.
Contact: Fellows and Grants Office, (202) 691-4100; fax: (202) 691-4100; email: krunkaitytee@wwic.si.edu
Internet: http://wwics.si.edu/index.cfm?fuseaction=news.item&news_id=8713
Sponsor: Kennan Institute for Advanced Russian Studies
1300 Pennsylvania NW
Washington, DC 20004-3027

Kennan Institute Research Scholarships 3075

Research scholarships are normally available to academic participants in the early stages of their careers (before tenure) or scholars whose careers have been interrupted or delayed. For nonacademics, an equivalent degree of professional achievement is expected. Research proposals examining neglected or underdeveloped disciplines or topics in Russian studies, such as regional issues within the Russian Federation or subjects concerning the successor states to the U.S.SR, are particularly welcome. The grant offers a stipend, research facilities, word processing support, and some research assistance.
Requirements: Eligibility is limited to the postdoctoral level for academic participants, although doctoral candidates in the process of completing dissertations may apply. U.S. citizens or permanent residents are eligible.
Samples: Kathleen Kuehnast, Research Associate, Institute of European, Russian, and Eurasian Studies, George Washington U (Washington, DC)—to support research on Women, Poverty, and Conflicting Ideologies in Post-Soviet Kyrgyzstan.; Willard Sunderland, Department of History, U of Cincinnati (OH)—to support research on Steppe-Building: Colonization and Empire in the Russian South, 1764-1850s.
Amount of Grant: $3000 per month
Date(s) Application is Due: Dec 1
Contact: Edita Krunkaityte, (202) 691-4100; fax: (202) 691-4247; email: krunkaitytee@wwic.si.edu
Internet: http://wwics.si.edu/index.cfm?fuseaction=news.item&news_id=8713&page=2&endPoints=5996%2C11990%2C13445
Sponsor: Kennan Institute for Advanced Russian Studies
1300 Pennsylvania NW
Washington, DC 20004-3772

Kennan Institute Short-Term Grants 3076

The institute offers short-term grants of up to one-month duration to scholars studying the former Soviet Union having a particular need to utilize the library, archival, and other specialized resources of the Washington, DC, area. Academic participants must either possess doctoral degrees or be doctoral candidates who have nearly completed their dissertations. For nonacademics, an equivalent degree of professional achievement is expected. Application form is not required for short-term grants.
Requirements: U.S. citizens or permanent residents may request application materials.
Samples: Nathan Brooks, Associate Professor, Department of History, New Mexico State U (NM)—to support research on D.J. Mendeleev and Russian Society.; Hafizullah Emadi (Seattle, WA)—to support research on The Question of State-Building in the former Soviet Republic of Tajikistan.
Amount of Grant: $100 per day for duration of grant
Date(s) Application is Due: Mar 1; Jun 1; Sep 1; Dec 1
Contact: Fellows and Grants Office, (202) 691-4100; fax: (202) 691-4247; email: krunkaitytee@wwic.si.edu
Internet: http://wwics.si.edu/index.cfm?fuseaction=news.item&news_id=8713&page=2&endPoints=5996%2C11990%2C13445
Sponsor: Kennan Institute for Advanced Russian Studies
1300 Pennsylvania NW
Washington, DC 20024-3772

Kennedy Center Fellowships 3077

This fellowship program is a comprehensive management-training program for up to 10 highly qualified and motivated individuals who aspire to manage performing arts institutions and arts service organizations in both the public and private sector. The fellowships include extensive coursework in contemporary business practices and practical management experience through the lens of planning, presenting, and producing performing arts programming at an internationally recognized performing arts institution. Fellows have departmental assignments in addition to ongoing course work. They also attend performances and additional educational events and complete significant projects within the context of the Kennedy Center. Selected fellows may be offered the opportunity to continue their fellowships for an additional year. Fellowships provide a bimonthly stipend and cover all courses, materials, health insurance, attendance at selected performances and other educational events at the Kennedy Center, and transportation to and from Washington, DC for international fellows only. U.S. citizens living abroad or U.S. citizens with dual citizenship are not eligible for travel assistance. Housing is the responsibility of the individual.
Requirements: The ideal candidate must demonstrate knowledge of an artistic discipline(s), an aptitude for management skills, proven organizational and communication skills, maturity and confidence, and the ability to work independently, as well as with diverse groups. A minimum of a bachelor's degree in the arts or a related discipline and at least two years of professional administration experience working with a performing arts organization, or more than five years experience performing with a professional arts company (theater, dance, music) is required.
Amount of Grant: $18,000
Date(s) Application is Due: Apr 1
Contact: Fellowships Administrator, (202) 416-8897
Internet: http://www.kennedy-center.org/education/vilarinstitute/fellowships/home.html
Sponsor: John F. Kennedy Center for the Performing Arts
2700 F Street NW
Washington, DC 20566

Kenneth King Foundation Grants 3078

The foundation awards grants to Colorado nonprofit organizations in its areas of interest, including agriculture, arts and culture, civic and public affairs, community development, disabled, disadvantaged (economically), education (elementary, secondary, and higher), environment, healthcare, international (Hungary, Russia), religion, social services, technology, and women's issues. Types of support include general operating support, matching/challenge grants, program development, research grants, and scholarship funds. Guidelines are available online.
Requirements: Colorado 501(c)3 nonprofit organizations are eligible. Organizations that received more than $1500 in the previous year should contact the office by phone or email before submitting a proposal.
Restrictions: Capital requests will not be considered.
Geographic Focus: Colorado
Amount of Grant: $300-$200,000; $1000-$10,000 average
Date(s) Application is Due: Mar 1
Contact: Janice Fritsch, Program Contact, (303) 832-3200; fax: (303) 832-4176; email: jfritsch@kennethkingfoundation.org
Internet: http://www.kennethkingfoundation.org
Sponsor: Kenneth King Foundation
900 Pennsylvania Street
Denver, CO 80203-3163

Kenneth T. and Eileen L. Norris Foundation Grants 3079

The foundation supports Los Angeles County nonprofits in the areas of medicine—to improve access to health care, increase knowledge through research, and provide facilities for those activities to take place; youth—to provide constructive activities, positive role models, and opportunities for disadvantaged, disabled, and misguided children; community—to support law enforcement agencies, good citizenship, and environmental conservation; arts and culture—to support museums, symphony orchestras, and dance and theater companies; and education and science—to focus on private education, especially secondary and college levels. Types of support include general operating support, continuing support, building construction/renovation, equipment acquisition, endowment funds, program development, professorships, scholarship funds, research, and matching funds. Education/science projects are due between May 1 and June 30; youth requests are due between February 15 and March 31; community and cultural (the arts) requests are due between December 1 and January 31; and medicine proposals are due between May 1 and June 30.
Requirements: Grants are awarded to organizations in southern California.
Geographic Focus: California

Samples: Los Angeles Regional Foodbank (Los Angeles, CA)—for general operating support, $10,000.
Amount of Grant: $5000-$25,000 average
Date(s) Application is Due: Jan 31; Mar 31; Jun 30
Contact: Ronald Barnes, Executive Director, (562) 435-8444; fax: (562) 436-0584; email: grants@ktn.org
Internet: http://www.norrisfoundation.org/grant_funding.html
Sponsor: Kenneth T. and Eileen L. Norris Foundation
11 Golden Shore, Suite 450
Long Beach, CA 90802

Kettering Fund Grants **3080**
The mission of the Kettering Fund is to support scientific, medical, social and educational studies and research conducted by nonprofit, charitable organizations that are located and/or provide service in Ohio. The Distribution Committee meets on a semi-annual basis to make funding decisions, generally in May and November. The committee will consider support for the following purposes: capital needs (such as the purchase of land, new construction, renovations and equipment); seed grants for new organizations or programs; project support; endowments; research; and scholarship funds.
Requirements: 501(c)3 nonprofit organizations that provide services in or are located in Ohio may apply.
Restrictions: The fund does not support: religious organizations for religious purposes; activities of specific individuals; public elementary or secondary schools; loans; travel; deficit reduction; benefit events; or efforts to carry on propaganda or otherwise attempt to influence legislation.
Geographic Focus: Ohio
Samples: Kettering U (Flint, MI)—to support the Kettering Scholars Program, which awards scholarships to Ohio residents, $2.3 million.
Amount of Grant: $5,000-$200,000 range
Contact: Judith M. Thompson, Executive Director; (973) 228-1021; fax (973) 228-2399; email: info@ketteringfund.org
Internet: http://www.ketteringfund.org/main.html
Sponsor: Kettering Fund
1480 Kettering Tower
Dayton, OH 45423

KFC Allied Health Research Grants **3081**
The foundation supports research into the incidence and cure of kidney and urinary tract disease, patient services, and public education. Preference will be given to allied health personnel undertaking research related to the kidney and urinary tract and to a new research project not supported by any other agency. Applications from investigators in related specialty areas also are considered. Types of support include capital grants, fellowships, general operating grants, program development grants, scholarships, and travel assistance. One- and two-year renewable grants are awarded.
Requirements: Application is open to individuals affiliated with Canadian universities for research conducted within Canada.
Restrictions: Grants will not be awarded for equipment only.
Amount of Grant: $C50,000 maximum
Date(s) Application is Due: Oct 15
Contact: Manager, National Research Programs, (514) 369-4806 ext 223 or (800) 361-7494 ext. 223; email: research@kidney.ca
Internet: http://www.kidney.ca/english/research/allied-funding-03.htm
Sponsor: Kidney Foundation of Canada
5165 Rue Sherbrooke Street, Suite 300
Montreal, QC H4A 1T6 Canada

KFC Biomedical Fellowships **3082**
Fellowships support full-time, postdoctoral research training in the renal and urinary tract field for training in Canada or abroad. The award amount is determined by level of experience and the holding of MD, DDS, DVM, and/or PhD degrees. Fellows are expected to devote 12 consecutive months and 80 percent of their time toward fulfilling the objectives of the fellowship. Awards are applied for through the National Office of the Kidney Foundation. Obtain application forms.
Requirements: Applicants must be Canadian citizens or landed immigrants and apply through a university department or affiliated institution.
Amount of Grant: $C45,000
Date(s) Application is Due: Oct 15
Contact: Manager, National Research Programs, (514) 369-4806 ext 225 or (800) 361-7494 ext 225; email: research@kidney.ca
Internet: http://www.kidney.ca/english/research/biomed-funding-03.htm
Sponsor: Kidney Foundation of Canada
5165 Rue Sherbrooke Street, Suite 300
Montreal, QC H4A 1T6 Canada

KFC Biomedical Research Grants **3083**
The program fosters and encourages research about the kidney and urinary tract and the understanding of kidney diseases. Types of support include general operating grants, research grants, materials and supplies, and travel assistance. One- and two-year renewable grants are offered.
Requirements: Canadian citizens or landed immigrants may apply. Research must be conducted in Canada by individuals holding staff appointments at Canadian universities or academic institutions.
Restrictions: Requests for equipment only are denied.
Amount of Grant: $C50,000 maximum
Date(s) Application is Due: Oct 15
Contact: Manager, National Research Programs, (514) 369-4806 ext 225 or (800) 361-7494 ext 225; fax: (514) 369-2472; email: research@kidney.ca
Internet: http://www.kidney.ca/english/research/biomed-funding-03.htm
Sponsor: Kidney Foundation of Canada
5165 Rue Sherbrooke Street, Suite 300
Montreal, PQ H4A 1T6 Canada

KFC Nephrology and Urology Scholarships **3084**
Nephrology and urology scholarships provide salary support for up to two years of an initial faculty appointment at an approved medical school in Canada. Successful applicants must use the scholarship within three years of the date of their first appointment as assistant professor and must commit to spending 75 percent or more time in research.
Amount of Grant: $C45,000 per year for two years
Date(s) Application is Due: Oct 15
Contact: Manager, National Research Programs, (514) 369-4806 ext 225 or (800) 361-7494 ext 225; fax: (514) 369-2472; email: research@kidney.ca
Internet: http://www.kidney.ca/english/research/biomed-funding-03.htm#bio-scholar
Sponsor: Kidney Foundation of Canada
5165 Rue Sherbrooke Street, Suite 300
Montreal, QC H4A 1T6 Canada

KFF Barbara Jordan Health Policy Scholars Program **3085**
The program brings talented African American, Latino, American Indian/Alaska Native, and Asian/Pacific Islander college seniors and recent graduates to Washington, DC, where they are placed in Congressional offices and learn about health policy. Through the nine-week program (May 23 through July 28), scholars gain knowledge about federal legislative procedure and health policy issues, while further developing their critical thinking and leadership skills. Scholars also participate in seminars and site visits to augment their knowledge of health care issues, and write and present a health policy research memo. Scholars receive a stipend upon completion of the program; a daily expense allowance for meals and local transportation; transportation/airfare to and from Washington, DC; and lodging at Howard University in Washington, DC.
Requirements: Eligible candidates must be U.S. citizens who will be seniors or recent graduates of an accredited U.S. college or university in the fall. Successful applicants will show evidence of academic achievement and will demonstrate the following: be a member of a population that is adversely affected by racial or ethnic disparities in health; and have experience working in or with programs that address health disparities (e.g., local health departments, community health centers, inner city/rural hospitals, or organizations that conduct research on or develop policy related to health or health care disparities) or have done previous academic work (e.g., taken courses, written papers, or participate in research projects) related to health disparities.
Restrictions: Currently enrolled law, medical, and graduate students are not eligible to apply.
Amount of Grant: $7000 approximately, including $1500 stipend
Date(s) Application is Due: Jan 9
Contact: Grants Administrator, (650) 854-9400; fax: (650) 854-4800
Internet: http://www.kff.org/about/jordanscholars.cfm
Sponsor: Henry J. Kaiser Family Foundation
2400 Sand Hill Road
Menlo Park, CA 94025

Kimmel Scholar Award **3086**
The foundation seeks to improve the basic understanding of cancer biology and to develop new cancer treatment and prevention methods. Grants support programs of accomplished young cancer researchers that emphasize basic cancer research, rapid translation of basic science concepts into potential therapeutic applications, and clinical research with novel or innovative treatment strategies. Up to 10 awards are made each year. Additional grants are awarded to physicians engaged in translational science. Application and guidelines are available online.
Requirements: Applicants must hold an MD, PhD, or equivalent graduate degree. Researchers who have achieved the equivalent rank of assistant professor on or after July 2001. Physicians who have both clinical and research responsibility are eligible if appointed on or after July 2000. Research must be conducted at U.S. nonprofit organizations.
Samples: Sheil Steart, Washington U (Saint Louis, MO)—for research on cancer, $200,000 over two years (2004); Brian Schaefer, Uniformed Services U of the Health Sciences (Bethesda, MD)—for research on cancer, $200,000 over two years (2004).
Amount of Grant: $100,000 per year for two years
Date(s) Application is Due: Dec 6
Contact: Gary Cohen, (443) 849-3729; fax: (443) 849-3719; email: gcohen@gbmc.org
Internet: http://www.kimmel.org/About41.html
Sponsor: Sidney Kimmel Foundation for Cancer Research
6569 N Charles Street, Suite 201
Baltimore, MD 21204

King Faisal Foundation Grants **3087**
The foundation operates in Saudi Arabia and internationally in the fields of science and medicine and arts and humanities. King Faisal International Prizes are awarded in the areas of Arabic literature, Islamic studies, service to Islam, and science and medicine. Nominations are due May 31. The foundation also sponsors a scholarship program

intended to offer opportunities to outstanding Muslim graduates in the Islamic World to complete their higher education in the most prestigious universities of the world.
Requirements: Institutions and organizations around the world nominate candidates for the prizes. Scholarship applicants must be Muslim; have completed university studies with an average of 86 percent or above; have acceptance from an accredited European or North American university that is highly regarded in the particular field of specialization; not exceed 40 years of age; be fluent in the language of his/her studies; have specialization in one of the fields of medicine, engineering, and sciences (physics, chemistry, and geology); and have a graduation date of not more than three years.
Amount of Grant: $200,000 prize; $400 per month scholarships
Date(s) Application is Due: May 31
Contact: Grants Administrator, 966-1-465-2255; fax: 966-1-465-6524; email: info@kff.com
Internet: http://www.kff.com
Sponsor: King Faisal Foundation
P.O. Box 352
Riyadh 11411 Saudi Arabia

King Foundation Grants 3088
The foundation awards grants to eligible nonprofit organizations in its areas of interest, including scholarships, education, youth organizations, the indigent, cultural and historic community resources, medical and scientific research, and the aging population. Priority is given to agencies with innovative programs, a proven record of quality service delivery, and policies consistent with the foundation's mission. Single-year projects, rather than multiyear projects, are favored. After an application has been received and reviewed, the foundation may request further information, request a meeting, or request a site visit.
Requirements: Texas and Arkansas 501(c)3 tax-exempt organizations are eligible.
Restrictions: Grants are not made to individuals; organizations outside of Texas and Arkansas; church or seminary construction or religious programs other than social service-based initiatives; ongoing operating expenses or funds to offset operating losses; loan financing; endowments; professional conferences or symposia; balls, events, or galas benefiting charitable organizations; or organizations without tax-exempt status.
Geographic Focus: Arkansas; Texas
Samples: Communities in Schools-Dallas Region (TX)—to expand services available to at-risk children and their families at Maple Lawn Elementary School, $15,000 (2005); Saint Philip's School and Community Ctr (Dallas, TX)—for student scholarships, $10,000 (2005); Genesis Women's Shelter (Dallas, TX)—to renovate its shelter facilities, $35,000 (2005); Goodwill Industries of Dallas (TX)—to purchase a trailer truck for use in its Contributed Goods Program, $35,000 (2005).
Contact: Michelle Monse, Executive Director, (214) 750-1884; fax: (214) 750-1651
Sponsor: Carl B. and Florence E. King Foundation
2929 Carlisle Street, Suite 222
Dallas, TX 75204

Kinsman Foundation Grants 3089
The foundation awards grants to eligible nonprofit organizations in its areas of interest, including historic preservation, animal/wildlife welfare, and health care policy. Types of support include annual campaigns, building construction/renovation, capital campaigns, challenge/matching grants, conferences and seminars, consulting services, continuing support, curriculum development, endowments, equipment acquisition, general operating support, internships, program development, publication, research, seed grants, and technical assistance. Contact the office for application forms and guidelines.
Requirements: Oregon and southern Washington nonprofit organizations are eligible.
Geographic Focus: Oregon; Washington
Samples: Kam Wah Chung and Company Museum (John Day, OR)—for restorations at this museum, a former trading post built in 1875, that documents the lives and culture of Chinese immigrants in eastern Oregon, $110,000 (2004).
Date(s) Application is Due: Aug 1
Contact: Grants Administrator, (503) 654-1668; fax: (503) 654-1759; email: grants@kinsmanfoundation.org
Sponsor: Kinsman Foundation
3727 SE Spaulding Avenue
Milwaukie, OR 97267-3938

Kiplinger Program in Public Affairs Journalism Fellowships 3090
The program in public affairs journalism is dedicated to giving journalists time to produce meaningful work and deepen their knowledge of public affairs issues so they can better inform citizens and contribute to public life. Fellows will have six months to work on an independent public affairs print project of their choice. Fellows who successfully complete the program will receive a Certificate in Public Affairs Journalism from the John Glenn Institute, as well as a transcript reflecting credits for the courses they have taken. Fellows must either be U.S. citizens with five years of relevant professional experience in the United States or abroad, or foreign nationals with five years of professional experience within the United States or working abroad for U.S. organizations. Application and guidelines are available online.
Requirements: Applicants must have at least five years of professional experience in journalism. Those eligible include reporters, editors, and employees from newspapers, wire services, magazines, online publications, and freelancers. Television and radio reporters will also be considered if they seek to produce a print project.
Amount of Grant: $20,000
Date(s) Application is Due: Jul 20
Contact: Betsy Hubbard, Program Manager, (614) 247-8845; fax: (614) 292-4868; email: hubbard.160@osu.edu
Internet: http://www.kiplingerprogram.org/kiplinger
Sponsor: Ohio State University
350 Page Hall, 1810 College Road
Columbus, OH 43210

Klingenstein Fellowship Awards in the Neurosciences 3091
The purpose of these awards is to support, in the early stages of their careers, young investigators engaged in basic or clinical research that may lead to a better understanding of epilepsy. The fund recognizes that to accomplish this goal it is necessary to encourage a variety of new approaches. Several areas within the neurosciences are of particular interest to the fund: Cellular and molecular neuroscience'Studies of the mechanisms of neuronal excitability and development, and of the genetic basis of seizure disorders. Neural systems'Studies of the integrative function of the nervous system. Clinical research'Studies designed to improve the prevention, diagnosis, treatment and our understanding of the causes of epilepsy. Application may be made at any time.
Requirements: Applicants must hold the PhD and/or MD degrees, and have completed all research training, including post-doctoral training. U.S. citizenship is not a requirement, but it is expected that candidates will be permanent residents of the U.S. and that their research will be carried out in U.S. institutions.
Restrictions: The fund does not contribute to endowments and rarely contributes to buildings or other kinds of capital projects.
Samples: Teachers College Columbia U (New York, NY)—for program support, $578,595.; American Jewish Committee (New York, NY)—for work on issues of separation of church and state, $40,000.; Foundation for Biomedical Research (Washington, DC)—for animal biomedical research, $25,000.
Date(s) Application is Due: Dec 10
Contact: Kathleen Pomerantz, Vice President, (212) 492-6181; email: kathleen.pomerantz@klingenstein.com
Internet: http://www.klingfund.org
Sponsor: Esther A. and Joseph Klingenstein Fund
787 Seventh Avenue, 6th Fl
New York, NY 10019

Klingenstein Neuroscience Fellowships 3092
The purpose of the fellowships is to support, in the early stages of their careers, young investigators engaged in basic or clinical research that may lead to a better understanding of epilepsy. Several areas within the neurosciences are of particular interest: cellular and molecular neuroscience, neural systems, and clinical research. Up to 10 fellows will be appointed. While U.S. citizenship is not a requirement, it is expected that candidates will be permanent residents of the United States and that their research will be carried out in U.S. institutions. The award may be used for salary support, research assistants, equipment, or for activities that promote the scientific advancement of the fellow. Payments will be made to the investigator's institution. Holding an award from another foundation concurrently with a Klingenstein award requires the approval of the advisory committee.
Requirements: Investigators must hold the PhD and/or MD degrees and have completed all research training, including postdoctoral training. The candidate should be an independent investigator, holding a tenure-track academic rank (but not yet tenured) in a university or medical school, or the equivalent standing in a research institute or medical center.
Amount of Grant: $150,000 over three years
Date(s) Application is Due: Dec 9
Contact: John Klingenstein, President, (212) 492-6181; fax: (212) 492-7007
Internet: http://www.klingfund.org
Sponsor: Esther A. and Joseph Klingenstein Fund
787 7th Avenue, 6th Fl
New York, NY 10019

Klingenstein Third Generation Foundation Grants in Depression 3093
The foundation strives to improve the lives of families afflicted by clinical depression, including childhood and adolescent depression, and Attention Deficit/Hyperactive Disorder (ADHD). Grants are awarded in the areas of intervention and referral, prevention, public education/training, and infrastructure. The foundation will consider making capital grants, operating grants, challenge grants, seed grants, and grants given over more than a one-year period. The foundation also sponsors fellowships for postdoctoral research in depression. Funding proposals will be accepted from any geographic area, but the board expects to fund primarily in the New York, Chicago, and District of Columbia metropolitan areas. There is no formal application procedure or form. Applicants should submit a brief letter that succinctly describes the project and the amount sought and explains why the project is significant. The foundation will request a more detailed proposal.
Requirements: U.S. tax-exempt organizations are eligible.
Restrictions: Grants do not support direct services, unless they include a research or program evaluation component, or unless they represent the application of research findings to a clinical setting.
Samples: American Academy of Child and Adolescent Psychiatry (Washington, DC)—for general support, $15,000 (2003); Columbia U, Carmel Hill Ctr for Early Diagnosis and Treatment (New York, NY)—to create brochures on mental-health disorders written in different languages and at a low reading level, $15,000 (2003); State U of New York (Buffalo, NY)—fellowship support for James Waxmonsky, for research on child and adolescent depression, $30,000 over two years (2003).
Amount of Grant: $5000-$35,000 average

Contact: Sally Klingenstein Martell, Executive Director, (212) 492-6179; fax: (212) 492-7007; email: sally@ktgf.org
Internet: http://www.ktgf.org/depress.html
Sponsor: Klingenstein Third Generation Foundation
787 Seventh Avenue, 6th Fl
New York, NY 10019-6016

Kluge Center David B. Larson Fellowship in Health and Spirituality 3094
The Library of Congress invites qualified scholars to apply for a postdoctoral fellowship in the field of health and spirituality. It seeks to encourage the pursuit of scholarly excellence in the scientific study of the relation of religiousness and spirituality to physical, mental, and social health. The fellowship provides an opportunity for a period of six to 12 months of concentrated use of the collections of the Library of Congress, through full-time residency in the Library's John W. Kluge Center. If necessary, special arrangements may be made with the National Library of Medicine for access to its materials as well. Applicants must submit a formal application packet, including an application form, a two-page curriculum vita that should indicate major prior scholarship, a one-paragraph project summary, a bibliography of basic sources, a research proposal of no more than 1500 words, and three letters of reference (in English) from people who have read the research proposal.
Requirements: Applicants must by U.S. citizens or permanent residents and must possess a doctoral degree (PhD, MD, ScD, DrPH, DSW, PPsy, DST, ThD, and JD) awarded by the deadline date.
Amount of Grant: $4,000 per month for up to twelve months
Date(s) Application is Due: Apr 17
Contact: Larson Fellowships, (202) 707-3302; fax: (202) 707-3595; email: scholarly@loc.gov
Internet: http://www.loc.gov/loc/kluge/kluge-fellow-larson.html
Sponsor: Library of Congress
101 Independence Avenue, SE
Washington, DC 20540-4860

Kluge Center Fellowships 3095
The fellowships are awarded to qualified scholars worldwide, and especially encourages humanistic and social science research that makes use of the Library's large and varied collections. Interdisciplinary, cross-cultural, or multi-lingual research is particularly welcome. Among the collections available to researchers are the world's largest law library and outstanding multi-lingual collections of books and periodicals. Deep special collections of manuscripts, maps, music, films, recorded sound, prints and photographs are also available.
Requirements: Scholars worldwide who have received a terminal advanced degree within the past seven years in the humanities, social sciences, or in a professional field such as architecture or law are eligible. Exceptions may be made for individuals without continuous academic careers. Applicants may be U.S. citizens or foreign nationals. Upon selection, and in accordance with relevant visa regulations, foreign nationals will be assisted in obtaining the appropriate visa. To meet the minimum eligibility requirements, the degree must be formally awarded by the deadline date.
Amount of Grant: $4,000 per month for up to twelve months
Date(s) Application is Due: Jul 15
Contact: Kluge Fellowships, (202) 707-3302; fax: (202) 707-3595; email: scholarly@loc.gov
Internet: http://www.loc.gov/loc/kluge/fellowships/kluge.html
Sponsor: Library of Congress
101 Independence Avenue, SE
Washington, DC 20540-4860

Kluge Center Kislak Fellowships in American Studies 3096
The Program supports scholarly research that contributes significantly to a greater understanding of the cultures and history of the Americas. It provides an opportunity for a period of up to 8 months of concentrated use of materials from the Kislak Collection and other collections of the Library of Congress, through full-time residency at the Library. The program supports research projects in the disciplines of archaeology, history, cartography, epigraphy, linguistics, ethno-history, ethnography, bibliography and sociology, with particular emphasis on Florida, the circum-Caribbean region and Mesoamerica. We encourage interdisciplinary projects that combine disciplines in novel and productive ways.
Requirements: Applicants may be of any nationality and must possess a PhD degree, or equivalent terminal degree, awarded by the application deadline date.
Amount of Grant: $4,000 per month for up to eight months
Date(s) Application is Due: Feb 28
Contact: Program Director, (202) 707-3302; fax: (202) 707-3595; email: scholarly@loc.gov
Internet: http://www.loc.gov/loc/kluge/fellowships/kislak html
Sponsor: Library of Congress
101 Independence Avenue, SE
Washington, DC 20540-4860

Kluge Center Kislak Short-Term Fellowships in American Studies 3097
The Program offers short-term fellowships for independent scholars, undergraduate and graduate students, and college and university faculty to conduct research based on items from the Kislak Collection. The Program supports scholarly research that contributes significantly to a greater understanding of the cultures and history of the Americas. It provides an opportunity for a period of up to four months of concentrated use of materials from the Kislak Collection and other collections of the Library of Congress, through full-time residency at the Library. The program supports research projects in the disciplines of archaeology, history, cartography, epigraphy, linguistics, ethno-history, ethnography, bibliography and sociology, with particular emphasis on Florida, the circum-Caribbean region and Mesoamerica. We encourage interdisciplinary projects that combine disciplines in novel and productive ways.
Requirements: Applicants may be of any nationality and must possess a PhD degree, or equivalent terminal degree, awarded by the application deadline date.
Amount of Grant: $4,000 per month for up to four months
Date(s) Application is Due: Aug 31
Contact: Program Director, (202) 707-3302; fax: (202) 707-3595; email: scholarly@loc.gov
Internet: http://www.loc.gov/loc/kluge/fellowships/kislakshort.html
Sponsor: Library of Congress
101 Independence Avenue, SE
Washington, DC 20540-4860

Kluge Center Kissinger Chair in Foreign Policy and International Relations 3098
The Kissinger Scholar is a distinguished senior research position in residence at the Library for a period of ten months. Using research facilities and services at the Library of Congress, the Scholar is expected to engage in research on foreign policy and international affairs that will lead to publication. The Scholar may be of any nationality. A stipend of $135,000 during the term of appointment supports the Scholar.
Samples: James M. Goldgeier—professor of political science and director of the Institute for European, Russian and Eurasian Studies at George Washington University (2005); Melvyn P. Leffler—dean of the faculty of arts and sciences and Edward Stettinius Professor of American History at the University of Virginia (2004); Xiang Lanxin—professor of international history and politics at the Institut universitaire de hautes etudes internationales in Geneva, Switzerland (2003).
Amount of Grant: $135,000
Date(s) Application is Due: Nov 1
Contact: Program Director, (202) 707-3302; fax: (202) 707-3595; email: scholarly@loc.gov
Internet: http://www.loc.gov/loc/kluge/fellowships/kissinger.html
Sponsor: Library of Congress
101 Independence Avenue, SE
Washington, DC 20540-4860

Kluge Center Kissinger Lecturer Award 3099
Another part of the Kissinger Program is the Kissinger Lecturer, who is appointed annually to deliver the Kissinger Lecture. Like the Kissinger Scholar, the Lecturer may be of any nationality. The Lecturer will have achieved distinction in the field of foreign affairs. Each lecture will be published and, every five years, the lectures will be aggregated into an edited volume that may contain additional material. The Lecturers, who will receive a $20,000 honorarium and round- trip transportation, will be appointed upon recommendation of the Steering Committee. There will be no direct applications, but nominations are welcome and may be submitted at any time. Selection will be based upon an achieved level of distinction.
Amount of Grant: $20,000
Contact: Program Director, (202) 707-3302; fax: (202) 707-3595; email: scholarly@loc.gov
Internet: http://www.loc.gov/loc/kluge/fellowships/kissinger.html
Sponsor: Library of Congress
101 Independence Avenue, SE
Washington, DC 20540-4860

Knight Public Health Journalism Fellowships at CDC 3100
The fellowships offer midcareer journalists a chance to break away from the day-to-day rigors of journalism to explore public health in-depth for a period of three months. Housed at CDC in Atlanta, with funding from the John S. and James L. Knight Foundation, the fellowships are a multi-faceted experience in research, field practice, and instruction. This year six Knight Fellows will train with CDC's Epidemic Intelligence Service officers, investigate a disease outbreak, work closely with other CDC scientists, attend seminars and Grand Rounds, and see how public health work is carried out by local and state authorities. Knight fellows receive a stipend to cover modest living expenses. Program guidelines are available online.
Requirements: Journalists with a track record of accomplishment, a minimum of five years of reporting experience, and a desire to strengthen the reporting of public health are invited to apply.
Date(s) Application is Due: Mar 15
Contact: Fellowship Administrator, (888) 880-4232 or (404) 653-0790; fax: (404) 653-0330
Internet: http://www.cdcfoundation.org/pages.html?page=296
Sponsor: Centers for Disease Control Foundation
50 Hurt Plz, Suite 765
Atlanta, GA 30303

Knight Science Journalism Fellowships 3101
The fellowships provide one academic year in residence at the Massachusetts Institute of Technology, auditing courses, visiting laboratories, and attending twice-weekly seminars with leading scientists. The goal of the program is to increase exposure of the fellows to front-rank researchers and to deepen the fellows' knowledge of current issues in science and technology.

Requirements: Full-time staff or freelance science and technology journalists working for U.S. newspapers, magazines, press services, radio, or television are eligible. Applicant must have a minimum of three years' coverage of science and technology.
Amount of Grant: $55,000 stipend
Date(s) Application is Due: Mar 1
Contact: Martha Henry, Program Coordinator, (617) 253-2336; fax: (617) 258-8100; email: mshenry@mit.edu
Internet: http://web.mit.edu/knight-science/fellowships.html
Sponsor: Massachusetts Institute of Technology
E32-300, 77 Massachusetts Avenue
Cambridge, MA 02139-4307

Knight-Bagehot Fellowship Program in Economics and Business Journalism 3102
Up to 10 fellowships are awarded annually to working journalists for the study of economics, finance, and business at Columbia for an academic year. Recipients of these fellowships pursue an intensive one-year program at Columbia's business, international, or law school, including a specially tailored seminar and dinner program. Stipend plus full tuition is provided. Application forms are available on the Web site.
Requirements: The program is open to full-time employees of newspapers, wire services, magazines, and broadcast stations whose work regularly appears in the United States or Canada. Journalists applying must have at least four years of professional experience. There are no age requirements.
Amount of Grant: $40,000 stipend plus tuition
Date(s) Application is Due: Mar 1
Contact: Terri Thompson, Knight-Bagehot Fellowships, (212) 854-2711; fax: (212) 854-7837
Internet: http://www.jrn.columbia.edu/knight-bagehot
Sponsor: Columbia University
Graduate School of Journalism, Room 500
New York, NY 10027

Knowles Science Teaching Foundation Fellowships 3103
The teaching fellowships support young men and women who have received a bachelor's or advanced degree in science, engineering, or mathematics and are committed to teaching high school science and/or mathematics in U.S. schools. The fellowship supports them professionally and financially for up to five years through a teacher preparation program to eligibility for tenure. During the summers, fellows have opportunities to engage in activities that will enhance their ability to teach science and mathematics. During the initial academic year, fellows may participate in a recognized teacher certification program. Each year, the fellows meet in summer, fall, and spring, occasionally in conjunction with a professional meeting for science or mathematics teachers. Fellows communicate electronically with each other regularly. Support is available for a local mentor and for instructional materials. Application and guidelines are available online.
Date(s) Application is Due: Jan 15
Contact: Nicole Gillespie, (856) 216-8080; fax: (856) 216-9987; email: ngillespie@kstf.org
Internet: http://www.kstf.org/teaching_fellowships
Sponsor: Janet H. and C. Harry Knowles Foundation
20 E Redman Avenue
Haddonfield, NJ 08033

Kobe College Corporation Graduate Fellowships 3104
The institute sponsors graduate fellowships for one year of study or research in Japan. The program is open to U.S. citizens and supports teaching or research at the master's or doctoral level. Graduate students who have a record of teaching effectively about Japan, or who show promise to do so in the future, are favorable candidates. Fellowship recipients must plan to teach in the U.S. after completing their degree, either at the secondary or higher education levels, and should have good public speaking skills. One fellowship will be awarded to a female graduate student. Currently, the Graduate Fellowship is only be offered in alternate years.
Requirements: Applicants must be U.S. citizens, enrolled, in good standing, in a graduate program at an accredited higher education institution in the U.S..
Restrictions: Students enrolled in professional graduate degree programs are not eligible (i.e., M.B.A., J.D., M.D., etc.).
Amount of Grant: $24,000
Date(s) Application is Due: Jan 10
Contact: Fellowship Administrator; (312) 364-9336; fax: (312) 364-9337; email: kobecollegecorp@prodigy.net
Internet: http://www.kccjee.org/pages/program_pages/gradfellowshippage.htm
Sponsor: Kobe College Corporation
188 W. Randolph, Suite 1809
Chicago, IL 60601

Kodak Community Relations Grants 3105
The company awards grants in Kodak communities to support community revitalization, education, and health and human services. Education grants support preschools, K-12, higher education, diversity initiatives, and umbrella education funds. University contributions are focused on recruitment, technology, and market development. Community revitalization grants support economic development, culture, and the arts. The company awards grants to international nonprofits for relief efforts. There are no application deadlines or forms.
Requirements: Nonprofit organizations in company-operating areas nationwide are eligible.
Restrictions: Kodak does not support individuals; a commitment beyond three to five years; endowed chairs; university capital campaigns; event sponsorships; operating costs of organizations that receive funds from a Kodak supported United Way; legislators, political organizations, or campaigns; or sectarian organizations whose programs are limited to members of one religious group.
Samples: American National Red Cross (Washington, DC)—for relief efforts in South Asia and Africa, $1 million (2005).
Amount of Grant: $11 million total
Date(s) Application is Due: Apr 30
Contact: Community Relations, (716) 724-3041; fax: (716) 724-1376
Internet: http://www.kodak.com/U.S./en/corp/community.shtml
Sponsor: Eastman Kodak Corporation
343 State Street
Rochester, NY 14650-0517

Koinonia Foundation Grants 3106
Support will be considered for projects that teach, research, or demonstrate some of the following ideas: respectful use of human potential and resources; respectful use of the resources of our planet; harmonious balance of humankind and nature; sensitive awareness of the needs of the environment; creative problem solving; innovative alternative living styles; sensitivity to all levels of consciousness or our planet; enriching ways of relating and sharing; global thinking coupled with local initiatives; deepened understanding of spirituality and spiritual living; and exploration and development of higher consciousness. The foundation gives priority to projects that work for systemic change, empower trickle-up activity that enables small group action to influence large movements, support grassroots programs and action, and articulate a clearly defined goal or activity as part of a larger project or organizational effort.
Requirements: 501(c)3 nonprofit organizations are eligible.
Amount of Grant: $2000 maximum
Contact: Grants Review Committee, (410) 828-1869
Internet: http://www.koinoniafoundation.org
Sponsor: Koinonia Foundation
P.O. Box 11050
Baltimore, MD 21212

Komen Greater NYC Clinical Trials Enrollment Grants 3107
The Greater New York City Affiliate of Susan G. Komen for the Cure (Komen Greater NYC) fulfills its mission of eradicating breast cancer as a life-threatening disease, in part, by offering grants to enhance the capacity of National Cancer Institute- and Department of Defense-approved breast cancer clinical research conducted in the Komen Greater NYC area to increase the number of women enrolled in breast cancer clinical research studies. Komen Greater NYC seeks to fund projects that implement evidence-based strategies to overcome barriers to enrollment in breast cancer clinical research. Collaborative projects between institutions and community-based organizations are encouraged.
Requirements: To apply, the organization must be: a not-for-profit organization (a charitable or educational tax-exempt organization), a government agency, or an Indian tribe; conducting National Cancer Institute (NCI) or Department of Defense (DOD) approved breast cancer clinical research (this includes studies through the Clinical Trials Cooperative Group Program including but not limited to ACOSOG, ECOG, SWOG, NSABP, ACRIN, and RTOG; and located in the Komen Greater NYC service area, which includes the five boroughs of New York City, Long Island, Westchester County, and Rockland County.
Geographic Focus: New York
Amount of Grant: $200,000 over two years
Date(s) Application is Due: Oct 9
Contact: Mary Haviland, Esq. Director of Grants and Public Policy; (212) 560-9590, ext. 230; email: mhaviland@komennyc.org
Internet: http://www.komennyc.org/site/PageServer?pagename=grants_clinicaltrials
Sponsor: Susan G. Komen for the Cure Foundation
341 West 38th Street, 10th Floor
New York, NY 10018-2918

Komen Greater NYC Community Breast Health Grants 3108
The mission of Komen Greater NYC is to eradicate breast cancer as a life-threatening disease by advancing research, education, screening, and treatment. The Greater New York City Affiliate is, therefore, soliciting applications from non-profit organizations that have current or planned breast health programs within the five boroughs of New York City, Long Island, Westchester County, and Rockland County. The purpose of the Community Breast Health Grants Program (formerly known as Screening, Treatment, and Education Program- STEP) is to support community-based programs that provide the medically under-served with access to breast health education, screening coordination, diagnostic, treatment, and support services.
Requirements: Applicant organizations must be: a not-for-profit organization (a charitable or educational tax-exempt organization), a government agency, or an Indian tribe; located in the Komen Greater NYC service area, which includes the five boroughs of New York City, Long Island, Westchester County, and Rockland County; and a current or planned provider of breast health services.
Geographic Focus: New York
Amount of Grant: $100,000 maximum
Date(s) Application is Due: Oct 9

Contact: Mary Haviland, Esq. Director of Grants and Public Policy; (212) 560-9590, ext. 230; email: mhaviland@komennyc.org
Internet: http://www.komennyc.org/site/PageServer?pagename=grants_breasthealthgrants
Sponsor: Susan G. Komen for the Cure Foundation
341 West 38th Street, 10th Floor
New York, NY 10018-2918

Komen Greater NYC Conference Grants **3109**
Komen Greater NYC provides grants to non-profit organizations to support educational conferences, symposiums or meetings. To be considered for funding, events must be specific to breast health or help further the body of knowledge related to breast cancer research, education, screening, diagnosis, treatment and/or support. Grants are considered three times per year, in May, August, and January. Awards will not exceed $5,000. Award notifications are generally made within 6 weeks after receipt of the application.
Requirements: To apply, the organization must be: a not-for-profit organization (a charitable or educational tax-exempt organization), a government agency, or an Indian tribe; conducting National Cancer Institute (NCI) or Department of Defense (DOD) approved breast cancer clinical research (this includes studies through the Clinical Trials Cooperative Group Program including but not limited to ACOSOG, ECOG, SWOG, NSABP, ACRIN, and RTOG; and located in the Komen Greater NYC service area, which includes the five boroughs of New York City, Long Island, Westchester County, and Rockland County.
Geographic Focus: New York
Amount of Grant: $5,000 maximum
Date(s) Application is Due: May 14; Jul 25; Dec 1
Contact: Cara Thunder, Grants Program Associate; (212) 461-6185; cthunder@komennyc.org
Internet: http://www.komennyc.org/site/PageServer?pagename=grants_conference
Sponsor: Susan G. Komen for the Cure Foundation
341 West 38th Street, 10th Floor
New York, NY 10018-2918

Kosciuszko Foundation Grants for Graduate and Postgraduate Study in Poland **3110**
This program is open to U.S. graduate and postgraduate students who wish to pursue courses of graduate or postgraduate study and/or research at institutions of higher learning in Poland. Also eligible are university faculty who wish to spend sabbaticals pursuing research in Poland. Individuals accepted into the program are eligible to apply for limited subsidies for tuition, transportation, and living expenses. A personal interview may be required. Contact the office for further information and an application.
Requirements: Applicants must have working knowledge of the Polish language and are reviewed based on academic background, motivation for pursuing graduate studies and/or research in Poland, and proposals for studies and research in Poland. Applicants must be U.S. citizens or permanent residents.
Date(s) Application is Due: Jan 14
Contact: Addy Tymczyszyn, (212) 734-2130 ext 210; email: addy@thekf.org
Internet: http://www.kosciuszkofoundation.org/EDScholarships.html
Sponsor: Kosciuszko Foundation
15 E 65th Street
New York, NY 10021-6595

Kosciuszko Foundation Tuition Scholarships **3111**
Domestic scholarships foster knowledge and appreciation of Polish culture, history, and traditions and serve to strengthen understanding and friendship between citizens of the United States and Poland through educational and cultural ties. Tuition scholarships support full-time study in law, medicine, education, Slavic studies, literature, political science, chemistry, marketing, international relations, journalism, dentistry, nursing, and many other fields. Certain scholarships are available to assist eligible students of law and engineering at DePaul University, Chicago; Bayonne, NJ, residents studying nursing, teaching or business; and to eligible students who are residents of Amsterdam, NY; Chicopee, MA; and the state of New Hampshire. The period of study is one academic with possible renewal for for one additional year. Applications are available on the Web site.
Requirements: Eligible students must be U.S. citizens of Polish descent, Poles who are permanent residents of the United States, and Americans of non-Polish descent who are pursuing courses of study or research concerning Polish subjects.
Restrictions: Members of the Massachusetts Federation of Polish Women's Clubs and the Polish American Club of North Jersey and their children are ineligible to receive these scholarships.
Amount of Grant: $3000 average; $1000-$7000
Date(s) Application is Due: Jan 15
Contact: Addy Tymczyszyn, Grant Officer, (212) 734-2130 ext 210; fax: (212) 628-4552; email: addy@thekf.org
Internet: http://www.kosciuszkofoundation.org/EDScholarships.html
Sponsor: Kosciuszko Foundation
15 E 65th Street
New York, NY 10021-6595

KPMG Foundation Grants **3112**
The foundation supports colleges and universities under the following basic programs: Employee Matching Gifts; Special Grants for specific innovative projects or programs that will create new educational opportunities; KPMG Professorships; Professional Development Program allowing academics to keep abreast of current professional practice by attending KPMG courses; Auditing Practice and Research Seminars; and support of educational organizations that aid the Beta Alpha Psi Scholarship Program and the Minority Accounting and Information Systems Doctoral Scholarship Program. Grants are renewable for up to five years of support.
Requirements: Grants must be used for educational purposes related to the firm's functional areas of practice—accounting and auditing, taxation, management consulting, and private business advisory services.
Samples: International Year of Volunteers (New York, NY)—to support events and activities related to the International Year of Volunteers 2001, $100,000.
Contact: Grants Administrator, (201) 307-7932; email: kpmgfoundation-GM@kpmg.com
Internet: http://www.kpmgfoundation.org
Sponsor: KPMG Foundation
3 Chestnut Ridge Road
Montvale, NJ 07645

KPMG Foundation Minority Accounting and Information Systems Doctoral Scholarships **3113**
The scholarships are available to African American, Hispanic American, and Native American accounting or information systems doctoral students. Scholarships are renewable for up to five years.
Requirements: Applicants must be U.S. citizens or U.S. residents enrolled in full-time, accredited, U.S. doctoral programs.
Samples: PhD Project (Montvale, NJ)—to provide 24 scholarships through the KPMG Minority Doctoral Scholarships program, $1.2 million over five years.
Amount of Grant: $10,000 per year
Contact: Anita English, (201) 307-7152; email: acenglish@kpmg.com
Internet: http://www.kpmgfoundation.org/foundinit.asp
Sponsor: KPMG Foundation
3 Chestnut Ridge Road
Montvale, NJ 07645

Kraft Cares Canada Grants **3114**
Kraft Canada has a long tradition of supporting charitable and nonprofit organizations across the country. Corporate citizenship is focused on fighting hunger—helping Canadians to achieve healthy lifestyles and building stronger communities; achieving healthy lifestyles—healthy eating and physical activity, including education for youth; and building stronger communities—contributing to local charitable organizations. Hunger and healthy living proposals are reviewed on an annual basis. Submit a letter of intents; full proposals are by invitation. Guidelines are available online.
Requirements: Nonprofit organizations in Canada ae eligible.
Restrictions: Kraft Canada will not consider funding for the following: general operating costs; organizations with a limited constituency, such as fraternal or veterans; organizations that restrict their services to members of a religious or ethnic group; organizations that discriminate on the basis of race, sex, religion, national origin, age handicap, or veteran status; individuals; travel, tuition, ad registration fees; special events (e.g., fundraisers, charitable golf tournaments, gala dinners); or benefit tickets or courtesy advertising.
Samples: Boys and Girls Clubs of Canada (Markham, Ontario), $100,000 (2005); Canadian Association of Food Banks (Tornto, Ontario), $150,000 (2005); Centraide (Montreal, Quebec), $90,000 for hunger relief (2005).
Date(s) Application is Due: May 18
Contact: Corporate Contributions Specialist, (847) 646-3332; fax: (847) 646-5241; email: adickerson@kraft.com
Internet: http://www.kraftcares@kraft.com
Sponsor: Kraft Foods Inc
3 Lakes Drive
Northfield, IL 60093

Kresge Foundation Science Initiative Grants **3115**
The program's current funding goal is to improve teaching and/or research laboratories by replacing and/or upgrading their equipment and, when necessary, the facilities where the equipment is housed. The foundation will provide funding in two equal installments, the first for one-third of the equipment cost, and the second for one-fifth of new endowment funds for future updating of such equipment and facilities. It is necessary to raise funds equal to two-thirds of the equipment prior to applying. Project plans should be specific at the time of application, including firm costs of both the equipment and space renovation. The foundation will waive the minimum unit cost restriction associated with its traditional equipment grants. Organizations may submit applications for traditional bricks-and-mortar challenge grants within the same 12-month period. For details, request a pamphlet—the Kresge Foundation Science Initiative. Send for the Policies and Application Procedures brochure before applying.
Requirements: Colleges and universities, teaching hospitals, medical schools, and research institutions are eligible.
Samples: U of Evansville (IN)—to purchase equipment for the Koch Center for Engineering and Science, $228,000 challenge grant (2005); Monell Chemical Senses Ctr (Philadelphia, PA)—to begin the renovation of its facilities, $600,000 challenge grant (2005); Methodist College (Fayetteville, NC)—to expand its science building, $550,000 challenge grant (2005); Clark U (Worcester, MA)—to purchase a magnetometer and diffractometer, $205,920 challenge grant (2005).
Amount of Grant: $200,000-$800,000 average
Date(s) Application is Due: May 15

Contact: Grants Administrator, (248) 643-9630; fax: (248) 643-0588
Internet: http://www.kresge.org
Sponsor: Kresge Foundation
P.O. Box 3151, 3215 W Big Beaver Road
Troy, MI 48007-3151

Kress Foundation Art History Pre-Doctoral Fellowships 3116
The predoctoral fellowship in the history of European art supports a two-year research appointment in association with one of the following institutes: Florence, Kunsthistorishches Institut; Jerusalem, Nelson Glueck School of Biblical Archaeology, and Hebrew Union College; Leiden, Prentenkabinet/Kunsthistorisch Instituut der Rijksuniversiteit; London, Courtauld Institute of Art, University of London Warburg Institute, and University of London; Munich, Zentralinstitut f'r Kunstgeschichte; Nicosia, Cyprus American Archaeological Research Institute; Paris, American University in Paris; Rome, Bibliotheca Hertziana; and Zurich, Swiss Institute for Art Research.
Requirements: Nominees must be U.S. citizens or individuals matriculated at an American university, and must be nominated by their art history departments.
Samples: Anna Sigridur, U of Chicago (Chicago, IL)—for dissertation-writing support, $10,000.; Susan Germond, Stanford U (CA)—for dissertation-writing support, $10,000.
Amount of Grant: $22,500
Date(s) Application is Due: Nov 30
Contact: Lisa Ackerman, Executive Vice President, (212) 861-4993; fax: (212) 628-3146
Internet: http://www.kressfoundation.org/prog.html
Sponsor: Samuel H. Kress Foundation
174 E 80th Street
New York, NY 10021

Kress Foundation Art History Travel Fellowships 3117
Predoctoral candidates at U.S. universities or U.S. students abroad may apply for travel fellowships to view materials essential for the completion of dissertation research. Fifteen to 20 fellowships are awarded each year. Contact the foundation for guidelines.
Requirements: Applicants must be nominated by their art history departments.
Amount of Grant: $3500-$10,000
Date(s) Application is Due: Nov 30
Contact: Lisa Ackerman, Executive Vice President, (212) 861-4993; fax: (212) 628-3146
Internet: http://www.kressfoundation.org/kressorg/travel.html
Sponsor: Samuel H. Kress Foundation
174 E 80th Street
New York, NY 10021

Kress/ARIT Predoctoral Fellowship in the History of Art and Archaeology 3118
The fellowship supports doctoral research in Turkey in art history and archeology for one academic year. Applicants should be engaged in advanced dissertation research that necessitates a period of study in Turkey. Eligible fields include the history of art and architecture from antiquity to the present, and archeology. Awards are made for tenure of one academic year; awards for shorter periods of time also are possible. Turkish law requires all foreign scholars, prior to entering the country, to obtain formal permission for any research to be carried out in Turkey. Forms should be obtained from the Turkish Embassy, 2525 Massachusetts Avenue NW, Washington, DC 20008. Applicants are urged to apply well in advance of the time they expect to carry out their research.
Requirements: Candidacy is open to U.S. or Canadian citizens and students matriculated at U.S. or Canadian universities.
Amount of Grant: $17,000 maximum for one academic year
Date(s) Application is Due: Nov 1
Contact: President, (215) 898-3474; fax: (215) 898-0657; email: leinwand@sas.upenn.edu
Internet: http://ccat.sas.upenn.edu/ARIT/Kress-ARITFellowship.htm
Sponsor: American Research Institute in Turkey
University Museum, 33rd & Spruce Sts
Philadelphia, PA 19104-6324

KSTF Young Scholars Fellowship Program 3119
The fellowship supports early career scholars engaged in critical research in education. Applications are encouraged from scholars in all disciplines that promise to make significant scholarly contributions to areas that are consistent with improving high school mathematics and science teaching in U.S. schools via the recruitment, preparation, induction, mentoring, and/or retention of high- quality mathematics and science teachers. The fellowship is intended to support research relevant to the foundation's mission in mathematics, science, education, psychology, sociology, or other fields. Up to three fellowships per year are awarded. Application and guidelines are available online.
Requirements: The fellowship is awarded to recent, pre-tenure PhDs in research or academic positions.
Amount of Grant: $110,000; $55,000 per year for two years
Date(s) Application is Due: Nov 1
Contact: Nicole Gillespie, (856) 216-8080; fax: (856) 216-9987; email: ngillespie@kstf.org
Internet: http://www.kstf.org/research_fellowships
Sponsor: Janet H. and C. Harry Knowles Foundation
20 E Redman Avenue
Haddonfield, NJ 08033

Kulas Foundation Grants 3120
The foundation awards annual grants in the greater Cleveland, OH, area to support arts and culture—performing arts institutions, visual arts, and museums; education—music schools and conservatories and arts programs benefiting low income and minority students in Cleveland public schools; social service organizations; community-based organizations; and musical therapy research. Applicants must demonstrate plans to disseminate grant results and set clear time limits for funded activities. Requests should be submitted approximately six weeks prior to quarterly board meetings (generally held in March, June, September, and December). Guidelines are available online.
Requirements: Ohio nonprofit organizations are eligible. Preference is given to requests from Cuyahoga County or the greater Cleveland area.
Geographic Focus: Ohio
Samples: Playhouse Square Foundation (Cleveland, OH)—to help construct the Idea Center, a downtown facility that will house Ideastream, Cleveland's public-broadcasting organization, and Playhouse Square Foundation's educational programs, $600,000.
Amount of Grant: $2.5 million total
Contact: Program Contact, (216) 623-4770; fax: (216) 623-4773
Internet: http://www.fdncenter.org/grantmaker/kulas
Sponsor: Kulas Foundation
50 Public Sq, Suite 924
Cleveland, OH 44113

Kurt Weill Foundation for Music Grants Program 3121
The foundation promotes public understanding and appreciation of the musical works by Kurt Weill and solicits proposals from individuals and nonprofit organizations for funding of projects related to the perpetuation of Weill's artistic legacy. The foundation is accepting proposals in one or more of the following categories: research grants, publication assistance, dissertation fellowships, travel grants, professional performance and production grants, college and university performance and production grants, recording projects, and broadcasts.
Requirements: Music and theater scholars and professionals, as well as nonprofit organizations, are eligible to apply.
Amount of Grant: $1500-$5000 typically
Date(s) Application is Due: Nov 1
Contact: Carolyn Weber, Director of Program Administration and Business Affairs, (212) 505-5240; fax: (212) 353-9663; email: cweber@kwf.org
Internet: http://www.kwf.org/pages/granthom.html
Sponsor: Kurt Weill Foundation for Music
7 E 20th Street, 3rd Fl
New York, NY 10003

L'Oreal U.S.A Women in Science Fellowship 3122
The program supports the aspirational goals of young women interested in pursuing studies and careers in the natural sciences (biological and physical), engineering, computer science, and mathematics. Applicants must be planning to pursue their field of study or advanced research at a U.S. nonprofit institution. Five one-year grants will be offered. The fellowships may not be deferred or renewed. Candidates may obtain matching grants or supplemental fellowship grants.
Requirements: The fellowship is available to postdoctoral associates only. U.S. citizens and individuals with permanent resident status are eligible.
Amount of Grant: $20,000
Date(s) Application is Due: Oct 15
Contact: L'Oreal-UNESCO for Women in Science U.S. Fellowships; email: InfoOffice@us.loreal.com
Internet: http://www.lorealusa.com/fwis/2006_fellowship.aspx
Sponsor: L'Oreal U.S.A
575 Fifth Avenue
New York, NY 10017

Lady Davis Fellowships 3123
Fellowships of three to nine months make the cultural heritage of ancient and modern Israel widely known to people from both technologically advanced and evolving societies. The program enables scholars of various disciplines and nationalities to study, perform research, or teach at the Hebrew University of Jerusalem or at the Technion-Israel Institute of Technology, Haifa. Types of support include postdoctoral fellowships visiting professorships, and graduate studies abroad. Prospective candidates for Hebrew University should contact the university to ensure that there is a faculty member willing to sponsor the application. Fellows will be selected on the basis of demonstrated excellence in their studies, promise of distinction in their chosen fields of specialization, as well as on qualities of mind, intellect, and character. The graduate fellowship includes tuition, travel, and a stipend to meet reasonable living expenses; the postdoctoral fellowship includes travel and a larger stipend to meet reasonable living expenses; and the visiting professorship includes travel and stipend.
Requirements: Citizenship is unrestricted. Only students who are enrolled in a PhD program overseas are eligible applicants for the graduate fellowship at the Hebrew University. Applicants for the Technion must have completed their studies with excellent marks.
Contact: M. Mark Sopher, 972-2-658-4723; fax: 972-2-566-3848; email: LDFT@vms.huji.ac.il
Internet: http://ldft.huji.ac.il
Sponsor: Lady Davis Fellowship Trust
Hebrew University, Givat Ram
Jerusalem 91904 Israel

Lady Gregory Postgraduate Fellowship 3124

The Faculty of Arts at NUI Galway awards three postgraduate fellowships to support PhD research within the Arts Faculty. Fellowships are awarded on a competitive basis, one in each of the general academic categories of humanities, languages and literatures, and social and behavioral sciences. Areas of study include: archeology, classics, education, English, French, Gaeilge, geography, German, history, Italian, philosophy, political science and sociology, psychology, Spanish and the Centers for Human Settlement and Historical Change, Irish studies, and the Houston School of Film and Digital Media. The fellowship, which begins in the fall, is tenable for a maximum of three years. Guidelines and application are available online.

Requirements: To be considered for a fellowship, the applicant must be the holder of, or expect to attain before July, a first-class honors degree (or equivalent) from a recognized institution in a relevant discipline. Fellowships will normally be awarded to individuals embarking on the first year of their PhD program. The applicant must also have an established proficiency in English or Irish languages. For applicants whose first language is not English or Irish, a language proficiency assessment may be applied.

Amount of Grant: Euro 20,000

Date(s) Application is Due: Mar 3

Contact: Mairead Ni Dhomhnaill, Office of Faculty of Arts; +353-91-493622; fax: +353-91-495508; email: mairead.nidhomhnaill@nuigalway.ie

Internet: http://www.nuigalway.ie/arts/ladygregory.html

Sponsor: National University of Ireland Galway
University Road
Galway
Ireland

Laidlaw Foundation Grants 3125

The foundation awards grants to registered charitable organizations in Canada in its areas of interest, including the development of health, well-being, and social justice services and the performing arts in Canada; to support child development programs and community services in improving the opportunity and life quality of children, particularly those who are socially and economically disadvantaged; to support the performing arts in enriching the quality, vitality, and diversity of cultural and artistic expression; and to balance the need for environmental preservation with economic development. Proposals are reviewed quarterly; applications are accepted at anytime.

Requirements: Canadian nonprofit organizations are eligible.

Contact: Grants Administrator, (416) 964-3614; fax: (416) 975-1428; email: mail@laidlawfdn.org

Internet: http://www.laidlawfdn.org

Sponsor: Laidlaw Foundation
365 Bloor Street E, Suite 2000
Toronto, ON M4W 3L4 Canada

Laird Norton Endowment Foundation Grants 3126

The foundation supports programs in conservation and forestry education. It welcomes proposals to develop sustainable forestry certification and audit standards, promote forest and owner education, study restoration forestry, encourage K-12 sustainable forestry education, provide forestry-related internships for college students, and foster ecosystem-based approaches to forestry. Types of support include curriculum development, equipment, internship funds, matching/challenge grants, program development, publication, research, and seed money. Fifty to 60 percent of recent grants have been made to organizations in the Pacific Northwest. Deadlines are for preliminary letters of inquiry.

Requirements: Nonprofits, including schools and other institutions, that serve K-12 teachers and students are eligible.

Restrictions: The foundation does not fund capital campaigns, annual drives, endowments, building programs, scholarship funds, general operating expenses, international projects, or efforts for specific animals.

Samples: Community Forestry Resource Center/IATP (Minneapolis, MN)—to help support the long-term health of private forests, $20,000. New England Forestry Foundation (Orange, MA)—Landowner Outreach Program in the North Quabbin region to make certification accessible to small private landowners and assist the community in developing market strategies for certified forest products, $14,700.; Ecotrust (Portland, OR), $10,000.; Southwest Ctr for Biological Diversity (Tucson, AZ)—for the Southwestern Forests Reforestation program, $8000.

Amount of Grant: $7500–$20,000; $15,000 average

Date(s) Application is Due: Feb 4; Sep 9

Contact: Bruce Reed, Program Director, (206) 464-5242; fax: (206) 464-5099; email: reed@lairdnorton.org

Internet: http://www.lairdnorton.org/#ApplicationInfo

Sponsor: Laird Norton Endowment Foundation Grants
801 Second Avenue, Suite 1300
Seattle, WA 98104-1581

Lalor Foundation Postdoctoral Research Grants 3127

Grants are awarded to institutions for basic postdoctoral research in mammalian reproductive physiology and biochemistry yielding improved methods of contraception, termination of pregnancy, and/or sterilization. Funds may be used to cover fellowship stipends, institution overhead, and laboratory and miscellaneous expenses. Less money is awarded for projects requiring less than one year.

Requirements: Applications must come from tax-exempt institutions; individuals nominated must have training and experience at least equal to the PhD, or MD level. People who have held the doctoral degree less than five years are preferred.

Amount of Grant: $35,000 maximum

Date(s) Application is Due: Jan 13

Contact: Pamela Labonte Maksy, (617) 426-7080 ext 314; fax (617) 426-5441; email: pmaksy@grantsmanagement.com

Internet: http://www.lalorfound.org/guidelines.html

Sponsor: Lalor Foundation
77 Summer Street
Boston, MA 02110-1006

LAM Foundation Established Investigator Awards 3128

The award supports research directly related to lymphangioleiomyomatosis (LAM) and is targeted at established investigators who focus on the abnormal proliferation of smooth muscle that occurs in the disease. Awards provide a maximum of $50,000 per year, renewable for up to two additional years. The structure and terms of this award are identical to the LAM Fellowship Award except that with the LAM Established Award, faculty level investigators are eligible to receive funding for technician support and supplies.

Requirements: U.S. citizens and foreign nationals with appropriate immigrant visa status are eligible. At the time of application, an applicant must hold a doctoral degree and faculty appointment at the level of assistant or associate professor and be undertaking a project related to LAM.

Date(s) Application is Due: Apr 15; Sep 15

Contact: Francis X. McCormack, MD, Director; (513) 558-4831; fax: (513) 558-4858; email: frank.mccormack@uc.edu

Internet: http://www.thelamfoundation.org/research1/research-funding.html&searched=Awards&highlight=ajaxSearch_highlight+ajaxSearch_highlight1

Sponsor: LAM (Lymphangioleiomyomatosis) Foundation
MSB Room 6053 231 Albert Sabin Way
Cincinnati, OH 45267-0564

LAM Foundation Pilot Project Awards 3129

Pilot project awards are granted for one year to enable investigators to gather sufficient preliminary data for more substantial LAM funding. Funding is intended to support faculty-level investigators, or postdoctoral-level investigators who will be performing LAM research in the laboratory of an established scientist who is an expert in areas that are directly pertinent to LAM. Letters of intent (LOI) are required prior to submission of grant proposals to the LAM Foundation. Application materials and details are available online.

Requirements: Applications are accepted from investigators of all nationalities who have MDs or PhDs, but candidates must possess visas that allow for completion of the proposed project in the original laboratory.

Amount of Grant: $25,000 maximum

Date(s) Application is Due: Apr 15; Sep 15

Contact: Francis X. McCormack, MD, Director; (513) 558-4831; fax: (513) 558-4858; email: frank.mccormack@uc.edu

Internet: http://www.thelamfoundation.org/research1/research-funding.html&searched=Awards&highlight=ajaxSearch_highlight+ajaxSearch_highlight1

Sponsor: LAM (Lymphangioleiomyomatosis) Foundation
MSB Room 6053 231 Albert Sabin Way
Cincinnati, OH 45267-0564

LAM Foundation Postoctoral Fellowships 3130

The foundation offers postdoctoral fellowships for the study of the cellular and molecular basis of the abnormal smooth muscle proliferation that occurs in the disease, Lymphangioleiomyomatosis (LAM). Fellowships are renewable for up to two additional years. The fellowship is intended to support postdoctoral level investigators who will be performing LAM research in the laboratory of an established scientist who is an expert in areas that are directly pertinent to LAM. Letters of intent (LOI) are required prior to submission of grant proposals to the LAM Foundation.

Requirements: Applications are accepted from investigators of all nationalities who have MDs or PhDs, but candidates must possess visas that allow for completion of the proposed project in the original laboratory.

Restrictions: The award is generally not available to individuals who have attained faculty rank, or who are seeking transition funds for the establishment of a laboratory in a new setting.

Amount of Grant: $50,000 maximum

Date(s) Application is Due: Apr 15; Sep 15

Contact: Francis X. McCormack, MD, Director; (513) 558-4831; fax: (513) 558-4858; email: frank.mccormack@uc.edu

Internet: http://www.thelamfoundation.org/research1/research-funding.html&searched=Awards&highlight=ajaxSearch_highlight+ajaxSearch_highlight1

Sponsor: LAM (Lymphangioleiomyomatosis) Foundation
MSB Room 6053 231 Albert Sabin Way
Cincinnati, OH 45267-0564

LAM Foundation Research Grants 3131

The program provides seed money to investigators seeking to understand the abnormal smooth muscle proliferation that occurs in lymphangioleiomyomatosis (LAM). The grant supports clinical, laboratory, epidemiological, or any other kind of research that directly relates to LAM. Grants are subject to annual review and may be granted for up to two years. Scientists who which to apply for LAM Foundation funding must submit a Letter of Intent (LOI). Letters of Intent must reach The LAM Foundation by July 30th.

Requirements: U.S. citizens and foreign nationals with appropriate immigrant visa status are eligible. At the time of application, an applicant must hold a doctoral degree and faculty appointment with an academic institution and have completed two years of research training.
Restrictions: Residents, interns, postdoctoral fellows, students enrolled in degree-granting programs, and established investigators are ineligible.
Amount of Grant: $40,000 maximum per year
Date(s) Application is Due: Sep 15
Contact: Francis X. McCormack, MD, Director; (513) 558-4831; fax: (513) 558-4858; email: frank.mccormack@uc.edu
Internet: http://www.thelamfoundation.org/research1/research-funding.html&searched=Awards&highlight=ajaxSearch_highlight+ajaxSearch_highlight1
Sponsor: LAM (Lymphangioleiomyomatosis) Foundation
MSB Room 6053 231 Albert Sabin Way
Cincinnati, OH 45267-0564

Lance Armstrong Foundation Research Grants 3132
The foundation funds and disseminates research that builds on the body of knowledge and services focused on improving the quality of life for cancer survivors across physical, emotional, and practical challenges. The research program also supports research focused on testicular cancer. The foundation funds research that is not readily fundable from traditional sources and that encourages and supports the efforts of both young and established investigators. Young investigators are defined as within eight years of completing a terminal degree or within five years of initiating independent research within a mentored laboratory, while general research awards support new research projects initiated by established investigators. Proposals are accepted from all countries. The listed application deadline is for letters of intent; full RFPs are available online.
Amount of Grant: $50,000 maximum young investigator research awards; $75,000 maximum research awards
Date(s) Application is Due: May 13
Contact: Amy Nunn, Grants Coordinator, (512) 236-8820 ext 181; email: amy.nunn@laf.org or research.grants@laf.org
Internet: http://www.livestrong.org/site/c.jvKZLbMRIsG/b.739079/k.9EDA/Research_Grants.htm
Sponsor: Lance Armstrong Foundation
P.O. Box 161150
Austin, TX 78716-1150

Landon Foundation-AACR Innovator Award for International Collaboration 3133
This Award will support highly meritorious research that is being conducted collaboratively by investigators in different countries around the world. The goals of the program are to: promote international cancer research collaboration as an effective means to accelerate progress against cancer; provide the support necessary to sustain and enhance highly meritorious international cancer research collaborations; foster interactions between and among cancer scientists and disseminate the scientific knowledge gained from international collaboration; and contribute to a global impact against cancer. The Award of $100,000 will be provided to the recipients' institutions in the form of a two-year grant to support the collaborative research project, and may be used for such purposes including, but not limited to: direct research expenses attributable to the proposed research, which may include the salary of research assistants or technicians; expenses for travel to the institutions of the collaborators relevant to the research project; expenses for training in new techniques that will benefit the research project, such as training received through a scholar exchange/visiting professor program; and expenses related to the presentation of research data at scientific meetings or through other means that will contribute to the dissemination of the scientific knowledge gained.
Requirements: Two or more independent researchers who working within an established international cancer research collaboration involving institutes in multiple countries may apply. Proposals to initiate a collaborative project will not be accepted. The co-investigators may be affiliated with any institution involved in cancer research, cancer medicine, or cancer-related biomedical science anywhere in the world. There is no limit to the number of co-investigators within each of the collaborating institutes. There are no national or residency status restrictions. Applicants need not be members of the AACR. Applications must be completed online using the proposalCENTRAL website, with one paper copy submitted to the AACR office. Online applications must be completed by 12:00 noon (United States Eastern Daylight Time) on November 20. The paper copy with original signatures and all supporting documents must be postmarked no later than November 25.
Restrictions: Scientific investigators or health professionals who are funded by the tobacco industry for any project, or whose named mentors in the case of mentored grants are funded by the tobacco industry for any project, may not apply and will not be eligible for AACR grants.
Amount of Grant: two-year grant $100,000 ($50,000 per year)
Date(s) Application is Due: Nov 20; Nov 25
Contact: Hanna Hopfinger, Grants Program Assistant; (267) 646-0665; fax: (215) 440-9372; email: hanna.hopfinger@aacr.org
Internet: http://www.aacr.org/home/scientists/research-funding—fellowships/landon-aacr-innovator-award-for-international-collaboration.aspx
Sponsor: American Association for Cancer Research
615 Chestnut Street, 17th Floor
Philadelphia, PA 19106-4404

Landon Foundation-AACR Innovator Award for Cancer Prevention Research 3134
This Award was established and first given in 2008 to recognize the outstanding achievement of an early career Assistant Professor, and to provide support for cancer prevention research of significant scientific merit in any discipline across the continuum of research. The goals of the program are to: encourage younger investigators to pursue cancer prevention research of significant scientific merit; provide the support necessary to sustain and enhance highly meritorious cancer prevention research; foster interactions between and among cancer scientists and disseminate the scientific knowledge about cancer prevention research; and contribute to a global impact against cancer. A two-year grant of $100,000 ($50,000 per year) will be provided to the recipient's institution to support direct research expenses; salary and benefits; and attendance at the AACR Annual Meeting, Cancer Prevention Conference, or other AACR meeting relevant to cancer prevention for the purpose of participating in scholarly exchange.
Requirements: Candidates must be scientists at the level of Assistant Professor who completed postdoctoral studies or clinical fellowships on or after July 1. Candidates cannot be tenured or under consideration for tenured academic positions at the time of application. A candidate whose title is not Assistant Professor but who believes that he or she meets the eligibility requirements for this Award must attach a letter from the institution's dean certifying that the position held is equivalent to Assistant Professor. Both AACR members and nonmembers are eligible to apply. However, nonmembers must submit a satisfactory application for AACR Active membership on or before October 31. Applications must be submitted by 12:00 noon (United States Eastern Daylight Time) on November 20, using the proposalCENTRAL website at http://proposalcentral.altum.com. One paper copy with original signatures and all supporting documents must be mailed to the AACR office, and be postmarked no later then November 25.
Restrictions: Employees or subcontractors of a national government or private industry are not eligible for this grant.
Amount of Grant: two-year grant $100,000 ($50,000 per year)
Date(s) Application is Due: Nov 20; Nov 25
Contact: Julia Laurence, Program Assistant; (267) 646-0655; fax: (215) 440-9372; email: julia.laurence@aacr.org or awards@aacr.org
Internet: http://www.aacr.org/home/scientists/research-funding—fellowships/landon-aacr-innovator-award-for-cancer-prevention-research.aspx
Sponsor: American Association for Cancer Research
615 Chestnut Street, 17th Floor
Philadelphia, PA 19106-4404

Larry L. Hillblom Foundation Grants 3135
The mission of the foundation is to provide philanthropic support exclusively for charitable, religious, scientific, literary, and educational purposes. Primarily, the foundation will support medical research, including basic scientific research, clinical research, and research related to patient self-care and management. Medical research priorities include the areas of diabetes mellitus, diseases associated with aging, and children's physical and neurological diseases. Specific grants include building/institute grants, start-up research grants, network grants, fellowships, endowed chairs, and clinical research grants.
Samples: U of California (Oakland, CA)—to endow a chair in geriatric medicine at the U of California, San Diego, in La Jolla, $2 million (2003); U of California at San Francisco (CA)—to create a center for research on the basic biology of aging, and to support research on diabetes, eye disorders, and neurodegenerative diseases, $12 million.
Amount of Grant: $250,000-$10 million over 7 years
Date(s) Application is Due: Jan 14
Contact: Administrator, (707) 762-6691; fax: (707) 762-6694; email: som@llhf.org
Internet: http://www.llhf.org
Sponsor: Larry L. Hillblom Foundation
755 Baywood Drive, Suite 180
Petaluma, CA 94954

LASPAU Grants and Fellowships 3136
LASPAU is a nonprofit organization affiliated with Harvard University that designs, develops, and implements academic and professional exchange programs on behalf of individuals and institutions in the United States, Antigua and Barbuda, Argentina, the Bahamas, Barbados, Belize, Bolivia, Brazil, Canada, Chile, Columbia, Costa Rica, Dominica, Dominican Republic, Ecuador, El Salvador, Grenada, Guatemala, Guyana, Haiti, Honduras, Jamaica, Mexico, Nicaragua, Panama, Paraguay, Peru, Saint Kitts/Nevis, Saint Lucia, Saint Vincent and the Grenadines, Suriname, Trinidad and Tobago, Uruguay, and Venezuela. LASPAU administers scholarship and loan programs for a wide variety of sponsoring agencies. However, the organization itself does not offer scholarships, fellowships, or loans. Interested parties should apply directly to funding agencies and may contact those agencies to request information on criteria for eligibility and other application-related information. Information is available online.
Contact: Program Officer for Ecology Programs, (617) 495-5255; fax: (617) 495-8990; email: laspau@harvard.edu
Internet: http://www.laspau.harvard.edu/grant-LASPAU.htm
Sponsor: LASPAU: Academic and Professional Programs for the Americas
25 Mount Auburn Street
Cambridge, MA 02138-6095

Laura B. Vogler Foundation Grants 3137
The foundation supports innovative programs and projects in New York City and Long Island in the areas of education, health care, and social services. The foundation is particularly interested in organizations that serve and support children, the elderly, and the disadvantaged. Types of support include general operating support, program

development, research grants, and seed money grants. Grants provide one-time, nonrenewable support.
Requirements: Nonprofit organizations in New York City and Long Island, NY, may submit proposals.
Restrictions: Grants are not awarded to support building or endowment funds, annual fund-raising campaigns, or matching gifts.
Geographic Focus: New York
Samples: Shield Institute (NY)—for a joint activities program for preschoolers and the elderly, $2500.; Mercy Ctr Ministries Inc (NY)—to support three shelters for homeless young women and parenting teens, $3500.
Amount of Grant: $1000-$5000
Date(s) Application is Due: Mar 1; Jul 1; Nov 1
Contact: Lawrence D'Amato, President, (718) 423-3000; fax: (718) 631-4808
Internet: http://fdncenter.org/grantmaker/vogler
Sponsor: Laura B. Vogler Foundation
P.O. Box 610508
Bayside, NY 11361-0508

Laura J. Niles Foundation Grants **3138**
The foundation's mission is to encourage and support efforts that offer learning and economic growth opportunities for the motivated poor. Of equal importance are charitable initiatives that foster life enrichment through canine and other types of animal companionship. The foundation has a particular interest in education, economic self-sufficiency, and programs that alleviate unhealthy dependencies. With regard to animals, most notably dogs, the foundation's areas of concentration center around canine health research, animal protection and adoption, search and rescue training, human assistance, and similar fields of interest. A most favorable outcome is one where both people and animals benefit simultaneously from an initiative. Applications are accepted throughout the year. Guidelines are available online.
Requirements: 501(c)3 nonprofit organizations are eligible. The majority of grants are awarded in the northeastern United States, although, occasionally, grants may be made in other regions of the country and/or abroad.
Restrictions: Grants do not support general fund-raising drives, endowments, government agencies, or organizations that subsist on third-party funding.
Amount of Grant: $2000-$50,000
Contact: Rose Anna Miller, (203) 629-7314; email: rmiller@fcsn.com
Internet: http://www.ljniles.org/index.htm
Sponsor: Laura J. Niles Foundation
88 Field Point Road
Greenwich, CT 06830

Law and Social Science Doctoral Dissertation Fellowships **3139**
The purpose of this fellowship program is to encourage original and significant research on law and legal institutions. Applications are invited from outstanding students who are candidates for PhD degrees in the social sciences. Proposed dissertation research must be in the general area of socio-legal studies or in social scientific approaches to law, the legal profession, or legal institutions. The dissertation must address critical issues in the field and show promise of a major contribution to social scientific understanding of law and legal processes.
Requirements: Candidates must have completed all doctoral requirements except the dissertation by September 1 of the application year.
Amount of Grant: $30,000 stipend for 12 months; $3500 research support including travel to conferences; $1000 relocation expenses
Contact: Grants Administrator, (312) 988-6513; email: fellowships@abfn.org
Internet: http://www.abf-sociolegal.org/abffellow.html
Sponsor: American Bar Foundation
750 N Lake Shore Drive
Chicago, IL 60611

Law and Social Science Postdoctoral Fellowships **3140**
The purpose of this fellowship program is to encourage original and significant research on law and legal institutions. Applications are invited from junior scholars who have completed all requirements for the PhD within the past two years, or who are currently in the final stages of completing the dissertation. In exceptional circumstances, candidates with a JD who have substantial social science training may also be considered. Minority candidates are especially encouraged to apply.
Requirements: Applications must include a letter of application; a statement describing research interests and achievements to date and plans for the fellowship period; two letters of reference; a curriculum vita; and a sample of written work (conference paper, dissertation chapter, or published article).
Amount of Grant: $30,000 for 12 months, plus $3500 maximum research support and $1000 relocation expenses
Date(s) Application is Due: Feb 1
Contact: Kristine Martens, Program Associate, (312) 988-6523; fax: (312) 988-6579; email: kmartens@abfn.org
Internet: http://www.abf-sociolegal.org/Fellowship/pre.html
Sponsor: American Bar Foundation
750 N Lake Shore Drive
Chicago, IL 60611

Lawrence C. Gallen Teaching Fellowship **3141**
Gallen Teaching fellows will teach three courses a semester of the Augustine and Culture Villanova Seminar in the College of Arts and Sciences Core Humanities Program. The Core Humanities Program is a writing-intensive interdisciplinary seminar program required of all first-year students. Besides teaching responsibilities, Gallen fellows are expected to share in the life of the department and participate in the department's many activities, including curriculum and faculty development, as well as other committee and/or administrative responsibilities. In addition, all Gallen fellows are expected to contribute to Villanova's residential learning communities. Candidates must submit unofficial undergraduate and graduate school transcripts, three letters of recommendation, a writing sample, and a curriculum vita. Each fellowship is a two-year renewable fellowship. Three fellowships are available. Guidelines are available online.
Requirements: Candidates for these positions must have a PhD in a Humanities discipline, a strong commitment to undergraduate teaching in the humanities, and an interest in teaching the thought of Augustine and its influence. Candidates with an interest in Hebrew Scriptures and the New Testament are also encouraged to apply.
Contact: John A. Doody, Robert M. Birmingham Chair; (610) 519-4691 or (610) 519-7325; email: john.doody@villanova.edu
Internet: http://www.villanova.edu/artsci/vcle/fellowships/gallen.htm
Sponsor: Villanova University
St Augustine Center Liberal Arts Room 104, 800 Lancaster Avenue
Villanova, PA 19085

Lawrence Fellows Program **3142**
The postdoctoral program is designed for recent PhDs (or M.Mus/MFAs) who seek to develop a record of excellence in teaching and research experience in a liberal arts college setting. Initial appointments will be for two years and will involve a reduced teaching load so that fellows have time to engage in tutorials and research projects with undergraduate students. The full-time fellowships provide an annual salary, plus benefits, and a program fund to support research, travel, and other costs. Applicants should send a letter of interest, curriculum vita, and three letters of recommendation.
Requirements: The program is designed for recent PhDs (or M.Mus/MFAs). Applications are accepted from qualified applicants in any discipline and interdisciplinary program offered at the university.
Amount of Grant: $32,000-$36,000 stipend plus $5000 for program activities
Date(s) Application is Due: Mar 15
Contact: Office of the President, (920) 832-7000
Internet: http://www.lawrence.edu
Sponsor: Lawrence University
P.O. Box 599
Appleton, WI 54912

Lawrence K. Cecil Award in Environmental Chemical Engineering **3143**
This award is given to recognize outstanding chemical engineering contributions and achievements toward the preservation or improvement of the natural environment. There is no age limit. The award may be based on new discoveries, research, or the development of new processes or equipment relating to chemical engineering applications in the protection of the environment; outstanding contribution of a chemical engineering nature in the design, construction, operation, or management of environmental protection facilities or enterprises; or distinguished service of a chemical engineering nature in the protection of the environment as a professional engineer, industrial leader, or educator.
Requirements: All members of AIChE as well as other interested persons are urged to nominate deserving candidates. All information and supporting documents are to be included with the completed form; nominating forms are available from the institute.
Amount of Grant: $2500
Date(s) Application is Due: Feb 15
Contact: Awards Program, (212) 591-7107; fax: (212) 591-8890; email: awards@aiche.org
Internet: http://www.aiche.org/awards/awarddtl.asp?AwardID=32
Sponsor: American Institute of Chemical Engineers
3 Park Avenue
New York, NY 10016-5901

LBI David Baumgardt Memorial Fellowships **3144**
The fellowship is intended to assist scholars whose research projects are connected with the writings of Professor David Baumgardt or his scholarly interests in ethics and the modern intellectual history of German-speaking Jewry. The fellowship, in principal, requires extensive use of the facilities of the LBI in New York and particularly the David Baumgardt Collection. The fellowship pays a stipend based on the requirements of the particular project being conducted. Support for travel or family members is not available.
Requirements: Doctoral candidates and academics affiliated with an accredited institution of higher education are eligible to apply.
Amount of Grant: $3000 maximum
Date(s) Application is Due: Nov 1
Contact: Ellen Musso, (212) 744-6400; fax: (212) 988-1305; email: emusso@lbi.cjh.org
Internet: http://www.lbi.org/baumgardt.html
Sponsor: Leo Baeck Institute
15 W 16th Street
New York, NY 10011

LBI Fred Grubel Fellowship **3145**
This is a paid summer internship program for a graduate student who will participate in work on a specific research topic (jointly determined by the candidate and the LBI) related to LBI collections, which can include archives, library, photo collection, and art collection. The research project should pertain to the lives of refugees of the 1930s and

1940s in New York. The fellow will be supervised by the director of research and will work on a day-to-day basis with archives and library staff. The deadline for applications is the fall of the year prior to the summer internship. Decisions will be made in early spring. The application form is available online.
Requirements: PhD candidates from history, sociology, literature, or Jewish studies programs are eligible.
Amount of Grant: $1500 per month
Date(s) Application is Due: Nov 1
Contact: Ellen Musso, Executive Director, (212) 744-6400; fax: (212) 988-1305; email: emusso@lbi.cjh.org.
Internet: http://www.lbi.org/grubel.html
Sponsor: Leo Baeck Institute
15 W 16th Street
New York, NY 10011

LBI Fritz Halbers Fellowship 3146
The Leo Baeck Institute awards this fellowship to assist scholars whose projects are connected with the culture and history of German-speaking Jewry. The amount of the award will be determined according to the requirements of the project and the availability of funds. The fellowship should normally be taken up by the end of the year it is granted. Support for travel or family members is not available. Contact the institute for applications.
Requirements: Applicants must be students enrolled in a PhD program at an accredited institution of higher education.
Amount of Grant: $3000 maximum
Date(s) Application is Due: Nov 1
Contact: Ellen Musso, Executive Director, (212) 744-6400; fax: (212) 988-1305; email: emusso@lbi..cjh.com
Internet: http://www.lbi.org/halbers.html
Sponsor: Leo Baeck Institute
15 W 16th Street
New York, NY 10011

LBI/DAAD Fellowships in German-Jewish History and Culture 3147
The German Academic Exchange Service (DAAD) and the Leo Baeck Institute (LBI) offer two fellowships per year for doctoral students or recent PhDs affiliated with accredited U.S. institutions of higher education to research the social, communal, and intellectual history of German-speaking Jewry. The intent of the fellowship is to assist in the preparation of a scholarly essay or book through the use of the LBI's resources in New York. Normally, the fellowship must be used within one calendar year. Support for travel or family members is not available. Fellowship applications must be submitted in writing. Application materials are available from the institute.
Requirements: Applicants must be U.S. citizens. PhD candidates may not be older than 32; individuals who recently received the PhD may not be older than 35. A complete application consists of a curriculum vita, official transcripts (graduate and undergraduate work), two letters of recommendation from scholars in the field, and a full description of the research project.
Amount of Grant: $2000 paid in two equal installments
Date(s) Application is Due: Nov 1
Contact: Ellen Musso, (212) 744-6400; fax: (212) 988-1305; email: emusso@lbi.cjh.org
Internet: http://www.lbi.org/daad.html
Sponsor: Leo Baeck Institute
129 E 73rd Street
New York, NY 10021

Leadership in Education in Neurodevelopmental Disabilities Postdoctoral Fellowships 3148
Munroe-Meyer Institute, an administrative unit of the Nebraska Medical Center, offers postdoctoral fellowships supported by the Maternal and Child Health Bureau of the Health Resources and Services Administration. Fellowships are available in administration, nursing, nutrition, occupational therapy, pediatrics, physical therapy, psychology, social work, and speech language pathology. The program provides interdisciplinary training for careers and leadership roles in developmental disabilities and children with special needs. The applicant is encouraged to contact the home department director prior to submitting materials to determine if applicant interest is in line with department mission, goals, and objectives. The recruitment/selection committee meets every two weeks, beginning in January, to review materials. Applicants offered an interview will be scheduled to provide a scholarly presentation to the personnel and students within the institute.
Amount of Grant: $18,600 minimum
Contact: Patricia Gromak, Munroe-Meyer Institute, (800) 656-3937 or (402) 559-6402; fax: (402) 559-5737
Internet: http://www.unmc.edu/dept/mmi/index.cfm?L2_ID=60&L1_ID=17&CONREF=14
Sponsor: University of Nebraska Medical Center
985450 Nebraska Medical Ctr
Omaha, NE 68198-5450

Leahi Fund Grants 3149
The purpose of the Fund is to support programs of research and education in, and the prevention of, pulmonary disease. Highest priority will be given to: research and education with foreseeable clinical application to pulmonary problems that exist in Hawai'i; scientific research in the prevention or treatment of pulmonary diseases and infirmities. Preference will be given to supporting new investigators and to new areas of investigation for established researchers. Application information is available online.
Requirements: Research grants can only be approved for individuals affiliated with tax-exempt organizations classified by the Internal Revenue Services as 501(c)3 type, or with units of government.
Restrictions: Funding is not available for: salary for the principal investigator; patient care; travel; major equipment purchases; printing or publication costs.
Geographic Focus: Hawaii
Amount of Grant: $30,000
Date(s) Application is Due: Jul 16
Contact: Amy Luersen, (808) 566-5550 or (888) 731-3863; email: aluersen@hcf-hawaii.org
Internet: http://www.hawaiicommunityfoundation.org/index.php?id=71&categoryID=23
Sponsor: Hawaii Community Foundation
1164 Bishop Street, Suite 800
Honolulu, HI 96813

Leakey Foundation Franklin Mosher Baldwin Fellowships 3150
The fellowships are intended for scholars and students with citizenship in an African country who wish to obtain an advanced degree or specialize training in an area of study related to human origins research. This award is for a program of approved special training and/or advanced training toward an MA, PhD, or equivalent and is limited to two years of support. Fellowships may be used to cover travel, living expenses, student fees, insurance, books, and other relevant categories of expenditure. Guidelines and application are available online.
Requirements: Applicants must be prepared to demonstrate affiliation and/or employment with an institution in their home country; provisional acceptance (or evidence of application) to the host institution; financial assistance from the host institution; and intention to return and work in the home country upon completion of training.
Amount of Grant: $12,000 per year
Date(s) Application is Due: Feb 15
Contact: Grants Officer, (415) 561-4646; fax: (415) 561-4647; email: grants@leakeyfoundation.org
Internet: http://www.leakeyfoundation.org/grants/g3.jsp
Sponsor: Leakey Foundation
P.O. Box 29346, 1002A O'Reilly Avenue
San Francisco, CA 94129-4646

Leakey Foundation General Grants 3151
The foundation supports further research into human origins, behavior, and survival, and research related to human evolution. In recent years, priority has been given to research into the ecology, archaeology, and human paleontology of the Miocene, Pliocene, and Pleistocene; the behavior of great apes and other Old World primate species; and the ecology and adaptations of living hunter-gatherer peoples. The foundation recognizes that not all research related to human evolution falls into these categories, and other areas of study have been funded occasionally. Priority is normally given to the exploratory phase of promising new projects enabling researchers to exploit novel opportunities in establishing programs that most closely meet the stated purpose of the foundation. If grants are likely to be required in future years for long-term projects, some estimate of future requirements should be included in the initial application. Where research is to be undertaken abroad, evidence must be submitted that permission from the appropriate government agencies has been secured or sought.
Requirements: Grants are normally made to scientists with professional qualifications and demonstrated capability. Applications will be accepted from graduate students and PhDs.
Restrictions: High school, college, and lower level graduate students are not eligible.
Samples: Maria Cords (Kenya)—to support research on the Blue Monkeys in the Kakamega Forest, $7200.; Walter Hartwig (Brazil)—to study large-bodied primates during the Pleistocene in Brazil, $3450.
Amount of Grant: $3000-$13,000
Date(s) Application is Due: Jul 15; Dec 15
Contact: Grants Officer, (415) 561-4646; fax: (415) 561-4647; email: grants@leakeyfoundation.org
Internet: http://www.leakeyfoundation.org/grants
Sponsor: L.S.B. Leakey Foundation
P.O. Box 29346, 1002A O'Reilly Avenue
San Francisco, CA 94129-0346

Leakey Foundation Human Origins Research Grants 3152
The foundation was formed to further research into human origins. Recent priorities include research into the environments, archeology, and human paleontology; into the behavior, morphology, and ecology of the great apes and other primate species; and into the behavioral ecology of contemporary hunter-gatherers. Other areas of study are rarely considered. General research grants are awarded twice annually and fund the exploratory phases of promising new research projects that meet the stated purpose of the foundation. The program makes awards for projects of one year or less in length.
Requirements: Advanced doctoral students as well as established scientists are eligible for general research grants. Eligible applicants must either hold a PhD or equivalent qualification in anthropology or a related discipline; or be enrolled in a doctoral program with all degree requirements fulfilled other than the dissertation. There are no citizenship restrictions.

Restrictions: High school and undergraduate college students are not eligible.
Amount of Grant: $3000-$13,500 average; $22,000 maximum postdoctoral and senior scientist level
Date(s) Application is Due: Jul 15; Dec 15
Contact: Grants Officer, (415) 561-4646; fax: (415) 561-4647; email: grants@leakeyfoundation.org
Internet: http://www.leakeyfoundation.org/grants/g2.jsp
Sponsor: Leakey Foundation
P.O. Box 29346, 1002A O'Reilly Avenue
San Francisco, CA 94129-4646

Leakey Foundation Paleoanthropology Award **3153**
The grant is intended for potentially long-term, multidisciplinary research programs that seek to recover the physical and/or cultural remains of early humans and their hominid ancestors. All applicants must make an explicit argument for how their research is related to understanding human evolution. Usually awarded to senior scientists, these grants are potentially renewable for up to three years.
Requirements: Usually awarded to established scientists.
Amount of Grant: $22,000 maximum per year
Date(s) Application is Due: Jul 15; Dec 15
Contact: Programs/Grants Officer, (415) 561-4646; fax: (415) 561-4647; email: grants@leakeyfoundation.org
Internet: http://www.leakeyfoundation.org/grants.htm
Sponsor: L.S.B. Leakey Foundation
P.O. Box 29346, 1002A O'Reilly Avenue
San Francisco, CA 94129-0346

Learning and Behavioral Disorders Research Grants **3154**
The foundation's purpose is to provide funds for research and dissemination of research information which targets the proving, diagnosing, or treating of biologically-based disordered behavior. Areas of special interest include nutrition, food and chemical intolerance, neurometrics, neurochemistry, and genetics. Studies ascertaining validity of innovative biomedical therapies are especially encouraged Initial contact from parties interested in funding an appropriate research project should be by letter which briefly explains the proposed project and states approximate cost. A researcher need not have a doctoral degree to apply.
Amount of Grant: $1000-$50,000
Contact: John Wacker, President, (214) 373-3308; fax: (214) 373-3308
Sponsor: Wacker Foundation
10848 Strait Ln
Dallas, TX 75229

Lee and Ramona Bass Foundation Grants **3155**
The foundation makes grants in Texas in the following categories: schools, colleges, and universities, with emphasis on faculty development and liberal arts programs; community programs and projects, particularly related to the arts and the environment, such as museums, zoos, and educational/research institutions; and national and regional conservation programs. Types of support include general operating support, building construction/renovation, endowment funds, and curriculum development. Preliminary inquiries are requested, in the form of a letter briefly describing the project or program, prior to the filing of formal applications. Proposals are by invitation. Guidelines are available online.
Requirements: Eligible organizations must have 501(c)3 status. Grant requests must be limited to programs and projects within the state of Texas.
Restrictions: No grants can be made to individuals.
Geographic Focus: Texas
Samples: All Saints' Episcopal School of Fort Worth (TX)—for its capital campaign, $1 million (2004).
Amount of Grant: $10,000-$350,000 average
Contact: Valleau Wilkie Jr., Executive Director; (817) 336-0494; fax: (817) 332-2176; email: cjohns@sidrichardson.org
Sponsor: Lee and Ramona Bass Foundation
309 Main Street
Fort Worth, TX 76102

LEF Foundation California Grants **3156**
The foundation's California office supports projects on a national scope. It is focused on supporting work of high creative merit in the fields of contemporary art and architecture, design, literature, the performing arts, film and new media. Through its efforts, LEF seeks to provide opportunities for artists, architects, writers and organizations to develop, produce and present new work; honor creative merit and foster critical discourse; encourage the dissemination of work by emerging and underrecognized artists, as well as to increase exposure of established artists in regions where they have not been widely represented; promote new concepts, technologies, and approaches that are experimental or innovative; support work that may be considered controversial or provocative; and enhance the voices of marginalized cultures. Funding is provided to assist direct project expenses for new work, exhibitions, commissions, installations, performances, publications, symposia; advocacy and cultural policy research; and collaborative or interdisciplinary projects that are deemed compatible with LEF's mission.
Requirements: Funding is provided to organizations with current tax-exempt status. Individuals are required to apply through a nonprofit fiscal sponsor. Prospective applicants are asked to submit a letter of intent of three pages or less that includes: the legal name of the applicant organization as acknowledged by the I.R.S.; the start and completion dates for the proposed project; A description of the proposed project that includes a statement addressing its significance or timeliness; and a budget for the project demonstrating anticipated expense and income.
Restrictions: Funding is not available for capital campaigns; general or operational support; academic or educational programs; festivals or juried exhibitions; or to effect legislation or support candidates for political office. Also, funding is not available for fellowships, scholarships, or to complete academic work.
Samples: American Indian Film Institute (San Francisco, CA)—to support the 27th Annual American Indian Film Festival, $10,000 (2002); California College of Arts and Crafts (Oakland, CA)—to support a four-month residency and presentation of new work by Shirley Tse at the Capp Street Project, a program at CCAC's San Francisco campus, $15,000 (2002).
Amount of Grant: $5000-$25,000
Contact: Marina Drummer, Grants Administrator, (617) 963-9591; fax: (707) 963-2109; email: jmisraje@lef-foundation.org or lyda@lef-foundation.org
Internet: http://www.lef-foundation.org/page.php/id/135
Sponsor: LEF Foundation
1095 Lodi Ln
Saint Helena, CA 94574

Legal Education Research Grants **3157**
The council funds empirical research on legal education in the United States and Canada. The board welcomes research proposals from a variety of methodologies and a broad range of topics. The proposed time frame for the research subject matter can range from a student's earliest consideration of law school to his or her first job experience in the legal market. Eligible investigators need not be members of law school faculties. Proposals from interdisciplinary research teams are encouraged. There are two reviewing cycles each year.
Requirements: Researchers from universities, law schools, or other institutions with experience in designing and conducting empirical research are eligible. Principal investigators need not be based in law schools; proposals are equally welcome from social, behavioral, and educational researchers of all kinds.
Restrictions: Not eligible to receive grants are members of the LSAC board of trustees, members of the LSAC test development and research committee or its grants subcommittee, and persons who had been members of one or more of those bodies within one year prior to applying for a grant. This program is not used to fund legal education tuition.
Date(s) Application is Due: Feb 1; Sep 1
Contact: Lillian Worthington, Grants and Contracts Coordinator, (215) 968-1198; fax: (215) 968-1169; email: lworthington@lsac.org
Internet: http://www.lsac.org/LSAC.asp?url=lsac/lsac-legal-education-grant-program.asp
Sponsor: Law School Admission Council
P.O. Box 40, 662 Penn Street
Newtown, PA 18940-0040

Legler Benbough Foundation Grants **3158**
The foundation awards grants to nonprofits serving San Diego County, CA, in the areas of health, education, arts, and community development. Grants have been awarded to/for museums, educational institutions, hospitals and health care agencies, law enforcement, children and youth services agencies, hospices, elder care centers, and programs serving underserved area residents (minorities, the disabled, and the aging). Types of support include building/renovation, materials acquisition, capital endowment campaigns, and research. There are no application forms; initial approach should be by letter.
Requirements: Nonprofit organizations serving San Diego County, CA, and its residents may apply.
Restrictions: Grants are not awarded to individuals.
Geographic Focus: California
Samples: San Diego Natural History Museum (San Diego, CA)—to construct its new hall for traveling exhibitions, $1 million.
Date(s) Application is Due: Mar 15; Sep 15
Contact: Peter Ellsworth, President, (619) 235-8099; fax: (619) 235-8077; email: peter@benboughfoundation.org
Internet: http://www.benboughfoundation.org/criteria.php
Sponsor: Legler Benbough Foundation
2550 5th Avenue, Suite 132
San Diego, CA 92103-6622

Leland Fikes Foundation Grants Program **3159**
Grants are awarded for medical research, mental health programs, social welfare, education, reproductive health care, and arts and culture. Building programs and equipment acquisition are supported in these areas. Support is rarely given to programs located outside of the Dallas area. Additional types of support include general operating support, continuing support, annual campaigns, capital campaigns, endowment funds, emergency funds, program development, professorships, seed money grants, scholarship funds, research, and matching funds. Contact the foundation for guidelines.
Requirements: Giving is primarily to Texas organizations.
Restrictions: Grants are not awarded to individuals.
Geographic Focus: Texas
Samples: Saint Marks School of Texas (Dallas, TX)—for capital campaign, $200,000 (2002); Texas Partnership for End-of-Life Care (Austin, TX)—for general support, $175,000 (2002); Planned Parenthood (TX)—for operating support.
Amount of Grant: $5000-$75,000 average

Contact: Nancy Solana, (214) 754-0144
Sponsor: Leland Fikes Foundation
3050 Lincoln Plaza, 500 N Akard Street
Dallas, TX 75201

Leland McSpadden Memorial Fund for Medical and Scientific Research Grants 3160
The fund invites research proposals aimed at achieving a better understanding of the metabolic basis of Lowe syndrome and/or developing a better understanding and better treatments for the major complications of Lowe syndrome, a genetic disorder that affects boys. Grants are awarded for one year with possible renewal for a second year.
Requirements: The program invites applications from researchers of all types, including but not limited to universities, hospitals, and other nonprofit research organizations.
Restrictions: The grant may not be used as salary for the principal investigator, for honoraria for consultants, for the purchase of standard, nonconsumable laboratory equipment, to purchase services from other research laboratories, or for overhead or indirect costs.
Amount of Grant: $25,000
Date(s) Application is Due: Oct 15
Contact: Mary Tietz, President, (612) 770-8438; email: info@lowesyndrome.org
Internet: http://www.lowesyndrome.org/Research/index.html
Sponsor: Lowe Syndrome Association
18919 Voss Road
Dallas, TX 75287

Lemelson-MIT Lifetime Achievement Award 3161
The award recognizes outstanding individuals whose pioneering spirit and demonstrated inventiveness throughout their careers has improved our society and inspired others. Candidates are living role models in the fields of science, engineering, medicine, and technology. They are U.S. citizens or permanent residents. Candidates are suggested to the program by the inventor community and other national leaders in science and technology. The winner is selected by the Lemelson-MIT board and notified via telephone by the director. Annual deadline dates may vary; contact program staff for exact dates.
Samples: Dr. Ruth Rogan Benerito—for her patented processes in textiles for easy-care cotton and her lifelong commitment to education.
Amount of Grant: $100,000 prize
Contact: Lemelson-MIT Program Office, (617) 253-3352 or (617) 253-3490; email: idudek@mit.edu
Internet: http://web.mit.edu/invent/a-lifetime.html
Sponsor: Massachusetts Institute of Technology
77 Massachusetts Avenue, Building E60-215
Cambridge, MA 02139-4307

Lemelson-MIT Prize 3162
The prize recognizes scientists, engineers, and inventors who improve life for all of us. It is the world's largest single cash prize for invention, and it recognizes individuals whose ingenuity and innovation have made society stronger, safer, and healthier. This prestigious award is bestowed upon an individual who has a rich history of inventiveness and creativity and a proven commitment to inspiring others.
Requirements: The program invites leaders from industry, research, academia, and government to nominate candidates for the award. Eligible candidates are U.S. citizens or permanent residents who hold at least two patents. One patent must be for a product or process that provides a significant benefit—potential or realized—to all of society.
Amount of Grant: $500,000
Contact: Lemelson-MIT Program, (617) 253-3352; fax: (617) 258-8276; email: kariukit@mit.edu
Internet: http://web.mit.edu/invent/a-prize.html
Sponsor: Massachusetts Institute of Technology
77 Massachusetts Avenue, Building E60-215
Cambridge, MA 02139-4307

Leo Goodwin Foundation Grants 3163
The foundation awards general operating grants, primarily in Florida, in the areas of cancer research, child welfare, education, hospitals, mental health, and youth. There are no application forms.
Requirements: Nonprofit organizations are eligible.
Restrictions: Individuals are not eligible.
Geographic Focus: Florida
Amount of Grant: $1000-$80,000 range
Date(s) Application is Due: Oct 31
Contact: Helen Furia, Secretary-Treasurer, (954) 491-2000
Sponsor: Leo Goodwin Foundation
800 Corporate Drive, Suite 510
Fort Lauderdale, FL 33334-3621

Leon Lowenstein Foundation Grants 3164
The foundation awards grants to eligible organizations in its areas of interest, including education, medical research, and disaster relief. Types of support include research grants, general operating support, program development, and seed funding. There are no application deadlines or forms. The board meets twice each year to consider requests.
Requirements: Nonprofit organizations are eligible. Preference will be given to requests from the New York City metropolitan area.
Samples: U of Pennsylvania (Philadelphia, PA)—to create a loan-forgiveness fund for MBA graduates of the Wharton School who pursue careers at public and nonprofit institutions, $2.5 million (2005).
Contact: Executive Director, (212) 319-0670; fax: (212) 688-0134
Sponsor: Leon Lowenstein Foundation
126 E 56th Street
New York, NY 10022

Leonard and Helen R. Stulman Charitable Foundation Grants 3165
The Leonard and Helen R. Stulman Charitable Foundation supports work in four areas of interest. Ninety five percent (95%) of funding from The Foundation is reserved for programs in greater Baltimore and the State of Maryland. The foundation will make a limited number of new grants each year to projects that address one or more of the following: research as to causes and treatment of mental illness and services for people with mental illness and their families; initiatives for an older adult and aging population and programs for geriatric medicine; programs engaged in health care treatment and prevention as well as those involved in medical research; and higher educational institutions. To apply to the Foundation for funding, applicants should submit a two-page letter of inquiry, with basic background on your organization, the identified needs your project proposes to address, an overview of the proposed project, and the amount you intend to request.
Requirements: Organizations (or their fiscal agents) that qualify as public charities under section 501(c)3 of the Internal Revenue Code and do not discriminate on the basis of race, creed, national origin, color, physical handicap, gender or sexual orientation are eligible to apply.
Amount of Grant: $50,000 average
Contact: Cathy Brill, Program Officer; (410) 332-4171; fax: (410) 837-4701; email: cathybrill@comcast.net
Internet: http://www.bcf.org/ourgrants/ourgrantsdetail.aspx?grid=4
Sponsor: Leonard and Helen R. Stulman Charitable Foundation
2 East Read Street, 9th Floor
Baltimore, MD 21202

Lesbian Health Fund Grants 3166
The fund supports research into lesbian health, including research with the following goals: determination or rates and risk factors for cancers and other diseases among lesbians; issues of access to medical care for lesbians; study of mental health concerns of lesbians and their families; definition and investigation of lesbian family issues; and methodologic exploration of diversity in the lesbian population. Recently funded projects have included the first-ever longitudinal study of lesbian mothers; a survey of lesbian health curricula in U.S. internal medicine residencies; and an effort to develop the phone-survey method as a way of identifying lesbians missed by other survey approaches. Applications are available online.
Restrictions: Grants are not awarded for completed research or for projects in progress.
Amount of Grant: $500-$10,000
Date(s) Application is Due: May 1; Oct 1
Contact: Manager, (415) 255-4547; fax: (415) 255-4784; email: info@glma.org
Internet: http://www.glma.org/programs/lhf/index.shtml
Sponsor: Lesbian Health Fund
459 Fulton Street, Suite 107
San Francisco, CA 94102

Leslie Center Humanities Institute Fellowships 3167
This institute will bring researchers to the Dartmouth campus to investigate, in weekly meetings, the psychological and biological bases of moral beliefs and the implications of such empirical findings for moral philosophy. A conference on these topics will be held, and essays by participants in the conference and the institute will be published. External fellows who stay for the duration of the institute receive a stipend. Internal full-time fellows from the Dartmouth faculty receive one course relief. Internal part-time fellows from the Dartmouth faculty receive research funding. Applicants for external fellowships should submit a curriculum vita, names of three referees, and a letter of interest with a short (200-500 word) description of a research project to be pursued during the institute. Applicants for internal fellowships should submit a curriculum vita and a letter of interest with a short (200-500 word) description of a research project to be pursued during the institute. Materials should arrive by the end of March for full consideration.
Requirements: Applications are encouraged not only from philosophers but also from psychologists, biologists, and scholars in other fields.
Amount of Grant: $7500 stipend
Contact: Grants Administrator, (603) 646-0896; fax: (603) 646-0998; email: Humanities.Center@Dartmouth.EDU
Internet: http://www.dartmouth.edu/~lhc/fellowships.html
Sponsor: Dartmouth College
Dartmouth College, Gerry 205, HB 6240
Hanover, NH 03755

Leukaemia Research Fund Grants 3168
The fund was founded to encourage, promote, and assist research into the leukemias, Hodgkin's disease, and other related conditions. The fund supports clinical fellowships and research projects at hospitals and university medical centers throughout the United Kingdom.
Amount of Grant: ?67,866
Contact: Grants Administrator, (020) 7405 0101; fax: (020) 7405 3139; email: info@lrf.org.uk

Internet: http://www.lrf.org.uk/en/1/science.html
Sponsor: Leukaemia Research Fund
43 Great Ormond Street
London WC1N 3JJ United Kingdom

Leukemia and Lymphoma Society Career Development Awards **3169**

The program grants help scientists in various stages of their careers and encourage them to investigate promising new directions and devote their careers to curing blood-related cancers. The four awards in the program—Scholar, Scholar in Clinical Research, Fellow, and Special Fellow—provide stipends that allow investigators to perform basic research into the cause and prevention of leukemia, lymphoma, Hodgkin's lymphoma and myeloma. A preliminary application is mandatory and must be completed online by September 15; full applications must be received at the office by October 1. Guidelines are available online.
Requirements: Scientists pursuing the investigation of promising new directions to curing blood-related cancers are eligible.
Amount of Grant: $50,000-$110,000 per year
Date(s) Application is Due: Sep 15; Oct 1
Contact: Grants Administrator, (914) 821-8843 or 821-8859; fax: (914) 821-8946; email: researchprograms@leukemia-lymphoma.org
Internet: http://www.leukemia-lymphoma.org/all_page?item_id=11618
Sponsor: Leukemia and Lymphoma Society
1311 Mamaroneck Avenue
White Plains, NY 10605

Leukemia and Lymphoma Society Research Grants **3170**

The society's research program is based on the belief that all scientifically sound approaches toward a cure or control of leukemia, lymphoma, and myeloma should be encouraged on a worldwide basis. Basic laboratory research and its application to a clinical setting are supported.
Samples: Baylor College of Medicine, Ctr for Cell and Gene Therapy (TX)—for blood-cancer research, through the Specialized Center of Research grant program, $5 million over five years (2003); To be divided among 32 research projects—to develop new treatments for leukemia, lymphoma, and myeloma, through the Translational Research Program, $11.3 million (2003); Fred Hutchinson Cancer Research Ctr (Seattle, WA)—to support the work of Irwin Bernstein to develop less toxic and more effective immune-based therapies for treating advanced leukemia and lymphoma, $7 million.
Contact: Grants Administrator, (914) 949-5213; fax: (914) 949-6691
Internet: http://www.leukemia-lymphoma.org/all_page?item_id=7427
Sponsor: Leukemia and Lymphoma Society
1311 Mamaroneck Avenue
White Plains, NY 10605

Leverhulme Trust Grants **3171**

The trust makes awards for the support of research and education. The trust emphasises individuals and encompasses all subject areas. The trustees place special weight on the originality of the projects; the significance of the proposed work; the ability to judge and take appropriate risk in the project; and the removal of barriers between traditional disciplines. Categories of support include study abroad studentships, emeritus fellowships, early-career fellowships, research project grants, academic collaboration—international networks, artists in residence, training and professional development, major research fellowships, Philip Leverhulme Prizes, and visiting professorships. Check Web site for deadlines.
Restrictions: The trust does not make awards for equipment or for the provision of research premises, or provide: core funding for institutions; contributions to appeals; funding of exhibitions; endowments; funding of equipment, sites, buildings, or other capital expenditure; support for conferences, workshops, or symposia that are not directly related to research projects; and the means for making good withdrawals or deficiencies in public finance. The tustees do not usually support research in the following fields for grants made to institutions: social policy and welfare (especially action research); medicine; and education.
Contact: Beryl Thomas, 44-171-020-7822-5227; fax: 44-171-020-7822-5084; email: bthomas@leverhulme.ac.uk
Internet: http://www.leverhulme.org.uk
Sponsor: Leverhulme Trust
1 Pemberton Row
London EC4A 3BG United Kingdom

LI American Religion Dissertation Fellowship **3172**

The program is designed to support the final year of writing on promising PhD and ThD dissertation projects dealing with aspects of American religious life that are related to the concerns of the Louisville Institute. Preference will be given to proposals that attempt to describe more fully how the Christian faith is actually lived by contemporary persons and to bring the resources of the Christian faith into closer relation to their daily lives; and to help us understand more adequately the institutional reconfiguration of American religion. Proposals on certain other issues of importance to the churches are also welcome. Proposed projects may employ a variety of methodological perspectives, including, but not limited to, history; ethics; the social sciences; and historical, systematic, and practical theology. They may also be interdisciplinary in nature. Guidelines and application forms are available online.
Requirements: Applicants must be candidates for the PhD or ThD degree who have fulfilled all predissertation requirements, including approval of the dissertation proposal, by January 15 of the award year and expect to complete the dissertation by the end of the following academic year.
Restrictions: The institute is interested in funding projects that can contribute in significant ways to the understanding of contemporary religious communities. Consequently, very few dissertation projects funded in recent years have dealt with American religion prior to the mid-nineteenth century.
Amount of Grant: $18,000 for 12 months
Date(s) Application is Due: Jan 15
Contact: Dissertation Fellowships, (502) 895-3411 ext 251; fax: (502) 894-2286; email: info@louisville-institute.org
Internet: http://www.louisville-institute.org/secondary/DFdetail.asp
Sponsor: Louisville Institute
1044 Alta Vista Road
Louisville, KY 40205-1798

LI Christian Faith and Life Sabbatical Grants **3173**

The initiative seeks to support efforts to bring the resources of the ethical, liturgical, and doctrinal wisdom of the Christian faith into closer relation to the daily lives of practicing Christians and to describe more fully how the Christian faith is actually lived by contemporary Christians of various ages, circumstances, and traditions. The sabbatical grants program seeks to identify scholars already working in this area; encourage other scholars to develop research projects that would advance thinking in this area; and generate a broader conversation in the American scholarly community about the nature of and challenges to Christian faith and life in the contemporary world.
Requirements: The program is open to both academic and pastoral leaders. Applicants must have earned the terminal degree in their chosen vocation (e.g., MDiv, PhD, ThD) and must demonstrate a capacity to complete the proposed project in a timely fashion.
Amount of Grant: $50,000
Date(s) Application is Due: Nov 15
Contact: Grants Administrator, (502) 895-3411 ext 487; fax: (502) 894-2286; email: info@louisville-institute.org
Internet: http://www.louisville-institute.org/secondary/CFLdetail.asp
Sponsor: Louisville Institute
1044 Alta Vista Road
Louisville, KY 40205-1798

LI General Grants **3174**

The institute offers a modest general grant program that supports a limited number of individual and collaborative research and leadership education projects related to the priorities of the Louisville Institute. Such grants, for example, might cover an individual scholar's research expenses or the costs of convening a research team or group of religious leaders. Applicants who believe their request fits within institute guidelines should submit a preliminary letter of no more than two pages, in which they tell about their project and the amount of support they are requesting. The board meets three times per year (April, August, and December) to evaluate proposals.
Requirements: Academics, pastors, and other religious leaders are encouraged to apply.
Amount of Grant: $25,000-$75,000
Date(s) Application is Due: Mar 1; Jul 1; Nov 1
Contact: Grants Administrator, (502) 992-5432; fax: (502) 894-2286; email: info@louisville-institute.org
Internet: http://www.louisville-institute.org/secondary/generalgrant.asp
Sponsor: Louisville Institute
1044 Alta Vista Road
Louisville, KY 40205-1798

LI Religious Institutions Sabbatical Grants **3175**

The program seeks to advance the understanding of the contemporary situation confronting religious institutions so that they and their leaders can respond in more informed and effective ways to the challenges they face. A maximum of five grants will be awarded for the academic year. Each award is designed to make possible a full academic year of sabbatical research and writing by providing up to one-half of the grantee's salary and benefits for that year. The institute will pay the grant directly to the institutions of those selected, but no indirect costs to the institution will be allowed.
Requirements: Applicants must have completed a PhD or ThD and be employed full-time by an accredited academic institution. This includes faculty in both tenured and tenure-track positions as well as persons in full-time, non-tenure-track appointments. Applicants must be eligible for a full-year sabbatical leave.
Restrictions: Awards for sabbatical leaves of less than a full academic year will not be made.
Amount of Grant: $50,000
Date(s) Application is Due: Dec 1
Contact: Grants Administrator, (502) 895-3411 ext 487; fax: (502) 894-2286; email: info@louisville-institute.org
Internet: http://www.louisville-institute.org/secondary/pastoralgrants.asp
Sponsor: Louisville Institute
1044 Alta Vista Road
Louisville, KY 40205-1798

LI Summer Stipend Program Grants **3176**

The program offers grants to faculty members engaged in summer research projects in areas of contemporary American religion. The award is intended to support 10 weeks of research and writing. Eligible proposals should promise a significant contribution to the study of American religion. Preference will be given to proposals that attempt to describe

more fully how the Christian faith is actually lived by contemporary persons and to bring the resources of the Christian faith into closer relation to their daily lives; and to help in understanding more adequately the institutional reconfiguration of American religion. Proposals on certain other issues are also welcome. Proposed projects may employ a variety of methodological perspectives, including, but not limited to, history, ethics, the social sciences; and historical, systematic, and practical theology. They may also be interdisciplinary in nature. Application and guidelines are available online.
Requirements: Applicants should have completed a PhD or ThD and be currently employed by an accredited academic institution.
Amount of Grant: $9000
Date(s) Application is Due: Oct 15
Contact: Grants Administrator, (502) 992-5432; fax: (502) 894-2286; email: info@louisville-institute.org
Internet: http://www.louisville-institute.org/secondary/SSdetail.asp
Sponsor: Louisville Institute
1044 Alta Vista Road
Louisville, KY 40205-1798

Liberal Arts Fellowships in Law **3177**
These fellowships are offered to enable teachers in the social sciences or humanities to study fundamental techniques, concepts, and aims of law so that in their teaching and research they will be better able to use legal materials and legal insights that are relevant to their own disciplines. Fellowship holders will presumably take at least two first-year courses in law, in addition to more advanced courses. Each year, four or five fellowships are awarded; the fellowships are for one year, nonrenewable. Tuition, health fees, and office space are provided. Notification is within one month of deadline date.
Requirements: Applicants should be college teachers in the arts and sciences. Candidate must submit a biographical resume including academic record and list of publications, a statement explaining what the applicant hopes to achieve through his/her year of study, and two letters of recommendation mailed directly to the chair by the referees.
Contact: Chair, Committee on Liberal Arts, (617) 495-3356; fax: (617) 495-1110
Internet: http://www.law.harvard.edu/academics/fellowships
Sponsor: Harvard Law School
1557 Massachusetts Avenue, LILC/331
Cambridge, MA 02138

Library Company of Philadelphia and The Historical Society of Pennsylvania Visiting Research Fellowships **3178**
The Library Company of Philadelphia and The Historical Society of Pennsylvania will jointly award approximately thirty one-month fellowships for research in residence in either or both collections during the academic year. These two independent research libraries, adjacent to each other in Center City Philadelphia, have complementary collections capable of supporting research in a variety of fields and disciplines relating to the history of America and the Atlantic world from the 17th through the 19th centuries, as well as Mid-Atlantic regional history to the present. The Stipend is $2,000. Fellowships are tenable for any one-month period between June and May. They support advanced, post doctoral, and dissertation research. Candidates are strongly encouraged to inquire about the appropriateness of the proposed topic. Deadline for receipt of applications is March 1 with a decision to be made by April 15.
Amount of Grant: $2,000
Date(s) Application is Due: Mar 1
Contact: James Green, Library Company of Philadelphia; (215) 546-3181; fax: (215) 546-5167; email: jgreen@librarycompany.org
Internet: http://www.librarycompany.org/fellowships/american.htm
Sponsor: Library Company of Philadelphia; The Historical Society of Pennsylvania
1314 Locust Street
Philadelphia, PA 19107

Library Company of Philadelphia Barra Foundation International Research Fellowships **3179**
The Library Company of Philadelphia and the Historical Society of Pennsylvania each year offer two one-month fellowships to support research in residence in their collections by foreign national scholars of early American history and culture living outside the United States. The fellowships support both post-doctoral and dissertation research. The project proposal should demonstrate that the Library Company and the Historical Society have primary sources central to the research topic. Candidates are encouraged to inquire about the appropriateness of a proposed topic before applying. The fellowships are tenable for one month at any time from June to May. The stipend is $2,500, plus an allowance for travel expenses. Deadline for receipt of application is March 1, and appointments will be made by April 15. To apply send a curriculum vitae, a two to four page description of the project, and one letter of reference.
Amount of Grant: $2,500
Date(s) Application is Due: Mar 1
Contact: James Green, Library Company of Philadelphia; (215) 546-3181; fax: (215) 546-5167; email: jgreen@librarycompany.org
Internet: http://www.librarycompany.org/fellowships/barra.htm
Sponsor: Library Company of Philadelphia
1314 Locust Street
Philadelphia, PA 19107

Library Company of Philadelphia NEH Long-term Post-Doctoral Fellowships **3180**
The fellowship supports research in residence at the Library Company on any subject relevant to its collections, which are capable of supporting research in a variety of fields and disciplines relating to the history of America and the Atlantic world from the 17th through the 19th centuries. The fellowships are tenable from September through May, but the awards may both be divided between two applicants, each of whom would spend a semester in residence. Candidates are strongly encouraged to inquire about the appropriateness of the proposed topic before applying.
Requirements: Applicants must hold a PhD. Midcareer and senior scholars are particularly encouraged to apply.
Amount of Grant: $40,000 or $20,000 per semester
Date(s) Application is Due: Nov 1
Contact: James Green, Library Company of Philadelphia; (215) 546-3181; fax: (215) 546-5167; email: jgreen@librarycompany.org
Internet: http://www.librarycompany.org/fellowships/postdoc.htm
Sponsor: Library Company of Philadelphia
1314 Locust Street
Philadelphia, PA 19107-5698

Library Company of Philadelphia Prizes **3181**
The Program in Early American Economy and Society makes two awards annually to recognize the best book and journal article published in the previous year relating to an aspect of early American economic history, broadly defined, up to 1850. Submissions are welcomed in such fields as the history of finance, business, commerce, agriculture, manufacturing, technology, and economic policymaking. The author of the winning book receives a cash award and presents his or her work during an awards ceremony. The author of the winning article receives a cash award. Complete entries consist of three copies of the work and a nominating letter explaining the work's importance to the field of early American economic history. Forward nominations and entries to the Program in Early American Economy and Society.
Requirements: Nominations should be made by someone other than the author.
Amount of Grant: $2,500 to author of winning book; $1,000 to author of winning article
Date(s) Application is Due: Mar 15
Contact: Cathy Matson, Program Director; (215) 546-3181; fax: (215) 546-5167; email: cmatson@udel.edu
Internet: http://www.librarycompany.org/economics/earticleprize.htm
Sponsor: Library Company of Philadelphia
1314 Locust Street
Philadelphia, PA 19107-5698

Library Company of Philadelphia Program in Early American Economy and Society (PEAES) Dissertation-Level Fellowships **3182**
The fellowships are designed to promote scholarship in early American economy and society, broadly defined, from its colonial beginnings to roughly the 1850s. Some of the possible topics of research include the history of commerce, finance, technology, manufacturing, agriculture, internal improvements, and political economy. One dissertation-level fellowship, carrying a stipend of $20,000, is tenable for nine consecutive months of residency from September 1 to May 31. It may also be divided between two scholars, who would each receive $10,000 for the periods between September 1 to December 15, or December 15 to May 31. Deadline for receipt of the applications is March 1, with reply date being March 31.
Amount of Grant: $20,000
Date(s) Application is Due: Mar 1
Contact: Cathy Matson, Program Director; (215) 546-3181; fax: (215) 546-5167; email: cmatson@udel.edu
Internet: http://www.librarycompany.org/Economics/EFellowships.htm
Sponsor: Library Company of Philadelphia
1314 Locust Street
Philadelphia, PA 19107-5698

Library Company of Philadelphia Program in Early American Economy and Society (PEAES) One-Month Fellowships **3183**
These fellowships are designed to promote scholarship in early American economy and society, broadly defined, from its colonial beginnings to roughly the 1850s. Some of the possible topics of research include the history of commerce, finance, technology, manufacturing, agriculture, internal improvements, and political economy. Available to scholars at all levels, four one-month fellowships, carrying stipends of $2,000 each, are tenable for a month of research at the Library Company between June 1 and May 31. Deadline for receipt of one-month and dissertation fellowship applications is March 1, and reply date is March 31.
Amount of Grant: $20,000
Date(s) Application is Due: Mar 1
Contact: Cathy Matson, Program Director; (215) 546-3181; fax: (215) 546-5167; email: cmatson@udel.edu
Internet: http://www.librarycompany.org/Economics/EFellowships.htm
Sponsor: Library Company of Philadelphia
1314 Locust Street
Philadelphia, PA 19107-5698

Library Company of Philadelphia Program in Early American Economy and Society (PEAES) Post-Doctoral Research Fellowships **3184**
This fellowship is designed to promote scholarship in early American economy and society, broadly defined, from its colonial beginnings to roughly the 1850s. Some of the possible topics of research include the history of commerce, finance, technology, manufacturing, agriculture, internal improvements, and political economy. One post-doctoral research

fellowship, carrying a stipend of $40,000, is tenable for nine consecutive months of residency from September 1 to May 31. It may also be divided between two scholars, who would each receive $20,000 for the periods between September 1 to December 15 or December 15 to May 31. Deadline for receipt of applications is November 1, and reply date is December 15.
Amount of Grant: $40,000
Date(s) Application is Due: Nov 1
Contact: Cathy Matson, Program Director; (215) 546-3181; fax: (215) 546-5167; email: cmatson@udel.edu
Internet: http://www.librarycompany.org/Economics/EFellowships.htm
Sponsor: Library Company of Philadelphia
1314 Locust Street
Philadelphia, PA 19107-5698

Library of Congress Coca-Cola Fellowship for the Study of Advertising and World Cultures 3185
Fellowships support researchers studying the interrelationships among advertising, culture, commerce, and the media, beginning in the 20th century. The award provides recipients with access to resources for an extended period of in-depth research into the Library's broadcast advertising and other audio-visual collections. Investigations of related collections in other divisions of the Library are encouraged; further information about the Library's holdings can be found on the Library's Website: www.loc.gov. A supporting stipend is provided for the duration of the fellowship, which must include three months residency in the Washington, DC, area to conduct research in the Library's collections. Scholars of all nationalities are invited to apply. Application guidelines are available online.
Amount of Grant: $20,000 stipend
Date(s) Application is Due: Mar 15
Contact: Coca-Cola Fellowship Committee, Motion Picture, Broadcasting, and Recorded Sound Division, (202) 707-8572; fax: (202) 707-2371; email: cokefellow@loc.gov
Internet: http://www.loc.gov/rr/mopic/cokefellowship
Sponsor: Library of Congress
101 Independence Avenue, SE
Washington, DC 20540-4690

Library of Congress Fellowships in International Studies 3186
The program supports postdoctoral research in all disciplines of the humanities and social sciences using the foreign language collections of the Library of Congress. Proposals in multidisciplinary and cross-disciplinary studies are also welcome, as are proposals focused on single or multiple geographical areas. Up to 10 fellowships will be available for four to nine months each. During the fellowship period, scholars will be expected to be engaged in full-time research at the Library. The fellowships may be initiated at any time between July 1 and February 1. The fellowships may be combined with sabbatical and other fellowship funds to extend the research period up to a total of 12 months. Completed applications must be submitted through the ACLS online fellowship application system.
Requirements: Applicants must hold a PhD conferred prior to the listed application deadline, and strong preference will be given to scholars at an early stage of their career—those within seven years of the degree. Applicants must be U.S. citizens or permanent residents as of the application deadline date, and may be affiliated with any academic institution. Independent scholars are also welcome to apply.
Amount of Grant: $4000 per month
Date(s) Application is Due: Nov 3
Contact: Mary Lou Reker, Manager, Fellowships and Grants, (202) 707-5027; email: www.loc.gov/kluge
Internet: http://www.loc.gov/kluge
Sponsor: Library of Congress, John W. Kluge Center
Library of Congress, LJ-120, 101 Independence Avenue SE
Washington, DC 20540-4860

Library of Congress Henry A. Kissinger Scholar Grants 3187
The program offers a senior, distinguished, 10-month scholarly appointment at the Library of Congress for sustained research on issues of foreign policy and international affairs related to the United States. In addition to a stipend, the scholar will be provided with research space and services at the Library of Congress. Applications require a brief statement of the proposed research, an indication of the holdings at the Library of Congress that will be used in the research, evidence of prior scholarly work and publication, and the names of three people who can be reached for letters of recommendation. Candidates may apply directly or be identified to the selection committee by nomination.
Amount of Grant: $135,000
Date(s) Application is Due: Oct 1
Contact: Kissinger Scholar Selection Committee, Office of Scholarly Programs, (202) 707-3302; fax: (202) 707-3595; email: scholarly@loc.gov
Internet: http://www.loc.gov/loc/kluge/kissinger.html
Sponsor: Library of Congress
101 Independence Avenue SE
Washington, DC 20540-1320

Library of Congress International Studies Fellowships 3188
The program supports postdoctoral research in all disciplines of the humanities and social sciences using the foreign language collections of the Library of Congress. Proposals in multidisciplinary and crossdisciplinary studies are also welcome, as are proposals focused on single or multiple geographical areas. Up to 10 fellowships will be available for four to nine months each. Fellowships may be combined with sabbatical and other fellowship funds to extend the research period up to a total of 12 months. Scholars will be expected to be engaged in full-time research at the library during the fellowship term. Fellowships may be initiated at any time between July 1 and February 1.
Requirements: Applicants must hold a PhD conferred prior to November 1, and preference will be given to scholars at an early stage of the career—those within seven years of the degree. Applicants must be U.S. citizens or permanent residents as of the application deadline date, and may be affiliated with any academic institution. Independent scholars are also welcome to apply.
Amount of Grant: $3500 per month
Date(s) Application is Due: Nov 1
Contact: Lester Vogel, Office of Scholarly Program, (202) 707-1673; fax: (212) 949-8058; email: grants@acls.org
Internet: http://www.loc.gov/rr/international/int-fellow.html
Sponsor: Library of Congress
101 Independence Avenue SE
Washington, DC 20540-4860

Library of Congress Mellon Foreign Area Research Fellowships 3189
The fellowships enable scholars to pursue research at the Library of Congress's foreign-language and area-studies collections. The fellowship duration is four to nine months and can begin no sooner than August 1.
Requirements: The program is open to U.S. citizens with PhD degrees who are proposing or working on major research projects with focus on foreign-language materials.
Amount of Grant: $3500 per month
Date(s) Application is Due: Nov 1
Contact: Office of Fellowships and Grants, fax: (202) 949-8058; email: grants@acls.org
Internet: http://www.loc.gov/loc/mellon/int-fellow.html
Sponsor: Library of Congress
101 Independence Avenue SE
Washington, DC 20540-4860

Library of Congress Rockefeller Islamic Studies Fellowships 3190
The program supports residential fellowships for postdoctoral scholars wishing to conduct research in the humanities on globalization and Muslim societies using the resources of the Library of Congress. The Library of Congress wishes to focus scholarly attention on the issues that emerge from the discourse on globalization's impact on Muslim societies and the resulting interactions. The library's holdings include resources for studying the full range of related issues, through materials published in all Muslim societies in their own languages as well as through materials about them from all over the world. Materials include books, serials, maps, manuscripts, photographs, music, film, sound recordings, and Internet resources. Two to four scholars each year, funded for periods of not less than five or more than 10 months duration, will conduct original research on the humanistic dimensions of globalization. Particularly welcome are historical and cross-disciplinary studies of the way intellectual issues have arisen in Muslim societies in response to globalization. Applicants may be U.S. citizens or foreign nationals. Upon selection, foreign nationals will be assisted in obtaining the appropriate visa.
Requirements: Scholars who have received a terminal advanced degree within the past seven years in the humanities, social sciences, or in a professional field such as architecture or law are eligible. Exceptions may be made for individuals without continuous academic careers.
Amount of Grant: $3500 per month
Date(s) Application is Due: Nov 1
Contact: Office of Scholarly Programs, (202) 707-3302; fax: (202) 707-3595; email: scholarly@loc.gov
Internet: http://www.loc.gov/loc/kluge/kluge-fellow.html
Sponsor: Library of Congress
101 Independence Avenue SE
Washington, DC 20540-4860

Life Sciences Research Foundation Postdoctoral Fellowships 3191
Each year approximately 12-15 postdoctoral fellowships are awarded in the medical and life sciences to provide training and support of the highest quality young scientists in the best research environments. Research must be performed at nonprofit institutions. The fellowships are awarded for three-year periods. Send a self-addressed envelope for information and an application, or visit the Web site.
Requirements: Applicant must have an MD, PhD, or equivalent degree. Persons working on a second postdoctoral are eligible only if they are transferring to a different supervisor's laboratory and embarking on a new project not connected to their previous research. All U.S. citizens are eligible with no geographic restrictions. Foreign applicants will be eligible for study in U.S. laboratories only.
Restrictions: Candidate must not hold a faculty/permanent appointment. The LSRF does not award an institutional allowance for overhead.
Amount of Grant: $51,000 maximum per year
Date(s) Application is Due: Oct 1
Contact: Susan DiRenzo, Assistant Director, (609) 258-3551; email: sdirenzo@molbio.princeton.edu
Internet: http://www.lsrf.org/geninfo.htm
Sponsor: Life Sciences Research Foundation
c/o Lewis Thomas Laboratory, Princeton University, Washington Road
Princeton, NJ 08544

Lillian S. Wells Foundation Grants 3192

The Lillian S. Wells Foundation established in 1976, primarily supports the Fort Lauderdale, Florida, and Chicago, Illinois regions with funding interests that include: medical research, with emphasis on brain cancer research, women's health, substance abuse, and at-risk youth. Additional funding for education and the arts. Contact the Foundation for an application form, the Board meets quarterly in January, April, July, and October.

Requirements: Nonprofit organizations in Chicago, Illinois and Fort Lauderdale, Florida area are eligible to apply.

Geographic Focus: Florida; Illinois

Amount of Grant: $1,000 - $1,900,000 range

Date(s) Application is Due: Mar 15; Jun 15; Sep 15; Dec 15

Contact: Patricia F. Mulvaney, Executive Director; email: patricia.mulvaney@thewellsfamily.net

Sponsor: Lillian S. Wells Foundation
600 Sagamore Road
Fort Lauderdale, FL 33301-2215

Lillian Sholtis Brunner Summer Fellowships for Historical Research in Nursing 3193

These summer fellowships for historical research in nursing are offered by the Center for the Study of the History of Nursing in the School of Nursing at the University of Pennsylvania. Selection of the scholars will be based on evidence of preparation and/or productivity in historical research related to nursing. It is expected that the research and new materials produced by Brunner scholars will help ensure the growth of scholarly work focused on the history of nursing. The fellowship will support up to eight weeks of residential study and use of the center's collections.

Amount of Grant: $2,500

Date(s) Application is Due: Dec 31

Contact: Dr. Karen Buhler-Wilkerson, Director; (215) 898-4725; fax: (215) 573-2168; email: karenwil@nursing.upenn.edu

Internet: http://www.nursing.upenn.edu/history/Pages/Brunner_Fellowship.aspx

Sponsor: University of Pennsylvania
School of Nursing, 418 Curie Boulevard
Philadelphia, PA 19104-6020

Lilly Endowment Community Alliances to Promote Education (CAPE) 3194

The Community Alliances to Promote Education (CAPE) is a statewide, competitive grants program aimed at the state's more than 90 community foundations. The grant will help community foundations to become effective institutions in building community, assessing and prioritizing local needs and problems, and mobilizing support to address them.

Requirements: Each community foundation may request funds to support the identification, research and planning of programs to improve its residents' educational attainment. Implementation proposals will be developed during this phase.

Restrictions: Limited to community foundations in Indiana counties.

Geographic Focus: Indiana

Amount of Grant: $50,000 - $5 million

Contact: Sarah Cobb, Vice President - Education, (317) 924-5471; fax: (317) 926-4431

Internet: http://www.lillyendowment.org/ed_cape.html

Sponsor: Lilly Endowment Inc.
2801 N. Meridian Street
Indianapolis, IN 46208

Lilly Endowment Grants 3195

Areas for which funds are granted include religion—denominational studies, congregational studies, urban ministry, etc.; leadership education—training for trustees of educational institutions, including community leadership and other community-based nonprofit organizations; education—higher education, faculty and institutional development, minority doctoral education at Big Ten and UNCF-affiliated schools, dropout prevention, etc.; youth—community guidance programs, support for African American males; and community development—improved low-income housing, the United Way, statewide expansion in support of community foundations, arts and culture, etc. Religion, public policy, and select research and pilot programs are national in scope. Community development and education grants are mainly restricted to Indiana and UNCF colleges and universities. Applications are accepted at any time.

Requirements: Applicant should send a letter describing credentials of organization, project proposed, and amount of funding requested.

Restrictions: The endowment does not consider applications for health care and biological science projects, mass media initiatives, endowments or endowed chairs, libraries, building campaigns, K-12 schools, or general operations.

Samples: Heartland Film Festival (Indianapolis, IN)—to expand its programs, strengthen its finances, and develop additional opportunities for filmmakers to showcase their work, $3.7 million (2005); U of Indianapolis (IN)—for efforts to recruit and retain outstanding faculty members and students, $1 million (2005); Indiana U at Bloomington (IN)—for life-sciences research, primarily for studies of genetic coding and the internal workings of cells, $53 million (2005).

Amount of Grant: $557.3 million total

Contact: Sue Ellen Walker, Communications Associate, (317) 924-5471; fax: (317) 926-4431

Internet: http://www.lillyendowment.org

Sponsor: Lilly Endowment Inc
2801 N Meridian Street
Indianapolis, IN 46208-0068

Lilly Endowment National Clergy Renewal Grants 3196

The program seeks to strengthen Christian congregations by providing an opportunity for pastors to step away briefly from the persistent obligations of daily parish life and to engage in a period of renewal and reflection. Renewal periods are not vacations, but times for intentional exploration and reflection, religious rejuvenation, and regaining enthusiasm and creativity for ministry. High-quality proposals should include a strong sense of purpose; a sense of balance; evidence that the congregation understands and affirms the value of a renewal program for its leader; and benefits for both the pastor and the congregation. It is anticipated that approximately 125 grants will be awarded directly to congregations for the support of a renewal program for their pastor. Up to $15,000 of the grant may be used for congregational expenses associated with the renewal program.

Requirements: A Christian congregation whose ordained pastor has a master of divinity degree is eligible.

Restrictions: The program for Indiana congregation continues, so Indiana pastors are ineligible for the national program.

Samples: To be distributed among 36 congregations in Indiana (IN)—to enable pastors to take short sabbaticals, $1.62 million (2005); To be distributed among 141 U.S. congregations—for sabbaticals for pastors, $15.024 million (2005).

Amount of Grant: $45,000 maximum, including $15,000 maximum for congregational expenses

Date(s) Application is Due: May 15

Contact: National Clergy Renewal Program, (317) 916-7302; fax: (317) 926-4431; email: clergyrenewal@yahoo.com

Internet: http://www.clergyrenewal.org

Sponsor: Lilly Endowment Inc
P.O. Box 88068, 2801 N Meridian Street
Indianapolis, IN 46208

Lilly Endowment Teacher Creativity Fellowships 3197

The program awards fellowships to support creative projects that are personally renewing and intellectually revitalizing to Indiana teachers and education professionals. In a second level of funding, the endowment also will provide funding for up to five creative renewal projects proposed by teachers that will require additional financial support and time away from the classroom or school. At this level, each individual teacher will receive a separate grant and a grant will be available to the teacher's school district to cover the costs of a replacement teacher, should the recipient need time away from the school to pursue the project. Guidelines are available online.

Requirements: Licensed teachers in Indian public, private, and parochial schools (including guidance counselors and library/media specialists), principals, and assistant principals in such schools (who have held one of these positions since Fall 2003 or before) are eligible. They must be currently employed full-time and must be Indiana-certified in grades K-12.

Restrictions: Past fellowship recipients who received their awards in 1999 or later are ineligible.

Geographic Focus: Indiana

Amount of Grant: $8000; $25,000 maximum to teacher and $25,000 maximum to school for creative renewal projects

Date(s) Application is Due: Dec 1

Contact: Fellowship Administrator, email: info@teachercreativity.org

Internet: http://www.teachercreativity.org

Sponsor: Lilly Endowment
2801 N Meridian Street
Indianapolis, IN 46208-0068

Lindbergh Grants 3198

Grants are awarded to individuals whose proposed projects represent a significant contribution toward the achievement of a better balance between technology and our human and natural environment. Nine to 10 grants are awarded each year. The current funding program is interested in increasing representation in the following areas: aviation/aerospace; agriculture; arts and humanities; biomedical research; conservation of natural resources; exploration; health and population sciences; intercultural communication; oceanography; water resources management; waSuite disposal management; wildlife preservation; and adaptive technology. Grant application can be obtained by writing the fund; enclose a self-addressed, stamped envelope with request. The Web site contains additional information.

Requirements: Grants are international in scope and open to citizens of all countries. Letters of request, proposals, endorser reports, and required progress and final reports must be submitted in English. The fund welcomes candidates who may or may not be affiliated with academic or nonprofit institutions. Grants are made to individuals, not to affiliated institutions for institutional programs. Note: When sending any materials by fax, seven hard copies via mail or delivery are requested. The deadline for grant applications is the second Thursday of June in the year preceding the awarding of funds.

Restrictions: The fund does not provide support for overhead costs of organizations, tuition, scholarships, fellowships, or travel related to such.

Samples: Dr. Ernest Diedrich, Saint John's College (Collegeville, MN)—for Balancing Energy Consumption with Resource Conservation: A Wind Power Study for Central Minnesota, an Educational Model for the World.;Dr. Ping Liu, Eastern Illinois U (Charleston, IL)—for Analysis of Recycled Plastic Composites to Aid in the Development and Design of Recycled Plastic Products.;Dr. Vivian Charles McAlister, Dalhousie U (Halifax, Nova Scotia)—for Use of Human Response to Parasites to Improve Organ Transplant Survival.;Dr. Carl Wulfman, U of the Pacific (Stockton,

CA)—for Clean Energy into Clean Energy: Toward a New Method of Using Sunlight to Produce Electricity.
Amount of Grant: $10,580 maximum
Date(s) Application is Due: Jun 9
Contact: Grants Coordinator, (763) 576-1596; fax: (763) 576-1664; email: info@lindberghfoundation.org
Internet: http://www.lindberghfoundation.org/grants
Sponsor: Charles A. and Anne Morrow Lindbergh Foundation
2150 Third Avenue N, Suite 310
Anoka, MN 55303-2200

Link Foundation Energy Doctoral Fellowships 3199
The objective of the program is to foster energy research, to enhance both the theoretical and practical knowledge and application of energy research, and to disseminate the results of that research through lectures, seminars, and publications. On the basis of applications to the foundation in the form of research proposals, awards will be made to doctoral students in academic institutions. Preference will be shown to proposals dealing directly with energy and exploring ideas not yet fully tested, rather than to developed programs already in progress. Ideas that can be implemented in the relatively near term will be given first priority. Application guidelines are available online.
Requirements: Preference will be given to candidates who have a well-defined thesis direction but who still have enough time remaining working toward their PhD that receipt of a Link Foundation fellowship could make a difference.
Restrictions: The foundation does not support student tuition, institutional overhead, or pooled costs.
Amount of Grant: $25,000 total: $21,500 stipend; $2500 for research-related expenses; and $1000 to attend one or more technical meetings and/or publication expenses
Date(s) Application is Due: Dec 1
Contact: Dr. Lee Lynd, Manager, Thayer School of Engineering, (603) 646-2674
Internet: http://www.linkenergy.org
Sponsor: Link Foundation
Dartmouth College
Hanover, NH 03755

Linn-Henley Charitable Trust Grants 3200
The charitable trust awards grants to Alabama nonprofit organizations to improve the quality of life in its areas of interest, including arts and culture, higher education, libraries and library science, religion, and hospitals. Types of support include capital campaigns, research grants, and matching funds.
Requirements: Jefferson County, AL, nonprofit organizations are eligible.
Geographic Focus: Alabama
Samples: Southern Museum of Flight (AL)—$10,000.; U of Alabama (AL)—to support the Linn-Henley International Awards banquet, $7000.
Amount of Grant: $2000-$120,000 range
Contact: Collins Compere, Trustee, c/o Regions Bank, Trust Dept, (205) 326-7224
Sponsor: Linn-Henley Charitable Trust
P.O. Box 10247
Birmingham, AL 35202

Lionel Pearson Fellowships in the Classics 3201
These fellowships contribute to the training of American/Canadian classicists by providing for a period of study at an English or Scottish university. Fellows may use the fellowship to support one year of a longer program of study. Candidates must first be nominated by faculty members (preferably teachers of the classics) at their undergraduate institutions. Candidates will be interviewed at the APA convention. Funding covers academic fees, living stipend, airfare, and other necessary expenses of travel to and from the place of study. Those who wish to nominate students for the fellowship should send the name and address of the student to the current Chair, Professor Kenneth Scott Morrell, who will send the nominator an application form and other relevant materials. The deadline for receiving nominations is Monday October 4, 2004 and for receiving completed applications (including letters of recommendation), Monday November 4, 2004. Note that these are not postmark deadlines.
Requirements: Competition is open to majors in Greek, Latin, or the classics. Applicant should be in the final year of undergraduate study at a U.S. or Canadian university, expect to obtain the BA degree by September of the following calendar year, and be able to begin an academic year of postgraduate work at that time.
Amount of Grant: $25,000 maximum
Date(s) Application is Due: Oct 4; Nov 4
Contact: Dr. Adam Blistein, Executive Director, (215) 898-4975; fax: (215) 573-7874; email: blistein@sas.upenn.edu
Internet: http://www.apaclassics.org/Administration/pearson html
Sponsor: American Philological Association
292 Logan Hall, 249 S 36th Street
Philadelphia, PA 19104-6304

LITA Frederick G. Kilgour Award for Research in Library and Information Technology 3202
The Frederick G. Kilgour Award for Research in Library and Information Technology is sponsored by OCLC Online Computer Center, Inc. and the Library and Information Technology Association (LITA), a division of the American Library Association. The purpose of this award is to bring attention to research relevant to the development of information technologies, especially work which shows promise of having a positive and substantive impact on any aspect of the publication, storage, retrieval and dissemination of information, or the processes by which information and data is manipulated and managed. The award will consist of $2,000 cash and an expense paid trip to the ALA conference (airfare and two nights lodging).
Requirements: Nominations will be accepted from any member of the American Library Association. Nominating letters should briefly address how the research is relevant to libraries; is creative in its design or methodology; builds on existing research or enhances potential for future exploration; and/or solves an important current problem in the provision of information resources. In addition, a complete curriculum vita and a copy of several seminal publications by the nominee must be included. Nominees may be asked to submit additional information to the Kilgour Committee . Preference will be given to completed research over works in progress.
Restrictions: Currently-serving officers and elected officials of LITA, members of Kilgour Committee and OCLC employees and their immediate family members are ineligible.
Samples: Dr. Jane Greenberg, University of North Carolina, NC .;2008.; Dr. Jane Greenberg, University of North Carolina, NC .;2008,;
Amount of Grant: $2,000
Contact: Mary Taylor, LITA Executive Director, (312) 280-4267; fax: (312) 280-3257; email: mtaylor@ala.org; Michael Gorman, Kilgour Award Chair; email: michaelg@csufresno.edu
Internet: http://www.ala.org/ala/mgrps/divs/lita/litaresources/litascholarships/litascholarships.cfm
Sponsor: American Library Association
50 East Huron Street
Chicago, IL 60611

LITA/Gaylord Award for Achievement in Library and Information Technology 3203
The purpose of the award is to recognize distinguished leadership in technology, notable developments of applications of technology, superior accomplishment in research or education or original contribution to literature in the field.
Requirements: Nominations for the award may be made by any member of the American Library Association. Nominations are invited and judged by the LITA/Gaylord Awards Committee.
Amount of Grant: $1,000
Date(s) Application is Due: Mar 1
Contact: Mary Carmen Chimato, LITA Representative, (919) 513-0869; fax: (919) 513-3653; email: mary_chimato@ncsu.edu or lita@ala.org; Megan Griffin, (800) 545-2433 ext. 4270; fax: (312) 280-3257; email: mgriffin@ala.org
Internet: http://www.ala.org/ala/mgrps/divs/lita/litaresources/litascholarships/litascholarships.cfm#oclchttp://www.ala.org/ala/lita/litaresources/litascholarships/litascholarships.htm#oclc
Sponsor: American Library Association
50 East Huron Street
Chicago, IL 60611

LITA/LSSI Minority Scholarship in Library and Information Technology 3204
This scholarship is awarded annually for work toward an MLS degree in an ALA-accredited program with emphasis on library automation. Application forms are available on online. Candidate must not have earned more than 12 hours towards a MLS degree.
Requirements: An applicant must be a U.S. or Canadian citizen enrolled at or accepted for admission to an ALA-accredited MLS program, with an emphasis on library automation, and be a member of one of the following minority groups: Native American, Alaskan Native, Asian or Pacific Islander, African American, or Hispanic.
Amount of Grant: $2500
Date(s) Application is Due: Mar 1
Contact: Scholarship Staff Liaison, Library and Information Technology Association, (312) 280-4281 or (800) 545-2433 ext 4281; fax: (312) 280-3257; email: lita@ala.org
Internet: http://www.ala.org/ala/lita/litaresources/litascholarships/litascholarships.htm#oclchttp://www.ala.org/ala/lita/litaresources/litascholarships/litascholarships.htm#oclc
Sponsor: American Library Association
50 E Huron Street
Chicago, IL 60611

Little River Foundation Grants 3205
The foundation supports nonprofit organizations in the Mid-Atlantic states, primarily New Jersey, New York, and Pennsylvania. Grants are awarded for hospitals, AIDS research, substance abuse services, higher education, secondary education, early childhood education, adult basic education, building funds, literacy programs, animal welfare, legal education, international studies in Africa, environmental conservation, legal services, religious organizations, and community funds. Grants are awarded for general operating support, annual campaigns, endowment funds, conferences and seminars, research, and building construction/renovations. Application forms are not required. The board meets each year in the fall.
Requirements: Only residents of Maryland, New Jersey, New York, Pennsylvania, and Virginia may apply.
Geographic Focus: California; New York; Virginia
Samples: Grace Episcopal Church (The Plains, VA)—for facilities improvement, $25,000.; Piedmont Environmental Council (Warrenton, VA)—for the building program, $87,000.; Johns Hopkins U (Baltimore, MD)—for brain tumor research, $25,000.; Princeton U (Princeton, NJ)—for annual support, $25,000.
Amount of Grant: $100-$175,000 range
Contact: Dale Hogoboom, Assistant Treasurer

Sponsor: Little River Foundation
101 Park Avenue, Suite 3500
New York, NY 10178-0061

Littlefield-AACR Grants in Metastatic Colon Cancer Research 3206
The grants will provide support for innovative cancer research projects designed to accelerate the discovery and development of new agents to treat metastatic colon cancer and/or for pre-clinical research with direct therapeutic intent. Special emphasis will be placed on research that holds promise for leading to individualized therapeutic options for treatment in the near future or for developing promising new cancer therapeutics for metastatic colon cancer, which will translate into clinical applications within a one- to two-year period.
Requirements: The prizes are open to all cancer researchers who are affiliated with any institution involved in cancer research, cancer medicine, or cancer-related biomedical science anywhere in the world. Such institutions include those in academia, industry, or government. Candidates must be active researchers and have a record of recent publications.
Amount of Grant: $500,000-$1,000,000
Date(s) Application is Due: Feb 1
Contact: Julia Laurence, Staff Assistant; (215) 440-9300, ext. 102 or (267) 646-0655; fax: (215) 440-9372; email: laurence@aacr.org
Internet: http://www.aacr.org/home/scientists/research-funding—training-grants/research-funding/littlefield-aacr-grants-for-metastatic-colon-cancer-research.aspx
Sponsor: Littlefield 2000 Trust and the American Association for Cancer Research
615 Chestnut Street, 17th Floor
Philadelphia, PA 19106-4404

Lloyd G. Balfour Foundation Grants 3207
The foundation funds programs that remove barriers to education for minority and disadvantaged youth. Educational programs that encourage youth to excel in grade school and high school and to continue on with a postsecondary education are of particular interest. Types of support include scholarship funds, fellowships, matching funds, operating budgets, seed money, and special projects.
Requirements: IRS 501(c)3 organizations in Massachusetts and the New England area eligible.
Geographic Focus: Massachusetts; New England
Amount of Grant: $75,000-$200,000 average
Date(s) Application is Due: Feb 3
Contact: Kerry Herlihy Sullivan, Senior Vice President, (617) 434-4846; email: kerry.h.sullivan@bankofamerica.com
Internet: http://www.bankofamerica.com/privatebank/index.cfm?template=charitable_management_services
Sponsor: Fleet Investment Management
100 Federal Street
Boston, MA 02110

Loeb Fellowships in Advanced Environmental Design Studies 3208
Loeb Fellowships provide part-time and full-time residencies in environmental studies in Harvard's Graduate School of Design. Part-time fellowships are limited to persons living within a two-hour airline distance of Cambridge, MA; it is expected that they will spend at least two days per week at Harvard during an academic semester or the academic year. Full-time fellows are expected to obtain a leave of absence for the duration of the appointment. Experienced professionals and academicians in the field are requested to nominate those they consider eligible for the fellowships.
Requirements: Midcareer professionals in design, planning, and related fields, normally with at least five years of professional experience, may apply. While residency/citizenship are not required, fellowships are limited to individuals whose work directly impacts the built environment of the United States.
Date(s) Application is Due: Jan 4
Contact: Loeb Fellowships, Harvard University Graduate School of Design; (617) 495-9345; fax: (617) 384-9529; email: loeb_fellowship@gsd.harvard.edu
Internet: http://www.gsd.harvard.edu/loebfell
Sponsor: Harvard University
48 Quincy Street, Room 420
Cambridge, MA 02138

Longview Foundation Grants 3209
The foundation awards grants to U.S. regional and state organizations working to enrich existing K-12 curricula with global perspectives. Priority is given to elementary and middle school education and teacher education for those grades. Grants support projects, research, seed money, and conferences and seminars. Listed deadlines are for letters of intent. Annual deadline dates may vary; contact program staff for exact dates.
Requirements: 501(c)3 nonprofit organizations are eligible.
Samples: Associates for Cultural Exchange (Seattle, WA)—for support of the Washington State Coalition for International Education including a leadership summit, a survey on world languages and technology, a report on statewide international education activities, $15,000 (2004); Brown University, Choices Program (Providence, RI)—for staffing support for the expansion of the Capitol Forum for America's Future, an experiential civic engagement program designed to give high school students a voice in public consideration of current international issues, $15,000 (2004).
Amount of Grant: $15,000-$25,000 average range
Date(s) Application is Due: Jan 5; Jul 1
Contact: Betsy Devlin-Foltz, Program Director, (301) 681-0899; fax: (301) 681-0925; email: globaled@longviewfdn.org
Internet: http://fdncenter.org/grantmaker/longview
Sponsor: Longview Foundation for Education in World Affairs and International Understanding
8639 B 16th Street, Box 211
Silver Spring, MD 20910

Longwood Foundation Grants 3210
Limited grants are made generally to educational institutions, to hospitals for construction purposes, and to social and youth agencies primarily in Delaware, with emphasis on Wilmington and with some giving also in Pennsylvania. Grants are given for annual campaigns, building funds, equipment (including computers), land acquisition, and research.
Requirements: Delaware nonprofit organizations and hospitals serving southern Chester County in Pennsylvania are eligible.
Restrictions: Grants are not given to individuals or for special projects. Grants generally will not be awarded to support K-12 schools or public high schools.
Geographic Focus: Delaware; Pennsylvania
Samples: Christiana Care Health System (Wilmington, DE)—$4 million (2004); Connections Community Support Programs (Wilmington, DE)—$1 million (2004); Exceptional Care (Wayne, PA)—$500,000 (2004).
Amount of Grant: $25,000-$500,000 average
Date(s) Application is Due: Mar 15; Sep 15
Contact: Peter Morrow, Executive Director, (302) 654-2477; fax: (302) 654-2323
Sponsor: Longwood Foundation
100 W 10th Street, Suite 1109
Wilmington, DE 19801

Lorain County Grants and Scholarships 3211
The foundation awards medical scholarships and educational grants for postgraduate studies in the field of medical arts to residents of Lorain County, OH, who intend to return to practice in Lorain County.
Requirements: Applicant must be a Lorain County, OH, resident for at least two years and demonstrate financial need.
Geographic Focus: Ohio
Amount of Grant: $2000 maximum
Contact: Lynne Hammes, Administrator, c/o Bank One Ohio Trust Co., (216) 781-2386 or (800) 945-2265; email: lynne.h.hammes@jpmorgan.com
Sponsor: James R. Nicholl Memorial Foundation
600 Superior Avenue
Lorain, OH 44114

Louis Calder Foundation Grants 3212
The foundation supports projects that benefit the children and youth of New York City in the areas of health, welfare, religion, environment, and education. General operating grants, project/program grants, challenge/matching grants, and scholarships are awarded. The foundation's small grants program awards one-time grants to organizations whose programming and budgets would ordinarily fall outside of funding parameters. Small grants are designed to increase the recipient organization's visibility enabling it to attract new potential funding sources.
Requirements: Nonprofit organizations serving the New York City area and its residents are eligible. Typical recipients include schools, hospitals and other medical agencies, youth groups, human service agencies, museums, libraries, and arts groups.
Restrictions: Awards to performing arts groups or private colleges are by invitation only.
Geographic Focus: New York
Samples: Saint John's U (Jamaica, NY)—to endow a scholarship fund, $333,000 challenge grant.; Canisius College (Buffalo, NY)—for scholarships for New York City students who demonstrate academic promise and acute financial need, $250,000 challenge grant.; Citizens Advice Bureau (New York, NY)—for support to implement a whole language-based after-school program for South Bronx children, $25,000.
Amount of Grant: $5000-$250,000 typically; $5000 maximum small grants
Contact: Holly Nuechterlein, Program Manager, (203) 966-8925; fax: (203) 966-5785; email: admin@calderfdn.org
Internet: http://www.louiscalderfdn.org
Sponsor: Louis Calder Foundation
175 Elm Street
New Canaan, CT 06840

Louise Wallace Hackney Fellowship for the Study of Chinese Art 3213
The scholarships are for the study of Chinese art, with special emphasis on painting and its reflection of Chinese culture, and for translation into English of works on Chinese art. Scholarships include funding for travel if necessary and are renewable.
Requirements: Applicants must be citizens of the United States and should have completed three years of study of the Chinese language or the equivalent and should be able to demonstrate commitment to the serious study of Chinese art. Doctoral and postdoctoral students may apply.
Restrictions: Fellowships are not awarded to scholars of well-recognized standing but are given to men or women who show aptitude or promise in the field of Chinese art.
Amount of Grant: $8000
Date(s) Application is Due: Mar 1

Contact: Dr. Jonathan Rodgers, Secretary-Treasurer, (734) 764-7555; fax: (734) 763-6743; email: jrodgers@umich.edu
Internet: http://www.umich.edu/~aos/hackney.htm
Sponsor: American Oriental Society
University of Michigan, Hatcher Library
Ann Arbor, MI 48109-1205

Louisiana State Arts Council Artist Fellowships **3214**
Nonmatching artist fellowships are awarded to artists of exceptional talent to enable them to pursue their artistic goals. Applicant must designate a specific discipline and subcategory in which to be considered: crafts (glass, ceramics, fiber, wood, metals, etc.); dance (performance or choreography); design arts (architecture, landscape architecture, urban design, historic preservation and planning, interior design, industrial design, graphic design, and fashion design); folklife (traditional arts and crafts and such performing arts as music, storytelling, and dance); literature (fiction, nonfiction, and poetry); media (film, audio, video, or photography); music (vocal performance, instrumental performance, or composition); theater (dramatic performance—acting, play writing, or screenplay writing); or visual arts (painting, printmaking, drawing, sculpture, etc.). Artists may submit only one fellowship application per grant deadline.
Requirements: Artists who are legal residents of Louisiana and who have lived in the state for at least 24 consecutive months prior to the application deadline may apply. Applicants should have demonstrated professional experience in their fields, with a serious career commitment and some degree of peer acceptance either as emerging artists of outstanding promise or established artists with a body of work.
Restrictions: Students or persons enrolled in arts-related degree or certificate-granting programs are not eligible to apply. Previous recipients of division fellowships also are not eligible. (Former applicants who have not received fellowships are encouraged to reapply.) Absences from the state for six or more months during the two years prior to the application deadline disqualify the artist.
Geographic Focus: Louisiana
Amount of Grant: $5000
Contact: Program Officer, (225) 342-8180; fax: (225) 342-8173; email: arts@crt.state.la.us
Internet: http://www.crt.state.la.us/arts/state_art_grant/fellowsh.htm
Sponsor: Louisiana State Arts Council and Division of the Arts
P.O. Box 44247
Baton Rouge, LA 70804

Louisiana State Arts Council Individual Artist Mini-Grants **3215**
Mini-grants are designed to encourage artistic development, to support the realization of specific artistic ideas, and to recognize the significant contribution artists make to the creative environment of the state of Louisiana. These grants are available for Louisiana individual artists with unique professional development needs. They are designed to support specific activities relevant to the artist's work and career. Proposals may include, but are not limited to, supplies, equipment, or time necessary for the development, completion, or presentation of new work; documentation of work; travel essential to artistic research or to present or complete work; and advanced workshops. Up to 20 grants will be awarded in each funding cycle.
Requirements: Emerging and established Louisiana artists in all disciplines may apply.
Restrictions: Support for mini-grants is not granted for proposals eligible for funding through other Division of the Arts programs; activities not directly related to the enhancement of an artist's career; activities that do not demonstrate a proven commitment to a particular discipline; costs associated with any degree, i.e., tuition, fees, materials, curriculum planning, teaching materials, or teaching programs; activities completed prior to the award; or expenses incurred before the grant start date.
Geographic Focus: Louisiana
Amount of Grant: $500
Contact: Program Officer, (225) 342-8180; email: arts@crt.state.la.us
Internet: http://www.crt.state.la.us/arts/state_art_grant/minigran.htm
Sponsor: Louisiana State Arts Council and Division of the Arts
P.O. Box 44247
Baton Rouge, LA 70804

Love Thy Neighbor Course Competition for Secondary School Faculty **3216**
The competition encourages academically rigorous secondary school courses that focus on unselfish love of neighbor as a spiritual and practical ideal. The program supports innovative courses that couple the study and practice of the spirituality of love with the sciences, such as the neurology of compassion, the evolution of altruism, cosmology, or the enhancement of psychological well-being and health. The winning courses must combine the study of unselfish love as understood within spiritual traditions and scientific frameworks, such as physics, cosmology, evolution, biology, political science, the social sciences, and health. The institute encourages a deeply integrative approach to the topic, which may also include experiential learning. Two hard copies of the application should be mailed to Dr. Post, and a copy must also be emailed to him as attachments. Course ideas, additional information, and guidelines are available online.
Requirements: Faculty in religious studies and/or theology, humanities, and the sciences are encouraged to apply. Courses co-taught by two faculty members representing religion and science might be especially compelling in some cases, but team teaching is by no means imperative. In addition to new individual course development, it is possible to apply on the basis of refocusing an already existing course on the theme of unselfish love with significant attention to religion and science. In some cases, several faculty teaching in an integrative core humanities course might work together to include the topic as a major and sustained theme.
Amount of Grant: $5000
Date(s) Application is Due: Jul 15
Contact: Dr. Stephen Post, School of Medicine, Room 214A, (216) 368-6205; email: Stephen.Post@Case.edu
Internet: http://www.unlimitedloveinstitute.org/competition.html
Sponsor: Institute for Research on Unlimited Love—Altruism, Compassion, Service
10900 Euclid Avenue
Cleveland, OH 44106-4976

Lowe Foundation Grants **3217**
The foundation awards grants to organizations in Texas, primarily for programs that support the critical needs of women and children. The arts and higher education also receive support. Grants are awarded to support capital campaigns, building construction/renovation, and general operations. The board meets in April of each year. The December 1 deadline listed is for pre-proposals, with full proposals by request due on December 31.
Requirements: 501(c)3 organizations in Texas may apply for grant support.
Geographic Focus: Texas
Samples: U of Texas Southwestern Medical Ctr at Dallas (Dallas, TX)—to endow a professorship in pediatric critical care, $100,000.
Amount of Grant: $5000-$225,000 range
Date(s) Application is Due: Dec 1; Dec 31
Contact: Clayton Maebius, Trustee; (512) 322-0041; fax: (512) 322-0061; email: info@thelowefoundation.org
Internet: http://www.thelowefoundation.org/guidelines.htm
Sponsor: Lowe Foundation
1005 Congress Avenue, Suite 895
Austin, TX 78701

Lown Cardiovascular Center Developing World Fellowship **3218**
The fellowship offers training in noninvasive and preventive cardiology and public health for physicians from developing countries. Participants receive training in a model of treatment that maximizes risk factor modification and medical management in order to minimize intervention. After completing training, fellowship participants return to their home countries prepared to address the mounting epidemic of cardiovascular disease. This two-year fellowship program program is conducted in the United States. Fellowship recipients travel to the United States on a J-1 visa. Guidelines are available online.
Requirements: The following criteria are essential and cannot be waived: valid ECFMG certificate; training in internal medicine and/or cardiology; formal affiliation with a supporting institution in the country of origin; and commitment to returning to medical practice in the home country after completing the fellowship program. Candidates from developing countries who meet the specified criteria are invited to apply.
Amount of Grant: $40,000 stipend
Contact: Dr. Shmuel Ravid, (617) 732-1318; fax: (617) 734-5763
Internet: http://www.lowncenter.org/fellowship.htm
Sponsor: Lown Cardiovascular Research Foundation
21 Longwood Avenue
Brookline MA 02446

LRF New Investigator Leukemia Research Grants **3219**
The foundation supports leukemia-related research in a broad sense, including such areas as biochemistry, microbiology, virology, molecular biology, genetics, and immunology. Projects related to significant aspects of leukemia, lymphoma, or myelodysplasia also are considered. The goal of the program is to support new investigators; preference will be given to applicants proposing new lines of investigation. Types of support include research grants and postdoctoral fellowships. Grants and fellowships are for a one-year period and may be renewable for a second year. Annual deadline dates may vary; contact program staff for exact dates.
Requirements: Grants are restricted to researchers who are within five years of the end of their training or of their first independent faculty appointment at the time of the proposed starting date of the grant.
Restrictions: The funds may not be used for salary support of the principal investigator.
Samples: Dr. Curt Horvath, Mount Sinai School of Medicine (New York, NY)—for research to determine how the STAT:IRF complexes function to mediate interferon biological responses, $50,000.; Dr. Alona Weiss, Massachusetts General Hospital (Boston, MA)—for research to further understand the mechanisms and processes involved in normal gene regulation and to contribute to understanding of the causes of diseases, such as leukemia, where these mechanisms fail to function, $25,000.
Amount of Grant: $75,000
Date(s) Application is Due: Feb 15
Contact: Kelli Fitzgerald, Grant Manager, (847) 424-0600; fax: (847) 424-0606; email: Kelli@LRFMail.org
Internet: http://www.leukemia-research.org/phd/application.asp
Sponsor: Leukemia Research Foundation
2700 Patriot Boulevard, Suite 100
Glenview, IL 60026-8021

LRF Physician-Scientist Postdoctoral Fellowships **3220**
The foundation awards physician-scientist postdoctoral fellowships to help physicians prepare for independent careers in research of leukemia and related disorders and obtain significant research results. Fellowships support two years of research. Annual deadline dates may vary; contact program staff for exact dates.

Requirements: Fellows (MD, DO, DVM, or equivalent) not yet established as independent investigators are eligible.
Amount of Grant: $30,000 per year
Date(s) Application is Due: Feb 15
Contact: Kelli Fitzgerald, Medical Grants Administrator, (847) 424-0600; fax: (847) 424-0606; email: Kelli@LRFMail.org
Internet: http://www.leukemia-research.org/phd/application.asp
Sponsor: Leukemia Research Foundation
2700 Patriot B;vd, Suite 100
Glenview, IL 60026-8021

LRF Postdoctoral Fellowships 3221
The program awards postdoctoral fellowship support for trainees who wish to work in a research area related to leukemia and related disorders. Fellowships are awarded for two years. Annual deadlines dates may vary; contact program staff for exact dates.
Requirements: Fellows (MD or PhD) with up to three years of training (excluding medical residency) prior to receiving the fellowship are eligible.
Amount of Grant: $30,000 maximum per year
Date(s) Application is Due: Feb 15
Contact: Kelli Fitzgerald, (847) 424-0600; fax: (847) 424-0606; email: Kelli@LRFMail.org
Internet: http://www.leukemia-research.org/phd/application.asp
Sponsor: Leukemia Research Foundation
2700 Patriot Boulevard, Suite 100
Glenview, IL 60026-8021

LSA Fellowships 3222
Fellows of the Leukemia Society are promising investigators with no or minimal prior postdoctoral experience working in fields relevant to leukemia. They will have had less than two years of postdoctoral research training at the time of review. Fellows are encouraged to embark on academic careers involving clinical or fundamental research in leukemia, Hodgkin's disease, the lymphomas, and multiple myeloma. Fellowships are awarded for three years and are nonrenewable; however, an application may be made for a special fellowship in the third year of the fellowship award. The preliminary, two-page application must meet the September application deadline; complete applications must be received by the October date.
Requirements: Applicant must hold a PhD, MD, or equivalent degree at the time a grant is to become effective. There are no restrictions as to citizenship, age, race, sex, or creed.
Restrictions: An award will not be made to an individual with full institutional salary support; applicants must demonstrate lack of or minimal/limited salary support as per institutional policies. Awards are payable to the academic institution on behalf of the investigator.
Amount of Grant: $50,000 (stipend $47,000 + $3000 institutional overhead) per year for three years
Date(s) Application is Due: Sep 15; Oct 1
Contact: Director of Research Administration, (914) 821-8843; fax: (914) 821-8946; email: researchprograms@leukemia-lymphoma.org
Internet: http://www.leukemia-lymphoma.org/all_page?item_id=11618
Sponsor: Leukemia Society of America
1311 Mamaroneck Avenue
White Plains, NY 10605

LSA Scholar Award for Clinical Research 3223
Applicants for this award have demonstrated their ability to design and conduct original research on leukemia, lymphoma, and myeloma. They are expected to hold an independent faculty-level or equivalent position for a period of not less than three years and to have obtained substantial research support from a national agency. This program is designed as a career development award and is not intended for the support of well-established or senior investigators. The award runs for five years; however, annual renewal is required and is based on a yearly noncompetitive progress report review.
Amount of Grant: $110,000 ($105,000 + $5000 institutional overhead) per year for five years
Date(s) Application is Due: Sep 15; Oct 1
Contact: Director of Research Administration, (914) 821-8843; fax: (914) 821-8946; email: researchprograms@lls.org
Internet: http://www.leukemia-lymphoma.org/all_page?item_id=11618
Sponsor: Leukemia Society of America
1311 Mamaroneck Avenue
White Plains, NY 10605

LSA Scholar Awards 3224
This is a career development award for highly qualified scholars who have demonstrated over a five-year period the ability to conduct original research on leukemia, Hodgkin's disease, the lymphomas, and multiple myeloma. Recipients are expected to hold independent faculty-level or equivalent positions and will usually have successfully competed for research grant support for their work. The program has been designed to aid the individual rather than the project and to encourage continued investigation while ensuring the investigator an income. The applicant should have shown the capacity for sustained, original investigation in abnormal cell growth as it relates to leukemia, lymphoma, and myeloma and must desire an investigative career in the field of leukemia, lymphoma, and myeloma. Awards are usually given to individuals in the clinical or basic science departments of universities or associated research institutions. Scholarships are awarded for terms of five years and are nonrenewable. Preliminary, two-page applications must meet the September deadline; complete applications must meet the October deadline.
Requirements: Applicant must hold a PhD, MD, or equivalent degree at the time the grant is to become effective. There are no restrictions as to citizenship, age, race, sex, or creed.
Restrictions: An award will not be made to an individual with full institutional salary support; applicants must demonstrate lack of or minimal/limited salary support as per institutional policies. All awards must be payable to an academic institution.
Amount of Grant: $110,000 ($105,000 + $5000 institutional overhead) per year for five years
Date(s) Application is Due: Sep 15; Oct 1
Contact: Director of Research Administration, (914) 821-8843; fax: (914) 821-8946; email: researchprograms@lls.org
Internet: http://www.leukemia-lymphoma.org/all_page?item_id=11618
Sponsor: Leukemia Society of America
1311 Mamaroneck Avenue
White Plains, NY 10605

LSA Special Fellowships 3225
Special Fellows of the Leukemia Society are scientists who have completed more basic training and education in the field than are required for a fellow but less than is necessary for a scholar. A special fellow has had at least two years of postdoctoral training at the time of review but does not yet hold an independent position and relies on the sponsor for research support. Investigators applying must be willing to dedicate their careers to research bearing on a cure or control of leukemia, Hodgkin's disease, the lymphomas, or multiple myeloma. The special fellowships are awarded for a period of three years and are nonrenewable; however, an application may be made for a scholarship in the third year of the special fellowship award. Proposed research projects must be related directly to the field of leukemia, lymphomas, and myeloma. Preliminary, two-page applications must meet the September deadline; complete applications must meet the October deadline.
Requirements: Applicant should hold a PhD, MD, or equivalent degree at the time a grant is to become effective. There are no restrictions as to citizenship, age, race, sex, or creed.
Restrictions: An award will not be made to an individual with full institutional salary support; applicants must demonstrate lack of or minimal/limited salary support as per institutional policies. Award must be payable to an academic institution.
Amount of Grant: $60,000 ($ 56,000 + $ 4,000 institutional overhead) per year for three years
Date(s) Application is Due: Sep 15; Oct 1
Contact: Director of Research Administration, (914) 821-8843; fax: (914) 821-8946; email: researchprograms@lls.org
Internet: http://www.leukemia-lymphoma.org/all_page?item_id=11618
Sponsor: Leukemia Society of America
1311 Mamaroneck Avenue
White Plains, NY 10605

LSA Specialized Center of Research Grants 3226
The purpose of the program is to increase the cure rate of or prevent leukemia, Hodgkin's and non-Hodgkin's lymphoma, and myeloma in order to foster interdisciplinary and synergistic research. The proposed center should be interdisciplinary, cohesive, and sharply focused and should be composed of at least three relevant scientific projects capable of interacting. The research may be fundamental or applied or an integrated combination of the two approaches. Basic research tied to a related translational research project is encouraged but not mandatory. The most promising ideas and the likelihood of a significant impact on the morbidity and mortality of these diseases will be viewed favorably regardless of the particular approach. The center grant will also support scientific core laboratories required by the component research programs. Preliminary applications are due November 1; full applications are due March 15. Annual deadlines may vary; contact program staff for exact dates.
Requirements: Applications may be submitted by individuals holding an MD, PhD, or equivalent, working in domestic or foreign nonprofit organizations, such as universities, colleges, hospitals, or laboratories. Applications may be multi-institutional in nature. Applicants need not be U.S. citizens.
Amount of Grant: $1.25 million maximum
Date(s) Application is Due: Mar 15; Nov 1
Contact: Director of Research Administration, (914) 821-8843; fax: (914) 821-8946; email: researchprograms@lls.org
Internet: http://www.leukemia.org/all_page?item_id=11627
Sponsor: Leukemia Society of America
1311 Mamaroneck Avenue
White Plains, NY 10605

LSA Translational Research Program Grants 3227
The society sponsors research toward a cure or control of leukemia, lymphoma, and myeloma. Grants fund research that shows high promise for translating basic biomedical knowledge to prevention or new treatments and, ultimately, to prolong and enhance life. The goal of the program is to accelerate transfer of findings from the laboratory to clinical application. A renewal of the grant may be funded for an additional two years if it includes an approved clinical trial. A principal intent of the program is to permit the acquisition of sufficient data to apply to an appropriate agency or sponsor for sustained support. Guidelines and sample application are available online.
Requirements: Applications may be submitted by individuals holding an MD or PhD, or equivalent degree, working in domestic or foreign nonprofit organizations, such as

universities, colleges, hospitals, laboratories, units of state and local governments, or eligible agencies of the federal government.
Amount of Grant: $200,000 maximum
Date(s) Application is Due: Mar 1
Contact: Director of Research Administration, (914) 821-8843; fax: (914) 821-8946; email: researchprograms@lls.org
Internet: http://www.leukemia-lymphoma.org/all_page?item_id=11620
Sponsor: Leukemia and Lymphoma Society of America
1311 Mamaroneck Avenue
White Plains, NY 10605

Lubbock Area Foundation, Inc. 3228
The foundation awards grants, primarily in the South Plains, TX, area, in its areas of interest, including education (elementary, secondary, higher, medical school, and nursing education), hospitals, human and social services, arts and culture, and religion. Types of support include general operating support, annual campaigns, capital campaigns, building construction/renovation, equipment acquisition, publication, curriculum development, scholarship funds, research, and matching funds There are no application forms. Submit a brief letter of inquiry that includes a organization background, project description and objectives, purpose of request, detailed budget, additional sources of funding, specific amount requested, proof of tax-exempt status, recently audited financial statement, and a statement of approval of the request signed by the chief administrator.
Requirements: Texas nonprofits are eligible.
Geographic Focus: Texas
Samples: Texas Tech U (Lubbock, TX)—for general support for museums, $175,000.; All Saints Episcopal School (Lubbock, TX)—for the building fund, $100,000.; Lubbock Symphony Orchestra (Lubbock, TX)—for general support, $25,000.
Amount of Grant: $5000-$50,000 average
Date(s) Application is Due: May 1
Contact: Kathleen Stocco, Executive Director, (806) 762-8061; fax: (806) 762-8551
Internet: http://www.lubbockareafoundation.org/contact.shtml
Sponsor: CH Foundation
1655 Main Street, Suite 202
Lubbock, TX 79401

Lubrizol Foundation Grants Program 3229
The foundation supports education, youth, health care, human services, the arts, environment, and civic affairs in Lubrizol Corporation operating locations. Funding in education supports higher education scholarships, fellowships in chemistry and chemical and mechanical engineering, adult basic education and literacy, and secondary education. Capital and operating grants are made to colleges and universities and also to private elementary and secondary schools. Health and human services grants support united campaigns and grants to hospitals and organizations providing clothing, rehabilitation, and hospice care. Civic and cultural affairs grants support the performing arts, museums, parks and nature centers, public broadcasting stations, and community special interest groups. Youth activities promoting leadership, citizenship, and recreation also are considered. The foundation operates a matching gift program, which matches gifts up to $5000 per year from Lubrizol Corporation employees on a one-to-one basis. Gifts can support education institutions and other community nonprofit organizations. Applications are accepted at any time.
Requirements: Tax-exempt Ohio and Texas organizations are eligible to apply.
Restrictions: Funds are not awarded directly to individuals or to endowments.
Geographic Focus: Ohio; Texas
Samples: Purdue U—to expand the Lubrizol HP Process Systems Engineering Laboratory, $15,000.; Cleveland Institute of Art—for a scholarship, $4000.
Amount of Grant: $1500-$30,000 average
Contact: Karen Lerchbacher, (440) 347-1797; fax: (440) 347-1858; email: kal@lubrizol.com
Internet: http://corporate.lubrizol.com/Foundation/default.asp
Sponsor: Lubrizol Corporation
29400 Lakeland Boulevard, 053A
Wickliffe, OH 44092

Lucile Horton Howe and Mitchell B. Howe Foundation Grants 3230
The foundation supports youth organizations, social services, medical research, hospitals, religion, child welfare, drug abuse, education, and family services. Types of support include continuing support, general operating funds, and research. Requests from qualified organizations must be received prior to the second Tuesday of March for the grant year. There is only one consideration meeting a year.
Requirements: Only nonprofit organizations in the Prox-Pasadena area and the San Gabriel Valley of California are eligible. A brief letter and a copy of the organizations 501(c)3 form are the requirements for application.
Restrictions: No restricted grants will be funded by the foundation.
Geographic Focus: California
Amount of Grant: $1,000-$300,000 range
Date(s) Application is Due: Feb 28
Contact: Mitchell B. Howe, Jr., President; (626) 792-2771; email: lhmbhowefoun@earthlink.net
Sponsor: Lucile Horton Howe and Mitchell B. Howe Foundation
180 South Lake Avenue
Pasadena, CA 91101-4932

Lucius N. Littauer Foundation Grants 3231
The foundation awards grants to U.S. tax-exempt organizations for charitable and educational purposes, with emphasis on publication and scholarly research at the postdoctorate level in the fields of Jewish and Israeli studies. Other current areas of interest are medical research with an emphasis on medical ethics and the environment. Types of support include challenge/matching grants, conferences and seminars, endowments, program development, research grants, travel grants, and publication. Applications are reviewed throughout the year.
Requirements: Giving primarily in NY for medical ethics and environmental projects.
Restrictions: Scholarships, loans, or grants are not awarded directly to individuals.
Samples: Columbia U (New York, NY)—to preserve the university's Language and Culture Archive of Ashkenazic Jewry, $50,000.; George Washington U (Washington, DC)—to establish the Littauer Judaica Book Fund in the Gelman Library, $15,000.; American Assoc of the International Ctr for the U Teaching of Jewish Civilization (New York, NY)—for general operating support, $5000.
Amount of Grant: $1000-$10,000 average
Contact: William Lee Frost, President, (212) 697-2677
Sponsor: Lucius N. Littauer Foundation
60 E 42nd Street, Suite 4600
New York, NY 10165

Ludwick Family Foundation Grants 3232
The foundation is a California nonprofit public benefit corporation established exclusively for charitable, scientific, literary, and educational purposes. The foundation tends to fund tangible types of items that will remain with and can be used repeatedly by the organization.
Requirements: Eligible applicants include U.S. or U.S.-based international organizations, 501(c)3 nonprofit public charities, and government agencies (any level). All grants are to be used exclusively for charitable, public benefit purposes.
Restrictions: The foundation does not grant requests for salaries, general operating expenses, scholarships, endowment funds, fundraising events or capital campaigns, feasibility studies, consulting fees, or advertising. The foundation will no longer accept any unsolicited requests for research or from public/private schools (K-12), universities/colleges, child daycare/development centers, hospitals, or libraries.
Samples: AbilityFirst (Pasadena, CA)—to purchase an American Disabilities Act accessible van to enable the transportation of people who are disabled (2003); Alaska Public Radio Network (AK)—to purchase recorder kits and professional stereo headphones to enable reports of news, current affairs, and cultural information for Alaska's Native People (2003).
Amount of Grant: $5000-$50,000 typically for one year
Date(s) Application is Due: Mar 31; Aug 31
Contact: Deanna Monaghan, (626) 852-0092; fax: (626) 852-0776; email: ludwickfndn@ludwick.org
Internet: http://www.ludwick.org
Sponsor: Ludwick Family Foundation
P.O. Box 1796
Glendora, CA 91740

Ludwig Vogelstein Foundation Grants 3233
The foundation provides one-time grants to individuals working in the fine arts (excluding photography, film, and video) and in the literary arts, including playwriting. It also supports certain unaffiliated scholarly and/or research projects. Grants ranging from $1,000 to $3,000 are awarded on the criteria of merit and need; the foundation is especially concerned with those who have no other source of funding.
Requirements: Send email or SASE to request current guidelines (email is preferable and quick). The foundation prefers not to receive phone calls. Currently, the sponsor does not have a website.
Amount of Grant: $1,000 - $3,000
Contact: Diana Braunschweig, Executive Director; email: lvf@earthlink.net
Sponsor: Ludwig Vogelstein Foundation
4001 Inglewood Avenue, Suite 101-309
Redondo Beach, CA 90278

Lumina Foundation for Education Grants 3234
The foundation focuses on three theme areas: addressing the multiple barriers that impede access to postsecondary education among traditionally underserved groups; supporting efforts that help these students attain sound educational objectives; and understanding and addressing the challenges faced by adult learners, particularly those who face barriers, such as low income or academic underpreparation. The following types of strategic activities are supported: research that casts new light on issues affecting postsecondary education access and success; programs that present innovative and practical approaches in overcoming challenges to access and attainment; leadership-development efforts to share the expertise of higher education executives and scholars working on issues of access and attainment; communication initiatives that bring about positive change by sharing the lessons of access and degree attainment with educational leaders, policy-makers, the news media, and the public; public policy initiatives that help policy-makers reach informed decisions on issues of postsecondary education access and attainment; and evaluation efforts that look closely at the work grantees and others have done to answer the questions: What have we learned? What works? What's next? The first step in seeking a grant is to write a preliminary letter of inquiry that is no more than three pages long. The letter should describe the organization, its mission, the project, and the amount requested. The foundation responds in writing to all preliminary inquiries. If the inquiry fits with funding priorities, the foundation will request additional information

about the project, and will provide a grant-application packet. Guidelines are available online.
Requirements: IRS 501(c)3 or 509(a)(1, 2, or 3) tax-exempt organizations and public school organizations are eligible.
Restrictions: The foundation will not award grants that support corporate sponsorships and fundraising events; fund partisan political or lobbying efforts; or provide direct support for individuals, including scholarships or institutional scholarship funds.
Samples: Excelencia in Education (Washington, DC)—for a study on the effects of institutional leadership on practices that improved Latino students' success at institutions that primarily serve Hispanics, $230,000 (2005); Ohio Assoc of Student Financial Aid Administrators (Delaware, OH)—for events put on by the College Goal Sunday program in Ohio, which helps students find ways to pay for college, $25,000 (2005); U of Michigan (Ann Arbor, MI)—to develop a model for dealing with local issues that affect access to and success in higher education, $769,200 (2005); Women in Government (Washington, DC)—to establish the Legislative Research Center and inform policy makers about access to higher education for women and minorities, $114,055 (2005).
Amount of Grant: $75,000-$200,000 typically
Contact: Coordinator of Grant Programs and Special Projects, (800) 834-5756 or (317) 951-5300; fax: (317) 951-5063
Internet: http://www.luminafoundation.org/grants/apply.html
Sponsor: Lumina Foundation for Education
30 S Meridian Street, Suite 700
Indianapolis, IN 46204

Lupus Foundation of America Research Grants **3235**
Grants are awarded annually to encourage research in lupus erythematosus. It is the foundation's hope that this investment in research will produce new information that may directly lead to much larger projects and substantially increased funding from other sources, particularly the National Institutes of Health, in the future. Grants are awarded for up to two years.
Requirements: Awards are for junior investigators only (assistant professors or below, if in academic medicine). Teachers and aides at teaching centers, hospitals, and colleges are eligible.
Restrictions: First-year fellows are not eligible.
Amount of Grant: $125,000 per year
Date(s) Application is Due: Feb 15
Contact: Grants Coordinator, (202) 349-1155; fax: (202) 349-1156; email: lupusinfo@aol.com
Internet: http://www.lupus.org/research/grantproginfo.html
Sponsor: Lupus Foundation of America
2000 L Street, NW, Suite 710
Washington, DC 20036-4916

Lustgarten Foundation for Pancreatic Cancer Research Grants **3236**
The foundation aims to make rapid advances in the battle against pancreatic cancer through a combination of cutting-edge technology and novel ideas. Grants can be submitted in one of three areas: novel technologies for pancreatic cancer genetics; screening for the early detection of pancreatic cancer; and novel therapies in pancreatic cancer. Proposals that exhibit leveraging of institutional resources and the attainment of major new programs are encouraged. This initiative encourages, but does not require, the development of integrative and collaborative teams of investigators from within a single institution or among several institutions. Letters of intent must be received on or before 5:00 p.m. Eastern Time on July 31, with full applications due no later than 5:00 p.m. Eastern Time on August 18. Grants will be awarded for a one-year period for a maximum amount of $100,000, of which no more than 10% can be used for indirect costs.
Requirements: Applications will be accepted from individual investigators as well as collaborating investigators. Applicants must have an MD or a PhD degree to become principal investigators. Eligible organizations include nonprofit organizations; public or private institutions, such as universities, colleges, hospitals, and laboratories; and domestic or foreign institutions/organizations.
Samples: Johns Hopkins University, Allison Klein, Ph.D. - Identification of Pancreatic Cancer Susceptibility Genes (2009); University of Michigan Medical School, Chandan Kumar, Ph.D. - Discovery of Recurrent Gene Fusions in Pancreatic Cancer using High-Throughput Sequencing (2009); Johns Hopkins University School of Medicine, Joshua Mendell, M.D., Ph.D. - Ras-regulated microRNAs in Pancreatic Cancer (2009).
Amount of Grant: $100,000 maximum per year
Contact: Kerri Kaplan, Executive Director; (516) 803-2304 or (866) 789-1000; fax: (516) 803-2303; email: kkaplan@cablevision.com or lsasso@cablevision.com
Internet: http://www.lustgarten.org/Page.aspx?pid=666
Sponsor: Lustgarten Foundation for Pancreatic Cancer Research
1111 Stewart Avenue
Bethpage, NY 11714

Luther I. Replogle Foundation Grants **3237**
The foundation focuses its giving on the following areas: programs addressing the needs of youth and children living in, or at risk of, long-term poverty (especially children of inner-city residents and migrant workers); programs to improve educational opportunities for inner- city children, including enrichment programs in the arts and sciences, alternative schools, after-school tutoring and mentoring, and scholarship programs; programs for affordable and supportive housing that reach groups of people frequently left out of traditional shelter programs, including single mothers and families with children, the elderly, ex-offenders, and youth; projects, lectures, and fellowships in classical archaeology; projects and institutions working for the conservation of maps and globes, and dissemination and education in the area of geography; and a modest budget for support of the arts. Types of support include direct service efforts; advocacy at the local, state, and/or federal levels; general operating support, new projects, and capital campaigns (limited). The board meets twice a year to make funding decisions, once in the autumn and once in the spring. Guidelines and application are available online.
Requirements: 501(c)3 nonprofits in Chicago, IL; Minneapolis, MN; Palm Beach County, FL; and Washington, DC, are eligible. Preference is given to organizations with small or modest operating budgets
Geographic Focus: District of Columbia; Florida; Illinois; Minnesota
Samples: Bright Beginnings (Washington, DC)—day care for homeless children, $5000 (2004); The Enterprising Kitchen, (Chicago, IL)—for jobs training for women, $5000 (2004); First Presbyterian Church of Chicago (IL)—social service programming, $25,000 (2004); Oriental Institute, U of Chicago (Chicago, IL)—to support a conservation internship, $10,000 (2004).
Amount of Grant: $500-$45,000 range
Date(s) Application is Due: Mar 15; Sep 15
Contact: Gwenn Gebhard, Executive Director, (202) 296-3686; fax (202) 296-3948; email: info@lirf.org
Internet: http://www.lirf.org
Sponsor: Luther I. Replogle Foundation
1111 19th Street NW, Suite 900
Washington, DC 20036

Luther Northwestern Scholarships **3238**
Scholarships are awarded to Lutheran graduate students who are majoring in theology/religion.
Requirements: Lutheran graduate students enrolled at Luther Northwestern are eligible.
Contact: Dean of Students, (651) 641-3463; fax: (651) 641-3497; email: faohouse@luthersem.edu
Internet: http://www.luthersem.edu/admissions/faohouse/scholarships.asp
Sponsor: Luther Northwestern
2481 Como Avenue
Saint Paul, MN 55108

Lutheran Foundation Grants **3239**
The foundation awards Regular, Partnership, and Emergency Grants. Regular grants are available to any congregation or nonprofit organization in northeast Indiana. Regular grants are awarded four times a year. The foundation will consider funding of one-time grants, single-year grants, and multi-year grants that do not exceed three years. Partnership grants are available to each of the foundation's member congregations. They are designed to assist congregations with their outreach ministries and also to enrich the internal life of their own membership. Emergency grants are designed to address unforeseen circumstances that disrupt the ongoing ministry or service provided by churches and organizations. The foundation also encourages, promotes, and supports Lutheran and Christian educational opportunities.
Requirements: The foundation awards grants to ministries located in 10 counties in northeast Indiana, including Adams, Allen, DeKalb, Huntington, LaGrange, Noble, Steuben, Wabash, Wells, and Whitley counties.
Restrictions: Proposals for projects that have been completed, deficit financing, fundraising, and promotional items are not considered. In addition, the foundation does not fund individuals, endowments, other funds, or the general operating expenses of congregations or medical procedures that are destructive of life.
Geographic Focus: Indiana
Samples: Allen County Non-Public School Assoc (Fort Wayne, IN)—to fund school nurse program expenses for parochial and independent schools $57,565 three-year grant (2003); Bethlehem Evangelical Lutheran Church and School (Fort Wayne, IN)—for renovations to cafeteria, basement classrooms, and Fellowship Hall for handicap accessibility $9870 matching grant (2003); Camp Lutherhave (Albion, IN)—to purchase multi-passenger van, $22,000 (2003); Cancer Services of Allen County (Fort Wayne, IN)—for operational support, $75,000 three-year grant (2003).
Amount of Grant: $10,000 maximum
Contact: Terri Kortokrax, (800) 677-2451 or (260) 458-2112 ext 302; fax: (260) 458-3069; email: tkortokrax@thelutheranfoundation.org
Internet: http://www.thelutheranfoundation.org/grants.htm
Sponsor: Lutheran Foundation
3024 Fairfield Avenue, Suite 200
Fort Wayne, IN 46807-1697

Lymphoma Research Foundation Fellowships **3240**
Two-year fellowships uniquely permit the uninterrupted development of promising leads and help attract the nation's best scientific talent to careers in lymphoma. Its emphasis upon translational as well as purely basic research assures that breakthroughs are quickly transformed into viable treatment options.
Amount of Grant: $45,000 and $50,000 salary support for year one and year two, respectively; $5000 annual research budget
Date(s) Application is Due: Sep 9
Contact: Director of Medical and Scientific Outreach, (212) 349-2910; fax: (212) 349-2886; email: researchgrants@lymphoma.org
Internet: http://www.lymphoma.org/site/pp.asp?c=bfIKIVMIG&b=38480
Sponsor: Lymphoma Research Foundation
111 Broadway, 19th Floor
New York, NY 10016

Lymphoma Research Foundation Grants 3241

The foundation supports high-quality research that results in safer, more effective treatments and a cure for lymphoma. Applications are accepted for two programs. Fellowship research grants support young investigators (third-year fellows) pursuing careers in lymphoma research. Lymphoma-specific projects submitted by individuals associated with accredited academic institutions, JCAHO-accredited research hospitals, and other research organizations that have national or international reputations for excellence will be considered. Junior faculty research grants support researchers at the assistant or associate professor level. Mantle cell research projects also are funded at both levels.
Samples: U Hospital Grosshadern (Germany)—for research on mantle-cell lymphoma, a rare form of blood cancer, $890,000 (2004); U of Texas MD Anderson Medical Ctr (Houston, TX)—for research on mantle-cell lymphoma, a rare form of blood cancer, $2.4 million (2003).
Amount of Grant: $45,000 maximum fellowships; $75,000 maximum junior faculty research grants
Date(s) Application is Due: Sep 15
Contact: Grants Administrator, (212) 349-2910 or (800) 235-6848; email: ResearchGrants@lymphoma.org
Internet: http://www.lymphoma.org/site/pp.asp?c=bfIKIVMIG&b=36878
Sponsor: Lymphoma Research Foundation
8800 Venice Boulevard, Suite 207
Los Angeles, CA 90034

Lynde and Harry Bradley Foundation Grants 3242

The foundation aims to encourage projects that focus on cultivating a renewed, healthier, and more vigorous sense of citizenship among the American people and among peoples of other nations. Projects likely to be supported will generally treat free people as self-governing, personally responsible citizens, not as victims or clients; aim to restore the intellectual and cultural legitimacy of common sense, the wisdom of experience, everyday morality, and personal character; seek to reinvigorate and re-empower the traditional, local institutions—families, schools, churches, and neighborhoods—that provide training in and room for the exercise of genuine citizenship, that pass on everyday morality to the next generation, and that cultivate personal character; and encourage decentralization of power and accountability away from centralized, bureaucratic, national institutions back to states, localities, and revitalized mediating structures. Eligible projects may address any arena of public life—economics, politics, culture, or civil society; the problem of citizenship at home or abroad; Milwaukee and Wisconsin community and state projects that aim to improve the life of the community through increasing cultural and educational opportunities, grassroots economic development, and social and health services; the resuscitation of citizenship in the economic, political, cultural, or social realms; policy research and writing about approaches encouraging that resuscitation; academic research and writing that explore the intellectual roots of citizenship; and popular writing and media projects that illustrate for a broader public audience the themes of citizenship. The foundation supports programs that research the needs of gifted children and techniques of providing education for students with superior skills and/or intelligence; research programs investigating how learning occurs in gifted children; and demonstration programs of instruction.
Requirements: Tax-exempt and nonprofit organizations should prepare a letter of inquiry presenting a concise description of the project, its objectives and significance, and the qualifications of the organizations and individuals involved. If the project appears to fall within the foundation's mandate, a brochure with detailed proposal requirements will be sent to the applicant.
Restrictions: The foundation favors projects that are normally not financed by public funds and will consider requests from religious organizations that are not denominational in character. Grants without significant importance to the foundation's areas of interest will only under special conditions be considered for endowment or deficit financing proposals. Grants will not be made to individuals, for overhead costs, or for fund-raising counsel.
Samples: Black Alliance for Educational Options, Washington, D.C., $250,000 (2007); Texas A&M University, College Station, Texas, $50,000 (2007); Evergreen Freedom Foundation, Olympia, Washington, $50,000 (2007).
Amount of Grant: $25,000-$250,000
Date(s) Application is Due: Feb 1; May 1; Aug 1, Nov 1
Contact: Daniel Schmidt, Vice President for Programs; (414) 291-9915; fax: (414) 291-9991
Internet: http://www.bradleyfdn.org/app.html
Sponsor: Lynde and Harry Bradley Foundation
1241 North Franklin Place
Milwaukee, WI 53202-2901

M. Bastian Family Foundation Grants 3243

The foundation supports nonprofit organizations in its areas of interest, including music, the arts, higher education, health care and health organizations, religion (Christian and Latter-day Saints), social services, and wildlife conservation. Types of support include general operating support and scholarship funds. Contact the office for application forms. There are no application deadlines.
Requirements: Funding focus is primarily in Utah.
Restrictions: Grants are not made to individuals.
Samples: Brigham Young U (Provo, UT)—for scholarship endowment, $200 000.; Pediatric AIDS Foundation (Los Angeles, CA)—for general support, $20 000.; American Diabetes Assoc (Alexandra, VA)—for research, $100,000.
Amount of Grant: $2000-$200,000 range
Contact: McKay Matthews, Program Contact, (801) 225-2455
Sponsor: M. Bastian Family Foundation
51 W Center Street, Suite 305
Orem, UT 84057

M.D. Anderson Foundation Grants 3244

The foundation supports Texas nonprofit organizations, including religious organizations, higher education institutions, parochial education institutions, hospitals, and religious welfare organizations. Types of support include capital grants, general support, program support, scholarships, and research grants. Applicants should send a brief letter describing the proposed project, along with an original and four copies of the proposal. If applying for a matching grant, indicate other sources of funding. There are no application deadlines. The board meets on the third Tuesday of each month. Applications should be received a month prior to the the meeting to be considered.
Requirements: Texas nonprofit organizations are eligible.
Geographic Focus: Texas
Samples: U of Texas (Austin, TX)—for the law school, $800,000 (2002); U of Saint Thomas (Houston, TX)—to construct a humanities and education building, $250,000.
Amount of Grant: $25,000-$100,000 average
Contact: Toloria Allen, Secretary-Treasurer, (713) 216-4513
Sponsor: M.D. Anderson Foundation
P.O. Box 2558
Houston, TX 77252-8037

M.J. Murdock Charitable Trust Grants 3245

The trust supports nonprofit organizations in the Pacific Northwest. Grantmaking priorities include health and human services—mental health services, early-childhood and youth programs, economic and cultural development, support for the disabled, and disaster relief; education—projects that emphasize innovative learning techniques both in and outside the classroom (e.g., educational programs at museums, nonprofit organizations, schools, universities, and zoos); scientific research in the natural sciences that emphasizes new knowledge or trains students in research techniques; and arts and culture. Prospective applicants should submit a letter of inquiry; guidelines are available on the Web site. A positive response to the letter of inquiry will include an application packet and directions for its completion.
Requirements: Pacific Northwest 501(c)3 tax-exempt organizations in Alaska, Idaho, Montana, Oregon, and Washington State are eligible.
Restrictions: The charitable trust does not support ongoing operations, debt reduction, general fund drives, annual appeals, or emergency funds.
Geographic Focus: Alaska; Idaho; Montana; Oregon; Pacific Northwest; Washington
Samples: Pacific Northwest Research Institute (Seattle, WA)—to purchase equipment for research on the subcellular mechanisms of disease, with a focus on diabetes, $67,500 (2004); Northwest Children's Theater and School (Portland, OR)—to develop a fundraising department, $115,000 over three years (2004); Seattle's Union Gospel Mission (WA)—to repair a shelter that was damaged by an earthquake, $50,000 challenge grant.; NetCorps (Eugene, OR)—to provide technology support to nonprofit organizations in rural Oregon, $90,000 partial challenge grant over three years.
Amount of Grant: $22.8 million total
Contact: Dr. John Van Zytveld, Senior Program Officer, (360) 694-8415
Internet: http://www.murdock-trust.org
Sponsor: M.J. Murdock Charitable Trust
P.O. Box 1618
Vancouver, WA 98668

MAA-Tensor Foundation Women and Mathematics Grants 3246

MAA and the Tensor Foundation award grants to advance the study of math among women and girls, especially those in middle school, high school, and postsecondary grades. Grants support project directors and encourage mathematics faculty to develop projects that increase the participation of women in mathematics. Grants are made to the institution of the project director. An institution is expected to supply matching funds or in-kind support as an indication of commitment to the project. Some grants may be renewed for a second year. Possible projects are to create a network of women professional mentors who will direct mathematics projects for girls, organize a club for women interested in math, bring high school women onto a college campus for a Math Day, and form partnerships with industry to acquaint women students with real-life applications of mathematics. Proposals can be up to three single-spaced pages.
Requirements: College, university, and high school mathematics faculty (in conjunction with college or university faculty) and their departments and institutions may submit proposals.
Amount of Grant: $5000 maximum
Date(s) Application is Due: Feb 25
Contact: Dr. Florence Fasanelli, (202) 966-5591; fax: (202) 966-5591; email: ffasanelli@juno.com
Internet: http://www.maa.org/projects/tensor_solic.html
Sponsor: Mathematical Association of America
1529 18th Street NW
Washington, DC 20036

Mabel H. Flory Charitable Trust Grants 3247

The Mabel H. Flory Charitable Trust was established in 1979 and funds organizations primarily in Washington, DC., and Baltimore, MD. Support is available for research designed to develop new cures and methods of treatment: for children with hearing and sight problems; arthritis in both children and adults; and aphasia related problems.

Requirements: Your initial approach should be in the form of a letter, prior to the deadline date. Mail letter to the following address: c/o BB&T Co. of Virginia, Attn.: Trustee, 6400 Arlington Boulevard, Falls Church, VA, 22042.
Restrictions: No grants to individuals.
Geographic Focus: District of Columbia; Maryland
Date(s) Application is Due: Aug 31
Contact: BB&T Co. Trustee; (703) 531-2053;
Sponsor: Mabel H. Flory Charitable Trust c/o BB&T Co. of Virginia
P.O. Box 2907
Wilson, NC 27894-2907

Mabel Y. Hughes Charitable Trust Grants 3248
The trust awards grants to Denver-area nonprofit organizations in its areas of interest, including community funds, cultural programs, education, health, hospitals, museums, social services, and youth. Types of support include operating support, continuing support, annual campaigns, seed money, emergency funds, building funds, equipment acquisition, endowment funds, research grants, and special projects.
Requirements: Colorado nonprofit organizations are eligible.
Restrictions: Grants do not support conferences, individuals, or ticket purchases for benefits.
Geographic Focus: Colorado
Amount of Grant: $5000-$60,000 range
Contact: Judy Dowling, c/o Wells Fargo Bank, (720) 947-6725
Sponsor: Mabel Y. Hughes Charitable Trust
1740 Broadway, MC 7300 #483
Denver, CO 80274

Macalester College Andrew W. Mellon Postdoctoral Fellowship 3249
The college invites applications for a two-year postdoctoral fellowship from recent PhDs in socio-cultural anthropology. Area of interest is open, but the department is interested in applicants with a commitment to and extensive experience in field research, and whose teaching interests do not duplicate existing faculty strengths. The fellowship will begin in the fall and will entail half-time teaching responsibilities. The fellow will offer one course each semester, including one section of the introductory course, Cultural Anthropology and a topics course in his or her field each year. Candidates should have expertise that contributes to the overall objectives of the department and should be able to contribute to one of more of the college's interdisciplinary programs. The fellowship includes a salary plus benefits and funds to support research and travel. Applicants should submit a letter of application, a curriculum vita, a graduate school transcript, names of three references, and a brief description of teaching interests by the listed application deadline.
Requirements: The primary qualifications for a successful applicant are recent (within the past five years) completion of a PhD, commitment to excellent teaching in a liberal arts environment, and outstanding promise in original scholarship.
Amount of Grant: $54,000 salary
Date(s) Application is Due: Oct 31
Contact: Dan Balik, Associate Provost
Internet: http://www.macalester.edu/anthropology/Jobs.html
Sponsor: Macalester College
1600 Grand Avenue
Saint Paul, MN 55105-1899

MacArthur Fellows Program Grants 3250
The program provides unrestricted fellowships to exceptionally talented and promising individuals who have shown evidence of originality, dedication to creative pursuits, and capacity for self-direction. They may be writers, scientists, artists, social scientists, humanists, activists, or workers in any other field or fields, with or without institutional affiliations. The foundation awards fellows an income in quarterly installments over five years so that they may have the time and the freedom to fulfill their potential by devoting themselves to their own endeavors at their own pace. The fellowships are intended to support individuals, not projects. Fellows are free to focus on more than one area through interdisciplinary work, to change fields if they wish, or even to alter the direction of their careers.
Requirements: Only an individual whose name is submitted by a designated nominator is eligible. Candidates must be U.S. citizens or residents. No other restrictions are placed on nominators regarding their nominees.
Samples: Ida Applebroog (NY)—fellowship recipient, $375,000.;Peter Miller (Berlin)—fellowship recipient, $220,000.
Amount of Grant: $500,000 stipend
Contact: Office of Grants Management, Research, and Information, (312) 726-8000; fax: (312) 920-6258; TDD: (312) 920-6285; email: 4answers@macfound.org
Internet: http://www.macfound.org/programs/fel/fel_overview.htm
Sponsor: John D. and Catherine T. MacArthur Foundation
140 S Dearborn Street, Suite 1100
Chicago, IL 60603

Mackenzie King Traveling Scholarship, University of British Columbia 3251
Scholarships support work in postgraduate studies in the field of international or industrial relations, including the international or industrial aspects of law, history, politics, and economics.
Requirements: Traveling Scholarships are open to graduates of any Canadian university who engage in (commence or continue) postgraduate study, in the United States or the United Kingdom, of international relations or industrial relations (including the international or industrial relations aspects of law, history, politics and economics). The Open Scholarship is open to graduates of any Canadian university who engage in (commence or continue) postgraduate study in any field, in Canada or elsewhere.
Amount of Grant: $C9000 for traveling scholarships; $C7000 for open scholarship
Date(s) Application is Due: Feb 1
Contact: Faculty of Graduate Studies, (604) 822-8501; fax: (604) 822-5802; email: blom@law.ubc.ca or falbo@law.ubc.ca
Internet: http://www.grad.ubc.ca/awards/index.asp?menu=013,004,000,000
Sponsor: University of British Columbia
180-6371 Crescent Road
Vancouver, BC V6T 1Z2 Canada

Macquarie Bank Foundation Grants 3252
The Foundation is one of Australia's oldest and largest corporate foundations contributing more than $A8.5 million last financial year to more than 400 community organizations. The Foundation focuses its resources in six core areas - the arts, education, environment, health care, health research and welfare. We are also committed to projects specifically aimed at supporting indigenous communities. The Foundation's funding criteria is flexible and open. It welcomes applications from a diverse range of community organizations that are working in innovative ways to provide long-term benefits. Funding levels are flexible and are dictated by the needs of the organization and funding availability. Each application is assessed on its individual merit, with priority given to programs which: support a broad section of the community at a regional, state or national level; have the involvement or potential for involvement of Macquarie Bank staff through volunteering, fundraising, pro bono work and board and/or management committee involvement; are located in cities/countries where Macquarie Bank staff are located; and deliver long-term benefits and build community sustainability.
Amount of Grant: $100-$500,000
Contact: Sally Shepherd, Grants Project Officer, (612) 8232-7554; fax: (612) 8232 0019; email: sally.shepherd@macquarie.com
Internet: http://www.macquarie.com/foundation/index.htm
Sponsor: Macquarie Bank Foundation
GP.O. Box 4294
Sydney, Australia NSW 1164 Australia

Madison and Lila Self Graduate Fellowships 3253
This program combines personal and professional development with rigorous academic training. The fellowships are awarded to support exemplary students pursuing doctoral degrees in the natural and physical sciences, behavioral and social sciences, humanities, engineering, and business. Fellowships provide four years of support.
Requirements: Application is open to students planning to pursue doctoral degrees in designated fields.
Amount of Grant: $23,000 yearly stipend
Date(s) Application is Due: Jan 30
Contact: Self Graduate Fellowship Program, (785) 864-7249; fax: (785) 864-0394; email: self@ku.edu
Internet: http://www.ku.edu/~selfpro
Sponsor: University of Kansas
1450 Jayhawk Boulevard, Strong Hall, Room 222
Lawrence, KS 66045

Mallinckrodt Medical Community Partnership Grants 3254
The company supports nonprofit organizations and public and private K-12 schools for programs and projects designed to strengthen the community. Areas of support include health, youth, science, and the environment. Grants are awarded to support new and existing programs, conferences, research, equipment, products, technical assistance, and annual or capital campaigns. There are no application deadlines. Applications are reviewed annually in March, September, and December. Applications may be submitted by fax as well as by mail.
Requirements: Nonprofit organizations are eligible.
Amount of Grant: $2 million total; $100-$250,000 average
Contact: Grants Administrator, (314) 895-2000
Internet: http://www.mallinckrodt.com/corpprofile/cp-citizen.html
Sponsor: Mallinckrodt Medical Inc
675 McDonnell Boulevard
Hazelwood, MO 63042

Malone Fellowship in Arab and Islamic Studies 3255
The council is a nonprofit educational organization that seeks to further Americans' understanding of the Arab world. The program enables university and community educators and leaders who do not have extensive prior experience traveling to or studying in the Arab world to visit one or more Arab countries on two-week study visits. Host countries include Bahrain, Egypt, Jordan, Kuwait, Oman, Qatar, Saudi Arabia, Syria, Tunisia, the United Arab Emirates, Yemen, and the West Bank and Gaza. Applications are accepted any time.
Requirements: Applicants must be U.S. citizens with valid passports.
Restrictions: Students are ineligible.
Contact: Jennifer Sewell-Zeytin, (202) 293-6466; fax: (202) 293-7770; email: jennifer@ncusar.org
Internet: http://www.ncusar.org/malone
Sponsor: National Council on U.S.-Arab Relations
1730 M Street NW, Suite 503
Washington, DC 20036

Manitoba Arts Council Aboriginal Arts Creative Development Pilot Projects Grants 3256

The Aboriginal Arts Creative Development Pilot Projects initiative is designed to enable Aboriginal artists in all arts disciplines, including dance/music, drumming, singing, carving, painting, storytelling, new media, writing, theatre or crafts, to create or produce new works over a three month period. Only one Creative Development grant may be received in any given fiscal year (April 1 to March 31).
Requirements: Professional Aboriginal artists are eligible who: are Canadian citizens or permanent residents of Canada; have lived in Manitoba for at least one year immediately prior to submitting an application; have completed basic training (not necessarily in a formal school, college or university); view the practice of their art as a career and are recognized as professionals in their art form by their peers; have completed at least three full years of professional activity; are able to devote a sufficient time over a six-month period to complete the project for which they are requesting assistance.
Amount of Grant: $7,500
Date(s) Application is Due: Jun 30; Nov 30
Contact: Tracey Longbottom, Program Consultant; (204) 945-3180; tlongbottom@artscouncil.mb.ca
Internet: http://www.artscouncil.mb.ca/english/abor_grantind.html
Sponsor: Manitoba Arts Council
525-93 Lombard Avenue
Winnipeg, MB R3B 3B1 Canada

MAP International Medical Fellowships 3257

The MAP International Medical Fellowship encourages lifelong involvement in global health issues by providing selected medical students firsthand exposure in a Christian context to the health, social and cultural characteristics of a developing world community. MAP International, through a grant from the founders of Reader's Digest, DeWitt and Lila Wallace, sponsors fourth year medical students by funding travel arrangements for medical students to their chosen destination. The field experience must be designed to provide the student exposure to community health in a Christian context. Students select a mission agency or hospital that has an outreach among the poor in a rural or urban setting. Students must spend at least eight weeks in the field (six weeks for residents and interns). The fellowship provides 100 percent of the approved round trip airfare to one destination. In most instances, students pay room and board as well as any in-country travel expenses. To obtain an application, check with the dean of your school or the Student Affairs Office. Applications are sent to all medical schools annually. You may also download the application in PDF or Microsoft Word format on the Map International website.
Requirements: Fourth-year medical students, residents and interns are eligible. Applications should be submitted during the academic year prior to travel.
Date(s) Application is Due: Mar 1
Contact: Margaret Stevenson, Medical Fellowship Coordinator; (404) 880-0540; fax: (912) 265-6170; email: mstevenson@map.org
Internet: http://www.map.org/site/PageServer?pagename=what_Medical_Fellowship
Sponsor: Map International
4700 Glynco Parkway
Brunswick, GA 31525-6800

MAPFRE Foundation for Medicine Grants 3258

The foundation operates nationally and internationally in the fields of education, medicine and health, science and technology, and social welfare. Types of support include research grants, grants to individuals and organizations, scholarships, fellowships, and prizes.
Contact: Grants Administrator, 34-1-6265704; fax: 34-1-6265825; email: infofmm@mapfremedicina.es
Internet: http://www.mapfre.com/pmapfre/paraguas_mapfre/en/06/0608.shtml
Sponsor: MAPFRE Foundation for Medicine
Carretera de Pozuelo a Majadahonda s/n
Madrid 28220 Spain

Marcia and Otto Koehler Foundation Grants 3259

Grants are made to support arts and culture, medical, education, and social organizations in San Antonio, Texas. Proposals must demonstrate leadership in effecting positive change; encourage collaborative effort; serve large and diverse sectors of the population; and demonstrate vision, effectiveness, and good fiscal management. Types of support include general operating support, building construction/renovation, and research. Grants are made for one time only. Applicants must send a letter of request for application by March 1 to receive an application.
Requirements: Grants are awarded only to organizations in Bexar County, Texas.
Restrictions: Grants will not be made to individuals or to support other foundations or endowments; salaries; operating deficits; political organizations; or churches, synagogues, or parishes.
Geographic Focus: Texas
Samples: Centro Alameda (San Antonio, TX)—to restore a marquee, canopy, and facade, $30,000.; McNay Art Museum (San Antonio, TX)—to sponsor an art exhibit, $50,000.
Amount of Grant: $2,500-$50,000 range
Date(s) Application is Due: Jun 1
Contact: Thomas K. Killion, Senior Vice President, Bank of America; (210) 270-5422; fax: (210) 270-5520; email: thomas.k.killion@ustrust.com
Sponsor: Marcia and Otto Koehler Foundation
P.O. Box 121
San Antonio, TX 78291-0121

Marcus and Theresa Levie Educational Fund Scholarships 3260

Graduate scholarships in the helping professions are awarded to individual students for the academic year; eligible applicants may receive application forms from the Jewish Vocational Service. Personal interviews are required.
Requirements: Applicants must be Jewish, legally domiciled in Cook County, IL, demonstrate financial need, and be enrolled or entering an accredited college or university in a full-time vocational or professional training program, preferably graduate level or above, in the helping professions; at least a C-plus average is required. The scholarship supervisor may be reached between December 1 and February 15 at the fund.
Geographic Focus: Illinois
Amount of Grant: $1000- $5000
Date(s) Application is Due: Mar 1
Contact: Marla Levine, Scholarship Supervisor, c/o Jewish Federation of Metropolitan Chicago, (312) 346-6700; fax: (312) 855-3282; TYY: (312) 444-2877c; email: jvschicago@jon.cjfny.org
Internet: http://www.free-4u.com/levie_educational_fund_scholarship1.htm
Sponsor: Marcus and Theresa Levie Educational Fund
1 S Franklin Street
Chicago, IL 60606

Margaret and James A. Elkins Jr. Foundation Grants 3261

The foundation supports primarily education (elementary education, secondary education, medical school education, and higher education), social services, religious programs, cultural programs, and programs in the biological sciences. Types of support include capital campaigns, building construction/renovation, equipment acquisition, program development, and research. There are no application forms.
Requirements: 501(c)3 nonprofits in Texas may apply for grant support. Emphasis is on the greater Houston, TX area.
Restrictions: Grants are not made to individuals or to support annual fund-raising campaigns or operating deficits.
Geographic Focus: Texas
Samples: United Way of the Texas Gulf Coast (Houston, TX)—$32,500.; Baylor College of Medicine, Department of Ophthalmology (Houston, TX)—$50,000.; Institute of Religion (Houston, TX)—$5000.
Amount of Grant: $5000-$250,000 average
Date(s) Application is Due: Aug 31
Contact: Melinda Myers, (713) 652-2052
Sponsor: Margaret and James A. Elkins Jr. Foundation
1001 Fannin Street, 1166 First City Tower
Houston, TX 77002

Margaret Drew Alpha Fellowship 3262

The fellowship supports graduate study in a field of family and consumer sciences with preference given to dietetics or food and nutrition majors.
Requirements: Applicant must be a member of Phi Upsilon Omicron.
Amount of Grant: $1000
Date(s) Application is Due: Feb 1
Contact: Fellowships Administrator, National Office, (304) 368-0612; email: rickards@mountain.net
Internet: http://phiu.unl.edu/webdoc3.htm#Graduate%20Fellowships
Sponsor: Phi Upsilon Omicron Educational Foundation, Inc.
P.O. Box 329
Fairmont, WV 26555

Margaret L. Wendt Foundation Grants 3263

The foundation awards grants to eligible New York nonprofit organizations in its areas of interest, including education, the arts, social services, churches and religious organizations, health associations, public interest organizations, and youth agencies. There are no application deadlines or forms. The board meets quarterly
Requirements: Western New York nonprofit organizations are eligible. Giving is primarily in Buffalo.
Restrictions: Grants are not awarded to individuals, including scholarships.
Geographic Focus: New York
Samples: D'Youville College (Buffalo, NY)—for its chiropractic program, including equipping a computerized diagnostic-imaging classroom and an off-campus community-based clinic, $350,000 (2005).
Contact: Grants Administrator, (716) 855-2146
Sponsor: Margaret L. Wendt Foundation
40 Fountain Plaza, Suite 277
Buffalo, NY 14202

Margaret T. Morris Foundation Grants 3264

The foundation awards grants, primarily in Arizona, in its areas of interest, including arts and culture, children and youth, higher education. family, environment and animal welfare, social services, mental health, medical research, marine sciences, music, and hospice. Types of support include general operating costs, building construction/renovation, scholarship funds, deficit financing, endowments, land acquisition, and program development. Proposals are accepted between May and November, and the board meets in August and December.
Restrictions: No support for religious organizations. Grants are not made to individuals.
Geographic Focus: Arizona
Amount of Grant: $350-$100,000 range

Contact: Eugene Polk, Trustee
Sponsor: Margaret T. Morris Foundation
P.O. Box 592
Prescott, AZ 86302

Margaret Yardley Fellowship **3265**
This fellowship is awarded toward payment of tuition only for a woman from New Jersey who is a college graduate and plans to continue her education. She must be enrolled as a full-time student in graduate work or a doctoral program for an uninterrupted year of study in an American college of her choice. Selection is based on scholastic record, financial need, ability, character, and potential service to the chosen field.
Requirements: Request an application form by the listed deadline date; a self-addressed, stamped envelope must be enclosed.
Geographic Focus: New Jersey
Amount of Grant: $1000 per year
Date(s) Application is Due: Mar 1
Contact: Deen Meloro, President, (732) 249-5485 or (732) 249-5474; fax: (732) 249-8822; email: dmeloro@njsfwc.org
Internet: http://www.njsfwc.org/MYFellowship.html
Sponsor: New Jersey State Federation of Women's Clubs
55 Labor Center Wy
New Brunswick, NJ 08901

Margoes Foundation Grants **3266**
The foundation awards grants to California nonprofits in its areas of interest, including cardiac research, rehabilitation of mentally ill patients, and scholarships for minority students and for Asian and African nationals studying in the United States. Types of funding include general operating support, continuing support, program development, fellowships, scholarship funds, research, and matching funds. There are no application forms. Applicants should call or submit a preliminary letter prior to submitting a full proposal.
Requirements: California nonprofits are eligible, with preference given to the San Francisco Bay area.
Restrictions: Grants are not awarded to agencies supported by federated campaigns or to individuals.
Geographic Focus: California
Samples: Community Music Center (San Francisco, CA)—for financial aid, $10,000.; U of the Pacific (Stockton, CA)—to increase minority enrollment in the predentistry undergraduate programs, $20,000 over two years.; Fred Finch Youth Ctr (Oakland, CA)—for salary support of a part-time counselor for a transitional independent living program for older adolescents and young adults with severe emotional problems, $15,000.
Amount of Grant: $10,000-$20,000 average
Date(s) Application is Due: Apr 1; Oct 1
Contact: John Blum, Principal Manager, (415) 981-2966; fax: (415) 981-5218; email: margoesfdn@aol.com
Sponsor: Margoes Foundation
57 Post Street, Suite 510
San Francisco, CA 94104

Maria and Eric Muhlmann Award **3267**
The award is given annually by the Astronomical Society of the Pacific for significant observational research made possible through advances in astronomical instrumentation, software, or support infrastructure. The award will be made to the scientist(s) who acquired the results, to the builder of the instrument/software (if a different individual), or to both as appropriate. Individuals as well as small teams of individuals are eligible. The award will be granted for significant advances in any areas of astronomy, without restriction to wavelength or space/ground-based observations.
Requirements: Nominations are accepted from major observatories and departments of astronomy worldwide.
Amount of Grant: $500
Date(s) Application is Due: Jan 15
Contact: Marilyn Delgado, (415) 337-1100; fax: (415) 337-5205; email: mdelgado@astrosociety.org
Internet: http://www.astrosociety.org/membership/awards/muhlmann.html
Sponsor: Astronomical Society of the Pacific
390 Ashton Avenue
San Francisco, CA 94112

Marijuana Policy Project Grants **3268**
The program supports efforts that promote the legalization and regulation of marijuana. Projects may encompass any aspect of marijuana-law reform, such as: ballot initiatives, grassroots organizing, litigating for the right to use marijuana for religious purposes, lobbying the American Medical Association, marijuana research, and revising the definition of marijuana abuse and dependence in the American Psychiatric Association's Diagnostic and Statistical Manual. Submit letters of inquiry; full proposals are by invitation.
Requirements: Preference is given to 501(c)3 tax-exempt organizations, but such status is not an eligibility requirement.
Restrictions: The grants program does not fund political parties or candidates for office, state ballot initiatives, or hemp-related projects.
Amount of Grant: $60,000 maximum
Date(s) Application is Due: Jan 2; May 1; Sep 1
Contact: Stephanie Vogel, (202) 462-5747 ext 104; fax: (202) 232-0442; email: stephanie@mpp.org
Internet: http://www.mpp.org/grants/index.html
Sponsor: Marijuana Policy Project
P.O. Box 77492
Washington, DC 20013

Marilyn Yarbrough Dissertation/Teaching Fellowship **3269**
One of the most significant challenges confronting small liberal arts colleges in the United States today is increasing the members of underrepresented groups teaching at such institutions. In order to encourage such scholars to consider college rather than university teaching, Kenyon College offers this fellowship, which is for scholars in the final stages of their doctoral work who need only to finish the dissertation to complete requirements for the PhD. It is hoped that the experience of living and working for a year at Kenyon will encourage these fellows to consider a career at a liberal arts college. The fellow is expected to write the dissertation and to teach one course each semester, usually in the fellow's general research area. Fellows are also expected to offer a college lecture or departmental seminar on the dissertation topic at some point during the academic year in residence. Send a letter of application, vita, and three letters of references (including one from the dissertation advisor) to the office. Information is available online.
Requirements: Individuals meeting all of the following criteria are eligible: U.S. citizens or nationals; members of underrepresented groups (e.g., ethnic minorities; women in fields that attract mostly men, or men in fields that attract mostly women; and persons who are first‚Äëgeneration college attendees); individuals who are enrolled in a research‚Äëbased PhD program in one of the following fields—African and African American studies, American studies, anthropology, art, art history, Asian studies, biology, biochemistry, chemistry, classics, dance, drama, economics, English, environmental studies, history, humanities, international studies, legal studies, mathematics, modern languages and literature, music, neuroscience, philosophy, physics, political science, public policy, psychology, religious studies, scientific computing, sociology, and women's and gender studies; individuals who aspire to a teaching and research career; and persons who have not yet earned a doctoral degree at any time and in any field.
Amount of Grant: $32,500 plus health benefits, housing, and a small moving allowance
Date(s) Application is Due: Dec 1
Contact: Office of the Associate Provost
Internet: http://www.kenyon.edu/x27112.xml
Sponsor: Kenyon College
Gambier, OH 43022

Marine Biological Laboratory Summer Research Fellowships **3270**
Each year a number of summer research support opportunities for young investigators are awarded for studies and research to be carried out at the Marine Biological Laboratory (MBL) at Woods Hole, MA. The Robert Dale Allen Fellowship supports research in cell motility and cytoarchitecture; and the Frederik B. Bang Fellowship funds research on marine invertebrate immunology, pathology, infectious diseases, and molecular biology. The M.G.F. Fuortes Fellowship sponsors research in the neurosciences, and the Stephen W. Kuffler Fellowship supports research in neurobiology. The H. Burr Steinbach Fellowship funds research in cell biology and physiology. In addition, the Porter Fellowships for Minority Students/Investigators are offered for research in the field of physiology with senior investigators at the MBL. The MBL Summer Fellowships are offered to investigators to conduct independent or collaborative research in any area. Awarded funds may be used to pay laboratory rental fees, housing costs, travel to and from Woods Hole, and appropriate personal expenses. Notification of selection of fellows is made by March 1.
Requirements: Postdoctoral students, instructors, and assistant professors may apply for these fellowships.
Date(s) Application is Due: Jan 15
Contact: Fellowship Coordinator, Office of Research Administration and Education, (508) 289-7441; fax: (508) 289-1924; email: skaufman@mbl.edu
Internet: http://www.mbl.edu/research/summer/fellowships.html
Sponsor: Marine Biological Laboratory
7 MBL Street
Woods Hole, MA 02543

Marine Corps Historical Center Beeler-Raider Fellowship **3271**
The annual fellowship encourages graduate-level and advanced study of the combat contributions of enlisted marines. Applicants may either suggest a study topic or ask the staff at the center for guidance in selecting an appropriate topic. Preference is given to projects covering the pre-1991 period, for which records are made available to scholars. Fellowship recipients are encouraged to do part of their research in the main federal archival centers in Washington, DC. Preliminary application involves submitting a letter that outlines the applicant's qualifications and the topic (or requesting a topic); applicants will be invited to submit formal applications.
Requirements: The program concentrates on graduate students but considers other qualified persons. Applicants must be U.S. citizens or nationals.
Amount of Grant: $2500
Date(s) Application is Due: May 1
Contact: Coordinator, Grants and Fellowships, (703) 432-4874
Internet: http://hqinet001.hqmc.usmc.mil/HD/General/Grants_Internships_Grants.htm#Grant3
Sponsor: United States Marine Corps
Marine Corps University, 3079 Moreell Avenue
Quantico, Virginia 22134

Marine Corps Historical Center Dissertation Fellowship 3272
One dissertation fellowship per academic year is awarded to a qualified graduate student working on a doctoral dissertation pertinent to Marine Corps history. Topics in U.S. military and naval history, as well as history and history-based studies in the social and behavioral sciences, with a direct relationship to the history of the U.S. Marine Corps will be considered. Topics may encompass wars, institutions, organization and administration, policy, biography, civil affairs and civic action, civil-military relations, weaponry and technology, manpower, training and education, strategy, tactics, and logistics; as well as diplomatic, political, economic, social, and intellectual trends affecting the Marine Corps during peace and war. The topic must have the approval of the applicant's dissertation director and contribute to the study and understanding of Marine Corps history. Fellows are expected to do a portion of their research in Washington, DC.
Requirements: The applicant must be a U.S. citizen who is enrolled in a recognized graduate school; have completed by September all requirements for the doctoral degrees except the dissertation; and have an approved, pertinent dissertation topic.
Restrictions: A fellowship will not be awarded to an applicant who has held or accepted an equivalent fellowship from any other Department of Defense agency; however, recipients of the Marine Corps's master's thesis fellowships may apply.
Amount of Grant: $10,000 stipend
Date(s) Application is Due: May 1
Contact: Coordinator, Grants and Fellowships, (703) 432-4874
Internet: http://hqinet001.hqmc.usmc.mil/HD/General/Grants_Internships_Grants.htm
Sponsor: United States Marine Corps
Marine Corps University, 3079 Moreell Avenue
Quantico, Virginia 22134

Marine Corps Historical Center Master's Thesis Fellowships 3273
A number of fellowships are awarded each academic year to qualified graduate students working on topics pertinent to Marine Corps history. Topics in U.S. military and naval history, as well as history and history-based studies in the social and behavioral sciences, that have a direct relationship to the history of the U.S. Marine Corps will be considered. Topics may encompass wars, institutions, organization and administration, policy, biography, civil affairs and civic action, civil-military relations, weaponry and technology, manpower, training and education, strategy, tactics, and logistics; as well as diplomatic, political, economic, social, and intellectual trends affecting the Marine Corps during peace and war. Preference is given to projects covering the pre-1991 period. In all cases the topic must have the approval of the graduate student's thesis advisor. Fellows are expected to do a portion of their research in Washington, DC.
Requirements: U.S. citizens or nationals actively enrolled in accredited master's degree programs that require master's theses are eligible.
Restrictions: A fellowship will not be awarded to anyone who has held or accepted an equivalent fellowship from any other Department of Defense agency.
Amount of Grant: $3500
Date(s) Application is Due: May 1
Contact: Coordinator, Grants and Fellowships, (202) 433-4244; fax: (202) 433-7265
Internet: http://hqinet001.hqmc.usmc.mil/HD/General/Grants_Internships_Grants.htm
Sponsor: United States Marine Corps
1254 Charles Morris Street, SE
Washington Navy Yard, DC 20374-5040

Marine Corps Historical Center Research Grants 3274
Research grants are awarded to encourage graduate-level and advanced study in Marine Corps history and related fields. Research may encompass wars, institutions, organization and administration, policy, biography, civil affairs and civic action, civil-military relations, weaponry and technology, personnel, training and education, strategy, tactics, and logistics; as well as the interaction of diplomatic, political, economic, social, and intellectual trends with military affairs during and between wars. Research may also relate to such museum curatorial fields as military art, design, ordnance, uniforms, equipment, and aviation. Preference is given to projects covering the pre-1991 period where records are declassified or can be most readily declassified and made available to scholars. Grant recipients perform research in the main federal archival centers in Washington, DC, and receive personalized assistance in gaining access to archival and library sources in the Washington, DC, area. Preapplication letters are accepted at any time. The letter should outline the applicant's qualifications and should either propose a specific topic or request a suggested topic based on the applicant's interests and qualifications. The center invites formal applications.
Requirements: The program concentrates on graduate students but also awards grants to other qualified persons. Applicants should have the ability to conduct advanced study in those aspects of American military history and museum activities directly related to the U.S. Marine Corps.
Amount of Grant: $400-$3000
Date(s) Application is Due: May 1
Contact: Coordinator, Grants and Fellowships, (202) 433-4244; fax: (202) 433-7265
Internet: http://hqinet001.hqmc.usmc.mil/HD/General/Grants_Internships_Grants.htm#Research
Sponsor: United States Marine Corps
1254 Charles Morris Street, SE
Washington Navy Yard, DC 20374-5040

Marion Park Deaver and Harry Gilbert Deaver Foundation Grants 3275
The mission of the foundation is to promote education, with major emphasis on financial support to parochial and private schools for facilities, scholarships, computing and network capability, technical support for the use of computers in education, development and presentation of a course for teachers in the use of network for education, and the Deaver Foundation Award for Excellence in Teaching. Other areas designated for financial assistance are religion, science, health, environment, literacy arts, and the indigent. The geographic areas for grants are Florida, Wisconsin, and Minnesota. There are no application forms or deadlines.
Requirements: Florida, Minnesota, and Wisconsin nonprofit organizations are eligible.
Geographic Focus: Florida; Minnesota; Wisconsin
Samples: SpanishNet College (Boynton Beach, FL)—to offer Spanish-language classes to U.S. teachers and other school employees and to nurses and other health-care providers, beginning in Forsyth County, NC, $225,000 (2005).
Contact: Grants Administrator, c/o Wachovia Bank
Internet: http://www.deaverfoundation.org/AboutTheDeaverFoundation.htm
Sponsor: Marion Park Deaver and Harry Gilbert Deaver Foundation
975 S Federal Hwy
Boca Raton, FL 33432

Marjorie C. Adams Charitable Trust Grants 3276
The Foundation, established in New York in 1987, is administered by the JP Morgan Chase Bank of Newark and supports the care and treatment of deaf and blind people, research into the causes and the treatment of deafness and blindness, and K-12 education. Types of support for these programs include: building and renovation projects, capital campaigns, and endowments. There are no specific application forms required or deadlines with which to adhere.
Geographic Focus: Maryland; New York
Samples: League for the Hard of Hearing, New York, New York, $95,000 (2007); Foundation for Fighting Blindness, Owings Mills, Maryland, $102,000—for an endowment (2007).
Amount of Grant: $150,000 maximum
Contact: Frank Lemma, VP of JP Morgan Chase Bank; (212) 464-2439
Sponsor: Marjorie C. Adams Charitable Trust
345 Park Avenue
New York, NY 10154

Mark and Catherine Winkler Foundation Grants 3277
The foundation awards grants in its areas of interest, including children and youth, community development, education/higher education, environmental conservation/programs, health care/surgery, human services, and social sciences. Types of support include annual campaigns, building construction/renovation, capital campaigns and endowments, challenge/matching grants, conferences and seminars, equipment acquisition, fellowships, general operating support, professorships, program development, research, and seed grants. There are no application deadlines. The board meets in November to consider requests.
Requirements: Nonprofit organizations are eligible.
Samples: Centre College of Kentucky (Danville, KY)—for program support, $500,000.; Colorado College (Colorado Springs, CO)—to create a professorship and to add a new junior faculty position in the psychology department, $1.5 million.
Amount of Grant: $700-$1 million range
Contact: Lynne Ball, Assistant Treasurer, (703) 998-0400
Sponsor: Mark and Catherine Winkler Foundation
4900 Seminary Road, Suite 900
Alexandria, VA 22311

Marlene F. Johnson Memorial Fund for Scholarly Research on Christian Science Grants 3278
The fund finances serious scholarly studies on Christian Science and the Church of Christ, Scientist, for the purpose of helping advance the understanding of those subjects in a manner fit for the introductory decades of the 21st century. The fund provides a small number of annual, renewable grants for a full-time writing project. Proposals are sought for high-level scholarly works that produce honest information and probing insights. An applicant does not need to have any connection with Christian Science for a proposal to be reviewed and potentially accepted. Applications and questions are not to be sent electronically but instead should be mailed.
Requirements: An interested applicant should submit the following papers: a proposal of not more than four pages, typed double-space, outlining the project, its potential for significantly contributing to the fund's purposes; photocopies of one or several published articles, or of short writing samples; two or three references, preferably from recognized scholars; and curriculum vita detailing academic background and experience, and listing published works.
Amount of Grant: $40,000 maximum
Date(s) Application is Due: Mar 31
Contact: Marlene F. Johnson Memorial Fund, c/o Palmer & Dodge, LLP, (617) 239-0699
Internet: http://www.johnsonfund.org
Sponsor: Marlene F. Johnson Memorial Fund for Scholarly Research on Christian Science
22 Concord Avenue, Suite 100
Cambridge, MA 02138

Marpat Foundation Grants 3279

The foundation supports nonprofit organizations in the Washington, DC, metropolitan area for programs and projects that support culture, education, health care, science, and religion. Grants are awarded to support organizations that advance international understanding; schools, universities, museums, and libraries for the advancement and diffusion of knowledge; organizations that encourage family planning; organizations promoting or conducting programs and research in the natural sciences; organizations preserving natural and historic resources, or those advancing knowledge of American history and culture; and organizations that provide food, shelter, health care, and counseling for individuals in need of such services. Grant support includes general operating support, building funds, equipment, programs/projects, and publications. Stage One summary sheets are due by the listed application deadline; full proposals are by invitation.
Requirements: Nonprofits in the Washington, DC, area are eligible.
Restrictions: The foundation does not make grants for medical research, to endowment funds, to individuals, or to organizations based outside the United States.
Geographic Focus: District of Columbia
Samples: Thelonius Monk Institute of Jazz (Washington, DC)—to support the Jazz in the Classroom program in inner-city DC public schools, $7500.; Capital City Public Charter School (Washington, DC)—to acquire books and materials that will enhance the schoolwide literacy and reading programs, $10,000.; Scenic America (Washington, DC)—for general operating support, $7500.; Civil War Preservation Trust (Washington, DC)—to support the purchase and preservation of Civil War Battlefield land at Briscoe Station, Prince William County, VA, $15,000.
Amount of Grant: $4000-$50,000 average
Date(s) Application is Due: Jun 1
Contact: Joan Koven, Secretary/Treasurer; email: jkoven@marpatfoundation.org
Internet: http://fdncenter.org/grantmaker/marpat
Sponsor: Marpat Foundation
P.O. Box 1769
Silver Spring, MD 20915-1769

Marquette University Arnold L. Mitchem Dissertation Fellowships 3280

The primary goal of the program is to help increase the presence of currently underrepresented racial and cultural groups in the U.S. professoriate by supporting doctoral candidates in completing the dissertation. Fellowships are invited for the following academic areas: education, English, foreign languages and literature, history, mathematics and mathematics education, statistics and computer science, philosophy, political science, psychology, social and cultural sciences, and theology/religious studies. Fellows teach one course in their area of specialization, interact with faculty and undergraduate students, and devote their primary energies to the completion of their dissertations. The program provides a stipend, fringe benefits, and research and travel monies. African American, Native American, and Hispanic American candidates are especially encouraged to apply. Application guidelines are available online. Annual deadlines may vary; contact program staff for exact dates.
Requirements: Persons are eligible to apply who have not earned a doctoral degree at any time or in any field, are U.S. citizens, have completed all other requirements for the PhD, are well into the writing stage of their dissertation work, and who belong to a racial-cultural group historically underrepresented in the U.S. professoriate.
Amount of Grant: $35,000 stipend
Date(s) Application is Due: Jan 10
Contact: Mitchem Dissertation Fellowship, (414) 288-7472; email: mitchem.fellowship@marquette.edu
Internet: http://www.marquette.edu/mitchem
Sponsor: Marquette University
P.O. Box 1881
Milwaukee, WI 53201-1881

Marshall Scholarships 3281

The scholarships enable U.S. citizens who are graduates of U.S. colleges and universities to study for degrees in any subject at a university in the United Kingdom for a period of at least two academic years, extendible for one more year. Up to 40 new awards are offered each year. Application guidelines may be obtained after June 1 each year from British Information Services in New York City; the British Consulate-General in Boston, Chicago, Atlanta, and San Francisco; or from the British Embassy (Cultural Department). The deadline for application falls in mid-October of the year preceding the scholarship tenure. The award's amount includes University fees, cost of living expenses, annual book grant, thesis grant, research and daily travel grants, fares to and from the United States and, where applicable, a contribution towards the support of a dependent spouse.
Requirements: Marshall scholarships are open only to citizens of the United States 25 years of age or under who must, by the time of taking up residence in a British university, have obtained a first degree requiring at least three years of study from an accredited degree-granting university or college in the United States. A minimum grade point average of 3.7 for academic courses after the freshman year will normally be required. Married persons are eligible for the scholarships, but among candidates of equal merit, preference will be given to those who intend to remain unmarried until at least the end of their first-year tenure.
Date(s) Application is Due: Oct 1
Contact: Marshall Scholarships, (202) 588-7800; fax: (202) 588-7918; email: enquiries@britishcouncil-usa.org
Internet: http://www.acu.ac.uk/marshall/index.html
Sponsor: British Council
3100 Massachusetts Avenue NW
Washington, DC 20008

Marta Lange/CQ Award 3282

This annual award, supported by Congressional Quarterly Inc, recognizes an academic or law librarian who, through research, other creative activity, or service to the profession, makes distinguished contributions to bibliography and information service in law or political science. Nominees should have achieved distinction in one or more of the following areas: planning and implementing a model bibliography/information services program in a law or political science library; history of contributions to the field through research, publications, and other activities displaying active participation in the advancement of law/political science librarianship; service to the profession through ACRL or related regional and national organizations; and promotion or development of an education program for law and political science librarianship that has served as a model for other courses and programs.
Requirements: Individuals may nominate themselves or others.
Samples: Graham R. Walden—award recipient (2006); Merle Slyhoff—award recipient, $1000 (2005); Lucia Snowhill—award recipient, $1000 (2004).
Amount of Grant: $1000
Date(s) Application is Due: Dec 1
Contact: Megan Bielefeld, (312) 280-2514; email: mbielefeld@ala.org
Internet: http://www.ala.org/Template.cfm?Section=awards&template=/ContentManagement/ContentDisplay.cfm&ContentID=93952
Sponsor: Association of College and Research Libraries
50 E Huron Steet
Chicago, IL 60611-2795

Martin E.P. Seligman Award for Outstanding Research in Positive Psychology 3283

The competition awards a cash grant to a scholar who has completed a PhD dissertation in any area of research related to positive psychology in the last three years and who plans to continue research in this area. Interested candidates should submit six copies of their curriculum vita and a summary of their dissertation research (not to exceed eight single-spaced pages, including tables and figures, but excluding references). The summary should contain a description of the problem investigated; an explanation of the methods used to investigate the problem; a summary of the findings; a discussion of the significance of the findings in relation to positive psychology; and an outline of how the work might be continued in the future. The award is presented at the APA convention. Travel to the convention and one day's lodging and expenses are also included.
Requirements: The dissertation must have been completed after March 30, 2002.
Amount of Grant: $1000
Date(s) Application is Due: Apr 1
Contact: Grants Administrator, (610) 941-2828; fax: (610) 825-1730; email: info@templeton.org
Internet: http://www.templeton.org/SeligmanAward/index.html
Sponsor: John Templeton Foundation
300 Conshohocken State Road, Suite 500
West Conshohocken, PA 19428

Martin Luther King Jr. Graduate Memorial Scholarships 3284

The scholarships provide graduate study aid to racial/ethnic minority group CTA members and/or their dependent children, and student CTA members. This scholarship is supported by voluntary contributions from the membership and has the goal of promoting professional growth for minority teachers. Scholarships vary each year depending on the amount of contributions and on the financial need of individual applicants. Applications are available in January each year and may be obtained by contacting CTA regional resource center offices or the CTA Human Rights Department in Burlingame.
Date(s) Application is Due: Mar 15
Contact: Human Rights Department, (650) 697-1400; fax: (650) 552-5001; email: scholarships@cta.org
Internet: http://www.cta.org/InsideCTA/TrainingHR/Scholarships/ScholarMLK.htm
Sponsor: California Teachers Association
P.O. Box 921, 1705 Murchison Drive
Burlingame, CA 94011-0921

Mary Baker Eddy Library Summer Research Fellowships 3285

The library awards fellowships to scholars and other researchers for research in its collections in the fields of women's history, spirituality and health, religious studies, 19th-century history, and journalism—themes for which the library's collections hold special relevance and value. Consideration will be given to applicants with outstanding writing skills, evidence of having previous publications, and a definite plan for disseminating the research through publication, lecturing, etc. Fellowship recipients are expected to be in regular and continuous use of the collections at the library in Boston during the period of the fellowship (between June 13 and September 2). Fellows are responsible for making their own arrangements for housing in the Boston area.
Requirements: Fellowships are open to doctoral and postdoctoral scholars, faculty members, and independent researchers.
Amount of Grant: $700 per scholar week
Date(s) Application is Due: Feb 13
Contact: Fellowships Administrator, (617) 450-7206; email: fellowships@marybakereddylibrary.org
Internet: http://www.marybakereddylibrary.org/programs/fellowshipapplications.jhtml

Sponsor: Mary Baker Eddy Library
200 Massachusetts Avenue
Boston, MA 02115

Mary Ingraham Bunting Institute Biomedical Research Fellowships 3286

Funded by the Burroughs Wellcome Fund, the fellowships support women scientists in all fields of biomedical research. Two awards will be made in the current year. Fellows may compete for a second year of funding. Fellowships provide office or studio space, auditing privileges, and access to libraries and most other resources of Radcliffe College and Harvard University. Residence in the Boston area and participation in the institute community are required during the fellowship appointment.
Requirements: Women scientists who are U.S. citizens or permanent residents and have held a doctorate (MD or PhD) degree for at least two years prior to appointment are eligible.
Amount of Grant: $55,000 maximum for one year with plus research expenses
Date(s) Application is Due: Oct 1
Contact: Fellowship Coordinator, (617) 496-1324; email: fellowships@radcliffe.edu
Internet: http://www.radcliffe.edu/fellowships/index.php
Sponsor: Bunting Institute of Radcliffe College
34 Concord Avenue
Cambridge, MA 02138

Mary Ingraham Bunting Institute Peace Fellowship 3287

One fellowship will be awarded to support a woman committed to domestic or international policy work, or actively engaged in finding peaceful and just solutions to conflict, either real or potential, among groups or nations. Involvement with peace issues may be of an activist or scholarly nature.
Requirements: Women who have demonstrated effectiveness in senior-level work related directly to peace and justice are eligible.
Amount of Grant: $32,000
Date(s) Application is Due: Oct 3
Contact: Fellowship Coordinator, (617) 495-8212; fax: (617) 495-8136; email: fellowships@radcliffe.edu
Internet: http://www.radcliffe.edu/fellowships/bunting.php
Sponsor: Bunting Institute of Radcliffe College
34 Concord Avenue
Cambridge, MA 02138

Mary Isabel Sibley Fellowship 3288

The fellowship is awarded annually for advanced research in Greek language, literature, history, or archaeology (odd-numbered years); or French language or literature (even-numbered years). Applicants must have shown ability to carry on original research and must be planning to devote full-time work to research during the fellowship year. Application forms may be obtained from the society.
Requirements: Applicants must be unmarried women between 25 and 35 years of age who hold doctorates or have completed all requirements except the dissertation.
Restrictions: Eligibility is not restricted to Phi Beta Kappa members or to U.S. citizens.
Samples: Pauline LeVen, Princeton University, $20,000 (2007); Alicia Levin, University of North Carolina, $20,000 (2006); Camille Robcis, Correll University, $20,000 (2005).
Amount of Grant: $20,000
Date(s) Application is Due: Jan 15
Contact: Sandra Beasley, Awards Coordinator, (202) 265-3808; fax: (202) 986-1601; email: SBeasley@pbk.org or info@pbk.org
Internet: http://staging.pbk.org/AM/Template.cfm?Section=Sibley_Fellowship&Template=/CM/HTMLDisplay.cfm&ContentID=2647
Sponsor: Phi Beta Kappa Society
1606 New Hampshire Avenue, NW
Washington, DC 20009

Mary S. and David C. Corbin Foundation Grants 3289

The foundation gives primary consideration to the city of Akron and Summit County, OH, charitable organizations and/or local chapters of national charities located in Akron and Summit County, OH, although extremely worthy causes outside of this preferred area may be considered. Areas of interest include arts and culture, civic and community, education, health care, housing, social services, medical research, and youth. Types of support include building construction/renovation, equipment acquisition, seed grants, research, and matching funds. There are no application forms. The foundation meets at least semiannually to consider requests. Only written applications will be considered. Guidelines are available online.
Requirements: Only residents of Ohio may apply.
Restrictions: Telephone and personal interviews are discouraged unless requested by the foundation. Grants are not made to individuals.
Geographic Focus: Ohio
Samples: U of Akron (OH)—for a common area, $800,000.; Children's Hospital Medical Ctr of Akron (OH)—for burn research, $236,000.
Amount of Grant: $882,500 total
Date(s) Application is Due: Mar 1; Sep 1
Contact: Program Contact, (330) 762-6427; fax: (330) 762-6428; email: corbin@nls.net
Internet: http://fdncenter.org/grantmaker/corbin
Sponsor: Mary S. and David C. Corbin Foundation
159 S Main Street, Room 910
Akron, OH 44308

Maryland State Archives Summer Internship 3290

The purpose of this internship is to introduce students to the problems and procedures of establishing archival control over permanently valuable historical records with an emphasis on modern research techniques using those records. The 10-week program is held at the Maryland State Archives with periodic travel to Baltimore and to Washington, DC.
Requirements: Applicants must be graduate or advanced undergraduate students who are either Maryland residents or attend a college or university in Maryland.
Geographic Focus: Maryland
Amount of Grant: $2625 approximately
Date(s) Application is Due: Mar 31
Contact: Program Contact, (410) 260-6400; fax: (410) 974-2525; email: emilyo@mdarchives.state.md.us
Internet: http://www.mdarchives.state.md.us/msa/educ/html/interns.html
Sponsor: Maryland State Archives
350 Rowe Boulevard
Annapolis, MD 21401

Massachusetts Bar Foundation Legal Intern Fellowships 3291

The program was established to encourage careers in the law that further the goals of social justice, while contributing valuable legal support to organizations serving the underrepresented in Massachusetts. To assist law students in gaining practical experience in the public sector, the foundation awards four stipends to law students who intern during the summer months at nonprofit organizations providing civil legal services to low-income clients in Massachusetts. Guidelines are available online.
Requirements: Applications are welcome from law students who have secured a volunteer internship with a qualified nonprofit organization in Massachusetts; are currently enrolled in U.S. law schools, with preference given to second- and third-year students; demonstrate a commitment to public interest law, including experience working with low-income clients; and commit to work no less than 10 continuous, full-time weeks between June 1 and October 1.
Geographic Focus: Massachusetts
Amount of Grant: $6000
Date(s) Application is Due: Mar 17
Contact: Fellowships Administrator, (617) 338-0534; fax: (617) 338-0550; email: foundation@massbar.org
Internet: http://www.massbarfoundation.org/grant_programs/legal_intern
Sponsor: Massachusetts Bar Foundation
20 West Street
Boston, MA 02111-1204

Massachusetts Foundation for the Humanities Project Grants 3292

The Foundation makes project grants to support lecture series, professional development for teachers, public forums, post or pre-performance discussions, film screenings with discussions, reading and discussion programs, oral history projects, and other public humanities activities in Massachusetts. In most cases, the maximum award for a grant for a public program is $5,000, and most proposals the Foundation receives fit under the general category of 'public program'. The current thematic initiative - Liberty and justice for all - gives priority to public humanities programs that explore two fundamental principles in American political life, and their interplay, past and present. Projects that address this theme are eligible for a maximum of $10,000.
Requirements: Proposals are due on or before the first business day of February, May, August, and November. Any non-profit organization, whether or not it has 501(c)3 status, is eligible for MFH funding.
Geographic Focus: Massachusetts
Samples: Coruway Film Institute (Portsmouth, NH) and film maker William Rogers—for production of Boys Home: The History of Fernald State School, a one-hour television documentary on the nation's oldest institution for people with developmental disabilities, $2000.; Wenham Museum(MA)—for a two-part program featuring Native American author and storyteller Margaret Bruchac, $2205.; Clark U Teachers Ctr for Global Studies (Worcester, MA)—to train high-school teachers to offer a year-long curriculum unit on contemporary world issues, which will culminate in a State House forum involving students, teachers, and legislators, $3875.; Korean War Veterans Assoc, Cape and Islands Chapter (Centerville, MA)—to fabricate and install a wayside exhibit at the Korean War Memorial in Hyannis, $2550.
Amount of Grant: $5000 maximum; $2,500 average
Contact: Hayley Wood, Program Officer; (413) 584-8440; fax: (413) 584-8454; email: hwood@mfh.org
Internet: http://www.mfh.org/grants/index.htm
Sponsor: Massachusetts Foundation for the Humanities
66 Bridge Street
Northampton, MA 01060

Massage Therapy Foundation Practitioner Case Report Contest 3293

The Foundation has chosen to encourage the writing of case reports to provide an opportunity for massage therapists and bodyworkers to develop research skills and enhance their ability to provide knowledge-based massage to the public. This contest is intended to enhance professional development skills of the practitioner: writing case reports help develop communication skills, critical thinking skills, and could contribute to future research and clinical practice. Cash and publication recognition will be awarded to practitioners submitting the top reports.
Requirements: The Foundation requires that HIPAA guidelines for confidentiality are followed.

Samples: Erika Larson—for Massage Therapy Effects in a Long-Time Prosthetic User with Fibular Hemimelia (2008); Glenda Keller—for The effects of massage therapy in treatment of chronic plantar fasciitis (2008); Robin B. Anderson—for Reduction and Stabilization in Parkinson-related Peripheral Edema with Therapeutic Massage (2008).
Date(s) Application is Due: Oct 8
Contact: Alison Pittas, Program Manager; (847) 869-5019, ext 167 or (847) 905-1667; fax: (847) 864-1178; email: apittas@massagetherapyfoundation.org
Internet: http://www.massagetherapyfoundation.org/practitionercontest.html
Sponsor: Massage Therapy Foundation
500 Davis Street, Suite 900
Evanston, IL 60201

Massage Therapy Foundation Research Grants **3294**
One of the primary purposes of the Foundation is to fund solid research studies investigating the many beneficial applications of massage therapy. Foundation research grants are awarded to individuals or teams conducting studies that promise to advance our understanding of specific therapeutic applications of massage, public perceptions of and attitudes toward massage therapy, and the role of massage therapy in health care delivery. The Research Grant deadline is normally March 1 annually (or the following Monday if this falls on a weekend). The normal award for 12 months is $1,000 to $30,000, and must be used in the specific time period for which it has been awarded. This grant supports high quality, independent research that contributes to the basic science of massage therapy application, including applied research investigating massage therapy as a health/mental health treatment and/or prevention modality.
Requirements: Research Grants are available to investigators who: have experience in the relevant field of research; are presently associated with, or have secured the cooperation of a university, independent research organization, health center, or other institution qualified and willing to function as a Sponsoring Organization for the purpose of this project; or new investigators without prior research experience must document support from an experience investigator willing to act as a collaborator.
Restrictions: Research Grants will not be awarded to spouses, domestic partners, children, descendant, spouses of descendants or any other individual related to any officers or trustees of the Massage Therapy Foundation, or to members of the Research Proposal Review Committee.
Amount of Grant: $30,000 maximum
Date(s) Application is Due: Mar 1
Contact: Alison Pittas, Manager of Research and Grants; (847) 869-5019, ext 167 or (847) 905-1667; fax: (847) 864-1178; email: apittas@massagetherapyfoundation.org
Internet: http://www.massagetherapyfoundation.org/grants_research.html
Sponsor: Massage Therapy Foundation
500 Davis Street, Suite 900
Evanston, IL 60201

Massage Therapy Foundation Student Case Report Contest **3295**
The program is a creative way for the foundation and massage therapy schools to work together to advance the development of research literacy and evidence-based practice habits in massage practitioners. The contest encourages students at massage-therapy schools to conduct research and produce case reports on their findings.
Amount of Grant: $2,500 gold award; $2,000 silver award; $1,500 bronze award
Date(s) Application is Due: Mar 1
Contact: Alison Pittas, Program Manager of Research and Grants; (847) 869-5019, ext 167 or (847) 905-1667; fax: (847) 864-1178; email: apittas@massagetherapyfoundation.org
Internet: http://www.massagetherapyfoundation.org/contest.html
Sponsor: Massage Therapy Foundation
500 Davis Street, Suite 900
Evanston, IL 60201

Massage Therapy Research Grants **3296**
This grant supports high quality, independent research that contributes to the basic science of massage therapy application, including applied research investigating massage therapy as a health/mental health treatment and/or prevention modality. Research grants are awarded to individuals or teams conducting studies that promise to advance understanding of specific therapeutic applications of massage, public perceptions of and attitudes toward massage therapy, and the role of massage therapy in health care delivery.
Requirements: The grant is available to investigators who have experience in the relevant field of research, and are presently associated with or have secured the cooperation of a university, independent research organization, or other institution qualified and willing to function as a sponsoring organization for the purpose of this project.
Samples: Steve Keir, Duke University, $20,000—for Feasibility Study of Massage Therapy within a Brain Tumor Setting (2008); Jennifer Rosen, MD, Boston Medical Center, $20,000—for Pilot RCT of Cancer Patients Undergoing Port-a-Catheter Placement (2008).
Amount of Grant: $1,000-$30,000
Date(s) Application is Due: Mar 1
Contact: Alison Pittas; (847) 869-5019, ext. 167; fax: (847) 864-1178; email: apittas@massagetherapyfoundation.org
Internet: http://www.massagetherapyfoundation.org/grants_research.html
Sponsor: Massage Therapy Foundation
500 Davis Street, Suite 900
Evanston, IL 60201

Materials Research Society Medalist Awards **3297**
This program constitutes public and professional recognition of outstanding recent achievements in materials research. A medal and cash prize will be awarded annually for a specific discovery or advancement that is expected to have a major impact on the progress of any materials-related field. A medal will normally be awarded to one individual for a cited achievement; however, it may be awarded to the collaboration of two or three individuals. In that case, each recipient will receive a personal medal trophy, registration, and travel, and the cash award will be divided equally. Recipients are selected from nominations prepared by peers of the recipients, not the recipients themselves. A nominee need not be a member of the society and may be of any nationality or citizenship. A nomination for the medal remains under consideration for three years, and may be updated by the nominator during that time. The award includes an engraved medal, registration fee for MRS meeting, and reasonable travel expenses reimbursement.
Requirements: Nominations are open to scientists and engineers who have in recent years been responsible for major advances in any materials-related field of research. Nomination package consists of statement by nominator supporting the candidate's suitability with particular reference to specific research achievement, impact of the achievement within the relevant materials field and its effect on other sciences, and innovative and perceptive quality of the candidate's work; supporting information and documents such as curriculum vita including relevant publications and copies of important papers relevant to the research achievements; letters of support from two established scientists familiar with the nominee's qualifications and area of research; and a list of supporting documents submitted. Package should not exceed 10 pages (excluding reprints).
Restrictions: Current members of the MRS awards committee and MRS officers are not eligible.
Amount of Grant: $5000 cash prize
Date(s) Application is Due: Jun 1
Contact: John Ballance, Executive Director, (724) 779-3003; fax: (724) 779-8313; email: ballance@mrs.org
Internet: http://www.mrs.org/awards/index.html
Sponsor: Materials Research Society
506 Keystone Drive
Warrendale, PA 15086-7573

Matilda R. Wilson Fund Grants **3298**
The fund awards grants, primarily in southeast Michigan, in support of the arts, youth agencies, higher education, hospitals, and social services. Types of support include building construction/renovation, endowments, equipment acquisition, general operating support, matching/challenge grants, program development, research, and scholarship funds. There are no application forms; initial approach should be a letter of request. The board considers requests at quarterly board meetings, in January, June, September, and December.
Requirements: Michigan tax-exempt organizations are eligible.
Restrictions: Grants are not made to individuals.
Geographic Focus: Michigan
Samples: Oakland U (MI)—to repair and restore Meadow Brook Hall, the former home of the university's founder, Matilda Dodge Wilson, $7 million (2004).
Amount of Grant: $10,000-$100,000 average
Contact: David Larsen, Grants Administrator, (313) 259-7777; fax: (313) 393-7579; email: roosterveen@bodmanllp.com
Sponsor: Matilda R. Wilson Fund
100 Renaissance Center, 34th Fl
Detroit, MI 48243

Matsumae International Foundation Grants **3299**
The foundation operates worldwide in the fields of medicine/health and science/technology. Fields of study such as natural science, engineering, and medicine are given first priority. Types of support include grants, scholarships, and fellowships, tenable for three to six months. Awards include a monthly stipend; lump sum on arrival to assist with the cost of lodging, travel expense, etc.; round-trip travel; and insurance.
Requirements: Applicants must apply from outside Japan, be under the age of 40, and be postdoctoral researchers or have a minimum of two years of research experience after receipt of a master's degree. Study in the areas of natural science, engineering, and medicine receive priority.
Amount of Grant: @Y200,000 monthly stipend; @Y300,000 lump sum on arrival
Date(s) Application is Due: Jul 31
Contact: Grants Administrator, 81-3-3214-7611; fax: 81-3-3301-7601; email: contact@matsumae-if.org
Internet: http://www.mars.dti.ne.jp/~mif
Sponsor: Matsumae International Foundation
4-14-46, Kamiogi
Suginami-ku, Tokyo 167-0043 Japan

Mattel Children's Foundation Grants **3300**
Charitable organizations that demonstrate that they directly serve children in need may be eligible for grants of $5,000 up to $20,000. Such organizations must have a mission that focuses on the direct service of children ages zero to 12 years. Applicants or programs must show creative and/or innovative methods to address locally defined need directly impacting children, and must be aligned with Mattel's philanthropic priorities: Learnin—increasing access to education for underserved children and in particular, innovative strategies to promote and address literacy; health—supporting the physical health and well-being of children, with particular emphasis on promoting healthy,

active lifestyles; and Girl Empowerment—promoting self-esteem in young girls, up to age 12. Organizations may submit under a fiscal agent as long as the program has been in existence for at least two years.
Requirements: Applicants must be a 501(c)3 Public Charity organizations. If an organization is a project under a fiscal sponsor, that sponsor must have a valid tax exemption status. Organizations must serve children in communities within the U.S. Organizations must not discriminate against a person or a group on the basis of age, political affiliation, race, national origin, ethnicity, gender, disability, sexual orientation or religious beliefs. Organizations must have an annual operating budget of less than $1,000,000.
Restrictions: Organizations may not be affiliated in any way with national organizations, regardless of whether funds are received from the national entity. Examples of organizations that are ineligible to apply include Boys & Girls Clubs, United Ways, YMCA's, and disease-affiliated organizations such as the American Heart Association.
Samples: Lower Eastside Girls Club, New York, NY (2006); MicheLee Puppets, Inc., Orlando, FL (2006); Quiet Creek Herb Farm & School Country Living, Brookville, PA (2006).
Amount of Grant: $5,000-$20,000
Date(s) Application is Due: Jun 15
Contact: Grants Administrator, (310) 252-2000; fax: (310) 252-2180; email: foundation@mattel.com
Internet: http://www.mattel.com/about_us/philanthropy/ci_mcf_philanthropy_grantmaking.asp
Sponsor: Mattel Children's Foundation
333 Continental Boulevard, M1-1418
El Segundo, CA 90245-5012

Mattel International Grants Program 3301
The corporation and its foundation award grants to charitable organizations that directly serve children in need. Funding priorities include international organizations or programs that creatively address a defined need directly impacting children in need, and international organizations or programs that align with Mattel's philanthropic priorities, including health—supporting the physical and mental health and well-being of children, and increasing access to health care services for children in need; education—increasing access to education, promoting literacy to children in need, and resources that promote after-school educational achievement; and girls empowerment—promoting self-esteem of girls and increasing access to education, health, and community resources for girls. Types of support include program-specific grants—funding for new programs or expansion of existing programs; and core operating support—support of organizations to sustain their programs. Applications must be submitted online. Applications will not be accepted by fax or mail. Submit questions via email.
Requirements: Organizations must serve children in non-U.S. communities; and must not discriminate against a person or a group on the basis of age, political affiliation, race, national origin, ethnicity, gender, disability, sexual orientation, or religious belief. Pilot projects and new organizations may be considered as long as all eligibility criteria are met; however, preference will be given to organizations that have at least two years of experience. Preference is given to organizations that have an annual operating budget of less than $1 million and are not affiliated with a national organization.
Restrictions: The program does not not fund capital funding for physical property purchase, renovation, or developments; individuals; political parties, candidates, or partisan political organizations; labor organizations, fraternal organizations, athletic clubs, or social clubs; sectarian or denominational religious organizations, except for programs that are available to anyone, broadly promoted, and free from religious orientation; schools and school districts; fundraising events (e.g., dinners, tournaments); advertising or marketing sponsorships; or research overhead and/or indirect costs (including fiscal sponsor fees) that exceed 15 percent of the direct project costs.
Amount of Grant: $5000-$25,000 typically
Contact: Grants Administrator, (310) 252-2000; fax: (310) 252-2180; email: foundation@mattel.com
Internet: http://www.mattel.com/About_Us/Philanthropy
Sponsor: Mattel Children's Foundation
333 Continental Boulevard, M1-1418
El Segundo, CA 90245-5012

Maurice Amado Foundation Grants 3302
The foundation awards grants to nonprofit organizations in the Los Angeles and New York City metropolitan areas primarily for activities that emphasize the Sephardic Jewish heritage, with special interest in educational and cultural projects. Types of support include general operating support, challenge grants, conferences and seminars, continuing support, endowment funds, program development, publication, curriculum development, fellowships, seed money grants, and research. Letters of request and proposals are accepted and reviewed throughout the year. The board meets in the fall and the spring, at which time funding decisions are made. Organizations interested in applying are asked to submit a letter of inquiry.
Requirements: 501(c)3 organizations in New York and California are eligible. Individuals applying for grants must be affiliated with a tax-exempt organization.
Restrictions: Grants are not made directly to individuals.
Samples: 92nd Street Y (NY)—for operating support, $15,000.; American Jewish Historical Society (NY)—for operating support, $10,000.
Amount of Grant: $500-$200,000
Date(s) Application is Due: Feb 15; Aug 15
Contact: Pam Kaizer, Executive Director, (818) 980-9190; fax: (818) 980-9190; email: pkaizer@mauriceamadofdn.org
Internet: http://www.mauriceamadofdn.org
Sponsor: Maurice Amado Foundation
3940 Laurel Canyon Boulevard, No 809
Studio City, CA 91604

Maurice J. Masserini Charitable Trust Grants 3303
The trust awards one-year grants to eligible San Diego County California nonprofit organizations in its areas of interest, including children and youth, aging, music, higher education, and marine sciences. Types of support include building construction/renovation, equipment acquisition, program development, research grants, matching grants, development grants, internships, and scholarships. Submit a letter of application, and include a copy of the IRS tax-exempt letter. Contact foundation by email for current guidelines.
Requirements: San Diego County, CA, 501(c)3 tax-exempt organizations are eligible.
Geographic Focus: California
Amount of Grant: $2500-$25,000
Date(s) Application is Due: Feb 15; May 15; Aug 15; Nov 15
Contact: Brian Murphy, c/o Wells Fargo Bank, San Diego Private Banking Group, (858) 597-4307; fax: (858) 597-4336; email: brian.j.murphy@wellsfargo.com
Sponsor: Maurice J. Masserini Charitable Trust
4475 Executive Drive, Suite 100
San Diego, CA 92121-2130

Max and Victoria Dreyfus Foundation Grants 3304
The Foundation awards grants for: museums; cultural, performing, and visual arts programs, schools; hospitals, educational and skills training projects and programs for youth, seniors, and the disabled. Areas of interest include arts, education, health care, hospitals, social services, and medical research. Submit an inquiry letter (three-page maximum) that describes the nonprofit organization, its annual operating budget, the project to be funded, and the project budget; include proof of tax-exempt status.
Requirements: U.S. nonprofit organizations are eligible.
Restrictions: Foreign charitable organizations and individuals are ineligible.
Amount of Grant: $5,000-$20,000 average
Date(s) Application is Due: Mar 10; Jul 10; Nov 10
Contact: John W. Hager; (202) 337-3300; fax: (202) 337-4925
Sponsor: Max and Victoria Dreyfus Foundation
2233 Wisconsin Avenue, NW, Suite 414
Washington, DC 20007-4122

Max Bell Foundation Grants 3305
The foundation focuses its grant making on innovative projects having substantial impact within the areas of health, Canada and the Asian Pacific, and veterinary sciences and education. Under health care, the foundation directs its support to the areas of health education and preventive health care. It is also interested in assisting health care professionals to update their education on a continuing basis. The foundation also supports research intended to build a better understanding of the Asian Pacific area on the part of Canadian policymakers and businesspeople and assists Canadian scholars to build ties to their Asian counterparts. Also funded are major programs at centers of Asian research in Canadian universities. Within the field of veterinary sciences and education, the foundation seeks to provide broad-based support, although it has developed a particular interest in equine research. Applications are reviewed three times annually, eight weeks before the board meets (usually in May September, and November).
Requirements: Funding is restricted to registered Canadian charities. Applicants should submit a brief letter describing the project for which they are seeking funds, an estimated budget, and the organization's charitable gift registration number. If the letter proves of interest to the foundation, a more detailed proposal will be invited.
Restrictions: Support will not be provided for capital projects or building campaigns; continuing operating expenses, funds to cover budget deficits, general fund-raising drives, or annual charitable appeals; academic or sabbatical leaves, internships, bursaries or scholarships; conferences, seminars or workshops, equipment purchases, publication costs, or travel costs.
Samples: AboutFace (Toronto, Ontario)—to provide support to children with facial disfigurements and their families, $C100,000.; U of Toronto (Toronto, Ontario)—to develop a summer study program on the campus of the U of Hong Kong, $C105,000.; U of Guelph (Guelph, Ontario)—to explore the biological and physical attributes of the Great Lakes and consider the continuing impact of humans upon them, $C205,000.
Amount of Grant: $C20,000-$C25,000 average
Contact: Alida White, Administrator, (403) 215-7310; fax: (403) 215-7319; email: apply@maxbell.org
Internet: http://www.maxbell.org
Sponsor: Max Bell Foundation
639 Fifth Avenue SW, Suite 1230
Calgary, AB T2P OM9 Canada

Max Goldenberg Foundation Grants 3306
The foundation awards grants to Illinois nonprofit organizations in its areas of interest, including healthcare, health organizations, and hospitals; higher education; Jewish services; medical research; and social services delivery.
Requirements: Illinois nonprofit organizations are eligible.
Geographic Focus: Illinois
Amount of Grant: $2500-$40,000 range
Contact: Christopher McClear, c/o Harris Trust and Savings Bank, (312) 461-5992

Sponsor: Max Goldenberg Foundation
P.O. Box 755
Chicago, IL 60690

Max Kade Foundation Grants **3307**

The foundation awards grants to higher education institutions in the areas of medicine, and natural/physical sciences. Postdoctoral research exchange programs between the United States and Europe in the following disciplines receive high priority: biological sciences, biomedicine, chemistry, engineering, higher education, language and literature, medical research, and the physical/earth sciences. Grants also support visiting faculty exchange programs and training of language teachers. There are no application deadlines.
Restrictions: Grants do not support capital funds, development campaigns, endowment funds, general operating expenses, or individuals.
Samples: Concordia College (Moorhead, MN)—for construction and furnishing of Erich H. Markel German House, $350,000.; University of Rostock (Rostock, Germany)—to renovate Max Kade student dormitory and reservations for U.S. students, $100,000.
Amount of Grant: $10,000-$100,000 average
Contact: Hans Hachmann, President, (646) 672-4354
Sponsor: Max Kade Foundation
6 E 87th Street
New York, NY 10128

Mayo Foundation Multidisciplinary Training in Digestive Diseases Research Grants **3308**

Postdoctoral fellowships in gastroenterological research are available for two or more years. Faculty members employ a multidisciplinary approach using biochemical, cell/molecular biological, immunological, physiological, and electrophysiological techniques. Applicants may focus on clinical and/or basic studies. A curriculum vita and the names of three references should be submitted to the program director.
Requirements: Applicant should have an MD or PhD degree and must be a U.S. citizen or have permanent visa status.
Contact: Dr. N.F. LaRusso, Program Director, (507) 284-1000
Internet: http://mayoresearch.mayo.edu/mayo/research/traininggrants
Sponsor: Mayo Foundation
200 First Street SW
Rochester, MN 55905

Maytag Corporation Foundation Grants **3309**

The foundation awards grants in Iowa, primarily in Des Moines and Newton, for higher and other education, arts, humanities, cultural programs, public affairs, social services, health, cancer research, the disabled, and family planning. Types of support include operating budgets, continuing support, annual campaigns, seed money, building funds, equipment, land acquisition, matching funds, research, special projects, and renovation projects.
Requirements: Nonprofit organizations in Iowa may apply.
Restrictions: Grants are not awarded to individuals or for emergency funds, deficit financing, endowments, scholarships, fellowships, demonstration projects, publications, or conferences.
Geographic Focus: Iowa
Amount of Grant: $1000-$25,000 average
Contact: Grants Administrator, (641) 792-7000
Internet: http://www.maytag.com/mths/our_company/default.jsp
Sponsor: Maytag Corporation Foundation
P.O. Box 39, 403 W Fourth Street N
Newton, IA 50208-0039

Mazda Foundation Grants **3310**

The foundation awards grants to nonprofit organizations for programs promoting education and literacy, environmental conservation, cross-cultural understanding, social welfare, and scientific research. Preference is given to nonprofit organizations that are national in scope.
Requirements: Nonprofit organizations are eligible.
Restrictions: Grants do not support individuals, political organizations, religious organizations, fundraising dinners or events, capital building, endowments, or deficit reduction.
Geographic Focus: California; District of Columbia; Louisiana; New Hampshire; North Carolina; Texas
Amount of Grant: $400,000 total
Date(s) Application is Due: Jul 1
Contact: Barbara Nocera, Program Director, (202) 467-5088; fax: (202) 223-6490; email: bnocera@mazdausa.com
Internet: http://www.mazdafoundation.org/grant.html
Sponsor: Mazda Foundation
1025 Connecticut Avenue NW, Suite 910
Washington, DC 20036

MAZON Grants: A Jewish Response to Hunger **3311**

MAZON's domestic grant program makes awards for general support or special projects in four categories. Advocacy Education/Research grants, ranging from $5000-$20,000, support programs to change public policy or opinion or the operation and administration of federal food assistance programs. Emergency/Direct Food Assistance grants, ranging from $5000-$7000, support food pantries, soup kitchens, and programs providing meals to special-needs populations. Food Bank grants, ranging from $5000-$15,000, are awarded to warehouses that collect and distribute food to nonprofit feeding programs; applicants must demonstrate that they participate in antihunger advocacy and education efforts. Multiservice Organization grants, ranging from $5000-$15,000, support organizations that provide food and a broad array of services to hungry and low-income people. Deadlines listed are for letters of inquiry.
Requirements: U.S. nonprofit organizations that meet program criteria may apply.
Restrictions: MAZON does not make grants to groups that proselytize or place any religious requirements on service; have national name recognition; are principally focused on homelessness; charge individuals for food; or are government entities, professional associations, job training programs, or grantmaking organizations. Capital grants for building projects are not eligible for support. Prepared and perishable food programs are not eligible unless they are in their first two years of operation. MAZON does not fund holiday meal programs.
Samples: 28 nonprofit organizations nationwide—for their work to relieve hunger in the United States, $110,500 divided.
Date(s) Application is Due: May 15; Nov 15
Contact: Grants Administrator, (310) 442-0020; fax: (310) 442-0030; email: mazonmail@mazon.org
Internet: http://www.mazon.org/What_We_Fund/Funding_Guidelines/
Sponsor: MAZON: A Jewish Response to Hunger
1990 S Bundy Drive, Suite 260
Los Angeles, CA 90025-5232

McCarthy Family Foundation Grants **3312**

The foundation is currently considering grants in four major areas: K-12 science education—to foster increased study of science and to promote further scientific study and careers by qualified students; HIV/AIDS—research (to determine the pathogenesis of HIV and AIDS and related diseases, and to develop new experimental therapies), and education and support (to provide support to HIV-positive individuals and public education regarding the disease); assistance to homeless people—to help with basic needs and support to improve opportunity for independence, health, and self-esteem (including youth and adult education); and child abuse prevention—for child abuse prevention programs and to provide necessary services for victims of child abuse and their families.
Requirements: Proposals will only be accepted for programs within San Diego County.
Restrictions: The foundation does not consider grants for individuals, scholarship funds, or sectarian religious activities. Normally the foundation does not consider requests for general fundraising drives. It does not make grants intended directly or indirectly to support political candidates or to influence legislation.
Geographic Focus: California
Samples: Shelter Partnership (Los Angeles, CA)—for a resource bank, $5000.; KPBS-TV (San Diego, CA)—for the San Diego Night Sky documentary, $25,000.
Amount of Grant: $5000-$15,000 generally
Date(s) Application is Due: Mar 15; Sep 15
Contact: Program Contact, fax: (858) 485-0172; email: mail@mccarthyfamilyfdn.org
Internet: http://www.mccarthyfamilyfdn.org
Sponsor: McCarthy Family Foundation
P.O. Box 27389
San Diego, CA 92198-1389

McCombs Foundation Grants **3313**

The foundation makes grants to Texas nonprofit charitable, philanthropic, educational, and benevolent organizations. Areas of interest include arts, education and higher education, historic preservation, medical research, philanthropy, recreation and sports, youth services, and voluntarism. There are no application deadlines or forms.
Requirements: Texas nonprofit organizations are eligible.
Geographic Focus: Texas
Samples: M.D. Anderson Cancer Ctr (Houston, TX)—for cancer research, $30 million (2005).
Contact: Grants Administrator, (210) 821-6523
Internet: http://www.redmccombs.net/philanthropy.html
Sponsor: McCombs Foundation
755 E Mulberry, Suite 600
San Antonio, TX 78212

McCune Foundation Grants **3314**

The foundation supports nonprofit organizations in southwestern Pennsylvania and emphasizes two major program areas: independent higher education and human services. Education grants also support adult basic education and literacy programs. The foundation recognizes and supports civic, cultural, and community-based organizations. Funding requests may be made in support of capital campaigns, building construction/renovation, equipment acquisition, program development, and seed money. The board meets quarterly. Application forms are not required.
Requirements: The foundation supports organizations in southwestern Pennsylvania, with emphasis on the Pittsburgh area.
Restrictions: Grants are not awarded to individuals or for general operating purposes or loans.
Geographic Focus: Pennsylvania
Samples: Carnegie-Mellon U (Pittsburgh, PA)—program support, $500,000.; Family Resources (Pittsburgh, PA)—program support, $350,000.
Amount of Grant: $50,000-$500,000 average
Contact: Henry Beukema, Executive Director, (412) 644-8779; fax: (412) 644-8059; email: info@mccune.org

Internet: http://www.mccune.org
Sponsor: McCune Foundation
750 Sixth PPG Pl
Pittsburgh, PA 15222

MCEAS Dissertation Fellowships 3315
The center was established to marshal the vast scholarly and archival resources of 10 area institutions in an interdisciplinary research program focused on the history and culture of the United States to 1850, to increase scholarly and public awareness of the heritage of the Delaware Valley. At least six new fellows will be appointed to nine-month terms beginning September 1st, with a stipend of at least $18,000 payable in monthly installments. While a limited number of shorter-term awards may be made, all candidates should apply for a nine-month appointment. Fellows receive office space in the Center's magnificent new building on the University of Pennsylvania's historic campus and library, computer, and other privileges at the University. Limited travel funds for research are also available. While no teaching is required for most fellowships, all McNeil Center fellows are expected to be in residence in Philadelphia during the academic year and to participate regularly in the Center's program of seminars and other activities.
Requirements: Fellowships are available to advanced graduate students from any PhD-granting institution who are in the dissertation research or writing stage.
Amount of Grant: $18,000 minimum
Date(s) Application is Due: Mar 1
Contact: Daniel K. Richter; (215) 898-9251; fax (215) 573-3391; email: drichter@history.upenn.edu
Internet: http://www.mceas.org/dissertationfellowships.htm
Sponsor: McNeil Center for Early American Studies
3355 Woodland Walk
Philadelphia, PA 19104-6213

McGraw-Hill Companies Corporate Contributions 3316
The foundation supports innovative programs and initiatives that increase the abilities of people around the world to learn; to grow intellectually; to master new skills; and to maximize their individual talents for school, work, and community. Projects are considered that promote and support excellence in education and learning, focus on initiatives that advance knowledge and work and home skills, contribute to the education of youth in the communities and markets where McGraw-Hill companies operate, utilize unique applications of new and developing technologies, and extend their reach globally. Types of support include matching, program, and research grants; and scholarship funds and employee-related scholarships.
Requirements: Qualified applicants are invited to submit preliminary proposal letters that should include a brief background of the organization, concise program description, copy of the most recent audited financial statement and annual report, current year's budget, budget of the program, and evidence of 501(c)3 status.
Restrictions: The foundation does not contribute to political activities or organizations established to influence legislation; individuals; publication of books, magazines, films, or videos; organizations included in United Way drives; sectarian or religious organizations; endowment funds; conferences, trips, or tours; nor loans of any kind.
Samples: Hofstra U (Hempstead, NY)—for a business-technology laboratory at the university's Business Development Ctr.; Friends of the National Journalism Library, National Press Club (Washington, DC)—for its news information center, as a research center for journalists and journalism students.
Amount of Grant: $500-$10,000; $2500 average
Contact: Susan Wallman, Manager, Corporate Contributions, (212) 512-6480; fax: (212) 512-3611; email: susan_wallman@mcgraw-hill.com
Internet: http://www.mcgraw-hill.com/community/guidelines shtml
Sponsor: McGraw-Hill Companies
1221 Avenue of the Americas, 20th floor
New York, NY 10020-1095

McGregor Fund Grants 3317
The fund makes grants to organizations in the following areas: human services—emphasis on activities in southeastern Michigan addressing emergency needs in housing, food, clothing, and other direct aid; education—early child education and development, colleges and universities in the Detroit metropolitan area, liberal arts colleges and universities in Michigan and Ohio; arts and culture in southeastern Michigan; health care—improving access to primary care for underserved populations in Detroit; and public benefit—civic and community organizations that improve the quality of life in southeastern Michigan. Types of support include project support, operating support, special projects, and capital support. Grant decisions are made by the board of trustees four times annually, in February, May, September, and November. Grant requests may be submitted at any time, but requests may take up to three months for staff review. Application guidelines are available online.
Requirements: Organizations in the metropolitan Detroit, MI area, are eligible. Requests will be considered from organizations located elsewhere for programs or projects that significantly benefit the metropolitan Detroit area (city of Detroit and Wayne, Oakland, and Macomb Counties).
Restrictions: The fund discourages proposals for student scholarships, travel, seminars, conferences, workshops, film or video projects, as well as disease-specific organizations and their local affiliates.
Geographic Focus: Michigan
Amount of Grant: $25,000-$50,000 average
Contact: C. David Campbell, President, (313) 963-3495; fax: (313) 963-3512; email: info@mcgregfund.org
Internet: http://www.mcgregorfund.org/guideline.html
Sponsor: McGregor Fund
333 W Fort Street, Suite 2090
Detroit, MI 48226

McKnight Doctoral Fellowships 3318
The fellowship is awarded only to African Americans who intend to seek the PhD degree in any field in the arts and sciences, mathematics, business, or engineering. The fellowship award includes tuition and a stipend for up to five years of support. Additional information and applications are available on the Web site.
Requirements: Applicants must be African American, U.S. citizens, and hold a minimum of a bachelor's degree from a regionally accredited college or university.
Amount of Grant: $17,000
Date(s) Application is Due: Jan 15
Contact: Program Officer, (813) 272-2772; fax: (813) 272-2784; email: mdf@fl-educ-fd.org
Internet: http://www.fefonline.org/mdf.html
Sponsor: Florida Education Fund
201 E Kennedy Boulevard, Suite 1525
Tampa, FL 33602

McKnight Foundation Grants 3319
The foundation awards grants to eligible nonprofit organizations in support of efforts to strengthen communities, families, and individuals, particularly those in need. Grants contribute to the arts, encourage preservation of the natural environment, and promote research in selected fields. Types of support include planning, project, and operating support. Application deadlines are for letters of inquiry; full proposals are by invitation.
Requirements: Eligible applicants must be tax-exempt, nonprofit organizations in Minnesota that are not private foundations.
Restrictions: Examples of areas not supported by grants include mental health or disabilities, except for services to enhance parents' capacity to nurture their children; chemical dependency treatment; services for seniors; and health. In addition to these program areas, grants do not support scholarships or other types of assistance for individuals; attendance at or travel to conferences or costs related to conferences; travel, except when related to other McKnight support of an organization; scientific research outside established McKnight research programs; endowments, except in rare cases; or activities that have a specific religious purpose.
Samples: Greater Metropolitan Housing Corp of the Twin Cities (Minneapolis, MN)—for general operating support, $400,000 over two years (2005); Loft (Minneapolis, MN)—for a literary-arts center, and for a national magazine for readers and writers, $160,000 (2005); Resource Inc (Minneapolis, MN)—for a program that trains and supports low-income mothers to help them advance in their jobs, $200,000 (2005); Midwest Environmental Advocates (Madison, WI)—for a campaign to implement and enforce the Clean Water Act, $80,000 (2005).
Amount of Grant: $23 million total annually; $5000-$500,000 typically
Date(s) Application is Due: Jan 15; Apr 15; Jul 15; Oct 15
Contact: Stephanie Duffy, Grants Administration Manager, (612) 333-4220; fax: (612) 332-3833; email: info@mcknight.org
Internet: http://www.mcknight.org
Sponsor: McKnight Foundation
710 S Second Street, Suite 400
Minneapolis, MN 55401

McLean Contributionship Grants 3320
The program awards grants primarily in Pennsylvania in its areas of interest, including aging, children and youth, education, environment, health care, historic preservation, hospitals, libraries, medical research, museums, nursing homes, performing arts, and youth services. Types of support include building construction/renovation, capital campaigns, conferences and seminars, endowments, equipment, internships, program development, publication, scholarship funds, and seed grants. Applicants can apply using the Delaware Valley Grantmakers Common Grant Application. The board meets quarterly to review requests.
Restrictions: The contributionship does not fund the costs or expenses of existing staff allocated to a project it is asked to support.
Samples: Wistar Institute of Anatomy and Biology (Philadelphia, PA)—to purchase a new tissue processor, $28,000.; Lankenau Hospital Foundation (Wynnewood, PA)—for renovation of the emergency room, $100,000.
Amount of Grant: $2000-$100,000
Contact: Sandra McLean, Executive Director, (610) 527-6330; fax: (610) 527-9733
Internet: http://fdncenter.org/grantmaker/mclean
Sponsor: McLean Contributionship
945 Haverford Road, Suite A
Bryn Mawr, PA 19010-3814

MDA Development Grant 3321
The association will consider an application for a research grant from a candidate who may be a member of a research team in the laboratory of a senior investigator under whose guidance the applicant will be given flexibility to conduct a neuromuscular disease research project. Awards are for one to three years. Applications must be requested by the listed application deadlines. Guidelines are available online.

Requirements: An applicant must hold a Doctor of Medicine, Doctor of Philosophy, Doctor of Science or equivalent degree; be a professional member at an appropriate institution; be qualified to conduct a program of original research; have an acceptable research plan for a specific disease in MDA's program; have access to institutional resources necessary to conduct the proposed research project; and have eighteen months of postdoctoral research laboratory training at the time of application.
Amount of Grant: $45,000 maximum per year
Date(s) Application is Due: Jun 15; Dec 15
Contact: Grants Program Manager, (520) 529-2000; fax: (520) 529-5454; email: grants@mdausa.org
Internet: http://www.mdausa.org/research/guidelines.html#developmentGrant
Sponsor: Muscular Dystrophy Association
3300 E Sunrise Drive
Tucson, AZ 85718

MDA Neuromuscular Disease Research Grants **3322**
The association supports research into 40 diseases of the neuromuscular system to identify the causes of and effective treatments for the muscular dystrophies and related diseases, which include spinal muscular atrophies and motor neuron diseases, peripheral neuropathies, inflammatory myopathies, metabolic myopathies, and diseases of the neuromuscular junction. Awards cover periods of up to three years. The listed deadlines are for requesting applications.
Requirements: Applicants must be professionals or faculty members at appropriate educational, medical, or research institutions and qualified to conduct and supervise programs of original research; must have access to institutional resources necessary to conduct the proposed research project; and must have the MD, PhD, DSc, or equivalent degree. Preliminary letter requesting an application is required and must include name and address of institution(s) where the research is to be conducted, the specific aim of the proposed research, statement of relevance to the purposes of MDA, and a statement on whether human subjects or tissues will be used.
Samples: Dr. J. Edwin Blalock, U of Alabama (Birmingham, AL)—for research on diagnosis and immunotherapy of myasthenia gravis, $100,000 over two years.; Dr. Terrone Rosenberry, Case Western Reserve U (Cleveland, OH)—for research on catalytic sites and subunity assembly of acetylcholinesterase, $51,500.
Date(s) Application is Due: Jun 15; Dec 15
Contact: Grants Program Manager, (520) 529-2000; fax: (520) 529-5454; email: grants@mdausa.org
Internet: http://www.mdausa.org/research/guidelines.html
Sponsor: Muscular Dystrophy Association
3300 E Sunrise Drive
Tucson, AZ 85718

MDBD Basic Research Grants **3323**
Grants are awarded to qualified U.S. institutions to support medical and scientific research in the field of birth defects. Research supported by the foundation on the etiology of birth defects takes three main directions. The first supports investigation of the hereditary influences of birth defects. Requests may be to investigate the structure and function of chromosomes, their subunits, genes, and their supporting substructures. The second investigates environmental effects ranging from obvious factors such as drugs, radiation, infections, and malnutrition to more subtle influences such as cytoplasmic influences. The third is the interaction between heredity and environment such as enzymatic feedback and an individual's resistance to disease. Applications will be sent to those whose abstracts are considered relevant to this program. Abstracts must be received with a postmark on or before the listed application deadline date.
Requirements: A request for a new grant should be initiated by a 300-word abstract to the director of grants administration from the principal investigator of the proposed research program. This letter of inquiry, containing a brief protocol of the study, must be submitted before an application form will be provided. Letters of intent may be submitted online at http://researchgrants.modimes.org/LOI.
Samples: Dr. Mary Porter, U of Minnesota (Minneapolis, MN)—for the research study Molecular and Genetic Analyses of the Mechanisms of Flagellar Motility, $37,001.; Dr. John Tamkun, U of California (Santa Cruz, CA)—for the research study Regulators of Homeotic Gene Expression in Drosophilia, $52,242.
Amount of Grant: $25,000 average
Date(s) Application is Due: Apr 30
Contact: Vice President for Research, (914) 997-4555; fax: (914) 997-4560; email: researchgrants@marchofdimes.com
Internet: http://www.marchofdimes.com/professionals/685_2160.asp
Sponsor: March of Dimes Birth Defects Foundation
1275 Mamaroneck Avenue
White Plains, NY 10605

MDBD Research Grants and Basil O'Connor Starter Scholar Award Program **3324**
The program is designed for young scientists at the level of instructor or assistant professor who are embarking on their independent research careers. Both awards support research on birth defects. Grants do not include the recipients' salaries, but provide support for technical help and supplies. Programs may be approved for two years only. Each potential applicant must submit a letter of nomination written by the chair of the department, director of the institute, dean, or other senior person; brief abstract of the proposed research; and curriculum vita. Candidates meeting the criteria of the program will receive application forms.
Restrictions: Applicants may not submit simultaneously any other applications for March of Dimes research programs.
Samples: Dr. Anne Calof, U of Iowa (Iowa City, IA)—for Strategies for Identifying Genes that Regulate Neurogenesis in Mammals, $30,000.; Dr. Piero Rinaldo, Yale U (New Haven, CT)—for research on sudden infant death syndrome, $10,000.
Amount of Grant: $75,000 maximum
Date(s) Application is Due: Mar 19
Contact: Grants Administrator, (914) 997-4564 or (888) 663-4637; fax: (914) 997-4560; email: researchgrants@marchofdimes.com
Internet: http://www.marchofdimes.com/professionals/691_1452.asp
Sponsor: March of Dimes Birth Defects Foundation
1275 Mamaroneck Avenue
White Plains, NY 10605

MDBDF Agnes Higgins Award **3325**
The award recognizes distinguished achievement in research, education, or clinical services in the field of maternal-fetal nutrition. In addition to an honorarium, winners are honored at a presentation and reception at the American Public Health Association Annual Meeting in the year of their award. The award is made every two years. Nominations are invited in even-numbered years. Nomination form and guidelines are available online.
Requirements: Candidates have been widely involved in maternal-fetal nutrition through teaching, research, or clinical practice for at least five years; shown a demonstrable effect in raising the quality of maternal-fetal nutritional care through scholarly pursuits, research, education, or practice; and demonstrated the ability to apply maternal-fetal nutritional standards of practice or facilitate their implementation by others.
Amount of Grant: $3000
Date(s) Application is Due: Mar 1
Contact: Grants Administrator, (914) 997-4564 or (888) 663-4637; fax: (914) 997-4560; email: ResearchGrants@modimes.org
Internet: http://www.marchofdimes.com/professionals/685_1370.asp
Sponsor: March of Dimes Birth Defects Foundation
1275 Mamaroneck Avenue
White Plains, NY 10605

Mead Johnson Nutritionals Medical Education Grants **3326**
Mead Johnson Nutritionals has established a Grant program to support medical education and philanthropic initiatives. The program will consider supporting Independent Medical Education (IME) programs and activities designed to enhance the professional skills and knowledge of health care professionals involved in pediatrics, when the program is independently developed and conducted by a qualified medical education provider. Supported programs must be objective, balanced, and scientifically rigorous. Health care provider educational programming can be given for: medical education programs; scientific conferences; creation of health care publications; and other educational and scientific activities.
Requirements: Eligible recipients for medical education grants include: academic medical centers; accredited medical education companies; community health centers; hospitals; managed care organizations; medical/professional societies; medical universities; patient advocacy groups; pharmacies; and similar entities.
Restrictions: Grants may not be tied, in any way, to past, present, or future prescribing, purchasing, or recommending (including formulary recommendations) of any product.
Samples: U.S.DA Children's Nutrition Research Center (CNRC) at Baylor College of Medicine, Texas Medical Center, Houston, $300,000 (2002-04); University of Toronto, Faculty of Medicine, Department of Nutritional Sciences, $500,000, (2001-06).
Contact: Grants Administrator, (812) 429-7831; fax: (812) 647-8674; email: meadjohnson.grants1@bms.com
Internet: http://www.meadjohnson.com/app/iwp/MJN/Content2.do?dm=mj&id=-8504
Sponsor: Mead Johnson Nutritionals
2400 W. Lloyd Expwy.
Evansville, IN 47721-0001

Mead Johnson Nutritionals Scholarships **3327**
Mead Johnson Nutritionals provides scholarship grants to medical students, residents, fellows, and other healthcare professionals in training to attend classes and major educational or scientific meetings of a national medical or professional association. The selection of the individuals to receive such support and the grant application must be made by the institution at which they are being trained or the organization hosting the event.
Requirements: Eligible recipients for such grants: academic medical centers; hospitals; medical/professional societies; and medical universities.
Contact: Grants Administrator, (812) 429-7831; fax: (812) 647-8674; email: meadjohnson.grants1@bms.com
Internet: http://www.meadjohnson.com/app/iwp/MJN/Content2.do?dm=mj&id=-8515
Sponsor: Mead Johnson Nutritionals
2400 W. Lloyd Expwy.
Evansville, IN 47721-0001

Meadows Foundation Grants **3328**
Grants are made in five areas of interest: arts and culture, human services, health, education, and civic and public affairs. The foundation looks for innovative, well-planned projects that promote better human relationships and a sense of community. Projects should show the ability to continue and expand after the grant period. Funds support programs, projects, capital projects, general operations, seed grants, equipment

acquisition, matching funds, scholarship funds, professorships, internships, fellowships, research, publications, conferences and seminars, and consulting services. The foundation tries to give at least one-half of the grants each year to organizations not previously receiving foundation support. Requests not exceeding $25,000 should be submitted between June 1 and August 31 to be eligible for a special review program.
Requirements: Grants are made to qualified organizations in Texas or programs benefiting Texas residents.
Restrictions: Grants are not made to individuals or for church or seminary construction, annual fund-raising events or drives, biomedical research, out-of-state performances or competition expenses, or professional conferences and symposia.
Geographic Focus: Texas
Samples: Christian Community Action (Lewisville, TX)—to expand the resale operations of this organization that provides clothing, food, job counseling, shelter and other services to needy people, $150,000 (2005); Phoenix House, Texas (Austin, TX)—for the Phoenix Academy of Austin, which helps adolescents with substance-abuse problems through long-term, residential treatment and schooling, $107,000 (2005); Head Start of Greater Dallas (TX)—to help construct a Head Start center on the campus of Brookhaven Community College, $135,000 matching grant (2005).
Amount of Grant: $10,000-$1 million
Contact: Program Contact, (800) 826-9431 or (214) 826-9431; fax: (214) 827-7042; email: grants@mfi.org
Internet: http://www.mfi.org
Sponsor: Meadows Foundation of Texas
3003 Swiss Avenue
Dallas, TX 75204-6090

Measey Foundation Medical Education Grants 3329
The purpose of this grant program is to support and advance the medical education of physicians. Grants are made to medical schools in the immediate Philadelphia, PA, area that will determine eligible recipients. Types of support include fellowships, professorships, and scholarship funds. Grantmaking decisions are made at the quarterly meetings held in March, June, September, and December each year; grant requests should be received at least two weeks prior to the meetings.
Restrictions: Individuals are ineligible.
Geographic Focus: Pennsylvania
Samples: Thomas Jefferson U, Jefferson Medical College (Philadelphia, PA)—to endow a professorship in surgery, $1 million.
Amount of Grant: $15,000-$400,000 average
Contact: James Brennan, Manager and Counsel, (610) 566-5800
Sponsor: Benjamin and Mary Siddons Measey Foundation
P.O. Box 258
Media, PA 19063

Medicaid Infrastructure Grant Program to Support Employment of People with Disabilities 3330
This program is designed to help states support the competitive employment of persons with disabilities by facilitating improvements to states' Medicaid programs. Request applications from Judith Norris,(410)786-5130 or jnorris@cms.hhs.gov.
Requirements: Single state Medicaid agencies and other state agencies in partnership with the single agency are eligible to apply.
Amount of Grant: $500,000 minimum
Date(s) Application is Due: Jul 31
Contact: Jeremy Silanskis, Project Manager, (410) 786-6126; email: jsilanskis@cms.hhs.gov
Internet: http://cms.hhs.gov/twwiia/infrastr.asp
Sponsor: Health Care Financing Administration
7500 Security Boulevard
Baltimore, MD 21244-1850

Medical Foundation Health Initiatives 3331
The program supports two major health initiatives in the Northeast. The Charles H. Hood Child Health Research Grant program supports New England medical or research institutions that initiate studies to diminish health problems affecting large numbers of children. Postdoctoral research fellowships in the biomedical sciences offer two-year grants to Massachusetts nonprofits addressing issues of health in the community. Proposals are judged on medical relevance, innovative ideas, background, and need for support.
Requirements: New England nonprofit organizations may apply for Charles H. Hood Child Health Research Grants. Massachusetts nonprofits addressing issues of health in the community may apply for postdoctoral research fellowships in the biomedical sciences.
Geographic Focus: Massachusetts; New England
Amount of Grant: $75,000 Charles H. Hood Child Health Research Grants; $39,000-$45,000 postdoctoral research fellowships
Contact: Grants Administrator, (617) 451-0049 ext 702; fax: (617) 423-4619; TDD: (617) 451-0007; email: research@tmfnet.org
Internet: http://www.tmfnet.org/grantmake.html
Sponsor: Medical Foundation
95 Berkeley Street, Suite 201
Boston, MA 02116

Medical Foundation Postdoctoral Research Fellowships 3332
The foundation and Fleet National Bank's Charles A. King Trust support research fellowships in the biomedical and health sciences. Candidates come from basic science research and clinical or community health research settings. Criteria include that proposed research builds on current knowledge, objectives are realistic, current ethical and safety standards are in compliance, potential limitations have been identified, and data collection and analysis are feasible. Annual deadline is in November. Guidelines and applications are available on the Web site.
Requirements: Eligible to apply are those who come from basic science research and clinical or community health research settings. The geographic area is limited to Massachusetts.
Geographic Focus: Massachusetts
Amount of Grant: $39,000-$45,000
Contact: Program Contact, (617) 451-0049 ext 702; fax: (617) 423-4619; TDD: (617) 451-0007; email: research@tmfnet.org
Internet: http://www.tmfnet.org/grantmake.html
Sponsor: Medical Foundation
95 Berkeley Street, Suite 201
Boston, MA 02116

Medical/Nursing Education, Research, and Health Care Exchanges with Taiwan 3333
This organization provides for the exchange of medical and scientific knowledge and fosters cooperation between health scientists in Taiwan and the United States in biomedical teaching, research, and health care. Three separate programs are sponsored. The fellowship program assists Republic of China graduates or postgraduates to study and do research in the United States. The visiting specialists program provides for U.S. health specialists to teach and lecture in medical schools in Taiwan. The nursing program recruits and sponsors U.S. nurses for positions in Taiwan and places and sponsors nurses from Taiwan in U.S. institutions. In Taiwan, contact Executive Director, Room 9001, 212 Chung Hsiao Tung Lu, Section IV, Taipei, Taiwan 10646, Republic of China, 02-721-6511 ext 367, fax: 02-751-6639.
Requirements: Program recruitment is primarily by invitation.
Contact: Annie Kang, c/o New York Academy of Medicine, (212) 860-1990
Sponsor: American Bureau for Medical Advancement in China
1216 Fifth Avenue, Room 585
New York, NY 10029

Medieval Academy of America John Nicholas Brown Prize 3334
The prize is for a first book or monograph on a medieval topic by a resident of North America. Three copies of the book bearing a publication date no more than four years previous to the award year should be sent to the academy office, together with any reviews that have appeared, and a statement by the author that it is his or her first book in the medieval field. The prize will be awarded at the academy's annual meeting. Books, letters, and three copies of reviews are due by the listed application deadline. There is no application form.
Amount of Grant: $1000
Date(s) Application is Due: Oct 15
Contact: Richard Emmerson, Executive Director, (617) 491-1622; fax: (617) 492-3303; email: Speculum@MedievalAcademy.org or maa@fas.harvard.edu
Internet: http://www.medievalacademy.org/grants/gradstudent_cara_brown.htm
Sponsor: Medieval Academy of America
104 Mt Auburn Street, 5th Fl
Cambridge, MA 02138

MedImmune Charitable Grants 3335
The program supports educational and charitable activities that ultimately improve patient healthcare and further understanding of disease states and treatment options within the community. While the sponsor will consider initiatives in many areas, priority is given to those in the following areas of interest: health and science education for young people; influenza; immunology; pediatric infectious disease; and pediatric respiratory disease.
Requirements: Eligible programs include: (a) Community outreach designed to increase public awareness (e.g, fundraising walk or run, nonprofit community hospital or hospital foundation, etc.); (b) Fellowships; (c) Independent educational/scientific programs, accredited and non-accredited; (d) Medical or Scientific education for Healthcare Professionals; (e) Scholarships; or, (f) Academic Chair Endowments. Applications must be completed online at https://web.medimmune.com/corporategrants/, and must be submitted at least 6 weeks prior to the program date or event.
Restrictions: Applications will not be considered from: (1) individuals, physicians, or physician practices; (2) retroactive support; (3) exhibit or display booths at conventions/conferences; (4) textbooks; (5) operating expenses; (6) fraternal, social, leisure, labor or political organizations; (7) corporate foundations; (8) public or private schools or scholarship funds, except where a grant is specific to a science education initiative.
Contact: Grants Office; (866) 396-6235; email: corporatefunding@medimmune.com
Internet: http://www.medimmune.com/culture_grants.aspx
Sponsor: MedImmune, Inc.
One MedImmune Way
Gaithersburg, MD 20878

Medtronic Foundation Education Grants 3336
The STAR program (Science and Technology Are Rewarding) is a K-12 science education program intended to stimulate and sustain the interest of young people in science. Grants focus on five key program elements: curriculum, instruction, expectations, peer support,

and assessment. Priority for available funds will be given to programs meeting the following guidelines: creative K-12 education programs that have a lasting impact on children; replication in a Medtronic community of the Teacher Outstanding Performance (TOP) program that provides awards to professional teachers for classroom innovation and excellence; and postsecondary programs that increase the number of women and people of color who study engineering, science, technology, or health. Application forms are available on the Web site.
Requirements: Proposals are considered from private and parochial nonprofit schools, school districts, and other nonprofit organizations working with kindergarten through grade 12 in the seven county metro area of Minneapolis/St. Paul. The foundation encourages cooperation among schools or schools and organizations.
Restrictions: Grants are not available for regular school district obligations such as teacher salaries, textbooks, endowments, multi-year commitments, or capital improvements.
Geographic Focus: Minnesota.
Amount of Grant: $2000-$60,000 typically
Date(s) Application is Due: Jan 15; Jul 15; Oct 15
Contact: Kathleen O'Keefe, STAR Senior Program Officer, (763) 505-2638; email: kathleen.okeefe@medtronic.com
Internet: http://www.medtronic.com/foundation/programs_star.html?
Sponsor: Medtronic Foundation
710 Medtronic Pkwy NE
Minneapolis, MN 55432-5604

Medtronic Foundation Health Care Grants **3337**
The foundation supports health education programs that work to help people develop and maintain active, healthy lives. Grants support programs and projects that work to prevent or reduce the effects of disabling and chronic disease, serve children and/or older adults, and provide low-income people with nonreimbursed, preventive health services. Also supported are education programs for service providers that build multicultural understanding; programs that address the relationship between physical, mental, and chemical health; and science and technology for grades K-12. Types of support include continuing support, annual campaigns, seed grants, matching funds, scholarship funds, fellowships, special projects, and employee matching gifts. The foundation discourages requests for general operating support, equipment, capital projects, endowments, and multiyear commitments. Within the Twin Cities area, grants of $5000 or less are generally reviewed within 60 days of receipt of application. Grants over $5000 must meet listed deadlines for review. Outside the Twin Cities area, grants of up to $10,000 are made within three months of receipt of the application. Contact the foundation for deadlines for grants over $10,000. Application forms are available on the Web site.
Requirements: Proposals are considered from nonprofit, tax-exempt organizations, largely from the communities in which the company has operations (Tempe, AZ; Anaheim, Irvine, and San Diego, CA; Parker, CO; Milaca and Minneapolis, MN; Danvers, MA; and Humacao and Villalba, PR).
Restrictions: The foundation does not awards grants to religious or political political groups or for fund-raising events, general support of educational institutions, long-term counseling, or personal development.
Geographic Focus: Minnesota
Samples: Beth Israel Deaconess Medical Ctr (Boston, MA), Scripps Ctr for Integrative Medicine (La Jolla, CA), and Stanford Ctr for Research in Disease Prevention/The Stanford Heart Network (Palo Alto, CA)—to develop integrated acute- and chronic-care models for chronically ill patients that emphasize patient-centered care, $2.4 million divided over three years.
Amount of Grant: $1000-$50,000 average
Date(s) Application is Due: Jan 15; Jul 15; Oct 15
Contact: Deb Anderson, Grants Administrator, (763) 514-4000; fax: (763) 505-2648
Internet: http://www.medtronic.com/foundation
Sponsor: Medtronic Foundation
710 Medtronic Pkwy NE, MS LC110
Minneapolis, MN 55432-5604

Meet the Composer Global Connections Grants **3338**
Grants are awarded to composers for the creation of work to share with the global community. The program fosters international exchange and provides support to composers actively participating in events featuring the live performance of their music; and planning research and development activities toward the creation of new work with collaborating artists abroad. Eligible expenses are composer's appearance fee, travel, accommodation, and per diem costs. Applications are submitted jointly by a composer and a hosting organization. Guidelines and application are available online.
Requirements: The composer and hosting organization must not be based in the same country and one of the two must be based in the United States.
Amount of Grant: $500-$5000
Date(s) Application is Due: Apr 1
Contact: Grants Administrator, (212) 645-6949; fax: (212) 645-9669
Internet: http://www.meetthecomposer.org/programs/globalconnections.html
Sponsor: Meet the Composer
75 Ninth Avenue, 3R Suite C
New York, NY 10011

Melanoma Research Foundation Grants **3339**
The grant supports promising medical research that will further the goal of developing an effective treatment and possible cure of malignant melanoma, while encouraging a new generation of scientists and clinicians to join in this mission. The program is directed primarily at young independent investigators, e.g., untenured assistant professors and the equivalent who have not yet established strong federal funding, who are working at U.S. institutions. In addition, emphasis will be placed on projects that explore innovative approaches to understanding melanoma and its treatment. Both basic and clinical research projects will be considered. Awards will be announced by December, to commence by January 1 of the following year. Complete guidelines are available online.
Requirements: Researchers who are either U.S. permanent residents or citizens, and hold a PhD or MD degree, and a title equivalent to assistant professor, or have at least four years of postdoctoral experience are eligible to apply. Applicants should have evidence indicating strong departmental or institutional support and commitment, and be within eight years of receiving their advanced degree. The proposed research must be conducted in a nonprofit research organization, medical organization, or educational institution located in the United States.
Amount of Grant: $50,000 per year for two years
Date(s) Application is Due: Jul 1
Contact: Grants Program Director, (800) 673-1290; fax: (732) 821-5955; email: research@melanoma.org
Internet: http://www.melanoma.org
Sponsor: Melanoma Research Foundation
24 Old Georgetown Road
Princeton, NJ 08540

Mellon Postdoctoral Fellowships in the Humanities at Cornell University **3340**
To encourage academic growth of promising, nontenured scholars and teachers in the humanities, the program awards one-year postdoctoral teaching/research fellowships. Fellows spend a year at Cornell teaching half-time in a department pursuing their research. Fellowships are offered in comparative literature; English; feminist, gender, and sexuality studies; and theater, film, and dance. Applications and letters of recommendation must be postmarked by the listed application deadline.
Requirements: Eligible candidates must have completed requirements for the PhD within the last five years and must be citizens or permanent residents of the U.S. or Canada.
Amount of Grant: $40,000 annually
Date(s) Application is Due: Oct 1
Contact: Program Administrator, Society for the Humanities, (607) 255-9274; email: humctr-mailbox@cornell.edu
Internet: http://www.arts.cornell.edu/sochum
Sponsor: Cornell University
AD White House, 27 East Avenue
Ithaca, NY 14853-1101

Mellon Postdoctoral Research Fellowships **3341**
The Huntington Library and Art Gallery is an independent research center with holdings in British and American history, literature, art history, the history of science, and photography. The fellowships support researchers pursuing scholarship in a field appropriate to the Huntington's collections. Preference will be given to scholars who have not held major awards in the three years preceding the year of the award. Recipients are expected to be in continuous residence at the Huntington during the nine- to 12-month fellowship tenure and to participate in the intellectual life of the center. There is no application form. Application consists of cover sheet, description of the project, curriculum vita, and three letters of recommendation that must be received by the application deadline.
Requirements: Application is open to individuals with the PhD or equivalent degree.
Amount of Grant: $40,000
Date(s) Application is Due: Dec 15
Contact: Robert Ritchie, Director of Research, (626) 405-2194; fax: (626) 449-5703; email: cpowell@hungtington.org
Internet: http://www.huntington.org/ResearchDiv/Fellowships.html
Sponsor: Huntington Library and Art Gallery
1151 Oxford Road
San Marino, CA 91108

Mellon Visual Culture and/or Film or Media Studies Postdoctoral Fellowship **3342**
Applications are invited for a two-year postdoctoral fellowship in visual culture and/or film or media studies. The successful candidate will advise student research and work with faculty in Africana Studies and in English and potentially other departments to develop courses. Mellon fellows are included in department activities as regular junior faculty members. The fellowship includes a stipend plus benefits and funds to support research and travel. Submit a letter of application, curriculum vita, graduate-school transcript, three letters of recommendation, and a brief description of teaching interests. Review of applications will begin on January 31 and continue until the search is completed.
Requirements: Applicants must be U.S. citizens or permanent residents or expect to pursue a teaching career in the United States. The successful candidate will have a joint appointment in Africana Studies and in English. New PhD's are encouraged to apply.
Amount of Grant: $35,000 salary
Contact: Joy James, Africana Studies or Karen Swann, English
Internet: http://www.williams.edu
Sponsor: Williams College
Stetson Hall
Williamstown, MA 01267

Melville Charitable Trust Grants **3343**
The Melville Trust was organized in 1990 with the desire to understand the growing problem of homelessness and help eradicate it. The foundation decided to focus its study on Connecticut and to create programs that would alleviate and reduce the impact of

homelessness on children, adults and communities. The trust also funds educational, research and advocacy initiatives in the state and on the national level.
Requirements: The Trust does not use a formal application form. Instead, applicants are asked to submit a concept paper of not more than three pages describing the project. If there is interest in pursuing the request after the concept paper has been reviewed, the applicant organization will be contacted to either submit a more formal proposal and/or provide more detailed information to the Board. Applications must be from an organization that has been determined to be tax-exempt under section 501(c)3 of the Internal Revenue Code and "not a private foundation" under section 509(a) of the Code.
Restrictions: Grants will not be awarded for: direct support to individuals or scholarships; budget deficits; religious organizations for religious purposes; or general fund raising drives or events.
Samples: Partnership for Strong Communities, Hartford, CT, $815,000 - to continue to build on core strength of affordable housing policy and advocacy (2007); National Public Radio, Washington, DC, $175,000 - to support coverage from Gulf Coast and ongoing reporting on issues of poverty and housing (2007);
Contact: Aimee Hendrigan, Senior Program Officer; (617) 236-2244; fax: (617) 482-5624; email: ahendrigan@melvilletrust.org
Internet: http://www.melvilletrust.org/funding_guidelines.aspx
Sponsor: The Melville Charitable Trust
240 Newbury Street, 2nd Floor
Boston, MA 02116-2580

Melville Medal 3344
The medal is the highest ASME honor for the best current original paper, not published elsewhere, which has been presented before ASME or approved for publication by ASME during the two calendar years immediately preceding the year of award. The paper may have more than one author but one of the authors must be an ASME corporate member (fellow, member, or associate member). The paper must be specifically recommended for the medal by a review committee or qualified individual. The medal may be awarded for a paper that has been selected for some other best paper award in the society if specifically recommended by the committee responsible for that specific award.
Requirements: One author must a corporate member of AMSE.
Amount of Grant: $2000 and expense supplement
Date(s) Application is Due: Mar 1
Contact: Gilda DiTullio, Manager, (212) 591-7736; fax: (212) 705-7739; email: awards@asme.org
Internet: http://www.asme.org/honors/ms71/dla/melville.html
Sponsor: American Society of Mechanical Engineers
3 Park Avenue
New York, NY 10016

Mental Retardation Distinguished Research Award 3345
The award recognizes an individual or individuals whose research has had a significant impact on the prevention or amelioration of mental retardation. The recipient will receive, in addition to the monetary award, an expense-paid trip to speak at the luncheon of the national convention. The location of the convention varies; the date of the convention is usually in the fall. Nominations must be received by the listed application deadline. Annual deadline dates may vary; contact program staff for exact dates. Nomination forms are available on the Web site.
Amount of Grant: $1000
Date(s) Application is Due: May 15
Contact: Dr. Sharon Davis, (301) 565-5456; fax: (301) 565-3843; email: davis@the arc.org or info@thearc.org
Internet: http://thearc.org
Sponsor: The Arc
1010 Wayne Avenue, Suite 650
Silver Spring, MD 20910

Merck Company Foundation Grants 3346
The foundation advances the company's philanthropic outreach in helping to meet important needs of local and international communities that are consistent with Merck's overall mission to enhance the health and well-being of people around the world. Such needs include improving health care, fostering biomedical and science education; and supporting the arts, social services, and civic, environmental and other charitable organizations.
Requirements: Operating support of health and social service agencies is reserved for Merck communities and generally directed through annual contributions to United Way.
Restrictions: Grants are not made to political, labor, fraternal, or veterans organizations; sectarian groups; or to individuals. Except within foundation programs, grants are not given for elementary/secondary education, scholarships, fellowships, research, publications, conferences/seminars/symposia, and travel.
Samples: Alliance of Community Health Plans (Washington, DC)—to help healthcare groups that use culturally and linguistically appropriate health services demonstrate that those services are financially sound options, $235,000 (2005); American National Red Cross (Washington, DC), UNICEF (NY), and several local relief organizations—for relief efforts in South Asia and Africa, $3 million divided (2005); Task Force for Child Survival and Development (Atlanta, GA)—to expand programs to treat people in Africa who are suffering from onchocerciasis, more commonly known as river blindness, as part of the Mectizan Donation Program, $1 million over five years (2004).
Contact: Dr. Shuang Ruy Huang, (908) 423-1000; fax: (908) 423-1987
Internet: http://www.merck.com/about/cr/policies_performance/social/philanthropy.html
Sponsor: Merck Company Foundation
P.O. Box 100 (WSIAF-35)
Whitehouse Station, NJ 08889-0100

Merck Institute of Aging and Health Grants 3347
The institute is dedicated to improving the health and independence of older people around the world by communicating health information, educating the public and health professionals, and encouraging research on health in the aging field. Proposals are encouraged for demonstration and pilot programs in the following priority areas: innovative education of primary care providers about the care of older adults in community practice; research and evaluation of innovative health services provided for older adults in the community; and development and evaluation of innovative interventions that encourage health promotion and disease prevention for older adults. The institute currently serves the United States, Europe, and Latin America and may eventually expand into other areas of the world. Application guidelines are available online.
Requirements: Eligibility criteria include 501(c)3 designation and incorporation as an accredited school of medicine, nursing, osteopathy, pharmacy, or public health; a teaching hospital or institution affiliated with an accredited school of medicine, nursing or osteopathy; or a national health or health-care related organization, health professional organization, nursing or medical association, or consumer health or health-care organization. All non-U.S. organizations are required to provide evidence of nonprofit or NGO status, equivalent to a U.S. public charity.
Restrictions: Ineligible organizations and programs include organizations without 501(c)3 or NGO status; political organizations, campaigns, and activities; fraternal/labor/veteran's organizations and activities; religious organizations or groups whose activities are primarily sectarian in purpose; organizations that discriminate on the basis of race, color, sexual orientation, religion, marital status, age, national origin, veteran's status, or disability; fundraising events; capital or building campaigns and endowments; media productions; basic or clinical research projects; meetings, conferences, or symposia that are not part of a larger health-related program efforts; unrestricted general support; fellowship/tuition support for training purposes; programs that directly support marketing and/or sales objectives of Merck and Company; and programs for pre-college education.
Amount of Grant: $50,000-$100,000
Contact: Julie Kennedy, Grants Manager, (202) 842-0525; fax: (202) 354-6599
Internet: http://www.miahonline.org/aboutMIAH/grants.html
Sponsor: Merck Institute of Aging and Health
1100 New York Avenue NW, Suite 350W
Washington, DC 20005

Mericos Foundation Grants 3348
The foundation awards grants to organizations based in Santa Barbara, CA, in support of the arts, medical research, child development and the elderly, and education at all levels. Types of support include program/project support, building/renovations, equipment, endowment funds, fellowships, and matching funds. Grants are usually initiated by the foundation.
Requirements: Organizations in California are eligible to apply.
Restrictions: Grants are not made to individuals.
Geographic Focus: California
Samples: Laguna Cottages for Seniors (Santa Barbara, CA)—to refurbish apartments, $11,000.; Goleta Boys and Girls Club (Goleta, CA)—for renovations, $150,000 challenge grant.; U of California at Santa Barbara (CA)—for a program that matches high school students with mentors who are attending graduate school, $100,000 challenge grant.; Girl Scouts of Tres Condades (Santa Barbara, CA)—for technology upgrades for the Stay Safe project, $25,000.
Amount of Grant: $5000-$380,000 range
Contact: Linda Blinkenberg, Secretary, (626) 441-5188; fax (626) 441-3672
Sponsor: Mericos Foundation
625 S Fair Oaks Avenue, Suite 360
South Pasadena, CA 91030-2630

Merrill Lynch Philanthropic Program Grants 3349
The foundation supports a variety of charitable causes throughout the world, with a focus on education. Through 2005, the foundation has adopted children and youth as its global cause. Programs should meet the educational needs and interests of uderserved children and youth. Programs should support an ethnically diverse population of children and youth; provide direct services from early childhood to postsecondary education, and incorporate one or more of the following elements: academic support, global exchange, community service, internships, cultural enrichment, life skills, digital divide, mentoring, entrepreneurship, scholarships, personal finance, technology skills, GED programs linked with four-year colleges, and workforce preparation. The foundation gives priority to specific programs and projects, rather than general operating support. Preference will be given to requests from the five boroughs of New York City and national organizations that reflect that focus. Requests outside of New York City should be submitted to managers of local branches or offices.
Requirements: Tax-exempt 501(c)3 organizations are eligible.
Restrictions: Grants will not be made to private foundations; individuals; fundraising activities related to individual sponsorships (e.g., walk-a-thons, marathons); seed money for new organizations; political causes, candidates, and campaigns as well as organizations designed specifically for lobbying; religious, fraternal, social, or other membership organizations providing services mainly to their own constituencies; athletic

events and sports tournaments (Special Olympics is eligible); and fundraising events (dinners, luncheons).
Samples: Sesame Workshop (New York, NY)—for the WorldwideKids program, which will use multimedia tools and resources to educate three- to six-year olds about diverse people and cultures, $5 million (2005); United Nations Assoc of the United States of America (Washington, DC)—to expand to 14 countries the Global Classrooms program, which supports Model United Nations activities in schools, and to develop a curriculum on global development, finance, and trade, $7.5 million (2005).
Contact: Westina Matthews, Vice President, Philanthropic Programs, (212) 236-4319; fax: (212) 236-8007; email: SDimaggio@exchange.ml.com
Internet: http://philanthropy.ml.com
Sponsor: Merrill Lynch & Company Foundation
2 World Financial Ctr, 5th Fl
New York, NY 10281-6100

Mervin Bovaird Foundation Grants 3350
The foundation awards grants, with a focus on Tulsa, OK, to nonprofit organizations such as churches, homeless shelters, medical centers, nursing homes, parochial schools, religious welfare organizations, the Salvation Army, and youth organizations. Areas of interest include arts, community development, education, environment, health care and health organizations, and social services. Types of support include general operating support, matching, project, and research grants. The foundation also funds a scholarship program for Tulsa County High School graduating seniors and graduates of Tulsa Community College at the University of Tulsa (not for graduate or professional study). Applicants should submit a brief letter of inquiry, including program and organization descriptions.
Requirements: Nonprofits of the Christian, Methodist, Roman Catholic, and United Methodist faiths in Tulsa, OK, are eligible.
Geographic Focus: Oklahoma
Samples: Salvation Army (Tulsa, OK)—for facilities and program expansion, $50,000.; John 3:16 Mission (Tulsa, OK)—for the family/youth center, $25,000.
Amount of Grant: $1000-$250,000 range
Contact: R. Casey Cooper, President and Trustee, c/o Boesche, McDermott, and Eskridge, (918) 592-3300
Sponsor: Mervin Bovaird Foundation
401 S Boston Avenue, Suite 3300
Tulsa, OK 74103-4070

Mesothelioma Applied Research Foundation Grants 3351
The program seeks to stimulate translational research for the treatment of pleural or peritoneal mesothelioma. Grants support projects that apply cutting-edge general cancer research concepts specifically to mesothelioma, and translate directly to improved mesothelioma treatment. Eligible projects may relate to either benchwork research or clinical research, must not be presently funded or pending review, and may be conducted through any not-for-profit academic, medical, or research institution, in the United States or abroad. Grants are awarded for 12 months, renewable for 12 more months.
Requirements: The investigator must be affiliated with a nonprofit academic, medical, or research institution.
Amount of Grant: $50,000 per year for two years
Date(s) Application is Due: Aug 15
Contact: Danele Mangone, Grants Director; (805) 563-8400, 7273; fax: (805) 563-8411; email: dmangone@curemeso.org
Internet: http://www.curemeso.org/site/c.kkLUJ7MPKtH/b.4097447/k.E6A3/Mesothelioma_Research_Grants__Call_for_Applications.htm
Sponsor: Mesothelioma Applied Research Foundation
P.O. Box 91840, 3944 State Street, Suite 340
Santa Barbara, CA 93105

Metanexus Institute Local Societies Initiative 3352
The initiative provides three-year grants to fund start-up costs for dialog groups exploring the dynamic interface between religion and science. While focused on colleges, universities, and seminaries, the program also welcomes applications from a wide spectrum of interested parties in other venues worldwide. The applicant must be able to match the grant dollar for dollar over the three years. There will be a number of supplemental grants awarded annually to those projects that are particularly innovative, creative, and effective. Funds may be used for society events, speakers, the purchase of books for group study, communications, publicity, outreach, public forums, and travel directly related to the project. All societies will send representatives to annual conferences each June in Philadelphia. Guidelines are available online.
Requirements: The proposal must come from a group (and not an individual) of qualified persons (scientists, theologians, philosophers, etc.); demonstrate both the means and the willingness to address issues at the intersection of science and religion with all due rigor; give evidence that the applicant society will enthusiastically pursue collaborative opportunities with other Metanexus LSI societies in the global network, their members, and host institutions; and, include proof of dollar-for-dollar matching funds.
Amount of Grant: $15,000 paid over three years; $10,000 supplemental grants
Date(s) Application is Due: Jan 1; Apr 1; Oct 1
Contact: Grants Administrator, (215) 789-2200; fax: (215) 789-2222; email: lsi@metanexus.net
Internet: http://www.metanexus.net/local%5Fsocieties/default.html
Sponsor: Metanexus Institute
3624 Market Street, Suite 301
Philadelphia, PA 19104

Metanexus Institute Templeton Research Lectures 3353
The goal of the program is to promote the constructive engagement and original research between the physical, biological, and human sciences and those modes of inquiry and understanding generally found within the domains of theology, religious studies, and philosophy. This is a three-year program, with the possibility of a fourth-year extension, that supports scholars who want to develop on-going, interdisciplinary programs within their universities. The program seeks to create vibrant and long-term networks of scholars engaged in vigorous dialog, careful research, and significant publication. The grant offers an honorarium for an individual scholar selected each year of the program to give the annual Templeton Research Lectures at the host institution. This is an opportunity to bring a distinguished scholar from another institution for sustained interaction or possibly in specialized situations the elevation of an internal faculty to give the annual lectures. These lectures must result in a book- length publication whose target audience is either the academic community or the educated layperson. Guidelines and application procedures are available online.
Requirements: Applications for the grant should be submitted by an interdisciplinary committee of scholars employed by their host university and forming the core of a governing committee to oversee the program. One person should be identified as both the chair of the group and the contact person for the grant applicant.
Amount of Grant: $90,000 to $130,000 per year
Date(s) Application is Due: Jan 1
Contact: Dr. William Grassie, (215) 789-2200; fax: (215) 789-2222; email: lectures@metanexus.net
Internet: http://www.metanexus.net/lectures/request/overview.html
Sponsor: Metanexus Institute on Religion and Science
3624 Market Street, Suite 301
Philadelphia, PA 19104

MetLife Foundation Community College Excellence Award 3354
Jobs for the Future administers the MetLife Foundation Community College Excellence Award. The award is given biennially to two colleges that are especially effective in helping students from underserved populations to succeed in postsecondary learning. The award recognizes community colleges that, in different ways and in very different communities, make significant institutional commitments to helping first-time college-goers, new immigrants, working adults, welfare recipients, high school dropouts, and other populations with limited college experience and success prepare for further education or for a family-supporting career.
Requirements: Any public, accredited community college can apply. All applications must be submitted online. Successful applicants will show how they combine strong leadership, effective programming, and attention to outcomes to create and institutionalize improvement, throughout the college, that meets the varied learning needs of harder-to-serve students.
Samples: 2006: Indian River Community College - Fort Pierce, FL, and LaGuardia Community College, - Long Island City, NY. 2004: City College of San Francisco - San Francisco, CA, and Community College of Denver - Denver, CO. 2002: West Hills Community College - San Joaquin Valley, CA, and Sinclair Community College - Dayton, OH.
Amount of Grant: $30,000
Date(s) Application is Due: Aug 31
Contact: Monique Sheen, Jobs for the Future; (617) 728-4446, ext. 137; email: msheen@jff.org
Internet: http://www.jff.org/Content/Current+Projects_Building+Economic+Opportunity+for+Adults_MetLife+Foundation+Community+College+Excellence+Award.html
Sponsor: MetLife Foundation / Jobs for the Future
88 Broad Street
Boston, MA 02110

MetLife/TCG Aha! Think it Do it Grants 3355
The program supports creative thinking and action in two ways: Think It grants (up to $25,000) give theater professionals the time and space for research and development, Do It grants (up to $50,000) support the implementation and testing of new ideas. The intent of the program is to enable theaters to dare to try new approaches to problem-solving artistic, managerial, production and/or technological challenges - to try things the organization doesn't and couldn't normally do. Proposals from organizations seeking to test an idea that is new to them, but not to the field will be accepted. Emphasis is on identifying strategies that will be major catalysts for change to the organization, rather than strategies which will bring minimal change.
Requirements: To be eligible, the applying theater must be a member theater of TCG in good standing and may apply in only one category. The Project Leader must be a key artistic, administrative or production staff member who holds a salaried position at the theater. Theaters receiving THINK IT grants are eligible to apply for DO IT grants in a future round, provided that their original grant was completed. Preference will be given to projects that break from the organization's previous practice and have the best chance at long-term, field-wide impact over a longer span of time.
Restrictions: Strategies and projects that have already been tested or implemented at the applicant organization are not eligible.
Amount of Grant: $25,000 - $50,000
Date(s) Application is Due: May 15
Contact: Sarah McLellan, Artistic Programs Associate;
Internet: http://www.tcg.org/grants/aha/aha_guidelines.cfm
Sponsor: Theatre Communications Group (TCG)
520 8th Avenue, 24th Floor
New York, NY 10018-4156

Meyer Memorial Trust Operating Funds Grants 3356
MMT has launched an Operating Funds grant program to help small- and mid-sized organizations that have been hit particularly hard by the current economic crisis. The purpose of this initiative is organizational stability. Organizations for which a cash infusion is not critical to stability, but wish to request funds to maintain or expand current programming, should continue to apply for support through MMT, Responsive and Grassroots Grant programs, as usual. Applications will be accepted beginning September 1. Application deadline is October 1. Awards announcement is anticipated in December. Applications will be accepted through MMT's Responsive Grant inquiry process.
Requirements: Any 501(c)3 nonprofit organizations the operates in Oregon and Clark County, Washington is eligible to apply.
Restrictions: Meyers does not support the following: direct grants, scholarships, or loans to individuals; endowments; general fund drives, annual appeals, special events, or sponsorships; elimination of operating deficits; projects of sectarian or religious organizations that principally benefit their own members or adherents; direct replacement funding for activities previously supported by federal, state, or local public sources; acquisition of land for conservation purposes (except through Program Related Investments); hospital capital construction projects (except through Program Related Investments); animal welfare organizations (both domestic and wild animals); projects that primarily benefit students of a single K-12 school, unless the school is an independent alternative school primarily serving low-income and/or special needs populations; MMT funds cannot be earmarked for purposes of influencing legislation, and the grantee cannot expend any part of the grant in a way that violates its tax-exempt status.
Geographic Focus: Oregon; Washington
Amount of Grant: $50,000 maximum
Date(s) Application is Due: Oct 1
Contact: Barbara Gibbs, Grants Director; (503) 228-5512; email: barbara@mmt.org
Internet: http://www.mmt.org/grants_programs/
Sponsor: Meyer Memorial Trust
425 NW 10th Avenue, Suite 400
Portland, OR 97209

Meyer Memorial Trust Responsive Grants 3357
Responsive Grants are awarded in the areas of social services, health, affordable housing, community development, conservation and environment, public affairs and social benefit, arts and culture, and education. They generally range from $30,000 to $500,000; grant periods are from one to three years in length. General purpose grants are given for a wide variety of projects in many fields of activity including program expansions, capacity building, capital projects, and start-ups. Final decisions on Responsive Grants proposals are made by trustees at program meetings every month except January, April and August. Final decisions on proposals for capital projects requesting more than $200,000 are made in May and November. Applicants should time their Initial Inquiry submission at least six months before they hope to have a final decision from MMT.
Requirements: General purpose and small grants are awarded to 501(c)3 nonprofit organizations in Oregon and Clark County, Washington.
Restrictions: Meyer does not fund general administrative overhead or indirect rate allocations.
Geographic Focus: Oregon; Washington
Samples: Confederated Tribes of Warm Springs, Warm Springs, Oregon, $150,000—to help construct a building to co-locate the newspaper office and KWSO radio station (2008); Oregon Ballet Theatre, Portland, Oregon, $200,000—for two years of operating support (2008); South Lane Mental Health Services Cottage Grove, Oregon, $150,000—to purchase and renovate a building as a mental health clinic (2008).
Amount of Grant: $30,000-$500,000 average
Contact: Barbara Gibbs, Grants Director; (503) 228-5512; email: barbara@mmt.org
Internet: http://www.mmt.org/grants_programs/
Sponsor: Meyer Memorial Trust
425 NW 10th Avenue, Suite 400
Portland, OR 97209

MFA Established Investigator Grants 3358
The foundation is committed to advancing animal health care, including dog, cat, horse, llama and alpaca, wildlife, and mountain gorilla. Grants provide one to three years of funding for projects addressing animal health topics. The listed application deadline is for preproposals. Full proposals are by invitation. Guidelines are available online.
Amount of Grant: $38,000 average per year
Date(s) Application is Due: Nov 1
Contact: Kristin Benjamin, Director of Grants, (800) 243-2345
Internet: http://morrisanimalfoundation.org/apply/preproposal.asp
Sponsor: Morris Animal Foundation
45 Inverness Dr E
Englewood, CO 80112-5480

MFA Fellowship Training Grant 3359
The goal of the program is to provide training opportunities that will produce biomedical scientists (veterinarian and/or PhD) committed to a career in companion animal and/or wildlife health research. The applicant must be paired with a mentor who has current funding and a proven track record in companion animal or wildlife research as well as in training (preferably PhD candidates/postdoctoral fellows). The mentor must demonstrate excellent training environment including: adequate laboratory space for the trainee; adequate laboratory facilities; adequate funding for laboratory supplies; and a well-described training plan including scheduled lab meetings, access to seminars, etc. The fellowship supports two years of research and includes salary and benefits. The listed deadline is for preproposals; full proposals are by invitation. Guidelines are available online.
Requirements: Candidates who are post-DVM, postresidency, or post-PhD are preferred, but exceptional predoctoral students will be considered. Applicants must document their institutional appointment in a graduate program or postdoctoral position and will be expected to provide documentation of appropriateness of salary request relative to standard levels at institution.
Restrictions: Applications for support of training positions with clinical service requirements or significant didactic coursework will not be considered.
Amount of Grant: $35,000 per year
Date(s) Application is Due: Nov 1
Contact: Kristin Benjamin, Director of Grants, (800) 243-2345
Internet: http://morrisanimalfoundation.org/apply/fellowshiptraining.asp?section=2,0
Sponsor: Morris Animal Foundation
45 Inverness Dr E
Englewood, CO 80112-5480

MFA First Award Grants 3360
The goal of the program is to offer veterinary and/or doctoral (PhD) investigators opportunities early in their career by providing research funding for their first project in companion animal or wildlife research; and pairing them with a seasoned researcher as mentor to facilitate launching a successful, long-term, scientific career in advancing companion animal and wildlife health. The mentor must have a track record in companion animal and or wildlife research. Applications are judged on the merits of the mentor, the merits of the applicant, the training environment, and the scope of research proposed. The mentor must have a track record in training (preferably PhD candidates/post doctoral fellows). Research topic can be basic or applied research or a combination, however, the topic must be directed toward animal health issues relevant to companion animals or wildlife species. The listed application deadline is for preproposals. Full proposals are by invitation. Guidelines are available online.
Requirements: Applicant must be post-DVM, post-PhD and/or postspecialty boards, excluding graduate school and post-doc positions. Applicant must have a post-DVM career appointment/faculty equivalent position (beyond training program) in a university, accredited zoo, or conservation organization.
Restrictions: MAF does not pay salaries for faculty appointed to regular/tenure track, full-time positions. Applicant must not have been funded previously as a principal investigator for more than $10,000 for any single, extramural award (previous role as a co-investigator is acceptable).
Amount of Grant: $50,000 maximum per year
Date(s) Application is Due: Nov 1
Contact: Kristin Benjamin, Director of Grants, (800) 243-2345
Internet: http://morrisanimalfoundation.org/apply/firstaward.asp?section=2,0
Sponsor: Morris Animal Foundation
45 Inverness Dr E
Englewood, CO 80112-5480

MfA Newton Fellowships 3361
The fellowships are available to mathematically sophisticated individuals who are interested in using their talents to make a difference in the lives of young people. The program seeks college graduates and mid-career professionals who can demonstrate a strong interest in teaching, have a bachelor's degree with substantial coursework in mathematics, and are willing to commit to a five-year program that includes one year of full-time graduate study and four years of teaching mathematics in New York City high schools. Fellows receive a stipend, full tuition to master's in education program, mentoring and coaching, and professional development. Applicants are required to take the ETS Praxis II Math subject tests. Application and guidelines are available online.
Requirements: U.S. citizens and permanent residents are eligible.
Restrictions: Individuals who are currently teaching, are certified to teach, or have completed an education degree program are not eligible.
Geographic Focus: New York
Amount of Grant: $90,000 over five years
Date(s) Application is Due: Feb 10
Contact: Fellowship Administrator, (212) 206-0053; fax: (212) 206-1622; email: info@mathforamerica.org
Internet: http://www.mathforamerica.org
Sponsor: Math for America
101 Fifth Avenue, 5th Fl
New York, NY 10003

MfA Newton Master Teacher Fellowships 3362
The fellowship is a four-year award available to outstanding teachers of mathematics in the New York City public high schools. The program's purpose is to recognize mathematics teachers who demonstrate solid mathematical knowledge and outstanding performance records. The program provides opportunities for professional development and collaboration with other teachers. Guidelines are available online.
Requirements: Candidates must be current New York City public high school math teachers, hold permanent certification in mathematics from New York State, and possess a license to teach in New York City.
Geographic Focus: New York
Amount of Grant: $50,000; $12,500 per year
Date(s) Application is Due: Mar 24

Contact: Fellowship Administrator, (212) 206-0053; fax: (212) 206-1622; email: info@mathforamerica.org
Internet: http://www.mathforamerica.org
Sponsor: Math for America
101 Fifth Avenue, 5th Fl
New York, NY 10003

MGM Mirage Community Affairs Grants 3363
It is the mission of MGM Mirage to strengthen communities in which the corporation operates. MGM Mirage's key community investment areas are: childhood development—(birth to 18 years old), with emphasis placed on at-risk/low income youth and early childhood development; community development—specific community issues and deliverable outcomes within targeted populations; diversity—including efforts that encourage economic development as well as professional and personal growth of our employees and programs that enhance community resources; and education—including efforts to strengthen public education, from kindergarten through higher education.
Requirements: Requests must benefit residents in communities which the Company operates (Nevada, Michigan, Mississippi).
Restrictions: The program does not support the following types of activities: requests which fall outside of communities in which MGM MIRAGE operates; requests for any type of gaming-casino items (new or previously used) for children; organizations that discriminate by race, color, creed, gender or national origin; grants or sponsorship to individuals; or athletic sponsorships of individuals and/or teams.
Geographic Focus: Michigan; Mississippi; Nevada
Date(s) Application is Due: Oct 1
Contact: Director of Corporate Philanthropy, (702) 650-7409; fax: (702) 650-7401
Internet: http://www.mgmmiragevoice.com/pages/cg_giving.asp
Sponsor: MGM Mirage Corporation
3260 Industrial Road
Las Vegas, NV 89109

MHRC Graduate Studentships 3364
A limited number of studentships are offered to qualified graduate students who are undertaking full-time master's or doctoral programs. Studentships are awarded on a competitive basis. Each candidate must arrange to have three individuals send supporting letters directly to the council. Studentships are initially awarded for a 12-month period and may be renewed. The maximum period of support by the council is two years for individuals enrolled in a master's program, three years for individuals holding a master's degree enrolled in a PhD program, and five years for individuals proceeding directly to a PhD from an honor's BS, BA, or equivalent degree. Recipients should begin their award period by July 1 at the earliest and September 1 at the latest.
Restrictions: In each competition a maximum of one studentship will be awarded per supervisor. Individuals who hold a faculty position are not eligible for graduate studentship awards.
Contact: Program Contact, (204) 775-1096; fax: (204) 786-5401; email: mhrc@cc.umanitoba.ca
Internet: http://www.umanitoba.ca/academic_support/MHRC/PROGRAMS.HTM#contents
Sponsor: Manitoba Health Research Council
770 Bannatyne Avenue
Winnipeg, MB R3E 0W3 Canada

MHRC Operating Grants 3365
Grants in aid of research are designed to defray the normal direct costs of research including, among others, personnel costs, supplies and expendable materials, experimental animals, equipment costing less than $C5000, computer costs, and publication costs. Up to five percent of the operating grant may be used for travel related to the investigator's research program. Grants are normally awarded for one year. The council gives priority to new investigators and new projects. The starting date for successful grants is July 1.
Requirements: Normally only investigators holding academic appointments or equivalent in the province of Manitoba are eligible to apply. Investigators who do not hold academic appointments should consult with the council to establish their eligibility before applying.
Restrictions: Individuals holding continuing operating grants from the council are not eligible to apply for other operating grants that would become effective during the tenure of their present grants. Expenditures for student and postdoctoral salaries will not be considered on operating grants.
Amount of Grant: $C100,000 maximum for one year
Contact: Program Contact, (204) 775-1096; fax: (204) 786-5401; email: mhrc@cc.umanitoba.ca
Internet: http://www.umanitoba.ca/academic_support/MHRC/PROGRAMS.HTM#contents
Sponsor: Manitoba Health Research Council
770 Bannatyne Avenue
Winnipeg, MB R3E 0W3 Canada

MHRC Postdoctoral Research Fellowships 3366
A limited number of fellowships are available to qualified candidates with doctoral degrees whose prior postdoctoral research training shall normally not exceed five years. Fellowships are awarded on a competitive basis and will carry the same monetary value as postdoctoral salaries on the council's operating grants. Only individuals licensed to practice in the province will be eligible for the stipend scale for professional degree holders; all others will receive the PhD stipend. Fellowships are awarded for a 12-month period and may be renewed for one additional year by reapplication and competitive review. Starting dates are between July 1 and September 1.
Requirements: Holders of professional doctoral degrees (MD, DDS, DVM) who will be registered as full-time master's or doctoral students may apply for either a postdoctoral fellowship or a graduate studentship; however, they are eligible for consideration in only one of these two categories and must indicate clearly for which category they wish to be considered. Each candidate must arrange to have three individuals send supporting letters to the council.
Restrictions: In each competition a maximum of one postdoctoral fellowship will be awarded per supervisor. Individuals who hold a faculty position are not eligible for postdoctoral fellowship awards.
Contact: Program Contact, (204) 775-1096; fax: (204) 786-5401; email: mhrc@cc.umanitoba.ca
Internet: http://www.umanitoba.ca/academic_support/MHRC/PROGRAMS.HTM#contents
Sponsor: Manitoba Health Research Council
770 Bannatyne Avenue
Winnipeg, MB R3E 0W3 Canada

MHS Long Term Research Fellowships 3367
The Society will award two long-term grants of six to twelve months, although either fellowship may be divided into two awards of a maximum of five months. The stipend, governed by an NEH formula, will be no more than $40,000 for a term of six to twelve months and smaller amounts for shorter terms. Tenure must be continuous. Within the constraints of the NEH's guidelines, the Society will also supplement each stipend with a housing allowance of up to $500 per month.
Requirements: MHS-NEH fellowships are open to U.S. citizens and to foreign nationals who have lived in the United States for at least the three years immediately preceding the application deadline. Applicants must have completed their professional training.
Restrictions: NEH-sponsored fellowships are not available to graduate students.
Amount of Grant: $40,000 maximum stipend; $500 per month maximum housing allowance
Date(s) Application is Due: Jan 15
Contact: Jane Becker; (617) 646-0518 or (617) 536-1608; fax: (617) 859-0074; email: fellowships@masshist.org
Internet: http://www.masshist.org/fellowships/long_term.cfm
Sponsor: Massachusetts Historical Society
1154 Boylston Street
Boston, MA 02215

MHS New England Regional Fellowships 3368
The New England Regional Fellowship Consortium, a collaboration of 18 major cultural agencies, provides a stipend for eight weeks of research at participating institutions. Participants include Baker Library, Harvard Business School; Baker Library; Boston Athenaeum; Bostonian Society; Colonial Society of Massachusetts; Connecticut Historical Society; Francis A. Countway Library of Medicine; Harvard Law School, Special Collections; Historic Deerfield; Houghton Library, Harvard University; Maine Historical Society; Massachusetts Historical Society; Mystic Seaport; New England Historic Genealogical Society; New Hampshire Historical Society; Rhode Island Historical Society; Schlesinger Library; and Vermont Historical Society. The grants are designed to encourage projects that draw on the resources of several agencies. Each award will be for research at a minimum of three different institutions. Fellows must work at each of these organizations for at least two weeks. Grants are for June of one year through May 31 of the next. Application and guidelines are available online.
Requirements: Applications are welcome from anyone with a serious need to use the collections and facilities of the organizations.
Amount of Grant: $5,000
Date(s) Application is Due: Feb 1
Contact: Jane Becker; (617) 646-0518 or (617) 536-1608; fax: (617) 859-0074; email: fellowships@masshist.org
Internet: http://www.nerfc.org/
Sponsor: Massachusetts Historical Society
1154 Boylston Street
Boston, MA 02215

MHS Short Term Research Fellowships 3369
The society offers approximately 20 fellowships each year for researchers who need to use the collections to complete a major project. Fellowships for the current cycle include: African American Studies Fellowship; Boston Marine Society Fellowship; W.B.H. Dowse Fellowships; Marc Friedlaender Fellowship; Massachusetts Society of the Cincinnati Fellowship; Andrew W. Mellon Fellowships; Ruth R. Miller Fellowship; Andrew Oliver Research Fellowship; Paul Revere Memorial Association Fellowship; and Benjamin F. Stevens Fellowship. Each fellowship includes a stipend for four weeks of research, which must be conducted in the society's archives sometime between July 1 of one year June 30 of the next. Candidates who live 50 or more miles from Boston receive preference. Guidelines and application are available online.
Requirements: Recipients must be U.S. citizens or foreign nationals holding the appropriate U.S. government documents. Typical applicants include independent scholars, advanced graduate students, and college and university faculty in all disciplines.
Amount of Grant: $1,500 stipend
Date(s) Application is Due: Mar 1
Contact: Jane Becker; (617) 646-0518 or (617) 536-1608; fax: (617) 859-0074; email: fellowships@masshist.org

Internet: http://www.masshist.org/fellowships/short_term.cfm
Sponsor: Massachusetts Historical Society
1154 Boylston Street
Boston, MA 02215

MHS Suzanne and Caleb Loring Research Fellowship 3370
The Massachusetts Historical Society and the Boston Athenaeum will award one Suzanne and Caleb Loring Fellowship on the Civil War, Its Origins, and Consequences this year. The recipient will conduct research for at least four weeks at each institution. The Athenaeum, Civil War collections are anchored by its holdings of Confederate states imprints, the largest in the nation, consisting of books, maps, broadsides, sheet music, government documental publications, and other materials organized according to the Parrish & Willingham bibliography. The Society, manuscript holdings on the Civil War are particularly strong. They include, for instance, diaries, photographs, correspondence from the battlefield and the home front, papers of political leaders, materials on black regiments raised in Massachusetts, and extensive holdings on the U.S. Sanitary Commission. The Athenaeum and the Society are especially interested in projects for which both repositories' resources are vital. The fellowship carries a stipend of $4,000. Each institution will automatically refer unsuccessful proposals to its short-term fellowship competition.
Amount of Grant: $4,000 stipend
Date(s) Application is Due: Feb 15
Contact: Jane Becker; (617) 646-0518 or (617) 536-1608; fax: (617) 859-0074; email: fellowships@masshist.org
Internet: http://www.masshist.org/fellowships/loring.cfm
Sponsor: Massachusetts Historical Society
1154 Boylston Street
Boston, MA 02215

Michael J. Fox Foundation Clinical Discovery Grants 3371
The goal of this program is to encourage translational research by stimulating well-designed clinical research projects focused on potentially high-impact approaches to the field of Parkinson's disease. Through the program, researchers will be invited to submit grant applications to conduct new or novel clinical research in humans relevant to the cure, cause, prevention, or improved treatment of Parkinson's disease and its complications. The program will support a variety of innovative research studies that are directly related to patients or patient populations affected by the disease. Appliciants may request a project period of up to three years. The foundation employs an expedited, abbreviated granting mechanism to accelerate discoveries and hopefully translate findings to Parkinson's disease patients more quickly. A mandatory letter of intent is due on the listed application deadline; full applications are by invitation.
Requirements: Both U.S. and non-U.S. organizations are eligible. Applications may be submitted by public and private nonprofit entities, units of state and local governments, eligible agencies of the federal government, and for- profit entities. Collaborative efforts are welcome.
Restrictions: Postdoctoral students or fellows are not eligible to apply as principal investigators.
Amount of Grant: $200,000 maximum per year
Date(s) Application is Due: Nov 21
Contact: Grants Administrator, email: research@michaeljfox.org
Internet: http://www.michaeljfox.org/research/index.php
Sponsor: Michael J. Fox Foundation for Parkinson's Research
P.O. Box 4777, Grand Central Sta
New York, NY 10163

Michael J. Fox Foundation for Parkinson's Research Community Fast Track Grants 3372
This program invites researchers to submit investigator-initiated grant applications to conduct research relevant to the cure, cause, prevention or improved treatment of Parkinson's disease and its complications. The intent of the program is to stimulate novel, innovative, and/or high-impact approaches to the field of Parkinson's disease as well as to fill funding gaps missed by more conventional funding sources. Responsibility for the planning, direction, and execution of the proposed project will be solely that of the principal investigator. The intent of this RFA is to stimulate work on potentially high-impact approaches to the field of Parkinson's disease research, as well as to fill funding gaps missed by more conventional funding sources. This program is restricted to projects and ideas that are conducted on a one-year timeframe; however, successful projects meeting all of their original specific aims may be allowed to submit proposals for an additional one to two years of funding. The listed application deadline is for mandatory letters of intent, which must be submitted via email.
Requirements: Applications may be submitted by U.S. and non-U.S. entities; public and private nonprofit entities, such as universities, colleges, hospitals, laboratories, units of state and local governments, and eligible agencies of the federal government; and for-profit entities. Collaborative efforts are welcome.
Restrictions: Postdoctoral students or fellows are not eligible to apply as principal investigators.
Amount of Grant: $100,000 maximum
Date(s) Application is Due: Jul 27
Contact: Grants Administrator, email: research@michaeljfox.org
Internet: http://www.michaeljfox.org/research/grantsunderreview.php
Sponsor: Michael J. Fox Foundation for Parkinson's Research
P.O. Box 4777
New York, NY 10163

Michael J. Fox Foundation for Parkinson's Research LEAPS Program 3373
Investigators are invited to apply for funding through a novel mechanism called LEAPS (Linked Efforts to Accelerate Parkinson's Solutions). LEAPS are multi-year, potentially multi-million dollar grant awards for collaborative, milestone-driven efforts to address questions that will have significant practical impact on the understanding and treatment of Parkinson's disease. LEAPS grants are goal-oriented and focused on answering a broad research question, one whose answer would represent a major, practical step forward in Parkinson's disease—a solution to a Parkinson's problem. Examples of such questions could be categorized under broad aims of: repairing the PD brain; slowing or halting the progression of PD; preventing or controlling levodopa-induced dyskinesias; and identifying the cause of PD. LEAPS are not intended to be conventional research grants. LEAPS investigators will be asked to interact regularly with foundation staff and advisors to discuss elements of the project, research directions, progress against stated milestones, and possible ways (other than through additional financial resources) the foundation may offer support or assistance. T
Requirements: Applications may be submitted by U.S. and non-U.S. entities; public and private nonprofit entities, such as universities, colleges, hospitals, laboratories, units of state and local governments, and eligible agencies of the federal government; and for-profit entities.
Restrictions: Postdoctoral students or fellows are not eligible to apply as coordinating or principal investigators.
Date(s) Application is Due: Apr 29
Contact: Grants Administrator, (800) 708-7644
Internet: http://www.michaeljfox.org
Sponsor: Michael J. Fox Foundation for Parkinson's Research
P.O. Box 4777
New York, NY 10163

Michael J. Fox Foundation Parkinson's Cell Line Research Grants 3374
Researchers are invited to submit investigator-initiated proposals using embryonic stem cells targeted at achieving optimized transplantation materials and protocols promoting graft survival, stability of dopaminergic phenotype, and long-term functional recovery in animal models of Parkinson's disease. The intent is to use the acquired knowledge to develop cell replacement therapy for PD patients. Work in animal or human cell lines is acceptable; however, a primary focus on human cell lines is required, and research on human embryonic stem cells is considered appropriate. Mandatory letters of intent are due August 9. Guidelines are available online.
Requirements: Investigators with scientific interests and demonstrated success in the development, culture, and manipulation of individual cell lines are invited to apply.
Samples: Ernest Arenas of the Karolinska Institute (Stockholm Sweden), Jun Takahashi of the Kyoto U Graduate School of Medicine (Japan), Xuejun Wen of Clemson U (SC), and Su-Chung Zhang of the U of Wisconsin (Madison, WI)—for stem-cell research related to finding a cure for Parkinson's disease, $1 million divided (2005).
Date(s) Application is Due: Aug 9; Sep 20
Contact: Grants Administrator, (800) 708-7644; email: research@michaeljfox.org
Internet: http://www.michaeljfox.org/research/fundingopps.php?id=9
Sponsor: Michael J. Fox Foundation for Parkinson's Research
P.O. Box 4777, Grand Central Sta
New York, NY 10163

Michael Reese Health Trust Grants 3375
The trust awards grants in the metropolitan Chicago area to nonprofits for community-based health services provided to vulnerable groups, including children and youth, people with mental and physical disabilities, the elderly, immigrants and refugees, uninsured, medically indigent, and other underserved populations. Grants provide general and project support of innovative approaches to delivering health services to the target populations, efforts to improve community health through health promotion and disease prevention, and education to improve healthcare providers' ability to treat the underserved. Some research is funded, but emphasis is on direct service delivery. To apply, submit an inquiry letter. Deadlines listed are for letters of inquiry.
Requirements: Nonprofit organizations operating in the metropolitan Chicago area are eligible.
Restrictions: Grants do not support capital needs (i.e., building construction/renovation, vehicles, and equipment), endowments, fund-raising events, individuals, scholarships or debt reduction.
Geographic Focus: Illinois
Date(s) Application is Due: Jun 15; Dec 15
Contact: Dorothy Gardner, President, (312) 726-1008; fax (312) 726-2797; email: programs@healthtrust.net
Internet: http://www.healthtrust.net/index.php?option=com_content&task=view&id=13&Itemid=42
Sponsor: Michael Reese Health Trust
20 N Wacker Drive, Suite 760
Chicago, IL 60606

Michele Clark Fellowship 3376
This fellowship is awarded annually to an exceptional young minority electronic journalist. Fellowships may be applied to any legitimate educational or professional purpose. Applications must include a video- or audiotape of the journalist's work in radio or television news done as a member of a news staff. Format and length are not criteria for consideration. In addition to the award winners of the Michele Clark Fellowships receive an expenses-paid trip to the RTNDA International Conference.

Requirements: Applicants must be employed in electronic journalism, have five years or less experience in the field, and belong to a minority group.
Amount of Grant: $1000
Date(s) Application is Due: May 9
Contact: Karen Jackson-Bullitt, (202) 467-5218; fax: (202) 223-4007; email: karenb@rtndf.org
Internet: http://www.rtnda.org/asfi/fellowships/minority.shtml
Sponsor: Radio and Television News Directors Foundation
1600 K StNW, Suite 700
Washington, DC 20006

Michigan Society of Fellows Postdoctoral Fellowships **3377**
The society offers three-year postdoctoral fellowships to individuals selected for outstanding achievement, professional promise, and interdisciplinary interests in the humanities and arts, sciences, and professions. The fields of study include all departments within schools and colleges at the University of Michigan. Fellows are appointed as assistant professors in appropriate departments and are expected to teach for the equivalent of one academic year, to participate in the informal intellectual life of the society, and to devote time to their independent productive scholarship.
Amount of Grant: $47,271 annual stipend
Date(s) Application is Due: Sep 30
Contact: Michigan Society of Fellows, (734) 763-1259; email: society.of.fellows@umich.edu
Internet: http://www.rackham.umich.edu/Faculty/society.html
Sponsor: Michigan Society of Fellows
University of Michigan, 3572 Rackham Building, 915 E Washington Street
Ann Arbor, MI 48109-1070

Microscopy Society of America Student Scholarships **3378**
The scholarship program is designed to further the interest in microscopy and facilitate information exchange for students working in the field of microscopy. Recipient must attend the annual meeting and present a research paper at the regular scientific session.
Requirements: Recipient must be a full-time student at a recognized college or university at the time of the annual meeting of MSA, must submit an abstract of research or project, and must be sponsored with a signed statement by a member of MSA, preferably a research advisor.
Amount of Grant: $3000 maximum
Date(s) Application is Due: Dec 15
Contact: Business Office, (800) 538-3672 or (312) 644-1527; fax: (312) 644-8557; email: BusinessOffice@MSA.Microscopy.com
Internet: http://www.microscopy.org
Sponsor: Microscopy Society of America
230 E Ohio, Suite 400
Chicago, IL 60611

Microsoft Research A. Richard Newton Breakthrough Research Awards **3379**
Microsoft invites proposals for breakthrough research projects. Breakthrough, in this sense, refers to projects which clearly leverage innovative computational techniques and advantages at the heart of the research, and which may integrate approaches from multiple disciplines. Additionally, projects considered breakthrough demonstrate potentially high impact by solving problems of great importance or by inventing promising new approaches for enabling their solution. The total amount available under this request for proposals (RFP) is U.S. $1,000,000. Microsoft Research anticipates making approximately ten to twelve awards, with a maximum of $100,000 for any single award. Awards are made for the purpose of seed-funding larger initiatives, proofs of concept, or demonstrations of feasibility. It is important to understand that funding will continue after the first year only in exceptional circumstances, and that the principal investigators should therefore make every effort to leverage Microsoft Research funds as one component of a diverse funding base in a larger or longer-running project.
Requirements: All qualifying institutions are eligible without regard for geographic location. The proposing institution must be either: an accredited degree-granting college or university (or international equivalent) with non-profit status and awarding degrees at the baccalaureate level or above; or a research institution with non-profit status.
Restrictions: For all awards, payment of indirect costs (overhead) is not permitted.
Amount of Grant: $100,000 maximum
Date(s) Application is Due: Dec 3
Contact: Rick Rashid, Senior Vice President; (800) 642-7676; fax: (425) 936-7329; email: erpinq@microsoft.com
Internet: http://research.microsoft.com/ur/us/fundingopps/rfps/ARichardNewtonAward.aspx
Sponsor: Microsoft Research
One Microsoft Way
Redmond, WA 98052-6399

Microsoft Research Beyond Search - Semantic Computing and Internet Economics Grants **3380**
Microsoft would like to enable cross-disciplinary research in the area of semantic computing, Internet economics and online advertising, and is looking for innovative proposals which explore the technical, research, societal, and commercial issues around the topics of semantic computing and Internet economics. The total amount available under this request for proposals (RFP) is $1,000,000. Microsoft Research anticipates making approximately 20 awards averaging $50,000, with a maximum of $100,000 for any single award. Awards are made for the purpose of seed-funding larger initiatives, proofs of concept, or demonstrations of feasibility.
Requirements: All qualifying institutions are eligible without regard for geographic location. The proposing institution must be either: an accredited degree-granting college or university (or international equivalent) with non-profit status and awarding degrees at the baccalaureate level or above; or a research institution with non-profit status.
Restrictions: For all awards, payment of indirect costs (overhead) is not permitted.
Amount of Grant: $100,000 maximum
Date(s) Application is Due: Nov 1
Contact: Rick Rashid, Senior Vice President; (800) 642-7676; fax: (425) 936-7329; email: erpinq@microsoft.com
Internet: http://research.microsoft.com/ur/us/fundingopps/rfps/BeyondSearch_RFP.aspx
Sponsor: Microsoft Researh
One Microsoft Way
Redmond, WA 98052-6399

Microsoft Research Biomedical Computing for Genome Wide Association Studies Grants **3381**
Microsoft Research encourages researchers to develop tools that can facilitate better data usage and analysis for genomewide association studies to provide a stronger framework for enabling personalized treatment methods. It anticipates making approximately 8-10 awards averaging $70,000, with a maximum of $100,000 for any single award. Awards are made for the purpose of seed-funding larger initiatives, proofs of concept, or demonstrations of feasibility. It is important to understand that funding will continue after the first year only in exceptional circumstances, and that the principal investigators should therefore make every effort to leverage Microsoft Research funds as one component of a diverse funding base in a larger or longer-running project.
Requirements: All qualifying institutions are eligible without regard for geographic location. The proposing institution must be either: an accredited degree-granting college or university (or international equivalent) with non-profit status and awarding degrees at the baccalaureate level or above; or a research institution with non-profit status.
Restrictions: For all awards, payment of indirect costs (overhead) is not permitted.
Amount of Grant: $100,000 maximum; $700,000 total
Contact: Rick Rashid, Senior Vice President; (800) 642-7676; fax: (425) 936-7329; email: erpinq@microsoft.com
Internet: http://www.microsoft.com/presspass/press/2007/jul07/07-16FacultySummitPR.mspx
Sponsor: Microsoft Research
One Microsoft Way
Redmond, WA 98052-6399

Microsoft Research Cell Phone as a Platform for Healthcare Grants **3382**
Microsoft Research will support selected academic research with the goal of advancing the state-of-the-art in smart cell phones for health care applications. Successful projects must seek to increase the capacity of cell phone-based health care solutions in under-served rural and urban communities. They must also take into consideration the social context of application deployment. Topics to consider include: medical applications that are relevant, worldwide, for smart mobile phones (application and Web-enabled) in rural, and urban, communities; and appropriate services and infrastructures needed to provide affordable and accessible health care services. The total amount available under this request for proposals (RFP) is $1,000,000. Microsoft Research anticipates making approximately 10-12 awards averaging $80,000, with a maximum of $100,000 for any single award. Awards are made for the purpose of seed-funding larger initiatives, proofs of concept, or demonstrations of feasibility. It is important to understand that funding will continue after the first year only in exceptional circumstances, and that the principal investigators should therefore make every effort to leverage Microsoft Research funds as one component of a diverse funding base in a larger or longer-running project.
Requirements: All qualifying institutions are eligible without regard for geographic location. The proposing institution must be either: an accredited degree-granting college or university (or international equivalent) with non-profit status and awarding degrees at the baccalaureate level or above; or a research institution with non-profit status.
Restrictions: For all awards, payment of indirect costs (overhead) is not permitted.
Amount of Grant: $100,000 maximum
Date(s) Application is Due: Oct 29
Contact: Rick Rashid, Senior Vice President; (800) 642-7676; fax: (425) 936-7329; email: erpinq@microsoft.com
Internet: http://research.microsoft.com/en-us/um/redmond/about/collaboration/awards/cellphone-healthcare_awards.aspx
Sponsor: Microsoft Research
One Microsoft Way
Redmond, WA 98052-6399

Microsoft Research Human-Robot Interaction (HRI) Grants **3383**
Microsoft Research encourages research to take human-robot interaction to the next level through development of tools and methods that lead to practical applications with realistic commercial potential within five to 10 years. It anticipates making approximately 8-10 awards averaging $50,000, with a maximum of $100,000 for any single award. Awards are made for the purpose of seed-funding larger initiatives, proofs of concept, or demonstrations of feasibility. It is important to understand that funding will continue after the first year only in exceptional circumstances, and that the principal investigators should therefore make every effort to leverage Microsoft Research funds as one component of a diverse funding base in a larger or longer-running project.

Requirements: All qualifying institutions are eligible without regard for geographic location. The proposing institution must be either: an accredited degree-granting college or university (or international equivalent) with non-profit status and awarding degrees at the baccalaureate level or above; or a research institution with non-profit status.
Restrictions: For all awards, payment of indirect costs (overhead) is not permitted.
Amount of Grant: $100,000 maximum; $500,000 total
Contact: Rick Rashid, Senior Vice President; (800) 642-7676; fax: (425) 936-7329; email: erpinq@microsoft.com
Internet: http://www.microsoft.com/presspass/press/2007/jul07/07-16FacultySummitPR.mspx
Sponsor: Microsoft Research
One Microsoft Way
Redmond, WA 98052-6399

Microsoft Research Intelligent Web 3.0 Grants 3384
This Program encourages research to help find, discover, extract, publish, and share information, at a desk or on the go, safely, making the Web meaningful (from string manipulation to meaning computation) and enabling a human-centric, context-aware model of information access. It anticipates making approximately 8-10 awards averaging $50,000, with a maximum of $100,000 for any single award. Awards are made for the purpose of seed-funding larger initiatives, proofs of concept, or demonstrations of feasibility. It is important to understand that funding will continue after the first year only in exceptional circumstances, and that the principal investigators should therefore make every effort to leverage Microsoft Research funds as one component of a diverse funding base in a larger or longer-running project.
Requirements: All qualifying institutions are eligible without regard for geographic location. The proposing institution must be either: an accredited degree-granting college or university (or international equivalent) with non-profit status and awarding degrees at the baccalaureate level or above; or a research institution with non-profit status.
Restrictions: For all awards, payment of indirect costs (overhead) is not permitted.
Amount of Grant: $100,000 maximum; $500,000 total
Contact: Rick Rashid, Senior Vice President; (800) 642-7676; fax: (425) 936-7329; email: erpinq@microsoft.com
Internet: http://www.microsoft.com/presspass/press/2007/jul07/07-16FacultySummitPR.mspx
Sponsor: Microsoft Research
One Microsoft Way
Redmond, WA 98052-6399

Microsoft Research Mechanisms for Safe and Scalable Multicore Computing Grants 3385
Microsoft Research will encourage research in how operating systems and runtimes can evolve to enable safe and scalable concurrent programs. It anticipates making approximately 8-10 awards averaging $50,000, with a maximum of $100,000 for any single award. Awards are made for the purpose of seed-funding larger initiatives, proofs of concept, or demonstrations of feasibility. It is important to understand that funding will continue after the first year only in exceptional circumstances, and that the principal investigators should therefore make every effort to leverage Microsoft Research funds as one component of a diverse funding base in a larger or longer-running project.
Requirements: All qualifying institutions are eligible without regard for geographic location. The proposing institution must be either: an accredited degree-granting college or university (or international equivalent) with non-profit status and awarding degrees at the baccalaureate level or above; or a research institution with non-profit status.
Restrictions: For all awards, payment of indirect costs (overhead) is not permitted.
Amount of Grant: $100,000 maximum
Contact: Rick Rashid, Senior Vice President; (800) 642-7676; fax: (425) 936-7329; email: erpinq@microsoft.com
Internet: http://www.microsoft.com/presspass/press/2007/jul07/07-16FacultySummitPR.mspx
Sponsor: Microsoft Research
One Microsoft Way
Redmond, WA 98052-6399

Microsoft Research New Faculty Fellowship Program 3386
The program seeks nominees who are advancing computing research in novel directions with the potential for high impact, and who demonstrate the likelihood of becoming thought leaders of the field. The program identifies, recognizes, and supports exceptional new faculty members engaged in innovative computing research. The objective is to stimulate and support creative research by promising researchers who have the potential of making a profound impact to the state of the art in their research discipline. Examples of research areas may include, but are not limited to, security, scientific computing, interdisciplinary research, bioinformatics, computational biology, and software engineering. Each fellowship award includes a cash gift as well as Microsoft resources. The awardees will have the option of exploring collaborations with MSR researchers.
Requirements: The potential fellow must be nominated by the university provost. No more than one nomination from the university will be accepted. The new faculty member must hold a tenure-track faculty position in a U.S. research university. The applicant must be in the first, second, or third year as a new faculty member in his/her first faculty appointment. Five letters of recommendation are required from established researchers familiar with the applicant's research; no more than two letters of recommendation can come from within the nominating institution. Applications may only be submitted through the online application system.
Restrictions: Direct applications are not accepted from a new faculty candidate. The applicant must be no more than six years from the completion of his/her most recent PhD.
Samples: Magdalena Balazinska, University of Washington, $200,000 (2007); Josh Bongard, University of Vermont, $200,000 (2007); Adam Siepel, Cornell University, $200,000 (2007).
Amount of Grant: $200,000
Date(s) Application is Due: Nov 19
Contact: Faculty Fellowship Program Director; (800) 642-7676
Internet: http://research.microsoft.com/ur/us/nff/default.aspx
Sponsor: Microsoft Research
One Microsoft Way
Redmond, WA 98052-6399

Microsoft Research Sustainable Computing Grants 3387
Microsoft Research encourages research in innovative approaches toward power-optimized system architectures, and adaptive power management solutions for maximizing the energy efficiency of computing infrastructure. It anticipates making approximately 8-10 awards averaging $50,000, with a maximum of $100,000 for any single award. Awards are made for the purpose of seed-funding larger initiatives, proofs of concept, or demonstrations of feasibility. It is important to understand that funding will continue after the first year only in exceptional circumstances, and that the principal investigators should therefore make every effort to leverage Microsoft Research funds as one component of a diverse funding base in a larger or longer-running project.
Requirements: All qualifying institutions are eligible without regard for geographic location. The proposing institution must be either: an accredited degree-granting college or university (or international equivalent) with non-profit status and awarding degrees at the baccalaureate level or above; or a research institution with non-profit status.
Restrictions: For all awards, payment of indirect costs (overhead) is not permitted.
Amount of Grant: $100,000 maximum
Contact: Rick Rashid, Senior Vice President; (800) 642-7676; fax: (425) 936-7329; email: erpinq@microsoft.com
Internet: http://www.microsoft.com/presspass/press/2007/jul07/07-16FacultySummitPR.mspx
Sponsor: Microsoft Research
One Microsoft Way
Redmond, WA 98052-6399

Miles of Hope Breast Cancer Foundation Grants 3388
The Fund provides programs and support services for women and families in the Hudson Valley, New York, affected by breast cancer. All funds raised for the Foundation are used in the Hudson Valley. The Miles of Hope Breast Cancer Foundation is currently offering grants for projects in the areas of breast health and breast cancer education, outreach, screening, treatment and support projects.
Requirements: Services must be provided in the eight counties served by the Miles of Hope Breast Cancer Foundation, which include: Columbia, Dutchess, Greene, Orange, Putnam, Rockland, Westchester and Ulster.
Restrictions: Project must be specific to breast health and/or breast cancer. Applicants must be a U.S. nonprofit (federally tax-exempt) organization. Nonprofit organizations, known educational institutions (i.e. college or university), government agencies, and Indian tribes are eligible. Equipment costs, if applicable, may not exceed 30% of direct costs and should be used exclusively on this project. Salaries, if requested, are for personnel related to this project only and not the general work of the employee. Funds will not be awarded to capital campaigns
Geographic Focus: New York
Samples: Healing Vibrations—providing a classical chamber ensemble at the Dyson Center for Cancer Care at Vassar Brothers Medical Center; Medical Gap Care Fund—for individuals who are faced with life emergencies and/or costs not covered by health insurance.
Contact: Program Administrator, (845) 527-6884 or (845) 264-2005; email: info@milesofhopebcf.org
Internet: http://www.milesofhopebcf.org/cms/index.php?module=webpage&id=8
Sponsor: Miles of Hope Breast Cancer Foundation
P.O. Box 405
La Grangeville, NY 12540

Milheim Foundation Cancer Research Grants 3389
Project grants provide support for research work for the prevention, treatment, and cure of cancer. In addition to a written proposal and completed application form, at least one letter of recommendation and one curriculum vita must be submitted. The grants are awarded for one year; renewal is possible but not encouraged.
Requirements: Organizations that are incorporated as nonprofit and operate within the continental limits of the United States are eligible. PI must hold MD, PhD, or the equivalent.
Restrictions: Grants do not support travel or presentation of papers, overhead expenses or support and maintenance of hospitals, salaries of academic personnel, or individuals.
Amount of Grant: $10,000-$22,000
Date(s) Application is Due: Mar 15
Contact: Barbara Cole, c/o U.S. Bank PFS/Trust, (303) 316-5945 or (303) 316-5948; fax: (303) 388-9387; email: jnelson@foundation.sdsu.edu
Sponsor: Milheim Foundation for Cancer Research
200 University Boulevard
Denver, CO 80206

Military Medical Research and Development Grants 3390

The objective of the program is to reduce illness and injury acquired by U.S. military personnel on the battlefield through basic and applied medical research executed largely through grants and contracts with civilian educational and research institutions. The maximum period acceptable for a research proposal is five years. Preproposals are encouraged prior to submission of a formal application. There are no overall deadline dates for applications. Individual research areas may initiate deadlines, which will be communicated upon review of the preproposal.

Requirements: Applicant must be a public, nonprofit organization; other public organization; private nonprofit organization; specialized group; or quasi-public nonprofit organization.

Samples: Characterization and Utilization of Opioid-Like Hibernation Factors, $621,000.; Analysis of Investigational Drugs in Biological Fluids, Method Development and Routine Assay, $2.4 million.; Studies of Altered Response to Infection Induced by Severe Injury, $1.1 million; and Mechanisms of Cutaneous Vesication, $862,000.

Amount of Grant: $100,000-$5 million; $650,000 average

Contact: Craig Lebo, U.S. Army Medical Research Acquisition Activity, (301) 619-2036

Internet: https://mrmc-www.army.mil/mrdindex.asp

Sponsor: U.S. Army Medical Research and Material Command
SCRD-RMA-RC, Fort Detrick
Frederick, MD 21702-5014

Milken Family Foundation Grants 3391

The purpose of the family foundation is to discover and advance inventive ways to build human resources and help people lead productive and satisfying lives, primarily in Los Angeles, CA. This mission is pursued in three areas: education, medical research and health care, and community welfare. Under education, the foundation seeks to recognize and reward outstanding educators, expand their professional leadership and policy influence, and encourage talented young people to become educators; stimulate creativity and productivity among educators; and foster the involvement of the family and the community in schools. The foundation supports basic and applied medical research, especially in the areas of cancer and epilepsy, and supports programs designed to improve and protect the health of people of all ages. Under community welfare, the foundation helps build vibrant communities and responds to basic human needs. Preventive programs with long-range goals receive the closest consideration. Types of support include general operations, research, conferences and seminars, scholarship funds, and building and renovation projects. Applicant should submit a brief written statement that includes a description of the project, goals, procedure, and personnel; brief background of the organization; and letter of exemption from the IRS. Approximately 80 percent of all funding is directed toward foundation initiatives. Requests are considered throughout the year.

Requirements: Grants are made to 501(c)3 tax-exempt organizations. Grant recipients must have the financial potential to sustain the program for which funding is sought following the period of foundation support.

Restrictions: Grants are not made directly to individuals.

Geographic Focus: California

Samples: Teacher Advancement Program Foundation (Santa Monica, CA)—for the work of this new organization, which uses a research-based strategy designed to attract, develop, and retain high-quality teachers, $5 million (2005); To be distributed among 100 educators nationwide—for the Milken Educator Awards, which honor outstanding elementary and secondary teachers, principals, and other education professionals, $2.5 million (2004).

Contact: Dr. Julius Lesner, Executive Vice President, (310) 570-4800; fax: (310) 570-4801; email: admin@mff.org

Internet: http://www.mff.org/about/about.taf?page=funding

Sponsor: Milken Family Foundation
1250 Fourth Street, 6th Fl
Santa Monica, CA 90401-1353

Millard Meiss Publication Fund Grants 3392

The grant is to be used to subsidize publication costs of book-length scholarly manuscripts on any period of art history. Publisher will need to submit a written statement giving technical specifications of the proposed publication and amount of subsidy required, and enclosing all reviews of the manuscript. Application forms are available on the Web site.

Requirements: Applicant may be a scholar of any nationality who is a member of the College Art Association and whose manuscript has been accepted by a commercial or university press but cannot be published without a subsidy.

Restrictions: Eligibility does not include museum catalogues, excavation reports, articles, collections of previously published essays, or multi-author volumes (such as anthologies, collections of discrete writings, or congress proceedings).

Date(s) Application is Due: Mar 15; Oct 1

Contact: Director of Publications, (212) 619-1051 ext 208 or 217; fax: (212) 627-2381; email: publications@collegeart.org

Internet: http://www.collegeart.org/caa/resources/meiss/index.html

Sponsor: College Art Association
275 Seventh Avenue
New York, NY 10001

Miller Brewing Corporate Contributions Program Grants 3393

The corporate contributions program funds programs of nonprofit organizations in corporate operating communities. Applicants should call the local Miller community affairs manager to discuss an idea before submitting a proposal. The corporation also supports the Thurgood Marshall Scholarship Fund, which provides four-year merit scholarships to students attending the 44 historically black public colleges and universities, and two-year scholarships to law schools at Texas Southern, North Carolina Central, Southern, and Howard Universities; and ?Adelante! U.S. Education Leadership Fund, which awards scholarships to Hispanic-American students.

Requirements: Nonprofits must be near Miller operating locations, in Irwindale, CA; Albany, GA; Trenton, OH; Eden, NC; Fort Worth, TX; and Milwaukee, WI.

Geographic Focus: California

Samples: Advocates of Ozaukee (West Bend, WI)—for a round-the-clock crisis line and other services for victims of domestic violence, $15,000 (2005); Goodwill Industries of Southeastern Wisconsin (Milwaukee, WI)—to purchase a vehicle for a program that provides elderly people with assistance with errands and shopping, $10,000 (2005); America's Second Harvest-Milwaukee (WI)—for a nutrition program for elderly people, $15,000 (2005); COA Youth and Family Ctrs (Milwaukee, WI)—for GED programs and job-readiness training and placement services, $15,000 (2005).

Amount of Grant: $1000-$15,000

Contact: Grants Administrator, (414) 931-3110; fax: (414) 931-6352

Internet: http://www.millerbrewing.com/inthecommunity/default.asp

Sponsor: Miller Brewing Corporation
3939 W Highland Boulevard
Milwaukee, WI 53201-0482

Miller Center of Public Affairs Fellowships 3394

The fellowships support PhD candidates and authors focusing on contemporary issues in 20th-century U.S. politics and history. Annual deadline dates may vary; contact program staff for exact dates.

Amount of Grant: $18,000

Date(s) Application is Due: Feb 1

Contact: Chi Lam, Managing Director, Miller Center Fellowships in Public Affairs; email: ckl2q@virginia.edu

Internet: http://millercenter.virginia.edu/programs/apd/fellowship.html

Sponsor: University of Virginia
P.O. Box 400406, 2201 Old Ivy Road
Charlottesville, VA 22904-4406

Millipore Foundation Grants 3395

The foundation supports nonprofit organizations in company operating communities in five main areas: education and research, social services, culture, health care, and public policy. Types of support include general operating support, employee matching gifts, and employee-related scholarships. Applications are accepted throughout the year.

Requirements: 501(c)3 tax-exempt organizations are eligible.

Amount of Grant: $2500-$5000 average

Contact: Charleen Johnson, Executive Director, (978) 715-1268; fax: (978) 715-1385; email: Charleen_Johnson@millipore.com

Internet: http://www.millipore.com/corporate/milliporefoundation.nsf/foundationhome

Sponsor: Millipore Foundation
290 Concord Road
Billerica, MA 01821-7037

Millipore/Charles P. Schaufus Parenteral-Processing Technology Research Grants 3396

The grant is intended to stimulate research in parenteral processing through the support of a graduate student. Parenteral processing includes the investigation of aspects in bulk manufacture of parenteral drugs, such as fermentation and scale-up; intermediate steps, such as crystallization and purification; equipment surfaces, such as their contribution to particulate matter and their resistance to corrosion; and computer-assisted manufacture, such as software and its validation. Emphasis of this research is to be on parenteral processing. The grant is for up to three years and includes cash and equipment each year. Applications are available on the Web site.

Requirements: Applicants must be associated with U.S. 501(c)3 and 509(a)(1)(2) or (3) tax-exempt organizations.

Amount of Grant: $22,500 for one year; $22,500 in equipment

Date(s) Application is Due: Sep 15

Contact: Deanna Morton, c/o Morton Communications/Infinitech, (516) 829-5501; fax: (516) 829-1008; email: DeeMorton@aol.com

Internet: http://www.pdafoundation.org/grant_offerings.html

Sponsor: Parenteral Drug Association Foundation for Pharmaceutical Sciences
111 Great Neck Road, Suite 308
Great Neck, NY 11021

Millstone Foundation Grants 3397

The foundation awards grants to eligible Missouri nonprofit organizations in its areas of interest, including education and higher education, Jewish services, and Israel. Types of support include annual campaings, continuing support, deficit financing, emergency funds, general operating support, and research grants. There are no application forms or deadlines.

Requirements: Missouri tax-exempt organizations are eligible. Preference is given to requests from Saint Louis.

Geographic Focus: Missouri

Amount of Grant: $1.15 million total

Contact: I.E. Millstone, President or Colleen Millstone, Director, (314) 961-8500

Sponsor: Millstone Foundation
7701 Forsyth Boulevard, Suite 925
Saint Louis, MO 63105-1842

Milton Wexler Postdoctoral Fellowships 3398
Support will be for research projects that contribute to identifying and understanding the basic defect of Huntington's disease. Areas of interest include trinucleotide expansions, animal models, gene therapy, neurobiology and development of the basal ganglia, cell survival and death, and intercellular signaling in striatal neurons. Awards may be renewed for a third year. To receive an application, submit a letter of intent. Letter of intent forms may be submitted electronically at any time via the Web site.
Amount of Grant: $43,000 starting salary, $12,000 for supplies, and $2000 travel allowances for two years
Date(s) Application is Due: Feb 15; Jun 15; Oct 15
Contact: Dr. Carl Johnson, (310) 458-4183; fax: (310) 458-3937; email: carljohnson@hdfoundation.org
Internet: http://www.hdfoundation.org/fundinga.htm
Sponsor: Hereditary Disease Foundation
3960 Broadway, 6th Fl
New York, NY 10032

Mineralogical Society of America Research Grants 3399
This is a grant for research in crystallography. The award selection will be based on the qualifications of the applicant, the quality, innovativeness, and scientific significance of the proposed research, and the likelihood of success of the project. There are no restrictions on how the grant funds may be spent, as long as they are used in support of research.
Requirements: The applicant must have reached his or her 25th birthday but not yet have reached his or her 36th birthday on the date the grant is awarded, and that the person is not a MSA Counsellor.
Amount of Grant: $5000
Date(s) Application is Due: Jun 1
Contact: J. Alex Speer, (703) 652-9950; fax: (703) 652-9951; email: j_a_speer@minsocam.org
Internet: http://www.minsocam.org/MSA/Awards/Crystallography_Award.html
Sponsor: Mineralogical Society of America
1015 18th Street NW, Suite 601
Washington, DC 20036-5212

Minerva Fellowships 3400
The program enables German and Israeli scientists and researchers to spend longer-term research residencies at institutions in the host country from six to thirty-six months (for graduates) or from six to twenty-four months (for post docs). Some fifty scientists receive fellowships each year. Young scientists and researchers (post docs) are given priority in the selection process. In addition to the monthly stipend, dependency, children and travel allowance will be given along with payment in kind for books, travel to symposia, etc.
Requirements: For the graduate fellowship, eligible candidates must have completed a university degree ('Diplom', MA, M.Sc., or equivalent qualification). For the post-doctoral fellowship applications will be accepted from candidates who provide proof that they have submitted their doctoral thesis prior to the start of the fellowship year, or who have completed their doctorate no more than seven years prior to the beginning of the fellowship. Maternal leave will be taken into account. Applicants who are presently residing in Israel/Germany but are not nationals of either of the two countries must show a proven record of integration into the Israeli or German scientific community. An affiliation to a research institution in Israel or Germany is required together with a residency of 5 years or longer in either of the two countries. Conversely, if applicants are citizens of one country but have been living in the other for an extended period of time (more than 4 years) a scholarship in the second country cannot be granted since it would not truly support German-Israeli exchange.
Restrictions: Scientists with tenure are not eligible to apply. BA students are not eligible to apply.
Amount of Grant: 1,128-2,300 EURO per month
Date(s) Application is Due: Jun 15
Contact: Frau Sieglinde Reichardt, Minerva Fellowship Office; +49-(0)89-2108-1242; fax: +49-(0)89-2108-1222; email: reichardt@gv.mpg.de
Internet: http://www.minerva.mpg.de/fellowships/fellowships.html
Sponsor: Minerva Foundation
Hofgartenstr. 8
Munich D-80539 Germany

Minerva Short-Term Research Grants 3401
The primary goal of grants is to give young German and Israeli scientists the opportunity to visit research partners or to establish contacts with scientists in the respective guest country. The grants also allow the scientists to participate in seminars and workshops, particularly as guest speakers. The short-term grants are for stays of a minimum of 1 week and up to 8 weeks. Grants are available for all subject areas and are open to applicants from all research facilities (in Israel - public or governmental research institutes) and universities in Germany and Israel. Graduate students will receive EUR 300/week, while Postdocs will receive EUR 425/week. Flights are covered by a lump sum of up to EUR 700.
Requirements: Applicants should not be older than 38 years. (Maternal leave will be taken into account with regard to the age limit). Application forms may be downloaded from http://www.minerva.mpg.de. Both German and Israeli scientists must submit their applications to the Minerva Office in Munich. In addition, Israeli applicants are requested to send copies of their applications by electronic mail to the Israeli Ministry of Science (nurit@most.gov.il). Applications must be written in English and sent by electronic mail. Selection criteria are: scientific excellence; scientifically promising and innovative research topics; relevance of the research topic to German-Israeli cooperation.
Restrictions: Scientists presently holding a Minerva Fellowship are not eligible to apply. In addition, principal investigators in projects within the German Federal Ministry for Education and Research and the Israeli Ministry of Science and Technology programs are not eligible to apply. Only complete applications and applications sent by electronic mail will be processed.
Amount of Grant: 300-425 EURO per week
Date(s) Application is Due: May 1
Contact: Michael Nagel, Head Office, Munich; +49-89-2108-1258; email: nagel@gv.mpg.de
Internet: http://www.minerva.mpg.de/fellowships/research.html
Sponsor: Minerva Foundation
Gesellschaft fur die Forschung mbH; HofgartenstraBe 8
Munich D-80539 Germany

Minnesota War Orphans Educational Assistance Grants 3402
This program serves children of Minnesota veterans who died while on active duty military service or as a result of injuries or disease incurred while on active duty military service. Funding may be used for vocational/occupational, community college, undergraduate, graduate, professional, or postdoctoral education. Funds cover tuition, fees room and board, or books and supplies. Awards are renewable. An application form must be obtained from the VA, which includes necessary instructions for completing the form and supporting documents.
Requirements: Applicant must be resident of Minnesota for at least two years prior to application; the deceased parent-veteran must have been a Minnesota resident at time of entry to active duty service.
Restrictions: Scholarships may not be used at the University of Minnesota.
Geographic Focus: Minnesota
Amount of Grant: $750 per year
Contact: County Veterans Service Officer, (612) 296-2562; fax: (612) 296-3954; email: lee.peterson@state.mn.us
Internet: http://www.mdva.state.mn.us/education.htm
Sponsor: Minnesota State Department of Veterans Affairs
20 W 12th Street, Veterans Service Building
Saint Paul, MN 55155-2079

Minnie Pearl Cancer Foundation Grants 3403
Nonprofit organizations in Tennessee are eligible for grants to support programs in cancer research, education, and patient and family services. The foundation funds programs that offer the greatest benefit to cancer patients and their families.
Requirements: Nonprofits in Tennessee are eligible.
Restrictions: Grants are not normally funded for annual appeals or membership drives, capital campaigns, endowment funds, operating expenses of established organizations or debt retirement, or grants to individuals.
Geographic Focus: Tennessee
Samples: Cool Springs YMCA (TN)—for a program to help women after breast cancer treatment, $5000.; Vanderbilt Ctr for Health Services (TN)—to support the Student Health Coalition, $10,000.; Wellness Community East Tennessee (TN)—to develop its Newly Diagnosed Breast Cancer Support Group, $7500.
Amount of Grant: $25,000 maximum
Contact: Patrick Clemens, (615) 467-1936; fax: (615) 342-1745; email: pat@minniepearl.org
Internet: http://www.minniepearl.org/pages/grants.html
Sponsor: Minnie Pearl Cancer Foundation
2410 Patterson Street, Suite 110
Nashville, TN 37203

Minority Doctoral Fellowships in Science and Engineering 3404
The university accepts applications from members of minority groups for doctoral fellowships in the life sciences, physical sciences, and engineering. Contact the program director for detailed information on application and amounts of fellowships offered. Annual deadline dates may vary; contact program staff for exact dates. Contact program staff for program availability.
Requirements: Citizens or permanent residents of the United States who are African American, Hispanic, or Native American may apply. The recipient must have an earned baccalaurate degree and be accepted for admission to a doctoral program in the fields of life sciences, physical sciences, engineering, or mathematics at a participating Illinois institution. Upon completion, recipients must agree to seek and accept a full-time faculty appointment at an Illinois institution of higher learning for a period equal to the number of years of the award.
Amount of Grant: $1100 monthly stipend
Date(s) Application is Due: Jan 23
Contact: Karen Jennings, Graduate School, (618) 453-4555; fax: (618) 453-4562; email: kjenning@siu.edu or gaoffice@siu.edu
Internet: http://www.siu.edu/gradschl/doctoral_fellowship.htm
Sponsor: Southern Illinois University at Carbondale
Graduate School, Woody Hall B130
Carbondale, IL 62901-4716

Minority Faculty Fellowship Program 3405

Indiana University (IU) annually seeks minorities for participation in this fellowship program. The purpose of the program is to introduce to the Bloomington campus outstanding African American, Hispanic American, and Native American scholars. Each year summer fellows are invited to teach one class during the first (six-week) or second (eight-week) summer session, and academic-year fellows teach one or more classes during the fall and spring semesters. This program provides fellows the opportunity to interact with the Bloomington faculty and to conduct research in the IU libraries and academic facilities. In the past, some IU departments and schools have offered fellows permanent tenure-track positions.

Requirements: U.S. citizens or permanent residents within one year of completion of the doctorate and those who have completed the PhD within the last four years are eligible to apply.

Amount of Grant: $4000 stipend and salary equivalent to that ordinarily paid to an IU faculty member of the same rank

Date(s) Application is Due: Nov 30

Contact: Rebecca Jones, Program Coordinator, IUB Faculty Fellows Program, (812) 855-0542; fax: (812) 856-5477; email: mffp@indiana.edu

Internet: http://www.iub.edu/~shs/ffp_overview.shtml

Sponsor: Indiana University

1021 E Third Street, Memorial Hall W 108

Bloomington, IN 47405-7005

Minority Humanities and Social Sciences Dissertation Scholar in Residence Grants 3406

The program awards residential grants to minority scholars for dissertation research appointments in African American studies, classics and world religions, economics, English, geography, history, linguistics, modern languages, philosophy, political science, psychology, social work, and sociology/anthropology. University faculty in the scholar's discipline serve as mentors and advisors while the scholar completes the doctoral dissertation. The scholar will teach one one-quarter course; make presentations of his/her research in department, college, and university colloquia; and participate in forums with undergraduates about succeeding in graduate study, etc. The scholar will receive a stipend plus benefits, office space, library privileges, and travel funds to attend a professional meeting in his/her field. Applicants should submit a full curriculum vita, a copy of the dissertation prospectus, an autobiographical sketch, a statement of purpose, graduate school transcript, and three letters of recommendation.

Requirements: U.S. citizens who are members of the following ethnic minority groups may apply: Native American Indian, Alaskan Native (Eskimo or Aleut), African American, Mexican American/Chicana/Chicano, Native Pacific Islander (Polynesian or Micronesian), and Puerto Rican. The successful candidate must be committed to a career of college or university teaching after receiving the doctorate and must be able to complete the doctoral dissertation during the academic year.

Amount of Grant: $27,500 minimum stipend plus $1000 to attend a professional meeting

Date(s) Application is Due: Jan 31

Contact: Dean, College of Arts and Sciences, (740) 593-1000

Internet: http://www.ohio.edu

Sponsor: Ohio University

Athens, OH 45701

Minority Scholar in Residence Program: Dissertation and Postdoctoral Fellowships 3407

The consortium, an association of selective undergraduate colleges, sponsors this program in an effort to encourage African American, Asian American, Hispanic American, and Native American scholars to consider teaching and research careers in small liberal arts colleges. Applications are invited for the dissertation fellowship program or the postdoctoral fellowship program in all disciplines of the liberal arts and engineering. Scholars will be expected to do some teaching and research and to participate in other activities at the host institution under the supervision of a mentor. It is hoped that the scholars will, upon completion of their residency, qualify for tenure-track appointments at the host institution or similar institution. The award consists of a stipend equal to the typical salary for beginning instructors or assistant professors. Among the member institutions are: Bowdoin College, Bryn Mawr College, Carleton College, Claremont McKenna College, Coe College, The College of Wooster, Colorado College, Denison University, DePauw University, Dickinson College, Gettysburg College, Grinnell College, Hamilton College, Haverford College, Hope college, Juniata College, Lewis and Clark College, Luther College, Macalester College, Mount Holyoke College, Oberlin College, Occidental College, Pitzer College, Pomona College, Rhodes College, Saint Olaf College, Skidmore College, Southwestern University, Swarthmore College, Trinity University the University of the South, Vassar College, Wellesley College, Wheaton College, Whitman College, and Williamette University. Guidelines and application are available online. Annual deadline dates may vary; contact program staff for exact dates.

Requirements: Scholars must be U.S. citizens working toward the PhD or the MFA degree or should have been awarded the degree no more than five years before application.

Date(s) Application is Due: Nov 15

Contact: Jean Cousins, (641) 269-3000; email: cousins@grinnell.edu

Internet: http://www.grinnell.edu/offices/dean/csmp

Sponsor: Grinnell College

1121 Park Street

Grinnell, IA 50112

Missouri Foundation for Health Grants 3408

The foundation makes grants within five priority areas: elimination of health disparities; promotion of improved access to care; support for disease prevention and health promotion; strengthening the health care safety net; and enhancing the health of children. Proposals also are accepted for activities that improve the health of the people in the foundation's service region, but are outside the scope of the five priority areas. All applicants are required to complete: a preliminary proposal application, a preliminary proposal summary, and a budget sheet (forms available online). Detailed instructions, including deadline dates, will be provided to all applicants invited to submit a full proposal.

Requirements: 501(c)3 nonprofit organizations and organizations applying through a 501(c)3 fiscal sponsor are eligible.

Restrictions: The foundation will not fund ongoing general operating expenses of established programs; unspecified capital campaigns; annual appeals or other fundraising events; existing deficits incurred outside any foundation-funded project; restoration of funding cuts by government or other organizations; endowment building or development campaigns; basic biomedical research; research on drug therapies or devices; direct support of an individual's medical care; direct support of an individual's education or training; lobbying of any kind directly supported by foundation grant funds; purchase of health insurance for individuals or groups (other than as a part of employee fringes on approved projects); religious purposes; use of foundation funds to expand existing public insurance programs; or acquisition of real estate.

Geographic Focus: Missouri

Samples: Area Resources for Community and Human Services (Saint Louis, MO)—to provide eye exams and glasses free of charge to adults, $134,883 (2005); U of Missouri at Columbia (MO)—to expand a program that recruits and trains medical students to work in rural settings, $954,598 (2005); Food Outreach (Saint Louis, MO)—for a meal program for adults and children with HIV/AIDS, $150,000 (2005); National Alliance for the Mentally Ill of Southwest Missouri (Springfield, MO)—to help low-income individuals obtain prescriptions following hospitalization, $35,992 over two years (2005).

Contact: Grants Administrator, (800) 655-5560 or (314) 345-5500; fax: (314) 345-5599; email: info@mffh.org

Internet: http://www.mffh.org/for_grantees.html

Sponsor: Missouri Foundation for Health

1000 Street Louis Union Sta, Grand Central Building, Suite 400

Saint Louis, MO 63103

Missouri Historical Society Research Fellowships 3409

The society awards one to two short-term fellowships each year to scholars wishing to research the history of Saint Louis, MO, and the settlement of the trans-Mississippi West. Scholars are expected to spend their fellowship periods in residence at the society and to make substantial use of the society's collections. The precise focus of the fellowships changes annually based on ongoing institutional priorities in research and programming. Information on the current year's fellowship is available each November. Annual deadline dates may vary; contact the society for exact dates.

Requirements: Applicant, as a minimum, must have an undergraduate degree. Graduate students and established scholars equally are encouraged to apply.

Samples: Jennifer Turner, U of Wisconsin, Madison—for the project Supporting St. Louis: Community and Agriculture in the French, Spanish, and American Regimes.; William Foley, Central Missouri State U, Warrensburg—for the project Stepping Out of the Shadow: William Clark, the Other Partner in Discovery.

Amount of Grant: $1700 monthly stipend

Date(s) Application is Due: Jan 31

Contact: John Wolford, (314) 746-4599; fax: (314) 746-4548; email: jbw@mohistory.org

Internet: http://www.mohistory.org/content/libraryandresearch/researchcenter.aspx

Sponsor: Missouri Historical Society

P.O. Box 11940

Saint Louis, MO 63112-0040

Mizuho U.S.A Foundation Grants 3410

The foundation supports education on global issues and revitalization of inner-city neighborhoods in IBJ communities. The foundation has two main funding interests: education of global citizens grants support projects and efforts to cultivate greater understanding of global issues among students in grades K-12, teachers, and families. Projects that encourage and enhance students' knowledge, as well as programs that provide professional development for educators, are considered. Strengthening urban communities grants aim to improve inner-city neighborhoods by supporting housing, economic development, and neighborhood revitalization projects. Funding is given to organizations and programs that offer volunteer opportunities to IBJ employees; demonstrate creative and collaborative efforts among the public, private, and nonprofit sectors; and serve as replicable models for other organizations and programs. Preliminary requests are reviewed throughout the year.

Requirements: 501(c)3 tax-exempt organizations in New York, NY; Los Angeles and San Francisco, CA; Chicago, IL; Atlanta, GA; and Houston, TX, are eligible.

Restrictions: The foundation does not support individuals; religious, fraternal, veterans, labor, or political organizations; fund-raising activities; journal advertisements or ticket purchases; capital campaigns; or building construction/renovation.

Geographic Focus: California; Georgia; Illinois; New York; Texas

Samples: Low Income Investment Fund (New York, NY)—for the New York Child Care Seed Fund, which works to increase access to high-quality child-care and early-childhood education programs, $20,000 (2005); Local Initiatives Support Corp (New York,

NY)—to provide technical and financial assistance to community development groups in New York seeking to upgrade their information-technology and telecommunications infrastructure, $30,000 (2005); Highbridge Community Life Ctr (New York, NY)—to expand and improve a program that trains low-income workers to become certified nurse aides and helps graduates of the program advance in their profession, $10,000 (2005); Common Ground Community (New York, NY)—for its Neighborhood-Based Homelessness Prevention program, $20,000 (2005).
Amount of Grant: $326,682 total
Contact: Lesley Harris Palmer, Executive Director, (212) 282-4192; fax: (212) 282-3250; email: mizuho.usa.foundation@mizuhocbus.com
Sponsor: Mizuho U.S.A Foundation
1251 Avenue of the Americas, 31st Fl
New York, NY 10020-1104

MLA Aldo and Jeanne Scaglione Literary Translation Prize **3411**
The prize is awarded in even-numbered years for an outstanding translation into English of a book-length literary work, and in odd-numbered years for a translation of a book-length work of literary history, literary criticism, philology, or literary theory. For each year's prize, books published during the two preceding years are eligible to compete. Translators need not be members of MLA.
Samples: Joseph Solodow, Southern Connecticut State U (CT)—for Latin Literature: A History, by Gian Biagio Conte, $1000.
Amount of Grant: $2000
Date(s) Application is Due: Apr 1
Contact: Office of Special Projects, (646) 576-5141; fax: (646) 458-0030; email: awards@mla.org
Internet: http://www.mla.org/prizeinfo_translit
Sponsor: Modern Language Association of America
26 Broadway, 3rd Fl
New York, NY 10004-1789

MLA Aldo and Jeanne Scaglione Prize for a Translation of a Scholarly Study of Literature **3412**
The prize is awarded each odd-numbered year for a translation of a book-length work of literary history, literary criticism, philology, or literary theory. For each year's prize, books published during the two preceding years are eligible. Translators need not be members of MLA.
Amount of Grant: $2000
Date(s) Application is Due: May 1
Contact: Office of Special Projects, (646) 576-5141; fax: (646) 458-0030; email: awards@mla.org
Internet: http://www.mla.org/resources/awards/awards_submissions/awards_competitions/biennial_prizes2003/prizeinfo_trans
Sponsor: Modern Language Association of America
26 Broadway, 3rd Fl
New York, NY 10004-1789

MLA Aldo and Jeanne Scaglione Prize for Comparative Literary Studies **3413**
The annual prize will be awarded for an outstanding scholarly work in the field of comparative literary studies involving at least two literatures Works of literary history, literary criticism, philology, and literary theory are eligible, as are works dealing with literature and other arts and disciplines, including cinema. Publishers may enter more than one title, but no book may compete for more than one MLA prize.
Requirements: The committee solicits entries of books written by current members of the association that were published in the preceding year. To enter a book into competition, send four copies and a letter identifying the work and confirming the author's membership in MLA. Entries may be sent at any time but must be received by the listed application deadline date.
Restrictions: Books that are primarily translations will not be considered.
Amount of Grant: $2000
Date(s) Application is Due May 1
Contact: Office of Special Projects, (646) 576-5141; fax: (646) 458-0030; email: awards@mla.org
Internet: http://www.mla.org/resources/awards/awards_submissions/awards_competitions/prizes_annual2003/prizeinfo_complit
Sponsor: Modern Language Association of America
26 Broadway, 3rd Fl
New York, NY 10004-1789

MLA Aldo and Jeanne Scaglione Prize for French and Francophone Studies **3414**
Nominations are invited for this prize, to be awarded for an outstanding scholarly work in the field of French or Francophone linguistic or literary studies. Works of literary history, literary criticism, philology, and literary theory are eligible for consideration. Publishers may enter more than one title, but no book may compete for more than one MLA prize.
Requirements: To enter a book into competition, send four copies and a letter identifying the work and confirming the author's membership in the MLA. Entries may be sent at any time, but all must be received by the listed application deadline date.
Restrictions: Books that are primarily translations will not be considered.
Amount of Grant: $2000
Date(s) Application is Due: May 1
Contact: Office of Special Projects, (646) 576-5141; fax: (646) 458-0030; email: awards@mla.org
Internet: http://www.mla.org/resources/awards/awards_submissions/awards_competitions/prizes_annual2003/prizeinfo_french
Sponsor: Modern Language Association of America
26 Broadway, 3rd Fl
New York, NY 10004-1789

MLA Aldo and Jeanne Scaglione Prize for Germanic Languages and Literatures **3415**
The prize is awarded biennially in even-numbered years for an outstanding scholarly work on the linguistics or literatures of the Germanic languages, including Danish, Dutch, German, Icelandic, Norwegian, Swedish, and Yiddish. Works of literary history, literary criticism, philology, and literary theory are eligible for consideration. Publishers may enter more than one title, but no book may compete for more than one MLA prize.
Requirements: The committee solicits entries of books written by current members of the association that were published in the last two years. To enter a book into competition, send four copies and a letter identifying the work and confirming the author's membership in the MLA. Entries may be sent at any time but must be received by the listed application deadline date.
Restrictions: Books that are primarily translations will not be considered.
Amount of Grant: $2000
Date(s) Application is Due: May 1
Contact: Office of Special Projects, (646) 576-5141; fax: (646) 458-0030; awards@mla.org
Internet: http://www.mla.org/resources/awards/awards_submissions/awards_competitions/prizes_biennial2004/prizeinfo_germanic
Sponsor: Modern Language Association of America
26 Broadway, 3rd Fl
New York, NY 10004-1789

MLA Aldo and Jeanne Scaglione Prize for Italian Studies **3416**
The Committee on Honors and Awards of MLA invites submissions for the joint competition for these biennial prizes for a scholarly study of book or essay length on any phase of Italian literature or comparative literature involving Italian, published during the two years preceding the even-numbered year in which the prizes are awarded. The awards will be announced and presented at the annual convention in December. Applications are accepted in even-numbered years.
Requirements: To enter a work into competition, send four copies to the Aldo and Jeanne Scaglione Prize for Italian Literary Studies at MLA. Under separate cover, send a letter indicating the title submitted, the author, the date of publication, and affirmation that the author is a member in good standing of the association.
Samples: Robert Dombroski, City U of New York and U of Connecticut, Storrs—for Properties of Writing Ideological Discourse in Modern Italian Fiction, $1000.
Amount of Grant: $2000
Date(s) Application is Due: May 1
Contact: Office of Special Projects, (646) 576-5141; fax: (646) 458-0030; email: awards@mla.org
Internet: http://www.mla.org/resources/awards/awards_submissions/awards_competitions/biennial_prizes2003/prizeinfo_italian
Sponsor: Modern Language Association of America
26 Broadway, 3rd Fl
New York, NY 10004-1789

MLA Aldo and Jeanne Scaglione Prize for Studies in Slavic Languages and Literatures **3417**
Authors are invited to compete for this prize, which is awarded biennially in odd-numbered years for an outstanding scholarly work on the linguistics or literatures of the Slavic languages. Works of literary history, literary criticism, philology, and literary theory are eligible for consideration. The prize will be presented to the winning author at the association's annual convention. To enter a book into competition, send four copies and a letter identifying the work. Publishers may enter more than one title, but no book may compete for more than one MLA prize. Entries may be submitted at any time but must be received by the listed application deadline date.
Requirements: Books are eligible if published in the two years preceding the year in which the award is given.
Amount of Grant: $2000
Date(s) Application is Due: May 1
Contact: Office of Special Projects, (646) 576-5141; fax: (646) 458-0030; email: awards@mla.org
Internet: http://www.mla.org/resources/awards/awards_submissions/awards_competitions/biennial_prizes2003/prizeinfo_slavic
Sponsor: Modern Language Association of America
26 Broadway, 3rd Fl
New York, NY 10004-1789

MLA Cunningham Memorial International Fellowship **3418**
The fellowship supports the education and training of a medical librarian from a country outside the United States or Canada and includes a stipend plus approved travel for four months in either country. The candidate must be guaranteed a position in a medical library upon return to his/her country. Contact the office for application forms.
Requirements: Candidates must have an undergraduate degree and a library degree (latter requirement may be waived) and must be working in a medical library or preparing to work in one in their country. In addition, a satisfactory score must be achieved on the TOEFL English competency examination. The fellowship recipient must agree to return to his or her country and work in a medical library for a period of two years.

Restrictions: Fellowship is not available to U.S. or Canadian citizens. Travel arrangements to and from the United States or Canada must be paid by the fellow.
Amount of Grant: $6000 plus $2000 for travel
Date(s) Application is Due: Dec 1
Contact: Lisa Fried, (312) 419-9094 ext 28; fax: (312) 419-8950; email: mlapd2@mlahq.org
Internet: http://www.mlanet.org/awards/grants/cunningham.html
Sponsor: Medical Library Association
65 E Wacker Pl, Suite 1900
Chicago, IL 60601-7298

MLA Doctoral Fellowships **3419**
The award is intended to foster and encourage superior students to conduct doctoral research in some area of medical librarianship or information science. The one-year, nonrenewable award supports research or travel applicable to the candidate's study within a 12-month period; it may not be used for tuition.
Requirements: The candidate must have a master's degree from an ALA-accredited school of library science in the United States or the equivalent elsewhere and must be a candidate for a PhD in a doctoral program with emphasis on biomedical and health-related information science.
Restrictions: A past recipient of any MLA doctoral fellowship is ineligible to apply.
Amount of Grant: $2000 maximum
Date(s) Application is Due: Dec 1
Contact: Lisa Fried, (312) 419-094 ext 28; email: mlapd2@mlahq.org
Internet: http://www.mlanet.org/awards/grants/doctoral.html
Sponsor: Medical Library Association
65 E Wacker Pl, Suite 1900
Chicago, IL 60601-7298

MLA Estelle Brodman Academic Medical Librarian of the Year Award **3420**
The award recognizes an academic medical librarian who at midcareer demonstrates both significant achievement and potential for leadership and continuing excellence.
Requirements: Nominees must be members of MLA and academic medical librarians who have worked in academic medical libraries for at least five years preceding the award.
Amount of Grant: $1000
Date(s) Application is Due: Nov 1
Contact: Lisa Fried, (312) 419-9094 ext 28; email: mlapd2@mlahq.org
Internet: http://www.mlanet.org/awards/honors/brodman.html
Sponsor: Medical Library Association
65 E Wacker Pl, Suite 1900
Chicago, IL 60601-7298

MLA First Book Prize **3421**
The prize is awarded annually for the first book-length publication of a current member of the association. To qualify for the prize, a book must be a literary or linguistic study, a critical edition of an important work, or a critical biography. Studies dealing with literary theory, media, cultural history, and interdisciplinary topics are eligible; books that are primarily translations will not be considered. The award, which consists of a monetary prize and a certificate, will be presented to the winning author at the association's annual convention. Six copies of each entry should be submitted.
Requirements: An author who is not a member of the MLA at the time of publication may become eligible to compete by joining the association prior to May 1 of the following year.
Samples: Deidre Shauna Lynch, State U of New York (Buffalo, NY)—for the Economy of Character: Novels Market Culture, and the Business of Inner Meaning, $1000 .; Katie Trumpener, U of Chicago (IL)—for Bardic Nationalism: The Romantic Novel and the British Empire, $1000 .; Marc Redfield, Claremont Graduate U—for Phantom Formations: Aesthetic Ideology and the Bildungsroman, $1000.; Elaine Hadley, U of Chicago (IL)—for Melodramatic Tactics: Theatricalized Dissent in the English Marketplace, 1800-1885, $1000.
Amount of Grant: $1000
Date(s) Application is Due: Apr 1
Contact: Office of Special Projects, (646) 576-5141; fax: (646) 458-0030; email: awards@mla.org
Internet: http://www.mla.org/resources/awards/awards_submissions/awards_competitions/prizes_annual2003/prizeinfo_firstbook
Sponsor: Modern Language Association of America
26 Broadway, 3rd Floor
New York, NY 10004-1789

MLA Frank Bradway Rogers Information Advancement Award **3422**
The award is presented to an MLA member (or small group of members) who has made an outstanding contribution to the application of technology in the delivery of health sciences information, to the science of information, or to facilitate the delivery of health sciences information.
Requirements: Applicants must be health science librarians.
Date(s) Application is Due: Nov 1
Contact: Lisa Fried, (312) 419.9094 ext 28; email: mlapd2@mlahq.org
Internet: http://www.mlanet.org/awards/honors/rogers.html
Sponsor: Medical Library Association
65 E Wacker Pl, Suite 1900
Chicago, IL 60601-7298

MLA Graduate Scholarships **3423**
Scholarships are granted to provide graduate students the opportunity to study or continue studies in graduate medical librarianship. The scholarship is announced at the annual meeting of the association, where the recipient will also receive a student membership in MLA.
Requirements: Applicant must be a U.S. or Canadian citizen and be entering an ALA-accredited graduate library school or have at least one-half of the academic requirements to complete during the year following the granting of the scholarship.
Amount of Grant: $5000 maximum
Date(s) Application is Due: Dec 1
Contact: Professional Development Department, (312) 419-9094 ext 28; fax: (312) 419-8950; email: mlapd2@mlahq.org
Internet: http://www.mlanet.org/awards/grants/scholar.html
Sponsor: Medical Library Association
65 E Wacker Pl, Suite 1900
Chicago, IL 60601-7298

MLA Graduate Scholarships for Minority Students **3424**
Scholarships are granted to provide students from minority groups the opportunity to study or continue studies in medical librarianship. The scholarship will be granted to a student entering or continuing the study of health sciences librarianship at the master's level.
Requirements: Applicant must be a member of a minority group (African American, Hispanic, Asian, Native American, or Pacific Islander), be a U.S. or Canadian citizen, and be entering an ALA-accredited graduate library school or have at least one-half of the academic requirements to complete during the year following the granting of the scholarship.
Amount of Grant: $5000 maximum
Date(s) Application is Due: Dec 1
Contact: Lisa Fried, (312) 419-9094 ext 28; email: mlapd2@mlahq.org
Internet: http://www.mlanet.org/awards/grants/minstud.html
Sponsor: Medical Library Association
65 E Wacker Pl, Suite 1900
Chicago, IL 60601-7298

MLA James Russell Lowell Prize **3425**
The Committee on Honors and Awards of MLA invites authors to compete for this annual prize. Eligible for consideration are books published within the previous year by members of the association that meet one of the following criteria: an outstanding literary or linguistic study, a critical edition of an important work, or a critical biography. Studies dealing with literary theory, media, cultural history, and interdisciplinary topics are eligible; books that are primarily translations are ineligible. The award will be announced and presented at the annual convention in December.
Requirements: To enter a work into competition, send six copies and a letter indicating the title submitted, the author, the date of publication, and affirmation that the author is a member in good standing of MLA.
Samples: Gauri Viswanathan, Columbia U (NY)—for Outside the Fold: Conversion, Modernity, and Belief, $1000.; David Wallace, U of Pennsylvania (PA)—for Chaucerian Polity: Absolutist Lineages and Associational Forms in England and Italy, $1000.
Amount of Grant: $1000
Date(s) Application is Due: Mar 1
Contact: Office of Special Projects, (646) 576-5141; fax: (646) 458-0030; email: awards@mla.org
Internet: http://www.mla.org/resources/awards/awards_submissions/awards_competitions/prizes_annual2003/awardinfo_lowell
Sponsor: Modern Language Association of America
26 Broadway, 3rd Fl
New York, NY 10004-1789

MLA Janet Doe Lectureship Award **3426**
The Janet Doe Lecturer is an individual chosen annually by MLA for his/her unique perspective on the history or philosophy of medical librarianship. The award recipient delivers a lecture at the MLA annual meeting.
Requirements: Nominees should be regular members of the association who have been active members of the profession a sufficient number of years to have acquired a broad perspective of medical librarianship and possess the qualities to communicate their experiences and ideas articulately.
Date(s) Application is Due: Nov 1
Contact: Lisa Fried, (312) 419.9094 ext 28; email: mlapd2@mlahq.org
Internet: http://www.mlanet.org/awards/honors/doe.html
Sponsor: Medical Library Association
65 E Wacker Pl, Suite 1900
Chicago, IL 60601-7298

MLA Kenneth W. Mildenberger Prize **3427**
The Committee on Honors and Awards of MLA invites submissions for this annual prize for an outstanding research publication in the field of teaching foreign language and literature. The prize will be awarded in even-numbered years for a book published the previous year and in odd-numbered years for an article published in a journal. The selection committee will look for evidence of fresh and effective approaches to teaching and for works that are likely to be useful on a wide scale. The award, which includes a one-year membership to MLA, will be announced and presented at the association's annual convention in December.

Requirements: To enter works into competition, send four copies of each work and a letter indicating the titles submitted, the authors, and the dates of publication. Authors of works submitted need not be members of the association.
Samples: Evelyn Hatch, U of California, Los Angeles, and Cheryl Brown, Brigham Young U—for Vocabulary, Semantics, and Language Education, $500.
Amount of Grant: $1000 award-winning book; $500 award-winning article
Date(s) Application is Due: May 1
Contact: Office of Special Projects, (646) 576-5141; email: awards@mla.org
Internet: http://www.mla.org/resources/awards/awards_submissions/awards_competitions/prizes_annual2003/prizeinfo_kwm
Sponsor: Modern Language Association of America
26 Broadway, 3rd Fl
New York, NY 10004-1789

MLA Mina P. Shaughnessy Prize **3428**
The Committee on Honors and Awards of MLA invites submissions for this annual award for an outstanding research publication in the field of teaching English language and literature. The prize will be awarded to an author for a book published the previous year. The selection committee will look for evidence of fresh and effective approaches to teaching and for works that are likely to be useful on a wide scale. The award will be presented at the annual convention in December.
Requirements: To enter works in the competition, send four copies of each work and a letter indicating the titles submitted, the authors, and the dates of publication. Authors of works submitted need not be members of the association.
Samples: Ross Talarico, National U—for Spreading the Word: Poetry and the Survival of Community in America, $1000.; A. Suresh Canagarajah, Oxford U—for Linguistic Imperialism in English Teaching Resisting Linguistic, $1000.
Amount of Grant: $1000
Date(s) Application is Due: May 1
Contact: Special Projects, (646) 576-5141; fax: (646) 458-0030; email: awards@mla.org
Internet: http://www.mla.org/resources/awards/awards_submissions/awards_competitions/prizes_annual2003/prizeinfo_mps
Sponsor: Modern Language Association of America
26 Broadway, 3rd Fl
New York, NY 10004-1789

MLA Murray Gottlieb Prize Essay Award **3429**
The prize is awarded annually for the best unpublished essay on the history of medicine and allied sciences written by a health sciences librarian. The purpose of the award is to recognize and stimulate the health sciences librarian's interest in the history of medicine.
Requirements: Authors of papers submitted must be health sciences librarians.
Restrictions: Papers may be under consideration for publication at the time of submission but cannot have been published.
Amount of Grant: $100
Date(s) Application is Due: Nov 1
Contact: Lisa Fried, (312) 419-9094 ext 28; email: mlapd2@mlahq.org
Internet: http://www.mlanet.org/awards/honors/gottlieb.html
Sponsor: Medical Library Association
65 E Wacker Pl, Suite 1900
Chicago, IL 60601-7298

MLA Prize for Independent Scholars **3430**
To recognize and further encourage the achievements and contributions of independent scholars, the MLA invites authors to compete for this annual prize for a scholarly book in the fields of modern languages and literatures, including English. The book must have been published in the preceding calendar year. The award is given to a person who, at the time of publication of the book for which the award is given, has received his/her terminal academic degree no fewer than four years earlier, and does not hold a tenured or tenure-accruing position in a postsecondary educational institution. Application forms are available upon request.
Requirements: To enter a book into competition, send six copies and a completed application form to MLA Prize for Independent Scholars. Authors of works submitted need not be members of the association.
Samples: Janet Galligani Casey, Wesleyan College—for Dos Passos and the Ideology of the Feminine, $1000.; Gary Schmidgall (New York, NY)—for Walt Whitman: A Gay Life, $1000.; Graham Robb (Oxford, England)—for Unlocking Mallarme, $1000.; Nora Sayre—for Previous Convictions: A Journey through the 1950s, $1000.
Amount of Grant: $1000 and one year of membership in MLA
Date(s) Application is Due: May 1
Contact: Office of Special Projects, (646) 576-5141; fax: (646) 458-0030; email: awards@mla.org
Internet: http://www.mla.org/resources/awards/awards_submissions/awards_competitions/prizes_annual2003/prizeinfo_indsch
Sponsor: Modern Language Association of America
26 Broadway, 3rd Fl
New York, NY 10004-1789

MMA Andrew W. Mellon Fellowships in Conservation **3431**
Annual conservation fellowships, funded by the Andrew W. Mellon Foundation, are awarded for training in one or more of the following departments of the Metropolitan Museum: paintings conservation, objects conservation (including sculpture, metalwork, glass, ceramics, furniture, and archaeological objects), musical instruments, arms and armor, paper conservation, textile conservation, the costume institute, and Asian art conservation. Each fellowship will typically be one year in duration with the possibility of renewal for up to two additional years. Fellowship recipients will receive stipends appropriate to their level of training or experience. Additional funds may be available for travel, books, photographs, and other reasonable or necessary expenses incident to the fellowship.
Requirements: Fellowship applicants should already have reached an advanced level of experience or training; recipients will be expected to spend the duration of the fellowships in residence in the departments with which they are affiliated. It is the responsibility of the applicant to discuss his/her application in detail with the department concerned and to obtain that department's approval prior to submitting an application. Such approval should not, however, be construed as assurance that a fellowship will be awarded.
Amount of Grant: $30,000; $5000 for travel
Date(s) Application is Due: Jan 6
Contact: Marcie Karp, Coordinator for Fellowships, fax: (212) 396-5168; email: marcie.karp@metmuseum.org or education.grants@metmuseum.org
Internet: http://www.metmuseum.org/education/conserve_application.html
Sponsor: Metropolitan Museum of Art
1000 Fifth Avenue
New York, NY 10028-0198

MMA Bothmer Fellowship **3432**
The annual fellowship is awarded to an outstanding graduate student who has been admitted to a doctoral program of a university in the United States and who has submitted an outline of a thesis dealing with Greek and Roman art. The thesis outline must have already been accepted by the applicant's thesis advisor at the time of the fellowship application. Preference will be given to the applicant who, in the opinion of the grants committee, will profit most from utilizing the resources of the Department of Greek and Roman Art, its collections, library, photographic and other archives, and the guidance of the curatorial staff.
Requirements: It is the responsibility of the applicant, in connection with any project that would require assistance from a particular department, to discuss the project with the department concerned and to obtain that department's approval prior to submitting an application; however, departmental approval should not be construed as assurance of a fellowship.
Restrictions: Fellowships generally cannot be given for projects proposing that an exhibition be organized and presented within the fellowship period.
Amount of Grant: $40,000 for senior fellows, $30,000 for predoctoral fellows; $5000 for travel
Date(s) Application is Due: Nov 4
Contact: Marcie Karp, Coordinator for Fellowships, Office of Academic Programs, (212) 650-2763; fax: (212) 396-5168; email: education.grants@metmuseum.org
Internet: http://www.metmuseum.org/education/fellowship.html
Sponsor: Metropolitan Museum of Art
1000 Fifth Avenue
New York, NY 10028-0198

MMA Charitable Frohlich Fellowship **3433**
This two-year fellowship is available to promote art conservation. The value of the award varies according to applicant's level of training or experience, with the possibility of additional funds available for travel, books, photographs, and other expenses. Fellowships begin on September 1.
Requirements: Competition is open to conservators, art historians, and scientists who are at advanced levels in their training and have demonstrated commitment to the physical examination and treatment of objects of art. It is the responsibility of the applicant to discuss her or his application in detail with the department and to obtain that department's approval to submit the application.
Date(s) Application is Due: Jan 6
Contact: Marcie Karp, Coordinator for Fellowships, Office of Academic Programs, (212) 650-2763; fax: (212) 396-5168; email: education.grants@metmuseum.org
Internet: http://www.metmuseum.org/education/conservation.html
Sponsor: Metropolitan Museum of Art
1000 Fifth Avenue
New York, NY 10028-0198

MMA Chester Dale Fellowships **3434**
Individuals whose fields of study are related to the fine arts of the Western world and who are, preferably, American citizens under the age of 40 are eligible for Chester Dale fellowships for research at the Metropolitan Museum of Art. Grants typically cover periods from three months to a year.
Requirements: It is the responsibility of the applicant, in connection with any project that may reasonably be expected to require assistance from a particular museum department, to discuss the project with the department concerned and to obtain that department's approval before submitting an application. However, such approval should not be construed as assurance of being awarded a fellowship.
Restrictions: Fellowships generally cannot be given for projects proposing that an exhibition be organized and presented within the fellowship period.
Amount of Grant: $40,000 for senior fellows, $30,000 for predoctoral fellows; plus $5000 for travel
Date(s) Application is Due: Nov 4
Contact: Marcie Karp, Coordinator for Fellowships, Office of Academic Programs, (212) 650-2763; fax: (212) 570-3972; email: education.grants@metmuseum.org
Internet: http://www.metmuseum.org/education/fellowship.html

Sponsor: Metropolitan Museum of Art
1000 Fifth Avenue
New York, NY 10028-0198

MMA J. Clawson Mills Scholarships 3435

These scholarships are awarded for one year of study or research at the Metropolitan Museum in any branch of the fine arts relating to the museum's collections and carry the possibility of renewal for a second year. It is the responsibility of the applicant to discuss the intended project with the department concerned and to obtain that department's approval before submitting an application.
Requirements: Mature scholars of demonstrated ability who are interested in pursuing research projects relevant to the museum's collections are eligible to apply.
Amount of Grant: $40,000 for senior fellows, $30,000 for predoctoral fellows; plus $5000 for travel
Date(s) Application is Due: Nov 4
Contact: Marcie Karp, Coordinator for Fellowships, Office of Academic Programs, (212) 650-2763; fax: (212) 570-3972; email: education.grants@metmuseum.org
Internet: http://www.metmuseum.org/education/fellowship.html#clawson
Sponsor: Metropolitan Museum of Art
1000 Fifth Avenue
New York, NY 10028-0198

MMA Jane and Morgan Whitney Fellowships 3436

These fellowships are awarded for study, work, or research to students of the fine arts whose fields are related to the museum's collections, with preference given to students in the decorative arts who are under 40 years of age. The fellowship carries the possibility of renewal for one additional year.
Requirements: Applicants must discuss the proposed project with the department concerned and obtain approval before submitting the application.
Amount of Grant: $40,000 for senior fellows, $30,000 for predoctoral fellows; $5000 for travel
Date(s) Application is Due: Nov 4
Contact: Marcie Karp, Coordinator for Fellowships, Office of Academic Programs, (212) 650-2763; fax: (212) 396-5168; email: education.grants@metmuseum.org
Internet: http://www.metmuseum.org/education/fellowship.html
Sponsor: Metropolitan Museum of Art
1000 Fifth Avenue
New York, NY 10028-0198

MMA Lifchez/Stronach Curatorial Internship 3437

This internship will be awarded to a student who has recently graduated from college or is enrolled in an art history MA program and would use the internship to continue in a career in art history. The student should come from a background of financial need or other disadvantage that might jeopardize her or his pursuing such a career without this support. The intern will be assigned to a curatorial department to work on projects appropriate to her/his academic background, professional skills, and career goals. The internship is full-time from September to June. Interns are invited to participate in orientation activities during the summer. Annual deadline dates vary; contact the program office for exact dates.
Amount of Grant: $15,000 honorarium
Date(s) Application is Due: Jan 21
Contact: Coordinator for Fellowships, Office of Academic Programs, (212) 570-3710; email: mmainterns@metmuseum.org
Internet: http://www.metmuseum.org/education/er_internship.asp
Sponsor: Metropolitan Museum of Art
1000 Fifth Avenue
New York, NY 10028-0198

MMA Polaire Weissman Fund Fellowships 3438

This fund provides fellowships to qualified graduate students who preferably will have completed graduate studies in the fine arts or studies in costume and who are interested in pursuing costume history in museum or teaching careers or other careers including conservation related to the field of costume. These fellowships, generally awarded for nine-month terms, are offered in alternate years.
Requirements: Applicants must discuss the proposed project with the department concerned and obtain approval before submitting the application.
Amount of Grant: $40,000 for senior fellows, $30,000 for predoctoral fellows; $5000 for travel
Date(s) Application is Due: Nov 4
Contact: Marcie Karp, Coordinator for Fellowships, Office of Academic Programs; fax: (212) 650-396-5168; email: education.grants@metmuseum.org
Internet: http://www.metmuseum.org/education/fellowship.html
Sponsor: Metropolitan Museum of Art
1000 Fifth Avenue
New York, NY 10028-0198

MMA Summer Internship Program 3439

The internship program is for students and recent graduates interested in careers in art museums. Based upon their academic training, interests, and the availability of projects, interns work in one of the museum's departments—curatorial, education, conservation, administration, or library. All internships are nine to 10 weeks, full-time, five-day, 35 hours per week, and include a one-week orientation for graduate students and a two-week orientation for college students. The application deadlines for the internship for College Students and the internship for Graduate Students fall in mid- to late-January, respectively; call for exact dates. The deadline for the Cloisters Summer internship program for college students is the first Friday in February. For more information on the Cloisters internship, contact College Internship Program, the Cloisters, Fort Tryon Park, New York, NY 10040; or call (212) 650-2280. Contact the museum for information and application deadline for the Roswell L. Gilpatric internship. Applicants of diverse backgrounds are encouraged to apply.
Requirements: Most projects require strong preparation in the history of art.
Amount of Grant: $3000 for college students and recent graduates; $3250 for graduate students
Date(s) Application is Due: Jan 21
Contact: Marcie Karp, Coordinator for Fellowships, Office of Academic Programs, (212) 650-2763; fax: (212) 570-3972; email: mmainterns@metmuseum.org
Internet: http://www.metmuseum.org/education/er_internship.asp#sum
Sponsor: Metropolitan Museum of Art
1000 Fifth Avenue
New York, NY 10028-0198

MMA Theodore Rousseau Fellowships 3440

Fellowships are for the training of students planning to enter museums as curators of painting, by enabling them to undertake related study in Europe. The purpose of these fellowships is to develop the skills of connoisseurship by supporting firsthand examination of paintings in major European collections, rather than to support library research for the completion of degree requirements. Short-term fellowships of at least three months will be considered along with 12-month requests.
Requirements: Applicants should have been enrolled for at least one year in advanced degree programs in the field of art history. It is the responsibility of the applicant to discuss the project with the department concerned and to obtain the department's approval before submitting the application.
Amount of Grant: $40,000 for senior fellows, $30,000 for predoctoral fellows; $5000 for travel
Date(s) Application is Due: Nov 4
Contact: Marcie Karp, Coordinator for Fellowships, Office of Academic Programs, (212) 650-2763; fax: (212) 396-5168; email: education.grants@metmuseum.org
Internet: http://www.metmuseum.org/education/fellowship.html
Sponsor: Metropolitan Museum of Art
1000 Fifth Avenue
New York, NY 10028-0198

MMRF Research Grants 3441

The foundation's mission is to accelerate a cure for multiple myeloma. The program supports three categories of research grants. Fellows awards support researchers entering the field of multiple myeloma working under the direction of a research sponsor (March 1 deadline). Senior research awards support investigators with an interest in myeloma who have been working in cancer research for a minimum of five years (May 2 deadline). Collaborative programs grants support multidisciplinary research in multiple myeloma. Cohesive and synergistic grant proposals that involve several research projects across one or more scientific core laboratories are sought (preliminary deadline for submitting a Collaborative Program Grant application is June 15, final application deadline is September 15).
Samples: Johns Hopkins U (Baltimore, MD)—for research on immunotherapy of myeloma at the Sidney Kimmel Comprehensive Cancer Ctr, $1.5 million (2004).
Amount of Grant: $50,000 fellows awards; $100,000 senior research awards; $1.5 million collaborative program grants
Contact: Jill Shook, Grants Administrator, (203) 972-1250; fax: (203) 972-1259; email: shookj@themmrf.org
Internet: http://www.multiplemyeloma.org/research
Sponsor: Multiple Myeloma Research Foundation
3 Forest Street
New Canaan, CT 06840

MNI Jeanne Timmins Fellowships 3442

The institute offers annual postdoctoral fellowships to support research and study in neurology, neurosurgery, or neuroscience for a period of one year at the institute. One renewal is possible. An application form may be requested in writing. Successful applicants will be notified by letter.
Requirements: Candidates must have MD or PhD degrees; those with MDs will ordinarily have completed clinical studies in neurology or neurosurgery.
Amount of Grant: $C25,000 maximum
Date(s) Application is Due: Oct 31
Contact: Assistant to the Director, (514) 398-1903; fax: (514) 398-8248; email: john@mni.lan.mcgill.ca
Internet: http://www.mni.mcgill.ca/MNIfellowships.html
Sponsor: Montreal Neurological Institute
3801 University Street, Room 636
Montreal, PQ H3A 2B4 Canada

Molson Companies Donations Fund Grants 3443

The corporation supports Canadian charitable organizations in its efforts to meet the changing needs of Canadian society. Areas of interest include the visual and performing arts, historic preservation, literacy, universities and community colleges, education research, basic science research, hospitals, the environment, and community and social services. Funding decisions are made throughout the year. Applicants are requested to

apply by letter and to include descriptive material outlining activities of the organization and reason for the request, copy of latest annual report and/or audited financial statements, and charitable registration number. Decisions are made throughout the year.
Requirements: Organizations must be registered and provide a valid Canadian charitable registration number to be eligible.
Contact: Program Contact, (416) 975-1786; fax: (416) 975-1981
Internet: http://www.molson.com/en/community/donation/index.php?sec=5&com=3&mdf=1
Sponsor: Molson Companies
33 Carlingview Drive
Toronto, ON M9W 5E4 Canada

Monfort Family Foundation Grants 3444
The foundation awards general operating grants to Colorado nonprofit organizations in its areas of interest, including education, health and medical research, arts and culture, agriculture, and general charitable giving. Application forms must be obtained.
Requirements: Colorado nonprofits, primarily in Weld County, are eligible.
Restrictions: Grants are not made to individuals.
Geographic Focus: Colorado
Samples: U of Northern Colorado, College of Business Administration (Greeley, CO)—to bring nationally prominent business leaders to campus to teach; to create a technology center; and to establish an entrepreneurship institute, a lecture series, and an instruction-improvement fund, $10.5 million.
Amount of Grant: $5000-$100,000 average range
Date(s) Application is Due: May 1; Oct 1
Contact: Dave Evans, Program Contact, (970) 454-1357
Sponsor: Monfort Family Foundation
Box 337300
Greeley, CO 80633

Montana Arts Council Cultural and Aesthetic Project Grants 3445
Applications must be for cultural and aesthetic projects in Montana, including, but not limited to, the visual, performing, literary, and media arts; history; archaeology; folklore; archives; collections; research; historic preservation; and the construction or renovation of cultural facilities. Applications are encouraged for projects serving rural communities, racial and ethnic groups, the disabled, institutionalized populations, youth and the aging, and projects addressing the role of women in Montana's cultural life. Grant categories include special project ($4500 or less), operational support, capital expenditure, and challenge grants for endowment development. Generally, each dollar in special projects grants is to be matched in cash or in-kind goods and services. Capital expenditures require three dollars or in-kind goods and services for each dollar granted. Annual deadline dates may vary; contact program staff for exact dates.
Requirements: Any person, association, or representative of a governing unit in Montana including, but not limited to, state, regional, county, city, town, or Indian tribal governments may submit an application. Examples of eligible applicants are county art or historical museums, public libraries, public educational institutions or school districts, state agencies, city arts commissions or parks and recreation departments, and tribal cultural or education committees.
Geographic Focus: Montana
Date(s) Application is Due: Aug 1
Contact: Program Contact, (406) 444-6430; fax: (406) 444-6548; email: mac@mt.gov
Internet: http://www.art.state.mt.us/orgs/orgs_ca.asp
Sponsor: Montana Arts Council
P.O. Box 202201, 316 N Park Avenue, Suite 252
Helena, MT 59620

Moody Foundation Grants 3446
Funds are awarded in the areas of education; arts, humanities, and religion; health and sciences; and community and social services, with primary emphasis on foundation-initiated projects in Galveston, TX. Grants are awarded for seed money, building funds, equipment, consulting services, technical assistance, matching funds, special projects, research, publications, conferences, and seminars. Applications are accepted at any time but must be received six weeks prior to each of four meetings every year.
Requirements: Grants are limited to Texas nonprofit organizations.
Restrictions: Grants are not awarded to individuals (except for students covered by one scholarship program in Galveston County), for continuing support, annual campaigns, deficit financing, or for operating budgets except for start-up purposes.
Geographic Focus: Texas
Samples: American Cancer Society, Inc./Cattle Baron's Ball (Dallas, TX)—to assist funding cancer research projects in Dallas County, $30,000 (2005); Communities in Schools (Dallas, TX)—for Bright Kids Reading Improvement Program for lower-elementary reading and financial literacy, $15,000 (2005); Galveston Independent School District, Educational Foundation, Inc. (Galveston, TX)—for assistance with the Good Start Early Childhood Preschool Program, $50,000 (2005).
Amount of Grant: $11 million total
Contact: Peter Moore, Director, Grants, (409) 763-5333; fax (409) 763-5564
Internet: http://www.moodyf.org/HTMLversion/grantappset.htm
Sponsor: Moody Foundation
2302 Post Office Street, Suite 704
Galveston, TX 77550

Morgan Adams Foundation Grants 3447
The Morgan Adams Foundation supports laboratory and clinical research in the area of pediatric cancer, with an emphasis on cancers of the brain and spine. The Foundation's particular interest is to provide seed and bridge grants for viable investigations not yet ripe enough for total funding by larger organizations. Additionally, the Foundation is able to partially fund translational studies, experimental therapeutics studies and/or Phase I and II clinical trials on a case by case basis. The Foundation is seeking out projects that are highly collaborative and/or multi-institutional in nature, hoping to facilitate through the strategic placement of funding, more rapid trials development in order to get the best possible treatment options into circulation with the greatest expediency possible. Through careful consideration and placement of its funding, the Foundation encourages and supports research intended to improve treatment effectiveness, improve treatment outcomes and improve the quality of life for children battling cancer.
Contact: Joan Slaughter, Executive Director; (303) 758-2130; fax: (303) 758-2134; email: joan@morganadamsfoundation.org or info@morganadamsfoundation.org
Internet: http://www.morganadamsfoundation.org/fundsBecomeHope.html
Sponsor: Morgan Adams Foundation
5303 E. Evans Avenue, Suite 202
Denver, CO 80222

Morris K. Udall Environmental Public Policy Dissertation Fellowships 3448
Dissertation fellowships are available to outstanding doctoral candidates in the area of environmental public policy and conflict resolution. The program is designed to support individuals who are committed to environmental public policy and conflict resolution and who have the potential to make significant contribution in the field. An application packet may be requested by mail, phone, or fax.
Requirements: Dissertation fellowship applicants must be United States citizens, permanent residents, or, in the case of applicants from American Samoa, a U.S. National.
Restrictions: Students already holding a fellowship for the purpose of writing the dissertation the year before the fellowship are not eligible to apply.
Amount of Grant: $24,000 maximum for one year
Date(s) Application is Due: Feb 3
Contact: Jane Curlin, (520) 670-5609; fax: (520) 670-5530; email: curlin@udall.gov
Internet: http://udall.gov/p_fellowships.asp
Sponsor: The Morris K. Udall Scholarship and Excellence in National Environmental Policy Foundation
130 South Scott Avenue
Tucson, AZ 85701-1922

Morris K. Udall Foundation Dissertation Fellowships 3449
The foundation awards fellowships to doctoral candidates entering the final year of writing dissertations concerning environmental public policy and/or environmental conflict resolution. Dissertation fellowships are intended to cover both academic and living expenses. The foundation anticipates awarding up to two fellowships annually. Awards cannot be deferred. Application and guidelines are available online.
Requirements: Dissertation fellowship applicants must be U.S. citizens, permanent residents, or, in the case of applicants from American Samoa, a U.S. national. Applicants must have completed all coursework, passed all preliminary exams, and had the dissertation research proposal or plan approved by the listed application deadline.
Restrictions: Funds may not be used for extended field research. Fellowships do not fund research assistants; laboratory supplies and equipment; purchase of computer equipment; publication costs; travel to professional meetings, conferences, or seminars; tuition for additional coursework; repayment of loans or other personal obligations; or tuition for a dependent's education.
Amount of Grant: $24,000
Date(s) Application is Due: Feb 3
Contact: Jane Curlin, (520) 670-5609; fax: (520) 670-5530; email: curlin@udall.gov
Internet: http://www.udall.gov/p_fell_guide.htm
Sponsor: Morris K. Udall Foundation
130 S Scott Avenue
Tucson, AZ 85701-1922

Morris K. Udall Native American Congressional Internship Program 3450
The internship program is financed by a permanent trust fund endowment to develop increased opportunities for young Americans to prepare for and pursue careers related to the environment, and to Native Americans and Alaska Natives who intend to pursue careers in health care and tribal public policy. Annual deadlines may vary; contact program staff for exact dates.
Requirements: Applicant must be an enrolled member of a recognized tribe; be a matriculated senior, graduate student, or law student in a college or university; have a minimum 3.2 GPA; and have an interest in tribal government and policy.
Amount of Grant: $1200 stipend
Date(s) Application is Due: Jan 31
Contact: Monica Nuvamsa, Program Coordinator, (520) 670-5187 or (520) 670-5529; fax: (520) 670-5530; email: nuvamsa@udall.gov
Internet: http://www.udall.gov/p_internships.asp
Sponsor: The Morris K. Udall Scholarship and Excellence in National Environmental Policy Foundation
130 S Scott Avenue
Tucson, AZ 85701-1922

Morton-Kelly Charitable Trust Grants 3451

The trust awards grants to Maine nonprofits in its areas of interest, including historic preservation, arts and culture, education and vocational education, and environment. Types of support include general operating support, continuing support, annual campaigns, capital campaigns, building construction/renovation, endowment funds, equipment acquisition, fellowships, capital campaigns, program development, seed grants, scholarship funds, and research. There are no application forms; submit a letter of request.
Requirements: Maine nonprofit organizations are eligible.
Amount of Grant: $209,000 total
Contact: Program Contact, (207) 775-7271; fax: (207) 775-7935
Sponsor: Morton-Kelly Charitable Trust
10 Free Street, Box 4510
Portland, ME 04112

Motorola Foundation Grants 3452

Motorola concentrates its giving in the following areas: engineering, technical, science or mathematics programs in K-12 schools or in universities and colleges; programs reaching traditionally under-represented groups in the areas of math, science, engineering and business; programs providing technical assistance, research and statistical information on the state of science and engineering education, or; progams that support the protection and preservation of the environment.
Requirements: All grants are made to 501(c)3 tax-exempt organizations.
Restrictions: Grants may not be made to individuals (scholarships, travel, business loans), religious groups, fraternity or sorority programs, political campaigns, private foundations, benefit events or ads, single disease causes, national health organizations or their local chapters, trade schools, product donations, media projects, capital fund drives, sports sponsorships, or endowment funds.
Geographic Focus: Arizona; California; Florida; Georgia; Illinois; Massachusetts; New Jersey; Pennsylvania; Texas
Amount of Grant: $7.1 million total; $10,000-$100,000 typically
Contact: Matthew Blakely, (847) 576-7895; fax: (847) 576-9440; email: giving@motorola.com
Internet: http://www.motorola.com/content.jsp?globalObjectId=8152
Sponsor: Motorola Foundation
1303 East Algonquin Road
Schaumburg, IL 60196

Mount Sinai Health Care Foundation Grants 3453

The foundation seeks to assist greater Cleveland's organizations and leaders to improve the health and well-being of the Jewish and general communities now and for generations to come. The foundation concentrates its efforts on several well-defined program areas, including health policy, health of the Jewish community, health of the urban community, and academic medicine and bioscience. In addition to these program areas, the foundation seeks to strengthen Cleveland's prominence as a national leader in medical education and research. Whenever possible, the foundation will encourage the active participation of other funders to increase the overall amount of support available.
Requirements: 501(c)3 nonprofits serving the greater Cleveland, OH, area may submit proposals for grant support.
Restrictions: In general, the foundation does not support general operating expenses, direct provision of health services, building or equipment expenses, fund-raising events, projects outside of greater Cleveland, endowment funds, lobbying, or program advertising.
Geographic Focus: Ohio
Samples: Jewish Community Federation of Cleveland (OH)—to support health and human services projects at Jewish-sponsored institutions, as part of its Centennial Initiative for Jewish Cleveland, $5 million (2005).
Date(s) Application is Due: Jan 1; Apr 1; Jul 1; Oct 1
Contact: Shelly Galvin, Program Officer, (216) 421-5500; fax: (216) 421-5633; email: sgg2@case.edu
Internet: http://www.mtsinaifoundation.org/grants.html
Sponsor: Mount Sinai Health Care Foundation
11000 Euclid Avenue, 3rd Fl
Cleveland, OH 44106-1714

Mount Zion Health Fund Grants 3454

The fund provides grants predominantly in the San Francisco Bay area to nonprofit organizations in its areas of interest, including UCSF at Mount Zion and its affiliates, the Institute on Aging, and the Bay Area Jewish community. MZHF will consider proposals for grants for up to three years in duration. For multiyear grants, grant distributions will be contingent upon annual review of the grantee's progress toward defined expectations. There are no deadlines for submission. Applicants will generally receive a response within three months after submitting a complete proposal.
Requirements: California 501(c)3 tax-exempt organizations and organizations with 501(c)3 fiscal sponsors are eligible.
Restrictions: Grants are not awarded to support the personal benefit of individuals; endowments, annual appeals, membership drives, and fundraising events; or the reducation of operating deficits.
Geographic Focus: California
Samples: Bay Area Jewish Healing Ctr (San Francisco, CA)—for Jewish spiritual care, outreach, and education, $145,000 (2004); Homeless Prenatal Program (San Francisco, CA)—to provide care to homeless pregnant women and mothers, $15,000 (2004); Jewish Home (San Francisco, CA)—for general support of this geriatric center, $100,000 (2004); Life Tapes Project (San Francisco, CA)—to produce videotaped interviews designed to help relatives of people diagnosed with cancer, $40,000 (2004).
Amount of Grant: $5000-$100,000 generally
Date(s) Application is Due: Feb 15; Sep 15
Contact: Grants Administrator, (415) 750-4100; fax: (415) 495-6635; email: information@mzhf.org
Internet: http://sandbox.sfjcf.org/mountzion/grantapp
Sponsor: Mount Zion Health Fund
P.O. Box 194065
San Francisco, CA 94119-4065

MPAAT Grants 3455

The organization's mission is to reduce the harm that tobacco causes the people of Minnesota. Requests for applications have been issued for intervention funding and for research funding. Intervention funding proposals are invited in the areas of Minnesota Tobacco Treatment Centers, and Minnesota Statewide Worksite Cessation Programs. In the research category, applications are sought for Research Funding for Projects to Reduce the Harm Tobacco Causes Minnesotans. Guidelines and application for research grants are available online.
Requirements: Minnesota nonprofit organizations are eligible.
Geographic Focus: Minnesota
Contact: Chris Tholkes, (952) 767-1409 or (952) 767-1400; fax: (952) 767-1422; email: ctholkes@mpaat.org
Internet: http://www.mpaat.org
Sponsor: Minnesota Partnership for Action Against Tobacco
8011 34th Avenue S, Suite 400
Minneapolis, MN 55425

MPI for Social Anthropology Department II Postdoctoral Grants 3456

The postdoctoral grants are awarded with a specific theme in mind (see the Society's website for the chosen theme). All studies will gather standardized data, and individuals will be expected to expand their research based on specific topics related to the theme. The postdoctoral grants are awarded initially for two years, with the possibility of a one-year extension. Individuals with disabilities will be given priority, assuming equal qualifications.
Requirements: Participants must have a PhD and will undertake comparative fieldwork in light of the project, theme. PhD must be in anthropology or a related discipline to be submitted before commencement. There is no application to fill out; individuals should send the following documentation either by mail or email: a cover letter; a CV including a list of publications; a project proposal (one page outline); photocopies of university degrees; names of 2-3 referees.
Restrictions: Postdoctoral grants can only be awarded to scientists who have received their doctorate within the last ten years.
Geographic Focus: Alabama; Alaska; Arizona; Arkansas; California; Colorado; Connecticut; Delaware; District of Columbia; Florida; Georgia; Guam; Hawaii; Idaho; Illinois; Indiana; Iowa; Kansas; Kentucky; Louisiana; Maine; Maryland; Massachusetts; Michigan; Midwest; Minnesota; Mississippi; Missouri; Montana; Nebraska; Nevada; New England; New Hampshire; New Jersey; New Mexico; New York; North Carolina; North Dakota; Northeast; Northwest; Ohio; Oklahoma; Oregon; Pacific Northwest; Pennsylvania; Puerto Rico; Rhode Island; South; South Carolina; South Dakota; Southwest; Tennessee; Texas; U.S. Virgin Islands; Utah; Vermont; Virginia; Washington; West; West Virginia; Wisconsin; WyomingAfrica; Argentina; Asia; Australia; Belgium; Bolivia; Brazil; Bulgaria; Canada; Caribbean; Central America; Central Europe; China; Colombia; Czech Republic; Eastern Europe; Ecuador; Estonia; Europe; Germany; Hungary; Israel; Japan; Kazakhstan; Kyrgyzstan; Marshall Islands; Mexico; Micronesia; New Independent States; New Zealand; Northern Mariana Islands; Peru; Philippines; Poland; Republic of Palau; Romania; Russia; Scandinavia; Slovakia; Slovenia; South America; Sweden; Ukraine; United Kingdom; Uzbekistan; Venezuela
Samples: Chris Hahn - Feudalism, Socialism and the Present Mixed Economy in Rural Eastern Xinjiang; Helena Obendiek - Educational Mobility and Social Support in Rural China; Friedrich Binder - Urban Spirit-Mediums and the Construction of Moralities in Modern Taiwan; Edyta Roszko - Buddhas, Gods, and Spirits: Dialogic Interplay and Moral Dimensions in Vietnamese Religion; Dr. Florian M??hlfried - The Landscape of Citizenship in Highland Georgia; Milena Baghdasaryan - Social Implications of Armenian Citizenship for Refugees from Azerbaijan
Date(s) Application is Due: Jun 30
Contact: Ms. Janka Diallo; +49 345/ 29 27 404; fax: +49 345/ 29 27 402; email: jdiallo@eth.mpg.de
Internet: http://www.mpg.de/english/careerOpportunities/jobOffers/extern/ethnologische_forschung/stellenangebot29/
Sponsor: Max Planck Institute for Social Anthropology
P.O. Box 11 03 51
Halle/Saale, Munich 06017 Germany

MPI for Social Anthropology Max Planck Fellow Research Group Law, Organizations, Science, and Technology (LOST) 3457

This fellowship is open to scientists who have completed their doctorate within the last ten years. The primary regional focus is on Africa. Special attention is given to medical practice and argumentation in juridical contexts such as in the control of epidemics, the legitimization and legalization of diagnostic and healing practices, intellectual property rights, medical evidence in various forms of courts of justice etc. (for more details see http://www.eth.mpg.de/). The PhD grants are generally awarded for 2 years, with the

possibility of two six-month extensions. The postdoctoral grants are for two years (a one-year extension is possible).
Requirements: Applicants should have an anthropological background with at least some experience in Science & Technology Studies, or an S&TS background with a least some anthropological experience. Good field research skills, the ability to work in a team, communication skills with good written and spoken English are required. PhD students are expected to complete a year-long period of fieldwork as part of their studies. No application form is required, however applicants should mail or email the following: a cover letter; a CV including a list of publications; a project resume (five pages only); photocopies of university degrees; names of 2-3 referees, who can be contacted.
Restrictions: Postdoctoral grants can only be awarded to scientists who have received their doctorate within the last ten years.
Date(s) Application is Due: Apr 30
Contact: Ms. Wagenbrett, Personnel Administration; +49 345/ 29 27 404; fax: +49 345/ 29 27 402; email: wagenbrett@eth.mpg.de
Internet: http://www.mpg.de/english/careerOpportunities/jobOffers/extern/ethnologische_forschung/stellenangebot23/
Sponsor: Max Planck Institute for Social Anthropology
P.O. Box 11 03 51
Halle (Saale), Munich 06017 Germany

MRC Career Establishment Grants 3458

Career establishment grants provide research support for scientists and clinicians newly appointed to academic posts in medical schools, teaching hospitals, and universities who are at an early stage in the development of their research career and who can show that they have the potential to become the research team leaders of the future. A career establishment grant is intended to help award holders to establish themselves as independent scientists capable of winning support in open competition by the time the grant ends.
Requirements: An applicant must be in an established academic post in a university, medical school, teaching hospital, or similar institution; have a PhD or DPhil (for scientifically qualified applicants) or an MD (for medically qualified applicants); have at least three years postdoctoral research experience; be planning to make a long-term career in biomedical or health-related research; and make their initial application for a career establishment grant within the first three years of appointment to their first established academic post.
Restrictions: Scientists who do not hold an established academic appointment are not eligible to apply. Grants are not intended for researchers whose career in research is already well established, as demonstrated by their publication track record and their success in winning research funding.
Samples: Dr. C.R. Byrne, School of Biological Science, U of Manchester—for research on the relationship between fetal skin patterning and epidermal differentiation, @ L310,000.; Dr. S.L. Dickson, Physiological Laboratory, U of Cambridge—for research on hypothalamic circuits involved in body weight homeostasis, @L276,000.; Dr. S.J. Talbot, Department of Medical Microbiology, U of Edinburgh—for research on characterization of HHV-8 latency associated genes and the development of an animal model for studying tumorigenesis, @L476,000.
Contact: Michele Michalczuk, Research Management Group, 020 7636 5422; fax: 020 7436 6179; email: michele.michalczuk@headoffice.mrc.ac.uk
Internet: http://www.mrc.ac.uk/index/funding/funding-specific_schemes/funding-revised_grant_schemes.htm
Sponsor: Medical Research Council
20 Park Crescent
London W1B 1AL United Kingdom

Ms. Foundation Public Voices, Public Policy Grants 3459

The foundation is actively seeking organizations that are led by women of color and that are engaged in public policy advocacy and action. Because the goals of this initiative are not issue specific but rather about broad issues of access and participation of women of color in the public policy process, grantee organizations can be engaged in either multi- or single-issue work. Types of support include general support grants, technical assistance, training, and peer-to-peer opportunities. Start-up organizations that have no track record or experience in advocating for policy change for their communities will be given priority. Preference will be given to applicants with a budget under $1 million, although organizations that fall outside of this range are encouraged to apply. The listed application is for letters of intent; full proposals are by invitation. Guidelines and application are available online.
Requirements: 501(c)3 nonprofit organizations in the United States or Puerto Rico that are implementing local, state, tribal, national, or collective approaches around organizing, research, advocacy, or communications for public policy change are eligible.
Amount of Grant: $35,000-$75,000
Date(s) Application is Due: Nov 12
Contact: Grants Administrator, (212) 742-2300; fax: (212) 742-1653; email: info@ms.foundation.org
Internet: http://ms.foundation.org/wmspage.cfm?parm1=5
Sponsor: Ms. Foundation
120 Wall Street, 33rd Fl
New York, NY 10005

MSI Alden G. Clayton Dissertation Proposal Competition 3460

This annual competition will grant up to five awards of $5,000 each for dissertation proposals. Some preference will be given to proposals that address the current MSI Research Priority Topics. (See the website for current research priorities.) Each winner will receive an unrestricted cash prize of $5,000 designed to aid in the completion of the dissertation. Winners are asked to submit the results to MSI's working paper series.
Requirements: Entrants must be candidates for the doctorate in marketing or a related field (e.g., organizational behavior, consumer psychology, economics, strategy, management, finance etc.) in a fully accredited university. The dissertation must be no more than fifty percent completed at the time of submission, and the primary thesis advisor must approve submission of the proposal for the competition. Submissions must be accompanied by a letter from the primary thesis advisor certifying that the dissertation is no more than fifty percent completed at the time of submission.
Restrictions: For blind reviewing purposes, please omit any identifying information (name, school, etc.) from the proposal itself. Please include with your submission a separate file/hard copy sheet containing your name, the title of the dissertation, school, and detailed contact information for both the entrant and his/her thesis advisor(s).
Samples: 2006 Winners: Brett R. Gordon, Carnegie Mellon University, 'Replacement Cycles and Innovation in the PC Processor Industry.' Advisors - Kannan Srinivasan and Ronald Goettler; Michael Trusov, UCLA, 'Your Members are also Your Customers: Marketing for Internet Social Networks.' Advisor - Randolph E. Bucklin
Amount of Grant: $5,000
Date(s) Application is Due: Jul 31
Contact: Ross Rizley, Research Director, (617) 491-2060; email: rizley@msi.org
Internet: http://www.msi.org/awards/index.cfm?id=60
Sponsor: Marketing Science Institute
1000 Massachusetts Avenue
Cambridge, MA 02138-5396

MSI Marketing Strategy Meets Wall Street Research Competition 3461

The Marketing Science Institute (MSI) and the Emory Brand Institute (EmoryBI) have allocated resources to fund high-quality empirical research on several critical questions in this area. These research topics were prioritized in a multi-disciplinary discussion session of leading accounting, finance, and marketing faculty, as well as representatives from corporations and investment banks. Priority Research Topics: How informed are stock prices relative to important customer, brand, and marketing developments?; Understanding brand valuation; Marketing spending and firm value; The investor community as a customer.
Requirements: The co-sponsors invite proposals for original, scientifically rigorous research projects on these priority topics. Proposals should follow the MSI guidelines (available at the website).
Restrictions: The following expense categories are acceptable for funding: research assistance, data acquisition and analysis, travel among co-authors. Awarded funds may be used only for out-of-pocket expenses of the research (i.e., they may not be used for university overhead, researcher, salary or course relief, tuition remission, etc.).
Amount of Grant: $5,000 - $15,000
Date(s) Application is Due: Sep 1
Contact: Ross Rizley, Research Director, (617) 491-2060; email: rizley@msi.org
Internet: http://www.msi.org/research/index.cfm?id=146
Sponsor: Marketing Science Institute
1000 Massachusetts Avenue
Cambridge, MA 02138-5396

MSI Research Support Standard Grants 3462

MSI funds high-quality research that deals with topics of importance to member companies. The organization supports research with the potential for application by managers as well as more basic or exploratory work. No one approach or methodology is favored over another as long as the form is appropriate to the objectives of the research. Studies may be conceptual or empirical and may involve literature reviews, comparative studies, field or laboratory experiments, model building, or theory development. Cross-disciplinary work building on theories, research results, and methods from disciplines of relevance to marketing are encouraged. MSI and its member companies strongly endorse using actual consumers, customers, and executives rather than student subjects in research projects.
Requirements: Academic researchers (faculty members, or doctoral students working collaboratively with faculty advisors) can apply for financial and/or non-financial support for research projects. Most MSI grants are made to cover researchers' out-of-pocket costs for data collection, respondent fees, research assistants, and similar expenses. Both proposals and pre-proposals are accepted, though there is no required standard format. MSI operates on a monthly review cycle, and proposals may be submitted at any time. See website for detailed guidelines.
Restrictions: Grants do not provide salary replacement for the principal researcher(s), funds for the purchase of equipment or software, university overhead, tuition, or funds for travel to non-MSI conferences.
Amount of Grant: $5,000 - $20,000
Contact: Research Director, (617) 491-2060; fax: (617) 491-2065; email: msi@msi.org
Internet: http://www.msi.org/research/index.cfm?id=48
Sponsor: Marketing Science Institute
1000 Massachusetts Avenue
Cambridge, MA 02138-5396

MSSC Career Development Awards 3463

The awards further research into the cause, prevention, treatment, and cure of multiple sclerosis. Awards are intended to support independent research by the young, beginning researcher who holds a doctoral degree and intends to work full time at a Canadian school of medicine on research relevant to multiple sclerosis. Awards are normally made for three years with possible renewal.

Date(s) Application is Due: Oct 1
Contact: Jackie Munroe, Research Grants Manager, (416) 967-3001; fax: (416) 967-3040; email: jackie.munroe@mssociety.ca or info@mssoc.ca
Internet: http://www.mssociety.ca/en/research/researchfunding.htm
Sponsor: Multiple Sclerosis Society of Canada
175 Bloor Street E, Suite 700, N Tower
Toronto, ON M4W 3R8 Canada

MSSC Postdoctoral Fellowships **3464**
The fellowships further research into the cause, prevention, treatment, and cure of multiple sclerosis. These fellowships are intended for qualified scientists who have MDs or PhDs and intend to pursue further training in research related to multiple sclerosis. Fellowships are awarded for one year, renewable for up to three years if progress is deemed satisfactory.
Date(s) Application is Due: Oct 1
Contact: Jackie Munroe, Research Grants Manager, (416) 967-3001; fax: (416) 922-7538; email: jackie.munroe@mssociety.ca or info@mssoc.ca
Internet: http://www.mssociety.ca/en/research/researchfunding.htm
Sponsor: Multiple Sclerosis Society of Canada
175 Bloor Street E, Suite 700, N Tower
Toronto, ON M4W 3R8 Canada

MSSC Research Grants **3465**
The grants further independent research into the cause, prevention, treatment, and cure of multiple sclerosis. Research grants are awarded for projects covering a one- to three-year period. Research must be conducted in or under the authority of an institution approved by the society. Grants are to be used to cover research costs, including funding of equipment up to $C5000.
Samples: Drs. Albert Aguayo and Garth Bray, McGill U—for Neuron-Glia Interactions: Experimental Manipulation of Trophic Factors in Demyelination, $88,100.; Dr. Luanne Metz, U of Calgary—for The Cost of Multiple Sclerosis, $5050.
Date(s) Application is Due: Oct 1
Contact: Jackie Munroe, Research Grants Manager, (416) 967-3001; fax: (416) 922-7538; email: jackie.munroe@mssociety.ca or info@mssoc.ca
Internet: http://www.mssociety.ca/en/research/researchfunding.htm
Sponsor: Multiple Sclerosis Society of Canada
175 Bloor Street E, Suite 700, N Tower
Toronto, ON M4W 3R8 Canada

MSSC Research Studentships **3466**
Studentships are for doctoral candidates who require further training. The awards offer candidates an opportunity to acquire this training in specialized areas related to research into the cause, prevention, treatment, and/or cure of multiple sclerosis. Awards are for a maximum period of four years, renewable annually.
Date(s) Application is Due: Oct 1
Contact: Jackie Munroe, Research Grants Manager, (416) 967-3001; fax: (416) 922-7538; email: jackie.munroe@mssociety.ca or info@mssoc.ca
Internet: http://www.mssociety.ca/en/research/researchfunding.htm
Sponsor: Multiple Sclerosis Society of Canada
175 Bloor Street E, Suite 700, N Tower
Toronto, ON M4W 3R8 Canada

Multicultural Advertising Intern Program **3467**
This program offers internships to college juniors, seniors, and graduate students for 10 weeks during the summer in member agencies across the country. Intern duties include training in account management, art direction, copywriting, research, or media. Stipend plus reimbursement for 60 percent of housing expenses and 60 percent round-trip travel to the internship site are provided. Each year, 40 to 50 awards are made. Annual deadline dates may vary; contact program staff for exact dates.
Requirements: African Americans, Hispanics, Asians, and Native Americans who are U.S. citizens and currently enrolled in undergraduate or graduate programs, with a minimum GPA of 3.0 out of 4.0, are eligible. Applicants must have completed at least their junior year by the time of the internship and be returning to school in the fall following the internship.
Date(s) Application is Due: Apr 15
Contact: Tiffany Warren, Diversity Programs, (212) 682-2028; email: tiffany@aaaa.org or maip@aaaa.org
Internet: http://www.aaaa.org/diversity/maip/agencies/internselection_copy.htm
Sponsor: American Association of Advertising Agencies
405 Lexington Avenue, 18th Fl
New York, NY 10147-1801

Munby Fellowship in Bibliography **3468**
One fellowship is offered annually to cover bibliographical research, normally at the postdoctoral level, based on the collections of the Cambridge University libraries. There is no restriction on the choice of topic within the fields of historical bibliography, the history of the book trade, and book collecting. Research may be concerned with printed or manuscript material, Western or Oriental, in any language. Application consists of a curriculum vita, a statement outlining the research proposed, a list of publications, and names and addresses of two references. There are no interviews. Election will take place in autumn for the fellowship beginning the following October and running until September 30 of the next year.
Requirements: Eligible to apply are graduates of any university, who may be of any nationality. Preference will be given to promising younger scholars already established in their careers.
Samples: Dr. S. Avery-Quash—to research the rise of color printing from wood block 1850-70.
Amount of Grant: @L20,540
Date(s) Application is Due: Sep 9
Contact: Deputy Librarian, 01223-333047; fax: 01223-333160; email: library@lib.cam.ac.uk
Internet: http://www.lib.cam.ac.uk/munby.html
Sponsor: Cambridge University Library
West Road
Cambridge, England CB3 9DR United Kingdom

Murphy Institute Center for Ethics and Public Affairs Faculty Fellowships **3469**
The Center for Ethics and Public Affairs supports residential faculty fellowships to support outstanding faculty whose teaching and research focus on questions of ethics and moral choice in such areas as architecture, business, government, law, medicine, urban design and planning, and engineering. Fellows will participate in conferences and seminars and devote the majority of time to conducting their own research. All Fellows will receive a stipend, an office, library privileges, a research allowance and a monthly housing supplement. The fellowships are open to all, regardless of citizenship. See website for application form.
Requirements: Applicants should hold a doctorate in philosophy, political theory, or related disciplines; or a professional degree in architecture, business, engineering, law, medicine, urban design and planning, or public policy. Preference will be given to applicants at an early stage of their careers, normally no more than 10 years from the terminal degree in their field.
Geographic Focus: Louisiana
Amount of Grant: $35,000 maximum
Date(s) Application is Due: Nov 17
Contact: Meg Keenan, Fellowship Coordinator; (504) 862-3236; fax: (504) 862-8360; email: mkeenan@tulane.edu
Internet: http://www.tulane.edu/~murphy/Faculty%20F.htm
Sponsor: The Murphy Institute, Tulane University
108 Tilton Hall
New Orleans, LA 70118

Murphy Institute Center for Ethics and Public Affairs Graduate Fellowships **3470**
The Center for Ethics and Public Affairs encourages teaching and research on ethics, in moral and political philosophy, and in public life and the professions. The Center, Graduate Fellowships support outstanding Tulane graduate students in these fields. Graduate Fellows receive a stipend for one academic year, which varies according to individual circumstance.
Requirements: Student applicants should be enrolled in a doctoral program at Tulane University or in one of Tulane, professional schools; and should have completed all course requirements and general examinations by September prior to the fellowship year. The Center seeks applicants who have excelled in their fields of specialization, have demonstrated an interest in ethical questions that cut across disciplinary boundaries, and who are likely to make significant contributions to teaching and scholarship in the fields of ethics, moral and political philosophy, political theory, or professional ethics.
Geographic Focus: Louisiana
Date(s) Application is Due: Mar 3
Contact: Meg Keenan, Fellowship Coordinator; (504) 862-3236; fax: (504) 862-8360; email: mkeenan@tulane.edu
Internet: http://murphy.tulane.edu/center/graduate-fellowships.php
Sponsor: The Murphy Institute, Tulane University
108 Tilton Hall
New Orleans, LA 70118

Musculoskeletal Rehabilitation Fellowships **3471**
The fellowship provides a unique opportunity for specialized training in clinical, education, and research aspects of musculoskeletal rehabilitation. Resources at the clinic are vast and include a multispecialty spine center, sports medicine center, biomechanics department, hand center, arthritis clinic, pain clinic, and orthopedic and general musculoskeletal rehabilitation. Fellows will have the opportunity to assist in medical coverage of sporting events in southeastern Minnesota to gain musculoskeletal management skills. Multiple educational opportunities are available, including the teaching of medical students and residents. Dedicated research time can be allotted with faculty mentorship and integrated use of other departments. Appointments and application deadlines vary each year.
Requirements: Candidate must have successfully completed an accredited residency program in physical medicine and rehabilitation and be board eligible or board certified.
Date(s) Application is Due: Aug 30
Contact: Cindy Weber, (507) 284-2946; fax: (507) 284-3431; email: pmreduc@mayo.edu
Internet: http://www.mayo.edu/msgme/physmed-musculo-rch.html
Sponsor: Mayo Clinic
200 First Street SW
Rochester, MN 55905

Musgrave Foundation Grants 3472
The foundation awards grants to eligible Missouri tax-exempt organizations in its areas of interest, including civic and public affairs, community development, elementary education, secondary education, higher education, and social services. Types of support include annual campaigns, building construction/renovation, capital campaigns, continuing support, equipment acquisition, general operating support, matching grants, and scholarship funds.
Requirements: 501(c)3 tax-exempt organizations and colleges, universities, and schools in Missouri may apply.
Restrictions: Grants are not made to individuals.
Geographic Focus: Missouri
Amount of Grant: $500-$61,000 range
Date(s) Application is Due: May 31
Contact: Jerry Redfern, Program Contact, (417) 841-4698 or (417) 882-9090; fax: (417) 882-2529; email: jredfern@musgravefoundation.org
Internet: http://www.musgravefoundation.org
Sponsor: Musgrave Foundation
P.O. Box 10327
Springfield, MO 65804

Myrna F. Bernath Fellowship Award 3473
This fellowship is awarded biannually (odd years) for a U.S. woman to do historically based research abroad or for a female citizen from a foreign country to do historically based research in the United States on U.S. foreign relations, transnational history, international history, peace studies, cultural interchange, or defense or strategic studies. Whenever possible, preference will be given to a graduate student.
Requirements: Individuals interested in applying are asked to submit three copies of each proposal, consisting of no more than 15 double-spaced pages and three references.
Amount of Grant: $2500
Date(s) Application is Due: Dec 1
Contact: Grants Administrator, (614) 292-1951 or (614) 292-7200; fax: (614) 292-2282; email: shafr@osu.edu
Internet: http://www.shafr.org/prizes.htm#myrnafellowship
Sponsor: Society for Historians of American Foreign Relations
Department of History, Ohio State University, 106 Dulles Hall, 230 W 17th Avenue
Columbus, OH 43210

Myrna L. Bernath Book Prize 3474
This prize is offered every two years for the best book published in the previous two years by a woman in the areas of U.S. foreign relations, transnational history, international history, peace studies, cultural interchange, and defense or strategic studies. Three copies of each book (or page proofs) should be submitted.
Amount of Grant: $2500
Date(s) Application is Due: Dec 1
Contact: Grants Administrator, (614) 292-1951 or (614) 292-7200; fax: (614) 292-2282; email: shafr@osu.edu
Internet: http://www.shafr.org/prizes.htm#myrnabook
Sponsor: Society for Historians of American Foreign Relations
Department of History, Ohio State University, 106 Dulles Hall, 230 W 17th Avenue
Columbus, OH 43210

N.S. Beinstock Fellowship 3475
This fellowship was created to recognize promising minority journalists in radio or television news. Applications are available on the Web site.
Amount of Grant: $2500 plus expenses-paid trip to annual conference
Date(s) Application is Due: May 9
Contact: Karen Jackson-Bullitt, (202) 467-5218; email: karenb@rtndf.org
Internet: http://www.rtndf.org/asfi/fellowships/minority.shtml
Sponsor: Radio and Television News Directors Foundation
1000 Connecticut Avenue NW, Suite 615
Washington, DC 20036

NAA Norwegian Marshall Fund Grants 3476
The program awards up to 15 grants each year to support Americans invited to conduct postgraduate study or research in Norway on areas of mutual importance to Norway and the United States. The goal of the program is to increase knowledge and understanding and strengthen the ties of friendship between the two countries.
Requirements: Citizens of the United States, who have arranged with a Norwegian sponsor or research institution to pursue a research project or program in Norway are encouraged to apply for an award. Under special circumstances, the awards can be extended to Norwegians for study or research in the United States.
Amount of Grant: $U.S.1500-$U.S.4500
Contact: Gunnar Gran, Secretary General, 47 22 447716; fax: 47 22 447831; email: namerika@online.no
Internet: http://www.noram.no/index_am.html
Sponsor: Norway-America Association
Drammensveien 20c
Oslo N-0255 Norway

NAAB Coordinated Research Grants Program 3477
This program is designed to sustain growth and improve efficiency of the artificial insemination industry by fostering research and development in animal breeding and reproductive physiology with special emphasis on artificial insemination. Research priorities include disease transmittability, improvement of health testing procedures as they relate to artificial insemination, male and female fertility, semen quality, indicators of genetic merit, genetic factors affecting semen production, genetic evaluation, and marketing methods and strategies. The program supports research primarily related to dairy and beef cattle. However, research on other species may be considered if applicable to the bovine species. Grants are awarded for one year, July 1 to June 30. Preproposals are due on the listed application deadline date.
Requirements: Applicants should hold the doctorate or be of equivalent professional level in applicable discipline fields.
Amount of Grant: $1000-$15,000 per year
Date(s) Application is Due: Oct 1
Contact: Jere Mitchell, Technical Director, (573) 445-4406; fax: (573) 446-2279; email: jmitchel@naab-css.org or naab-css@naab-css.org
Internet: http://www.naab-css.org/membership/research_grant.html
Sponsor: National Association of Animal Breeders
P.O. Box 1033
Columbia, MO 65205

NAACP Agnes Jones Jackson Graduate Scholarships 3478
This scholarship is awarded annually to NAACP members who are or plan to be full-time graduate students at accredited colleges or universities and who can demonstrate financial need. The scholarship is awarded for one year and a recipient may apply for renewal at deadline date. Application forms and required documents must be postmarked on or before the last Friday in the month of March.
Requirements: Current regular members of the NAACP who are U.S. citizens under the age of 25 with a 2.5 (undergraduate) or 3.0 (graduate) GPA are eligible to apply.
Amount of Grant: $1500 undergraduate; $2500 graduate
Date(s) Application is Due: Mar 31
Contact: Jackson Graduate Scholarships, Education Department, (410) 358-8900; fax: (410) 764-7357
Internet: http://www.naacp.org/departments/education/scholarship_jones.html
Sponsor: National Association for the Advancement of Colored People
8260 Willow Oaks Corporate Drive
Fairfax, VA 22031

NAACP Willems Scholarships 3479
The scholarship is awarded to U.S. citizens who is male and majoring in engineering, chemistry, physics, or computer and mathematical sciences. Applicant must possess a cumulative grade point average of at least 3.0. Both undergraduate and graduate students may apply. Membership and participation in NAACP is highly desired. To request a scholarship packet, your request must be in writing. Application forms and required documents must be postmarked on or before the last Friday in the month of March.
Amount of Grant: $3000 for graduates; $2000 for undergraduates
Date(s) Application is Due: Mar 31
Contact: National Education Coordinator, (410) 580-5760
Internet: http://www.naacp.org/inc/docs/education/education_scholarship_instruction_app.pdf
Sponsor: National Association for the Advancement of Colored People
4805 Mount Hope Drive
Baltimore, MD 21215

NABT Four-Year College Biology Teaching Award 3480
The award is made to a faculty member at a four-year college who demonstrates creativity and innovation in his or her teaching of biology. Innovation may address curriculum design, teaching strategies, and laboratory utilization, and the strategy must have been implemented in the classroom and demonstrated to be effective. The award includes a Leica microscope, a recognition plaque to be presented at the NABT National Convention, and $500 in travel funds to attend the Annual NABT meeting to receive the award. Application forms are available on the Web site.
Requirements: NABT members who teach at four-year institutions are eligible.
Date(s) Application is Due: Mar 15
Contact: Louise Pittack, Awards Manager, (703) 264-9696 or (800) 406-0775; fax: (703) 264-7778; email: lpittack@nabt.org or office@nabt.org
Internet: http://www.nabt.org/sup/education/awards.asp
Sponsor: National Association of Biology Teachers
12030 Sunrise Valley Drive, Suite 110
Reston, VA 20190

NACC David Stevenson Fellowships 3481
The goals of this fellowship are to acknowledge David Stevenson's contribution toward knowledge of the nonprofit sector and his role to teach and work with undergraduate students; and to encourage junior faculty members of color to provide leadership for the field and to assist them in gaining tenure in their respective academic institutions. Two fellowships are awarded annually. Fellowships may be used to support research, travel to present research at conferences (other than ARNOVA), and professional education seminars related to research and teaching. The fellowship must be completed between August 1 and July 31. The award will be made payable to the award recipient's affiliate institution. Guidelines are available online.
Requirements: Applicants must be: junior faculty in a tenure track position; affiliated with a U.S. or Canadian accredited institution of higher education; members of the African American, Latino, Asian/Pacific Islander, or American Indian/Alaskan Native communities; committed to excellence in research and teaching in the field of philanthropic or nonprofit sector studies or nonprofit management from any one

of a variety of academic disciplines; and willing and able to participate in the annual ARNOVA conference.
Amount of Grant: $15,000
Date(s) Application is Due: Mar 31
Contact: Amy McClellan, Executive Director, (216) 368-0969; email: amcclellan-nacc@cwru.edu
Internet: http://www.naccouncil.org/stevenson.asp
Sponsor: Nonprofit Academic Centers Council
10900 Euclid Avenue
Cleveland, OH 44106-7167

NACC William Diaz Fellowships **3482**
The goals of this fellowship are: to acknowledge William Diaz's contributions to the development of minority leadership for nonprofit organizations, foundations, and the field of philanthropic and nonprofit management studies; and to encourage a better understanding of philanthropy from diverse communities. Two fellowships will be awarded annually. Funds from these fellowships may be used for support of research, travel to present research at conferences (other than ARNOVA), professional education seminars related to research and teaching on philanthropic and nonprofit studies, and course development. All funds must be expended and accounted for within one year after the end of the fellowship year. The award will be made payable to the award recipient's affiliate institution. Guidelines are available online.
Requirements: Applicants must be: faculty of any rank, including senior faculty new to philanthropic and nonprofit management studies whose work focuses on philanthropy, nonprofit organizations and diversity; affiliated with a U.S. or Canadian accredited institution of higher education; members of the African American, Latino, Asian/Pacific Islander, or American Indian/Alaskan Native communities; committed to excellence in research and teaching in the field of philanthropic or nonprofit sector studies or nonprofit management from any one of a variety of academic disciplines; and willing and able to participate in the annual ARNOVA Conference. (Travel expenses will be paid by NACC. Travel to attend the ARNOVA conference is limited to the year of the award and up to two years after the award to present fellowship-related research.)
Amount of Grant: $15,000
Date(s) Application is Due: Mar 30
Contact: Amy McClellan, Executive Director, (216) 368-0969; email: amcclellan-nacc@cwru.edu
Internet: http://www.naccouncil.org/diaz.asp
Sponsor: Nonprofit Academic Centers Council
10900 Euclid Avenue
Cleveland, OH 44106-7167

NAF Fellowships **3483**
The foundation supports new and innovative studies that are relevant to the cause, pathogenesis, or treatment of the hereditary or sporadic ataxias. Postdoctoral fellowship awards serve as a bridge from postdoctoral positions to junior faculty positions. The award will permit individuals to spend an additional third year in a postdoctoral position and increase chances to establish an independent ataxia research program. Letters of Intent should arrive by August 15, with full proposals due by September 15. Application and guidelines are available online.
Requirements: Applicants should have completed at least one year of postdoctoral training, but not more than two at the time of application, and should have shown a commitment to research in the field of ataxia.
Restrictions: Indirect costs are not allowed.
Amount of Grant: $35,000 average
Date(s) Application is Due: Aug 15; Sep 15
Contact: Grants Administrator; (763) 553-0020; fax: (763) 553-0167; email: naf@ataxia.org
Internet: http://www.ataxia.org/research/ataxia-research-grants.aspx
Sponsor: National Ataxia Foundation
2600 Fernbrook Lane, Suite 119
Minneapolis, MN 55447-4752

NAF Kyle Bryant Translational Research Award **3484**
The Friedreich's Ataxia Research Alliance (FARA) and the National Ataxia Foundation (NAF) invite proposals, under a competitive Request for Applications (RFA) process, to award a grant focusing on pre-clinical investigations that will facilitate clinical trials for Friedreich's Ataxia. Funding will be for one year. The total award is limited to $120,000 (direct costs only). The Foundation is requesting that applicants submit Letters of Intent by July 15, and full applications by July 31.
Requirements: Each application must include a project objectives and lay summary that will be used as FARA and NAF wishes and should not contain confidential information. The other parts of the grant application are considered confidential and will only be released to review committee, FARA and NAF staff and Board of Directors. Applications will be accepted from for-profit organizations, non-profit organizations, public or private institutions, and foreign institutions.
Amount of Grant: $120,000 maximum in direct costs
Date(s) Application is Due: Jul 15; Jul 31
Contact: Jennifer Farmer, Grant Manager; (610) 246-9017; email: jen.farmer@cureFA.org; Susan Hagen; email: (763) 553-0020; susan@ataxia.org
Internet: http://www.ataxia.org/research/ataxia-research-grants.aspx
Sponsor: National Ataxia Foundation
2600 Fernbrook Lane, Suite 119
Minneapolis, MN 55447-4752

NAF Research Grants **3485**
NAF research grants are for new and innovative studies that are relevant to the cause, pathogenesis or treatment of the hereditary or sporadic ataxias. Due to the larger availability of funding for Ataxia-Telangiectasia (A-T), those research proposals will receive a lower priority. However, a higher ranking will be given to those Ataxia-Telangiectasia research studies which lend themselves to an overall better understanding of the ataxia process. Research grants are offered primarily as seed monies to assist investigators in the early or pilot phase of their studies and as additional support for ongoing investigations on demonstration of need. It is hoped that these studies will be further developed to attract future funding from other sources. Grants are awarded for one year only. If funding for a second year is desired then another grant application should be submitted for that funding period. The average funds granted will be up to $15,000, but funding may be considered for up to $30,000 for projects deserving special consideration.
Restrictions: Indirect costs will not be funded.
Amount of Grant: $5,000-$15,000 average; $30,000 for special projects
Date(s) Application is Due: Aug 15
Contact: Grants Administrator; (763) 553-0020; fax: (763) 553-0167; email: naf@ataxia.org
Internet: http://www.ataxia.org/research/ataxia-research-grants.aspx
Sponsor: National Ataxia Foundation
2600 Fernbrook Lane, Suite 119
Minneapolis, MN 55447-4752

NAF Young Investigator Awards **3486**
The purpose of this award is to encourage young clinical and scientific investigators to pursue a career in the field of ataxia research. It is the foundation's hope that ataxia research will be invigorated by the work of young, talented individuals supported by stable multi-year funding. This award will be made to the sponsoring institution and will be used as salary support or for direct research expenses. The sponsoring institution must agree in writing to the following provision: a clinical investigator (MD) must be free to allocate a minimum of half time for a project; a non-clinician investigator (PhD) must allocate a minimum of three-fourths time for the project. Periodic progress reports will be required. Applicants are asked to submit a brief email with the intentions to apply for NAF funding, including a full title of the research proposal, on or before August 1. Deadline for full application is September 1.
Requirements: Candidates must have attained the MD or PhD degree, and have an appointment as a junior faculty member. Clinicians must have finished their residency no more than five years prior to applying, and investigators must have completed their postdoctoral training no more than five years prior to applying.
Restrictions: Individuals at the associate professor level are not eligible. Awards are for direct costs only, not for indirect costs or institutional overhead.
Amount of Grant: $50,000 per year for two years
Date(s) Application is Due: Aug 1; Sep 1
Contact: Grants Administrator; (763) 553-0020; fax: (763) 553-0167; email: naf@ataxia.org
Internet: http://www.ataxia.org/research/ataxia-research-grants.aspx
Sponsor: National Ataxia Foundation
2600 Fernbrook Lane, Suite 119
Minneapolis, MN 55447-4752

NAGC Hollingworth Award **3487**
The annual Hollingworth Award competition is sponsored by the National Association for Gifted Children (NAGC) to encourage educational and psychological research studies of potential benefit to gifted and talented children. The award is named for Leta Stetter Hollingworth (1886 - 1939), a pioneer in the field of gifted education. Long associated with Teachers College of Columbia University.
Requirements: This international competition is open to individuals, educational institutions, and other organizations that present proposals for publishable research projects concerning gifted and talented youth. You may be a professor, educational administrator, psychologist, teacher, graduate student, or other professional individual. The research projects may be sponsored by universities, schools systems, individual schools, public agencies, or private nonprofit organizations.
Amount of Grant: $2500
Date(s) Application is Due: Apr 20
Contact: Jeff Danielian, NAGC, Gifted Resource Specialist, (202) 785-4268; fax: (202) 785-4248; email: danielian@nagc.org
Internet: http://www.nagc.org/index.aspx?id=1120
Sponsor: National Association for Gifted Children
1707 L Street NW, Suite 550
Washington, DC 20036

NAGC Masters and Specialists Award **3488**
The National Association for Gifted Children (NAGC) presents a maximum of three awards to those who have demonstrated at least four of the following: teaching; writing; advocacy; curriculum development; community service in gifted education; juried presentations (local, state, or national); facilitating the social and emotional development of gifted and talented students; mentoring/tutoring of children and adolescents; publication of papers in juried and/or non-juried publications; participation in graduate student research colloquia; teacher in-service/parent workshop.
Requirements: Viable candidates will have a proven record of at least four of the following: teaching; writing; advocacy; curriculum development; community service in gifted education; juried presentations (local, state, or national); facilitating the social and

emotional development of gifted and talented students; mentoring/tutoring of children and adolescents; publication of papers in juried and/or non-juried publications; participation in graduate student research colloquia; teacher in-service/parent workshop
Restrictions: An individual may receive this award once. The award must be given within a year of the candidate's graduation. A maximum of one candidate and award can be given to one institution.
Amount of Grant: $200 and a plaque
Date(s) Application is Due: Apr 20
Contact: Jeff Danielian, NAGC, Gifted Resource Specialist, (202) 785-4268; fax: (202) 785-4248; email: danielian@nagc.org
Internet: http://www.nagc.org/index.aspx?id=696
Sponsor: National Association for Gifted Children
1707 L Street NW, Suite 550
Washington, DC 20036

NAGPRA Museum Documentation Grants **3489**
The Native American Graves Protection and Repatriation Act (NAGPRA) addresses the rights of lineal descendants, Indian tribes, Alaska Native villages and corporations, and Native Hawaiian organizations to certain Native American human remains and cultural items with which they are affiliated. Section 10 of the statute authorizes the Secretary of the Interior to make grants to museums and to Indian tribes, Alaska Native villages and corporations, and Native Hawaiian organizations for the purposes of assisting in the inventory, documentation, and repatriation of Native American human remains and cultural items.
Requirements: NAGPRA grants are available to any institution or state or local government agency (including any institution of higher learning) that has possession of, or control over, Native American human remains or cultural items.
Restrictions: Museums that have previously received a NAGPRA grant, but whose grant expired without successfully completing major elements of the proposed work, or without meeting the conditions of the grant award will not be considered for funding.
Amount of Grant: $5000-$75,000
Date(s) Application is Due: Dec 30; Mar 1
Contact: Paula Molloy, NAGPRA Consultant, (202) 354-2207; fax: (202) 371-5197; email: Paula_Molloy@nps.gov
Internet: http://www.cr.nps.gov/nagpra/GRANTS/INDEX.HTM
Sponsor: National Park Service
1849 C Street NW, (2253)
Washington, DC 20240

NAGPRA Museum Repatriation Grants **3490**
The Native American Graves Protection and Repatriation Act (NAGPRA) addresses the rights of lineal descendants, Indian tribes, Alaska Native villages and corporations, and Native Hawaiian organizations to certain Native American human remains and cultural items with which they are affiliated. Section 10 of the statute authorizes the Secretary of the Interior to make grants to museums and to Indian tribes, Alaska Native villages and corporations, and Native Hawaiian organizations for the purposes of assisting in the inventory, documentation, and repatriation of Native American human remains and cultural items. Repatriation awards will be granted to museums to repatriate culturally affiliated Native American human remains and cultural items.
Requirements: NAGPRA grants are available to any institution or state or local government agency (including any institution of higher learning) that has possession of, or control over, Native American human remains or cultural items.
Restrictions: Applicants for repatriation awards must collaborate with the tribes(s), Alaska Native village(s) and corporation(s), or Native Hawaiian organization(s) to whom the human remains or cultural items are being repatriated.
Amount of Grant: $15,000 maximum
Date(s) Application is Due: Jun 30
Contact: Paula Molloy, (202) 354-2207; fax: (202) 371-5197; email: Paula_Molloy@nps.gov or NAGPRA_Info@nps.gov
Internet: http://www.cr.nps.gov/nagpra/GRANTS/INDEX.HTM
Sponsor: National Park Service
1849 C Street NW, (2253)
Washington, DC 20240

Nakanishi Prize **3491**
To Prize is intended to recognize and stimulate significant work that extends chemical and spectroscopic methods to the study of important biological phenomena, and was established by the students and colleagues of Koji Nakanishi in the United States and Japan and by the members of the scientific community of both countries. Consequently, the prize is given in the United States and Japan in alternate years through the designated societies and separate endowments administered accordingly. Presentations in odd-numbered years are administered by ACS and in even-numbered years by the Kanagawa Academy of Science and Technology. The prize shall be awarded to an individual without regard to age or nationality for outstanding scientific achievement. Special consideration will be given for work that has contributed broadly on an international scope. In addition to the prize, a bronze medal and certificate are awarded.
Restrictions: Self-nominations are not accepted.
Samples: Hung-wen Liu—Prize winner, $3,000 (2007); Stephen Benkovic—Prize winner, $3,000 (2005); Isao Kitagawa—Prize winner, $3,000 (2004).
Amount of Grant: $3,000 and travel allowance
Date(s) Application is Due: Nov 1
Contact: Felicia Dixon, Awards Administrator; (800) 227-5558 or (202) 872-4408; fax: (202) 776-8008; email: f_dixon@acs.org or awards@acs.org
Internet: http://portal.acs.org/portal/acs/corg/content?_nfpb=true&_pageLabel=PP_ARTICLEMAIN&node_id=1319&content_id=CTP_004540&use_sec=true&sec_url_var=region1
Sponsor: American Chemical Society
1155 Sixteenth Street, NW
Washington, DC 20036

NAMM Foundation Music Research Grants **3492**
The Foundation supports scientific research exploring the relationship between music and physical and emotional wellness, with particular attention to the elderly population, the impact of music making on at-risk youth, and music education and the effects of music and music making. The research division also convenes scientists, educators and others around critical issues in music research and disseminates research through various on-line research referral services and archives. Interested applicants should submit a pre-proposal by email to info@nammfoundation.org. Pre-proposal information is available online.
Samples: Frank Porter Graham Child Development Institute (Chapel Hill, NC)—to support the work of Noreen Yazejian and Ellen Peisner-Feinberg, $80,000 over two years (2003); Harvard Medical School (MA)—to support the work of Gottfried Schlaug and Ellen Winner, for research entitled The Effects of Music Training on Brain and Cognitive Development, $50,000 over two years (2003); U of Texas Health Ctr, Research Imaging Ctr (San Antonio, TX)—to support the work of Steven Brown, for music-research projects, including a study entitled Analysis of Brain Systems Underlying Music and Dance Improvisations, $25,000 (2003).
Contact: Grants Administrator, (760) 438-8001; fax: (760) 438-7327; email: info@nammfoundation.org
Internet: http://www.music-research.org/Grants/guidelines.html
Sponsor: NAMM Foundation
5790 Armada Drive
Carlsbad, CA 92008

NAMME-McCormick Tribune Fellowship **3493**
The association, in partnership with the Robert R. McCormick Tribune Foundation, awards fellowships to high-performing senior managers and executives of color in the news media. Eight fellows, four from the newspaper industry and four from the television industry, are selected to attend leadership programs at the Media Management Center at Northwestern University in Evanston, IL. The program's duration is four weeks for senior executives in the newspaper industry and seven days for television executives. Fellows also participate in other programs and activities. Information and guidelines are available on the Web site.
Requirements: African American, Latino, American Indian, and Asian American senior managers and executives in the news business (newspapers, broadcast stations, and news companies) are eligible. Editorial candidates must have 10 or more years of experience in the media and should be at the level of assistant managing editor or higher at newspapers, news companies, or the equivalent at television stations and groups. Business applicants should have at least 10 years of professional experience and have reached the level of assistant director. Guidelines are available online.
Date(s) Application is Due: Oct 1
Contact: McCormick Fellowships, (888) 968-7658 or (703) 288-6501; fax: (703) 893-2414; email: gjasik@namme.org
Internet: http://www.namme.org/programs/mccormick
Sponsor: National Association of Minority Media Executives Foundation
1921 Gallows Road, Suite 600
Vienna, VA 22182

Nanyang Technological University Lee Kuan Yew Postdoctoral Fellowship **3494**
The three-year fellowship will be held concurrently with the candidate's appointment as a research staff in the school working on any of the following identified research areas: College of Engineering, School of Biological Sciences, School of Civil and Environmental Engineering, School of Computer Engineering, School of Electrical and Electronic Engineering, School of Mechanical and Aerospace and Engineering, School of Materials Science and Engineering, and School of Natural Sciences. The program is open to nationals of all countries. Applications must be made on the NTU application form, available at http://www.ntu.edu.sg/personnel/adminres/StaffRec/Researchform.doc. Application and guidelines are available online.
Requirements: Applications are invited from young candidates (below 35 years of age) who have obtained their PhD degrees in the last few years and with excellent records and research potential.
Amount of Grant: S$72,800-S$123,900
Date(s) Application is Due: Mar 31
Contact: Vice President, Human Resources, fax: (65) 67919340; email: lklim@ntu.edu.sg
Internet: http://www.ntu.edu.sg/personnel/adminres/StaffRec/AdLKY2005.html
Sponsor: Nanyang Technological University
50 Nanyang Avenue
Singapore 639798 Singapore

NAPM Doctoral Dissertation Grant **3495**
Four yearly fellowships are available to produce useful research that can be applied to the purchasing and materials management functions, and to help develop more high-potential academicians who will teach and conduct research in the purchasing and materials management fields. Grants are to be paid directly to the doctoral candidate in three equal installments: upon receipt of the final research proposal approved by

the candidate's dissertation committee; upon receipt of a written statement from the dissertation committee verifying that the research is approximately 50 percent completed; and upon receipt of completed dissertation.
Requirements: Doctoral candidates who are pursuing PhDs or DBAs in purchasing, business, management, logistics, economics, industrial engineering, or related fields in accredited U.S. universities are eligible for the grant.
Amount of Grant: $12,000 maximum
Date(s) Application is Due: Jan 31
Contact: Dr. Joseph Cavinato, Senior Vice President, (480) 752-6276 ext 3029 or (800) 888-6276; fax: (480) 752-7890; email: jcavinato@ism.ws
Internet: http://www.napm.org/OnlineGuides/doctoralgrant.cfm
Sponsor: National Association of Purchasing Management
P.O. Box 22160, 2055 E Centennial Cir
Tempe, AZ 85285-2160

NAPNAP Foundation Graduate Student Research Grant **3496**
The program supports research by graduate students currently enrolled in a pediatric nurse practitioner program. The purposes of the award are to: encourage graduate students to participate in the research process; and stimulate interest in research regarding children and their families. The recipient of the award will be invited to participate at the research presentation session at the annual conference in the spring following the award. Application and guidelines are available online.
Requirements: The graduate student must be a current member of NAPNAP.
Samples: Cathy Woodward—for Routine Chest Radiographs Post Chest Tube Removal in Children, $1,000 (2006-07); Sheri Vlam—for ADHD Assessment Methods Used by APRNs, $500 (2003-04); Jacqueline Rychnovsky—for A Pediatric Perspective of Maternal Fatigue Across the First Ten Weeks Postpartum, $1,000 (2002-03).
Amount of Grant: $1,000
Date(s) Application is Due: Apr 1
Contact: Dolores Jones, Director; (856) 857-9700; fax (856) 857-1600; email: djones@napnap.org or info@napnap.org
Internet: http://www.napnap.org/PNPResources/research/AvailGrantsScholarships.aspx
Sponsor: National Association of Pediatric Nurse Associates and Practitioners
20 Brace Road, Suite 200
Cherry Hill, NJ 08034-2633

NAPNAP Foundation Nursing Research Grants **3497**
The foundation supports nursing research efforts that contribute to the improvement of the quality of life for children and their families. The foundation is primarily interested in the following areas of research: impact of nurse practitioner role in healthcare delivery; development/implementation of innovative methods for delivery of health care services in the areas of school health, community health, behavior and development, adolescent health, and chronic disease management; health care problems of children and their families; advanced nursing practice problems and issues; exploratory/evaluative research of PNP education and credentialing; and educational programs and materials for healthcare providers and/or children and families. Up to two grants will be awarded annually.
Amount of Grant: $2,500
Date(s) Application is Due: Apr 1
Contact: Dolores Jones, Director; (856) 857-9700; fax (856) 857-1600; email: djones@napnap.org or info@napnap.org
Internet: http://www.napnap.org/index.cfm?page=13&sec=102
Sponsor: National Association of Pediatric Nurse Associates and Practitioners
20 Brace Road, Suite 200
Cherry Hill, NJ 08034-2633

Narcolepsy Network Seed Grants **3498**
The network supports medical/scientific research of narcolepsy and awards seed/pilot grants in the following categories: tissue banks—facilitation of acquisition, storage, or utilization of tissue; subjects—addition of subjects with narcolepsy to an ongoing or recently completed study of some other condition; food—as regards understanding of short-term adverse effects in cataplectic-narcoleptic persons of some aspect of food consumption; and comorbidities—the sesarch for coexsisting structural or physiological alterations in persons with narcolepsy, which along with alterations in sleep reflect narcolepsy susceptibility gene(s). The network anticipates awarding one grant in each category. Complete application information is available online. Annual deadline dates may vary; contact program staff for exact dates.
Requirements: 501(c)3 tax-exempt organizations are eligible. A principal investigator with the MD, PhD, or DO degree must originate or be responsible for the projectt.
Amount of Grant: $8000
Date(s) Application is Due: Sep 15
Contact: Grants Administrator, (401) 667-2523; fax: (401) 633-6567; email: narnet@narcolepsynetwork.org
Internet: http://www.narcolepsynetwork.org
Sponsor: Narcolepsy Network
P.O. Box 294
Pleasantville, NY 10570

NARSAD Distinguished Investigator Award **3499**
The award supports established scientists who maintain peer-reviewed competitively funded scientific programs. This program is intended to stimulate new approaches and projects from senior investigators, rather than extending current work of established investigators. Research must be relevant to schizophrenia; major affective disorders; or other serious mental illnesses, including bipolar disease, borderline disorders with depression or suicide, or research with children. Applicants must demonstrate that, without new funds, the proposed work is not feasible.
Requirements: The researcher must be at the full professor academic rank (or equivalent), maintain a laboratory, and be affiliated with a university or research institution.
Amount of Grant: $100,000 maximum
Date(s) Application is Due: May 15
Contact: Audra Moran, Director, Research Grants Program, (800) 829-8289 or (516) 829-5576; email: amoran@narsad.org
Internet: http://www.narsad.org/research/apply
Sponsor: National Alliance for Research on Schizophrenia and Depression
60 Cutter Mill Road, Suite 404
Great Neck, NY 11021

NARSAD Independent Investigator Award **3500**
The program provides up to two years of support to enable the scientist at the academic level of associate professor or equivalent, who has won national competitive support as a principal investigator, to facilitate innovative research opportunities. Research must be relevant to schizophrenia, major affective disorders, or other serious mental illnesses, including bipolar disease, borderline disorders with depression or suicide, and research with children. The award is not renewable; applicants may reapply for a second award immediately following the completion of the first award, pending they have fulfilled all requirements for the first award. A maximum of two NARSAD Independent Investigator Awards may be received.
Requirements: The researcher must be at the associate professor academic rank (or equivalent), have received national competitive support as a principal investigator, and be affiliated with a university or research institution.
Amount of Grant: $50,000 maximum per year; up to $100,000 for two years
Date(s) Application is Due: Mar 6
Contact: Audra Moran, (800) 829-8289 or (516) 829-5576; fax: (516) 487-6930; email: amoran@narsad.org or grants@narsad.org
Internet: http://www.narsad.org/research/apply/independent
Sponsor: National Alliance for Research on Schizophrenia and Depression
60 Cutter Mill Road, Suite 404
Great Neck, NY 11021

NARSAD Young Investigator Award **3501**
The award supports basic and/or clinical research relevant to schizophrenia, major affective disorders, or other serious mental illnesses, including bipolar disease, borderline disorders with depression or suicide, and research with children. The program provides up to two years of support to enable promising investigators to either extend their research fellowship training or to begin careers as independent research faculty. Applicants must have an on-site mentor who is an established investigator in areas relevant to schizophrenia, depression, or other serious mental illness. Applications will not be accepted before June 1.
Requirements: Applicants must have a doctoral-level degree (e.g., MD, PhD, PsyD, etc.) by the application deadline and be employed in research training or be in a faculty or independent research position. The program is intended to support postdoctoral fellows and assistant professors (or equivalent).
Restrictions: Predoctoral students are ineligible.
Samples: To be distributed among 190 researchers in the United States and elsewhere—for research on anxiety disorders, bipolar illness, depression, schizophrenia, and other brain disorders, $11.9 million (2003).
Amount of Grant: $30,000 maximum per year up to two years
Date(s) Application is Due: Jul 25
Contact: Audra Moran, Director, Research Grants Program, (516) 829-5576; fax: (516) 487-6930; email: amoran@narsad.org
Internet: http://www.narsad.org/research/apply/young
Sponsor: National Alliance for Research on Schizophrenia and Depression
60 Cutter Mill Road, Suite 404
Great Neck, NY 11021

NASA Cooperative Agreement Notice for Technology Development Grants **3502**
The objective of this opportunity is to establish partnerships with the goal of developing a technology to meet a specific NASA need. This goal will be accomplished by selecting partners who will cooperatively share in the development cost of the technology that meets the specified NASA need. The partnership selection is a two step process with an initial Notice of Intent followed with the request for the Project Proposal on selected technology concepts. Specific areas of interest include: Rocket Propulsion Test Technologies and Integrated System Health Management. NASA plans to fund projects at approximately $20,000 to $500,000 per project with a period of performance of approximately 12 months.
Requirements: The partner is expected to contribute at least 50 percent of the total resources required to accomplish the objectives of the project. However, special situations may arise in which NASA will allow the partner contribution to be less than 50 percent. Partner contributions may be cash, non-cash or both.
Amount of Grant: $20,000 - $500,000
Date(s) Application is Due: Sep 30
Contact: Jake Jacobs, Contract Specialist; (228) 688-3931; fax: (228) 688-1141; email: james.e.jacobs@nasa.gov; James D. Huk, Contracting Officer; (228) 688-1045; fax: (228) 688-1141; email: james.huk@ssc.nasa.gov

Internet: http://nspires.nasaprs.com/external/solicitations/summary.do?method=init&solId={993EDEC9-8D62-6A05-EEB7-DF2EDF97F902}&path=open
Sponsor: National Aeronautics and Space Administration (NASA)
NASA Business Management Division, Acquisition Management Office, Code BA30
Stennis Space Center, MS 39529-6000

NASA Goddard Visiting Fellowships 3503
The program seeks highly qualified PhD scientists for collaboration in scientific research in one of the following research theme areas of the Division: aerosols; carbon cycle; clouds and precipitation; polar climate change; solar impacts on climate; terrestrial water; or weather and short-term climate forecasting. Selected candidates will be offered appointments as Goddard Visiting Fellows to pursue independent research in collaboration with scientists in the Laboratories within the Earth Sciences Division, and will have full access to NASA computing facilities and other resources at either the Goddard Space Flight Center in Greenbelt, Maryland, or at the Goddard Institute for Space Studies in New York, New York. Candidates will be selected, by competition, amongst the above theme areas of research but awards may not be made in all theme areas in any one competition year.
Requirements: This solicitation is aimed at exceptional post-doctoral candidates and university faculty interested in sabbatical arrangements.
Date(s) Application is Due: Mar 1
Contact: Dr. Raymond M. Hoff, Director; (410) 455-1610; fax: (410) 455-1291; email: hoff@umbc.edu
Internet: http://gest.umbc.edu/employment_opp/goddard_visiting_fellows.html
Sponsor: Goddard Earth Sciences and Technology Center (GEST) / NASA Goddard Earth Sciences Division
5523 Research Park Drive, Suite 320
Baltimore, MD 21228

NASA Graduate Student Researchers Program (GSRP) Fellowships 3504
The Program awards fellowships for graduate study leading to masters or doctoral degrees in the fields of science, mathematics, and engineering related to NASA research and development. This twelve month award includes a required internship at the NASA Center affiliated with the NASA sponsored research. The goal of the Program is to cultivate research ties to the academic community, to help to meet the continuing needs of the Nation's aeronautics and space requirements by increasing the number of highly trained scientists and engineers in aeronautics and space-related disciplines, and to broaden the base of students pursuing advanced degrees in science, mathematics, and engineering. Research opportunities are assessed and updated annually to complement the mission requirements of NASA. Research areas are in disciplines that lead to aeronautics and space careers. Training Grants are awarded for one in the amount of $30,000. This amount includes a $21,000 student stipend, a student travel allowance of $4,000, up to $1,000 for health insurance, and a $4,000 university allowance, which typically goes to the Research Adviser, who becomes the Principal Investigator for the Training Grant. The student stipend of $21,000 may cover tuition, room and board, books, software, meal plans, school and laboratory supplies, and other related expenses. The $4,000 student allowance may be used for national and international conferences and data collection. Awards are renewable up to three years based on satisfactory academic advancement, research progress, and available funding. The program supports approximately 300 graduate students annually.
Requirements: Applicants must be currently enrolled or accepted as a full-time graduate student in an accredited U.S. college or university. Applicants must be U.S. citizens. Students may apply at any time during their graduate program, or prior to receiving their baccalaureate degree, provided they have been accepted to an accredited graduate program at a U.S. college or university. All applicants must have a Faculty Adviser (who will be the Principal Investigator of the GSRP Training Grant) from the institution where they plan to receive their graduate degree.
Amount of Grant: $30,000
Date(s) Application is Due: Feb 1
Contact: Dr. Katie Blanding; (202) 358-0402; fax: (202) 606-0122; email: katie.blanding@nasa.gov
Internet: http://fellowships.hq.nasa.gov/gsrp/program/index.cfm
Sponsor: National Aeronautics and Space Administration (NASA)
Office of Education, NASA Headquarters
Washington, DC 20546-0001

NASA Lewis Educational Research and Collaborative Internships 3505
This is an educational program that provides internships for students in science, engineering, professional administrative, and technical areas. Internships are available with various starting dates during the summer at the NASA Glenn Research Center in Cleveland, Ohio. The internships provide students with introductory professional experiences to complement their academic programs. Interns receive assignments in research and development as well as technical and administrative projects under the personal guidance of NASA professional staff members. Each assignment is commensurate with the academic level and field of study of the student. Interns are integrated into the day-to-day activities of the Center to the greatest extent possible. Time is devoted to a program of scheduled educational, professional, and social events. Internships are available for college students, secondary teachers, and high school students.
Requirements: Teachers must be U.S. citizens employed full-time as secondary school teachers. College students must be U.S. citizens and full-time students pursuing a baccalaureate or higher degree in science, engineering, technology, or professional administration at an accredited college or university in the United States. College-bound high school graduates or two-year college students are also eligible. Students must have cumulative GPA of 3.0 on a 4.0 scale.
Date(s) Application is Due: Jan 15; Jan 31
Contact: Mary Roberts, Program Manager; (440) 962-3170, ext. 5006; email: internships@oai.org
Internet: http://www.oai.org/education/student/lercip.html
Sponsor: Ohio Aerospace Institute
22800 Cedar Point Road
Cleveland, OH 44142

NASA Marslink Initiative 3506
This program is an effort to bring one of the most current NASA missions into the classrooms of various U.S. schools across the country. This initiative is designed to impact one school per congressional district. Schools will be chosen on a first-come, first-served basis. The interactive program allows students to simulate a mission to Mars in three phases. The project is funded by $75,000 in matching grant money from NASA. Schools must be able to provide $225 toward the cost of the $500 curriculum.
Requirements: K-12 public schools with sufficient technological capability and ability to meet the required match are eligible to apply.
Amount of Grant: $500
Contact: Tim Lewaren; (800) 965-3763, ext. 262; email: tim@space-explorers.com
Internet: http://www.state.nj.us/cgi-bin/education/grants/grants_active2.pl?string=recnum=00050&maxhits=500&fed=1
Sponsor: National Aeronautics and Space Administration (NASA)
1825 Nimitz Drive
De Pere, WI 54115

NASA New Aeronautics Competition for University Students 3507
University students are invited to design the next generation, 21st Century, DC-3 type aircraft. Some of the design considerations include reduced environmental impact, reduced noise, daily operations on short runways, cost analysis for production and operation, passenger and cargo limits, structure and materials, engines. Students should also briefly describe three valid scenarios for potential use of this vehicle in the year 2058. Entries will be limited to a 25-page research paper, excluding ancillary material, and adhere to a standard format. Teams or individuals may enter. Winning entries may be invited to a student forum sponsored by ARMD and/or industry, may receive offers of student internships, and may receive other prizes, including cash, depending on available funds. Cash prize amounts will be on the order of $5000 for first place, $3500 for second place, $2000 for third place. In the case of a tie, prizes will be determined based on funds available.
Requirements: Any U.S. citizen enrolled in an accredited college or university in the U.S. or its territories is eligible to enter the competition. International students, either in the U.S. or elsewhere, may enter but will not be eligible for certain prizes. Trophies and certificates will be awarded to each winner, regardless of citizenship. Students who are receiving NASA funds for research projects or who are working on a NASA funded project for a faculty member should disclose this information in their entry packet.
Restrictions: Only U.S. citizens are eligible for cash prizes or NASA funded internships.
Samples: Heavy MEDL, Team Georgia Institute of Technology, $5,000 (2007); Caelus Team, Subsonic Transport, University of Virginia, $3,500 (2007); Mars Heavy-Payload Entry, Descent, and Landing Team, Georgia Institute of Technology, $2,000 (2007); Adaptable Airfoil, Supersonic Flight, Montana State University, $2,000 (2007).
Amount of Grant: $2,000-$5,000
Date(s) Application is Due: Apr 30
Contact: Dr. Elizabeth Ward, High School Competition; (757) 864-7638; email: elizabeth.b.ward@nasa.gov
Internet: http://aero.larc.nasa.gov/competitions_univ.htm
Sponsor: National Aeronautics and Space Administration (NASA)
Mail Stop 218, Building 1212, Room 128, NASA Langley Research Center
Hampton, VA 23681

NASA Undergraduate Student Researchers Project (U.S.RP) Grants 3508
The Program offers undergraduate students across the United States internship experiences at NASA Centers. Students may apply for a 10-week summer session or a 15-week spring or fall session. The purpose of the Program is three-fold: to attract undergraduate students from the widest array of backgrounds, who are fully representative of America's racial, ethnic, and cultural diversity and to provide them with hands-on, challenging research experiences that stimulate continued student interest in the fields/disciplines aligned with NASA's research and development mission; to build a national program bridge - from existing NASA K-12 Education Program activities to NASA Higher Education Program options - that encourages and facilitates student interest in future professional opportunities with NASA and its partner organizations; and to extend and strengthen NASA's commitment to educational excellence and university research, highlighting the critical need to increase the Nation's undergraduate and graduate science, engineering, mathematics, and technology skill base. Students will receive a $6,000 (10-week summer session) or $9,000 (15-week spring or fall session) stipend for the research experience plus one round-trip airfare or ground transportation allowance to and from the NASA Host Center.
Requirements: Applicants must meet the following eligibility requirements to participate: U.S. citizenship; classified as a Junior or Senior after Spring semester/quarter of the previous year; enrolled full-time in an accredited U.S. college or university (community college students may apply, however placement is contingent on matriculation into a four-year institution as a rising Junior); academic major or demonstrated coursework

concentration in engineering, mathematics, computer science, or physical/life sciences; demonstrated strong interest/commitment to one of the above career fields or disciplines (e.g., science/math/engineering fairs, clubs, or awards; tutoring/mentoring; internships or other related experience); and minimum cumulative grade point average of 3.0 on a scale of 4.0.
Amount of Grant: $6,000-$9,000, plus transportation allowance
Date(s) Application is Due: Jan 31; Feb 28
Contact: Nick Koltun, Educational Program Specialist; (757) 766-5210; fax: (757) 766-5205; email: nkoltun@odu.edu
Internet: http://www.vsgc.odu.edu/Menu3_1_3.htm
Sponsor: National Aeronautics and Space Administration (NASA)
Virginia Space Grant Consortium; 600 Butler Farm Road, Suite 200
Hampton, VA 23666

NASDAQ Educational Foundation Grants **3509**
The foundation seeks grants proposals that advance its mission of financial markets literacy through: educational projects or programs, university campus trading labs, curriculum development, academic study or research, and PhD dissertation fellowships (examples of appropriate topics—capital formation, free market economy, the function and operation of securities markets, investing/finance, and business/entrepreneurship). Targeted audiences include high school students, college and university communities, middle school students, and adults. Guidelines and instructions are available online.
Requirements: Proposals are accepted from 501(c)3 educational institutions and organizations and specifically qualified individuals, for the purpose of conducting independent academic study or research on financial markets.
Amount of Grant: $15,000 maximum dissertation fellowships
Date(s) Application is Due: Mar 1; Sep 1
Contact: Grants Administrator, (800) 842-0356; email: Foundation@NASDAQ.com
Internet: http://www.nasdaq.com/services/education_initiatives.stm
Sponsor: NASDAQ Educational Foundation
1801 K Street NW, 8th Fl
Washington, DC 20006

NASM Guggenheim Fellowships **3510**
Through support of the Guggenheim Foundation, the National Air and Space Museum sponsors one-year resident fellowships for historical and scientific research related to aviation and space. Fellowships begin between July 1 and October 1. Requests for applications should be mailed to the address listed.
Requirements: The fellowship is open to predoctoral and postdoctoral applicants. Predoctoral applicants should have completed preliminary coursework and examinations and be engaged in dissertation research. Postdoctoral applicants preferably should have received the PhD within the past seven years.
Amount of Grant: $20,000 predoctoral stipend; $30,000 postdoctoral stipend
Date(s) Application is Due: Jan 15
Contact: Collette Williams, Fellowship Program Coordinator, (202) 633-2648; fax: (202) 786-2447; email: collette.williams@nasm.si.edu
Internet: http://www.nasm.si.edu/getinvolved/fellow/gvrfellowinfo.cfm
Sponsor: National Air and Space Museum
P.O. Box 37012, Sixth and Independence Avenue SW, Room 3313, MRC 312
Washington, DC 20560

NASM Ramsey Fellowship in Aviation History **3511**
The Ramsey fellowship is a 12-month, in-residence fellowship focused on the history of aviation at sea and in naval service, particularly the U.S. Navy. The fellowship is open to all interested candidates with demonstrated skills in research and writing. A stipend will be awarded for a 12-month fellowship, with limited additional funds for travel and miscellaneous expenses.
Requirements: The fellowship is open to all interested candidates with demonstrated skills in research and writing. An advanced degree is not required. Residence in the Washington, DC, metropolitan area during the fellowship term is a requirement of the fellowship.
Amount of Grant: $45,000
Date(s) Application is Due: Jan 15
Contact: Collette Williams, Fellowship Coordinator, (202) 633-2648; fax: (202) 786-2447; email: collette.williams@nasm.si.edu
Internet: http://www.nasm.si.edu/getinvolved/fellow/fellowap2004-5.pdf
Sponsor: National Air and Space Museum
955 L'Enfant Plaza, Suite 7000
Washington, DC 20560

NASM Verville Fellowships **3512**
This is a 12-month fellowship intended for the analysis of major trends, developments, and accomplishments in the history of aviation or space studies. Candidate need not hold an advanced degree in history, engineering, or related field to qualify. Applications will be mailed November 1. Notification will be made by April 15 with the program beginning between July 1 and October 1. Request an application packet from the address listed.
Requirements: The fellowship is open to all interested candidates with demonstrated skills in research and writing.
Amount of Grant: $45,000
Date(s) Application is Due: Jan 15
Contact: Collette Williams, Fellowship Program Coordinator, (202) 357-2515; fax: (202) 786-2447; email: collette.williams@nasm.si.edu
Internet: http://www.nasm.si.edu/getinvolved/fellow/vfellow.cfm
Sponsor: National Air and Space Museum
P.O. Box 37012, Sixth and Independence Avenue SW, Room 3313, MRC 312
Washington, DC 20560

Nathan Burkan Memorial Competition **3513**
Each law school holds a competition for the best papers on the subject of copyright law. Information and rules governing the competition are forwarded to ABA-accredited law schools. All winning entries are published.
Requirements: Competition is open to third-year law school students at schools accredited by the American Bar Association. Second-year students may also be eligible at the discretion of the dean. Second-year students who receive awards may not compete in their third year.
Restrictions: As papers are presumed to represent individual study and research, collaboration with others in their preparation is not permitted.
Amount of Grant: $600 first prize; $250 second prize law school competition
Date(s) Application is Due: Jun 30
Contact: Director, Nathan Burkan Memorial Competition, (212) 621-6000; fax: (212) 724-9064; email: nbmc@ascap.com
Internet: http://www.ascap.com/burkan/burkanrules.html
Sponsor: American Society of Composers, Authors, and Publishers
1 Lincoln Plaza
New York, NY 10023

National Academies Postdoctoral and Senior Research Awards **3514**
The program administers postdoctoral and senior research awards through its associateship programs, part of the Policy and Global Affairs Division. The Research Associateship Programs (RAP) are sponsored by federal laboratories and NASA research centers at over 100 locations in the United States and overseas. Awards are given for the purpose of conducting research at a specific laboratory chosen by the applicant; and are made to doctoral level scientists and engineers who can apply their special knowledge and research talents to research areas that are of interest to them and to the host laboratories and centers. Guidelines and application are available online.
Requirements: Postdoctoral research associateships are awarded to persons who have held the doctorate less than five years at the time of application and are made initially for one year. Senior research associateships are awarded to applicants who have held the doctorate five years or more at the time of application. Senior research associateship applicants should have research experience that has resulted in significant contributions and recognition as established investigators in their specialized fields.
Amount of Grant: $45,000 minimum yearly stipend
Date(s) Application is Due: Feb 1; May 1; Aug 1; Nov 1
Contact: Postdoctoral Research Awards, (202) 334-2760; fax: (202) 334-2759; email: rap@nas.edu
Internet: http://www4.nationalacademies.org/pga/rap.nsf/WebDocuments/Home+Page
Sponsor: National Academy of Sciences
500 Fifth Street NW, GR 322A
Washington, DC 20001

National Academy of Education Spencer Postdoctoral Fellowships **3515**
Fellowships are designed to promote scholarly inquiry and discussion concerning the ends and means of education, in all of its forms, in the United States and abroad. Each year the National Academy of Education solicits applicants in education, the humanities, and the social and the behavioral sciences; their interests must describe research whose relevance to education is apparent. Fellows are expected to devote one academic year to full-time research or two contiguous academic years to half-time research. Applications may be made directly by the individual. Up to 30 fellowships will be awarded each year. This fellowship is open to citizens of all countries; the grant is expected to be used for salary replacement or to pay related research costs. Guidelines and application packet are available online.
Requirements: Applicants must have received the PhD, EdD, or equivalent degrees between January 1, 2000, and December 31 of the current year.
Restrictions: Concurrent funding for the proposed project is not permitted.
Amount of Grant: $55,000 for one academic year of research or $27,500 for each of two contiguous years, working part-time
Date(s) Application is Due: Nov 10
Contact: Spencer Fellowship Program, 202-334-2341; email: nae.info@nyu.edu
Internet: http://www.nae.nyu.edu
Sponsor: National Academy of Education
500 Fifth Street NW, No 1049
Washington, DC 20001

National Assessment of Educational Progress Visiting Scholar Program **3516**
ETS sponsors the visiting scholar program to enable scholars to conduct their own independent research using the NAEP database with access to senior NAEP staff and ETS research staff and have an opportunity to contribute to the effectiveness of the program, especially as it pertains to African American, Hispanic, and other minority students by identifying productive areas of investigation for future NAEP activities relevant to the education of those minority students. The goals of this program are to provide research opportunities for recent awardees of the doctorate and to increase the number of women and minority professionals working in the areas specified. The term of visit is normally from September 1 through June 30. Stipend is set in relation to compensation at the guest scholar's home institution. In addition, limited relocation

expenses consistent with ETS guidelines will be reimbursed upon presentation of receipts.
Requirements: The program is open to any individual who holds a doctorate in a relevant discipline and provides evidence of prior research relevant to the education of minorities.
Contact: Linda DeLauro, Fellowship Programs, (609) 734-1806; fax: (609) 683-2800; email: internfellowships@ets.org
Internet: http://www.ets.org/portal/site/ets/menuitem.22f30af61d34e9c39a77b13bc3921509/?vgnextoid=49f5be3a864f4010VgnVCM10000022f95190RCRD
Sponsor: Educational Testing Service
MS 18N, Rosedale Road
Princeton, NJ 08541

National Association of Black Journalists Scholarship 3517
The scholarship will be awarded to a college undergraduate or graduate student. Applicants must have completed at least one full semester of college to be eligible. Eligible students must be planning to major or currently be majoring in journalism—print, photography, radio, or television—or planning a career in one of those fields. Annual deadline dates may vary; contact program staff for exact dates.
Requirements: African American college students enrolled in graduate school are eligible.
Restrictions: Previous winners are ineligible.
Amount of Grant: $5000 maximum
Date(s) Application is Due: Apr 1
Contact: Scholarship Coordinator, (301) 445-7100; fax: (301) 445-7101; email: nabj@nabj.org
Internet: http://www.nabj.org/programs/scholarships/index.html
Sponsor: National Association of Black Journalists
University of Maryland, 8701-A Adelphi Road
Adelphi, Md 20783-1716

National Association of Broadcasters Grants for Research in Broadcasting 3518
The purpose of this program is to stimulate interest in broadcast research, especially on economic, social, cultural, or policy issues of importance to the U.S. commercial broadcast industry. The goal is to make high-quality academic research available to industry practitioners as well as other academics. Application forms are available on the Web site. Annual deadline dates may vary; contact the program office for exact dates.
Requirements: Individuals eligible to apply include academic personnel working in fields that relate to the social, cultural, political, or economic aspects of broadcasting. Undergraduate seniors and graduate students in the field of broadcasting also are eligible.
Restrictions: Research proposals concerning foreign broadcasting, and instructional or public radio and television, are not eligible.
Amount of Grant: $25,000 total
Date(s) Application is Due: Jan 30
Contact: Brian Baxter, (202) 429-5377; email bbaxter@nab.org
Internet: http://www.nab.org/Research/Grants/grants.asp
Sponsor: National Association of Broadcasters
1771 N Street NW
Washington, DC 20036

National Black MBA Association Annual Scholarships 3519
NBMBAA is a nonprofit organization of minorities that hold or are currently pursuing graduate business degrees. Students applying for the national scholarships (27 awarded each year) must complete the official NBMBAA scholarship application, provide verification of grade point average and enrollment, complete a financial need assessment provided, and write a two-page essay that responds to a topic specified by the association. There are also 30 state chapters, many of which offer local scholarships in addition to supporting the national scholarships. Scholarships are offered at the undergraduate, master's, and doctoral levels. Students wishing to pursue chapter scholarships must contact the chapter president in their area for details; a list of chapters, presidents, and their mailing addresses is available upon request.
Requirements: Minority students actively pursuing graduate degrees in business are eligible to apply. They must be enrolled full time in accredited graduate business programs in either the fall or winter term of the year in which they apply.
Amount of Grant: $7000-$12,500
Date(s) Application is Due: Apr 30
Contact: Lori Johnson, NBMBAA National Office, (312) 236-2622 ext 8086; fax: (312) 236-4131; scholarship@nbmbaa.org
Internet: http://nbmbaa.org/scholarship.cfm
Sponsor: National Black MBA Association
180 N Michigan Avenue
Chicago, IL 60601

National Black Programming Program Development Grants 3520
The consortium funds and promotes programs from the African American community and offers them to the public broadcasting system. NBPC funds projects in various stages, including production, postproduction, distribution, and acquisition. Proposals should include the completed application form, tape of work in progress and treatment summary, goal statement, program format and length, production personnel list with brief bios, itemized budget, timeline, list of consultants, and additional materials. The complete proposal should be no more than 10 pages in length, including supplemental information. Annual deadline dates may vary; contact program staff for exact dates.
Samples: Mississippi America (MS)—postproduction support of a one-hour documentary, $10,000.
Amount of Grant: $1000-$80,000 range
Date(s) Application is Due: Jun 3
Contact: Program Development Fund, (212) 234-8200; fax: (212) 234-7032; email: info@nbpc.tv
Internet: http://www.nbpc.tv/grants.php
Sponsor: National Black Programming Consortium
68 E 131th Street, 7th Fl
Harlem, NY 10037

National Blood Foundation Research Grant Program 3521
Research grants support investigator-initiated original research in blood banking and transfusion medicine. Projects dealing with all aspects of blood banking, including transfusion sciences, blood banking administration, and education will be considered. Priority will be given to the following areas: transfusion medicine and immunology, hematology, immunohematology, infectious diseases, and medicine administration. Preference will be given to innovative new projects; applications from junior and less well-established investigators are encouraged. Grants will be awarded for periods of one or two years. Applications are available on the Web site.
Requirements: Preference is given to new investigators in the field. There is a $150 application fee.
Amount of Grant: $65,000 maximum
Date(s) Application is Due: Dec 13
Contact: Program Contact, (301) 215-6552; fax: (301) 907-6895; email: nbf@aabb.org
Internet: http://www.aabb.org/about_the_aabb/nbf/nbf2.htm#3
Sponsor: National Blood Foundation
8101 Glenbrook Road
Bethesda, MD 20814-2749

National Book Scholarship Fund 3522
The purpose of the National Book Scholarship Fund (NBSF) is to distribute direct in-kind assistance to local literacy providers throughout the United States. NBSF grants are intended to provide the New Readers Press books and other materials essential to begin a new literacy outreach program or to significantly expand an existing effort. First priority will be given to programs that focus on women and intergenerational or family literacy concerns. Other organizations working in economically disadvantaged urban and rural areas, among homeless populations, in refugee resettlement efforts, in workplace literacy efforts, and English-as-a-second-language initiatives will also receive consideration for support. Types of support include program development grants, research grants, scholarships, technical assistance, and training grants. Applications are accepted from September until December. applicants are encouraged to apply early because NBSF applications are reviewed and awards are made on a first-come, first-served basis.
Requirements: Programs applying for a grant must agree to provide, from local sources, a cash payment equal to 20% of the total amount being requested in the grant application to defray the costs of NBSF administration.
Amount of Grant: $500-$8,000 range
Date(s) Application is Due: Apr 14
Contact: Michele Diecuch, Program Administrator; (888) 528-2224 or (315) 422-9121, ext. 321; fax: (315) 422-6369; email: mdiecuch@proliteracy.org
Internet: http://www.servicelearning.org/funding-sources#category310
Sponsor: ProLiteracy Worldwide
1320 Jamesville Avenue
Syracuse, NY 13210

National Council of Farmer Cooperatives Graduate Awards 3523
The awards support one doctoral dissertation and two master's theses with themes dealing with some aspect of economics, finance, operation, or structure of American agricultural cooperatives. The Edwin G. Nourse award is for the best doctoral dissertation. The Kenneth D. Naden award and E.A. Stokdyk award are for the two best master's theses. Application forms are required.
Requirements: Any graduate student in economics, business, communications, sociology, or other relevant field who has an agricultural vocational objective is eligible. All entries must be completed between April 1 of the previous year and March 31 of the year of the award and must be accompanied by a registration form available from NCFC.
Amount of Grant: $1500 Kenneth D. Naden Award; $1000 Edwin G. Nourse Award; and $600 E.A. Stokdyk Award
Date(s) Application is Due: Apr 15
Contact: Catherine Wilson, Executive Assistant, (202) 626-8700; fax: (202) 626-8722
Internet: http://www.ncfc.org
Sponsor: National Council of Farmer Cooperatives Education Foundation
50 F Street NW, Suite 900
Washington, DC 20001

National Dairy Promotion and Research Board Scholarships 3524
Scholarships are awarded annually to up to 19 students enrolled in college/university programs that emphasize dairy food science, technology, or marketing. Scholarships are awarded based upon outstanding academic achievement; a commitment to a career in a dairy food-related discipline; and demonstrated leadership, initiative, and integrity. Application forms are available on the Web site.
Requirements: Applications must be accompanied by an official transcript of all completed university credits and a letter of recommendation from a faculty member familiar with the applicant's scholastic performance. Sophomore, junior, and/or senior undergraduate

students enrolled in college/university programs that emphasize dairy and majoring in communications, public relations, journalism, marketing, business, economics, nutrition, food science or ag education are eligible to apply.
Amount of Grant: $1500
Date(s) Application is Due: May 31
Contact: Scholarship Administrator, (847) 803-2000 ext 255 or (800) 853-2479
Internet: http://www.dairycheckoff.com/DairyCheckoff/about/scholarship.htm
Sponsor: National Dairy Council
10255 W Higgins Road, Suite 900
Rosemont, IL 60018-5616

National Education Reporting Fellowships 3525
The association offers a study/travel fellowship for a two-month period for reporters to pursue stories of their own design. Fellows will receive half-salary for the two-month study period; travel expenses; access to expert sources; editing assistance; the opportunity to share work with other reporters at EWA regional meetings and the national seminar; and consultation with EWA staff about progress and problems during the project. A distinguished panel of journalists will select the fellows on the basis of the quality of their proposals. Annual deadlines fall in May; contact the office for specific dates.
Requirements: The fellowships are open to full-time print or broadcast journalists who have been covering education for at least two years, have the endorsement of their employers, and can show that they have likely outlets for the products of their studies. Freelance writers may apply if they write about education for a substantial portion of their time and if they have agreements with media organizations to publish or air stories resulting from the fellowships.
Contact: Lisa Walker, Executive Director, (202) 452-9830; fax: (202) 637-9707; email: lwalker@ewa.or
Internet: http://www.ewa.org/desktopdefault.aspx?page_id=12
Sponsor: Education Writers Association
2122 P Street NW, Suite 201
Washington, DC 20037

National Emergency Medicine Association Grants 3526
The association supports projects that foster the development and delivery of quality emergency services, with special emphasis given to heart attack and stroke. Proposals should address education or direct services to victims. Also to receive consideration are childrens' emergency issues and projects teaching good parenting skills. Programs must be related to health and emergency issues and prevention or diminishing of trauma events. Types of support include capital, program development, research, and training grants. Applications are accepted on an ongoing basis.
Samples: Beebe Medical Ctr (Lewes, DE)—to improve its trauma care services for people in southern Delaware, $83,778.; National Stroke Assoc (Englewood, CO)—to update a publication that educates emergency room physicians about emergency evaluation and treatment of stroke victims, $17,000.
Contact: Grants Administrator, (410) 494-0300; fax: (410) 494-0725; email: info@nemahealth.org
Internet: http://www.nemahealth.org/organization/grants.htm
Sponsor: National Emergency Medicine Association
306 W Joppa Road
Baltimore, MD 21204

National Endowment for Democracy Discretionary Grants 3527
The objective of the worldwide grant program is to strengthen democratic institutions throughout the world through private, nongovernmental efforts. The endowment is currently funding programs abroad in the areas of pluralism—development of strong, independent, private sector organizations, especially trade unions and business associations; democratic governance and political processes—promotion of strong, stable political parties committed to the democratic process; education, culture, and communications—programs that nourish a strong democratic civic culture including support for publications and other communications media and training programs; the production and dissemination of books and other materials to strengthen popular understanding and intellectual advocacy of democracy; and programs of democratic education. The Endowment is especially interested in proposals that originate with indigenous democratic groups. It is also interested in nonpartisan programs seeking to strengthen democratic values among all sectors of the democratic political spectrum. To a much lesser extent, the endowment also funds programs in international cooperation and research. The endowment staff is willing to discuss proposal ideas and welcomes preliminary letters of inquiry prior to submission of formal proposals. Applications are accepted any time, but are due six weeks prior to the board meeting. The board meets quarterly to approve grants; call to confirm deadlines.
Restrictions: The endowment does not make grants to individuals or for projects within the domestic United States.
Date(s) Application is Due: Jan 18; Mar 28; Jun 13; Sep 12
Contact: Public Information Officer; (202) 378-9700; fax: (202) 378-9407; email: info@ned.org
Internet: http://www.ned.org/grants/grants.html
Sponsor: National Endowment for Democracy
1025 F Street, NW, Suite 800
Washington, DC 20004

National Endowment for Democracy Reagan-Fascell Democracy Fellowships 3528
The program enables democracy activists, practitioners, scholars, and journalists from around the world to deepen their understanding of democracy and enhance their ability to promote democratic change. Fellows maintain full-time residence at the International Forum for Democratic Studies, the research arm of the Endowment. The Forum hosts 12 to 15 Reagan-Fascell fellows per year for periods ranging from three to ten months. The program offers two tracks, a practitioner track and a scholarly track. Fellowships on the practitioner track tend to be short-term (three to five months), typically culminating in a strategy memorandum, short article, or op-ed and a presentation of the fellow's analysis and ideas. The scholarly track is principally for professors and researchers from emerging and aspiring democracies. Accomplished scholars from established democracies are also eligible to apply. During their three- to 10-month stay at the Forum, scholars make at least one presentation and complete a substantial piece of writing (a monograph or book) for publication. Each fellow receives a monthly stipend for living expenses, plus health insurance and reimbursement for travel to and from Washington, DC. Fellows also receive a fully equipped office and support services. Application and guidelines are available online.
Requirements: There are no specific degree requirements for applicants selecting the practitioner track; candidates are expected to have substantial practical experience working on behalf of some aspect of democracy building—such as human rights advocacy, political party activism, voter education, election monitoring, journalism and press freedom, ethnic pluralism, civic education, etc. Individuals applying in the scholarly track are expected to possess a PhD or academic equivalent at the time of application, to have published in their field of expertise, and to have developed a detailed research outline for their fellowship project. Examples of eligible candidates for the scholarly track include college and university professors, scholars, analysts at research centers, and independent writers.
Date(s) Application is Due: Nov 10
Contact: Program Assistant, Fellowship Programs; (202) 378-9700; fax: (202) 378-9407; email: fellowships@ned.org
Internet: http://www.ned.org/forum/reagan-fascell.html
Sponsor: National Endowment for Democracy
1025 F Street, NW, Suite 800
Washington, DC 20004

National Foundation for Catholic Youth Grants 3529
In an effort to provide support for the field of Catholic youth ministry, the foundation awards financial grants to parish, diocesan, and national initiatives. Grant projects are activities, events, strategies, or campaigns designed to train youth and coordinators of youth ministry, to develop new approaches to comprehensive youth ministry, and to respond to the needs of today's young church. Guidelines are available online.
Requirements: The foundation awards financial grants to parish, diocesan, and national initiatives.
Restrictions: The foundation does not provide scholarships for undergraduate or postsecondary education.
Samples: Los Embajadores in the Archdiocese of Portland (OR)—for group-leader training for the missionary outreach program to underdeveloped communities in Mexico, $1250.; Diocese of El Paso (TX) for youth leadership development training, $1000.; Diocese of Knoxville (TN)—for youth leadership training, $2000.; European Catholic Youth Conference for American youth living on military bases $3000.; Good Shepherd Parish (Tallahassee, FL)—for the God's House ecumenical youth outreach, $500.
Amount of Grant: $500-$4000
Date(s) Application is Due: Dec 9
Contact: Grants Administrator, (202) 636-3825; fax: (202) 526-7544; email: info@cyfusa.org
Internet: http://www.cyfusa.org/grants.html
Sponsor: National Foundation for Catholic Youth
415 Michigan Avenue NE, Suite 40
Washington, DC 20017-4502

National Geographic Society Conservation Trust Grants 3530
The objective of the trust is to support conservation activities around the world. Grants support projects that contribute significantly to the preservation and sustainable use of the Earth's biological, cultural, and historical resources. Researchers planning work in foreign countries should include at least one local collaborator as part of their research teams. The strength of the program is to identify cutting edge programs that might be overlooked due to inherent risks associated with new investigators and fields. Funds are intended to function as complementary support; the trust strongly encourages applicants to seek additional, concurrent funding from other funding agencies. Before receiving an application form, each principal investigator must submit a pre-application form online. If the pre-application is approved, the principal investigator will be sent an electronic application and instructions.
Requirements: Applicants are not expected to have advanced degrees (PhD or equivalent). However, applicants must provide a record of prior research or conservation action as it pertains to the proposed project.
Restrictions: Society grants may not be used for indirect costs, overhead, and other expenses not directly related to the project. Fringe benefits are also excluded, as are salaries. Funds may not be used for travel to scientific/professional meetings or conferences, legal actions, land acquisition, endowments, construction of permanent field stations, or to publish research results.
Amount of Grant: $15,000-$20,000 average
Contact: Grants Administrator; (800) 647-5463; email: conservationtrust@ngs.org
Internet: http://www.nationalgeographic.com/conservation
Sponsor: National Geographic Society
1145 17th Street NW
Washington, DC 20036-4688

National Geographic Society Expeditions Council (EC) Grants 3531

The Expeditions Council consists of representatives from National Geographic editorial divisions (magazines, television, books, and so on) who review and vote on grant applications and an advisory board of external consultants. The Expeditions Council is editorially driven; projects must have the potential to yield compelling stories and images. Applications are also judged on the qualifications of applicants and their teams and on the merit and uniqueness of the project. In addition to financial support, the Expeditions Council offers its grantees the opportunity to work effectively with the National Geographic's many divisions. Grantees are therefore able to share the results of their expeditions with the National Geographic global audience. Grants generally range from $5,000 to $35,000 and are to be used for direct field expenses: transportation, supplies, subsistence, and permit costs as well as other related fees.
Requirements: International applicants are encouraged. However, submissions must be made in English if they are to receive timely consideration by the Expeditions Council. Applicants are expected to have qualifications and experience pertinent to the expedition or project they propose, and advanced academic degrees are not required. Those planning work in countries other than their own should consider including at least one local collaborator as part of their expedition team.
Restrictions: The Expeditions Council does not provide fees for photography, videography, and writing. Society grants may not be used for indirect costs, overhead, and other expenses not directly related to the project. Fringe benefits are also excluded, as are salaries. Funds may not be used for travel to scientific/professional meetings or conferences, legal actions, land acquisition, endowments, construction of permanent field stations, or to publish research results.
Amount of Grant: $5,000-$35,000
Contact: Grants Administrator; (800) 647-5463; email: ecouncil@ngs.org
Internet: http://www.nationalgeographic.com/field/grants-programs/expeditions-council.html
Sponsor: National Geographic Society
1145 Seventeenth Street, NW
Washington, DC 20036-4688

National Geographic Society Field Research and Exploration Grants 3532

The National Geographic Society awards grants for scientific field research and exploration through its Committee for Research and Exploration. The program provides grants throughout the world for basic research in the sciences pertinent to geography including, but not limited to, anthropology, archaeology, astronomy, biology, botany, geography, geology, oceanography, paleontology, and zoology. Particular emphasis is placed on multidisciplinary projects of an environmental nature. Priorities favor research that addresses pressing environmental concerns. All projects must have geographic relevance or significance. Types of support include grants in aid, research grants, and seed money grants. Prior to applying, a potential investigator must submit a pre-proposal (500 words maximum length) outlining the significance of the research and describing who will conduct the research and where it will occur.
Requirements: Applicants are expected to have advanced degrees (Ph.D. or equivalent) and be associated with an educational organization or institution Independent researchers or those pursuing a Ph.D.-level degree may apply, but awards to non-Ph.D. applicants are rare. As a general rule, all applicants are expected to have published a minimum of three articles in peer-reviewed scientific journals.
Restrictions: Society grants may not be used for expenses that are not directly related to the project. Fringe benefits for any individual or institution are excluded, as are salaries for those with faculty status, or any other paid professional affiliation. Funds may not be used for travel to scientific/professional meetings or conferences, nor to publish research results.
Amount of Grant: $15,000-$20,000 average
Contact: Grants Administrator; (800) 647-5463; email: cre@ngs.org
Internet: http://www.nationalgeographic.com/research/grant/rg1.html
Sponsor: National Geographic Society
1145 Seventeenth Street, NW
Washington, DC 20036-4688

National Geographic Society Young Explorers Grants 3533

Young Explorers Grants (YEG) offer opportunities to individuals between the ages of 18 and 25 to pursue research, conservation, and exploration-related projects consistent with National Geographic's existing grant programs, including: the Committee for Research and Exploration (CRE), the Expeditions Council (EC), and the Conservation Trust (CT). Grants range between $2,000 and $5,000.
Requirements: Applicants are not expected to have advanced degrees (PhD or equivalent). However, applicants must provide a record of prior research or conservation action as it pertains to the proposed project.
Restrictions: Society grants may not be used for indirect costs, overhead, and other expenses not directly related to the project. Fringe benefits are also excluded, as are salaries. Funds may not be used for travel to scientific/professional meetings or conferences, legal actions, land acquisition, endowments, construction of permanent field stations, or to publish research results.
Amount of Grant: $2,000-$5,000 average
Contact: Grants Administrator; (800) 647-5463; email: cre@ngs.org or conservationtrust@ngs.org
Internet: http://www.nationalgeographic.com/field/grants-programs/yeg-application.html
Sponsor: National Geographic Society
1145 Seventeenth Street, NW
Washington, DC 20036-4688

National Geographic Society/Waitt Grants 3534

The National Geographic Society/Waitt Grants Program funds projects that require venture capital, supporting exceptional projects while foregoing a time-consuming peer-review process. NGS/Waitt grants are able to fund proof of concept research for applicants at an earlier stage in their careers than other NGS grant programs. Special emphasis is placed on expedited grant processing and turnaround. The selection committee endeavors to have funding decisions made within eight weeks of application submission. Applications are accepted on a rolling basis.
Requirements: Funding is not restricted to United States citizens. Applicants planning work in foreign countries should include at least one local collaborator as part of their research teams. Applicants are not required to have advanced degrees to be eligible for funding, though they will be required to show a commensurate level of expertise and experience. Applicants must have an affiliation with an educational organization or other institution.
Restrictions: This grant program does not pay educational tuition, nor does it offer scholarships or fellowships of any kind. Society grants may not be used for indirect costs, overhead, and other expenses not directly related to the project. Fringe benefits are also excluded, as are salaries. Funds may not be used for travel to scientific/professional meetings or conferences, legal actions, land acquisition, endowments, construction of permanent field stations, or to publish research results.
Amount of Grant: $5,000-$15,000
Contact: Grants Administrator; (800) 647-5463; email: waitt@ngs.org
Internet: http://www.nationalgeographic.com/field/grants-programs/waitt-grants.html
Sponsor: National Geographic Society / Waitt Family Foundation
1145 Seventeenth Street, NW
Washington, DC 20036-4688

National Headache Foundation Research Grants 3535

The National Headache Foundation is interested in research protocols that are objectively sound and whose results can, when published in the medical literature, contribute to the better understanding and treatment of headache and pain. Grants may be used for the purchasing of supplies needed for the study (chemical or pharmacological reagents, laboratory animals, tissue culture materials. forms, etc.) and for costs related to the recruiting of patients, data analysis, interpreting or reading results, etc. Applications are available online, and are due by December 1.
Requirements: Funds for a research project approved by the National Headache Foundation are awarded to the individual who is the principal investigator of the study. The funds are sent to the institution that employs the PI but designated for the PI and the approved study.
Restrictions: Grants from the NHF will not cover costs including overhead, salaries (payment for technicians, others collecting data, administrative services, etc.), purchasing of equipment, rent or other indirect expenses.
Date(s) Application is Due: Dec 1
Contact: Carolyn Smith, Executive Assistant; (312) 274-2652 or (312) 274-2650; fax: (312) 640-9049; email: info@headaches.org
Internet: http://www.headaches.org/For_Professionals/Research_Grants
Sponsor: National Headache Foundation
820 N. Orleans, Suite 217
Chicago, IL 60610-3132

National Headache Foundation Seymour Diamond Clinical Fellowship in Headache Education 3536

The one-year fellowship is intended to encourage, provide for, and engage physicians in the clinical management of headache patients. The program is an individualized hands-on approach at a headache clinic mutually selected by the committee and the applicant. The objectives are to develop competence in the management of common and complex headache problems, to publish a research paper keyed to the treatment of headache, and to prepare the applicant for a career in treating headache patients. The participating institution will be a specialized facility dedicated primarily to headache management, with facilities for inpatient and outpatient treatment. Up to 80 percent of the fellow's time will be clinical, 20 percent study/research.
Requirements: Applicant will be a graduate of an approved U.S. medical school and be four years or more postgraduate.
Amount of Grant: $40,000 stipend
Date(s) Application is Due: Feb 1
Contact: Fellowship Selection Committee; (312) 274-2650; fax: (312) 640-9049; email: info@headaches.org
Internet: http://www.headaches.org/For_Professionals/Fellowship_Grants
Sponsor: National Headache Foundation
820 N. Orleans, Suite 217
Chicago, IL 60610-3132

National Heart Foundation of New Zealand Grants 3537

The foundation is dedicated to the reduction of premature death and disability from heart and circulatory disease, including stroke. Types of support include research project grants, training fellowships enabling New Zealand cardiology trainees to study abroad, senior fellowships, and a chair in cardiovascular studies.
Contact: Program Contact, (09) 571 9191; fax: (09) 571 9190; email: info@nhf.org.nz
Internet: http://www.nhf.org.nz/index.asp?PageID=2145820407
Sponsor: National Heart Foundation of New Zealand
P.O. Box 17160
Greenlane, Auckland 1130 New Zealand

National History Day Brown v Board of Education Essay Contest **3538**
The program invites students in grades 6 to 12 to participate in a national essay contest in conjunction with the National History Day competition. As part of the National History Day program, students are encouraged to write an essay with supporting research that examines the Brown decision and its ongoing impact. The essay must relate to the theme Exploration, Encounter, Exchange in History and should follow the National History Day rules for the paper category, which can be found on the Website. Papers selected at district or regional National History Day contests to advance to the state contest are eligible to compete in the special Brown v Board essay contest. For information on deadlines for district contests contact the state coordinator.
Amount of Grant: $1000 first place; $500 second place; $350 third place
Contact: Bill Lickiss, Contest and Program Manager, (301) 314-9739; email: bill@nhd.org
Internet: http://www.nationalhistoryday.org/brownvboard.htm
Sponsor: University of Maryland
0119 Cecil Hall
College Park, MD 20742

National Housing Endowment Challenge/Build/Grow Grant **3539**
The National Housing Endowment is the philanthropic arm of the National Association of Home Builders (NAHB). Working with organizations within the NAHB federation, the National Housing Endowment is offering eligible applicants up to $5,000 in matching challenge grant funds (dollar for dollar) for programs and projects addressing NAHB, highest priority areas of Education, Land Use and Labor Shortage/Worker Training.
Requirements: The program/project must address the National Housing Endowment mission of furthering the education, training, and research goals of the home building industry. The program/project can initially be at the local level, but should have the potential to be used as a national model to assist in addressing the education, land use and workforce training/recruitment issue(s). In addition, the program/project should adhere to the IRS regulation definition of ????serving the public interest???? and not exist to serve the sole interests of the home building industry.
Restrictions: Programs and projects are not eligible for support over a multi-year period. HBA projects that have already received funding in 2007 may not apply. Preference will be given to newly created programs/projects especially those that create partnerships in their communities.
Amount of Grant: up to $3,000 matching funds
Date(s) Application is Due: Apr 16
Contact: Tracy Hensler, (800) 368-5242, ext. 8483; fax: (202) 266-8177; email: nhe@nahb.com
Internet: http://www.nationalhousingendowment.com/CBG.htm
Sponsor: National Housing Endowment
1201 15th Street NW
Washington, DC 20005

National Leadership Grant Competition **3540**
The Sons of Italy awards merit-based grants for Italian American students enrolled full time in undergraduate or graduate programs at accredited colleges or universities. Criteria for the award include academic excellence and leadership potential.
Requirements: Applicants must be U.S. citizens who are full-time undergraduate or graduate students of Italian heritage.
Amount of Grant: $4000-$25,000
Date(s) Application is Due: Feb 28
Contact: Scholarship Administrator, (202) 547-5106; fax: (202) 547-8168; email: scholarships@osia.org
Internet: http://www.osia.org/public/scholarships/grants.asp
Sponsor: Sons of Italy Foundation
219 E Street NE
Washington, DC 20002

National LGBT Scholarship Fund **3541**
The mission of the foundation is to provide financial support, mentoring, and hope to meritorious gay, lesbian, bisexual, and transgender students, who are marginalized because of their sexual orientation or gender identity. Submit applications online by the listed application deadline.
Requirements: The application process is open to all LGBT students nationwide regardless of level of education.
Samples: To be distributed among 10 individuals—for scholarships for gay, lesbian, bisexual, and transgender college students, $132,000 (2003).
Date(s) Application is Due: Mar 1
Contact: Scholarship Administrator, (866) 337-6468; fax: (866) 397-6468; email: info@thepointfoundation.org
Internet: http://www.thepointfoundation.org/scholarships.html
Sponsor: Point Foundation
P.O. Box 11210
Chicago, IL 60611

National Lottery Community Fund Grants **3542**
The organization operates in the United Kingdom and internationally and awards grants to charitable, benevolent, and philanthropic organizations. The organization particularly wants to support projects that help people who face disadvantage that is severe, long-term, difficult to tackle, or who face a combination of needs. Grants support projects that try to prevent or reduce, as far as possible, future disadvantage.
Contact: Gerald Oppenheim, Director, 44-020-7-747-5299; fax: 44-020-7-747-5214; email: enquiries.res@biglotteryfund.org.uk
Internet: http://www.community-fund.org.uk/about-us/our-grants-programmes/index.html
Sponsor: National Lottery Community Fund
Saint Vincent House, 16 Suffolk Street
London SW1Y 4NL United Kingdom

National MPS Society Grants **3543**
The society awards research grants and fellowships to qualified individuals to promote medical research in the fields of mucopolyssaccharidosis (MPS) and mucolipidosis (ML). The program supports MPS (I, II, III IV, V, and VI) research, ML II/III research, and general MPS research. Grants and fellowships are awarded for a two-year period. Application forms and guidelines are available online.
Requirements: Qualified medical researchers are eligible.
Samples: Mount Sinai School of Medicine (New York, NY)—to support the work of Caleogera Simanaro, for medical research on bone and joint diseases in mucopolysaccharide diseases, $100,000 (2005); Iowa State U (IA)—to support the work of Matthew Ellinwood, for medical research on Sanfilippo syndrome, a genetic disease that prevents the normal degeneration of cells and progressively complicates cognitive and physical function, $60,000 (2005).
Amount of Grant: $30,000 per year
Date(s) Application is Due: May 1
Contact: Barbara Wedehase, Executive Director, (207) 947-1445; fax: (207) 990-3074; email: barbara@mpssociety.org
Internet: http://www.mpssociety.org/grants.html
Sponsor: National MPS Society
P.O. Box 736
Bangor, ME 04402-0736

National Patient Safety Foundation Grants **3544**
The foundation awards grants in the United States and Canada to health, medical, and research institutions to promote studies that prevent human errors, system errors, and preventable injuries in health care. Top priority will be given to proposals that apply to a broad spectrum of problems, are relevant to patient safety, and promise to improve conditions with a direct path to implementation into medical care. Of particular interest are projects that address how the relationship between physicians and patients contributes to or inoculate against error and how enhanced patient involvement in care may reduce errors. The foundation also encourages innovation, creative approaches, and cross-disciplinary research teams. Requests may be made for up to two years of support. The listed deadline date is for letters of intent.
Requirements: U.S. and Canadian health, medical, and research institutions are eligible.
Samples: Bruce Lambert, U of Illinois (Chicago, IL)—for research on the mechanisms that lead to patient injuries sustained in hospitals and on ways of reducing preventable healthcare errors, $99,923 over two years.
Amount of Grant: $100,000 maximum plus 15 percent maximum indirect cost
Date(s) Application is Due: Jul 1
Contact: Emily Wilson, Research and Special Projects Manager, (703) 506-3280; email: info@npsf.org
Internet: http://www.npsf.org/html/research/rfp.html
Sponsor: National Patient Safety Foundation
8405 Greensboro Drive, Suite 800
McLean, Virginia 22102-5120

National Patient Safety Foundation Patient Safety Research Grants **3545**
NPSF seeks to stimulate new, innovative projects directed toward enhancing patient safety in the United States. Its objective is to promote studies leading to prevention of human errors, system errors, and any forms of preventable injuries in health care and the adverse consequences to patients that may result. In the first stage of a two-stage application process, letters of intent (LOIs) are solicited for research and development that is broadly related to identifying the causes of preventable injuries and errors and/or developing prevention strategies and methods to implement them. From these LOIs, a subset will be selected for preparation of a more detailed proposal. Highest priority is given to investigations that apply to a broad spectrum of problems and patients and that are clearly relevant to patient safety. Priority is also given to projects with the promise of improved methods of patient safety with a defined and direct path to implementation into health care. Areas of research include, but are not limited to: identifying, reporting on, analyzing, or otherwise assisting in the understanding of the causes of preventable injuries in health care; enhancing understanding of the impact of the changes in health care on the system, on human error, and on safety; developing methods for avoiding preventable injuries; assessing the cost effectiveness of techniques designed to increase patient safety; and developing educational tools and/or training materials designed to alter behaviors to prevent errors and increase safety. Innovation and creativity are strongly encouraged, as are cross-disciplinary research teams. Priority will be given to topics that do not have other available sources for funding. LOIs are due July 1; full applications are due December 2.
Requirements: Awards are made to a sponsoring institution, not to individuals or to departments. Any official member of a nonprofit institution in the United States may apply. Only one person should be listed as the principal investigator. All coinvestigators, collaborators, and consultants should be listed on the LOI.
Amount of Grant: $100,000 maximum for up to two years
Date(s) Application is Due: Jul 1; Dec 2

Contact: NPSF at the AMA, (703) 506-3280; fax: (703) 506-3266; email: info@npsf.org.
Internet: http://www.npsf.org/html/research
Sponsor: American Medical Association Foundation
8405 Greensboro Drive, Suite 800
McLean, VA 22102

National Psoriasis Foundation Research Grants 3546
The foundation will award up to five $50,000 one year pilot project grants to support advancement of both basic and clinical research into the cause and cure of psoriasis and psoriatic arthritis. The purpose of pilot project grants is to provide support for innovative early stage ideas. These grants are designed to allow the investigator to determine whether their idea is worth pursuing and then to secure longer-term funding.
Requirements: Applicants should use their NIH biographical sketches. Download the required application form from the website.
Amount of Grant: up to $50,000
Date(s) Application is Due: Oct 20
Contact: Bruce F. Bebo, Jr., Ph.D.; Director of Research and Medical Programs; (503) 244-7404 or (800) 723-9166; email: research@psoriasis.org
Internet: http://www.psoriasis.org/research/grants/index.php
Sponsor: National Psoriasis Foundation
6600 SW 92nd Avenue, Suite 300
Portland, OR 97223-7195

National Rising Scholars Award to Advance Research on Higher Education for the Public Good 3547
The program makes awards for pretenured faculty, young practitioners, and advanced graduate students in any discipline in U.S. higher education institutions who engage in research that explores higher education's role in serving the public good. The award will be made to individuals in order to support outstanding research and encourage publication. Recipients receive waiver for conference registration fees to attend one of the national conferences of a collaborating association (Association for Institutional Research, American Association of Higher Education, American Education Research Association, Association for the Study of Higher Education, and American College Personnel Association); participation in the Kellogg Forum Higher Educational Intergenerational Scholars Symposium; and inclusion of the manuscript from their project in an annual Kellogg Forum edited monograph and consideration for publication in the peer review journals of the collaborating associations. Applicants may apply online.
Requirements: All recipients should hold a doctorate (no more than five years post degree) or be at the dissertation stage of their doctorate studies; they must also be affiliated with a higher educational institution.
Amount of Grant: $3000
Date(s) Application is Due: Apr 1
Contact: Kellogg Forum, (734) 615-8882; fax: (734) 615-9777; email: Rising.Scholars@umich.edu
Internet: http://www.thenationalforum.org/projects_risingscholars.shtml
Sponsor: National Forum on Higher Education for the Public Good
610 E University Avenue, 2339 SEB
Ann Arbor, MI 48109-1259

National Security Education Program David L. Boren Graduate Fellowships 3548
Each year, NSEP publishes a list of geographic areas, languages, and fields of study identified as critical to U.S. national security. In addition, it emphasizes a diverse list of fields of study, including business, economics, history, international affairs, law, applied sciences and engineering, health and biomedical sciences, political science, and other social sciences. Applications will be considered for study in other countries and world regions where increased understanding and appreciation represents an important contribution to U.S. national security. Fellowship recipients are required to seek employment with an agency or office of the federal government involved in national security affairs. Awards are made for a minimum of one and a maximum of six academic semesters (24 months). Applicants design their own programs and may combine domestic language and cultural study with overseas study. All fellowships must include study of a modern language other than English and the study of an area and culture.
Requirements: NSEP welcomes applications from U.S. citizens enrolled in or applying to a graduate degree program in an accredited U.S. college or university located within the United States.
Restrictions: Support may not be used for study of Western Europe, Canada, Australia, or New Zealand.
Amount of Grant: $12,000 maximum per semester for domestic or overseas study; $30,000 maximum for overseas and domestic program
Date(s) Application is Due: Jan 30
Contact: Program Contact, (800) 498-9360 or (202) 884-8285; fax: (202) 884-8407; email: nsep@aed.org
Internet: http://www.aed.org/nsep/application.html
Sponsor: Academy for Educational Development
1825 Connecticut Avenue NW
Washington, DC 20009

National Security Education Grants to Institutions of Higher Education 3549
The National Security Education Program (NSEP) encourages proposals that address programs in specific foreign languages, countries, and areas generally underrepresented in U.S. programs; and/or address professional, disciplinary, and/or interdisciplinary opportunities to incorporate international education into the curriculum. Proposals are sought that improve language acquisition and cultural knowledge; improve study and work opportunities abroad, stimulate faculty interest and involvement in international aspects of curriculum, improve information dissemination and linkages across institutions, and/or broaden the base of interdisciplinary and institutional relationships. Evaluation of proposals will be based on significance of the problem, responsiveness to identified needs, cost effectiveness, prospects for wider impact, applicant's capacity, and evaluation plans. Prospective applicants should obtain the complete program announcement, which contains final deadline dates and proposal requirements. The grant solicitation may be downloaded from the listed NSEP Web site.
Requirements: Accredited two-year and four-year U.S. institutions of higher education.
Restrictions: Federal government schools are ineligible.
Samples: Johnson C. Smith U (Charlotte, NC)—for a program in Russian language and culture, $419,000.; Metropolitan Community College (Omaha, NE)—for studies in the language and culture of Brazil, $50,000.
Amount of Grant: $25,000-$450,000 for projects of one, two, three, or four years
Contact: Institutional Grants Manager, (202) 326-7697; fax: (202) 326-7698; email: nsep@iie.org
Internet: http://www.ndu.edu/nsep
Sponsor: National Security Education Program
1400 K Street NW, Suite 650
Washington, DC 20005-2403

National Sleep Foundation Pickwick Postdoctoral Fellowship 3550
The fellowships enable young researchers to devote full-time professional effort to research related to the study of sleep or sleep disorders. Fellows receive funding for one or two years. The second year of support is dependent upon satisfactory progress in the first year. Applicants must conduct research in recognized U.S. or Canadian programs of study or laboratories with strong mentorship in the appropriate area. Fellowships are available for basic, applied, or clinical research. Minorities are encouraged to apply.
Requirements: Candidates must hold either an MD, DVM, PhD, or DO degree in which the degree or subsequent training has been completed within the past five years.
Restrictions: Individuals in a faculty position or who have received an NIH grant are ineligible.
Amount of Grant: $45,048 maximum per year; $5500 maximum benefit allowance
Date(s) Application is Due: Dec 1
Contact: Jessica Steinitz, (202) 347-3471 ext 209; fax: (202) 347-3472; email: jsteinitz@sleepfoundation.org
Internet: http://www.sleepfoundation.org/nsfalert/index.php?id=31&secid=159#item159
Sponsor: National Sleep Foundation
1522 K Street NW, Suite 500
Washington, DC 20005

National Turkey Federation Research Award 3551
This award is given each even-numbered year for the outstanding record of turkey research in the United States during the past five years. The work must be original and deal specifically with turkeys. The recipient will be announced at the annual meeting of the Poultry Science Association. The award includes a cash amount and a plaque.
Requirements: The work, specifically with turkeys, must be original in one of a wide range of fields such as, but not limited to, economics, food technology, genetics, merchandising, pathology, physiology, and nutrition. Any administrator or researcher may nominate a qualified candidate. Nominations can be made by any individual knowledgeable about the nominee, accomplishments. The nomination should include the following items: nomination letter, biography, pertinent bibliography, reprints of appropriate and supporting data. Material must be submitted electronically via the PSA awards website no later than March 1 prior to the annual meeting at which the award will be given.
Samples: (2006) Sandra Velleman, Ohio Agricultural Research and Development Center, Ohio State University; (2004) T. D. Siopes; (2002) M. S. Lilburn.
Amount of Grant: $1,000
Date(s) Application is Due: Mar 1
Contact: Mary Swenson, Membership Services; (217) 356-5285; fax: (217) 398-4119; email: marys@assochq.org
Internet: http://www.poultryscience.org/awards.asp
Sponsor: Poultry Science Association
1111 North Dunlap Avenue
Savoy, IL 61874

National Urban/Rural Fellows Program 3552
A limited number of fellowships are awarded to men and women who aspire to positions of administrative and community leadership in the public, private, and nonprofit sectors of the community. The goal of the program is to further the careers and talents of highly skilled, midcareer men and women, especially members of minority groups, for a new diverse leadership in the private and public sectors across the nation. The 14-month program encompasses a nine-month mentorship with a prominent public administrator and academic graduate studies leading to a Master of Public Administration at Bernard M. Baruch College, CUNY, New York City. The fellowship begins and ends with two summer sessions at the college. The first summer session is followed by the nine-month mentorship assignment as a special assistant to an outstanding urban or rural administrator. Upon successful completion of the academic program, the fellow is awarded the MPA degree. Application materials are available on the Web site. Annual deadline dates may vary; contact the program office for exact dates.

Requirements: U.S. citizens who have bachelor's degrees; have three to five years' full-time work experience in an administrative or management capacity; have demonstrated exceptional ability, maturity, leadership potential, and commitment to the solutions of urban and rural problems; and are willing to relocate for the duration of the fellowship year are eligible. An official transcript of all undergraduate and graduate education will be required.
Amount of Grant: $25,000 stipend over 14 months; $500 relocation allowance; $500 book allowance
Date(s) Application is Due: Feb 24
Contact: Luis Benitez, Program Director, National Urban/Rural Fellows, (212) 349-6200; fax: (212) 921-9572; email: lbenitez@nuf.org
Internet: http://www.nuf.org/nufprogram.asp
Sponsor: National Urban Fellows Inc
59 John Street, Suite 310
New York, NY 10038

National Wetlands Awards **3553**
Cosponsored by the Environmental Law Institute, U.S. Environmental Protection Agency, U.S. Fish and Wildlife Service, U.S.DA Forest Service, Natural Resources Conservation Service, and National Marine Fisheries Service, the awards recognize individuals who have demonstrated excellence in wetland conservation. Awards will be made in the the following categories: education and outreach; science research; conservation and restoration; landowner stewardship; state, tribal, and local program development; and wetland community leader. Awardees are chosen by a committee of wetland experts representing the conservation and business communities, as well as federal, state, and local governments. Guidelines and nomination form are available online.
Requirements: Nominations must be for individuals—not organizations or groups of individuals—whose programs or projects operate at the regional, state, or local level.
Restrictions: Federal employees are not eligible. Projects not currently in compliance with the Clean Water Act or similar state or local statutes, as well as projects compelled by enforcement actions under these statutes, are not eligible.
Date(s) Application is Due: Dec 15
Contact: Roxanne Thomas, Wetlands Program, (202) 939-3827; fax: (202) 939-3868; email: thomas@eli.org or wetlandsawards@eli.org
Internet: http://www2.eli.org/nwa/nwaprogram.htm
Sponsor: Environmental Law Institute
2000 L Street NW, Suite 620
Washington, DC 20036

National Wild Turkey Federation Research Grant Program **3554**
The federation awards grants to promote research that will maximize wild turkey populations in all suitable habitats and assure the optimum sustained consumptive and nonconsumptive recreational use. The research program is also intended to promote public awareness and support for the conservation and wise management of the American wild turkey. State Super Fund grants deal with priority projects as determined by the NWTF state chapters and their respective wildlife agencies. These projects may include habitat management (planting wildlife openings), education (creating private landowner awareness of turkey management concepts), hunter safety, and restoration. National projects, which are broader in scope than state projects, may take the form of research grants to support university research or as a publication that can be used by government and corporate biologists as a management reference tool.
Contact: Director of Research and Management, (800) 843-6983 or (803) 637-3106; fax: (803) 637-0034; email: kmorris@nwtf.net
Internet: http://www.nwtf.com/conservation
Sponsor: National Wild Turkey Federation
P.O. Box 530, 770 Augusta Road
Edgefield, SC 29824

National Wildlife Federation Internships **3555**
The federation offers an internship program for college graduates and graduate students with an interest in environmental issues to work in its resources conservation and international affairs departments. These departments work on a wide range of issues, seeking agency compliance with environmental laws and regulations, identifying and filling legislative gaps in protection of critical natural resources, and litigating issues that cannot be resolved by other means. Two interns are assigned to teams comprising a scientist, lawyer, and lobbyist in each program area, including biodiversity conservation, environmental quality, and international. Much of an intern's time is spent researching environmental policy issues and covering congressional activity. Internships are for 24 weeks.
Requirements: Applicants must be undergraduate juniors or seniors, recent graduates and graduate students with academic background and experience in areas of wildlife biology or management, wetlands ecology, fisheries science, natural resource management, land management, forestry, environmental science/policy, biology, chemistry, toxicology, energy, international environmental/development issues, political science, grassroots organizing, or journalism. No application form is required. An application consists of a cover letter, indicating areas of interest; resume; names and telephone numbers of three to five academic or professional references; and a two- to four-page sample of nontechnical academic or professional writing. Applicants for the journalism internships must submit four writing samples of any length.
Amount of Grant: $500 per semester approximately plus some benefits
Contact: Internship Coordinator, (202) 797-6800; fax: (202) 797-6646; email: internopp@nwf.org
Internet: http://www.nwf.org/careergateway/internships.cfm
Sponsor: National Wildlife Federation
1400 16th Street NW, Suite 501
Washington, DC 20036

National Wildlife Federation Species Recovery Fund **3556**
The fund was created to spur the implementation of habitat restoration efforts, species reintroduction projects, private landowner habitat protection activities, and other creative endeavors that directly improve conditions for one or more of 25 specified species. Grants support innovative local and regional species conservation efforts that improve on-the-ground conditions for these species. The list of the 25 species is available on the Web site.
Requirements: Applications are accepted from any organization, agency, tribe, for-profit company, university, or individual working to improve on-the-ground conditions for one or more of the 25 highlighted species.
Restrictions: Grants will not be given for research that does not directly contribute to the conservation of an endangered species.
Amount of Grant: $3000-$7000
Contact: Ron Ohrel, (202) 797-6639; fax: (202) 797-6646; email: ohrel@nwf.org
Internet: http://www.nwf.org/wildlife/grants
Sponsor: National Wildlife Federation
1400 16th Street NW, Suite 501
Washington, DC 20036

National Wildlife Refuge System Centennial Scholarship Program **3557**
The scholarship supports students whose research will contribute toward improved management and understanding of the diverse fish, wildlife, and plant resources found on U.S. National Wildlife Refuges. Under this competitive program, scholarships will be awarded to support students' research expenses, tuition, fees, books, and room and board. The one-time, one-year awards are based on merit and will take into consideration the value of the research toward improving management and understanding of fish, wildlife, and plant resources on or related to National Wildlife Refuges as well as the student's academic achievements.
Requirements: To be eligible for consideration, a student must be a U.S. citizen or legal resident enrolled in an accredited institution of higher education in the United States pursuing a graduate or undergraduate degree (sophomores and juniors in the current academic year only) in environmental science, natural resource management, biology, public policy, education, geography, political science, or related disciplines. Only those students working on a National Wildlife Refuge or whose work is directly applicable to resource management on a National Wildlife Refuge(s) are eligible.
Amount of Grant: $5000 awards to undergraduate, master's, and juris doctorate students; $15,000 awards to PhD and DVM students
Date(s) Application is Due: Apr 15
Contact: Kevin Kilcullen, (703) 358-2029; email: Kevin_Kilcullen@fws.gov
Internet: http://www.nfwf.org/programs/centennialscholarship/index.cfm
Sponsor: National Fish and Wildlife Foundation
1120 Connecticut Avenue NW, Suite 900
Washington, DC 20036

Nationwide Foundation Grants **3558**
Grants are given to tax-exempt organizations located in central Ohio or recommended for funding by a state office or subsidiary/affiliate, In order of emphasis, grants support: health and human services, higher education, culture and arts, and civic and community. Funds are provided for arts programs, including fine arts, dance, music, museum, and theater programs; business and economics; children and youth; health programs, mental health, and programs dealing with alcoholism and drug abuse; assistance for and care of the aged; educational programs for minorities, persons with disabilities, and the disadvantaged; social service and social welfare programs; and urban affairs. Types of support include operating budgets, continuing support, annual campaigns, seed grants, projects and programs, employee matching gifts, and capital campaigns. Preference is given to organizations located in central Ohio.
Requirements: Seventy percent of grants are awarded to nonprofit organizations in the greater Columbus, OH, area, with the rest going to organizations in which Nationwide has a large number of employees and agents.
Restrictions: Individuals, organizations that support political candidates or political philosophies, public elementary and secondary schools, lobbying groups, and fraternal or veterans organizations are not eligible for these grants. Research is not funded.
Geographic Focus: Ohio
Samples: To be divided among more than 1000 organizations nationwide—to benefit these nonprofit groups, which include more than 700 United Way chapters, $14 million (2005).
Amount of Grant: $11.8 million total; $5000-$50,000 average
Date(s) Application is Due: Sep 1
Contact: Grants Administrator, (614) 249-4310 or (614) 249-0039; fax: (614) 249-3147
Internet: http://www.nationwide.com/aboutus/community/fndatn.htm
Sponsor: Nationwide Foundation
1 Nationwide Plaza, MD 1-22-05
Columbus, OH 43215-2220

NATO Advanced Research Workshops (ARW) Program **3559**
The aims of an ARW are to contribute to the critical assessment of existing knowledge on new important topics, to identify directions for future research, and to promote close working relations between scientists from different countries and with different

professional experience. Workshops host from 20 to 50 experts and last from two to five days. Funding is provided to ARW directors to cover organizational expenses and the travel and living expenses of key speakers, and to contribute to the travel and living expenses of other participants of whom 40-50 percent are from partner countries. ARWs are codirected by a scientist from a partner country and a scientist from a NATO country; ARWs may be held in either a NATO or a partner country. Areas of support and deadlines are as follows: physical and engineering science, March 1, August 1, November 1; life sciences and environmental and earth sciences, April 1, September 1, December 1; and security-related civil sciences, February 1, May 1, October 1. Applications are available on the Web site.
Requirements: Scientists and engineers working permanently in NATO member countries and cooperating Euro-Atlantic Partnership Council (EAPC) countries are eligible to apply.
Date(s) Application is Due: Feb 1; Mar 1; Apr 1; May 1; Aug 1; Sep 1; Oct 1; Nov 1; Dec 1
Contact: Scientific and Environmental Affairs Division, 011 322 707 41 11; fax: 011 322 707 42 32; email: science@hq.nato.int
Internet: http://www.nato.int/science-old/e/arw.htm
Sponsor: North Atlantic Treaty Organization
Bd Leopold III
Brussels B-1110 Belgium

NATO Advanced Study Institutes (ASI) Program **3560**
Grants are offered to organize Advanced Study Institutes, which are high-level tutorial activities where a carefully defined subject is treated in depth by lecturers of international standing, and which include new advances in the subject not yet in university courses. Guidelines are available online.
Requirements: Support is available for ASIs only in priority research topics in the areas of defense against terrorism or countering other threats to security and/or in a partner-country priority.
Date(s) Application is Due: Mar 1; Sep 1; Dec 1
Contact: Scientific and Environmental Affairs Division, 011 322 707 41 11; fax: 011 322 707 42 32; email: science@hq.nato.int
Internet: http://www.nato.int/science/support/asi/asi-nfa.htm
Sponsor: North Atlantic Treaty Organization
Blvd Leopold III
Brussels B-1110 Belgium

NATO Collaborative Linkage Grants (CLG) **3561**
The aim of the program is to facilitate collaboration between research scientists or research teams in order to stimulate scientific research through the pooling of capabilities and resources. CLGs are available for collaboration between researchers in partner and NATO countries. Funding is available for reciprocal visits abroad of up to five members of each research team collaborating on a joint research project. Awards are for one or two years. Computer networking supplements are available to upgrade the computer networking capabilities of the partner-country laboratories, upon additional application once a CLG has been awarded. Grants are given for projects that rely for basic costs on national funding but where the costs for the international collaboration cannot be met from other sources. The grants support travel and living expenses of investigators for short visits to partner institutions abroad. Areas of support and deadlines are as follows: physical and engineering science, March 1, August 1, November 1; life sciences and environmental and earth sciences, April 1, September 1, December 1; and security-related civil sciences, February 1, May 1, October 1. Applications are available on the Web site. Grants will be in Euro. Payment of the grant, however, may be in any currency, as specified by the grantholder at the time of payment.
Requirements: To be eligible for support a project must be specific and the collaboration should be between scientists in NATO countries and NATO's Partner countries of the Euro-Atlantic Partnership Council. Scientists of the Mediterranean Dialogue countries may also apply for support of a CLG, and should draw up applications in cooperation with scientists from NATO countries. Applicants should follow the guidelines given for NATO Partner countries.
Restrictions: No salaries or stipends for any of the collaborators, students, technical assistants, etc., will be offered under these grants. No institutional overhead expenses may be charged to the grant. No support can be supplied for scientists while on sabbatical or other extended leave abroad, or to allow attendance at conferences, symposia, workshops, etc. Scientists may not be involved in more than one collaborative linkage grant at the same time.
Amount of Grant: $5000-$25,000 depending on how many years and how many collaborators
Date(s) Application is Due: Feb 1; Mar 1; Apr 1; May 1; Aug 1; Sep 1; Oct 1; Nov 1; Dec 1
Contact: Scientific Affairs Division, NATO, 011 322 707 41 11; fax: 011 322 707 42 32; email: natodoc@hq.nato.int
Internet: http://www.nato.int/science-old/e/clg.htm
Sponsor: North Atlantic Treaty Organization
Bd Leopold III
Brussels B-1110 Belgium

NATO Expert Visits (EV) Program Grants **3562**
The objective of expert visits program is to allow high-level specialists from NATO and partner countries to benefit from one another's expertise. Grants support the travel and living expenses of experts from NATO countries visiting research laboratories in partner countries, and vice versa, for consultation on current research projects. The duration of visits may range from a few days to a maximum of one month. Applications are available on the Web site.
Requirements: To be eligible for support the visits must enable senior scientists to provide expertise on a specific project, and must be for visits either of a partner-country expert to institutions in a NATO country or a NATO-country expert to institutions in a Euro-Atlantic Partnership Council (EAPC) country.
Date(s) Application is Due: Mar 1; Sep 1; Dec 1
Contact: Scientific and Environmental Affairs Division, 011 322 707 41 11; fax: 011 322 707 42 32; email: science@hq.nato.int
Internet: http://www.nato.int/science/support/ev/ev-info.htm
Sponsor: North Atlantic Treaty Organization
Bd Leopold III
Brussels B-1110 Belgium

NATO Science for Peace Grants **3563**
The objective of the Science for Peace subprogram is to offer support to Euro-Atlantic Partnership Council (EAPC) partner countries in their transition towards a market-oriented, environmentally sound economy. This subprogram supports applied science and technology projects in partner countries that relate to industrial or environmental problems, when such problems involve collaboration between science and industry or between science and other end-users. Science for Peace helps partner scientists to increase contacts in the NATO science community, while building a stronger science infrastructure in their home countries. Support is for three- to five-year funding of applied R&D projects having well-defined objectives, well-defined work plans, and well-defined budgets. Projects are of a cooperative nature, jointly carried out among NATO and partner countries; they take place in partner countries. Projects involving more than one partner country are encouraged. Funds are provided for partner-country participants (and Greece and Turkey) to cover mainly scientific equipment, computers, software, travel, training of project personnel and project-specific consumables. Funds are provided for NATO-country participants mainly to cover travel and project-specific consumables. Calls for proposals, application forms, and guidelines are posted on the Web site when available, generally in March.
Contact: Scientific and Environmental Affairs Division, 011 322 707 41 11; fax: 011 322 707 42 32; email: natodoc@hq.nato.int
Internet: http://www.nato.int/science/sfp/index.htm
Sponsor: North Atlantic Treaty Organization
Blvd Leopold III
Brussels 1110 Belgium

NATSO Foundation Grants **3564**
The foundation accepts proposals for research from qualified individuals for projects focusing on the travel plaza and truckstop industry. Projects should be no more than six months. Topics and guidelines are available on the Web site.
Samples: Kimberly Barden (Cambria, WI)—scholarship winner, $2500 (2004); Lauren Chrystal (Vancouver, WA)—scholarship winner, $2500 (2004).
Amount of Grant: $2500
Contact: Sharon Corigliano, Executive Director, (703) 549-2100 ext 856; fax: (703) 684-9667; email: scorigliano@natso.com or foundation@natso.com
Internet: http://www.natsofoundation.org/web/education/billmoon.html
Sponsor: NATSO Foundation
1737 King Street, Suite 200
Alexandria, VA 22314

Naval Historical Center Internship Program **3565**
The center offers internships to students who wish to have professional work experience in areas related to their educational programs and career goals. In addition to undertaking historical research, writing and editing, the center operates the Navy Department Library and the Navy Museum and maintains collections of naval archives, photographs, artifacts, and art. Internships are available in all of these areas. There is no set length for an internship. The minimum period is three weeks, full-time. Internships can last an academic quarter or semester, during which time a student may serve on either a full- or part-time basis. Applications are available on the Web site.
Requirements: Registered students of colleges or universities and graduates thereof are eligible for this program. A writing sample, transcript, and letter of recommendation must be submitted with the application form.
Amount of Grant: $400 maximum total depending on available funds
Contact: Dr. Edward Furgol, Curator, The Navy Museum, (202) 433-6901; fax: (202) 433-8200; email: edward.furgol@navy.mil
Internet: http://www.history.navy.mil/prizes/intern.htm
Sponsor: Naval Historical Center
805 Kidder Bresse SE
Washington, DC 20374-5060

NBME Stemmler Medical Education Research Fund Grants **3566**
The purpose of the fund is to provide support for research or development of innovative evaluation methodology or techniques with the potential to advance assessment in medical education or practice, at eligible medical schools. Collaborative investigations within or among institutions are eligible, particularly as they strengthen the likelihood of the project's contribution and success. Proposed projects may request up to two years of support. Applicants for both new proposals and competing renewals may request up to $150,000 of NBME funding support for a project period of up to two years.
Requirements: Medical schools accredited by the Liaison Committee on Medical Education or the American Osteopathic Association are eligible.

Amount of Grant: $150,000 maximum for two years
Date(s) Application is Due: Oct 3
Contact: Officer of Research Support; (215) 590-9657; fax: (215) 590-9488; email: stemmlerfund@nbme.org
Internet: http://www.nbme.org/research/stemmler/index.html
Sponsor: National Board of Medical Examiners
3750 Market Street
Philadelphia, PA 19104

NBNA Lauranne Sams Scholarships **3567**
NBNA awards undergraduate and graduate nursing scholarships primarily to African Americans enrolled at any level in a nursing program. Each chapter must submit completed NBNA scholarship application forms filled out by each candidate it recommends. Scholarship winners are notified by July 1, and the awards are presented at the annual August conference.
Requirements: Candidate must be enrolled in a nursing program (AD, BSN, or LPN/LVN) and be in good scholastic standing at the time of application. Applications must be accompanied by two letters of recommendation (one from the school of nursing and one from the local chapter). Letters from the school of nursing and a nurse in the area will suffice for candidates who have no local chapter in the area.
Amount of Grant: $500-$2000 range
Date(s) Application is Due: Apr 15
Contact: Scholarships Administrator, (800) 575-6298 or (301) 589-3200; fax: (301) 589-3223; email: NBNA@erols.com
Internet: http://www.nbna.org/scholarship.htm
Sponsor: National Black Nurses Association
8630 Fenton Street, Suite 330
Silver Spring, MD 20910-3803

NCAA Postgraduate Scholarships **3568**
The NCAA awards up to 174 postgraduate scholarships annually, 87 for men and 87 for women. The scholarships are awarded to student-athletes who excel academically and athletically and who are in their last year of intercollegiate athletics competition. The one-time grants are awarded for fall sports, winter sports, and spring sports. Each sports season (fall, winter and spring), there are 29 scholarships available for men and 29 scholarships available for women. The scholarships are one-time, nonrenewable grants.
Requirements: Eligibility is restricted to student-athletes at NCAA member institutions who, in their final year of eligibility, have performed with distinction as members of varsity teams in the sport for which they are nominated. Nominees must have a minimum grade-point average of 3.0 (based on a 4.0 scale), be nominated by the faculty athletics representative or athletics director, and intend to continue academic work beyond the baccalaureate degree as a full-time graduate student.
Amount of Grant: $7500 one-time grant
Date(s) Application is Due: Feb 24; May 5; Dec 9
Contact: Committee Coordinator, (317) 917-6222; fax: (317) 917-6888; email: info@ncaa.org
Internet: http://www1.ncaa.org/membership/ed_outreach/prof_development/postgrad_scholarships.html
Sponsor: National Collegiate Athletic Association
P.O. Box 6222
Indianapolis, IN 46206-6222

NCAR Graduate Research Fellowships **3569**
The NCAR Advanced Study Program offers a small number of graduate fellowships that support university PhD thesis research conducted in collaboration with NCAR scientists. Usually, these fellowships are restricted to research projects that require facilities or expertise available at NCAR but not at the collaborating university. Fellowships may also be supported when the NCAR scientist contributes scientific or technical expertise in an area outside the research interests of the faculty at the collaborating university. Research topics at NCAR include atmospheric dynamics, climate, mesoscale weather, atmospheric chemistry, studies of the upper atmosphere, solar physics, turbulence, cloud physics, oceanography, the planetary boundary layer, societal impacts of atmospheric processes, mathematical and numerical methods applied to modeling of the atmosphere or sun, development of instrumentation for field and laboratory investigations, and many other areas of atmospheric research. The fellowship may support benefits and some travel costs. Fellowships are not available every year. Contact the organization or check the Web site for more information.
Requirements: The student must intend to work on a PhD thesis in cooperation with an NCAR program and must be in residence at NCAR for the thesis research.
Amount of Grant: $16,870 plus $1200 maximum travel costs
Contact: Barbara Hansford, Coordinator, Advanced Study Program, (303) 497-1601; fax: (303) 497-1646; email: barbm@ucar.edu
Internet: http://www.asp.ucar.edu/asp/arnaud.html
Sponsor: National Center for Atmospheric Research
P.O. Box 3000
Boulder, CO 80307-3000

NCAR Postdoctoral Fellowships in the Atmospheric Sciences **3570**
The program provides an opportunity for talented atmospheric scientists to continue to pursue their research interests and to offer an opportunity to scientists from other disciplines to apply their training to research in the atmospheric sciences. Each fellow is encouraged to pursue his/her own research objectives while taking maximum advantage of the opportunities available at NCAR, including collaboration with staff and use of research facilities. NCAR concentrates most of its research efforts on coordinated projects in atmospheric dynamics (on all scales) and models, climate science, cloud physics, atmospheric chemistry and radiation, turbulence, upper-atmosphere physics, solar and solar-terrestrial physics (including ionospheric studies and aeronomy), oceanography, and atmospheric technology. The NCAR program also includes studies of the interaction of the atmosphere with the oceans, the cryosphere, the earth's surface, and human society. Scientists interested in applying such skills as biology, ecology, geology, economics, and political science to atmospheric issues are encouraged to apply.
Requirements: An applicant must have a PhD in an area related to NCAR science during the first four years preceding the application date or must be able to complete such a PhD before starting the NCAR appointment.
Restrictions: Scientists with more experience should apply to the director of the division in which they would like to work for positions as visiting scientists.
Amount of Grant: $47,000 first year; $49,000 second year
Date(s) Application is Due: Jan 5
Contact: Paula Fisher, Administrator, fax: (303) 497-1646; email: asp-apply@ucar.edu
Internet: http://www.asp.ucar.edu/asp/pdann.html
Sponsor: National Center for Atmospheric Research
P.O. Box 3000, 1850
Boulder, CO 80307-3000

NCCAM Exploratory Developmental Grants for Complementary and Alternative Medicine (CAM) Studies of Humans **3571**
The National Center for Complementary and Alternative Medicine (NCCAM), the National Cancer Institute (NCI), the National Institute on Aging (NIA), and the Office of Dietary Supplements (ODS) invites high quality exploratory/developmental research grant applications of humans in all domains of complementary and alternative medicine (CAM), in order to obtain preliminary data that can be used as a foundation for a larger clinical study. NCCAM groups CAM practices into four domains: biologically-based practices; energy medicine; manipulative and body-based therapies, and mind-body medicine. In addition, NCCAM funds studies of whole medical systems that employ practices drawn from the four domains. The total project period for an application submitted in response to this funding opportunity may not exceed three years. Direct costs are limited to $400,000 over a three-year period, with no more than $250,000 in direct costs allowed in any single year.
Requirements: The following organizations and institutions may apply: public/state controlled institutions of higher education; private institutions of higher education; Hispanic-serving institutions; historically Black colleges and universities (HBCUs); tribally controlled colleges and universities (TCCUs); Alaska native and native Hawaiian serving institutions; nonprofits with 501(c)3 IRS status; nonprofits without 501(c)3 IRS status; small businesses; for-profit organizations; State governments; regional organizations; U.S. territories or possessions; Indian/Native American tribal governments; Indian/Native American tribally designated organizations; county governments; city or township governments; special district governments; independent school districts; public housing authorities/Indian housing authorities; eligible agencies of the Federal government; and faith-based or community based organizations.
Amount of Grant: $250,000 maximum for single entity; $400,000 maximum for 3-year project
Date(s) Application is Due: Feb 16; Jun 16; Oct 16
Contact: April Bower; (301) 451-3560; fax: (301) 480-3621; email: bowera@mail.nih.gov
Internet: http://grants.nih.gov/grants/guide/pa-files/PAR-08-135.html#SectionVII
Sponsor: National Center for Complementary and Alternative Medicine (NCCAM)
6707 Democracy Boulevard, Suite 401
Bethesda, MD 20892

NCCAM Predoctoral Research Training in Complimentary and Alternative Medicine **3572**
NCCAM provides awards to individuals for research training in specified areas of biomedical and behavioral research related to complementary and alternative medicine (CAM) to help ensure that highly trained scientists will be available in adequate numbers and in appropriate research areas and fields to meet the nation's health research needs. As such, this program is intended to provide a mechanism to train future generations of outstanding scientists committed to pursuing a research career in CAM.
Requirements: Applicants must be citizens or noncitizen nationals of the United States, or have been lawfully admitted to the United States for permanent residence and have in their possession an Alien Registration Receipt Card at the time of award. Applicants must have received, as of the activation date of the award, a baccalaureate degree and must be enrolled in a program leading to a research doctorate such as the PhD or DSc, or a combined clinical and research degree such as DC, DO, MD, or ND/PhD by the proposed fellowship activation date.
Date(s) Application is Due: Apr 5; Aug 5; Dec 5
Contact: Dr. Neal West, Program Officer, (301) 402-5867; fax: (301) 402-4741; email: westn@od.nih.gov
Internet: http://grants.nih.gov/grants/guide/pa-files/PAR-00-023.html
Sponsor: National Center for Complementary and Alternative Medicine
Building 31, Room 5B58
Bethesda, MD 20892

NCCAM Ruth L. Kirschstein National Research Service Awards for Postdoctoral Training in Complementary and Alternative Medicine **3573**
The National Center for Complementary and Alternative Medicine (NCCAM) awards individual postdoctoral research training fellowships to promising applicants with the

desire and potential to become productive, independent investigators in the field of complementary and alternative medicine (CAM) research. The primary objective of this funding opportunity is to help ensure that diverse pools of highly trained scientists will be available in adequate numbers and in appropriate research areas to carry out the Nation, biomedical, behavioral, and clinical CAM research agendas. Applicants may come from a range of career backgrounds, including biomedical research, conventional allopathic medicine, or CAM healing practices. The number of awards and the total amount of funding that NCCAM expects to award through this announcement will depend on the scientific merit of applications received, relevance to NCCAM's program priorities, and the availability of funds.
Requirements: Eligible organizations include for-profit and non-profit organizations, public or private institutions such as universities, colleges, hospitals and laboratories, units of State and local governments, eligible agencies and labs of the Federal government including NIH intramural labs, and domestic or foreign institutions/organizations. Applicant fellows must be citizens or non-citizen nationals of the United States, or have been lawfully admitted to the United States for permanent residence and have in their possession an Alien Registration Receipt Card.
Restrictions: Individuals on temporary or student visas are not eligible.
Date(s) Application is Due: Apr 8; Aug 8; Dec 8
Contact: Nancy J. Pearson; (301) 594-0519; fax: (301) 480-3621; email: pearsonn@mail.nih.gov
Internet: http://grants.nih.gov/grants/guide/pa-files/PAR-07-319.html
Sponsor: National Center for Complementary and Alternative Medicine (NCCAM)
6707 Democracy Boulevard, Suite 401, MSC 5475
Bethesda, MD 20892-5475

NCCAM Translational Tools for Clinical Studies of CAM Interventions Grants 3574
This Funding Opportunity Announcement (FOA) issued by the National Center for Complementary and Alternative Medicine (NCCAM) of the National Institutes of Health (NIH) encourages investigator(s)-initiated applications that propose to develop, enhance, and validate translational tools to facilitate rigorous study of complementary and alternative medicine (CAM) approaches that are in wide use by the public. Multidisciplinary studies and collaboration among investigators with expertise in appropriate disciplines are encouraged. NCCAM intends to award up to $3 million (total costs) per year, to support approximately 4 to 6 new grants. A maximum of $300,000 (direct costs) per year may be requested and the total project period for an application submitted in response to this FOA will be five years.
Requirements: The following organizations and institutions are eligible to apply: individuals; public/state controlled institutions of higher education; private institutions of higher education; Hispanic-serving institutions; historically Black colleges and universities (HBCUs); tribally controlled colleges and universities (TCCUs); Alaska Native and Native Hawaiian serving institutions; nonprofits with 501(c)3 IRS status; nonprofits without 501(c)3 IRS status; for-profit organizations; state governments; Indian/Native American tribal governments (federally recognized); Indian/Native American tribally designated organizations; county governments; city or township governments; special district governments; independent school districts; public housing authorities/Indian housing authorities; U.S. territory or possession; Indian/Native American tribal governments (other than federally recognized); regional organizations; non-domestic (non-U.S.) entities (foreign organizations); other eligible agencies of the federal government; and faith-based or community-based organizations.
Amount of Grant: $300,000 per year
Date(s) Application is Due: Jul 17
Contact: Dr. Partap S. Khalsa, Division of Extramural Research; (301) 594-3462; fax: (301) 480-1587; email: khalsap@mail.nih.gov
Internet: http://grants.nih.gov/grants/guide/rfa-files/RFA-AT-09-002.html
Sponsor: National Center for Complementary and Alternative Medicine (NCCAM)
6707 Democracy Boulevard, Suite 401
Bethesda, MD 20892-5475

NCEEER Collaborative Research Program 3575
NCEEER offers grants over a two-year period to teams of researchers who conduct both campus-based and field-based collaborative research on topics in the humanities and social sciences that focus on understanding current developments in the former Soviet Union and Central and Eastern Europe and analysis of their significance for the present and future. Studies of earlier periods must also be relevant for understanding contemporary history, culture, politics, and social trends in the region.
Requirements: The program is limited to collaborative research projects conducted by at least two U.S.-based postdoctoral scholars.
Restrictions: Research on the following countries will not be supported: Czech Republic, Estonia, Hungary, Latvia, Lithuania, Poland, Slovakia, and Slovenia.
Amount of Grant: $40,000 maximum
Date(s) Application is Due: Feb 15
Contact: Program Contact, (202) 822-6950; fax: (202) 822-6955; email: dc@nceeer.org
Internet: http://www.nceeer.org/Programs/national_research_competition.htm
Sponsor: National Council for Eurasian and East European Research
910 17th Street NW, Suite 300
Washington, DC 20006

NCEEER Individual Research Programs 3576
Grants are awarded for a two-year period to individual U.S.-based scholars for research on topics in the humanities and social sciences that focus on understanding current developments in the former Soviet Union and Central and Eastern Europe and analysis of their significance for the present and future. Studies of earlier periods must also be relevant for understanding contemporary history, culture, politics, and social trends in the region. Applications and guidelines are available on the Web site.
Amount of Grant: $70,000 maximum
Date(s) Application is Due: Feb 15
Contact: Robert Huber, President, (202) 822-6950; fax: (202) 822-6955; email: dc@nceeer.org
Internet: http://www.nceeer.org/Programs/national_research_competition.htm
Sponsor: National Council for Eurasian and East European Research
910 17th Street NW, Suite 300
Washington, DC 20006

NCI Academic-Industrial Partnerships for Development and Validation of In Vivo Imaging Systems and Methods for Cancer Investigations 3577
The overall goal of this Funding Opportunity Announcement (FOA) is to accelerate the transition of in vivo spectroscopic and imaging systems and methods into cancer research, clinical trials, and/or clinical practice. A related goal is to advance new technologies and methods for animal imaging. To promote these goals, the Cancer Imaging Program (CIP) of the National Cancer Institute (NCI) solicits applications from multi-institutional research partnerships formed by academic, industrial, and other investigators. The partners will establish an inter-disciplinary, multi-institutional research team for the purpose of developing imaging technology that is applicable to cancer research involving human subjects and/or animal models. The academic-industrial partnerships must be dedicated to the complete research and development cycle, culminating in the product prototype stage. The partnership model is expected to more readily overcome various barriers to the development of new imaging devices and technologies faced by either academia or industry working alone.
Requirements: Eligible institutions include: public and state controlled institutions of higher education; private institutions of higher education; nonprofits with 501(c)3 IRS status (other than institutions of higher education); nonprofits without 501(c)3 IRS status (other than institutions of higher education); small businesses; for-profit organizations (other than small businesses); state governments; U.S. territories or possessions; non-domestic (non-U.S.) entities (foreign organizations); and eligible agencies of the federal government.
Date(s) Application is Due: Feb 5; Jun 5; Oct 5
Contact: Guoying Liu, Ph.D.; (301) 496-9531; fax: (301) 480-3507; email: guoyingl@mail.nih.gov
Internet: http://grants.nih.gov/grants/guide/pa-files/PAR-07-214.html
Sponsor: National Cancer Institute (NCI)
6130 Executive Plaza, EPN Suite 6000, MSC 7412
Bethesda, MD 20892-7412

NCI Application and Use of Transformative Emerging Technologies in Cancer Research 3578
This Funding Opportunity Announcement (FOA) solicits grant applications proposing exploratory research projects on evaluating the performance of emerging molecular and cellular analysis technologies utilizing an appropriate cancer-relevant biological system. Technologies proposed for evaluation may have broad applicability but must ultimately be suitable for cancer-relevant analyses such as the detection (at various molecular and cellular levels) of cancer-related characteristics and/or structural and/or functional alterations. Letters of intent are requested by January 23, April 27, and August 30; full proposal deadlines are February 23, May 27, and September 30. Annual deadline dates may vary; contact program staff for exact dates.
Requirements: Eligible institutions include: public and state controlled institutions of higher education; private institutions of higher education; nonprofits with 501(c)3 IRS status (other than institutions of higher education); nonprofits without 501(c)3 IRS status (other than institutions of higher education); small businesses; for-profit organizations (other than small businesses); state governments; U.S. territories or possessions; non-domestic (non-U.S.) entities (foreign organizations); and eligible agencies of the federal government.
Amount of Grant: $275,000 maximum request per two-year period
Date(s) Application is Due: Jan 23; Feb 23; Apr 27; May 27; Aug 30; Sep 30
Contact: Richard Aragon, Ph.D.; (301) 496-1550; fax: (301) 496-7807; email: raragon@mail.nih.gov
Internet: http://grants.nih.gov/grants/guide/rfa-files/RFA-CA-09-006.html
Sponsor: National Cancer Institute
Building 31, Room 10A52, MSC 2580
Bethesda, MD 20892-2580

NCI Application of Metabolomics for Translational and Biological Research Grants 3579
The purpose of this Funding Opportunity Announcement (FOA), sponsored by multiple NIH Institutes/Centers (ICs), is to promote the use of metabolomic technologies in translational research in human health and disease for the purposes of enabling and improving disease detection, diagnosis, risk assessment, prognosis, and prediction of therapeutic responses. In addition to this overall goal, specific scopes and objectives of the participating ICs are presented below. Using the NIH Research Project Grant (R01) funding mechanism, this FOA focuses on well developed projects supported by preliminary data.
Requirements: Application can be made by the following institutions and organizations: public and state controlled institutions of higher education; private institutions of higher education; small businesses; for-profit organizations (other than small businesses); and non-domestic (non-U.S.) entities (foreign organizations).
Date(s) Application is Due: Feb 5; Feb 16; Jun 5; Jun 16; Oct 5; Oct 16

Contact: Padma Maruvada, Ph.D.; (301) 496-3893; fax: (301) 402-8990; email: maruvadp@mail.nih.gov
Internet: http://grants.nih.gov/grants/guide/pa-files/PA-07-301.html
Sponsor: National Cancer Institute (NCI)
6130 Executive Boulevard, EPN Suite 3144, MSC 7362
Bethesda, MD 20892-7362

NCI Basic Cancer Research in Cancer Health Disparities Grants 3580
The Through this Funding Opportunity Announcement (FOA), the Center to Reduce Cancer Health Disparities (CRCHD) and the Division of Cancer Biology (DCB), at the National Cancer Institute (NCI), invite cooperative agreement research (U01) grant applications from investigators interested in conducting basic research studies into the causes and mechanisms of cancer health disparities. These awards will support pilot and feasibility studies, development and testing of new methodologies, secondary data analyses, and innovative mechanistic studies that investigate biological/genetic bases of cancer health disparities. This FOA is also designed to aid and facilitate the growth of a nationwide cohort of scientists with a high level of basic research expertise in cancer health disparities research who can develop resources and tools, such as biospecimens, cell lines and methods that are necessary to conduct basic research in cancer health disparities. The total project period for an application submitted in response to this funding opportunity may not exceed five years. Direct costs are limited to $250,000/year and may not exceed 5 years.
Requirements: The following organizations and institutions are eligible to apply: public/state controlled institutions of higher education; private institutions of higher education; Hispanic-serving institutions; historically Black colleges and universities (HBCUs); tribally controlled colleges and universities (TCCUs); Alaska Native and Native Hawaiian serving institutions; nonprofits with 501(c)3 IRS status; nonprofits without 501(c)3 IRS status; small businesses; for-profit organizations; state governments; Indian/Native American tribal governments (federally recognized); Indian/Native American tribally designated organizations; U.S. territories or possessions; Indian/Native American tribal governments (other than federally recognized); regional organizations; eligible agencies of the federal government; and units of local governments.
Amount of Grant: $250,000 per year for up to 5 years
Date(s) Application is Due: Jun 23; Nov 23
Contact: Phillip J. Daschner; (301) 496-1951; fax: (301) 496-2025; email: PD93U@nih.gov; Kenneth C. Chu; (301) 435-9213; fax: (301) 435-9225; email: kc10d@nih.gov
Internet: http://grants.nih.gov/grants/guide/pa-files/PAR-09-161.html
Sponsor: National Cancer Institute (NCI)
Executive Plaza North (EPN), Room 5014, 6130 Executive Boulevard, MSC 7344
Bethesda, MD 20892-7344

NCI Cancer Centers Support Grants (CCSG) 3581
The purpose of this program is to provide an organizational focus and stimulus for the highest quality cancer research that effectively promotes interdisciplinary cancer research aimed toward the ultimate goal of reducing cancer incidence, mortality and morbidity. The Cancer Center Support Grant (CCSG) provides the resources and infrastructure to facilitate the coordination of interdisciplinary programs across a broad spectrum of research from basic laboratory research to clinical investigation to population science. The CCSG supports salaries for scientific leadership of the Center, shared resources for funded center investigators, certain administrative costs, planning and evaluation, and developmental funds for new recruitments and feasibility studies.Annual deadline dates may vary; contact program staff for exact dates.
Requirements: Universities, colleges, hospitals, public agencies, or nonprofit research institutions are eligible to apply.
Amount of Grant: $15,950-$10.37 million; $2.9 million average
Date(s) Application is Due: Feb 1; Jun 1; Oct 1
Contact: Dr. Linda Weiss, Chief, Cancer Centers Branch, Division of Cancer Treatment, Diagnosis, and Centers, (301) 496-8531; Leo Buscher Jr, Grants Management Officer, (301) 496-7753
Internet: http://12.46.245.173/pls/portal30/CATALOG.PROGRAM_TEXT_RPT.show
Sponsor: National Cancer Institute
6116 Executive Boulevard, Suite 700
Bethesda, MD 20892

NCI Cancer Communications System Research Grants 3582
The Division of Cancer Prevention and Control of NCI invites applications for studies to initiate cancer communications research activities. This project proposes to enable qualified investigators to identify a cancer communications issue or problem, and develop, implement, and evaluate a research project to address the identified issue or problem. The project should be targeted at specific audiences and may or may not utilize the resources of the Cancer Communications System. Investigators may propose new data collection activities and/or design studies comparing public knowledge, attitudes, and/or practices by location, ethnicity, age, and so forth. A specific group such as women, smokers, African Americans, or people over age 50 may be chosen for study. Projects may be proposed for durations of 18 to 36 months. Shared resources on the part of the applicant, such as contribution of staff time, should be described in the application. Annual deadline dates may vary contact program staff for exact dates. PA: PAR-99-141
Requirements: The awardee will be a university, college, hospital, public agency, nonprofit research institution, or for-profit organization that submits an application and receives a grant or cooperative agreement for support of research by a named principal investigator.
Restrictions: Restrictions are imposed against the use of funds for entertainment, foreign travel (unless specifically authorized), office equipment, and other items not normally necessary for the effective prosecution of such research.
Amount of Grant: $100,000 maximum
Contact: Connie Dresser, Division of Cancer Control and Population Sciences, (301) 496-8520; fax: (301) 480-6637; email: cd34b@nih.gov
Internet: http://www.grants.nih.gov/grants/guide/pa-files/PAR-99-141.html
Sponsor: National Cancer Institute
6120 Executive Boulevard, Executive Plaza S
Bethesda, MD 20892

NCI Cancer Construction Grants 3583
The objective of this program is to renovate existing cancer research facilities and build new cancer research facilities to meet basic and clinical space requirements, laboratory safety, biohazard containment, and animal care standards in order to support the facility demands of NCI research programs.
Requirements: Except where otherwise prohibited by law, any nonprofit organizations competent to carry out cancer research, such as universities, colleges, hospitals, public agencies, or research institutions, are eligible to apply for support.
Date(s) Application is Due: Feb 1
Contact: Daryl Paunil, Deputy Director, Office of Space and Facilities Planning, (301) 496-8534; Leo Buscher Jr., Grants Management Officer, (301) 496-7753
Internet: http://12.46.245.173/pls/portal30/CATALOG.PROGRAM_TEXT_RPT.show
Sponsor: National Cancer Institute
6120 Executive Boulevard, EPS-638
Bethesda, MD 20892

NCI Cancer Control Research Grants 3584
Grants are awarded to programs that will reduce cancer incidence, morbidity, and mortality through an orderly sequence from research on interventions and their impact in defined populations to the broad, systematic application of the research results. Primary emphasis is on the inclusion of a cancer control intervention in any proposed study. Cancer control research studies are classified into one of five phases that represent the orderly progression noted in the definition: hypothesis development; methods development and testing; controlled intervention trials to establish cause-and-effect relationships; research in defined populations; and demonstration and implementation studies. Primary interest is in research on cancer control interventions in phases two through five. Cancer control programs include those in the following areas: chemoprevention; nutrition and diet; screening and early detection; community oncology, rehabilitation, and pain management; special populations research; public health applications; biobehavioral mechanisms; and surveillance research. Annual deadline dates may vary; contact program staff for exact dates.
Requirements: University, college, hospital, public agency, nonprofit research institutions or for-profit organization may apply.
Amount of Grant: $15,000-$14.9 million; $1.08 million average
Date(s) Application is Due: Feb 1; Jun 1; Oct 1
Contact: Leo Buscher Jr., Grants Management Officer, (301) 496-7753
Internet: http://12.46.245.173/pls/portal30/CATALOG.PROGRAM_TEXT_RPT.show
Sponsor: National Cancer Institute
6120 Executive Boulevard, Executive Plaza S
Bethesda, MD 20892

NCI Cancer Detection and Diagnosis Research Grants 3585
The purpose of this program is to improve screening and early detection strategies and to develop accurate diagnostic techniques and methods for predicting the course of disease in cancer patients. Screening and Early Detection research includes development of strategies to decrease cancer mortality by finding tumors early when they are more amenable to treatment. Diagnosis research focuses on methods to determine the presence of a specific type of cancer; to predict its course and response to therapy; and to monitor the effect of the therapy and the appearance of disease recurrence. These methods include diagnostic imaging and direct analyses of specimens from tumor or other tissues. Support is also provided for establishing and maintaining resources of human tissue to facilitate research. Annual deadline dates may vary; contact program staff for exact dates.
Requirements: The awardee will be a university, college, hospital, public agency, nonprofit research institution or organization, or a for-profit organization that submits an application and receives a grant or cooperative agreement for support of research by a named principal investigator.
Amount of Grant: $60,000-$3.8 million; $438,522 average
Date(s) Application is Due: Feb 1; Jun 1; Oct 1
Contact: Dr. Sheila Taube, Associate Director, (301) 496-8639; Leo Buscher Jr., Grants Management Officer, (301) 496-7753
Internet: http://12.46.245.173/pls/portal30/CATALOG.PROGRAM_TEXT_RPT.show
Sponsor: National Cancer Institute
6120 Executive Boulevard, Executive Plaza S
Bethesda, MD 20892

NCI Cancer Education and Career Development Program 3586
The purpose of the program is to support the development and implementation of curriculum-dependent programs to train predoctoral and postdoctoral candidates in cancer research settings that are highly interdisciplinary and collaborative. This program

requires sustained leadership, dedicated faculty time, specialized curriculum development and implementation, interdisciplinary research environments, and more than one mentor per trainee to achieve career development research and education objectives. Areas of research particularly applicable but not all inclusive to interdisciplinary training are cancer prevention and control, nutrition, population sciences, behavioral sciences, imaging and molecular diagnostics. PA: PAR-03-148
Requirements: Applications may be submitted on behalf of the principal investigator by domestic, nonfederal, public or private organizations. Applications may include more than one institution to create a program through consortium agreements. Predoctoral and postdoctoral candidates must be citizens or noncitizen nationals of the U.S., or must have been lawfully admitted to the U.S. for permanent residence.
Restrictions: No application may request more than $500,000 in direct costs per annum without written approval from NCI.
Amount of Grant: $500,000 maximum
Contact: Catherine Blount, Grants Administration Branch, (301) 496-3179; fax: (301) 496-8601; email: blountc@mail.nih.gov
Internet: http://grants2.nih.gov/grants/guide/pa-files/PAR-03-148.html
Sponsor: National Cancer Institute
6120 Executive Boulevard
Bethesda, MD 20892

NCI Cancer Intervention and Surveillance Modeling Network (CISNET) 3587
The purpose of the project is to support collaborative research using simulation and other modeling techniques to describe the impact of interventions (i.e., primary prevention, screening, and treatment) in population-based settings in the United States or in non-U.S. settings that will shed light on U.S. population-based trends. It is well known that great progress can be made in the war against cancer by the complete use and adequate delivery of existing modalities of cancer control. A letter of intent is due by September 14; full proposal is due October 14. Annual deadline dates may vary; contact program staff for exact dates. PA: CA-05-018
Requirements: Applications may be submitted by domestic and foreign for-profit and nonprofit organizations, public and private, such as universities, colleges, medical centers, units of state and local governments, and eligible agencies of the federal government.
Amount of Grant: $1.5 million total
Date(s) Application is Due: Sep 14; Oct 14
Contact: Crystal Wolfrey, Grants Administration Branch, (301) 496-8634; fax: (301) 496-8601; email: crystal.wolfrey@nih.gov
Internet: http://grants.nih.gov/grants/guide/rfa-files/RFA-CA-05-018.html
Sponsor: National Cancer Institute
6701 Rockledge Drive, Room 1040, MSC 7710
Bethesda, MD 20892

NCI Cancer NRSAs—Individual 3588
The awards offer health scientists with a doctoral degree the opportunity to receive full-time research training in cancer research. These fellowships may not be used for study leading to the MD, DO, DDS, or similar professional degree, or for the support of residency training. Awardees are subject to the payback provisions of the NRSA by which one year of research or teaching, or acceptable combination thereof, is required for each year of support (except for the first 12 months, which are free of the payback obligation). Annual deadline dates may vary; contact program staff for exact dates.
Requirements: Universities, colleges, hospitals, public agencies, or research institutions may apply for institutional grants, and individuals are eligible for fellowships.
Date(s) Application is Due: Apr 5 Aug 5; Dec 5
Contact: Program Contact, (301) 435-0714; email: GrantsInfo@nih.gov
Internet: http://grants.nih.gov/grants/policy/nihgps_2001/part_iib_3.htm
Sponsor: National Cancer Institute
6120 Executive Boulevard, Executive Plaza S
Bethesda, MD 20892

NCI Cancer NRSAs—Institutions 3589
These grants are made to eligible institutions to develop or enhance pre- and post-PhD research training opportunities for individuals who are interested in preparing for careers in cancer research. The program is to ensure that an adequate number of highly competent basic and clinical cancer research specialists will be trained to meet needs in the following areas of cancer research: etiology, prevention, detection, diagnosis, treatment, and restorative care. Applications in surgical, radiation, and preventive oncology are of special interest. Proposed institutional training programs may be multidisciplinary. The training program director at the grantee institution will be responsible for selection and appointment of trainees and for the overall direction of the program. Trainees are subject to payback provisions of the NRSA by which one year of research or teaching, or acceptable combination thereof, is required for each year of support, except for the first 12 months, which are free from payback. Annual deadline dates may vary; contact program staff for exact dates.
Requirements: Universities, colleges, hospitals, public agencies, or nonprofit research institutions may apply for institutional grants, and individuals are eligible for fellowships.
Date(s) Application is Due: Jan 10; May 10; Sep 10
Contact: Program Contact, (301)-435-0714; email: GrantsInfo@nih.gov
Internet: http://grants.nih.gov/grants/policy/nihgps_2001/part_iib_3.htm
Sponsor: National Cancer Institute
6120 Executive Boulevard, Executive Plaza S
Bethesda, MD 20892

NCI Cancer Prevention Research Small Grant Program 3590
NCI invites applications that address developmental research in chemoprevention agent development, biomarkers, early detection, and nutrition science. The program is designed to aid and facilitate the growth of a nationwide cohort of scientists with a high level of research expertise in cancer prevention research. Annual deadline dates may vary; contact program staff for exact dates. PA: PAR-04-147
Requirements: Applications may be submitted by domestic, for-profit and nonprofit organizations, public and private, such as universities, colleges, hospitals, laboratories, units of state and local governments, eligible agencies of federal government, and small businesses.
Amount of Grant: $100,000 maximum for indirect costs; $50,000 maximum for direct costs
Date(s) Application is Due: Mar 21; Jul 21; Nov 21; Dec 20
Contact: Shane Woodward, Grants Administration Branch, (301) 846-1017; fax: (301) 846-5720; email: woodwars@mail.nih.gov
Internet: http://grants.nih.gov/grants/guide/pa-files/PAR-04-147.html
Sponsor: National Cancer Institute
6116 Executive Boulevard
Bethesda, MD 20892

NCI Cancer Treatment Research Grants 3591
The objective of this program is to develop the means to cure as many cancer patients as possible and to control the disease in those patients who are not cured. Research includes the development and evaluation of improved methods of cancer treatment through the support and performance of both fundamental and applied laboratory and clinical research. Research is supported in the discovery, development, and clinical testing of all modes of therapy including: surgery, radiotherapy, chemotherapy, and biological therapy, both individually and in combination. In addition, research is carried out in areas of nutritional support, stem cell and bone marrow transplantation, studies to reduce toxicity of cytotoxic therapies, and other methods of supportive care that may supplement and enhance primary treatment. Annual deadline dates may vary; contact program staff for exact dates.
Requirements: Awardee will be a university, college hospital, public agency, nonprofit research institution, or for-profit organization that submits an application and receives a grant or cooperative agreement for support of research by a named principal investigator.
Amount of Grant: $59,416-$6.24 million; $467,739 average
Date(s) Application is Due: Feb 1; Jun 1; Oct 1
Contact: Dr. James Doroshow, Director, Cancer Treatment Division, (301) 496-4291; Leo Buscher Jr., Grants Management Officer, (301) 496-7753
Internet: http://aspe.os.dhhs.gov/cfda/p93395.htm
Sponsor: National Cancer Institute
6120 Executive Boulevard, Executive Plaza S
Bethesda, MD 20892

NCI Centers of Excellence in Cancer Communications Research 3592
NCI invites applications for Centers of Excellence in Cancer Communications Research (CECCRs). The purpose is to identify areas of discovery that build upon important recent developments in knowledge and technology and that hold promise for making significant progress against all cancers. The centers must include three or more individual hypothesis-driven research projects, pilot or developmental research projects, shared resources, and a plan for career development. To be effective, the centers' research should integrate cancer communications appropriately into one or more contexts of the cancer continuum—from prevention through treatment to survivorship and end-of-life research. Communications research also is needed about challenging topics such as cancer information seeking, decision making under uncertainty, and genetic testing. Centers' research also should provide insight into mechanisms underlying how people process information. Annual deadline dates may vary; contact program staff for exact dates. RFA: CA-03-007
Requirements: Applications may be submitted by domestic for-profit and nonprofit organizations, both public and private; units of state and local governments; and eligible agencies of the federal government.
Amount of Grant: $1.5 million maximum
Contact: Crystal Wolfrey, Grants Administration Branch, (301) 496-8634; email: crystal.wolfrey@nih.gov
Internet: http://grants2.nih.gov/grants/guide/rfa-files/RFA-CA-03-007.html
Sponsor: National Cancer Institute
6130 Executive Boulevard
Bethesda, MD 20892

NCI Community Networks to Reduce Cancer Health Grants 3593
Cooperative agreement awards will be made to eligible institutions to develop and implement a variety of community-based cancer control and prevention activities. The overall goals of this program are to significantly improve access to and utilization of beneficial cancer interventions in communities with cancer health disparities, thereby reducing these disparities. The CNP will be implemented in three phases. The goal of Phase I is to develop and increase capacity building to support community-based participatory education, research, and training to reduce cancer health disparities. The goal of Phase II is to develop community-based participatory research and training programs to reduce cancer health disparities. The goal of Phase III is to establish credibility and sustainability of the CNP. Annual deadline dates may vary; contact program staff for exact dates. RFA: CA-05-012

Requirements: Applications may be submitted by domestic organizations, public and private entities such as colleges, universities, hospitals, medical/nursing and sociological associations, biomedical research centers, and eligible agencies or units of federal, state, and local government. Applications are encouraged from existing Leadership Initiatives projects, and from subgroups throughout the country that focus on America's minority and underserved populations. Applicants should have a history of involvement with the community. Collaborative applications are encouraged; among collaborators, one must be designated as the lead applicant.
Restrictions: Foreign organizations are not eligible, nor are organizations that must rely primarily on consultants to conduct the project.
Amount of Grant: $1.4 million approximately per year for large, multisite projects; $350,000 maximum per year for small, single-site projects
Date(s) Application is Due: Jun 14; Jul 13
Contact: Crystal Wolfrey, Grants Administration Branch, (301) 496-8634; fax: (301) 496-8601; email: Crystal.Wolfrey@nih.gov
Internet: http://grants.nih.gov/grants/guide/rfa-files/RFA-CA-05-012.html
Sponsor: National Cancer Institute
6120 Executive Boulevard, Executive Plaza S
Bethesda, MD 20892

NCI Correlative Studies Using Specimens from Multi-Institutional Prevention and Treatment Trials 3594
The objectives of this funding opportunity are to foster collaborations and interactions between basic researchers, scientists working in private industry, and clinical investigators to perform clinical translational research on promising predictive and prognostic markers. These studies should focus on clinical correlative or mechanistic studies that will be useful for cancer risk assessment, early detection, and prognosis, and the predicting responses to therapy and to prevention interventions. These studies should focus on correlations between biologic features of tissue specimens (collected from the NCI Cooperative Groups or other large multi-institutional clinical trials) and patient outcomes. Annual deadline dates may vary; contact program staff for exact dates. PA: PA-05-062
Requirements: Applications may be submitted by foreign and domestic, for-profit and nonprofit organizations, public and private. Applications may be from a single institution or may include arrangements with one or more institutions (e.g., consortia, NCI groups) if appropriate.
Date(s) Application is Due: Feb 1; Jun 1; Oct 1
Contact: Jill Rogers, Grants Management Branch, (301) 496-8699; fax: (301) 496-8601; email: jr261m@nih.gov
Internet: http://grants.nih.gov/grants/guide/pa-files/PA-05-062.html
Sponsor: National Cancer Institute
6130 Executive Boulevard, Executive Plaza N
Bethesda, MD 20892

NCI Development and Application of Imaging in Therapeutic Studies 3595
NCI invites applications to apply imaging technologies in the assessment of investigational cancer therapeutic agents. This request for applications is designed to support research projects that address the development and application of labeled therapeutic agents as compounds for imaging studies, and/or the development and application of imaging agents as metabolic markers of response to newly developed therapeutic agents. The total project period for an application submitted in response to this RFA may not exceed four years. RFA: CA-98-024
Requirements: Applications may be submitted by domestic and foreign, for-profit and nonprofit organizations, public and private.
Amount of Grant: $2.8 million per year total
Contact: Dr. Anne Menkens, Division of Cancer Treatment and Diagnosis, (301) 496-9531; fax: (301) 480-5785; email: am187k@nih.gov
Internet: http://www.grants.nih.gov/grants/guide/rfa-files/RFA-CA-98-024.html
Sponsor: National Cancer Institute
6130 Executive Boulevard, Executive Plaza N
Bethesda, MD 20852

NCI Development of Digital Mammography Displays and Workstations (SBIR/STTR) Grants 3596
The purpose of the program is to alert the investigator community to the need for a concerted effort to overcome the problems of display for digital mammograms. Integration of advanced hardware, software, and psychophysics research data is needed to optimize the early diagnosis of breast cancer using digital mammography. This program will utilize the Small Business Innovation Research (SBIR) and Small Business Technology Transfer (STTR) mechanisms, but will run in parallel with a program of identical scientific scope utilizing the R01 research project grant mechanism (PA-99-082). This PA encourages the small business community to develop displays for digital mammography. Funding will be based on the quality of the proposed project as determined by peer review, availability of funds, and program priority. Contact program staff for deadline dates. PA: PA-99-083
Requirements: Any small business, independently owned by U.S. citizens and located in the U.S., may apply.
Contact: Dr. Barbara Croft, Diagnostic Imaging Program, (301) 496-9531; fax: (301) 480-5785; email: bc129b@nih.gov
Internet: http://www.grants.nih.gov/grants/guide/pa-files/PA-99-083.html
Sponsor: National Cancer Institute
6130 Executive Boulevard
Bethesda, MD 20892

NCI Development of Novel Imaging Technologies Grants 3597
The NCI invites applications for the development and delivery of novel image acquisition or enhancement technologies and methods for biomedical imaging and image-guided interventions and therapy, and which may incorporate limited pilot or clinical feasibility evaluations using either pre-clinical models or clinical studies. This initiative is primarily intended to facilitate the proof-of-feasibility, development, and delivery of novel imaging technologies for early detection, screening, diagnosis, image-guided interventions, and treatment of various diseases, and, secondarily, to facilitate limited evaluation studies to show proof-of-concept and functionality. Annual deadline dates may vary; contact program staff for exact dates. PA: PA-04-095
Requirements: Applications may be submitted by foreign and domestic, for-profit and nonprofit organizations, both public and private; units of state and local governments; and eligible agencies of the federal government.
Contact: Houston Baker, Biomedical Imaging Program, (301) 496-9531; fax: (301) 480-3507; email: bakerhou@mail.nih.gov
Internet: http://grants.nih.gov/grants/guide/pa-files/PA-04-095.html
Sponsor: National Cancer Institute
6130 Executive Plaza
Bethesda, MD 20892

NCI Diagnostic Imaging and Guided Therapy in Prostate Cancer Award 3598
NCI invites applications on the development, risk assessment, and application of improved imaging methods for the localization, biopsy, and image guided biopsy or therapy of prostate cancer. Relevant investigations could include technology development, in vitro laboratory work, preclinical animal studies, or early feasibility testing in humans depending on the maturity of the methods proposed, or evaluation of the effects of age-associated changes and co-morbid conditions as they affect imaging diagnosis and treatment techniques. The development of several methodologies and their optimization for this particular organ system is required. The specific goals include the development and application of one or more of the following interrelated components: means for measuring local extent of disease using anatomic, metabolic or alternative novel imaging methods, means for improved image guided biopsy, staging or identification of aggressive cancers by metabolic or alternative novel imaging methods, and means for navigation, control of image guided therapy or measurement of early biological effects of therapy. Research is also encouraged on how age-associated differences in tumor characteristics and age-related changes in the prostate and adjacent tissues may affect the sensitivity, specificity, prognostic value, or the efficacy of imaging techniques in guiding therapy. The development of methods to increase sensitivity, specificity, prognostic value, and therapeutic applicability of these techniques across the full range of ages in which prostate cancer most frequently occurs, and in the presence of age-related co-morbid conditions in the prostate, other organs, and systems, is of particular interest. Annual deadline dates may vary; contact program staff for exact dates. CA-99-015
Requirements: Applications may be submitted by foreign and domestic, for-profit and nonprofit organizations, public and private, such as universities, colleges, hospitals, laboratories, industries, units of state and local governments, and eligible agencies of the federal government.
Contact: Barbara Croft, Diagnostic Imaging Program, (301) 496-9531; fax: (301) 480-5785; email: bc129b@nih.gov
Internet: http://grants2.nih.gov/grants/guide/rfa-files/RFA-CA-99-015.html
Sponsor: National Cancer Institute
6701 Rockledge Drive
Bethesda, MD 20892

NCI Diet, Epigenetic Events, and Cancer Prevention Grants 3599
The purpose of this Funding Opportunity Announcement (FOA) is to promote innovative preclinical and clinical research to determine how diet and dietary factors, including dietary supplements, impact epigenetic processes involved in cancer prevention. Although much evidence exists that dietary components are linked to cancer prevention, the specific nutrients and sites of action remain elusive. Diet, in fact, has been implicated in many of the pathways of cancer, including apoptosis, cell cycle control, differentiation, inflammation, angiogenesis, DNA repair, and carcinogen metabolism. These are also processes that are likely regulated by epigenetic events to impact gene function and chromatin stability. Thus, research supported by this initiative could address one of the following issues: How bioactive food components regulate epigenetic events for cancer prevention, how bioactive food components might alter aberrant epigenetic patterns or events and restore gene function, and how these components might circumvent or compensate for genes and pathways that are altered by epigenetic events. Another important aim of this FOA is to encourage collaborations between nutrition scientists and experts in epigenetics to study bioactive food components with cancer-preventative properties and to examine key epigenetic events in cancer processes (e.g., carcinogen metabolism, cell division, differentiation, and apoptosis) in order to begin to establish linkages between epigenetic events and patterns, and tumor incidences/behaviors. NIDDK is particularly interested in receiving applications that address normal and neoplastic conditions focusing on the impact of epigenetics on cell division, differentiation, and/or apoptosis in the gastrointestinal (GI) tract and related organs. It is also expected that each application demonstrate experience in nutrition and cancer prevention as well as in epigenetic research. Because the nature and scope of the proposed research will vary from application to application, it is anticipated that the size and duration of each award will also vary.
Requirements: Application can be made by the following institutions and organizations: public/state controlled institutions of higher education; private institutions of higher education; nonprofits with 501(c)3 IRS status (other than institutions of higher education); nonprofits without 501(c)3 IRS status (other than institutions of higher

education); small businesses; for-profit organizations (other than small businesses); state governments; U.S. territories or possessions; Indian/Native American tribal governments (federally recognized); Indian/Native American tribal governments (other than Federally recognized); Indian/Native American tribally designated organizations; non-domestic (non-U.S.) entities (foreign organizations); Hispanic-serving institutions; historically Black colleges and universities (HBCUs); tribally controlled colleges and universities (TCCUs); Alaska Native and Native Hawaiian serving institutions; regional organization; eligible agencies of the Federal government; and faith-based or community based organizations.
Date(s) Application is Due: Feb 5; Jun 5; Oct 5
Contact: Sharon Ross, Ph.D.; (301) 594-7547; fax: (301) 480-3925; email: rossshа@mail.nih.gov; Vishnudutt Purohit, Ph.D.; (301) 443-2689; fax: (301) 594-0673; email: vpurohit@mail.nih.gov
Internet: http://grants.nih.gov/grants/guide/pa-files/PA-07-175.html
Sponsor: National Cancer Institute (NCI)
6130 Executive Boulevard, EPN Suite 3157, MSC 7328
Bethesda, MD 20892-7328

NCI Division of Cancer Treatment Clinical Trials Cooperative Groups Grants 3600
NCI is accepting applications from institutions interested in conducting multi-institutional clinical trials in a cooperative group setting. Awards will be made using the cooperative agreement mechanism (U10). Potential applicants are encouraged to contact program staff to discuss and/or clarify any issues or questions regarding this announcement. The program announcement, which describes the research objectives, application procedures, review considerations, and award criteria for this solicitation, may be obtained through the NIH grant line (301-402-2221), and the NIH Web site, and by mail and email. Annual deadline dates may vary; contact program staff for exact dates. PA: PA-99-058
Date(s) Application is Due: Feb 1; Jun 1; Oct 1
Contact: Dr. Richard Ungerleider, Division of Cancer Treatment and Diagnosis, (301) 496-2522; fax: (301) 402-0557; email: ru4m@nih.gov
Internet: http://www.grants.nih.gov/grants/guide/pa-files/PA-99-058.html
Sponsor: National Cancer Institute
6130 Executive Boulevard, Executive Plaza N
Bethesda, MD 20892

NCI Early Detection Research Network: Biomarkers Validation Laboratories Research Grants 3601
The Division of Cancer Prevention invites applications for cooperative agreements to establish a national network that will have responsibility for the development, evaluation, and validation of biomarkers for earlier cancer detection and risk assessment. Biomarkers are defined as cellular, biochemical, molecular, or genetic alterations by which a normal, abnormal, or simply biologic process can be recognized or monitored. Biomarkers are measurable in biological media, such as in tissues, cells, or fluids. The purpose of the network is to establish a scientific consortium of investigators, academic as well as industrial, with resources for basic, translational, and clinical research. The consortium will have three main components—biomarkers developmental laboratories, biomarkers validation laboratories, and clinical/epidemiologic centers. The biomarkers developmental laboratories will have responsibility for the development and characterization of new, or refinement of existing, biomarkers; the biomarkers validation laboratories will serve as a network resource for clinical and laboratory validation of biomarkers, which include technological development and refinement; and the clinical/epidemiology centers will conduct clinical and epidemiological research regarding the wide clinical application of biomarkers. A steering committee composed of the principal investigators in the network and appropriate NCI staff will coordinate the work of the consortium. Logistic support and informatics will be provided through an auxiliary data management and coordinating center. Approximately two awards will be made. Letter of intent is requested by July 15; full application is due August 16. Annual deadline dates may vary; contact program staff for exact dates. RFA: CA-05-009
Requirements: Domestic public and private for-profit and nonprofit organizations are eligible. Domestic institutions may propose collaborations/consortia with foreign institutions.
Restrictions: Applications will not be accepted from foreign institutions.
Amount of Grant: $1 million total
Date(s) Application is Due: Jul 16; Aug 16
Contact: Karen Chuang, Grants Administration Branch, (301) 496-2784; fax: (301) 496-8601; email: chuangk@mail.nih.gov; Dr. Sudhir Srivastava, Division of Cancer Prevention, (301) 435-1594; fax: (301)402-8990; email: srivasts@mail.nih.gov
Internet: http://grants.nih.gov/grants/guide/rfa-files/RFA-CA-05-009.html
Sponsor: National Cancer Institute
6130 Executive Boulevard, Executive Plaza N
Bethesda, MD 20892

NCI Early Detection Research Network: Clinical and Epidemiologic Centers Grants 3602
The Division of Cancer Prevention invites applications for cooperative agreements to establish a national network that will have responsibility for the development, evaluation, and validation of biomarkers for earlier cancer detection and risk assessment. Biomarkers are defined as cellular, biochemical, molecular, or genetic alterations by which a normal, abnormal, or simply biologic process can be recognized or monitored. Biomarkers are measurable in biological media, such as in tissues, cells, or fluids. The purpose of the network is to establish a scientific consortium of investigators, academic as well as industrial, with resources for basic, translational, and clinical research. The consortium will have three main components—biomarkers developmental laboratories, biomarkers validation laboratories, and clinical/epidemiologic centers. The biomarkers developmental laboratories will have responsibility for the development and characterization of new, or refinement of existing, biomarkers; the biomarkers validation laboratories will serve as a network resource for clinical and laboratory validation of biomarkers, which include technological development and refinement; and the clinical/epidemiology centers will conduct clinical and epidemiologic research regarding the wide clinical application of biomarkers. A steering committee composed of the principal investigators in the network and appropriate NCI staff will coordinate the work of the consortium. Logistic support and informatics will be provided through an auxiliary data management and coordinating center. Letter of intent is requested by May 14; full application is due June 14. RFA: CA-05-005
Requirements: Domestic public and private for-profit and nonprofit organizations are eligible. Domestic institutions may propose collaborations/consortia with foreign institutions.
Restrictions: Applications will not be accepted from foreign institutions.
Date(s) Application is Due: May 14; Jun 14
Contact: Karen Chuang, Grants Administration Branch, (301) 496-2784; fax: (301) 496-8601; email: chuangk@mail.nih.gov; Dr. Sudhir Srivastava, Division of Cancer Prevention, (301) 435-1594; fax: (301) 402-8990; email: srivasts@mail.nih.gov
Internet: http://grants.nih.gov/grants/guide/rfa-files/RFA-CA-05-005.html
Sponsor: National Cancer Institute
6130 Executive Boulevard, Executive Plaza N
Bethesda, MD 20892

NCI Engineered Isogenic Cell Lines with Relevant Cancer Targets Research Grants 3603
This PA encourages the small business community to develop isogenic cell lines in which functions of a molecular activity or pathway found in human tumors can be analyzed in pair-wise fashion. Such cell lines could be used for drug screening with the intent of identification of compounds that influence the action of the identified target. The end result of the PA-supported projects would be a series of genetically engineered screening models that the small businesses could market directly or utilize in a license agreement or other type of collaboration. PAR-98-068 may be obtained through the NIH grant line (301-402-2221), the NIH Web site, mail, and email.
Contact: Dr. George Johnson, Division of Cancer Treatment and Diagnosis, (301) 496-8783; fax: (301) 402-5200; email: gj16m@nih.gov; Toby Friedberg, Division of Extramural Activities, (301) 496-3428; fax: (301) 402-0275; email: friedbet&7dea.nci.nih.gov
Internet: http://www.grants.nih.gov/grants/guide/pa-files/PAR-98-068.html
Sponsor: National Cancer Institute
6130 Executive Boulevard, Executive Plz N
Bethesda, MD 20892

NCI Epidemiology Small Grants Program 3604
This is a short-term award, not to exceed three years, intended to provide support for pilot projects, testing of new techniques, or innovative or high-risk projects that could provide a basis for more extended research. Examples of research proposals that would apply to this PAR include, but are not limited to validating measurements in body fluids and tissues of exogenous exposures associated with the development of cancer; applying epigenetic approaches in cancer epidemiology; analyzing existing data that otherwise may have gone unexplored, such as pooled analyses of data from multiple studies coordinated into consortia; explore relationship between reported tobacco use and actual delivered amounts of tobacco constituents in epidemiologic studies and other large datasets in order to develop standardized markers of exposure in cancer risk; planning an epidemiological study to apply validated relevant carcinogenesis markers to better understand the natural history of neoplasms and identify individuals at high risk. Annual deadline dates may vary; contact program staff for exact dates. PA: PAR-04-159
Requirements: Applications may be submitted by domestic and foreign, for-profit and nonprofit organizations, both public and private; units of state and local governments; eligible agencies of federal government; and small businesses.
Amount of Grant: $100,000 maximum
Date(s) Application is Due: Mar 20; Jul 20; Nov 21
Contact: Crystal Wolfrey, Program Contact, (301) 496-8634; fax: (301) 496-8601; email: wolfreyc@mail.nih.gov
Internet: http://grants2.nih.gov/grants/guide/pa-files/PAR-04-159.html
Sponsor: National Cancer Institute
6120 Executive Boulevard, Executive Plaza S
Bethesda, MD 20892

NCI Established Investigator Award in Cancer Prevention, Control, Behavioral, and Population Sciences 3605
The purpose of this program is to provide established investigators protected time to devote to research and to act as mentors for new investigators and junior faculty members. The target candidates are outstanding established scientists who have demonstrated a sustained, high level of research productivity and significant contributions to cancer prevention, control, behavioral and/or population cancer research. They must demonstrate the need to develop and enhance their own research and a commitment to serve as mentors to new scientists. The award provides salary support for up to five years and it is renewable for one additional five-year period. The NCI is especially interested in training and retaining scientists who conduct high-quality research in cancer prevention, control, behavioral, and population sciences. Examples of relevant disciplines for this PA include any aspect of human cancer prevention (modifiable risk

factors, new animal models, and extrapolation of these models to human cancer; the development of preventive interventions to decrease cancer risk behaviors and/or increase health-promoting behaviors; chemoprevention from studies of the identification and early-phase characterization of candidate agents to preventive intervention trials; studies of nutritional supplements or complementary/alternative interventions in relation to cancer prevention and control; health services research, including patient outcome studies, practice research, and medical decision analyses related to cancer prevention or to cancer care and patient outcomes; Palliative care studies, including interventions to improve quality of life; survivorship research, including studies of outcomes and quality of life as related to cancer course, treatments, and treatment side effects; and studies of the effectiveness of cancer health communications in reducing high risk behaviors or in increasing participation in screening activities. Annual deadline dates may vary; contact program staff for exact dates. PA: PAR-05-145
Requirements: Eligible applicants must be established cancer scientists with a strong track record of publications and successful competition for cancer research support; must have peer-reviewed, independent research support at the time of, and during the entire duration of, the award; must have a record of supervising junior researchers; must demonstrate the need for protected time to advance their research careers and mentoring activities; must be U.S. citizens or non-citizen nationals, or must have been lawfully admitted for permanent residence and possess an Alien Registration Receipt Card (I-151 or I-551) or some other verification of legal admission as a permanent resident.
Contact: Bill Wells, Grants Administration Branch, (301) 496-8796; email: wellsw@mail.nih.gov
Internet: http://grants2.nih.gov/grants/guide/pa-files/PAR-05-145.html
Sponsor: National Cancer Institute
6120 Executive Boulevard, Executive Plaza S
Bethesda, MD 20892

NCI Exploratory Grants for Behavioral Research in Cancer Control **3606**
This FOA will support innovative pilot projects or feasibility studies, which will facilitate the growth of research science in the cancer control continuum from a behavioral perspective. This FOA includes and incorporates the research interests of the Behavioral Research program, the Office of Cancer Survivorship, and the Community Oncology and Prevention Trials Research Group. This FOA is appropriate for testing timely interventions in pilot studies for feasibility or using rigorous qualitative research methods to assess the potential efficacy of an intervention. It is also appropriate for the psychometric evaluation of new measures or culturally appropriate ones to be adapted for use in populations where measures have not yet been developed or validated. This FOA encourages applications that include small cross-disciplinary teams of investigators who bring perspectives from the behavioral and social sciences, as well as other fields of public health.
Requirements: Eligible institutions and agencies include: public/state controlled institutions of higher education; private institutions of higher education; nonprofits with 501(c)3 IRS status (other than institutions of higher education); nonprofits without 501(c)3 IRS status (other than institutions of higher education); small businesses; for-profit organizations (other than small businesses); state governments; U.S. territories or possessions; regional organizations; non-domestic (non-U.S.) entities (foreign organizations); eligible agencies of the federal government; and units of local governments.
Amount of Grant: $275,000 maximum request per two-year period
Date(s) Application is Due: Feb 16; Jun 16; Oct 16
Contact: Sabra F. Woolley, Ph.D.; (301) 435-4589; fax: (301) 594-0673; email: woolleys@mail.nih.gov; Barbara C. Sorkin, Ph.D.; (301) 594-8018; email: sorkinb@mail.nih.gov
Internet: http://grants.nih.gov/grants/guide/pa-files/PA-09-130.html
Sponsor: National Cancer Institute (NCI)
6130 Executive Boulevard, Room 4084, MSC 7365
Bethesda, MD 20850-7365

NCI Exploratory/Developmental Grants for Diagnostic Cancer Imaging **3607**
The Cancer Imaging Program (CIP) of the Division of Cancer Treatment and Diagnosis solicits exploratory/developmental (R21) grants that articulate highly innovative research concepts in diagnostic cancer imaging. Within each area of importance in imaging, there exists a need for innovative and creative approaches leading to new avenues of research. One way to encourage research into high-risk/high-impact areas is to provide investigators with the initial resources required to accomplish feasibility and pilot testing of innovative ideas. Contact program staff for deadline dates. PA-04-045
Requirements: Applications may be submitted by foreign and domestic, for-profit and nonprofit organizations, public and private.
Amount of Grant: $275,000 maximum for the two year period
Contact: Kathryn Dunn, Grants Management Specialist, National Cancer Institute, (301) 846-6829; fax: (301) 846-5720; email: dunnkath@mail.nih.gov
Internet: http://grants.nih.gov/grants/guide/pa-files/PA-04-045.html
Sponsor: National Cancer Institute
6120 Executive Boulevard, Executive Plz S
Bethesda, MD 20892

NCI Flexible System to Advance Innovative Research for Cancer Drug Discovery by Small Businesses Grants **3608**
The objective of this RFA is to support the research activities required to enable small businesses to bring their research efforts for drug discovery and development to clinical evaluation. Annual deadline dates may vary; contact program staff for exact dates. PA: PAR-03-074
Requirements: Any small business, independently owned and operated by United States citizens or permanent resident aliens may apply. An eligible business must be organized for-profit, cannot be dominant in its field of expertise, and must have its principal place of business in the United States. Including any affiliates, the company can be the employer of no more than 500 people.
Amount of Grant: $1.8 million
Contact: Shane Woodward, Grants Administration Branch, (301) 496-8649; fax: (301) 496-8601; email: woodwars@mail.nih.gov
Internet: http://grants2.nih.gov/grants/guide/pa-files/PAR-03-074.html
Sponsor: National Cancer Institute
6120 Executive Boulevard, Executive Plaza S
Bethesda, MD 20892

NCI Howard Temin Award **3609**
The goal of this award is to bridge the transition from a mentored research environment to an independent research career for scientists who have demonstrated unusually high potential during their initial stages of training and development. This special award is aimed at fostering the research careers of outstanding junior scientists in basic research who are committed to developing research programs highly relevant to the understanding of human biology and human disease as it relates to the etiology, pathogenesis, prevention, diagnosis, and treatment of cancer. Candidates are offered up to five years to develop knowledge in the basic sciences and research skills, with up to three of the initial years (at least one year required) in a mentored environment followed by a transition to an unmentored independent investigator phase for the remaining time. More information may be obtained by mail and email or through the NIH grant line (301-402-2221) and the NIH Web site. PA: PAR-03-104
Requirements: Candidates must be U.S. citizens or noncitizen nationals, or must have been lawfully admitted for permanent residence by the time of award.
Restrictions: Former principal investigators on any NIH research grants or non-PHS peer reviewed grants over $70,000 direct costs per year are not eligible.
Amount of Grant: $75,000 maximum salary plus fringe benefits and $30,000 per year for expenses during mentored phase; $50,000 per year for expenses after mentored phase
Contact: Catherine Blount, Grants Administration Branch, (301) 496-3179; fax: (301) 496-8601; email: blountc@mail.nih.gov
Internet: http://grants2.nih.gov/grants/guide/pa-files/PAR-03-104.html
Sponsor: National Cancer Institute
6120 Executive Boulevard, Executive Plaza S
Bethesda, MD 20892

NCI Improving Diet and Physical Activity Assessment Grants **3610**
This Funding Opportunity Announcement (FOA) is aimed at advancing the quality of measurements of dietary intake and physical activity pertinent to cancer and/or other pathologies through support of research on improved instruments, technologies, and/or statistical/analytical techniques. Studies proposed in the grant applications should be aimed at optimizing the combined use of objective and self-report measures of physical activity and/or dietary intake for testing in both general and diverse populations. Applicants responding to this FOA, which uses the R01 grant mechanism, should have already tested and validated new approaches and have obtained feasibility data in the context of cancer or other relevant diseases. Because the nature and scope of the proposed research will vary from application to application, it is anticipated that the size and duration of each award will also vary.
Requirements: Application can be made by the following institutions and organizations: public/state controlled institutions of higher education; private institutions of higher education; nonprofits with 501(c)3 IRS status (other than institutions of higher education); nonprofits without 501(c)3 IRS status (other than institutions of higher education); small businesses; for-profit organizations (other than small businesses); state governments; U.S. territories or possessions; Indian/Native American tribal governments (federally recognized); Indian/Native American tribal governments (other than Federally recognized); Indian/Native American tribally designated organizations; non-domestic (non-U.S.) entities (foreign organizations); Hispanic-serving institutions; historically Black colleges and universities (HBCUs); tribally controlled colleges and universities (TCCUs); Alaska Native and Native Hawaiian serving institutions; regional organization; eligible agencies of the Federal government; and faith-based or community based organizations.
Restrictions: Applicants without sufficiently extensive preliminary data or those who wish to explore the utility of new untested dietary or physical activity assessment methods in pilot studies are advised to consider submitting applications under the parallel FOA, PAR-06-103, which uses the Exploratory/Developmental Grant (R21) award mechanism.
Date(s) Application is Due: Feb 5; Jun 5; Oct 5
Contact: Amy Subar, Ph.D.; (301) 594-0831; fax: (301) 435-3710; email: subara@mail.nih.gov; Richard Troiano, Ph.D.; (301) 435-6822; fax: (301) 435-3710; email: troianor@mail.nih.gov
Internet: http://grants.nih.gov/grants/guide/pa-files/PAR-09-225.html
Sponsor: National Cancer Institute (NCI)
6130 Executive Boulevard, EPN Room 4005, MSC 7344
Bethesda, MD 20892-7344

NCI In Vivo Cellular and Molecular Imaging Centers Planning Grants **3611**
Applications are invited for three-year planning grants which will lead to the establishment of in vivo cellular and molecular imaging centers. Planning grants are designed to assist institutions that have most of the separate scientific components necessary for productive interaction but have no track record of performing multidisciplinary scientific research. The initiative is designed to capitalize on the extraordinary opportunity for studying cancer noninvasively, and in many cases, quantitatively due to recent advances

in molecular imaging modalities and in molecular and cellular biology. Awards will facilitate the interaction of scientists from a variety of fields such as, but not limited to, imaging sciences, chemistry, radiopharmaceutical chemistry, cellular and molecular biology, pharmacology, computer science, biomedical engineering, immunology, and neuroscience. The program will also provide resources to conduct multidisciplinary research. A letter of intent is requested by July 14; full application is due August 18. Annual deadline dates may vary; contact program staff for exact dates.
Requirements: Applications may be submitted by domestic public and private nonprofit and for-profit organizations including universities, colleges, hospitals, and laboratories.
Restrictions: Foreign institutions are not eligible to apply; foreign components of applications from domestic organizations will be accepted, however, with adequate justification.
Amount of Grant: $439,000 average
Date(s) Application is Due: Feb 1; Jun 1; Oct 1
Contact: Dr. Sheila Taube, Associate Director, Cancer Diagnosis Program, (301) 496-8639
Internet: http://grants.nih.gov/grants/guide/rfa-files/RFA-CA-01-010.html
Sponsor: National Cancer Institute
6130 Executive Boulevard, Executive Plaza N
Bethesda, MD 20892

NCI Innovative Technologies for the Molecular Analysis of Cancer: Phased Innovation Award (R21/R33) Grants 3612
NCI invites applications for research projects to develop novel technologies that will support the molecular analysis of cancers and their host environment in support of basic, clinical, and epidemiologic research. Technology encompasses methods and tools that enable research including, but not limited to, instrumentation, techniques, devices, and analysis tools (e.g., computer software). Technology is distinct from resources such as databases and tissue repositories. Applications for support of such resources will not be considered to be responsive to this announcement. Technologies solicited include those that are suitable for the detection of alterations and instabilities of genomic DNA; measurement of the expression of genes and gene products; analysis and detection of gene and or cellular products including post-translational modification, and function of proteins; identification and characterization of exogenous infectious agents in cancer; and assaying the function of major signal transduction networks involved in cancer. This PA is intended to support the development of all required components of fully integrated systems for analysis including front end preparation of sample materials from cells, bodily fluids, and tumor specimens; novel chemistries or contrast agents; molecular detection systems; data acquisition methods; and data analysis tools. Technologies under consideration include those that will support molecular analysis either in vitro, in situ, or in vivo (by imaging or other methods) in the discovery process, as well as in preclinical models and clinical research. This solicitation will utilize the Phased Innovation Award mechanism (R21/R33). Letters of intent are requested by February, June, and October; full applications are due in March, July, and November. Annual deadline dates may vary; contact program staff for exact dates. PA: PAR-01-104
Amount of Grant: $100,000 maximum per year
Contact: Dr. Jay George, Office of Technology and Industrial Relations, (301) 496-1550; fax: (301) 496-7807; email: jgeorge@mail.nih.gov
Internet: http://grants.nih.gov/grants/guide/pa-files/PAR-01-104.html
Sponsor: National Cancer Institute
31 Center Drive
Bethesda, MD 20892

NCI Mentored Career Development Award 3613
The Comprehensive Minority Biomedical Program, Division of Extramural Activities, invites underrepresented minority research scientists who have been recipients of NIH Research Supplements for Underrepresented Minority Individuals in Postdoctoral Training (MIPT) or Minority Investigator Supplements (MIS), funded by NCI, who need extended periods of sponsored research as a way to gain scientific expertise while bridging the transition from mentored research environments to independent research/academic careers to submit applications. This award offers opportunities for a mentored peer review experience in cancer research which will enhance the candidate's knowledge and understanding of the peer review process with the intended purpose of developing skills with the expectation that the candidate will submit a grant application for nontargeted mechanisms (R01, R03, R21). Applicants are encouraged to contact NCI program staff regarding their eligibility. PA: PAR-03-016 may be obtained from the Web site and by mail and email.
Amount of Grant: $75,000 maximum salary plus related fringe benefits; $30,000 for research development during first phase, $50,000 during second phase
Contact: Barbara Fisher, Grants Administration Branch, (301) 846-1015; fax: (301) 846-5720; email: fisherb@gab.nci.nih.gov
Internet: http://grants2.nih.gov/grants/guide/pa-files/PAR-03-016.html
Sponsor: National Cancer Institute
6130 Executive Boulevard, Executive Plaza N
Bethesda, MD 20892

NCI Mentored Career Development Award to Promote Diversity 3614
This mechanism establishes a pathway of recruiting, training, and retaining underrepresented investigators in research fields that address problems pertinent to the biology, etiology, pathogenesis, prevention, diagnosis, control, and treatment of human cancer and who can conduct independent competitive cancer research programs. The Mentored Career Development Award to Promote Diversity (K01) provides support for a sustained period of protected time for intensive research career development under the guidance of an experienced mentor, or sponsor, in the biomedical, behavioral or clinical sciences leading to research independence. The expectation is that through this sustained period of research career development and training, awardees will launch independent research careers and become competitive for new research project grant funding.
Requirements: The following organizations and institutions may apply: for-profit and non-profit organizations; public or private institutions, such as universities, colleges, hospitals, and laboratories; and domestic institutions.
Restrictions: Foreign institutions and organizations are not eligible to apply.
Amount of Grant: $75,000 salary plus fringe benefits
Date(s) Application is Due: Feb 12; Mar 12; Jun 12; Jul 12; Oct 12; Nov 12
Contact: Belinda Locke; (301) 496-7344; fax: (301) 402-4551; email: Lockeb@mail.nih.gov
Internet: http://grants.nih.gov/grants/guide/pa-files/PAR-06-220.html
Sponsor: National Cancer Institute (NCI)
6116 Executive Boulevard, Suite 7031, MSC 8350
Bethesda, MD 20892-8350

NCI Minorities in Clinical Oncology Grants 3615
The Comprehensive Minority Biomedical Program of NCI announces the availability of minority medical oncology awards. The purpose of these grants is to encourage recently trained underrepresented minority clinicians to acquire clinical training and research experience in medical oncology and to increase representation of minorities in medical oncology. More information may be obtained by mail or email or through the NIH grant line (301-402-2221), or the NIH Web site. Annual deadline dates may vary; contact program staff for exact dates. PA: PAR-03-002
Requirements: The award is intended for underrepresented minorities who have a health professional degree or its equivalent, or are doctorally trained oncology nurses to acquire research experience in clinical oncology. Individuals with the PhD or other doctoral degrees in clinical disciplines such as clinical psychology, nursing, clinical genetics, speech-language pathology, audiology and rehabilitation are also eligible.
Restrictions: Current and former PI,s on NIH research project grants, FIRST Awards, comparable career development awards, sub-projects of program project or center grants, and the equivalent are not eligible to apply.
Contact: Barbara Fisher, Grants Administration Branch, National Cancer Institute, (301) 846-1015; fax: (301) 846-5720; email: fisherb@gab.nci.nih.gov
Internet: http://grants.nih.gov/grants/guide/pa-files/PAR-03-002.html
Sponsor: National Cancer Institute
6120 Executive Boulevard, Room 243
Bethesda, MD 20892

NCI Molecular Target Drug Discovery for Cancer 3616
The goal of this program is the discovery and validation of molecular targets for cancer prevention or treatment intervention which can be developed into drug discovery assays. PA: PAR-01-046
Requirements: Applications may be submitted by domestic for-profit and nonprofit organizations, both public and private; units of state and local governments; and eligible agencies of the federal government.
Amount of Grant: $250,000 maximum
Contact: Toby Friedberg, Extramural Activities, (301) 496-3428; fax: (301) 402-0275; email: tf12w@nih.gov
Internet: http://grants.nih.gov/grants/guide/pa-files/PAR-01-046.html
Sponsor: National Cancer Institute
6116 Executive Boulevard
Bethesda, MD 20892

NCI National Cooperative Drug Discovery Research on Opportunistic Infections 3617
Investigator-initiated research project grant applications are invited to support projects with adequate preliminary data for serious development of new therapies to treat AIDS- and cancer-associated opportunistic infections. Studies may include research on identified targets for rational design of inhibitors or projects with chemically identified candidate compounds with suitable efficacy and selectivity for exploration as candidate drugs. Applications that include collaborations with the private sector (e.g., pharmaceutical, chemical or biotechnological companies) are strongly encouraged. The opportunistic pathogens emphasized in this program announcement are human cytomegalovirus, Mycobacterium tuberculosis, Mycobacterium avium, Cryptosporidium parvum, systemic candidiasis, Aspergillus, and the Microsporida. Annual deadline dates may vary; contact program staff for exact dates. PA: PA-98-100
Requirements: Applications may be submitted by public and private, domestic and foreign, for-profit and nonprofit organizations.
Contact: Dr. Mary Wolpert, Division of Cancer Treatment and Diagnosis, (301) 496-8783; fax: (301) 402-5200; email: mw8u@nih.gov
Internet: http://www.grants.nih.gov/grants/guide/pa-files/PA-98-100.html
Sponsor: National Cancer Institute
6130 Executive Boulevard, Executive Plaza N
Bethesda, MD 20892

NCI Nonmammalian Organisms as Models for Anticancer Drug Discovery Research Grants 3618
This PA encourages the use of nonmammalian organisms to discover new cancer therapies. The goal is to identify key genes, enzymatic activities, components of signaling pathways, or cellular processes which are altered in human cancer, as potential intervention points that could be used in the design of new cancer drugs. Projects could

focus on validating inferences already drawn from comparative study of cancerous and normal cells, exploring promising leads from the nature of mutations known to be associated with cancer development, or testing hypotheses generated by the identification of new and unknown gene products such as those sequenced through the NCI's Cancer Genome Project (CGAP). For example, projects may include development of organisms in which genes in key pathways or processes are mutated or in which human transgenes are expressed to simulate changes known to occur in human cancers. These genetic manipulations could present a proof of principle or validation of the importance of the target genes to the development of cancer. Some examples of genes that are well studied in model organisms and known or strongly suspected of involvement in cancer include oncogenes, proto-oncogenes, some cell cycle and signal transduction genes, and DNA repair genes. Investigations in response to this PA should provide insight into which genes or gene products in cancer-relevant pathways are potential targets for human cancer treatment. Projects of a general nature not specifically related to new cancer target or drug discovery will not be considered responsive. A search for small molecule inhibitors through screening is appropriate. Although this PA is being issued by NCI, NIGMS has an interest in supporting projects related to this topic. Studies of growth regulation, cell cycle control, or other basic studies that are not explicitly focused on tumor target cell systems may be assigned to NIGMS or another NIH institute. Annual deadline dates may vary; contact program staff for exact dates. PA: PA-01-042
Requirements: Applications may be submitted by foreign and domestic, public and private, for-profit and nonprofit organizations, units of state and local governments, and eligible agencies of the federal government. Small businesses are encouraged to consider a parallel program announcement of identical scientific scope that utilizes the SBIR and STTR mechanisms (PAR-99-020). It should be noted that a search for small molecule inhibitors through screening is appropriate for projects submitted for the SBIR program announcement.
Contact: Dr. George Johnson, Division of Cancer Treatment and Diagnosis, (301) 496-8783; fax: (301) 402-5200; email: johnsong@exchange.nih.gov
Internet: http://grants.nih.gov/grants/guide/pa-files/PA-01-042.html
Sponsor: National Cancer Institute
6130 Executive Boulevard, Executive Plaza N
Bethesda, MD 20892

NCI Paul Calabresi Award for Clinical Oncology 3619
The purpose of this program is to increase the number of medical doctors (MD's, DO's), and doctorally-degreed nurses (PhD's or equivalents) and basic scientists (PhD's DVM's, or equivalents) who are highly-motivated and trained to: perform clinical oncology therapeutic research that develops and tests scientific hypotheses based on fundamental and clinical research findings; design and test hypothesis-based, clinical therapeutic protocols and adjunct biological analyses and for clinician candidates to administer all phases (i.e., pilot/Phase I, Phase II, and Phase III) of cancer therapeutic clinical trials, and conduct cancer therapeutic research in team research settings in which basic and clinical scientists collaborate and interact to expedite the translation of basic research discoveries into patient-oriented therapeutic cancer research. The PCACO is not intended to train laboratory-based scientists whose research will emphasize the use of animal or other model systems. Letters of intent are requested by May 2; full application is due by June 1. Annual deadline dates may vary; contact program staff for exact dates. PA: PAR-04-096
Requirements: Applications may be submitted by domestic, for-profit and nonprofit organizations, both public and private; units of state and local governments; and eligible agencies of the federal government. The principal investigator and all clinician candidates must currently be physicians holding the MD or DO degrees, or be doctorally prepared oncology registered nurses, and must have completed their clinical training.
Amount of Grant: $750,000 maximum direct costs per year
Date(s) Application is Due: May 2; Jun 1
Contact: Kimery Griffin, Grants Administration Branch, (301) 496-3196; fax: (301) 496-8601; email: griffink3@mail.nih.gov; Dr. Lester Gorelic, Training and Resources Program, (301) 496-8580; fax: (301) 402-4472; email: gorelicl@mail.nih.gov
Internet: http://grants.nih.gov/grants/guide/pa-files/PAR-04-096.html
Sponsor: National Cancer Institute
6116 Executive Boulevard, Suite 7025, MSC 8346
Bethesda, MD 20892

NCI Planning Grants for NCI Cancer Research Centers 3620
The Cancer Centers Branch of NCI invites planning grant applications for the development of Cancer Research Centers in a variety of organizational settings. The purpose is to expand the scientific, geographic, and demographic diversity of the Cancer Centers Program by encouraging research-oriented organizations to develop the qualities of a strong cancer research center and become competitive for a Cancer Center Support grant. Cancer center planning strategies may focus on a specific research theme (e.g., diagnosis, therapy, epidemiology) or integrate a broad spectrum of research to include the basic, clinical, prevention and control, and population sciences (i.e., an NCI-designated comprehensive cancer center). Information may be obtained electronically through the NIH Web site, email, and by the NIH grant line (301-402-2221) and mail. Annual deadline dates may vary; contact program staff for exact dates. PA: PAR-03-022
Requirements: Applications may be submitted by domestic for-profit and nonprofit organizations, both public and private; units of state and local governments; and eligible agencies of the federal government.
Amount of Grant: $250,000 maximum in direct costs
Contact: Eileen Natoli, Grants Administration Branch, (301) 496-8791; fax: (301) 496-8601; email: cm113g@nih.gov
Internet: http://grants2.nih.gov/grants/guide/pa-files/PAR-03-022.html
Sponsor: National Cancer Institute
6130 Executive Boulevard, Executive Plaza N
Bethesda, MD 20892

NCI Priorities in Behavioral Research in Cancer Prevention & Control Grants 3621
NCI and NIDR invite researchers to submit research project grant (R01) applications that address behavioral research issues in cancer prevention and control. The PA addresses recommendations made by a special Behavioral Research in Cancer Prevention and Control Working Group in 1996 that consisted of leading national experts whose role was to identify behavioral research needs in cancer prevention and control during the coming years. PA-97-055 may be obtained electronically through the NIH grant line (301-402-2221), the NIH Web site, and by mail and email.
Contact: Veronica Chollett, (301) 435-2837; email: vc24a@nih.gov
Internet: http://www.grants.nih.gov/grants/guide/pa-files/PA-97-055.html
Sponsor: National Cancer Institute
6130 Executive Boulevard, Executive Plaza N
Bethesda, MD 20892

NCI Prostate Cancer Aging, Race, and Ethnicity Research Grants 3622
NIA, NCI, and NIEHS invite applications to expand the understanding of biological and clinical factors leading to the development, progression, and treatment of prostate cancer in aging men. The increased risk of prostate cancer with advancing age and its prominence in older-aged men are well-known characteristics of this tumor. The unusually high incidence and mortality rates of prostate cancer for older white and black American men and, by contrast, the much lower rates in men of Hispanic and Asian descent, provide the need for research that emphasizes the role of race and ethnic factors, as well as age, in early diagnosis, management, and etiology of this tumor. This PA is intended to stimulate research that applies the expanding scientific knowledge gained on prostate cancer to older men and to extend the knowledge base on age-related aspects of the etiology of this malignancy. The PA, which describes the research objectives, application procedures, review considerations, and award criteria, may be obtained electronically through the NIH grant line (301-402-2221), the NIH Web site, and by mail and email. PA: PA-97-019
Contact: Dr. Andrew Chiarodo, Organ System Coordinating Branch, (301) 496-5278; fax: (301) 402-0181; email: ac53a@nih.gov
Internet: http://www.grants.nih.gov/grants/guide/pa-files/PA-97-019.html
Sponsor: National Cancer Institute
6130 Executive Boulevard, Executive Plaza N
Bethesda, MD 20892

NCI Review and Analysis of Tobacco Industry Documents Grants 3623
This program announcement is intended to stimulate research on a wide variety of scientific, technical, marketing, and tactical undertakings by the tobacco industry that were documented in papers, memos, and other records. The systematic, comprehensive analysis and evaluation of these documents will greatly contribute to the understanding of what the tobacco industry knew and will help researchers and the public health community identify effective strategies to reduce tobacco use. Annual deadline dates may vary; contact the program office for exact dates. PA: PAR-01-063
Requirements: Applications may be submitted by foreign and domestic, for-profit and nonprofit organizations, public and private. Applicants requesting budgets greater than $500,000 in direct costs are required to contact program staff prior to submitting applications.
Contact: Dr. Michele Bloch, Division of Cancer Control and Population Sciences, (301) 496-8584; fax: (301) 496-8675; email: blochm@mail.nih.gov
Internet: http://grants.nih.gov/grants/guide/pa-files/PAR-01-063.html
Sponsor: National Cancer Institute
6130 Executive Boulevard, Executive Plaza N
Bethesda, MD 20892

NCI Ruth L. Kirschstein National Research Service Award (NRSA) Institutional Research Training Grants 3624
The National Institutes of Health (NIH) will award Ruth L. Kirschstein National Research Service Award (NRSA) Institutional Research Training Grants (T32) to eligible institutions as the primary means of supporting predoctoral and postdoctoral research training to help ensure that a diverse and highly trained workforce is available to assume leadership roles related to the Nation, biomedical, behavioral and clinical research agenda. The primary objective of the T32 program is to prepare qualified individuals for careers that have a significant impact on the health-related research needs of the Nation. This program supports predoctoral, postdoctoral and short term research training programs at domestic institutions of higher education with the T32 funding mechanism. Stipends are provided as a subsistence allowance for trainees to help defray living expenses during the research training experience and are based on a 12-month appointment period.
Requirements: The following organizations are eligible to apply: public/state controlled institutions of higher education; private institutions of higher education; nonprofits with 501(c)3 IRS status (other than institutions of higher education); nonprofits without 501(c)3 IRS status (other than institutions of higher education); for-profit organizations (other than small businesses); state governments; U.S. territories or possessions; Indian/Native American tribal governments (Federally recognized); Hispanic-serving institutions; historically Black colleges and universities (HBCUs); tribally controlled colleges and universities (TCCUs); Alaska Natives and Native Hawaiians serving institutions; regional organizations; eligible agencies of the federal government; faith-based or community based organizations.

Date(s) Application is Due: Apr 8; Aug 8; Dec 8
Contact: Crystal Wolfrey; (301) 496-8634; (301) 496-8601; email: wolfreyc@mail.nih.gov
Internet: http://grants.nih.gov/grants/guide/pa-files/PA-08-226.html
Sponsor: National Cancer Institute (NCI)
6120 Executive Boulevard, Room 243, MSC 7150
Bethesda, MD 20892-7150

NCI Short-Term Scientist Exchange Program **3625**
The program promotes collaborative research between established U.S. and foreign scientists by supporting, in part, exchange visits of U.S. cancer researchers to foreign laboratories, or foreign cancer researchers to U.S. laboratories. The visits may be from one week to six months in duration. The Office of International Affairs (OIA) funds an APEX fare airline ticket through an OIA contractor. The foreign host institution covers the in-country expenses of the U.S. visitor. In the case of foreign visitors, a foreign source pays for the airfare, while OIA provides a subsistence allowance. The subsistence allowance is generally the monthly equivalent of $32,004 per year (split equally by OIA and the sponsor). Cost sharing of the subsistence allowance by the sponsoring American laboratory is required.
Requirements: Candidates must have a minimum of three years postdoctoral experience in cancer research and have an invitation from a qualified sponsor. Sponsors or exchangees from the United States must be NCI-supported principal investigators, in either the intramural or extramural research programs. Foreign sponsors must be established scientists recognized in the international peer reviewed literature. Foreign candidates must be proficient in spoken and written English. Preference is given to applicants from not yet fully developed countries.
Amount of Grant: $37,100 per year
Contact: Office of International Affairs, (301) 496-4761; fax: (301) 496-3945; email: NC6@cu.nih.gov or cancergovstaff@mail.nih.gov
Internet: http://www.cancer.gov/about_nci/doc_img.aspx?viewid=8fc72aff-ab84-40b5-8fe9-20da67662417&docid=775a930b-99ec-40ff-ae9d-59beb7175527
Sponsor: National Cancer Institute
6130 Executive Boulevard, Executive Plaza N
Bethesda, MD 20892

NCI Small Grants Program for Behavioral Research in Cancer Control **3626**
The small grants program is designed to aid and facilitate the growth of a nationwide cohort of scientists with a high level of research expertise in behavioral cancer control research. Small grants are short-term awards to provide support for pilot projects, development and testing of new methodologies, secondary data analyses, or innovative projects that provide a basis for more extended research. New investigators in relevant fields and disciplines (e.g., medicine, public health, health promotion, health communications and informatics, epidemiology, anthropology, social work, nursing research, nutrition, health policy, health services research, and behavioral sciences, such as psychology, health education, and sociology) may apply for small grants to test ideas or do pilot studies. The small grant award is not renewable. PA: PAR-04-020
Requirements: Applications may be submitted by public and private, domestic, for-profit and nonprofit organizations, units of state and local governments, and eligible agencies of the federal government. Eligible applicants include new investigators who have not previously been a principal investigator on an NCI-funded cancer control research grant (R03, R01, P01), or established scientists refocusing their research interests to behavioral research in cancer. Predoctoral investigators currently enrolled in an accredited doctoral degree program also are eligible to apply. All applicants should identify a mentor or sponsor from whom they will receive guidance regarding the proposed research. Small research grants may not be used to supplement research projects currently supported by federal or nonfederal funds, or to provide interim support of research project applications under review by the Public Health Service.
Restrictions: Foreign organizations are ineligible to apply.
Amount of Grant: $50,000 maximum for one year for up to two years
Date(s) Application is Due: Apr 20; Aug 22; Dec 22
Contact: Veronica Chollette, Division of Cancer Control and Population Sciences, (301) 435-2837; email: vc24a@nih.gov
Internet: http://www.grants.nih.gov/grants/guide/pa-files/PAR-04-020.html
Sponsor: National Cancer Institute
6130 Executive Boulevard, Executive Plaza N
Bethesda, MD 20892

NCI Specialized Program of Research Excellence in Human Cancer **3627**
A SP.O.RE must address the scientific information base and provide focal points for sustaining and maintaining state-of-the-art research that will contribute to improved detection, diagnosis, treatment, and prevention of an organ-specific cancer. SP.O.RES not only are expected to conduct a wide spectrum of research activities, but also to contribute significantly to the development of specialized research resources, the development of improved research model systems and the expansion of the research base through collaborative research with scientists and clinicians in other institutions locally and nationwide. A letter of intent is due at least 60 days prior to the specific due date. Application receipt dates are February 1 for breast cancer SP.O.RES, June 1 for lung cancer SP.O.RES, and October 1 for prostate and genitourinary cancer SP.O.RES. PA: PAR-99-167
Requirements: Applications may be submitted by domestic for-profit and nonprofit organizations, both public and private; units of state and local governments; and eligible agencies of the federal government.
Date(s) Application is Due: Feb 1; Jun 1; Oct 1
Contact: Dr. Jorge Gomez, Organ Systems Branch, (301) 496-8528; email: jg1w@nih.gov
Internet: http://grants.nih.gov/grants/guide/pa-files/PAR-99-167.html
Sponsor: National Cancer Institute
6130 Executive Boulevard
Rockville, MD 20852

NCI Stages of Breast Development: Normal to Metastatic Disease Grants **3628**
The National Cancer Institute (NCI), National Institute of Child Health and Human Development (NICHD), National Institute of Diabetes and Digestive and Kidney Diseases (NIDDK), National Institute on Aging (NIA), and National Institute of Environmental Health Sciences (NIEHS) invite investigator-initiated research grant applications to study the molecular, cellular, endocrine, and other physiological influences on the development and maturation of the normal mammary gland and alterations involved in early malignant and metastatic breast cancer. Multi-disciplinary collaborations, for example between cell biologists, molecular endocrinologists, bioengineers, geneticists, and mammary pathologists, are encouraged. Appropriate studies include, but are not limited to, documenting the role of dynamic hormonal influences and determining the role of cell growth, apoptosis, and differentiation in mammary gland maturation; integrating knowledge of cell signaling in breast tissue with whole organ biology; developing models of breast differentiation; and studies focusing on the characteristics of breast tumor physiology particularly in relation to metastasis. It is expected that a complete understanding of mammary gland development will form critical underpinnings for continued advances in detecting, preventing and treating breast cancer.
Requirements: Applications may be submitted by domestic and foreign, for-profit and non-profit organizations, public and private, such as universities, colleges, hospitals, laboratories, units of State or local governments, and eligible agencies of the Federal government.
Amount of Grant: $250,000 maximum per year
Date(s) Application is Due: Feb 5; Jun 5; Oct 5
Contact: Padma Maruvada, Ph.D.; (301) 496-3893; fax: (301) 402-8990; email: maruvadp@mail.nih.gov
Internet: http://grants.nih.gov/grants/guide/pa-files/PA-99-162.html
Sponsor: National Cancer Institute (NCI)
6130 Executive Boulevard, EPN Suite 3144, MSC 7362
Bethesda, MD 20892-7362

NCI Technologies and Software to Support Integrative Cancer Biology Research (SBIR) Grants **3629**
This Funding Opportunity Announcement (FOA), issued by the National Cancer Institute (NCI), invites Small Business Innovation Research (SBIR) grant applications from small business concerns (SBCs) that propose the development of commercial software tools, computational/mathematical methods, and technologies that will enable integrative cancer biology research. Integrative cancer biology focuses on understanding cancer as a complex biological system by utilizing both computational and experimental biology to integrate heterogeneous data sources and ultimately to generate predictive computational models of cancer processes. Phase I applications may request up to 2 years of support with total costs of up to $300,000 for the duration of the award (not to exceed $200,000 total costs in any one year). Budgetary requests of Phase II applications will conform to standard SBIR guidelines (up to $750,000 total costs for up to two years).
Requirements: Only United States small business concerns (SBCs) are eligible to submit SBIR applications.
Amount of Grant: $300,000 maximum for Phase 1; $750,000 maximum for Phase 2
Date(s) Application is Due: Apr 5; Aug 5; Dec 5
Contact: Greg Evans, Ph.D.; (301) 594-8807; fax: (301) 480-4082; email: evansgl@mail.nih.gov
Internet: http://grants.nih.gov/grants/guide/pa-files/PA-09-188.html
Sponsor: National Cancer Institute (NCI)
Building 31, Room 10A19, 31 Center Drive, MSC 2580
Bethesda, MD 20892-2580

NCI Transition Career Development Award **3630**
The purpose of the program is to facilitate the transition of investigators from the mentored to the independent stage of their careers in cancer research. It applies to clinicians who are pursuing basic science careers; clinicians who are pursuing careers in patient-oriented research; and to individuals pursuing careers in the prevention, control, and population sciences. To apply, a candidate must have completed two years or more of postdoctoral, mentored research or have been in an independent position for less than one year at the time of the application. The unique feature of this award is that individuals may apply without a sponsoring institution while they are still in a mentored position. Successful postdoctoral applicants will be given up to 12 months to identify an independent, preferably tenure-track, position at a sponsoring institution before an award can be activated. Awardees must apply for an R01 research grant or equivalent prior to the end of the second year of the award. Annual deadline dates may vary; contact program staff for exact dates. PA: PAR-01-134
Requirements: Eligible candidates are doctorally degreed individuals who have been educated as clinicians (e.g., MDs, oncology nurses) or as prevention, control, and population scientists (e.g., PhDs, DPHs, MDs) and are ready to pursue independent careers in cancer research.
Restrictions: The award is not for basic scientists following a traditional basic science research career track (e.g., PhDs in molecular biology).
Date(s) Application is Due: Feb 1; Jun 1; Oct 1

Contact: Kimery Griffin, Grants Administration Branch, (301) 496-3196; fax: (301) 496-8601; email: griffink3@mail.nih.gov
Internet: http://grants.nih.gov/grants/guide/pa-files/PAR-04-040.html
Sponsor: National Cancer Institute
6130 Executive Boulevard, Executive Plaza N
Bethesda, MD 20892

NCI/NCCAM Quick-Trials for Novel Cancer Therapies 3631
This PA sponsored by NCI and NCCAM expands the initiative to all cancer sites and provides investigators with rapid access to support for pilot, phase I, and phase II cancer clinical trials testing new agents and patient monitoring and laboratory studies to ensure the timely development of new therapeutic approaches. QUICK-TRIAL will provide a new approach designed to simplify the grant application process and provide a rapid turnaround from application to funding. Features include a modular grant application and award process, inclusion of the clinical protocol within the grant application, and accelerated peer review with the goal of issuing new awards within five months of application receipt. Inclusion of the complete clinical protocol within the PHS 398 grant application is intended to simplify the application process by eliminating the need to duplicate protocol details in the research plan section. Investigators may apply for a maximum of two years of funding support using the exploratory/developmental (R21) grant mechanism for up to $250,000 direct costs per year. Therapeutic trials in human subjects employing new agents and therapeutic approaches, including CAM treatments, whether used as a single agent/modality or in combination, are appropriate. Preference will be given to clinical trials that include laboratory studies to validate mechanistic hypotheses or clinical correlates that can meaningfully guide further clinical development. A letter of intent is due one month prior to application receipt date. PA: PAR-04-155
Requirements: Applications may be submitted by foreign and domestic, for-profit and nonprofit organizations, both public and private; units of state and local governments; and eligible agencies of the federal government.
Amount of Grant: $250,000 maximum
Date(s) Application is Due: Apr 9; Aug 9; Dec 9
Contact: Amy Connolly, Grants Management Branch, (301) 496-8786; fax: (301) 496-8601; email: amyconnolly@nih.gov
Internet: http://grants.nih.gov/grants/guide/pa-files/par-04-155.html
Sponsor: National Cancer Institute
6130 Executive Boulevard, Executive Plaza N
Bethesda, MD 20892

NCI/NIDA Transdisciplinary Tobacco Use Research Centers Grants 3632
Transdisciplinary tobacco use research centers (TTURCs) are expected to conduct research that will lead to major scientific advances in knowledge of tobacco use, nicotine addiction, prevention and treatment interventions, and the translation of this knowledge into practice. There are several overarching goals: increase the number of investigators from relevant disciplines who focus on the study of tobacco use as part of transdisciplinary teams; generate basic research evidence to improve understanding of the etiology and natural history of tobacco use; produce evidence-based tobacco use interventions that can translate to the community and specific understudied or underserved populations; increase the number of evidence-based interventions that are novel, including the development, testing and dissemination of innovative behavioral treatments and prevention strategies based upon findings from basic research; train transdisciplinary investigators capable of conducting cutting-edge tobacco use research; and to increase the number of peer-reviewed publications in the area of tobacco use, nicotine addiction, and treatment. RFA-CA-04-012
Requirements: Applications may be submitted by public and private, domestic, for-profit and nonprofit organizations, units of state and local governments, and eligible agencies of the federal government. Domestic applications with foreign components are permitted. To be considered, applicant organizations must have: (1) an established investigator as a strong principal investigator who can oversee and conduct planning activities, provide direction to the TTURC, ensure a transdisciplinary research emphasis, and build a career development program; full institutional commitment from the applicant institution; clearly delineated and formally confirmed, subcontracted, collaborative scientific arrangements with scientists from other institutions, including foreign institutions, may be included.
Contact: Crystal Wolfrey, Grants Administration Branch, NCI, (301) 496-8634; email: crystal.wolfrey@nih.gov
Internet: http://grants1.nih.gov/grants/guide/rfa-files/RFA-CA-04-012.html
Sponsor: National Cancer Institute
6130 Executive Boulevard, Executive Plz N
Rockville, MD 20852

NCIC Canadian Breast Cancer Research Initiative Grants 3633
The initiative is a broad alliance of partners, comprising the Canadian Cancer Society, Health Canada through the National Health Research and Development Program, the Medical Research Council, the National Cancer Institute of Canada, the Breast Cancer International Centre, the Canadian Breast Cancer Foundation, and the Canadian Breast Cancer Network. The grants provide research support for up to three years to individuals formally affiliated with Canadian universities or other recognized Canadian institutions. Proposals must clearly state their relevance to the problem of breast cancer. Grantees are required to submit annual progress reports. Obtain application form 801 from the institute.
Restrictions: Applications may not be submitted in duplicate to sponsoring agencies. Applications for support of infrastructure will not be accepted.
Contact: Dr. Pascale Macgregor, Research Program Director, (416) 596-6598 ext 360; fax; (416) 596-1714; email: pfmacgregor@cbcf.org
Internet: http://www.breast.cancer.ca/cbcra_research_program
Sponsor: National Cancer Institute of Canada
10 Alcorn Avenue, Suite 200
Toronto, ON M4V 3B1 Canada

NCIC Canadian Cancer Society Research Grants 3634
The grants are intended to stimulate Canadian investigators in the design and execution of clinical and psychosocial/behavioral research in cancer and in the exploration of all the areas of the basic sciences having relevance to cancer. Grants for research will be awarded to projects deemed worthy of support, provided that the basic equipment and research facilities are available in the institution concerned and that it will provide the necessary administrative services. Such awards will be granted for the purchase and maintenance of animals, for expendable supplies, minor items of equipment, payment of graduate students and other assistants, and for research travel. Researchers focusing on breast cancer are strongly encouraged to submit applications to the Canadian Breast Cancer Research Initiative.
Requirements: A researcher who is designated as the principal investigator must be based in, or formally affiliated with (but not necessarily receive salary support from), a Canadian host institution such as a university, research institute, or healthcare agency.
Restrictions: Funds may not be used for personal salary support of the grantee.
Samples: Ralph Durand, British Columbia Cancer Research Centre—for a three-year project beginning in 1998 entitled Modifying Tumour Physiology to Improve Response to Radiation, $C149,544 per year.; Christopher Overall, U of British Columbia—for a three-year project beginning in 1998 entitled Cellular Activation of Human Gelatinase A, $C121,500 per year.; Heather Sutherland, U of British Columbia—for a three-year project beginning in 1998 entitled Purification and Characterization of Human Acute Myeloid Leukemia Stem Cells, $C97,600 per year.; Kevin Kane, U of Alberta—for a three year project beginning in 1998 entitled Ly-49 Interactions with Class I Major Histocompatibility Complex Molecules, $C95,390 per year.
Amount of Grant: $C300,000 maximum per year
Date(s) Application is Due: Oct 17
Contact: Director of Research Programs, (416) 961-7223; fax: (416) 961-4189; email: research@cancer.ca
Internet: http://www.ncic.cancer.ca/ncic/internet/standard/0,3621,84658243_85805597__langId-en,00.html
Sponsor: National Cancer Institute of Canada
10 Alcorn Avenue, Suite 200
Toronto, ON M4V 3B1 Canada

NCIC Canadian Cancer Society Travel Award for Senior Level PhD Students 3635
The purpose of this program is to provide financial assistance to senior-level PhD or MD/PhD students by helping to defray the travel costs associated with making a scientific presentation at a conference, symposium, or other appropriate professional gathering. The travel award will be made to defray actual costs of travel, registration, accommodation, etc. Three competitions are held annually. Conference dates for which funding is being requested must occur within six months of the competition deadline. Individuals may receive this award only once. A maximum of one award per supervisor will be given in the current calendar year. Applications must be completed using form 405/96. Letters announcing the results of eligible applications will be issued approximately one month following the deadline date.
Requirements: Applicants must be students enrolled in a PhD or MD/PhD program at a Canadian institution and be in their final phase of study. Candidates must be attending a conference for the purpose of presenting data (paper or poster presentation) from a cancer-related project on a first-author basis. It is not a requirement that the student and/or the supervisor be supported by an NCIC grant.
Amount of Grant: $C1500 maximum
Date(s) Application is Due: Apr 1; Sep 1; Dec 1
Contact: Program Contact, (416) 961-7223; fax: (416) 961-4189; email: research@cancer.ca
Internet: http://www.ncic.cancer.ca/ncic/internet/standard/0,3621,84658243_85813951_91005239_langId-en,00.html
Sponsor: National Cancer Institute of Canada
10 Alcorn Avenue, Suite 200
Toronto, ON M4V 3B1 Canada

NCIC Equipment Grants 3636
The grants provide for the purchase of items of permanent equipment necessary for research projects supported by the institute, if such equipment is not otherwise available. Application for minor items of apparatus and equipment with a total cost for each item of less than $C10,000 should be included as part of the application for the research grant. Application for funds to purchase each item with a total cost of less than $C100,000 but more than $C5000 may be made either as part of an application for a grant for research or as a separate request if the applicant is the recipient of a continuing award. For items of specialized major equipment having a value of $C100,000 or more, the institute may have to contemplate alternative approaches to funding; therefore, a prior consultation with the institute is required.
Amount of Grant: $C75,000 maximum
Date(s) Application is Due: Oct 17
Contact: Grants Administrator, (416) 961-7223; fax: (416) 961-4189; email: research@cancer.ca

Internet: http://www.ncic.cancer.ca/ncic/internet/standard/0,3621,84658243_85808828__langId-en,00.html
Sponsor: National Cancer Institute of Canada
10 Alcorn Avenue, Suite 200
Toronto, ON M4V 3B1 Canada

NCIC Terry Fox Postdoctoral Research Fellowships 3637
Fellowships are designed to provide training in cancer research for outstanding candidates who plan careers in Canada in this field of investigation. Post-MD fellowships are made to provide support for specialized training in the research aspects of the broad field of oncology (including epidemiology, medical statistics, clinical trials, and other forms of biomedical research). Initial tenure of a fellowship will normally be two years. One or more years of additional support may be available to individuals who demonstrate a serious commitment to cancer research, e.g., by registration in a degree program. Post-MD fellows may wish to maintain their clinical skills by means of service to the host institution during the period of the award. Such activities are acceptable provided the service is directly related to the cancer-oriented plan of the fellow and neither exceeds 10 working hours per week nor adversely affects the quantity or quality of the research training.
Requirements: Candidates must be graduates of universities recognized by NCIC and must be accepted for postdoctoral training in the field of cancer research in an academic environment also recognized by NCIC.
Restrictions: In general, applicants will not be eligible for funding beyond five years of postdoctoral training.
Amount of Grant: $C35,000-$C47,500 per year based on number of years of postdoctoral experience; Plus for up to $C1,200 per year to use towards the cost of attending and/or travelling to scientific meetings or conferences
Date(s) Application is Due: Feb 1
Contact: Director of Research Programs, (416) 961-7223; fax: (416) 961-4189; email: research@cancer.ca
Internet: http://www.ncic.cancer.ca/ncic/internet/standard/0,3621,84658243_85811512_90904221_langId-en,00.html
Sponsor: National Cancer Institute of Canada
10 Alcorn Avenue, Suite 200
Toronto, ON M4V 3B1 Canada

NCIC Terry Fox Research Studentships 3638
Research studentships provide training to PhD students in cancer research and are offered to outstanding individuals who plan a career in cancer research in Canada. Awards are made for up to a maximum of four years and are not renewable. Research studentship award recipients must devote full time to the research and training program during the tenure of the studentship. If the supervisor considers a certain amount of teaching to be part of the training, such teaching and duties must be limited to not more than 10 hours per week. At the completion of each award year, a letter summarizing the progress to date, signed by the research student and the research supervisor, must be submitted to the institute.
Requirements: Applicants must be residents of Canada and be enrolled in a PhD program at a Canadian institution at the time of application. Candidates must have completed at least two years of research training at the graduate level or have equivalent research experience.
Amount of Grant: $C21,500; Plus up to $C1200 per year to use towards the cost of attending and/or travelling to scientific meetings or conferences
Contact: Michael Wosnick, Executive Director, (416) 934-5672; fax: (416) 961-4189; email: mwosnick@cancer.ca
Internet: http://www.ncic.cancer.ca/ncic/internet/standard/0,3621,84658243_85813949_90911828_langId-en,00.html
Sponsor: National Cancer Institute of Canada
10 Alcorn Avenue, Suite 200
Toronto, ON M4V 3B1 Canada

NCRR Development of Nonmammalian Models and Related Biological Materials Research Grants 3639
The Comparative Medicine area, NCRR, encourages the submission of applications for the development of nonmammalian models and associated biological materials for biomedical research. Models to be considered must be applicable to the research interests of two or more categorical disease institutes of NIH. The overall goal of these studies is to develop new nonmammalian research models of broad interest and expand on the usefulness of existing nonmammalian model systems. Accordingly, the mechanisms for research projects (R01), exploratory/developmental (R21) and resource-related research projects (R24) grants apply. PA: PAR-99-115
Requirements: Applications may be submitted by foreign and domestic, for-profit and nonprofit organizations, public and private; units of state or local governments; and eligible agencies of the federal government.
Amount of Grant: $100,000 maximum direct costs per year
Contact: Dr. Jill Carrington, Comparative Medicine Area, (301) 435-0744; fax: (301) 480-3819; email: jillc@ncrr.nih.gov
Internet: http://www.grants.nih.gov/grants/guide/pa-files/PAR-99-115.html
Sponsor: National Center for Research Resources
6705 Rockledge Drive
Bethesda, MD 20892

NCRR Innovative Approaches to Developing New Technologies 3640
The purpose of this PA is to encourage submission of new exploratory/developmental grant applications to support innovative approaches to the development of technologies exploring new research paradigms in engineering, instrumentation, physical sciences, mathematics, or computer science applicable to all areas of biomedical research. The projects should provide the opportunity to develop new technologies, methods, devices, and materials that can be used in research aimed at gaining a greater understanding of fundamental elements of biological phenomena as well as potentially providing a basis for the development of new products useful in all aspects of biomedical research and health care. Annual deadline dates may vary; contact program staff for exact dates. PA: PAR-98-047
Requirements: Applications may be submitted by foreign and domestic, nonprofit and for-profit organizations, public and private, such as universities, colleges, hospitals, laboratories, units of state and local governments, and eligible agencies of the federal government.
Amount of Grant: $75,000 maximum
Date(s) Application is Due: Jun 1; Oct 1
Contact: Dr. Richard Dubois, Biomedical Technology, (301) 435-0755; fax: (301) 480-3659
Internet: http://www.grants.nih.gov/grants/guide/pa-files/PAR-98-047.html
Sponsor: National Center for Research Resources
6705 Rockledge Drive
Bethesda, MD 20892

NCRR Institutional Animal Resources Developing and Improving Grants 3641
The grants assist biomedical and behavioral research institutions in upgrading and developing their animal facilities. Support is limited to alterations and renovations to improve laboratory animal facilities and to the purchase of major equipment items for animal resources, diagnostic laboratories, transgenic animal resources, or similar associated activities. Facilities and administrative costs are not provided. Matching funds from nonfederal sources are required, equal to or exceeding one-half of the total allowable costs of the requested project. These matching funds must be applied to the specific project described in the application and cannot be met by citing other expenditures. The initial budget period may be either two or three years in length; the total project period may not exceed five years. PA: PAR-03-077
Requirements: Any domestic public or private nonprofit institution, organization, or association with one or more projects supported by the Public Health Service involving the use of animals is eligible to apply.
Restrictions: The program does not support requests for equipment used for teaching purposes and for housing non-research animals.
Amount of Grant: $700,000 maximum ($500,000 maximum for alterations and renovations; $200,000 maximum for movable equipment)
Contact: Paul Karadbil, Office of Grants Management, (301) 435-0844; email: paulk@ncrr.nih.gov
Internet: http://grants2.nih.gov/grants/guide/pa-files/PAR-03-077.html
Sponsor: National Center for Research Resources
6705 Rockledge Drive
Bethesda, MD 20892

NCRR National Stem Cell Resource Grants 3642
The purpose of this RFA is to solicit applications to establish a national stem cell resource. The resource will be for deposit, maintenance, preservation, and distribution of nonhuman-derived embryonic and other stem cells from a variety of species, as well as for deposit and distribution of related reagents and information in order to make these critical research tools available for biomedical research. Annual deadline dates may vary; contact program staff for exact dates. RFA: RR-00-001
Requirements: Applications may be submitted by domestic, public and private, for-profit and nonprofit organizations, units of state and local governments, and eligible agencies of the federal government.
Restrictions: Foreign institutions are not eligible to apply.
Amount of Grant: $800,000 approximately each year
Contact: Dr. Jill Carrington, Comparative Medicine, (301) 435-0744; fax: (301) 480-3819; email: jillc@ncrr.nih.gov
Internet: http://grants.nih.gov/grants/guide/rfa-files/RFA-RR-00-001.html
Sponsor: National Center for Research Resources
6705 Rockledge Drive
Bethesda, MD 20892

NCRR Networks and Pathways Collaborative Research Projects Grants 3643
This announcement solicits applications for research project grants that will leverage and complement the ongoing technology development being pursued in the National Technology Centers for Networks and Pathways (TCNPs), a program of the NIH Roadmap for Medical Research. These collaborative projects should focus either on addressing a challenging biological problem using the technology developed in one or more of the TCNPs, or on the development of technology that will complement that which is being developed in the centers. Applicants may request support for their own work as well as supplemental support for components pursued in the participating TCNP. While the nature and scope of these applications will vary, it is anticipated that the annual direct costs for these projects will typically range from $150,000 to $250,000.
Requirements: Application can be made by the following institutions and organizations: public/state controlled institutions of higher education; private institutions of higher education; nonprofits with 501(c)3 IRS status (other than institutions of higher education); nonprofits without 501(c)3 IRS status (other than institutions of higher

education); small businesses; for-profit organizations (other than small businesses); state governments; U.S. territories or possessions; non-domestic (non-U.S.) entities (foreign organizations); historically Black colleges and universities (HBCUs); Tribally controlled colleges and universities (TCCUs); and other eligible agencies of the Federal government.
Amount of Grant: $150,000 to $250,000 range
Date(s) Application is Due: Feb 5; Jun 5; Oct 5
Contact: Douglas M. Sheeley, Sc.D.; (301) 594-9762; email: sheeleyd@mail.nih.gov; David M. Balshaw, Ph.D.; (919) 541-2448; email: balshaw@niehs.nih.gov
Internet: http://grants.nih.gov/grants/guide/pa-files/PA-07-266.html
Sponsor: National Center for Research Resources (NCRR)
6701 Democracy Boulevard, MSC 4874
Bethesda, MD 20892-4874

NCRR Novel Approaches to Enhance Animal Stem Cell Research **3644**
The purpose of this funding opportunity announcement (FOA) is to encourage the submission of applications for research to enhance animal stem cells as model biological systems. Innovative approaches to isolate, characterize and identify toti-potent and multi-potent stem cells from nonhuman biomedical research animal models, as well as to generate reagents and techniques to characterize and separate those stem cells from other cell types is encouraged.
Requirements: The following organizations and institutions may apply: public/state controlled institutions of higher education; private institutions of higher education; Hispanic-serving institutions; historically Black colleges and universities (HBCUs); tribally controlled colleges and universities (TCCUs); Alaska Native and Native Hawaiian serving institutions; nonprofits with 501(c)3 IRS status (other than institution of higher education); nonprofits without 501(c)3 IRS status (other than institution of higher education); non-domestic (non-U.S.) entities (foreign organizations); small businesses; for-profit organizations (other than small businesses).
Restrictions: Studies involving human subjects are not allowed.
Date(s) Application is Due: Feb 5; Jun 5; Oct 5
Contact: John D. Harding, PhD; (301) 435-0744; fax: (301) 480-3819; email: hardingj@mail.nih.gov; Donald Blair, Ph.D.; (301) 496-7028; fax: (301) 402-1037; email: blaird@mail.nih.gov
Internet: http://grants.nih.gov/grants/guide/pa-files/PA-07-303.html
Sponsor: National Center for Research Resources (NCRR)
6701 Democracy Boulevard, Room 950, MSC 4874
Bethesda, MD 20892-4874

NCRR Shared Instrumentation Grants **3645**
The objective of the program is to make available to institutions expensive research instruments that can only be justified on a shared-use basis and for which meritorious research projects are described. This program is designed to provide for the acquisition or updating of expensive shared-use instrumentation not generally available through other NIH mechanisms, such as the regular research project, program project, or center grant programs. Proposals for research on advancing the design or for the development of new instrumentation will not be considered. Annual deadline dates may vary; contact the NCRR for exact dates. PA: PAR-05-028
Requirements: Nonprofit organizations, public or private institutions, and domestic institutions/organizations are eligible to apply.
Restrictions: Federal institutions, foreign institutions, and for-profit institutions are not eligible to apply.
Amount of Grant: $500,000 maximum
Date(s) Application is Due: Mar 22
Contact: Program Contact, (301) 435-0714; email: GrantsInfo@nih.gov
Internet: http://grants.nih.gov/grants/guide/pa-files/PAR-05-028.html
Sponsor: National Center for Research Resources
6705 Rockledge Drive
Bethesda, MD 20892

NCSS Exemplary Reserach in Social Studies Award **3646**
It is the mission of National Council for the Social Studies to engage and support educators in strengthening and advocating social studies, where social studies is defined as the integrated study of the social sciences and humanities to promote civic competence. In support of this mission, NCSS and the Research Community sponsor this annual research award designed to recognize an exemplary single-study research conception, model, design, procedure and analysis in social studies and social education. For the purpose of this award, social education includes work in the individual social sciences; women's studies and gender studies; multicultural, citizenship, global, economic, and geographic education; decision making; reflective inquiry; moral education; critical theory; teacher education; teacher thinking; and teaching-learning phenomena. Research is broadly constructed to mean any and all forms of recognized, formal research methodologies, paradigms, and perspectives.
Restrictions: No self-nominations are accepted.
Date(s) Application is Due: May 30
Contact: Prema Parmar, Program Manager; (301) 588-1800, ext. 106; fax: (301) 588-2049; email: pparmar@ncss.org
Internet: http://www.socialstudies.org/awards/research/exemplary/
Sponsor: National Council for the Social Studies
8555 Sixteenth Street, Suite 500
Silver Spring, MD 20910

NCSS Jean Dresden Grambs Distinguished Career Research in Social Studies Award **3647**
The Jean Dresden Grambs Distinguished Career Research Award recognizes professionals who have made extensive contributions to knowledge concerning significant areas of social studies education through meritorious research. The use of the term research in reference to this award shall be broadly construed to mean any and all forms of recognized formal research - methodologies, paradigms. and perspectives. Eligible research is that which has: added significantly to the understanding of the status, impact, characteristics, and activities of social studies education or social education, including several separate and distinct contributions related to curriculum, instruction, evaluation, and the learning and teaching of and in the social studies; stimulated a sense of responsibility among and fostered other professionals to engage in additional research efforts; added significantly to the information bases to those working in the social studies/social education field as evidenced by citations of their work within the social studies literature; and been used by professionals as the basis for significant decisions, rationales, and practices related to one or more areas of social studies curriculum instruction, evaluation, and teaching or learning.
Restrictions: No self-nominations are accepted.
Date(s) Application is Due: May 1
Contact: Prema Parmar, Program Manager; (301) 588-1800, ext. 106; fax: (301) 588-2049; email: pparmar@ncss.org
Internet: http://www.socialstudies.org/awards/research/grambs/
Sponsor: National Council for the Social Studies
8555 Sixteenth Street, Suite 500
Silver Spring, MD 20910

NCSS Larry Metcalf Exemplary Dissertation Award **3648**
The award, which is conferred in odd-numbered years, is designed to recognize substantive scholarly inquiry in social education completed in pursuit of the doctoral degree. Social education includes work in the individual social sciences; women studies and gender studies; multicultural, citizenship, global, economic, and geographic education; decision making; reflective theory; moral education; critical theory; teacher education; teacher thinking; and teaching-learning phenomena. Research is broadly defined to include experimental, qualitative, historical, and philosophical work. The award recipient must be willing to present the research at a one-hour session during the annual conference.
Requirements: Research in pursuit of the doctoral degree must be completed between March of the year preceding the award and February of the year in which the award is presented.
Amount of Grant: $250
Date(s) Application is Due: Mar 21
Contact: Prema Parmar, Program Manager; (301) 588-1800, ext. 106; fax: (301) 588-2049; email: pparmar@ncss.org
Internet: http://www.socialstudies.org/awards/research/metcalf/
Sponsor: National Council for the Social Studies
8555 Sixteenth Street, Suite 500
Silver Spring, MD 20910

NCSS/FASSE/CUFA Inquiry Grant **3649**
The Fund for the Advancement of Social Studies Education (FASSE) and the College and University Faculty Assembly (CUFA) of the National Council for the Social Studies (NCSS) have established a grant to support inquiry in citizenship education. Grant proposals should affirm social, cultural, and racial diversity and address issues of equality, equity, and social justice. Proposals that address aims for citizen action are preferred. Proposals should be relevant to school, university or community-based educational settings. FASSE and CUFA will award a $10,000 grant to the successful applicant who presents a proposal for a research project that demonstrates potential to inform the educational field about justice oriented, citizenship education. The original and four (4) copies of the proposal must be post-marked no later than June 30.
Requirements: Assistant, Associate or a full professor, or a graduate student with the demonstrated support of a university mentor/advisor are eligible to apply. Applicant(s) must be an NCSS member. If a graduate student, then the mentor must also be an NCSS member. Researchers from all groups, including under-represented groups, are encouraged to apply.
Restrictions: No fax or email submissions will be accepted.
Amount of Grant: $10,000
Date(s) Application is Due: Jun 30
Contact: Prema Parmar, Program Manager; (301) 588-1800, ext. 106; fax: (301) 588-2049; email: pparmar@ncss.org
Internet: http://www.socialstudies.org/fasse/rfp/
Sponsor: National Council for the Social Studies / Fund for the Advancement of Social Studies Education / College and University Faculty Assembly
8555 Sixteenth Street, Suite 500
Silver Spring, MD 20910

NCTE Teacher Researcher Award **3650**
The council regards as appropriate research any organized inquiry, theoretical or applied, into the subject of English, including the language arts or related fields, that has significance for the teaching or learning of English. Proposals are ecouraged that focus on underrepresented populations, curriculum changes and the effect the changes have on students, school policies, home-school literacy reationships, after-school programs, student literacy practices, etc. Applications are available online.
Requirements: An applicant must be a member of NCTE. Applicant is urged to define the purpose clearly and limit the project so that it may be completed in a reasonable time,

include a valid method of analyzing data, demonstrate that both the question and related research have been explored, and select persons as references who can speak specifically of applicant's ability to complete the project.
Restrictions: Award may not be used for travel costs or subsidies for loss of salary, nor for indirect or overhead costs.
Amount of Grant: $500-$12,500
Date(s) Application is Due: Aug 15
Contact: Dale Allender, Program Officer, (510) 642-5791; fax: (510) 642-4799
Internet: http://www.ncte.org/groups/research/awards/120596.htm
Sponsor: National Council of Teachers of English
1111 W Kenyon Road
Urbana, IL 61801

NEA Arts on Radio and Television **3651**
NEA seeks to make the excellence and diversity of the arts widely available to the American public through nationally distributed television and radio programs. Grants are available to support the development, production, and national distribution of significant radio and television programs on the arts. Priority will be given to artistically excellent programs that have the potential to reach a significant national audience. Only projects of artistic excellence and merit, in terms of both the content and the use of the medium, will be funded. Projects may include high profile multi-part television and radio series, single documentaries, performance programs, or arts segments for use within an existing series. Programs may deal with any art form (e.g., music, dance, literature, design, theater, musical theater, opera, visual arts, film/video/audio art, folk and traditional arts, etc.). All phases of a project are eligible for support, including research and development, production, and completion costs. For radio projects, grants generally range from $20,000 to $100,000. For television projects: research and development grants generally range from $20,000 to $50,000; grants for single documentaries generally range from $25,000 to $50,000; and grants for multipart series will vary according to the complexity and scope of the project. In rare instances, NEA may recommend an award over $200,000 for a project with demonstrated significance and impact. All grants require a match of at least one-to-one.
Requirements: The 501(c)3 applicant organization must meet the legal requirements listed on the Web site; have staff, paid or volunteer, who can devote the time and effort that are required to accomplish the project; and must have, if a former grantee, submitted acceptable final report packages on any completed arts endowment grant(s) by the final report due date(s).
Restrictions: Funding is not available for direct grants to individuals, programs that are intended primarily for local or regional distribution, films/tapes that are produced primarily for non-broadcast distribution (e.g. an instructional video, a film to accompany an exhibition, etc.), or documentation or simple recording of performances or events primarily for archival purposes.
Amount of Grant: $20,000-$200,000
Date(s) Application is Due: Sep 9
Contact: Program Staff, (202) 682-5738
Internet: http://arts.endow.gov/grants/apply/Media.html
Sponsor: National Endowment for the Arts
Nancy Hanks Ctr, 1100 Pennsylvania Avenue NW
Washington, DC 20506

NEA Heritage and Preservation Grants **3652**
This theme has three purposes: to honor, assist, and make visible those artists and forms of artistic expression that are rooted in the reflective or the many cultural groups that make up our nation; to preserve our most significant artistic accomplishments for future generations; and to conserve important works of art. Projects may include, but are not limited to, the documentation and/or presentation of artists and art forms that reflect our diverse cultural traditions; projects that pass on artistic repertoire, techniques, aesthetic principles, and oral traditions to future generations through apprenticeships or other forms of instruction; the documentation and preservation of significant artistic works, styles, techniques, and aesthetic/cultural perspectives; projects that provide technical assistance to traditional/folk artists; publications that document and/or disseminate artistic works, models of preservation, or other material that is aimed at strengthening our artistic heritage; the conservation treatment of highly significant works of art, artifacts, and collections wholly owned by the applicant; and innovative uses of modern technology as a means of preserving and strengthening our artistic heritage. Annual deadline dates may vary; contact the program office for exact dates. For guidelines, contact the Public Information Office at (202) 682-5400. Do not send applications through U.S. Postal Service—send applications by an alternate carrier.
Requirements: Arts institutes, state and local arts agencies, and other nonprofit or public organizations are eligible to apply. Consortia of such organizations are also eligible. Except for folk and traditional arts, applicant organization must have a four-year history of arts programming.
Restrictions: Funding is not available for general operating support; seasonal support; construction or renovation of facilities; marketing expenses that are not directly related to the proposed project; commercial enterprises or activities; fund raising or development; debt and deficit reduction; cash reserves and endowments; social activities, entertainment costs, receptions, etc.; lobbying expenses; individuals directly; elementary and secondary schools directly; avocational or student groups; generally, professional training in degree-granting institutions; work toward academic degrees; literary projects, programming, or publishing that does not focus primarily on contemporary literature and/or writers; research that is directed primarily to academic purposes or scholarly projects; publication of books or exhibition of works by the applicant organization's staff, board members, faculty, or trustees; exhibitions of, and other projects involving, single, privately owned collections; purchase of major equipment, except in limited cases where the equipment is an integral component of the project and where rental costs would exceed the purchase price, and subgranting or regranting, except for state arts agencies and regional arts organizations.
Samples: Cornerstones Community Foundation (Santa Fe, NM)—for technical assistance and training to rural communities to help them preserve historic buildings and building practices that are central to the Hispanic and Native American culture, $75,000. Evansville Philharmonic (Evansville, IN)—to support two performances of orchestral gospel music, $5000.
Contact: Heritage and Preservation Program, (202) 682-5428; fax: (202) 682-5512/5669; email: leniartk@arts.endow.gov
Internet: http://www.arts.endow.gov/honors/heritage/index.html
Sponsor: National Endowment for the Arts
Nancy Hanks Center, 1100 Pennsylvania Avenue NW, Room 720
Washington, DC 20506

NEA Literature Fellowships **3653**
Fellowships are available to published writers of exceptional talent and enable recipients to set aside time for writing, research, travel, and general career advancement. The program operates on a two-year cycle with fellowships in prose available one cycle and fellowships in poetry available in the next. The current cycle supports fiction/creative nonfiction and translation of prose or poetry into English. The January deadline is for translation projects; the March deadline is for fiction and creative nonfiction. Complete information and applications are available online.
Restrictions: An individual may submit only one application per year.
Amount of Grant: $10,000 -$20,000 translation projects; $20,000 maximum grants
Date(s) Application is Due: Jan 10; Mar 1
Contact: Literature Fellowships, (202) 682-5034; TDD: (202) 682-5496; email: webmgr@arts.endow.gov
Internet: http://arts.endow.gov/grants/apply/index_ind.html
Sponsor: National Endowment for the Arts - Office of Information and Technology - Creative Writing Fellowship
Nancy Hanks Ctr, 1100 Pennsylvania Avenue NW
Washington, DC 20506

NEA Public Works Grants **3654**
NEA makes a limited number of grants for design competitions to stimulate excellence in design in the public realm. This initiative is a pilot program with a different emphasis each year. Previous emphases have been on planning, architecture and interior design, landscape architecture, industrial design, and graphic design. Projects may include, but are not limited to, competitions for masterplans; public building design; landscape design, reclamation, or restoration. Subsequent years will be dedicated to specific types of work, such as parks and landscape, housing, or civic structures. A letter of interest is requested; a limited number of organizations will be invited to submit formal applications.
Requirements: To be eligible for this category, a nonprofit, tax-exempt organization must have staff, paid or volunteer, who can devote the time and effort that are required to accomplish the project; have a three-year history of programming prior to the application deadline; have evidence of financial resources substantial enough to implement the results of the competition; and for former grantees, have submitted acceptable final report packages on any completed Arts Endowment grant(s) by the final report due date(s). Applicant must also meet the legal requirements listed on the Web page.
Restrictions: This initiative is to fund design competitions only. Under this initiative, funding is not available for construction costs; costs of purchasing or renovating buildings or land; commercial (for-profit) enterprises or activities; direct grants to individuals; or subgranting or regranting, except for state arts agencies, their regional arts organizations, and local arts agencies that are designated to operate on behalf of their local governments or are operating units of city or county government.
Samples: Archeworks (on behalf of the City of Chicago) Chicago, IL. $75,000 was awarded to support a design competition for a proto-typical living and learning center that combines intergenerational housing and early-childhood educational programs in Chicago. The design competition will emphasize universal design principles to provide accessible facilities for all users.
Amount of Grant: $75,000
Contact: Nicki Jacobs, Grants Officer, (202) 682-5403; fax: (202) 682-5609; TDD/TTY: (202) 682-5496; email: jacobsn@arts.endow.gov
Internet: http://arts.endow.gov/grants/apply/Design.html
Sponsor: National Endowment for the Arts
1100 Pennsylvania Avenue NW, Suite 726
Washington, DC 20506

NEA/TCG Career Development Grants for Theatre Directors **3655**
The program offers six exceptionally talented, early-career theatre directors the opportunity to spend six months over a two-year period expanding their artistic boundaries and increasing their knowledge of the field. Each experience is hand-tailored, matching recipients' goals to appropriate and challenging assignments. Under the program, recipient directors will receive $25,000, which represents six stipend payments of $3,900 and a final payment of $1,600. these stipends cannot be used to underwrite fees for directing jobs, unless those assignments are related to the goals the recipient articulates in their application and interview. In addition, each recipient will receive up to $2,000 to support ongoing life needs such as health insurance, child care or elder care and up to $1,500 in supplementary travel.
Requirements: Applicants of diverse backgrounds representing the full spectrum of theatrical activity, including those who have worked in non-traditional venues, are

encouraged to apply to this program. All applicants: must have submitted online registration and mailed completed application packet by the postmark deadline; must be a citizen or permanent resident of the United States at the time of the application postmark deadline; must have directed three fully staged professional productions. Productions are considered professional if all artists are compensated at least a minimal amount for their work. Productions at educational institutions or community theaters do not meet this requirement. It is not necessary for the production to have operated on a union contract.
Restrictions: Applicants must not be enrolled in, or on leave from, a university or conservatory training program at the commencement of the program period; and, must not work exclusively in other disciplines (e.g. opera, dance). Previous recipients of this program, the National Endowment for the Arts Director Fellowship or the NEA/TCG Director Fellowship are not eligible to apply.
Amount of Grant: $25,000
Date(s) Application is Due: Jan 30
Contact: Casey Baltes, Assistant Director of Artistic Programs; (212) 609-5900, ext.281; email: cbaltes@tcg.org
Internet: http://www.tcg.org/grants/cdpdirect/cdp_dir_index.cfm
Sponsor: Theatre Communications Group (TCG)
520 8th Avenue, 24th Floor
New York, NY 10018-4156

NEBC Graduate Student Research Awards **3656**
Grants support botanical research to be conducted by graduate students on the New England flora, and make possible visits to the New England region by those who would not otherwise be able to do so. The awards are given to the graduate student(s) submitting the best research proposal dealing with systematic botany, biosystematics, plant ecology, or plant conservation biology. Typically two awards are given, although the actual number of awards and amount depend upon the proposals received.
Requirements: Graduate students must submit a proposal no longer than three pages, a budget, a curriculum vita, and two letters of support.
Samples: Krissa Skogen, University of Connecticut—award winner (2004); Julie Ellis, Brown University—to study the role of nesting seabirds in structuring New England coastal plant communities.
Amount of Grant: $2000 maximum
Date(s) Application is Due: Mar 1
Contact: Awards Committee, (617) 308-3656; fax: (617) 495-9484
Internet: http://blodeuwedd.huh.harvard.edu/nebc/Graduate.html
Sponsor: New England Botanical Club
22 Divinity Avenue
Cambridge, MA 02138

NEC Foundation of America Education Grants **3657**
The foundation makes cash grants to nonprofit organizations in support of the development, application, and use of technology by and for people with disabilities. Eligible proposals must demonstrate national reach and impact. Science and technology education proposals are not accepted unless they address innovative technology for people with disabilities. One-page preliminary proposals can be sent in advance of a formal proposal. A preliminary proposal must include the amount requested, purpose of grant and how the project matches the funding interests of the foundation. Preliminary proposals may be submitted on-line; formal proposals are not accepted on-line. The listed deadlines apply to formal proposals. Submit preliminary proposals well in advance of these deadlines so that the formal proposal, if invited, can be received in advance of these deadlines.
Restrictions: Grants do not support individual elementary and secondary schools or school districts or regional consortia; local chapters of national organizations; individuals (scholarships, stipends, fellowships, health care, etc.); capital campaigns/building funds; endowment campaigns; sports teams or athletic competitions; fundraising events or advertising; sectarian or religious activities; political campaigns or causes; organizations outside of the United States; or projects that inure to the commercial benefit of a private sector manufacturer, service provider, or individual.
Samples: Disability Funders Network (Falls Church, VA)—to market a database that encourages financing for disability issues and helps nonprofit disability groups prepare and submit grant solicitations, $15,000 (2005); Institute for Innovative Blind Navigation (Saginaw, MI)—for the First World Congress on Blind Navigation Technologies, $10,000 (2005); San Francisco State U Foundation (CA)—for an Internet-based tool designed to measure the sign-language abilities of deaf students and improve methods of teaching American sign language, $50,000 (2005); TERC (Cambridge, MA)—to help create an interactive sign-language dictionary, $25,000 (2005).
Amount of Grant: $1500-$75,000 average; $30,000 median
Date(s) Application is Due: Mar 1; Sep 1
Contact: Sylvia Clark, Executive Director, (631) 232 2212; fax: (631) 753-7096; email: foundation@necfoundation.org
Internet: http://www.necfoundation.org
Sponsor: NEC Foundation of America
2950 Express Dr South, Suite 102
Islandia, NY 11749-1412

NEDA/AED Charron Family Research Grant **3658**
Committed to promoting study into the causes, cures and prevention of eating disorders, NEDA first established the Research Grants Program in 2002 in partnership with the Academy for Eating Disorders (AED). The aim of these grants is to expand eating disorders research while drawing promising new scientists into the field. In 2004 Shelley Charron-Bahniuk died from complications of her eating disorder while she was in recovery. The Charron Family Research Grant was created as a direct result of this illness that impacted both her immediate and extended family.
Requirements: Individuals working in the field of eating disorders who have completed their training, and have the doctorate degree or beyond are eligible.
Amount of Grant: $10,000
Contact: Dee Christoff, Development Director; (206) 382-3587, ext. 15; fax: (206) 829-8501; email: info@nationaleatingdisorders.org
Internet: http://www.nationaleatingdisorders.org/p.asp?WebPage_ID=712
Sponsor: National Eating Disorders Association / Academy for Eating Disorders
603 Stewart Street, Suite 803
Seattle, WA 98101

NEDA/AED Joan Wismer Research Grant **3659**
Committed to promoting study into the causes, cures and prevention of eating disorders, NEDA first established the Research Grants Program in 2002 in partnership with the Academy for Eating Disorders (AED). The aim of these grants is to expand eating disorders research while drawing promising new scientists into the field.
Requirements: Individuals working in the field of eating disorders who have completed their training, and have the doctorate degree or beyond are eligible.
Amount of Grant: $10,000
Contact: Dee Christoff, Development Director; (206) 382-3587, ext. 15; fax: (206) 829-8501; email: info@nationaleatingdisorders.org
Internet: http://www.nationaleatingdisorders.org/p.asp?WebPage_ID=818
Sponsor: National Eating Disorders Association / Academy for Eating Disorders
603 Stewart Street, Suite 803
Seattle, WA 98101

NEDA/AED Tampa Bay Eating Disorders Task Force Award **3660**
The National Eating Disorders Association sponsors a small grants program to support research in the areas of eating disorders etiology, prevention, and treatment. Through the grants, the organization seeks to expand innovative eating disorders research while supporting investigators in the early stages of their careers in the eating disorders field. Support can be included for: salary for technicians, research assistants and other ancillary personnel; laboratory, radiological and pharmaceutical costs in clinical research projects; rating instruments and questionnaires; equipment and supplies; acquisition and maintenance of laboratory animals; subject payment; and advertising.
Requirements: Any junior investigator worldwide who works in the field of eating disorders and has completed their terminal graduate level degree and training is eligible.
Restrictions: Trainees (e.g. students, grad students, residents) are not eligible and investigators must be within seven years following the end of their training. Individuals who previously served or currently serve as a principal investigator on a substantial externally funded grant (e.g. RO1, RO3, or K award in the U.S.) are ineligible. Funds cannot be used for the following purposes: salary support for residents, fellows or other trainees; faculty or investigator salary; secretarial salary and benefits; tuition for residents, graduate students or post-doctoral fellows; publication costs or purchase of reprints or books; room and board for clinical subjects; studies that are also sponsored by pharmaceutical firms; computers; or travel.
Amount of Grant: $10,000
Contact: Dee Christoff, Development Director; (206) 382-3587, ext. 15; fax: (206) 829-8501; email: info@nationaleatingdisorders.org
Internet: http://www.nationaleatingdisorders.org/p.asp?WebPage_ID=816
Sponsor: National Eating Disorders Association / Academy for Eating Disorders
603 Stewart Street, Suite 803
Seattle, WA 98101

NEETF Competitive Grants **3661**
NEETF supports environmental education grants that explicitly connect environment, economy, and equity. Because of the immediate need to make this people/community/environment connection, NEETF is most interested in projects that educate adults through workplace programs, community initiatives, and in formal education settings. Youth-focused programs will be considered only where need can be clearly shown. Competitive grants are awarded in three program areas: health and environment, safe water, and environmental education innovation. The foundation's one-year challenge grants require a cash match; the preferred match ratio is two or more nonfederal dollars for each NEETF dollar awarded. To screen potential grant projects, NEETF uses a two-page preproposal application. Applicants will be invited to submit full proposals. Preproposals must be received by the listed application deadline date.
Requirements: Proposals may be submitted by public and private agencies, including nonprofit organizations such as schools and community-based organizations. Application forms are required.
Restrictions: Funds will not be provided for general administrative expenses, capital projects, scholarships or fellowships, basic research, political advocacy, or litigation projects.
Samples: Schuylkill River Development Council (Philadelphia, PA)—to create and test a middle- and high-school curriculum that links historical water-quality data with urban river-focused exploration and study, $15,000.; Headwaters Institute (San Francisco, CA)—to provide a series of environmental seminars for river guides on nine major recreational rivers, $15,000.; Arm-of-the-Sea Productions (Malden-on-Hudson, NY)—for a program that uses theater to convey to young people the relationship between watershed protection and New York City's drinking water, $15,000.
Amount of Grant: $5000-$15,000

Contact: Samantha Blodgett, Grants Coordinator, (202) 833-2933; fax: (202) 261-6464; email: blodgett@neetf.org
Internet: http://www.neetf.org/Education/index.htm
Sponsor: National Environmental Education and Training Foundation
1707 H Street NW, Suite 900
Washington, DC 20006-3915

NEFE Fellowships 3662
The fellowships are awarded to recent college graduates interested in gaining practical, meaningful professional experience in the personal finance and nonprofit sectors. Fellows work closely with executive staff at the foundation's headquarters near Denver. Fellowships are funded for up to two years and include competitive salary and benefits. Promising students are invited to apply. At the Web site, click on Innovative Thinking.
Requirements: Applicants must hold a bachelor's or master's degree, preferably with an emphasis in family and consumer economics, applied economics, business or personal finance, or journalism or communications.
Restrictions: faxed and email applications are not accepted.
Amount of Grant: $27,150 per year
Date(s) Application is Due: Apr 15
Contact: Fellows Program, (303) 224-3503 or (303) 741-6333
Internet: http://www.nefe.org/pages/innovativefellows.html
Sponsor: National Endowment for Financial Education
5299 DTC Boulevard, Suite 1300
Greenwood Village, CO 80111

NEFE Grants 3663
The foundation's mission is to help Americans acquire the information and gain the skills necessary to take control of their financial lives. NEFE accomplishes its mission by partnering with other concerned organizations and individuals to provide financial education to young people; underserved segments of society whose financial concerns are not being addressed by others, or who have special needs; and the general public. The NEFE Grants Program was established to encourage those outside the foundation to pursue activities that will help further the mission of NEFE. Submit applications by the first Tuesday in February, April, and August. At the Web site, refer to the Innovative Thinking section.
Requirements: 501(c)3 tax-exempt organizations are eligible.
Restrictions: NEFE does not fund local financial education programs, workshops, etc., or organizational operating expenses.
Samples: Economic Opportunity Agency of Washington County (Springdale, Ark)—to develop a culturally sensitive Spanish translation of a financial education program, $20,000.; Consumer Federation of America Foundation (Washington, DC)—to encourage low- and moderate-income individuals to save money and convince policymakers that low-income wealth building should be a national priority, $90,000.
Amount of Grant: $20,000-$137,538; $60,750 average
Contact: Mare Canfield, (303) 224-3534; fax: (303) 220-0838; email: mhc@nefe.org
Internet: http://www.nefe.org/pages/innovative.html
Sponsor: National Endowment for Financial Education
5299 DTC Boulevard, Suite 1300
Greenwood Village, CO 80111

NEH Documenting Endangered Languages Grants 3664
Documenting Endangered Languages is a joint, multi-year funding program of the National Science Foundation and the National Endowment for the Humanities to develop and advance scientific and scholarly knowledge concerning endangered human languages. Made urgent by the imminent death of approximately half of the 6000-7000 currently used human languages, DEL seeks not only to document these endangered languages but to integrate, systematize, and make knowledge concerning them widely available by exploiting advances in information technology. Approximately 6-10 Standard or Continuing Grants ranging from $12,000 to $150,000 per year for one to three years are available. Approximately 12 Fellowships of $4,200 per month for awards lasting from six to twelve months are available; the maximum stipend is $50,400 for a twelve-month tenure period.
Requirements: Academic institutions and non-profit, non-academic organizations located in the United States are eligible. U.S. citizens are eligible to apply for fellowships. Foreign nationals who have been living in the United States or its jurisdictions for at least the three years prior to the proposal deadline are also eligible to apply for fellowships.
Restrictions: For-profit organizations are not eligible to apply.
Amount of Grant: $100,000 average; $2,000,000 total ($1 million from NSF and $1 million from NEH)
Date(s) Application is Due: Sep 15
Contact: Joan Maling, Linguistics Program Director; (703) 292-8046, fax: (703) 292-9068, email: jmaling@nsf.gov; Douglas Whalen; (703) 292-7321; email: dwhalen@nsf.gov
Internet: http://www.nsf.gov/pubs/2006/nsf06577/nsf06577.htm
Sponsor: National Endowment for the Humanities / National Science Foundation
1100 Pennsylvania Avenue, NW, Room 426
Washington, DC 20506

NEH Advancing Knowledge Grants: The IMLS/NEH Digital Partnership 3665
The partnership invites proposals for innovative, collaborative humanities projects using the latest digital technologies for the benefit of the American public, humanities scholarship, and the nation's cultural institutions. These grants will support collaborations among libraries, museums, archives, universities, and other cultural organizations that may serve as models for the field. The program encourages projects that explore new ways to share, examine, and interpret humanities collections in a digital environment and to develop new uses and audiences for existing digital resources. Awards normally are for two years and typically range from $50,000 to a maximum of $350,000. In special circumstances an award may exceed this amount, depending on the scope and complexity of the project. Successful applicants will be awarded a grant in outright funds, federal matching funds, or a combination of the two, depending on the applicant's preference and the availability of NEH funds.
Requirements: Any U.S. nonprofit organization with 501(c)3 tax exempt status is eligible, as are state and local governmental agencies and tribal governments. Collaboration is a requirement of the Advancing Knowledge grant program, and to be eligible an application must include at least one library, museum, or archive as an integral member of the project team. Potential partners could also include universities, state humanities councils and other cultural organizations, or private sector organizations, as appropriate.
Restrictions: Individuals are not eligible to apply. Funds may not be used for: projects that focus primarily on digitizing existing collections; recurring or established conferences or professional meetings; creative or performing arts; empirical social scientific research; specific policy studies; the preparation or publication of textbooks; projects that focus on general pedagogical theory, research on educational methods, tests, or measurements; projects that focus on cognitive psychology; projects devoted to political, religious, commercial, or social advocacy; purchase of artifacts or collections; professional development; or general operations, renovation, or construction.
Amount of Grant: $50,000-$350,000 per year for up to two years
Date(s) Application is Due: Mar 18
Contact: Division of Preservation and Access; (202) 606-8570; email: preservation@neh.gov
Internet: http://www.neh.gov/grants/guidelines/Digital_Partnership.html
Sponsor: National Endowment for the Humanities
1100 Pennsylvania Avenue, NW
Washington, DC 20506

NEH Challenge Grants 3666
NEH challenge grants help institutions and organizations secure long-term improvements in and support for their humanities programs and resources. Awards are made to museums, public libraries, colleges, research institutions, historical societies and historic sites, public television and radio stations, universities, scholarly associations, state humanities councils, and other nonprofit entities. Because of the matching requirements, these NEH awards also strengthen the humanities by encouraging non-federal sources of support. Challenge grants are offered only when NEH funds will help institutions carry out long-term plans and enhance their financial stability. The grants may support ongoing humanities activities in education, public programming, scholarly research, and preservation. Examples include faculty and staff positions, fellowships, lecture or exhibition series, visiting scholars, publishing subventions, consultants, maintenance of facilities, faculty development, acquisitions, and preservation/conservation programs. Types of support include building construction and renovation and computer technology for the humanities. Guideline and application are available online.
Requirements: With the exception of elementary and secondary schools or school districts, any U.S. nonprofit institution (public agency, private nonprofit organization, or tribal government) working wholly or in part with the humanities may apply for a challenge grant. Affiliated institutions (e.g., university museums) should consult with NEH staff on questions of separate eligibility.
Restrictions: Challenge grant funds, federal or non-federal, may not be used for direct expenditures for operations or programs; recovery of indirect costs; awards or stipends for students below the graduate level; or support for short-term projects eligible for grants from other NEH programs.
Samples: Chicago Historical Society, Chicago, IL—campaign to endow the research collections, $150,000.; Amarillo College (Amarillo, TX)—for an endowment to continue and expand the college's public lecture series and faculty development seminars, $60,000.; U of San Francisco (CA)—for an endowment for a professorship in American Jewish History, faculty development, and a lecture series to strengthen an interdisciplinary Jewish Studies program, $375,000.
Amount of Grant: $30,000-$800,000 range
Date(s) Application is Due: May 1
Contact: Stephen M. Ross, Director; (202) 606-8309; fax: (202) 606-8204; email: sross@neh.gov or challenge@neh.gov
Internet: http://www.neh.gov/grants/guidelines/challenge.html
Sponsor: National Endowment for the Humanities
1100 Pennsylvania Avenue, NW
Washington, DC 20506

NEH Collaborative Research Grants 3667
The grants support original research undertaken by a team of two or more scholars; or research coordinated by an individual scholar that, because of its scope or complexity, requires additional staff and resources beyond the individual's salary. Eligible projects include research that significantly adds to knowledge and understanding in the humanities; conferences on topics of major importance in the humanities that will benefit ongoing research; archaeology projects that interpret and communicate the results of archaeological fieldwork. Projects may encompass excavation, materials analysis, laboratory work, field reports, and preparation of interpretive monographs; translations into English of works that provide insight into the history, literature, philosophy, and artistic achievements of other cultures; and research that uses the knowledge, methods, and perspectives of the humanities to enhance understanding of science, technology, and medicine. These grants support full-time or part-time activities for periods of one to three

years. Support is available for various combinations of scholars, consultants, and research assistants; project-related travel; field work; applications of information technology; and technical support and services. All grantees are expected to communicate the results of their work to the appropriate scholarly and public audiences. Annual deadline dates may vary; contact program staff for exact dates.
Requirements: Eligible applicants include nonprofit, IRS tax-exempt institutions and organizations in the United States; state and local governmental agencies; and U.S. citizens or foreign nationals who have been living in the United States or its jurisdictions for at least the three years prior to the time of application. Applicants affiliated with an eligible institution must apply through their institution.
Restrictions: Grants cannot be used for specific policy studies or educational or technical impact assessments; support for the preparation or publication of textbooks; archaeological materials analysis whose primary goal is refinement of a method; inventories of collections; research in educational methods, tests, or measurements; or recurrent meetings of professional organizations, societies, and scholarly organizations. Project directors cannot be degree candidates. NEH generally does not award grants to other federal entities or to applicants whose projects are so closely intertwined with a federal entity that the project takes on characteristics of the federal entity's own authorized activities.
Samples: Portland State University, Portland, OR—for Chinookan Households on the Lower Columbia River: Contact and Complexity, $150,000 over three years (2006); Harriet Beecher Stowe Center, Hartford, CT—for Uncle Tom's Cabin in the Web of Culture: A Multi disciplinary Conference, $48,496 (2006).
Amount of Grant: $25,000-$100,000 per year
Date(s) Application is Due: Nov 1
Contact: Division of Research; (202) 606-8200; email: collaborative@neh.gov
Internet: http://www.neh.gov/grants/guidelines/collaborative.html#poc
Sponsor: National Endowment for the Humanities
1100 Pennsylvania Avenue, NW
Washington, DC 20506

NEH Digital Humanities Start-Up Grants **3668**

The program is designed to encourage innovations in the digital humanities. By awarding relatively low-dollar grants during the planning stages, the goal is to identify projects that are particularly innovative and have the potential to make a positive impact on the humanities. Proposals should be for the planning or initial stages of digital initiatives in any area of the humanities. Digital Humanities Start-Up Grants may involve: research that brings new approaches or documents best practices in the study of the digital humanities; planning and prototyping new digital tools for preserving, analyzing, and making accessible digital resources, including libraries' and museums' digital assets; scholarship that examines the philosophical implications and impact of the use of emerging technologies; innovative uses of technology for public programming and education utilizing both traditional and/or new media; and new digital modes of publication facilitating the dissemination of humanities scholarship in advanced academic as well as informal or formal educational settings at all academic levels. Two levels of awards will be made in this program. Level I awards are small grants designed to fund brainstorming sessions, workshops, early alpha-level prototypes, and initial planning. Level II awards are larger grants that can be used for more fully-formed projects that are ready to start the first stage of implementation or the creation of working prototypes.
Requirements: Eligible applicants include: any U.S. nonprofit organization or institution with IRS 501(c)3 tax exempt status; state and local governmental agencies and Native American tribal organizations; and U.S. citizens or foreign nationals who have been living in the United States or its jurisdictions for at least the three years immediately prior to the application deadline. Individuals affiliated with an eligible institution must apply through an institution, ordinarily their own institution. Adjunct faculty may apply as individuals.
Restrictions: Grants cannot be used for: projects that mainly involve digitization, unless the applicant is proposing an innovative new method for digitization; the implementation or assessment of existing digital applications in the humanities (however, exploration of or planning for a new direction or tool for an established project is allowed); recurring or established conferences or professional meetings; acquisition of computer equipment or software in excess of 20% of the grant total; creative or performing arts; empirical social scientific research; work undertaken in the pursuit of an academic degree; the preparation or publication of textbooks; or projects devoted to political, religious, commercial, or social advocacy.
Amount of Grant: $5,000-$25,000 (level 1); $25,001-$50,000 (level 2)
Date(s) Application is Due: Apr 2; Oct 8
Contact: Digital Humanities Initiative (DHI) Director; (202) 606-8364; email: dhi@neh.gov
Internet: http://www.neh.gov/grants/guidelines/digitalhumanitiesstartup.html#poc
Sponsor: National Endowment for the Humanities
1100 Pennsylvania Avenue, NW, Room 426
Washington, DC 20506

NEH Faculty Humanities Workshops **3669**

Grants support local and regional professional development programs for K-12 teachers and faculty at postsecondary institutions. Workshops should extend and deepen knowledge of the humanities by fostering collaborative study of significant topics, texts, and issues; provide faculty with the opportunity to engage in rigorous intellectual inquiry, including reading, reflection, and discussion; involve scholars from outside the institution(s) who are experts in the topic of the workshop; use creative formats and programs to engage faculty members; and advance the study and teaching of the humanities at the participating institution(s). Proposals to provide workshops for teachers with limited access to professional development in the humanities are encouraged. These workshops may include teachers at charter schools, parents who homeschool, private license school faculty, and community college faculty. If the workshop is designed for K-12 educators, project directors are encouraged to make arrangements with the appropriate state agency for participants to receive continuing education units (CEUs) or in-service credit. Funds may be used to pay for guest scholars and visiting consultants, books and other materials, logistical support, and appropriate release time for project staff. Guidelines and application are available online.
Requirements: U.S. nonprofit, IRS tax-exempt organizations or institutions are eligible.
Restrictions: Grants are not awarded to individuals. NEH generally does not award grants to other federal entities or to applicants whose projects are so closely intertwined with a federal entity that the project takes on characteristics of the federal entity's own authorized activities.
Amount of Grant: $30,000-$75,000
Date(s) Application is Due: Apr 7
Contact: Division of Education Programs, (202) 606-8380; TDD: (866) 372-2930; email: info@neh.gov
Internet: http://www.neh.gov/grants/guidelines/facworkshops.html
Sponsor: National Endowment for the Humanities
1100 Pennsylvania Avenue NW, Room 302
Washington, DC 20506

NEH Faculty Research Awards for Historically Black, Tribal Colleges, Universities, and Institutions with High Hispanic Enrollment **3670**

Grants awarded under this program support faculty research in the humanities at historically black, Hispanic-serving, and tribal colleges and universities. Projects are welcome that intend to make contributions to research and synthesis in the humanities generally and which may result in furthering the educational missions of the institution by enriching the humanities content of classroom teaching.
Requirements: Applicants must hold a full-time tenured, tenure-track, or annual contract faculty position at a historically black, Hispanic-serving, or tribal college or university. Applicants must be U.S. citizens, native residents of U.S. jurisdictions, or foreign nationals who have resided in the United States or its jurisdictions for at least three years immediately preceding the application deadline.
Amount of Grant: $4,200 per month; $50,400 maximum
Date(s) Application is Due: May 1
Contact: Division of Research Programs; (202) 606-8200; email: facultyresearch@neh.gov
Internet: http://www.neh.gov/grants/guidelines/facultyresearch.html
Sponsor: National Endowment for the Humanities
1100 Pennsylvania Avenue, NW
Washington, DC 20506

NEH Fellowship Programs at Independent Research Institutions **3671**

Grants for Fellowship Programs at Independent Research Institutions (FPIRI) support fellowships at institutions devoted to advanced study and research in the humanities. NEH fellowships provide scholars with research time and access to resources that might not be available at their home institutions. Fellowship programs may be administered by independent centers for advanced study, libraries, and museums in the United States; American overseas research centers; or organizations that have expertise in promoting research on foreign cultures. Individual scholars must apply directly to the institutions themselves.
Requirements: Any U.S. nonprofit organization or institution with IRS 501(c)3 status is eligible, as are state and local governmental agencies and tribal governments. In addition, to apply for a grant, institutions must: be financed, governed, and administered independently of institutions of higher education; and have established and maintained fellowship programs with their own or other funding.
Restrictions: Individuals are not eligible to apply.
Amount of Grant: $81,600-$468,000 range
Date(s) Application is Due: Aug 19
Contact: Program Staff, Research Programs; (202) 606-8200; email: fpiri@neh.gov
Internet: http://www.neh.gov/grants/guidelines/fpiri.html
Sponsor: National Endowment for the Humanities
1100 Pennsylvania Avenue, NW, Room 426
Washington, DC 20506

NEH Fellowships **3672**

The fellowships provide opportunities for individuals to pursue advanced work that will enhance their capacities as interpreters of the humanities and enable them to make significant contributions to thought and knowledge in the humanities. Recipients usually produce articles, monographs, books, digital materials, archaeological site reports, translations, editions, and other scholarly tools. Applicants may be faculty or staff members of colleges or universities, or of primary or secondary schools. However, the fellowships are not limited to individuals in educational institutions, and scholars and writers in various other positions may apply. Projects supported by fellowships may contribute to scholarly knowledge, to the advancement of teaching, or to the general public's understanding of the humanities. All applicants, particularly first-time applicants, are encouraged to contact the division six weeks prior to submitting an application. Application materials are available on the Web site. Applications are now accepted online.

Requirements: Applicants may be faculty or staff members of colleges or universities, or of primary or secondary schools, or independent scholars or writers. Applicants must be citizens or permanent residents of the United States for three years prior to applying.
Restrictions: Persons seeking support for work leading toward a degree are not eligible to apply nor are active candidates for degrees. Applications for the planning of institutional curricula, empirical educational research projects, educational surveys, work in the creative or performing arts, and studies whose content or methods are not those of the humanities are not eligible for consideration.
Amount of Grant: $4,200 per month, $50,400 for a twelve-month tenure
Date(s) Application is Due: May 1
Contact: Division of Research: (202) 606-8200; fax: (202) 606-8204; email: fellowships@neh.gov
Internet: http://www.neh.gov/grants/guidelines/fellowships.html
Sponsor: National Endowment for the Humanities
1100 Pennsylvania Avenue, NW, Room 318
Washington, DC 20506

NEH Fellowships at Digital Humanities Centers **3673**
Fellowships at Digital Humanities Centers (FDHC) support collaboration between digital centers and individual scholars. An award provides funding for both a stipend for the fellow while in residence at the center and a portion of the center's costs for hosting a fellow. Awards are for periods of six to twelve months. The intellectual cooperation between the visiting scholar and the center may take many different forms and may involve humanities scholars of any level of digital expertise. Fellows may work exclusively on their own projects in consultation with center staff, collaborate on projects with other scholars affiliated with the center, function as apprentices on existing digital center projects, or any combination of these. The results of the collaboration may range from proof of concept to finished product. The maximum award for a twelve month grant, therefore, is $100,800: $50,400 for the fellow's stipend and $50,400 for the direct and indirect costs of administering the fellowship.
Requirements: All individual scholars must have completed their formal education by the application deadline. While applicants need not have advanced degrees, individuals currently enrolled in a degree-granting program are ineligible to apply. Centers may submit one application per deadline; individual scholars may apply in collaboration with only one digital center per deadline.
Restrictions: Center eligibility is limited to: digital humanities centers affiliated with either American institutions of higher education or U.S. nonprofit organizations or institutions with IRS 501(c)3 tax exempt status. Individual scholar eligibility is limited to: U.S. citizens; and foreign nationals who have been living in the United States or its jurisdictions for at least the three years immediately prior to the application deadline.
Amount of Grant: $100,800 maximum
Date(s) Application is Due: Sep 15
Contact: Division of Research; (202) 606-8200; fax: (202) 606-8204; email: fellowships@neh.gov
Internet: http://www.neh.gov/grants/guidelines/fdhc.html#poc
Sponsor: National Endowment for the Humanities
1100 Pennsylvania Avenue, NW, Room 318
Washington, DC 20506

NEH Humanities Fellowships at Winterthur **3674**
Supported by NEH, Winterthur awards long-term fellowships for four to 12 months of continuous advanced research. The program encourages the exchange of ideas among the research fellows, the Winterthur staff, and graduate students through colloquia, conferences, discussions, and informal gatherings. Rental housing is available on the Winterthur grounds in the research fellowship residence. Winterthur is a research center for American studies; American cultural, social, and intellectual history; art and design history; and material culture. Application materials are available on the Web site or from the office.
Requirements: Academic, museum, and independent scholars who are U.S. citizens or foreign residents in the U.S. for three years prior to application are invited to apply.
Restrictions: Candidates for degrees are ineligible to apply.
Amount of Grant: $40,000 maximum stipend
Date(s) Application is Due: Jan 15
Contact: Program Contact, (800) 448-3883; fax: (302) 888-4870; email: kcollins@winterthur.org
Internet: http://www.winterthur.org/research/fellowship.asp
Sponsor: Winterthur Museum, Garden, and Library
Winterthur Museum
Wilmington, DE 19735

NEH Institutional Grants to Historically Black Minority-Serving and Tribal Colleges and Universities **3675**
The purpose of this program is to enrich the humanities at historically black, Hispanic-serving, tribal colleges and universities by improving their ability to teach, conduct research, and support learning at the undergraduate level and throughout life. The grants may also better prepare recipients to gain future access to federal funds available for the support of these activities.
Requirements: Historically black, Hispanic-serving, and tribal colleges and universities are eligible to apply.
Amount of Grant: $25,000 maximum
Date(s) Application is Due: Jun 1
Contact: Division of Education Programs, (202) 606-8500; TDD: (866) 372-2930; email: education@neh.gov
Internet: http://www.neh.gov/grants/guidelines/institutional.html
Sponsor: National Endowment for the Humanities
1100 Pennsylvania Avenue NW
Washington, DC 20506

NEH Preservation and Access Research and Development Grants **3676**
Grants are given to support research and development projects that advance the nation's capacity to preserve and provide access to humanities resources. The endowment particularly encourages applications for projects that would enhance the use of digital technology for broadening access to and ensuring the long-term availability of cultural resources. Eligible activities include projects that would develop technical standards and best practices for preserving and providing access to humanities collections; devise more effective physical methods of preserving humanities collections; develop new or improved procedures for the creation of reference works; and address issues involving the policies and practices of humanities research institutions that affect preservation and access on a national level. Applicants should discuss potential projects with other professionals in their field to ensure that the proposed research will address important preservation and access problems. Guidelines are available online.
Requirements: Any U.S. nonprofit organization or institution is eligible. State and local government agencies are also eligible. When two or more institutions or organizations collaborate on a project, one of them must serve as the lead applicant and administer the grant on behalf of all the participating entities.
Amount of Grant: $150,000-$390,000
Date(s) Application is Due: Jul 1
Contact: Division of Preservation and Access, (202) 606-8570; TDD: (866) 372-2930; email: preservation@neh.gov
Internet: http://www.neh.gov/grants/guidelines/researchdevelopment.html
Sponsor: National Endowment for the Humanities
1100 Pennsylvania Avenue NW
Washington, DC 20506

NEH Preservation Assistance Grants **3677**
The grants help small and mid-size institutions—libraries, museums, and historical societies, town and county records offices, archival repositories, community colleges, and underserved departments and units within colleges and universities and other larger institutions—improve their ability to preserve and care for their humanities collections. These collections may include books, journals, manuscript and archival materials, maps, moving images, sound recordings, decorative and fine arts, prints and photographs, textiles, archaeological and ethnographic artifacts, furniture, and historical objects. Grants may be used for general preservation assessments; consultations with professionals to address a specific preservation issue, need, or problem; purchase of storage furniture and preservation supplies; purchase of environmental monitoring equipment for humanities collections; and education and training. Institutions that have never received an NEH grant are encouraged to apply. The grant period normally ranges from six months to one year. These grants do not require an applicant to provide cost sharing. An institution may submit only one application per deadline. Complete guidelines are available online.
Requirements: Any U.S. nonprofit, tax-exempt organization, institution, or consortium is eligible to apply.
Restrictions: Grants cannot be used for projects that focus on collections that fall outside of the humanities; projects that seek to catalog, index, or arrange and describe collections; projects that focus on preserving or restoring buildings or other structures; capital improvements to buildings and building systems, including the purchase of equipment such as air conditioners, dehumidifiers, and security and fire protection systems; salaries and fringe benefits for the staff of an institution; reformatting of collections (e.g., digitizing, microfilming, or copying to another medium), or the purchase of equipment for reformatting (e.g., computers, scanners, digital cameras, cassette decks, and CD-ROM drives); library binding; graduate-level conservation training, or training related to advanced conservation treatment; conservation or restoration treatments or the purchase of conservation or restoration treatment supplies and equipment; or attendance at regular meetings of museum, library, archives, or preservation organizations.
Samples: Atlanta U Ctr (Atlanta, GA)—to arrange and describe archival collections that document the African American experience in the arts, religion, education, race relations, and civil rights, $90,000.
Amount of Grant: $2000, $3500, or $5000
Date(s) Application is Due: May 16
Contact: Division of Preservation and Access, (202) 606-8570; fax: (202) 606-8369; email: preservation@neh.gov
Internet: http://www.neh.gov/grants/guidelines/pag.html#program
Sponsor: National Endowment for the Humanities
1100 Pennsylvania Avenue NW
Washington, DC 20506

NEH Reference Materials Grants **3678**
Grants are given to support projects that will create reference works and research tools. Eligible projects include print and online encyclopedias that encompass various humanities fields or that focus on a particular area or subject; databases and electronic archives that codify and integrate humanities materials, or provide bibliographical control of a subject or field; historical, etymological, and bilingual dictionaries that document a language, as well as reference grammars and other linguistic tools; tools for spatial analysis and representation of humanities data, such as atlases and geographical information systems (GIS); and descriptive catalogs that provide detailed information about humanities materials.

Requirements: Any U.S. nonprofit organization or institution is eligible. State and local government agencies are also eligible. When two or more institutions or organizations collaborate on a project, one of them must serve as the lead applicant and administer the grant on behalf of all the participating entities.
Restrictions: Grants cannot be used for tools and works designed solely for classroom instruction; printing costs that are the responsibility of the publisher; and scholarly editions.
Amount of Grant: $50,000-$325,000
Date(s) Application is Due: Jul 15
Contact: Division of Preservation and Access, (202) 606-8570; TDD: (866) 372-2930; email: preservation@neh.gov
Internet: http://www.neh.gov/grants/guidelines/referencematerials.html
Sponsor: National Endowment for the Humanities
1100 Pennsylvania Avenue NW
Washington, DC 20506

NEH Summer Stipends **3679**
Stipends provide opportunities for individuals to pursue advanced work that will enhance their capacities as interpreters of the humanities and enable them to make significant contributions to thought and knowledge in the humanities. Applicants may be faculty or staff members of colleges or universities, or of primary or secondary schools. However, stipends are not limited to individuals in educational institutions, and scholars and writers in various other positions may apply. The stipends support projects that can be completed during the tenure of an award, but they may also be part of a long-term endeavor. Projects supported by stipends may contribute to scholarly knowledge to the advancement of teaching, or to the general public's understanding of the humanities. Recipients may hold other research grants during the tenure of their awards, but they must work full-time on their projects during the two months of their tenure. All applicants, particularly first-time applicants, are encouraged to contact the division six weeks prior to application.
Requirements: U.S. citizens or foreign nationals who have been living in the United States or its jurisdictions for at least the three years prior to the time of application are eligible. Faculty members teaching full-time at colleges and universities must be nominated by their institutions to apply. Applications should be submitted online by faculty only after they have been nominated. Each college and university in the United States and its jurisdictions may nominate two faculty members. Of the two, at least one should be a junior nominee. Applicants who hold the rank of instructor or assistant professor or who are at comparably early stages of their careers will be considered junior nominees. Those holding the rank of associate professor or professor will be considered senior nominees. The following individuals may apply online without a nomination: independent scholars not affiliated with a college or university; nonfaculty college and university staff members who will not be teaching during the academic year preceding the grant tenure; and adjunct faculty, part-time faculty, and applicants with academic appointments that terminate by the summer of the grant tenure.
Restrictions: Summer stipends cannot be used for research for doctoral dissertations or theses by students enrolled in a degree program; specific policy studies or educational or technical impact assessments; preparing or publishing textbooks; studying teaching methods or theories, surveys of courses and programs, or preparing institutional curricula; or creating inventories of collections. Former recipients must wait five years before they can reapply for this program.
Amount of Grant: $5000 per stipend
Date(s) Application is Due: Oct 1
Contact: Fellowships and Summer Stipends, Division of Research and Education Programs, (202) 606-8200; fax: (202) 606-8204; email: stipends@neh.gov
Internet: http://www.neh.gov/grants/guidelines/stipends.html
Sponsor: National Endowment for the Humanities
1100 Pennsylvania Avenue NW
Washington, DC 20506

NEH/AIAR Postdoctoral Fellowships **3680**
Fellowships are open to those in ancient Near Eastern studies, including the fields of archaeology, anthropology, art history, Bible, epigraphy, historical geography, history, language, literature, philology, and religion or related disciplines from Pre-history, through the early Islamic period. The research period should be continuous, without frequent trips outside the country. Residence at the Albright is required. Grants are awarded for six to 12 months (up to four awards from four to 12 months). Guidelines are available online.
Requirements: Postdoctoral scholars who are U.S. citizens (or foreign nationals for at least three years) are eligible.
Amount of Grant: $40,000; $20,000 six months
Date(s) Application is Due: Oct 15
Contact: Dr. Joan Branham, (401) 865-1789; fax: (401) 865-1036; email: jbranham@providence.edu
Internet: http://www.aiar.org/deadlines.html
Sponsor: W.F. Albright Institute of Archaeological Research (AIAR) Jerusalem
Providence College
Providence, RI 02918

NEI Clinical Study Planning Grant **3681**
This program provides support for the development of a comprehensive research protocol for a large-scale clinical trial or epidemiologic study. This grant provides early peer review of the proposed clinical study in terms of its rationale, design, organizational structure, and implementation plan. The planning grant is used to support the development of a detailed MOP and may include preliminary studies to refine study procedures or to document recruitment potential. PAR-05-115
Requirements: Applications may be submitted by domestic for-profit and nonprofit organizations, both public and private; units of state and local governments; and eligible agencies of the federal government; domestic and foreign institutions.
Restrictions: Foreign institutions are not eligible.
Amount of Grant: $150,000 maximum for two years
Internet: http://grants.nih.gov/grants/guide/pa-files/PAR-05-115.html
Sponsor: National Eye Institute
5635 Fishers Lane, Suite 1300, MSC 9300
Bethesda, MD 20892-9300

NEI Clinical Vision Research Development Award **3682**
This award is intended to strengthen interactions among clinicians, biostatisticians, epidemiologists, statistical geneticists, and other clinical trial specialists to facilitate the design and conduct of clinical research projects, such as the development of coordinating center capabilities. The intent of this projects is to advance the understanding, prevention, or clinical management of visual system disorders. A number of experienced investigators and research groups are needed to provide leadership in the design and implementation of such clinical vision research projects. Therefore, NEI wishes to help institutions with NEI-funded investigators augment staff expertise and acquire the resources necessary to enhance their clinical vision research programs. PA: PAR-00-050
Requirements: Applications may be submitted by domestic for-profit and non-profit organizations, public and private, such as universities, colleges, hospitals, laboratories, units of State and local governments, and eligible agencies of the Federal government.
Restrictions: Foreign institutions and institutions having current or past NEI Core Grant for Vision Research (P30) support for a biostatistics module are not eligible to apply.
Amount of Grant: $100,000 maximum
Contact: Mary Frances Cotch, (301) 496-5983; fax: (301) 402-0528; email: mfcotch@nei.nih.gov or deverett@nei.nih.gov
Internet: http://grants.nih.gov/grants/guide/pa-files/PAR-00-050.html
Sponsor: National Eye Institute
6120 Executive Boulevard, MSC 7164
Bethesda, MD 20892-7164

NEI Mentored Clinical Scientist Development Program Award (K12) **3683**
The NEI Mentored Clinician Scientist Development Award (K12) is an award to an educational institution or professional organization to facilitate and support career development experiences which lead clinician scientists to research independence. It is expected that scholars appointed to the program will subsequently apply for their own individual research support. The program should be designed to accommodate candidates with varying levels of research experience who require an individualized career development program ranging from two to five years. Applications for this award should propose a structured, phased development plan overseen by a Program Director and a Program Advisory Committee. A designated period of didactic training should be followed by a period of supervised research experience. Candidates should be exposed to both research that is typically undertaken with a few investigators and to research approaches that require a multidisciplinary team to address complex research questions as appropriate. It is expected that at the end of the career development period, appointees will transition successfully into positions as independent investigators. This award provides five years of potentially renewable support. The subsequent continuation will depend in part on the progress made by the Scholars in achieving independent research support. Renewal applications must contain a progress report documenting the accomplishments and current career activities of former Scholars. PA# PAR-09-083
Requirements: The following organizations and institutions are eligible to apply: public/state controlled institutions of higher education; private institutions of higher education; hispanic-serving Institutions; historically black colleges and universities; tribally controlled colleges and universities; Alaska Native and Native Hawaiian serving institutions; nonprofits with 501(c)3 IRS Status (other than institutions of higher education); nonprofits without 501(c)3 IRS Status (other than institutions of higher education). Any individual with the skills, knowledge, and resources necessary to carry out the proposed research is invited to work with his/her institution to develop an application for support. Individuals from underrepresented racial and ethnic groups as well as individuals with disabilities are always encouraged to apply for NIH support.
Restrictions: Foreign institutions are not eligible to apply.
Amount of Grant: $1,125,000
Date(s) Application is Due: May 13
Contact: Neeraj Agarwal, (301) 451-2020; fax: (301) 402-0528; email: agarwalnee@nei.nih.gov
Internet: http://grants.nih.gov/grants/guide/pa-files/PAR-09-083.html
Sponsor: National Eye Institute
5635 Fishers Lane, Suite 1300
Bethesda, MD 20892-9300

NEI Research Grant For Secondary Data Analysis Grants **3684**
The NEI Research Grant for Secondary Data Analysis (R21) program is designed to provide investigators with the support necessary to conduct secondary data analyses utilizing existing database resources. Applications may be related to, but must be distinct from, the specific aims of the original data collection. The total project period for an application submitted in response to this funding opportunity may not exceed two years. Direct costs are limited to $275,000 over an R21 two-year period, with no more than $200,000 in direct costs allowed in any single year. PA # PAR-06-326

Requirements: Eligible organizations: for-profit and non-profit organizations; public and private institutions (such as universities, colleges, hospitals, and laboratories); units of state and local governments; eligible agencies of the federal government; units of state; local tribal governments; domestic and foreign institutions. Eligible Project Directors/ Principal Investigators (PD/PIs): any individual with the skills, knowledge, and resources necessary to carry out the proposed research. Individuals from underrepresented racial and ethnic groups as well as individuals with disabilities are always encouraged to apply for NIH support.
Amount of Grant: $275,000 over a two year period
Date(s) Application is Due: May 8
Contact: Natalie Kurinij, (301) 451-2020; fax: (301) 402-0528; email: kurinijn@mail.nih.gov
Internet: http://grants.nih.gov/grants/guide/pa-files/PAR-06-326.html
Sponsor: National Eye Institute
5635 Fishers Lane, Suite 1300, MSC 9300
Bethesda, MD 20892-9300

NEI Ruth L. Kirschstein National Research Service Award Short-Term Institutional Research Training Grants 3685

The National Institutes of Health (NIH) will award Ruth L. Kirschstein National Research Service Award (NRSA) Short-Term Institutional Research Training Grants (T35) to eligible institutions to develop or enhance research training opportunities for individuals interested in careers in biomedical, behavioral and clinical research. Many of the NIH Institutes and Centers (ICs) use this grant mechanism exclusively to support intensive, short-term research training experiences for students in health professional schools during the summer. In addition, the Short-Term Institutional Research Training Grant may be used to support other types of predoctoral and postdoctoral training in focused, often emerging scientific areas relevant to the mission of the funding IC. The proposed training must be in either basic, behavioral or clinical research aspects of the health-related sciences. This program is intended to encourage graduate and/or health professional students to pursue research careers by exposure to and short-term involvement in the health- related sciences. The training should be of sufficient depth to enable the trainees, upon completion of the program, to have a thorough exposure to the principles underlying the conduct of research. PA # PA-08-227
Requirements: The following organizations/institutions are eligible to apply: public/ state controlled institutions of higher education; private institutions of higher education; hispanic-serving institutions; historically black colleges and universities (HBCUs); tribally controlled colleges and universities (TCCUs); Alaska Native and Native Hawaiian serving institutions; nonprofits with 501(c)3 IRS Status (Other than Institutions of Higher Education); nonprofits without 501(c)3 IRS Status (Other than institutions of higher education); indian/native american tribal governments (federally recognized); indian/native american tribally designated organizations; U.S. Territory or possession; indian/native american tribal governments (other than federally recognized); faith-based or community-based organizations. Only domestic, non-profit, private or public institutions may apply for grants to support National Research Service Award (NRSA) short-term research training programs. Foreign institutions are not eligible to apply. The applicant institution must have a strong and high quality research program in the area(s) proposed for the research training and must have the requisite staff and facilities on site to conduct the proposed research training program. Any individual with the skills, knowledge, and resources necessary to organize and implement a high-quality research training program is invited to work with their institution to develop an application for support. Individuals from underrepresented racial and ethnic groups, individuals with disabilities, and individuals from disadvantaged backgrounds are always encouraged to apply for NIH support.
Restrictions: Foreign institutions are not eligible to apply.
Date(s) Application is Due: Sep 8
Contact: Neeraj Agarwal, (301) 451-2020; fax: (301) 402-0528; email: agarwalnee@mail.nih.gov
Internet: http://grants.nih.gov/grants/guide/pa-files/PA-08-227.html
Sponsor: National Eye Institute
5635 Fishers Lane, Suite 1300
Rockville, MD 20892-9300

NEI Scholars Program 3686

This program provides an opportunity for outstanding individuals to obtain laboratory or clinical research training within NEI intramural environment and to facilitate the successful transition to continue their research career at an extramural institution as independent vision researchers. This is accomplished by first providing individuals with the necessary resources to receive high quality research training for three to four years at NEI. This is followed by providing extramural funding to support the research program for two years at the extramural institution to which NEI scholar is recruited. It is anticipated that NEI scholars will subsequently compete for independent funding to continue their research. During the intramural phase, the scholar is expected to spend full time on research. During the extramural phase, the scholar must spend a minimum of 75 percent of a full-time professional effort conducting research and engaging in research career development activities for the two years of the award. Applicants are encouraged to contact NEI extramural staff regarding their eligibility for this award and NEI intramural laboratory scientist regarding sponsorship information, prior to submitting an application. PA: PAR-98-107
Requirements: Individuals must have a research or health professional doctoral level degree or its equivalent and must have demonstrated the potential for a highly productive research career during predoctoral training and the immediate postdoctoral period, if applicable. Individuals at NEI or other NIH intramural laboratories who meet the eligibility requirements are eligible to apply. Before submitting an application for the program, the candidate must identify an individual in the NEI Division of Intramural Research who will serve as the sponsor and will be committed to supervise the training and research project. The sponsor must be an active investigator in the area of the proposed research and have research training experience and resources needed to support the NEI scholar.
Restrictions: Individuals who have had more than five years of postdoctoral research training at the time of application are not eligible to apply; however, clinical training does not count against the five years. Former principal investigators on NIH research project grants R01; SBIR/STTR awards R43, R44/R41, R42; subprojects of program project grants P01; center grants P30, P50; K08, K23, K24, or the equivalent are not eligible.
Amount of Grant: $175,000 maximum per year or $350,000 for two years
Date(s) Application is Due: Feb 1; Jun 1; Oct 1
Contact: Dr. Maria Giovanni, Division of Extramural Research, (301) 496-0484; fax: (301) 402-0528; email: myg@nei.nih.gov
Internet: http://www.grants.nih.gov/grants/guide/pa-files/PAR-98-107.html
Sponsor: National Eye Institute
9000 Rockville Pike, Building 31
Bethesda, MD 20892

NEI Translational Research Program On Therapy For Visual Disorders 3687

This program continues support of collaborative, multidisciplinary research programs focused on new therapeutic approaches to restore or prevent the loss of function due to visual system diseases and disorders. The rapid and efficient translation of laboratory research findings into clinical development requires a comprehensive and highly integrated approach involving collaborative teams of scientists and clinicians with expertise in multiple disciplines. Such a collaborative approach would be particularly appropriate for research focused on pathways that will likely be targeted by biological intervention, such as gene therapy, cell-based therapy, pharmacological approaches, or development of appropriate delivery systems. The intention of this program is to make resources available to scientists from several disciplines to form research teams to address scientific and technical questions that would be beyond the capabilities of any one research group. It is anticipated that applications funded under this program will lead to therapies that can be tested in a clinical trial.
Requirements: Eligible Institutions, you may submit an application (s) if your organization has any of the following characteristics: for-profit organizations; non-profit organizations; public or private institutions, such as universities, colleges, hospitals, and laboratories; units of State government; units of local government; eligible agencies of the Federal government; domestic Institutions Foreign components to domestic applications will be allowed if the foreign component provides significant scientific input that is critical to the program and for which there is no similar resource in the United States. Eligible Individuals: any individual with the skills, knowledge, and resources necessary to carry out the proposed research is invited to work with their institution to develop an application for support. Individuals from underrepresented racial and ethnic groups as well as individuals with disabilities are always encouraged to apply for NIH programs.
Amount of Grant: $2 million maximum
Contact: Dr. Peter Dudley, (301) 451-2020; fax: (301) 402-0528; email: pad@nei.nih.gov
Internet: http://grants.nih.gov/grants/guide/pa-files/PAR-05-110.html
Sponsor: National Eye Institute
5635 Fishers Lane, Suite 1300, MSC 9300
Bethesda, MD 20892

NEI Vision Research Core Grants 3688

The primary objective of Core Grants for Vision Research is to provide groups of investigators who have achieved independent National Eye Institute (NEI) funding with additional, shared support to enhance their own and their institution's capability for conducting vision research. Secondary objectives of this program include facilitating collaborative studies and attracting other scientists to research on the visual system.
Requirements: Institutions applying for a Core Grant must hold, on the receipt date, a minimum of eight NEI research project grants (R01), Bioengineering Research Partnership Grants (R24), Mentored Clinical Scientist Development Awards (K08), Mentored Patient-Oriented Research Career Development Award (K23), or Midcareer Investigator Award in Patient-Oriented Research (K24), including any noncompeting extensions of these awards made with or without additional funds. No other mechanisms or source of research support will be considered in determining eligibility. Only one Core Grant will be made to any single applicant organization. For multicampus institutions, no more than one Core Grant will be made to each campus. Joint applications may be submitted by investigators at neighboring, independent institutions.
Amount of Grant: $2,000,000 - $2,500,000 range over a five-year period
Date(s) Application is Due: Sep 25
Contact: Hemin R. Chin, (301) 451-2020; fax: (301) 402-0528; email: hemin@nei.nih.gov
Internet: http://www.nei.nih.gov/funding/p30.asp
Sponsor: National Eye Institute
5635 Fishers Lane, Suite 1300 MSC 9300
Bethesda, MD 20892-9300

Nelda C. and H.J. Lutcher Stark Foundation Grants 3689

The foundation awards grants in its areas of interest, including education, health and social services, medical and dental, community enrichment, arts and culture. Giving is primarily in Texas, with preference to programs that directly impact Southeast Texas and limited giving in Southwest Louisiana. There are no application forms. Submit a letter of

proposal containing the following information: description of the project signed by the president or CEO; brief history, description, programs, and mission of the organization; project time line; specific amount requested; list of other funding sources and amounts; list of board of directors and occupations; form 990 for most recent year-end; most recent audited financial statements or organization prepared statements; and current copy of IRS determination letter certifying tax-exempt status. Guidelines are available online. All grants from the third quarter of 2008 and for 2009 have been suspended due to recovery efforts in which the Foundation must engage for its own operations as a result of damages from Hurricane Ike, together with the negative impact of the downturn in financial markets on the Foundation's corpus, the Stark Foundation will continue the prior suspension of the grant-making aspect of its operations through the entirety of Year 2009. Therefore, the Stark Foundation will not accept any grant applications for Year 2009, nor will be able to consider any grant applications that may be submitted notwithstanding this suspension.
Requirements: To be eligible for a grant, an organization must: be exempt from taxation under Section 501(c)3 of the Internal Revenue Code; not be a private foundation within the meaning of Section 509(a) of the Code. No grants are made for endowment purposes, to individuals, or to supporting organizations as described in Section 509(a)(3) of the Internal Revenue Code.
Samples: Lamar State College, Orange, Texas $128,269 - Funding to establish a regional EMT training program in cooperation with Acadian Ambulance Service (2007); Texas Energy Museum, Beaumont, Texas $2,500 - Matching funds for an interactive science exhibit (2007); Orange Community Band, Orange, Texas $12,340 - equipment and program support (2007); Orange County Historical Commission, Orange, Texas $2,800 - matching funds to capture in digital format and place on-line the large collection of historical photographs belonging to the Historical Commission (2007); The Conservation Fund, Austin, Texas $375,000 - funding for purchase of strategic conservation lands in Orange County, Texas for donation to the Big Thicket National Preserve for permanent protection (2007);
Date(s) Application is Due: Mar 1; Jun 1; Oct 1
Contact: Grant Department
Internet: http://www.starkfoundation.org/Grants/Grant-Guidelines.aspx
Sponsor: Nelda C. and H.J. Lutcher Stark Foundation
P.O. Box 909
Orange, TX 77631-0909

Nell J. Redfield Foundation Grants **3690**
The foundation awards grants to eligible Nevada organizations in its areas of interest, including biomedical research, health care and health organizations, elderly, children with disabilities, education, human services, and religion. Types of support include building construction/renovation, capital campaigns, equipment acquisition, program development, and scholarship funds. Applications are accepted from January 1 to the listed application deadline. Contact the office for application form and guidelines.
Requirements: Nevada nonprofit organizations are eligible.
Geographic Focus: Nevada
Samples: Truckee Meadows Community College (Reno, NV)—to renovate the performing-arts center, $250,000 (2004).
Amount of Grant: $1000-$500,000 range
Date(s) Application is Due: Jun 1
Contact: Grants Administrator, (775) 323-1373
Sponsor: Nell J. Redfield Foundation
P.O. Box 61
Reno, NV 89502

Nell Warren Elkin and William Simpson Elkin Foundation **3691**
The foundation awards grants to charitable organizations in its areas of interest, including universities, cancer clinics, specific disease associations, religious groups, and Bible studies. Application form and guidelines are available online.
Requirements: Nonprofit, charitable organizations are eligible.
Amount of Grant: $2000-$10,000 average
Date(s) Application is Due: Feb 28; Jul 31
Contact: Dale Welch, c/o SunTrust Bank, Atlanta, (404) 724-3773; email: rs.foundationinfo@suntrust.com
Internet: http://www.elkinfoundation.org
Sponsor: Nell Warren Elkin and William Simpson Elkin Foundation
P.O. Box 4655, MC 221
Atlanta, GA 30302

Nemmers Prizes in Economics and in Mathematics **3692**
The prizes are awarded for distinguished achievement in economics and mathematics. The Erwin Plein Nemmers Prize in Economics and the Frederic Esser Nemmers Prize in Mathematics are awarded in even-numbered years, with nominations due in odd-numbered years. Prize recipients will deliver a public lecture and participate in other scholarly activities at Northwestern University for 10 weeks. Nominations from experts in the field are preferred to institutional nominations. Nominations for both prizes will be accepted until the listed deadline date for applications. Direct applications will not be accepted. Send nominations to the selection committee secretary; nominations may be sent by email.
Requirements: Candidacy is open to individuals of all nationalities for outstanding achievement in their disciplines, as demonstrated by major contributions to new knowledge or the development of significant new modes of analysis. A nominating letter of no more than three pages should describe the nominee's professional experience, accomplishments, and qualifications for the award.
Restrictions: Current or former members of the Northwestern University faculty and recipients of the Nobel Prize are ineligible.
Samples: Ariel Rubinstein, Tel Aviv U and New York U—for a broad series of highly original contributions to game theory ranging from analyses of bargaining and repeated games to models of bounded nationality, $150,000 (2004); Mikhael Gromov, Institut des Hautes Etudes Scientifiques and Courant Institute of Mathematical Sciences, New York U—for his work in Riemannian geometry, which revolutionized this subject, his theory of pseudoholomorphic curves in symplectic manifolds, his solution of the problem of groups of polynomial growth, and his construction of the theory of hyperbolic groups, $150,000 (2004).
Amount of Grant: $150,000 awards
Date(s) Application is Due: Dec 1
Contact: Secretary, Selection Committee, Office of the Provost, email: nemmers@northwestern.edu
Internet: http://www.northwestern.edu/provost/awards/nemmers
Sponsor: Northwestern University
633 Clark Street
Evanston, IL 60208

Nessa Cohen Memorial Fund Awards for Advanced Study **3693**
Awards cover advanced study and/or research, including living and travel expenses, for students of the Art Students League, who must also be members of the league in good standing. The award is to be used by the recipient to further his/her interest or abilities in the specified areas. Student should submit to the board of control an outline of his/her plan for future study; projects involving further study at the Art Students League will be encouraged. Application forms are available from the league office.
Requirements: Applicant must be a member of the league in good standing and must be a full-time student for a full school year prior to receiving the award, out of which four of the months must have been in the same year as the granting of the award. Applicant must be registered in classes in fine arts drawing, fine arts printmaking, sculpture, and watercolor painting.
Restrictions: Students registered only in oil painting or pastel classes are not eligible.
Contact: Gaetano Scognamillo, Scholarship Coordinator, (212) 247-4510; fax: (212) 541-7024
Internet: http://www.theartstudentsleague.org/Navigation/Home/HP-FRAME.html
Sponsor: Art Students League of New York
215 W 57th Street
New York, NY 10019

Nestle Foundation International Fellowship Program in Human Nutrition Research **3694**
The foundation awards fellowships in human nutritional science. Fellowships in the areas of public health and epidemiology are considered as long as the study program places a heavy emphasis on problems of human nutrition. The training should take place in approved training or research centers. The foundation welcomes inquiries from institutions wishing to benefit from this program. Letters of intent may be submitted at any time; application forms will be sent and must be received by the listed deadlines.
Restrictions: The foundation does not fund studies in the areas of food policy, food production, and food technology.
Date(s) Application is Due: Jan 10; May 10
Contact: Program Contact, 41-21-3203351; fax: 41-21-3203392; email: nestle.foundation@vtxnet.ch
Internet: http://www.nestlefoundation.org/e/apply.htm
Sponsor: Nestle Foundation for the Study of the Problems of Nutrition in the World
4 place de la Gare
Lausanne 1001 Switzerland

Netherlands Institute for Advanced Study in the Humanities and Social Sciences Fellowships **3695**
Fellowships are awarded to stimulate research in the humanities and social sciences and to promote work at an interdisciplinary level. Twenty scholars from abroad are invited to join approximately 20 scholars from the Netherlands for a full academic year (September 1 to June 30) to work on research of their own choosing in a research setting that facilitates interaction and cooperation among scholars. Three multidisciplinary research theme groups are formed each academic year. A theme group consists of approximately eight fellows whose fields of study coincide with the theme. Fellowships are awarded for periods of not less than 10 months, which includes five weeks' leave. Monthly stipend plus refund for round-trip fare from place of residence to Wassenaar and free housing or housing at subsidized rent are provided. Scholars who wish to apply for fellowships are invited to write for the necessary forms. Applications for a particular year should be submitted at least 18 months in advance. All correspondence should be addressed to the rector.
Requirements: Postdoctoral candidates should have made substantial contributions to their fields through publications and other activities. In addition to the scholarly accomplishments, a candidate's research must fit in the multidisciplinary setting and research pattern of a particular year.
Contact: Jos Hooghuis, Secretary to the Scholarship Committee, 31-70-512-27-00; fax: 31-70-511-71-62; email: Hooghuis@NIAS.KNAW.NL
Internet: http://www.nias.knaw.nl
Sponsor: Netherlands Institute for Advanced Study in the Humanities and Social Sciences
Meijboomlaan 1
Wassenaar 2242 PR Netherlands

Neuroscience Research on Problem Gaming Grants 3696

The center requests applications for projects on underage gambling and the neurobiology of disordered gambling. Funding will support pioneering research on pathological gambling. Proposed projects should produce original research about disordered gambling behavior and advance the development of methodologies used in this field. Preference is given to research activities that inform and guide current prevention and treatment programs. Behavioral social sciences projects should research the role that cognitive, personality, emotional, and psychosocial processes play in the etiology, onset, and maintenance of disordered gambling. Priority is placed on research on youth gambling and the development of effective prevention strategies. Neuroscience research efforts should investigate the causes and consequences of addictive gambling by identifying neurobiologic mechanisms associated with gambling disorders and the potential treatments for these conditions, including effective pharmacotherapy.

Requirements: U.S. 501(c)3 nonprofits are eligible. Canadian organizations must have proof of equivalent tax-exempt status.

Samples: U of Connecticut Health Ctr (Storrs, CT)—for a project that examines how altered brain functioning affects risk taking, $172,056.; U of Iowa College of Medicine (Ames, IA)—for a study based on interviews with a family of pathological gamblers, $169,929.

Contact: Christine Reilly, Executive Director, (617) 384-9028; fax: (617) 384-9023; email: Christine_Reilly@hms.harvard.edu

Internet: http://www.ncrg.org/research/institute.cfm

Sponsor: National Center for Responsible Gaming
401 Park Drive, 2nd Fl E
Boston, MA 02215

New Hampshire Charitable Foundation Grants 3697

The foundation seeks to improve the quality of life for New Hampshire citizens by distributing grants to qualifying organizations and scholarship assistance to the state's residents. Areas of support include arts, humanities, environment and conservation, health, and social and community services. Types of support include seed grants, scholarships and scholarship funds, program development, technical assistance, demonstration grants, development grants, endowments, and training grants. Four regional divisions of the foundation exist to benefit particular areas of the state: Greater Piscataqua Community Foundation, Lakes Region Charitable Foundation, Northern New Hampshire Foundation, and Upper Valley Community Foundation. Although a limited number of multi-year grants are made, awards are not usually repeated or renewed. Grantmaking meetings are held three times each year by the foundation.

Requirements: New Hampshire 501(c)3 tax-exempt organizations are eligible.

Restrictions: Grants are not awarded for capital campaigns, the purchase of major equipment, as operating support for ongoing or annual programs, to replace public funding, for sectarian or religious purposes, for deficit reduction, to support academic research, or for travel out of state by school or youth groups or by individuals for scholarly or professional purposes.

Geographic Focus: New Hampshire

Amount of Grant: $5,000 maximum for Project Grants; $15,000 maximum for Strategic Focus Grants

Contact: Jennifer P. Hopkins, Director; (800) 464-6641, ext. 1239; fax: (603) 225-1700; email: jph@nhcf.org

Internet: http://www.nhcf.org/page16898.cfm

Sponsor: New Hampshire Charitable Foundation
37 Pleasant Street
Concord, NH 03301-4005

New Jersey Historical Commission Grants Program 3698

To be considered for funding, the project or organization must serve the public directly and accomplish at least one of the following: help strengthen and further develop existing New Jersey history programs and operations as well as initiate new programs; improve management and interpretation of historic sites and historical collections; expand public understanding and awareness of historic resources; increase public and organization participation in historical programs and activities; increase the body and quality of information on New Jersey history available to the public; and preserve materials for the study or preservation of New Jersey history. Organizations may apply for special projects grants; research, publication, and media grants; general operating support; and mini-grants. General operating support applications are due April 1; projects are due September 15; mini-grant applications are reviewed bimonthly. Applications are available on the Web site. Applications will not be accepted by fax or email.

Requirements: Applications for projects on research, publication, and films may be made by all institutions, organizations, and individuals. Applications for projects on conservation of historical materials or production of public programs or exhibitions may be made only by New Jersey institutions or organizations. Applications for teaching projects may be made by New Jersey educational institutions or their teachers, administrators, or librarians. Students must be sanctioned by their schools, and grant funds awarded to students will be paid to their schools.

Restrictions: Projects are ineligible if they do not deal with New Jersey history, are not accessible to the general public, or are for the restoration or preservation of historic structures. Only organizations or institutions with annual operating budgets of at least $500,000 may apply for more than one project at a time.

Geographic Focus: New Jersey

Samples: Asbury Park Public Library (NJ)—to assist microfilming of city directories in Monmouth and Middlesex Counties, $6000.; Monmouth County Board of Recreation Commissioners (NJ)—for a series of programs about the history of African Americans in the county, $2786.; Sea Isle City Historical Museum (NJ)—for conservation work on the museum's collections, $4120.

Amount of Grant: $10,000 general operating support for organizations with budgets under $30,000; 33 percent of operating budget for organizations with budgets over $30,000; $3000-$20,000 research, publications, and media grants; $3000-$20,000 special projects grants

Contact: Mary Murrin, Director Grants Program, (609) 984-0954; email: mary.murrin@sos.state.nj.us

Internet: http://www.state.nj.us/state/history/hisidx.html

Sponsor: New Jersey Historical Commission
225 W State Street, 4th Fl
Trenton, NJ 08625-0325

New York Public Library Fellowships 3699

The fellowships are open to scholars, nonacademic research professionals, scientists engaged with the humanities, and creative writers regardless of nationality, whose proposed subjects will benefit directly from access to the New York Public Library's collections at the Humanities and Social Sciences Library. Fellows will be required to be in continuous residence for the length of the award and to participate as much as possible in center activities including daily lunches, readings, lectures, colloquia, symposia, and conferences. Each fellow will be responsible for a public presentation—a reading, a paper, a lecture—of publishable quality. The fellows receive stipends and, when necessary, housing allowances. Write or email to the address listed for an application form.

Requirements: Eligible fields of study include anthropology, art, geography, history, languages and literatures of the world, philosophy, politics, popular culture, psychology, religion, sociology, and sports.

Amount of Grant: $50,000

Date(s) Application is Due: Sep 30

Contact: Dorothy and Lewis B. Cullman Center for Scholars and Writers;email: csw@nypl.org

Internet: http://www.nypl.org/research/chss/scholars/scholars1.html

Sponsor: New York Public Library
Fifth Avenue and 42nd Street
New York, NY 10018-2788

New York State Archives Hackman Research Residency Program 3700

Residencies support advanced work in New York State history, government, or public policy using the holdings of the New York State Archives. It also encourages public dissemination of research products. At the end of the residency, awardee is expected to submit a written report on the research or a draft of the written product and a statement of publication plans or other public dissemination, as well as one public presentation in New York State on the results of the project. Application forms are available on the Web site.

Requirements: Applicants working on doctoral dissertations and those at the postdoctoral level are particularly encouraged to apply, but any proposal for advanced research will be considered. Topics or area of study must draw on the holdings of the New York State Archives. Preference will be given to projects that have application to enduring public policy issues, particularly in New York State, and that have a high probability of publication or other public dissemination. Other considerations aside, preference will also be given to proposals for research in the archives' holdings that have been little used or have only recently become available.

Amount of Grant: $100-$4500

Date(s) Application is Due: Jan 15

Contact: Trust Administrator, (518) 473-7091; fax: (518) 473-7058; email: hackmanres@mail.nysed.gov

Internet: http://www.nysarchivestrust.org/apt/hackman

Sponsor: New York State Archives Partnership Trust
Cultural Education Ctr, Suite 9C49
New York, NY 12230

New York State Science, Technology and Academic Research Grants 3701

This agency will concentrate its resources on high-technology research and development projects that have the greatest potential for industrial application and for contributing to the state's economic growth and development. Project grants and technical assistance are awarded to provide hands-on technical assistance to small and medium-sized manufacturing companies through a network of technical specialists or field agents based in regions across the state. Capital grants, challenge/matching grants, and seed money grants also are available. Cooperation of industry in these projects is strongly encouraged.

Requirements: Investigators at colleges, universities, and nonprofit research laboratories in New York State are eligible to submit proposals.

Geographic Focus: New York

Amount of Grant: $125 million total

Contact: Program Contact, (518) 292-5700; fax: (518) 292-5798; email: contact@nystar.state.ny.us

Internet: http://www.nystar.state.ny.us/initiatives.htm

Sponsor: NYSTAR
30 S Pearl Street, 11th Fl
Albany, NY 12207

New-Land Foundation Grants 3702

The foundation supports U.S. nonprofit organizations for programs that emphasize the areas of civil rights, justice, peace, and arms control. The foundation also focuses on the

environment, population control, and leadership development in these areas. Types of support include general operating expenses, annual campaigns, seed grants, research, continuing support, internships, matching funds, and programs/projects. The foundation board generally meets in the fall and spring of each year to consider grant requests. Application form and guidelines should be requested in writing.
Restrictions: Individuals are ineligible.
Samples: Ms. Foundation for Women (New York, NY)—$60,000.; Institute for Policy Studies (Washington, DC)—$40,000.; Land and Water Fund of the Rockies (Boulder, CO)—$25,000.
Amount of Grant: $5000-$25,000 average
Date(s) Application is Due: Feb 1; Aug 1
Contact: Program Contact, (212) 479-6162
Sponsor: New-Land Foundation
1114 Avenue of the Americas, 46th Fl
New York, NY 10036-7798

New-York Historical Society Research Fellowships **3703**
The society awards short-term visiting fellowships to encourage research in the collections by scholars who reside beyond a 75-mile radius of New York City, and to enable research by New York area scholars who can demonstrate financial need. The one-month fellowship must be undertaken for consecutive weeks between June 1 and May 31. Fellows will be able to conduct research in the library and may also undertake research by appointment in the library's department of prints, photographs and architectural collections, and in the museum collections, which are housed in the Henry Luce III Center for the Study of American Culture on the society's fourth floor. There is no application form. Applicants must submit: a cover sheet with name, permanent address and email, current employer/affiliation, title of project, proposed dates of residency, and signature; a letter of two single-spaced pages maximum describing the project and its relation to specifically cited collections at the society and to previous work on the same theme, and describing the projected outcome of the work. If residents of the New York City metropolitan area are applying, they must explain their financial need for the stipend; a resume; and three confidential letters of reference. Graduate students must include their thesis advisor.
Requirements: U.S. citizens and foreign nationals may apply. Applicants must hold an undergraduate degree. The applicant's field of research must demonstrably and specifically relate to society's collections.
Amount of Grant: $2000
Date(s) Application is Due: Apr 1
Contact: Fellowships Administrator, (212) 873-3400 ext 221; email: fellows@nyhistory.org
Internet: http://www.nyhistory.org/fellowships.html
Sponsor: New-York Historical Society
170 Central Park W
New York, NY 10024

NewsBank/Readex/GODORT/ALA Catharine J. Reynolds Research Grants **3704**
The NewsBank/Readex/GODORT/ALA Catharine J. Reynolds Award provides funding for research in the field of documents librarianship, or in a related area that would benefit the individual's performance as a documents librarian or make a contribution to the field. This award, established in 1987, is named for Catharine J. Reynolds, former Head of Government Publications at the University of Colorado, Boulder. It is supported by an annual contribution of $2,000 from the NewsBank/Readex Corporation.
Requirements: Applications for the NewsBank/Readex/GODORT/ALA Catharine J. Reynolds Awards are accepted from documents librarians. The awards support research in all areas of government information, U.S., foreign, international, state, regional, or local. The grant can be used to finance research costs, such as computer time or including travel. Recipients are selected on the basis of a well-defined proposal, potential ability to complete the project, promise of future contributions to the profession, financial need, and activity to the profession. Preference is given to GODORT members. Two confidential letters of recommendation should accompany the application form.
Samples: Judith Downie, California State University, CA, $2,000 .;2008.;
Amount of Grant: $2,000
Contact: Stephen Woods, ALA GODORT Awards Committee Chair: (814) 865-0665; fax: (814) 863-1403; email: swoods@psu.edu
Internet: http://www.ala.org/ala/mgrps/rts/godort/godortcommittees/godortawards/cjrform.cfm
Sponsor: American Library Association
50 East Huron Street
Chicago, IL 60611

NFID Advanced Vaccinology Course Travel Grant **3705**
These grants are intended to assist recent postdoctoral graduates who have a demonstrated interest in a career in vaccinology. This program provides participants with a comprehensive overview of vaccinology, from immunology to the social, economic, political, and ethical issues of vaccination. Practical approaches to the implementation of effective vaccination programs suitable for particular population groups will be discussed. To be eligible for this award the applicant must have made separate application to the course. Further information about the course can be obtained from: Betty Dodet, Fondation Merieux, 17, rue Bourgelat, BP 2021, Lyon Cedex 02, France. Annual deadline dates vary; contact program staff for exact dates. Note: Please do not contact NFID for information about the course. Contact the listed contact person.
Requirements: Individuals who are U.S. citizens who reside in the United States at the time of the course and who are within three years of obtaining a doctoral degree, or in the case of physicians, within two years of completing specialty training or a subspecialty fellowship, are eligible to apply. Candidates must have a demonstrated commitment to vaccinology as a career, and must be conducting research or working in a recognized and accredited U.S. institution of higher learning or in a governmental agency.
Amount of Grant: $4000
Date(s) Application is Due: Nov 14
Contact: Betty Dodet, Fondation Merieux; fax: (33-4) 72 40 79 50; email: betty.dodet@fond-merieux.org
Internet: http://www.nfid.org/fellow/grantsfellow.pdf
Sponsor: National Foundation for Infectious Diseases
4733 Bethesda Avenue, Suite 750
Bethesda, MD 20814

NFID Colin L. Powell Minority Postdoctoral Fellowship in Tropical Disease Research **3706**
The program, sponsored by NFID and GlaxoSmithKline, invites applications from qualified minority researchers to become specialists and investigators in the field of tropical diseases. The one-year fellowship includes a stipend, $3000 of which may be used for travel and supplies. Fellowships commence on or about July 1, with notification on or about April 15. Annual deadline dates may vary; contact program staff for exact dates.
Requirements: The applicant must hold a doctorate from a recognized U.S. university and be a U.S. citizen or permanent resident or spouse of a U.S. citizen or permanent resident.
Amount of Grant: $30,000 stipend
Date(s) Application is Due: Jan 6
Contact: Senior Executive Director, (301) 656-0003; fax: (301) 907-0878; email: info@nfid.org
Internet: http://www.nfid.org/fellow/grantsfellow.pdf
Sponsor: National Foundation for Infectious Diseases
4733 Bethesda Avenue, Suite 750
Bethesda, MD 20814

NFID John P. Utz Medical Mycology Postdoctoral Fellowship **3707**
The fellowship is intended to encourage and assist a qualified physician to become a specialist and investigator in medical mycology. Selection criteria for the one-year fellowship will include, but not be limited to, the applicant's record of scholarship performance and professional qualifications; scientific merit of the clinical research project proposed; and validity of research rationale and adequacy of facilities available to the applicant, host department, and host institution. The recipient will be required to submit two single-spaced, one-page progress reports during his/her year of clinical research. One report will be due after six months and the second after 12 months or upon completion of the fellowship year. Notification of award will be made on or about April 15.
Requirements: The applicant must be a physician who is a U.S. citizen. He/she must demonstrate aptitude and accomplishment in research and must confirm arrangements for conduct of the proposed research in a recognized host laboratory. A letter from the chairperson of the department of infectious diseases expressing a willingness to assume responsibility for training the applicant must accompany the application.
Restrictions: The fellowship may not be awarded if the applicant has received or will receive a major fellowship, research grant, or traineeship from the federal government or another foundation in excess of the amount of this award.
Amount of Grant: $40,000 stipend which includes $1000 for travel and supplies
Date(s) Application is Due: Jan 6
Contact: Senior Executive Director, (301) 656-0003; fax: (301) 907-0878; email: info@nfid.org
Internet: http://nfid.org/fellow/grantsfellow.pdf
Sponsor: National Foundation for Infectious Diseases
4733 Bethesda Avenue, Suite 750
Bethesda, MD 20814-5278

NFID New Investigator Matching Grants **3708**
Matching grants are intended to assist new investigators who are beginning their research. The proposal must be based on research relevant to infectious diseases, microbiology, clinical medicine, epidemiology, nursing, etc. Priority will be given to applications proposing research on HIV and AIDS; tuberculosis, influenza, and other respiratory diseases; hospital-acquired and opportunistic infections; viral hepatitis; gastrointestinal, diarrheal, and food-borne diseases; sexually transmitted diseases; antibiotic resistance; tropical, travelers', and parasitic diseases; diseases transferred from animals to humans; and new and reemerging infectious diseases. Each application must be accompanied by an agreement from the applicant's sponsoring institution to match the award with equal funds.
Requirements: A new investigator is defined as an individual with full-time junior faculty (instructor or assistant professor) status at a recognized and accredited institution of higher learning. Priority will be given to those who do not have research grants and whose studies represent pilot work with intent to further develop their research.
Samples: Dr. Nancy Connell, UMDNJ/New Jersey Medical School (NJ)—to support experiments to study the cell wall and membranes of mycobacteria, $2000.; Dr. Christine Wanke, Harvard U Medical School (MA)—to support research on E. coli in adult patients infected with HIV with chronic diarrhea, $2000.
Amount of Grant: $2000
Date(s) Application is Due: Feb 14
Contact: Senior Executive Director, (301) 656-0003; fax: (301) 907-0878; email: info@nfid.org

Internet: http://www.nfid.org/fellow/grantsfellow.pdf
Sponsor: National Foundation for Infectious Diseases
4733 Bethesda Avenue, Suite 750
Bethesda, MD 20814

NFID Nosocomial Infection Research and Training Postdoctoral Fellowship 3709
The fellowship is intended to encourage and assist a qualified physician researcher to become a specialist and investigator in the field of nosocomial infections. Priority will be given to the study of gram positive nosocomial infections. Each application must be accompanied by a letter from the host laboratory and signed by the laboratory director or appropriate department chairperson indicating a willingness to accept the applicant and to provide guidance, research space, and necessary research equipment. Overhead to the sponsoring institution will not be paid. One thousand dollars of the stipend may be used for travel and supplies.
Requirements: The applicant must be a physician and a citizen of the United States. He/she must demonstrate aptitude and accomplishment in research and must confirm arrangements for conduct of the proposed research in a recognized host laboratory. An applicant with no grant support will receive priority consideration.
Restrictions: Applicants receiving or to be awarded grants in the same academic year in excess of the amount of this award are ineligible.
Amount of Grant: $40,000 total
Date(s) Application is Due: Jan 6
Contact: Senior Executive Director, (301) 656-0003; fax: (301) 907-0878; email: info@nfid.org
Internet: http://nfid.org/fellow/grantsfellow.pdf
Sponsor: National Foundation for Infectious Diseases
4733 Bethesda Avenue, Suite 750
Bethesda, MD 20814-5278

NFID Postdoctoral Fellowships in Emerging Infectious Diseases 3710
This one-year fellowship is intended to encourage and assist a qualified physician to become a recognized authority on emerging infectious diseases and epidemiology. The fellow will be assigned to the National Center for Infectious Diseases, CDC, in Atlanta, GA, and will participate in analysis and reporting of the national burden of infectious diseases and potential threats of emerging infectious diseases.
Requirements: The applicant must be a physician who is a U.S. citizen who preferably will have completed an infectious diseases fellowship.
Restrictions: The fellowship may not be awarded if the applicant has received or will receive a major fellowship, research grant, or traineeship from the federal government or another foundation in excess of $20,000.
Amount of Grant: $50,000 stipend
Date(s) Application is Due: Jan 6
Contact: Senior Executive Director, (301) 656-0003; fax: (301) 907-0878; email: info@nfid.org
Internet: http://nfid.org/fellow/grantsfellow.pdf
Sponsor: National Foundation for Infectious Diseases
4733 Bethesda Avenue, Suite 750
Bethesda, MD 20814

NFIE Foundation Learning and Leadership Grants 3711
Grants will fund recipients to participate in a high-quality professional development experience or organize a collegial study group that leads to improvements in practice, curriculum, and student achievement in the school or university. One-shot professional growth experiences such as attending a national conference or engaging a professional speaker are discouraged. Decisions regarding the content of the professional growth activities must be based on an assessment of student work undertaken with colleagues and must be embedded in the institutional planning process. Grant funds may be used for fees, travel expenses, books, or other materials that enable applicants to learn subject matter, instructional approaches, and skills. Recipients are expected to exercise professional leadership by sharing their new learning with their colleagues. Applications may be submitted at any time; notification will be made within seven months of receipt of the application.
Requirements: Practicing public school teachers, public school education support personnel, and public higher education faculty and staff are eligible to apply. Preference will be given to members of the National Education Association.
Restrictions: Grant funds may not be used to pursue degrees, to pay indirect costs or grant administration fees, or to pay salaries. With the exception of study groups, funds may not be used by applicants to pay themselves stipends. Grant funds may not be used to serve K-12 students in after-school or weekend settings. Funds may not be used for lobbying or religious purposes. Identical applications will not be considered.
Samples: Ruth Allmand (Westport, WA)—to attend a summer institute and to purchase training materials on new approaches to teaching gifted children, $1000.; Carrie Brennan (Tucson, AZ)—to attend the annual colloquium of the Annenberg Institute for School Reform, $1000.
Amount of Grant: $2000-$5000
Date(s) Application is Due: Feb 1; Jun 1; Sep 15
Contact: Learning and Leadership Grants, (202) 822-7840; fax: (202) 822-7779
Internet: http://www.nfie.org/grants.htm
Sponsor: NEA Foundation for the Improvement of Education
1201 16th Street NW, Suite 416
Washington, DC 20036

NFJC Gantz-Zahler Grant 3712
The grant is awarded every other year to support the publication of a nonfiction work reflecting on Jewish life, history, tradition, thought, or culture. Proposals must include a contract or letter from an established publisher indicating intent to publish the completed work. Grants can cover any publication cost, including research, editorial, printing, marketing, or distribution. Guidelines and application are available online.
Requirements: The author must be a U.S. citizen or permanent resident of the United States. The work must be nonfiction. The work must have a commitment to be published from a publisher.
Amount of Grant: $2500
Date(s) Application is Due: Apr 21
Contact: Kristen Runk, (212) 629-0500 ext 215; fax: (212) 629-0508; email: krunk@jewishculture.org or nfjc@jewishculture.org
Internet: http://www.jewishculture.org/grants/gantz_zahler_grant.html
Sponsor: National Foundation for Jewish Culture
330 Seventh Avenue, 21st Fl
New York, NY 10001

NFJC Maurice and Marilyn Cohen Fund for Doctoral Dissertation Fellowships in Jewish Studies 3713
Fellowships for Jewish studies are granted for one academic year and are normally given for the final stages of completing the dissertation. Application and guidelines are available online.
Requirements: Student must be a U.S. citizen or permanent resident; have completed all requirements for the PhD degree, except the dissertation itself; and have proficiency in a Jewish language adequate for pursuing an academic career in their chosen field.
Amount of Grant: $8000-$10,000
Date(s) Application is Due: Jan 20
Contact: Nancy Schwartzman, (212) 629-0500 ext 219; fax: (212) 629-0508; email: Nschwartzman@jewishculture.org
Internet: http://www.jewishculture.org/doc_diss/jewish_scholarship_dd.html
Sponsor: National Foundation for Jewish Culture
330 7th Avenue, 21st Fl
New York, NY 10001

NFPW Member Graduate Scholarships 3714
The scholarship supports one NFPW member who wishes to continue her education by returning to school for courses leading to a graduate degree at an accredited college or university. The recipient is selected on the basis of career potential, scholarship, and financial need. Application forms are available upon request.
Requirements: Applicant must have a BA or BS degree, have been a member of NFPW for at least two years, and have worked in the communications profession for at least two years.
Amount of Grant: $1000
Date(s) Application is Due: May 1
Contact: Executive Director, (800) 780-2715; fax: (703) 812-4555; email: presswomen@aol.com
Internet: http://www.nfpw.org/about/Grants.htm
Sponsor: National Federation of Press Women
P.O. Box 5556
Arlington, VA 22205

NFWF Bird Conservancy Initiative Grants 3715
Through its Bird Conservation Initiative, the Foundation and its partners seek to stabilize and increase declining (but still common) populations of select bird species or suites of species and accelerate recovery of imperiled species. The Foundation seeks to make a significant impact on bird conservation by focusing on five key issues: demonstrating that implementation of actions within existing conservation strategies will have a positive impact on bird populations; evaluating which species conservation plans are succeeding, and why; stimulating development of effective plans for those species for which no strategy is available, and revising those that need improvement;; building international capacity for conservation along migratory paths; and accelerating the recovery of imperiled species, particularly those designated as candidate, threatened or endangered at the federal level. The preliminary list of priority species includes, but is not limited to: black-capped vireo & golden cheeked warbler; northern bobwhite quail; California condor; red cockaded woodpecker; cerulean warbler; sage grouse (greater and Gunnison); golden winged warbler; shorebirds; lesser prairie chicken; and whooping crane. Grants generally range from $50,000-$300,000. Preproposals are due April 1, and the deadline for full proposal submission is June 1.
Requirements: The program will fund projects that directly address conservation of priority bird habitats in the Western Hemisphere. A minimum of 1:1 non-federal match is required; however, most competitive projects have at least 2:1 in non-federal match.
Amount of Grant: $50,000-$300,000 range
Date(s) Application is Due: Apr 1; Jun 1
Contact: Daniel Petit, Ph.D., Director, Bird Conservation; (202) 857-0166; fax: (202) 857-0162; email: daniel.petit@nfwf.org
Internet: http://www.nfwf.org/AM/Template.cfm?Section=Bird_Conservation2
Sponsor: National Fish and Wildlife Foundation
1120 Connecticut Avenue NW, Suite 900
Washington, DC 20036

NFWF Bring Back the Natives Grants 3716

The program requests pre-proposals (due December 1) from nonprofit organizations; universities; Native American tribes; and local, state, and federal agencies interested in restoring native populations of sensitive or listed aquatic species. Grants fund on-the-ground efforts to restore native aquatic species to their historic range. Projects should involve partnerships between communities, agencies, private landowners, and organizations that seek to rehabilitate streamside and watershed habitats. Projects should focus on habitat needs of species such as fish, invertebrates, and amphibians that originally inhabited the waterways across the country. Special emphasis is placed on cutthroat trout restoration with specific preference given to projects that will protect or reestablish migration corridors between breeding populations such as Lahontan cutthroat trout and Yellowstone cutthroat trout. Other eligible species include bull trout, redband trout, Apache trout, steelhead trout, desert pupfish, Pacific tree frog, and several species of salamanders and salmon, that are keystone or indicator species of the health of aquatic communities. Preference will also be given to support projects that will keep sensitive or declining species off of the Endangered Species List, or for listed species that have a chance for recovery and de-listing within five to 10 years. The program will award between 12 and 15 matching grants annually evaluated on a national competitive basis. Pre-proposals should be submitted electronically through the online application form available on the NFWF website.

Restrictions: No part of a matching grant, either the federal funds from NFWF or the non-federal match, may be used to: support litigation expenses or lobbying activities; cover general operating expenses, overhead, or other indirect expenses without NFWF's prior written approval; supplement shortfalls in government agency budgets; support multi-year grants due to the nature of NFWF's annual appropriations; support basic research; or support basic planning or school education projects without an on-the-ground component to the project.

Amount of Grant: $60,000 average

Date(s) Application is Due: Dec 1; Feb 11

Contact: Barrett Bohnengel, Program Coordinator; (503) 417-8700; fax: (503) 417-8787; email: Barrett.Bohnengel@nfwf.org

Internet: http://www.nfwf.org/AM/Template.cfm?Section=Browse_All_Programs&Template=/CM/ContentDisplay.cfm&ContentID=3677

Sponsor: National Fish and Wildlife Foundation
806 SW Broadway, Suite 750
Portland, OR 97205

NFWF Budweiser Renewable Energy and Wildlife Conservation Research Prize 3717

This research prize will be awarded under a competitive process and will recognize and support a single innovative project that advances new technologies or practices that contribute to making renewable energy a practical energy alternative. The $100,000 total prize will be awarded in consideration of a project's ability to develop or evaluate new cost-effective renewable energy technologies for industrial application while demonstrating the measurable benefits such practices can have in advancing wildlife conservation. Strong preference will be given to projects that demonstrate additional financial commitment in the form of cash or in-kind matching funds at a ratio equal to or exceeding 1:1. Funds may be used for items such as: salaries and benefits, student stipends, travel, publication expenses, and supplies. Applications will be accepted through August 15.

Requirements: Eligibility for the research prize is limited to accredited post-secondary academic institutions with preference given to those working collaboratively among different departments within an institution(s), and with public agencies and nonprofit organizations.

Restrictions: Funds may not be used for: political advocacy or litigation of any kind, shortfalls in government agency budgets, or general administrative overhead or indirect costs.

Amount of Grant: $100,000

Date(s) Application is Due: Aug 15

Contact: Ellen Gabel; (202) 857-0166; fax: (202) 857-0162; email: ellen.gabel@nfwf.org or info@nfwf.org

Internet: http://www.nfwf.org/AM/Template.cfm?Section=Grants&Template=/CM/ContentDisplay.cfm&ContentID=3634

Sponsor: National Fish and Wildlife Foundation
1120 Connecticut Avenue NW, Suite 900
Washington, DC 20036

NFWF Freshwater Fish Conservation Initiative Grants 3718

Through its Freshwater Fish Conservation Initiative, the Foundation and its partners seek to increase populations of flagship freshwater, anadromous and diadramous fish species and other aquatic species through the protection, restoration and enhancement of spawning and rearing habitat in the upper reaches of targeted watersheds, removal of barriers to fish passage and maintenance of critical in-stream flows. Funding priorities include implementing projects that meet the goals to: protect and maintain intact and healthy aquatic systems; prevent further degradation of fish habitats that have been already affected; reverse declines in the quality and quantity of aquatic habitats to improve overall health of fish and aquatic organisms, and increase the quality and quantity of fish habitats that support a broad natural diversity of fish and other aquatic species. Grants generally range from $50,000-$300,000. Preproposals are due April 1, and the deadline for full proposal submission is June 1.

Requirements: A minimum of 1:1 non-federal match is required; however, most competitive projects have at least 2:1 in non-federal match.

Amount of Grant: $50,000-$300,000 range

Date(s) Application is Due: Apr 1; Jun 1

Contact: Krystyna Wolniakowski, Director; (503) 417-8700; fax: (503) 417-8787; email: wolniakowski@nfwf.org

Internet: http://www.nfwf.org/AM/Template.cfm?Section=Fish_Conservation2

Sponsor: National Fish and Wildlife Foundation
1120 Connecticut Avenue NW, Suite 900
Washington, DC 20036

NFWF International Sea Turtle Conservation Fund Grants 3719

The National Fish and Wildlife Foundation announces the availability of matching grant funding for international sea turtle conservation projects in the Western Hemisphere. Priority will be given to projects along the Atlantic and Pacific Coasts of Central and South America and in the wider Caribbean (including the U.S. Virgin Islands and Puerto Rico) and Canada. Priority consideration will be given to projects demonstrating clear and immediate benefit to the conservation of hawksbill (Eretmochelys imbricata) and leatherback (Dermochelys coriacea) turtles in the wider Caribbean and East Pacific, and east Pacific green turtles (Chelonia mydas). A limited amount of funds are available to support single or multi-year projects focused on addressing the incidental capture of sea turtles in artisanal and commercial gillnet fisheries. Proposals must be submitted in English to the International Sea Turtle Conservation Fund via the Foundation's online system by 11:59PM EST on February 4.

Amount of Grant: $25,000 maximum

Date(s) Application is Due: Feb 4

Contact: Tom Barry, Assistant Program Director; (202) 857-0166; fax: (202) 857-0162; email: Tom.Barry@nfwf.org

Internet: http://www.nfwf.org/AM/Template.cfm?Section=Browse_All_Programs&TEMPLATE=/CM/ContentDisplay.cfm&CONTENTID=7639

Sponsor: National Fish and Wildlife Foundation
1120 Connecticut Avenue NW, Suite 900
Washington, DC 20036

NFWF Marine & Coastal Conservation Initiative Grants 3720

The Foundation seeks to help reduce unintended human impacts to marine and coastal habitats and species. Primary Conservation objectives are to: increase and sustain productivity of key spawning grounds and unique marine habitats by reducing human-generated impacts to the environment and establishing and improving management of Marine Protected Areas; increase populations of select federally-listed marine species (e.g., whales, sea turtles) to levels needed for recovery to facilitate change in species listing; and improve and sustain the health of the nation, major estuaries and the Great Lakes by restoring and protecting critical coastal habitat, improving water quality in tributaries, and enhancing populations of keystone species. Grants generally range from $50,000-$300,000. Preproposals are due April 1, and the deadline for full proposal submission is June 1.

Requirements: A minimum of 1:1 non-federal match is required; however, most competitive projects have at least 2:1 in non-federal match.

Amount of Grant: $50,000-$300,000 range

Date(s) Application is Due: Apr 1; Jun 1

Contact: Anthony C. Chatwin, Ph.D., Director; (202) 857-0166; fax: (202) 857-0162; email: anthony.chatwin@nfwf.org

Internet: http://www.nfwf.org/AM/Template.cfm?Section=Coastal_and_Marine

Sponsor: National Fish and Wildlife Foundation
1120 Connecticut Avenue NW, Suite 900
Washington, DC 20036

NFWF Marine Debris Research and Technology Grants 3721

The Marine Debris Research and Technology Grants Program (Program) is a partnership between the NOAA Marine Debris Program and National Fish and Wildlife Foundation. The Program seeks to support projects that have a strong likelihood of reducing marine debris caused by derelict fishing gear (both commercial and recreational), to increase our understanding of the impacts, location and best methods for removal, and projects that engage marinas, ports and the fishing industry. The Program provides competitive grants to finance innovative proposals that seek to work with marinas, ports, and the fishing industry to significantly reduce the occurrence of debris in these areas. Technology proposals to test new innovations in gear development to prevent loss or make the gear stop fishing or entangling if lost are also of high priority and strongly encouraged. In addition, the Program is accepting research proposals that address the biological, social, or economic impact of marine debris on species, habitat, and coastal businesses. Approximately $800,000 is available for calendar year. The majority of grants are anticipated to fall between $20,000 and $200,000. Larger grants will be reserved for those projects that are structured to definitively answers key research questions or implement projects over a broad scale. Pre-proposals must be received by May 5, and final proposals are due June 27.

Requirements: Eligible applicants are institutions of higher education, other non-profits, commercial organizations, and state (including Commissions), local, and tribal governments whose projects have the potential to benefit NOAA trust resources through marine debris research and prevention projects.

Restrictions: Applications from federal agencies or employees of federal agencies will not be considered. Funds cannot be used for political advocacy, boycotts, litigation expenses, terrorist activities or activities conducted in violation of the Foreign Corrupt Practices Act.

Amount of Grant: $20,000-$200,000 average range

Date(s) Application is Due: May 5; June 27

Contact: Tom Barry, Assistant Program Director; (202) 857-0166; fax: (202) 857-0162; email: Tom.Barry@nfwf.org

Internet: http://www.nfwf.org/AM/Template.cfm?Section=Home&Template=/TaggedPage/TaggedPageDisplay.cfm&TPLID=32&ContentID=9272
Sponsor: National Fish and Wildlife Foundation
1120 Connecticut Avenue NW, Suite 900
Washington, DC 20036

NFWF More Fish Partnership Fund Grants **3722**

The goal of the Fund is to improve awareness about aquatic resources by supporting both grassroots capacity building with the ability to measure that performance and successfully plan and implement habitat conservation projects. Under this program, grants of up to $50,000 will be awarded competitively. Proposed activities eligible for funding will include: aquatic habitat restoration; enhancement and protection that protect and improve water quantity/quality, aquatic vegetation; landowner/water user conservation, outreach and stewardship training; and conservation watershed planning and management. The program's primary focus is smaller, community-based projects for up to $50,000, but larger requests for funds will be considered on a competitive basis. The following costs are eligible: protection, enhancement, and restoration of fish habitat within and along rivers and streams, lakes, reservoirs, ponds, coastal areas, etc. and the upland areas affecting them; volunteer training workshops, stakeholder roundtables, and community outreach; and project design and development that is anticipated to lead to an on-the-ground project within 18 months. Matching funds will be required on a 1:1 basis or higher.
Requirements: Applicants representing B.A.S.S. Clubs, local non-profits, educational institutions, angling clubs, tribes, conservation districts, or local governments are eligible. Community groups without non-profit status are encouraged to seek an eligible sponsor/fiscal agent.
Restrictions: Full fee acquisition, less-than-fee acquisition, free-standing educational projects, mitigation or corrective actions, general planning and studies, and legal or lobbying efforts are ineligible.
Amount of Grant: $50,000 maximum primarily
Date(s) Application is Due: Nov 1
Contact: Barrett Bohnengel, Coordinator; (503) 417-8700; fax: (503) 417-8787
Internet: http://www.nfwf.org/AM/Template.cfm?Section=Home&Template=/CM/ContentDisplay.cfm&ContentID=3937
Sponsor: National Fish and Wildlife Foundation
806 SW Broadway, Suite 750
Portland, OR 97205

NFWF National Whale Conservation Fund Grants **3723**

The fund supports research, management, conservation, and education/outreach activities related to the conservation and recovery of whales, particularly projects relating to great whales that are the most endangered (e.g., North Atlantic and North Pacific right whales). Priority will be given to projects that address anthropogenic threats to endangered whales of U.S. waters (i.e., fishing gear entanglement, ship/whale collisions, noise, habitat degradation, contaminants, etc.). Proposals that address tasks or needs identified in endangered whale recovery plans or conservation plans adopted by the National Marine Fisheries Service are encouraged. Applicants are encouraged to show leverage of a one-to-one or greater match to the proposal request. NWCF actively seeks to form new partnerships with corporations, organizations, and individuals to leverage NWCF??s resources sufficiently to meet its aggressive conservation goals. Applications should clearly describe anticipated benefits to whale conservation; project goals, objectives, and methodology, including provisions for monitoring and evaluation; qualifications of key project personnel; detailed budget illustrating use of NWCF grant; cooperation and coordination with other appropriate federal, state, and private organizations; and ability to leverage NWCF grant award.
Restrictions: NWCF will not fund political advocacy, boycotts, litigation, or indirect or unallocated expenses.
Contact: Stephanie Pendergrass, Assistant Director; (202) 857-0166; fax: (202) 857-0162; email: Stephanie.Pendergrass@nfwf.org
Internet: http://www.nfwf.org/AM/Template.cfm?Section=Who_We_Are&Template=/TaggedPage/TaggedPageDisplay.cfm&TPLID=30&ContentID=9725
Sponsor: National Fish and Wildlife Foundation
1120 Connecticut Avenue NW, Suite 900
Washington, DC 20036

NFWF Native Plant Conservation Initiative Grants **3724**

This program supports projects that protect, enhance, and/or restore native plants on public and private lands. Project categories include species, communities (of plants), and/or habitat protection and restoration; public education and information; and species and/or communities inventory and assessment. Applicants must match the full amount of the grant. The listed deadlines are for pre-proposals only.
Requirements: Government agencies and nongovernmental partners are eligible to apply. Preference will be given to projects that involve local communities and volunteers in the conservation efforts.
Amount of Grant: $5,000-$40,000; $15,000 average
Date(s) Application is Due: Feb 15; Aug 15
Contact: Ellen Gabel, Program Director; (202) 857-0166; fax: (202) 857-0162; email: Ellen.Gabel@nfwf.org
Internet: http://www.nfwf.org/AM/Template.cfm?Section=Who_We_Are&CONTENTID=9578&TEMPLATE=/CM/ContentDisplay.cfm
Sponsor: National Fish and Wildlife Foundation
1120 Connecticut Avenue NW, Suite 900
Washington, DC 20036

NFWF Nature Restoration Trust Grants **3725**

The Nature Restoration Trust seeks to fund projects that benefit fish, wildlife, and plants, as well as provide educational opportunities and encourage community involvement in conservation efforts. The Nature Restoration Trust will fund conservation projects within the Pacific Gas and Electric Company, service area in northern and central California through the Five Star Restoration Program. Note that the Nature Restoration Trust program especially encourages schools and school-based projects to apply.
Requirements: Partners, funders, and/or participants necessary to complete the project include schools or youth organizations; local or tribal governments; local businesses or corporations; conservation organizations or local citizens groups; state and federal resource management agencies; and foundations or other funders. Conservation projects within the Pacific Gas and Electric Company, service area in northern and central California are eligible.
Restrictions: Projects involving only research, monitoring, or planning; and projects that are part of a mitigation requirement are not eligible for funding.
Geographic Focus: California
Amount of Grant: $24,000 average
Date(s) Application is Due: Feb 15
Contact: Claire Thorp, Assistant Director; (415)243-3104 or (415) 778-0999; fax: (415) 778-0998; email: Claire.Thorp@nfwf.org
Internet: http://www.nfwf.org/AM/Template.cfm?Section=Home&Template=/TaggedPage/TaggedPageDisplay.cfm&TPLID=30&ContentID=8301
Sponsor: National Fish and Wildlife Foundation
90 New Montgomery Street, Suite 700
San Francisco, CA 94105

NFWF Northern Coastal California Restoration Fund Grants **3726**

The fund invests in conservation projects in the area of environmental impact. The program awards grants to projects for the purposes of: on the ground habitat conservation and protection projects in and near the Port of San Francisco, including the estuary of San Francisco Bay (SF Bay) and its tributaries and watershed and the northern coast of California to benefit the fish and wildlife species and the habitats on which they depend; improvement of public access areas and other public information/education projects that benefit the U.S. Fish and Wildlife National Wildlife Refuge System; projects that encourage collaboration and support local communities seeking environmentally and economically sustainable solutions for natural resource management; and applied research that is directly related to improvement of the SF Bay's and northern coastal California's natural resources management. Proposals must be submitted in electronic word format, and may not be longer than 10 pages.
Requirements: Nonprofit conservation organizations, counties and cities, state government agencies, tribes, federal agencies, and schools are eligible.
Restrictions: Funds cannot be used for political advocacy, litigation expenses, mitigation, or terrorist activities; equipment purchases not primarily related to the direct completion of the project; or website development or videos.
Amount of Grant: $200,000 maximum
Date(s) Application is Due: Sep 1
Contact: Cara Rose, Assistant Director; (503) 417-8700, ext. 21; fax: (503) 417-8787; email: cara.rose@nfwf.org
Internet: http://www.nfwf.org/
Sponsor: National Fish and Wildlife Foundation
806 SW Broadway, Suite 750
Portland, OR 97205

NFWF Puget Sound Marine Conservation Fund Grants **3727**

Together with the U.S. Fish and Wildlife Service, the Foundation will use funds to establish the Puget Sound Marine Conservation Fund to receive monies and begin a grant-making program to finance projects in the State of Washington within a geographic region that includes Puget Sound, and the Straits of Juan de Fuca, and waters subject to tidal influence within that geographic area for any of the following purposes: habitat conservation, protection, restoration and management projects to benefit the fish and wildlife resources and the habitats upon which they depend; landowner outreach and incentive programs for restoration and management of natural resources; public use and natural/cultural projects that benefit the National Wildlife Refuge System and other federal and state refuges, reserves and protected natural areas; and collaborative projects from local coastal communities seeking environmentally and economically sustainable solutions to natural resource problems. Pre-proposals can address a multitude of program purposes (listed at the web site). The people living in or near such habitats are intended to be the beneficiaries of the Fund, on behalf of such species and habitats.
Requirements: The following organizations are eligible to apply: 501(c)3 registered non-profit conservation organizations; counties; local, state or federal government agencies; tribes; schools; soil and water conservation districts; and other special districts.
Restrictions: Funds cannot be used for: political advocacy, litigation expenses, mitigation or terrorist activities; equipment purchases not primarily related to the direct completion of the project; videos; improvement of natural resources for commercial or private profit required mitigation; required Habitat Conservation Plan activities; or straight percentage of administrative overhead/indirect costs (a portion of the grant may be used for direct project-related administrative expenses such as salaries, travel, supplies, telephone, but these must be identified specifically in the budget as line items).
Amount of Grant: $20,000-$250,000
Contact: Cara Rose, Assistant Director; (503) 417-8700, ext. 21; fax: (503) 417-8787; email: cara.rose@nfwf.org
Internet: http://www.nfwf.org/AM/Template.cfm?Section=Who_We_Are&Template=/CM/ContentDisplay.cfm&ContentID=3998

Sponsor: National Fish and Wildlife Foundation
806 SW Broadway, Suite 750
Portland, OR 97205

NFWF Pulling Together Initiative Grants **3728**
The program aims to create partnerships between local agencies, private landowners, and other interested parties in developing long-term weed management projects as part of an integrated pest management strategy. The goals of PTI are: to prevent, manage, or eradicate invasive and noxious plants through a coordinated program of public/private partnerships; and to increase public awareness of the adverse impacts of invasive and noxious plants. Preproposals should be submitted inline by the October 26 due date, with full proposals due in January.
Requirements: Eligible applicants are partnerships of federal, state, and/or local government agencies, nonprofit organizations, private businesses, and landowners. The foundation makes challenge grants based on a two-to-one nonfederal funding match; awardees are required to supply remaining funds.
Amount of Grant: $10,000-$100,000; $35,000-$45,000 average
Date(s) Application is Due: Oct 26; Jan 11
Contact: Ellen Gabel, Program Director; (202) 857-0166; fax: (202) 857-0162; email: Ellen.Gabel@nfwf.org
Internet: http://www.nfwf.org/AM/Template.cfm?Section=Who_We_Are&Template=/CM/ContentDisplay.cfm&ContentID=4790
Sponsor: National Fish and Wildlife Foundation
1120 Connecticut Avenue NW, Suite 900
Washington, DC 20036

NFWF Shell Marine Habitat Program Grants **3729**
NFWF and Shell Oil Company Foundation fund natural resource conservation projects in the Gulf of Mexico. Projects will be funded that protect, conserve, restore, or enhance the Gulf coasts of Florida, from Tampa north; Mobile Bay, AL; coastal Louisiana, with an emphasis on the La Branche wetlands; and Port Arthur, Galveston Bay, and the Deer Park Ship Channel in Texas. Projects of interest include those that protect or restore estuarine habitat in the above geographic areas. Projects that include opportunities for hands-on participation by volunteers are of special interest. Focal species include brown pelicans, with an emphasis on habitat restoration and conservation of breeding sites. In addition, education programs intended to minimize harm from monofilament entanglement and other human impacts are of particular interest. Priority will be given to projects that build new or enhance existing partnerships and that address conservation needs of species characteristic of southeastern U.S. marine ecosystems. Preproposals must be submitted online by the listed application deadline. Application and guidelines are available online.
Requirements: Nonprofit IRS 501(c)3 tax-exempt organizations are eligible. Entities that are subdivisions of government (federal, state, county/parish, or city) are also eligible. Applicants must be based in the United States.
Restrictions: Individuals are not eligible for grants under this program. The program will not fund international projects, with very limited exceptions. The foundation will not fund political advocacy, lobbying, or litigation of any kind; shortfalls in government agency budgets; general administrative overhead or indirect costs of any kind; multiyear grants (applicant may reapply); or basic research.
Samples: Cameron Creole Watershed Marsh Project (LA)—to construct a protective system of terraces along 30 miles of marsh in southwest Louisiana, $50,000.; Gulf Coast/Mississippi Basin Hypoxia Outreach (Gulf Coast/Mississippi River Basin)—to initiate a targeted outreach effort as a preliminary step in addressing the hypoxia problems in the Gulf of Mexico, $50,000.; National Fish and Wildlife Foundation Minority Landowner Outreach in South Texas (Laguna Madre, TX)—to hire a local, Spanish-speaking individual to perform outreach work with local farmers and ranchers in the Laguna Madre ecosystem and along the Lower Rio Grande River, $40,000.
Amount of Grant: $50,000-$150,000
Date(s) Application is Due: Apr 1; Jun 1
Contact: Suzanne Sessine, Assistant Director, Southeast; (202) 857-0166; fax: (202) 857-0162; email: Suzanne.Sessine@nfwf.org
Internet: http://www.nfwf.org/AM/Template.cfm?Section=Home&Template=/TaggedPage/TaggedPageDisplay.cfm&TPLID=32&ContentID=9173
Sponsor: National Fish and Wildlife Foundation
1120 Connecticut Avenue NW, Suite 900
Washington, DC 20036

NFWF Wildlife & Habitat Conservation Initiative Grants **3730**
The Foundation seeks to make a significant impact on increasing populations of imperiled wildlife and plant species and to protect and restore priority wildlife habitats at a landscape scale. The Foundation will target investments towards the following key activities: * Development and implementation of conservation plans or strategies that contribute to targeted landscape-scale habitat objectives, in particular on or around federal lands or other conserved lands; investments will support land conservation and stewardship activities within these landscapes to achieve targeted habitat objectives: development and implementation of incentive-based tools that engage private landowners in imperiled species recovery; support private, voluntary or incentive-based land conservation models (e.g., rangeland trusts) to promote conservation of pivotal private lands adjacent to important public wildlife habitats; and accelerate the recovery of imperiled wildlife and plant species, particularly those designated as candidate, threatened or endangered at the federal level. Grants generally range from $50,000-$300,000. Preproposals are due April 1, and the deadline for full proposal submission is June 1.
Requirements: The program will fund projects that directly address conservation of priority in the Western Hemisphere. A minimum of 1:1 non-federal match is required; however, most competitive projects have at least 2:1 in non-federal match.
Amount of Grant: $50,000-$300,000 range
Date(s) Application is Due: Apr 1; Jun 1
Contact: Tim Male, Ph.D., Director, Wildlife and Habitat Conservation; (202) 857-0166; fax: (202) 857-0162; email: timothy.male@nfwf.org
Internet: http://www.nfwf.org/AM/Template.cfm?Section=Wildlife_and_Habitat
Sponsor: National Fish and Wildlife Foundation
1120 Connecticut Avenue NW, Suite 900
Washington, DC 20036

NFWF/Exxon Save the Tiger Fund Grants **3731**
The fund supports a variety of efforts aimed at the conservation of tigers in the wild, including on-the-ground tiger conservation and habitat/prey protection projects in tiger-range countries, as well as education activities both in tiger-range countries and internationally. Applications should demonstrate: anticipated benefits to wild tiger populations; improved management/protection of occupied and/or suitable tiger habitats; enhanced prey protection measures for in situ efforts; cooperation and coordination with other tiger conservation organizations; integration of tiger-range governments and local communities into project; lack of duplication and conflict with other ongoing conservation efforts; and methods for resolving conflicts between tiger conservation and the economic health and day-to-day lives of surrounding communities.
Requirements: U.S. or international nonprofit organizations, academic institutions, and government agencies, excluding U.S. federal agencies, are eligible.
Restrictions: The program will not fund political advocacy or litigation activities; captive tiger projects that are not presently part of approved Species Survival Plan; and general administrative overhead of sponsoring agency.
Date(s) Application is Due: Nov 5; Jan 21
Contact: Mahendra Shrestha, Director; (202) 857-0166; fax: (202) 857-0162; email: Mahendra.Shrestha@nfwf.org
Internet: http://www.savethetigerfund.org/Content/NavigationMenu2/Research/Grantees/Applyforagrant/default.htm
Sponsor: National Fish and Wildlife Foundation
1120 Connecticut Avenue NW, Suite 900
Washington, DC 20036

NFX Patient-Oriented Research Grants **3732**
The foundation supports a broad range of research endeavors that will lead to better recognition, treatment, and an eventual cure of children and adults with fragile X syndrome. The following forms of investigation are potentially appropriate for the current round of funding: educational or therapy interventions, medical treatment interventions, studies of family support and counseling, and diagnostic and reproductive studies. Basic research and patient-oriented research grants are awarded on a six-month alternating cycle. The January deadline is for basic science and clinical research applications. Applications should be submitted both as email attachments and as hard-copy documents.
Requirements: The program is open to all investigators with demonstrated ability to carry out research on fragile X syndrome.
Samples: Dr. Emily Osterweil—for research, Investigating FMRP in mGluR-mediated Protein Synthesis and Synaptic Plasticity, $35,000 (2005); To be distributed between two researchers—for research on fragile X syndrome, a hereditary condition that causes mental impairment, $60,000 (2004).
Amount of Grant: $35,000 maximum for junior investigators; $50,000 maximum for senior investigators
Date(s) Application is Due: Jan 31
Contact: Grants Administrator, (800) 688-8765 or (925) 938-9300; fax: (925) 938-9315; email: NATLFX@FragileX.org
Internet: http://www.fragilex.org/html/request_for_application.htm
Sponsor: National Fragile X Foundation
P.O. Box 190488
San Francisco, CA 94119

NFX Summer Research Fellowships **3733**
The foundation awards summer research fellowships in the area of fragile X syndrome. Proposals should be no longer than four pages, and should describe the research project in depth and the specific roles and activities of the summer fellow and the mentor. The curriculum vitas of the mentor and the fellow should be included as an addendum. Guidelines are available online.
Requirements: The fellow can be either an undergraduate or graduate student, and the mentor should be experienced in research concerning fragile X syndrome.
Amount of Grant: $2500
Date(s) Application is Due: Apr 17
Contact: Grants Administrator, (800) 688-8765 or (925) 938-9300; fax: (925) 938-9315; email: NATLFX@FragileX.org
Internet: http://www.fragilex.org/html/request_for_application.htm
Sponsor: National Fragile X Foundation
P.O. Box 190488
San Francisco, CA 94119

NHC Fellowships **3734**
Up to 40 fellowships are available to scholars from all fields of the humanities and humanistically inclined individuals from the arts, the natural and social sciences, and other professions for advanced study at the center. Most of the center's fellowships are

intended for any humanistic research, but several will be specifically available to scholars whose work concerns religion; art history or visual culture; Asian studies; French history or culture; and American art history. Scholars from all nations may apply. Stipends are determined individually, depending on the needs of fellows and the center's ability to meet them. Round-trip travel for fellows and their immediate families is also provided. Application and guidelines are available online.
Requirements: Doctoral degree holders or those who have equivalent professional accomplishments as well as younger scholars engaged in work significantly beyond the revision of a doctoral dissertation are eligible. Applicant must complete the center's application form and financial request form and must submit a current curriculum vita and a description of the project that would be undertaken at the center. Three letters of recommendation also are required by the deadline date.
Amount of Grant: $50,000 maximum
Date(s) Application is Due: Oct 15
Contact: Fellowship Program Coordinator, (919) 549-0661; fax: (919) 990-8535; email: nhc@ga.unc.edu
Internet: http://www.nhc.rtp.nc.us/fellowships/fellshipapinfo.htm
Sponsor: National Humanities Center
P.O. Box 12256, 7 Alexander Drive
Research Triangle Park, NC 27709

NHC Richard W. Lyman Award **3735**
The center welcomes nominations for this award, which recognizes humanities scholars who make imaginative use of information technology to advance scholarship and teaching. In recent years, scholars in the classics, English and American literature, history, and other humanistic disciplines have increasingly used computers and the World Wide Web to make available facsimiles of rare manuscripts; to archive, index, and annotate literary, artistic, and scholarly materials; to link text, visual images, and sound in new ways; and to create a new social structure that will bring scholars and students together to break down boundaries among learning, teaching, and research. The award is presented to individuals who break new ground by exploiting information technology toward these ends. Complete information and a nomination form are available online.
Samples: John M. Unsworth, University of Illinois at Urbana-Champaign—award recipient, $25,000 (2005); Robert K. Englund, University of California at Los Angeles—award recipient, $25,000 (2004).
Amount of Grant: $25,000 prize
Date(s) Application is Due: Dec 15
Contact: Award Administrator, (919) 549-0661; fax: (919) 990-8535
Internet: http://www.nhc.rtp.nc.us/lymanaward/lymanaward.htm
Sponsor: National Humanities Center
P.O. Box 12256, 7 Alexander Drive
Research Triangle Park, NC 27709-2256

NHC Summer Institutes in Literary Studies Grants **3736**
The center offers three summer institutes designed to provide junior faculty members in departments of literature and other disciplines the opportunity for intensive critical study of a single text or small group of texts. The institutes will seek to engage literary works deeply through close reading. Each institute will include only 12 participants and will be led by a distinguished scholar. The center will supply all texts and cover travel, meal, and lodging costs. Participants also receive a stipend. Information and application are available online.
Requirements: The institutes are open to scholars who have received a PhD within the last 10 years and who teach in departments of literature or other relevant disciplines at U.S. colleges or universities.
Amount of Grant: $1500 stipend
Date(s) Application is Due: Feb 25
Contact: Richard Schramm, (877) 271-7444; email: rschramm@ga.unc.edu
Internet: http://www.nhc.rtp.nc.us/siliterarystudies/index.htm
Sponsor: National Humanities Center
P.O. Box 12256, 7 Alexander Drive
Research Triangle Park, NC 27709

NHGRI Courses in Genomic Analysis and Interpretation and Ethical, Legal, and Social Implications (ELSI) Research Grants **3737**
NHGRI invites applications for support of short, advanced-level courses in genomic analysis and interpretation and in ethical, legal, and social implications (ELSI) research. Of particular interest are courses that emphasize new laboratory techniques in genetic and physical mapping; technology and theory related to studying sequence variation; DNA sequencing technology applicable to large-scale projects, informatics as it relates to the Human Genome Project (HGP); new technologies for interpretation of the genome; interdisciplinary training in principles of genomic analysis for nonbiologists; and principles and methods of studying the ethical, legal, and social issues relevant to HGP for biologists. These courses are meant to enhance the skills of U.S. scientists and scholars who are interested in pursuing laboratory research that utilizes the data and materials generated by the HGP or scholarly research relevant to the goals of HGP. Faculty should consist of established investigators or scholars actively working in the area of instruction. Plans for the inclusion of individuals who are currently underrepresented in the field of genomic research, such as women, minorities, and persons with disabilities should be addressed. Courses may range from a few days to a few weeks and may be offered annually, although other terms will be acceptable. PA: PAR-99-059
Requirements: Applications may be submitted by domestic public and private organizations, such as universities, colleges, and research institutions. Applicants may initially request support for two years. Course offerers are expected to be academic or research institutions experienced in training; however, applications from for-profit institutions will also be considered.
Restrictions: Foreign institutions are not eligible for support under this mechanism.
Contact: Dr. Bettie Graham, Division of Extramural Research, (301) 496-7531; email: Bettie_Graham@nih.gov
Internet: http://www.grants.nih.gov/grants/guide/pa-files/PAR-99-059.html
Sponsor: National Human Genome Research Institute
9000 Rockville Pk
Bethesda, MD 20892

NHGRI Curriculum Development Award in Genomic Research and Analysis **3738**
NHGRI, through this initiative, supports the development of courses and curricula designed to train interdisciplinary scientists who combine knowledge of genomics and genetics research with expertise in computer sciences, mathematics, chemistry, physics, engineering, or closely related sciences. It is anticipated that these courses or curricula would be useful to students and scientists who wish to develop new conceptual approaches to genome research and analysis or organize, analyze, or interpret large data sets resulting from genomic and genetics research. Development of courses at the graduate and undergraduate levels is encouraged. As part of this program, awardees will be expected to develop and implement the courses or curricula in their institution. The courses or curricula are expected to be models that could be transferable to other institutions in whole or in part. PAR-98-063 may be obtained through the NIH grant line (301-402-2221), the NIH Web site, mail, and email.
Requirements: The principal investigator must be engaged in genomics/genetics research or research in one of the following areas: computer sciences, mathematics, chemistry, physics, engineering, or a related field, such as bioinformatics, statistics, biomathematics, computational biology, bioengineering with genomics research or genetics research.
Date(s) Application is Due: Feb 1; Jun 1; Oct 1
Contact: Dr. Bettie Graham, (301) 496-7531; email: Bettie_Graham@nih.gov; Jean Cahill, Grants and Contracts Management Section, (301) 402-0733; email: Jean_Cahill@nih.gov
Internet: http://grants.nih.gov/grants/guide/pa-files/PAR-98-063.html
Sponsor: National Human Genome Research Institute
9000 Rockville Pike
Bethesda, MD 20892

NHGRI Genome Scholar Development and Faculty Transition Award **3739**
The purpose of the program is to enable promising new genome researchers to establish an independent research program in genomic research and analysis and to secure a tenure-track appointment in an academic institution in the United States. This is accomplished by providing the necessary resources to initiate an independent research program as a Genome Scholar in an extramural or an NIH intramural laboratory for up to two years followed by funding of their extramural research program for up to four years at an extramural institution to which she/he has been recruited as tenure-track faculty. The total number of years for both phases of the award cannot exceed five. This award has two phases—a Genome Scholar Development Phase and a Faculty Transition Phase. Applicants may apply for both phases or for the Faculty Transition Phase only. It is anticipated that awardees will subsequently obtain a research project grant (R01) to support the continuation of their research. PAR: PA-99-022
Requirements: At the time of application, individuals who apply to this program must have no more than five years postdoctoral experience, if applying for a Genome Scholar Development and Faculty Transition award; have no more than six years postdoctoral experience, if applying for a Faculty Transition award only; be in a postdoctoral or equivalent position; be engaged in fundamental or applied genomics research or technology development research as it relates to genomics; be able to devote at least 75 percent of time to research-related activities; not be a principal investigator on R01, R21, R29 or subproject of P grant, or recipient of a K award; and be a U.S. citizen or permanent resident of the United States.
Date(s) Application is Due: Feb 1; Jun 1; Oct 1
Contact: Dr. Bettie Graham, (301) 496-7531; email: Bettie_Graham@nih.gov; Jean Cahill, Grants and Contracts Management, (301) 402-0733; email: Jean_Cahill@nih.gov
Internet: http://www.grants.nih.gov/grants/guide/pa-files/PAR-99-022.html
Sponsor: National Human Genome Research Institute
9000 Rockville Pike
Bethesda, MD 20892

NHGRI Genomic Mapping, Sequencing, and Analysis Pilot Project Grants **3740**
NHGRI invites applications for exploratory/developmental grants (R21) to develop new and/or significantly improve existing technologies that will accelerate the genome mapping, sequencing, and analysis goals for the Human Genome Project in the most expeditious and economical manner. The purpose of this program announcement is to encourage high-risk/potential high-payoff applications that are not yet developed fully enough to successfully compete for a standard R01 grant. PA-97-045 may be obtained electronically through the NIH grant line (301-402-2221), the NIH Web site, and by mail and email.
Requirements: Applications may be submitted by domestic, for-profit and nonprofit organizations, public and private, such as universities, colleges, hospitals, laboratories, units of state and local governments, and eligible agencies of the federal government.
Amount of Grant: $100,000 maximum
Contact: Dr. Bettie Graham, Genetic Mapping and Sequence Variation, (301) 496-7531; fax: (301) 480-2770; email: bettie_graham@nih.gov
Internet: http://grants.nih.gov/grants/guide/pa-files/PA-97-045.html

Sponsor: National Human Genome Research Institute
9000 Rockville Pike
Bethesda, MD 20892

NHGRI Individual Mentored Research Scientist Awards in Genomic Research and Analysis 3741

These awards are intended to foster the career development of individuals with expertise in scientific disciplines that would further technological developments critical to the success of the Human Genome Project and the understanding of the genetic basis of diseases. Applications for this award should propose a development plan that has both intrinsic research importance and will also serve as a suitable vehicle for learning the methodology, theories, and conceptualizations necessary for well-trained independent, interdisciplinary researchers. PA: PAR-98-061
Requirements: Applicants must hold a PhD or equivalent professional degree in a scientific discipline other than biology, such as engineering, mathematics, computer sciences, physics, chemistry, or a closely related scientific discipline (such as bioinformatics, computational biology, statistics, biomathematics, or bioengineering), and show evidence of expertise in their discipline.
Restrictions: Applicants who are or have been former principal investigators on NIH research projects, FIRST Awards, sub-projects of program project or center grants, or the equivalent, are not eligible to apply for a K01 award. Applicants may not concurrently hold any other PHS award that duplicates the provisions of this award.
Date(s) Application is Due: Feb 1; Jun 1; Oct 1
Contact: Dr. Bettie Graham, (301) 496-7531; email: Bettie_Graham@nih.gov; Jean Cahill, Grants and Contracts Management Section, (301) 402-0733; email: Jean_Cahill@nih.gov
Internet: http://grants.nih.gov/grants/guide/pa-files/PAR-98-061.html
Sponsor: National Human Genome Research Institute
9000 Rockville Pike
Bethesda, MD 20892

NHGRI Mentored Research Scientist Development Award 3742

These awards are intended to foster the career development of individuals with expertise in scientific disciplines that would further technological developments critical to the success of the Human Genome Project and the understanding of the genetic basis of diseases. Up to three scientists may be selected and appointed to this program by the grantee institutions. Up to the Federal salary cap, plus fringe benefits, is available. Up to $20,000 of research support per year is also available.
Requirements: Eligibility is limited to individuals with degrees in computer sciences, mathematics, chemistry, engineering, physics, and closely related scientific disciplines, such as bioinformatics, computational biology, statistics, biomathematics, bioengineering. Applications will be accepted from domestic, nonfederal organizations of higher education that have strong, well-established research and training programs. The following organizations and institutions may apply: public/state controlled institutions of higher education; private institutions of higher education; Hispanic-serving institutions; historically Black colleges and universities (HBCUs); tribally controlled colleges and universities (TCCUs); Alaska native and native Hawaiian serving institutions; nonprofits with 501(c)3 IRS status; nonprofits without 501(c)3 IRS status; small businesses; for-profit organizations; State governments; regional organizations; U.S. territories or possessions; Indian/Native American tribal governments; Indian/Native American tribally designated organizations; eligible agencies of the Federal government; and faith-based or community based organizations.
Date(s) Application is Due: Feb 12; Mar 12; Jun 12; Jul 12; Oct 12; Nov 12
Contact: Bettie J. Graham; (301) 496-7531; email: bettie_graham@nih.gov
Internet: http://grants.nih.gov/grants/guide/pa-files/PAR-98-062.html
Sponsor: National Human Genome Research Institute (NHGRI)
5635 Fishers Lane, Suite 4076
Rockville, MD 20852

NHGRI Technologies for Genomic Mapping, Sequencing, and Analysis Research Grants 3743

The NHGRI solicits applications for projects to develop new technologies, and/or significantly improve existing technologies, that will facilitate and accelerate the genome mapping, sequencing, and analysis goals of the Human Genome Project in the most expeditious and economical manner. The resources produced will be used to further studies of diseases and other biological phenomena. PA-97-044 may be obtained electronically through the NIH grant line (301-402-2221), the NIH Web site, and by mail and email.
Requirements: Applications may be submitted by domestic and foreign, for-profit and nonprofit organizations, public and private, such as universities, colleges, hospitals, laboratories, units of state and local governments, and eligible agencies of the federal government.
Contact: Dr. Bettie Graham, Genetic Mapping and Sequence Variation, (301) 496-7531; fax: (301) 480-2770; email: bettie_graham@nih.gov
Internet: http://grants2.nih.gov/grants/guide/pa-files/PA-97-044.html
Sponsor: National Human Genome Research Institute
9000 Rockville Pike
Bethesda, MD 20892

NHLBI Airway Smooth Muscle Function and Targeted Therapeutics in Human Asthma Grants 3744

The purpose of this funding opportunity announcement, issued by the NHLBI, NIH, is to solicit Research Project Grant applications from institutions and organizations that propose to investigate the complex role of airway smooth muscle (ASM) functions in the development of human asthma and identify novel therapeutic targets. Budgets for direct costs of up to $350,000 per year (exclusive of indirect costs associated with consortia) and a project duration of up to four years may be requested for a maximum of $1,400,000 direct costs over a four-year project period. Letters of Intent should be received by December 8, and full proposals are due by January 6.
Requirements: The following organizations and institutions are eligible to apply: public/state controlled institutions of higher education; private institutions of higher education; Hispanic-serving institutions; historically Black colleges and universities (HBCUs); tribally controlled colleges and universities (TCCUs); Alaska Native and Native Hawaiian serving institutions; nonprofits with 501(c)3 IRS status; nonprofits without 501(c)3 IRS status; for-profit organizations; state governments; Indian/Native American tribal governments (federally recognized); Indian/Native American tribally designated organizations; county governments; city or township governments; special district governments; independent school districts; public housing authorities/Indian housing authorities; U.S. territory or possession; Indian/Native American tribal governments (other than federally recognized); regional organizations; non-domestic (non-U.S.) entities (foreign organizations); other eligible agencies of the federal government; and faith-based or community-based organizations.
Amount of Grant: $350,000 per year maximum
Date(s) Application is Due: Dec 8; Jan 6
Contact: Susan Banks-Schlegel, Ph.D.; (301) 435-0202; fax: (301) 480-3557; email: Schleges@nhlbi.nih.gov; Dorothy Gail, Ph.D.; (301) 435-0222; fax: (301) 480-3557; email: Gaild@nhlbi.nih.gov
Internet: http://grants1.nih.gov/grants/guide/rfa-files/RFA-HL-09-007.html
Sponsor: National Heart, Lung, and Blood Institute (NHLBI)
6701 Rockledge Drive
Bethesda, MD 20892-7952

NHLBI Ancillary Studies in Clinical Trials 3745

The purpose of this Funding Opportunity Announcement (FOA) is to solicit research grant applications to conduct time-sensitive ancillary studies related to heart, lung, and blood diseases and sleep disorders in conjunction with ongoing NIH- or non-NIH-supported clinical trials. An applicant may request a project period of up to four years and a budget for direct costs up to $250,000 (10 modules) per year. Dates for Letters of Intent are: December 30, April 29, and August 31. Respective due dates fot full applications are: January 30, May 29, and September 30.
Requirements: The following organizations and institutions are eligible to apply: individuals; public/state controlled institutions of higher education; private institutions of higher education; Hispanic-serving institutions; historically Black colleges and universities (HBCUs); tribally controlled colleges and universities (TCCUs); Alaska Native and Native Hawaiian serving institutions; nonprofits with 501(c)3 IRS status; nonprofits without 501(c)3 IRS status; for-profit organizations; state governments; Indian/Native American tribal governments (federally recognized); Indian/Native American tribally designated organizations; county governments; city or township governments; special district governments; independent school districts; public housing authorities/Indian housing authorities; U.S. territory or possession; Indian/Native American tribal governments (other than federally recognized); regional organizations; non-domestic (non-U.S.) entities (foreign organizations); other eligible agencies of the federal government; and faith-based or community-based organizations.
Amount of Grant: $250,000 per year for up to 4 years
Date(s) Application is Due: Jan 30; May 29; Sep 30
Contact: Gail Weinmann, M.D.; (301) 435-0233; fax: (301) 480-3547; email: weinmang@nhlbi.nih.gov; Peter Kaufmann, Ph.D.; (301) 435-0422; fax: (301) 480-1773; email: kaufmannp@nhlbi.nih.gov
Internet: http://grants.nih.gov/grants/guide/rfa-files/RFA-HL-09-001.html
Sponsor: National Heart, Lung, and Blood Institute
6701 Rockledge Drive, Room 10188
Bethesda, MD 20892-7952

NHLBI Bioengineering and Obesity Grants 3746

The purpose of this FOA is to solicit applications to develop and validate new and innovative engineering approaches to address clinical problems related to energy balance, intake, and expenditure. Novel sensors, devices, imaging, and other technologies, including technologies to detect biochemical markers of energy balance are expected to be developed and evaluated by collaborating engineers, physical scientists, mathematicians, and scientists from other relevant disciplines with expertise in obesity and nutrition. Basic and applied scientific applications with rigorous hypothesis-testing designs or for design-directed research are encouraged. The goal is to increase the number of useful technologies and tools available to scientists to facilitate their research in energy balance and health. Eventually these research tools should facilitate therapeutic advances and behavioral changes to address such problems as weight control and obesity.
Requirements: The following organizations and institutions are eligible to apply: public/state controlled institutions of higher education; private institutions of higher education; Hispanic-serving institutions; historically Black colleges and universities (HBCUs); tribally controlled colleges and universities (TCCUs); Alaska Native and Native Hawaiian serving institutions; nonprofits with 501(c)3 IRS status; nonprofits without 501(c)3 IRS status; for-profit organizations; state governments; Indian/Native American tribal governments (federally recognized); Indian/Native American tribally designated organizations; county governments; city or township governments; special district governments; independent school districts; public housing authorities/Indian housing authorities; U.S. territory or possession; Indian/Native American tribal governments (other than federally recognized); regional organizations; non-domestic (non-U.S.)

entities (foreign organizations); other eligible agencies of the federal government; and faith-based or community-based organizations.
Date(s) Application is Due: Feb 5; Jun 5; Oct 5
Contact: Abby G. Ershow; (301) 435-0550; fax: (301) 480-1338; email: ErshowA@mail.nih.gov; J. Timothy Baldwin; (301) 435-0513; fax: (301) 480-1335; email: BaldwinT@mail.nih.gov
Internet: http://grants.nih.gov/grants/guide/pa-files/PA-07-354.html
Sponsor: National Heart, Lung, and Blood Institute (NHLBI)
6701 Rockledge Drive, Suite, MSC 7956
Bethesda, MD 20892-7956

NHLBI Bioengineering Approaches to Energy Balance and Obesity Grants for SBIR 3747

This funding opportunity will develop and validate new and innovative bioengineering technology to address clinical problems related to energy balance, intake, and expenditure. Novel sensors, devices, imaging, and other approaches are expected to be developed and evaluated by collaborating engineers, physical scientists, and scientists from other relevant disciplines with expertise in obesity and nutrition. The goal is to increase the number of useful technologies and tools available to scientists to facilitate their research in energy balance and health. Eventually these research tools should facilitate therapeutic advances and behavioral changes to address such problems as weight control and obesity. Because the nature and scope of the proposed research will vary from application to application, it is anticipated that the size and duration of each award will also vary. The total amount awarded and the number of awards will depend upon the mechanism numbers, quality, duration, and costs of the applications received.
Requirements: Only United States small business concerns (SBCs) are eligible to submit SBIR applications.
Date(s) Application is Due: Apr 5; Aug 5; Dec 5
Contact: Abby G. Ershow; (301) 435-0550; fax: (301) 480-1338; email: ErshowA@mail.nih.gov; J. Timothy Baldwin; (301) 435-0513; fax: (301) 480-1335; email: BaldwinT@mail.nih.gov
Internet: http://grants.nih.gov/grants/guide/pa-files/PA-07-435.html
Sponsor: National Heart, Lung, and Blood Institute (NHLBI)
Two Rockledge Center, Suite 9152, MSC 7956, 6701 Rockledge Drive
Bethesda, MD 20892-7956

NHLBI Bioengineering Approaches to Energy Balance and Obesity Grants for STTR 3748

This funding opportunity will develop and validate new and innovative bioengineering technology to address clinical problems related to energy balance, intake, and expenditure. Novel sensors, devices, imaging, and other approaches are expected to be developed and evaluated by collaborating engineers, physical scientists, and scientists from other relevant disciplines with expertise in obesity and nutrition. The goal is to increase the number of useful technologies and tools available to scientists to facilitate their research in energy balance and health. Eventually these research tools should facilitate therapeutic advances and behavioral changes to address such problems as weight control and obesity. Because the nature and scope of the proposed research will vary from application to application, it is anticipated that the size and duration of each award will also vary. The total amount awarded and the number of awards will depend upon the mechanism numbers, quality, duration, and costs of the applications received.
Requirements: Only United States small business concerns (SBCs) are eligible to submit STTR applications.
Date(s) Application is Due: Apr 5; Aug 5; Dec 5
Contact: Abby G. Ershow; (301) 435-0550; fax: (301) 480-1338; email: ErshowA@mail.nih.gov; J. Timothy Baldwin; (301) 435-0513; fax: (301) 480-1335; email: BaldwinT@mail.nih.gov
Internet: http://grants.nih.gov/grants/guide/pa-files/PA-07-436.html
Sponsor: National Heart, Lung, and Blood Institute (NHLBI)
Two Rockledge Center, Suite 9152, MSC 7956, 6701 Rockledge Drive
Bethesda, MD 20892-7956

NHLBI Biomedical Research Training Program for Individuals from Underrepresented Groups 3749

The NHLBI established the Biomedical Research Training Program for Individuals from Underrepresented Groups (BRTPUG) to offer opportunities for underrepresented post-baccalaureate individuals, undergraduate, and graduate students to receive training in fundamental biomedical sciences and clinical research disciplines. The purpose of the program is to enhance career opportunities in biomedical and behavioral research, including clinical and laboratory medicine, epidemiology, and biostatistics as applied to the etiology and treatment of heart, blood vessel, lung, and blood diseases. BRTPUG offers each participant the opportunity to work closely with leading research scientists in the Division of Intramural Research and extramural scientists in the Division of Prevention and Population Sciences. The program provides students with hands-on training in a research environment, which will prepare them to continue their studies and advance their careers in clinical and basic research. The trainee's appointment is a one-time appointment of 6 (2 summers) to 24 months over a 2 year period beginning the summer of selection. Each trainee is assigned to a mentor who is responsible for designing a carefully planned training program. For placements beginning in June of the calendar year or summer opportunities, NHLBI strongly encourages applicants to submit your application by January 15. For Post-Baccalaureate 1-2 year research internships beginning in July-September of the calendar year, applications should be submitted by March 1.
Requirements: Underrepresented post-baccalaureate individuals, and undergraduate and graduate students planning to pursue advanced degrees in the biomedical and behavioral sciences and meeting all other program criteria are eligible. Applicants must have a cumulative grade point average (GPA) or science course GPA of 3.3 or better on a 4.0 scale, or 4.3 or better on a 5.0 scale, and be U.S. citizens or permanent residents.
Date(s) Application is Due: Jan 15; Mar 1
Contact: Helena O. Mishoe, PhD; (301) 451-5081; fax: (301) 480-0862; email: mishoeh@nhlbi.nih.gov
Internet: http://www.nhlbi.nih.gov/funding/training/redbook/undbrtum.htm
Sponsor: National Heart, Lung, and Blood Institute (NHLBI)
6701 Rockledge Drive, Room 8188
Bethesda, MD 20892-7913

NHLBI Blood Diseases and Resources Research Grants 3750

To foster research for the prevention, improved diagnosis, and treatment of non-malignant blood diseases such as Coolie's anemia, sickle cell disease, hemophilia, hemochromatosis, and disorders of hemostasis and thrombosis; to support research on hematopoietic stem cell biology and transplantation and on the use of transplantation to treat blood diseases; and to promote research in transfusion medicine for improving the availability, safety, and use of blood and blood products; and to develop new scientists for blood disease and resources research. Grants may support salaries, equipment, supplies, travel, and patient hospitalization as required to perform the research effort. National Research Service Awards (NRSAs) are made directly to individuals for research training in specified biomedical shortage areas. Grants may be made to institutions to enable them to make NRSAs to individuals selected by them.
Requirements: Any nonprofit organization engaged in biomedical research and institutions (or companies) organized for profit may apply for grants, with the exception of NRSAs. An individual may apply for a NRSA or, in some cases, may qualify for a research grant if adequate facilities in which to perform the research are available. SBIR grants can be awarded only to domestic small businesses (entities that are independently owned and operated for profit, are not dominant in the field in which research is proposed, and have no more than 500 employees).
Restrictions: Restrictions or limitations are imposed against the use of funds for entertainment, foreign travel, general-purpose equipment, alterations and renovations, and other items not regularly required for the performance of research.
Amount of Grant: $67,988-$2.9 million; $470,787 average
Date(s) Application is Due: Feb 1; Jun 1; Oct 1
Contact: Director, Division of Blood Diseases and Resources; (301) 435-0080; Suzanne White, Grants Management Officer, (301) 435-0144
Internet: http://www.federalgrantswire.com/blood-diseases-and-resources-research.html
Sponsor: National Heart, Lung, and Blood Institute
6701 Rockledge Drive
Bethesda, MD 20892

NHLBI Cardiac Development Consortium Grants 3751

The National Heart, Lung, and Blood Institute (NHLBI) and the Canadian Institutes of Health Research (CIHR) invite applications to participate in a new NHLBI Cardiac Development Consortium, a cooperative investigative group that will drive an integrated approach to the investigation of cardiovascular development. The purpose of this initiative is to support basic collaborative research leading to a comprehensive understanding of the regulatory networks controlling cardiovascular development. The consortium will consist of up to four Research Centers, scientific Cores as necessary, and a Steering Committee. Research Center applicants will propose projects that will select key regulatory pathways, identify the components of the pathways and targets, and rapidly disseminate data to the scientific community. The Research Centers will collaborate to maximize interoperability of the data. This Consortium is part of a new NHLBI translational program in pediatric cardiovascular disease, which includes the companion NHLBI Pediatric Cardiac Genomics Consortium and Administrative Coordinating Center. These consortia will interact with each other and the NHLBI Pediatric Heart Network to encourage translation of results from basic science to clinical research, and to provide clinical input on pressing needs for basic research. Maximum allowable direct costs for the research centers are $160,000 in the Year 1 planning phase, and up to $1,070,000 per year in years two through six. Letters of Intent should be received by January 6, and full proposals are due by February 6.
Requirements: The following organizations and institutions are eligible to apply: public/state controlled institutions of higher education; private institutions of higher education; Hispanic-serving institutions; historically Black colleges and universities (HBCUs); tribally controlled colleges and universities (TCCUs); Alaska Native and Native Hawaiian serving institutions; nonprofits with 501(c)3 IRS status; nonprofits without 501(c)3 IRS status; small businesses; for-profit organizations; state governments; and non-domestic (non-U.S.) entities in North America.
Amount of Grant: $160,000 maximum in the Year 1 planning phase; $1,070,000 maximum per year in years 2-6
Date(s) Application is Due: Jan 6; Feb 6
Contact: Charlene Schramm; (301) 435-0510; email: SchrammC@nhlbi.nih.gov
Internet: http://grants.nih.gov/grants/guide/rfa-files/RFA-HL-09-002.html
Sponsor: National Heart, Lung, and Blood Institute (NHLBI)
Rockledge 2, Room 8100, 6701 Rockledge Drive, MSC 7940
Bethesda, MD 20892

NHLBI Cardiac Translational Research Implementation Program Grants 3752

The National Heart, Lung, and Blood Institute (NHLBI) solicits Stage 1 exploratory planning (P20) applications to the Cardiac Translational Research Implementation Program (C-TRIP). The overall objective of the entire two-staged C-TRIP is to accelerate translation of promising new therapeutic interventions derived from fundamental research discoveries for the treatment and prevention of heart failure or arrhythmias through the planning and execution of well-designed clinical trials to demonstrate safety and efficacy. Stage 1 exploratory planning (P20) applications will support the planning and development of Stage 2 specialized center (P50) applications that support the conduct of the safety and efficacy trials recommended for support following a second phase of peer review. Subsequent Stage 2 (P50) applications for the conduct of the clinical trials planned and developed during Stage 1 will be solicited at a later time. The to-be-released solicitation for Stage 2 specialized center (P50) applications will be limited to successful Stage 1 applicants. Stage 2 (P50) applications will be anticipated 18 months following the initial award of Stage 1 applications. Support during Stage 2 of C-TRIP is expected to provide up to $1,515,000 direct costs per year for up to five years. During this period, applicants may request no more than $500,000 per year in direct costs for a maximum of $1,000,000 direct costs.

Requirements: The following organizations and institutions are eligible to apply: individuals; public/state controlled institutions of higher education; private institutions of higher education; Hispanic-serving institutions; historically Black colleges and universities (HBCUs); tribally controlled colleges and universities (TCCUs); Alaska Native and Native Hawaiian serving institutions; nonprofits with 501(c)3 IRS status; nonprofits without 501(c)3 IRS status; for-profit organizations; state governments; Indian/Native American tribal governments (federally recognized); Indian/Native American tribally designated organizations; county governments; city or township governments; special district governments; independent school districts; public housing authorities/Indian housing authorities; U.S. territory or possession; Indian/Native American tribal governments (other than federally recognized); regional organizations; non-domestic (non-U.S.) entities (foreign organizations); other eligible agencies of the federal government; and faith-based or community-based organizations.

Amount of Grant: $500,000

Date(s) Application is Due: Apr 28; May 28

Contact: David A. Lathrop, Ph.D.; (301) 435-0507; fax: (301) 451-5458; email: lathropd@nhlbi.nih.gov

Internet: http://grants.nih.gov/grants/guide/rfa-files/RFA-HL-10-001.html

Sponsor: National Heart, Lung, and Blood Institute
6701 Rockledge Drive, MSC 7956
Bethesda, MD 20892-7956

NHLBI Career Enhancement Awards for Stem Cell Research 3753

The Career Enhancement Award for Stem Cell Research enables investigators to acquire new research capabilities in the use of human or animal embryonic, adult, or cord blood stem cells. Candidates must have a clinical or research doctoral degree and be actively engaged in research of interest to the NHLBI. Candidates must meet the NIH career development requirements. Actual salary is based on candidate's full-time, 12-month institutional salary, and the level of effort requested. Most recently, for mid-career and senior investigators, the maximum allowable salary was $186,600 for full-time effort. For junior faculty, maximum allowable salary was $75,000. Salary adjustments are announced in the NIH Guide. Up to $50,000 per year is allowed for research expenses, tuition, technical support, personnel consultant fees for mentor, and travel.

Requirements: Investigators at U.S. institutions with current or previous independent research support are eligible.

Amount of Grant: $200,000 maximum

Date(s) Application is Due: Feb 12; Mar 12; Jun 12; Jul 12; Oct 12; Nov 12

Contact: Traci Heath Mondoro, Ph.D.; (301) 435-0052; fax: (301) 480-1060; email: mondorot@nhlbi.nih.gov

Internet: http://www.nhlbi.nih.gov/funding/training/redbook/k18est-junior.htm

Sponsor: National Heart, Lung, and Blood Institute (NHLBI)
Two Rockledge Center, Room 9174, 6701 Rockledge Drive, MSC 7950
Bethesda, MD 20892-7950

NHLBI Career Transition Award 3754

The goal of the program is to provide highly qualified new investigators with an opportunity to receive research and research training experience in the NHLBI Division of Intramural Research and then to provide them with bridge funding to facilitate the transition of their research programs to extramural institutions. Total direct costs for the extramural phase cannot exceed $150,000 including fringe benefits per year in direct costs or $300,000 for the two-year period.

Requirements: The candidate must have a research or a health-professional doctorate, or its equivalent, and must have completed at least three and no more than five years of postdoctoral training, during which the potential for highly productive basic or clinical research was demonstrated.

Restrictions: Individuals on temporary or student visas are not eligible.

Amount of Grant: $150,000 total for intramural phase; $300,000 total direct costs for extramural phase

Date(s) Application is Due: Feb 5; Jun 5; Oct 5

Contact: Herbert M. Geller; (301) 451-9440; fax: (301) 594-8133; email: direducation@nhlbi.nih.gov

Internet: http://grants.nih.gov/grants/guide/pa-files/PAR-07-317.html#SectionVII

Sponsor: National Heart, Lung, and Blood Institute
10 Center Drive, Room 2N242, MSC 1754
Bethesda, MD 20892-1754

NHLBI Characterizing the Blood Stem Cell Niche Grants 3755

This FOA issued by the National Heart, Lung, and Blood Institute, National Institutes of Health, solicits Research Project Grant applications in a specific area of stem cell research, the blood stem cell niche, an area critical to advancing stem cell biology and its applications to cellular therapeutics including hematopoietic stem cell transplantation. This FOA is being initiated to foster collaborative research projects and insightful approaches to dissect the cellular components and factors involved in the hematopoietic stem cell niche. Development of conditional genetic knock-out models to test the role of factors in specific cell lineages and imaging technology to facilitate following stem cell engraftment in the niche in vivo are integral to this initiative. An individual research grant application may not request more than $250,000 direct cost a year unless collaborative in vivo cell imaging studies are included, in which case up to $350,000 direct cost may be requested in a year. An applicant may request a project period of up to 4 years. Letters of Intent should be received by December 8, and full proposals are due by January 6.

Requirements: The following organizations and institutions are eligible to apply: public/state controlled institutions of higher education; private institutions of higher education; Hispanic-serving institutions; historically Black colleges and universities (HBCUs); tribally controlled colleges and universities (TCCUs); Alaska Native and Native Hawaiian serving institutions; nonprofits with 501(c)3 IRS status; nonprofits without 501(c)3 IRS status; for-profit organizations; state governments; Indian/Native American tribal governments (federally recognized); Indian/Native American tribally designated organizations; county governments; city or township governments; special district governments; independent school districts; public housing authorities/Indian housing authorities; U.S. territory or possession; Indian/Native American tribal governments (other than federally recognized); regional organizations; non-domestic (non-U.S.) entities (foreign organizations); other eligible agencies of the federal government; and faith-based or community-based organizations.

Amount of Grant: $250,000 maximum for single entity; $350,000 maximum for collaboratives

Date(s) Application is Due: Dec 8; Jan 6

Contact: John Thomas, Ph.D.; (301) 435-0065; fax: (301) 451-5453; email: ThomasJ@NHLBI.NIH.Gov

Internet: http://grants.nih.gov/grants/guide/rfa-files/RFA-HL-09-010.html

Sponsor: National Heart, Lung, and Blood Institute (NHLBI)
Rockledge II, Suite 9030, 6701 Rockledge Drive
Bethesda, MD 20892-7950

NHLBI Childhood Obesity Prevention and Treatment Research Consortium Grants 3756

The goal of this Funding Opportunity Announcement (FOA) is to fund randomized trials testing intervention approaches to prevent excess weight gain in non-overweight youth and in those already overweight, and/or to reduce weight in obese and severely obese youth. The whole program is designed to promote collaboration, coordination and exchange of scientific information among institutions, and partnerships with local, state and national agencies. The ultimate goal is to improve children's health by preventing obesity-related morbidity and mortality from cardiovascular and other chronic diseases. Letters of Intent are due by September 8, with full proposals received by October 6.

Requirements: The following organizations and institutions are eligible to apply: individuals; public/state controlled institutions of higher education; private institutions of higher education; Hispanic-serving institutions; historically Black colleges and universities (HBCUs); tribally controlled colleges and universities (TCCUs); Alaska Native and Native Hawaiian serving institutions; nonprofits with 501(c)3 IRS status; nonprofits without 501(c)3 IRS status; for-profit organizations; state governments; Indian/Native American tribal governments (federally recognized); Indian/Native American tribally designated organizations; county governments; city or township governments; special district governments; independent school districts; public housing authorities/Indian housing authorities; U.S. territory or possession; Indian/Native American tribal governments (other than federally recognized); regional organizations; non-domestic (non-U.S.) entities (foreign organizations); other eligible agencies of the federal government; and faith-based or community-based organizations.

Amount of Grant: $700,000 in years 1 and 2; $1,500,000 per year in years 3-6; $700,000 in year 7

Date(s) Application is Due: Sep 8; Oct 6

Contact: Charlotte A. Pratt, Program Director; (301) 435-0382; fax: (301) 480-5158; email: prattc@nhlbi.nih.gov; Terry T-K Huang, Director, Obesity Research Strategic Core; (301) 594-1846; fax: (301) 480-9791; email: huangter@mail.nih.gov

Internet: http://grants.nih.gov/grants/guide/rfa-files/RFA-HL-10-004.html

Sponsor: National Heart, Lung, and Blood Institute (NHLBI)
6701 Rockledge Drive, MSC 7936, Room 10018
Bethesda, MD 20892

NHLBI Childhood Obesity Prevention and Treatment Research Coordinating Unit Grants 3757

This funding opportunity announcement (FOA) invites applications from eligible institutions to participate as the Research Coordinating Unit (RCU) for the childhood obesity prevention and treatment research consortium, which runs in parallel to this FOA. Applicants for the clinical trial will conduct randomized controlled trials that are innovative and that would test innovative interventions to prevent overweight and obesity in youth and to treat existing overweight, obesity and severe obesity in youth. The participating organizations intend to award up to $5.25 million over seven years under this FOA. It is anticipated that one Research Coordinating Unit (RCU) award will be made. Budgets for direct costs of up to $500,000 per year and a project duration

of up to seven years may be requested for a maximum of $3.5 million direct costs over a seven year project period. Letters of Intent are due by September 8, with full proposals received by October 6.
Requirements: The following organizations and institutions are eligible to apply: individuals; public/state controlled institutions of higher education; private institutions of higher education; Hispanic-serving institutions; historically Black colleges and universities (HBCUs); tribally controlled colleges and universities (TCCUs); Alaska Native and Native Hawaiian serving institutions; nonprofits with 501(c)3 IRS status; nonprofits without 501(c)3 IRS status; for-profit organizations; state governments; Indian/Native American tribal governments (federally recognized); Indian/Native American tribally designated organizations; county governments; city or township governments; special district governments; independent school districts; public housing authorities/Indian housing authorities; U.S. territory or possession; Indian/Native American tribal governments (other than federally recognized); regional organizations; non-domestic (non-U.S.) entities (foreign organizations); other eligible agencies of the federal government; and faith-based or community-based organizations.
Amount of Grant: $500,000 per year for 7 years
Date(s) Application is Due: Sep 8; Oct 6
Contact: Charlotte A. Pratt, Program Director; (301) 435-0332; fax: (301) 480-5158; email: prattc@nhlbi.nih.gov
Internet: http://grants.nih.gov/grants/guide/rfa-files/RFA-HL-10-005.html
Sponsor: National Heart, Lung, and Blood Institute (NHLBI)
6701 Rockledge Drive, MSC 7936, Room 10018
Bethesda, MD 20892

NHLBI Circadian-Coupled Cellular Function in Heart, Lung, and Blood Tissue Grants 3758
The purpose of this FOA is to stimulate Phase I translation (T1) of fundamental genome-based circadian biology to improve our understanding of heart, lung, and blood disease pathogenesis in non-neural tissues. For the purpose of this initiative, T1 research starts after gene discovery and has as its goal the development of molecular models with application to an improved understanding of human health and disease pathogenesis, predictive testing for disease susceptibility, prognostic testing, or the selection of therapeutic strategies. Budget for direct costs up to $325,000 per year for project duration of up to four years may be requested. Letters of Intent should be received by December 8, and full proposals are due by January 6.
Requirements: The following organizations and institutions are eligible to apply: public/state controlled institutions of higher education; private institutions of higher education; Hispanic-serving institutions; historically Black colleges and universities (HBCUs); tribally controlled colleges and universities (TCCUs); Alaska Native and Native Hawaiian serving institutions; nonprofits with 501(c)3 IRS status; nonprofits without 501(c)3 IRS status; for-profit organizations; state governments; Indian/Native American tribal governments (federally recognized); Indian/Native American tribally designated organizations; county governments; city or township governments; special district governments; independent school districts; public housing authorities/Indian housing authorities; U.S. territory or possession; Indian/Native American tribal governments (other than federally recognized); regional organizations; non-domestic (non-U.S.) entities (foreign organizations); other eligible agencies of the federal government; and faith-based or community-based organizations.
Amount of Grant: $325,000 per year maximum
Date(s) Application is Due: Dec 8; Jan 6
Contact: Aaron Laposky, Ph.D.; (301) 435-0199; fax: (301) 480-3451; email: laposkya@nhlbi.nih.gov
Internet: http://grants.nih.gov/grants/guide/rfa-files/RFA-HL-09-012.html
Sponsor: National Heart, Lung, and Blood Institute (NHLBI)
6701 Rockledge Dr. MSC 7952
Bethesda, MD 20892-7952

NHLBI Clinical Centers for the NHLBI Asthma Network Grants 3759
The purpose of this funding opportunity, issued by NHLBI, NIH, is to invite applications to participate in the NHLBI Asthma Network (AsthmaNet), a clinical research network that will develop and conduct multiple clinical trials to address the most important asthma management questions and new treatment approaches in pediatric and adult populations. AsthmaNet is designed to promote cooperation and coordination, facilitate scientific exchange, provides training opportunities, and leverage resources. AsthmaNet will include multiple Clinical Centers and one Data Coordinating Center. The protocols will include clinical trials to evaluate and/or compare existing or new therapeutic approaches to asthma management as well as a limited number of proof-of-concept studies to advance the development of novel therapies and studies that investigate the mechanistic bases for interventions examined in AsthmaNet. Letters of Intent should be received by December 30, and full proposals are due by January 30.
Requirements: The following organizations and institutions are eligible to apply: public/state controlled institutions of higher education; private institutions of higher education; Hispanic-serving institutions; historically Black colleges and universities (HBCUs); tribally controlled colleges and universities (TCCUs); Alaska Native and Native Hawaiian serving institutions; nonprofits with 501(c)3 IRS status; nonprofits without 501(c)3 IRS status; for-profit organizations; state governments; Indian/Native American tribal governments (federally recognized); Indian/Native American tribally designated organizations; county governments; city or township governments; special district governments; independent school districts; public housing authorities/Indian housing authorities; U.S. territory or possession; Indian/Native American tribal governments (other than federally recognized); regional organizations; non-domestic (non-U.S.) entities (foreign organizations); other eligible agencies of the federal government; and faith-based or community-based organizations.
Date(s) Application is Due: Dec 30; Jan 30
Contact: Virginia S. Taggart; (301) 435-0202; email: taggartv@nhlbi.nih.gov; Robert Smith, Ph.D.; (301) 435-0202; email: smithra3@nhlbi.nih.gov
Internet: http://grants.nih.gov/grants/guide/rfa-files/RFA-HL-08-010.html
Sponsor: National Heart, Lung, and Blood Institute (NHLBI)
6701 Rockledge Drive
Bethesda, MD 20892-7952

NHLBI Developmental Origins of Altered Lung Physiology and Immune Function Grants 3760
This FOA issued by the NHLBI, National Institutes of Health, solicits Research Project Grant applications from institutions and organizations that propose to perform research that will enhance the understanding of how the pre- and postnatal environments affect the interplay of the lung and immune system during development resulting in sustained changes in lung physiology and immune function that compromise respiratory health and outcomes. Budgets for direct costs of up to $350,000 per year (exclusive of indirect costs associated with consortia) and a project duration of up to five years may be requested for a maximum of $1,750,000 direct costs over a five-year project period. Letters of Intent should be received by September 19, and full proposals are due by October 21.
Requirements: The following organizations and institutions are eligible to apply: public/state controlled institutions of higher education; private institutions of higher education; Hispanic-serving institutions; historically Black colleges and universities (HBCUs); tribally controlled colleges and universities (TCCUs); Alaska Native and Native Hawaiian serving institutions; nonprofits with 501(c)3 IRS status; nonprofits without 501(c)3 IRS status; for-profit organizations; state governments; Indian/Native American tribal governments (federally recognized); Indian/Native American tribally designated organizations; county governments; city or township governments; special district governments; independent school districts; public housing authorities/Indian housing authorities; U.S. territory or possession; Indian/Native American tribal governments (other than federally recognized); regional organizations; non-domestic (non-U.S.) entities (foreign organizations); other eligible agencies of the federal government; and faith-based or community-based organizations.
Amount of Grant: $350,000 per year maximum
Date(s) Application is Due: Sep 19; Oct 21
Contact: Patricia Noel, Ph.D.; (301) 435-0202; email: noelp@nhlbi.nih.gov
Internet: http://grants.nih.gov/grants/guide/rfa-files/RFA-HL-08-009.html
Sponsor: National Heart, Lung, and Blood Institute (NHLBI)
6701 Rockledge Drive
Bethesda, MD 20892-7952

NHLBI Elucidating Nature, Solutions to Heart, Lung, and Blood Diseases and Sleep Disorder Processes Research Grants 3761
This Funding Opportunity Announcement (FOA) issued by the National Heart, Lung, and Blood Institute (NHLBI), National Institutes of Health (NIH), solicits grant applications proposing studies that elucidate the natural molecular and cellular adaptations of mammalian species to extreme environmental conditions that would rapidly evoke life-threatening cardiovascular or respiratory responses in other species, including humans. Extreme environmental conditions may include food scarcity, limited access to oxygen, or temperature extremes. The overall objective of the funding opportunity is to identify new targets for therapeutic interventions to treat and prevent heart, blood, and lung diseases and sleep disorders. There are no set aside specific dollars for this funding opportunity, and the amount of each award will be limited to $250,000 in direct costs for up to four years. Check the website for various deadlines.
Requirements: The following organizations and institutions are eligible to apply: individuals; public/state controlled institutions of higher education; private institutions of higher education; Hispanic-serving institutions; historically Black colleges and universities (HBCUs); tribally controlled colleges and universities (TCCUs); Alaska Native and Native Hawaiian serving institutions; nonprofits with 501(c)3 IRS status; nonprofits without 501(c)3 IRS status; for-profit organizations; state governments; Indian/Native American tribal governments (federally recognized); Indian/Native American tribally designated organizations; county governments; city or township governments; special district governments; independent school districts; public housing authorities/Indian housing authorities; U.S. territory or possession; Indian/Native American tribal governments (other than federally recognized); regional organizations; non-domestic (non-U.S.) entities (foreign organizations); other eligible agencies of the federal government; and faith-based or community-based organizations.
Restrictions: This FOA will not support the examination of animal models that simply mimic the progression of human cardiovascular and respiratory diseases.
Amount of Grant: $250,000 in direct costs for up to four years
Date(s) Application is Due: Jan 16; May 15
Contact: David Lathrop; (301) 435-0504; fax: (301) 451-5458; email: LathropD@nhlbi.nih.gov; David Reiter, Grants Operations Branch; (301) 480-3310; email: reiterd@nhlbi.nih.gov
Internet: http://grants.nih.gov/grants/guide/pa-files/PAR-07-102.html
Sponsor: National Heart, Lung, and Blood Institute
6701 Rockledge Drive, MSC 7940
Bethesda, MD 20892-7940

NHLBI Exploratory Studies in the Neurobiology of Pain in Sickle Cell Disease 3762
The goal of this initiative is to foster novel basic and translational research of the neurobiology of pain in sickle cell disease. Priority will be given to the application of

investigational techniques that have been utilized in other pain syndromes in both human and nonhuman studies. A multidisciplinary collaboration, between neurobiologists, clinical specialists in pain, and hematologists will be necessary to maximize the research effort.
Requirements: The following organizations and institutions are eligible to apply: public/state controlled institutions of higher education; private institutions of higher education; Hispanic-serving institutions; historically Black colleges and universities (HBCUs); tribally controlled colleges and universities (TCCUs); Alaska Native and Native Hawaiian serving institutions; nonprofits with 501(c)3 IRS status; nonprofits without 501(c)3 IRS status; for-profit organizations; state governments; Indian/Native American tribal governments (federally recognized); Indian/Native American tribally designated organizations; county governments; city or township governments; special district governments; independent school districts; public housing authorities/Indian housing authorities; U.S. territory or possession; Indian/Native American tribal governments (other than federally recognized); regional organizations; non-domestic (non-U.S.) entities (foreign organizations); other eligible agencies of the federal government; and faith-based or community-based organizations.
Amount of Grant: $300,000 per year maximum
Date(s) Application is Due: Jan 5; Feb 3
Contact: Harvey Luksenburg, M.D.; (301) 435-0050; fax: (301) 480-0868; email: luksenburgh@mail.nih.gov
Internet: http://grants.nih.gov/grants/guide/rfa-files/RFA-HL-09-008.html
Sponsor: National Heart, Lung, and Blood Institute (NHLBI)
6701 Rockledge Drive, MSC 7950
Bethesda, MD 20892-7950

NHLBI Hyperglycemic Exacerbation of Ischemia/Reperfusion Injury after Ischemic Insult Grants **3763**

This funding opportunity announcement (FOA) will stimulate a focus on the causal relationship between hyperglycemia and impaired cardioprotection in a cohesive, integrated manner. The concept of this initiative evolved from a survey of the grant portfolio in ischemic heart disease and from the literature and consultation with investigators in the scientific community. Appropriate topics for investigation under this FOA include but are not limited to the following: to elucidate mechanisms of acute hyperglycemic exacerbation of myocardial ischemic injury following an ischemic insult to identify new therapeutic targets; to examine the relationship between severity of hyperglycemia and increases in myocardial infarct size mediated by hyperglycemia-induced impairment of cardioprotective signal transduction; to determine the role of inflammatory cells and oxidative stress in the hyperglycemic exacerbation of myocardial injury in infarction; to examine the interaction of metabolic and sex specific hormonal effects on endothelial and coronary microvascular function with acute hyperglycemia; to elucidate the specific mechanisms by which hyperglycemia affects glucose and lipid metabolism in the myocardium; and to study the pathogenesis of cardiovascular disease during acute hyperglycemia in suitable animal models.
Requirements: The following organizations and institutions are eligible to apply: individuals; public/state controlled institutions of higher education; private institutions of higher education; Hispanic-serving institutions; historically Black colleges and universities (HBCUs); tribally controlled colleges and universities (TCCUs); Alaska Native and Native Hawaiian serving institutions; nonprofits with 501(c)3 IRS status; nonprofits without 501(c)3 IRS status; for-profit organizations; state governments; Indian/Native American tribal governments (federally recognized); Indian/Native American tribally designated organizations; county governments; city or township governments; special district governments; independent school districts; public housing authorities/Indian housing authorities; U.S. territory or possession; Indian/Native American tribal governments (other than federally recognized); regional organizations; non-domestic (non-U.S.) entities (foreign organizations); other eligible agencies of the federal government; and faith-based or community-based organizations.
Date(s) Application is Due: Feb 5; Jun 5; Oct 5
Contact: Isabella Liang, Ph.D.; (301) 435-0504; fax: (301) 451-5458; email: Liangi@mail.nih.gov
Internet: http://grants.nih.gov/grants/guide/pa-files/PA-07-323.html
Sponsor: National Heart, Lung, and Blood Institute (NHLBI)
6701 Rockledge Drive, MSC 7940
Bethesda, MD 20892-7940

NHLBI Hypersensitivity Pneumonitis - Grants for Early Identification and Mechanisms of Disease **3764**

The objectives of this funding opportunity announcement (FOA) are to investigate basic mechanisms and early clinical identification of hypersensitivity pneumonitis (HP) interstitial lung disease, in order to elucidate how chronic disease progresses and how it can evolve into either interstitial fibrosis or airway obstructive disease.
Requirements: The following organizations and institutions are eligible to apply: individuals; public/state controlled institutions of higher education; private institutions of higher education; Hispanic-serving institutions; historically Black colleges and universities (HBCUs); tribally controlled colleges and universities (TCCUs); Alaska Native and Native Hawaiian serving institutions; nonprofits with 501(c)3 IRS status; nonprofits without 501(c)3 IRS status; for-profit organizations; state governments; Indian/Native American tribal governments (federally recognized); Indian/Native American tribally designated organizations; county governments; city or township governments; special district governments; independent school districts; public housing authorities/Indian housing authorities; U.S. territory or possession; Indian/Native American tribal governments (other than federally recognized); regional organizations; non-domestic (non-U.S.) entities (foreign organizations); other eligible agencies of the federal government; and faith-based or community-based organizations.
Date(s) Application is Due: Feb 5; Jun 5; Oct 5
Contact: Herbert Y. Reynolds, M.D.; (301) 435-0222; fax: (301) 480-3557; email: Reynoldh@nhlbi.nih.gov
Internet: http://grants.nih.gov/grants/guide/pa-files/PA-07-300.html
Sponsor: National Heart, Lung, and Blood Institute (NHLBI)
6701 Rockledge Drive, 2 Rockledge Center, Suite 10042
Bethesda, MD 20892-7952

NHLBI Immunomodulatory, Inflammatory, and Vasoregulatory Properties of Transfused Red Blood Cell Units as a Function of Preparation and Storage Grants **3765**

The goal of this Funding Opportunity Announcement (FOA) in blood banking and transfusion medicine is to solicit novel basic and translational research projects which will identify the molecular and cellular changes that occur during red blood cell unit preparation and storage; and evaluate the immunomodulatory, inflammatory, and vasoactive effects of storage lesion elements from red blood cell units on the blood vessel wall, host cells, and tissue oxygenation. Only research applications involving human red blood cell units (and their components) suitable for transfusion in the U.S. and that are collected and stored according to methods approved by the U.S. FDA will be considered responsive to this FOA. Applications are encouraged that propose collaborations between investigators from different disciplines such as hematology, blood banking, transfusion medicine, immunology, cell biology, rheology, nanotechnologies, molecular biology, and other relevant disciplines. This program will promote interactions and resource sharing among the award recipients to maximize their research efforts. Letters of Intent should be received by April 21 (1st cycle) or December 19 (2nd cycle), and full proposals are due by either May 21 (1st cycle) or January 21 (2nd cycle).
Requirements: The following organizations and institutions are eligible to apply: public/state controlled institutions of higher education; private institutions of higher education; Hispanic-serving institutions; historically Black colleges and universities (HBCUs); tribally controlled colleges and universities (TCCUs); Alaska Native and Native Hawaiian serving institutions; nonprofits with 501(c)3 IRS status; nonprofits without 501(c)3 IRS status; small businesses; for-profit organizations; state governments; Indian/Native American tribal governments; Indian/Native American tribally designated organizations; and U.S. territories or possessions.
Date(s) Application is Due: Apr 21; May 21; Dec 19; Jan 21
Contact: Simone Glynn, M.D.; (301) 435-0065; email: glynnsa@nhlbi.nih.gov
Internet: http://grants.nih.gov/grants/guide/rfa-files/RFA-HL-08-005.html
Sponsor: National Heart, Lung, and Blood Institute (NHLBI)
Rockledge II, Room 9142, 6701 Rockledge Drive
Bethesda, MD 20892-7950

NHLBI Improving Heart Failure Disease Management Grants **3766**

The purpose of this initiative is to stimulate innovative research to address inadequately answered questions about disease management in clinical practice for chronic heart failure (HF). The broader goal is to identify and improve, better implement, and disseminate clinically effective disease management tools into clinical practice to reduce HF morbidity and mortality and improve patient outcomes. Because the nature and scope of the proposed research will vary from application to application, it is anticipated that the size and duration of each award will also vary. The total amount awarded and the number of awards will depend upon the mechanism numbers, quality, duration, and costs of the applications received.
Requirements: The following organizations and institutions are eligible to apply: individuals; public/state controlled institutions of higher education; private institutions of higher education; Hispanic-serving institutions; historically Black colleges and universities (HBCUs); tribally controlled colleges and universities (TCCUs); Alaska Native and Native Hawaiian serving institutions; nonprofits with 501(c)3 IRS status; nonprofits without 501(c)3 IRS status; for-profit organizations; state governments; Indian/Native American tribal governments (federally recognized); Indian/Native American tribally designated organizations; county governments; city or township governments; special district governments; independent school districts; public housing authorities/Indian housing authorities; U.S. territory or possession; Indian/Native American tribal governments (other than federally recognized); regional organizations; non-domestic (non-U.S.) entities (foreign organizations); other eligible agencies of the federal government; and faith-based or community-based organizations.
Date(s) Application is Due: Feb 5; Jun 5; Oct 5
Contact: Lawton S. Cooper, M.D., M.P.H.; (301) 435-0419; fax: (301) 480-1773; email: CooperLS@nhlbi.nih.gov; Patrice Desvigne-Nickens, M.D.; (301) 435-0504; fax: (301) 480-1336; email: DesvignP@mail.nih.gov
Internet: http://grants.nih.gov/grants/guide/pa-files/PA-07-355.html
Sponsor: National Heart, Lung, and Blood Institute
6701 Rockledge Drive, Room 10108, MSC 7936
Bethesda, MD 20892-7936

NHLBI Independent Scientist Award **3767**

The award is a special salary-only grant designed to provide protected time for newly independent scientists who currently have non-research obligations such as heavy teaching loads, clinical work, committee assignments, service, and administrative duties that prevent them from having a period of intensive research focus. The award is targeted to persons with doctoral degrees who have completed their research training, have independent peer-reviewed research support, and who need a period of protected research time in order to foster their research career development. New applications are due February 12, June 12, and October 12, while renewal applications are due March

12, July 12, and November 12. Up to $75,000 per year plus fringe benefits is available. No other research development support funds are provided.
Requirements: The following organizations and institutions are eligible to apply: public/state controlled institutions of higher education; private institutions of higher education; nonprofits with 501(c)3 IRS status; nonprofits without 501(c)3 IRS status; small businesses; for-profit organizations; State governments; U.S. territories or possessions; Indian/Native American tribal governments (Federally recognized and other than Federally recognized); Indian/Native American tribally designated organizations; Hispanic-serving institutions; historically Black colleges and universities (HBCUs); tribally controlled colleges and universities (TCCUs); Alaska Native and Native Hawaiian serving institutions; regional organizations; and faith-based or community based organizations.
Restrictions: The award is not intended for investigators who already have full time to perform research, or have substantial publication records or considerable research support indicating that they are well established in their fields. Foreign institutions are not eligible to apply.
Amount of Grant: $75,000 per year maximum, plus fringe benefits
Date(s) Application is Due: Feb 12; Mar 12; Jun 12; Jul 12; Oct 12; Nov 12
Contact: Traci Heath Mondoro, Ph.D.; (301) 435-0052; fax: (301) 480-1060; email mondorot@nhlbi.nih.gov
Internet: http://grants.nih.gov/grants/guide/pa-files/PA-09-038.html
Sponsor: National Heart, Lung, and Blood Institute (NHLBI)
6701 Rockledge Drive, MSC 7950
Bethesda, MD 20892-7950

NHLBI Individual Postdoctoral National Research Service Awards 3768
This program offers health scientists the opportunity to receive full-time research training for up to 3 years in areas that reflect the national need for biomedical, clinical and behavioral research in cardiovascular, pulmonary, hematologic, and sleep disorders.
Requirements: When the fellowship begins, the applicant must have received a doctoral degree and have arranged to work with a sponsor affiliated with an institution that has the staff and facilities needed for the proposed training. Applicants must be U.S. citizens, noncitizen nationals, or legal permanent residents of the U.S. Training can be conducted abroad if the site provides opportunities that are not available in this country.
Restrictions: These grants are not intended for study leading to the M.D., D.O., D.D.S., or equivalent professional degrees, nor do they support residency training. One month of payback for each month of training, up to a maximum of 12 months, is required. This requirement can be fulfilled by teaching or research (a minimum of 20 hours per week) on a continuous basis beginning within 2 years after support ends.
Date(s) Application is Due: Apr 8; Aug; Dec 8
Contact: Sandra Colombini Hatch, M.D.; (301) 435-0222; fax: (301) 480-3557; email: Hatchs@nhlbi.nih.gov; Hannah H. Peavy, M.D.; (301) 435-0222; fax: (301) 480-3557; email: peavyh@nhlbi.nih.gov
Internet: http://www.nhlbi.nih.gov/funding/training/redbook/phdf32.htm
Sponsor: National Heart, Lung, and Blood Institute
Two Rockledge Center, Room 10162, 6701 Rockledge Drive, MSC 7952
Bethesda, MD 20892-7952

NHLBI Individual Predoctoral Fellowships to Promote Diversity in Health-Related Research 3769
This program encourages students from underrepresented racial and ethnic groups, individuals with disabilities, and individuals from disadvantaged backgrounds to seek graduate degrees in the biomedical and behavioral sciences to help increase the number of well-trained scientists from underrepresented groups. The fellowship provides up to 5 years of support for research training leading to the Ph.D. or equivalent research degree, the combined M.D./Ph.D. degree, or other combined degrees in the biomedical or behavioral sciences.
Requirements: At the time of appointment, students must be U.S. citizens, noncitizen nationals, or lawfully admitted to the U.S. for permanent residence. Individuals on temporary or student visas are not eligible. Individuals with disabilities, or from racial and ethnic groups, or individuals from disadvantaged backgrounds pursuing advanced degrees in the biomedical and behavioral sciences are eligible.
Date(s) Application is Due: Apr 13; Aug 13; Dec 13
Contact: Ann Rothgeb; (301) 435-0202; fax: (301) 480-3557; email: rothgeba@mail.nih.gov
Internet: http://www.nhlbi.nih.gov/funding/training/redbook/gradf31.htm
Sponsor: National Heart, Lung, and Blood Institute
Two Rockledge Center, Room 10158, 6701 Rockledge Drive, MSC 7952
Bethesda, MD 20892-7952

NHLBI Integrating Lung Genetics and Genomics in Human Populations Grants 3770
The purpose of this Funding Opportunity Announcement (FOA) is to stimulate innovative approaches to merge genetic and genomic techniques to find genes associated with human complex traits for lung disease. Because the nature and scope of the proposed research will vary from application to application, it is anticipated that the size and duration of each award will also vary. The total amount awarded and the number of awards will depend upon the mechanism numbers, quality, duration, and costs of the applications received.
Requirements: The following organizations and institutions are eligible to apply: individuals; public/state controlled institutions of higher education; private institutions of higher education; Hispanic-serving institutions; historically Black colleges and universities (HBCUs); tribally controlled colleges and universities (TCCUs); Alaska Native and Native Hawaiian serving institutions; nonprofits with 501(c)3 IRS status; nonprofits without 501(c)3 IRS status; for-profit organizations; state governments; Indian/Native American tribal governments (federally recognized); Indian/Native American tribally designated organizations; county governments; city or township governments; special district governments; independent school districts; public housing authorities/Indian housing authorities; U.S. territory or possession; Indian/Native American tribal governments (other than federally recognized); regional organizations; non-domestic (non-U.S.) entities (foreign organizations); other eligible agencies of the federal government; and faith-based or community-based organizations.
Date(s) Application is Due: Feb 5; Jun 5; Oct 5
Contact: Susan Banks-Schlegel, Ph.D.; (301) 435-0202; fax: (301) 480-3557; email: Schleges@nhlbi.nih.gov; Dorothy Gail, Ph.D.; (301) 435-0222; fax: (301) 480-3557; email: Gaild@nhlbi.nih.gov
Internet: http://grants.nih.gov/grants/guide/pa-files/PA-07-135.html
Sponsor: National Heart, Lung, and Blood Institute
Two Rockledge Center, Suite 10018, 6701 Rockledge Drive
Bethesda, MD 20892-7952

NHLBI Intramural Research Training Awards 3771
The Office of Education of the Division of Intramural Research coordinates and assists the training program and provides administrative support and resources for all research training activities within the Division. Programs are available within the Laboratories and Branches within the Division to provide specialized research training for high school students, college students, graduate and medical students, postdoctoral fellows and medical residents and fellows. The Office of Education provides assistance to applicants at each of these levels. NHLBI is committed to improving the representation of underrepresented groups, including minorities, women, and scientists with disabilities in the mainstream of basic and clinical research.
Contact: Herbert M. Geller; (301) 451-9440; fax: (301) 594-8133; email: direducation@nhlbi.nih.gov
Internet: http://www.training.nih.gov/student/pre-irta/previewpostbac.asp
Sponsor: National Heart, Lung, and Blood Institute (NHLBI)
Building 10, Room 2N242, 10 Center Drive MSC 1754
Bethesda, MD 20892-1754

NHLBI Investigator Initiated Multi-Site Clinical Trials 3772
The purpose of this Funding Opportunity Announcement is to provide a vehicle for submitting grant applications for investigator-initiated multi-site Phase II or Phase III randomized, controlled clinical trials. The trials may address any research question related to the mission and goals of NHLBI and may test clinical or behavioral interventions. The trials may randomize at the individual (patient) level or at a group or cluster level (e.g., randomization of clinics, schools, worksites, etc.).
Requirements: The following organizations and institutions are eligible to apply: public/state controlled institutions of higher education; private institutions of higher education; Hispanic-serving institutions; historically Black colleges and universities (HBCUs); tribally controlled colleges and universities (TCCUs); Alaska Native and Native Hawaiian serving institutions; nonprofits with 501(c)3 IRS status; nonprofits without 501(c)3 IRS status; for-profit organizations; state governments; Indian/Native American tribal governments (federally recognized); Indian/Native American tribally designated organizations; county governments; city or township governments; special district governments; independent school districts; public housing authorities/Indian housing authorities; U.S. territory or possession; Indian/Native American tribal governments (other than federally recognized); regional organizations; non-domestic (non-U.S.) entities (foreign organizations); other eligible agencies of the federal government; and faith-based or community-based organizations.
Restrictions: This Funding Opportunity Announcement is not intended for support of single-center studies or multi-center observational studies that are not testing an intervention. Foreign institutions are not eligible to apply.
Date(s) Application is Due: Feb 5; Jun 5; Oct 5
Contact: David Gordon; (301) 435-0564; fax: (301) 480-7971; email: gordond@nhlbi.nih.gov
Internet: http://grants.nih.gov/grants/guide/pa-files/PAR-07-373.html
Sponsor: National Heart, Lung, and Blood Institute (NHLBI)
6701 Rockledge Drive, MSC 7940
Bethesda, MD 20892-7940

NHLBI Lymphatic Biology in Health and Disease Grants 3773
This Funding Opportunity Announcement (FOA) issued by the National Heart, Lung, and Blood Institute (NHLBI) solicits Research Project Grant (R01) applications from qualified researchers to investigate the biology of the lymphatic system and potential new therapeutic implications. The purpose of this program announcement is to stimulate research on the biology of the lymphatic system at all biologic levels: molecular, cellular, tissue, organ, and whole body levels. Further, it includes research on innovations for identifying and intervening in lymphatic diseases across all age groups and disease states. Individuals with the skills, knowledge, and resources necessary to carry out the proposed research are invited to work with their institution/organization to develop an application for support. Individuals from underrepresented racial and ethnic groups as well as individuals with disabilities are always encouraged to apply for NIH support. Letters of Intent are due on August 17, while full ptoposals must arrive by September 18.
Requirements: The following organizations and institutions are eligible to apply: public/state controlled institutions of higher education; private institutions of higher education; Hispanic-serving institutions; historically Black colleges and universities (HBCUs); tribally controlled colleges and universities (TCCUs); Alaska Native and Native

Hawaiian serving institutions; nonprofits with 501(c)3 IRS status; nonprofits without 501(c)3 IRS status; for-profit organizations; state governments; Indian/Native American tribal governments (federally recognized); Indian/Native American tribally designated organizations; county governments; city or township governments; special district governments; independent school districts; public housing authorities/Indian housing authorities; U.S. territory or possession; Indian/Native American tribal governments (other than federally recognized); regional organizations; non-domestic (non-U.S.) entities (foreign organizations); other eligible agencies of the federal government; and faith-based or community-based organizations.
Restrictions: Foreign institutions are not eligible to apply.
Date(s) Application is Due: Aug 17; Sep 18
Contact: Stephen Goldman, Ph.D.; (301) 435-0560; fax: (301) 480-7971; email: goldmans@nhlbi.nih.gov; A. Tyl Hewitt, Ph.D.; (301) 496-5541; fax: (301) 480-0303; email: th119v@nih.gov
Internet: http://grants.nih.gov/grants/guide/pa-files/PAR-07-420.html
Sponsor: National Heart, Lung, and Blood Institute (NHLBI)
6701 Rockledge Drive, Rockledge II, Room 8108
Bethesda, MD 20892-7940

NHLBI Mentored Career Award For Faculty At Minority Institutions 3774
The NHLBI Mentored Career Award for Faculty at Minority Institutions is a three-to-five-year non-renewable career development award made to faculty at a minority institution. The candidates are faculty members at a minority institution, must have research experience (length of time may vary), and be committed to developing into independent biomedical investigators in research areas relevant to the mission of the NHLBI (i.e, cardiovascular, pulmonary, hematologic, or sleep disorders research). The award will enable suitable faculty members holding doctoral degrees, such as the Ph.D., M.D., D.O., D.V.M., or an equivalent degree, to undertake special study and supervised research under a mentor who is an accomplished investigator in the research area proposed and has experience in developing independent investigators. Letters of Intent are due on June 21, while full proposals are due by July 19.
Requirements: Minority scientists and physicians with limited research experience needing guided coursework and supervised laboratory experiences, as well as minority faculty needing only an intensive research experience under the guidance of an established scientist, are eligible to apply. At the time of award, it is required that at least two years have elapsed since the receipt of the doctoral degree and that the candidate have at least one year of prior documented research experience.
Restrictions: Current or past principal investigators of an NIH grant or its equivalent, including the Clinical Investigator Award, Physician Scientist Award, Clinical Investigator Development Award, or Mentored Clinical Scientist Development Award are not eligible for the Mentored Research Scientist Development Award for Minority Faculty. Similarly, individuals serving as responsible investigators or project leaders on large grants, such as a Program Project Grant, are not eligible for the Mentored Research Scientist Development Award for Minority Faculty.
Amount of Grant: $1.2 million total
Date(s) Application is Due: Jun 21; Jul 19
Contact: Traci Heath Mondoro, Ph.D.; (301) 435-0052; fax: (301) 480-1060; email: mondorot@nhlbi.nih.gov
Internet: http://grants.nih.gov/grants/guide/rfa-files/RFA-HL-05-016.html#SectionVII
Sponsor: National Heart, Lung, and Blood Institute
6701 Rockledge Drive, MSC 7950
Bethesda, MD 20892-0052

NHLBI Mentored Career Award for Faculty at Minority Institutions 3775
This program provides research support to faculty members at minority institutions who have the potential to conduct high quality research in the areas of cardiovascular, pulmonary, hematologic, or sleep disorders. Important NHLBI program goals are to enhance the minority institution's science infrastructure by developing the research career and teaching skills of the candidate. All faculty members at domestic institutions with student enrollment drawn substantially from minority racial and ethnic groups (such as African Americans/Blacks, Hispanics, American Indians, Alaska Natives, Native Hawaiians, and non-Asian Pacific Islanders) currently underrepresented in biomedical and behavioral research are eligible. Up to $75,000 per year is available for salary, plus fringe benefits, and up to $36,000 for research development support.
Requirements: Candidates for this award must be minority school faculty members who: are U.S. citizens, noncitizen nationals, or permanent residents by the time of the award; have a doctoral degree or equivalent in a biomedical or behavioral science at least two years prior to the award; and wish to receive specialized training in cardiovascular, pulmonary, hematologic, or sleep disorders research. Each candidate must identify and complete arrangements with a mentor (at the same institution or at a collaborating research center) who is recognized as an accomplished investigator in the proposed research area and who will provide guidance for the candidate's career development and research plan.
Amount of Grant: $75,000 per year for salary; $36,000 maximum for research development
Date(s) Application is Due: Aug 15
Contact: Traci Heath Mondoro, Ph.D.; (301) 435-0052; fax: (301) 480-1060; email: mondorot@nhlbi.nih.gov
Internet: http://www.nhlbi.nih.gov/funding/training/redbook/newrek01.htm
Sponsor: National Heart, Lung, and Blood Institute (NHLBI)
Two Rockledge Center, Room 9174, 6701 Rockledge Drive, MSC 7950
Bethesda, MD 20892-7950

NHLBI Mentored Career Development Award to Promote Faculty Diversity/Re-Entry in Biomedical Research 3776
This award is offered to encourage underrepresented faculty members at U.S. institutions to enhance their research skills in areas of interest to the NHLBI and to increase the number of highly trained investigators from diverse backgrounds or who have experienced an interruption in their research careers. It is geared toward individuals whose basic and clinical research interests are grounded in the advanced methods and experimental approaches needed to solve problems related to cardiovascular, pulmonary, hematological, and sleep disorders research. Salary support up to $75,000 per year plus fringe benefits is available, with research support up to $30,000 per year.
Requirements: Faculty members who have been determined by the grantee institution to be underrepresented on faculty in biomedical and behavioral research on a national or institutional basis, such as individuals from racial and ethnic minority groups, individuals with disabilities, individuals from disadvantaged backgrounds or individuals who have experienced an interruption in their research careers are eligible.
Amount of Grant: $105,000 maximum
Date(s) Application is Due: Aug 15
Contact: Lorraine M. Silsbee; (301) 435-0709; fax: (301) 480-1667; email: Lorraine_Silsbee@nih.gov
Internet: http://www.nhlbi.nih.gov/funding/training/redbook/newsck01.htm
Sponsor: National Heart, Lung, and Blood Institute (NHLBI)
Two Rockledge Center, Room 10018, 6701 Rockledge Drive, MSC 7936
Bethesda, MD 20892-7936

NHLBI Mentored Clinical Scientist Research Career Development Awards 3777
This award enables candidates holding professional degrees (e.g., M.D., D.O., D.V.M., or equivalent degrees) to undertake 3 to 5 years of special study and supervised research with the goal of becoming independent investigators. The award also allows awardees to pursue a research career development program suited to their experience and capabilities under a mentor who is competent to provide guidance in the chosen research area. Institutions may submit applications on behalf of candidates who hold professional degrees. At least 2 years must have elapsed since the health professional degree was granted. Candidates can have varying levels of clinical training and research experience. New applications are due February 12, June 12, and October 12, while resubmitted applications are due March 12, July 12, and November 12. Up to $75,000 per year plus fringe benefits is available, and up to $25,000 per year for research support for new awards.
Requirements: All candidates must be U.S. citizens, non-citizen nationals, or legal permanent residents of the U.S. Persons with temporary or student visas are not eligible. The grantee institution must have a strong, well-established research and research training program in the chosen area, accomplished faculty in the basic and clinical sciences, and a commitment to the candidate's research development. The proposed program should include an appropriate mentor.
Amount of Grant: $100,000 maximum
Date(s) Application is Due: Feb 12; Mar 12; Jun 12; Jul 12; Oct 12; Nov 12
Contact: Lorraine M. Silsbee; (301) 435-0709; fax: (301) 480-1667; email: Lorraine_Silsbee@nih.gov
Internet: http://www.nhlbi.nih.gov/funding/training/redbook/newrek08.htm
Sponsor: National Heart, Lung, and Blood Institute (NHLBI)
6701 Rockledge Drive, MSC 7936, Two Rockledge Center, Room 10202
Bethesda, MD 20892-7936

NHLBI Mentored Patient-Oriented Research Career Development Awards 3778
This award supports the career development of investigators who are committed to patient-oriented research. It provides support for supervised study and research for clinically trained professionals who have the potential to develop into productive, clinical investigators focusing on patient-oriented research. This program provides research development opportunities for clinicians with varying levels of research experience. Support is provided for a minimum of 3 years and a maximum of 5 years. Up to $75,000 per year in salary, plus fringe benefits, is available. In addition, $30,000 for research development support per year is available (with sufficient justification up to $50,000 per year can be requested). New applications are due February 12, June 12, and October 12, while resubmitted applications are due March 12, July 12, and November 12.
Requirements: All candidates must be U.S. citizens, noncitizen nationals, or legal permanent residents of the U.S. Candidates must have an M.D., D.D.S., D.M.D., D.O., D.C., Pharm.D. or the Ph.D. or other doctoral degrees in disciplines such as clinical psychology, nursing, clinical genetics, and rehabilitation, or equivalent degree from an accredited domestic or foreign institution at the time of application. Certification by an authorized individual at the degree-granting institution that all degree requirements have been met is acceptable. Candidates must have completed their clinical training, including specialty and, if applicable, subspecialty training prior to receiving an award.
Restrictions: Persons with temporary or student visas are not eligible.
Amount of Grant: $125,000 maximum
Date(s) Application is Due: Feb 12; Mar 12; Jun 12; Jul 12; Oct 12; Nov 12
Contact: Ann Rothgeb; (301) 435-0202; fax: (301) 480-3557; email: rothgeba@mail.nih.gov
Internet: http://www.nhlbi.nih.gov/funding/training/redbook/newk23.htm
Sponsor: National Heart, Lung, and Blood Institute (NHLBI)
Two Rockledge Center, Room 10158, 6701 Rockledge Drive, MSC 7952
Bethesda, MD 20892-7952

NHLBI Mentored Quantitative Research Career Development Awards 3779
The goal of this program is to foster interdisciplinary collaboration in biomedical and behavioral research by supporting supervised research experiences for scientists with

quantitative and engineering backgrounds. This award provides research and career development opportunities for scientists and engineers with little or no biomedical or behavioral research experience who are committed to establishing themselves in careers as independent biomedical or behavioral investigators. Examples of quantitative scientific and technical backgrounds outside of biology and medicine considered appropriate for this award include, but are not limited to: mathematics, statistics, computer science, informatics, physics, chemistry, and engineering. This mechanism is intended for research-oriented investigators from the postdoctoral level to the level of senior faculty. Institutions may submit applications on behalf of candidates who hold an advanced degree in a quantitative area of science or engineering. Up to $75,000 per year plus fringe benefits and up to $40,000 per year for research development support is available. New applications are due February 12, June 12, and October 12, while resubmitted applications are due March 12, July 12, and November 12.
Requirements: All candidates must be U.S. citizens, noncitizen nationals, or legal permanent residents of the U.S. The grantee institution must have a strong, well-established research and research training program in the chosen area, accomplished faculty in the basic and clinical sciences, and a commitment to the candidate's research development. The proposed program should include an appropriate mentor.
Amount of Grant: $115,000 maximum
Date(s) Application is Due: Feb 12; Mar 12; Jun 12; Jul 12; Oct 12; Nov 12
Contact: Michael Commarato, Ph.D.; (301) 435-0535; fax: (301) 480-7971; email: CommaraM@nhlbi.nih.gov
Internet: http://www.nhlbi.nih.gov/funding/training/redbook/newk25.htm
Sponsor: National Heart, Lung, and Blood Institute (NHLBI)
Two Rockledge Center, Room 8140, 6701 Rockledge Drive, MSC 7940
Bethesda, MD 20892-7940

NHLBI Microbiome of the Lung and Respiratory Tract in HIV-Infected Individuals and HIV-Uninfected Controls 3780
The National Heart, Lung, and Blood Institute (NHLBI) solicits grant applications under this Funding Opportunity Announcement (FOA) to characterize the microbiome of the lung (the airways and airspaces below the glottis and the lung parenchyma) alone or in combination with the nasal and/or oropharyngeal cavities in HIV-infected individuals and matched HIV-uninfected controls, including normal healthy controls, using molecular techniques to identify bacteria and if possible other organisms, e.g., viruses, cell-wall deficient organisms, protozoa, and fungi. Investigators should use high-throughput technology platforms to create a data set of sufficient quality and depth to allow analysis of how changes of microbiota relate to HIV lung disease progression/complications. These data will be used to examine the impact of changes in the respiratory microbiome on the pathogenesis and progression of HIV disease, on HIV-related respiratory complications, and the effects of anti-HIV therapies. The NHLBI also invites applications for four Clinical/Sequencing research sites and one supporting Data Coordinating Center (DCC). Each application for a clinical/sequencing research site should propose approaches to characterize the lung microbiome and applicants are encouraged to include at least one hypothesis-driven mechanistic aim. Direct costs up to $525,000 per year for project duration of up to five years may be requested. Letters of Intent are due by February 25, with full proposals to be received by March 25.
Requirements: The following organizations and institutions are eligible to apply: individuals; public/state controlled institutions of higher education; private institutions of higher education; Hispanic-serving institutions; historically Black colleges and universities (HBCUs); tribally controlled colleges and universities (TCCUs); Alaska Native and Native Hawaiian serving institutions; nonprofits with 501(c)3 IRS status; nonprofits without 501(c)3 IRS status; for-profit organizations; state governments; Indian/Native American tribal governments (federally recognized); Indian/Native American tribally designated organizations; county governments; city or township governments; special district governments; independent school districts; public housing authorities/Indian housing authorities; U.S. territory or possession; Indian/Native American tribal governments (other than federally recognized); regional organizations; non-domestic (non-U.S.) entities (foreign organizations); other eligible agencies of the federal government; and faith-based or community-based organizations.
Amount of Grant: $525,000 maximum per year for up to 5 years
Date(s) Application is Due: Feb 25; Mar 25
Contact: Sandra Colombini Hatch, M.D.; (301) 435-0222; fax: (301) 480-3557; email: Hatchs@nhlbi.nih.gov; Hannah H. Peavy, M.D.; (301) 435-0222; fax: (301) 480-3557; email: peavyh@nhlbi.nih.gov
Internet: http://grants.nih.gov/grants/guide/rfa-files/RFA-HL-09-006.html
Sponsor: National Heart, Lung, and Blood Institute (NHLBI)
6701 Rockledge Dr. MSC 7952
Bethesda, MD 20892-7952

NHLBI Midcareer Investigator Award in Patient-Oriented Research 3781
This award provides support for established clinicians within 15 years of their specialty training to further their research and mentoring of outstanding patient-oriented investigators. It will enable them to expand their potential for significant contributions to their field, and to act as mentors for beginning clinical researchers. This award provides protected time to enable the clinician to carry out these activities. The salary that may be requested and awarded is 25% to 50% of the recipient's institutional salary, exclusive of fringe benefits, up to the maximum allowable annual salary rate in effect at the time of award. New applications are due February 12, June 12, and October 12, while resubmitted applications are due March 12, July 12, and November 12.
Requirements: Patient-oriented investigators at U.S. institutions with independent research support are eligible.
Date(s) Application is Due: Feb 12; Mar 12; Jun 12; Jul 12; Oct 12; Nov 12
Contact: Sandra Colombini Hatch, M.D.; (301) 435-0222; fax: (301) 480-3557; email: Hatchs@nhlbi.nih.gov; Hannah H. Peavy, M.D.; (301) 435-0222; fax: (301) 480-3557; email: peavyh@nhlbi.nih.gov
Internet: http://www.nhlbi.nih.gov/funding/training/redbook/estk24.htm
Sponsor: National Heart, Lung, and Blood Institute (NHLBI)
Two Rockledge Center, 6701 Rockledge Drive, MSC 7952
Bethesda, MD 20892-7952

NHLBI Minority Institution Research Scientist Development Award 3782
This program provides research support to faculty members at minority institutions who have the interest and potential to conduct high quality research in the areas of cardiovascular, pulmonary, hematologic, or sleep disorders. Important goals are to enhance the institution's science programs and to assist in the acquisition of hands-on research opportunities for minority students at the applicant's institution. Annual deadline dates may vary; contact the program office for exact dates.
Requirements: The institution must be a domestic medical or non-medical college, university, or equivalent school in which students of underrepresented minority ethnic groups including African Americans, Hispanics, American Indians, Alaska Natives, and Pacific Islanders compose a majority (more than 50 percent) of the school's enrollment. It must have ongoing staff and facilities required for the proposed program.
Amount of Grant: $400,000 maximum
Date(s) Application is Due: Jul 21; Sep 12
Contact: Joyce Creamer, Division of Blood Diseases and Resources; (301) 435-0064; fax: (301) 480-0867; email: CreamerJ@gwgate.nhlbi.nih.gov
Internet: http://www.grants.nih.gov/grants/guide/rfa-files/RFA-HL-00-017.html
Sponsor: National Heart, Lung, and Blood Institute
6701 Rockledge Drive, MSC 7952
Bethesda, MD 20892-7952

NHLBI Minority Institutional Research Training Program Grants 3783
The NHLBI, Minority Institutional Research Training Program is a Ruth L. Kirschstein National Research Service Award Program intended to support training of predoctoral and health professional students and individuals in postdoctoral training at minority schools having the potential to develop meritorious training programs in cardiovascular, pulmonary, hematologic, and sleep disorders. Predoctoral and health professional students and individuals in postdoctoral training in minority schools need further opportunities to develop biomedical and behavioral research skills. The NHLBI Minority Institutional Research Training Program is designed to attract students in their developmental stages; to increase their awareness of cardiovascular, pulmonary, hematologic, and sleep disorders research; and to encourage them to pursue research career opportunities in these areas. Trainee appointments will be determined by the institution and carried out in a manner that will include the recruitment of racial and ethnic minorities, individuals from disadvantaged backgrounds, individuals with disabilities, and women into heart, lung, blood, and sleep disorders scientific research areas. Letters of Intent are due on July 15, with full proposals due by August 15.
Requirements: The institution must be a domestic medical or non-medical college, university, or equivalent school in which students from underrepresented minority groups including African Americans, Hispanics, American Indians, Alaska Natives, and Pacific Islanders compose a majority or a significant proportion of the school enrollment. The individual to be trained must be a citizen of the United States, a noncitizen national, or have been lawfully admitted to the United States for permanent residence at the time of appointment for training, and have a baccalaureate degree. Trainees must be training at the post-baccalaureate level (i.e., predoctoral level) in a relevant biomedical or behavioral science and have made a strong commitment to completing a doctoral degree, be enrolled in a minority health professional school, or have a doctoral degree or equivalent in a biomedical or behavioral science. The program may not support studies leading to a health professional degree.
Amount of Grant: $250,000 total
Date(s) Application is Due: Jul 15; Aug 15
Contact: Sandra Colombini Hatch, M.D.; (301) 435-0222; fax: (301) 480-3557; email: Hatchs@nhlbi.nih.gov; Hannah H. Peavy, M.D.; (301) 435-0222; fax: (301) 480-3557; email: peavyh@nhlbi.nih.gov
Internet: http://grants.nih.gov/grants/guide/rfa-files/RFA-HL-08-017.html
Sponsor: National Heart, Lung, and Blood Institute
Two Rockledge Center, Suite 10162, MSC 7952, 6701 Rockledge Drive
Bethesda, MD 20892-7952

NHLBI Multidisciplinary Translational Research in Critical Care Grants 3784
The purpose of this funding opportunity is to promote multidisciplinary translational research that will improve the treatment, diagnosis, and outcome of patients with critical illness and increase the understanding of fundamental processes causing critical illness in humans of all ages. The research team must include scientists from at least two medical or surgical specialties with direct relevance to critical care. Individuals with the skills, knowledge, and resources necessary to carry out the proposed research are invited to work with their institution/organization to develop an application for support.
Requirements: The following organizations and institutions are eligible to apply: individuals; public/state controlled institutions of higher education; private institutions of higher education; Hispanic-serving institutions; historically Black colleges and universities (HBCUs); tribally controlled colleges and universities (TCCUs); Alaska Native and Native Hawaiian serving institutions; nonprofits with 501(c)3 IRS status; nonprofits without 501(c)3 IRS status; for-profit organizations; state governments; Indian/Native American tribal governments (federally recognized); Indian/Native American tribally designated organizations; county governments; city or township

governments; special district governments; independent school districts; public housing authorities/Indian housing authorities; U.S. territory or possession; Indian/Native American tribal governments (other than federally recognized); regional organizations; non-domestic (non-U.S.) entities (foreign organizations); other eligible agencies of the federal government; and faith-based or community-based organizations.
Date(s) Application is Due: Feb 5; Jun 5; Oct 5
Contact: Andrea L Harabin, Ph.D.; (301) 435-0222; email: harabin@nih.gov; Basil Eldadah, M.D., Ph.D.; (301) 496-4996; email: eldadahb@mail.nih.gov
Internet: http://grants.nih.gov/grants/guide/pa-files/PA-07-233.html
Sponsor: National Heart, Lung, and Blood Institute
6701 Rockledge Drive, MSC 7952
Bethesda, MD 20892-7952

NHLBI National Research Service Award Programs in Cardiovascular Epidemiology and Biostatistics **3785**

This program enables research institutions to support pre- and post-doctoral training in areas of particular interest to the Institute and to prepare individuals for careers in biomedical and behavioral research. The cardiovascular epidemiology and biostatistics research training programs offer opportunities in the fields of epidemiology of cardiovascular diseases and statistical analysis. Trainees are selected through review procedures established by the program director at the grantee institution. Attention is given to recruiting individuals from minority groups that are under-represented nationally in the biomedical sciences (African Americans, Hispanics, Native Americans, Alaskan Natives, and Pacific Islanders).
Requirements: Students must be U.S. citizens, noncitizen nationals, or legal permanent U.S. residents. Trainees must have received a baccalaureate degree and must be training at the postbaccalauareate level in a program leading to a PhD, ScD, or equivalent degree.
Restrictions: These awards do not support study leading to the MD, DDS, or other similar professional degrees, nor do they support residency training. Persons on temporary or student visas are not eligible.
Date(s) Application is Due: Apr 5; Aug 5; Dec 5
Contact: Lorraine M. Silsbee; (301) 435-0709; fax: (301) 480-1667; email: Lorraine_Silsbee@nih.gov
Internet: http://www.nhlbi.nih.gov/funding/training/epi-bio/nrsa.htm
Sponsor: National Heart, Lung, and Blood Institute
6701 Rockledge Drive
Bethesda, MD 20892-7934

NHLBI National Research Service Awards for Senior Fellowships **3786**

This program enables experienced scientists to change the direction of their research careers, broaden their scientific background, acquire new research capabilities, or enlarge their command of an allied research field. Selection is by national competition. All experienced scientists interested in changing the direction of their careers are eligible to apply.
Requirements: When the fellowship begins, an applicant must have received the Ph.D., M.D., D.O., D.D.S., D.V.M., O.D., D.P.M., Sc.D., D.Eng., DNSc., or equivalent degree. The applicant also must have at least 7 subsequent years of relevant postdoctoral research or professional training, and must have a sponsor affiliated with a public or private institution with appropriate facilities for the proposed training.
Restrictions: One month of payback for each month of training, up to a maximum of 12 months, is required. This requirement can be fulfilled by teaching or research (a minimum of 20 hours per week) on a continuous basis beginning within 2 years after support ends.
Date(s) Application is Due: Apr 8; Aug; Dec 8
Contact: Charlotte A. Pratt, Program Director; (301) 435-0382; fax: (301) 480-5158; email: prattc@nhlbi.nih.gov
Internet: http://www.nhlbi.nih.gov/funding/training/redbook/estf33.htm
Sponsor: National Heart, Lung, and Blood Institute (NHLBI)
Two Rockledge Center, Room 10118, 6701 Rockledge Drive, MSC 7936
Bethesda, MD 20892-7936

NHLBI New Approaches to Arrhythmia Detection and Treatment Grants for SBIR **3787**

The purpose of this initiative is to improve our ability to detect, treat and prevent cardiac arrhythmias using new or improved methods, tools, and technologies. Applications are invited for the development and significant improvements of innovative monitoring tools, diagnostic methods, and therapeutic approaches for arrhythmia detection, treatment and prevention.
Requirements: Only United States small business concerns (SBCs) are eligible to submit SBIR applications.
Date(s) Application is Due: Apr 5; Aug 5; Dec 5
Contact: Bishow B. Adhikari, Ph.D.; (301) 435-0507; fax: (301) 480-7971; email: AdhikarB@NHLBI.NIH.GOV; David A. Lathrop, Ph.D.; (301) 435-0507; fax: (301) 451-5458; email: LathropD@NHLBI.NIH.GOV
Internet: http://grants.nih.gov/grants/guide/pa-files/PA-07-032.html
Sponsor: National Heart, Lung, and Blood Institute (NHLBI)
Rockledge Two, Room 9192, 6701 Rockledge Drive
Bethesda, MD 20892

NHLBI New Approaches to Arrhythmia Detection and Treatment Grants for STTR **3788**

The purpose of this initiative is to improve our ability to detect, treat and prevent cardiac arrhythmias using new or improved methods, tools, and technologies. Applications are invited for the development and significant improvements of innovative monitoring tools, diagnostic methods, and therapeutic approaches for arrhythmia detection, treatment and prevention.
Requirements: Only United States small business concerns (SBCs) are eligible.
Date(s) Application is Due: Apr 5; Aug 5; Dec 5
Contact: Bishow B. Adhikari, Ph.D.; (301) 435-0507; fax: (301) 480-7971; email: AdhikarB@NHLBI.NIH.GOV; David A. Lathrop, Ph.D.; (301) 435-0507; fax: (301) 451-5458; email: LathropD@NHLBI.NIH.GOV
Internet: http://grants.nih.gov/grants/guide/pa-files/PA-07-031.html
Sponsor: National Heart, Lung, and Blood Institute (NHLBI)
Rockledge Two, Room 9192, 6701 Rockledge Drive
Bethesda, MD 20892

NHLBI Nutrition and Diet in the Causation, Prevention, and Management of Heart Failure Research Grants **3789**

The purpose of this Funding Opportunity Announcement (FOA) is to encourage submission of investigator-initiated research applications on the role of nutrition and diet in the causation, prevention, and treatment of cardiomyopathies and heart failure. Basic, translational, and applied interdisciplinary research applications with rigorous hypothesis-testing designs for projects in animals or humans are of interest. The overall goal is to develop a satisfactory science base for preventive approaches in high-risk individuals and for rational nutritional management of patients in various stages of heart failure.
Requirements: The following groups are eligible to apply: for-profit organizations on-profit organizations; public or private institutions, such as universities, colleges, hospitals, and laboratories; units of State government; units of local government; eligible agencies of the Federal government; foreign institutions; domestic institutions; faith-based or community-based organizations; units of State tribal government; and units of local Tribal government.
Amount of Grant: $200,000 maximum annually
Date(s) Application is Due: Feb 16; Jun 16; Oct 16
Contact: Abby G. Ershow; (301) 435-0550; fax: (301) 480-1338; email: ErshowA@mail.nih.gov
Internet: http://grants.nih.gov/grants/guide/pa-files/PA-06-136.html
Sponsor: National Heart, Lung, and Blood Institute (NHLBI)
6701 Rockledge Drive, Suite, MSC 7956
Bethesda, MD 20892-7956

NHLBI Pathogenesis and Treatment Of Lymphedema and Lymphatic Diseases Grants **3790**

The purpose of this Funding Opportunity Announcement (FOA) is to stimulate research on the lymphatic system, to characterize its function on the molecular, cellular, tissue, organ, and intact organism levels and pathophysiologic mechanisms that cause disease, to develop new methods for imaging and/or quantitating lymph flow, and to discover new therapeutic interventions, including nursing, complementary and alternative treatments. The scope of this research includes developmental biology and genetics of the lymphatic system to identify and characterize processes important for its organization and regulation. Such knowledge will help to improve early diagnosis of affected individuals, the choice and timing of treatment, and genetic counseling. Research is also needed on disorders of skin and subcutaneous tissue due to chronic lymphedema, and lymphedema which results from cancers or cancer treatment, with a goal to develop better therapies.
Requirements: The following organizations and institutions are eligible to apply: public/state controlled institutions of higher education; private institutions of higher education; Hispanic-serving institutions; historically Black colleges and universities (HBCUs); tribally controlled colleges and universities (TCCUs); Alaska Native and Native Hawaiian serving institutions; nonprofits with 501(c)3 IRS status; nonprofits without 501(c)3 IRS status; for-profit organizations; state governments; Indian/Native American tribal governments (federally recognized); Indian/Native American tribally designated organizations; non-domestic (non-U.S.) entities (foreign organizations); Hispanic-serving institutions; and regional organization.
Date(s) Application is Due: Feb 5; Jun 5; Oct 5
Contact: Henry Chang, M.D.; (301) 435-0067; fax: (301) 480-1060; email: changh@nih.gov; Stephen S. Goldman, Ph.D.; (301) 435-0565; fax: (301) 480-2849; email: goldmans@nih.gov
Internet: http://grants.nih.gov/grants/guide/pa-files/PA-07-165.html
Sponsor: National Heart, Lung, and Blood Institute (NHLBI)
6701 Rockledge Drive, MSC 7950
Bethesda, MD 20892-7950

NHLBI Pathway to Independence Award **3791**

The Pathway to Independence (PI) Award is designed to facilitate a timely transition from a mentored postdoctoral research position to a stable independent research position with independent NIH or other independent research support at an earlier stage than is currently the norm. The PI award will provide up to 5 years of support consisting of two phases. The initial phase will provide 1-2 years of mentored support for postdoctoral research scientists. This phase will be followed by up to 3 years of independent support contingent on securing an independent tenure-track or equivalent research position. A typical successful candidate will have an impressive record of research productivity and publications, and will have identified a unique research direction to pursue after the mentored phase of the award. Domestic for-profit or non-profit institutions and organizations, or public or private institutions, such as universities, colleges, hospitals and laboratories, eligible agencies of the federal government, and NIH intramural laboratories are eligible (all institution and organization types listed above are eligible

for both the mentored and independent phase, with one exception: ineligible agencies of the Federal government, such as the NIH intramural program, are eligible only for the mentored phase). For the Mentored Phase, salary up to $75,000 plus fringe benefits per year is available, with up to $25,000 additional per year for support (F&A costs will be provided at 8% of modified direct costs). During the Independent investigator phase, total cost may not exceed $249,000 per year. New applications are due February 12, June 12, and October 12, while resubmitted applications are due March 12, July 12, and November 12.
Requirements: At the time of application submission (or resubmission) candidates for this award must have earned a terminal clinical or research doctorate (including Ph.D., M.D., D. O., D.C., N.D., D.D.S., D.V.M., Sc.D., D.N.S., Pharm. D., or equivalent doctoral degree, or a combined clinical and research doctoral degree); and have no more than 5 years (60 months) of research experience since completing the requirements of the doctoral degree (resubmissions must also comply with this requirement).
Restrictions: Foreign institutions are not eligible.
Amount of Grant: $100,000 maximum for mentored phase; $249,000 for Independent Investigator phase
Date(s) Application is Due: Feb 12; Mar 12; Jun 12; Jul 12; Oct 12; Nov 12
Contact: Herbert Geller, Ph.D.; (301) 451-9440; fax: (301) 594-8133; email direducation@nhlbi.nih.gov
Internet: http://www.nhlbi.nih.gov/funding/training/redbook/phdk99r00.htm
Sponsor: National Heart, Lung, and Blood Institute (NHLBI)
Building 10, Room 2N242, 10 Center Drive, MSC 1754
Bethesda, MD 20892-1754

NHLBI Pediatric Cardiac Genomics Consortium Grants **3792**
The purpose of this initiative is to support collaborative genetic and genomic research leading to a comprehensive understanding of the causes and outcomes of congenital heart disease in the human population. The Consortium will consist of up to 6 Research Centers, Cores, and a Steering Committee. Research Center applicants will propose detailed investigation of the genetics and genomics of one or more human cardiac malformations, with the goal of identifying genetic influences on outcome, as well as causative genes. Individual Research Centers will collaborate in recruitment of individuals with congenital heart disease and their families for all Research Center projects. Maximum allowable direct costs for the Research Centers per year are $160,000 in Year 1, and $535,000 per year in years 2-6. Letters of Intent should be received by January 6, and full proposals are due by February 6.
Requirements: The following organizations and institutions are eligible to apply: public/state controlled institutions of higher education; private institutions of higher education; Hispanic-serving institutions; historically Black colleges and universities (HBCUs); tribally controlled colleges and universities (TCCUs); Alaska Native and Native Hawaiian serving institutions; nonprofits with 501(c)3 IRS status; nonprofits without 501(c)3 IRS status; for-profit organizations; small businesses; state governments; non-domestic (non-U.S.) entities in North America; and other eligible agencies of the Federal Government.
Amount of Grant: $160,000 maximum in year 1 planning; $535,000 in years 2-6
Date(s) Application is Due: Jan 6; Feb 6
Contact: Gail D. Pearson; (301) 435-0510; email: pearsong@mail.nih.gov
Internet: http://grants.nih.gov/grants/guide/rfa-files/RFA-HL-09-003.html
Sponsor: National Heart, Lung, and Blood Institute (NHLBI)
6701 Rockledge Drive, Room 8104
Bethesda, MD 20892

NHLBI Phase II Clinical Trials of Novel Therapies for Lung Diseases **3793**
The purpose of this FOA is to solicit research grant applications to conduct Phase II clinical therapeutic trials that have the potential to advance development of novel therapies for a lung disease or a cardiopulmonary disorder of sleep. Each application will propose one Phase II interventional trial that will most likely use physiological or biochemical rather than clinical endpoints along with at least one smaller basic ancillary research study that is tightly related to the clinical question. Although definitive Phase III trials will not be supported, the proposed studies must provide proof of concept for a novel intervention that has high potential for modifying current treatments and could be disease modifying.
Requirements: The following organizations and institutions are eligible to apply: public/state controlled institutions of higher education; private institutions of higher education; Hispanic-serving institutions; historically Black colleges and universities (HBCUs); tribally controlled colleges and universities (TCCUs); Alaska Native and Native Hawaiian serving institutions; nonprofits with 501(c)3 IRS status; nonprofits without 501(c)3 IRS status; for-profit organizations; state governments; Indian/Native American tribal governments (federally recognized); Indian/Native American tribally designated organizations; county governments; city or township governments; special district governments; independent school districts; public housing authorities/Indian housing authorities; U.S. territories or possessions; Indian/Native American tribal governments (other than federally recognized); regional organizations; non-domestic (non-U.S.) entities (foreign organizations); other eligible agencies of the federal government; and faith-based or community-based organizations.
Amount of Grant: $800,000 maximum 1st year; $1,515,000 maximum 2nd and 3rd year
Date(s) Application is Due: May 18; Jun 18; Dec 30; Jan 29; Aug 15; Sep 15
Contact: Andrea L. Harabin, Ph.D.; (301) 435-0222; email: harabina@nhlbi.nih.gov
Internet: http://grants.nih.gov/grants/guide/rfa-files/RFA-HL-10-003.html
Sponsor: National Heart, Lung and Blood Institute (NHLBI)
6701 Rockledge Drive
Bethesda, MD 20892-7952

NHLBI Prematurity and Respiratory Outcomes Program (PROP) Grants **3794**
This RFA intends to fill the need for understanding the underlying pathophysiology by supporting the investigation of the molecular mechanisms that contribute to acute and chronic respiratory morbidity of the premature infant. The well phenotyped populations developed in this program will significantly increase the chance of studying groups with common pathophysiology and lay the groundwork for novel biomarker discovery. The program will use the cooperative agreement mechanism and require multidisciplinary expertise including neonatologists/pulmonologists, and physiological, molecular, biological, or bioinformatics scientists. Participating sites will plan and coordinate their own research and will develop and implement shared protocols for respiratory phenotyping and respiratory outcomes of the premature infant NICU graduate. Letters of Intent should be received by May 4, and full proposals are due by June 2 each year.
Requirements: The following organizations and institutions are eligible to apply: public/state controlled institutions of higher education; private institutions of higher education; Hispanic-serving institutions; historically Black colleges and universities (HBCUs); tribally controlled colleges and universities (TCCUs); Alaska Native and Native Hawaiian serving institutions; nonprofits with 501(c)3 IRS status; nonprofits without 501(c)3 IRS status; for-profit organizations; state governments; Indian/Native American tribal governments (federally recognized); Indian/Native American tribally designated organizations; county governments; city or township governments; special district governments; independent school districts; public housing authorities/Indian housing authorities; U.S. territories or possessions; Indian/Native American tribal governments (other than federally recognized); regional organizations; non-domestic (non-U.S.) entities (foreign organizations); other eligible agencies of the federal government; and faith-based or community-based organizations.
Amount of Grant: $150,000 maximum in year 1; $350,000 maximum in years 2-5.
Date(s) Application is Due: May 4; Jun 2
Contact: Carol J. Blaisdell, M.D.; (301) 435-0222; email: blaisdellcj@nhlbi.nih.gov
Internet: http://grants.nih.gov/grants/guide/rfa-files/RFA-HL-10-007.html
Sponsor: National Heart, Lung, and Blood Institute (NHLBI)
6701 Rockledge Drive
Bethesda, MD 20892

NHLBI Progenitor Cell Biology Consortium Administrative Coordinating Center Grants **3795**
The goal of the Consortium is to identify and characterize progenitor cell lineages, to direct the differentiation of stem and progenitor cells to desired cell fates, and to develop new strategies to address the unique challenges presented by the transplantation of these cells. The Consortium will assemble multiple independent research projects, each with a multi-disciplinary team of Principal Investigators, core research support facilities, and skills development components to establish synergistic virtual hubs focused on progenitor cell biology. The total project period for an application submitted in response to this FOA is 7 years. Maximum allowable direct costs are $600,000 per year for ACC operations. Letters of Intent are dur by April 1, while full applications should arrive by May 1.
Requirements: The following organizations and institutions are eligible to apply: public/state controlled institutions of higher education; private institutions of higher education; Hispanic-serving institutions; historically Black colleges and universities (HBCUs); tribally controlled colleges and universities (TCCUs); Alaska Native and Native Hawaiian serving institutions; nonprofits with 501(c)3 IRS status; nonprofits without 501(c)3 IRS status; small businesses; for-profit organizations; and state governments.
Amount of Grant: $600,000 per year maximum for 7 years
Date(s) Application is Due: Apr 1; May 1
Contact: Denis Buxton, Ph.D.; (301) 435-0513; email: Buxtond@nhlbi.nih.gov; Carol J. Blaisdell, M.D.; (301) 435-0222; email: blaisdellcj@nhlbi.nih.gov
Internet: http://grants.nih.gov/grants/guide/rfa-files/RFA-HL-09-005.html
Sponsor: National Heart, Lung, and Blood Institute (NHLBI)
Rockledge 2, Room 8216, 6701 Rockledge Drive, MSC 7940
Bethesda, MD 20892

NHLBI Progenitor Cell Biology Consortium Research Hubs **3796**
The National Heart, Lung, and Blood Institute (NHLBI) invites applications to participate in the NHLBI Progenitor Cell Biology Consortium. The goal of the Consortium is to identify and characterize progenitor cell lineages, to direct the differentiation of stem and progenitor cells to desired cell fates, and to develop new strategies to address the unique challenges presented by the transplantation of these cells. The Consortium will assemble clusters of synergistic research projects, each with a multi-disciplinary team of Principal Investigators, to establish virtual Research Hubs focused on progenitor cell biology. The Consortium will also contain core research support facilities and skills development components. The expected direct cost amount for the Consortium awards will be up to $750,000 per year. Letters of Intent are due by April 1, with full proposals received by May 1.
Requirements: The following organizations and institutions are eligible to apply: public/state controlled institutions of higher education; private institutions of higher education; Hispanic-serving institutions; historically Black colleges and universities (HBCUs); tribally controlled colleges and universities (TCCUs); Alaska Native and Native Hawaiian serving institutions; nonprofits with 501(c)3 IRS status; nonprofits without 501(c)3 IRS status; small businesses; for-profit organizations; and state governments.
Amount of Grant: $750,000 per year
Date(s) Application is Due: Apr 1; May 1
Contact: Denis Buxton, Ph.D.; (301) 435-0513; email: Buxtond@nhlbi.nih.gov; Carol J. Blaisdell, M.D.; (301) 435-0222; email: blaisdellcj@nhlbi.nih.gov
Internet: http://grants.nih.gov/grants/guide/rfa-files/RFA-HL-09-004.html

Sponsor: National Heart, Lung, and Blood Institute (NHLBI)
Rockledge 2, Room 8216, 6701 Rockledge Drive, MSC 7940
Bethesda, MD 20892

NHLBI Protein Interactions Governing Membrane Transport in Pulmonary Health and Disease Grants 3797

This Funding Opportunity Announcement (FOA) issued by the National Heart, Lung, and Blood Institute, National Institutes of Health (NIH), solicits research grant applications to delineate the protein interactions and pathways governing membrane trafficking pathways operative in pulmonary health and disease and develop novel therapeutic interventions. Because the nature and scope of the proposed research will vary from application to application, it is anticipated that the size and duration of each award will also vary. The total amount awarded and the number of awards will depend upon the mechanism numbers, quality, duration, and costs of the applications received.
Requirements: The following organizations and institutions are eligible to apply: public/state controlled institutions of higher education; private institutions of higher education; Hispanic-serving institutions; historically Black colleges and universities (HBCUs); tribally controlled colleges and universities (TCCUs); Alaska Native and Native Hawaiian serving institutions; nonprofits with 501(c)3 IRS status; nonprofits without 501(c)3 IRS status; for-profit organizations; state governments; Indian/Native American tribal governments (federally recognized); Indian/Native American tribally designated organizations; county governments; city or township governments; special district governments; independent school districts; public housing authorities/Indian housing authorities; U.S. territory or possession; Indian/Native American tribal governments (other than federally recognized); regional organizations; non-domestic (non-U.S.) entities (foreign organizations); other eligible agencies of the federal government; and faith-based or community-based organizations.
Date(s) Application is Due: Feb 5; Jun 5; Oct 5
Contact: Susan Banks-Schlegel, Ph.D.; (301) 435-0202; fax: (301) 480-3557; email: Schleges@nhlbi.nih.gov; Dorothy Gail, Ph.D.; (301) 435-0222; fax: (301) 480-3557; email: Gaild@nhlbi.nih.gov
Internet: http://grants.nih.gov/grants/guide/pa-files/PA-07-137.html
Sponsor: National Heart, Lung, and Blood Institute (NHLBI)
Two Rockledge Center, Suite 10018, 6701 Rockledge Drive
Bethesda, MD 20892-7952

NHLBI Research Demonstration and Dissemination Grants 3798

The purpose of this Funding Opportunity Announcement (FOA) is to encourage the scientific community to conduct Demonstration and Dissemination (D&D) studies to test the effectiveness of interventions in children, adolescents, and/or adults to: promote healthful behaviors; reduce risk factors for heart, lung, and blood diseases, and sleep disorders; improve the prevention or management of heart, lung, and blood diseases, and sleep disorders, including the delivery of health care services; and enhance understanding of the processes of intervention implementation and diffusion, or sustainability in a defined population, or defined clinical or community setting. The total project period for an application submitted in response to this funding opportunity may not exceed 5 years. Direct costs are limited to less than $500,000 in any single year, unless pre-approval is obtained from the Institute.
Requirements: The following organizations and institutions are eligible to apply: public/state controlled institutions of higher education; private institutions of higher education; Hispanic-serving institutions; historically Black colleges and universities (HBCUs); tribally controlled colleges and universities (TCCUs); Alaska Native and Native Hawaiian serving institutions; nonprofits with 501(c)3 IRS status; nonprofits without 501(c)3 IRS status; for-profit organizations; state governments; Indian/Native American tribal governments (federally recognized); Indian/Native American tribally designated organizations; county governments; city or township governments; special district governments; independent school districts; public housing authorities/Indian housing authorities; U.S. territory or possession; Indian/Native American tribal governments (other than federally recognized); regional organizations; non-domestic (non-U.S.) entities (foreign organizations); other eligible agencies of the federal government; and faith-based or community-based organizations.
Amount of Grant: $500,000 per year maximum for 5 years
Date(s) Application is Due: Jan 25; May 25; Sep 25
Contact: Charlotte Pratt, Ph.D.; (301) 435-0382; fax: (301) 480-5158; email: PrattC@nhlbi.nih.gov
Internet: http://grants.nih.gov/grants/guide/pa-files/PA-07-017.html
Sponsor: National Heart, Lung, and Blood Institute (NHLBI)
6701 Rockledge Drive, MSC 7936, Room 8134
Bethesda, MD 20892

NHLBI Research Grants on the Relationship Between Hypertension and Inflammation 3799

This Funding Opportunity Announcement (FOA) solicits grant applications that propose the study of the sequence of events in which the vascular inflammatory state contributes to the development and maintenance of hypertension. Ample evidence suggests that inflammation may play a role in the pathogenesis of hypertension or that it may characterize a functional state of the vessel wall as a consequence of high blood pressure. Angiotensin II (Ang II), a widely recognized vasoconstrictor and anti-natriuretic involved in blood pressure regulation, also acts as a pro-inflammatory factor in the cardiovascular system. Ang II stimulates the expression of several inflammatory cytokines, which in turn affect blood pressure. A potential linkage between Ang II and immuno-cytokines is their shared ability to induce the level of reactive oxygen species (ROS), which serve as second messengers for many intracellular signaling pathways. The production of ROS not only decreases bioavailability of nitric oxide (NO), a vasodilator, but also initiates the functional and morphological alterations, such as remodeling, in the vascular wall that accompany the hypertensive state over time. This FOA would provide an opportunity to bring focus on the potential causal relationship between hypertension and inflammation in a cohesive, integrated manner. A new understanding of hypertension and inflammation would provide novel opportunities to prevent and treat the disease.
Requirements: The following organizations and institutions are eligible to apply: public/state controlled institutions of higher education; private institutions of higher education; Hispanic-serving institutions; historically Black colleges and universities (HBCUs); tribally controlled colleges and universities (TCCUs); Alaska Native and Native Hawaiian serving institutions; nonprofits with 501(c)3 IRS status; nonprofits without 501(c)3 IRS status; for-profit organizations; state governments; Indian/Native American tribal governments (federally recognized); Indian/Native American tribally designated organizations; county governments; city or township governments; special district governments; independent school districts; public housing authorities/Indian housing authorities; U.S. territory or possession; Indian/Native American tribal governments (other than federally recognized); regional organizations; non-domestic (non-U.S.) entities (foreign organizations); other eligible agencies of the federal government; and faith-based or community-based organizations.
Date(s) Application is Due: Feb 5; Jun 5; Oct 5
Contact: Eser Tolunay, Ph.D.; (301) 435-0560; email: tolunaye@nhlbi.nih.gov
Internet: http://grants.nih.gov/grants/guide/pa-files/PA-07-138.html
Sponsor: National Heart, Lung, and Blood Institute (NHLBI)
6701 Rockledge Drive
Bethesda, MD 20892

NHLBI Research on the Role of Cardiomyocyte Mitochondria in Heart Disease: An Integrated Approach 3800

The National Heart, Lung, and Blood Institute (NHLBI) invites applications for collaborative research projects to develop an integrated understanding of cardiomyocyte mitochondria and its contributions to myocardial adaptations and heart disease progression by combining functional data with information derived from powerful new technologies. Budgets for direct costs of up to $500,000 per year and a project duration of up to four years may be requested for a maximum of $2,000,000 direct costs over a four-year project period. Because the nature and scope of the proposed research will vary from application to application, it is anticipated that the size and duration of each award will also vary.
Requirements: The following organizations and institutions are eligible to apply: individuals; public/state controlled institutions of higher education; private institutions of higher education; Hispanic-serving institutions; historically Black colleges and universities (HBCUs); tribally controlled colleges and universities (TCCUs); Alaska Native and Native Hawaiian serving institutions; nonprofits with 501(c)3 IRS status; nonprofits without 501(c)3 IRS status; for-profit organizations; state governments; Indian/Native American tribal governments (federally recognized); Indian/Native American tribally designated organizations; county governments; city or township governments; special district governments; independent school districts; public housing authorities/Indian housing authorities; U.S. territory or possession; Indian/Native American tribal governments (other than federally recognized); regional organizations; non-domestic (non-U.S.) entities (foreign organizations); other eligible agencies of the federal government; and faith-based or community-based organizations.
Amount of Grant: $500,000 direct costs per year for no more than four years
Date(s) Application is Due: Apr 20; May 19
Contact: Isabella Liang, Ph.D.; (301) 435-0504; fax: (301) 451-5458; email: Liangi@mail.nih.gov
Internet: http://grants.nih.gov/grants/guide/rfa-files/RFA-HL-10-002.html
Sponsor: National Heart, Lung, and Blood Institute (NHLBI)
Two Rockledge Center, Room 8178, 6701 Rockledge Drive, MSC 7956
Bethesda, MD 20892-7956

NHLBI Research Supplements to Promote Diversity in Health-Related Research (Postdoctoral Individuals/New Independent Researchers) 3801

Supplement enables principal investigators with eligible NHLBI research grants to include recently appointed junior faculty from underrepresented racial and ethnic groups and individuals with disabilities in their projects. Nationally, underrepresented groups in biomedical research careers include but are not limited to, African Americans, Hispanic Americans, American Indians/Alaska Natives, Native Hawaiians and Pacific Islanders. The research proposed for the investigator must be a logical extension of the parent grant. The investigator must be given opportunities to interact with other individuals on the grant, to contribute intellectually to the study, and to enhance his or her research skills and knowledge. Funds may be requested to make changes or adjustments in the laboratory setting that will enable the investigator with disabilities to perform the essential functions associated with his or her research project. Accommodation requests must be reasonable in relationship to the direct costs of the parent grant. There is no application receipt date. However, awards are generally made four times per year (December, March, June, and September). Principal Investigators are encouraged to synchronize the supplement requested start date with the parent grant budget period start date. A maximum of $75,000 per year in salary is available (plus fringe benefits), and a maximum of $10,000 per year for supplies and travel.
Requirements: All principal investigators at U.S. institutions with eligible NHLBI research grants are eligible to apply.
Amount of Grant: $85,000 maximum

Contact: Helena O. Mishoe, PhD; (301) 451-5081; fax: (301) 480-0862; email: mishoeh@nhlbi.nih.gov
Internet: http://www.nhlbi.nih.gov/funding/training/redbook/newmirs.htm
Sponsor: National Heart, Lung, and Blood Institute (NHLBI)
6701 Rockledge Drive, Room 8188
Bethesda, MD 20892-7913

NHLBI Research Supplements to Promote Diversity in Health-Related Research for Established Investigators Who Become Disabled 3802
This supplement enables principal investigators with eligible NHLBI research grants who become disabled to obtain additional support to continue their research. Funds may be requested to make changes or adjustments in the laboratory setting that will enable the disabled investigator to perform the essential functions associated with his or her research project and/or to hire an assistant. Accommodation requests must be reasonable in relationship to the direct costs of the parent grant. Applications can be submitted at any time, but no later than 3 months before the last 2 years remaining on the NHLBI research grant. Applicants may request support for special equipment, an assistant, or other modifications to facilitate reasonable accommodation to a disabling injury or illness that has occurred during the current project period.
Requirements: All principal investigators at U.S. institutions with eligible NHLBI research grants who become disabled are eligible to apply.
Contact: Helena O. Mishoe, PhD; (301) 451-5081; fax: (301) 480-0862; email: mishoeh@nhlbi.nih.gov
Internet: http://www.nhlbi.nih.gov/funding/training/redbook/estdissu.htm
Sponsor: National Heart, Lung, and Blood Institute (NHLBI)
6701 Rockledge Drive, Room 8188
Bethesda, MD 20892-7913

NHLBI Research Supplements to Promote Diversity in Health-Related Research for Graduate Research Assistants 3803
This supplement enables principal investigators with eligible NHLBI research grants to include graduate students in their projects. Priority is given to applications requesting support for students with disabilities and/or from underrepresented racial and ethnic groups. The research activity proposed for the student must be part of the approved research for the parent grant or a logical extension. The student must be given opportunities to interact with other individuals on the grant and to contribute intellectually to the study. Applications can be submitted at any time, but no later than 3 months before the last 2 years remaining on the NHLBI research grant. Stipends include: salary (including fringe benefits) up to a maximum of $26,000 per year; and a maximum of $4,000 a year for supplies, travel (up to $1,000 for travel to scientific meetings), and other.
Requirements: All principal investigators at U.S. institutions with eligible NHLBI research grants who wish to offer research experiences to students with disabilities and/or students from underrepresented racial and ethnic groups who are pursuing a masters or doctoral degree in one of the biomedical or behavioral sciences or enrolled in a master's degree program in nursing are eligible to apply.
Amount of Grant: $30,000 maximum
Contact: Helena O. Mishoe, PhD; (301) 451-5081; fax: (301) 480-0862; email: mishoeh@nhlbi.nih.gov
Internet: http://www.nhlbi.nih.gov/funding/training/redbook/gradsup.htm
Sponsor: National Heart, Lung, and Blood Institute
6701 Rockledge Drive, Room 8188
Bethesda, MD 20892-7913

NHLBI Research Supplements to Promote Diversity in Health-Related Research for High School Students 3804
This supplement enables principal investigators with eligible NHLBI research grants or contracts to include high school students in their projects. Priority is given to applications requesting support for students with disabilities and/or from underrepresented racial and ethnic groups. The research activity proposed for the student must be part of the approved research for the parent grant or contract, or a logical extension. The student must be given opportunities to interact with other individuals on the study and to contribute to the study. The program is designed to support high school students who express an interest in the biomedical or behavioral sciences and who will devote at least 2 years (two 3-month periods) to the program. Support should be requested for at least 3 months during any year. Applications are due any time, but no later than 3 months before the last 2 years remaining on the NHLBI research grant or contract. Up to $3 000 per year is available for salary. Supplies, travel (scientific meetings, home, and/or research site) and other are allocated up to $1,000 per year.
Requirements: All principal investigators at U.S. institutions with eligible NHLBI research grants or contracts may apply.
Amount of Grant: $3,000 per year for salary; $1,000 per year for suppliments
Contact: Helena O. Mishoe, PhD; (301) 451-5081; fax: (301) 480-0862; email: mishoeh@nhlbi.nih.gov
Internet: http://www.nhlbi.nih.gov/funding/training/redbook/hsminsup.htm
Sponsor: National Heart, Lung, and Blood Institute
6701 Rockledge Drive, Room 8188
Bethesda, MD 20892-7913

NHLBI Research Supplements to Promote Diversity in Health-Related Research for Individuals in Postdoctoral Training 3805
This supplement enables principal investigators with eligible NHLBI research grants to include postdoctoral individuals from underrepresented racial and ethnic groups and individuals with disabilities in their projects. Nationally, underrepresented groups in biomedical research careers include but are not limited to African Americans, Hispanic Americans, American Indians/Alaska Natives, Native Hawaiians and Pacific Islanders. The research proposed for the individual must be a logical extension of the parent grant. The individual must be given opportunities to interact with other researchers on the grant, to contribute intellectually to the study, and to enhance his or her research skills and knowledge. Funds may be requested to make changes or adjustments in the laboratory setting that will enable the investigator with disabilities to perform the essential functions associated with his or her research project. Accommodation requests must be reasonable in relationship to the direct costs of the parent grant. There is no application receipt date. However, awards are generally made four times per year (December, March, June, and September). Principal Investigators are encouraged to synchronize the supplement requested start date with the parent grant budget period start date. Up to $40,000 per year is provided for salary and fringe benefits, plus up to $10,000 per year for supplies, travel, and other.
Requirements: The individual may be affiliated with the principal investigator's institution or another institution. Supplements are available for long-term research support for individuals with a postdoctoral degree. A minimum of two years and a maximum of four years full-time research effort (at least 40 hours a week) is required.
Restrictions: Normally, funding will not be provided for postdoctoral individuals to work with their former predoctoral mentors. If the postdoctoral mentor is the individual's predoctoral mentor, a justification for continued mentorship must be provided. Cumulative Postdoctoral research experience must not exceed 6 years.
Amount of Grant: $50,000 maximum
Contact: Helena O. Mishoe, PhD; (301) 451-5081; fax: (301) 480-0862; email: mishoeh@nhlbi.nih.gov
Internet: http://www.nhlbi.nih.gov/funding/training/redbook/phdressu.htm
Sponsor: National Heart, Lung, and Blood Institute
6701 Rockledge Drive, Room 8188
Bethesda, MD 20892-7913

NHLBI Research Supplements to Promote Diversity in Health-Related Research for Undergraduate Students 3806
This supplement enables principal investigators with eligible NHLBI research grants to include undergraduate students in their projects. Priority is given to applications requesting support for students with disabilities and/or from underrepresented racial and ethnic groups. The research proposed for the student must be part of the approved research for the parent grant or a logical extension. The student must be given opportunities to interact with other individuals on the study, to contribute intellectually, and to enhance his or her research skills and knowledge. The program is designed to support undergraduate students who are interested in biomedical or behavioral research and who will devote at least 2 years (two 3-month periods) to the program. Support should be requested for at least 3 months during any 1 year. Applications can be submitted any time, but no later than 3 months before the last 2 years remaining on the NHLBI research grant. Stipend include: $10 per hour for salary; and supplies, travel (up to $1,000 for travel to scientific meeting(s) or between home/school and worksite), and other up to $2,400.
Requirements: All principal investigators at U.S. institutions with eligible NHLBI research grants are eligible.
Contact: Helena O. Mishoe, PhD; (301) 451-5081; fax: (301) 480-0862; email: mishoeh@nhlbi.nih.gov
Internet: http://www.nhlbi.nih.gov/funding/training/redbook/undersup.htm
Sponsor: National Heart, Lung, and Blood Institute
6701 Rockledge Drive, Room 8188
Bethesda, MD 20892-7913

NHLBI Right Heart Function in Health and Chronic Lung Diseases Grants 3807
This program will promote research on cellular, molecular and physiological determinants of right ventricular function in health, and function/ dysfunction in association with chronic lung diseases. The goal is to encourage research which concentrates on building a foundation of basic knowledge of right ventricular physiology and the basic molecular and biomechanical factors contributing to right ventricular dysfunction associated with pulmonary diseases. This knowledge base will be used in determining the best methods of preventing, diagnosing and treating right heart failure. This FOA seeks to fill the current gaps that exist in our understanding of the right heart and how right heart function affects lung function in both health and disease. The emphasis on interaction of the heart and the lung during chronic lung disease in this program is intended to promote collaboration between the heart and lung research communities.
Requirements: The following organizations and institutions are eligible to apply: public/state controlled institutions of higher education; private institutions of higher education; Hispanic-serving institutions; historically Black colleges and universities (HBCUs); tribally controlled colleges and universities (TCCUs); Alaska Native and Native Hawaiian serving institutions; nonprofits with 501(c)3 IRS status; nonprofits without 501(c)3 IRS status; for-profit organizations; state governments; Indian/Native American tribal governments (federally recognized); Indian/Native American tribally designated organizations; county governments; city or township governments; special district governments; independent school districts; public housing authorities/Indian housing authorities; U.S. territory or possession; Indian/Native American tribal governments (other than federally recognized); regional organizations; non-domestic (non-U.S.) entities (foreign organizations); other eligible agencies of the federal government; and faith-based or community-based organizations.
Date(s) Application is Due: Feb 5; Jun 5; Oct 5

Contact: Elizabeth M. Denholm, Ph.D.; (301) 435-0222; fax: (301) 480-3557; email: denholme@mail.nih.gov
Internet: http://grants.nih.gov/grants/guide/pa-files/PA-07-043.html
Sponsor: National Heart, Lung, and Blood Institute (NHLBI)
6701 Rockledge Drive, MSC 7952
Bethesda, MD 20892-7952

NHLBI Ruth L. Kirschstein National Research Service Award (NRSA) Institutional Research Training Grants 3808

This award enables research institutions to support predoctoral research training in specific areas and fields of shortage. Selection of institutions is by national competition; trainees are selected through local review procedures established by the program director at the grantee institution. Institutional grants are renewable, through competition, for up to 5 years. The NHLBI will accept all types of competing T32 applications (new, renewal/competing continuation, resubmission/amended) until January 25. Only resubmission/amended T32 applications will be accepted until September 25.
Requirements: Trainees are required to pursue their research training full-time, and trainees in clinical areas are expected to confine their clinical duties to those that are part of their research training. Trainees must be U.S. citizens, noncitizen nationals, or legal permanent residents of the U.S. Persons on temporary or student visas are not eligible.
Date(s) Application is Due: Jan 25; Sep 25
Contact: Michael Commarato, Ph.D.; (301) 435-0535; fax: (301) 480-7971; email: CommaraM@nhlbi.nih.gov
Internet: http://www.nhlbi.nih.gov/funding/training/redbook/gradt32.htm
Sponsor: National Heart, Lung, and Blood Institute
Two Rockledge Center, 6701 Rockledge Drive, Room 8140, MSC 7940
Bethesda, MD 20892-7940

NHLBI Ruth L. Kirschstein National Research Service Award Short-Term Institutional Research Training Grants 3809

This program provides funds to research institutions to make awards to individuals in health professional schools for research opportunities that would not be available through their regular course of study. Awards are made to training institutions by national competition. The applicant institution may request the proportion of the NIH standard annual NRSA Training Related Expenses to help defray other costs of the short-term training experience, such as research supplies, tuition, fees, certain types of travel, and other expenses. Amounts for TRE are $4,200 annually ($350/month) for each predoctoral trainee and $7,850 annually ($654/month) for each postdoctoral trainee. Health insurance is an allowable expense that may be charged to the Trainee Related Expenses budget category but only to the extent that the same health insurance fees are charged to non-Federally students and postdoctoral individuals.
Requirements: Trainees are selected by the grantee institution and must be U.S. citizens, noncitizen nationals, or legal permanent residents of the U.S. Trainees should have successfully completed at least 1 semester at an accredited school of medicine, optometry, osteopathy, dentistry, veterinary medicine, pharmacy, or public health before entering the program. The award cannot be used to support courses that are required for the M.D., D.O., D.D.S., D.V.M., or similar professional degrees.
Date(s) Application is Due: Jan 25
Contact: Charlotte Pratt, Ph.D.; (301) 435-0382; fax: (301) 480-5158; email: PrattC@nhlbi.nih.gov
Internet: http://www.nhlbi.nih.gov/funding/training/redbook/gradt35.htm
Sponsor: National Heart, Lung, and Blood Institute (NHLBI)
6701 Rockledge Drive, MSC 7936
Bethesda, MD 20892-7936

NHLBI Ruth L. Kirschstein National Research Service Awards for Individual Predoctoral MD/PhD Fellows 3810

The National Heart, Lung, and Blood Institute (NHLBI) provides Ruth L. Kirschstein National Research Service Awards (Kirschstein-NRSA) to individuals for combined MD/PhD fellowship training. The NHLBI awards this Kirschstein-NRSA individual fellowship to promising applicants with the potential to become productive, independent, highly trained physician-scientists, including patient-oriented physician-scientists in its scientific mission area. The Institute plans, conducts, fosters, and supports an integrated and coordinated program of basic research, clinical investigations and trials, observational studies, and demonstration and education projects. Research is related to the causes, prevention, diagnosis, and treatment of heart, blood vessel, lung, and blood diseases; and sleep disorders. The NIH will reimburse 60 percent of the cost of tuition up to $21,000.
Requirements: Applications may be submitted by: for-profit organizations; non-profit organizations; public or private institutions, such as universities, colleges, hospitals, and laboratories; eligible agencies of the Federal government; and domestic Institutions
Date(s) Application is Due: Apr 8; Aug 8; Dec 8
Contact: Jane D. Scott, Sc.D., M.S.N.; (301) 435-0535; fax: (301) 480-7404; email: scottj2@nhlbi.nih.gov
Internet: http://grants.nih.gov/grants/guide/pa-files/PA-08-021.html
Sponsor: National Heart, Lung, and Blood Institute
6701 Rockledge Drive, MSC 7940
Bethesda, MD 20892-7940

NHLBI Sarcoidosis: Research Grants into the Cause of Multi-Organ Disease and Clinical Strategies for Therapy 3811

The purpose of this Funding Opportunity Announcement (FOA) is to stimulate research on the etiology and management of sarcoidosis, an immune,Äìmediated granulomatous inflammatory disorder. Studies supported by this FOA would include investigations to find the etiology(ies) and related host factors that might enhance susceptibility to sarcoidosis, especially development of symptomatic multi-organ disease that has a propensity to involve critical organs (such as lungs, heart, eyes, central/peripheral nervous system, liver, kidneys and other abdominal viscera) that create serious illness. Study of the still problematic management and the biological, behavioral, or psychosocial burden on individuals, families, and community is encouraged. Moreover, there is interest for approaches to risk reduction, psychological coping, and management of complications and side effects of treatment. Because the nature and scope of the proposed research will vary from application to application, it is anticipated that the size and duration of each award will also vary. The total amount awarded and the number of awards will depend upon the mechanism numbers, quality, duration, and costs of the applications received.
Requirements: The following organizations and institutions are eligible to apply: public/state controlled institutions of higher education; private institutions of higher education; Hispanic-serving institutions; historically Black colleges and universities (HBCUs); tribally controlled colleges and universities (TCCUs); Alaska Native and Native Hawaiian serving institutions; nonprofits with 501(c)3 IRS status; nonprofits without 501(c)3 IRS status; for-profit organizations; state governments; Indian/Native American tribal governments (federally recognized); Indian/Native American tribally designated organizations; county governments; city or township governments; special district governments; independent school districts; public housing authorities/Indian housing authorities; U.S. territory or possession; Indian/Native American tribal governments (other than federally recognized); regional organizations; non-domestic (non-U.S.) entities (foreign organizations); other eligible agencies of the federal government; and faith-based or community-based organizations.
Date(s) Application is Due: Feb 5; Jun 5; Oct 5
Contact: Herbert Y. Reynolds, M.D.; (301) 435-0222; fax: (301) 480-3557; email: Reynoldh@nhlbi.nih.gov
Internet: http://grants.nih.gov/grants/guide/pa-files/PA-07-136.html
Sponsor: National Heart, Lung, and Blood Institute (NHLBI)
Two Rockledge Center, Suite 10018, 6701 Rockledge Drive, MSC 7952
Bethesda, MD 20892-7952

NHLBI Short-Term Research Education Program to Increase Diversity in Health-Related Research 3812

This FOA issued by the National Heart, Lung, and Blood Institute, National Institutes of Health, solicits Research Education (R25) applications from institutions/organizations to promote diversity in undergraduate and health professional student populations by providing short-term research education support to stimulate career development in cardiovascular, pulmonary, hematologic, and sleep disorders research. The overall goal of the program is to provide research opportunities for individuals from disadvantaged backgrounds, racial and ethnic minorities, and others who are underrepresented in biomedical and behavioral research (e.g., individuals with disabilities), that will significantly contribute to a diverse research workforce in the future. The estimated funds (total costs) available for the first year of support for the entire program are expected to be $850,000. Up to eight new awards are anticipated, depending on the number of student participants requested and the availability of funds.
Requirements: The following organizations and institutions are eligible to apply: domestic institutions; non-profit organizations; public/state controlled institutions of higher education; private institutions of higher education; nonprofits with 501(c)3 IRS status; for-profit organizations (other than small businesses); state governments; county governments; city or township governments; special district governments; and faith-based or community-based organizations.
Restrictions: Applications from foreign institutions will not be accepted. Individuals on temporary or student visas and individuals holding PhD, MD, DVM, or equivalent doctoral degrees in the health sciences are not eligible.
Amount of Grant: $850,000 total
Date(s) Application is Due: Jul 15; Aug 15
Contact: Dr. Michael Commarato; (301) 435-0530; fax: (301) 480-1454; email: michael_commarato@nih.gov
Internet: http://grants.nih.gov/grants/guide/rfa-files/RFA-HL-08-016.html
Sponsor: National Heart, Lung, and Blood Institute
6701 Rockledge Drive, MSC 7940
Bethesda, MD 20892-7924

NHLBI Stress Management Interventions to Reduce Risk of Coronary Artery Disease 3813

The purpose of this program announcement is to encourage applications that evaluate the efficacy of stress management interventions on intermediate outcomes purported to be in the biological or behavioral mechanistic pathway through which mental stress (MS) is thought to influence major adverse cardiac events (MACE) in patients with confirmed coronary artery disease (CAD). These include subclinical measures of atherosclerosis, measures of cardiac or endothelial function, or measures such as cardiovascular reactivity or neurohumoral response to a provocative mental stress test. The ultimate goal is to determine whether stress management interventions can slow progression of CAD or reduce MACE and, if so, to translate effective stress management interventions to clinical practice.
Requirements: The following organizations and institutions are eligible to apply: individuals; public/state controlled institutions of higher education; private institutions of higher education; Hispanic-serving institutions; historically Black colleges and universities (HBCUs); tribally controlled colleges and universities (TCCUs); Alaska Native and Native Hawaiian serving institutions; nonprofits with 501(c)3 IRS status;

nonprofits without 501(c)3 IRS status; for-profit organizations; state governments; Indian/Native American tribal governments (federally recognized); Indian/Native American tribally designated organizations; county governments; city or township governments; special district governments; independent school districts; public housing authorities/Indian housing authorities; U.S. territory or possession; Indian/Native American tribal governments (other than federally recognized); regional organizations; non-domestic (non-U.S.) entities (foreign organizations); other eligible agencies of the federal government; and faith-based or community-based organizations.
Date(s) Application is Due: Feb 5; Jun 5; Oct 5
Contact: Peter G. Kaufmann, Ph.D.; (301) 435-2467; fax: (301) 480-1773; email: KaufmannP@mail.nih.gov; Cristina Rabadan-Diehl, Ph.D.; (301) 435-0550; fax: (301) 480-2858
Internet: http://grants.nih.gov/grants/guide/pa-files/PA-07-322.html
Sponsor: National Heart, Lung, and Blood Institute (NHLBI)
6701 Rockledge Drive, Room 8118, MSC 7936
Bethesda, MD 20892-7936

NHLBI Summer Institute for Training in Biostatistics II **3814**
The National Heart, Lung, and Blood Institute (NHLBI) and the National Center for Research Resources (NCRR) invite applications for grants to develop, conduct and evaluate summer courses in the basic principles and methods of biostatistics as employed in biomedical research. The courses would introduce participants to the field of biostatistics for the purpose of attracting new students into the field. The course will attract students from the entire U.S.A and will cover the fundamental concepts of probability, statistical reasoning and inferential methods motivated, in part, by examples that include data collected in studies of heart, lung, blood, and sleep disorders. The course will be taught during the summer, with appropriate modifications or refinements following each of the first two summers' sessions. Programs may choose an area of emphasis, such as clinical trials, statistical genetics, bioinformatics or epidemiology or take a general approach. Recent college graduates as well as advanced undergraduates should be eligible to participate provided they have not participated in SIBS previously. Direct costs are limited to $248,000 per year for a three-year period. Letters of Intent should be received by December 5, and full proposals are due by January 6.
Requirements: The following organizations and institutions are eligible to apply: public/state controlled institutions of higher education; private institutions of higher education; Hispanic-serving institutions; historically Black colleges and universities (HBCUs); tribally controlled colleges and universities (TCCUs); Alaska Native and Native Hawaiian serving institutions; nonprofits with 501(c)3 IRS status; nonprofits without 501(c)3 IRS status; for-profit organizations; state governments; Indian/Native American tribal governments (federally recognized); Indian/Native American tribally designated organizations; county governments; city or township governments; special district governments; independent school districts; public housing authorities/Indian housing authorities; U.S. territory or possession; Indian/Native American tribal governments (other than federally recognized); regional organizations; non-domestic (non-U.S.) entities (foreign organizations); other eligible agencies of the federal government; and faith-based or community-based organizations.
Amount of Grant: $248,000 maximum per year for a three-year period
Date(s) Application is Due: Dec 5; Jan 6
Contact: Song Yang, Ph.D.; (301) 435-0431; fax: (301) 480-1862; email: yangso@nhlbi.nih.gov
Internet: http://grants.nih.gov/grants/guide/rfa-files/RFA-HL-09-009.html
Sponsor: National Heart, Lung, and Blood Institute (NHLBI)
6701 Rockledge Drive, Room 9210
Bethesda, MD 20892

NHLBI Summer Internship Program in Biomedical Research **3815**
The goal of the Summer Internship Program in Biomedical Research at NHLBI is to expose students to research investigation in a highly enriched environment that is devoted exclusively to biomedical research and training. The Program is open to high school and college, graduate or Medical and Dental Students. Participants join a research laboratory for a minimum of ten weeks between June and August and conduct research in selected areas of investigation under the guidance of an NHLBI intramural research scientist. To augment the training experience, students participate in a Seminar Series at the NHLBI, as well as the National Institutes of Health Summer Seminar Series and Summer Research Program Poster Day. While applications may be filed between Nov. 15 and March 1, early applications are encouraged.
Requirements: Applicants must be either U.S. citizens or permanent residents and be currently enrolled full-time in high school, college, professional or graduate school with a minimum GPA of 3.0 on a 4.0 scale. College graduates are eligible for the summer between graduation and beginning either graduate or professional school. High school students should be at least 16 years of age prior to June 1. Applications must be submitted online using the link for provided for Summer Internship Applicants.
Date(s) Application is Due: Mar 1
Contact: Llewellyn, Jessica, MBA, Program Coordinator; (301) 451-9440; fax: (301) 594-8133; email: llewellj@mail.nih.gov or direducation@nhlbi.nih.gov
Internet: http://dir.nhlbi.nih.gov/oe/summerprogs.asp
Sponsor: National Heart, Lung, and Blood Institute (NHLBI)
6701 Rockledge Drive, Building 10, Room 2N242
Bethesda, MD 20892-1754

NHLBI Targeted Approaches to Weight Control for Young Adults Grants **3816**
The National Heart, Lung, and Blood Institute (NHLBI) of the National Institutes of Health (NIH) invites applications for cooperative agreements (U01) to conduct two-phase clinical research studies to develop, refine, and test innovative behavioral and/or environmental approaches for weight control in young adults at high risk for weight gain. Weight control interventions can address weight loss, prevention of weight gain, or prevention of excessive weight gain during pregnancy. The first phase will consist of formative research to refine the proposed intervention, recruitment, retention, and adherence strategies targeted to young adults. The second phase will consist of a randomized controlled trial to test the efficacy of the intervention. For the purpose of this FOA, young adults are defined as 18-35 years of age. The maximum award for the research component will be approximately $3,000,000 in direct costs, including direct costs of up to $500,000 for the first year, $650,000 for the second year, $650,000 for the third year, $650,000 for the fourth year, and $550,000 for the fifth year. Letters of Intent should be received by September 10, and full proposals are due by October 10.
Requirements: The following organizations and institutions are eligible to apply: public/state controlled institutions of higher education; private institutions of higher education; Hispanic-serving institutions; historically Black colleges and universities (HBCUs); tribally controlled colleges and universities (TCCUs); Alaska Native and Native Hawaiian serving institutions; nonprofits with 501(c)3 IRS status; nonprofits without 501(c)3 IRS status; for-profit organizations; Indian/Native American tribal governments; Indian/Native American Tribally Designated Organizations; eligible agencies of the federal government; faith-based or community-based organizations; and regional organizations.
Amount of Grant: $500,000-$650,000 maximum
Date(s) Application is Due: Sep 10; Oct 10
Contact: Catherine (Cay) Loria, Ph.D.; (301) 435-0702; fax: (301) 480-5158; email: loriac@nhlbi.nih.gov
Internet: http://grants.nih.gov/grants/guide/rfa-files/RFA-HL-08-007.html
Sponsor: National Heart, Lung, and Blood Institute (NHLBI)
6701 Rockledge Drive, Suite 10118, MSC 7936
Bethesda, MD 20892-7936

NHLBI Targeting Calcium Regulatory Molecules for Arrhythmia Prevention Grants **3817**
The purpose of this Funding Opportunity Announcement (FOA) is to stimulate the development of therapeutic strategies for cardiac arrhythmias that target Ca2+ handling and Ca2+ signaling molecules. The overall goal is to exploit present knowledge of Ca2+ regulatory molecules and the role they play in arrhythmogenesis to develop novel therapeutic targets. Critical components of successful responses to this program announcement are: preclinical or early clinical testing of promising new therapeutic strategies to terminate and prevent cardiac arrhythmias in established models of human arrhythmogenic disease; and the feasibility to move positive findings to Phase I or II clinical trials. Grantees must budget for and attend two Project Director/Principal Investigator (PD/PI) meetings in Bethesda, MD, usually scheduled during the second and fourth year of the award.
Requirements: The following organizations and institutions are eligible to apply: public/state controlled institutions of higher education; private institutions of higher education; Hispanic-serving institutions; historically Black colleges and universities (HBCUs); tribally controlled colleges and universities (TCCUs); Alaska Native and Native Hawaiian serving institutions; nonprofits with 501(c)3 IRS status; nonprofits without 501(c)3 IRS status; for-profit organizations; state governments; Indian/Native American tribal governments (federally recognized); Indian/Native American tribally designated organizations; county governments; city or township governments; special district governments; independent school districts; public housing authorities/Indian housing authorities; U.S. territory or possession; Indian/Native American tribal governments (other than federally recognized); regional organizations; non-domestic (non-U.S.) entities (foreign organizations); other eligible agencies of the federal government; and faith-based or community-based organizations.
Date(s) Application is Due: Feb 5; Jun 5; Oct 5
Contact: Dennis A. Przywara, Ph.D.; (301) 435-0506; fax: (301) 480-1454; email: przywarad@nhlbi.nih.gov
Internet: http://grants.nih.gov/grants/guide/pa-files/PA-07-101.html
Sponsor: National Heart, Lung, and Blood Institute (NHLBI)
6701 Rockledge Drive, Room 9186, MSC 7940
Bethesda, MD 20892-7940

NHLBI Transfusion-Related Acute Lung Injury Basic Research Grants **3818**
This Funding Opportunity Announcement (FOA) issued by the National Heart, Lung, and Blood Institute (NHLBI), National Institutes of Health, solicits Research Project Grant (R01) applications from institutions/organizations that propose to study the mechanisms responsible for acute lung injury after blood transfusion and stimulate the development of strategies to improve the prevention, diagnosis, and treatment of the disorder for the four million transfusion recipients per year in the U.S. who may be at risk for developing Transfusion-Related Acute Lung Injury (TRALI).
Requirements: The following organizations and institutions are eligible to apply: individuals; public/state controlled institutions of higher education; private institutions of higher education; nonprofits with 501(c)3 IRS status; nonprofits without 501(c)3 IRS status; small businesses; for-profit organizations; and non-domestic entities.
Date(s) Application is Due: Feb 5; Jun 5; Oct 5
Contact: Traci Heath Mondoro, Ph.D.; (301) 435-0052; fax: (301) 480-1060; email: mondorot@nhlbi.nih.gov
Internet: http://grants.nih.gov/grants/guide/pa-files/PA-07-385.html
Sponsor: National Heart, Lung, and Blood Institute
6701 Rockledge Drive, MSC 7950
Bethesda, MD 20892-0052

NHLBI Translating Basic Behavioral and Social Science Discoveries into Interventions to Reduce Obesity: Centers for Behavioral Intervention Development Grants 3819

This FOA solicits cooperative agreement (U01) applications from institutions/organizations that propose to translate findings from basic research on human behavior into more effective clinical, community, and population interventions to reduce obesity and improve obesity-related behaviors. This FOA will support Centers for Behavioral Intervention Development (CBIDs) in which interdisciplinary teams of basic and applied behavioral and social science researchers develop and refine novel interventions to reduce obesity and alter obesity-related health behaviors (e.g., diet, physical activity). The interventions to be developed include any of a wide range of strategies aimed at promoting weight loss and/or preventing weight gain. Interventions can be targeted to the individual, family, social network, community, environmental, clinical or population level, or combinations of these, and to any age group. Investigators at each CBID will conduct several types of studies during the funding period, including laboratory experimental studies, formative research, early phase trials and pilot/feasibility studies, in order to create promising new avenues for reducing obesity and improving obesity-related behaviors. The maximum award for the research component will be approximately $3,750,000 in direct costs across five years, with a limit of $750,000 in direct cost for any single year. Letters of Intent should be received by December 16, and full proposals are due by January 13.

Requirements: The following organizations and institutions are eligible to apply: public/state controlled institutions of higher education; private institutions of higher education; Hispanic-serving institutions; historically Black colleges and universities (HBCUs); tribally controlled colleges and universities (TCCUs); Alaska Native and Native Hawaiian serving institutions; nonprofits with 501(c)3 IRS status; nonprofits without 501(c)3 IRS status; for-profit organizations; state governments; Indian/Native American tribal governments (federally recognized); Indian/Native American tribally designated organizations; county governments; city or township governments; special district governments; independent school districts; public housing authorities/Indian housing authorities; U.S. territory or possession; Indian/Native American tribal governments (other than federally recognized); regional organizations; non-domestic (non-U.S.) entities (foreign organizations); other eligible agencies of the federal government; and faith-based or community-based organizations.

Amount of Grant: $750,000 maximum per year

Date(s) Application is Due: Dec 16; Jan 13

Contact: Susan M. Czajkowski, Ph.D.; (301) 435-0406; fax: (301) 480-1773; email: czajkows@mail.nih.gov

Internet: http://grants.nih.gov/grants/guide/rfa-files/RFA-HL-08-013.html

Sponsor: National Heart, Lung, and Blood Institute (NHLBI)
6701 Rockledge Drive, Suite 10118, MSC 7936
Bethesda, MD 20892-7936

NHLBI Translational Programs in Lung Diseases 3820

The purpose of this program is to encourage interdisciplinary approaches to lung translational research, allowing for potentially high-impact research, which will translate basic observations to applied clinical research. An application to NHLBI requires a minimum of three interrelated research projects that focus a number of scientific disciplines on investigations of a complex biomedical theme or research question. Applicants will be required to propose research that will become progressively more clinical and translational during the life of the program project, and they are expected to assemble interactive, multidisciplinary teams that have the combined expertise to formulate a plan for successful translation. Research teams are not required to have prior collaborative experience but must be able to demonstrate an integrated, practical approach that will result in the effective progression of mechanistic/basic concepts toward application in the clinic. This program allows for a maximum of two 5-year awards (a new application and one competing renewal). There are two application cycles. Letters of Intent should be received by April 25 and August 25 (respectively), and full proposals are due by May 25 and September 25 (respectively).

Requirements: The following organizations and institutions are eligible to apply: public/state controlled institutions of higher education; private institutions of higher education; Hispanic-serving institutions; historically Black colleges and universities (HBCUs); tribally controlled colleges and universities (TCCUs); Alaska Native and Native Hawaiian serving institutions; nonprofits with 501(c)3 IRS status; nonprofits without 501(c)3 IRS status; for-profit organizations; state governments; Indian/Native American tribal governments (federally recognized); Indian/Native American tribally designated organizations; county governments; city or township governments; special district governments; independent school districts; public housing authorities/Indian housing authorities; U.S. territories or possessions; Indian/Native American tribal governments (other than federally recognized); regional organizations; non-domestic (non-U.S.) entities (foreign organizations); other eligible agencies of the federal government; and faith-based or community-based organizations.

Date(s) Application is Due: Apr 25; May 25; Aug 25; Sep 25

Contact: Patricia Noel, Ph.D.; (301) 435-0202; email: noelp@nhlbi.nih.gov

Internet: http://grants.nih.gov/grants/guide/pa-files/PAR-09-185.html

Sponsor: National Heart, Lung and Blood Institute (NHLBI)
6701 Rockledge Drive
Bethesda, MD 20892-7952

NHPRC Archives - Basic Projects Grants 3821

The National Historical Publications and Records Commission seeks proposals for archives and records projects to carry out fundamental archival activities that promote the preservation and use of America's documentary heritage essential to understanding our democracy, history, and culture. The Commission is particularly concerned that some historical collections in archives and repositories are difficult for most researchers to find because they are not processed or represented in national catalogs. It hopes to encourage repositories to reveal these hidden collections by concentrating on materials in their backlogs. In addition, the NHPRC wishes to continue support for planning for archives, collection development, and appropriate preservation. As a result, the Commission seeks proposals for projects that support backlog processing and collections cataloging, phased preservation, and/or collections development. A grant normally is for one or two years and for up to $150,000. The Commission expects to make up to 20 grants in this category for a total of up to $1,000,000.

Requirements: Although projects focused exclusively on professional development are not eligible in this grant category, the Commission strongly encourages each application to include professional development components necessary for the success of the project.

Amount of Grant: $150,000 maximum

Date(s) Application is Due: Oct 6

Contact: Daniel A. Stokes, Program Officer; (202) 357-5487 or (202) 357-5010; fax: (202) 357-5914; email: daniel.stokes@nara.gov

Internet: http://www.archives.gov/nhprc/announcement/basic.html

Sponsor: National Historical Publications and Records Commission
700 Pennsylvania Avenue NW, Room 106
Washington, DC 20408-0001

NHPRC Archives - Detailed Processing Projects Grants 3822

The National Historical Publications and Records Commission seeks proposals for detailed processing and preservation of collections of national significance. The collections should also have high research demand or substantial preservation challenges. Detailed Processing Projects may include reformatting records onto microfilm or other media in order to ensure their permanence in the face of high demand. Applicants may process and create detailed descriptions at the series or file level. Such descriptions will improve user access to historical records and help preserve collections. If parts of collections deserve processing to the item level, applicants must provide specific justifications for this detailed degree of work and provide estimates of the percentage of collections to be processed to the item level. A grant normally is for one to three years and ranges between $40,000-$200,000. The Commission expects to make up to 5 grants in this category, for a total of up to $500,000.

Requirements: Eligible projects include archives and other repositories of historical documents that are part of: nonprofit organizations or institutions; colleges and universities and other academic institutions; state or local government agencies; and federally-acknowledged or state-recognized Native American tribes or groups.

Amount of Grant: $40,000-$200,000

Date(s) Application is Due: Oct 6

Contact: Daniel A. Stokes, Program Officer; (202) 357-5487 or (202) 357-5010; fax: (202) 357-5914; email: daniel.stokes@nara.gov

Internet: http://www.archives.gov/nhprc/announcement/detailed.html

Sponsor: National Historical Publications and Records Commission
700 Pennsylvania Avenue NW, Room 106
Washington, DC 20408-0001

NHPRC Digitizing Historical Records Grants 3823

The Commission seeks proposals that use cost-effective methods to digitize nationally significant historical record collections and make the digital versions freely available online. Projects must make use of existing holdings of historical repositories and consist of entire collections or series. The materials should already be available to the public at the archives and described so that projects can re-use existing information to serve as metadata for the digitized collection. A grant normally is for 1 to 3 years and up to $150,000. The Commission expects to make up to 5 grants in this category, for a total of up to $400,000.

Requirements: Eligible applicants include: nonprofit organizations or institutions with IRS 501(c)3 tax exempt status; colleges, universities, and other academic institutions; state or local government agencies; and federally-acknowledged or state-recognized Native American tribes or groups. Cost sharing is required. It is the financial contribution the applicant pledges to the cost of a project. Cost sharing can include both direct and indirect expenses, in-kind contributions, non-Federal third-party contributions, and any income earned directly by the project. The NHPRC will provide up to 50 percent of the total project costs.

Restrictions: Projects may not use grant funds to create descriptive metadata or edited transcriptions of the digitized materials.

Amount of Grant: $150,000 maximum over three years

Date(s) Application is Due: Jun 5

Contact: Daniel A. Stokes, Program Officer; (202) 357-5487 or (202) 357-5010; fax: (202) 357-5914; email: daniel.stokes@nara.gov

Internet: http://www.archives.gov/nhprc/announcement/digitizing.html

Sponsor: National Historical Publications and Records Commission
700 Pennsylvania Avenue NW, Room 106
Washington, DC 20408-0001

NHPRC Electronic Records Projects Grants 3824

The National Historical Publications and Records Commission seeks proposals that will increase the capacity of archival repositories to create electronic records archives that preserve records of enduring historical value. The NHPRC supports efforts by archivists and records managers to meet the challenges of electronic records. Projects must involve institutions that have already established archives and records management programs. A grant normally is for 1 to 3 years and up to $300,000. The Commission expects to make

up to 6 grants in this category, for a total of up to $600,000. The NHPRC will provide up to 50 percent of the total project costs for Electronic Records Archives projects.
Requirements: Eligible applicants include: nonprofit organizations or institutions with IRS 501(c)3 tax exempt status; colleges, universities, and other academic institutions; state or local government agencies; and federally-acknowledged or state-recognized Native American tribes or groups. Cost sharing is required. It is the financial contribution the applicant pledges to the cost of a project. Cost sharing can include both direct and indirect expenses, in-kind contributions, non-Federal third-party contributions, and any income earned directly by the project.
Restrictions: Projects in this category cannot digitize historical records.
Amount of Grant: $300,000 maximum for up to three years
Date(s) Application is Due: Jun 5
Contact: Daniel A. Stokes, Program Officer; (202) 357-5487 or (202) 357-5010; fax: (202) 357-5914; email: daniel.stokes@nara.gov
Internet: http://www.archives.gov/nhprc/announcement/electronic.html
Sponsor: National Historical Publications and Records Commission
700 Pennsylvania Avenue NW, Room 106
Washington, DC 20408-0001

NHPRC Historical Editing Fellowships 3825
The National Historical Publications and Records Commission seeks proposals from active NHPRC-supported publications projects to serve as a host institution for a Historical Documentary Editing Fellowship. Applicants should demonstrate the capability to provide strong post-graduate training in documentary editing, including document collection, accessioning, and control; selection; transcription; annotation; proofreading; indexing; and project management. Awards are for one year grants of $55,000 each. The basic stipend may be up to $42,000. A 25 percent allowance for fringe benefits supplements the stipend. In addition, up to $2 500 may be used by the institution to cover the costs of interviewing individual applicants and/or sending fellows to professional conferences or training institutes. Depending on the quality of proposals and the availability of funding, the Commission expects to fund one to two fellowship projects, each with a single fellow.
Requirements: Proposals may be submitted only by active NHPRC-supported historical documentary editing projects that are affiliated with: nonprofit organizations and institutions; colleges, universities, and other academic institutions; state or local government agencies; federally-acknowledged or state-recognized Native American tribes or groups.
Amount of Grant: $55,000 maximum
Date(s) Application is Due: Oct 6
Contact: Daniel A. Stokes, Program Officer; (202) 357-5487 or (202) 357-5010; fax: (202) 357-5914; email: daniel.stokes@nara.gov
Internet: http://www.archives.gov/nhprc/announcement/fellowship.html
Sponsor: National Historical Publications and Records Commission
700 Pennsylvania Avenue NW, Room 106
Washington, DC 20408-0001

NHPRC Publishing Historical Records Grants 3826
The National Historical Publications and Records Commission seeks proposals to publish historical records of national significance. Projects may focus on the papers of major figures from American life or cover broad historical movements in politics, military, business, reform movements, the arts, and other aspects of the national experience. The historical value of the records and their expected usefulness to broad audiences must justify the costs of the project. Grants are awarded for collecting, describing, preserving, compiling, editing, and publishing documentary source materials. Eligible Activities Include: scholarly documentary editions in printed and bound volumes; scholarly documentary editions in online, compact disc, and other formats; image editions in online, compact disc, microfilm, and other formats; conversion of existing print and microfilm editions to electronic publications; and combinations of the above. Award amounts ordinarily range from $20,000 to $250,000 annually. The Commission expects to make as many as 35-40 grants in this category. Cost sharing can include both direct and indirect expenses, in-kind contributions, non-Federal third-party contributions, and any income earned directly by the project.
Requirements: Eligibility includes: individuals; nonprofit organizations or institutions with IRS 501(c)3 tax exempt status; colleges, universities, and other academic institutions; state or local government agencies; and federally-acknowledged or state-recognized Native American tribes or groups. Cost sharing is required.
Restrictions: The NHPRC does not fund proposals to purchase historical records; it also does not fund proposals to publish the papers of anyone who has been deceased for fewer than ten years.
Amount of Grant: $20,000-$250,000 annually
Date(s) Application is Due: Jun 5
Contact: Daniel A. Stokes, Program Officer; (202) 357-5487 or (202) 357-5010; fax: (202) 357-5914; email: daniel.stokes@nara.gov
Internet: http://www.archives.gov/nhprc/announcement/publishing.html
Sponsor: National Historical Publications and Records Commission
700 Pennsylvania Avenue NW, Room 106
Washington, DC 20408-0001

NHPRC State and National Archival Partnership Grants 3827
The National Historical Publications and Records Commission seeks proposals to strengthen archives and historical records programs in each of the states and build a national archival network. The Commission awards grants to State Historical Records Advisory Boards to: provide statewide archival services, including professional education, public information about records and archival programs, and other activities to implement state plans; operate grant programs for eligible archives, manuscript repositories, and other organizations within a state; collaborate on projects with other organizations to address common problems or shared opportunities within a state or among a consortium of state archives; assess the health of archival and records programs, the conditions of records, and the challenges and opportunities facing historical records keepers and users; and to implement strategies and programs to address pressing archival issues; hold meetings and public forums on statewide or national archival issues; and participate in national archival organizations. The NHPRC expects to award up to 56 grants in this program. Applicants may apply for outright basic grants of between $5,000 and $20,000 for 18 month projects. For State and National Archival Partnership Projects, the Commission will supply up to 50 percent of the total project costs.
Requirements: These grants are awarded only to State Historical Records Advisory Boards (SHRABs) in each state, or to the state agency responsible for the SHRAB, ordinarily the state archives. Another state agency, or a non-profit organization, such as a foundation or university, acting on behalf of the designated state agency may apply.
Amount of Grant: $5,000-$20,000 for 18 months
Date(s) Application is Due: Mar 6
Contact: Daniel A. Stokes, Program Officer; (202) 357-5487 or (202) 357-5010; fax: (202) 357-5914; email: daniel.stokes@nara.gov
Internet: http://www.archives.gov/nhprc/announcement/partnership.html
Sponsor: National Historical Publications and Records Commission
700 Pennsylvania Avenue NW, Room 106
Washington, DC 20408-0001

NHPRC Strategies and Tools for Archives and Historical Publishing Projects Grants 3828
The National Historical Publications and Records Commission seeks proposals to develop new strategies and tools that can improve the preservation, public discovery, or use of historical records. Projects may also focus on techniques and tools that will improve the professional performance and effectiveness of those who work with such records, such as archivists, documentary editors, and records managers. Projects concerning records may focus on methods of working with any format including born-digital records. Publishing of historical records must focus on methods of presenting archival records as primary sources. The Commission does not fund projects focused on artifacts or books. A grant normally is for one to three years. The Commission expects to make one to three grants of between $50,000 and $150,000. The total amount allocated to this category is up to $350,000 during the fiscal year.
Requirements: Eligibility includes: nonprofit organizations or institutions with IRS 501(c)3 tax exempt status; colleges, universities, and other academic institutions; state or local government agencies; and federally-acknowledged or state-recognized Native American tribes or groups. Cost sharing is required.
Amount of Grant: $50,000-$150,000
Date(s) Application is Due: Jun 5
Contact: Daniel A. Stokes, Program Officer; (202) 357-5487 or (202) 357-5010; fax: (202) 357-5914; email: daniel.stokes@nara.gov
Internet: http://www.archives.gov/nhprc/announcement/strategies.html
Sponsor: National Historical Publications and Records Commission
700 Pennsylvania Avenue NW, Room 106
Washington, DC 20408-0001

NIA Aging Research Dissertation Awards to Increase Diversity 3829
This Funding Opportunity Announcement (FOA) issued by the National Institute on Aging (NIA) of the National Institutes of Health (NIH) publicizes the re-issuance of an FOA that publicizes the availability of dissertation awards in aging research to increase the diversity of the research-on-aging workforce. These awards are available to all qualified Predoctoral students in accredited research doctoral programs in the United States (including Puerto Rico and other U.S. territories or possessions).
Requirements: Public or private institutions, such as universities, colleges, hospitals, and laboratories, that offer accredited research doctoral programs may submit an application.
Restrictions: Domestic institutions only are eligible.
Amount of Grant: $50,000 average
Date(s) Application is Due: Feb 16; Jun 16; Oct 16
Contact: Michael-David A.R.R. Kerns; (301) 402-7713; fax: (301) 402-2945; email: michael-david.kerns@nih.hhs.gov
Internet: http://grants.nih.gov/grants/guide/pa-files/PAR-08-250.html
Sponsor: National Institute on Aging (NIA)
7201 Wisconsin Avenue, Suite 2C-218
Bethesda, MD 20814

NIA Aging, Oxidative Stress and Cell Death Grants 3830
The purpose of this Program Announcement is to encourage the submission of applications to support research on the relationship between oxidative stress and apoptosis, and how these biological processes are involved in aging and/or change with age. Although the National Institute of General Medical Sciences is not participating in this program announcement, that Institute continues to be interested in receiving applications on basic mechanisms of programmed cell death and oxidative stress when these applications do not focus on aging mechanisms. Modular Grant applications will request direct costs in $25,000 modules, up to a total direct cost request of $250,000 per year.
Requirements: The following organizations and institutions may apply: public/state controlled institutions of higher education; private institutions of higher education; Hispanic-serving institutions; historically Black colleges and universities (HBCUs);

tribally controlled colleges and universities (TCCUs); Alaska native and native Hawaiian serving institutions; nonprofits with 501(c)3 IRS status; nonprofits without 501(c)3 IRS status; small businesses; for-profit organizations; State governments; regional organizations; U.S. territories or possessions; Indian/Native American tribal governments; Indian/Native American tribally designated organizations; county governments; city or township governments; special district governments; independent school districts; public housing authorities/Indian housing authorities; eligible agencies of the Federal government; and faith-based or community based organizations.
Amount of Grant: $250,000 maximum per year
Date(s) Application is Due: Feb 5; Jun 5; Oct 5
Contact: Chyren Hunter, Ph.D.; (301) 496-9322; fax: (301) 402-0528; email: Hunterc@nia.nih.gov
Internet: http://grants.nih.gov/grants/guide/pa-files/PA-00-081.html
Sponsor: National Institute on Aging (NIA)
7201 Wisconsin Avenue, MSC 9205, Gateway Building, Room 2C218
Bethesda, MD 20892-9205

NIA AIDS and Aging: Behavioral Sciences Prevention Research Grants 3831
NIA, NIMH, and NINR invite qualified researchers to submit applications to investigate prevention issues relevant to AIDS in middle-aged and older populations. This announcement solicits AIDS prevention research proposals to study primary prevention of disease transmission as well as secondary and tertiary prevention of negative behavioral and social consequences of HIV/AIDS for persons with AIDS, their families, and communities. Thus, the primary goals are to identify social and behavioral factors associated with HIV transmission and disease progression in later life; examine behavioral and social consequences of HIV infection/AIDS across the life course; develop and evaluate age-appropriate behavioral and social interventions for preventing AIDS in middle-aged and older adults and/or ameliorating problems associated with older adult caregiver responsibilities and burdens; explore health care issues surrounding AIDS care; and strengthen existing research and evaluation methods.
Requirements: Applications may be submitted by domestic and foreign for-profit and non-profit organizations, public and private, such as universities, colleges, hospitals, laboratories, units of State and local governments, and eligible agencies of the Federal government. Applications may be submitted by single institutions or by a consortia of institutions.
Date(s) Application is Due: Feb 5; Feb 16; Jun 5; Jun 16; Oct 5; Oct 16
Contact: Ying Tian; (301) 496-6761; fax: (301) 402-1784; email: tiany@nia.nih.gov; John W. Kusek; (301) 594-7735; fax: (301) 480-3510; email: kusekj@extra.niddk.nih.gov
Internet: http://www.grants.nih.gov/grants/guide/pa-files/PA-97-069.html
Sponsor: National Institute on Aging (NIA)
7201 Wisconsin Avenue, Gateway Building, Suite 3C-307
Bethesda, MD 20892-9205

NIA Alzheimer's Disease Core Centers Grants 3832
The National Institute on Aging (NIA) invites applications from qualified institutions for support of Alzheimer's Disease Core Centers (ADCCs). These centers are designed to support and conduct research on Alzheimer's disease (AD), to serve as shared research resources that will facilitate research in AD and related disorders, distinguish them from the processes of normal brain aging and mild cognitive impairment (MCI), provide a platform for training, develop novel techniques and methodologies, and translate these research findings into better diagnostic, prevention and treatment strategies. New applications should request a project period of five years and a budget for direct costs of up to $750,000 per year. Competing renewal applications may request direct costs for all cores (both required and optional), and the other listed functions, i.e., satellites and pilot grants at a level not exceeding the combined direct costs of all funded activities awarded during the final year of the present funding period plus a 3% increase, or $750,000 per year, whichever is larger. Letters of Intent should arrive by the April 6 deadline, with full proposals due by May 5
Requirements: The following organizations and institutions are eligible to apply: public/state controlled institutions of higher education; private institutions of higher education; Hispanic-serving institutions; historically Black colleges and universities (HBCUs); tribally controlled colleges and universities (TCCUs); Alaska Native and Native Hawaiian serving institutions; nonprofits with 501(c)3 IRS status (other than institutions of higher education); and nonprofits without 501(c)3 IRS status (other than institutions of higher education).
Restrictions: At the time of award, the applicant institution cannot have another Alzheimer's Disease Center funded by the NIA.
Amount of Grant: $750,000 per year for five years
Date(s) Application is Due: Apr 6; May 5
Contact: Creighton H. Phelps, (301) 496-9350; fax: (301) 496-1494; email: phelpsc@nia.nih.gov
Internet: http://grants.nih.gov/grants/guide/rfa-files/RFA-AG-10-001.html
Sponsor: National Institute on Aging (NIA)
7201 Wisconsin Avenue, Suite 350
Bethesda, MD 20892-9205

NIA Alzheimer, Disease Drug Development Program Grants 3833
The NIA and other NIH Institutes currently support research projects for the study of the epidemiology, etiology, diagnosis, and treatment of Alzheimer's disease (AD). Investigators in all relevant scientific disciplines and organizations are invited to develop and pre-clinically test compounds which can slow the disease progression, reverse the disease process, or delay the onset of or prevent AD, mild cognitive impairment (MCI), or age-related cognitive decline.
Requirements: The following organizations and institutions may apply: public/state controlled institutions of higher education; private institutions of higher education; nonprofits with 501(c)3 IRS status; nonprofits without 501(c)3 IRS status; small businesses; for-profit organizations; State governments; and non-domestic (non-U.S.) entities (foreign organizations).
Amount of Grant: $300,000-$800,000 annually in direct costs
Date(s) Application is Due: Jan 2; Feb 2; Sep 1; Oct
Contact: Suzana Petanceska, Ph.D.; (301) 496-9350; fax: (301) 496-1494; email: petanceskas@nia.nih.gov
Internet: http://grants.nih.gov/grants/guide/pa-files/PAR-08-266.html
Sponsor: National Institute on Aging (NIA)
7201 Wisconsin Avenue, Suite 350, MSC 9205
Bethesda, MD 20892-9205

NIA Alzheimer, Disease Pilot Clinical Trials 3834
The objective of the Alzheimer's Disease Pilot Clinical Trials initiative is to improve the quality of clinical research designed to evaluate interventions for the prevention and treatment of Alzheimer's disease (AD), mild cognitive impairment (MCI), and age-associated cognitive decline by stimulating applications for pilot clinical trials to test interventions aimed at delaying the onset of or preventing AD, MCI, and age-associated cognitive decline; slowing, halting, or, if possible, reversing the progressive decline in cognitive function; and modifying the cognitive and behavioral symptoms in AD and MCI. The goal is not to duplicate or compete with pharmaceutical companies but to encourage, complement, and accelerate the process of testing new, innovative, and effective treatments. The pilot research project application should directly address how it will inform and advance the design of a subsequent full-scale clinical trial. The application should also address the intrinsic scientific merit of the pilot trial itself. The application may include the following: studies to refine the intervention strategy (e.g. drug dosage, duration, delivery system; behavioral intensity and duration); studies to define and refine the target population and ensure adequate enrollment, protocol adherence and subject retention; collection of preliminary data for establishing measures of clinical efficacy including clinical/neuropsychological/behavioral measures, neuroimaging measures, and other biological measures in blood and cerebrospinal fluid; assessment of efficacy (including primary and secondary outcome and assessment measures for both patients and caregivers); and studies of safety.
Requirements: The following organizations and institutions may apply: public/state controlled institutions of higher education; private institutions of higher education; Hispanic-serving institutions; historically Black colleges and universities (HBCUs); tribally controlled colleges and universities (TCCUs); Alaska native and native Hawaiian serving institutions; nonprofits with 501(c)3 IRS status; nonprofits without 501(c)3 IRS status; small businesses; for-profit organizations; State governments; regional organizations; U.S. territories or possessions; Indian/Native American tribal governments; Indian/Native American tribally designated organizations; county governments; city or township governments; special district governments; independent school districts; public housing authorities/Indian housing authorities; eligible agencies of the Federal government; and faith-based or community based organizations.
Amount of Grant: $450,000 per year for up to 3 years
Date(s) Application is Due: Feb 5; Jun 5; Oct 5
Contact: Laurie M. Ryan, PhD; (301) 496-9350; fax: (301) 496-1494; email: ryanl@nia.nih.gov; April Bower, RN; (301) 451-3560; fax: (301) 480-3621; email: bowera@mail.nih.gov
Internet: http://grants.nih.gov/grants/guide/pa-files/PAR-08-062.html
Sponsor: National Institute on Aging (NIA)
7201 Wisconsin Avenue, Suite 350, MSC 9205
Bethesda, MD 20892-9205

NIA Alzheimer, Disease Research Centers Grants 3835
The National Institute on Aging (NIA) invites applications from qualified institutions for support of Alzheimer's Disease Research Centers (ADRCs). These centers are designed to support and conduct research on Alzheimer's disease (AD), to serve as shared research resources that will facilitate research in AD and related disorders, distinguish them from the processes of normal brain aging and mild cognitive impairment (MCI), provide a platform for training, develop novel techniques and methodologies, and translate these research findings into better diagnostic, prevention and treatment strategies. New applications should request a project period of five years and a budget for direct costs of up to $1 million per year. Letters of Intent are due by April 6 annually, with full applications received by May 5.
Requirements: The following organizations and institutions are eligible to apply: public/state controlled institutions of higher education; private institutions of higher education; Hispanic-serving institutions; historically Black colleges and universities (HBCUs); tribally controlled colleges and universities (TCCUs); Alaska native and native Hawaiian serving institutions; nonprofits with 501(c)3 IRS status (other than institutions of higher education); and nonprofits without 501(c)3 IRS status (other than institutions of higher education).
Amount of Grant: $1 million maximum per year for five years
Date(s) Application is Due: Apr 5; May 6
Contact: Creighton H. Phelps, Ph.D.; (301) 496-9350; fax: (301) 496-1494; email: phelpsc@nia.nih.gov
Internet: http://grants.nih.gov/grants/guide/rfa-files/RFA-AG-10-002.html
Sponsor: National Institute on Aging (NIA)
7201 Wisconsin Avenue, Suite 350
Bethesda, MD 20892-9205

NIA Archiving and Development of Socialbehavioral Datasets in Aging Related Studies Grants 3836

This Funding Opportunity Announcement (FOA), issued by the National Institute on Aging (NIA) is seeking small grant applications to stimulate and facilitate data archiving and development related to cognitive psychology, behavioral interventions in the context of randomized controlled trials (RCTs), demography, economics, epidemiology, behavioral genetics and other behavioral research on aging for secondary analysis.
Requirements: The following organizations and institutions may apply: public/state controlled institutions of higher education; private institutions of higher education; Hispanic-serving institutions; historically Black colleges and universities (HBCUs); tribally controlled colleges and universities (TCCUs); Alaska native and native Hawaiian serving institutions; nonprofits with 501(c)3 IRS status; nonprofits without 501(c)3 IRS status; small businesses; for-profit organizations; State governments; regional organizations; U.S. territories or possessions; Indian/Native American tribal governments; Indian/Native American tribally designated organizations; county governments; city or township governments; special district governments; independent school districts; public housing authorities/Indian housing authorities; eligible agencies of the Federal government; and faith-based or community based organizations.
Amount of Grant: $100,000 maximum over two years
Date(s) Application is Due: Feb 16; Jun 16; Oct 16
Contact: Partha Bhattacharyya, Ph.D.; (301) 496-3136; fax: (301) 402-0051; email: bhattacharyyap@mail.nih.gov
Internet: http://grants.nih.gov/grants/guide/pa-files/PA-08-252.html
Sponsor: National Institute on Aging (NIA)
7201 Wisconsin Avenue, Gateway 533
Bethesda, MD 20892

NIA Awards to Support Research on the Biology of Aging in Invertebrates Grants 3837

This Funding Opportunity Announcement (FOA) issued by the National Institute on Aging (NIA), National Institutes of Health, invites applications that propose the identification and development of new invertebrate models for use in biomedical research that is highly relevant to the biology of aging. Applications should focus on the identification, development and characterization of new invertebrate models that have short lifespans. Invertebrate models with tractable genetics and genome sequence available or in progress, negligible or induced senescence, and tissue regeneration in adults are of particular interest to NIA. Examples of invertebrate models of interest include, but are not limited to planaria, hydra, rotifers, and tunicates. These invertebrate models are currently not well developed for aging research. Within the scope of this FOA, invertebrate models other than those mentioned above currently used by the biomedical research community, but not yet fully developed or realized for aging research, will be considered responsive to this FOA. In particular, applications that focus on the characterization of these novel models for aging research utilizing molecular, genetic, cellular or physiological tools and approaches to identify genes involved in longevity, senescence and cellular pathways in tissue homeostasis during aging are encouraged. Letters of Intent are due by September 30, with final applications due by October 30.
Requirements: The following organizations and institutions are eligible to apply for support: public/state controlled institutions of higher education; private institutions of higher education; Hispanic-serving institutions; historically Black colleges and universities (HBCUs); tribally controlled colleges and universities (TCCUs); Alaska Native and Native Hawaiian serving institutions; nonprofits with 501(c)3 IRS status (other than institutions of higher education); nonprofits without 501(c)3 IRS status (other than institutions of higher education); Indian/Native American tribal governments (federally recognized); Indian/Native American tribally designated organizations; U.S. territories or possessions; and eligible agencies of the federal government.
Amount of Grant: $250,000 maximum
Date(s) Application is Due: Sep 30; Oct 30
Contact: Anna M. McCormick, Ph.D.; (301) 496-6402; email: mccormia@nia.nih.gov
Internet: http://grants.nih.gov/grants/guide/rfa-files/RFA-AG-10-004.html
Sponsor: National Institute on Aging (NIA)
Gateway Building, Suite 2C231, 7201 Wisconsin Avenue
Bethesda, MD 20892-9205

NIA Behavioral and Social Research Grants on Disasters and Health 3838

The purpose of this FOA is to stimulate research in the behavioral and social sciences on the consequences of natural and man-made disasters for the health of children, the elderly, and vulnerable groups, with an ultimate goal of preventing and mitigating harmful consequences and health disparities. Disasters include severe weather-related events, earthquakes, large-scale attacks on civilian populations, technological catastrophes or perceived catastrophes, and influenza pandemics. For the elderly and for children and youth, the health outcomes of greatest interest include mortality, disability and resilience, severe distress and clinically significant morbidity (as opposed to mild or transient symptoms and dysphoria), and economic hardship sufficient to harm health. For children and youth, long-term effects on development are also of interest.
Requirements: The following organizations and institutions may apply: for-profit organizations; non-profit organizations; public or private institutions, such as universities, colleges, hospitals, and laboratories; units of State government; units of Local government; units of State Tribal government; units of Local Tribal government; eligible agencies of the Federal government; foreign institutions; domestic institutions; faith-based or community-based organizations; Indian/Native American tribal governments (federally recognized); Indian/Native American tribal governments (other than federally recognized); and Indian/Native American tribally designated organizations.
Date(s) Application is Due: Feb 16; Jun 16; Oct 16
Contact: John G. Haaga, Ph.D.; (301) 496-3131; email: HaagaJ@mail.nih.gov; Rebecca L. Clark, Ph.D.; (301) 496-1175; email: rclark@mail.nih.gov
Internet: http://grants.nih.gov/grants/guide/pa-files/PA-06-452.html
Sponsor: National Institute on Aging (NIA)
7201 Wisconsin Avenue, Gateway Building, Room 533
Bethesda, MD 20892-9205

NIA Bioenergetics, Fatigability, and Activity Limitations in Aging 3839

This FOA encourages Research Project Grant applications proposing to study bioenergetic factors underlying increased fatigability and activity limitations in aging. Increased fatigability is a significant cause of restricted physical and cognitive activity in older adults. Alterations in bioenergetics - the production and utilization of energy, and the regulation of these processes - may contribute significantly to increased fatigability. This FOA encourages applications that propose to: elucidate specific alterations in bioenergetics related to increased fatigability and activity limitations; develop and evaluate improved measures of fatigability related to bioenergetics; and evaluate interventions for increased fatigability and activity limitations that target alterations in bioenergetics and lead to improved quality of life.
Requirements: The following organizations and institutions may apply: public/state controlled institutions of higher education; private institutions of higher education; Hispanic-serving institutions; historically Black colleges and universities (HBCUs); tribally controlled colleges and universities (TCCUs); Alaska native and native Hawaiian serving institutions; nonprofits with 501(c)3 IRS status; nonprofits without 501(c)3 IRS status; small businesses; for-profit organizations; State governments; regional organizations; U.S. territories or possessions; Indian/Native American tribal governments; Indian/Native American tribally designated organizations; county governments; city or township governments; special district governments; independent school districts; public housing authorities/Indian housing authorities; eligible agencies of the Federal government; and faith-based or community based organizations.
Date(s) Application is Due: Feb 5; Feb 16; Jun 5; Jun 16; Oct 5; Oct 16
Contact: Basil Eldadah; (301) 496-6761; fax: (301) 402-1784; email: eldadahb@nia.nih.gov; Cheryl L. McDonald; (301) 435-0560; fax: (301) 480-2858; email: mcdonalc@nhlbi nih.gov
Internet: http://grants.nih.gov/grants/guide/pa-files/PA-09-190.html
Sponsor: National Institute on Aging (NIA)
7201 Wisconsin Avenue, 3C-307
Bethesda, MD 20892-9205

NIA Diversity in Medication Use and Outcomes in Aging Populations Grants 3840

The National Institute on Aging (NIA) in collaboration with the National Center for Complementary and Alternative Medicine, the Office of Research on Minority Health and the Office of Research on Women"s Health invites qualified researchers to submit applications to investigate issues relevant to medication use and outcomes among older people. This program announcement solicits both basic and applied research proposals to investigate the multiple factors that influence medication use by the elderly. Goals of this announcement are to: identify the epidemiological, psychosocial, health care and clinical factors associated with medication use by older people, including the use of unconventional medical products; assess social, behavioral, psychological and cognitive factors that play a role in older people's understanding of and adherence to medication regimens; determine the role of medical and pharmaceutical professionals in facilitating or hindering proper use of prescribed and over-the-counter medications and unconventional medications including herbs and dietary supplements; refine methodologies to link these factors to medical or psychosocial outcomes; investigate interventions to improve medication adherence; and increase our understanding of biological factors which contribute to therapeutic outcomes in the use of medications by the elderly.
Requirements: Applications may be submitted by foreign and domestic for-profit and non-profit organizations, public and private, such as universities, colleges, hospitals, laboratories, units of state and local governments, and eligible agencies of the Federal government.
Contact: Laurie M. Ryan, PhD; (301) 496-9350; fax: (301) 496-1494; email: ryanl@nia.nih.gov; April Bower, RN; (301) 451-3560; fax: (301) 480-3621; email: bowera@mail.nih.gov
Internet: http://grants.nih.gov/grants/guide/pa-files/PA-99-097.html
Sponsor: National Institute on Aging (NIA)
7201 Wisconsin Avenue, Suite 350, MSC 9205
Bethesda, MD 20892-9205

NIA Effects of Gene-Social Environment Interplay on Health and Behavior in Later Life Grants 3841

The National Institute on Aging (NIA) invites applications for the development of multidisciplinary collaborations among existing longitudinal twin and family studies, with a focus on social and behavioral factors associated with aging outcomes. This FOA is intended to lay the foundation for future studies of the role of gene-environment interplay in accounting for links between social experiences and physical health, functionality, and psychological well-being in midlife and older age. Of particular interest are applications that can embed this foundation within a lifespan perspective. Applicants are encouraged to use this FOA as a starting point for: developing collaborations among existing data sets; harmonizing social phenotypes across a number of twin and family studies; conducting analyses aimed at further refining social phenotypes for genetic analysis; using animal models of gene-social environment interplay to inform work using twin studies; and/or establishing preliminary data on the feasibility of genetic/genomic approaches that could be applied to questions of how social experiences are biologically embedded and influence middle and later life outcomes. Use of existing data is strongly encouraged; funds from

this award are not intended for new data collection, but a small percentage of funds may be used for adding modules to ongoing data collection efforts to augment phenotypic or genetic data. NIA expects to make 1-2 awards totaling $750,000 total costs in the first year, and $3,750,000 in total costs over the 5 year project period. Direct costs are expected to range from $200,000 -$500,000 in the first year. Letters of Intent are due by October 9, with final applications due by November 9.
Requirements: The following organizations and institutions may apply: public/state controlled institutions of higher education; private institutions of higher education; Hispanic-serving institutions; historically Black colleges and universities (HBCUs); tribally controlled colleges and universities (TCCUs); Alaska native and native Hawaiian serving institutions; nonprofits with 501(c)3 IRS status; nonprofits without 501(c)3 IRS status; small businesses; for-profit organizations; State governments; regional organizations; U.S. territories or possessions; Indian/Native American tribal governments; Indian/Native American tribally designated organizations; county governments; city or township governments; special district governments; independent school districts; public housing authorities/Indian housing authorities; eligible agencies of the Federal government; and faith-based or community based organizations.
Amount of Grant: $200,000-$500,000 in the first year
Date(s) Application is Due: Oct 9; Nov 9
Contact: Erica L. Spotts, Ph.D.; (301) 451-4503; fax: (301) 402-0051; email: spottse@mail.nih.gov
Internet: http://grants.nih.gov/grants/guide/rfa-files/RFA-AG-10-006.html
Sponsor: National Institute on Aging (NIA)
7201 Wisconsin Avenue, Suite 533
Bethesda, MD 20892-9201

NIA Factors Affecting Cognitive Function in Adults with Down Syndrome Grants 3842

The purpose of this solicitation is to encourage investigator initiated research to focus on various factors that maximize and maintain cognitive function in adults (>21 years of age) with Down syndrome. This solicitation encourages applications from investigators from very diverse backgrounds and interdisciplinary and transdisciplinary proposals, as well as those that involve systematic longitudinal study rather than study of samples of convenience. These studies could serve to form an evidence base that will inform not only health care management and treatment, but also issues such as quality of life, access to services, education, and ultimately affect public policy. This solicitation encourages applications that would inform several areas of inquiry: the natural history of medical and mental health issues in adults with Down syndrome, their incidence and onset; health disparities that occur, and their prevention or effective treatment; the development of better diagnostic tools to assess functional and cognitive impairment in adults with Down syndrome; improved quality of life measures; community-based interventions that improve individual educational, occupational, and social outcomes; the impact of early entry to services, type of intervention, and/or family history on quality of life for adults with Down Syndrome; and the impact of family structure and ancestry on critical life transitions, co-occurring conditions, and developmental changes of adults with Down syndrome through longitudinal studies rather than studies of samples of convenience. The requested budget is limited to $500,000 total costs per year for a maximum of five years. Letters of Intent are due May 29, and full applications should arrive by June 30.
Requirements: The following organizations and institutions are eligible to apply: public/state controlled institutions of higher education; private institutions of higher education; nonprofits with 501(c)3 IRS status (other than institutions of higher education); nonprofits without 501(c)3 IRS status (other than institutions of higher education); for-profit organizations (other than small businesses); State governments; county governments; eligible agencies of the Federal government; and faith-based or community-based organizations.
Restrictions: Foreign institutions may not submit applications, but collaborators at foreign institutions are allowed.
Amount of Grant: $500,000 maximum per year for up to five years
Date(s) Application is Due: May 29; Jun 30
Contact: Mary Lou Oster-Granite; (301) 435-6866; fax: (301) 496-3791; email: mo96o@nih.gov
Internet: http://grants.nih.gov/grants/guide/rfa-files/RFA-HD-09-028.html
Sponsor: Eunice Kennedy Shriver National Institute of Child Health and Human Development (NICHD) / National Institute on Aging (NIA)
6100 Executive Boulevard, Room 4B05L, MSC 7510
Bethesda, MD 20892-7510

NIA Grants for Alzheimer's Disease Drug Discovery 3843

The objective of this solicitation is to stimulate preclinical research in the discovery, design, development and testing of novel compounds aimed at slowing, halting, or, if possible, reversing the progressive decline in cognitive function and modifying the behavioral symptoms in Alzheimer's disease as well as delaying the onset of or preventing AD. This initiative is intended to stimulate basic research and development efforts. The goal is not to duplicate or compete with pharmaceutical companies but to encourage, complement, and accelerate the process of discovering new, innovative, and effective compounds for the prevention and treatment of the cognitive impairment and behavioral symptoms associated with Alzheimer's disease. The development of compounds for ameliorating, modifying, or improving potential aberrations in neuronal cellular communication mechanisms is encouraged. These compounds should be designed to affect fundamental processes such as the neuronal dysfunction, death, and loss of connectivity associated with the disease by targeting molecules and mechanisms such as neurotransmitters, neuromodulators, and neurotrophins; receptors and ion channels; second and third messenger systems; all facets of relevant amyloid precursor protein, amyloid beta protein, and tau neurobiology; protein synthesis, aggregation, and degradation; energy utilization; and oxidative, immunological, and inflammatory mechanisms.
Requirements: The following organizations are eligible to apply: for-profit organizations; non-profit organizations; public or private institutions, such as universities, colleges, hospitals, and laboratories; units of State government; units of local government; eligible agencies of the Federal government; foreign institutions; and domestic institutions.
Amount of Grant: $200,000 maximum per year; $275,000 maximum per two-year cycle
Date(s) Application is Due: Feb 16; Jun 16; Oct 16
Contact: Neil S. Buckholtz, Ph.D.; (301) 496-9350; fax: (301) 496-1494; email: Buckholn@nia.nih.gov
Internet: http://grants.nih.gov/grants/guide/pa-files/PAS-06-261.html
Sponsor: National Institute on Aging (NIA)
7201 Wisconsin Avenue, Suite 350, MSC 9205
Bethesda, MD 20892-9205

NIA Harmonization of Longitudinal Cross-National Surveys of Aging Grants 3844

The National Institute on Aging (NIA) will provide support through the R21 mechanism to facilitate the ex-ante and ex-post harmonization of international longitudinal surveys of aging populations to the Health and Retirement Study to maximize cross-national comparability of data. Facilitation will include, but is not limited to, meetings, pilots, and the development of methods to improve cross-national comparability of international aging surveys to the Health and Retirement Study and existing international data files that are harmonized with the HRS: the English Longitudinal Study on Ageing (ELSA); and the Survey of Health, Ageing, and Retirement in Europe (SHARE).
Requirements: The following organizations and institutions may apply: public/state controlled institutions of higher education; private institutions of higher education; nonprofits with 501(c)3 IRS status; nonprofits without 501(c)3 IRS status; for-profit organizations; State governments; and non-domestic (non-U.S.) entities (foreign organizations).
Amount of Grant: $200,000 maximum per year; $275,000 maximum per two-year cycle
Date(s) Application is Due: Feb 16; Jun 16; Oct 16
Contact: John W. R. Phillips, Ph.D.; (301) 496-3138; fax: (301) 402-0051; email: phillipj@mail.nih.gov
Internet: http://grants.nih.gov/grants/guide/pa-files/PAS-07-387.html
Sponsor: National Institute on Aging (NIA)
7201 Wisconsin Avenue, Gateway Building, Room 533
Bethesda, MD 20892

NIA Health Behaviors and Aging Grants 3845

The National Institute on Aging (NIA) invites qualified researchers to submit applications for research and research training on those health-related behaviors and attitudes of older adults, their families, and significant others, that can affect health and functioning as people grow older. Studies are sought that extend scientific understanding of how older adults health behaviors and attitudes develop under varying social conditions; how they relate to health promotion and disease prevention, care and treatment of disease, rehabilitation or death; and how they can be modified as relevant new scientific knowledge is developed.
Requirements: Applications may be submitted by foreign and domestic, for-profit and non-profit organizations, public and private, such as universities, colleges, hospitals, laboratories, units of State and local governments, and eligible agencies of the Federal government. Applications from minority individuals and women are encouraged.
Amount of Grant: $150,000 maximum per year
Date(s) Application is Due: Feb 5; Jun 5; Oct 5
Contact: Chyren Hunter, Ph.D.; (301) 496-9322; fax: (301) 402-0528; email: Hunterc@nia.nih.gov
Internet: http://grants.nih.gov/grants/guide/pa-files/PA-93-064.html
Sponsor: National Institute on Aging (NIA)
7201 Wisconsin Avenue, MSC 9205, Gateway Building, Room 2C218
Bethesda, MD 20892-9205

NIA Healthy Aging through Behavioral Economic Analyses of Situations Grants 3846

The National Institute on Aging (NIA), National Institutes of Health, solicits Research Project Grant applications that propose to translate basic findings from Behavioral Economics into behavior change interventions targeting health behaviors associated with chronic health conditions of mid-life and older age. Applications should propose small pilot clinical trials or demonstration projects, ideally based on collaborations between individuals with expertise in behavioral economics and psychologists, psychiatrists, clinicians, or others with expertise in aging or implementing behavioral interventions. Depending on the availability of funding, the NIA expects to make 4 or 5 awards totaling $1,250,000 total costs in the first year, and $3,750,000 over the 3 year project period. The direct cost amount for each individual award will be no greater than $525,000 in total for three years, and no more than $175,000 in any year. Letters of Intent are due by October 2, with final applications due by November 2.
Requirements: The following organizations and institutions may apply: public/state controlled institutions of higher education; private institutions of higher education; Hispanic-serving institutions; historically Black colleges and universities (HBCUs); tribally controlled colleges and universities (TCCUs); Alaska native and native Hawaiian serving institutions; nonprofits with 501(c)3 IRS status; nonprofits without 501(c)3 IRS status; small businesses; for-profit organizations; State governments; regional organizations; U.S. territories or possessions; Indian/Native American tribal

governments; Indian/Native American tribally designated organizations; county governments; city or township governments; special district governments; independent school districts; public housing authorities/Indian housing authorities; eligible agencies of the Federal government; and faith-based or community based organizations.
Amount of Grant: $525,000 maximum for three years; $175,000 maximum per year
Date(s) Application is Due: Oct 2; Nov 2
Contact: Dr. Jonathan W. King; (301) 402-4156; fax: (301) 402-0051; email: kingjo@nia.nih.gov
Internet: http://grants.nih.gov/grants/guide/rfa-files/RFA-AG-10-008.html
Sponsor: National Institute on Aging (NIA)
7201 Wisconsin Avenue, Room 533
Bethesda, MD 20892-9205

NIA Higher-Order Cognitive Functioning and Aging Grants 3847
The National Institute on Aging (NIA) invites qualified researchers to submit new applications for research projects that focus on adulthood and aging related changes in the higher-order processes and strategies required for judgment, decision-making, reasoning, problem-solving, and processing complex information. As in earlier periods of life older adults continue to make decisions related to everyday life, but with advanced age, new, and sometimes even more complex, decision-making is required of them. Recent research indicates that age-related limitations in cognitive processing resources (e.g., speed and working memory) may impact decision-making. Research also indicates that some older adults experience growth in specific areas of cognitive functioning (e.g., expertise, semantic knowledge, emotional regulation) and continue to use adaptive intelligence, demonstrating multi-directionality in adult cognitive change. It is generally recognized that research on higher-order processing is underdeveloped in the field of aging. Research proposals are needed that examine the actual processes that are engaged when older adults make important decisions, how these processes change with age and context, and what environmental supports, interventions, and training may be necessary for optimal functioning. Research may investigate either individual or collaborative, or social processes. Modular Grant applications will request direct costs in $25,000 modules, up to a total direct cost request of $250,000 per year.
Requirements: Applications may be submitted by foreign and domestic, for-profit and non-profit, public and private organizations such as universities, colleges, hospitals, laboratories, units of State and local governments, and eligible agencies of the Federal government. Racial/ethnic minority individuals, women, and persons with disabilities are encouraged to apply as principal investigators.
Amount of Grant: $250,000 maximum per year
Date(s) Application is Due: Feb 16; Jun 16; Oct 16
Contact: Chyren Hunter, Ph.D.; (301) 496-9322; fax: (301) 402-0528; email: Hunterc@nia.nih.gov
Internet: http://grants.nih.gov/grants/guide/pa-files/PA-00-052.html
Sponsor: National Institute on Aging (NIA)
7201 Wisconsin Avenue, MSC 9205, Gateway Building, Room 2C218
Bethesda, MD 20892-9205

NIA Human Biospecimen Resources for Aging Research Grants 3848
The purpose of this announcement is to disseminate information about available human biospecimen resources for age-related studies, and promote independent use of existing human biospecimen resources through collaborative and other arrangements with the studies that collect biospecimens. Websites for the participating studies are accessed via the Virtual Repository. NIA is interested in aging-related research that integrates biomarker analyses with the study of normal aging, age-associated diseases, and behavioral and psychosocial data. Budgets for direct costs of up to $50,000 per year and a project duration of up to two years may be requested for a maximum of $100,000 direct costs over a two-year project period.
Requirements: The following organizations and institutions may apply: for-profit organizations; non-profit organizations; public or private institutions, such as universities, colleges, hospitals, and laboratories; units of State government; units of Local government; units of State Tribal government; units of Local Tribal government; eligible agencies of the Federal government; foreign institutions; domestic institutions; faith-based or community-based organizations; Indian/Native American tribal governments (federally recognized); Indian/Native American tribal governments (other than federally recognized); and Indian/Native American tribally designated organizations.
Amount of Grant: $50,000 maximum per year for up to 2 years
Date(s) Application is Due: Feb 16; Jun 16; Oct 16
Contact: Nancy L. Nadon, Ph.D.; (301) 402-7744; fax: (301) 402-0010; email: nadonn@nia.nih.gov
Internet: http://grants.nih.gov/grants/guide/pa-files/PA-06-443.html
Sponsor: National Institute on Aging (NIA)
7201 Wisconsin Avenue, GW 2C231
Bethesda, MD 20892

NIA Independent Scientist Award 3849
The award is a special salary-only grant designed to provide protected time for newly independent scientists who currently have non-research obligations such as heavy teaching loads, clinical work, committee assignments, service, and administrative duties that prevent them from having a period of intensive research focus. The award is targeted to persons with doctoral degrees who have completed their research training, have independent peer-reviewed research support, and who need a period of protected research time in order to foster their research career development. New applications are due February 12, June 12, and October 12, while renewal applications are due March 12, July 12, and November 12. Up to 75% per year of base salary plus fringe benefits is available. No other research development support funds are provided.
Requirements: The following organizations and institutions are eligible to apply: public/state controlled institutions of higher education; private institutions of higher education; nonprofits with 501(c)3 IRS status; nonprofits without 501(c)3 IRS status; small businesses; for-profit organizations; State governments; U.S. territories or possessions; Indian/Native American tribal governments (Federally recognized and other than Federally recognized); Indian/Native American tribally designated organizations; Hispanic-serving institutions; historically Black colleges and universities (HBCUs); tribally controlled colleges and universities (TCCUs); Alaska Native and Native Hawaiian serving institutions; regional organizations; and faith-based or community based organizations. The candidate must have a doctoral degree and peer-reviewed, independent research support at the time the award is made. The candidate must spend a minimum of 75 percent effort conducting research during the period of the award.
Restrictions: The award is not intended for investigators who already have full time to perform research, or have substantial publication records or considerable research support indicating that they are well established in their fields. Foreign institutions are not eligible to apply.
Date(s) Application is Due: Feb 12; Mar 12; Jun 12; Jul 12; Oct 12; Nov 12
Contact: Chyren Hunter, Ph.D.; (301) 496-9322; fax: (301) 402-0528; email: Hunterc@nia.nih.gov
Internet: http://grants.nih.gov/grants/guide/pa-files/PA-09-038.html
Sponsor: National Institute on Aging (NIA)
7201 Wisconsin Avenue, MSC 9205, Gateway Building, Room 2C218
Bethesda, MD 20892-9205

NIA Malnutrition in Older Persons Research Grants 3850
NIA continues to encourage research on the full spectrum of issues related to the causes, prevention, and treatment as well as sociobehavioral and economic aspects related to malnutrition in older persons. Support for this program will be through the research project grant and mentored research scientist development award. Applications will compete for available funds on the basis of scientific merit with all other applications assigned to the institute.
Requirements: Applications may be submitted by foreign and domestic, for-profit and non-profit organizations, public and private, such as universities, colleges, hospitals, laboratories, unit of State and local governments, and eligible agencies of the Federal government. Applications from minority individuals and women are encouraged.
Date(s) Application is Due: Feb 12; Mar 12; Jun 12; Jul 12; Oct 12; Nov 12
Contact: Chyren Hunter, Ph.D.; (301) 496-9322; fax: (301) 402-0528; email: Hunterc@nia.nih.gov
Internet: http://grants.nih.gov/grants/guide/pa-files/PA-94-088.html
Sponsor: National Institute on Aging (NIA)
7201 Wisconsin Avenue, Gateway 533
Bethesda, MD 20892

NIA Mechanisms Underlying the Links between Psychosocial Stress, Aging, the Brain and the Body Grants 3851
This FOA encourages multidisciplinary and interdisciplinary research to elucidate the mechanistic links between psychosocial stress and health in aging, as well as how the aging process and age-related diseases affect the responses to psychosocial stressors. Generally, research should be focused on: aging and how neural mechanisms respond to psychosocial stress and affect other body systems; characterizing the behavioral, psychological and social mechanisms and pathways involved in transducing psychosocial stressors into health outcomes; how stressors modulate physiological process underlying life-span, immune mechanisms, and metabolism; and how psychosocial stress contributes to the development or progression of geriatric syndromes, chronic medical conditions, and disabilities in later life. Research is strongly encouraged that aims to identify appropriate targets for intervention, at any level of analysis, from societal to molecular. Research spanning multiple levels of analysis is particularly encouraged.
Requirements: The following organizations and institutions are eligible to apply: public/state controlled institutions of higher education; private institutions of higher education; Hispanic-serving institutions; historically Black colleges and universities (HBCUs); tribally controlled colleges and universities (TCCUs); Alaska native and native Hawaiian serving institutions; nonprofits with 501(c)3 IRS status (other than institutions of higher education); nonprofits without 501(c)3 IRS status (other than institutions of higher education); small businesses; for-profit organizations (other than small businesses); state governments; Indian/Native American tribal governments (federally recognized); Indian/Native American tribally designated organizations; county governments; city or township governments; special district governments; independent school districts; public housing authorities/Indian housing authorities; U.S. territories or possessions; Indian/native American tribal governments (other than federally recognized); regional organizations; and non-domestic (non-U.S.) entities (foreign organizations).
Restrictions: Research focused on oxidative stress or on environmental or physical stressors of a non-psychosocial nature is not appropriate for this FOA.
Date(s) Application is Due: Mar 5; Jul 5; Nov 5
Contact: Suzana S. Petanceska, Ph.D.; (301) 496-9350; fax: (301) 496-1494; email: petanceskas@mail.nih.gov
Internet: http://grants.nih.gov/grants/guide/pa-files/PA-09-216.html
Sponsor: National Institute on Aging (NIA)
Gateway Building, Suite 350 7201, Wisconsin Avenue, MSC 9205
Bethesda, MD 20892-9205

NIA Mechanisms, Measurement, and Management of Pain in Aging: from Molecular to Clinical 3852

This FOA encourages Research Project Grant applications from institutions and organizations that propose to: study biological, neurobiological, psychosocial, and clinical mechanisms and processes by which aging and/or age-related diseases affect the experience of pain; examine biological, neurobiological, psychosocial, and clinical factors that impact pain experience and prevalence in older people; evaluate existing pain assessment and/or management approaches in older adults; or develop new assessment methods and/or management strategies for pain with particular attention to the needs of older adults. Studies involving animal models or human subjects are appropriate under this program announcement.

Requirements: The following organizations and institutions may apply: public/state controlled institutions of higher education; private institutions of higher education; Hispanic-serving institutions; historically Black colleges and universities (HBCUs); tribally controlled colleges and universities (TCCUs); Alaska native and native Hawaiian serving institutions; nonprofits with 501(c)3 IRS status; nonprofits without 501(c)3 IRS status; small businesses; for-profit organizations; State governments; regional organizations; U.S. territories or possessions; Indian/Native American tribal governments; Indian/Native American tribally designated organizations; county governments; city or township governments; special district governments; independent school districts; public housing authorities/Indian housing authorities; eligible agencies of the Federal government; and faith-based or community based organizations.

Date(s) Application is Due: Feb 5; Feb 16; Jun 5; Jun 16; Oct 5; Oct 16

Contact: Wen G. Chen; (301) 496-9350; fax: (301) 496-1494; email: chenw@nia.nih.gov; Basil Eldadah; (301) 496-6761; fax: (301) 402-1784; email: eldadahb@nia.nih.gov

Internet: http://grants.nih.gov/grants/guide/pa-files/PA-09-193.html

Sponsor: National Institute on Aging (NIA)

7201 Wisconsin Avenue, Gateway Building, Suite 350

Bethesda, MD 20892

NIA Medical Management of Older Patients with HIV/AIDS Grants 3853

The National Institute on Aging (NIA), the National Institute of Allergy and Infectious Diseases (NIAID), the Center for Mental Health Research on AIDS of the National Institute of Mental Health (NIMH), and the National Institute of Nursing Research (NINR), invite Research Project Grant applications addressing clinical and translational health issues in the diagnosis and/or management of HIV infection and its consequences in older persons. The goal of this funding opportunity is to improve medical outcomes, functional status and quality of life in older patients with HIV/AIDS through improved understanding of interactions among aging processes, HIV viral infection, treatment effects and toxicities, and multiple morbidities commonly occurring in older persons.

Requirements: The following organizations and institutions may apply: public/state controlled institutions of higher education; private institutions of higher education; Hispanic-serving institutions; historically Black colleges and universities (HBCUs); tribally controlled colleges and universities (TCCUs); Alaska native and native Hawaiian serving institutions; nonprofits with 501(c)3 IRS status; nonprofits without 501(c)3 IRS status; small businesses; for-profit organizations; State governments; regional organizations; U.S. territories or possessions; Indian/Native American tribal governments; Indian/Native American tribally designated organizations; county governments; city or township governments; special district governments; independent school districts; public housing authorities/Indian housing authorities; eligible agencies of the Federal government; and faith-based or community based organizations.

Date(s) Application is Due: May 7; Sep 7; Jan 7

Contact: Susan G. Nayfield, M.D.; (301) 496-6761; fax: (301) 402-1784; email: nayfiels@nia.nih.gov

Internet: http://grants.nih.gov/grants/guide/pa-files/PA-09-017.html

Sponsor: National Institute on Aging (NIA)

7201 Wisconsin Avenue, Gateway Building, Suite 3C-307

Bethesda, MD 20892-9205

NIA Mentored Clinical Scientist Research Career Development Awards 3854

The overall objective of the NIH Research Career Development Award program is to prepare qualified individuals for careers that have a significant impact on the health-related research needs of the Nation. The objective of the NIH Mentored Clinical Scientist Research Career Development Award (K08) program is to support didactic study and mentored research for individuals with clinical doctoral degrees (e.g., M.D., D.D.S., D.M.D., D.O., D.C., O.D., N.D., D.V.M., Pharm.D., or Ph.D. in clinical disciplines). The purpose of this Funding Opportunity Announcement (FOA) is to continue the long-standing NIH support of the K08 program. This award provides support and protected time for an intensive, mentored research career development experience in biomedical or behavioral research, including translational research. For the purpose of this award, translational research is defined as application of basic research discoveries toward the diagnosis, management, and prevention of human disease.

Requirements: The following organizations and institutions may apply: public/state controlled institutions of higher education; private institutions of higher education; Hispanic-serving institutions; historically Black colleges and universities (HBCUs); tribally controlled colleges and universities (TCCUs); Alaska native and native Hawaiian serving institutions; nonprofits with 501(c)3 IRS status; nonprofits without 501(c)3 IRS status; small businesses; for-profit organizations; State governments; regional organizations; U.S. territories or possessions; Indian/Native American tribal governments; Indian/Native American tribally designated organizations; county governments; city or township governments; special district governments; independent school districts; public housing authorities/Indian housing authorities; eligible agencies of the Federal government; and faith-based or community based organizations.

Restrictions: Individuals on temporary or student visas and organizations described in section 501(c)4 of the Internal Revenue Code that engage in lobbying are not eligible.

Amount of Grant: $100,000 maximum

Date(s) Application is Due: Feb 12; Mar 12; Jun 12; Jul 12; Oct 12; Nov 12

Contact: Chyren Hunter, Ph.D.; (301) 496-9322; fax: (301) 402-0528; email: Hunterc@nia.nih.gov

Internet: http://grants.nih.gov/grants/guide/pa-files/PA-09-042.html

Sponsor: National Institute on Aging (NIA)

7201 Wisconsin Avenue, MSC 9205, Gateway Building, Room 2C218

Bethesda, MD 20892-9205

NIA Mentored Research Scientist Development Award 3855

The award is for research scientists who have established careers in biomedical, behavioral, or social research and wish to change career direction toward aging research; for more junior researchers with training in aging research who need additional periods of mentored research experience prior to becoming fully independent; or for researchers with training and experience in some aspect of aging research who wish to gain complementary training to expand their research interests in aging. The primary purpose of the program is to provide support and protected time (three, four, or five years) for an intensive, supervised career development experience in the biomedical, behavioral, or clinical sciences leading to research independence.

Requirements: The candidate must have a research or health-professional doctorate or its equivalent and must have demonstrated the capacity or potential for highly productive independent research in the period after the doctorate. The candidate must identify a mentor with extensive research experience and must spend a minimum of 75 percent of full-time professional effort conducting research for the period of the award. The following organizations and institutions may apply: public/state controlled institutions of higher education; private institutions of higher education; Hispanic-serving institutions; historically Black colleges and universities (HBCUs); tribally controlled colleges and universities (TCCUs); Alaska native and native Hawaiian serving institutions; nonprofits with 501(c)3 IRS status; nonprofits without 501(c)3 IRS status; small businesses; for-profit organizations; State governments; regional organizations; U.S. territories or possessions; Indian/Native American tribal governments; Indian/Native American tribally designated organizations; eligible agencies of the Federal government; and faith-based or community based organizations.

Amount of Grant: $75,000 salary plus fringe benefits

Date(s) Application is Due: Feb 12; Mar 12; Jun 12; Jul 12; Oct 12; Nov 12

Contact: Chyren Hunter, Ph.D.; (301) 496-9322; fax: (301) 402-0528; email: Hunterc@nia.nih.gov

Internet: http://grants.nih.gov/grants/guide/pa-files/PAR-09-060.html

Sponsor: National Institute on Aging (NIA)

7201 Wisconsin Avenue, MSC 9205, Gateway Building, Room 2C218

Bethesda, MD 20892-9205

NIA Minority Dissertation Research Grants in Aging 3856

Small grants to support doctoral dissertation research will be available for underrepresented minority doctoral candidates. Grant support is designed to aid the research of such investigators and to encourage underrepresented minority individuals from a variety of academic disciplines and programs to conduct research related to aging. It is anticipated that a total of five to six new applications will be funded in each year. The maximum budget request should be limited to $25,000 in direct costs for the initial budget period. No more than $30,000 in total direct costs will be provided across the two year period.

Requirements: This research initiative is to provide underrepresented minority students assistance to complete their dissertation research on an aging-related topic and thereby increase their representation in aging research.

Amount of Grant: $25,000 maximum per year; $30,000 maximum for 2-year period

Date(s) Application is Due: Mar 15; Nov 15

Contact: Laurie M. Ryan, PhD; (301) 496-9350; fax: (301) 496-1494; email: ryanl@nia.nih.gov; April Bower, RN; (301) 451-3560; fax: (301) 480-3621; email: bowera@mail.nih.gov

Internet: http://grants.nih.gov/grants/guide/pa-files/PAR-98-110.html

Sponsor: National Institute on Aging (NIA)

7201 Wisconsin Avenue, Suite 350, MSC 9205

Bethesda, MD 20892-9205

NIA Multicenter Study on Exceptional Survival in Families: The Long Life Family Study (LLFS) Grants 3857

This limited competition FOA is to continue the Long Life Family Study (LLFS). The LLFS was designed to determine the degree and patterns of familial transmission and aggregation of exceptional longevity and healthy survival to advanced age as characterized by a variety of phenotypic measures, with a further goal of also understanding potential genetic factors that contribute to exceptional survival. Five awards, to four study centers and one data management and coordinating center, were made in response to RFA-AG-03-004.

Date(s) Application is Due: Aug 30; Sep 30

Contact: Winifred K. Rossi; (301) 496-3836; fax: (301) 402-1784; email: winnie_rossi@nih.gov

Internet: http://grants.nih.gov/grants/guide/rfa-files/RFA-AG-10-005.html

Sponsor: National Institute on Aging (NIA)

7201 Wisconsin Avenue, Room 3C307

Bethesda, MD 20892-9205

NIA Nathan Shock Centers of Excellence in Basic Biology of Aging Grants 3858
The NIA invites applications for support of centers, known as Nathan Shock Centers of Excellence in Basic Biology of Aging. These center grants will provide funding for research and training activities related to basic biology of aging. They are intended for institutions with a substantial investment in and commitment to aging research, but they are not intended to directly support clinical research or clinical trials. The goal of this program is to enhance the ability of institutions with well-developed research programs in basic research on aging to utilize state-of-the-art research resources to provide the strongest environment for the conduct of research on aging. Thus this RFA is intended to enhance the quality of research in the basic biology of aging, facilitate the planning and coordination of aging research activities, provide support and a suitable environment for investigators new to aging research to acquire research skills and experience at institutions that have demonstrated commitment to, and expertise in, basic biology of aging research, and to develop potential regional and/or national resource centers. Annual deadline dates may vary.
Requirements: Applications may be submitted by domestic nonprofit organizations and institutions, state and local governments and their agencies, and authorized federal institutions. The applicant institution must currently support a minimum of 15 peer-reviewed, externally funded research projects on aging to be eligible. In the case of currently funded program projects (P01s), or similar multiproject research grants, each research project will be deemed to be a separate project. (Core components do not qualify.)
Date(s) Application is Due: Apr 16; May 20
Contact: Dr. Huber Warner, Ph.D., Associate Director; (301) 496-4996; fax: (301) 402-0010; email: warnerh@nia.nih.gov
Internet: http://grants.nih.gov/grants/guide/rfa-files/RFA-AG-04-010.html
Sponsor: National Institute on Aging
7201 Wisconsin Avenue, Gateway Building, Suite 2C231
Bethesda, MD 20892-9205

NIA NRSAs—Individual Postdoctoral Fellowships 3859
The fellowships support training for basic and clinical research on aging and the problems of the aging and the aged. The training should be directed toward long-term involvement in research in aging and gerontology. The institute will accept proposals in the behavioral and social sciences related to aging and in the biological and biomedical sciences related to aging. Support may be requested for up to three years. PA: PA-03-067
Requirements: Applicants must be citizens of the United States or permanent resident aliens who have received doctoral degrees before the beginning date of the fellowship.
Amount of Grant: $26,256-$41,268
Date(s) Application is Due: Apr 5; Aug 5; Dec 5
Contact: Dr. Robin Barr, Extramural Affairs, (301) 496-9322; fax: (301) 402-2945 email: rb42h@nih.gov
Internet: http://grants.nih.gov/grants/guide/pa-files/PA-03-067.html
Sponsor: National Institute on Aging
6701 Rockledge Drive, Room 1040, MSC 7710
Bethesda, MD 20892-7710

NIA NRSAs—Institutional Training Awards 3860
NIA will award new and competing renewal NRSA (NRSA) Institutional Training Grants (T32) to eligible institutions to develop or enhance research training opportunities for individuals, selected by the institution, who are training for careers in specified areas of aging research. The purpose of this announcement is to help ensure a broad cadre of researchers trained in the content and methods of aging research and in the major biomedical, behavioral, and social areas of research most closely related to their field of study within aging. NIA supports both predoctoral and postdoctoral training in aging. The institute also supports short-term research training for students in health-professional programs as part of an overall T32 program. The NIA has a single receipt date each year for applications. PA: PA-00-057
Requirements: Applicant must be a citizen or noncitizen national of the United States or must have been lawfully admitted for permanent residence. Predoctoral trainees must have received a baccalaureate degree by the beginning date of their NRSA appointment, and must be training at the postbaccalaureate level and enrolled in a program leading to a PhD in science or in an equivalent research doctoral degree program. Postdoctoral trainees must have received, as of the beginning date of the NRSA appointment, a PhD, MD, or comparable doctoral degree from an accredited domestic or foreign institution. To be eligible for short-term predoctoral research training positions, health-professional students must have completed at least one quarter in a program leading to a clinical doctorate prior to participating in the program.
Contact: Dr. Robin Barr, Extramural Affairs, (301) 496-9322, fax: (301) 402-2945; email: rb42h@nih.gov
Internet: http://grants.nih.gov/grants/guide/pa-files/PA-00-057 html
Sponsor: National Institute on Aging
7201 Wisconsin Avenue, Gateway Building
Bethesda, MD 20892

NIA Paul Beeson Career Development Awards in Aging 3861
The program will foster the independent research careers of clinically trained investigators whose research will enhance the health and quality of life of Americans, particularly older people. The aims of the program are to encourage and assist the development of future leaders in the field of aging by supporting junior and midcareer faculty members to gain additional research training as needed and to establish independent programs in aging research; deepen the commitment of research institutions to academic research in aging and to translating research outcomes to geriatric medicine by involving mentor and recipient in establishing and advancing the recipient's career in aging research; expand medical research on aging broadly defined as including the biology of aging, maintenance of health and independence in old age, diseases and disabilities of old age and issues in their clinical management, and systems of care for the elderly. Project periods of three, four, or five years are supported. A letter of intent is requested but not required by October 24. Program information and guidelines are available online. RFA-AG-06-005
Requirements: Eligible institutions are for-profit or nonprofit, public or private, units of state and local governments, and eligible agencies of the federal government. Individuals eligible to become principal investigators have a clinical doctoral degree (e.g., MD, DO, DDS) or its equivalent. At the time of award, applicants must be U.S. citizens or noncitizen nationals, or have been lawfully admitted to the United States for permanent residence.
Restrictions: Foreign institutions are not eligible to apply. Ineligible individuals include current and former principal investigators on NIH research projects, FIRST Awards, sub-projects of program project, or center grants. Individuals on temporary or student visas are not eligible.
Amount of Grant: $200,000 maximum per year
Date(s) Application is Due: Nov 23
Contact: Office of Extramural Affairs, (301) 496-9322; fax: (301) 402-2945; email: BarrR@nia.nih.gov
Internet: http://grants1.nih.gov/grants/guide/rfa-files/RFA-AG-06-005.html
Sponsor: National Institutes of Health
7201 Wisconsin Avenue, Room 2C218, MSC 9205
Bethesda, MD 20892-9205

NIA Pilot Research Grant Program 3862
NIA is seeking small grant applications in specific areas to stimulate and facilitate the entry of promising new investigators into aging research or encourage established investigators to enter new targeted, high-priority areas in this research field. This small grant program provides support for pilot research that is likely to lead to subsequent individual research project grants and/or a significant advancement of aging research. Annual deadline dates may vary; contact program staff for exact dates.
Requirements: Applications may be submitted by domestic for-profit and nonprofit organizations, both public or private; units of state and local governments; and eligible agencies of the federal government. The program supports new and established investigators. For a new investigator to be eligible, the individual should be in the first five years of his or her independent research career. If the applicant is in the final stages of training it is permissible to apply for an R03, but no award will be made to individuals who are still in training or fellowship status at the time of award. For an established investigator to be eligible, the individual must propose research that is unrelated to a currently funded research project in which the investigator participates.
Restrictions: Foreign organizations and institutions are not eligible.
Amount of Grant: $50,000 maximum direct costs for one year
Contact: Dr. David Finkelstein, Biology of Aging Program; (301) 496-6402; fax: (301) 402-0010; email: BAPquery@extramur.nia.nih.gov
Internet: http://grants.nih.gov/grants/guide/pa-files/PA-99-049.html
Sponsor: National Institute on Aging (NIA)
7201 Wisconsin Avenue, Gateway Building, Suite 2C231, MSC 9205
Bethesda, MD 20892

NIA Postdoctoral Training in Research on Aging in Canada 3863
The objective of the National Research Service Award (NRSA) F32 program is to provide support to promising postdoctoral applicants who have the potential to become productive and successful independent research investigators in scientific health-related fields relevant to the missions of the participating NIH Institutes and Centers (ICs). A primary purpose of this initiative is to offer training in research on aging that exposes developing American postdoctoral investigators to diverse combinations of expertise and resources singularly available in Canada through the Canadian Institutes of Health-Institute on Aging (CIHR-IA), combinations that also take into account the unique cultural and ethnic composition of Canada and its aging population.
Requirements: Applicants must choose Canadian institutions and mentors selected by CIHR-IA. These institutions and mentors can be located through the CIHR Funding Database. Any individual with the skills, knowledge, and resources necessary to carry out the proposed research training is invited to work with his/her sponsor and institution to develop an application for support. Individuals from underrepresented racial and ethnic groups as well as individuals with disabilities are always encouraged to apply for NIH support.
Restrictions: Candidates for the postdoctoral fellowship award must be citizens or non-citizen nationals of the United States, or must have been lawfully admitted to the United States for Permanent Residence.
Date(s) Application is Due: Apr 8; Aug 8; Dec 8
Contact: Michael-David A.R.R. Kerns; (301) 402-7713; fax: (301) 402-2945; email: michael-david.kerns@nih.hhs.gov
Internet: http://grants.nih.gov/grants/guide/pa-files/PA-06-469.html
Sponsor: National Institute on Aging (NIA)
7201 Wisconsin Avenue, Suite 2C-218
Bethesda, MD 20892 -9205

NIA Promoting Careers in Aging and Health Disparities Research Grants 3864
The Mentored Research Scientist Development Award provides support for a sustained period of protected time for intensive research career development under the guidance of an experienced mentor, or sponsor, in the biomedical, behavioral or clinical sciences

leading to research independence. The expectation is that through this sustained period of research career development and training, awardees will launch independent research careers and become competitive for new research project grant funding.
Requirements: The following organizations and institutions may apply: public/state controlled institutions of higher education; private institutions of higher education; Hispanic-serving institutions; historically Black colleges and universities (HBCUs); tribally controlled colleges and universities (TCCUs); Alaska native and native Hawaiian serving institutions; nonprofits with 501(c)3 IRS status; nonprofits without 501(c)3 IRS status; small businesses; for-profit organizations; State governments; regional organizations; U.S. territories or possessions; Indian/Native American tribal governments; Indian/Native American tribally designated organizations; county governments; city or township governments; special district governments; independent school districts; public housing authorities/Indian housing authorities; eligible agencies of the Federal government; and faith-based or community based organizations.
Amount of Grant: $150,000 dollars per year for a project period of up to 3-5 years
Date(s) Application is Due: Feb 12; Jun 12; Oct 12
Contact: Dr. J. Taylor Harden; (301) 496-0765; fax: (301) 496-2525; email: Taylor_Harden@nih.gov
Internet: http://grants.nih.gov/grants/guide/pa-files/PAR-09-136.html
Sponsor: National Institute on Aging (NIA)
31 Center Drive
Bethesda, MD 20892-2292

NIA Prostate Cancer Aging, Race, and Ethnicity Research Grants **3865**
NIA invites research project grant applications to expand the understanding of biological and clinical factors leading to the development, progression, and treatment of prostate cancer in aging men. The increased risk of prostate cancer with advancing age and its prominence in older-aged men are well-known characteristics of this tumor. The unusually high incidence and mortality rates of prostate cancer for older white and black American men and, by contrast, the much lower rates in men of Hispanic and Asian descent, provide the need for research that emphasizes the role of race and ethnic factors, as well as age, in early diagnosis, management, and etiology of this malignancy. The purpose of this RFA is to support research that will provide information leading to better prevention, diagnosis, prognosis, and treatment of prostate cancer in the age range in which prostate cancer most frequently occurs in the diverse population groups at risk (i.e., in men 65 years and older). Proposals to address the issues indicated above may be focused on genetic and environmental risk factors, pre-malignant changes, tumorigenesis, detection of localized and advanced disease, prognostic indicators, disease progression, and response to treatment. Studies to evaluate or refine the ability of existing diagnostic, screening, or follow-up methods of persons of different age, race, and ethnicity are also responsive to this RFA. The PA, which describes the research objectives, application procedures, review considerations, and award criteria may be obtained electronically through the NIH grant line (301-402-2221), the NIH Web site, and by mail and email. RFA: AG-02-003
Contact: Dr. Rosemary Yancik, Geriatrics Program, (301) 496-5278; fax: (301) 496-2793; email: YancikR@31.nia.nih.gov
Internet: http://grants2.nih.gov/grants/guide/rfa-files/RFA-AG-02-003.html
Sponsor: National Institute on Aging
7201 Wisconsin Avenue, Gateway Building
Bethesda, MD 20892

NIA Renal Function and Chronic Kidney Disease in Aging Grants **3866**
This Funding Opportunity Announcement (FOA) issued by the National Institute on Aging (NIA) and the National Institute of Diabetes and Digestive and Kidney Diseases (NIDDK), National Institutes of Health, invites applications that propose basic, clinical, and translational research on chronic kidney disease (CKD) and its consequences in aging and in older persons. Applications should focus on: biology and pathophysiology of CKD in animal models; etiology and pathophysiology of CKD in the elderly; epidemiology and risk factors for the development of CKD with advancing age; and/or diagnosis, medical management and clinical outcomes of CKD in this population. Research supported by this initiative should enhance knowledge of CKD and its consequences in the elderly and provide evidence-based guidance in the diagnosis, prevention, and treatment of CKD in older persons.
Requirements: The following organizations and institutions may apply: public/state controlled institutions of higher education; private institutions of higher education; Hispanic-serving institutions; historically Black colleges and universities (HBCUs); tribally controlled colleges and universities (TCCUs); Alaska native and native Hawaiian serving institutions; nonprofits with 501(c)3 IRS status; nonprofits without 501(c)3 IRS status; small businesses; for-profit organizations; State governments; regional organizations; U.S. territories or possessions; Indian/Native American tribal governments; Indian/Native American tribally designated organizations; county governments; city or township governments; special district governments; independent school districts; public housing authorities/Indian housing authorities; eligible agencies of the Federal government; and faith-based or community based organizations.
Date(s) Application is Due: Feb 5; Feb 16; Jun 5; Jun 16; Oct 5; Oct 16
Contact: Susan G. Nayfield; (301) 496-6761; fax: (301) 402-1784; email: nayfiels@mail.nih.gov; John W. Kusek, Ph.D.; (301) 594-7735; fax: (301) 480-3510; email: kusekj@extra.niddk.nih.gov
Internet: http://grants.nih.gov/grants/guide/pa-files/PA-09-165.html
Sponsor: National Institute on Aging (NIA)
7201 Wisconsin Avenue, Gateway Building, Suite 3C-307
Bethesda, MD 20892-9205

NIA Research Grants on Chemical Senses: Neurotransmitters and Neuromodulators **3867**
NIDCD and NIA sponsor research grants that support research fundamental to understanding the neurochemistry of pathways in the olfactory and gustatory systems throughout the life span. This research includes the identification and characterization of the neurotransmitters, neuromodulators, receptors, and secondary messengers throughout the chemosensory systems. More information may be obtained by mail and email, or through the NIH grant line (301-402-2221), and the NIH Web site. PA: PA-96-035
Requirements: Applications may be submitted by domestic and foreign, for-profit and non-profit organizations, public and private, such as universities, colleges, hospitals, laboratories, units of state and local governments, and eligible agencies of the federal government.
Contact: Dr. Judy Finkelstein, (301) 496-9350; fax: (301) 496-1494; email: FinkelsJ@gw.nia.nih.gov
Internet: http://www.grants.nih.gov/grants/guide/pa-files/PA-96-035.html
Sponsor: National Institute on Aging
7201 Wisconsin Avenue, Gateway Building
Bethesda, MD 20892

NIA Role of Apolipoprotein E, Lipoprotein Receptors and CNS Lipid Homeostasis in Brain Aging and Alzheimer, Disease **3868**
This FOA encouarges multidisciplinary and interdisciplinary research to elucidate how Apolipoprotein E, lipoprotein receptors and CNS lipid homeostasis influence brain aging and the transition to neurodegeneration in Alzheimer, disease (AD). The ultimate goal is to gain an in depth understanding of the mechanisms by which the Apolipoprotein E e4 allele confers increased AD risk for the purpose of advancing the overall search for efficacious AD treatments and Apolipoprotein E e4-directed therapeutics in particular. To this end we encourage research spanning multiple levels of analysis in multiple species (from mice to man). Studies aimed at identifying new therapeutic targets for the treatment of AD and other age-related neurodegenerative conditions associated with lipid neurobiology are strongly encouraged. Also of great interest are projects that aim to use lipidomics, various types of imaging and other cutting edge technologies to identify and develop early biomarkers of neurodegeneration associated lipid dyshomeostasis. The estimated amount of funds available to support projects under this FOA is $2,500,000.
Requirements: The following organizations and institutions are eligible to apply: public/state controlled institutions of higher education; private institutions of higher education; Hispanic-serving institutions; historically Black colleges and universities (HBCUs); tribally controlled colleges and universities (TCCUs); Alaska native and native Hawaiian serving institutions; nonprofits with 501(c)3 IRS status (other than institutions of higher education); nonprofits without 501(c)3 IRS status (other than institutions of higher education); small businesses; for-profit organizations (other than small businesses); state governments; Indian/Native American tribal governments (federally recognized); Indian/Native American tribally designated organizations; county governments; city or township governments; special district governments; independent school districts; public housing authorities/Indian housing authorities; U.S. territories or possessions; Indian/native American tribal governments (other than federally recognized); regional organizations; and non-domestic (non-U.S.) entities (foreign organizations).
Date(s) Application is Due: Mar 5; Jul 5; Nov 5
Contact: Suzana S. Petanceska, Ph.D.; (301) 496-9350; fax: (301) 496-1494; email: petanceskas@mail.nih.gov
Internet: http://grants.nih.gov/grants/guide/pa-files/PA-09-217.html
Sponsor: National Institute on Aging (NIA)
7201 Wisconsin Avenue, Suite 350
Bethesda, MD 20892-9205

NIA Role of Nuclear Receptors in Tissue and Organismal Aging Grants **3869**
This Funding Opportunity Announcement (FOA) solicits grant applications proposing to conduct research into underlying biologic mechanisms involving nuclear receptors, their co-regulators and intracellular signaling systems in the process of aging and the connections of the aging process with pathophysiology in middle- and old-age. The focus of the proposed research must be on processes of aging and/or age-related changes, i.e., research proposed in applications responding to this FOA must have clear relevance to aging. Potential topic areas of interest to the NIA include, but are not limited to, the following: age-changes in levels, sensitivity, and gene expression activation profiles of NRs and co-regulators, and their effects on processes of aging (e.g., age-changes in tissue function, inflammation, glucose and lipid metabolism, adipogenesis, muscle, and brain function) and connections with disease processes for diseases common in older individuals; modification of lifespan in model organisms (e.g., the NR homologue DAF-12 in the insulin-IGF-like pathway in the nematode, and mammalian homologues) and the health-protecting, life-extending effects of caloric restriction or other life-span altering genetic mutations or transgenic models in mammalian organisms; and age-changes in steroid metabolism or in tissue sensitivity and/or response to exogenous sex steroid (or tissue-selective sex steroid receptor modulators) therapy in middle- and old-age.
Requirements: The following organizations and institutions may apply: public/state controlled institutions of higher education; private institutions of higher education; Hispanic-serving institutions; historically Black colleges and universities (HBCUs); tribally controlled colleges and universities (TCCUs); Alaska native and native Hawaiian serving institutions; nonprofits with 501(c)3 IRS status; nonprofits without 501(c)3 IRS status; small businesses; for-profit organizations; State governments; regional organizations; U.S. territories or possessions; Indian/Native American tribal

governments; Indian/Native American tribally designated organizations; eligible agencies of the Federal government; and faith-based or community based organizations.
Date(s) Application is Due: Feb 5; Jun 5; Oct 5
Contact: Jill Carrington, Ph.D.; (301) 496-6402; fax: (301) 402-0010; email: carringj@mail.nih.gov; Ronald Margolis, Ph.D.; (301) 594-8819; fax: (301) 435-6047; email: rm76f@nih.gov
Internet: http://grants.nih.gov/grants/guide/pa-files/PAS-07-257.html
Sponsor: National Institute on Aging (NIA)
Gateway Building, Suite 2C231, 7201 Wisconsin Avenue
Bethesda, MD 20892-9205

NIA Significance of Sarcopenia in Old Age Grants 3870
NIA and NIAMS invite research project applications to elucidate the metabolic/physiological and functional consequences of sarcopenia in old age. Sarcopenia is defined as the loss of skeletal muscle mass, quality, and strength. More information may be obtained by mail and email, or through the NIH grant line (301-402-2221) and the NIH Web site. PA: PA-96-038
Requirements: Applications may be submitted by foreign and domestic, for-profit and nonprofit organizations, both public and private; units of state and local governments; and eligible agencies of the Federal government.
Contact: Dr. Chhanda Dutta, Geriatrics Program, (301) 496-6761; fax: (301) 402-1784; email: DuttaC@gw.nia.nih.gov
Internet: http://www.grants.nih.gov/grants/guide/pa-files/PA-96-038.html
Sponsor: National Institute on Aging
7201 Wisconsin Avenue, Gateway Building
Bethesda, MD 20892

NIA Sociobehavorial Data Analysis and Archiving in Aging 3871
NIA) is seeking small grant (R03) applications to stimulate and facilitate data archiving and secondary analyses of data related to caregiving, cognition, demography, economics, epidemiology, behavioral genetics and other behavioral research on aging; provide support for preliminary projects using secondary analysis that could lead to subsequent applications for other research project grant award mechanisms; provide support for rapid analyses of new databases and experimental modules for purposes such as informing the design and content of future study waves; provide support for the development, enhancement and assembly of new databases from existing data; and provide support for pilot research on under-utilized databases. PA: PA-04-123
Requirements: Applications may be submitted by domestic for-profit and nonprofit organizations, public and private, such as universities, colleges, hospitals, laboratories, units of state and local governments, and eligible agencies of the federal government
Restrictions: Foreign organizations and institutions are not eligible.
Amount of Grant: $850,000 maximum
Contact: Linda Whipp, Grants and Contracts Management Office, (301) 496-1472; fax: (301) 402-3672; email: lw17m@nih.gov
Internet: http://grants.nih.gov/grants/guide/pa-files/PA-04-123.html
Sponsor: National Institute on Aging
7201 Wisconsin Avenue, Gateway Building
Bethesda, MD 20892

NIA Support of Scientific Meetings as Cooperative Agreements 3872
The purpose of the program announcement is to inform the scientific community that the National Institute on Aging (NIA) will now support scientific meetings as cooperative agreements in addition to the current practice of supporting them through the traditional grant mechanism. This program provides guidelines for when it is appropriate to request support of a meeting as a cooperative agreement and explains procedures for preparing and submitting such applications. The Principal Investigator will have the primary authority and responsibility to define objectives and approaches; plan, publicize, and conduct the scientific meeting; and publish the results of the meeting.
Requirements: Applications may be submitted by U.S. institutions, including scientific or professional societies eligible to receive grants from Public Health Service (PHS) agencies. In the case of an international conference, the U.S. representative organization of an established international scientific or professional society is the eligible applicant.
Restrictions: Foreign institutions are not eligible.
Contact: Chyren Hunter, Ph.D.; (301) 496-9322; fax: (301) 402-0528; email: Hunterc@nia.nih.gov
Internet: http://grants.nih.gov/grants/guide/pa-files/PA-00-128.html
Sponsor: National Institute on Aging (NIA)
Gateway Building, Room 2C218, 7201 Wisconsin Avenue, MSC 9205
Bethesda, MD 20892-9205

NIA Symptom Management for Chronic Neurological Conditions Research Grants 3873
NINR, NIA, NICHD, NIDCD, and NINDS are interested in facilitating investigator-initiated research designed to deal with symptoms associated with chronic neurological conditions including stroke, epilepsy, brain and spinal cord injury, multiple sclerosis, and Parkinson's disease. The overall goals are to manage symptoms effectively, to prevent secondary effects of the condition when possible, to maximize the health potential, and to improve the health-related quality of life for patients, care givers, and families. The mechanism of support is the research project grant (R01). PA-98-030 may be obtained through the NIH grant line (301-402-2221), the NIH Web site, mail, and email.
Requirements: Applications may be submitted by domestic and foreign for-profit and nonprofit organizations, public or private.
Contact: Dr. Judith Finkelstein, Sensory/Motor Disorders of Aging Program, (301) 496-9350; fax: (301) 496-1494; email: jf119k@nih.gov; Joseph Ellis, Grants Management Officer, (301) 496-1472; fax: (301) 402-3672; email: ellisj@exmur.nia.nih.gov
Internet: http://www.grants.nih.gov/grants/guide/pa-files/PA-98-030.html
Sponsor: National Institute on Aging
7201 Wisconsin Avenue, Gateway Building
Bethesda, MD 20892

NIA Thyroid in Aging Grants 3874
The purpose of this funding opportunity announcement (FOA) is to encourage submission of investigator-initiated research applications on the thyroid in aging. This FOA is intended to promote basic, translational, and clinical studies leading to increased understanding of the physiology of the aging thyroid and improved diagnosis and management of thyroid disease in the elderly.
Requirements: The following organizations and institutions may apply: public/state controlled institutions of higher education; private institutions of higher education; Hispanic-serving institutions; nonprofits with 501(c)3 IRS status; nonprofits without 501(c)3 IRS status; small businesses; for-profit organizations; State governments; regional organizations; U.S. territories or possessions; Indian/Native American tribal governments; Indian/Native American tribally designated organizations; non-domestic (non-U.S.) entities (foreign organizations); Hispanic-serving institutions; historically Black colleges and universities (HBCUs); tribally controlled colleges and universities (TCCUs); Alaska native and native Hawaiian serving institutions; and regional organizations.
Date(s) Application is Due: Feb 5; Feb 16; Jun 5; Jun 16; Oct 5; Oct 16
Contact: Basil A. Eldadah, MD; (301) 496-6761; fax: (301) 402-1784; email: eldadahb@nia.nih.gov; Nancy J. Emenaker, PhD; (301) 496-0116; fax: (301) 480-3925; email: emenaken@mail.nih.gov
Internet: http://grants.nih.gov/grants/guide/pa-files/PA-08-037.html
Sponsor: National Institute on Aging (NIA)
7201 Wisconsin Avenue, Gateway Building, Suite 3C-307
Bethesda, MD 20892-9205

NIA Transdisciplinary Research on Fatigue and Fatigability in Aging Grants 3875
The purpose of this Funding Opportunity Announcement (FOA) is to encourage submission of research applications on fatigue and fatigability in aging. This FOA is intended to promote research studies employing transdisciplinary approaches that could lead to increased understanding of mechanisms contributing to, assessment of, or potential interventions for, increased fatigue or fatigability in older persons. Both animal models and humans are appropriate for study under this FOA.
Requirements: The following organizations and institutions may apply: public/state controlled institutions of higher education; private institutions of higher education; Hispanic-serving institutions; historically Black colleges and universities (HBCUs); tribally controlled colleges and universities (TCCUs); Alaska native and native Hawaiian serving institutions; nonprofits with 501(c)3 IRS status; nonprofits without 501(c)3 IRS status; small businesses; for-profit organizations; State governments; regional organizations; U.S. territories or possessions; Indian/Native American tribal governments; Indian/Native American tribally designated organizations; county governments; city or township governments; special district governments; independent school districts; public housing authorities/Indian housing authorities; eligible agencies of the Federal government; and faith-based or community based organizations.
Date(s) Application is Due: Feb 5; Feb 16; Jun 5; Jun 16; Oct 5; Oct 16
Contact: Basil A. Eldadah, MD; (301) 496-6761; fax: (301) 402-1784; email: eldadahb@nia.nih.gov; Kathleen Jett, PhD; (301) 594-2154; fax: (301) 480-8260; email: jettk@mail.nih.gov
Internet: http://grants.nih.gov/grants/guide/pa-files/PA-08-161.html
Sponsor: National Institute on Aging (NIA)
7201 Wisconsin Avenue, Suite 3C307
Bethesda, MD 20892-9205

NIA Translational Research at the Aging/Cancer Interface Grants 3876
The purpose of this Funding Opportunity Announcement (FOA) is to enhance translational research in the overlapping areas of human aging and cancer by: integrating knowledge of basic processes in cancer biology and aging into clinical care of older patients with cancer (bench to bedside); and exploring clinical observations from the patient care setting at more basic and molecular levels (bedside to bench). Ultimately, information from the research supported by this initiative should lead to further improvements in prevention, diagnosis and disease management, improving the health and well-being of elderly patients at risk for, or diagnosed with, cancer and decreasing the functional impairment and morbidity associated with cancer in this population.
Requirements: The following organizations and institutions may apply: public/state controlled institutions of higher education; private institutions of higher education; Hispanic-serving institutions; historically Black colleges and universities (HBCUs); tribally controlled colleges and universities (TCCUs); Alaska native and native Hawaiian serving institutions; nonprofits with 501(c)3 IRS status; nonprofits without 501(c)3 IRS status; small businesses; for-profit organizations; State governments; regional organizations; U.S. territories or possessions; Indian/Native American tribal governments; Indian/Native American tribally designated organizations; county governments; city or township governments; special district governments; independent school districts; public housing authorities/Indian housing authorities; eligible agencies of the Federal government; and faith-based or community based organizations.
Date(s) Application is Due: Feb 5; Feb 16; Jun 5; Jun 16; Oct 5; Oct 16

Contact: Susan G. Nayfield, M.D.; (301) 496-6761; fax: (301) 402-1784; email: nayfiels@nia.nih.gov
Internet: http://grants.nih.gov/grants/guide/pa-files/PA-08-230.html
Sponsor: National Institute on Aging (NIA)
7201 Wisconsin Avenue, Gateway Building, Suite 3C-307
Bethesda, MD 20892-9205

NIA Vulnerable Dendrites and Synapses in Aging and Alzheimer, Disease Grants 3877
The purpose of this initiative is to solicit research proposals that will investigate the factors regulating synaptic plasticity and dysfunction with a particular emphasis on the age-dependent changes in the functions of dendrites, spines and synapses of key cell types in regions of brain that are selectively vulnerable in certain neurodegenerative disorders such as AD, or with aging using, for example, in vitro and in vivo models.
Requirements: The following organizations and institutions may apply: public/state controlled institutions of higher education; private institutions of higher education; Hispanic-serving institutions; historically Black colleges and universities (HBCUs); tribally controlled colleges and universities (TCCUs); Alaska native and native Hawaiian serving institutions; nonprofits with 501(c)3 IRS status; nonprofits without 501(c)3 IRS status; small businesses; for-profit organizations; State governments; regional organizations; U.S. territories or possessions; Indian/Native American tribal governments; Indian/Native American tribally designated organizations; county governments; city or township governments; special district governments; independent school districts; public housing authorities/Indian housing authorities; eligible agencies of the Federal government; and faith-based or community based organizations.
Date(s) Application is Due: Feb 5; Jun 5; Oct 5
Contact: D. Stephen Snyder, Ph.D.; (301) 496-9350; fax; (301) 496-1494; email: ss82f@nih.gov
Internet: http://grants.nih.gov/grants/guide/pa-files/PA-09-061.html
Sponsor: National Institute on Aging (NIA)
7201 Wisconsin Avenue, Gateway Building, Room 350
Bethesda, MD 20892-9205

NIA/NIEHS Biomedical Epidemiologic and Intervention Studies Planning Grants 3878
This program provides grant support for planning and protocol development of biomedical epidemiologic and intervention studies in research areas supported by NIA's geriatrics program. PA: PA-99-145
Requirements: Applications may be submitted by foreign and domestic for-profit and nonprofit organizations, both public and private; units of state and local governments; and eligible agencies of the federal government.
Amount of Grant: $150,000 maximum
Contact: Dr. Evan Hadley, Associate Director, NIA; fax: (301) 402-1784; email: HadleyE@exmur.nia.nih.gov; Dr. Gwen Collman, Program Administrator, Environmental and Molecular Epidemiology, NIEHS, (301) 541-4980; fax: (919) 541-4937; email: Collman@niehs.nih.go
Internet: http://grants.nih.gov/grants/guide/pa-files/PA-99-145.html
Sponsor: National Institute on Aging
7201 Wisconsin Avenue
Bethesda, MD 20892

NIAAA Alcohol Research Centers Grants 3879
The program's objective is to provide long-term support for a concerted and multidisciplinary research effort into the problems of alcohol use and alcoholism by coordinating the activities of investigators from biomedical, behavioral, and social science disciplines around a basic research theme. Funds may be used for support of core resources including, for example, salaries of center personnel, services and facilities shared by center personnel, and program enrichment activities such as symposia. Support for individual components may include staff salaries, supplies, publication, travel, and equipment. Support will be provided for up to five years (renewable for subsequent periods) to qualifying centers selected for funding.
Requirements: State and local governments and any domestic (nonfederal) public or private nonprofit and for-profit institution may apply for a center grant. The proposed center must be affiliated with an institution, such as a university, medical center, or research center, that has the resources to sustain a long-term coordinated research program around a central theme related to problems of alcohol use and alcoholism. An applicant institution must demonstrate the ability to attract high-quality scientists, from biomedical, behavioral, and social science disciplines, who are willing to make a long-term commitment to alcohol research. In addition, the applicant must assure that alcoholism related research and clinical training opportunities will be available.
Restrictions: Funds may not be used for training, although the applicant is required to demonstrate an institutional training capacity related to alcohol. Funds may not be used for purchase of any land or the purchase, construction, preservation, or repair of any building, and grantees must have sufficient laboratory facilities and reference services to carry out the grant.
Amount of Grant: $2 million maximum total costs
Contact: Dr. Ernestine Vanderveen, (301) 443-2531; fax: (301) 584-0673; email: tvander@willco.niaaa.nih.gov
Internet: http://grants.nih.gov/grants/guide/rfa-files/rfa-aa-00-003.html
Sponsor: National Institute on Alcohol Abuse and Alcoholism
6000 Executive Boulevard, MSC 7003
Bethesda, MD 20892-7003

NIAAA Hepatitis C Infection and Alcoholic Liver Disease 3880
NIAAA invites basic and clinical research grant applications that focus on the synergistic interactions between alcohol consumption and hepatitis C virus (HCV) infection to produce chronic liver disease. This announcement solicits grant applications for research that elucidate mechanism(s) in the development and progression of liver disease in a setting of combined alcohol use and HCV infection. Studies that focus on cellular and molecular mechanisms of viral pathogenesis, immune responses and liver disease, in the context of gender, race, and age are encouraged. Development of animal models and in vitro experimental systems to study the interactive effects of alcohol exposure and HCV pathogenesis are highly encouraged. Annual deadline dates may vary; contact program staff for exact dates. PA: PAS-99-155
Requirements: Applications may be submitted by domestic and foreign, for-profit and nonprofit organizations, both public and private; units of state and local governments; and eligible agencies of the federal government.
Amount of Grant: $2 million total
Contact: Thomas Kresina, Division of Basic Research, (301) 443-6537; fax: (301) 594-0673; email: tk13v@nih.gov; Raye Litten, Division of Clinical and Prevention Research, (301) 443-0796; fax: (301) 443-8774; email: rlitten@willco.niaa.nih.gov
Internet: http://grants.nih.gov/grants/guide/pa-files/PAS-99-155.html
Sponsor: National Institute on Alcohol Abuse and Alcoholism
6000 Executive Boulevard
Bethesda, MD 20892

NIAAA Independent Scientist Award 3881
The award is a special salary-only grant designed to provide protected time for newly independent scientists who currently have non-research obligations such as heavy teaching loads, clinical work, committee assignments, service, and administrative duties that prevent them from having a period of intensive research focus. The award is targeted to persons with doctoral degrees who have completed their research training, have independent peer-reviewed research support, and who need a period of protected research time in order to foster their research career development. New applications are due February 12, June 12, and October 12, while renewal applications are due March 12, July 12, and November 12. Up to 75% per year of base salary plus fringe benefits is available. Research support is not provided in most cases. Individuals engaged in predominantly theoretical work, such as modeling or computer simulation, may request up to $25,000 per year for research expenses.
Requirements: The following organizations and institutions are eligible to apply: public/state controlled institutions of higher education; private institutions of higher education; nonprofits with 501(c)3 IRS status; nonprofits without 501(c)3 IRS status; small businesses; for-profit organizations; State governments; U.S. territories or possessions; Indian/Native American tribal governments (Federally recognized and other than Federally recognized); Indian/Native American tribally designated organizations; Hispanic-serving institutions; historically Black colleges and universities (HBCUs); tribally controlled colleges and universities (TCCUs); Alaska Native and Native Hawaiian serving institutions; regional organizations; and faith-based or community based organizations. The candidate must have a doctoral degree and peer-reviewed, independent research support at the time the award is made. The candidate must spend a minimum of 75 percent effort conducting research during the period of the award.
Restrictions: The award is not intended for investigators who already have full time to perform research, or have substantial publication records or considerable research support indicating that they are well established in their fields. Foreign institutions are not eligible to apply.
Amount of Grant: $100,000 maximum
Date(s) Application is Due: Feb 12; Mar 12; Jun 12; Jul 12; Oct 12; Nov 12
Contact: Michael Hilton, Ph.D.; (301) 402- 9402; email: mhilton@mail.nih.gov
Internet: http://grants.nih.gov/grants/guide/pa-files/PA-09-038.html
Sponsor: National Institute on Alcohol Abuse and Alcoholism (NIAAA)
5635 Fishers Lane
Bethesda, MD 20892-9304

NIAAA Mechanisms of Alcohol and Drug-Induced Pancreatitis Grants 3882
This Funding Opportunity Announcement (FOA) issued by the National Institute on Alcohol Abuse and Alcoholism (NIAAA), the National Institute of Diabetes and Digestive and Kidney Diseases (NIDDK), and the National Institute on Drug Abuse (NIDA), National Institutes of Health (NIH), solicits grant applications from institutions/organizations that propose to investigate the underlying molecular, biochemical, and cellular mechanisms by which long-term alcohol ingestion and drugs lead to the development of pancreatitis. Research is also encouraged to understand the role of various predisposing factors, including substance abuse, that make the pancreas susceptible to alcoholic injury. Understanding the mechanisms as well as the role of predisposing factors may help in developing strategies for the prevention or treatment of the disease. Because the nature and scope of the proposed research will vary from application to application, it is anticipated that the size and duration of each award will also vary.
Requirements: Application can be made by the following institutions and organizations: public/state controlled institutions of higher education; private institutions of higher education; nonprofits with 501(c)3 IRS status (other than institutions of higher education); nonprofits without 501(c)3 IRS status (other than institutions of higher education); small businesses; for-profit organizations (other than small businesses); state governments; U.S. territories or possessions; Indian/Native American tribal governments (federally recognized); Indian/Native American tribal governments (other than Federally recognized); Indian/Native American tribally designated organizations; non-domestic (non-U.S.) entities (foreign organizations); Hispanic-serving institutions;

historically Black colleges and universities (HBCUs); tribally controlled colleges and universities (TCCUs); Alaska Native and Native Hawaiian serving institutions; regional organization; eligible agencies of the Federal government; and faith-based or community based organizations.
Date(s) Application is Due: Feb 5; Jun 5; Oct 5
Contact: Vishnudutt Purohit, Ph.D.; (301) 443-2689; fax: (301) 443-0673; email: vpurohit@mail.nih.gov
Internet: http://grants.nih.gov/grants/guide/pa-files/PA-07-067.html
Sponsor: National Institute on Alcohol Abuse and Alcoholism (NIAAA)
5635 Fishers Lane, Room 2035, MSC 9304
Bethesda, MD 20892-9304

NIAAA Mechanisms of Alcohol-Induced Hepatic Fibrosis Research Grants 3883
The Division of Basic Research of NIAAA invites grant applications that focus on the developmental mechanisms of hepatic fibrosis that occur in alcohol-induced liver disease. PA: PA-99-110
Requirements: Applications may be submitted by domestic and foreign, for-profit and nonprofit organizations, public and private. Applicants requesting $500,000 or more in direct costs for any year must contact program staff before submitting the application.
Amount of Grant: $50,000 maximum direct costs per year
Contact: Dr. Vishnudutt Purohit, Division of Basic Research, (301) 443-2689; email: vpurohit@willco.niaaa.nih.gov; Dr. Leslie Isaki, Division of Basic Research, (301) 594-6228; fax: (301) 594-0673; email: lisaki@willco.niaaa.nih.gov
Internet: http://www.grants.nih.gov/grants/guide/pa-files/PA-99-110.html
Sponsor: National Institute on Alcohol Abuse and Alcoholism
6000 Executive Boulevard, Willco Bldg.
Bethesda, MD 20892

NIAAA Mentored Clinical Scientist Research Career Development Awards 3884
This award enables candidates holding professional degrees (e.g., M.D., D.O., D.V.M., or equivalent degrees) to undertake 3 to 5 years of special study and supervised research with the goal of becoming independent investigators. The award also allows awardees to pursue a research career development program suited to their experience and capabilities under a mentor who is competent to provide guidance in the chosen research area. Institutions may submit applications on behalf of candidates who hold professional degrees. At least 2 years must have elapsed since the health professional degree was granted. Candidates can have varying levels of clinical training and research experience. New applications are due February 12, June 12, and October 12, while resubmitted applications are due March 12, July 12, and November 12. Up to Up to 75 percent of base salary plus fringe benefits is available, and up to $50,000 per year for research support for new awards.
Requirements: All candidates must be U.S. citizens, non-citizen nationals, or legal permanent residents of the U.S. Persons with temporary or student visas are not eligible The grantee institution must have a strong, well-established research and research training program in the chosen area, accomplished faculty in the basic and clinical sciences, and a commitment to the candidate's research development. The proposed program should include an appropriate mentor.
Amount of Grant: $100,000 average
Date(s) Application is Due: Feb 12; Mar 12; Jun 12; Jul 12; Oct 12; Nov 12
Contact: Judith A. Arroyo, Ph.D.; (301) 402-0717; fax: (301) 443-6077; email: jarroyo@mail.nih.gov
Internet: http://grants.nih.gov/grants/guide/pa-files/pa-06-512.html
Sponsor: National Institute on Alcohol Abuse and Alcoholism (NIAAA)
5635 Fishers Lane, Room 2079
Bethesda, MD 20892 -9304

NIAAA Peptide Regulation of Alcohol Intake 3885
The purpose of this program is to identify peptides and their receptors that regulate alcohol consumption, ultimately leading to an understanding of their mechanism(s) of action. Although not exclusive, several areas have potential for further investigation including the hypothalamic-pituitary-adrenal axis, opioid peptides, the renin-angiotensin system, plus the group of peptides regulating food intake and energy balance, exemplified by leptin and neuropeptide Y. The goal of this initiative is to elucidate the role of peptides in pathological alcohol consumption, suggesting novel pharmacotherapeutic approaches for alcoholism. Annual deadlines may vary; contact program staff for exact dates. PA: PAS-99-156
Requirements: Applications may be submitted by domestic and foreign, for-profit and nonprofit organizations, both public and private; units of state and local governments; and eligible agencies of the federal government.
Amount of Grant: $2 million total
Contact: Jules Selden, Division of Basic Research, (301) 443-2673; fax: (301) 594-0673; email: jselden@willco.niaaa.nih.gov
Internet: http://grants.nih.gov/grants/guide/pa-files/PAS-99-155.html
Sponsor: National Institute on Alcohol Abuse and Alcoholism
6000 Executive Boulevard
Bethesda, MD 20892

NIAAA Research Grants on Violence against Women and within the Family 3886
The NIH Office of Behavioral and Social Sciences Research, the National Institute of Justice, the NIH Office of Research on Women's Health, NIDA NIAAA, NIMH, the NIA, the National Center on Child Abuse and Neglect, and the CDC invite applications for a three-year research grant program to conduct investigator-initiated research on the causes, course, treatment, management, and prevention of violence against women and violence within the family, as well as the health and legal consequences of this violence for victims. One of the goals of this joint program is to bring together perspectives of these different agencies to advance knowledge of violence against women and family violence. The estimated total funds available for the first year of support for awards is $1.435 million. Further information may be obtained through the NIH grant line (301-402-2221), and the NIH Web site, and by mail and email. RFA: OD-96-002
Amount of Grant: $10,000-$873,000; $240,374 average
Contact: Division of Basic Research, (301) 443-2530
Internet: http://grants2.nih.gov/grants/guide/rfa-files/RFA-OD-96-002.html
Sponsor: National Institute on Alcohol Abuse and Alcoholism
6000 Executive Boulevard, Willco Building
Bethesda, MD 20892-7003

NIAAA Secondary Analysis of Existing Health Services Data Sets Grants 3887
The NIAAA is soliciting exploratory/developmental research grant applications (R21) to support the secondary analysis of existing data sets relevant to developing a knowledge base to improve the delivery of services for alcohol-related problems, including both treatment and preventive interventions. Research objectives of this PA include, but are not limited to, five major areas: determining the effects of financing and reimbursement mechanisms on alcohol-related healthcare program availability, accessibility, delivery, organization, content, quality, and outcomes; assessing sources of variation in access to and utilization of treatment and prevention interventions for alcohol-related problems; identifying organizational and managerial factors that influence the delivery of treatment and prevention services for alcohol-related problems, either within or across regions, populations, and settings; evaluating the cost, cost-effectiveness, cost-benefit, and cost-utility of alcohol-related treatment and prevention services; and identifying and assessing the effectiveness and outcomes of alcohol-related treatment and preventive services.
Requirements: Applications may be submitted by foreign and domestic, for-profit and nonprofit organizations, both public and private; units of state and local governments; and eligible agencies of the federal government.
Amount of Grant: $100,000 maximum in direct costs per year for two years
Contact: Dr. Harold Perl, Division of Clinical and Prevention Research, (301) 443-0788; fax: (301) 443-8774; email: hperl@willco.niaaa.nih.gov
Internet: http://grants.nih.gov/grants/guide/pa-files/PA-00-100.html
Sponsor: National Institute on Alcohol Abuse and Alcoholism
6000 Executive Boulevard, Willco Building
Bethesda, MD 20892

NIAAA Strategies for the Prevention of DUI Recidivism Research Grants 3888
The purpose of this program announcement is to stimulate research that addresses the problem of recidivism among people convicted of drinking and driving offenses. This includes individuals who are apprehended for driving under the influence (DUI), driving while intoxicated (DWI), or related offenses such as driving after suspension for a DUI and violation of zero tolerance laws, and are either convicted or otherwise sanctioned for such illegal behavior. Applicants who respond to this program announcement are encouraged to develop research applications that design, develop, implement, and/or test interventions that have the potential of preventing or reducing DUI recidivism. Such interventions may be investigator-initiated and implemented in collaboration with appropriate members of judicial, administrative, and policy-making bodies specifically to test or compare their efficacy in preventing or reducing DUI recidivism. Alternatively, the interventions to be studied may occur naturally in society as laws, regulations, policies, and practices implemented by legally constituted bodies and may be tested for their effectiveness by researchers who use the established methodologies of studying natural experiments. PA PAS-99-023
Requirements: Applications may be submitted by domestic and foreign, for-profit and nonprofit, both public and private; units of state and local governments; and eligible agencies of the federal government.
Amount of Grant: $56,000-$2.64 million; $320,000 average
Contact: Dr. Ernestine Vanderveen, (301) 443-2531
Internet: http://grants2.nih.gov/grants/guide/pa-files/PAS-99-023.html
Sponsor: National Institute on Alcohol Abuse and Alcoholism
6000 Executive Boulevard, Willco Building
Bethesda, MD 20892

NIAF Italian Culture and Heritage Grants 3889
The grants are available to individuals or organizations pursuing projects to promote, research, educate, or preserve Italian culture, history, or heritage. Examples include documentaries, doctoral research, exhibits, conferences, books, media stereotyping/anti-defamation surveys, campaigns, plays, etc., and course syllabi. The foundation requests One-page letters of inquiry describing the project are invited. Guidelines are available online.
Amount of Grant: $2000-$15,000
Date(s) Application is Due: May 31; Nov 30
Contact: Grants Administrator, (202) 387-0600; fax: (202) 387-0800; email: grants@niaf.org
Internet: https://www.niaf.org/grants/index.asp
Sponsor: National Italian American Foundation Scholarship Program
1860 19th Street NW
Washington, DC 20009-5501

NIAF Scholarships 3890
The NIAF will award scholarships and grants to outstanding students in the summer for use during the following academic year. The awards will be made on the basis of academic merit and divided between two groups of students: general category 1—Italian American

students who demonstrate outstanding potential and high academic achievements. Area of study is open; and general category 2— students from any ethnic background majoring or minoring in Italian language, Italian studies, Italian American studies or a related field, who demonstrate outstanding potential and high academic achievements. Each scholarship award can only cover tuition and university-provided room and board. Scholarship recipients are eligible for one year of scholarship support. Scholarship monies not used during one academic year are not transferable to the following academic year. Scholarship applicants and winners can and are encouraged to reapply in subsequent years. Apply online.
Requirements: A student must meet the following criteria: be enrolled in a U.S. accredited institution of higher education for the fall; have a grade-point average of at least 3.25 out of 4.0 (or the equivalent); be a U.S. citizen or a permanent resident; and fit into one of the two NIAF scholarship categories. For the purposes of this program, to be Italian American, the student must have at least one ancestor who has immigrated from Italy.
Amount of Grant: $2500-$10,000
Date(s) Application is Due: Mar 1
Contact: Scholarships Administrator, (202) 387-0600; fax: (202) 387-0800; email: scholarships@niaf.org
Internet: https://www.niaf.org/scholarships/index.asp
Sponsor: National Italian American Foundation Scholarship Program
1860 19th Street NW
Washington, DC 20009-5501

NIAID Allergy, Immunology, and Transplantation Program Research Grants 3891
This program concerns the immune system as it functions in the maintenance of health and as it malfunctions in the production of disease. Because of this dual focus, the program encompasses both basic research (immunology) and clinical research (allergic and immunologic diseases). Basic research is supported under immunobiology and immunochemistry to provide the information necessary for understanding immune system structure and function and under genetics and transplantation biology, directed more specifically at the histocompatibility antigens and their role in graft rejection, autoimmunity, immune responsiveness, and a variety of diseases. Clinical research is supported under asthma and allergic diseases, directed at determining the cause and improving the diagnosis, prevention, and treatment of a variety of common allergic diseases, and under immunologic diseases covering disorders of the human immune system and their underlying pathophysiologic inflammation and immunopathologic mechanisms and consequences. Clinical aspects of transplantation and genetic predisposition to disease are supported under genetics and transplantation biology. The small instrumentation program supports the purchase of relatively low-cost pieces of research equipment that generally are not funded in research project grants and that do not qualify for support under the NIH larger shared instrumentation program.
Requirements: Universities, colleges, hospitals, laboratories, and other public or private nonprofit domestic institutions, including state and local units of government, and individuals are eligible for grant support of research by a named principal investigator or a research career development candidate. Also, for-profit organizations are eligible.
Amount of Grant: $9.24 million-$4.72 million; $373,726 average
Date(s) Application is Due: Feb 1; Jun 1; Oct 1
Contact: Dr. John McGowan, Acting Director, Division of Extramural Activities, (301) 496-7291; Mary Kirker, Grants Management, (301) 496-7075; email: mk35h@nih.gov
Internet: http://12.46.245.173/pls/portal30/CATALOG.PROGRAM_TEXT_RPT.show
Sponsor: National Institute of Allergy and Infectious Diseases
6003 Executive Boulevard, Solar Building
Bethesda, MD 20892

NIAID Allergy, Immunology, Transplantation, Infectious Diseases, and AIDS NRSAs—Individual 3892
NRSAs are offered to increase the number of biomedical investigators who are conducting high-quality research in the areas of allergy, immunology, transplantation, infectious diseases, and AIDS. Predoctoral fellowships for minority and disabled students provide up to five years of support. NRSA postdoctoral fellowships provide support for up to three years. NRSA senior fellowships provide support for two years. Deadlines for postdoctoral and senior fellowships are April 5, August 5, and December 5. Annual deadline dates may vary; contact program, staff for exact dates.
Requirements: Awards are given only to U.S. citizens and noncitizen nationals. Minority predoctoral fellowships are available to underrepresented minority students (African Americans, Hispanic, Native Americans, and Pacific Islanders) or scientists with disabilities who will be entering a PhD program. Postdoctoral fellowship applicants must have a doctoral degree (PhD, MD, DVM, DDS, OD, etc.). Senior fellowship applicants should plan to do research that will provide training the candidate does not have and will become an asset to her or his academic career.
Amount of Grant: $7586-$743,606; $113,228 average
Date(s) Application is Due: Apr 5; Aug 5; Dec 5
Contact: Mary Kirker, Grants Management Officer, (301) 496-7075
Internet: http://12.46.245.173/pls/portal30/CATALOG.PROGRAM_TEXT_RPT.show
Sponsor: National Institute of Allergy and Infectious Diseases
6003 Executive Boulevard, Solar Building
Bethesda, MD 20892

NIAID Challenge Grants: Joint Ventures in Biomedicine and Biotechnology 3893
The purpose of this initiative is to support research and development efforts whose outcomes could significantly reduce the impact of infectious diseases nationally and worldwide. NIAID has identified areas of high importance where it believes successful product research and development combined with existing infrastructures and federal challenge grant funding could significantly impact a major health or medical problem. Applications in response to this RFA for challenge grants are limited to these areas: Medicines for Malaria; Medicines for Tuberculosis; Vaccines for Influenza; Vaccines for Emerging and Resistant Infections; and Therapeutics for Emerging and Resistant Infections. Awards will be made in two Phases: Phase I awards will be short-term (two month), small dollar ($25,000) research grants for development of detailed plans for implementation of research and development efforts (i.e., a Phase II grant application); and Phase II awards will be cooperative agreement awards to those Phase I awardees whose research plans present the best opportunities, based on peer review and potential public health significance, for products that will benefit the public health. Annual deadline dates may vary; contact program staff for exact dates. RFA: AI-00-010
Requirements: Applications may be submitted by domestic for-profit and nonprofit organizations such as biotechnology, bioengineering and pharmaceutical companies with collaborators as necessary from universities, colleges, hospitals, and laboratories.
Restrictions: Phase I applications must directly address one or more of the research topics listed. Only Phase I awardees will be eligible for Phase II awards.
Amount of Grant: $20 million total for Phases I and II
Contact: Program Officer; email: niaidchallenge@niaid.nih.gov
Internet: http://grants.nih.gov/grants/guide/rfa-files/RFA-AI-00-010.html
Sponsor: National Institute of Allergy and Infectious Diseases
6700-B Rockledge Drive
Bethesda, MD 20892

NIAID Enhancing Vaccine-Elicited Protective Immunity in Malaria Research Grants 3894
NIAID invites applications for research project (R01) grants to conduct innovative research to advance fundamental understanding of eliciting optimal protective immune responses with candidate malaria vaccines. As part of its Malaria Vaccine Development Plan, NIAID seeks to identify and validate methods and strategies for improved antigen delivery, processing, and presentation or enhancement of protective immunity with existing candidate antigens. Identification, characterization, and evaluation of new candidate vaccines, especially combination vaccines based on a rational strategy for antigen inclusion, are also encouraged. Because many of the poorest nations of the world are afflicted with malaria, to be effective any intervention must be inexpensive, cost-effective, and relatively easy to administer and maintain. Annual deadline dates may vary; contact program staff for exact dates. RFA: AI-99-006
Requirements: Applications may be submitted by domestic and foreign for-profit and nonprofit organizations and public and private institutions.
Amount of Grant: $250,000 maximum per year
Contact: Dr. B.F. Hall, Division of Microbiology and Infectious Diseases, (301) 496-2544; fax: (301) 402-0659; email: bh24q@nih.gov
Internet: http://www.grants.nih.gov/grants/guide/rfa-files/RFA-AI-99-006.html
Sponsor: National Institute of Allergy and Infectious Diseases
6003 Executive Boulevard, Solar Building
Bethesda, MD 20892

NIAID HIV Prevention Trial Units Grants 3895
NIAID invites applications for HIV Prevention Trial Units (HPTUs), which will be integral parts of an HIV Prevention Trials Network (HPTN) designed to carry out a comprehensive scientific agenda on HIV prevention strategies. The HPTN will conduct domestic and international research on promising biomedical and behavioral strategies for the prevention of HIV transmission among adult, pediatric, and adolescent populations. Modalities of interventions may include, but are not limited to, antiretroviral drugs, microbicides, behavioral approaches, barrier and other contraceptive/STD prevention methods, vaccines (in perinatal settings), chemoprophylaxis, treatment of sexually transmitted diseases, and combined approaches. The HPTUs will contribute to the HPTN research agenda; conduct/participate in Phase I, II, and/or III prevention intervention clinical trials domestically and/or internationally; and expand for multiple prevention efficacy trials. Potential applicants should refer to the following Web site for further information on this initiative: http://www.niaid.nih.gov/daids/vtn-ptn. Annual deadline dates may vary; contact program staff for exact dates. RFA: AI-99-010
Requirements: Applications may be submitted by domestic and foreign for-profit and nonprofit organizations, public and private institutions, and eligible agencies of the federal government.
Contact: Rodney Hoff, Division of AIDS, (301) 496-6177; fax: (301) 402-3684; email: rh25v@nih.gov
Internet: http://www.grants.nih.gov/grants/guide/rfa-files/RFA-AI-99-010.html
Sponsor: National Institute of Allergy and Infectious Diseases
6003 Executive Boulevard, Solar Building
Bethesda, MD 20892

NIAID HIV Vaccine Clinical Trial Units Grants 3896
The Division of AIDS (DAIDS) of NIAID solicits applications for HIV Vaccine Trial Clinical Units (HVTUs) that are integral parts of an HIV Vaccine Trials Network (HVTN) designed to carry out a comprehensive scientific agenda on HIV vaccines. The HVTUs will contribute to the HVTN research agenda, participate in domestic and international Phase I and II clinical trials of candidate HIV vaccines, and participate in domestic and international Phase III efficacy trials of HIV vaccines. Potential applicants should refer to the following Web site for further information on this initiative: http://www.niaid.nih.gov/daids/vtn-ptn. In addition to the HVTN, NIAID supports the creation of the HIV Prevention Trials Network (HPTN) to perform Phase I, II, and

III trials of other prevention strategies to prevent HIV-1 transmission, domestically and internationally. Clinical trials of vaccines focused on questions around maternal-infant transmission will be conducted within the HPTN in collaboration with the HVTN and, when desirable, the Pediatric AIDS Clinical Trials Network. Sites funded under the HPTN may serve as effective venues for the eventual implementation of larger HIV vaccine trials; therefore, linkages between HPTN and HVTN, in addition to other scientific organizations, are strongly encouraged. Annual deadline dates may vary; contact program staff for exact dates. RFA: AI-99-009
Requirements: Applications may be submitted by domestic and foreign for-profit and nonprofit organizations, public and private institutions, and eligible agencies of the federal government.
Contact: Dr. Jorge Flores, Division of AIDS, (301) 435-3578; fax: (301) 402-3684; email: jf30t@nih.gov
Internet: http://www.grants.nih.gov/grants/guide/rfa-files/RFA-AI-99-009.html
Sponsor: National Institute of Allergy and Infectious Diseases
6003 Executive Boulevard, Solar Building
Bethesda, MD 20892

NIAID HIV Vaccine Research and Design—Program Project Grants **3897**
NIAID invites applications for research aimed at obtaining an efficacious vaccine against HIV or AIDS. NIAID supports all stages of AIDS vaccine research and development—from basic research to iterative product development through clinical trials. Applications may target any area of AIDS vaccine research. These areas include, but are not limited to, HIV antigen processing, correlates of immunity, animal model development, DNA vaccination, studies of the structure of HIV immunogens, development of virus and bacterial vaccine vectors, studies of existing vaccine cohorts, studies targeting initial infection (dendritic cells), and extensive modeling of vaccine concepts in macaques or chimps, etc. Clinical studies involving humans would not be considered appropriate for this announcement.
Requirements: Applications may be submitted by domestic and foreign, for-profit and nonprofit organizations, public and private.
Date(s) Application is Due: Jun 10; Nov 10
Contact: Dr. Michael Pensiero, Division of AIDS, (301) 435-3749; fax: (301) 402-3684; email: mpensiero@niaid.nih.gov
Internet: http://grants.nih.gov/grants/guide/pa-files/PAR-00-093.html
Sponsor: National Institute of Allergy and Infectious Diseases
6003 Executive Boulevard, Solar Building
Bethesda, MD 20892

NIAID Independent Scientist Award **3898**
The award is a special salary-only grant designed to provide protected time for newly independent scientists who currently have non-research obligations such as heavy teaching loads, clinical work, committee assignments, service, and administrative duties that prevent them from having a period of intensive research focus. The award is targeted to persons with doctoral degrees who have completed their research training, have independent peer-reviewed research support, and who need a period of protected research time in order to foster their research career development. New applications are due February 12, June 12, and October 12, while renewal applications are due March 12, July 12, and November 12. Up to $75,000 plus fringe benefits per year is available. Additional research is not generally provided.
Requirements: The following organizations and institutions are eligible to apply: public/state controlled institutions of higher education; private institutions of higher education; nonprofits with 501(c)3 IRS status; nonprofits without 501(c)3 IRS status; small businesses; for-profit organizations; State governments; U.S. territories or possessions; Indian/Native American tribal governments (Federally recognized and other than Federally recognized); Indian/Native American tribally designated organizations; Hispanic-serving institutions; historically Black colleges and universities (HBCUs); tribally controlled colleges and universities (TCCUs); Alaska Native and Native Hawaiian serving institutions; regional organizations; and faith-based or community based organizations. The candidate must have a doctoral degree and peer-reviewed, independent research support at the time the award is made. The candidate must spend a minimum of 75 percent effort conducting research during the period of the award.
Restrictions: The award is not intended for investigators who already have full time to perform research, or have substantial publication records or considerable research support indicating that they are well established in their fields. Foreign institutions are not eligible to apply.
Amount of Grant: $75,000 plus fringe benefits
Date(s) Application is Due: Feb 12; Mar 12; Jun 12; Jul 12; Oct 12; Nov 12
Contact: Milton J. Hernandez, Ph.D.; (301) 496-3775; email: mh35c@nih.gov
Internet: http://grants.nih.gov/grants/guide/pa-files/PA-09-038.html
Sponsor: National Institute of Allergy and Infectious Diseases (NIAID)
6700B Rockledge Drive, Room 2149
Bethesda, MD 20886

NIAID Integrated Preclinical/Clinical AIDS Vaccine Development Grants **3899**
NIAID invites applications for research aimed at obtaining a safe and efficacious vaccine against HIV or AIDS. The overall goal of this program is to support multidisciplinary research that will lead to a safe and protective AIDS vaccine. As such, IPCAVD applications must include human clinical studies within the term of the award. Applicants must address the following issues in their application regarding human studies: design of the clinical trial, plans for recruitment, enrollment, education and follow-up of clinical subjects, including attention to the possibility of social harms and discrimination arising from vaccine- induced positive screening tests that may be interpreted incorrectly as HIV infection, plus documentation of compliance with all applicable federal (FDA), state and local laws and regulations governing clinical research. Annual deadlines may vary; contact program staff for exact dates. PA: PAR-00-072
Requirements: Applications may be submitted by domestic for-profit and nonprofit organizations, both public and private; units of state and local governments; and eligible agencies of the federal government.
Contact: Dr. Michael Pensiero, Division of AIDS, (301)-435-3749; fax: (301) 402-3684; email: mpensiero@niaid.nih.gov
Internet: http://grants.nih.gov/grants/guide/pa-files/PAR-00-072.html
Sponsor: National Institute of Allergy and Infectious Diseases
6003 Executive Boulevard, Solar Building
Bethesda, MD 20892

NIAID Laboratory Methods to Assess Responses to HIV Vaccine Candidates Research Grants **3900**
NIAID invites applications to conduct innovative research to develop and improve technologies for evaluation of immune responses to HIV vaccines and to expand the range and scope of immune functions currently measured in vaccine trials. This PA intends to complement the scope of vaccine research carried out in human volunteers and animals receiving experimental HIV/SIV vaccines in prophylactic and/or therapeutic trials funded through other programs. Development of new techniques that detect and quantitate changes in innate (nonspecific) and HIV-specific responses will be fostered. Research may include comparative analysis of samples acquired from individuals infected with HIV or who have been exposed to, but not infected with, HIV. Studies in animals, particularly primates that have received SIV/SHIV vaccines, are also encouraged. Applications to develop reagents (including non-clade B based materials), to adapt assays to field conditions and to improve on cell cryopreservation methods are also appropriate. The PA is not intended to directly support the conduct of human or animal clinical trials. Contact the program office for deadline dates. PA: PA-99-044
Requirements: Applications may be submitted by domestic and foreign, for-profit and nonprofit organizations, public and private.
Contact: Dr. Jorge Flores, (301) 435-3758; fax: (301) 402-3684; email: jf30t@nih.gov
Internet: http://www.grants.nih.gov/grants/guide/pa-files/PA-99-044.html
Sponsor: National Institute of Allergy and Infectious Diseases
6003 Executive Boulevard, Solar Building
Bethesda, MD 20892

NIAID Microbiology and Infectious Diseases Research Grants and Contracts **3901**
This program supports research with the broad aim of improving health by controlling diseases caused by infectious or parasitic agents. Investigator-originated and grant-supported research focuses on all aspects of microbiology, with particular emphasis on human pathogens. Projects range from studies of microbial physiology and antigenic structure to collaborative trials of experimental drugs and vaccines. NIAID also supports studies on the mechanisms of resistance to antibiotics as well as research dealing with epidemiologic observations in hospitalized patients or community populations. Program project grants and research contracts are used to fund selected projects of high institute relevance in the branches covering molecular microbiology; bacteriology, mycobacteriology, and mycology; virology; and parasitology.
Requirements: Universities, colleges, hospitals, laboratories, and other public or private nonprofit domestic institutions, including state and local units of government, are eligible. Individuals are eligible to make application for grant support of research by a named principal investigator or a research career development candidate.
Amount of Grant: $2000-$18.07 million; $389,416 average
Date(s) Application is Due: Feb 1; Jun 1; Oct 1
Contact: Dr. John McGowan, Acting Director, Division of Extramural Activities, (301) 496-7291; Mary Kirker, Grants Management Officer, (301) 496-7075
Internet: http://12.46.245.173/pls/portal30/CATALOG.PROGRAM_TEXT_RPT.show
Sponsor: National Institute of Allergy and Infectious Diseases
6003 Executive Boulevard, Solar Building
Bethesda, MD 20892

NIAID Pediatric Clinical Trials Program for AIDS **3902**
The purpose of this RFA is to solicit applications from institutions interested in establishing cooperative clinical trials groups to plan and direct pediatric therapeutics and prevention research. The awardees will have the capacity and expertise to conduct all phases of clinical trials involving children, adolescents, and pregnant women with all stages of infection. Studies focus on high priority research questions on the treatment and pathogenesis of HIV disease and its sequelae. The scope of activities ranges from studies requiring in-depth laboratory support to research in long-term clinical outcomes. Annual deadline dates may vary: contact program staff for exact dates. RFA: AI-00-015
Requirements: Applications may be submitted by domestic for-profit and nonprofit organizations, both public and private; units of state and local governments; and eligible agencies of the federal government.
Restrictions: Foreign institutions are not eligible to apply as primary grantees; however, foreign institutions may participate as part of a domestic application.
Amount of Grant: $36 million total
Contact: Dr. Frederick Batzold, Division of AIDS, (301) 402-0143; fax: (301) 480-4582; email: fb10c@nih.gov
Internet: http://grants.nih.gov/grants/guide/rfa-files/RFA-AI-00-015.html
Sponsor: National Institute of Allergy and Infectious Diseases
6700-B Rockledge Drive
Bethesda, MD 20892-7624

NIAID Preparedness against Illegitimate Use of Bacterial Pathogens 3903

The purpose of this program is to support basic research on molecular and genetic aspects of pathogenesis of, and host immune/defense responses to, the following bacterial pathogens: Bacillus anthracis, Yersinia pestis, Francisella tularensis, Coxiella burnetii, or Brucella species. It is hoped that research from funded projects will be used to devise appropriate defense strategies that can be implemented shortly before or soon after bioterrorist acts.

Requirements: Applications may be submitted by domestic for-profit and nonprofit organizations, both public and private; units of state and local governments; and eligible agencies of the federal government.

Amount of Grant: $1.5 million total

Contact: Dr. Phillip Baker, Division of Microbiology and Infectious Diseases, (301) 435-2855; fax: (301) 402-2508; email: pb26o@nih.gov

Internet: http://grants.nih.gov/grants/guide/rfa-files/RFA-AI-01-002.html

Sponsor: National Institute of Allergy and Infectious Diseases
6700-B Rockledge Drive
Bethesda, MD 20892

NIAID Research Scholar Development Award 3904

NIAID invites applications from outstanding intramural and extramural postdoctoral fellows for the Research Scholar Development Award (RSDA). The RSDA provides support for postdoctoral fellows who are moving to assistant professor positions in academic institutions. The purpose of the RSDA is to ease the transition to an academic position by enabling the awardee to focus on the establishment of his or her research laboratory prior to submitting applications for grant support. PAR-98-072 may be obtained through the NIH grant line, the NIH Web site, mail, and email.

Requirements: The candidate must have a health professional doctoral degree and have demonstrated the capacity for highly productive, independent postdoctoral research in her or his chosen field prior to applying for this award.

Contact: Milton Hernandez, (301) 496-3775; fax: (301) 402-0369; email: mh35c@nih.gov; Barbara Huffman, (301) 496-3821; fax: (301) 402-0369; email: bh23g@nih.gov

Internet: http://www.grants.nih.gov/grants/guide/pa-files/PAR-98-072.html

Sponsor: National Institute of Allergy and Infectious Diseases
6003 Executive Boulevard, Solar Building
Bethesda, MD 20892

NIAID Small Business Innovation Research Advanced Technology Grants 3905

This PA invites grant applications for Small Business Innovation Research (SBIR) projects with award duration and amounts greater than those routinely allowed under the SBIR program. This PA provides examples of advanced technology areas in which NIAID will entertain SBIR applications and provides NIAID's interpretation and clarification of instructions relating to total cost, dates of project period, and page limitations for these applications. Information may be obtained through the NIH grant line (301-402-2221), the NIH Web site, mail, and email. PA: PA-04-127

Date(s) Application is Due: Apr 1; Aug 1; Dec 1

Contact: Gregory Milman, Division of Extramural Activities, (301) 496-8666; fax: (301) 402-0369; email: gm16s@nih.gov

Internet: http://grants.nih.gov/grants/guide/pa-files/PA-04-127.html

Sponsor: National Institute of Allergy and Infectious Diseases
6003 Executive Boulevard, Solar Building
Bethesda, MD 20892

NIAID Statistical Methods in HIV and AIDS Research Grants 3906

NIAID and NIMH invite applications for the development of original statistical methods to advance the understanding, treatment, and prevention of HIV and AIDS. Traditional research project grant (R01) applications may be submitted in response to this PA. PA-98-054 may be obtained through the NIH grant line (301-402-2221), the NIH Web site, mail, and email.

Requirements: Applications may be submitted by domestic and foreign, for-profit and nonprofit organizations, public or private.

Contact: Dr. Jeffrey Albert, Division of AIDS, (301) 435-3752; fax: (301) 480-5703; email: ja24o@nih.gov; Mary Ledford, Grants Management, (301) 402-6446; fax: (301) 480-3780; email: m128g@nih.gov

Internet: http://www.grants.nih.gov/grants/guide/pa-files/PA-98-054.html

Sponsor: National Institute of Allergy and Infectious Diseases
6003 Executive Boulevard, Solar Building
Bethesda, MD 20892

NIAID Therapeutics Research on AIDS-Associated Opportunistic Infections 3907

The goal of this program is the discovery of agents for treatment of AIDS-associated opportunistic infections. Given present advances in molecular biology, the principal objectives of the program are to stimulate original and innovative preclinical research to identify and characterize new, pathogen-specific molecular targets and to develop promising therapeutic approaches based on new targets. Program announcement is available on the Web site along with an addendum (http://grants2.nih.gov/grants/guide/notice-files/NOT-AI-00-001.html). PA: PA-99-124

Requirements: Applications may be submitted by domestic and foreign, for-profit and nonprofit organizations, both public and private; units of state and local governments; and eligible agencies of the federal government. Racial/ethnic minority individuals, women, and persons with disabilities are encouraged to apply.

Amount of Grant: $25,000-$250,000 R01 modules; $50,000 R03 modules

Contact: Dr. Chris Lambros, (301) 435-3769; fax: (301) 402-3171; email: clambros@niaid.nih.gov

Internet: http://www.grants.nih.gov/grants/guide/pa-files/PA-99-124.html

Sponsor: National Institute of Allergy and Infectious Diseases
6003 Executive Boulevard, Solar Building
Bethesda, MD 20892

NIAMS Clinical Trial Planning Grants 3908

The goal of this program is to provide support for planning activities critical for the successful implementation of clinical trials in areas of interest to NIAMS. In addition, the grant is intended to allow for early peer review of the rationale and design for complex or large-scale clinical trials and studies in those areas that are in the NIAMS mission; provide support for the development of a detailed clinical trial research plan, including a manual of operations and procedures, as a means of decreasing the long start-up time often needed for initiating large trials after award; and provide support to refine critical components of a clinical trial protocol, such as outcome measures, recruitment strategies, inclusion and exclusion criteria, etc. Annual deadline dates may vary; contact program staff for exact dates. RFA: AR-00-010

Requirements: Applications may be submitted by domestic for-profit and nonprofit organizations, both public and private; units of state and local governments; and eligible agencies of the federal government.

Amount of Grant: $500,000 total

Contact: Dr. Joan McGowan, Bone Diseases, (301) 594-5055; fax: (301) 480-4543; email: Joan_Mcgowan@nih.gov; Dr. Alan Moshell, Skin Diseases, (301) 594-5017; fax: (301) 480-4543; email: Alan_N_Moshell@nih.gov; Dr. James Panagis, Orthopaedics, (301) 594-5055; fa

Internet: http://grants.nih.gov/grants/guide/rfa-files/RFA-AR-00-010.html

Sponsor: National Institute of Arthritis and Musculoskeletal and Skin Diseases
45 Center Drive
Bethesda, MD 20892

NIAMS Collaborative Arthritis and Musculoskeletal and Skin Diseases Science Award 3909

The objective of the Collaborative Arthritis and Musculoskeletal and Skin Diseases Science Award (CAMSSA) program is to develop and promote competitive scientific research programs in areas within the mission of the National Institute of Arthritis and Musculoskeletal and Skin Diseases at institutions with substantial minority enrollment. The CAMSSA program is designed to encourage collaborative investigations among scientists at institutions with substantial minority enrollment with grantees from research-intensive institutions who have grant support to conduct research in arthritis, musculoskeletal diseases or skin diseases. The nature of the collaborations will include joint research efforts, specialized training in research techniques, and participation in research seminars. The CAMSSA program should develop and expand scientific opportunities among the participating institutions for research in arthritis, musculoskeletal diseases or skin diseases. PAR 04-003

Requirements: Applicant institution must be an institution with substantial minority enrollment that may be a non-Federal, public or private nonprofit institution located in the United States, its possessions or territories. The collaborating institution must also be a non-Federal, public or private nonprofit institution located in the United States, its possessions or territories. Individuals are eligible to become principal investigators. Applicant investigator must have a doctoral degree in a basic or clinical science area, and have completed one or more years of postdoctoral research experience and must be a full-time employee of the applicant institution.

Amount of Grant: $750,000 maximum direct costs

Date(s) Application is Due: Feb 1; Jun 1; Oct 1

Contact: Melinda Nelson, Chief Grants Management Officer, (301) 594-3535; fax: (301) 480-5450; email: nelsonm@mail.nih.gov

Internet: http://grants2.nih.gov/grants/guide/pa-files/PAR-04-003.html

Sponsor: National Institute of Arthritis and Musculoskeletal and Skin Diseases
Building 31, Room 4C02, 31 Center Dr MSC 2350
Bethesda, MD 20892-2350

NIAMS Core Centers for Musculoskeletal Disorders Grants 3910

NIAMS invites applications for research core centers in musculoskeletal disorders. The centers will provide the resources for a number of established, currently funded investigators, often from different disciplines, to adopt a multidisciplinary approach to common research problems in musculoskeletal disorders and to ensure greater productivity than from each of the separate projects. Annual deadline dates may vary; contact program staff for exact dates. RFA: AR-98-004

Requirements: Applications may be submitted by domestic for-profit and nonprofit organizations, public or private.

Restrictions: Foreign institutions are not eligible for center grants.

Amount of Grant: $400,000 maximum

Contact: Dr. Julia Freeman, Centers Program, (301) 594-5052; fax: (301) 480-4543; email: Julia_B_Freeman@nih.gov

Internet: http://grants.nih.gov/grants/guide/rfa-files/RFA-AR-98-004.html

Sponsor: National Institute of Arthritis and Musculoskeletal and Skin Diseases
45 Center Drive, Natcher Building
Bethesda, MD 20892

NIAMS Exploratory Developmental Grants for Multidisciplinary Clinical Studies of Sarcopenia 3911

NIAMS, with NIA, invites exploratory developmental research grant (R21) applications for multidisciplinary clinical studies on the mechanisms underlying the physical functional consequences of sarcopenia in older persons. This exploratory grants program is intended

to facilitate new collaborative efforts in the development of novel scientific hypotheses on the causes of changes in skeletal muscle morphology and/or metabolism that lead to physical disabilities in old age by providing research support for the appropriate pilot/feasibility studies to refine these hypotheses. Multidisciplinary approaches may include collaborations among various types of clinical research expertise or between basic and clinical researchers, with or without previous aging research experience. PA-97-060 may be obtained electronically through the NIH grant line (301-402-2221), the NIH Web site, and by mail and email.
Contact: Dr. Richard Lymn, Muscle Biology and Musculoskeletal Fitness Program Director, (301) 594-5128; fax: (301) 480-4543; email: lymnr@ep.niams.nih.gov
Internet: http://www.grants.nih.gov/grants/guide/pa-files/PA-97-060.html
Sponsor: National Institute of Arthritis and Musculoskeletal and Skin Diseases
45 Center Drive, Natcher Building
Bethesda, MD 20892

NIAMS High Risk Arthritis and Musculoskeletal and Skin Diseases Research 3912
This program seeks to broaden the base of inquiry in fundamental biomedical, biobehavioral, and biomedical technology research by encouraging research projects that involve a high degree of innovation and novelty. It supports research in rheumatic diseases; cartilage biology and diseases; bone biology and diseases (e.g., osteoporosis, Paget's disease); skin biology and skin diseases; autoimmune diseases (e.g., lupus); connective tissue diseases; musculoskeletal diseases, injuries, and disorders; muscle biology and diseases (e.g., muscular dystrophy); exercise physiology and musculoskeletal fitness; sports injuries; occupational diseases and injuries; and orthopedics and bioengineering. Annual deadline dates may vary; contact program staff for exact dates.
Requirements: Applications may be submitted by domestic for-profit and nonprofit organizations, both public and private; units of state and local governments; and eligible agencies of the federal government.
Amount of Grant: $50,000 maximum
Contact: Dr. Bernadette Tyree, (301) 594-5032; email: bt16w@nih.gov; Dr. Elizabeth Gretz, Rheumatic Diseases, (301) 594-5032; email: gretze@mail.nih.gov; Dr. Deborah Ader, Behavioral and Prevention Research, (301) 594-5032; e-m
Internet: http://grants.nih.gov/grants/guide/rfa-files/RFA-AR-0?-004.html
Sponsor: National Institute of Arthritis and Musculoskeletal and Skin Diseases
45 Center Drive, Natcher Building
Bethesda, MD 20892

NIAMS Low-Back Pain and Common Spinal Disorders Research Grants 3913
The program is sponsored by NIAMS, NICHD, NINDS, and NIOSH at the CDC. This initiative invites applications directed to the study of the pathogenesis, epidemiology, prevention, and treatment of three common causes of low-back pain: herniated nucleus pulposus, spinal stenosis, and idiopathic low-back pain. The support mechanism for grants in this area will be the individual investigator-initiated research grant (R01). PA-97-058 may be obtained electronically through the NIH grant line (301-402-2221), the NIH Web site, and by mail and email.
Contact: Dr. James Panagis, Orthopaedics Program, (301) 594-5055; fax: (301) 480-4543; email: panagisj@ep.niams.nih.gov
Internet: http://www.grants.nih.gov/grants/guide/pa-files/PA-97-058.html
Sponsor: National Institute of Arthritis and Musculoskeletal and Skin Diseases
45 Center Drive, Natcher Building
Bethesda, MD 20892

NIAMS Mechanisms of Chondroprotection Research Grants 3914
Approximately six to eight awards will be made for research on mechanisms of chondroprotection. Applications may be for individual research projects or for exploratory/development grants. The objective of the exploratory/developmental mechanism is to encourage applications from individuals who are interested in testing innovative or conceptually creative ideas that are scientifically sound and may advance our understanding of chondroprotective mechanisms. Research should be specifically targeted towards identification and evaluation of chondroprotective agents to prevent cartilage destruction and/or facilitate its repair in such conditions as rheumatoid arthritis, juvenile rheumatoid arthritis, and osteoarthritis, where there is a decline in the structural integrity of the articular cartilage. This request for applications is for basic and applied research projects, but not epidemiological treatment projects. Annual deadline dates may vary; contact program staff for exact dates. RFA: AR-99-004
Requirements: Applications may be submitted by domestic and foreign public and private for-profit and nonprofit organizations, including universities, colleges, hospitals, laboratories, units of state and local government, and eligible agencies of the federal government.
Amount of Grant: $1.5 million total
Contact: Dr. Bernadette Tyree, Cartilage and Connective Tissue Program, or Dr. Susana Serrate-Sztein, Arthritis Branch, (301) 594-5032; fax: (301) 480-4543; email: tyreeb@ep.niams.nih.gov or szteins@ep.niams.nih.gov
Internet: http://www.grants.nih.gov/grants/guide/rfa-files/RFA-AR-99-004.html
Sponsor: National Institute of Arthritis and Musculoskeletal and Skin Diseases
45 Center Drive, Natcher Building
Bethesda, MD 20892

NIAMS Molecular Pathogenesis and New Interventions in Scleroderma 3915
The goal of this RFA is to foster developmental and traditional research projects to advance understanding of the pathogenesis of scleroderma and to promote the design, development, and pilot testing of hypothesis-driven innovative therapeutic approaches. Annual deadline dates may vary; contact program staff for exact dates.
Requirements: Applications may be submitted by domestic for-profit and nonprofit organizations, both public and private; units of state and local governments; and eligible agencies of the federal government.
Restrictions: Foreign institutions are not eligible to apply for these grants, however, applicants may collaborate, through consultation or contractual agreements, with investigators at foreign institutions.
Amount of Grant: $1.5 million total
Contact: Dr. Susana Serrate-Sztein, Rheumatic Diseases Branch, (301) 594-5032; fax: (301) 480-4543; email: szteins@mail.nih.gov
Internet: http://grants.nih.gov/grants/guide/rfa-files/RFA-AR-00-007.html
Sponsor: National Institute of Arthritis and Musculoskeletal and Skin Diseases
45 Center Drive
Bethesda, MD 20892

NIAMS Pathogenesis and Therapy of the Muscular Dystrophies Research Grants 3916
NINDS and NIAMS encourage investigator-initiated research grant applications to study the pathogenesis and therapy of the various forms of muscular dystrophy in children and adults. Responses to this PA may include studies in appropriate animal models or preclinical or clinical studies in patients with facioscapulohumeral dystrophy (FSH), limb-girdle muscular dystrophy (LGMD), myotonic dystrophy, congenital muscular dystrophy (CMD), Emery-Dreifuss muscular dystrophy (EMD), Duchenne muscular dystrophy (DMD), Becker muscular dystrophy (BMD), or other forms of muscular dystrophy. Support mechanisms for grants are research project grant (R01) and program project grant (P01). PA-98-044 may be obtained through the NIH grant line (301-402-2221), the NIH Web site, mail, and email.
Requirements: Applications may be submitted by domestic and foreign, for-profit and nonprofit organizations, public or private.
Contact: Dr. Richard Lymn, Muscle Biology Program, (301) 594-4128; fax: (301) 480-4543; email: r128b@nih.gov; Sally Nichols, Grants Management Officer, (301) 594-3535; fax: (301) 480-5450; email: nicholss@ep.niams.nih.gov
Internet: http://www.grants.nih.gov/grants/guide/pa-files/PA-98-044.html
Sponsor: National Institute of Arthritis and Musculoskeletal and Skin Diseases
45 Center Drive, Natcher Building
Bethesda, MD 20892

NIAMS Repetitive Motion Disorders Research Grants 3917
The program is sponsored by NIAMS, NICHD, NINDS, and NIOSH at the CDC. This initiative invites applications directed to the study of the pathogenesis, epidemiology, prevention, and treatment of repetitive motion disorders. The support mechanism for grants in this area will be the individual investigator-initiated research grant (R01). PA-97-059 may be obtained electronically through the NIH grant line (301-402-2221), the NIH Web site, and by mail and email.
Contact: Dr. James Panagis, Orthopaedics Program, (301) 594-5055; fax: (301) 480-4543; email: panagisj@ep.niams.nih.gov
Internet: http://www.grants.nih.gov/grants/guide/pa-files/PA-97-059.html
Sponsor: National Institute of Arthritis and Musculoskeletal and Skin Diseases
45 Center Drive, Natcher Building
Bethesda, MD 20892

NIAMS Research on Musculoskeletal Fitness and Sports Medicine Grants 3918
NIAMS and NINR invite investigator-initiated research grant applications to study a broad range of basic and clinical topics related to musculoskeletal fitness, exercise physiology, and sports medicine. The National Center for Medical Rehabilitation Research of NICHD encourages applications for both basic and clinical studies of musculoskeletal fitness and exercise physiology of persons with physical disabilities. The principal investigator and participating investigators will plan, direct, and perform the research. Applicants for program project grants are requested to contact NIAMS representative as early as possible in the planning stages. NINR will support individual research project grants. PA-97-025 may be obtained electronically through the NIH grant line (301-402-2221), the NIH Web site, and by mail and email.
Contact: Dr. Richard Lymn, Muscle Biology and Musculoskeletal Fitness Program Director, (301) 594-5128; fax: (301) 480-4543; email: lymnr@ep.niams.nih.gov
Internet: http://www.grants.nih.gov/grants/guide/pa-files/PA-97-025.html
Sponsor: National Institute of Arthritis and Musculoskeletal and Skin Diseases
45 Center Drive, Natcher Building
Bethesda, MD 20892

NIAMS Significance of Sarcopenia in Old Age Grants 3919
NIA and NIAMS invite research project applications to elucidate the metabolic/physiological and functional consequences of sarcopenia in old age. Sarcopenia is defined as the loss of skeletal muscle mass, quality, and strength. More information may be obtained by mail and email, or through the NIH grant line (301-402-2221) or the NIH Web site. PA: PA-96-038
Amount of Grant: $10,000-$1.5 million; $239,000 average
Date(s) Application is Due: Feb 1; Jun 1; Oct 1
Contact: Dr. S. Hausman, Director, Extramural Program, (301) 594-2463; Sally Nichols, Grants Management Officer, Extramural Program, (301) 594-3535
Internet: http://www.grants.nih.gov/grants/guide/pa-files/PA-96-038.html
Sponsor: National Institute of Arthritis and Musculoskeletal and Skin Diseases
45 Center Drive, Natcher Building
Bethesda, MD 20892

NIAMS Skin Diseases Research Core Centers Grants 3920

NIAMS is accepting applications for research core centers in skin diseases. The centers will provide the resources for a number of established, currently funded investigators, often from different disciplines, to adopt a multidisciplinary approach to common research problems in skin diseases and to ensure greater productivity through interaction among the investigators of individually funded research projects. New research directions are encouraged through a pilot and feasibility component. Broad advances in skin disease research can be effectively fostered by research core centers. Examples of advances include, but are not limited to, stratum corneum—biochemistry, structure, function; epidermis—differentiation, keratinization, cellular constituents; dermal-epidermal junction—structure, functions, diseases; skin and an immunological organ; and autoimmune skin diseases. Annual deadline dates may vary; contact program staff for exact dates. RFA: AR-99-009

Requirements: Applications may be submitted by domestic for-profit and nonprofit organizations, both public and private; units of state and local governments; and eligible agencies of the federal government. An established clinical and basic research program in skin diseases must be present.

Restrictions: Foreign institutions are not eligible for center grants.

Amount of Grant: $400,000 maximum

Contact: Dr. Julia Freeman, Centers Program, Extramural Program, (301) 594-5052; fax: (301) 480-4543; email: Julia_B_Freeman@nih.gov

Internet: http://grants.nih.gov/grants/guide/rfa-files/RFA-AR-99-009.html

Sponsor: National Institute of Arthritis and Musculoskeletal and Skin Diseases
45 Center Drive, Natcher Building
Bethesda, MD 20892

NIAMS Specialized Centers of Research in Osteoarthritis and Systemic Lupus Erythematosus 3921

NIAMS invites applications for specialized centers of research in osteoarthritis and systemic lupus erythematosus. These centers should foster a coordinated research effort that strongly emphasizes basic disciplines but also involves significant interaction between basic research and clinical investigations. Annual deadline dates may vary; contact program staff for exact dates. RFA: AR-01-005

Requirements: Applications may be submitted by domestic for-profit and nonprofit organizations, both public and private; units of state and local governments; and eligible agencies of the federal government.

Amount of Grant: $750,000 maximum

Contact: Dr. Julia Freeman, Centers Program, (301) 594-5052; fax: (301) 480-4543; email: Julia_B_Freeman@nih.gov; Melinda Nelson, Grants Management Officer, (301) 594-3535; fax: (301) 480-5450; email: nelsonm@exchange.nih.gov

Internet: http://grants.nih.gov/grants/guide/rfa-files/RFA-AR-01-005.html

Sponsor: National Institute of Arthritis and Musculoskeletal and Skin Diseases
45 Center Drive, Natcher Building
Bethesda, MD 20892

NIAMS Specialized Centers of Research in Osteoporosis, Rheumatoid Arthritis, and Scleroderma 3922

NIAMS invites applications for specialized centers of research in the following disease areas: osteoporosis, rheumatoid arthritis, and scleroderma. A center should foster a coordinated research effort that strongly emphasizes basic disciplines but also involves significant interaction between basic research and clinical investigations. More information may be obtained by mail and email or through the NIH grant line (301-402-2221), and the NIH Web site. Annual deadline dates may vary; contact program staff for exact dates. RFA: AR-98-001

Requirements: Domestic for-profit and nonprofit organizations, both public and private; units of state and local governments; and eligible agencies of the federal government may apply.

Amount of Grant: $750,000 maximum

Contact: Dr. Julia Freeman, Centers Program, (301) 594-5052; fax: (301) 480-4543; email: freemanj@ep.niams.nih.gov; Sally Nichols, Grants Management Officer, (301) 594-3535; fax: (301) 480-5450; email: nicholss@exchange.nih.gov

Internet: http://grants.nih.gov/grants/guide/rfa-files/RFA-AR-98-001.html

Sponsor: National Institute of Arthritis and Musculoskeletal and Skin Diseases
45 Center Drive, Natcher Building
Bethesda, MD 20892

NIAMS/NICHD Deep Infections of Total Joint Replacements 3923

This PA seeks to stimulate the receipt of a broad range of basic science and clinical studies to better understand the pathophysiology, diagnosis, and treatment of deep infections around total joint replacement implants. PA: PA-00-014

Requirements: Applications may be submitted by domestic and foreign, for-profit and nonprofit organizations, both public and private; units of state and local governments; and eligible agencies of the federal government.

Contact: Dr. James Panagis, Director, Orthopaedics Program, NIAMS, (301) 594-5055; fax: (301) 480-4543; email: jp149d@nih.gov; Dr. Ralph Nitkin, NICHD, (301) 402-2242; fax: (301) 402-0832; email: rn21e@nih.gov

Internet: http://grants.nih.gov/grants/guide/pa-files/PA-00-014.html

Sponsor: National Institute of Arthritis and Musculoskeletal and Skin Diseases
6500 Center Drive, Natcher Building
Bethesda, MD 20892

NICHD Central Nervous System Effects of HIV Infection: Neurobiological, Neurovirological, and Neurobehavioral Studies Grants 3924

NIMH, NINDS, NIAAA, and NIDA, invites applications for support of research focused on determining the pathogenic mechanisms involved in Human Immunodeficiency Virus (HIV)-1 associated neurobehavioral and neurological dysfunction in domestic and international settings. The objective of this cooperative effort is to foster investigations that will provide the foundation for the rapid development of therapeutic interventions to prevent and treat the effects of HIV-1 on the central nervous system (CNS). Applications ranging from basic research to clinical diagnosis and treatment are of interest. Multidisciplinary research teams and collaborative alliances are encouraged but not required. Applications will be accepted under PHS receipt dates for AIDS applications. PA: PA-04-154

Requirements: Applications may be submitted by domestic and foreign, for-profit and nonprofit organizations, both public and private; units of state and local governments; and eligible agencies of the federal government.

Amount of Grant: $375,000 maximum

Contact: Dr. Jeymohan Joseph, (301) 443-3012; fax: (301) 443-9719; email: jjeymoha@mail.nih.gov

Internet: http://grants.nih.gov/grants/guide/pa-files/PA-04-154.html

Sponsor: National Institute of Child Health and Human Development
6100 Executive Boulevard
Bethesda, MD 20892

NICHD Clinical Trial Planning Grants for Pediatric Rehabilitation 3925

The purpose of this grant is to provide support for the organization of an effective research group and development of elements essential for a successful clinical trial. Applications may address the development of activities relevant to such elements as plans for recruitment of patients, experimental design and protocols, testing of procedures, data management, analysis techniques, facilities, administrative procedures, and collaborative arrangements. Detailed information regarding the rationale for the clinical trial based on preclinical and preliminary clinical research must be included in the application. Annual deadline dates may change; contact program staff for exact dates. RFA: HD-01-006

Requirements: Applications may be submitted by foreign and domestic, for-profit and nonprofit organizations, both public and private; units of state and local governments; and eligible agencies of the federal government.

Restrictions: Clinical trial planning support is for new projects and may not be used to supplement research projects already being supported or to provide interim support of projects under review.

Amount of Grant: $75,000 maximum

Contact: Dr. Beth Ansel, Director, Clinical Practices Research Program, (301) 402-2242; fax: (301) 402-0832; email: Beth_Ansel@nih.gov

Internet: http://grants.nih.gov/grants/guide/rfa-files/RFA-HD-01-006.html

Sponsor: National Institute of Child Health and Human Development
6100 Executive Boulevard
Bethesda, MD 20892

NICHD Cooperative Specialized Infertility Research Centers 3926

The objective of this program is to expedite the development of new approaches for the diagnosis and treatment of human male or female infertility. Annual deadline dates may vary; contact program staff for exact dates. RFA: HD-01-001

Requirements: Applications may be submitted by domestic for-profit and nonprofit organizations, both public and private; units of state and local governments; and eligible agencies of the federal government.

Amount of Grant: $3 million total

Contact: Dr. Louis DePaolo, Reproductive Sciences Branch, (301) 435-6970; fax: (301) 496-0962; email: ld38p@nih.gov

Internet: http://grants.nih.gov/grants/guide/rfa-files/RFA-HD-01-001.html

Sponsor: National Institute of Child Health and Human Development
6100 Executive Boulevard
Bethesda, MD 20892

NICHD Innovative Therapies and Clinical Studies for Screenable Disorders Grants 3927

This Funding Opportunity Announcement (FOA) issued by the National Institute of Child Health and Human Development (NICHD), the National Institute on Deafness and Other Communication Disorders (NIDCD), and the National Institute of Diabetes and Digestive and Kidney Diseases (NIDDK), National Institutes of Health (NIH), solicit applications for research relevant to the basic understanding and development of therapeutic interventions for currently screened conditions and high priority genetic conditions for which screening could be possible in the near future. Because the nature and scope of the proposed research will vary from application to application, it is anticipated that the size and duration of each award will also vary. The total amount awarded and the number of awards will depend upon the numbers, quality, duration, and costs of the applications received.

Requirements: Application can be made by the following institutions and organizations: public/state controlled institutions of higher education; private institutions of higher education; nonprofits with 501(c)3 IRS status (other than institutions of higher education); nonprofits without 501(c)3 IRS status (other than institutions of higher education); small businesses; for-profit organizations (other than small businesses); state governments; U.S. territories or possessions; Indian/Native American tribal governments (federally recognized); Indian/Native American tribal governments (other than Federally recognized); Indian/Native American tribally designated organizations; non-domestic (non-U.S.) entities (foreign organizations); Hispanic-serving institutions;

historically Black colleges and universities (HBCUs); tribally controlled colleges and universities (TCCUs); Alaska Native and Native Hawaiian serving institutions; regional organization; eligible agencies of the Federal government; and faith-based or community based organizations.
Date(s) Application is Due: Feb 5; Mar 5; Jun 5; Jul 5; Oct 5; Nov 5
Contact: Gilian Engelson; (301) 451-2137; fax: (301) 451-5512; email: engelsong@mail.nih.gov
Internet: http://grants.nih.gov/grants/guide/pa-files/PAR-07-184.html
Sponsor: National Institute of Child Health and Human Development (NICHD)
6100 Executive Boulevard, Room 4A05E, MSC 7510
Bethesda, MD 20892-7510

NICHD Low-Back Pain and Common Spinal Disorders Research Grants 3928
The program is sponsored by NIAMS, NICHD, NINDS, and NIOSH at the CDC. This initiative invites applications directed to the study of the pathogenesis, epidemiology, prevention, and treatment of three common causes of low-back pain: herniated nucleus pulposus, spinal stenosis, and idiopathic low-back pain. The support mechanism for grants in this area will be the individual investigator-initiated research grant (R01). Information may be obtained electronically through the NIH grant line (301-402-2221), the NIH Web site, and by mail and email. PA-97-058
Requirements: Applications may be submitted by foreign and domestic, for-profit and nonprofit organizations, both public and private; units of state and local governments; and eligible agencies of the federal government.
Restrictions: Foreign institutions or organizations in foreign countries are not eligible.
Contact: Dr. Stephen Tuel, National Center for Medical and Rehabilitation Research, (301) 402-2242; fax: (301) 402-0832; email: tuels@hd01.nichd nih.gov
Internet: http://www.grants.nih.gov/grants/guide/pa-files/PA-97-058.html
Sponsor: National Institute of Child Health and Human Development
6100 Executive Boulevard
Bethesda, MD 20892

NICHD Medical Rehabilitation Research 3929
This program announcement invites qualified researchers to submit grant applications for research on restoring, replacing, or enhancing the function of adults and children with disabilities. Medical rehabilitation research is directed towards restoration and improvement of functional capability lost as a consequence of injury, disease, or congenital disorders. The mission of the NCMRR is to increase the effectiveness of medical rehabilitation practices through research on improving functional mobility; promoting behavioral adaptation to functional losses; assessing the efficacy and outcomes of medical rehabilitation therapies and practices; developing improved assistive technology; understanding whole body system responses to physical impairments and functional changes; developing more precise methods to measure impairments, disabilities, and societal and functional limitations; and training research scientists in the field of medical rehabilitation. PA: PA-99-041
Requirements: Applications for research project grants (R01) may be submitted by domestic and foreign, for-profit and nonprofit organizations, both public and private; units of state or local governments; and eligible agencies of the federal government.
Contact: Dr. Louis Quatrano, Behavioral Sciences and Rehabilitation Engineering Program, (301) 402-2242; email: lq2n@nih.gov; Dr. Stephen Tuel, Clinical Practices Program, (301) 402-2242; email: st100v@nih.gov
Internet: http://www.grants.nih.gov/grants/guide/pa-files/PA-99-041.html
Sponsor: National Institute of Child Health and Human Development
6100 Executive Boulevard
Rockville, MD 20852

NICHD NRSAs—Individual 3930
The program offers health scientists with doctoral degrees the opportunity to receive full-time research training in areas that reflect the national need for biomedical and behavioral research on specific programs of mothers and children and population and reproductive research. Areas of concentration include pregnancy and infancy, developmental biology and nutrition, learning and behavior, mental retardation, fertility-infertility, fertility regulation, nutrition and reproduction, social and behavioral aspects of reproduction, and population change. These fellowships may not be used for study leading to an MD, DO, DDS, or similar degree, or for the support of residency training. Awardees are subject to the payback provisions of the NRSA by which one year of research or teaching, or acceptable combination thereof, is required for each year of support, except for the first 12 months, which are free of the payback obligation.
Requirements: Universities, colleges, medical, dental, and nursing schools and schools of public health, laboratories, hospitals, state and local health departments, other for-profit public or private institutions, and individuals may apply for research grants. NRSA support is provided for academic and research training only, in health and health-related areas that are periodically specified by the NIH. SBIR grants can be awarded only to domestic small businesses. STTR grants can be awarded only to domestic small business concerns that partner with research institutions in cooperative research and development.
Date(s) Application is Due: Apr 5; Aug 5; Dec 5
Contact: NICHD Office of Program and Public Liaison, (301) 496-0536 or (301) 435-3447
Internet: http://www.nichd.nih.gov/funding/funding.htm
Sponsor: National Institute of Child Health and Human Development
Building 31 Room 2A32, MSC 2425, 31 Center Drive
Bethesda, MD 20892-2425

NICHD NRSAs—Institutional 3931
Awards provide opportunities for research training in the biological and/or behavioral science aspects of child health and human development. Primary concern in awarding institution awards will be given to multidisciplinary or interdisciplinary programs that cannot be provided through individual fellowships. Although major concern is for postdoctoral training, predoctoral training will be considered where a special case for support can be justified. In the Center for Research for Mothers and Children, areas covered include pregnancy and infancy, developmental biology and nutrition, learning and behavior, and mental retardation. In the Center for Population Research, areas covered include fertility-infertility, fertility regulation, nutrition and reproduction, social and behavioral aspects of reproduction, and population change. The training program director at the grantee institution will be responsible for the selection and appointment of trainees and for the overall direction of the program. These grants are not made for study leading to the MD, DDS, DVM, or other similar health professional degrees. Trainees are subject to the payback provisions of the NRSA by which one year of research or teaching, or acceptable combination thereof, is required for each year of support, except for the first 12 months, which are free of payback obligation.
Requirements: Universities; colleges; medical, dental, and nursing schools and schools of public health; laboratories; hospitals; state and local health departments; other for-profit public or private institutions; and individuals may apply for research grants. NRSA support is provided for academic and research training only, in health and health-related areas that are periodically specified by the National Institutes of Health. SBIR grants can be awarded only to domestic small businesses. STTR grants can be awarded only to domestic small business concerns that partner with research institutions in cooperative research and development.
Date(s) Application is Due: May 10
Contact: NICHD Office of Program and Public Liaison, (301) 496-0536 or (301) 435-3447
Internet: http://www.nichd.nih.gov/funding/funding.htm
Sponsor: National Institute of Child Health and Human Development
Building 31 Room 2A32, MSC 2425, 31 Center Drive
Bethesda, MD 20892-2425

NICHD Population Research Grants 3932
The objective of the program is to seek solutions to the fundamental problems of the reproductive processes; to strive to develop and evaluate safer and more effective contraceptives; and to understand how population structure and change affect the health and well-being of individuals and society. The Center for Population Research supports biomedical research in the reproductive sciences, clinical and epidemiological studies of contraceptives and the risk of sexually transmitted diseases (STDs) including HIV, and behavioral and demographic studies on population processes and structure. Mechanisms are projects grants, small business innovation research (SBIR) grants, small business technology transfer (STIR) grants, and individual fellowships (NRSAs).
Requirements: Project grants: Universities, colleges, medical, dental, and nursing schools, and schools of public health; laboratories, hospitals, state and local health departments, and other nonprofit or for-profit public or private institutions; and individuals. NRSAs: Individuals with a professional or scientific degree are eligible (MD, PhD, DDS, DO, DVM, ScD, DEng, or equivalent domestic or foreign degree). SBIR and STIR grants can be awarded only to domestic small businesses (entities that are independently owned and operated for profit, are not dominant in the field in which research is proposed, and have no more than 500 employees).
Amount of Grant: $50,000-$5 million range; $318,200 average
Date(s) Application is Due: Feb 1; Jun 1; Oct 1
Contact: Dr. Florence Haseltine, Director, Center for Population Research, (301) 496-1101; fax: (301) 496-0962; email: haseltif@mail.nih.gov
Internet: http://www.nichd.nih.gov/cpr
Sponsor: National Institute of Child Health and Human Development
6100 Executive Boulevard, Executive Building
Bethesda, MD 20892

NICHD Repetitive Motion Disorders Research Grants 3933
The program is sponsored by NIAMS, NICHD, NINDS, and NIOSH at the CDC. This initiative invites applications directed to the study of the pathogenesis, epidemiology, prevention, and treatment of repetitive motion disorders. The support mechanism for grants in this area will be the individual investigator-initiated research grant (R01). Information may be obtained electronically through the NIH grant line (301-402-2221), the NIH Web site, and by mail and email. PA-97-059
Requirements: Applications may be submitted by foreign and domestic, for-profit and nonprofit organizations, both public and private; units of state and local governments; and eligible agencies of the federal government.
Restrictions: Foreign institutions or organizations in foreign countries are not eligible.
Contact: Dr. Stephen Tuel, National Center for Medical and Rehabilitation Research, (301) 402-2242; fax: (301) 402-0832; email: tuels@hd01.nichd.nih.gov
Internet: http://www.grants.nih.gov/grants/guide/pa-files/PA-97-059.html
Sponsor: National Institute of Child Health and Human Development
6100 Executive Boulevard, Executive Building
Bethesda, MD 20892

NICHD Research on Musculoskeletal Fitness and Sports Medicine Grants 3934
NIAMS and NINR invite investigator-initiated research grant applications to study a broad range of basic and clinical topics related to musculoskeletal fitness, exercise physiology, and sports medicine. The National Center for Medical Rehabilitation Research of NICHD encourages applications for both basic and clinical studies of

musculoskeletal fitness and exercise physiology of persons with physical disabilities. PA: PA-97-025

Requirements: Applications may be submitted by domestic and foreign, for-profit organizations, both public and private; units of state or local governments; and eligible agencies of the federal government.

Contact: Dr. Danuta Krotoski, National Center for Medical Rehabilitation Research, (301) 402-2242; email: krotoskd@hd01.nichd.nih.gov

Internet: http://www.grants.nih.gov/grants/guide/pa-files/PA-97-025.html

Sponsor: National Institute of Child Health and Human Development
6100 Executive Boulevard
Bethesda, MD 20892

NICHD Ruth L. Kirschstein National Research Service Award (NRSA) Institutional Predoctoral Training Program in Systems Biology of Developmental Biology & Birth Defects 3935

The objective of this NRSA T32 program is to provide research training to predoctoral students interested in establishing research careers that use systems biology approaches to study developmental biology and the formation of structural birth defects. Stipends are provided as a subsistence allowance to help defray trainees' living expenses during the research training experience. Funds are provided to offset the cost of tuition and fees. The NIH provides 60% of the combined requested costs, up to $16,000 annually per Ph.D. student, and up to $21,000 annually for students in formal programs leading to combined professional and Ph.D. degrees. Letters of Intent are due by April 25, with full applications due May 25.

Requirements: The following organizations are eligible to apply: public/state controlled institutions of higher education; private institutions of higher education; nonprofits with 501(c)3 IRS status (other than institutions of higher education); nonprofits without 501(c)3 IRS status (other than institutions of higher education); for-profit organizations (other than small businesses); state governments; U.S. territories or possessions; Indian/Native American tribal governments (Federally recognized); Hispanic-serving institutions; historically Black colleges and universities (HBCUs); tribally controlled colleges and universities (TCCUs); Alaska Natives and Native Hawaiians serving institutions; regional organizations; eligible agencies of the federal government; faith-based or community based organizations.

Date(s) Application is Due: Apr 25; May 25

Contact: James Coulombe, Ph.D.; (301) 496-5541; fax: (301) 480-0303; email: coulombej@mail.nih.gov

Internet: http://grants.nih.gov/grants/guide/pa-files/PAR-08-054.html

Sponsor: National Institute of Child Health and Human Development (NICHD)
6100 Executive Boulevard, 4B01 MSC 7510
Bethesda, MD 20892-7510

NICHD Small Grants Program 3936

Applications for these small grants may be submitted by new or more established investigators for support of research in any scientific area relevant to the institute's programs in the Center for Population Research, the Center for Research for Mothers and Children, or the National Center for Medical Rehabilitation Research. Although proposals may involve a wide variety of biomedical, biobehavioral, or clinical disciplines, relevance to the mission of NICHD must be clear. Deadline dates are February 1, June 1, or October 1 for new applications; March 1, or July 1, or November 1 for revised applications. Additional information may be obtained by mail and email, or through the NIH grant line (301-402-2221), and the NIH Web site. PA: PAR-99-126

Requirements: Applications may be submitted by domestic, for-profit and nonprofit organizations, both public and private; units of state and local governments; and eligible agencies of the federal government.

Restrictions: Foreign institutions and organizations are not eligible.

Amount of Grant: $50,000 for one year; $100,000 for two years

Date(s) Application is Due: Feb 1; Jun 1; Oct 1

Contact: Program Contact, (301) 435-0714; email: grantsinfo@nih.gov

Internet: http://grants.nih.gov/grants/guide/pa-files/PAR-99-126.html

Sponsor: National Institute of Child Health and Human Development
6100 Executive Boulevard, Executive Building
Bethesda, MD 20892

NICHD Socioeconomic Status and Health Across the Life Course 3937

NHLBI, NIA, NICHD, NIEHS, and NIMH seek research grant applications on the cumulative and contemporaneous relationships between socioeconomic status (SES) and physical and mental health and functioning over the life course and across generations. Given that the relationships between SES and physical and mental health, morbidity, disability, and mortality have been long and extensively documented, additional studies aimed at merely describing or demonstrating these relationships are outside the scope of this program announcement. Encouraged are studies relating to appropriate conceptualization and measurement of SES over the life course, across generations, and in various population groups; specification of the processes through which SES influences cumulatively and contemporaneously physical and mental health, disability, morbidity, and mortality outcomes over the life course, and how these outcomes, in turn, impact on SES; the relationship between SES and physical and mental health, disability, morbidity, and mortality over the life course in various population groups. PA: PA-98-098

Requirements: Applications may be submitted by public and private, foreign and domestic for-profit and nonprofit organizations, and eligible agencies of the federal government.

Contact: Dr. Rose Maria Li, Center for Population Research, (301) 496-1174; fax: (301) 496-0962; email: Rose_Li@nih.gov

Internet: http://www.grants.nih.gov/grants/guide/pa-files/PA-98-098.html

Sponsor: National Institute of Child Health and Human Development
6100 Executive Boulevard
Bethesda, MD 20892

NICHD Symptom Management for Chronic Neurological Conditions Research Grants 3938

NINR, NIA, NICHD, NIDCD, and NINDS are interested in facilitating investigator-initiated research designed to deal with symptoms associated with chronic neurological conditions including stroke, epilepsy, brain and spinal cord injury, multiple sclerosis, and Parkinson's disease. The overall goals are to manage symptoms effectively, to prevent secondary effects of the condition when possible, to maximize the health potential, and to improve the health-related quality of life for patients, caregivers, and families. The mechanism of support is the research project grant (R01).

Requirements: Research applications may be submitted by domestic and foreign, for-profit and nonprofit organizations, both public and private; units of state and local governments; and eligible agencies of the federal government.

Contact: Dr. Louis Quatrano, National Center for Medical Rehabilitation Research, (301) 402-2242; fax: (301) 402-0832; email: LQ2N@nih.gov; Mary Ellen Colvin, Grants Management Branch, (301) 496-1303; fax: (301) 402-0915; email: MC113B@nih.gov

Internet: http://www.grants.nih.gov/grants/guide/pa-files/PA-98-030.html

Sponsor: National Institute of Child Health and Human Development
6100 Executive Boulevard, Executive Building
Bethesda, MD 20892

Nick Traina Foundation Grants 3939

The foundation supports organizations that are involved in the diagnosis, research, treatment, and/or family support of manic-depression, suicide prevention, child abuse, and children in jeopardy. Grants also support struggling musicians in the areas of health and mental illness. The foundation gives special consideration to proposals that address manic-depression in children and young adults. There are no application deadlines; proposals are reviewed four times each year.

Contact: Grants Administrator, (415) 771-4224; email: info@nicktrainafoundation.org

Internet: http://www.nicktrainafoundation.org/main.htm

Sponsor: Nick Traina Foundation
P.O. Box 470427
San Francisco, CA 94147-0427

NIDA Behavioral Science Track Awards for Rapid Transition 3940

NIDA invites newly independent investigators to submit applications for small-scale, exploratory (i.e., pilot) research projects related to NIDA's behavioral sciences mission. The award (B/START-NIDA) provides rapid review and funding decisions of applications. Experimentally based research applications are encouraged across a wide variety of behavioral factors in drug abuse, including neurocognitive, cognitive, and perceptual processes; psychosocial; and more broadly motivational, social, and community factors in drug abuse. Given the role that drug abuse plays in HIV/AIDS transmission, studies applying basic behavioral science models and methods to address this issue are especially encouraged. More information may be obtained by mail and email or through the NIH grant line (301-402-2221), and the NIH Web site. PAR-03-146

Amount of Grant: $50,000 maximum

Contact: Gary Fleming, Grants Management Branch, (301) 443-6710

Internet: http://grants2.nih.gov/grants/guide/pa-files/PAR-03-146.html

Sponsor: National Institute on Drug Abuse
5600 Fishers Ln, Parklawn Building
Rockville, MD 20857

NIDA Behavioral Therapies Development Program Grants 3941

The purpose of the program is to reaffirm NIDA's continued and ongoing commitment to a major program of research on behavioral therapies for drug abuse and dependence. The term behavioral therapy is used here in a broad sense and includes various forms of psychotherapy, behavior therapy, cognitive therapy, family therapy, skills training, counseling, and other rehabilitative therapies. Behavioral therapy research has been conceptualized, for the purposes of this initiative, to consist of three stages. Stage I, or early therapy development, involves research on the development, refinement, and pilot testing of behavioral interventions. Stage I includes research aimed at translating basic behavioral and cognitive research into novel behavioral therapies. Stage II involves the efficacy testing of therapies that show promise and includes research on the hypothesized mechanisms through which these therapies may have their effect. Stage III is research aimed at determining if and how efficacious behavioral therapies may be transported to community settings, including primary care sites. Stage III includes research examining methods of training therapists and counselors to administer new behavioral therapies. PA: PA-03-066

Requirements: Applications may be submitted by domestic and foreign, for-profit and nonprofit organizations, both public and private; units of state and local government; and eligible agencies of the federal government.

Amount of Grant: $100,000 maximum per year

Contact: Gary Fleming, NIDA, Grants Management Branch, (301) 443-6710; fax: (301) 594-6849; email: GF6S@NIH.GOV

Internet: http://grants2.nih.gov/grants/guide/pa-files/PA-03-066.html

Sponsor: National Institute on Drug Abuse
6001 Executive Boulevard, Executive Building
Bethesda, MD 20892

NIDA Clinical Research Education Programs in Drug Abuse and Addiction 3942

Grants support educational programs developing the research skills of investigators who wish to focus on clinical, prevention, health services, or treatment issues in drug abuse; or developing a cadre of clinical drug abuse experts proficient in the use of research findings (for example, in public health, behavioral science, or medicine) to develop or implement early detection, diagnosis, treatment, or prevention interventions. This latter category would also include applications of health services research findings. Programs that address both of these aims are particularly desired. NIDA will grant awards to eligible institutions to develop or enhance research training opportunities for individuals, selected by the institution, who are training for careers as researchers, as clinicians (broadly defined), or (ideally) as a combination of the two. Educational partnerships between research institutions and drug abuse treatment organizations are highly encouraged in order to broaden research training and educational opportunities and as a result, increase the number of well-trained clinical and health services researchers in the field. Contact program staff for deadline dates. PA: PAR-04-054
Requirements: Applications may be submitted by domestic, for-profit and nonprofit organizations, both public and private; units of state and local governments; and eligible agencies of the federal government; All students must be U.S. citizens, nationals, or lawfully admitted permanent residents of the United States. Predoctoral students must have an MS, MSW, or equivalent degree in biostatistics, nursing, psychology, or other social sciences as a prerequisite to pursuing a doctorate degree.
Restrictions: This award will not support a master's level degree for a predoctoral student.
Amount of Grant: $350,000 maximum
Contact: Gary Fleming, Chief, Grants Management Officer, (301) 443-6710; fax: (301) 594-6847; email: gf6s@nih.gov
Internet: http://grants.nih.gov/grants/guide/pa-files/PAR-04-054.html
Sponsor: National Institute on Drug Abuse
6101 Executive Boulevard, Suite 242, MSC 8403
Bethesda, MD 20892-8403

NIDA Cutting-Edge Basic Research Awards (CEBRA) 3943

The goal of this program is to foster highly innovative or conceptually creative research that advances our understanding of drug abuse and addiction and how to prevent and treat them. It is specifically designed to support research that is high-risk and potentially high-impact and that is underrepresented or not included in NIDA's current portfolio. The program is aimed at experienced drug abuse research investigators who wish to develop or adapt new methods or techniques and at new investigators or scientists with expertise in fields other than drug abuse who wish to establish new programs in drug abuse research. PA: PAR-01-047
Requirements: Applications may be submitted by domestic and foreign, for-profit and nonprofit organizations, both public and private; units of state and local governments; and eligible agencies of the federal government.
Restrictions: Amended R21 applications for CEBRA will not be accepted.
Amount of Grant: $1 million total; $100,000 maximum for Stage I awards
Contact: Dr. Rebekah Rasooly, (301) 443-6300; fax: (301) 594-6043; email: rr185i@nih.gov
Internet: http://grants.nih.gov/grants/guide/pa-files/PAR-01-047.html
Sponsor: National Institute on Drug Abuse
6001 Executive Boulevard
Bethesda, MD 20892

NIDA Drug Abuse Dissertation Research: Epidemiology, Prevention, Treatment, Services, and Women and Gender Differences 3944

NIDA announces the availability of small grants (R03) to support drug abuse doctoral dissertation research in epidemiology, prevention, treatment, services, and women and sex/gender differences, areas in which there is a significant need for new investigators. Grant support is designed to encourage doctoral candidates from a variety of academic disciplines and programs to conduct research in these areas of interest to NIDA. It is hoped that this program will ultimately facilitate the entry of promising new investigators into the field of drug abuse research. PA: PAR-05-083
Requirements: The applicant investigator must be enrolled in an accredited doctoral degree program in the behavioral, biomedical, or social sciences and must be conducting or intending to conduct research in one of the areas specified. The applicant must be a registered doctoral candidate student. All requirements for the doctoral degree other than the dissertation and clinical internship, if applicable, must be completed by the time of the award. This information and the approval of the dissertation topic by a named committee must be verified in a letter of certification from the thesis chairperson and submitted with the grant application. The applicant institution administering the grant on behalf of the proposed applicant must be domestic. Applications may be submitted by any public or private nonprofit university, college, or professional school. The doctoral candidate must be a citizen or noncitizen national of the United States or hold a permanent residence visa.
Restrictions: An applicant who receives support for dissertation research under a grant from NIDA may not hold concurrently another federally sponsored fellowship or similar Federal award, such as a National Research Service Award (NRSA) that provides a stipend or otherwise duplicates provisions of this grant program.
Contact: Gary Fleming, (301) 443-6710; fax: (301) 594-6849; email: gfleming@nida.nih.gov
Internet: http://grants.nih.gov/grants/guide/pa-files/PAR-05-083.html
Sponsor: National Institute on Drug Abuse
5600 Fishers Ln, Parklawn Building
Rockville, MD 20857

NIDA Drug Addiction Neuroscience Research Grants 3945

The intent of this program announcement is to continue to encourage investigator interest in the wide range of neuroscience research relevant to drug abuse, drug dependence, and drug addiction supported by NIDA. The research relevant to this announcement does not involve human subjects; for information about the neuroscience research program involving human subjects, contact Dr. Joseph Frascella, (301) 443-4877, jf80t@nih.gov. The scientific understanding gained by this research is anticipated ultimately to be applicable to improved treatment and prevention of drug abuse and drug addiction, and their consequences. The specific research topics emphasized in this program announcement are animal models to study the neurobiology of addiction; vulnerability to drug addiction; neuroadaptation and neurotoxicity; developmental neurobiology; drugs and learning, memory, and cognition; pain and analgesia; drug effects on sensory and perceptual systems; drugs of abuse; sleep and circadian rhythms; neuropsychopharmacology of drugs of abuse; and central nervous system (CNS) interaction with other systems. Contact the program officer for deadline dates. PA: PA-02-085
Requirements: Applications may be submitted by public and private, domestic and foreign, for-profit and nonprofit organizations, units of state and local governments, and eligible agencies of the federal government. Racial/ethnic minority individuals, women, and persons with disabilities are encouraged to apply as principal investigators.
Contact: Dr. Roger Brown, Division of Basic Research, (301) 443-1887; fax: (301) 594-6043; email: rb99w@nih.gov
Internet: http://grants2.nih.gov/grants/guide/pa-files/PA-02-085.html
Sponsor: National Institute on Drug Abuse
6100 Executive Boulevard, Executive Building
Rockville, MD 20892

NIDA HIV Drug Therapy for HIV Users: Access, Adherence, Effectiveness 3946

NIDA invites applications for research on access, adherence, and effectiveness relevant to the treatment of drug users with HIV. This initiative supports research on a broad range of HIV care-related issues, including, but not limited to, access issues related to receipt of HIV-related treatment and outreach strategies to improve access; factors influencing initiation of HIV treatment and long-term adherence to HIV treatment regimens; and the development of intervention strategies to improve use of and adherence to complex medical regimens in a variety of settings. Annual deadline dates may vary; contact program staff for exact dates. RFA: DA-00-007
Requirements: Applications may be submitted by domestic and foreign for-profit and nonprofit organizations, both public and private; units of state and local governments; and eligible agencies of the federal government.
Amount of Grant: $3 million total
Contact: Dr. Elizabeth Lambert, (301) 402-1933; fax: (301) 443-4100; email: el46i@nih.gov
Internet: http://grants.nih.gov/grants/guide/rfa-files/RFA-DA-00-007.html
Sponsor: National Institute on Drug Abuse
5600 Fishers Ln, Parklawn Building
Rockville, MD 20857

NIDA International Studies on Drug Abuse and HIV/AIDS 3947

This program invites applications for international, cross-disciplinary research on HIV/AIDS, other blood-borne infections [i.e., hepatitis B virus (HBV), hepatitis C virus (HCV)], tuberculosis (TB), and comorbid conditions and consequences in injecting and noninjecting drug users, their sexual partners, and their children. This RFA is intended to build on lessons learned in developed countries in response to the intertwined epidemics of drug abuse and the spread of HIV/AIDS and other infectious diseases. It seeks to foster cross-national and international research collaborations that, through both independent research and the recruitment, training, and mentoring of new, multidisciplinary researchers, lead to the development, adaptation, replication, and evaluation of effective interventions and approaches to slow or reverse the spread of HIV and other infections in vulnerable drug-using populations. Annual deadline dates may vary; contact program staff for exact dates. RFA: DA-01-006
Requirements: Applications may be submitted by domestic and foreign for-profit and nonprofit organizations, both public and private; units of state and local governments; and eligible agencies of the federal government.
Amount of Grant: $1.5 million total
Contact: Helen Cesari, Center on AIDS and Other Medical Consequences of Drug Abuse, (301) 402-1918; fax: (301) 443-4100; email: hc30x@nih.gov
Internet: http://grants.nih.gov/grants/guide/rfa-files/RFA-DA-01-006.html
Sponsor: National Institute on Drug Abuse
6001 Executive Boulevard
Bethesda, MD 20892

NIDA Medications Development for Cannabis-Related Disorder 3948

This program seeks grant applications focusing on the identification, evaluation, and development of safe and effective pharmacological treatments for cannabis-related disorders (CRDs), such as cannabis abuse or dependence, and cannabis-induced disorders (e.g., intoxication, delirium, psychotic disorder, and anxiety disorder), and their comorbidity with other medical and psychiatric disorders (e.g., depression), with special interest in the treatment of children and adolescents. Cannabis use includes marijuana, hashish, and other tetrahydrocannabinol (THC) containing substances. This RFA will support clinical trials to test potential medications in humans, and the preclinical development of new chemical entities (NCEs) for advancement to the clinical development stage. Applications may focus on the pharmacotherapy of cannabis abuse

and they may target one or various CRDs or clinical manifestations of the disorders. Annual deadline dates may vary; contact program staff for exact dates.
Requirements: Applications may be submitted by domestic and foreign, for-profit and nonprofit organizations, both public and private; units of state and local governments; and eligible agencies of the federal government.
Contact: Gary Fleming, Grants Management Branch, (301) 443-6710; fax: (301) 594-6847; email: gf6s@nih.gov
Internet: http://www.grants.nih.gov/grants/guide/rfa-files/RFA-DA-04-014.html
Sponsor: National Institute on Drug Abuse
6101 Executive Boulevard
Bethesda, MD 20892-8403

NIDA Mentored Clinical Scientist Research Career Development Awards **3949**
The overall objective of the NIH Research Career Development Award program is to prepare qualified individuals for careers that have a significant impact on the health-related research needs of the Nation. The objective of the NIH Mentored Clinical Scientist Research Career Development Award (K08) program is to support didactic study and mentored research for individuals with clinical doctoral degrees (e.g., M.D., D.D.S., D.M.D., D.O., D.C., O.D., N.D., D.V.M., Pharm.D., or Ph.D. in clinical disciplines). The purpose of this Funding Opportunity Announcement (FOA) is to continue the long-standing NIH support of the K08 program. This award provides support and protected time for an intensive, mentored research career development experience in biomedical or behavioral research, including translational research. For the purpose of this award, translational research is defined as application of basic research discoveries toward the diagnosis, management, and prevention of human disease.
Requirements: The following organizations and institutions may apply: public/state controlled institutions of higher education; private institutions of higher education; Hispanic-serving institutions; historically Black colleges and universities (HBCUs); tribally controlled colleges and universities (TCCUs); Alaska native and native Hawaiian serving institutions; nonprofits with 501(c)3 IRS status; nonprofits without 501(c)3 IRS status; small businesses; for-profit organizations; State governments; regional organizations; U.S. territories or possessions; Indian/Native American tribal governments; Indian/Native American tribally designated organizations; county governments; city or township governments; special district governments; independent school districts; public housing authorities/Indian housing authorities; eligible agencies of the Federal government; and faith-based or community based organizations.
Amount of Grant: $90,000 salary maximum per yer plus fringe benefits; $50,000 for research support
Date(s) Application is Due: Feb 12; Mar 12; Jun 12; Jul 12; Oct 12; Nov 12
Contact: Mimi M. Ghim, Ph.D.; (301) 443-6071; email: ghimm@mail.nih.gov; Carol Alderson; (301) 594-5614; email: ca10h@nih.gov
Internet: http://grants.nih.gov/grants/guide/pa-files/PA-09-042.html#SectionIII
Sponsor: National Institute on Drug Abuse (NIDA)
6001 Executive Boulevard, Room 5232, MSC 9591
Bethesda, MD 20892

NIDA Microarray-Based Research on Drug Abuse **3950**
With this announcement, NIDA plans to develop a strong, multidisciplinary basic research program—drawing from neuroscience, genetics, behavioral science, bioengineering, and computational biology—aimed at adapting and applying microarray techniques to the general problem of drug abuse. Eventually, it is anticipated that the identification of genes and gene products that may underlie the process of addiction will provide new targets for the development of new medications and therapies for drug abuse. Annual deadline dates may vary; contact program staff for exact dates. RFA: DA-00-003
Requirements: Applications may be submitted by domestic and foreign, for-profit and nonprofit organizations, both public and private; units of state and local governments; and eligible agencies of the federal government.
Amount of Grant: $2 million total
Contact: Dr. Rebekah Rasooly, (301) 443-6300; fax: (301) 594-6043; email: rr185i@nih.gov
Internet: http://grants.nih.gov/grants/guide/rfa-files/RFA-DA-00-003.html
Sponsor: National Institute on Drug Abuse
6001 Executive Boulevard
Bethesda, MD 20892

NIDA Minority Institutions Drug Abuse Research Development Grants **3951**
In its efforts to broaden and strengthen drug abuse research in minority institutions, NIDA is initiating MIDARP. Grants will be provided to develop the capacity of minority institutions and their faculty, staff, and students to conduct rigorous drug abuse research in all areas of research supported by NIDA including neuroscience, behavioral, clinical, social science, public health, biological, HIV/AIDS, and health services areas. In addition, this PA seeks to broaden the scientific knowledge base in drug abuse in those areas where minority institutions may have particular interest, knowledge, and commitment. Awards under this PA will use the R24 grant mechanism. Applicants are responsible for the planning, direction, and execution of the proposed projects. The total project period for applications submitted in response to this PA may not exceed five years. PAR-98-001 may be obtained electronically through the NIH Web site, email, and by mail.
Requirements: An applicant institution must be a public or private nonprofit university, four-year college, or other institution with a traditionally high (more than 50 percent) minority student enrollment; an institution with a substantial enrollment of minority students, which provides evidence of efforts to recruit into scientific careers members of ethnic or racial groups who have been found to be underrepresented in biomedical and behavioral research; or a Native American tribe that, in conjunction with one or more institutions of higher learning, offers undergraduate and graduate degrees in ADM disciplines.
Contact: Dr. Lula Beatty, Special Populations Office, (301) 443-0441; fax: (301) 480-8179; email: lb75x@nih.gov; Dr. Pushpa Thadani, Division of Basic Research, (301) 443-6300; fax: (301) 443-6043; email: pt24e@nih.gov; Dr. C. Jamie Biswas, Medications Develop
Internet: http://www.grants.nih.gov/grants/guide/pa-files/PAR-98-001.html
Sponsor: National Institute on Drug Abuse
5600 Fishers Ln, Parklawn Building
Rockville, MD 20857

NIDA National Drug Abuse Treatment Clinical Trials Network **3952**
NIDA invites cooperative agreement applications from established clinical investigators to participate in the National Drug Abuse Treatment Clinical Trials Network (CTN). Awardees will conduct and participate in coordinated multisite clinical trials of behavioral, pharmacological, and combined behavioral/pharmacological therapies for drug abuse and addiction, and conduct research on practices, i.e., studies of factors that impact upon successful adoption of new treatments. The overall goal of the program is to improve the quality of drug abuse treatment throughout the nation, thereby reducing the prevalence, morbidity, and mortality of drug abuse and addiction. Some of the program objectives include supporting studies of behavioral, pharmacological, and combined behavioral and pharmacological treatment interventions of established efficacy in rigorous, multisite clinical trials to determine effectiveness across a broad range of treatment settings and patient populations; encouraging research on the translation of science-based, effective treatment interventions into clinical practice by establishing effectiveness in community-based treatment programs and by determining effective strategies for transporting effective treatments; furthering the development of effective treatments by integrating behavioral, pharmacological, and practice research; investigating the impact of advances in treatment research on community-level treatment practices; and ensuring that treatment research in drug abuse and addiction meets the needs of the wider community, including minorities, women, children, and underserved populations. The complete RFA is available on the Web site. Annual deadline dates may vary; contact a program officer for exact dates. RFA: DA-00-002
Requirements: Applications may be submitted by domestic for-profit and nonprofit organizations, both private and public. Principal investigators should be able to document a substantial history of managing complex research initiatives in clinical sites and have an extensive publication record in clinical research.
Restrictions: Foreign applicants are not eligible, and domestic applications may not include a foreign site.
Amount of Grant: $11 million total
Contact: Dr. Jack Blaine, Division of Clinical and Services Research, (301) 443-0107; fax: (301) 443-8674; email: jb167n@nih.gov; Dr. Betty Tai, Medications Development Division, (301) 443-2397; fax: (301) 443-2599; email: bt26r@nih.gov
Internet: http://grants.nih.gov/grants/guide/rfa-files/RFA-DA-00-002.html
Sponsor: National Institute on Drug Abuse
6001 Executive Boulevard, Executive Building
Bethesda, MD 20892

NIDA Neurobiological and Behavioral Research on Nicotine and Tobacco Products **3953**
The purpose of this PA is to expand the basic science knowledge base on the neurobiological and behavioral effects of nicotine and associated tobacco chemicals, as part of continuing efforts to explain and prevent their use and to develop effective treatments for nicotine addiction. This PA encourages research on any aspect of the effects of nicotine and other tobacco components, using neurobiological, behavioral, or other methods in humans, animals, or in vitro systems, that seeks to explain nicotine use, addiction, or other effects in humans. This PA solicits research in two general categories, neurobiology and behavioral sciences. Neurobiological, cognitive, behavioral, and social processes, methods, and models are all relevant to this announcement. For both neurobiological and behavioral studies, approaches are encouraged that consider differences in gender, race, ethnicity, and age or developmental stage. PA: PA-00-045
Requirements: Applications may be submitted by domestic and foreign, for-profit and nonprofit organizations, both public and private; units of state and local governments; and eligible agencies of the federal government.
Contact: Dr. Thomas Aigner, Neurobiological Studies, (301) 443-6975; fax: (301) 594-6043; email: ta17r@nih.gov; Dr. Jaylan Turkkan, Cognitive or Behavioral Studies, (301) 443-1263; fax: (301)-594-6043; email: jt58t@nih.gov
Internet: http://grants.nih.gov/grants/guide/pa-files/PA-00-045.html
Sponsor: National Institute on Drug Abuse
6001 Executive Boulevard
Bethesda, MD 20892

NIDA Origins and Pathways to Drug Abuse Research Grants **3954**
The main goal of this PA is to stimulate innovative investigations that enhance our understanding of: drug use patterns and trends within and across populations; interplay of social interactions, social environment, structural context with individual behavioral characteristics and genetic vulnerability; the phenotypic heterogeneity of drug abuse; causal mechanisms leading to onset, maintenance, and remittance of drug abuse, as well as protective mechanisms that reduce the risk of drug abuse; and drug abuse over the life course, including developmental processes that influence drug use trajectories and behavioral, health, and social consequences of drug abuse. In addition, research

is encouraged to develop methodologies to improve the accuracy, efficiency, scope, timeliness, and analytic yield of drug abuse epidemiologic data. PA: PA-04-100
Requirements: Applications may be submitted by foreign and domestic, for-profit and nonprofit organizations, both public and private; units of state or local governments; and eligible agencies of the federal government.
Restrictions: Foreign institutions are not eligible for program project or centers (P-Series) awards.
Contact: Gary Fleming, Grants Management, (301) 443-6710; fax: (301) 594-6849; email: gfleming@nida.nih.gov
Internet: http://grants.nih.gov/grants/guide/pa-files/PA-04-100.html
Sponsor: National Institute on Drug Abuse
6001 Executive Boulevard
Bethesda, MD 20892

NIDA Pilot and Feasibility Studies in Preparation for Drug Abuse Prevention Trials 3955
This FOA for R34 applications seeks to support: pilot and/or feasibility testing of new, revised, or adapted preventive intervention approaches targeting the initiation of drug use, the progression to abuse or dependence, and the acquisition or transmission of HIV infection among diverse populations and settings; and pre-trial feasibility testing for prevention services and systems research.
Requirements: The following organizations and institutions may apply: public/state controlled institutions of higher education; private institutions of higher education; Hispanic-serving institutions; historically Black colleges and universities (HBCUs); tribally controlled colleges and universities (TCCUs); Alaska native and native Hawaiian serving institutions; nonprofits with 501(c)3 IRS status; nonprofits without 501(c)3 IRS status; small businesses; for-profit organizations; State governments; regional organizations; U.S. territories or possessions; Indian/Native American tribal governments; Indian/Native American tribally designated organizations; county governments; city or township governments; special district governments; independent school districts; public housing authorities/Indian housing authorities; eligible agencies of the Federal government; and faith-based or community based organizations.
Restrictions: The NIDA R34 mechanism does not support the development of intervention protocols, manuals, or the standardization of protocols.
Amount of Grant: $225,000 annually; $450,00 maximum for three years
Date(s) Application is Due: Feb 16; Jun 16; Oct 16
Contact: Aleta Meyer, Ph.D.; (301) 402-1725; fax: (301) 443-2636; email: meyera2@nida.nih.gov
Internet: http://grants.nih.gov/grants/guide/pa-files/PA-09-146.html
Sponsor: National Institute on Drug Abuse (NIDA)
6001 Executive Boulevard, Room 5170
Bethesda, MD 20892

NIDA Research on Adolescent Drug Abuse 3956
NIDA is firmly committed to support of research in the area of adolescent drug abuse. The purpose of this program announcement is to encourage further investigations in this area, particularly with regard to gaps in current knowledge. Additional information can be obtained electronically through the NIH grant line (301-402-2221), the NIH Web site, and by mail or email from the program contact listed. PA: PA-97-005
Requirements: Applications may be submitted by foreign and domestic, for-profit and nonprofit organizations, both public and private; units of state and local governments; and eligible agencies of the federal government.
Date(s) Application is Due: Feb 1; Jun 1; Oct 1
Contact: Dr. Elizabeth Rahdert, Division of Clinical and Services Research, (301) 443-0107; fax: (301) 443-8674; email: er34g@nih.gov
Internet: http://www.grants.nih.gov/grants/guide/pa-files/PA-97-005.html
Sponsor: National Institute on Drug Abuse
5600 Fishers Ln, Parklawn Building
Rockville, MD 20857

NIDA Role of Limbic System and Brain Ontogeny in Drug Abuse Research Grants 3957
This initiative is designed to support basic research into the fundamental mechanisms of development of the midbrain and basal forebrain structures that mediate the euphoric properties of drugs as well as understanding how drugs of abuse affect the cellular and molecular mechanism underlying nervous system development. In the past, vertebrate model systems (such as chick, frog, mouse, and zebrafish) and invertebrate systems (such as Drosophila and C. Elegans) have provided insights into mechanisms of development and are also likely to provide new information about the formation and specification of the limbic system. Approaches using these or other model systems both in vitro and in vivo are applicable to this PA. PA-02-015 may be obtained through the NIH grant line (301-402-2221), the NIH Web site, mail, and email.
Requirements: Applications may be submitted by foreign and domestic for-profit and nonprofit organizations, public or private.
Amount of Grant: $90,000 maximum direct costs per year
Contact: Dr. Jonathan Pollock, Division of Basic Research, (301) 443-6300; fax: (301) 594-6043; email: jp183r@nih.gov; Gary Fleming, Grants Management Branch, (301) 443-6710; fax: (301) 594-6847; email: gf6s@nih.gov
Internet: http://grants2.nih.gov/grants/guide/pa-files/PA-02-015.html
Sponsor: National Institute on Drug Abuse
5600 Fishers Ln, Parklawn Building
Rockville, MD 20857

NIDA Senior Scientist Award 3958
The Senior Scientist Award provides support to outstanding scientists who have demonstrated sustained, high levels of productivity and whose expertise, research accomplishments, and contributions to the field have been and will continue to be critical to the mission of the particular NIH center or institute. The award provides salary support for award periods of five years. Contact the program office for more information. PA: PA-95-051
Requirements: The candidate must be a senior scientist and a recognized leader in the field with a distinguished record of original contributions, must have long-term support from a funding institute or center, and must have peer-reviewed grant support at the time of the award. Applications may be submitted on behalf of candidates by domestic, nonfederal organizations, both public or private; or other institutions. Candidates must be U.S. citizens, noncitizen nationals, or permanent residents.
Amount of Grant: $1000-$1.235 million; $190,000 average
Date(s) Application is Due: Feb 1; Jun 1; Oct 1
Contact: Dr. Timothy Condon, Office of Science Policy, Education, and Legislation, (301) 443-6072; fax: (301) 443-6277; email: tc52x@nih.gov; Dr. Charles Sharp, Division of Basic Research, (301) 443-1887; fax: (301) 594-6043; email: cs107m@nih.gov
Internet: http://www.grants.nih.gov/grants/guide/pa-files/PA-95-051.html
Sponsor: National Institute on Drug Abuse
5600 Fishers Ln, Parklawn Building
Rockville, MD 20857

NIDA Transition from Drug Abuse to Drug Addiction 3959
The intent of this program is to stimulate research on the processes and mechanisms underlying the transition from drug use to drug addiction from a wide variety of academic disciplines and approaches. Research from epidemiologic, basic, and clinical disciplines, including research on other transitional states with relevance to addiction, will be supported under this program. Annual deadline dates may vary; contact program staff for exact dates. RFA: DA-01-004
Requirements: Applications may be submitted by domestic, for-profit and nonprofit organizations, both public and private; units of state and local governments; and eligible agencies of the federal government.
Amount of Grant: $400,000 maximum
Contact: Dr. Steven Gust, (301) 443-6480; fax: (301) 443-9127; email: sgust@mail.nih.gov
Internet: http://grants.nih.gov/grants/guide/rfa-files/RFA-DA-01-004.html
Sponsor: National Institute on Drug Abuse
6001 Executive Boulevard
Bethesda, MD 20892

NIDCD Clinical Trial Planning Grant 3960
NIDCD provides funding to those who are seeking to improve health by supporting well-executed clinical trials in its mission areas of hearing, balance, smell, taste, voice, speech, and language. Investigators who submit applications for large-scale clinical trials for consideration are expected to provide detailed information regarding the study rationale, design, protocols and procedures, analytical techniques, facilities and environment, administrative procedures, and collaborative arrangements. PA: PAR-02-158
Requirements: Applications may be submitted by domestic for-profit and nonprofit organizations, both public and private; units of state and local governments; and eligible agencies of the federal government.
Restrictions: Foreign organizations are not eligible to apply.
Amount of Grant: $75,000 maximum
Contact: A. Julianna Gulya, Chief, Clinical Trials Branch, (301) 435-4085; fax: (301) 402-6251; email: julie_gulya@nih.gov
Internet: http://grants2.nih.gov/grants/guide/pa-files/PAR-02-158.html
Sponsor: National Institute on Deafness and Other Communication Disorders
6120 Executive Boulevard
Bethesda, MD 20892

NIDCD Communicative Competence of Users of Augmentative and Alternative Communication Systems Research Grants 3961
Awards are given for the support of research studies on issues related to the communicative competence of users of augmentative and alternative communication (AAC) systems. Such studies promise to lead to a better understanding of the impact of AAC on the development of or restoration of communicative competence of individuals with severe communication difficulties commonly seen in association with developmental or acquired disorders such as brain injury and other diseases or disorders affecting the oral motor system. It will also define the ways in which augmentative and alternative strategies can influence all phases of communication development and activity. Findings from these studies will direct the establishment and refinement of tools and intervention strategies enabling the development or recovery of communicative function. Additional information may be obtained by mail and email or through the NIH grant line (301-402-2221), and the NIH Web site. PA: PA-96-032
Amount of Grant: $72,486-$458,342; $200,000 average
Date(s) Application is Due: Feb 1; Jun 1; Oct 1
Contact: Sara Stone, Grants Management, (301) 402-0909
Internet: http://www.grants.nih.gov/grants/guide/pa-files/PA-96-032.html
Sponsor: National Institute on Deafness and Other Communication Disorders
6120 Executive Boulevard, Executive Plaza S
Bethesda, MD 20892

NIDCD Exploratory Research: Feasibility Pilot Studies Grants 3962

The NIDCD invites applications for exploratory, initial feasibility pilot studies focused on hearing, balance, smell, taste, voice, speech, and language, the organization's scientific mission areas. Exploratory research involves initial feasibility pilot studies in which the technological, methodological, or theoretical approach to a problem lacks sufficient preliminary/baseline data and a body of peer-reviewed publications, but whose successful outcome would make a significant contribution to a scientific area or field. It is anticipated that most applications will be submitted by investigators with ongoing research programs who wish to change the focus of their current research effort or move into a new area of research within the communication sciences but need additional funds to complete initial pilot studies. The NIDCD encourages applications from investigators conducting research outside the communication sciences who possess expertise with methodological or technological approaches which have not been typically applied to but which would greatly promote scientific progress within the communication sciences. Annual deadline dates may vary; contact program staff for exact dates. PA: PAR-01-103

Requirements: Applications may be submitted by domestic for-profit and nonprofit organizations, both public and private; units of state or local governments; and eligible agencies of the federal government.

Contact: Dr. Amy Donahue, Hearing, (301) 402-3458; email: amy_donahue@nih.gov; Dr. Lynn Luethke, Hearing, (301) 402-3461; email: lynn_huerta@nih.gov; Dr. Daniel Sklare, Balance/Vestibular, (301) 496-1804; email: daniel_sklare@nih.gov

Internet: http://grants2.nih.gov/grants/guide/pa-files/PAR-01-103.html

Sponsor: National Institute on Deafness and Other Communication Disorders
6120 Executive Boulevard, Executive Plaza S
Bethesda, MD 20892

NIDCD Exploratory/Developmental Research Grants 3963

NIDCD invites grant applications from basic and clinical investigators who are interested in pursuing high-risk/high-impact (HR/HI) research focused on hearing, balance, smell, taste, voice, speech, or language related to the specific mission of the institute. HR/HI research involves pilot/feasibility studies in which the technological, methodological, or theoretical approach to the problem lacks a traditional historical basis or pilot data, but which could have a major impact on a scientific area or field. Additional information may be obtained electronically through the NIH grant line (301-402-2221), and by mail and email from the program contact listed. Annual deadline dates may vary; contact program staff for exact dates. PA: PAR-96-073

Amount of Grant: $72,486-$458,342; $200,000 average

Date(s) Application is Due: Feb 1; Jun 1; Oct 1

Contact: Sharon Hunt, Grants Management, (301) 402-0909

Internet: http://www.grants.nih.gov/grants/guide/pa-files/PAR-96-073.html

Sponsor: National Institute on Deafness and Other Communication Disorders
6120 Executive Boulevard, Executive Plaza S
Bethesda, MD 20892

NIDCD Health Promotion and Disease Prevention: Health Communication, Development, and Dissemination 3964

The purpose of this RFA is to encourage the submission of research grant applications that create, develop, and evaluate health communication information. Projects should be based in current health communication theory and directed to address important problems of disease prevention, health promotion, and improved understanding of the basis for human communication disorders through public information. Annual deadline dates may vary; contact program staff for exact dates. RFA: DC-00-003

Requirements: Applications may be submitted by domestic and foreign, for-profit and nonprofit organizations, both public and private; units of state and local governments; and eligible agencies of the federal government.

Amount of Grant: $600,000 total

Contact: Dr. Amy Donahue, (301) 402-3458; fax: (301) 402-6251; email: Amy_Donahue@nih.gov

Internet: http://grants.nih.gov/grants/guide/rfa-files/RFA-DC-00-003.html

Sponsor: National Institute on Deafness and Other Communication Disorders
6120 Executive Boulevard, Executive Plaza S
Bethesda, MD 20892

NIDCD Mentored Clinical Scientist Research Career Development Award 3965

This award enables candidates holding professional degrees (e.g., M.D., D.O., D.V.M., or equivalent degrees) to undertake 3 to 5 years of special study and supervised research with the goal of becoming independent investigators. The award also allows awardees to pursue a research career development program suited to their experience and capabilities under a mentor who is competent to provide guidance in the chosen research area. Institutions may submit applications on behalf of candidates who hold professional degrees. At least 2 years must have elapsed since the health professional degree was granted. Candidates can have varying levels of clinical training and research experience. New applications are due February 12, June 12, and October 12, while resubmitted applications are due March 12, July 12, and November 12. Up to $105,000 per year plus fringe benefits is available, and up to $85,000 per year for research support for new awards.

Requirements: A candidate must have a clinical degree or its equivalent, must have initiated postgraduate clinical training, must identify a mentor with extensive research experience, and must be willing to spend a minimum of 75 percent of full-time professional effort conducting research. Applications may be submitted on behalf of candidates by domestic, nonfederal organizations, public or private, such as medical, dental, or nursing schools or other institutions of higher education. Candidates must be U.S. citizens, noncitizen nationals, or permanent residents.

Amount of Grant: $185,000 maximum

Date(s) Application is Due: Feb 12; Mar 12; Jun 12; Jul 12; Oct 12; Nov 12

Contact: Dr. Daniel Sklare; (301) 496-1804; fax: (301) 402-6251; email: sklared@nidcd.nih.gov

Internet: http://grants.nih.gov/grants/guide/pa-files/pa-06-512.html

Sponsor: National Institute on Deafness and Other Communication Disorders (NIDCD)
6120 Executive Blvd. , MSC-7180
Bethesda, MD 20892-7180

NIDCD Mentored Research Scientist Development Award 3966

The NIDCD utilizes the K01 Program for the retooling and transition of junior- and midcareer-level scientists and clinicians (generally below the rank of full professor, or the equivalent in nonacademic settings) into two research domains: translational research, i.e., the application of basic research discoveries toward advancing the diagnosis, management and prevention of deafness and other communication disorders; and clinical research. Junior-level clinically-trained individuals should apply for the Mentored Clinical Scientist Development Award or the Mentored Patient-Oriented Research Career Development Award, as appropriate.

Requirements: The candidate must have a research or health-professional doctorate or its equivalent, and must have demonstrated the capacity or potential for highly productive independent research in the period after the doctorate. The candidate must identify a mentor with extensive research experience, and must be willing to spend a minimum of 75 percent of full-time professional effort conducting research and research career development activities for the period of the award. Applications may be submitted on behalf of candidates by domestic, public or private, nonfederal organizations. Candidates must be U.S. citizens or noncitizen nationals.

Amount of Grant: $105,000 maximum salary per year plus fringe benefits; $50,000 per year for research support

Date(s) Application is Due: Feb 12; Mar 12; Jun 12; Jul 12; Oct 12; Nov 12

Contact: Daniel A. Sklare; (301) 496-1804; email: sklared@nidcd.nih.gov

Internet: http://www.grants.nih.gov/grants/guide/pa-files/PAR-99-027.html

Sponsor: National Institute on Deafness and Other Communication Disorders (NIDCD)
6120 Executive Blvd. , MSC-7180, Executive Plaza South, Room 400C
Bethesda, MD 20892-7180

NIDCD Minority Dissertation Research Grants in Human Communication 3967

NIDCD invites applications for small grants to support doctoral dissertation research in human communication for doctoral candidates who are members of racial and ethnic minority groups that are under-represented in biomedical and behavioral research. The goals of this program are to aid the research of new minority investigators and to encourage minority individuals from a variety of academic disciplines and programs to conduct research in hearing, balance, smell, taste, voice, speech, and language. These grants are not eligible for competitive renewal. Annual deadline dates may vary; contact the program office for exact dates. More information may be obtained by mail and email, or through the NIH grant line (301-402-2221) and the NIH Web site. RFA: DC-00-001

Requirements: Candidates for this award must be graduate students in good standing and must be identified by the grantee institution as a member of an ethnic or racial group that is underrepresented in biomedical or behavioral research.

Amount of Grant: $36,000 maximum direct costs for up to two years, $30,000 maximum direct costs for any one year

Contact: Daniel Sklare, Division of Extramural Research, (301) 496-1804; fax: (301) 402-6251; email: daniel_sklare@nih.gov

Internet: http://grants2.nih.gov/grants/guide/rfa-files/RFA-DC-00-001.html

Sponsor: National Institute on Deafness and Other Communication Disorders
6120 Executive Boulevard, Executive Plaza S
Bethesda, MD 20892

NIDCD Molecular and Cellular Pathology of the Human Auditory and Vestibular Systems Grants 3968

These grants support cellular biologic, molecular biologic, and molecular genetic studies of newly acquired and archival human temporal bones and associated brain tissues of individuals who had well-documented evidence of disorders of hearing and/or balance before death. It is believed that studies of this kind can lead to a better understanding of the pathogenesis and pathophysiology of disorders of the human auditory and vestibular systems. Further information may be obtained by mail and email or through the NIH grant line (301-402-2221), and the NIH Web site. Annual deadline dates may vary; contact program staff for exact dates. RFA: DC-96-005

Requirements: Applications may be submitted by domestic and foreign for-profit and non-profit organizations, public and private; and eligible agencies of the federal government.

Amount of Grant: $72,486-$458,342; $200,000 average

Contact: Sharon Hunt, Grants Management, (301) 402-0909

Internet: http://www.grants.nih.gov/grants/guide/rfa-files/RFA-DC-96-005.html

Sponsor: National Institute on Deafness and Other Communication Disorders
6120 Executive Boulevard, Executive Plaza S
Bethesda, MD 20892

NIDCD Neurobiology and Genetics of Autism Grants 3969

NICHD and NIDCD fund grants for research that will elucidate the biological basis of autism and contribute to the discovery of a biological and/or diagnostic marker for autism. The purpose of these grants is to provide support for collaborative, multidisciplinary,

and methodologically rigorous programs of research that will use advanced techniques in the biological and behavioral sciences to study the diagnosis, biological etiology, pathophysiology, and developmental course of autism. Annual deadline dates may vary; contact program staff for exact dates. More information may be obtained by mail and email, or through the NIH grant line (301-402-2221), and the NIH Web site. RFA: HD-96-004
Amount of Grant: $36,000-$2.3 million; $282,409 average
Date(s) Application is Due: Feb 1; Jun 1; Oct 1
Contact: Marie Bristol, Center for Research for Mothers and Children, (301) 496-1383; fax: (301) 496-3791; email: BRISTOLM@HD01.NICHD.NIH.GOV
Internet: http://www.grants.nih.gov/grants/guide/rfa-files/RFA-HD-96-004.html
Sponsor: National Institute on Deafness and Other Communication Disorders
6120 Executive Boulevard, Executive Plaza S
Bethesda, MD 20892

NIDCD Physiologic and Molecular Bases of Tinnitus Research Grants 3970
Three to five exploratory/developmental grants will be awarded to support research addressing the physiologic and molecular bases of tinnitus. The goal is to stimulate research on pathologic processes in the auditory system that produce tinnitus. More information may be obtained by mail and email, or through the NIH grant line (301-402-2221), and the NIH Web site. Annual deadline dates vary; contact program staff for exact dates. RFA: DC-96-003
Amount of Grant: $72,486-$458,342; $200,000 average
Contact: Sharon Hunt, Grants Management, (301) 402-0909
Internet: http://www.grants.nih.gov/grants/guide/rfa-files/RFA-DC-96-003.html
Sponsor: National Institute on Deafness and Other Communication Disorders
6120 Executive Boulevard, Executive Plaza S
Bethesda, MD 20892

NIDCD Program for Postdoctoral Research Training in Clinical Trials 3971
Grants support programs of postdoctoral research training in the design and conduct of clinical trials for individuals trained in the clinical disciplines of human communications. The purpose of this initiative is to foster the development of trained professionals into independent investigators capable of conducting clinical trials of the efficacy of treatment for diseases and disorders affecting hearing, balance, smell, taste, voice, speech, or language. More information may be obtained by mail and email or through the NIH grant line (301-402-2221), and the NIH Web site. Annual deadline dates may vary; contact program staff for exact dates. PA: PA-96-024
Amount of Grant: $72,486-$458,342; $200,000 average
Date(s) Application is Due: Feb 1; Jun 1; Oct 1
Contact: Daniel Sklare, Division of Human Communication, (301) 496-1804; fax: (301) 402-6251; email: daniel_sklare@nih.gov; Sharon Hunt, Grants Management, (301) 402-0909
Internet: http://www.grants.nih.gov/grants/guide/pa-files/PA-96-024.html
Sponsor: National Institute on Deafness and Other Communication Disorders
6120 Executive Boulevard, Executive Plaza S
Bethesda, MD 20892

NIDCD Research Core Centers 3972
NIDCD invites grant applications for research core centers. The overall goal of a core center is to promote a cooperative interaction among basic science and/or clinical investigators in a manner that will enrich the effectiveness of ongoing research and promote new research directions. Grants support centralized resources and facilities shared by investigators with existing research projects. A letter of intent is requested by April 1, August 1, and December 1; full application is due by February 1, June 1, and October 1. Annual deadline dates may vary; contact program staff for exact dates. PA: PA-01-011
Requirements: Applications may be submitted by domestic for-profit and nonprofit organizations, both public and private; units of state or local governments; and eligible agencies of the federal government.
Restrictions: Foreign organizations or organizations in foreign countries are not eligible subcontracts to foreign organizations are not allowed.
Date(s) Application is Due: Feb 1; Apr 1; Jun 1; Aug 1; Oct 1; Dec 1
Contact: Dr. Amy Donahue, (301) 402-3458; fax: (301) 402-6251; email: amy_donahue@nih.gov
Internet: http://grants.nih.gov/grants/guide/pa-files/PA-01-011.html
Sponsor: National Institute on Deafness and Other Communication Disorders
6120 Executive Boulevard, Executive Plaza S
Bethesda, MD 20892

NIDCD Research Grants on Chemical Senses: Neurotransmitters and Neuromodulators 3973
NIDCD and NIA sponsor research grants that support research fundamental to understanding the neurochemistry of pathways in the olfactory and gustatory systems throughout the lifespan. This research includes the identification and characterization of the neurotransmitters, neuromodulators, receptors, and secondary messengers throughout the chemosensory systems. More information may be obtained by mail and email, or through the NIH grant line (301-402-2221), and the NIH Web site. Annual deadline dates may vary; contact program staff for exact dates. PA: PA-96-035
Amount of Grant: $72,486-$458,342; $200,000 average
Date(s) Application is Due: Feb 1; Jun 1; Oct 1
Contact: Program Contact, Division of Human Communication, (301) 496-5061; Sara Stone, Grants Management, (301) 402-0909
Internet: http://www.grants.nih.gov/grants/guide/pa-files/PA-96-035.html
Sponsor: National Institute on Deafness and Other Communication Disorders
6120 Executive Boulevard, Executive Plaza S
Bethesda, MD 20892

NIDCD Small Grants Program 3974
NIDCD provides support for pilot research that is likely to lead to subsequent application for individual research project grants or first independent research support and transition grants. The research must be focused on one or more of the areas within the biomedical and behavioral scientific mission of NIDCD: hearing, balance, smell, taste, voice, speech, or language. Applicants may request support for up to two years, nonrenewable. For additional information, investigators are encouraged to contact NIDCD staff responsible for grants in the investigator's particular area of interest. PA: PAR-99-018
Requirements: This program is designed solely to support basic and clinical scientists with limited research experience who are at the beginning stages of their research careers. Former recipients of federally funded Training or Career Development awards are eligible for this program.
Restrictions: Current and previous principal investigators of NIH research grants or other federal funding agencies are ineligible.
Amount of Grant: $35,000 maximum per year
Contact: Dr. Beth Ansel, Voice/Speech, (301) 402-3461; fax: (301) 402-6251; email: Beth_Ansel@nih.gov; Dr. Chyren Hunter, Hearing, (301) 402-3461; email: Chyren_Hunter@nih.gov; Dr. Daniel Sklare, Balance/Vestibular, (301) 496-1804; email: Daniel_Sklare@nih.g
Internet: http://www.grants.nih.gov/grants/guide/pa-files/PAR-99-018.html
Sponsor: National Institute on Deafness and Other Communication Disorders
6120 Executive Boulevard, Executive Plaza S
Bethesda, MD 20892

NIDCD Symptom Management for Chronic Neurological Conditions Research Grants 3975
NINR, NIA, NICHD, NIDCD, and NINDS are interested in facilitating investigator-initiated research designed to deal with symptoms associated with chronic neurological conditions including stroke, epilepsy, brain and spinal cord injury, multiple sclerosis, and Parkinson's disease. The overall goals are to manage symptoms effectively, to prevent secondary effects of the condition when possible, to maximize the health potential, and to improve the health-related quality of life for patients, caregivers, and families. The mechanism of support is the research project grant (R01). PA-98-030 may be obtained through the NIH grant line (301-402-2221), the NIH Web site, mail, and email.
Requirements: Applications may be submitted by domestic and foreign for-profit and nonprofit organizations, public or private.
Contact: Dr. Beth Ansel, Division of Human Communication, (301) 402-3461; fax: (301) 402-6251; email: Beth_Ansel@nih.gov
Internet: http://www.grants.nih.gov/grants/guide/pa-files/PA-98-030.html
Sponsor: National Institute on Deafness and Other Communication Disorders
6120 Executive Boulevard, Executive Plaza S
Bethesda, MD 20892

NIDCR Clinical Trial Pilot Grant 3976
The goal of this program is to identify effective preventive, diagnostic, and treatment approaches to craniofacial, oral and dental diseases, and disorders that will improve health by supporting well-executed clinical trials and to support pilot studies to obtain preliminary data necessary to address preliminary design and methodological issues for clinical trials. PA: PAR-99-158
Requirements: Applications may be submitted by foreign and domestic, for-profit and nonprofit organizations, both public and private; units of state and local governments; and eligible agencies of the federal government.
Amount of Grant: $250,000 maximum for enrollment centers; $350,000 maximum for coordinating centers
Contact: Dr. Ann Sandberg, Division of Extramural Research, (301) 594-2419; email: Ann.Sandberg@nih.gov
Internet: http://grants.nih.gov/grants/guide/pa-files/PAR-99-158.html
Sponsor: National Institute of Dental and Craniofacial Research
45 Center Drive
Bethesda, MD 20892

NIDCR Clinical Trial Planning Grant 3977
The goal of this program is to identify effective preventive, diagnostic, and treatment approaches to craniofacial, oral and dental diseases, and disorders that will improve health by supporting well-executed clinical trials. PA: PAR-03-042
Requirements: Applications may be submitted by domestic and foreign, for-profit and nonprofit organizations, both public and private; units of state and local governments; and eligible agencies of the federal government.
Amount of Grant: $250,000 maximum
Contact: Dr. Bruce Pihlstrom, Acting Director, Division of Clinical Research and Health Promotion, (301) 594-4830; fax: (301) 480-8322; email: bruce.Pihlstrom@nih.gov
Internet: http://grants.nih.gov/grants/guide/pa-files/PAR-03-042.html
Sponsor: National Institute of Dental and Craniofacial Research
45 Center Drive
Bethesda, MD 20892

NIDCR HIV/AIDS Associated Oral Viral Infections 3978

NIDCR invites new and experienced scientists to submit research grant applications for basic, behavioral, epidemiological, or clinical investigations to study the pathogenesis, infectivity, transmission, reactivation, and pathotropic spread of viral infections in the oral cavity of the immunocompromised host. This RFA is intended to solicit a broad range of research proposals on oral opportunistic infections with viruses, including human herpes virus, Epstein Barr virus, cytomegalovirus varicella-zoster virus, and human papillomavirus in HIV-infected/AIDS individuals. The institute encourages collaborative projects that bring together investigators from all relevant scientific disciplines including microbiology, immunology, biochemistry, and pathology interested in studying viral infections of the oral cavity. Applications that include collaboration with foreign scientists conducting unique research on this topic are also encouraged. The goal of this RFA is to increase our understanding of how virulence, infection, and host response is modified in the immune compromised patient to allow infection, activation, or reactivation of opportunistic pathogens. This understanding can lead to improved diagnosis, prevention, and treatment strategies, for the various opportunistic pathogens.

Requirements: Applications may be submitted by domestic and foreign, for-profit and nonprofit organizations, both public and private; units of state and local governments; and eligible agencies of the federal government. Although an application must be submitted from a single institution, collaborative arrangements with other institutions are encouraged.

Amount of Grant: $200,000 maximum; $225,000 maximum for collaborative agreements

Contact: Dr. Denise Anne Russo, Infectious Diseases and Immunity, (301) 594-2617; fax: (301) 480-8318; email: Denise.Russo@nih.gov

Internet: http://grants.nih.gov/grants/guide/rfa-files/RFA-DE-00-002.html

Sponsor: National Institute of Dental and Craniofacial Research
45 Center Drive
Bethesda, MD 20892

NIDCR International Collaborative Oral Health Research Planning Grant 3979

The objective of this scientific program is to promote research that will advance understanding of the underlying causes of craniofacial dysmorphologies such as ectodermal dysplasia, enamel dysplasia, cleft lip and palate, and osteogenesis imperfecta. International collaboration in relevant areas such as genetic research, epidemiology, genetic registries, and treatment modalities can help speed progress toward the goal of improved prevention, diagnosis, and treatment by combining scientific talents from different countries to focus on selected populations at high risk. An international approach that includes all disciplines relevant to specific craniofacial anomalies might dramatically enhance the possibility of resolving the many complex issues surrounding these malformations. PA: PAR-00-012

Requirements: Applications may be submitted by foreign and domestic, for-profit and nonprofit organizations, both public and private; units of state and local governments; and eligible agencies of the federal government.

Amount of Grant: $75,000 maximum

Contact: Dr. Judy Small, Craniofacial Anomalies and Injuries Branch, (301) 594-2425; Dr. Dennis Mangan, Infectious Diseases Branch, (301) 594-2421; Dr. Ann Sandberg, Neoplastic Diseases Branch, (301) 594-2419

Internet: http://grants.nih.gov/grants/guide/pa-files/PAR-00-012.html

Sponsor: National Institute of Dental and Craniofacial Research
45 Center Drive, Natcher Building
Bethesda, MD 20892

NIDCR Oral Microbiology/Immunology of Type 1 Diabetes 3980

The purpose of this program is to encourage submission of applications for two-year exploratory grants that will broaden research on the oral microbiology and immunology of Type 1 diabetes. In particular, research is needed on the oral microflora, cytokines, and antibodies in the subgingival crevice. Annual deadline dates may vary; contact program staff for exact dates. RFA: DE-01-001

Requirements: Applications may be submitted by domestic and foreign, for-profit and nonprofit organizations, both public and private; units of state and local governments; and eligible agencies of the federal government.

Amount of Grant: $100,000 maximum

Contact: Dr. Dennis Mangan, Infectious Diseases and Immunity Branch, (301) 594-2421; fax: (301) 480-8318; email: Dennis.Mangan@nih.gov

Internet: http://grants.nih.gov/grants/guide/rfa-files/RFA-DE-01-001.html

Sponsor: National Institute of Dental and Craniofacial Research
45 Center Drive, Natcher Building
Bethesda, MD 20892

NIDCR Priorities in Behavioral Research in Cancer Prevention and Control Grants 3981

NCI and NIDCR invite researchers to submit research project grant (R01) applications that address behavioral research issues in cancer prevention and control. The PA addresses recommendations made by a special Behavioral Research in Cancer Prevention and Control Working Group in 1996 that consisted of leading national experts whose role was to identify behavioral research needs in cancer prevention and control during the coming years. PA-97-055 may be obtained electronically through the NIH grant line (301-402-2221), the Web site, and by mail and email.

Contact: Dr. Patricia Bryant, Behavior, Health Promotion, and Environment Program, (301) 594-2095; email: bryantp@de45.nidr.nih.gov

Internet: http://grants.nih.gov/grants/guide/pa-files/PA-97-055.html

Sponsor: National Institute of Dental and Craniofacial Research
45 Center Drive, Natcher Building
Bethesda, MD 20892

NIDDK Advances in Polycystic Kidney Disease Grants 3982

The National Institute of Diabetes and Digestive and Kidney Diseases (NIDDK) through its Division of Kidney, Urologic and Hematologic Diseases (DKUHD) invites experienced and new investigators to submit research grant applications to pursue basic and applied investigations in order to better understand the etiology and pathogenesis of Polycystic Kidney Disease (PKD), in both its autosomal dominant and autosomal recessive forms. Such applications may examine the genetic determinants, and cellular and molecular mechanisms, which disrupt normal kidney function; mechanisms of cyst formation and growth; development of experimental model systems; development of markers of disease progression; and the identification of innovative therapeutic interventions and gene targeted strategies to prevent progressive renal insufficiency due to this disorder. The intent of this funding opportunity is to intensify investigator-initiated research, to attract new investigators to the field, and to increase interdisciplinary research. The ultimate aim is to facilitate PKD-related research studies, which will provide the basis for new therapeutic approaches.

Requirements: The following organizations may apply: for-profit organizations; non-profit organizations; public or private institutions, such as universities, colleges, hospitals, and laboratories; units of State government; units of local government; eligible agencies of the Federal government; foreign institutions; domestic institutions; and faith-based or community-based organizations.

Date(s) Application is Due: Feb 5; Jun 5; Oct 5

Contact: Catherine M. Meyers, M.D.; (301) 594-7717; fax: (301) 480-3510; email: cm420i@nih.gov

Internet: http://grants.nih.gov/grants/guide/pa-files/PA-06-158.html

Sponsor: National Institute of Diabetes and Digestive and Kidney Diseases (NIDDK)
6707 Democracy Boulevard, Room 641
Bethesda, MD 20892-5458

NIDDK Ancillary Studies of Kidney Disease Accessing Information from Clinical Trials, Epidemiological Studies, and Databases Grants 3983

The National Institute of Diabetes and Digestive and Kidney Diseases (NIDDK) invites investigator-initiated research project applications for ancillary studies to ongoing or completed clinical trials and epidemiological studies of kidney disease as well as clinical trials and epidemiological studies for other diseases or populations that lend themselves to the study of kidney disease. These studies may range from new analyses of existing datasets of completed studies to additional collection of data and biological specimens in ongoing investigations. The goal of these studies should be to extend our understanding of the risk factors for developing kidney disease and their associated co-morbid illnesses such as malnutrition and cardiovascular disease, factors associated with rapid decline in kidney function among persons with chronic kidney disease, and the impact of these diseases on quality of life and mental and physical functioning. Studies of biomarkers for patients with acute renal failure and other kidney diseases, for example, are also appropriate topics for further investigation. Studies ancillary to both government and non-government supported clinical trials and epidemiological studies are encouraged. Analysis of large public access databases and other databases is also encouraged.

Requirements: The following organizations and institutions may apply: public/state controlled institutions of higher education; private institutions of higher education; Hispanic-serving institutions; historically Black colleges and universities (HBCUs); tribally controlled colleges and universities (TCCUs); Alaska native and native Hawaiian serving institutions; nonprofits with 501(c)3 IRS status; nonprofits without 501(c)3 IRS status; small businesses; for-profit organizations; State governments; regional organizations; U.S. territories or possessions; Indian/Native American tribal governments; Indian/Native American tribally designated organizations; eligible agencies of the Federal government; and faith-based or community based organizations.

Date(s) Application is Due: Feb 5; Jun 5; Oct 5

Contact: John W. Kusek, Ph.D.; (301) 594-7735; fax: (301) 480-3510; email: jk61x@nih.gov

Internet: http://grants.nih.gov/grants/guide/pa-files/PA-07-050.html

Sponsor: National Institute of Diabetes and Digestive and Kidney Diseases (NIDDK)
6707 Democracy Boulevard, Room 617
Bethesda, MD 20892-5458

NIDDK Ancillary Studies to Major Ongoing NIDDK and NHLBI Clinical Research Studies 3984

This Funding Opportunity Announcement (FOA) solicits grant applications from qualified investigators to conduct ancillary studies of selected ongoing major clinical research studies, including clinical trials, epidemiological studies and disease databases, supported by the National Institute of Diabetes and Digestive and Kidney Diseases and the National Heart, Lung, and Blood Institute, National Institutes of Health (NIH). The objective is to introduce NIDDK-oriented ancillary studies into either NIDDK or NHLBI-funded parent studies, or NHLBI-oriented ancillary studies into NIDDK-funded parent studies.

Requirements: The following organizations and institutions may apply: public/state controlled institutions of higher education; private institutions of higher education; Hispanic-serving institutions; historically Black colleges and universities (HBCUs); tribally controlled colleges and universities (TCCUs); Alaska native and native Hawaiian serving institutions; nonprofits with 501(c)3 IRS status; nonprofits without 501(c)3 IRS status; small businesses; for-profit organizations; State governments; regional organizations; U.S. territories or possessions; Indian/Native American tribal

governments; Indian/Native American tribally designated organizations; eligible agencies of the Federal government; and faith-based or community based organizations.
Date(s) Application is Due: Feb 5; Jun 5; Oct 5
Contact: John W. Kusek, Ph.D.; (301) 594-7735; fax: (301) 480-3510; email: jk61x@nih.gov
Internet: http://grants.nih.gov/grants/guide/pa-files/PAR-07-024.html
Sponsor: National Institute of Diabetes and Digestive and Kidney Diseases (NIDDK)
Two Democracy Plaza, Room 617, 6707 Democracy Boulevard
Bethesda, MD 20892-5458

NIDDK Basic and Clinical Studies of Congenital Urinary Tract Obstruction Grants 3985

The purpose of this funding opportunity is to address the numerous scientific and clinical uncertainties related to the development, treatment and prognosis of congenital obstructive uropathy, by encouraging and facilitating research in diverse areas. These areas include: the development of objective prognostic markers; the genetic determinants of this congenital disorder; the development of reliable animal models of the disorder; and, evaluation of the long-term effectiveness of various treatment strategies. Because the nature and scope of the proposed research will vary from application to application, it is anticipated that the size and duration of each award will also vary. The total amount awarded and the number of awards will depend upon the mechanism numbers, quality, duration, and costs of the applications received.
Requirements: The following organizations and institutions may apply: public/state controlled institutions of higher education; private institutions of higher education; Hispanic-serving institutions; historically Black colleges and universities (HBCUs); tribally controlled colleges and universities (TCCUs); Alaska native and native Hawaiian serving institutions; nonprofits with 501(c)3 IRS status; nonprofits without 501(c)3 IRS status; small businesses; for-profit organizations; State governments; regional organizations; U.S. territories or possessions; Indian/Native American tribal governments; Indian/Native American tribally designated organizations; eligible agencies of the Federal government; and faith-based or community based organizations.
Date(s) Application is Due: Feb 5; Jun 5; Oct 5
Contact: Marva Moxey-Mims, M.D.; (301) 594-7717; fax: (301) 480-3510; email: mm726k@nih.gov
Internet: http://grants.nih.gov/grants/guide/pa-files/PA-07-059.html
Sponsor: National Institute of Diabetes and Digestive and Kidney Diseases (NIDDK)
6707 Democracy Boulevard, Room 639
Bethesda, MD 20892-5458

NIDDK Basic Research Grants in the Bladder and Lower Urinary Tract 3986

This funding opportunity encourages basic cellular, molecular, developmental and genetic research relevant to the bladder and lower urinary tract. Basic research studies that address age and gender differences in bladder and lower urinary tract function are also encouraged. New and established investigators from related fields of study are encouraged to apply their expertise to these problem areas. Investigators with diverse basic science and clinical backgrounds are encouraged to develop collaborative research relationships. Discoveries resulting from these studies may serve as the basis for the development of new agents, techniques, and strategies for detecting, preventing, and treating diseases of the bladder and lower urinary tract, as well as dealing with bladder complications resulting from other diseases, such as diabetes.
Requirements: Application can be made by the following institutions and organizations: for-profit organizations; non-profit organizations; public or private institutions (such as universities, colleges, hospitals, and laboratories); units of State government; units of local government; eligible agencies of the Federal government; foreign institutions; and domestic institutions.
Date(s) Application is Due: Feb 5; Jun 5; Oct 5
Contact: Chris Mullins, Ph.D.; (301) 594-7717; fax: (301) 480-3510; email: cm419z@nih.gov
Internet: http://grants.nih.gov/grants/guide/pa-files/PA-06-254.html
Sponsor: National Institute of Diabetes and Digestive and Kidney Diseases (NIDDK)
6707 Democracy Boulevard, Room 637
Bethesda, MD 20892-5458

NIDDK Calcium Oxalate Stone Diseases Grants 3987

The purpose of this Funding Opportunity Announcement (FOA) is to increase investigator interest in research into the genetics and heritability of oxalate regulation and the oxalate stone diseases, and to develop new treatments for the disorder. This initiative solicits research as well as pilot and feasibility studies that utilize new and innovative approaches to study the diagnosis, treatment and prevention of these disorders. Because the nature and scope of the proposed research will vary from application to application, it is anticipated that the size and duration of each award will also vary. The total amount awarded and the number of awards will depend upon the mechanism numbers, quality, duration, and costs of the applications received.
Requirements: The following organizations and institutions may apply: public/state controlled institutions of higher education; private institutions of higher education; Hispanic-serving institutions; historically Black colleges and universities (HBCUs); tribally controlled colleges and universities (TCCUs); Alaska native and native Hawaiian serving institutions; nonprofits with 501(c)3 IRS status; nonprofits without 501(c)3 IRS status; small businesses; for-profit organizations; State governments; regional organizations; U.S. territories or possessions; Indian/Native American tribal governments; Indian/Native American tribally designated organizations; eligible agencies of the Federal government; and faith-based or community based organizations.
Date(s) Application is Due: Feb 5; Jun 5; Oct 5
Contact: Rebekah S. Rasooly, Ph.D.; (301) 594-6007; fax: (301) 480-3510; email: rr185i@nih.gov
Internet: http://grants.nih.gov/grants/guide/pa-files/PA-07-051.html
Sponsor: National Institute of Diabetes and Digestive and Kidney Diseases (NIDDK)
6707 Democracy Boulevard, Room 643
Bethesda, MD 20892-5458

NIDDK Cell-Specific Delineation of Prostate and Genitourinary Development 3988

This RFA is intended to encourage the development of new research tools and methods to study the development and biology of the prostate and genitourinary tract. Strategies to be supported include systematic assessment of gene expression in specific cell types in the developing prostate and genitourinary tract; methods to tag individual cell types for purification, analysis, and characterization; tools to manipulate gene expression in vivo in individual cell types; and development of clinically relevant cell lines. The major goal of this initiative is to ensure that research tools are available so that a broad range of innovative methods can be applied to studying prostate development. Annual deadline dates may vary; contact program staff for exact dates.
Requirements: Application can be made by the following institutions and organizations: public/state controlled institutions of higher education; private institutions of higher education; nonprofits with 501(c)3 IRS status (other than institutions of higher education); nonprofits without 501(c)3 IRS status (other than institutions of higher education); small businesses; for-profit organizations (other than small businesses); state governments; U.S. territories or possessions; Indian/Native American tribal governments (federally recognized); Indian/Native American tribal governments (other than Federally recognized); Indian/Native American tribally designated organizations; non-domestic (non-U.S.) entities (foreign organizations); Hispanic-serving institutions; historically Black colleges and universities (HBCUs); tribally controlled colleges and universities (TCCUs); Alaska Native and Native Hawaiian serving institutions; regional organization; eligible agencies of the Federal government; and faith-based or community based organizations.
Amount of Grant: $350,000 maximum for RO1 grants; $100,000 maximum for R21 grants
Date(s) Application is Due: Feb 5; Mar 5; Jun 5; Jul 5; Oct 5; Nov 5
Contact: Dr. Robert Star; (301) 594-7715; fax: (301) 480-3510; email: StarR@extra.niddk.nih.gov; Dr. Michael McClure, Organs and Systems Toxicology, (919) 541-5327; fax: (919) 541-5064; email: mm461n@nih.gov
Internet: http://grants.nih.gov/grants/guide/rfa-files/RFA-DK-00-015.html
Sponsor: National Institute of Diabetes and Digestive and Kidney Diseases (NIDDK)
Room 9A-35, 31 Center Drive, MSC 2560
Bethesda, MD 20892-2560

NIDDK Co-Activators and Co-Repressors in Gene Expression Grants 3989

This initiative stems from an NIDDK workshop in molecular endocrinology and is designed to stimulate research that addresses the fundamental underlying mechanisms by which nuclear accessory proteins mediate signaling through hormone receptors at the level of the regulation of gene expression. Staff contact prior to submission is highly recommended.
Requirements: Applications may be submitted by domestic and foreign, for-profit and nonprofit organizations, both public and private; units of state and local governments; and eligible agencies of the federal government.
Amount of Grant: $100,000 maximum per year R21 grants, $250,000 maximum per year modular R01 grants
Date(s) Application is Due: Feb 16; Jun 16; Oct 16
Contact: Dr. Ronald Margolis; (301) 594-8819; fax: (301) 435-6047; email: rm76f@nih.gov
Internet: http://grants.nih.gov/grants/guide/pa-files/PA-99-111.html
Sponsor: National Institutes of Health
6707 Democracy Boulevard, Building 2DEM, Room 693
Bethesda, MD 20892

NIDDK Collaborative Interdisciplinary Team Science in Diabetes, Endocrinology and Metabolic Diseases Grants 3990

The purpose of the Collaborative Interdisciplinary Team Science Program described in this announcement is to provide support to enable strong investigative teams to do inter- and/or trans-disciplinary research on a complex problem in biomedical science relevant to Diabetes, Endocrinology and Metabolic Diseases. It is anticipated that 1-3 projects will be funded per year, but because the nature and scope of the proposed research will vary from application to application, it is anticipated that the size and duration of each award will also vary. The total amount awarded and the number of awards will depend upon the quality, duration, and costs of the applications received, as well as on the availability of funds.
Requirements: The following organizations and institutions are eligible to apply: public/state controlled institutions of higher education; private institutions of higher education; Hispanic-serving institutions; historically Black colleges and universities (HBCUs); tribally controlled colleges and universities (TCCUs); Alaska Native and Native Hawaiian serving institutions; and non-domestic (non-U.S.) entities (foreign organizations).
Date(s) Application is Due: Feb 24; Mar 24
Contact: Karen Salomon; (301) 594-7733; fax: (301) 435-6047; email: ks495c@nih.gov
Internet: http://grants.nih.gov/grants/guide/pa-files/PAR-08-182.html
Sponsor: National Institute of Diabetes and Digestive and Kidney Diseases (NIDDK)
6707 Democracy Boulevard, Room 687
Bethesda, MD 20892-5460

NIDDK Development of Assays for High-Throughput Drug Screening Grants 3991
The overall goal of the Molecular Libraries Initiative, part of the NIH Roadmap for Medical Research, is to offer public sector researchers opportunities to screen small molecules with the high-throughput chemical screening (HTS) methods commonly used by the private sector to develop therapeutic agents. The purpose of this FOA is to support the development of innovative assays that may ultimately be adapted for automated screening. The assay should aim to identify new tools for basic research or promising new avenues for therapeutics development, especially in areas related to the missions of the participating institutes. Awards issued under this FOA are contingent upon the availability of funds and the submission of a sufficient number of meritorious applications. The total amount awarded and the number of awards will depend upon the quality and costs of the applications received.
Requirements: The following organizations and institutions may apply: public/state controlled institutions of higher education; private institutions of higher education; Hispanic-serving institutions; historically Black colleges and universities (HBCUs); tribally controlled colleges and universities (TCCUs); Alaska native and native Hawaiian serving institutions; nonprofits with 501(c)3 IRS status; nonprofits without 501(c)3 IRS status; small businesses; for-profit organizations; State governments; regional organizations; U.S. territories or possessions; Indian/Native American tribal governments; Indian/Native American tribally designated organizations; eligible agencies of the Federal government; and faith-based or community based organizations.
Date(s) Application is Due: Feb 5; Jun 5; Oct 5
Contact: Laura K. Moen, Ph.D.; (301) 594-4748; fax: (301) 480-3510; email: lm232f@nih.gov
Internet: http://grants.nih.gov/grants/guide/pa-files/PA-07-320.html
Sponsor: National Institute of Diabetes and Digestive and Kidney Diseases (NIDDK)
6707 Democracy Boulevard, Room 623
Bethesda, MD 20892-5458

NIDDK Development of Disease Biomarkers Grants 3992
The goal of this Funding Opportunity Announcement (FOA) is to validate biomarkers for well-defined human diseases of liver, kidney, urological tract, digestive and hematologic systems, and endocrine and metabolic disorders, diabetes and its complications, and obesity, for which there are no or very few biomarkers, or for which standard biomarkers are currently prohibitively invasive or expensive. New biomarkers will stimulate bench to bedside translation by providing measures of the biological effects of potential new treatments. The ideal biomarker can be measured in a minimally invasive way, can be measured repeatedly over time, identifies early stages of disease, is indicative of disease prognosis, and correlates well with progression and response to therapy. Especially of interest would be studies designed to test the validity of candidate biomarkers or new technologies to monitor candidate biomarkers in patient tissue samples or small groups of well-characterized patients. For biomarkers already validated in human subjects, the larger effort might be to establish a reliable assay prior to pursuing larger patient studies. Priority will be given to those projects with high promise for improving clinical care in the relatively near future. Studies of potential biomarkers conducted in animal models of disease, and extended prospective clinical trials for marker validation against hard clinical endpoints, are outside the scope of this initiative. Because the nature and scope of the proposed research will vary from application to application, it is anticipated that the size and duration of each award will also vary.
Requirements: Application can be made by the following institutions and organizations: public/state controlled institutions of higher education; private institutions of higher education; nonprofits with 501(c)3 IRS status (other than institutions of higher education); nonprofits without 501(c)3 IRS status (other than institutions of higher education); small businesses; for-profit organizations (other than small businesses); state governments; U.S. territories or possessions; Indian/Native American tribal governments (federally recognized); Indian/Native American tribal governments (other than Federally recognized); Indian/Native American tribally designated organizations; non-domestic (non-U.S.) entities (foreign organizations); Hispanic-serving institutions; historically Black colleges and universities (HBCUs); tribally controlled colleges and universities (TCCUs); Alaska Native and Native Hawaiian serving institutions; regional organization; eligible agencies of the Federal government; and faith-based or community based organizations.
Date(s) Application is Due: Feb 5; Jun 5; Oct 5
Contact: Maren R. Laughlin, Ph.D.; (301) 594-8802; fax: (301) 480-0475; email: ml33q@nih.gov
Internet: http://grants.nih.gov/grants/guide/pa-files/PA-07-052.html
Sponsor: National Institute of Diabetes and Digestive and Kidney Diseases (NIDDK)
6707 Democracy Boulevard, Room 787
Bethesda, MD 20892-5460

NIDDK Developmental Biology and Regeneration of the Liver Grants 3993
The purpose of this FOA is to invite qualified scientific investigators to submit applications on liver development and regeneration to fully define the molecular and cellular mechanisms underlying these processes in health and disease and to apply these findings to developing improved therapies for liver disease. Because the nature and scope of the proposed research will vary from application to application, it is anticipated that the size and duration of each award will also vary. The total amount awarded and the number of awards will depend upon the mechanism numbers, quality, duration, and costs of the applications received.
Requirements: The following organizations and institutions may apply: public/state controlled institutions of higher education; private institutions of higher education; Hispanic-serving institutions; historically Black colleges and universities (HBCUs); tribally controlled colleges and universities (TCCUs); Alaska native and native Hawaiian serving institutions; nonprofits with 501(c)3 IRS status; nonprofits without 501(c)3 IRS status; small businesses; for-profit organizations; State governments; regional organizations; U.S. territories or possessions; Indian/Native American tribal governments; Indian/Native American tribally designated organizations; eligible agencies of the Federal government; and faith-based or community based organizations.
Date(s) Application is Due: Feb 5; Jun 5; Oct 5
Contact: Jose Serrano; (301) 594-8871; fax: (301) 480-8300; email: js362q@nih.gov; Samir Zakhari, Ph.D.; (301) 4 43-0799; fax: (301) 594-0673; email: szakhari@mail.nih.gov
Internet: http://grants.nih.gov/grants/guide/pa-files/PA-07-026.html
Sponsor: National Institute of Diabetes and Digestive and Kidney Diseases (NIDDK)
6707 Democracy Boulevard, Room 657
Bethesda, MD 20892-5450

NIDDK Diabetes Research Centers 3994
The mission of the Diabetes Centers is to serve as a key component of the NIDDK-supported research effort to develop new therapies and improve the health of Americans with, or at risk for, diabetes and related endocrine and metabolic disorders. The Centers promote new discoveries and enhance scientific progress through support of cutting-edge basic and clinical research related to the etiology and complications of diabetes, with the goal of rapidly translating research findings into novel strategies for the prevention, treatment and cure of diabetes and related conditions. This Funding Opportunity Announcement (FOA) solicits new and competing continuation applications for Diabetes Endocrinology Research Centers (DERCs) and Diabetes Research and Training Centers (DRTCs). Both types of Centers are designed to support and enhance the national research effort in diabetes and related endocrine and metabolic diseases. DERCs support three primary research-related activities: biomedical research cores, a Pilot and Feasibility (P&F) program, and an Enrichment program. DRTCs possess all elements of a DERC, with additional dedicated core services and P&F awards to support research in diabetes prevention and control. All activities pursued by Diabetes Centers are designed to enhance the efficiency, productivity, effectiveness and multidisciplinary nature of research in Diabetes Center topic areas.
Requirements: The following organizations and institutions are eligible to apply: public/state controlled institutions of higher education; private institutions of higher education; Hispanic-serving institutions; historically Black colleges and universities (HBCUs); tribally controlled colleges and universities (TCCUs); Alaska Native and Native Hawaiian serving institutions; nonprofits with 501(c)3 IRS status (other than institutions of higher education); nonprofits without 501(c)3 IRS status (other than institutions of higher education); for-profit organizations (other than small businesses); and state governments.
Date(s) Application is Due: Jun 22; Jul 15
Contact: James F. Hyde, Ph.D.; (301) 594-7692; fax: (301) 480-3503; email: jh486z@nih.gov
Internet: http://grants.nih.gov/grants/guide/rfa-files/RFA-DK-08-008.html
Sponsor: National Institute of Diabetes and Digestive and Kidney Diseases (NIDDK)
6707 Democracy Boulevard, Room 789
Bethesda, MD 20892-5460

NIDDK Diabetes, Endocrinology, and Metabolic Diseases NRSAs—Individual 3995
Individual National Research Service Awards (NRSAs) are given only at the postdoctoral level for research training in diabetes, endocrine, or metabolic diseases. To receive one, an individual must submit an application that describes a specific research project, which is guided and sponsored by a specific preceptor at a particular institution. The award may be for up to 36 months of training and consists of a stipend and an allowance to help meet expenses incurred during the course of fellowship.
Requirements: All awardees must be citizens or have been admitted to the United States for permanent residence. Individuals and public and private institutions, nonprofit and for-profit, that propose to establish, expand, and improve research activities in health sciences and related fields may apply for research grants. Individuals must be nominated and sponsored by public or nonprofit private institutions having staff and facilities appropriate to the proposed research training programs. SBIR grants can be awarded only to domestic small businesses. STTR grants can be awarded only to domestic small business concerns that partner with research institutions in cooperative research and development.
Amount of Grant: $16,600-$2.09 million; $180,100 average
Date(s) Application is Due: Apr 5; Aug 5; Dec 5
Contact: Dr. Judith Fradkin, Director; (301) 496-7348
Internet: http://grants.nih.gov/grants/guide/pa-files/PA-07-107.html
Sponsor: National Institute of Diabetes and Digestive and Kidney Diseases
45 Center Drive, Natcher Bldg.
Bethesda, MD 20892-2560

NIDDK Diet Composition and Energy Balance Grants 3996
The goal of this funding opportunity announcement (FOA) is to invite Research Project Grant applications investigating the role of diet composition in energy balance, including studies in both animals and humans. Both short and longer-term studies are encouraged, ranging from basic studies investigating the impact of micro-or macronutrient composition on appetite, metabolism, and energy expenditure through clinical studies evaluating the efficacy of diets differing in micro- or macronutrient composition, absorption, dietary variety, or energy density for weight loss or weight maintenance. Because the nature and scope of the proposed research will vary from application to application, it is anticipated that the size and duration of each award will

also vary. The total amount awarded and the number of awards will depend upon the mechanism numbers, quality, duration, and costs of the applications received.
Requirements: The following organizations and institutions may apply: public/state controlled institutions of higher education; private institutions of higher education; Hispanic-serving institutions; historically Black colleges and universities (HBCUs); tribally controlled colleges and universities (TCCUs); Alaska native and native Hawaiian serving institutions; nonprofits with 501(c)3 IRS status; nonprofits without 501(c)3 IRS status; small businesses; for-profit organizations; State governments; regional organizations; U.S. territories or possessions; Indian/Native American tribal governments; Indian/Native American tribally designated organizations; eligible agencies of the Federal government; and faith-based or community based organizations.
Date(s) Application is Due: Feb 5; Jun 5; Oct 5
Contact: Susan Z. Yanovski, M.D.; (301) 594-8882; fax: (301) 480-8300; email: sy29f@nih.gov
Internet: http://grants.nih.gov/grants/guide/pa-files/PA-07-218.html
Sponsor: National Institute of Diabetes and Digestive and Kidney Diseases (NIDDK)
6707 Democracy Boulevard, Room 665
Bethesda, MD 20892-5450

NIDDK Education Program Grants 3997

This funding opportunity announcement (FOA) solicits Research Education grant applications from applicant organizations that propose to create educational opportunities to attract undergraduate students, graduate students, and postdoctoral fellows to careers in areas of biomedical or behavioral research of particular interest to the NIDDK while fostering the career development of these students and fellows. The NIDDK is especially interested in attracting students and postdoctoral fellows from scientific disciplines underrepresented in disease-oriented biomedical research such as engineering, informatics, computer science, and computational sciences, to encourage them to apply their expertise to research relevant to diabetes and other endocrine and metabolic diseases, digestive and liver diseases, nutrition, obesity research and prevention, and kidney, urologic and hematologic diseases. The total project period for an application submitted in response to this funding opportunity may not exceed five years. Direct costs are limited to $100,000 per year.
Requirements: The following organizations and institutions may apply: public/state controlled institutions of higher education; private institutions of higher education; Hispanic-serving institutions; historically Black colleges and universities (HBCUs); tribally controlled colleges and universities (TCCUs); Alaska native and native Hawaiian serving institutions; nonprofits with 501(c)3 IRS status; nonprofits without 501(c)3 IRS status; and Indian/Native American tribally designated organizations.
Amount of Grant: $100,000 maximum per year
Date(s) Application is Due: Jan 25; May 25; Sep 25
Contact: James F. Hyde; (301) 594-7692; fax: (301) 480-3503; email: jh486z@nih.gov
Internet: http://grants.nih.gov/grants/guide/pa-files/PAR-06-55-.html
Sponsor: National Institute of Diabetes and Digestive and Kidney Diseases (NIDDK)
6707 Democracy Boulevard, Room 785
Bethesda, MD 20892-5460

NIDDK Endoscopic Clinical Research Grants In Pancreatic and Biliary Diseases 3998

These small grants (R03) may be used as planning grants for full-scale multi-center clinical trials or for pilot studies that could lead to full-scale multi-center clinical trials designed to provide evidence for or against changes in the current standard of care. Pilot epidemiological studies are encouraged that could lead to more extended research that would provide evidence for or against changes in health policy, especially as related to disease and cancer prevention. It is expected that these R03 grants will serve as a basis for planning future multi-center research project grant applications (R01) or cooperative agreement (U01) awards. New and experienced investigators in relevant fields and disciplines may apply for these small grants. Investigators are encouraged to take advantage of recent endoscopic and laboratory developments. In addition, the small grant is a good mechanism for new and experienced investigators to become better equipped to perform clinical and epidemiological research. A project period of up to two years and a budget for direct costs of up to two $25,000 modules or $50,000 per year may be requested.
Requirements: The following organizations and institutions may apply: for-profit organizations; non-profit organizations; public or private institutions, such as universities, colleges, hospitals, and laboratories; units of State government; units of local government; eligible agencies of the Federal government; domestic institutions and foreign institutions.
Date(s) Application is Due: Feb 16; Jun 16; Oct 16
Contact: Frank Hamilton M.D.; (301) 594-8877; fax: (301) 480-8300; email: fh14e@nih gov; Jose Serrano; (301) 594-8871; fax: (301) 480-8300; email: js362q@nih.gov
Internet: http://grants.nih.gov/grants/guide/pa-files/PAR-06-171.html
Sponsor: National Institute of Diabetes and Digestive and Kidney Diseases (NIDDK)
6707 Democracy Boulevard, Room 669
Bethesda, MD 20892-5450

NIDDK Enhancing Zebrafish Research with Research Tools and Techniques 3999

This FOA encourages investigator-initiated applications designed to exploit the power of the zebrafish as a vertebrate model for biomedical and behavioral research. Applications proposing to develop new research tools or techniques that are of high priority to the zebrafish community and that will advance the detection and characterization of genes, pathways, and phenotypes of interest in development and aging, organ formation, neural processes, behavior, sensory processing, physiological processes, and disease processes are welcome. Budgets for direct costs under $500,000 (direct costs) per year and a project duration of up to five years may be requested. Letters of Intent should arrive by August 17, with full proposals due by September 17.
Requirements: The following organizations and institutions may apply: public/state controlled institutions of higher education; private institutions of higher education; Hispanic-serving institutions; historically Black colleges and universities (HBCUs); tribally controlled colleges and universities (TCCUs); Alaska native and native Hawaiian serving institutions; nonprofits with 501(c)3 IRS status; nonprofits without 501(c)3 IRS status; small businesses; for-profit organizations; State governments; regional organizations; U.S. territories or possessions; Indian/Native American tribal governments; Indian/Native American tribally designated organizations; eligible agencies of the Federal government; and faith-based or community based organizations.
Date(s) Application is Due: Aug 17; Sep 17
Contact: Dr. Rebekah S. Rasooly; (301) 594-6007; fax: (301) 480-3510; email: rr185i@nih.gov
Internet: http://grants.nih.gov/grants/guide/pa-files/PAR-08-139.html
Sponsor: National Institute of Diabetes and Digestive and Kidney Diseases (NIDDK)
6707 Democracy Boulevard, MSC 5458
Bethesda, MD 20892-5458

NIDDK Erythroid Lineage Molecular Toolbox Grants 4000

The ultimate goals of this Funding Opportunity Announcement (FOA) will be to use erythroid cells at different stages of development and differentiation to assemble the complete collection of genes expressed, describe how they are expressed and elucidate physical parameters, including subcellular localization of the proteins translated during erythropoiesis. These reagents may then be used to discover the structure-function relationships that exist in erythroid cells with possible application to other cell types. Because the nature and scope of the proposed research will vary from application to application, it is anticipated that the size and duration of each award will also vary.
Requirements: Application can be made by the following institutions and organizations: public/state controlled institutions of higher education; private institutions of higher education; nonprofits with 501(c)3 IRS status (other than institutions of higher education); nonprofits without 501(c)3 IRS status (other than institutions of higher education); small businesses; for-profit organizations (other than small businesses); state governments; U.S. territories or possessions; Indian/Native American tribal governments (federally recognized); Indian/Native American tribal governments (other than Federally recognized); Indian/Native American tribally designated organizations; non-domestic (non-U.S.) entities (foreign organizations); Hispanic-serving institutions; historically Black colleges and universities (HBCUs); tribally controlled colleges and universities (TCCUs); Alaska Native and Native Hawaiian serving institutions; regional organization; eligible agencies of the Federal government; and faith-based or community based organizations.
Date(s) Application is Due: Feb 5; Jun 5; Oct 5
Contact: Terry Rogers Bishop, Ph.D.; (301) 594-7726; email: tb232j@nih.gov
Internet: http://grants.nih.gov/grants/guide/pa-files/PA-07-011.html
Sponsor: National Institute of Diabetes and Digestive and Kidney Diseases (NIDDK)
6707 Democracy Boulevard, Room 619
Bethesda, MD 20892-5458

NIDDK Erythropoiesis: Components and Mechanisms Grants 4001

This FOA issued by the National Institute of Diabetes and Digestive and Kidney Diseases (NIDDK) the National Institute of Aging (NIA), and the National Heart, Lung, and Blood Institute (NHLBI), National Institutes of Health, encourages investigator-initiated R01 applications that propose hypothesis-driven research using erythroid cells. The aim of this program is to support research efforts towards a complete description of the molecular and cellular components of erythropoiesis and how these components contribute to erythropoiesis. Components include genes that are expressed (transcriptome) in erythroid cells, either during development or during differentiation, and the proteins (proteome) that are translated in erythroid cells, especially with post-translational modifications or subcellular localizations that are unique to erythroid cells. A long range goal of this program is to generate a concise description of erythropoiesis that unifies genetics, molecular processes and cytokine determinants in the erythroid lineages so that new therapeutics may be developed to measure and combat anemia. Budgets for direct costs of up to $500,000 per year and a project duration of up to five years may be requested.
Requirements: The following organizations and institutions may apply: public/state controlled institutions of higher education; private institutions of higher education; Hispanic-serving institutions; historically Black colleges and universities (HBCUs); tribally controlled colleges and universities (TCCUs); Alaska native and native Hawaiian serving institutions; nonprofits with 501(c)3 IRS status; nonprofits without 501(c)3 IRS status; small businesses; for-profit organizations; State governments; regional organizations; U.S. territories or possessions; Indian/Native American tribal governments; Indian/Native American tribally designated organizations; county governments; city or township governments; special district governments; independent school districts; public housing authorities/Indian housing authorities; eligible agencies of the Federal government; and faith-based or community based organizations.
Amount of Grant: $500,000 maximum per year for five years
Date(s) Application is Due: Feb 5; Jun 5; Oct 5
Contact: Terry Rogers Bishop, Ph.D.; (301) 594-7726; email: tb232j@nih.gov
Internet: http://grants.nih.gov/grants/guide/pa-files/PA-09-255.html
Sponsor: National Institute of Diabetes and Digestive and Kidney Diseases (NIDDK)
6707 Democracy Boulevard, Room 619
Bethesda, MD 20892-5458

NIDDK Exploratory/Developmental Clinical Research Grants in Obesity 4002

The goal of this initiative is to encourage exploratory/developmental clinical research that will accelerate the development of effective interventions for prevention or treatment of overweight or obesity in adults and/or children. The goal of this mechanism is to provide flexibility for initiating exploratory, short-term studies, thus allowing new ideas to be investigated in a more expeditious manner without stringent requirements for preliminary data. Such support is needed to encourage investigators to pursue new approaches, underdeveloped topics, or more creative avenues for research including new partnerships. Epidemiological research with a goal of informing translational/clinical research on prevention or treatment of obesity or overweight in adults and/or children is encouraged. The emphasis is thus on the development of exploratory clinical studies, pilot and feasibility studies, or small randomized clinical trials that will provide preliminary data for intervention and epidemiological studies that will inform translational/clinical research.

Requirements: The following organizations and institutions may apply: public/state controlled institutions of higher education; private institutions of higher education; Hispanic-serving institutions; historically Black colleges and universities (HBCUs); tribally controlled colleges and universities (TCCUs); Alaska native and native Hawaiian serving institutions; nonprofits with 501(c)3 IRS status; nonprofits without 501(c)3 IRS status; small businesses; for-profit organizations; State governments; regional organizations; U.S. territories or possessions; Indian/Native American tribal governments; Indian/Native American tribally designated organizations; eligible agencies of the Federal government; and faith-based or community based organizations.

Amount of Grant: $200,000 in direct costs allowed in any single year; $275,000 for the combined two-year award

Date(s) Application is Due: Feb 16; Jun 16; Oct 16

Contact: Carolyn W. Miles, Ph.D.; (301)451-3759; fax: (301) 480-8300; email: cm294e@nih.gov

Internet: http://grants.nih.gov/grants/guide/pa-files/PA-09-124.html

Sponsor: National Institute of Diabetes and Digestive and Kidney Diseases (NIDDK)
6707 Democracy Boulevard, Room 665
Bethesda, MD 20817

NIDDK Grants for Basic Research in Glomerular Diseases 4003

This FOA invites applications from investigators with diverse scientific interests to apply their expertise in basic research to enhance the understanding of the pathogenesis of the various primary or secondary forms of glomerular disease. Recent observations regarding intrinsic glomerular cell biology, particularly in the podocyte, have provided exciting new insights into potential pathogenic mechanisms of human glomerular disease. Although both immune and nonimmune mechanisms of glomerular injury have been studied previously, experimental models of disease and recent techniques that provide tools for molecular and proteomic profiling show great promise for identifying glomerular disease biomarkers. Despite these recent advances, additional basic studies are needed to facilitate a better understanding of glomerular disease pathogenesis and ultimately impact on the treatment of human glomerulopathies. It is anticipated that applications submitted in response to this FOA could address a number of different aspects concerning the pathogenesis, natural history, therapy pre-emption or prevention of the various morphologic forms of experimental glomerular disease.

Requirements: The following organizations and institutions may apply: public/state controlled institutions of higher education; private institutions of higher education; Hispanic-serving institutions; historically Black colleges and universities (HBCUs); tribally controlled colleges and universities (TCCUs); Alaska native and native Hawaiian serving institutions; nonprofits with 501(c)3 IRS status; nonprofits without 501(c)3 IRS status; small businesses; for-profit organizations; State governments; regional organizations; U.S. territories or possessions; Indian/Native American tribal governments; Indian/Native American tribally designated organizations; eligible agencies of the Federal government; and faith-based or community based organizations.

Date(s) Application is Due: Feb 5; Jun 5; Oct 5

Contact: Marva Moxey-Mims, M.D.; (301) 594-7717; fax: (301) 480-3510; email: mm726k@nih.gov

Internet: http://grants.nih.gov/grants/guide/pa-files/PA-07-367.html

Sponsor: National Institute of Diabetes and Digestive and Kidney Diseases (NIDDK)
6707 Democracy Boulevard, Room 639
Bethesda, MD 20892-5458

NIDDK Identifying and Reducing Diabetes and Obesity Related Health Disparities within Healthcare Systems 4004

This FOA requests applications designed to identify or address factors or barriers that result in disparate outcomes within a healthcare system. All applications should measure the impact of identified factors or interventions on health outcomes. Research is sought that examines at least one of the following factors in a healthcare system and/or the interaction between these factors; healthcare professionals, healthcare organizations, and the patients and communities they serve.

Requirements: The following organizations and institutions may apply: public/state controlled institutions of higher education; private institutions of higher education; Hispanic-serving institutions; historically Black colleges and universities (HBCUs); tribally controlled colleges and universities (TCCUs); Alaska native and native Hawaiian serving institutions; nonprofits with 501(c)3 IRS status; nonprofits without 501(c)3 IRS status; small businesses; for-profit organizations; State governments; regional organizations; U.S. territories or possessions; Indian/Native American tribal governments; Indian/Native American tribally designated organizations; eligible agencies of the Federal government; and faith-based or community based organizations.

Date(s) Application is Due: Feb 5; Jun 5; Oct 5

Contact: Christine Hunter, Ph.D.; (301) 594-4728; fax: (301) 480-3503; fax: ch514c@.nih.gov

Internet: http://grants.nih.gov/grants/guide/pa-files/PA-07-388.html

Sponsor: National Institute of Diabetes and Digestive and Kidney Diseases (NIDDK)
6707 Democracy Boulevard, Room 605
Bethesda, MD 20892-5460

NIDDK Insulin Signaling and Receptor Cross Talk Grants 4005

This Funding Opportunity Announcement (FOA) solicits investigator-initiated research projects that will define the mechanistic basis of crosstalk between the insulin receptor signaling pathway and other signaling pathways, with the goal of defining how insulin action is influenced by the complex microenvironment found in insulin responsive tissues.

Requirements: The following organizations and institutions may apply: public/state controlled institutions of higher education; private institutions of higher education; Hispanic-serving institutions; historically Black colleges and universities (HBCUs); tribally controlled colleges and universities (TCCUs); Alaska native and native Hawaiian serving institutions; nonprofits with 501(c)3 IRS status; nonprofits without 501(c)3 IRS status; small businesses; for-profit organizations; State governments; regional organizations; U.S. territories or possessions; Indian/Native American tribal governments; Indian/Native American tribally designated organizations; eligible agencies of the Federal government; and faith-based or community based organizations.

Date(s) Application is Due: Feb 5; Jun 5; Oct 5

Contact: Kristin Abraham, Ph.D.; (301) 451-8048; fax: (301) 480-3503; email: ka136s@nih.gov

Internet: http://grants.nih.gov/grants/guide/pa-files/PA-07-058.html

Sponsor: National Institute of Diabetes and Digestive and Kidney Diseases (NIDDK)
6707 Democracy Boulevard, Room 795
Bethesda, MD 20892-5460

NIDDK Intestinal Failure, Short Gut Syndrome and Small Bowel Transplantation Grants 4006

The National Institute of Diabetes and Digestive and Kidney Diseases (NIDDK) seeks grant applications to study the pathogenesis, natural history, treatment and complications of intestinal failure and its therapies, including parenteral nutrition and small bowel transplantation. The overall objective of this funding opportunity is to encourage basic and clinical research into intestinal failure, short gut syndrome and intestinal transplantation. Because the nature and scope of the proposed research will vary from application to application, it is anticipated that the size and duration of each award will also vary.

Requirements: Application can be made by the following institutions and organizations: public/state controlled institutions of higher education; private institutions of higher education; nonprofits with 501(c)3 IRS status (other than institutions of higher education); nonprofits without 501(c)3 IRS status (other than institutions of higher education); small businesses; for-profit organizations (other than small businesses); state governments; U.S. territories or possessions; Indian/Native American tribal governments (federally recognized); Indian/Native American tribal governments (other than Federally recognized); Indian/Native American tribally designated organizations; non-domestic (non-U.S.) entities (foreign organizations); Hispanic-serving institutions; historically Black colleges and universities (HBCUs); tribally controlled colleges and universities (TCCUs); Alaska Native and Native Hawaiian serving institutions; regional organization; eligible agencies of the Federal government; and faith-based or community based organizations.

Date(s) Application is Due: Feb 5; Jun 5; Oct 5

Contact: Dr. Michael K. May; (301) 594-8884; fax: (301) 480-8300; email: mm102i@nih.gov

Internet: http://grants.nih.gov/grants/guide/pa-files/PA-07-010.html

Sponsor: National Institute of Diabetes and Digestive and Kidney Diseases (NIDDK)
6707 Democracy Boulevard, Room 663, MSC 5450
Bethesda, MD 20892-5450

NIDDK Intestinal Stem Cell Consortium Grants 4007

This FOA invites new applications to participate in the Intestinal Stem Cell Consortium (ISCC). The goals of this initiative are: to establish a network of individual research projects and a Coordinating Center focused on research on stem cells of the small intestine; to isolate, culture, characterize, functionally validate, and compare stem cell populations from the small intestine epithelium in vivo and in vitro; to share data, biomaterials, models, reagents, resources and methods between projects in the ISCC; and make these publicly available through a web site to be developed by the Coordinating Center. The objective of support for this consortium is to accelerate research on stem cells of the intestinal epithelium in order to facilitate understanding of the biology and function of the intestine and aid development of therapies for intestinal diseases and conditions where damage and replacement of this epithelium are important components. Direct costs are limited to $250,000 per year for a maximum five-year period for research projects and $350,000 per year for a maximum five-year period for the Coordinating Center.

Requirements: Application can be made by the following institutions and organizations: public/state controlled institutions of higher education; private institutions of higher education; nonprofits with 501(c)3 IRS status (other than institutions of higher education); nonprofits without 501(c)3 IRS status (other than institutions of higher education); small businesses; for-profit organizations (other than small businesses); state governments; U.S. territories or possessions; Indian/Native American tribal governments (federally recognized); Indian/Native American tribal governments (other

than Federally recognized); Indian/Native American tribally designated organizations; non-domestic (non-U.S.) entities (foreign organizations); Hispanic-serving institutions; historically Black colleges and universities (HBCUs); tribally controlled colleges and universities (TCCUs); Alaska Native and Native Hawaiian serving institutions; regional organization; eligible agencies of the Federal government; and faith-based or community based organizations.
Amount of Grant: $250,000 maximum per year for five years
Date(s) Application is Due: Mar 18
Contact: Jill Carrington, Ph.D.; (301) 496-6402; fax: (301) 402-0010; email: carringj@mail.nih.gov; Ronald Margolis, Ph.D.; (301) 594-8819; fax: (301) 435-6047; email: rm76f@nih.gov
Internet: http://grants.nih.gov/grants/guide/rfa-files/RFA-DK-08-010.html
Sponsor: National Institute of Diabetes and Digestive and Kidney Diseases (NIDDK)
Democracy 2, Room 661, 6707 Democracy Boulevard, MSC 5450
Bethesda, MD 20892

NIDDK Mentored Research Scientist Development Award 4008
The overall objective of the NIH Research Career Development Award program is to prepare qualified individuals for careers that have a significant impact on the health-related research needs of the Nation. The objective of the NIDDK Mentored Research Scientist Development Award (K01) is to provide support for a sustained period of protected time for intensive research career development under the guidance of an experienced mentor, or sponsor, in the biomedical, behavioral or clinical sciences leading to research independence, thereby ensuring a future cadre of well-trained Ph.D. scientists working in research areas supported by NIDDK. The expectation is that through this sustained period of research career development and training, awardees will launch independent research careers and become competitive for new research project grant (R01) funding. By providing support for the critical transition period between postdoctoral training and independent R01 funding for non-clinical investigators, the NIDDK hopes to foster the careers of these investigators who are vital for the future excellence of the NIDDK research endeavor. Applicants must justify the need for a period of mentored research experience and provide a convincing case that the proposed period of support will substantially enhance their careers as independent investigators.
Requirements: The following organizations and institutions may apply: public/state controlled institutions of higher education; private institutions of higher education; Hispanic-serving institutions; historically Black colleges and universities (HBCUs); tribally controlled colleges and universities (TCCUs); Alaska native and native Hawaiian serving institutions; nonprofits with 501(c)3 IRS status; nonprofits without 501(c)3 IRS status; small businesses; for-profit organizations; State governments; regional organizations; U.S. territories or possessions; Indian/Native American tribal governments; Indian/Native American tribally designated organizations; eligible agencies of the Federal government; and faith-based or community based organizations.
Date(s) Application is Due: Mar 12; Jul 12; Nov 12
Contact: James F. Hyde, Ph.D.; (301) 594-7692; fax: (301) 480-3503; email: jh486z@nih.gov
Internet: http://grants.nih.gov/grants/guide/pa-files/PAR-09-060.html
Sponsor: National Institute of Diabetes and Digestive and Kidney Diseases (NIDDK)
6707 Democracy Boulevard, Room 789
Bethesda, MD 20892-5460

NIDDK Multi-Center Clinical Study Cooperative Agreements 4009
This PAR provides for grant applications for investigator-initiated, multi-center clinical studies. NIDDK supports investigator-initiated, multi-center clinical studies through a two-part process that includes an implementation planning (U34) grant. Beginning with the receipt date, NIDDK will accept, peer review, and consider for funding applications for investigator-initiated, multi-center clinical studies only from NIDDK Multi-Center Clinical Study Implementation Planning (U34) grant awardees. Because the nature and scope of the proposed research will vary from application to application, it is anticipated that the size and duration of each award will also vary. The total amount awarded and the number of awards will depend upon the mechanism numbers, quality, duration, and costs of the applications received.
Requirements: The following organizations and institutions may apply: public/state controlled institutions of higher education; private institutions of higher education; Hispanic-serving institutions; historically Black colleges and universities (HBCUs); tribally controlled colleges and universities (TCCUs); Alaska native and native Hawaiian serving institutions; nonprofits with 501(c)3 IRS status; nonprofits without 501(c)3 IRS status; small businesses; for-profit organizations; State governments; regional organizations; U.S. territories or possessions; Indian/Native American tribal governments; Indian/Native American tribally designated organizations; eligible agencies of the Federal government; and faith-based or community based organizations.
Date(s) Application is Due: Feb 5; Jun 5; Oct 5
Contact: Catherine Meyers, M.D.; (301) 451-4901; fax: (301) 480-3510; email: meyersc@mail.nih.gov
Internet: http://grants.nih.gov/grants/guide/pa-files/PAR-08-058.html
Sponsor: National Institute of Diabetes and Digestive and Kidney Diseases (NIDDK)
Two Democracy Plaza, Room 641, 6707 Democracy Boulevard
Bethesda, MD 20892-5458

NIDDK Multi-Center Clinical Study Implementation Planning Grants 4010
NIDDK supports investigator-initiated, multi-center clinical studies through a two-part process that includes an implementation planning grant. The planning grant is designed to: permit early peer review of the rationale for the proposed clinical study; permit assessment of the design/protocol of the proposed study; provide support for the development of a complete study protocol and associated documents including a manual of operations; and support the development of other essential elements required for the conduct of a clinical study. Completion of the required products of a U34 grant is a prerequisite for submission of a multi-center clinical study cooperative agreement application, which will support the actual conduct of the study. Because the nature and scope of the proposed research will vary from application to application, it is anticipated that the size and duration of each award will also vary. The total project period for an application submitted in response to this funding opportunity may not exceed two years. Applicants may request up to $250,000 in direct costs per year.
Requirements: The following organizations and institutions may apply: public/state controlled institutions of higher education; private institutions of higher education; Hispanic-serving institutions; historically Black colleges and universities (HBCUs); tribally controlled colleges and universities (TCCUs); Alaska native and native Hawaiian serving institutions; nonprofits with 501(c)3 IRS status; nonprofits without 501(c)3 IRS status; small businesses; for-profit organizations; State governments; regional organizations; U.S. territories or possessions; Indian/Native American tribal governments; Indian/Native American tribally designated organizations; eligible agencies of the Federal government; and faith-based or community based organizations.
Date(s) Application is Due: Jun 23; Mar 18
Contact: Catherine Meyers, M.D.; (301) 451-4901; fax: (301) 480-3510; email: meyersc@mail.nih.gov
Internet: http://grants.nih.gov/grants/guide/pa-files/PAR-08-057.html
Sponsor: National Institute of Diabetes and Digestive and Kidney Diseases (NIDDK)
Two Democracy Plaza, Room 641, 6707 Democracy Boulevard
Bethesda, MD 20892-5458

NIDDK Neurobiology of Diabetic Complications Grants 4011
NIDDK invites investigator-initiated research grant applications to study the mechanisms by which diabetes results in painful and disabling neuropathies and other neurological complications and to apply this information to the development of interventions to prevent, limit, or reverse these complications. The intent of this RFA is to attract basic neuroscientists to the study of diabetic neuropathy and neurobiology relevant to diabetes, and enhance interdisciplinary approaches to research in this area. Annual deadline dates may vary; contact the appropriate NIH institute for exact dates. Letters of intent are due by March 24, with full applications due by April 25 each year.
Requirements: Applications may be submitted by domestic and foreign, for-profit and nonprofit organizations, both public and private; units of state and local governments; and eligible agencies of the federal government.
Amount of Grant: $3 million total
Date(s) Application is Due: Mar 24; Apr 25
Contact: Dr. Barbara Linder; (301) 594-0021; fax: (301) 480-3503; email: bl99n@nih.gov. Dr. Paul Nichols; (301) 496-9964; fax: (301) 402-2060; email: pn13w@nih.gov
Internet: http://grants.nih.gov/grants/guide/rfa-files/RFA-NS-00-002.html
Sponsor: National Institute of Diabetes and Digestive and Kidney Diseases (NIDDK)
45 Center Drive, MSC 6600
Bethesda, MD 20892-6600

NIDDK Neuroimaging in Obesity Research Grants 4012
This FOA issued by the National Institute of Diabetes and Digestive and Kidney Diseases, the National Institute on Drug Abuse, and the National Institute of Biomedical Imaging and Bioengineering, of the National Institutes of Health, solicits Research Project Grant (R01) applications from institutions/ organizations that propose to use neuroimaging approaches in obesity research in human subjects and animal models. Many areas of the brain interact or communicate with other organs to control eating behavior, physical activity and energy metabolism, and functional neuroimaging holds enormous promise for expanding our understanding of how food intake and energy expenditure are mismatched in a setting of abundantly available nutrients, leading to excessive fat storage. The total amount of funding that the ICs expect to award through this announcement is up to $4.5 Million, to fund approximately 9-15 awards. The duration and requested direct costs will vary depending on the scope of the proposed research, but cannot exceed 5 years and $500,000 per year. Letters of intent are due by February 18, with full applications due by 5:00 p.m. on March 18.
Requirements: Application can be made by the following institutions and organizations: public/state controlled institutions of higher education; private institutions of higher education; nonprofits with 501(c)3 IRS status (other than institutions of higher education); nonprofits without 501(c)3 IRS status (other than institutions of higher education); small businesses; for-profit organizations (other than small businesses); state governments; U.S. territories or possessions; Indian/Native American tribal governments (federally recognized); Indian/Native American tribal governments (other than Federally recognized); Indian/Native American tribally designated organizations; non-domestic (non-U.S.) entities (foreign organizations); Hispanic-serving institutions; historically Black colleges and universities (HBCUs); tribally controlled colleges and universities (TCCUs); Alaska Native and Native Hawaiian serving institutions; regional organization; eligible agencies of the Federal government; and faith-based or community based organizations.
Amount of Grant: $500,000 maximum in direct costs per year, for up to 5 years
Date(s) Application is Due: Feb 18; Mar 18
Contact: Maren R. Laughlin, Ph.D.; (301) 594-8802; fax: (301) 480-0475; email: ml33q@nih.gov
Internet: http://grants.nih.gov/grants/guide/rfa-files/RFA-DK-08-009.html
Sponsor: National Institute of Diabetes and Digestive and Kidney Diseases (NIDDK)
6707 Democracy Boulevard Room 787, MSC 5460
Bethesda, MD 20892-5460

NIDDK New Technologies for Liver Disease SBIR **4013**

The purpose of this Funding Opportunity Announcement (FOA) is to solicit Small Business Innovation Research (SBIR) grant applications from small business concerns (SBCs) that propose to develop resources, research tools, instrumentations, biomarkers, devices, drugs or new and innovative approaches to diagnosis, monitoring, management, treatment and prevention of liver diseases. Areas of interest include development of reliable and practical means of diagnosis of liver diseases; biomarkers for disease activity and stage; noninvasive tests for inflammation, fibrosis and fat in the liver; and drugs, complementary and alternative modalities, biologics or molecular reagents for the therapy or prevention of liver diseases. The goal of this announcement is to enlist members of the small business research community in advancing means of diagnosis, treatment and prevention of liver disease and facilitate the goals outlined in the trans-NIH Action Plan for Liver Disease Research.
Requirements: Only United States small business concerns (SBCs) are eligible to submit SBIR applications.
Amount of Grant: $100,000 maximum for Phase 1; $750,000 maximum for Phase 2
Date(s) Application is Due: Apr 5; Aug 5; Dec 5
Contact: Christine L. Densmore, M.S.; (301) 402-8714; fax: (301) 480-8300; email: cd121z@nih.gov
Internet: http://grants.nih.gov/grants/guide/pa-files/PA-09-095.html
Sponsor: National Institute of Diabetes and Digestive and Kidney Diseases (NIDDK)
6707 Democracy Boulevard, Room 649
Bethesda, MD 20892-5450

NIDDK New Technologies for Liver Disease STTR **4014**

The objective of this FOA is to encourage and enable scientists at small businesses to develop and evaluate new technologies, drugs, devices, and approaches to diagnosis, management, and prevention of liver disease. Development of new technologies as well as application of existing technologies may be proposed. Studies may include use of animal models or human participants or both. If appropriate, plans for manufacturing and clinical evaluation of developed technologies, drugs, devices and innovative approaches should be included in the application. However, clinical trials beyond Phase I studies will not be considered appropriate to this announcement.
Requirements: Only United States small business concerns (SBCs) are eligible to submit STTR applications.
Amount of Grant: $100,000 maximum for Phase 1; $750,000 maximum for Phase 2
Date(s) Application is Due: Apr 5; Aug 5; Dec 5
Contact: Christine L. Densmore, M.S.; (301) 402-8714; fax: (301) 480-8300; email: cd121z@nih.gov
Internet: http://grants.nih.gov/grants/guide/pa-files/PA-09-094.html
Sponsor: National Institute of Diabetes and Digestive and Kidney Diseases (NIDDK)
6707 Democracy Boulevard, Room 649
Bethesda, MD 20892-5450

NIDDK Non-Invasive Methods for Diagnosis and Progression of Diabetes, Kidney, Urological, Hematological and Digestive Diseases **4015**

This funding opportunity is a call for the application of imaging and other non- or minimally-invasive technologies to detect, characterize, diagnose, identify persons with predisposition to, or monitor treatment of diseases of interest to the National Institute of Diabetes and Digestive and Kidney Diseases. Also needed are new, robust surrogate markers for clinical trial endpoints, and new ways to characterize normal and pathological tissues in vivo. Diseases of interest include type 1 and 2 diabetes, obesity, and kidney, liver, urologic, hematologic, digestive, endocrine and metabolic diseases and their complications. Applicable techniques include molecular imaging and functional imaging approaches, imaging methods with high spatial, chemical or time resolution, and new spectroscopic or sensor array technologies for monitoring metabolic or physiological events.
Requirements: The following organizations and institutions may apply: for-profit organizations; non-profit organizations; public or private institutions, such as universities, colleges, hospitals, and laboratories; units of State government; units of local government; eligible agencies of the Federal government; foreign institutions; domestic institutions; and faith-based or community-based organizations.
Date(s) Application is Due: Feb 5; Jun 5; Oct 5
Contact: Maren R. Laughlin, Ph.D.; (301) 594-8802; fax: (301) 480-0475; email: ml33q@nih.gov
Internet: http://grants.nih.gov/grants/guide/pa-files/PA-06-143.html
Sponsor: National Institute of Diabetes and Digestive and Kidney Diseases (NIDDK)
6707 Democracy Boulevard, Room 787
Bethesda, MD 20892-5460

NIDDK Pancreatic Development and Regeneration Grants: Toward Cellular Therapies for Diabetes **4016**

This Funding Opportunity Announcement (FOA) is intended to promote basic research in pancreatic development, stem cell biology and regeneration that will provide the foundation for the rationale design of future cell replacement therapies for diabetes. Appropriate topics for investigation include, but are not limited to: prospective isolation, purification and characterization of pancreatic stem/progenitor cells; identification of the signals, signaling pathway components and transcriptional factors that could direct the formation of pancreatic endoderm and progenitors, islet cell differentiation, morphogenesis, growth, function and survival in vitro; use of existing NIH-approved human ES cell lines or mouse embryonic cells, pancreatic stem cells, non-pancreatic adult stem cells such as those from liver or gut, human amniotic cells or bone marrow-derived cells to efficiently generate endoderm, pancreatic progenitor cells, or differentiated beta cells in culture; development of methods and reagents to assess quantitatively, the survival, migration, fate and function of transplanted animal or human pancreatic progenitors and their progeny in the living animal; determination of the factors important for beta cell regeneration and designing strategies to foster regeneration; development of new animal models that can used to identify exogenous or endogenous factors that might promote beta cell regeneration or neogenesis; characterization the lineage of cells that may contribute or participate in beta cell regeneration by genetic marking; and development of assays to screen chemical libraries for modifiers of beta cell regeneration, differentiation, and or expansion.
Requirements: The following organizations and institutions may apply: public/state controlled institutions of higher education; private institutions of higher education; Hispanic-serving institutions; historically Black colleges and universities (HBCUs); tribally controlled colleges and universities (TCCUs); Alaska native and native Hawaiian serving institutions; nonprofits with 501(c)3 IRS status; nonprofits without 501(c)3 IRS status; small businesses; for-profit organizations; State governments; regional organizations; U.S. territories or possessions; Indian/Native American tribal governments; Indian/Native American tribally designated organizations; eligible agencies of the Federal government; and faith-based or community based organizations.
Date(s) Application is Due: Feb 5; Jun 5; Oct 5
Contact: Sheryl M. Sato, Ph.D.; (301) 594-8811; fax: (301) 480-0475; email: ss68z@nih.gov
Internet: http://grants.nih.gov/grants/guide/pa-files/PA-07-056.html
Sponsor: National Institute of Diabetes and Digestive and Kidney Diseases (NIDDK)
2 Democracy Plaza, Room 790, 6707 Democracy Boulevard
Bethesda, MD 20892-5460

NIDDK Pilot and Feasibility Clinical Research Grants in Diabetes, Endocrine and Metabolic Diseases **4017**

The goal of this initiative is to encourage exploratory/developmental clinical research related to the prevention or treatment of diabetes, obesity and endocrine and genetic metabolic diseases. The Pilot and Feasibility Clinical Research Grants Program is designed to allow initiation of exploratory, short-term clinical studies, so that new ideas may be investigated without stringent requirements for preliminary data. Such support can be used by experienced investigators, as well as new investigators, to pursue new approaches and investigate underdeveloped research areas. The short-term studies should focus on research questions that are likely to have high clinical impact. They can include testing a new prevention strategy, a new intervention, or unique combinations of therapies. A high priority is the use of such studies to help stimulate the translation of promising research developments from the laboratory into clinical practice in diabetes, endocrine diseases and genetic metabolic diseases, including cystic fibrosis.
Requirements: The following organizations and institutions may apply: public/state controlled institutions of higher education; private institutions of higher education; Hispanic-serving institutions; historically Black colleges and universities (HBCUs); tribally controlled colleges and universities (TCCUs); Alaska native and native Hawaiian serving institutions; nonprofits with 501(c)3 IRS status; nonprofits without 501(c)3 IRS status; small businesses; for-profit organizations; State governments; regional organizations; U.S. territories or possessions; Indian/Native American tribal governments; Indian/Native American tribally designated organizations; eligible agencies of the Federal government; and faith-based or community based organizations.
Amount of Grant: $200,000 in direct costs allowed in any single year; $275,000 for the combined two-year award
Date(s) Application is Due: Feb 16; Jun 16; Oct 16
Contact: Karen Salomon; (301) 594-7733; fax: (301) 435-6047; email: ks495c@nih.gov
Internet: http://grants.nih.gov/grants/guide/pa-files/PA-09-133.html
Sponsor: National Institute of Diabetes and Digestive and Kidney Diseases (NIDDK)
6707 Democracy Boulevard, Democracy 2, Room 687
Bethesda, MD 20892-5460

NIDDK Pilot and Feasibility Clinical Research Grants in Kidney or Urologic Diseases **4018**

This funding opportunity announcement (FOA) is a re-issuance of PAR-06-113 and specifically encourages the submission of applications for small scale or pilot and feasibility clinical and translational research studies, including epidemiological studies or clinical trials related to kidney or urologic disease research that address important clinical and translational questions and are potentially of high impact. It is anticipated that some applications for pilot and feasibility studies may lead to full-scale clinical studies including evaluation of diagnostic strategies, epidemiologic studies, or trials in the diagnosis, prevention, or treatment of kidney or urologic diseases. These grants may be used to plan, evaluate the feasibility, or implement clinical trials that assess pharmacological, dietary, surgical, or behavioral interventions for the prevention or treatment of kidney or urologic disease. Small scale or pilot epidemiological studies are also encouraged.
Requirements: The following organizations and institutions may apply: public/state controlled institutions of higher education; private institutions of higher education; Hispanic-serving institutions; historically Black colleges and universities (HBCUs); tribally controlled colleges and universities (TCCUs); Alaska native and native Hawaiian serving institutions; nonprofits with 501(c)3 IRS status; nonprofits without 501(c)3 IRS status; small businesses; for-profit organizations; State governments; regional organizations; U.S. territories or possessions; Indian/Native American tribal governments; Indian/Native American tribally designated organizations; eligible agencies of the Federal government; and faith-based or community based organizations.
Amount of Grant: $200,000 in direct costs allowed in any single year; $275,000 for the combined two-year award

Date(s) Application is Due: Feb 16; Jun 16; Oct 16
Contact: Catherine M. Meyers, M.D.; (301) 594-7717; fax: (301) 480-3510; email: cm420i@nih.gov
Internet: http://grants.nih.gov/grants/guide/pa-files/PAR-09-077.html
Sponsor: National Institute of Diabetes and Digestive and Kidney Diseases (NIDDK)
6707 Democracy Boulevard, Room 641
Bethesda, MD 20892-5458

NIDDK Pilot and Feasibility Clinical Research Studies in Digestive Diseases and Nutrition 4019

The goal of this pilot and feasibility mechanism is to provide flexibility for initiating preliminary, short-term clinical studies, thus allowing new ideas to be investigated in a more expeditious manner without stringent requirements for preliminary data. Such support is needed to encourage experienced investigators as well as new investigators to pursue new approaches, underdeveloped topics, or more creative avenues of research. If successful, these awards should lead to significant scientific advances in digestive disease research and should facilitate translation into clinical practice to improve patient outcomes.
Requirements: The following organizations and institutions are eligible to apply: public/state controlled institutions of higher education; private institutions of higher education; Hispanic-serving institutions; historically Black colleges and universities (HBCUs); tribally controlled colleges and universities (TCCUs); Alaska Native and Native Hawaiian serving institutions; nonprofits with 501(c)3 IRS status (other than institutions of higher education); nonprofits without 501(c)3 IRS status (other than institutions of higher education); small businesses; for-profit organizations (other than small businesses); state governments; Indian/Native American tribal governments (federally recognized); Indian/Native American tribally designated organizations; county governments; city or township governments; special district governments; independent school districts; public housing authorities; Indian housing authorities; U.S. territory or possession; Indian/Native American tribal governments (other than federally recognized); regional organizations; non-domestic (non-U.S.) entities (foreign organizations); other eligible agencies of the federal government; and faith-based or community-based organizations.
Date(s) Application is Due: Feb 16; Jun 16; Oct 16
Contact: Mary Evans, Ph.D.; (301) 594-4578; fax: (301) 480-8300; email: me189d@nih.gov
Internet: http://grants.nih.gov/grants/guide/pa-files/PA-09-151.html
Sponsor: National Institute of Diabetes and Digestive and Kidney Diseases (NIDDK)
6707 Democracy Boulevard, Room 681
Bethesda, MD 20892-5450

NIDDK Planning Grants For Translational Research For The Prevention and Control Of Diabetes and Obesity 4020

The National Institute of Diabetes and Digestive and Kidney Diseases (NIDDK) seeks to develop cost effective and sustainable interventions, that can be adopted in real world settings, for the prevention and control of diabetes and obesity. Research should be based on interventions already proven efficacious in clinical trials to prevent and reverse obesity and type 2 diabetes, to improve care of type 1 and type 2 diabetes and to prevent or delay its complications. The interventions proposed under this solicitation should have the potential to be widely disseminated to clinical practice, individuals and communities at risk. Relevant topics include: methods to improve health care delivery to patients with or at risk of diabetes; to enhance diabetes self management; methods to develop strategies to promote healthy lifestyles to reduce the risk of diabetes; and methods to promote lifestyle change that will prevent or reverse overweight and obesity, and 5) new cost effective ways to identify people with pre-diabetes and undiagnosed diabetes. This FOA will use the National Institutes of Health (NIH) clinical trial planning grant (R34) award mechanism, which will provide up to $150,000 in direct costs per year.
Requirements: Applications may be submitted by: for-profit organizations; non-profit organizations; public or private institutions, such as universities, colleges, hospitals, and laboratories; units of State government; units of local government; eligible agencies of the Federal government; domestic institutions; faith-based or community-based organizations; Indian/Native American tribal governments and Indian/Native American tribally designated organizations.
Amount of Grant: $150,000 maximum per year
Date(s) Application is Due: Feb 16; Jun 16; Oct 16
Contact: Sanford Garfield, Ph.D.; (301) 594-8803; fax: (301) 480-3503; email: sg50o@nih.gov
Internet: http://grants.nih.gov/grants/guide/pa-files/PAR-06-353.html
Sponsor: National Institute of Diabetes and Digestive and Kidney Diseases
6707 Democracy Boulevard, Room 685
Bethesda, MD 20892-5460

NIDDK Proteomics: Diabetes, Obesity, and Endocrine, Digestive, Kidney, Urologic, and Hematologic Diseases 4021

This Funding Opportunity Announcement (FOA) issued by the National Institute of Diabetes and Digestive and Kidney Diseases (NIDDK), National Institutes of Health (NIH), solicits projects that advance research to identify and quantitate protein expression patterns, post-translational modification of proteins, and protein-protein interactions on cells, tissues, organ systems relevant to diabetes, obesity, endocrine and metabolic diseases, nutritional function and diseases of the alimentary tract, exocrine pancreas, liver, kidney, bladder and prostate and normal biological processes related to the function of these systems. The development and improvement of innovative proteomic technologies is also encouraged through their application to relevant biological questions related to the pathophysiology of endocrine glands, gastrointestinal tract, liver and kidney, bladder and prostate.
Requirements: The following organizations and institutions may apply: public/state controlled institutions of higher education; private institutions of higher education; Hispanic-serving institutions; historically Black colleges and universities (HBCUs); tribally controlled colleges and universities (TCCUs); Alaska native and native Hawaiian serving institutions; nonprofits with 501(c)3 IRS status; nonprofits without 501(c)3 IRS status; small businesses; for-profit organizations; State governments; regional organizations; U.S. territories or possessions; Indian/Native American tribal governments; Indian/Native American tribally designated organizations; eligible agencies of the Federal government; and faith-based or community based organizations.
Date(s) Application is Due: Feb 5; Jun 5; Oct 5
Contact: Salvatore Sechi, Ph.D.; (301) 594-8814; fax: (301) 480-2688; email: ss24q@nih.gov
Internet: http://grants.nih.gov/grants/guide/pa-files/PA-07-016.html
Sponsor: National Institute of Diabetes and Digestive and Kidney Diseases (NIDDK)
6707 Democracy Blvd. Room 611
Bethesda, MD 20892-5460

NIDDK Race/Ethnic Disparities in the Incidence of Diabetes Complications 4022

This PA solicits research to investigate differences among contemporary populations in the United States, categorized by race-ethnicity and other factors, in risk factors for complications of diabetes and in rates of these complications; and the extent to which factors, including inherent metabolic and genetic variations, medical care, socioeconomic status, and behavioral factors, account for these differences. PA: PA-02-165
Requirements: Applications may be submitted by domestic for-profit and nonprofit organizations, both public and private; units of state and local governments; and eligible agencies of the federal government.
Contact: Dr. Kristin Abraham, (301) 451-8048; fax: (301) 480-3503; email: ka136s@nih.gov
Internet: http://grants2.nih.gov/grants/guide/pa-files/PA-02-165.html
Sponsor: National Institute of Diabetes and Digestive and Kidney Diseases
45 Center Drive, Natcher Building
Bethesda, MD 20892

NIDDK Research Grants for Studies of Hepatitis C in the Setting of Renal Disease 4023

It is the intent of this Funding Opportunity Announcement (FOA) to invite applications from investigators with diverse scientific interests, who wish to apply their expertise into basic and applied research to enhance the understanding of the pathogenesis, natural history, therapy,and prevention of hepatitis C in the setting of renal disease or renal transplantation. Because the nature and scope of the proposed research will vary from application to application, it is anticipated that the size and duration of each award will also vary. The total amount awarded and the number of awards will depend upon the mechanism numbers, quality, duration, and costs of the applications received.
Requirements: The following organizations and institutions may apply: public/state controlled institutions of higher education; private institutions of higher education; Hispanic-serving institutions; historically Black colleges and universities (HBCUs); tribally controlled colleges and universities (TCCUs); Alaska native and native Hawaiian serving institutions; nonprofits with 501(c)3 IRS status; nonprofits without 501(c)3 IRS status; small businesses; for-profit organizations; State governments; regional organizations; U.S. territories or possessions; Indian/Native American tribal governments; Indian/Native American tribally designated organizations; eligible agencies of the Federal government; and faith-based or community based organizations.
Date(s) Application is Due: Feb 5; Jun 5; Oct 5
Contact: Catherine M. Meyers, M.D.; (301) 594-7717; fax: (301) 480-3510; email: cm420i@nih.gov
Internet: http://grants.nih.gov/grants/guide/pa-files/PA-07-015.html
Sponsor: National Institute of Diabetes and Digestive and Kidney Diseases (NIDDK)
6707 Democracy Boulevard, Room 641
Bethesda, MD 20892-5458

NIDDK Research Grants on Improving Health Care for Obese Patients 4024

This funding opportunity announcement (FOA) solicits Research Project Grant applications from institutions and organizations that propose to conduct research to determine the barriers to optimal health care for obese patients, and to test innovations or modifications in care delivery to improve health outcomes for obese patients independent of weight loss. Awards issued under this FOA are contingent upon the availability of funds and the submission of a sufficient number of meritorious applications. Applications can be renewed by competing for additional project periods.
Requirements: The following organizations and institutions may apply: public/state controlled institutions of higher education; private institutions of higher education; Hispanic-serving institutions; historically Black colleges and universities (HBCUs); tribally controlled colleges and universities (TCCUs); Alaska native and native Hawaiian serving institutions; nonprofits with 501(c)3 IRS status; nonprofits without 501(c)3 IRS status; small businesses; for-profit organizations; State governments; regional organizations; U.S. territories or possessions; Indian/Native American tribal governments; Indian/Native American tribally designated organizations; eligible agencies of the Federal government; and faith-based or community based organizations.
Date(s) Application is Due: Feb 5; Jun 5; Oct 5
Contact: Susan Z. Yanovski, M.D.; (301) 594-8882; fax: (301) 480-8300; email: sy29f@nih.gov
Internet: http://grants.nih.gov/grants/guide/pa-files/PA-07-013.html

Sponsor: National Institute of Diabetes and Digestive and Kidney Diseases (NIDDK)
6707 Democracy Boulevard, Room 675
Bethesda, MD 20892-5450

NIDDK Role of Gastrointestinal Surgical Procedures in Amelioration of Obesity-Related Insulin Resistance & Diabetes Independent of Weight Loss 4025

This Funding Opportunity Announcement (FOA) issued by the National Institute of Diabetes and Digestive and Kidney Diseases, National Institutes of Health encourages Research Project Grant (R01) applications from institutions and organizations that propose to explain the underlying mechanism(s) by which various gastrointestinal surgical procedures ameliorate obesity-related insulin resistance and diabetes independent of the resultant weight loss. Awards issued under this FOA are contingent upon the availability of funds and the submission of a sufficient number of meritorious applications.
Requirements: The following organizations and institutions may apply: public/state controlled institutions of higher education; private institutions of higher education; Hispanic-serving institutions; historically Black colleges and universities (HBCUs); tribally controlled colleges and universities (TCCUs); Alaska native and native Hawaiian serving institutions; nonprofits with 501(c)3 IRS status; nonprofits without 501(c)3 IRS status; small businesses; for-profit organizations; State governments; regional organizations; U.S. territories or possessions; Indian/Native American tribal governments; Indian/Native American tribally designated organizations; eligible agencies of the Federal government; and faith-based or community based organizations.
Date(s) Application is Due: Feb 5; Jun 5; Oct 5
Contact: Carolyn W. Miles, Ph.D.; (301)451-3759; fax: (301) 480-8300; email: cm294e@nih.gov
Internet: http://grants.nih.gov/grants/guide/pa-files/PA-08-014.html
Sponsor: National Institute of Diabetes and Digestive and Kidney Diseases (NIDDK)
6707 Democracy Boulevard, Room 665
Bethesda, MD 20892-5450

NIDDK Secondary Analyses in Obesity, Diabetes and Digestive and Kidney Diseases Grants 4026

The specific objectives of this announcement on Secondary Analyses In Obesity, Diabetes, Digestive and Kidney Diseases are to support the following: secondary analyses of data related to the epidemiology of disease areas of NIDDK; important and/or innovative hypotheses explored through analysis of existing data sets; secondary analyses designed to inform and support subsequent applications for individual research awards; rapid analyses of new databases and experimental modules to inform the design and content of future studies; the archiving of datasets to be made publicly available for research purposes related to disease areas of NIDDK, including both epidemiological studies and multi-center clinical trials. Research that employs analytic techniques that demonstrate or promote methodological advances in patient oriented and epidemiologic research is also of interest. International comparative analyses are encouraged. Applications that are innovative and high risk with the likelihood for high impact are especially encouraged.
Requirements: The following organizations and institutions may apply: public/state controlled institutions of higher education; private institutions of higher education; Hispanic-serving institutions; historically Black colleges and universities (HBCUs); tribally controlled colleges and universities (TCCUs); Alaska native and native Hawaiian serving institutions; nonprofits with 501(c)3 IRS status; nonprofits without 501(c)3 IRS status; small businesses; for-profit organizations; State governments; regional organizations; U.S. territories or possessions; Indian/Native American tribal governments; Indian/Native American tribally designated organizations; eligible agencies of the Federal government; and faith-based or community based organizations.
Amount of Grant: $200,000 in direct costs allowed in any single year; $275,000 for the combined two-year award
Date(s) Application is Due: Feb 16; Jun 16; Oct 16
Contact: James E. Everhart, MD, MPH; (301) 594-8878; fax: (301) 480-8300; email: je17g@nih.gov
Internet: http://grants.nih.gov/grants/guide/pa-files/PA-09-131.html
Sponsor: National Institute of Diabetes and Digestive and Kidney Diseases (NIDDK)
6707 Democracy Boulevard, Room 655
Bethesda, MD 20892-5450

NIDDK Seeding Collaborative Interdisciplinary Team Science in Diabetes, Endocrinology and Metabolic Diseases Grants 4027

The purpose of the Seeding Collaborative Interdisciplinary Team Science Program described in this announcement is to provide initial support to enable strong new investigative teams to form, and to foster preliminary research activities. It is anticipated that research teams receiving seeding support under this program will be well positioned to compete for funds through the parent NIDDK Collaborative Interdisciplinary Research Program in Diabetes, Endocrinology and Metabolic Diseases solicitation. It is anticipated that 5-10 projects will be funded, but because the nature and scope of the proposed research will vary from application to application, it is anticipated that the size and duration of each award will also vary. Budgets for direct costs of up to $300,000 per year with a project duration of up to one year may be requested.
Requirements: The following organizations and institutions are eligible to apply: public/state controlled institutions of higher education; private institutions of higher education; Hispanic-serving institutions; historically Black colleges and universities (HBCUs); tribally controlled colleges and universities (TCCUs); Alaska Native and Native Hawaiian serving institutions; and non-domestic (non-U.S.) entities (foreign organizations).
Amount of Grant: $375,000 maximum
Date(s) Application is Due: Jan 25; May 25; Sep 25
Contact: Karen Salomon; (301) 594-7733; fax: (301) 435-6047; email: ks495c@nih.gov
Internet: http://grants.nih.gov/grants/guide/pa-files/PAR-08-181.html
Sponsor: National Institute of Diabetes and Digestive and Kidney Diseases (NIDDK)
6707 Democracy Boulevard, Room 687
Bethesda, MD 20892-5460

NIDDK Silvio O. Conte Digestive Diseases Research Core Centers Grants 4028

The objective of the Silvio O. Conte Digestive Diseases Research Core Centers is to bring together, on a cooperative basis, basic science and clinical investigators to enhance the effectiveness of their research related to digestive and/or liver diseases and their complications. A core center must be an identifiable unit within a single university medical center or a consortium of cooperating institutions, including an affiliated university. An existing program of excellence in biomedical research in the area of digestive and/or liver diseases is a prerequisite for applying. This research must be in the form of NIH research projects, program projects, or other peer-reviewed research that is already funded at the time of submission of a center grant application. Close cooperation, communication, and collaboration among all involved personnel of various professional disciplines are the ultimate objectives.
Requirements: The following organizations and institutions are eligible to apply: public/state controlled institutions of higher education; private institutions of higher education; nonprofits with 501(c)3 IRS status (other than institutions of higher education); nonprofits without 501(c)3 IRS status (other than institutions of higher education); and for-profit organizations (other than small businesses).
Amount of Grant: $750,000 per year for five years
Date(s) Application is Due: Oct 12; Nov 9
Contact: Judith Podskalny; (301) 594-8876; fax: (301) 480-8300; email: jp53s@nih.gov
Internet: http://grants.nih.gov/grants/guide/rfa-files/RFA-DK-09-005.html
Sponsor: National Institute of Diabetes and Digestive and Kidney Diseases (NIDDK)
6707 Democracy Boulevard, MSC 5450
Bethesda , MD 20892

NIDDK Small Business Innovation Research to Develop New Therapeutics and Monitoring Technologies for Type 1 Diabetes (T1D)—Towards an Artificial Pancreas (SBIR) 4029

A promising therapeutic option for the treatment of diabetes is a system (termed an artificial pancreas or closed-loop) that can mimic normal pancreatic beta cell function thereby restoring normal metabolic homeostasis without causing hypoglycemia. The design of any system capable of achieving this goal is complex and raises scientific, clinical and regulatory challenges. This FOA solicits Small Business Innovation Research (SBIR) grant applications from small business concerns (SBCs) for funding to perform research leading to the development of innovative technologies that may advance progress toward integrated long term glucose regulated insulin delivery system (artificial pancreas). Letters of Intent are due by March 16, with full proposals to be received by April 15.
Requirements: Only United States small business concerns (SBCs) are eligible to submit SBIR applications.
Amount of Grant: $500,000 maximum for phase I; $1,000,000 maximum for phase II
Date(s) Application is Due: Mar 16; Apr 15
Contact: Dr. Guillermo Arreaza; (301) 594-4724; fax: (301) 480-3503; email: ga96b@nih.gov
Internet: http://grants.nih.gov/grants/guide/rfa-files/RFA-DK-09-001.html
Sponsor: National Institute of Diabetes and Digestive and Kidney Diseases (NIDDK)
6707 Democracy Boulevard, Room 6101
Bethesda, MD 20892

NIDDK Small Grant Program for K08/K23 Recipients 4030

The NIDDK invites recipients of its K08/K23 awards to apply for grant support during the final two years of their awards through the NIDDK Small Grant Program. This additional support will enable K awardees to either expand their current research objectives or to branch out to a closely related pilot study, thus demonstrating their growth as investigators and their independence. This should facilitate the transition to fully independent investigator status. The added grant support in the latter years of an award is expected to have the following benefits: increased fiscal independence for the K08/K23 award recipient as a precursor to complete independence; an opportunity for the recipient to generate additional publications and data to form the basis for an R01 application; an opportunity for a Scientific Review Group (SRG) to evaluate accomplishments made during the first two years of the K08/K23 award; an opportunity for the applicant to provide more detailed research plans for the last two years of the K award; and an opportunity for the applicant to demonstrate additional success in the peer review process during the course of their K08/K23 award. A project period of up to two years and a budget for direct costs of up to two $25,000 modules or $50,000 per year may be requested.
Requirements: This award is available only to those currently holding the K08 award from NIDDK. Details of the K08 award are given in NIH Program Announcement PA-00-003. Applications may be submitted on behalf of candidates by domestic non-federal nonprofit organizations, public or private.
Amount of Grant: $50,000 per year for two years
Date(s) Application is Due: Feb 16; Jun 16; Oct 16
Contact: Dr. James Hyde; (301) 594-7692; fax: (301) 480-0475; email: jh486z@nih.gov; Dr. Judith Podskalny; (301) 594-8876; fax: (301) 480-8300; email: jp53s@nih.gov
Internet: http://grants.nih.gov/grants/guide/pa-files/PAR-06-172.html
Sponsor: National Institute of Diabetes and Digestive and Kidney Diseases
6707 Democracy Boulevard, Room 786
Bethesda, MD 20892-5460

NIDDK Small Grants for Underrepresented Minority Scientists in Diabetes and Digestive and Kidney Diseases 4031

The goal of this program is to enable the applicant to accept a tenure-earning position, gain additional research experience and obtain preliminary data on which to base a subsequent research grant application in an area of diabetes, endocrinology, metabolism, digestive diseases, and nutrition, kidney, urology, or hematology. Annual deadline dates may vary; contact program staff for exact dates.

Requirements: Awards will be limited to minority citizens or non-citizen nationals of the United States or to individuals who have been lawfully admitted for permanent residence at the time of application. Applicants must also have a doctoral degree.

Restrictions: Applicants may not hold, nor apply concurrently for any other PHS research project grant at the time of this application.

Amount of Grant: $250,000 maximum

Date(s) Application is Due: Mar 22

Contact: Dr. Charles H. Rodgers; (301) 594-7717; email: rodgersc@extra.niddk.nih.gov

Internet: http://grants.nih.gov/grants/guide/rfa-files/RFA-DK-00-007.html

Sponsor: National Institute of Diabetes and Digestive and Kidney Diseases (NIDDK)
45 Center Drive, Room 6AS 19J, MSC 6600
Bethesda, MD 20892-6600

NIDDK Traditional Conference Grants 4032

Conference Grants are awarded to institutions and organizations (not individuals) to provide partial support for international or domestic meetings, conferences, and workshops to coordinate, exchange, and disseminate research information. NIDDK is especially interested in conference grants used to cover travel expenses for young and minority investigators, although meeting publications, salaries, consultant services, equipment rental, and supplies may be requested. Indirect costs normally are not allowed.

Requirements: Applicants must submit a letter of intent to apply at least 8 weeks before the application receipt date. This letter should include the names of the organizing committee, a preliminary agenda, a list of accepted or invited speakers, a proposed budget and other planned sources of funding, and a description of how the conference will address each of the above criteria.

Contact: Dr. M. Ken May, Nutrient Metabolism Program Director; (301) 594-8884; fax: (301) 480-8300; email: maym@mail.nih.gov

Internet: http://www2.niddk.nih.gov/Funding/Grants/ApplicantGuidelines/ConferenceSupport.htm

Sponsor: National Institute of Diabetes and Digestive and Kidney Diseases (NIDDK)
Building 2DEM, Room 663, 6707 Democracy Boulevard
Bethesda, MD 20892-1754

NIDDK Training in Clinical Investigation in Kidney and Urology Grants 4033

The Division of Kidney, Urology, and Hematology (DKUH) has long been interested in stimulating more training in clinical research methods in programs related to nephrology and urology. At the same time, there is a dearth of clinicians that have advanced training in clinical trial methodology. The purpose of this program announcement is to encourage applications for new training programs and to encourage program directors of existing Institutional Training Grants supported by DKUH to collaborate with schools of public health and departments of epidemiology, statistics, and biostatistics to offer didactic and practical training in clinical research methodologies. The goal of this program announcement is to increase the number of well-trained investigators who are capable of conceiving, designing, conducting, analyzing, and presenting the results of clinical research in areas pertaining to nephrology or urology.

Requirements: For new training programs, only domestic, nonprofit, private or public institutions may apply. The applicant institution must have the staff and facilities required for the proposed program. The research training program director at the institution will be responsible for the selection and appointment of trainees to receive National Research Service Award (NRSA) support and for the overall direction of the program.

Restrictions: Positions on NRSA institutional grants may not be used for study leading to the MD, DDS or other clinical health professional degrees except when those studies are a part of a formal combined research degree program such as the MD/PhD. Similarly, trainees may not accept NRSA support for studies which are a part of residency training leading to a medical specialty or subspecialty except when the residency program credits a period of full-time, postdoctoral research training toward board certification and the trainee intends to pursue a research career.

Amount of Grant: $2000 maximum for tuition and fees plus 60 percent of costs exceeding that amount

Date(s) Application is Due: Mar 10

Contact: Dr. Charles Rodgers; (301) 594-7717; fax: (301) 480-3510; email: rodgersc@extra.niddk.nih.gov

Internet: http://grants.nih.gov/grants/guide/pa-files/PAR-99-012.html

Sponsor: National Institute of Diabetes and Digestive and Kidney Diseases
45 Center Drive, MSC 6600
Bethesda, MD 20892-6600

NIDDK Translational Research for the Prevention and Control of Diabetes and Obesity 4034

The National Institute of Diabetes and Digestive and Kidney Diseases (NIDDK), the National Institute of Nursing Research and the Office of Behavior and Social Sciences Research seek to develop cost effective and sustainable interventions that can be adopted in real world settings, for the prevention and control of diabetes and obesity. Research should be based on interventions already proven efficacious in clinical trials to prevent and reverse obesity and type 2 diabetes, to improve care of type 1 and type 2 diabetes and to prevent or delay its complications. The interventions proposed under this solicitation should have the potential to be widely disseminated to clinical practice, individuals and communities at risk. Relevant topics include: methods to improve health care delivery to patients with or at risk of diabetes; strategies to enhance diabetes self management; methods to develop strategies to promote healthy lifestyles to reduce the risk of diabetes; methods to promote lifestyle change that will prevent or reverse overweight and obesity; and new cost effective ways to identify people with pre-diabetes and undiagnosed diabetes.

Requirements: The following organizations and institutions may apply: public/state controlled institutions of higher education; private institutions of higher education; Hispanic-serving institutions; historically Black colleges and universities (HBCUs); tribally controlled colleges and universities (TCCUs); Alaska native and native Hawaiian serving institutions; nonprofits with 501(c)3 IRS status; nonprofits without 501(c)3 IRS status; small businesses; for-profit organizations; State governments; regional organizations; U.S. territories or possessions; Indian/Native American tribal governments; Indian/Native American tribally designated organizations; eligible agencies of the Federal government; and faith-based or community based organizations.

Date(s) Application is Due: Jan 25; May 25; Sep 25

Contact: Sanford Garfield, Ph.D.; (301) 594-8803; fax: (301) 480-3503; email: sg50o@nih.gov

Internet: http://grants.nih.gov/grants/guide/pa-files/PAR-06-532.html

Sponsor: National Institute of Diabetes and Digestive and Kidney Diseases (NIDDK)
6707 Democracy Boulevard, Room 685
Bethesda, MD 20892-5460

NIDDK Transmission of Human Immunodeficiency Virus In Semen 4035

The National Institute of Diabetes and Digestive and Kidney Diseases (NIDDK) through its Division of Kidney, Urologic and Hematologic Diseases (DKUHD) and the National Institute of Child Health and Human Development, Center for Population Research invite investigators to submit research grant applications that will increase the basic and clinical knowledge of the biology of HIV in semen. Direct exposure to semen of HIV seropositive men is a major route for transmission of HIV type I. This Funding Opportunity Announcement (FOA) focuses on studies that will elucidate the factors that determine HIV shedding in the male genital tract. This includes studies that elucidate the infectivity of HIV in semen fractions, the effect of antiretroviral therapy on HIV infectivity in semen fractions, the relationship between the immunobiology of the male genital tract and HIV replication and infectivity, and factors such as genital tract inflammation which influence HIV transmission through semen. Of particular interest are studies of the anatomic sites of virus infected cells and virus production in the male genital tract. The intent of this Funding opportunity is to encourage investigator-initiated research, to attract new investigators to the field, and to increase interdisciplinary research to enhance the scope and effectiveness of research in this area. Because the nature and scope of the proposed research will vary from application to application, it is anticipated that the size and duration of each award will also vary. The total amount awarded and the number of awards will depend upon the mechanism numbers, quality, duration, and costs of the applications received.

Requirements: The following organizations and institutions are eligible to apply: public/state controlled institutions of higher education; private institutions of higher education; Hispanic-serving institutions; historically Black colleges and universities (HBCUs); tribally controlled colleges and universities (TCCUs); Alaska Native and Native Hawaiian serving institutions; nonprofits with 501(c)3 IRS status (other than institutions of higher education); nonprofits without 501(c)3 IRS status (other than institutions of higher education); small businesses; for-profit organizations (other than small businesses); state governments; Indian/Native American tribal governments (federally recognized); Indian/Native American tribally designated organizations; county governments; city or township governments; special district governments; independent school districts; public housing authorities; Indian housing authorities; U.S. territory or possession; Indian/Native American tribal governments (other than federally recognized); regional organizations; non-domestic (non-U.S.) entities (foreign organizations); other eligible agencies of the federal government; and faith-based or community-based organizations.

Date(s) Application is Due: Feb 5; Jun 5; Oct 5

Contact: Leroy M. Nyberg, Jr.; (301) 594-7717; email: ln10f@nih.gov

Internet: http://grants.nih.gov/grants/guide/pa-files/PA-07-044.html

Sponsor: National Institute of Diabetes and Digestive and Kidney Diseases (NIDDK)
6707 Democracy Boulevard, Room 627
Bethesda, MD 20892

NIDDK Type 2 Diabetes in the Pediatric Population Research Grants 4036

NIDDK and NICHD invite investigator-initiated research grant applications to study the epidemiology, natural history, pathophysiology, prevention, and treatment of type 2 diabetes in children in the United States. Type 2 diabetes has traditionally been viewed as a disease of adults. More recently, however, it has become apparent that an increasing number of cases of type 2 diabetes are being reported in children. New hypothesis-driven studies are needed to describe the epidemiology, refine the diagnosis, define the metabolic abnormalities, and formulate treatment options for type 2 diabetes in children. Annual deadline dates may vary; contact program staff for exact dates.

Requirements: Applications may be submitted by domestic for-profit and nonprofit organizations, both public and private; units of state and local governments; and eligible agencies of the federal government.

Contact: Dr. Barbara Linder; (301) 594-0021; fax: (301) 480-3503; email: linderb@mail.nih.gov; Dr. Gilman Grave; (301) 496-5593; fax: (301) 480-9791

Internet: http://www2.niddk.nih.gov/Research/ScientificAreas/Diabetes/Type2Diabetes/DM2K.htm
Sponsor: National Institute of Diabetes and Digestive and Kidney Diseases (NIDDK)
Building 2DEM, Room 699, 6707 Democracy Boulevard
Bethesda, MD 20892

NIDRR Field-Initiated Projects **4037**
The goals of this project include: developing methods and technologies that maximize the integration of individuals with disabilities into society; employment and family environments; and improving the effectiveness of services authorized under the Rehabilitation Act. Applicants who receive awards under this program may carry out either research or development activities, and applications for each type of project will be reviewed separately. The applicant is responsible for indicating which type of project is being proposed. Projects may last up to three years. Annual deadline dates may vary; contact program staff for exact dates.
Requirements: Proposals will be accepted from any public or private agency or organization.
Contact: Donna Nangle; (202) 245-7462; fax: (202) 245-7323; email: Donna.Nangle@ed.gov
Internet: http://www.ed.gov/fund/grant/apply/nidrr/index.html
Sponsor: National Institute on Disability and Rehabilitation Research
550 12th Street SW, 6th Floor, Switzer Building
Washington, DC 20202-6510

NIEHS Applied Toxicological Research and Testing Grants **4038**
A small number of grants are supported to further the primary aims of the program to determine the toxicity and health hazards of chemicals by the best means available and to develop and validate better test methods. Toward the latter objective, the conventional two-year bioassays previously used to detect and assess carcinogenesis alone have been modified to assess other toxic responses in the same animals. Chemicals of interest are selected on the basis of recommendations from public and private sources including state and local government agencies, colleges and universities, trade unions, and industry. Testing priorities are established by an executive committee made up of personnel from key federal regulatory and research agencies. Testing of selected chemicals has been broadened and improved with new quality assessment standards, expanded screening for effects in various organs, on fertility and fetal development, and on distribution, metabolism, and excretion of the chemicals. Program emphases include development of reliable test methods, improvement of existing bioassay systems, and development of model test systems for xenobiotics.
Requirements: A university, college, hospital, state or local government, nonprofit research institution, or for-profit organization may submit an application and receive a grant for support or research by a named principal investigator.
Amount of Grant: $141,500-$1.6 million; $332,897 average
Date(s) Application is Due: Feb 1; Jun 1; Oct 1
Contact: Dr. William Suk, Chief, Chemical Exposures and Molecular Biology Branch, (919) 541-0797; email: suk@niehs.nih.gov
Internet: http://12.46.245.173/pls/portal30/CATALOG.PROGRAM_TEXT_RPT.show
Sponsor: National Institute of Environmental Health Sciences
P.O. Box 12233
Research Triangle Park, NC 27709

NIEHS Biological Response to Environmental Health Hazards Research Grants **4039**
The aim of this program is to develop a better understanding of the mechanisms of injury and disease caused by environmental factors. Research focuses on perturbations by chemical and physical agents at the molecular, cellular, tissue, and organ levels. This is a broad area of inquiry that integrates the traditional disciplines of anatomy, physiology, cell biology, endocrinology, genetics, biochemistry, and reproduction. Problem areas include, but are not limited to, neurological and behavioral disorders, respiratory diseases, cardiovascular disease, biochemical and endocrine disturbances, kidney dysfunction, carcinogenesis, mutagenesis, reproductive disorders, and birth defects.
Requirements: A university, college, hospital, state or local government, nonprofit research institution, or for-profit organization may submit an application and receive a grant for support of research by a named principal investigator.
Amount of Grant: $38,500-$2.03 million; $350,490 average
Date(s) Application is Due: Feb 1; Jun 1; Oct 1
Contact: Dr. William Suk, Director, Office of Program Development, (919) 541-0797; email: suk@niehs.nih.gov
Internet: http://12.46.245.173/pls/portal30/CATALOG.PROGRAM_TEXT_RPT.show
Sponsor: National Institute of Environmental Health Sciences
P.O. Box 12233
Research Triangle Park, NC 27709

NIEHS Carcinogenicity of Drinking Water Disinfection By-Products **4040**
NIEHS, in collaboration with EPA, is conducting short-term and long-term toxicity studies of the disinfection by-products found in drinking water, on traditional rodent models with additional studies in genetically modified (transgenic) mice. Emphasis should be on whether lower concentrations, which may not cause cancer in rodents, can be either expected to pose a human health hazard or may be considered of minimal risk for the human population, including sensitive subpopulations. Annual deadline dates may vary; contact program staff for exact dates. RFA: ES-99-007
Requirements: Applications may be submitted by domestic, for-profit and nonprofit organizations, public and private, such as universities, colleges, hospitals, laboratories, units of state or local governments and eligible agencies of the federal government. Applications from minority individuals and women are encouraged.
Restrictions: Foreign institutions and organizations are not eligible.
Amount of Grant: $450,000 total
Contact: Dr. Jerrold Heindel, (919) 541-0781; fax: (919) 541-5064; email: heindelj@niehs.nih.gov
Internet: http://grants.nih.gov/grants/guide/rfa-files/RFA-ES-99-007.html
Sponsor: National Institute of Environmental Health Sciences
P.O. Box 12233
Research Triangle Park, NC 27709

NIEHS Community-Based Prevention Intervention Research in Environmental Health Sciences Grants **4041**
NIEHS, NINR, and the Office of Behavioral and Social Sciences invite research grant applications addressing development of community-based strategies aimed at prevention and intervention activities in economically disadvantaged and/or underserved populations adversely impacted by an environmental contaminant. These funds support research activities that develop and implement improved prevention and intervention strategies related to underserved populations. In community-based research, active cooperation and participation of organizations within the community that is the focus of the study are essential components of the research. Hence, applicants must describe an existing or proposed involvement with one or more community-based organizations. These grants provide support to develop, test, and evaluate health service activities and to enhance the application or transfer of existing knowledge to efforts designed to prevent or control dysfunction or disease. Annual deadline dates may vary; contact program staff for exact dates. More information may be obtained by mail and email, or through the NIH grant line (301-402-2221), and the NIH Web site. RFA: ES-99-012
Requirements: Applications may be submitted by domestic and foreign, for-profit and nonprofit organizations, both public and private; units of state and local governments; and eligible agencies of the federal government.
Amount of Grant: $300,000 maximum
Contact: Frederick Tyson, Chemical Exposures and Molecular Biology Branch, (919) 541-0176; fax: (919) 316-4606; email: tyson2@niehs.nih.gov
Internet: http://grants.nih.gov/grants/guide/rfa-files/RFA-ES-99-012.html
Sponsor: National Institute of Environmental Health Sciences
P.O. Box 12233
Research Triangle Park, NC 27709

NIEHS Environmental Justice: Partnerships for Communication **4042**
The purpose of this program is to strengthen the NIEHS effort that supports research aimed at achieving environmental equity for socioeconomically disadvantaged and medically underserved populations in the United States. One goal of the institute is to stimulate investigative efforts that attempt to address questions related to the influence of economic and social factors on the health status of individuals exposed to environmental toxicants. The main objective of this announcement is to establish a new model for linking members of a community who are directly affected by adverse environmental conditions with researchers and health care providers. The mechanism of support for this program will be the NIH Education Grant. The total project period for applications submitted in response to the announcement may not exceed four years, and projects are not renewable. Annual deadline dates may vary; contact the program office for exact dates. RFA: ES-99-005
Requirements: Applications may be submitted by domestic for-profit, nonprofit, public, and private institutions.
Amount of Grant: $200,000 maximum per year
Contact: Dr. Frederick Tyson, Chemical Exposures and Molecular Biology Branch, (919) 541-0176; fax: (919) 541-4937; email: tyson2@niehs.nih.gov
Internet: http://www.grants.nih.gov/grants/guide/rfa-files/RFA-ES-99-005.html
Sponsor: National Institute of Environmental Health Sciences
P.O. Box 12233
Research Triangle Park, NC 27709

NIEHS Hazardous Materials Worker Health and Safety Training Grants **4043**
This FOA issued by the National Institute of Environmental Health Sciences (NIEHS) invites applications for cooperative agreements to support the development of model programs for the training and education of workers engaged in activities related to hazardous materials and waSuite generation, removal, containment, transportation and emergency response. The major objective of this solicitation is to prevent work-related harm by assisting in the training of workers in how best to protect themselves and their communities from exposure to hazardous materials encountered during hazardous waSuite operations, hazardous materials transportation, environmental restoration of contaminated facilities or chemical emergency response. A variety of sites, such as those involved with chemical waSuite clean up and remedial action and transportation-related chemical emergency response may pose severe health and safety concerns to workers and the surrounding communities. These sites contain a multiplicity of hazardous substances, sometimes unknown substances, and often the site is uncontrolled. A major goal of the Worker Education and Training Program (WETP) is to provide assistance to organizations in developing their institutional competency to provide appropriate model training and education programs. The total project period for an application submitted in response to this funding opportunity must be 5 years. A new applicant may request a budget for direct costs of up to $700,000 for the first year. Letters of Intent (LOIs) are due by October 23, with full proposals due by November 23.

Requirements: The following organizations and institutions are eligible to apply: public/state controlled institutions of higher education; private institutions of higher education; Hispanic-serving institutions; historically Black colleges and universities (HBCUs); tribally controlled colleges and universities (TCCUs); Alaska Native and Native Hawaiian serving institutions; and nonprofits with 501(c)3 IRS status (other than institutions of higher education).
Amount of Grant: $700,000 maximum for the first year
Date(s) Application is Due: Oct 23; Nov 23
Contact: Joseph Hughes, Director; (919) 541-0217; fax: (919) 541-0462; email: hughes3@niehs.nih.gov
Internet: http://grants.nih.gov/grants/guide/rfa-files/RFA-ES-09-004.html
Sponsor: National Institute of Environmental Health Sciences (NIEHS)
Keystone Building, Room 3039, P.O. Box 12233, MD K3-14
Research Triangle Park, NC 27709-2233

NIEHS Hazmat Training at Doe Nuclear Weapons Complex Grants **4044**
The National Institute of Environmental Health Sciences (NIEHS) invites applications for cooperative agreements to support the development of model programs for the training and education of workers engaged in activities related to hazardous materials and waSuite generation, removal, containment, transportation and emergency response within the Department of Energy (DOE) Nuclear Weapons Complex. The major objective of this solicitation is to prevent work related harm by assisting in the training and education of workers in the DOE nuclear weapons complex. Safety and health training will transmit skills and knowledge to workers in how best to protect themselves and their communities from exposure to hazardous materials encountered during hazardous waSuite operations, facility decommissioning and decontamination, hazardous materials transportation, environmental restoration of contaminated facilities or chemical emergency response. The total project period for an application submitted in response to this funding opportunity must be 5 years. A new applicant may request a budget for direct costs of up to $700,000 for the first year. Letters of Intent (LOIs) are due by October 23, with full proposals due by November 23.
Requirements: The following organizations and institutions are eligible to apply: public/state controlled institutions of higher education; private institutions of higher education; Hispanic-serving institutions; historically Black colleges and universities (HBCUs); tribally controlled colleges and universities (TCCUs); Alaska Native and Native Hawaiian serving institutions; and nonprofits with 501(c)3 IRS status (other than institutions of higher education).
Amount of Grant: $700,000 maximum for the first year
Date(s) Application is Due: Oct 23; Nov 23
Contact: Joseph Hughes, Director; (919) 541-0217; fax: (919) 541-0462; email: hughes3@niehs.nih.gov
Internet: http://grants.nih.gov/grants/guide/rfa-files/RFA-ES-09-003.html
Sponsor: National Institute of Environmental Health Sciences (NIEHS)
Keystone Building, Room 3039, P.O. Box 12233, MD K3-14
Research Triangle Park, NC 27709-2233

NIEHS Independent Scientist Award **4045**
The award is a special salary-only grant designed to provide protected time for newly independent scientists who currently have non-research obligations such as heavy teaching loads, clinical work, committee assignments, service, and administrative duties that prevent them from having a period of intensive research focus. The award is targeted to persons with doctoral degrees who have completed their research training, have independent peer-reviewed research support, and who need a period of protected research time in order to foster their research career development. Strong preference is given to candidates with active research project grant support from the NIEHS. New applications are due February 12, June 12, and October 12, while renewal applications are due March 12, July 12, and November 12. Up to $75,000 per year of base salary plus fringe benefits is available. Research support of up to $25,000 is provided in some cases. Individuals engaged in predominantly theoretical work, such as modeling or computer simulation, may request up to $25,000 per year for research expenses.
Requirements: The following organizations and institutions are eligible to apply: public/state controlled institutions of higher education; private institutions of higher education; nonprofits with 501(c)3 IRS status; nonprofits without 501(c)3 IRS status; small businesses; for-profit organizations; State governments; U.S. territories or possessions; Indian/Native American tribal governments (Federally recognized and other than Federally recognized); Indian/Native American tribally designated organizations; Hispanic-serving institutions; historically Black colleges and universities (HBCUs); tribally controlled colleges and universities (TCCUs); Alaska Native and Native Hawaiian serving institutions; regional organizations; and faith-based or community based organizations. The candidate must have a doctoral degree and peer-reviewed, independent research support at the time the award is made. The candidate must spend a minimum of 75 percent effort conducting research during the period of the award.
Restrictions: The award is not intended for investigators who already have full time to perform research, or have substantial publication records or considerable research support indicating that they are well established in their fields. Foreign institutions are not eligible to apply.
Contact: Carol Shreffler; (919) 541-1445; fax: (919) 541-5064; email: shreffl1@niehs nih.gov
Internet: http://grants.nih.gov/grants/guide/pa-files/PA-09-038.html
Sponsor: National Institute of Environmental Health Sciences (NIEHS)
P.O. Box 12233
Research Triangle Park, NC 27709

NIEHS Mentored Clinical Scientist Research Career Development Awards **4046**
This award enables candidates holding professional degrees (e.g., M.D., D.O., D.V.M., or equivalent degrees) to undertake 3 to 5 years of special study and supervised research with the goal of becoming independent investigators. The award also allows awardees to pursue a research career development program suited to their experience and capabilities under a mentor who is competent to provide guidance in the chosen research area. Institutions may submit applications on behalf of candidates who hold professional degrees. At least 2 years must have elapsed since the health professional degree was granted. Candidates can have varying levels of clinical training and research experience. New applications are due February 12, June 12, and October 12, while resubmitted applications are due March 12, July 12, and November 12. Up to $75,000 per year plus fringe benefits is available, and up to $25,000 per year for research support for new awards.
Requirements: All candidates must be U.S. citizens, non-citizen nationals, or legal permanent residents of the U.S. Persons with temporary or student visas are not eligible. The grantee institution must have a strong, well-established research and research training program in the chosen area, accomplished faculty in the basic and clinical sciences, and a commitment to the candidate's research development. The proposed program should include an appropriate mentor.
Amount of Grant: $100,000 maximum
Date(s) Application is Due: Feb 12; Mar 12; Jun 12; Jul 12; Oct 12; Nov 12
Contact: Dr. Carol Shreffler; (919) 541-1445; fax: (919) 541-5064; email: shreffl1@niehs.nih.gov
Internet: http://grants.nih.gov/grants/guide/contacts/pa-06-512_contacts.htm
Sponsor: National Institute of Environmental Health Sciences (NIEHS)
P.O. Box 12233
Research Triangle Park, NC 27709

NIEHS Mentored Research Scientist Development Award **4047**
The MRSDA is for a research scientist who needs an additional period of sponsored research experience as a way to gain expertise in a research area new to the candidate or in an area that would demonstrably enhance the candidate's scientific career. Research projects proposed in response to this Announcement will be expected to have a defined impact on the environmental health sciences and be directly responsive to the mission of the NIEHS, which is distinguished from that of other Institutes by its focus on research programs seeking to link the effects of environmental exposures to the cause, mechanisms, moderation, or prevention of a human disease or disorder or relevant pathophysiologic process. Applications must include as part of the justification the specific human disease, dysfunction, pathophysiologic condition, or relevant human biologic process altered, and the specific environmentally relevant toxicant. Examples of environmentally relevant toxicants include industrial chemicals or manufacturing byproducts, metals, pesticides, herbicides, air pollutants and other inhaled toxicants, particulates or fibers, fungal, bacterial, or biologically derived toxins. Agents considered non-responsive to this announcement include, but are not limited to: alcohol, chemotherapeutic agents, radiation which is not the result of an ambient environmental exposure, drugs of abuse, and infectious or parasitic agents. Ecologic, biomonitoring, biotransformation or biodegradation studies are also not responsive. Applicants must demonstrate a significant career shift or career enhancement activity which will achieve research independence upon completion of the program. Preference will be given to candidates whose mentor has NEIHS research grant support.
Requirements: The candidate must have a research or health-professional doctorate or its equivalent and must have demonstrated the capacity or potential for highly productive independent research in the period after the doctorate. The candidate must identify a mentor with extensive research experience and must spend a minimum of 75 percent of full-time professional effort conducting research for the period of the award. Applications may be submitted on behalf of candidates by domestic, nonfederal organizations, public or private, such as medical, dental, or nursing schools or other institutions of higher education. Candidates must be U.S. citizens, noncitizen nationals, or permanent residents.
Amount of Grant: $75,000 plus fringe benefits
Date(s) Application is Due: Feb 12; Mar 12; Jun 12; Jul 12; Oct 12; Nov 12
Contact: Carol Shreffler; (919) 541-1445; fax: (919) 541-5064; email: shreffl@niehs.nih.gov
Internet: http://grants.nih.gov/grants/guide/pa-files/PA-09-040.html
Sponsor: National Institute of Environmental Health Sciences (NIEHS)
P.O. Box 12233
Research Triangle Park, NC 27709

NIEHS Small Business Innovation Research (SBIR) Program Grants **4048**
The focus of these grants is on understanding how chemical and physical agents cause pathological changes in molecules, cells, tissues, and organs and become manifested as respiratory disease; neurological, behavioral, and developmental abnormalities; cancer; and other disorders. By understanding the relationship between environmental exposures and the subsequent development of disease or biological injury, human health may be protected. SBIR program objectives are to stimulate technological innovation, use small businesses to meet federal research and development needs, increase private sector commercialization of innovations derived from federal research and development, and foster and encourage participation by minority and disadvantaged persons in technological innovation. Phase I awards, generally for six months, establish the technical merit and feasibility of proposed research efforts that may lead to commercial products or processes. Phase II awards are for the continuation of research initiated in Phase I.
Requirements: SBIR grants can be awarded only to domestic small businesses (entities that are independently owned and operated for profit, are not dominant in the field in which research is proposed, and have no more than 500 employees). Primary employment

(more than one-half) of the principal investigator must be with the small business at the time of award and during the conducting of the proposed research. In both Phase I and II, the research must be performed in the United States.
Amount of Grant: $100,000 Phase I grants; $750,000 maximum Phase II grants
Date(s) Application is Due: Apr 15; Aug 15; Dec 15
Contact: Jo Anne Goodnight, SBIR/STTR, (301) 435-2688; fax: (301) 480-0146; email: jg128w@nih.gov or sbir@od.nih.gov
Internet: http://grants.nih.gov/grants/funding/sbir.htm#sbir
Sponsor: National Institute of Environmental Health Sciences
P.O. Box 12233
Research Triangle Park, NC 27709

NIEHS Structure-Function Relationships of Environmentally Relevant Genetic Variants Research Grants **4049**
The objective of this RFA is to encourage multidisciplinary projects to identify and characterize the gene products coded by allelic variants of environmental responsive genes. NIEHS is the principal funding agency that supports this RFA and NIA is the cosponsor. Data generated from the studies will allow for rapid developments of innovative approaches to elucidate the mechanisms of the xenobiotic-induced molecular alterations in specific target cells and organs. Collaborative research efforts between investigators in environmental health sciences and in molecular and structural biology, and/or molecular modeling disciplines are especially encouraged. The major research objectives for this RFA include: biochemical and structural biological studies of the proteins coded by the allelic variants of candidate genes; evaluation of the functional significance of single nucleotide polymorphisms or other allelic variations in candidate genes and their regulatory regions; characterization of the molecular mechanisms controlling temporal and cellular regulation of expression of the allelic variants of a specific class of genes; functional genomic studies to assess the expression patterns and functions of these classes of environmental responsive genes at the organ/tissue/cell levels; and studies on protein-protein interactions. RFA: ES-98-007
Requirements: Applications may be submitted by domestic and foreign, for-profit and nonprofit organizations, both public and private; units of state or local governments; and eligible agencies of the federal government. Applications from minority individuals and women are encouraged.
Amount of Grant: $3.5 million
Contact: Dr. Jose Velazquez, Division of Extramural Research and Training, (919) 541-4998; fax: (919) 541-4937; email: velazqu1@niehs.nih.gov
Internet: http://www.grants.nih.gov/grants/guide/rfa-files/RFA-ES-98-007.html
Sponsor: National Institute of Environmental Health Sciences
P.O. Box 12233
Research Triangle Park, NC 27709

NIEHS Superfund Hazardous Substances Basic Research and Education Grants **4050**
Grants support a wide range of research to address the broad public health concerns arising from the release of hazardous chemical wastes and other hazardous substances into the environment from waste-disposal sites. The primary objective is to expand the base of scientific knowledge needed for adequate assessment of exposure and health risks from environmental contamination, reduction in the amount and toxicity of hazardous substances, and ultimately to prevent adverse human health effects. Grants fund university-based research programs to develop advanced techniques for the detection, assessment, and evaluation of the effects of hazardous substances on human health; methods to assess the risks to human health presented by hazardous substances; methods and technologies to detect hazardous substances in the environment; and basic biological, chemical, and physical methods to reduce the amount and toxicity of hazardous substances. Advanced or graduate training should be integrated into the second year of the multidisciplinary research program to provide for training in environmental and occupational health and safety and the engineering aspects of hazardous wastes control, and graduate training in the geosciences including hydrogeology, geological engineering, geophysics, geochemistry, and related fields. Contact the program office for deadline dates. RFA: ES-99-001
Requirements: An accredited institution of higher education may submit an application and receive a grant for support of research by a named principal investigator. Subcontracts may be made with public and private organizations including generators of hazardous wastes; persons involved in the detection, assessment, evaluation, and treatment of hazardous substances; owners and operators of facilities at which hazardous substances are located; and state and local governments. Strong biomedical research is a requisite of this program. A minimum of three approved biomedical projects is required.
Restrictions: Single-project applications will not be accepted. This program is not intended to support individual research project grants.
Amount of Grant: $182,787$3.81 million; $1.31 million average
Contact: Dr. William Suk, Director, Superfund Hazardous Substance Basic Research and Training, (919) 541-0797; fax: (919) 541-2843; email: suk@niehs.nih.gov;
Internet: http://www.grants.nih.gov/grants/guide/rfa-files/RFA-ES-99-001.html
Sponsor: National Institute of Environmental Health Sciences
P.O. Box 12233
Research Triangle Park, NC 27709

NIEHS/NINDS Role of the Environment in Parkinson's Disease **4051**
The objective of this program is to stimulate research on the relative roles of environmental, endogenous neurochemical and modifying genetic factors in the cause of Parkinson's disease. Annual deadline dates may vary; contact program staff for exact dates.
Requirements: Applications may be submitted by domestic and foreign, for-profit and nonprofit organizations, public and private, such as universities, colleges, hospitals, laboratories, units of state and local governments, and eligible agencies of the federal government.
Amount of Grant: $4 million total
Contact: Dr. Annette Kirshner, Division of Extramural Research and Training, (919) 541-0488; fax: (919) 541-5064; email: kirshner@niehs.nih.gov; Dr. Eugene Oliver, Division of Stroke, Trauma and Neurodegenerative Disorders, (301) 496-5680; fax: (301) 480-1080; e
Internet: http://grants.nih.gov/grants/guide/rfa-files/RFA-ES-00-002.html
Sponsor: National Institute of Environmental Health Sciences
P.O. Box 12233
Research Triangle Park, NC 27709

Nieman Fellowships for Journalists at Harvard University **4052**
The foundation annually awards 10-month academic fellowships to approximately 12 midcareer journalists from the United States and approximately 10 to 12 journalists from abroad. Fellows design their own courses of study, and all parts of Harvard University are open to them. Academic credit or degree is not granted. U.S. journalists receive stipends for living expenses. Foreign applicants must obtain funding by successfully competing for restricted grants available to the Nieman Foundation or by securing their own financial backing from outside sources.
Requirements: Applicants must be full-time staff or freelance journalists working for the news or editorial department of newspapers, news services, radio, television, or magazines of general public interest; must have had at least three years of professional work in the media; and must obtain their employers' consent for leaves of absence for the academic year, September to June.
Amount of Grant: $55,000 stipend
Date(s) Application is Due: Jan 31; Mar 1
Contact: Rose Moss, Program Officer, (617) 495-2237; fax: (617) 495-8976; email: rosemoss@fas.harvard.edu or nieman@harvard.edu
Internet: http://www.nieman.harvard.edu
Sponsor: Nieman Foundation
One Francis Avenue, Harvard University
Cambridge, MA 02138

NIGMS Exploratory Studies for High Risk/High Impact Research Grants **4053**
This program attempts to broaden the base of inquiry in fundamental biomedical research by encouraging applications for research projects that involve an especially high degree of innovation and novelty, such that their potential for highly significant outcomes may be difficult to judge by the standard criteria used in evaluating R01 proposals. Research projects proposed under this program may lack preliminary data establishing feasibility, but should present the opportunity for conceptual or technological breakthroughs. PA-03-100 may be obtained electronically through the NIH grant line (301-402-2221), the NIH Web site, and by mail and email.
Amount of Grant: $275,000 maximum for two years
Contact: Joe Ellis, Grants Management Officer, (301) 594-5135; fax: (301) 480-1969; email: ellisj@nigms.nih.gov; Dr. James Cassatt, Division of Cell Biology and Biophysics, (301) 594-0828; fax: (301) 480-2004; email: cassattj@nigms.nih.gov
Internet: http://grants2.nih.gov/grants/guide/pa-files/PA-03-100.html
Sponsor: National Institute of General Medical Sciences
45 Center Drive, Natcher Building
Bethesda, MD 20892

NIGMS Fellowships in Quantitative Biology **4054**
The NIGMS encourages applications for individual postdoctoral and senior fellowships from highly qualified individuals with doctoral training in both the traditional quantitative disciplines (such as mathematics, physics, engineering, and computer science) and biology. NIGMS supports individual postdoctoral and senior fellowships in those areas covered by its research grant portfolio. These areas include biomedical engineering, cell biology, molecular biophysics, genetics and developmental biology, pharmacology, physiology, especially burn and trauma research, and biological chemistry. PA 98-082
Requirements: Individuals must be citizens or noncitizen nationals of the United States, or have been lawfully admitted to the United States for permanent residence (i.e., possess a currently valid Alien Registration Receipt Card I-551, or other legal verification of such status) at the time of application. Applicants for individual postdoctoral fellowships must have completed all of the requirements for the doctoral degree by the time of award. Applicants for senior fellowships must have had at least seven years of relevant research or professional experience beyond the doctoral degree. In addition, a significant amount of retraining is required for senior fellowship applicants. Applications from women, underrepresented minorities, and individuals with disabilities are especially encouraged.
Restrictions: Individuals on temporary or student visas are not eligible.
Amount of Grant: $21,000-$33,012 stipend; $4000 allowance per 12-month period for sponsoring institutions (nonfederal and nonprofit organizations)
Date(s) Application is Due: Apr 5; Aug 5; Dec 5
Contact: Dr. James Cassatt, Division of Cell Biology and Biophysics, (301) 594-0828; fax: (301) 402-2004; email: cassattj@nigms.nih.gov; Dr. James Anderson, Division of Genetics and Developmental Biology, (301) 594-0943; fax: (301) 480-2228; email: andersoj&,ãä
Internet: http://www.grants.nih.gov/grants/guide/pa-files/PA-98-082.html
Sponsor: National Institute of General Medical Sciences
45 Center Drive, Natcher Building
Bethesda, MD 20892

NIGMS Individual Postdoctoral NRSA Fellowships in Clinical Pharmacology 4055
Fellowships support high-quality physicians seeking research training in the field of clinical pharmacology. These fellowships are intended to support up to three years of advanced training to individuals with the MD degree to qualify them to perform basic laboratory and/or clinical investigations of the effects and mechanisms of drug actions in humans, in the scientific areas consistent with the mission of NIGMS. One goal of this initiative is to promote high-quality clinical pharmacology research training at academic health centers. Therefore, a special feature of this award is to encourage applicants at institutions with a General Clinical Research Center (GCRC) supported by the NCRR to submit a letter of endorsement from the director of the training institution's GCRC. The intent is to promote interaction between the applicant and the staff and director of the GCRC, expose the applicant to the extensive resources and rich research environment of the GCRC, and raise the visibility of this important area of translational research within the institution. PA: PA-99-078
Requirements: The candidate must have an MD degree or its equivalent and must meet the criteria for the NIH National Research Service Awards for Individual Postdoctoral Fellows (F32) described in program announcement PA-99-025, available on the Web site at http://www.nih.gov/grants/guide/pa-files/PA-99-025.html.
Contact: Dr. Alison Cole, Division of Pharmacology, Physiology, and Biological Chemistry, (301) 594-1826; fax: (301) 480-2802; email: colea@nigms.nih.gov
Internet: http://www.grants.nih.gov/grants/guide/pa-files/PA-99-078.html
Sponsor: National Institute of General Medical Sciences
45 Center Drive, Natcher Building
Bethesda, MD 20892

NIGMS Initiative for Minority Students: Bridges to the Baccalaureate Degree 4056
NIGMS and the Office of Research on Minority Health (ORMH) reannounce two research initiatives directed at increasing the number of underrepresented minorities entering careers in biomedical research. The programs target two different underrepresented minority student populations: those in colleges and universities offering only master of science (MS) degree programs in biomedically-related sciences and those in two-year junior or community colleges. These have been identified as two key transition points for students considering careers in biomedical research. This RFA solicits new applications for a partnership program involving two-year colleges awarding the associate's degree and institutions awarding the baccalaureate degree. Annual deadline dates may change; contact program staff for exact dates. RFA: GM-01-002
Requirements: Applications may be submitted by domestic, private or public, educational institutions and state or local systems of higher education.
Amount of Grant: $600,000 maximum
Contact: Dr. Irene Eckstrand, (301) 594-5402; fax: (301) 480-2228; email: eckstrai@nigms.nih.gov
Internet: http://grants.nih.gov/grants/guide/rfa-files/RFA-GM-01-002.html
Sponsor: National Institute of General Medical Sciences
45 Center Drive, Natcher Building
Bethesda, MD 20892

NIGMS Initiative for Minority Students: Bridges to the Doctoral Degree 4057
The program develops partnerships between educational institutions to support underrepresented minorities in the transition to doctoral programs. Besides reflecting the plans and priorities of the individual institutions, the partnership must address the needs of the minority graduate students. Support activities the institutions may implement include, but are not limited to, enhancing the curriculum of the master's institution; establishing a mentor program for master's students with faculty at the doctoral institution; enabling and encouraging students from either institution to take classes at the other institution; and providing research opportunities for master's students at the doctoral institution or in private industrial laboratories. Annual deadline dates may vary; contact program staff for exact dates. RFA: GM-01-003
Requirements: Partnerships involve a minimum of two educational institutions. One institution must have a significant enrollment of underrepresented minorities, and the highest postgraduate degree offered in the related sciences must be the master's degree. The other participating institution must be a research university and offer doctoral degrees in biomedical relevant areas. Applications also may represent several institutions within a single state, with one institution designated as the applicant institution.
Amount of Grant: $20 million total
Contact: Dr. Irene Eckstrand, (301) 594-5402; fax: (301) 480-2228; email: eckstrai@nigms.nih.gov
Internet: http://grants.nih.gov/grants/guide/rfa-files/RFA-GM-01-003.html
Sponsor: National Institute of General Medical Sciences
45 Center Drive, Natcher Building
Bethesda, MD 20892

NIGMS Institutional Research and Academic Career Development Award 4058
The goal of this program is to prepare qualified individuals for careers that significantly impact the nation's research agenda. The objectives of this initiative are: to enhance research-oriented teaching at minority serving institutions (MSIs); to increase the research and other skills needed by developing scientists to conduct high quality research in an academic environment; and to promote linkages between research-intensive institutions and MSIs that can lead to further collaborations in research and teaching; and to increase the number of well-qualified underrepresented minority students entering competitive careers in biomedical research. Grant funds may be used to support the following expenses: salary/wages and benefits for a full-time, 12-month staff appointment; tuition, fees, and books; research expenses; supplies; travel to one training or scientific meeting per year; and statistical services. PAR-98-085.
Requirements: An application must be submitted on behalf of a consortium of domestic private or public educational institutions; the applicant institution should be the primary site of the postdoctoral research experience. An institution may not submit more than one application. A consortium must include one or more partners that are MSIs. Candidates appointed under this program award must be U.S. citizens or noncitizen nationals, or have been lawfully admitted for permanent residence and possess Alien Registration Receipt Cards (I-151 or I-551) or some other verification of legal status as permanent residents. Postdoctoral candidates must have received, as of the beginning date of the appointment to the program, the PhD, MD, or comparable doctoral degree from an accredited domestic or foreign institution.
Restrictions: Individuals on temporary or student visas are not eligible.
Contact: Dr. Clifton Poodry, Division of Minority Opportunities in Research, (301) 594-3900; fax: (301) 480-2753; email: poodryc@nigms.nih.gov
Internet: http://www.grants.nih.gov/grants/guide/pa-files/PAR-98-085.html
Sponsor: National Institute of General Medical Sciences
45 Center Drive, Natcher Building
Bethesda, MD 20892

NIGMS Integrative and Collaborative Approaches to Research Grants 4059
The purpose of this PA is to encourage collaborative and integrative approaches to research on multifaceted biological problems. While the traditional R01 research project grant remains the backbone of NIH and NIGMS support for biomedical science, individual R01 grants may be inadequate to support larger efforts that require collaborative approaches. In some cases, individual investigators with existing R01 or R37 support, or other sources of independent funding, will need to attract and coordinate expertise in different disciplines and approaches and may also require access to specialized resources, such as computational facilities, high-throughput technologies, and equipment. In order to provide a new mechanism to support such collaborative and interactive activities, this initiative will use the R24 consortium grant mechanism, designed for groups of currently funded investigators. This PA also strongly encourages consideration of existing and previously announced mechanisms whenever appropriate, to enhance collaborative and integrative approaches to research problems. These include administrative supplements to NIGMS-funded research grants, competitive supplements to NIGMS-funded research grants, individual R01 grants containing a consortium arrangement for a collaborative component, program project (P01) and center (P50) grants, and Interactive Research Project Grants (IRPG). PA: PA-00-099
Requirements: Applications may be submitted by domestic and foreign, for-profit and nonprofit organizations, both public and private; units of state and local governments; and eligible agencies of the federal government.
Amount of Grant: $300,000 maximum direct costs per year
Contact: Dr. Marion Zatz, Division of Genetics and Developmental Biology, (301) 594-0943; fax: (301) 480-2228; email: zatzm@nigms.nih.gov
Internet: http://grants.nih.gov/grants/guide/pa-files/PA-00-099.html
Sponsor: National Institute of General Medical Sciences
45 Center Drive, Natcher Building
Bethesda, MD 20892

NIGMS Large-Scale Collaborative Project Awards 4060
The purpose of this program is to enable the solution of major problems in biomedical research and to facilitate the next evolutionary stage of integrative biomedical science. The intention is to make resources available for independently funded scientists to form research teams to solve a complex biological problem that would be beyond the means of any one research group, and that is of central importance to biomedical science and to the mission of NIGMS. It is expected that the participating investigators will already hold externally peer reviewed and funded research grants in the area of the proposal, and, in general, support of new individual research projects will not be a part of these large-scale project awards. A high level of resources may be requested to allow participating investigators to extend their research efforts to form a consortium to approach a research problem of overarching importance in a comprehensive and highly integrated fashion. Grants for this initiative will be awarded in two phases. The purpose of the Phase I award is to provide resources for detailed planning to applicants who have demonstrated the selection of an appropriate complex biological problem, an innovative plan, and appropriate commitments to its solution from participating investigators and institutions. Phase I applicants will submit an overview of the proposed large-scale project for peer review. Successful Phase I applicants will receive a $25,000 (direct costs) planning grant, and those applicants who receive awards will be eligible to submit a more extensively planned and detailed application for a Phase II award to support the large-scale project itself. Annual deadline dates may vary; contact program staff for exact dates. RFA: GM-02-007
Requirements: Applications may be submitted by domestic, nonprofit organizations, both public and private; units of state and local governments; and eligible agencies of the federal government. Multi-institutional applications are encouraged. A collaborative project may include participating investigators from foreign and/or for-profit organizations. Participation by industry is also encouraged.
Contact: Dr. Michael Rogers, Division of Pharmacology, Physiology, and Biological Chemistry, (301) 594-3827; fax: (301) 480-2802; email: rogersm@nigms.nih.gov; Dr. James Cassatt, Division of Cell Biology and Biophysics, (301) 594-0828; fax: (301) 480-2004; e-mai
Internet: http://grants2.nih.gov/grants/guide/rfa-files/RFA-GM-02-007.html
Sponsor: National Institute of General Medical Sciences
45 Center Drive, Natcher Building
Bethesda, MD 20892

NIGMS MARC Undergraduate Student Training in Academic Research 4061
The Minority Access to Research Careers (MARC) Program Branch of the Division of Minority Opportunity in Research of NIGMS provides research training opportunities that target students and faculty from minority groups who are underrepresented in the biomedical sciences, including mathematics. As part of its ongoing commitment to the development of prospective underrepresented minority researchers and the enhancement of their training environment, the MARC Program provides support directed at underrepresented minority students to improve their preparation for graduate training in the biomedical sciences. PA: PAR-02-033
Requirements: Only domestic, nonprofit, private or public institutions which offer the baccalaureate degree and in which student enrollments are drawn substantially from ethnic minority groups that are underrepresented in the biomedical sciences, including mathematics, may apply. Only one grant per eligible institution will be awarded. The program is open to qualified undergraduate honors students majoring in the sciences with expressed interest in careers in biomedical research and intentions to pursue graduate education leading to the PhD, MD/PhD or other combined professional degree/PhD.
Contact: Dr. Adolphus Toliver, Division of Minority Opportunities in Research, (301) 594-3900; fax: (301) 480-2753; email: tolivera@nigms.nih.gov
Internet: http://grants2.nih.gov/grants/guide/pa-files/PAR-02-033.html
Sponsor: National Institute of General Medical Sciences
45 Center Drive, Natcher Building
Bethesda, MD 20892

NIGMS Mentored Clinical Scientist Research Career Development Awards 4062
This award enables candidates holding professional degrees (e.g., M.D., D.O., D.V.M., or equivalent degrees) to undertake 3 to 5 years of special study and supervised research with the goal of becoming independent investigators. The award also allows awardees to pursue a research career development program suited to their experience and capabilities under a mentor who is competent to provide guidance in the chosen research area. Institutions may submit applications on behalf of candidates who hold professional degrees. At least 2 years must have elapsed since the health professional degree was granted. Candidates can have varying levels of clinical training and research experience. New applications are due February 12, June 12, and October 12, while resubmitted applications are due March 12, July 12, and November 12. Up to $75,000 per year plus fringe benefits is available, and up to $20,000 per year for research support for new awards.
Requirements: The candidate must have an MD degree or its equivalent, must have completed postgraduate clinical training and have secured a faculty appointment in an appropriate research-intensive environment, must identify a mentor with extensive research experience, and must be willing to spend a minimum of 75 percent of full-time professional effort conducting research and research career development. Applications may be submitted on behalf of candidates by domestic, nonfederal organizations, public or private, such as medical schools. Minorities and women are encouraged to apply. Candidates must be U.S. citizens or noncitizen nationals, or must have been lawfully admitted for permanent residence and possess Alien Registration Receipt Card (I-551) or some other verification of legal admission as permanent residents.
Restrictions: Former principal investigators on NIH research project (R01), sub-projects of program project (P01), or center grants (P50), or the equivalent, are not eligible. A candidate for the award may not concurrently apply for any other PHS award that duplicates the provisions of this award nor have another application pending award. K08 recipients are encouraged to apply for independent research grant support during the period of this award. The K08 recipient would be allowed to maintain the award if other PHS support is procured, as long as the new support does not interfere with the ability to meet the MCSDA requirements.
Amount of Grant: $95,000 salary per year maximum
Date(s) Application is Due: Feb 12; Mar 12; Jun 12; Jul 12; Oct 12; Nov 12
Contact: Dr. Alison Cole, Anesthesiology Program; (301) 594-1826; fax: (301) 480-2802; email: colea@nigms.nih.gov
Internet: http://www.grants.nih.gov/grants/guide/pa-files/PAR-98-084.html
Sponsor: National Institute of General Medical Sciences (NIGMS)
45 Center Drive, Natcher Building, Room 2AS-49, MSC6200
Bethesda, MD 20892-6200

NIGMS Minority Access to Research Careers (MARC) Ancillary Training Activities Grants 4063
The MARC program provides research training opportunities for students and faculty from minority groups underrepresented in the biomedical and behavioral sciences relevant to biomedicine, including mathematics and engineering. Under the ancillary training initiative, the program provides support for ancillary research training activities that are harmonious with its overall objective of increasing the number of underrepresented minorities who participate fully in the biomedical enterprise of this country. Program-related scientific meetings, conferences, technical assistance workshops, and other ancillary training activities will be funded. Funds may be used for salaries, equipment, supplies, travel expenses, per diem costs, conference services, publication costs, registration fees, and speaker fees. Grant amounts are variable, based on the nature and scope of proposed activities. A letter of intent is requested by January 1, May 1, and September 1; full application is due February 1, June 1, and October 1. Annual deadline dates may vary; contact program officer for exact dates. PA: PAR-99-091
Requirements: Applications may be submitted by U.S. institutions, including scientific or professional societies, eligible to receive grants from the Department of Health and Human Services.
Restrictions: Foreign organizations are not eligible.
Date(s) Application is Due: Jan 1; Feb 1; May 1; Jun 1; Sep 1; Oct 1
Contact: Dr. Hinda Zlotnick, MARC Program Director, (301) 594-3900; fax: (301) 480-2573; email: hindaz@nigms.nih.gov
Internet: http://www.grants.nih.gov/grants/guide/pa-files/PAR-99-091.html
Sponsor: National Institute of General Medical Sciences
45 Center Drive, Natcher Building
Bethesda, MD 20892

NIGMS Minority Access to Research Careers (MARC) Program 4064
The program's goals are to increase the number and capabilities of scientists from underrepresented minority groups who are engaged in biomedical research and to strengthen science curricula and student research opportunities at institutions with substantial minority enrollments in order to prepare minority students for research careers. The program offers postdoctoral and predoctoral fellowships, and undergraduate student training. PA: PAR-03-140
Requirements: Applications are accepted from domestic private and public universities, and research institutions or centers with graduate programs that offer a solid research environment as evidenced by a high level of faculty involvement in biomedical and/or behavioral research. To be eligible for participation in the program as a PREP scholar, individuals from underrepresented minority groups must have graduated with a baccalaureate degree in a biomedically relevant science from an accredited U.S. college or university no more than 36 months prior to the date of submission of the application. PREP scholars must be U.S. citizens or permanent residents.
Restrictions: An applicant institution may submit only one application for this program announcement.
Contact: Dr. Adolphus Toliver, Chief, MARC Branch, (301) 594-3900; fax: (301) 480-8506; email: rm63f@nih.gov
Internet: http://grants.nih.gov/grants/guide/pa-files/PAR-03-140.html
Sponsor: National Institute of General Medical Sciences
45 Center Drive, Natcher Building
Bethesda, MD 20892

NIGMS Molecular Pharmacology of Anesthetic Action Research Grants 4065
NIGMS invites applications for the study of the molecular mechanisms of action of general and local anesthetics. The objective of this program is to encourage the submission of applications from individuals who have been trained either in basic or clinical research and who possess experience in research disciplines that are pertinent to answering questions germane to the understanding of anesthetic action such as the neurosciences, the pharmacological sciences, biophysics, biochemistry, and cell biology. Presumably, a wide variety of approaches and model systems could be employed to study anesthetic action at the molecular level. These model systems would include, but are not limited to, artificial membranes; ion channels, receptors, and other macromolecules in reconstituted membranes; electrophysiological measurements and biochemical determinations in isolated whole membrane systems; studies of whole cell responses to anesthetics; and actions of anesthetics on integrated systems such as brain slices and the spinal cord. Additionally, experimental designs utilizing recombinant DNA and molecular biological techniques, immunological approaches, and biophysical procedures whose aims are to elucidate structural features of anesthetic interactions with components of biological systems would also be of interest. There are no limits to the types of approaches that may be proposed to answer questions relevant to anesthetic action, so long as those approaches emphasize clarification of molecular aspects of anesthetic pharmacodynamics. Ideally, the ultimate goal of this research would be to develop hypotheses that might explain the relationship between molecular actions and physiological effects of general and local anesthetics.
Requirements: Research project awards can be made to any public or private, for-profit or nonprofit university, college, hospital, laboratory, or other institution, including state and local units of government, qualifying small businesses, and individuals. SBIR grants can be awarded only to domestic small businesses. STTR grants can be awarded only to domestic small business concerns that partner with research institutions in cooperative research and development.
Date(s) Application is Due: Feb 1; Jun 1; Oct 1
Contact: Dr. Alison Cole, (301) 594-1826; fax: (301) 480-2802; email: colea@gm1.nigms.nih.gov
Internet: http://www.grants.nih.gov/grants/guide/pa-files/PA-96-026.html
Sponsor: National Institute of General Medical Sciences
45 Center Drive, Natcher Building
Bethesda, MD 20892

NIGMS MORE Faculty Development Awards 4066
The MORE program provides support for eligible faculty to spend a summer or one academic term every year for two to five years in full-time research in a research-intensive laboratory. Contact the program office for deadline dates. PA: PAR-95-048
Date(s) Application is Due: Feb 1; Jun 1; Oct 1
Contact: Dr. Adolphus Toliver, Predoctoral NRSA Awards, (301) 594-3900; Carol Tippery, Grants Management Officer, (301) 594-5135
Internet: http://grants.nih.gov/grants/guide/pa-files/PAR-95-048.html
Sponsor: National Institute of General Medical Sciences
45 Center Drive, Natcher Building
Bethesda, MD 20892

NIGMS Pharmacogenomics Research Network Grants 4067
The purpose of this funding opportunity announcement is to allow new and renewal applications for research groups to compete to become part of the Pharmacogenomics Research Network (PGRN), and to create network resources that serve the entire PGRN.

The PGRN is a consortium of interdisciplinary groups with a core mission to conduct research into understanding the genetic basis of variable drug responses, both therapeutic and adverse. The scientific objectives of the research network include definition of the pressing problems for safe and effective use of medications, establishment of truly cross-disciplinary research groups, application of the most appropriate genetic and genomic methods to important sample collections, discovery and replication studies that identify variation in critical genes and biological pathways, elucidation of the functional consequences of genetic variation for the drugs being studied, and understanding the clinical implications of genotyping to predict drug responses. The expected direct costs amount for each individual research group award is $1 to $1.5 million per year, and for each add-on network resource is $0.1 to $0.5 million per year. Letters of Intent are due by May 2, with full proposals received by June 2.
Requirements: The following organizations and institutions are eligible to apply: public/state controlled institutions of higher education; private institutions of higher education; Hispanic-serving institutions; historically Black colleges and universities (HBCUs); nonprofits with 501(c)3 IRS status (other than institutions of higher education); nonprofits without 501(c)3 IRS status (other than institutions of higher education); small businesses; and for-profit organizations (other than small businesses).
Date(s) Application is Due: May 2; Jun 2
Contact: Rochelle M. Long, PhD; (301) 594-3827; email: rochelle.long@nih.gov; Richard A. Anderson, MD, PhD; (301) 594-0943; email: anderscr@nigms.nih.gov
Internet: http://grants.nih.gov/grants/guide/rfa-files/RFA-GM-10-001.html
Sponsor: National Institute of General Medical Sciences (NIGMS)
45 Center Drive, Building 45
Bethesda, MD 20892

NIGMS Pilot Projects for the Protein Structure Initiative 4068

This program funds research centers that serve as pilots to examine the best approach for developing subsequent integrated, large-scale research networks in structural genomics. This is part of the institute's Protein Structure Initiative (PSI), whose goal is the understanding of protein structural families, structural folds, and the relationship of structure and function. These research centers should include all the constituent tasks of structural genomics and should test strategies for large-scale high throughput structure determination by X-ray crystallography and/or NMR. A letter of intent is requested by November 3; full application is due February 12. Annual deadline dates may vary; contact program staff for exact dates. RFA: GM-00-006
Requirements: Applications may be submitted by domestic for-profit and nonprofit organizations, both public and private; units of state and local governments; and eligible agencies of the federal government.
Restrictions: Foreign institutions are not eligible for research center grants.
Amount of Grant: $3 million maximum
Date(s) Application is Due: Feb 12; Nov 3
Contact: Dr. John Norvell, Division of Cell Biology and Biophysics, (301) 594-0533; fax: (301) 480-2004; email: norvellj@nigms.nih.gov
Internet: http://grants.nih.gov/grants/guide/rfa-files/RFA-GM-00-006.html
Sponsor: National Institute of General Medical Sciences
45 Center Drive
Bethesda, MD 20892

NIGMS Predoctoral Training in Bioinformatics and Computation Biology Grants 4069

The purpose of this program is to develop a mechanism to train a cadre of scientists whose primary professional identification and disciplinary affiliation is bioinformatics or computational biology. For the purpose of this announcement, bioinformatics and computational biology are defined broadly to include the use of theory, computer implementation, and application to the full spectrum of basic research in the biomedical sciences. The terms thus include analysis of molecular sequence and structure, molecular function, cellular function, physiology, genomics, and genetics, as well as computational modeling of complex phenomena such as neural circuits and equilibrium phenomena, population biology, theoretical and mathematical biology, and the analysis of complex systems. The goal of this program is to train PhD students in the background theory, computational implementation, and biological application of information sciences (including computer science, statistics, mathematics and others), and to use this training to study problems relevant to biomedical research. PA: PAR-99-146
Requirements: Applications may be submitted by domestic public and private institutions with established programs leading to a PhD degree.
Amount of Grant: $14,688 annual stipend, $1500 total direct support costs, $300 per year for travel
Contact: Dr. James Cassatt, Director, Division of Cell Biology and Biophysics, (301) 594-0828; fax: (301) 402-2004; email: cassattj@nigms.nih.gov; Dr. Marion Zatz, Division of Genetics and Developmental Biology, (301) 594-0943; fax: (301) 480-2228; email: zatzm
Internet: http://grants.nih.gov/grants/guide/pa-files/PAR-99-146.html
Sponsor: National Institute of General Medical Sciences
45 Center Drive, Natcher Building
Bethesda, MD 20892

NIGMS Quantitative Approaches to the Analysis of Complex Biological Systems 4070

The purpose of this program announcement is to advertise NIGMS and NIMH interest in supporting research projects that develop quantitative approaches to describe, analyze, and predict the behavior of complex biological systems, especially those requiring the integration of potentially large amounts of molecular, biochemical, cell biological, and physiological data. Such studies, adapted to the analysis of complex systems in humans, will ultimately have an impact on the treatment of human disorders and disease. These projects are expected to require the participation of individuals with diverse expertise and therefore to be of a collaborative and cross disciplinary nature. Additional information for this program may be obtained by phone, fax, email, or the Internet. PA: PA-98-077
Requirements: Applications may be submitted by foreign and domestic, for-profit and nonprofit organizations, both public and private. Racial/ethnic minority individuals, women, and persons with disabilities are encouraged to apply.
Restrictions: Foreign institutions are not eligible for program project (P01) grants.
Contact: Dr. James Cassatt, Division of Cell Biology and Biophysics, (301) 594-0828; fax: (301) 480-2004; email: czj@cu.nih.gov; Dr. Judith Greenberg, Division of Genetics and Developmental Biology, (301) 594-0943; fax: (301) 480-2228; email: greenbej@nigms.ni
Internet: http://www.grants.nih.gov/grants/guide/pa-files/PA-98-077.html
Sponsor: National Institute of General Medical Sciences
45 Center Drive, Natcher Building
Bethesda, MD 20892

NIGMS Ruth L. Kirschstein National Research Service Awards for Individual Predoctoral Fellows in PharmD/PhD Grants 4071

NIGMS is committed to increasing the number of pharmacy students who wish to pursue doctoral research in the pharmaceutical sciences. It recognizes that current career paths for Doctor of Pharmacy graduates have reduced the number of individuals who pursue academic research training. NIGMS, intent in issuing this Funding Opportunity Announcement (FOA) is to provide individual research training fellowships to promising predoctoral applicants who are enrolled in PharmD/PhD programs in Schools of Pharmacy who desire to become productive and successful independent investigators whose research meets the mission of NIGMS. This fellowship program will provide support for the dissertation stage of predoctoral students enrolled in joint PharmD/PhD programs. It is expected that the length of time for an award will be 3 years or less.
Requirements: The following organizations and institutions are eligible to apply: for-profit organizations; non-profit organizations; domestic institutions; public or private institutions, such as universities, colleges, and hospitals; units of State and local governments; eligible agencies of the Federal government.
Restrictions: Individuals on temporary or student visas are not eligible.
Date(s) Application is Due: Apr 8; Aug 8; Dec 8
Contact: Richard Okita, Ph.D.; (301) 594-3827; fax: (301) 480-2802; email: okitar@nigms.nih.gov
Internet: http://grants.nih.gov/grants/guide/pa-files/PA-09-029.html
Sponsor: National Institute of General Medical Sciences (NIGMS)
45 Center Drive, 2AS.49A, MSC 6200
Bethesda, MD 20892-6200

NIGMS Short Courses on Mathematical and Statistical Tools for the Study of Complex Phenotypes and Complex Systems 4072

The purpose of this program is to provide support for short courses or workshops to assist scientists in preparing for research on complex phenotypes and complex systems. This program announcement solicits applications for education grants (R25), which provide for one to five years of support for short courses. Support is limited to courses that are highly focused and that reach a wide audience of scientists. The program announcement is not intended for university course or curriculum development. A variety of approaches will be considered. The nature of the activity, the needs of the target community, and the course goals should determine the instructional approach. These might include new approaches, such as Web-based instruction, or more traditional approaches, such as summer workshops or on-site instruction. PA: PA-98-083
Requirements: Applications may be submitted by domestic nonprofit and for-profit organizations, public and private.
Restrictions: Foreign organizations are not eligible.
Contact: Dr. Irene Anne Eckstrand, (301) 594-0943; fax: (301) 480-2228; email: Irene_Eckstrand@nih.gov
Internet: http://www.grants.nih.gov/grants/guide/pa-files/PA-98-083.html
Sponsor: National Institute of General Medical Sciences
45 Center Drive, Natcher Building
Bethesda, MD 20892

NIGMS Short Courses on Mathematical, Statistical, and Computational Tools for Studying Biological Systems 4073

The goal of this FOA is to prepare researchers at all professional levels for multidisciplinary research on complex biological systems and phenotypes by improving integration of mathematical, statistical, and computational approaches into biological and/or behavioral research. Progress in contemporary biological and behavioral disciplines depends heavily on investigators who are skilled in the use of mathematics, computation, and statistics (collectively referred to as quantitative methods). This FOA issued by the National Institute of General Medical Sciences, the National Center for Complementary and Alternative Medicine, the National Institute on Aging and the National Institute of Mental Health encourages applications for Research Education Grants (R25) from institutions and organizations to conduct workshops and short courses to improve integration of mathematical, statistical, and computational approaches into biological and/or behavioral research. Support will be limited to activities that reach a wide audience of researchers.
Requirements: Application can be made by the following institutions and organizations: public/state controlled institutions of higher education; private institutions of higher education; nonprofits with 501(c)3 IRS status (other than institutions of higher

education); nonprofits without 501(c)3 IRS status (other than institutions of higher education); small businesses; for-profit organizations (other than small businesses); state governments; U.S. territories or possessions; Indian/Native American tribal governments (federally recognized); Indian/Native American tribal governments (other than Federally recognized); Indian/Native American tribally designated organizations; non-domestic (non-U.S.) entities (foreign organizations); Hispanic-serving institutions; historically Black colleges and universities (HBCUs); tribally controlled colleges and universities (TCCUs); Alaska Native and Native Hawaiian serving institutions; regional organization; eligible agencies of the Federal government; and faith-based or community based organizations.
Restrictions: The program announcement is not intended for university course or curriculum development.
Amount of Grant: $200,000 maximum annually
Date(s) Application is Due: Jan 25; May 25; Sep 25
Contact: Matthew E. Portnoy, Ph.D.; (301) 594-0943; fax: (301) 480-2228; email: mportnoy@nigms.nih.gov
Internet: http://grants.nih.gov/grants/guide/pa-files/PA-09-002.html
Sponsor: National Institute of General Medical Sciences (NIGMS)
Building 45, 2AS.25P, 45 Center Drive
Bethesda, MD 20892-6200

NIGMS Supplements for the Study of Complex Biological Systems **4074**
The purpose of this initiative is to support new quantitative approaches to the study of complex, fundamental biological processes by encouraging nontraditional collaborations across disciplinary lines. NIGMS will provide supplements to existing NIGMS grants to support the salary and expenses of collaborating investigators such as physicists, engineers, mathematicians, and other experts with quantitative skills relevant to the analysis of complex systems. It is expected that the collaboration will result in new directions for the parent project, or new research projects that will compete for independent funding. PA-98-024 may be obtained through the NIH grant line (301-402-2221), the NIH Web site, mail, and email.
Requirements: Applications may be submitted by foreign, domestic, for-profit and nonprofit organizations, public or private.
Contact: Dr. James Cassatt, Division of Cell Biology and Biophysics, (301) 594-0828; fax: (301) 480-2004; email: czj@cu.nih.gov; Dr. Judith Greenberg, Division of Genetics and Developmental Biology, (301) 594-0943; fax: (301) 480-2228; email: greenbej@nigms.ni
Internet: http://www.grants.nih.gov/grants/guide/pa-files/PA-98-024.html
Sponsor: National Institute of General Medical Sciences
45 Center Drive, Natcher Building
Bethesda, MD 20892

NIGMS Support of Competitive Research (SCORE) Pilot Project Awards **4075**
The objective of the SCORE program is to foster the development of faculty at minority serving institutions (MSIs) in order to increase their research competitiveness and promote their transition to non-SCORE external sources of funding. This objective is expected to translate into an increase in the number of individuals from groups underrepresented in biomedical and behavioral research professionally engaged in these areas of research, and an enhancement of an institution, research base. SCORE SC grants are offered to eligible MSIs. For the purposes of this program, eligible MSIs are those with more than 50% student enrollment of individuals from groups underrepresented in biomedical and behavioral research. The SC2 awards may not exceed $300,000 (direct costs) for the entire length of the award which may be one to three years maximum.
Requirements: Eligible applicants include: public or private post secondary educational institution, such as a university or college awarding associate, undergraduate or graduate degrees located in the United States of America or its territories including Puerto Rico, Guam and Virgin Islands with more than 50 percent student enrollment from groups underrepresented in the biomedical and behavioral sciences (such as African American, Hispanic American, Native American, Alaska Natives or natives of the U.S. Pacific Islands); an Indian tribe that has a recognized governing body and that performs substantial governmental functions, or an Alaska Regional Corporation (ARC), as defined in the Alaska Native Claims Settlement Act (43 U.S.C. 1601 et. Seq.); or an institution that has a significantly high student enrollment from groups underrepresented in the biomedical or behavioral sciences but less than 50% (if the Secretary of the Department of Health and Human Services, through the MBRS Chief, determines that the institution has demonstrated special commitment to the retention and graduation of students from groups underrepresented in the biomedical and behavioral sciences and to the hiring and retention of science faculty from underrepresented groups).
Restrictions: Foreign institutions are not eligible to apply.
Amount of Grant: $50,000 to $100,000 maximum
Date(s) Application is Due: Jan 25; May 25; Sep 25
Contact: Hinda Zlotnik, Ph.D.; (301) 594-3900; fax: (301) 480-2753; email: zlotnikh@nigms.nih.gov
Internet: http://grants.nih.gov/grants/guide/pa-files/PAR-08-027.html
Sponsor: National Institute of General Medical Sciences (NIGMS)
45 Center Drive, Room 2AS.37, MSC 6200
Bethesda, MD 20892-6200

NIGMS Support of Competitive Research (SCORE) Research Advancement Awards **4076**
The objective of the SCORE program is to foster the development of faculty at minority serving institutions (MSIs) in order to increase their research competitiveness and promote their transition to non-SCORE external sources of funding. This objective is expected to translate into an increase in the number of individuals from groups underrepresented in biomedical and behavioral research professionally engaged in these areas of research, and an enhancement of an institution's research base. SCORE SC grants are offered to eligible MSIs. For the purposes of this program, eligible MSIs are those with more than 50% student enrollment of individuals from groups underrepresented in biomedical and behavioral research. The anticipated amount of individual SC1 awards is expected to range between $125,000 to $250,000 maximum direct costs per year.
Requirements: Eligible applicants include: public or private post secondary educational institution, such as a university or college awarding associate, undergraduate or graduate degrees located in the United States of America or its territories including Puerto Rico, Guam and Virgin Islands with more than 50 percent student enrollment from groups underrepresented in the biomedical and behavioral sciences (such as African American, Hispanic American, Native American, Alaska Natives or natives of the U.S. Pacific Islands); an Indian tribe that has a recognized governing body and that performs substantial governmental functions, or an Alaska Regional Corporation (ARC), as defined in the Alaska Native Claims Settlement Act (43 U.S.C. 1601 et. Seq.); or an institution that has a significantly high student enrollment from groups underrepresented in the biomedical or behavioral sciences but less than 50% (if the Secretary of the Department of Health and Human Services, through the MBRS Chief, determines that the institution has demonstrated special commitment to the retention and graduation of students from groups underrepresented in the biomedical and behavioral sciences and to the hiring and retention of science faculty from underrepresented groups).
Restrictions: Foreign institutions are not eligible to apply.
Amount of Grant: $125,000-$250,000
Date(s) Application is Due: Jan 25; May 25; Sep 25
Contact: Hinda Zlotnik, Ph.D.; (301) 594-3900; fax: (301) 480-2753; email: zlotnikh@nigms.nih.gov
Internet: http://grants1.nih.gov/grants/guide/pa-files/PAR-08-026.html
Sponsor: National Institute of General Medical Sciences (NIGMS)
45 Center Drive, Room 2AS.37, MSC 6200
Bethesda, MD 20892-6200

NIGMS/NIDCR Evolution of Infectious Diseases Research Grants **4077**
The goal of this joint NIGMS/NIDCR request for applications is to encourage development of a predictive science of infectious diseases by applying the perspectives, theories, and methods of population and evolutionary biology to important issues of disease emergence, prevention, and treatment. To achieve this goal, this RFA seeks collaborations between two major groups of scientists: those with expertise in population and evolutionary biology and molecular phylogenetics, including mathematical modeling and complexity theory; and infectious disease experts such as clinicians, epidemiologists, immunologists, microbiologists, veterinarians, or plant pathologists. PA: PA-99-147
Requirements: Applications may be submitted by domestic and foreign, for-profit and nonprofit organizations, units of state and local governments, and eligible agencies of the federal government.
Restrictions: Foreign institutions are not eligible for program project grants.
Amount of Grant: $250,000 total for direct costs per year
Contact: Dr. Irene Anne Eckstrand, Division of Genetics and Developmental Biology, (301) 594-0943; fax: (301) 480-2228; email: Irene_Eckstrand@nih.gov; Dennis Mangan, Division of Extramural Research, (301) 594-2421; fax: (301) 480-8318; email: Dennis.Mangan&,àã
Internet: http://grants2.nih.gov/grants/guide/pa-files/PA-99-147.html
Sponsor: National Institute of General Medical Sciences
45 Center Drive, Natcher Building
Bethesda, MD 20892

NIH Academic Career Awards **4078**
The awards support individuals interested in introducing or improving curriculum in a particular scientific field as a means of enhancing the educational or research capacity at the grantee institution. The award supports two types of activities. Development—support for more junior candidates who are interested in developing academic and research expertise in particular fields as a way to increase the overall pool of individuals capable of research or teaching in the identified area. Leadership—supports more senior individuals with acknowledged scientific expertise and leadership skills who are interested in improving the curricula and enhancing the research capacity within an academic institution. Inquiries may be directed to program staff at the appropriate NIH institute: Dr. Robin Barr, Office of Extramural Affairs, (301) 496-9322, NIA, 7201 Wisconsin Avenue, Bethesda, MD 20892; Dr. Ernestine Vanderveen, Division of Basic Research, (301) 443-1273, NIAAA, 6000 Executive Boulevard, Bethesda, MD 20892; Dr. Richard Lymn, Extramural Programs, (301) 594-5128, NIAMS, Natcher Building, Bethesda, MD 20892; Dr. John Schneider, (301) 496-8580, NCI, Executive Plaza North, Bethesda, MD 20892; Dr. Annette Kirshner, (919) 541-0488, NIEHS, P.O. Box 12233, Research Triangle Park, NC 27709; and Dr. George Niederehe, (301) 443-3264, NIMH, 5600 Fishers Ln, Rockville, MD 20857. Annual deadline dates may vary; contact program staff for exact dates. PA: PA-95-052
Requirements: Candidate for this award must have a clinical or doctoral degree. Candidate for the development award must demonstrate the potential to develop into an excellent academician and must be able to identify a mentor who is an expert in the research field of interest. Candidate must also be able to devote 75 percent of full-time professional effort to the research program. For the leadership award, candidate must have sufficient clinical training, research, or teaching experience in the academic area of interest to implement a program of curriculum development within the applicant institution; must have an academic appointment at a level sufficient to exert an influence on the coordination of

research, teaching, and clinical practice in an emerging field; and must be able to devote at least 25 percent effort to the program.
Date(s) Application is Due: Feb 1; Jun 1; Oct 1
Contact: Program Officer
Internet: http://www.grants.nih.gov/grants/guide/pa-files/PA-95-052.html
Sponsor: National Institutes of Health
6701 Rockledge Drive
Bethesda, MD 20892

NIH Academic Research Enhancement Awards (AREA) 4079
NIH makes a special effort to stimulate research in educational institutions that provide baccalaureate training for a significant number of the nation's research scientists but that historically have not been major recipients of NIH support. The AREA funds are intended to support new research projects or expand ongoing research activities proposed by faculty members of these institutions in areas related to the health sciences. Funding decisions will be based on the proposed research project's scientific merit and relevance to NIH programs and the institution's contribution to the undergraduate preparation of doctoral-level health professionals. Annual deadline dates vary; contact program staff for exact dates. PA: PA-99-062
Requirements: All health professional schools/colleges and other academic components of domestic institutions offering baccalaureate or advanced degrees in the sciences related to health are eligible to apply.
Restrictions: Applicant principal investigators must not have active research grant support from either NIH or ADAMHA at the applicant institution at the time of award of an AREA grant; may not submit a regular NIH or ADAMHA research grant application for essentially the same project as a pending AREA application; are expected to conduct the majority of their research at their own institutions, although limited access to special facilities or equipment other institutions is permitted; and may not be awarded more than one AREA grant at a time nor be awarded a second AREA grant to continue research initiated under the first AREA grant.
Amount of Grant: $150,000 maximum in direct costs
Date(s) Application is Due: May 7; Jun 25; Sep 7
Contact: Denise Russo, Ph.D., Office of Extramural Research; (301) 451-7972; fax: (301) 480-0146; email: drusso1@mail.nih.gov
Internet: http://grants.nih.gov/grants/guide/pa-files/PA-06-042.html
Sponsor: National Institutes of Health
6705 Rockledge Drive, Rockledge I Building, Room 3524
Bethesda, MD 20892

NIH Alzheimer's Disease Clinical Trial Planning Grant 4080
NIA, NIMH, and NINDS invite qualified investigators to submit research grant applications for clinical trial planning grants directed toward the drug treatment of the cognitive and behavioral symptoms of Alzheimer's disease (AD). The purpose of the grant is to allow for early peer review of the rationale and design for Alzheimer's disease clinical drug trials and to provide support for the development of a detailed clinical trial research plan, including a manual of operations and procedures. Note: There is a companion program announcement available for Alzheimer's Disease Pilot Clinical Trials (PA-99-038). Inquiries may be directed to program staff at the appropriate NIH institute: Dr. Neil Buckholtz, (301) 496-9350, Neuroscience and Neuropsychology of Aging Program, NIA, 7201 Wisconsin Avenue, Bethesda, MD 20892; Dr. Eugene Oliver, (301) 496-5680, Division of Stroke, Trauma, and Neurodegenerative Disorders, NINDS, 7550 Wisconsin Avenue, Bethesda, MD 20892; Dr. George Niederehe, (301) 443-1185, NIMH, 5600 Fishers Ln, Rockville, MD 20857. PA: PA-99-039
Requirements: Applications may be submitted by foreign and domestic, for-profit and nonprofit organizations, both public and private.
Restrictions: The award cannot be renewed.
Amount of Grant: $150,000 maximum direct costs for one year
Contact: Program Officer
Internet: http://www.grants.nih.gov/grants/guide/pa-files/PA-99-039.html
Sponsor: National Institutes of Health
6701 Rockledge Drive
Bethesda, MD 20892

NIH Autism and Autism Spectrum Disorders Research 4081
The purpose of this program announcement is to encourage grant applications for the support of research designed to elucidate the diagnosis, epidemiology, etiology, genetics, treatment, and optimal means of service delivery in relation to autistic disorder (autism) and autism spectrum disorders. Inquiries may be directed to program staff at the appropriate NIH institute: Dr. Judith Rumsey, (301) 443-9264, NIMH, 5600 Fishers Ln, Rockville, MD 20857; Dr. Marie Bristol, (301) 496-1383, NICHD, 6100 Executive Boulevard, Bethesda, MD 20892; Dr. Judith Cooper, (301) 496-5061, NIDCD, Executive Plaza South, Bethesda, MD 20892; and Dr. Giovanna Spinella, (301) 496-5821, NINDS, 7550 Wisconsin Avenue, Bethesda, MD 20892. PA: PA-98-108
Requirements: Applications may be submitted by foreign and domestic, for-profit organizations, both public and private; units of state and local governments; and eligible agencies of the federal government.
Restrictions: Foreign institutions are not eligible to apply.
Contact: Program Officer
Internet: http://www.grants.nih.gov/grants/guide/pa-files/PA-98-108.html
Sponsor: National Institutes of Health
6701 Rockledge Drive
Bethesda, MD 20892

NIH Biobehavioral Pain Research Grants 4082
The purpose of this PA is to inform the scientific community of the interests of the various institutes at the NIH and to stimulate and foster a wide range of basic and clinical studies on pain as they relate to the missions of these institutes. Applicants are encouraged to study individual differences in pain responses which may be due to factors such as genetic differences, endocrine activity, neural activity, immune function, psychological state, disability state, age, gender, and cultural background. Research is also needed in areas such as understanding the neuroanatomical pathways and the neurophysiological mechanisms in pain. The pain experience needs to be examined at all levels of research including the gene, molecule, cell, organ, and individual with the goal of developing biobehavioral interventions to manage or prevent pain. Inquiries may be directed to the appropriate NIH institute: Dr. Martha Hare, (301) 451-3874, Office of Extramural Programs, NINR, 6701 Democracy Boulevard, Room 710, Bethesda, MD 20892; Dr. Patricia Bryant, (301) 594-2095, Division of Extramural Research, NIDCR, Building 45, Bethesda, MD 20892; Dr. Judith Finkelstein, (301) 496-9350, Neuroscience and Neuropsychology of Aging Program, NIA, 7201 Wisconsin Avenue, Bethesda, MD 20892; Dr. Deborah Ader, (301) 594-5032, Behavior and Prevention Research Program Director, NIAMS, Building 45, Bethesda, MD 20892; Dr. Louis Quatrano, (301) 402-4221, Behavioral Sciences and Rehabilitation Engineering, NICHD, 6100 Executive Boulevard, Bethesda MD 20892; Dr. David Thomas, (301) 443-6975, Behavioral Neurobiology Research Branch, NIDA, 5600 Fishers Ln, Rockville, MD 20857; Dr. Peter Muehrer, (301) 443-4708, Prevention, Early Intervention, and Epidemiology Research Branch, NIMH, 5600 Fishers Ln, Rockville, MD 20857; Dr. Linda Porter, (301) 496-9964, Systems and Cognitive Neuroscience, NINDS, 6001 Executive Boulevard, Bethesda, MD 20892; and Dr. Nancy Pearson, (301) 594-0519, NCCAM, 6707 Democracy Boulevard, Bethesda, MD 20892. PA: PA-03-152
Requirements: Applications may be submitted by foreign and domestic, for-profit and nonprofit, both public and private organizations; units of state and local governments; and eligible agencies of the federal government.
Contact: Program Officer
Internet: http://grants2.nih.gov/grants/guide/pa-files/PA-03-152.html
Sponsor: National Institutes of Health
6701 Rockledge Drive
Bethesda, MD 20892

NIH Biobehavioral Research for Effective Sleep 4083
The goal of this program announcement is to stimulate clinical and applied research on behavioral, psychosocial, and physiological consequences of acute and chronic partial sleep deprivation in either chronically ill or healthy individuals and to develop environmental, clinical management, and other interventions with the potential to reduce sleep disturbances and significantly improve the health of large numbers of people. Although sleep disorders are a cause of sleep loss in affected individuals, the questions to be addressed under this solicitation should focus on causes and consequences of sleep deprivation, apart from any sleep pathology. Inquiries may be directed to program staff at the appropriate NIH institute.
Requirements: Applications may be submitted by domestic and foreign, for-profit and non-profit organizations, public and private, such as universities, colleges, hospitals, laboratories, units of State and local governments, and eligible agencies of the Federal government. Racial/ethnic minority individuals, women, and persons with disabilities are encouraged to apply as Principal Investigators.
Amount of Grant: $250,000 maximum per year
Date(s) Application is Due: Feb 16; Jun 16; Oct 16
Contact: Susan Grove, Program Officer; (301) 435-0714; email: sg16d@nih.gov or grantsinfo@od.nih.gov
Internet: http://grants.nih.gov/grants/guide/pa-files/PA-00-046.html
Sponsor: National Institutes of Health (NIH)
6701 Rockledge Drive
Bethesda, MD 20892

NIH Bioengineering Nanotechnology Initiative 4084
NIH invites grant applications for projects on nanotechnologies useful to biomedicine. Nanotechnology is defined as the creation of functional materials, devices, and systems through control of matter at the scale of one to 100 nanometers, and the exploitation of novel properties and phenomena at the same scale. Inquiries may be directed to program staff at the appropriate NIH institute. See web site for listing of program staff. PA PA-02-125
Amount of Grant: $400,000 maximum
Date(s) Application is Due: Apr 1; Aug 1; Dec 1
Contact: Program Officer
Internet: http://grants2.nih.gov/grants/guide/pa-files/PA-02-125.html
Sponsor: National Institutes of Health
6701 Rockledge Drive
Bethesda, MD 20892

NIH Biology, Development, and Progression of Malignant Prostate Disease Research Grants 4085
Investigator-initiated research grants will be made to examine a range of fundamental biological issues considered critical for progress in defeating prostate cancer. The purpose of funding is to encourage new projects focusing on the biology that underlies the development and progression of malignant prostatic disease. Multidisciplinary collaborations between basic and clinical scientists in projects ranging from exploratory basic biology studies to disease progression in appropriate biological systems or models are encouraged.

Requirements: Applications may be submitted by foreign or domestic for-profit and nonprofit organizations, both public and private; units of state and local government; and eligible agencies of the federal government.
Amount of Grant: $250,000 maximum per year
Date(s) Application is Due: Feb 5; Feb 16; Jun 5; Jun 16; Oct 5; Oct 16
Contact: Frank Bellino; (301) 496-6402; fax: (301) 402-0010; email: fb12a@nih.gov
Internet: http://grants.nih.gov/grants/guide/pa-files/PA-99-081.html
Sponsor: National Institutes of Health (NIH)
Gateway Building, Suite 2C231
Bethesda, MD 20892

NIH Biomedical Research on the International Space Station 4086
The National Institutes of Health (NIH) and the National Aeronautics and Space Administration (NASA) are cooperating to facilitate biomedical research in space for better understanding of human physiology and human health on Earth. Applications to this FOA should propose innovative biomedical research on the molecular or cellular level that is directly relevant to the NIH mission and can be carried out on the ISS. Awards made through this FOA will initially support milestone-driven, ground based preparatory studies (UH2 ground feasibility phase), with possible rapid transition to the second, ISS-based research phase (UH3 ISS experimental phase). The ground feasibility phase (UH2) will allow investigators to focus on ground-based preparatory work to meet scientific milestones and technical requirements leading to the ISS experimental phase (UH3). The UH3 phase will include preparing the experiments for launch, conducting them on the ISS, and the subsequent data analyses on Earth. UH3s will be awarded after administrative review of the eligible UH2s that have met the scientific milestones and feasibility requirements necessary to conduct research on the ISS. Applicants should contact individuals at the specific departments listed on the web site. Letters of Intent are due by August 31, and full applications should arrive by September 30.
Requirements: The UH2/UH3 application must be submitted as a single application, and applicants should note specific instructions for each phase in this FOA. The following organizations and institutions may apply: public/state controlled institutions of higher education; private institutions of higher education; Hispanic-serving institutions; historically Black colleges and universities (HBCUs); tribally controlled colleges and universities (TCCUs); Alaska native and native Hawaiian serving institutions; nonprofits with 501(c)3 IRS status; nonprofits without 501(c)3 IRS status; small businesses; for-profit organizations; State governments; regional organizations; U.S. territories or possessions; Indian/Native American tribal governments; Indian/Native American tribally designated organizations; county governments; city or township governments; special district governments; independent school districts; public housing authorities/ Indian housing authorities; eligible agencies of the Federal government; and faith-based or community based organizations.
Amount of Grant: $300,000 in direct costs per year
Date(s) Application is Due: Aug 31; Sep 30
Contact: Susan Grove, Program Officer; (301) 435-0714; email: sg16d@nih.gov or grantsinfo@od.nih.gov
Internet: http://grants.nih.gov/grants/guide/pa-files/PAR-09-120.html
Sponsor: National Institutes of Health (NIH)
6701 Rockledge Drive
Bethesda, MD 20892-0052

NIH Bone and the Hematopoietic and Immune Systems Research Grants 4087
The participating institutes seek investigator-initiated projects that have the potential to illuminate functional interactions between bone and the hematopoietic and immune systems. Recent observations underscore the linkage between endochondral bone formation and the establishment of hematopoietic marrow, and suggest that interactions between bone, marrow, and the immune system persist in the mature skeleton. Marrow stromal cells include the precursors of the osteochondrogenic lineage, exert important influences on osteoclastogenesis and lymphopoiesis, and mediate the effects of some systemic factors on bone turnover. Recent evidence indicates that hematopoietic cells can influence the differentiation of osteogenic cells, and suggests that mature lymphocytes can influence osteoclastic and osteoblastic functions. In order to explore the mechanisms that underlie these interactions, the participating institutes support outstanding projects that have the potential to either clarify the importance of specific cell types and effector molecules or identify previously unrecognized cellular and molecular agents that influence bone physiology. Collaborations among bone biologists, hematologists, and immunologists, and between basic scientists and clinical investigators, are particularly encouraged. Inquiries may be directed to program staff at the appropriate NIH institute; see web site for staff listings. PA: PA-99-085
Requirements: Applications may be submitted by domestic and foreign, for-profit and nonprofit, public and private organizations, units of state and local governments, and eligible agencies of the federal government.
Restrictions: Foreign institutions are not eligible for program project grant support.
Amount of Grant: $250,000 maximum per year
Contact: Frank Bellino; (301) 496-6402; fax: (301) 402-0010; email: fb12a@nih.gov
Internet: http://grants.nih.gov/grants/guide/pa-files/PA-99-085.html
Sponsor: National Institutes of Health (NIH)
Gateway Building, Suite 2C231
Bethesda, MD 20892

NIH Building Interdisciplinary Research Careers in Women, Health Grants 4088
The NIH Office of Research on Women's Health (ORWH) and its cosponsors invite institutional career development award applications for Building Interdisciplinary Research Careers in Women's Health (BIRCWH) Career Development Programs, hereafter termed Programs. Programs will support mentored research career development of junior faculty members, known as BIRCWH Scholars, who have recently completed clinical training or postdoctoral fellowships, and who will be engaged in interdisciplinary basic, translational, behavioral, clinical, and/or health services research relevant to women's health or sex/gender factors. Direct costs are limited to $500,000 total costs for a five-year period. Letters of Intent are due by September 23, with final applications due by October 22.
Requirements: The following organizations and institutions are eligible to apply for support: public/state controlled institutions of higher education; private institutions of higher education; Hispanic-serving institutions; historically Black colleges and universities (HBCUs); tribally controlled colleges and universities (TCCUs); Alaska Native and Native Hawaiian serving institutions; nonprofits with 501(c)3 IRS status (other than institutions of higher education); nonprofits without 501(c)3 IRS status (other than institutions of higher education); Indian/Native American tribal governments (federally recognized); Indian/Native American tribally designated organizations; U.S. territories or possessions; and eligible agencies of the federal government.
Amount of Grant: $500,000 maximum for a five-year period
Date(s) Application is Due: Sep 23; Oct 22
Contact: Joan Davis Nagel; (301) 496-9186; fax: (301) 402-1798; email: joandav@mail.nih.gov
Internet: http://grants.nih.gov/grants/guide/rfa-files/RFA-OD-09-006.html
Sponsor: National Institutes of Health (NIH)
6707 Democracy Boulevard, Suite 400
Bethesda, MD 20892

NIH Cancer Surveillance Using Health Claims-Based Data System Grants 4089
Three institutes of the NIH invite investigator-initiated grant applications for research to investigate the utility of health claims information as a reporting source for assessing the national cancer burden. Health claims include secondary data sources, for example fee-for-service insurance bills, managed care encounter data, and discharge summaries. Responses to this program announcement would initiate mechanisms to expand understanding of the capability of and methods needed to use claims data for cancer surveillance as well as applying these findings to the claims data in order to assess all aspects of cancer surveillance, such as cancer screening, incidence, treatment, and outcomes. The NIDCR's specific interests focus on health claims-based data as related to oral and pharyngeal cancers or the oral complications of cancer therapies. Inquiries may be directed to program staff at the appropriate NIH institute; see web site for staff listings. Annual deadline dates vary; contact program staff for exact dates.
Requirements: Applications may be submitted by public and private, foreign and domestic, for-profit and nonprofit organizations.
Contact: Crystal Wolfrey, Grants Administration Branch, (301) 496-8634; fax: (301) 496-8601; email: wolfreyc@il.nih.gov
Internet: http://grants2.nih.gov/grants/guide/pa-files/PA-04-012.html
Sponsor: National Institutes of Health
6701 Rockledge Drive
Bethesda, MD 20892

NIH Career Development Awards: Child Abuse and Neglect Research 4090
The five sponsoring institutes invite applications in response to this announcement for the career development of investigators who have made a commitment to focus their research endeavors on child abuse and neglect through research career enhancement to conduct high-quality, multidisciplinary, clinically relevant research on basic biological, behavioral, and social aspects of child and adolescent abuse and neglect. While the focus of the career development program is on child abuse and neglect in human populations, the award may include complementary, appropriate laboratory and animal research related to the child abuse and neglect research proposed in the application. Inquiries may be directed to program staff at the appropriate NIH institute; see web site for staff listings. PA: PA-99-133
Requirements: Prospective candidates should contact the program staff most relevant to the applicant's research interests to ascertain the conditions of the award supported by each institute.
Date(s) Application is Due: Feb 1; Jun 1; Oct 1
Contact: Program Officer
Internet: http://grants.nih.gov/grants/guide/pa-files/PA-99-133.html
Sponsor: National Institutes of Health
6701 Rockledge Drive, Room 1040
Bethesda, MD 20892

NIH Centers for AIDS Research (CFARs) 4091
Participating institutes of NIH invite applications for center core grants (P30) to support Centers for AIDS Research (CFARs). CFAR cores provide infrastructure and promote basic, clinical, behavioral and translational AIDS research activities at institutions that receive significant AIDS funding from multiple NIH institutes or centers. CFARs foster synergy and improve coordination of research, support emerging research opportunities, and promote economy of scale through resources shared by multiple independent laboratories. Inquiries may be directed to program staff at the appropriate NIH institute; see web site for staff listings. PA: PAR-03-089
Requirements: Applications may be submitted by domestic for-profit and nonprofit organizations, both public and private; units of state or local government; and eligible agencies of the federal government.
Restrictions: Foreign institutions are not eligible to apply.
Date(s) Application is Due: Jun 16
Contact: Lori Zimand, Program Contact, (301) 451-4580; fax; (301) 493-0597; email: lz36k@nih.gov

Internet: http://grants2.nih.gov/grants/guide/pa-files/PAR-03-089.html
Sponsor: National Institutes of Health
6701 Rockledge Drive
Bethesda, MD 20892

NIH Centers for Dietary Supplements Research: Botanicals **4092**
Applications are invited to establish specialized research centers to investigate the biological effects of botanicals including, but not limited to, botanicals available as dietary supplements. The creation of such centers is needed to advance the quality and quantity of scientific information on botanicals and to promote further research in this area. The major goal of this RFA is to foster interdisciplinary research to develop systematic evaluation of the safety and efficacy of botanicals. Inquiries may be directed to program staff at the appropriate NIH institute; see web site for staff listings. Annual deadline dates may vary; contact program staff for exact dates. RFA: OD-00-004
Requirements: Applications may be submitted by domestic for-profit and nonprofit organizations, both public and private; units of state and local governments; and eligible agencies of the federal government. Foreign organizations may participate if they are components of domestic U.S. organizations or via contractual or consortium agreements with domestic U.S. organizations.
Amount of Grant: $1.5 million maximum for up to five years
Contact: Program Officer
Internet: http://grants.nih.gov/grants/guide/rfa-files/RFA-OD-00-004.html
Sponsor: National Institutes of Health
6701 Rockledge Drive
Bethesda, MD 20892

NIH Clinical Bioethics Postdoctoral Fellowships **4093**
The Department of Clinical Bioethics of the Warren Grant Magnuson Clinical Center, NIH, offers a limited number of postdoctoral fellowship positions in bioethics. Fellows will participate in the activities and the intellectual life of the department and study ethical issues related to conduct of research, clinical practice, and health policy. Fellows will conduct their research under the guidance of senior faculty, participate in weekly bioethics seminars, case conferences, ethics consultations, and IRB deliberations, and have access to multiple educational opportunities at the NIH. Fellowships are two years in length, and stipends are based on the applicant's previous experience and the current U.S. government schedule. Guidelines are available online.
Requirements: Applicants should have a PhD, MD, JD, or other advanced degree in a relevant field. Fellows will be selected on the basis of their previous academic or professional achievements, commitment to scholarship, and the contribution they are likely to make in the field of bioethics. No prior experience in bioethics is necessary.
Date(s) Application is Due: Dec 30
Contact: Becky Chen, Department of Clinical Bioethics, (301) 496-2429; fax: (301) 496-0760; email: bioethics-inquiries@mail.nih.gov
Internet: http://www.bioethics.nih.gov/opportunities/post-doc.html
Sponsor: National Institutes of Health
10 Center Drive, Building 10, Room 1C118
Bethesda, MD 20892-1156

NIH Clinical Bioethics Predoctoral Fellowships **4094**
The Department of Clinical Bioethics of the Warren Grant Magnuson Clinical Center, NIH, supports a limited number of predoctoral fellowship positions in bioethics. Fellows will participate in the activities and intellectual life of the department, and study ethical issues related to the conduct of research, clinical practice, and health policy. Fellows will conduct their research under the guidance of the senior faculty, participate in weekly bioethics seminars, case conferences, ethics consultations, and IRB deliberations, and have access to multiple educational opportunities at NIH. Fellowships are two years in length. Stipends are based on the current U.S. government schedule. Application guidelines are available online.
Requirements: Applicants should have an undergraduate degree and should be planning postgraduate work. Fellows will be selected on the basis of their previous academic achievements, commitment to scholarship, and the contribution they are likely to make in the field of bioethics. No prior experience in bioethics is necessary.
Date(s) Application is Due: Jan 15
Contact: Program Contact, Department of Clinical Bioethics, (301) 496-2429; fax: (301) 496-0760; email: bioethics-inquiries@mail.nih.gov
Internet: http://www.bioethics.nih.gov/opportunities/pre-doc.html
Sponsor: National Institutes of Health
10 Center Drive, Building 10, Room 1C118
Bethesda, MD 20892-1156

NIH Collaborations for Advanced Strategies in Complication of HIV Infection **4095**
NIDA, NIAAA, NIDDK, and NIMH invite applicants to expand the clinical knowledge of the complications of HIV by expediting the translation of promising innovative preclinical findings into clinical research applications. The focus of the program is to support focused collaborative research between preclinical and clinical scientists to test, refine, and improve diagnostic, pathogenic, or therapeutic concepts. It anticipates such research will lead to innovative approaches for the prevention and management of complications causing significant morbidity or mortality in HIV infected persons including the recently recognized metabolic complications, coincident pathogens such as hepatitis C virus, and the opportunistic pathogens. Inquiries may be directed to program staff at the appropriate NIH institute: Dr. Barbara Laughon, (301) 402-2304, NIAID 6700 B Rockledge Drive, Bethesda, MD 20892; Dr. Jag Khalsa, Health Scientist Administrator, (301) 443-1801, NIDA, 6001 Executive Boulevard, Bethesda, MD 20892; Dr. Thomas Kresina, Chief, Biomedical Research Branch, (301) 443-6537, NIAAA, 6000 Executive Boulevard, Bethesda, MD 20892; Dr. Philip Smith, Senior Advisor, (301) 594-8816, NIDDK, 45 Center Drive, Bethesda, MD 20892; and Dr. Dianne Rausch, (301) 443-7281, NIMH, 6001 Executive Boulevard, Bethesda, MD 20892. Annual deadline dates vary; contact program staff for exact dates.
Requirements: Applications may be submitted by domestic and foreign, for-profit and nonprofit organizations, both public and private; units of state and local governments; and eligible agencies of the federal government.
Contact: Program Officer
Internet: http://grants.nih.gov/grants/guide/pa-files/PA-00-048.html
Sponsor: National Institutes of Health
6701 Rockledge Drive
Bethesda, MD 20892

NIH Complementary and Alternative Medicine Postdoctoral Training Grants **4096**
The Office of Alternative Medicine funds individual postdoctoral training awards using the National Research Service Award (NRSA) mechanism. The purpose is to provide a cadre of investigators capable of conducting systematic studies on safety, efficacy, cost effectiveness, or mechanisms of action of unconventional methods for treating major diseases and promoting well-being. This training is expected to attract postdoctoral candidates who are in the early stages of their careers. They will have obtained expertise in conventional research methodology and some familiarity with/or interest in alternative medical procedures. Prospective trainees will be expected to form alliances with established researchers to provide mutual learning experiences. Individuals may request up to three years of aggregate NRSA support at the postdoctoral level. Written and telephone inquiries are invited. PA: PA-01-088
Requirements: U.S. citizens or noncitizen nationals may apply if, prior to beginning the award, they have received the PhD, MD, DO, DDS, DVM, OD, DPM, ScD, D.Eng, DrPH, DNS, DPharm, or equivalent doctoral degree from an accredited domestic or foreign institution. Before submitting an application, the applicant must arrange for appointment to an appropriate accredited university, hospital, or other institution with facilities including staff for postdoctoral training. The candidate should be accepted by a sponsor who will actively supervise the training. The sponsor should have research experience in clinical medicine and/or basic preclinical research along with an involvement in the evaluation of alternative medicine. The sponsor should be qualified to supervise in the application of rigorous study design to the assessment of individual alternative therapies.
Amount of Grant: $28,260-$44,412 depending on years of experience
Date(s) Application is Due: Apr 5; Aug 5; Dec 5
Contact: Dr. Nancy Pearson, (301) 594-0519; fax: (301) 480-3621; email: pearsonn@mail.nih.gov
Internet: http://grants2.nih.gov/grants/guide/pa-files/PA-01-088.html
Sponsor: National Institutes of Health
6120 Executive Boulevard, Executive Plaza S
Rockville, MD 20892-9904

NIH Demographic Research on Sexual Behaviors Related to HIV **4097**
NICHD, NIDR and NIMH invite qualified researchers to submit applications to study the social and behavioral aspects of the transmission of HIV through sexual intercourse, including oral sexual practices. Inquiries may be directed to program staff at the appropriate NIH institute: Dr. Susan Newcomer, (301) 496-1174, NICHD, 6100 Executive Boulevard, Bethesda, MD 20892; Dr. Willo Pequegnat, (301) 443-6100, NIMH, Parklawn Building, 5600 Fishers Ln, Rockville, MD 20857. Annual deadline dates may vary; contact program staff for exact dates. PAS-00-136
Requirements: Applications may be submitted by domestic for-profit and nonprofit organizations, both public and private; units of state and local governments; and eligible agencies of the federal government.
Amount of Grant: $2.7 million total
Contact: Program Officer
Internet: http://grants2.nih.gov/grants/guide/pa-files/PAS-00-136.html
Sponsor: National Institutes of Health
6701 Rockledge Drive
Bethesda, MD 20892

NIH Dietary Supplement Research Centers: Botanicals Grants **4098**
ODS, NCCAM and NCI invite new and renewal applications to support research centers which will: promote collaborative integrated interdisciplinary study of botanicals, particularly those found as ingredients in dietary supplements; and conduct research of high potential for being translated into practical benefits for human health. This initiative is intended to advance the spectrum of botanical research activities, ranging from plant identification and characterization to early phase clinical studies. Preclinical research that will inform future clinical studies is encouraged as the primary research focus of this program. ODS has a particular interest in botanicals as part of health care for conditions relating to health maintenance or primary prevention. NCCAM is interested in the study of botanicals broadly used by the American public. The NCI Division of Cancer Prevention is interested in supporting research focused on mechanisms by which botanically derived bioactive food components might influence cancer risk and tumor behavior. An applicant may request a project period of up to five years; an annual budget in direct costs up to $1.0 million.
Requirements: The following organizations and institutions are eligible to apply: individuals; public/state controlled institutions of higher education; private institutions of higher education; Hispanic-serving institutions; historically Black colleges and universities (HBCUs); tribally controlled colleges and universities (TCCUs); Alaska

Native and Native Hawaiian serving institutions; nonprofits with 501(c)3 IRS status; nonprofits without 501(c)3 IRS status; for-profit organizations; state governments; Indian/Native American tribal governments (federally recognized); Indian/Native American tribally designated organizations; county governments; city or township governments; special district governments; independent school districts; public housing authorities/Indian housing authorities; U.S. territory or possession; Indian/Native American tribal governments (other than federally recognized); regional organizations; non-domestic (non-U.S.) entities (foreign organizations); other eligible agencies of the federal government; and faith-based or community-based organizations.
Amount of Grant: $1,000,000 per year
Date(s) Application is Due: Dec 3
Contact: Christine A. Swanson, Ph.D., Director, Botanical Research Centers Program; (301) 435-2920; fax: (301) 480-1845; email: Swansonc@od.nih.gov
Internet: http://grants.nih.gov/grants/guide/pa-files/PAR-09-091.html
Sponsor: National Institutes of Health (NIH)
6100 Executive Boulevard, Room 3B01, MSC 7517
Bethesda, MD 20892

NIH Earth-Based Research Relevant to the Space Environment **4099**
The purpose of this Program Announcement (PA) is to stimulate ground-based research on basic, applied, and clinical biomedical and behavioral problems that are relevant to human space flight or that could use the space environment as a laboratory. Although none of the research supported under this initiative would be conducted in space, it is anticipated that it would form a basis for future competitively reviewed studies which could be conducted on the International Space Station, or other space flight opportunities, by skilled on-board specialists. Modular Grant applications will request direct costs in $25,000 modules, up to a total direct cost request of $250,000 per year.
Requirements: The following organizations and institutions may apply: public/state controlled institutions of higher education; private institutions of higher education; Hispanic-serving institutions; historically Black colleges and universities (HBCUs); tribally controlled colleges and universities (TCCUs); Alaska native and native Hawaiian serving institutions; nonprofits with 501(c)3 IRS status; nonprofits without 501(c)3 IRS status; small businesses; for-profit organizations; State governments; regional organizations; U.S. territories or possessions; Indian/Native American tribal governments; Indian/Native American tribally designated organizations; county governments; city or township governments; special district governments; independent school districts; public housing authorities/Indian housing authorities; eligible agencies of the Federal government; and faith-based or community based organizations.
Amount of Grant: $250,000 maximum per year
Date(s) Application is Due: Feb 5; Jun 5; Oct 5
Contact: Susan Grove, Program Officer; (301) 435-0714; email: sg16d@nih.gov or grantsinfo@od.nih.gov
Internet: http://grants.nih.gov/grants/guide/pa-files/PA-00-088.html
Sponsor: National Institutes of Health (NIH)
6701 Rockledge Drive
Bethesda, MD 20892

NIH Ecology of Infectious Diseases **4100**
NIH, NSF, NASA, U.S.GS, and the Agricultural Research Service (ARS) invite applications for the establishment of research programs to elucidate the underlying mechanisms that govern the relationships between anthropogenic environmental changes and the transmission dynamics of infectious diseases. This RFA calls for the development of interdisciplinary research programs on the ecology of infectious diseases in the context of anthropogenic environmental changes such as biodiversity loss, habitat transformation, environmental contamination, climate change, and other influences. The focus of this RFA is on discovery of basic ecological and biological mechanisms and development of predictive models for the emergence and transmission of diseases in humans and other animals, and ultimately the development of strategies to prevent or control them. Annual deadline dates may vary; contact program staff for exact dates.
Requirements: Applications may be submitted by domestic and foreign, for-profit and nonprofit organizations, both public and private (including domestic institutions with foreign collaborators); units of state and local governments; and eligible agencies of the federal government.
Amount of Grant: $4 million total
Contact: Susan Grove, Program Officer; (301) 435-0714; email: sg16d@nih.gov or grantsinfo@od.nih.gov
Internet: http://grants.nih.gov/grants/guide/rfa-files/RFA-TW-00-002.html
Sponsor: National Institutes of Health
6701 Rockledge Drive
Bethesda, MD 20892

NIH Economic Studies in Cancer Prevention, Screening, and Care **4101**
NCI and AHCPR invite investigator-initiated grant applications for research directed at increasing the knowledge base in the area of the economic aspects of cancer prevention, screening, and care. The goal of this program announcement is to generate new economic knowledge that will promote the optimal design of cancer prevention and control trial studies and interventions and will facilitate the formulation of effective health care policy related to cancer prevention and control. NIDCR has an interest in economic and health services studies as these relate specifically to oral and pharyngeal cancers and health care policies related to prevention and control of oral cancers. This initiative requests research applications on new methods development, the synthesis and extension of existing methods, and innovative data gathering strategies. Applications that propose to implement actual data collection on a pilot or full-scale basis as well as analytical studies that use existing data and methodology will be entertained. PA: PA-04-017
Requirements: Foreign and domestic, for-profit and nonprofit organizations, both public and private; units of state and local governments; and eligible agencies of the federal government may apply.
Contact: Crystal Wolfrey, Grants Administration Branch, NCI, (301) 496-8634; fax: (301) 496-4801; email: wolfreyc@mail.nih.gov
Internet: http://grants1.nih.gov/grants/guide/pa-files/PA-04-017.html
Sponsor: National Institutes of Health
6701 Rockledge Drive
Bethesda, MD 20892

NIH Endocrine Disruptors: Epidemiologic Approaches Grants **4102**
Through this program NIOSH, EPA, NIEHS, and NCI support research on the relationship between exposure to endocrine disruptors and adverse health effects in humans, particularly reproductive and developmental, with a focus on epidemiologic approaches. Synthetic chemicals released in the workplace and environment collectively called endocrine disruptors (EDs) or endocrine disrupting chemicals (EDCs) are thought to interfere with one or more aspects of the normal endocrine functions that are responsible for the maintenance of homeostasis and the regulation of developmental processes. The cooperating agencies support research that examines human health consequences of exposure to physical and chemical toxicants in the environment and in occupational settings as well as the underlying mechanisms involved in environmentally related disease and dysfunction. Research applications are sought that will investigate the relationships between exposure to endocrine disruptors and adverse health effects, particularly reproductive/developmental effects in humans. Annual deadline dates may vary; contact program staff for exact dates. RFA: OH-01-001
Requirements: Applications may be submitted by domestic and foreign, public and private nonprofit and for-profit organizations, and by governments and their agencies; state and local governments or their bona fide agents; and federally recognized Indian tribal governments, Indian tribes, or Indian tribal organizations.
Amount of Grant: $6 million total
Contact: Susan Grove, Program Officer; (301) 435-0714; email: sg16d@nih.gov or grantsinfo@od.nih.gov
Internet: http://www.grants.nih.gov/grants/guide/rfa-files/RFA-OH-01-001.html
Sponsor: National Institutes of Health
6701 Rockledge Drive
Bethesda, MD 20892

NIH Enhancing Adherence to Diabetes Self-Management Behaviors **4103**
This PA solicits applications for investigator-initiated research related to sociocultural, environmental, and behavioral mechanisms and biological/technological factors that contribute to successful and ongoing self-management in diabetes. Self-management is defined as client strategies and behaviors that contribute to blood glucose normalization, improved health, and prevention or reduction of complications. It is broader than adherence to specific regimen components and incorporates deliberate problem solving and decision-making processes. Applications are encouraged for both type 1 and type 2 diabetes; representative and minority populations; and all age groups. Inquiries may be directed to program staff at the appropriate NIH institute. Modular Grant applications will request direct costs in $25,000 modules, up to a total direct cost request of $250,000 per year.
Requirements: Applications may be submitted by domestic and foreign, for-profit and non-profit organizations, public and private, such as universities, colleges, hospitals, laboratories, units of State and local governments, and eligible agencies of the Federal government. Racial/ethnic minority individuals, women, and persons with disabilities are encouraged to apply as principal investigators.
Amount of Grant: $250,000 maximum
Date(s) Application is Due: Feb 16; Jun 16; Oct 16
Contact: Susan Grove, Program Officer; (301) 435-0714; email: sg16d@nih.gov or grantsinfo@od.nih.gov
Internet: http://grants.nih.gov/grants/guide/pa-files/PA-00-049.html
Sponsor: National Institutes of Health (NIH)
6701 Rockledge Drive
Bethesda, MD 20892

NIH Epidemiological and Behavioral Research in Oral Health **4104**
This program seeks to stimulate epidemiological and behavioral research to improve oral health or increase the understanding of the determinants of oral health. It supports studies that expand and enhance the integration of population-based, genetic, social, and behavioral research. Specific research activities highlighted in the strategic plan include, but are not limited to, furthering the understanding of how genetic, behavioral, and environmental factors interact to contribute to disease susceptibility or resistance; assessing and bridging gaps in the knowledge, opinions, and practices of the public, educators, and health care professionals about oral diseases and their prevention; identifying risk factors and markers for the progression of disease to oral, dental, and craniofacial diseases and disorders; improving oral health literacy; and developing more effective strategies to encourage the appropriate and timely adoption of research findings into clinical practice. PA Number: PA-04-022
Requirements: Applications may be submitted by domestic and foreign, for-profit and nonprofit organizations, both public and private; units of state and local governments; and eligible agencies of the federal government.
Contact: Mary Daley, Grants Management Branch, (301) 594-4808; (301) 480-3562; email: daleym@mail.nih.gov

Internet: http://www.grants.nih.gov/grants/guide/pa-files/PA-04-022.html
Sponsor: National Institutes of Health
6701 Rockledge Drive
Bethesda, MD 20892

NIH Ethical Issues in Human Studies Research Grants 4105

NIH and CDC invite research grant applications (R01) for support to conduct research on ethical issues that arise with research involving human participants. Grants may be requested in $25,000 direct cost modules. Annual deadline dates may vary; contact program staff for exact dates. PA: PA-02-103
Requirements: Applications may be submitted by any domestic and foreign, for-profit and nonprofit, public and private organizations; units of state and local governments; and eligible agencies of the federal government. Applicants are encouraged to visit the Web sites of the participating institutes and centers to learn more about their specific interests.
Restrictions: The PA does not solicit essays, editorials, literature or historical reviews of science, conferences, or speculative, theoretical, or policy analyses.
Amount of Grant: $250,000 maximum per year
Date(s) Application is Due: Feb 1; Jun 1; Oct 1
Contact: Dr. Della Hann, Program Contact, (301) 402-2725; email: hannd@od.nih.gov
Internet: http://grants2.nih.gov/grants/guide/pa-files/PA-02-103.html
Sponsor: National Institutes of Health
6701 Rockledge Drive
Bethesda, MD 20892

NIH Exceptional, Unconventional Research Enabling Knowledge Acceleration (EUREKA) Grants 4106

The purpose of the EUREKA (Exceptional Unconventional Research Enabling Knowledge Acceleration) initiative is to foster exceptionally innovative research that, if successful, will have an unusually high impact on the areas of science that are germane to the mission of one or more of the participating NIH Institutes. Before submitting an application, it is extremely important to verify that the proposed research is of interest to at least one of the NIH Institutes that is participating in the EUREKA FOA, since applications that are not germane to the mission of one or more of the participating Institutes will be withdrawn without review.
Requirements: The following organizations and institutions are eligible to apply: public/state controlled institutions of higher education; private institutions of higher education; Hispanic-serving institutions; historically Black colleges and universities (HBCUs); tribally controlled colleges and universities (TCCUs); Alaska Native and Native Hawaiian serving institutions; nonprofits with 501(c)3 IRS status (other than institutions of higher education); nonprofits without 501(c)3 IRS status (other than institutions of higher education); small businesses; for-profit organizations (other than small businesses); non-domestic (non-U.S.) entities (foreign organizations).
Restrictions: EUREKA is for new projects, not for continuation of existing projects. Nor is EUREKA for support of pilot projects, i.e., projects of limited scope that are designed primarily to generate data that will enable the PI to seek other funding opportunities.
Amount of Grant: $800,000 maximum in direct costs
Date(s) Application is Due: Nov 24
Contact: Laurie Tompkins, Ph.D.; (301) 594-0943; fax: (301) 480-2228; email: tompkinL@nigms.nih.gov; Judy Mietz, Ph.D.; (301) 496-9326; fax: (301) 496-1224; email: mietzj@mail.nih.gov
Internet: http://grants.nih.gov/grants/guide/rfa-files/RFA-GM-10-009.html
Sponsor: National Institutes of Health (NIH)
45 Center Drive, MSC 6200
Bethesda, MD 20892-6200

NIH Exploratory Centers for Alternative Medicine Research Grants 4107

The objective of the program is to develop exploratory centers for the purpose of investigating and validating alternative therapies. These centers will support new interdisciplinary programs involving experienced investigators from conventional medicine with clinicians and investigators from alternative medicine and should provide clinical/scientific/technical assistance to alternative medicine clinicians who may have potentially useful accumulations of patient data that could be evaluated for treatment outcome. It will be important that the research faculty at these centers have expertise in broad areas such as biostatistics, computer processing, data management, and protocol design. Additionally, through a peer review process that will be established by the centers, small prospective pilot studies will be started. Program areas to be developed and evaluated can include nutrition, diet, and lifestyle/behavioral health changes; mind/body control therapies; traditional and ethnomedicine therapies; structural manipulations and energetic therapies; pharmacological and biological therapies; and bioelectromagnetic therapies. Annual deadline dates vary; contact program staff for exact dates. RFA: OD-94-004
Requirements: Applications may be submitted by domestic for-profit, nonprofit, public, and private organizations.
Amount of Grant: $30,000 planning grants, $20,000-$30,000 training grants, $350,000 cooperative agreements, $115,000 investigator-initiated grants
Contact: Dr. John Spencer, Office of Alternative Medicine, (301) 402-2466; fax: (301) 402-4741
Internet: http://www.grants.nih.gov/grants/guide/rfa-files/RFA-OD-94-004.html
Sponsor: National Institutes of Health
6120 Executive Boulevard, Executive Plaza S
Rockville, MD 20892

NIH Exploratory Innovations in Biomedical Computational Science and Technology (BISTI) Grants 4108

The NIH is interested in promoting research and developments in computational science and technology that will support rapid progress in areas of scientific opportunity in biomedical research. As defined here, biomedical computing or biomedical information science and technology includes database design, graphical interfaces, querying approaches, data retrieval, data visualization and manipulation, data integration through the development of integrated analytical tools, and tools for electronic collaboration, as well as computational and mathematical research including the development of structural, functional, integrative, and analytical models and simulations. This particular FOA is intended to support exploratory biomedical informatics and computational biology research - applications should be innovative, with high risk/high impact in new areas that are lacking preliminary data or development. Because the nature and scope of the proposed research will vary from application to application, it is anticipated that the size and duration of each award will also vary.
Requirements: Application can be made by the following institutions and organizations: for-profit organizations; non-profit organizations; public or private institutions (such as universities, colleges, hospitals, and laboratories); units of state government; units of local government; eligible agencies of the Federal government; foreign institutions; domestic institutions; faith-based or community-based organizations; Indian/Native American tribal governments (federally recognized); Indian/Native American tribal governments (other than federally recognized); and Indian/Native American tribally designated organizations.
Date(s) Application is Due: Feb 16; Jun 16; Oct 16
Contact: Peter Lyster, Ph.D.; (301) 451-6446; email: lysterp@mail.nih.gov
Internet: http://grants.nih.gov/grants/guide/pa-files/PAR-06-411.html
Sponsor: National Institutes of Health (NIH)
45 Center Drive, Room 2AS.55k, MSC 6200
Bethesda, MD 20892-6200

NIH Exploratory/Development Grants for Mental Health Intervention Research 4109

NIMH is requesting research grant applications to develop somatic, pharmacologic, psychosocial, or combination interventions for the prevention, rehabilitation, treatment, and/or management of mental disorders and emotional or behavioral problems; research the co-occurrence of mental, physical, and substance abuse problems and the functional consequences of these problems across the life span; and to perform preliminary tests of feasibility, acceptability, and efficacy/effectiveness of interventions. PA-03-078
Requirements: Applications may be submitted by domestic and foreign, for-profit and nonprofit organizations, both public and private; units of state and local governments; and eligible agencies of the federal government.
Amount of Grant: $125,000 average
Contact: Joy Knipple, Grants Management Branch, National Institute of Mental Health, (301) 443-8811; fax: (301) 443-6885; email: jk173@nih.gov
Internet: http://grants.nih.gov/grants/guide/pa-files/PAR-03-078.html
Sponsor: National Institute of Mental Health
6701 Rockledge Drive, Room 1040, MSC 7710
Bethesda, MD 20892

NIH Exploratory/Developmental Research Grant 4110

The Exploratory/Developmental Grant (R21) mechanism is intended to encourage exploratory and developmental research projects by providing support for the early and conceptual stages of these projects. These studies may involve considerable risk but may lead to a breakthrough in a particular area, or to the development of novel techniques, agents, methodologies, models, or applications that could have a major impact on a field of biomedical, behavioral, or clinical research.
Requirements: Eligible organizations: for profit organizations; non-profit organizations; public or private institutions, such as universities, colleges, hospitals and laboratories; units of state government; units of local government; eligible institutions of the federal government; domestic institutions; foreign institutions; faith-based or community-based organizations; units of state tribal government; and units of local tribal government. Eligible Project Directors/Principal Investigators: any individual with the skills, knowledge and resources necessary to carry out the proposed research, is invited to work with his/her institution to develop an application for support; individuals from underrepresented racial and ethnic groups; individuals with disabilities
Amount of Grant: $275,000 over the two years of the R21 award
Date(s) Application is Due: Feb 16, Jun 16, Oct 16
Contact: Ellen S. Liberman, Ph.D., (301) 451-2020; email: esl@nei.nih.gov
Internet: http://grants.nih.gov/grants/guide/pa-files/PA-06-181.html
Sponsor: National Eye Institute
5635 Fishers Lane, Suite 1300
Bethesda, MD 20892-9300

NIH Fine Mapping and Function of Genes for Type 1 Diabetes Grants 4111

The purpose of this initiative is to bring together investigators with experience in genetics, immunology, and biochemistry to perform fine mapping of the loci discovered and study the function of the genes being identified and to identify the mechanisms by which newly discovered genes predispose to T1D. This would lead to elucidation of the mechanisms whereby changes in the function or regulation of these genes are likely to provide crucial new insights into disease pathogenesis. The discovery of the genes would be relevant to developing a predictive strategy for individuals who may develop diabetes. In the future, genetic testing may lead to personalized treatment regimens by identifying the most appropriate class of drugs for particular patients. Total funding

available for this five year program is $20 million to fund 4-10 applications, contingent upon the availability of funds and the submission of a sufficient number of meritorious applications. The maximum direct costs are $5 million that can be spent on a project period of up to 5 years. Letters of Intent are due by February 27, with final applications due by March 30. The office strongly urges communication from applicants via email.
Requirements: The following organizations and institutions are eligible to apply: public/state controlled institutions of higher education; private institutions of higher education; Hispanic-serving institutions; historically Black colleges and universities (HBCUs); tribally controlled colleges and universities (TCCUs); Alaska Native and Native Hawaiian serving institutions; nonprofits with 501(c)3 IRS status (other than institutions of higher education); nonprofits without 501(c)3 IRS status (other than institutions of higher education); small businesses; for-profit organizations (other than small businesses); non-domestic (non-U.S.) entities (foreign organizations); and other eligible agencies of the federal government.
Amount of Grant: $1,500,000-$5,000,000 over five years
Date(s) Application is Due: Feb 27; Mar 30
Contact: Beena Akolkar, PhD.; (301) 594-8812; email: ba92i@nih.gov
Internet: http://grants.nih.gov/grants/guide/rfa-files/rfa-dk-08-006.html
Sponsor: National Institutes of Health (NIH)
6707 Democracy Boulevard, Room 6105
Bethesda, MD 20892-5460

NIH Genetic Architecture of Complex Phenotypes **4112**
The purpose of this initiative is to support new studies on the architecture of complex phenotypes, including research using human and model systems as well as research using theoretical approaches.The studies targeted by this program announcement are expected to expand our understanding of the roles of genetic and environmental variation and their interactions in causing phenotypic variation in populations; increase the quantity and quality of population-based data; lead to the development of mathematical and statistical tools for analyzing measured genotype data; lead to improvements in study design; and create biologically relevant models for understanding the origins, roles, and implications of genetic variation in causing variation in phenotypes. This program announcement encourages efforts including organization of DNA sequence variation in populations and subpopulations with the goal of understanding the causal forces that shape that organization; studies of the relationship of genotypic and phenotypic variation; data analysis; research on study design, including comparisons of different strategies for studying complex phenotypes; development of statistical and analytical tools; and construction of meaningful tests of relationships between genotypic and phenotypic data. Further information for this program is available on the Web site. Inquiries may be directed to program staff at the appropriate NIH institute; Dr. Irene Anne Eckstrand, (301) 594-0943, NIGMS, 45 Center Drive, Bethesda, MD 20892; Dr. Jose Velazquez, (919) 541-4998, NIEHS, P.O. Box 12233, Research Triangle Park, NC 27709; Dr. William Sharrock, (301) 594-5055, NIAMS, 45 Center Drive, Bethesda, MD 20892; Dr. Evan Hadley, (301) 435-3048, NIA, 7201 Wisconsin Avenue, Bethesda, MD 20892; Dr. Steven Moldin, (301) 443-2037, NIMH, 5600 Fishers Ln, Rockville, MD 20857; Dr. Harold Gordon, (301) 443-4877, NIDA, 5600 Fishers Ln, Rockville, MD 20857; Dr. Robert Karp, (301) 443-4223, NIAAA, 6000 Executive Boulevard, Bethesda, MD 20892; Dr. Ralph Nitkin, (301) 496-1383, NICHD, 6100 Executive Boulevard, Bethesda, MD 20892; Dr. Lisa Brooks, (301) 496-7531, NHGRI, 38 Library Drive, Bethesda, MD 20892; and Dr. Cashell Jaquish, (301) 435-0447, NHLBI, 6701 Rockledge Drive, Bethesda, MD 20892. PA: PA-02-110
Requirements: Applications may be submitted by domestic and foreign, for-profit and nonprofit organizations, both public and private. Racial/ethnic minority individuals, women, and persons with disabilities are encouraged to apply.
Restrictions: Foreign institutions are not eligible for program projects (P01) grants.
Contact: Program Officer
Internet: http://grants2.nih.gov/grants/guide/pa-files/PA-02-110.html
Sponsor: National Institutes of Health
6701 Rockledge Drive
Bethesda, MD 20892

NIH Genetic Susceptibility and Variability of Human Malformations Research Grants **4113**
NICHD, NIDCR, NIEHS, and EPA invite research grant applications to study the genetic susceptibility and mechanisms of human congenital malformations including research on the contribution of potential genetic and environmental factors, identified at the molecular level, to the etiology, distribution, and prevention of disease within families and across populations. The research conducted as a result of this RFA will provide the direction and focus for future initiatives to dissect the molecular and genetic mechanisms underlying specific human malformations. The current initiative is designed to encourage collaborative, interdisciplinary, and innovative studies that integrate the latest advances in developmental genetics, functional genomics, epidemiology, and high throughput biotechnology with the new mathematical, methodological, and statistical tools for analyzing and evaluating data related to the etiology and distribution of birth defects in populations. The primary objective of this initiative is to develop and use integrated multidisciplinary approaches to obtain information that will enhance our knowledge of and provide a basis for future initiatives focusing on the basic mechanisms and etiologies of human malformations. This RFA is part of a larger birth defects program which will include an upcoming initiative related to fundamental developmental mechanisms underlying the formation of structural anomalies. Inquiries may be directed to the staff at the appropriate institute/organization: Dr. Deborah Henken, (301) 496-5541, Center for Research for Mothers and Children, NICHD, Executive Building, 6100 Executive Boulevard, Bethesda, MD 20892; Dr. Judy Small, (301) 594-2425, Craniofacial Anomalies and Injury Branch, NIDCR, Building 45, Bethesda, MD 20892; Dr. Gwen Collman, (919) 541-4980, Chemical Exposures and Molecular Biology Branch, Division of Extramural Research and Training, NIEHS, P.O. Box 12233 Research Triangle Park, NC 27709; and Dr. Chris Saint, (202) 564-6909, National Center for Environmental Research & Quality Assurance, EPA, 1300 Pennsylvania Avenue, Washington, DC 20004. Annual deadline dates may vary; contact program staff for exact dates. RFA: HD-99-002
Requirements: Applications may be submitted by domestic and foreign, for-profit and nonprofit, public and private organizations; units of state and local governments; and eligible agencies of the federal government.
Restrictions: Applications limited to descriptive or traditional epidemiological approaches will be considered nonresponsive. The EPA will not be able to consider applications from for-profit organizations or federal agencies.
Amount of Grant: $700,000 maximum
Contact: Program Officer
Internet: http://www.grants.nih.gov/grants/guide/rfa-files/RFA-HD-99-002.html
Sponsor: National Institutes of Health
6701 Rockledge Drive
Bethesda, MD 20892

NIH Genetics of Neurobehavioral Disorders in Existing Samples Grants **4114**
The purpose of the program is to encourage applications for multidisciplinary, collaborative genetic research projects, employing state-of-the-art diagnostic procedures and analytic methods, to integrate different pedigree data sets that have already been collected for the study of a given neurobehavioral disorder. Applications will implement the assembling of data from multiple research groups for combined analysis, and will focus on one of the following: attention-deficit hyperactivity disorder, bipolar disorder or other related mood disorders, recurrent early-onset depression, eating disorders, obsessive-compulsive disorder or other anxiety disorders, panic disorder, schizophrenia or other psychotic disorders, personality disorders, or Tourette's syndrome. The creation and analysis of clinically well-characterized samples of sufficient size for linkage analyses and for linkage disequilibrium mapping studies in genetically isolated populations studied previously may be facilitated by the establishment of international consortia. Full collaborations between American scientists and scientists at international institutions are encouraged, when scientifically appropriate. In these cases, awards may be made to international institutions or to domestic applications that include international components. Inquiries may be directed to program staff at the appropriate NIH institute: Dr. Steven Moldin, Division of Neuroscience and Basic Behavioral Science, NIMH, (301) 443-2037, 6001 Executive Boulevard, Bethesda, MD 20892; Dr. Lisa Freund, Child Development and Behavior Branch, NICHD, (301) 435-6879, Building 6100, 9000 Rockville Pike, Bethesda, MD 20892. Annual deadline dates may vary; contact program staff for exact dates. PA: PAR-99-120
Requirements: Applications may be submitted by domestic, for-profit and nonprofit organizations, both public and private; units of state and local governments; and eligible agencies of the federal government.
Amount of Grant: $250,000 maximum per year
Date(s) Application is Due: Feb 1; Jun 1; Oct 1
Contact: Program Officer
Internet: http://www.grants.nih.gov/grants/guide/pa-files/PA-99-120.html
Sponsor: National Institutes of Health
6701 Rockledge Drive
Bethesda, MD 20892

NIH Health Promotion Among Racial and Ethnic Minority Males **4115**
This Funding Opportunity Announcement solicits Exploratory/Developmental grant applications from applicants that propose to stimulate and expand research in the health of minority men. Specifically, this initiative is intended to: enhance our understanding of the numerous factors (e.g., sociodemographic, community, societal, personal) influencing the health promoting behaviors of racial and ethnic minority males and their sub-populations across the life cycle; and solicit applications focusing on the development and testing of culturally and linguistically appropriate health-promoting interventions designed to reduce health disparities among racially and ethnically diverse males and their sub-populations age 21 and older. The total project period for an application submitted in response to this funding opportunity may not exceed two years. Direct costs are limited to $275,000 over an R21 two-year period, with no more than $200,000 in direct costs allowed in any single year.
Requirements: Application can be made by the following institutions and organizations: public/state controlled institutions of higher education; private institutions of higher education; nonprofits with 501(c)3 IRS status (other than institutions of higher education); nonprofits without 501(c)3 IRS status (other than institutions of higher education); small businesses; for-profit organizations (other than small businesses); state governments; U.S. territories or possessions; Indian/Native American tribal governments (federally recognized); Indian/Native American tribal governments (other than Federally recognized); Indian/Native American tribally designated organizations; non-domestic (non-U.S.) entities (foreign organizations); Hispanic-serving institutions; historically Black colleges and universities (HBCUs); tribally controlled colleges and universities (TCCUs); Alaska Native and Native Hawaiian serving institutions; regional organization; eligible agencies of the Federal government; and faith-based or community based organizations.
Amount of Grant: $200,000 maximum per year; $275,000 maximum per two-year cycle
Date(s) Application is Due: Feb 16; Jun 16; Oct 16

Contact: Dr. Charlotte Pratt, Program Director; (301) 435-0382; fax: (301) 480-5158; email: prattc@nhlbi.nih.gov
Internet: http://grants.nih.gov/grants/guide/pa-files/PA-07-421.html
Sponsor: National Institutes of Health (NIH)
6701 Rockledge Drive, Suite 10118, MSC 7936
Bethesda, MD 20892-7936

NIH Human Brain Project (Neuroinformatics) Phase I & II 4116

The purpose of this initiative is to encourage and support investigator- initiated research on neuroscience informatics (neuroinformatics). This research will lead to the development of new web based databases, analytical tools, and knowledge management systems to foster sharing of data for all domains of neuroscience research. This program combines neuroscience and informatics (neuroinformatics) research to develop and apply advanced tools and approaches essential for efficient understanding of the structure, function and development in health and disorders of the nervous system, from the genetic to whole systems level. Annual deadline dates may vary; contact program staff for exact dates. PA: PAR-03-035
Requirements: Applications may be submitted by domestic and foreign, for-profit and nonprofit organizations, both public and private; units of state and local governments; and eligible agencies of the federal government.
Restrictions: Foreign institutions are not eligible to apply.
Date(s) Application is Due: Jan 21; May 20; Sep 22
Contact: Stephen Koslow, Office on Neuroinformatics, NIMH, (301) 443-1815; fax: (301) 443-1867; email: koz@helix.nih.gov
Internet: http://grants2.nih.gov/grants/guide/pa-files/PAR-03-035.html
Sponsor: National Institutes of Health
6701 Rockledge Drive, Room 1040
Bethesda, MD 20892

NIH Human Connectome Project Grants 4117

This Funding Opportunity Announcement is issued as an initiative of the NIH Blueprint for Neuroscience Research. The Neuroscience Blueprint is a collaborative framework through which 16 NIH Institutes, Centers and Offices jointly support neuroscience-related research, with the aim of accelerating discoveries and reducing the burden of nervous system disorders. The overall purpose of this five year Human Connectome Project (HCP) is to develop and share knowledge about the structural and functional connectivity of the human brain. This purpose will be pursued through the following specific efforts: existing, but cutting-edge, non-invasive imaging technologies will be optimized and combined to acquire structural and functional in vivo data about axonal projections and neural connections from brains of hundreds of healthy adults. Demographic data and data regarding sensory, motor, cognitive, emotional, and social function will also be collected for each subject, as will DNA samples and blood (to establish cell lines). Models to better understand and use these data will be developed. Connectivity patterns will be linked to existing architectonic data. Data and models will be made available to the research community immediately via a user-friendly system to include tools to query, organize, visualize and analyze data. Outreach activities will be conducted to engage and educate the research community about the imaging tools, data, models, and informatics tools. After five years, these specific efforts are expected to deliver: a set of integrated, non-invasive imaging tools to obtain connectivity data from humans in vivo; a high quality and well characterized, quantitative set of human connectivity data linked to behavioral and genetic data as well as to general, existing architectonic data, and associated models, from up to hundreds of healthy adult female and male subjects; and rapid, user-friendly dissemination of connectivity data, models, and tools to the research community via outreach activities and an informatics platform. Letters of Intent are due by October 24, with final applications due by November 24.
Requirements: The following organizations and institutions are eligible to apply for support: public/state controlled institutions of higher education; private institutions of higher education; Hispanic-serving institutions; historically Black colleges and universities (HBCUs); tribally controlled colleges and universities (TCCUs); Alaska Native and Native Hawaiian serving institutions; nonprofits with 501(c)3 IRS status (other than institutions of higher education); nonprofits without 501(c)3 IRS status (other than institutions of higher education); Indian/Native American tribal governments (federally recognized); Indian/Native American tribally designated organizations; U.S. territories or possessions; and eligible agencies of the federal government.
Amount of Grant: $6 million maximum per year for five years
Date(s) Application is Due: Oct 24; Nov 24
Contact: Michael F. Huerta, Ph.D.; (301) 443-1815; fax: (301) 443-1731; email: mhuert1@mail.nih.gov
Internet: http://grants.nih.gov/grants/guide/rfa-files/RFA-MH-10-020.html
Sponsor: National Institutes of Health (NIH)
6001 Executive Boulevard, Room 7202, MSC 9645
Bethesda, MD 20892-9645

NIH Human Models and Markers of Skeletal Aging Grants 4118

NIA, NIAMS, NIDR, and NIDDK invite research grant applications to develop or refine models and markers using human bone, bone marrow, or bone cell constituents that more accurately reflect those age-related and/or pathophysiological processes occurring within the mature human skeleton that lead to osteoporosis and other age-related skeletal diseases. It is expected that improved human models and markers of age-related changes in skeletal tissue structure and function will enhance the accuracy and reliability of diagnostic and prognostic capabilities and serve to expedite the development of more effective, targeted preventative, and therapeutic strategies. Annual deadlines vary; contact program staff for exact dates. Additional information may be obtained by mail and email, or through the NIH grant line (301-402-2221), and the NIH Web site. PA: PA-96-022
Requirements: Applications may be submitted by domestic and foreign, for-profit and nonprofit organizations, both public and private; units of state and local governments; and eligible agencies of the federal government.
Date(s) Application is Due: Feb 1; Jun 1; Oct 1
Contact: Dr. Richard Sprott, Biology of Aging, (301) 496-4996; Dr. Evan Hadley, Geriatrics and Clinical Research, (301) 496-6761
Internet: http://www.grants.nih.gov/grants/guide/pa-files/PA-96-022.html
Sponsor: National Institutes of Health
6701 Rockledge Drive
Bethesda, MD 20892

NIH Hyperaccelerated Award/Mechanisms in Immunomodulation Trials 4119

NIAID, NIA, NIAMS, NIDCR, NIDDK, and NINDS invite investigator-initiated research applications for mechanistic studies in clinical trials of: immunomodulatory interventions for immune system mediated diseases, including, but not limited to: asthma and allergic diseases; graft failure in solid organ, cell, tissue and stem cell transplantation; and chronic inflammatory, autoimmune, and immunodeficiency diseases; and preventative and therapeutic, vaccines for non-HIV/AIDS infectious diseases, including NIAID Category A, B, and C agents of bioterrorism and emerging/re-emerging infectious diseases. Applications will be accepted monthly on the ninth of each month; a letter of intent is due one month prior to application receipt date. RFA: AI-01-001
Requirements: Applications may be submitted by domestic and foreign, for-profit and nonprofit organizations, both public and private; units of state and local governments; and eligible agencies of the federal government.
Amount of Grant: $2 million total
Contact: Ann White-Devine, NIAID, (301) 402-5601; fax: (301) 480-3780; email: ad22x@nih.gov
Internet: http://grants.nih.gov/grants/guide/rfa-files/RFA-AI-01-001.html
Sponsor: National Institutes of Health
6701 Rockledge Drive
Bethesda, MD 20892

NIH Impact of HAART on HIV/CNS Disease 4120

This RFA solicits applications to study the effects of HAART on HIV/CNS disease with respect to the clinical, neurological, neuropsychological, virological, immunological, and neuropathological response to the drugs. This solicitation also includes studies to develop drug formulations with improved ability to cross the blood-brain barrier, studies to develop improved methods of delivering drugs to the CNS, and the interaction of anti-retrovirals with psychotropic drugs. Annual deadline dates vary; contact program staff for exact dates. Inquiries may be directed to program staff at the appropriate NIH institute: Dr. Dianne Rausch, (301) 443-7281, NIMH, 6001 Executive Boulevard, Bethesda, MD 20892; Dr. A.P. Kerza-Kwiatecki, (301) 496-1431, NINDS, 6001 Executive Boulevard, Bethesda, MD 20892; and Dr. Thomas Johnson, (301) 402-3461, NIDCD, 6120 Executive Boulevard, Bethesda, MD 20892. RFA: MH-00-003
Requirements: Applications may be submitted by domestic and foreign, for-profit and nonprofit organizations, both public and private; units of state and local governments; and eligible agencies of the federal government.
Amount of Grant: $3.4 million total
Contact: Program Officer
Internet: http://grants.nih.gov/grants/guide/rfa-files/RFA-MH-00-003.html
Sponsor: National Institutes of Health
6701 Rockledge Drive
Bethesda, MD 20892

NIH Independent Scientist Award 4121

The award is a special salary-only grant designed to provide protected time for newly independent scientists who currently have non-research obligations such as heavy teaching loads, clinical work, committee assignments, service, and administrative duties that prevent them from having a period of intensive research focus. The award is targeted to persons with doctoral degrees who have completed their research training, have independent peer-reviewed research support, and who need a period of protected research time in order to foster their research career development. New applications are due February 12, June 12, and October 12, while renewal applications are due March 12, July 12, and November 12.
Requirements: The following organizations and institutions are eligible to apply: public/state controlled institutions of higher education; private institutions of higher education; nonprofits with 501(c)3 IRS status; nonprofits without 501(c)3 IRS status; small businesses; for-profit organizations; State governments; U.S. territories or possessions; Indian/Native American tribal governments (Federally recognized and other than Federally recognized); Indian/Native American tribally designated organizations; Hispanic-serving institutions; historically Black colleges and universities (HBCUs); tribally controlled colleges and universities (TCCUs); Alaska Native and Native Hawaiian serving institutions; regional organizations; and faith-based or community based organizations. The candidate must have a doctoral degree and peer-reviewed, independent research support at the time the award is made. The candidate must spend a minimum of 75 percent effort conducting research during the period of the award.
Restrictions: The award is not intended for investigators who already have full time to perform research, or have substantial publication records or considerable research support indicating that they are well established in their fields. Foreign institutions are not eligible to apply.
Amount of Grant: $75,000-$100,000 average

Date(s) Application is Due: Feb 12; Mar 12; Jun 12; Jul 12; Oct 12; Nov 12
Contact: Susan Grove, Program Officer; (301) 435-0714; email: sg16d@nih.gov or grantsinfo@od.nih.gov
Internet: http://grants.nih.gov/grants/guide/contacts/PA-09-038_contacts.html
Sponsor: National Institutes of Health (NIH)
6701 Rockledge Drive
Bethesda, MD 20892

NIH Individual Postdoctoral Fellowships in Genomics and Related ELSI Topics 4122
The purpose of these fellowships is to train scientists who will have the multidisciplinary skills that will enable them to engage in research to accomplish the short- and long-term objectives of the HGP and similar genomic projects and to take full advantage of the resulting genomic data and resources to solve biomedical and bioethical problems. Broad areas of research that are relevant include genomic analysis (including technology development) and the ethical, legal, and social implications of human genetics research. The goal of the fellowship program in the area of genomic analysis and technology is to train highly skilled scientists who will use the expertise gained to develop research programs in the mapping and sequencing of the human genome and the genomes of other organisms, in sequence variation, in the analysis and interpretation of the resulting data, and in the development of biological, medical, or biotechnological applications based on the data, or in the development of technologies to support any of this research. Inquiries may be directed to program staff at the appropriate NIH institute. Annual deadline dates vary; contact program staff for exact dates. PA: PA-99-122
Requirements: Individuals must meet the criteria for the NIH National Research Service Awards for Individual Postdoctoral Fellows (F32) described in program announcement PA-99-025, available in the NIH Guide for Grants and Contracts. By the time of award, individuals must be citizens or noncitizen nationals of the United States, or have been lawfully admitted to the United States for permanent residence. Before an NRSA award can be activated, the individual must have received a doctoral degree or its equivalent to be eligible for an individual postdoctoral fellowship (F32). To be eligible for a senior fellowship (F33), individuals must have received a doctoral degree and must have had at least seven subsequent years of relevant research or professional experience.
Restrictions: Individuals on temporary or student visas are not eligible.
Amount of Grant: $26,256-$41,268 depending on experience
Date(s) Application is Due: Apr 5; Aug 5; Dec 5
Contact: Program Officer
Internet: http://www.grants.nih.gov/grants/guide/pa-files/PA-99-122.html
Sponsor: National Institutes of Health
6701 Rockledge Drive
Bethesda, MD 20892

NIH Innovations in Biomedical Computational Science and Technology Grants 4123
The NIH is interested in promoting research and developments in computational science and technology that will support rapid progress in areas of scientific opportunity in biomedical research. As defined here, biomedical computing or biomedical information science and technology includes database design, graphical interfaces, querying approaches, data retrieval, data visualization and manipulation, data integration through the development of integrated analytical tools, and tools for electronic collaboration, as well as computational and mathematical research including the development of structural, functional, integrative, and analytical models and simulations. Because the nature and scope of the proposed research will vary from application to application, it is anticipated that the size and duration of each award will also vary.
Requirements: Application can be made by the following institutions and organizations: for-profit organizations; non-profit organizations; public or private institutions (such as universities, colleges, hospitals, and laboratories); units of state government; units of local government; eligible agencies of the Federal government; foreign institutions; domestic institutions; faith-based or community-based organizations; Indian/Native American tribal governments (federally recognized); Indian/Native American tribal governments (other than federally recognized); and Indian/Native American tribally designated organizations.
Amount of Grant: $500,000 maximum per year
Date(s) Application is Due: Feb 5; Jun 5; Oct 5
Contact: Peter Lyster, Ph.D.; (301) 451-6446; email: lysterp@mail.nih.gov
Internet: http://grants.nih.gov/grants/guide/pa-files/PAR-09-218.html
Sponsor: National Institutes of Health (NIH)
45 Center Drive, Room 2AS.55k, MSC 6200
Bethesda, MD 20892-6200

NIH Innovations in Biomedical Computational Science and Technology Initiative Grants for SBIR 4124
This funding opportunity announcement (FOA) solicits Small Business Innovation Research (SBIR) grant applications from small business concerns (SBCs) that propose innovative research in biomedical computational science and technology to promote the progress of biomedical research. There exists an expanding need to speed the progress of biomedical research through the power of computing to manage and analyze data and to model biological processes. The NIH is interested in promoting research and developments in biomedical computational science and technology that will support rapid progress in areas of scientific opportunity in biomedical research. As defined here biomedical computing or biomedical information science and technology includes database design, graphical interfaces, querying approaches, data retrieval, data visualization and manipulation, data integration through the development of integrated analytical tools, and tools for electronic collaboration, as well as computational research including the development of structural, functional, integrative, and analytical models and simulations.
Requirements: Only United States small business concerns (SBCs) are eligible to submit SBIR applications.
Amount of Grant: $150,000 maximum per year in phase 1; $750,000 maximum per year in phase 2
Date(s) Application is Due: Apr 5; Aug 5; Dec 5
Contact: Peter Lyster, Ph.D.; (301) 451-6446; email: lysterp@mail.nih.gov
Internet: http://grants.nih.gov/grants/guide/pa-files/PAR-07-160.html
Sponsor: National Institutes of Health (NIH)
45 Center Drive, Room 2AS.55k, MSC 6200
Bethesda, MD 20892-6200

NIH Innovations in Biomedical Computational Science and Technology Initiative Grants for STTR 4125
This funding opportunity announcement (FOA) solicits Small Business Technology Transfer (STTR) grant applications from small business concerns (SBCs) that propose innovative research in biomedical computational science and technology to promote the progress of biomedical research. There exists an expanding need to speed the progress of biomedical research through the power of computing to manage and analyze data and to model biological processes. The NIH is interested in promoting research and developments in biomedical computational science and technology that will support rapid progress in areas of scientific opportunity in biomedical research. As defined here biomedical computing or biomedical information science and technology includes database design, graphical interfaces, querying approaches, data retrieval, data visualization and manipulation, data integration through the development of integrated analytical tools, and tools for electronic collaboration, as well as computational research including the development of structural, functional, integrative, and analytical models and simulations.
Requirements: Only United States small business concerns (SBCs) are eligible to submit STTR applications.
Amount of Grant: $150,000 maximum per year in phase 1; $750,000 maximum per year in phase 2
Date(s) Application is Due: Apr 5; Aug 5; Dec 5
Contact: Peter Lyster, Ph.D.; (301) 451-6446; email: lysterp@mail.nih.gov
Internet: http://grants.nih.gov/grants/guide/pa-files/PAR-07-161.html
Sponsor: National Institutes of Health (NIH)
45 Center Drive, Room 2AS.55k, MSC 6200
Bethesda, MD 20892-6200

NIH Investigator-Initiated Interactive Research Project Grants 4126
This program provides support for formal, investigator-initiated, collaborative relationships. An IRPG group could be used constructively to support collaborative efforts to accelerate the development of fundamental knowledge and/or enhance the clinical application of that knowledge. The IRPG program may provide an effective means to fund applications for focused clinical research and related, correlative laboratory studies. Annual deadline dates may vary; contact program staff for exact dates.
Requirements: Applications may be submitted by foreign and domestic, for-profit and nonprofit organizations, both public and private; state and local governments; and eligible agencies of the federal government.
Date(s) Application is Due: Feb 15; Jun 15; Oct 15
Contact: Division of Research Grants, (301) 435-0715
Internet: http://www.grants.nih.gov/grants/guide/pa-files/PA-96-001.html
Sponsor: National Institutes of Health
6701 Rockledge Drive
Bethesda, MD 20892

NIH Low Birthweight in Minority Populations Research Grants 4127
This PA solicits research to expand our understanding of the underlying mechanisms that contribute to the ethnic variations in low birth weight (LBW) and strategies for prevention. The goals of this PA are the development of innovative strategies to prevent LBW in minority populations and to expand our understanding of how psychosocial and environmental factors affect or interact with the biologic mechanisms that influence pregnancy outcomes. Inquiries may be directed to the program staff at the appropriate NIH institute: Dr. Cara Krulewitch, (301) 594-2542, Division of Extramural Activities, NINR, Building 45, Bethesda, MD 20892; Dr. Dennis Mangan, (301) 594-2421, Division of Extramural Research, NIDCR, 45 Center Drive, Bethesda, MD 20892; Dr. Gwen Collman, (919) 541-4980, Division of Extramural Research and Training, NIEHS, Box 12233 Research Triangle Park, NC 27709; and Dr. Jonelle Rowe, (301) 496-5575, Center for Research for Mothers and Children, NICHD, 6100 Executive Boulevard, Rockville, MD 20852. PA: PA-99-045
Requirements: Applications may be submitted by foreign and domestic for-profit and nonprofit organizations, public and private.
Contact: Program Officer
Internet: http://www.grants.nih.gov/grants/guide/pa-files/PA-99-045.html
Sponsor: National Institutes of Health
6701 Rockledge Drive
Bethesda, MD 20892

NIH Mechanisms of Orofacial Pain: Anatomy, Genomics, and Proteomics 4128
NINDS and NIDCR invite applications to stimulate and support innovative, interdisciplinary research studies to elucidate the molecular mechanisms underlying orofacial pain, particularly the discovery of proteins and protein networks critical to

processing nociceptive information. The purpose of this RFA is to encourage the use of genomic and proteomic approaches and imaging techniques to clarify the molecular events involved in acute orofacial pain, the transition from unrelieved acute pain to chronic pain (i.e. neuroplasticity), neuronal hyperexcitability as manifested by hyperalgesia and allodynia, and chronic orofacial pain disorders of an inflammatory and neuropathic origin. This improved understanding could lead to new therapeutic interventions to effectively treat chronic pain conditions. A letter of intent is requested by April 19; full proposal is due May 14. Annual deadline dates may vary; contact program staff for exact dates. RFA: RFA-DE-05-004
Requirements: Applications may be submitted by domestic and foreign, for-profit and nonprofit organizations, both public and private; units of state and local governments; and eligible agencies of the federal government.
Amount of Grant: $250,000 maximum
Date(s) Application is Due: Apr 19; May 14
Contact: Mary Daley, Chief Grants Management Officer, (301) 594-4808; fax: (301) 480-3562; email: daleym@mail.nih.gov
Internet: http://www.grants.nih.gov/grants/guide/rfa-files/RFA-DE-05-004.html
Sponsor: National Institutes of Health
45 Center Drive, Building 45, Room 4AN-44F
Bethesda, MD 20892-6402

NIH Mechanisms Underlying Individual Variations in Drug Responses 4129
The purpose of this program announcement is to stimulate research into identifying the critical candidate proteins and/or genes that play essential roles in determining individual variations in drug responses. The study of pharmacogenetic/pharmacogenomic variation presents opportunities to a wide range of researchers, working at levels ranging from the most molecular to the most clinical, in the fields of pharmacology, genetics, genomics, medicine, epidemiology, statistics, and computer science. Although NIGMS is leading the initiative, a number of other institutes of NIH are participating, and they are particularly interested in studies in the following areas: NCI—drug response proteins and/or genes involved in cancer, genetic/molecular epidemiologic studies of polymorphisms associated with cancer, and studies of nicotine susceptibility; NHLBI—drug response proteins and/or genes involved in cardiovascular, pulmonary, and hematologic systems; NICHD—development of drug response proteins and/or genes in children; NIEHS—drug response proteins and/or genes affected by environmental exposure; NIMH—drug response proteins and/or genes involved in the treatment of mental disorders and cognitive disorders; NIA—drug response proteins and/or genes involved in susceptibility to adverse drug reactions and drug-drug interactions in older persons; NIAAA—drug response proteins and/or genes that mediate the effects of alcohol and/or drugs that modify alcohol's effects. Applicants may contact the appropriate institute for information: Dr. Kumiko Iwamoto, (301) 496-9600, Division of Cancer Control and Population Sciences, NCI, Executive Plaza N, Rockville, MD 20852; Dr. Mary Wolpert, (301) 496-8783, Division of Cancer Treatment and Diagnosis, NCI, Executive Plaza N, Rockville, MD 20852; Dr. Peter Savage, (301) 435-0421, Division of Epidemiology and Clinical Applications, NHLBI, 2 Rockledge Center, Bethesda, MD 20817; Dr. George Giacoia, (301) 496-5589, Endocrinology, Nutrition and Growth Branch, NICHD, Building 62, Bethesda, MD 20892; Dr. Jose Velazquez, (919) 542-4998, Division of Extramural Research and Training, NIEHS, P.O. Box 12233, Research Triangle Park, NC 27709; Dr. Linda Brady, (301) 443-9875, Division of Basic and Clinical Science Research, NIMH, Parklawn Building, Rockville, MD 20857; Dr. Stanley Slater, (301) 496-6761, Geriatrics Program, NIA, Gateway Building, Bethesda, MD 20892; or Dr. Samir Zakhari, (301) 443-0799, Division of Basic Research, Willco Building, NIAAA, Bethesda, MD 20892. PA: PA-99-016
Requirements: Applications may be submitted by domestic and foreign, for-profit and nonprofit organizations, units of state and local governments, and eligible agencies of the federal government
Contact: Program Officer
Internet: http://www.grants.nih.gov/grants/guide/pa-files/PA-99-016.html
Sponsor: National Institutes of Health
45 Center Drive, Natcher Building
Bethesda, MD 20892

NIH Mechanisms Underlying Secondary Conditions in Mobility Disorders 4130
Primary disabling conditions are often associated with secondary conditions that further inhibit full participation, cause additional negative health consequences, and dramatically increase the burden of the disorder. Treatment or prevention of such secondary conditions (e.g., spasticity, pain, ulceration, infection, weakness and limited motion) will help to create a more enabling environment for persons with limited mobility, so that functioning of the whole person is enhanced. To accomplish these goals, additional research is required to investigate the underlying mechanisms that exacerbate secondary conditions in mobility disorders. Inquiries may be directed to program staff at the appropriate NIH institute: Dr. Mary Ellen Cheung, (301) 402-2242, NICHD, 6100 Executive Boulevard, Bethesda, MD 20892; Dr. Chhanda Dutta, (301) 435-3048 NIA, 7201 Wisconsin Avenue, Bethesda, MD 20892; Dr. James Panagis, (301) 594-5055, NIAMS, 45 Center Drive, Bethesda, MD 20892; Dr. Sanford Garfield, (301) 594-8803, NIDDK, 45 Center Drive, Bethesda, MD 20892; Dr. Scott Somers, (301) 594-5560, NIGMS, 45 Center Drive, Bethesda, MD 20892; and Dr. Karin Helmers, Program Director, (301) 594-2177, NINR, Building 45, Bethesda, MD 20892. Annual deadline dates vary; contact program staff for exact dates. PA: PA-99-096
Requirements: Applications may be submitted by domestic and foreign, for-profit and nonprofit organizations, both public and private; units of state or local governments; and eligible agencies of the federal government.
Amount of Grant: $250,000 maximum per year
Contact: Program Officer
Internet: http://www.grants.nih.gov/grants/guide/pa-files/PA-99-096.html
Sponsor: National Institutes of Health
6701 Rockledge Drive
Bethesda, MD 20892

NIH Mentored Clinical Scientist Development Award 4131
This award enables candidates holding professional degrees (e.g., M.D., D.O., D.V.M., or equivalent degrees) to undertake 3 to 5 years of special study and supervised research with the goal of becoming independent investigators. The award also allows awardees to pursue a research career development program suited to their experience and capabilities under a mentor who is competent to provide guidance in the chosen research area. Institutions may submit applications on behalf of candidates who hold professional degrees. At least 2 years must have elapsed since the health professional degree was granted. Candidates can have varying levels of clinical training and research experience. New applications are due February 12, June 12, and October 12, while resubmitted applications are due March 12, July 12, and November 12.
Requirements: All candidates must be U.S. citizens, non-citizen nationals, or legal permanent residents of the U.S. Persons with temporary or student visas are not eligible. The grantee institution must have a strong, well-established research and research training program in the chosen area, accomplished faculty in the basic and clinical sciences, and a commitment to the candidate's research development. The proposed program should include an appropriate mentor.
Date(s) Application is Due: Feb 12; Mar 12; Jun 12; Jul 12; Oct 12; Nov 12
Contact: Susan Grove, Program Officer; (301) 435-0714; email: sg16d@nih.gov or grantsinfo@od.nih.gov
Internet: http://grants.nih.gov/grants/guide/pa-files/pa-06-512.html
Sponsor: National Institutes of Health
6701 Rockledge Drive
Bethesda, MD 20892

NIH Mentored Patient-Oriented Research Career Development Award 4132
The purpose of this program is to support the career development of investigators who have made a commitment to focus their research endeavors on patient-oriented research. This mechanism provides support for three to five years of supervised study and research for clinically trained professionals who have the potential to develop into productive, clinical investigators focusing on patient-oriented research. Inquiries may be directed to program staff at the appropriate NIH institute. See the following web page for specific contact information: http://grants.nih.gov/grants/guide/contacts/PA-09-043_contacts.html
Requirements: Applicants must have a clinical doctoral degree or its equivalent and must have completed their clinical training, specialty and, if applicable, sub-specialty training prior to receiving an award. However, candidates may submit an application prior to the completion of clinical training. Candidates must identify a mentor with extensive research experience, and must be willing to spend a minimum of 75 percent of full-time professional effort conducting research career development and clinical research.
Restrictions: Current and former principal investigators on NIH research project (R01), FIRST Awards (R29), comparable career development awards (K01, K07, or K08), sub-projects of program project (P01) or center grants (P50), and the equivalent are not eligible for funding under this grant.
Contact: Susan Grove, Program Officer; (301) 435-0714; email: sg16d@nih.gov or grantsinfo@od.nih.gov
Internet: http://grants.nih.gov/grants/guide/pa-files/PA-09-043.html#SectionVII
Sponsor: National Institutes of Health (NIH)
6701 Rockledge Drive
Bethesda, MD 20892

NIH Mentored Research Scientist Development Awards 4133
The overall objective of the NIH Research Career Development Award program is to prepare qualified individuals for careers that have a significant impact on the health-related research needs of the Nation. The objective of the NIH Mentored Research Scientist Development Award is to provide support for a sustained period of protected time for intensive research career development under the guidance of an experienced mentor, or sponsor, in the biomedical, behavioral or clinical sciences leading to research independence. The expectation is that through this sustained period of research career development and training, awardees will launch independent research careers and become competitive for new research project grant funding. Inquiries should be directed to program staff at the appropriate NIH institute. See the website for specific contact information.
Requirements: Applications may be submitted, on behalf of candidates, by domestic, non-federal organizations, public or private, such as medical, dental, or nursing schools or other institutions of higher education. Candidates must be U.S. citizens or non-citizen nationals, or must have been lawfully admitted for permanent residence by the time of award. Candidates must have a research or a health-professional doctorate or its equivalent; have demonstrated the capacity or potential for highly productive independent research in the period after the doctorate; must identify a mentor with extensive research experience; and must be willing to spend a minimum of 75 percent of full-time professional effort conducting research and research career development during the entire award period.
Restrictions: Individuals on temporary or student visas are not eligible for this award.
Amount of Grant: $75,000-$100,000 average
Date(s) Application is Due: Feb 12; Mar 12; Jun 12; Jul 12; Oct 12; Nov 12

Contact: Susan Grove, Program Officer; (301) 435-0714; email: sg16d@nih.gov or grantsinfo@od.nih.gov
Internet: http://grants.nih.gov/grants/guide/pa-files/PA-09-040.html
Sponsor: National Institutes of Health (NIH)
6701 Rockledge Drive
Bethesda, MD 20892

NIH Midcareer Investigator Award in Mouse Pathobiology Research Grants 4134
The purpose of the program is to provide support for established pathobiologists to allow them protected time to devote to mouse pathobiology research and to act as mentors for beginning investigators. The target candidates are outstanding scientists engaged in pathobiology research who are within 15 years of their specialty training, who can demonstrate the need for a period of intensive research focus as a means of enhancing their research careers, and who are committed to mentoring the next generation of mouse pathobiologists. The award is intended to relieve mouse pathobiologists from time consuming service obligations and administrative responsibilities, thereby increasing the opportunities for their own research and for mentoring. Direct inquiries regarding programmatic issues to: Dr. Franziska Grieder, Comparative Medicine Area, (301) 435-0744, NCRR, 6705 Rockledge Drive, Bethesda, MD 20892. Inquiries regarding research and training related to age-related pathology may be directed to: Dr. Huber Warner, (301) 496-6402, Biology of Aging Program, NIA, Gateway Building, Bethesda, MD 20892. PA: PAR-01-064
Requirements: Most candidates for this award will have a Doctor of Veterinary Medicine degree (or equivalent) from an institution recognized by the American Veterinary Medical Association (AVMA). In addition, individuals holding other clinical (such as the MD) or research (such as the PhD) degree(s) may apply for the award if they have been certified or have demonstrated the necessary expertise to perform high quality, funded research in mouse pathobiology. Candidates must have completed their specialty or research training within 15 years of submitting the application, and there is no age limit for candidates. In exceptional circumstances, the period of eligibility may be extended if it can be demonstrated that candidates had an interruption in their career progression due to family or personal circumstances. Applications may be submitted on behalf of candidates by domestic, nonfederal organizations, public or private. At the time of award, candidates must be U.S. citizens or noncitizen nationals.
Restrictions: Individuals on temporary visas are not eligible.
Contact: Program Officer
Internet: http://grants.nih.gov/grants/guide/pa-files/PAR-01-064.html
Sponsor: National Institutes of Health
6701 Rockledge Drive
Bethesda, MD 20892

NIH Molecular Epidemiology of Prostate Carcinogenesis Grants 4135
NCI, NIDDK, and NIEHS invite investigator-initiated research grant applications of molecular epidemiologic studies for understanding prostate cancer development and progression. The purpose of this initiative is to stimulate development and application of biological markers of prostate cancer risk and tumor aggressiveness in developing strategies for prostate cancer prevention and control. Of special interest are studies of markers to elucidate multiethnic differences in prostate cancer susceptibility. Interdisciplinary collaborations, which may include multicenter study sites as well as joint effort between NIH/NCI intramural and extramural scientists, are encouraged. These collaborations may include utilization of shared laboratory and specimen resources whenever possible. Investigations in understudied populations and in populations of contrasting risk, including those who are residing outside of the United States, are strongly encouraged. This PA invites a range of interdisciplinary investigations, including molecular epidemiologic, family, or population-based studies of prostate cancer, utilizing available or new/novel biomarkers (e.g., biochemical, molecular, cellular, genetic) and sources of specimens (e.g., prostate tissue, prostatic fluid, blood components) whenever possible. Annual deadline dates vary; contact program staff for exact dates. PA: PA-00-080
Requirements: Applications may be submitted by foreign and domestic, for-profit and nonprofit organizations, public and private.
Contact: Dr. Kumiko Iwamoto, Division of Cancer Control and Population Sciences, NCI (Executive Plaza N, Bethesda, MD 20892), (301) 435-4911; fax: (301) 402-4279; email: ki6n@nih.gov; Dr. Leroy Nyberg, Urology Program, NIDDK (Natcher Building, Bethesda, MD 20892), (
Internet: http://grants2.nih.gov/grants/guide/pa-files/PA-00-080.html
Sponsor: National Institutes of Health
6701 Rockledge Drive
Bethesda, MD 20892

NIH Mouse Mutagenesis and Phenotyping: Developmental Defects Research Grants 4136
Funds will be awarded to establish a facility for large-scale mutagenesis and phenotyping of developmental defects in the laboratory mouse. The immediate objective of the facility is to produce and characterize mouse strains harboring mutations that affect normal developmental processes. Ultimately, the mutant mice produced are expected to help elucidate the basic cellular, molecular, and genetic mechanisms that direct embryonic and post-embryonic growth and function, as well as yield insights into the mechanisms of human disease. It is anticipated that this facility will devise and perform efficient genome-wide mutagenesis, devise and perform high-throughput phenotyping to screen for mutations that disrupt normal developmental processes, and devise and perform detailed characterization of mutants that display defects in development. Mutant mice, protocols, assays, assessment criteria, and other materials and information generated in projects will be made available to the wider biomedical community. The activities of this facility will be coordinated with the facility(s) established in response to RFA MH-99-007, Mouse Mutagenesis and Phenotyping: Nervous System and Behavior. Inquiries may be directed to program staff at the appropriate NIH institute: Dr. Steven Klein, (301) 496-5541, Developmental Biology, Genetics and Teratology Branch, NICHD, 6100 Executive Boulevard, Rockville, MD 20852; Dr. Judith Greenberg, (301) 594-0943, Division of Genetics and Developmental Biology, NIGMS, Building 45, Bethesda, MD 20892; Dr. Huber Warner, (301) 496-6402, Biology of Aging Program, NIA, Gateway Building, Bethesda, MD 20892; Dr. William Sharrock, (301) 594-5055, Musculoskeletal Diseases Branch, NIAMS, Building 45, Bethesda, MD 20892; Dr. Judy Small, (301) 594-2425, Inherited Craniofacial Diseases and Injury Program, NIDCR, Building 45, Bethesda, MD 20892; Dr. Josie Briggs, (301) 496-6325, Division of Kidney, Urology and Hemotology, NIDDK, Building 31, 31 Center Drive, Bethesda, MD 20892; Dr. John Fakunding, (301) 435-0544, Heart Research Program, Division of Heart and Vascular Disease, NHLBI, 6701 Rockledge Drive, Bethesda, MD 20892. A letter of intent is requested by August 2; full application is due October 14. Annual deadline dates may vary; contact program staff for exact dates. RFA: HD-99-007
Requirements: Applications may be submitted by domestic public and private for-profit and nonprofit organizations such as universities, colleges, hospitals, laboratories, units of state and local government, and eligible agencies of the federal government.
Restrictions: Foreign institutions are not eligible to apply.
Amount of Grant: $2.8 million approximate total funding available
Date(s) Application is Due: Aug 2; Oct 14
Contact: Diane Watson, Grants Management Branch, National Institute of Child Health and Human Development, (301) 496-5001; fax: (301) 402-0915; email: watsond@exchange.nih.gov
Internet: http://www.grants.nih.gov/grants/guide/rfa-files/RFA-HD-99-007.html
Sponsor: National Institutes of Health
6701 Rockledge Drive
Bethesda, MD 20892

NIH Mouse Mutagenesis and Phenotyping: Nervous System and Behavior Research Grants 4137
Funds are awarded to establish facilities for large-scale mutagenesis and phenotyping of nervous system functions and complex behaviors in the laboratory mouse. Neuroscience-focused mutagenesis and phenotyping facilities established are expected to serve as a national resource by producing a bank of mouse strains that harbor a wide range of mutations affecting murine nervous system function and behavior. Data and biomaterials produced in projects supported will be made widely available to the scientific community. Projects may be funded for up to five years. Inquiries may be directed to program staff at the appropriate NIH institute: Dr. Hemin Chin, (301) 443-1706, Division of Neuroscience and Basic Behavioral Science, NIMH, 6001 Executive Boulevard, Bethesda, MD 20892; Dr. Jonathan Pollock, (301) 435-1309, Division of Basic Research, NIDA, 6001 Executive Boulevard, Bethesda, MD 20892; Dr. Robert Karp, (301) 443-4223, Division of Basic Research, NIAAA, 6000 Executive Boulevard, Bethesda, MD 20892; Dr.Bradley Wise, (301) 496-9350, Neuroscience and Neuropsychology of Aging Program, NIA, 7201 Wisconsin Avenue, Bethesda, MD 20892; Dr. Maria Giovanni, (301) 496-0484, Fundamental Retinal Processes, NEI, 6120 Executive Boulevard, Bethesda, MD 20892; Dr. Gabrielle Leblanc, (301) 496-5745, Division of Fundamental Neuroscience and Developmental Disorders, NINDS, 6001 Executive Boulevard, Bethesda, MD 20892; Dr. Rochelle Small, (301) 402-3464, Division of Human Communication, NIDCD, 6120 Executive Boulevard, Bethesda, MD 20892. A letter of intent is requested by August 2; full application is due October 14. Annual deadline dates vary; contact program staff for exact dates. RFA: MH-99-007
Requirements: Applications may be submitted by domestic public and private for-profit and nonprofit organizations such as universities, colleges, hospitals, laboratories, units of state and local government, and eligible agencies of the federal government.
Restrictions: Foreign institutions are not eligible.
Amount of Grant: $5.5 million total approximately
Date(s) Application is Due: Aug 2; Oct 14
Contact: Robert Barnie, (301) 443-4116; fax: (301) 443-0501; email: rb245s@nih.gov
Internet: http://www.grants.nih.gov/grants/guide/rfa-files/RFA-MH-99-007.html
Sponsor: National Institutes of Health
6701 Rockledge Drive
Bethesda, MD 20892

NIH Multidisciplinary Approach for Research on Oral Complications of HIV Infection 4138
This initiative is designed to encourage multidisciplinary research on the oral complications of HIV infection and their prevention that crosses several academic disciplines (e.g., virology, immunology, cell biology, pathology, epidemiology, biochemistry, pharmacology, medical imaging, etc.) It is anticipated that this research will increase the knowledge of the basic mechanisms involved in the pathogenesis of the oral disorders associated with AIDS and that it will identify novel strategies for prevention, treatment, and diagnosis of the oral manifestations of AIDS. A letter of intent is due by August 17; full applications are due September 14. Annual deadline dates may vary; contact program staff for exact dates. RFA Number: RFA-DE-05-003
Requirements: Applications may be submitted by domestic and foreign, for-profit and nonprofit organizations, both public and private; units of state and local governments; and eligible agencies of the federal government.
Amount of Grant: $1.1 million maximum per year
Date(s) Application is Due: Aug 17; Sep 14

Contact: Mary Daley, Grants Management Branch, NIDCR, (301) 594-4808; fax: (301) 480-3562; email: md74u@nih.gov
Internet: http://www.grants.nih.gov/grants/guide/rfa-files/RFA-DE-05-003.html
Sponsor: National Institutes of Health
Building 45, Room 4AN-44F
Bethesda, MD 20892-6402

NIH National Centers for Neurofibromatosis Research 4139
NINDS and NIDCD invite qualified investigators to submit grant applications for the establishment of National Centers for Neurofibromatosis Research. Recent discoveries have created important opportunities for basic, translational, and clinical research on the neurofibromatoses. The purpose of this Program Announcement (PA) is to encourage the formation and development of research centers that can capitalize on these opportunities, and ultimately develop therapeutic interventions for neurofibromatosis patients. These centers are intended to provide focused expertise and resources, and establish a multi-disciplinary environment that will accelerate research progress. PA Number: PAR-04-018
Requirements: Applications may be submitted by domestic and foreign, for-profit and nonprofit organizations, both public and private; units of state and local governments; and eligible agencies of the federal government.
Contact: Kathleen Howe, Grants Management Branch, NINDS, (301) 496-9231; fax: (301) 402-0219; email: kh52x@nih.gov; Sara Stone, Division of Extramural Activities, NIDCD, (301) 402-0909; fax: (301) 402-1758; email: stones@nidcd.nih.gov
Internet: http://grants1.nih.gov/grants/guide/pa-files/PAR-04-018.html
Sponsor: National Institutes of Health
6701 Rockledge Drive, Room 1040, MSC 7710
Bethesda, MD 20892-7710

NIH National Occupational Research Agenda Grants 4140
The purpose of the grants program is to develop knowledge that can be used in preventing occupational diseases and injuries and to better understand their underlying pathophysiology. The following types of research projects will be supported: causal research to identify and investigate the relationships between hazardous working conditions and associated occupational diseases and injury; the nature and magnitude of special risk factors experienced by older and/or minority workers; methods research to develop more sensitive means of evaluating hazards at work sites; and evaluations of the effectiveness of prevention and intervention programs, including new approaches or combinations of techniques such as control technologies, personal protective equipment, and changes in work organization factors which have been developed and implemented in workplaces. Inquiries may be directed to program staff at the appropriate NIH institute: Dr. Roy Fleming, (404) 639-3343, Research Grants, NIOSH, 1600 Clifton Rd NE, Atlanta, GA 30333; Dr. Kumiko Iwamoto, (301) 435-4911, Epidemiology and Genetics Research, NCI, 6130 Executive Boulevard, Bethesda, MD 20892; Dr. Gail Weinmann, (301) 435-0202, Division of Lung Diseases, NHLBI, 6701 Rockledge Drive, Bethesda, MD 20892; Dr. Sidney Stahl, (301) 402-4156, Behavioral and Social Research, NIA, 7201 Wisconsin Avenue, Bethesda, MD 20892; Dr. Amy Donhaue, (301) 402-3458, Hearing and Balance/Vestibular Sciences, NIDCD, 6120 Executive Boulevard, Bethesda, MD 20892; Dr. Gwen Collman, (919) 541-4980, Division of Extramural Research and Training, NIEHS, P.O. Box 12233, Research Triangle Park, NC 27709. Awards will be made for projects lasting one to three years. Annual deadline dates may vary; contact program staff for exact dates. RFA: OH-99-002
Requirements: Applications may be submitted by domestic and foreign public and private nonprofit and for-profit organizations; state and local governments or their bona fide agents; and federally recognized Indian tribal governments, Indian tribes, or Indian tribal organizations.
Restrictions: Any organization involved in lobbying activities is not eligible.
Amount of Grant: $250,000 maximum per year
Contact: Program Officer
Internet: http://www.grants.nih.gov/grants/guide/rfa-files/RFA-OH-99-002.html
Sponsor: National Institutes of Health
6701 Rockledge Drive
Bethesda, MD 20892

NIH National Research Service Awards—Senior Fellows 4141
This program awards senior fellowships to experienced scientists who wish to make major changes in the direction of their research careers or who wish to broaden their scientific background by acquiring new research capabilities. These awards will enable individuals with at least seven years of research experience beyond the doctorate, and who have progressed to the stage of independent investigator, to take time from regular professional responsibilities for the purpose of receiving training to increase their scientific capabilities. In most cases, this award is used to support sabbatical experiences for established independent scientists. This program is not designed for postdoctoral- level investigators seeking to prove their research potential prior to independence. Inquiries may be directed to program staff at the appropriate NIH institute: Dr. Robin Barr, (301) 496-9322, NIA; Dr. Ernestine Vanderveen, (301) 443-2530, NIAAA; Dr. Milton Hernandez, (301) 496-3775, NIAID; Dr. Richard Lymn, (301) 594-5128, NIAMS; Dr. Lisa Begg, (301) 496-8580, NCI; Dr. Steven Klein, (301) 496-5541, NICHD; Dr. Daniel Sklare, (301) 496-1804, NIDCD; Dr. James Lipton, (301) 594-2618, NIDR; Dr. Judith Podskalny, (301) 594-8876, NIDDK; Dr. Carol Shreffler, (919) 541-1445, NIEHS; Dr. Chyren Hunter, (301) 496-5301, NEI; Dr. John Norvell, (301) 594-0533, NIGMS; Dr. Sarah Knox, (301) 435-0409, NHLBI; Dr. Henry Khachaturian, (301) 496-4188, NINDS; Dr. Nell Armstrong, (301) 594-5973, NINR; Dr. Bettie Graham, (301) 496-7531, NCHGR; Dr. David Wilde, (301) 435-0799, NCRR; and Dr. Neal West, (301) 402-5867, NCCAM. Annual deadline dates may vary; contact program staff for exact dates. PA: PA-00-131
Requirements: Applicants must have at least seven years of postdoctoral research or professional experience in the field. By the time of award, individuals must be citizens or noncitizen nationals of the United States or have been lawfully admitted to the United States for permanent residence. The sponsoring institution may be private (profit or nonprofit) or public.
Date(s) Application is Due: Apr 5; Aug 5; Dec 5
Contact: Program Officer; email: grantsinfo@nih.gov
Internet: http://grants.nih.gov/grants/guide/pa-files/PA-00-131.html
Sponsor: National Institutes of Health
6701 Rockledge Drive
Bethesda, MD 20892

NIH New Directions in Pain Research Grants 4142
This program encourages investigator-initiated research project grant applications to study mechanisms underlying analgesic response and pain to advance the development of novel pain interventions, treatments, and management strategies. The purpose of this PA is to inform the scientific community of broad, shared interests in pain research encompassing the various components of the NIH, and to stimulate and foster a wide range of basic, translational, and patient-oriented clinical studies on pain. Applications are particularly encouraged to study pain throughout the lifespan from the perspectives of molecular genetics, transcriptional controls, signal transduction, including cellular/molecular mechanisms, innovative imaging technologies, plasticity, and from hormonal or gender influences. The pain experience needs to be examined at all levels of analysis from the gene, molecule, cell, tissue, and organ, to the individual, family, and community, with the ultimate goal of developing new insights into pain intervention, treatment, and management. Inquiries may be directed to program staff at the appropriate NIH institute: Dr. Cheryl Kitt, (301) 496-1431, Division of Convulsive, Infectious, and Immune Disorders, NINDS, Federal Building, Bethesda, MD 20892; Dr. Patricia Bryant, (301) 594-2095, Temporomandibular Joint Disorders and Neuropathies and Neurodegenerative Diseases, NIDRC, 45 Center Drive, Bethesda, MD 20892; Claudette Varricchio, (301) 496-8541, Division of Cancer Prevention, NCI, 6130 Executive Boulevard, Bethesda, MD, 20892; Dr. Judith Finkelstein, (301) 496-9350, Neuroscience and Neuropsychology of Aging Programs, NIA, 7201 Wisconsin Avenue, Bethesda, MD, 20892; Dr. John Killen, (301) 496-0545, Division of Acquired Immunodeficiency Syndrome, NIAID, 6003 Executive Boulevard, Rockville, MD 20892; Dr. James Panagis, (301) 594-5055, Orthopedics Program, NIAMS, 45 Center Drive, Bethesda, MD 20892; Dr. Rochelle Small, Division of Human Communication, (301) 402-3464, NIDCD, 6120 Executive Boulevard, Bethesda, MD 20892; Dr. David Thomas, (301) 443-4975, Division of Basic Research, NIDA, 5600 Fishers Ln, Rockville, MD 20857; Dr. Alison Cole, (301) 594-1826, Division of Pharmacology, Physiology, and Biological Chemistry, NIGMS, 45 Center Drive, Bethesda, MD 20892; Dr. Mary Leveck, (301) 594-5963, Program Director, NINR, Building 45, Bethesda, MD 20892; and Joyce Rudick, (301) 402-1770, Office of Research on Women's Health, NIH, Building 1, Bethesda, MD 20892. PA: PA-98-102
Requirements: Applications may be submitted by foreign and domestic, for-profit and nonprofit organizations, public and private.
Contact: Program Officer
Internet: http://www.grants.nih.gov/grants/guide/pa-files/PA-98-102.html
Sponsor: National Institutes of Health
9000 Rockville Pike
Bethesda, MD 20892

NIH Nonhuman Primate Immune Tolerance Cooperative Study Group Grants 4143
NIAID and NIDDK invite applications from single institutions or consortia of institutions to participate in a multi-center, cooperative research program to evaluate existing and new tolerance induction treatment regimens and to elucidate the underlying mechanisms of the induction, maintenance and/or loss of tolerance in nonhuman primate models of kidney and islet transplantation. The goals of this research program are to evaluate further the safety, toxicity, and efficacy of existing tolerance induction regimens; define the underlying mechanisms of action of the therapeutic approaches under investigation; and develop and validate immune and/or surrogate markers of the induction, maintenance, and loss of tolerance, graft function, graft acceptance, and graft survival. The deadline for Letters of Intent is August 22, with full applications due September 22.
Requirements: Applications may be submitted by domestic for-profit and nonprofit organizations and public and private institutions.
Restrictions: Foreign institutions are not eligible to apply. This RFA will not support: studies in animal models other than non-human primates; human studies or trials; xenotransplantation; preliminary development of a non-human primate transplantation model; studies to improve the isolation, preservation, or supply of organs, tissues or cells; hematopoietic stem cell transplantation, unless in the context of kidney or islet transplantation (i.e. bone marrow chimerism experiments); heart or lung transplantation models; or studies of non-immunological side effects of an immuno-modulatory agent.
Amount of Grant: $6.7 million approximate total
Date(s) Application is Due: Aug 22; Sep 22
Contact: Kristy Kraemer, Ph.D., Division of Allergy, Immunology and Transplantation; (301) 496-5598; fax: (301) 480-0693; email: kkraemer@niaid.nih.gov
Internet: http://grants.nih.gov/grants/guide/rfa-files/RFA-AI-06-018.html
Sponsor: National Institutes of Health
6610 Rockledge Drive, Room 3043, MSC-6601
Bethesda, MD 20892

NIH NRSAs for Individual Predoctoral Fellowships 4144

The awards are provided to individuals for research training in specified areas of biomedical and behavioral research to help ensure that highly trained scientists will be available in adequate numbers and in appropriate research areas and fields to meet the nation's mental health, drug abuse, alcohol abuse, and alcoholism research needs. As such, this program is intended to provide a mechanism to train future generations of outstanding scientists committed to pursuing a career in mental health, drug abuse, and/or alcohol abuse and alcoholism research. Inquiries may be directed to program staff at the appropriate NIH institute: Dr. Henry Khachaturian, (301) 443-4335, NIMH, 5600 Fishers Ln, Rockville, MD 20857; Dr. Cindy Miner, (301) 443-6071, NIDA, 5600 Fishers Ln, Rockville, MD 20850; and Dr. Tina Vanderveen, (301) 443-2531, NIAAA, 6000 Executive Boulevard, Bethesda, MD 20892. Annual deadline dates vary; contact program staff for exact dates. PA: PA-00-125

Requirements: Applicants must be citizens or non-citizen nationals of the United States, or have been lawfully admitted to the United States for permanent residence and have in their possession an Alien Registration Receipt Card at the time of award. Applicants must have received, as of the activation date of the award, a baccalaureate degree and must be enrolled in a program leading to a research doctorate such as the PhD or DSc, or a combined clinical and research degree such as MD/PhD, by the proposed fellowship activation date.

Contact: Program Officer

Internet: http://grants.nih.gov/grants/guide/pa-files/PA-00-125.html

Sponsor: National Institutes of Health
6701 Rockledge Drive
Bethesda, MD 20892

NIH NRSAs Predoctoral Fellowship Awards for Students with Disabilities 4145

The intent of this program is to encourage students with disabilities to seek graduate degrees and thus further the goal of increasing the number of scientists with disabilities who are prepared to pursue careers in biomedical and behavioral research. Inquiries may be directed to program staff at the appropriate NIH institute: Dr. Robin Barr, (301) 496-9322, NIA; Dr. Ernestine Vanderveen, (301) 443-1273, NIAAA; Dr. Milton Hernandez, (301) 496-7291, NIAID; Dr. Richard Lymn, (301) 594-5128, NIAMS; Dr. John Schneider, (301) 496-8580, NCI; Hildegard Topper, (301) 496-0104, NICHD; Dr. Daniel Sklare, (301) 496-1804, NIDCD; Dr. James Lipton, (301) 594-2618, NIDR; Dr. Ronald Margolis, (301) 594-8819, NIDDK; Dr. Charles Sharp, (301) 443-1887, NIDA; Dr. Michael Galvin Jr., (919) 541-7825, NIEHS; and Dr. Maria Giovanni, (301) 496-0484, NEI. Annual deadline dates may vary; contact program staff for exact dates. PA: PA-95-028

Requirements: Applicants must be citizens or non-citizen nationals of the United States, or have been lawfully admitted to the United States for permanent residence.

Restrictions: Individuals on temporary or student visas are not eligible.

Amount of Grant: $10,008 annual stipend plus a $2000 institutional allowance

Contact: Program Officer

Internet: http://www.grants.nih.gov/grants/guide/pa-files/PA-95-028.html

Sponsor: National Institutes of Health
6701 Rockledge Drive
Bethesda, MD 20892

NIH Pathological Gambling: Basic, Clinical, and Services Research Grants 4146

NIMH, NIAAA, and NIDA invite investigator-initiated research relevant to their programs that will advance basic, clinical, and services research concerning pathological gambling, as well as enable more effective clinical assessment, prevention, and treatment. This program announcement encourages research that builds on extant research findings concerning pathological gambling and research that expands the breadth and depth of scientific knowledge through increased involvement of various disciplines, e.g., epidemiology; genetics; neuroscience; developmental psychopathology; and behavioral, cognitive, and social science. Multidisciplinary research is especially encouraged. Major questions include factors that predispose for pathological gambling and that distinguish types of pathological gamblers; models for explaining pathological gambling, e.g., social learning, addiction, impulse dyscontrol; developmental pathways and course; gender, age, and sociocultural factors; mechanisms involved in onset, continuation, and cessation; efficacious clinical interventions for the prevention of onset, for treatment, and for prevention and management of relapse; effectiveness of interventions in general use; and effects of the organization and financing of services. Inquiries may be directed to the appropriate NIH institute: Dr. James Breiling, (301) 443-3527, Division of Mental Disorders, Behavior, and AIDS Research, NIMH, 5600 Fishers Ln, Rockville, MD 20857; Darryl Bertolucci, (301) 443-4898, Epidemiology Branch, NIAAA, 6000 Executive Boulevard, Bethesda, MD 20892; and Dr. Meyer Glantz, Division of Epidemiology and Prevention Research, NIDA, 5600 Fishers Ln, Rockville, MD 20857. Annual deadline dates may vary; contact program staff for exact dates. PA: PA-98-106

Requirements: Applications may be submitted by public and private, foreign and domestic, for-profit and nonprofit organizations.

Date(s) Application is Due: Oct 19; Nov 16

Contact: Program Officer

Internet: http://grants.nih.gov/grants/guide/pa-files/PA-98-106.html

Sponsor: National Institutes of Health
6003 Executive Boulevard
Bethesda, MD 20892

NIH Predoctoral Fellowship Awards for Minority Students 4147

These fellowships are for well-qualified students from minority groups found to be underrepresented in the biomedical and behavioral sciences in the United States. The intent of this program is to encourage students from minority groups that are underrepresented in the biomedical and behavioral sciences to seek graduate degrees and thus further the goal of increasing the number of minority scientists who are prepared to pursue careers in biomedical and behavioral research. Inquiries may be directed to program staff at the appropriate NIH institute: Dr. Robin Barr, (301) 496-9322, NIA; Dr. Ernestine Vanderveen, (301) 443-1273, NIAAA; Dr. Milton Hernandez, (301) 496-7291, NIAID; Dr. Richard Lymn, (301) 594-5128, NIAMS; Dr. John Schneider, (301) 496-8580, NCI; Hildegard Topper, (301) 496-0104, NICHD; Dr. Daniel Sklare, (301) 496-1804, NIDCD; Dr. James Lipton, (301) 594-2618, NIDR; Dr. Ronald Margolis, (301) 594-8819, NIDDK; Dr. Charles Sharp, (301) 443-1887, NIDA; Dr. Michael Galvin Jr., (919) 541-7825, NIEHS; and Dr. Maria Giovanni, (301) 496-0484, NEI. Annual deadline dates may vary; contact program staff for exact dates. PA: PA-95-029

Requirements: Applicants must be citizens or non-citizen nationals of the United States, or have been lawfully admitted to the United States for permanent residence

Restrictions: Individuals on temporary or student visas are not eligible.

Amount of Grant: $10,008 annual stipend plus a $2000 institutional allowance

Contact: Program Officer

Internet: http://www.grants.nih.gov/grants/guide/pa-files/PA-95-029.html

Sponsor: National Institutes of Health
6701 Rockledge Drive
Bethesda, MD 20892

NIH Protein Structure Initiative (Structural Genomics) Research Grants 4148

The purpose of this PA is to encourage research on the development of methodology and technology underpinning the emerging field of structural genomics, whose goal is the understanding of protein structural families, structural folds, and the relation of structure and function. Projects related to high throughput structure determination by x-ray crystallography and/or NMR, as well as those addressing other constituent tasks of structural genomics, are relevant to this PA. Inquiries may be directed to the staff at the appropriate NIH institute: Dr. John Norvell, (301) 594-0533, Division of Cell Biology and Biophysics, NIGMS, 45 Center Drive, Bethesda, MD 20892; Dr. Abraham Levy, (301) 435-0755, Biomedical Technology, NCRR, 6705 Rockledge Drive, Bethesda, MD 20892; and Dr. Jose Velazquez, (919) 541-4998, Division of Extramural Research and Training, NIEHS, P.O. Box 12233, Research Triangle Park, NC 27709. Annual deadlne dates vary; contact program staff for exact dates. PA: PA-99-116

Requirements: Applications may be submitted by domestic and foreign, for-profit and nonprofit organizations, public and private.

Restrictions: Foreign institutions are not eligible for program project (P01) grants.

Date(s) Application is Due: Feb 1; Jun 1; Oct 1

Contact: Program Officer

Internet: http://www.grants.nih.gov/grants/guide/pa-files/PA-99-116.html

Sponsor: National Institutes of Health
6701 Rockledge Drive
Bethesda, MD 20892

NIH Protein Structure Initiative (Structural Genomics)—SBIR/STTR Grants 4149

The purpose of this SBIR/STTR program announcement is to encourage research on the development of methodology and technology underpinning the emerging field of structural genomics, whose goal is the understanding of protein structural families, structural folds, and the relation of structure and function. Projects related to high-throughput structure determination by x-ray crystallography and/or NMR, as well as those addressing other constituent tasks of structural genomics, are relevant to this PA. PA-99-117

Requirements: Any small business concern, independently owned and operated by U.S. citizens or lawfully admitted permanent resident aliens, which is located in the United States and is organized for profit, may apply.

Date(s) Application is Due: Apr 1; Aug 1; Dec 1

Contact: Dr. John Norvell, Division of Cell Biology and Biophysics, (301) 594-0533; fax: (301) 480-2004; email: norvellj@nigms.nih.gov

Internet: http://grants2.nih.gov/grants/guide/pa-files/PA-99-117.html

Sponsor: National Institutes of Health
6701 Rockledge Drive
Bethesda, MD 20892

NIH Receptors and Signaling in Bone in Health and Disease 4150

The objective of this initiative is to elicit grant submissions that focus on systemic hormones, local growth factors, and bone-active cytokines, their receptors and mechanisms of signaling in bone. While the primary focus is on basic research, the long-term emphasis is on identifying mechanisms or processes related to hormone action with potential applicability as targets for therapeutic agents that may have efficacy in the treatment of diseases that adversely affect bone, such as osteoporosis and primary hyperparathyroidism. Modular Grant applications will request direct costs in $25,000 modules, up to a total direct cost request of $250,000 per year.

Requirements: Applications may be submitted by domestic and foreign for-profit and nonprofit organizations, public and private, such as universities, colleges, hospitals, laboratories, units of State and local governments, and eligible agencies of the Federal Government. Racial/ethnic minority individuals, women, and persons with disabilities are encouraged to apply as principal investigators.

Amount of Grant: $250,000 maximum per year

Date(s) Application is Due: Feb 5; Jun 5; Oct 5

Contact: Susan Grove, Program Officer; (301) 435-0714; email: sg16d@nih.gov or grantsinfo@od.nih.gov

Internet: http://grants.nih.gov/grants/guide/pa-files/PA-00-017.html

Sponsor: National Institutes of Health (NIH)
6701 Rockledge Drive
Bethesda, MD 20892

NIH Recovery Act Limited Competition: Academic Research Enhancement Awards 4151

The purpose of the Academic Research Enhancement Award (AREA) program is to stimulate research in educational institutions that provide baccalaureate or advanced degrees for a significant number of the Nation's research scientists, but that have not been major recipients of NIH support. These AREA grants create opportunities for scientists and institutions otherwise unlikely to participate extensively in NIH programs, to contribute to the Nation's biomedical and behavioral research effort. AREA grants are intended to support small-scale health-related research projects proposed by faculty members of eligible, domestic institutions. The requested budget is limited to $300,000 total costs for the entire three-year project (maximum time period). Applicants should contact individuals at the specific departments listed on the web site. No Letters of Intent are necessary, though full applications should arrive by September 24.
Requirements: The following institutions and organizations may apply: public/state controlled institutions of higher education; private institutions of higher education; Hispanic-serving institutions; historically Black colleges and universities (HBCUs); tribally controlled colleges and universities (TCCUs); Alaska Native and Native Hawaiian serving institutions; public or private institutions and components of institutions defined below as health professional schools or colleges; and other academic components.
Restrictions: Foreign organizations/institutions are not permitted as the applicant organization.
Amount of Grant: $300,000 maximum total direct costs for a three-year period
Date(s) Application is Due: Sep 24
Contact: Susan Grove, Program Officer; (301) 435-0714; email: sg16d@nih.gov or grantsinfo@od.nih.gov
Internet: http://grants.nih.gov/grants/guide/rfa-files/RFA-OD-09-007.html
Sponsor: National Institutes of Health (NIH)
6701 Rockledge Drive
Bethesda, MD 20892

NIH Recovery Act Limited Competition: Biomedical Research, Development, and Growth to Spur the Acceleration of New Technologies (BRDG-SPAN) Pilot Program Grants 4152

This Funding Opportunity Announcement (FOA), supported by funds provided to the NIH under the American Recovery & Reinvestment Act of 2009 (Recovery Act or ARRA), Public Law 111-5, solicits grant applications for a new initiative called Biomedical Research, Development, and Growth to Spur the Acceleration of New Technologies (BRDG-SPAN) Pilot Program. The purpose of this pilot program is to address the funding gap between promising research and development (R&D) and transitioning to the market — often called the Valley of Death — by contributing to the critical funding needed by applicants to pursue the next appropriate milestone(s) toward ultimate commercialization; i.e., to carry out later stage research activities necessary to that end. This program aims to accelerate the transition of research innovations and technologies toward the development of products or services that will improve human health, help advance the mission of NIH and its Institutes and Centers (ICs), and create significant value and economic stimulus. This program also aims to foster partnerships among a variety of research and development (R&D) collaborators working toward these aims. The requested budget is limited to $1 million total costs per year for a maximum of three years. Applicants should contact individuals at the specific departments listed on the web site. Letters of Intent are due by August 3, and full applications should arrive by September 1.
Requirements: The applicant United States institution and organization must be located in the 50 states, territories and possessions of the U.S., Commonwealth of Puerto Rico, Trust Territory of the Pacific Islands, or District of Columbia.
Restrictions: Foreign organizations/institutions are not permitted as the applicant organization.
Amount of Grant: $1,000,000 total costs per year over three years
Date(s) Application is Due: Aug 3; Sep 1
Contact: Susan Grove, Program Officer; (301) 435-0714; email: sg16d@nih.gov or grantsinfo@od.nih.gov
Internet: http://grants.nih.gov/grants/guide/rfa-files/RFA-OD-09-008.html
Sponsor: National Institutes of Health (NIH)
6701 Rockledge Drive
Bethesda, MD 20892

NIH Recovery Act Limited Competition: Building Sustainable Community-Linked Infrastructure to Enable Health Science Research 4153

Applications are invited from domestic (United States) institutions/organizations proposing to develop or expand needed infrastructures that will fundamentally transform collaboration and communication between academic health centers and local communities. Such collaborative infrastructures are essential to advance the health science research enterprise while ensuring that important research findings are effectively disseminated and implemented to improve public health and advance health care delivery. The primary intent is to fund projects to develop infrastructure for productive and sustainable academic-community research partnerships that can be leveraged in the future for efficiently conducting research that includes and is relevant to affected communities, and through which research findings can be disseminated in a manner that maximizes impact on public health. It is expected that each grantee will be able to leverage this Recovery Act funding into future research grants using the infrastructure partnership between itself and the participating community. Letters of Intent are due by November 12, with final applications due by December 11.
Requirements: The following organizations and institutions may apply: public/state controlled institutions of higher education; private institutions of higher education; Hispanic-serving institutions; historically Black colleges and universities (HBCUs); tribally controlled colleges and universities (TCCUs); Alaska native and native Hawaiian serving institutions; nonprofits with 501(c)3 IRS status; nonprofits without 501(c)3 IRS status; small businesses; for-profit organizations; State governments; regional organizations; U.S. territories or possessions; Indian/Native American tribal governments; Indian/Native American tribally designated organizations; county governments; city or township governments; special district governments; independent school districts; public housing authorities/Indian housing authorities; eligible agencies of the Federal government; and faith-based or community based organizations.
Restrictions: This RFA is not intended to fund research or evaluation projects, clinical trials, or public health campaigns. Foreign organizations/institutions are not permitted as the applicant organization.
Date(s) Application is Due: Nov 12; Dec 11
Contact: Dr. Suzanne Heurtin-Roberts; (301) 402-2277; email: sheurtin@mail.nih.gov
Internet: http://grants.nih.gov/grants/guide/rfa-files/RFA-OD-09-010.html
Sponsor: National Institutes of Health (NIH)
1 Center Drive, Room B3-11
Bethesda, MD 20892

NIH Recovery Act Limited Competition: Small Business Catalyst Awards for Accelerating Innovative Research Grants 4154

This NIH Funding Opportunity Announcement, supported by funds provided to the NIH under the American Recovery and Reinvestment Act of 2009 (Recovery Act or ARRA), Public Law 111-5, invites grant applications from small business concerns that propose to accelerate innovation through high risk, high reward research and development (R&D) that has commercial potential and is relevant to the mission of the NIH. The Small Business Catalyst Award is further expected to support entrepreneurs of exceptional creativity, drawn from scientific and technological environments beyond NIH, who propose pioneering and possibly transformative approaches to addressing major biomedical or behavioral challenges with the potential for downstream commercial development. The Small Business Catalyst Award for Accelerating Innovative Research funding opportunity seeks to encourage fresh research perspectives and approaches to serve the mission of NIH. In particular, applications from small business concerns without a history of NIH Small Business Innovation Research (SBIR) or Small Business Technology Transfer (STTR) support may receive funding priority. Solicited are applications for support for projects that have the potential to generate high impact results (e.g., products, processes or services) and/or innovative research applications, research tools, techniques, devices, inventions, or methodologies. The outcomes of the research supported should have potential to lead to products that will improve public health and create significant value and economic stimulus. This FOA solicits early-stage ideas that promise to lead to major leaps forward in capabilities important to serving the mission of NIH rather than incremental improvements of existing technologies. In accord with the funding priority of this initiative to attract applicants without a history of SBIR/STTR support from NIH, the focus of the projects solicited by this FOA is on early stage technology development. Budget requests are limited to $200,000 total costs for a maximum project period of one year. Applicants should contact individuals at the specific departments listed on the web site. Letters of Intent are due by August 3, and full applications should arrive by September 1.
Requirements: Only United States small business concerns (SBCs) are eligible to submit SBIR applications. A small business concern is one that, at the time of award of SBIR Phase I and Phase II, meets all of the following criteria: is organized for profit, with a place of business located in the United States, which operates primarily within the United States or which makes a significant contribution to the United States economy through payment of taxes or use of American products, materials or labor; is in the legal form of an individual proprietorship, partnership, limited liability company, corporation, joint venture, association, trust or cooperative, except that where the form is a joint venture, there can be no more than 49 percent participation by foreign business entities in the joint venture; is at least 51 percent owned and controlled by one or more individuals who are citizens of, or permanent resident aliens in, the United States, or it must be a for-profit business concern that is at least 51% owned and controlled by another for-profit business concern that is at least 51% owned and controlled by one or more individuals who are citizens of, or permanent resident aliens in, the United States, except in the case of a joint venture, where each entity to the venture must be 51 percent owned and controlled by one or more individuals who are citizens of, or permanent resident aliens in, the United States; and has, including its affiliates, not more than 500 employees.
Amount of Grant: $200,000 over a one-year period
Date(s) Application is Due: Aug 3; Sep 1
Contact: Susan Grove, Program Officer; (301) 435-0714; email: sg16d@nih.gov or grantsinfo@od.nih.gov
Internet: http://grants.nih.gov/grants/guide/rfa-files/RFA-OD-09-009.html
Sponsor: National Institutes of Health (NIH)
6701 Rockledge Drive
Bethesda, MD 20892

NIH Recovery Act Limited Competition: Supporting New Faculty Recruitment to Enhance Research Resources through Biomedical Research Core Centers Grants 4155

This initiative, which is supported by funds provided to the NIH under the Recovery Act, is designed to provide the necessary resources for U.S. academic institutions/

organizations to enhance their biomedical research efforts through the development of Biomedical Research Core Centers. For this announcement, Biomedical Core Centers are defined as a community of multidisciplinary researchers focusing on areas of biomedical research relevant to NIH, such as centers, departments, programs, and/or trans-departmental collaborations or consortia. These Core Centers are designed to provide scientific and programmatic support for promising research faculty and their areas of research. The NIH invites applications that include plans to recruit and hire investigators to conduct biomedical research in all scientific disciplines, including the field of bioethics. Applicants should contact individuals at the specific departments listed on the web site. Letters of Intent are due by April 29, and full applications should arrive by May 29.
Requirements: The following institutions and organizations may apply: public/state controlled institutions of higher education; private institutions of higher education; Hispanic-serving institutions; historically Black colleges and universities (HBCUs); tribally controlled colleges and universities (TCCUs); Alaska Native and Native Hawaiian serving institutions; nonprofits with 501(c)3 IRS status (other than institutions of higher education); nonprofits without 501(c)3 IRS status (other than institutions of higher education); and Indian/Native American tribally designated organizations .
Restrictions: Foreign organizations/institutions are not permitted as the applicant organization.
Date(s) Application is Due: Apr 29; May 29
Contact: Susan Grove, Program Officer; (301) 435-0714; email: sg16d@nih.gov or grantsinfo@od.nih.gov
Internet: http://grants.nih.gov/grants/guide/rfa-files/RFA-OD-09-005.html
Sponsor: National Institutes of Health (NIH)
6701 Rockledge Drive
Bethesda, MD 20892

NIH Research On Ethical Issues In Human Subjects Research **4156**
The purpose of this funding opportunity announcement is to solicit research addressing the ethical challenges of human subjects research in order to optimize the protection of human subjects and enhance the ethical conduct of human subjects research. The research design for studies on ethical issues in human subjects research should be appropriate to the nature of the project(s) proposed and the disciplines involved. Given the conceptual and methodological complexity of many of these research questions, interdisciplinary and collaborative projects are encouraged, particularly those involving clinical researchers, ethicists, and behavioral/social scientists.
Requirements: The following organizations and institutions may apply: public/state controlled institutions of higher education; private institutions of higher education; nonprofits with 501(c)3 IRS status (other than institution of higher education); nonprofits without 501(c)3 IRS status (other than institution of higher education); small businesses; for-profit organizations (other than small business); State governments; U.S. territories or possessions; Indian/Native American tribal governments (Federally recognized); Indian/Native American tribal governments (other than Federally recognized); Indian/Native American tribally designated organizations; non-domestic (non-U.S.) entities (foreign organization); eligible agencies of the Federal government; and faith-based or community based organizations.
Date(s) Application is Due: Feb 5; Jun 5; Oct 5
Contact: Kim Witherspoon; (301) 496-8866; fax: (301) 480-4663; email: withersk@ctep.nci.nih.gov; Ellen M. Werner, Ph.D.; (301) 435-0050; fax: (301) 480-0868; email: wernere@nhlbi.nih.gov
Internet: http://grants.nih.gov/grants/guide/pa-files/PA-07-277.html
Sponsor: National Institutes of Health (NIH)
6130 Executive Blvd. (EPN), Room 7009, MSC 7432
Bethesda, MD 20892-7432

NIH Research on Sleep and Sleep Disorders **4157**
This Funding Opportunity Announcement (FOA) solicits grant applications proposing research to advance biomedical knowledge related to sleep or sleep disorders, improve understanding of the neurobiology or functions of sleep over the life-span, enhance timely diagnosis and effective treatment for individuals affected by sleep-related disorders, or implement and evaluate innovative community-based public health education and intervention programs. Because the nature and scope of the proposed research will vary from application to application, it is anticipated that the size and duration of each award will also vary. The total amount awarded and the number of awards will depend upon the mechanism numbers, quality, duration, and costs of the applications received.
Requirements: The following organizations and institutions may apply: public/state controlled institutions of higher education; private institutions of higher education; nonprofits with 501(c)3 IRS status; nonprofits without 501(c)3 IRS status; small businesses; State governments; non-domestic (non-U.S.) entities (foreign organizations); eligible agencies of the Federal government; and faith-based or community based organizations.
Date(s) Application is Due: Feb 5; Jun 5; Oct 5
Contact: Susan Grove, Program Officer; (301) 435-0714; email: sg16d@nih.gov or grantsinfo@od.nih.gov
Internet: http://grants.nih.gov/grants/guide/pa-files/pa-07-140.html
Sponsor: National Institutes of Health (NIH)
6701 Rockledge Drive
Bethesda, MD 20892

NIH Research Project Grants **4158**
The Research Project Grant (R01) is an award to support a discrete, specified, circumscribed project to be performed by named Project Directors/Principal Investigators (PDs/PIs) in areas representing the investigators' specific interests and competencies, based on the mission of the NIH. The R01 is the original and historically the oldest grant mechanism used by the NIH to support health-related research and development. The NIH awards R01 grants to institutions/organizations of all types. This mechanism allows the PDs/PIs to define the scientific focus or objective of the research based on particular areas of interest and competence. Although the PDs/PIs write the grant application and are responsible for conducting and supervising the research, the actual applicant is the research institution/organization. Research grant applications are assigned to an NIH IC based on receipt and referral guidelines and many applications are assigned to multiple ICs with related research interests. Each IC maintains a Web site with funding opportunities and areas of interest. Contact with an IC representative may help focus the research plan based on an understanding of the mission of the IC. For specific information about the mission of each NIH IC, see http://www.nih.gov/icd, which provides a brief summary of the research interests in each IC and access to individual IC home pages.
Requirements: The following organizations and institutions may apply: public/state controlled institutions of higher education; private institutions of higher education; Hispanic-serving institutions; historically Black colleges and universities (HBCUs); tribally controlled colleges and universities (TCCUs); Alaska native and native Hawaiian serving institutions; nonprofits with 501(c)3 IRS status; nonprofits without 501(c)3 IRS status; small businesses; for-profit organizations; State governments; regional organizations; U.S. territories or possessions; Indian/Native American tribal governments; Indian/Native American tribally designated organizations; eligible agencies of the Federal government; and faith-based or community based organizations.
Date(s) Application is Due: Feb 5; Jun 5; Oct 5
Contact: Visit the web site for departmental contact information
Internet: http://grants.nih.gov/grants/guide/pa-files/PA-07-070.html
Sponsor: National Institutes of Health (NIH)
6100 Executive Boulevard
Bethesda, MD 20892

NIH Rett Syndrome: Genetics, Pathophysiology, and Biomarkers Research Grants **4159**
The Mental Retardation and Developmental Disabilities (MRDD) Branch and the Center for Research for Mothers and Children (CRMC) of NICHD and the Division of Fundamental Neuroscience and Developmental Disorders (DFNDD) of NINDS invite research grant applications that address the genetics, pathophysiological mechanisms, and the development of biomarkers for Rett syndrome. Areas of interest include developmental neurobiology, pathophysiology, genetics, epidemiology, structure-function correlations, and clinical studies that have a direct link to Rett syndrome. Specific funds will be set aside to support investigator-initiated applications responsive to this grant solicitation. PA: PAS-99-037
Requirements: Applications may be submitted by domestic, for-profit and nonprofit organizations, both public and private; units of state or local governments; and eligible agencies of the federal government.
Amount of Grant: $1.5 million total
Contact: Dr. Mary Lou Oster-Granite, NICHD, 6100 Executive Boulevard, Bethesda, MD 20892; (301) 496-1383; fax: (301) 496-3791; email: mo96o@nih.gov; Dr. Giovanna Spinella, NINDS, (301) 496-5821; (301) 402-1501;email: gs41b@nih.gov
Internet: http://www.grants.nih.gov/grants/guide/pa-files/PAS-99-037.html
Sponsor: National Institutes of Health
6701 Rockledge Drive
Bethesda, MD 20892

NIH Role of Microglia in Normal and Abnormal Immune Responses of the Nervous System **4160**
NINDS, NIMH, NCRR, NIDCD, and NIDA invite applications to promote research into the role of microglia in the initiation and expansion of autoimmune processes of the central nervous system (CNS) and the resulting injury to CNS components. The intent of this PA is to intensify interest and investigator-initiated research to attract new investigators to this field and to mobilize interdisciplinary approaches. Inquiries may be directed to program staff at the appropriate NIH institute: Dr. A.P. Kerza-Kwiatecki, Neural Environment Team, (301) 496-1431, NINDS, 6001 Executive Boulevard, Bethesda, MD 20892; Dr. John Strandberg, Comparative Medicine, (301) 435-0744, NCRR, 1 Rockledge Ctr, Bethesda, MD 20892; Dr. Dianne Rausch, (301) 443-7281, NIMH, 6001 Executive Boulevard, Bethesda, MD 20892; Dr. Thomas Johnson, Program Officer, (301) 402-3461, NIDCD, 6120 Executive Boulevard, Bethesda, MD 20892; and Dr. Charles William Sharp, Division of Basic Research, (301) 443-1887, NIDA, 6001 Executive Boulevard, Bethesda, MD 20892. PA: PA-00-029
Requirements: Applications may be submitted by foreign and domestic, for-profit and nonprofit organizations, both public and private; units of state and local governments; and eligible agencies of the federal government.
Contact: Program Officer
Internet: http://grants.nih.gov/grants/guide/pa-files/PA-00-029.html
Sponsor: National Institutes of Health
6701 Rockledge Drive
Bethesda, MD 20892

NIH Roybal Centers for Translational Research on Aging **4161**
The long-range objectives of the Roybal Centers are to improve the health, quality of life, and productivity of middle-aged and older people, through: facilitating the translation from the basic behavioral and social sciences (including human factors) to practical outcomes, including new technologies, for the benefit of the aged, and if possible also

stimulating new "use-inspired" basic research in the behavioral and social sciences. A letter of intent is due by December 22; full application is due January 22. Annual deadline dates mayb vary; contact program staff for exact dates.
Requirements: An Edward R. Roybal Center Grant (P30) requires relevant pre-existing research activity at the institution. Any individual with the skills, knowledge, and resources necessary to carry out the proposed research is invited to work with their institution to develop an application for support.
Restrictions: Foreign organizations are not eligible to apply for P30 grants.
Amount of Grant: $360,000 maximum allowable annual costs
Date(s) Application is Due: Jan 22; Dec 22
Contact: Jean Richelsen, Grants and Contracts Management Office, (301) 496-1472; email: richelsj@nia.nih.gov
Internet: http://grants1.nih.gov/grants/guide/rfa-files/RFA-AG-04-007.html
Sponsor: National Institutes of Health
6701 Rockledge Drive, Room 1040, MSC 7701
Bethesda, MD 20892-7710

NIH Ruth L. Kirschstein National Research Service Awards (NRSA) for Individual Senior Fellows 4162

The objective of the National Research Service Award (NRSA) F33 program is to provide senior fellowship support to experienced scientists who wish to make major changes in the direction of their research careers or who wish to broaden their scientific background by acquiring new research capabilities as independent research investigators in scientific health-related fields relevant to the missions of the participating NIH Institutes and Centers (ICs). Senior fellowship applicants for the F33 award must include a research training proposal that offers an opportunity for individuals to broaden their scientific background or to extend their potential for research in health-related areas as independent researchers. These awards will enable individuals with at least seven years of research experience beyond the doctorate, and who have progressed to the stage of independent investigator, to take time from regular professional responsibilities for the purpose of receiving training to increase their scientific capabilities. In most cases, this award is used to support sabbatical experiences for established independent scientists seeking support for retraining or additional career development.
Requirements: The following organizations are eligible to apply: for-profit organizations; non-profit organizations; public or private institutions, such as universities, colleges, hospitals, and laboratories; eligible agencies and labs of the Federal government, including NIH intramural labs; domestic institutions; and foreign institutions.
Restrictions: This program is not designed for postdoctoral level investigators seeking to enhance their research experience prior to independence.
Date(s) Application is Due: Apr 8; Aug 8; Dec 8
Contact: Nancy C. Lohrey; (301) 496-8580; fax: (301) 402-4472; email: lohreyn@mail.nih.gov
Internet: http://grants.nih.gov/grants/guide/pa-files/PA-07-172.html
Sponsor: National Institutes of Health (NIH)
6116 Executive Boulevard, Suite 7019, MSC 8346
Bethesda, MD 20892-8346

NIH Ruth L. Kirschstein National Research Service Awards for Individual Postdoctoral Fellows 4163

The purpose of the National Research Service Award Act (NRSA) is to help ensure that diverse pools of highly trained scientists will be available in adequate numbers and in appropriate research areas to carry out the Nation's biomedical, behavioral and clinical research agendas. The National Institutes of Health (NIH) awards individual postdoctoral fellowships (F32) to promising applicants who have the potential to become productive, independent investigators in fields related to the mission of participating NIH Institutes and Centers. The proposed postdoctoral training must be within the broad scope of biomedical, behavioral, or clinical research or other specific disciplines relevant to the research mission of the participating NIH Institutes and Centers. The proposed training must offer an opportunity to enhance the fellow's understanding of the health-related sciences and extend his/her potential for a productive research career.
Requirements: Applicant organizations must have any of the following characteristics: for-profit organizations; non-profit organizations; public or private institutions, such as universities, colleges, hospitals, and laboratories; units of State and local governments; eligible agencies and labs of the Federal government including NIH intramural labs; domestic institutions; and foreign institutions.
Date(s) Application is Due: Apr 8; Aug 8; Dec 8
Contact: Dr. Walter Schaffer; (301) 435-2687; fax: (301) 480-0146; email: ws11q@nih.gov
Internet: http://grants.nih.gov/grants/guide/pa-files/PA-06-373.html
Sponsor: National Institutes of Health (NIH)
45 Center Drive, Natcher Bldg.
Bethesda, MD 20892

NIH Ruth L. Kirschstein National Research Service Awards for Individual Predoctoral Fellows 4164

The objective of this funding opportunity announcement is to help ensure that highly trained scientists will be available in adequate numbers and in appropriate research areas to carry out the Nation, biomedical, behavioral, and clinical research agenda. The participating Institutes of the National Institutes of Health (NIH) provide individual predoctoral research training fellowship awards to promising doctoral candidates who have the potential to become productive, independent investigators in research fields relevant to the missions of these participating NIH Institutes and Centers. Each participating NIH Institute has a unique scientific purview and different program goals and initiatives that evolve over time. Prior to preparing an application, it is critical that all applicants consult the appropriate Institute website for details of research areas supported by that Institute. Applicants should also contact the appropriate Institute scientific/research contact to obtain current information about specific program priorities and policies.
Requirements: The following organizations are eligible to apply: for-profit organizations; non-profit organizations; public or private institutions, such as universities, colleges, hospitals, and laboratories; eligible agencies and labs of the Federal government, including NIH intramural labs; domestic institutions; and foreign institutions.
Date(s) Application is Due: Apr 8; Aug 8; Dec 8
Contact: Michael-David Kerns, Ph.D.; (301) 496-9322; email: kernsmd@mail.nih.gov; Dennis Twombly, Ph.D.; (301) 443-9334; fax: (301) 443-1650; email: dtwombly@mail.nih.gov
Internet: http://grants.nih.gov/grants/guide/pa-files/PA-07-002.html
Sponsor: National Institutes of Health (NIH)
6120 Executive Boulevard
Bethesda, MD 20892-7180

NIH Ruth L. Kirschstein National Research Service Awards for Individual Predoctoral Fellowships to Promote Diversity in Health Research 4165

The purpose of the F31 predoctoral fellowship to promote diversity in health-related research is to provide up to five years of support for research training leading to the PhD or equivalent research degree, the combined MD/PhD degree; or another formally combined professional degree and research doctoral degree in biomedical, behavioral, health services, or clinical sciences. These fellowships will enhance the diversity of the biomedical, behavioral, health services, and clinical research labor force in the United States by providing opportunities for academic institutions to identify and recruit students from diverse population groups to seek graduate degrees in health-related research and apply for this fellowship. The goal of this program is to increase the number of scientists from diverse population groups who are prepared to pursue careers in biomedical, behavioral, social, clinical, or health services research.
Requirements: The following organizations and institutions are eligible to apply: public/state controlled institutions of higher education; private institutions of higher education; Hispanic-serving institutions; historically Black colleges and universities (HBCUs); tribally controlled colleges and universities (TCCUs); Alaska Native and Native Hawaiian serving institutions; nonprofits with 501(c)3 IRS status (other than institutions of higher education); nonprofits without 501(c)3 IRS status (other than institutions of higher education); for-profit organizations (other than small businesses); non-domestic (non-U.S.) entities (foreign organizations); and other eligible agencies of the Federal Government.
Date(s) Application is Due: Apr 8; Aug 8; Dec 8
Contact: Susan Grove, Program Officer; (301) 435-0714; email: sg16d@nih.gov or grantsinfo@od.nih.gov
Internet: http://grants.nih.gov/grants/guide/pa-files/PA-09-209.html
Sponsor: National Institutes of Health (NIH)
6701 Rockledge Drive
Bethesda, MD 20892

NIH Ruth L. Kirschstein National Research Service Awards for Individual Predoctoral MD/PhD and Other Dual Doctoral Degree Fellows 4166

The purpose of the Ruth L. Kirschstein National Research Service Awards (Kirschstein-NRSA) is to provide support to individuals for combined MD/PhD and other dual doctoral degree training (e.g. DO/PhD, DDS/PhD, AuD/PhD). The participating Institutes award this Kirschstein-NRSA individual fellowship to qualified applicants with the potential to become productive, independent, highly trained physician-scientists and other clinician-scientists, including patient-oriented researchers in their scientific mission areas. This funding opportunity supports individual predoctoral F30 fellowships with the expectation that these training opportunities will increase the number of future investigators with both clinical knowledge and skills in basic, translational or clinical research.
Requirements: The following organizations and institutions are eligible to apply: public/state controlled institutions of higher education; private institutions of higher education; Hispanic-serving institutions; historically Black colleges and universities (HBCUs); tribally controlled colleges and universities (TCCUs); Alaska Native and Native Hawaiian serving institutions; nonprofits with 501(c)3 IRS status (other than institutions of higher education); nonprofits without 501(c)3 IRS status (other than institutions of higher education); for-profit organizations (other than small businesses); and other eligible agencies of the Federal Government.
Restrictions: Non-domestic (non-U.S.) entities (foreign organizations) are not eligible to apply.
Date(s) Application is Due: Apr 8; Aug 8; Dec 8
Contact: Nancy C. Lohrey; (301) 496-8580; fax: (301) 402-4472; email: lohreyn@mail.nih.gov
Internet: http://grants.nih.gov/grants/guide/pa-files/PA-09-207.html
Sponsor: National Institutes of Health (NIH)
6116 Executive Boulevard, Suite 7019, MSC 8346
Bethesda, MD 20892-8346

NIH Sarcoidosis: Research into the Cause of Multi-Organ Disease and Clinical Strategies for Therapy Grants 4167

The purpose of this Funding Opportunity Announcement (FOA) is to stimulate research on the etiology and management of sarcoidosis, an immune,Ämediated granulomatous inflammatory disorder. Studies supported by this FOA would include investigations

to find the etiology(ies) and related host factors that might enhance susceptibility to sarcoidosis, especially development of symptomatic multi-organ disease that has a propensity to involve critical organs (such as lungs, heart, eyes, central/peripheral nervous system, liver, kidneys and other abdominal viscera) that create serious illness. Study of the still problematic management and the biological, behavioral, or psychosocial burden on individuals, families, and community is encouraged. Moreover, there is interest for approaches to risk reduction, psychological coping, and management of complications and side effects of treatment.
Requirements: The following organizations and institutions may apply: public/state controlled institutions of higher education; private institutions of higher education; Hispanic-serving institutions; historically Black colleges and universities (HBCUs); tribally controlled colleges and universities (TCCUs); Alaska native and native Hawaiian serving institutions; nonprofits with 501(c)3 IRS status; nonprofits without 501(c)3 IRS status; small businesses; for-profit organizations; State governments; regional organizations; U.S. territories or possessions; Indian/Native American tribal governments; Indian/Native American tribally designated organizations; eligible agencies of the Federal government; and faith-based or community based organizations.
Date(s) Application is Due: Feb 5; Jun 5; Oct 5
Contact: Herbert Y. Reynolds, M.D.; (301) 435-0222; fax: (301) 480-3557; email: Reynoldh@nhlbi.nih.gov
Internet: http://grants.nih.gov/grants/guide/pa-files/PA-07-136.html
Sponsor: National Institutes of Health (NIH)
Two Rockledge Center, Suite 10018, 6701 Rockledge Drive, MSC 7952
Bethesda, MD 20892-7952

NIH School-based Interventions to Prevent Obesity **4168**
This Funding Opportunity Announcement (FOA) encourages the formation of partnerships between academic institutions and school systems in order to develop and implement controlled, school-based intervention strategies designed to reduce the prevalence of obesity in childhood. This FOA also encourages evaluative comparisons of different intervention strategies, as well as the use of methods to detect synergistic interactions between different types of interventions. Because the nature and scope of the proposed research will vary from application to application, it is anticipated that the size and duration of each award will also vary. The total amount awarded and the number of awards will depend upon the mechanism numbers, quality, duration, and costs of the applications received.
Requirements: The following organizations and institutions may apply: public/state controlled institutions of higher education; private institutions of higher education; Hispanic-serving institutions; historically Black colleges and universities (HBCUs); tribally controlled colleges and universities (TCCUs); Alaska native and native Hawaiian serving institutions; nonprofits with 501(c)3 IRS status; nonprofits without 501(c)3 IRS status; small businesses; for-profit organizations; State governments; regional organizations; U.S. territories or possessions; Indian/Native American tribal governments; Indian/Native American tribally designated organizations; eligible agencies of the Federal government; and faith-based or community based organizations.
Date(s) Application is Due: Feb 5; Jun 5; Oct 5
Contact: Gilman Grave, M.D.; (301) 496-5593; fax: (301) 480-9791; email: graveg@mail.nih.gov
Internet: http://grants.nih.gov/grants/guide/pa-files/PA-07-180.html
Sponsor: National Institutes of Health (NIH)
6100 Executive Boulevard, 4B-11, MSC 7510
Bethesda, MD 20892-7510

NIH Self-Management Strategies Across Chronic Diseases Grants **4169**
The purpose of this Program Announcement (PA) is to solicit applications to expand research on established self-management interventions to multiple chronic diseases across the life-course. Interventions aimed at chronic disease self-management are numerous and many are well described in the literature. They are often presented as specific to a particular chronic disease. This PA encourages applicants to investigate the applicability of effective self-management interventions to a broader spectrum of chronic diseases. Chronic disease, for this announcement, is defined as illnesses that are prolonged, are rarely cured completely, and require self-management behaviors by affected individuals and/or their caretakers. Modular Grant applications will request direct costs in $25,000 modules, up to a total direct cost request of $250,000 per year.
Requirements: The following organizations and institutions may apply: public/state controlled institutions of higher education; private institutions of higher education; Hispanic-serving institutions; historically Black colleges and universities (HBCUs); tribally controlled colleges and universities (TCCUs); Alaska native and native Hawaiian serving institutions; nonprofits with 501(c)3 IRS status; nonprofits without 501(c)3 IRS status; small businesses; for-profit organizations; State governments; regional organizations; U.S. territories or possessions; Indian/Native American tribal governments; Indian/Native American tribally designated organizations; county governments; city or township governments; special district governments; independent school districts; public housing authorities/Indian housing authorities; eligible agencies of the Federal government; and faith-based or community based organizations.
Amount of Grant: $250,000 maximum per year
Date(s) Application is Due: Feb 5; Jun 5; Oct 5
Contact: Susan Grove, Program Officer; (301) 435-0714; email: sg16d@nih.gov or grantsinfo@od.nih.gov
Internet: http://grants.nih.gov/grants/guide/pa-files/PA-00-109.html
Sponsor: National Institutes of Health (NIH)
6701 Rockledge Drive
Bethesda, MD 20892

NIH Senior Scientist Award **4170**
This award provides stability of support to outstanding scientists who have demonstrated a sustained, high level of productivity and whose expertise, research accomplishments, and contributions to the field have been and will continue to be critical to the mission of the particular NIH center or institute. The award provides salary support for award periods of up to five years as a means of enhancing the individual recipient's skills and dedication to his/her area of research. It permits NIH institutes and centers to identify and support exceptionally talented investigators who are well established in their field of research. Inquiries may be directed to program staff at the appropriate NIH institute: Dr. Ernestine Vanderveen, (301) 443-2531, NIAAA; Dr. Lester Gorelic, (301) 496-8580, NCI; Dr. Andrea Baruchin, (301) 443-6071, NIDA; Dr. Henry Khachaturian, (301) 443-4335, NIMH; and Dr. Neal West, (301) 402-5867, NCCAM. PA: PA-00-021
Requirements: Applications may be submitted on behalf of candidates by domestic, nonfederal organizations, public or private, such as medical, dental, or nursing schools or other institutions of higher education. At the time of award, candidates must be citizens or noncitizen nationals of the United States or must have been lawfully admitted for permanent residence. The candidate must be a senior scientist and a recognized leader in the field with a distinguished record of original contributions; must have a record of support from a funding institute or center; and must have peer-reviewed grant support at the time of the award to be eligible.
Restrictions: Individuals on temporary or student visas are not eligible.
Contact: Program Officer
Internet: http://grants.nih.gov/grants/guide/pa-files/PA-00-021.html
Sponsor: National Institutes of Health
6701 Rockledge Drive
Bethesda, MD 20892

NIH Small Business Innovation Research (SBIR): Animal Models of HCV Infection Research Grants **4171**
This is a multi-institute solicitation targeting the development/identification of one or more small animal models of hepatitis C infection and disease progression including acute and chronic states, fibrosis/cirrhosis, and liver tumor development—not necessarily all in the same model. Identification of a small, inexpensive, reproducible, well-defined, and nonendangered mammalian animal model for hepatitis C research is critical to the progress of exploration of this pathogen. Such a model is needed to study the pathogenesis, immunology, natural history, and viral replication of hepatitis C, as well as to act as a model for the testing of new therapies and vaccine candidates. This RFA invites grant applications for SBIR projects with award duration and amounts greater than those routinely allowed under the SBIR program. Inquiries may be directed to the appropriate NIH institute: Dr. Leslye Johnson, (301) 496-7051, Division of Microbiology and Infectious Diseases, NIAID, 6003 Executive Boulevard, Bethesda, MD 20892; Dr. John Cole III, (301) 496-1718, Division of Cancer Biology, NCI, Executive Plaza N, Bethesda, MD 20892; Dr. Raymond O'Neil, (301) 435-0744, Comparative Medicine, NCRR, 6701 Rockledge Drive, Bethesda, MD 20892; and Dr. Thomas Kresina, (301) 443-6537, Biomedical Research Branch, NIAAA, 6000 Executive Boulevard, Bethesda, MD 20892. Annual deadline dates may vary; contact program staff for exact dates. RFA: AI-00-008
Requirements: Any small business, independently owned and operated by U.S. citizens or permanent resident aliens may apply. Such a business must be organized for profit, cannot be dominant in its field of expertise, and must have its principal place of business in the United States. Including any affiliates, the company can be the employer of no more than 500 people. Full eligibility details may be found in the NIH Omnibus Solicitation.
Amount of Grant: $3.2 million
Contact: Program Officer
Internet: http://grants.nih.gov/grants/guide/rfa-files/RFA-AI-00-008.html
Sponsor: National Institutes of Health
6701 Rockledge Drive
Bethesda, MD 20892

NIH Solicitation of Assays for High Throughput Screening (HTS) in the Molecular Libraries Probe Production Centers Network **4172**
he NIH Molecular Libraries Roadmap Initiative wishes to encourage HTS assay applications from investigators who have the interest and capability to work with the Molecular Libraries Probe Production Centers Network (MLPCN) for chemical probe development. This Funding Opportunity Announcement (FOA) promotes discovery and development of new chemical probes as research tools for use by scientists in both the public and private sectors to advance the understanding of biological functions and disease mechanisms. This initiative is one of the integrated components of the NIH Molecular Libraries Roadmap initiative that offers biomedical researchers access to large-scale automated high throughput screening (HTS) centers in the MLPCN, diverse compound libraries in the Small Molecule Repository (MLSMR) and information on biological activities of small molecules in the PubChem BioAssay public database. Letters of Intent (LOIs) are due on December 4, April 4, and August 3 respectively. Full application must be received by January 4, May 4, and September 3 respectively.
Requirements: The following organizations and institutions are eligible to apply: public/state controlled institutions of higher education; private institutions of higher education; Hispanic-serving institutions; historically Black colleges and universities (HBCUs); tribally controlled colleges and universities (TCCUs); Alaska native and native Hawaiian serving institutions; nonprofits with 501(c)3 IRS status (other than institutions of higher education); nonprofits without 501(c)3 IRS status (other than institutions of higher education); small businesses; for-profit organizations (other than small businesses); state governments; Indian/Native American tribal governments (federally recognized); Indian/Native American tribally designated organizations; county governments; city

or township governments; special district governments; independent school districts; public housing authorities/Indian housing authorities; U.S. territories or possessions; Indian/native American tribal governments (other than federally recognized); regional organizations; non-domestic (non-U.S.) entities (foreign organizations); eligible agencies of the federal government; and faith-based or community-based organizations.
Amount of Grant: $50,000 maximum
Date(s) Application is Due: Jan 4; May 4; Sep 3
Contact: Yong Yao, Ph.D.; (301) 443-6102; email: yyao@mail.nih.gov
Internet: http://grants.nih.gov/grants/guide/pa-files/PAR-09-129.html#SectionIII
Sponsor: National Institutes of Health (NIH)
6001 Executive Boulevard, Room 7175, MSC 9641
Bethesda, MD 20892-9641

NIH Structural Biology of Membrane Proteins Grants 4173
This PA solicits applications to develop research and methods to enhance the rate of membrane protein structure determination and to determine specific membrane protein structures. Innovative methods for expression, oligomerization, solubilization, stabilization, purification, characterization, crystallization, isotopic labeling, and structure determination of unique and biologically significant membrane proteins by x-ray diffraction, nuclear magnetic resonance (NMR), electron microscopic, mass spectrometry, and other biophysical techniques are encouraged. Projects that will lead in the near term to determining the structures of biologically important membrane proteins are also encouraged.
Requirements: Applicant organizations must be: for-profit organizations; non-profit organizations; public or private institutions, such as universities, colleges, hospitals, and laboratories; units of State government; units of local government; eligible agencies of the Federal government; foreign institutions; domestic institutions; faith-based agencies; or community-based organizations.
Date(s) Application is Due: Feb 5; Jun 5; Oct 5
Contact: Peter C. Preusch, Ph.D.; (301) 594-3827; email: preuschp@nigms.nih.gov; Jean Chin, Ph.D.; (301) 594-0828; email: chinj@nigms.nih.gov
Internet: http://grants.nih.gov/grants/guide/pa-files/PA-06-119.html
Sponsor: National Institutes of Health (NIH)
Building 45, 2AS.55E 45 Center Drive, MSC 6200
Bethesda, MD 20892

NIH Studies of Ethical, Legal, and Social Implications of Research into Human Genetic Variation Grants 4174
The program solicits projects that examine the ethical, legal, and social implications of the study of human DNA sequence variation. Of particular interest are studies that explore how research on human genetic variation can be conducted in an ethically and culturally sensitive way; how the information that results from this research will interact with current concepts of race and ethnicity; how cultural and socioeconomic factors influence the interpretation and use of this information; and how this information may influence access to and use of genetic health services by various groups. The study of sequence variation and the exploration of related ethical, legal, and social issues have been identified as two of the new five-year goals of the U.S. Human Genome Project and will be of continuing interest to the NHGRI program over the next several years. Inquiries may be directed to program staff at the appropriate NIH institute: Jean Cahill, Division of Extramural Research, NHGRI, (301) 402-1733, 38 Library Drive, Bethesda, MD 20892; Sherry Dennison, Grants Management Branch, NIDCD, (301) 402-0909, 6120 Executive Boulevard, Bethesda, MD 20892; David Mineo Division of Extramural Research and Training, NIEHS, (919) 541-1373, P.O. Box 12233, T.W. Alexander Drive, E Campus, Research Triangle Park, NC 27709; Antoinette Holland, Grants Administration Branch, NIGMS, (301) 594-5132, Building 45, Bethesda, MD 20892. Annual deadline dates vary; contact program staff for exact dates. RFA: HG-99-002
Requirements: Applications may be submitted by domestic and foreign, for-profit and nonprofit, public and private organizations; units of state and local governments; and eligible agencies of the federal government.
Amount of Grant: $2 million per year for up to three years
Contact: Program Officer
Internet: http://www.grants.nih.gov/grants/guide/rfa-files/RFA-HG-99-002.html
Sponsor: National Institutes of Health
6701 Rockledge Drive
Bethesda, MD 20892

NIH Summer Internship Program in Biomedical Research 4175
Summer programs at the National Institutes of Health (NIH) provide an opportunity to spend a summer working side-by-side with some of the leading scientists in the world, in an environment devoted exclusively to biomedical research. The NIH consists of the 240-bed Mark O. Hatfield Clinical Research Center and more than 1,200 laboratories located on the main campus in Bethesda, Maryland, as well as in Baltimore and Frederick, Maryland; Research Triangle Park, North Carolina; Phoenix, Arizona; Hamilton, Momtana; and Detroit, Michigan. Awards cover a minimum of eight weeks, with students generally arriving at the NIH in May or June. The stipends for trainees are adjusted yearly, with supplements for prior experience. Prospective candidates must apply online. The application is available from mid-November to March 1.
Requirements: The Summer Internship Program is for students who will be sixteen years of age or older at the time they begin the program and who are currently enrolled at least half-time in high school or an accredited U.S. college or university. Students who have been accepted into a college or university program may also apply.
Restrictions: To be eligible, candidates must be U.S. citizens or permanent residents.
Amount of Grant: $1,700 - $3,100 per month
Date(s) Application is Due: Mar 1
Contact: Arnetta Courtney; (301) 827-3724; email: arnetta.courtney@fda.hhs.gov
Internet: http://www.training.nih.gov/student/sip/info.asp
Sponsor: National Institutes of Health (NIH)
Building 31, Room 9A06, 31 Center Drive, MSC 2560
Bethesda, MD 20892-2560

NIH Supplements to Promote Reentry into Biomedical and Behavioral Research Careers 4176
NIH provides administrative supplements to research grants to support individuals with high potential to reenter an active research career after taking time off to care for children or parents or to attend to other family responsibilities. This program will provide administrative supplements to existing NIH research grants for the purpose of supporting full-time or part-time research by these individuals in a program geared to bring their existing research skills and knowledge up to date. For more information, contact the program officer at the appropriate institute. PA: PA-99-106
Requirements: Candidates must have a doctoral degree, such as MD, DDS, PhD, OD, DVM, or equivalent; at least two years of postdoctoral research experience; and must have had sufficient prior research experience to have qualified for a faculty appointment at the assistant professor or equivalent level at the time of their leaving active research.
Restrictions: Candidates who have begun the reentry process through a fellowship, traineeship, or similar mechanism are not eligible.
Amount of Grant: $40,000 maximum
Contact: NIAAA, (301) 443-2531; NIA, (301) 496-9322; NIAID, (301) 496-8697; NIAMS, (301) 594-5052; NCI, (301) 496-7344; NICHD, (301) 496-1848; NIDCD, (301) 496-5061; NIDR, (301) 594-2089; NIDDK, (301) 594-8842; NIDA, (301) 443-0441; NIEHS, (919) 541-7723
Internet: http://www.grants.nih.gov/grants/guide/pa-files/PA-99-106.html
Sponsor: National Institutes of Health
6701 Rockledge Drive
Bethesda, MD 20892

NIH Targeting Research Gaps in HIV Therapeutics Research Grants 4177
NIAID, NIDDK, and NIMH invite applications to conduct applied studies in specific gap areas identified in current HIV therapeutics. Study areas include new delivery or formulation methods to enhance the clinical potential of anti-HIV drugs (FDA-approved or those undergoing clinical testing) in infected adults and children; validating new viral and cellular targets for HIV inhibition and for developing new drugs (or developing new agents against existing targets); and defining and validating the role of innate (nonacquired) immune responses in susceptibility to HIV infection, modulation of disease progression, and their potential use in therapeutic application in infected adults and children. Studies targeted in this PA should generate preclinical proof-of-concept on the therapeutic feasibility of the specific target or approach, and engage in applied preclinical studies. When relevant, applications with clinical studies (e.g., evaluating a drug delivery method) will be considered. Inquiries regarding eligibility and research scope may contact Dr. Nava Sarver, (301) 496-2970, Division of AIDS, NIAID, 6003 Executive Boulevard, Bethesda, MD 20892, email: ns18p@nih.gov. Inquiries regarding other programmatic issues may contact Dr. Dianne Rausch, (301) 443-7281, Office of AIDS Research, NIMH, 6001 Executive Boulevard, Bethesda, MD 20892; or Dr. Philip Smith, (301) 594-8816, Division of Diabetes, Endocrinology, and Metabolic Diseases, NIDDK, 45 Center Drive, Bethesda, MD 20892. PA-99-067
Requirements: Applications may be submitted by domestic and foreign for-profit and nonprofit organizations, both public and private; units of state and local governments; and eligible agencies of the federal government.
Contact: Program Officer
Internet: http://www.grants.nih.gov/grants/guide/pa-files/PA-99-067.html
Sponsor: National Institutes of Health
6701 Rockledge Drive
Bethesda, MD 20892

NIH Technologies to Improve the Utility of Animal Models Research Grants 4178
This PA encourages the small business community to develop technologies, reagents, and equipment to improve the utility of animal models for biomedical research. The results of projects supported under this PA are expected to be improvement of laboratory animal welfare, improved long-term maintenance and preservation of laboratory animal models, and improved technologies for generation of animal models of human diseases. Cost-effective methods, reagents, and equipment to improve animal health and to characterize, culture, and preserve biomedically important organisms are critical needs for biomedical research. Reagents to diagnose and control laboratory animal diseases, including those of aquatic animals, are needed. Equipment and techniques for cost-effective husbandry of animal models in controlled environments are needed for a variety of aquatic and nonaquatic laboratory animals. Improved technologies for the cryopreservation of embryos and gametes are needed for several scientific and economic reasons. Projects supported under the PA are to be directed along one or more of these objectives. This PA must be read in conjunction with the Omnibus Solicitations for SBIR and STTR applications (available on the Web site). Inquiries may be directed to the staff at the appropriate NIH institute: Director, Comparative Medicine, (301) 435-0744, NCRR, 6705 Rockledge Drive, Bethesda, MD 20892; Jo Anne Goodnight, (301) 496-5307, Division of Cancer Biology, NCI, Executive Plaza N, Bethesda, MD 20892; Dr. Christine Kelley, (301) 435-0565, Division of Heart and Vascular Diseases, NHLBI, 2 Rockledge Center, Bethesda, MD 20892; Dr. Joyce Reese, (301) 594-2088, Division of Extramural Research, NIDCR, Building 45, Bethesda, MD 20892; and Dr. Michael

Galvin Jr., (910) 541-7825, Division of Extramural Research and Training, NIEHS, P.O. Box 12233, Research Triangle Park, NC 27709. PA: PA-99-048
Requirements: Any small business, independently owned by U.S. citizens and located in the United States, may apply.
Contact: Program Officer
Internet: http://www.grants.nih.gov/grants/guide/pa-files/PA-99-048.html
Sponsor: National Institutes of Health
6701 Rockledge Drive
Bethesda, MD 20892

NIH Transitional Career Development Award in Women's Health Research 4179
This award is designed to support career development experiences leading to independence for clinical investigators interested in patient-oriented or population-based research related to women's health. The career development program will provide an opportunity for investigators to develop solid clinical research skills during two years of study and research within the environment of the NIH Intramural Research Programs (IRP). The award will also include a follow-on two-year period of salary and research support at an academic institution of the candidate's choice. This period of extramural support will facilitate the transition to independence as a clinical researcher specializing in research issues related to women's health. Inquiries may be directed to program staff at the appropriate NIH institute. PA: OD-00-003
Requirements: Individuals must have a health-professional doctoral degree or its equivalent and must be able to demonstrate the potential for highly productive patient-oriented research. Candidates should possess aptitudes for independent, patient-oriented, or population-based research (e.g. epidemiology, behavior, or disease prevention) and should be able to demonstrate a commitment to the study of diseases that occur in women. Candidates must be U.S. citizens or noncitizen nationals, or must have been lawfully admitted for permanent residence and possess an Alien Registration Card.
Restrictions: Individuals on temporary or student visas are not eligible to apply.
Contact: Program Officer
Internet: http://grants.nih.gov/grants/guide/rfa-files/RFA-OD-00-003.html
Sponsor: National Institutes of Health
6701 Rockledge Drive
Bethesda, MD 20892

NIH Undergraduate Scholarships 4180
The program offers competitive one-year scholarships to students from disadvantaged backgrounds who are committed to careers in biomedical, behavioral, and social science health-related research. Scholarships support tuition, educational expenses, and reasonable living expenses. Recipients may reapply for additional years (not to exceed four scholarship years).The program is designed to improve access to education leading to research careers for those who have had fewer opportunities than others, and to provide an incentive for exceptional scholars to pursue research careers at the NIH. For each full or partial scholarship year, the recipient will incur two types of service obligation: 10-week summer laboratory experience; and employment at the NIH after graduation. Guidelines are available online.
Requirements: The applicant must be a U.S. citizen, national, or qualified noncitizen; enrolled or accepted for enrollment as a full-time student for the academic year at an accredited undergraduate institution; from a disadvantaged background; and must have a grade point average of 3.5 (on a 4.0 scale) or be within the top 5 percent of his or her class.
Amount of Grant: $20,000 maximum per year
Date(s) Application is Due: Feb 28
Contact: Office of Loan Repayment and Scholarship, (800) 528-7689; fax: (301) 480-3123; email: ugsp@nih.gov
Internet: http://ugsp.info.nih.gov/exesumfaq.htm
Sponsor: National Institutes of Health
2 Center Drive, MSC 0230
Bethesda, MD 20892-0230

NIH Urinary Incontinence Treatment Network 4181
NIDDK and NICHD invite cooperative agreement applications for investigators to design and implement a multicenter, prospective cohort study to clarify the benefits of the different surgical interventions for treatment of patients with urinary incontinence. This study will assess the long-term outcomes of the most commonly utilized surgical interventions to correct urinary incontinence in adult women and the utilization of concomitant medical and behavioral therapy. These cooperative agreements will support a collaborative network of continence treatment centers and a biostatistical coordinating center that will prospectively assess the outcomes of different surgical interventions for the treatment of urinary incontinence in women over a three- to four-year period. This group of investigators will develop standardized diagnostic and outcome measures that can be applied uniformly to all of the different surgical cohorts to be evaluated by this initiative. This includes the establishment of diagnostic entry criteria, identification of clinically and patient relevant outcomes, and the development of questionnaires to measure quality of life and other patient-reported factors. Inquiries may be directed to program staff at the appropriate NIH institute: Dr. John Kusek, Program Director, (301) 594-7717, NIDDK, 45 Center Drive, Bethesda, MD 20892; and Dr. Trent MacKay, (301) 435-6972, NICHD, 6100 Executive Boulevard, Bethesda, MD 20892. Annual deadline dates may vary; contact program staff for exact dates. RFA: DK-99-001
Requirements: Applications may be submitted by domestic, for-profit and nonprofit institutions, both public and private organizations; units of state and local government; and eligible agencies of the federal government. Racial/ethnic minorities, women, and persons with disabilities are encouraged to apply.
Restrictions: Foreign institutions are not eligible to apply.
Amount of Grant: $250,000-$750,000
Contact: Program Officer
Internet: http://www.grants.nih.gov/grants/guide/rfa-files/RFA-DK-99-001.html
Sponsor: National Institutes of Health
6701 Rockledge Drive
Bethesda, MD 20892

NIH Xenobiotics and Cell Death/Injury in Neurodegenerative Disease Research Grants 4182
The objective of this program announcement is to stimulate research on the relative roles of environmental, endogenous, neurochemical, and genetic factors in the cause of neurodegenerative diseases (e.g., Parkinson's disease, Alzheimer's disease, and amyotrophic lateral sclerosis). Attempts are needed to understand the contribution of environmental exposure, endogenous susceptibility factors, and increasing age in the etiopathology of neurodegenerative disease and will require a concurrent advancement and refinement of methodologies and sciences. Of particular significance may be those approaches that can be used across species from lower animals to humans. Such approaches permit a precise characterization in animal models of alterations arising from defined environmental exposures which can serve as a cogent guide to underlying cellular and molecular mechanisms. The results of these investigations should help clarify the part xenobiotics play in the etiology of these diseases. Applications that are collaborations between basic and clinical neuroscientists and/or are interdisciplinary in nature are encouraged. Inquiries may be directed to the staff of the appropriate NIH institute: Dr. Annette Kirshner, (919) 541-0488, Division of Extramural Research and Training, NIEHS, Research Triangle Park, NC 27709; Dr. Eugene Oliver, (301) 496-5680, Division of Stroke, Trauma, and Neurodegenerative Disorders, NINDS, 7550 Wisconsin Avenue, Bethesda, MD 20892; Dr. Bradley Wise, (301) 496-9350, Neuroscience and Neuropsychology of Aging Program, NIA, 7201 Wisconsin Avenue, Bethesda, MD 20892; and Dr. Linda Brady, Preclinical and Clinical, Therapeutics Research Branch, NIMH, 5600 Fishers Ln, Rockville, MD 20857. PA: PAS-99-054
Requirements: Applications may be submitted by domestic and foreign, for-profit and nonprofit organizations, public and private.
Amount of Grant: $2 million total per year
Contact: Program Officer
Internet: http://www.grants.nih.gov/grants/guide/pa-files/PAS-99-054.html
Sponsor: National Institutes of Health
6701 Rockledge Drive
Bethesda, MD 20892

NIH Zebrafish as an Animal Model for Development and Disease Research Grants 4183
The purpose of this PA is to solicit applications as part of an NIH initiative to increase support of the zebrafish as an animal model for research. The goals of this PA are to encourage new and innovative research and approaches using the zebrafish to identify the genes and elucidate the molecular and genetic mechanisms responsible for normal and defective development and disease. The mechanism of support for this PA will be the investigator-initiated research project grant (R01) award. Applications from new investigators are particularly encouraged. Inquiries may be directed to program staff at the appropriate NIH institute. PA: PA-01-095
Requirements: Applications may be submitted by domestic and foreign, for-profit and nonprofit organizations, public or private. The total project period for an application submitted in response to this PA may not exceed five years.
Contact: Program Officer
Internet: http://grants.nih.gov/grants/guide/pa-files/PA-01-095.html
Sponsor: National Institutes of Health
6705 Rockledge Drive
Bethesda, MD 20892

NIH/EPA Mechanistic-Based Cancer Risk Assessment Methods Research Grants 4184
This RFA solicits studies that focus on the development of cancer risk assessment methods and practices. Emphasis should be on the development of new methods and practices that reduce the uncertainties associated with cancer risk assessment, especially uncertainties that are encountered in extrapolating from animal species to humans. Studies that propose refinement, validation, or invalidation of existing methods and practices are also acceptable if they will result in a significant reduction in uncertainty. Methods of development can be based on in vitro and/or in vivo animal model systems, in vitro human cell model systems, and exploratory epidemiologic approaches. Short-term, mechanistic-based methods that can be used to augment long-term bioassays are also encouraged. A clear and unambiguous demonstration of mechanistic commonality with humans is a requirement of any method based on animal model systems. Inquiries may be directed to the staff at the appropriate organization/institute. Annual deadline dates may vary; contact program staff for exact dates. RFA: OH-99-003
Requirements: Applications may be submitted by domestic and foreign, public and private, nonprofit and for-profit organizations; governments and their agencies; and federally recognized Indian tribal governments, Indian tribes, or Indian tribal organizations.
Amount of Grant: $250,000 maximum average
Contact: Program Officer
Internet: http://www.grants.nih.gov/grants/guide/rfa-files/RFA-OH-99-003.html
Sponsor: National Institutes of Health
6701 Rockledge Drive
Bethesda, MD 20892

NIHCM Foundation Health Care Research Awards 4185

To encourage outstanding work from researchers furthering innovation in health care policy and management, the National Institute for Health Care Management (NIHCM) Research and Educational Foundation recognizes excellence in original and creative research with a $10,000 award. Selection criteria for the award includes the originality and creativity of the research. Impact that the research may have in advancing the public policy debate, improving health care delivery, management and the relevance to one or more of the subject areas. The award will be presented to the winner at a dinner in Washington, DC in May.

Requirements: Articles must have been published in a peer-reviewed journal or similar quality publication during calendar year 2009. Applications are encouraged from authors in diverse fields, including economics, health policy, political science, public health, etc. and from individuals in non-academic settings, such as research firms and policy organizations. All research articles must address some aspect of health care financing, delivery, organization or the implementation of health care policy.

Amount of Grant: $10,000 prize

Date(s) Application is Due: Feb 26

Contact: Nancy Chockley, President and CEO; (202) 296-4426; fax: (202) 296-4319; email: nihcm@nihem.org

Internet: http://nihcm.org/awards/research

Sponsor: National Institute for Health Care Management Foundation
1225 19th Street, NW, Suite 710
Washington, DC 20036

NIHCM Foundation Health Care Television and Radio Journalism Awards 4186

The Third Annual NIHCM Foundation Health Care Television and Radio Journalism Award recognizes excellence in television and radio reporting on health care issues and policy. Selection criteria will include, new insights generated by the reporting, impact on health care policy and quality of writing and production. A $10,000 prize will be presented to the winner at a dinner in Washington, DC in May.

Requirements: Entries must originally have aired during calendar year 2009. Entries will be accepted from staff employees of national or local news organizations, nationally-distributed cable programs, or from freelance journalists who have produced reports disseminated by same. Online stories will be accepted if produced for and disseminated originally on the web in affiliation with a television or radio news organization. Material repurposed from other media will not be accepted.

Amount of Grant: $10,000

Date(s) Application is Due: Feb 26

Contact: Nancy Chockley, President and CEO; (202) 296-4426; fax: (202) 296-4319; email: nihcm@nihem.org

Internet: http://nihcm.org/awards/televison_and_radio_journalism

Sponsor: National Institute for Health Care Management Foundation
1225 19th Street, NW, Suite 710
Washington, DC 20036

NIJ Graduate Research Fellowship Program 4187

This program supports independent research fellowships for advanced-level doctoral students. The research undertaken should be the subject of the fellow's doctoral dissertation, and must focus on a topic related to one of NIJ's substantive long-term goals or be directly related to the concerns of an operating agency. NIJ's long-term goals are to reduce violent crime, reduce drug- and alcohol-related crime, reduce the consequences of crime, and improve the effectiveness of crime prevention programs. NIJ's current priority topic areas are law enforcement/policing; justice systems; courts, prosecution, defense; corrections; offender programs and treatment; crime prevention/causes of crime; violence and victimization; drugs, alcohol, and crime; international crime and justice; and evaluation research. Fellows are required to spend 10 weeks in Washington, DC, at the NIJ offices. Awards are made through the student's sponsoring university or through a criminal justice agency for up to 18 months.

Requirements: To be eligible to administer a graduate research fellowship grant on behalf of a doctoral candidate, an institution must be fully accredited by one of the regional institutional accreditation commissions recognized by the U.S. Commissioner of Education and the Council on Postsecondary Accreditation.

Restrictions: Individuals may not receive awards directly; grant awards will be made only to educational institutions.

Amount of Grant: $20,000 maximum

Date(s) Application is Due: Sep 15

Contact: Program Contact, (800) 421-6770

Internet: http://www.ojp.usdoj.gov/nij/funding.htm

Sponsor: National Institute of Justice
810 Seventh Street NW
Washington, DC 20531

Nils Klim Prize 4188

Bergen-born Ludvig Holberg held the chairs of Metaphysics and Logic, Latin Rhetoric and History at the University of Copenhagen, and played an important part in bringing the Enlightenment to the Nordic countries. Through his interdisciplinary and internationally-oriented efforts, Holberg sought to modernize academic subjects and teaching methods. This prize of NOK 250,000 (approximately EUR 32,000) is awarded to a young Nordic researcher in one of the chosen academic fields. Recommendations for the Prize are made by a committee of outstanding Nordic researchers in the relevant fields.

Requirements: Nominees for the Prize must be aged under 35.

Samples: Carina Keskitalo, Associate Professor at the Department of Political Science, Umea University in Sweden (2007).

Amount of Grant: NOK 250,000

Contact: Kare Rommetveit, Head of Secretariat; +47 55 58 20 00 or +47 55 58 69 92; email: Kare.Rommetveit@bfs.uib.no

Internet: http://www.holbergprisen.no/en_minnefondet.html

Sponsor: University of Bergen
Post Box 7800, Thormohlensgate 51 (2. floor)
Bergen, Norway N-5020 Norway

NIMH AIDS and Aging: Behavioral Sciences Prevention Research Grants 4189

NIA, NIMH, and NINR invite qualified researchers to submit applications to investigate prevention issues relevant to AIDS in middle-aged and older populations. This announcement solicits AIDS prevention research proposals to study primary prevention of disease transmission as well as secondary and tertiary prevention of negative behavioral and social consequences of HIV/AIDS for persons with AIDS, their families, and communities. Thus, the primary goals are to identify social and behavioral factors associated with HIV transmission and disease progression in later life; examine behavioral and social consequences of HIV infection/AIDS across the life course; develop and evaluate age-appropriate behavioral and social interventions for preventing AIDS in middle-aged and older adults and/or ameliorating problems associated with older adult caregiver responsibilities and burdens; explore health care issues surrounding AIDS care; and strengthen existing research and evaluation methods.

Requirements: Applications may be submitted by domestic and foreign for-profit and non-profit organizations, public and private, such as universities, colleges, hospitals, laboratories, units of State and local governments, and eligible agencies of the Federal government. Applications may be submitted by single institutions or by a consortia of institutions.

Date(s) Application is Due: Feb 5; Feb 16; Jun 5; Jun 16; Oct 5; Oct 16

Contact: Dr. Willo Pequegnat; (301) 443-6100; fax: (301) 443-9719; email: wpequegn@nih.gov

Internet: http://www.grants.nih.gov/grants/guide/pa-files/PA-97-069.html

Sponsor: National Institute of Mental Health (NIMH)
5600 Fishers Lane, Parklawn Building, Room 18-101
Rockville, MD 20857

NIMH Basic and Translational Research in Emotion Grants 4190

This Funding Opportunity Announcement encourages Research Project Grant applications to expand basic and translational research on the processes and mechanisms involved in the experience, expression, and regulation of emotion. The study of emotion encompasses a wide range of physiological, psychological, social, cognitive, and developmental phenomena. Emotional processes can be studied in human or animal subjects. Important objects of study include, but are not limited to: central and peripheral nervous system activity in the origins, expression, regulation and modulation of emotion; the contribution of emotional and motivational systems to cognitive faculties such as perception, attention, learning, and memory; investigations of overt behaviors, interpersonal relationships, communication and decision making; and the environmental circumstances and experiences that evoke and modulate different emotions.

Requirements: The following organizations and institutions may apply: public/state controlled institutions of higher education; private institutions of higher education; Hispanic-serving institutions; historically Black colleges and universities (HBCUs); tribally controlled colleges and universities (TCCUs); Alaska native and native Hawaiian serving institutions; nonprofits with 501(c)3 IRS status; nonprofits without 501(c)3 IRS status; small businesses; for-profit organizations; State governments; regional organizations; U.S. territories or possessions; Indian/Native American tribal governments; Indian/Native American tribally designated organizations; county governments; city or township governments; special district governments; independent school districts; public housing authorities/Indian housing authorities; eligible agencies of the Federal government; and faith-based or community based organizations.

Date(s) Application is Due: Feb 5; Jun 5; Oct 5

Contact: Janine Simmons; (301) 443-6652; fax: (301) 402-4740; email: simmonsj@mail.nih.gov; Wendy Nelson; (301) 435 4590; fax: (301) 435 7547; email: nelsonw@mail.nih.gov

Internet: http://grants.nih.gov/grants/guide/pa-files/PA-09-137.html

Sponsor: National Institute of Mental Health (NIMH)
6001 Executive Boulevard, Room 7179, MSC 9637
Bethesda, MD 20892-9637

NIMH Behavior Genetics in Adulthood and Old Age Research Grants 4191

NIA and NIMH invite qualified researchers to submit research grant applications on behavior genetics in adulthood and old age. The goal of this PA is to further scientific knowledge about the relative contributions of hereditary and environmental factors in the health, well-being, and quality of life of adults (age 21 and over) and older persons. This PA encourages research that will address one or more of the following: behavior genetics and cognitive functioning/intellectual ability and behavior genetics and personality.

Requirements: Applications may be submitted by foreign and domestic for-profit and nonprofit organizations, public or private.

Contact: Dr. Marilyn Miller, National Institute on Aging,(301) 496-9350; fax: (301) 496-1494; email: Millerm@nia.nih.gov

Internet: http://www.grants.nih.gov/grants/guide/pa-files/PAS-03-128.html

Sponsor: National Institute of Mental Health
5600 Fishers Ln, Parklawn Building
Rockville, MD 20857

NIMH Career Opportunities in Research (COR) Education and Training Honors High School Research Education Grant 4192

This program encourages applications for training activities designed to stimulate interest among minority youth in research careers in biomedical and behavioral sciences relevant to mental health. This program is specifically designed to increase the participation of groups that are underrepresented in biomedical and behavioral sciences, including African Americans, Hispanics, Native Indians and Alaska Natives, and Asians/Pacific Islanders. The program is intended to support activities that allow hands-on experiences in research or science education and encourage high school students to make the choice of a research career in a mental health-related field. PA: PAR-01-009
Requirements: Applicant institutions must have a currently funded NIMH Career Opportunities in Research Education and Training Program Honors Undergraduate Research Training Grant. Such institutions are baccalaureate granting, four-year domestic, public or private, nonprofit colleges, universities, or health professional schools with at least 50 percent minority students.
Restrictions: Foreign organizations are not eligible to apply.
Amount of Grant: $25,000-$35,000
Date(s) Application is Due: May 10
Contact: Dr. Carolyn Strete, Office for Special Populations, (301) 443-2847; fax: (301) 443-8552; email: cstrete@nih.gov
Internet: http://grants.nih.gov/grants/guide/pa-files/PAR-01-009.html
Sponsor: National Institute of Mental Health
6001 Executive Boulevard
Bethesda, MD 20892

NIMH Career Opportunities in Research (COR) Education and Training Honors Undergraduate Research Training Grant 4193

This program is intended to strengthen research and research training experiences of undergraduate minority students in scientific disciplines related to mental health. It is directed specifically to four-year colleges and universities that serve substantial numbers of students of one or more racial/ethnic minority groups, including African Americans, Hispanics, Native Americans and Alaska Natives, and Asians/Pacific Islanders. An applicant institution must propose a two-year COR Honors Undergraduate program for which six to 10 highly talented third-and fourth-year undergraduate students are selected. Proposed training programs should provide students with special research training experiences designed to improve their qualifications for entry into advanced research career training programs leading to doctoral-level or MD research career degrees. PA: PAR-01-008
Requirements: Applicant institutions must be four-year public or private, nonprofit colleges, universities, or heath professional schools with at least 50 percent racial/ethnic minority students; and be able to develop the faculty and institutional facilities in order to provide a suitable research environment and high-quality research training in areas relevant to mental health.
Restrictions: Foreign organizations are not eligible to apply.
Amount of Grant: $300,000 total
Contact: Robert Mays, Office for Special Populations, (301) 443-3017; fax: (301) 443-8022; email: rmays@mail.nih.gov
Internet: http://grants.nih.gov/grants/guide/pa-files/PAR-01-008.html
Sponsor: National Institute of Mental Health
6001 Executive Boulevard
Bethesda, MD 20892

NIMH Centers for AIDS Research (CFARs) 4194

NIMH is interested in establishing AIDS research centers to provide support for coordinated, multidisciplinary research programs on the mental health and drug abuse aspects of AIDS, ARC, and HTLV-III infection. AIDS research centers support may be requested for up to five years. The overall aims of the center must be clearly defined for the requested period of support. Areas of future development should be indicated and justified. Annual deadline dates may vary; contact program staff for exact dates. PA: PAR-03-089
Contact: Dianne Rausch, (301) 443-7281; fax: (301) 443-9719; email: dr89b@nih.gov
Internet: http://grants2.nih.gov/grants/guide/pa-files/PAR-03-089.html
Sponsor: National Institute of Mental Health
5600 Fishers Ln, Parklawn Building
Rockville, MD 20857

NIMH Central Nervous System Effects of HIV Infection: Neurobiological, Neurovirological, and Neurobehavioral Studies Grants 4195

NIMH, with NINDS, NICHD, and NIDA, invites applications for support of research on the role of HIV in the central nervous system and its consequent effects on behavior. Applicants may request up to five years of support. Applications will be accepted under PHS receipt dates for AIDS applications. Potential applicants should contact the AIDS coordinator. PA: PA-04-154
Contact: Brian Albertini, NIMH Grants Management Branch, (301) 443-0004; fax: (301) 443-0219; email: albertib2@mail.nih.gov
Internet: http://grants.nih.gov/grants/guide/pa-files/PA-04-154.html
Sponsor: National Institute of Mental Health
5600 Fishers Ln, Parklawn Building
Rockville, MD 20857

NIMH Collaborative R01s for Clinical Studies of Mental Disorders 4196

NIMH invites applications for collaborative research efforts on clinical studies of mental disorders (CSMD), including schizophrenia, mood disorders, anxiety disorders, and other severe mental illnesses. This PA provides guidelines for the conceptualization, preparation, review, and administration of collaborative CSMD applications submitted to the NIMH. PA-01-123 may be obtained through the NIH grant line (301-402-2221), the NIH Web site, mail, and email.
Requirements: This mechanism requires more than one applicant organization, although multiple campuses of a single institution are also eligible. Applications may be submitted by domestic and foreign, for-profit and nonprofit organizations, public and private.
Contact: Dr. Barry Lebowitz, Division of Services and Intervention Research, (301) 443-1185; fax: (301) 443-4045; email: blebowit@nih.gov; Diana Trunnell, Grants Management Branch, (301) 443-2805; fax: (301) 443-6885; email: Diana_Trunnell@nih.gov
Internet: http://grants2.nih.gov/grants/guide/pa-files/PA-01-123.html
Sponsor: National Institute of Mental Health
5600 Fishers Ln, Parklawn Building
Rockville, MD 20857

NIMH Core Support Program for Mental Health/AIDS Research Grants 4197

The purpose of this program is to provide a mechanism to facilitate currently funded investigators in improving and expanding research infrastructure, including administrative coordination, subject recruitment, equipment, laboratories, statistical analysis, and database management. A letter of intent is requested by December 1; full application is due by January 2. Annual deadline dates may vary; contact program staff for exact dates. PA: PAR-03-142
Requirements: Applications may be submitted by domestic, for-profit and nonprofit organizations, both public and private; units of state and local governments; and eligible agencies of the federal government.
Amount of Grant: $2 million total
Date(s) Application is Due: Jan 2; Dec 1
Contact: Dr. Dianne Rausch, Office of AIDS Research, (301) 443-7281; fax: (301) 443-9719; email: dr89b@nih.gov
Internet: http://grants.nih.gov/grants/guide/pa-files/PAR-03-142.html
Sponsor: National Institute of Mental Health
6001 Executive Boulevard
Bethesda, MD 20892

NIMH Curriculum Development Award in Neuroinformatics Research and Analysis 4198

The purpose of this program is to encourage and support applications from individuals, with the requisite scientific expertise and leadership, for the development of courses and curricula designed to train interdisciplinary neuroinformatics scientists at U.S. educational institutions. The field of neuroinformatics combines neuroscience research with informatics research developed from the computer sciences, mathematics, physics, engineering or closely related sciences. It is anticipated that these courses or curricula would be useful to students and scientists who wish: to develop new conceptual approaches to basic and/or clinical neuroscientific research and analysis; or to acquire, store, retrieve, organize, manage, analyze, visualize, manipulate, integrate, synthesize, disseminate, and share data about the brain and behavior. Development of courses at the graduate and undergraduate level is encouraged. As part of this program, awardees will be expected to develop and implement the courses or curricula in their institution. It is expected that such courses and curricula will be models that could be transferable to other institutions in whole or in part.
Requirements: The principal investigator must be engaged in neuroscience research or research in computer science, mathematics, physics, engineering, or a related informatics field. Collaborator(s) who will contribute to the interdisciplinary nature of the courses or curricula must be identified. The principal investigator must be willing to spend at least 20 percent of full-time professional effort on course(s) and curricula development during the period of the award. The principal investigator must also identify appropriate researcher(s) who will agree to collaborate on the development of course(s) and curricula.
Date(s) Application is Due: Feb 16; Jun 16; Oct 16
Contact: Stephen H. Koslow; (301) 443-1815; fax: (301) 443-1867; email: koz@helix.nih.gov
Internet: http://grants.nih.gov/grants/guide/pa-files/PAR-99-135.html
Sponsor: National Institute of Mental Health (NIMH)
6001 Executive Boulevard, Room 6167, MSC 9613
Bethesda, MD 20892

NIMH Early Detection of HIV Grants 4199

NIMH offers funds to conduct research on ways to promote early detection of HIV, and the impact of these findings on prevention. Research should look at new strategies to increase HIV testing and communication of test results to at-risk individuals, as well as develop innovative prevention strategies associated with emerging technologies in the early detection of HIV. A letter of intent is requested by February 10; full proposal is due March 10. Annual deadline dates may vary; contact program staff for exact dates. RFA: MH-00-008
Requirements: Public and private nonprofits may apply.
Amount of Grant: $1 million total
Date(s) Application is Due: Feb 10; Mar 10
Contact: Louis Steinberg, (301) 443-6100; fax: (301) 443-9719; email: lsteinbe@nih.gov
Internet: http://grants.nih.gov/grants/guide/rfa-files/RFA-MH-00-008.html
Sponsor: National Institute of Mental Health
5600 Fishers Ln, Parklawn Building
Rockville, MD 20857

NIMH Early Identification and Treatment of Mental Disorders in Children and Adolescents Grants 4200

This program supports research on early identification and treatment of mental disorders in children and adolescents. In particular, this announcement intends to encourage research on disorders such as schizophrenia, schizoaffective disorder, bipolar disorder, major depression, obsessive-compulsive disorder, and anorexia nervosa, alone or comorbid with other common mental or substance abuse disorders. PA: PA-00-094
Requirements: Applications may be submitted by domestic for-profit and nonprofit organizations, both public and private; units of state or local governments; and eligible agencies of the federal government.
Amount of Grant: $125,000; $250,000
Contact: Rebecca Claycamp, Grants Management Officer, (301) 443-2811
Internet: http://www.nimh.nih.gov/researchfunding/grants.cfm
Sponsor: National Institute of Mental Health
6001 Executive Boulevard
Bethesda, MD 20892

NIMH HIV/AIDS, Severe Mental Illness, and Homelessness Research Grants 4201

This PA is sponsored by the NIMH to stimulate investigator-initiated research that targets persons with severe mental illness (SMI) either before or after HIV infection and to expand HIV-related research to homeless persons. It solicits studies on the SMI population and/or homeless persons with special attention to the development, implementation, and evaluation of effective HIV-prevention interventions and their dissemination and translation to the community and public health service organizations. also solicit studies on the SMI population and/or homeless persons that propose improved methodologies to study descriptive/analytic epidemiology of HIV infection and epidemiology of the behavior associated with risk and the spread of HIV infection. An important objective of this PA is to encourage integration both across and within the different research areas by establishing multidisciplinary research teams and collaborative alliances. PA: PA-04-024
Requirements: Applications may be submitted by domestic and foreign, for-profit and nonprofit organizations, both public and private; units of state and local governments; and eligible agencies of the federal government.
Contact: Dr. David Stoff, (301) 443-4625; fax: (301) 443-9719; email: dstoff@nih.gov; Brian Albertini, NIMH, (301) 443-0004; fax: (301) 443-6885; email: albertib2@mail.nih.gov
Internet: http://www.grants.nih.gov/grants/guide/pa-files/PA-04-024.html
Sponsor: National Institute of Mental Health
6701 Rockledge Drive
Bethesda, MD 20817

NIMH Informed Consent in Research Grants 4202

This program invites applications for a three-year research grant to stimulate investigations into the informed consent process in scientific research involving human participants. The grant is being offered because voluntary informed consent is the defining aspect of interactions between researchers and participants and is integral to the conduct of scientific research. One of the goals of this project is to bring together perspectives of different agencies, because various areas of research reflect a diversity of issues relating to informed consent. The focus of the program will be to document the degree of understanding achieved by research participants regarding the identity of the sponsoring federal agency or agencies; the purposes for which the research is being conducted; the comprehension of a study's methods and procedures; the relative risks and benefits of deciding to consent or refuse participation; the confidentiality and any exceptions to confidentiality; the implications of withdrawal from a study; and the planned and other possible use of the data. The goal of the present initiative is to develop and test alternative strategies for obtaining informed consent in diverse populations and to determine optimal ways to obtain informed consent for research participation. Additional information may be obtained electronically through the NIH grant line (301-402-2221), the NIH Web site, and by mail and email from the program contact listed. Annual deadline dates may vary; contact program, staff for exact dates. RFA: OD-97-001
Requirements: Applications may be submitted by any domestic or foreign, for-profit or nonprofit organizations, both public or private; units of state and local governments; or eligible agencies of the federal government.
Contact: Dr. John Hsiao, Chief, Special Projects Research Program, (301) 443-1185; fax: (301) 594-6784; email: jhsiao@heilix.nih.gov or jh23f@nih.gov
Internet: http://www.grants.nih.gov/grants/guide/rfa-files/RFA-OD-97-001.html
Sponsor: National Institute of Mental Health
5600 Fishers Ln, Parklawn Building
Rockville, MD 20857

NIMH Interdisciplinary Behavioral Science Centers for Mental Health 4203

The purpose of these centers is to support collaborative, hypothesis-driven basic research activities that will extend the most cutting-edge theories and approaches in basic behavioral science to incorporate current approaches in neuroscience. A letter of intent is requested by January 18; full application is due February 18. Annual deadline dates may vary; contact program staff for exact dates. PA: PAR-04-004
Requirements: Applications may be submitted by domestic, for-profit and nonprofit organizations, both public and private; units of state and local governments; and eligible agencies of the federal government.
Amount of Grant: $1.5 million maximum for full-scale centers; $600,000 maximum for start-up centers
Date(s) Application is Due: Jan 18; Feb 18
Contact: Dr. Mary Ellen Oliveri, (301) 443-3942; fax: (301) 443-9876; email: moliveri@nih.gov
Internet: http://grants.nih.gov/grants/guide/pa-files/PAR-04-004.html
Sponsor: National Institute of Mental Health
6001 Executive Boulevard, Room 7220, MSC 9651
Bethesda, MD 20892-9651

NIMH Interventions Infrastructure Program Research Grants 4204

NIMH is seeking to expand the capacity of the field to support state-of-the-art public health-oriented studies in the treatment, rehabilitation, and prevention of mental disorders in children, adolescents, and in mid- and late-life adults. The Interventions Infrastructure Program (IIP) encourages the expansion of intervention research collaborations between academic health centers and health care provider organizations that are oriented to community care, and have access to large, stable, and diverse patient populations. The purpose of the IIP is to provide the infrastructure for the support of research that has sufficient power to define standards of appropriate and cost-effective care for the diverse populations of patients with mental disorders seen in all health care settings. Applications are accepted annually. Annual deadline date may vary; contact program staff for exact dates. PA: PAR-04-015
Requirements: Applications may be submitted by domestic, for-profit and nonprofit organizations, public and private.
Restrictions: This program is nonrenewable.
Amount of Grant: $250,000 maximum, plus negotiated indirect costs, for up to five years
Date(s) Application is Due: Jun 1
Contact: Dr. Junius Gonzales, NIMH, (301) 443-3364; fax: (301) 443-4045; email: jgonzale@mail.nih.gov
Internet: http://grants2.nih.gov/grants/guide/pa-files/PAR-04-015.html
Sponsor: National Institute of Mental Health
6001 Executive Boulevard
Bethesda, MD 20892

NIMH Jointly Sponsored Ruth L. Kirschstein National Research Service Award Institutional Predoctoral Training Program in the Neurosciences 4205

The Jointly Sponsored NIH Predoctoral Training Program in the Neurosciences supports broad and fundamental, early-stage graduate research training in the neurosciences via institutional NRSA research training grants (T32) at domestic institutions of higher education. Trainees are supported during years 1 and 2 of their graduate training when they are typically not committed to a dissertation laboratory. The primary objective is to prepare qualified individuals for careers in neuroscience that have a significant impact on the health-related research needs of the Nation. Because the nature and scope of the proposed research training will vary from application to application, it is anticipated that the size and duration of each award will also vary. The total amount awarded and the number of awards will depend upon the number, quality, duration, and costs of the applications received.
Requirements: The following organizations are eligible to apply: non-profit organizations; public or private institutions, such as universities, colleges, hospitals, and laboratories; and domestic institutions.
Date(s) Application is Due: Apr 25; May 25
Contact: Lindsey Grandison, Ph.D.; (301) 443-0606; fax: (301) 443-1650; email: lgrandis@mail.nih.gov
Internet: http://grants.nih.gov/grants/guide/pa-files/PAR-08-101.html
Sponsor: National Institute of Mental Health (NIMH)
5635 Fishers Lane, MSC 9304
Bethesda, MD 20892-9304

NIMH Mental Health Dissemination Research Grants 4206

NIMH invites applications for studies addressing factors influencing the dissemination of mental health research knowledge across mental health service settings. The purpose of this announcement is to stimulate research on the array of influences, and their interaction, that beneficially or adversely affect the adoption of valid mental health research findings into clinical practice; and to promote the development of a sound knowledge base on the effectiveness of knowledge dissemination interventions. PA: PA-02-131
Requirements: Applications may be submitted by domestic and foreign, for-profit and nonprofit organizations, public and private.
Restrictions: Foreign institutions are not eligible for small (R03) awards.
Contact: Dr. David Chambers, Division of Services and Intervention Research, (301) 443-3747; fax: (301) 443-4045; email: dchamber@mail.nih.gov; Joy Knipple, Grants Management Branch, (301) 443-8811; fax: (301) 443-6885; email: jk173r@nih.gov
Internet: http://grants2.nih.gov/grants/guide/pa-files/PA-02-131.html
Sponsor: National Institute of Mental Health
6001 Executive Boulevard
Bethesda, MD 20892

NIMH Minority Research Infrastructure Support Program Grants 4207

The purpose of this program is to increase the capacity of minority institutions and their faculty to conduct rigorous mental health-relevant research. The intent of the research infrastructure program is to strengthen the research environments of minority institutions through grant support to develop and/or expand existing capacities for conducting behavioral and neuroscience research in all fields related to mental health. The program provides for two types of activities: institutional development support to strengthen the institutional infrastructure and enhance the capability of individual

faculty members to undertake mental health-related research; and individual investigator research support for developing research scientists to conduct small grant research activities that can lead to successful applications for funding under regular research grant mechanisms. PA: PAR-01-029
Requirements: Applications may be submitted by domestic for-profit and nonprofit organizations, public and private; units of state and local governments; and eligible agencies of state and local government.
Amount of Grant: $400,000 maximum
Date(s) Application is Due: Jun 1
Contact: Dr. Robert Mays Jr., Office for Special Populations, (301) 443-3017; fax: (301) 443-8022; email: rmays@mail.nih.gov
Internet: http://grants.nih.gov/grants/guide/pa-files/PAR-01-029.html
Sponsor: National Institute of Mental Health
5600 Fishers Ln, Parklawn Building
Rockville, MD 20857

NIMH National Research Service Awards (NRSA)—Individual **4208**
A National Research Service Award to an individual supports research training experiences in mental health areas of biomedical and behavioral research. The NIMH research objective is to increase the knowledge of, and improve research methods on, mental and behavioral disorders; to generate information regarding basic biological and behavioral processes underlying these disorders and the maintenance of mental health; and to improve mental health services. PA: PA-99-025
Requirements: Applicants must be citizens or noncitizen nationals of the United States. Predoctoral applicants must have completed two or more years of graduate work and be enrolled in a doctoral degree program by the proposed activation date of the fellowship. Postdoctoral applicants must have received a doctoral or equivalent degree as of the proposed activation date, or submit certification by an authorized official of the degree-granting institution that all requirements have been met.
Amount of Grant: $14,688 predoctoral stipend; $26,256-$41-268 postdoctoral stipend
Date(s) Application is Due: Apr 5; Aug 5; Dec 5
Contact: Dr. Robin Barr, (301) 443-2530; email: rb42h@nih.gov
Internet: http://www.grants.nih.gov/grants/guide/pa-files/PA-99-025.html
Sponsor: National Institute of Mental Health
5600 Fishers Ln, Parklawn Building
Rockville, MD 20857

NIMH New Minority Faculty Scientist Development Award **4209**
The purpose of this program is to enable new minority faculty members to have the necessary time and assistance early in their academic careers to initiate a program of research and to help them to become outstanding independent investigators in mental health research. PA: PAR-99-169
Requirements: Applications may be submitted on behalf of the proposed candidate by domestic, nonprofit, public and private universities, colleges, and professional schools that are engaged in mental health research. The candidate must be a citizen or non-citizen national of the United States or hold a permanent residence visa. Candidates for this award must be new, non-tenured minority faculty in their first or second faculty position who have earned a doctoral degree (PhD, MD, DSc, etc.) by the time the award is made.
Contact: Henry Khachaturian, Office of Science Policy and Program Planning, (301) 443-4335; fax: (301) 443-3225; email: hkhachat@mail.nih.gov
Internet: http://grants.nih.gov/grants/guide/pa-files/PAR-99-169.html
Sponsor: National Institute for Mental Health
6001 Executive Boulevard
Bethesda, MD 20892

NIMH Pilot Effectiveness Trials for Mental Disorders Grants **4210**
The purpose of the program is to obtain needed preliminary data and to conduct studies to support subsequent effectiveness trials of interventions for mental disorders of children and adults. This is part of a series of four program announcements making up a broad-scale public health-oriented intervention research program. Projects may be funded for up to three years.
Requirements: Applications may be submitted by domestic, for-profit and nonprofit organizations, both public and private; units of state and local governments; and eligible agencies of the federal government.
Amount of Grant: $125,000 maximum per year
Date(s) Application is Due: Feb 1; Jun 1; Oct 1
Contact: Dr. Barry Lebowitz, Adult and Geriatric Division, (301) 443-1419; fax: (301) 594-6784; email: blebowit@mail.nih.gov; Joy Knipple, Grants Management Division, (301) 443-8811; fax: (301) 443-6885; email: jk173@nih.gov
Internet: http://www.grants.nih.gov/grants/guide/pa-files/PAR-03-078.html
Sponsor: National Institute of Mental Health
6001 Executive Boulevard
Bethesda, MD 20892

NIMH Quantitative Approaches to the Analysis of Complex Biological Systems **4211**
The purpose of this program announcement is to advertise the interest of NIGMS and NIMH in supporting research projects that develop quantitative approaches to describe, analyze, and predict the behavior of complex biological systems, especially those requiring the integration of potentially large amounts of molecular, biochemical, cell biological, and physiological data. These projects are expected to require the participation of individuals with diverse expertise and therefore to be of a collaborative and cross-disciplinary nature. NIMH expresses particular interest in studies using mathematical, computational, or theoretical approaches to understanding the fundamental biological mechanisms underlying behavior. PA: PA-98-077
Requirements: Applications may be submitted by foreign and domestic, for-profit and nonprofit organizations, both public and private. Racial/ethnic minority individuals, women, and persons with disabilities are encouraged to apply.
Restrictions: Foreign institutions are not eligible for program project (P01) grants.
Contact: Dr. Dennis Glanzman, Division of Basic and Clinical Neuroscience Research, (301) 443-1576; fax: (301) 443-4822; email: glanzman@helix.nih.gov; Diana Trunnell, Grants Management Branch, (301) 443-2805; fax: (301) 433-6885; email: Diana_Trunnel@nih.gov
Internet: http://www.grants.nih.gov/grants/guide/pa-files/PA-98-077.html
Sponsor: National Institute of Mental Health
5600 Fishers Ln, Parklawn Building
Rockville, MD 20857

NIMH Risk Factors for Psychopathology Using Existing Data Sets Grants **4212**
The program encourages extensive and innovative use of existing data sets to study the development of psychopathology or resilience in order to guide the development of prevention and early intervention strategies. The aims of the program are to mine the full potential of public use and other extant data sets to increase our knowledge of risk and protective factors for the development of psychopathology or resilience in community-based and clinical populations, and to encourage applications from new investigators to examine these research areas using state-of-the-art data analytic procedures. NIMH encourages proposals to conduct analyses using mental health research data that is in public-use format or that is privately held by a principal investigator. Annual deadline dates may vary; contact program staff for exact dates. PA: PA-03-044
Requirements: Applications may be submitted by domestic and foreign, for-profit and nonprofit organizations, both public and private; units of state and local governments; and eligible agencies of the federal government.
Restrictions: Foreign institutions are not eligible for NIMH R03 (small grant) awards.
Amount of Grant: $250,000 maximum
Date(s) Application is Due: Feb 1; Jun 1; Oct 1
Contact: Karen Bourdon, (301) 443-5944; fax: (301) 480-4415; email: kbourdon@nih.gov
Internet: http://www.grants.nih.gov/grants/guide/pa-files/PA-03-044.html
Sponsor: National Institute of Mental Health
6701 Executive Boulevard
Bethesda, MD 20892

NIMH Short Courses in Neuroinformatics **4213**
The purpose of this program is to encourage and support short-term research for the development of short courses, seminars, and workshops on interdisciplinary neuroinformatics education . This short-term training will be provided to scientists seeking to combine knowledge about the various subdisciplines of neuroscience and behavioral science research with expertise in informatics research. It is anticipated that these short courses will allow the participants to acquire new conceptual approaches to basic neuroscience research and analyses; and learn to develop unique strategies for acquiring, storing, retrieving, organizing, managing, analyzing, visualizing, manipulating, integrating, synthesizing, disseminating, and sharing data about the brain and behavior.
Requirements: Any nonprofit or for-profit organization engaged in health-related education or research and located in the United States, its possessions, or territories is eligible to apply.
Amount of Grant: $150,000 maximum
Date(s) Application is Due: Feb 16; Jun 16; Oct 16
Contact: Stephen H. Koslow; (301) 443-1815; fax: (301) 443-1867; email: koz@helix.nih.gov
Internet: http://grants.nih.gov/grants/guide/pa-files/PAR-99-137.html
Sponsor: National Institute of Mental Health (NIMH)
6001 Executive Boulevard, Room 6167, MSC 9613
Bethesda, MD 20892

NIMH Short Courses on Mathematical and Statistical Tools for the Study of Complex Phenotypes and Complex Systems **4214**
The purpose of this program announcement is to provide support for short courses or workshops to assist scientists in preparing for research on complex phenotypes and complex systems. This program announcement solicits applications for education grants (R25), which provide for one to five years of support for short courses. Support is limited to courses that are highly focused and that reach a wide audience of scientists. The program announcement is not intended for university course or curriculum development. A variety of approaches will be considered. The nature of the activity, the needs of the target community, and the course goals should determine the instructional approach. These might include new approaches, such as Web-based instruction, or more traditional approaches, such as summer workshops or on-site instruction. PA: PA-98-083
Requirements: Applications may be submitted by domestic nonprofit and for-profit organizations, both public and private.
Restrictions: Foreign organizations are not eligible.
Contact: Dr. Walter Goldschmidts, Division of Basic and Clinical Neurosciences Research, (301) 443-3563; fax: (301) 443-1731; email: wg8u@nih.gov
Internet: http://www.grants.nih.gov/grants/guide/pa-files/PA-98-083.html
Sponsor: National Institute of Mental Health
5600 Fishers Ln, Parklawn Building
Rockville, MD 20857

NIMH Silvio O. Conte Centers for Neuroscience Research Grants 4215
NIMH invites research grant applications for centers. The primary goal of this initiative is to encourage investigators representing a variety of disciplines and/or approaches to bring to the field of mental health research the full range of expertise and advanced technologies available in the basic sciences. Centers for the feasibility of neuroscience research, centers for neuroscience research, and centers for the neuroscience of mental disorders will be represented. The letter of intent receipt date is July 1, and the application receipt date is October 20. PAR-02-122 may be obtained through the NIH grant line (301-402-2221), the NIH Web site, mail, and email.
Requirements: Applications may be submitted by domestic, for-profit and nonprofit organizations, public or private.
Contact: Carol Robinson, Grants Management Branch, (301) 443-3858; fax: (301) 443-6885; email: crobinso@mail.nih.gov
Internet: http://grants2.nih.gov/grants/guide/pa-files/PAR-02-122.html
Sponsor: National Institute of Mental Health
5600 Fishers Ln, Parklawn Building
Rockville, MD 20857

NIMH Supplements for Underrepresented Minorities 4216
NIMH recognizes the need to increase the number of underrepresented minority scientists participating in mental health research. Therefore, funds are made available for administrative supplements designed to provide a continuum of support for underrepresented minority scientists and students. These supplements may support research experience for minority undergraduate students, graduate research assistants, and minority investigators. The aim of these programs is to attract and encourage these minority individuals to pursue biomedical and behavioral research careers in areas within the mission of NIMH by providing supplemental funding to ongoing research grants. The minority individual must be given the opportunity to interact with individuals on the parent grant, to contribute intellectually to the research, and to enhance her/his research skills and knowledge regarding the particular area of biomedical or behavioral science. Applications are accepted at any time. Copies of the specific guidelines may be obtained upon request. PA: PA-99-104
Requirements: Any principal investigator at a domestic institution holding an active ADAMHA grant that has a minimum of two years of research support remaining at the time of a requested supplemental award is eligible. In making requests, the grantee institution, on behalf of the principal investigator of the parent grant and in cooperation with the minority individual, should submit the request for supplemental funds directly to the project officer for the parent grant.
Contact: Deputy Associate Director for Special Populations, (301) 443-2847; fax: (301) 443-8552; email: sr14j@nih.gov
Internet: http://grants.nih.gov/grants/guide/pa-files/PA-99-104.html
Sponsor: National Institute of Mental Health
5600 Fishers Ln, Parklawn Building
Rockville, MD 20857

NIMH Therapeutic Effectiveness Protocol Development Program Grants 4217
The program is directed toward the development of infrastructure and resources for investigation of effectiveness of therapeutic interventions through support of protocol design. This program has been developed in response to a recommendation of the Clinical Treatment and Services Research Workgroup of the National Advisory Mental Health Council. The program will support effectiveness trial development that will lead to large-scale studies focusing on diseases and populations of major public health significance that are consistent with the mission of NIMH. This is part of a series of four program announcements. Projects may receive funding for up to three years. Annual deadline dates may vary; contact program staff for exact dates. PA: PAR-03-078
Requirements: Applications may be submitted by domestic, for-profit and nonprofit organizations, both public and private; units of state and local governments; and eligible agencies of the federal government.
Amount of Grant: $125,000 maximum per year
Date(s) Application is Due: Feb 1; Jun 1; Oct 1
Contact: Dr. Linda Street, Division of Services and Intervention Research, (301) 443-0651; fax: (301) 594-6784; email: Lstreet@nih.gov; Joy Knipple, Grants Management Branch, (301) 443-8811; fax: (301) 443-6885; email: jk173@nih.gov
Internet: http://www.grants.nih.gov/grants/guide/pa-files/PAR-03-078.html
Sponsor: National Institute of Mental Health
6001 Executive Boulevard
Bethesda, MD 20892

NIMH/NIDA Children's Services Program Grants 4218
The purpose of this program is to encourage research grant applications on services delivered to children, adolescents, and their families. Conducting the proposed research at one of the Children's Services Program sites (either existing sites or graduated sites) is a prerequisite for funding under this PA. This PA encourages studies of the effectiveness of treatments or services delivered at these sites, the nature and impact of routine clinical practice, and factors related to successful implementation of treatments or services. PA: PA-04-019
Requirements: Applications may be submitted by domestic and foreign, for-profit and nonprofit organizations, both public and private; units of state and local governments; and eligible agencies of the federal government.
Restrictions: Research grant funds may not be used to operate treatment, rehabilitation, or service programs.
Amount of Grant: $25,000-$250,000
Contact: Dr. Heather Ringeisen, Chief, Child and Adolescent Services Research Program, (301) 496-7227; fax: (301) 443-4045; email: hringeis@mail.nih.gov
Internet: http://grants.nih.gov/grants/guide/pa-files/PA-04-019.html
Sponsor: National Institute of Mental Health
6001 Executive Boulevard
Bethesda, MD 20892

NINDS Central Nervous System Effects of HIV Infection: Neurobiological, Neurovirological, and Neurobehavioral Studies Grants 4219
NINDS, with NIMH, NICHD, and NIDA, invites applications for support of research on the role of the HIV virus in the central nervous system and its consequent effects on behavior. Applicants may request up to five years of support. Applications will be accepted under PHS receipt dates for AIDS applications. PA: PA-04-154
Requirements: Any public, private, nonprofit, or for-profit institution is eligible to apply for research grants. For-profit institutions are not eligible for institutional NRSAs but are eligible for individual NRSAs. SBIR grants can be awarded only to domestic small businesses. STTR grants can be awarded only to domestic small business concerns that partner with a research institution in cooperative research and development.
Date(s) Application is Due: Feb 1; Jun 1; Oct 1
Contact: Dr. Jeymohan Joseph, Chief, HIV Neurovirology, (301) 443-3012; fax: (301) 443-9719; email: jjeymoha@mail.nih.gov
Internet: http://grants.nih.gov/grants/guide/pa-files/PA-04-154.html
Sponsor: National Institute of Neurological Disorders and Stroke
6701 Rockledge Drive, Room 1040, MSC 7710
Bethesda, MD 20892-7710

NINDS Cerebral Radiobiology and Neuroimaging of Brain Tumors Research Grants 4220
This program supports research that will increase our knowledge of the genetic, molecular, cellular, and physiological mechanisms of radiation-induced cell injury and recovery, and the sensitizing and protective mechanisms in the central nervous system under radiation treatment conditions for brain tumors. The intent of this program announcement is to encourage investigator-initiated applications to study tumor and normal brain cell injury and repair mechanisms induced by brain tumor radiotherapy including stereotactic radiosurgery procedures such as the Gamma Knife, altered fractionation and/or radioenhancing agents, using state-of-the-art neurobiological and neuroimaging approaches. An applicant planning to submit a new (Type 1) investigator-initiated grant application requesting $500,000 or more in direct costs for any year is advised that he or she must contact the program staff before submitting the application.
Requirements: Applications may be submitted by domestic and foreign, for-profit and nonprofit organizations, both public and private; units of state and local governments; and eligible agencies of the federal government.
Restrictions: Foreign institutions are not eligible for P01 grants.
Contact: Dr. Thomas Jacobs, Division of Stroke, Trauma, and Neurodegenerative Disorders, (301) 496-4226; fax: (301) 480-1080; email: TJ12G@nih.gov; Dr. Francis Mahoney, Division of Cancer Treatment and Diagnosis, NCI, (301) 496-9360; fax: (301) 480-5785; email:
Internet: http://www.grants.nih.gov/grants/guide/pa-files/PA-98-094.html
Sponsor: National Institute of Neurological Disorders and Stroke
6001 Executive Boulevard, Suite 2209
Bethesda, MD 20892

NINDS Low-Back Pain and Common Spinal Disorders Research Grants 4221
The program is sponsored by NIAMS, NICHD, NINDS, and NIOSH at CDC. This initiative invites applications directed to the study of the pathogenesis, epidemiology, prevention, and treatment of three common causes of low-back pain: herniated nucleus pulposus, spinal stenosis, and idiopathic low-back pain. The support mechanism for grants in this area will be the individual investigator-initiated research grant (R01). PA-97-058 may be obtained electronically through the NIH grant line (301-402-2221), the NIH Web site, and by mail and email.
Requirements: Applications may be submitted by foreign and domestic, for-profit and nonprofit organizations, both public and private; units of state and local governments; and eligible agencies of the federal government.
Restrictions: Foreign institutions or organizations in foreign countries are not eligible for First Independent Research Support and Transition (R29) awards.
Contact: Dr. Mary Ellen Cheung, Division of Stroke, Trauma, and Neurodegenerative Disorders, (301) 496-4226; fax: (301) 480-1080; email: mm108w@nih.gov
Internet: http://www.grants.nih.gov/grants/guide/pa-files/PA-97-058.html
Sponsor: National Institute of Neurological Disorders and Stroke
6001 Executive Boulevard, Suite 2209
Bethesda, MD 20892

NINDS Optimization of Small Molecule Probes for the Nervous System (SBIR) Grants 4222
This Funding Opportunity Announcement (FOA) encourages Small Business Innovation Research (SBIR) grant applications from small business concerns (SBCs) that propose to develop new small molecule probes for investigating biological function in the nervous system via the application of advanced medicinal chemistry and the biological testing of compounds. Eligible SBCs will have identified probe candidates via screening of small molecule collections, using in vitro assays of biological activity developed to interrogate these collections, and be able to show that the structural features of these small molecules are related to their biological activity. Applications should nominate small molecule probe candidates from distinct structural series for the further, iterative

design and testing of analogues in structure-activity relationship studies, using in vitro assays of biological function adapted to the medium throughput screening requirements of this work. These studies should have the goal of developing a small molecule probe possessing the attributes (eg: affinity, selectivity, activity) required for its use in future pharmacological studies proposed by the SBC. Applicants are strongly encouraged to utilize publicly available cheminformatic capabilities for the acquisition of compounds, and semi-custom synthesis of analogues.
Requirements: Only United States small business concerns (SBCs) are eligible to submit SBIR applications.
Amount of Grant: $150,000 maximum per year
Date(s) Application is Due: Apr 5; Aug 5; Dec 5
Contact: Mark Scheideler, Ph.D.; (301) 496-1779; fax: (301) 402-1501; email: scheidelerm@ninds.nih.gov
Internet: http://grants.nih.gov/grants/guide/pa-files/PAR-09-260.html
Sponsor: National Institute of Neurological Disorders and Stroke (NINDS)
6001 Executive Boulevard, NSC 2107
Bethesda, MD 20892-9527

NINDS Optimization of Small Molecule Probes for the Nervous System (STTR) Grants 4223
This Funding Opportunity Announcement (FOA) encourages Small Business Technology Transfer (STTR) grant applications from small business concerns (SBCs) that propose to develop new small molecule probes for investigating biological function in the nervous system via the application of advanced medicinal chemistry and the biological testing of compounds. Eligible SBCs will have identified probe candidates via screening of small molecule collections, using in vitro assays of biological activity developed to interrogate these collections, and be able to show that the structural features of these small molecules are related to their biological activity. Applications should nominate small molecule probe candidates from distinct structural series for the further, iterative design and testing of analogues in structure-activity relationship studies, using in vitro assays of biological function adapted to the medium throughput screening requirements of this work. These studies should have the goal of developing a small molecule probe possessing the attributes (eg: affinity, selectivity, activity) required for its use in future pharmacological studies proposed by the SBC. Applicants are strongly encouraged to utilize publicly available cheminformatic capabilities for the acquisition of compounds, and semi-custom synthesis of analogues, which is required of these studies.
Requirements: Only United States small business concerns (SBCs) are eligible to submit STTR applications.
Amount of Grant: $150,000 maximum per year
Date(s) Application is Due: Apr 5; Aug 5; Dec 5
Contact: Mark Scheideler, Ph.D.; (301) 496-1779; fax: (301) 402-1501; email: scheidelerm@ninds.nih.gov
Internet: http://grants.nih.gov/grants/guide/pa-files/PAR-09-260.html
Sponsor: National Institute of Neurological Disorders and Stroke (NINDS)
6001 Executive Boulevard, NSC 2107
Bethesda, MD 20892-9527

NINDS Optimization of Small Molecule Probes for the Nervous System Grants 4224
This FOA issued by participating institutes of the National Institutes of Health, encourages research grant applications from institutions/ organizations that propose to develop new small molecule probes for investigating biological function in the nervous system via the application of advanced medicinal chemistry and the biological testing of compounds. Eligible investigators will have identified probe candidates via screening of small molecule collections, using in vitro assays of biological activity developed to interrogate these collections, and be able to show that the structural features of these small molecules are related to their biological activity. Proposals should nominate small molecule probe candidates from distinct structural series for the further, iterative design and testing of analogues in structure-activity relationship studies, using in vitro assays of biological function adapted to the medium throughput screening requirements of this work. These studies should have the goal of developing a small molecule probe possessing the attributes (eg: affinity, selectivity, activity) required for its use in future pharmacological studies proposed by the investigator. Applicants are strongly encouraged to utilize publicly available cheminformatic capabilities for the acquisition of compounds, and semi-custom synthesis of analogues.
Requirements: The following organizations and institutions are eligible to apply: public/state controlled institutions of higher education; private institutions of higher education; Hispanic-serving institutions; historically Black colleges and universities (HBCUs); tribally controlled colleges and universities (TCCUs); Alaska native and native Hawaiian serving institutions; nonprofits with 501(c)3 IRS status (other than institutions of higher education); nonprofits without 501(c)3 IRS status (other than institutions of higher education); small businesses; for-profit organizations (other than small businesses); state governments; Indian/Native American tribal governments (federally recognized); Indian/Native American tribally designated organizations; county governments; city or township governments; special district governments; independent school districts; public housing authorities/Indian housing authorities; U.S. territories or possessions; Indian/native American tribal governments (other than federally recognized); regional organizations; non-domestic (non-U.S.) entities (foreign organizations); eligible agencies of the federal government; and faith-based or community-based organizations.
Amount of Grant: $150,000 maximum per year
Date(s) Application is Due: Feb 16; Jun 16; Oct 16
Contact: Mark Scheideler, Ph.D.; (301) 496-1779; fax: (301) 402-1501; email: scheidelerm@ninds.nih.gov
Internet: http://grants.nih.gov/grants/guide/pa-files/PAR-09-251.html
Sponsor: National Institute of Neurological Disorders and Stroke (NINDS)
6001 Executive Boulevard, NSC 2107
Bethesda, MD 20892-9527

NINDS Pathogenesis and Therapy of the Muscular Dystrophies Research Grants 4225
NINDS and NIAMS encourage investigator-initiated research grant applications to study the pathogenesis and therapy of the various forms of muscular dystrophy in children and adults. Responses to this PA may include studies in appropriate animal models or preclinical or clinical studies in patients with facioscapulohumeral dystrophy (FSH), limb-girdle muscular dystrophy (LGMD), myotonic dystrophy, congenital muscular dystrophy (CMD), Emery-Dreifuss muscular dystrophy (EMD), Duchenne muscular dystrophy (DMD), Becker muscular dystrophy (BMD), or other forms of muscular dystrophy. Support mechanisms for grants are research project grant (R01) and program project grant (P01). Information may be obtained through the NIH grant line (301-402-2221), the NIH Web site, mail, and email. PA: PA-98-044
Requirements: Applications may be submitted by domestic and foreign, for-profit and nonprofit organizations, both public and private; units of state and local governments; and eligible agencies of the federal government.
Contact: Dr. Paul Nichols, Division of Convulsive, Infectious, and Immune Disorders, (301) 496-1431; fax: (301) 402-2060; email: pn13w@nih.gov; Dawn Richardson, Grants Management Branch, (301) 496-9231; fax: (301) 402-0219; email: r128b@nih.gov
Internet: http://www.grants.nih.gov/grants/guide/pa-files/PA-98-044.html
Sponsor: National Institute of Neurological Disorders and Stroke
6001 Executive Boulevard, Suite 2209
Bethesda, MD 20892

NINDS Repetitive Motion Disorders Research Grants 4226
The program is sponsored by NIAMS, NICHD, NINDS, and NIOSH at CDC. This initiative invites applications directed to the study of the pathogenesis, epidemiology, prevention, and treatment of repetitive motion disorders. The support mechanism for grants in this area will be the individual investigator-initiated research grant (R01). PA-97-059 may be obtained electronically through the NIH grant line (301-402-2221), the NIH Web site, and by mail and email.
Requirements: Applications may be submitted by foreign and domestic, for-profit and nonprofit organizations, both public and private; units of state and local governments; and eligible agencies of the federal government.
Restrictions: Foreign institutions or organizations in foreign countries are not eligible for First Independent Research Support and Transition (FIRST) (R29) awards.
Contact: Dr. Paul Nichols, (301) 496-1431; fax: (301) 402-2060; email: pn13w@nih.gov
Internet: http://www.grants.nih.gov/grants/guide/pa-files/PA-97-059.html
Sponsor: National Institute of Neurological Disorders and Stroke
6001 Executive Boulevard, Suite 2209
Bethesda, MD 20892

NINDS Specialized Neuroscience Research Programs (SNRP) on Health Disparity: HIV and the Nervous System 4227
The purpose of this program is to augment and strengthen the research capabilities of faculty, students, and fellows at minority institutions by supporting the development of new, and/or the enhancement of ongoing basic and clinical neuro-AIDS related research projects and programs. Applicants planning to submit an investigator-initiated new, competing continuation, competing supplement, or any amended/revised version of the preceding grant application types requesting $500,000 or more in direct costs for any year are advised that he or she must contact institute program staff before submitting the application. Annual deadline dates may vary; contact program staff for exact dates. RFA: NS-01-009
Requirements: Applications will only be accepted from domestic, public or private, nonprofit, academic institutions that offer PhD, MD, and/or equivalent health professional degrees, at which more than 50 percent of the students enrolled are from ethnic or racial minority groups. Each applicant institution must identify a collaborating institution that can be a domestic federal or nonfederal, public or private, nonprofit organization. The principal investigator, who serves as SNRP director, must be a U.S. citizen, permanent resident, or noncitizen national.
Amount of Grant: $1 million maximum
Contact: Dr. Joana Rosario, Office of Special Programs in Neuroscience, (301) 496-5489; fax: (301) 594-5929; email: jr69z@nih.gov; Dr. A.P. Kerza-Kwiatecki, (301) 496-1431; fax: (301) 402-4020; email: ak45w@nih.gov
Internet: http://grants.nih.gov/grants/guide/rfa-files/RFA-NS-01-009.html
Sponsor: National Institute of Neurological Disorders and Stroke
6001 Executive Boulevard, Suite 2209
Bethesda, MD 20892

NINDS Support of Scientific Meetings as Cooperative Agreements 4228
The purpose of the program announcement is to inform the scientific community that the National Institute of Neurological Disorders and Stroke (NINDS) will now support scientific meetings as cooperative agreements in addition to the current practice of supporting them through the traditional grant mechanism. This program provides guidelines for when it is appropriate to request support of a meeting as a cooperative agreement and explains procedures for preparing and submitting such applications. The Principal Investigator will have the primary authority and responsibility to define objectives and approaches; plan, publicize, and conduct the scientific meeting; and publish the results of the meeting.

Requirements: Applications may be submitted by U.S. institutions, including scientific or professional societies eligible to receive grants from Public Health Service (PHS) agencies. In the case of an international conference, the U.S. representative organization of an established international scientific or professional society is the eligible applicant.
Restrictions: Foreign institutions are not eligible.
Contact: Danilo A. Tagle, Ph.D.; (301) 496-5745; fax: (301) 402-2060; email: tagled@ninds.nih.gov
Internet: http://grants.nih.gov/grants/guide/notice-files/NOT-NS-02-003.html
Sponsor: National Institute of Neurological Disorders and Stroke (NINDS)
6001 Executive Boulevard, Neuroscience Center, Room 2114
Bethesda, MD 20892-9525

NINDS Symptom Management for Chronic Neurological Conditions Research Grants 4229
NINR, NIA, NICHD, NIDCD, and NINDS are interested in facilitating investigator-initiated research designed to deal with symptoms associated with chronic neurological conditions including stroke, epilepsy, brain and spinal cord injury, multiple sclerosis, and Parkinson's disease. The overall goals are to manage symptoms effectively, to prevent secondary effects of the condition when possible, to maximize the health potential, and to improve the health-related quality of life for patients, care givers, and families. The mechanism of support is the research project grant (R01). More information may be obtained through the NIH grant line (301-402-2221), the NIH Web site, mail, and email. PA: PA-98-030
Requirements: Research applications may be submitted by domestic and foreign, for-profit and nonprofit organizations, both public and private; units of state and local governments; and eligible agencies of the federal government.
Contact: Dr. Mary Ellen Cheung, Division of Stroke, Trauma, and Neurodegenerative Disorders, (301) 496-4226; fax: (301) 480-1080; email: MM108W@nih.gov; Gladys Melendez-Bohler, Grants Management Branch, (301) 496-9231; fax: (301) 402-0219; email: GB13Y@nih.gov
Internet: http://www.grants.nih.gov/grants/guide/pa-files/PA-98-030.html
Sponsor: National Institute of Neurological Disorders and Stroke
6001 Executive Boulevard, Suite 2209
Bethesda, MD 20892

NINR Acute and Chronic Care During Mechanical Ventilation Research Grants 4230
NINR seeks research applications for investigator-initiated research to improve the clinical management and care delivery of patients who require mechanical ventilation. For the purpose of this program announcement, the time course of mechanical ventilation is defined as beginning with endotracheal or tracheal tube intubation, maintenance of full and/or partial ventilatory support, discontinuation of mechanical ventilation via weaning techniques, and concluding with care of patients who require chronic mechanical ventilation. Two different, but related, types of basic science and/or clinical research on acute and/or chronic mechanical ventilation are encouraged: studies examining the effects of physiological, environmental, and/or psychological approaches to facilitate the care of patients during any phase of mechanical ventilation, and studies directed toward examining the physical and psychological ramifications of technology dependency on patients and their families/significant others. Priority research areas include optimum ways to deliver patient care during mechanical ventilatory support; ethical issues related to initiation, maintenance, and withdrawal of life support technology; emotional support and communication during mechanical ventilation; methods of optimally managing the artificial airway; impact of care on family dynamics; and decision making regarding withdrawal of ventilatory support. Research is also needed to describe and intervene with care-giving resources, family functioning, and quality of life for patients on mechanical ventilation particularly chronically ill children and adults being maintained on home ventilators; optimum methods and techniques to deliver nursing care while critically ill patients receive mechanical ventilation; and determine optimal measures for promoting recovery as patients move through levels of care in acute, intermediate, and long-term care. This PA covers a wide range of topics relevant to clinical management and clinical problem solving issues for patients who require mechanical ventilation.
Requirements: Applications may be submitted by public and private, domestic and foreign, for-profit and nonprofit organizations; units of state and local governments; and eligible agencies of the federal government.
Restrictions: The total project period for an application may not exceed five years.
Date(s) Application is Due: Feb 5; Feb 16; Jun 5; Jun 16; Oct 5; Oct 16
Contact: Josephine Boyington; (301) 594-2542; fax: (301) 480-8260; email: boyingtonje@mail.nih.gov
Internet: http://grants.nih.gov/grants/guide/pa-files/PA-99-003.html
Sponsor: National Institute of Nursing Research (NINR)
6701 Democracy Boulevard, Suite 710, Room 715
Bethesda, MD 20892-4870

NINR Aging Women and Breast Cancer Grants 4231
NINR, NCI, and NIA are supporting grants that focus on the unique problems of older women with breast cancer. Their purpose is to inform the scientific community of the interests of NIA and NCI and to expand the knowledge base on breast cancer in older women through studies in the fields of biology, clinical medicine, epidemiology, and the behavioral and social sciences. More information may be obtained by mail and email or through the NIH grant line (301-402-2221), and the NIH Web site. PA: PA-00-001
Requirements: Foreign and domestic for-profit and nonprofit organizations, both public and private; units of state and local governments; and eligible agencies of the federal government are eligible to apply.
Amount of Grant: $29,175-$627,861; $272,777 average
Date(s) Application is Due: Feb 1; Jun 1; Oct 1
Contact: Dr. Janice Phillips, Program Director, (301) 594-6152; fax: (301) 480-8260; email: Janice_Phillips@nih.gov
Internet: http://grants.nih.gov/grants/guide/pa-files/PA-00-001.html
Sponsor: National Institute of Nursing Research
45 Center Drive, Natcher Building
Bethesda, MD 20892

NINR AIDS and Aging: Behavioral Sciences Prevention Research Grants 4232
NIA, NIMH, and NINR invite qualified researchers to submit applications to investigate prevention issues relevant to AIDS in middle-aged and older populations. This announcement solicits AIDS prevention research proposals to study primary prevention of disease transmission as well as secondary and tertiary prevention of negative behavioral and social consequences of HIV/AIDS for persons with AIDS, their families, and communities. Thus, the primary goals are to identify social and behavioral factors associated with HIV transmission and disease progression in later life; examine behavioral and social consequences of HIV infection/AIDS across the life course; develop and evaluate age-appropriate behavioral and social interventions for preventing AIDS in middle-aged and older adults and/or ameliorating problems associated with older adult caregiver responsibilities and burdens; explore health care issues surrounding AIDS care; and strengthen existing research and evaluation methods.
Requirements: Applications may be submitted by domestic and foreign for-profit and non-profit organizations, public and private, such as universities, colleges, hospitals, laboratories, units of State and local governments, and eligible agencies of the Federal government. Applications may be submitted by single institutions or by a consortia of institutions.
Date(s) Application is Due: Feb 5; Feb 16; Jun 5; Jun 16; Oct 5; Oct 16
Contact: Josephine Boyington; (301) 594-2542; fax: (301) 480-8260; email: boyingtonje@mail.nih.gov
Internet: http://www.grants.nih.gov/grants/guide/pa-files/PA-97-069.html
Sponsor: National Institute of Nursing Research (NINR)
6701 Democracy Boulevard, Suite 710, Room 715
Bethesda, MD 20892-4870

NINR Career Transition Award 4233
The overall goal of this program is to support outstanding individuals to obtain a research training experience at the NIH in the intramural research laboratories and to facilitate their successful transition to an extramural environment as independent researchers. Annual deadline dates may vary; contact program staff for exact dates. RFA: NR-01-003
Requirements: Registered Nurses (RN) who are recent recipients of a doctoral degree or its equivalent and who have demonstrated the potential for a highly productive research career are eligible to apply.
Amount of Grant: $125,000 maximum
Contact: Hilary Sigmon, NINR, (301) 594-5970; fax: (301) 480-8260; email: hilary_sigmon@nih.gov
Internet: http://www.grants.nih.gov/grants/guide/rfa-files/RFA-NR-01-003.html
Sponsor: National Institute of Nursing Research
Building 45
Bethesda, MD 20892

NINR Chronic Illness Self-Management in Children and Adolescents Grants 4234
The purpose of this Funding Opportunity Announcement (FOA) issued by the NINR, NICHD, NIDDK, NCI, and the ODS is to solicit research to improve self-management and quality of life in children and adolescents with chronic illnesses. Biobehavioral studies of children within the context of family and family-community dynamics are encouraged. Children with a chronic disease and their families have a long-term responsibility for self-management. The child with the illness will have a life-long responsibility to maintain and promote health and prevent complications. Research related to sociocultural, environmental, and behavioral mechanisms as well as biological/technological factors that contribute to successful and ongoing self-management of particular chronic illnesses in children is encouraged. Because the nature and scope of the proposed research will vary from application to application, it is anticipated that the size and duration of each award will also vary.
Requirements: Application can be made by the following institutions and organizations: public/state controlled institutions of higher education; private institutions of higher education; nonprofits with 501(c)3 IRS status (other than institutions of higher education); nonprofits without 501(c)3 IRS status (other than institutions of higher education); small businesses; for-profit organizations (other than small businesses); state governments; U.S. territories or possessions; Indian/Native American tribal governments (federally recognized); Indian/Native American tribal governments (other than Federally recognized); Indian/Native American tribally designated organizations; non-domestic (non-U.S.) entities (foreign organizations); Hispanic-serving institutions; historically Black colleges and universities (HBCUs); tribally controlled colleges and universities (TCCUs); Alaska Native and Native Hawaiian serving institutions; regional organization; eligible agencies of the Federal government; and faith-based or community based organizations.
Restrictions: This FOA is restricted to studies of chronic illnesses in children and adolescents ages 8 to 21 grouped by developmental stages according to the discretion of the investigator. Studies of chronic mental illness or serious cognitive disability are beyond the scope of this FOA.
Date(s) Application is Due: Feb 5; Jun 5; Oct 5

Contact: Dr. Linda S. Weglicki, PhD; (301) 594-6908; fax: (301) 480-8260; email: weglickils@mail.nih.gov
Internet: http://grants.nih.gov/grants/guide/pa-files/PA-07-097.html
Sponsor: National Institute of Nursing Research (NINR)
6701 Democracy Boulevard, Suite 710, One Democracy Plaza
Bethesda, MD 20892-4870

NINR Clinical Interventions for Managing the Symptoms of Stroke Research Grants 4235

NINR, NINDS, NIA, NIDCD, and NICHD seek research applications that address effective nonpharmacological approaches to managing the initial events and subsequent symptoms of a stroke. This research would focus on reducing the impairments, preventing secondary complications, and improving the functional independence and the quality of life for the individual following a stroke. Research areas that need to be explored include, but are not limited to: techniques to improve assessment during the initial stroke event and at each stage of recovery to guide nonpharmacologic treatment decisions and monitor patient progress; interventions to manage common symptoms, such as dysphagia, aphasia, bowel and bladder problems, sensorimotor deficits, depression, cognitive and perceptual deficits as well as difficulties in nutrition, hydration, skin integrity, and sleep; strategies to determine optimal timing, intensity, and duration of nonpharmacologic treatments; methods to prevent secondary complications, improve function, and address the sequelae of common comorbid conditions; strategies to promote self-care in both activities of daily living and instrumental activities of daily living and to enhance or maintain health promotion activities; measures to enhance family caregiving during all stages of recovery; methods to improve care during transition from one setting to another and care in specialized nurse managed units; and studies on the extent to which age and pre-existing conditions, such as specific diseases, sensory, motor, and cognitive declines, influence treatment outcome and management after stroke.
Requirements: Applications may be submitted by domestic and foreign, for-profit and nonprofit organizations, public and private.
Amount of Grant: $250,000 maximum per year
Contact: Karin F. Helmers; (301) 594-2177; email: karin_helmers@nih.gov
Internet: http://www.grants.nih.gov/grants/guide/pa-files/PA-99-088.html
Sponsor: National Institute of Nursing Research (NINR)
6701 Rockledge Drive, Building 45, Room 3AN-12
Bethesda, MD 20892

NINR Clinical Trials: Collaborations for Nursing Research 4236

NINR invites nurse investigators to participate in an initiative that capitalizes on the substantial NIH resource of available study populations in clinical trials from the unique perspective of nursing research and special expertise of nurse researchers. The purpose of this RFA is to link substudies posed by nurse researchers to currently funded clinical research studies, or to link substudies by non-nurse researchers with ongoing clinical studies where nurse researchers are the principal investigators. Collaborations and consortia promoting the cross-fertilization of ideas are encouraged. The rationale for this RFA is that substudies are cost effective ways to stimulate ongoing research while capitalizing on resources expended on a parent clinical trial. Nurse researchers are to collaborate with non-nurse PIs and/or the steering committees on currently funded clinical trials where resources can be accessed to answer a clinical question or solve a nursing problem. Likewise, non-nurse researchers in various fields are encouraged to collaborate with an NIH-funded nurse researcher to access a study population in a scientific field of their choice.
Requirements: Registered nurses with doctoral preparation who pose a research question from her/his ongoing program of research that can be answered by accessing a currently funded clinical trial, or, a non-nurse doctorally prepared scientist who seeks to answer a clinical question or solve a nursing problem using the clinical study resources where the PI is a nurse researcher are eligible to apply. Applications may be submitted by domestic and foreign for-profit and nonprofit organizations, both public and private; units of state or local governments; and eligible agencies of the federal government.
Amount of Grant: $1.5 million total
Contact: Dr. Hilary Sigmon, Program Director, (301) 594-5970; fax: (301) 480-8260; email: Hilary_Sigmon@nih.gov
Internet: http://www.grants.nih.gov/grants/guide/rfa-files/RFA-NR-01-004.html
Sponsor: National Institute of Nursing Research
45 Center Drive, Natcher Building
Bethesda, MD 20892

NINR Community-Based Prevention Intervention Research in Environmental Health Sciences Grants 4237

NINR, NIEHS, and the Office of Behavioral and Social Sciences invite research grant applications addressing development of community-based strategies aimed at prevention and intervention activities in economically disadvantaged and/or underserved populations adversely impacted by an environmental contaminant. These funds support research activities that develop and implement improved prevention and intervention strategies related to underserved populations. In community-based research, active cooperation and participation of organizations within the community that is the focus of the study are essential components of the research. Hence, applicants must describe an existing or proposed involvement with one or more community-based organizations. These grants provide support to develop, test, and evaluate health service activities and to enhance the application or transfer of existing knowledge to efforts designed to prevent or control dysfunction or disease. Annual deadline dates may vary; contact program staff for exact dates. More information may be obtained by mail and email, or through the NIH grant line (301-402-2221), and the NIH Web site. RFA: ES-96-005
Amount of Grant: $64,658-$1.184 million; $218,000 average
Date(s) Application is Due: Feb 1; Jun 1; Oct 1
Contact: Dr. Allen Dearry, NIEHS, (919) 541-4500; fax: (919) 541-2843; email: DEARRY@NIEHS.NIH.GOV; J. Taylor Harden, Division of Extramural Programs, (301) 594-5976; fax: (301) 480-8260; email: THARDEN@EP.NINR.NIH.GOV
Internet: http://www.grants.nih.gov/grants/guide/rfa-files/RFA-ES-96-005.html
Sponsor: National Institute of Nursing Research
45 Center Drive, Natcher Building
Bethesda, MD 20892

NINR Diabetes Self-Management in Minority Populations Grants 4238

This Program Announcement (PA) solicits applications for investigator-initiated research related to sociocultural, environmental, and behavioral mechanisms and biological/technological factors that contribute to successful and ongoing self-management of diabetes in minority populations. Applications that expand accepted intervention strategies in majority populations to minority populations are encouraged. Testing new interventions designed to promote self-management in minority diabetes populations will also be responsive to the PA. Self-management is defined as client strategies and behaviors that contribute to blood glucose normalization, improved health, and prevention or reduction of complications. The concept is broader than adherence to specific regimen components and incorporates deliberate problem solving and decision making processes. Applications are encouraged for both type 1 and type 2 diabetes and all age groups. Modular Grant applications will request direct costs in $25,000 modules, up to a total direct cost request of $250,000 per year.
Requirements: Applications may be submitted by domestic and foreign, for-profit and non-profit organizations, public and private, such as universities, colleges, hospitals, laboratories, units of State and local governments, and eligible agencies of the Federal government. Racial/ethnic minority individuals, women, and persons with disabilities are encouraged to apply as principal investigators.
Amount of Grant: $250,000 maximum per year
Date(s) Application is Due: Feb 5; Jun 5; Oct 5
Contact: Dr. Karen Huss, Ph.D.; (301) 594-5970; fax: (301) 451-5649; email: husk@mail.nih.gov
Internet: http://grants.nih.gov/grants/guide/pa-files/PA-00-113.html
Sponsor: National Institute of Nursing Research (NINR)
6701 Democracy Boulevard, Room 710
Bethesda, MD 20892-4870

NINR Health Risk Reduction: Community-Based Strategies 4239

NINR is interested in facilitating investigator-initiated research into the clinical application of intervention strategies designed to reduce health risks at the community level. Community-based strategies targeting health problems of rural residents and of underserved minority groups are of particular interest. More information may be obtained by mail and email, or through the NIH grant line, and the NIH Web site.
Date(s) Application is Due: Feb 1; Jun 1; Oct 1
Contact: Grants Administrator, (301) 594-6906; fax: (301) 480-8260; email: info@ninr.nih.gov
Internet: http://grants2.nih.gov/grants/guide/pa-files/PA-96-037.html
Sponsor: National Institute of Nursing Research
45 Center Drive, Natcher Building
Bethesda, MD 20892

NINR Management of Symptoms at the End of Life Grants 4240

NINR, NCI, NIAID, NIMH, and the Office of Alternative Medicine, NIH, seek research grant applications concerning the clinical management of symptoms and syndromes that are associated with life-limiting illness, such as pain, dyspnea, delirium, cachexia, nausea, fatigue, and depression. The purpose of this initiative is to stimulate research that will lead to improved quality of life for those at the end of life and decreased distress for their caregivers. PA-98-019 may be obtained through the NIH grant line (301-402-2221), the NIH Web site, mail, and email.
Requirements: Applications may be submitted by domestic and foreign, for-profit and nonprofit organizations, public and private.
Contact: Dr. June Lunney, Division of Extramural Activities, (301) 594-6908; fax: (301) 480-8260; email: Lunneyj@ninr.nih.gov; Jeff Carow, Grants Management Branch, (301) 496-7800 ext 250; fax: (301) 496-8601; email: Jeff_Carow@nih.gov
Internet: http://grants.nih.gov/grants/guide/pa-files/PA-98-019.html
Sponsor: National Institute of Nursing Research
6300 Center Drive
Bethesda, MD 20892

NINR Mechanisms, Models, Measurement, & Management in Pain Research Grants 4241

The purpose of this Funding Opportunity Announcement (FOA), Mechanisms, Models, Measurement, and Management in Pain Research, is to inform the scientific community of the pain research interests of the various Institutes and Centers (ICs) at the National Institutes of Health (NIH) and to stimulate and foster a wide range of basic, clinical, and translational studies on pain as they relate to the missions of these ICs. New advances are needed in every area of pain research, from the micro perspective of molecular sciences to the macro perspective of behavioral and social sciences. Although great strides have been made in some areas, such as the identification of neural pathways of pain, the experience of pain and the challenge of treatment have remained uniquely individual and unsolved. Furthermore, our understanding of how and why individuals transition to a chronic pain state after an acute insult is limited. Research to address these issues conducted

by interdisciplinary and multidisciplinary research teams is strongly encouraged, as is research from underrepresented, minority, disabled, or female investigators. Direct costs are limited to $275,000 over a two-year period, with no more than $200,000 in direct costs allowed in any single year.
Requirements: The following organizations and institutions are eligible to apply: public/state controlled institutions of higher education; private institutions of higher education; Hispanic-serving institutions; historically Black colleges and universities (HBCUs); tribally controlled colleges and universities (TCCUs); Alaska native and native Hawaiian serving institutions; nonprofits with 501(c)3 IRS status (other than institutions of higher education); nonprofits without 501(c)3 IRS status (other than institutions of higher education); small businesses; for-profit organizations (other than small businesses); state governments; Indian/Native American tribal governments (federally recognized); Indian/Native American tribally designated organizations; county governments; city or township governments; special district governments; independent school districts; public housing authorities/Indian housing authorities; U.S. territories or possessions; Indian/native American tribal governments (other than federally recognized); regional organizations; non-domestic (non-U.S.) entities (foreign organizations); eligible agencies of the federal government; and faith-based or community-based organizations.
Amount of Grant: $200,000 maximum annually; $275,000 maximum request per two-year period
Date(s) Application is Due: Feb 16; Jun 16; Oct 16
Contact: Dr. Susan Marden, PhD; (301) 496-9623; fax: (301) 480-8260; email: mardens@mail.nih.gov
Internet: http://grants.nih.gov/grants/guide/pa-files/PA-10-007.html
Sponsor: National Institute of Nursing Research (NINR)
6701 Democracy Boulevard, Suite 710, One Democracy Plaza
Bethesda, MD 20892-4870

NINR Mentored Research Scientist Development Award 4242

The objective of the NIH Mentored Research Scientist Development Award is to provide support for a sustained period of protected time for intensive research career development under the guidance of an experienced mentor, or sponsor, in the biomedical, behavioral or clinical sciences leading to research independence. The expectation is that through this sustained period of research career development and training, awardees will launch independent research careers and become competitive for new research project grant funding. NINR encourages mentored research scientist development in areas of symptom management, pulmonary, critical care, trauma, reproductive health, end-of-life and palliative care. NINR limits the length of the K01 award to a 3-year period and generally considers career development applications only from doctorally prepared applicants who will advance the science of nursing.
Requirements: The candidate must have a research or health-professional doctorate or its equivalent and must have demonstrated the capacity or potential for highly productive independent research in the period after the doctorate. The candidate must identify a mentor with extensive research experience and must spend a minimum of 75 percent of full-time professional effort conducting research for the period of the award. Applications may be submitted on behalf of candidates by domestic, nonfederal organizations, public or private, such as medical, dental, or nursing schools or other institutions of higher education. Candidates must be U.S. citizens, noncitizen nationals, or permanent residents.
Amount of Grant: $50,000 maximum salary plus fringe benefits
Date(s) Application is Due: Feb 12; Mar 12; Jun 12; Jul 12; Oct 12; Nov 12
Contact: David Banks, Ph.D.; (301) 496-9558; fax: (301) 480-8260; email: banksd@mail.nih.gov
Internet: http://grants.nih.gov/grants/guide/pa-files/PA-09-040.html
Sponsor: National Institute of Nursing Research (NINR)
6701 Democracy Boulevard, Suite 710, One Democracy Plaza
Bethesda, MD 20892-4870

NINR Nursing National Research Service Awards (NRSA)—Individuals 4243

NRSAs are given to increase the opportunities for qualified nurses to engage in full-time graduate study and research training; to prepare professional nurses to conduct independent research, collaborate in interdisciplinary research, and stimulate and guide others in nursing research; to promote the availability and utilization of nurses with research training in nursing and/or the basic sciences to function as faculty in schools of nursing at undergraduate and graduate levels; and to prepare nurses to conduct scientific inquiry in disciplines that have significance for nursing theory and practice. PA: PA-03-067
Requirements: Applicant must be a registered professional nurse with either a baccalaureate and/or a master's degree in nursing, must be a U.S. citizen or lawfully admitted for permanent residence. Applicants for postdoctoral study must have received PhD, DN, ScD, or equivalent prior to beginning date of proposed fellowship.
Restrictions: Those on temporary or student visas are ineligible.
Date(s) Application is Due: Apr 5; Aug 5; Dec 5
Contact: Dr. Nell Armstrong, (301) 594-5973; email: armstrongn@nih.gov
Internet: http://grants.nih.gov/grants/guide/pa-files/PA-03-067.html
Sponsor: National Institute of Nursing Research
45 Center Drive, Natcher Building
Bethesda, MD 20892

NINR Nursing Research Exploratory Center Grants 4244

NINR invites applications for exploratory center grants in key clinical and basic areas of nursing research that establish a scientific basis for the care of individuals across the life span. NINR seeks to increase the numbers of research-intensive schools of nursing through this program. These grants augment the research resources in schools of nursing by facilitating the growth of infrastructure to centralize resources and facilities in a specific area of inquiry that has a beginning base of research funding. Annual deadline dates may vary; contact program staff for exact dates. RFA: NR-01-001
Requirements: Applications may be submitted by domestic, for-profit and nonprofit organizations both public and private; units of state and local governments; and eligible agencies of the federal government.
Restrictions: Foreign institutions are not eligible for the center program grants.
Contact: Carole Hudgings, (301) 594-5976; fax: (301) 480-8260; email: carole_hudgings@nih.gov
Internet: http://grants.nih.gov/grants/guide/rfa-files/RFA-NR-01-001.html
Sponsor: National Institute of Nursing Research
45 Center Drive, Natcher Building
Bethesda, MD 20892

NINR Opportunities in Genetics and Nursing Research Grants 4245

The purpose of the PA is to integrate genetics and nursing research. Currently PHS-funded investigators with at least two years remaining on an NIH-funded research project grant (R01) are eligible. Nurse researchers are encouraged to collaborate with an ethicist, geneticist, molecular biologist, or an equivalent professional trained in genetic research. Likewise, PHS-funded genetic researchers in fields of ethics, genetics, molecular biology, or equivalent fields are encouraged to collaborate with a nurse researcher. Collaborations and consortia promoting the cross-fertilization of ideas are welcome. In such cases, each participant's contribution should be identified and well-integrated into the overall design. PA-97-047 may be obtained electronically through the NIH grant line (301-402-2221), the NIH Web site, and by mail and email.
Contact: Dr. Hilary Sigmon, Division of Extramural Activities, (301) 594-5970; fax: (301) 480-8260; email: hsigmon@ep.ninr.nih.gov
Internet: http://www.grants.nih.gov/grants/guide/pa-files/PA-97-047.html
Sponsor: National Institute of Nursing Research
45 Center Drive, Natcher Building
Bethesda, MD 20892

NINR Quality of Life for Individuals at the End of Life Grants 4246

The National Institute of Nursing Research (NINR) and 6 other ICs seek research grant applications that will generate scientific knowledge to improve the quality of life for individuals who are facing end-of-life issues and for their families. Research applications may include basic, clinical or care delivery studies focused on management of physical and psychological symptoms, patient-provider and patient-family communication, ethics and clinical decision-making, caregiver support, or the context of care delivery for those facing life-limiting illnesses. In a broad sense the purpose of this program announcement is to enhance the quality of life remaining for individuals who are nearing the end of their lives. Modular Grant applications will request direct costs in $25,000 modules, up to a total direct cost request of $250,000 per year.
Requirements: Applications may be submitted by domestic and foreign, for-profit and non-profit organizations, public and private, such as universities, colleges, hospitals, laboratories, units of State and local governments, and eligible agencies of the Federal government. Racial/ethnic minority individuals, women, and persons with disabilities are encouraged to apply as principal investigators.
Amount of Grant: $250,000 maximum per year
Date(s) Application is Due: Feb 5; Jun 5; Oct 5
Contact: Nancy C. Lohrey; (301) 496-8580; fax: (301) 402-4472; email: lohreyn@mail.nih.gov
Internet: http://grants.nih.gov/grants/guide/pa-files/PA-00-127.html
Sponsor: National Institute of Nursing Research (NINR)
6116 Executive Boulevard, Suite 7019, MSC 8346
Bethesda, MD 20892-8346

NINR Research on Musculoskeletal Fitness and Sports Medicine Grants 4247

NIAMS and NINR invite investigator-initiated research grant applications to study a broad range of basic and clinical topics related to musculoskeletal fitness, exercise physiology, and sports medicine. The National Center for Medical Rehabilitation Research of NICHD encourages applications for both basic and clinical studies of musculoskeletal fitness and exercise physiology of persons with physical disabilities. The principal investigator and participating investigators will plan, direct, and perform the research. Applicants for program project grants are requested to contact the NIAMS representative as early as possible in the planning stages. PA-97-025 may be obtained electronically through the NIH grant line (301-402-2221), the NIH Web site, and by mail and email.
Contact: Dr. Hilary Sigmon, Division of Extramural Activities, (301) 594-5970; fax: (301) 480-8260; email: hsigmon@ep.ninr.nih.gov
Internet: http://www.grants.nih.gov/grants/guide/pa-files/PA-97-025.html
Sponsor: National Institute of Nursing Research
45 Center Drive, Natcher Building
Bethesda, MD 20892

NINR Ruth L. Kirschstein National Research Service Award (NRSA) for Individual Predoctoral Fellows in Nursing Research 4248

This grant program will provide predoctoral training support for doctoral students. The applicant must propose a research training program and dissertation research that is consistent with the scientific mission of the NINR. Research topics and skills that will serve as a foundation for an ongoing program of research are of particular interest. The research training experience must enhance the applicant's conceptualization of research

problems and research skills, under the guidance and supervision of a committed mentor who is an active and established investigator in the area of the applicant's proposed research. The research training program should be carried out in a research environment that includes appropriate human and technical resources and is demonstrably committed to the research training of the applicant in the program he/she proposes in the application. Up to five years of aggregate NRSA support may be provided. Fellowship awardees are required to pursue their research training on a full-time basis, devoting at least 40 hours per week to the training program.
Requirements: The following organizations are eligible to apply: for-profit organizations; non-profit organizations; public or private institutions, such as universities, colleges, hospitals, and laboratories; and domestic institutions.
Date(s) Application is Due: Apr 8; Aug 8; Dec 8
Contact: David Banks, Ph.D.; (301) 496-9558; fax: (301) 480-8260; email: banksd@mail.nih.gov
Internet: http://grants.nih.gov/grants/guide/pa-files/PAR-09-227.html
Sponsor: National Institute of Nursing Research (NINR)
6701 Democracy Boulevard, Suite 710, One Democracy Plaza
Bethesda, MD 20892-4870

NINR Symptom Management for Chronic Neurological Conditions Research Grants 4249
NINR, NIA, NICHD, NIDCD, and NINDS are interested in facilitating investigator-initiated research designed to deal with symptoms associated with chronic neurological conditions, including stroke, epilepsy, brain and spinal cord injury, multiple sclerosis, and Parkinson's disease. The overall goals are to manage symptoms effectively, to prevent secondary effects of the condition when possible, to maximize the health potential, and to improve the health-related quality of life for patients, care givers, and families. The mechanism of support is the research project grant (R01). PA-98-030 may be obtained through the NIH grant line (301-402-2221), the NIH Web site, mail, and email.
Requirements: Applications may be submitted by domestic and foreign for-profit and nonprofit organizations, public or private.
Contact: Dr. Mary Leveck, Scientific Program Administrator, (301) 594-5963; fax: (301) 480-8260; email: mary_leveck@nih.gov; Jeff Carow, Grants Management Officer, (301) 594-6869; fax: (301) 480-8260; email: jcarow@ep.ninr.nih.gov
Internet: http://grants.nih.gov/grants/guide/pa-files/PA-98-030.html
Sponsor: National Institute of Nursing Research
45 Center Drive, Natcher Building
Bethesda, MD 20892

NINR/NHGRI Nursing and Genetics Research Individual Postdoctoral and Senior Fellowships 4250
NINR and NHGRI announce the availability of long- and short-term support for individual postdoctoral and senior fellowships to scientists who are seeking training that will enable them to investigate the clinical implications of human genetics research. The three-fold purpose of this initiative is to incorporate genetic research into programs of research and research training for nurse scientists; integrate nursing research with research training programs for genetic scientists; and provide opportunities for addressing the ethical, legal, and social implications of genetic research training. PA-97-039 may be obtained electronically through the NIH grant line (301-402-2221), the NIH Web site, and by mail and email.
Contact: Dr. Hilary Sigmon, Scientific Program Administrator, (301) 594-5970; fax: (301) 480-8260; email: hsigmon@ep.ninr.nih.gov; Dr. Elizabeth Thomson, NHGRI, (301) 402-4997; fax: (301) 402-083; email: Elizabeth_Thomson@nih.gov
Internet: http://grants.nih.gov/grants/guide/pa-files/PA-97-039.html
Sponsor: National Institute of Nursing Research
45 Center Drive, Natcher Building
Bethesda, MD 20892

NIOSH Occupational Safety and Health Research and Demonstration Grants 4251
This program aims to increase knowledge about the underlying characteristics of occupational safety and health problems to prevent occupational diseases and injuries. Research projects supported include causal research to identify and investigate relationships between hazardous working conditions and associated occupational diseases and injuries; methods research to develop more sensitive means of evaluating hazards at work sites, as well as methods for measuring early markers of adverse health effects and injuries; control research to develop new protective equipment, engineering control technology, and work practices to reduce the risks of occupational hazards; and demonstrations to evaluate the technical feasibility or application of a new or improved occupational safety and health procedure, method, technique, or system. Funds support career development, small grants, and competing renewal and revised applications. PA: PA-04-038
Requirements: Eligible applicants include for-profit and nonprofit organizations, universities, colleges, research institutions, and other public and private organizations, including state and local governments, and small minority- and/or women-owned businesses.
Samples: Epidemiologic Study of Isocyanate Asthma in Autobody Shops—to assess the impact of isocyanate exposures on the development of adult asthma and to assess potential biomarkers for exposure to isocyanates, and genetic markers of susceptibility to isocyanate-induced asthma.; Trunk Stability and Spinal Load during MMH Lifting—to use biomechanical modeling to quantify spinal loads and musculoskeletal behavior in response to unstable events.; Electrostatic Sampling of Airborne Microorganisms—to use electrostatic mechanisms to collect the airborne microorganisms.
Amount of Grant: $30,000-$400,000; $220,000 average
Date(s) Application is Due: Mar 1; Jul 1; Nov 1
Contact: Susan Board, Office of Extramural Programs, (404) 498-2512; fax: (404) 498-2517; email: sboard@cdc.gov
Internet: http://grants.nih.gov/grants/guide/pa-files/PA-04-038.html
Sponsor: National Institute for Occupational Safety and Health
1600 Clifton Rd NE
Atlanta, GA 30333

NIOSH Repetitive Motion Disorders Research Grants 4252
The program is sponsored by NIAMS, NICHD, NINDS, and NIOSH at CDC. This initiative invites applications directed to the study of the pathogenesis, epidemiology, prevention, and treatment of repetitive motion disorders. PA-97-059 may be obtained electronically through the NIH grant line (301-402-2221), the NIH Web site, and by mail and email.
Contact: Roy Fleming, Grants Management Branch, (404) 639-2810; fax: (404) 639-2196; email: rmf2@cdc.gov
Internet: http://www.grants.nih.gov/grants/guide/pa-files/PA-97-059.html
Sponsor: National Institute for Occupational Safety and Health
1600 Clifton Rd NE
Atlanta, GA 30333

NIST Fire Research Grants 4253
This program supports research into the dynamics of fire, causes of fire, and a better understanding of fire prevention and safety measures. It aims to advance fire science and engineering practices by developing a better understanding of dynamic fire phenomena. This includes performing research into the heat and mass transfer processes of fires, in order to improve predictions of growth, suppression, and emissions from fires. This program also seeks to develop a more definitive understanding of the behavior, prevention, and control of large fires.
Requirements: The Program is open to institutions of higher education; hospitals; non-profit organizations; commercial organizations; state, local, and Indian tribal governments; foreign governments; organizations under the jurisdiction of foreign governments; and international organizations. Joint programs with participation by more than one eligible entity are possible.
Samples: Harvey Mudd College, Claremont, California, $287,000—for Real-Time Monitoring of Burning Structures (2006); University of Texas, San Antonio, Texas, $95,000—for The Integration of Biometric Technologies with Fire Fighting Information Systems (2006); Purdue University, West Lafayette, Indiana, $149,000—for Realistic Fire Behavior and Stability of Steel Building Structures (2006).
Amount of Grant: $10,000-$100,000
Contact: Wanda J. Duffin-Ricks, Fire Grants Assistant; (301) 975-6863; email: wanda.duffin-ricks@nist.gov
Internet: http://www.nist.gov/public_affairs/grants.htm
Sponsor: National Institute of Standards and Technology
100 Bureau Drive, Stop 8660
Gaithersburg, MD 20899-8660

NIST Hollings Manufacturing Extension Partnerships 4254
The purpose of the program is to establish, maintain, and support manufacturing extension centers and services to accelerate the usage of appropriate manufacturing technology by smaller U.S.-based manufacturing firms; and partner with the states in developing such technical assistance programs and services for their manufacturing base. Funds shall be used for the creation and support of regional manufacturing extension services, or used by the states to plan for and pilot test statewide extension services. Funds may also be used to plan for and pilot test services within a multi-state region which has sufficient regional linkages to justify such services. Extension service providers shall be affiliated with a U.S.-based nonprofit institution or organization or group thereof. Funds may be used for the purpose of demonstrations, technology deployment, active transfer and dissemination of research findings and extension service expertise to a wide range of companies and enterprises, especially small manufacturers with fewer than 500 employees. Extension service planning and pilot testing agreements require a state to provide cost share by an amount at least equal to the amount of federal assistance requested. Deadline dates are published in the Federal Register.
Requirements: Eligible to apply for extension services are U.S.-based nonprofit institutions or organizations or groups thereof. Applicants must contribute 50 percent or more of the proposed service's capital, annual operating and maintenance costs. At least 50 percent of the applicant's share must be as cash or full-time personnel.
Amount of Grant: $25,000-$100,000 individual grants; $200,000-$6.5 million centers grants annually
Contact: Roger D. Kilmer, Director; (301) 975-4676; fax: (301) 963-6556; email: roger.kilmer@nist.gov
Internet: http://www.nist.gov/public_affairs/guide/mep.htm
Sponsor: National Institute of Standards and Technology
100 Bureau Drive, Stop 4800
Gaithersburg, MD 20899-4800

NIST Materials Science and Engineering Grants 4255
The Program supports work in polymers, ceramics, metallurgy, and neutron scattering and spectroscopy research at academic, industrial, and other non-federal institutions. Projects are usually funded for one year and are subject to availability of funds.
Requirements: Universities, colleges, professional institutes and associations, nonprofit organizations, and state and local governments are eligible for these grants.
Contact: Tanya Burke; (301) 975-4711; email: tanya.burke@nist.gov

Internet: http://www.nist.gov/public_affairs/grants.htm
Sponsor: National Institute of Standards and Technology
100 Bureau Drive, Room A120, MS 3580
Gaithersburg, MD 20899-3580

NIST National Standard Reference Data System Research Grants 4256
These grants support research for data evaluation that will aid the scientist or engineer by providing reliable evaluated information on the chemical and physical properties of substances. Evaluations are carried out through a network of data centers and projects. Data compilations published and databases established by the program are for sale without restriction. Contact the program for deadline information. Project support is provided for one to two years.
Requirements: Research grants, contracts, and cooperative agreements for data evaluation are made to technical groups in universities, industrial laboratories, and other research establishments.
Contact: Office of Standard Reference Data, (301) 975-2208 or (301) 975-2200; fax: (301) 926-0416
Internet: http://www.nist.gov/srd/
Sponsor: National Institute of Standards and Technology
100 Bureau Drive, Stop 2300
Gaithersburg, MD 20899-2300

NIST Precision Measurement Grants 4257
The Program supports researchers in U.S. colleges and universities for experimental and theoretical studies of fundamental physical phenomena. It is seeking proposals for significant, primarily experimental, research in the field of fundamental measurement or the determination of fundamental constants. The purpose of the program is to encourage basic, measurement-related research and to foster contacts between NIST scientists and those faculty members of U.S. academic institutions who are actively engaged in such work. Applicants should propose multi-year projects for up to three years at no more than $50,000 per year. Applicants should begin by submitting an abbreviated proposal, which must be received by the end of business on February 2. Finalists will be selected by March 23, and will be required to submit full proposals to NIST by close of business on May 4.
Requirements: Universities and colleges in the United States are eligible to apply.
Samples: J.C. Seamus Davis and Hiro Miyake, Cornell University, Ithaca, New York—Precision Measurement of Gravitational Forces at the Submicron Length Scale (2006); Krzysztof Szalewicz, University of Delaware, Newark, Delaware—High-precision calculations of density and dielectric virial coefficients of helium for new pressure and temperature standards (2006).
Amount of Grant: $50,000 maximum per year
Date(s) Application is Due: Feb 2; May 4
Contact: Dr. Peter J. Mohr; (301) 975-3217; email: mohr@nist.gov
Internet: http://physics.nist.gov/ResOpp/grants/grants.html
Sponsor: National Institute of Standards and Technology
100 Bureau Drive, Stop 8420
Gaithersburg, MD 20899-8420

NIST Small Business Innovation Research (SBIR) Grants 4258
Ten federal agencies set aside a portion of their extramural research and development budget each year to fund research proposals from small science and technology-based firms. This Program supports creative advanced research in important scientific and engineering areas and is designed to encourage the conversion of government-funded R&D into technological innovation and commercial application. The Program goals are: to increase private sector commercialization of innovations derived from federal R&D; to use small business to meet federal research and development (R&D) needs; to stimulate small business innovation in technology; and to foster and encourage participation by minority and disadvantaged persons in technological innovation. The program consists of three phases: In Phase 1, a small business can receive up to $75,000 to establish the scientific or technical merit or feasibility of ideas that appear to have commercial potential. The successful completion of Phase 1 enables the awardee to compete for a Phase 2I contract for up to $300,000 to continue research in the same technical area. Phase 1I awardees have 6 months to complete their work; Phase 2 awardees have up to 2 years. Phase 2 is the principal R&D effort. Under Phase 3, no SBIR funds are available to pursue commercial applications of the technology.
Requirements: Each organization submitting a proposal for both Phase 1 and Phase 2 must qualify as a small business concern (Section 2.10) for research or R&D purposes (Section 2.7) at the time of award. In addition, the primary employment of the principal investigator must be with the small business at the time of the award and during the conduct of the proposed research. More than one-half of the principal investigator's time must be spent with the small business for the period covered by the award. Primary employment with a small business precludes full-time employment with another organization.
Amount of Grant: $50,000-$200,000
Contact: Program Director; (301) 975-3085; fax: (301) 975-3482; email: sbir@nist.gov
Internet: http://patapsco.nist.gov/ts_sbir/
Sponsor: National Institute of Standards and Technology
100 Bureau Drive, Stop 2200
Gaithersburg, MD 20899-2200

NIU Horatio Alger Jr. Fellowship 4259
The fellowships supports scholars who will be using materials from the libraries' major holdings in American popular culture, including the Albert Johannsen Collection of more than 50,000 dime novels, and the nation's preeminent collections related to Horatio Alger Jr. and Edward Stratemeyer. Topics that could draw on the collections' strengths might include the plight of urban children, image of the American West in popular literature, widespread use of pseudonyms, and stereotypical portrayals. Preference will be given to applicants who signify an interest in research related to Horatio Alger Jr. The fellowship award may be used between July 1 and December 31. Candidates should submit a letter of interest, curriculum vita, brief proposal for their research, and two letters of recommendation.
Amount of Grant: $2000 stipend
Date(s) Application is Due: Jun 1
Contact: University Libraries, Rare Books and Special Collections, (815) 753-9838
Sponsor: Northern Illinois University
Founders' Memorial Library, Rare Books and Special Collections
DeKalb, IL 60115-2868

NKF Clinical Scientist Awards 4260
The award supports investigators who have demonstrated outstanding clinical or basic research potential to promote their continued success as independent investigators. Scientists are expected to devote a minimum of 75 percent of their time to research. This undertaking includes the establishment of a new Study Section which will review all proposals involving clinical research for all funding mechanisms, based on the following definition from the Nathan Report, which describes three categories of clinical research: Patient-Oriented Research, defined as research conducted with human subjects (or on material of human origin such as tissues, specimens and cognitive phenomena) for which an investigator (or colleague) directly interacts with human subjects in either an outpatient or inpatient setting; Epidemiologic and Biobehavioral Studies; and Outcomes Research and Health Services Research. Applications must be submitted electronically no later than November 30.
Requirements: Applicants must be citizens or permanent residents of the United States at the time of application. Applicants must hold an M.D. or equivalent domestic or foreign degree and must be full-time staff members of a department within an institution in the United States at the time of the award. Applicants will generally have at least five, but not more than 10 years of postdoctoral research experience in nephrology by the beginning of the award date.
Amount of Grant: $50,000 per year
Date(s) Application is Due: Nov 30
Contact: Emily B. Newell, Director; (212) 889-2210, ext. 288 or (800) 622-9010; fax: (212) 689-9261; email: emilyn@kidney.org or research@kidney.org
Internet: http://www.kidney.org/professionals/research/index.cfm
Sponsor: National Kidney Foundation
30 East 33rd Street
New York, NY 10016

NKF Franklin McDonald/Fresenius Medical Care Clinical Research Grants 4261
The National Kidney Foundation is soliciting applications for the annual Franklin McDonald/Fresenius Medical Care Clinical Research Grant. Applicants should be able to spend a minimum of 50% of their time conducting research and 25% of effort in the direct care of maintenance dialysis patients and/or in teaching medical students, residents or nephrology fellows the pathophysiology and treatment of CKD, ESRD, or maintenance dialysis. Proposals are due on December 1.
Requirements: Applicants must hold a full-time faculty appointment at a university or an equivalent position as a scientist on the staff of a research organization.
Date(s) Application is Due: Dec 1
Contact: Emily B. Newell, Director; (212) 889-2210, ext. 288 or (800) 622-9010; fax: (212) 689-9261; email: emilyn@kidney.org or research@kidney.org
Internet: http://www.kidney.org/professionals/research/index.cfm
Sponsor: National Kidney Foundation
30 East 33rd Street
New York, NY 10016

NKF Professional Councils Research Grants 4262
The purpose of the CNNT Research Grants Program is to further knowledge of nursing and technician issues in the management of kidney failure. The grants may cover basic or applied research on nursing and technical issues in the area of kidney failure; early intervention and treatment of chronic kidney disease; development and evaluation of education programs to enhance patient/family understanding of kidney failure treatment; demonstration projects related to kidney failure and rehabilitation. Letter of intent due on October 15, and full proposals must be received by December 1.
Requirements: Applicants must hold a regular membership in CNNT, a minimum of two years work experience as a nephrology nurse or technician, residence in the U.S. or its territories, written approval from the department head or facility director at the institution/facility where the research is to be conducted, and have prior research training or evidence of support from an individual with research experience.
Restrictions: Project support and stipend not to exceed 20% of requested amount; consultant fees not to exceed 15% of total requested funds.
Amount of Grant: $20,000 maximum
Date(s) Application is Due: Oct 15; Dec 1
Contact: Beverly Sneed, Grant Contact; (214) 514-3846 or (800) 622-9010; fax: (212) 689-9261; email: bevyann1@sbcglobal.net or research@kidney.org
Internet: http://www.kidney.org/professionals/research/profcouncil.cfm
Sponsor: National Kidney Foundation
30 East 33rd Street
New York, NY 10016

NKF Research Fellowships 4263

The purpose of National Kidney Foundation (NKF) Research Fellowships is: to foster training of young and new investigators with the potential of making contributions to the understanding and cure of kidney diseases; to foster, through training fellows, kidney-related research that is of high merit; and to encourage high quality applicants who want to make a career change into academic Nephrology. The fellowships are open to qualified investigators of any nationality who are interested in a career in kidney research and who will not have had more than four and one-half years of research training prior to the commencement of the award. The one-year awards may be renewed.
Requirements: Applicants must have completed no more than 4.5 years of post-doctoral research training (after receipt of M.D., Ph.D, D.O., or equivalent degree) at time of the activation of the award. For those applying for a second year (competitive renewal), applicants must have completed no more than 5.5 years of post-doctoral training.
Restrictions: Anyone who has ever held a faculty appointment at the level of Assistant Professor is ineligible.
Amount of Grant: $50,000 per year for a maximum of two years
Date(s) Application is Due: Nov 30
Contact: Emily B. Newell, Director; (212) 889-2210, ext. 288 or (800) 622-9010; fax: (212) 689-9261; email: emilyn@kidney.org or research@kidney.org
Internet: http://www.kidney.org/professionals/research/awards.cfm
Sponsor: National Kidney Foundation
30 East 33rd Street
New York, NY 10016

NKF Young Investigator Grants 4264

The purpose of this program is to support research in the fields of nephrology, urology, and related disciplines by junior faculty now starting their investigative work, who are often hampered in their progress by difficulty in obtaining suitable financial support for their research during the formative periods of their careers. The applicant's research and career goals must be directed to the study of normal or abnormal kidney function or of diseases of the kidney or urinary tract.
Requirements: Applicants must have completed research fellowship training in nephrology, urology or closely related field prior to the start of the grant award, intend to pursue directly related research, and hold a full-time appointment to a faculty position at a university or an equivalent position as a scientist on the staff of a research oriented institution for no more than four years at the start of the award. Recipients typically hold an appointment as Assistant Professor.
Amount of Grant: $50,000 maximum
Date(s) Application is Due: Nov 30
Contact: Emily B. Newell, Scientific Communications Director; (212) 889-2210, ext. 288 or (800) 622-9010; fax: (212) 689-9261; email: emilyn@kidney.org or research@kidney.org
Internet: http://www.kidney.org/professionals/research/awards.cfm
Sponsor: National Kidney Foundation
30 East 33rd Street
New York, NY 10016

NL American Society for Eighteenth-Century Studies (ASECS) Fellowships 4265

ASECS Fellowships are available for one to three months in residence at the Newberry for studies in the period 1660-1815. Application materials are available on the Web site.
Requirements: Applicants must be postdoctoral scholars.
Amount of Grant: $1200 per month
Date(s) Application is Due: Mar 1
Contact: Program Contact, (312) 255-3666; email: research@newberry.org
Internet: http://www.newberry.org/nl/research/L3rfellowships.html
Sponsor: Newberry Library
60 W Walton Street
Chicago, IL 60610-3380

NL Arthur Weinberg Independent Scholars Fellowships 4266

The one-month fellowships are for scholars working outside the academy who need to use particular Newberry collections. Applicants need not have the PhD, but must have demonstrated, through their publications, particular excellence in a field appropriate to the Newberry. Preference is given to scholars working on historical issues related to social justice and/or reform. Applications are available on the Web site.
Restrictions: Scholars whose principal residence or place of employment is within the Chicago area are not eligible.
Amount of Grant: $1200 per month
Date(s) Application is Due: Mar 1
Contact: Program Contact, (312) 255-3666; email: research@newberry.org
Internet: http://www.newberry.org/nl/research/L3rfellowships.html
Sponsor: Newberry Library
60 W Walton Street
Chicago, IL 60610-3380

NL Associated Colleges of the Midwest/Great Lakes College Association Faculty Fellowships 4267

These fellowships support faculty from the colleges of the Associated Colleges of the Midwest and the Great Lakes Colleges Association, Inc, as they teach a small group of select undergraduate students in an advanced research seminar, while doing their own research. Fellows can come from any of the colleges in ACM or GLCA, from any discipline.
Requirements: Applicants must hold the PhD at the time of application.
Amount of Grant: $40,000 maximum
Date(s) Application is Due: Mar 1
Contact: Daniel Sack, (312) 263-5000; email: dsack@acm.edu
Internet: http://www.newberry.org/nl/research/L3rfellowships.html
Sponsor: Newberry Library
60 Walton Street
Chicago, IL 60610-3380

NL Audrey Lumsden-Kouvel Fellowship 4268

The Audrey Lumsden-Kouvel Fellowship is for post-doctoral scholars who wish to use the Newberry's extensive holdings in late medieval and Renaissance history and literature. Preference will be given to projects focusing on Romance cultures. The fellowship is intended to encourage scholars to pursue research at the Newberry during sabbaticals; it could also help a scholar to extend a leave. Applicants must plan to be in residence for a minimum of three months. This fellowship may be combined with non-residential fellowships. Applications are available on the Web site. Applicants for this fellowship need not be from outside the Chicago area.
Amount of Grant: $4000
Contact: Program Contact, (312) 255-3666; email: research@newberry.org
Internet: http://www.newberry.org/nl/renaissance/L3rrenaissance.html
Sponsor: Newberry Library
60 W Walton Street
Chicago, IL 60610-3380

NL British Academy Fellowship for Study in the United Kingdom 4269

In cooperation with the British Academy, the Newberry Library offers an exchange fellowship for three months of study in the United Kingdom in any field in humanities in which the Newberry's collections are strong.
Requirements: Applicants must be established scholars at the postdoctoral level or its equivalent who are on the staffs of universities, museums, or libraries. Preference is given to readers and staff of the Newberry and to scholars who have previously used the Newberry.
Amount of Grant: @L1350 per month while the fellow is in the United Kingdom; the fellow's home institution is expected to continue to pay his/her salary
Date(s) Application is Due: Jan 10
Contact: Committee on Awards, (312) 255-3666; email: research@newberry.org
Internet: http://www.newberry.org/nl/research/L3rfellowships.html
Sponsor: Newberry Library
60 W Walton Street
Chicago, IL 60610-3380

NL Center for Great Lakes Culture/Michigan State University Fellowships 4270

Two one-month residential fellowships are available for projects using the library's collections to understand and interpet the cultural history and expressions of the diverse people of the Great Lakes/Ohio Valley region (Michigan, Illinois, Wisconsin, Minnesota, Ohio, Indiana, West Virginia, Kentucky, and Ontario).
Requirements: Scholars with the PhD or equivalent or an established record of scholarly research are available.
Amount of Grant: $1250 per month
Date(s) Application is Due: Mar 1
Contact: Jaes Grossman, Vice president of Research and Education, (312) 255-3535; email: research@newberry.org
Internet: http://www.newberry.org/nl/research/L3rfellowships.html
Sponsor: Newberry Library
60 Walton Street
Chicago, IL 60610-3380

NL Frances C. Allen Fellowships 4271

These fellowships are available to women of Native American heritage who are pursuing an academic program at any stage beyond the undergraduate degree. Candidates may be working in any graduate or preprofessional field, but the particular purpose of the fellowship is to encourage study in the humanities and social sciences. The length of the award may vary from one month to a year. Financial support varies according to need and may include travel expenses. The majority of the research should be carried out at the D'Arcy McNickle Center for the History of the American Indian. Applications are available on the Web site.
Restrictions: Scholars whose principal residence or place of employment is within the Chicago area are not eligible.
Amount of Grant: $8000 maximum
Date(s) Application is Due: Mar 1
Contact: James Grossman, Vice president of Research and Education, (312) 255-3535;email: research@newberry.org
Internet: http://www.newberry.org/nl/research/L3rfellowships.html
Sponsor: Newberry Library
60 W Walton Street
Chicago, IL 60610-3380

NL Herzog August Bibliothek Wolfenbuttel Fellowship 4272

The Newberry Library and the Herzog August Bibliothek in Wolfenbuttel, Germany, awards a joint fellowship intended to support related research at the two libraries. This fellowship is awarded to complement a long- or short-term Newberry fellowship. The proposed project must link the collections of both libraries. Applicants should plan to

hold the fellowships sequentially to ensure continuity of research. Application deadlines are January 10 for long-term fellowships and March 1 for short-term fellowships. Contact the Herzog August Bibliothek Wolfenbuttel at http://www.hab.de for additional information.
Amount of Grant: DM1200 travel expenses and three-month stipend of DM2000 per month
Date(s) Application is Due: Jan 10; Mar 1
Contact: Program Contact, (312) 255-3666; email: research@newberry.org
Internet: http://www.newberry.org/research/L3rfellowships.html
Sponsor: Newberry Library
60 W Walton Street
Chicago, IL 60610-3380

NL Lester J. Cappon Fellowship in Documentary Editing 4273
The fellowship supports editing projects based on Newberry materials and carried out while in residence for two weeks to two months at the library. Funds can be used to defray other costs related to the project's preparation and publication, including travel, photocopying, digitization, copyright fees, publication subventions, etc. Application materials are available on the Web site.
Requirements: Applicants must have PhDs or equivalent degrees.
Restrictions: Scholars whose principal residence or place of employment is within the Chicago area are not eligible.
Amount of Grant: $5000 maximum
Date(s) Application is Due: Mar 1
Contact: Program Contact, (312) 255-3666; email: research@newberry.org
Internet: http://www.newberry.org/nl/research/L3rfellowships.html
Sponsor: Newberry Library
60 W Walton Street
Chicago, IL 60610-3380

NL Lloyd Lewis Fellowships in American History 4274
These fellowships are available for established scholars in any field of American history appropriate to the Newberry's collections, for periods of six to 11 months; two or three awards will be made each year. Fellows will be expected to participate actively in the Newberry's scholarly community. Applicants may combine grants with sabbaticals or other stipendiary support. Application forms are available on the Web site.
Requirements: Applicants must have the PhD or its equivalent and must have demonstrated through their publications particular excellence in the field.
Restrictions: faxed applications will not be accepted.
Amount of Grant: $40,000 maximum
Date(s) Application is Due: Jan 10
Contact: Program Contact, (312) 255-3666; email: research@newberry.org
Internet: http://www.newberry.org/research/felshp/long-term.html
Sponsor: Newberry Library
60 W Walton Street
Chicago, IL 60610-3380

NL Mellon Postdoctoral Research Fellowships 4275
Applications are invited from postdoctoral scholars in any field relevant to the Newberry's collections for awards of six to 10 months of residential research and writing. Fellows will become part of the Newberry's community of scholars, participating in biweekly fellows' seminars, colloquia, and other events. Applicants may combine this award with sabbatical or other stipendiary support. Individuals applying for NEH and Lloyd Lewis fellowships will also be considered for Mellon Research fellowships and need not submit a second application. Application forms are available on the Web site.
Requirements: Postdoctoral scholars at any stage in their careers are eligible.
Amount of Grant: $40,000 maximum
Date(s) Application is Due: Jan 10
Contact: Program Contact, (312) 255-3666;email: research@newberry.org
Internet: http://www.newberry.org/research/felshp/long-term.html
Sponsor: Newberry Library
60 W Walton Street
Chicago, IL 60610-3380

NL Monticello College Foundation Fellowships for Women 4276
With the support of a grant from the Monticello College Foundation, the Newberry Library offers a six-month fellowship for work in residence at the Newberry by a woman scholar. The fellowship is designed primarily for women at early stages in their professional careers whose work gives clear promise of scholarly productivity and whose careers would be significantly enhanced by six months of research and writing. The applicant's topic should be related to the Newberry's collections; preference will be given to proposals particularly concerned with the study of women. Application materials are available on the Web site.
Requirements: Applicants must have the PhD at the time of application and must be U.S. citizens or permanent residents.
Amount of Grant: $15,000
Date(s) Application is Due: Jan 10
Contact: James Grossman, Vice president of Research and Education, (312) 255-3535;email: Grossman@newberry.org
Internet: http://www.newberry.org/research/felshp/long-term.html
Sponsor: Newberry Library
60 W Walton Street
Chicago, IL 60610-3380

NL National Endowment for the Humanities Fellowships 4277
Fellowships are available to scholars who wish to spend from six to 11 months in residence to pursue their own research and writing, while contributing to and benefiting from the exchange of ideas among the community of scholars at the Newberry. Four to six awards will be made each year. Applicants may combine this award with sabbatical or other stipendiary support. Completed applications must include all letters of reference. Scholars conducting research in American history may also ask to be considered for the Lloyd Lewis fellowship at the time of their application. Application materials are available on the Web site.
Requirements: Applicants must be established scholars at the postdoctoral level or its equivalent; must be U.S. citizens, nationals, or foreign nationals who have been living in the United States or its territories for at least three years; and must be pursuing research in fields appropriate to the Newberry's collections. Preference is given to applicants who have not held major fellowships or grants in the three years preceding proposed period of residency.
Amount of Grant: $40,000 maximum
Date(s) Application is Due: Jan 10
Contact: Committee on Awards, (312) 255-3666; email: research@newberry.org
Internet: http://www.newberry.org/research/felshp/long-term.html
Sponsor: Newberry Library
60 W Walton Street
Chicago, IL 60610-3380

NL Rockefeller Foundation Residential Fellowships in the Humanities 4278
These short-term, residential fellowships support research in any aspect of American Indian studies supported by Newberry collections. Applicants' projects may culminate in a variety of formats, including but not limited to curriculum development projects, artistic works, or publications. The fellowships support one to three months of research at the Newberry and carry a monthly stipend plus travel expenses. This award has a different application form from other Newberry awards. Application and guidelines are available online.
Requirements: Application is open to historians working in reservation-based communities, tribal college faculty, and librarians or curators at American Indian cultural centers or museums.
Amount of Grant: $3000 per month; $1000 travel expenses
Date(s) Application is Due: Jan 15; Apr 15
Contact: Committee on Awards, (312) 255-3666; email: research@newberry.org
Internet: http://www.newberry.org/nl/research/L3rfellowships.html
Sponsor: Newberry Library
60 W Walton Street
Chicago, IL 60610-3380

NL Short-Term Fellowships in the History of Cartography 4279
The short-term fellowships support two weeks to two months of research in residence at the Newberry on projects related to the history of cartography and focused on cartographic materials in the library's collection. Application materials are available on the Web site.
Requirements: Applicants must have the PhD or equivalent degree.
Restrictions: Scholars whose principal residence or place of employment is within the Chicago area are not eligible.
Amount of Grant: $1200 per month
Date(s) Application is Due: Mar 1
Contact: Program Contact, (312) 255-3666; email: research@newberry.org
Internet: http://www.newberry.org/nl/research/L3rfellowships.html
Sponsor: Newberry Library
60 W Walton Street
Chicago, IL 60610-3380

NL Short-Term Resident Fellowships for Individual Research 4280
These short-term fellowships provide access to the Newberry's collections for scholars who live or work outside the Chicago area. The tenure of short-term fellowships varies from one week to two months, unless otherwise noted under the award description. A majority of fellowships will be for one month or less.
Requirements: Applicants must have the PhD or have completed all requirements except the dissertation.
Restrictions: Scholars whose principal residence or place of employment is within the Chicago area are not eligible.
Amount of Grant: $1200 per month
Date(s) Application is Due: Mar 1
Contact: James Grossman, Vice President of Research and Education, (312) 255-3666; email: research@newberry.org
Internet: http://www.newberry.org/nl/research/L3rfellowships.html
Sponsor: Newberry Library
60 W Walton Street
Chicago, IL 60610-3380

NL South Central Modern Languages Association Fellowship 4281
A one-month fellowship is available for postdoctoral scholars who are members of the SCMLA for work in residence at the Newberry. The awards committee gives preference to applicants who need the Newberry's collections. Application forms are available on the Web site.
Requirements: PhD candidates and postdoctoral scholars are eligible.
Amount of Grant: $2000

Date(s) Application is Due: Mar 1
Contact: Program Contact, (312) 255-3666; email: research@newberry.org
Internet: http://www.newberry.org/nl/research/L3rfellowships.html
Sponsor: Newberry Library
60 W Walton Street
Chicago, IL 60610-3380

NLM Academic Research Enhancement Awards (AREA) 4282
The purpose of the Academic Research Enhancement Award (AREA) program is to stimulate research in educational institutions that provide baccalaureate or advanced degrees for a significant number of the Nation's research scientists, but that have not been major recipients of support. NLM's participation is confined to applications that address research in biomedical informatics and bioinformatics. Only electronic applications are accepted. Deadlines for new applications are February 25, June 25 and October 25 each year.
Date(s) Application is Due: Feb 25; Jun 25; Oct 25
Contact: Dr. Hua-Chuan Sim, Clinical and Public Health Informatics; (301) 594-5983; fax: (301) 402-1384; email: simh@mail.nih.gov
Internet: http://www.nlm.nih.gov/ep/area.html
Sponsor: National Library of Medicine
6705 Rockledge Drive, Suite 301
Bethesda, MD 20892

NLM Advancing Novel Science in Women's Health Research (ANSWHR) Exploratory/Developmental Research Grants 4283
The purpose of this funding opportunity announcement (FOA), issued by the Office of Research on Women's Health (ORWH), is to promote innovative, interdisciplinary research that will advance new concepts in women's health research and the study of sex/gender differences. Recent research reports have established the importance of studying issues specific to women, including the scientific and clinical importance of analyzing data separately for females and males. ORWH is particularly interested in encouraging extramural investigators to undertake new interdisciplinary research to advance studies on how sex and gender factors affect women's health; however, applications in all areas of women's health and/or sex/gender research are invited. The deadline for new applications is October 16, and the deadline for revised applications is November 16.
Date(s) Application is Due: Oct 16; Nov 16
Contact: Dr. Hua-Chuan Sim, Clinical and Public Health Informatics; (301) 594-5983; fax: (301) 402-1384; email: simh@mail.nih.gov
Internet: http://www.nlm.nih.gov/ep/answhr.html
Sponsor: National Library of Medicine
6705 Rockledge Drive, Suite 301
Bethesda, MD 20892

NLM Advancing Novel Science in Women's Health Research (ANSWHR) Small Project Grants 4284
The purpose of this funding opportunity announcement (FOA), issued by the Office of Research on Women's Health (ORWH), is to promote innovative, interdisciplinary research that will advance new concepts in women's health research and the study of sex/gender differences. Recent research reports have established the importance of studying issues specific to women, including the scientific and clinical importance of analyzing data separately for females and males. ORWH is particularly interested in encouraging extramural investigators to undertake new interdisciplinary research to advance studies on how sex and gender factors affect women's health; however, applications in all areas of women's health and/or sex/gender research are invited. The deadline for new applications is October 16, and the deadline for revised applications is November 16.
Date(s) Application is Due: Oct 16; Nov 16
Contact: Dr. Hua-Chuan Sim, Clinical and Public Health Informatics; (301) 594-5983; fax: (301) 402-1384; email: simh@mail.nih.gov
Internet: http://www.nlm.nih.gov/ep/answhr.html
Sponsor: National Library of Medicine
6705 Rockledge Drive, Suite 301
Bethesda, MD 20892

NLM Behavioral and Social Science Research on Understanding and Reducing Health Disparities Research Grants 4285
The goal of this funding opportunity announcement (FOA) is to encourage behavioral and social science research on the causes and solutions to health and disability disparities in the U.S. population. Health disparities between, on the one hand, racial/ethnic populations, lower socioeconomic classes, and rural residents and, on the other hand, the overall U.S. population are major public health concerns.
Date(s) Application is Due: Sep 18
Contact: Dr. Hua-Chuan Sim, Clinical and Public Health Informatics; (301) 594-5983; fax: (301) 402-1384; email: simh@mail.nih.gov
Internet: http://www.nlm.nih.gov/ep/healthdisp.html
Sponsor: National Library of Medicine
6705 Rockledge Drive, Suite 301
Bethesda, MD 20892

NLM Exceptional, Unconventional Research Grants Enabling Knowledge Acceleration 4286
This RFA is for exceptionally innovative research on novel hypotheses or difficult problems, solutions to which would have an extremely high impact on biomedical or biobehavioral research that is germane to the mission of the National Library of Medicine. The RFA supports new projects, not continuation of projects that have already been initiated. For the EUREKA program, NLM seeks innovative applications in the following two discovery science areas: integrated discovery mining for biology and medicine; and integrated hypothesis testing for biology and medicine. Only electronic applications are accepted. The deadline for Letter of Intent is September 29, and the deadline for new applications is October 28.
Restrictions: This grant does not support pilot projects, i.e., projects of limited scope that are designed primarily to generate data that will enable the PI to seek other funding opportunities.
Date(s) Application is Due: Sep 29; Oct 28
Contact: Dr. Valerie Florance; (301) 594-5983; fax: (301) 402-1384; email: florancev@mail.nih.gov
Internet: http://www.nlm.nih.gov/ep/eureka.html
Sponsor: National Library of Medicine
6705 Rockledge Drive, Suite 301
Bethesda, MD 20892

NLM Exploratory/Developmental Grants 4287
The purpose of this program is to support for early stage conceptual work and feasibility tests in biomedical informatics and bioinformatics. These studies may involve considerable risk but may lead to a breakthrough in a particular area, or to the development of novel techniques or approaches that could have major impact on a field of biomedical, behavioral, or clinical research. Preliminary data are not required, but applicants are urged to provide milestones by which progress on the project can be measured. Only electronic applications are accepted. Deadlines for new applications are February 16, June 16 and October 16 each year. Deadlines for revised applications are March 16, July 16 and November 16 each year.
Date(s) Application is Due: Feb 16; Mar 16; Jun 16; Jul 16; Oct 16; Nov 16
Contact: Dr. Hua-Chuan Sim, Clinical and Public Health Informatics; (301) 594-5983; fax: (301) 402-1384; email: simh@mail.nih.gov; Dr. Jane Ye, Bioinformatics and Computational Biology; (301) 594-5983; fax: (301) 402-1384; email: yej@mail.nih.gov
Internet: http://www.nlm.nih.gov/ep/GrantExDev.html
Sponsor: National Library of Medicine
6705 Rockledge Drive, Suite 301
Bethesda, MD 20892

NLM Express Research Grants in Biomedical Informatics and Bioinformatics 4288
The purpose of the program is to provide support for rigorous scientific research in biomedical informatics and bioinformatics. NLM Express Research Grant applications are limited to a 15-page research plan and to a modular budget. The scope of interests is outlined in the funding opportunity announcement. Investigators whose needs are not met by this announcement can also apply through the NIH Parent R01 (PA 07-070), with a cover letter requesting assignment to NLM. Those who do so should use PAR 08-080 as a source of guidance about NLM's research grant priorities and scope of interest. Deadlines for new applications are February 5, June 5 and October 5 each year. Deadlines for revised applications are March 5, July 5 and November 5 each year.
Date(s) Application is Due: Feb 5; Mar 5; Jun 5; Jul 5; Oct 5; Nov 5
Contact: Dr. Hua-Chuan Sim, Clinical and Public Health Informatics; (301) 594-5983; fax: (301) 402-1384; email: simh@mail.nih.gov; Dr. Jane Ye, Bioinformatics and Computational Biology; (301) 594-5983; fax: (301) 402-1384; email: yej@mail.nih.gov
Internet: http://www.nlm.nih.gov/ep/GrantResearch.html
Sponsor: National Library of Medicine
6705 Rockledge Drive, Suite 301
Bethesda, MD 20892

NLM Grants for Scholarly Works in Biomedicine and Health 4289
Small grants are awarded for up to three years to provide short-term assistance for the preparation of book-length manuscripts and, in some cases, the publication of important scientific information needed by U.S. health professionals. Grants are awarded for major critical reviews and analyses of current developments in important areas of the health sciences, historical studies, works about health sciences informatics, librarianship, and certain kinds of secondary information and literature tools in the health sciences. Publication may be in formats other than print-on-paper (e.g., electronic, film, etc.), and may involve new and innovative ways of organizing and presenting information. Only electronic applications are accepted.
Requirements: Grants may be made to public or private, nonprofit institutions on behalf of a principal investigator; or to unaffiliated individuals to support salaries, consultant fees, equipment and supplies, travel, and other justified costs.
Restrictions: The program does not support textbooks, the production of curriculum materials, the initial reporting of original biomedical scientific research findings, proceedings of annual meetings, projects highly likely to have substantial commercial sales (best-seller type publications), or projects of local interest only. The program does not provide support for the free distribution of publications resulting from grants.
Amount of Grant: $50,000 maximum direct annual costs for up to three years
Date(s) Application is Due: Feb 2
Contact: Dr. Hua-Chuan Sim, Clinical and Public Health Informatics; (301) 594-5983; fax: (301) 402-1384; email: simh@mail.nih.gov
Internet: http://www.nlm.nih.gov/ep/GrantPubs.html
Sponsor: National Library of Medicine
6705 Rockledge Drive, Suite 301
Bethesda, MD 20892

NLM Internet Connection for Medical Institutions Grants 4290
NLM encourages the development of a communications infrastructure to promote the rapid interchange of medical information nationally and throughout the world. Internet access provides health professionals engaged in education, research, clinical care, and administration with a means of accessing remote databases and libraries, transferring files and images, and interacting with colleagues throughout the world. To accelerate the pace with which health-related institutions become part of the electronic information web, NLM offers grants to support Internet connections. Funds may be used for the purchase and installation of a gateway system and associated connection hardware. Funds may also be used to defray the cost of installation and leasing of communication circuits to connect the network. Annual deadline dates may vary; contact the program office for exact dates. RFA: LM-00-001
Requirements: Domestic public and private nonprofit institutions working in health sciences administration, education research, and/or clinical care services can apply for these grants. Groups that are health-related institutions are also encouraged to apply.
Amount of Grant: $30,000 maximum for a single institution; $50,000 maximum for a group of institutions
Contact: Dr. Hua-Chuan Sim, Clinical and Public Health Informatics; (301) 594-5983; fax: (301) 402-1384; email: simh@mail.nih.gov; Dr. Jane Ye, Bioinformatics and Computational Biology; (301) 594-5983; fax: (301) 402-1384; email: yej@mail.nih.gov
Internet: http://grants.nih.gov/grants/guide/rfa-files/RFA-LM-00-001.html
Sponsor: National Library of Medicine
8600 Rockville Pike, Building 38A
Bethesda, MD 20894

NLM Limited Competition for Continuation of Biomedical Informatics Bioinformatics Resource Grants 4291
This limited competition funding opportunity announcement (FOA) solicits competitive renewal applications from investigators currently supported under PAR-04-142, NLM Resource Grant in Biomedical Informatics/Bioinformatics (P41) or its predecessors. NLM, Biomedical Informatics/Bioinformatics Resource Grant program provides support for five (5) resource centers that support activities, databases or software tools that advance research or practice in the biomedical sciences, clinical medicine or health services research. Direct costs for the first year may not exceed 5% above the current award of the active year of NLM funding at the time of application, or a maximum of $500,000, whichever is smaller. Applicants may include cost-of-living increases of up to 3% per year for direct costs in future years. The deadline for Letters of Intent is July 23, while the deadline for Renewal Applications is August 27.
Date(s) Application is Due: Jul 28; Aug 27
Contact: Dr. Valerie Florance; (301) 594-5983; fax: (301) 402-1384; email: florancev@mail.nih.gov
Internet: http://www.nlm.nih.gov/ep/grantp41.html
Sponsor: National Library of Medicine
6705 Rockledge Drive, Suite 301
Bethesda, MD 20892

NLM Manufacturing Processes of Medical, Dental, and Biological Technologies (SBIR) Grants 4292
NLM is encouraging eligible United States small business concerns to submit SBIR and STTR Phase I, Phase II, and Fast-Track grant applications whose biomedical research is related to advanced processing, manufacturing processes, equipment and systems, and manufacturing workforce skills and protection. Only electronic applications are accepted.
Date(s) Application is Due: Apr 5; Aug 5; Dec 5
Contact: Dr. Jane Ye, Bioinformatics and Computational Biology; (301) 594-5983; fax: (301) 402-1384; email: yej@mail.nih.gov
Internet: http://www.nlm.nih.gov/ep/grantsbmanu.html
Sponsor: National Library of Medicine
6705 Rockledge Drive, Suite 301
Bethesda, MD 20892

NLM New Technologies for Liver Disease Grants 4293
The goal of this announcement is to enlist members of the small business research community in advancing means of diagnosis, treatment and prevention of liver disease and facilitate the goals outlined in the trans-NIH Action Plan for Liver Disease Research. Annual deadline dates are April 5, August 5 and December 5.
Date(s) Application is Due: Apr 5; Aug 5; Dec 5
Contact: Dr. Jane Ye, Bioinformatics and Computational Biology; (301) 594-5983; fax: (301) 402-1384; email: yej@mail.nih.gov
Internet: http://www.nlm.nih.gov/ep/grantsbliver.html
Sponsor: National Library of Medicine
6705 Rockledge Drive, Suite 301
Bethesda, MD 20892

NLM Predictive Multiscale Models of the Physiome in Health and Disease Grants 4294
The goal of this funding opportunity announcement (FOA) is to move the field of biomedical computational modeling forward through the development of more realistic and predictive models of health and disease. NIH recognizes the need for sophisticated, predictive, computational models of development and disease that encompass multiple biological scales. These models may be designed to uncover biological mechanisms or to make predictions about clinical outcome and may draw on a variety of data sources including relevant clinical data. Ultimately the models and the information derived from their use will enable researchers and clinicians to better understand, prevent, diagnose and treat the diseases or aberrations in normal development. Specifically this FOA solicits the development of predictive multiscale models of health and disease states that must include higher scales of the physiome. The specific objectives are to develop multiscale models that are physiologically mechanistic and biomedically relevant, to bring together modeling and biomedical expertise to collaborate on building models, to validate and test models with standard datasets, and to develop models that can be explicitly shared with other modelers.
Date(s) Application is Due: Jan 14; May 14; Sep 15
Contact: Dr. Jane Ye, Bioinformatics and Computational Biology; (301) 594-5983; fax: (301) 402-1384; email: yej@mail.nih.gov
Internet: http://www.nlm.nih.gov/ep/predictive.html
Sponsor: National Library of Medicine
6705 Rockledge Drive, Suite 301
Bethesda, MD 20892

NLM Research Project Grants 4295
The primary purpose of this program is to provide support for rigorous scientific research in biomedical informatics and bioinformatics. Medical informatics is an interdisciplinary field that combines health sciences specialty areas with the information and computer sciences, computational linguistics, and decision theory, among others. NLM's program emphasizes knowledge issues without restriction as to specific categorical diseases. Proposed research must be clearly health relevant. Areas of interest include knowledge representation, more effective human-machine interfaces, and more efficient information storage and retrieval. The cognitive aspects of information-related behaviors are of particular interest. This area includes problems of medical decision analysis, medical decision support systems, and retrieval from image databases. Deadlines for new applications are February 5, June 5 and October 5 each year. Deadlines for revised applications are March 5, July 5 and November 5 each year.
Requirements: Institutions or organizations with research capabilities in the health information fields or in medical informatics are eligible to apply.
Amount of Grant: $25,000-$2 million; $225,000 average
Date(s) Application is Due: Feb 5; Mar 5; Jun 5; Jul 5; Oct 5; Nov 5
Contact: Dr. Hua-Chuan Sim, Clinical and Public Health Informatics; (301) 594-5983; fax: (301) 402-1384; email: simh@mail.nih.gov; Dr. Jane Ye, Bioinformatics and Computational Biology; (301) 594-5983; fax: (301) 402-1384; email: yej@mail.nih.gov
Internet: http://www.nlm.nih.gov/ep/GrantResearchParent.html
Sponsor: National Library of Medicine
6705 Rockledge Drive, Suite 301
Bethesda, MD 20892

NLM Research Supplements to Promote Diversity in Health-Related Research 4296
Principal Investigators holding NIH research grants of the specified types (see announcement) can apply for administrative supplements to improve the diversity of the research workforce by supporting and recruiting students, postdoctoral candidates, and eligible investigators from groups that have been shown to be underrepresented in science. A currently funded Principal Investigator can submit one application on one topic. The grantee institution must submit the application for supplemental funds directly to the awarding component that supports the parent grant. The application must not be submitted to the NIH Center for Scientific Review.
Contact: Dr. Hua-Chuan Sim, Clinical and Public Health Informatics; (301) 594-5983; fax: (301) 402-1384; email: simh@mail.nih.gov
Internet: http://www.nlm.nih.gov/ep/grantsupdiversity.html
Sponsor: National Library of Medicine
6705 Rockledge Drive, Suite 301
Bethesda, MD 20892

NLM Research Supplements to Promote Reentry in Health-Related Research 4297
Principal Investigators holding NIH research grants of the specified types (see announcement) can apply for administrative supplements to support individuals with high potential to reenter an active research career after taking time off to care for children or attend to other family responsibilities. It is anticipated that at the completion of the supplement, the reentry scientist will be in a position to apply for a career development award, a research award, or some other form of independent research support. The grantee institution must submit the application for supplemental funds directly to the awarding component that supports the parent grant. The application must not be submitted to the NIH Center for Scientific Review.
Contact: Dr. Hua-Chuan Sim, Clinical and Public Health Informatics; (301) 594-5983; fax: (301) 402-1384; email: simh@mail.nih.gov
Internet: http://www.nlm.nih.gov/ep/grantsupreentry.html
Sponsor: National Library of Medicine
6705 Rockledge Drive, Suite 301
Bethesda, MD 20892

NLM Small Business Innovation Research (SBIR) Grants 4298
The NLM supports research and development projects in biomedical informatics and bioinformatics. NLM defines biomedical informatics as the intersection of basic informational and computing sciences with a wide range of application domains in biomedicine and public health. Bioinformatics is the intersection of basic informational and computing sciences with the biological sciences. Innovative technologies and methodologies fuel progress in biomedical and behavioral research and represent an increasingly important area of the economy. The SBIR and STTR programs provide support for research or development of new or improved technologies and methodologies

that have the potential to succeed as commercial products or services. Only electronic applications are accepted.
Date(s) Application is Due: Apr 5; Aug 5; Dec 5
Contact: Dr. Jane Ye, Bioinformatics and Computational Biology; (301) 594-5983; fax: (301) 402-1384; email: yej@mail.nih.gov
Internet: http://www.nlm.nih.gov/ep/grantsbir.html
Sponsor: National Library of Medicine
6705 Rockledge Drive, Suite 301
Bethesda, MD 20892

NLM Technological Innovations for Interdisciplinary Research Incorporating the Behavioral and Social Sciences Grants **4299**
The intent of the program is to solicit Small Business Innovation Research (SBIR) and Small Business Technology Transfer (STTR) grant applications from small business concerns (SBCs) for development of new, innovative technologies for research integrating human social and/or behavioral science with other disciplines. These announcements cover Phase I, Phase II, and Fast-Track applications. Only electronic applications are accepted.
Date(s) Application is Due: Jan 15; May 1; Sep 17
Contact: Dr. Jane Ye, Bioinformatics and Computational Biology; (301) 594-5983; fax: (301) 402-1384; email: yej@mail.nih.gov
Internet: http://www.nlm.nih.gov/ep/grantsbsoc.html
Sponsor: National Library of Medicine
6705 Rockledge Drive, Suite 301
Bethesda, MD 20892

NMF California Community Service Scholarships **4300**
The program offers scholarships to third- and fourth-year students attending medical schools in California. Scholarships are awarded based on demonstrated commitment to practice in California, interest in community-based primary care, satisfactory academic performance, leadership, and financial need. Students are encouraged to apply for the scholarship program during their second or third year of medical school. The goal of the program is to expose students who are interested in receiving first-hand exposure to community-based primary care for the equivalent of six weeks; or conducting research for the equivalent of at least eight to 12 weeks in areas of critical need (e.g., HIV/AIDS care and research, hypertension, tuberculosis, cardiovascular disease, diabetes, asthma, substance abuse, women's health research, etc.). It also fosters their commitment to practicing in California. The scholarships will be awarded on the basis of personal commitment to practice in California, demonstrated interest in community-based primary care or research, satisfactory academic performance, and leadership. Priority will be given to NMF scholars.
Requirements: African-Americans, mainland Puerto Ricans, Mexican-Americans, Native Hawaiians, Alaska Natives, and American Indians who are U.S. citizens attending California medical schools. Eligibility is limited to rising third-year or fourth-year students. The required application is available at the sponsor's website.
Restrictions: Only students attending California medical schools are eligible for the program.
Geographic Focus: California
Amount of Grant: $15,000
Contact: Mary Wade, CSSP Interim Program Coordinator; (415) 397-2526; email: marywade@nmfonline.org
Internet: http://www.nmfonline.org/Programs/Scholarships/CaliforniaCSSP/Overview.htm
Sponsor: National Medical Fellowships, Inc.
5 Hanover Square
New York, NY 10004

NMF General Electric Medical Scholars Fellowships **4301**
The sponsor has established a partnership with General Electric (GE) to provide need-based scholarship support to historically underrepresented minority medical students, and to allow them to engage in clinical work in underserved communities. The program also offers eligible students an opportunity to complete a two-month elective in Ghana, West Africa focusing on areas of critical need such as HIV/AIDS, diabetes, tuberculosis, heart disease, cancer and infant mortality. Five fellowships are awarded annually. Students will be matched with a mentor for the period of the program. Travel, hosing and a stipend will be provided for students.
Requirements: Students must be nominated by their medical school deans. Eligible candidates must be underrepresented minority students enrolled in their fourth year in an accredited U.S. medical school. Candidates must demonstrate outstanding academic achievement, leadership, and potential for community-based work or initiation of innovative projects in school or community. Additionally, candidates must also demonstrate potential for distinguished contributions to medicine. Complete application requirements can be found at the sponsor's website.
Geographic Focus: Alabama; Alaska; Arizona; Arkansas; California; Colorado; Connecticut; Delaware; District of Columbia; Florida; Georgia; Hawaii; Idaho; Illinois; Indiana; Iowa; Kansas; Kentucky; Louisiana; Maine; Maryland; Massachusetts; Michigan; Minnesota; Mississippi; Missouri; Montana; Nebraska; Nevada; New Hampshire; New Jersey; New Mexico; New York; North Carolina; North Dakota; Ohio; Oklahoma; Oregon; Pennsylvania; Puerto Rico; Rhode Island; South Carolina; South Dakota; Tennessee; Texas; Utah; Vermont; Virginia; Washington; West Virginia; Wisconsin; Wyoming
Contact: Dr. Stephen Keith, President and COO of Panacea Pharmaceuticals, Inc.; (240) 243-8000; fax: (240) 465-0450
Internet: http://www.nmfonline.org/index.html
Sponsor: National Medical Fellowships, Inc.
5 Hanover Square
New York, NY 10004

NMF National Medical Association Awards for Medical Journalism **4302**
In general, these awards recognize academic achievement, leadership, and potential for distinguished contributions to medicine. Specifically, these awards recognize demonstrated skills in journalism. Deadlines vary - contact the sponsor for dates.
Requirements: Eligibility is limited to African-American medical students. Each medical school may submit up to two nominations. Candidates must be in the third or fourth year of medical school and must have published articles and photographs in, or been writers, editors, photographers on the staffs of: medical school newspapers; medical students' journals; recognized professional journals; or, other accredited scientific journal. Students who have written, produced or directed health-related films, commercials or videos are also encouraged to submit their applications and samples of their work. Applicants must also submit samples of their journalistic work (articles, photographs, audio tapes, video tapes, etc.) for review by the selection committee. Entries must have been published or produced within the last three years. The application form and more detailed guidelines can be found at the sponsor's website.
Geographic Focus: Alabama; Alaska; Arizona; Arkansas; California; Colorado; Connecticut; Delaware; District of Columbia; Florida; Georgia; Hawaii; Idaho; Illinois; Indiana; Iowa; Kansas; Kentucky; Louisiana; Maine; Maryland; Massachusetts; Michigan; Minnesota; Mississippi; Missouri; Montana; Nebraska; Nevada; New Hampshire; New Jersey; New Mexico; New York; North Carolina; North Dakota; Ohio; Oklahoma; Oregon; Pennsylvania; Puerto Rico; Rhode Island; South Carolina; South Dakota; Tennessee; Texas; Utah; Vermont; Virginia; Washington; West Virginia; Wisconsin; Wyoming
Amount of Grant: $2,500
Contact: Programs Department; (212) 483-8880; fax: (212) 239-9718; email: nmf2@nmfonline.org
Internet: http://www.nmfonline.org/index.html
Sponsor: National Medical Fellowships, Inc.
5 Hanover Square
New York, NY 10004

NMF National Medical Association Emerging Scholar Awards **4303**
The National Medical Association and NMF sponsor need-based award programs that recognize and reward African American medical students for special achievements, academic excellence, leadership, and potential for outstanding contributions to medicine. The NMA Emerging Scholar Awards are presented to current first, second, and third-year medical students.
Requirements: Eligibility for these awards is limited to African-American medical students. Each school may submit two nominations. Applicants must demonstrate and document financial need. Deadlines vary - contact the sponsor for specific dates.
Geographic Focus: Alabama; Alaska; Arizona; Arkansas; California; Colorado; Connecticut; Delaware; District of Columbia; Florida; Georgia; Hawaii; Idaho; Illinois; Indiana; Iowa; Kansas; Kentucky; Louisiana; Maine; Maryland; Massachusetts; Michigan; Minnesota; Mississippi; Missouri; Montana; Nebraska; Nevada; New Hampshire; New Jersey; New Mexico; New York; North Carolina; North Dakota; Ohio; Oklahoma; Oregon; Pennsylvania; Puerto Rico; Rhode Island; South Carolina; South Dakota; Tennessee; Texas; Utah; Vermont; Virginia; Washington; West Virginia; Wisconsin; Wyoming
Amount of Grant: $2,250
Contact: Programs Department; (212) 483-8880; email: nmf1@nmfonline.org
Internet: http://www.nmfonline.org/index.html
Sponsor: National Medical Fellowships, Inc.
5 Hanover Square
New York, NY 10004

NMF National Medical Association Patti LaBelle Award **4304**
This award is one of the highest honors presented to students by the National Medical Association. It recognizes academic achievement, leadership and potential for distinguished contributions to medicine. This award is presented to current first-, second-, and third-year medical students with demonstrated financial need.
Requirements: Eligibility is limited to African-American medical students. Application materials must include a letter of recommendation that fully discusses the candidate's academic performance, extracurricular activities and other qualifications; a complete scholarship application form; an essay of at least 500 words discussing the applicant's motivation for a career in medicine and plans over the next ten years; and, an official academic transcript.
Geographic Focus: Alabama; Alaska; Arizona; Arkansas; California; Colorado; Connecticut; Delaware; District of Columbia; Florida; Georgia; Hawaii; Idaho; Illinois; Indiana; Iowa; Kansas; Kentucky; Louisiana; Maine; Maryland; Massachusetts; Michigan; Minnesota; Mississippi; Missouri; Montana; Nebraska; Nevada; New Hampshire; New Jersey; New Mexico; New York; North Carolina; North Dakota; Ohio; Oklahoma; Oregon; Pennsylvania; Puerto Rico; Rhode Island; South Carolina; South Dakota; Tennessee; Texas; Utah; Vermont; Virginia; Washington; West Virginia; Wisconsin; Wyoming
Amount of Grant: $5,000
Contact: Programs Department; (212) 483-8880; fax: (212) 483-8897; email: nmf1@nmfonline.org
Internet: http://www.nmfonline.org/index.html

Sponsor: National Medical Fellowships, Inc.
5 Hanover Square
New York, NY 10004

NMF Need-Based Scholarships 4305

The program provides financial assistance to United States citizens from groups currently underrepresetned in the medical profession; specifically, African-Americans, Mexican-Americans, Native Americans, Alaska Natives, Native Hawaiians, and mainland Puerto Ricans who permanently reside within the 50 U.S. states. These awards are offered to first- and second-year medical students. Scholarships are granted on the basis of financial need as determined by the student's total resources (including parental and spousal support), cost of education, and receipt of other scholarships and grants. Awards have ranged from $500 to $10,000.
Requirements: Applicants must be U.S. citizens by the application deadline and must have been accepted to AAMC or AOA-accredited U.S. medical schools for study leading to M.D. or D.O. degrees. The application form and FAQs can be found at the sponsor's website. Applicants must demonstrate and document financial need by submitted complete copies of their parents', spouse's and their own most recent 1040, 1040A or 1040EZ tax forms. All nontaxable income (e.g. AFDC, ADC, Social Security benefits, etc.) must also be documented by the appropriate agency. Applicants with neither taxable nor nontaxable income must provide verification of means of support. All parental financial data must be submitted in order to evaluate dependent status.
Geographic Focus: Alabama; Alaska; Arizona; Arkansas; California; Colorado; Connecticut; Delaware; District of Columbia; Florida; Georgia; Hawaii; Idaho; Illinois; Indiana; Iowa; Kansas; Kentucky; Louisiana; Maine; Maryland; Massachusetts; Michigan; Minnesota; Mississippi; Missouri; Montana; Nebraska; Nevada; New Hampshire; New Jersey; New Mexico; New York; North Carolina; North Dakota; Ohio; Oklahoma; Oregon; Pennsylvania; Puerto Rico; Rhode Island; South Carolina; South Dakota; Tennessee; Texas; Utah; Vermont; Virginia; Washington; West Virginia; Wisconsin; Wyoming
Amount of Grant: $500 -$10,000
Date(s) Application is Due: Aug 31
Contact: Programs Department; (212) 483-8880 ext 304; fax: (212) 483-8897; email: NMF1@NMFonline.org or info@NMFonline.org
Internet: http://www.nmfonline.org/programs.html
Sponsor: National Medical Fellowships, Inc.
5 Hanover Square
New York, NY 10004

NMF Ralph W. Ellison Prize 4306

The prize is presented to a graduating underrepresented medical student who has demonstrated outstanding academic achievement, leadership, and potential to make significant contributions to medicine. One prize is presented each year (nonrenewable) and includes a certificate of merit and a stipend. Annual deadline date may vary; contact program staff for exact dates.
Requirements: African-Americans, mainland Puerto Ricans, Mexican-Americans, Native Hawaiians, Alaska Natives, and American Indians who are U.S. citizens attending M.D. degree-granting institutions accredited by the Liaison Committee on Medical Education of the Association of American Medical Colleges, or in D.O. degree-granting colleges of osteopathic medicine accredited by the Bureau of Professional Education of the American Osteopathic Association are eligible.
Geographic Focus: Alabama; Alaska; Arizona; Arkansas; California; Colorado; Connecticut; Delaware; District of Columbia; Florida; Georgia; Hawaii; Idaho; Illinois; Indiana; Iowa; Kansas; Kentucky; Louisiana; Maine; Maryland; Massachusetts; Michigan; Minnesota; Mississippi; Missouri; Montana; Nebraska; Nevada; New Hampshire; New Jersey; New Mexico; New York; North Carolina; North Dakota; Ohio; Oklahoma; Oregon; Pennsylvania; Puerto Rico; Rhode Island; South Carolina; South Dakota; Tennessee; Texas; Utah; Vermont; Virginia; Washington; West Virginia; Wisconsin; Wyoming
Amount of Grant: $500
Contact: Programs Department; (212) 483-8880; fax: (212) 483-8897; email: nmf1@nmfonline.org
Internet: http://www.nmfonline.org/index.html
Sponsor: National Medical Fellowships, Inc.
5 Hanover Square
New York, NY 10004

NMSF Ernest F. Hollings Ocean Awareness Trust Fund Grants 4307

The Foundation builds partnerships to promote conservation and develop an ocean literate public who is able to make informed and responsible decisions regarding the ocean and its resources. This fund was established in 2005 as a legacy to the longstanding commitment of former U.S. Senator Ernest F. Hollings to promoting public involvement and understanding of the benefits of the oceans to current and future generations. The Ocean Awareness Trust Fund seeks to continue this important work by ensuring that all Americans understand and value these precious resources, and by promoting opportunities for the people of all ages to become actively interested and involved in ocean stewardship. Special consideration will be given to projects that address marine debris, human interaction with marine mammals, runoff and pollution or public health issues such as red tides or marine pharmaceuticals.
Requirements: Eligible applicants are formal or informal education institutions, nonprofits, commercial organizations, and state, local and Indian tribal governments.
Amount of Grant: $10,000-$25,000
Date(s) Application is Due: Aug 17
Contact: Allison Alexander, Grants Manager; (301) 608-3040; fax: (301) 608-3044; email: Allison@nmsfocean.org or grants@NMSFocean.org
Internet: http://nmsfocean.org/grants.asp
Sponsor: National Marine Sanctuary Foundation
8601 Georgia Avenue, Suite 501
Silver Spring, MD 20910

NMSS Daniel Haughton Senior Faculty Awards 4308

The award is granted to established investigators seeking support for specialized training in a field which they are not currently expert, inasmuch as such training will enhance their capacity to conduct research related to multiple sclerosis. Awards may be used for sabbatical support.
Date(s) Application is Due: Feb 13
Contact: Dr. Timothy Coetzee, Director, Research Training Programs, (212) 476-0478; email: timothy.coetzee@nmss.org
Internet: http://www.nationalmssociety.org/Research-srfaculty.asp
Sponsor: National Multiple Sclerosis Society
733 Third Avenue
New York, NY 10017

NMSS Harry Weaver Neuroscience Scholar Awards 4309

These junior faculty awards provide higher support for qualified candidates who have concluded their research training and begun academic careers as independent investigators in areas of the neurosciences related to multiple sclerosis. The awards are designed to provide salary and grant support for a five-year period, thus permitting the awardee to establish competence in his/her chosen research area. Candidates may request up to 75 percent annual salary support, not to exceed $50,000 in the first year of the award, with a 5 percent annual increase in each of the remaining four years. Research support of $25,000 may be requested in the first year and up to $35,000 in the last year.
Requirements: Candidate must hold the MD, PhD, or equivalent with no more than five years of independent research experience and must have been offered a position equivalent to assistant professor at an approved institution.
Date(s) Application is Due: Feb 1
Contact: Timothy Coetzee, Director, Research Training Programs, (212) 476-0478; email: timothy.coetzee@nmss.org
Internet: http://www.nationalmssociety.org/Research-jrfaculty.asp
Sponsor: National Multiple Sclerosis Society
733 Third Avenue
New York, NY 10017-3288

NMSS Research Grants 4310

Grants are awarded for fundamental or applied research in scientific areas pertinent to multiple sclerosis. Grants are available for clinical or nonclinical studies, providing they show a reasonable relevance to the society's interests. Annual deadline dates may vary; contact program staff for exact dates.
Requirements: Qualified investigators affiliated with appropriate institutions are eligible to apply. Institutions are the official recipients of the grants.
Amount of Grant: $35,568-$51,036
Date(s) Application is Due: Feb 1; Aug 3
Contact: Dr. Patricia O'Looney, Research Programs Department, (212) 476-0413; fax: (212) 986-7981; email: patricia.olooney@nmss.org
Internet: http://www.nationalmssociety.org/Research-Grants.asp
Sponsor: National Multiple Sclerosis Society
733 Third Avenue
New York, NY 10017-3288

NMSS Sylvia Lawry Physician Fellowships 4311

The society welcomes applications for support of physicians for training in key elements associated with conducting multiple sclerosis clinical trials. The purpose of this award is to help maintain and foster high quality clinical trials in multiple sclerosis and to ensure that there is a continuing pool of physicians who are trained in this area. In this program the individual with an MD or equivalent medical degree will acquire formal training, under the tutelage of an established investigator, in key elements associated with conducting clinical trials in MS. Annual deadline dates may vary; contact program staff for exact dates.
Requirements: Individuals must have received the MD or equivalent medical degree from an accredited institution and must be licensed to practice medicine in the United States.
Amount of Grant: $40,000 maximum
Date(s) Application is Due: Feb 1
Contact: Dr. Timothy Coetzee, Research Programs Department, (212) 476-0478; email: timothy.coetzee@nmss.org
Internet: http://www.nationalmssociety.org/Research-physfellow.asp
Sponsor: National Multiple Sclerosis Society
733 Third Avenue
New York, NY 10017-3288

NOAA Ballast Water Management Demonstration Program (Technologies and Practices) Grants 4312

The Ballast Water Management Demonstration Program supports projects to develop, test, and demonstrate ballast water technologies and practices that reduce the threat of introduction of aquatic invasive species to U.S. waters through the discharge of ballast water. The ultimate mission of the program is the implementation and use of effective

treatments. This grant program supports projects to develop, test, and demonstrate ballast water technologies and practices in order to reduce the threat of introduction of aquatic invasive species to U.S. waters through the discharge of ballast water. This FFO is for projects dealing with technologies and practices necessary for successful ballast water management. For example, this could include technologies or practices that treat ballast water, or that make it possible to test, evaluate, regulate or use ballast water treatment technologies or practices. Letters of Intent must be received by the National Sea Grant Office by 4:00 PM EST February 21. Full proposals must be received by 4:00 PM EDT April 3. Proposals selected for funding from non-Federal applicants will be funded through project grants or cooperative agreements.
Requirements: Only those applicants who submit Letters of Intent by the preliminary proposal deadline and who submit LOIs that meet the other requirements are eligible to submit full proposals. Eligible applicants are individuals, institutions of higher education, nonprofit organizations, for-profit organizations, Federal, State, local and Indian tribal governments, foreign governments, organizations under the jurisdiction of foreign governments, and international organizations.
Restrictions: Applications from non-Federal and eligible Federal applicants (including NOAA and MARAD employees) will be evaluated in the same selection process.
Amount of Grant: $200,000 average; $1,000,000 total
Date(s) Application is Due: Apr 3
Contact: Dorn Carlson, NOAA National Sea Grant Office; (301) 734-1080; email: ballast.water@noaa.gov; Carolyn Junemann, U.S. Maritime Administration; (202) 366-1920; email: carolyn.junemann@dot.gov
Internet: http://www07.grants.gov/search/search.do;jsessionid=HWbpMDzmZ104MfvCGSbHhsGGPnb8dGTRG2hM0vfdTvpFRTZdkPyK!-796138783?oppId=16385&flag2006=true&mode=VIEW
Sponsor: National Oceanic and Atmospheric Administration
1315 East West Highway, 12th Floor
Silver Spring, MD 20910

NOAA Ballast Water Technology Demonstration Program (Treatment Technology Demonstration Project) Grants 4313

The Program supports projects to develop, test, and demonstrate technologies that treat ships' ballast water in order to reduce the threat of introduction of aquatic invasive species to U.S. waters through the discharge of ballast water. This Program is only for the technology demonstration grants competition described. There is a separate FFO for the competition to develop and operate a regional ballast water technology research, development, testing and evaluation facility. Depending on annual appropriations, NOAA and the U.S. Fish and Wildlife Service expect to make available up to about $2 million, and the U.S. Maritime Administration (MARAD) expects to make available several vessels for use as test platforms, to support ballast water treatment technology demonstration projects. The maximum amount of award will vary with the scale of the proposed project. Depending on the funding available and the number and quality of proposals received, approximately 10 grants with a median value of about $200,000 are anticipated to be awarded. Preliminary proposals must be received by 4 p.m. EDT on August 26, with full proposals due by 4 p.m. EST on December 2. Only those who submit preliminary proposals meeting the preliminary proposal deadline and other requirements are eligible to submit full proposals.
Requirements: Eligible applicants include: individuals; institutions of higher education; nonprofit organizations; for-profit organizations; Federal, State, local and Indian tribal governments; foreign governments; organizations under the jurisdiction of foreign governments; and international organizations. Applications from non-Federal and eligible Federal applicants (including NOAA employees) will be evaluated in the same selection process.
Amount of Grant: $200,000 maximum; $2,000,000 total funding
Date(s) Application is Due: Aug 26; Dec 2
Contact: Dorn Carlson, NOAA National Sea Grant Office; (301) 734-1080; email: ballast.water@noaa.gov; Carolyn Junemann, U.S. Maritime Administration; (202) 366-1920; email: carolyn.junemann@dot.gov
Internet: http://www07.grants.gov/search/search.do;jsessionid=HXXNv3Mt6rGXD7y5CX8dvQqGkVgGF8H6Hw5BGwTw1BFBfBQs3fdn!1282164016?oppId=1414&flag2006=true&mode=VIEW
Sponsor: National Oceanic and Atmospheric Administration
1315 East West Highway, 12th Floor
Silver Spring, MD 20910

NOAA Abrupt Climate Change Studies (ARCHES) Grants 4314

The office invites proposals that build on the ARCHES program model - a multidisciplinary, single-institution effort that advances NOAA's abrupt climate change research interests (including paleo-drought) along three timescales: the era of anthropogenic influence, which includes the modern period of instrumental data as well as the future; the last 2000 years; and the Holocene. It is expected that proposals will describe an integrated research program that incorporates a variety of paleoproxy and modern datasets, has strong modeling, diagnostics and synthesis components and also incorporates a research thrust to study the past and future impacts of abrupt climate change. Due to the level of funding available, new observations are not encouraged, unless they are highly-leveraged and greatly enhance the overall effort. It is anticipated that the successful proposal will be from a single institution that can demonstrate a strong, collocated multi-disciplinary research program. Consideration will be given to an institution wishing to link its effort with one of the several NOAA Cooperative Institutes as well as the NOAA Regional Integrated Science and Assessments program. While it is in the best interest of the applicants and their institutions to submit a Letter of Intent (LOI), it is not a requirement; applicants who do not submit an LOI are allowed to submit a full proposal. For those that do submit an LOI, full proposals will be encouraged only for LOIs deemed relevant.
Requirements: Eligible applicants are institutions of higher education, other nonprofits, commercial organizations, international organizations, and state, local and Indian tribal governments.
Restrictions: Federal agencies or institutions are not eligible to receive Federal assistance under this notice.
Amount of Grant: $2,000,000 per year for five years
Date(s) Application is Due: Sep 24
Contact: James Todd, Grants Manager; (301) 734-1258; fax: (301) 713-0518; email: james.todd@noaa.gov
Internet: http://www.climate.noaa.gov/index.jsp?pg=/opportunities/opp_index.jsp&opp=2008/acchange_info.jsp
Sponsor: National Oceanic and Atmospheric Administration
1315 East West Highway, 12th Floor
Silver Spring, MD 20910

NOAA Anadromous Fish Conservation Act Grants 4315

These projects are conducted for the conservation, development, and enhancement of anadromous fishery resources (those that migrate from salt to fresh water for spawning) including similar species in the Great Lakes and Lake Champlain. All projects must be cleared with the State fishery agency of the State that the work is carried out in. Many projects funded under this program are critical elements of larger multi-funded programs to manage, restore or enhance United States anadromous species such as Pacific and Atlantic salmon, Atlantic and shortnose sturgeon, and American Shad and river herrings. Funds may be used for spawning area improvement, installment of fishways, data collection, construction of fish protection devices and hatcheries, and research to improve management and increase anadromous fish resources. This program is also supported by the Fish and Wildlife Service of the DOI. Applications are accepted at any time for one- to three-year projects. The project application should be submitted 90 days in advance of desired effective date.
Requirements: Any interested person or organization may propose a cooperative undertaking; however, all proposals must be coordinated with the state fishery agency having responsibility for the resource to be affected by the proposal.
Restrictions: Funds may not be used for law enforcement, public relations, or construction of facilities and vessels, whose primary purpose is to commercially harvest, handle, and process fishery products and may not be used for projects in the Columbia River Basin with the exception of the state of Idaho. The federal share of a project cost shall usually not exceed 50 percent. Federal funds for projects supporting an interstate fishery management plan may be up to 90 percent.
Samples: Delaware—monitor Atlantic sturgeon abundance and movements in the Delaware River.; Wisconsin—research on the influence of woody debris on salmon in Wisconsin tributaries of Lake Superior.; Louisiana—studies on striped bass in the lower Mississippi and Atchafalaya rivers.
Amount of Grant: $2,000-$400,000; average $40,000
Contact: Anne Lange, Chief; (301) 713-2334; fax: (301) 713-0596; email: anne.lange@noaa.gov or grants.information@noaa.gov
Internet: http://www.nmfs.noaa.gov/sfa/state_federal/State-Federal-WEB/afca.htm
Sponsor: National Oceanic and Atmospheric Administration
1315 East-West Highway
Silver Spring, MD 20910

NOAA Arctic Research Program Grants 4316

The goal of the Arctic Research Program is to provide climate-relevant observations and analysis of the broader Arctic region, with an emphasis on the Pacific sector of the Arctic, to support improved climate projections and assessment of impacts in the Arctic and regions influenced by the Arctic. The Program is soliciting proposals in the following area: participation in the Russian-American Long-term Census of the Arctic (RU.S.ALCA) program in FY2008 and beyond. All proposals should display proposed start dates of April 1, 2008, and extend to March 31, 2013 for a total of 5 years.
Requirements: Eligible applicants are institutions of higher education, other nonprofits, commercial organizations, international organizations, and state, local and Indian tribal governments. Proposals will be accepted only from investigators that responded to the previous call for proposals for RU.S.ALCA issued by CIFAR and who were selected for future funding, or from the existing RU.S.ALCA logistics providers.
Restrictions: Federal agencies or institutions are not eligible to receive Federal assistance under this notice.
Date(s) Application is Due: Sep 24
Contact: John Calder; (301) 734-1207; fax: (301) 713-0517; email: John.Calder@noaa.gov; Kathy Crane; (301) 734-1212; fax: (301) 713-0517; email: Kathy.Crane@noaa.gov
Internet: http://www.climate.noaa.gov/index.jsp?pg=/opportunities/opp_index.jsp&opp=2008/arp_info.jsp
Sponsor: National Oceanic and Atmospheric Administration
1315 East West Highway, 12th Floor
Silver Spring, MD 20910

NOAA Assessing Atlantic Meridional Overturning Circulation Grants 4317

The AMOC near-term priority is designed to take advantage of rapidly advancing observing and assimilation capabilities as well as leverage substantial international investment. The program will include activities in the following areas: fundamental research (e.g. description of the AMOC, its variability, critical processes); observing and monitoring the AMOC (e.g. design and development, stewardship of data); ocean analyses (e.g. improving ocean models and historical ocean analyses); development

of nowcasting capabilities and experimental products; and assessment of potential impacts of rapid AMOC changes. Letters of Intent (LOI) will be due November 2, with responses from program management by November 7. Full proposals will be due December 7, with funding to begin August 1.
Requirements: Eligible applicants are institutions of higher education, other nonprofits, commercial organizations, international organizations, and state, local and Indian tribal governments.
Restrictions: Federal agencies or institutions are not eligible to receive Federal assistance under this notice.
Amount of Grant: $50,000 - $200,000 per year
Date(s) Application is Due: Nov 2; Dec 7
Contact: James Todd, Grants Manager; (301) 734-1258; fax: (301) 713-0518; email: james.todd@noaa.gov
Internet: http://www.climate.noaa.gov/index.jsp?pg=/opportunities/opp_index.jsp&opp=2008/amoc_info.jsp
Sponsor: National Oceanic and Atmospheric Administration
1315 East West Highway, 12th Floor
Silver Spring, MD 20910

NOAA Atlantic Salmon Conservation Grants 4318
NOAA Atlantic Salmon Conservation Grants provide funding to catalyze the implementation of locally driven projects that address priority needs for Atlantic salmon recovery and restoration. Through these grants, NOAA provides funding and technical assistance to support restoration of habitat connectivity and function for the benefit of Atlantic salmon within their current and historical range in New England. Funded projects have strong on-the-ground habitat restoration or protection components that lead to long-term ecological habitat improvements for Atlantic salmon, and also provide educational and social benefits related to Atlantic salmon conservation for citizens and their communities. Proposals selected for funding through this solicitation will be implemented through a cooperative agreement. Funding of up to $2,500,000 is expected to be available for Atlantic Salmon Conservation Grants. Typical awards will range from $50,000 to $250,000. Although a select few may fall outside of this typical award range, project proposals requesting less than $30,000 or more than $500,000 will not be accepted.
Requirements: Eligible applicants are institutions of higher education, other non-profits, industry and commercial (for profit) organizations, organizations under the jurisdiction of foreign governments, international organizations, and state, local and Indian tribal governments whose projects have the potential to benefit NOAA trust resources.
Restrictions: Applications from federal agencies or employees of federal agencies will not be considered.
Amount of Grant: $30,000-$500,000
Date(s) Application is Due: Nov 14
Contact: Cathy Bozek; (301) 713-0174; email: Cathy.Bozek@noaa.gov; Craig Woolcott; (301) 713-0174; email: Craig.Woolcott@noaa.gov
Internet: http://www.grants.gov/search/search.do;jsessionid=LP0hqWfp11PVPKSr9yfL1LLhNQTn2gsNBmYzZtzGKTQhJRdRBYxW!-2005900814?oppId=42303&flag2006=false&mode=VIEW
Sponsor: National Oceanic and Atmospheric Administration
1315 East-West Highway
Silver Spring, MD 20910

NOAA Atmospheric Composition and Climate Grants 4319
The program pursues two overall research objectives: to improve the predictive understanding of the radiative forcing of the climate system by aerosols and by chemically active greenhouse gases, and to better characterize the recovery of the stratospheric ozone layer, including its role in climate change. The integrated research activities that address these objectives involve instrument 'developmentegional' to global observations, laboratory studies, and theoretical modeling by NOAA and extramural partners. While it is in the best interest of the applicants and their institutions to submit a Letter of Intent (LOI), it is not a requirement; applicants who do not submit an LOI are allowed to submit a full proposal. For those that do submit an LOI, full proposals will be encouraged only for LOIs deemed relevant.
Requirements: Eligible applicants are institutions of higher education, other nonprofits, commercial organizations, international organizations, and state, local and Indian tribal governments.
Restrictions: Federal agencies or institutions are not eligible to receive Federal assistance under this notice.
Amount of Grant: $50,000-$200,000 per year
Date(s) Application is Due: Sep 24
Contact: Kea Duckenfield, Program Manager; (301) 734-1216; fax: (301) 713-0518; email: kea.duckenfield@noaa.gov
Internet: http://www.climate.noaa.gov/index.jsp?pg=/opportunities/opp_index.jsp&opp=2008/acc_info.jsp
Sponsor: National Oceanic and Atmospheric Administration
1315 East West Highway, 12th Floor, Room 12707
Silver Spring, MD 20910

NOAA Ballast Water Management Demonstration (RDTE Facility) Grants 4320
The Program supports projects to develop, test, and demonstrate ballast water technologies and practices that reduce the threat of introduction of aquatic invasive species to U.S. waters through the discharge of ballast water. The ultimate mission of the program is the implementation and use of effective treatments. The Program's priority is to support establishment of RDTE facilities that demonstrate: a clear statement of the purpose and goals of the facility; appropriate physical and organizational infrastructure to accomplish their purpose; access to appropriate expertise in multiple disciplines; ability to provide fair and unbiased decisions and advice concerning what technologies or ballast water management strategies will be addressed and what activities will be undertaken; active support of state regulatory and maritime private sector interests; adequate quality control and accountability procedures; and the financial ability to maintain the facility even in the event that NOAA funding is less than expected or unavailable after the grant period.
Requirements: Only those applicants who submit Letters of Intent by the preliminary proposal deadline and who submit LOIs that meet the other requirements are eligible to submit full proposals. Eligible applicants are individuals, institutions of higher education, nonprofit organizations, for-profit organizations, Federal, State, local and Indian tribal governments, foreign governments, organizations under the jurisdiction of foreign governments, and international organizations.
Restrictions: Applications from non-Federal and eligible Federal applicants (including NOAA and MARAD employees) will be evaluated in the same selection process.
Amount of Grant: $500,000 average; $1,000,000 total
Date(s) Application is Due: Apr 24
Contact: Dorn Carlson, NOAA National Sea Grant Office; (301) 734-1080; email: ballast.water@noaa.gov
Internet: http://www07.grants.gov/search/search.do;jsessionid=HWbpMDzmZ104MfvCCSbHhsGGPnb8dGTRG2hM0vfdTvpFRTZdkPyK!-796138783?oppId=16384&flag2006=true&mode=VIEW
Sponsor: National Oceanic and Atmospheric Administration
1315 East West Highway, 12th Floor
Silver Spring, MD 20910

NOAA Ballast Water Technology Demonstration Program (Research, Development, Testing and Evaluation Facility) Grants 4321
The Program supports projects to develop, test, and demonstrate technologies that treat ships' ballast water in order to reduce the threat of introduction of aquatic invasive species to U.S. waters through the discharge of ballast water. The purpose of the Program is to foster the development of technologies that: are effective at removing, inactivating, or preventing the transfer of aquatic organisms in the ballast tanks of ships, including in the ballast water or associated sediments; are practicable from the standpoint of ship operations, safety, and environmental protection; have the ability to meet all applicable regulatory requirements; and have the potential to be a commercially viable product. The Program has funded, and continues to fund, grants for one- and two-year ballast water treatment technology demonstration projects. Preliminary proposals are due on October 27th, with final proposals due on February 10th.
Requirements: Eligible applicants include: individuals; institutions of higher education; nonprofit organizations; for-profit organizations; Federal, State, local and Indian tribal governments; foreign governments; organizations under the jurisdiction of foreign governments; and international organizations. Applications from non-Federal and eligible Federal applicants (including NOAA employees) will be evaluated in the same selection process.
Amount of Grant: $950,000
Date(s) Application is Due: Oct 27; Feb 10
Contact: Dorn Carlson, NOAA National Sea Grant Office; (301) 734-1080; email: ballast.water@noaa.gov; Carolyn Junemann, U.S. Maritime Administration; (202) 366-1920; email: carolyn.junemann@dot.gov
Internet: http://www07.grants.gov/search/search.do;jsessionid=HXXNv3Mt6rGXD7y5CX8dvQqGkVgGF8H6Hw5BGwTw1BFBfBQs3fdn!1282164016?oppId=1562&flag2006=true&mode=VIEW
Sponsor: National Oceanic and Atmospheric Administration
1315 East West Highway, 12th Floor
Silver Spring, MD 20910

NOAA California Bay Watershed Education and Training (B-WET) Grants 4322
The California B-WET grant program is a competitively based program that supports existing environmental education programs, fosters the growth of new programs, and encourages the development of partnerships among environmental education programs throughout the San Francisco Bay, Monterey Bay, and Santa Barbara Channel watersheds. Projects support organizations that provide students meaningful watershed educational experiences and teachers professional development opportunities in the area of environmental education related watersheds.
Requirements: Eligible applicants are K-through-12 public and independent schools and school systems, institutions of higher education, nonprofit organizations, state or local government agencies, and Indian tribal governments. No cost sharing is required under this program; however, the National Marine Sanctuary Program strongly encourages applicants applying for either area of interest to share as much of the costs of the award as possible.
Amount of Grant: $50,000-$60,000 average
Date(s) Application is Due: Oct 6
Contact: Seaberry Nachbar; (831) 647-4201; fax: (831) 647-4250; seaberry.nachbar@noaa.gov
Internet: http://www.grants.gov/search/search.do;jsessionid=LT5QC6LgrvSTzGVTkBfmSkJyDJJs2yy1yRRxhSQppp55y4bm5Hbl!-1724161176?oppId=42316&flag2006=false&mode=VIEW
Sponsor: National Oceanic and Atmospheric Administration
299 Foam Street
Monterey, CA 93940

NOAA Chesapeake Bay Office B-WET Grants 4323

The grant is a competitively based program that supports existing environmental education programs, fosters the growth of new programs, and encourages the development of partnerships among environmental education programs throughout the entire Chesapeake Bay watershed. Funded projects assist in meeting the Stewardship and Community Engagement goals of the Chesapeake 2000 Agreement. Specifically, projects support organizations that provide meaningful watershed educational experiences for students or related professional development for teachers. NCBO is encouraging applications that include innovative technologies in the delivery of these experiences.
Requirements: Eligible applicants are K-12 public and independent schools and school systems, institutions of higher education, community-based and nonprofit organizations, state or local government agencies, interstate agencies, and Indian tribal governments in the Chesapeake Bay watershed. The Department is strongly committed to broadening the participation of historically black colleges and universities, Hispanic serving institutions, tribal colleges and universities, and institutions that work in undeserved areas.
Restrictions: Federal agencies or institutions are not eligible to receive federal assistance under this funding program.
Geographic Focus: Delaware; District of Columbia; Maryland; New York; Pennsylvania; Virginia; West Virginia
Samples: National Environmental Education and Training Foundation, Washington, DC, $292,042—75 high school students with mentors in the natural resources field conduct research projects on environmental issues.; Highland School, Warrenton, VA, $5,000—for Hawks Get Wet, designed to improve the local watershed by converting retention ponds in Warrenton into active wetlands (2003).
Date(s) Application is Due: Oct 19
Contact: Peyton Robertson, Acting Director; (410) 267-5652 or (410) 267-5660; fax: (410) 267-5666; email: Peyton.Robertson@noaa.gov; Shannon Sprague, Education Coordinator; (410) 267-5664 or (410) 267-5660; fax: (410) 267-5666; email: Shannon.Sprague@noaa.gov
Internet: http://noaa.chesapeakebay.net/educationgrants.aspx
Sponsor: National Oceanic and Atmospheric Administration / NOAA Chesapeake Bay Office (NCBO)
410 Severn Avenue
Annapolis, MD 21403

NOAA Chesapeake Bay Office Cooperative Research Program Grants 4324

The Program is intended to formalize and expand collaborative research among the Chesapeake Bay's commercial fishing industry, marine scientists, and fishery management communities. The goal of this initiative is to enhance the data upon which fishery management decisions are made as well as to facilitate communication and collaboration among commercial and collaboration among commercial fishermen, scientists, and fishery managers. Through this initiative, the NCBO will develop a collaborative and cooperative program to set research priorities that meet management and fishing industry needs. Priority areas are modified slightly each year but may include stock assessment research and management, and multi-species research, management and fisheries ecosystem plan implementation, including ecosystem understanding, retrospective analyses, and ecosystem modeling. $100,000 to $150,000 in funding may be available for awards; there are no restrictions on minimum or maximum funding requests. Award periods may be up to a maximum of three years. Letters of Intent (LOI) are requested within 30 days and full proposals 30-45 days after LOI. Applicants must submit a full proposal by the deadline, regardless of whether they have heard back from NOAA following the LOI submission. The deadline for proposal submission is usually within the March/April time frame.
Requirements: Eligible applicants are institutions of higher education, other nonprofits, commercial organizations, foreign governments, organizations under the jurisdiction of foreign governments, international organizations, state, local and Indian tribal governments.
Restrictions: Federal agencies or institutions are not eligible to receive federal assistance under this funding program.
Geographic Focus: Delaware; District of Columbia; Maryland; New York; Pennsylvania; Virginia; West Virginia
Samples: University of Maryland, Center for Environmental Science, Cambridge, MD, $45,914—to measure age composition of summertime samples of blue crabs in support of growth rate estimates for juvenile age 0-1 bue crabs, a temperature-specific growth model for juvenile blue crabs, and estimation of age (2004); University of Maryland, College Park, MD, $147,300—for Striped Bass Stock Health Assessment: Mycobacteriosis Prevelance and Distribution (2003-05).
Contact: Derek Orner, Fishery Biologist; (410) 267-5676 or (410) 267-5660; fax: (410) 267-5666; email: Derek.Orner@noaa.gov; Shannon Sprague, Education Coordinator; (410) 267-5664 or (410) 267-5660; fax: (410) 267-5666; email: Shannon.Sprague@noaa.gov
Internet: http://noaa.chesapeakebay.net/CooperativeResearch.aspx
Sponsor: National Oceanic and Atmospheric Administration / NOAA Chesapeake Bay Office (NCBO)
410 Severn Avenue
Annapolis, MD 21403

NOAA Chesapeake Bay Office Fisheries Science Program Grants 4325

The Fisheries component of the Program is competitively based and supports research, monitoring, modeling and management addressing various aspects of Chesapeake Bay fisheries. Funded projects foster knowledge and understanding of the Chesapeake Bay ecosystem by providing biological information and life history characteristics for many individual Chesapeake Bay fisheries stocks, and by broadening the multispecies knowledge base for development of Fisheries Ecosystem Planning. Priority areas are modified slightly each year but may include stock assessment research and management, and multi-species research, management and fisheries ecosystem plan implementation, including ecosystem understanding, retrospective analyses, and ecosystem modeling. Letters of Intent (LOI) are requested within 30 days and full proposals 30-45 days after LOI. Applicants must submit a full proposal by the deadline, regardless of whether they have heard back from NOAA following the LOI submission. The deadline for proposal submission is usually within the March/April time frame.
Requirements: Eligible applicants are institutions of higher education, other nonprofits, commercial organizations, foreign governments, organizations under the jurisdiction of foreign governments, international organizations, state, local and Indian tribal governments.
Restrictions: Federal agencies or institutions are not eligible to receive federal assistance under this funding program.
Geographic Focus: Delaware; District of Columbia; Maryland; New York; Pennsylvania; Virginia; West Virginia
Samples: Chesterfield County Public Schools, Chesterfield, VA, $42,330—for intensive and sustained staff development that combined a meaningful experience for Chesterfield County teachers involved as well as specific guidance in the integration of outdoor learning experiences into the classroom curriculum (2003-05); National Aquarium Society, Washington, DC, $573,000 (2004).
Contact: Derek Orner, Fishery Biologist; (410) 267-5676 or (410) 267-5660; fax: (410) 267-5666; email: Derek.Orner@noaa.gov; Shannon Sprague, Education Coordinator; (410) 267-5664 or (410) 267-5660; fax: (410) 267-5666; email: Shannon.Sprague@noaa.gov
Internet: http://noaa.chesapeakebay.net/fisheriesgrants.aspx
Sponsor: National Oceanic and Atmospheric Administration / NOAA Chesapeake Bay Office (NCBO)
410 Severn Avenue
Annapolis, MD 21403

NOAA Chesapeake Bay Office Non-Native Oyster Research Grants 4326

The NCBO seeks proposals for projects that will provide scientific information to support a programmatic Environmental Impact Statement (EIS) on the proposed introduction of a non-native oyster species. The EIS is being prepared by federal and state agencies in response to a proposal by the states of Maryland and Virginia to introduce the non-native oyster, Crassostrea ariakensis, to the Chesapeake Bay and other tidal waters of Maryland and Virginia. It is expected that approximately $1.8M will be available through this research program. Funds will be used to support multi-year projects currently underway, as well as new projects that can be completed within a 12-18 month project period to fill remaining information gaps. The NCBO requests proposals for projects that will provide data and information in the following three areas: biological research and analysis; economic research and analysis; and evaluation of EIS alternatives. Proposals should exhibit familiarity with related work that is completed or ongoing. Proposals may be submitted for up to 3 years.
Requirements: Eligible applicants are institutions of higher education, other nonprofits, commercial organizations, foreign governments, organizations under the jurisdiction of foreign governments, international organizations, state, local and Indian tribal governments. No cost sharing is required under this program, however, the NCBO strongly encourages applicants to share as much of the project costs as possible. Funds from other Federal awards may not be considered matching funds.
Restrictions: Federal agencies or institutions are not eligible to receive Federal assistance under this notice.
Geographic Focus: Maryland; Virginia
Date(s) Application is Due: Mar 15
Contact: Derek Orner, Fishery Biologist; (410) 267-5676 or (410) 267-5660; fax: (410) 267-5666; email: Derek.Orner@noaa.gov; Jamie King, Specialist; (410) 267-5655; fax: (410) 267-5666; email: Jamie.King@noaa.gov
Internet: http://noaa.chesapeakebay.net/nonnativeoystergrants.aspx
Sponsor: National Oceanic and Atmospheric Administration / NOAA Chesapeake Bay Office (NCBO)
410 Severn Avenue
Annapolis, MD 21403

NOAA Chesapeake Bay Office Submerged Aquatic Vegetation (SAV) Culture and Restoration Grants 4327

The Submerged Aquatic Vegetation (SAV) component of the Chesapeake Bay Integrated Science Program is competitively-based and designed to enhance and increase the important fisheries habitat in the Chesapeake Bay and its tidal tributaries. This program focuses on areas including large-scale SAV planting, site assessments for future SAV planting, applied research on SAV seed ecology, enhanced capacity for SAV propagation, and integrated restoration (e.g., with oysters). Availability of funding for this program is included in the December NOAA Omnibus notice published in the Federal Register. Pre-proposals are typically due in January or February.
Requirements: Eligible applicants are institutions of higher education, other nonprofits, commercial organizations, foreign governments, organizations under the jurisdiction of foreign governments, international organizations, state, local and Indian tribal governments.
Restrictions: Federal agencies or institutions are not eligible to receive federal assistance under this funding program.
Geographic Focus: Delaware; District of Columbia; Maryland; New York; Pennsylvania; Virginia; West Virginia

Samples: Maryland Department of Natural Resources, Annapolis, MD, $472,859—for Large Scale Restoration of Eelgrass (Zostera marina) in the Patuxent River, Maryland (2003-06); Virginia Institute of Marine Science, Gloucester Point, VA, $24,718—A Comparative Test of Mechanized and Manual Planting of Eelgrass Seeds (2005).
Date(s) Application is Due: Mar 26
Contact: Derek Orner, Fishery Biologist; (410) 267-5676 or (410) 267-5660; fax: (410) 267-5666; email: Derek.Orner@noaa.gov; Shannon Sprague, Education Coordinator; (410) 267-5664 or (410) 267-5660; fax: (410) 267-5666; email: Shannon.Sprague@noaa.gov
Internet: http://noaa.chesapeakebay.net/aquaticvegetationgrants.aspx
Sponsor: National Oceanic and Atmospheric Administration / NOAA Chesapeake Bay Office (NCBO)
410 Severn Avenue
Annapolis, MD 21403

NOAA Climate Change Data and Detection (CCDD) Program Grants 4328
The goal of the program is to provide data and information management support to assure the availability of critical data sets for a variety of international programs and assessments. The data and resulting products extend the existing long-term climate record and, also, serve as essential input to predictive models. For the upcoming fiscal year, the CCDD program is soliciting proposals for research in the following areas: development of climate reference data sets; climate change detection and attribution studies, in joint sponsorship with the Department of Energy's Climate Change Prediction Program; and paleoclimate. While it is in the best interest of the applicants and their institutions to submit a Letter of Intent (LOI) by the July 23 due date, it is not a requirement; applicants who do not submit an LOI are allowed to submit a full proposal. For those that do submit an LOI, full proposals will be encouraged only for LOIs deemed relevant.
Requirements: Eligible applicants are institutions of higher education, other nonprofits, commercial organizations, international organizations, and state, local and Indian tribal governments.
Restrictions: Federal agencies or institutions are not eligible to receive Federal assistance under this notice.
Amount of Grant: $50,000-$200,000 per year
Date(s) Application is Due: Sep 24
Contact: Chris Miller; (301) 734-1241; fax: (301) 713-0518; email: christopher.d.miller@noaa.gov; Bill Murray; (301) 734-1243; fax: (301) 713-0518; email: william.l.murray@noaa.gov
Internet: http://www.climate.noaa.gov/cpo_pa/ccdd/
Sponsor: National Oceanic and Atmospheric Administration
1315 East West Highway, 12th Floor
Silver Spring, MD 20910

NOAA Climate Dynamics and Experimental Prediction Grants 4329
NOAA is soliciting proposals to initiate Climate Test Bed (CTB) projects under the Climate Dynamics and Experimental Prediction (CDEP) program. Due to limited available funding, the priority areas are focused on enhancing multi-model ensemble based forecasts, anchored by the NCEP Climate Forecast System, and on improving monthly-to-seasonal climate forecast products and applications, especially for drought. Priority areas include: climate forecast system improvements; generation and evaluation of multi-model ensemble forecasts; improving drought prediction; and enhancing operational drought forecast products and applications. Prospective investigator is strongly encouraged to submit a Letter of Intent (LOI) by the published deadline of July 23.
Requirements: Eligible applicants are institutions of higher education, other nonprofits, commercial organizations, international organizations, and state, local and Indian tribal governments.
Restrictions: Federal agencies or institutions are not eligible to receive Federal assistance under this notice.
Amount of Grant: $50,000-$200,000 per year
Date(s) Application is Due: Sep 24
Contact: Kenneth Mooney; (301) 734-1242; fax: (301) 713-0517; email: Kenneth.Mooney@noaa.gov
Internet: http://www.climate.noaa.gov/index.jsp?pg=/opportunities/opp_index.jsp&opp=2008/cdep_info.jsp
Sponsor: National Oceanic and Atmospheric Administration
1315 East West Highway, 12th Floor
Silver Spring, MD 20910

NOAA Climate Prediction Program for the Americas (CPPA) Grants 4330
The Program is a competitive research program with a goal to improve operational intraseasonal to interannual climate and hydrologic predictions for the Americas with quantified uncertainties sufficient for making informed decisions. To achieve its goal, CPPA has the following major objectives: to improve the predictive understanding and model simulation of ocean, atmosphere and land-surface processes; to quantify the sources and limits of predictability of climate variations on intra-seasonal to interannual time scale; to advance NOAA's operational climate forecasts, monitoring; and analysis systems and to develop climate-based hydrologic forecasting capabilities for decision support and water resource applications. Currently, CPPA is soliciting proposals under the following research areas: VOCALS (VAMOS Ocean-Cloud-Atmosphere-Land-Studies) Field Experiment; Hydrologic applications of NCEP multi-model seasonal hindcasts; and predictability and mechanisms of drought. While it is in the best interest of the applicants and their institutions to submit a Letter of Intent (LOI) by the July 23 due date, it is not a requirement; applicants who do not submit an LOI are allowed to submit a full proposal. For those that do submit an LOI, full proposals will be encouraged only for LOIs deemed relevant.
Requirements: Eligible applicants are institutions of higher education, other nonprofits, commercial organizations, international organizations, and state, local and Indian tribal governments.
Restrictions: Federal agencies or institutions are not eligible to receive Federal assistance under this notice.
Samples: Arthur V. Douglas, Creighton University—The warm water pools of the western Caribbean and eastern tropical Pacific: Their influence on the predictability of intraseasonal rainfall regimes in Mexico and basin wide tropical storm activity (2007); Michael W. Douglas, National Severe Storms Laboratory—Explaining the spatial variability of the mid-summer drought over the Inter-American Seas region (2007).
Amount of Grant: $50,000-$200,000 per year
Date(s) Application is Due: Sep 24
Contact: Jin Huang; (301) 734-1226; fax: (301) 713-0517; email: Jin.Huang@noaa.gov; Annarita Mariotti; (301) 734-1237; fax: (301) 713-0517; email: Annarita.Mariotti@noaa.gov
Internet: http://www.climate.noaa.gov/cpo_pa/cppa/
Sponsor: National Oceanic and Atmospheric Administration
1315 East West Highway, 12th Floor
Silver Spring, MD 20910

NOAA Climate Variability and Human Health Program (CVHH) Grants 4331
The Program is designed to take a problem-oriented approach to the use of climate and environmental information for social benefit. The goal of the program is to provide scientifically sound, socially relevant and technically usable climate and environmental information for the public health community. The CVHH program is comprised of four main components: research; application; capacity building and training; and community building. In collaboration with other agencies, academia, international institutions, and regional partners, the CVHH program supports training and other activities that build capacity to use climate information for public health purposes. The program also focuses on building a robust climate and health community, helping to bring together various communities and disciplines to foster an iterative definition of the problem and collective development of approaches and solutions.
Requirements: All institutions eligible to receive support from NOAA, EPA, NASA, and NSF are able to apply. Universities, nonprofit organizations, for profit organizations, state and local governments, and Indian tribes are included among entities eligible for funding under this announcement.
Restrictions: Funding for foreign institutions is not available. Federal agencies or institutions are not eligible to receive Federal assistance under this notice.
Samples: Laura C. Harrington and Arthur T. DeGaetano, Cornell University—Climate effects, West Nile virus vector development, and transmission risk; Patrick L. Kinney, Columbia University—Climate Variability, Air Quality and Human Health: Measuring Regional Vulnerability for Improved Decision-Making.
Amount of Grant: $50,000-$250,000
Date(s) Application is Due: Feb 27; Apr 23
Contact: Diane Brown, Grants Administration Team Leader; (301) 734-1206; fax: (301) 713-0518; email: diane.brown@noaa.gov
Internet: http://www.ogp.noaa.gov/mpe/csi/cvhh/index.htm
Sponsor: National Oceanic and Atmospheric Administration
1315 East West Highway, 12th Floor
Silver Spring, MD 20910

NOAA Climate Variability and Predictability (CVP) Program Grants 4332
The program seeks to understand, through modeling and diagnostic studies, large-scale patterns of climate variability and their teleconnections on seasonal to decadal time scales. The program aims to address climate variability and predictability over the United States, recognizing that this may require knowledge of climate variability in other regions of the globe. Goals include: identifying and understanding the major patterns of climate variability on seasonal, decadal and longer time scales and evaluating their predictability; expanding capacity in short term (seasonal to interannual) climate predictability and searching for ways to predict decadal variability; better documenting the record of rapid climate changes and the mechanisms for these events, and evaluating the potential for abrupt climate changes in the future; evaluating and enhancing the models used to project climate change due to human activity, including anthropogenically induced changes in atmospheric composition; and detecting and describing any global climate changes that may occur. Currently, CVP is soliciting proposals that specifically address strategies for correcting tropical biases in coupled climate models to advance the understanding and prediction of climate. While it is in the best interest of the applicants and their institutions to submit a Letter of Intent (LOI) by the July 23 due date, it is not a requirement; applicants who do not submit an LOI are allowed to submit a full proposal. For those that do submit an LOI, full proposals will be encouraged only for LOIs deemed relevant.
Requirements: Eligible applicants are institutions of higher education, other nonprofits, commercial organizations, international organizations, and state, local and Indian tribal governments.
Restrictions: Federal agencies or institutions are not eligible to receive Federal assistance under this notice.
Amount of Grant: $50,000 - $200,000 per year
Date(s) Application is Due: Sep 24
Contact: James Todd, Grants Manager; (301) 734-1258; fax: (301) 713-0518; email: james.todd@noaa.gov

Internet: http://www.climate.noaa.gov/index.jsp?pg=./opportunities/opp_index.jsp&opp=info_cvp.jsp
Sponsor: National Oceanic and Atmospheric Administration
1315 East West Highway, 12th Floor
Silver Spring, MD 20910

NOAA Coastal and Estuarine Land Conservation Program Grants 4333
The purpose of this document is to advise eligible coastal states and territories (requirements described below) that OCRM is soliciting coastal and estuarine land conservation project proposals for competitive funding under the CELCP. This announcement solicits proposals for land acquisition projects (fee simple interest or conservation easements) that can be completed within 18 months from the start date of the award and that have the purpose of protecting important coastal and estuarine areas. NOAA may extend project grants up to an additional 18 months (for a total project period of 3 years maximum) if circumstances warrant.
Requirements: Only states and territories with Coastal Zone Management Programs or National Estuarine Research Reserves approved under the CZMA that have submitted a draft CELCP plan to NOAA on or before May 1, 2008, are eligible to participate in this funding competition. The eligible states and territories are: Alabama, Alaska, American Samoa, California, Connecticut, Delaware, Florida, Georgia, Guam, Hawaii, Indiana, Louisiana, Maine, Maryland, Massachusetts, Michigan, Mississippi, New Hampshire, New Jersey, New York, North Carolina, Northern Mariana Islands, Ohio, Oregon, Pennsylvania, Puerto Rico, Rhode Island, South Carolina, Texas, Virginia, Washington, and Wisconsin.
Amount of Grant: $3,000,000 maximum
Date(s) Application is Due: Jul 23
Contact: Elaine Vaudreuil; (301) 713-3155, ext. 103; email: Elaine.Vaudreuil@noaa.gov; Elisabeth Morgan; (301) 713-3155, ext. 166; email: Elisabeth.Morgan@noaa.gov
Internet: http://www07.grants.gov/search/search.do;jsessionid=LQCZGLyvwpgGPznL6jnLRyjK1vpGyTRsfyCJKLZtfJnqhLHphMnJ!280595828?oppId=41800&flag2006=false&mode=VIEW
Sponsor: National Oceanic and Atmospheric Administration
1315 East-West Highway
Silver Spring, MD 20910

NOAA Coastal Ecosystem Research Project 4334
This program solicits 1-year and 2-year proposals for modeling, monitoring, and retrospective studies of coastal ecosystem research in the Northern Gulf of Mexico. Modeling studies are needed to provide a framework for studies in the northern Gulf of Mexico. Modeling activities will be used to guide further program development and identify important processes for the extensive fieldwork anticipated to follow this preliminary phase. Modeling studies may include: models that simulate impacts of varying nutrient flux on productivity and trophic response in the northern Gulf of Mexico ecosystem, including impacts to and responses of commercially and recreationally important fisheries; NPZF models; physical-biological coupled models of processes in the Gulf ecosystem influenced by the Mississippi River discharge, including transport and population dynamics of key zooplankton and fishery populations; models of oceanographic and climate influences on nutrients and their impact on Gulf productivity; models of biogeochemical cycling of nutrients within the Gulf and its relationship to the dynamics of organic carbon flux in the Gulf; and models of water column stability and hypoxic zone dynamics. This announcement requests full proposals only.
Amount of Grant: $25,000-$1 million range
Date(s) Application is Due: Mar 14
Contact: Rob Magnien, Director; (301) 713-3338 or (301) 713-3020; fax: (301) 713-4044; email: rob.magnien@noaa.gov
Internet: http://frwebgate2.access.gpo.gov/cgi-bin/getdoc.cgi?dbname=2001_register&docid=fr17ja01-56
Sponsor: National Oceanic and Atmospheric Administration
1305 East-West Highway, N/SCI2 SSMC4, Room 8243
Silver Spring, MD 20910

NOAA Coastal Hypoxia Research Program (CHRP) Grants 4335
The purpose of this document is to advise the public that NOAA is soliciting proposals for projects of 2 to 5 years in duration that advance understanding, predicting, and managing the causes and ecological and economic impacts of hypoxia in representative coastal ecosystems. Proposals for funding under this announcement will be accepted for hypoxia research in U.S. Great Lakes, estuaries, and coastal ocean regions, exclusive of the dead zone in the Gulf of Mexico over the Louisiana continental shelf. Funding is contingent upon the availability of annual Federal appropriations.
Requirements: Institutions of higher education, not-for- profit institutions, State, commercial organizations, local and Indian tribal governments and Federal agencies are eligible to apply.
Date(s) Application is Due: Sep 11
Contact: Rob Magnien, Director; (301) 713-3338 or (301) 713-3020; fax: (301) 713-4044; email: rob.magnien@noaa.gov
Internet: http://www.cop.noaa.gov/opportunities/grants/fundingarchive/fy2007/CHRP_FFO_0606.html
Sponsor: National Oceanic and Atmospheric Administration
1305 East-West Highway, N/SCI2 SSMC4, Room 8243
Silver Spring, MD 20910

NOAA Coastal Management Fellowships 4336
The Coastal Management Fellowship was established in 1996 to provide on-the-job education and training opportunities in coastal resource management and policy for postgraduate students and to provide project assistance to state coastal zone management programs. The program matches postgraduate students with state coastal zone programs to work on projects proposed by the state and selected by the fellowship sponsor, the National Oceanic and Atmospheric Administration (NOAA) Coastal Services Center. This two-year opportunity offers a competitive salary, medical benefits, and travel and relocation expense reimbursement. Proposals due from the states on October 17, and applications are due from candidates on January 26.
Date(s) Application is Due: Jan 26
Contact: Margaret VanderWilt , Fellowship Coordinator; (843) 740-1273; fax: (843) 740-1224; email: margaret.vanderwilt@noaa.gov or csc.fellowships@noaa.gov
Internet: http://www.csc.noaa.gov/cms/fellows.html
Sponsor: National Oceanic and Atmospheric Administration
2234 South Hobson Avenue
Charleston, SC 29405

NOAA Coastal Ocean Program Grants 4337
This program is designed to provide predictive capability for managing coastal ecosystems. Grants may be used to fund research and inter-agency initiatives under specific COP programs related to NOAA's mission in coastal ecosystem oceanography, cumulative coastal impacts, and coastal forecasting/natural hazards. The program supports research on critical issues which exist in the nation's estuaries, coastal waters, and the Great Lakes; and translates its findings into accessible information for coastal managers, planners, lawmakers, and the public. Grant periods may last from one to three years with funding approved annually. Deadlines are announced each year.
Requirements: Institutions of higher education, not-for- profit institutions, State, commercial organizations, local and Indian tribal governments and Federal agencies are eligible to apply.
Amount of Grant: $25,000-$1 million range
Contact: Rob Magnien, Director; (301) 713-3338 or (301) 713-3020; fax: (301) 713-4044; email: rob.magnien@noaa.gov
Internet: http://12.46.245.173/cfda/CFDA_Static/Programs/11_478.html
Sponsor: National Oceanic and Atmospheric Administration
1305 East-West Highway, Room 8243
Silver Spring, MD 20910

NOAA Collaborative Science, Technology, and Applied Research Grants 4338
Through the Collaborative Science, Technology, and Applied Research (CSTAR) program, the NWS provides funding to the university community for highly collaborative applied research activities. The CSTAR program creates a cost-effective transition from basic and applied research to operations and services. This service-science linkage provides a foundation for the ongoing infusion of science and technology into NWS operations. The long-term objective of the CSTAR program is to improve the overall forecast and warning capabilities of the operational hydrometeorological community by addressing the following national science priorities: quantitative precipitation estimation (QPE) and forecasting (QPF), including precipitation type and probabilistic QPF (PQPF); flash flood and probabilistic river prediction; prediction of seasonal-to-interannual and decadal climate variability, and the impacts of these variabilities on extreme weather events; prediction of tropical cyclones near landfall, including track, intensity, and associated precipitation, and hazardous weather; prediction of marine conditions, including fog, winds, coastal ocean, and open ocean waves; the effect of topography and other surface forcing on local weather regimes; locally hazardous weather, especially severe convection, winter weather, and phenomena that affect aviation; and conditions conducive for the rapid development of wildfires and the dispersion of smoke and other air-quality hazards. Individual NWS regions and centers have a subset of these science priorities due to differences in factors such as topography, weather regimes, and mission.
Requirements: All accredited U.S. colleges and universities, including federally funded educational institutions, such as the Naval Postgraduate School, are eligible.
Amount of Grant: $100,000-$125,000
Contact: Sam Contorno, Federal Program Officer; (301) 713-3557, ext. 150; fax: (301) 713-1253; email: samuel.contorno@noaa.gov
Internet: http://www.epa.gov/EPA-IMPACT/2003/January/Day-07/i224.htm
Sponsor: National Oceanic and Atmospheric Administration / National Weather Service
1315 East-West Highway, Room 15330
Silver Spring, MD 20910

NOAA Communicating Hurricane Information Grants 4339
This solicitation aims to advance basic research in the social, behavioral and economic sciences related to the communication of hurricane outlooks, forecasts, watches, and warnings both to decision makers (i.e., emergency managers, elected officials) and to the general public. Although all proposals addressing this topic will be considered, investigators are urged to consider specific communication concerns: how messages are received and understood; and what the linkages are between communications and behavior. The role of the physical scientist or engineer on each project is to provide expertise about the state of hurricane forecast technologies and what is likely to be available in the near future. This interaction between the social scientists and the physical scientist or engineer should continue throughout the duration of the project. The Program anticipates making approximately a total of 6-12 awards for research projects. These awards will be for 1 to 3 years.

Requirements: Eligible research proposals must have at least one investigator from the social, behavioral, or economic sciences, and at least one from the physical sciences or engineering.
Amount of Grant: $400,000 per project maximum
Date(s) Application is Due: Jun 3
Contact: Robert E. O'Connor, Directorate for Social, Behavioral and Economic Sciences; (703) 292-7263; fax: (703) 292-9068; email: roconnor@nsf.gov
Internet: http://www07.grants.gov/search/synopsis.do;jsessionid=HYTSYzGB611pwv pt7Dx3cqnZjD2lJsFhsGLMd6v2XS8TJcVsWY6j!1282164016
Sponsor: National Oceanic and Atmospheric Administration / National Science Foundation
4201 Wilson Boulevard
Arlington, VA 22230

NOAA Community-Based Habitat Restoration Program (CRP) Grants 4340
The Program funds grass-roots marine and coastal habitat restoration projects that will benefit anadromous fish species, commercial and recreational resources, and endangered and threatened species. The program invites the public to submit proposals for available funding to implement grass-roots habitat restoration projects that will benefit living marine resources, including anadromous fish. The Federal Funding Opportunity describes the conditions under which applications (project proposals) will be accepted under the CRP and describes criteria under which applications will be evaluated for funding consideration. Projects funded through the CRP will be expected to have strong on-the-ground habitat restoration components that provide educational and social benefits for people and the communities in addition to long-term ecological habitat improvements for NOAA trust resources. Proposals selected for funding through this solicitation will be implemented through a project grant or cooperative agreement mechanism.
Requirements: Eligible applicants are institutions of higher education, hospitals, other non-profits, commercial (for profit) organizations, organizations under the jurisdiction of foreign governments, international organizations, and state, local and Indian tribal governments whose projects have the potential to benefit NOAA trust resources.
Restrictions: Applications from federal agencies or employees of Federal agencies will not be considered.
Geographic Focus: Alabama; Alaska; California; Connecticut; Delaware; District of Columbia; Florida; Georgia; Idaho; Illinois; Indiana; Louisiana; Maine; Maryland; Massachusetts; Michigan; Mississippi; New England; New Hampshire; New Jersey; New York; North Carolina; Ohio; Oregon; Pennsylvania; Rhode Island; South Carolina; Texas; Vermont; Virginia; Washington; West Virginia; Wisconsin
Amount of Grant: $30,000-$250,000
Date(s) Application is Due: Sep 27
Contact: Cathy Bozek; (301) 713-0174; fax: (301) 713-0184; email: Cathy.Bozek@noaa.gov
Internet: http://www.nmfs.noaa.gov/habitat/restoration/projects_programs/crp/partners_funding/callforprojects.html
Sponsor: National Oceanic and Atmospheric Administration
1315 East-West Highway, 12th Floor
Silver Spring, MD 20910

NOAA Community-Based Marine Debris Prevention and Removal Grants 4341
The Program provides funding to catalyze the implementation of locally driven, community-based marine debris prevention and removal projects that will benefit coastal habitat, waterways, and NOAA trust resources including diadromous fish. Projects funded through the MDP have strong on-the-ground habitat components involving the removal of marine debris and derelict fishing gear that will provide educational and social benefits for people and their communities in addition to long-term ecological habitat improvements for NOAA trust resources. The principal objective of the Program is to provide federal financial and technical assistance to grass-roots, community-based activities that improve living marine resource habitats through the removal of marine debris and the prevention of debris entering the marine environment, and that promote stewardship and a conservation ethic for NOAA trust resources. Generally, costs that are allowable include salaries and equipment and supplies (as long as these are necessary and reasonable specifically for the purpose of the award).
Requirements: Eligible applicants are institutions of higher education, other non-profits, commercial (for profit) organizations, organizations under the jurisdiction of foreign governments, international organizations, and state, local and Indian tribal governments whose projects have the potential to benefit NOAA trust resources. Applicants are encouraged to demonstrate a minimum 1:1 non-Federal match for MDP funds requested to conduct the proposed project.
Restrictions: Applications from federal agencies or employees of Federal agencies will not be considered.
Amount of Grant: $15,000 to $250,000
Date(s) Application is Due: Oct 31
Contact: David Landsman; (301) 713-0174; email: David.Landsman@noaa.gov
Internet: http://www.nmfs.noaa.gov/habitat/restoration/projects_programs/crp/partners_funding/callforprojects2.html
Sponsor: National Oceanic and Atmospheric Administration
1315 East-West Highway, 12th Floor
Silver Spring, MD 20910

NOAA Cooperative Institute for Applied Meteorological Studies Grants 4342
CIAMS is affiliated with the Fort Worth based headquarters office of the Southern Region of the National Weather Service. The Institute employs research scientists and graduate students in a broad program of applied research and service to Texas and surrounding states, in agricultural meteorology, marine meteorology and air-sea interactions over the Gulf of Mexico, lightning and severe weather, and Doppler radar studies from the installations in the Texas A&M Department of Atmospheric Sciences and at the Houston Weather Forecast Office. The NWS is responsible for the overall identification of tasks needed to be addressed and funds (through cooperative agreements) for faculty and student work towards those goals. Texas A&M University also provides some funding, as well as computer and observational facilities, to help support faculty and students' work on projects. Awards are for a 12-month period. In some instances, award periods may extend beyond 12 months. There are no deadline dates.
Requirements: Eligible applicants are Sole Source programs with Texas A&M University and public- and state-supported institutions.
Amount of Grant: $100,000-$125,000
Contact: Sam Contorno, Federal Program Officer; (301) 713-3557, ext. 150; fax: (301) 713-1253; email: samuel.contorno@noaa.gov
Internet: http://atmo.tamu.edu/ciams/
Sponsor: National Oceanic and Atmospheric Administration
1315 East-West Highway, Room 15330
Silver Spring, MD 20910

NOAA Cooperative Institute for Tropical Meteorology Research Grants 4343
The NOAA Cooperative Institute for Tropical Meteorology (CITM) is affiliated with the Department of Meteorology at Florida State University. The Program offers cooperative research opportunities for graduate level students. This cooperative research has an emphasis on hurricane studies, flash floods, tropical convection, and local weather. Student participation is a large component of the program. Florida State University also provides some funding, as well as computer and observational facilities, to help support faculty and students' work on projects. Awards are for a 12-month period. In some instances, award periods may extend beyond 12 months. There are no deadline dates.
Requirements: Eligible applicants are Sole Source programs with Florida State University and public- and state-supported institutions.
Amount of Grant: $100,000-$125,000
Contact: Sam Contorno, Federal Program Officer; (301) 713-3557, ext. 150; fax: (301) 713-1253; email: samuel.contorno@noaa.gov
Internet: http://www.met.fsu.edu/index.pl/acprograms/gradprogram
Sponsor: National Oceanic and Atmospheric Administration / National Weather Service
1315 East-West Highway, Room 15330
Silver Spring, MD 20910

NOAA Cooperative Institute: Eastern U.S. Continental Shelf Frontier Exploration, Research, and Technology Development Grants 4344
The purpose of this announcement is to invite the submission of proposals to establish a CI for the eastern U.S. Continental Shelf frontier exploration, research, and technology development, and to provide details on the application, review, and selection process. This CI will give NOAA the benefit of working with complementary capabilities at one or more research institutions that contribute to NOAA's ocean exploration, research, and technology priorities on the eastern U.S. Continental Shelf. CI Concept/Program Background A CI is a NOAA-supported, non-Federal organization that has established an outstanding research program in one or more areas that are relevant to the NOAA mission to understand and predict changes in the Earth's environment and conserve and manage coastal and marine resources to meet our Nation's economic, social, and environmental needs.
Requirements: Eligibility is limited to non-Federal public and private non-profit universities, colleges and research institutions that offer accredited graduate level degree-granting programs in NOAA- related sciences.
Amount of Grant: $150,000-$2.5 million
Date(s) Application is Due: Oct 6
Contact: Dr. John Cortinas; (301) 734-1090; email: John.Cortinas@noaa.gov
Internet: http://www.grants.gov/search/search.do;jsessionid=LP0hqWfp11PVPKSr9y fL1LLhNQTn2gsNBmYzZtzGKTQhJRdRBYxW!-2005900814?oppId=42235&flag 2006=false&mode=VIEW
Sponsor: National Oceanic and Atmospheric Administration
1315 East-West Highway, Room 11326
Silver Spring, MD 20910

NOAA Cooperative Research Program Grants 4345
The CRP program provides financial assistance for projects that seek to increase and improve the working relationship between researchers from the NMFS, state fishery agencies, universities, and fishermen in the Gulf of Mexico and off the South Atlantic states of North Carolina, South Carolina, Georgia, and Florida involving the U.S. fishing industry (recreational and commercial). The program is a means of involving commercial and recreational fishermen in the collection of fundamental fisheries information. Collection efforts support the development and evaluation of management and regulatory options.
Requirements: Eligible applicants may be institutions of higher education, nonprofits, commercial organizations, individuals, and state, local, and Indian tribal governments.
Restrictions: Federal agencies or institutions are not eligible. Foreign governments, organizations under the jurisdiction of foreign governments, and international organizations are also excluded for purposes of this solicitation.
Amount of Grant: $25,000-$400,000
Date(s) Application is Due: Sep 9

Contact: Robert Sadler, State/Federal Liaison Office; (727) 824-5324; Robert.Sadler@noaa.gov
Internet: http://www07.grants.gov/search/search.do;jsessionid=LQCZGLyvwpgGPznL6jnLRyjK1vpGyTRsfyCJKLZtfJnqhLHphMnJ!280595828?oppId=42299&flag2006=false&mode=VIEW
Sponsor: National Oceanic and Atmospheric Administration
263 13th Avenue South
Saint Petersburg, FL 33702

NOAA Cooperative Unit for Fisheries Education and Research Internships **4346**
The Cooperative Unit for Fisheries Education and Research (CUFER) was established in 1992 within the Cooperative Institute for Marine and Atmospheric Studies (CIMAS) of the Rosenstiel School of Marine and Atmospheric Sciences (RSMAS) of the University of Miami (UM) in response to a need for the development of methods for improved quantitative assessment of fish populations and as a source of advice for resource sustainability. These activities are envisioned to complement NOAA Fisheries mission of resource stewardship. CUFER is designed to attract bright and skilled people to a program that encourages students and scientists with quantitative interests to apply themselves to collaborative research projects on quantitative fisheries science and marine resource modeling. Check the web site for available internships.
Contact: David Die; (305) 421-4607; fax: (305) 361-4457; email: ddie@rsmas.miami.edu
Internet: http://cufer.rsmas.miami.edu/wordpress/?page_id=77
Sponsor: National Oceanic and Atmospheric Administration
4600 Rickenbacker Causeway
Miami, FL 33149

NOAA Coral Reef Ecosystem Studies (CRES) Grants **4347**
The purpose of this program is to advise the public that NOAA/NOS/NCCOS/CSCOR is soliciting proposals for projects of 3-5 years in duration for the Coral Reef Ecosystem Studies Program (CRES) and up to 3 years in duration for the Deep Coral Reef Ecosystem Studies Program (Deep-CRES). The development of predictive models (e.g., bio-physical models to investigate larval transport of reef organisms and their recruitment to reef systems in the context of variable oceanographic conditions; water quality models to investigate the relationship between watershed-based pollutant inputs and effects on reef ecosystems; economic models to investigate the relationship between coral reef health and local economies) as a component of these studies is required. Scientific information, syntheses, models, and ecosystem forecasting capabilities from this multi-disciplinary, long-term effort should be clearly usable by resource managers to make more informed decisions on managing U.S. coral reef ecosystems. Funding is contingent upon the availability of Fiscal Year Federal appropriations.
Requirements: Institutions of higher education, not-for- profit institutions, State, commercial organizations, local and Indian tribal governments and Federal agencies are eligible to apply.
Date(s) Application is Due: Nov 13
Contact: Rob Magnien, Director; (301) 713-3338 or (301) 713-3020; fax: (301) 713-4044; email: rob.magnien@noaa.gov
Internet: http://www.cop.noaa.gov/opportunities/grants/fundingarchive/fy2007/CRES_FFO_0606.html
Sponsor: National Oceanic and Atmospheric Administration
1305 East-West Highway, N/SCI2 SSMC4, Room 8243
Silver Spring, MD 20910

NOAA CRCP Coral Reef Ecosystem Monitoring Grants **4348**
The Grant Program, as authorized under the Coral Reef Conservation Act of 2000, provides matching grants to Governor-appointed point of contact agencies for the jurisdictions of Puerto Rico, the U.S. Virgin Islands, Florida, Hawaii, American Samoa, Guam, the Commonwealth of the Northern Mariana Islands, the Republic of Palau, the Federated States of Micronesia (including Chuuk, Yap, Kosrae, and Pohnpei), and the Republic of the Marshall Islands to support State and territory coral reef monitoring activities. The current closing date for applications is February 22, with pre-applications due on November 6. The expected number of awards is 15, with total funding equaling $1,100,000.
Requirements: The island jurisdictions of Puerto Rico, the U.S. Virgin Islands, Florida, Hawaii, American Samoa, Guam, the Commonwealth of the Northern Mariana Islands, the Republic of Palau, the Federated States of Micronesia (including Chuuk, Yap, Kosrae, and Pohnpei), and the Republic of the Marshall Islands are eligible to apply.
Amount of Grant: $73,333 average; $1,100,000 total
Date(s) Application is Due: Nov 6; Feb 22
Contact: Steve Drescher, Policy Advisor; (301) 713-0942, ext.122; fax: (301) 713-0947; email: Steve.J.Drescher@noaa.gov
Internet: http://www07.grants.gov/search/search.do;jsessionid=HWbpMDzmZ104MfvCGSbHhsGGPnb8dGTRG2hM0vfdTvpFRTZdkPyK!-796138783?oppId=14663&flag2006=true&mode=VIEW
Sponsor: National Oceanic and Atmospheric Administration
1315 East West Highway, 12th Floor
Silver Spring, MD 20910

NOAA CSCOR Regional Ecosystem Prediction Program: Cumulative Impacts of Stressors at the Land-Water Interface in the Mid-Atlantic **4349**
Proposals are solicited for projects of up to 5 years in duration for a regional-scale ecosystem research study in the Mid-Atlantic investigating the cumulative impacts of multiple stressors at the land-water interface of estuaries and bays on recreationally, economically or ecologically important living resource populations and communities. This research is needed to support management efforts to protect and restore submerged and wetland plant communities at the land-water interface of Mid-Atlantic estuaries and bays and species that depend upon these critical habitats including finfish, shellfish, amphibians, reptiles and birds. Generally, costs that are allowable include salaries, equipment, supplies, and training, as long as these are necessary and reasonable.
Requirements: Eligible applicants are: institutions of higher education; other non-profits; states; local governments; commercial organizations; and Federal agencies that possess the statutory authority to receive financial assistance.
Restrictions: The Program will not fund any Federal Full Time Employee (FTE) salaries.
Amount of Grant: $500,000 - $1,000,000 per year
Date(s) Application is Due: Oct 15
Contact: Elizabeth Turner, Program Manager; (603) 862-4680; email: elizabeth.turner@noaa.gov
Internet: http://www07.grants.gov/search/search.do;jsessionid=HXXNv3Mt6rGXD7y5CX8dvQqGkVgGF8H6Hw5BGwTw1BFBfBQs3fdn!1282164016?oppId=14667&flag2006=true&mode=VIEW
Sponsor: National Oceanic and Atmospheric Administration
1315 East West Highway, 12th Floor
Silver Spring, MD 20910

NOAA CSCOR Regional Ecosystem Prediction Program: Ecosystem Goal-Setting in Coastal Waters and Reefs of South Florida **4350**
Two- to three-year proposals are solicited to develop, undertake and conclude a consensus-building process that results in scientifically-based quantifiable goals for aquatic resources and habitats of the Florida Bay and Keys that are of concern to or fall within the legal mandates of management agencies at the federal, state and local level. Funding instruments available are project grants and cooperative agreements. Generally, costs that are allowable include salaries, equipment, supplies, and training, as long as these are necessary and reasonable.
Requirements: Eligible applicants are: institutions of higher education; other non-profits; states; local governments; commercial organizations; and Federal agencies that possess the statutory authority to receive financial assistance.
Restrictions: The Program will not fund any Federal Full Time Employee (FTE) salaries.
Amount of Grant: $500,000 per year maximum
Date(s) Application is Due: Oct 29
Contact: Larry Pugh, Program Manager; (301) 713-3338, ext. 160; email: larry.pugh@noaa.gov
Internet: http://www07.grants.gov/search/search.do;jsessionid=HXXNv3Mt6rGXD7y5CX8dvQqGkVgGF8H6Hw5BGwTw1BFBfBQs3fdn!1282164016?oppId=14667&flag2006=true&mode=VIEW
Sponsor: National Oceanic and Atmospheric Administration
1315 East West Highway, 12th Floor
Silver Spring, MD 20910

NOAA CSCOR Regional Ecosystem Prediction Program: Invasive Species in the Great Lakes—A Regional Scale Approach **4351**
Within an ecosystem framework impacted by invasive species, proposals are solicited for projects up to 5 years in duration for a regional-scale ecosystem research study to develop environmental forecasting tools for assessing and predicting changes in water quality, habitat, living resources and consequences to the food web of the Great Lakes. Proposals must address this issue in a regional context (all or most of the Great Lakes) and be of obvious utility to managing natural resources in the face of current invasive populations. Proposals must include justification for the questions addressed, in terms of environmental management importance, linkage to invasive species and need for a regional ecosystem construct. Proposals must also provide detailed descriptions of the improved quantitative understanding to be achieved and predictive capability to be developed, identify management issue(s) that are being addressed and specify how results could change current management practices. Funding instruments available are project grants and cooperative agreements. Generally, costs that are allowable include salaries, equipment, supplies, and training, as long as these are necessary and reasonable.
Requirements: Eligible applicants are: institutions of higher education; other non-profits; states; local governments; commercial organizations; and Federal agencies that possess the statutory authority to receive financial assistance.
Restrictions: The Program will not fund any Federal Full Time Employee (FTE) salaries.
Amount of Grant: $500,000 - $1,000,000 per year
Date(s) Application is Due: Oct 1
Contact: Felix Martinez, Program Manager; (301) 713-3338, ext. 153; email: felix.martinez@noaa.gov
Internet: http://www07.grants.gov/search/search.do;jsessionid=HXXNv3Mt6rGXD7y5CX8dvQqGkVgGF8H6Hw5BGwTw1BFBfBQs3fdn!1282164016?oppId=14667&flag2006=true&mode=VIEW
Sponsor: National Oceanic and Atmospheric Administration
1315 East West Highway
Silver Spring, MD 20910

NOAA Cumulative Impacts of Multiple Stressors (MultiStress) Grants **4352**
The purpose of this program is to advise the public that NOAA/NOS/NCCOS/CSCOR is soliciting proposals for projects of up to 5 years in duration to investigate the impacts of multiple stressors in coastal ocean ecosystems, including estuaries and the Great Lakes.

These projects should be interdisciplinary, multiple investigator, and well integrated studies designed to develop capabilities for understanding, predicting, and managing the effects of multiple stressors (both anthropogenic and natural) in coastal ecosystems. Funding is contingent upon the availability of annual Federal appropriations.
Requirements: Institutions of higher education, not-for- profit institutions, State, commercial organizations, local and Indian tribal governments and Federal agencies are eligible to apply.
Date(s) Application is Due: Oct 23
Contact: Rob Magnien, Director; (301) 713-3338 or (301) 713-3020; fax: (301) 713-4044; email: rob.magnien@noaa.gov
Internet: http://www.cop.noaa.gov/opportunities/grants/fundingarchive/fy2007/Multistress_FFO_0606.html
Sponsor: National Oceanic and Atmospheric Administration
1305 East-West Highway, N/SCI2 SSMC4, Room 8243
Silver Spring, MD 20910

NOAA Dr. Nancy Foster Scholarships 4353
The Scholarship Program provides support for outstanding scholarship and encourages independent graduate-level research in oceanography, marine biology, or maritime archaeology, particularly by women and members of minority groups. Dr. Nancy Foster Scholarships carry a 12-month stipend for each student of $20,000 and an annual cost-of-education allowance of up to $12,000 and up to $20,000 support for a four to six week research collaboration at a NOAA facility. Masters students may be supported for up to two years, and doctoral students for up to four years. About four scholarships are awarded each year. The annual stipend is paid directly to the scholar.
Requirements: Those eligible to apply are United States citizens currently pursuing or intending to pursue a masters or doctoral level degree in oceanography, maritime archaeology (including the curation, preservation, and display of maritime artifacts) or marine biology, (including all science, engineering, and resource management of ocean and coastal areas). Studies must be conducted on a full-time basis.
Amount of Grant: $52,000 maximum per year
Date(s) Application is Due: Feb 8
Contact: Pritti Brahma; (301) 713-9437, ext. 118; email: priti.brahma@noaa.gov
Internet: http://fosterscholars.noaa.gov/aboutscholarship.html
Sponsor: National Oceanic and Atmospheric Administration
1315 East-West Highway, Room 10725
Silver Spring, MD 20910

NOAA Ecological Forecasting (ECOFORE) Grants 4354
The NOAA/NOS/NCCOS/CSCOR is soliciting proposals for projects of 2 to 5 years in duration for the development of new ecological forecasting capabilities and the transition of existing ecological forecasts to operational status. These ecological forecasts are to support critical management decisions for the management of the Nation's Great Lake , estuarine, coastal and ocean ecosystems. The EcoFore program seeks to develop capabilities to provide coastal managers with state-of-the-art science in a form that they can utilize to make proactive management decisions in an ecosystem context. Proposals should provide detailed descriptions of the predictive/forecasting capability to be developed, management issue(s) to be addressed, the target audience or users of developed forecasts, and how forecasts will be provided to users. Explicit identification of the end user group(s) is required.
Requirements: Institutions of higher education, not-for- profit institutions, State, commercial organizations, local and Indian tribal governments and Federal agencies are eligible to apply.
Date(s) Application is Due: Nov 18
Contact: Elizabeth Turner; (603) 862-4680 or (301) 713-3020; fax: (301) 713-4044; email: Elizabeth.Turner@noaa.gov
Internet: http://www.cop.noaa.gov/opportunities/grants/fundingarchive/fy2006.html
Sponsor: National Oceanic and Atmospheric Administration
1305 East-West Highway, N/SCI2 SSMC4
Silver Spring, MD 20910

NOAA Ecology and Oceanography of Harmful Algal Blooms Grants 4355
The Program is seeking applications proposing targeted research projects of up to 3 years duration and multi-disciplinary regional studies for 3 to 5 years duration for the Ecology and Oceanography of Harmful Algal Blooms (ECOHAB) program. This program provides support for research on algal species whose populations may cause or result in deleterious effects on ecosystems, coastal economies, and human health. Studies of the causes of such blooms, their detection, effects, mitigation, and control in U.S. coastal waters (including estuaries and Great Lakes) are solicited.
Requirements: Institutions of higher education, not-for- profit institutions, State, commercial organizations, local and Indian tribal governments and Federal agencies are eligible to apply.
Date(s) Application is Due: Oct 4
Contact: Rob Magnien, Director; (301) 713-3338 or (301) 713-3020; fax: (301) 713-4044; email: rob.magnien@noaa.gov
Internet: http://www.cop.noaa.gov/opportunities/grants/fundingarchive/fy2008/Ecohab_08.html
Sponsor: National Oceanic and Atmospheric Administration
1305 East-West Highway, N/SCI2 SSMC4, Room 8243
Silver Spring, MD 20910

NOAA Ernest F. Hollings Scholarships 4356
The purposes of the Scholarship are to: increase undergraduate training in oceanic and atmospheric science, research, technology, and education and foster multidisciplinary training opportunities; increase public understanding and support for stewardship of the ocean and atmosphere and improve environmental literacy; recruit and prepare students for public service careers with the National Oceanic and Atmospheric Administration and other natural resource and science agencies at the federal, state and local levels of government; and recruit and prepare students for careers as teachers and educators in oceanic and atmospheric science and to improve scientific and environmental education in the United States. The award provides financial assistance (up to a maximum of $8,000) for two academic years of full-time study and a 10-week, full-time internship position ($650/week) during the summer at a NOAA facility.
Requirements: Applicants must be U.S. citizens; have full-time junior academic status in an accredited college or university (fall semester) within the United States or U.S. territories; have a cumulative GPA of 3.0 (based on a 4.0 scale) in all completed undergraduate courses and major field of study; be majoring in a discipline area related to oceanic and atmospheric science, research, technology, and education; and be supportive of the purposes of NOAA???s programs and mission.
Amount of Grant: $22,500 maximum per student
Date(s) Application is Due: Feb 28
Contact: Chantell Haskins; (301) 713-9437, ext. 125; email: chantell.haskins@noaa.gov or StudentScholarshipPrograms@noaa.gov
Internet: http://www.oesd.noaa.gov/Hollings_info.html
Sponsor: Oak Ridge Institute for Science and Education
1315 East-West Highway
Silver Spring, MD 20910

NOAA Faculty and Student Intern Research Program (FSIRP) 4357
The primary focus of the program is to provide work experiences and research opportunities to faculty, as well as graduate and undergraduate students who are pursuing degrees in the NOAA,Äërelated sciences. Appointments are also available for high school students planning to enroll in and attend an institution of higher education upon graduation. Disciplines include: business, computer sciences, engineering, life sciences, physical sciences. Appointments are made available on a year-round basis. Summer internships are generally for ten weeks. A weekly stipend of $440 is provided for undergraduate students, $550 for graduate students, and $1,200 for faculty. Limited travel reimbursement (round-trip transportation expenses between facility and home or campus)and housing allowance are also available.
Contact: Alfred A. Corea; (301) 713-0500, ext. 114; email: Alfred.A.Corea@noaa.gov; Victoria Dancy; (301) 713-0500, Ext. 136; email: victoria.g.dancy@noaa.gov
Internet: http://see.orau.org/ProgramDescription.aspx?Program=10049
Sponsor: National Oceanic and Atmospheric Administration
1315 East-West Highway
Silver Spring, MD 20910

NOAA Financial Assistance for National Centers for Central Coastal Ocean Science 4358
Research grants are awarded to determine the long-term consequences of human activities that adversely affect the coastal and marine environments; to assess the consequences of these activities in terms of ecological, economic, and social impacts; and to define and evaluate management alternatives that minimize adverse consequences of human use of the coastal and marine environments and resources. There are no specific deadlines for proposal submission. Proposals will be reviewed several times a year. Support for projects may be provided for periods up to five years.
Requirements: Universities, colleges, technical schools, institutes, laboratories, state and local government agencies, profit or nonprofit entities or individuals are eligible to apply.
Amount of Grant: $20,000-$550,000; $250,000
Contact: Chief, National Ocean Service, Office of Ocean Resources Conservation and Assessment, (301) 713-3102
Internet: http://www.nccos.noaa.gov
Sponsor: National Oceanic and Atmospheric Administration
1305 East-West Hwy
Silver Spring, MD 20910

NOAA General Coral Reef Conservation Grants 4359
The NOAA Coral Reef Conservation Program/ General Coral Reef Conservation Grants Program (GCRCGP) provides funding to institutions of higher education, non-profit organizations, commercial organizations, Freely Associated State government agencies, and local and Indian tribal governments to support coral reef conservation projects in the United States and the Freely Associated States in the Pacific, as authorized under the Coral Reef Conservation Act of 2000. Projects funded through the GCRCGP support on-the ground efforts that: help preserve, sustain and restore the condition of coral reef ecosystems; promote the wise management and sustainable use of coral reef resources; increase public knowledge and awareness of coral reef ecosystems and issues regarding their conservation; and develop sound scientific information on the condition of coral reef ecosystems and the threats to such ecosystems. Projects should complement and fill gaps in state, territorial and commonwealth coral reef programs, emphasize community- based conservation, or address local action strategy priorities.
Requirements: Institutions of higher education, non-profit organizations, commercial organizations, local and Indian tribal governments and Freely Associated State Government Agencies can apply for funding under the GCRCGP.

Restrictions: U.S. federal, state, territory, and commonwealth governments and Regional Fishery Management Councils are not eligible for this program.
Amount of Grant: $15,000-$50,000
Date(s) Application is Due: Nov 3
Contact: Liz Fairey; (301) 713-3459; email: Liz.fairey@noaa.gov; Jennifer Koss; (301) 713-3459, ext. 195; email: Jennifer.koss@noaa.gov
Internet: http://www07.grants.gov/search/search.do;jsessionid=LQCZGLyvwpgGPznL6jnLRyjK1vpGyTRsfyCJKLZtfJnqhLHphMnJ!280595828?oppId=42306&flag2006=false&mode=VIEW
Sponsor: National Oceanic and Atmospheric Administration
1315 East-West Highway
Silver Spring, MD 20910

NOAA Global Carbon Cycle (GCC) Program Grants **4360**
The goal of the program is to improve the ability to predict the fate of anthropogenic carbon dioxide and future atmospheric carbon dioxide concentrations using a combination of atmospheric and oceanic global observations, process-oriented field studies and modeling. GCC research is integrated across four over-riding themes: global location, magnitude and dynamics of carbon sources and sinks; northern hemisphere carbon sinks; causes of variability in sources and sinks; and future atmospheric carbon dioxide concentrations.
Requirements: Eligible applicants are institutions of higher education, other nonprofits, commercial organizations, international organizations, and state, local and Indian tribal governments.
Restrictions: Federal agencies or institutions are not eligible to receive Federal assistance under this notice.
Amount of Grant: $50,000 - $200,000 per year
Contact: Kathy Tedesco; (301) 427-2382; fax: (301) 713-0518; email: kathy.tedesco@noaa.gov
Internet: http://www.climate.noaa.gov/index.jsp?pg=./cpo_pa/cpo_pa_index.jsp&pa=gcc,äÇ=1
Sponsor: National Oceanic and Atmospheric Administration
1315 East West Highway, 12th Floor
Silver Spring, MD 20910

NOAA Graduate Sciences Program (GSP) Grants **4361**
The Graduate Sciences Program (GSP) is aimed primarily at increasing opportunities for students in NOAA-related fields to pursue research and educational training in atmospheric, environmental, remote sensing and oceanic sciences at minority serving institutions (MSI) when possible. The GSP offers at least two years of NOAA-related research and training opportunities. The program's primary objective is to provide college graduates with degrees in math, science, economics, law, social science, and engineering entry-level employment and hands-on research and work experience at NOAA. The program is designed to improve NOAA's outreach and recruitment efforts of under-represented individuals in the scientific mission-related occupational fields, i.e., fishery biology, environmental law, meteorology, cartography, oceanography, hydrology, computer science, environmental economics, and engineering. The award is equal to 640 hours paid employment, full tuition, housing allowance, meal allowance, and travel.
Requirements: Qualified candidates must have and maintain a minimum GPA of 3.0 or higher every semester/quarter.
Date(s) Application is Due: Jan 31
Contact: Pritti Brahma; (301) 713-9437, ext. 118; email: priti.brahma@noaa.gov
Internet: http://epp.noaa.gov/grad_science/welcome.html
Sponsor: National Oceanic and Atmospheric Administration
1315 East-West Highway, Room 10725
Silver Spring, MD 20910

NOAA Great Lakes Habitat Restoration Partnership Grants **4362**
The NOAA Great Lakes Habitat Restoration Program invites applications requesting funding to establish one or more regional habitat restoration partnership(s) for 1 to 3 years. The partnership(s) is expected to catalyze the implementation of habitat restoration projects that will benefit coastal resources through improved Great Lakes habitat quality. The centerpiece of the program will be one or more restoration projects in an Area of Concern that: are based on strong science and data availability; are ecosystem focused; and, involve significant problems and lake-wide improvements. Project areas should include locations where: maximum use can be made of on-going restoration efforts and partnerships; availability of matching funds are met; the problem is significant to the Great Lakes region, NOAA's mission and established priorities; and there is a scientific merit in restoration. NOAA envisions working jointly on such a partnership(s) through its Great Lakes Habitat Restoration Program (GLHRP) to fund and administer projects that support community-identified priorities such as: restoring and enhancing critical, nearshore areas, tributaries and connecting channels; remediating basin-wide sources of stress; protecting healthy functioning areas; and monitoring ecosystem health.
Requirements: Eligible applicants are institutions of higher education, hospitals, other non-profits, commercial (for-profit) organizations, organizations under the jurisdiction of foreign governments, international organizations, state, local and Indian tribal governments whose applications propose to benefit Great Lakes coastal and open-lake habitats.
Restrictions: Applications from federal agencies or employees of federal agencies will not be considered.
Amount of Grant: $250,000 average; $1,000,000 total
Date(s) Application is Due: Aug 31
Contact: Jenni Wallace; (301) 713-0174, ext. 191; email: GLHRP.GLERL@NOAA.gov; David Landsman; (301) 713-0174 ext. 151; email: GLHRP.GLERL@NOAA.gov
Internet: http://www07.grants.gov/search/search.do;jsessionid=LQCZGLyvwpgGPznL6jnLRyjK1vpGyTRsfyCJKLZtfJnqhLHphMnJ!280595828?oppId=14637&flag2006=true&mode=VIEW
Sponsor: National Oceanic and Atmospheric Administration
1315 East-West Highway
Silver Spring, MD 20910

NOAA International Coral Reef Conservation Project Grants **4363**
The NOAA Coral Reef Conservation Grant Program, as authorized under the Coral Reef Conservation Act of 2000, provides matching grants of financial assistance for international coral reef conservation projects. The Program solicits proposals under four funding categories: Promote Watershed Management in the Wider Caribbean, Brazil, and Bermuda; Support Planning for Effective Marine Protected Area Management; Encourage the Development of National Networks of Marine Protected Areas in the Wider Caribbean, Bermuda, Brazil, Southeast Asia, and the South Pacific; and Promote Regional Socio-Economic Training and Monitoring in Coral Reef Management in the Wider Caribbean, Brazil, Bermuda, the Western Indian Ocean, the Red Sea, the South Pacific, South Asia, and Southeast Asia. Each funding category has specific applicant and project eligibility criteria. For specific country eligibility per category, refer to individual category descriptions online. Pre-applications are due on November 3, with full applications received by February 13.
Requirements: Eligible applicants include institutions of higher education, U.S. and international non-profit organizations, and commercial organizations. The proposed work must be conducted at a non-U.S. site.
Restrictions: U.S. federal agencies and individuals are not eligible.
Amount of Grant: $15,000-$80,000
Date(s) Application is Due: Nov 3; Feb 13
Contact: Scot Frew; (301) 713-3078, ext. 220; email: Scot.Frew@noaa.gov
Internet: http://www07.grants.gov/search/search.do;jsessionid=LQCZGLyvwpgGPznL6jnLRyjK1vpGyTRsfyCJKLZtfJnqhLHphMnJ!280595828?oppId=42317&flag2006=false&mode=VIEW
Sponsor: National Oceanic and Atmospheric Administration
1315 East-West Highway
Silver Spring, MD 20910

NOAA Joint Hurricane Testbed Grants **4364**
The Office of Oceanic and Atmospheric Research (OAR), National Oceanic and Atmospheric Administration (NOAA), is soliciting LOIs (Letters of Intent) under the United States Weather Research Program (U.S.WRP), as administrated by the U.S.WRP Joint Hurricane Testbed (JHT). This notice: provides guidelines for the submission of full proposals; describes opportunities and application procedures for the transfer of relevant research and technology advances into tropical cyclone analysis and forecast operations; and calls for researchers to submit proposals to test and evaluate, and modify if necessary, in a quasi-operational environment, their own scientific and technological research applications. Projects satisfying metrics for success and operational constraints may be selected for operational implementation by the operational center(s) after the completion of the JHT-funded work. The period of the award is from one to two years. LOIs must be submitted by July 31, and full proposals are due by October 30.
Requirements: Eligible applicants are institutions of higher education; other nonprofits; commercial organizations; foreign governments; organizations under the jurisdiction of foreign governments; international organizations; state, local and Indian tribal governments; and Federal agencies. Applications from non-Federal and Federal agencies will be competing against each other.
Amount of Grant: $50,000-$200,000 averagw
Date(s) Application is Due: Jul 31; Oct 30
Contact: Dorothy Fryar, Office of Weather & Air Quality Research; (301) 734-1179; email: Dorothy.Fryar@noaa.gov
Internet: http://www07.grants.gov/search/search.do;jsessionid=LQCZGLyvwpgGPznL6jnLRyjK1vpGyTRsfyCJKLZtfJnqhLHphMnJ!280595828?oppId=42328&flag2006=false&mode=VIEW
Sponsor: National Oceanic and Atmospheric Administration
1315 East-West Highway, Room 11209
Silver Spring, MD 20910

NOAA Joint Research Institutes Grants **4365**
The objective of this grants program is to promote collaboration between university scientists and NOAA scientists so that academic and government researchers can work together to learn about and apply their knowledge to environmental processes. The NOAA Joint Research Institutes are centers of excellence in research relevant to understanding the Earth's oceans, atmosphere, Great Lakes, Arctic regions, and regions similar to the U.S. intermountain west, as well as the solar-terrestrial environment. Cutting-edge research is being conducted under a broad umbrella of environmental themes, such as climate variability, air quality, sea-ice dynamics, coastal ecosystems, fisheries oceanography, hydrology, and weather research and forecasting. Contact the office for further details.
Requirements: Major state universities located near the NOAA environmental research laboratories are eligible to apply.
Amount of Grant: $3,000-$60,000 individual proposals, $25,000 average; $60,000-$150,000 group proposals, $80,000 average
Contact: Dr. John Cortinas Jr., Joint Institute Program Manager; (301) 713-2465, ext. 206; fax: (301) 713-0158; email: john.cortinas@noaa.gov

Internet: http://www.oar.noaa.gov/organization/backgrounders/joint.html
Sponsor: National Oceanic and Atmospheric Administration
1315 East-West Highway, Room 11554
Silver Spring, MD 20910

NOAA Living Marine Resources Cooperative Science Center Internships/Fellowships 4366
The Mission of the Center is to conduct research congruent with the interests of NOAA Fisheries and to prepare students for careers in research, management, and public policy that support the sustainable harvest and conservation of our nation's living marine resources. Research activities of the Center fall into four thematic areas and is a collaborative effort among researchers from the partner institutions and NOAA Fisheries staff. The thematic areas of research are: Fisheries Socioeconomics, Quantitative Fisheries, Essential Fish Habitat, and Aquaculture. The Internship/Fellowship program is a NOAA partnership with the University of Maryland Eastern Shore, Delaware State University, Savannah State University, Hampton University, the UMBI Center of Marine Biotechnology, and the University of Miami.
Contact: Dionne Hoskins, LMRCSC Program Director; (312) 352-8925; email: dionne.hoskins@noaa.gov
Internet: http://www.ssufisheries.com/
Sponsor: National Oceanic and Atmospheric Administration
P.O. Box 20228
Savannah, GA 31404

NOAA Living Marine Resources Cooperative Science Center Mini Grants 4367
The program has been established by the Living Marine Resources Cooperative Science Center (LMRCSC) to award and distribute funds up to $1,000.00 for research projects and/or travel directly related to the goals, objectives, and Mission of the LMRCSC. Requests for fee payment are accepted up to $500. Suggested activities include travel to NOAA laboratories or partner institutions, student centered projects, resource enhancement that encourages and sustains research and collaboration. Travel or projects with clear NOAA partnerships will be given preferential consideration. Applications are received and awarded on a continuing basis as funding allows.
Requirements: Eligible research and travel applicants are SSU faculty or fulltime staff, and students enrolled at Savannah State University. Fee applicants must be graduate students matriculating at SSU. All students must have and maintain a 3.0 GPA.
Geographic Focus: Georgia
Amount of Grant: $1,000 maximum
Contact: Dionne Hoskins, LMRCSC Program Director; (312) 352-8925; email: dionne.hoskins@noaa.gov
Internet: http://www.ssufisheries.com/
Sponsor: National Oceanic and Atmospheric Administration
P.O. Box 20228
Savannah, GA 31404

NOAA Marine Fisheries Initiative (MARFIN) Grants 4368
The National Marine Fisheries Service (NMFS), Southeast Region, is seeking proposals under the Marine Fisheries Initiative Program (MARFIN), for research and development projects that optimize the use of fisheries in the Gulf of Mexico and off the South Atlantic states of North Carolina, South Carolina, Georgia, and Florida involving the U.S. fishing industry (recreational and commercial), including fishery biology, resource assessment, socioeconomic assessment, management and conservation, selected harvesting methods, and fish handling and processing. The NMFS Southeast Regional Office anticipates awarding projects that will range from $25,000 to $300,000. The average award is $150,000.
Requirements: Eligible applicants may be institutions of higher education, nonprofits, commercial organizations, individuals, state, local and Indian tribal governments.
Restrictions: Federal agencies or institutions are not eligible. Foreign governments, organizations under the jurisdiction of foreign governments, and international organizations are excluded for purposes of this solicitation
Amount of Grant: $25,000-$300,000; $150,000 average
Date(s) Application is Due: Aug 11
Contact: Robert Sadler, Liaison Office; (727) 824-5324; Robert.Sadler@noaa.gov
Internet: http://sero.nmfs.noaa.gov/grants/marfin.htm
Sponsor: National Oceanic and Atmospheric Administration
263 13th Avenue South
Saint Petersburg, FL 33702

NOAA Miami Lab Summer Internships 4369
The Miami Lab Summer Internship Program provides educational opportunities for both high school and undergraduate level students with research positions offering on-the-job training experience. Students receive direct supervision from faculty members at the Rosenstiel School of Marine and Atmospheric Science (RSMAS) and from staff scientists at the Atlantic Oceanographic Marine Laboratory (AOML) and the Southeast Fisheries Center (SEFC). This program provides students an opportunity to work in a scientific environment and to appraise career opportunities in oceanographic science. The annual deadline is the end of May, and grant awards vary depending upon the experience of the individual student. Interested applicants should contact the office directly.
Date(s) Application is Due: May 31
Contact: Essie C. Duffie; (305) 361-4237; email: essie.c.duffie@noaa.gov
Sponsor: National Oceanic and Atmospheric Administration
1315 East-West Highway
Silver Spring, MD 20910

NOAA Mid-Atlantic Bight National Undersea Research Center Grants 4370
The MAB Center supports undersea research to improve knowledge of processes governing change and stability in ecosystems of coastal and oceanic waters and seabed south of Montauk Point, New York and off New York, New Jersey, Pennsylvania, Delaware, Maryland and Virginia. The Center provides undersea technology with an emphasis on the REMU.S. Autonomous Undersea Vehicle (AUV) and the LEO-15 Observatory and associated sensors in order to address research questions at a range of scales, and is now requesting research proposals related to these priorities. Qualified proposals are eligible to receive support in the way AUV and LEO-15 Node access, including operators and divers, data download and processing, and AUV shipment and operator travel costs. Modest funds will be available for development project supplies. Proposals should provide a summary of the proposed research demonstration or development, describe research goals and facilities/equipment requirements, outline time or logistic constraints, specify area of operations including depths, and estimate the level of support required. Demonstration proposals have a limit of $15,000 per project, to include platform costs, travel, transportation, and salaries for NURP technical support (such as REMU.S. operators). Proposals received after May 21 will be reviewed on a rotational basis dependent on funds.
Requirements: Demonstration or development proposals are required and must be submitted by email by May 21.
Restrictions: Proposals for support of science beyond technical demonstrations or developments are discouraged during this funding cycle.
Amount of Grant: $15,000
Date(s) Application is Due: May 21
Contact: Michael P. De Luca, Director; (732) 932-6555, ext. 512; fax: (732) 932 8578; email: deluca@marine.rutgers.edu; Karen Kohanowich, Acting Director; (301) 734-1006 or (301) 713-1000; fax: (301) 713-1967; email: karen.kohanowich@noaa.gov
Internet: http://marine.rutgers.edu/nurp/proposals.html
Sponsor: National Oceanic and Atmospheric Administration
1315 East-West Highway
Silver Spring, MD 20910

NOAA Mid-Atlantic Research Set-Aside Program Grants 4371
The RSA program provides a mechanism to fund research and compensate vessel owners through the sale of fish harvested under the research quota. Vessels participating in an approved research project to harvest and land fish in excess of any imposed trip limit or during fishery closures. Landings from such trips are sold to generate funds that help defray the costs associated with the approved research projects. No Federal funds are provided for research under this notification. NMFS is soliciting proposals for research activities concerning the summer flounder, scup, black sea bass, Loligo squid, Illex squid, Atlantic mackerel, butterfish, Atlantic bluefish, and tilefish fisheries. The closing date for applications is March 24.
Requirements: Eligible applicants include institutions of higher education, hospitals, other nonprofits, commercial organizations, individuals, and state, local, and Native American tribal governments.
Restrictions: Federal agencies and institutions are not eligible to receive Federal assistance under this notice. Additionally, employees of any Federal agency or Regional Fishery Management Council are ineligible to submit an application under this program.
Date(s) Application is Due: Mar 24
Contact: Cheryl A. Corbett, Cooperative Programs Specialist; (508) 495-2070; fax: (508) 495-2004; email: cheryl.corbett@noaa.gov
Internet: http://www07.grants.gov/search/search.do;jsessionid=HWbpMDzmZ104MfvCGSbHhsGGPnb8dGTRG2hM0vfdTvpFRTZdkPyK!-796138783?oppId=16824&flag2006=true&mode=VIEW
Sponsor: National Oceanic and Atmospheric Administration
166 Water Street
Woods Hole, MA 02543

NOAA Monitoring and Event Response for Harmful Algal Blooms Grants 4372
The purpose of this document is to advise the public that NCCOS/CSCOR is soliciting proposals for two types of research projects: MERHAB-targeted and MERHAB-regional. MERHAB-targeted proposals will incorporate tools, approaches and technologies from HAB research programs into existing harmful algal bloom (HAB) monitoring programs. MERHAB regional proposals will create partnerships to enhance existing and initiate new HAB monitoring capabilities that will provide managers with timely information needed to mitigate HAB impacts on coastal communities.
Requirements: Institutions of higher education, not-for- profit institutions, State, commercial organizations, local and Indian tribal governments and Federal agencies are eligible to apply.
Date(s) Application is Due: Oct 2
Contact: Rob Magnien, Director; (301) 713-3338 or (301) 713-3020; fax: (301) 713-4044; email: rob.magnien@noaa.gov
Internet: http://www.cop.noaa.gov/opportunities/grants/fundingarchive/fy2007/MERHAB_FFO_0606.html
Sponsor: National Oceanic and Atmospheric Administration
1305 East-West Highway, N/SCI2 SSMC4, Room 8243
Silver Spring, MD 20910

NOAA Monkfish Research Set-Aside Program Grants 4373
The intent of this program is for fishing vessels to utilize these research DAS to conduct monkfish related research, rather than their allocated monkfish DAS, thereby eliminating any cost to the vessel associated with using a monkfish DAS. Landings from such research trips may be sold to generate funds to help defray research costs. No Federal

funds are provided for research under this notification. Rather, projects funded under the Program would be provided with additional opportunity to harvest monkfish, and the catch sold to generate income to offset research costs. Projects funded under an RSA DAS award must enhance the knowledge of the monkfish fishery resource or contribute to the body of information on which management decisions are made. The Councils and NMFS will give priority to funding research proposals in the following general areas: research concerning monkfish bycatch and discards; research to minimize bycatch and interactions with sea turtles and other protected species; research to minimize impacts of monkfish fishery on essential fish habitat (EFH), or other sensitive habitats; research to establish an exempted monkfish trawl fishery in the Northern Fishery Management Area; research on the biology or population structure and dynamics of monkfish; tagging studies; mesh and gear selectivity studies, including studies on gear efficiency; and cooperative stock assessment surveys. Note that the research subject areas listed above are not listed in order of priority.
Requirements: Eligible applicants are institutions of higher education, hospitals, other nonprofits, commercial organizations, individuals, state, local and Indian tribal governments.
Restrictions: Federal agencies or institutions are not eligible to receive Federal assistance under this notice. Also, a person is not eligible to submit an application under this program if he/she is an employee of any Federal agency. Fishery Management Council members who are not Federal employees may submit an application.
Date(s) Application is Due: Aug 31
Contact: Steve Drescher, Policy Advisor; (301) 713-0942, ext.122; fax: (301) 713-0947; email: Steve.J.Drescher@noaa.gov
Internet: http://www07.grants.gov/search/search.do;jsessionid=HV8GpydVhXQylKWtj8sRt2JWS1dB9H223h3VT8t1y7LpfYB52vmB!2066172073?oppId=14650&flag2006=true&mode=VIEW
Sponsor: National Oceanic and Atmospheric Administration
1315 East West Highway, 12th Floor
Silver Spring, MD 20910

NOAA National Climatic Data Center (NCDC) Internships 4374
Students from across the Nation are invited to participate in the NCDC Internship Program. Topics for investigation will cover a variety of scientific areas. Previous student research has ranged from hurricanes to configuration management to precipitation trends. Any climate variable holds possibilities for an opportunity of study. Each student is expected to prepare a presentation. The presentation can be completed using a team approach. Several students are permitted to work together if the research topics are related. All students hired will be employees of The National Oceanic and Atmospheric Administration or the designated program. Atmospheric Administration or the designated program. Salary is commensurate with experience.
Date(s) Application is Due: Feb 28
Contact: Carmella Davis Watkins; (828) 271-4105; fax: (828) 271-4876; email: carmella.davis.watkins@noaa.gov
Internet: http://www.ncdc.noaa.gov/oa/about/about.html
Sponsor: National Oceanic and Atmospheric Administration
151 Patton Avenue
Asheville, NC 28801

NOAA National Coastal and Estuarine Research and Technology Grants 4375
The National Oceanic and Atmospheric Administration (NOAA) seeks to establish a national estuarine research and technology program which operates in partnership with the National Estuarine Research Reserve System (NERRS). Funds will be used to conduct collaborative research and transform the best available science into practical innovative tools that coastal managers can use to detect, prevent, and reverse the impacts of coastal pollution and habitat degradation. Additionally, the program will provide coastal and estuarine managers a better understanding of what tools are available, how well they work, and how best to apply them to detect, prevent, and reverse the impacts of coastal pollution and habitat degradation.
Requirements: Eligible applicants are non-Federal institutions of higher education, other non-profits, commercial organizations, and state and local governments that possess the statutory authority to receive financial assistance.
Amount of Grant: $5,232,000 total
Date(s) Application is Due: Sep 9
Contact: Erica Seiden, OCRM/ERD Project Manager; (301) 563-1172; email: Erica.Seiden@noaa.gov
Internet: http://www07.grants.gov/search/search.do;jsessionid=LQCZGLyvwpgGPznL6jnLRyjK1vpGyTRsfyCJKLZtfJnqhLHphMnJ!280595828?oppId=42319&flag2006=false&mode=VIEW
Sponsor: National Oceanic and Atmospheric Administration
1315 East-West Highway
Silver Spring, MD 20910

NOAA National Estuarine Research Reserve Land Acquisition and Construction Program Grants 4376
he National Estuarine Research Reserve System consists of estuarine areas of the United States and its territories which are designated and managed for research and educational purposes. Each reserve within the system is chosen to represent different bio-geographic regions and to include a variety of ecosystem types in accordance with the classification scheme of the national program. Through the funding of designated reserve agencies and universities to undertake land acquisition and construction projects that support the NERRS purpose, NOAA will strengthen protection of key land and water areas; enhance long-term protection of the area for research and education; and provide for facility and exhibit construction. It is anticipated that the awards will run for up to two years. In the past, funding for land acquisition/construction awards has ranged in amount from approximately $50,000 to $3 million.
Requirements: Eligible applicants are NERR lead state agencies or universities in coastal states.
Amount of Grant: $50,000-$3,000,000 average
Date(s) Application is Due: Nov 30
Contact: Nina Garfield; (301) 563-1171, ext. 171; fax: (301) 713-4012; email: Nina.Garfield@noaa.gov
Internet: http://www07.grants.gov/search/search.do;jsessionid=LQCZGLyvwpgGPznL6jnLRyjK1vpGyTRsfyCJKLZtfJnqhLHphMnJ!280595828?oppId=42321&flag2006=false&mode=VIEW
Sponsor: National Oceanic and Atmospheric Administration
1315 East-West Highway
Silver Spring, MD 20910

NOAA National Estuarine Research Reserve System Graduate Research Fellowships 4377
The National Estuarine Research Reserve System's Graduate Research Fellowship provides master's degree students and Ph.D. candidates with an opportunity to conduct research of local and national significance that focuses on enhancing coastal zone management. Fellows conduct their research within a National Estuarine Research Reserve and gain hands-on experience by participating in the host reserve's research and monitoring programs. Graduate Research Fellowship projects are based on the reserves' local needs, the reserve system's national priorities and the students' interest. Fellows receive a stipend of $20,000 per year, and Fellowships may be funded for up to three years.
Requirements: Students admitted to or enrolled in a full-time master's or doctoral program at U.S. accredited universities are eligible to apply.
Restrictions: Electronically transmitted letters of support are not acceptable.
Amount of Grant: $20,000 stipend per year
Date(s) Application is Due: Nov 1
Contact: Alison Krepp; (301) 713-3155, ext. 105; fax: (301) 713-4367; email: Alison.Krepp@noaa.gov
Internet: http://nerrs.noaa.gov/Fellowship/welcome.html
Sponsor: National Oceanic and Atmospheric Administration
1305 East-West Highway
Silver Spring, MD 20910

NOAA National Height Modernization Program Grants 4378
The purpose of this notice is to solicit proposals for cooperative agreements and/or grants between NOAA and partnering entities in the United States, implementing NOAA's National Height Modernization Program (NHMP) Plan. Proposals submitted in response to this announcement should contribute to the beneficial public outcomes associated with the five priority issues in this plan which consists of enhancing the vertical component of the National Spatial Reference System (NSRS); enabling users to access the vertical component of the NSRS; outreach and education regarding geospatial issues and activities as they relate to NHMP; capacity building and technology transfer as they relate to NHMP; coordination, cooperation, and collaboration with other entities to accomplish common goals as they relate to NHMP. This competition is focused on the geography of the United States and its territories in response to NOAA's NHMP Regional Implementation Plan and subsequent congressional appropriations. Pre-proposals are required and due by August 1. Full proposals are due by September 2.
Requirements: Eligible funding applicants are institutions of higher education, state, local and Indian tribal governments.
Amount of Grant: $50,000-$1,200,000
Date(s) Application is Due: Aug 1; Sep 2
Contact: Gilbert Mitchell; (301) 713-3228, ext. 114; email: Glbert.Mitchell@NOAA.gov
Internet: http://www.grants.gov/search/search.do;jsessionid=LP0hqWfp11PVPKSr9yfL1LLhNQTn2gsNBmYzZtzGKTQhJRdRBYxW!-2005900814?oppId=42318&flag2006=false&mode=VIEW
Sponsor: National Oceanic and Atmospheric Administration
1305 East-West Highway, N/NGS1, Room 9356, SSMC3
Silver Spring, MD 20910

NOAA National Marine Fisheries Service - Sea Grant Joint Graduate Fellowships 4379
NOAA National Sea Grant Office and NOAA Fisheries established a Graduate Fellowship Program in two specialized areas: population dynamics and marine resource economics. Population dynamics is the study of fish populations as affected by fishing mortality, growth, recruitment and natural mortality. Ph.D. candidates interested in the population dynamics of living marine resources and the development and implementation of quantitative methods for assessing their status can receive up to three years of funding. Ph.D. students in marine resource economics, concentrating on the conservation and management of living marine resources, can receive two years of funding. The four main goals of the NOAA Fisheries/Sea Grant Fellowship Program are: to encourage qualified applicants to pursue careers in either population dynamics and stock assessment or in marine resource economics; to increase available expertise related to these fields; to foster closer relationships between academic scientist and NOAA Fisheries; and to provide real-world experience to graduate students and accelerate their career development.
Requirements: The Fisheries Fellowship is available to U.S. citizens who are graduate students enrolled in PhD degree programs in academic institutions in the United States and its territories.

Date(s) Application is Due: Jan 20
Contact: Miguel Lugo, Program Manager; (301) 734-1077, ext. 1075; fax: (301) 713-0799; email: miguel.lugo@noaa.gov
Internet: http://www.seagrant.noaa.gov/funding/rfp.html#nmfs
Sponsor: National Oceanic and Atmospheric Administration
1315 East-West Highway, Room 8243
Silver Spring, MD 20910

NOAA National Sea Grant Gulf of Mexico Oyster Industry Program Grants 4380
The goal of this program is to encourage multi-disciplinary research and extension projects that contribute directly to the efficiency and profitability of oyster-related businesses and to the safety of oyster products. This competition generally runs every two years. Annual deadlines vary. Applicants should visit the web site for further details.
Requirements: Institutions of higher education, not-for- profit institutions, State, commercial organizations, local and Indian tribal governments and Federal agencies are eligible to apply.
Contact: Megan Agy, Program Coordinator; (301) 734-1066 or (301) 734-1077; fax: (301) 713-0799; email: Megan Agy@noaa.gov
Internet: http://www.seagrant.noaa.gov/funding/rfp.html#Gulf%20Oyster
Sponsor: National Oceanic and Atmospheric Administration
1305 East-West Highway
Silver Spring, MD 20910

NOAA National Sea Grant Industry Fellowships 4381
The Sea Grant Industry Fellowship provides support for graduate students who are pursuing research and development projects in topics of interest to a particular industry/company. In a true partnership, the student, the faculty advisor, the Sea Grant college or institute, and the industry representative work together on a project from beginning to end.
Requirements: The Fellowship is available to U.S. citizens who are graduate students enrolled in PhD degree programs in academic institutions in the United States and its territories.
Date(s) Application is Due: Jan 20
Contact: Miguel Lugo, Program Manager; (301) 734-1077, ext. 1075; fax: (301) 713-0799; email: miguel.lugo@noaa.gov
Internet: http://www.seagrant.noaa.gov/funding/fellowships.html#industry
Sponsor: National Oceanic and Atmospheric Administration
1305 East-West Highway
Silver Spring, MD 20910

NOAA National Sea Grant John A. Knauss Marine Policy Fellowships 4382
The program is designed to provide a unique educational experience to students who have an interest in marine/ocean/Great Lakes resources and in the national policy decisions affecting those resources. The program matches highly qualified graduate students with hosts in the legislative branch, the executive branch, or appropriate associations/institutions located in the Washington, DC, area for a one-year paid fellowship.
Requirements: A student who, at the time of application, is in a graduate or professional degree program in a marine-related field at an accredited institution of higher education in the United States may apply. All applicants must submit an application to one of the Sea Grant College Programs in their state. If there is no program in the applicant's state, he or she should apply through the closest state Sea Grant College Program. A list of the programs may be obtained from the Web site http://www.mdsg.umd.edu/NSGO/SGDirectors.html.
Amount of Grant: $41,500
Date(s) Application is Due: Feb 20
Contact: Miguel Lugo, Knauss Program Manager; (301) 734-1077, ext. 1075; fax: (301) 713-0799; email: miguel.lugo@noaa.gov
Internet: http://www.seagrant.noaa.gov/funding/rfp.html#knauss
Sponsor: National Oceanic and Atmospheric Administration
1315 East-West Highway, Room 8243
Silver Spring, MD 20910

NOAA National Sea Grant Oyster Disease Research Program Grants 4383
This extensive program of ongoing research coupled with outreach and management efforts aims to better serve the restoration of healthy populations of oysters in the nation's coastal waters through research related to oyster disease. This competition generally runs every two years. Annual deadlines vary. Applicants should visit the web site for further details.
Requirements: Institutions of higher education, not-for- profit institutions, State, commercial organizations, local and Indian tribal governments and Federal agencies are eligible to apply.
Contact: Megan Agy, Program Coordinator; (301) 734-1066 or (301) 734-1077; fax: (301) 713-0799; email: Megan.Agy@noaa.gov
Internet: http://www.seagrant.noaa.gov/funding/rfp.html#oyster
Sponsor: National Oceanic and Atmospheric Administration
1305 East-West Highway
Silver Spring, MD 20910

NOAA National Sea Grant Support Program Grants 4384
The program objective is to support the establishment and operation of major university centers for marine resources, research, education, and training, and to support marine advisory services. Grant and fellowship funds may be used for research and development education and training, and advisory services. At least one-third of total cost must be obtained from non-federal sources; however, projects designed to meet specified national needs may receive 100 percent support.
Requirements: Universities, colleges, junior colleges, technical schools, institutes, laboratories, any public or private corporation, partnership, or other association or entity may apply for grants under this program. Also, any individual, state, political subdivision of a state or agency, or officer thereof is eligible to apply.
Restrictions: Grant money may not be used to purchase or construct ships or facilities.
Amount of Grant: $5,000-$3.9 million
Contact: Leon Cammen, Director; (301) 734-1088 or (301) 734-1077; fax: (301) 713-0799; email: leon.cammen@noaa.gov
Internet: http://www.nsgo.seagrant.org/funding/funding.html
Sponsor: National Oceanic and Atmospheric Administration
1315 East-West Hwy
Silver Spring, MD 20910

NOAA National Sea Grant Technology Program Grants 4385
The Sea Grant program accepts proposals for a technology transfer and development program to fulfill its broad responsibilities in fostering economic competitiveness through the transfer of technology pertaining to the development and utilization of ocean, coastal, and Great Lakes resources. It is desirable that proposals, which must be submitted through state Sea Grant Programs, involve industrial partners. The program provides support for applied research and development projects that ultimately facilitate the transfer of new products and processes related to the development of marine resources, including cost reductions for processes and product safety. Goals of the program are to conduct focused projects that can lead to the development and utilization of marine resources and related technological innovations and their acceptance in the marketplace (both in the United States and abroad), to increase interactions between the nation's academic scientists and engineers and their industrial counterparts, to accelerate the transfer of research-based marine science from universities to new technologies in industry, to provide a mechanism for industry to influence Sea Grant research priorities and solve problems of importance to industry, to increase cost-effectiveness of seafood production through aquaculture, and to forge long-term relationships between Sea Grant colleges and industrial firms.
Requirements: Matching funds equal to a minimum of 50 percent of the federal request must be provided.
Amount of Grant: $5,000-$3.6 million
Contact: Dr. Vijay Panchang, Program Manager; (301) 713-2435, ext. 142; fax: fax: (301) 713-0799; email: Vijay.Panchang@noaa.gov
Internet: http://www.aquanic.org/news/1999/ustechprog.htm
Sponsor: National Oceanic and Atmospheric Administration
1315 East-West Highway, Room 8243
Silver Spring, MD 20910

NOAA Northern Gulf of Mexico Ecosystems and Hypoxia Assessment Grants 4386
NGOMEX has two distinct components. The Modeling the Causes of Hypoxia component takes a regional ecosystem prediction approach to advance model development assessing the association between the northern Gulf hypoxic zone and causative factors. The proposed research for this competition should be 3-5 years in duration. The Modeling the Impacts of Hypoxia component takes an ecosystem stressors approach to advance understanding of hypoxia on ecologically and commercially important living resource populations of the northern Gulf of Mexico ecosystem. These projects should be 3 to 5 years of duration. Funding is contingent upon the availability of annual Federal appropriations. It is anticipated that final recommendations for funding under this announcement will be made by February and that projects funded under this announcement will have a July 1 through August 1 start date.
Date(s) Application is Due: Oct 20
Contact: Rob Magnien, Director; (301) 713-3338 or (301) 713-3020; fax: (301) 713-4044; email: rob.magnien@noaa.gov
Internet: http://www.cop.noaa.gov/opportunities/grants/fundingarchive/fy2009.html
Sponsor: National Oceanic and Atmospheric Administration
1315 East-West Highway, Room 8243
Silver Spring, MD 20910

NOAA Ocean Exploration Omnibus: Marine Archaeology Grants 4387
Proposals for exploration and discovery of significant maritime heritage sites are the priority. Submerged, previously subaerial, landscapes, shipwrecks, and other maritime cultural sites are typical focus subjects of the program. Proposals will emphasize the early phases of field archaeology: searching, locating, evaluating or inventorying sites. All applicants must convincingly describe: the archaeological significance of their site(s), and their importance to human history; how their research fits within the realm of exploration; and why their methodologies are innovative and make the most economical use of current marine technology. Pre-proposals are required and due by August 29. Full proposals are due by October 21.
Requirements: Eligible applicants are: institutions of higher education; other nonprofits; commercial organizations; foreign governments; organizations under the jurisdiction of foreign governments; international organizations; state, local and Indian tribal governments; and Federal agencies.
Restrictions: The Announcement does not invite proposals to support later phases of archaeological research, such as intensive site excavations, and major conservation projects.
Amount of Grant: $400,000 total
Date(s) Application is Due: Aug 29; Oct 21

Contact: Nicolas Alvarado, Proposal Manager; (301) 734-1015; email: OAR.OE.FAQ@noaa.gov
Internet: http://www.grants.gov/search/search.do;jsessionid=LP0hqWfp11PVPKSr9yfL1LLhNQTn2gsNBmYzZtzGKTQhJRdRBYxW!-2005900814?oppId=42332&flag2006=false&mode=VIEW
Sponsor: National Oceanic and Atmospheric Administration
1315 East West Highway, SSMC 3, Room 10244
Silver Spring, MD 20910

NOAA Ocean Exploration Omnibus: Ocean Exploration Grants **4388**
OE is seeking pre-proposals and full proposals to support its mission, consistent with NOAA's Strategic Plan, to search, investigate, and document poorly-known and unknown areas of the ocean and Great Lakes through interdisciplinary exploration, and to advance and disseminate knowledge of the ocean environment and its physical, chemical, and biological resources. Competitive OE proposals will be bold, innovative and interdisciplinary in their approach to Ocean Exploration. NOAA OE anticipates a total of approximately $1,400,000 including costs for ship and submersible assets will be available through this announcement. Pre-proposals are required and due by August 29. Full proposals are due by October 21.
Requirements: Eligible applicants are: institutions of higher education; other nonprofits; commercial organizations; foreign governments; organizations under the jurisdiction of foreign governments; international organizations; state, local and Indian tribal governments; and Federal agencies.
Restrictions: Only exploratory proposals will be funded, any other kind of project will not be reviewed.
Amount of Grant: $350,000 average
Date(s) Application is Due: Aug 29; Oct 21
Contact: Nicolas Alvarado, Proposal Manager; (301) 734-1015; email: OAR.OE.FAQ@noaa.gov
Internet: http://www.grants.gov/search/search.do;jsessionid=LP0hqWfp11PVPKSr9yfL1LLhNQTn2gsNBmYzZtzGKTQhJRdRBYxW!-2005900814?oppId=42333&flag2006=false&mode=VIEW
Sponsor: National Oceanic and Atmospheric Administration
1315 East West Highway, SSMC 3, Room 10244
Silver Spring, MD 20910

NOAA Pacific Islands Assistantships **4389**
The NOAA Pacific Services Center (PSC) and the NOAA Coastal Services Center have supported the Pacific Islands Assistantship (PIA) since 2001. The PIA program matches highly qualified recipients of professional degrees with hosts from the four Pacific Island coastal resource management organizations of American Samoa, the Commonwealth of the Northern Mariana Islands, Guam, and Hawaii. New assistants are placed every two years for two-year terms. The focus of each assistant depends on the needs of the programs and partners.
Date(s) Application is Due: Jan 26
Contact: Margaret VanderWilt , Fellowship Coordinator; (843) 740-1273; fax: (843) 740-1224; email: margaret.vanderwilt@noaa.gov or csc.fellowships@noaa.gov; NOAA Pacific Services Center; (808) 532-3202; fax: (808) 532-3224; email: psc@noaa.gov
Internet: http://www.csc.noaa.gov/cms/fellows/pacific-island.html
Sponsor: National Oceanic and Atmospheric Administration
737 Bishop Street, Suite 2250
Honolulu, HI 96813

NOAA Population Dynamics Recruiting Program Grants **4390**
The Population Dynamics Recruiting Program (PDRP) was developed to increase the number of highly qualified applicants for population dynamics positions with the Southeast Fisheries Science Center. The PDRP provides unique opportunities for top undergraduate students in the southeastern U.S. to learn about the field of marine resources population dynamics first-hand, all expenses paid, from some of the top experts in the field. Students interested must enter the PDRP through a one week Undergraduate Workshop, held in early January, each year, in the Florida Keys . The highly competitive application process seeks students with a strong quantitative background and a basic understanding of ecology. The second phase of the PDRP is the Summer Program where 4-8 of the top, PDRP Workshop participants spend part of their summers learning about current, important marine resource issues. The top students participating in the summer program are offered M.S. Graduate Assistantships with the NMFS RTR Unit at Virginia Tech.
Date(s) Application is Due: Oct 15
Contact: Dr. Jim Berkson; (540) 231-5910 or (540) 552-2510; email: Jim.Berkson@noaa.gov or nmfs@vt.edu
Internet: http://www.nmfs.vt.edu/sumprograms.htm
Sponsor: National Oceanic and Atmospheric Administration
100 Cheatham Hall (0321)
Blacksburg, VA 24061

NOAA Regional Ecosystem Prediction Program on Coral Reef Ecosystem Studies (CRES) Grants **4391**
The goal of this funding opportunity is to utilize existing scientific tools and approaches (e.g., biophysical models; coupled watershed and hydrodynamic models) within a social, cultural, and economic framework to develop and implement effective coastal ecosystem management practices in the Pacific Islands. Proposals should be regional in scale, interdisciplinary, comprehensive, integrated, and include multiple investigators to develop capabilities for innovative forecasts and predictions for improved management and control capabilities. Projects under this program will be 3 to 5 years in duration. Funding is contingent upon the availability of annual Federal appropriations. It is anticipated that final recommendations for funding under this announcement will be made by February and that projects funded under this announcement will have a July 1 through August 1 start date.
Date(s) Application is Due: Oct 9
Contact: Rob Magnien, Director; (301) 713-3338 or (301) 713-3020; fax: (301) 713-4044; email: rob.magnien@noaa.gov
Internet: http://www.cop.noaa.gov/opportunities/grants/fundingarchive/fy2009.html
Sponsor: National Oceanic and Atmospheric Administration
1305 East-West Hwy., N/SCI2 SSMC4, Room 8243
Silver Spring, MD 20910

NOAA Regional Integrated Sciences and Assessments (RISA) Grants **4392**
The program supports integrated, place-based research across a range of social, natural, and physical science disciplines to expand decision-makers' options in the face of climate change and variability at the regional level. RISA teams are comprised of researchers from the physical, natural, engineering and social sciences who work together and partner with stakeholders in a region to determine how climate impacts key resources and how climate information could aid in decision making and planning for those stakeholders. This effort often includes analyses of adaptation options in the face of a varying and changing climate. Currently, the RISA program is soliciting proposals to launch one new RISA effort in a region of the U.S. not currently covered by the program. The region needs to be one where drought is a serious concern to decision makers and public policy officials. Proposed projects should build on progress already achieved by integrated climate-society research and assessments projects in the region in terms of working with stakeholders and advancing regional climate impacts science. NOAA plans to have funding on the order of up to $500,000 to $600,000 per year available for this new RISA. While it is in the best interest of the applicants and their institutions to submit a Letter of Intent (LOI) by the July 23 due date, it is not a requirement; applicants who do not submit an LOI are allowed to submit a full proposal. For those that do submit an LOI, full proposals will be encouraged only for LOIs deemed relevant.
Requirements: Eligible applicants are institutions of higher education, other nonprofits, commercial organizations, international organizations, and state, local and Indian tribal governments.
Restrictions: Federal agencies or institutions are not eligible to receive Federal assistance under this notice.
Amount of Grant: $500,000 - $600,000 per year
Date(s) Application is Due: Sep 24
Contact: Hannah Campbell; (301) 734-1208; fax: (301) 713-0516; email: hannah.campbell@noaa.gov; Caitlin Simpson; (301) 734-1251; email: caitlin.simpson@noaa.gov
Internet: http://www.climate.noaa.gov/index.jsp?pg=/opportunities/opp_index.jsp&opp=2008/risa_info.jsp
Sponsor: National Oceanic and Atmospheric Administration
1315 East West Highway, 12th Floor
Silver Spring, MD 20910

NOAA Saltonstall-Kennedy Grants **4393**
The Saltonstall-Kennedy Act established a fund (known as the S-K fund) that the Secretary of Commerce uses to provide grants or cooperative agreements for fisheries research and development projects addressing aspects of U.S. fisheries, including, but not limited to, harvesting, processing, marketing, and associated infrastructures. U.S. fisheries include any fishery, commercial or recreational, that is, or may be, engaged in by citizens or nationals of the United States, or citizens of the Northern Mariana Islands (NMI), the Republic of the Marshall Islands, Republic of Palau, and the Federated States of Micronesia.
Requirements: Any U.S. citizen or national, or any citizen of the Northern Mariana Islands, the Republic of the Marshall Islands, Republic of Palau, or the Federated States of Micronesia; or group including state and local governments may apply.
Restrictions: Employees of federal agencies and regional fishery management councils and their employees are not eligible.
Amount of Grant: $200,000 average
Date(s) Application is Due: Oct 1
Contact: Steve Aguzin, Program Manager; (301) 713-2358; email: Stephen.Aguzin@noaa.gov
Internet: http://www.nmfs.noaa.gov/mb/financial_services/skhome.htm
Sponsor: National Oceanic and Atmospheric Administration
1315 East-West Highway
Silver Spring, MD 20910

NOAA Scientific Data Stewardship (SDS) Grants **4394**
The Program seeks to support the routine, operational production of Climate Data Records (CDRs) for data from the atmosphere, oceans, and land surface. This production requires collaboration between experts in the climate community and experts in data management, and must be informed by user feedback on the accessibility and usability of the produced CDRs. Applicants to this program must demonstrate that proposed CDRs have reached a critical level of maturity in science and data preservation and are of demonstrable societal benefit to be considered for funding under this Program.
Requirements: Eligible applicants are institutions of higher education, other nonprofits, commercial organizations, international organizations, and state, local and Indian tribal governments.

Restrictions: Federal agencies or institutions are not eligible to receive Federal assistance under this notice.
Amount of Grant: $50,000-$200,000 per year
Date(s) Application is Due: Sep 24
Contact: Diane Brown, Grants Administration Team Leader; (301) 734-1206; fax: (301) 713-0518; email: diane.brown@noaa.gov
Internet: http://www.climate.noaa.gov/index.jsp?pg=/opportunities/opp_index.jsp&opp=2008/sds_info.jsp
Sponsor: National Oceanic and Atmospheric Administration
1315 East West Highway, 12th Floor
Silver Spring, MD 20910

NOAA Sea Scallop Research Set-Aside Program Grants 4395
NMFS requests research proposals for fishing to utilize portions of the total allowable catch (TAC) and Days-at-Sea (DAS) allowance in the Atlantic sea scallop fishery that have been set aside by the New England Fishery Management Council (Council) for sea scallop research endeavors under a research set-aside (RSA) program. The program provides a mechanism to fund research and compensate vessel owners through the sale of fish harvested under the research quota. Vessels participating in an approved research project may be authorized by the Administrator, Northeast Region, NMFS (Regional Administrator), to harvest and land species in excess of any imposed trip limit or to harvest and land during fishery closures. Landings from such trips would be sold to generate funds that would help defray the costs associated with research projects. The DOC/NOAA encourages applications from members of the fishing community and applications that involve fishing community cooperation and participation.
Requirements: Eligible applicants are institutions of higher education, hospitals, other nonprofits, commercial organizations, individuals, state, local and Indian tribal governments.
Restrictions: Federal agencies or institutions are not eligible to receive Federal assistance under this notice. Also, a person is not eligible to submit an application under this program if he/she is an employee of any Federal agency. Fishery Management Council members who are not Federal employees may submit an application.
Date(s) Application is Due: Nov 18
Contact: Steve Drescher, Policy Advisor; (301) 713-0942, ext.122; fax: (301) 713-0947; email: Steve.J.Drescher@noaa.gov
Internet: http://www07.grants.gov/search/search.do;jsessionid=HV8GpydVhXQvlKWtj8sRt2JWS1dB9H223h3VT8t1y7LpfYB52vmB!2066172073?oppId=16903&flag2006=true&mode=VIEW
Sponsor: National Oceanic and Atmospheric Administration
1315 East West Highway, 12th Floor
Silver Spring, MD 20910

NOAA Sector Applications Research Program (SARP) Grants 4396
The Program supports the overarching goals of the NOAA Climate Program by developing the knowledge base, decision support tools, capacities and partnerships in sectors affected by climate in a substantial and increasingly visible way. SARP is designed to catalyze and support interdisciplinary research, innovative outreach and education activities that enhance the capacity of key socio-economic sectors to respond to and plan for climate variability and change through the use of climate information and related decision support resources. The program is designed to systematically build an interdisciplinary and expressly applicable knowledge base and mechanism for the creation, dissemination and exchange of climate-related research findings and decision support resources critical for understanding and addressing resource management challenges in vital social and economic sectors (e.g., coastal resources, water, agriculture, health, etc.). While it is in the best interest of the applicants and their institutions to submit a Letter of Intent (LOI) by the July 23rd due date, it is not a requirement; applicants who do not submit an LOI are allowed to submit a full proposal. For those that do submit an LOI, full proposals will be encouraged only for LOIs deemed relevant.
Requirements: Eligible applicants are institutions of higher education, other nonprofits, commercial organizations, international organizations, and state, local and Indian tribal governments.
Restrictions: Federal agencies or institutions are not eligible to receive Federal assistance under this notice.
Amount of Grant: $50,000-$200,000 per year
Date(s) Application is Due: Sep 24
Contact: Lisa Vaughan; (301) 734-1266; email: Lisa.Vaughan@noaa.gov; Nancy Beller-Simms; (301) 734-1205; email: Nancy.Beller-Simms@noaa.gov
Internet: http://www.climate.noaa.gov/index.jsp?pg=/opportunities/opp_index.jsp&opp=2008/sarp_info.jsp
Sponsor: National Oceanic and Atmospheric Administration
1315 East West Highway, 12th Floor
Silver Spring, MD 20910

NOAA South Florida Program Grants 4397
NCCOS/CSCOR is soliciting 1-year and 2-year proposals to support coastal ecosystem studies in South Florida including Florida Bay , Florida Keys, the Florida Keys National Marine Sanctuary (FKNMS), and adjacent coastal waters. It will provide support for the NOAA South Florida Program (SFP) and the FKNMS. The overall goal of this announcement is to fund high priority research and long term observational data collection needed to model and predict the impacts of Everglades restoration on the South Florida coastal ecosystem and to fulfill NOAA commitments to the South Florida Ecosystem Restoration effort and the Comprehensive Everglades Restoration Plan (CERP).
Requirements: Institutions of higher education, not-for- profit institutions, State, commercial organizations, local and Indian tribal governments and Federal agencies are eligible to apply.
Date(s) Application is Due: Sep 29
Contact: Rob Magnien, Director; (301) 713-3338 or (301) 713-3020; fax: (301) 713-4044; email: rob.magnien@noaa.gov
Internet: http://www.cop.noaa.gov/opportunities/grants/fundingarchive/fy2006/sfl06.html
Sponsor: National Oceanic and Atmospheric Administration
1305 East-West Highway, N/SCI2 SSMC4, Room 8243
Silver Spring, MD 20910

NOAA Spherical Display Systems for Earth System Science - Installations and Content Development Grants 4398
NOAA seeks partners to further promote the inclusion of Earth system science concepts in informal science education venues through the development of exhibits involving spherical display systems. There are two priorities for this competition: priority 1 installations of spherical display systems into public exhibits and, priority 2 development and evaluation of Earth system science modules. A single institution may apply to both priorities, but with separate applications, one to each priority. Applications will be reviewed independently. Funds awarded cannot necessarily pay for all the costs that the recipient might incur in the course of carrying out the project. NOAA anticipates the availability of approximately $4,000,000 of Federal financial assistance
Requirements: Eligible applicants are institutions of higher education, other nonprofits, and state, local and Indian tribal governments in the United States.
Restrictions: Applications to priority 1 requesting Federal support from NOAA of more than $100,000 total will not be considered for funding through this announcement. Applications for priority 2 requesting Federal support from NOAA of more than $300,000 total will not be considered for funding. For profit organizations, foreign institutions, foreign organizations and foreign government agencies are not eligible to apply.
Amount of Grant: $100,000 maximum for Priority 1; $300,000 maximum for Priority 2
Date(s) Application is Due: Oct 30
Contact: Carrie McDougall; (202) 482-0875; email: carrie.mcdougall@noaa.gov
Internet: http://www.oesd.noaa.gov/Spherical_Display_awards.html
Sponsor: National Oceanic and Atmospheric Administration
1401 Constitution Avenue NW, Room 6863
Washington, DC 20230

NOAA Transition of Research Applications to Climate Services Grants 4399
The TRACS Program transitions experimentally mature climate information tools, methods, and processes, including computer related applications (e.g. web interfaces, visualization tools), from research mode into settings where they may be applied in an operational and sustained manner. TRACS primary goal is to generate sustained delivery of useful climate information products and services to local, regional, national, and international decision and policy makers. TRACS seeks not only to support implementation of these transitions, but also to learn from partners how to better accomplish technology transition processes for public goods applications and improved risk management. The title TRACS, besides evoking a focus of being on-track, should also call attention to three key elements and their interplay. These activities include, transitions (i.e. a focus on partnerships where technology hand-offs occur), research applications (i.e. experimentally developed and tested, end-user-friendly information to support decision making), and climate services (i.e. the routine and timely delivery of that information, including via partnerships).
Requirements: Eligible applicants are institutions of higher education, other nonprofits, commercial organizations, international organizations, and state, local and Indian tribal governments.
Restrictions: Federal agencies or institutions are not eligible to receive Federal assistance under this notice.
Contact: Josh Foster, Climate Transition Coordinator; (301) 734-1218; fax: (301) 713-0518; email: josh.foster@noaa.gov
Internet: http://www.climate.noaa.gov/cpo_pa/nctp/
Sponsor: National Oceanic and Atmospheric Administration
1315 East West Highway, 12th Floor
Silver Spring, MD 20910

NOAA Undersea Research Center Coral Reef Conservation Program (CRCP) Coral Research Grants 4400
The NOAA Undersea Research Center at the University of North Carolina Wilmington seeks proposals for undersea research projects to be conducted off the southeast U.S. from North Carolina to Texas. This funding opportunity is for one to two year targeted research projects that address one or more of the following threats facing U.S. coral reefs in Florida (includes the Florida Keys, Southeast Florida, and the Eastern Gulf of Mexico) and the Flower Garden Banks: fishing, pollution, coastal uses (including invasive species), climate change, and extreme events. Proposals should be geared towards management-oriented research to: develop a better understanding of the role of the threats listed above in coral reef ecosystem decline; predict the impacts from and responses of reef communities to anthropogenic and natural stressors; and design, test, and evaluate the effectiveness of specific reef management measures, including marine protected areas and habitat restoration techniques. Pre-proposals are required and must be submitted by email by January 14 to: pottst@uncw.edu.

Requirements: Proposals submitted under the CRCP funding opportunity require 100% non-federal match. Non-federal matching funds may be comprised of a variety of public and private sources and may include in-kind contributions and other non-cash support.
Amount of Grant: $75,000
Date(s) Application is Due: Jan 14
Contact: Thomas Potts, Associate Director; (910) 962-2442; fax: (910) 962-2444; email: pottst@uncw.edu; Karen Kohanowich, Acting Director; (301) 734-1006 or (301) 713-1000; fax: (301) 713-1967; email: karen.kohanowich@noaa.go
Internet: http://www.uncw.edu/nurc/funding/opportunities.htm
Sponsor: National Oceanic and Atmospheric Administration
1315 East-West Highway
Silver Spring, MD 20910

NOAA Undersea Research Center Ocean Acidification Research at Aquarius Reef Base Grants **4401**
The NOAA Undersea Research Center at the University of North Carolina Wilmington seeks proposals for undersea research projects to be conducted off the southeast U.S. from North Carolina to Texas. Private foundation funds in the amount of $75,000 for direct science support are available for a research mission focused on the impacts of ocean acidification on coral reefs at Aquarius Reef Base. Proposals may be written for one year (opportunity 1) or two years (opportunity 2) of direct science support of $75,000 per year and should contain an in situ field component. Undersea system and support vessel time is provided by the Center at no cost to the principal investigator. Pre-proposals are required and must be submitted by email by January 14 to: pottst@uncw.edu.
Requirements: Proposals submitted under the CRCP funding opportunity require 100% non-federal match. Non-federal matching funds may be comprised of a variety of public and private sources and may include in-kind contributions and other non-cash support.
Amount of Grant: $75,000
Date(s) Application is Due: Jan 14
Contact: Ellen Prager, Chief Scientist; (305) 451-0233, ext. 206; fax: (305) 453-9719; email: pragere@earthlink.net; Karen Kohanowich, Acting Director; (301) 734-1006 or (301) 713-1000; fax: (301) 713-1967; email: karen.kohanowich@noaa.gov
Internet: http://www.uncw.edu/nurc/funding/opportunities.htm
Sponsor: National Oceanic and Atmospheric Administration
1315 East-West Highway
Silver Spring, MD 20910

NOAA Undersea Research Program Grants **4402**
NOAA's Undersea Research Program (NURP) is a unique comprehensive research program that provides scientists with the tools and expertise to conduct cutting-edge underwater research using advanced underwater technologies (e.g., human occupied submersibles, remotely operated vehicles, an undersea laboratory, mixed gas and technical diving). Research priorities are identified on an annual basis by NURP Headquarters in its Science Guidance. The research priorities are based upon the NOAA Strategic Plan and research needs as identified by NOAA resource managers. Highest priority is given to proposals for studies within the U.S. Exclusive Economic Zone and waters under U.S. jurisdiction, including the Great Lakes, territorial seas, and adjacent waters of the United States; U.S. Territories; and the Freely-Associated States. Funding opportunities will be announced through the individual NURP Center or NIU.S.T.
Requirements: Eligible applicants are U.S. Institutions of higher education, not-for-profit institutions, and federal, state, and local governments.
Restrictions: Federal agencies may not charge salary or overhead.
Amount of Grant: $15,000-$2.225 million
Contact: Karen Kohanowich, Acting Director; (301) 734-1006 or (301) 713-1000; fax: (301) 713-1967; email: karen.kohanowich@noaa.gov
Internet: http://www.nurp.noaa.gov/Funding.htm
Sponsor: National Oceanic and Atmospheric Administration
1315 East-West Highway
Silver Spring, MD 20910

NOAA Undersea Research Program Project Grants **4403**
NOAA's Undersea Research Program (NURP) is a unique comprehensive research program that provides scientists with the tools and expertise to conduct cutting-edge underwater research using advanced underwater technologies (e.g., human occupied submersibles, remotely operated vehicles, an undersea laboratory, mixed gas and technical diving). Some funds (less than $10,000 per project) are available for developmental projects or small exploratory research grants. These funds are reserved for late-breaking, high priority issues and may be awarded at the discretion of the Center Director. Funding opportunities will be announced through the individual NURP Center or NIU.S.T.
Requirements: Eligible applicants are U.S. Institutions of higher education, not-for-profit institutions, and federal, state, and local governments.
Restrictions: Federal agencies may not charge salary or overhead.
Amount of Grant: $10,000 maximum
Contact: Karen Kohanowich, Acting Director; (301) 734-1006 or (301) 713-1000; fax: (301) 713-1967; email: karen.kohanowich@noaa.gov
Internet: http://www.nurp.noaa.gov/Funding.htm
Sponsor: National Oceanic and Atmospheric Administration
1315 East-West Highway
Silver Spring, MD 20910

Nobel Laureate Signature Award for Graduate Education in Chemistry **4404**
This award, sponsored by Mallinckrodt Baker Inc. and administered by the American Chemical Society, is awarded annually to recognize an outstanding graduate student and his or her preceptor(s) in the field of chemistry as broadly defined. The graduate student nominee must have completed a PhD dissertation in chemistry within the 12-month period before the deadline for receipt of nominations. The award will recognize only work done while the nominee was a graduate student; it is granted without regard to age or nationality. The award includes a stipend, plaque, and travel expenses for the student and preceptor(s).
Requirements: A nominating document shall consist of a letter of nomination, a brief biographical sketch of the graduate student nominee and the preceptor(s), a synopsis of the nominee's PhD dissertation not longer than 10 typewritten pages (double-spaced), and one or two letters from experts in the field of the dissertation independently appraising its significance. Nominees may subsequently be asked to submit five copies of the complete dissertation. All documents must be in English. Six copies of all items to be included in the nomination must be submitted.
Restrictions: Self-nominations are not accepted.
Samples: (2008) Rebecca Anne Nelson and David Eisenberg; (2007) Anirban Banerjee and Gregory L. Verdine; (2006) Raquel L. Lieberman and Amy C. Rosenzweig; (2005) Christy L. Haynes and Richard P. Van Duyne
Amount of Grant: $3000 for student; $3000 for preceptor(s)
Date(s) Application is Due: Feb 1
Contact: Awards Administrator, (202) 872-4408; fax: (202) 872-6317; email: awards@acs.org
Internet: http://portal.acs.org/portal/acs/corg/content?_nfpb=true&_pageLabel=PP_ARTICLEMAIN&node_id=1319&content_id=CTP_004541&use_sec=true&sec_url_var=region1
Sponsor: American Chemical Society
1155 16th St. NW
Washington, DC 20036

Nokomis Foundation Grants **4405**
The foundation seeks to make a difference in the lives of women and girls and primarily funds grants in the West Michigan area. However, proposals with a state or national focus, provided they include a West Michigan component, are accepted. The foundation is particularly interested in pilot programs and gives priority to organizations that do not have access to traditional funding sources. In reviewing proposals, Nokomis looks for these characteristics: evidence of innovative thinking and creative problem-solving, organizational values similar to the foundation's (e.g., respect for diversity), intent to foster collaboration among groups sharing similar goals, and intent to accomplish long-term social change (multiyear projects are welcomed). The best way to begin the process is to send the executive director a concise letter or email (no longer than two pages) that describes the proposed project and need for funding. The foundation will review the letter promptly, and indicate whether a detailed proposal may be submitted to the foundation's Board of Advisors. In some cases, more information or a site visit may be required to be able to evaluate the request. There are no deadline dates; however, the board meets quarterly to consider requests. Details are available on the Web site. Unsolicited proposals are not currently being accepted.
Requirements: Giving primarily in the greater Grand Rapids, MI, area; limited support for national organizations.
Restrictions: The foundation does not provide funding to individuals or to religious organizations for religious purposes, and does not normally fund scholarships, fellowships, medical research, capital requests, endowments, or attendance at conferences. Support is rarely provided for capital expenditures (e.g., building renovations, new equipment, or endowments).
Samples: Abundant Life Ministries (Grand Rapids, MI)—to support Mr. B's Basketball and Reading Camp, $2000 (2003); Ctr for Women in Transition (Holland, MI)—to support an infrastructure project, $25,000 (2003); Grand Rapids Civic Theatre (MI)—for the summer stock program, The Glass Slipper, $1000 (2003); Jewish Federation of Grand Rapids (MI)—for Rachel Simmon's lecture, $2000 (2003).
Contact: Program Contact, (616) 451-0267; fax: (616) 451-9914; email: info@nokomisfoundation.org
Internet: http://www.nokomisfoundation.org/whathow.htm
Sponsor: Nokomis Foundation
161 Ottawa NW, Suite 305-C
Grand Rapids, MI 49503

Norcliffe Foundation Grants **4406**
Grants are awarded to organizations in the Puget Sound Region in and around Seattle, WA, with emphasis on higher and secondary education, cultural programs, religious associations, hospitals, and historic preservation. Support is also given for medical research, youth agencies, adult basic education, vocational education, literacy, the aged, and conservation. Grants supply funds for operating budgets, capital campaigns, conferences and seminars, scholarship funds, continuing support, annual campaigns, seed money, building funds, equipment, land acquisition, research, and special projects. Application guidelines are available upon request; applications are accepted at any time.
Requirements: Washington 501(c)3 organizations may apply.
Restrictions: Grants are not awarded to individuals or for deficit financing.
Geographic Focus: Washington
Samples: Seattle Center Foundation (Seattle, WA)—to construct Marion Oliver McCaw Hall, $1 million.
Amount of Grant: $1000-$25,000 average
Contact: Arline Hefferline, Foundation Manager, (206) 682-4820; fax: (206) 682-4821; email: arline@thenorcliffefoundation.com
Internet: http://www.thenorcliffefoundation.com

Sponsor: Norcliffe Foundation
999 Third Avenue, First Interstate Ctr, Suite 1006
Seattle, WA 98104

NORD Orphan Diseases Research Grants 4407
The program provides seed money in small grants to academic scientists studying new treatments or diagnostics for rare diseases. One-year clinical-research projects should focus on the early detection, diagnosis, or treatment of one of the following disorders: adenoid cystic carcinoma, alveolar capillary dysplasia, Cat Eye Syndrome, Ehlers-Danlos Syndrome, erythromelalgia, giant hypertrophic gastritis, and olivopontocerebellar atrophy or closely related diseases. The clinical researchers supported by the research grants provide preliminary data indicating that a treatment (drug, device, or medical food) may be safe and effective when used for a larger number of patients. Researchers can then use the preliminary data to apply for larger multiyear government grants or to attract a commercial sponsor who will manufacture an orphan product and get it approved for marketing by the U.S. Food and Drug Administration.
Requirements: Individuals who have earned an MD or MD/PhD in the past 10 years are eligible.
Samples: Dr. Cynthia Jackson and Dr. Douglas Gnepp, Rhode Island Hospital (RI)—to focus on the identification of genes as potentially new prognostic markers, as well as to study their contribution to the development of adenoid cystic carcinoma.; Dr. Eli Chang, Doheny Eye Institute, Keck School of Medicine of the U of Southern California (CA)—to research the establishment of an animal model for adenoid cystic carcinoma of the lacrimal gland, which may lead to the development of new and improved treatment modalities.
Contact: Grants Administrator, (203) 744-0100; fax: (203) 798-2291; email: orphan@rarediseases.org
Internet: http://www.rarediseases.org/research/requests
Sponsor: National Organization for Rare Disorders
55 Kenosia Avenue
Danbury, CT 06813-1968

NORD Roscoe Brady Lysosomal Storage Disease Fellowships 4408
The fellowships assist physicians who desire to establish careers in lysosomal storage diseases (LSDs) clinical medicine. Clinical research funded by the fellowships will be collaborative, involve multidisciplinary teams, and focus on genetics, new treatments and diagnostics, and/or epidemiology of LSDs in general, or for a specific LSD. The program's primary goal is to develop and support clinicians beginning their medical career that will specialize and focus on research, treatment, and coordination of multidisciplinary care for people suffering with LSDs. The research will be conducted in the United States, countries of the European Union, Canada, Australia, New Zealand, Brazil, Argentina, Chile, South Africa, Japan, or Israel, and where supervision of grant administration is possible. Complete guidelines and application are available online.
Requirements: Applicants should have earned the degree(s) of MD or MD/PhD within the past 10 years.
Amount of Grant: $50,000-$70,000
Date(s) Application is Due: Apr 13
Contact: Fellowships Administrator, (203) 744-0100 or (800) 999-6673 (voicemail only);TDD: (203) 797-9590; fax: (203)798-2291; email: orphan@rarediseases.org
Internet: http://www.rarediseases.org/research/roscoe
Sponsor: National Organization for Rare Disorders
55 Kenosia Avenue
Danbury, CT 06813-1968

Nordson Corporation Foundation Grants 4409
The corporate foundation awards grants to nonprofit organizations in geographic areas where Nordson facilities and employees are located: Cuyahoga and Lorain counties in Ohio; the Greater Atlanta, Georgia area; California, Rhode Island and Southeastern Massachusetts. The foundation provides a source of stable funding for community programs and projects in the areas of education, human welfare, civic, and arts and culture. Educational support is generally limited to improving elementary and secondary schools and certain programs for public and private higher education. Human welfare grants focus on children and youth. The foundation considers other funding areas, including urban affairs, volunteerism, public policy, health and health organizations, and literacy. Grants are awarded for general operating support, continuing support, annual campaigns, capital campaigns, building construction/renovation, equipment acquisition, emergency funds, seed money, scholarship funds, employee matching gifts, and technical assistance. The board meets four times each year. Deadlines vary by state; contact program staff for appropriate deadlines. Application forms are available on the website and may only be submitted online.
Requirements: Private, nonprofit organizations in California, Georgia, Ohio, Rhode Island, and southeastern Massachusetts may apply.
Restrictions: Funding is not provided for organizations whose services are not provided within the foundation's geographic areas of interest; direct grants or scholarships to individuals; organizations not eligible for tax-deductible support; organizations not exempt under Section 501(c)3 of the Internal Revenue Code; political causes, candidates, organizations or campaigns; organizations that discriminate on the basis of race, sex, or religion; or special occasion, goodwill advertising, i.e., journals or dinner programs.
Geographic Focus: California; Georgia; Massachusetts; Ohio; Rhode Island
Samples: Ctr for Leadership in Education (Lorain, OH)—for general operating support, $250,000.; Salvation Army of Lorain (Lorain, OH)—for the capital campaign, $50,000.; Neighborhood House Assoc of Lorain County (Lorain, OH)—for rejuvenation of the Cityview Ctr, $39,625.
Amount of Grant: $1.5 million total
Date(s) Application is Due: Feb 15; May 15; Aug 15; Nov 15
Contact: Cecilia H. Render, Director; (440) 892-1580, ext. 5172; email: crender@nordson.com
Internet: http://www.nordson.com/Corporate/Community/Foundation
Sponsor: Nordson Corporation Foundation
28601 Clemens Road
Westlake, OH 44145-1119

North Carolina Arts Council Artist Fellowships 4410
The program operates on a two-year cycle. Composers of jazz, chamber, electronic, experimental, and symphonic music; songwriters; playwrights and screenwriters; spoken-word artists; and writers of fiction, poetry, literary nonfiction, and literary translation are eligible to apply during even years. Choreographers, film/video artists, craft artists, and visual artists are eligible to apply in odd years.
Requirements: Applicants must: be at least 18 years old; be a U.S. citizen or holder of permanent resident alien status; have been a year-round resident of North Carolina for at least one year immediately prior to the application deadline; remain a North Carolina resident during the grant period and be physically present in the state for the majority of the grant period; not be enrolled at the time of application and will not be enrolled during the grant period in an academic or degree-granting program.
Geographic Focus: North Carolina
Amount of Grant: $10,000
Date(s) Application is Due: Nov 3
Contact: Wayne Martin, Senior Program Director; (919) 807-6506; email: wayne.martin@ncmail.net
Internet: http://www.ncarts.org/grants_category.cfm?ID=35
Sponsor: North Carolina Arts Council
MSC #4632, Department of Cultural Resources
Raleigh, NC 27601-2807

North Carolina Arts Council Arts Folklife Internship 4411
This three-month internship provides intensive, directed training with the Arts Council's Folklife staff. The intern receives a $4,000 stipend for the three-month period. The internship takes place for twelve weeks between April and August. Applicants must have completed at least one year of graduate study in folklife or related fields (ethnomusicology, anthropology, history, etc.) or have experience with traditional arts and culture. The next application deadline is March 2.
Requirements: Applicants must have completed at least one year of graduate study in folklife or related fields (ethnomusicology, anthropology, history, etc.) or have otherwise gained significant experience with traditional arts and culture. Applicants must be North Carolina residents.
Geographic Focus: North Carolina
Amount of Grant: $4,000 stipend
Date(s) Application is Due: Mar 2
Contact: Wayne Martin, Senior Program Director; (919) 807-6506; email: wayne.martin@ncmail.net
Internet: http://www.ncarts.org/freeform_scrn_template.cfm?ffscrn_id=119
Sponsor: North Carolina Arts Council
MSC #4632, Department of Cultural Resources
Raleigh, NC 27601-2807

North Carolina Arts Council Folklife Public Program Grants 4412
Public programs support: exhibitions, performances and workshops, publications, and radio and television programs that promote public knowledge and appreciation of the state's traditional arts; or salary assistance to organizations that create a staff position for a professional folklorist. Grant funds also support expenses directly related to the project including production costs, stipends, honoraria, marketing, travel, supplies, telephone, and postage.
Requirements: North Carolina nonprofit organizations are eligible.
Geographic Focus: North Carolina
Amount of Grant: $10,000 maximum
Date(s) Application is Due: Mar 2
Contact: Sally Peterson, Folklife Specialist; (919) 807-6507; email: sally.peterson@ncdcr.gov
Internet: http://www.ncarts.org/grants_programs.cfm?ID=6
Sponsor: North Carolina Arts Council
MSC #4632, Department of Cultural Resources
Raleigh, NC 27601-2807

North Carolina Arts Council Residency Center Grants 4413
The N.C. Arts Council sponsors one- and two-month residencies for artists in a variety of disciplines at residency centers in California and Vermont. The purpose of the N.C. Arts Council's residency program is to provide both emerging and established North Carolina artists with time for their work. Two artists will be selected for residencies at Headlands Center for the Arts and for residencies at Vermont Studio Center as part of this program. The deadline for Headlands is June 6, and the deadline for Vermont is June 16.
Requirements: North Carolina poets, fiction writers, playwrights, storytellers, and visual artists may apply.
Geographic Focus: North Carolina
Amount of Grant: $7,500 award; $1,000 stipend
Date(s) Application is Due: Jun 6 (Headlands); Jun 16 (Vermont)

Contact: Jeff Pettus, Visual Arts Director; (919) 807-6513; email: jeff.pettus@ncmail.net; Debbie McGill, Literature Director; (919) 807-6512; email: debbie.mcgill@ncmail.net
Internet: http://www.ncarts.org/grants_category.cfm?ID=36&CFID=4457402&CFTOKEN=25653586&jsessionid=28302898591231777988333
Sponsor: North Carolina Arts Council
MSC #4632, Department of Cultural Resources
Raleigh, NC 27601-2807

North Carolina Biotechnology Center Collaborative Funding Grants **4414**
The grant financially supports a postdoctoral fellow or technician in a university research laboratory who will perform a project of commercial interest under the guidance of a principal investigator. The Kenan Institute for Engineering, Technology and Science and the Biotechnology Center jointly sponsor this grant. The Grant aims to: encourage collaboration between academia and industry; help to move technology from the lab to commercialization; and leverage Biotechnology Center money with the Kenan Institute.
Requirements: Proposed programs and activities must clearly relate to biotechnology research, development, commercialization, education, or training. Priority will be giving to proposals requesting support for innovative programs or activities in biotechnology. Only North Carolina academic institutions, non profit research organizations, or businesses may apply.
Restrictions: Awards are not made to individuals.
Geographic Focus: North Carolina
Amount of Grant: $50,000 per year maximum
Date(s) Application is Due: Feb 25; Aug 6
Contact: Ginny DeLuca; (919) 549-8842; fax: (919) 549-9710; email: virginia_deluca@ncbiotech.org
Internet: http://www.ncbiotech.org/services_and_programs/grants_and_loans/collaborative_funding/index.html
Sponsor: North Carolina Biotechnology Center
P.O. Box, 13547, 15 T.W. Alexander Drive
Research Triangle Park, NC 27709-3547

North Carolina Biotechnology Center Institutional Development Grants **4415**
The purpose of the Institutional Development Grant program is to provide research equipment or core facilities that serve multiple investigators. Research Intensive universities require at least six investigators; Non-research intensive universities require at least three. If the proposal has any of the following features, it will be more competitive: shows that the request is occurring within the context of an institutional initiative; involves new ideas that may lead to biotechnology breakthroughs; involves collaboration between academic and industrial scientists; and strengthens regional and statewide biotechnology capabilities.
Requirements: Awards are made to any North Carolina academic or nonprofit agency.
Geographic Focus: North Carolina
Samples: University of North Carolina at Charlotte (Charlotte, North Carolina)—for the purchase of equipment that characterizes the thermal properties of polymers, $96,000.; Duke University (Durham, North Carolina)—for the development of a recruitment package, $150,000.
Amount of Grant: $250,000 maximum
Date(s) Application is Due: Sep 10
Contact: Ginny DeLuca; (919) 549-8842; fax: (919) 549-9710; email: virginia_deluca@ncbiotech.org
Internet: http://www.ncbiotech.org/services_and_programs/grants_and_loans/institutional_development/index.html
Sponsor: North Carolina Biotechnology Center
P.O. Box, 13547, 15 T.W. Alexander Drive
Research Triangle Park, NC 27709-3547

North Carolina Biotechnology Center Oliver Smithies Faculty Recruitment Grant **4416**
The goal of this program is to aid recruitment of outstanding research faculty into tenure-track positions at the senior level to non-Research Extensive colleges and universities. It is expected that the recruit will significantly impact an academic institution's overall biotechnology research initiative. These awards are intended to increase an institution's competitive advantage in negotiations with the recruit.
Requirements: Proposed programs and activities must clearly relate to biotechnology research, development, commercialization, education, or training. Priority will be giving to proposals requesting support for innovative programs or activities in biotechnology. Only North Carolina academic institutions, non profit research organizations, or businesses may apply.
Restrictions: Awards are not made to individuals. Duke, North Carolina State University and University of North Carolina at Chapel Hill are ineligible for this program.
Geographic Focus: North Carolina
Amount of Grant: $250,000
Contact: Ginny DeLuca; (919) 549-8842; fax: (919) 549-9710; email: virginia_deluca@ncbiotech.org
Internet: http://www.ncbiotech.org/services_and_programs/grants_and_loans/faculty_recruitment/index.html
Sponsor: North Carolina Biotechnology Center
P.O. Box, 13547, 15 T.W. Alexander Drive
Research Triangle Park, NC 27709-3547

North Carolina Biotechnology Center Research Grants **4417**
Biotechnology Research Grants support novel research projects that have the potential to strengthen biotechnology research and development at academic and non-profit research institutions in the early stages of building their research capacity. Awards are intended to support investigators at these institutions allowing for the generation of preliminary data required to attract funding from other sources, with the ultimate goal of technology commercialization. The preproposal application deadline is October 1. For applicants invited, the deadline for full proposals is February 4.
Requirements: Proposed programs and activities must clearly relate to biotechnology research, development, commercialization, education, or training. Priority will be giving to proposals requesting support for innovative programs or activities in biotechnology. Only North Carolina academic institutions, non profit research organizations, or businesses may apply.
Restrictions: Awards are not made to individuals. Duke, North Carolina State University and University of North Carolina at Chapel Hill are ineligible for this program.
Geographic Focus: North Carolina
Amount of Grant: $75,000 maximum
Date(s) Application is Due: Oct 1; Feb 4
Contact: Ginny DeLuca; (919) 549-8842; fax: (919) 549-9710; email: virginia_deluca@ncbiotech.org
Internet: http://www.ncbiotech.org/services_and_programs/grants_and_loans/biotech_research.html
Sponsor: North Carolina Biotechnology Center
P.O. Box, 13547, 15 T.W. Alexander Drive
Research Triangle Park, NC 27709-3547

North Carolina Biotechnology Centers of Innovation Grants **4418**
The Centers of Innovation Program is bringing together North Carolina, best scientific and technical minds in the life sciences. This program is designed to focus the state, efforts in biotechnology research, development and commercialization in targeted industrial sectors important to economic development and job creation. The North Carolina Biotechnology Center (NCBC) will work with university researchers, technology transfer offices, industrial partners, non-profit stake holders as well as regional and state-wide community leaders to establish nine Centers of Innovation (COI). Initial Centers of Innovation will complement efforts already under way in the state to align academic and industrial resources. The Phase I award is a 12-month $100,000 planning grant that will allow the development of a business plan and organization plan. The Phase II award is funding for the implementation of the business plan. Only invited applicants may apply. Potential applicants will need to demonstrate that they have established a cohesive academic-industry consortium in order to receive an invitation to submit an application.
Geographic Focus: North Carolina
Amount of Grant: $100,000 maximum (phase I); $2,400,000 maximum (phase II)
Contact: Mary Beth Thomas, Ph.D.; (919) 541-9366; fax: (919) 549-9710; email: marybeth_thomas@ncbiotech.org
Internet: http://www.ncbiotech.org/services_and_programs/COI/index.html
Sponsor: North Carolina Biotechnology Center
P.O. Box, 13547, 15 T.W. Alexander Drive
Research Triangle Park, NC 27709-3547

North Carolina Community Development Grants **4419**
The foundation supports nonprofit organizations in North Carolina to enrich the quality of life in the state, with special concern for rural regions and Cabarrus County. Grants are awarded primarily for healthcare, higher education, and community service. Other more limited areas of interest are arts, culture, historic preservation, religion, and the environment. Grants generally will be awarded for capital improvements, challenge opportunities, or special purposes rather than for operating purposes. Grant requests are considered on a quarterly basis.
Requirements: IRS 501(c)3 organizations serving North Carolina communities are eligible. Churches and governmental agencies also may apply.
Restrictions: Normally, the foundation does not make multiyear grants or frequent grants to the same organization (except for select, historic relationships); or provide operating funds, start-up funding, or lead gifts. The foundation does not make grants to individuals, to endowment funds, for loans, or for scholarships or fellowships.
Geographic Focus: North Carolina
Samples: Lenoir-Rhyne College (Hickory, NC)—for equipment and technology upgrades, $156,000 (2003); Mars Hill College (Mars Hill, NC)—to purchase two boilers and a fire-alarm system, $150,000.
Amount of Grant: $10,000-$30,000 average
Date(s) Application is Due: Jan 5; Apr 5; Jul 5; Oct 5
Contact: Frank Davis, Executive Director, (704) 786-8216; fax: (704) 785-2051
Internet: http://www.thecannonfoundationinc.org/guidelines.asp
Sponsor: Cannon Foundation
P.O. Box 548
Concord, NC 28026-0548

North Dakota Community Foundation (NDCF) Grants **4420**
Unrestricted funds are awarded in North Dakota for aid to the elderly and disadvantaged. Support is also given for health services, youth agencies, parks and recreation, arts and cultural programs, and mental health. Grants support operating budgets, continuing support, annual campaigns, seed money, building funds, equipment, matching funds, scholarship funds, special projects, research, publications, conferences, and seminars. Initial contact should be by letter. If a project appears to be appropriate to NDCF

guidelines, formal application materials will be sent to the inquiring organization. Students seeking scholarships should not contact the foundation directly.
Requirements: North Dakota nonprofit organizations are eligible.
Restrictions: Grants are not made to individuals.
Geographic Focus: North Dakota
Amount of Grant: $2000 typically
Date(s) Application is Due: Aug 15
Contact: Kevin Dvorak, President, (701) 222-8349; fax: (701) 222-8257; email: kdvorak@ndcf.net
Internet: http://www.ndcf.net/Grants/grantsindex.html
Sponsor: North Dakota Community Foundation
P.O. Box 387, 1025 N Third Street
Bismarck, ND 58502

North Dakota Indian Scholarships **4421**
The intent of the program is to assist Native American students in obtaining a basic college education. To that end, priority in making awards will be given to undergraduate full-time students. In order to provide a reward as well as an incentive to students with high academic achievements, those applicants with a cumulative grade point average of 3.50 or above, shall be given priority in funding.
Requirements: Applicants must be an enrolled member of a federally recognized Indian tribe and a resident of North Dakota. Residency for each student shall be determined by the institution that student is attending. All applicants must have been accepted for admission at an institution of higher learning or state vocational education program within North Dakota. Scholarship recipients must be enrolled full-time and may not have a grade point average below 2.00. Students participating in internships, student teaching, teaching assistance, or cooperative education programs may be eligible for a scholarship award only if participation in that program is required for the degree and only if tuition must be paid for the credits earned.
Geographic Focus: North Dakota
Amount of Grant: $500 - $2,000
Date(s) Application is Due: Jul 15
Contact: Rhonda Schauer, Coordinator of Multicultural Education, (701) 224-2497; fax: (701) 328-2961; email: rhonda.schauer@ndus.nodak.edu
Internet: http://www.ndus.nodak.edu/students/financial-aid/details.asp?id=111
Sponsor: North Dakota University System
10th Floor, State Capitol 600 East Boulevard Avenue, Dept. 215
Bismarck, ND 58505-0230

Northwest Fund for the Environment Grants **4422**
The fund's mission is to protect Washington's fish, native wildlife, natural forests, wetlands and shorelines; and preserve pure and free-flowing waters. Grant programs include aquatic ecosystem protection—freshwater ecosystems, saltwater ecosystems, and salmon; growth management—prevent damage to environmentally sensitive areas, limit rural growth, support the use of smart growth policies within urban growth areas, and support environmentally beneficial use of Washington State's land and natural resources; and capacity building—develop and increase the capacity of organizations to achieve their programmatic objectives in an effective, efficient, and sustainable manner. Sixteen-month grants are awarded. Guidelines are available online.
Requirements: Nonprofit organizations serving Washington state are eligible.
Restrictions: The fund does not support endowment or capital campaigns, land acquisition, structural restoration, political activities, capital projects, scholarships, purely educational activities, public conferences, youth groups, videos and publications, Web page development, museum displays, art projects, research, government agencies, or debt reduction.
Geographic Focus: Washington
Samples: Friends of Gray's Harbor (WA)—for a program to protect the Westport Interdunal and Aquatic Ecosystem, $18,000 (2004); People for Puget Sound (WA)—for a program to protect and restore Puget Sound's near shore habitats, $12,000 (2004); Western Environmental Law Ctr (WA)—for general operating support, $3000 (2004).
Amount of Grant: $1000-$40,000
Date(s) Application is Due: Feb 1
Contact: Pam Fujita-Yuhas or Zoe Rothchild, (206) 386-7220; fax: (206) 386-7223; email: staff@nwfund.org
Internet: http://www.nwfund.org
Sponsor: Northwest Fund for the Environment
1904 Third Avenue, Suite 615
Seattle, WA 98101

Northwestern Mutual Foundation Grants **4423**
The people of Northwestern Mutual have a long history of community involvement. To carry out its corporate contributions program, Northwestern Mutual created a private foundation, The Northwestern Mutual Foundation. The Northwestern Mutual Foundation focuses its funding in Milwaukee, where the company is headquartered. Therefore, applying for a grant from the Northwestern Mutual Foundation is limited to Milwaukee-area nonprofit organizations with programs centered on Education, Health & Human Services or Arts & Culture.
Requirements: Northwestern Mutual Foundation does not accept unsolicited grant requests. Groups new to the Foundation are asked to submit a written letter of intent, describing the organization, funding request, and provide information about the program for which you are seeking support. The Foundations team will review your letter of intent and, if they are interested in learning more about your project, you will receive an application form. Groups that have an active relationship with the Foundation and currently receive funding are allowed to apply annually for support. Once an organization submits an application, the review process takes about 90 days. The Foundation review's Milwaukee-area grants on a monthly basis.
Restrictions: Milwaukee-area nonprofits are eligible to apply. The Foundation defines the Milwaukee-area as Milwaukee County, with limited programmatic grants in Waukesha, Racine and Ozaukee Counties.
Geographic Focus: Wisconsin
Samples: Boys and Girls Clubs of Greater Milwaukee, Milwaukee, WI, $800,000 (2008); United Performing Arts Fund, Milwaukee, WI, $605,000 (2008); Scholarship America, Saint Peter, MN, $205,425 (2008);
Amount of Grant: $10,000- $800,000 range
Contact: Edward J. Zore President and CEO; (414) 271-1444;
Internet: http://www.nmfn.com/tn/aboutus—fd_intro
Sponsor: Northwestern Mutual Foundation
720 East Wisconsin Avenue
Milwaukee, WI 53202-4797

Northwood University/Alden B. Dow Creativity Center Summer Residency Grants **4424**
The program provides a 10-week residency at the university from mid-June through mid-August to allow each successful applicant to pursue innovative ideas having the potential for impact in his or her field. Project ideas must be specific with accomplishable goals for the 10-week period. Selection is based on newness and uniqueness of idea. Fellows are expected to be capable of independent research on their proposed projects. From education and literature to health care and chemistry, physics, engineering, and architecture and the arts, no creative area will be overlooked when a project idea may result in a significant advancement. Foreign citizens are also invited, provided they can communicate in written and spoken English. Project budgets are approved on an individual basis; however, ideas involving heavy developmental costs are discouraged. The grant includes travel within the United States (foreign applicants must provide transportation to the United States), room and board, and a stipend for personal expenses and/or project materials. Application fee is $10; forms and instructional brochures are available upon request.
Amount of Grant: $750 stipend
Date(s) Application is Due: Dec 31
Contact: Dr. Grover Proctor, Jr, Executive Director, (989) 837-4478; fax: (989) 837-4468; email: creativity@northwood.edu
Internet: http://www.northwood.edu/abd
Sponsor: Northwood University
4000 Whiting Drive
Midland, MI 48640-2398

Norwegian Academy of Science and Letters Abel Prize **4425**
The Abel Prize is an international prize for outstanding scientific work in the field of mathematics, including mathematical aspects of information technology, mathematical physics, probability theory, numerical analysis and computational science, statistics and applications of mathematics in other sciences. The prize is meant to recognize contributions of extraordinary depth and influence to pure and applied mathematics. Such work may have resolved fundamental problems, created powerful new techniques, introduced unifying principles or opened up major new fields of research. The prize amount is 6 million NOK (about 750,000 Euro) and was awarded for the first time in 2003.
Samples: Lennart Carleson, Royal Institute of Technology, Sweden—for his profound and seminal contributions to harmonic analysis and the theory of smooth dynamical systems (2006); Peter D. Lax, Courant Institute of Mathematical Sciences, New York University—for his groundbreaking contributions to the theory and application of partial differential equations and to the computation of their solutions (2005).
Amount of Grant: NOK 6,000,000
Contact: Prize Coordinator; + 47 22 12 10 90; fax: + 47 22 12 10 99; email: dnva@online.no
Internet: http://www.dnva.no/c26893/artikkel/vis.html?tid=26898
Sponsor: Norwegian Academy of Science and Letters
Drammensveien 78
Oslo, Norway N-0271 Norway

Norwich Jubilee Esperanto Foundation Grants **4426**
Grants are made to help young people become more competent speakers of the international language Esperanto. Grants are for travel and subsistence in the United Kingdom and to enable recipients to use their knowledge in a different language community than their own and for research into the teaching of Esperanto. Travel to and from the United Kingdom is not normally included in the award. Applications are accepted at any time.
Requirements: Candidates must not be over 25 years of age and must be in need of some financial assistance. Candidates who are not citizens of the United Kingdom or Ireland must use their grants to come to the British Isles and speak or lecture in Esperanto to local clubs or groups. Application must be made by letter in Esperanto.
Amount of Grant: @L40-@L1000
Contact: Secretary, 01865-245509
Internet: http://esperanto.org/uk/nojef
Sponsor: Norwich Jubilee Esperanto Foundation
37 Granville Ct, Cheney Ln
Oxford, England OX3 0HS United Kingdom

NPF Investigator Initiated Research Grants 4427

The foundation awards investigator-initiated research grants worldwide for projects that are directly relevant to the study of causes of and a cure for Parkinson disease; and are complementary to, not duplicative of, other research in the field. High-risk, high yield projects are encouraged, provided that they are thoughtfully composed and reasonably likely to advance the state of the knowledge about Parkinson disease. Both basic-research and clinical-research proposals are eligible for support. Preference will be given to scientists who are at an early stage of their professional careers. Awards will be made for one year, but second- and third-year awards may be sought, contingent upon successful competition with new proposals. Funding is effective July 1 through June 30. The foundation encourages applicants to submit grant applications via email.
Amount of Grant: $40,000 maximum
Date(s) Application is Due: Mar 14
Contact: Pam Olmo, (305) 243-4830; fax: (305) 243-8144; email: iiresearchgrants@parkinson.org
Internet: http://www.parkinson.org/site/pp.asp?c=9dJFJLPwB&b=71127
Sponsor: National Parkinson's Foundation
1501 NW Ninth Avenue
Miami, FL 33136-1494

NPF Large Grants 4428

The program funds significant Parkinson-related projects through large research grants. Eligible research projects should be targeted, time-limited preclinical or clinical studies—directly relevant to studying the cause and cure for Parkinson disease—in which applicants identify promising and exciting hypotheses, with the goal of obtaining sufficient data to establish or refute the hypotheses and perhaps qualify for funding from more conventional sources such as the National Institutes of Health. Grantees will be required to present a progress report and an application for further funding at the end of each year. Guidelines are available online.
Restrictions: No portion of the grant monies may be used for overhead or administrative purposes.
Amount of Grant: $100,000-$150,000 per year for up to three years, $450,000 maximum
Date(s) Application is Due: Mar 14
Contact: Pam Olmo, (305) 243-4830; fax: (305) 243-8144; email: iiresearchgrants@parkinson.org
Internet: http://www.parkinson.org/site/pp.asp?c=9dJFJLPwB&b=278606
Sponsor: National Parkinson's Foundation
1501 NW Ninth Avenue
Miami, FL 33136-1494

NPPF-Kit C. King Graduate Scholarship 4429

The scholarship is open to anyone pursuing an advanced degree in journalism with an emphasis in photojournalism. The applicant must present a portfolio that demonstrates talent and initiative in documentary photojournalism. The portfolio should be accompanied by a statement of goals and philosophy relating to documentary photojournalism. Entries should be submitted to Scott Sines, Managing Editor, Memphis Commercial-Appeal, (901) 529-5843, email: sines@commercialappeal.com.
Requirements: The applicant must have indication of acceptance to an accredited graduate program in photojournalism.
Amount of Grant: $1000
Date(s) Application is Due: Mar 1
Contact: Program Contact, (800) 289-6772 or (305) 376-3750; fax: (305) 376-0287; email: nppa@mindspring.com
Internet: http://www.nppa.org/professional_development/students/scholarships/king.html
Sponsor: National Press Photographers Foundation
3200 Croasdaile Drive, Suite 306
Durham, NC 27702

NPSC Graduate Fellowships in the Physical Sciences 4430

The consortium awards graduate fellowships in the physical sciences (astronomy, chemistry, computer science, geology, materials science, mathematical sciences, physics, and their subdisciplines, and related engineering fields—chemical, computer, electrical, environmental, and mechanical) at any participating NPSC member university or college (list available on request). This is a full-time study, six-year, renewable, doctoral track fellowship program. The fellowship covers tuition and fees and provides a stipend. NPSC welcomes applications from all qualified students with continued emphasis on recruiting underrepresented minorities: African American; Hispanic; Native American Indian, Eskimo, Aleut; Pacific Islander (Polynesian); and/or women, the underrepresented groups in the physical sciences and related engineering fields. Application and guidelines are available online.
Requirements: Applicant must be: a U.S. citizen and have the ability to pursue graduate work at an NPSC member institution; in the senior year with at least a 3.0/4.0 grade-point average, or in the first year of a graduate program pursuing a PhD, or in a terminal master's program, or returning from the workforce with no more than a master's degree.
Restrictions: Persons who already possess a doctoral degree in any field are ineligible.
Amount of Grant: $16,000 annual stipend
Date(s) Application is Due: Nov 5
Contact: Executive Director, (800) 854-6772 or (213) 821-2409; email: npschq@npsc.org
Internet: http://www.npsc.org/students/info.html
Sponsor: National Physical Science Consortium
3716 S Hope, Suite 348
Los Angeles, CA 90007

NRA Foundation National and State Fund Grants 4431

The foundation's grants focus on youth programs, range improvement and development, public safety, education and training, wildlife and natural resource conservation, and constitutional research and education. The board considers state fund applications monthly, with submission deadlines set by the field representative, and meets annually to review national grants applications (deadline listed). Applicants are required to complete a formal application form. Contact NRA field representatives or the foundation secretary for additional application information and procedures.
Restrictions: The following are not eligible for grants: deficit financing, political candidates or organizations, labor organizations, and lobby efforts.
Amount of Grant: $13.7 million total
Date(s) Application is Due: Aug 1
Contact: Sandy Elkin, Grants Manager, (800) 423-6894 ext 1131 or (703) 267-1131; email: selkin@nrahq.org
Internet: http://www.nrafoundation.org/foundation/grants.asp
Sponsor: National Rifle Association Foundation
11250 Waples Mill Road
Fairfax, VA 22030

NRC-Ford Foundation Dissertation Fellowships for Minorities 4432

Dissertation fellowships support students in research-based doctoral programs (PhD or ScD), who are in their final year of dissertation writing, and who intend to pursue an academic career in teaching and research in the behavioral and social sciences, humanities, engineering, mathematics, physical sciences, and life sciences.
Requirements: Fellowships are open to U.S. citizens or native residents of a possession of the United States who are members of one of the following minority groups: Native Americans, Alaskan Natives (Eskimo or Aleut), African Americans, Mexican Americans/Chicanas/Chicanos, Native Pacific Islanders (Micronesian or Polynesian), and Puerto Ricans. Applicants must have finished all requirements for their degree except for the writing and defense of the dissertation.
Restrictions: These fellowships are not made for work leading to degrees in practice-based disciplines. A fellow may not accept remuneration from another fellowship or similar external award while on Ford Foundation Fellowship tenure.
Amount of Grant: $21,000 stipend for one year
Date(s) Application is Due: Dec 1
Contact: Fellowship Office, (202) 334-2872; fax: (202) 334-3419; email: infofell@nas.edu
Internet: http://www7.nationalacademies.org/fellowships/forddiss.html
Sponsor: National Research Council
500 Fifth Street NW
Washington, DC 20001

NRC-Ford Foundation Diversity Fellowships 4433

This program allows fellowship recipients to engage in a year of postdoctoral research and scholarship in an environment free from the interference of their normal professional responsibilities. Supported fields are behavioral and social sciences, humanities, engineering, mathematics, physical sciences, life sciences, education, and interdisciplinary programs composed of two or more eligible research-based disciplines.
Requirements: Fellowships are open to U.S. citizens or native residents of a possession of the United States who are members of one of the following minority groups: Native Americans, Alaskan Natives (Eskimo or Aleut), African Americans, Mexican Americans/Chicanas/Chicanos, Native Pacific Islanders (Micronesian or Polynesian), and Puerto Ricans. Applicants may not have held their degree for more than seven years at time of application.
Amount of Grant: $40,000 stipend; $1500 employing institution allowance
Date(s) Application is Due: Dec 15
Contact: Fellowship Office, (202) 334-2872; fax: (202) 334-3419; email: infofell@nas.edu
Internet: http://www7.nationalacademies.org/fellowships/fordpost.html
Sponsor: National Research Council
550 Fifth Street NW
Washington, DC 20001

NRC-Ford Foundation Predoctoral Fellowships for Minorities 4434

Fellowships promote diversity in the faculty of colleges and universities by supporting students from underrepresented minority groups who are at or near the beginning of their doctoral study. Awards are made for three years of support in research-based programs (PhD or ScD) that lead to an academic career in teaching and research in the behavioral and social sciences, humanities, engineering, mathematics, physical sciences, and life sciences, education, and an interdisciplinary program composed of two or more eligible disciplines.
Requirements: Fellowships are open to U.S. citizens or native residents of a possession of the United States who are members of one of the following minority groups: Native Americans, Alaskan Natives (Eskimo or Aleut), African Americans, Mexican Americans/Chicanas/Chicanos, Native Pacific Islanders (Micronesian or Polynesian), and Puerto Ricans. Applicants must have become a citizen by the application deadline date.
Restrictions: Awards will not be made for work leading to degrees in areas related to education, business, management and administration, health sciences, nursing, home

economics, library science, speech pathology, audiology, personnel, guidance, social work, fine arts, and performing arts. In addition, awards will not be made for work leading to terminal master's degrees, doctorates in education (PhD or EdD), Doctor of Fine Arts (DFA) degrees, professional degrees in such areas as medicine, law, or public health, or for study in joint degree programs such as the MD/PhD, JD/PhD, and MFA/PhD.
Amount of Grant: $19,000 stipend, $3000 institution fee)
Date(s) Application is Due: Nov 17
Contact: Fellowship Office, (202) 334-2872; fax: (202) 334-3419; email: infofell@nas.edu
Internet: http://www7.nationalacademies.org/fellowships/fordpredoc.html
Sponsor: National Research Council
550 Fifth Str NW
Washington, DC 20001

NSERC Brockhouse Canada Prize for Interdisciplinary Research in Science and Engineering Grant 4435
The Brockhouse Canada Prize for Interdisciplinary Research in Science and Engineering recognizes outstanding Canadian teams of researchers from different disciplines who have combined their expertise to produce achievements of outstanding international significance in the natural sciences and engineering. The prize, accompanied by a team research grant of $250,000, reflects NSERC, commitment to supporting Canadian research through strategic investments in people, discovery and innovation. The grant supports the direct costs of university-based research and/or the enhancement of research facilities. The grant may be distributed in one lump sum or up to five instalments, depending on the needs of the recipients.
Requirements: Research teams nominated for the Brockhouse Canada Prize must have at least two members who are independent researchers, one of whom must hold an NSERC grant. The team can be part of an international effort, but the majority of the nominated team members must be employed at a Canadian university or public or private organization. NSERC recognizes that teams may change between the time of the specific research achievements and the time of nomination. Nominations will be accepted when changes have occurred but only as long as the core of the team remains intact. Contributions must be primarily in the natural sciences and engineering, and of an interdisciplinary and collaborative nature.
Restrictions: Any Canadian citizen may nominate a research team for the Brockhouse Canada Prize. Self nominations will not be accepted. Current NSERC Council members cannot be nominated. Since the prize is for interdisciplinary research, nominators should consult NSERC's Guidelines for the Preparation and Review of Applications in Interdisciplinary Research. Nominators should also consult NSERC's Guidelines for the Preparation and Review of Applications in Engineering and the Applied Sciences. See website for deadline & hyper links to above mentioned guidelines.
Amount of Grant: $250,000
Date(s) Application is Due: Jun 1
Contact: Program Contact; (613) 995-5829; fax: (613) 992-5337; email: brockhouse@nserc-crsng.gc.ca
Internet: http://www.nserc-crsng.gc.ca/Prizes-Prix/Brockhouse-Brockhouse/Index-Index_eng.asp
Sponsor: Natural Sciences and Engineering Research Council of Canada
350 Albert Street
Ottawa, Ontario K1A 1H5 Canada

NSERC E.W.R. Steacie Memorial Fellowships 4436
Every year, NSERC awards up to six Steacie Fellowships that are held for a two-year period. Successful Fellows are relieved of teaching and administrative duties, so that they can devote all their time and energy to research. The Fellowships are held at a Canadian university or affiliated research institution.
Requirements: Candidates should have obtained their doctorate within the last 12 years and hold a grant from NSERC.
Restrictions: Candidates may not apply on their own behalf. They must be nominated by senior members of the Canadian science and engineering community, and the nominations must be endorsed by the executive head of the candidate's university.
Amount of Grant: $90,000
Date(s) Application is Due: Jun 30
Contact: Program Contact; (613) 996-7042; fax: (613) 992-5337; email: inquiries.steacie@nserc-crsng.gc.ca
Sponsor: Natural Sciences and Engineering Research Council of Canada
350 Albert Street
Ottawa, Ontario K1A 1H5 Canada

NSERC Gerhard Herzberg Canada Gold Medal for Science and Engineering Grants 4437
The NSERC Herzberg Medal is awarded annually to an individual who has demonstrated sustained excellence and influence in research, for a body of work conducted in Canada that has substantially advanced the natural sciences or engineering fields. The award is meant to celebrate Canada's most outstanding scientists and engineers and to raise public awareness about the major contributions that Canada's top researchers make to international science and technology, and to bettering people's lives. In addition to the medal, the winner is guaranteed $1 million to use for personal university-based research or to direct in some related way such as the establishment of research scholarships fellowships or chairs in his or her name at Canadian universities. The monetary award will be distributed over a five-year period.
Requirements: To be eligible, a candidate must be a scientist or engineer from a Canadian university, government lab or private firm. Candidates may be nominated by any individual or group; however, the nominators must be Canadian citizens or permanent residents of Canada. Posthumous or self nominations will not be accepted. Current NSERC Council members are not eligible for nomination.
Restrictions: An individual may win the NSERC Herzberg Medal only once.
Samples: John C. Polanyi, University of Toronto; Chemistry (2008)
Amount of Grant: $1,000,000
Date(s) Application is Due: May 1
Contact: Program Contact; (613) 995-5829; fax: (613) 995-7753; email: inquiries.herzberg@nserc-crsng.gc.ca
Internet: http://www.nserc-crsng.gc.ca/Prizes-Prix/Herzberg-Herzberg/About-Apropos_eng.asp
Sponsor: Natural Sciences and Engineering Research Council of Canada
350 Albert Street
Ottawa, Ontario K1A 1H5 Canada

NSERC Howard Alper Postdoctoral Prize 4438
The NSERC Howard Alper Postdoctoral Prize is awarded to the most outstanding candidate in NSERC, postdoctoral fellowship competition. Valued at $20,000, the prize recognizes academic excellence, existing and potential research contributions, interpersonal and communication skills, and leadership abilities. It was established by Howard Alper, recipient of the 2000 Gerhard Herzberg Canada Gold Medal for Science and Engineering.
Requirements: Award candidates must be in the NSERC, postdoctoral fellowship competition.
Amount of Grant: $20,000
Contact: Program Contact; (613) 947-3789; email: scholfin@nserc-crsng.gc.ca
Internet: http://www.nserc-crsng.gc.ca/Prizes-Prix/Alper-Alper/Index-Index_eng.asp
Sponsor: Natural Sciences and Engineering Research Council of Canada
350 Albert Street
Ottawa, Ontario K1A 1H5 Canada

NSERC Industrial R&D Fellowships 4439
The IRDF Program provides financial contributions that support the most promising recent doctoral graduates to engage in research and development in the private sector and with not-for-profit and non-governmental organizations. Its objective is to: encourage excellent recent Ph.D. graduates in science and engineering to gain experience and seek careers in the Canadian private, not-for-profit and non-governmental sectors; promote awareness in the Canadian private, not-for-profit and non-governmental sectors of the capabilities of Canadian universities and university research; facilitate the transfer of expertise and technology; and provide an opportunity for Ph.D. holders seeking university careers to gain experience in the private, not-for-profit and non-governmental sectors; research and development. The value is $30,000 per year for two years plus a minimum contribution of $10,000 per year from the host organization. See website for further detail.
Requirements: Industrial R&D Fellowships are intended primarily for recent doctoral graduates seeking employment in the Canadian private, not-for-profit and non-governmental sectors for the first time. To be nominated for a fellowship, you must: be a Canadian citizen or permanent resident of Canada at the time of your nomination;hold a doctoral degree in a discipline of science or engineering that NSERC supports or expect to have completed all the requirements for such a degree, including your thesis defense, by the proposed date of appointment. In addition, candidates must have completed their doctoral degree within the last five years. Awards to candidates who completed their doctoral degree more than five years but no more than seven years prior to the nomination will be made only in exceptional circumstances and will have to be justified by the nominating organization. If you have withdrawn from the workforce and active research for at least one year after you received your doctorate for maternity leave or to raise a child, NSERC will extend this period to nine years.
Amount of Grant: $30,000
Contact: Director of Programs; (613) 947-0428; email: schol@nserc-crsng.gc.ca
Internet: http://www.nserc-crsng.gc.ca/Students-Etudiants/PD-NP/Industrial-Industrielle_eng.asp
Sponsor: Natural Sciences and Engineering Research Council of Canada
350 Albert Street
Ottawa, ON K1A 1H5 Canada

NSERC Industrial Research Chairs 4440
Industrial Research Chairs (IRC) are intended to assist universities in building on existing strengths to achieve the critical mass required for a major research endeavour in science and engineering of interest to industry and/or assist in the development of research efforts in fields that have not yet been developed in Canadian universities but for which there is an important industrial need. Description NSERC offers three types of Industrial Research Chairs: Senior Industrial Research Chairs (for distinguished senior researchers), Associate Industrial Research Chairs (for promising junior faculty researchers), and Executive Industrial Research Chairs (for outstanding R&D professionals). Emphasis is on the development and transfer of new knowledge from the university to industry, and on the training of highly qualified personnel in fields of importance to industry. A reasonable level of industry participation in terms of a monetary, equipment, or personnel commitment is a prerequisite for consideration of NSERC funding. University authorities should contact NSERC concerning application procedures. Applications are accepted at any time.

Requirements: Candidates must be researchers of international stature who are at or approaching the most productive period of their careers, who have the necessary qualifications to hold the rank of professor or associate professor and are recognized experts in their fields. The applicant must hold a senior administrative position in a Canadian university, such as department head, dean, vice president or president, with direct or ultimate responsibility for the proposed chair.
Restrictions: The applicant cannot be the candidate for the chair.
Contact: Canada Research Secretary: (613) 943-3087; fax: (613) 943-3282; email: information@chairs.gc.ca
Internet: http://www.nserc-crsng.gc.ca/Professors-Professeurs/CFS-PCP/IRC-PCI_eng.asp
Sponsor: Natural Sciences and Engineering Research Council of Canada
350 Albert Street, 16th Floor, P.O. Box 1610
Ottawa, ON K1A 1H5 Canada

NSERC Industrial Undergraduate Student Research Awards **4441**
These awards are meant to stimulate your interest in research in the natural sciences and engineering. They are also meant to encourage you to undertake graduate studies and to pursue a research career in these fields. If you would like to gain research work experience in a non-university setting that complements your studies, these awards can provide you with financial support through your host organization.
Requirements: To be eligible to apply for an award, you must: be a Canadian citizen or permanent resident of Canada; be registered at the time you apply in a bachelor's degree program at an eligible university; and have obtained, over the previous years of study, a cumulative average of at least second class (a grade of "B" or "B-," if applicable) as defined by your university.
Amount of Grant: $4,500 plus organization contribution
Contact: Undergraduate Cordinator: (613) 947-0428; email: schol@nserc-crsng.gc.ca
Internet: http://www.nserc-crsng.gc.ca/Students-Etudiants/UG-PC/U.S.RAI-BRPCI_eng.asp
Sponsor: Natural Sciences and Engineering Research Council of Canada
350 Albert Street
Ottawa, ON K1A 1H5 Canada

NSERC Innovation Challenge Awards **4442**
The Innovation Challenge Awards honor graduate students in the natural sciences or engineering who have demonstrated an entrepreneurial spirit by identifying ways in which their research thesis results can be developed into products and processes to benefit Canadians. Since 2004, these awards have recognized and rewarded Canada, brightest minds. The awards are aimed at fostering, in master, or PhD students who are in their final year or have recently graduated, an appreciation of real-world applications of their research. The students review their thesis work and submit an essay identifying potential products or services that could result from applying their research. Since universities play a crucial role in the innovation cycle in Canada, university Industrial Liaison Offices (ILOs), or their equivalent, are charged with selecting and submitting their institution, nominations. There are three awards: one $10,000 grand prize and two $5,000 runner-up prizes.
Requirements: Eligible nominees must have conducted research within the natural sciences or engineering at the masters or PhD level at an eligible Canadian university & be Canadian citizens or permanent residents in Canada at the time of nomination.
Samples: Darren Kraemer; University of Toronto; $10,000 (2008) Jiang Liu; University of Toronto; $5,000 (2008)
Amount of Grant: $5,000 - $10,000
Date(s) Application is Due: May 15
Contact: Program Contact; (613) 947-9452; fax: (613) 992-5337; email: innovation@nserc-crsng.gc.ca
Internet: http://www.nserc-crsng.gc.ca/Prizes-Prix/Ica-Pdi/About-Apropos_eng.asp
Sponsor: Natural Sciences and Engineering Research Council of Canada
350 Albert Street
Ottawa, Ontario K1A 1H5 Canada

NSERC John C. Polanyi Awards **4443**
In tribute to the excellence in research that Dr. John C. Polanyi continues to exemplify, NSERC offers a prestigious research prize in his name. Although Dr. Polanyi, most significant research accomplishments have been in chemistry, the NSERC John C. Polanyi Award recognizes a recent outstanding advance in any field of the natural sciences or engineering.
Requirements: To be considered for the NSERC John C. Polanyi Award, candidates must: be nominated by two active researchers from the Canadian science and engineering research community; currently hold an NSERC grant. If a team is being nominated, one team member must hold an NSERC grant and have received NSERC support for the research that led to the advance for which the individual or team is being nominated.
Amount of Grant: $250,000
Date(s) Application is Due: Mar 2
Contact: Award Contact; (613) 995-5829; fax: (613) 992-5337; email: polanyi@nserc-crsng.gc.ca
Internet: http://www.nserc-crsng.gc.ca/Prizes-Prix/Polanyi-Polanyi/Index-Index_eng.asp
Sponsor: Natural Sciences and Engineering Research Council of Canada
350 Albert Street
Ottawa, Ontario K1A 1H5 Canada

NSERC Northern Research Internships **4444**
Northern Research Internships Program supports senior undergraduate students, graduate students and postdoctoral fellows pursuing studies in the Canadian Northl by facilitating extended stays and subsidizing the costs associated with activities in the North. Up to $10,000 for a 16-week internship, including a logistics allowance of up to $4,000.
Requirements: See website for extensive list of requirements.
Amount of Grant: $10,000
Date(s) Application is Due: no deadline date
Contact: Scholarships and Fellowships Division; (613) 947-0428; fax: (613) 996-2589; email: schol@nserc-crsng.gc.ca
Internet: http://www.nserc-crsng.gc.ca/Students-Etudiants/NorthernIntern-NordiqueStage_eng.asp
Sponsor: Natural Sciences and Engineering Research Council of Canada
350 Albert Street
Ottawa, ON K1A 1H5 Canada

NSERC Postdoctoral Fellowships **4445**
The Postdoctoral Fellowships (PDF) Program provides support to a core of the most promising researchers at a pivotal time in their careers. The fellowships are also intended to secure a supply of highly qualified Canadians with leading-edge scientific and research skills for Canadian industry, government, and universities. The value of his fellowship is $40,000 per year for two years.
Requirements: You must be a Canadian citizen or a permanent resident of Canada and hold or expect to hold a doctorate in one of the fields of research that NSERC supports. In addition: You must expect to complete all your doctoral requirements by January 1 of the year following the award announcement. You must have completed all requirements of your first doctoral degree no more than two years before the deadline date of the year in which you apply. If you have acquired at least six months of full-time relevant work experience in industry or government (experience in a university or its affiliated hospitals, research centers, and other laboratories will not be considered) after you received your doctorate, NSERC will extend this eligibility period to three years. If you have withdrawn from the workforce and active research for maternity leave or to raise a child for at least one year after you received your doctorate, NSERC will extend this eligibility period to six years. NSERC uses the date of degree completion to determine applicant eligibility. This is the date on which all requirements of your degree have been met, including successful defense and submission of the corrected copy of your thesis. You can hold only one NSERC Postdoctoral Fellowship. If NSERC does not award you a fellowship after your first application, you may apply again, provided you meet the eligibility criteria. You can apply only twice for a Postdoctoral Fellowship. You may submit only one application per year to either NSERC or CIHR or SSHRC.
Amount of Grant: $40,000
Date(s) Application is Due: Oct 15
Contact: Director of Programs; (613) 947-0428; email: schol@nserc-crsng.gc.ca
Internet: http://www.nserc-crsng.gc.ca/Students-Etudiants/PD-NP/PDF-BP_eng.asp
Sponsor: Natural Sciences and Engineering Research Council of Canada
350 Albert Street
Ottawa, ON K1A 1H5 Canada

NSERC Research Tools and Instruments (RTI) Grants **4446**
Research Tools and Instruments (RTI) Grants foster and enhance the discovery, innovation and training capability of university researchers in the natural sciences and engineering by supporting the purchase of research equipment and installations. RTI Grants are one-year awards that assist in buying or developing research equipment that costs more than $7,000. In exceptional cases, multiple-year requests will be accepted. The onus is on the applicant to clearly explain the rationale. Items costing less than $7,000 each can be purchased with Discovery Grant funds or be included as a complement to a main piece of equipment being requested.RTI grants are divided into three categories according to the total net cost of the equipment, i.e., excluding taxes, shipping, and handling. These categories are: RTI - Category 1: $7,001 to $150,000; RTI - Category 2: $150,001 to $325,000; and RTI - Category 3: more than $325,000.
Requirements: To apply, submit an Application for a Grant (Form 101) and a Personal Data Form (Form 100) for the applicant and each co-applicant. Only applications from applicants and co-applicants who currently hold or are applying for an NSERC research grant (e.g., Discovery, Strategic, Research Partnership, Canada Research Chairs) at the time of application will be accepted.
Amount of Grant: $7,000 minimum
Date(s) Application is Due: Oct 1, Oct 25
Contact: Grants Administrator, (613) 995-5829; fax: (613) 992-5337; email: resgrant@nserc.ca
Internet: http://www.nserc-crsng.gc.ca/Professors-Professeurs/RTII-OIRI/RTI-OIR_eng.asp
Sponsor: Natural Sciences and Engineering Research Council of Canada
350 Albert Street
Ottawa, ON K1A 1H5 Canada

NSERC Synergy Awards for Innovation **4447**
The annual Synergy Awards for Innovation recognize examples of university-industry collaboration that stand as a model of effective partnership. By working together, award-winning companies and universities have proven that effective partnerships are the foundation of achievement. Their success has enriched the academic and research programs within Canadian universities while providing tangible benefits to Canadians. Synergy Awards are offered in four categories: small and Medium-Sized Companies

- a partnership with a single company of up to 500 employees; large Companies - a partnership with a single company of more than 500 employees; two or More Companies - A partnership with two or more companies of any size; Leo Derikx Award - An established innovative model of long-standing university-industry partnership in pre-competitive R&D that has improved the general well-being of an industry. Up to four awards may be presented annually, one in each category.
Requirements: The nominated partnership must be between a Canadian university professor or group of professors and a Canadian-based company or companies. Nominations must be submitted by a senior official from one of the participating organizations. Nominations will be accepted only from the President, CEO or Vice-President of Research (or equivalent) of the industrial partner, or from the President, Rector, Vice-President of Research or Vice-Rector of Research (or equivalent) from the university. Letters of support from the other partner organization(s) must be from a senior executive in the company or executive head of the university, and must be included with the nomination. Refer to Nomination Checklist to ensure that your nomination package is complete. The checklist will be found on the NSERC website.
Amount of Grant: $200,000
Date(s) Application is Due: May 1
Contact: (613) 947-9452; fax: (613) 992-5337; email: synergy@nserc-crsng.gc.ca
Internet: http://www.nserc-crsng.gc.ca/Prizes-Prix/Synergy-Synergie/Index-Index_eng.asp
Sponsor: Natural Sciences and Engineering Research Council of Canada
350 Albert Street
Ottawa, ON K1A 1H5 Canada

NSERC Undergraduate Student Research Awards **4448**
Undergraduate Student Research Awards (U.S.RA) are meant to stimulate your interest in research in the natural sciences and engineering. They are also meant to encourage you to undertake graduate studies and pursue a research career in these fields. If you would like to gain research work experience that complements your studies in an academic setting, these awards can provide you with financial support through your host university.
Requirements: To be eligible to apply for an award, you must: be a Canadian citizen or permanent resident of Canada; be registered at the time you apply in a bachelor's degree program at an eligible university and have obtained, over the previous years of study, a cumulative average of at least second class (a grade of "B" or "B-," if applicable) as defined by your university.
Amount of Grant: $4,500 plus organization contribution
Date(s) Application is Due: Minimum of four weeks before start date of work term
Contact: Undergraduate Cordinator: (613) 947-0428; email: schol@nserc-crsng.gc.ca
Internet: http://www.nserc-crsng.gc.ca/Students-Etudiants/UG-PC/U.S.RA-BRPC_eng.asp
Sponsor: Natural Sciences and Engineering Research Council of Canada
350 Albert Street
Ottawa, ON K1A 1H5 Canada

NSF Conferences, Workshops, and Special Meetings in the Mathematical Sciences **4449**
NSF supports research, education, and infrastructure for the mathematical sciences in a variety of modes, including support for symposia, workshops, regional conferences, and special research years. These provide exposure for scholarly work, encourage graduate students and junior scientists early in their careers, and they influence the sense of participation, the concept of a role-model, and the perception of current opportunities in a given scientific field. proposals normally request funding in the range of $5,000 to $25,000, although awards of up to $50,000 have occasionally been made. Their duration is normally for one year. NSF 05-540
Requirements: Proposals may be submitted by universities, colleges, and nonprofit, nonacademic organizations in support of individual investigators or small groups.
Amount of Grant: $5,000 - $25,000
Date(s) Application is Due: Aug 23
Contact: Appropriate Program Officer (see DMS Homepage http://www.nsf.gov/mps/dms/start.htm)
Internet: http://www.nsf.gov/publications/pub_summ.jsp?ods_key=nsf05540
Sponsor: National Science Foundation
4201 Wilson Boulevard
Arlington, VA 22230

NSF Accelerating Discovery In Science and Engineering through PETASCALE Simulations and Analysis (PetaApps) **4450**
The primary purpose of this program is to support projects that will enable researchers to capitalize on emerging petascale computing architectures and thus catalyze progress in science and engineering beyond the current state-of-the-art. The emphasis is on implementation and exploitation of forefront techniques. Proposals from or including junior researchers are encouraged as one of the goals of this solicitation is to build a community capable of using petascale computing. Proposers must be prepared to demonstrate that they have a science or engineering research challenge requiring petascale computing and that they have a path to exploiting such computation capabilities effectively.
Requirements: Proposals are encouraged in the following areas although work in other areas will also be considered: enhancing algorithmic scalability using techniques that better exploit multi-threaded, highly parallel, hierarchical architectures; improving and creating data sampling, analysis, and clustering algorithms for massive data sets, including scalable interactive visualization tools; developing innovative modeling, simulation or optimization algorithms suitable for petascale systems; developing platform-independent software to solve forefront scientific problems on petascale systems; optimizing software for specific petascale hardware or predicted 'best guess' extrapolations to future hardware; exploring innovative computational techniques that were not previously considered viable because of limited hardware capability; conducting performance analysis and profiling of software that is heavily used but may never have been analyzed for scalability, bottle-necks, and optimization; changing functioning problem solvers by substituting algorithmic implementations known from CS research to scale more effectively to computing systems of very large scale; adapting a research code that currently uses MPI on supercomputers, to one or more of Co-Array Fortran, UPC or Titanium, in a form that will be useful for specific petascale problems. Estimated program budget, number of awards and average award size/duration are subject to the availability of funds. Proposal budgets should not exceed $2M in size and the requested duration should not exceed five years.
Restrictions: An individual may be the Primary Investigator in no more than one proposal that responds to this solicitation.
Amount of Grant: $900,000 - $2 million
Date(s) Application is Due: Oct 30
Contact: Daryl Hess; (703) 292-4942; fax: (703) 292-9035; email: dhess@nsf.gov; Hans Kaper; (703) 292-4859; email: hkaper@nsf.gov
Internet: http://nsf.gov/funding/pgm_summ.jsp?pims_id=501015&org=NSF&sel_org=NSF&from=fund
Sponsor: National Science Foundation
4201 Wilson Boulevard
Arlington, VA 22230

NSF Active Nanostructures and Nanosystems Grants **4451**
The National Science Foundation announces a program on collaborative research and education in the area of active nanostructures, nanosystems, and on the long-term societal change associated with these innovations. Active nanostructures change or evolve their structure, property, or function during their operation. The goal of this program is to support fundamental research and catalyze synergistic science and engineering research and education in several emerging areas of nanoscale science and engineering, including: active nanostructures; nanosystems with improved functionality and new architectures; hierarchical nanomanufacturing; and long-term societal and educational implications of scientific and technological advances on the nanoscale. This solicitation provides support for Nanoscale Interdisciplinary Research Teams (NIRT) and Nanoscale Exploratory Research (NER). Other research and education projects in nanoscale science and engineering will continue to be supported in the relevant programs and divisions. The anticipated funding amount is $29,000,000 total, of which $24,500,000 is for approximately 20 NIRT awards and $4,500,000 is for approximately 40 NER awards. Full proposals should be submitted via FastLane or Grants.gov.
Requirements: Proposals may be submitted by a single organization or a group of organizations consisting of a lead organization in partnership with one or more partner organizations. Principal investigators are encouraged to form synergistic collaborations among researchers.
Restrictions: Only U.S. academic institutions with significant research and degree-granting education programs in disciplines normally supported by NSF are eligible.
Amount of Grant: $29,000,000 total
Date(s) Application is Due: Nov 15
Contact: Kevin Lyons, Program Director; (703) 292-5365; fax: (703) 292-9056; email: klyons@nsf.gov; Matthew Realff, Program Director; (703) 292-7081; fax: (703) 292-9055; email: mrealff@nsf.gov
Internet: http://www.nsf.gov/funding/pgm_summ.jsp?pims_id=13633
Sponsor: National Science Foundation
4201 Wilson Boulevard
Arlington, VA 22230

NSF Activities in Science, Engineering, and Mathematics for Persons with Disabilities **4452**
The program funds projects that enhance the academic and career development of persons with disabilities in fields related to science, math, technology, and engineering (SMET). Demonstration and intervention projects may last from one to eight years, and funding for this category is to help grantees design and carry out innovative projects that reduce barriers to science, mathematics, engineering, and technology education for persons with disabilities. Information dissemination projects support innovative activities designed to disseminate and promote use of program results, to promote replication of successful strategies used to increase the recruitment, education, and retention in SMET education and career development of persons with disabilities, and to provide the Nation's SMET teaching force information on strategies that permit full and successful participation of students with disabilities in SMET educational environments. Capacity building within community colleges projects encourage proposals designed to establish and test models by which students with disabilities are recruited and trained in SMET disciplines in community college environments. All applicants must submit a preliminary proposal and are encouraged to discuss their proposals with NSF staff before doing so. Full proposals should be submitted via the FastLane system at NSF or Grants.gov. Refer to NSF brochure NSF 00-69.
Requirements: Organizations eligible to submit proposals under this program include universities and colleges; nonprofit, nonacademic institutions, including museums, professional associations, and community-based organizations; state and local governments, including school districts; and for-profit research institutions and businesses.
Amount of Grant: $4.5 million total

Date(s) Application is Due: May 15; Aug 15
Contact: Lawrence Scadden, Program Director; (703) 306-1635, ext 6865; email: lscadden@nsf.gov
Internet: http://www.nsf.gov/pubs/2000/nsf0069/nsf0069.htm
Sponsor: National Science Foundation
4201 Wilson Boulevard
Arlington, VA 22230

NSF ADVANCE: Increasing the Participation and Advancement of Women in Academic Science and Engineering Careers Grants **4453**
The goal of the ADVANCE program is to develop systemic approaches to increase the representation and advancement of women in academic science, technology, engineering and mathematics (STEM) careers, thereby contributing to the development of a more diverse science and engineering workforce. Creative strategies to realize this goal are sought from women and men. Members of underrepresented minority groups and individuals with disabilities are especially encouraged to apply. Proposals that address the participation and advancement of women with disabilities and women from underrepresented minority groups are particularly encouraged. Creative strategies to realize this goal are sought from women and men. Members of underrepresented minority groups and individuals with disabilities are especially encouraged to apply. Proposals that address the participation and advancement of women with disabilities and women from underrepresented minority groups are particularly encouraged. Letters of Intent are due by January 20, with full proposals due by February 24.
Requirements: Proposals from primarily undergraduate institutions, teaching intensive colleges, community colleges, minority-serving institutions (e.g. Tribal Colleges and Universities, Historically Black Colleges and Universities, Hispanic-Serving Institutions), women's colleges, and institutions primarily serving persons with disabilities are encouraged. Institutional Transformation proposals may be submitted by academic institutions of higher learning in the U.S., its territories or possessions, or the Commonwealth of Puerto Rico, that award degrees in a field supported by NSF. Partnerships involving industry, government, professional societies and other not-for-profit organizations are encouraged but not required; however, in the case of partnerships, the lead partner organization must be an academic institution of higher learning and must accept overall management responsibility for the activity. Government organizations (other than academic institutions of higher learning) are not eligible to apply, but may participate in partnerships when the lead partner organization is an academic institution of higher learning with overall management responsibility for the activity. Partner organizations and institutions must be based in the U.S., its territories or possessions, or the Commonwealth of Puerto Rico. Partnerships for Adaptation, Implementation and Dissemination proposals may be submitted by: academic institutions of higher learning that award degrees in a field supported by NSF; professional societies; or other not-for-profit organizations. Submitting institutions and organizations must be based in the U.S., its territories or possessions, or the Commonwealth of Puerto Rico. Government organizations (other than academic institutions of higher learning) are not eligible to apply, but may participate in partnerships when the lead partner organization is an academic institution of higher learning with overall management responsibility for the activity. Partner institutions and organizations must be based in the U.S., its territories or possessions, or the Commonwealth of Puerto Rico.
Restrictions: Organizations that have received NSF ADVANCE Institutional Transformation awards are not eligible to apply for another institutional transformation award or for an IT-Start award. Organizations may submit only one Institutional Transformation proposal or one IT-Start proposal. There is no limitation on the number of PAID proposals.
Samples: CUNY Hunter College, NY - $499,901; University of Missouri-Columbia, MO - $499,993; University of Wisconsin-Madison - $499,991; Wesleyan University, CT - $259,593
Amount of Grant: $500,000
Date(s) Application is Due: Jan 20; Feb 24
Contact: Jessie DeAro, Program Director; (703) 292-5350; fax: (703) 292-9018; email: jdearo@nsf.gov
Internet: http://www.nsf.gov/funding/pgm_summ.jsp?pims_id=5383
Sponsor: National Science Foundation
4201 Wilson Boulevard
Arlington, VA 22230

NSF Advanced Learning Technologies (ALT) Grants **4454**
Through this program, the NSF supports research that: enables radical improvements in learning through innovative computer and information technologies; and advances research in computer science, information technology, learning, and cognitive science through the unique challenges posed by learning environments and learning technology platforms. Integrative research approaches that build across disciplines and establish tight linkages among theory, experiment, and design are strongly encouraged. Technology goals may include systems for tutoring or assessment, modeling and sensing of cognitive or emotional states, context awareness, natural language interfaces, collaboration, knowledge management, and non-traditional goals that redefine the roles of technology in learning. Educational foci for ALT projects must include an area of science, technology, engineering, or mathematics (STEM), or general cross-cutting skills directly relevant to STEM. The anticipated funding amount is $2,800,000 total, to be divided into approximately 6 to 10 awards annually. Full proposals should be submitted via FastLane or Grants.gov.
Requirements: In response to this solicitation, an investigator may participate as PI or Co-PI in no more than one proposal.
Amount of Grant: $280,000-$470,000 average; $2,800,000 total
Date(s) Application is Due: Apr 25
Contact: Amy Baylor, Program Director; (703) 292-8930; email: abaylor@nsf.gov; John C. Cherniavsky, Senior EHR Advisor for Research; (703) 292-5136; fax: (703) 292-9046; email: jchernia@nsf.gov
Internet: http://www.nsf.gov/funding/pgm_summ.jsp?pims_id=12834
Sponsor: National Science Foundation
4201 Wilson Boulevard
Arlington, VA 22230

NSF Advanced Technological Education (ATE) Grants **4455**
With an emphasis on two-year colleges, the ATE program focuses on the education of technicians for the high-technology fields that drive our nation's economy. The program involves partnerships between academic institutions and employers to promote improvement in the education of science and engineering technicians at the undergraduate and secondary school levels. The ATE program supports curriculum development; professional development of college faculty and secondary school teachers; career pathways to two-year colleges from secondary schools and from two-year colleges to four-year institutions; and other activities. A secondary goal is articulation between two-year and four-year programs for K-12 prospective teachers that focus on technological education. The program also invites proposals focusing on applied research relating to technician education. This program provides educational opportunities for Undergraduate Students, K-12 Educators. This program supports institutions which may provide support to individuals at those institutions. Optional preliminary proposals are due by the April deadline; full proposals are due by in October. Full proposals should be submitted via FastLane or Grants.gov. NSF 07-530.
Requirements: Proposers may opt to submit proposals via Grants.gov or via the NSF FastLane system. In determining which method to utilize in the electronic preparation and submission of the proposal, please note that all collaborative proposals submitted as separate submissions from multiple organizations must be submitted via the NSF FastLane system. Chapter II, Section D.3 of the Grant Proposal Guide provides additional information on collaborative proposals.
Amount of Grant: $600,000-$700,000 range
Date(s) Application is Due: Oct 15
Contact: Eileen L. Lewis, Lead Program Director; (703) 292-4627; fax: (703) 292-9015; email: ellewis@nsf.gov; Gerhard L. Salinger, Lead Program Director; (703) 292-5116; email: gsalinge@nsf.gov
Internet: http://www.nsf.gov/funding/pgm_summ.jsp?pims_id=5464
Sponsor: National Science Foundation
4201 Wilson Boulevard, Room 8
Arlington, VA 22230

NSF Advanced Technologies and Instrumentation (ATI) Grants **4456**
The Advanced Technologies and Instrumentation (ATI) program provides grants to support the development and construction of state-of-the-art detectors and instruments for the visible, infrared, and radio regions of the spectrum, including interferometric imaging instrumentation and adaptive optics, and the application of new hardware and software technology and innovative techniques in astronomical research. Instrumentation projects should be driven by scientific considerations and clearly articulated scientific goals. Proposals to the ATI program should include a brief task implementation plan with milestones, schedules, and costs. Full proposals should be submitted via the FastLane system or Grants.gov.
Samples: Richard Dekany, California Institute of Technology, for 2007 Adaptive Optics Roadmap Update Meeting, $49,507 (2007); Gaspar Bakos, Smithsonian Institution Astrophysical Observatory, for MRI: Acquisition of a 3-Station Global Network of Automated Telescopes to Detect a Large Number of Nearby Transiting Extra-solar Planets, $719,533 (2007).
Date(s) Application is Due: Nov 1
Contact: Jeffrey R. Pier, Program Director; (703) 292-2977; fax: (703) 292-9034; email: jpier@nsf.gov
Internet: http://www.nsf.gov/funding/pgm_summ.jsp?pims_id=5660
Sponsor: National Science Foundation
4201 Wilson Boulevard
Arlington, VA 22230

NSF Advancing Theory in Biology (ATB) Grants **4457**
The Biological Sciences Directorate invites submission of proposals that advance our conceptual and theoretical understanding of living systems. The Advancing Theory in Biology (ATB) solicitation supports the development of new theoretical approaches that will improve our understanding of fundamental biological principles that integrate phenomena across levels of biological organization. The anticipated funding amount is $5,000,000 total, of which 10 to 15 grants will be awarded. Full proposals should be submitted via the FastLane system at NSF or Grants.gov.
Amount of Grant: $330,000 - $500,000 range; $5,000,000 toal funding
Date(s) Application is Due: Feb 17
Contact: Kamal Shukla, Program Director; (703) 292-7131; fax: (703) 292-9061; email: kshukla@nsf.gov; Irwin Forseth, Program Director; (703) 292-8413; fax: (703) 292-9153; email: iforseth@nsf.gov
Internet: http://www.nsf.gov/funding/pgm_summ.jsp?pims_id=501066
Sponsor: National Science Foundation
4201 Wilson Boulevard
Arlington, VA 22230

NSF Aeronomy Grants 4458

The Aeronomy program supports research on upper and middle atmosphere phenomena of ionization, recombination, chemical reaction, photo emission, and transport; the transport of energy, and momentum. This program also supports research into mass in the mesosphere-thermosphere-ionosphere system, including the processes involved and the coupling of this global system to the stratosphere below and magnetosphere above and the plasma physics of phenomena manifested in the coupled ionosphere-magnetosphere system, including the effects of high-power radio wave modification. Full proposals should be submitted via the FastLane system at NSF or Grants.gov.
Contact: Cassandra Fesen, Program Director; (703) 292-8529; fax: (703) 292-9022; email: cfesen@nsf.gov
Internet: http://www.nsf.gov/funding/pgm_summ.jsp?pims_id=11686
Sponsor: National Science Foundation
4201 Wilson Boulevard
Arlington, VA 22230

NSF Alan T. Waterman Award 4459

The annual award recognizes an outstanding young researcher in any field of science or engineering supported by the National Science Foundation. In addition to a medal, the awardee receives a grant of $500,000 over a three year period for scientific research or advanced study in the mathematical, physical, biological, engineering, social, or other sciences at the institution of the recipient's choice.
Requirements: Candidates must be U.S. citizens, normally 35 years of age or younger, or not more than seven years beyond receipt of the PhD degree by December 31 of the year nominated. Candidates should have demonstrated exceptional individual achievements in scientific or engineering research of sufficient quality to place them at the forefront of their peers. Criteria include originality, innovation, and significant impact on the field.
Samples: David Charbonneau, Professor of Astronomy, Harvard University (2009); Terence Tao, Professor of Mathematics, University of California, Los Angeles (2008).
Amount of Grant: $500,000 over a three-year period
Date(s) Application is Due: Dec 31
Contact: Mayra N. Montrose, Program Contact; (703) 292-8040; fax: (703) 292-9040; email: mmontros@nsf
Internet: http://www.nsf.gov/od/waterman/waterman.jsp
Sponsor: National Science Foundation
4201 Wilson Boulevard
Arlington, VA 22230

NSF Algebra, Number Theory and Combinatorics (ANTC) Grants 4460

The Program supports research in algebra, including algebraic structures, general algebra, and linear algebra; number theory, including algebraic, analytic number theory, arithmetic geometry, quadratic forms, and automorphic forms; combinatorics, including graph theory; and algebraic geometry. Full proposals should be submitted via the FastLane system at NSF or Grants.gov.
Date(s) Application is Due: Oct 6
Contact: Tie Luo, Program Director; (703) 292-8448; fax: (703) 292-9032; email: tluo@nsf.gov; Andrew D. Pollington; (703) 292-4878; fax: (703) 292-9032; email: adpollin@nsf.gov
Internet: http://www.nsf.gov/funding/pgm_summ.jsp?pims_id=5431&org=DMS&from=home
Sponsor: National Science Foundation
4201 Wilson Boulevard
Arlington, VA 22230

NSF Analysis Grants 4461

The Program supports basic research in that area of mathematics whose roots can be traced to the calculus of Newton and Leibniz. Given its centuries-old ties to physics, analysis has influenced developments from Newton's mechanics to quantum mechanics and from Fourier's study of heat conduction to Maxwell's equations of electromagnetism to Witten's theory of supersymmetry. More generally, research supported by analysis provides the theoretical underpinning for the majority of applications of the mathematical sciences to other scientific disciplines. Current areas of significant activity include: nonlinear partial differential equations; dynamical systems and ergodic theory; real, complex and harmonic analysis; operator theory and algebras of operators on Hilbert space; mathematical physics; and representation theory of Lie groups/algebras. Emerging areas include random matrix theory and its ties to classical analysis, number theory, quantum mechanics, and coding theory, and development of noncommutative geometry with its applications to modeling physical phenomena. It should be stressed, however, that the underlying role of the Analysis Program is to provide support for research in mathematics at the most fundamental level. Although this is often done with the expectation that the research will generate a payoff in applications at some point down the road, the principal mission of the Program is to tend and replenish an important reservoir of mathematical knowledge, maintaining it as a dependable resource to be drawn upon by engineers, life and physical scientists, and other mathematical scientists, as need arises. Full proposals should be submitted via the FastLane system at NSF or Grants.gov.
Date(s) Application is Due: Oct 6
Contact: Joe W. Jenkins, Program Director; (703) 292-4870; fax: (703) 292-9032; email: jjenkins@nsf.gov
Internet: http://www.nsf.gov/funding/pgm_summ.jsp?pims_id=5434&org=DMS&from=home
Sponsor: National Science Foundation
4201 Wilson Boulevard
Arlington, VA 22230

NSF Analytical & Surface Chemistry (ASC) Grants 4462

Supports fundamental chemical research directed toward the characterization and analysis of all forms of matter. Studies of elemental and molecular composition and of the microstructure of both bulk and surface domains are included. The program supports projects that develop the fundamentals of measurement science, new sensors and new instruments, and innovative approaches to data processing and interpretation. Investigations designed to probe the chemical structure and reactivity of the interface between different forms of matter also are supported. The program is linked to several other chemistry research programs within NSF, including Solid State Chemistry (Materials Research Division, MPS Directorate); Biochemistry and Biophysics (Molecular and Cellular Biosciences Division, BIO Directorate); and Chemical Reaction Processes and Interfacial, Transport, and Separation Processes (Chemical and Transport Systems Division, ENG Directorate). Full proposals should be submitted via the FastLane system at NSF or Grants.gov.
Date(s) Application is Due: Jul 31; Nov 30
Contact: Kelsey D. Cook, Program Officer; (703) 292-7490; fax: (703) 292-9037; email: kcook@nsf.gov; Zeev Rosenzweig, Program Officer; (703) 292-7719; fax: (703) 292-9037; email: zrosenzw@nsf.gov
Internet: http://www.nsf.gov/funding/pgm_summ.jsp?pims_id=5677
Sponsor: National Science Foundation
4201 Wilson Boulevard
Arlington, VA 22230

NSF Animal Developmental Mechanisms Grants 4463

Animal Developmental Systems programmatic area supports research that seeks to understand the processes that result in the complex phenotype of animals. Because different organisms may be more amenable to certain approaches than others, analyses of development in a wide range of different species are encouraged. The Developmental Systems Cluster is also particularly interested in understanding how emergent properties result in the development of complex phenotypes and lead to the evolution of developmental mechanisms. Proposals may be submitted at any time; however, receipt by target dates will ensure review at the regular panel meeting. Full proposals should be submitted via the FastLane system at NSF or Grants.gov.
Requirements: The most frequent recipients of support are academic institutions and nonprofit research groups. In special circumstances, grants also are awarded to other types of institutions and to individuals.
Date(s) Application is Due: Jan 12; Jul 12
Contact: James Deshler, Program Director; (703) 292-8417; email: jdeshler@nsf.gov
Internet: http://www.nsf.gov/funding/pgm_summ.jsp?pims_id=501087
Sponsor: National Science Foundation
4201 Wilson Boulevard
Arlington, VA 22230

NSF Antarctic Aeronomy and Astrophysics Program Grants 4464

The Program supports studies of three major domains: the stratosphere and the mesosphere; the thermosphere, the ionosphere, and the magnetosphere; and astronomy and astrophysical studies of the Universe, including solar astronomy and cosmic ray physics. Major goals are to sponsor research that requires or would benefit from the unique conditions of the Antarctic, to contribute to the understanding of the role of the Antarctic in global environmental change, to participate in interdisciplinary studies of the middle and upper atmosphere, and to improve understanding of the coupling of the Earth's polar atmosphere with the magnetosphere and of the ways in which both are affected by solar activity. A total of $20,000,000 is available, plus up to $35,000,000 in outyear increments for continuing awards. Deadlines are: May 2 for research proposals on the Oden during November through December; and June 6 for all proposals other than work proposed on Oden for November through December. Full proposals should be submitted via the FastLane system at NSF or Grants.gov.
Amount of Grant: $400,000 average
Contact: Vladimir Papitashvili, Program Director; (703) 292-7425; email: vpapita@nsf.gov
Internet: http://www.nsf.gov/funding/pgm_summ.jsp?pims_id=13420
Sponsor: National Science Foundation
4201 Wilson Boulevard
Arlington, VA 22230

NSF Antarctic Artists and Writers Program Grants 4465

The purpose of the Program is to enable serious writings and works of art that exemplify the Antarctic heritage of humankind. In particular, the program seeks to increase public understanding of the Antarctic region, including the continent and the surrounding oceans, as well as the associated research and education endeavors. The Program provides opportunities for professional artists and writers to travel to Antarctica — at research stations, field camps, and aboard ships — to make the observations necessary to complete their proposed projects. While the majority of award recipients are established artists and writers, the program also seeks to support early career artists and writers in an effort to broaden participation. Full proposals should be submitted via the FastLane system at NSF or Grants.gov. *Requirements*:
Date(s) Application is Due: Jun 8
Contact: Winifred M. Reuning; (703) 292-8033; fax: (703) 292-9079; email: wreuning@nsf.gov
Internet: http://www.nsf.gov/funding/pgm_summ.jsp?pims_id=12783
Sponsor: National Science Foundation
4201 Wilson Boulevard
Arlington, VA 22230

NSF Antarctic Earth Sciences Program Grants **4466**
The Program supports research in both terrestrial and marine settings to interpret Antarctica's rich history and to understand the processes that shape it today. Current projects address such diverse topics as: understanding the ice sheets using sediment records to reconstruct their history and determine the geologic controls of their formation and stability; deciphering paleoenvironmental and paleobiological records to understand global climate, ocean circulation, and the evolution of life; exploring Antarctica's tectonic evolution, from its central role in Gondwana's breakup to the present-day deformation driving volcanism, rifting, and orogenesis; and investigating unique processes, such as the formation of subglacial lakes or the aeolian and permafrost sculpting of the Dry Valleys. All of these problems require a better understanding of Antarctica's geology, including the structure and composition of its crust and underlying mantle, and its connection to global earth science questions. The program therefore supports and encourages field, laboratory, and theoretical work in geology, geophysics, and any other area of earth sciences, including the intersections between these fields and biology, glaciology, and oceanography. The program also strongly encourages investigators new to the program, international collaboration, and research-related education and outreach. Work on previously collected samples and data is also encouraged. A total of $20,000,000 is available, plus up to $35,000,000 in outyear increments for continuing awards. Deadlines are: May 2 for research proposals on the Oden during November through December; and June 6 for all proposals other than work proposed on Oden for November through December. Full proposals should be submitted via the FastLane system at NSF or Grants.gov.
Amount of Grant: $400,000 average
Date(s) Application is Due: May 2; Jun 6
Contact: Thomas P. Wagner, Program Director; (703) 292-4746; fax: (703) 292-9079; email: twagner@nsf.gov
Internet: http://www.nsf.gov/funding/pgm_summ.jsp?pims_id=8173
Sponsor: National Science Foundation
4201 Wilson Boulevard
Arlington, VA 22230

NSF Antarctic Glaciology Program Grants **4467**
The Program is concerned with the study of the history and dynamics of all naturally occurring forms of snow and glacial ice, including floating ice shelves, glaciers, and continental and marine ice sheets. Program emphases include paleoenvironments from ice cores, ice dynamics, numerical modeling, glacial geology, and remote sensing of ice sheets. Some specific objectives are: correlation of climatic fluctuations evident in antarctic ice cores with data from arctic and lower-latitude ice cores, and integration of the ice record with the terrestrial and marine record; documentation of the geographic extent of climatic events noted in paleoclimatic records; extension of the ice core time series to provide information on astronomical forcing of climate; establishment of more precise dating methodologies for deep ice cores; determination of the Cenozoic history of antarctic ice sheets and their interaction with global climate and uplift of the Transantarctic Mountains; response of the antarctic ice sheets to the Pliocene warming; investigation of the physics of fast glacier flow with emphasis on processes at glacier beds.; investigation of ice-shelf stability; and identification and quantification of the feedback between ice dynamics and climate change. A total of $20,000,000 is available, plus up to $35,000,000 in outyear increments for continuing awards. Deadlines are: May 2 for research proposals on the Oden during November through December; and June 6 for all proposals other than work proposed on Oden for November through December. Full proposals should be submitted via the FastLane system at NSF or Grants.gov.
Amount of Grant: $400,000 average
Date(s) Application is Due: May 2; Jun 6
Contact: Julie M. Palais, Program Manager; (703) 292-8033; fax: (703) 292-9079; email: jpalais@nsf.gov
Internet: http://www.nsf.gov/funding/pgm_summ.jsp?pims_id=12798
Sponsor: National Science Foundation
4201 Wilson Boulevard
Arlington, VA 22230

NSF Antarctic Integrated System Science (AISS) Program Grants **4468**
In general terms, the AISS program administers projects that transcend disciplinary boundaries, are highly integrated and address questions broader in scope than those typically supported by the disciplinary programs. AISS projects must have compelling intellectual merit, broad impact and expand the frontiers of our knowledge. Projects must not be so broad in scope that tractable research strategies are not practical. It is recognized that integrated system proposals can be challenging to review. At this juncture in accordance with existing NSF guidelines, proposers may choose to submit single collaborative proposals or multiple related proposals that share some common text. Proposals will be reviewed by both ad-hoc mail reviews and a combination of panelists from the disciplinary panels as appropriate. Those considering submission to AISS are encouraged to contact the program director in advance. A total of $20,000,000 is available, plus up to $35,000,000 in outyear increments for continuing awards. Deadlines are: May 2 for research proposals on the Oden during November through December; and June 6 for all proposals other than work proposed on Oden for November through December. Full proposals should be submitted via the FastLane system at NSF or Grants.gov.
Restrictions: AISS does not fund programs that recast disciplinary questions into a form requiring minimal expertise from other disciplines when progress is possible within a discipline.
Amount of Grant: $400,000 average
Date(s) Application is Due: May 2; Jun 6
Contact: Kelly Falkner, Program Director; (703) 292-8033; fax: (703) 292-9079; email: kfalkner@nsf.gov
Internet: http://www.nsf.gov/pubs/2008/nsf08535/nsf08535.htm
Sponsor: National Science Foundation
4201 Wilson Boulevard
Arlington, VA 22230

NSF Antarctic Ocean and Atmospheric Sciences Grants **4469**
The Program is intended to foster advances in understanding of the oceanic and tropospheric environments at high southern latitudes and their links to each other and global scale processes. Innovative approaches involving surface and/or satellite-based field observations and/or modeling studies are particularly encouraged. Major program elements include but are not limited to: physical oceanographic studies that probe the dynamics and kinematics of the polar oceans, such as the effects of interface driving forces like wind, solar radiation, tides, and heat exchange, water mass production and modification processes, ocean dynamics at the pack ice edge, and the effect of polynyas on ventilation and relationships to the distribution of marine biota; chemical oceanographic studies that investigate the chemical composition of and reactions in sea water, track fluxes of material within ocean basins and at their boundaries, and apply chemical tracers to study time and space scales of physical and biogeochemical oceanic processes; sea ice studies that encompass material characteristics at the individual crystal level to large-scale patterns of freezing, deformation, and melting and the relationships of sea ice dynamics with the ocean and atmosphere over a variety of temporal and spatial scales; and meteorological studies that investigate atmospheric circulation systems and dynamics entailing aspects such as local and/or global exchange of heat, momentum and trace constituents and atmospheric chemistry. A total of $20,000,000 is available, plus up to $35,000,000 in outyear increments for continuing awards. Deadlines are: May 2 for research proposals on the Oden during November through December; and June 6 for all proposals other than work proposed on Oden for November through December. Full proposals should be submitted via the FastLane system at NSF or Grants.gov.
Amount of Grant: $400,000 average
Date(s) Application is Due: May 2; Jun 6
Contact: Kelly Falkner, Program Director; (703) 292-8033; fax: (703) 292-9079; email: kfalkner@nsf.gov
Internet: http://www.nsf.gov/funding/pgm_summ.jsp?pims_id=13422
Sponsor: National Science Foundation
4201 Wilson Boulevard
Arlington, VA 22230

NSF Antarctic Organisms and Ecosystems Grants **4470**
The goal of the Program is to improve understanding of antarctic organisms and their interactions within the biosphere and geosphere. The program supports projects directed at all levels of biological organization from molecular, cellular, and organismal, to communities and ecosystems over regional and global scales. Investigators are encouraged to develop and apply theory and innovative technologies to understand how organisms adapt to and live in high-latitude environments and how populations and ecosystems may respond to global change. Particular emphases include: marine ecosystems; terrestrial and freshwater ecosystems; population dynamics physiological ecology, and adaptation; and genomics. A total of $20,000,000 is available, plus up to $35,000,000 in outyear increments for continuing awards. Deadlines are: May 2 for research proposals on the Oden during November through December; and June 6 for all proposals other than work proposed on Oden for November through December. Full proposals should be submitted via the FastLane system at NSF or Grants.gov.
Amount of Grant: $400,000 average
Date(s) Application is Due: May 2; Jun 6
Contact: Roberta L. Marinelli, Program Director; (703) 292-7448; fax: (703) 292-9079; email: rmarinel@nsf.gov
Internet: http://www.nsf.gov/funding/pgm_summ.jsp?pims_id=13421
Sponsor: National Science Foundation
4201 Wilson Boulevard
Arlington, VA 22230

NSF Antarctic Research Grants **4471**
Scientific research and operational support of that research are the principal activities supported by the United States Government in Antarctica. The goals are to expand fundamental knowledge of the region, to foster research on global and regional problems of current scientific importance, and to use Antarctica as a platform from which to support research. For projects involving fieldwork, the U.S. Antarctic Program supports only research that can be done exclusively in Antarctica or that is best done from Antarctica. The program also supports antarctic-related analytical research performed at home organizations. A total of $20,000,000 is available, plus up to $35,000,000 in outyear increments for continuing awards. Deadlines are: May 2 for research proposals on the Oden during November through December; and June 6 for all proposals other than work proposed on Oden for November through December. Full proposals should be submitted via the FastLane system at NSF or Grants.gov.
Amount of Grant: $400,000 average; 20,000,000 total
Date(s) Application is Due: May 2; Jun 6
Contact: Kelly Falkner, Program Director; (703) 292-8033; fax: (703) 292-9079; email: kfalkner@nsf.gov
Internet: http://www.nsf.gov/funding/pgm_summ.jsp?pims_id=5519
Sponsor: National Science Foundation
4201 Wilson Boulevard
Arlington, VA 22230

NSF Antarctic Special Opportunity for Science aboard the Swedish Ice Breaker Oden Grants 4472

The 5-year U.S.-Swedish cooperative agreement for use of the icebreaker Oden provides the opportunity for research proposals to support participation on cruises aboard the Class 1A icebreaker Oden in the Southern Ocean. The Oden will lead the annual break-in of the McMurdo Ship Channel in the Ross Sea in the January-to-February time period. Prior to that time, and in conjunction with the transit of the Oden from South America to the Ross Sea, there will be approximately 20 dedicated science days co-sponsored by Sweden and the United States. The expeditions provide the possibility to collect data in sectors of Antarctic marine regions that are rarely visited and data-sparse.
Requirements: Proposals that include well-developed collaborations among U.S. and Swedish scientists are particularly encouraged. Scientists from Sweden may apply for support from the Swedish Research Council (http://www.vr.se/) The Swedish Research Council is offering a similar schedule for cruise participation, with review and decisions compatible with the U.S. timetable outlined above. Scientists from third countries can be part of U.S. or Swedish Oden-programs.
Date(s) Application is Due: May 2
Contact: Kelly Falkner, Program Director; (703) 292-8033; fax: (703) 292-9079; email: kfalkner@nsf.gov; Roberta L. Marinelli, Program Director; (703) 292-7448; fax: (703) 292-9079; email: rmarinel@nsf.gov
Internet: http://www.nsf.gov/pubs/2008/nsf08535/nsf08535.htm
Sponsor: National Science Foundation
4201 Wilson Boulevard
Arlington, VA 22230

NSF Applied Mathematics Grants 4473

These grants support research motivated by or having an effect on problems arising in science and engineering. Mathematical merit and novelty, as well as breadth and quality of impact on applications, are important factors. Proposals to develop critical mathematical techniques from individual investigators as well as interdisciplinary teams are encouraged. The proposal window is November 1 through November 15 each year. Full proposals should be submitted via the FastLane system at NSF or Grants.gov. PD 07-1266.
Date(s) Application is Due: Nov 15
Contact: Henry A. Warchall, Program Director; (703) 292-4861; fax: (703) 292-9032; email: hwarchal@nsf.gov; Michael Steuerwalt, Program Director; (703) 292-4860; fax: (703) 292-9032; email: msteuerw@nsf.gov
Internet: http://www.nsf.gov/funding/pgm_summ.jsp?pims_id=5664&org=DMS&from=home
Sponsor: National Science Foundation
4201 Wilson Boulevard
Arlington, VA 22230

NSF Archaeology and Archeaometry Research Grants 4474

The program provides support for anthropologically relevant archaeological research at both a senior and doctoral dissertation level. It also funds anthropologically significant archaeometric research. High risk exploratory research proposals are accepted for consideration and a description of these competitions is provided in the Archaeology Program Overview. Deadline dates for senior research grants are July 1 and December 1; other archaeometry grants are due October 31. Doctoral dissertation research and high risk research in anthropology proposals are accepted at any time. Full proposals should be submitted via the FastLane system at NSF or Grants.gov. Refer to brochure PD 98-1391 for program announcement.
Date(s) Application is Due: Jul 1; Oct 31; Dec 1
Contact: Dr. John E. Yellen, Program Director; (703) 292-8759; fax: (703) 292-9068; email: jyellen@nsf.gov
Internet: http://www.nsf.gov/funding/pgm_summ.jsp?pims_id=11690
Sponsor: National Science Foundation
4201 Wilson Boulevard
Arlington, VA 22230

NSF Arctic Natural Science Grants in Atmospheric Sciences 4475

Research in arctic atmospheric sciences focuses on arctic climate and meteorology, including atmosphere-sea and atmosphere-ice interactions. Research on past climates and atmospheric gases preserved in snow and ice is encouraged. The program also supports research on stratospheric and tropospheric processes. For upper atmospheric and space physics, proposals often are considered jointly with the Division of Atmospheric Sciences in the Geosciences Directorate and OPP, Antarctic Aeronomy and Astrophysics Program when conjugate studies are proposed. Full proposals should be submitted via the FastLane system at NSF or Grants.gov.
Requirements: Only U.S. organizations are eligible to submit proposals under this solicitation. There is no limit on PI eligibility, and there is no limit on the number of proposals that may be submitted.
Restrictions: Proposals that treat generic processes that could be adequately studied outside the Arctic are more appropriate to other programs within the Foundation.
Amount of Grant: $400,000 average
Date(s) Application is Due: Nov 10
Contact: Jane Dionne, Arctic Natural Sciences Program Director; (703) 292-7427; fax: (703) 292-9082; email: jdionne@nsf.gov
Internet: http://www.nsf.gov/funding/pgm_summ.jsp?pims_id=13424
Sponsor: National Science Foundation
4201 Wilson Boulevard
Arlington, VA 22230

NSF Arctic Natural Science Grants in Biological Sciences 4476

The Program supports projects that emphasize understanding of the adaptation of organisms to the arctic environment. Biological studies in the Arctic include research in freshwater, marine, and terrestrial biology, and ecology. Full proposals should be submitted via the FastLane system at NSF or Grants.gov.
Requirements: Only U.S. organizations are eligible to submit proposals under this solicitation. There is no limit on PI eligibility, and there is no limit on the number of proposals that may be submitted.
Restrictions: Proposals that treat generic processes that could be adequately studied outside the Arctic are more appropriate to other programs within the Foundation.
Amount of Grant: $400,000 average
Date(s) Application is Due: Nov 10
Contact: Jane Dionne, Arctic Natural Sciences Program Director; (703) 292-7427; fax: (703) 292-9082; email: jdionne@nsf.gov
Internet: http://www.nsf.gov/funding/pgm_summ.jsp?pims_id=13424
Sponsor: National Science Foundation
4201 Wilson Boulevard
Arlington, VA 22230

NSF Arctic Natural Science Grants in Earth Sciences 4477

This Program supports a range of terrestrial and marine geology and geophysics project, but of greatest interest are projects that will bring a better understanding of the arctic geological processes, which will improve our ability to interpret the geologic record of environmental change in the polar regions, particularly in the Late Cretaceous and Cenozoic with special emphasis on the Quaternary. Also a priority is understanding the processes responsible for the evolution of permafrost and its consequent effects and a better understanding and reconstruction of the plate tectonic history of the Arctic Ocean. Full proposals should be submitted via the FastLane system at NSF or Grants.gov. NSF 06-603.
Requirements: Only U.S. organizations are eligible to submit proposals under this solicitation. There is no limit on PI eligibility, and there is no limit on the number of proposals that may be submitted.
Restrictions: Proposals that treat generic processes that could be adequately studied outside the Arctic are more appropriate to other programs within the Foundation.
Amount of Grant: $400,000 average
Date(s) Application is Due: Nov 10
Contact: Jane Dionne, Arctic Natural Sciences Program Director; (703) 292-7427; fax: (703) 292-9082; email: jdionne@nsf.gov
Internet: http://www.nsf.gov/funding/pgm_summ.jsp?pims_id=13424
Sponsor: National Science Foundation
4201 Wilson Boulevard
Arlington, VA 22230

NSF Arctic Natural Science Grants in Glaciology 4478

Glaciology projects generally focus on the history and dynamics of all naturally-occurring forms of snow and ice, including seasonal snow, glaciers, and the Greenland ice sheet. Modeling of mass balance, glacial geology, and remote sensing studies of ice sheets are also supported. Full proposals should be submitted via the FastLane system at NSF or Grants.gov. NSF 06-603
Requirements: Only U.S. organizations are eligible to submit proposals under this solicitation. There is no limit on PI eligibility, and there is no limit on the number of proposals that may be submitted.
Restrictions: Graduate students are not encouraged to initiate proposals but are encouraged to participate as research assistants to principal investigators. Proposals that treat generic processes that could be adequately studied outside the Arctic are more appropriate to other programs within the Foundation.
Amount of Grant: $400,000 average
Date(s) Application is Due: Nov 10
Contact: Jane Dionne, Arctic Natural Sciences Program Director; (703) 292-7427; fax: (703) 292-9082; email: jdionne@nsf.gov
Internet: http://www.nsf.gov/funding/pgm_summ.jsp?pims_id=13424
Sponsor: National Science Foundation
4201 Wilson Boulevard
Arlington, VA 22230

NSF Arctic Natural Science Grants in Ocean Sciences 4479

The Program supports ocean science projects that advance knowledge of the structure and processes of the Arctic Ocean and adjacent seas and their interactions with their boundaries, including the arctic sea-ice cover. Areas of special interest are: low-temperature life processes; the formation, movement, and mixing of arctic water masses; the dynamics of sea ice; and the role of the Arctic Ocean and adjacent seas in global climate. Proposals concerned with the interdependencies of chemical and physical processes and marine organisms and productivity are encouraged. Full proposals should be submitted via the FastLane system at NSF or Grants.gov. Brochure NSF 06-603
Requirements: Only U.S. organizations are eligible to submit proposals under this solicitation. There is no limit on PI eligibility, and there is no limit on the number of proposals that may be submitted.
Restrictions: Graduate students are not encouraged to initiate proposals but are encouraged to participate as research assistants to principal investigators. Proposals that treat generic processes that could be adequately studied outside the Arctic are more appropriate to other programs within the Foundation.
Amount of Grant: $400,000 average
Date(s) Application is Due: Nov 10

Contact: Jane Dionne, Arctic Natural Sciences Program Director; (703) 292-7427; fax: (703) 292-9082; email: jdionne@nsf.gov
Internet: http://www.nsf.gov/funding/pgm_summ.jsp?pims_id=13424
Sponsor: National Science Foundation
4201 Wilson Boulevard
Arlington, VA 22230

NSF Arctic Natural Sciences Program Grants **4480**
The Program provides core support for disciplinary and interdisciplinary research on arctic processes and coordinates its support of arctic research with the Directorates for Geosciences, Mathematical and Physical Sciences, Social, Behavioral, and Economic Sciences, and Biological Sciences. Upper atmospheric and space physics proposals often are considered jointly with the Division of Atmospheric Sciences in the Geosciences Directorate and with OPP's Antarctic Aeronomy and Astrophysics Program when conjugate studies are proposed. Areas of special interest include marine and terrestrial ecosystems, arctic atmospheric and oceanic dynamics and climatology, arctic geological and glaciological processes, and their connectivity to lower latitudes. The Program particularly encourages proposals that treat uniquely arctic processes and that provide hypothesis testing required to produce the understanding needed to develop predictive tools based on first principles. Proposals to perform monitoring per se are discouraged. ANS supports projects that emphasize understanding the adaptation of organisms to the unique aspects of the arctic environment. Terrestrial and marine geology and geophysics projects of greatest interest are those that will improve our ability to interpret the geologic record of environmental change in the Arctic, particularly during the Quaternary. Understanding the processes responsible for the evolution of permafrost and consequences of changing permafrost remains a priority, as well. Projects that focus on the history and dynamics of all naturally occurring forms of arctic snow and ice, including seasonal snow, glaciers, and the Greenland ice sheet, are encouraged. The Program supports ocean science projects that advance knowledge of the processes of the Arctic Ocean and adjacent seas and their interactions with their boundaries. The development of sensors necessary to observe these processes is also supported by ANS. Full proposals should be submitted via the FastLane system at NSF or Grants.gov.
Requirements: Only U.S. organizations are eligible to submit proposals under this solicitation. There is no limit on PI eligibility, and there is no limit on the number of proposals that may be submitted.
Restrictions: Proposals that treat generic processes that could be adequately studied outside the Arctic are more appropriate to other programs within the Foundation.
Amount of Grant: $400,000 average
Date(s) Application is Due: Nov 10
Contact: William J. Wiseman, Program Manager; (703) 292-4750; fax: (703) 292-9082; email: wwiseman@nsf.gov
Internet: http://www.nsf.gov/funding/pgm_summ.jsp?pims_id=13424
Sponsor: National Science Foundation
4201 Wilson Boulevard
Arlington, VA 22230

NSF Arctic Research and Education Grants **4481**
This program supports activities that bridge research and education in concert with funded research grants and agreements through supplement requests or as separate proposal requests to support new ventures. Proposals to this program may include formal or informal education or outreach for students K-12 and higher or to the broader public. Most commonly awards are made as supplements to research grants or as small grants. The program seeks to collaborate with other directorates at NSF to promote the integration of research and education. Arctic research spans the major fields of science, technology, engineering and mathematics and is often multi- or interdisciplinary. Research in the Arctic has clear applications for education and outreach at many levels. The region itself is an interesting hook for teaching about life, physical, and social sciences, and concepts such as ocean and atmosphere circulation, climate, earth system science, animal migrations, and life in extreme environments. Projects bridging research and education that have received support from this program include: teacher enhancement experiences through field research and mentoring by researchers; curriculum development infusing inquiry-based research into K-12 science; travel for arctic experts (e.g., researcher, Alaska Natives, educators) to share their knowledge through seminars and lectures to K-12 and university classrooms, community groups, and other audiences; opportunities for minorities and underrepresented groups to participate in science; projects that involve people of the arctic in science and education, and journalist/science writer immersion in arctic science. Full proposals should be submitted via the FastLane system at NSF or Grants.gov.
Amount of Grant: $400,000 average
Date(s) Application is Due: Nov 10
Contact: Renee D. Crain, Research Support and Logistics Manage; (703) 292-4482; fax: (703) 292-9082; email: rcrain@nsf.gov
Internet: http://www.nsf.gov/funding/pgm_summ.jsp?pims_id=13448
Sponsor: National Science Foundation
4201 Wilson Boulevard
Arlington, VA 22230

NSF Arctic Research Opportunities **4482**
The goal of the NSF Arctic Sciences Section is to gain a better understanding of the Earth's physical, biological, geological, chemical, social and cultural processes, and the interactions of ocean, land, atmosphere, biological, and human systems in the Arctic. The Arctic Sciences Section and other NSF programs support projects that contribute to the development of the next generation of researchers and scientific literacy for all ages through education, outreach and broadening participation in science, technology, engineering and mathematics. Program representatives from the Office of Polar Programs (OPP) and other NSF programs that support arctic research coordinate across NSF, including joint review and funding of arctic proposals and mutual support of special projects with high logistical costs. Pending availability of funds, $16,000,000 may be available for proposals to this program. This does not include logistics support that may be provided through the Arctic Research Support and Logistics program. NSF estimates 40 awards per year as standard or continuing grants, or cooperative agreements. The number of awards and average award size and duration are subject to the availability of funds. NSF 06-603
Requirements: Only U.S. organizations are eligible to submit proposals to this program. There is no limit on PI eligibility, and there is no limit on the number of proposals that may be submitted. Emphasis areas and special research opportunities include: (1) Arctic Natural Sciences Program; (2) Arctic Social Sciences Program; (3) Arctic System Science Program; (4) Arctic Research Support and Logistics Program; (5) Arctic Research and Education Program; (6) International Polar Year (IPY). Detailed descriptions for each category can be found in the program solicitation.
Samples: Lee Cronk, Rutgers University New Brunswick (NJ), $15,968 - Doctoral Dissertation Research: Cooperation and Collective Action in the Russian Far East; Peter Schweitzer, University of Alaska Fairbanks Campus (AK), $16,000 - Dissertation Research: Hunting as Being and Knowing in Northwest Alaska; George Spangler, University of Minnesota-Twin Cities (MN), $12,468 - Doctoral Dissertation Improvement Grant: A Collaborative Exploration of Environmental Change, Fisheries, and Fishing Communities in Old Harbor, AK
Date(s) Application is Due: Nov 10
Contact: Renee D. Crain, Arctic Research and Education Assistant Program Director, (703) 292-4482, fax: (703) 292-9082, email: rcrain@nsf.gov; Jane V. Dionne, Arctic Natural Sciences Program Director, (703) 292-7427, fax: (703) 292-9082, email: jdionne@n
Internet: http://nsf.gov/funding/pgm_summ.jsp?pims_id=5521
Sponsor: National Science Foundation
4201 Wilson Boulevard
Arlington, VA 22230

NSF Arctic Research Support and Logistics Program Grants **4483**
The Program supports the field component of research projects funded through science programs in the Arctic Sciences Section of the Office of Polar Programs and through other programs at National Science Foundation (NSF). The program accepts proposals that: support long-term observations of the Arctic; support the acquisition of data sets useful to a broad segment of the arctic research community; will lead to Cooperative Agreements to operate multi-use arctic research facilities; or provide services that broadly support the arctic research community, such as facilitating communication, developing research ideas in an arctic-wide community setting, and cooperating with arctic communities. The types of proposals considered by RSL are: long-term observations; data acquisitions or collection; facility operations; and field work. Full proposals should be submitted via the FastLane system at NSF or Grants.gov. Detailed information pertaining to proposal submission is included in the Arctic Research Opportunities solicitation (NSF 05-514).
Requirements: Proposals may only be submitted by U.S. organizations. Academic institutions and academically related nonprofit research organizations may submit proposals on behalf of any U.S. scientist; nonprofit, nonacademic research institutions; private profit organizations; and state and local governments are also eligible for support. Under certain circumstances unaffiliated scientists may be eligible. All proposals involving fieldwork must include sufficient information to assure safety and feasibility. Proposals must identify all proposed field personnel and describe their qualifications for the work they are to do.
Amount of Grant: $400,000 average
Date(s) Application is Due: Nov 10
Contact: Renee D. Crain, Research Support and Logistics Manager; (703) 292-4482; fax: (703) 292-9082; email: rcrain@nsf.gov
Internet: http://www.nsf.gov/funding/pgm_summ.jsp?pims_id=13437
Sponsor: National Science Foundation
4201 Wilson Boulevard
Arlington, VA 22230

NSF Arctic Social Sciences Grants **4484**
The Arctic Social Sciences Program (ASSP) encompasses all social sciences. These include, but are not limited to anthropology, archaeology, economics, geography, linguistics, political science, psychology, science and technology studies, sociology, traditional knowledge and related subjects. Although unsolicited proposals in any of the social sciences mentioned above are welcome, areas of particular interest include culture and environment, resources and economic change, development of social and political institutions, ethnic (cultural) and regional identities, and knowledge systems. Full proposals should be submitted via the FastLane system at NSF or Grants.gov.
Requirements: Proposals may only be submitted by U.S. organizations. Academic institutions and academically related nonprofit research organizations may submit proposals on behalf of any U.S. scientist; nonprofit, nonacademic research institutions; private profit organizations; and state and local governments are also eligible for support. Under certain circumstances unaffiliated scientists may be eligible. All proposals involving fieldwork must include sufficient information to assure safety and feasibility. Proposals must identify all proposed field personnel and describe their qualifications for the work they are to do.
Amount of Grant: $400,000 average

Date(s) Application is Due: Nov 10
Contact: Anna M. Kerttula de Echave, Program Director; (703) 292-7432; fax: (703) 292-9082; email: akerttul@nsf.gov
Internet: http://www.nsf.gov/funding/pgm_summ.jsp?pims_id=13425
Sponsor: National Science Foundation
4201 Wilson Boulevard
Arlington, VA 22230

NSF Arctic System Science (ARCSS) Program Grants **4485**
The program goals are to understand the physical, chemical, biological, and sociocultural processes of the arctic system that interact with the total earth system and thus contribute to or are influenced by global change. ARCSS has four linked and ongoing components: ocean-ice-atmosphere interactions; land-atmosphere-ice interactions; human dimensions of the arctic system; and paleoenvironmental arctic sciences. Specific program opportunities are announced annually for single and multidisciplinary investigations.
Requirements: Any U.S. scientist interested in doing research may initiate a proposal, which is submitted formally by his or her institution. Nonprofit, nonacademic research institutions; private profit-making organizations; and local and state governments also are eligible for support.
Restrictions: Graduate students are not encouraged to initiate proposals but are encouraged to participate as research assistants to principal investigators.
Amount of Grant: $400,000 average
Date(s) Application is Due: Nov 10
Contact: Neil R. Swanberg, Program Manager; (703) 292-8029; fax: (703) 292-9082; email: nswanber@nsf.gov
Internet: http://www.nsf.gov/funding/pgm_summ.jsp?pims_id=13426
Sponsor: National Science Foundation
4201 Wilson Boulevard
Arlington, VA 22230

NSF Astronomy and Astrophysics Postdoctoral Fellowships **4486**
The Fellowships provide an opportunity for highly qualified young investigators within 3 years of obtaining their PhD to carry out an integrated program of independent research and education. Fellows may engage in research of observational, instrumental, or theoretical nature, in combination with a coherent educational plan for the duration of the fellowship. The program supports researchers for a period of up to 3 years with fellowships that can be taken to the institution or national facility of their choice. The program is intended to recognize young investigators of significant potential, and provide them with experience in research and education that will establish them in positions of distinction and leadership in the community.
Samples: Tamara Rogers, $67,000 (2006); Marcel Agueros, $67,000 (2006); Matthew Browning, $134,000 (2005).
Date(s) Application is Due: Oct 10
Contact: Dana E. Lehr, Associate Program Director, (703) 292-7456; fax: (703) 292-9034; email: dlehr@nsf.gov
Internet: http://www.nsf.gov/funding/pgm_summ.jsp?pims_id=5291
Sponsor: National Science Foundation
4201 Wilson Boulevard
Arlington, VA 22230

NSF Astronomy and Astrophysics Research Grants (AAG) **4487**
The Astronomy and Astrophysics Research Grants (AAG) Program provides individual investigator and collaborative research grants for observational, theoretical, laboratory and archival data studies in all areas of astronomy and astrophysics, including but not limited to the following areas of study: (1) Planetary Astronomy - Studies of the detailed structure and composition of the surfaces, interiors and atmospheres of the planets and satellites in the Solar System; the nature of small bodies (asteroids and comets); the inter-planetary medium; and the origin and development of the Solar System. (2) Stellar Astronomy and Astrophysics - Studies of the structure and activity of the Sun and other stars; the physical properties and composition of all types of single and multiple stars; compact objects and their interactions; extra-solar system planet formation and detection; star formation and stellar evolution; stellar nucleosynthesis; and the properties of atoms and molecules of relevance to stellar astronomy. (3) Galactic Astronomy - Studies on the composition, structure and evolution of the Milky Way galaxy and nearby galaxies. Research may focus on the stellar populations in these galaxies; the characteristics of star clusters; the interstellar medium; and the properties of atomic and molecular constituents of the interstellar medium. (4) Extragalactic Astronomy and Cosmology - Studies of the more distant Universe. Research topics include galaxy formation, evolution and interaction; active galaxies; quasars; large-scale structure; and all areas of cosmology. NSF 05-608
Requirements: Proposals submitted to the AAG Program do not require categorization into one of the study areas identified above. Proposals may span multiple disciplines and/or areas of study and may utilize multiple techniques. Principal Investigators are encouraged to contact one of the Program Officers listed in this announcement prior to submitting a proposal to the AAG Program, particularly if the proposal will include investigators at multiple institutions. Standard NSF eligibility requirements apply.
Date(s) Application is Due: Nov 15
Contact: Nigel A. Sharp, Program Director; (703) 292-4905; fax: (703) 292-9034; email: nsharp@nsf.gov
Internet: http://www.nsf.gov/funding/pgm_summ.jsp?pims_id=13630
Sponsor: National Science Foundation
4201 Wilson Boulevard
Arlington, VA 22230

NSF Atmospheric Chemistry Grants **4488**
This program supports research to measure and model the concentration and distribution of gases and aerosols in the lower and middle atmosphere. Also supports research on the chemical reactions among atmospheric species; the sources and sinks of important trace gases and aerosols; the aqueous-phase atmospheric chemistry; the transport of gases and aerosols throughout the atmosphere; and the improved methods for measuring the concentrations of trace species and their fluxes into and out of the atmosphere.
Requirements: Individuals and organizations in the following categories may submit proposals: (1) Universities and Colleges - Universities and two- and four-year colleges (including community colleges) located and accredited in the U.S., acting on behalf of their faculty members. Such organizations also are referred to as academic institutions; (2) Non-profit, non-academic organizations - Independent museums, observatories, research laboratories, professional societies and similar organizations in the U.S. that are directly associated with educational or research activities; (3) For-profit organizations - U.S. commercial organizations, especially small businesses with strong capabilities in scientific or engineering research or education. An unsolicited proposal from a commercial organization may be funded when the project is of special concern from a national point of view, special resources are available for the work, or the proposed project is especially meritorious. NSF is interested in supporting projects that couple industrial research resources and perspectives with those of universities; therefore, it especially welcomes proposals for cooperative projects involving both universities and the private commercial sector; (4) State and Local Governments - State educational offices or organizations and local school districts may submit proposals intended to broaden the impact, accelerate the pace, and increase the effectiveness of improvements in science, mathematics and engineering education in both K-12 and post-secondary levels; (5) Unaffiliated Individuals - Scientists, engineers or educators in the U.S. and U.S. citizens may be eligible for support, provided that the individual is not employed by, or affiliated with, an organization, and: (a) the proposer has demonstrated the capability and has access to any necessary facilities to carry out the project; and, (b) the proposer agrees to fiscal arrangements that, in the opinion of the NSF Division of Grants & Agreements, ensure responsible management of Federal funds. Unaffiliated individuals should contact the appropriate program before preparing a proposal for submission.
Restrictions: NSF rarely provides support to foreign organizations. NSF will consider proposals for cooperative projects involving U.S. and foreign organizations, provided support is requested only for the U.S. portion of the collaborative effort. Also, NSF does not normally support research or education activities by scientists, engineers or educators employed by Federal agencies or Federally Funded Research and Development Centers (FFRDCs). Under unusual circumstances, other Federal agencies and FFRDCs may submit proposals directly to NSF.
Samples: Measurements of Aerosol Precursor Trace Gases Over the Tropical and Southeast Pacific Ocean - Rainer Volkamer, University of Colorado at Boulder ($85,000); Determination of Nitrous Acid and Nitrogen Dioxide Flux and Concentration Over Forest Canopies Using Quantum Cascade Laser Absorption Spectroscopy - Mark Zahniser, Aerodyne Research, Inc. ($90,999) and William J. Munger, Harvard University ($107,117); A Particle-Resolved Aerosol Model of Soot Aging - Nicole Riemer, University of Illinois at Urbana-Champaign ($161,928); Line Shape Parameters and Their Temperature Dependence for Molecules of Importance in Earth's Atmosphere - Robert Gamache, University of massachusetts Lowell ($97,176).
Amount of Grant: $30,000 - $475,000
Contact: Eric Hintsa, (703) 292-8522, email: ehintsa@nsf.gov; John E. Mak, (703) 292-8524, email: jmak@nsf.gov
Internet: http://www.nsf.gov/funding/pgm_summ.jsp?pims_id=11692&org=ATM
Sponsor: National Science Foundation
4201 Wilson Boulevard
Arlington, VA 22230

NSF Atmospheric Sciences Mid-Size Infrastructure Opportunity Grants **4489**
The Division of Atmospheric Sciences (ATM) at the National Science Foundation recognizes the need for new experimental infrastructure that directly enables areas of research traditionally supported by the Division. For the purpose of this program, mid-size infrastructure is defined as research equipment costing between $4M and $25M. The development of mid-size infrastructure is intended to be a long-term effort in ATM to phase in high priority science-enabling tools consistent with community needs, ATM goals, and the NSF strategic vision. The potential for broad community usage and support is essential. Projects to be funded as mid-size infrastructure must be multi-user facilities that have demonstrable broad community support and directly enable research activities traditionally supported by ATM. Broad community advocacy for proposed projects is essential, as evidenced by the results of decadal surveys, focused workshops, and relevant steering or advisory committee recommendations. Selected projects will be funded for one to five years. NSF 07-602
Requirements: Proposals may be submitted by: (a) Academic Institutions located in the U.S. - U.S. universities and colleges located in the U.S.; for-profit organizations - U.S. commercial organizations, especially small businesses with strong capabilities in scientific or engineering research or education; non-profit, non-academic organizations - Independent museums, observatories, research labs, professional societies and similar organizations in the U.S. associated with educational or research activities. Selected projects will be funded for one or more years, not to exceed five years. The development of mid-size infrastructure is intended to be a long-term effort in ATM to phase in high priority science-enabling tools consistent with community needs, ATM goals, and the NSF strategic vision. Design, engineering, and prototyping of mid-size infrastructure will be considered for funding only if there is valid justification for why this activity cannot be funded at the program level.

Restrictions: FFRDC's (Federally Funded Research and Development Centers) affiliated with agencies other than the National Science Foundation are NOT eligible to submit a proposal to this competition.
Amount of Grant: $4,000,000 - $25,000,000
Date(s) Application is Due: Jan 11; Jun 2
Contact: Robert M. Robinson; (703) 292-8529; email: rmrobins@nsf.gov
Internet: http://www.nsf.gov/publications/pub_summ.jsp?ods_key=nsf07602
Sponsor: National Science Foundation
4201 Wilson Boulevard
Arlington, VA 22230

NSF Atomic Molecular and Optical Physics Grants **4490**
The Atomic Molecular and Optical Physics program encompasses four sub-areas of this broad discipline: Precision Measurements, Atomic and Molecular Dynamics, Atomic and Molecular Structure, and Optical Physics. Research supported in the first three sub-areas includes activities in quantum control, cooling and trapping of atoms and ions, low-temperature collision dynamics, the collective behavior of atoms in weakly interacting gases (Bose-Einstein Condensates and dilute Fermi degenerate systems), precision measurements of fundamental constants, and the effects of electron correlation on structure and dynamics. In Optical Physics, support is provided in areas such as nonlinear response of isolated atoms to intense, ultra-short electromagnetic fields, the atom-cavity interaction at high fields, and quantum properties of the electromagnetic field.
Date(s) Application is Due: Sep 30
Contact: Robert Dunford, Program Director; (703) 292-7173; fax: (703) 292-9078; email: rdunford@nsf.gov
Internet: http://www.nsf.gov/funding/pgm_summ.jsp?pims_id=13622
Sponsor: National Science Foundation
4201 Wilson Boulevard
Arlington, VA 22230

NSF Behavioral Neurosciences Research Grants **4491**
This program focuses on the neural regulation of behavioral events ranging from simple movements to complex adaptive and interactive responses. Major activities are sensorimotor integration, biological rhythms, and cognitive functions such as attention, spatial representation, and learning and memory. Approaches that use novel techniques to study behavior within an evolutionary and ecological context are encouraged. Proposals may be submitted at any time. For review within six months, proposals should be submitted in line with the listed target dates for applications. Refer to NSF brochure PD 98-1191
Requirements: The most frequent recipients of support are academic institutions and nonprofit research groups. In special circumstances, grants also are awarded to other types of institutions and to individuals.
Date(s) Application is Due: Jan 12; Jul 12
Contact: Judith Verbeke, Acting Division Director, (703) 292-8420; fax: (703) 292-9153; email: jverbeke@nsf.gov
Internet: http://www.nsf.gov/pubs/progdesc/1998/bio/ibn/1191.htm
Sponsor: National Science Foundation
4201 Wilson Boulevard
Arlington, VA 22230

NSF BIO Minority Postdoctoral Research Fellowships **4492**
These fellowships seek to provide opportunities for recent minority doctoral scientists to obtain additional training, to gain research experience under the sponsorship of established scientists, and to broaden their scientific horizons beyond their research experiences during their undergraduate or graduate training. They are further designed to assist new scientists to direct their research efforts across traditional disciplinary lines and to avail themselves of unique research resources, sites, and facilities, including foreign locations. Fellows must affiliate with an appropriate research institution and are expected to devote themselves full time to the fellowship activities for the duration of the fellowship. At the conclusion of the fellowship, a fellow who accepts a tenure-track appointment at a U.S. institution eligible to receive NSF funds may apply for a research starter grant. Application materials are available on request. The deadline for FastLane submission of new postdoctoral fellowships is 5 p.m. local time on the first Monday in December each year. NSF 00-139
Requirements: Applicants must be U.S. nationals from underrepresented minority groups. Graduate students who will complete the PhD within one year or have completed it within the previous four years and who have not completed more than two years of postdoctoral support are eligible to apply.
Amount of Grant: $1.8 million total
Date(s) Application is Due: Dec 5
Contact: Carter Kimsey, Program Officer, (703) 292-8470; email: ckimsey@nsf.gov
Internet: http://www.nsf.gov/pubs/2001/nsf00139/nsf00139.txt
Sponsor: National Science Foundation
4201 Wilson Boulevard
Arlington, VA 22230

NSF Biogeosciences Research Grants **4493**
Within the oceans and on land, microbes exert their impact through active interactions with earth materials (including minerals, rocks, hydrates, soils and dust). Proposers should seek: to increase our understanding of microbial processes at the interface with such materials; to elucidate the geologic record of microbial activity at that interface at any time scale including the co-evolution of the geosphere and biosphere through such interactions; to better understand the impact of microbes on biogeochemical processes operating at that interface; or to develop new methods and research techniques for the study of geomicrobial processes. Annual deadline dates may vary; contact program staff for exact dates. NSF 05-511
Requirements: U.S. organizations and individuals in the following categories may submit proposals: universities and two-and four-year colleges (including community colleges) acting on behalf of their faculty members; nonprofit, non-academic organizations; for-profit organizations; state and local governments; unaffiliated individuals; and other federal agencies.
Amount of Grant: $100,000 per year for up to five years
Date(s) Application is Due: Feb 10
Contact: Rachael Craig, Program Director, Directorate for Geosciences, Earth Sciences, (703) 292-8233; fax: (703) 292-9025; email: rcraig@nsf.gov
Internet: http://www.nsf.gov/pubs/2005/nsf05511/nsf05511.txt
Sponsor: National Science Foundation
4201 Wilson Boulevard, Room 775
Arlington, VA 22230

NSF Biological Databases and Informatics (BDI) Program Grants **4494**
This program seeks to encourage new approaches to the management, analysis, and dissemination of biological knowledge for the benefit of both the scientific community and the broader public. The BD&I program is especially interested in the development of informatics tools and resources that have the potential to advance all fields of biology under the purview of the Directorate for Biological Sciences at the NSF. Annual deadline date is the second Monday in July. Refer to brochure NSF 05-577.
Requirements: Consortia of eligible individuals or organizations may apply, but a single individual or organization must accept overall management responsibility. International collaboration is encouraged; however, financial support for any non-U.S. participant organization must be provided from within the participant's country or other non-U.S. sources.
Amount of Grant: $50,000-$500,000 average per year for up to five years
Date(s) Application is Due: Jul 9
Contact: Manfred Zorn, Division of Biological Infrastructure, (703) 292-8470; email: dbidba@nsf.gov
Internet: http://www.nsf.gov/pubs/2005/nsf05577/nsf05577.txt
Sponsor: National Science Foundation
4201 Wilson Boulevard
Arlington, VA 22230

NSF Biological Field Stations and Marine Laboratories (FSML) Grants **4495**
This program supports refurbishment, rehabilitation, enhancement, and construction, including laboratory renovations, at these research resources. Sites are required to have a plan for archiving (in electronic form) and making shareable scientific databases available on the unique data they obtain. Support can be provided for necessary computer equipment, connectivity, and development of these scientific database. Deadline is the first Friday in March each year. NSF 01-59
Requirements: U.S. colleges and universities, free-standing research and education institutions, and U.S. chartered corporations with formally constituted research and education programs at field stations or marine laboratories are eligible to apply.
Amount of Grant: $500,000 maximum
Contact: Dr. Gerald Selzer, Program Director, (703) 292-8470; email: gselzer@nsf.gov
Internet: http://www.nsf.gov/pubs/2001/nsf0159/nsf0159.htm
Sponsor: National Science Foundation
4201 Wilson Boulevard
Arlington, VA 22230

NSF Biological Information Technology and Systems **4496**
This program supports high risk/high return research at the interface of biology and information technology. Biological systems have enormous capabilities as powerful and agile control systems for robotic and regulatory systems, and for pattern recognition, adaptability, information storage, retrieval and processing, sensor fusion and other information-handling tasks. Biology often performs orders of magnitude better than systems based on today's silicon device technologies. Determining what needs to and what can be learned about information processing in biological systems should lead to important new information systems (algorithms, software and systems) and technologies (computer platforms, sensors, robotic devices, etc.). The initial phase of this program focuses on developing computational models and theories for the information processing mechanisms encountered in biological systems that will lead to new information technology systems and hardware platforms. While these new information technology systems will not necessarily or exclusively be implemented in biological matter, the program emphasizes hybrid (bio-silical) systems, particularly as a means for experimenting with and validating new theories of biological information technologies and systems. Annual deadline is the second Friday in February; optional letter of intent is requested 4 weeks in advance of the due date. NSF 01-102
Amount of Grant: $100,000-$200,000 per year for theoretical proposals; $100,000-$500,000 per year for hybrid systems
Contact: Frederica Darema, Senior Science and Technology Advisor, (703) 292-8980; email: fdarema@nsf.gov; James Hickman, Special Advisor to the Division Director, (703) 292-8980; email: jhickman@nsf.gov
Internet: http://www.nsf.gov/pubs/2001/nsf01102/nsf01102.htm
Sponsor: National Science Foundation
4201 Wilson Boulevard
Arlington, VA 22230

NSF Biological Oceanography Research Grants 4497
The program supports research in marine ecology which encompasses relationships among marine organisms and their interactions with the environment of the sea. Funded projects focus on understanding variability within marine ecosystems, including ocean margins, continental shelves, central gyres, basins, as well as the Great Lakes. Major studies aim to determine the roles of organisms in global-scale processes. Areas of research include ecosystem and biogeochemical processes, community and population ecology, behavioral, reproductive and life-history ecology, physiological and chemical ecology, molecular, cellular and biochemical studies, evolutionary ecology, and sometimes systematic biology and paleoecology. Submit proposals at any time; allow approximately six months for review and processing. Target dates indicate proximity of review panel meetings.
Date(s) Application is Due: Feb 15; Aug 15
Contact: David L. Garrison; (703) 292-7588; fax: (703) 292-9085; email: dgarriso@nsf.gov; Cynthia Suchman, Program Director; (703) 292-8143; email: csuchman@nsf.gov
Internet: http://www.nsf.gov/funding/pgm_summ.jsp?pims_id=11696&org=OCE&from=home
Sponsor: National Science Foundation
4201 Wilson Boulevard
Arlington, VA 22230

NSF Biological Physics (BP) 4498
The program supports projects in which the analytical and experimental tools of physics are applied to the study of problems originating in the living world. Both experimental and theoretical projects will be considered, although the main focus of the program is in the experimental area. Of particular interest are projects in which new experimental approaches are brought to bear on a well-identified problem. These approaches should at the same time have the potential for broad applicability to a set of similar problems, thereby adding to the set of tools the scientist has for addressing biological problems in general. While the problems under study must be important to advancing understanding of the living world in a meaningful way, particular emphasis will be placed on those projects in which the lessons learned from the application serve to foster new concepts and ideas that expand the intellectual basis of physics. The program funds individual investigators, although collaborative proposals between physicists and biologists are welcome.
Requirements: Unrestricted (i.e., open to any type of entity)
Samples: Trustees of Boston University - ($195,000) Electronic Recognition of Gene Regulatory Proteins Bound to DNA; University of California-Berkeley - ($300,000) In vitro and in vivo Single Molecular Experiments of Biological Systems; University of Illinois at Chicago - ($180,000) Study of the Dynamics of Protein-DNA Interactions to Probe Site-Specific Recognition; University of California-Davis - ($135,000) Computational study of DNA repair enzyme photolyase.
Amount of Grant: $135,000 - $500,000
Date(s) Application is Due: Jul 31
Contact: C. D. Caldwell, dcaldwel@nsf.gov, (703) 292-7371; James M. Olsen, jolsen@nsf.gov, (703) 292-7388; Ramona Winkelbauer, winkelb@nsf.gov, (703) 292-7390
Internet: http://nsf.gov/funding/pgm_summ.jsp?pims_id=6673&org=NSF&sel_org=NSF&from=fund
Sponsor: National Science Foundation
4201 Wilson Boulevard
Arlington, VA 22230

NSF Biological Research Collections (BRC) 4499
The program provides support for biological collection enhancement, computerization of specimen-related data, research to develop better methods for specimen curation and collection management, and activities such as symposia and workshops to investigate support and management of biological collections. Biological collections supported include those housing natural history specimens and jointly curated collections such as frozen tissues and other physical samples, e.g. DNA libraries and digital images. Such collections provide the materials necessary for research in a broad area of biological sciences. NSF 06-569
Requirements: Proposals are accepted from U.S. organizations including colleges and universities that maintain research collections, natural history museums including herbaria, and other collections administered by independent organizations or by state, county, or local governments; non-federal and non-profit research organizations that maintain collections; and field stations, marine laboratories, botanical gardens, zoological parks, and aquaria that maintain research collections that document biological diversity. The size of an organization is not a factor in determining eligibility. The maximum that may be requested in a proposal is $500,000. Numbers of awards and average award size/duration are subject to the availability of funds.
Restrictions: In any single round of the BRC competition, only one proposal may be submitted from any individual collection within an organization.
Samples: (2007 Awards) University of Colorado at Boulder - Conservation, Curation, and Website Development for the Karl Hirsch Fossil Eggshell Collection at the University of Colorado Museum; University of California-Berkeley - Ornithological Collection Improvement Project (Phase I), Museum of Vertebrate Zoology, University of California, Berkeley; Michigan State University - RCN: Building a National Community of Natural History Collections; University of Alaska Fairbanks Campus - SGER: Rescuing an important collection of Cambodian mammal specimens for curation and archiving at the University of Alaska Museum.
Amount of Grant: $500,000 maximum
Date(s) Application is Due: Jul 28
Contact: W. Carl Taylor, BRC Program Director, Directorate for Biological Sciences; (703) 292-8470; fax: (703) 292-9063; email: dbibrc@nsf.gov
Internet: http://www.nsf.gov/publications/pub_summ.jsp?ods_key=nsf06569
Sponsor: National Science Foundation
4201 Wilson Boulevard
Arlington, VA 22230

NSF Biomedical Engineering (BME) Grants 4500
The mission of the Biomedical Engineering (BME) Program is to: provide opportunities to develop novel ideas into discovery-level and transformative projects that integrate engineering and life science principles in solving biomedical problems that serve humanity in the long-term; and advance both engineering and life sciences with biomedical engineering projects that are at the interface of engineering and biomedical sciences. The BME program supports fundamental, transformative, and discovery research applied to biological systems.
Date(s) Application is Due: Mar 1; Sep 15
Contact: Semahat S. Demir, Program Director; (703) 292-7950; fax: (703) 292-9098; email: sdemir@nsf.gov
Internet: http://www.nsf.gov/funding/pgm_summ.jsp?pims_id=501023
Sponsor: National Science Foundation
4201 Wilson Boulevard
Arlington, VA 22230

NSF Biomedical Engineering and Engineering Healthcare Grants 4501
The mission of the Biomedical Engineering and Engineering Healthcare cluster is to provide opportunities to develop novel ideas into projects that integrate engineering and life science principles in solving biomedical problems that serve humanity. The cluster focuses on high impact transforming technologies for deriving information from cells, tissues, organs, and organ systems, extraction of useful information from complex biomedical signals, new approaches to the design of structures and materials for eventual medical use, biophotonics, and new methods of controlling living systems. This cluster is also directed toward the characterization, restoration, and/or substitution of normal functions in humans.
Contact: Marcia Rawlings, Information Technology Specialist; (703) 292-7956; fax: (703) 292-9013; email: mrawling@nsf.gov
Internet: http://www.nsf.gov/funding/pgm_summ.jsp?pims_id=501032
Sponsor: National Science Foundation
4201 Wilson Boulevard
Arlington, VA 22230

NSF Biomolecular Systems Cluster Grants 4502
The Biomolecular Systems Cluster emphasizes the structure, function, dynamics, interactions, and interconversions of biological molecules. The context for such studies can range from investigations of individual macromolecules to the large-scale integration of metabolic and energetic processes. Research supported by this cluster includes development of cutting-edge technologies integrating theoretical, computational, and experimental approaches to the study of biological molecules and their functional complexes; mechanistic studies of the regulation and catalysis of enzymes and RNA, and higher-order characterization of the biochemical processes by which all organisms acquire, transform, and utilize energy from substrates. This cluster emphasizes the importance of multi-disciplinary research carried out at the interfaces of biology, physics, chemistry, mathematics and computer science, and engineering.
Requirements: Standard NSF Grant Proposal Guidelines apply (see http://www.nsf.gov/publications/pub_summ.jsp?ods_key=gpg).
Date(s) Application is Due: Jan 12; Jul 12
Contact: Kamal Shukla, Program Director; (703) 292-7131; email: kshukla@nsf.gov
Internet: http://www.nsf.gov/funding/pgm_summ.jsp?pims_id=12771
Sponsor: National Science Foundation
4201 Wilson Boulevard
Arlington, VA 22230

NSF Biophotonics, Advanced Imaging, and Sensing for Human Health Grants 4503
Innovative basic research in photonics, imaging, and sensing that is very fundamental in science and engineering is needed to lay the foundation for new technologies beyond those that are mature and ready for application in medical diagnostics and therapies. Developing molecularly specific sensing (molecular photonics), imaging, and monitoring systems with high sensitivity and resolution would be an enormous accomplishment with powerful applications to both biology and medicine. Low cost diagnostics will require novel integration of photonics, molecular biology, and material science. Complex biosensors capable of detecting and discriminating among large classes of biomolecules could be important not only to biology and medicine, but also to environmental sensing and homeland security.
Date(s) Application is Due: Mar 1; Sep 15
Contact: Leon Esterowitz, Program Director; (703) 292-7942; fax: (703) 292-9098; email: lesterow@nsf.gov
Internet: http://www.nsf.gov/funding/pgm_summ.jsp?pims_id=501025
Sponsor: National Science Foundation
4201 Wilson Boulevard
Arlington, VA 22230

NSF Biotechnology, Biochemical, and Biomass Engineering (BBBE) Grants 4504
The program deals with fundamental problems involved in the processing and manufacturing of products of economic importance by effectively utilizing renewable

resources of biological origin and bioinformatics originating from genomic and proteomic information. The BBBE program emphasizes basic engineering and biological research that advances the fundamental knowledge base that contributes to a better understanding of cellular and biomolecular processes (in vivo, in vitro, and/or ex vivo) and eventually to the development of generic enabling technology and practical application. Quantitative assessments of bioprocesses and their rates at the levels of gene regulation and expression, signal transduction pathways, posttranslational protein processing, enzymes in reaction systems, metabolic pathways, cells and tissues in cultivation, and biological systems including animal, plant, microbial and insect cells, etc. are considered vital to the successful research projects in the BBBE program. Research projects supported through the BBBE program include, but are not limited to: fermentation technology; enzyme technology; recombinant DNA technology; cell culture technology; ex vivo and therapeutic stem cell culture technology; metabolic engineering; tissue engineering; nanobiotechnology; quantitative systems biotechnology; biosensor development; and food processing with special focus on the safety of the nation's food supply.
Date(s) Application is Due: Mar 1; Sep 15
Contact: Frederick G. Heineken, Program Director; (703) 292-7944; fax: (703) 292-9098; email: fheineke@nsf.gov
Internet: http://www.nsf.gov/funding/pgm_summ.jsp?pims_id=501024
Sponsor: National Science Foundation
4201 Wilson Boulevard
Arlington, VA 22230

NSF Biotic Surveys and Inventories Grants **4505**
The program invites proposals to document levels of diversity among living and fossil species throughout the world, to include viruses, prokaryotes, and eukaryotes from all marine and terrestrial habitats. The program supports research to record and document the diversity of life on earth, both past and present, as a prologue to investigations of patterns and processes and the development of plans for conservation of that diversity. Well-curated, permanent collections and computerized databases are strongly encouraged as products of such support. Proposals should involve collecting specimens of extant and extinct organisms as samples of taxa of geographic regions and/or geologic horizons; conducting inventories of existing collections; developing and disseminating electronic databases of the collected or inventoried specimens and taxa; and producing biotic treatments, authenticated species lists, catalogs, manuals, keys, expert identification systems, and/or other documents in electronic and printed form. Applications are due the first Friday in November each year. All proposals must be submitted by Fastlane.
Restrictions: The program does not provide support for phylogenetic, monographic, biogeographic, genetic, or ecological studies. The program does not accept proposals to fund assays of genetic diversity within single species, projects to monitor diversity over time, ecological comparison of the diversity in two or more sites, or studies of the functional significance of diversity at the ecosystem level.
Amount of Grant: $30,000-$150,00; $70,000 average
Contact: Program Director, Biotic Surveys and Inventories, Division of Environmental Biology, Directorate for Biological Sciences, (703) 306-1481; fax: (703) 306-0817; email: bsirev@nsf.gov
Internet: http://www.nsf.gov/pubs/1998/nsf98158/nsf98158.htm
Sponsor: National Science Foundation
4201 Wilson Boulevard
Arlington, VA 22230

NSF Catalysis and Biocatalysis Grants **4506**
The Catalysis and Biocatalysis program primarily supports fundamental and applied research on: kinetics and mechanisms of important catalyzed chemical reactions as they relate to the production of chemicals, fuels, and specialized materials; characterization of chemical and biochemical phenomena occurring at or near solid surfaces and interfaces; electrocatalytic processes having engineering significance or commercial potential; sustainability, environmental catalysis, and basic research related to green chemistry or utilization of biorenewable resources; kinetic modeling and theory of heterogeneous, homogeneous, and biocatalysis; fundamental aspects of reactive deposition and processing for thin film materials; interactions between chemical reactions and transport processes in reactive systems, and the use of this information in the design or control of complex chemical reactors. This program promotes multidisciplinary research in: chemical kinetics of heterogeneous, homogeneous, and biocatalysis; synthesis and processing of chemically functional materials; interfacial, electrochemical, and photochemical reaction processes; theory and modeling of reaction processes at surfaces; synthesis and characterization of catalysts that function at the nanoscale.
Date(s) Application is Due: Mar 1; Sep 15
Contact: John Regalbuto; (703) 292-8371; email: jregalbu@nsf.gov
Internet: http://www.nsf.gov/funding/pgm_summ.jsp?pims_id=13360
Sponsor: National Science Foundation
4201 Wilson Boulevard
Arlington, VA 22230

NSF Cell Biology Research Grants **4507**
Through programs in cellular organization and signal transduction, this program focuses on the structure, function, and regulation of plant, animal and microbial cells, and their interactions with the environment and with one another. Areas supported include studies of the structure, function, and assembly of cellular elements, such as the cytoskeleton, membranes, organelles, intracellular compartments, intranuclear structures, and extracellular matrix, including eukaryotic and prokaryotic cell walls and envelopes. Support is also provided for the study of intracellular and transmembrane signal transduction mechanisms and cell-cell signaling processes, including those that occur in biofilms. Research on cellular recognition and self defense mechanisms is included. Research utilizing both traditional and innovative methodologies , multidisciplinary approaches, technique development, computation and modeling, and approaches that exploit genomic information is encouraged. Multi-disciplinary approaches to the study of cellular systems, including research carried out at the interfaces of biology, physics, chemistry, mathematics and computer science, and engineering, are also encouraged. The Microbial Observatories and Microbial Interactions and Processes competition is also housed in this cluster. This expanded activity supports integrative studies that explore novel microorganisms, their interactions in consortia and communities, and aspects of their physiology, biochemistry and genomics in relationship to the processes that they carry out in the environments. Apply to PD 04-1114
Date(s) Application is Due: Jan 12; Jul 12
Contact: Eve Barak, Program Director, (703) 292-7113; email: ebarak@nsf.gov
Internet: http://www.nsf.gov/funding/pgm_summ.jsp?pims_id=12772
Sponsor: National Science Foundation
4201 Wilson Boulevard
Arlington, VA 22230

NSF Center for Synthesis in Biological Evolution **4508**
NSF requests proposals to establish a Center for Synthesis in Biological Evolution. This center will serve the needs of the evolutionary biology community by providing mechanisms to foster synthetic, collaborative, cross-disciplinary studies. It will play a pivotal role in the further unification of the biological sciences as it draws together knowledge from disparate biological fields to increase our general understanding of biological design and function. Finally, the center will play a critical role in organizing and synthesizing evolutionary knowledge that will be useful to policy makers, government agencies, educators and society. A preliminary proposal is due by September 17; full proposal, by request only, is due February 1. Annual deadline dates change; contact program staff for exact dates. NSF 03-570
Amount of Grant: $3 million per year
Date(s) Application is Due: Feb 1; Sep 17
Contact: Samuel Scheiner, Program Director, (703) 292-8481; fax: (703) 292-9064; email: sscheine@nsf.gov; Gerald Guala, Program Director, Division of Biological Infrastructure, (703) 292-8470; fax: (703) 292-9063; email: ggual
Internet: http://www.nsf.gov/pubs/2003/nsf03570/nsf03570.htm
Sponsor: National Science Foundation
4201 Wilson Boulevard
Arlington, VA 22230

NSF Centers for Learning and Teaching (CLT) Grants **4509**
This program is a comprehensive, research-based effort that will address critical issues and national needs of the science, technology, engineering, and mathematics (STEM) instructional workforce. Centers will provide a rich environment that melds research, teacher professional development, and education practice. Individual Centers may have specific foci (e.g., the nature of learning, teaching strategies, and reform policies and outcomes), but each will address the following three equally important components: renewing and diversifying the cadre of leaders in STEM education, particularly the higher education faculty who educate STEM teachers and professionals in leadership positions at state and district-levels and other education organizations; enhancing the content knowledge and pedagogical skills of current and future elementary and secondary teachers; and supporting research into STEM education issues of national import. The CLT effort builds previous activities in the preparation and professional development of teachers and provide opportunities for doctoral and postdoctoral students in the disciplines and in STEM education to acquire the knowledge and skills to educate the next generation of K-12 teachers. Awards will be made in two categories: full Center proposals and developmental proposals. A letter of intent is requested by December 2; full application is due February 20. Annual deadline dates may vary; contact program staff for exact dates. NSF 04-501
Requirements: Educational partnerships that include at least one K-12 school district and one institution of higher education, that awards doctoral degrees, are eligible to apply.
Amount of Grant: $16 million approximate total
Date(s) Application is Due: Feb 20; Dec 2
Contact: John Bradley, (703) 292-5091; email: jbradley@nsf.gov; Michael Haney, (703) 292-5102; email: mhaney@nsf.gov
Internet: http://www.nsf.gov/funding/pgm_summ.jsp?pims_id=5465
Sponsor: National Science Foundation
4201 Wilson Boulevard
Arlington, VA 22230

NSF Centers for Ocean Education Excellence (COSEE) Grants **4510**
Centers for Ocean Education Excellence (COSEE), supports partnerships between ocean science researchers, educators, and informal science organizations, providing the public a deeper understanding of the ocean and its influence on each person's quality of life and our national prosperity. Efforts to support recruitment of minorities to ocean science are also supported through OCE Education and Geo Diversity (OEDG).
Contact: Elizabeth L. Rom, Program Director; (703) 292-7709; fax: (703) 292-9085; email: elrom@nsf.gov; Donald Elthon; (703) 292-8580; fax: (703) 292-9085; email: delthon@nsf.gov
Internet: http://www.nsf.gov/funding/pgm_summ.jsp?pims_id=13548&org=OCE&from=home
Sponsor: National Science Foundation
4201 Wilson Boulevard
Arlington, VA 22230

NSF Centers of Research Excellence in Science and Technology and HBCU Research Infrastructure for Science and Engineering (RISE) Grants 4511
The Centers of Research Excellence in Science and Technology (CREST) program makes resources available to enhance the research capabilities of minority-serving institutions through the establishment of centers that effectively integrate education and research. CREST promotes the development of new knowledge, enhancements of the research productivity of individual faculty, and an expanded presence of students historically underrepresented in STEM disciplines. This solicitation requests proposals for: CREST centers; supplements for partnerships applied to existing CREST awards; HBCU Research Infrastructure for Science & Engineering (HBCU-RISE) proposals; supplements for diversity collaboration for projects co-funded with NSF's Small Business Innovation Research and Small Business Technology Transfer (SBIR/STTR) programs, which are administered by NSF's Directorate for Engineering; and new projects in the Directorate for Education and Human Resources track: Innovation through Institutional Integration.
Date(s) Application is Due: Feb 24; Feb 27; Mar 13; Aug 25
Contact: Toni Edquist, Program Specialist; (703) 292-4649; email: tedquist@nsf.gov
Internet: http://www.nsf.gov/funding/pgm_summ.jsp?pims_id=6668&org=ENG&from=home
Sponsor: National Science Foundation
4201 Wilson Boulevard
Arlington, VA 22230

NSF Chemical and Biological Separations (CBS) Grants 4512
The Chemical and Biological Separations (CBS) program supports fundamental research on novel methods and materials for separation processes. These processes are central to the chemical, biochemical, materials, energy, and pharmaceutical industries. A fundamental understanding of the interfacial, transport, and thermodynamic behavior of multiphase chemical systems as well as quantitative descriptions of processing characteristics in the process-oriented industries is critical for efficient resource management and effective environmental protection. The program encourages proposals that address emerging research areas and technologies, have a high degree of interdisciplinary thought coupled with knowledge creation, and integrate education and research. Areas of emphasis include separation of biological molecules and separations that lead to environmentally benign processing.
Date(s) Application is Due: Mar 1; Sep 15
Contact: Rosemarie D. Wesson, Program Director; (703) 292-7070; fax: (703) 292-9098; email: rwesson@nsf.gov
Internet: http://www.nsf.gov/funding/pgm_summ.jsp?pims_id=13363
Sponsor: National Science Foundation
4201 Wilson Boulevard
Arlington, VA 22230

NSF Chemical Oceanography Research Grants 4513
The Chemical Oceanography Program supports research into the chemical components, reaction mechanisms, and geochemical pathways within the ocean and at its interfaces with the solid earth and the atmosphere. Major emphases include: studies of material inputs to and outputs from marine waters; orthochemical and biological production and transformation of chemical compounds and phases within the marine system; and the determination of reaction rates and study of equilibria. The Program encourages research into the chemistry, distribution, and fate of inorganic and organic substances introduced into or produced within marine environments including those from estuarine waters to the deep sea.
Date(s) Application is Due: Feb 15; Aug 15
Contact: Donald Rice, Program Director; (703) 292-8582; fax: (703) 292-9085; email: drice@nsf.gov
Internet: http://www.nsf.gov/funding/pgm_summ.jsp?pims_id=11698&org=OCE&from=home
Sponsor: National Science Foundation
4201 Wilson Boulevard
Arlington, VA 22230

NSF Chemical, Biochemical, and Biotechnology Systems Research Grants 4514
The Chemical, Biochemical, and Biotechnology cluster deals with problems involved in the processing and manufacture of products of economic importance by effectively utilizing chemical resources and renewable resources of biological origin, often with the aid of bioinformatics originating from genomic and proteomic information. This cluster supports fundamental and applied research on catalytic processes and understanding the kinetics and mechanisms of pertinent reactions. Research in control, optimization, and process modeling are also supported. Chemical and Biological Separations, Biochemical Engineering, Biotechnology, Catalysis and Biocatalysis, and Process and Reaction Engineering are funded as part of this activity.
Amount of Grant: $400,000 maximum
Contact: Marcia Rawlings, Information Technology Specialist; (703) 292-7956; fax: (703) 292-9013; email: mrawling@nsf.gov
Internet: http://www.nsf.gov/funding/pgm_summ.jsp?pims_id=13481
Sponsor: National Science Foundation
4201 Wilson Boulevard, Room 565
Arlington, VA 22230

NSF Chemistry Research Experiences for Undergraduates (REU) 4515
The program supports active research participation by undergraduate students in any of the areas of research funded by the National Science Foundation. The NSF Division of Chemistry funds about 20-25 REU Sites a year (depending on the availability of funds) as part of the NSF-wide REU activity. Projects involve students in meaningful ways in ongoing research programs or in research projects specifically designed for the program. Two mechanisms are featured for support of student research: (1) REU Sites are based on independent proposals to initiate and conduct projects that engage a number of students in research. REU Sites may be based in a single discipline or academic department, or on interdisciplinary or multi-department research opportunities with a coherent intellectual theme. Proposals with an international dimension are welcome. A partnership with the Department of Defense supports REU Sites in DoD-relevant research areas. (2) REU Supplements may be requested for ongoing NSF-funded research projects or may be included as a component of proposals for new or renewal NSF grants or cooperative agreements. NSF 07-569.
Requirements: For REU Site proposals, a single individual should be designated as the Principal Investigator. This individual will be responsible for overseeing all aspects of the award. However, one additional person may be designated as Co-Principal Investigator, if developing and operating the REU Site would involve such shared responsibility. Other anticipated research supervisors should be listed as Non-Co-PI Senior Personnel. Undergraduate student participants in either Sites or Supplements must be citizens or permanent residents of the United States or its possessions. An undergraduate student is a student who is enrolled in a degree program (part-time or full-time) leading to a baccalaureate or associate degree. Students who are transferring from one college or university to another and are enrolled at neither institution during the intervening summer may participate. High school graduates who have been accepted at an undergraduate institution but who have not yet started their undergraduate study are also eligible to participate. Students who have received their bachelor's degrees and are no longer enrolled as undergraduates are generally not eligible to participate. For REU Sites, a significant fraction of the student participants should come from outside the host institution or organization. Some NSF directorates encourage inclusion in the REU program of K-12 teachers of science, technology, engineering, and mathematics. Please contact the appropriate disciplinary program officer for guidance.
Restrictions: Students may not apply to NSF to participate in REU activities. Students apply directly to REU Sites and should consult the directory of active REU Sites.
Contact: Wilfredo Colon, Program Director; (703) 292-8171; fax: (703) 292-9061; email: wcolon@nsf.gov
Internet: http://www.nsf.gov/funding/pgm_summ.jsp?pims_id=503210
Sponsor: National Science Foundation
4201 Wilson Boulevard
Arlington, VA 22230

NSF Chemistry Research Instrumentation and Facilities Program Grants 4516
This program provides funds to research institutions and consortia thereof for the purchase of multi-user instruments, for major instrumentation development and construction, and for the establishment and support of multi-user research facilities in the chemical sciences. This program is structured to enable the NSF, through its Division of Chemistry, to respond to a variety of needs for infrastructure to undergird advanced research and education in chemistry. Proposal deadline is the fourth Monday in June annually.
Requirements: Proposals may be submitted by U.S. institutions of higher education and independent nonprofit research institutions.
Amount of Grant: $6 million total
Contact: Wade Sisk, (703) 292-4454; fax: (703) 292-9037; email: wsisk@nsf.gov
Internet: http://www.nsf.gov/staff/staff_bio.jsp?lan=wsisk&org=NSF
Sponsor: National Science Foundation
4201 Wilson Boulevard
Arlington, VA 22230

NSF Chemistry Research Instrumentation and Facilities: Departmental Multi-User Instrumentation (CRIF:MU) 4517
This program is structured to enable the NSF's Division of Chemistry to respond to a variety of needs for infrastructure that promotes research and education in areas traditionally supported by the Division. The Departmental Multi-User Instrumentation component of CRIF (CRIF:MU) provides funds to universities, colleges, and consortia thereof for the purchase of multi-user instruments. The maximum request is $500,000 for instrumentation. Additional funds may be requested for personnel who are needed to support cyber-enhanced projects. The deadline is the fourth Monday of June for full proposals. Refer to NSF brochure NSF 05-578.
Amount of Grant: $500,000 maximum
Contact: Wade Sisk, Program Director, (703) 292-8840; fax: (703) 292-9037; email: wsisk@nsf.gov
Internet: http://www.nsf.gov/pubs/2005/nsf05578/nsf05578.htm
Sponsor: National Science Foundation
4201 Wilson Boulevard
Arlington, VA 22230

NSF CISE Increasing the Participation and Advancement of Women in Academic Science and Engineering Careers (ADVANCE) Grants 4518
The goal of the ADVANCE Program is to increase the participation of women in the scientific and engineering workforce through the increased representation and advancement of women in academic science and engineering careers. Through these awards, NSF seeks to support new and systemic approaches to improving the climate for women in U.S. academic institutions and to facilitate women's retention and advancement to the highest ranks of academic leadership. Creative approaches to realize the goal of this program are sought from women and men. Proposals may include international

activities that directly support the goals of ADVANCE. This program will support the following types of ADVANCE Projects: Partnerships for Adaptation, Implementation, and Dissemination (PAID) Awards (deadline January 17); Institutional Transformation (IT) Awards (deadline December 6); and Institutional Transformation Planning Grants (deadline December 6).
Requirements: Institutional Transformation proposals may be submitted by academic institutions of higher learning in the U.S., its territories or possessions, or the Commonwealth of Puerto Rico, that award degrees in a field supported by NSF. Partnerships for Adaptation, Implementation and Dissemination proposals may be submitted by: academic institutions of higher learning that award degrees in a field supported by NSF; professional societies; or other not-for-profit organizations.
Date(s) Application is Due: Jan 17; Dec 6
Contact: Jessie A. DeAro, Program Director; (703) 292-5350; fax: (703) 292-9018; email: jdearo@nsf.gov
Internet: http://www.nsf.gov/funding/pgm_summ.jsp?pims_id=5383&org=CISE&sel_org=CISE&from=fund
Sponsor: National Science Foundation
4201 Wilson Boulevard
Arlington, VA 22230

NSF CISE Accelerating Discovery in Science and Engineering Through Petascale Simulations and Analysis (PetaApps) Grants **4519**
This solicitation seeks proposals to develop the future simulation and analysis tools that can use petascale computing to advance the frontiers of scientific and engineering research. Proposals are sought from researchers aiming to capitalize on emerging petascale computing architectures, catalyzing potentially transformative research. NSF's emphasis is on implementation and exploitation of forefront techniques. Proposers must be prepared to demonstrate that they have a research problem that requires and can exploit petascale computing capabilities. Proposals from or including junior researchers are encouraged as one of the goals of this solicitation is to build a community capable of using petascale computing. The focus of the proposal may be in any of the following areas: Atmospheric Sciences; Astronomy and Astrophysics; Chemistry; Computer Science; Earth Sciences; Engineering; Materials Science; Mathematics and Statistics; Ocean Sciences; Physics; Social, Behavioral and Economic Sciences.
Requirements: An individual may be the PI or co-PI in no more than two proposal that responds to this solicitation. There is no limit on the number of proposals with which an individual may be associated as senior personnel to provide specific expertise.
Amount of Grant: $1,125,000-$2,000,000 range
Date(s) Application is Due: Oct 30
Contact: Abani Patra, OCI Program Director; (703) 292-8970, email: apatra@nsf.gov; Almadena Chtchelkanova, Program Director, CISE; (703) 292-8910, email: achtchel@nsf.gov
Internet: http://www.nsf.gov/pubs/2008/nsf08592/nsf08592.htm
Sponsor: National Science Foundation
4201 Wilson Boulevard
Arlington, VA 22230

NSF CISE Advanced Learning Technologies (ALT) Grants **4520**
Through the Advanced Learning Technologies (ALT) program, CISE and EHR support learning technologies research that enables radical improvements in learning through innovative computer and information technologies, and advances research in computer science, information technology, learning, and cognitive science through the unique challenges posed by learning environments and learning technology platforms. Integrative research approaches that build across disciplines and establish tight linkages among theory, experiment, and design are strongly encouraged. Under this program, CISE and EHR will support research and prototyping work that is built around well defined learning and technology goals, and leads to generalizable advances in knowledge and technology. Large-scale studies, infrastructure, large-scale system development, or educational materials development will not be supported under this program. Both learning and technology goals are expected to be well articulated and significant: educational applications of off-the-shelf technologies or projects that neither model nor measure learning will not be supported under this program.
Requirements: Proposals that do not address both learning and technology goals may be returned without review by the program.
Amount of Grant: $100,000-$200,000 per year over three years
Date(s) Application is Due: Apr 25
Contact: Amy L. Baylor, Program Director; (703) 292-8930; fax: (703) 292-9073; email: abaylor@nsf.gov; Kenneth Whang, Program Director; (703) 292-8930; fax: (703) 292-9073; email: kwhang@nsf.gov
Internet: http://www.nsf.gov/funding/pgm_summ.jsp?pims_id=12834&org=CISE&sel_org=CISE&from=fund
Sponsor: National Science Foundation
4201 Wilson Boulevard
Arlington, VA 22230

NSF CISE Assembling the Tree of Life (ATOL) Grants **4521**
Projects for Assembling the Tree of Life are expected to be ambitious, large scale, and when appropriate to involve multiple investigators from multiple disciplines, likely from multiple organizations, and to include training, outreach, and dissemination components. Tree of Life projects that are taxon-oriented will focus on phylogenetic resolution of large lineages or clades; this taxon focus is not intended to deflect interest in and attention to theoretical or analytical issues, particularly when the clade under study raises critical questions about the suitability or power of current phylogenetic methods of analysis, such as complexities caused by reticulate evolution and lateral gene transfer. In addition to hypothesis-driven work, Tree of Life projects may also be method or theory-oriented, in which case they will address major analytical or computational problems in phylogenetic research and phyloinformatics. The taxon-focus and method-focus approaches described here are intended for guidance only, and not as constraints on innovative projects for Assembling the Tree of Life.
Requirements: Individuals and organizations in the following categories may submit proposals: universities and colleges; non-profit, non-academic organizations; for-profit organizations; state and local governments; unaffiliated individuals; foreign organizations; and other federal agencies.
Amount of Grant: $3,000,000 maximum
Date(s) Application is Due: Mar 14
Contact: Patrick Herendeen; (703) 292-7184; email: pherende@nsf.gov
Internet: http://www.nsf.gov/pubs/2008/nsf08515/nsf08515.htm
Sponsor: National Science Foundation
4201 Wilson Boulevard
Arlington, VA 22230

NSF CISE Broadening Participation in Computing (BPC) Grants **4522**
The Broadening Participation in Computing (BPC) program is intended to significantly increase the number of domestic students receiving post secondary degrees in the computing disciplines, with an emphasis on those communities that have a longstanding underrepresentation in computing: persons with disabilities, women, and minorities. Included minorities are African Americans, Hispanics, American Indians, Alaska Natives, Native Hawaiians, and Pacific Islanders. Members of these groups participate in computing at rates well below their proportionate representation in the population. Specifically, the BPC program aims to develop and implement innovative models for recruiting, mentoring, and retaining students in post secondary programs in the computing disciplines, with particular attention to students from underrepresented groups. Since the lack of role models in the professoriate can be a barrier to participation, the program also aims to develop effective strategies for encouraging students who want to pursue academic careers in computing and can serve as these role models. While the focus is on implementation, a project may include complementary research that informs the design of its activities. PIs are encouraged to include social scientists in any research, evaluation, and assessment activities.
Requirements: There is no organizational limit. Alliance proposals must include at least one partner that is a degree-granting, academic institution of higher learning located in the U.S., its territories or possessions, or a consortium of such institutions. A person can be a PI or CoPI on at most one Alliance or Alliance Extension proposal (although he or she may serve in other capacities on additional proposals). The PI on an Extension proposal must be a PI or CoPI on the original BPC Alliance award.
Amount of Grant: $200,000 to 750,000 per year for three years;$2,000,000 maximum
Date(s) Application is Due: May 21
Contact: Janice Cuny, Program Director; (703) 292-8489; fax: (703) 292-9010; email: jcuny@nsf.gov; Harriet G. Taylor, Program Manager; (703) 292-8950; fax: (703) 292-9010; email: htaylor@nsf.gov
Internet: http://www.nsf.gov/pubs/2007/nsf07548/nsf07548.htm
Sponsor: National Science Foundation
4201 Wilson Boulevard
Arlington, VA 22230

NSF CISE Cluster Exploratory (CluE) Grants **4523**
The CluE program invites researchers to submit proposals that describe new, innovative use of the cluster, and/or that probe the possibilities and fundamental limits of the computing paradigm it enables. Since this program provides researchers with access to the Google-IBM resource, proposals must clearly describe why access to the cluster will significantly advance the line of research inquiry.
Requirements: Proposals may only be submitted by the following: non-profit, non-academic organizations—independent museums, observatories, research labs, professional societies and similar organizations in the U.S. associated with educational or research activities; and universities and colleges—universities and two- and four-year colleges (including community colleges) located and accredited in the U.S., acting on behalf of their faculty members. Such organizations also are referred to as academic institutions.
Amount of Grant: $500,000 maximum for up to 2 years
Date(s) Application is Due: Jul 17
Contact: Jim French; 703 292 8930; email: jfrench@nsf.gov
Internet: http://www.nsf.gov/pubs/2008/nsf08560/nsf08560.htm
Sponsor: National Science Foundation
4201 Wilson Boulevard
Arlington, VA 22230

NSF CISE Collaborative Research in Computational Neuroscience Grants **4524**
Computational neuroscience provides a theoretical foundation and a rich set of technical approaches for understanding the principles and dynamics of the nervous system. Building on the theory, methods, and findings of computer science, neuroscience, biology, psychology, the mathematical and physical sciences, engineering, and other fields, computational neuroscience employs a broad spectrum of analytical and experimental approaches to study function, organization, and computational strategies across all levels of the nervous system. Two classes of proposals will be considered in response to this solicitation: research proposals describing new collaborative research projects, and data sharing proposals to enable sharing of data and other resources. In general, appropriate scientific areas of investigations may be related to any of the participating funding organizations.

Restrictions: In response to this solicitation, an investigator may participate as PI or Co-PI in no more than one research proposal per review cycle, and in no more than one data sharing proposal per review cycle.
Amount of Grant: $100,000 to $250,000 per year for three to five years
Date(s) Application is Due: Feb 26; Oct 30
Contact: Kenneth Whang, CRCNS Program Coordinator; (703) 292-5149; fax: (703) 292-9073; email: kwhang@nsf.gov
Internet: http://www.nsf.gov/pubs/2008/nsf08514/nsf08514.htm
Sponsor: National Science Foundation
4201 Wilson Boulevard
Arlington, VA 22230

NSF CISE Communicating Research to Public Audiences Grants 4525
The purpose of Communicating Research to Public Audiences grants is to promote the discovery, integration, dissemination, and employment of new knowledge in service to society and to achieve excellence in U.S. science, technology, engineering, and mathematics (STEM) education at all levels. These grants will provide an opportunity for Principal Investigators (PIs) to explain in non-technical terms the methods and/or results of their research to a broad and diverse audience. Projects with funding levels up to $75,000 will be supported to communicate the processes and results of current research awards from any NSF directorate or the Office of Polar Programs to public audiences in order to assist in the broader dissemination of research results and to promote science and technological literacy for the general public in an out-of-school setting. The grant can be used for any activity that falls within the definition of an informal science education activity such as media presentations, exhibits, or youth-based activities, in order to disseminate research results, research in progress, or research methods.
Requirements: Individuals and organizations in the following categories may submit proposals: U.S. universities and two-and four-year colleges (including community colleges) acting on behalf of their faculty members; independent museums, observatories, research laboratories, professional societies and similar organizations in the U.S. that are directly associated with educational or research activities; U.S. commercial organizations (especially small businesses with strong capabilities in scientific or engineering research or education); state and local governments; scientists, engineers or educators in the U.S. and U.S. citizens; foreign organizations; and a limited number of other types of Federal agencies.
Restrictions: NSF rarely provides support to foreign organizations. However, NSF will consider proposals for cooperative projects involving U.S. and foreign organizations, provided support is requested only for the U.S. portion of the collaborative effort.
Amount of Grant: $75,000 maximum
Contact: Alphonse T. Desena, Program Director; (703) 292-5106; fax: (703) 292-9044; email: adesena@nsf.gov; Sylvia M. James, Program Director; (703) 292-5333; fax: (703) 292-9044; email: sjames@nsf.gov
Internet: http://www.nsf.gov/pubs/2003/nsf03509/nsf03509.html
Sponsor: National Science Foundation
4201 Wilson Boulevard
Arlington, VA 22230

NSF CISE Community-Based Data Interoperability Networks Grants 4526
This NSF crosscutting program supports community efforts to provide for broad interoperability with the goal of enhancing interaction and information sharing across all of the areas of science and engineering research and education represented at the National Science Foundation. The program supports the formation of community-based Data Interoperability Networks groups that enable communities to work together in the development of effective data interoperability strategies. Each Network holds dual responsibilities for: enabling broad community engagement in the development of consensus and agreement on strategies, priorities, and best approaches for achieving broad interoperability; and providing the technical expertise necessary to turn consensus and agreement into robust interoperability frameworks along with the appropriate tools and resources for their broad use and implementation.
Requirements: Individuals and organizations in the following categories may submit proposals: U.S. universities and two-and four-year colleges (including community colleges) acting on behalf of their faculty members; independent museums, observatories, research laboratories, professional societies and similar organizations in the U.S. that are directly associated with educational or research activities; U.S. commercial organizations (especially small businesses with strong capabilities in scientific or engineering research or education); state and local governments; scientists, engineers or educators in the U.S. and U.S. citizens; foreign organizations; and a limited number of other types of Federal agencies.
Restrictions: NSF rarely provides support to foreign organizations. However, NSF will consider proposals for cooperative projects involving U.S. and foreign organizations, provided support is requested only for the U.S. portion of the collaborative effort.
Amount of Grant: $250,000 total costs per year for three to five years
Date(s) Application is Due: Jul 23
Contact: Lucille Nowell; (703) 292-8970; email: lnowell@nsf.gov; Maria Burka; (703) 292-7030; email: mburka@nsf.gov
Internet: http://www.nsf.gov/pubs/2007/nsf07565/nsf07565.htm
Sponsor: National Science Foundation
4201 Wilson Boulevard
Arlington, VA 22230

NSF CISE Computer and Network Systems (CNS): Core Programs Grants 4527
CISE's Division of Computer and Network Systems (CNS) supports research and education projects that develop new knowledge in two core programs: the Computer Systems Research (CSR) program; and the Networking Technology and Systems (NeTS) program. Proposers are invited to submit proposals in three project classes, which are defined as follows: Small Projects—up to $500,000 total budget with durations up to three years; Medium Projects—$500,001 to $1,200,000 total budget with durations up to four years; and Large Projects—$1,200,001 to $3,000,000 total budget with durations up to five years. Small Project proposals are due by December 17, Medium Project proposals are due by August 31, and Large Project proposals are due by November 28.
Requirements: Proposals may only be submitted by the following: non-profit, non-academic organizations—independent museums, observatories, research labs, professional societies and similar organizations in the U.S. associated with educational or research activities; and universities and two- and four-year colleges (including community colleges) located and accredited in the U.S., acting on behalf of their faculty members.
Amount of Grant: $500,000 average
Date(s) Application is Due: Aug 31; Nov 28; Dec 17
Contact: Darleen Fisher; (703) 292-8950, email: dlfisher@nsf.gov; Krishna Kant; (703) 292-4776, email: kkant@nsf.gov
Internet: http://www.nsf.gov/pubs/2008/nsf08576/nsf08576.htm
Sponsor: National Science Foundation
4201 Wilson Boulevard
Arlington, VA 22230

NSF CISE Computing and Communication Foundations: Core Grants 4528
CISE's Division of Computing and Communication Foundations (CCF) supports research and education projects that develop new knowledge in three core programs: the Algorithmic Foundations program; the Communications and Information Foundations program; and the Software and Hardware Foundations program. Proposers are invited to submit proposals in three project classes, which are defined as follows: Small Projects—up to $500,000 total budget with durations up to three years; Medium Projects—$500,001 to $1,200,000 total budget with durations up to four years; and Large Projects—$1,200,001 to $3,000,000 total budget with durations up to five years. Small Project proposals are due by December 17, Medium Project proposals are due by August 31 (except for October 31 in 2008), and Large Project proposals are due by November 28.
Requirements: Proposals may only be submitted by the following: non-profit, non-academic organizations—independent museums, observatories, research labs, professional societies and similar organizations in the U.S. associated with educational or research activities; and universities and two- and four-year colleges (including community colleges) located and accredited in the U.S., acting on behalf of their faculty members.
Amount of Grant: $500,000 average
Date(s) Application is Due: Aug 30; Nov 28; Dec 17
Contact: Richard Beigel; (703) 292-8910, email: rbeigel@nsf.gov; John H. Cozzens; (703) 292-8910, email: jcozzens@nsf.gov
Internet: http://www.nsf.gov/pubs/2008/nsf08577/nsf08577.htm
Sponsor: National Science Foundation
4201 Wilson Boulevard
Arlington, VA 22230

NSF CISE Computing Research Infrastructure (CRI) Grants 4529
The CISE Computing Research Infrastructure (CRI) program drives discovery and learning in the computing disciplines by supporting the creation, enhancement and operation of world-class computing research infrastructure. Further, through the CRI program CISE seeks to ensure that individuals from a diverse range of academic institutions, including minority-serving and predominantly undergraduate institutions, have access to such infrastructure. The CRI program supports two classes of awards: (1) Institutional Infrastructure awards support either the creation of new computing research infrastructure or the enhancement of existing computing research infrastructure to enable world-class research and education opportunities at the awardee and collaborating institutions. (2) Community Infrastructure awards support the planning for computing research infrastructure, or the creation of new computing infrastructure, or the enhancement of existing computing research infrastructure to enable world-class research and education opportunities for broadly-based communities of researchers and educators that extend well beyond the awardee institutions. Furthermore, CI awards support the operation of such infrastructure, ensuring that awardee institutions are well-positioned to provide a high quality of service to community researchers and educators expected to use the infrastructure to realize their research and education goals. NSF 08-570
Requirements: Individuals and organizations in the following categories may submit proposals: U.S. universities and two-and four-year colleges (including community colleges) acting on behalf of their faculty members; independent museums, observatories, research laboratories, professional societies and similar organizations in the U.S. that are directly associated with educational or research activities; U.S. commercial organizations (especially small businesses with strong capabilities in scientific or engineering research or education); state and local governments; scientists, engineers or educators in the U.S. and U.S. citizens; foreign organizations; and a limited number of other types of Federal agencies.
Restrictions: NSF rarely provides support to foreign organizations. NSF will consider proposals for cooperative projects involving U.S. and foreign organizations, provided support is requested only for the U.S. portion of the collaborative effort. NSF does not normally support research or education activities by scientists, engineers or educators employed by Federal agencies or Federally Funded Research and Development Centers (FFRDCs).
Amount of Grant: $200,000 - $750,000
Date(s) Application is Due: Sep 22; Aug 5

Contact: Tanya Korelsky, Program Director; (703) 292-8930; email: tkorelsk@nsf.gov
Internet: http://www.nsf.gov/funding/pgm_summ.jsp?pims_id=12810&govDel=U.S.NSF_39
Sponsor: National Science Foundation
4201 Wilson Boulevard
Arlington, VA 22230

NSF CISE Cross-Cutting Program Grants **4530**
This solicitation seeks proposals in three programs of interest to all three CISE divisions, and which pose important and timely research and education challenges optimally addressed by researchers from a range of computing fields or sub-fields and beyond, including: data-intensive computing; network science and engineering; and trustworthy computing. Proposals submitted to each of the cross-cutting programs will be reviewed collaboratively by the three CISE divisions, with the goal of creating a comprehensive portfolio of awards with high-impact potential. CISE expects that, over time, these cross-cutting programs will evolve or be absorbed in our core programs, and that new cross-cutting programs will be introduced. Proposers are invited to submit proposals in three project classes, which are defined as follows: small projects—up to $500,000 total budget with durations up to three years; medium projects—$500,001 to $1,200,000 total budget with durations up to four years; and large projects—$1,200,001 to $3,000,000 total budget with durations up to five years.
Requirements: Proposals may only be submitted by the following: non-profit, non-academic organizations—independent museums, observatories, research labs, professional societies and similar organizations in the U.S. associated with educational or research activities; and universities and two- and four-year colleges (including community colleges) located and accredited in the U.S., acting on behalf of their faculty members. Such organizations also are referred to as academic institutions.
Date(s) Application is Due: Aug 30; Nov 28; Dec 17
Contact: Darleen L. Fisher, Network Science and Engineering Program; (703) 292-8950; email: dlfisher@nsf.gov; Jim French, Data-intensive Computing Program; (703) 292-8930; fax: (703) 292-9073; email: jfrench@nsf.gov
Internet: http://www.nsf.gov/pubs/2008/nsf08578/nsf08578.htm
Sponsor: National Science Foundation
4201 Wilson Boulevard
Arlington, VA 22230

NSF CISE Distributed Systems Grants **4531**
To achieve efficient, flexible, and robust computing in parallel/multiprocessor/network execution environment, this program sponsors research and education projects on methodologies and designs of system software and runtime support for distributed applications. The program focuses on novel concepts for the design of distributed systems to capture the dynamic and open nature of the underlying system platforms. Work that extends and integrates operating system, and network technologies is emphasized. The program encompasses subject areas ranging from fundamental operating systems, cluster and grid computing, middleware, to emerging pervasive agent/mobile systems, and peer-to-peer and web-based computing. Experimental designs and formal methods and analytical/simulation tools for such systems are also encouraged. In line with the NSF mission, proposals are expected to integrate research topics with educational activities. Estimated program budget, number of awards and average award size/duration are subject to the availability of funds.
Date(s) Application is Due: Dec 4
Contact: D. Helen Gill, Program Director; (703) 292-8950; fax: (703) 292-9010; email: hgill@nsf.gov
Internet: http://www.nsf.gov/pubs/progdesc/2003/cise/2876.html
Sponsor: National Science Foundation
4201 Wilson Boulevard, Room 1175 N
Arlington, VA 22230

NSF CISE Educational Innovation Program Grants **4532**
Grants support innovative educational activities at the undergraduate level in computer and information science and engineering that transfer research results into the undergraduate curriculum. Projects supported are expected to act as national models of excellence by being prototypes of educational experiences for use by a broader segment of the CISE community. Proposals may address a variety of educational activities, including, but not limited to, the development of courses, instructional technologies, software, and other educational materials. Institutions must match at least one-third of the total cost. Refer to brochure NSF 00-33 for program announcement. Deadline is generally the second Tuesday in March each year.
Requirements: Proposals may be submitted from any U.S. college or university that currently offers baccalaureate degree programs in any discipline supported by the CISE Directorate.
Amount of Grant: $250,000 average
Date(s) Application is Due: Apr 25
Contact: Dr. Anthony Maddox; (703) 306-1980; email: eipd@nsf.gov; Program Contact; (703) 306-1980; fax: (703) 292-9030; email: eipd@nsf.gov
Internet: http://www.nsf.gov/pubs/2000/nsf0033/nsf0033.htm
Sponsor: National Science Foundation
4201 Wilson Boulevard
Arlington, VA 22230

NSF Climate and Large-Scale Dynamics (CLD) Grants **4533**
The goals of the Program are to: advance knowledge about the processes that force and regulate the atmosphere, synoptic and planetary circulation, weather and climate; and sustain the pool of human resources required for excellence in synoptic and global atmospheric dynamics and climate research. Research topics include: theoretical, observational and modeling studies of the general circulation of the stratosphere and troposphere; synoptic scale weather phenomena; processes that govern climate; the causes of climate variability and change; methods to predict climate variations; extended weather and climate predictability; development and testing of parameterization of physical processes; numerical methods for use in large-scale weather and climate models; the assembly and analysis of instrumental and/or modeled weather and climate data; data assimilation studies; development and use of climate models to diagnose and simulate climate and its variations and change. PD 06-5740
Requirements: Individuals and organizations in the following categories may submit proposals: universities and colleges; non-profit, non-academic organizations; and for-profit organizations (an unsolicited proposal from a commercial organization may be funded when the project is of special concern from a national point of view, special resources are available for the work, or the proposed project is especially meritorious. NSF is interested in supporting projects that couple industrial research resources and perspectives with those of universities; therefore, it especially welcomes proposals for cooperative projects involving both universities and the private commercial sector; state and local governments—state educational offices or organizations and local school districts may submit proposals intended to broaden the impact, accelerate the pace, and increase the effectiveness of improvements in science, mathematics and engineering education in both K-12 and post-secondary levels; unaffiliated individuals—scientists, engineers or educators in the U.S. and U.S. citizens may be eligible for support, provided that the individual is not employed by, or affiliated with, an organization, and: (a) the proposer has demonstrated the capability and has access to any necessary facilities to carry out the project; and, (b) the proposer agrees to fiscal arrangements that, in the opinion of the NSF Division of Grants and Agreements, ensure responsible management of Federal funds. Unaffiliated individuals should contact the appropriate program before preparing a proposal for submission.
Restrictions: NSF rarely provides support to foreign organizations. NSF will consider proposals for cooperative projects involving U.S. and foreign organizations, provided support is requested only for the U.S. portion of the collaborative effort. NSF does not normally support research or education activities by scientists, engineers or educators employed by Federal agencies or Federally Funded Research and Development Centers (FFRDCs). Under unusual circumstances, other Federal agencies and FFRDCs may submit proposals directly to NSF.
Samples: Atmospheric and Planetary Sciences - William Paterson, Hampton University ($92,513); VOCALS—The Influence of Continental Aerosols on Stratocumulus Clouds - Gilberto Fochesatto, University of Alaska Fairbanks ($100,000); Testing Climate Models by Feedback Analysis Using Atmospheric Infrared Sounder (AIRS) and Global Positioning System (GPS) Radio Occultation Data - Richard Goody, Harvard University ($172,383); Testing Climate Models by Feedback Analysis Using Atmospheric Infrared Sounder (AIRS) and Global Positioning System (GPS) Radio Occultation Data - Xianglei Huang, University of Michigan Ann Arbor ($72,211).
Contact: Jay S. Fein, Program Director; (703) 292-8527, fax: (703) 292-9022; email: jfein@nsf.gov
Internet: http://nsf.gov/funding/pgm_summ.jsp?pims_id=11699&org=ATM
Sponsor: National Science Foundation
4201 Wilson Boulevard
Arlington, VA 22230

NSF Collaborations in Mathematical Geosciences (CMG) **4534**
The primary purposes of the CMG activity are to promote cutting-edge research in areas that require the collaboration of experts in both the geosciences and the mathematical sciences and to promote the training of researchers with skills in both the mathematical sciences and geosciences. We are particularly interested in supporting collaborations in areas where such interdisciplinary collaboration is presently uncommon. We are also particularly interested in supporting collaborations between young investigators. It is not the intent of this activity to provide general support for infrastructure. NSF 05-535
Amount of Grant: $8.3 million total funding
Date(s) Application is Due: Feb 1
Contact: Xuming He, Division of Mathematical Sciences, (703) 292-4876; fax: (703) 292-9032; email: xhe@nsf.gov; Stephen Meacham, Division of Atmospheric Sciences, (703) 292-8527; fax: (703) 292-9022; email: smeacham@nsf.gov
Internet: http://www.nsf.gov/pubs/2005/nsf05535/nsf05535.htm
Sponsor: National Science Foundation
4201 Wilson Boulevard
Arlington, VA 22230

NSF Collaborative Research in Chemistry **4535**
This program is designed to promote interdisciplinary collaborative research in a coherent, defined project at the forefront of the chemical sciences. CRC proposals will involve three or more investigators with complementary expertise. Co-investigators may include researchers with backgrounds in diverse areas of chemistry and other science and engineering disciplines appropriate to the proposed research. The use of cyber-infrastructure to enable and enhance collaborations is encouraged. Projects should be scientifically focused in areas supported by the NSF Division of Chemistry, limited in duration, and substantial in their scope and impact. Listed deadlines are for letters of internet; full proposals are by invitation only. Annual deadline dates may vary; contact program staff for exact dates. NSF 05-606
Requirements: Only U.S. academic institutions and nonprofit, nonacademic organizations may submit proposals. The PI must be affiliated with a U.S. academic institution or nonprofit, non-academic organization. Other investigators may be affiliated with U.S.

academic institutions, nonprofit, nonacademic organizations, industry, government laboratories, or international institutions.
Restrictions: No CRC award funds may go directly to industry, government laboratories or international institutions.
Amount of Grant: $2.5 million total
Date(s) Application is Due: Aug 11; Nov 14
Contact: Katharine Covert, Program Director, Directorate for Mathematical & Physical Sciences, (703) 292-4950, fax: (703) 292-9037; email: kcovert@nsf.gov
Internet: http://www.nsf.gov/pubs/2005/nsf05606/nsf05606.htm
Sponsor: National Science Foundation
4201 Wilson Boulevard
Arlington, VA 22230

NSF Communicating Research to Public Audiences 4536
Communicating Research to Public Audiences is a component of the Informal Science Education program (ISE) in the Division of Elementary, Secondary, and Informal Education. ISE projects provide rich and stimulating contexts and experiences for individuals of all ages, interests, and backgrounds to increase their appreciation for, and understanding of, science, technology, engineering, and mathematics (STEM) in out-of-school settings. Requests for up to $75,000 will be considered to support projects that communicate to public audiences the process and results of current research that is being supported by any NSF directorate through informal science education activities, such as media presentations, exhibits, or youth-based activities. The purpose of these efforts is to disseminate research results, research in progress, or research methods. NSF 03-509
Requirements: The Principal Investigator (PI) must have an active NSF research award; a letter of support from the cognizant Program Officer for the research award is required. (NSF research awards do not include Small Grants for Exploratory Research Awards; Conference, Symposia, and Workshops grants; Dissertation Improvement Awards; or Post-doc Fellowships.) There is no fixed deadline. Proposals may be submitted at anytime, but at least six months prior to anticipated start date. The maximum award amount is $75,000. The maximum award period is 24 months. It is expected that the product (e.g. exhibit, film, program) will have a life beyond the expiration of the award. Estimated program budget, number of awards, and average award size/duration are subject to the availability of funds.
Restrictions: One proposal per active research award.
Samples: Neil F. Humphrey, W. Tad Pfeffer, B. Amadei (University of Wyoming and University of Colorado). - Direct Measure of the In Situ Stress Tensor at Depth in Glacier Ice: A Collaborative Study Glaciological Interpretation at Worthington Glacier State Park.; Ceil Lucas (Gallaudet University). - The Socio-linguistic Variation in American Sign Language, Phase II.; Veerabhadran Ramanathan (University of California San Diego). - Catch A Cloud.
Amount of Grant: up to $75,000
Contact: Sylvia M. James, Program Director; (703) 292-5333; fax: (703) 292-9044; email: sjames@nsf.gov
Internet: http://www.nsf.gov/funding/pgm_summ.jsp?pims_id=5362&org=DGE&sel_org=DGE&from=fund
Sponsor: National Science Foundation
4201 Wilson Boulevard
Arlington, VA 22230

NSF Computational Mathematics Research Grants 4537
This program supports mathematical research in areas of science where computing plays a central and essential role, emphasizing algorithms design, numerical methods and their analysis, and symbolic methods. The prominence of computation in the research is a hallmark of the program. Proposals ranging from single-investigator projects that develop and analyze innovative computational methods to interdisciplinary team projects that not only create new mathematical and computational techniques but use them to model, study, and solve important application problems are encouraged.
Contact: Junping Wang, Program Director; (703) 292-4488; fax: (703) 292-9032; email: msteuerw@nsf.gov
Internet: http://www.nsf.gov/funding/pgm_summ.jsp?pims_id=5390&org=DMS&from=home
Sponsor: National Science Foundation
4201 Wilson Boulevard
Arlington, VA 22230

NSF Condensed Matter Physics Research Grants 4538
This program supports fundamental, experimental, and combined experiment and theory projects on the physics of solid, liquid, and amorphous systems. Phenomena of interest include phase transitions; localization; electronic, magnetic, and lattice structure; superconductivity; elementary excitations, including electronic, magnetic, plasma, and lattice; transport, magnetic, and optical properties; and nonlinear dynamics. Low-temperature physics is represented by research on quantum fluids and solids as well as two-dimensional electron systems. Soft condensed matter research includes partially ordered fluids, colloid physics. Characterization and analysis of new materials by novel methods and research on condensed matter under extreme conditions-such as low temperatures, high pressures, and high magnetic fields-are of interest. Development of new experimental techniques to carry out proposed projects is encouraged. Apply to PD 03-1710
Date(s) Application is Due: Nov 2
Contact: Wendy Fuller-Mora, Program Director, Division of Materials Research. Directorate for Mathematics and Physical Sciences, (703) 292-4931; fax: (703) 292-9035; email: wfullerm@nsf.gov
Internet: http://www.nsf.gov/funding/pgm_summ.jsp?pims_id=5666
Sponsor: National Science Foundation
4201 Wilson Boulevard
Arlington, VA 22230

NSF Conferences, Workshops, and Special Meetings in the Mathematical Sciences Grants 4539
For conferences, workshops, and special meetings, the Division of Mathematical Sciences (DMS) invites proposals of two types: regular conference, symposia, and workshop proposals; and proposals for special meetings, which comprise longer-term or larger-scale activities that more widely engage and connect the mathematical sciences community, such as special research years or semesters, multi-institutional regional meetings, and summer schools. Regular proposals are submitted to the cognizant DMS programs according to those programs' usual deadlines or target dates. These proposals normally request funding in the range of $5,000 to $25,000, although awards of up to $50,000 have occasionally been made. Their duration is normally for one year. Proposals for special meetings are submitted to the cognizant DMS programs but at the common deadline stated in this solicitation. Special meetings proposals may request funding of any amount and for durations of up to three years, but most awards are expected to be in the range of $50,000 to $150,000 per year. This is not a change in the sort of proposals that DMS is willing to accept or to fund; it is simply a reminder to the mathematical sciences research community that this opportunity is available.
Date(s) Application is Due: Aug 27
Contact: Program Director; (202) 293-1170; fax: (202) 293-3412; email: rosier@math.georgetown.edu
Internet: http://www.nsf.gov/funding/pgm_summ.jsp?pims_id=11701
Sponsor: National Science Foundation
4201 Wilson Boulevard
Arlington, VA 22230

NSF Continental Dynamics Research Grants 4540
Grants support multidisciplinary research that focuses on an improved understanding of the processes governing the origin, structure, composition, and dynamical evolution of the continents and continental building blocks. The program is particularly oriented toward projects whose scope and complexity require a cooperative or multi-institutional approach and multiyear planning and execution. It is intended that the program would fund only relatively large projects that do not fit easily within earth sciences project support and that have broad support of major sections of the earth science community. Preliminary proposals are required by April 1; full proposals must be received by November 15. NSF 04-512
Requirements: Proposals may be submitted by academic institutions and nonprofit research organizations.
Date(s) Application is Due: Apr 1; Nov 15
Contact: Leonard Johnson, Program Director, Directorate for Geosciences, Division of Earth Sciences, (703) 292-8559; fax: (703) 292-9025; email: lejohnso@nsf.gov
Internet: http://www.nsf.gov/pubs/2004/nsf04512/nsf04512.htm
Sponsor: National Science Foundation
4201 Wilson Boulevard
Arlington, VA 22230

NSF Cooperative Studies Of The Earth's Deep Interior (CSEDI) Grants 4541
The Division of Earth Sciences (EAR) invites the submission of proposals for collaborative, interdisciplinary studies of the Earth's interior within the framework of the community-based initiative known as Cooperative Studies of the Earth's Deep Interior (CSEDI). Funding will support basic research on the character and dynamics of the Earth's mantle and core, their influence on the evolution of the Earth as a whole, and on processes operating within the deep interior that affect or are expressed on the Earth's surface. Projects may employ any combination of field, laboratory, and computational studies with observational, theoretical, or experimental approaches. Support is available for research and research infrastructure through grants and cooperative agreements awarded in response to investigator-initiated proposals from U.S. universities and other eligible institutions. Multidisciplinary work is required. EAR will consider co-funding of projects with other agencies and supports international work and collaborations.
Requirements: Individuals and organizations in the following categories may submit proposals: (1) Universities and Colleges - Universities and two- and four-year colleges (including community colleges) located and accredited in the U.S., acting on behalf of their faculty members. (2) Non-profit, non-academic organizations - Independent museums, observatories, research laboratories, professional societies and similar organizations in the U.S. that are directly associated with educational or research activities (3.) For-profit organizations - U.S. commercial organizations, especially small businesses with strong capabilities in scientific or engineering research or education. An unsolicited proposal from a commercial organization may be funded when the project is of special concern from a national point of view, special resources are available for the work, or the proposed project is especially meritorious. NSF is interested in supporting projects that couple industrial research resources and perspectives with those of universities; therefore, it especially welcomes proposals for cooperative projects involving both universities and the private commercial sector. (4) State and Local Governments - State educational offices or organizations and local school districts may submit proposals intended to broaden the impact, accelerate the pace, and increase the effectiveness of improvements in science, mathematics and engineering education in both K-12 and post-secondary levels. (5) Unaffiliated Individuals - Scientists, engineers or educators in the U.S. and U.S. citizens may be eligible for support, provided that the individual is not employed by, or affiliated with, an organization, and: (a) the proposer has demonstrated the capability and has

access to any necessary facilities to carry out the project; and, (b) the proposer agrees to fiscal arrangements that, in the opinion of the NSF Division of Grants & Agreements, ensure responsible management of Federal funds. Unaffiliated individuals should contact the appropriate program before preparing a proposal for submission.
Restrictions: NSF rarely provides support to foreign organizations. NSF will consider proposals for cooperative projects involving U.S. and foreign organizations, provided support is requested only for the U.S. portion of the collaborative effort. NSF does not normally support research or education activities by scientists, engineers or educators employed by Federal agencies or Federally Funded Research and Development Centers (FFRDCs).
Amount of Grant: $200,000 - $400,000
Date(s) Application is Due: Sep 25
Internet: http://www.nsf.gov/funding/pgm_summ.jsp?pims_id=11702&govDel=U.S.NSF_39
Sponsor: National Science Foundation
4201 Wilson Boulevard
Arlington, VA 22230

NSF Coupling, Energetics, and Dynamics of Atmospheric Regions (CEDAR) Research Grants **4542**
CEDAR is a broad-based, community-initiated, upper atmospheric research program. The goal is to understand the behavior of atmospheric regions from the middle atmosphere upward through the thermosphere and ionosphere into the exosphere in terms of coupling, energetics, chemistry, and dynamics on regional and global scales. These processes are related to the sources of perturbations that propagate upward from the lower atmosphere as well as to solar radiation and particle inputs from above. The activities within this program combine observations, theory and modeling. Estimated program budget, number of awards and average award size/duration are subject to the availability of funds. Normally, CEDAR awards are made for a duration of three years, but proposers may request from one to five years of funding provided the requested duration is adequately justified. NSF estimates making 10 to 15 continuing grant awards. The typical award size will be about $85,000 per year. The maximum award size will be about $150,000 per year. NSF 06-561 (replaces Document NSF 05-554)
Requirements: Individuals and organizations in the following categories may submit proposals: (1) Universities and Colleges - Universities and two- and four-year colleges (including community colleges) located and accredited in the U.S., acting on behalf of their faculty members. Such organizations also are referred to as academic institutions; (2) Non-profit, non-academic organizations - Independent museums, observatories, research laboratories, professional societies and similar organizations in the U.S. that are directly associated with educational or research activities; (3) For-profit organizations - U.S. commercial organizations, especially small businesses with strong capabilities in scientific or engineering research or education. An unsolicited proposal from a commercial organization may be funded when the project is of special concern from a national point of view, special resources are available for the work, or the proposed project is especially meritorious. NSF is interested in supporting projects that couple industrial research resources and perspectives with those of universities; therefore, it especially welcomes proposals for cooperative projects involving both universities and the private commercial sector; (4) State and Local Governments - State educational offices or organizations and local school districts may submit proposals intended to broaden the impact, accelerate the pace, and increase the effectiveness of improvements in science, mathematics and engineering education in both K-12 and post-secondary levels; (5) Unaffiliated Individuals - Scientists, engineers or educators in the U.S. and U.S. citizens may be eligible for support, provided that the individual is not employed by, or affiliated with, an organization, and: (a) the proposer has demonstrated the capability and has access to any necessary facilities to carry out the project; and, (b) the proposer agrees to fiscal arrangements that, in the opinion of the NSF Division of Grants & Agreements, ensure responsible management of Federal funds. Unaffiliated individuals should contact the appropriate program before preparing a proposal for submission.
Restrictions: NSF rarely provides support to foreign organizations. NSF will consider proposals for cooperative projects involving U.S. and foreign organizations, provided support is requested only for the U.S. portion of the collaborative effort. Also, NSF does not normally support research or education activities by scientists, engineers or educators employed by Federal agencies or Federally Funded Research and Development Centers (FFRDCs). Under unusual circumstances, other Federal agencies and FFRDCs may submit proposals directly to NSF.
Amount of Grant: $85,000 typically
Date(s) Application is Due: May 10
Contact: Robert Kerr, Program Director, (703) 292-8529, email: rkerr@nsf.gov; Robert M. Robinson, Program Manager, (703) 292-8529, email: rmrobins@nsf.gov; Shaun Young, Program Assistant, (703) 292-8521, email: slyoung@nsf.gov
Internet: http://www.nsf.gov/publications/pub_summ.jsp?ods_key=nsf06561
Sponsor: National Science Foundation
4201 Wilson Boulevard
Arlington, VA 22230

NSF Course, Curriculum, and Laboratory Improvement Research Grants **4543**
The goal of this program is improvement in the quality of science, technology, engineering, and mathematics (STEM) education for all undergraduate students. The program supports efforts to create new learning materials and teaching strategies, develop faculty expertise, implement educational innovations, assess learning and evaluate innovations, and conduct research on STEM teaching and learning. The program supports three types of projects representing three different phases of development, ranging from small, exploratory investigations to large comprehensive projects. NSF 06-536
Date(s) Application is Due: Jan 10
Contact: Division of Undergraduate Education, (703) 292-8666; email: undergrad@nsf.gov
Internet: http://www.nsf.gov/funding/pgm_summ.jsp?pims_id=5741
Sponsor: National Science Foundation
4201 Wilson Boulevard
Arlington, VA 22230

NSF CreativeIT Grants **4544**
The goal of the CreativeIT Program is to fund research that focuses on creativity to produce simultaneous advances in both computer science and creative cognition, creativity support tools, engineering design or science. The CreativeIT Program solicits proposals for projects that explore the synergies of cross disciplinary research in creativity and computer science and information technology. Information technology is playing an increasing role in extending the capability of human creative thinking and problem solving. The study of creativity as a way to advance computer science and information technology can lead to new models of creative computational processes, innovative approaches to education that encourage creativity, innovative modes of research that include creative professionals, and new technology to support human creativity. A better understanding of creativity and it's role in computer science research, encouraging creativity in education, and supporting creativity with new information technology will improve American competitiveness and innovation.
Requirements: All collaborative proposals submitted as separate submissions from multiple organizations must be submitted via the NSF FastLane system. Individuals and organizations in the following categories may submit proposals: (1) Universities and Colleges - Universities and two- and four-year colleges (including community colleges) located and accredited in the U.S., acting on behalf of their faculty members. (2) Non-profit, non-academic organizations - Independent museums, observatories, research laboratories, professional societies and similar organizations in the U.S. that are directly associated with educational or research activities (3.) For-profit organizations - U.S. commercial organizations, especially small businesses with strong capabilities in scientific or engineering research or education. An unsolicited proposal from a commercial organization may be funded when the project is of special concern from a national point of view, special resources are available for the work, or the proposed project is especially meritorious. NSF is interested in supporting projects that couple industrial research resources and perspectives with those of universities; therefore, it especially welcomes proposals for cooperative projects involving both universities and the private commercial sector. (4) State and Local Governments - State educational offices or organizations and local school districts may submit proposals intended to broaden the impact, accelerate the pace, and increase the effectiveness of improvements in science, mathematics and engineering education in both K-12 and post-secondary levels. (5) Unaffiliated Individuals - Scientists, engineers or educators in the U.S. and U.S. citizens may be eligible for support, provided that the individual is not employed by, or affiliated with, an organization, and: (a) the proposer has demonstrated the capability and has access to any necessary facilities to carry out the project; and, (b) the proposer agrees to fiscal arrangements that, in the opinion of the NSF Division of Grants & Agreements, ensure responsible management of Federal funds. Unaffiliated individuals should contact the appropriate program before preparing a proposal for submission.
Restrictions: NSF rarely provides support to foreign organizations. NSF will consider proposals for cooperative projects involving U.S. and foreign organizations, provided support is requested only for the U.S. portion of the collaborative effort. NSF does not normally support research or education activities by scientists, engineers or educators employed by Federal agencies or Federally Funded Research and Development Centers (FFRDCs).
Amount of Grant: $200,000-$800,000
Date(s) Application is Due: Sep 26
Contact: Mary Lou Maher, Lead Program Director for CreativeIT; (703) 292-7242; fax: (703) 292-9073; email: mmaher@nsf.gov
Internet: http://www.nsf.gov/funding/pgm_summ.jsp?pims_id=501096&govDel=U.S.NSF_39
Sponsor: National Science Foundation
4201 Wilson Boulevard
Arlington, VA 22230

NSF Cultural Anthropology Research Grants **4545**
The program promotes basic scientific research on the causes and consequences of human social and cultural variation. The program solicits research proposals of theoretical importance in all substantive and theoretical subfields within the discipline of cultural anthropology. Types of support include dissertation improvement grants, research experiences for graduates, supplements; and research experiences for undergraduates, supplements. Proposals submitted through Fastlane should use PD 98-1390 Cultural Anthropology when selecting a program announcement/solicitation number. Submission and program guidelines are available online.
Date(s) Application is Due: Jan 15; Aug 15
Contact: Deborah Winslow, Program Director; (703) 292-7315; fax: (703) 292-9068; email: dwinslow@nsf.gov
Internet: http://www.nsf.gov/funding/pgm_summ.jsp?pims_id=5388
Sponsor: National Science Foundation
4201 Wilson Boulevard
Arlington, VA 22230

NSF Cyber-Enabled Discovery and Innovation (CDI) 4546

Cyber-Enabled Discovery and Innovation (CDI) is NSF, bold five-year initiative to create revolutionary science and engineering research outcomes made possible by innovations and advances in computational thinking. Computational thinking is defined comprehensively to encompass computational concepts, methods, models, algorithms, and tools. Applied in challenging science and engineering research and education contexts, computational thinking promises a profound impact on the Nation, ability to generate and apply new knowledge. Collectively, CDI research outcomes are expected to produce paradigm shifts in our understanding of a wide range of science and engineering phenomena and socio-technical innovations that create new wealth and enhance the national quality of life. NSF 08-604

Requirements: Proposals may only be submitted by the following: (1) Non-profit, non-academic organizations: Independent museums, observatories, research labs, professional societies and similar organizations in the U.S. associated with educational or research activities; or (2) Universities and Colleges: Universities and two- and four-year colleges (including community colleges) located and accredited in the U.S., acting on behalf of their faculty members. Such organizations also are referred to as academic institutions. Preliminary proposals are required and must be submitted via the NSF FastLane system, even if full proposals will be submitted via Grants.gov.

Restrictions: Limit of 2 proposals per Principal Investigator (PI). Projects that make use of existing computational concepts, methods, models, algorithms, and tools to significantly advance only one field of science or engineering should be submitted to an appropriate NSF program in that field instead of to CDI.

Date(s) Application is Due: Dec 8

Contact: Mary Lou Maher, Program Director; (703) 292-8930; fax: (703) 292-9073; email: mmaher@nsf.gov

Internet: http://www.nsf.gov/funding/pgm_summ.jsp?pims_id=503163&org=DGE&sel_org=DGE&from=fund

Sponsor: National Science Foundation
4201 Wilson Boulevard
Arlington, VA 22230

NSF Decision, Risk, and Management Science Research Grants 4547

This program supports scientific research directed at increasing the understanding and effectiveness of decision making by individuals, groups, organizations, and society. Disciplinary and interdisciplinary research, doctoral dissertation research, and workshops are funded in the areas of judgment and decision making; decision analysis and decision aids; risk analysis; perception, and communication; societal and public policy decision making; management science and organizational design. The program also supports small grants for exploratory research of a time-critical or high-risk, potentially transformative nature. Funded research must be relevant to an operational or applied context, grounded in theory, based on empirical observation, or subject to empirical validation, and generalizable. Proposals are accepted at any time from investigators in all fields of science. The target dates indicate the proximity of panel review meetings. Refer to NSF brochure PD 98-1321 for program announcement. Full Proposal Target dates are January 18 and August 18.

Amount of Grant: $1000-$1 million; $94,826 average

Date(s) Application is Due: Jan 11; Aug 11

Contact: Robert O'Connor, Program Director, (703) 292-7263; fax: (703) 292-9068; email: roconnor@nsf.gov

Internet: http://www.nsf.gov/funding/pgm_summ.jsp?pims_id=5423

Sponsor: National Science Foundation
4201 Wilson Boulevard
Arlington, VA 22230

NSF Deep Underground Science and Engineering Laboratory Grants 4548

The particle physics, cosmology/astrophysics and nuclear physics communities have identified the need for a deep underground laboratory infrastructure in order to address some of the most compelling, transformational questions at the frontier of their disciplines. While such an underground infrastructure is envisioned to specifically address forefront physics research, it may also provide cost effective opportunities for other communities to address important new research areas. The funds awarded for the proposals selected in S4 will allow the proposing team to complete the design phase(s) through Preliminary Design or beyond, as defined in the LFM, within the three year solicitation funding period.

Date(s) Application is Due: Jan 9

Contact: Richard J. Fragaszy, Program Director; (703) 292-7011; email: rfragasz@nsf.gov

Internet: http://www.nsf.gov/funding/pgm_summ.jsp?pims_id=503136

Sponsor: National Science Foundation
4201 Wilson Boulevard
Arlington, VA 22230

NSF Developing Global Scientists and Engineers (International Research Experiences for Students (IRES) and Doctoral Dissertation Enhancement Projects (DDEP) 4549

The United States needs to educate a globally-engaged science and engineering workforce capable of performing in an international research environment in order to remain at the forefront of world science and technology. To support this aim, the Developing Global Scientists and Engineers program provides highest quality international research experiences for U.S. students. Whereas the International Research Experiences for Students (IRES) component of the program supports groups of U.S. undergraduate or graduate students conducting research abroad in collaboration with foreign investigators, the Doctoral Dissertation Enhancement Projects (DDEP) component supports the dissertation research abroad of one doctoral student in collaboration with a foreign investigator. In addition to the activities described in this solicitation, the Office of International Science and Engineering (OISE) supports other targeted international research and education experiences for early-career scientists and engineers via the Research Experience for Undergraduates program, the East Asia and Pacific Summer Institutes for U.S. Graduate Students, the Pan-American Advanced Studies Institutes (for advanced graduate students and post-doctoral fellows), and the International Research Fellowship Program (for post-doctoral fellows or new faculty). Anticipated annual funding is estimated as: International Research Experiences for Students (IRES): $2,000,000 to fund up to 14 awards; Doctoral Dissertation Enhancement Projects (DDEP): $100,000 to fund up to 10 awards. Estimated program budget, number of awards and average award size/duration are subject to the availability of funds.

Requirements: This program provides educational opportunities for Undergraduate Students and Graduate Students, however proposals must be submitted by a U.S. institution, organization, or professional society. Doctoral Dissertation Enhancement Project (DDEP) proposals must be submitted by the faculty advisor of the graduate student whose dissertation project is the subject of the proposal. On International Research Experiences for Students (IRES) proposals, indirect costs are allowable, consistent with NSF, general policy. Student stipends, travel, and subsistence costs for this training experience are typically listed as participant support costs and as such are not subject to indirect costs. Off-campus indirect rates are applicable to activities at foreign sites.

Restrictions: No indirect costs are allowable on Doctoral Dissertation Enhancement Projects (DDEP) proposals.

Samples: Elizabeth Catlos, Oklahoma State University, $150,000 - IRES: Research Opportunities in Extensional Dynamics for U.S. Undergraduate and Graduate Geosciences Students in Western Turkey; Amy McCune, Cornell University, $18,962 - U.S.-Tanzania Dissertation Enhancement: Diversification in Lake Tanganyika Endemics: Gastropods and Cichlid Fishes Compared; Robert Jackson, Duke University, $15,000 - U.S.-Argentina Dissertation Enhancement: Afforestation Induced Shifts in Soil Microbial Community Composition and Function in Southern South America

Amount of Grant: $87,500 average

Date(s) Application is Due: Feb 15; Sep 15

Contact: Contact the Program Officer listed for the country/economy where your primary international activity will occur - go to http://www.nsf.gov/od/oise/country-list.jsp

Internet: http://www.nsf.gov/funding/pgm_summ.jsp?pims_id=12831

Sponsor: National Science Foundation
4201 Wilson Boulevard
Arlington, VA 22230

NSF Director's Award for Distinguished Teaching Scholars (DTS) 4550

The award recognizes and rewards individuals who have contributed significantly to the scholarship of their discipline and to the education of students in science, technology, engineering, and mathematics, and exemplify the ability to engage productively in both research and education. The program is part of NSF's efforts to foster an academic culture that values a scholarly approach to both research and education. The award is the highest honor bestowed by the NSF for excellence in both teaching and research in science, technology, education, and mathematics fields, or in educational research related to these disciplines. The awards will be conferred at a special ceremony recognizing the awardees. Preliminary proposals (required) are due in October; full proposals are due in February. Contact a program officer for updated deadline dates. Guidelines are available online. NSF 04-594

Requirements: NSF is most interested in supporting projects that promote scholarly work in STEM education; enhance its value in the higher education community; have an impact beyond the Awardees' institutions; and that maintain the prestige of the award. Scientists, engineers, and educators usually initiate proposals that are officially submitted by their employing organization.

Restrictions: Although there is no limit on the number of candidates nominated from an institution, no more than one award will be made to a single institution in a fiscal year's competition.

Amount of Grant: $300,000 maximum, over 4 years

Date(s) Application is Due: Feb 25; Nov 19

Contact: Myles G. Boylan, Program Director, Room 875 N, (703) 292-4617, fax: (703) 292-9048, email: mboylan@nsf.gov; Daniel P. Maki, Program Director, Room 835 N, (703) 292-4620, email: dmaki@nsf.gov

Internet: http://www.nsf.gov/funding/pgm_summ.jsp?pims_id=8170

Sponsor: National Science Foundation
4201 Wilson Boulevard
Arlington, VA 22230

NSF Doctoral Dissertation Improvement Grants in the Directorate for Biological Sciences (DDIG) 4551

This program awards grants in selected areas of the biological sciences. These grants provide partial support of doctoral dissertation research to improve the overall quality of research. Allowed are costs for doctoral candidates to participate in scientific meetings, to conduct research in specialized facilities or field settings, and to expand an existing body of dissertation research. NSF 08-564.

Requirements: U.S. institutions and organizations that are eligible for awards from the National Science Foundation, including colleges, universities, and other nonprofit research organizations such as botanical gardens, marine and freshwater institutes, and natural history museums may submit proposals. See Chapter I, Section E of the NSF Grant Proposal Guide (GPG) for specific definitions of these categories of proposers.

The NSF encourages collaborations with scientists at foreign organizations; however, primary support for any foreign participants' activities must be secured through their own national sources. A student must have advanced to candidacy for a Ph.D. degree before the submission deadline to be eligible to submit a proposal. A statement that the student has advanced to candidacy for a Ph.D., signed and dated by the department chairperson, graduate dean, or similar administrative official is required. The proposal must be submitted through regular organizational channels by the dissertation advisor(s) on behalf of a graduate student who is at the point of initiating or is already conducting dissertation research. The student must be enrolled at a U.S. institution, but need not be a U.S. citizen. Organizations should limit applications to outstanding dissertation proposals with unusual financial requirements that cannot be met otherwise. Preference may be given to projects that are underway and for which feasibility is demonstrated.
Restrictions: Indirect costs not allowed.
Amount of Grant: up to $15,000
Date(s) Application is Due: Nov 20
Contact: DEB Program Officer, (703) 292-8480, ddig-deb@nsf.gov; IOS Program Officer, (703) 292-8423, ddig-ios@nsf.gov
Internet: http://www.nsf.gov/funding/pgm_summ.jsp?pims_id=5234&org=BIO&sel_org=BIO&from=fund
Sponsor: National Science Foundation
4201 Wilson Boulevard
Arlington, VA 22230

NSF East Asia and Pacific Summer Institutes for U.S. Graduate Students 4552
The East Asia and Pacific Summer Institutes provide U.S. graduate students in science and engineering: 1) first-hand research experience in Australia, China, Japan, Korea, New Zealand, Singapore or Taiwan; 2) an introduction to the science and science policy infrastructure of the respective location; and 3) orientation to the society, culture and language. The primary goals of EAPSI are to introduce students to East Asia and Pacific science and engineering in the context of a research setting, and to help students initiate scientific relationships that will better enable future collaboration with foreign counterparts. The institutes last approximately eight weeks from June to August. Awardees will receive a $5,000 stipend, an allowance of up to $2,500 for international travel, and will be supported to attend a pre-departure orientation in the Washington, D.C. area. Foreign co-sponsoring organizations will provide additional support to cover EAPSI students' living expenses abroad during the period of the summer institutes, and will provide an orientation to the science environment and culture(s) of each location.
Requirements: Applicants: (1) Must be a citizen or permanent resident of the United States by the deadline; (2) Must be enrolled in a research-oriented master's or Ph.D. degree program at a U.S. institution located in the United States; (3) Pursuing studies in fields of science and engineering research and education supported by the National Science Foundation; and, (4) Pursuing studies in fields that are supported by the foreign co-sponsoring organization. Applications are submitted directly by the individual graduate student, unlike standard NSF proposals that are submitted through the principal investigator's U.S. authorized organizational representative. In the EAPSI FastLane application process (Section V.D.), the applicant acts as the Authorized Organizational Representative (AOR).
Restrictions: One application per student per year. Previous EAPSI awardees may not receive funding to the same location (Australia, China, Japan, Korea, New Zealand, Singapore or Taiwan).
Samples: (2008) Jina Lee - $5,637.; Todd Otanicar - $5,637.; (2007) Brendan O'Donnell - $4,579.; Abigail Watrous - $4,579.; Stephen Obrochta - $4,579.; Lisa Adams - $956.; Jon Danielson - $4,579.; Jason Funk - $4,579
Amount of Grant: $5,000 stipend plus travel expenses
Date(s) Application is Due: Dec 12
Contact: EASI Program Manager, Office of International Science and Engineering; (703) 292-8704; fax: (703) 292-9175; email: eapinfo@nsf.gov
Internet: http://www.nsf.gov/publications/pub_summ.jsp?ods_key=nsf07584
Sponsor: National Science Foundation
4201 Wilson Boulevard
Arlington, VA 22230

NSF East Asia and Pacific Summer Institutes for U.S. Host Institutions 4553
The East Asia and Pacific Summer Institutes provide U.S. host institutions in science and engineering: 1) first-hand research experience in Australia, China, Japan, Korea, New Zealand or Taiwan; 2) an introduction to the science and science policy infrastructure of the respective location; and 3) orientation to the society, culture and language. The primary goals are to introduce students to East Asia and Pacific science and engineering in the context of a research laboratory, and to initiate personal relationships that will better enable them to collaborate with foreign counterparts in the future. The institutes last approximately eight weeks from June to August. The National Institutes of Health (NIH) co-sponsor the Summer Institute in Japan.
Requirements: As of the deadline date of the application year, applicants must be: U.S. citizens or permanent residents; Enrolled at U.S. institutions in a research-oriented master's, M.D. or Ph.D. degree program; Pursuing studies in fields of science and engineering research and education supported by the National Science Foundation (Biology; Computer and Information Science and Engineering; Cyberinfrastructure; Education; Engineering; Environmental Research and Education; Geosciences; Mathematical and Physical Sciences; Polar Research; and Social, Behavioral, and Economic Sciences). See http://www.nsf.gov/funding/browse_all_funding.jsp for descriptions of these fields. For Japan, fields of study may also include those supported by the National Institutes of Health (http://www.nih.gov/science/); and, Pursuing studies in fields of science and engineering research and education that are represented among the potential host institutions at the desired location. Applications must be submitted directly by the individual graduate student, unlike standard NSF proposals that are submitted through the principal investigator's U.S. authorized organizational representative.
Restrictions: If a desired host is not listed, that host will be considered on a case-by-case basis. Students who have identified a potential host at an institution not listed are encouraged to coordinate with EAPSI staff (e-mail: eapsi@asee.org) to determine the possibility of receiving support.
Amount of Grant: $4,000 stipend plus travel expenses
Date(s) Application is Due: Dec 11
Contact: EAPSI Program Manager, Office of International Science and Engineering; phone: 703-292-8704; fax: 703-292-9175; email: eapinfo@nsf.gov
Internet: http://nsfsi.org/how_to_locate_potential_hosts_
Sponsor: National Science Foundation
4201 Wilson Boulevard
Arlington, VA 22230

NSF Ecosystem Science Grants 4554
This program supports mechanistic or empirical investigations of whole-system ecological processes and relationships in the following areas: biogeochemistry, such as studies of decomposition, global and regional elemental budgets, and biotic versus abiotic controls of nutrient cycles; primary productivity, particularly ecophysiology within an ecosystem framework; and landscape dynamics with an emphasis on quantitative models of disturbances, ecosystem resilience, and successional patterns. Research on natural, managed and disturbed ecosystems is supported, including terrestrial, freshwater, wetland, coastal (including salt marsh and mangrove), and human-dominated environments.
Requirements: Proposals will be considered that focus on advancing ecosystem science through either the pursuit of new theoretical paradigms or novel modeling efforts. Proposals that, in whole or part, strive to develop new techniques can be supported when a compelling argument exists that there is the potential for a major advance in ecosystem research. Inter- and multi-disciplinary proposals that fall across traditional programmatic boundaries are welcomed and encouraged. Proposals may focus on the cycling of non-nutrient elements, but proposals that are ecotoxicological in orientation will not be considered. Apply to PD 04-7381.
Samples: (2008) Collaborative Research: Testing Paleo to Present Tree Carbon Capture Across Ontogeny, Atmospheric CO2 and Productivity Gradients, Richard Guyette - University of Missouri-Columbia and Frederick Meinzer - Oregon State University.; Dissertation Research: The Spatial Dynamics of Lead Levels in Urban Soil and Correlations with Land Cover, Stewart T. Pickett - Institute of Ecosystem Studies.; Dissertation Research: The Role of Anthropogenic Acidification in Stream Nitrogen Cycling: Understanding Stress Effects on Ecosystem Function, Maurice H. Valett - Virginia Polytechnic Institute and State University.; Dissertation Research: Secondary Production in Urban Stream Food Webs: From Pattern to Process, Margaret Carreiro - University of Louisville Research Foundation.; Dissertation Research: Plant species and functional trait effects on methane and nitrous oxide fluxes from a North Carolina restored wetland, Robert Jackson - Duke University
Date(s) Application is Due: Jan 9; Jul 9
Contact: Todd A. Crowl, Program Director; (703) 292-7870; fax: (703) 292-9064; email: tcrowl@nsf.gov; Ann Russell; (703) 292-7909; email: arussell@nsf.gov
Internet: http://www.nsf.gov/funding/pgm_summ.jsp?pims_id=12822&org=DEB&sel_org=DEB&from=fund
Sponsor: National Science Foundation
4201 Wilson Boulevard
Arlington, VA 22230

NSF Education and Human Resources Project Grants 4555
The ultimate goal of the program is to provide leadership and ensure the vitality of the nation's science, mathematics, engineering, and technology education from K-16, and through graduate levels. Grants support research and educational studies or activities, such as salaries and expenses, permanent equipment, expendable equipment and supplies, travel, publication costs, and other direct and indirect costs. Fellowships provide for stipends and allowances to be paid to the awardee through her/his institution.
Requirements: Public and private colleges (two-year and four-year) and universities, state and local educational agencies, nonprofit and private organizations, professional societies, science academies and centers, science museums and zoological parks, research laboratories, and other institutions with an educational mission may apply.
Amount of Grant: $2500-$7 mil; $128,125 average
Contact: Program Contact, (703) 292-8600; email: jramaley@nsf.gov
Internet: http://www.ehr.nsf.gov
Sponsor: National Science Foundation
4201 Wilson Boulevard, Room 805
Arlington, VA 22230

NSF Education and Interdisciplinary Research (EIR) Grants 4556
The program supports activities in conjunction with NSF-wide programs such as Faculty Early Career Development (CAREER), Research Experiences for Undergraduates (REU), and programs aimed at women, minorities, and persons with disabilities. Further information about all of these programs and activities is available in the Crosscutting Investment Strategies section of the NSF Guide to Programs. The program also supports activities that seek to improve the education and training of physics students (both undergraduate and graduate), such as curriculum development for upper-level physics courses, and activities that are not included in specific programs elsewhere within NSF.

The program supports research at the interface between physics and other disciplines and extending to emerging areas. Broadening activities related to research at the interface with other fields, possibly not normally associated with physics, also may be considered.
Date(s) Application is Due: Sep 30
Contact: Marvin Goldberg, Program Director; (703) 292-7374; email: mgoldber@nsf.gov
Internet: http://www.nsf.gov/funding/pgm_summ.jsp?pims_id=5610
Sponsor: National Science Foundation
4201 Wilson Boulevard
Arlington, VA 22230

NSF Elementary Particle Physics (EPP) Grants 4557
The program supports research on the properties and interactions of elementary particles, the most fundamental building blocks of matter, at the frontiers of energy and sensitivity. Research includes the exploration of quarks and leptons and the interactions among these elementary constituents. The program supports university groups working at major accelerator laboratories, including those operated by the Department of Energy, and university groups involved in the construction of detectors for the Large Hadron Collider at the European Organization for Nuclear Research (CERN). The program also supports the Cornell Electron Storage Ring (CESR), which produces electron and positron colliding beams that allow detailed studies by university groups of b-meson physics and upsilon physics, and facilitates an aggressive program of synchrotron radiation research at the Cornell High-Energy Synchrotron Source, which is operated by the Division of Materials Research. CESR is among the highest luminosity electron-positron colliders in the world in this energy range. CESR also maintains a vigorous program of accelerator research and development. The annual deadline is the last Wednesday in September.
Requirements: It is recommended that proposals come in near the general Physics Division target date. Individuals and organizations in the following categories may submit proposals: (1) Universities and Colleges - Universities and two- and four-year colleges (including community colleges) located and accredited in the U.S., acting on behalf of their faculty members. Such organizations also are referred to as academic institutions. Institutions located outside the U.S. fall under (6) below. (2) Non-profit, non-academic organizations - Independent museums, observatories, research laboratories, professional societies and similar organizations in the U.S. that are directly associated with educational or research activities. (3) For-profit organizations - U.S. commercial organizations, especially small businesses with strong capabilities in scientific or engineering research or education. An unsolicited proposal from a commercial organization may be funded when the project is of special concern from a national point of view, special resources are available for the work, or the proposed project is especially meritorious. NSF is interested in supporting projects that couple industrial research resources and perspectives with those of universities; therefore, it especially welcomes proposals for cooperative projects involving both universities and the private commercial sector. (4) State and Local Governments - State educational offices or organizations and local school districts may submit proposals intended to broaden the impact, accelerate the pace, and increase the effectiveness of improvements in science, mathematics and engineering education in both K-12 and post-secondary levels. (5) Unaffiliated Individuals - Scientists, engineers or educators in the U.S. and U.S. citizens may be eligible for support, provided that the individual is not employed by, or affiliated with, an organization and (a) the proposer has demonstrated the capability and has access to any necessary facilities to carry out the project; and (b) the proposer agrees to fiscal arrangements that, in the opinion of the NSF Division of Grants & Agreements, ensure responsible management of Federal funds. Unaffiliated individuals should contact the appropriate program before preparing a proposal for submission. (6) Foreign organizations - NSF rarely provides support to foreign organizations. NSF will consider proposals for cooperative projects involving U.S. and foreign organizations, provided support is requested only for the U.S. portion of the collaborative effort.
Restrictions: Other Federal agencies - NSF does not normally support research or education activities by scientists, engineers or educators employed by Federal agencies or Federally Funded Research and Development Centers (FFRDCs). Under unusual circumstances, other Federal agencies and FFRDCs may submit proposals directly to NSF. A proposed project is only eligible for support if it meets one or more of the following exceptions, as determined by a cognizant NSF Program Officer: (a) Special Projects. Under exceptional circumstances, research or education projects at other Federal agencies or FFRDCs that can make unique contributions to the needs of researchers elsewhere or to other specific NSF objectives may receive NSF support. (b) National and International Programs. The Foundation may fund research and logistic support activities of other Government agencies or FFRDCs directed at meeting the goals of special national and international research programs for which the Foundation bears special responsibility, such as the U.S. Antarctic Research Program. (c) International Travel Awards. In order to ensure appropriate representation or availability of a particular expertise at an international conference, staff researchers of other Federal agencies may receive NSF international travel awards. Proposers who think their project may meet one of the exceptions listed above should contact a cognizant NSF Program Officer before preparing a proposal for submission. In addition, a scientist, engineer or educator who has a joint appointment with a university and a Federal agency (such as a Veterans Administration Hospital, or with a university and a FFRDC) may submit proposals through the university and may receive support if he/she is a bona fide faculty member of the university, although part of his/her salary may be provided by the Federal agency. Preliminary inquiry should be made to the appropriate program before preparing a proposal for submission.
Samples: Flavor Physics and CP Violation 2009 ($8,000) - Sheldon Stone, Syracuse University, NY.; Ringside Seats for U.S. CMS Graduate Students During the Startup of the LHC ($125,212.00) - Daniel Marlow, Princeton University, NJ.; LHC Experimental Experience at CER ($120,000.00) - Michael P. Tuts, Columbia University, NY.; Support for the 2008 Hadron Collider Summer School; August 2008, Batavia, IL ($7,000) - Emanuela Barberis, Northeastern University, MA.
Date(s) Application is Due: Sep 30
Contact: Marvin Goldberg, Program Director; (703) 292-7374; email: mgoldber@nsf.gov
Internet: http://www.nsf.gov/funding/pgm_summ.jsp?pims_id=5624&org=PHY&from=home
Sponsor: National Science Foundation
4201 Wilson Boulevard
Arlington, VA 22230

NSF Elementary, Secondary, and Informal Education Grants 4558
Programs of the Division of Elementary, Secondary, and Informal Education (ESIE) support projects designed to enable all students, pre-K-12, to succeed in science, mathematics, and technology (SMT) and to increase the scientific and technological literacy of students and adults of all ages. Its programs provide the educational foundation for future generations of scientists, engineers, and technologists; for those pursuing postsecondary education in other disciplines; and for those who enter the workforce directly from secondary school. ESIE programs promote the achievement, competency, and literacy considered essential to inform citizens about the SMT issues that impact their lives, as well as promote effective advocacy for high-quality education in these fields.
Requirements: Any institution or organization with a scientific or educational mission may submit a proposal.
Samples: Franklin Institute of Science Museum (Philadelphia, PA)—to improve family science learning in museums, $616,248.; Desert Botanical Garden (Phoenix, AZ)—to create a science learning center focusing on deserts, $634,776.
Contact: Division of Elementary, Secondary, and Informal Education, (703) 292-8620; fax: (703) 292-9044; email: ehr-esi-info@nsf.gov
Internet: http://www.nsf.gov/div/index.jsp?div=ESIE
Sponsor: National Science Foundation
4201 Wilson Boulevard
Arlington, VA 22230

NSF Emerging Frontiers in Research and Innovation (EFRI) Grants 4559
The Directorate for Engineering at the National Science Foundation has established the Office of Emerging Frontiers in Research and Innovation (EFRI) to serve a critical role in focusing on important emerging areas in a timely manner. The EFRI Office is launching a new funding opportunity for interdisciplinary teams of researchers to embark on rapidly advancing frontiers of fundamental engineering research. For this program, they will consider proposals that aim to investigate emerging frontiers in the following two specific research areas: (1) BioSensing & BioActuation: Interface of Living and Engineered Systems (BSBA), and (2) Hydrocarbons from Biomass (HyBi). EFRI seeks proposals with transformative ideas that represent an opportunity for a significant shift in fundamental engineering knowledge with a strong potential for long term impact on national needs or a grand challenge. It is anticipated that 11 or more standard grants will be made (4-year awards).
Requirements: Proposals may be submitted by a single organization or a group of organizations consisting of a lead organization in partnership with one or more partner organizations. Only U.S. academic institutions with significant research and degree-granting education programs in disciplines normally supported by NSF are eligible to be the lead organization. Principal investigators are encouraged to form synergistic collaborations among researchers and with private and public sector organizations, government laboratories, and scientists and engineers at foreign organizations where appropriate, though no NSF funds will be provided to those organizations. Submission of Letters of Intent is required, due by October 14. Submission of Preliminary Proposals is required, due by December 2. Full proposals will be due by April 30. Each project team may receive support of up to a total of $500,000 per year for up to four years, pending the availability of funds. It is not expected that all awards will receive the maximum amount; the size of awards will depend upon the type of research program proposed. See the website for the detailed solicitation, which includes guidelines, requirements, and application instructions.
Restrictions: The principal investigator and co-principal investigators may participate in one proposal submitted to this program.
Amount of Grant: up to $500,000 per year
Date(s) Application is Due: Oct 14; Dec 2; Apr 30
Contact: Sohi Rastegar, Director, Office of Emerging Frontiers in Research and Innovation (EFRI); (703) 292-8305; email: srastega@nsf.gov
Internet: http://www.nsf.gov/funding/pgm_summ.jsp?pims_id=13708
Sponsor: National Science Foundation
4201 Wilson Boulevard
Arlington, VA 22230

NSF Engineering Research Centers (ERC) Grants 4560
The goal of the Generation Three (Gen-3) Engineering Research Centers (ERC) Program is to create a culture of innovation in engineering research and education that links scientific discovery to technological innovation through transformational engineered systems research in order to advance technology and produce engineering graduates who will be creative innovators in a global economy. These ERCs will be at the forefront as the U.S. competes in the 21st century global economy where R&D resources and engineering talent are internationally and domestically distributed. Recognizing that optimizing efficiency and product quality is no longer sufficient for U.S. industry to remain competitive, these ERCs will optimize academic engineering research and

education to stimulate increased innovation. They will develop this culture of discovery and innovation through a symbiotic relationship between academic researchers, small innovative firms, and larger industrial and practitioner partners. These ERCs will build bridges from science-based discovery to technological innovation by focusing on research needed to realize transforming engineered systems. They will have the opportunity to partner with foreign universities and provide unique opportunities for research and learning collaboration that will prepare U.S. engineering graduates for leadership in innovation in a global economy. Their faculty will be diverse and talented individuals who will prepare diverse and talented domestic and international graduates who can function in a global world where design and production efforts cross national borders. Their transforming engineering education programs will strategically impart the capacity to create and exploit knowledge for technological innovation. Awards will be made as cooperative agreements between NSF and the lead university, with subawards to the partner institutions as appropriate for their role in the ERC. ERCs are expected to compensate pre-college teachers for time spent carrying out research in the ERCs laboratories and the materials needed to develop the courseware. 07-521
Requirements: Only U.S. academic institutions with undergraduate, masters, and doctoral engineering programs of sufficient depth and breadth appropriate to support the vision may submit proposals as the lead institution. Proposals must be submitted by the lead institution with subaward budgets for partner domestic academic institutions. As a multi-university effort, the lead and partner institutions share the responsibility for the ERC. The other university partner institutions must have graduate or undergraduate engineering programs or both. At least one of these academic partners will be a university or college that serves students predominantly underrepresented in engineering. These partners also must include a few middle and high schools commited to long-term sustained partnerships to bring engineering concepts into the classroom and engage talented older pre-college students in the ERC's research programs as Young Scholars in order to raise the enrollment and diversity of domestic students in engineering and science degree program. The partners must also include an organization devoted to speeding entrepreneurship or innovation as discussed above. Full proposals must include letters from firms/agencies indicating that they are committed to the ERC's industrial/practitioner partnership program. Proposals may include a foreign university as a long-term partner in research and education. Only universities able to provide headquarters space for the proposed ERC may submit preliminary and invited full proposals as the lead institution. This headquarters should provide space for the administrative staff, the director, the deputy director, and physical and virtual conferencing. It may also provide space for facilities to support the ERC, research needs. The center director must be a tenured faculty member in an engineering department at the lead institution. The director's doctoral degree must be in engineering or an associated field of science; if the latter, she/he must have substantial career experience in engineering and a joint appointment in an engineering department. A one to two-page letter of intent is required to facilitate the NSF review process. The letter should be submitted via FastLane no later than the date specified in this solicitation.
Restrictions: Separately submitted collaborative proposals are not acceptable and will be returned without review.
Date(s) Application is Due: Feb 2; May 3
Contact: Lynn Preston, Deputy Division Director; (703) 292-5358; fax: (703) 292-9051; email: lpreston@nsf.gov
Internet: http://www.nsf.gov/funding/pgm_summ.jsp?pims_id=5502&org=ENG&sel_org=ENG&from=fund
Sponsor: National Science Foundation
4201 Wilson Boulevard
Arlington, VA 22230

NSF Engineering Research Grants **4561**
The Directorate for Engineering makes catalytic investments in knowledge creation and intellectual growth that can in turn spur technological innovation and economic growth in an increasingly swift, complex, and interconnected world. Although the research supported by the directorate is fundamental in nature, much of it focuses on societal needs. Over the long term, the directorate's investments contribute to an innovation system that enables the creation of valuable new products, new services, and new and more productive enterprises that enhance future economic strength, security, and quality of life. The directorate supports the following research activity programs: bioengineering and environmental systems (BES), chemical and transport systems (CTS), civil and mechanical systems (CMS), design, manufacture, and industrial innovation (DMII), electrical and communications systems (ECS), and engineering education and centers (EEC).
Requirements: Public and private colleges and universities; nonprofit institutions; and profit-making institutions including small businesses, federal, state, and local government agencies are eligible to apply.
Amount of Grant: $5000-$4 million; average $119,000
Contact: Joanne Culbertson; (703) 292-4602; fax: (703) 292-9013; email: jculbert@nsf.gov
Internet: http://www.eng.nsf.gov
Sponsor: National Science Foundation
4201 Wilson Boulevard
Arlington, VA 22230

NSF Ethics and Values Studies and Research on Science and Technology Grants **4562**
This program supports research and educational projects that contribute to developing and transmitting knowledge about ethics, values, and the conduct of science and engineering. The program is interested in research that examines the intellectual, ethical, value, and policy contexts that influence and that are influenced by development, use, and effects of science and technology. Projects are expected to contribute to understanding the ethical and social values, policies, obligations, and responsibilities that arise in these interactions. Refer to brochure NSF 01-152 for program announcement.
Requirements: Proposals may be submitted by qualified individual investigators who do not have institution affiliations or through normal institutional channels in accordance with NSF procedures. Projects may involve a single investigator or several investigators with different disciplinary backgrounds.
Restrictions: In general, SDEST does not consider proposals from individual academic institutions to support lectureships or conference activities. It does consider proposals where conferences or workshops are part of a research or education project plan, and proposals for workshops to develop research agendas on topics important to program goals. Research focused primarily on ethical, value or policy issues for clinical research or practice or resource allocation in health care is not normally considered by the program.
Date(s) Application is Due: Feb 1; Aug 1
Contact: Dr. Rachelle Hollander, Division of Social, Behavioral, and Economic Research, (703) 292-7272; fax: (703) 292-9068; email: rholland@nsf.gov
Internet: http://www.nsf.gov/pubs/2001/nsf01152/nsf01152.html
Sponsor: National Science Foundation
4201 Wilson Boulevard
Arlington, VA 22230

NSF Evolution of Developmental Systems Grants **4563**
The Developmental Systems Cluster supports research aimed at understanding how interacting developmental processes give rise to the emergent properties of organisms. A systems level approach to understanding these processes, at the molecular, cellular, and organismal levels of organization, requires the use of molecular, genetic, biochemical, and physiological techniques as well as techniques from outside biology. The Evolution of Developmental Systems programmatic area supports research to discover the developmental processes shared by all organisms and those singular ones that produce diversity (phenotypic variation within a species and/or between species).
Requirements: The most frequent recipients of support are academic institutions and nonprofit research groups. In special circumstances, grants also are awarded to other types of institutions and to individuals.
Date(s) Application is Due: Jan 12; Jul 12
Contact: Thomas Jack, Program Director; (703) 292-8417; fax: (703) 292-9153; email: tjack@nsf.gov
Internet: http://www.nsf.gov/funding/pgm_summ.jsp?pims_id=501087
Sponsor: National Science Foundation
4201 Wilson Boulevard
Arlington, VA 22230

NSF Extragalactic Astronomy and Cosmology Research Grants **4564**
The program supports studies of the more distant Universe. Research topics include galaxy formation, evolution and interaction; active galaxies; quasars; large-scale structure; and all areas of cosmology. Proposals may span multiple disciplines and/or areas of study and may utilize multiple techniques. Principal Investigators are encouraged to contact one of the Program Officers listed in this announcement prior to submitting a proposal to the AAG Program, particularly if the proposal will include investigators at multiple institutions.
Requirements: Scientists initiate research proposals, which are usually submitted on their behalf by their employing institutions.
Date(s) Application is Due: Nov 15
Contact: Nigel A. Sharp, Program Director; (703) 292-4905; fax: (703) 292-9034; email: nsharp@nsf.gov
Internet: http://www.nsf.gov/funding/pgm_summ.jsp?pims_id=13630
Sponsor: National Science Foundation
4201 Wilson Boulevard
Arlington, VA 22230

NSF Faculty Early Career Development (CAREER) Grants **4565**
The Faculty Early Career Development (CAREER) Program is a Foundation-wide activity that offers the National Science Foundation's most prestigious awards in support of junior faculty who exemplify the role of teacher-scholars through outstanding research, excellent education and the integration of education and research within the context of the mission of their organizations. Such activities should build a firm foundation for a lifetime of leadership in integrating education and research. NSF encourages submission of CAREER proposals from junior faculty members at all CAREER-eligible organizations and especially encourages women, members of underrepresented minority groups, and persons with disabilities to apply. This program also includes the NSF component of the Presidential Early Career Awards for Scientists and Engineers (PECASE). Deadlines vary according to the field of study. NSF 08-557
Requirements: Proposals may be submitted by (a) Academic institutions in the U.S., its territories or possessions, and the Commonwealth of Puerto Rico that award degrees in fields supported by NSF; or, (b) Non-profit, non-degree-granting organizations such as museums, observatories or research labs may also be eligible to submit proposals, if the eligibility requirements of the PI's position are satisfied. There is no limit on the number of proposals that an organization may submit. Each Principal Investigator may only submit one CAREER proposal per annual competition.
Restrictions: A Principal Investigator may not participate in more than three CAREER competitions.
Amount of Grant: $150,000 - $200,000
Date(s) Application is Due: Jul 21; Jul 22; Jul 23

Contact: Elizabeth L. Rom, Program Director; (703) 292-7709; fax: (703) 292-9085; email: elrom@nsf.gov; Donald Elthon; (703) 292-8580; fax: (703) 292-9085; email: delthon@nsf.gov
Internet: http://www.nsf.gov/funding/pgm_summ.jsp?pims_id=503214&from=fund
Sponsor: National Science Foundation
4201 Wilson Boulevard
Arlington, VA 22230

NSF Focused Research Groups in the Mathematical Sciences Grants **4566**
The purpose of this program is to allow groups of researchers to respond to recognized scientific needs of pressing importance, to take advantage of current scientific opportunities, or to prepare the ground for anticipated significant scientific developments in the mathematical sciences. It is expected that some of these groups will include, in addition to mathematical scientists, researchers from other science and engineering disciplines appropriate to the proposed research. The activity will support projects for which the collective effort by a group of researchers is necessary to reach the scientific goals. Projects should also be timely, limited in duration to up to three years, and substantial in their scope and impact. Fastlane use is required. A letter of intent is requested by August 21; full application is due by September 18.
Requirements: Proposals may be submitted by U.S. academic institutions and other U.S. nonprofit organizations in support of small groups.
Amount of Grant: $800,000 average; $12 million approximate total
Date(s) Application is Due: Aug 21; Sep 18
Contact: Henry A. Warchall, Program Director; (703) 292-4861; fax: (703) 292-9032; email: hwarchal@nsf.gov
Internet: http://www.nsf.gov/funding/pgm_summ.jsp?pims_id=5671
Sponsor: National Science Foundation
4201 Wilson Boulevard
Arlington, VA 22230

NSF Foundations Grants **4567**
The program in Foundations supports research in mathematical logic and the foundations of mathematics, including proof theory, recursion theory, model theory, set theory, and infinitary combinatorics. For full proposals submitted via FastLane, standard Grant Proposal Guidelines apply. The full proposal target date is October 6.
Date(s) Application is Due: Oct 6
Contact: Tomek Bartoszynski; (703) 292-4885; fax: (703) 292-9032; email: tbartosz@nsf.gov
Internet: http://www.nsf.gov/funding/pgm_summ.jsp?pims_id=5548&org=DMS&from=home
Sponsor: National Science Foundation
4201 Wilson Boulevard
Arlington, VA 22230

NSF Galactic Astronomy Research Grants **4568**
This program supports studies on the composition, structure and evolution of the Milky Way galaxy and nearby galaxies. Research may focus on the stellar populations in these galaxies; the characteristics of star clusters; the interstellar medium; and the properties of atomic and molecular constituents of the interstellar medium.
Requirements: Scientists initiate research proposals, which are usually submitted on their behalf by their employing institutions.
Date(s) Application is Due: Sep 30
Contact: Nigel A. Sharp, Program Director; (703) 292-4905; fax: (703) 292-9034; email: nsharp@nsf.gov
Internet: http://www.nsf.gov/funding/pgm_summ.jsp?pims_id=13630
Sponsor: National Science Foundation
4201 Wilson Boulevard
Arlington, VA 22230

NSF Genes and Genome Systems Cluster Grants **4569**
Grants support studies on genomes and genetic mechanisms in all organisms, whether prokaryote, eukaryote, phage, or virus. Proposals on the structure, maintenance, expression, transfer, and stability of genetic information in DNA, RNA, and proteins and how those processes are regulated are appropriate. Areas of interest include genome organization, molecular and cellular evolution, replication, recombination, repair, and vertical and lateral transmission of heritable information. Of equal interest are the processes that mediate and regulate gene expression, such as chromatin structure, epigenetic phenomena, transcription, RNA processing, editing and degradation, and translation. The use of innovative in vivo and/or in vitro approaches, including biochemical, physiological, genetic, genomic, and/or computational methods, is encouraged, as is research at the interfaces of biology, physics, chemistry, mathematics and computer science, and engineering.
Requirements: Standard NSF Grant Proposal Guidelines apply (see http://www.nsf.gov/publications/pub_summ.jsp?ods_key=gpg).
Date(s) Application is Due: Jan 12; Jul 12
Contact: Patrick P. Dennis, Program Director; (703) 292-7145; fax: (703) 292-9061; email: pdennis@nsf.gov
Internet: http://www.nsf.gov/funding/pgm_summ.jsp?pims_id=12780
Sponsor: National Science Foundation
4201 Wilson Boulevard
Arlington, VA 22230

NSF Geometric Analysis Research Grants **4570**
The program supports research on differential geometry and its relation to partial differential equations and variational principles; aspects of global analysis, including the differential geometry of complex manifolds and geometric Lie group theory; geometric methods in modern mathematical physics; and geometry of convex sets, integral geometry, and related geometric topics.
Requirements: Apply to PD 07-1265 as follows: For full proposals submitted via FastLane: standard Grant Proposal Guidelines apply. For full proposals submitted via Grants.gov: NSF Grants.gov Application Guide; A Guide for the Preparation and Submission of NSF Applications via Grants.gov Guidelines apply (Note: The NSF Grants.gov Application Guide is available on the Grants.gov website and on the NSF website at: http://www.nsf.gov/bfa/dias/policy/docs/grantsgovguide.pdf)
Restrictions: All collaborative proposals submitted as separate submissions from multiple organizations must be submitted via the NSF FastLane system.
Contact: Joanna Kania-Bartoszynska, Program Director; (703) 292-4881; fax: (703) 292-9032; email: jkaniaba@nsf.gov
Internet: http://www.nsf.gov/funding/pgm_summ.jsp?pims_id=5549
Sponsor: National Science Foundation
4201 Wilson Boulevard
Arlington, VA 22230

NSF Geophysics Grants **4571**
The Geophysics Program supports basic research in the physics of the solid earth to explore its composition, structure, and processes. Laboratory, field, theoretical, and computational studies are supported. Topics include seismicity, seismic wave propagation, and the nature and occurrence of earthquakes; the earth's magnetic, gravity, and electrical fields; the earth's thermal structure; and geodynamics. Supported research also includes geophysical studies of active deformation, including geodesy, and studies of the properties and behavior of earth materials in support of geophysical observation and theory.
Requirements: Individuals and organizations in the following categories may submit proposals: (1) Universities and Colleges - Universities and two- and four-year colleges (including community colleges) located and accredited in the U.S., acting on behalf of their faculty members. Such organizations also are referred to as academic institutions. Institutions located outside the U.S. fall under (6) below. (2) Non-profit, non-academic organizations - Independent museums, observatories, research laboratories, professional societies and similar organizations in the U.S. that are directly associated with educational or research activities. (3) For-profit organizations - U.S. commercial organizations, especially small businesses with strong capabilities in scientific or engineering research or education. An unsolicited proposal from a commercial organization may be funded when the project is of special concern from a national point of view, special resources are available for the work, or the proposed project is especially meritorious. NSF is interested in supporting projects that couple industrial research resources and perspectives with those of universities; therefore, it especially welcomes proposals for cooperative projects involving both universities and the private commercial sector. (4) State and Local Governments - State educational offices or organizations and local school districts may submit proposals intended to broaden the impact, accelerate the pace, and increase the effectiveness of improvements in science, mathematics and engineering education in both K-12 and post-secondary levels. (5) Unaffiliated Individuals - Scientists, engineers or educators in the U.S. and U.S. citizens may be eligible for support, provided that the individual is not employed by, or affiliated with, an organization, and (a) the proposer has demonstrated the capability and has access to any necessary facilities to carry out the project; and (b) the proposer agrees to fiscal arrangements that, in the opinion of the NSF Division of Grants & Agreements, ensure responsible management of Federal funds. Unaffiliated individuals should contact the appropriate program before preparing a proposal for submission. (6) Foreign organizations - NSF rarely provides support to foreign organizations. NSF will consider proposals for cooperative projects involving U.S. and foreign organizations, provided support is requested only for the U.S. portion of the collaborative effort.
Restrictions: Other Federal agencies - NSF does not normally support research or education activities by scientists, engineers or educators employed by Federal agencies or Federally Funded Research and Development Centers (FFRDCs). Under unusual circumstances, other Federal agencies and FFRDCs may submit proposals directly to NSF. A proposed project is only eligible for support if it meets one or more of the following exceptions, as determined by a cognizant NSF Program Officer: (a) Special Projects. Under exceptional circumstances, research or education projects at other Federal agencies or FFRDCs that can make unique contributions to the needs of researchers elsewhere or to other specific NSF objectives may receive NSF support. (b) National and International Programs. The Foundation may fund research and logistic support activities of other Government agencies or FFRDCs directed at meeting the goals of special national and international research programs for which the Foundation bears special responsibility, such as the U.S. Antarctic Research Program. (c) International Travel Awards. In order to ensure appropriate representation or availability of a particular expertise at an international conference, staff researchers of other Federal agencies may receive NSF international travel awards. Proposers who think their project may meet one of the exceptions listed above should contact a cognizant NSF Program Officer before preparing a proposal for submission. In addition, a scientist, engineer or educator who has a joint appointment with a university and a Federal agency (such as a Veterans Administration Hospital, or with a university and a FFRDC) may submit proposals through the university and may receive support if he/she is a bona fide faculty member of the university, although part of his/her salary may be provided by the Federal agency. Preliminary inquiry should be made to the appropriate program before preparing a proposal for submission. Proposals that have been declined are not eligible for resubmission for one year from the original

date of submission and must be substantially revised to be considered. A proposal that has not been substantially revised will be returned without review as per the NSF Grant Proposal Guide.
Date(s) Application is Due: Dec 1
Contact: Robin Reichlin, Program Director; (703) 292-8556; fax: (703) 292-9025; email: rreichli@nsf.gov
Internet: http://www.nsf.gov/funding/pgm_summ.jsp?pims_id=13682
Sponsor: National Science Foundation
4201 Wilson Boulevard
Arlington, VA 22230

NSF George E. Brown Jr. Network for Earthquake Engineering Simulation Research (NEESR) **4572**
The Division of Civil, Mechanical and Manufacturing Innovation (CMMI) in the Directorate for Engineering (ENG) of the National Science Foundation (NSF) invites proposals for research that uses the George E. Brown, Jr. Network for Earthquake Engineering Simulation (NEES) to advance knowledge discovery and innovation for: earthquake and tsunami loss reduction of our nation's civil infrastructure; and new experimental simulation techniques and instrumentation for NEES. NEES comprises a network of earthquake engineering experimental equipment sites available for experimentation on-site or in the field and through telepresence. NEES equipment sites include shake tables, geotechnical centrifuges, a tsunami wave basin, unique large-scale testing laboratory facilities, and mobile and permanently installed field equipment. The NEES networking cyberinfrastructure connects, via Internet2, the equipment sites as well as provides telepresence, a curated central data repository, simulation tools, and collaborative tools for facilitating on-line planning, execution, and post-processing of experiments. Projects proposed and supported under this solicitation must use one or more of the NEES equipment sites.
Requirements: U.S. universities and colleges and U.S. nonprofit, nonacademic organizations may submit NEESR proposals as the lead organization.
Amount of Grant: $9 million total
Date(s) Application is Due: Mar 27
Contact: Joy M. Pauschke, Program Director; (703) 292-7024; fax: (703) 292-9053; email: jpauschk@nsf.gov
Internet: http://www.nsf.gov/funding/pgm_summ.jsp?pims_id=6192
Sponsor: National Science Foundation
4201 Wilson Boulevard
Arlington, VA 22230

NSF Geoscience Education **4573**
There are two elements to this NSF program. Awards to Facilitate Geoscience Education (AFGE) is the first element. Proposals may target any educational level: graduate and postdoctoral education and training (outside the framework of normal NSF research grants), undergraduate education, elementary and secondary education, and education outside the classroom. Awards made under this element are intended to facilitate the initiation or piloting of highly innovative educational activities that involve leading geoscience researchers where support may not otherwise be available. The second element is Application of Digital Libraries to Undergraduate Earth Systems Education. This element supports a process leading to the establishment of a national digital library which would support and promote high-quality undergraduate education in the geosciences, in particular innovative Earth systems curricula, associated archive data sets, and tools for handling real-time data. In addition to individual focused proposals, larger-scale collaborative proposals are invited that specifically seek to implement the recommendations of the community workshop Portal to the Future: A Digital Library for Earth System Education. FastLane submittal is required. NSF 00-38
Amount of Grant: $2 million total
Date(s) Application is Due: Apr 10
Contact: Michael Mayhew, Directorate for Geosciences, (703) 306-1557; email: mmayhew@nsf.gov; Dorothy Stout, Division of Undergraduate Education, (703) 306-1557; email: mmayhew@nsf.gov
Internet: http://www.nsf.gov/pubs/2000/nsf0038/nsf0038.htm
Sponsor: National Science Foundation
4201 Wilson Boulevard
Arlington, VA 22230

NSF Geosciences Research Grants **4574**
The objective of the program is to strengthen and enhance the national scientific enterprise through the expansion of fundamental knowledge and increased understanding of the integrated earth system through the support of basic research in the atmospheric, earth, and ocean sciences. Major objectives include new knowledge of the atmosphere from the sun to the earth's surface over the entire spectrum of physical and chemical phenomena; a better understanding of the physical, chemical, and biological character of the earth and the processes that govern its evolution; and increased insight into the world's oceans, their composition, structure, behavior, and tectonics. Support is also provided for science and technology centers, undergraduate student research, facility enhancement, instrumentation, and laboratory equipment; and for research opportunities for women, minority, and scientists and engineers with disabilities.
Requirements: Proposals may be submitted by academic institutions, nonacademic and nonprofit research organizations, profit-making and private research organizations, and individuals.
Amount of Grant: $1000-$60 million; $101,000 average
Date(s) Application is Due: Jun 1; Dec 1
Contact: Margaret Leinen, Geosciences Assistant Director, (703) 292-8500; fax: (703) 292-9022; Program Director, Earth Sciences, (703) 292-8550; fax: (703) 292-9025; Dr. G. Michael Purdy, Ocean Sciences, (703) 292-8580; fax: (703) 292-9042; email: mleinen@nsf.go
Internet: http://www.nsf.gov/home/programs/geo.cfm
Sponsor: National Science Foundation
4201 Wilson Boulevard
Arlington, VA 22230

NSF Geospace Environment Modeling (GEM) **4575**
GEM is a broad-based, community-initiated research program on the physics of the Earth's magnetosphere and the coupling of the magnetosphere to the atmosphere and to the solar wind. The purpose of the GEM program is to support basic research into the dynamical and structural properties of geospace, leading to the construction of a global Geospace General Circulation Model (GGCM) with predictive capability. This GGCM model will be modularized and will complement parallel developments of magnetohydrodynamic models. The strategy for achieving GEM goals is to undertake a series of campaigns, in both theory and observational modes, each focusing on particular aspects of the geospace environment. Estimated program budget, number of awards and average award size/duration are subject to the availability of funds. The typical award size is approximately $85,000 per year with a duration of three years. Proposers may request from one to five years of funding provided the requested duration is adequately justified. The maximum award size is $175,000 per year. The estimated number of awards is 8 to 12 new awards in each year. NSF 04-576 (replaces Document 02-122)
Requirements: Individuals and organizations in the following categories may submit proposals: (1) Universities and Colleges - Universities and two- and four-year colleges (including community colleges) located and accredited in the U.S., acting on behalf of their faculty members. Such organizations also are referred to as academic institutions; (2) Non-profit, non-academic organizations - Independent museums, observatories, research laboratories, professional societies and similar organizations in the U.S. that are directly associated with educational or research activities; (3) For-profit organizations - U.S. commercial organizations, especially small businesses with strong capabilities in scientific or engineering research or education. An unsolicited proposal from a commercial organization may be funded when the project is of special concern from a national point of view, special resources are available for the work, or the proposed project is especially meritorious. NSF is interested in supporting projects that couple industrial research resources and perspectives with those of universities; therefore, it especially welcomes proposals for cooperative projects involving both universities and the private commercial sector; (4) State and Local Governments - State educational offices or organizations and local school districts may submit proposals intended to broaden the impact, accelerate the pace, and increase the effectiveness of improvements in science, mathematics and engineering education in both K-12 and post-secondary levels; (5) Unaffiliated Individuals - Scientists, engineers or educators in the U.S. and U.S. citizens may be eligible for support, provided that the individual is not employed by, or affiliated with, an organization, and: (a) the proposer has demonstrated the capability and has access to any necessary facilities to carry out the project; and, (b) the proposer agrees to fiscal arrangements that, in the opinion of the NSF Division of Grants & Agreements, ensure responsible management of Federal funds. Unaffiliated individuals should contact the appropriate program before preparing a proposal for submission.
Restrictions: NSF rarely provides support to foreign organizations. NSF will consider proposals for cooperative projects involving U.S. and foreign organizations, provided support is requested only for the U.S. portion of the collaborative effort. Also, NSF does not normally support research or education activities by scientists, engineers or educators employed by Federal agencies or Federally Funded Research and Development Centers (FFRDCs). Under unusual circumstances, other Federal agencies and FFRDCs may submit proposals directly to NSF.
Amount of Grant: $80,000 average per year
Date(s) Application is Due: Oct 15
Contact: Kile B. Baker, Program Director, (703) 292-8519, fax: (703) 292-9022, email: kbaker@nsf.gov; Robert M. Robinson, Program Manager, (703) 292-8529, fax: (703) 292-9022, email: rmrobins@nsf.gov
Internet: http://www.nsf.gov/publications/pub_summ.jsp?ods_key=nsf04576
Sponsor: National Science Foundation
4201 Wilson Boulevard
Arlington, VA 22230

NSF Graduate Teaching Fellows in K-12 Education (GK-12) Grants **4576**
This program supports fellowships and associated training that will enable graduate students and advanced undergraduates in the sciences, mathematics, engineering, and technology to serve in K-12 schools as resources knowledgeable about both the content and applications of science, mathematics, engineering, and technology. Academic institutions apply for awards to support fellowship activities, and are responsible for selecting fellows. The fellows will serve as resources for teachers in science and mathematics instruction. Expected outcomes include improved communication and teaching skills for the fellows, enriched learning by K-12 students, professional development opportunities for K-12 teachers, and strengthened partnerships between institutions of higher education and local school districts. Fastlane submission is required. Annual deadline dates may vary; contact program staff for exact dates. NSF 05-553
Requirements: Proposals may be submitted only by academic institutions that grant master's or doctoral degrees in the sciences, mathematics, engineering, or technology. An institution may submit no more than two proposals in which it is the only participating institution in the prospective award, and no more than one proposal in which it is the lead institution in a multi-institution collaboration, per round of competition. The principal

investigator must be a faculty member in the sciences, mathematics, engineering, or technology.
Amount of Grant: $19 million total
Date(s) Application is Due: May 4; Jun 2
Contact: Sonia Ortega, Program Director, Division of Graduate Education, (703) 292-8697; fax: (703) 292-9048; email: sortega@nsf.gov
Internet: http://www.nsf.gov/publications/pub_summ.jsp?ods_key=nsf05553&org=NSF
Sponsor: National Science Foundation
4201 Wilson Boulevard
Arlington, VA 22230

NSF Grant Opportunities for Academic Liaison with Industry (GOALI) 4577
Grant Opportunities for Academic Liaison with Industry (GOALI) promotes university-industry partnerships by making project funds or fellowships/traineeships available to support an eclectic mix of industry-university linkages. Special interest is focused on affording the opportunity for: faculty, postdoctoral fellows, and students to conduct research and gain experience in an industrial setting; industrial scientists and engineers to bring industry's perspective and integrative skills to academe; and interdisciplinary university-industry teams to conduct research projects. This solicitation targets high-risk/high-gain research with a focus on fundamental research, new approaches to solving generic problems, development of innovative collaborative industry-university educational programs, and direct transfer of new knowledge between academe and industry. GOALI seeks to fund transformative research that lies beyond that which industry would normally fund.
Contact: Donald Senich, Program Director; (703) 292-7082; fax: (703) 292-9056; email: dsenich@nsf.gov
Internet: http://www.nsf.gov/funding/pgm_summ.jsp?pims_id=13706
Sponsor: National Science Foundation
4201 Wilson Boulevard
Arlington, VA 22230

NSF Grants for the Analysis of Science and Technology (S&T) Resources 4578
NSF welcomes proposals for research, workshops, and studies leading to improved approaches to indicator development and presentation, new S&T indicator development, and better understanding of the S&T enterprise in the United States and globally under this program. The development of new approaches to the analysis and presentation of data as indicators, with a particular focus on effective methods of presenting complex information as indicators is a priority. Although it is not required, PIs are encouraged to submit a letter of intent one month before submitting a proposal.
Requirements: Proposals may be submitted by colleges, universities, and nonprofit or commercial organizations, or combinations of such organizations.
Amount of Grant: $400,000 total
Date(s) Application is Due: Sep 18
Contact: Derek Hill, Science Resources; (703) 292-7805; email: dhill@nsf.gov
Internet: http://www.nsf.gov/pubs/2000/nsf00111/nsf00111.htm
Sponsor: National Science Foundation
4201 Wilson Boulevard
Arlington, VA 22230

NSF Gravitational Physics Research Grants 4579
This funding program emphasizes the theory of strong gravitational fields and their application to astrophysics and cosmology, computer simulations of strong and gravitational fields, and gravitational radiation; and construction of a quantum theory of gravity. The program oversees the management of the construction, commissioning, and operation of the Laser Interferometer Gravity Wave Observatory (LIGO), and provides support for LIGO users and other experimental investigations in gravitational physics and related areas.
Requirements: Individuals and organizations in the following categories may submit proposals: (1) Universities and Colleges - Universities and two- and four-year colleges (including community colleges) located and accredited in the U.S., acting on behalf of their faculty members. Such organizations also are referred to as academic institutions. Institutions located outside the U.S. fall under (6) below. (2) Non-profit, non-academic organizations - Independent museums, observatories, research laboratories, professional societies and similar organizations in the U.S. that are directly associated with educational or research activities. (3) For-profit organizations - U.S. commercial organizations, especially small businesses with strong capabilities in scientific or engineering research or education. An unsolicited proposal from a commercial organization may be funded when the project is of special concern from a national point of view, special resources are available for the work, or the proposed project is especially meritorious. NSF is interested in supporting projects that couple industrial research resources and perspectives with those of universities; therefore, it especially welcomes proposals for cooperative projects involving both universities and the private commercial sector. (4) State and Local Governments - State educational offices or organizations and local school districts may submit proposals intended to broaden the impact, accelerate the pace, and increase the effectiveness of improvements in science, mathematics and engineering education in both K-12 and post-secondary levels. (5) Unaffiliated Individuals - Scientists, engineers or educators in the U.S. and U.S. citizens may be eligible for support, provided that the individual is not employed by, or affiliated with, an organization, and (a) the proposer has demonstrated the capability and has access to any necessary facilities to carry out the project; and (b) the proposer agrees to fiscal arrangements that, in the opinion of the NSF Division of Grants & Agreements, ensure responsible management of Federal funds. Unaffiliated individuals should contact the appropriate program before preparing a proposal for submission. (6) Foreign organizations - NSF rarely provides support to foreign organizations. NSF will consider proposals for cooperative projects involving U.S. and foreign organizations, provided support is requested only for the U.S. portion of the collaborative effort.
Restrictions: Other Federal agencies - NSF does not normally support research or education activities by scientists, engineers or educators employed by Federal agencies or Federally Funded Research and Development Centers (FFRDCs). Under unusual circumstances, other Federal agencies and FFRDCs may submit proposals directly to NSF. A proposed project is only eligible for support if it meets one or more of the following exceptions, as determined by a cognizant NSF Program Officer: (a) Special Projects. Under exceptional circumstances, research or education projects at other Federal agencies or FFRDCs that can make unique contributions to the needs of researchers elsewhere or to other specific NSF objectives may receive NSF support. (b) National and International Programs. The Foundation may fund research and logistic support activities of other Government agencies or FFRDCs directed at meeting the goals of special national and international research programs for which the Foundation bears special responsibility, such as the U.S. Antarctic Research Program. (c) International Travel Awards. In order to ensure appropriate representation or availability of a particular expertise at an international conference, staff researchers of other Federal agencies may receive NSF international travel awards. Proposers who think their project may meet one of the exceptions listed above should contact a cognizant NSF Program Officer before preparing a proposal for submission. In addition, a scientist, engineer or educator who has a joint appointment with a university and a Federal agency (such as a Veterans Administration Hospital, or with a university and a FFRDC) may submit proposals through the university and may receive support if he/she is a bona fide faculty member of the university, although part of his/her salary may be provided by the Federal agency. Preliminary inquiry should be made to the appropriate program before preparing a proposal for submission.
Samples: Searching for Stochastic Gravitational-Wave Background with LIGO: Data Analysis and Detector Development - Vuk Mandic, University of Minnesota-Twin Cities ($100,000); Numerical simulations of Einstein's equations - Manuel Tiglio, University of Maryland College Park ($40,000); Student Travel for Numerical Relativity and Data Analysis 2008 Conference - Duncan Brown, Syracuse University ($5,000); Investigations in Theoretical and Numerical Relativity - John Brown, North Carolina State University ($40,000).
Date(s) Application is Due: Sep 30
Contact: Beverly Berger, Program Director; (703) 292-7372; fax: (703) 292-9078; email: bberger@nsf.gov
Internet: http://www.nsf.gov/funding/pgm_summ.jsp?pims_id=5628
Sponsor: National Science Foundation
4201 Wilson Boulevard
Arlington, VA 22230

NSF Historically Black Colleges and Universities (HBCU) Program Grants 4580
This program is committed to enhancing the quality of undergraduate science, technology, engineering, and mathematics (STEM) education and research at Historically Black Colleges and Universities (HBCUs). Support is available for Implementation Projects, Planning Grants, Education Research Projects, and Targeted Infusion Projects. Additional information is available on the internet. This program provides educational opportunities for Undergraduate Students. This program provides indirect funding for students at this level or focuses on educational developments for this group such as curricula development, training or retention. Fastlane submission is required for collaborative proposals. NSF 06-606 provides additional information.
Requirements: The organizational eligibility has been broadened to include accredited two-year HBCUs that do not currently have undergraduate science, technology, engineering and mathematics (STEM) degree programs. These HBCUs are encouraged to aply for funding to enhance the quality of their general STEM courses and programs and/or to establish new undergraduate programs.
Amount of Grant: $7 million total anticipated
Date(s) Application is Due: Nov 14; Dec 15
Contact: Jessie A. DeAro, Assistant Program Director, (703) 292-5350; fax: (703) 292-9018; email: jdearo@nsf.gov
Internet: http://www.nsf.gov/funding/pgm_summ.jsp?pims_id=5481
Sponsor: National Science Foundation
4201 Wilson Boulevard
Arlington, VA 22230

NSF Human Cognition and Perception Research Grants 4581
This program supports research on perception, action and cognition including the development of these capacities. Emphasis is on research strongly grounded in theory. Research topics include vision, audition, haptics, attention, memory reasoning, written and spoken discourse, motor control and developmental issues in all topic areas. The program encompasses a wide range of theoretical perspectives, such as symbolic computation, connectionism, ecological, nonlinear dynamics, and complex systems, and a variety of methodologies including both experimental studies and modeling. Research involving acquired or developmental deficits is appropriate if the results speak to basic issues of perception, action, and cognition. Apply to PD 03-7252.
Amount of Grant: $1100-$4.9 million range; $85,395 average
Date(s) Application is Due: Jan 15; Jul 15
Contact: Christopher Kello, Program Director, (703) 292-7337; email: ckello@nsf.gov
Internet: http://www.nsf.gov/funding/pgm_summ.jsp?pims_id=5686
Sponsor: National Science Foundation
4201 Wilson Boulevard
Arlington, VA 22230

NSF Hydrologic Sciences Research Grants 4582

Hydrologic Sciences focuses on the flow of water and transport processes within streams, soils, and aquifers. Particular attention is given to spatial and temporal heterogeneity of fluxes and storages of water and chemicals over a wide range of scales, to geolimnology and to interfaces with the landscape, microbial communities, and coastal areas. Studies may also deal with processes in aqueous geochemistry and with the physical, chemical, and biological processes within water bodies. Study of these processes requires expertise from many basic sciences and mathematics, and proposals often require joint review with related programs.
Contact: L. Douglas James, Program Director; (703) 292-8549; fax: (703) 292-9025; email: ldjames@nsf.gov
Internet: http://www.nsf.gov/funding/pgm_summ.jsp?pims_id=13684
Sponsor: National Science Foundation
4201 Wilson Boulevard
Arlington, VA 22230

NSF Industry/University Cooperative Research Centers (I/UCRC) Grants 4583

The Industry/University Cooperative Research Centers (I/UCRCs) program develops long-term partnerships among industry, academe, and government. The centers are catalyzed by a small investment from the National Science Foundation (NSF) and are primarily supported by industry center members, with NSF taking a supporting role in their development and evolution. Each center is established to conduct research that is of interest to both the industry and the center. An I/UCRC contributes to the Nation's research infrastructure base and enhances the intellectual capacity of the engineering and science workforce through the integration of research and education. 2 to 8 full center awards and 4-12 planning grants are awarded annually. NSF 07-537
Requirements: U.S. academic institutions with graduate research programs may apply. The center director must be the PI and come from the lead institution. A PI must be a tenured faculty member (or equivalent) at an institution eligible to submit an I/UCRC proposal. A Letter of Intent must be approved by the NSF Program Director before the institution may submit a planning grant proposal. Institutions must have been awarded a planning grant before they are eligible to submit a full center proposal. Letters of Intent are due by the first Friday in January or the fourth Friday in June. Planning Grant and Full Center Proposals are due by the fourth Friday in March or by the fourth Friday in September.
Restrictions: Each institution is limited to a maximum of two single institution I/UCRC awards. Institutions which currently have two single institution awards are not eligible to submit another single institution proposal. There is no limit on the number of proposals that may be submitted when collaborating with another institution. It is permissible for an institution to have two single institution center awards and also submit a proposal to be a site in a multi-institutional center.
Contact: Alex Schwarzkopf, Lead I/UCRC Program Director; (703) 292-5359; fax: (703) 292-9051; email: aschwarz@nsf.gov
Internet: http://www.nsf.gov/funding/pgm_summ.jsp?pims_id=5501
Sponsor: National Science Foundation
4201 Wilson Boulevard
Arlington, VA 22230

NSF Information and Intelligent Systems: Core Programs Grants 4584

CISE's Division of Information and Intelligent Systems (IIS) supports research and education projects that develop new knowledge in three core programs: the Human-Centered Computing program; the Information Integration and Informatics program; and the Robust Intelligence program. The Division of Information and Intelligent Systems (IIS) studies the inter-related roles of people, computers, and information. IIS supports research and education activities that: develop new knowledge about the role of people in the design and use of information technology; increase our capability to create, manage, and understand data and information in circumstances ranging from personal computers to globally-distributed systems; and advance our understanding of how computational systems can exhibit the hallmarks of intelligence. Proposers are invited to submit proposals in three project classes, which are defined as follows: Small Projects—up to $500,000 total budget with durations up to three years; Medium Projects—$500,001 to $1,200,000 total budget with durations up to four years; and Large Projects—$1,200,001 to $3,000,000 total budget with durations up to five years.
Date(s) Application is Due: Aug 30; Nov 28; Dec 17
Contact: William S. Bainbridge, Program Director; (703) 292-8930; fax: (703) 292-9073; email: wbainbri@nsf.gov
Internet: http://www.nsf.gov/funding/pgm_summ.jsp?pims_id=13707
Sponsor: National Science Foundation
4201 Wilson Boulevard
Arlington, VA 22230

NSF Innovation and Organizational Change (IOC) 4585

This program program supports scientific research directed at advancing understanding of innovation and organizational phenomena. Levels of analysis may include (but are not limited to) individuals, groups and/or institutional arrangements. Disciplinary perspectives may include (but are not limited to) organization theory, organizational behavior, organizational sociology, social and industrial psychology, public administration, computer and information sciences, complexity sciences, decision and management sciences. Research methods may span a broad variety of qualitative and quantitative methods, including (but not limited to) archival analyses, surveys, simulation studies, experiments, comparative case studies, and network analyses. Research may involve industrial, educational, service, government, not-for-profits, voluntary organizations or interorganizational arrangements.
Requirements: The most frequent recipients of support are academic institutions, although awards are occasionally made to profit-making organizations, individuals, and government agencies at all levels. Research must be grounded in theory and generalizable. It must advance our scientific understanding of innovation and organizations. Scientific inquiries that are relevant to real problems and organizations in generalizable ways are encouraged.
Restrictions: Proposals that aim to implement or evaluate innovations or particular organizational changes rather than to advance fundamental, generalizable knowledge about innovation and organizations are not appropriate for IOC.
Amount of Grant: $2.2 million total
Date(s) Application is Due: Feb 2
Contact: Jacqueline Meszaros, (703) 292-7261, jmeszaro@nsf.gov; Crystal Aikens, (703) 292-4562, caikens@nsf.gov; Robbie Brown, (703) 292-7264, rbrown@nsf.gov
Internet: http://nsf.gov/funding/pgm_summ.jsp?pims_id=5378
Sponsor: National Science Foundation
4201 Wilson Boulevard
Arlington, VA 22230

NSF Instrument Development for Biological Research (IDBR) 4586

The IDBR program provides support for development of the following: (1) New instrumentation that addresses emerging biological research needs with the capacity to transform biological research; (2) Novel instrument configurations (hybrids, etc.) that demonstrably address a fundamental biological research challenge; (3) Instrumentation that leverages advances in micro- and nano-fabrication and device design to provide new capabilities in measurement and observation of biological phenomena; (4) Concept and proof-of-concept of novel instruments for biological research; (5) New instruments that provide new capabilities for detection, measurement, and/or observation of biological phenomena, or that greatly extend currently achievable sensitivity, accuracy or resolution; (6) Sensors that meet emerging biological research needs and that have the potential to transform biological research at any level of organization (7) Data acquisition and analysis tools that, through the development of novel devices/instruments, meet biological research needs. The program is not limited to the above mentioned activities. Any proposal that is designed to meet the goals of the Program will be considered. Proposals must, in all cases, demonstrate a strong connection to biological research needs through, for example, collaboration with biological researchers as well as demonstrate that the need for the instrumentation is firmly based on biological research drivers. NSF Program Solicitation 08-566.
Requirements: The program encourages proposals which conduct collaborative and planning activities such as workshops and the development of virtual organization frameworks. Those activities which promote interactions between the instrument development community and biological researchers as well as innovative networking strategies that foster research collaborations or enable new instrument development directions are especially encouraged. Activities which increase participation of colleagues at smaller institutions, minority-serving institutions, community colleges, and K-12 students and teachers are also recommended. Recognizing that the development and use of novel instrumentation have become increasingly integral to activities supported by all BIO programs, the IDBR program will place a higher priority on proposals that create novel instrumentation that enables biological research in a new and potentially transformative way and that serves a broad user community. Check the NSF Grant Proposal Guide (GPG) available electronically on the NSF website at: http://www.nsf.gov/publications/pub_summ.jsp?ods_key=gpg for other general eligibility guidelines.
Restrictions: The program does not support research or technique development activities, except to the extent these are required as part of the development of the new or improved instrument, or for the testing of its utility. Projects emphasizing the development of new research techniques should be addressed to an appropriate research program. The anticipated uses of the instrumentation to be developed or improved should include areas of research that fall within the scope of the Directorate for Biological Sciences (see BIO Home Page at http://www.nsf.gov/bio). Projects aimed at instrumentation whose primary use will be in studies of the etiology, diagnosis or treatment of physical or mental disease, abnormality, or malfunction in human beings or animals, is not supported by IDBR. Similarly, the development or testing of drugs or of instruments whose primary application is in pharmaceutical chemistry are not eligible for support. Such projects should be addressed to an appropriate program in another NSF Directorate or to another agency. Projects in which the main portion of the instrument development activity will be subcontracted to a Federally Funded Research and Development Center (FFRDC) or a commercial (for profit) organization are also not eligible for support by IDBR, and should be addressed to an appropriate NSF program or to another agency.
Date(s) Application is Due: Aug 28
Contact: Robyn Hannigan; (703) 292-8470; fax: (703) 292-9063; email: dbiiid@nsf.gov
Internet: http://www.nsf.gov/publications/pub_summ.jsp?ods_key=nsf08566&org=NSF
Sponsor: National Science Foundation
4201 Wilson Boulevard
Arlington, VA 22230

NSF Integrative Graduate Education and Research Traineeship (IGERT) 4587

The Integrative Graduate Education and Research Traineeship (IGERT) program has been developed to meet the challenges of educating U.S. Ph.D. scientists and engineers who will pursue careers in research and education, with the interdisciplinary backgrounds, deep knowledge in chosen disciplines, and technical, professional, and personal skills to become, in their own careers, leaders and creative agents for change. The program is intended to catalyze a cultural change in graduate education, for students, faculty, and institutions, by establishing innovative new models for graduate education

and training in a fertile environment for collaborative research that transcends traditional disciplinary boundaries. It is also intended to facilitate diversity in student participation and preparation, and to contribute to a world-class, broadly inclusive, and globally engaged science and engineering workforce.
Requirements: Academic institutions in the United States and its territories that grant the PhD degree and have research and training programs in the sciences and engineering are invited to submit preproposals. Projects involving more than one U.S. institution are eligible, but a single institution must accept overall management responsibility. Collaborating institutions need not be academic.
Amount of Grant: $600,000 per year maximum plus up to $200,000 for research instrumentation and special purpose research materials during the first year
Date(s) Application is Due: Mar 13; Sep 14
Contact: Carol Van Hartesveldt, Lead Program Director; (703) 292-8696; fax: (703) 292-9048; email: cvanhart@nsf.gov
Internet: http://www.nsf.gov/funding/pgm_summ.jsp?pims_id=12759
Sponsor: National Science Foundation
4201 Wilson Boulevard
Arlington, VA 22230

NSF Interagency Education Research Initiative (IERI) Program Grants **4588**
The goal of the IERI is to improve preK-12 student learning and achievement in reading, mathematics, and science by supporting rigorous, interdisciplinary research on large-scale implementations of promising educational practices and technologies in complex and varied learning environments. To this end, the Initiative will support an evolving, cumulative, and integrated portfolio of research projects that, when taken together, will provide a substantive corpus of effective instructional practices and a body of knowledge that informs the ways in which these practices can be implemented in real, complex, and varied educational environments and lead to enhanced student learning. Annual deadline dates may vary; contact program staff for exact dates. NSF 04-553
Requirements: Proposals may be submitted only by individual investigators or by small groups of investigators from colleges, universities, or other nonprofit organizations including local and state agencies.
Amount of Grant: $2 million total for contextual projects; $6 million total for scaling projects
Date(s) Application is Due: Mar 1; Apr 7
Contact: Finbarr Sloane, (703) 292-5146; email: fsloane@nsf.gov
Internet: http://www.nsf.gov/funding/pgm_summ.jsp?pims_id=5486
Sponsor: National Science Foundation
4201 Wilson Boulevard
Arlington, VA 22230

NSF Interagency Opportunities in Metabolic Engineering Research Grants **4589**
This solicitation describes a collaborative effort among the Department of Agriculture, Department of Commerce, Department of Defense, Department of Energy, Environmental Protection Agency, National Aeronautics and Space Administration, National Institute of General Medical Sciences (National Institutes of Health), and the National Science Foundation. The intent of this interagency solicitation is to provide an opportunity for an interagency granting activity in the area of metabolic engineering (ME). The eight participating agencies or departments are providing research funding and agency in-kind support such as equipment, laboratory space, personnel time, and materials in support of this solicitation. Upon conclusion of the review process, meritorious proposals may be recommended for funding by a participating agency or department. Each participating agency will make its own awards and the subsequent grant administration procedures will be in accordance with the individual policies of the awarding agency or department.
Requirements: Academic and nonprofit institutions, industrial organizations, and government (federal, state, and local) laboratories, are eligible. Investigators are strongly encouraged to discuss their idea for a preproposal with agency staff member(s).
Contact: Frederick G. Heineken, Program Director; (703) 292-7944; fax: (703) 292-9098; email: fheineke@nsf.gov
Internet: http://www.nsf.gov/funding/pgm_summ.jsp?pims_id=5494
Sponsor: National Science Foundation
4201 Wilson Boulevard
Arlington, VA 22230

NSF Interfacial Processes and Thermodynamics Research Grants **4590**
The Interfacial Processes and Thermodynamics program supports fundamental research in engineering areas related to: interfacial phenomena; mass transport phenomena; and solution phase equilibrium thermodynamics. Currently, emphasis is placed on molecular engineering approaches at interfaces, especially as applied to the processing of soft materials. Molecules at interfaces with functional interfacial properties are of special interest. These interfacial molecules may have biomolecular functions at the micro and nanoscale. Interfacial materials are generally formed through molecular self-directed, -templated, and/or -assembly, and they are driven primarily by thermodynamic intermolecular forces. In some cases, these interfacial processes may also be supplemented by weak chemical reactions.
Date(s) Application is Due: Mar 1; Sep 15
Contact: Robert M. Wellek, Program Director; (703) 292-8370; fax: (703) 292-9054; email: rwellek@nsf.gov
Internet: http://www.nsf.gov/funding/pgm_summ.jsp?pims_id=13362
Sponsor: National Science Foundation
4201 Wilson Boulevard
Arlington, VA 22230

NSF Linguistics Research Grants **4591**
This program supports research into the syntactic, semantic, phonological, and phonetic properties of individual languages and of language in general; psychological processes involved in the use of language; the development of linguistic capacities in children; social and cultural factors in language use, variation, and change; the acoustics of speech and the physiological and psychological processes involved in the production and perception of speech; and the biological bases of language in the central nervous system. Proposals are accepted at any time; target dates indicate approximate time of review panel meetings. Refer to brochure PD 98-1311 for program announcement.
Amount of Grant: $60,000 average
Date(s) Application is Due: Jan 15; Jul 15
Contact: Dr. Joan Maling, Program Director, (703) 292-8046; fax: (703) 292-9068; email: jmaling@nsf.gov
Internet: http://www.nsf.gov/funding/pgm_summ.jsp?pims_id=5408
Sponsor: National Science Foundation
4201 Wilson Boulevard
Arlington, VA 22230

NSF Living Stock Collections (LSC) Grants **4592**
The program supports operation of and improvements in outstanding collections of living organisms used in basic biological research. In addition to the collected organisms, biological reagents relevant to their study (such as isolated DNA, DNA libraries, antibodies, etc.) may be included in the collection when appropriate. To facilitate public access to collections supported on an ongoing basis by the LSC Program, all such collections must have an up-to-date web site that provides information about the contents of the collection and about procedures for ordering strains. The program provides support for two types of projects. Short-term projects are one-time awards (up to 36 months) leading to innovative handling of living stocks or to well-defined improvements in existing collections, including those not otherwise supported by LSC. Long-term projects (up to 60 months) support ongoing operation of significant collections. Collections receiving long-term support are expected to receive support from user fees. NSF 06-574
Requirements: The LSC Program accepts proposals from U.S. colleges, universities and non-profit organizations. Submission of proposals that are similar or identical to proposals submitted to another federal agency is allowed with the prior approval of NSF and of the other agency. Proposers planning such dual submission must contact the LSC Program Director in advance. Full proposals must be submitted via Grants.gov.
Restrictions: The program does not support the establishment of new collections.
Samples: University of Texas at Austin, $187,849 - Operation of UTEX Culture Collecton of Algae; West Virginia University Research Corporation, $103,656 - INVAM, An International Culture Collection of Arbuscular Mycorrhizal Fungi; American Type Culture Collection, $245,810.00 - ATCC Collection of Protists; Ohio State University Research Foundation, $1,199,330 - The Arabidopsis Biological Resource Center at The Ohio State University; Duke University, $764,137 - Continued Support of the Duke University Primate Center for the Study of Primate Biology and History
Date(s) Application is Due: Jul 2
Contact: Gerald B. Selzer, LSC Program Director, 615 N; (703) 292-8470; fax: (703) 292-9063; email: gselzer@nsf.gov
Internet: http://www.nsf.gov/funding/pgm_summ.jsp?pims_id=9189
Sponsor: National Science Foundation
4201 Wilson Boulevard
Arlington, VA 22230

NSF Louis Stokes Alliances for Minority Participation (LSAMP) Grants **4593**
This program is aimed at increasing the quality and quantity of students successfully completing science, technology, engineering, and mathematics (STEM) baccalaureate degree programs, and increasing the number of students interested in, academically qualified for and matriculated into programs of graduate study. LSAMP supports sustained and comprehensive approaches that facilitate achievement of the long-term goal of increasing the number of students who earn doctorates in SMET fields, particularly those from populations underrepresented in SMET fields. The program goals are accomplished through the formation of alliances. Phase I awards place emphasis on aggregate baccalaureate production. Phase II awards augment the Phase I emphasis with attention to individual student retention and progression to baccalaureate degrees. Phase III awards augment the Phase I and Phase II emphases with attention to aggregate student progression to graduate school entry. NSF 03-520
Requirements: The program is limited to alliances of academic institutions which have exemplary records over several years of enrolling and retaining significant numbers of undergraduate students underrepresented in SMET disciplines. The PI must be either the Chief Executive Officer or the Chief Academic Officer of the lead institution. Proposers are required to prepare and submit all proposals for this announcement/solicitation through the FastLane system. Victoria Smoot, Fastlane, (703) 292-4677; email: vsmoot@nsf.com
Amount of Grant: $6 million approximate total funding
Date(s) Application is Due: Oct 15
Contact: A. James Hicks, Program Director, Division of Human Resource Development, Education and Human Resources, (703) 292-4668; fax: (703) 292-9018; email: ahicks@nsf.gov
Internet: http://www.nsf.gov/funding/pgm_summ.jsp?pims_id=13646
Sponsor: National Science Foundation
4201 Wilson Boulevard
Arlington, VA 22230

NSF Magnetospheric Physics Research Grants 4594

Supports research on the magnetized plasma envelope of the outer atmosphere, including energization by the solar wind; the origin of geomagnetic storms and substorms; the population by solar and ionospheric sources; the origin of electric fields; the coupling among the magnetosphere, ionosphere, and atmosphere; and waves and instabilities in the natural plasma. Also supported are ground-based observational programs at high latitudes and laboratory experiments applicable to the geospace environment. Theoretical research programs may include numerical simulations using a variety of MHD, hybrid and particle codes. The analysis of data from all sources, whether ground-based or from spacecraft, is also supported.

Requirements: Proposals may be submitted by academic institutions, nonacademic and nonprofit research organizations, profit-making and private research organizations, and individuals. Occasionally, NSF sponsors efforts by other government agencies, particularly for field programs. Apply to PD 98-5750 as follows: (a) For full proposals submitted via FastLane: standard Grant Proposal Guidelines apply. (b) For full proposals submitted via Grants.gov: NSF Grants.gov Application Guide; A Guide for the Preparation and Submission of NSF Applications via Grants.gov Guidelines apply.

Samples: Johns Hopkins University (MD) $99,942 - NSWP: Sustained Sub-auroral Storm Electric Fields: Polarization or Minimum Dissipation?; Utah State University (UT) $79,004 - NSWP: Space Weather Prediction Using Hybrid Physics/Black-Box Methods; University of California-Los Angeles (CA) $170,212 - Renewal of the South American Meridional B-field Array: An American-Chilean magnetometer chain

Date(s) Application is Due: Proposals accepted anytime

Contact: Kile Baker, Program Director; Magnetospheric Physics, Division of Atmospheric Science; (703) 292-8519; fax: (703) 292-9023; email: kbaker@nsf.gov

Internet: http://www.nsf.gov/funding/pgm_summ.jsp?pims_id=11725

Sponsor: National Science Foundation
4201 Wilson Boulevard
Arlington, VA 22230

NSF Major Research Instrumentation Program (MRI) Grants 4595

The MRI program assists in the acquisition or development of major research instrumentation that is, in general, too costly for support through other NSF programs. This program seeks to improve the quality and expand the scope of research and research training in science and engineering, and to foster the integration of research and education by providing instrumentation for research-intensive learning environments. For proposals over $2 million, requests must be for the acquisition of a single instrument. For proposals requesting $2 million or less, investigators may seek support for instrument development or for acquisition of a single instrument, a large system of instruments, or multiple instruments that share a common or specific research focus. Submission of Letters of Intent is required (only for acquisition requests between $2 million and $4 million). NSF 08-503

Requirements: Proposals may only be submitted by the following: (1) U.S. colleges, universities and organizations of higher education located in the U.S., its territories and possessions; (2) U.S. independent research museums located in the U.S., its territories and possessions; (3) U.S. independent nonprofit research organizations located in the U.S., its territories and possessions, including consortia whose members consist only of organizations described in items (1) and (2). (Requests for instrumentation that will be located at a Federally Funded Research and Development Center (FFRDC) must be submitted as consortium proposals. This is the only mechanism by which instrumentation, whether purchased or developed, can be placed at an FFRDC.); (4) U.S. small businesses located in the U.S., its territories and possessions are eligible for instrument development support as private sector partners with submitting organizations; they may not submit proposals as a lead organization. The deadline is the fourth Thursday in January. Proposers are required to prepare and submit all proposals for this program solicitation through use of the NSF FastLane system.

Restrictions: The MRI program will NOT support proposal requests for: Computer networks used for general-purposes; General laboratory equipment or assorted instruments that do not share a common or specific research or research training focus; Instrumentation used primarily for standard science and engineering courses; Renovation or modernization of research facilities, fixed equipment, or facilities such as research vessels, airplanes, large telescopes, and supercomputing centers. The term 'research facilities' refers to the bricks-and-mortar physical plant in which sponsored or unsponsored research activities (including research training) take place, including related infrastructure, systems (e.g., HVAC and power systems, toxic waSuite removal systems), and fixed equipment. The term 'fixed equipment' refers to the permanent components of a research facility that are integral (i.e., built in, rather than affixed) to the facility (e.g., clean rooms, fume hoods, elevators, laboratory casework); their removal would affect the integrity or basic operation of the facility.

Amount of Grant: up to $4,000,000

Date(s) Application is Due: Jan 22

Contact: Joan M. Frye, (703) 292-8040, email: jfrye@nsf.gov; Randy Phelps, (703) 292-8040, email: rphelps@nsf.gov

Internet: http://nsf.gov/funding/pgm_summ.jsp?pims_id=5260

Sponsor: National Science Foundation
4201 Wilson Boulevard
Arlington, VA 22230

NSF Marine Geology and Geophysics Research Grants 4596

This program supports research on all aspects of geology and geophysics of the ocean basins and margins, as well as the Great Lakes. The program includes: (a) Structure, tectonic evolution and volcanic activity of the ocean basins, the continental margins, the mid-ocean ridges, and island arc systems; (b) Processes controlling exchange of heat and chemical species between seawater and ocean rocks; (c) Genesis, chemistry, and mineralogic evolution of marine sediments; (d) Processes controlling deposition, erosion and transport of marine sediments; (e) Past ocean circulation patterns and climates; (f) and, Interactions of continental and marine geologic processes.

Requirements: Individuals and organizations in the following categories may submit proposals: (1) Universities and Colleges - Universities and two- and four-year colleges (including community colleges) located and accredited in the U.S., acting on behalf of their faculty members. Such organizations also are referred to as academic institutions. Institutions located outside the U.S. fall under (6) below. (2) Non-profit, non-academic organizations - Independent museums, observatories, research laboratories, professional societies and similar organizations in the U.S. that are directly associated with educational or research activities. (3) For-profit organizations - U.S. commercial organizations, especially small businesses with strong capabilities in scientific or engineering research or education. An unsolicited proposal from a commercial organization may be funded when the project is of special concern from a national point of view, special resources are available for the work, or the proposed project is especially meritorious. NSF is interested in supporting projects that couple industrial research resources and perspectives with those of universities; therefore, it especially welcomes proposals for cooperative projects involving both universities and the private commercial sector. (4) State and Local Governments - State educational offices or organizations and local school districts may submit proposals intended to broaden the impact, accelerate the pace, and increase the effectiveness of improvements in science, mathematics and engineering education in both K-12 and post-secondary levels. (5) Unaffiliated Individuals - Scientists, engineers or educators in the U.S. and U.S. citizens may be eligible for support, provided that the individual is not employed by, or affiliated with, an organization, and (a) the proposer has demonstrated the capability and has access to any necessary facilities to carry out the project; and (b) the proposer agrees to fiscal arrangements that, in the opinion of the NSF Division of Grants & Agreements, ensure responsible management of Federal funds. Unaffiliated individuals should contact the appropriate program before preparing a proposal for submission. (6) Foreign organizations - NSF rarely provides support to foreign organizations. NSF will consider proposals for cooperative projects involving U.S. and foreign organizations, provided support is requested only for the U.S. portion of the collaborative effort.

Restrictions: Other Federal agencies - NSF does not normally support research or education activities by scientists, engineers or educators employed by Federal agencies or Federally Funded Research and Development Centers (FFRDCs). Under unusual circumstances, other Federal agencies and FFRDCs may submit proposals directly to NSF. A proposed project is only eligible for support if it meets one or more of the following exceptions, as determined by a cognizant NSF Program Officer: (a) Special Projects. Under exceptional circumstances, research or education projects at other Federal agencies or FFRDCs that can make unique contributions to the needs of researchers elsewhere or to other specific NSF objectives may receive NSF support. (b) National and International Programs. The Foundation may fund research and logistic support activities of other Government agencies or FFRDCs directed at meeting the goals of special national and international research programs for which the Foundation bears special responsibility, such as the U.S. Antarctic Research Program. (c) International Travel Awards. In order to ensure appropriate representation or availability of a particular expertise at an international conference, staff researchers of other Federal agencies may receive NSF international travel awards. Proposers who think their project may meet one of the exceptions listed above should contact a cognizant NSF Program Officer before preparing a proposal for submission. In addition, a scientist, engineer or educator who has a joint appointment with a university and a Federal agency (such as a Veterans Administration Hospital, or with a university and a FFRDC) may submit proposals through the university and may receive support if he/she is a bonafide faculty member of the university, although part of his/her salary may be provided by the Federal agency. Preliminary inquiry should be made to the appropriate program before preparing a proposal for submission.

Date(s) Application is Due: Feb 15; Aug 15

Contact: Bilal Haq, Program Director; (703) 292-8581; fax: (703) 292-9085; email: bhaq@nsf.gov

Internet: http://www.nsf.gov/funding/pgm_summ.jsp?pims_id=11726&org=OCE&from=home

Sponsor: National Science Foundation
4201 Wilson Boulevard
Arlington, VA 22230

NSF Materials Research Science and Engineering Centers (MRSEC) 4597

Grants support interdisciplinary materials research and education while addressing fundamental problems in science and engineering. MRSECs require outstanding research quality and intellectual breadth, provide support for research infrastructure and flexibility in responding to new opportunities, and strongly emphasize the integration of research and education. These centers foster active collaboration between universities and other sectors, including industry, and they constitute a national network of university-based centers in materials research. MRSECs address problems of a scope or complexity requiring the advantages of scale and interdisciplinary interaction provided by a center. A MRSEC may be located at a single institution, or may involve two or more institutions in partnership. Preliminary proposals are due September 5; full proposals are due January 18. Annual deadline dates may vary; contact program staff for exact dates. NSF 01-109

Requirements: Proposals may be submitted by universities in support of center-type activities. FastLane Submission is available.

Restrictions: Only two pre-proposals may be submitted by any one institution. Any one individual may be the Principal Investigator for only one pre-proposal. The same limitations apply to full proposals.

Amount of Grant: $24 million anticipated total

Date(s) Application is Due: Jan 18
Contact: Maija M. Kukla, Program Director, (703) 292-4940; email: mkukla@nsf.gov; Thomas P. Rieker, Program Director, (703) 292-4914; email: trieker@nsf.gov
Internet: http://www.nsf.gov/funding/pgm_summ.jsp?pims_id=5295&org=NSF&sel_org=NSF&from=fund
Sponsor: National Science Foundation
4201 Wilson Boulevard
Arlington, VA 22230

NSF Materials World Network: Cooperative Activity in Materials Research Between U.S. Investigators and their Counterparts Abroad (MWN) 4598
The program is intended to foster collaboration in materials and condensed matter research between investigators in the U.S. and their counterparts abroad. It includes joint activities between NSF and funding organizations in Africa, through the U.S.-Africa Materials Collaboration; Asia, Europe and other countries or regions. The estimated number of awards is 20 to 40, depending on quality of proposals and availability of funds. The total anticipated funding amount is $2,500,000 to $4,500,000. Estimated total funding, number of awards and average award size/duration are subject to quality of proposals and availability of funds. Awards may be standard or continuing grants. Proposals submitted must have clear relevance to fundamental materials and condensed matter research supported by the NSF Division of Materials Research (DMR). Projects not having this focus will not be considered for funding. NSF 07-574
Requirements: NSF will accept proposals from U.S. universities and two- and four-year colleges (including community colleges) addressing collaborations between researchers from the U.S. and participating countries or regions. Concurrently, investigators at non-U.S. research institutions should submit to the counterpart funding organization in their country or region a request for support of their side of the collaboration. NSF will consider support of all appropriate research costs for the U.S. side of such collaborations, with the expectation that funding or research organizations from the appropriate countries or regions will consider supporting the costs of the non-U.S. participants. In addition to proposals for collaborative materials research and education, proposals to interface networks of U.S. investigators and counterpart networks abroad will also be considered. Proposals from networks of U.S. researchers to NSF in support of activities such as workshops, meetings, brief personnel exchanges, etc., may be linked to counterpart networks abroad supported, for example, by the European Science Foundation. Proposed projects are expected to offer students and junior researchers the opportunity to participate in an international research and education experience and, more generally, for integrating research and training in an international environment. Proposals that include exchange of students and post-doctoral research associates between the U.S. and abroad are strongly encouraged, as well as proposals from junior faculty and members of underrepresented groups in science and engineering. Of special interest to NSF are proposals including activities that build upon and expand current cyber infrastructure capabilities, such as remote use of instrumentation, database creation and use, visualization and virtual experimentation, virtual networking, etc, to enhance and advance the international collaboration.
Restrictions: Proposals from investigators at non-U.S. institutions will not be accepted. An investigator may be Principal Investigator in only one proposal submitted to this program.
Samples: University of California-Santa Barbara, $191,000 - Materials World Network for the Discovery of Low Conductivity Oxides by Integrated Simulation and Experimentation; Kent State University, $122,000 - Materials World Network: Lyotropic Chromonic Liquid Crystals; Michigan Technological University, $120,000 - Materials World Network: Nonlinear Magnetophotonic Crystals
Date(s) Application is Due: Nov 13
Contact: Dr. Carmen Huber, Office of Special Programs; (703) 292-4939; fax: 703-292-9036; email: chuber@nsf.gov
Internet: http://www.nsf.gov/funding/pgm_summ.jsp?pims_id=12820
Sponsor: National Science Foundation
4201 Wilson Boulevard
Arlington, VA 22230

NSF Mathematical Biology Grants 4599
The Mathematical Biology Program supports research in areas of applied and computational mathematics with relevance to the biological sciences. Successful proposals are mathematically innovative and address challenging problems of interest to members of the biological community. Projects may include development of mathematical concepts and tools traditionally seen in other disciplinary programs within the Division of Mathematical Sciences, e.g., topology, probability, statistics, and computation, etc. To receive appropriate and timely review, such proposals should be submitted directly to the relevant disciplinary program, but will be considered for co-review by the Mathematical Biology program which may be selected as a secondary program.
Date(s) Application is Due: Jan 13
Contact: Mary Ann Horn, Program Director; (703) 292-4879; fax: (703) 292-9032; email: mhorn@nsf.gov
Internet: http://www.nsf.gov/funding/pgm_summ.jsp?pims_id=5690
Sponsor: National Science Foundation
4201 Wilson Boulevard
Arlington, VA 22230

NSF Microbial Observatories (MO) and Microbial Interactions and Processes 4600
Program activities will support research to discover and characterize novel microorganisms microbial consortia, communities, activities and other novel properties, and to study their roles in diverse environments. The long-term goal of this activity is to develop a network of sites or microbial observatories in different habitats to study and understand microbial diversity over time and across environmental gradients. Projects supported are expected to establish or participate in an established, Internet-accessible knowledge network to disseminate information resulting from these activities. In addition, educational and outreach activities such as formal or informal training in microbial biology, and activities that will broaden the participation of underrepresented groups in microbial research and education are expected. Fastlane submission is required. NSF 03-571
Requirements: Proposals may be submitted by universities or other NSF eligible institutions in support of individual investigators or small groups.
Restrictions: A PI may submit only one proposal and he/she may only collaborate in one other proposal as a co-PI.
Amount of Grant: $5,000,000 total
Date(s) Application is Due: Jul 23
Contact: Matthew D. Kane, Division of Molecular & Cellular Biosciences; (703) 292-7186; fax: (703) 292-9061; email: mkane@nsf. gov; Ronald Weiner, Program Director; (703) 292-8442; email: rweiner@nsf.gov
Internet: http://www.nsf.gov/pubs/2003/nsf03571/nsf03571.pdf
Sponsor: National Science Foundation
4201 Wilson Boulevard
Arlington, VA 22230

NSF MPS Internships in Public Science Education (MPS-IPSE) 4601
This program is intended to bring the most recent science research results from MPS disciplines to the public by promoting partnerships between the MPS research community and professionals in public science education. Science centers and museums play an important role in informing and educating society by providing an effective mechanism that brings current, accurate, and compelling scientific concepts and discoveries to the public. IPSE activities should be designed to increase the museum or science center's effectiveness in outreach and public science education through specific projects or exhibits that bring new scientific ideas to museum programming. Full proposal deadlines will be the second Monday in February. NSF 02-064
Requirements: Proposals will be accepted from academic institutions in the U.S. and its territories, science centers or museums, and MPS-funded centers, facilities, and institutes.
Amount of Grant: $2 million total
Contact: Henry Blount, Head, MPS/Office of Multidisciplinary Activities, (703) 292-8803; email: hblount@nsf.gov
Internet: http://www.nsf.gov/pubs/2002/nsf02064/nsf02064.htm
Sponsor: National Science Foundation
4201 Wilson Boulevard
Arlington, VA 22230

NSF National Astronomy and Ionosphere Center (NAIC) Grants 4602
NSF supports the National Astronomy and Ionosphere Center (NAIC), a visitor-oriented national research center devoted to scientific investigations in radio and radar astronomy and atmospheric sciences. NAIC provides telescope users with a wide range of research and observing instrumentation including receivers, transmitters, movable line feeds, and digital data acquisition and processing equipment. The center has a permanent staff of scientists, engineers, and technicians who are available to help visiting investigators with their observing programs. The NAIC facilities and instrumentation are available on a competitive basis to qualified scientists from all over the world. NAIC is operated and managed for NSF by Cornell University.
Contact: Program Director, Division of Astronomical Sciences, Directorate for Mathematical and Physical Sciences; (703) 292-8820; fax: (703) 292-9034
Internet: http://www.nsf.gov/funding/pgm_summ.jsp?pims_id=5652&org=AST&from=home
Sponsor: National Science Foundation
4201 Wilson Boulevard
Arlington, VA 22230

NSF National Science, Mathematics, Engineering, and Technology Education Digital Library (NSDL) 4603
This program aims to found a national digital library that will constitute an online network of learning environments and resources for science, technology, engineering, and mathematics (STEM) education at all levels. The program the program will accept proposals in three tracks: (1) Pathways projects are expected to provide stewardship for the content and services needed by major communities of learners. (2) Services projects are expected to develop services that support users, resource collection providers, and the Core Integration effort and that enhance the impact, efficiency, and value of the library. (3) Targeted Research projects are expected to explore specific topics that have immediate applicability to collections, services, and other aspects of the development of the digital library. Annual deadline dates may vary; contact program staff for exact dates. NSF 05-545
Requirements: Graduate students, public and private colleges (2-year and 4-year) and universities, State and local education agencies, tribal entities, nonprofit and private organizations, professional societies, science academies and centers, science museums and zoological parks, other informal science education institutions, research laboratories, and other institutions with an educational mission may apply.
Amount of Grant: $9 million total
Date(s) Application is Due: Mar 9; Apr 11
Contact: Dr. Lee Zia, Lead Program Director, NSDL Program, (703) 292-5140; fax: (703) 292-9046; email: lzia@nsf.gov; FastLane Help Desk, (800) 673-6188; email: fastlane@nsf.gov

Internet: http://www.nsf.gov/funding/pgm_summ.jsp?pims_id=5487
Sponsor: National Science Foundation
4201 Wilson Boulevard
Arlington, VA 22230

NSF National Space Weather Program **4604**
Proposals should benefit and advance the National Space Weather Program's goal to mitigate the adverse effects of space weather by providing timely, accurate, and reliable space environment observations, specifications, and forecasts. An effective program requires a strong commitment to basic research in many areas of space-related science. Proposals may be submitted for basic research in solar, heliospheric, magnetospheric, ionospheric, and thermospheric physics. NSF 07-520
Amount of Grant: $100,000 maximum
Date(s) Application is Due: Jan 16
Contact: Richard Behnke , Section Head, Upper Atmosphere Research , (703) 292-8518; fax: (703) 292-9023; email: rbehnke@nsf.gov
Internet: http://www.nsf.gov/funding/pgm_summ.jsp?pims_id=5575
Sponsor: National Science Foundation
4201 Wilson Boulevard
Arlington, VA 22230

NSF NATO Postdoctoral Fellowships in Science and Engineering **4605**
NSF/NATO invites proposals for 12-month postdoctoral research fellowships from beginning scientists, mathematicians, and engineers. Approximately 20 awards will be made to U.S. institutions that would like to host scientists from NATO Partner Countries (Albania, Armenia, Azerbaijan, Belarus, Bulgaria, Estonia, Georgia, Kazakhstan, Kyrgyz Republic, Latvia, Lithuania, Moldova, Romania, Russian Federation, Slovak Republic, Slovenia, Tajikistan, Turkmenistan, the former Yugoslav Republic of Macedonia, Ukraine, and Uzbekistan). Eligible fields of research are: mathematics, engineering, computer and information science, geosciences, the physical, biological, social, behavioral, and economic sciences, the history and philosophy of science, and interdisciplinary areas comprised of two or more of these fields. Contact NSF for deadline dates and additional eligibility requirements. Fellows are also provided with dependency allowances of $200 per month for a dependent spouse and for each of not more than two dependent children for 12 months. Annual deadline dates may vary; contact program staff for exact dates. Refer to brochure NSF 01-163 for program announcement.
Requirements: U.S. citizens who have received their doctorates within the past five years or who will have done so by the start of their tenure may study or conduct research in other NATO countries. Proposers are required to prepare and submit all proposals for this program solicitation through the FastLane system. For questions about FastLane, contact Sheryl Balke-Smith, Program Analyst, Education and Human Resources, Division of Graduate Education, (703)292-8630, email: sbalke@nsf.gov.
Amount of Grant: $2750 monthly stipend for 12 months; $1200 travel and research
Date(s) Application is Due: Dec 13
Contact: Dr. Terry Woodin, Program Director, Division of Graduate Education, Directorate For Education and Human Resources, (703) 292-8697; email: nsf-nato@nsf.gov
Internet: http://www.nsf.gov/pubs/2001/nsf01163/nsf01163.html
Sponsor: National Science Foundation
4201 Wilson Boulevard
Arlington, VA 22230

NSF Natural, Social, and System Sciences and Research Support and Logistics Grants **4606**
The Program supports research in glaciology and in the atmospheric, biological, earth, contaminants, and ocean sciences. This program provides core support for disciplinary research in the Arctic and coordinates its support of arctic research with the Directorates for Geosciences, Mathematical and Physical Sciences, and Biological Sciences. Areas of special interest include marine and terrestrial ecosystems, arctic adaptations, atmospheric chemistry, exploration of the Arctic Ocean, as well as arctic geological and glaciological processes. The goal of this program is to gain a better understanding of the Earth's biological, geological, chemical, and social processes, and the interactions of ocean, land, atmosphere, biological, and human systems. Approximately 130-160 awards are expected to be made, with a combination of standard and continuing awards. NSF 00-96
Amount of Grant: $150,000 to $5 million; $143,000 average
Date(s) Application is Due: Feb 15; Aug 8
Contact: Dr. Jane Dionne, Arctic Natural Sciences; (703) 306-1029; email: jdionne@nsf.gov; Dr. Neil R. Swanberg, Program Officer; (703) 306-1029; email: nswanber@nsf.gov
Internet: http://www.nsf.gov/pubs/2000/nsf0096/nsf0096.htm
Sponsor: National Science Foundation
4201 Wilson Boulevard
Arlington, VA 22230

NSF Nuclear Physics Research Grants **4607**
The program supports research at the frontiers of nuclear science, including: properties and behavior of nuclei and nuclear matter under extreme conditions, and/or as they relate to astrophysical phenomena; the quark-gluon basis for the structure and dynamics of nuclear matter; phase transitions of nuclear matter from normal nuclear density and temperature to the predicted high-temperature quark-gluon plasma; and basic interactions and fundamental symmetries. This research involves many venues, including low-energy to multi-GeV electrons and photons; intermediate-energy light ions; low-energy to relativistic heavy ions, including radioactive beams; cold and ultra-cold neutrons; as well as non-accelerator-based experiments. The program supports university user groups executing experiments at a large number of laboratories in the United States and abroad, and a national user facility: the National Superconducting Cyclotron Laboratory, a superconducting, heavy-ion cyclotron facility at Michigan State University. The program also supports smaller accelerator facilities, such as those at Florida State University and the University of Notre Dame.
Requirements: Individuals and organizations in the following categories may submit proposals: (1) Universities and Colleges - Universities and two- and four-year colleges (including community colleges) located and accredited in the U.S., acting on behalf of their faculty members. Such organizations also are referred to as academic institutions. Institutions located outside the U.S. fall under (6) below. (2) Non-profit, non-academic organizations - Independent museums, observatories, research laboratories, professional societies and similar organizations in the U.S. that are directly associated with educational or research activities. (3) For-profit organizations - U.S. commercial organizations, especially small businesses with strong capabilities in scientific or engineering research or education. An unsolicited proposal from a commercial organization may be funded when the project is of special concern from a national point of view, special resources are available for the work, or the proposed project is especially meritorious. NSF is interested in supporting projects that couple industrial research resources and perspectives with those of universities; therefore, it especially welcomes proposals for cooperative projects involving both universities and the private commercial sector. (4) State and Local Governments - State educational offices or organizations and local school districts may submit proposals intended to broaden the impact, accelerate the pace, and increase the effectiveness of improvements in science, mathematics and engineering education in both K-12 and post-secondary levels. (5) Unaffiliated Individuals - Scientists, engineers or educators in the U.S. and U.S. citizens may be eligible for support, provided that the individual is not employed by, or affiliated with, an organization, and (a) the proposer has demonstrated the capability and has access to any necessary facilities to carry out the project; and (b) the proposer agrees to fiscal arrangements that, in the opinion of the NSF Division of Grants & Agreements, ensure responsible management of Federal funds. Unaffiliated individuals should contact the appropriate program before preparing a proposal for submission. (6) Foreign organizations - NSF rarely provides support to foreign organizations. NSF will consider proposals for cooperative projects involving U.S. and foreign organizations, provided support is requested only for the U.S. portion of the collaborative effort.
Restrictions: Other Federal agencies - NSF does not normally support research or education activities by scientists, engineers or educators employed by Federal agencies or Federally Funded Research and Development Centers (FFRDCs). Under unusual circumstances, other Federal agencies and FFRDCs may submit proposals directly to NSF. A proposed project is only eligible for support if it meets one or more of the following exceptions, as determined by a cognizant NSF Program Officer: (a) Special Projects. Under exceptional circumstances, research or education projects at other Federal agencies or FFRDCs that can make unique contributions to the needs of researchers elsewhere or to other specific NSF objectives may receive NSF support. (b) National and International Programs. The Foundation may fund research and logistic support activities of other Government agencies or FFRDCs directed at meeting the goals of special national and international research programs for which the Foundation bears special responsibility, such as the U.S. Antarctic Research Program. (c) International Travel Awards. In order to ensure appropriate representation or availability of a particular expertise at an international conference, staff researchers of other Federal agencies may receive NSF international travel awards. Proposers who think their project may meet one of the exceptions listed above should contact a cognizant NSF Program Officer before preparing a proposal for submission. In addition, a scientist, engineer or educator who has a joint appointment with a university and a Federal agency (such as a Veterans Administration Hospital, or with a university and a FFRDC) may submit proposals through the university and may receive support if he/she is a bonafide faculty member of the university, although part of his/her salary may be provided by the Federal agency. Preliminary inquiry should be made to the appropriate program before preparing a proposal for submission.
Samples: Investigations in Intermediate Energy Spin Physics - Samuel Danagoulian, North Carolina Agricultural & Technical State University ($183,476); Strange Form Factors and Precision Standard Model Tests - Mark Pitt, Virginia Polytechnic Institute and State University ($90,000); Parity-Violating Electron Scattering - David Armstrong, College of William and Mary ($330,000); Studies in Nuclear Physics and Fundamental Interactions at Indiana University - William Snow, Indiana University ($1,000,000); Study of spin structure of hadrons using polarized photons - Daniel Sober, Catholic University of America ($424,937)
Date(s) Application is Due: Sep 30
Contact: Bradley D. Keister, Program Director; (703) 292-7377; fax: (703) 292-9078; email: bkeister@nsf.gov
Internet: http://www.nsf.gov/funding/pgm_summ.jsp?pims_id=5627&org=PHY&from=home
Sponsor: National Science Foundation
4201 Wilson Boulevard
Arlington, VA 22230

NSF OCE Education Grants **4608**
OCE Education supports efforts to integrate ocean research and education via three main program areas, which are: Research Experiences for Undergraduates (REU) Site Program—this program provides funding to Universities and Marine Laboratories that allows them to offer summer internships to undergraduate students who would like to participate in ocean-related research efforts. Proposals may be submitted annually (August deadline); Faculty Early Career Development (CAREER)—this program

supports pre-tenure researchers who would like to combine their research efforts with excellent educational programs. Proposals may be submitted annually (July deadline); and Centers for Ocean Education Excellence (COSEE)—this program supports partnerships between ocean science researchers, educators, and informal science organizations, providing the public a deeper understanding of the ocean and its influence on each person's quality of life and our national prosperity.
Contact: Elizabeth L. Rom, Program Director; (703) 292-7709; fax: (703) 292-9085; email: elrom@nsf.gov; Donald Elthon; (703) 292-8580; fax: (703) 292-9085; email: delthon@nsf.gov
Internet: http://www.nsf.gov/funding/pgm_summ.jsp?pims_id=13548&org=OCE&from=home
Sponsor: National Science Foundation
4201 Wilson Boulevard
Arlington, VA 22230

NSF Ocean Drilling (OD) Program Grants **4609**
ODP activities are a unique, international effort to explore the earth's crust beneath oceans for the purpose of revealing the composition, structure, and history of the submerged portion of the earth's surface. Ocean drilling involves logging and collecting geologic samples from the floor of the deep ocean basins through rotary coring and hydraulic piston coring. The logs and samples of the cores are made available to qualified scientists throughout the world for individual research projects. Drilling programs have sampled sites in the North Atlantic, Norwegian Sea, Mediterranean Sea, southern and equatorial Atlantic, Pacific Ocean off the west coast of South America, Weddell Sea off Antarctica, Indian Ocean and western and equitorial Pacific Ocean. Drilling operations are managed by Texas A&M University; logging is managed by Lamont-Doherty Geological Observatory. The general contractor for the overall management and operation of ODP is Joint Oceanographic Institutions. Support for participation and drilling-related research performed by U.S. scientists is provided by NSF. This support focuses on investigations of potential drilling regions, especially by means of regional geophysical field studies; feasibility and initial development of downhole instruments and techniques; and downhole geophysical and geochemical experiments. In addition, NSF will consider proposals for studies leading to long-range definition of future drilling objectives. To be considered for support, proposed projects should be clearly relevant to the drilling plans of the international drilling community and focus on predrilling or concurrent drilling activities. Additional support for U.S. scientists may be obtained via the Joint U.S. Science Advisory Committee (U.S.SAC), which includes support for U.S. scientists participating on the drillship and for necessary follow-up studies related to initial publication of drilling results; and planning activities, such as workshops, to define concepts and develop problem-related drilling programs, including U.S. participation in Joint Oceanographic Institutions for Deep Earth Sampling (JOIDES). Also, requests for proposals may be issued for other surveys, regional and topical synthesis of existing data, and development of downhole tools and instrumentation as these tasks are identified. Contact the NSF for further information and list of addresses to which specific proposals should be sent.
Date(s) Application is Due: Feb 15; Aug 15
Contact: Deborah Smith, Program Director; (703) 292-7478; fax: (703) 292-9085; email: dksmith@nsf.gov; James F. Allan; Program Director; (703) 292-8144; fax: (703) 292-9085; email: jallan@nsf.gov
Internet: http://www.nsf.gov/funding/pgm_summ.jsp?pims_id=13524&org=NSF&sel_org=NSF&from=fund
Sponsor: National Science Foundation
4201 Wilson Boulevard
Arlington, VA 22230

NSF Ocean Sciences Division Research Grants **4610**
This division supports research to improve understanding of the ocean and ocean basins. Basic research programs support individual scientists, small groups of cooperating scientists, and some large coordinated projects. Ocean sciences also supports efforts to develop, acquire, and operate the instrumentation and facilities needed to carry out these research programs. Programs include ocean sciences instrumentation development, oceanographic instrumentation, shipboard scientific support equipment, and ocean sciences research support and its subprograms of physical oceanography, chemical oceanography, marine geology and geophysics, and biological oceanography; oceanographic centers and facilities; and the ocean drilling program.
Date(s) Application is Due: Feb 15; Aug 15
Contact: David L. Garrison; (703) 292-7588; fax: (703) 292-9085; email: dgarriso@nsf.gov; Cynthia Suchman, Associate Program Director; (703) 292-3143; email: csuchman@nsf.gov
Internet: http://www.geo.nsf.gov/oce/ocefund.htm
Sponsor: National Science Foundation
4201 Wilson Boulevard, Room 725
Arlington, VA 22230

NSF Ocean Technology and Interdisciplinary Coordination Grants **4611**
The Oceanographic Technology and Interdisciplinary Coordination (OTIC) Program supports a broad range of research and technology development activities. Unsolicited proposals are accepted for instrumentation development that has broad applicability to ocean science research projects and that enhance observational, experimental or analytical capabilities of the ocean science research community. Specific announcements for funding opportunities are made for additional projects involving Improvements in Facilities, Communications, and Equipment at Biological Field Stations and Marine Laboratories (FSML) and the National Ocean Partnership Program.
Date(s) Application is Due: Feb 15; Aug 15
Contact: Kandace S. Binkley; (703) 292-7577; fax: (703) 292-9085; email: kbinkley@nsf.gov
Internet: http://www.nsf.gov/funding/pgm_summ.jsp?pims_id=12724&org=OCE&from=home
Sponsor: National Science Foundation
4201 Wilson Boulevard
Arlington, VA 22230

NSF Oceanographic Centers and Facilities Grants **4612**
NSF supports construction, conversion, acquisition, and operation of major shared-use oceanographic facilities. The University-National Oceanographic Laboratory System (UNOLS) schedules these facilities and expeditionary programs. This program supports expensive facilities that are necessary for NSF-funded research and training of oceanographers. Examples of these facilities are ships, submersibles, large shipboard equipment, and shared-use instruments to collect and analyze data. NSF encourages local contributions from nonfederal funds; however, there is no fixed requirement for institutional contributions.
Contact: Linda Goad, Program Director; (703) 292-7706; fax: (703) 292-9085; email: lgoad@nsf.gov
Internet: http://www.nsf.gov/funding/pgm_summ.jsp?pims_id=12725&org=OCE&from=home
Sponsor: National Science Foundation
4201 Wilson Boulevard
Arlington, VA 22230

NSF Paleo Perspectives on Climate Change (P2C2) Grants **4613**
The goal of research funded under the interdisciplinary P2C2 solicitation is to utilize key geological, chemical, and biological records of climate system variability to provide insights into the mechanisms and rate of change that characterized Earth's past climate variability, the sensitivity of Earth's climate system to changes in forcing, and the response of key components of the Earth system to these changes. Important scientific objectives of P2C2 are to: 1) provide comprehensive paleoclimate data sets that can serve as model test data sets analogous to instrumental observations; and 2) enable transformative syntheses of paleoclimate data and modeling outcomes to understand the response of the longer-term and higher magnitude variability of the climate system that is observed in the geological record. Collaborative Proposals. All collaborative proposals submitted as separate submissions from multiple organizations must be submitted via the NSF FastLane system. Solicitation: 08-505
Amount of Grant: $101,000 average
Date(s) Application is Due: Feb 4; Oct 15
Contact: Dr. David Verardo, Program Officer; (703) 292-8527; fax: (703) 292-9022; email: dverardo@nsf.gov; Paul E. Filmer, Program Officer; (703) 292-8551; fax: (703) 292-9025; email: pfilmer@nsf.gov
Internet: http://www.nsf.gov/funding/pgm_summ.jsp?pims_id=5750
Sponsor: National Science Foundation
4201 Wilson Boulevard
Arlington, VA 22230

NSF Paleoclimate Grants **4614**
This program supports research on the natural evolution of Earth's climate with the goal of providing a baseline for present variability and future trends through improved understanding of the physical, chemical, and biological processes that influence climate over the long-term. The Geosciences Directorate and the Office of Polar Programs have joined in creating the annual Paleo Perspectives on Climate Change (P2C2) competition in paleoclimate global change research. Researchers are encouraged to consider the P2C2 competition as a possible source of support for their global change research. Proposals are welcome at any time during the year for the Paleoclimate Program but, investigators are encouraged to submit proposals early in the fiscal year. PD 98-1530 Since proposals eligible for funding in the P2C2 competition are not eligible for funding in the Paleoclimate Program, researchers are strongly advised to contact the Director of the Paleoclimate Program for guidance as to the suitability of their proposed research for either program.
Requirements: Individuals and organizations in the following categories may submit proposals: (1) Universities and Colleges - Universities and two- and four-year colleges (including community colleges) located and accredited in the U.S., acting on behalf of their faculty members. Such organizations also are referred to as academic institutions; (2) Non-profit, non-academic organizations - Independent museums, observatories, research laboratories, professional societies and similar organizations in the U.S. that are directly associated with educational or research activities; (3) For-profit organizations - U.S. commercial organizations, especially small businesses with strong capabilities in scientific or engineering research or education. An unsolicited proposal from a commercial organization may be funded when the project is of special concern from a national point of view, special resources are available for the work, or the proposed project is especially meritorious. NSF is interested in supporting projects that couple industrial research resources and perspectives with those of universities; therefore, it especially welcomes proposals for cooperative projects involving both universities and the private commercial sector; (4) State and Local Governments - State educational offices or organizations and local school districts may submit proposals intended to broaden the impact, accelerate the pace, and increase the effectiveness of improvements in science, mathematics and engineering education in both K-12 and post-secondary levels; (5) Unaffiliated Individuals - Scientists, engineers or educators in the U.S. and U.S. citizens may be eligible for support, provided that the individual is not employed by, or affiliated

with, an organization, and: (a) the proposer has demonstrated the capability and has access to any necessary facilities to carry out the project; and, (b) the proposer agrees to fiscal arrangements that, in the opinion of the NSF Division of Grants & Agreements, ensure responsible management of Federal funds. Unaffiliated individuals should contact the appropriate program before preparing a proposal for submission.
Restrictions: NSF rarely provides support to foreign organizations. NSF will consider proposals for cooperative projects involving U.S. and foreign organizations, provided support is requested only for the U.S. portion of the collaborative effort. Also, NSF does not normally support research or education activities by scientists, engineers or educators employed by Federal agencies or Federally Funded Research and Development Centers (FFRDCs). Under unusual circumstances, other Federal agencies and FFRDCs may submit proposals directly to NSF.
Samples: Mesoamerican Dendroclimatology: Climate Dynamics and Socioeconomic Applications - David Stahle, University of Arkansas ($327,273); Holocene Sea Surface Temperature (SST) and El Nino-Southern Oscillation (ENSO) Reconstructions from Peruvian Mollusks - Julian Sachs, University of Washington ($98,898); Workshop: Paleoclimate Modeling Intercomparison Project on Coupled Simulations - Bette Otto-Bliesner, University Corporation for Atmospheric Research ($20,000); Workshop on Synthesizing Transient Climate Evolution of the Last 21,000 Years (SynTraCE-21000) - Zhengyu Liu, University of Wisconsin-Madison ($24,995); Reducing and Representing Uncertainties in High-Resolution Proxy Climate Data Workshop - Jonathan Overpeck, University of Arizona ($15,000); Millennial-scale Tropical Rainfall Variability From 100 to 20 ka: Testing Cross-Isthmian Water Vapor Transport and Feedbacks on Thermohaline Circulation - Matthew Lachniet, University of Nevada Las Vegas ($95,064).
Contact: David J. Verardo; (703) 292-8527; email: dverardo@nsf.gov
Internet: http://www.nsf.gov/funding/pgm_summ.jsp?pims_id=12727&org=ATM
Sponsor: National Science Foundation
4201 Wilson Boulevard
Arlington, VA 22230

NSF Particle and Nuclear Astrophysics (PNA) Grants **4615**
The program supports university research in particle and nuclear astrophysics, including experiments or research and design projects in underground facilities. Current supported activities are high-energy cosmic ray studies, solar and high-energy neutrino astrophysics, the study of gamma ray bursts, and searches for dark matter. Funding is also provided for accelerator-based nuclear astrophysics studies of stellar processes, nucleosynthesis, and processes related to cosmology and the early universe. Currently, the Deep Underground Science and Engineering Laboratory (DU.S.EL) program is funding Research and Development studies for experiments that need to be located in an underground laboratory to reduce backgrounds. NSF PD 06-1643
Requirements: It should be noted that proposals that are submitted to the PNA program and are requesting in excess of $1,000,000 may, at the discretion of the Program Officer, be subjected to an additional level of scrutiny in the form of a cost review which would take place before the annual PNA panel that meets to discuss all of the submitted proposals. Individuals and organizations in the following categories may submit proposals: (1) Universities and Colleges - Universities and two- and four-year colleges (including community colleges) located and accredited in the U.S., acting on behalf of their faculty members. (2) Non-profit, non-academic organizations - Independent museums, observatories, research laboratories, professional societies and similar organizations in the U.S. that are directly associated with educational or research activities (3.) For-profit organizations - U.S. commercial organizations, especially small businesses with strong capabilities in scientific or engineering research or education. An unsolicited proposal from a commercial organization may be funded when the project is of special concern from a national point of view, special resources are available for the work, or the proposed project is especially meritorious. NSF is interested in supporting projects that couple industrial research resources and perspectives with those of universities; therefore, it especially welcomes proposals for cooperative projects involving both universities and the private commercial sector. (4) State and Local Governments - State educational offices or organizations and local school districts may submit proposals intended to broaden the impact, accelerate the pace, and increase the effectiveness of improvements in science, mathematics and engineering education in both K-12 and post-secondary levels. (5) Unaffiliated Individuals - Scientists, engineers or educators in the U.S. and U.S. citizens may be eligible for support, provided that the individual is not employed by, or affiliated with, an organization, and: (a) the proposer has demonstrated the capability and has access to any necessary facilities to carry out the project; and, (b) the proposer agrees to fiscal arrangements that, in the opinion of the NSF Division of Grants & Agreements, ensure responsible management of Federal funds. Unaffiliated individuals should contact the appropriate program before preparing a proposal for submission.
Restrictions: NSF rarely provides support to foreign organizations. NSF will consider proposals for cooperative projects involving U.S. and foreign organizations, provided support is requested only for the U.S. portion of the collaborative effort. NSF does not normally support research or education activities by scientists, engineers or educators employed by Federal agencies or Federally Funded Research and Development Centers (FFRDCs).
Date(s) Application is Due: Sep 24
Contact: Marvin Goldberg, Program Director; (703) 292-7374; email: mgoldber@nsf.gov
Internet: http://www.nsf.gov/funding/pgm_summ.jsp?pims_id=5633&govDel=U.S.NSF_39
Sponsor: National Science Foundation
4201 Wilson Boulevard
Arlington, VA 22230

NSF Partnership for Advancing Technologies in Housing Grants **4616**
The objective of this program is to support the objectives of the PATH program, which are to: improve the quality, affordability, durability, environmental performance, and energy efficiency of today's new and existing homes; strengthen the technology infrastructure of the United States; and help create the next generation of American housing. In addition to proposals in the above areas of technology and technology management, NSF and PATH are also interested in proposals involving partnerships between research institutions, industrial enterprises, local government, and/or other broadly-defined research and development participants relevant to the home building industry. NSF 01-45
Requirements: Proposals may be submitted by U.S. academic institutions in support of individual investigators or small groups.
Amount of Grant: $150,000 maximum
Date(s) Application is Due: Apr 17
Contact: John Scalzi, Program Director, Division of Civil and Mechanical Systems; (703) 292-7020; email: jscalzi@nsf.gov; Robert Wellek, Deputy Division Director, Chemical and Transport Systems; (703) 292-8370; email: rwellek@nsf.gov
Internet: http://www.nsf.gov/pubs/2001/nsf0145/nsf0145.htm
Sponsor: National Science Foundation
4201 Wilson Boulevard
Arlington, VA 22230

NSF Partnerships for Innovation (PFI) Grants **4617**
The program is intended to forge connections between new knowledge created in the discovery process to learning and innovation, while broadening the participation of people and institutions in NSF activities. For the purposes of this program, innovation is defined as the transformation of knowledge into the products, processes, systems and services that fuel economic development, create wealth and generate improvements in the national standard of living. The goals of this program are to stimulate the transformation of knowledge created by the national research and education enterprise into innovations that create new wealth, build strong local, regional and national economies, and improve the national well-being; broaden the participation of all types of academic institutions and all citizens in NSF activities to more fully meet the broad workforce needs of the national innovation enterprise; and create or enhance enabling infrastructure necessary to foster and sustain innovation in the long-term. Cost sharing is required at a level of 10 percent of the requested NSF budget. An optional letter of intent is due March 10; full proposal is due May 17. Annual deadline dates may vary; contact program staff for exact dates. NSF 04-556
Requirements: Proposals must be submitted by U.S. universities and colleges located in the U.S., its territories and possessions, including universities, two-and four-year colleges and community colleges, hereafter referred to as universities and colleges, or academic institutions. A Partnership proposal must be submitted as a single administrative package by the lead institution. The submitting institution must serve as the lead institution or partner in the proposed partnership, that is, as the institution responsible for overall management of the partnership activity with administrative responsibility for the proposed effort. Partners may include, but are not limited to, other academic institutions, not-for-profit organizations, private sector firms, entrepreneurs, venture capitalists, state and local government entities, trade and professional associations, and Federal laboratories. At a minimum, proposed partnerships must include academic institutions and private sector organizations. Partnerships that include state/local government entities are strongly encouraged.
Amount of Grant: $36,000 - $600,000
Date(s) Application is Due: Jun 3
Contact: John Hurt, Program Director; (703) 292-5332; fax: (703) 292-9051; email: jhurt@nsf.gov
Internet: http://www.nsf.gov/pubsys/ods/getpub.cfm?ods_key=nsf04556
Sponsor: National Science Foundation
4201 Wilson Boulevard
Arlington, VA 22230

NSF Partnerships for International Research and Education (PIRE) **4618**
The Partnerships for International Research and Education (PIRE) program seeks to catalyze a higher level of international engagement in the U.S. science and engineering community by supporting innovative, international research and education collaborations. The program will enable U.S. scientists and engineers to establish collaborative relationships with international colleagues in order to advance new knowledge and discoveries at the frontiers of science and engineering and to promote the development of a diverse, globally-engaged U.S. scientific and engineering workforce. The PIRE program will support bold, forward-looking research whose successful outcome results from all partners,ÄiU.S. and foreign,Äìproviding unique contributions to the research endeavor. It is also intended to facilitate greater student preparation for and participation in international research collaboration, and to contribute to the development of a diverse, globally-engaged U.S. science and engineering workforce. The program aims to support partnerships that will strengthen the capacity of institutions, multi-institutional consortia, and networks to engage in and benefit from international research and education collaborations. This program provides educational opportunities for undergraduate students, graduate students, and postdoctoral fellows. This program supports institutions which may provide support to individuals at those institutions.
Requirements: Submission of Preliminary Proposals is required. Full proposals will be by invitation only. Proposals may only be submitted by universities that granted one (1) or more PhD in a science or engineering field since January 1, 2006, accredited in, and having a campus located in the United States. Data on science and engineering PhD production is available from The Survey of Earned Doctorates for 2006. Institutions that

are not on this list but that have granted one (1) or more PhD in a science or engineering field since 2006 should contact PIRE Program staff regarding eligibility. Proposers are strongly encouraged to develop synergistic collaborations with a wide range of U.S. academic institutions, including four-year and two-year institutions, in order to take advantage of complementary strengths in international research and/or education and broaden the impact of the proposed PIRE program. They are also encouraged to develop synergistic linkages with other domestic and international partners that will enhance the PIRE project, including research laboratories, professional societies, corporations, institutions of informal learning (e.g., museums, zoos), etc.
Restrictions: Eligible institutions may submit up to three (3) preliminary proposals maximum.
Date(s) Application is Due: Feb 26
Contact: Elizabeth Lyons, Program Coordinator; (703) 292-7256; fax: (703) 292-9067; email: PIRE-info@nsf.gov
Internet: http://www.nsf.gov/funding/pgm_summ.jsp?pims_id=12819&org=OISE&from=home
Sponsor: National Science Foundation
4201 Wilson Boulevard
Arlington, VA 22230

NSF Partnerships in Astronomy & Astrophysics Research and Education **4619**
The objective of the program is to enhance diversity in astronomy and astrophysics research and education by stimulating the development of formal, potentially long-term, collaborative research and education relationships between minority-serving colleges and universities and the NSF-supported facilities, projects or faculty members at research institutions (including private observatories). As retention of traditionally underrepresented students is key to the success of this program, a solid mentoring scheme is vital. PAARE awards are expected to help build a foundation for students at an early point in their careers and help these students make the next step in their careers. The partnerships are intended to provide education and mentoring as a form of professional development to help these students succeed long after their work on the PAARE research. Funded activities might include, but are not limited to, the development of collaborative and mutually beneficial astronomy-related research and education projects, support for graduate and undergraduate students and exchanges of faculty and students. Well-prepared high-school students may also participate. Of special interest are activities based on research and education connections between the participants, and designed to increase recruitment, retention and degree attainment by members of groups underrepresented in astronomy and astrophysics research. Proposers are encouraged to contact NSF program staff to discuss the appropriateness of the planned activities.
Requirements: Proposals may only be submitted by minority-serving colleges and universities, including Historically Black Colleges and Universities (HBCUs), Hispanic-serving institutions (HSIs), and Tribally Controlled Colleges (TCCs). Each proposal must be submitted in collaboration with (and may include a subaward to) one or more NSF-supported facilities, projects or partnering faculty members at research institutions. Colleges and universities eligible to participate in this activity must offer baccalaureate degrees in physics or astronomy and meet at least one of the following criteria: 1) Be designated by the Department of Higher Education in FY 1994 as a Hispanic-serving Institution (HSI) under Title III of the Higher Education Act of 1965, as amended [See 20 U.S.C 1059 (copyright symbol); Public Law 102-325, Section 316, July 22, 1992]. 2) Be designated by the Department of Education as a Historically Black College or University (HBCU) under Title III of the Higher Education Act of 1965, as amended (see 34 CFR 608.2). 3) Be cited as a Tribal College or University in Section 532 of the Equity in Education Land-Grant Status October of 1964; Tribally Controlled Community College Assistance Act of 1978, Public Law 95-471. 4) Be a college or university whose undergraduate enrollment consists of at least 50% one or more ethnic minorities underrepresented in science and engineering in the U.S., and not a Doctoral/Research University-Extensive institution in the Carnegie classification (for the Carnegie classification see http://www.carnegiefoundation.org/classifications/index.asp?key=785). African American, Hispanic Americans, Native Americans or Native Alaskans and Hawaiian/Pacific islanders are ethnic minorities underrepresented in science and engineering in the U.S.. Partnering institutions or organizations may be any institution or organization eligible for NSF funding that supports a research program in astronomy and astrophysics including AST-funded facilities and also private observatories. Research groups and individual scientists at research institutions are also eligible partners, provided their institution shows evidence of commitment to the proposed program as demonstrated in a letter from the institution describing its support of the partnership. Proposals may include subawards to the partnering institution, provided it has demonstrated a commitment to the activity at least for the duration of the PAARE grant.
Amount of Grant: $2,000,000 total funding, 3-5 awards estimated.
Date(s) Application is Due: Aug 3
Contact: Brian Patten, (703) 292-4910, bpatten@nsf.gov; Dana Lehr, (703) 292-7456, dlehr@nsf.gov; Tammy Bosler, AAAS Fellow, (703) 292-8248, fax: (703) 292-9034, tbosler@nsf.gov
Internet: http://nsf.gov/funding/pgm_summ.jsp?pims_id=501046&org=NSF&sel_org=NSF&from=fund
Sponsor: National Science Foundation
4201 Wilson Boulevard
Arlington, VA 22230

NSF Physical and Dynamic Meteorology (PDM) Grants **4620**
Physical and Dynamic Meteorology supports research involving studies of cloud physics; atmospheric electricity; radiation; boundary layer and turbulence; the initiation, growth, and propagation of gravity waves; all aspects of mesoscale meteorological phenomena, including their morphological, thermodynamic, and kinematic structure; development of mesoscale systems and precipitation processes; and transfer of energy between scales. The program also sponsors the development of new techniques and devices for atmospheric measurements. PD 98-1522
Requirements: Individuals and organizations in the following categories may submit proposals: (1) Universities and Colleges - Universities and two- and four-year colleges (including community colleges) located and accredited in the U.S., acting on behalf of their faculty members. Such organizations also are referred to as academic institutions; (2) Non-profit, non-academic organizations - Independent museums, observatories, research laboratories, professional societies and similar organizations in the U.S. that are directly associated with educational or research activities; (3) For-profit organizations - U.S. commercial organizations, especially small businesses with strong capabilities in scientific or engineering research or education. An unsolicited proposal from a commercial organization may be funded when the project is of special concern from a national point of view, special resources are available for the work, or the proposed project is especially meritorious. NSF is interested in supporting projects that couple industrial research resources and perspectives with those of universities; therefore, it especially welcomes proposals for cooperative projects involving both universities and the private commercial sector; (4) State and Local Governments - State educational offices or organizations and local school districts may submit proposals intended to broaden the impact, accelerate the pace, and increase the effectiveness of improvements in science, mathematics and engineering education in both K-12 and post-secondary levels; (5) Unaffiliated Individuals - Scientists, engineers or educators in the U.S. and U.S. citizens may be eligible for support, provided that the individual is not employed by, or affiliated with, an organization, and: (a) the proposer has demonstrated the capability and has access to any necessary facilities to carry out the project; and, (b) the proposer agrees to fiscal arrangements that, in the opinion of the NSF Division of Grants & Agreements, ensure responsible management of Federal funds. Unaffiliated individuals should contact the appropriate program before preparing a proposal for submission.
Restrictions: NSF rarely provides support to foreign organizations. NSF will consider proposals for cooperative projects involving U.S. and foreign organizations, provided support is requested only for the U.S. portion of the collaborative effort. Also, NSF does not normally support research or education activities by scientists, engineers or educators employed by Federal agencies or Federally Funded Research and Development Centers (FFRDCs). Under unusual circumstances, other Federal agencies and FFRDCs may submit proposals directly to NSF.
Samples: VORTEX 2: Storm-scale Baroclinity Using Fine-scale Observations and Numerical Models - Christopher Weiss, Texas Tech University ($317,261); Observations of the Spatial-Statistical Structures of Precipitation - Arthur Jameson, RJH Scientific Inc. ($148,280); Mechanisms of Convection-Wave Interactions - Stefan Tulich, University of Colorado at Boulder ($120,158); High Temporal-Resolution Analysis of Electrical Activity in a Severe Hailstorm Scanned Using Phased Array Radar - Pamela Heinselman, University of Oklahoma Norman Campus ($56,783).
Amount of Grant: $50,000-$2,000,000
Contact: Andrew G. Detwiler, Program Director; (703) 292-8524; fax: (703) 292-9022; email: adetwile@nsf.gov
Internet: http://www.nsf.gov/funding/pgm_summ.jsp?pims_id=12728&org=ATM
Sponsor: National Science Foundation
4201 Wilson Boulevard
Arlington, VA 22230

NSF Physical Anthropology Research Grants **4621**
The program supports basic research in areas related to human evolution and contemporary human biological variation. Research areas supported by the program include, but are not limited to, human genetic variation, human adaptation, human osteology and bone biology, human and nonhuman primate paleontology, functional anatomy, and primate socioecology. Grants supported in these areas are united by an underlying evolutionary framework, and often a consideration of adaptation as a central theoretical theme. Many proposals also have a biocultural orientation. The program frequently serves as a bridge within NSF between the social and behavioral sciences and the natural and physical sciences, and proposals are commonly jointly reviewed and funded with other programs. Refer to NSF brochure PD 98-1392.
Date(s) Application is Due: Jul 1; Dec 3
Contact: Dr. Mark Weiss, Director, Physical Anthropology Program, (703) 292-7321; fax: (703) 292-9068; email: mweiss@nsf.gov
Internet: http://www.nsf.gov/funding/pgm_summ.jsp?pims_id=5407
Sponsor: National Science Foundation
4201 Wilson Boulevard
Arlington, VA 22230

NSF Physical Chemistry (PCHEM) Grants **4622**
This program supports 2 broad areas: (1) Experimental Physical Chemistry - Supports experimental research directed at the molecular level of understanding of the physical properties of chemical systems. Experimental methodologies employed include frequency domain and time domain spectroscopic techniques covering the entire range of the electromagnetic spectrum, time-resolved dynamical studies of state-selected and mass-selected systems, and reactive scattering in molecular beams. Chemical systems studied range from single isolated molecules or ions to clusters, liquids, and solids. Chemical properties of interest include molecular structure and the shape of the ground and

excited electronic-state potential energy surfaces, chemical dynamics of unimolecular and bimolecular chemical processes, time-resolved internal energy redistribution and state-to-state dynamics in molecular systems, and solute/solvent interactions in clusters and liquids. (2) Theoretical and Computational Chemistry - Supports the development of theoretical and computational methods for chemical problems. Methods of special interest include, but are not limited to, electronic structure techniques, statistical mechanics, molecular dynamics, and simulation methods for molecular or supramolecular systems. Areas of application span the full range of chemical systems from small molecules to macromolecules and degrees of aggregation from single molecules or small clusters to macroscopic systems. While application areas may involve any chemical system, including biological systems or materials, the goal of the program is to support the development of new theoretical and computational methodologies that will be broadly applicable to a range of chemical problems.
Requirements: Proposals to the Physical Chemistry Program should be submitted during the Division of Chemistry's submission windows that are that are open between July 1 and July 31 and then between November 1 and November 30. Proposals may be submitted at any time the windows are open.
Date(s) Application is Due: Jul 31; Nov 30
Contact: Tanja Pietrass; (703) 292-2170; email: tpietras@nsf.gov
Internet: http://www.nsf.gov/funding/pgm_summ.jsp?pims_id=5674&org=NSF&more=Y
Sponsor: National Science Foundation
4201 Wilson Boulevard
Arlington, VA 22230

NSF Physical Oceanography Research Grants **4623**
The Physical Oceanography Program supports research on a wide range of topics associated with the structure and movement of the ocean, with the way in which it transports various quantities, with the way the ocean's physical structure interacts with the biological and chemical processes within it, and with interactions between the ocean and the atmosphere, solid earth and ice that surround it.
Date(s) Application is Due: Feb 15; Aug 15
Contact: Eric Itsweire, Program Director; (703) 292-8582; fax: (703) 292-9085; email: eitsweir@nsf.gov
Internet: http://www.nsf.gov/funding/pgm_summ.jsp?pims_id=12729&org=OCE&from=home
Sponsor: National Science Foundation
4201 Wilson Boulevard
Arlington, VA 22230

NSF Physics at the Information Frontier (PIF) Grants **4624**
This program (PIF) is intended to provide support for physics proposals in three subareas: computational physics; information intensive physics; and quantum information science and revolutionary computing. The computational physics subarea focuses on computational problems in physics requiring significant long-term code development, and/or medium to large collaboratories involving physicists or physicists interacting with applied mathematicians and computer scientists. Priority will be given to proposals which: advance computational capabilities crossing intellectual boundaries in physics; result in dramatic new computational advances within a specific sub-area of physics; or provide generally useful software tools for the physics community. Information intensive physics seeks proposals to develop rapid, secure and efficient access to physics data stores rising from Petabytes (today) to Exabytes (in 10 years) via heterogeneous and distributed computing resources and networks of varying capability and reliability and to develop internally consistent approaches to the usage of common resources required in the multiple collaborations and serving virtual science organizations on a global scale. Priority will be given to proposals which: develop tools which can serve a broad community within physics or reach out to other communities in need of rapid, secure access to large data stores; or which bring dramatic new capabilities to a specific sub-area of physics.
Date(s) Application is Due: Oct 28
Contact: Barry I. Schneider, Program Director; (703) 292-7383; fax: (703) 292-9060; email: bschneid@nsf.gov
Internet: http://www.nsf.gov/funding/pgm_summ.jsp?pims_id=6681
Sponsor: National Science Foundation
4201 Wilson Boulevard
Arlington, VA 22230

NSF Physics Frontiers Centers (PFC) Grants **4625**
This program supports university-based centers and institutes where the collective efforts of a larger group of individuals can enable transformational advances in the most promising research areas. The program is designed to foster major breakthroughs at the intellectual frontiers of physics by providing needed resources such as combinations of talents, skills, disciplines, and/or specialized infrastructure, not usually available to individual investigators or small groups, in an environment in which the collective efforts of the larger group can be shown to be seminal to promoting significant progress in the science and the education of students. Activities supported are in all sub-fields of physics within the purview of the Division of Physics: atomic, molecular, optical, plasma, elementary particle, nuclear, astro-, gravitational, and biological physics. Interdisciplinary projects at the interface between these physics areas and other disciplines and physics sub-fields, e.g. biology, quantum information science, mathematical physics, condensed matter physics, and emerging areas of physics are also included. The successful PFC activity will demonstrate: the potential for a profound advance in physics; creative, substantive activities aimed at enhancing education, diversity, and public outreach; potential for broader impacts, e.g., impacts on other field(s) and benefits to society; and a synergy or value-added rationale that justifies a center- or institute-like approach.
Requirements: Proposals may be submitted by universities and colleges in the United States. In cases involving multi-institutional consortia, a single entity must accept overall management responsibility.
Restrictions: No more than two pre-proposals may be submitted by any one institution.
Amount of Grant: $500,000-$4,000,000
Date(s) Application is Due: Jan 30
Contact: Dr. C.D. Caldwell, Atomic, Molecular, Optical, and Plasma Physics; (703) 292-7371; fax: (703) 292-9078; email: dcaldwel@nsf.gov; Ramona Winkelbauer, Computer Specialist; (703) 292-7390; fax: (703) 292-9078; email: rwinkelb@nsf.gov
Internet: http://www.nsf.gov/funding/pgm_summ.jsp?pims_id=5305
Sponsor: National Science Foundation
4201 Wilson Boulevard
Arlington, VA 22230

NSF Physics of Living Systems (PoLS) Grants **4626**
PoLS will encourage research that will emphasize the physical principles of organization and function of living systems, including the exploration of artificial life forms. While the problems under study must be important to advancing our understanding of the living world in a quantitative way, particular emphasis will be placed on those projects in which lessons learned from the biological application also expand the intellectual range of physics. Awards will cover a broad spectrum of physics approaches in biology, ranging from the physical principles and mechanisms at the single cell level such as cellular organization (e.g. cytoskeleton), energy metabolism, gene regulation and intracellular and intercellular communication, to collective behavior and evolution of complexity in life forms and living populations of organisms. This systems approach in physics has been very successful in understanding inanimate systems, and has the potential to bring deep understanding of the world of animated, replicating systems, through falsifiable phenomenological theories. In vitro systems, if necessary for understanding of the physical principles in a living organism, will be supported as well. The program funds individual investigators, although collaborative proposals between physicists and biological researchers are welcome.
Date(s) Application is Due: Jul 31
Contact: Krastan B. Blagoev, Program Director; (703) 292-4666; fax: (703) 292-9078; email: kblagoev@nsf.gov
Internet: http://www.nsf.gov/funding/pgm_summ.jsp?pims_id=6673
Sponsor: National Science Foundation
4201 Wilson Boulevard
Arlington, VA 22230

NSF Physics Research Grants **4627**
The Physics Division of the Mathematical and Physical Sciences Directorate (MPS) supports research projects in the areas of atomic, molecular, optical, and plasma physics, elementary particle physics, nuclear physics, and gravitational physics. The division also supports a significant level of activity in nuclear and particle astrophysics, which involves substantial linkages between the NSF nuclear and particle physics programs and with the astronomy division. Much of the research supported by the division is heavily involved in computational physics and nonlinear science. Research in theoretical physics is supported in a separate program within the division, and the theoretical physics program has strong linkages to programs in all of the other divisions in the MPS Directorate. The research activities of the division are linked to education. Research programs support and involve students at the postdoctoral, graduate, and undergraduate levels.
Contact: Joseph L. Dehmer, Division Director; (703) 292-7370; fax: (703) 292-9078; email: jdehmer@nsf.gov
Internet: http://www.nsf.gov/mps/divisions/phy/about/c_programs.htm
Sponsor: National Science Foundation
4201 Wilson Boulevard
Arlington, VA 22230

NSF Physiology and Ethology Research Grants **4628**
The Physiology and Ethology Cluster of thematic areas is located within the Division of Integrative Biology and Neuroscience (IBN) and supports integrative studies of physiological functions at the genomic, cellular, systemic, and organismal levels, and animal behavior in both field and laboratory settings. Also considered are Long-Term Research in Environmental Biology (LTREB) proposals. The cluster supports research on: the mechanism, development, function, and evolution of all animal behavior, including behavioral ecology and evolution; nonhuman learning and cognition; behavioral genetics; development of behavior; and behavioral physiology and motivation, including behavioral endocrinology, animal communication, and animal orientation. Also included are studies that address ecological or evolutionary questions in the areas of morphology, comparative physiology, physiological ecology, and biomechanics of plants, animals, protists, fungi, and bacteria, with emphasis on the study of whole organisms, living or extinct.
Requirements: Scientists affiliated with U.S. institutions are eligible to apply.
Restrictions: Clinical research is not considered for support.
Contact: William E. Zamer, Program Director; (703) 292-7894; fax: (703) 292-9154; email: wzamer@nsf.gov
Internet: http://www.nsf.gov/funding/pgm_summ.jsp?pims_id=12730
Sponsor: National Science Foundation
4201 Wilson Boulevard
Arlington, VA 22230

NSF Planetary Astronomy (PLA) Research Grants 4629

The grants award research support for theoretical and observational studies of the detailed structure and composition of planetary surfaces, interiors, atmospheres, and satellites; the nature of small bodies (asteroids and comets); and the origin and development of the solar system. All applications must be submitted electronically via FastLane.
Requirements: Scientists initiate research proposals, which are usually submitted on their behalf by their employing institutions.
Date(s) Application is Due: Nov 15
Contact: Nigel A. Sharp, Program Director; (703) 292-4905; fax: (703) 292-9034; email: nsharp@nsf.gov
Internet: http://www.nsf.gov/funding/pgm_summ.jsp?pims_id=13630
Sponsor: National Science Foundation
4201 Wilson Boulevard
Arlington, VA 22230

NSF Plant Genome Research Program (PGRP) 4630

The goals of this program are to support basic research in plant genomics, and to accelerate the acquisition and utilization of new knowledge and innovative approaches to elucidate fundamental biological processes in plants. The focus is on plants of economic importance and plant processes of potential economic value. NSF 06-581
Requirements: Proposals are invited from U.S. academic institutions, U.S. non-profit research organizations including museums, research laboratories, professional societies and similar organizations in the U.S. that are directly associated with educational or research activities, and consortia of only the eligible organizations listed here. When a consortium of eligible organizations submits a proposal, it must be submitted as a single proposal with one organization serving as the lead and all other organizations as subawardees. Separately submitted collaborative proposals will not be accepted and will be returned without review.
Restrictions: Organizations ineligible to submit to this program solicitation may not receive subawards. There continues to be a limit on the number of proposals in which an investigator can be included. An investigator may only submit one proposal per fiscal year as a principal investigator or a co-principal investigator in response to this Program Solicitation. Proposals received in excess of this limit may be returned without review.
Date(s) Application is Due: Oct 26; Jan 21
Contact: Jane Silverthorne, Program Director, Plant Genome Research Program; (703) 292-8470; email: dbipgr@nsf.gov
Internet: http://www.nsf.gov/pubs/2006/nsf06581/nsf06581.htm
Sponsor: National Science Foundation
4201 Wilson Boulevard
Arlington, VA 22230

NSF Plant, Fungal and Microbial Developmental Systems Grants 4631

The Developmental Systems Cluster supports research aimed at understanding how interacting developmental processes give rise to the emergent properties of organisms. A systems level approach to understanding these processes, at the molecular, cellular, and organismal levels of organization, requires the use of molecular, genetic, biochemical, and physiological techniques as well as techniques from outside biology. The Plant, Fungal and Microbial Development programmatic area supports research that addresses developmental processes in plants from algae to angiosperms, microbes and fungi.
Requirements: The most frequent recipients of support are academic institutions and nonprofit research groups. In special circumstances, grants also are awarded to other types of institutions and to individuals.
Date(s) Application is Due: Jan 12; Jul 12
Contact: Thomas Jack, Program Director; (703) 292-8417; fax: (703) 292-9153; email: tjack@nsf.gov
Internet: http://www.nsf.gov/funding/pgm_summ.jsp?pims_id=501087
Sponsor: National Science Foundation
4201 Wilson Boulevard
Arlington, VA 22230

NSF Plasma Physics Grants 4632

The plasma physics program jointly with DoE funds research in the fundamental physics of plasmas. Research areas include plasma turbulence and shocks, turbulent and nonlocal, collisional transport with and without strong magnetic fields, non-neutral plasmas, cold plasmas, strongly-coupled and dusty plasmas, laser-plasma interactions, ultra-short pulse and/or ultra-intense laser plasma interactions, high-energy-density plasmas, and low temperature plasmas. Both theoretical and experimental research is included. Note, that there is often coordination between the Physics Division and other organizations within the Mathematical and Physical Science Directorate as well as in the Engineering and Geophysical Sciences Directorates to co-review proposals that are plasma related but cross disciplinary boundaries.
Contact: Richard L. Berger; (703) 292-2183; fax: (703) 292-9078; email: rlberger@nsf.gov
Internet: http://www.nsf.gov/funding/pgm_summ.jsp?pims_id=503252
Sponsor: National Science Foundation
4201 Wilson Boulevard
Arlington, VA 22230

NSF Polar Programs Division Research Grants 4633

The program seeks to strengthen and enhance the national scientific enterprise through the expansion of fundamental knowledge and increased understanding of the polar regions; to encourage and support basic research in the arctic and antarctic focused on the solid earth, glacial and sea ice, terrestrial ecosystems, the oceans, the atmosphere, and beyond. Major objectives include understanding of the natural phenomena and processes in the Antarctic and Arctic regions. Support is also provided for science and technology centers, undergraduate student research, facility enhancement, instrumentation, and laboratory equipment.
Requirements: Eligible applicants include public and private colleges and universities; nonacademic research institutions; private profit organizations; and, under special circumstances, unaffiliated scientists.
Amount of Grant: $1000-$5 million; average $143,000
Contact: Karl A. Erb, Director, Office of Polar Programs; (703) 292-8030; fax: (703) 292-9081; email: kerb@nsf.gov
Internet: http://www.nsf.gov/funding/pgm_list.jsp?org=OPP
Sponsor: National Science Foundation
4201 Wilson Boulevard
Arlington, VA 22230

NSF Polymers Research Grants 4634

The program supports basic research and education on polymeric materials. The program portfolio is mainly experimental and highly diverse with components of materials science, chemistry, physics, and related disciplines. Polymers are studied across the nano-to-macro continuum through fundamental materials-focused scientific approaches. Areas addressed include synthesis, molecular assembly, characterization, phase behavior, structure, morphology, and properties. Particular focus is on new materials or materials with novel and/or superior properties, as well as on fundamental understanding of structure-property relationships. Proposals of high potential impact for research and education in the above areas are welcome.
Contact: Andrew Lovinger, Program Director; (703) 292-4933; email: alovinge@nsf.gov
Internet: http://www.nsf.gov/funding/pgm_summ.jsp?pims_id=5357
Sponsor: National Science Foundation
4201 Wilson Boulevard
Arlington, VA 22230

NSF Population and Evolutionary Processes Grants 4635

The Population and Evolutionary Processes Cluster supports research on population properties that lead to variation within and among populations and among species. Approaches include empirical and theoretical studies of microevolution, organismal adaptation, geographical differentiation, natural hybridization and speciation, as well as processes that lead to macroevolutionary patterns of trait evolution. Please note that studies focusing on interactions among species should be directed to the Ecological Biology Cluster.The Population and Evolutionary Processes Cluster funds projects within the Population and Evolutionary Processes Program as well as CAREER, OPU.S. and LTREB.
Date(s) Application is Due: Jan 9; Jul 9
Contact: Nancy J. Huntly, Program Director; (703) 292-8980; email: nhuntly@nsf.gov
Internet: http://www.nsf.gov/funding/pgm_summ.jsp?pims_id=12824
Sponsor: National Science Foundation
4201 Wilson Boulevard
Arlington, VA 22230

NSF Postdoctoral Research Fellowships in Biology (PRFB) 4636

The Directorate for Biological Sciences (BIO) awards Postdoctoral Research Fellowships in Biology to recent recipients of the doctoral degree for research and training in selected areas of biology supported by BIO to encourage independence early in their research careers and to permit them to pursue their research and training goals in the most appropriate research locations regardless of the availability of funding for the Fellows at that site. The fellowships are also designed to provide active mentoring of the Fellows by the sponsoring scientists who will benefit from having additional members in their research groups. The research and training plan of each fellowship must address important scientific questions in contemporary biology within the scope of the BIO Directorate and the specific guidelines in this fellowship program solicitation. Because the fellowships are offered only to postdoctoral scientists early in their careers, doctoral advisors are encouraged to discuss the availability of BIO fellowships with their graduate students early in their doctoral programs. Fellowships are awards to individuals, not institutions, and are administered by the Fellows. Fellowship stipends are being adjusted on a yearly basis. NSF 07-580.
Requirements: Only individuals may apply. NSF postdoctoral fellowships are awards to individuals, and applications are submitted directly by applicants to NSF. However, applications must include sponsoring scientists' statements and the applicants must affiliate with institutions (e.g., colleges and universities, and privately-sponsored nonprofit institutes and museums, government agencies and laboratories, and, under special conditions, for-profit organizations) anywhere in the world. Applicants must: (1) be U.S. citizens (or nationals) or legally admitted permanent residents of the United States (i.e., have a 'green card') at the time of application; (2) earn or plan to earn the doctoral degree in a scientific or engineering field prior to the requested start date of the fellowship; (3) either currently be a graduate student or, at the deadline date, have served in a position requiring the doctoral degree for no more than 12 full time months since earning the degree. There is a one time exception to this criterion for Biological Informatics in November, 2007 only, as described below; (4) must present a research and training plan that falls within the purview of BIO and includes the required information for the specific competition as described below; (5) select a host institution and sponsoring scientist different from the doctoral degree and current position or provide compelling justification for why such a change is not being proposed; (6) not have received Federal funding of more than $20,000 as PI or co-PI (except graduate fellowships and doctoral

dissertation improvement grants); (7) not have submitted concurrently the same project to another NSF program; (8) not be a named participant on any other proposal submitted to NSF, including regular research proposals, concurrent with the fellowship application, regardless of who is the named principal investigator; AND (9) meet other limitations on eligibility imposed within the specific competitive areas, if any. There is no limit on the number of applicants that an institution may host. For now, submission can only be made using NSF's FastLane system, given that Grants.gov does not support all required forms. Applications are due the first Monday in November by 5 p.m. proposer's local time.
Amount of Grant: $60,000 - $66,000 per year
Date(s) Application is Due: Nov
Contact: Carter Kimsey, Program Manager, Broadening Participation, (703) 292-8470, email: ckimsey@nsf.gov; Peter McCartney, Program Director, Informatics, (703) 292-8470, email: pmccartn@nsf.gov
Internet: http://www.nsf.gov/publications/pub_summ.jsp?ods_key=nsf07580&org=BIO
Sponsor: National Science Foundation
4201 Wilson Boulevard
Arlington, VA 22230

NSF Presidential Awards for Excellence in Mathematics and Science Teaching 4637
This program provides national recognition of outstanding elementary, middle, and secondary school teachers. Every year, two mathematics teachers and two science teachers are selected for the honor from each state and each of the four jurisdictions listed. The program seeks to enhance the status and visibility of the teaching profession by identifying and rewarding exemplary teachers. Awardees are brought to Washington, DC, for a week of activities and professional interaction. The continuation of this interaction is encouraged, and opportunities for individual and collective leadership are facilitated. The deadline for nomination within each state is usually mid-March. Twelve state/jurisdiction finalists are forwarded to a national section committee that selects the four presidential awardees. In addition to the award, a grant from NSF is included to be used at the teacher's discretion to improve science and/or mathematics education at the awardee's home institution. Annual deadline dates may vary; contact program staff for exact dates. NSF 04-578
Requirements: Eligible applicants are teachers of K-12 science and/or mathematics in a public or private school who plan to teach for the following year. Teachers must also be full-time employees and have at least five years of K-12 teaching experience in science and/or mathematics.
Amount of Grant: $10,000
Date(s) Application is Due: May 1
Contact: Mark Saul, Program Director; (703) 292-5092; fax: (703) 292-9044; email: msaul@nsf.gov
Internet: http://www.nsf.gov/funding/pgm_summ.jsp?pims_id=12802
Sponsor: National Science Foundation
4201 Wilson Boulevard
Arlington, VA 22230

NSF Presidential Awards for Excellence in Science, Math, and Engineering Mentoring 4638
The PAESMEM Program seeks to identify outstanding mentoring efforts that enhance the participation of groups (i.e., women, minorities, and persons with disabilities) that are underrepresented in science, technology, engineering, and mathematics. This program provides educational opportunities for Undergraduate Students, Graduate Students, Postdoctoral Fellows, K-12 Educators. The awardees serve as leaders in the national effort to develop fully the Nation's human resources in science, technology, engineering, and mathematics.
Requirements: Individuals and institutions must be nominated by a colleague, student, administrator, or another organization.
Amount of Grant: $10,000
Date(s) Application is Due: May 2
Contact: David L. Temple; (703) 292-5111; fax: (703) 292-9018; email: dtemple@nsf.gov
Internet: http://www.nsf.gov/funding/pgm_summ.jsp?pims_id=5473&org=EHR&from=fund
Sponsor: National Science Foundation
4201 Wilson Boulevard
Arlington, VA 22230

NSF Presidential Early Career Awards for Scientists and Engineers Grants 4639
Each year NSF selects nominees for the Presidential Early Career Awards for Scientists and Engineers (PECASE) from among the most meritorious new CAREER awardees. Selection for this award is based on two important criteria: innovative research at the frontiers of science and technology that is relevant to the mission of the sponsoring organization or agency, and community service demonstrated through scientific leadership, education or community outreach. These awards foster innovative developments in science and technology, increase awareness of careers in science and engineering, give recognition to the scientific missions of the participating agencies, enhance connections between fundamental research and national goals, and highlight the importance of science and technology for the Nation, future. Individuals cannot apply for PECASE. These awards are initiated by the participating federal agencies. At NSF, up to twenty nominees for this award are selected each year from among the PECASE-eligible CAREER awardees who are most likely to become the leaders of academic research and education in the twenty-first century. The White House Office of Science and Technology Policy makes the final selection and announcement of the awardees.
Requirements: Applicant must be employed at an institution in the United States, its territories or possessions, or Puerto Rico, that awards a baccalaureate or advanced degree in a field supported by NSF; be in their first or second full-time tenure-track academic appointment and have begun the first tenure-track or appointment (at any institution) on or after October 1, 1996, and before October 1, 2001; not be tenured or have held tenure on or before October 1, 2001; and not be a previous recipient of a CAREER or PECASE award. Contact program officer for exact requirements.
Restrictions: No salary support for other Senior Personnel is permitted, in either the primary budget or in any sub awards.
Amount of Grant: $400,000 - $500,000 for a 5-yr period
Date(s) Application is Due: Jul 21; Jul 22; Jul 23
Contact: Elizabeth L. Rom, Program Director; (703) 292-7709; fax: (703) 292-9085; email: elrom@nsf.gov; Donald Elthon; (703) 292-8580; fax: (703) 292-9085; email: delthon@nsf.gov
Internet: http://www.nsf.gov/funding/pgm_summ.jsp?pims_id=5262
Sponsor: National Science Foundation
4201 Wilson Boulevard
Arlington, VA 22230

NSF Probability Research Grants 4640
The Probability Program supports research on the theory and applications of probability. Subfields include discrete probability, stochastic processes, limit theory, interacting particle systems, stochastic differential and partial differential equations, and Markov processes. Research in probability which involves applications to other areas of science and engineering is especially encouraged.
Date(s) Application is Due: Nov 7
Contact: Tomek Bartoszynski; (703) 292-4885; fax: (703) 292-9032; email: tbartosz@nsf.gov
Internet: http://www.nsf.gov/funding/pgm_summ.jsp?pims_id=5555&org=DMS&from=home
Sponsor: National Science Foundation
4201 Wilson Boulevard
Arlington, VA 22230

NSF Process and Reaction Engineering Grants 4641
The Process and Reaction Engineering program supports fundamental and applied research on: rates and mechanisms of important classes of catalyzed and uncatalyzed chemical reactions as they relate to the design, production, and application of catalysts, chemical processes, biochemical processes, and specialized materials; chemical and biochemical phenomena occurring at or near solid surfaces and interfaces; electrochemical and photochemical processes of engineering significance or with commercial potential; design and optimization of complex chemical and biochemical processes; dynamic modeling and control of process systems and individual process units; reactive processing of polymers, ceramics, and thin films; and interactions between chemical reactions and transport processes in reactive systems, and the use of this information in the design of complex chemical and biochemical reactors.
Date(s) Application is Due: Mar 1; Sep 15
Contact: Maria Burka; (703) 292-7030; email: mburka@nsf.gov
Internet: http://www.nsf.gov/funding/pgm_summ.jsp?pims_id=13361
Sponsor: National Science Foundation
4201 Wilson Boulevard
Arlington, VA 22230

NSF Project 2010 Grants 4642
The Directorate for Biological Sciences (BIO) announces its intention to continue support of research to determine the functions of all genes in the model plant Arabidopsis thaliana by the year 2010. Individual investigators or groups of investigators will be supported to conduct creative and innovative, genome-wide or systems-level research designed to determine, using all available means, the functions of Arabidopsis genes. This year, as in the past, the Program will focus on: research on exemplary networks using high throughput methods and integrating modeling with experimental data to understand the gene circuitry underlying basic plant processes; projects that will develop experimental and computational methods, tools, and resources for enabling a broad community of scientists to conduct functional genomics research on Arabidopsis; and projects to perform genome-wide analyses of the gene function. Approximately 12 grants totaling $1,000,000 will be awarded. Full proposals should be submitted via FastLane or Grants.gov.
Requirements: Proposals are accepted only from U.S. academic institutions, U.S. non-profit research organizations including museums, research laboratories, professional societies and similar organizations in the U.S. that are directly associated with educational or research activities, and consortia of such organizations with appropriate research and educational facilities. The eligibility criteria apply to both the main and sub-awardees.
Amount of Grant: $800,000-$1,000,000 average
Date(s) Application is Due: Jan 21
Contact: Parag R. Chitnis, Acting Deputy Division Director; (703) 292-8443; fax: (703) 292-9061; email: bio-2010@nsf.gov; Diane Jofuku Okamuro, Program Director; (703) 292-8470; fax: (703) 292-9063; email: bio-2010@nsf.gov
Internet: http://www.nsf.gov/funding/pgm_summ.jsp?pims_id=5337
Sponsor: National Science Foundation
4201 Wilson Boulevard
Arlington, VA 22230

NSF Research Coordination Networks in Biological Sciences Grants 4643

The goal of this program is to encourage and foster communications and collaborations among scientists with common goals and interests. Groups of investigators are supported to communicate and coordinate their research efforts across disciplinary, organizational, institutional, and geographical boundaries. The proposed networking activities should have a theme as a focus of its collaboration. The focus could be on a broad research question, a specific group of organisms, or particular technologies or approaches. Innovative ideas for implementing novel networking strategies to promote research coordination and collaboration that enable new research directions or advancement of a field are especially encouraged. Although the research coordination networks are expected to be multi-institutional, a single institution must serve as the submitting institution.

Requirements: U.S. academic institutions and U.S. nonprofit research institutions are eligible.

Amount of Grant: $50,000-$100,000 average

Contact: Peter H. McCartney, Program Director; (703) 292-8470; email: pmccartn@nsf.gov

Internet: http://www.nsf.gov/funding/pgm_summ.jsp?pims_id=11691

Sponsor: National Science Foundation
4201 Wilson Boulevard
Arlington, VA 22230

NSF Research Experiences for Undergraduates (REU) Grants 4644

The Research Experiences for Undergraduates (REU) program supports active research participation by undergraduate students in any of the areas of research funded by the National Science Foundation. REU projects involve students in meaningful ways in ongoing research programs or in research projects specifically designed for the REU program. This solicitation features two mechanisms for support of student research: (1) REU Sites are based on independent proposals to initiate and conduct projects that engage a number of students in research. REU Sites may be based in a single discipline or academic department, or on interdisciplinary or multi-department research opportunities with a coherent intellectual theme. Proposals with an international dimension are welcome. A partnership with the Department of Defense supports REU Sites in DoD-relevant research areas. (2) REU Supplements may be requested for ongoing NSF-funded research projects or may be included as a component of proposals for new or renewal NSF grants or cooperative agreements. NSF is particularly interested in increasing the numbers of women, underrepresented minorities, and persons with disabilities in research. REU projects are strongly encouraged to involve students who are members of these groups. (Underrepresented minorities are African Americans, Hispanics, American Indians, Alaska Natives, and Native Hawaiians or Other Pacific Islanders.) Historically, the vast majority of REU participants have been junior- or senior-level undergraduates—students who have typically already committed to a major in science or engineering. So that the REU program can succeed in attracting students into science and engineering who might not otherwise consider these majors and careers, projects are also encouraged, when appropriate, to involve students at earlier stages in their college experience. NSF anticipates investing approximately $57 million (pending availability of funds) in approximately 150 new Site awards and 1,600 new Supplement awards. REU Site awards have durations of one to five years, with a three-year duration being typical in most NSF directorates. The term of REU Supplements may not exceed that of the underlying research project. NSF 05-592

Requirements: NSF welcomes proposals on behalf of all qualified scientists, engineers and educators. The Foundation strongly encourages women, minorities and persons with disabilities to participate fully in its programs. Undergraduate student participants in either Sites or Supplements must be citizens or permanent residents of the United States or its possessions. An undergraduate student is a student who is enrolled in a degree program (part-time or full-time) leading to a baccalaureate or associate degree. Students who are transferring from one college or university to another and are enrolled at neither institution during the intervening summer may participate. High school graduates who have been accepted at an undergraduate institution but who have not yet started their undergraduate study are also eligible to participate. For REU Sites, a significant fraction of the student participants should come from outside the host institution or organization. Some NSF directorates encourage inclusion in the REU program of K-12 teachers of science, technology, engineering, and mathematics.

Restrictions: Students who have received their bachelor's degrees and are no longer enrolled as undergraduates are generally not eligible to participate.

Samples: University of California - Merced (CA) $234,206 - REU Site: Summer Yosemite Research Training in Environmental Science; Ohio State University Research Foundation (OH) $80,000 - REU Site: Interdisciplinary Research at the Chemistry/Biology Interface for Undergraduates; Texas Engineering Experiment Station (TX) $76,860 - REU Site: Ecology and Environment of the Southern High Plains; George Washington University (DC) $224,010 - REU Site: Molecular Biology, Bioinformatics and Biomaterials

Date(s) Application is Due: Sep 13; Jun 5; Aug 18

Contact: Elizabeth L. Rom, Program Director; (703) 292-7709; fax: (703) 292-9085; email: elrom@nsf.gov; Donald Elthon; (703) 292-8580; fax: (703) 292-9085; email: delthon@nsf.gov

Internet: http://www.nsf.gov/funding/pgm_summ.jsp?pims_id=5517

Sponsor: National Science Foundation
4201 Wilson Boulevard
Arlington, VA 22230

NSF Research on Gender in Science and Engineering (GSE) Grants 4645

The program seeks to broaden the participation of girls and women in all fields of science, technology, engineering, and mathematics (STEM) education by supporting research, the diffusion of research-based innovations, and extension services in education that will lead to a larger and more diverse domestic science and engineering workforce. Typical projects will contribute to the knowledge base addressing gender-related differences in learning and in the educational experiences that affect student interest, performance, and choice of careers; how pedagogical approaches and teaching styles, curriculum, student services, and institutional culture contribute to causing or closing gender gaps that persist in certain fields. Projects will communicate and apply findings, evaluation results, and proven good practices and products to a wider community. It is estimated that 15 to 22 grants will be awarded per year; a mix of Research Awards, Diffusion of Research-based Innovations Awards, and Extension Services Awards. Research grants will be up to 3 years. Diffusion of Research-based Innovations grants will be up to 3 years. Extension Services grants are for five years, with years 4 and 5 depending on performance. The total anticipated funding amount is $5,000,000 for new grants in all tracks, pending availability of funds. NSF 07-578

Requirements: The categories of proposers eligible to submit proposals to the National Science Foundation are identified in the Grant Proposal Guide, Chapter I, Section E. A Preliminary Proposal is required prior to submitting a Full Proposal. (Deadlines for Preliminary Proposals: Research Proposals - Nov. 5, Extension Services Proposals - Jan. 7.) The program seeks to build resources - developing the nation's knowledge capital, social capital, and human capital — toward the goal of broadening the participation of girls and young women in STEM education from kindergarten through undergraduate education. The following types of projects will be considered: (1) Research projects: investigate factors behind the underrepresentation of girls and women in STEM education; societal, formal and informal educational systems' interaction with individuals that encourage or discourage interest and persistence in study or careers in certain STEM fields. (2) Diffusion of Research-Based Innovation projects: provide a mechanism for informing a wider audience (e.g., teachers, faculty, guidance counselors, parents, etc.) about issues, research findings, and strategies for changing educational practice. (3) Extension Services: provide training and consulting services to educators and institutions, to enable them to adopt and embed proven gender-inclusive policies and practices in pedagogy, the design of curriculum materials, student support programs, educator and faculty development. Extension services employ a 'train-the-trainer' model and are based on a 'unified program of change.' The program does not currently fund intervention or education projects that directly serve students as their primary purpose. Research projects may involve students as subjects only if the intervention is an integral part of gathering data and if the findings from the intervention would substantially answer the research questions posed within the context of theory and hypotheses. There should be meaningful control groups also included in the design. Those wishing to undertake direct intervention or education service projects are encouraged to search the NSF web site and other publications for appropriate funding programs. All proposals submitted to GSE are already intended to broaden participation in STEM. Proposals that, in addition, include a partnership with institutions serving underrepresented populations may request up to $100,000 more (for a total of $350,000) for Diffusion of Research-Based Innovation activities. The partnership must be documented with, at a minimum, strong letters of support from the partnering institutions.

Samples: The Research on Gender in Science and Engineering program has been funding these objectives since 1993, under the prior names 'Program for Women and Girls' (PWG), 'Program for Gender Equity in Science, Mathematics, Engineering and Technology' (PGE), and 'Gender Diversity in STEM Education' (GDSE).

Date(s) Application is Due: Nov 5; Jan 7; Feb 25; Apr 7

Contact: Jolene Jesse, Program Director, 815 N, phone: (703) 292-7303, fax: (703) 292-9018, email: jjesse@nsf.gov; Tayana Casseus, Science Assistant, NSF, 815N, telephone: (703) 292-4684, email: tcasseus@nsf.gov

Internet: http://www.nsf.gov/funding/pgm_summ.jsp?pims_id=5475

Sponsor: National Science Foundation
4201 Wilson Boulevard
Arlington, VA 22230

NSF Research to Aid Persons with Disabilities (RAPD) Grants 4646

The Research to Aid Persons with Disabilities (RAPD) program supports research that will lead to the development of new technologies, devices, or software for persons with disabilities. Research may be supported that is directed to the characterization, restoration, and/or substitution of human functional ability or cognition, or to the interaction of persons with disabilities and their environment. Areas of particular recent interest are disability-related research in neuroscience/neuroengineering and rehabilitation robotics. Emphasis is placed on significant advancement of fundamental engineering and scientific knowledge and not on incremental improvements. Proposals should advance discovery or innovation beyond the frontiers of current knowledge in disability-related research. Applicants are encouraged to contact a program director prior to submitting a proposal.

Date(s) Application is Due: Mar 1; Sep 15

Contact: Ted A. Conway; (703) 292-7091; email: tconway@nsf.gov

Internet: http://www.nsf.gov/funding/pgm_summ.jsp?pims_id=501021

Sponsor: National Science Foundation
4201 Wilson Boulevard
Arlington, VA 22230

NSF SBE Doctoral Dissertation Research Improvement Grants 4647

This program award grants to doctoral students to improve the quality of dissertation research. These grants provide funds for items not normally available through the

student's university. Additionally, these grants allow doctoral students to undertake significant data-gathering projects and to conduct field research in settings away from their campus that would not otherwise be possible. Proposals are judged on the basis of their scientific merit, including the theoretical importance of the research question and the appropriateness of the proposed data and methodology to be used in addressing the question. NSF 06-605
Requirements: The student applying must be enrolled at a U.S. academic institution, but need not be a U.S. citizen. The proposal must be submitted by the dissertation advisor(s) on behalf of the graduate student who is at the point of initiating or already conducting dissertation research. Proposals from women, minorities, and persons with disabilities are strongly encouraged. Items such as budget limitations, target dates and/or deadlines, page length restrictions, and review procedures vary widely across programs. Please consult the relevant program's webpage for specific information and contact the program director if necessary.
Restrictions: This program does not support research with disease-related goals, including research on the etiology, diagnosis, or treatment of physical or mental disease, abnormality, or malfunction of human beings, animals or plants. Funds may not be used as a stipend for the student, for tuition, textbooks, journals, or for the typing, reproduction, or publication costs of the student's dissertation.
Samples: James Biles (Indiana University, IN) - Doctoral Dissertation Research: Intercountry Adoption in the Philippines and the United States: Global Networks and Local Processes; Christine Padoch (CUNY Graduate School University Center, NY) - Doctoral Dissertation Research: Shihuahuaco (Dipteryx spp.) and the Chinese Timber Trade; Kathleen Morrison (University of Chicago, IL) - Doctoral Dissertation Improvement Grant: Socio-Political Differentiation and Land Use in Early South India; Karen Barkey (Columbia University, NY) - Doctoral Dissertation Research: Free Agents: The Civilian Patrol of America's Borders.
Amount of Grant: $2.5 million total; 200 to 300 estimated number of awards.
Date(s) Application is Due: Due dates vary across Divisions.
Contact: See http://www.nsf.gov/sbe/ddrig_contacts.jsp for the appropriate contact person in your field of interest.
Internet: http://nsf.gov/funding/pgm_summ.jsp?pims_id=13453&org=NSF&sel_org=NSF&from=fund
Sponsor: National Science Foundation
4201 Wilson Boulevard
Arlington, VA 22230

NSF Scholarships in Science, Technology, Engineering, and Mathematics **4648**
This program supports scholarships for academically talented but financially needy students, enabling them to pursue associate, baccalaureate, or graduate-level degrees in science, technology, engineering, or mathematics. Academic institutions apply for awards to support scholarship activities, and are responsible for selecting scholarship recipients, reporting demographic information about student scholars, and managing the S-STEM project at the institution. An optional letter of intent is requested by January 15, October 10 or July 10, depending on the competition period; full proposal is due January 28. Annual deadline dates may vary; contact program staff for exact dates.
Requirements: Proposals may only be submitted by Institutions of higher education (as defined in section 101 (a) of the Higher Education Act of 1965) in the United States and its territories that grant associate, baccalaureate, or graduate degrees in the following disciplines: biological sciences (except medicine and other clinical fields); physical sciences, including physics, chemistry, astronomy, and materials science; mathematical sciences; computer and information sciences; geosciences; engineering; technology areas associated with the preceding fields (for example, biotechnology, chemical technology, engineering technology, information technology, etc.). The Principal Investigator must be a faculty member currently teaching within one of the S-STEM disciplines who can provide the leadership required to ensure the success of the project. Projects involving more than one department within an institution are eligible, but a single Principal Investigator must accept overall management responsibility. Other members of the S-STEM project management team may be listed as Co-Principal investigators.
Restrictions: An institution may submit no more than one proposal per competition.
Amount of Grant: $600,000 maximum
Date(s) Application is Due: Jan 15; Oct 10; Jul 10
Contact: Duncan McBride, Section Head, (703) 292-4630, email: dmcbride@nsf.gov; Bevlee Waterford, (703) 292-5323, email: bwatford@nsf.gov; Pratibha Varma-Nelson, (703) 292-4653, email: pvarmane@nsf.gov
Internet: http://www.nsf.gov/funding/pgm_summ.jsp?pims_id=5257
Sponsor: National Science Foundation
4201 Wilson Boulevard
Arlington, VA 22230

NSF Science and Society (S&S) **4649**
The program seeks proposals that examine questions that arise in the interactions of engineering, science, technology, and society. There are four components: Ethics and Values in Science, Engineering and Technology (EVS); History and Philosophy of Science, Engineering and Technology (HPS); Social Studies of Science, Engineering and Technology (SSS); and Studies of Policy, Science, Engineering and Technology (SPS). The components overlap, but are distinguished by the different scientific and scholarly orientations they take to the subject matter, as well as by different focuses within the subject area. This program offers the following modes of support: Scholars Awards; Standard Research Grants and Grants for Collaborative Research; Postdoctoral Fellowships; Professional Development Fellowships; Doctoral Dissertation Research Improvement Grants; Small Grants for Training and Research; Conference and Workshop Awards; Other Funding Opportunities.
Requirements: 1) Scholar Awards: Normally made to U.S. academic institutions, although an individual who is not affiliated with an appropriate U.S. academic institution may submit a proposal as an independent scholar. In that case, the scholar must be a U.S. citizen or national, or have permanent resident status. Awards may provide support for full-time academic year (nine months) research, including salary, fringe benefits, and other direct costs, up to a ceiling that is ordinarily $70,000 for total direct costs. 2) Standard Research Grants and Grants for Collaborative Research: These awards are made to U.S. academic institutions. Awards can include proposals for research, infrastructure or education projects. These proposals ordinarily do not require full-time investigator support like that for Scholars' Awards. These grants can also support projects that require several investigators, advisors, or collaboration among Principal Investigators, including investigators at different institutions. They may also involve postdoctoral researchers, or graduate or undergraduate student assistants. 3) Postdoctoral Fellowships: Available for S&S researchers within 5 years of receipt of the Ph.D. degree. Fellows must be U.S. citizens or nationals, or have permanent resident status. Proposals must describe both a training and a research component, and the site for the Fellowship must be different from the institution where the Fellow received the Ph.D. degree. 4) Professional Development Fellowships: All Fellows must be U.S. citizens or nationals, or have permanent resident status. These fellowships are available for researchers trained in all areas of Science and Society who wish to improve and expand their skills in the areas of science or engineering, and conversely for physical and natural scientists and engineers who desire training in S&S disciplines. 5) Doctoral Dissertation Research Improvement Grants: Doctoral students who are enrolled in U.S. graduate programs are eligible to apply. The dissertation advisor is the principal investigator. Doctoral students must have passed the qualifying exams, completed all course work required for the degree, and had the dissertation topic approved prior to receiving the award. These awards provide funds for dissertation research expenses not normally available through the student's university. The dissertation advisor is the principal investigator on these proposals; the doctoral student should be listed as co-principal investigator. 6) Small Grants for Training and Research (SGTR): Available for postdoctoral researchers within 5 years of receipt of the Ph.D. and for graduate students who are regularly admitted students in S&S graduate programs. All Postdoctoral Fellows must be U.S. citizens or nationals, or have permanent resident status. 7) Conference and Workshop Support: Can help to support national and international conferences, symposia, and research workshops that enable scientists, engineers, researchers in S&S areas of support, policy makers, and representatives of interested groups to develop, evaluate, and share new research findings. Also supports projects on the interactions of engineering, science, technology and society that emphasize capacity building. Such activities can include national summer workshops for graduate students or faculty, or projects by professional societies to develop concentrations in the ethical, philosophical, historical and social context of science and engineering for undergraduate or graduate level science and engineering students. 8) Other Funding: may provide supplemental funding to existing awards in order to create research experiences for undergraduates (REU). See the REU Announcement in the listings of NSF funding opportunities.
Amount of Grant: $6,000,000 total funds; estimated 40 awards
Date(s) Application is Due: Feb 1; Aug 1
Contact: Frederick Kronz, (703) 292-7283, fkronz@nsf.gov; Stephen Zehr, (703) 292-7318, szehr@nsf.gov; Hilary Harding, (703) 292-4927, hharding@nsf.gov
Internet: http://nsf.gov/funding/pgm_summ.jsp?pims_id=5324&org=NSF&sel_org=NSF&from=fund
Sponsor: National Science Foundation
4201 Wilson Boulevard
Arlington, VA 22230

NSF Science and Technology Studies (STS) Grants **4650**
This program supports research and related activities that contribute to systematic understanding of the character and development of science and technology, including their cultural, intellectual, material,and social dimensions. The program supports research on the nature and development of technology and science, both in the past and present; on differences in the nature of theory and evidence in various fields of science and engineering; and on the interactions among science, technology and society. Proposals are welcomed from various disciplinary perspectives, including history of science, history of technology, philosophy of science, and various social sciences, including sociology, anthropology, and political science. Fastlane submission is required. NSF-01-159
Requirements: Scientists, engineers and educators usually initiate proposals that are officially submitted by their employing organization. Before formal submission, the proposal may be discussed with appropriate NSF program staff. Graduate students are not encouraged to submit research proposals, but should arrange to serve as research assistants to faculty members. Some NSF divisions accept proposals for Doctoral Dissertation Research Grants when submitted by a faculty member on behalf of the graduate student. The Foundation also provides support specifically for women and minority scientists and engineers, scientists and engineers with disabilities, and faculty at primarily undergraduate academic institutions.
Amount of Grant: $57,000 average
Date(s) Application is Due: Feb 1; Aug 1
Contact: Bruce E. Seely, STS, Program Director; (703) 292-7283; email: bseely@nsf.gov; John Perhonis, STS-Dissertation Proposals, Program Director; (703) 292-7279; email: jperhoni@nsf.gov
Internet: http://www.nsf.gov/pubs/2001/nsf01159/nsf01159.html
Sponsor: National Science Foundation
4201 Wilson Boulevard
Arlington, VA 22230

NSF Scientific Computing Research Environments for the Mathematical Sciences (SCREMS) Grants 4651

The Division of Mathematical Sciences (DMS) of the National Science Foundation plans a limited number of awards for the support of computing environments for research in the mathematical sciences. Scientific Computing Research Environments for the Mathematical Sciences (SCREMS) proposals are for computing environments dedicated to research in the mathematical sciences. Proposals may request support for the purchase of computing equipment and limited support for professional systems administrators or programmer personnel for research computing needs. These grants are intended to support research projects of high quality that require access to advanced computing resources. Requests for routine upgrades of standard desk-environment workstations or laptop computers are not appropriate for this program. Awards are made to provide support for specific research projects rather than to provide general computing capacity. Proposers are encouraged to include projects involving symbolic and algebraic computations, numerical computations and simulations, and graphical representations (visualization) in aid of the research. The total number of awards is estimated between 8 and 12, and the total funding amount is approximately $1,000,000. Estimated program budget, number of awards and average award size/duration are subject to the availability of funds. NSF 07-502

Requirements: Proposals may be submitted by U.S. educational institutions with ongoing research programs in mathematics, applied mathematics, or statistics. Proposals involving inter-institutional or inter-departmental sharing arrangements are welcome. SCREMS proposals are primarily for funding for equipment. If appropriate justification is made, a limited amount of funding may be requested for certain personnel costs. When appropriate and cost-effective, requests for group or departmental servers may be suitable for the SCREMS program. The net costs (after discounts) of the equipment portion should be at least $50,000. Requests to SCREMS may be as high as $200,000, provided a case is made for substantial impact and cost-effectiveness.

Restrictions: Identical proposals may not be submitted to both the SCREMS and MRI (Major Research Instrumentation) solicitations.

Date(s) Application is Due: Jan 24

Contact: Dean Evasius, Program Director, telephone: (703) 292-8132, email: devasius@nsf.gov; MPS FastLane *Contact*: Florence Rabanal , Electronic Business Coordinator, Directorate for Mathematical & Physical Sciences, 1005 N, (703) 292-8808

Internet: http://www.nsf.gov/funding/pgm_summ.jsp?pims_id=5616

Sponsor: National Science Foundation
4201 Wilson Boulevard
Arlington, VA 22230

NSF Sedimentary Geology and Paleobiology Grants 4652

Sedimentary Geology and Paleobiology supports studies of: (1) the changing aspects of life, ecology, environments, and biogeography in past geologic time based on fossil plants, animals, and microbes; (2) all aspects of the Earth, sedimentary carapace ,Äî insights into geological processes recorded in its historical records and rich organic and inorganic resources locked in rock sequences; (3) the science of dating and measuring the time sequence of events and rates of geological processes of the Earth, past sedimentary and biological record; (4) the geologic record of the production, transportation, and deposition of physical and chemical sediments; and (5) understanding the complexities of Earth's deep time climate systems. The Sedimentary Geology and Paleobiology Program especially encourages integrative studies at the national and international levels that seek to link subdisciplines, such as paleoclimatology, paleogeography, and paleoenvironmental and paleoecologic reconstructions. The program is committed to supporting the most meritorious research in any relevant area, including interdisciplinary and multidisciplinary research, as well as research involving international collaboration. The program is especially interested in proposals in emerging fields.

Requirements: Individuals and organizations in the following categories may submit proposals: (1) Universities and Colleges - Universities and two- and four-year colleges (including community colleges) located and accredited in the U.S., acting on behalf of their faculty members. Such organizations also are referred to as academic institutions. Institutions located outside the U.S. fall under (6) below. (2) Non-profit, non-academic organizations - Independent museums, observatories, research laboratories, professional societies and similar organizations in the U.S. that are directly associated with educational or research activities. (3) For-profit organizations - U.S. commercial organizations, especially small businesses with strong capabilities in scientific or engineering research or education. An unsolicited proposal from a commercial organization may be funded when the project is of special concern from a national point of view, special resources are available for the work, or the proposed project is especially meritorious. NSF is interested in supporting projects that couple industrial research resources and perspectives with those of universities; therefore, it especially welcomes proposals for cooperative projects involving both universities and the private commercial sector. (4) State and Local Governments - State educational offices or organizations and local school districts may submit proposals intended to broaden the impact, accelerate the pace, and increase the effectiveness of improvements in science, mathematics and engineering education in both K-12 and post-secondary levels. (5) Unaffiliated Individuals - Scientists, engineers or educators in the U.S. and U.S. citizens may be eligible for support, provided that the individual is not employed by, or affiliated with, an organization, and (a) the proposer has demonstrated the capability and has access to any necessary facilities to carry out the project; and (b) the proposer agrees to fiscal arrangements that, in the opinion of the NSF Division of Grants & Agreements, ensure responsible management of Federal funds. Unaffiliated individuals should contact the appropriate program before preparing a proposal for submission. (6) Foreign organizations - NSF rarely provides support to foreign organizations. NSF will consider proposals for cooperative projects involving U.S. and foreign organizations, provided support is requested only for the U.S. portion of the collaborative effort.

Restrictions: Proposals that have been declined are not eligible for resubmission for one year from the original date of submission and must be substantially revised to be considered. A proposal that has not been substantially revised will be returned without review as per the NSF Grant Proposal Guide.

Date(s) Application is Due: Jan 16; Jul 16

Contact: Paul E. Filmer, Program Director, (703) 292-7858, email: pfilmer@nsf.gov; H. Richard Lane, Program Director, (703) 292-8551, email: hlane@nsf.gov

Internet: http://www.nsf.gov/funding/pgm_summ.jsp?pims_id=13691

Sponsor: National Science Foundation
4201 Wilson Boulevard
Arlington, VA 22230

NSF Small Business Innovation Research (SBIR) Grants 4653

The Small Business Innovation Research (SBIR) Program stimulates technological innovation in the private sector by strengthening the role of small business concerns in meeting Federal research and development needs, increasing the commercial application of federally supported research results, and fostering and encouraging participation by socially and economically disadvantaged and women-owned small businesses.

Requirements: Eligible to apply are small business firms—those organized for profit, individually owned or operated, not dominant in the field in which they are bidding, and with an average of no more than 500 employees in all affiliated firms. In addition, the primary employment of the principal investigator must be with the small business firm at the time of the award.

Samples: Inrad Inc (Northvale, NJ)—Growth of potassium iodate single crystals for nonlinear applications (Phase I SBIR), $65,000.; Metcal Inc (Menlo Park, CA)—Computer design of a new class of magnetic alloys (Phase I SBIR), $65,000.

Amount of Grant: $5000-$4 million; $119,000 average

Contact: Gregory T. Baxter, Program Director; (703) 292-7795; fax: (703) 292-9057; email: gbaxter@nsf.gov

Internet: http://www.nsf.gov/funding/pgm_summ.jsp?pims_id=503361

Sponsor: National Science Foundation
4201 Wilson Boulevard
Arlington, VA 22230

NSF Small Business Technology Transfer Programs Phase I Grants 4654

The small business programs stimulate technological innovation in the private sector by strengthening the role of small business concerns in meeting Federal research and development needs, increasing the commercial application of federally supported research results, and fostering and encouraging participation by socially and economically disadvantaged and women-owned small businesses. The Small Business Technology Transfer Program (STTR) requires researchers at universities and other research institutions to play a significant intellectual role in the conduct of each STTR project. These university-based researchers, by joining forces with a small company, can spin-off their commercially promising ideas while they remain primarily employed at the research institution. Letters of Intent are due January 14, with the full proposal deadline February 25.

Requirements: Primary employment must be with the small business concern at the time of an award.

Amount of Grant: $100,000 maximum

Date(s) Application is Due: Jan 14; Feb 25

Contact: Cheryl F. Albus, Program Director; (703) 292-7051; fax: (703) 292-9057; email: calbus@nsf.gov

Internet: http://www.nsf.gov/funding/pgm_summ.jsp?pims_id=5527

Sponsor: National Science Foundation
4201 Wilson Boulevard
Arlington, VA 22230

NSF Social Psychology Research Grants 4655

Grants support laboratory and field research in all areas of human social behavior, including attitude formation and change, social cognition, personality processes, interpersonal relations and group processes, the self, emotion, social comparison and social influence, the social psychology of health, and the psychophysiological correlates of social behavior. This program also supports research on social development over the human life span. Proposals are accepted at any time; target dates indicate the approximate time of review panel meetings. PD 98-1332

Date(s) Application is Due: Jan 15; Jul 15

Contact: Amber Story, Program Director, (703) 292-7249; email: astory@nsf.gov

Internet: http://www.nsf.gov/funding/pgm_summ.jsp?pims_id=5712

Sponsor: National Science Foundation
4201 Wilson Boulevard
Arlington, VA 22230

NSF Solar Terrestrial Research Grants 4656

This program supports research into processes by which energy in diverse forms is generated by the sun, transported to earth, and ultimately deposited in the terrestrial environment. Major topics include space weather impacts, helioseismology, the solar dynamo, the solar activity cycle, magnetic flux emergence, solar flares and eruptive activity, coronal mass ejections, solar wind heating, solar energetic particles, interactions with cosmic rays, and solar wind/magnetosphere boundary problems. Studies on terrestrial influences include solar spectral irradiance changes, solar constant changes and climactic impacts, C14 and sun/climate connections, solar activity and its effects on

the terrestrial environment, and sun/weather variations on various time scales. Proposals may be submitted at any time during the year for all programs except those involving the allocation of observational and computing facilities.
Contact: Paul Bellaire, Program Director; (703) 292-8519; fax: (703) 292-9023; email: pbellair@nsf.gov
Internet: http://www.nsf.gov/funding/pgm_summ.jsp?pims_id=12741&org=NSF&sel_org=NSF&from=fund
Sponsor: National Science Foundation
4201 Wilson Boulevard
Arlington, VA 22230

NSF Solar, Heliospheric, and INterplanetary Environment (SHINE) Grants 4657
The solar physics program in the Upper Atmospheric Research Section of the Division of Atmospheric Sciences, National Science Foundation (NSF), solicits proposals to investigate the connections between eruptive events and magnetic phenomena on the Sun and the corresponding solar wind structures in the inner heliosphere. SHINE research focuses upon the connections between eruptive events and magnetic phenomena on the Sun and the corresponding solar wind structures in the inner heliosphere. SHINE fosters research on those processes by which magnetic fields and particles produced by the Sun permeate interplanetary space and on the mechanisms by which these fields and particles are transported to geospace through the inner heliosphere. Proposals may be submitted for any funding amount up to $200,000 per year. A maximum duration of five years will be considered in those exceptional cases where the justification and promise are compelling. Typically, SHINE awards are for three years, but proposers may request from one to five years of funding, provided the requested duration is justified. NSF expects to fund approximately 6 to 8 standard or continuing grants annually, depending on the quality of submissions and the availability of funds. NSF 04-585 (replaces Document NSF 02-189)
Requirements: Individuals and organizations in the following categories may submit proposals: (1) Universities and Colleges - Universities and two- and four-year colleges (including community colleges) located and accredited in the U.S., acting on behalf of their faculty members. Such organizations also are referred to as academic institutions; (2) Non-profit, non-academic organizations - Independent museums, observatories, research laboratories, professional societies and similar organizations in the U.S. that are directly associated with educational or research activities; (3) For-profit organizations - U.S. commercial organizations, especially small businesses with strong capabilities in scientific or engineering research or education. An unsolicited proposal from a commercial organization may be funded when the project is of special concern from a national point of view, special resources are available for the work, or the proposed project is especially meritorious. NSF is interested in supporting projects that couple industrial research resources and perspectives with those of universities; therefore, it especially welcomes proposals for cooperative projects involving both universities and the private commercial sector; (4) State and Local Governments - State educational offices or organizations and local school districts may submit proposals intended to broaden the impact, accelerate the pace, and increase the effectiveness of improvements in science, mathematics and engineering education in both K-12 and post-secondary levels; (5) Unaffiliated Individuals - Scientists, engineers or educators in the U.S. and U.S. citizens may be eligible for support, provided that the individual is not employed by, or affiliated with, an organization, and: (a) the proposer has demonstrated the capability and has access to any necessary facilities to carry out the project; and, (b) the proposer agrees to fiscal arrangements that, in the opinion of the NSF Division of Grants & Agreements, ensure responsible management of Federal funds. Unaffiliated individuals should contact the appropriate program before preparing a proposal for submission.
Restrictions: NSF rarely provides support to foreign organizations. NSF will consider proposals for cooperative projects involving U.S. and foreign organizations, provided support is requested only for the U.S. portion of the collaborative effort. Also, NSF does not normally support research or education activities by scientists, engineers or educators employed by Federal agencies or Federally Funded Research and Development Centers (FFRDCs). Under unusual circumstances, other Federal agencies and FFRDCs may submit proposals directly to NSF.
Amount of Grant: up to $200,000 per year
Date(s) Application is Due: Aug 27
Contact: Paul Bellaire, Program Director; (703) 292-8529; fax: (703) 292-9023; email: pbellair@nsf.gov
Internet: http://www.nsf.gov/publications/pub_summ.jsp?ods_key=nsf04585
Sponsor: National Science Foundation
4201 Wilson Boulevard
Arlington, VA 22230

NSF Starter Grants for Postdoctoral Fellows in Microbial Biology 4658
The purpose of this program is to support training and research in a host institution on the basic biology of protozoan, microalgal, fungal, archaeal, bacterial, and viral species that are not generally considered to be model organisms (e.g., E. coli, Saccharomyces cerevesiae, and TMV). The use of model organisms in comparative studies with nonmodel organisms is not excluded. Deadline is the first Monday in October. NSF 99-142
Requirements: Applicants must be: U.S. citizens, nationals, or legally admitted permanent resident aliens; may not have earned the doctoral degree more than 18 months prior to the deadline; must propose a research and training plan in microbial biology; may not have previously received a federal research grant; may not submit a research and training plan duplicated in another NSF proposal; and may not have been involved in the research before June 1 of the deadline year. Fastlane submission is required. NSF 99-142
Samples: Maria Elana Pereyra, University of Tulsa $50,000 (OK) - Neuroendocrine modulation of avian breeding schedules; Gerald Downes, University of Massachusetts Amherst (MA) $50,000 - Analysis of Glycinergic Neurotransmission in the Early Zebrafish Spinal Cord; Ignacio Moore, Virginia Polytechnic Institute and State University (VA) $50,000 - Role of Sex Steroids in Reproductive Timing in Rufous-Collared Sparrows; Kathleen Treseder, University of California-Irvine (CA) $6,849 - Minority Postdoctoral Research Fellowships and Supporting Activities
Amount of Grant: $50,000 per year for two or three years
Date(s) Application is Due: Oct 1
Contact: Ms. Carter Kimsey, BIO Postdoctoral Fellowship Program Officer; Room 615, Division of Biological Infrastructure; phone: (703) 292-8470; email: ckimsey@nsf.gov
Internet: http://www.nsf.gov/funding/pgm_summ.jsp?pims_id=13441
Sponsor: National Science Foundation
4201 Wilson Boulevard
Arlington, VA 22230

NSF Statistics Research Grants 4659
The Statistics Program supports research in statistical theory and methods, including research in statistical methods for applications to any domain of science and engineering. The theory forms the base for statistical science. The methods are used for stochastic modeling, and the collection, analysis and interpretation of data. The methods characterize uncertainty in the data and facilitate advancement in science and engineering. The Program encourages proposals ranging from single-investigator projects to interdisciplinary team projects.
Date(s) Application is Due: Nov 7
Contact: Gabor Szekely, Program Director; (703) 292-8869; fax: (703) 292-9032; email: wrundell@nsf.gov
Internet: http://www.nsf.gov/funding/pgm_summ.jsp?pims_id=5556&org=DMS&from=home
Sponsor: National Science Foundation
4201 Wilson Boulevard
Arlington, VA 22230

NSF Stellar Astronomy and Astrophysics (SAA) Research Grants 4660
This program supports theoretical and observational studies of the structure and activity of the sun; the physical properties of all types of stars; all aspects of stellar formation and evolution; the effects of mass loss, rotation, and magnetic fields; and the properties of stellar atoms and molecules of relevance to stellar astronomy.
Requirements: Scientists initiate research proposals, which are usually submitted on their behalf by their employing institutions.
Date(s) Application is Due: Nov 15
Contact: Nigel A. Sharp, Program Director; (703) 292-4905; fax: (703) 292-9034; email: nsharp@nsf.gov
Internet: http://www.nsf.gov/funding/pgm_summ.jsp?pims_id=13630
Sponsor: National Science Foundation
4201 Wilson Boulevard
Arlington, VA 22230

NSF Synthesis and Modeling Project of the U.S. Joint Global Ocean Flux Study 4661
The purpose of this announcement is to invite innovative research proposals to augment the ongoing Synthesis and Modeling Project (SMP) of the U.S. Joint Global Ocean Flux Study (U.S.JGOFS). As part of a multinational oceanographic effort, U.S.JGOFS has as its primary objectives the determination and understanding of processes controlling the biogeochemical cycling of carbon in the ocean and the prediction of response of marine biogeochemical processes to climate change. Now in its fourth year, the SMP is focused on the resolution of major unanswered questions during the course of U.S.JGOFS and related activities and on the modeling of the global ocean carbon cycle. NSF 01-103
Requirements: Scientists, engineers, and educators usually initiate proposals that are officially submitted by their employing organization.
Restrictions: Graduate students are not encouraged to submit research proposals, but should arrange to serve as research assistants to faculty members.
Amount of Grant: $300,000-$430,000 average
Date(s) Application is Due: Sep 13
Contact: Donald L. Rice, Chemical Oceanography Program; (703) 306-1589; email: drice@nsf.gov; Phillip Taylor, Biological Oceanography Program; (703) 306-1587; email: prtaylor@nsf.gov
Internet: http://www.nsf.gov/pubs/2000/nsf00103/nsf00103.htm
Sponsor: National Science Foundation
4201 Wilson Boulevard
Arlington, VA 22230

NSF Systematic Biology and Biodiversity Inventories Grants 4662
The Systematic Biology and Biodiversity Inventories Cluster supports research in taxonomy and systematics that contributes to: 1) using phylogenetic methods to understand the evolution of life in time and space, 2) discovery, description, and cataloguing global species diversity, and 3) organizing information from the above in efficiently retrievable forms that best meet the needs of science and society. The Systematic Biology and Biodiversity Inventories Cluster funds projects within the two Programs, Systematic Biology and Biodiversity Inventories, in addition to the PEET and PBI solicitations listed below. In addition, the cluster participates in AToL and other related funding opportunities. The SBBI Cluster continues to encourage and support studies that seek to synthesize available and new species-level taxonomic information in the context of providing revisionary treatments and predictive classifications for particular groups of organisms. NSF PD 047374.

Requirements: The most frequent recipients are academic institutions and nonprofit research groups. In special circumstances, grants also are awarded to other types of institutions and to individuals. Standard NSF Grant Proposal Guidelines apply (see http://www.nsf.gov/publications/pub_summ.jsp?ods_key=gpg).
Date(s) Application is Due: Jan 9; Jul 9
Contact: Maureen Kearney, Program Director; (703) 292-7187; email: mkearney@nsf.gov
Internet: http://www.nsf.gov/publications/pub_summ.jsp?ods_key=pd047374&org=NSF
Sponsor: National Science Foundation
4201 Wilson Boulevard
Arlington, VA 22230

NSF Tectonics Research Grants **4663**
The Tectonics Program supports a broad range of field, laboratory, computational, and theoretical investigations aimed at understanding the evolution and deformation of continental lithosphere through time. Proposals to elucidate the processes that act on the lithosphere at various time-scales and length-scales, either at depth or the surface, are encouraged. Because understanding such large-scale phenomena commonly requires a variety of expertise and methods, the Tectonics Program supports integrated research involving the disciplines of structural geology, petrology, geochronology, sedimentology, stratigraphy, geomorphology, rock mechanics, paleomagnetics, geodesy, and other geophysical techniques.
Requirements: Proposals will be accepted from colleges, universities, and other U.S. institutions with formal research programs in the earth sciences.
Date(s) Application is Due: Jun 1; Dec 1
Contact: David M. Fountain, Program Officer; (703) 292-8552; fax: (703) 292-9025; email: dfountai@nsf.gov
Internet: http://www.nsf.gov/funding/pgm_summ.jsp?pims_id=13673&org=NSF&sel_org=NSF&from=fund
Sponsor: National Science Foundation
4201 Wilson Boulevard
Arlington, VA 22230

NSF Theoretical Atomic, Molecular, and Optical Physics (TAMOP) Grants **4664**
The Theoretical Atomic, Molecular, and Optical Physics (TAMOP) program supports theoretical and computational research in all areas of atomic structure, the molecular structure of small molecules, electron, and atomic collisions, photoionization and photodetachment of electrons from atoms and small molecules, time-dependent interactions with atoms and small molecules, quantum optics, ultracold phenomena in Bose and Fermi gases, and quantum information. Investigations primarily directed toward a theoretical understanding of larger molecules or condensed matter systems should be directed toward the appropriate programs in the Division of Chemistry or the Division of Materials Research.
Date(s) Application is Due: Sep 30
Contact: Barry I. Schneider, Program Director; (703) 292-7383; fax: (703) 292-9060; email: bschneid@nsf.gov
Internet: http://www.nsf.gov/funding/pgm_summ.jsp?pims_id=503273
Sponsor: National Science Foundation
4201 Wilson Boulevard
Arlington, VA 22230

NSF Theoretical Physics Research Grants **4665**
The program supports the development of qualitative and quantitative understanding of fundamental physical systems, ranging from the most elementary constituents of matter through nuclei and atoms to astrophysical objects and the cosmos. This includes formulating new approaches for theoretical, computational, and experimental research that explore the fundamental laws of physics and the behavior of physical systems; formulating quantitative hypotheses; exploring and analyzing the implications of such hypotheses analytically and computationally; and, in some cases, interpreting the results of experiments. Support is given for research and conferences in the following categories: elementary particle physics; nuclear physics; atomic, molecular, optical, and plasma physics; astrophysics and cosmology; gravitational physics, and a broad spectrum of topics in mathematical physics, described below. The effort also includes a considerable number of interdisciplinary grants.
Date(s) Application is Due: Sep 30
Contact: Beverly Berger, Program Director; (703) 292-7372; fax: (703) 292-9078; email: bberger@nsf.gov
Internet: http://www.nsf.gov/funding/pgm_summ.jsp?pims_id=5626&org=PHY&from=home
Sponsor: National Science Foundation
4201 Wilson Boulevard, Room 1015
Arlington, VA 22230

NSF Thermal Transport Processes Research Grants **4666**
The Thermal Transport Processes program supports engineering research aimed at gaining a basic understanding of the microscopic and macroscopic levels of thermal transport phenomena (heat and mass transfer) underlying energy conversion and conservation, the synthesis and processing of materials, cooling and heating of infrastructure and equipment, the interaction of industrial processes with the environment, the propulsion of air and land-based vehicles, and thermal phenomena in biological and environmental systems. The program supports fundamental research and engineering education in transport processes that are driven by thermal gradients, and their manipulation to achieve engineering goals. This engineering science forms an important part of the intellectual infrastructure of a wide array of modern technologies.
Date(s) Application is Due: Mar 1; Sep 15
Contact: Theodore L. Bergman; (703) 292-7494; email: tbergman@nsf.gov
Internet: http://www.nsf.gov/funding/pgm_summ.jsp?pims_id=13367
Sponsor: National Science Foundation
4201 Wilson Boulevard
Arlington, VA 22230

NSF Topology Research Grants **4667**
The program supports research on algebraic topology, including homotopy theory, ordinary and extraordinary homology and cohomology, cobordism theory, and K-theory, topological manifolds and cell complexes, fiberings, knots, and links; differential topology and actions of groups of transformations; geometric group theory; and general topology and continua theory.
Date(s) Application is Due: Nov 3
Contact: Mark F. Feshbach, Program Director; (703) 292-5371; fax: (703) 292-9032; email: mfeshbac@nsf.gov
Internet: http://www.nsf.gov/funding/pgm_summ.jsp?pims_id=5551&org=DMS&from=home
Sponsor: National Science Foundation
4201 Wilson Boulevard
Arlington, VA 22230

NSF Tribal Colleges and Universities Program (TCUP) **4668**
This program provides awards to enhance the quality of science, technology, engineering and mathematics(STEM) instructional and outreach programs, with an emphasis on the leveraged use of information technologies at tribal colleges and universities, Alaskan native-serving institutions and native Hawaiian-serving institutions. Support is available for the implementation of comprehensive institutional approaches to strengthen STEM teaching and learning in ways that improve access to, retention within, and graduation from STEM programs, particularly those that have a strong technological foundation. Typical project implementation strategies include curriculum enhancement, faculty professional development, undergraduate research and community service, academic enrichment, infusion of technology to enhance STEM instruction, collaborations, and other activities that meet institutional and community needs. Annual deadline dates may vary; contact program staff for exact dates. Approximately $6 million available funds - 9 awards total - approximately 6 implementation awards resulting from Full Proposals will be awarded as cooperative agreements annually, and approximately 3 planning grants resulting from Planning Proposals will be awarded annually. NSF 04-602
Requirements: Tribal colleges and universities, Alaskan Native-serving institutions, and Native Hawaiian-serving institutions are eligible to apply. The Principal Investigator is expected to be the chief academic officer of the institution, or other senior academic officer responsible for oversight and management of curriculum and instructional policies for the institution.
Restrictions: Only one proposal (for either a planning grant or an implementation award) per institution in any one year.
Date(s) Application is Due: Feb 5; Nov 20
Contact: Lura Chase, Program Director, Room 815 N; (703) 292-8682; fax: (703) 292-9018; email: lchase@nsf.gov
Internet: http://www.nsf.gov/funding/pgm_summ.jsp?pims_id=5483
Sponsor: National Science Foundation
4201 Wilson Boulevard
Arlington, VA 22230

NSF Undergraduate Research and Mentoring in the Biological Sciences **4669**
This program will fund projects that have strong research and mentoring activities designed to prepare students for successful entry into graduate programs. URM will support projects involving the recruitment, retention and development of undergraduate students, especially those from underrepresented groups, for the purpose of preparing them for graduate study in the biological sciences. Proposed projects are expected to create a URM program that will actively engage students in interesting and exciting research ideas, provide hands-on research experience, and develop their academic skills. URM will enable institutions to create innovative programs that will increase the number and diversity of students who enter graduate research programs in the biological sciences. URM will support well-defined research and mentoring activities that will enable students, especially underrepresented minorities, to become independent thinkers and effective communicators, as well as professional development activities that will expose students to exciting careers in biology. Students are expected to be recruited early in their academic career. The project should frame all student activities around an effective mentoring strategy. Factors that are known to increase the likelihood of success for students to enter, and complete, graduate research programs should be emphasized. Approximately $4,000,000 will be available for this program, subject to the availability of funds. Under the solicitation, requests may be submitted for funding amounts up to a total of $1,000,000 for up to 5 years. However, funding beyond the first year is contingent upon satisfactory progress in the program. The program expects to make at least eight awards, depending upon the quality of submissions and the availability of funds. NSF 06-591
Requirements: Preliminary Proposals (deadline: Sep. 18) are required prior to submitting a Full Proposal (deadline: Mar. 4). Proposals may only be submitted by academic institutions located in the United States. Collaborative Proposals from two or more institutions are allowed. The proposing institution is expected to foster an enriched and intellectually stimulating research and educational environment. URM proposals should involve year-

round mentoring and include a major emphasis on direct student participation in research leading to publishable data. URM projects may be based in a single discipline or academic department, or on interdisciplinary or multi-department research opportunities. A proposal should reflect the unique combination of the proposing organization's interests and capabilities and those of any partnering organizations. Partnerships between minority-serving institutions, especially Historically Black Colleges and Universities (HBCU), Tribal Colleges and Universities (TCU), and Hispanic-Serving Institutions (HSI), community colleges, and research intensive institutions are encouraged, if such partnerships will result in a diverse pool of qualified participants as well as an enriched research environment for the students. Undergraduate student participants supported with NSF funds must be citizens or permanent residents of the United States or its possessions. An undergraduate student is one who is enrolled in a degree program (part-time or full-time) leading to a baccalaureate or associates degree. High school graduates who have not yet enrolled, and students who have received their bachelor's degrees and are no longer enrolled as undergraduates, are not eligible.
Restrictions: All collaborative proposals submitted as separate submissions from multiple organizations must be submitted via the NSF FastLane system.
Samples: University of Arkansas (AR) $149,233 - RIG: Biophysical Characterization of Cdc42Hs Forms Associated with Cell Transformation; University of Montana (MT) $170,382 - Collaborative UMEB: Training American Indians in Environmental Biology; Salish Kootenai College (MT) $132,715 - Collaborative UMEB: Training American Indians in Environmental Biology; University of Maryland Eastern Shore (MD) $150,000 - Discovery of Novel Transcription Factor III Proteins in Trypanosomes
Amount of Grant: $1,000,000 maximum
Date(s) Application is Due: Sep 18; Mar 4
Contact: URM Program Director; phone: (703) 292-8470; email: biourm@nsf.gov
Internet: http://www.nsf.gov/funding/pgm_summ.jsp?pims_id=500036
Sponsor: National Science Foundation
4201 Wilson Boulevard
Arlington, VA 22230

NSF University Radio Observatories Program (AST-URO) 4670

The URO program provides support for operations, technical enhancements, scientific investigations, and student training at university-based radio facilities. A primary function of funded University Radio Observatories is to provide focused research environments that foster hands-on student access to state-of-the-art radio astronomy instrumentation. UROs also serve the general astronomical community by providing open access to 30% - 50% of available telescope time. NSF Solicitation 08-543
Requirements: URO awards are for three-year intervals, and are renewable under the proposal review procedures described below. UROs fully funded by NSF are expected to provide the general astronomical community with access to between one-third and one-half of all telescope observing time. UROs should also develop and maintain a strong program of student education and training in radio astronomy, and should conduct public outreach activities appropriate to the location and nature of the facility. Proposals may only be submitted by universities and two- and four-year colleges (including community colleges) located and accredited in the U.S., acting on behalf of their faculty members. New applicants are strongly encouraged to contact the Program Director prior to preparation of a full proposal for an evaluation of the suitability of the proposed project to the URO program.
Date(s) Application is Due: May 30
Contact: Richard E. Barvainis, Program Director; (703) 292-4891; fax: (703) 292-9034; email: rbarvai@nsf.gov
Internet: http://www.nsf.gov/funding/pgm_summ.jsp?pims_id=5292
Sponsor: National Science Foundation
4201 Wilson Boulevard
Arlington, VA 22230

NSF Upper Atmospheric Facilities (UAF) Grants 4671

The major goal of the Upper Atmospheric Facilities (UAF) is to promote basic research on the structure and dynamics of the Earth's upper atmosphere. Research efforts utilizing these facilities have strong links to the Aeronomy Program and the Magnetospheric Physics Program. The National Science Foundation supports four large incoherent-scatter radar facilities and the SuperDARN coherent scatter radar system. The incoherent-scatter radars are located along a longitudinal chain from Greenland to Peru. Each of the incoherent-scatter facilities is also equipped with powerful optical diagnostic instruments. The SuperDARN consists of a number of coherent-scatter HF radars in both the northern and southern hemispheres.
Requirements: Standard NSF Grant Proposal Guidelines apply (see http://www.nsf.gov/publications/pub_summ.jsp?ods_key=gpg). There are no deadlines or target dates for proposals sent in to any of the Upper Atmospheric Research Section (UARS) base programs. However, it is recommended that PIs try to send in proposals early in the fiscal year. Certain special programs within UARS do have deadlines for proposal submissions.
Contact: Robert Robinson, Program Director; (703) 292-8529; fax: (703) 292-9023; email: rmrobins@nsf.gov
Internet: http://www.nsf.gov/funding/pgm_summ.jsp?pims_id=12808
Sponsor: National Science Foundation
4201 Wilson Boulevard
Arlington, VA 22230

NSF Western European Cooperative Program Grants 4672

Proposals for cooperative research and joint workshops with all countries in western Europe, including Iceland, will be considered. Projects that provide research experiences for scientists and engineers in the early stages of their careers, and projects that involve significant participation of advanced graduate students and postdoctoral researchers are encouraged. Preference will be given to applicants who have not had extensive experience in international research or prior support from the division. Workshops and seminars should be designed to encourage new collaborations. Regional collaborations, including projects involving participants from more than one western European country, are encouraged. Cooperation with western European multilateral organizations and center-to-center cooperation is encouraged as well. Program focus may differ from year to year. Refer to brochure NSF 00-138 for program announcement.
Requirements: Proposals may be submitted by U.S. academic institutions; public or private nonprofit, nonacademic scientific organizations; private for-profit organizations; and by unaffiliated individual scientists and engineers. Except for Austria and France, parallel proposals from foreign counterparts are not required; however, all proposals must include endorsement of the project from the cooperating institution.
Contact: Rose Gombay, Program Manager, (703) 292-8702; fax: (703) 292-9177; email: rgombay@nsf.gov
Internet: http://www.nsf.gov/sbe/int/w_europe/start.htm
Sponsor: National Science Foundation
4201 Wilson Boulevard
Arlington, VA 22230

NSF-CONACyT Collaborative Research Opportunities 4673

The purpose of the initiative is to advance scientific and engineering knowledge through joint research efforts of investigators from the United States of America and Mexico having complementary talents and interests, and capitalizing on the international character of modern scientific research and its support by computer network infrastructures. Projects supported in this collaborative initiative must include researchers and educators in institutions from both countries in areas relating to either CISE or ENG. Projects may range in scope from supplements to an existing award through joint efforts to a new joint-effort proposal up to $100,000 with a duration from one to three years; larger grants may be awarded under special circumstances if appropriate. Refer to brochure NSF 96-145.
Requirements: Scientists and engineers from universities and other nonprofit research institutions in the United States of America and Mexico conducting research in areas relating to either CISE or ENG may submit proposals for support under this initiative. All research proposals submitted under this initiative must be bilateral, collaborative, and related to a corresponding proposal from the other country.Proposals must be received by NSF no later than the first Wednesday in February.
Restrictions: Research proposals submitted to NSF under this collaborative initiative should avoid requests for equipment. Nonetheless, under special circumstances, the latter may be requested with appropriate justification. CONACyT has its own guidelines with respect to equipment.
Geographic Focus: Alabama; Alaska; Arizona; Arkansas; California; Colorado; Connecticut; Delaware; District of Columbia; Florida; Georgia; Guam; Hawaii; Idaho; Illinois; Indiana; Iowa; Kansas; Kentucky; Louisiana; Maine; Maryland; Massachusetts; Michigan; Midwest; Minnesota; Mississippi; Missouri; Montana; Nebraska; Nevada; New England; New Hampshire; New Jersey; New Mexico; New York; North Carolina; North Dakota; Northeast; Northwest; Ohio; Oklahoma; Oregon; Pacific Northwest; Pennsylvania; Puerto Rico; Rhode Island; South; South Carolina; South Dakota; Southwest; Tennessee; Texas; U.S. Virgin Islands; Utah; Vermont; Virginia; Washington; West; West Virginia; Wisconsin; WyomingMexico
Amount of Grant: up to $100,000
Contact: Jeannette M. Wing, Assistant Director; Directorate for Computer & Information Science & Engineering; (703) 292-8900; email: jwing@nsf.gov
Internet: http://www.nsf.gov/pubs/1998/nsf96145/nsf96145.htm
Sponsor: National Science Foundation
4201 Wilson Boulevard
Arlington, VA 22230

NSF-NIST Interaction in Chemistry, Materials Research, Molecular Biosciences, Bioengineering, and Chemical Engineering 4674

This program is intended to facilitate interactions between faculty and students supported by the National Science Foundation (NSF) and scientists at the National Institute of Standards and Technology's (NIST) Chemical Science and Technology Laboratory (CSTL) and Materials Science and Engineering Laboratory (MSEL), including the NIST Center for Neutron Research (NCNR). Chemistry, materials research, molecular biology, bioengineering, and chemical engineering are centralized at NIST in these laboratories. Support may be requested for supplements to existing NSF awards to provide the opportunity for faculty and students to participate in research at NIST facilities. Support may be requested as supplements to existing NSF awards for travel expenses and per diem associated with work at these NIST laboratories for faculty, students and other personnel associated with the NSF/NIST activity. It is estimated that 10-20 supplements will be made annually, up to $20,000 each.
Requirements: Support may be requested only by organizations that have currently active awards in any of the participating NSF Divisions, viz. the Division of Bioengineering and Environmental Systems (BES), the Division of Chemical and Transport Systems (CTS), the Division of Chemistry (CHE), the Division of Materials Research (DMR), and the Division of Molecular and Cellular Biosciences (MCB). Only faculty who are principal investigators on current NSF awards from BES, CHE, CTS, DMR, or MCB are eligible to apply for supplements. Standard NSF Grant Proposal Guidelines apply (see http://www.nsf.gov/publications/pub_summ.jsp?ods_key=gpg).
Restrictions: Indirect costs (F&A) are not allowed.
Amount of Grant: up to $20,000

Contact: Dr. William F. Koch, Deputy Director, (301) 975-8301, fax: (301) 975-3845, william.koch@nist.gov; Dr. Stephen W. .Freiman, Deputy Director; (301) 975-5658, fax: (301) 975-5012, stephen freiman@nist.gov
Internet: http://www.nsf.gov/publications/pub_summ.jsp?ods_key=nsf03568
Sponsor: National Science Foundation
4201 Wilson Boulevard
Arlington, VA 22230

NSF/DOE Partnership in Basic Plasma Science and Engineering Grants **4675**
The goal of this three year (FY09-FY11) program initiative is to enhance plasma research and education in this broad, multidisciplinary field by coordinating efforts and combining resources of the two agencies. The initiative will address fundamental issues in plasma science and engineering that can have impact in other areas or disciplines in which improved basic understanding of the plasma state is needed.
Requirements: Proposals may be submitted by individual investigators or small groups from universities and nonprofit organizations.
Restrictions: An applicant may submit only one proposal as principal or co-principal investigator.
Amount of Grant: $25,000-$200,000 average
Date(s) Application is Due: Oct 22
Contact: Richard L. Berger, (703) 292-2183; fax: (703) 292-9078; email: rlberger@nsf.gov
Internet: http://www.nsf.gov/funding/pgm_summ.jsp?pims_id=5602
Sponsor: National Science Foundation
4201 Wilson Boulevard
Arlington, Va 22230

NSF/NEH Documenting Endangered Languages Project Grants and Fellowships **4676**
The program supports projects to develop and advance knowledge concerning endangered human languages. Funding will support fieldwork and other activities relevant to recording, documenting, and archiving endangered languages, including the preparation of lexicons, grammars, text samples, and databases. Types of support include one- to three-year project grants and fellowships for up to 12 months. At least half the available funding will be awarded to projects involving fieldwork.
Requirements: U.S. academic institutions and other nonprofit organizations are eligible for project grants. Personnel in for-profit organizations may participate as co-investigators. Principal investigators worldwide are eligible to apply for project grants. U.S. citizens are eligible to apply for fellowships. Foreign nationals who have been living in the United States or its jurisdictions for at least the three years prior to the proposal deadline also are eligible to apply for fellowships.
Amount of Grant: $2 million total
Date(s) Application is Due: Sep 15
Contact: Joan Maling, Linguistics Director; (703) 292-8046; fax (703) 292-9068; email: jmaling@nsf.gov; Anna Kerttula; (703) 292-7432; email: akerttul@nsf.gov
Internet: http://www.nsf.gov/pubs/2006/nsf06577/nsf06577.htm
Sponsor: National Science Foundation/ National Endowment for the Humanities
4201 Wilson Boulevard
Arlington, VA 22230

NSU Fischler School of Education and Human ServicesDoctoral Fellowships **4677**
The doctoral fellowship program is open to prospective and current doctoral students at FGSEHS. Fellowships will be awarded to persons pursuing doctoral degrees in one of the six programs offered by FGSEHS: EdD in Child, Youth, and Human Services; EdD in Educational Leadership; EdD in Higher Educational Leadership; EdD in Instructional Technology and Distance Education; EdD in Organizational Leadership; and Doctor of Speech-Language Pathology. Each fellowship includes a stipend; full tuition, books, and fees; office space at FGSEHS in North Miami Beach, FL; computer support; and all expenses paid for the Conference on Global Leadership, Learning, and Research. Program application criteria and procedures are available online.
Requirements: Applicants must be fully accepted or currently enrolled in one of the doctoral programs before the fellowship will be awarded.
Restrictions: Former fellows are not eligible for the program
Amount of Grant: $25,000 annual stipend
Date(s) Application is Due: Sep 1
Contact: Dr. Nelson Diaz, Executive Dean , (800) 986-3223; fax: (954) 262-3903; email: ndiaz@nova.edu
Internet: http://www.fgse.nova.edu/gradfellow/application.htm
Sponsor: Nova Southeastern University
1750 NE 167th Street
North Miami Beach, FL 33162-3017

Nuffield Foundation Children and Families Grants **4678**
The foundation was founded for the advancement of health through medical research and teaching and for the care of the elderly through scientific research and education. The Children and Families program supports work to help ensure that the legal and institutional framework is best adapted to meet the needs of children and families. At present, particular interests include (but are not limited to): work that links education and child development, either in the case of adolescent mental health or younger children; work that considers policies relevant to child welfare in a broader institutional context: parents' paid working patterns; childcare and early years provision; work that considers especially the well-being of children growing up in adverse conditions, and what institutional responses may be appropriate; work in family law including cohabitation, child contact, child support; and work in child protection and placement (adoption and fostering) but only when it raises significant issues. Where a proposal is for a research study, Trustees are interested in the dispassionate examination of evidence. It notes that evidence is likely to be different in different cases, for different types of children and families, and are more likely to support work that takes this approach.
Requirements: Usually grants are made only to United Kingdom organizations, and support work that will be mainly based in the UK, although the Trustees welcome proposals for collaborative projects involving partners in European or Commonwealth countries.
Restrictions: The Foundation does not make grants for the running costs of voluntary bodies but will consider making a contribution to voluntary sector overheads on funded projects.
Samples: Gillian Douglas, Cardiff Law School—for Attitudes towards the law of inheritance in the context of changing family structures, 193,542 GBP (2009); Ingrid Schoon, Department of Quantitative Social Sciences, Institute of Education, 152,771 GBP (2009).
Amount of Grant: 5,000-150,000 GBP
Date(s) Application is Due: Jan 8; May 7; Sep 3
Contact: Sharon Witherspoon; 44-171-631-0566; fax: 44-171-323-4877; email: sfwpa@nuffieldfoundation.org or info@nuffieldfoundation.org
Internet: http://www.nuffieldfoundation.org/go/grants/cplfj/page_30.html
Sponsor: Nuffield Foundation
28 Bedford Square
London WC1B 3JS United Kingdom

Nuffield Foundation Law and Society Grants **4679**
The foundation was founded for the advancement of health through medical research and teaching and for the care of the elderly through scientific research and education. The Foundation has long had an interest in the area of law. Its current objectives in this area are: to promote developments in the legal system that will improve its accessibility to all people; to promote wider access to legal services and advice and a better understanding of the obstacles to access to justice; to fund research and promote developments in alternative dispute resolution; to help promote a greater knowledge of the rights and duties of the individual, including those of the European citizen; to examine the implications of new human rights' obligations on civil (not criminal) justice; and to help promote a greater public understanding of the role of law in society and of the legal system.
Requirements: Usually grants are made only to United Kingdom organizations, and support work that will be mainly based in the UK, although the Trustees welcome proposals for collaborative projects involving partners in European or Commonwealth countries.
Samples: Jo Shaw, School of Law, University of Edinburgh—for Friction and overlap between EU free movement rules and immigration law in the United Kingdom, 65,171 GBP (2009); Nigel Blackwood, Forensic Mental Health Science, King's College London, 193,398 GBP, (2008).
Amount of Grant: 5,000-150,000 GBP
Date(s) Application is Due: Jan 8; May 7; Sep 3
Contact: Sharon Witherspoon; 44-171-631-0566; fax: 44-171-323-4877; email: sfwpa@nuffieldfoundation.org or info@nuffieldfoundation.org
Internet: http://www.nuffieldfoundation.org/go/grants/accesstojustice/page_57.html
Sponsor: Nuffield Foundation
28 Bedford Square
London WC1B 3JS United Kingdom

Nuffield Foundation Oliver Bird Rheumatism Program Grants **4680**
The Fund will offer grants to up to five United Kingdom Higher Education Institutions where students of outstanding potential will be recruited into doctoral training. The institutions will provide multidisciplinary environments for researching problems in rheumatic disease, and be able to develop both the students' awareness of the broader context of these diseases and the professional skills required of future scientists. For each studentship the grant will provide: a stipend of not less than 13,500GBP per annum plus United Kingdom fees; research expenses at a rate of 8,000GBP per annum; and a meeting and travel allowance of 1,000GBP per annum. In addition the Fund will provide a grant to each institution of 25,000GBP over the five years of the program to support such activities as: meetings within the Collaborative Center to promote the field and assist the recruitment and training of doctoral students; the hosting of a program meeting to be attended by all program participants from the other Collaborative Centers; and occasional bursaries for interested undergraduates to work alongside the doctoral students and potential supervisors or to attend meetings in the Collaborative Centers.
Amount of Grant: 27,500GBP per year for four years
Date(s) Application is Due: Feb 28
Contact: Vicki Hughes; 44-171-631-0566; fax: 44-171-323-4877; email: oliverbird@nuffieldfoundation.org
Internet: http://www.nuffieldfoundation.org/go/grants/oliverbird/page_86.html
Sponsor: Nuffield Foundation
28 Bedford Square
London WC1B 3JS United Kingdom

Nuffield Foundation Open Door Grants **4681**
The Foundation keeps an open door to proposals of exceptional merit for research projects or practical innovations that lie outside our main program areas, but that meet Trustees' wider interests. These must have some bearing on the Foundation's widest charitable object. Particular interest is given to projects which identify change or interventions which will have a practical implications for policy or practice, or that will improve the

quality of research evidence in areas of public debate. Through the Open Door, the Foundation may also identify emerging areas that justify more sustained attention.
Requirements: Usually grants are made only to United Kingdom organizations, and support work that will be mainly based in the UK, although the Trustees welcome proposals for collaborative projects involving partners in European or Commonwealth countries.
Samples: Dr Alex Bryson, National Institute of Economic & Social Research—for British Workplaces in the shadow of recession, 219,167 GBP (2009); Colin Aitken, School of Mathematics, University of Edinburgh, 75,869 GBP (2009).
Amount of Grant: 5,000-150,000 GBP
Date(s) Application is Due: Jan 8; May 7; Sep 3
Contact: Sharon Witherspoon; 44-171-631-0566; fax: 44-171-323-4877; email: sfwpa@nuffieldfoundation.org or info@nuffieldfoundation.org
Internet: http://www.nuffieldfoundation.org/go/grants/opendoor/page_115.html
Sponsor: Nuffield Foundation
28 Bedford Square
London WC1B 3JS United Kingdom

NWHF Community-Based Participatory Research Grants **4682**
The Northwest Health Foundation supports community-based participatory research as a tool to generate meaningful information about community health and build the capacity of groups united by common challenges to organize and advocate for change. The RFP will seek to support community-based participatory research (CBPR) projects that: build community capacity for policy advocacy; and apply research findings to support policy change that improves community health.
Requirements: 501(c)3 educational institutions and governmental entities are eligible. Generally speaking, foundation programs serve the entire state of Oregon, and the following Washington counties: Clark, Cowlitz, Pacific, Skamania, and Wahkiakum. Specific programs and funds may have additional geographic restrictions. Applying organizations must sign a statement of nondiscrimination in leadership, staffing, and services.
Restrictions: Support will not be considered for basic operations, or for salaries or other expenses associated with ongoing programs.
Geographic Focus: Oregon; Washington
Date(s) Application is Due: Sep 8
Contact: Judith Woodruff, Program Director; (503) 220-1955; fax: (503) 220-1355; email: judith@nwhf.org; Chris Kabel, Program Officer; (503) 220-1955; fax: (503) 220-1355; email: ckabel@nwhf.org
Internet: http://www.nwhf.org/index.php?/apply/cbpr
Sponsor: Northwest Health Foundation / Robert Wood Johnson Foundation
221 NW Second Avenue, Suite 300
Portland, OR 97209

NWHF Mark O. Hatfield Research Fellowship **4683**
The Mark O. Hatfield Research Fellowship Program was established in 2005 to honor the legacy of Senator Mark O. Hatfield, founding Board Chairman of the Northwest Health Foundation (NWHF). The purpose of the Fellowship program is to advance the research career of health practitioners who serve communities in Oregon and southwest Washington. The fellowship is open to health professionals from a variety of disciplines who wish to launch, renew, or expand their research activities in biomedical, applied, or translational research. A secondary purpose is to increase the intellectual resources of research-based institutions. One two-year fellowship of up to $150,000 will be awarded. Funds will be distributed to the fellow's sponsoring institution in two annual disbursements of half of the amount awarded.
Requirements: Eligible candidates must: be a clinical, allied health, public health, or social service provider (early-career researchers specializing in translational research are also encouraged to apply); possess (or be in the process of completing) a doctoral degree or equivalent in an appropriate field related to medicine, nursing, health care, public health, or biomedical sciences; obtain sponsorship from a research-based institution that can act as fiscal manager of the award; and demonstrate a commitment to communities in Oregon and southwest Washington.
Amount of Grant: $150,000 maximum
Date(s) Application is Due: Mar 16
Contact: R. David Rebanal, Program Officer; (503) 220-1955; email: drebanal@nwhf.org
Internet: http://www.nwhf.org/index.php?/apply/hatfield
Sponsor: Northwest Health Foundation
221 NW Second Avenue, Suite 300
Portland, OR 97209

NWSA Abafazi-Africana Women's Studies Essay Award **4684**
The purpose of the writing awards is to discover, encourage, and promote the intellectual development of emerging black women scholars who engage in critical theoretical discussions and/or analyses about feminist/womanist issues among women of color. Scholarly essays may be on any subject relevant to black girl children, women's issues, and/or experiences in the United States or throughout the diaspora.
Requirements: For most scholarships, NWSA members will be given preference over nonmembers.
Amount of Grant: $400
Date(s) Application is Due: Feb 15
Contact: Program Contact, (301) 403-0524; fax: (301) 403-4137; email: nwsaoffice@nwsa.org
Internet: http://www.nwsa.org/wocscol.html
Sponsor: National Women's Studies Association
7100 Baltimore Avenue, Suite 502
College Park, MD 20740

NWSA Graduate Scholarship in Jewish Women's Studies **4685**
The scholarship will be awarded to a student who, in the fall, will be engaged in the research or writing stages of a master's thesis or PhD dissertation in the interdisciplinary field of women's studies. The research project must be on women and must enhance the NWSA mission.
Requirements: Applicants must be NWSA members at the time of application.
Amount of Grant: $1000
Date(s) Application is Due: Mar 1
Contact: National Executive Administrator, (301) 403-0525; fax: (301) 403-4137; email: nwsaoffice@nwsa.org
Internet: http://www.nwsa.org
Sponsor: National Women's Studies Association
7100 Baltimore Avenue, Suite 502
College Park, MD 20740

NWSA Graduate Scholarships **4686**
NWSA will award the scholarship to a student who, in the fall of the year of the award, will be engaged in the research or writing stages of a master's Thesis or PhD dissertation in the interdisciplinary field of women's studies. The research project must be on women and must enhance the NWSA mission.
Requirements: NWSA members are eligible.
Amount of Grant: $1000
Date(s) Application is Due: Feb 15
Contact: National Executive Administrator, (301) 403-0525; fax: (301) 403-4137; email: nwsaoffice@nwsa.org
Internet: http://www.nwsa.org/gradscol.html
Sponsor: National Women's Studies Association
7100 Baltimore Avenue, Suite 502
College Park, MD 20740

NWSA Graduate Scholarships in Lesbian Studies **4687**
One scholarship is awarded per year to support a graduate student doing research for, or writing a master's thesis or PhD dissertation in, lesbian studies. Applicants need not be enrolled in a women's studies program. Preference is given to candidates who are NWSA members.
Amount of Grant: $500
Date(s) Application is Due: Feb 15
Contact: National Executive Administrator, (301) 403-0525; fax: (301) 403-4137; email: nwsaoffice@nwsa.org
Internet: http://www.nwsa.org/sslesbian.html
Sponsor: National Women's Studies Association
7100 Baltimore Avenue, Suite 502
College Park, MD 20740

NYAM Audrey and William H. Helfand Fellowship **4688**
The Fellowship supports work in the humanities as they apply to medicine and health, including works of non-fiction, visual or performing arts, biography and memoir, as well as scholarly research in a humanistic discipline other than the history of medicine. Although residence is not obligatory, preference will be given applicants whose projects require use of the resources of the Academy Library and who plan to spend time at the Academy. Besides completing a research or creative project, each fellow will be expected to make a public presentation at the Academy and submit a final report.
Requirements: Applications are invited from anyone, regardless of citizenship, academic discipline, or academic status. Preference will be given to (1) those whose research will take advantage of resources that are uniquely available at the Academy, and (2) scholars or creative artists in the early stages of their careers.
Samples: Sabine Arnaud (2005-06); Bryan Waterman (2004-05); Angus Fletcher (2003-04).
Amount of Grant: $5,000 maximum
Date(s) Application is Due: Mar 6
Contact: Program Administrator, (212) 822-7314; fax: (212) 822-7338; email: history@nyam.org
Internet: http://www.nyam.org/grants/history.shtml
Sponsor: New York Academy of Medicine
1216 Fifth Avenue
New York, NY 10029-5202

NYAM Bowen-Brooks Fellowship for Advanced Studies Abroad **4689**
The Fellowship provides full tuition, a travel allowance, and a $50,000 stipend to enable the Fellow to spend one year in Oxford, England at the Oxford University Program in Evidence-Based Health Care, in affiliation with Green College, Oxford University, and the John Radcliffe Hospital. Depending on the Fellow's level of academic commitment, either a Diploma or MSc degree can be awarded on completion of the course requirements.
Requirements: Eligible candidates are physicians who have recently completed residency training in internal medicine, pediatrics, emergency medicine, or family practice in a New York area program, who will return to New York to practice medicine at the completion of their studies.
Amount of Grant: $50,000 plus tuition and travel allowance

Date(s) Application is Due: Feb 6
Contact: Program Coordinator, (212) 822-7204; fax: (212) 822-7338; email: bowen-brooks@nyam.org
Internet: http://www.nyam.org/grants/bowenbrooks.shtml
Sponsor: New York Academy of Medicine
1216 Fifth Avenue
New York, NY 10029-5202

NYAM Charles A. Elsberg Fellowship **4690**
The Fellowship in Neurological Surgery was established to support research training in the specialty of neurological surgery for individuals who have completed, or will shortly complete, accredited residency training in neurological surgery, and who intend to use research training for continued development of academic careers in this field. The Elsberg Fellowship program seeks to support individuals in supervised programs that will develop the candidate's capacity to perform independent clinical or laboratory research.
Requirements: Candidates who have, or will have completed an accredited neurological surgery residency program by June 30, and whose career direction suggests eligibility for subsequent research support are invited to apply. In general, applicants are expected to be at the start of their academic careers, having completed their residency training within the past year or two. Candidates must be board certified or eligible for certification by the beginning of the grant period. The proposed research must be conducted at an institution located in the United States under the mentorship of a senior researcher.
Samples: Jason H. Huang, MD, University of Rochester School of Medicine and Dentistry (2006-07); John A. Boockvar, MD, Weill Medical College of Cornell University (2005-06); Saadi Ghatan, MD, Columbia University Medical Center (2004-05).
Amount of Grant: $50,000
Date(s) Application is Due: Feb 27
Contact: Program Administrator, (212) 822-7204; fax: (212) 822-7338; email: elsberg@nyam.org
Internet: http://www.nyam.org/grants/elsberg.shtml
Sponsor: New York Academy of Medicine
1216 Fifth Avenue
New York, NY 10029-5202

NYAM David E. Rogers Fellowships **4691**
The Fellowship is meant to enrich the educational experiences of medical and dental students through projects that bear on medicine and dentistry as social enterprises, such as enterprises devoted to the capacity of these professions in any and all of their expressions to serve human needs. Particularly, the focus is on the needs of underserved or disadvantaged patients or populations. The content of the Fellowship might include clinical investigation, health policy analysis, activities linking biomedicine, the social infrastructure and human need, or community activities. Half the fellowships awarded are dedicated to projects focused on HIV/AIDS prevention or care. The union of service or research activities and a mentoring program is intended to contribute to the maturation of the student and to promote consideration of the societal responsibilities of medicine and dentistry and exploration of new career directions, including linkages between health policy and administrative elements on the one hand and clinical care on the other.
Requirements: Available to first-year medical and dental students for support of projects to be executed during the summer between the first and second years of medical or dental school.
Samples: Christy Boraas Alsleben, University of Minnesota Medical School (2005); Ashita Satish Batavia, Weill Medical College of Cornell University (2005); Emma Ester Bendana, Albany Medical College (2005).
Amount of Grant: $3,500
Date(s) Application is Due: Mar 23
Contact: Laura Van Wyk, (212) 822-7220; fax: (212) 423-0274; email: rogers@nyam.org
Internet: http://www.nyam.org/grants/rogers.shtml
Sponsor: New York Academy of Medicine
1216 Fifth Avenue
New York, NY 10029-5202

NYAM Edwin Beer Research Fellowship **4692**
The Edwin Beer Research Fellowship seeks to increase the number of investigators in urology and urology-related fields by providing transitional funding in support of research conducted by qualified candidates. The program provides two-year grants of up to $80,000 for research projects commencing July 1 of the award year. The grants are made to the awardees' institutions for the direct support of their salary and research activities. Fellows are required to devote at least fifty percent of their time to the research funded by the program.
Requirements: Candidates must hold an MD, PhD, or equivalent degree and are expected to conduct research in a supervised program in the United States. Applicants should have completed residency and fellowship or postdoctoral training, having demonstrated their ability to perform and direct research, and should currently be under the mentorship of a senior researcher. Fellowships are awarded to qualified candidates who are at a stage of transition prior to receiving major independent funding. Because the program seeks to increase the number of investigators in urology and urology-related fields, applicants must have a direct association with the department of urology at their sponsoring institution. Candidates are not required to be United States citizens; however, non-citizen applicants are required to provide assurances of employment eligibility to perform the proposed research.
Amount of Grant: $80,000 maximum
Date(s) Application is Due: Mar 26
Contact: Program Director, (212) 822-7204; fax: (212) 822-7338; email: beer@nyam.org
Internet: http://www.nyam.org/grants/beer.shtml
Sponsor: New York Academy of Medicine
1216 Fifth Avenue
New York, NY 10029-5202

NYAM Ferdinand C. Valentine Fellowship **4693**
The Fellowship offers one-year, $50,000 fellowships in support of research by individuals who have completed, or will shortly complete, residency training acceptable to the American Board of Urology, and who intend to pursue academic research careers in urology.
Requirements: Fellows must commit no less than fifty percent of their time to the supported research and preference is given to investigators conducting their research at institutions in the greater New York area.
Restrictions: Indirect costs and fringe benefits will not be paid.
Samples: George J. Huang, MD, University of Southern California, Keck School of Medicine (2006-07); Ricardo R. Gonzalez, MD, Weill Medical College of Cornell University (2005-06); Richard Lee, MD, Children's Hospital Boston (2005-06).
Amount of Grant: $50,000
Date(s) Application is Due: Mar 5
Contact: Program Administrator, (212) 822-7204; fax: (212) 822-7338; email: valentine@nyam.org
Internet: http://www.nyam.org/grants/valentine.shtml
Sponsor: New York Academy of Medicine
1216 Fifth Avenue
New York, NY 10029-5202

NYAM Ferdinand C. Valentine Student Research Grants **4694**
Up to four $3,500 grants for students pursuing mentored research projects in urology will be awarded to MD candidates who are either enrolled in a New York-area medical school or plan to conduct research at a New York-area institution. The grants typically support 8-12 week urology research projects initiated during the summer between the first and second years of medical school, but medical school students in any year of their education may apply. Students are encouraged to pursue research projects that extend beyond this startup period.
Requirements: Applicants must provide a one-page research proposal, as well as a letter of support from the faculty mentor supervising the research.
Samples: Michelle Diana Wilson (2006); Stephen David Marshall (2006); Stuart Lance Cohen (2006).
Amount of Grant: $3,500
Date(s) Application is Due: Mar 16
Contact: Program Administrator, (212) 822-7204; fax: (212) 996-7338; email: valentine@nyam.org
Internet: http://www.nyam.org/grants/valentine_student.shtml
Sponsor: New York Academy of Medicine
1216 Fifth Avenue
New York, NY 10029-5202

NYAM Gladys Brooks Internship **4695**
The Internship at the New York Academy of Medicine is designed to provide an opportunity for training in book and paper conservation. Paid internships will last for 2-3 months, depending on the number of days spent in the lab. The internship will include: examining and applying book and paper conservation principles; understanding the steps in determining conservation needs; learning which materials and structures are appropriate for use in repairing, binding and storing rare or fragile materials; and executing the agreed-upon conservation treatments. Interns will become familiar with conservation activities through extensive hands-on experience and observations in a working book and paper conservation laboratory.
Requirements: Applicants for this internship must: have some experience in the field of bookbinding and/or book conservation, and possess a desire to gain expertise through hands-on work in a functioning conservation laboratory; have aesthetic appreciation and the manual dexterity to work with delicate and fragile materials; submit a detailed letter of interest, a current resume, and three professional recommendations - all to the address below; have student or working visas if they are not U.S. citizens.
Amount of Grant: $5,000 stipend
Date(s) Application is Due: Dec 15
Contact: Susan Martin, (212) 822-7363; Anne Hillam, (212) 822-7365
Internet: http://www.nyam.org/initiatives/im-histc-i.shtml
Sponsor: New York Academy of Medicine
1216 Fifth Avenue, Room 550
New York, NY 10029-5202

NYAM Glorney-Raisbeck Fellowships **4696**
The Fellowships in Cardiovascular Diseases are one-year grants funded by the Corlette Glorney Foundation that are awarded for basic science, clinical, or epidemiological research in cardiovascular diseases. The program seeks to enhance the potential of exceptional candidates to become physician scientists by supporting individuals in supervised programs that develop the candidate, ability to perform independent research.
Requirements: Candidates must hold an MD or equivalent degree, and are expected to have completed residency by the start of the grant period. The applicant pool includes physicians in training throughout the United States, but in order to be considered for

this program candidates must pursue the funded research in a supervised program at an institution located in New York City, Long Island, or Westchester County. Candidates must be United States citizens or permanent visa holders. Renewal of awards to previous recipients will be considered on a competitive basis.
Amount of Grant: $60,000 maximum
Date(s) Application is Due: Feb 5
Contact: Program Administrator, (212) 822-7204; fax: (212) 822-7338; email: glorney-raisbeck@nyam.org
Internet: http://www.nyam.org/grants/glorneyraisbeck.shtml
Sponsor: New York Academy of Medicine
1216 Fifth Avenue
New York, NY 10029-5202

NYAM Glorney-Raisbeck Medical Student Grants **4697**
Up to four $4,000 grants, consisting of a $3,500 student stipend plus $500 in support of the laboratory or department at which the student will conduct research, will be awarded to MD candidates who are either enrolled in a New York-area medical school or plan to conduct research at a New York-area institution. The grants typically support 8-12 week mentored cardiovascular research projects initiated during the summer between the first and second years of medical school, but medical school students in any year of their education may apply. Students are encouraged to pursue research projects that extend beyond this startup period.
Requirements: Applicants must provide a one-page research proposal, as well as a letter of support from the faculty mentor supervising the research.
Samples: Amy Chao (2006); Rachel Sharon Edlin (2006); Matthew Robert Greives (2006).
Amount of Grant: $4,000
Date(s) Application is Due: Mar 16
Contact: Program Administrator, (212) 822-7204; fax: (212) 822-7338; email: glorney_raisbeck@nyam.org
Internet: http://www.nyam.org/grants/raisbeckstudent.shtml
Sponsor: New York Academy of Medicine
1216 Fifth Avenue
New York, NY 10029-5202

NYAM Louis L. Seaman Medical Student Research Grants **4698**
Up to four $3,500 grants for students pursuing mentored research projects in microbiology will be awarded in 2007 to MD candidates who are either enrolled in a New York-area medical school or plan to conduct research at a New York-area institution. The grants typically support 8-12 week microbiology research projects initiated during the summer between the first and second years of medical school, but medical school students in any year of their education may apply. Students are encouraged to pursue research projects that extend beyond the start up period.
Requirements: Applicants must provide a one-page research proposal, as well as a letter of support from the faculty mentor supervising the research.
Samples: Ronald Damani Howell (2006); Stefan Kantrowitz (2006); Tanvi Nitin Parikh (2006); Roseann I. Wu (2006).
Amount of Grant: $3,500
Date(s) Application is Due: Mar 16
Contact: Program Administrator, (212) 822-7204; fax: (212) 996-7338; email: seaman@nyam.org
Internet: http://www.nyam.org/grants/seaman.shtml
Sponsor: New York Academy of Medicine
1216 Fifth Avenue
New York, NY 10029-5202

NYAM Mary and David Hoar Fellowship **4699**
The Mary and David Hoar Fund was established to promote research in the prevention and treatment of hip fractures. The Fund provides a two-year grant of $100,000 to support pilot programs in clinical, epidemiologic and health services research in the following areas related to hip fractures: clinical care and treatment; prevention; rehabilitation; and model program development.
Requirements: Candidates must hold an MD, PhD, or equivalent degree and are expected to conduct research in a supervised program in the greater New York area. Preferential consideration will be given to new investigators for whom this award is likely to provide interim support pending future funding, or to more senior investigators who are making a change in career or research direction for whom this award will facilitate the transition. Candidates are not required to be United States citizens; however, non-citizen applicants are required to provide assurances, such as permanent resident status, of legal eligibility for employment to perform the proposed research throughout the award period, regardless of the disposition of this or any other pending grants.
Amount of Grant: $100,000
Date(s) Application is Due: Feb 26
Contact: Program Administrator, (212) 822-7204; fax: (212) 822-7338; email: hoar@nyam.org
Internet: http://www.nyam.org/grants/hoar.shtml
Sponsor: New York Academy of Medicine
1216 Fifth Avenue
New York, NY 10029-5202

NYAM Paul Klemperer Fellowship **4700**
The Fellowship supports research using the Academy Library's resources for scholarly study of the history of medicine. It is intended specifically for a scholar in residence at the Academy Library. Although residence is not obligatory, preference will be given applicants whose projects require use of the resources of the Academy Library and who plan to spend time at the Academy. Besides completing a research or creative project, each fellow will be expected to make a public presentation at the Academy and submit a final report.
Requirements: Applications are invited from anyone, regardless of citizenship, academic discipline, or academic status. Preference will be given to (1) those whose research will take advantage of resources that are uniquely available at the Academy, and (2) scholars or creative artists in the early stages of their careers.
Samples: Daniel Margocsy (2005-06); Britta McEwen (2004-05); Sarah Tracy (2003-04).
Amount of Grant: $5,000 maximum
Date(s) Application is Due: Mar 6
Contact: Program Administrator, (212) 822-7314; fax: (212) 822-7338; email: history@nyam.org
Internet: http://www.nyam.org/grants/history.shtml
Sponsor: New York Academy of Medicine
1216 Fifth Avenue
New York, NY 10029-5202

NYCT Biomedical Research Grants **4701**
The Trust supports projects that strengthen preventive health care, improve access to services, promote the efficient use of health resources, and develop the skills and independence of people with special needs. The Trust provides funding for biomedical research projects in four areas: blood diseases, cancer, leprosy and tuberculosis. Grants for these projects are awarded in response to requests for proposals.
Requirements: Grants are made primarily to colleges, universities, and research organizations located in the five boroughs of New York City. Grants for programs outside the area are generally from funds designated for specific charities or have been made at the suggestion of donors.
Restrictions: The trust does not make grants to individuals, offer general or capital funding, endowments, building construction/renovation, deficit financing, films, or religion.
Geographic Focus: New York
Amount of Grant: $5,000-$100,000
Contact: Joyce M. Bove, Senior Vice President; (212) 686-0010, ext. 552; fax: (212) 534-8528; email: info@nycommunitytrust.org or grants@nycommunitytrust.org
Internet: http://www.nycommunitytrust.org/ForGrantSeekers/GrantmakingGuidelines/HealthandPeoplewithSpecialNeeds/tabid/207/Default.aspx
Sponsor: New York Community Trust
909 Third Avenue, 22nd Floor
New York, NY 10022

NYCT Blindness and Visual Disabilities Grants **4702**
The Trust supports projects that strengthen preventive health care, improve access to services, promote the efficient use of health resources, and develop the skills and independence of people with special needs. The goals of this program are to support program innovation and reform and eliminate service gaps. Specifically, grants are made to projects that: improve services to those people with visual disabilities who are presently underserved, such as the elderly, minorities, people with multiple disabilities, people in institutions, and youth in transition from school to work (projects need not be limited to direct service, and may include advocacy and organizational improvement); involve people with visual disabilities more fully in community activities; expand programs that identify people with vision problems at an early stage and link them with appropriate resources; empower persons with visual disabilities by enabling them to participate in planning programs that affect them; and support research in the prevention and treatment of blindness (this may include clinical, epidemiological, and applied studies). The grant review process takes from two to six months. Grants range from $5,000 to $100,000; an average grant is around $60,000.
Requirements: Grants are made primarily to nonprofit organizations located in the five boroughs of New York City. Grants for programs outside the area are generally from funds designated for specific charities or have been made at the suggestion of donors.
Restrictions: The trust does not make grants to individuals, offer general or capital funding, endowments, building construction/renovation, deficit financing, films, or religion.
Geographic Focus: New York
Amount of Grant: $5,000-$100,000
Contact: Joyce M. Bove, Senior Vice President; (212) 686-0010, ext. 552; fax: (212) 534-8528; email: info@nycommunitytrust.org or grants@nycommunitytrust.org
Internet: http://www.nycommunitytrust.org/ForGrantSeekers/GrantmakingGuidelines/HealthandPeoplewithSpecialNeeds/tabid/207/Default.aspx
Sponsor: New York Community Trust
909 Third Avenue, 22nd Floor
New York, NY 10022

NYCT Blood Disease Research Grants **4703**
The New York Community Trust is a charitable foundation concerned with the needs of the Greater New York area. The NYCT has a special fund to be used in support of research in the field of blood diseases. A panel of recognized hematologists will convene to review proposals solicited from the major medical research institutions on a competitive basis. It is estimated that about $700,000 will be available this year to support such research projects. A total of no more than two projects may be submitted from any one organization. Basic laboratory research. Candidates may apply for support

totaling $100,000 for up to 2 years, but funds for any one project year may not exceed $50,000. Second-year funding is contingent upon submission of a satisfactory progress report at the end of the first year. Research that may have application to New York City residents is of particular interest to The Trust. Young investigators will be given special consideration.
Requirements: Proposals are due no later than 5:00 p.m. on September 25th. Additional guidelines and forms are available online at the NYCT website.
Geographic Focus: New York
Amount of Grant: $100,000 over two years
Date(s) Application is Due: Sep 25
Contact: Joyce M. Bove, Senior Vice President; (212) 686-0010, ext. 552; fax: (212) 534-8528; email: info@nycommunitytrust.org or grants@nycommunitytrust.org
Internet: http://www.nycommunitytrust.org/BloodDiseaseResearch/tabid/504/Default.aspx
Sponsor: New York Community Trust
909 Third Avenue, 22nd Floor
New York, NY 10022

NYCT New York City Environment Grants 4704
The New York City program focuses on four urban environmental issues: promoting a more effective and sustainable solid waSuite management program through consumer education, research, and advocacy, with an emphasis on waSuite reduction and equitable distribution of solid waSuite facilities; expanding open space and parks through advocacy, constituency building, and support of local planning and greening activities; reclaiming the waterfront and brownfields through technical assistance, advocacy, and partnerships between community development and environmental groups; reducing air pollutants and other environmental toxins through cleaner fuels and engines, mass transit improvements, and elimination of indoor pollutants and the concentration of polluting facilities in poor communities. In the New York metropolitan region, the NYCT aims are, to preserve and restore habitat, promote a more sustainable transportation system, protect open space and drinking water, and redevelop brownfields.
Requirements: Grants are made primarily to nonprofit organizations located in the five boroughs of New York City. Grants for programs outside the area are generally from funds designated for specific charities or have been made at the suggestion of donors.
Restrictions: The trust does not make grants to individuals, offer general or capital funding, endowments, building construction/renovation, deficit financing, films, or religion.
Geographic Focus: New York
Contact: Joyce M. Bove, Senior Vice President; (212) 686-0010, ext. 552; fax: (212) 534-8528; email: info@nycommunitytrust.org or grants@nycommunitytrust.org
Internet: http://www.nycommunitytrust.org/ForGrantSeekers/GrantmakingGuidelines/CommunityDevelopmentandtheEnvironment/tabid/204/Default.aspx
Sponsor: New York Community Trust
909 Third Avenue, 22nd Floor
New York, NY 10022

NYSCA Architecture, Planning, and Design: Design and Planning Studies Grants 4705
Funding is available for any nonprofit organization or local governmental agency in New York State to engage the services of an architect, or planning, design, or historic preservation professional for a wide variety of planning and design studies. Previously funded projects have included the design of an all new, universally designed live/work facility for an artist colony; a seven-county regional planning study; urban design and Main Street revitalization studies; open-space planning projects; site-specific architect/artist collaborations; conversion of a former train depot into a regional visitor center; and the development of a graphic sign system for a museum. Jointly sponsored projects and those that actively engage the applicant's community - for example, through a design charrette - are also encouraged. The registration deadline is March 3rd.
Requirements: New York State organizations that present documented proof of nonprofit status are eligible to apply.
Restrictions: The Program does not support bid negotiation, construction administration, purchase of real estate or capital construction costs, out-of-state travel, salaries or overhead of public agencies, or student projects.
Geographic Focus: New York
Date(s) Application is Due: Apr 1; Jul 1; Sep 5
Contact: Anne van Ingen, Director; (212) 741-7013; email: avaningen@nysca.org; Marcy Shaffer, Associate; (212) 741-5258; email: mshaffer@nysca.org
Internet: http://www.nysca.org/public/guidelines/architecture/design_planning.htm
Sponsor: New York State Council on the Arts
175 Varick Street
New York, NY 10014-4604

NYZS/Wildlife Conservation Society Research Grants 4706
The society operates an international program of research and action that has a direct bearing on the conservation of threatened wild animals and habitats. The research development program is designed to support individual research projects that lead to concrete advances in the understanding and conservation of wildlife. Each project must incorporate clearly identifiable, specific conservation components. To obtain an application, send a one- or two-page cover letter containing a summary of the proposed research. The RFP will not limit any individual from applying, however, most of our grantees are either professional conservationists from the country of research, and/or postgraduates pursuing a higher degree.
Requirements: Any individual is eligible to apply. Applications will be accepted for field research in Africa, Asia, and Latin America (including Mexico.)
Restrictions: Organizations are not eligible for funding. Previous research fellows are not eligible for funding for the same project. Faculty and/or research advisors should not be listed as principal investigators unless they plan to conduct and achieve the majority of the field work. The principal researcher must write the proposal. Applications written on behalf of another individual are not allowed. Translation by a third party is permitted, however, applications are accepted in English, French and Spanish. Support will not be provided for conferences, airfare to scientific meetings, legal actions, building of permanent field stations, salaries at institutions, overhead costs, or expensive laboratory analyses. Stipends will be considered only when the investigator has no other source of support. Research will not be supported in North America (except Mexico), Europe, or Australia.
Amount of Grant: $1000-$25,000; $11,000 average
Date(s) Application is Due: Mar 1; Sep 1
Contact: Research Fellowship Coordinator, (718) 220-6828; fax: (718) 364-4275; email: fellowship@wcs.org
Internet: http://www.wcs.org/international/rfp
Sponsor: New York Zoological Society/The Wildlife Conservation Society
2300 Southern Boulevard
Bronx, NY 10460

O.H. Ammann Research Fellowship in Structural Engineering 4707
Nominations are being solicited for the annual O.H. Ammann Research Fellowship in Structural Engineering, which is bestowed annually to a member for the purpose of encouraging the creation of new knowledge in the field of structural design and construction. The recipient may apply for a renewal, additional stipend to be determined annually. Applications must be postmarked by the listed application deadline date with notification being made in May for a fellowship beginning that same year.
Requirements: Applicants must be members of the society or applicants for membership; citizens of countries other than the United States may apply if they are eligible under this rule. Each application should include completed application form, official transcripts, a description of proposed research and its objectives, a statement from the institution at which the research is to be done that the applicant and proposed research are acceptable to that institution, and a statement in general terms of the purposes for which the funds are expected to be used.
Restrictions: During the tenure of the fellowship, the recipient may not work on research projects other than that for which this fellowship has been granted, but recipients may accept other awards if the conditions of such awards are the same as those for this fellowship.
Samples: Dragana Jankovic, Delft University of Technologies, Delft, Netherlands, $5,000 (2006).
Amount of Grant: $5,000
Date(s) Application is Due: Feb 19
Contact: Debbie Smith, Senior Program Coordinator; (703) 295-6095; fax: (703) 295-6132; email: dsmith@asce.org
Internet: http://www.seinstitute.org/inside/ammann.cfm
Sponsor: American Society of Civil Engineers
1801 Alexander Bell Drive
Reston, VA 20191-4400

OAH ABC-CLIO America: History and Life Award 4708
The biennial award recognizes and encourages scholarship in American history in the journal literature advancing new perspectives on accepted interpretations or previously unconsidered topics. The award is not confined to any particular subject area or chronological period but instead seeks to recognize journal articles that display a pathbreaking character, articles that do not simply elaborate previous interpretations but that propose new perspectives or examine new areas and in so doing extend historical knowledge and raise new avenues of investigation. The award will be presented at the appropriate annual meeting of the Organization of American Historians. One copy of each entry, clearly labeled ABC-CLIO America: History and Life Award Entry, must be mailed directly to committee members. Guidelines are available online.
Requirements: Individuals as well as editors are encouraged to submit nominations. Each entry must be published during the period November 16 of last year through November 15 of next year.
Samples: Joon K. Kim, Colorado State University—The Political-Economy of the Mexican Farm Labor Program, 1942-1964 (Aztlan: A Journal of Chicano Studies), 2007.; Francois Furstenberg, Universite de Montreal—Beyond Freedom and Slavery: Autonomy, Virtue, and Resistance in Early American Political Discourse (Journal of American History, March 2003), 2005.
Amount of Grant: $750
Date(s) Application is Due: Dec 1
Contact: Kara Hamm, Award and Prize Coordinator, (812) 855-9852; fax: (812) 855-0696; email: kara@oah.org
Internet: http://www.oah.org/activities/awards/abcclio/index.html
Sponsor: Organization of American Historians
P.C. Box 5457, 112 N. Bryan Avenue
Bloomington, IN 47408-4199

OAH Avery O. Craven Award 4709
The award is given each year for the most original book on the coming of the Civil War, the Civil War years, or the era of Reconstruction, with the exception of works of purely

military history. One copy of each entry, clearly labeled Avery O. Craven Award Entry, must be mailed directly to committee members. Guidelines are available online.
Requirements: Each entry must be published during the period January 1 through December 31 of the current year.
Samples: Mark Elliott, Wagner College—Color-Blind Justice: Albion Tourgee and the Quest for Racial Equality from the Civil War to Plessy v. Ferguson (Oxford University Press, 2007); Anne Sarah Rubin, University of Maryland—Baltimore County, A Shattered Nation: The Rise and Fall of the Confederacy, 1861-1868 (University of North Carolina Press, 2006); C. Wyatt Evans, Drew University—The Legend of John Wilkes Booth: Myth, Memory, and a Mummy (University Press of Kansas, 2005).
Amount of Grant: $500
Date(s) Application is Due: Oct 1
Contact: Kara Hamm, Award and Prize Coordinator, (812) 855-9852; fax: (812) 855-0696; email: kara@oah.org
Internet: http://www.oah.org/activities/awards/craven/index.html
Sponsor: Organization of American Historians
P.O. Box 5457, 112 N. Bryan Avenue
Bloomington, IN 47408-4199

OAH Binkley-Stephenson Award 4710
The Binkley-Stephenson Award has honored scholarly work since 1967. The award recognizes the best article published in the Journal of American History during the preceding calendar year. The journal is published quarterly in March, June, September, and December. It is widely recognized as an important source of research on all aspects of American history. William C. Binkley was president of the OAH 1944-1946 and editor of the Journal of American History 1956-1963.
Samples: Laura McEnaney, Whittier College—Nightmares on Elm Street: Demobilizing in Chicago, 1945-1953 (JAH, March 2006), $500 (2007); Jamie J. Fader, Michael B. Katz, and Mark J. Stern, University of Pennsylvania—The New African American Inequality (JAH, June 2005), $500 (2006); Jason Opal, Colby College—Exciting Emulation: Academies and the Transformation of the Rural North, 1780s-1820s (JAH, September 2004), $500 (2005).
Amount of Grant: $500
Contact: Kara Hamm, Award and Prize Coordinator, (812) 855-9852; fax: (812) 855-0696; email: kara@oah.org
Internet: http://www.oah.org/activities/awards/binkleystephenson/index.html
Sponsor: Organization of American Historians
P.O. Box 5457, 112 N. Bryan Avenue
Bloomington, IN 47408-4199

OAH David Thelen Award 4711
The OAH sponsors this biennial award (formerly the Foreign Language Article Prize through 1998) for the best article on American history published in a foreign language. The winning article will be published in the Journal of American History. An article should be concerned with the past (recent or distant) or with issues of continuity and change. It should also be concerned with events or processes that began, developed, or ended in what is now the United States. It should make a significant and original contribution to the understanding of U.S. history. Comparative and international studies that fall within these guidelines are welcome.
Requirements: Entries must have been published during the preceding two years.
Samples: Manfred Berg, Ruprecht-Karls-Universitat—Black Civil Rights and Liberal Anticommunism: The NAACP during the McCarthy Era (2006); Simone Cinotto, University of Turin—Leonard Covello, the Covello Collection, and the History of Eating Habits among Italian Immigrants in New York (2004).
Amount of Grant: $500
Date(s) Application is Due: May 1
Contact: Kara Hamm, Award and Prize Coordinator, (812) 855-9852; fax: (812) 855-0696; email: kara@oah.org
Internet: http://www.oah.org/activities/awards/thelen/index.html
Sponsor: Organization of American Historians
P.O. Box 5457, 112 N. Bryan Avenue
Bloomington, IN 47408-4199

OAH Ellis W. Hawley Prize 4712
The prize is awarded annually for the best book-length historical study of the political economy, politics, or institutions of the United States, in its domestic or international affairs, from the Civil War to the present. Eligible works include book-length historical studies, written in English, published during a given calendar year. Dissertations completed by October 1 also are eligible. One copy of each entry, clearly labeled Ellis W. Hawley Prize Entry, must be mailed directly to committee members. Guidelines are available online.
Requirements: Each entry must be published during the period January 1 through December 31 of the current year,
Samples: Marie Gottschalk, University of Pennsylvania—The Prison and the Gallows: The Politics of Mass Incarceration in America (Cambridge University Press, 2007); Meg Jacobs, Massachusetts Institute of Technology—Pocketbook Politics: Economic Citizenship in Twentieth-Century America (Princeton University Press, 2006); Alison Isenberg, Rutgers University—New Brunswick, Downtown America: A History of the Place and the People Who Made It (The University of Chicago Press, 2005).
Amount of Grant: $500
Date(s) Application is Due: Oct 1
Contact: Kara Hamm, Award and Prize Coordinator, (812) 855-9852; fax: (812) 855-0696; email: kara@oah.org
Internet: http://www.oah.org/activities/awards/hawley/index.html
Sponsor: Organization of American Historians
P.O. Box 5457, 112 N. Bryan Avenue
Bloomington, IN 47408-4199

OAH Erik Barnouw Award 4713
One or two awards are given annually in recognition of outstanding reporting or programming on network or cable television, or in documentary film, concerned with American history, the study of American history, and/or the promotion of history. The winning film(s) or video program(s) will be screened, and the award(s) will be presented at the annual meeting in March. Films and videos selected as honorable mentions will be shown at the annual meeting, along with the Barnouw winner, as part of the OAH Screening History film series. One copy of each entry, clearly labeled Erik Barnouw Award Entry, must be mailed directly to committee members.
Requirements: Films and video programs released between January 1 and December 31 of the current year are eligible for entry.
Samples: Randall MacLowry—The Gold Rush (a Yellow Jersey Films, Inc. production for American Experience, produced in association with KQED television; Randall MacLowry and Laura Longsworth producers; directed by Randall MacLowry, 2007); The Documentary Institute, College of Journalism and Communications—Negroes with Guns: Rob Williams and Black Power (Sandra Dickson & Churchill Roberts, co-directors; Cindy Hill & Cara Pilson, associate directors; California Newsreel, distributor, 2006).
Amount of Grant: $1,000 (or $500 should two films be selected)
Date(s) Application is Due: Dec 1
Contact: Kara Hamm, Award and Prize Coordinator, (812) 855-9852; fax: (812) 855-0696; email: kara@oah.org
Internet: http://www.oah.org/activities/awards/barnouw/index.html
Sponsor: Organization of American Historians
P.O. Box 5457, 112 N. Bryan Avenue
Bloomington, IN 47408-4199

OAH Frederick Jackson Turner Award 4714
The award is given annually for an author's first book dealing with some significant phase of American history. The work must be the first book-length study of history published by the author and must have been published or scheduled for publication in the calendar year before the award is given. A copy of each entry should be submitted. Final page proofs may be used for books to be published after September 1 and before January 1 of the following year. The publisher of the winning book will receive a complimentary ad for the book in the Journal of American History and a certificate and the author will receive a medal.
Requirements: If the author is a PhD, she/he must have received the degree no earlier than seven years prior to submitting the manuscript for publication. Each entry must be published during the period January 1 through December 31 of the current year.
Restrictions: Each press may submit one entry per year.
Samples: Ned Blackhawk, University of Wisconsin, Madison—Violence over the Land: Indians and Empires in the Early American West (Harvard University Press, 2007); Tiya Alicia Miles, University of Michigan—Ties that Bind: The Story of an Afro-Cherokee Family in Slavery and Freedom (University of California Press, 2006); Mae M. Ngai, University of Chicago—Impossible Subjects: Illegal Aliens and the Making of Modern America (Princeton University Press, 2005).
Amount of Grant: $1000
Date(s) Application is Due: Oct 1
Contact: Kara Hamm, Award and Prize Coordinator, (812) 855-9852; fax: (812) 855-0696; email: kara@oah.org
Internet: http://www.oah.org/activities/awards/turner/index.html
Sponsor: Organization of American Historians
P.O. Box 5457, 112 N. Bryan Avenue
Bloomington, IN 47408-4199

OAH Huggins-Quarles Award 4715
The award is given annually to one or two minority graduate students at the dissertation research stage of their PhD program. To apply, a student should submit a brief two-page abstract of his/her dissertation project, along with a one-page budget explaining travel and research plans. Each application must be accompanied by a letter from the dissertation adviser attesting to the student's status and the ways in which the award will facilitate the completion of the dissertation project. Include email addresses for both the applicant and the adviser, if available. One complete copy of each application (including cover letter, abstract, budget, and reference letter) should be clearly labeled Huggins-Quarles Award Entry and received, by the listed deadline date, by each member of the committee. Guidelines are available online.
Samples: Geraldo Lujan Cadava, Yale University—Ethnic Histories of Tucson and the Arizona-Sonora Borderlands, 1941-1981, $1,000 (2007); Julian S. Lim, Cornell University—Race, Nations, Borders, and Freedoms: Multiracial Societies in the U.S.-Mexico Borderlands, 1880-1930, $1,000 (2007).
Amount of Grant: $1,000 award ($2,000 if only one is awarded)
Date(s) Application is Due: Dec 1
Contact: Kara Hamm, Award and Prize Coordinator, (812) 855-9852; fax: (812) 855-0696; email: kara@oah.org
Internet: http://www.oah.org/activities/awards/hugginsquarles/index.html
Sponsor: Organization of American Historians
P.O. Box 5457, 112 N. Bryan Avenue
Bloomington, IN 47408-4199

OAH James A. Rawley Prize 4716

This prize is given each year for a book dealing with the history of race relations in the United States. Each entry must be published during the period January 1 through December 31 in the year of deadline date. One copy of each entry, clearly labeled James A. Rawley Prize Entry, must be mailed directly to each committee member. The award will be presented at the OAH annual meeting. The publisher will receive a certificate of merit.

Samples: Paul A. Kramer, University of Michigan, Ann Arbor MI—The Blood of Government: Race, Empire, the United States, and the Phillippines (The University of North Carolina Press, 2007); James Edward Smethurst, University of Massachusetts, Amherst, MA—The Black Arts Movement: Literary Nationalism in the 1960s and 1970s (University of North Carolina Press, 2006); Robert O. Self, Brown University—American Babylon: Race and the Struggle for Postwar Oakland (Princeton University Press, 2005).

Amount of Grant: $1,000

Date(s) Application is Due: Oct 1

Contact: Award and Prize Coordinator, (812) 855-9852; fax: (812) 855-0696; email: awards@oah.org

Internet: http://www.oah.org/activities/awards/rawley/index.html

Sponsor: Organization of American Historians
P.O. Box 5457, 112 N. Bryan Avenue
Bloomington, IN 47408-4199

OAH Japanese Residencies Program Grants 4717

The OAH sends two Ameri–can scholars to Japanese universities annually for two-week residencies. There, in English, the American historians give lectures and seminars in their specialty and provide individual consultation to Japanese schol–ars, graduate students and sometimes undergraduates studying American history and culture. Visitors also participate in the collegial life of their host institutions and help expand personal scholarly networks between Japan and the U.S. The aim of the program is to contribute to the expansion of personal scholarly networks between the two countries. Round-trip airfare to Japan, housing, and modest daily expenses are covered. Awardees are also encouraged to explore Japan before or after their two-week residency at their own expense.

Requirements: Applicants must be members of the OAH, have a Ph.D., and be scholars of American history.

Samples: Kathryn Kish Sklar, State University of New York, Binghamton, cisiting Tokyo Woman's Christian University (2007); Thomas J. Sugrue, University of Pennsylvania, visiting Nanzan University (2007); Donald L. Fixico, Arizona State University, visiting Rikkyo University (2006); Leo P. Ribuffo, George Washington University, visiting Sophia University (2006).

Date(s) Application is Due: Oct 15

Contact: Kara Hamm, Award and Prize Coordinator, (812) 855-9852; fax: (812) 855-0696; email: kara@oah.org

Internet: http://www.oah.org/activities/awards/japan/index.html

Sponsor: Organization of American Historians
P.O. Box 5457, 112 N. Bryan Avenue
Bloomington, IN 47408-4199

OAH Lerner-Scott Dissertation Prize 4718

This prize is given annually to the best doctoral dissertation in women's history. Each application must contain a letter of support from a faculty member at the degree-granting institution, along with an abstract, table of contents and sample chapter from the dissertation. Also include email addresses for both the applicant and the adviser, if available. One complete copy of each entry must be received by each member of the prize committee by the listed deadline. Finalists will be asked to submit a complete copy of the dissertation at a later date. One copy of each entry, clearly labeled Lerner-Scott Dissertation Prize Entry, must be mailed directly to committee members. Guidelines are available online.

Requirements: A dissertation must be completed during the period July 1 of last year through June 30 of the current year.

Samples: Serena Mayeri, University of Pennsylvania Law School—Reasoning from Race: The Civil Rights Paradigm and American Legal Feminism, 1960-1979 (2007); Margot Canaday, Princeton University—The Straight State: Sexuality and American Citizenship, 1900-1969 (2006); Vasantha Lynn Kennedy, University of Lethbridge—Partus Sequitur Ventrem: Narratives of Childbirth and Motherhood in the Antebellum South (2005).

Amount of Grant: $1,000

Date(s) Application is Due: Oct 1

Contact: Kara Hamm, Award and Prize Coordinator, (812) 855-9852; fax: (812) 855-0696; email: kara@oah.org

Internet: http://www.oah.org/activities/awards/lernerscott/index.html

Sponsor: Organization of American Historians
P.O. Box 5457, 112 N. Bryan Avenue
Bloomington, IN 47408-4199

OAH Liberty Legacy Foundation Award 4719

The award will be given for the best book on any historical aspect of the struggle for civil rights in the United States from the nation's founding to the present. One copy of each entry must be received by each committee member by the listed application deadline date. Final page proofs may be used for books to be published after the deadline and before 1 January. If a final page proof is submitted, a bound copy of the entry must be received no later than 7 January. The author of the winning book will receive a monetary prize and a certificate. The publisher of the book will be honored with a certificate of merit. One copy of each entry, clearly labeled 2006 Liberty Legacy Foundation Award, must be mailed directly to committee members. Guidelines are avaialable onlin.

Requirements: Each entry must be published during the period 1 January through 31 December.

Samples: Thomas F. Jackson, University of North Carolina Greensboro—From Civil Rights to Human Rights: Martin Luther King, Jr. and the Struggle for Economic Justice (University of Pennsylvania Press, 2007); Matthew J. Countryman, University of Michigan—Up South: Civil Rights and Black Power in Philadelphia (University of Pennsylvania Press, 2006); Nikhil Pal Singh, University of Washington, Seattle—Black is a Country: Race and the Unfinished Struggle for Democracy (Harvard University Press, 2005).

Amount of Grant: $2,000

Date(s) Application is Due: Oct 1

Contact: Kara Hamm, Award and Prize Coordinator, (812) 855-9852; fax: (812) 855-0696; email: kara@oah.org

Internet: http://www.oah.org/activities/awards/liberty/index.html

Sponsor: Organization of American Historians
P.O. Box 5457, 112 N. Bryan Avenue
Bloomington, IN 47408-4199

OAH Louis Pelzer Memorial Award 4720

Candidates for graduate degrees are invited to submit essays for the Louis Pelzer Memorial Award competition. Essays may deal with any period or topic in the history of the United States. The winning essay will be published in the Journal of American History. Significance of the subject matter, literary craftsmanship and competence in the handling of evidence are some of the factors that will be considered in judging the essays. Manuscripts should be clearly labeled (list current year) Louis Pelzer Memorial Award Entry and and addressed to the committee chair). The winner will be notified by the organization and furnished with details of the annual meeting and the awards presentation. Guidelines are available online.

Requirements: Essays should be submitted in quintuplicate and should not exceed 7000 words (including endnotes) in length. The endnotes should be triple-spaced. Because manuscripts are judged anonymously, the author's name and graduate program should appear only on a separate cover page.

Samples: Andrew W. Kahrl, Indiana University—Why the Police at No. 4 Get Busy When They Hear the Whistle of the Razor Beach Boat: Steamboat Excursions, Pleasure Resorts, and the Emergence of Segregation Culture on the Potomac River, 1890-1920 (2007); Wendy Anne Warren, Yale University—The Cause of Her Grief: The Rape of a Slave Woman in Early New England (2006); Kevin Dawson, University of South Carolina—Enslaved Swimmers and Divers in the Atlantic World (2005).

Amount of Grant: $500

Date(s) Application is Due: Nov 30

Contact: Kara Hamm, Award and Prize Coordinator, (812) 855-9852; fax: (812) 855-0696; email: kara@oah.org

Internet: http://www.oah.org/activities/awards/pelzer/index.html

Sponsor: Organization of American Historians
P.O. Box 5457, 112 N. Bryan Avenue
Bloomington, IN 47408-4199

OAH Merle Curti Award 4721

The award is given annually to the best work published in the last two years in the field of American social history (even-numbered years) and intellectual history (odd-numbered years). The author of the winning book will receive a monetary prize, a certificate, and a medal. The publisher(s) of the book(s) will be honored with a certificate(s) of merit. The award will be presented at the annual meeting of the OAH. One copy of each entry, clearly labeled Merle Curti Award Entry, must be mailed directly to the members of the chair. Guidelines are available online.

Samples: Scott Reynolds Nelson, College of William & Mary—Steel Drivin' Man: John Henry: the Untold Story of an American Legend (Oxford University Press), $2,000 (2007); Steven Mintz, U of Houston (TX)—Huck's Raft, A History of American Childhood (Harvard University Press), $1000 (2005); Michael O'Brien, University of Cambridge (MA)—Conjectures of Order, Intellectual Life and the American South, 1810-1860, $1000 (2005).

Amount of Grant: $2,000 (or $1,000 should two books be selected)

Date(s) Application is Due: Oct 1

Contact: Kara Hamm, Award and Prize Coordinator, (812) 855-9852; fax: (812) 855-0696; email: kara@oah.org

Internet: http://www.oah.org/activities/awards/curti/index.html

Sponsor: Organization of American Historians
P.O. Box 5457, 112 N. Bryan Avenue
Bloomington, IN 47408-4199

OAH Ray A. Billington Prize 4722

The award is given biennially for the best book on American frontier history published during the last two years. American frontier history is defined broadly so as to include the pioneer periods of all geographical areas and comparisons between American frontiers and others. The deadline is for even-numbered years. One copy of each entry, clearly labeled Ray Allen Billington Prize Entry, must be mailed directly to each committee member. Guidelines are available online.

Samples: Pablo R. Mitchell, Oberlin College—Coyote Nation: Sexuality, Race, and Conquest in Modernizing New Mexico, 1880-1920 (The University of Chicago Press), $1,000 (2007); Colin G. Calloway, Dartmouth College—for One Vast Winter Count:

The Native American West before Lewis and Clark (University of Nebraska Press), $1,000 (2005).
Amount of Grant: $1,000
Date(s) Application is Due: Oct 1
Contact: Kara Hamm, Award and Prize Coordinator, (812) 855-9852;fax: (812) 855-0696;email: kara@oah.org
Internet: http://www.oah.org/activities/awards/billington/index.html
Sponsor: Organization of American Historians
P.O. Box 5457, 112 N. Bryan Avenue
Bloomington, IN 47408-4199

OAH Richard W. Leopold Prize **4723**
This award is designed to improve contacts and interrelationships within the historical profession where an increasing number of history-trained scholars hold distinguished positions in governmental agencies. This prize recognizes the significant historical work being done by historians outside academe. The award is given every two years for the best book written by a historian connected with federal, state, or municipal government, in the areas of foreign policy, military affairs broadly construed, the historical activities of the federal government, or biography in one of the foregoing areas. One copy of each entry, clearly labeled Richard W. Leopold Prize Entry, must be mailed directly to committee members.
Requirements: Each entry must be published during the two-year period between January 1 of the last year and December 31 of the current year. The winner must have been employed in a government position for at least five years.
Samples: Robert J. Schneller Jr., Naval Historical Center—Breaking the Color Barrier: The U.S. Naval Academy's First Black Midshipmen and the Struggle for Racial Equality (New York University Press), $1,500 (2006); Peter S. Kindsvatter, U.S. Army Ordnance Center and School—American Soldiers: Ground Combat in the World Wars, Korea, and Vietnam (University Press of Kansas), $1,500 (2004).
Amount of Grant: $1,500
Date(s) Application is Due: Oct 1
Contact: Kara Hamm, Award and Prize Coordinator, (812) 855-9852; fax: (812) 855-0696; email: kara@oah.org
Internet: http://www.oah.org/activities/awards/leopold/index.html
Sponsor: Organization of American Historians
P.O. Box 5457, 112 N. Bryan Avenue
Bloomington, IN 47408-4199

OAH White House Historical Association Fellowships **4724**
The program seeks proposals for projects shedding light on the roles of the White House as home, workplace, museum, structure, and symbol. Fellowships include: The White House History Fellowship in Precollegiate Education—for initiatives that reach the K-12 classroom; the White House History Research Fellowship—for forwarding or completing dissertation, postdoctoral, or advanced academic work; and the White House History Fellowship in Public History—for public presentation in the form of exhibits, multimedia projects, films, etc., or for other projects that make historical collections available to broad audiences. Fellowships lasting one to six months will be considered. To apply, mail eight (8) complete copies of the application to each committee member. Guidelines are available online.
Requirements: Teachers and scholars whose work enhances understanding of how the White House functions in its several capacities and of life and work at all levels within the walls of the President's House are encouraged to apply. (Studies that deal primarily with political or governmental policy issues would not be appropriate for this program, but ones concerning the operation of the White House as a political institution would be considered.)
Samples: Glenda Armand Sheppard, Los Angeles Unified School District, California—Frederick Douglass: From Slave Cabin to the White House (2007); Michelle L. Pearson, Hulstrom Options at Rocky Top Middle School, Thornton, Colorado, and Christopher T. Jennings, Library of Congress, Washington, D.C.—Using Google Earth to Track the Mobile White House (2007).
Amount of Grant: $2,000 per month for up to six months
Date(s) Application is Due: Dec 1
Contact: Kara Hamm, Award and Prize Coordinator, (812) 855-9852; fax: (812) 855-0696; email: kara@oah.org
Internet: http://www.oah.org/activities/awards/whha/index.html
Sponsor: Organization of American Historians
P.O. Box 5457, 112 N. Bryan Avenue
Bloomington, IN 47408-4199

OAH Willi Paul Adams Award **4725**
The OAH sponsors a biennial award for the best book on American history published in a foreign language. The award was formerly known as the Foreign Language Book Prize. The OAH defines both "history" and "American" broadly. To be eligible, a book should be concerned with the past (recent or distant) or with issues of continuity and change. It should also be concerned with events or processes that began, developed, or ended in what is now the United States. Comparative and international studies that fall within these guidelines are welcomed. Authors of eligible books are invited to nominate their work. Under unusual circumstances unpublished manuscripts will be considered.
Requirements: Each entry must have been published during the two-year period July 1 through June 30 of the current year.
Samples: Pierre Gervais, Departement d'Histoire, Universite Paris 8/CENA-Mascipo UMR 8168, Ecole des Hautes Etudes en Sciences Sociales-CNRS—The Origins of the Industrial Revolution in the United States: From Mercantile Economy to Industrial Capitalism, 1800-1850 (2007); Michel Cordillot, Universite Paris—The Democratic and Social Republic In America: French-Speaking Radicals in the U.S.A, a Biographical Dictionary, 1848-1922 (2005).
Amount of Grant: $1,000
Date(s) Application is Due: May 1
Contact: Kara Hamm, Award and Prize Coordinator, (812) 855-9852; fax: (812) 855-0696; email: kara@oah.org
Internet: http://www.oah.org/activities/awards/Adams/index.html
Sponsor: Organization of American Historians
P.O. Box 5457, 112 N. Bryan Avenue
Bloomington, IN 47408-4199

OAH/IEHS Higham Travel Grants **4726**
Travel grants of $500 are awarded to three (3) graduate students each year. Funds are to be used by graduate students toward costs of attending the OAH/IEHS annual meeting. The successful candidates will have a preferred area of concentration in American Immigration and/or American Ethnic and/or American Intellectual history. The committee will seek some balance by gender, region of country, and type of university (e.g., major research university, and second tier).
Samples: Maddalena Marinari, University of Kansas—Toward a New Era: World War II and the Fight Against Immigration Restriction (2007); Eric R. Schlereth, Brandeis University—Creating a Disenchanted Republic: American Political Independence and the Problem of Religion (2007); Stephen Seng-hua Mak, Northwestern University—The Other Internment: The United States, Latin America and Enemy Aliens During the Second World War (2007).
Amount of Grant: $500
Date(s) Application is Due: Dec 1
Contact: Kara Hamm, Award and Prize Coordinator, (812) 855-9852; fax: (812) 855-0696; email: kara@oah.org
Internet: http://www.oah.org/activities/awards/higham/index.html
Sponsor: Organization of American Historians / Immigration and Ethnic History Society
P.O. Box 5457, 112 N. Bryan Avenue
Bloomington, IN 47408-4199

OAS Regular Training Program Fellowships **4727**
Fellowships contribute to the economic, social, technical, and cultural development of OAS member countries and bring about a closer understanding among the peoples of the Americas. These fellowships (PRA) are awarded for advanced study at the graduate level and for research in any field except the medical sciences. The member countries in which a candidate may apply are Antigua and Barbuda, Argentina, Bahamas, Barbados, Bolivia, Brazil, Chile, Colombia, Costa Rica, Dominica, Dominican Republic, Ecuador, El Salvador, Grenada, Guatemala, Haiti, Honduras, Jamaica, Mexico, Nicaragua, Panama, Paraguay, Peru, Saint Lucia, Saint Vincent and the Grenadines, Suriname, Trinidad and Tobago, the United States, Uruguay, and Venezuela, with the exception of the country of which the candidate is a citizen or in which he/she maintains a permanent residence. Fellowships are granted for periods of not less than three months nor more than two years. Travel expenses, living allowance, tuition fees, study materials, and health insurance that varies from country to country are included; no benefits are provided for the family.
Requirements: Those who may apply are citizens or permanent residents of an OAS member country with university degrees or who have demonstrated ability to pursue advanced studies in the field chosen. Candidates must know the language of the study country and must choose the university, study center, or research site and make the necessary contacts to secure acceptance or show evidence that the facilities needed to complete a research project will be provided.
Restrictions: Fellowships are not intended for research in medicine or for language training.
Amount of Grant: $20,000
Contact: Inter-American Agency for Cooperation and Development, (202) 458-3902; fax: (202) 458-3527; email: fellcon4@iacd.oas.org
Internet: http://www.educoas.org/portal/en/becas/acerca.aspx?culture=en&tabindex=44&childindex=45
Sponsor: Organization of American States
1889 F Street NW
Washington, DC 20006

Oberly Award for Bibliography in the Agricultural Sciences **4728**
This is a biennial award given in odd-numbered years to a U.S. citizen who compiles the best bibliography in the field of agriculture or one of the related sciences in the two-year period preceding the year in which the award is made. The bibliographies are judged on accuracy, scope, usefulness, format, and special features such as explanatory introductions, annotations, and indexes. The award consists of a cash award and a citation. Nominations should be submitted to Paul Kelsey, Louisiana State University Libraries, Baton Rouge, LA 70803, Phone: (225)578-9433. If you have questions, contact Paul via email at pkelsey@lsu.edu.
Requirements: Nomination may be made in the form of a letter and should point out the reasons the bibliography should be considered for the award; a copy of the bibliography should accompany the nomination.
Samples: David Frodin—award winner for Guide to Standard Floras of the World. Second Edition (2003).
Amount of Grant: $350
Date(s) Application is Due: Dec 31
Contact: Paul Kelsey, (225) 578-9433; email: pkelsey@lsu.edu

Internet: http://www.ala.org/Template.cfm?Section=Awards17&template=/ContentManagement/ContentDisplay.cfm&ContentID=44129
Sponsor: Association of College and Research Libraries
50 E. Huron Street
Chicago, IL 60611

Obsessive-Compulsive Foundation Research Awards 4729
The foundation invites proposals for research projects on any aspect of obsessive-compulsive disorders. Applicants should submit a proposal limited to five pages and a budget. The foundation will request more detailed proposals following initial review.
Amount of Grant: $39,000-$49,000
Contact: Jeannette Cole, (888) 302-6662 or (203) 401-2069; fax: (203) 315-2196; email: cole@ocfoundation.org
Internet: http://www.ocfoundation.org/ocf1140a.htm
Sponsor: Obsessive-Compulsive Foundation
337 Notch Hill Road
North Branford, CT 06471

OBSSR Behavioral and Social Science Research on Understanding and Reducing Health Disparities Grants 4730
The purpose of this program is to encourage behavioral and social science research on the causes and solutions to health and disabilities disparities in the U. S. population. Health disparities between, on the one hand, racial/ethnic populations, lower socioeconomic classes, and rural residents and, on the other hand, the overall U.S. population are major public health concerns. Emphasis is placed on research in and among three broad areas of action: public policy; health care; and disease/disability prevention. Particular attention is given to reducing health gaps among groups. Proposals that utilize an interdisciplinary approach, investigate multiple levels of analysis, incorporate a life-course perspective, and/or employ innovative methods such as system science or community-based participatory research are particularly encouraged. The NIH anticipates supporting 20 to 30 awards. Because the nature and scope of the proposed research will vary from application to application, it is anticipated that the size and duration of each award will also vary.
Requirements: Application can be made by the following institutions and organizations: public/state controlled institutions of higher education; private institutions of higher education; nonprofits with 501(c)3 IRS status (other than institutions of higher education); nonprofits without 501(c)3 IRS status (other than institutions of higher education); small businesses; for-profit organizations (other than small businesses); state governments; U.S. territories or possessions; Indian/Native American tribal governments (federally recognized); Indian/Native American tribal governments (other than Federally recognized); Indian/Native American tribally designated organizations; non-domestic (non-U.S.) entities (foreign organizations); Hispanic-serving institutions; historically Black colleges and universities (HBCUs); tribally controlled colleges and universities (TCCUs); Alaska Native and Native Hawaiian serving institutions; regional organization; eligible agencies of the Federal government; and faith-based or community based organizations.
Date(s) Application is Due: Feb 5; Jun 5; Oct 5
Contact: Ronald P. Abeles, Ph.D.; (301) 496-7859; fax: (301) 435-8779; email: abeles@nih.gov
Internet: http://grants.nih.gov/grants/guide/pa-files/PAR-07-379.html
Sponsor: Office of Behavioral and Social Sciences Research (OBSSR)
31 Center Drive, Building 31, Room B1C19
Bethesda, MD 20892-2027

OCLC Library and Information Sciences Research Grants 4731
In recognition of the importance of research to the advancement of librarianship and information science, the Office of Research of OCLC promotes independent research that helps librarians integrate new technologies into areas of traditional competence and contributes to a better understanding of the environment in which OCLC services exist. Grant funds may be applied to the costs of release time from teaching for the principal investigator; research assistants; project-related travel; equipment, if integral to the research; and other research expenses. Staff training, general operating or overhead expenses, and other indirect costs are not funded. Projects are expected to be conducted within approximately one year from the date of the award. As a condition of the grant, the researcher must furnish OCLC a final project report, which may be published and distributed all or in part by OCLC.
Requirements: Faculty in schools of library and information science in the United States and Canada are eligible to apply. OCLC encourages collaborative projects.
Restrictions: Staff training, general operating or overhead expenses, and other indirect costs are not funded.
Samples: Francis Miksa, Professor, U of Texas at Austin (TX)—for Examining the Attributes of Information Resources on the World Wide Web and Testing for Their Usefulness as Metadata.; Jian Qin, Assistant Professor, U of Southern Mississippi (MS)—for Computational Representation of Web Objects in an Interdisciplinary Digital Library: A Survey and an Experiment in Polymer Science.
Amount of Grant: $15,000 maximum
Date(s) Application is Due: Sep 15
Contact: Administrator, (614) 764-6000; fax: (614) 764-6096; email: oclc@oclc.org
Internet: http://www.oclc.org/research/grants
Sponsor: Online Computer Library Center
6565 Frantz Road
Dublin, OH 43017-3395

Ohio Learning Network Grants 4732
The network funds programs to help Ohioans increase work-related knowledge and skills through college credits; assist faculty in teaching using emerging technologies; and better understand the role of technology in learning. Grants support emerging needs, learning communities, distance learning research, partnership grants, and technology initiatives/innovations.
Geographic Focus: Ohio
Amount of Grant: $100,000 maximum
Contact: Judith M. Leach, Assistant Director; (614)485-6738 or (614) 995-3240; email: jleach@oln.org
Internet: http://www.oln.org/about_oln/grants.php
Sponsor: Ohio Learning Network
35 E. Chestnut Street, 8th Floor
Columbus, OH 43215-2541

OIEAHC Jamestown Prize 4733
This prize is offered to the author of the best book-length, scholarly manuscript on early American history or culture submitted to the institute for publication. The subject of the manuscript must pertain to America before ca. 1820 or to the related history of the British Isles, Europe, West Africa, or the Caribbean. Manuscripts may be submitted for the prize at any time. A separate letter should indicate that the manuscript is being submitted as a Jamestown Prize entry.
Requirements: The prize competition is open only to authors who have not already published scholarly books.
Restrictions: Past or present holders of the institute fellowship are ineligible.
Samples: David Waldstreicher—prize winner, for In the Midst of Perpetual Fetes: The Making of American Nationalism, 1776-1820, $1500.
Amount of Grant: $3000 plus publication and royalties
Contact: Ronald Hoffman, Director, (757) 221-1110; fax: (757) 221-1047; email: ieahc1@wm.edu
Internet: http://www.wm.edu/oieahc/pub/jp.html
Sponsor: Omohundro Institute of Early American History and Culture
Box 8781
Williamsburg, VA 23187-8781

OIEAHC Postdoctoral Fellowships 4734
The institute annually offers one two-year postdoctoral fellowship and one one-year Andrew W. Mellon postdoctoral research fellowship, both in any area of early American studies. A principal criterion for the selection is that the candidate's dissertation or other manuscript have significant potential as a distinguished, book-length contribution to scholarship. The institute holds first claim on publishing the appointed fellow's completed manuscript. A substantial portion of the work must be submitted with the application.
Amount of Grant: $30,000 per year
Date(s) Application is Due: Nov 1
Contact: Ronald Hoffman, Director, (757) 221-1110; fax: (757) 221-1047; email: ieahc1@wm.edu
Internet: http://www.wm.edu/oieahc/Fellowships/NEH.html
Sponsor: Omohundro Institute of Early American History and Culture
Box 8781
Williamsburg, VA 23187-8781

OISE Graduate Scholarships 4735
OISE/UT graduate scholarships are awarded on the basis of academic merit to students who will be registered in a graduate degree program at the institute. Scholarships are renewable for required academic residence.
Restrictions: Student must be registered full-time.
Date(s) Application is Due: Dec 2
Contact: M.A. Brennan, Administrative Officer, (416) 923-6641; fax: (416) 926-4765; email: stuserv@oise.utoronto.ca
Internet: http://ro.oise.utoronto.ca/bulPage175.htm
Sponsor: Ontario Institute for Studies in Education
252 Bloor Street W
Toronto, ON M5S 1V6 Canada

Oklahoma Arts Council Artists in Residence Program 4736
This program offers a roster of professional artists, and financial assistance to fund residencies in schools and communities throughout the state.
Requirements: Applications for financial assistance from the council are accepted only from Oklahoma nonreligious, nonprofit, tax-exempt organizations.
Geographic Focus: Oklahoma
Date(s) Application is Due: 60 days prior to the beginning date of the project
Contact: Program Officer, Arts in Education, (405) 521-2931; fax: (405) 521-6418; TDD: (405) 521-2931; email: okarts@arts.state.ok.us
Internet: http://www.state.ok.us/~arts/Pages/rosters/rosters.html
Sponsor: Oklahoma Arts Council
P.O. Box 52001-2001
Oklahoma City, OK 73152-2001

Oklahoma City Community Programs and Grants 4737
The community foundation functions to enhance the quality of life in Oklahoma by awarding grants to nonprofits in its areas of interest, including arts and cultural programs, education, health care, health associations, and human services. Types of support include general operating support, continuing support, annual campaigns, building/renovation,

equipment, emergency funds, program development, seed grants, fellowships, scholarship funds, research, and matching funds. The grants assist organizations with activities that provide new or expanded services to the community in response to new opportunities and changing circumstances. The community foundation views its role as a partner with other organizations and regards contributions and participation by other individuals and groups as an essential element in grant activities. Especially encouraged are programs and services that are oriented toward the entire population and are not restricted only to those people who live in a particular geographic location or who are members of a particular organization. The foundation encourages nonprofit organizations to cooperate and pool their resources to meet community needs. The use of volunteers as a significant program component is considered favorably as is active participation of the agency's board of directors and a broad base of financial support in the community. Grants are generally made for a specified time period and are nonrenewable.
Requirements: 501(c)3 nonprofit organizations serving the greater Oklahoma City area, and programs that benefit persons living in the greater Oklahoma City metropolitan area, are eligible.
Restrictions: Grants are not awarded to individuals or to benefit specific individuals. No faxed or emailed proposals are accepted.
Geographic Focus: Oklahoma
Samples: Arts Council of Oklahoma City (OK)—to support a collaborative summer program of the Oklahoma City Parks and Recreation Department, After School Options, and four local school districts, $20,000.; Mesta Park Neighborhood Assoc (OK)—to create a new median strip to reduce the volume of traffic around NW 18th Street and N Shartel Boulevard, $10,000.; Neighborhood Alliance (OK)—for a computer-resource center, $22,788.
Contact: Grants Manager, (405) 235-5603; fax: (405) 235-5612; email: info@occf.org
Internet: http://www.occf.org/occf/community_programs/index.php
Sponsor: Oklahoma City Community Foundation
P.O. Box 1146
Oklahoma City, OK 73103-1146

Olfactory Research Fund Grants **4738**
The fund sponsors innovative scientific research in the olfactory arts and sciences. Grants promote the study of olfaction and also support educational and outreach programs about the sense of smell. The fund is especially interested in supporting research that integrates the study of olfaction with current issues in developmental, perceptual, social and cognitive psychology, and related disciplines. The fund is currently seeking proposals for research that focuses on the influence of olfaction in the mental and physical aspects of performance. Contact organization for submission deadlines.
Requirements: Applicants must have a doctoral degree.
Restrictions: Generally, the following types of research are ineligible for support: studies not directly related to olfaction and/or the human reaction to odors; studies exclusively on animal olfaction without direct relevance to humans; or invasive human studies.
Contact: Theresa Molnar, Executive Administrator, (212) 725-2755; fax: (212) 779-9072; email: tmolnar@senseofsmell.org or info@senseofsmell.org
Internet: http://www.senseofsmell.org/resources/research.php
Sponsor: Olfactory Research Fund
145 E 32nd Street
New York, NY 10016

Olin Corporation Charitable Trust Grants **4739**
The charitable trust supports nonprofits in company-operating areas. Grants focus on three priority categories: education, with a science and technology emphasis; programs that include Olin employees as volunteers and that impact communities in which Olin employees work and live; and conservation, environmental education, and environmental research. Types of support include general operating budgets, continuing support, annual campaigns, seed money, emergency funds, building funds, equipment, land acquisition, special projects, research, publications, conferences and seminars, internships, scholarship funds, employee-related scholarships, fellowships, and employee matching gifts. U.S.-based nonprofits with an international focus are supported primarily through matching gifts. There are no application deadlines. Applicants should submit a letter of request. The board meets in January to consider requests.
Requirements: Support is directed primarily to communities where Olin has a major employee presence.
Restrictions: Grants are not awarded to individuals or for endowments.
Samples: Thirteen communities in Olin operating areas—for education programs, with a focus on computer technology and science education, $242,137 divided.
Amount of Grant: $50-$50,000
Contact: Program Contact, (314) 480-1400
Internet: http://www.olin.com/about/charitable.asp
Sponsor: Olin Corporation Charitable Trust
190 Carondelet Plz, Suite 1530
Clayton, MO 63105-3443

Olin Fellowships **4740**
The fellowships are offered to North American residents seeking to improve their knowledge or skills in advanced salmon-related fields while also seeking solutions to current problems in Atlantic salmon biology, management, or conservation. The fellowships may be applied toward a wide range of endeavors, including salmon management, graduate study, and research. Applicants need not be enrolled in a degree program. The awards are tenable at any accredited university or research laboratory or in an active management program. Application forms may be obtained from the Atlantic Salmon Federation, P.O. Box 807, Calais, ME, 04619, U.S.A; however, the completed form must be received at the federation's Saint Andrews office by the deadline date. Applicants will be advised of awards by May 15.
Requirements: Applicants must be legal residents of the United States or Canada and must complete the formal application form to which should be attached a statement of qualifications, transcript of grades if applicable, description of the program/project for which the fellowship is sought, and all supporting documentation required.
Amount of Grant: $C1000-$C3000
Date(s) Application is Due: Mar 15
Contact: Muriel Ferguson, Program Contact, (506) 529-1033
Internet: http://www.asf.ca/Awards/Awards.html
Sponsor: Atlantic Salmon Federation
P.O. Box 5200
Saint Andrews, NB E0G 2X0 Canada

Olivia James Traveling Fellowship **4741**
One fellowship will be awarded to an American student who desires to travel and study in Greece, the Aegean Islands, Sicily, southern Italy, Asia Minor, or Mesopotamia. The most suitable areas of study are the classics, sculpture, architecture, archaeology, and history. The term student shall not be taken to restrict awards to persons registered in academic institutions; however, preference will be given to individuals engaged in dissertation research or to recent recipients of the PhD. Official application materials are available upon request from the institute in the spring of each year. Preference will be given to projects of at least a half-year in duration.
Requirements: Applicants must be citizens or permanent residents of the United States engaged in dissertation research or recent recipients of the PhD.
Restrictions: The award is not intended to support field excavation projects. Recipients may not hold other major fellowships during the requested tenure of the Olivia James award.
Amount of Grant: $22,000
Date(s) Application is Due: Nov 1
Contact: Elizabeth Gilgan, Program Administrator, (617) 353-8705; email: egilgan@aia.bu.edu
Internet: http://www.archaeological.org/webinfo.php?page=10008
Sponsor: Archaeological Institute of America
656 Beacon Street, 4th Fl
Boston, MA 02215-2006

OM Association Outstanding Dissertation Awards Program **4742**
The association honors doctoral candidates whose dissertations significantly advance the knowledge and understanding of education theories, concepts, and practices related to OM's mission to develop creative problem-solving skills important to a changing world. OM's mission includes the encouragement and development of cooperation, self-respect, and the appreciation and understanding of others. Winners will be notified by September 1.
Requirements: Any doctoral candidate may enter if his/her dissertation was completed and approved over the last three academic years from an institution of higher education that is sanctioned by a state, regional, or provincial accrediting agency.
Amount of Grant: $1500 first place; $750 second place
Date(s) Application is Due: Jun 1
Contact: Dissertation Awards Program, fax: (609) 881-3596
Internet: http://www.isop.ucla.edu/eas/fellowships/om-assoc.htm
Sponsor: OM Association
P.O. Box 547
Glassboro, NJ 08028

Onassis Grants and Scholarships for Foreign Scholars and Students **4743**
The foundation awards grants and scholarships to foreign scholars, researchers, and students of humanistic and political sciences; Hellenists; and teachers, artists, and translators of Greek literature. The program awards research grants—for a one-month stay in Greece, awarded to university professors whose scientific work has been widely acclaimed and who wish to visit Greece in order to conduct research or to collaborate with scientific institutions/organizations; postgraduate scholarships for Greek language teachers—for a maximum duration of two months, awarded to elementary or high school teachers from abroad who teach Greek language, literature, history, and civilization; and postgraduate research scholarships—for a maximum duration of 12 months, awarded to foreign postgraduate and PhD students who pursue theoretical or artistic studies in universities, scientific centers, or fine arts schools. The subject of the thesis or PhD dissertation should be approved in advance by the university or institution, and the supervising professor must concur in the necessity of conduction the study in Greece. Awards include round-trip airline tickets (business class) for the interested party and his or her spouse (if the spouse accompanies the applicant); single or double room in a hotel with bath and breakfast; and monthly stipend to cover moving expenses and other necessities. Guidelines are available online.
Requirements: Non-Greek citizens are eligible. Foreigners of Greek origin (second generation and on) may also be eligible. Foreign scholars in university or research institutions (maximum age 50 years old) who wish to perform research in Greece in cooperation with a Greek university, research venture, institute, or other scientific organization are eligible for research grants. Applicants for postgraduate scholarships for Greek language teachers must receive a recommendation from the director of the school where the applicant teaches. Foreign postgraduate and PhD students not more than 40 years old are eligible for postgraduate research scholarships.
Amount of Grant: $900-$1600
Date(s) Application is Due: Feb 28

Contact: Secretariat, (+30) 1 311-0900-2 ext 140, 143
Internet: http://www.yasou.org/education/onassis_scholarships.htm
Sponsor: Alexander S. Onassis Public Benefit Foundation
7 Aischinou Street
Athens 10558 Greece

OneSight Research Foundation Block Grants **4744**
Formerly known as the Pearle Vision Foundation, the OneSight Research Foundation is dedicated to sight preservation through research and education. A significant portion of their grant dollars will be awarded to fund research and treatment for diabetic eye diseases, hoping to find a cure for diabetic retinopathy, the number one cause of adult blindness in the United States.
Requirements: U.S. 501(c)3 nonprofits are eligible for grant support. Requests for funding should cover a period of time not to exceed one year. Exceptions to this policy will be considered on an individual basis.
Restrictions: Grants will not be considered for endowments or operating expenses.
Samples: Retina Foundation of the Southwest (Dallas, TX)—to equip and furnish a new visual-testing laboratory at the Morchower Pediatric Eye Research Laboratory, $16,342.; Lighthouse International (New York, NY)—to study a new reading method for people with macular degeneration and other diseases that result in central-vision loss, $10,000.; New Eyes for the Needy (Short Hills, NV)—for eyeglass vouchers to serve individuals nationwide, $25,000.
Amount of Grant: $5,000-$50,000
Date(s) Application is Due: Jun 30; Dec 31
Contact: Trina Parasiliti, Administrator; (972) 277-6191; fax: (972) 277-6414; email: tparasil@onesight.org
Internet: http://www.onesight.org/northamerica/na/
Sponsor: OneSight Research Foundation
2465 Joe Field Road
Dallas, TX 75229

ONF Career Development Awards **4745**
The awards support the career development of professional nurses in the field of oncology nursing. The awards allow attendance in an educational program focusing on clinical ethics that will further the nurse's professional goals and potential to contribute to oncology nursing. This award is not intended to supplement tuition in a graduate degree program. Preference will be given to applicants who are seeking beginning level knowledge in clinical ethics.
Requirements: All applicants must be registered nurses or nurse administrators actively involved in some aspect of cancer patient care, education, or research.
Amount of Grant: $1000
Contact: ONS Foundation Office, (412) 859-6100 ext 8503; email: customer.service@ons.org or member@ons.org
Internet: http://www.ons.org/awards/foundawards/ethics.shtml
Sponsor: Oncology Nursing Foundation
125 Enterprise Drive
Pittsburgh, PA 15275-1214

ONF Small Grants **4746**
These grant awards are designed to support oncology nursing research conducted by beginning and experienced researchers. The majority of the small grants are for general investigator-initiated oncology topics. Funding preference is given to research that addresses the ONS Research Priorities and Research Agenda. A letter of intent is requested by October 1; full proposal is due November 1.
Requirements: The principal investigator must be actively involved in some aspect of cancer patient care, education, or research. Funding preference is given to projects that involve nurses in the design and conduct of the research activity and that promote oncology nursing theory and a scientific basis for practice.
Samples: Kimberly Kahn, U of Virginia Health Science Ctr (Charlottesville, VA)—nursing education research grant, The Informed Consent Process in Oncology Clinical Trials, $7500.; Dr. June Eilers, U of Nebraska Medical Ctr (Omaha, NE)—research grant, Longitudinal Study of the Impact of BMT on Family Members, $5000.
Amount of Grant: $10,000
Date(s) Application is Due: Oct 1; Nov 1
Contact: Grants Administrator, email: research@ons.org
Internet: http://www.ons.org/awards/foundawards/smallGrants.shtml
Sponsor: Oncology Nursing Foundation
125 Enterprise Drive, RIDC Park W
Pittsburgh, PA 15275-1214

ONR 100 GHz Logic Grants **4747**
This work seeks new approaches to achieving 100 GHZ logic. This includes investigations in devices, architecture, power dissipation, materials and related all-encompassing aspects. This work is expected to eventually lead to the ability to synthesize RF signals at frequencies up to at least 25 GHz as well as to amplitude, phase and frequency modulate these signals. Additional objectives of this work are expected to lead to digital beamforming and steering of electronic antenna arrays and the issue of pre-compensating signals to reduce distortion at the output of power amplifiers. Techniques to control phase jitter are also relevant. Though the proposal submission date is ongoing, proposals are preferred by April 1 of each year.
Requirements: Grants are awarded to academic, industrial, and nonprofit organizations with strong emphasis on university research. Investigators are encouraged to contact the staff prior to formally submitting a proposal.
Date(s) Application is Due: Apr 1
Contact: Dr. Daniel Purdy; (703) 588-1920 or (703) 696-5031; fax: (703) 696-5940; email: dan.purdy@navy.mil or 312_GHz@onr.navy.mil
Internet: http://www.onr.navy.mil/sci_tech/31/312/100ghz.asp
Sponsor: Office of Naval Research
One Liberty Center, 875 North Randolph Street, Suite 1425
Arlington, VA 22203-1995

ONR Aerospace Science Research Grants **4748**
This division is structured into two major programs, weapons technology and aircraft technology. These programs directly support Navy Science and Technology requirements for Joint Strike Warfare involving Air Superiority and Precision Attack, and Air and Surface Battlespace requirements of Joint Littoral Warfare involving Aircraft, Naval Surface Fire Support, and Ship Self Defense. In support of these mission area requirements, the division conducts exploratory and advanced development programs in the areas of aeronautics, avionics, air vehicle propulsion, missile solid and air breathing propulsion, gun propulsion, missile and gun projectile guidance and control, warheads, fuze and safe and arm devices, fire control and targeting.
Requirements: Grants are awarded to academic, industrial, and nonprofit organizations, with strong emphasis on university research. Investigators are encouraged to contact the staff prior to formally submitting a proposal.
Contact: Dr. Katherine Stevens; (703) 696-3798 or (703) 696-5031; fax: (703) 696-5940; email: katherine.stevens@navy.mil
Internet: http://www.onr.navy.mil/sci_tech/35/351/
Sponsor: Office of Naval Research
One Liberty Center, 875 North Randolph Street, Suite 1425
Arlington, VA 22203-1995

ONR Agile and Reconfigurable Organizational Structures for Command and Control Grants **4749**
This program thrust focuses on developing a principled basis for organizational design of flexible Command and Control Systems. Commanders and their staffs must deal with a variety of task environments that change rapidly in response to changes in the battle rhythm, threat conditions and uncertain contingencies. Military operations are increasingly knowledge driven, mediated by four structures: the command structure, the information structure, the communication structure and the resource structure. New adversaries, rapidly changing, high-tempo and ambiguous environments require flexible and adaptive organizational designs to engage the dynamic task environment without losing mission effectiveness, speed, and efficiency. This research thrust supports the achievement of Sea Power 21 and FORCEnet goals concerning human centric design, decision support systems for distributed decision making and agile and adaptive command and control structures. Submission of the Research Proposal should always be preceded by submission of a Preproposal and a discussion of that Preproposal with the assigned Program Officer.
Requirements: Grants are awarded to academic, industrial, and nonprofit organizations, with strong emphasis on university research. Investigators are encouraged to contact the staff prior to formally submitting a proposal. New grants or contracts are typically started near the beginning of the fiscal year (1 October). New proposals for the next fiscal year should be submitted no later than June 1, unless otherwise directed by the BAA. Competitive renewals should be submitted no later than six months prior to the award termination date after contacting the Program Officer. Conference proposals should be submitted at least four months prior to the conference.
Contact: Gerald Malecki; (703) 696-4044 or (703) 696-5031; fax: (703) 696-5940; email: gerald.malecki@navy.mil
Internet: http://www.onr.navy.mil/sci_tech/34/341/hp_agile.asp
Sponsor: Office of Naval Research
875 North Randolph Street, Suite 1425
Arlington, VA 22203-1995

ONR Air Warfare and Naval Weapons Applications Division Grants **4750**
Naval unique or essential projects in this program are involved with applied research and advanced technology aligned with current and future Naval capability (FNC) gaps and the Electro-Magnetic Railgun Innovative Naval Prototype (INP). The Portfolio of current FNC projects include intelligent autonomy, unmanned vehicle systems, precision strike power generation advanced gun barrel, and turbine engine propulsion. Submission of the Research Proposal should always be preceded by submission of a Preproposal and a discussion of that Preproposal with the assigned Program Officer.
Requirements: Grants are awarded to academic, industrial, and nonprofit organizations, with strong emphasis on university research.
Restrictions: New grants or contracts are typically started near the beginning of the fiscal year (October 1). New proposals for the next fiscal year should be submitted no later than June 1, unless otherwise directed by the BAA. Competitive renewals should be submitted no later than six months prior to the award termination date after contacting the Program Officer. Conference proposals should be submitted at least four months prior to the conference.
Contact: Roger Ellis; (703) 696-9504 or (703) 588-2427; fax: (703) 696-5940; email: roger.ellis@navy.mil
Internet: http://www.onr.navy.mil/sci_tech/35/352/
Sponsor: Office of Naval Research
875 North Randolph Street, Suite 1425
Arlington, VA 22203-1995

ONR Antennas, Propagation and Electromagnetic Interactions Grants 4751
This program has as its primary interests: electromagnetic propagation and scattering with emphasis on antennas; target, clutter and background modeling; and sensor performance prediction. Submission of the Research Proposal should always be preceded by submission of a Preproposal and a discussion of that Preproposal with the assigned Program Officer.
Requirements: Grants are awarded to academic, industrial, and nonprofit organizations, with strong emphasis on university research. Investigators are encouraged to contact the staff prior to formally submitting a proposal.
Restrictions: Grants are awarded to academic, industrial, and nonprofit organizations, with strong emphasis on university research. New grants or contracts are typically started near the beginning of the fiscal year (1 October). New proposals for the next fiscal year should be submitted no later than June 1, unless otherwise directed by the BAA. Competitive renewals should be submitted no later than six months prior to the award termination date after contacting the Program Officer. Conference proposals should be submitted at least four months prior to the conference.
Contact: Dr. Michael Pollock; (703) 696-4202 or (703) 696-1438; fax: (703) 696-5940; email: michael.a.pollock@navy.mil or 312_APEI@onr.navy.mil
Internet: http://www.onr.navy.mil/sci_tech/31/312/propagation_interactions.asp
Sponsor: Office of Naval Research
875 North Randolph Street, Suite 1425
Arlington, VA 22203-1995

ONR Applied and Computational Analysis Grants 4752
Program emphases are probabilistic analysis and statistical inference for models of the temporal and spatial evolution of physical phenomena and engineering systems subject to stochastic variability. Thrust areas include optimal nonlinear filtering for target tracking, object modeling under uncertainty for image analysis and automatic target recognition, minefield detection and mine countermeasures, and command and control. Proposals are accepted at any time. Submission of the Research Proposal should always be preceded by submission of a Preproposal and a discussion of that Preproposal with the assigned Program Officer.
Requirements: Grants are awarded to academic, industrial, and nonprofit organizations, with strong emphasis on university research.
Restrictions: New grants or contracts are typically started near the beginning of the fiscal year (1 October). New proposals for the next fiscal year should be submitted no later than June 1, unless otherwise directed by the BAA. Competitive renewals should be submitted no later than six months prior to the award termination date after contacting the Program Officer. Conference proposals should be submitted at least four months prior to the conference.
Contact: Dr. Tristan Nguyen, Program Officer; (703) 588-2360 or (703) 696-5031; fax: (703) 696-5940; email: tristan.nguyen@navy.mil
Internet: http://www.onr.navy.mil/media/poc/default.asp?pocid=270
Sponsor: Office of Naval Research
875 North Randolph Street, Suite 1425
Arlington, VA 22203-1995

ONR Atomic and Molecular Physics Grants 4753
The research supported under this program seeks primarily to understand and exploit the interactions of light, atoms, ions, and molecules. Areas supported may include laser cooling and trapping of both ions and neutral atoms; Bose-Einstein condensation; atom laser concepts; coherent control, e.g., of chemical reactions; femtosecond pulse shaping; quantum computing; and quantum state teleportation. Also supported are novel theoretical methods for computational electronic structure. Challenges addressed through the atomic and molecular physics program include: obtaining ultra-high precision spectroscopy; overcoming limitations in atomic time and frequency standards, leading, for example, to the pursuit of the physics of ultra-cold collisions; the ability to control chemical dynamics and branching ratios; and the ability to predict properties of materials, and devise novel new materials (e.g., through a better understanding of strong electron correlations). Many of these subprograms and goals are interrelated and interdependent. Submission of the Research Proposal should always be preceded by submission of a Preproposal and a discussion of that Preproposal with the assigned Program Officer. Proposals are preferred by 1 April of each year.
Requirements: Grants are awarded to academic, industrial, and nonprofit organizations, with strong emphasis on university research.
Restrictions: New grants or contracts are typically started near the beginning of the fiscal year (1 October). New proposals for the next fiscal year should be submitted no later than June 1, unless otherwise directed. Competitive renewals should be submitted no later than six months prior to the award termination date after contacting the Program Officer. Conference proposals should be submitted at least four months prior to the conference.
Date(s) Application is Due: Apr 1
Contact: Dr. Charles W. Clark, Physics Division; (703) 696-5267 or (703) 696-5031; fax: (703) 696-5940; email: charles.clark3@navy.mil or 312_AM@onr.navy.mil
Internet: http://www.onr.navy.mil/sci_tech/31/312/atomic_molecular_physics.asp
Sponsor: Office of Naval Research
875 North Randolph Street, Suite 1425
Arlington, VA 22203-1995

ONR Autonomous Systems Grants 4754
Projects in robotics pursue the understanding of how to design, build, and utilize mobile, intelligent agents. Research is conducted in novel actuator design using muscle-like materials, tactile information processing, motion planning algorithms, artificial vision, and behavior of teams of agents. Submission of the Research Proposal should always be preceded by submission of a Preproposal and a discussion of that Preproposal with the assigned Program Officer.
Requirements: Grants are awarded to academic, industrial, and nonprofit organizations, with strong emphasis on university research.
Restrictions: New grants or contracts are typically started near the beginning of the fiscal year (October 1). New proposals for the next fiscal year should be submitted no later than June 1, unless otherwise directed by the BAA. Competitive renewals should be submitted no later than six months prior to the award termination date after contacting the Program Officer. Conference proposals should be submitted at least four months prior to the conference.
Contact: Dr. Behzad Kamgar-Parsi; (703) 696-5754 or (703) 588-2427; fax: (703) 696-5940; email: behzad.kamgar-parsi@.navy.mil or 311_IS@onr.navy.mil
Internet: http://www.onr.navy.mil/sci_tech/31/311/autonomous_systems.asp
Sponsor: Office of Naval Research
875 North Randolph Street, Suite 1425
Arlington, VA 22203-1995

ONR Biological & Biomedical Division Grants 4755
The Biomedical and Biological Division's mission is to conduct research and technology demonstration programs directed at maintaining the survival, health and performance of Navy and Marine Corps personnel during training, routine operations, special operations, and in time of war. The Division's program is organized into five research programs: Casualty Care & Management, Casualty Prevention, Undersea Medicine, Stress Physiology and Naval Biosciences and Biocentric Technology. Each research program supports a focused initiative, the Force Health Protection Future Naval Capability, that will deliver specific technology products in the FY08-FY12 timeframe. Submission of the Research Proposal should always be preceded by submission of a Preproposal and a discussion of that Preproposal with the assigned Program Officer.
Requirements: Grants are awarded to academic, industrial, and nonprofit organizations, with strong emphasis on university research.
Restrictions: New grants or contracts are typically started near the beginning of the fiscal year (October 1). New proposals for the next fiscal year should be submitted no later than June 1, unless otherwise directed by the BAA. Competitive renewals should be submitted no later than six months prior to the award termination date after contacting the Program Officer. Conference proposals should be submitted at least four months prior to the conference.
Contact: Dr. Igor Vodyanoy; (703) 696-4109 or (703) 588-2427; fax: (703) 696-5940; email: igor.vodyanoy@navy.mil
Internet: http://www.onr.navy.mil/sci_tech/34/342/
Sponsor: Office of Naval Research
875 North Randolph Street, Suite 1425
Arlington, VA 22203-1995

ONR Bisson Prize for Naval Technology Achievement 4756
This award is presented to recognize and reward an individual's successes in technology achievement. It is granted annually to a current or former navy scientist, engineer, or S&T program manager that has a significant, direct transitional impact from S&T to naval operations or systems. Such impact shall be so outstanding as to be widely recognized in the navy and the relevant technical community nationally. While the impact must have been realized in the year prior to nomination, it is recognized that the efforts leading to transition may have been ongoing for many years. The award consists of an individual plaque, a certificate, and a cash award. Additionally, the recipient's name is placed on the master plaque located on the Honorary Wall of Fame at ONR Headquarters.
Requirements: High-quality science and/or technology accomplishment, well-conducted transition of that accomplishment from S&T to naval acquisition or operations, and a significant impact upon the war fighting and/or peacekeeping capabilities of naval forces are criteria for the award. At the time of the achievement cited as the basis for the prize, nominees must have been military or civilian employees of the navy in positions managing or performing S&T.
Amount of Grant: $3,000
Date(s) Application is Due: Feb 15
Contact: Awards Administrator; (703) 696-5031; fax: (703) 696-5940; email: onrpao@onr.navy.mil
Internet: http://www.onr.navy.mil/sci_tech/34/341/training/transitions/bisson.htm
Sponsor: Office of Naval Research
875 North Randolph Street, Suite 1425
Arlington, VA 22203-1995

ONR Casualty Care and Management Research Grants 4757
This program addresses applied and advanced research and technology efforts in trauma medicine for use in austere environments with a focus on control of internatl bleeding without surgical access. Program interests include treatment of hemorrhagic shock, agents that control acute pain without impairing cognitive performance, therapeutics or other strategies that effect resuscitation without the use of fluids, and automated care (closed loop-control) of casualties during transport. Products of this research must ultimately be stable to ambient temperatures, small in size, light in weight, function in small volumes, be affordable, be patentable, and have the ability to be administered by a minimally trained individual. Submission of the Research Proposal should always be preceded by submission of a Preproposal and a discussion of that Preproposal with the assigned Program Officer.
Requirements: Grants are awarded to academic, industrial, and nonprofit organizations, with strong emphasis on university research.

Restrictions: New grants or contracts are typically started near the beginning of the fiscal year (October 1). New proposals for the next fiscal year should be submitted no later than June 1, unless otherwise directed by the BAA. Competitive renewals should be submitted no later than six months prior to the award termination date after contacting the Program Officer. Conference proposals should be submitted at least four months prior to the conference.
Contact: Dr. Michael Given; (703) 696-4055 or (703) 588-2427; fax: (703) 696-5940; email: michael.given@navy.mil
Internet: http://www.onr.navy.mil/sci_tech/34/342/cas_care.asp
Sponsor: Office of Naval Research
875 North Randolph Street, Suite 1425
Arlington, VA 22203-1995

ONR Casualty Prevention Grants 4758

This program addresses research and technology efforts focused on reducing casualties by enhancing an individual's situational awareness and countering threats from disease, battle and non-battle injuries. Solutions can be targeted by diverse disciplines such as biological, engineering and behavioral sciences, with the objective of developing improved methods, models, and techniques for understanding, preventing or mitigating sources of military casualties. Program interests include, but are not limited to: developing non-invasive or minimally invasive procedures for identifying infection by pathogenic organisms; characterizing blunt trauma effects of body armor; mitigating thermal effects during diving; and protecting personnel from lethal and non-lethal effects from lasers. Products of this program must function with minimal negative impact on operational capability. Submission of the Research Proposal should always be preceded by submission of a Preproposal and a discussion of that Preproposal with the assigned Program Officer.
Requirements: Grants are awarded to academic, industrial, and nonprofit organizations, with strong emphasis on university research.
Restrictions: New grants or contracts are typically started near the beginning of the fiscal year (October 1). New proposals for the next fiscal year should be submitted no later than June 1, unless otherwise directed by the BAA. Competitive renewals should be submitted no later than six months prior to the award termination date after contacting the Program Officer. Conference proposals should be submitted at least four months prior to the conference.
Contact: Elizabeth A. Montcalm-Smith; (703) 696-2580 or (703) 588-2427; fax: (703) 696-5940; email: elizabeth.montcalmsm@navy.mil
Internet: http://www.onr.navy.mil/sci_tech/34/342/cas_prevent.asp
Sponsor: Office of Naval Research
875 North Randolph Street, Suite 1425
Arlington, VA 22203-1995

ONR Circuits/Architectures Grants 4759

This program seeks innovation in mixed signal circuits, architectures and related all-encompassing aspects with emphasis on higher operating frequency, greater signal dynamic range. Innovations are expected to eventually lead to improvements in complex circuits such as analog-to-digital converters (ADCs), digital-to-analog converters (DACs), and direct high power conversion DACs at frequencies of microwave and millimeter wave. Examples are distributed sigma-delta and switched-mode amplifier based circuits and architectures which can be advantageously exploited for generation of multiple simultaneous RF/Microwave beams in electronically scanned arrays. Techniques that show potential to operate efficiently at 25 GHz and beyond are highly sought after. Also of interest are methods for rigorous electromagnetic modeling of complex mixed signal circuits with a potential to provide 30 dB improvement in dynamic range over current state of art. Submission of the Research Proposal should always be preceded by submission of a Preproposal and a discussion of that Preproposal with the assigned Program Officer. Proposals are preferred by April 1 of each year.
Requirements: Grants are awarded to academic, industrial, and nonprofit organizations, with strong emphasis on university research.
Amount of Grant: $115,000 average
Date(s) Application is Due: Apr 1
Contact: Dr. Daniel Purdy; (703) 588-1920 or (703) 696-5031; fax: (703) 696-5940; email: dan.purdy@navy.mil or 312_GHz@onr.navy.mil
Internet: http://www.onr.navy.mil/sci_tech/31/312/circuits_architechtures.asp
Sponsor: Office of Naval Research
875 North Randolph Street, Suite 1425
Arlington, VA 22203-1995

ONR Coastal Geosciences Grants 4760

The Coastal Geosciences program funds basic and applied research in two main areas: predicting the 4D coastal environment and predicting seafloor characteristics in the global ocean margins. Areas of interest are motivated by on-going and future Navy and Marine Corps needs, including all littoral operations in nearshore, estuarine, and riverine environments; prediction of sediment transport phenomena across a broad spectrum of coastal environments; and assessment of geological sources of sonar clutter, among many others. Basic research interests include: nearshore, estuarine and riverine processes; sediment transport processes; seafloor and subsurface geological characteristics, including bed geometry, geoacoustical and geotechnical properties; and remote sensing of coastal environments. Studies by individual investigators and collaborative teams are often undertaken in conjunction with other programs at ONR, including the Physical Oceanography, Ocean Optics, Ocean Engineering and Marine Systems, and Ocean Acoustics programs. Focused, field-oriented research programs are undertaken as Departmental Research Initiatives (DRIs) or as Multidisciplinary University Research Initiatives (MURIs) each of which typically has a five-year timeframe and substantial funding for field studies. Submission of the Research Proposal should always be preceded by submission of a Preproposal and a discussion of that Preproposal with the assigned Program Officer.
Requirements: Grants are awarded to academic, industrial, and nonprofit organizations, with strong emphasis on university research.
Restrictions: New grants or contracts are typically started near the beginning of the fiscal year (October 1). New proposals for the next fiscal year should be submitted no later than June 1, unless otherwise directed by the BAA. Competitive renewals should be submitted no later than six months prior to the award termination date after contacting the Program Officer. Conference proposals should be submitted at least four months prior to the conference.
Contact: Dr. Thomas Drake; (703) 696-1206 or (703) 588-2427; fax: (703) 696-5940; email: tom.drake@navy.mil or 321_CG@onr.navy.mil
Internet: http://www.onr.navy.mil/sci_tech/32/321/coastal_geosciences.asp
Sponsor: Office of Naval Research
875 North Randolph Street, Suite 1425
Arlington, VA 22203-1995

ONR Cognitive Science Basic Research Grants 4761

The overall technical objectives of this program are to provide formal, computational theories of the fundamental information processing characteristics of the human cognitive architecture. Within this context, four distinct areas are attractive loci for future research efforts. They include: theoretical foundations for socio-cognitive architectures; perception, metacognition and cognitive control; representing and reasoning about uncertainty; and skill acquisition and learning. The application of much of the basic research generated through investigations of cognitive architectures is toward the development of artificially intelligent tutoring technology for training. Submission of the Research Proposal should always be preceded by submission of a Preproposal and a discussion of that Preproposal with the assigned Program Officer.
Requirements: Grants are awarded to academic, industrial, and nonprofit organizations, with strong emphasis on university research. Investigators are encouraged to contact the staff prior to formally submitting a proposal.
Restrictions: New grants or contracts are typically started near the beginning of the fiscal year (1 October). New proposals for the next fiscal year should be submitted no later than June 1, unless otherwise directed. Competitive renewals should be submitted no later than six months prior to the award termination date after contacting the Program Officer. Conference proposals should be submitted at least four months prior to the conference.
Contact: Dr. Paul Bello; (703) 696-4318 or (703) 696-5031; fax: (703) 696-5940; email: paul.bello@navy.mil
Internet: http://www.onr.navy.mil/sci_tech/34/341/training_cog.asp
Sponsor: Office of Naval Research
875 North Randolph Street, Suite 1425
Arlington, VA 22203-1995

ONR Command and Control Grants 4762

The program focuses on developing new approaches to support tactical decision makers. This is performed by producing distributed situation assessment of a commanders environment through development of a common tactical picture; developing collaborative, distributed mission plans; monitoring and dynamically replanning mission execution as required; understanding their environment by being able to access distributed data bases intelligently through use of software agents; effectively using sensor information and making resource allocation decisions. Specific research involves real time computing, decision aids, and collaborative workspace; virtual reality displays and secure distributed architecture; and information warfare techniques to protect secure information. The deadline for submission is ongoing. Submission of the Research Proposal should always be preceded by submission of a Preproposal and a discussion of that Preproposal with the assigned Program Officer.
Requirements: Grants are awarded to academic, industrial, and nonprofit organizations, with strong emphasis on university research.
Restrictions: New grants or contracts are typically started near the beginning of the fiscal year (October 1). New proposals for the next fiscal year should be submitted no later than June 1, unless otherwise directed by the BAA. Competitive renewals should be submitted no later than six months prior to the award termination date after contacting the Program Officer. Conference proposals should be submitted at least four months prior to the conference.
Contact: Dr. Gary Toth; (703) 696-4961 or (703) 588-2427; fax: (703) 696-5940; email: gary.toth@navy.mil
Internet: http://www.onr.navy.mil/sci_tech/31/311/command_control.asp
Sponsor: Office of Naval Research
875 North Randolph Street, Suite 1425
Arlington, VA 22203-1995

ONR Communications and Networking Grants 4763

This program supports the research and development of improved radio communications and wireless networking. Areas of interest in radio communications include anti-jam and low-probability-of-intercept techniques, adaptive equalization, bandwidth efficient modulation, cognitive radio for dynamic spectrum management, and high data rate tactical communications techniques including communications at speed and depth (for submarines). Areas of interest in wireless networks include mobile ad-hoc wireless networking algorithms/protocols, end-to-end Quality-of-Service, joint/coalition interoperability, IPv6, adaptation of commercial standards, network management and security. Interest areas in antenna technology include wideband multifunction

antennas, compatibility of phased array antennas with naval platforms and marine environments, directional beam forming/steering techniques, and special-purpose submarine communication antenna systems. Submission of the Research Proposal should always be preceded by submission of a Preproposal and a discussion of that Preproposal with the assigned Program Officer.
Requirements: Grants are awarded to academic, industrial, and nonprofit organizations, with strong emphasis on university research.
Restrictions: New grants or contracts are typically started near the beginning of the fiscal year (October 1). New proposals for the next fiscal year should be submitted no later than June 1, unless otherwise directed by the BAA. Competitive renewals should be submitted no later than six months prior to the award termination date after contacting the Program Officer. Conference proposals should be submitted at least four months prior to the conference.
Contact: Dr. Santanu Das; (703) 588-1036 or (703) 588-2427; fax: (703) 696-5940; email: santanu.das@navy.mil
Internet: http://www.onr.navy.mil/sci_tech/31/312/communications.asp
Sponsor: Office of Naval Research
875 North Randolph Street, Suite 1425
Arlington, VA 22203-1995

ONR Communications Systems Grants **4764**
This division supports programs in surface and aerospace surveillance, communications, and electronic combat. Research areas in surface and aerospace surveillance include sensors, primarily radar and electro- optic/infrared, and associated signal and image processing methods. Areas of interest in communications include military radio communications with an emphasis on anti-jam and low-probability-of-intercept techniques, radio networks, and dynamic internetworking. Electronic combat research areas include threat warning systems, electronic support measures, decoys, and electronic countermeasures. Navigation research includes GPS anti-jam and associated inertial navigation techniques. Submission of the Research Proposal should always be preceded by submission of a Preproposal and a discussion of that Preproposal with the assigned Program Officer.
Requirements: Grants are awarded to academic, industrial, and nonprofit organizations, with strong emphasis on university research.
Restrictions: New grants or contracts are typically started near the beginning of the fiscal year (October 1). New proposals for the next fiscal year should be submitted no later than June 1, unless otherwise directed by the BAA. Competitive renewals should be submitted no later than six months prior to the award termination date after contacting the Program Officer. Conference proposals should be submitted at least four months prior to the conference.
Contact: Dr. Santanu Das; (703) 588-1036 or (703) 588-2427; fax: (703) 696-5940; email: santanu.das@navy.mil
Internet: http://www.onr.navy.mil/sci_tech/31/313/default.asp
Sponsor: Office of Naval Research
875 North Randolph Street, Suite 1425
Arlington, VA 22203-1995

ONR Computational Analysis Grants **4765**
The Program develops and exploits methods in modern and classical mathematical analysis with emphasis on mathematical and computational models of physical phenomena. Naval interest in waves, flows, materials, structures, and information processing motivates research in the areas of multi-phase, multi-physics, multi-material problems; fluid transport and mixing; predictability of models for nonlinear dynamics; motion and still image analysis, processing, and enhancement; and inverse problems arising from acoustic and electromagnetic wave propagation and scattering. Submission of the Research Proposal should always be preceded by submission of a Preproposal and a discussion of that Preproposal with the assigned Program Officer.
Requirements: Grants are awarded to academic, industrial, and nonprofit organizations, with strong emphasis on university research.
Restrictions: New grants or contracts are typically started near the beginning of the fiscal year (October 1). New proposals for the next fiscal year should be submitted no later than June 1, unless otherwise directed by the BAA. Competitive renewals should be submitted no later than six months prior to the award termination date after contacting the Program Officer. Conference proposals should be submitted at least four months prior to the conference.
Contact: Dr. Reza Malek-Madani; (703) 696-4314 or (703) 588-2427; fax: (703) 696-5940; email: reza.malekmadani@navy.mil
Internet: http://www.onr.navy.mil/sci_tech/31/311/computational_analysis.asp
Sponsor: Office of Naval Research
875 North Randolph Street, Suite 1425
Arlington, VA 22203-1995

ONR Computational Neuroscience & Biorobotics Grants **4766**
This program aims to discover the principles of control, sensorimotor integration and biomechanics that underlie the agility, quietness, dexterity, and intelligence of swimming and flying biological systems. We seek to understand the robust dynamically stable locomotion of legged animals in unknown terrestrial and aquatic environments, and to formalize these principles to facilitate engineering invention and design. The program seeks to build a bridge between the science of biology and the art of Naval engineering by laying emphasis on the distillation of science in biology and the implementation of the kernel of it on existing Navy products to make significant gains in performance via novel inventions. Research proposals are expected to meld biology and engineering. Of particular interest are the following areas: interdisciplinary research on biology-based high-lift mechanisms and their incorporation on to existing Navy products like propulsors; principles of neural control and hybrid systems for biological and biology-based high-lift devices and system conceptualization, analysis and integration of biology-inspired high-lift mechanisms, artificial-muscle based actuators and their neural control for the development of advanced underwater vehicles and microflying vehicles; research on the neurobiological mechanisms and principles of control of fish pectoral fins, including closed loop control from biosensors; and visual guidance and sensorimotor integration for legged (especially bipedal robots) or wheeled robots that support human robot interaction. Submission of the Research Proposal should always be preceded by submission of a Preproposal and a discussion of that Preproposal with the assigned Program Officer.
Requirements: Grants are awarded to academic, industrial, and nonprofit organizations, with strong emphasis on university research. Investigators are encouraged to contact the staff prior to formally submitting a proposal.
Restrictions: New grants or contracts are typically started near the beginning of the fiscal year (1 October). New proposals for the next fiscal year should be submitted no later than June 1, unless otherwise directed. Competitive renewals should be submitted no later than six months prior to the award termination date after contacting the Program Officer. Conference proposals should be submitted at least four months prior to the conference.
Contact: Dr. Thomas McKenna; (703) 696-4503 or (703) 696-5031; fax: (703) 696-5940; email: tom.mckenna@navy.mil
Internet: http://www.onr.navy.mil/sci_tech/34/341/ne_biorobotics.asp
Sponsor: Office of Naval Research
875 North Randolph Street, Suite 1425
Arlington, VA 22203-1995

ONR Defense University Research Instrumentation Grants **4767**
DURIP is a multi-agency DoD program within the University Research Initiative designed to improve the capabilities of U.S. institutions of higher education to conduct research and to educate scientists and engineers in areas important to national defense by providing funds for the acquisition of research equipment. Each year that DURIP is in effect, a complete announcement will provide detailed program information, including the deadline date for submitting proposals. DURIP proposals submitted to ONR should facilitate research in an area of interest to ONR, as described in the Science and Technology section of the ONR Website. Potential proposers may contact the appropriate program managers, listed with each area of interest, to explore possible mutual interests before submitting proposals. Proposals are evaluated competitively. The evaluation criteria are listed in the current program announcement.
Requirements: Grants are awarded to academic, industrial, and nonprofit organizations, with strong emphasis on university research.
Contact: Dr. William Lukens; (703) 696-4668 or (703) 588-2427; fax: (703) 696-5940; email: william.luckens1@navy.mil
Internet: http://www.onr.navy.mil/sci_tech/3t/corporate/durip.asp
Sponsor: Office of Naval Research
875 North Randolph Street, Suite 1425
Arlington, VA 22203-1995

ONR Device and Circuit Cooling Grants **4768**
This program investigates fundamental properties associated with heat transport and cooling of semiconductor and superconductor materials, devices, and circuits. Of particular attention are interfaces between dissimilar materials exhibiting differing characteristic acoustic impedances. Submission of the Research Proposal should always be preceded by submission of a Preproposal and a discussion of that Preproposal with the assigned Program Officer.
Requirements: Grants are awarded to academic, industrial, and nonprofit organizations, with strong emphasis on university research.
Restrictions: New grants or contracts are typically started near the beginning of the fiscal year (October 1). New proposals for the next fiscal year should be submitted no later than June 1, unless otherwise directed by the BAA. Competitive renewals should be submitted no later than six months prior to the award termination date after contacting the Program Officer. Conference proposals should be submitted at least four months prior to the conference.
Contact: Dr. Deborah Van Vechten; (703) 696-4219 or (703) 588-2427; fax: (703) 696-5940; email: deborah.vanvechten@navy.mil
Internet: http://www.onr.navy.mil/sci_tech/31/312/device_circuit_cooling.asp
Sponsor: Office of Naval Research
875 North Randolph Street, Suite 1425
Arlington, VA 22203-1995

ONR DOD National Defense Science & Engineering Graduate Fellowships **4769**
The NDSEG Fellowship Program is a joint program of the United States Army, Navy and Air Force within the University Research Initiative (URI), designed to increase the number of U.S. citizens trained in disciplines of science and engineering important to defense goals. DoD awards approximately 100-150 new three-year graduate fellowships each year to individuals for study and research leading to doctoral degrees in, or closely related to, the disciplines of aeronautical and astronautical engineering; biosciences, chemical engineering; chemistry; cognitive, neural, and behavioral sciences; electrical engineering; geosciences; civil engineering; computer and computational sciences; materials science and engineering; mathematics; mechanical engineering; naval architecture and ocean engineering; oceanography; and physics. The Fellowships are intended for students at or near the beginning of graduate study. The tenure of an NDSEG Fellowship is 36 months. Consult the current program announcement for

stipend levels. In addition to stipends, the DoD pays the Fellow's full tuition and required fees. The deadline for application is in January.
Requirements: Applicants must be citizens or nationals of the United States who intend to pursue a Ph.D. in one of the designated fields. Applications are encouraged from women, underrepresented minorities and persons with disabilities. Recipients of new three-year fellowships must begin tenure not later than the fall following award.
Contact: Dr. Anthony V. Junior; (703) 696-0453 or (703) 588-2427; fax: (703) 696-5940; email: anthony.junior@navy.mil
Internet: http://www.onr.navy.mil/sci_tech/3t/corporate/gradfe.asp
Sponsor: Office of Naval Research
875 North Randolph Street, Suite 1425
Arlington, VA 22203-1995

ONR Electro-Optical (EO)/Infrared (IR) Sensor Processing Research Grants 4770
The primary goal of this Program is to detect, classify/identify, and localize/geolocate air, sea-surface, and ground targets by improving the performance of the signal and image processing techniques associated with electro-optical (EO) and infrared (IR) sensors (passive and active) that operate from the visible through longwave infrared bands. Submission of the Research Proposal should always be preceded by submission of a Preproposal and a discussion of that Preproposal with the assigned Program Officer.
Requirements: Grants are awarded to academic, industrial, and nonprofit organizations, with strong emphasis on university research.
Restrictions: New grants or contracts are typically started near the beginning of the fiscal year (October 1). New proposals for the next fiscal year should be submitted no later than June 1, unless otherwise directed by the BAA. Competitive renewals should be submitted no later than six months prior to the award termination date after contacting the Program Officer. Conference proposals should be submitted at least four months prior to the conference.
Contact: Dr. Preston Grounds; (703) 696-0331 or (703) 588-2427; fax: (703) 696-5940; email: chip.grounds@navy.mil
Internet: http://www.onr.navy.mil/sci_tech/31/312/processing_eo_ir.asp
Sponsor: Office of Naval Research
875 North Randolph Street, Suite 1425
Arlington, VA 22203-1995

ONR Electronic and Optical Ceramics Research Grants 4771
The goal of the program is to further the understanding of piezoelectric crystals and ceramics. Lying at the heart of most SONAR transducers, these materials and structures perform the essential roles of producing strong interrogating sound pulse and also detecting very weak incoming sound waves. As looking through the sea for navigation, threat detection and weapon guidance is a critical Navy need, ONR has designated this area as a National Naval Responsibility. Submission of the Research Proposal should always be preceded by submission of a Preproposal and a discussion of that Preproposal with the assigned Program Officer.
Requirements: Grants are awarded to academic, industrial, and nonprofit organizations, with strong emphasis on university research.
Restrictions: New grants or contracts are typically started near the beginning of the fiscal year (October 1). New proposals for the next fiscal year should be submitted no later than June 1, unless otherwise directed by the BAA. Competitive renewals should be submitted no later than six months prior to the award termination date after contacting the Program Officer. Conference proposals should be submitted at least four months prior to the conference.
Contact: Dr. Wallace Smith; (703) 696-0284 or (703) 588-2427; fax: (703) 696-5940; email: wallace.smith1@navy.mil
Internet: http://www.onr.navy.mil/sci_tech/ocean/322_processes/prog_op.htm
Sponsor: Office of Naval Research
875 North Randolph Street, Suite 1425
Arlington, VA 22203-1995

ONR Electronic Combat Technology Program Grants 4772
The goal of electronic combat is to disrupt, deny, and exploit enemy use of the electromagnetic (EM) spectrum while ensuring its secure friendly use. To that end, ONR's Electronic Combat Technology Program provides improved threat warning systems, electronic support measures, decoys, and electronic countermeasures (radio frequency and EO/IR) as well as new and improved communication and navigation countermeasures. Submission of the Research Proposal should always be preceded by submission of a Preproposal and a discussion of that Preproposal with the assigned Program Officer.
Requirements: Grants are awarded to academic, industrial, and nonprofit organizations, with strong emphasis on university research.
Restrictions: New grants or contracts are typically started near the beginning of the fiscal year (October 1). New proposals for the next fiscal year should be submitted no later than June 1, unless otherwise directed by the BAA. Competitive renewals should be submitted no later than six months prior to the award termination date after contacting the Program Officer. Conference proposals should be submitted at least four months prior to the conference.
Contact: Dr. Peter E. Craig; (703) 696-0114 or (703) 588-2427; fax: (703) 696-5940; email: peter.craig@navy.mil
Internet: http://www.onr.navy.mil/sci_tech/31/312/electronic_combat.asp
Sponsor: Office of Naval Research
875 North Randolph Street, Suite 1425
Arlington, VA 22203-1995

ONR Electronics Division Research Grants 4773
Projects within this division cover all areas of electronics, including the growth of electronic materials, understanding of surfaces and interfaces, semiconductor and magnetic material theory, microelectronics, nanoscale electronic devices, nanolithography, novel electronic device concepts, memory, high-power RF and switching components, vacuum electronics, high-temperature superconductor electronics, optoelectronics, solid-state lasers and light emitters, multiband IR and UV sensors, ionospheric effects, and innovative electronic circuits. Submission of the Research Proposal should always be preceded by submission of a Preproposal and a discussion of that Preproposal with the assigned Program Officer.
Requirements: Grants are awarded to academic, industrial, and nonprofit organizations, with strong emphasis on university research.
Restrictions: New grants or contracts are typically started near the beginning of the fiscal year (October 1). New proposals for the next fiscal year should be submitted no later than June 1, unless otherwise directed by the BAA. Competitive renewals should be submitted no later than six months prior to the award termination date after contacting the Program Officer. Conference proposals should be submitted at least four months prior to the conference.
Contact: Christopher Rouin; (703) 696-0107 or (703) 588-2427; fax: (703) 696-5940; email: chris.rouin@navy.mil
Internet: http://www.onr.navy.mil/sci_tech/31/default.asp
Sponsor: Office of Naval Research
875 North Randolph Street, Suite 1425
Arlington, VA 22203-1995

ONR Electronics, Sensors, and Network Research Grants 4774
This division conducts an integrated program of 6.1, 6.2, and 6.3 research into the uses of the electromagnetic spectrum in support on Navy and Marine Corps needs. Specific applied programs include research in the areas of surface and aerospace surveillance, communications, electronic combat, and navigation. All these areas are supported by a broad research program in electronics, focused on the reducing the cost weight and size of transmit and receive systems. Two over arching foci of this division are: the development of technologies and techniques to support adaptive persistent surveillance; and the development of digital/RF technologies and techniques to support active aperture phased arrays capable of performing multiple functions simultaneously. Submission of the Research Proposal should always be preceded by submission of a Preproposal and a discussion of that Preproposal with the assigned Program Officer.
Requirements: Grants are awarded to academic, industrial, and nonprofit organizations, with strong emphasis on university research.
Restrictions: New grants or contracts are typically started near the beginning of the fiscal year (October 1). New proposals for the next fiscal year should be submitted no later than June 1, unless otherwise directed by the BAA. Competitive renewals should be submitted no later than six months prior to the award termination date after contacting the Program Officer. Conference proposals should be submitted at least four months prior to the conference.
Contact: Dr. Harry Dietrich; (703) 696-0240 or (703) 588-2427; fax: (703) 696-5940; email: harry.dietrich@navy.mil
Internet: http://www.onr.navy.mil/sci_tech/31/312/
Sponsor: Office of Naval Research
875 North Randolph Street, Suite 1425
Arlington, VA 22203-1995

ONR Environmental Quality Research Grants 4775
The program seeks to develop technology that leads to affordable environmental compliance and pollution prevention afloat and ashore. The program focuses on technologies that are directed toward environmentally sound ships, shore-related facilities, and sediment issues. Specific research interests includes sensors and improved cleaning methods. The program supports basic research, applied research, and advanced technology development. Currently, there are two primary focus areas: Environmentally Benign Marine Antifouling Coatings; and WaSuite Treatment/Reduction. Submission of the Research Proposal should always be preceded by submission of a Preproposal and a discussion of that Preproposal with the assigned Program Officer.
Requirements: Grants are awarded to academic, industrial, and nonprofit organizations, with strong emphasis on university research.
Restrictions: New grants or contracts are typically started near the beginning of the fiscal year (October 1). New proposals for the next fiscal year should be submitted no later than June 1, unless otherwise directed by the BAA. Competitive renewals should be submitted no later than six months prior to the award termination date after contacting the Program Officer. Conference proposals should be submitted at least four months prior to the conference.
Contact: Dr. Stephen McElvany; (703) 696-1449 or (703) 588-2427; fax: (703) 696-5940; email steve.mcelvany.navy.mil
Internet: http://www.onr.navy.mil/sci_tech/33/332/
Sponsor: Office of Naval Research
875 North Randolph Street, Suite 1425
Arlington, VA 22203-1995

ONR Expeditionary Maneuver Warfare & Combating Terrorism Department Grants 4776
The primary focus of this Program is to lead the Department of the Navy's Science and Technology effort that develops future combat capabilities for Naval Expeditionary Maneuver Warfare and the Department's role in Combating Terrorism through the exploitation and subsequent application of Science and Technology in order to enhance

the ability of the Navy-Marine Corps team to achieve assured access and conduct decisive operations as the naval portion of a Joint campaign. Submission of the Research Proposal should always be preceded by submission of a Preproposal and a discussion of that Preproposal with the assigned Program Officer.
Requirements: Grants are awarded to academic, industrial, and nonprofit organizations, with strong emphasis on university research.
Restrictions: New grants or contracts are typically started near the beginning of the fiscal year (October 1). New proposals for the next fiscal year should be submitted no later than June 1, unless otherwise directed by the BAA. Competitive renewals should be submitted no later than six months prior to the award termination date after contacting the Program Officer. Conference proposals should be submitted at least four months prior to the conference.
Contact: John Moniz; (703) 696-2492 or (703) 588-2427; fax: (703) 696-5940; email: john.moniz@navy.mil
Internet: http://www.onr.navy.mil/sci_tech/30/
Sponsor: Office of Naval Research
875 North Randolph Street, Suite 1425
Arlington, VA 22203-1995

ONR Faculty Sabbatical Leave Program Grants **4777**
The Faculty Sabbatical Leave Program provides science and engineering faculty the opportunity to conduct research at Navy laboratories while on sabbatical leave. Participants in the Sabbatical Leave Program will receive a monthly stipend making up the difference between salary and sabbatical leave pay from their home institution. In addition, participants will receive reimbursement for travel to and from the laboratory site and a relocation allowance for those who must relocate their residence during their sabbatical leave tenure. Sabbatical leave appointments are contingent on the availability of sabbatical leave positions. Appointments are for a minimum of one semester and a maximum of one year. This is a residential sabbatical — participants must conduct research on site.
Requirements: Faculty should submit applications at least 6 months prior to their proposed sabbatical leave starting date.
Contact: Dr. Anthony Junior; (703) 696-0453 or (703) 588-2427; fax: (703) 696-5940; email: anthony.junior@navy.mil
Internet: http://www.onr.navy.mil/sci_tech/3t/corporate/sabbat.asp
Sponsor: Office of Naval Research
875 North Randolph Street, Suite 1425
Arlington, VA 22203-1995

ONR Historically Black Colleges and Universities (HBCU) Future Engineering Faculty Fellowships **4778**
This fellowship program is designed to develop and attract qualified engineering faculty to Historically Black Colleges and Universities (HBCUs) with engineering programs. Each year, three recipients who have agreed to join the engineering faculty of an HBCU after receiving their degrees are competitively selected for study and research support leading to doctoral degrees in engineering. Disciplines include the following: aerospace engineering; electrical engineering; mechanical engineering; manufacturing science and engineering; civil engineering; ocean engineering; and chemical engineering. The tenure of an ONR-HBCU Future Engineering Faculty Fellowship is generally 36 months. In addition to stipends, the program pays the Fellow's full tuition and fees (not to include room and board).
Requirements: Applicants must be citizens or nationals of the United States who intend to pursue a Ph.D. in the designated fields and, in return for the support, agree to join the faculty of an HBCU. Applicants must be nominated by one of the participating HBCUs. Applications are especially encouraged from women, underrepresented minorities and persons with disabilities.
Contact: Dr. Anthony Junior; (703) 696-0453 or (703) 588-2427; fax: (703) 696-5940; email: anthony.junior@navy.mil
Internet: http://www.onr.navy.mil/sci_tech/3t/corporate/hbec.asp
Sponsor: Office of Naval Research
875 North Randolph Street, Suite 1425
Arlington, VA 22203-1995

ONR Human Performance Division Research Grants **4779**
The mission of the Human Performance S&T Division is to direct, plan, foster and encourage Science and Technology in cognitive science, computational neuroscience, bio-inspired systems, social/organizational science, manpower & personnel, training, human factors, decision making and human systems integration as related to Naval needs. Major goals include: to sustain warfighter performance and enhanced decision making in all environments through training technologies, manpower management, and human systems integration; to create options for future (perhaps unanticipated) naval decisions, based upon fundamental understanding of the cognitive and neurosciences; to support integrated interdisciplinary research programs; and to cultivate transition of findings to government and commercial funding of applied research, advanced technology development, small business and acquisition projects. Submission of the Research Proposal should always be preceded by submission of a Preproposal and a discussion of that Preproposal with the assigned Program Officer.
Requirements: Grants are awarded to academic, industrial, and nonprofit organizations, with strong emphasis on university research.
Restrictions: New grants or contracts are typically started near the beginning of the fiscal year (October 1). New proposals for the next fiscal year should be submitted no later than June 1, unless otherwise directed by the BAA. Competitive renewals should be submitted no later than six months prior to the award termination date after contacting the Program Officer. Conference proposals should be submitted at least four months prior to the conference.
Amount of Grant: $70,000-$150,000 per year
Contact: Dr. John F. Tangney; (703) 696-4501 or (703) 588-2427; fax: (703) 696-5940; email: john.f.tangney@navy.mil
Internet: http://www.onr.navy.mil/sci_tech/34/341/
Sponsor: Office of Naval Research
875 North Randolph Street, Suite 1425
Arlington, VA 22203-1995

ONR Human Performance: Collaboration and Knowledge Interoperability Grants **4780**
This program seeks to develop cognitive science-based tools, models, computational methods, and human-agent interfaces to help attain common situational awareness among multidisciplinary, distributed team members engaged in collaboration for issue resolution or decision making. Collaboration is defined as joint activity for attaining a shared understanding. It examines the cognitive aspects of joint analysis or problem solving for the purpose of attaining shared understanding sufficient to achieve situational awareness for decision-making or creation of a product. Knowledge interoperability is the identification, collection, fusion and presentation of data for the purpose of attaining comprehension of a situation at both the individual and team level. The program technical approach addresses two principal issues: understanding knowledge building; and understanding team cognition. Research issues include: individual mental model construction; developing knowledge interoperability; developing shared understanding; and achieving consensus. Submission of the Research Proposal should always be preceded by submission of a Preproposal and a discussion of that Preproposal with the assigned Program Officer.
Requirements: Grants are awarded to academic, industrial, and nonprofit organizations, with strong emphasis on university research.
Restrictions: New grants or contracts are typically started near the beginning of the fiscal year (October 1). New proposals for the next fiscal year should be submitted no later than June 1, unless otherwise directed by the BAA. Competitive renewals should be submitted no later than six months prior to the award termination date after contacting the Program Officer. Conference proposals should be submitted at least four months prior to the conference.
Contact: Mike Letsky; (703) 696-4251 or (703) 588-2427; fax: (703) 696-5940; email: Michael.Letsky@navy.mil; Gerald Malecki; (703) 696-4044 or (703) 588-2427; fax: (703) 696-5940; email: gerald.malecki@navy.mil
Internet: http://www.onr.navy.mil/sci_tech/34/341/hp_ckm.asp
Sponsor: Office of Naval Research
875 North Randolph Street, Suite 1425
Arlington, VA 22203-1995

ONR Independent Research & Development Grants **4781**
Independent Research & Development (IR & D) is a DOD sponsored program that encourages contractors to pursue independent research and development projects that are of potential interest to DOD. Through this program, DOD seeks to encourage maintenance of a strong national technology base. The IR & D program allows contractors to recover some of the costs incurred in doing independent research and development as a necessary cost of doing business in a high technology environment. In addition, the Navy benefits from this program by being able to avoid duplication of effort and being able to tailor its R & D efforts to meet future needs that industry R & D efforts are not directed towards. Section 824 of P.L. 101-510 defines "potential interest " to include activities intended to accomplish any of the following: enabling superior performance of future United States' weapon systems and components; reducing acquisition costs and life-cycle costs of military systems; strengthening the defense industrial base and the technology base of the United States; enhancing the industrial competitiveness of the United States; promoting the development of technologies identified as critical under section 2522 of Title 10, the United States Code; increasing the development and promotion of efficient and effective applications of dual-use technologies; and providing efficient and effective technologies for achieving such environmental benefits as improved environmental data gathering, environmental cleanup and restoration, pollution reduction in manufacturing, environmental conservation, and environmentally safe management of facilities. Submission of the Research Proposal should always be preceded by submission of a Preproposal and a discussion of that Preproposal with the assigned Program Officer.
Requirements: Grants are awarded to academic, industrial, and nonprofit organizations, with strong emphasis on university research.
Restrictions: New grants or contracts are typically started near the beginning of the fiscal year (October 1). New proposals for the next fiscal year should be submitted no later than June 1, unless otherwise directed by the BAA. Competitive renewals should be submitted no later than six months prior to the award termination date after contacting the Program Officer. Conference proposals should be submitted at least four months prior to the conference.
Contact: Dorothy Vincent; (703) 696-4792 or (703) 588-2427; fax: (703) 696-5940; email: dorothy.vincent@navy.mil
Internet: http://www.onr.navy.mil/sci_tech/3t/transition/irad.asp
Sponsor: Office of Naval Research
875 North Randolph Street, Suite 1425
Arlington, VA 22203-1995

ONR Intelligent Systems Research Grants 4782

The major thrusts in the Intelligent Systems program are: intelligent agents; planning and reasoning; machine learning; natural language and speech; and computer vision. Minor thrusts include: knowledge representation; information management; and theoretical aspects of AI. Other topics are considered on a case by case basis. Submission of the Research Proposal should always be preceded by submission of a Preproposal and a discussion of that Preproposal with the assigned Program Officer.

Requirements: Grants are awarded to academic, industrial, and nonprofit organizations, with strong emphasis on university research.

Restrictions: New grants or contracts are typically started near the beginning of the fiscal year (October 1). New proposals for the next fiscal year should be submitted no later than June 1, unless otherwise directed by the BAA. Competitive renewals should be submitted no later than six months prior to the award termination date after contacting the Program Officer. Conference proposals should be submitted at least four months prior to the conference.

Contact: Dr. Behzad Kamgar-Parsi; (703) 696-5754 or (703) 588-2427; fax: (703) 696-5940; email: behzad.kamgar-parsi@.navy.mil or 311_IS@onr.navy.mil

Internet: http://www.onr.navy.mil/sci_tech/31/311/intelligent_systems.asp

Sponsor: Office of Naval Research
875 North Randolph Street, Suite 1425
Arlington, VA 22203-1995

ONR Ionosphere Interactions (Space) Research Grants 4783

This program investigates physical and chemical processes that govern active and passive electromagnetic spectrum scattering from the Earth's surface, and propagation through the upper atmosphere and the near space environment. Of particular interest for surface effects is short water wave roughness modulation mechanisms; surfactant effects; intermittency in wave breaking and non-linear water waves. Research is directed toward improving the knowledge base for development of mechanistic EO/EM clutter models and automatic target recognition, and to investigate techniques that invert sensor information for the development of algorithms for assimilation into environmental models. Additional interests include electromagnetic scattering theory, microwave properties, scattering surface characterization, and wave and flux modulation mechanisms. Submission of the Research Proposal should always be preceded by submission of a Preproposal and a discussion of that Preproposal with the assigned Program Officer.

Requirements: Grants are awarded to academic, industrial, and nonprofit organizations, with strong emphasis on university research.

Restrictions: New grants or contracts are typically started near the beginning of the fiscal year (October 1). New proposals for the next fiscal year should be submitted no later than June 1, unless otherwise directed by the BAA. Competitive renewals should be submitted no later than six months prior to the award termination date after contacting the Program Officer. Conference proposals should be submitted at least four months prior to the conference.

Contact: Dr. Robert McCoy; (703) 696-8699 or (703) 588-2427; fax: (703) 696-5940; email: bob.mccoy@navy.mil

Internet: http://www.onr.navy.mil/sci_tech/32/321/space.asp

Sponsor: Office of Naval Research
875 North Randolph Street, Suite 1425
Arlington, VA 22203-1995

ONR Joint Capability Technology Demonstration (JCTD) Program Grants 4784

The Joint Capability Technology Demonstration (JCTD) Program is dedicated to: supporting the combatant commands (CoComs) by addressing their joint and coalition challenges that are specific to their theater or mission; bridging the gaps in joint and coalition warfighting capabilities that have been identified by the Joint Capabilities Integration and Development System (JCIDS); and bringing new technology-based capabilities to the joint and coalition warfighter. Proposals competing for project starts should submit all documentation by January 15 each year.

Requirements: Grants are awarded to academic, industrial, and nonprofit organizations, with strong emphasis on university research. Investigators are encouraged to contact the staff prior to formally submitting a proposal.

Date(s) Application is Due: Jan 15

Contact: John Barkley; (703) 696-3158 or (703) 696-5031; fax: (703) 696-5940; email: john.barkley@navy.mil or jctdcoord@onr.navy.mil

Internet: http://www.onr.navy.mil/sci_tech/3t/jctd/

Sponsor: Office of Naval Research
One Liberty Center, 875 North Randolph Street, Suite 1425
Arlington, VA 22203-1995

ONR Magnetics Grants 4785

This program investigates magnetic films and electronic devices made therefrom. The scope of this program encompasses materials synthesis and device/circuit design ranging from non-volatile memory through microwave circulators and isolators. Submission of the Research Proposal should always be preceded by submission of a Preproposal and a discussion of that Preproposal with the assigned Program Officer.

Requirements: Grants are awarded to academic, industrial, and nonprofit organizations, with strong emphasis on university research.

Restrictions: New grants or contracts are typically started near the beginning of the fiscal year (October 1). New proposals for the next fiscal year should be submitted no later than June 1, unless otherwise directed by the BAA. Competitive renewals should be submitted no later than six months prior to the award termination date after contacting the Program Officer. Conference proposals should be submitted at least four months prior to the conference.

Contact: Dr. Chagaan Baatar; (703) 696-0483 or (703) 588-2427; fax: (703) 696-5940; email: chagaan.baatar@navy.mil

Internet: http://www.onr.navy.mil/sci_tech/31/312/magnetics.asp

Sponsor: Office of Naval Research
875 North Randolph Street, Suite 1425
Arlington, VA 22203-1995

ONR Manufacturing Technology Grants 4786

Managed within the Office of Naval Research (ONR), the Navy ManTech Program provides for the development of enabling manufacturing technology and the transition of this technology for the production and sustainment of Navy weapon systems to support the Fleet. Navy ManTech is currently focused on shipbuilding affordability. Reducing the acquisition cost of current and future platforms is a critical goal of the Navy, and ManTech aids in achieving this goal by developing and transitioning key manufacturing technology. Transition of technology is key to Navy ManTech success. Advances in manufacturing technology are useful to the Navy only if they result in implementation in the production of weapon systems. Transition of manufacturing processes to private and government industrial entities that manufacture and repair systems and components for the Fleet is the goal of every ManTech project. Submission of the Research Proposal should always be preceded by submission of a Preproposal and a discussion of that Preproposal with the assigned Program Officer.

Requirements: Grants are awarded to academic, industrial, and nonprofit organizations, with strong emphasis on university research.

Restrictions: New grants or contracts are typically started near the beginning of the fiscal year (October 1). New proposals for the next fiscal year should be submitted no later than June 1, unless otherwise directed by the BAA. Competitive renewals should be submitted no later than six months prior to the award termination date after contacting the Program Officer. Conference proposals should be submitted at least four months prior to the conference.

Contact: John Carney; (703) 696-0352 or (703) 588-2427; fax: (703) 696-5940; email: john.u.carney@navy.mil

Internet: http://www.onr.navy.mil/sci_tech/3t/mantech/default.asp

Sponsor: Office of Naval Research
875 North Randolph Street, Suite 1425
Arlington, VA 22203-1995

ONR Marine Mammals & Biological Oceanography Grants 4787

ONR's Marine Mammals and Biological Oceanography Program was created late in 2007 and replaces two previous ONR programs (the Marine Mammals and Biology Program in Code 34 and the biological component of the Ocean Optics & Biology Program in Code 32). The Marine Mammals and Biological Oceanography Program supports research on a wide range of topics related to oceanic biota - ranging from the largest whales to the fish, zooplankton, phytoplankton and benthos. The program is guided by, though not limited to, research and technology development related to: impacts of sound on marine mammals; and the interactions between biota and sound or light in the ocean. Submission of the Research Proposal should always be preceded by submission of a Pre-proposal and a discussion of that Pre-proposal with the assigned Program Officer. Pre-proposals are requested by April 1.

Requirements: Grants are awarded to academic, industrial, and nonprofit organizations, with strong emphasis on university research.

Date(s) Application is Due: Apr 1

Contact: Dr. Frank Herr; (703) 696-4125 or (703) 588-2427; fax: (703) 696-5940; email: frank.herr@navy.mil

Internet: http://www.onr.navy.mil/sci_tech/32/322/marine_mammals_biology.asp

Sponsor: Office of Naval Research
875 North Randolph Street, Suite 1425
Arlington, VA 22203-1995

ONR Marine Meteorology and Atmospheric Effects Research Grants 4788

This program sponsors integrated basic, applied, and developmental research with emphasis on improving the modeling and prediction of environmental parameters critical to Navy and Marine Corps platform, sensor, and weapons performance. The program includes research and development, leading to enhanced environmental support for operations, training, mission planning, and systems development, and focuses on addressing Navy and Marine Corps real-time, high-resolution environmental parameter requirements to support tactical sensors and operations in littoral zones worldwide. Topics of interest include meso-scale coastal phenomena, such as coastally trapped disturbances; data assimilation incorporating high data rate, asynchronous sensors (radar, lidar, etc.); global, mesoscale, and on-scene modeling focusing on the marine atmosphere and/or coastal zone; simulation and visualization of weather phenomena; marine boundary layer process, including aerosols and marine convective and nonconvective clouds; environmental effects on electromagnetic and electrooptic propagation; and western and southern Pacific tropical cyclone behavior and evolution in motion and structure. The encouraged submission date is June 15, though unsolicited programs are accepted at any time. Submission of the Research Proposal should always be preceded by submission of a Preproposal and a discussion of that Preproposal with the assigned Program Officer.

Requirements: Grants are awarded to academic, industrial, and nonprofit organizations, with strong emphasis on university research.

Date(s) Application is Due: Jun 15

Contact: Dr. Ronald Ferek; (703) 696-0518 or (703) 588-2427; fax: (703) 696-5940; email: ron.ferek@navy.mil

Internet: http://www.onr.navy.mil/sci_tech/32/322/meteorology_atmospheric.asp
Sponsor: Office of Naval Research
875 North Randolph Street, Suite 1425
Arlington, VA 22203-1995

ONR Maritime Sensing Grants 4789

This program conducts multidisciplinary science and technology development of acoustic and non-acoustic sensors systems as a means to automatically develop and maintain comprehensive tactical and strategic awareness in the maritime undersea battle space. These systems may be carried as on-board equipment for Navy surface ships, submarines, aircraft, and unmanned vehicles or may deployed from those platforms as mobile, drifting, or fixed ocean surveillance systems operating under autonomous or remote control. The intent of this program is to provide next-generation acoustic and non-acoustic sensing technology for the tactical, surveillance, and undersea warfare missions of the Navy. Project scopes may range from component-level research and development to system-level technology demonstrations. Submission of the Research Proposal should always be preceded by submission of a Preproposal and a discussion of that Preproposal with the assigned Program Officer. A planning letter must be received by close of business on May 23rd in order to receive full consideration.
Requirements: Grants are awarded to academic, industrial, and nonprofit organizations, with strong emphasis on university research.
Restrictions: New grants or contracts are typically started near the beginning of the fiscal year (1 October). New proposals for the next fiscal year should be submitted no later than June 1, unless otherwise directed by the BAA. Competitive renewals should be submitted no later than six months prior to the award termination date after contacting the Program Officer. Conference proposals should be submitted at least four months prior to the conference.
Date(s) Application is Due: May 23
Contact: David Johnson; (703) 696-0807 or (703) 588-2427; fax: (703) 696-5940; email: dave.h.johnson@navy.mil or 321_MSTL@onr.navy.mil
Internet: http://www.onr.navy.mil/sci_tech/32/321/maritime_sensing.asp
Sponsor: Office of Naval Research
875 North Randolph Street, Suite 1425
Arlington, VA 22203-1995

ONR Mathematical, Computer, and Information Sciences Research Grants 4790

Sponsored programs in this area focus on experimental and theoretical research, as well as technology in these areas with applications across near-, mid-, and far-term applications. The research programs range from fundamental investigations into mathematical foundations for models, computability, and algorithms; solid state materials and electronic devices; and image and signal analysis in the far-term to C2 applications and infrastructure, sensor technology, and robust navigation technology in the near-term. The technology programs seek to exploit the results of research opportunities to enhance surveillance capabilities, communications, command and control, new concepts for electronic devices, application of information sciences to complex problems including human-computer interaction, navigation, and electronic warfare. Submission of the Research Proposal should always be preceded by submission of a Preproposal and a discussion of that Preproposal with the assigned Program Officer.
Requirements: Grants are awarded to academic, industrial, and nonprofit organizations, with strong emphasis on university research.
Restrictions: New grants or contracts are typically started near the beginning of the fiscal year (October 1). New proposals for the next fiscal year should be submitted no later than June 1, unless otherwise directed by the BAA. Competitive renewals should be submitted no later than six months prior to the award termination date after contacting the Program Officer. Conference proposals should be submitted at least four months prior to the conference.
Contact: Christopher Rouin; (703) 696-0107 or (703) 588-2427; fax: (703) 696-5940; email: chris.rouin@navy.mil
Internet: http://www.onr.navy.mil/sci_tech/31/default.asp
Sponsor: Office of Naval Research
875 North Randolph Street, Suite 1425
Arlington, VA 22203-1995

ONR Medical Science and Technology Division Grants 4791

The Biomedical & Biological Division's mission is to conduct research and technology demonstration programs directed at maintaining the survival, health and performance of Navy and Marine Corps personnel during training, routine operations, special operations, and in time of war. Its primary goals include: increasing the survival of casualties through immediate, life-saving treatment and stabilization; preventing personnel injury caused by the stresses of demanding Naval occupations and environments; and ensuring the fitness and physical readiness of Navy and Marine Corps personnel. Submission of the Research Proposal should always be preceded by submission of a Preproposal and a discussion of that Preproposal with the assigned Program Officer.
Requirements: Grants are awarded to academic, industrial, and nonprofit organizations, with strong emphasis on university research.
Restrictions: New grants or contracts are typically started near the beginning of the fiscal year (October 1). New proposals for the next fiscal year should be submitted no later than June 1, unless otherwise directed by the BAA. Competitive renewals should be submitted no later than six months prior to the award termination date after contacting the Program Officer. Conference proposals should be submitted at least four months prior to the conference.
Contact: Dr. Michael Given; (703) 696-4055 or (703) 588-2427; fax: (703) 696-5940; email: michael.given@navy.mil
Internet: http://www.onr.navy.mil/sci_tech/34/342/
Sponsor: Office of Naval Research
875 North Randolph Street, Suite 1425
Arlington, VA 22203-1995

ONR Microelectronics Applied Research Grants 4792

This program addresses analog and digital VLSI, radiation hardened electronics, high speed A/D converters, advanced microelectronic and nanoelectronic circuits, as well as special purpose architectures including non-volatile memory, non-destructive read-out memory, and chip-to-chip optical connections. Also included are smart switching systems integrating efficient high power switches and their associated control components. Semiconductor material. Proposals are preferred by April 1 of each year. Submission of the Research Proposal should always be preceded by submission of a Preproposal and a discussion of that Preproposal with the assigned Program Officer.
Requirements: Grants are awarded to academic, industrial, and nonprofit organizations, with strong emphasis on university research.
Restrictions: New grants or contracts are typically started near the beginning of the fiscal year (October 1). New proposals for the next fiscal year should be submitted no later than June 1, unless otherwise directed by the BAA. Competitive renewals should be submitted no later than six months prior to the award termination date after contacting the Program Officer. Conference proposals should be submitted at least four months prior to the conference.
Contact: Dr. Ingham Mack; (703) 696-4825 or (703) 588-2427; fax: (703) 696-5940; email: ingham.mack@navy.mil
Internet: http://www.onr.navy.mil/sci_tech/31/312/microelectronics.asp
Sponsor: Office of Naval Research
875 North Randolph Street, Suite 1425
Arlington, VA 22203-1995

ONR Multidisciplinary Research Program of the URI (MURI) Grants 4793

The Multidisciplinary Research Program of the University Research Initiative (MURI) is a multi-agency DoD program that supports research teams whose efforts intersect more than one traditional science and engineering discipline. Multidisciplinary team effort can accelerate research progress in areas particularly suited to this approach. Multidisciplinary research also can help to hasten the transition of research findings to practical application. MURI awards are made in research topics specified by the participating defense agencies each year that the program is in force. Specified topics change each year. Awards are typically for a period of three years (funded incrementally or as options) with two additional years possible as options to bring the total award to five-years, and at a funding level ranging from half a million to about a million dollars per year, with the size of the award dependent upon the topic, technical goals, and availability of appropriations. White papers are due on October 31, 4:00 p.m. Eastern Daylight Time, and full proposals are due on January 9, 4:00 p.m. Eastern Standard Time. Submission of the Research Proposal should always be preceded by submission of a Preproposal and a discussion of that Preproposal with the assigned Program Officer.
Requirements: Grants are awarded to academic, industrial, and nonprofit organizations, with strong emphasis on university research.
Date(s) Application is Due: Oct 31; Jan 9
Contact: Dr. William Lukens; (703) 696-4668 or (703) 588-2427; fax: (703) 696-5940; email: william.luckens1@navy.mil
Internet: http://www.onr.navy.mil/sci_tech/3t/corporate/muri.asp
Sponsor: Office of Naval Research
875 North Randolph Street, Suite 1425
Arlington, VA 22203-1995

ONR Nanoscale Electronics Grants 4794

The objective of this program is to study the electronic, optical and magnetic properties of semiconductor and magnetic structures with nanometer scale dimensions. Layered materials of semiconductors and magnetic metals are developed and nanolithographic methods used to fabricate device structures dominated by quantum or spin phenomena. Devices of interest include quantum wells, quantum dots, single electron transistors, spin transistors, magneto-resistance devices, and ultrashort channel FETs. Novel circuit architectures based on quantum effect devices are investigated. Submission of the Research Proposal should always be preceded by submission of a Preproposal and a discussion of that Preproposal with the assigned Program Officer. Proposals are preferred by April 1 of each year.
Requirements: Grants are awarded to academic, industrial, and nonprofit organizations, with strong emphasis on university research.
Restrictions: Competitive renewals should be submitted no later than six months prior to the award termination date after contacting the Program Officer.
Amount of Grant: $50,000-$400,000
Date(s) Application is Due: Apr 1
Contact: Dr. Chagaan Baatar; (703) 696-0483 or (703) 588-2427; fax: (703) 696-5940; email: chagaan.baatar@navy.mil
Internet: http://www.onr.navy.mil/sci_tech/31/312/nanoscale_electronics.asp
Sponsor: Office of Naval Research
875 North Randolph Street, Suite 1425
Arlington, VA 22203-1995

ONR Naval Materials Division Grants 4795

The Naval Materials Division supports basic and applied research and advanced development in the physical sciences and engineering, materials and processing, and environmental quality that enable enhanced performance, affordability and reliability for

future and legacy Navy and Marine Corps systems and platforms. Programs cover a full spectrum of activities from long-range, fundamental scientific and engineering research in the design and realization of new materials and systems to fulfilling the unique requirements of marine and military applications. For example, functional materials efforts emphasize naval needs for advanced power, energy and weapons systems with volume and weight efficiencies beyond current, materials-limited capabilities. Submission of the Research Proposal should always be preceded by submission of a Preproposal and a discussion of that Preproposal with the assigned Program Officer.
Requirements: Grants are awarded to academic, industrial, and nonprofit organizations, with strong emphasis on university research. Investigators are encouraged to discuss their project with program staff prior to submitting a proposal.
Restrictions: New grants or contracts are typically started near the beginning of the fiscal year (1 October). New proposals for the next fiscal year should be submitted no later than June 1, unless otherwise directed. Competitive renewals should be submitted no later than six months prior to the award termination date after contacting the Program Officer. Conference proposals should be submitted at least four months prior to the conference.
Contact: Dr. Michele Anderson; (703) 696-1938 or (703) 696-5031; fax: (703) 696-5940; email: michele.anderson1@navy.mil
Internet: http://www.onr.navy.mil/sci_tech/33/332/
Sponsor: Office of Naval Research
875 North Randolph Street, Suite 1425
Arlington, VA 22203-1995

ONR Naval Research Enterprise Intern Program (NREIP) Grants 4796
This ten-week intern program is designed to provide opportunities for undergraduate and graduate students to participate in research, under the guidance of an appropriate research mentor, at a participating Navy laboratory. Interns will be selected based upon academic achievement, personal statements, recommendation, and career and research interests. NREIP stipends are $5,500 for undergraduates and $6,500 for graduate students.
Requirements: Applicants must be: U.S. citizens (Permanent Resident Alien status is considered at some labs); junior, senior, or graduate students; and enrolled at a participating university.
Amount of Grant: $5,500 for undergraduates; $6,500 for graduates
Date(s) Application is Due: Jan 12
Contact: Dr. Anthony Junior; (703) 696-0453 or (703) 588-2427; fax: (703) 696-5940; email: anthony.junior@navy.mil
Internet: http://www.onr.navy.mil/sci_tech/3t/corporate/nreip.asp
Sponsor: Office of Naval Research
875 North Randolph Street, Suite 1425
Arlington, VA 22203-1995

ONR Navy Postdoctoral Fellowships 4797
The Navy, through the Naval Research Laboratory (NRL) , sponsors a Postdoctoral Fellowship Program at NRL and a number of Naval R&D centers and laboratories. The objective of this program is to encourage the involvement of creative, capable and highly trained scientists and engineers who have received a Ph.D. or equivalent within the prior seven years in research areas of interest and relevance to the Navy. The program is administered for NRL by the American Society for Engineering Education. Information about the program is also available at the ASEE website.
Contact: Dr. Anthony Junior; (703) 696-0453 or (703) 588-2427; fax: (703) 696-5940; email: anthony.junior@navy.mil
Internet: http://www.onr.navy.mil/sci_tech/3t/corporate/postfe.asp
Sponsor: Office of Naval Research
875 North Randolph Street, Suite 1425
Arlington, VA 22203-1995

ONR Navy Research Facilities Grants 4798
The Research Facilities program manages, under a charter party agreement with selected universities, the operation and maintenance of all oceanographic research vessels owned by ONR for Department of the Navy. As part of the management of the ONR fleet, the program actively supports fleet modernization to maintain the level of excellence that is required to support cutting edge oceanographic research. Other oceanographic platforms such as unmanned underwater vehicles, aerial unmanned vehicles may be used to collect field data through the Research Facilities Program. Submission of the Research Proposal should always be preceded by submission of a Preproposal and a discussion of that Preproposal with the assigned Program Officer.
Requirements: Grants are awarded to academic, industrial, and nonprofit organizations, with strong emphasis on university research. Investigators are encouraged to contact the staff prior to formally submitting a proposal.
Restrictions: New grants or contracts are typically started near the beginning of the fiscal year (October 1). New proposals for the next fiscal year should be submitted no later than June 1, unless otherwise directed by the BAA. Competitive renewals should be submitted no later than six months prior to the award termination date after contacting the Program Officer. Conference proposals should be submitted at least four months prior to the conference.
Contact: Bob Houtman; (703) 696-5123 or (703) 588-2427; fax: (703) 696-5940; email: bauke.houtman@navy.mil
Internet: http://www.onr.navy.mil/sci_tech/32/321/navy_research_facilities.asp
Sponsor: Office of Naval Research
875 North Randolph Street, Suite 1425
Arlington, VA 22203-1995

ONR Ocean Acoustics Research Grants 4799
The program supports research that addresses an understanding of the physics of the generation, propagation, and scattering of narrowband and broadband acoustic (and elastic) waves in the temporally and spatially varying ocean environment. Though research in acoustical signal processing is no longer a significant thrust of this program, support is granted to research in collaboration with the Undersea Signal Processing Program, which also manages development and demonstration efforts. The Ocean Acoustics program contains three primary thrusts and a miscellaneous thrust. The Shallow-Water Acoustics thrust seeks to understand the propagation and scattering of low frequency (10Hz to a few kHz) acoustic energy within the shallow-water ocean environment. Areas of research include investigations of the dominant shallow-water scattering mechanisms, the conversion of seafloor-incident acoustic energy into elastic body waves and interface waves, and acoustic propagation through linear and nonlinear internal waves. The High-Frequency Acoustics thrust seeks to understand the interaction of high frequency (few kHz to thousand kHz) sound with the ocean environment. Components of this thrust include the propagation of sound through an intervening turbulent or stochastic medium; scattering from rough surfaces, biologics, and bubbles; and penetration/propagation within the porous seafloor. The Long-Range Propagation thrust seeks to understand the behavior of sound as it propagates over very long ranges (several hundred kilometers to several thousand kilometers) in the ocean. The main area of interest is understanding the effect of ocean internal waves and mesoscale on transmitted broadband acoustic signals. Submission of the Research Proposal should always be preceded by submission of a Preproposal and a discussion of that Preproposal with the assigned Program Officer.
Requirements: Grants are awarded to academic, industrial, and nonprofit organizations, with strong emphasis on university research. Investigators are encouraged to contact the staff prior to formally submitting a proposal.
Restrictions: New grants or contracts are typically started near the beginning of the fiscal year (1 October). New proposals for the next fiscal year should be submitted no later than June 1, unless otherwise directed. Competitive renewals should be submitted no later than six months prior to the award termination date after contacting the Program Officer. Conference proposals should be submitted at least four months prior to the conference.
Contact: Dr. Ellen Livingston; (703) 696-4203 or (703) 696-5031; fax: (703) 696-5940; email: ellen.livingston@navy.mil
Internet: http://www.onr.navy.mil/sci_tech/32/321/ocean_acoustics.asp
Sponsor: Office of Naval Research
875 North Randolph Street, Suite 1425
Arlington, VA 22203-1995

ONR Ocean Engineering and Marine Systems Research Grants 4800
The goals of this program are to improve the knowledge base of fluid-structure interactions for engineering designs and to develop and demonstrate new technologies for expeditionary operations and special warfare. The investigators conduct multidisciplinary science and technology efforts in the behavior of ocean systems in general as well as specifically for the Marine Corps, Navy, Special Forces, and Navy Explosive Ordnance Disposal. The basic research program focuses on fluid forcing and structural response mechanisms and the coupled nonlinear interactions of fluids with generic structural components and unmanned platform systems. Technology development efforts are primarily in the areas of detection and imaging technologies applicable to underwater and buried objects, technologies for neutralization of explosive devices, underwater life support technologies, surface and subsurface transport systems for Navy Special Forces, mobile autonomous platform system technologies for very shallow water, surf zone and beach zone reconnaisance, and technologies leading to capability for rapid wide-area clearance of mines and obstacles from the surf zone. Calls for preproposals are posted on the Web site. Submission of the Research Proposal should always be preceded by submission of a Preproposal and a discussion of that Preproposal with the assigned Program Officer.
Requirements: Grants are awarded to academic, industrial, and nonprofit organizations, with strong emphasis on university research. Investigators are encouraged to contact the staff prior to formally submitting a proposal.
Restrictions: New grants or contracts are typically started near the beginning of the fiscal year (1 October). New proposals for the next fiscal year should be submitted no later than June 1, unless otherwise directed. Competitive renewals should be submitted no later than six months prior to the award termination date after contacting the Program Officer. Conference proposals should be submitted at least four months prior to the conference.
Contact: Dr. Thomas Swean; (703) 696-4025 or (703) 696-5031; fax: (703) 696-5940; email: tom.swean@navy.mil
Internet: http://www.onr.navy.mil/sci_tech/32/321/ocean_engineering_marine_systems_applied.asp#app
Sponsor: Office of Naval Research
875 North Randolph Street, Suite 1425
Arlington, VA 22203-1995

ONR Optical and Biological Oceanography Research Grants 4801
Funded projects focus on elucidating the optical and biological processes in the world ocean. Investments in optics support basic and applied long-term research leading to a more complete understanding of how light interacts with the ocean environment, including the surface, volume, and sea floor. Biological oceanographic investments are directed primarily at understanding the distribution, growth, and abundance of biota in the coastal ocean and shallow-water sediments, focusing especially on those organisms that affect or control the bulk optical and acoustical properties of the ocean. Overarching, long-term investments within the program are directed at developing better observational capabilities, in situ sensors, ocean color algorithms, and autonomous

systems, and distilling the accumulated knowledge into multidisciplinary, predictive models. Planning letters are requested by March 30, prior to submitting a proposal. The department encourages telephone inquiries to discuss ideas.
Requirements: Grants are awarded to academic, industrial, and nonprofit organizations, with strong emphasis on university research.
Date(s) Application is Due: Mar 30
Contact: Dr. Steve Ackleson; (703) 696-4732 or (703) 588-2427; fax: (703) 696-5940; email: steve.ackleson@navy.mil
Internet: http://www.onr.navy.mil/sci_tech/32/322/ocean_optics_biology.asp
Sponsor: Office of Naval Research
875 North Randolph Street, Suite 1425
Arlington, VA 22203-1995

ONR Physical Oceanography Research Grants **4802**
This program supports process-oriented and hypothesis-driven science and technology in the area of physical oceanography and in the applications of oceanic research tools and techniques for fleet usage. In response to post-Cold War naval strategy and tactics, increased emphasis is given to the littoral, defined here as the oceanographic region encompassing the continental shelf and slope and the adjacent deep water. The approach is principally on field observations with theory and modeling expected to be closely integrated. The following thrusts are supported: air-sea interaction: impact on mixed layer dynamics for improved parameterizations in oceanic and atmospheric models; internal waves/turbulence: development of a global littoral model and a low level of open ocean work; marginal seas/straits: identify key physics, develop archetypes, generalize results; shelf/slope dynamics: shelf/slope coupling, parameterizations for improved slope/shelf modeling; boundary current/slope interaction: quantify key physics for enhancement of improved regional modeling; fleet meteorological and oceanographic support: develop tools, techniques, and observations of direct impact to activities at sea. An overarching objective of the program is to foster transition of research products such as numerical and theoretical models, analysis algorithms, in situ data, sea-going instrumentation, and platforms into operational naval systems. In addition, support is also available for critical evaluation of the impact of the environment on fleet exercises that test and evaluate operational naval systems.
Requirements: Grants are awarded to academic, industrial, and nonprofit organizations, with strong emphasis on university research. A proposal may be submitted at any time; however, it is recommended that a planning letter or preproposal be submitted by May 1, prior to a submission of a proposal. A call for planning letters occurs annually.
Date(s) Application is Due: May 1
Contact: Dr. C. Linwood Vincent; (703) 696-4118 or (703) 588-2427; fax: (703) 696-5940; email: linwood.vincent@navy.mil
Internet: http://www.onr.navy.mil/sci_tech/32/322/physical_oceanography.asp
Sponsor: Office of Naval Research
875 North Randolph Street, Suite 1425
Arlington, VA 22203-1995

ONR Radio Frequency (RF) Sensor Processing Research Grants **4803**
This program investigates physical and chemical processes that govern active and passive electromagnetic spectrum scattering from the earth's surface, and propagation through the upper atmosphere and the near space environment. Algorithms and hardware (including photonic) for storing, relaying, processing, extracting and exploiting target information from radio frequency (RF) sensors and arrays in real-time are of interest. Submission of the Research Proposal should always be preceded by submission of a Preproposal and a discussion of that Preproposal with the assigned Program Officer.
Requirements: Grants are awarded to academic, industrial, and nonprofit organizations, with strong emphasis on university research.
Restrictions: New grants or contracts are typically started near the beginning of the fiscal year (October 1). New proposals for the next fiscal year should be submitted no later than June 1, unless otherwise directed by the BAA. Competitive renewals should be submitted no later than six months prior to the award termination date after contacting the Program Officer. Conference proposals should be submitted at least four months prior to the conference.
Contact: George McNeal; (703) 696-4845 or (703) 588-2427; fax: (703) 696-5940; email: george.mcneal@navy.mil
Internet: http://www.onr.navy.mil/sci_tech/31/312/processing_rf.asp
Sponsor: Office of Naval Research
875 North Randolph Street, Suite 1425
Arlington, VA 22203-1995

ONR Radio Frequency Solid State Devices Applied Research Grants **4804**
The program objective is to develop solid state (e.g., wide bandgap semiconductor) low noise and power devices and circuits for use in power supplies, communications, fire control, surveillance, EW, and multifunctional RF systems. Emphasis is on decade wide instantaneous bandwidth, excellent linearity, high (e.g., 60%) power added efficiency, and power output of at least 5 times that of GaAs-based devices. Control components such as limiters, circulators, and filters are also included. Submission of the Research Proposal should always be preceded by submission of a Preproposal and a discussion of that Preproposal with the assigned Program Officer. Proposals are preferred by April 1 of each year.
Requirements: Grants are awarded to academic, industrial, and nonprofit organizations, with strong emphasis on university research.
Date(s) Application is Due: Apr 1
Contact: Dr. Ingham Mack; (703) 696-4825 or (703) 588-2427; fax: (703) 696-5940; email: ingham.mack@navy.mil
Internet: http://www.onr.navy.mil/sci_tech/31/312/solid_state_devices.asp
Sponsor: Office of Naval Research
875 North Randolph Street, Suite 1425
Arlington, VA 22203-1995

ONR Semiconductor Devices Grants **4805**
This program investigates electronic properties of wide bandgap and silicon-based semiconductors and novel devices. The program is focussed on novel high power microwave and switching devices for broadband radar, EW, communications, and multifunctional systems applications and for high power control applications. Submission of the Research Proposal should always be preceded by submission of a Preproposal and a discussion of that Preproposal with the assigned Program Officer.
Requirements: Grants are awarded to academic, industrial, and nonprofit organizations, with strong emphasis on university research.
Restrictions: New grants or contracts are typically started near the beginning of the fiscal year (October 1). New proposals for the next fiscal year should be submitted no later than June 1, unless otherwise directed by the BAA. Competitive renewals should be submitted no later than six months prior to the award termination date after contacting the Program Officer. Conference proposals should be submitted at least four months prior to the conference.
Amount of Grant: $115,000 average
Contact: Dr. Harry Dietrich; (703) 696-0240 or (703) 588-2427; fax: (703) 696-5940; email: harry.dietrich@navy.mil
Internet: http://www.onr.navy.mil/sci_tech/31/312/semiconductor_devices.asp
Sponsor: Office of Naval Research
875 North Randolph Street, Suite 1425
Arlington, VA 22203-1995

ONR Semiconductor Materials Grants **4806**
This program has two major thrust areas: methods for the growth, characterization, and application of high perfection bulk and epitaxial semiconductors with particular emphasis on wide band-gap materials such as diamond, silicon carbide, and the group III nitrides are supported; and methods for the growth and integration of single crystal non-semiconductor epitaxial functional materials and hetero-structures with wide band-gap semiconductors, and their novel functional device/application fabrication and performance evaluation. Fundamental theory necessary for understanding the electronic properties of these epitaxial structures required to understand co-operative and transductive behavior of functional multi-layer hetero-structures are also encouraged. White papers are due on March 2, with final proposals encouraged by April 1. Submission of the Research Proposal should always be preceded by submission of a Preproposal and a discussion of that Preproposal with the assigned Program Officer.
Requirements: Grants are awarded to academic, industrial, and nonprofit organizations, with strong emphasis on university research.
Date(s) Application is Due: Mar 2; Apr 1
Contact: Dr. Paul Maki; (703) 696-3060 or (703) 588-2427; fax: (703) 696-5940; email: paul.maki@navy.mil
Internet: http://www.onr.navy.mil/sci_tech/31/312/semiconductor_materials.asp
Sponsor: Office of Naval Research
875 North Randolph Street, Suite 1425
Arlington, VA 22203-1995

ONR Ship Systems and Engineering Research Division Grants **4807**
The division is focused on providing technologically superior warfighting capabilities at reduced total ownership costs for surface and subsurface platforms through investments in basic and applied research, and advanced technology development of programs in the following areas: Advanced Naval Power; Advanced Platforms; Platform Survivability; and Sea Basing Systems. Research topic areas of interest that contribute to these focus areas include: electrical power generation systems and equipment; energy conversion, distribution, and storage; electro-mechanics; thermoelectrics; phototonics; thermal energy management; stability and control of electrical networks; ship design and analysis; systems engineering; ship systems integration; hydromechanics; maneuvering and seakeeping; hydroacoustics; structural mechanics; solid mechanics, thermo-mechanics; structural acoustics; structural dynamics; computational mechanics; advanced structural systems; protection systems and design methodology; automation; control of dynamic systems; damage control and firefighting; control of acoustic and non-acoustic signatures; high speed platforms; sense and respond logistics; and inter- and intra-ship material movement and tracking. Submission of the Research Proposal should always be preceded by submission of a Preproposal and a discussion of that Preproposal with the assigned Program Officer.
Requirements: Grants are awarded to academic, industrial, and nonprofit organizations, with strong emphasis on university research. Investigators are encouraged to contact the staff prior to formally submitting a proposal.
Restrictions: New grants or contracts are typically started near the beginning of the fiscal year (1 October). New proposals for the next fiscal year should be submitted no later than June 1, unless otherwise directed. Competitive renewals should be submitted no later than six months prior to the award termination date after contacting the Program Officer. Conference proposals should be submitted at least four months prior to the conference.
Contact: Dr. Richard Carlin; (703) 696-5075 or (703) 696-5031; fax: (703) 696-5940; email: richard.carlin1@navy.mil
Internet: http://www.onr.navy.mil/sci_tech/33/331/default.asp
Sponsor: Office of Naval Research
875 North Randolph Street, Suite 1425
Arlington, VA 22203-1995

ONR Signal and Image Processing Grants 4808

In a complex and dynamically changing environment that war fighters operate, there are many essential and important disparate sources of information such as speech, images from different sensors - visual camera, IR, LADAR, SAR, etc., signals from chemical/biological sensors and text. The main objectives of this program are to develop underlying mathematical framework, algorithms and tools for intelligent and efficient way of extracting, representing, retrieving and understanding information from these disparate sources based on the context. Submission of the Research Proposal should always be preceded by submission of a Preproposal and a discussion of that Preproposal with the assigned Program Officer.
Requirements: Grants are awarded to academic, industrial, and nonprofit organizations, with strong emphasis on university research.
Restrictions: New grants or contracts are typically started near the beginning of the fiscal year (October 1). New proposals for the next fiscal year should be submitted no later than June 1, unless otherwise directed by the BAA. Competitive renewals should be submitted no later than six months prior to the award termination date after contacting the Program Officer. Conference proposals should be submitted at least four months prior to the conference.
Contact: Dr. Behzad Kamgar-Parsi; (703) 696-5754 or (703) 588-2427; fax: (703) 696-5940; email: behzad.kamgar-parsi@.navy.mil or 311_IS@onr.navy.mil
Internet: http://www.onr.navy.mil/sci_tech/31/311/signal_image_processing.asp
Sponsor: Office of Naval Research
875 North Randolph Street, Suite 1425
Arlington, VA 22203-1995

ONR Small Business Innovation Research (SBIR/STTR) Grants 4809

The ONR SBIR program is an integral part of the Naval Science and Technology program. ONR SBIR topics address naval needs and should ultimately transition to the fleet through enabling technologies for future naval capabilities or directly to naval acquisition programs. Topics are derived from ONR Program Officers via periodic calls which are issued by the ONR SBIR Program Manager to the ONR Department Heads. ONR SBIR topics are published and solicited with those of other Naval components in the Department of Defense SBIR Program Solicitations. Qualified small businesses may submit proposals for these topics which are then competitively evaluated by ONR Program Officers. Awards are made to the best proposing companies within allotted funding. Funding for this program is not appropriated, but provided annually under the statutory 2-1/2% set-aside from ONR extramural research, development, testing and evaluation (RDT&E) funds. DoD goals are for all Phase Is to be under contract within 4 months of receipt and for all Phase IIs to be under contract within 6 months following the end of the Phase I contract. Timeline/calendar of events is enclosed for planning purposes and to help meet these goals.
Requirements: A business may apply that is independently owned and operated; is not dominant in the field of operation in which it is proposing; has its principal place of business located in the United States; is organized for profit; is at least 51 percent owned, or in the case of a publicly owned business, at least 51 percent of its voting stock is owned by United States citizens or lawfully admitted permanent resident aliens; and including its affiliates, has a number of employees amounting to 500 or less.
Contact: John Williams, (703) 696-0342 or (703) 588-2427; fax (703) 696-5940; email: john.williams6@navy.mil
Internet: http://www.onr.navy.mil/sci_tech/3t/sbir_sttr/sbir_onr.asp
Sponsor: Office of Naval Research
875 North Randolph Street, Suite 1425
Arlington, VA 22203-1995

ONR Small Business Technology Transfer (STTR) Program Grants 4810

This mission-oriented science and technology program seeks to produce commercializable products through scientific research areas. The program requires that small businesses and non-profit research institutions submit a joint proposal responsive to broad topics defined in the Navy section of the annual DOD STTR solicitation. Unlike the Navy's SBIR program, the Navy STTR program focuses on a different science or technology area each year. You may register to receive the DoD Solicitation on-line via the DoD SBIR Program.
Requirements: A business may apply that is independently owned and operated; is not dominant in the field of operation in which it is proposing; has its principal place of business located in the United States; is organized for profit; is at least 51 percent owned, or in the case of a publicly owned business, at least 51 percent of its voting stock is owned by United States citizens or lawfully admitted permanent resident aliens; and including its affiliates, has a number of employees amounting to 500 or less.
Contact: John Williams, (703) 696-0342 or (703) 588-2427; fax: (703) 696-5940; email: john.williams6@navy.mil
Internet: http://www.onr.navy.mil/sci_tech/3t/sbir_sttr/sttr_pilot.asp
Sponsor: Office of Naval Research
875 North Randolph Street, Suite 1425
Arlington, VA 22203-1995

ONR Software and Computer Systems Grants 4811

The software program is developing the foundations for the verifiably correct design and construction of complex software systems. Research is supported in linear logic and related proof systems, software testing, formal algorithm derivation, and formal proof of correctness, among other areas. Submission of the Research Proposal should always be preceded by submission of a Preproposal and a discussion of that Preproposal with the assigned Program Officer.
Requirements: Grants are awarded to academic, industrial, and nonprofit organizations, with strong emphasis on university research.
Restrictions: New grants or contracts are typically started near the beginning of the fiscal year (October 1). New proposals for the next fiscal year should be submitted no later than June 1, unless otherwise directed by the BAA. Competitive renewals should be submitted no later than six months prior to the award termination date after contacting the Program Officer. Conference proposals should be submitted at least four months prior to the conference.
Contact: Dr. Ralph Wachter; (703) 696-4304 or (703) 588-2427; fax: (703) 696-5940; email: ralph.wachter@navy.mil
Internet: http://www.onr.navy.mil/sci_tech/31/311/software_computer_systems.asp
Sponsor: Office of Naval Research
875 North Randolph Street, Suite 1425
Arlington, VA 22203-1995

ONR Special Research Awards in Ocean Acoustics 4812

Three types of awards are available for conducting research in the area of ocean acoustics: Graduate Traineeship Awards (5 available); Postdoctoral Fellowship Awards (3 available); and Entry-Level Faculty Awards (2 available). The deadline for awards starting next calendar year (January 1 to December 31) is August 1. Notification of awards will be made no later than the following October and grants will begin after January.
Date(s) Application is Due: Aug 1
Contact: Dr. Ellen Livingston; (703) 696-4203 or (703) 696-5031; fax: (703) 696-5940; email: ellen.livingston@navy.mil
Internet: http://www.onr.navy.mil/sci_tech/32/321/ocean_acoustics_awards.asp
Sponsor: Office of Naval Research
875 North Randolph Street, Suite 1425
Arlington, VA 22203-1995

ONR Summer Faculty Research Program Grants 4813

The Summer Faculty Research Program provides science and engineering faculty members from institutions of higher education the opportunity to participate in research at Navy laboratories for a ten week period during the summer. Participants work with professional peers in the Navy laboratories on research tasks of mutual interest. Information about the research interests of the participating laboratories is available via the ASEE website. The research for the summer is defined in advance through correspondence and an optional pre-program visit to the research site, for which funds are provided. Three levels of appointment are available: Summer Faculty Fellow, Senior Summer Faculty Fellow, and Distinguished Summer Faculty Fellow. Stipends vary by level of appointment and are set each year. In addition, each Fellow will be reimbursed for his/her personal travel for an optional pre-program visit to the research site.
Requirements: The ONR Summer Faculty Research Program is open to United States citizens or nationals and to legal permanent residents holding a teaching or research position at a U.S. college or university. Some restrictions may apply to permanent residents regarding eligibility for a DoD security clearance.
Contact: Dr. Anthony Junior; (703) 696-0453 or (703) 588-2427; fax: (703) 696-5940; email: anthony.junior@navy.mil
Internet: http://www.onr.navy.mil/sci_tech/3t/corporate/summfac.asp
Sponsor: Office of Naval Research
875 North Randolph Street, Suite 1425
Arlington, VA 22203-1995

ONR Surface/Aerospace Surveillance Grants 4814

This program is devoted to the development of enhanced sensor technology, primarily radar and electro-optic/infrared (EO/IR) and associated signal, image and array processing methods. With respect to radar, areas of emphasis include wideband, frequency-agile, multifunction radar; phased-array affordability; the enhancement of radar functionality through the integration of new operating modes such as synthetic aperture radar (SAR), inverse synthetic aperture radar (ISAR), moving target indication (MTI) and moving target imaging (MTIm); and automatic target recognition (ATR). With respect to EO/IR, areas of emphasis include EO/IR shared apertures; multispectral sensing; clutter, background and target modeling; dim target detection; and active/passive classification/discrimination techniques. Additionally, there are efforts underway that seek to develop techniques for target tracking, sensor/data fusion, sensor resource management, and multi-sensor integration. Submission of the Research Proposal should always be preceded by submission of a Preproposal and a discussion of that Preproposal with the assigned Program Officer.
Requirements: Grants are awarded to academic, industrial, and nonprofit organizations, with strong emphasis on university research.
Restrictions: New grants or contracts are typically started near the beginning of the fiscal year (October 1). New proposals for the next fiscal year should be submitted no later than June 1, unless otherwise directed by the BAA. Competitive renewals should be submitted no later than six months prior to the award termination date after contacting the Program Officer. Conference proposals should be submitted at least four months prior to the conference.
Contact: Dr. Michael Pollock; (703) 696-4202 or (703) 696-1438; fax: (703) 696-5940; email: michael.a.pollock@navy.mil or 312_APEI@onr.navy.mil
Internet: http://www.onr.navy.mil/sci_tech/31/312/surveillance.asp
Sponsor: Office of Naval Research
875 North Randolph Street, Suite 1425
Arlington, VA 22203-1995

ONR Target Tracking and Sensor Fusion Grants **4815**

This program's primary purpose is to support research programs in the development of target tracking, sensor/data fusion, sensor resource management, and multi-sensor integration. Submission of the Research Proposal should always be preceded by submission of a Preproposal and a discussion of that Preproposal with the assigned Program Officer.

Requirements: Grants are awarded to academic, industrial, and nonprofit organizations, with strong emphasis on university research.

Restrictions: New grants or contracts are typically started near the beginning of the fiscal year (October 1). New proposals for the next fiscal year should be submitted no later than June 1, unless otherwise directed by the BAA. Competitive renewals should be submitted no later than six months prior to the award termination date after contacting the Program Officer. Conference proposals should be submitted at least four months prior to the conference.

Contact: Dr. Rabinder Madan; (703) 696-4217 or (703) 588-2427; fax: (703) 696-5940; email: rabinder.madan@navy.mil

Internet: http://www.onr.navy.mil/sci_tech/31/311/tracking_fusion.asp

Sponsor: Office of Naval Research
875 North Randolph Street, Suite 1425
Arlington, VA 22203-1995

ONR Undersea Signal Processing Grants **4816**

The goal is to develop signal processing algorithms that improve the Navy's ability to rapidly detect, identify and locate submarines in shallow and deep ocean environments. To this end, the program funds basic and applied research projects at universities, Federal laboratories and industry. The program is organized broadly into the following technical thrusts: Active Sonar Signal Processing—projects focus on active sonar systems for wide area surveillance and battle group defense; Passive Sonar Signal Processing—projects focus on signal processing algorithms that detect, classify and locate modern diesel-electric submarines at tactically useful ranges; Data Fusion Technology—development of new ways to fuse sonar data from multiple sensors; Fundamental Research Initiatives—basic research in statistical and ocean acoustic signal processing projects conducted at universities throughout North America; and Acoustic Communications—to develop and demonstrate signal processing systems that improve the performance of underwater acoustic communication networks.

Requirements: Grants are awarded to academic, industrial, and nonprofit organizations, with strong emphasis on university research.

Date(s) Application is Due: May 9

Contact: Dr. John Tague; (703) 696-4399 or (703) 588-2427; fax: (703) 696-5940; email: john.a.tague@navy.mil

Internet: http://www.onr.navy.mil/sci_tech/32/321/undersea_signal_processing.asp

Sponsor: Office of Naval Research
875 North Randolph Street, Suite 1425
Arlington, VA 22203-1995

ONR Vacuum Electronics Research Grants **4817**

This program develops techniques and technologies for compact high power, high frequency vacuum electronic amplifiers exhibiting high gain, bandwidth, and efficiency. Of particular interest is the development of compact high power amplifiers that can be applied to the millimeter and sub-millimeter wave frequency range. Geometry and physics-based computationally efficient algorithms, models and simulation based design. Submission of the Research Proposal should always be preceded by submission of a Preproposal and a discussion of that Preproposal with the assigned Program Officer.

Requirements: Grants are awarded to academic, industrial, and nonprofit organizations, with strong emphasis on university research.

Restrictions: New grants or contracts are typically started near the beginning of the fiscal year (October 1). New proposals for the next fiscal year should be submitted no later than June 1, unless otherwise directed by the BAA. Competitive renewals should be submitted no later than six months prior to the award termination date after contacting the Program Officer. Conference proposals should be submitted at least four months prior to the conference.

Contact: Dr. Ingham Mack; (703) 696-4825 or (703) 588-2427; fax: (703) 696-5940; email: ingham.mack@navy.mil

Internet: http://www.onr.navy.mil/sci_tech/31/312/vacuum_electronics.asp

Sponsor: Office of Naval Research
875 North Randolph Street, Suite 1425
Arlington, VA 22203-1995

ONR Young Investigator Program Grants **4818**

The objectives of the Young Investigator Program (YIP) are to attract to naval research outstanding new faculty members at institutions of higher education, to support their research, and to encourage their teaching and research careers. ONR anticipates making about 24 new awards annually. Awards of up to $170,000 per year for three years, with the possibility of additional support for capital equipment or collaborative research with a Navy laboratory, are made, based on research proposals and supporting materials. Special attention will be given to proposals in naval priority research areas, listed in the announcement.

Requirements: This program is open to U.S. citizens, nationals, and permanent residents holding tenure track or permanent faculty positions at U.S. institutions of higher education who have received graduate degrees (Ph.D. or equivalent) on or after the date published in the full announcement.

Amount of Grant: $170,000 per year for three years

Date(s) Application is Due: Jan 12

Contact: Dr. William Lukens; (703) 696-4668 or (703) 588-2427; fax: (703) 696-5940; email: william.luckens1@navy.mil

Internet: http://www.onr.navy.mil/sci_tech/3t/corporate/yip.asp

Sponsor: Office of Naval Research
875 North Randolph Street, Suite 1425
Arlington, VA 22203-1995

ONR-ASEE Faculty Sabbatical Leave Program Fellowships **4819**

The Sabbatical Leave Program provides fellowship appointments for a minimum of one semester to a maximum of one year in length. The purpose of sabbatical leave is to provide an opportunity for faculty members to engage in scholarly, creative, professional, research, or other academic activities that will enhance the faculty member's further contributions to their institution. Participants in the Sabbatical Leave Program receive a monthly stipend making up the difference between salary and sabbatical leave pay from their home institution. Relocation and travel assistance are provided to qualifying participants.

Requirements: Sabbatical leave appointments are contingent on the availability of sabbatical leave positions. Participants must conduct research on-site. Prospective applicants should contact the research facility at which they are interested in working to jointly develop a suitable research proposal. Also required is an estimate of base salary and percentage of salary provided by the home institution while on sabbatical leave.

Restrictions: Faculty members from Historically Black Colleges and Universities, Minority Institutions, American Indian Tribal Colleges and Universities, and Hispanic Serving Institutions (HBCU/MI/TCU/HIS), as designated by the U.S. Department of Education, are especially encouraged to apply.

Contact: Artis Hicks, Program Coordinator; (202) 331-3558; fax: (202) 265-8504; email: a.hicks@asee.org

Internet: http://onr.asee.org/about_the_sabbatical_leave_program

Sponsor: Office of Naval Research
1818 N Street, NW, Suite 600
Washington, DC 20036

ONR-ASEE Summer Faculty Research Program Fellowships **4820**

The program provides science and engineering faculty members from colleges and universities the opportunity to participate in research at navy laboratories for a 10-week period during the summer break. Participants work with professional peers in the navy laboratories on research tasks of mutual interest. The research for the summer is defined in advance through correspondence and an optional pre-program visit to the research site, for which funds are provided. Three levels of appointment are available: Summer Faculty Fellow, Senior Summer Faculty Fellow, and Distinguished Summer Faculty Fellow. Applications may be obtained by emailing the American Society for Engineering Education (ASEE).

Requirements: Applicants must be U.S. citizens and hold a teaching or research position at a U.S. college or university. Senior Summer Faculty Fellow applicants must have at least six years (preferably more) of research experience in his/her field of expertise since earning a PhD or equivalent degree. Applicants must present a substantial, significant documented record of research accomplishments and publications, particularly in refereed journals and books. A limited number of appointments as a Distinguished Summer Faculty Fellow are available for faculty members who are preeminent in their field of research (e.g., hold a senior appointment at a leading research university and are internationally recognized for their research achievements).

Amount of Grant: $1,400 per week Summer Faculty Fellow, $1,650 per week Senior Summer Faculty Fellow, $1,900 per week at the Distinguished Summer Faculty Fellow

Date(s) Application is Due: Dec 1

Contact: Artis Hicks, Program Coordinator; (202) 331-3558; fax: (202) 265-8504; email: a.hicks@asee.org

Internet: http://onr.asee.org/about_the_summer_faculty_program

Sponsor: Office of Naval Research
1818 N Street, NW, Suite 600
Washington DC 20036

ONS Major Research Grants **4821**

Applications are invited for research proposals which focus on increasing understanding of and management of understudied symptoms and/or symptom clusters in diverse populations and across settings. Priority will be given to proposals that address management of symptoms and/or symptom clusters that include dermatologic problems in cancer patients. Examples of symptom clusters include dermatitis, pain, and sleep/wake disturbances or fatigue, depression, and sleep/wake disturbances. Other symptom clusters and symptoms will be considered. Justification must be provided for selecting the cluster and/or the rationale for the research needed for the symptom. Membership in ONS is not required.

Requirements: The principal investigator must be actively involved in some aspect of cancer patient care, education, or research.

Amount of Grant: $50,000

Contact: Research Team, (412) 859-6298; fax: (877) 369-5497; email: research@ons.org

Internet: http://www.ons.org/awards/foundawards

Sponsor: Oncology Nursing Society
125 Enterprise Drive
Pittsburgh, PA 15275-1214

ONS Scholarships 4822

Nursing scholarships are awarded each year to enable registered nurses to continue their education in the area of oncology nursing. Each year, undergraduate scholarships, graduate scholarships, and doctoral scholarships are awarded. In addition Congress Scholarships (December 1 deadline) may be applied toward registration, travel to and from Congress, and per diem Congress expenses. Guidelines are available online.
Requirements: An applicant must have a current license to practice as a registered nurse and be enrolled in (or accepted in) a nursing-degree program in an NLN-accredited school of nursing.
Amount of Grant: $2000 undergraduate, $3000 master's, $3000-$5000 doctoral; $1000 Congress Scholarship
Date(s) Application is Due: Feb 1
Contact: Research Team, (412) 859-6100; email: foundation@ons.org
Internet: http://www.ons.org
Sponsor: Oncology Nursing Society
125 Enterprise Drive
Pittsburgh, PA 15275-1214

Ontario Arts Council Integrated Arts Grants 4823

The Program supports a wide range of integrated arts practice and interdisciplinary work through the research, development, creation, production and presentation of original work. This includes new practices, hybrid forms and performance. Integrated arts are creative art forms, artistic processes or practices that use two or more artistic disciplines including visual arts, music, theater, literature, film and dance. Applicants may also choose to combine these above-mentioned disciplines with design, architecture or disciplines outside the arts. Each of these disciplines has expanded over the years to include aspects of the others. The program seeks to fund innovative integration of disciplines where there is a contribution to the development of each discipline and the artist(s). This program funds: artistic research; exploration and experimentation (including workshops/residencies where the artist(s) are integrating disciplines); collaborations fusing film making, visual/installation art, literature, dance, theater, music, architecture, design or any other disciplines; creation/presentation/production of new artistic work integrating two or more art forms; interdisciplinary performance pieces; and projects that combine artificial and natural processes. Aboriginal, culturally diverse, francophone and regional organizations receiving operating grant support are eligible to apply for additional project funding in any program outside their granting offices. Grants in research and development are generally awarded in the $3,000 to $8,000 range depending on the stage of development. Production or Presentation grants are generally awarded in the $5,000 to $15,000 range.
Requirements: This program is open to professional Ontario-based artists, collectives, teams of artists and not-for-profit organizations engaged in integrated arts development, production and presentation. Music theater and opera projects are no longer eligible in this program.
Restrictions: The following are not eligible for project funding: events/activities that take place before the deadline date; events/activities that have been completed before the results of the competition have been announced, which will be approximately 4 months after the deadline date; capital expenses;
Samples: Jill Battson, Elora, Ontario, Canada, $5,000 (2006); Carey Dodge, St. Catharines, Ontario, Canada, $8,500 (2006),; Alexandra Garrison, Toronto, Ontario, Canada, $10,000 (2006); Jac-Tak Projects, Toronto, Ontario, Canada, $7,000 (2006).
Amount of Grant: $20,000 maximum
Date(s) Application is Due: Nov 1; Mar 17
Contact: Bushra Junaid, Community and Multidisciplinary Arts Officer; (416) 969-7425; toll free: (800) 387-0058, ext. 7425; email: bjunaid@arts.on.ca; Georgina Braoudakis, Francophone (French services); (416) 969-7417; (800) 387-0058, ext. 7417
Internet: http://www.arts.on.ca/Page120.aspx
Sponsor: Ontario Arts Council
151 Bloor Street West, 5th Floor
Toronto, Ontario M5S 1T6 Canada

Opera America Fellowships 4824

The program is designed to identify and encourage individuals who are committed to a career in opera administration or production in North America, and to offer them hands-on, specialized training. During the 12-month program, fellows work in-residence with up to four opera companies in Opera America's professional company membership. Under the guidance of advisors and host company supervisors, fellows gain a broad overview of opera and the professional field, and acquire the knowledge and skills necessary to advance toward leadership positions. Up to four fellowships will be awarded in the areas of production/stage management, technical direction, artistic administration, general management, and other administrative specialty areas. Each fellow receives a monthly stipend, optional health insurance, travel expenses, and local housing. Fellowships begin in September.
Requirements: Fellowships are available to individuals who are one of the following opera personnel with limited experience who wish to enhance or augment their skills: individuals entering opera administration or production from other fields or disciplines who show exceptional promise; and graduates of arts administration or technical/production training programs. In addition to having a commitment to a career in opera in North America and the ability to devise and articulate clear goals for the fellowship applicants should manifest a high degree of personal maturity and motivation. Candidates must possess a strong background in the fundamental technical skills necessary to pursue advanced-level training with host companies. Further, candidates must be willing and able to travel as warranted for the duration of the fellowship.
Amount of Grant: $1200 per month
Contact: Fellowships Administrator, (202) 293-4466; fax: (202) 393-0735; email: fellowship@operaamerica.org
Internet: http://www.operaam.org/development/fellowhm.htm
Sponsor: Opera America
1156 15th Street NW, Suite 810
Washington, DC 20005-1704

Oppenstein Brothers Foundation Grants 4825

Grants are awarded in the Kansas City, MO, area primarily for social services and early childhood, elementary, secondary, adult basic, vocational, and higher education; family planning and services; social services and welfare agencies; Jewish welfare organizations; and programs for youth, the handicapped, disadvantaged, mentally ill, homeless, minorities, and elderly. Additional areas of interest include arts/cultural programs, the performing arts, museums, health care and health organizations, and AIDS research. The foundation considers requests for general operating support, capital campaigns, building and renovation, emergency funds, conferences and seminars, seed money, curriculum development, consulting services, and matching funds. Proposals may be submitted at any time.
Requirements: Nonprofit organizations in the metropolitan Kansas City, MO, area may submit grant proposals.
Restrictions: Grants are not awarded to support individuals, annual campaigns, scholarships, fellowships, medical equipment, or endowment funds.
Geographic Focus: Missouri
Samples: Arts Partners Fund (Kansas City, MO)—for program support.; Children's Mercy Hospital (Kansas City, MO)—for program support.
Amount of Grant: $5000-$15,000 average
Contact: Sheila Rice, Program Officer, (816) 234-8671
Sponsor: Oppenstein Brothers Foundation
922 Walnut Street, Suite 200
Kansas City, MO 64106-1809

Oracle Corporate Contributions Grants 4826

The corporation awards grants for medical research in the areas of cancer, AIDS, and neuroscience; endangered animal protection; environmental protection, specifically for education and preservation of important open space; and K-12 education, with emphasis on mathematics, science, and technology programs. Grants are awarded in February and August.
Requirements: Nonprofits nationwide and in operating communities are eligible.
Restrictions: Grants generally do not support fundraising benefits, charitable dinners, sporting events, and marketing brochures. The corporation does not award multiyear grants and rarely approves funds for more than three consecutive years.
Samples: U of Arkansas, Sam M. Walton College of Business (Fayetteville, AR)—for the Information Technology Research Ctr, software valued at $7.078 million (2003); Boys and Girls Club of the Peninsula (Menlo Park, CA)—for the Academic Program, which offers homework assistance, tutoring, and technology training, $1.4 million (2003); United Negro College Fund (Fairfax, VA)—to strengthen the technical capacities of its 39 member institutions, $10 million, including cash donation, scholarships, internships, and curricular training, certification, and technology resources.
Contact: Corporate Giving, (650) 506-7000
Internet: http://www.oracle.com/corporate/giving/community/index.html?giveform.html
Sponsor: Oracle Corporation
500 Oracle Pkwy, MS 5OP11
Redwood City, CA 94065

Orchard Foundation Grants 4827

The foundation makes grants to eligible New York and New England nonprofit organizations in its areas of interest, including environment (air quality, biodiversity, fresh and coastal waters, forests, toxic substances, and pollution prevention); children, youth, and families (child welfare systems, literacy, and pregnancy prevention); and campaign finance reform. The foundation will occasionally provide seed money, operating support, and start-up capital to smaller home-grown organizations provided that strong local support has been obtained for a project in the form of membership and contributions, and that the entire mission of the organization coincides with a field of interest to the foundation. The listed deadlines are for concept letters.
Requirements: The foundation focuses on Massachusetts, Maine, New Hampshire, New York, Vermont, Connecticut, and Rhode Island. The foundation accepts letters of inquiry from groups in those seven states as well as from national groups with regional offices or projects in the area.
Restrictions: Grants are not made to individuals or for endowments, annual or capital campaigns, religious programs, any religion-affiliated organization, conference participation/travel unrelated to current foundation grant, research efforts unrelated to advocacy interests of the foundation, scholarships, fellowships, building projects, equipment needs, film and video projects, land acquisition, animal hospitals/rehabilitation centers, or groups that focus on specific diseases or conditions. Loans are not made.
Geographic Focus: Connecticut; Maine; Massachusetts; New England; New Hampshire; New York; Rhode Island; Vermont
Samples: New England Forestry Foundation (Groton, MA)—to purchase the largest forestland easement in U.S. history, which will protect 762,192 acres in Maine from development, $35,000.
Amount of Grant: $400,000 total; $3000-$15,000 average
Date(s) Application is Due: Mar 1; Sep 1
Contact: Executive Director, (207) 799-0686; email: orchard@maine.rr.com

Internet: http://home.maine.rr.com/orchard
Sponsor: Orchard Foundation
P.O. Box 2587
Portland, ME 04116

Orchid Research Fellowships **4828**

The purpose of this fellowship program is to encourage doctoral candidates in their pursuit of fundamental and applied research on orchids; to attract future generations of scientists to orchidology; and to address critical research needs in the biology of orchids. Fellows spend up to three years working on orchid-related dissertation projects that lead to the PhD degree from accredited institutions in the United States. The candidate's dissertation project must deal with any aspect of orchid biology in the disciplines of physiology, molecular biology, structure, systematics, cytology, ecology, and/or evolution. The fellowship is limited to three consecutive years. Contact the American Orchid Society for application deadline.
Requirements: A candidate must be enrolled full time in a doctoral program of an accredited academic institution in the United States and should submit a current curriculum vita, transcripts of all college coursework, synopsis of the proposed research, and three letters of recommendation. Women, minorities, and persons with disabilities are especially encouraged to apply.
Restrictions: Fellowships will be awarded to the institution sponsoring the candidate. Institutional indirect costs are not to be paid from the awards.
Amount of Grant: $9000 per year for up to three years
Contact: Program Contact, (561) 404-2050; fax: (561) 404-2100; email: pgiust@aos.org
Internet: http://orchidweb.org/aos/research/page07.aspx
Sponsor: American Orchid Society
16700 AOS Ln
Delray Beach, FL 33446-4351

OREF Educational Awards **4829**

The awards support continuing medical education programs and other projects of relevance to orthopedic surgeons of all ages developed in conjunction with a national, recognized organization. Types of support include conferences and seminars, visiting scholars, and program grants. Honorariums support visiting scholars for lectureships; program grants vary in amount.
Requirements: Departments and state orthopedic societies may apply for honorariums. Orthopedic departments and regional and national specialty societies may apply for special symposia, conferences, and workshops.
Amount of Grant: $25,000 maximum
Date(s) Application is Due: Oct 3
Contact: Jean Maguire, VP Grants, (847) 698-9980; fax: (847) 698-7806
Internet: http://www.oref.org/rga/rga.html#educational
Sponsor: Orthopaedic Research and Education Foundation
6300 N River Road, Suite 700
Rosemont, IL 60018

OREF Orthopedic Career Development Awards **4830**

Career development awards are granted to candidates who have demonstrated a commitment to scientific research as well as excellence in clinical training in the field of orthopedic surgery. Program requirements specify that recipients spend a minimum of 20 percent of a five-day week on the research project. A minimum of 50 percent of grant funds may be used for the candidate's salary. Applications must note budget requirements. Awards for up to two years, conditional upon annual review, will be considered. The grant period each year runs from July 1 through June 30 following the year of application. Recipients will be notified by letter after the annual board meeting held in the first quarter of each year.
Requirements: Candidates must have completed a residency in orthopaedic surgery and demonstrated a sustained interest in research and excellence in clinical training. Letters of nomination and support must offer convincing evidence of the candidate's potential to develop as an investigator.
Restrictions: Recipients of current or past NIH RO1 grants are ineligible.
Amount of Grant: $75,000 maximum per year for three years
Date(s) Application is Due: Oct 3
Contact: Jean McGuire, Vice President of Grants, (847) 698-9980; fax: (847) 698-7806; email: mcguire@oref.org
Internet: http://www.oref.org/rga/rga.html#career
Sponsor: Orthopaedic Research and Education Foundation
6300 N River Road, Suite 700
Rosemont, IL 60018

OREF Orthopedic Resident Research Awards **4831**

Applications will be considered for one year of laboratory or clinical research by orthopedic surgery residents in approved orthopedic programs at U.S. medical centers. The objective of these awards is to encourage young investigators by providing funding for research projects that would not ordinarily be funded. Grants provide funds for supplies and expenses but not for a resident's salary. The foundation will notify each applicant by letter after the annual meeting of the board, held in the first quarter of each year.
Requirements: The principal investigator must be an orthopedic surgery resident in an approved U.S. or Canadian orthopedic program.
Amount of Grant: $15,000 maximum for one year
Date(s) Application is Due: Oct 3
Contact: Jean McGuire, Vice President of Grants, (847) 698-9980; fax: (847) 698-7806; email: mcguire@oref.org
Internet: http://www.oref.org/rga/rga.html#resident
Sponsor: Orthopaedic Research and Education Foundation
6300 N River Road, Suite 700
Rosemont, IL 60018

OREF Prospective Clinical Research Proposal Grants **4832**

The objective of these grants is to provide funding for promising prospective clinical research projects in areas of high clinical importance in orthopedic surgery. Subject to available funds, at least four applications will be considered each year for funding up to three years, conditional on annual review. Annual deadline dates may vary; contact program staff for exact dates.
Requirements: The principal or coprincipal investigator must be a suitably qualified orthopedic surgeon working at a U.S. institution. Investigators with an MD, PhD, or DVM degree may serve with the orthopedic surgeon.
Amount of Grant: $50,000 maximum per year for up to three years
Date(s) Application is Due: Oct 3
Contact: Jean McGuire, Vice President of Grants, (847) 698-9980; fax: (847) 698-7806; email: mcguire@oref.org
Internet: http://www.oref.org/rga/rga.html#prospective
Sponsor: Orthopaedic Research and Education Foundation
6300 N River Road, Suite 700
Rosemont, IL 60018

OREF Research Grants **4833**

The objective of the program is to encourage research into the causes, cure, and prevention of orthopedic-related injuries, illness, and conditions, and to encourage research by new investigators in the field of orthopedic musculoskeletal surgery. Grants provide seed and start-up funding for promising research projects for no more than two years, conditional upon annual review. Grant period runs from July 1st through June 30th for two years. Successful applicants will be notified by letter after the annual board meeting, which is held in the first quarter of each year.
Requirements: The principal or coprincipal investigator must be an orthopedic surgeon at a U.S. or Canadian institution. In addition, the candidate must have adequate protected time to accomplish the proposed grant within a 24-month time period.
Restrictions: Recipients of current or past NIH RO1 grants are ineligible.
Amount of Grant: $50,000 per year for up to two years
Date(s) Application is Due: Oct 3
Contact: Jean McGuire, Vice President of Grants, (847) 698-9980; fax: (847) 698-7806; email: mcguire@oref.org
Internet: http://www.oref.org/rga/rga.html#research
Sponsor: Orthopaedic Research and Education Foundation
6300 N River Road, Suite 700
Rosemont, IL 60018

OREF/AAOS Clinical Research Training Fellowship **4834**

The goal of this program is to encourage the development of a cadre of orthopaedic surgeons who have the research skills necessary to undertake and manage various types of health services and outcomes research. The candidate is encouraged to earn a public health degree. Candidates who have completed an accredited North American orthopedic residency and are recommended by their department chairs may apply.
Requirements: Applicants must have completed an orthopedic residency at an accredited North American institution and be recommended by their department chairs.
Amount of Grant: $70,000 per year
Date(s) Application is Due: Oct 3
Contact: Jean McGuire, Vice President of Grants, (847) 698-9980; email: mcguire@oref.org
Internet: http://www.oref.org/rga/rga.html#aaos
Sponsor: Orthopaedic Research and Education Foundation
6300 N River Road, Suite 700
Rosemont, IL 60018

Oregon Arts Commission Individual Artist Fellowships **4835**

The fellowships honor Oregon's professional artists and their artistic achievements and supports their efforts in advanceing their careers. The commission chooses Oregon artists of outstanding talen, demonstrated ability, and commitment to the creation of new work(s) to receive fellowships. Fellowship are awarded to artists demonstrating excellence in literature, the performing arts, and the visual arts.
Requirements: Professional Oregon artists working in any media (with the exception of film/video) may apply. Fellowship applicants must be Oregon residents for one year prior to the application deadline date to be eligible.
Restrictions: Full-time students, nonprofessionals who produce art as a hobby, and previous recipients of an individual artist fellowship are ineligible.
Geographic Focus: Oregon
Amount of Grant: $3000 maximum
Date(s) Application is Due: Jan 26
Contact: Shannon Planchon, Assistant Director, (503) 986-0086; email: shannon.planchon@state.or.us
Internet: http://www.oregonartscommission.org
Sponsor: Oregon Arts Commission
775 Summer Street NE
Salem, OR 97310

Oregon State University Humanities Resident Fellowships 4836

Each year, the center accepts applications for resident fellowships. The fellowship program is primarily concerned with offering research opportunities both for members of humanities departments as traditionally defined and for other scholars seriously interested in humanistic issues. Application includes a prospectus for a research project related to the center's annual research theme, plus reference letters.
Requirements: Applicants must have the PhD or equivalent terminal degree.
Contact: Director, Center for Humanities, (541) 737-2450; email: cfth@orst.edu
Internet: http://osu.orst.edu/dept/humanities/fellowship.html
Sponsor: Oregon State University
811 SW Jefferson Avenue
Corvallis, OR 97333-4506

Orentreich Foundation for the Advancement of Science Grants 4837

Occasional and limited grants are made for joint or collaborative biomedical research, usually initiated by OFAS, with emphasis in dermatology, aging, endocrinology, and serum markers for human diseases. The amount of the grants varies with joint interest and need. Unsolicited applications may be submitted at any time.
Requirements: Applicants must be at or above postgraduate level in science or medicine at accredited universities and research institutions in the United States. There is no restriction on citizenship. Submit an outline of proposed program, curriculum vita of principal investigator, and estimated budget.
Amount of Grant: $50,000 total
Contact: Dr. Rozlyn Krajcik, fax: (914) 265-4210; email: ofas@orentreich.org
Internet: http://www.orentreich.org/research.htm
Sponsor: Orentreich Foundation for the Advancement of Science Inc
Cold Spring-on-Hudson
New York, NY 10516

Organic Farming Research Foundation Grants 4838

The mission of the foundation is to sponsor research related to organic farming practices. Grants support research into organic farming methods, dissemination of research results to organic farmers and to growers interested in making the transition to organic production systems, and education of the general public about organic farming issues.
Requirements: Farmers, researchers, organic advocacy organizations, and other individuals or entities may apply. Projects should involve farmers in both design and execution and take place on working organic farms whenever possible and appropriate.
Samples: Carol Miles, Washington State Cooperative Extension (Chehalis, WA)—to study cover crops for weed management on organic farms, $7500.; Mark Schonbeck, Virginia Assoc for Biological Farming (Floyd, VA)—to support examination of soil nutrient balancing in sustainable vegetable production, $6000.; Terence Robinson, Cornell U (Geneva, NY)—for development of a commercial organic apple production system for New York, $10,000.
Amount of Grant: $15,000 maximum; $8000 average
Date(s) Application is Due: Jul 15; Dec 15
Contact: Jane Sooby, (831) 426-6606; fax: (831) 426-6670; email: jane@ofrf.org
Internet: http://www.ofrf.org/research/index.html
Sponsor: Organic Farming Research Foundation
P.O. Box 440, 303 Potrero Street, Suite 29-203
Santa Cruz, CA 95061

Orgone Biophysical Research Lab Lou Hochberg Awards 4839

Hochberg Awards provide cash awards and recognition to high school and university students in several categories, and to published authors and journalists, for research and open-theme essays that are topically focused upon Wilhelm Reich's sociological discoveries. The awards are open to students and journalists worldwide, but must be submitted in the English language. Awards are made in the following categories: thesis and dissertation awards—for the best accepted and approved university level thesis or dissertation addressing Wilhelm Reich's sociological discoveries; university/college essay awards— given for the best university/college-level student research paper (maximum 50 pages) addressing Wilhelm Reich's sociological discoveries; university thesis/dissertation improvement and implementation grants—for undertaking original graduate-level research addressing Wilhelm Reich's sociological discoveries; and outstanding research and journalism appearing in popular or academic publications— for research and writing on Wilhelm Reich's discoveries, with subsequent publication of accurate articles in the popular press and/or academic journals. The journalist award is open to any article focused upon Reich's sociological and natural scientific discoveries that is published in either a regional and/or national publication (newspaper, magazine, journal). Listed deadlines are for university/college essay awards and high school essay awards.
Requirements: Competitions are open to students worldwide. Submissions must be in English, typed, and presented in triplicate no later than May 15 for the spring/summer competition or December 15 for the fall/winter competition.
Amount of Grant: $1000 Thesis and Dissertation Award; $500 University/College Essay Awards; $1500 University Thesis/Dissertation Improvement and Implementation Grants; $1000 Outstanding Research and Journalism
Date(s) Application is Due: May 15; Dec 15
Contact: Award Administrator, (541) 552-0118; fax: (541) 552-0118; email: info@orgonelab.org
Internet: http://www.orgonelab.org/hochberg.htm
Sponsor: Orgone Biophysical Research Lab
P.O. Box 1148
Ashland, OR 97520

ORISE Civilian Radioactive WaSuite Management Fellowship Program 4840

These one-year, renewable fellowships are offered for graduate work in fields related to the management of spent nuclear fuel and high-level radioactive wastes. Application is usually made before or during the first year of graduate school. Benefits include payment of tuition and fees, monthly stipend, and one required summer practicum at a U.S. DOE national laboratory. Applications are due the last Monday in January.
Requirements: Graduate students at participating universities in the earth sciences, engineering, materials science, or radiation sciences are eligible.
Amount of Grant: $22,000 minimum stipend, additional $300 per month during practicum, some travel expenses, and tuition and fees up to $12,000 per year
Contact: M. Carl Wheeler, Program Manager, (865) 241-6704; email: Wheelerc@orau.gov
Internet: http://www.orau.gov/orise/edu/DOE/gi-gCRWM.htm
Sponsor: Oak Ridge Institute for Science and Education
P.O. Box 117, MS 44
Oak Ridge, TN 37831-0117

ORISE Department of Commerce Internship for Postsecondary Students 4841

This program offers opportunities to participate in hands-on education and training related to the Department of Commerce mission. Deadlines are the third Tuesday in January for summer; July 30 for fall semester; and December 31 for spring semester.
Requirements: Undergraduate and graduate students who are U.S. citizens are eligible to apply.
Amount of Grant: $550 for graduate students; $450 weekly stipend; $125 weekly dislocation allowance
Date(s) Application is Due: Jul 30; Dec 31
Contact: Michael Hubbard, Group Manager, (865) 576-3937; email: hubbardm@orau.gov; Pat Pressley, Program Specialist , (865) 576-3409; email: presslep@orau.gov
Internet: http://www.orau.gov/orise/edu/DOC/gi-gugIPS.htm
Sponsor: Oak Ridge Institute for Science and Education
P.O. Box 117
Oak Ridge, TN 37831-0117

ORISE Environmental Management Participation for the U.S. Army Environmental Center 4842

An interagency agreement between DOE and the U.S. Army Environmental Center makes it possible for ORISE to offer research grants in environmental programs involving restoration, compliance, conservation, pollution prevention, validation, demonstration, transfer, quality assurance and quality control, training, information management and reporting, and related resource management and planning programming. The duration of the program is 10 weeks to three months; some sabbatical appointments are available, as are full-time and part-time appointments. A negotiable stipend is based on regular salary; reimbursement for travel is included. Applications are accepted at any time.
Requirements: Applicants must be college or university full-time faculty, K-12 educators, and U.S. citizens.
Contact: Joanne Rasnake, Project Manager, (410) 436-7257; email: Joanne.Rasnake@us.army.mil; Kim Myers, Recruiter, (410) 436-7258; email: kim.myers3@us.army.mil
Internet: http://www.orau.gov/orise/edu/U.S.ARMY/gi-fEMP.htm
Sponsor: Oak Ridge Institute for Science and Education
P.O. Box 117
Oak Ridge, TN 37831-0117

ORISE Faculty Internship at the U.S. Army Center for Health Promotion and Preventive Medicine 4843

The program is available through an interagency agreement between DOE and the U.S. Army Center for Health Promotion and Preventive Medicine. It provides opportunities and support for applied clinical research and training activities in such areas as environmental health engineering, entomology, ionizing and nonionizing radiation, occupational and environmental health, industrial hygiene and worksite hazards, environmental sanitation and hygiene, and laboratory services. The duration of the program is 10 weeks to one year; some sabbatical appointments are available, as are full-time and part-time appointments. The stipend is negotiable based on regular salary, and reimbursement for travel is included. Applications are accepted at any time.
Requirements: Applicants must be college or university full-time faculty, K-12 educators, and U.S. citizens.
Contact: Joanne Rasnake, Project Manager, (410) 436-7257; email: Joanne.Rasnake@us.army.mil; Kim Myers, Recruiter, (410) 436-7258; email: kim.myers3@us.army.mil
Internet: http://www.orau.gov/orise/edu/U.S.ARMY/gi-fHPPM.htm
Sponsor: Oak Ridge Institute for Science and Education
P.O. Box 117
Oak Ridge, TN 37831-0117

ORISE Faculty Research Participation at the Centers for Disease Control and Prevention 4844

This program provides opportunities and support to conduct research on infectious diseases, environmental health, and occupational safety and health. One program provides support to conduct projects and studies in areas supportive of epidemiological programs. The duration of the programs is 10 weeks to three months; some sabbatical appointments are offered up to one year, and full-time and part-time appointments are available. The stipends are negotiable based on regular salary. Applications are accepted at any time.
Requirements: Applicants must be college or university full-time faculty members.

Contact: Tom Richmond, Program Manager, (865) 576-2194; email: richmont@orau.gov; Debbie Alcorn, Program Specialist, (865) 576-3428; email: alcornd@orau.gov
Internet: http://www.orau.gov/orise/edu/CDC/gi-fRP.htm
Sponsor: Oak Ridge Institute for Science and Education
P.O. Box 117
Oak Ridge, TN 37831-0117

ORISE Faculty Research Participation at the National Center for Toxicological Research **4845**
The program provides opportunities and support for research in biological effects of potentially toxic chemicals and deriving solutions to problems in toxicology that have a major impact on human health and the environment. Benefits include a negotiable stipend based on regular salary and reimbursement for travel and moving. Applications are accepted at any time.
Requirements: U.S. college and university full-time faculty members in the areas of toxicology, pharmacology, chemistry, biological sciences, mathematics, computer science, medicine, or other related scientific disciplines are eligible to apply.
Contact: Linda McCamant, Group Manager, (865) 576-1089; email: mccamanl@orau.gov; Sherry Edwards, Program Specialist, (865) 576-7385; email: warrens@orau.gov
Internet: http://www.orau.gov/orise/edu/FDA/gi-fNCTR.htm
Sponsor: Oak Ridge Institute for Science and Education
P.O. Box 117
Oak Ridge, TN 37831-0117

ORISE Faculty Research Participation at the U.S. Army Joint P.O.W/MIA Accounting Command, Central Identification Laboratory **4846**
This project offers opportunities to participate in research to locate, exhume, and identify the remains of individuals killed during military service. The program can last from 10 weeks to one year. Applications are accepted on a year-round basis.
Requirements: College or university full-time faculty are eligible to apply.
Contact: Jeffrey Johnson, Project Manager, (865) 576-6489; email: johnsonj@orau.gov; Norma Ward, Program Assistant, (865) 241-2890; email: wardn@orau.gov
Internet: http://www.orau.gov/orise/edu/U.S.ARMY/gi-fCIL.htm
Sponsor: Oak Ridge Institute for Science and Education
P.O. Box 117
Oak Ridge, TN 37831-0117

ORISE Fusion Energy Postdoctoral Research **4847**
Magnetic fusion energy collaborative research opportunities are available on fusion energy research and development projects at DOE Office of Fusion Energy-supported laboratories and contractor sites for appropriate science and engineering disciplines. Awards cover one year of support and are renewable. Limited reimbursement for health insurance costs and inbound travel and moving are included.
Requirements: Applicants must have completed the doctoral degree within the last three years.
Amount of Grant: $50,400
Date(s) Application is Due: Mar 1
Contact: Tom Richmond, Program Manager, (865) 576-2194; email: richmont@orau.gov; Debbie Alcorn, Program Specialist, (865) 576-3428; email: alcornd@orau.gov
Internet: http://www.orau.gov/orise/edu/DOE/gi-pdFEPR.htm
Sponsor: Oak Ridge Institute for Science and Education
P.O. Box 117
Oak Ridge, TN 37831-0117

ORISE Fusion Energy Sciences Fellowships **4848**
DOE and ORISE offer fellowships for graduate work in magnetic fusion science and engineering in the following disciplines: physical sciences; engineering; and mathematics. Application is usually made before or during the first year of graduate school. The deadline is the last Monday in January, and the program duration is one year. Tuition and fees are provided.
Requirements: Doctoral students are eligible.
Amount of Grant: $19,800 stipend and some travel expenses; $750 per month practicum allowance; additional $200 per month during practicum
Date(s) Application is Due: Jan 31
Contact: Tom Richmond, Program Manager, (865) 576-2194; email: richmont@orau.gov; Debbie Alcorn, Program Specialist, (865) 576-3428; email: alcornd@orau.gov
Internet: http://www.orau.gov/orise/edu/DOE/gi-gFES.htm
Sponsor: Oak Ridge Institute for Science and Education
P.O. Box 117
Oak Ridge, TN 37831-0117

ORISE Higher Education Research Experiences at Oak Ridge National Laboratory for Faculty **4849**
Opportunities exist to participate in ongoing energy research in the following areas: computer sciences; engineering; mathematics; and physical and natural sciences. The stipend is based on regular salary; limited travel reimbursement is provided. The duration of the program is 10 weeks to three months; some sabbatical appointments up to 12 months are offered. Proposals are due the first Monday in February.
Requirements: College or university full-time faculty who are U.S. citizens or permanent residents are eligible to apply.
Contact: Ruth Keller, Project Manager, (865) 574-7798; email: kellerr@orau.gov; Norma Patterson, Program Specialist, (865) 241-2877; email: pattersn@orau.gov
Internet: http://www.orau.gov/orise/edu/ornl/gi-fHERE.htm
Sponsor: Oak Ridge Institute for Science and Education
P.O. Box 117
Oak Ridge, TN 37831-0117

ORISE Laboratory Technology Program **4850**
opportunities exist to receive hands-on experience in technical areas. Current opportunities are in mechanical engineering technology. Stipends are based on academic level or degree. Applications are accepted on a year-round basis.
Requirements: Undergraduate and graduate students who are U.S. citizens or legal permanent residents are eligible.
Contact: Sam Held, Project Manager, (865) 576-8223; email: helds@orau.gov; Marie Woodby, Program Specialist, (865) 576-9934; email: woodbym@orau.gov
Internet: http://www.orau.gov/orise/edu/ornl/gi-ugIPLT.htm
Sponsor: Oak Ridge Institute for Science and Education
P.O. Box 117
Oak Ridge, TN 37831-0117

ORISE National Library of Medicine, Center for Biotechnology Information Scientific Visitors Program **4851**
This program offers opportunities to participate in informatics research and training programs under the discipline(s) of biochemistry, computer science, mathematics, medical research, molecular biology, and structural biology.
Requirements: Faculty, postdoctorals, and research scientists are eligible to apply.
Contact: Elizabeth Kittrell, Project Manager, (865) 241-3319; email: kittrell@orau.gov; Barbara Dorsey, Program Specialist, (865) 576-9975; email: dorseyb@orau.gov
Internet: http://www.orau.gov/orise/edu/nlm/gi-fpdgNCBI.htm
Sponsor: Oak Ridge Institute for Science and Education
P.O. Box 117
Oak Ridge, TN 37831-0117

ORISE National Oceanic and Atmospheric Administration Faculty Research Participation **4852**
Opportunities exist to participate in research and development relating to science, math, and engineering in the following disciplines: life sciences and physical sciences. Limited travel reimbursement is provided. The duration of the program is 10 weeks in the summer; some on-campus support to continue research begun during the summer is provided. Applications are accepted at any time.
Requirements: College or university full-time faculty who are U.S. citizens or permanent residents are eligible.
Contact: Harold Rider, Project Manager, (865) 576-8158; email: riderh@orau.gov; Jennifer Garren, Program Specialist, (865) 576-5300; email: garrenj@orau.gov
Internet: http://www.orau.gov/orise/edu/NOAA/gi-fRP.htm
Sponsor: Oak Ridge Institute for Science and Education
P.O. Box 117
Oak Ridge, TN 37831-0117

ORISE National Oceanic and Atmospheric Administration Student Research Participation **4853**
Opportunities exist to participate in research and development relating to science, math, and engineering in the following disciplines: computer science, engineering, life sciences, and physical sciences. Applications are accepted at any time. The duration of the program is 10 weeks in the summer.
Requirements: Undergraduate and graduate students who are U.S. citizens or permanent residents are eligible.
Amount of Grant: $420-$515 weekly stipend, depending on classification, and limited travel reimbursement
Contact: Harold Rider, Project Manager, (865) 576-8158; email: riderh@orau.gov; Jennifer Garren, Program Specialist, (865) 576-5300; email: garrenj@orau.gov
Internet: http://www.orau.gov/orise/edu/NOAA/gi-gugSRP.htm
Sponsor: Oak Ridge Institute for Science and Education
P.O. Box 117
Oak Ridge, TN 37831-0117

ORISE Nuclear Regulatory Commission Historically Black Colleges and Universities Faculty Research Participation **4854**
The U.S. Nuclear Regulatory Commission affords faculty members from historically black colleges and universities the opportunity to participate in research at various federal facilities across the United States. The duration of the program varies: 10 weeks to three months for summer appointments, and one year for sabbatical appointments and on-campus appointments during the academic year. The deadline for applications is the third Tuesday in January. Travel reimbursement is provided.
Requirements: Applicants must be full-time faculty members from historically black institutions and U.S. citizens or permanent residents.
Amount of Grant: $3800 per month summer stipend or normal monthly salary rate, whichever is greater
Contact: Michael Hubbard, Project Manager, (865) 576-3937; email: hubbardm@orau.gov; Pat Pressley, Program Specialist, (865) 576-3409; email: presslep@orau.gov
Internet: http://www.orau.gov/orise/edu/NRC/gi-fHBCU.htm
Sponsor: Oak Ridge Institute for Science and Education
P.O. Box 117
Oak Ridge, TN 37831-0117

ORISE Nuclear Regulatory Commission Historically Black Colleges and Universities Student Research Participation 4855

Opportunities exist to participate in research relating to science and engineering in laboratories where NRC research is being conducted in the following disciplines: computer sciences, engineering, geosciences, health physics, human factors, materials science, mathematics, molecular/radiation biology, physical sciences, and statistics-related nuclear material control and accounting. The duration of the program is 10 to 12 weeks in the summer; some one-year appointments are available. Annual deadline dates may vary; contact program staff for exact dates.

Requirements: Undergraduate and graduate students from historically black colleges and universities, who are U.S. citizens or permanent residents, are eligible.

Amount of Grant: $500-$600 weekly stipend and travel reimbursement

Contact: Michael Hubbard, Group Manager, (865) 576-3937; email: hubbardm@orau.gov; Pat Pressley, Program Specialist, (865) 576-3409; email: presslep@orau.gov

Internet: http://www.orau.gov/orise/edu/NRC/gi-gugHBCU.htm

Sponsor: Oak Ridge Institute for Science and Education
P.O. Box 117
Oak Ridge, TN 37831-0117

ORISE Oak Ridge National Laboratory Postdoctoral Research Associates Grants 4856

This program provides opportunities and support for research and training in a broad range of science and engineering activities related to basic sciences, energy, and the environment, as well as access to modern and extensive research facilities. Appointments are for one year and are renewable. Applications are accepted at any time. Stipends are based on research area and degree; reimbursement for inbound travel and moving is available. In addition, full-time temporary employees of Oak Ridge Associated Universities receive benefits.

Requirements: Applicants must have completed the doctoral degree within the last five years; others are considered on a case-by-case basis.

Contact: Bonnie DeJarnette, Program Specialist, (865) 241-4852; email: dejarneb@orau.gov

Internet: http://www.orau.gov/orise/edu/ornl/gi-pdRA.htm

Sponsor: Oak Ridge Institute for Science and Education
P.O. Box 117
Oak Ridge, TN 37831-0117

ORISE Postgraduate Environmental Management Research Participation at the U.S. Army Environmental Center 4857

An interagency agreement between DOE and the U.S. Army Environmental Center makes the program possible. It provides research in environmental programs involving restoration, compliance, conservation, pollution prevention, validation, demonstration, transfer, quality assurance and quality control, training, information management and reporting, and related resource management and planning programming. The duration of the program is one year, and is renewable. Stipends are based on research area and degree, and limited reimbursement for inbound moving and travel is included. Applications are accepted at any time.

Requirements: Applicants must be U.S. citizens and must have completed a bachelor's or undergraduate degree within the last three years; others are considered on a case-by-case basis.

Contact: Joanne Rasnake, Project Manager, (410) 436-7257; email: Joanne.Rasnake@apg.amedd.army.mil; Kim Myers, Recruiter, (410) 436-7258; email: kim.myers3@us.army.mil

Internet: http://www.orau.gov/orise/edu/U.S.ARMY/gi-pdEMF.htm

Sponsor: Oak Ridge Institute for Science and Education
P.O. Box 117
Oak Ridge, TN 37831-0117

ORISE Postgraduate Internship at the Office of Water 4858

These one-year renewable grants provide opportunities and support for studies related to development and implementation of drinking water regulations. Internships are located at the Office of Ground Water and Drinking Water in Cincinnati, OH and Washington, DC. Applications are accepted at any time. Awards are granted for one year and are renewable. Stipends are based on research area and degree; limited reimbursement for inbound travel and moving is available.

Requirements: Applicants must have completed a bachelor's, master's, or doctoral degree within the last three years. Others are considered on a case-by-case basis.

Contact: Sharon Kern, Project Manager, (865) 576-9361; email: kerns@orau.gov; Betty Bowling, Program Specialist, (865) 576-8503; email: bowlingb@orau.gov

Internet: http://www.orau.gov/orise/edu/EPA/gi-rgpdIOW.htm

Sponsor: Oak Ridge Institute for Science and Education
P.O. Box 117
Oak Ridge, TN 37831-0117

ORISE Postgraduate Internship at the U.S. Army Center for Health Promotion and Preventive Medicine 4859

An interagency agreement between DOE and the U.S. Army Center for Health Promotion and Preventive Medicine provides opportunities and support for applied clinical research and training activities in such areas as environmental health engineering, entomology ionizing and nonionizing radiation, occupational and environmental health, industrial hygiene and worksite hazards, environmental sanitation and hygiene, and laboratory services. The duration of the program is one year, and is renewable. Stipends are based on research area and degree, and limited reimbursement for inbound travel and moving is included. Applications are accepted at any time.

Requirements: Applicants must be U.S. citizens and must have completed a bachelor's, master's, or doctoral degree within the last three years; others are considered on a case-by-case basis.

Contact: Joanne Rasnake, Project Manager, (410) 436-7257; email: Joanne.Rasnake@us.army.mil; Kim Myers, Recruiter, (410) 436-7258; email:kim.myers3@us.army.mil

Internet: http://www.orau.gov/orise/edu/U.S.ARMY/gi-pdCHPPM.htm

Sponsor: Oak Ridge Institute for Science and Education
P.O. Box 117
Oak Ridge, TN 37831-0117

ORISE Postgraduate Research Participation at the Agency for Toxic Substances and Disease Registry 4860

An interagency agreement between DOE and ATSDR enables ORISE to provide opportunities and support for research in exposure and disease registries, health investigations, public health assessments, toxicological profiles, emergency response, and health education. The one-year, renewable grants are tenable at the ATSDR in Atlanta, GA. Stipends are based on research area and degree, and include reimbursement for inbound travel and moving. Applications are accepted at any time.

Requirements: Applicants should have completed graduate degree requirements within the last three years; others are considered on a case-by-case basis.

Date(s) Application is Due: Feb 1

Contact: Linda McCamant, Group Manager, (865) 576-9361; email: mccamanl@orau.gov; Tasha Powell, Program Specialist, (865) 574-3911; email: powellt@orau.gov

Internet: http://www.orau.gov/orise/edu/ATSDR/gi-gugsrp.htm

Sponsor: Oak Ridge Institute for Science and Education
P.O. Box 117
Oak Ridge, TN 37831-0117

ORISE Postgraduate Research Participation at the Center for Biologics Evaluation and Research 4861

An interagency agreement between DOE and FDA provides opportunities and support for research concerning biologics. The duration of the program is one year, and is renewable. Stipends are based on research area and degree, and limited reimbursement for inbound travel and moving is included. Applications are accepted at any time.

Requirements: Applicants must have completed a master's or doctoral degree within the last three years; others are considered on a case-by-case basis.

Contact: Linda McCamant, Group Manager, (865) 576-1089; email: mccamanl@orau.gov; Sherry Edwards, Program Specialist, (865) 576-7385; email: warrens@orau.gov

Internet: http://www.orau.gov/orise/edu/FDA/gi-rgpdCBER.htm

Sponsor: Oak Ridge Institute for Science and Education
P.O. Box 117
Oak Ridge, TN 37831-0117

ORISE Postgraduate Research Participation at the Centers for Disease Control and Prevention 4862

An interagency agreement between DOE and CDC provides opportunities and support to conduct research on infectious diseases, environmental health, and occupational safety and health; and for projects and studies in areas supportive of epidemiological programs. Programs are offered by the Division of AIDS, Sexually Transmitted Diseases, and Tuberculosis Laboratory Research; Division of Environmental Health Laboratory Sciences; Division of Mycotic and Bacterial Diseases; Division of Prevention Research and Analytic Methods; Division of Viral and Rickettsial Control and Prevention; and NIOSH. The programs last up to one year, and are renewable. Stipends are based on research area and degree. Applications are accepted at any time.

Requirements: Applicants must have completed a bachelor's, master's, or doctoral degree within the last three years; others are considered on a case-by-case basis.

Contact: Tom Richmond, Program Manager, (865) 576-2194; email: richmont@orau.gov; Debbie Alcorn, Program Specialist, (865) 576-3428; email: alcorndd@orau.gov

Internet: http://www.orau.gov/orise/edu/CDC/gi-rgpdCDC.htm

Sponsor: Oak Ridge Institute for Science and Education
P.O. Box 117
Oak Ridge, TN 37831-0117

ORISE Postgraduate Research Participation at the National Center for Toxicological Research 4863

An interagency agreement between DOE and FDA enables ORISE to administer this program, which provides opportunities and support for research in biological effects of potentially toxic chemicals and deriving solutions to problems in toxicology that have a major impact on human health and the environment. The program is tenable at the NCTR in Jefferson, AR. Stipends are based on research area and degree. Applications are accepted at any time for the one-year, renewable grants. Applications for summer appointments due March 1.

Requirements: Candidates should have completed graduate degree requirements within the last three years; others are considered on a case-by-case basis.

Contact: Linda McCamant, Group Manager, (865) 576-1089; email: mccamanl@orau.gov; Tasha Powell, Program Specialist, (865) 574-3911; email: powellt@orau.gov

Internet: http://www.orau.gov/orise/edu/FDA/gi-gugNCTR.htm

Sponsor: Oak Ridge Institute for Science and Education
P.O. Box 117
Oak Ridge, TN 37831-0117

ORISE Postgraduate Research Participation at the National Exposure Research Laboratory 4864

An interagency agreement between DOE and EPA provides research opportunities in environmental areas in the following disciplines: physical sciences, environmental sciences, life sciences, engineering, and related sciences. Laboratories are located in Athens, GA, and Cincinnati, OH. The duration of the program is one year, and is renewable. Stipends are based on research area and degree, and limited reimbursement for inbound moving and travel is included. Applications are accepted at any time.
Requirements: Applicants must have completed a bachelor's or graduate degree within the last three years; others are considered on a case-by-case basis.
Contact: Sharon Kern, Project Manager, (865) 576-9361; email: kerns@orau.gov; Betty Bowling, Program Specialist, (865) 576-8503; email: bowlingb@orau.gov
Internet: http://www.orau.gov/orise/edu/EPA/gi-rgpdNERL.htm
Sponsor: Oak Ridge Institute for Science and Education
P.O. Box 117
Oak Ridge, TN 37831-0117

ORISE Postgraduate Research Participation at the U.S. Army Central Identification Laboratory, Hawaii 4865

This program offers opportunities to participate in research to locate, exhume, and identify the remains of individuals killed during military service.
Requirements: Eligible applicants must have completed a doctoral degree degree within the last three years; others considered on a case-by-case basis. U.S. citizens preferred.
Contact: Jeffrey Johnson, Project Manager, (865) 576-6489; email: johnsonj@orau.gov; Norma Ward, Program Assistant, (865) 241-2890; email: wardn@orau.gov
Internet: http://www.orau.gov/orise/edu/U.S.ARMY/gi-pdCIL.htm
Sponsor: Oak Ridge Institute for Science and Education
P.O. Box 117
Oak Ridge, TN 37831-0117

ORISE Postgraduate Research Program at the Center for Devices and Radiological Health 4866

These one-year renewable grants provide opportunities and support for research concerning medical devices and radiation emissions from electronic products. This is an interagency agreement between the DOE and FDA, with the positions located at the Center for Devices and Radiological Health in Rockville, MD. Stipends are based on research area and degree and include limited reimbursement for inbound travel and moving.
Requirements: Applicants must have completed bachelor's or graduate degree within last three years; others are considered on a case-by-case basis.
Contact: Linda McCamant, Project Manager, (865) 576-1089; email: mccamanL@orau.gov; Sherry Edwards, Program Specialist, (865) 576-7385; email: warrens@orau.gov
Internet: http://www.orau.gov/orise/edu/FDA/gi-rgpdCDRH.htm
Sponsor: Oak Ridge Institute for Science and Education
P.O. Box 117
Oak Ridge, TN 37831-0117

ORISE Savannah River Site Faculty Research Participation Program 4867

Opportunities exist to participate in ongoing energy research in the following areas: computer sciences; engineering; mathematics; and physical and natural sciences. The stipend is based on regular salary; limited travel reimbursement is provided. The duration of the program is 10 weeks to three months; some sabbatical appointments of up to 12 months are offered. Proposals due the first Monday in February for summer appointments; other applications are accepted on a year-round basis.
Requirements: College or university full-time faculty who are U.S. citizens are eligible.
Contact: Ruth Keller, Project Manager, (865) 574-7798; email: kellerr@orau.gov; Marie Woodby, Program Specialist, (865) 576-9934; email: woodbym@orau.gov
Internet: http://www.orau.gov/orise/edu/SRS/gi-fRP.htm
Sponsor: Oak Ridge Institute for Science and Education
P.O. Box 117
Oak Ridge, TN 37831-0117

ORISE Savannah River Site Postgraduate Research Program 4868

DOE and the Savannah River Site provide opportunities and support to conduct collaborative research that complements ongoing Savannah River research projects. The duration of the program is one year, and is renewable. Stipends are based on research area and degree, and limited reimbursement for inbound travel and moving is included. Applications are accepted at any time.
Requirements: Students must have completed a graduate degree within the last three years and be U.S. citizens.
Contact: Ruth Keller, Project Manager, (865) (865) 574-7798; email: kellerr@orau.gov; Marie Woodby, Program Specialist, (865) 576-9934; email: woodbym@orau.gov
Internet: http://www.orau.gov/orise/edu/SRS/gi-rgpdRPP.htm
Sponsor: Oak Ridge Institute for Science and Education
P.O. Box 117
Oak Ridge, TN 37831-0117

ORISE Savannah River Site Professional Internship Program 4869

The Savannah River Site provides full-time and part-time opportunities for students enrolled in advanced degree programs to participate in research projects for their theses or dissertations. The duration of the program is three to 12 consecutive months. Travel (one round trip) to center/campus and off-campus tuition fees are required by the home institution.
Requirements: Applicants must be U.S. citizens and be undergraduate students, postbaccalaureates, or graduate students.
Date(s) Application is Due: Feb 15; Jun 1; Oct 1
Contact: Ruth Keller, Project Manager, (865) 574-7798; email: kellerr@orau.gov; Marie Woodby, Program Specialist, (865) 576-9934; email: woodbym@orau.gov
Internet: http://www.orau.gov/orise/edu/SRS/gi-gugrgPIP.htm
Sponsor: Oak Ridge Institute for Science and Education
P.O. Box 117
Oak Ridge, TN 37831-0117

ORISE Student Environmental Management Participation Program for the U.S. Army Environmental Center 4870

An interagency agreement between DOE and the U.S. Army Environmental Center provides research in environmental programs involving restoration, compliance, conservation, pollution prevention, validation, demonstration, transfer, quality assurance and quality control, training, information management and reporting, and related resource management and planning programming. The duration of the program is three months to one year, and full-time or part-time appointments are available. Stipends are based on research areas and academic classification.
Requirements: Students currently enrolled in undergraduate or graduate degree programs and who are U.S. citizens are eligible.
Contact: Joanne Rasnake, Project Manager, (410) 436-7257; email: Joanne.Rasnake@us.army.mil; Kim Myers, Recruiter, (410) 436-7258; email: kim.myers3@us.army.mil
Internet: http://www.orau.gov/orise/edu/U.S.ARMY/gi-gugEMP.htm
Sponsor: Oak Ridge Institute for Science and Education
P.O. Box 117
Oak Ridge, TN 37831-0117

ORISE Student Internship Program at the U.S. Army Center for Health Promotion and Preventive Medicine 4871

An interagency agreement between DOE and the U.S. Army Center for Health Promotion and Preventive Medicine provides opportunities and support for applied research and development in the areas of environmental health engineering programs, projects, and activities. The duration of the program is three months to one year, and full-time or part-time appointments are available. Applications are accepted at any time.
Requirements: Students who are U.S. citizens and are currently enrolled in undergraduate or graduate degree programs are eligible.
Contact: Joanne Rasnake, Group Manager, (410) 436-7257; email: Joanne.Rasnake@us.army.mil; Kim Myers, Recruiter, (410) 436-7258; email: kim.myers3@us.army.mil
Internet: http://www.orau.gov/orise/edu/U.S.ARMY/gi-gugHPPM.htm
Sponsor: Oak Ridge Institute for Science and Education
P.O. Box 117
Oak Ridge, TN 37831-0117

ORISE Student Research Participation at the Agency for Toxic Substances and Disease Registry 4872

An interagency agreement between DOE and the Agency for Toxic Substances and Disease Registry (Public Health Service) provides opportunities and support for research in exposure and disease registries, health investigations, public health assessments, toxicological profiles, emergency response, and health education. The duration of the program is 10 weeks to 12 months, and part-time appointments are available. A stipend based on academic status and research area is provided, plus limited travel reimbursement. Off-campus tuition and fees are required by the home institution. Applications are accepted at any time.
Requirements: Undergraduate and graduate students are eligible.
Date(s) Application is Due: Feb 1
Contact: Linda McCamant, Group Manager, (865) 576-9361; email: mccamanl@orau.gov; Tasha Powell, Program Specialist, (865) 574-3911; email: powellt@orau.gov
Internet: http://www.orau.gov/orise/edu/ATSDR/gi-gugsrp.htm
Sponsor: Oak Ridge Institute for Science and Education
P.O. Box 117
Oak Ridge, TN 37831-0117

ORISE Student Research Participation at the Centers for Disease Control and Prevention 4873

An interagency agreement between DOE and CDC provides opportunities and support to conduct research on infectious diseases, environmental health, and occupational safety and health; and to conduct projects and studies in areas supportive of epidemiological programs. The program is offered by the Division of AIDS, Sexually Transmitted Diseases, and Tuberculosis Laboratory Research; Division of Environmental Health Laboratory Sciences; Division of Mycotic and Bacterial Diseases; Division of Prevention Research and Analytic Methods; and NIOSH. The duration of the programs is one month to one year, and full-time or part-time appointments are available. Stipends are based on research area and academic classification. Applications are accepted at any time.
Requirements: Undergraduate and graduate students are eligible.
Contact: Tom Richmond, Group Manager, (865) 576-2194; email: richmont@orau.gov; Debbie Alcorn, Program Specialist, (865) 576-3428; email: alcornd@orau.gov
Internet: http://www.orau.gov/orise/edu/CDC/gi-gugRP.htm
Sponsor: Oak Ridge Institute for Science and Education
P.O. Box 117
Oak Ridge, TN 37831-0117

ORISE Student Research Participation at the National Center for Toxicological Research 4874

Full-time or part-time appointments, made for ten weeks to one year, provide opportunities and support for research participation in biological effects of potentially toxic chemicals and deriving solutions to problems in toxicology that have a major impact on human health and the environment. Stipends are based on research areas and academic classification. This is an interagency agreement between the U.S. Department of Energy and the National Center for Toxicological Research, with positions located in Jefferson, AR, at the research center. Applications are accepted at any time for academic-year appointments, March 1 for summer appointments.
Requirements: Applicants must be students currently enrolled in undergraduate or graduate degree programs.
Date(s) Application is Due: Mar 1
Contact: Linda McCamant, Group Manager, (8653) 576-1089; email: mccamanl@orau.gov; Tasha Powell, Program Specialist, (865) 574-3911; email: powellt@orau.gov
Internet: http://www.orau.gov/orise/edu/FDA/gi-gugNCTR.htm
Sponsor: Oak Ridge Institute for Science and Education
P.O. Box 117
Oak Ridge, TN 37831-0117

ORISE University Radioactive Ion Beam (UNIRIB) Consortium 4875

This program provides opportunities for basic research on rare, short-lived atomic nuclei at the Holifield Radioactive Ion Beam Facility. For best results, apply by February 1 for summer appointments.
Requirements: Faculty members, graduate and postgraduate students, and research fellows are eligible to apply.
Contact: H. Kennon Carter, Program Manager, (865) 576-2642; email: carterk@orau.gov
Internet: http://www.orau.gov/orise/edu/DOE/gi-fgrgpdURIB.htm
Sponsor: Oak Ridge Institute for Science and Education
P.O. Box 117
Oak Ridge, TN 37831-0117

Oscar Rennebohm Foundation Grants 4876

The foundation awards grants to Wisconsin nonprofit organizations in its areas of interest, including higher education, the arts, environmental conservation, and health and social services. Types of support include building construction/renovation, equipment acquisition, and research. There are no application forms or deadlines.
Geographic Focus: Wisconsin
Samples: U of Wisconsin at Madison (WI)—to construct an interdisciplinary-research complex that will focus on cancer, cardiovascular medicine, imaging science, neuroscience, regenerative medicine, and stem cells, $15 million (2005).
Contact: Grants Administrator, (608) 274-5991
Sponsor: Oscar Rennebohm Foundation
P.O. Box 5187
Madison, WI 53705

OSI Budapest Open Access Institutional Memberships to Public Library of Science Grants 4877

The program, part of the Budapest Open Access Initiative, will support open access to international research and scholarship. OSI will provide funding to support 50 Public Library of Science Institutional memberships for institutions in developing and transition countries. Authors from the least developed countries will receive a waiver of the publication charge upon request, a policy the Public Library of Science has had in place since its inception. The Public Library of Science currently publishes PLoS Biology and will accept papers for PLoS Medicine and other discipline-based journals. The Institutional memberships will provide the following benefits to the recipient institutions: the right for any of the institution's research staff or students to automatically qualify for a publication-charge waiver when their manuscript submitted to any of the Public Library of Science journals is accepted for publication; and the right to list all the relevant publications generated at the institution on a special page on the Public Library of Science web site, customized with the institution's name. The Institutional membership grants will be for twelve months.
Amount of Grant: $50,000
Date(s) Application is Due: Apr 15; Jun 15; Sep 15; Dec 15
Contact: Melissa Hagemann; (212) 548-0600; fax: (212) 548-4679; email: mhagemann@sorosny.org or openaccess@soros.org
Internet: http://www.soros.org/openaccess/grants-available.shtml#oaj
Sponsor: Open Society Institute
400 W. 59th Street, 3rd Floor
New York, NY 10019

OSI Harm Reduction and Drug Use Grants 4878

Founded in 1995, the International Harm Reduction Development Program (IHRD) works to reduce HIV and other harms related to injecting drug use, and to press for policies that reduce stigmatization of illicit drug users and protect their human rights. IHRD, which has supported more than 200 programs in Central and Eastern Europe, the former Soviet Union, and Asia, bases its activities on the philosophy that people unable or unwilling to abstain from drug use can make positive changes to protect their health and that of their families and communities. Since 2001, IHRD has prioritized advocacy to expand availability of needle exchange, opiate substitution treatment, and treatment for HIV; to reform discriminatory policies and practices; and to increase the political participation of people who use drugs and those living with HIV.
Contact: Janna Ataiants, Program Coordinator; (212) 548-0600; fax: (212) 548-4679; email: jataiants@sorosny.org
Internet: http://www.soros.org/initiatives/health/focus/ihrd/about
Sponsor: Open Society Institute
400 West 59th Street, 3rd Floor
New York, NY 10019

OSI International Pain Policy Fellowships 4879

The International Pain Policy Fellowship seeks to provide candidates with the knowledge and skills necessary to develop and implement a project to evaluate national policy and improve the availability of pain medications for pain relief and palliative care in their respective countries. The fellowship is intended for health professionals (for example, oncologists, AIDS clinicians, pharmacists, pain and palliative care physicians), health care administrators, policy experts, social workers, or lawyers from low- or middle-income countries who have an interest in drug policy advocacy to improve availability of opioid analgesics for pain relief and palliative care.
Contact: Francoise Girard; (212) 548-0600; fax: (212) 548-4679; email: fgirard@sorosny.org
Internet: http://www.soros.org/initiatives/health/focus/ipci/grants/palliative
Sponsor: Open Society Institute
400 W. 59th Street, 3rd Floor
New York, NY 10019

OSI Soros Justice Advocacy Fellowship 4880

The fellowships fund outstanding individuals including lawyers, advocates, grassroots organizers, and activist academics to initiate innovative projects that will have a measurable impact on one or more of the U.S. Justice Fund's criminal justice priorities. The program, which offers fellowships in two distinct tracks, seeks to identify and nurture emerging (Track 1) and seasoned (Track 2) leaders at either the local or national level. Projects may range from litigation to public education to coalition-building to grassroots mobilization to action research, and must identify a clear policy goal. Fellowships are 18 months in duration and may be implemented in conjunction with large or small nonprofit organizations. Fellows are expected to work full-time on their projects during the term of the fellowship. The fellowship funds a stipend, graduate student loan repayment, health insurance, professional development, and to attend fellows conference.
Requirements: Advocacy Track 1: Applicants must have substantive first-hand knowledge of and/or two to five years of advocacy experience with the issues and communities with which they propose to work. Advocacy Track 2: Applicants must have demonstrated expertise and a minimum of 10 years experience in their fields and five years of advocacy experience in criminal justice reform.
Restrictions: The institute cannot support lobbying activities. Projects that include lobbying activities will not be funded.
Samples: Alina Das, New York, NY—to develop reentry and reintegration strategies for immigrants facing criminal charges and convictions (2006); Angela Chan, San Francisco, CA—a juvenile justice project at the Asian Law Caucus (ALC) by utilizing litigation, education, and policy advocacy to obtain equal access for Asian Pacific Islander (API) families to participate in the rehabilitation of their youth (2006); Katherine Beckett, Seattle, WA—examine changes to the implementation of trespass law and its implications for access to public space and the expansion of the criminal justice system (2006).
Amount of Grant: $65,000-$85,000
Date(s) Application is Due: Sep 24
Contact: Christina Voight, Program Assistant; (212) 548-0600; fax: (212) 548-4666; email: cvoight@sorosny.org
Internet: http://www.soros.org/initiatives/usprograms/focus/justice/programs/justice_fellows/guidelines/advocacy
Sponsor: Open Society Institute
400 W. 59th Street, 3rd Floor
New York, NY 10019

OSI Soros Justice Media Fellowship 4881

The Fellowships support mid-career and veteran print and radio journalists, filmmakers and individuals with unique voices proposing to write books or complete other writing projects. The program seeks to improve the quality of media coverage and representation of issues at the core of the U.S. Justice Fund's criminal justice priorities. The program intends to mitigate the time, space, and market constraints that often discourage journalists from pursuing in-depth stories. Media Fellowships are one year in duration and support print and radio journalism, film and video post-production and dissemination, and book projects. Fellows are expected to make their projects the major focus of their work during the term of the fellowship. The fellowship includes: $45,000 stipend; $2,500 for health insurance; $1,000 to attend the OSI Fellows conference.
Requirements: Applicants must be mid-career or veteran print journalists, experienced filmmakers, or individuals with unique voices. Applicants must have well-established records of publication or broadcast in regional or national markets. All fellows are expected to publish articles or books, broadcast or exhibit-ready pieces, with the number of works to be determined at the outset of the fellowship. Book proposals must include plans to write companion magazine and newspaper articles.
Restrictions: Documentary film or video projects must be in the post-production or distribution stages; research and pre-production proposals are ineligible.
Samples: Elizabeth Amon, staff reporter at the National Law Journal (New York, NY)—to write articles that explore the nexus between immigration policy and criminal justice after September 11.; Amy Bach, freelance journalist (New York, NY)—to write a book about how the courts treat ordinary people and the structural and systemic reasons why

justice fails.; Nathan Blakeslee, co-editor of the Texas Observer (Austin, TX)—to write a book about the controversial drug sting in Tulia, TX.; Slawomir Grunberg, Spencer, New York: filmmaker, to complete a documentary film, Borderline, about a borderline mentally retarded woman serving 15 years to life for murder, despite evidence that the death was accidental.
Amount of Grant: $48,500 average
Date(s) Application is Due: Sep 24
Contact: Christina Voight, Program Assistant; (212) 548-0600; fax: (212) 548-4666; email: cvoight@sorosny.org
Internet: http://www.soros.org/initiatives/usprograms/focus/justice/programs/justice_fellows/guidelines/media
Sponsor: Open Society Institute
400 W. 59th Street, 3rd Floor
New York, NY 10019

OSI-Baltimore Community Fellowships **4882**
The program will award up to ten (10) individual fellowships to implement innovative projects that seek to improve the circumstances and capacity of a marginalized or disadvantaged community in Baltimore. Applicants may either apply for a fellowship: to work under the auspices of a nonprofit organization in Baltimore; or, to work independently. In cases where the fellowship takes place at an organization in Baltimore, applicants must secure sponsorship from that organization. The goals of these fellowships are to encourage public and community service careers, expand the number of mentors and role models available to youth in inner-city neighborhoods, and promote entrepreneurial initiatives that empower communities to increase opportunities and improve the quality of life for their residents.
Requirements: Applicants may come from any field, including, but not limited to, business management, law, medicine, education, architecture, and engineering. OSI strongly encourages individuals from disadvantaged communities and people of color to apply. Applicants need not be from Baltimore City, but should be knowledgeable about social and economic justice issues affecting Baltimore communities.
Restrictions: Individuals already receiving wages or a salary for the proposed project from a host organization are not eligible.
Samples: Luisa C. Bieri de Rios—a teacher and artist, $48,750 (2006); Tonya Featherston—a school principal, $48,750 (2006); Helen Keith—a day care provider, $48,750 (2006); Kenya Lee—a parent advocate, $48,750 (2006); Aisling McGuckin—a nurse, $48,750 (2006); Philip J. Merrill—a historian, $48,750 (2006); Galen Sampson—an executive chef, $48,750 (2006); and Beth Wacks—a master bicycle mechanic, $48,750 (2006).
Amount of Grant: $48,750
Date(s) Application is Due: Mar 16
Contact: Pamela King; (410) 234-1091, ext. 210; fax: (410) 234-2816; email: pking@sorosny.org
Internet: http://www.soros.org/initiatives/baltimore/focus_areas/community_fellowship/guidelines
Sponsor: Open Society Institute
201 North Charles Street, Suite 1300
Baltimore, MD 21201

OSRI Biological Monitoring: Intertidal Resources at Risk to Oil Spills **4883**
OSRI is seeking proposals to conduct surveys of unique, dominant, and/or previously undescribed, intertidal resources at risk to oil spills in Prince William Sound (PWS) and the Copper River Delta (CRD). OSRI anticipates surveys staging out of Cordova, Alaska. Professional services sought through this RFP include the survey prelogistics (design, equipment preparation, etc.), data acquisition, data analysis (including preparing a GIS of the data), and reporting of results. All sampling shall be conducted synoptic with GPS measurements, ensuring the acquisition of data that is both temporally and geographically encoded. The contractor is responsible for providing an electronic copy of the second survey in ARCINFO file format with the final report.
Requirements: Qualified scientists, engineers, and educators are eligible to apply.
Restrictions: Proposals will not be accepted electronically or by fax.
Amount of Grant: $100,000 maximum
Contact: Penny Oswalt, Finance Director; (907) 424-5800 ext 224; fax: (907) 424-5820; email: poswalt@pwssc.org
Internet: http://www.pws-osri.org
Sponsor: Oil Spill Recovery Institute
P.O. Box 705
Cordova, AK 99574

Osserman-Sosin Postdoctoral Myasthenia Gravis Research Fellowships **4884**
These postdoctoral fellowships are intended to provide some support to physicians, PhD scientists, and allied health professionals for conducting research into myasthenia gravis (MG) or related conditions. Clinical investigations as well as laboratory studies concerned with research on MG and with underlying mechanisms of neuromuscular transmission and the pathogenesis of myasthenia gravis are supported. A qualified candidate desiring training in MG-related research first finds a scientist with an established research program who will agree to supervise her/his training and the basic or clinical research to be carried out. The number of fellowships awarded per year is dependent on the amount of funds available.
Requirements: applicant must be a permanent resident of the United States or Canada who has been accepted to work in the laboratory of an established investigator at an institution in the United States, Canada, or abroad; or a foreign national who has been accepted to work in the laboratory of an established investigator a an institution in the United States or Canada.
Samples: Dr. Marika Falcone, U of Arizona—to study the role of B-lymphocytes as antigen-presenting cells in the induction of the pathogenic T-cell responses against nicotinic acetylcholine receptor, $30,000.; Dr. Brian Hare, Harvard Medical School—to study the solution structure determination of a single chain T-cell receptor and design of T-cell ligands with therapeutic potential, $30,000.; Drs. Anthony Auerbach and Claudio Grosman, State U of New York at Buffalo—to study binding, gating, and desensitization of the acetylcholine receptor channel.
Amount of Grant: $50,000 maximum
Date(s) Application is Due: Oct 1
Contact: Fellowships Administrator, (651) 917-6256 or (800) 541-5454; fax: (651) 917-1835; email: mgfa@myasthenia.org
Internet: http://www.myasthenia.org/research/MGFA_PostDoctoralFellowship.pdf
Sponsor: Myasthenia Gravis Foundation of America
1821 University Avenue W, Suite S256
St. Paul, MN 55104

Osteogenesis Imperfecta Foundation Research Grants **4885**
Funding is available for fellowships for young investigators working to develop expertise in osteogenesis imperfecta (OI) research or seed grants for either basic or clinical studies with relevance to OI. Michael Geisman Research Fellowships provide support for investigator's salary and for supplies. (Second year funding may be available.) Seed grants may be funded for one year. (Grants do not support indirect costs or principal investigator salaries.) Clinical Seed Grants may be funded for two years. (Grants do not support indirect costs or principal investigator salaries.)
Amount of Grant: $50,000 per year Michael Geisman Research Fellowships; $60,000 maximum seed grants; $120,000 clinical seed grants for two years
Date(s) Application is Due: Nov 1
Contact: Research Department, (301) 947-0083; fax: (301) 947-0456; email: bonelink@oif.org
Internet: http://www.oif.org/site/PageServer?pagename=OIFGrantApp
Sponsor: Osteogenesis Imperfecta Foundation
804 W Diamond Avenue, Suite 210
Gaithersburg, MD 20878

Otto Klineberg Intercultural and International Relations Award **4886**
This award honors the best paper or article of the year on intercultural or international relations—a field about which Professor Klineberg cared deeply from his social psychological commitment. Originality of the contribution, whether theoretical or empirical, will be given special weight. Please note that an individual or group may only submit one paper to one SPSSI paper award (from amongst the Allport, Klineberg, and Dissertation Awards) per fiscal year (July 1-June 30).
Requirements: Entries can be either unpublished manuscripts, in press papers or books, or papers or books published no more than 18 months prior to the submission deadline. Entries cannot be returned. The competition is open to nonmembers, as well as members of SPSSI, and graduate students are especially urged to submit papers.
Amount of Grant: $1000
Date(s) Application is Due: Feb 15
Contact: Klineberg Award Committee, (202)-675-6956; fax: (202)-675-6902; email: awards@spssi.org
Internet: http://www.spssi.org/awards.html
Sponsor: Society for the Psychological Study of Social Issues
208 I Street NE
Washington, DC 20002

Outstanding Junior Investigator Program in High Energy Physics **4887**
The purpose of this program is to support the development of individual research programs of outstanding scientists early in their careers. Awards made under this program will help to maintain the vitality of university research and assure continued excellence in the teaching of physics.
Requirements: Applications should be from tenure-track faculty investigators who are currently involved in experimental or theoretical high energy physics or accelerator physics research, and should be submitted through a U.S. academic institution.
Amount of Grant: $70,000 average
Date(s) Application is Due: Nov 15
Contact: Dr. Jeffrey Mandula, Division of High Energy Physics, (301) 903-4829; fax (301) 903-2597; email: jeffrey.mandula@science.doe.gov
Internet: http://www.science.doe.gov/grants/FAPN05-26.html
Sponsor: Department of Energy
19901 Germantown Road
Germantown, MD 20874

Overbrook Foundation Grants **4888**
The foundation strives to improve the lives of people by supporting projects that protect human and civil rights, advance the self-sufficiency and well being of individuals and their communities, and conserve the natural environment. The foundation's environment program supports organizations working to develop better consumption and production habits in the United States and in Latin America (currently Brazil, Mexico and Ecuador only). In Latin America the primary objective is to conserve the planet's dwindling biodiversity. The foundation's human and civil rights program is multifaceted and provides support to the following: reproductive health rights; domestic criminal system reform efforts; lesbian, gay, bisexual, and transgender rights; gun violence prevention

and international human rights. An emphasis is placed on work that protects the rights of children and women. The foundation supports projects both domestically and internationally (with a particular focus in Latin America and South Africa). Generally speaking, the foundation's domestic programs support policy development, advocacy, coalition building, research, legal and other strategies likely to impact large classes of people or shape issues important to the foundation's mission. In its international funding, the foundation also supports organizations providing direct services. The first step in the application process is to Submit a letter of inquiry; full application is by invitation. Guidelines are available online.
Requirements: 501(c)3 tax-exempt organizations are eligible.
Restrictions: The foundation does not make grants to individuals and rarely for endowments, building campaigns, deficit financing, or religious purposes. Grant funds may not be used to participate in or intervene in any political campaign on behalf of or in opposition to any candidate for public office or to conduct, directly or indirectly, any voter registration drive (within 4945(d)2 of the IRS code). Grants are not made for debt reduction. Funding for conferences, publications, and media is limited to projects directly related to priorities in the foundation's program areas.
Samples: Rainforest Alliance (New York, NY)—for efforts to ensure the conservation of biological diversity within the tourism industry in Ecuador, $70,000 (2005); Gender Public Advocacy Coalition (Washington, DC)—for operating or program support, $40,000 (2005); National Public Radio (Washington, DC)—to support coverage of environmental issues on its news program, $50,000 (2005); Educational Fund to Stop Gun Violence (Washington, DC)—for its work to spur congressional legislation designed to curb gun-related violence, $50,000 (2005).
Contact: Grants Administrator, (212) 661-8710; fax: (212) 661-8664; email: apply@overbrookfoundation.org
Internet: http://www.overbrook.org
Sponsor: Overbrook Foundation
122 E 42nd Street, Suite 2500
New York, NY 10168

Oxalosis and Hyperoxaluria Foundation Grants 4889
Grants are awarded for research on hyperoxaluria, the presence of excess oxalic acid or oxalates in the urine, and the related conditions oxalosis and calcium-oxalate kidney stones. The foundation seeks to fund grants that will increase the understanding of hyperoxaluria and oxalosis and improve the clinical management and treatment of these two orphan diseases. Investigation into alternative methods of treatment that will alleviate the need for transplantation therapy are encouraged. Applications for up to two years of support will be considered. These funds may be used for salaries for technical assistance, special equipment, animals, supplies, and travel. Other organizations that have been approached about supporting the same project should be identified.
Requirements: Applicants must hold an MD, PhD, or equivalent degree by the time the funding period begins.
Amount of Grant: $160,000 maximum
Date(s) Application is Due: Oct 15
Contact: Program Contact, (212) 777-0470; fax: (212) 777-0471; email: execdirector@ohf.org
Internet: http://www.ohf.org/prof_application.html
Sponsor: Oxalosis and Hyperoxaluria Foundation
201 E 19th Street, 12E
New York, NY 10003

Pace University Center for Downtown New York Visiting Scholar Fellowships 4890
Applications for visiting scholar fellowships are invited from journalists, researchers, scholars, and students who are studying the 9/11/01 terrorist attacks and their aftermath in New York City, Washington, and Pennsylvania. The program is interested in receiving proposals from journalists/ scholars/students from a variety of disciplines and fields, such as research focusing on the economic, social, and political impact of the attacks; policy effects (both local, national, and international); specific case studies of aspects of the rebuilding effort; and the implications of the attacks for democratic rule and international governance. Proposals outside of these areas also will be considered. A fellowship will provide the researcher with office space for three to six months, information technology access, library services, basic support services (photocopying, etc.), and limited secretarial service. (The fellowship does not provide a stipend or housing.) Fellows will be expected to give one public lecture related to their work at Pace University, are encouraged to interact with the other fellows in residence as well as present their research to the Pace University research community, and acknowledge Pace support in subsequent publications. Submit all application materials, including resume/vita, description of the project (research topic and focus, methodology to be used, timeline for project completion, plan for dissemination of research, and its outcome) and two letters of recommendation.
Date(s) Application is Due: May 1
Contact: Dan Slippen, Director, CDNY, (212) 346-1025
Internet: http://appserv.pace.edu/execute/page.cfm?doc_id=8294
Sponsor: Pace University Center for Downtown New York
1 Pace Plaza
New York, NY 10038

Pacific Cultural Foundation Grants 4891
The foundation promotes international cultural exchange. Types of support include grants for students outside of China to research Chinese culture, publication grants in Chinese studies to support the publication of research by foreign students on contemporary Chinese problems, and research and writing grants to foreign students conducting research on contemporary Chinese problems.
Requirements: Application is open to citizens of any country except for the Republic of China. Research grant applicants must hold at least a master's degree. For a publication grant, the book to be published must be recommended by two prominent scholars in the field and must be published within 10 months. Schools, academic organizations, or research institutions are eligible for seminar grants, the subject of which must be relevant to the foundation and to Chinese studies.
Restrictions: The program does not support individuals from the Republic of China.
Amount of Grant: $5000 maximum
Date(s) Application is Due: Mar 1; Sep 1
Contact: Yu-sheng Chang, 886-2-237-77155; fax: 886-2-233-77167; email: arts@pcf.org.tw
Internet: http://www.comnet.ca/~moecult/english/pcf.htm
Sponsor: Pacific Cultural Foundation
38 Chung King S Road, Sec 3
Taipei 10567 Taiwan

Pacific Islanders in Communications Grants 4892
The Media Fund is the prime vehicle to spark the development of Pacific Islander programming for national public television. The Media Fund is comprised of the following programs: Open Call, an annual call that invites proposals for television projects in all stages of production; the Completion Fund, that is designed to quickly respond to urgent post-production needs; the Short Film Initiative, that encourages and builds narrative storytelling; Travel Awards to support producers' professional development; Acquisitions; and Executive Productions. All stages of production, from research and development through completion, are supported. Projects selected under the Media Fund must present the Pacific Islander experience in an authentic and educational manner, as well as meet PBS' high standards of journalistic balance, storytelling, and technical excellence. A training component is an essential part of each project to help develop future Pacific Islander filmmakers.
Requirements: Applicants must be U.S. citizens, have primary artistic and editorial control, and have a nonprofit fiscal sponsor to receive and administer PIC funds.
Restrictions: Industrial films and videos, instructional videos, student productions, and similar non-broadcast categories are not eligible. An applicant may not apply for both a research and development grant and a production grant for the same project in the same grant cycle.
Amount of Grant: $15,000 maximum for research/development and scripting; $50,000 maximum for works in progress, including production, postproduction, and completion
Contact: Grants Administrator, (808) 591-0059; fax: (808) 591-1114; email: info@piccom.org
Internet: http://www.piccom.org/services.php
Sponsor: Pacific Islanders in Communications
1221 Kapiolani Boulevard, Suite 6A-4
Honolulu, HI 96814

PacifiCorp Foundation for Learning Grants 4893
The foundation awards grants to eligible nonprofit organizations in four broad areas, including educational and research institutions, both public and private, from early childhood through university level; youth organizations; cultural enrichment organizations dedicated to the performing arts, the visual arts, historic preservation, cross-cultural education, and other such activities; and health and human services organizations, including chemical dependency treatment and prevention programs, senior citizen centers, runaway youth services, and domestic violence treatment and prevention programs. Deadline for education organizations is March 15; for civic, community, and organizations not covered in other categories, June 15; for culture and arts organizations, September 15; for health, welfare, and social service organizations, December 15. Guidelines and application are available online.
Requirements: 501(c)3 community organizations in Idaho, Oregon, Utah, California, Washington, and Wyoming are eligible.
Restrictions: Organizations and activities not eligible for foundation support include religious groups for religious purposes; political organizations, ballot measure campaigns, or candidates for public office; organizations that discriminate against individuals on the basis of creed, color, sex, age, national origin, or veteran status; individuals; contributions to or memberships in chambers of commerce, service clubs, taxpayer associations, and other similar bodies; or any non-charitable purpose. As a general rule, the foundation also will not support establishment or support of endowments; coverage of operating deficits; capital campaigns; conferences, conventions, and events; projects that do not have a bearing on communities within the six-state region served by PacifiCorp subsidiaries or communities served by PacifiCorp Power Marketing.
Geographic Focus: California; Idaho; Oregon; Utah; Washington; Wyoming
Amount of Grant: $1000-$10,000 average
Date(s) Application is Due: Mar 15; Jun 15; Sep 15; Dec 15
Contact: Grants Manager, (503) 813-7257; fax: (503) 813-5272; email: pacificorpfoundation@pacificorp.com
Internet: http://www.pacificorpfoundation.org/Article/Article25110.html
Sponsor: PacifiCorp Foundation for Learning
825 NE Multnomah, Suite 2000
Portland, OR 97232-4116

Packard Foundation Children, Families, and Communities Program—Child Health Care Access Grants 4894

The foundation's goal in this area is to ensure that all children have health insurance that provides them with access to health care appropriate to their needs. Grants are awarded to: sustain and improve existing public health insurance programs for children; develop sustainable systems to enroll and retain children eligible for existing programs; and grow county-based programs across California to cover all children in families with incomes up to at least 300 percent of the federal poverty level. Most grants will be awarded to California organizations. Some nationally focused grants will be awarded for research, advocacy, technical assistance, and policy development aligned with our programmatic goals. Staff solicits proposals from specific applicants. A limited pool of funding is available to respond to unsolicited letters of inquiry for innovative projects that fall within the areas listed.
Requirements: Eligible applicants are 501(c)3 nonprofit organizations and institutions.
Samples: Community Health Councils (Los Angeles, CA)—for the 2004 Covering Kids and Families Health Coverage Conference and Summit, $20,000 (2005); Labor Project for Working Families (Berkeley, CA)—for outreach and education related to paid family leave, $50,000 (2005); Edison Consulting (Atlanta, GA)— for strategic communications work regarding health insurance for children, $65,000 (2004); California Economic Development Lending Initiative (Oakland, CA)—for initial staffing costs for the statewide Working Capital Initiative for Community Needs, to support licensed nonprofit child care and community health organizations in California, $150,000 (2004).
Amount of Grant: $4 billion total
Date(s) Application is Due: Mar 15; Jun 15; Sep 15; Dec 15
Contact: Program Director, (650) 948-7658; email: cfc@packard.org
Internet: http://www.packard.org/index.cgi?page=cfc-chi
Sponsor: David and Lucile Packard Foundation
300 Second Street, Suite 200
Los Altos, CA 94022

Packard Foundation Children, Families, and Communities Program—Preschool and Early Education Grants 4895

The program's central focus is to achieve access to voluntary preschool for all three- and four-year-olds by funding research, analysis, leadership, and constituency-building that promote this goal. Grants are awarded to enhance and expand statewide advocacy efforts and strengthen the coalition of constituency groups in support of preschool for all; refine preschool for all policy proposals and recommendations and support further research on the benefits of preschool for all in California; and provide ongoing support to local flagship preschool for all efforts that demonstrate the promise of high-quality preschool when implemented on a large scale. Most grants will be awarded in California. Staff will solicit proposals for local flagship programs from specific applicants. A limited pool of funding will be available to respond to unsolicited letters of inquiry for innovative projects that fall within the areas listed.
Requirements: Eligible applicants are 501(c)3 nonprofit organizations.
Samples: California Head Start Assoc (Sacramento, CA)—to mobilize Head Start parents and program employees in advocating preschool for all children, $50,000 (2004); Child Care Law Ctr (San Francisco, CA)—to provide technical assistance related to advocating preschool for all children and to the Affordable Buildings for Children's Development program, and to analyze and develop policies related to California's child-care system, $320,000 (2005); First 5 Alameda County (San Leandro, CA)—to create workforce-development models in the fields of early-childhood education, $100,000 (2005); Mexican American Legal Defense and Educational Fund (Los Angeles, CA)—to mobilize Latino parents, to advocate preschool for all children, and to identify solutions to implementation challenges, $155,000 (2005).
Amount of Grant: $428.9 million
Contact: Program Officer, (650) 948-7658
Internet: http://www.packard.org/index.cgi?page=cfc-upe
Sponsor: David and Lucile Packard Foundation
300 Second Street, Suite 200
Los Altos, CA 94022

Padilla Bay Foundation Grants 4896

The Padilla Bay Foundation is a membership-based non-profit organization formed in 1987. Its mission is to help preserve the Padilla Bay estuary through support of the Padilla Bay National Estuarine Research Reserve. Foundation programs include: support for education; support for research; and special projects involving the environment. Interested applicants should contact the Foundation directly.
Geographic Focus: Washington
Contact: Kay Reinhardt, Administrative Officer; (360) 757-3234; email: foundation@padillabay.gov
Internet: http://padillabay.gov/npfriends.asp
Sponsor: Padilla Bay Foundation
P.O. Box 1305
Mount Vernon, WA 98273

Padilla Bay National Estuarine Research Reserve Graduate Research Fellowships 4897

Grants are available on a competitive basis to students admitted to or enrolled in a full-time masters or doctoral program at U.S. accredited colleges and universities. Fellowships may be funded for up to three years to support thesis research. The Graduate Research Fellowship provides students the opportunity for hands-on training in ecological monitoring. The award amount is $20,000 per year and may be used for research, tuition, and other expenses. The application deadline is November 1.
Amount of Grant: $20,000 per annum
Contact: Susan White, Research Coordinator; (360) 428-1558; email: susan.white@noaa.gov
Internet: http://padillabay.gov/researchfellowships.asp#RA
Sponsor: National Oceanic and Atmospheric Administration / Padilla Bay National Estuarine Research Reserve
10441 Bayview-Edison Road
Mt. Vernon, WA 98273-9668

Padilla Bay National Estuarine Research Reserve Research Assistantships 4898

Research assistantships will be awarded to graduate students seeking to fund thesis work in Padilla Bay or work comparing Padilla Bay with other Washington bays or estuaries or coastal zone management issues anywhere in Washington. These assistantships require the sponsorship of a college/university faculty member who will supervise the student's thesis in coordination with the Reserve's research staff. A contract with the student totaling $5,000 will be provided over the period July 1 to December 31 of the following year (18 months). The Padilla Bay NERR is particularly interested in proposals addressing: priorities of the Puget Sound Partnership; either historical connections of Padilla Bay to the Skagit and Samish flats, or restoration of estuarine habitats and processes adjacent to Padilla Bay; the potential impacts on eelgrasses and on other plants and animals in Padilla Bay that may result from diversion of Skagit River floodwater to Swinomish Channel; the interaction and ecology of native eelgrass, Zostera marina, and Japanese eelgrass, Z. japonica; or research and resource management issues on Hat Island in Padilla Bay. First consideration will be given to applications received by May 16 each year.
Requirements: Applicants should usually be accepted into a Master, or Doctoral program in marine science, biology, botany, zoology, or one of the related natural or physical sciences.
Amount of Grant: $5,000
Contact: Dr. Douglas Bulthuis, Research Coordinator; (360) 428-1558; email: bulthuis@padillabay.gov
Internet: http://padillabay.gov/researchfellowships.asp#RA
Sponsor: National Oceanic and Atmospheric Administration / Padilla Bay National Estuarine Research Reserve
10441 Bayview-Edison Road
Mt. Vernon, WA 98273-9668

PAHO/WHO Fellowships 4899

Fellowship is a specially tailored training activity for an individual or a group with the purpose of fulfilling specific learning objectives. Fellowships may be of short or long duration and take place in a training institution or in the field inside or outside of the fellow's country. Each of the six regional WHO offices share responsibility for management and administration of the fellowship program within its region: regional office for Europe in Copenhagen, Denmark; regional office for Eastern Mediterranean in Alexandria, Egypt; regional office for the Western Pacific in Manila, the Philippines; regional office for Africa in Brazzaville, Congo; regional office for South-East Asia in New Delhi, India; and regional office for the Americas in Washington, DC. The Pan American Health Organization (PAHO) serves as the Regional WHO Office for all of the Americas.
Requirements: Fellowships are open to nationals of member states of the PAHO who are public health workers in all fields, especially deans or faculty members of schools of public health, medicine, dentistry, engineering, veterinary medicine, and nursing. Individual must be recommended by his/her own government and return to his/her home country for at least three years service on completion of studies. Contact program staff for specific eligibility.
Restrictions: Candidates normally should not be more than 55 years of age.
Contact: Regional Office for the Americas, Pan American Health Organization, (202) 974-3000; fax: (202) 974-3663; email: postmaster@paho.org
Internet: http://www.paho.org/English/HSP/HSR/fellowships.htm
Sponsor: Pan American Health Organization
525 23rd Street NW
Washington, DC 20037

Paired Fellowships for Research in Conservation and the History of Art and Archaeology 4900

Two paired fellowships are awarded annually. The Samuel H. Kress Foundation Paired Research Fellowship supports research teams working on European visual arts prior to the early 19th century. The J. Paul Getty Trust Paired Research Fellowship supports research teams working in any field of art history, archaeology, or architecture. The fellowship includes a two- to three-month period for field, collections, and/or laboratory research, followed by a two-month residency at the Center. Applications will be considered for study in the history and conservation of the visual arts (painting, sculpture, architecture, landscape architecture, urbanism, prints and drawings, film, photography, decorative arts, industrial design, and other arts) of any geographical area and of any period. Each team is required to submit an application.
Requirements: Applications are invited from teams consisting of two scholars: one in the field of art history, archaeology, or another related discipline in the humanities or social sciences, and one in the field of conservation or materials science. Fellowships are open to those who have held the appropriate terminal degree for five years or more. An equivalent record of professional accomplishment, in exceptional cases, may fulfill this requirement.

Amount of Grant: $12,000 stipend per team member, $5500 maximum for project-related research and travel expenses per segment of fellowship; $3000 maximum for research-related travel; $1500 for round-trip travel to Washington, if necessary
Date(s) Application is Due: Mar 21
Contact: Fellowships Program, Center for Advanced Study in the Visual Arts, (202) 842-6482; fax: (202) 789-3026; email: advstudy@nga.gov
Internet: http://www.nga.gov/resources/casvacon.htm
Sponsor: National Gallery of Art
2000B S Club Drive
Landover, MD 20785

Palm Beach and Martin Counties Grants 4901
The community foundation supports nonprofit organizations serving Florida's Palm Beach and Martin Counties in eight program areas: arts and culture, education, environment, health, human services, community development, the conservation and preservation of historical and cultural resources, and human and race relations. Guidelines and forms are available online.
Requirements: Florida organizations based in and serving Palm Beach and Martin Counties are eligible to apply.
Restrictions: Proposals will not be funded for routine operating expenses, endowments, building campaigns, computers, fund-raising events, celebration functions, feasibility studies, religious organizations for religious purposes, or recurring requests for the same purpose.
Geographic Focus: Florida
Samples: Pleasant City Family Reunion Committee (West Palm Beach, FL)—to support artists and instructors who work with disadvantaged youths, $15,000 (2003); TAB Development Corp (West Palm Beach, FL)—for classes and training courses for first-time homebuyers, $15,000 (2003); FAI/FIU Joint Ctr for Environmental and Urban Problems (Jupiter, FL)—to integrate economic, environmental, and social concerns in community design, $15,000 (2003); Florida Resource Ctr for Women and Children (Riviera Beach, FL)—for salary and program expenses of an antiviolence program for teenagers, $25,000 (2003).
Amount of Grant: $25,000 maximum
Contact: Linda Raybin, Program Contact, (561) 659-6800; fax: (561) 832-6542; email: lraybin@cfpdmc.org
Internet: http://www.cfpbmc.org/grant/Guidelines.asp
Sponsor: Community Foundation for Palm Beach and Martin Counties
700 S Dixie Hwy, Suite 200
West Palm Beach, FL 33401

Pancreatic Cancer Action Network-AACR Fellowship 4902
The Pancreatic Cancer Action Network-AACR Fellowship represents a joint effort to attract young scientist to careers in pancreatic cancer research. The Fellowship is open to Postdoctoral Fellows and Clinical Research Fellows working at an academic, medical, or research institution who will be in the 1st, 2nd, or 3rd year of their postdoctoral training at the start of the grant term on July 1. The Fellowship provides a one-year grant of $45,000 to support the salary and benefits of the Fellow while working on a mentored pancreatic cancer research project. A partial amount of funds, up to 25% of the total grant, may be designated for direct research support. The research proposed for funding may be basic, translational, clinical or epidemiological in nature and must have direct application and relevance to pancreatic cancer. It is anticipated that one Fellowship will be funded. AACR requires applicants submit both an online and a paper application. The online application must be submitted by 12:00 noon (United Sates Eastern Time) on October 28, using the proposalCENTRAL website at http://proposalcentral.altum.com. The paper copy with original signatures and all required documents must be postmarked and sent no later then October 30.
Requirements: Applicants must have acquired a doctoral degree (including Ph.D., M.D., D.O., D.C., N.D., D.D.S., D.V.M., Sc.D., D.N.S., Pharm.D., or equivalent doctoral degree, or a combined clinical and research doctoral degree) in a related field and may not currently be candidate for another degree. Applicants must be in the 1st, 2nd, or 3rd year of Postdoctoral or Clinical Research Fellowship at an academic, medical, or research institution within the United States at the start of the Fellowship term on July 1.
Restrictions: Scientific investigators or health professionals who are funded by the tobacco industry for any project, or whose named mentors in the case of mentored grants are funded by the tobacco industry for any project, may not apply and will not be eligible for AACR grants.
Amount of Grant: $45,000
Date(s) Application is Due: Oct 28; Oct 30
Contact: Hanna Hopfinger, Grants Program Assistant; (267) 646-0665; fax: (215) 440-9372; email: hanna.hopfinger@aacr.org
Internet: http://www.aacr.org/home/scientists/research-funding—fellowships/postdoctoral-fellowships.aspx
Sponsor: American Association for Cancer Research
615 Chestnut Street, 17th Floor
Philadelphia, PA 19106-4404

Pancreatic Cancer Action Network-AACR Pathway to Leadership Grants 4903
The Pancreatic Cancer Action Network-AACR Pathway to Leadership Grant represents a joint effort to ensure the future leadership of pancreatic cancer research by supporting outstanding early career investigators beginning in their postdoctoral research positions and continuing through their successful transition to independence. Applicants must be in the first five years of their postdoctoral or clinical research fellowships at the time of application and not already have a full-time tenure track assistant professor position (or equivalent). The Pathway to Leadership Grant will provide up to five years of support, for a total of $600,000, consisting of two phases. The initial Mentored Phase lasts up to two years, during which time recipients will receive $75,000 per year and are expected to work closely with mentors to develop their research. During the subsequent three years, recipients are expected to be independent research positions and will be funded at $150,000 per year. AACR requires applicants to submit both an online and a paper application. Online applications must be completed by 12:00 noon (United States Eastern Daylight Time) on October 28, using the proposalCENTRAL website. The paper copy with original signatures and all supporting documents must be postmarked and sent no later than October 30.
Requirements: Applicants must be in mentored training positions at an academic, medical, or research institution within the United States. There are no citizenship requirements. The applicant institution will be the mentored phase institution. Employees or subcontractors of a government entity or for-profit private industry are not eligible. Exceptions may apply if an applicant holds a full-time position at a Veterans' Hospital or national laboratory in the U.S. Contact AACR before submitting an application to determine your eligibility. Applicants must be Postdoctoral or Clinical Research Fellows who completed all doctoral degree requirements no more then five years prior to the start of the grant term.
Amount of Grant: $600,000
Date(s) Application is Due: Oct 28; Oct 30
Contact: Elizabeth Martin, Program Assistant; (267) 646-0664; fax: (215) 440-9372; email: grants@aacr.org
Internet: http://www.aacr.org/home/scientists/research-funding—fellowships/postdoctoral-fellowships.aspx
Sponsor: American Association for Cancer Research
615 Chestnut Street, 17th Floor
Philadelphia, PA 19106-4404

Parapsychology Foundation Grants 4904
The grant program is designed to advance the understanding of parapsychology. Grants are given to scientists, universities, and laboratories for original study, research, and experiments in extrasensory perception, psychokinesis, and precognition.
Restrictions: The foundation does not consider proposals for travel or for general graduate and undergraduate studies.
Amount of Grant: $3000-$5000
Contact: Program Contact, (212) 628-1550; fax: (212) 628-1559; email: info@parapsychology.org
Internet: http://www.parapsychology.org/dynamic/040100.html
Sponsor: Parapsychology Foundation
P.O Box 1562
New York, NY 10021-0043

Park Foundation Grants 4905
The foundation awards grants to nonprofit organizations in central New York and areas in the U.S. Southeast. Areas of interest include education and higher education, animal welfare, environment, television, and general charitable giving. Types of support include general operating support, program development, professorships, seed money, fellowships, scholarship funds, research grants, and matching funds. There are no application deadlines. The board meets in March, June, August, October, and December. Obtain application forms.
Requirements: 501(c)3 nonprofits in he eastern U.S., primarily in central NY, Washington, DC, and North Carolina are eligible.
Restrictions: For-profit organizations and individuals are not eligible.
Samples: North Carolina State University (Raleigh, NC)—for Park Scholarships, $647,710.; Beach Preservation Assoc of Pine Knoll Shores (Atlantic Beach, NC)—for general operating support, $150,000.; North Carolina State U (Raleigh, NC)—for the Park Scholarship program, $226,018.
Amount of Grant: $5000-$100,000 average
Contact: Linda Madeo, Executive Director, (607) 272-9124; fax: (607) 272-6057
Sponsor: Park Foundation
P.O. Box 550
Ithaca, NY 14851

Parker B. Francis Fellowships in Pulmonary Research 4906
Grants are awarded to institutions for the purpose of providing stipends, fringe benefits, and modest travel expenses in support of qualified post-doctoral fellows or newly appointed assistant professors. Award recipients will thereby be enabled to devote the major part of their professional effort to research related to pulmonary disease and lung biology. Sponsorship of the fellow by an established investigator is required. There can be only two fellows in a department at any one time. Only one application will be accepted from a department annually. Fellowships are awarded for three years.
Requirements: Fellows must be U.S. or Canadian citizens, or intend to make permanent residence in the United States or Canada, must hold doctoral degrees, and must be sponsored by established investigators. Foreign nationals must document intent to become permanent residents or citizens of North America.
Samples: Dr. Zissis Chroneos, Pediatrics, Children's Hospital Medical Ctr—for research entitled Cloning of the SP-A Receptor Gene.; Dr. Emily DiMango, Medicine, Columbia U—for research entitled Signal Transduction of Inflammation in Cystic Fibrosis.; Dr. Michael DuVall, Anesthesiology, U of Alabama, Birmingham—for research entitled Effects of Reactive Nitrogen Species on C1-secretion in Airway Epithelium.
Amount of Grant: $44,000 first year, $46,000 second year, $48,000 third year; plus $1500 maximum for travel and fringe benefits

Date(s) Application is Due: Oct 17
Contact: Dr. Joseph Brain, Director of the Fellowship Program, (617) 432-4099; fax: (617) 277-2382; email: brain@hsph.harvard.edu
Internet: http://www.francisfellowships.org
Sponsor: Francis Families Foundation
Harvard School of Public Health, Room 1411, Building 1
Boston, MA 02115

Parker Foundation Grants **4907**
Grants are given to nonprofit organizations operating in San Diego County, CA, for purposes leading to the betterment of all aspects of life of the people of the county. Grants may be in support of adult and youth services, performing and visual arts, museums and zoos, AIDS, education, and community activities. Types of support include seed money, matching funds, operating grants, annual campaigns, emergency funds, building construction/renovation, equipment acquisition, land acquisition, research, program-related investments, continuing support, special projects, and publications. Applications in writing are accepted at any time.
Requirements: 501(c)3 San Diego County, CA, nonprofit organizations are eligible.
Restrictions: Grants are not made to individuals nor is grant support given to religious organizations for denominational programs.
Geographic Focus: California
Samples: Neurosciences Institute (La Jolla, CA)—for its campaign for endowment and research operations, $50,000.
Amount of Grant: $5000-$50,000 average
Contact: Program Contact, (858) 677-1431; fax: (858) 677-1477
Internet: http://www.TheParkerFoundation.org
Sponsor: Parker Foundation
4365 Executive Drive, Suite 1100
San Diego, CA 92121-2133

Partnership for Cures Charles E. Culpeper Scholarships in Medical Science **4908**
The program awards scholarships in medical science to support the career development of academic physicians. In direct costs, $100,000 per year to the sponsoring institution will be committed for up to three consecutive years. It is expected that a significant portion of the award will be used as a contribution to the Scholar's salary. Up to $50,000 (plus fringe benefits) of this sum may be used for this purpose; the rest for support of research including personnel and some travel costs. Eight percent will be added to each award for indirect costs. All scientific research relevant to human health is eligible for consideration.
Requirements: Candidates must: be U.S. citizens or aliens who have been granted permanent U.S. residence (proof required); hold the M.D. degree from a U.S. medical school or the equivalent of an M.D. degree from an educational institution equivalent to a United States medical school; and be judged worthy of support by virtue of the quality of their research proposals and their potential for successful careers in academic medicine. Applicants must have at least one year of post-doctoral clinical training.
Restrictions: Individuals who have achieved a rank above assistant professor are ineligible for this program, and an awardee who is promoted above the rank of assistant professor before the initiation of the scholarship will be ineligible to receive the award.
Amount of Grant: $108,000 per year for three years
Date(s) Application is Due: Aug 16
Contact: Dr. Bruce E. Bloom; (312) 696-1366; fax: (312) 263-0939; email: Bruce@4cures.org
Internet: http://www.4cures.org/home/culpeper_scholar_program_research
Sponsor: Partnership for Cures
70 West Madison Street, Suite 1500
Chicago, IL 60602

Partnership for Cures Two Years To Cures Grants **4909**
The Initiative is intended to bring researchers, funders and institutions together to produce better treatments and cures for patients with disease right now. The Program works to quickly bring together: breakthrough scientific discoveries about diseases and treatments; FDA approved drugs and other medical therapies (Eastern and Western) with new potential uses; research teams at top institutions ready to proceed with urgency and focus; and independent validation, compressed time lines, and financial leverage. Projects must be able to create a direct patient impact in 24 months from the date of initiation of the project. Budgets can range from $35,000 to $500,000. The Program is accepting research projects in any catastrophic disease, but has special interest in the following diseases: lupus, Crohns disease, MS, ALS, protein intolerances in children, ischemic colitis, autism, type I diabetes, lung cancer, osteosarcoma, Fragile X syndrome, Jewish genetic diseases, and Parkinsons disease. There is no deadline for submissions, but funders are looking for projects right now.
Requirements: Each project must come with a minimum of 30% matching funds. Those funds can come from the institution itself, from government or other public funders, or from individuals, foundations or other private funders.
Restrictions: No more than 10% of the budget can be allocated to overhead, travel and publications.
Amount of Grant: $35,000 to $500,000
Contact: Bruce E. Bloom; (312) 696-1366; fax: (312) 263-0939; email: Bruce@4cures.org
Internet: http://www.4cures.org/home/funding_opportunities
Sponsor: Partnership for Cures
70 West Madison Street, Suite 1500
Chicago, IL 60602

Partnership for Pediatric Epilepsy Research Grants **4910**
This program stimulates epilepsy research by providing funding for investigators in the early stages of their careers. Seed grants are awarded to clinical investigators or basic scientists for support of biological or behavioral research which will advance the understanding, treatment, and prevention of epilepsy. All research must be conducted in the United States. The partnership is a consortium of professional organizations and individuals with the common objective of researching treatments and cures to pediatric epilepsy.
Requirements: Proposals are invited from investigators who hold an advanced degree (MD or PhD) and have completed all research training. Investigators may be based at corporations as well as academic/university settings.
Amount of Grant: $100,000 maximum for two years
Date(s) Application is Due: Sep 1
Contact: Grants Administrator, (301) 459-3700; fax: (301) 577-2684; email: grants@efa.org
Internet: http://www.epilepsyfoundation.org/research/grants.cfm
Sponsor: Partnership for Pediatric Epilepsy
4351 Garden City Drive
Landover, MD 20785

Pasadena Foundation Average Grants **4911**
The community foundation serves the greater Pasadena, CA, area and awards grants for capital improvements to agencies serving children, the elderly, and the disabled. Fields of interest include child development, the elderly, disabled, youth, drug abuse, family planning, education, arts and the humanities, mental health, community development, and the environment. Challenge/matching grants also are awarded. The deadline is the first business day in March.
Restrictions: Schools, religious institutions, and political organizations are ineligible.
Geographic Focus: California
Samples: Women at Work—for the acquisition of three computer systems, $6000.; Pasadena Boys Choir—to purchase concert uniforms, $5725.; All Saints Children's Ctr—to purchase cots and floor coverings, $2610.
Amount of Grant: $8000 maximum
Contact: Program Contact, (626) 796-2097; fax: (626) 583-4738; email: pcfstaff@pasadenacf.org
Internet: http://www.pasadenacf.org/grant_reg.htm
Sponsor: Pasadena Foundation
260 S Los Robles Avenue, #119
Pasadena, CA 91101

Patron Saints Foundation Grants **4912**
The foundation awards one-year grants to California nonprofit organizations to improve health care in the West San Gabriel Valley. Areas of interest include medical care and medical research. Types of support include capital grants, matching gifts, project grants, research grants, and seed money grants. Applications must be postmarked by the first Friday in March and October.
Requirements: California 501(c)3 tax-exempt organizations serving the West San Gabriel Valley are eligible.
Geographic Focus: California
Amount of Grant: $3500-$15,000 range
Contact: Program Contact, (626) 564-0444; fax: (626) 564-0444; email: patronsaintsfdn@sbcglobal.net
Internet: http://www.patronsaintsfoundation.org
Sponsor: Patron Saints Foundation
260 S Los Robles Avenue, Suite 201
Pasadena, CA 91101

Paul and Daisy Soros Fellowships for New Americans **4913**
The program offers support for graduate study by New Americans, individuals who hold Green Cards or who have been naturalized as U.S. citizens or are children of two parents who are both naturalized citizens. The program offers maintenance and tuition grants for up to two years of graduate study in the United States. Fellows must have shown potential in the fields for which they seek further education; the capacity for creativity, initiative, originality, and work; and the commitment to the values of the U.S. Constitution and Bill of Rights. Applications are available from college academic advisors, the program's Web site, or by contacting the program directly.
Requirements: Applicants must have either the bachelor's degree or be in the final year of undergraduate study and must not be older than 30 years of age as of the listed application deadline.
Samples: 2007: Eunice Cho, second-year JD student at Stanford Law School; Yahonnes Cleary, a second-year student at Yale Law School; Wei Lien Dang, PhD candidate in Physical Biology and Engineering at Harvard University; Mary Farag, MA candidate in Religion at Yale University; Shelli Farhadian, fourth year student at the Tri-Institutional MD/PhD program with Weill Cornell Medical College, Rockefeller University and Sloan Kettering Institute.
Amount of Grant: $20,000
Date(s) Application is Due: Nov 1
Contact: Fellowships Administrator, (212) 547-6926; fax: (212) 548-4623; email: pdsoros_fellows@sorosny.org
Internet: http://www.pdsoros.org
Sponsor: Paul and Daisy Soros Fellowships for New Americans
400 W. 59th Street
New York, NY 10019

Paul Cuffe Memorial Fellowships 4914

The museum awards these annual fellowships to encourage research that considers the participation of Native Americans and African Americans in the maritime activities of southeastern New England. Applicants are judged on the merits of the proposed research project, their qualifications to do the research, and the use to be made of the scholarly resources of southeastern New England. Application should be in the form of a letter including resume, outline of project, budget, and names and addresses of three references. Applications are available on the Web site.
Amount of Grant: $2400 maximum
Date(s) Application is Due: Mar 30
Contact: John Hattendorf, Munson Institute of American Maritime Studies, (860) 572-5359; fax: (860) 572-5329; email: munson@mysticseaport.org
Internet: http://www.mysticseaport.org/index.cfm?fuseaction=home.viewPage&page_id=BA0F6A5D-D0B2-1CEA-5778CABEA131F76E
Sponsor: Mystic Seaport Museum
P.O. Box 6000, 75 Greenmanville Avenue
Mystic, CT 06355-0990

Paul G. Allen Family Foundation Grants 4915

The single foundation, created through the consolidation of Allen's six previous foundations (The Allen Foundation for the Arts, The Paul G. Allen Charitable Foundation, The Paul G. Allen Foundation for Medical Research, The Paul G. Allen Forest Protection Foundation, The Allen Foundation for Music, and The Paul G. Allen Virtual Education Foundation), will continue to focus on the Allen family's philanthropic interests in the areas of arts and culture, youth engagement, community development and social change, and scientific and technological innovation. The Arts and Culture Program fosters creativity and promotes critical thinking by helping strong arts organizations become sustainable and supporting projects that feature innovative and diverse artistic forms. The Youth Engagement Program improves the way young people learn by supporting organizations that use innovative teaching strategies and provide opportunities for children to address issues relevant to their lives. The Community Development and Social Change Program promotes individual and community development by supporting initiatives and organizations that provide access to resources and opportunities. The Scientific and Technological Innovation Program advances promising scientific and technology research that has the potential to enhance understanding and stewardship of the world in which we live. Organizations may only receive one grant per year. Organizations must not have any delinquent final reports due to any of the Paul G. Allen Foundations for previous grants. Grantseekers are encouraged to apply through the online application process, where basic organizational and project information will be requested. Guidelines are available online.
Requirements: 501(c)3 tax-exempt organizations, status from the Internal Revenue government entities, and IRS-recognized tribes are eligible. Eligible organizations must be located in, or serving populations of, the Pacific Northwest, which includes Alaska, Idaho, Montana, Oregon, and Washington.
Restrictions: In general, the foundation will not consider requests for general fund drives, annual appeals, or federated campaigns; special events or sponsorships; direct grants, scholarships, or loans for the benefit of specific individuals; projects of organizations whose policies or practices discriminate on the basis of race, ethnic origin, sex, creed, or sexual orientation; contributions to sectarian or religious organizations whose principle activity is for the benefit of their own members or adherents; loans or debt retirement; projects that will benefit the students of a single school; general operating support for ongoing activities; or projects not aligned with the foundation's specified program areas. 509(a) private foundations are ineligible.
Geographic Focus: Alaska; Idaho; Montana; Northwest; Oregon; Pacific Northwest; Washington
Samples: Koahnic Broadcast Corp (Anchorage, AK)—to produce a weekly public-radio program of contemporary American Indian music, $225,000 (2005); Boys and Girls Club of Yellowstone County (Billings, MT)—for a new club at Castle Rock Middle School, $150,000 (2005); Housing Hope (Everett, WA)—for its campaign to build 270 low-cost housing units in Snohomish County, WA, $350,000 (2005); Mercy Corps (Portland, OR), Northwest Medical Teams (Tigard, OR), and World Vision (Federal Way, WA)—to assist tsunami-relief efforts, $500,000 each (2005).
Date(s) Application is Due: Mar 31; Sep 30
Contact: Grants Administrator, (206) 342-2030; fax: (206) 342-3030; email: info@pgafamilyfoundation.org
Internet: http://www.pgafamilyfoundation.org
Sponsor: Paul G. Allen Family Foundation
505 Fifth Avenue S, Suite 900
Seattle, WA 98104

Paul Marks Prizes for Cancer Research 4916

The prizes recognize outstanding young investigators who have made significant contributions to increase the understanding of cancer or improve the treatment of the disease through basic or clinical research. The winners will present their work at the center, be honored at a dinner, and share a cash award of $150,000. Nomination packets must include a letter from the nominator outlining the significance of the accomplishments for which the candidate should be recognized, a one-page scientific biography of the candidate; a list of up to eight of the candidate's significant published papers with a brief (fewer than 100 words) explanation of the importance of each one and the candidate's curriculum vita. Up to three supporting letters may also be submitted Annual deadline date may vary; contact program staff for exact dates.
Requirements: Nominees are required to be age 45 or younger at the time of the submission deadline.
Amount of Grant: $150,000
Date(s) Application is Due: Apr 29
Contact: Paizhe Pressley, Program Contact, (212) 639-6561; email: presslep@mskcc.org
Internet: http://www.mskcc.org/mskcc/html/53983.cfm
Sponsor: Memorial Sloan-Kettering Cancer Center
1275 York Avenue
New York, NY 10021

Paul Mellon Centre for British Art Junior Fellowships 4917

Three-month fellowships in the United Kingdom (based at the Paul Mellon Centre in London) or the United States (based at the Yale Center for British Art in New Haven) are offered to scholars already engaged in doctoral research on a British art-historical topic. Applications outlining the field of research, projected completion date, and proposed publication date should be submitted in triplicate, along with a curriculum vita and the names of three referees.
Requirements: Candidates may be of any nationality but normally must be enrolled in a graduate program at a U.S. university (for study in the United Kingdom) or at a non-U.S. university (for study in the United States).
Amount of Grant: @L2000 monthly stipend and return airfare; up to 3 months
Date(s) Application is Due: Jan 15
Contact: Grants Administrator, 44 (0) 20 7580 0311; fax: 44 (0) 20 7636 6730; email: grants@paul-mellon-centre.ac.uk
Internet: http://www.paul-mellon-centre.ac.uk/support.html
Sponsor: Paul Mellon Centre for British Art
16 Bedford Sq
London, England WC1B 3JA United Kingdom

Paul Mellon Centre for British Art Postdoctoral Fellowships 4918

Postdoctoral fellowships of three- to six-months' duration are offered to enable the fellow to transform doctoral research into publishable form as a book, series of articles, exhibition catalog, etc.; and to support new research arising out of a successfully submitted doctoral dissertation where that research may readily lead to publication. The scholar's field must be British art or architecture. Grants are made either as a stipend to the fellow or to fund a temporary replacement at the fellow's institution. Fellows may choose, if appropriate, to be affiliated with either the Paul Mellon Centre in London or the Yale Center for British Art in New Haven. Applications outlining the field of research, projected completion date, and proposed publication date should be submitted in triplicate, along with a curriculum vita and the names of three referees.
Requirements: Normally the postdoctoral fellowship is awarded within four years of an applicant's doctoral award. In exceptional cases, where the likelihood of publication is evident, consideration will be given to postgraduate students holding a master's degree or equivalent.
Amount of Grant: @L1200 monthly stipend and return airfare; up to 6 months
Date(s) Application is Due: Jan 15
Contact: Grants Administrator, 44 (0) 20 7580 0311; fax: 44 (0) 20 7636 6730; email: grants@paul-mellon-centre.ac.uk
Internet: http://www.paul-mellon-centre.ac.uk/support.html
Sponsor: Paul Mellon Centre for British Art
16 Bedford Square
London, England WC1B 3JA United Kingdom

Paul Mellon Centre for British Art Senior Fellowships 4919

The fellowship supports an established scholar in the field of British art and architecture in completing a manuscript or book for publication. The grant is made either as a stipend to the fellow or to fund a temporary replacement at the fellow's institution and includes a traveling allowance.
Amount of Grant: @L27,500 or $42,000, plus a small travelling allowance
Date(s) Application is Due: Jan 15
Contact: Grants Administrator, 44 (0) 20 7580 0311; fax: 44 (0) 20 7636 6730; email: grants@paul-mellon-centre.ac.uk
Internet: http://www.paul-mellon-centre.ac.uk/support.html
Sponsor: Paul Mellon Centre for British Art
16 Bedford Sq
London, England WC1B 3JA United Kingdom

Paul Mellon Centre Rome Fellowship 4920

In conjunction with the British School at Rome, the Paul Mellon Centre offers a four-month fellowship open to scholars working on Grand Tour subjects or in the field of Anglo-Italian cultural and artistic relations. The fellowship provides full residential accommodation at the British School at Rome and offers a monthly stipend plus travel to and from Rome. Applications outlining the field of research and the importance to this study of residence in Rome should be submitted in triplicate, along with a curriculum vita and the names of three referees.
Amount of Grant: @L2000, plus travel to and from Rome
Date(s) Application is Due: Jan 15
Contact: Grants Administrator, 44 (0) 20 7580 0311; fax: 44 (0) 20 7636 6730; email: grants@paul-mellon-centre.ac.uk
Internet: http://www.paul-mellon-centre.ac.uk/support.html
Sponsor: Paul Mellon Centre for British Art
16 Bedford Square
London, England WC1B 3JA United Kingdom

Paul Mellon Predoctoral Fellowship **4921**

The fellowship is awarded annually for three years of research in Western art to be held partly in residence at the National Gallery of Art, Center for Advanced Study in Visual Arts, and a two-year period of research in Europe or elsewhere. The Mellon fellow is expected to spend the third year of the fellowship at the center to complete the dissertation. Application must be made through the chair of the graduate department of art history or other appropriate department. Departments should limit nominations to one candidate. Fellowships begin September 1 and are not renewable. Nomination forms are available from the chairs of graduate departments with PhD programs.
Requirements: Applicants must have completed their residence requirements and coursework for the PhD and general or preliminary examinations before the date of application. Students must know two foreign languages related to the topic of the dissertation. Applicants must be either U.S. citizens or enrolled in a university in the United States.
Samples: Irina Oryshkevich, Columbia U—for dissertation, Christian Reclamation: Rediscovery of the Catacombs in Counter-Reformation Rome.; Leila Whittemore, Columbia U—for dissertation, Theory, Practice and the Architectural Imagination: Filarete's Treatise on Architecture and Early Sforza Urbanism, 1450-1466.
Amount of Grant: $24,000 annually for up to three years
Date(s) Application is Due: Nov 15
Contact: Fellowships Program, Center for Advanced Study in the Visual Arts, (202) 842-6482; fax: (202) 789-3026; email: advstudy@nga.gov
Internet: http://www.nga.gov/resources/casvapre.htm
Sponsor: National Gallery of Art
Fourth Street and Constitution Avenue NW
Washington, DC 20565

Paul P. Fidler Research Grants **4922**

The grant is designed to encourage the development and dissemination of knowledge that has potential to improve the experiences of college students in transition. The award includes a cash stipend, travel to two national conferences, a presentation at a national conference, and priority consideration for publication. The competition is open to faculty, staff, and graduate students who plan to conduct research on issues of college student transitions. Cross-institutional research teams are encouraged to apply.
Requirements: Applicants for the grant must complete a research proposal as outlined by the National Resource Center. Only complete proposals will be considered.
Restrictions: The following individuals are not eligible: National Resource Center staff and fellows; current members of the Center's National Advisory Board; and students, staff, and faculty of the University of South Carolina system.
Samples: Barbara Hofer, Associate Professor, Middlebury College, Middlebury, VT—The Electronic Tether: Parental Regulation, Self-Regulation, and the Role of Technology in College Transitions (2006-07); Christine and Michael Kirk-Kuwaye, University of Hawaii at Manoa, Hawaii—First-Semester Patterns of Student Engagement: Lateral and Community College Transfer Students at a Four-Year Institution (2005-06).
Amount of Grant: $5,000 stipend plus travel expenses
Date(s) Application is Due: Jul 16
Contact: Angela Griffin, Coordinator of Research, Grants, and Assessment; (803) 777-2134; email: AngelaG@mailbox.sc.edu
Internet: http://www.sc.edu/fye/research/grant/index.html
Sponsor: National Resource Center
University of South Carolina, 1728 College Street
Columbia, SC 29208

Paul Robeson Fund for Independent Media Grants **4923**

The fund supports media activism and grassroots organizing. Grants support local, state, national, and international organizations and individual media producers working in radio, film, and video production and distribution. The fund supports independent media productions that offer compelling political and artistic messages to promote progressive social change activism and organizing. Funding for radio projects supports the efforts of audio producers and activists to create a pilot or demo for a series; or to produce a special or package of special programming, a series of short features, an ongoing series. Funding for film and video projects is awarded in the categories of preproduction and distribution. Preproduction grants support such efforts as researching a topic, research-related travel and administrative costs, recruitment of production crew or talent, preparation of fund-raising materials, and script development. Distribution grants support creative initiatives that encourage and expand the distribution of media as resources for organizing and empowerment among broadcast-based groups working for social change. Applications can be downloaded from the Web site.
Requirements: 501(c)3 nonprofits are eligible. Individuals must have a 501(c)3 fiscal sponsor.
Restrictions: Grants do not support media projects other than film , video, and radio production; film, video and audio installations, festivals, conferences, exhibits, or special events; organizational projects for internal or promotional use; purely personal works or human interest stories with little or no social change relevance or impact; sociological and anthropological explorations that do not provide a strong progressive political analysis; documentation that is primarily of cultural events, personalities, or performances with little or no political relvance; public television or radio station productions; general operating expenses of distribution companies, nonprofit media organizations, or an artist's office, studio, or home; script development for dramatic features or radio dramas; student productions or other projects associated with a degree program; or productions originating in countries outside the United States unless the distribution strategy incorporates (in part) organizing in the United States.
Samples: American Socrates: The Life of Bayard Rustin, Nancy Kates, producer, $8000.; Justice For All, Michael Moore, producer, $2500.; Blink, Elizabeth Thompson, producer, $2500.
Amount of Grant: $20,000 maximum; $5000-$15,000 average
Date(s) Application is Due: May 16
Contact: Trinh Duong, (212) 529-5300; fax: (212) 982-9272; email: trinhh.duong@fex.org
Internet: http://www.fex.org/grantmaking.shtml#robeson
Sponsor: Funding Exchange
666 Broadway, Suite 500
New York, NY 10012

Paul Stock Foundation Grants **4924**

The foundation awards grants in Park County, WY, for higher education, university support, recreation, and human services. Types of support include annual campaigns, building construction/renovation, research grants, and scholarships to individuals. The board meets in July and December. There are no application forms.
Requirements: Giving primarily in Park County, WY; student aid is limited to those who have resided in WY for one year or more.
Geographic Focus: Wyoming
Contact: Charles Kepler, President, (307) 587-5275
Sponsor: Paul Stock Foundation
P.O. Box 2020
Cody, WY 82414-2020

PC Reproductive Biomedicine Postdoctoral Fellowships **4925**

The council awards fellowships to persons who wish to pursue advanced study in the physiology and biochemistry of reproduction. Fellows train in the laboratories of the council's Center for Biomedical Research. Training in any aspect of reproductive biology may be considered. Priority research topics include, but are not limited to, the following areas of reproductive biomedicine: physiology of hypothalamo-pituitary-gonadal axis; leydig cell development; cell and molecular biology of Sertoli cells; cell-cell communication in the seminiferous tubule; molecular biology of steroid-hormone action; biosynthesis and secretion of reproductive hormones; transmission of immunodeficiency viruses; genetics of male infertility; and sperm maturation. Fellowship awards are for full-time research training and study, and are made initially for one year. The aim of the program is to support two years of training, provided the preceptor and fellow agree on the second year. Fellowships can begin any time during the year. There are no deadlines for submission of applications. Guidelines are available online.
Requirements: Candidates must have successfully completed an advanced degree—MD, PhD, or equivalent—and be proficient in English. Applications may be made in anticipation of receiving the doctoral degree, but the fellowship cannot begin until after the degree has been awarded. Fellowships are open to individuals of all countries. If the applicant is other than a United States citizen and enters this country on a Population Council-sponsored J-1 visa, she/he must have a strong commitment to return to her/his own country upon completion of the fellowship.
Amount of Grant: $36,000 starting stipend
Contact: Fellowships Coordinator, (212) 339-0500; fax: (212) 755-6052; email: fellowships@popcouncil.org
Internet: http://www.popcouncil.org/slr/biofellowships.html
Sponsor: Population Council
1230 York Avenue, Box 273
New York, NY 10021

Peabody Foundation Grants **4926**

The giving program of this foundation consists of grants primarily to provide care, treatment, rehabilitation, education, and assistance to children and for medical research relating to disabled children. The majority of recipient organizations will be located in the Boston, MA, area. Proposals may be submitted at any time and are reviewed in the spring.
Restrictions: Grants are not made to individuals.
Geographic Focus: Massachusetts
Amount of Grant: $3000-$275,000
Date(s) Application is Due: Feb 1
Contact: William Tripp III, c/o Choate, Hall & Stewart, (617) 523-2700
Sponsor: Peabody Foundation
53 State Street, Exchange Pl
Boston, MA 02109-2804

Peacock Foundation Grants **4927**

Established by Henry B. Peacock, Jr. in 1947, the mission of Peacock Foundation, Inc. is to enhance and promote the good health and well being of children, families, and underprivileged persons in Southeast Florida, through contributions, gifts, and grants to eligible nonprofit organizations. The priorities of Peacock Foundation, Inc. include: making grants to human services providers that promote youth development, assist abused or neglected children, women, and the elderly, and seek to reduce abuse, prevent homelessness, and end hunger in our community; supporting educational programs in the arts and the environment, as well as special education for disabled persons; contributing to medical research, health care organizations, and hospitals.
Requirements: All applicants must be IRS recognized 501(c)3 public charities classified as not a private foundation, registered with the Department of Agriculture to solicit funds in Florida, when applicable, and located in and/or of significant benefit to residents to the Southeast Florida counties of Miami-Dade, Broward, or Monroe. In order for a proposal

to be considered for funding, the applicant first must send a brief letter of inquiry. See Foundations website for letter of inquiry guidelines: http://www.peacockfoundationinc.org/review_progress.html.
Restrictions: Peacock Foundation, Inc. does not fund: capital campaigns, construction, or renovation projects; deficit financing or debt reduction; conferences or festivals; fundraising events or advertising; special events or athletic events; individuals; lobbying to influence legislation; religious organizations, unless engaged in a significant project benefiting the entire community.
Geographic Focus: Florida
Contact: Joelle Allen, Executive Director; (305) 373-1386
Internet: http://www.peacockfoundationinc.org/eligibility.html
Sponsor: Peacock Foundation, Inc.
100 South East Second Street, Suite 2370
Miami, FL 33131

Pediatric Brain Tumor Research Grants 4928
The foundation awards grants in the area of basic pediatric brain tumor research. The laboratory-based, hypothesis driven project should demonstrate the potential to advance basic scientific research in brain tumors. Funds may be used for start-up projects or as supplementary funding. Guidelines and application are available online.
Requirements: Grant applications will be accepted from institutions in the United States, Canada, and Australia.
Amount of Grant: $50,000 per year
Date(s) Application is Due: Sep 30
Contact: Dianne Traynor, Grants Administrator, (800) 253-6530 or (828) 665-6891; fax: (828) 665-6894; email: dtraynor@pbtfus.org
Internet: http://www.pbtfus.org/grants/grntfund.htm
Sponsor: Pediatric Brain Tumor Foundation of the 302 Ridgefield Ct
Asheville, NC 28806

Pediatric Cancer Research Foundation Grants 4929
The foundation focuses its efforts on improving the care, quality of life, and survival rate of children with malignant diseases. One-year grants support laboratory research that translates into immediate treatment for children with cancer. The foundation funds research that brings innovative approaches to the treatment, diagnosis, or prevention of pediatric cancer. Applications proposing novel approaches to the prevention, management, or cure of pediatric cancer are sought. Proposals should be based on molecular, cellular, or integrated systems and be conceptually innovative. The application should have a clear plan for study and future clinical use of the facutal data obtained.
Requirements: Applications may be submitted by individuals holding a MD, PhD, or equivalent degree working in domestic nonprofit organizations such as universities, colleges, hospitals, or laboratories.
Samples: Colubmia U, College of Physicians and Surgeons, Division of Pediatric Oncology (New York, NY) and New York-Presbyterian Children's Hospital (New York, NY)—to establish a state-of-the-art laboratory for research on pediatric cancer, $3 million jointly over three years.
Amount of Grant: $25,000-$100,000
Date(s) Application is Due: Sep 30
Contact: Grants Administrator, (949) 859-6312; fax: (949) 859-6323; email: admin@pcrf-kids.com
Internet: http://www.pcrf-kids.com
Sponsor: Pediatric Cancer Research Foundation
9272 Jeronimo Road, Suite A-107A
Irvine, CA 92618

Pedro Barrie de la Maza, Count of Fenosa Foundation Grants 4930
The foundation operates mainly in the Galician region and awards grants in its areas of interest, including education, social welfare and studies, science, medicine, and arts and humanities. Types of support include research grants to individuals and institutions, scholarships, and prizes.
Contact: Vicente Arias Mosquera, Secretary, 34-81-221525; fax: 34-81-224448; email: info@fbarrie.org
Internet: http://www.fbarrie.org/fundacion/ingles/home.htm
Sponsor: Pedro Barrie de la Maza, Count of Fenosa Foundation
Canton Grande 9
La Coruna
Spain

Pell Grant Program 4931
This program provides grants to students attending eligible institutions of higher education to help them pay for their educational costs.
Requirements: Undergraduate students enrolled as regular students in an eligible program at an eligible institution of higher education and making satisfactory academic progress are eligible to apply. The applicants must be U.S. citizens or eligible noncitizens and have a high school diploma, a GED, or demonstrate the ability to benefit from the program offered.
Amount of Grant: $400-$4050 range
Contact: Program Specialist, (800) 433-3243
Internet: http://www.ifap.ed.gov/IFAPWebApp/index.jsp
Sponsor: Department of Education
400 Maryland Avenue SW
Washington, DC 20202

Pennsylvania Council on the Arts Individual Artist Fellowship Program 4932
The council supports outstanding Pennsylvania artists by awarding fellowships on an annual basis. Categories include dance, folk arts, interdisciplinary arts, literature, media, music, and visual arts. Not all categories are offered annually. Contact the council for information on arts organizations that administer the fellowships. Annual deadline dates may vary; contact the program director for exact dates.
Requirements: Any individual artist who established residency in Pennsylvania for at least two years prior to the August 1 deadline.
Restrictions: Individuals enrolled in high school, or in a bachelor's, master's or PhD degree program in the arts are ineligible.
Geographic Focus: Pennsylvania
Amount of Grant: $5000-$10,000 average
Date(s) Application is Due: Aug 1
Contact: Program Contact, email: keswartz@state.pa.us
Internet: http://www.pacouncilonthearts.org/pca.cfm?id=9&level=Third&sid=13
Sponsor: Pennsylvania Council on the Arts
216 Finance Building
Harrisburg, PA 17120

Pennsylvania State University Education Graduate Assistantships 4933
The program prepares graduates for education careers in research, administration, and policy analysis. Students work closely with established scholars in research or teaching assistantships. Graduate assistantships include full tuition remission and a monthly stipend. Application and guidelines are available online.
Date(s) Application is Due: Feb 1
Contact: Department of Education Policy Studies, fax: (814) 865-1480; email: Highered@psu.edu
Internet: http://www.ed.psu.edu/hied/financial2.asp
Sponsor: Pennsylvania State University
400 Rackley Building
University Park, PA 16802

Perl 6 Microgrants 4934
The program was established by a donation from Best Practical Solutions to help support Perl 6 development. Grants of $500 will be awarded to a range of Perl 6-related projects over the life of the grant program. Proposals will be accepted on a rolling schedule.
Requirements: Accepted grants might be for coding, documentation, testing or even writing articles about Perl 6. The program isn't tied to any one implementation of Perl 6 — they are interested in seeing proposals related to Pugs, Perl 6 on Parrot, Perl 6 on Perl 5 or any other Perl 6 implementation, and in general, projects that can be completed in 4-6 calendar weeks. To submit a grant proposal, send an email to perl6- microgrants@perl.org with the following information: a) A two to three paragraph summary of the work you intend to do; b) A quick bio; c) A brief description of what 'success' will mean for your project (How will we know you're done?); d) Where (if anywhere) you've discussed your project in the past; e) Where you'll be blogging about your progress. (Twice-weekly blog posts are a requirement for getting your grant money.)
Amount of Grant: $500
Contact: Jesse Vincent, Grant Manager; email: jesse@bestpractical.com
Internet: http://www.perlfoundation.org/patrick_michaud_awarded_perl_6_development_grant_as_a_joint_initiative_by_the_perl_foundation_and_mozilla_foundation
Sponsor: Perl Foundation
6832 Mulderstraat
Grand Ledge, MI 48837

Perpetual Trust for Charitable Giving Grants 4935
The Foundation, established in 1957, supports organizations involved with education, health care, and family services. Giving is limited to Massachusetts. Contact the Foundation for further application information and guidelines.
Requirements: Must be a non-profit in Massachusetts to be eligible.
Restrictions: No grants to individuals.
Geographic Focus: Massachusetts
Samples: Boston Urban Youth Foundation, Boston, MA, $20,000 (2004); Pine Street Inn, Boston, MA, $30,000 - for program support (2004); Education Resources Institute, Boston, MA, $35,000 - for CKP program (2004);
Amount of Grant: $1,000-$50,000 range
Date(s) Application is Due: Sep 1
Contact: Kerry Sullivan. Director, Bank of America, N.A. Foundation & Philanthropic Services; (617) 4334-4846; email: kerry.h.sullivan@bankofamerica.com
Sponsor: Perpetual Trust for Charitable Giving
100 Federal Street, MA5-100-16-01
Boston, MA 02110

Peter C. Craigie Memorial Scholarship in Humanities 4936
One scholarship is offered every second year to a student with an outstanding scholastic record in a department of the Faculty of Humanities and who must have been or currently is involved in activities contributing to the general welfare of the university community.
Requirements: The scholarship is open to full-time students who are registered in and have completed one term of study in a program of studies leading to a master's degree in a department of the Faculty of Humanities at the University of Calgary. Students must first apply to this faculty.
Amount of Grant: $C4000

Date(s) Application is Due: Apr 15
Contact: Connie Busch, Graduate Scholarship Officer, (403) 220-5690; fax: (403) 289-7635; email: cbusch@ucalgary.ca
Internet: http://www.grad.ucalgary.ca/funding/internal_scholarships/lev_4/craigie.htm
Sponsor: University of Calgary
2500 University Dr NW, Earth Sciences Building, Room 720
Calgary, AB T2N 1N4 Canada

Peter Debye Award in Physical Chemistry **4937**
This award is given annually to encourage and reward outstanding research in physical chemistry. A nominee must have accomplished outstanding research of a theoretical or experimental nature in the field. Some consideration should be given to the nominee's success as a mentor and colleague. The award will be granted without regard to age or nationality. The award consists of $5,000 and a certificate.
Requirements: Any individual, except a member of the award committee, may submit one nomination or seconding letter for the award in any given year. The nominating documents consist of a letter of not more than 1000 words containing an evaluation of the nominee's accomplishments and a specific identification of the work to be recognized, a biographical sketch including date of birth, and a list of publications and patents authored by the nominee.
Restrictions: Self-nominations are not accepted.
Samples: Michael L. Klein—award winner, $5,000 (2008); John T. Yates, Jr.—award winner, $5,000 (2007); Donald G. Truhlar—award winner, $5,000 (2006); Stephen R. Leone—award winner, $5,000 (2005).
Amount of Grant: $5,000 and travel expenses to award meeting
Date(s) Application is Due: Nov 1
Contact: Felicia Dixon, Awards Administrator; (800) 227-5558 or (202) 872-4408; fax: (202) 776-8008; email: f_dixon@acs.org or awards@acs.org
Internet: http://portal.acs.org/portal/acs/corg/content?_nfpb=true&_pageLabel=PP_ARTICLEMAIN&node_id=1319&content_id=CTP_004543&use_sec=true&sec_url_var=region1
Sponsor: American Chemical Society
1155 Sixteenth Street, NW
Washington, DC 20036

Peter F. Drucker Award for Nonprofit Innovation **4938**
The award is given each November to three nonprofit organizations in recognition of an innovative, existing program that has made a difference in the lives of the people it serves. The annual awards (one first place, two runners up) are accompanied by unrestricted cash prizes designed to further the work of the recipient organizations. Application and guidelines are available online.
Requirements: Organization must be a registered 501(c)3 nonprofit organization and must have been launched no earlier than January 1, 2000, to be eligible.
Amount of Grant: $25,000 first place; $2500 runners up
Date(s) Application is Due: Jun 8
Contact: Awards Coordinator; (212) 224-1174; fax: (212) 224-2508; email: info@leadertoleader.org
Internet: http://www.leadertoleader.org/events/index.html
Sponsor: Leader to Leader Institute
320 Park Avenue, 3rd Floor
New York, NY 10022

Peter Krueger Summer Internship **4939**
The internship was established to encourage promising students of art history, architectural history, museum studies, museum education, and design to explore careers in the museum profession. Interns are assigned to specific curatorial, education, or administrative departments where they will assist on special research or exhibition projects, as well as participate in daily museum activities. Interns are for 10 weeks. Housing is not provided.
Requirements: The internship is open to undergraduate students considering a career in art history, design, museum studies, and museum education, as well as graduate students who have not yet completed their MA degree.
Amount of Grant: $2500
Date(s) Application is Due: Feb 1
Contact: Intern Coordinator, (212) 860-6871; email: edu@si.edu
Internet: http://ndm.si.edu/EDUCATION/index.html
Sponsor: Cooper-Hewitt National Design Museum
2 E 91st Street
New York, NY 10128-9990

Petroleum Geology and Marine Geoscience Regional Graduate Scholarships **4940**
Three scholarships are awarded each year to promote advanced education and research in those fields of geology and marine geoscience having application to the petroleum industry. One scholarship will be awarded in each of the three regions of Canada (Atlantic Provinces, Ontario/Quebec, and Western Canada). Petroleum geology related disciplines include structural geology, sedimentology, stratigraphic studies involving clastic or carbonate rocks, paleontology, geochemistry, and coal geology. A fourth scholarship is awarded across Canada in the field of marine geoscience. Marine geoscience disciplines include all earth sciences actively directed toward the research, evaluation, development, and protection of the offshore regions of Canada and the Great Lakes. Scholarships are awarded for one year. Application forms are required.
Requirements: Candidates must hold Canadian citizenship or landed immigrant status and be accepted on the basis of merit and academic standing. The scholarships are open to MS or PhD students attending Canadian universities engaged in their second year of study in petroleum geology or marine geoscience-related disciplines.
Amount of Grant: $C1500
Contact: Shawna Christensen, Scholarships Program, (403) 218-1625 or (403) 264-5610; fax: (403) 264-5898; email: schristensen@petrelrob.com
Internet: http://www.cspg.org/body_award_descriptions.html
Sponsor: Canadian Society of Petroleum Geologists
540 Fifth Avenue SW, #160
Calgary, AB T2P 0M2 Canada

Pew Charitable Trusts Biomedical Research Grants **4941**
Pew invests in an array of programs related to science that are aimed at improving the quality of scientific research as well as making data widely available. These projects are working toward solutions to environmental, health and safety dilemmas. Pew also has a decades-long commitment to support groundbreaking research by biomedical researchers early in their careers. All of these initiatives are part of our broader effort to study and promote nonpartisan policy solutions for pressing and emerging problems affecting the American public and the global community. Pew, experts partner with leading authorities in science, the environment and associated fields to conduct research and advance fact-based solutions to compelling problems. Pew targets the frontiers of scientific knowledge where it can make a difference.
Requirements: Grants are made only to colleges and universities, as well as 501(c)3 tax-exempt organizations that are not private foundations.
Restrictions: Grants are not made to individuals or for endowments, capital campaigns, unsolicited construction requests, debt reduction, or scholarships or fellowships that are not part of a program initiated by the Trusts.
Contact: Susan A. Magill, Managing Director; (202) 552-2129 or (202) 552-2000; fax: (202) 552-2299; email: smagill@pewtrusts.org or info@pewtrusts.com
Internet: http://www.pewtrusts.org/our_work_category.aspx?id=336
Sponsor: Pew Charitable Trusts
2005 Market Street, Suite 1700
Philadelphia, PA 19103-7077

Pew Fellowships in the Arts **4942**
Fellowships support artists working in a wide variety of performing, visual, and literary disciplines. Awards are made to artists working in 12 different discipline categories, which rotate on a four-year cycle. Applications are accepted annually for the three discipline categories under consideration in that year. Fields of application for the current year are performance art; poetry; and sculpture and installation. The grants provide financial support directly to the artists so that they may have the opportunity to dedicate themselves to creative pursuits exclusively. The program aims to provide such support at moments in artists' careers when a concentration on artistic growth and exploration is most likely to have the greatest impact on an artist's long-term professional development. Guidelines and application are available online. Annual deadline dates may vary; contact program staff for exact dates.
Requirements: All applicants must be practicing artists of demonstrated commitment and professional accomplishment, and be able to provide evidence of such. Refer to specific eligibility requirements under appropriate field of application. Applicants must be verifiable Pennsylvania residents of Bucks, Chester, Delaware, Montgomery, or Philadelphia County for two years or longer immediately prior to the application deadline; and maintain permanent residence in the five-county Philadelphia area through the selection process. Artists who are granted fellowships must maintain permanent residence in the five-county Philadelphia area during their fellowship period.
Restrictions: Artists are not eligible if they are: not able to provide evidence of their professional achievements within their field of application; a matriculated student (full-time or part-time) in a degree-granting program or equivalent at the time of application or during the selection process; a current employee of PFA or The Pew Charitable Trusts, or immediate family member of such an individual; or a previous recipient of a Pew Fellowship in the Arts.
Geographic Focus: Pennsylvania
Samples: Divided among 13 artists—for fellowships to artists working in creative nonfiction, fiction, the media arts, and works on paper, $600,000 (2005).
Amount of Grant: $50,000
Date(s) Application is Due: Dec 2
Contact: Fellowships Administrator, (215) 875-2285; fax: (215) 875-2276; email: pewarts@mindspring.com
Internet: http://www.pewarts.org/apply.html
Sponsor: Pew Charitable Trusts
230 S Broad Street, Suite 1003
Philadelphia, PA 19102

Pew Scholars in the Biomedical Sciences **4943**
This program is designed to support young investigators of outstanding promise in the basic and clinical sciences relevant to the advancement of human health. The funding of the awards is provided by The Pew Charitable Trusts. The award is intended to provide assured support, during their earlier years, for junior members of the faculty as they establish their laboratories. It is hoped that the assurance provided through the program will encourage successful applicants to be more venturesome in their research and future applications for support than would otherwise be likely. ALL applicants must be nominated by their institution and complete the online application.

Requirements: The program is open to individuals with a doctorate in medicine, other health related professions, or biomedical sciences. Candidates must hold full- time appointments at the rank of assistant professor or independent researcher on the faculty of the sponsoring institution.
Restrictions: A biomedical scholar cannot apply on his/her own for this scholarship but must apply through a selected institution.
Samples: U of California at San Francisco (CA)—for the Pew Scholars Program in the Biomedical Sciences, $3.6 million over four years (2004).
Amount of Grant: $60,000 per year
Date(s) Application is Due: Nov 1
Contact: Leslie Tucker, Program Officer, (215) 575-4850; email: shans@pewtrusts.com or info@pewtrust.com
Internet: http://futurehealth.ucsf.edu/biomed/scholdes.html
Sponsor: Pew Charitable Trusts
2005 Market Street, Suite 1700
Philadelphia, PA 19103-7077

Pezcoller Foundation-AACR International Award for Cancer Research 4944
The Award for Cancer Research was established in 1997 to annually recognize a scientist: who has made a major scientific discovery in basic cancer research OR who has made significant contributions to translational cancer research; who continues to be active in cancer research and has a record of recent, noteworthy publications; and whose ongoing work holds promise for continued substantive contributions to progress in the field of cancer. More than one scientist may be co-nominated and selected to share the Award when their investigations are closely related in subject matter and have resulted in work that is worthy of the Award. The Award consists of an unrestricted grant of Euro 75,000 and a commemorative plaque.
Requirements: Eligible candidates are cancer researchers affiliated with institutions in academia, industry, or government that are involved in cancer research, cancer medicine, or cancer-related biomedical science anywhere in the world.
Restrictions: Institutions or organizations are not eligible for the Award.
Amount of Grant: Euro 75,000
Date(s) Application is Due: Sep 15
Contact: Monique P. Eversley, Staff Associate; (267) 646-0576 or (215) 440-9300, ext. 1400; fax: (215) 440-9372; email: monique.eversley@aacr.org or awards@aacr.org
Internet: http://www.aacr.org/home/scientists/scientific-achievement-awards/pezcoller-aacr-international-award.aspx
Sponsor: Pezcoller Foundation / American Association for Cancer Research
615 Chestnut Street, 17th Floor
Philadelphia, PA 19106-4404

Pfizer Foundation Epilepsy Scholarship Award 4945
The scholarship honors 16 outstanding students who have overcome the challenges of epilepsy, been successful in school, done well in activities out of the classroom, and shown the ability to make the most out of college or graduate school. Experts in education and medicine will choose the winners. Information and application are available online.
Requirements: An applicant must be under a doctor's care for epilepsy; a high school senior who has applied to college; a freshman, sophomore, or junior in college; or a college senior who has applied to graduate school.
Amount of Grant: $3000
Date(s) Application is Due: Mar 1
Contact: Pfizer Epilepsy Scholarship, (800) 292-7373; email: Scholarship@edencomgroup.com
Internet: http://www.epilepsy-scholarship.com
Sponsor: Eden Communications Group
515 Valley Street, Suite 200
Maplewood, NJ 07040

Pfizer Foundation Grants 4946
The foundation's charitable giving programs provide grants through targeted initiatives focused primarily on health care and science education. Areas of interest include: global health—health literacy programs to improve the ability of low-income, low-literate populations to effectively comprehend and act on health information; HIV/AIDS—programs to provide Pfizer's antifungal medicine (Diflucan) to low-income HIV/AIDS patients in the world's least developed countries where the disease is most prevalent, the training of health-care workers in opportunistic infections related to HIV, and health-promotion programs that target those at high-risk of HIV infection living in countries where Pfizer is donating Diflucan; international trachoma initiative—programs to provide Pfizer's key antibiotic (Zithromax) to countries implementing public-health programs to combat trachoma, including community-based initiatives to provide surgery, antibiotics, promotion of clean faces, and education about the disease; disaster relief—requests for product donations following floods, earthquakes, droughts, wars, and other disasters domestically and abroad; K-12 science education—partnerships between Pfizer sites and local schools, with an emphasis on training teachers, developing curriculum, constructing science labs, and support of exhibits and educational outreach programs at science-rich institutions across the country; and community and cultural—programs in Pfizer communities, with a particular focus in the New York City for a range of organizations that address urgent community needs, seek long-term solutions to community issues, and strengthen the cultural vitality of diverse communities.
Requirements: Nonprofit 501(c)3 organizations and schools are eligible.
Restrictions: The following types of organizations or activities are excluded from support fundraising events or activities such as telethons, walkathons, and races; specific performances or concerts; sporting events; endowment campaigns; building fund drives and capital campaigns (except science lab renovations in partner schools and universities); film, video, television, or radio projects (except as part of a health literacy program); requests for loans or debt retirement; trips, tours, or cultural exchange programs; conferences, seminars, briefing programs, and similar activities (except in instances where the event evolves from Pfizer programs in health care and science education); organizations already supported through United Way contributions; operating expenses of United Way local agencies (except through annual United Way campaigns); grants to individuals; political causes or candidates; anti-business organizations; or organizations that practice discrimination or limit membership on the basis of race, creed, gender, age, sexual preference, or national origin.
Contact: Grants Administrator, (800) 733-4717 or (212) 573-2323
Internet: http://www.pfizer.com/subsites/philanthropy/caring/index.html
Sponsor: Pfizer Inc
235 E 42nd Street
New York, NY 10017-5755

Pfizer Foundation Health Literacy Grant Program 4947
The program invites applications for research to advance the understanding of health literacy and for an intervention that overcomes problems associated with low levels of health literacy. Examples of proposed projects include developing new measures or new ways of looking at the causes or impact of health literacy; developing programs that demonstrate improved health literacy; or creating policy initiatives that promote improved health outcomes. Proposals for community-based research and interventions are encouraged.
Requirements: Applications are encouraged from investigators in diverse fields such as health literacy, adult education, health economics, medicine, nursing, pharmacy, public health, anthropology, health policy, social work, management sciences, and ethics.
Amount of Grant: $65,000 per year
Contact: Programs Administrator, (212) 733-4250
Internet: http://www.pfizerhealthliteracy.com/partnerships_awards.html
Sponsor: Pfizer Foundation
235 E 42nd Street
New York, NY 10017-5755

Pharmacia-ASPET Award for Experimental Therapeutics 4948
The Pharmacia-ASPET Award in Experimental Therapeutics is given annually to recognize and stimulate outstanding research in pharmacology and experimental therapeutics,Äìbasic laboratory or clinical research that has had, or potentially will have, a major impact on the pharmacological treatment of disease. The winner will receive a $2,500 honorarium, a plaque, hotel and economy airfare for the winner and spouse to the award ceremony at the ASPET annual meeting.
Requirements: Nominations must be made by ASPET members; however, the nominee need not be a member. No restrictions exist as to age or institutional affiliation of nominee.
Amount of Grant: $2,500
Date(s) Application is Due: Sep 15
Contact: Christine Carrico, Ph.D., Executive Officer; (301) 634-7060; fax: (301) 634-7061; email: ccarrico@aspet.org or info@aspet.org
Internet: http://www.aspet.org/public/awards/exp_ther_award.html
Sponsor: Pfizer / American Society for Pharmacology and Experimental Therapeutics
9650 Rockville Pike
Bethesda, MD 20814-3995

PHF Andrew W. Mellon Postdoctoral Fellowships in the Humanities 4949
Five one-year residential postdoctoral fellowships are available for untenured junior scholars who are no more than eight years out of their doctorate. The programs of the Penn Humanities Forum are conceived through yearly topics that invite broad interdisciplinary collaboration. Research proposals are invited on the annual topic in all areas of humanistic study, except educational curriculum-building and the performing arts. Candidates from all humanistic disciplines are eligible, as well as those in allied areas such as Anthropology and History of Science. Fellows teach one undergraduate course each of two terms in addition to conducting their research. The fellowship is open to all scholars, national and international.
Requirements: The program is open to scholars who are not yet tenured (may not be tenured during the fellowship year) and who are no more than eight years out of their doctorate. The PhD is the only terminal degree eligible.
Samples: Kinga Araya, PhD, Concordia University, Independent Scholar, Quebec, Art History, Visual Arts—for Walking in Exile (2006-07); John Ghazvinian, PhD, Oxford University, Contract Author, Harcourt Trade Publishers, History—for A certaine tickling humour: Passport Records and the Re-Fashioning of English Foreign Travel, 1560,Äì1660 (2006-07).
Amount of Grant: $42,000 annual stipend plus health insurance
Date(s) Application is Due: Oct 15
Contact: Jennifer Conway, Penn Humanities Forum, (215) 898-8220; email: conwayj@sas.upenn.edu
Internet: http://humanities.sas.upenn.edu/mellon_description.shtml
Sponsor: Penn Humanities Forum, University of Pennsylvania
3619 Locust Walk
Philadelphia, PA 19104-6213

PHF Penn Faculty Research Fellowships 4950
Applications are solicited from standing faculty members in the humanities and allied departments at Penn for the Penn Humanities Forum's Mellon Faculty Research

Fellowships. Available are junior faculty awards of $5000 disbursed as salary, as well as senior faculty awards of $10,000 paid directly to departments to offset the cost of a course release. Fellows must be engaged in research directly related to the Forum's theme for the year in which the award is granted, and are expected to participate in the Forum's weekly Mellon Research Seminars.
Requirements: Applicants must be Penn State University faculty in the humanities or allied departments.
Samples: Aditya Behl, Associate Professor & Chair, South Asia Studies, Pennsylvania University (2006-07); Phoebe Kropp, Assistant Professor, History, Pennsylvania University (2006-07); Simon Richter, Associate Professor and Chair, German, Pennsylvania University (2006-07).
Amount of Grant: $5,000 for junior faculty (assistant professors) and $10,000 for senior faculty
Date(s) Application is Due: Feb 2
Contact: Jennifer Conway, Penn Humanities Forum, (215) 898-8220; email: conwayj@sas.upenn.edu
Internet: http://humanities.sas.upenn.edu/faculty_description.shtml
Sponsor: Penn Humanities Forum, University of Pennsylvania
3619 Locust Walk
Philadelphia, PA 19104-6213

PHF Regional Mellon Faculty Research and Cultural Fellowships **4951**
The PHF research stipends of $3,000 are intended for faculty in the humanities from Philadelphia-area colleges and universities and for department directors and curators of Philadelphia-area arts and cultural organizations. The stipends are designed to encourage interdisciplinary exchange among the region's academies and cultural institutions. Fellows must be engaged in research directly related to the Forum's theme for the year in which the award is granted, and are expected to participate in the Forum's weekly Mellon Research Seminars.
Requirements: Applicants must be faculty members of Philadelphia-area colleges or universities, or department directors and curators of Philadelphia-area arts and cultural organizations.
Geographic Focus: Pennsylvania
Samples: Edmund Campos, Assistant Professor, English, Swarthmore College (2006-07); Ann Gardiner, Assistant Professor, Literature, Philadelphia University (2006-07); Keally McBride, Senior Fellow, Political Science, University of Pennsylvania (2006-07).
Amount of Grant: $3,000
Date(s) Application is Due: Feb 2
Contact: Jennifer Conway, Penn Humanities Forum, (215) 898-8220; email: conwayj@sas.upenn.edu
Internet: http://humanities.sas.upenn.edu/regional_description.shtml
Sponsor: Penn Humanities Forum, University of Pennsylvania
3619 Locust Walk
Philadelphia, PA 19104-6213

PHF Undergraduate Humanities Forum Research Fellowships **4952**
Each year, a number of Fellowships are offered to promote undergraduate research in the humanities and cultivate the importance of humanistic thought across all disciplines. Interested candidates should be conducting work related to the Forum's theme for the year in which the award is granted. Calls for applications are solicited in February of the year preceding the fellowship. One $1500 PHF Coordinating Research Fellowship and approximately eighteen PHF Research Fellowships at $600 each are available each year. Applicants are drawn from students across Penn, as well as from the university's twelve undergraduate College Houses.
Samples: Gisele Aris—The Power and Politics of Dress in Africa (2007); Sarah Breger—Sentenced to Marriage: Chained Women in Wartime (2007); Sharon Cantor—Treading the Abyss: The Distressing Journey in Kierkegaardian Faith (2007).
Amount of Grant: $600-$1,500
Date(s) Application is Due: Mar 15
Contact: Jennifer Conway, Penn Humanities Forum, (215) 898-8220; email: conwayj@sas.upenn.edu
Internet: http://humanities.sas.upenn.edu/undergraduate_description.shtml
Sponsor: Penn Humanities Forum, University of Pennsylvania
3619 Locust Walk
Philadelphia, PA 19104-6213

Phi Beta Kappa Award in Science **4953**
The award is offered for outstanding contributions by scientists to the literature of science. The intent of the award is to encourage literate and scholarly interpretations of the physical and biological sciences and mathematics. Entries must be submitted by the publisher who should first obtain a copy of conditions of eligibility from Phi Beta Kappa. By March 31, applicants should ship one copy of the book to the appropriate awards committee. By April 10, if the submission is eligible, applicants will receive the shipping list for judges, and should ship one book to each judge by April 30.
Requirements: To be eligible, biographies of scientists must have a substantial critical emphasis on their scientific research. Entries must be original works published in the United States during the 12-month period between May 1 (of the past year) and the April 30 (of the current year) deadline date. Entries ordinarily will be the work of a single writer. Exceptions may be made for books written by small teams of scholars working in close collaboration.
Restrictions: Entries should not be of a technical or highly specialized character. Monographs and compendiums are not eligible.
Samples: Sean B. Carroll—for The Making of the Fittest: DNA and the Ultimate Forensic Record of Evolution (2007); William F. Ruddiman—for Plows, Plagues, and Petroleum: How Humans Took Control of Climate (2006); Chris Beard—for The Hunt for the Dawn Monkey: Unearthing the Origins of Monkeys, Apes and Humans (2005).
Amount of Grant: $2,500
Date(s) Application is Due: Apr 30
Contact: Sandra Beasley, Awards Coordinator, (202) 265-3808; fax: (202) 986-1601; email: SBeasley@pbk.org or info@pbk.org
Internet: http://staging.pbk.org/AM/Template.cfm?Section=Book_Awards&Template=/CM/HTMLDisplay.cfm&ContentID=2646
Sponsor: Phi Beta Kappa Society
1606 New Hampshire Avenue, NW
Washington, DC 20009

Phi Delta Kappa Outstanding Doctoral Dissertation Awards **4954**
These biennial awards recognize scholars who show promise for noteworthy contributions to educational research. One award is given in each of Phi Delta Kappa's regions in the fall of even-numbered years. Winners receive a monetary award and a certificate of recognition. Each entry must include a summary of the dissertation and a completed application form. Application forms, nominating guidelines, and criteria for judging are available online.
Requirements: Competition is open to both Phi Delta Kappa members and nonmembers, but all applicants must be nominated by a Phi Delta Kappa chapter. The nominee must be a recent graduate of an accredited college or university. The nominee's major field must be in education, or the dissertation must focus on an educational topic. The nominee's dissertation must have been completed (approved by the doctoral committee and accepted by the graduate school or school of education) between September 1, 2002, and August of the current year.
Amount of Grant: $500
Date(s) Application is Due: Aug 31
Contact: Carol Langdon, (800) 766-1156 or (812) 339-1156; fax: (812) 339-0018; email: clangdon@pdkintl.org
Internet: http://pdkintl.org/edres/cedr_p2.htm
Sponsor: Phi Delta Kappa International
P.O. Box 789
Bloomington, IN 47402

Phi Upsilon Omicron Candle Fellowships **4955**
Two fellowships are awarded annually for study leading to a master's degree in family and consumer sciences. Selection will be made on the basis of applicant's scholastic record, participation in Phi Upsilon and other home economics-related organizations and community activities, and a statement of professional aims and goals. Application materials for undergraduate and graduate fellowships/scholarships/awards and nomination forms for the alumni awards are available from advisors and presidents of local collegiate chapters, presidents of alumni chapters, and the national office. Forms are also available on the Web site.
Requirements: Applicant must be a member of Phi Upsilon Omicron who is enrolled in or planning to enroll in graduate school.
Amount of Grant: $1000
Date(s) Application is Due: Feb 1
Contact: Fellowships Administrator, National Office, (304) 368-0612; email: rickards@access.mountain.net
Internet: http://www.ianr.unl.edu/phiu/Graduate_Fellowship_.htm
Sponsor: Phi Upsilon Omicron
P.O. Box 329
Fairmont, WV 26555

Phi Upsilon Omicron Diamond Anniversary Fellowships **4956**
Two fellowships will be awarded per year for study at the master's or doctoral level in home economics. Selection will be made on the basis of the applicant's scholastic record, participation in Phi Upsilon and other home economics-related organizations and community activities, and statement of professional aims and goals. Application materials for undergraduate and graduate fellowships/scholarships/awards and nomination forms for the alumni awards are available from advisors and presidents of local collegiate chapters, presidents of alumni chapters, and the national office.
Requirements: Applicant must be a member of Phi Upsilon Omicron.
Amount of Grant: $1000
Date(s) Application is Due: Feb 1
Contact: Fellowships Administrator, National Office, (304) 368-0612; email: rickards@mountain.net
Internet: http://www.ianr.unl.edu/phiu/Graduate_Fellowship_.htm
Sponsor: Phi Upsilon Omicron
P.O. Box 329
Fairmont, WV 26555

Phi Upsilon Omicron Founders Fellowship **4957**
One fellowship is awarded per year to be used toward doctoral degree work in the field of home economics. Applicants must be Phi Upsilon members who have completed at least half the hour requirements toward the doctorate in some phase of home economics and who have had professional employment in home economics or a related field. Application materials for undergraduate and graduate fellowships/scholarships/awards and nomination forms for the alumni awards are available from advisors and presidents of local collegiate chapters, presidents of alumni chapters, and the national office.

Requirements: Applicant must be a member of Phi Upsilon Omicron.
Amount of Grant: $1500
Date(s) Application is Due: Feb 1
Contact: Fellowships Administrator, National Office, (304) 368-0612; email: rickards@access.mountain.net
Internet: http://www.ianr.unl.edu/phiu/Graduate_Fellowship_.htm
Sponsor: Phi Upsilon Omicron
P.O. Box 329
Fairmont, WV 26555

Phi Upsilon Omicron Presidents Research Fellowship **4958**
The national home economics honor society awards this graduate research fellowship at the master's, doctoral, or postdoctoral level. Research prospectus must accompany the application and show organization and demonstrate need for the research. Selection will be based on applicant's scholastic record; honors and recognitions; participation in honor society; professional, community, and religious organizations; scholarly work; statement of professional aims and goals; and recommendations. Application materials for undergraduate and graduate fellowships/scholarships/awards and nomination forms for the alumni awards are available from advisors and presidents of local collegiate chapters, presidents of alumni chapters, and the national office.
Requirements: Applicant must be a member of Phi Upsilon Omicron.
Amount of Grant: $1000
Date(s) Application is Due: Feb 1
Contact: Fellowships Administrator, National Office, (304) 368-0612; email: rickards@mountain.net
Internet: http://www.ianr.unl.edu/phiu/Graduate_Fellowship_ htm
Sponsor: Phi Upsilon Omicron
P.O. Box 329
Fairmont, WV 26555

Phil Hardin Foundation Grants **4959**
The foundation targets its funding to education. Specific priorities are to strengthen the capacity of communities to nurture and educate young children; the capacity of higher education institutions to renew communities and their economies; the capacity of communities for locally initiated educational improvement and economic development; and state- and local-level policy and leadership initiatives that fit with foundation goals. Types of support include general operating support, continuing support, building construction/renovation, equipment acquisition, endowment funds, program development, conferences and seminars, professorships, publication, seed grants, fellowships, scholarship funds, research, and matching funds. The foundation also operates four K-12 fellowship programs. Deadlines vary between programs; contact program staff for specific fellowship deadlines.
Requirements: Applicants must either be based in Mississippi or the project must benefit Mississippi, depending on the program. Contact program staff for eligibility.
Samples: Meridian Foundation (MS)—to endow the Riley Education and Performing Arts Center, $2.5 million.
Amount of Grant: $6 million total
Contact: C. Thompson Wacaster, (601) 483-4282; fax: (601) 483-5665; email: info@philhardin.org
Internet: http://www.philhardin.org
Sponsor: Phil Hardin Foundation
1921 24th Avenue
Meridian, MS 39301

PhRMA Foundation Health Outcomes Post Doctoral Fellowships **4960**
The program provides stipend support for individuals engaged in a research training program that will create or extend their credentials in health outcomes. The intent of this program is to support post doctoral career development activities of individuals prepared (or preparing) to engage in research that will strengthen representation of health outcomes in schools of pharmacy, medicine, nursing and public health. To accomplish these goals, support will be provided for a two-year period to selected individuals who are beginning careers in health outcomes research and who give promise of outstanding development as researchers. The award, consisting of a $40,000 annual stipend, is made to the institution on behalf of the fellow. The award is intended solely as a stipend and may not be used otherwise.
Requirements: Eligible applicants include well-trained graduates from Pharm. D., M.D., and Ph.D. programs who seek to further develop and refine their research skills through formal postdoctoral training. They must also have a firm commitment from an accredited U.S. university and be a U.S. citizen or permanent resident. The application must include a research plan written by the applicant, the mentor's research record, and a description of how the mentored experience will enhance the applicant's career development in health outcomes research. The sponsor (mentor) of the post doctoral program must describe how the goals of the research training program will be accomplished and provide assurance that key collaborating mentors endorse and are willing to support the training plan.
Restrictions: Applications, however good, which do not meet the aims of the program will be disapproved.
Amount of Grant: $40,000
Contact: Eileen M. Cannon, Executive Director; (202) 572-7756; fax: (202) 572-7799; email: info@phrmafoundation.org
Internet: http://www.phrmafoundation.org/awards/outcomes/postdoc.php
Sponsor: Pharmaceutical Research and Manufacturers of America Foundation
950 F Street, N.W., Suite 300
Washington, DC 20004

PhRMA Foundation Health Outcomes Research Starter Grants **4961**
The purpose of the program is to offer financial support to individuals beginning their independent research careers at the faculty level. The program provides a research grant of $30,000 per year for up to two years. This program supports individuals beginning independent research careers in academia who do not have other substantial sources of research.
Requirements: Those holding the academic rank of instructor or assistant professor and investigators at the doctoral level with equivalent positions are eligible to apply, providing their proposed research is neither directly nor indirectly subsidized to any significant degree by an extramural support mechanism. The program is not intended for those in postdoctoral training programs. However, individuals in postdoctoral training scheduled to conclude and who will hold an academic appointment by January 1 may apply. Applicants must be sponsored by the department or unit within which the proposed research is to be undertaken. The grant is made to the university on behalf of the applicant and with the understanding that the university will administer the funds. Schools of medicine, pharmacy, public health, nursing, dentistry and schools of other areas where appropriate are eligible for this award. Applications must be submitted by an accredited school in the U.S., and all applicants must be U.S. citizens or permanent residents. Before an individual is eligible to apply for a PhRMA Foundation award, the applicant must first have a firm commitment from a U.S. university.
Restrictions: This program is not offered as a means to augment substantially funded research efforts. It is intended to offer support for researchers who are starting their independent research efforts. This award is granted in part based on need. If an individual currently has or is guaranteed substantial funding, they should not apply.
Amount of Grant: $30,000
Contact: Eileen M. Cannon, Executive Director; (202) 572-7756; fax: (202) 572-7799; email: info@phrmafoundation.org
Internet: http://www.phrmafoundation.org/awards/outcomes/starter.php
Sponsor: Pharmaceutical Research and Manufacturers of America (PhRMA) Foundation
950 F Street, N.W., Suite 300
Washington, DC 20004

PhRMA Foundation Health Outcomes Sabbatical Fellowships **4962**
The fellowship provides stipend support for individuals engaged in a multidisciplinary research training program that will create or extend their credentials in health outcomes. The program enables faculty with active research programs to work outside of their home institution for periods of six months to one year to learn new skills or develop new collaborations that will enhance their research and research training capabilities in health outcomes. The applicant and mentor of the program must describe how the multidisciplinary goals of the research experiential program will be accomplished and provide assurance that key collaborating mentors endorse and are willing to support the plan. Matching funds must be provided by the home institution. Awards may be activated beginning January 1 or on the first day of any month thereafter, up to and including December 1. Guidelines and application are available online.
Requirements: Eligible applicants must (1) hold a PhD, Pharm.D., M.D. or Sc.D. degree in a field of study logically or functionally related to the proposed post doctoral activities, (2) hold a faculty appointment that imparts eligibility for a sabbatical leave from their home institution, (3) have institutional approval of a sabbatical plan that includes partial salary that matches the PhRMA stipend, (4) hold an endorsement from a mentor who agrees to sponsor the applicant's visiting scientist activity, and (5) be a U.S. citizen or permanent resident. Matching funds must be provided by the home institution.
Restrictions: The program provides no other subsidies (travel, tuition, fringe benefit costs, etc.), and indirect costs to the institution are not provided.
Amount of Grant: $40,000
Contact: Eileen M. Cannon, Executive Director; (202) 572-7756; fax: (202) 572-7799; email: info@phrmafoundation.org
Internet: http://www.phrmafoundation.org/awards/outcomes/sabbatical.php
Sponsor: Pharmaceutical Research and Manufacturers of America (PhRMA) Foundation
950 F Street, N.W., Suite 300
Washington, DC 20004

PhRMA Foundation Informatics Sabbatical Fellowships **4963**
The fellowship supports individuals engaged in a multidisciplinary research training program that will create or extend their credentials in informatics. The program enables faculty with active research programs to work outside of their home institution for periods of six months to one year to learn new skills or develop new collaborations that will enhance their research and research training capabilities in informatics. Applicants are expected to engage in research that supports career development of scientists engaged in computational and experimental research to integrate cutting edge information technology with advanced biological, chemical, and pharmacological sciences in the fields of genetics, genomics, and proteomics biological pathways. Emphasis will be placed on the development of new informatics technologies that demonstrate the translation of genomic data into an elucidation and understanding of biological and disease processes. The applicant and mentor of the program must describe how the multidisciplinary goals of the research experiential program will be accomplished and provide assurance that key collaborating mentors endorse and are willing to support the plan. The award provides up to $40,000 stipend support for mid-career scientists to engage in an academic year or calendar year experiential program intended to redirect their core research focus to an area of emerging importance to pharmaceutical research and development.
Requirements: Eligible applicants must hold a PhD or MD degree in a field of study logically or functionally related to the proposed postdoctoral activities; hold a faculty

appointment that imparts eligibility for a sabbatical leave from their home institution; have institutional approval of a sabbatical plan that includes partial salary that matches the PhRMA stipend; and be a U.S. citizen or permanent resident. Preference will be given to those individuals whose research combines novel computational methods with experimental validation. Emphasis will be placed on the development of new informatics technologies that demonstrate the translation of genomic data into an elucidation and understanding of biological and disease processes. Matching funds must be provided by the home institution.
Restrictions: The program provides no other subsidies (travel, tuition, fringe benefit costs, etc.), and indirect costs to the institution are not provided.
Amount of Grant: $40,000
Contact: Eileen M. Cannon, Executive Director; (202) 572-7756; fax: (202) 572-7799; email: info@phrmafoundation.org
Internet: http://www.phrmafoundation.org/awards/informatics/sabbatical.php
Sponsor: Pharmaceutical Research and Manufacturers of America (PhRMA) Foundation
950 F Street, N.W., Suite 300
Washington, DC 20004

PhRMA Foundation Informatics Post Doctoral Fellowships **4964**
The postdoctoral program provides stipend support for individuals engaged in a multidisciplinary research training program that will create or extend their credentials in informatics. The intent of this program is to support postdoctoral career development activities of individuals preparing to engage in research that will bridge the gap between experimental and computational approaches in genomic and biomedical studies. It is anticipated that this research training will be accomplished in academic and/or industrial laboratory settings where multidisciplinary teams are organized to address problems that span the range of biological complexity rather than focus on the application of single technologies. The award, consisting of a $40,000 annual stipend, is made to the institution on behalf of the fellow. The award is intended solely as a stipend and may not be used otherwise.
Requirements: Eligible applicants must either hold a PhD degree in a field of study logically or functionally related to the proposed postdoctoral activities, or expect to receive the PhD before activating the award. They must also have a firm commitment from an accredited U.S. university and be a U.S. citizen or permanent resident. The sponsor (mentor) of the postdoctoral program must describe how the multidisciplinary goals of the research training program will be accomplished and provide assurance that key collaborating mentors endorse and are willing to support the training plan.
Restrictions: The program provides no other subsidies (travel, tuition, fringe benefit costs, etc.) and indirect costs to the institution are not provided. The second year of this award is contingent upon a progress report approved by the Foundation and submission of a financial report.
Amount of Grant: $40,000
Contact: Eileen M. Cannon, Executive Director; (202) 572-7756; fax: (202) 572-7799; email: info@phrmafoundation.org
Internet: http://www.phrmafoundation.org/awards/informatics/postdoc.php
Sponsor: Pharmaceutical Research and Manufacturers of America (PhRMA) Foundation
950 F Street, N.W., Suite 300
Washington, DC 20004

PhRMA Foundation Informatics Pre Doctoral Fellowships **4965**
The program is designed for candidates who expect to complete the requirements for the Ph.D. in the fields of informatics in two years or less from the time the fellowship begins. It aims at supporting promising students during their thesis research. The award is made to the university on behalf of the fellow. The fellowship provides a stipend of $20,000 a year payable monthly for a minimum of one year and a maximum of two years. Of the $20,000 awarded annually, up to $500 a year may be used for incidentals directly associated with the thesis research preparation (e.g., secretarial help, artwork, books, travel, etc.). Applications will be accepted for a minimum of one year and a maximum of two years. The program assumes that the fellows will devote full-time (including summers) to their research.
Requirements: The fellowship program of pre doctoral support is designed to assist full-time, in-residence Ph.D. candidates in the fields of informatics who are enrolled in U.S.A schools of medicine, pharmacy, dentistry or veterinary medicine. The program supports full-time advanced students who will have completed the bulk of their pre-thesis requirements (at least two years of study) and are engaged in thesis research as Ph.D. Candidates by the time the award is activated. Students just starting in graduate school should not apply. Highest priority will be given to applications who support the goal of the Informatics Programs. The goal of the program is to promote the use of informatics in an integrative approach to the understanding of biological and disease processes. Informatics awards will support career development of scientists engaged in computational and experimental research to integrate cutting-edge information technology with advanced biological, chemical, and pharmacological sciences. Applications must be submitted by an accredited U.S. school, and all applicants must be U.S. citizens or permanent residents.
Restrictions: The fellowship may not be used to supplement funds from other fellowships, traineeships or assistantships, unless necessary to make stipend levels compliant with institutional policy. The focus of the program is to assist in the predoctoral training of the candidate. It should not be viewed as a program to fund a research project. Indirect costs are not provided to the institution and PhRMA Foundation grant funds may not be used for this purpose.
Amount of Grant: $20,000
Contact: Eileen M. Cannon, Executive Director; (202) 572-7756; fax: (202) 572-7799; email: info@phrmafoundation.org
Internet: http://www.phrmafoundation.org/awards/informatics/predoc.php
Sponsor: Pharmaceutical Research and Manufacturers of America (PhRMA) Foundation
950 F Street, N.W., Suite 300
Washington, DC 20004

PhRMA Foundation Informatics Research Starter Grants **4966**
This program supports individuals beginning independent research careers in academia who do not have other substantial sources of research. The areas of interest within this program consists of research that supports career development of scientists engaged in computational and experimental research to integrate cutting edge-edge information technology with advanced biological, chemical, and pharmacological sciences in: Genetics (Molecular, Medical-human, Pharmaco, Population); Genomics (Function, Structural, Toxico, Pharmaco, Comparative); Proteomics; Biological pathways. The program provides a research grant of $30,000 per year for up to two years. The 'starter' aspect of the program strives to assist those individuals who are establishing careers as independent investigators in the field of informatics. The program is not offered as a means to augment an ongoing research effort or is the grant intended to be used for any direct effort to obtain further extramural funding. The funds are to be used to conduct the proposed research.
Requirements: Those holding the academic rank of instructor or assistant professor and investigators at the doctoral level with equivalent positions are eligible to apply for these research starter grants, providing their proposed research is neither directly nor indirectly subsidized to any significant degree by an extramural support mechanism. The program is not intended for those in postdoctoral training programs. However, individuals in postdoctoral training scheduled to conclude and who will hold an academic appointment by January 1 may apply. Applicants must be sponsored by the department or unit in which the proposed research is to be undertaken. The grant is made to the university on behalf of the applicant and with the understanding that the university will administer the funds. Applications must be submitted by an accredited U.S. school, and all applicants must be U.S. citizens or permanent residents.
Restrictions: Research projects that extend or develop the proprietary value of specific drug products are not acceptable in this program. This exclusion does not preclude research in which specific drug products are used to test hypotheses that have a general applicability. The funds may not be used for salary support of the grantee, but may be used to support technical assistance. No more than $500 a year may be used for travel to professional meetings by the grantee. Indirect costs are not provided to the institution, and grant funds may not be used for this purpose.
Amount of Grant: $30,000
Contact: Eileen M. Cannon, Executive Director; (202) 572-7756; fax: (202) 572-7799; email: info@phrmafoundation.org
Internet: http://www.phrmafoundation.org/awards/informatics/starter.php
Sponsor: Pharmaceutical Research and Manufacturers of America (PhRMA) Foundation
950 F Street, N.W., Suite 300
Washington, DC 20004

PhRMA Foundation Paul Calabresi Medical Student Research Fellowship **4967**
The fellowship is offered to medical or dental students who have substantial interests in research and teaching careers in pharmacology - clinical pharmacology and who are willing to spend full-time in a specific research effort within a pharmacology or clinical pharmacology unit. Fellowships are available for a minimum period of three months or any period of time up to 24 months. The commitment must be full-time. The student may undertake this investigative effort at his/her own school or at another institution. Requests for a fellowship are to be submitted online by the appropriate representative of the school or university. The fellowship is made to the institution on behalf of the student and the institution must agree to administer the funds. The program offers funding of up to $1,500 per month. It is expected that the funds will be distributed to the fellow in regular payments over the period of the fellowship by the recipient school. The funds may be used towards the stipend, tuition or affiliated fees for the fellow. Fellowships are available for a minimum period of six months or any period of time up to 24 months. However, if the student's award period is 24 months, the maximum stipend is still $18,000.
Requirements: A candidate must be enrolled in a U.S. medical/dental school and have finished at least one year of the school curriculum. Priority consideration will be given to those candidates who project strong commitments to careers in the field of clinical pharmacology. The applicant must be sponsored by the pharmacology or clinical pharmacology program in which the investigative project is to be undertaken. The sponsoring unit must have a demonstrated commitment to training in the fields of pharmacology or clinical pharmacology. Applications must be submitted by an accredited U.S. school, and all applicants must be U.S. citizens or permanent residents.
Restrictions: There are no indirect costs provided to the institution nor may any of the funds be used for this purpose. Once an individual has received a Medical Student Fellowship, that individual may not reapply.
Amount of Grant: up to $18,000
Contact: Eileen M. Cannon, Executive Director; (202) 572-7756; fax: (202) 572-7799; email: info@phrmafoundation.org
Internet: http://www.phrmafoundation.org/awards/clinical/medstudent.php
Sponsor: Pharmaceutical Research and Manufacturers of America Foundation
950 F Street, N.W., Suite 300
Washington, DC 20004

PhRMA Foundation Pharmaceutics Postdoctoral Fellowships 4968
The purpose of the fellowships is to encourage graduates from PhD programs in pharmaceutics to continue to refine their research skills through formal postdoctoral training. Applicants should indicate a strong determination to continue their research careers in pharmaceutics following completion of the fellowship. Applications will be accepted for postdoctoral research training extending over a minimum of one year and a maximum of two years. Each fellowship carries an annual stipend of $40,000 which can be supplemented by the sponsor and/or his/her institution.
Requirements: To be eligible, applicants must either hold a PhD degree in pharmaceutics from a school of pharmacy accredited by the American Council on Pharmaceutical Education or expect to receive such a degree before activating the fellowship. The mentor of the fellow must be in an accredited school of pharmacy. Alternatively, a student with a Ph.D. from a school of pharmacy may use the fellowship to study with a mentor in any science or engineering department. An equally important aspect of eligibility is that suitable facilities for the necessary training and research must be available to the applicant. The program with which the applicant is, or will be, associated must provide an environment where the applicant's potential can be developed to the fullest extent possible.
Restrictions: The award is intended solely as a stipend and may not be used otherwise. If necessary, the institution may supplement the award to a level that is consistent with other postdoctoral fellowships it currently offers. Research projects that extend or develop the proprietary value of specific drug products are not acceptable in this program. The program provides no other subsidies (travel, tuition, fringe benefit costs, etc.) and indirect costs to the institution are not provided.
Amount of Grant: $40,000
Contact: Eileen M. Cannon, Executive Director; (202) 572-7756; fax: (202) 572-7799; email: info@phrmafoundation.org
Internet: http://www.phrmafoundation.org/awards/pharmaceutics/postdoc.php
Sponsor: Pharmaceutical Research and Manufacturers of America (PhRMA) Foundation
950 F Street, N.W., Suite 300
Washington, DC 20004

PhRMA Foundation Pharmaceutics Pre Doctoral Fellowships 4969
To encourage the entry of more qualified individuals into this area, the program aims at supporting promising students during their thesis research by providing assistance in the form of stipend and funds to cover costs incidental to the training. Applications will be accepted for a minimum of one year and a maximum of two years. The program is designed for applicants who expect to complete the requirements for the Ph.D. in pharmaceutics in two years or less from the time the fellowship begins. For the purposes of this program, pharmaceutics includes basic pharmaceutics, biopharmaceutics and pharmaceutical technology. The fundamental aspects of pharmacokinetics are not included since these are covered by PhRMA Foundation awards in pharmacology and toxicology.
Requirements: To be eligible, applicants must hold a B.S., M.S., or Pharm.D. degree in pharmacy or a related area such as chemistry or biology from an accredited school in the U.S. A candidate enrolled in a Pharm.D./Ph.D. program should not be taking required clinical course work or clinical clerkships during the tenure of the fellowship. Also, before an individual is eligible to apply for a PhRMA Foundation award, the applicant must first have a firm commitment from a U.S. university, and must be a U.S. citizen or permanent resident. The program assumes that the fellows will devote full-time (including summers) to their research. The fellowship may not be used as a supplement to funds from other fellowships or assistantships. However, this does not preclude institutions supplementing a fellowship with their own funds in order to bring the stipend within the normal range of the institution.
Restrictions: Research projects which extend or develop the proprietary value of specific drug products are not acceptable in this program. This exclusion does not preclude research in which specific drug products are used to test hypotheses which have a general applicability. Applications, however good, which do not meet the aims of the program will be disapproved.
Contact: Eileen M. Cannon, Executive Director; (202) 572-7756; fax: (202) 572-7799; email: info@phrmafoundation.org
Internet: http://www.phrmafoundation.org/awards/pharmaceutics/predoc.php
Sponsor: Pharmaceutical Research and Manufacturers of America (PhRMA) Foundation
950 F Street, N.W., Suite 300
Washington, DC 20004

PhRMA Foundation Pharmaceutics Research Starter Grants 4970
The purpose of the starter grants is to offer financial support to individuals beginning their independent research careers at the faculty level. The areas of interest within this program are research efforts in the fields of pharmaceutics. The program provides a research grant of $30,000 per year for up to two years. For the purposes of this program, pharmaceutics includes basic pharmaceutics, biopharmaceutics, pharmaceutical technology and pharmaceutical biotechnology. The fundamental aspects of pharmacokinetics are not included since they are covered by PhRMA Foundation programs in pharmacology and toxicology. The 'starter' aspect of the program strives to assist those individuals who are establishing careers as independent investigators in the fields listed above.
Requirements: Those holding the academic, tenure track, rank of instructor or assistant professor and investigators at the doctoral level with equivalent positions are eligible to apply for these research starter grants, providing their proposed research is neither directly nor indirectly subsidized to any significant degree by an extramural support mechanism. It is anticipated that this research experience will occur in an academic and/or industrial laboratory setting. Applications must be submitted by a school of pharmacy accredited by the American Council on Pharmaceutical Education, and all applicants must be U.S. citizens or permanent residents. The grant is made to the university on behalf of the applicant and with the understanding that the university will administer the funds. Only schools of pharmacy are eligible for this award.
Restrictions: The program is not offered as a means to augment an ongoing research effort. Nor is the grant intended to be used for any direct effort to obtain further extramural funding. The funds are to be used to conduct the proposed research. The funds may not be used for salary support of the grantee, but may be used to support technical assistance. No more than $500 a year may be used for travel to professional meetings by the grantee. Indirect costs are not provided to the institution, and grant funds may not be used for this purpose.
Amount of Grant: $30,000
Contact: Eileen M. Cannon, Executive Director; (202) 572-7756; fax: (202) 572-7799; email: info@phrmafoundation.org
Internet: http://www.phrmafoundation.org/awards/pharmaceutics/starter.php
Sponsor: Pharmaceutical Research and Manufacturers of America (PhRMA) Foundation
950 F Street, N.W., Suite 300
Washington, DC 20004

PhRMA Foundation Pharmaceutics Sabbatical Fellowships 4971
The purpose of this program is to enable faculty with active research programs to work outside of their home institution for periods of 6 months to one year to learn new skills or develop new collaborations that will enhance their research and research training capabilities in pharmaceutics. For the purposes of this program, pharmaceutics includes basic pharmaceutics, biopharmaceutics, pharmaceutical technology and pharmaceutical biotechnology. The fundamental aspects of pharmacokinetics are not included since they are covered by PhRMA Foundation programs in pharmacology and toxicology. It is anticipated that this research experience will occur in an academic and/or industrial laboratory setting. The award provides up to $40,000 stipend support for mid-career scientists to engage in an academic year or calendar year experiential program intended to redirect their core research focus to an area of emerging importance to pharmaceutical research and development.
Requirements: Eligible applicants must hold a PhD degree in pharmaceutics from a School of Pharmacy accredited by the American Council on Pharmaceutical Education; hold a faculty appointment that imparts eligibility for a sabbatical leave from their home institution; have institutional approval of a sabbatical plan that includes partial salary that matches the PhRMA stipend; hold an endorsement from a mentor who agrees to sponsor the applicants visiting scientist activity; and be a U.S. citizen or permanent resident. The applicant and mentor of the program must describe how the multidisciplinary goals of the research experiential program will be accomplished and provide assurance that key collaborating mentors endorse and are willing to support the plan. Matching funds must be provided by the home institution.
Restrictions: The program provides no other subsidies (travel, tuition, fringe benefit costs, etc.) and indirect costs to the institution are not provided.
Amount of Grant: $40,000
Contact: Eileen M. Cannon, Executive Director; (202) 572-7756; fax: (202) 572-7799; email: info@phrmafoundation.org
Internet: http://www.phrmafoundation.org/awards/pharmaceutics/sabbatical.php
Sponsor: Pharmaceutical Research and Manufacturers of America (PhRMA) Foundation
950 F Street, N.W., Suite 300
Washington, DC 20004

PhRMA Foundation Pharmacology/Toxicology Post Doctoral Fellowships 4972
The postdoctoral program provides stipend support for individuals engaged in a multidisciplinary research training program that will create or extend their credentials in pharmacology or toxicology. The intent of this program is to support postdoctoral career development activities of individuals prepared (or preparing) to engage in research that integrates information on molecular or cellular mechanisms of action with information on the effect of an agent in the intact organism. It is anticipated that this research training will be accomplished in academic and/or industrial laboratory settings where multidisciplinary teams are organized to integrate informatics, molecular, cell, and systems biology with pharmacology/toxicology research. Funding is for up to 2 years but not beyond the 3rd year of the existing Post Doc.
Requirements: Eligible applicants must either hold a PhD degree or appropriate terminal research doctorate in a field of study logically or functionally related to the proposed post doctoral activities; or expect to receive the PhD before activating the award. They must also have a firm commitment from an accredited U.S. university and be a U.S. citizen or permanent resident. The sponsor (mentor) must describe how the multidisciplinary goals of the research training program will be accomplished and provide assurance that key collaborating mentors endorse and are willing to support the training plan. The award, consisting of a $40,000 annual stipend, is made to the institution on behalf of the fellow. The award is intended solely as a stipend and may not be used otherwise. If necessary, the institution may supplement the award to a level that is consistent with other postdoctoral fellowships it currently offers.
Restrictions: The program provides no other subsidies (travel, tuition, fringe benefit costs, etc.) and indirect costs to the institution are not provided. The second year of this award is contingent upon a progress report approved by the foundation and submission of a financial report.
Amount of Grant: $40,000

Contact: Eileen M. Cannon, Executive Director; (202) 572-7756; fax: (202) 572-7799; email: info@phrmafoundation.org
Internet: http://www.phrmafoundation.org/awards/pharmacology/postdoc.php
Sponsor: Pharmaceutical Research and Manufacturers of America (PhRMA) Foundation
950 F Street, N.W., Suite 300
Washington, DC 20004

PhRMA Foundation Pharmacology/Toxicology Pre Doctoral Fellowships 4973

The program is designed for candidates who expect to complete the requirements for the Ph.D. in the fields of pharmacology or toxicology in two years or less from the time the fellowship begins. The award is made to the university on behalf of the fellow. The fellowship provides a stipend of $20,000 a year payable monthly for a minimum of one year and a maximum of two years. Of the $20,000 awarded annually, up to $500 a year may be used for incidentals directly associated with the thesis research preparation (e.g., secretarial help, artwork, books, travel, etc.). Awards may be activated beginning January 1 or on the first day of any month thereafter, up to and including August 1. The second year of a fellowship is contingent upon certification by the thesis advisor that satisfactory progress is being achieved.
Requirements: The fellowship program of pre doctoral support is designed to assist full-time, in-residence Ph.D. candidates in the fields of pharmacology or toxicology who are enrolled in U.S.A schools of medicine, pharmacy, dentistry or veterinary medicine. The program supports full-time advanced students who will have completed the bulk of their pre-thesis requirements (at least two years of study) and are engaged in thesis research as Ph.D. Candidates by the time the award is activated. Students just starting in graduate school should not apply. In the field of pharmacology or toxicology, highest priority will be given to applications for research that attempts to integrate information on the mechanism of action of a drug or chemical at a molecular or cellular level with the drug's effects in the intact laboratory animal or human, encompassing the potential influences of biochemical, physiological or behavioral systems. The pharmacology/toxicology research and teaching mission of the unit awarding the degree must be apparent as a core mission component on their web site. Applications must be submitted by an accredited U.S. school, and all applicants must be U.S. citizens or permanent residents. The program assumes that the fellows will devote full-time (including summers) to their research. The PhRMA Foundation fellowship may not be used to supplement funds from other fellowships, traineeships or assistantships, unless necessary to make stipend levels compliant with institutional policy. The focus of the program is to assist in the predoctoral training of the candidate. It should not be viewed as a program to fund a research project.
Restrictions: A candidate enrolled in an M.D./Ph.D. program should not be taking required clinical course work or clerkships during the tenure of the fellowship. The foundation will accept only one application per academic institution. Exceptions may be considered for institutions with multiple campuses. Indirect costs are not provided to the institution and PhRMA Foundation grant funds may not be used for this purpose.
Amount of Grant: $20,000
Contact: Eileen M. Cannon, Executive Director; (202) 572-7756; fax: (202) 572-7799; email: info@phrmafoundation.org
Internet: http://www.phrmafoundation.org/awards/pharmacology/predoc.php
Sponsor: Pharmaceutical Research and Manufacturers of America (PhRMA) Foundation
950 F Street, N.W., Suite 300
Washington, DC 20004

PhRMA Foundation Pharmacology/Toxicology Research Starter Grants 4974

The purpose of the program is to offer financial support to individuals beginning their independent research careers at the faculty level. The areas of interest within this program are research efforts in the fields of pharmacology and drug toxicology. The program provides a research grant of $30,000 per year for up to two years. The 'starter' aspect of the program strives to assist those individuals who are establishing careers as independent investigators in the fields listed above. The program is not offered as a means to augment an ongoing research effort. The funds are to be used to conduct the proposed research. This program supports individuals beginning independent research careers in academia who do not have other substantial sources of research. Applicants will be judged on the scientific merit of the proposed research, and on the degree of financial need.
Requirements: Those holding the academic rank of instructor or assistant professor and investigators at the doctoral level with equivalent positions are eligible to apply for these research starter grants, providing their proposed research is neither directly nor indirectly subsidized to any significant degree by a competitive extramural grant. The program is not intended for those in postdoctoral training programs. However, individuals in postdoctoral training scheduled to conclude and who will hold an academic appointment by January 1 may apply. Applicants must be sponsored by the department or unit in which the proposed research is to be undertaken. The sponsoring unit must have responsibility for pharmacology/toxicology teaching and research as part of its core mission. If the pharmacology/toxicology mission is not apparent on the sponsoring unit's web site, the sponsor letter must describe how the academic appointment will support the applicant's career development in pharmacology/toxicology. The grant is made to the university on behalf of the applicant, with the understanding that the university will administer the funds. Schools of medicine (human or veterinary), pharmacy and dentistry are eligible for this award. In the field of pharmacology or toxicology, highest priority will be given to applications for research that attempts to integrate information on the mechanism of action of a drug or chemical at a molecular or cellular level with the drug's effects in the intact laboratory animal or human, encompassing the potential influences of biochemical, physiological or behavioral systems. Applications must be submitted by an accredited U.S. school, and all applicants must be U.S. citizens or permanent residents.
Restrictions: Research projects that extend or develop the proprietary value of specific drug products are not acceptable in this program. This exclusion does not preclude research in which specific drug products are used to test hypotheses that have a general applicability. The funds may not be used for salary support of the grantee, but may be used to support technical assistance. No more than $500 a year may be used for travel to professional meetings by the grantee. Indirect costs are not provided to the institution, and grant funds may not be used for this purpose. Applicants holding multi-year national competitive awards or grants supporting their independent research will not be considered. If a national competitive grant is obtained after the application has been submitted, the Foundation must be so informed.
Amount of Grant: $30,000
Contact: Eileen M. Cannon, Executive Director; (202) 572-7756; fax: (202) 572-7799; email: info@phrmafoundation.org
Internet: http://www.phrmafoundation.org/awards/pharmacology/starter.php
Sponsor: Pharmaceutical Research and Manufacturers of America (PhRMA) Foundation
950 F Street, N.W., Suite 300
Washington, DC 20004

PhRMA Foundation Pharmacology/Toxicology Sabbatical Fellowships 4975

This program provides stipend funding to enable faculty members at all levels with active research programs an opportunity to work at other institutions for periods of six months to one year to learn new skills or develop new collaborations that will enhance their research and research training activities in pharmacology/toxicology. Applicants are expected to engage in multidisciplinary research that integrates information on molecular or cellular mechanisms of action with information on the effect of an agent in the intact organism. It is anticipated that this research experience will occur in academic and/or industrial laboratory settings as part of a multidisciplinary team organized to integrate informatics, molecular, cell, and systems biology with pharmacology/toxicology research. The award provides up to $40,000 stipend support for mid-career scientists to engage in an academic year or calendar year experiential program intended to redirect their core research focus to an area of emerging importance to pharmaceutical research and development.
Requirements: Eligible applicants must hold a PhD degree or appropriate terminal doctorate and record of research accomplishment in a field of study logically or functionally related to the proposed post doctoral activities; hold a faculty appointment that imparts eligibility for a sabbatical leave from their home institution; have institutional approval of a sabbatical plan that includes partial salary that matches the PhRMA stipend; hold an endorsement from a mentor who agrees to sponsor the applicants visiting scientist activity; and be a U.S. citizen or permanent resident. The applicant and mentor of the program must describe how the multidisciplinary goals of the research experiential program will be accomplished and provide assurance that key collaborating mentors endorse and are willing to support the plan. Matching funds must be provided through the university.
Restrictions: The program provides no other subsidies (travel, tuition, fringe benefit costs, etc.) and indirect costs to the institution are not provided.
Amount of Grant: $40,000
Contact: Eileen M. Cannon, Executive Director; (202) 572-7756; fax: (202) 572-7799; email: info@phrmafoundation.org
Internet: http://www.phrmafoundation.org/awards/pharmacology/sabbatical.php
Sponsor: Pharmaceutical Research and Manufacturers of America (PhRMA) Foundation
950 F Street, N.W., Suite 300
Washington, DC 20004

PHSC Publication Grant 4976

The society offers a grant to a PHSC member to aid the publication, in book or monograph form, of original research into Canada's photographic history. A proposal must be submitted outlining the topic, format, print run, cost, and timing, together with a copy of the material and a sample of any previously published work. Proposals may be submitted at any time.
Requirements: Applicants must be PHSC members at the time of application and at the time of publication.
Restrictions: Material should not have been previously published.
Amount of Grant: $C1000 maximum
Contact: Mike Robinson, First Vice-President; email: mike@centurydarkroom.com
Internet: http://www.phsc.ca/awards.html
Sponsor: The Photographic Historical Society of Canada
P.O. Box 54620, RP.O. Avenue, Fairlawn
Toronto, ONT M5M 4N5 Canada

PHSC Research Grant 4977

The society offers a research grant to a PHSC member to promote research into Canada's photographic history. A one-page proposal outlining the research must be submitted. The selection criteria include the extent to which the proposed research expands existing knowledge and the relevance to Canada.
Requirements: Applicants must be PHSC members at the time of application and at the time of publication.
Amount of Grant: $C500 first place; $C250 runner-up
Date(s) Application is Due: Oct 31
Contact: Mike Robinson, First Vice-President; email: mike@centurydarkroom.com

Internet: http://www.phsc.ca/awards.html
Sponsor: The Photographic Historical Society of Canada
P.O. Box 54620, RP.O. Avenue, Fairlawn
Toronto, ONT M5M 4N5 Canada

Physicians Foundation for Health Systems Excellence Grants **4978**
PFHSE seeks to fund projects that match the foundation's funding priorities: practice management—to create and increase efficiency within the practice for delivery of high-quality health care; physician education—to enhance practice-based delivery of health care through the application of technological and clinical developments; and patient safety/disease management—to develop and implement innovative practice strategies aimed at improving the health status and well being of patients. Concept papers, which may be submitted electronically, may be submitted at any time; full proposals are by invitation. Application and guidelines are available online.
Requirements: 501(c)3 tax-exempt organizations are eligible.
Restrictions: Grants are not awarded for unrestricted general operating expenses; payment for services of a fiscal agent; endowment funds; religious purposes; fundraising activities or events; capital expenditures by the grantee; lobbying or political activities; activities related to litigation, arbitration, or other dispute resolution; medical research or animal research; research and/or development of drugs or medical devices; or any activity inconsistent with the foundation's status as a 501(c)3 charitable organization.
Amount of Grant: $100,000-$1 million
Contact: Tracy Little, c/o Grants Management Associates, (617) 399-0417; email: physicians@grantsmanagement.com
Internet: http://www.physiciansfoundation.org/e_intro.htm#funding
Sponsor: Physicians Foundation for Health Systems Excellence
77 Summer Street, Suite 800
Boston, MA 02110-1006

Pi Gamma Mu Graduate Scholarships **4979**
The annual scholarships are awarded to Pi Gamma Mu members for the first or second year of graduate study in all social science fields. Law school is excluded. Application consists of completed form, personal statement, transcripts, and three letters of recommendation.
Requirements: Applicants must be members of the international honor society in social science.
Amount of Grant: $1000-$2000
Date(s) Application is Due: Jan 30
Contact: Executive Director, (316) 221-3128; fax: (316) 221-7124; email: pgm@jinx.sckans.edu
Internet: http://www.pigammamu.org/scholarship.html
Sponsor: Pi Gamma Mu
1001 Millington, Suite B
Winfield, KS 67156

Pi Lambda Theta Research Grants **4980**
Research Grants are awarded in amounts up to $2,500 for clearly conceptualized research proposals concerning education. A research grant proposal can be for an independent project or a part of a larger project, such as a dissertation. Research grants are funded by the Pi Lambda Theta Educational Endowment and the Maxine Northrop Stephanian Fund.
Requirements: Grant available to Pi Lambda Theta members only.
Amount of Grant: $2500 maximum
Date(s) Application is Due: Apr 1; Oct 1
Contact: Pam Todd, Manager; (812) 339-3411; fax: (812) 339-3462; email: pam@pilambda.org
Internet: http://www.pilambda.org/benefits/awards/ScholarshipsAwards.html
Sponsor: Pi Lambda Theta
P.O. Box 6626, 4101 E Third Street
Bloomington, IN 47407-6626

PILF Fellows Program **4981**
The Constitutional and Legislative Policy Institute (COLPI), an affiliate of the Open Society Institute (OSI-Budapest), in collaboration with the Public Interest Law Initiative in Transitional Societies (PILI) at Columbia Law School, invite applications for the fellowship program. Three lawyers will be selected from Central and Eastern Europe and the former Soviet Union for two years of study and practical work experience. Criteria for selection will include the experience of the applicant, the applicant's potential to contribute to the development of the human rights or public interest law field in Eastern Europe and the former Soviet Union, and the suitability of the applicant's proposed role in the nominating NGO. Fellows will reside a total of one year in the United States: one semester of study at Columbia University and two three-month internships. Fellows will return to their home countries after the first year, where they will spend at least one year working with their nominating NGO on human rights/public interest advocacy on a nonprofit basis (providing legal services, litigating test cases, training/educating in human rights, etc.). Upon their selection, fellows will be required to sign an agreement with COLPI/OSI-Budapest according to which he/she will commit to two years in the program (first year in the United States, second year in fellow's home country working with the nominating NGO).
Requirements: Applicants with a strong commitment to human rights or public interest law, a law degree, eligibility for legal practice in his/her country, and proficiency in English may apply. Preference will be given to applicants under 35 years of age.
Date(s) Application is Due: Apr 4
Contact: David Caughlin, Fellowship Program Manager, (212) 851-1060; fax: (212) 851-1064; email: dcaugh@law.columbia.edu
Internet: http://www.pili.org/aboutpili/fellowship.html
Sponsor: Public Interest Law Foundation
435 W 116th Street
New York, NY 10027

Pioneer Fund Grants **4982**
The fund was established to advance the scientific study of heredity and human differences. The research supported looks at the human evolutionary past (human origins), present (individual and group differences), and future (the impact of technology and globalization on human ecology and demography). Grants support specialized, niche projects, which have difficulty attracting funds from government sources or from larger foundations. There are no application forms or deadlines. The fund requires letters from scholars with institutional support, briefly summarizing proposals and the researchers' qualifications. Guidelines are available online.
Requirements: Research institutions are eligible. Researchers are evaluated based on their scholastic background, extent of field research, published writings and citations of those writings, and reputation among peers.
Restrictions: Individuals are not eligible.
Contact: Grants Administrator, (212) 459-4084; email: info@pioneerfund.org
Internet: http://www.pioneerfund.org
Sponsor: Pioneer Fund, Inc
954 Lexington Avenue, Suite 211
New York, NY 10021

Pioneer Hi-Bred Honorary Plant Breeding Fellowships **4983**
The Pioneer Hi-Bred Honorary Plant Breeding Fellowships honor the contributions of 10 outstanding Pioneer plant breeders to the progress of science and agriculture. These fellowships are intended to promote graduate education in plant breeding and genetics, and are expected to attract high-quality students whose future careers will continue to advance the science of plant breeding. A Fellowship Committee established by Pioneer will review applications and select recipients based on their qualifications, including academic excellence; proposed study program area, with consideration given to agronomic crops such as corn, soybeans, canola, sorghum, sunflower, wheat and alfalfa; future professional plans; and leadership and other experiences predicting future success. The funding limit is $25,000 per year for the duration of the fellowship.
Requirements: Scholarship candidates must be nominated and sponsored by a Pioneer Hi-Bred employee.
Amount of Grant: $25,000 maximum per year
Contact: Tabare Abadie; (515) 270-3200; fax: (515) 334-4415; email: Tabare.Abadie@pioneer.com or community.investment@pioneer.com
Internet: http://www.pioneer.com/web/site/portal/menuitem.07196fe7964ae318bc0c0a03d10093a0/
Sponsor: Pioneer Hi-Bred International, Inc.
6900 NW 62nd Avenue, P.O. Box 246
Johnston, IA 50131

Pioneer Hi-Bred International Scholarships **4984**
The Pioneer Hi-Bred International Scholarship is designed to contribute to agricultural research and education by providing financial support for outstanding students studying for graduate level degrees in plant breeding overseas. Funding is considered to be an unrestricted gift. Funding decisions will be reviewed annually and any changes will be communicated by September 30 of each year. For specific details about the program, contact the listed individual by email.
Contact: Peter Freymark; (515) 270-3200 or (800) 247-6803, ext. 3915; fax: (515) 334-4415; email: Peter.Freymark@pioneer.com
Internet: http://www.pioneer.com/web/site/portal/menuitem.07196fe7964ae318bc0c0a03d10093a0/
Sponsor: Pioneer Hi-Bred International, Inc.
6900 NW 62nd Avenue, P.O. Box 246
Johnston, IA 50131

Pioneer Hi-Bred Society Fellowships **4985**
The Pioneer Hi-Bred Society Fellowship is designed to contribute to agricultural research and education by providing financial support for outstanding students studying for graduate level degrees in plant breeding. Leading societies that support plant improvement graduate programs in crops of interest to Pioneer have been selected by the Fellowship Sub-committee. The institutions are responsible for student selection and administering the program. Funded societies include: Agronomic Science Foundation; American Phytopathological Society; American Society of Plant Biologists; the Consultative Group on International Agricultural Research (CGIAR); and the Entomological Foundation.
Requirements: Pioneer prefers this grant to be used solely to fund a student(s) progressing towards a graduate degree in plant breeding from the funded institution. No university overhead will be paid from these funds.
Contact: David Harwood; (515) 270-3200 or (800) 247-6803, ext. 3915; fax: (515) 334-4415; email: David.Harwood@pioneer.com
Internet: http://www.pioneer.com/web/site/portal/menuitem.07196fe7964ae318bc0c0a03d10093a0/
Sponsor: Pioneer Hi-Bred International, Inc.
6900 NW 62nd Avenue, P.O. Box 246
Johnston, IA 50131

Pioneer Hi-Bred University Fellowships 4986

The Pioneer Hi-Bred University Plant Breeding Fellowships are designed to contribute to agricultural research and education by providing financial support for outstanding students working toward graduate level degrees in plant breeding. Leading universities that support plant improvement graduate programs in crops of interest to Pioneer have been selected by the Fellowship Subcommittee. The institutions are responsible for student selection and administering the program. Funded institutions currently include: Cornell University; North Carolina State University; University of Illinois; University of Minnesota; and University of Wisconsin-Madison.
Contact: Stephanie Raber; (515) 270-3200 or (800) 247-6803, ext. 3915; fax: (515) 334-4415; email: Stephanie.Raber@pioneer.com
Internet: http://www.pioneer.com/web/site/portal/menuitem.07196fe7964ae318bc0c0a03d10093a0/
Sponsor: Pioneer Hi-Bred International, Inc.
6900 NW 62nd Avenue, P.O. Box 246
Johnston, IA 50131

Pittsburgh Foundation Medical Research Grants 4987

Through the generosity of individuals and families who have established medical research funds with The Pittsburgh Foundation, the Foundation provides support for this cutting edge research. Based on the donors' fields of interest, the Foundation provides grants for the following: Richard S. Caliguiri Fund - for research in the field of amyloidosis; The Walter L. Copeland Fund - for cranial research (disease of the brain/neurological disease), to be conducted within the Department of Neurological Surgery at the University of Pittsburgh; John F. and Nancy A. Emmerling Fund - for research in the field of neuropsychobiology (with emphasis on biochemistry) of mental disorders, with the goal of discovering the causes and cure, of amelioration, of mental disorders. The principal investigator must have full time status at the University of Pittsburgh; Albert B. Ferguson, M.D. Orthopaedic Fund - for orthopaedic research; GLBT Health and Wellness Fund - to assess the incidence, cause or cure of certain diseases (i.e., cancer, heart disease, diabetes, arthritis, AIDS and Alzheimer's disease) in the Gay, Lesbian, Bisexual and Transgender (GLBT) community; Medical Research Funding - for research in the field of cancer, heart disease, arthritis, diabetes, multiple sclerosis and geriatric diseases (with focuses particularly on Alzheimer's disease and other related diseases of the elderly).
Requirements: Pennsylvania not-for-profit institutions with 501(c)3 tax-exempt status are invited to apply for funding. Support is given to investigators in Allegheny County with a proven record of performance in the area under investigation. Guidelines and applications vary, see: http://www.pittsburghfoundation.org/page32428.cfm for a complete listing of each grant.
Geographic Focus: Pennsylvania
Contact: Judy Powell, Administrative Support; (412) 394-4294; email: powellj@pghfdn.org
Internet: http://www.pittsburghfoundation.org/page32428.cfm
Sponsor: Pittsburgh Foundation
5 PPG Place, Suite 250
Pittsburgh, PA 15222-5414

PKD Foundation Lillian Kaplan International Prize for the Advancement in the Understanding of Polycystic Kidney Disease 4988

The prize is a joint venture between the foundation and the International Society of Nephrology. The winning nominee of the initial award received the prize in Berlin; it has been awarded every other year since. Each award consists of a monetary award, a framed certificate with an appropriate citation, and a designated sculpture to clinicians and scientists who have made outstanding discoveries in basic science or who have made outstanding clinical contributions to the care of individuals. This is the largest cash prize for a clinical medical specialty in the world signaling the prestige and importance it carries. PKD experts from both organizations judge nominations.
Requirements: Awards are made to residents of any country without restriction to gender, race, religion, creed, or nationality. Nomination letters should describe the key role of the nominee(s) in a biomedical advance relevant to PKD and should be accompanied by a full curriculum vita with a complete mailing address and at least two supporting letters that detail a major discovery or sustained advance. If two individuals are nominated to share the prize, the reasons for why the prize would be shared must be clearly delineated.
Samples: Dr. Corinne Antignac, Necker Hospital, Paris, France, $50,000 (2009); Dr. Lisa Guay-Woodford, University of Alabama at Birmingham, $50,000 (2009); Dr. Friedhelm Hildebrandt, University of Michigan, $50,000 (2009).
Amount of Grant: $50,000
Date(s) Application is Due: Feb 1
Contact: Dan Lara, Government Relations Manager; (816) 268-8480; fax: (816) 931-8655; email: larad@pkdcure.org or pkdcure@pkdcure.org
Internet: http://www.pkdcure.org
Sponsor: PKD Foundation
9221 Ward Parkway, Suite 400
Kansas City, MO 64114-3367

PKD Foundation Research Grants 4989

The PKD Foundation, established in 1982 by Joseph H. Bruening and Jared J. Grantham, leads the fight against polycistic kidney disease through vital research funding and patient education. Annually, the PKD Foundation funds more than $4.2 million in vital PKD research. Most recently, clinical trials in cardiovascular health, Rapamycin, Somatostatin, Tolvaptan, and more have offered help and hope to the 12.5 million worldwide with the disease.
Amount of Grant: $25,000-$1,000,000
Contact: Dan Lara, Government Relations Manager; (816) 268-8480; fax: (816) 931-8655; email: larad@pkdcure.org or pkdcure@pkdcure.org
Internet: http://www.pkdcure.org/Research/tabid/82/Default.aspx
Sponsor: PKD Foundation
9221 Ward Parkway, Suite 400
Kansas City, MO 64114-3367

Plan for Social Excellence Grants 4990

Projects supported by the foundation take place at various levels including early childhood education, technology, after-school mentoring, and higher education. Additionally, the foundation supports programs that are fluid and responsive to the needs of individual schools and communities rather than programs that attempt to address these needs through a system-wide process of reform. Fields of interest include: Grants to schools or other forms of educational institutions for innovative projects; Publication of monographs and videotapes related to the work of the foundation; Advocacy on behalf of children in selected issues; and, Support of research in specific education areas.
Requirements: Projects desiring consideration for funding must be for either the replication of Plan's projects or new initiatives. If not previously invited to apply via an RFP, submit a proposal for the funding of a project or program that is in line with the vision and the mission of the foundation. Review the guidelines posted on the website for clarification. NOTE: The deadline for proposals to be funded during the following school year is by 5:00 P.M. EST on the first Friday in May.
Restrictions: The foundation does not make operating grants.
Date(s) Application is Due: May 4
Contact: Dr. Mario J. Pena, Executive Director; (813) 282-1966; fax: 813-282-1679; email: mariopena@pfse.org
Sponsor: Plan for Social Excellence
2502 Rocky Point Drive, Suite 880
Tampa, FL 33607

Plastic Surgery Educational Foundation Basic Research Grants 4991

The Plastic Surgery Educational Foundation and the American Society for Aesthetic Plastic Surgery offer annual research grants to provide seed money to U.S. and Canadian surgeons for research projects in the field of plastic and reconstructive surgery. Nonmembers of PSEF must submit a letter of sponsorship from a foundation member.
Requirements: Applicant must be a plastic surgeon, MD, or PhD working in plastic surgery.
Samples: Dr. James Anthony, U of California (San Francisco, CA)—for research on wound cytokine levels, $4600. Dr. Steven Buchman, U of Michigan (MI)—for research on cranial suture closure, $4998. Dr. James Doyle, U of Massachusetts Medical Ctr (MA)—for research questioning IL-10 as an immunosuppressive factor in chronic wound fluid, $5000.
Amount of Grant: $5,000
Date(s) Application is Due: Jan 17
Contact: Carol V. Wargo, Director of Corporate Relations; (847) 228-3358 or (847) 228-9900; fax: (847) 228-9131 or (847) 228-0628; email: cvw@plasticsurgery.org or cschmieden@plasticsurgery.org
Internet: http://www.plasticsurgery.org/medical_professionals/research/Grant-Applications.cfm
Sponsor: Plastic Surgery Educational Foundation
444 E. Algonquin Road
Arlington Heights, IL 60005

Plastic Surgery Educational Foundation Research Fellowship Grants 4992

The Research Fellowship Grants are awarded for the purpose of encouraging research and academic career development in Plastic Surgery. The PSEF Research Fellowship is for training in any area of Plastic Surgery and is designed for those who wish to supplement one year of clinical training for a research experience. The PSEF/KCI Wound Care Research Fellowship is open to plastic surgery residents and recent plastic surgery resident graduates focusing on a project in wound care management.
Samples: Dr. Emmanuella Joseph (Boston, MA)—senior award winner for clinical research, $3000. Dr. Jayesh Panchal (Oklahoma City, OK)—junior award winner for clinical research, $3000. Dr. W.P. Andrew Leww (Boston, MA)—senior award winner for basic science, $3000. Dr. Vaishali Doolabh (Saint Louis, MO)—junior award winner for basic science, $3000.
Contact: Carol V. Wargo, Director of Corporate Relations; (847) 228-3358 or (847) 228-9900; fax: (847) 228-9131 or (847) 228-0628; email: cvw@plasticsurgery.org or cschmieden@plasticsurgery.org
Internet: http://www.plasticsurgery.org/medical_professionals/research/Grant-Applications.cfm
Sponsor: Plastic Surgery Educational Foundation
444 E. Algonquin Road
Arlington Heights, IL 60005

Plastic Surgery Educational Foundation/AAO-HNSF Combined Grant 4993

The PSEF/AAO-HNSF Combined Grant is designed to promote interdisciplinary research by encouraging a collaborative research effort between the specialties of Plastic Surgery and Otolaryngology. Recipients of this grant must demonstrate a potential for excellence in research and training and a commitment to an academic research career in either Plastic Surgery or Otolaryngology.
Requirements: Applicants must have demonstrated a potential for excellence in research and teaching and serious commitment to an academic research career in otolaryngology

or plastic surgery. Priority will be given to senior residents, fellows or faculty who have completed residencies or fellowships within four years of the application receipt date.
Amount of Grant: $20,000 maximum
Date(s) Application is Due: Dec 15
Contact: Carol V. Wargo, Director of Corporate Relations; (847) 228-3358 or (847) 228-9900; fax: (847) 228-9131 or (847) 228-0628; email: cvw@plasticsurgery.org or cschmieden@plasticsurgery.org
Internet: http://www.plasticsurgery.org/medical_professionals/research/Grant-Applications.cfm
Sponsor: Plastic Surgery Educational Foundation
444 E. Algonquin Road
Arlington Heights, IL 60005

Ploughshares Fund Grants **4994**
Grants address the significant proliferation risks in emerging nuclear technologies through research projects and education efforts directed to policymakers, the media, and the public. Types of support include conferences and seminars, continuing support, emergency support, general operating support, project development, publication grants, research grants, and seed grants. Specific grant programs are directed to ending the threat from nuclear, chemical, and biological weapons; stopping the spread of weapons of war; addressing the environmental legacy of the nuclear age; promoting public understanding and participation; and preventing global and regional conflict. Ploughshares has initiated media programs including radio talk shows, op-ed writing services, and seminars for reporters to inform the public and influence policymakers on related issues. In an effort to prevent conflict, the foundation supports peacekeeping efforts and research, public education, and advocacy on nuclear weapons and security issues.
Requirements: U.S. and international nonprofit organizations and individuals may apply.
Amount of Grant: $4.5 million total; $500-$50,000
Date(s) Application is Due: Feb 15; Jul 15
Contact: Executive Director, (415) 775-2244; fax: (415) 775-4529; email: proposals@ploughshares.org
Internet: http://www.ploughshares.org/grants.php?a=2&b=0&c=0
Sponsor: Ploughshares Fund
Fort Mason Ctr, Building B, Suite 330
San Francisco, CA 94123

Pollock-Krasner Foundation Grants **4995**
The foundation seeks to find artists of recognized merit throughout the world who have current financial need, whether professional or personal, for whom grants will assist in pursuing their careers. The amount of the grant is determined by the individual circumstances of the artist. The foundation's mission is to aid individuals who have worked as professional artists over a significant period of time. The foundation cannot respond to application requests by telephone or in person. Write to the office for guidelines. Applications are accepted anytime.
Requirements: The foundation welcomes applications from painters, sculptors, printmakers, and artists who work on paper.
Restrictions: The foundation will not accept applications from commercial artists, photographers, video artists, filmmakers, craftmakers, or any artist whose work primarily falls into these categories. Grants are not made to students nor to support academic The Foundation does not make grants to pay for past debts, legal fees, the purchase of real estate, moves to other cities, or to pay for the costs of installations, commissions, or projects ordered by others. With very few exceptions, the foundation will not fund travel expenses. study.
Samples: To be divided among 107 visual artists worldwide—for fellowships to support their personal and professional expenses, $2.069 million (2004); To be divided among 154 artists worldwide—for fellowships for visual artists, $2.23 million (2003).
Contact: Program Officer, (212) 517-5400; fax: (212) 288-2836; email: grants@pkf.org
Internet: http://www.arts.state.va.us/pkfgrants.htm
Sponsor: Pollock-Krasner Foundation
863 Park Avenue
New York, NY 10021

Polytechnic University International Research Studentships **4996**
Research Studentships support full-time PhD or MPhil students. The period of the studentship is decided by the department according to the student's ability, academic background, or experience. Recipients spend up to six hours of training each week within their department. The training involves assisting teaching and scholarly activities and is relevant to students' programs of study.
Requirements: MPhil and PhD students admitted to the university are eligible.
Restrictions: Part-time students are not eligible.
Amount of Grant: HK$12,800 monthly stipend
Contact: Research, 852-2766-5025; fax: 852-2355-7651; email: roro@inet.polyu.edu.hk
Internet: http://www.me.polyu.edu.hk/virtual/Acad_prog/mphil.htm#Studentships
Sponsor: Polytechnic University
Chung Sze Yuen Building, Room AG328
Hung Hom, Kowloon Hong Kong

Posen Foundation Grants for the Study of Secular Judaism **4997**
These grants are intended to cultivate and support the study of secular Judaism within already well-established university programs and departments of Judaic Studies, and other departments or programs in the humanities, arts or the social sciences (e.g., philosophy, literature, sociology, anthropology, etc.) Grants will be awarded to support the curricular development and teaching of two to four courses per year in the history, texts, philosophy and practices of secular Judaism. Upon review, these grants are renewable for up to two years. Guidelines are available online.
Samples: U of Miami (Coral Gables, FL)—for new courses of study in secular Judaism, $50,000 (2004); Temple U (Philadelphia, PA)—for new courses of study in secular Judaism, $50,000 (2004).
Amount of Grant: $50,000 per year
Date(s) Application is Due: Nov 28
Contact: Myrna Baron, (212) 564-6711 ext 301; fax: (212) 564-6721; email myrna@culturaljudaism.org
Internet: http://www.culturaljudaism.org/ccj/grants
Sponsor: Center for Cultural Judaism
224 W 35th Street, Suite 410
New York, NY 10001

Postdoctoral Research Fellowships in Gerontological Public Health **4998**
The Center for Research on Health and Aging (CRHA) at the university offers postdoctoral fellowships in gerontological public health. These are two- to three-year appointments sponsored by the National Institute on Aging. This program focuses on developing research skills through mentoring, a comprehensive curriculum, hands-on research experience, presentations and publications, and opportunities for grantsmanship. The goal of the program is to foster and hone the research skills of future faculty and academic researchers in health and aging. This is an integrated multidisciplinary program with academic faculty and researchers in public health, gerontology, geriatrics, epidemiology, biostatistics, psychology, sociology, disability and human development, medicine and occupational therapy. Each postdoc fellow will be matched with a mentor from the Training Program Core Faculty. This primary mentor will work with the fellow to guide and collaborate on research projects. The fellowship includes a salary, with reimbursement for health insurance and yearly travel support. Additional financial support is also available. All fellows will have office space, a computer, phone, and access to university resources through the CRHA. Applications are accepted at any time for open positions. Guidelines are available online.
Requirements: Candidates should have received (or be very close to receiving) a doctoral degree in public health, gerontology, sociology, medicine, disability studies, or a related discipline.
Amount of Grant: $35,568 per year
Contact: Dr. Thomas Prohaska, Center for Research on Health and Aging, (312) 413-9830; fax: (312) 413-9835; email: prohaska@uic.edu
Internet: http://www.uic.edu/depts/ovcr/hrpc/centers/fellow_gphpost.html
Sponsor: University of Illinois at Chicago
1747 W Roosevelt Road
Chicago, IL 60608

Pott Foundation Grants **4999**
Established in 1963 by Herman T. Pott, The Pott Foundation is a non-profit organization supporting children, education and health and human services. Most of the grant recipients are located in Herman Pott's adopted city of St. Louis, Missouri. The Foundation distributes approximately $1.5 million annually to over 100 charities.
Requirements: Non-profit 501(c)3 organization are eligible to apply in Missouri. Mail grant applications to: The Pott Foundation, C/O U.S. Bank, N.A., The Private Client Reserve, Attn: Carol Eaves, Mail Loc: SL-MO-CTCS, 10 North Hanley, Clayton, MO 63105.
Geographic Focus: Missouri
Date(s) Application is Due: Apr 1
Contact: Carol Eaves, U.S. Bank, N.A.; (314) 418-8317;
Sponsor: The Pott Foundation
10 North Hanley
Clayton, MO 63105

Potts Memorial Foundation Grants **5000**
The Potts Memorial Foundation is a private foundation, incorporated in 1922 in New York. The Foundation gives on a national and international basis. The foundation was established to provide for the care, treatment, and rehabilitation of persons afflicted with tuberculosis; support for tuberculosis eradication, including fellowship programs for physicians. Grants given for tuberculosis and related disorders.
Requirements: Application form not required, Applicants should submit a detailed description of project and amount of funding requested. Proposal must be submitted via mail.
Restrictions: No grants to individuals, or for endowment funds or matching gifts; no loans.
Samples: Johns Hopkins University, Baltimore, Maryland, $41,800 (2007); Columbia Memorial Hospital, Hudson, New York, $20,000 (2007);
Amount of Grant: $15,000 - $45,000 range
Date(s) Application is Due: Apr 15; Sep 15
Contact: Charles Inman, Secretary, (518) 828-3365
Sponsor: Potts Memorial Foundation
444 Warren Street
Hudson, NY 12534-2415

Poverty and Race Research Action Council Grants **5001**
The council, a national organization convened by major civil rights, civil liberties, and antipoverty groups, awards grants to link social science research to advocacy work to address successfully problems at the intersection of race and poverty. PRRAC funds

and commissions research tied to local, state, and national advocacy strategies; enhances communications between advocates and social science researchers in order to coordinate strategy in the race and poverty area; and disseminates ideas and materials focusing on issues of race and poverty. There is no application form. To apply, submit a title for and one-paragraph abstract of the proposal, detailed description of the research and research methodology, resumes(s) of researcher(s), time line for the work, detailed budget, detailed description of the advocacy work the research will support, how the project focuses on race and poverty, and statement as to the replicability of the research and/or advocacy work.
Requirements: Applications are welcome from advocate/social science research teams; advocacy groups—those with the capacity to carry out the research themselves, as well as those that need outside research help; and social science researchers at universities, research centers, etc.
Samples: Legal Services for New York City (New York, NY)—to establish a right to counsel for low-income tenants involved in eviction proceedings, $9000.; Vietnamese Community of Orange County (Santa Ana, CA)—to conduct a survey to advocate for services for its largely immigrant clientele, $3600.; Food Research and Action Ctr (Washington, DC)—to redefine the federal poverty line and advocate for changes in the food stamp program that will produce higher benefits for poor people, $10,000.
Amount of Grant: $10,000 maximum
Contact: Chester Hartman, (202) 387-8023; fax: (202) 387-0764; email: chartman@prrac.org
Internet: http://www.prrac.org/grants.php
Sponsor: Poverty and Race Research Action Council
3000 Connecticut Avenue NW, Suite 200
Washington, DC 20008

Poverty, the Underclass, and Public Policy Research and Training Postdoctoral Fellowships **5002**
This program, supported by the Ford Foundation and administered by the University of Michigan School of Social Work, enables outstanding American minority scholars in the social sciences the opportunity to spend a year in Ann Arbor at the university conducting research and pursuing extensive training. Fellows will participate in a year-long seminar on poverty, the underclass, and public policy, and conduct their own research while doing so. Annual deadline dates may vary; contact program staff for exact dates.
Requirements: Applicants must have completed their PhD by August 31 of the current year.
Amount of Grant: $44,000 stipend per year
Date(s) Application is Due: Jan 13
Contact: Program on Poverty, the Underclass, and Public Policy, Poverty Research and Training Center, (734) 998-8515; fax: (734) 998-8516; email: pswp@umich.edu
Internet: http://www.fordschool.umich.edu/poverty/fellowship.htm
Sponsor: University of Michigan
1015 E Huron Street, University of Michigan
Ann Arbor, MI 48104-1689

Prader Willi Syndrome Association Research Grants **5003**
The program awards grants to scientific researchers for projects aimed at discovering and developing treatments, cures, and technologies benefiting individuals with Prader-Willi syndrome (PWS). The foundation is open to insightful grants regarding PWS of any sort, but has identified health issues of meritorious research benefit, including obesity, respiratory, and gastrointestinal health issues. Funding may be used for technical support, supplies, equipment, and appropriate travel. Only minimal salary support for faculty level investigators will be provided. The listed application deadline is for letters of intent. Guidelines are available online.
Restrictions: Projects with limited direct connection to therapies that could make a difference for the PWS community will not be considered.
Amount of Grant: $50,000 per year for two years
Date(s) Application is Due: Jun 1
Contact: Grants Administrator, (800) 926-4797 or (941) 312-0400; fax: (941) 312-0142; email: national@pwsausa.org
Internet: http://www.pwsausa.org/research/research_funding_opportunities.htm
Sponsor: Prader Willi Syndrome Association U.S.A
5700 Midnight Pass Road, Suite 6
Sarasota, FL 34242

Praxair Foundation Grants **5004**
The foundation supports nonprofit organizations primarily in the communities where Praxair operates. Preference is given to projects in the following areas: higher education—science, engineering, and environmental studies and programs that encourage minority students to pursue studies in these areas; community service—United Way (matching employee donations), public Libraries (technology upgrading, acquiring new materials, or initiating innovative programs), employee volunteer grants (grants to nonprofit organizations where employees volunteer); public policy and the environment; and international organizations operating in countries where Praxair is expanding its presence, primarily in Asia. All proposals should describe projects that serve the needs of the community at large. Special attention will be given to innovative programs that can be started and completed within one year of the award. Goals should be clearly defined and measurable. The level of support varies depending upon the nature of the project and how closely it fits with foundation guidelines. To apply for a grant, complete the Praxair Foundation Funding Request form.
Requirements: 501(c)3 tax-exempt organizations are eligible.
Amount of Grant: $1,000-$50,000 average
Contact: Grants Administrator; (800) 772-9247; fax: (716) 879-2040
Internet: http://www.praxair.com/praxair.nsf/AllContent/1E607B9456292F1185256BF800779733?OpenDocument
Sponsor: Praxair Foundation
39 Old Ridgebury Road
Danbury, CT 06810

Press Ganey Best Practices Research Program Grants **5005**
The purpose of the program is to increase the systematic and rigorous study of satisfaction in the healthcare setting, through the funding of applied research that will identify best practices to be used throughout the healthcare industry.
Amount of Grant: $10,000
Date(s) Application is Due: Mar 14; Oct 1
Contact: Kelly Leddy, Research Specialist, (800) 232-8032; fax: (574) 232-3485; email: kleddy@pressganey.com
Internet: http://www.pressganey.com/news/grant_program.php
Sponsor: Press Ganey Associates
404 Columbia Pl
South Bend, IN 46601

Presser Foundation Grants **5006**
The foundation awards grants to college and university music departments to support scholarships, fellowships, and programs or projects. The program seeks to increase music education in institutions of learning and to popularize the teaching of music as a profession. Types of support include building construction/renovation, equipment acquisition, fellowships, matching/challenge grants, program development, scholarship funds, and seed grants. Grants to individuals are limited to emergency aid to worthy music teachers in need. Application forms are available for financial aid to needy music teachers and for scholarships. There are no application deadlines.
Requirements: Institution must have a minimum of 40 undergraduate music majors. Candidate must take at least one-third course credits in subjects not related to music.
Samples: Philadelphia Orchestra (PA)—to support education and community-outreach projects designed to engage and expand its audience, $500,000 over five years (2005).
Contact: Edith Reinhardt, President, (610) 658-9030
Sponsor: Presser Foundation
385 Lancaster Avenue, No 205
Haverford, PA 19041

Princess Grace Foundation Theater, Dance, and Film Grants **5007**
Grants are awarded to individuals, through their affiliation with nonprofit organizations, in the areas of dance, theater, and film. The awards are made on an annual basis by application for scholarships, fellowships, and apprenticeships. Scholarships for theater and dance cover tuition costs. Scholarships for film cover production costs. Emerging artists, fellowships, and apprenticeships are for salary assistance. The March 31 application deadline is for theater and playwriting, April 30 for dance, and June 1 for film. Application and guidelines are available online.
Requirements: Applicants must be studying with a U.S. company and must be U.S. citizens or have permanent resident status.
Samples: Terese Capucilli, dance fellowship recipient (2003); Stephen Hillenburg, film scholarship recipient (2003); Tony Kushner, theater fellowship recipient (2003).
Amount of Grant: $5000-$25,000 average
Date(s) Application is Due: Mar 31; Apr 30; Jun 1
Contact: Toby Boshak, Executive Director, (212) 317-1470; fax: (212) 317-1473; email: pgfusa@pgfusa.com
Internet: http://www.pgfusa.com/awards/grants/index.html
Sponsor: Princess Grace Foundation—U.S.A
150 E 58th Street, 21st Fl
New York, NY 10155

Princeton Society of Fellows in the Liberal Arts Postdoctoral Fellowships **5008**
The Princeton Society of Fellows invites applications for three postdoctoral fellowships. Open Fellowship—open to all disciplines represented in the Society of Fellows. The successful candidate will teach one semester of the first year in an interdisciplinary, team-taught course devoted to Western culture; Fellowship in the History of the Book—fellow will teach and pursue research in the history of books, their making, their diffusion or their interpretation, in any language or period; and Fellowship in the Study of Race and/or Ethnicity—fellow will pursue research that will make a significant contribution to the study of race and/or ethnicity in one or more selected disciplines in the humanities and social sciences, and contribute to public discourse on issues of race and/or ethnicity. Guidelines are available online.
Requirements: U.S. citizens and noncitizens, regardless of race, national origin, religion, gender, age, disability, or sexual orientation, are eligible to apply. Candidates must have received their PhD degree between January 1, 2003, and October 3, 2005.
Restrictions: Recipients of doctorates in Education and Jurisprudence, and holders of graduate degrees from Princeton University, are not eligible.
Amount of Grant: $62,000 approximately
Date(s) Application is Due: Oct 3
Contact: Fellowships Administrator, (609)258-4717; fax: (609) 258-2783; email: fellows@princeton.edu
Internet: http://www.princeton.edu/~sf
Sponsor: Princeton University
Joseph Henry House
Princeton, NJ 08544

Princeton University Whiting Postdoctoral Fellowships 5009

The program awards two postdoctoral fellowships to humanities scholars early in their careers to undertake and/or complete a major project, interact with a group of diverse postdoctoral fellows and regular faculty, and contribute to the undergraduate teaching program at the university. Fellows are appointed to a home department during the one-year tenure and will be expected to be in residence spending most of their time on their own research and to present a seminar on their work at least once during the year in their department of appointment. In addition, fellows are expected to teach one undergraduate course during the academic year. The fellowship may be renewable for one additional year. Interested applicants should send, in addition to a letter of application, a current curriculum vita, including educational background, professional experience, and prior publications; an abstract of their completed PhD dissertation, a two-page statement of their proposed research and writing project for the fellowship; and two confidential letters of recommendation, one from their dissertation supervisor and one from another established scholar in the field.

Requirements: Scholars who have been awarded a PhD from a U.S. institution of higher education between January 1, 2001, and January 1 2003, in the fields of architecture (history, theory, and criticism), art history, language and literature, history (including history of science), East Asian Studies, Near Eastern Studies, philosophy (including political philosophy and classical philosophy), and religious studies are eligible.

Amount of Grant: $37,000 plus faculty benefits

Date(s) Application is Due: Feb 15

Contact: Sandy Sussman, (609) 258-3035; email: ssussman@princeton.edu

Internet: http://gradschool.princeton.edu/financial/fellowships/competitive/honorifics/#comp000044bf59cc0000001c223c1d

Sponsor: Princeton University

The Graduate School, Clio Hall

Princeton, NJ 08544

Program in Law and Public Affairs Fellows Program 5010

The program will appoint teachers, scholars, lawyers, and judges as fellows for the upcoming academic year. Fellows devote an academic year or a semester in residence at Princeton to research, discussions, and scholarly collaborations concerned with when and how legal systems, practices, and concepts contribute to justice, order, individual well-being, and the common good. Some fellows will have the opportunity to teach. Fellows from academic institutions normally receive up to one-half their academic year salaries for the appointment period. A supplement may be paid to fellows who teach a course.

Requirements: Applicants should have a doctorate or a professional postgraduate degree.

Date(s) Application is Due: Nov 3

Contact: Program in Law and Public Affairs; (609) 258-5626; fax: (609) 258-0922; email: lapa@princeton.edu

Internet: http://lapa.princeton.edu/fellowships.php

Sponsor: Princeton University

Wallace Hall

Princeton, NJ 08544

Project AWARE Foundation Grants 5011

The foundation's grant program supports worthwhile aquatic conservation projects. Grants are awarded to projects that have a direct benefit to the aquatic environment such as: public education (formal and informal); grassroots conservation and enhancement projects; environmentally focused research that leads to conservation measures; public awareness initiatives; environmental assessment and monitoring projects; and volunteer-supported community activism. Micro grants provide funding for localized grassroots efforts, including but not limited to, beach and underwater cleanups, mooring buoy programs, research projects, local education, and public awareness programs. Macro grants support large-scale regional, national, or international efforts, including but not limited to, broad-based education and public awareness programs and high profile conservation projects. Quarterly board meetings are typically scheduled during the third week of the month. The program operates in the United States, Australia, and the United Kingdom.

Restrictions: Grants do not support legislative advocacy to influence policy; political campaigns; projects whose methods are not environmentally accepted; overhead expenses including salaries, stipends, benefits or tuition; capital expenses including computer hardware/software or office furnishings; travel and living expenses; dive equipment or instruction; or products designed for resale.

Amount of Grant: $1000 maximum for micro grants; $10,000 maximum for macro grants

Date(s) Application is Due: Feb 15; May 15; Aug 15; Nov 15

Contact: Grants Administrator, email: information@projectaware.org

Internet: http://www.projectaware.org/americas/english/grants.asp

Sponsor: Project AWARE Foundation

30151 Tomas Street

Rancho Santa Margarita, CA 92688

Prostate Cancer Foundation Research Grants 5012

The heart of the Prostate Cancer Foundation's effort to to find better treatments and a cure for recurrent prostate cancer is its research awards programs. Each year, the PCF invests millions of dollars in research projects that have the greatest potential to advance the fight against prostate cancer. The PCF's research awards program has three components: Competitive Awards, Special Awards and Donor-Directed Awards. The foundation favors research applications focused on translational and clinical programs with potential for near term benefit for patients with recurrent prostate cancer. The listed application deadline is for competitive awards. Application guidelines and form are available online.

Requirements: Applications from individuals with or without institutional affiliations will be considered.

Restrictions: For-profit organizations or government agencies are not eligible for consideration.

Amount of Grant: $25,000-$100,000

Date(s) Application is Due: Dec 2

Contact: Grants Administrator, (310) 570-4700; fax: (310) 570-4701

Internet: http://www.prostatecancerfoundation.org/site/pp.asp?c=itIWK2OSG&b=6879C

Sponsor: Prostate Cancer Foundation

1250 Fourth Street, Suite 360

Santa Monica, CA 90401

Province of Alberta Queen Elizabeth II Graduate Scholarships 5013

Administered by Alberta Scholarship Programs, this scholarship commemorates the Queen's visit to Alberta in May of 2005. Applicants are selected by their Alberta institution. Contact the faculty of graduate studies at your institution for more information about the nomination process.

Requirements: Candidates for the scholarships or fellowships must be enrolled full-time in a faculty of graduate studies in Alberta. Candidates must be either Canadian citizens or landed immigrants at the time of taking up the award.

Amount of Grant: $10,800 per year masters level; $15,000 per year doctoral level

Contact: Scholarships and Bursaries; (780) 427-8640 (dial 310-0000 to connect toll-free in Alberta); fax: (780) 427-1288; email: scholarships@gov.ab.ca

Internet: http://www.alis.gov.ab.ca/scholarships/info.asp?EK=39

Sponsor: Alberta Heritage Scholarship Endowment Fund

Capital Health Centre, South Tower 12th Floor, 10030-107 Street

Edmonton, AB T5J 3E4 Canada

Pulliam Journalism Fellowships 5014

Each year, CNI grants 20 fellowships covering a 10-week specialized internship program, either at the Indianapolis Star or the Arizona Republic. Fellows work as reporters or copy editors. Newsroom experience is enhanced by a series of individual and group sessions conducted by a writing critic. The program is balanced with seminar sessions led by reporters, editors, and executives of the host newspapers; noted journalists from other newspapers; wire services and syndicates; as well as specialists on economics, law, and newspaper design. Applications are accepted between November and February.

Requirements: Applicants must have received their undergraduate degree between August and June prior to the summer of the work-study program and must be newspaper journalism majors or liberal arts majors with part-time or summer newspaper experience. Graduates who plan to continue academic work toward advanced degrees are welcome, but evidence must be shown that the advanced degree will be applied in a newspaper career.

Amount of Grant: $6500

Date(s) Application is Due: Nov 15

Contact: Russell Pulliam, Editor, The Indianapolis Star, (317) 444-6001; email: russell.pulliam@indystar.com; Bill Hill (Phoenix), (602) 444-4368

Internet: http://www2.indystar.com/help/jobs/pjf

Sponsor: Central Newspapers Inc

P.O. Box 145

Indianapolis, IN 46206-0145

QLT Corporate Sponsorships 5015

Each year, corporate sponsorships support requests that are aligned with corporate objectives and fall into the following priority areas: local, national, and global health organizations or service agencies— ophthalmology, oncology, and dermatology research; science education—programs that encourage British Columbian K-12 students to explore science as a career, and university science education; community—in Vancouver, improving quality of life for the underprivileged or disadvantaged through social, athletic, artistic or cultural endeavors; and United Way—employee fundraising campaign and matching funds raised dollar for dollar. Guidelines and application are available online.

Samples: Johns Hopkins U, School of Medicine (Baltimore, MD)—to endow a professorship in ophthalmology at the Wilmer Eye Institute, $2 million (2004).

Contact: Corporate Communications , (800) 663-5486 or (604) 707-7000; fax: (604) 707-7179; email: corpcomm@qltinc.com

Internet: http://www.qltinc.com/Qltinc/main/mainpages.cfm?InternetPageID=4

Sponsor: QLT Inc

887 Great Northern Wy

Vancouver, BC V5T 4T5 Canada

Quest for Truth Foundation Grants 5016

Grants are awarded through organizations to support research and publication of papers on history, geography, politics, economics, medical, sociology, and related subjects, without restriction as to political jurisdiction, for the education of the reading public. An application form is required and should be requested from the address shown. Applications are accepted at any time.

Requirements: Pacific Northwest 501(c)3 tax-exempt organizations or organizations with private foundation status are eligible.

Restrictions: The foundation does not award scholarships or internships.

Geographic Focus: Idaho; Montana; Oregon; Utah; Washington

Amount of Grant: $5000-$50,000 range
Contact: DeLancey Lewis, Secretary, (206) 284-4424
Sponsor: Quest for Truth Foundation
10507 SE 28th Street
Beaux Arts, WA 98004

Quixote Foundation Grants **5017**

The foundation awards grants in its areas of interest, including restoring democracy in the United States through voter engagement and voter-verified paper ballots; safeguarding the world's natural environment through science, conservation, and advocacy; promoting sustainable energy use and technologies; and protecting civil liberties and reproductive rights in the United States. Support takes the form of operating and program grants, technical assistance, research, convening, and activism. Grants do not support endowment or capital purposes. Proposals are by invitation only. Currently letters of inquiry are accepted in the following areas: developing effective message strategies to safeguard the world's natural environment and reforming media coverage and promoting public awareness of economic inequality. Unless you are a current grant recipient, please do not telephone or email the foundation with project briefings or requests for meetings. Guidelines are available online.
Restrictions: Letters of request may not be submitted before November 15.
Samples: Electronic Frontier Foundation (San Francisco, CA)—to develop litigation strategies to respond to problems that arise from computerized voting, $21,000 (2004).
Contact: Program Contact, (206) 783-5554; fax: (206) 783-1815; email: lenore@quixotefoundation.org
Internet: http://www.iinet.com/~quixoteman/index.htm
Sponsor: Quixote Foundation
5703'20th Avenue NW
Seattle, WA 98107

R.B. Terry Charitable Foundation Grants **5018**

The foundation supports higher education institutions primarily in Raleigh, NC, and Woodberry Forest, VA. There are no application deadlines of forms. Submit a letter of inquiry to begin the application process.
Requirements: Higher education institutions in Raleigh, NC, and Woodberry Forest, VA, are eligible.
Restrictions: Individuals are not eligible.
Samples: North Carolina State U (Raleigh, NC)—to establish a new veterinary hospital at the College of Veterinary Medicine, $20 million over 10 years (2005).
Contact: Grants Administrator
Sponsor: R.B. Terry Charitable Foundation
P.O. Box 1009
High Point, NC 27261

R.C. Baker Foundation Grants **5019**

The foundation makes grants to U.S. nonprofit organizations for projects and programs that support social services for youth and the elderly, crime prevention, education, religion (Christian, Episcopal, Friends, Jewish, Methodist, and Presbyterian), scientific research, culture, and health. Support will be provided for fellowships and scholarships, general operating grants, challenge/matching grants, emergency funds, building funds, equipment, continuing support, annual campaigns, capital campaigns, renovation projects, and special projects. Submit cover letter with proposal. The board meets in June and November to consider requests.
Requirements: Nonprofit organizations are eligible. $25,000 to
Restrictions: Anaheim Memorial Medical Center, Anaheim, CA - $40,000 (2006); Harvey Mudd College, Claremont, CA - $25,000 (2006);
Geographic Focus: West
Samples: Anaheim Memorial Medical Center, Anaheim, CA - $40,000 (2006); Harvey Mudd College, Claremont, CA - $25,000 (2006);
Amount of Grant: $1000-$280,000 average
Date(s) Application is Due: May 1; Oct 1
Contact: Frank Scott, Chairman, (714) 750-8987
Sponsor: R.C. Baker Foundation
P.O. Box 6150
Orange, CA 92863-6150

R.H. Wilhelm Award in Chemical Reaction Engineering **5020**

Sponsored by ExxonMobil Research and Engineering Company and administered by AIChE, the award is given to recognize significant and new contributions in chemical reaction engineering. Recipients shall have been judged as advancing the frontiers of chemical reaction engineering with an emphasis on originality, creativity, and novelty of concept application. The recipient need not be a member of the institute.
Requirements: All members of AIChE, as well as other interested persons, are urged to nominate deserving candidates. All information and supporting documents are to be included with the completed form; nomination forms and instructions may be obtained from the institute.
Amount of Grant: $3000, up to $500 travel allowance
Date(s) Application is Due: Feb 15
Contact: Gordon Ellis, Awards Administrator; (646) 495-1348; fax: (646) 495-1503; email: gorde@aiche.org or awards@aiche.org
Internet: http://www.aiche.org/About/Awards/RHWilhelmAward.aspx
Sponsor: American Institute of Chemical Engineers
3 Park Avenue
New York, NY 10016

R.J. McElroy Trust Grants **5021**

The trust gives primarily in the KWWL viewing area in northeast Iowa. Grants are awarded to support higher education, especially scholarship and loan programs; public secondary education, particularly for the disadvantaged; early childhood and elementary education and programs for minorities; and youth, including internships. Support also is given for the cultural and performing arts, recreation, and the environment. Types of support include equipment acquisition, building construction/renovation, emergency funds, program development, fellowships, scholarships, professorships, seed money, internships, and matching funds.
Samples: Wartburg College (Waverly, IA)—to construct the Science Ctr, $1 million (2004).
Amount of Grant: $1000-$50,000
Date(s) Application is Due: Mar 1; Jun 1; Sep 1; Dec 1
Contact: Executive Director, (319) 287-9102; fax: (319) 287-9105; email: klinger@mcelroytrust.org
Internet: http://www.cedarnet.org/mcelroy
Sponsor: R.J. McElroy Trust
425 Cedar Street, Suite 312
Waterloo, IA 50701

RA Kleinhans Fellowship for Research in Tropical Nontimber Forest Products **5022**

These fellowships are offered to further sustainable use of tropical forest resources for the benefit of indigenous peoples, concurrent with conservation of biodiversity of forest ecosystems. The program promotes research into practical means for managing and using tropical forest resources without destroying the integrity of the ecosystem. The geographical focus of research is currently restricted to Latin America. However, projects with potential for other parts of the world are encouraged. Research should merge forest conservation and business prospects and must lead to the development of a product or marketing technique that will provide income for community groups living within or near a tropical forest. Research should culminate in an economically sound strategy that encourages sustainable extraction of tropical forest resources for food, fiber, medicine, and other products for which there is an existing market. Research on products encouraging the reforestation of degraded areas is encouraged. Annual deadline dates may vary; contact the alliance or visit the Web site for exact dates. Please do not fax applications. Please do not send videos. Photos, maps and other visual materials are encouraged but cannot be returned.
Requirements: Citizenship is unrestricted. Applicants with master's degrees in forestry, ecology, environmental science, or a related discipline may apply. However, doctoral candidates and postdoctoral researchers are preferred. Applicants may substitute relevant experience for degrees. Applicants are required to submit their completed application in digital (electronic) form by email to SRodriguez@RA.org. Check Web site for application requirements.
Samples: Past fellow Silvia E. Purata researched the Conservation and Use of Bursera spp in the Tropical Dry Forests of Oaxaca: The Role of Markets and Certification in the Management of Nontimber Forest Products,
Amount of Grant: $15,000 per year for two years
Date(s) Application is Due: Dec 31
Contact: Deanna Newsom, (212) 677-1900 or (888) 693-2784; fax: (212) 677-2187; email: dnewsom@ra.org
Internet: http://www.rainforest-alliance.org/kleinhans.cfm?id=main
Sponsor: Rainforest Alliance
665 Broadway, Suite 500
New York, NY 10012

RACGP Alan Chancellor Award **5023**

The Alan Chancellor Award is presented each year to the GP considered to be the best first time presenter of a research paper at the RACGP Annual Scientific Convention. The prize is $500 and an engraved medallion and is given in memory of Dr Chancellor.
Requirements: To be eligible for the Award, an individual must be: a GP or GP registrar; a member of the RACGP; in the early stages of their research career; a first-time presenter of a research paper at the RACGP Annual Scientific Conference, and; a major contributor to the research study being presented.
Restrictions: By applying for the award each applicant will be taken to have consented to permit RACGP to publicise their name, their institution, and the title of their research project.
Amount of Grant: $500
Date(s) Application is Due: Sep 1
Contact: Samantha Fernandes, Research and Grants Program Administrator, phone: 03-8699-0496; fax: 03-8699-0400; email: research@racgp.org.au
Internet: http://www.racgp.org.au/researchfoundation/awards/chancellor
Sponsor: Royal Australian College of General Practitioners
College House, 1 Palmerston Crescent
South Melbourne VIC 3205 Australia

RACGP Cardiovascular Research Grants in General Practice **5024**

Individual grants of up to $25,000 will be offered for a period of one year to fund projects investigating various aspects of cardiovascular disease including hypertension, stroke, lipid disorder, health promotion and lifestyle changes. The Cardiovascular Research Grants (CVRGs) in General Practice are a joint initiative of the RACGP and the Australian Association of Academic General Practice (AAAGP).
Requirements: To be eligible for the Cardiovascular Research Grants, the chief investigator must be a GP or GP registrar who is a member of the RACGP or the AAAGP. New and emerging researchers are particularly encouraged to apply.

Restrictions: Unless otherwise indicated, Grants must be used in the year for which they are awarded. Grants will be paid only to incorporated bodies that have an ABN and ACN and not to individuals. By applying for a Grant each applicant will be taken to have consented to permit the RACGP to publicise their name, institution, and title of their research project.
Amount of Grant: $25,000 maximum
Date(s) Application is Due: May 8
Contact: Samantha Fernandes, Research and Grants Program Administrator, phone: 03-8699-0496; fax: 03-8699-0400; email: research@racgp.org.au
Internet: http://www.racgp.org.au/researchfoundation/awards/cvrg
Sponsor: Royal Australian College of General Practitioners
College House, 1 Palmerston Crescent
South Melbourne VIC 3205 Australia

RACGP Chris Silagy Research Scholarship **5025**
The Chris Silagy Research Scholarship is designed to support a twelve month research project into an aspect of evidence-based primary health care by a general practitioner researcher. The scholarship is to the value of $20,000 per annum which includes a return economy airfare (booked by the College) to either attend a meeting of the Cochrane Collaboration or visit a Cochrane Research Centre during the year of their scholarship. Scholarships are awarded by the Chris Silagy Research Scholarship Committee, a subcommittee of the RACGP National Research Awards Committee.
Requirements: Applicants for the Chris Silagy Research Scholarship must be: a General Practitioner or General Practice Registrar; a member of the RACGP, and; currently undertaking a higher degree (Masters, PHD, MD) or have completed such a course in the past 5 years.
Restrictions: Unless otherwise indicated, the Scholarship must be used in the year for which it is awarded. The Scholarship will be paid only to incorporated bodies that have an ABN and ACN and not to individuals. By applying for the Scholarship each applicant will be taken to have consented to permit the RACGP to publicise their name, institution, and title of their research project.
Amount of Grant: $20,000
Date(s) Application is Due: May 8
Contact: Samantha Fernandes, Research and Grants Program Administrator, phone: 03-8699-0496; fax: 03-8699-0400; email: research@racgp.org.au
Internet: http://www.racgp.org.au/researchfoundation/awards/silagy
Sponsor: Royal Australian College of General Practitioners
College House, 1 Palmerston Crescent
South Melbourne VIC 3205 Australia

RACGP Family Medical Care, Education and Research Grants **5026**
The aim of the grant is to encourage and support general practitioners in the early stages of their research career to conduct research into an aspect of primary health care. The objective of the grant is to advance education in, and research into, medical knowledge and science. The grants are available for a 12 month period to provide new and emerging GP researchers with an opportunity to conduct research.
Requirements: To be eligible for the FMCER grants, the chief investigator must be a GP or GP registrar who is a member of the RACGP and must have a project that relates clearly to general practice.
Restrictions: Unless otherwise indicated, the Grant must be used in the year for which it is awarded. The Grant will be paid only to incorporated bodies that have an ABN and ACN and not to individuals. By applying for the Grant each applicant will be taken to have consented to permit the RACGP to publicise their name, institution, and title of their research project.
Amount of Grant: $10,000 maximum
Date(s) Application is Due: May 8
Contact: Samantha Fernandes, Research and Grants Program Administrator, phone: 03-8699-0496; fax: 03-8699-0400; email: research@racgp.org.au
Internet: http://www.racgp.org.au/researchfoundation/awards/fmcer
Sponsor: Royal Australian College of General Practitioners
College House, 1 Palmerston Crescent
South Melbourne VIC 3205 Australia

RACGP Grieve Memorial Award **5027**
The award, which is to the value of $2,500, is designed to support a plan of study in medical education for the general practitioner or for research in general practice and must be used with 12 months of being awarded. The award is administered by the RACGP National Research Awards Committee and funded by the RACGP Council.
Requirements: Applicants for the award must be a GP or GP registrar who is a member of the RACGP.
Restrictions: Unless otherwise indicated, the Award must be used in the year for which it is awarded. The Award will be paid only to incorporated bodies that have an ABN and ACN and not to individuals. By applying for the Award each applicant will be taken to have consented to permit the RACGP to publicise their name, institution, and title of their research project.
Amount of Grant: $2,500
Date(s) Application is Due: May 8
Contact: Samantha Fernandes, Research and Grants Program Administrator, phone 03-8699-0496; fax: 03-8699-0400; email: research@racgp.org.au
Internet: http://www.racgp.org.au/researchfoundation/awards/pwhgrieve
Sponsor: Royal Australian College of General Practitioners
College House, 1 Palmerston Crescent
South Melbourne VIC 3205 Australia

RACGP National Asthma Council Research Award **5028**
The Award is awarded annually by the National Asthma Council to the best presenter of an asthma related abstract or poster at the RACGP Annual Scientific Convention. The award is to the value of $500 and is presented at the closing ceremony.
Requirements: To be eligible for the award, an individual must be: a GP or GP registrar; a member of the RACGP; in the early stages of their research career; presenting an asthma-related paper or poster at the RACGP Annual Scientific Conference, and; a major contributor to the research study being presented.
Restrictions: By applying for the award each applicant will be taken to have consented to permit RACGP to publicise their name, their institution, and the title of their research project.
Amount of Grant: $500
Date(s) Application is Due: Sep 1
Contact: Samantha Fernandes, Research and Grants Program Administrator, phone: 03-8699-0496; fax: 03-8699-0400; email: research@racgp.org.au
Internet: http://www.racgp.org.au/researchfoundation/awards/nac
Sponsor: Royal Australian College of General Practitioners
College House, 1 Palmerston Crescent
South Melbourne VIC 3205

RACGP Rex Walpole Travelling Fellowship **5029**
The fellowship, which is to the value of $2500, is designed to support a plan of travel for a general practitioner and must be used within 12 months of being awarded. The fellowship is administered by the RACGP National Research Awards Committee and funded by the RACGP Victorian Faculty. Applications will be judged on benefits gained from travel to the individual, the profession, and the College and value for money.
Requirements: Applicants for the Fellowship must be a GP or GP registrar who is a member of the RACGP.
Restrictions: Unless otherwise indicated, the Fellowship must be used in the year for which it is awarded. The Fellowship will be paid only to incorporated bodies that have an ABN and ACN and not to individuals. By applying for the Fellowship each applicant will be taken to have consented to permit the RACGP to publicise their name, institution, and title of their research project.
Amount of Grant: $2,500
Date(s) Application is Due: May 8
Contact: Samantha Fernandes, Research and Grants Program Administrator, phone: 03-8699-0496; fax: 03-8699-0400; email: research@racgp.org.au
Internet: http://www.racgp.org.au/researchfoundation/awards/rexwalpole
Sponsor: Royal Australian College of General Practitioners
College House, 1 Palmerston Crescent
South Melbourne VIC 3205 Australia

RACGP Vicki Kotsirilos Integrative Medicine Grants **5030**
The Vicki Kotsirilos Integrative Medicine Grant is to encourage research that assists the integration of safe and ethical forms of natural and complementary medicines or therapies into mainstream general practice. The Grant is designed to encourage research that assists the integration of safe and ethical forms of natural and complementary medicines or therapies into mainstream general practice. A Grant of up $5,000 is offered to applicants conducting general practice research in the following areas: clinical research that contributes to the knowledge base concerning the efficacy of natural and complementary medicines or therapies; any project that supports research in behavioural and lifestyle factors impacting on health eg. diet, exercise, stress management, and; research that contributes to the efficient dissemination of the evidence regarding integrative medicine.
Requirements: Applicants must be: a General Practitioner or General Practice Registrar, and; a member of the RACGP.
Restrictions: Unless otherwise indicated, the Grant must be used in the year for which it is awarded. The Grant will be paid only to incorporated bodies that have an ABN and ACN and not to individuals. By applying for the Grant each applicant will be taken to have consented to permit the RACGP to publicise their name, institution, and title of their research project.
Amount of Grant: $5,000
Date(s) Application is Due: May 8
Contact: Samantha Fernandes, Research and Grants Program Administrator, phone: 03-8699-0496; fax: 03-8699-0400; email: research@racgp.org.au
Internet: http://www.racgp.org.au/researchfoundation/awards/kotsirilos
Sponsor: Royal Australian College of General Practitioners
College House, 1 Palmerston Crescent
South Melbourne VIC 3205 Australia

RACGP/APHCRI Indigenous Health Award **5031**
The College supports research which makes a difference to health outcomes, policy and practice, and the capacity of general practice to engage Aboriginal and Torres Strait Islanders' health issues. It particularly encourages a systematic approach to strategic and policy-driven research focusing on primary care, developing collaborative approaches and the building of research capacity within Indigenous populations and communities. The Award is valued at $2500 and is available for a 12 month period to provide support for: indigenous medical students; medical students to gain experience in indigenous health; indigenous medical graduates undertaking GP training; registrars and GPs undertaking research and educational projects in indigenous health, or; other activities supporting the provision of high quality health care to indigenous people.
Requirements: Applicants for the award must be either a medical student or a GP or GP registrar who is a member of the RACGP.

Restrictions: Unless otherwise indicated, the Award must be used in the year for which it is awarded. The Award will be paid only to incorporated bodies that have an ABN and ACN and not to individuals. By applying for the Award each applicant will be taken to have consented to permit the RACGP and APHCRI to publicise their name, institution, and title of their research project.
Amount of Grant: $2,500
Date(s) Application is Due: May 8
Contact: Samantha Fernandes, Research and Grants Program Administrator, phone: 03-8699-0496; fax: 03-8699-0400; email: research@racgp.org.au
Internet: http://www.racgp.org.au/researchfoundation/awards/indigenous
Sponsor: Royal Australian College of General Practitioners
College House, 1 Palmerston Crescent
South Melbourne VIC 3205 Australia

RACGP/CONROD Research Fellowship 5032
Fellowships are available to support research into general practice patient populations who have experienced motor vehicle related trauma. Fellowships encompass all types of motor-vehicle related trauma including (but not limited to) a particular interest in whiplash. Research grants of up $20,000 per year may be awarded to applicants who conduct general practice research in any of the following areas: social research on risk taking behavior in young males, the effects of the compensation process on patients with motor vehicle related injuries, and prevention of the effects of motor vehicle trauma at a community level; clinical research on short and/or long term management of patients who have had motor vehicle accidents, GP management of whiplash, and the influence of psychotherapeutic agents on driving safety; medical education and medical informatics research relating to implementing guidelines for dealing with impaired drivers, or; other areas of research that could involve the role of the GP in the rehabilitation needs of persons who have sustained motor vehicle related injuries, the assessment of the elderly driver, supporting carers of patients with motor vehicle related trauma, and assessing young people at high risk of motor vehicle accidents.
Requirements: Applicants for the RACGP / CONROD Research Fellowship must be a GP or GP registrar and a member of the RACGP.
Restrictions: Unless otherwise indicated, the Fellowship must be used in the year for which it is awarded. The Fellowship will be paid only to incorporated bodies that have an ABN and ACN and not to individuals. By applying for the Fellowship each applicant will be taken to have consented to permit the RACGP to publicise their name, institution, and title of their research project.
Amount of Grant: $20,000 maximum
Date(s) Application is Due: May 8
Contact: Samantha Fernandes, Research and Grants Program Administrator, phone: 03-8699-0496; fax: 03-8699-0400; email: research@racgp.org.au
Internet: http://www.racgp.org.au/researchfoundation/awards/conrod
Sponsor: Royal Australian College of General Practitioners
College House, 1 Palmerston Crescent
South Melbourne VIC 3205 Australia

Radcliffe Institute Carol K. Pforzheimer Student Fellowships 5033
The Arthur and Elizabeth Schlesinger Library on the History of Women in America invites Harvard undergraduates to make use of the library's collections with competitive awards of amounts up to $2,500 for relevant research projects. Preference will be given to applicants pursuing research in the history of work and the family, community service and volunteerism, the culinary arts, or women's health. The research may be, but is not required to be, in connection with a project for academic credit. Applications will be evaluated on the significance of the research and the project's potential contribution to the advancement of knowledge as well as its creativity in drawing on the library's holdings. The awards may be used to cover photocopying, other incidental research expenses, or living expenses in lieu of term time or summer employment, but not the purchase of durable equipment.
Requirements: Harvard undergraduates are eligible.
Amount of Grant: $2,500 maximum
Date(s) Application is Due: Apr 3
Contact: Program Officer; (617) 496-1324 or (617) 496-3048 (Science Office); email: fellowships@radcliffe.edu or science@radcliffe.edu
Internet: http://www.radcliffe.edu/schles/pforzheimer_grant.aspx
Sponsor: Radcliffe Institute for Advanced Study
10 Garden Street
Cambridge, MA 02138

Radcliffe Institute Dissertation Grants 5034
The Arthur and Elizabeth Schlesinger Library on the History of Women in America invites scholars whose dissertation research requires use of the library's collections to apply for research support. Applications will be evaluated on the significance of the research and the project's potential contribution to the advancement of knowledge as well as its creativity in drawing on the library's holdings. Grants of up to $3,000 will be given on a competitive basis to cover travel expenses, living expenses, photocopying, and other incidental research expenses, but not the purchase of durable equipment. A brief report about work accomplished under the grant will be due no later than one year from the date of the award.
Requirements: Applicants must have advanced to candidacy in a doctoral program in a relevant field and have an approved dissertation topic. Priority will be given to those whose projects require use of materials available only but the Schlesinger Library. Non-U.S. citizens are eligible but should contact the library regarding visas and other required paperwork prior to applying.
Amount of Grant: $3,000 maximum
Date(s) Application is Due: Apr 3
Contact: Grants Administrator; (617) 495-8647; fax: (617) 496-4640; email: slgrants@radcliffe.edu
Internet: http://www.radcliffe.edu/schles/dissertation_grant.aspx
Sponsor: Radcliffe Institute for Advanced Study
10 Garden Street
Cambridge, MA 02138

Radcliffe Institute Fellowships for Creative Artists 5035
The Radcliffe Institute is unique among centers for advanced study across the country in its commitment to encouraging and developing the creative arts along with all other disciplines. Through fellowships, public lectures and exhibits, the Institute offers artists not only a quiet place to create or compose, but also an opportunity to showcase their new work.
Requirements: Scholars in any field with a doctorate or appropriate terminal degree at least two years prior to appointment (by December of the prior year) in the area of the proposed project are eligible to apply. Only scholars who have published at least two articles or monographs are eligible to apply. Artists and writers need not have a PHD or an MFA to apply; however, they must meet other specific eligibility requirements. Fellows are expected to be free of their regular commitments so they may devote themselves full time to the work outlined in their proposal. Since this is a residential fellowship, fellows are expected to reside in the Boston area during that period and to have their primary office at the Institute so that they can participate fully in the life of the community. Applicants are strongly encouraged to use the online application system.
Restrictions: Former fellows of the Radcliffe Institute Fellowship Program (1999 to present) are not eligible to apply.
Amount of Grant: $70,000
Date(s) Application is Due: Oct 1
Contact: Program Officer; (617) 496-1324 or (617) 496-3048 (Science Office); email: fellowships@radcliffe.edu or science@radcliffe.edu
Internet: http://www.radcliffe.edu/research/arts.aspx
Sponsor: Radcliffe Institute for Advanced Study
8 Garden Street
Cambridge, MA 02138

Radcliffe Institute Fellowships for Humanists 5036
The Radcliffe Fellowship Program and its predecessor, the Bunting Institute, have a long tradition of supporting writers and scholars in the humanities. Many scholars and students have used the Arthur and Elizabeth Schlesinger Library on the History of Women in America, the premier women's history archive in the United States, to further their work. The Radcliffe Institute's interdisciplinary mission aims to advance understanding of the world through the humanities, and allows scholars, writers, and poets to work in tandem and in dialogue with each other and with researchers in other disciplines.
Requirements: Scholars in any field with a doctorate or appropriate terminal degree at least two years prior to appointment (by December of the prior year) in the area of the proposed project are eligible to apply. Only scholars who have published at least two articles or monographs are eligible to apply. Artists and writers need not have a PHD or an MFA to apply; however, they must meet other specific eligibility requirements. Fellows are expected to be free of their regular commitments so they may devote themselves full time to the work outlined in their proposal. Since this is a residential fellowship, fellows are expected to reside in the Boston area during that period and to have their primary office at the Institute so that they can participate fully in the life of the community. Applicants are strongly encouraged to use the online application system.
Restrictions: Former fellows of the Radcliffe Institute Fellowship Program (1999 to present) are not eligible to apply.
Amount of Grant: $70,000 plus project expenses
Date(s) Application is Due: Oct 1
Contact: Program Officer; (617) 496-1324 or (617) 496-3048 (Science Office); email: fellowships@radcliffe.edu or science@radcliffe.edu
Internet: http://www.radcliffe.edu/research/humanities.aspx
Sponsor: Radcliffe Institute for Advanced Study
8 Garden Street
Cambridge, MA 02138

Radcliffe Institute Fellowships for Natural Scientists and Mathematicians 5037
Radcliffe Institute Fellowships are designed to support scholars, scientists, artists, and writers of exceptional promise and demonstrated accomplishment who wish to pursue independent work in academic and professional fields and in the creative arts. Applicants must have received their doctorate or appropriate terminal degree at least two years prior to appointment or have achieved comparable professional achievement. In recognition of Radcliffe's historic contributions to the education of women and to the study of issues related to women, the Radcliffe Institute sustains a continuing commitment to the study of women, gender, and society. Applicants' projects need not focus on gender, however.
Requirements: Scholars in any field with a doctorate in the area of the proposed project by December of the prior year are eligible to apply. Only scholars who have published at least two articles or monographs are eligible to apply. Fellows are expected to be free of their regular commitments so they may devote themselves full time to the work outlined in their proposal. Since this is a residential fellowship, fellows are expected to reside in the Boston area during that period and to have their primary office at the Institute so that they can participate fully in the life of the community. Applicants are strongly encouraged to use the online application system.

Restrictions: Former fellows of the Radcliffe Institute Fellowship Program (1999 to present) are not eligible to apply.
Amount of Grant: $70,000 plus project expenses
Date(s) Application is Due: Dec 3
Contact: Program Officer; (617) 496-1324 or (617) 496-3048 (Science Office); email: fellowships@radcliffe.edu or science@radcliffe.edu
Internet: http://www.radcliffe.edu/research/science_fellowships.aspx
Sponsor: Radcliffe Institute for Advanced Study
8 Garden Street
Cambridge, MA 02138

Radcliffe Institute Fellowships for Social Scientists **5038**
Research clusters at the Radcliffe Institute for Advanced Study use the Radcliffe Fellowship Program to draw together scholars to focus on particular themes. A research cluster might have a longer life than a one-year fellowship, but it always includes a year when the scholars in the cluster are in residence at the Institute together. During the residential year, fellowship applicants with similar interests are encouraged to apply and to indicate the way in which their project will contribute to the topic. Previous cluster groups have examined evolution of career and family trajectories among highly educated men and women, new logical foundations for theoretical linguistics, unconscious prejudice and the law, and the social and political impact of immigration on the U.S. and Western Europe.
Requirements: Scholars in any field with a doctorate or appropriate terminal degree at least two years prior to appointment (by December of the prior year) in the area of the proposed project are eligible to apply. Only scholars who have published at least two articles or monographs are eligible to apply. Artists and writers need not have a PHD or an MFA to apply; however, they must meet other specific eligibility requirements. Fellows are expected to be free of their regular commitments so they may devote themselves full time to the work outlined in their proposal. Since this is a residential fellowship, fellows are expected to reside in the Boston area during that period and to have their primary office at the Institute so that they can participate fully in the life of the community. Applicants are strongly encouraged to use the online application system.
Restrictions: Former fellows of the Radcliffe Institute Fellowship Program (1999 to present) are not eligible to apply.
Amount of Grant: $70,000 plus project expenses
Date(s) Application is Due: Oct 1
Contact: Program Officer; (617) 496-1324 or (617) 496-3048 (Science Office); email: fellowships@radcliffe.edu or science@radcliffe.edu
Internet: http://www.radcliffe.edu/research/social_sciences.aspx
Sponsor: Radcliffe Institute for Advanced Study
8 Garden Street
Cambridge, MA 02138

Radcliffe Institute Oral History Grants **5039**
The Arthur and Elizabeth Schlesinger Library on the History of Women in America invites scholars who are conducting oral history interviews relevant to the history of women or gender in the United States to apply for support of up to $3,000. This grant stipulates that the interviews take place in accordance with guidelines of the Oral History Association, that consent is obtained from interviewees for their words to be viewed by researchers worldwide, and that true copies or transcripts of the original recording of the oral interviews, as well as copies of the consent forms, be deposited in the Schlesinger Library. Applicants may request support to cover travel expenses, living expenses, transcription services, and incidental research materials, but not the purchase of durable equipment. Applications will be evaluated on the significance of the research and documentation planned and the project's potential contribution to the advancement of knowledge. A brief report about work accomplished under the grant will be due no later than one year from the date of the award.
Requirements: Non-U.S. citizens are eligible but should contact the library regarding visas and other required paperwork prior to applying.
Amount of Grant: $3,000 maximum
Date(s) Application is Due: Apr 3
Contact: Grants Administrator; (617) 495-8647; fax: (617) 496-4640; email: slgrants@radcliffe.edu
Internet: http://www.radcliffe.edu/schles/oral_history_grant.aspx
Sponsor: Radcliffe Institute for Advanced Study
10 Garden Street
Cambridge, MA 02138

Radcliffe Institute Research Support Grants **5040**
The Arthur and Elizabeth Schlesinger Library on the History of Women in America invites scholars whose research requires use of the library's collections to apply for research support. Applications will be evaluated on the significance of the research and the project's potential contribution to the advancement of knowledge as well as its creativity in drawing on the library's holdings. Grants of up to $3,000 will be given on a competitive basis to cover travel expenses, living expenses, photocopying, and other incidental research expenses, but not the purchase of durable equipment).
Requirements: Applicants must have a doctoral degree or equivalent research and writing experience. Priority will be given to those who have demonstrated research productivity and whose projects require use of materials available only at the Schlesinger Library. Non-U.S. citizens are eligible but should contact the library regarding visas and other required paperwork prior to applying.
Amount of Grant: $3,000 maximum
Date(s) Application is Due: Apr 3
Contact: Grants Administrator; (617) 495-8647; fax: (617) 496-4640; email: slgrants@radcliffe.edu
Internet: http://www.radcliffe.edu/schles/research_grant.aspx
Sponsor: Radcliffe Institute for Advanced Study
10 Garden Street
Cambridge, MA 02138

Rajiv Gandhi Foundation Grants **5041**
The foundation conducts activities in India in the areas of the arts and humanities, conservation and the environment, economic affairs, education, international affairs, law and human rights, medicine and health, science and technology, and social welfare. Main projects are concerned with literacy and primary education, empowering women and children, aid for the disabled, health care services in rural and poor urban areas, and decentralized government. Types of support include research grants to institutions and scholarships, fellowships, and prizes to individuals.
Contact: Jawahar Bhawan, Director; (091-11) 23755117 or (091-11) 23312456; fax: (091-11) 23755119; email: info@rgfindia.com
Internet: http://www.rgfindia.com
Sponsor: Rajiv Gandhi Foundation
Dr. Rajendra Prasad Road
New Delhi 110 001 India

Ralph F. Hirschmann Award in Peptide Chemistry **5042**
This award, supported by Merck Research Laboratories, is given annually to recognize and encourage outstanding achievements in the chemistry, biochemistry, and biophysics of peptides. The award is granted without regard to age or nationality of the recipient. The award consists of $5,000 and a certificate. Up to $2,500 for travel expenses to the meeting at which the award will be presented will be reimbursed.
Requirements: Any individual, except a member of the award committee, may submit one nomination or seconding letter for each award in any given year. Nominating documents consist of a letter of not more than 1000 words containing an evaluation of the nominee's accomplishments and a specific identification of the work to be recognized, a biographical sketch including date of birth, and a list of publications and patents authored by the nominee.
Restrictions: Self-nominations are not accepted.
Samples: Morten P. Meldal—award winner, $5,000 (2009); William F. DeGrado—award winner, $5,000 (2008); Samuel H. Gellman—award winner, $5,000 (2007).
Amount of Grant: $5,000 and reimbursement of travel expenses to award meeting
Date(s) Application is Due: Nov 1
Contact: Felicia Dixon, Awards Administrator; (800) 227-5558 or (202) 872-4408; fax: (202) 776-8008; email: f_dixon@acs.org or awards@acs.org
Internet: http://portal.acs.org/portal/acs/corg/content?_nfpb=true&_pageLabel=PP_ARTICLEMAIN&node_id=1319&content_id=CTP_004547&use_sec=true&sec_url_var=region1
Sponsor: American Chemical Society
1155 Sixteenth Street, NW
Washington, DC 20036

Ralph M. Parsons Foundation Grants **5043**
The foundation encourages and supports projects and programs deemed beneficial in several major areas of interest. Higher education grants will be awarded with a strong emphasis on the fields of engineering, technology, and the sciences. University-based fellowships, developmental research, and equipment and building modifications for special projects or programs will be considered. Social impact grants focus on assistance to socially or economically disadvantaged groups, including women and children, families, youth, and the elderly. Civic and cultural grants extend and enhance the cultural life of all people by providing opportunities for access to cultural experiences where few or none have existed. Health grants target services to the disadvantaged, elderly, families, AIDS victims, and youth through support of programs and medical equipment associated with these services. Special projects grants respond to emerging societal needs or projects that fall outside the foundation's established areas of interest. Types of support include seed money, equipment, matching funds, special projects, internships, renovation projects, research, scholarship funds, fellowships, and operating budgets. The foundation staff submits applications to the board of directors six times per year, in alternate months, beginning in January. The first step in the application process is to mail a brief letter to the executive director, including a description of the project to be funded, the amount requested, justification for such a request, and background information about the applying organization.
Requirements: IRS 501(c)3 tax-exempt organizations located in Los Angeles County are eligible. (Higher education institutions are not limited to this geographic area.)
Restrictions: IRS 509(a) private foundations are ineligible. Funding generally is not favored for general fund-raising events, dinners, or mass mailings; direct aid to individuals; conferences, seminars, workshops, etc.; sectarian, religious, or fraternal purposes; endowment; federated fund-raising appeals; programs for which substantial support from government or other sources is available; nor support of candidates for political office or to influence legislation.
Samples: Pitzer College (Claremont, CA)—to develop the Science Learning Community, $250,000 (2005).
Contact: Wendy Hoppe, Executive Director, (213) 482-3185; fax: (213) 482-8878
Internet: http://www.rmpf.org
Sponsor: Ralph M. Parsons Foundation
1055 Wilshire Boulevard, Suite 1701
Los Angeles, CA 90017

Ralph Waldo Emerson Award **5044**
This award is offered annually for studies of the intellectual and cultural condition of the human race. The purpose of the award is to recognize comprehensive studies that contribute significantly to historical, philosophical, or religious interpretations of the human condition. History, philosophy, and religion are conceived in sufficiently broad terms to permit the inclusion of appropriate work in related fields. Biographies of public figures may be eligible if their critical emphasis is primarily on the intellectual and cultural condition of humanity. Entries must be submitted by publishers who have first obtained the conditions of eligibility from Phi Beta Kappa. By March 31, applicants should ship one copy of the book to the appropriate awards committee. By April 10, if the submission is eligible, applicants will receive the shipping list for judges, and should ship one book to each judge by April 30.
Requirements: Entries must be original works published in the United States during the 12-month period between May 1 (of the past year) and the April 30 (of the current year) deadline date. Entries will ordinarily be the work of a single writer. Exceptions may be made for books written by small teams of scholars working in close collaboration. Authors must be U.S. citizens or residents.
Restrictions: Narrowly limited or purely technical studies will not be considered. In general, reports on research and monographs as such are not eligible. Works of fiction and unpublished manuscripts are not eligible.
Samples: David Brion Davis—for Inhuman Bondage: The Rise and Fall of Slavery in the New World (2007); Susan Scott Parrish—for American Curiosity: Cultures of Natural History in the Colonial British Atlantic World (2006); Isabel Hull—for Absolute Destruction: Military Culture and the Practices of War in Imperial Germany (2005).
Amount of Grant: $2,500 maximum
Date(s) Application is Due: Apr 30
Contact: Sandra Beasley, Awards Coordinator, (202) 265-3808; fax: (202) 986-1601; email: SBeasley@pbk.org or info@pbk.org
Internet: http://staging.pbk.org/AM/Template.cfm?Section=Book_Awards&Template=/CM/HTMLDisplay.cfm&ContentID=2646
Sponsor: Phi Beta Kappa Society
1606 New Hampshire Avenue, NW
Washington, DC 20009

RAND Corporation Fellowships in Population Studies and the Study of Aging **5045**
The RAND Postdoctoral Training Program in the Study of Aging enables outstanding junior scholars in demographic and aging research to sharpen their analytic skills, learn to communicate research results effectively, and advance their research agenda. Housed within the Labor and Population Program, the program blends formal and informal training and extensive collaboration with distinguished researchers in a variety of disciplines. Fellowships are for one year, renewable for a second. Each fellow receives a competitive annual stipend, travel stipend and health insurance. Each fellow, in regular consultation with a staff adviser, will design a program that reflects his or her research needs and interests, and stage in their research career.
Requirements: Fellows must be U.S. citizens or permanent residents, and must have completed a Ph.D. (or its equivalent) in a relevant discipline before they begin the program. The program is open to new scholars in the field of demographic and aging research, as well as individuals who already have research experience on these topics. For example, in some cases, fellows are on leave from an academic position so they can engage in research and writing full time, and extend their research agenda and funding base. Applicants should submit a completed application form, curriculum vitae, two or more letters of recommendation, a sample of written work, and a statement of their research goals for the Fellows program. Application forms are available online (in PDF or in interactive format) or by writing to the address below. Members of the program's steering committee will rank applications on the basis of academic performance, the match between the applicant's research interests and those of the RAND staff, and the applicant's ability to benefit from the program. Members of minorities are encouraged to apply.
Date(s) Application is Due: Feb 9
Contact: Diana Malouf; (310) 393-0411, ext. 6462; email: diana_malouf@rand.org
Internet: http://www.rand.org/labor/fellows/aging.html
Sponsor: RAND Corporation
1776 Main Street
Santa Monica, CA 90407-2138

RAND Corporation Graduate Student Summer Associateships **5046**
The program introduces outstanding graduate students to RAND, an institution that conducts research on a wide range of national security problems and domestic and international social policy issues. Students receive a stipend and are given the opportunity to conduct research that can be completed during the three months they are at RAND. The program receives about 500 applications each year for the 20+ positions. The associates who end up finding matches with a research project do so because their skill sets match the needs of the project. The program runs in the summer months only. Summer associates work at RAND full time for a 12-week period. Positions are available in RAND's three major U.S. offices ‚Äì Santa Monica, Arlington, and Pittsburgh. All summer associates are collocated with project mentors. The location of the project mentor determines the location of the summer associate. The summer stipend in 2008 was about $12,000 (before taxes) for the 12 weeks of full time research.
Requirements: The program is designed for full time students who have completed at least two years of graduate work leading to a doctorate or professional degree (e.g., law or medical degree, professional engineer certificate). Students must be enrolled full time in a graduate degree program during the spring and fall of the current school year to be considered for the program. U.S. citizenship is not necessary except for positions that require security clearances. Most of RAND's professional hiring is at the PhD level, so the summer program is oriented toward individuals who are generally within a year or two of completing their doctorates. However, sometimes there are projects that match with the skills of graduate students in other stages of their education who also have significant work experience. To apply, go to the website at http://www.rand.org/jobs/. Click Find a Job, click Advanced Search, select Research-Summer Associate Prog under Job Families, and click the Search button. Click on the position and click the Apply Now button. Follow the prompts for application submission.
Restrictions: The program does not hire undergraduates for the summer. Also, those students graduating prior to the summer of employment are not eligible to apply. The program typically does not consider Master's degree students without significant work or research experience. Postdoctoral fellows are not eligible for the summer program.
Amount of Grant: $12,000
Date(s) Application is Due: Jan 31
Contact: Summer Director; (310) 393-0411; email: Summer_Director@rand.org
Internet: http://www.rand.org/about/edu_op/fellowships/gsap/
Sponsor: RAND Corporation
P.O. Box 2138
Santa Monica, CA 90407-2138

RAND Corporation Military Fellowships **5047**
RAND acts as a host to active-duty service members from all branches of our nation's Armed Forces through the Military Fellows Program. Typically, fellows are stationed at RAND's Santa Monica location for one year. Fellows are given access to all aspects of RAND's research community and are encouraged to participate actively in RAND research projects, seminars, and discussion groups.
Requirements: Requirements and procedures for the Military Fellows Program differ by branch of service: Army, Navy, Air Force, Marines, Coast Guard. Additionally, RAND conducts a fellowship with the Department of Homeland Security. Guidelines and contact information for each branch are available at the website.
Contact: Gene Gritton, NSRD Vice President and NDRI Director; (310) 393-0411 x6933; email: Gene_Gritton@rand.org
Internet: http://www.rand.org/natsec_area/fellowships.html
Sponsor: RAND Corporation
1776 Main Street
Santa Monica, CA 90407-2138

RAND Corporation Transatlantic Postdoctoral Fellowships for International Relations and Security (TAPIR) **5048**
The program is open to candidates who have recently received their doctorate in social and political sciences or economics and whose research focuses on topics of international relations and security. Fellowships are granted for a duration of 24 months to prepare Fellows for a career in policy-oriented and international research at renowned think tanks and political consulting research institutes. Fellows spend three eight-month stays at research institutions or think tanks participating in the program - at least one on the Eastern, and one on the Western side of the Atlantic. Participating institutes include: Carnegie Endowment for International Peace, Institut fran?ais des relations internationales (IFRI), European Union Institute for Security Studies (EUISS), RAND Corporation, Chatham House, SAIS Center for Transatlantic Relations, the Institute for Strategic and International Studies (IEEI), the Center for International Relations (CIR), the Center for Security Studies (CSS) and the German Institute for International and Security Affairs/Stiftung Wissenschaft und Politik (SWP). Fellows will be supervised by a researcher at each host institute, while being integrated into the relevant research units or projects. Fellows will have the opportunity to work on issues of major and immediate political significance, and to publish their work in journals or paper series that will be read by decision-makers at national and international levels. The fellowship aims to involve participants actively in the work and research of leading international think tanks. It is not intended to primarily fund the completion of major individual publications or of academic peer-reviewed journal articles. The fellowship consists of a monthly stipend of Euro 2,000 and a one-time travel allowance of Euro 3,500 for travel to and from the host institutions. Health insurance and fees for residence permits are not included.
Requirements: To be eligible, candidates must have successfully defended their Ph.D. by the start of the program (October). An exception can be made for candidates who can prove that their thesis has been accepted by the jury and requires no further work. Candidates must have a strong interest in working in a policy-oriented international environment and must demonstrate their ability to do so by previous international academic experience. U.S., Canadian and European nationals can apply to the program. Applications should contain: (1) five-page research statement outlining the candidate's current as well as longer-term academic interests. This does not have to be limited to trans-atlantic issues; (2) Curriculum Vitae, including a list of publications; (3) Two letters of recommendation. All files, including letters of recommendation, should be sent in PDF format to transatlanticfellowship@swp-berlin.org by the published deadline date.
Geographic Focus: Alabama; Alaska; Arizona; Arkansas; California; Colorado; Connecticut; Delaware; District of Columbia; Florida; Georgia; Guam; Hawaii; Idaho; Illinois; Indiana; Iowa; Kansas; Kentucky; Louisiana; Maine; Maryland; Massachusetts; Michigan; Midwest; Minnesota; Mississippi; Missouri; Montana; Nebraska; Nevada; New England; New Hampshire; New Jersey; New Mexico; New York; North Carolina; North Dakota; Northeast; Northwest; Ohio; Oklahoma; Oregon; Pacific Northwest; Pennsylvania; Puerto Rico; Rhode Island; South; South Carolina; South Dakota; Southwest; Tennessee; Texas; U.S. Virgin Islands; Utah; Vermont; Virginia; Washington; West; West Virginia; WisconsinCanada; Europe

Date(s) Application is Due: Feb 16
Contact: Tobias von Lossow, Stiftung Wissenschaft und Politik; +49 (0)30 88007-106; fax: +49 (0)30 88007-5321 / -100; email: tobias.vonlossow@swp-berlin.org
Internet: http://www.rand.org/about/edu_op/fellowships/transatlantic/
Sponsor: RAND Corporation
1200 South Hayes Street
Arlington, VA 22202

Rasmuson Foundation Individual Artists Awards **5049**
The foundation supports three award programs for individual artists living and working in Alaska: project awards—for emerging, midcareer, and mature artists for specific, short-term projects that have a clear benefit to the artist and the development of his/her work; artist fellowships—for midcareer or mature artists to focus their energy and attention for a one-year period on developing their creative work; and distinguished artist award—for an artist of recognized stature with a history of creative excellence and accomplishment in the arts. The foundation will fund individual Alaskan artists at various stages of their career whose work is defined by excellence and reflects any of the diverse cultural and aesthetic communities in the state. These awards will provide artists the resources to concentrate and reflect on their work; to immerse themselves in a creative endeavor; and to experiment, explore, and develop their artistry more fully. Deadlines for project awards are March 1 and September 1; fellowship award applications, and distinguished artist award applications are due September 1. Guidelines and application cover sheet are available online.
Requirements: Artists may apply if they meet the following eligibility *Requirements*: full-time Alaskan resident at the time of application and for the duration of the grant period; 18 years of age or older; and an artist who is currently producing work.
Restrictions: The following are ineligible: student, either part-time or full-time, enrolled in any degree-seeking program at any educational level; current board member, committee member, staff member, consultant, current panelist, or member of their immediate families; individuals whose work is primarily of a research or scholarly nature.
Geographic Focus: Alaska
Amount of Grant: $5000 project awards; $12,000 artist fellowships; $25,000 distinguished artist award
Date(s) Application is Due: Mar 1; Sep 1
Contact: Grants Administrator, (907) 297-2700 or (877) 366-2700 (within Alaska); fax: (907) 297-2770; email: rasmusonfdn@rasmuson.org
Internet: http://www.rasmuson.org/index.php?switch=viewpage&pageid=92
Sponsor: Rasmuson Foundation
301 W Northern Lights Boulevard, Suite 400
Anchorage, AK 99503

Rathmann Family Foundation Grants **5050**
The foundation's main funding areas are education, with priority given to science and math; the arts; children and youth health organizations; and preservation of the environment. Types of support include general operating support, continuing support, capital campaigns, equipment acquisition, endowment funds, program development, conferences and seminars, seed grants, curriculum development, fellowships, internships, scholarship funds, research, and matching funds.
Requirements: Grants are awarded to organizations in the San Francisco Bay, CA area; Annapolis, MD; Seattle, WA; Philadelphia, PA: and metropolitan Minneapolis/Saint Paul, MN.
Restrictions: Grants are not awarded to/for private foundations; religious organizations for religious activities; civil rights; social action; advocacy organizations; fraternal groups; political purposes; mental health counseling; individuals; or fundraisers, media events, public relations, annual appeals, or propaganda.
Geographic Focus: California; Maryland; Minnesota; Pennsylvania; Washington
Samples: U of Washington, School of Medicine (Seattle, WA)—to endow a chair in patient-centered clinical education, $1.5 million.
Amount of Grant: $1000-$100,000 average
Contact: Rick Rathmann, (410) 349-2376; fax: (410) 349-2377
Sponsor: Rathmann Family Foundation
1290 Bay Dale Drive, PMB No 352
Arnold, MD 21012

Ray Foundation Mental Health and Substance Abuse Prevention Grants **5051**
The purpose of this program is to support youth agencies with programs in mental health and substance abuse prevention, with particular interest in promoting sports and aviation as alternatives to drug abuse. Additional areas of interest include fine arts, historic preservation, higher and other education, and science and technology. Types of support include building construction/renovation, consulting services, continuing support, emergency funds, equipment acquisition, general operating grants, see grants, publications, research grants, and technical assistance.
Requirements: Youth agencies in the Pacific Northwest are eligible to apply.
Restrictions: Grants are not made to individuals or for deficit funding, and no support is given to religious organizations unless they are requesting funds for a community project.
Geographic Focus: Oregon; Pacific Northwest; Washington
Contact: James Ray, President , (775) 782-8337
Sponsor: Ray Foundation
2241 Park Place, Suite A1
Minden, NV 89423

Rayonier Foundation Grants **5052**
The foundation awards grants to eligible nonprofit organizations in its areas of interest, including children and youth, community development, disadvantaged (economically), education, engineering, environmental programs, families, health care and hospitals, libraries, minorities, performing arts, recreation science, social services, technology, and volunteerism. Types of support include annual campaigns, building construction/renovation, continuing support, employee matching gifts, employee-related scholarships, endowments, equipment acquisition, general operating support, matching/challenge grants, performing arts, program development, recreation, science, research grants, scholarship funds, and seed grants.
Requirements: Nonprofit organizations in Florida, Georgia, and Washington are eligible.
Geographic Focus: Florida; Georgia; Washington
Samples: United Way of Clallam County, Port Angeles, WA - $4,243 (2006); Florida Community College at Jacksonville Foundation, Jacksonville, FL - $25,000 (2006); University of Virginia, Charlottesville, VA - $7,500 (2006);
Amount of Grant: $250 - $25,000 range
Contact: Charles H. Hood, President; (904) 357-9100; fax: (904) 357-9101; email: info@rayonier.com
Sponsor: Rayonier Foundation
50 North Laura Street, Stuite 1900
Jacksonville, FL 32202

RCPSC Detweiler Traveling Fellowships **5053**
Fellowships are made to improve the quality of medical and surgical practice in Canada. The fellowships will enable the recipients to visit medical centers in Canada or abroad to study or gain experience in the use or application of new knowledge or techniques in their fields or to further the pursuit of a fundamental or clinical research problem. Five or more fellowships offered annually to a maximum of $21,000 will be reserved for junior applicants. Two additional fellowships offered annually will be reserved for senior applicants: one will be granted to a Fellow who has held Royal College certification for five years or more and has been in private practice for at least three years, and one will be granted to a Fellow who has held Royal College certification for five years or more and has been in academic practice for at least three years. Each award will be to a maximum of $10,500.
Requirements: Applicants must be Fellows of the Royal College in good standing, or residents in the final year of their training who will be admitted to Fellowship in the Royal College prior to December 31st.
Restrictions: Travel fellowships will not be awarded retroactively.
Amount of Grant: $1,750 per month to a maximum of 12 months (up to $21,000)
Date(s) Application is Due: Sep 30
Contact: Allison Shaw, Awards and Grants Section; (800) 668-3740, ext. 355 or (613) 730-8177; fax: (613) 730-2410; email: awards@rcpsc.edu
Internet: http://rcpsc.medical.org/awards/detweiler_e.php
Sponsor: Royal College of Physicians and Surgeons of Canada
774 Echo Drive
Ottawa, ON K1S 5N8 Canada

RCPSC International Traveling Fellowship **5054**
The objective of the fellowship is to enable fellows residing outside of Canada to study in a Canadian medical center in order to acquire a new knowledge and skills applicable to clinical practice, undertake clinical research, or to enable fellows residing in Canada to practice and teach in a less developed country. The fellowship is applicable for visits of six to 12 months duration. Visits may be broken into separate time blocks but must be completed within two years. An amount equivalent to 75% of the award will be given at the beginning of the fellowship and the remainder upon receipt of the final report at the end of the fellowship. The fellowship may be used to supplement financial assistance granted from other sources.
Requirements: Applicants must be Fellows of the Royal College in good standing, or residents in the final year of their training who will be admitted to Fellowship in the Royal College prior to December 31.
Amount of Grant: $1,750 per month to a maximum of 12 months (up to $21,000)
Date(s) Application is Due: Sep 30
Contact: Allison Shaw, Awards and Grants Section; (800) 668-3740, ext. 355 or (613) 730-8177; fax: (613) 730-2410; email: awards@rcpsc.edu
Internet: http://rcpsc.medical.org/awards/international_e.php
Sponsor: Royal College of Physicians and Surgeons of Canada
774 Echo Drive
Ottawa, ON K1S 5N8 Canada

RCPSC Medical Education Research Grants **5055**
The primary objective of the grant is to support quality research in Canada relevant to the field of postgraduate medical education or continuing professional development. Applications may be individual or joint, involving one or more departments, disciplines, or faculties of medicine. It is not necessary to be a Fellow of the RCPSC or a physician; however, a Fellow of the Royal College must contribute to the research project. The following areas of research are encouraged: curriculum design and implementation; instructional methods; assessment; program evaluation; continuing education; professional development; and faculty development. S
Requirements: Applicants must be involved in teaching and/or medical education research in Canada. Applications must meet contemporary standards of educational research, which include the relevance of the research question, the quality of the research design, ethical review of the submission, accountability for funds received, and the ability to

complete the project during the funding period. One or more grants, of up to $25,000 per year of study, will be awarded annually.
Restrictions: Ineligible costs include tuition fees, capital equipment costs (e.g. computer hardware, office equipment), overhead costs, university or other organization administrative fees.
Amount of Grant: $25,000 maximum per year of study
Date(s) Application is Due: Mar 2
Contact: Allison Shaw, Awards and Grants Section; (800) 668-3740, ext. 355 or (613) 730-8177; fax: (613) 730-2410; email: awards@rcpsc.edu
Internet: http://rcpsc.medical.org/awards/MERG_e.php
Sponsor: Royal College of Physicians and Surgeons of Canada
774 Echo Drive
Ottawa, ON K1S 5N8 Canada

RCPSC Medical Education Traveling Fellowships **5056**
The objective of the fellowship is to enable the recipient to acquire knowledge and expertise in the field of medical education. Upon completion, the recipient will be able to use education principles to develop educational programs and proceed to the evaluation of learning experiences at the undergraduate, postgraduate, or continuing education level. Fellowships are applicable for visits of six to 12 months duration. Up to 75 percent of the fellowship stipend will be given at the beginning of the fellowship and the remainder upon receipt of the final report at the end of the fellowship. The fellowship may be used to supplement financial assistance granted from other sources.
Requirements: Applicants must be Fellows of the Royal College in good standing, or residents in the final year of their training who will be admitted to Fellowship in the Royal College prior to December 31st.
Amount of Grant: $2,500 per month to a maximum of $21,000
Date(s) Application is Due: Sep 30
Contact: Allison Shaw, Awards and Grants Section; (800) 668-3740, ext. 355 or (613) 730-8177; fax: (613) 730-2410; email: awards@rcpsc.edu
Internet: http://rcpsc.medical.org/awards/medical_e.php
Sponsor: Royal College of Physicians and Surgeons of Canada
774 Echo Drive
Ottawa, ON K1S 5N8 Canada

RCPSC Royal College Fellowship for Studies in Medical Education **5057**
The goal of this fellowship is to increase the number and quality of professionally trained medical educators in Canada by providing training in the science of medical education to selected promising candidates. The program is to help specialists acquire knowledge and skills in the field of medical education in order to develop educational programs and evaluation methods that apply to any phase of the continuum of medical education. This fellowship funding is for one year. Applicants may reapply for one additional year of funding; however, repeat or renewal applications will be considered on a competitive basis with all other applications received in the same year, and will require an outline of the progress of the previous year, work.
Requirements: Applicants must register in a Masters or PhD level university program leading to a degree in Education, or a subject field/program of study closely related to medical education. The candidate must be registered and attending university as a full time or part time student, either within Canada or abroad. Applicants must be Canadian citizens or permanent residents who, are Fellows of the Royal College in good standing or residents who, at the time of the application, are registered in a program accredited by the Royal College.
Amount of Grant: $45,000 for residents and Fellows per year of study
Date(s) Application is Due: Sep 30
Contact: Allison Shaw, Awards and Grants Section; (800) 668-3740, ext. 355 or (613) 730-8177; fax: (613) 730-2410; email: awards@rcpsc.edu
Internet: http://rcpsc.medical.org/awards/fellow_e.php
Sponsor: Royal College of Physicians and Surgeons of Canada
774 Echo Drive
Ottawa, ON K1S 5N8 Canada

RCPSC Royal College Visiting Professorships in Medical Research **5058**
The intent of this grant is to provide medical students and residents with the opportunity to meet established researchers from other centers in order to promote medical research as a career choice. It is intended that the visiting professor have the opportunity to interact with both undergraduate medical students and residents through rounds and other teaching sessions, and that the visit not be department-specific. Such interaction will provide the visiting professor with opportunities to expound on the challenges and rewards of a career in medical research. A $3,000 grant is available for each medical faculty.
Amount of Grant: $3,000
Date(s) Application is Due: Sep 30
Contact: Ines Martin; (800) 668-3740, ext. 212 or (613) 730-6212; fax (613) 730-2410; email: imartin@rcpsc.edu
Internet: http://rcpsc.medical.org/awards/visiting-medresearch_e.php
Sponsor: Royal College of Physicians and Surgeons of Canada
774 Echo Drive
Ottawa, ON K1S 5N8 Canada

RCPSC Walter C. MacKenzie Fellowship in Surgery **5059**
The fellowship will enable the recipient to visit one or more centers to acquire surgical research or clinical surgical training. Knowledge gained should have a direct application to the recipient's clinical surgical practice. One or more fellowship are offered annually, depending on budget. Tenable for visits of two weeks to 12 months duration; visits may be broken into separate time blocks but must be completed within two years. An amount equivalent to 75% of the award will be given at the beginning of the fellowship, and the remainder upon receipt of a final report at the end of the fellowship. The fellowship may be used to supplement financial assistance granted from other sources.
Requirements: Applications are accepted from practicing surgeons who are fellows of the Royal College in good standing at the time of application and who reside in Canada. Practicing surgeons are those who have spent at least 50 percent of their professional time in active clinical surgery over the last five years.
Amount of Grant: $5,000 per month (up to $15,000 prorated)
Date(s) Application is Due: Sep 30
Contact: Allison Shaw, Awards and Grants Section; (800) 668-3740, ext. 355 or (613) 730-8177; fax: (613) 730-2410; email: awards@rcpsc.edu
Internet: http://rcpsc.medical.org/awards/mackenzie_e.php
Sponsor: Royal College of Physicians and Surgeons of Canada
774 Echo Drive
Ottawa, ON K1S 5N8 Canada

RCPSC/AMS CanMEDs Research and Development Grants **5060**
The primary objective of the grant is to support research, development, and/or implementation of projects to enhance specialty education in the promotion of the key roles for specialists as defined by CanMEDS. Applications may be individual or joint, involving one or more departments, disciplines, or faculties of medicine. It is not necessary to be a Fellow of the RCPSC or a physician; however, a Fellow of the Royal College must contribute to the project. Eligible projects may include faculty development, course development and/or medical education research that both apply and promote the CanMEDS roles. Budgets may include requests to defray the costs for travel, accommodation and meals for investigators to disseminate the project findings. One or more grants of up to $25,000 will be awarded each year.
Requirements: Applicants must be involved in teaching and/or medical education research in Canada. Applications that include research must meet contemporary standards of educational research, which include the relevance of the research question, the quality of the research design, ethical review of the submission, accountability for funds received, and the ability to complete the project during the funding period. Applications that include project development must include an evaluative component assessing the impact of the project.
Restrictions: Budgets should not include salary or honoraria for investigators. Ineligible costs include tuition fees, capital equipment costs (e.g. computer hardware, office equipment), overhead costs, university or other organization administrative fees.
Amount of Grant: $25,000 maximum
Date(s) Application is Due: Mar 2
Contact: Allison Shaw, Awards and Grants Section; (800) 668-3740, ext. 355 or (613) 730-8177; fax: (613) 730-2410; email: awards@rcpsc.edu
Internet: http://rcpsc.medical.org/awards/ams_e.php
Sponsor: Royal College of Physicians and Surgeons of Canada
774 Echo Drive
Ottawa, ON K1S 5N8 Canada

RealNetworks Helix Community Grants **5061**
The grant program will ensure the Helix DNA platform incorporates cutting edge research advances and has the widest extensibility. Ground-breaking research proposals, inventive implementations, and creative project ideas from independent developers, the academic and research communities, or any nonprofit or commercial enterprise engaged in digital media research and development are welcome. Proposals are encouraged to develop solutions for IP transmission of streaming data over terrestrial and wireless networks. Proposal guidelines are available online. Submit proposals by email.
Requirements: Independent open source developers, faculty members and the students under their purview from U.S.-based academic institutions recognized by the Council of the Higher Education Accreditation and the U.S. Department of Education, industry-recognized research organizations, 501(c)3 nonprofit organizations, and publicly or privately held companies are eligible.
Amount of Grant: $75,000 total
Date(s) Application is Due: Oct 3
Contact: Helix Community Manager; (206) 674-2700; fax: 206-674-2698; email: grants@helixcommunity.org
Internet: https://www.helixcommunity.org
Sponsor: RealNetworks, Inc
P.O. Box 91123
Seattle, WA 98111-9223

Rear Admiral John D. Hayes Predoctoral Fellowship in U.S. Naval History **5062**
The center offers predoctoral fellowships in U.S. naval history that provide financial and scholarly aid for dissertation research and writing. The subject matter of dissertations supported should significantly enhance knowledge of U.S. naval history. Topics in all periods of U.S. naval history will be considered, as well as all aspects, including operations, biography, technology and science, strategy and tactics, administration, social, diplomacy, etc. Applications are available on the Web site.
Requirements: Applicants must be U.S. citizens, be enrolled in a recognized graduate school, have completed requirements for the PhD except the dissertation by June 30 of the award year, and have an approved dissertation topic in the field of U.S. naval history.
Samples: Robert Mullins, Kings College, U of London—for research on how organizational cultures of the U.S. Navy and the Royal Navy influenced decisions in

1889 to create large, modern, steel navies.; Jeffrey Seiken, Ohio State U—for research on American naval policy, planning, and administration during the War of 1812.
Amount of Grant: $10,000 stipend
Date(s) Application is Due: Feb 28
Contact: Senior Historian, Washington Navy Yard, (202) 433-7230
Internet: http://www.history.navy.mil/prizes/prize1.htm
Sponsor: Naval Historical Center
805 Kidder Bresse SE
Washington, DC 20374-5060

Reference Service Press Fellowship 5063
The fellowship is intended to encourage college seniors, college graduates, or beginning library school students to prepare for careers in reference/information service librarianship. Selection is based on academic record (transcripts required), applicant essay (up to 1000 words, typed, double-spaced) on career plans and preparation, and two letters of recommendation that support the applicant's interest in and potential for reference or information service. The fellowship recipient must agree to take at least three classes specifically dealing with reference or information service. Funds will be distributed in three equal payments upon completion of each of the three required classes with a grade B or better. One fellowship is awarded annually. The recipient has four years from the date of the award to complete the fellowship requirements. Application forms are available upon request.
Requirements: Applicant must be a college senior or graduate who has been accepted in an accredited MLS program or a graduate student currently enrolled in an accredited MLS program, be either a California resident attending or planning to attend library school in any state or a resident of any state pursuing or planning to pursue an MLS at a California institution, and be interested in preparing for a career in reference or information service librarianship.
Geographic Focus: California
Amount of Grant: $3000
Date(s) Application is Due: Jun 15
Contact: Reference Service Press Fellowship, Chair, (916) 447-8541; fax: (916) 447-8394; email: info@cla-net.org
Internet: http://www.cla-net.org/awards/rspf.php
Sponsor: California Library Association
717 20th Street, Suite 200
Sacramento, CA 95814-3477

Regenstrief Geriatrics Fellowship 5064
The Department of Medicine and the Division of General Internal Medicine and Geriatrics offer a 2-year fellowship in Geriatric Medicine (with an optional third year). Residents who have completed three years of training in internal medicine or family practice and who are board eligible or board certified may apply for the fellowship. Fellows will be trained in clinical geriatrics, research, teaching and/or administration in preparation for a successful career in academic geriatric medicine. Formal instruction will be provided by a structured curriculum with goals and objectives that will be met by clinical experiences in a variety of settings, didactic seminars to augment the clinical experiences, and seminars in research, teaching and administration. For further information on the Geriatric Fellowship contact the Program Director.
Contact: Glenda Westmorland, Program Director: (317) 630-6906; fax: (317) 630-2667; email: GWestmorland@regenstrief.org
Internet: http://www.regenstrief.org/training/geriatrics-fellowship-program
Sponsor: Regenstrief Institute
410 West 10th Street, Suite 2000
Indianapolis, IN 46202-3012

Regenstrief General Internal Medicine - General Pediatrics Research Fellowship Program 5065
The Regenstrief Institute and the Division of General Internal Medicine offers a 2-year research oriented fellowship (with an optional third year) to prepare physicians for academic careers in general internal medicine and general pediatrics. Fellows receive formal didactic training through classes and seminars taught by faculty from both the Institute as well as Indiana University.
Requirements: Candidates for the Fellowship must hold the M.D. or equivalent degree, and must be U.S. citizens or permanent residents. Minority applicants are encouraged to apply.
Contact: Kurt Kroenke, Program Director: (317) 630-7447; fax: (317) 630-6611; email: KKroenke@regenstrief.org
Internet: http://www.regenstrief.org/training/general-internal-medicine-pediatric-research-fellowship-program
Sponsor: Regenstrief Institute
410 West 10th Street, Suite 2000
Indianapolis, IN 46202-3012

Regenstrief Master Of Science In Clinical Research/CITE Program 5066
Indiana University has received funding from the National Institute of Health through a K-30 grant to develop and implement the Clinical Investigation and Translational Education (CITE) program. The purpose of this program is to prepare health care professionals for a career in clinical research. Following completion of the program, graduates can embark on a career in clinical research with the skills necessary to successfully compete for grant funding, conduct and analyze research findings, and publish their work in scientific journals. By participating in the program, CITE trainees will complete a two-year formal clinical research curriculum, at the end of which they will receive a Master of Science in Clinical Research degree. The program is designed for participants that already have or are in training for a doctoral degree (e.g., MD, PhD, DNS, DDS, DPT).
Requirements: Most participants will already have a doctoral degree (e.g., MD, PhD, DNS, DDS, or DPT) or be in training for a doctoral degree. Potential applicants include, but are not limited to, the following: fellows or other health care professional trainees who have substantial protected time for clinical research; junior faculty who have career awards or support from their division or department head for participation in the CITE program; other faculty who previously have focused on basic research or other academic activities but now are committed to pursuing a career in clinical research CITE constitutes the formal didactic requirements for certain types of federal training grants (such as K-23's) and other career awards. Individuals in these programs would be expected to participate in CITE unless they were enrolled in some other type of formal degree program.
Contact: Kurt Kroenke, Program Director: (317) 630-7447; fax: (317) 630-6611; email: KKroenke@regenstrief.org
Internet: https://www.indianactsi.org/cite/citeapplication
Sponsor: Regenstrief Institute
1050 Wishard Boulevard, RG-6
Indianapolis, IN 46202-2872

Regenstrief Medical Informatics NLM Research Fellowships 5067
The Regenstrief Institute, an internationally recognized leader in medical informatics and health services research, offers National Library of Medicine supported fellowships in the use of computer technology to improve the quality and delivery of healthcare. Interacting closely with faculty mentors, fellows participate in a two-year (with optional third year) comprehensive training program to prepare for roles in developing, implementing and studying clinical information systems. The medical informatics fellowship is designed for physicians, other clinicians holding doctorates, and PhDs in relevant fields, with previous computer experience who desire to become part of a community of scholars and clinicians at the forefront of medical and technological innovations. The primary intent of this fellowship is to train fellows for careers in medical informatics research positions in academic or state and federal government programs, although graduates are also sought after by the health care and the medical systems development industry.
Requirements: Applicants must either hold an M.D. or equivalent degree and we prefer that M.D. applicants have completed their medical residency program or have a Ph.D. in a field related to medicine or technology. Applicants must have significant training or experience with computers. Due to limitations imposed by our NLM funded training programs, applicants must be U.S. citizens or permanent residents. Applications from women and minorities are encouraged.
Contact: Medical Informatics Research Fellowship Office; (317] 423-5670; fax: (317) 423-5695; Cindy Fox Aisen, Media Inquires: (317) 274-7722; email: caisen@iupui.edu
Internet: http://www.regenstrief.org/medinformatics/fellowship
Sponsor: Regenstrief Institute
410 West 10th Street, Suite 2000
Indianapolis, IN 46202-3012

Regents Professional Opportunity Scholarships 5068
Students pursuing a career in one of several professions may be eligible to apply for the scholarships. Degree programs include: associate degree, available at two-year community colleges and some four-year colleges—dental hygiene, dietetics/nutrition, massage therapy assistant, ophthalmic dispensing, physical therapy assistant, registered physician assistant, and veterinary technician; baccalaureate degree, available at four-year colleges only—certified public accountancy, architecture, athletic trainer, dietetics/nutrition, professional engineering, interior design, landscape architecture, registered nurse, occupational therapy, pharmacy, physical therapy, registered physician assistant, and veterinary technician; master's degree, graduate programs—acupuncture, architecture, audiology, landscape architecture, midwifery, occupational therapy, physical therapy, registered physician assistant, social work, speech language, and pathology; doctoral degree (Drive, PhD, PharmD)—chiropractic, optometry, pharmacy, podiatry, psychology, and veterinary medicine; and doctor of law (JD). Priority is given to individuals who are economically disadvantaged and/or who are members of a minority group that is historically underrepresented in the chosen profession; and enrolled in or graduate from SEEK, College Discovery, EOP, or HEOP opportunity programs.
Requirements: Students must study full time and be matriculated in an approved program of study in New York State, be New York State residents, and be U.S. citizens or qualifying noncitizens. Upon completion of study, the student must work as a licensed professional one year for each annual payment received. Employment must be in the studied profession and must be in New York State.
Geographic Focus: New York
Amount of Grant: $1,000-$5,000 per year based on income
Contact: Program Coordinator; (518) 486-1319 or (518) 473-1574
Internet: http://www.hesc.com/Content.nsf/SFC/2/NYS_Regents_Professional_Opportunity_Scholarships
Sponsor: New York State Higher Education Services Corporation
Education Building Addition, Room 1071
Albany, NY 12255

Regional Diver of the Year Awards 5069
The awards are intended to encourage and give local recognition to those having demonstrated leadership in the field of the underwater arts, science, sports, or distinguished service. Every club and council of clubs has a special person who has these qualities that should be recognized. Each award is based within each of the

society's geographical regions and administered by that regional vice president or her/his delegates. The program complements the society's national New Orleans Grand Isle (NOGI) award, whose rules serve as the guideline for administering the area awards. Those given the award may be given further recognition with their nomination for the NOGI award; the national rules must therefore apply. The application is to consist of no more than 500 words, typewritten in single-spaced form, and includes the nominee's name and the area in which she/he is nominated; listing of all the activities of the nominee, particularly as they relate to the underwater or water-related subject; listing of all endeavors the nominee has arranged, founded, originated, or inspired; and listing of all other activities that may be of interest. Each regional vice president may add specific regional rules to these requirements. Annual deadline dates may vary; contact the program director for exact dates.
Requirements: Nominations are accepted only from the membership of the Underwater Society or previous recipients of the NOGI award.
Contact: Program Contact, (650) 583-8492 or (650) 583-0614
Internet: http://www.underwater-society.org
Sponsor: Underwater Society of America
P.O. Box 628
Daly City, CA 94017

Rehabilitation Nursing Foundation Research Grants 5070
The Foundation makes available research grants totaling $44,500. RNF offers funding of up to $30,000 for rehabilitation nursing research which is awarded in the form of multiple grants. A New Investigator Award, offered to encourage nurses who are novice researchers, will be awarded for up to $10,000. Up to two grants from the remaining amount will be awarded to recipients, who will be designated RNF Research Fellows. Research proposals that address clinical practice or educational or administrative dimensions of rehabilitation nursing are accepted. Quantitative and qualitative research projects will be reviewed. Funding may be awarded in the form of multiple grants. The Research Grant Committee reviews proposals once a year. Proposals must be postmarked by February 1. Grant recipients will be notified in September, and funding begins January of the following year for the selected proposals.
Requirements: The principal investigator for the research project must be a registered nurse who is active in rehabilitation or who demonstrates interest in and significant contributions to rehabilitation nursing. Membership in the Association of Rehabilitation Nurses is not required. Graduate student researchers are eligible for funding.
Samples: Maryann Abendroth, Research Title: Understanding Caregivers' Decision Process When Choosing to Institutionalize a Relative with Parkinson's Disease (2008); Cheryl Gies, Research Title: Web-based Psychosocial Assessment for Caregivers of Persons with Dementia: A Feasibility Study (2008); Susan Newman, Research Title: Community Integration after Spinal Cord Injury: Identifying Barriers and Facilitators through Photovoice (2007).
Amount of Grant: $44,500 maximum
Date(s) Application is Due: Feb 1
Contact: Program Director; (847) 375-4710 or (800) 229-7530; email: info@rehabnurse.org
Internet: http://www.rehabnurse.org/research/researchgrants.html
Sponsor: Rehabilitation Nursing Foundation
4700 W. Lake Avenue
Glenview, IL 60025-1485

Reinberger Foundation Grants 5071
The foundation supports medical research, colleges and universities, cultural organizations, environmental organizations, and social services agencies. Types of support include building construction/renovation, equipment acquisition, general operating support, continuing support, annual campaigns, capital campaigns, endowment funds, debt reduction, program development, publication, scholarship funds, research, and matching funds.
Requirements: Most organizations receiving assistance are in the Cleveland and Columbus, OH, areas.
Geographic Focus: Ohio
Samples: Ideastream (Cleveland, OH) and Playhouse Square Foundation (Cleveland, OH)—for the Idea Center at Playhouse Square, which will offer educational programs that incorporate the performing arts, radio, television, and other technologies, $1 million jointly (2005).
Contact: Robert Reinberger, President, (216) 292-2790; fax: (216) 292-4466
Sponsor: Reinberger Foundation
27600 Chagrin Boulevard
Cleveland, OH 44122

Reinhard Mohn Fellowship 5072
Fellowships are awarded to up to five individuals worldwide. For 12 months, fellows are given the opportunity to gather valuable experiences and build networks in a variety of divisions and countries within Bertelsmann's global group of companies. The fellowship allows fellows to gain experiences in one of the biggest global media companies; work with inspiring personalities; obtain factual and professional knowledge about business and media, and build upon fellows' personal qualifications; and extend fellows' personal network, with other fellows and through the alumni network. Information and application are available online. Applications are accepted at any time.
Requirements: Individuals who have initiatied innovative projects in any of the following areas are eligible: business, culture, public and nonprofit organizations, science, social services, or sports.
Amount of Grant: 60,000 euro
Contact: Anette Bickmeyer, (+49) (52) 418-0892-14; fax: (+49) (52) 414-6970; email: info@reinhard-mohn-fellowship.de
Internet: http://reinhard-mohn-fellowship.com/wms/mohn/fset.php3?language=2
Sponsor: Bertelsmann AG
Postfach 111
Gutersloh D-33311 Germany

Religion and Health: Effects, Mechanisms, and Interpretation RFP 5073
The Center seeks proposals to elucidate how religious involvement (religious attendance, worship, altruistic and caring activities) influences individual and community health. The goals of this research program are to: quantitatively document effects on health (where health is broadly defined); clarify the biological, social, psychological mechanisms involved; and interpret what the findings mean for individual, congregational, and community health through trans-disciplinary collaboration and deliberation. This RFP focuses on research that examines the contribution of context and caring as part of the mechanism that accounts for the effects of religion on health, in particular the effects of membership and practices within a faith community. Letters of Intent are due July 15, 2007. Thirty finalists will be invited to submit full proposals; seven $200,000 grants will be awarded to outstanding research proposals. Finalists will be selected by Aug 5, 2007, with full proposals due by November 1, 2007, and awards announced on January 1, 2008.
Amount of Grant: $200,000
Date(s) Application is Due: Jul 15
Contact: Catherine Craver, Assistant Director, (919) 660-7578; fax: (919) 684-8569; email: catherine.craver@duke.edu
Internet: http://www.dukespiritualityandhealth.org/proposals/
Sponsor: Center for Spirituality, Theology and Health at Duke University Medical Center
Box 3825, Duke University Medical Center
Durham, NC 27710

Remarque Institute Fellowships 5074
The Remarque Institute invites applications for one-year fellowships from candidates with interests in contemporary Europe. Fellows receive a stipend and are provided with an office, studio housing, and assistance with travel and other expenses. Write to the director for further information.
Amount of Grant: $25,000 stipend
Date(s) Application is Due: Jan 15
Contact: Dr. Tony Judt, Director, (212) 998-3660; fax: (212) 995-4805; email: remarque.institute@nyu.edu
Internet: http://www.nyu.edu/pages/remarque/program.html#fellowships
Sponsor: New York University
53 Washington Square S
New York, NY 10012

Rennie Taylor/Alton Blakeslee Fellowships for Graduate Study in Science Writing 5075
The fellowships are intended to increase public understanding of science by upgrading the quality and quantity of science and medical writing and by improving relationships between scientists and the press. Journalists with two years of experience who wish to specialize in science writing will receive priority selection. Such applicants should be employed by daily newspapers, wire services, news magazines, radio stations, or television stations or networks. Students must have undergraduate degrees in science or journalism and prove to the satisfaction of a selection committee that they have the motivation and ability to pursue careers in science writing. Fellows may attend school full- or part-time. Application materials are available upon request.
Requirements: Eligible to apply are graduate students in science journalism and experienced reporters who wish to study science writing at the graduate level.
Amount of Grant: $2000 maximum per year
Date(s) Application is Due: Jul 1
Contact: Director, (304) 754-5077; email: diane@nasw.org
Internet: http://www.nasw.org/users/casw/fellowsh.htm
Sponsor: Council for the Advancement of Science Writing
P.O. Box 910
Hedgesville, WV 25427

Research Corporation Cotrell College Science Awards 5076
The program's objective is to support significant research that will enhance the professional and scholarly development of faculty at predominantly undergraduate institutions, along with their students. Faculty are challenged to explore new areas of science, to make new discoveries that contribute to their discipline, and to initiate new research programs that can be sustained by other extramural funding sources, as well as with institutional support. Successful programs are not summer research projects but, rather, those undertaken with continuity through the academic year. A key element in all such research programs is that they involve undergraduate students in meaningful ways. It is also expected that these endeavors will spill over into teaching, thereby raising the quality of undergraduate education.
Requirements: Institutional *Requirements*: applications are accepted from faculty members at public and private institutions of higher education in the United States and Canada. The principal investigator must have a faculty appointment in a department of astronomy, chemistry, or physics. A qualifying department must offer at least the baccalaureate, but not doctoral, degrees. Faculty *Requirements*: applications are accepted and encouraged from investigators at all stages of their professional careers. For both beginning and

established faculty, program emphasis is on initiating an area of research. Proposals from established faculty beginning a new program of research are strongly encouraged. Proposals for funding to continue ongoing projects (those that have extended beyond preliminary results) and requests for supplemental support of projects that have received support from other sources are not encouraged.
Samples: To be distributed among 62 scientists at U.S. and Canadian colleges and universities, $2.25 million (2005); To be distributed among 33 scientists at U.S. and Canadian colleges and universities—to support faculty research in astronomy, chemistry, and physics that has the potential for undergraduate participation, $1.2 million (2005).
Amount of Grant: $35,000 for one to two years; $3500 for 10-week student stipends; $7500 for 8-week faculty stipends
Date(s) Application is Due: May 15; Nov 15
Contact: Grants Administrator, (520) 571-1111; fax: (520) 571-1119; email: awards@rescorp.org
Internet: http://www.rescorp.org/ccsa.htm
Sponsor: Research Corporation
4703 E Camp Lowell Drive, Suite 201
Tucson, AZ 85712

Research Corporation Cottrell Scholars Awards **5077**
The program goal is to assist beginning faculty members who wish to excel at both teaching and research. Both research and teaching plans are required. Originality, the prospect for significant fundamental advances to science, and feasibility will be the main criteria for the research plan. Contributions to and success in undergraduate teaching, aspirations for development as a teacher, and the clarity of future plans will be considered in assessing the teaching plan. Guidelines and applications are available on the Web site.
Requirements: Awards will be made to faculty in PhD-granting astronomy, chemistry, and physics departments to further their teaching and research. Applicants must be in the third year of their first tenure-track position and must provide both a research and teaching plan for peer review.
Samples: To be distributed among 11 U.S. scientists—to recognize excellence in research and teaching in astronomy, chemistry, or physics, $825,000 (2004); Richard Taylor, U of Oregon (Eugene, OR)—award recipient, $75,000 (2003); Brian Stoltz, California Institute of Technology (Pasadena, CA)—award recipient, $75,000 (2003); Michael Hildreth, U of Notre Dame (South Bend, IN)—award recipient, $75,000 (2003).
Amount of Grant: $100,000
Date(s) Application is Due: Sep 1
Contact: Pamela George, Program Assistant, Cottrell Scholars, (520) 571-1111; fax: (520) 571-1119; email: pamela@rescorp.org
Internet: http://www.rescorp.org/csa/csa.htm
Sponsor: Research Corporation
101 N Wilmot Road, Suite 250
Tucson, AZ 85711-3332

Research Corporation Research Opportunity Awards **5078**
The awards are for scientists of demonstrated productivity and creativity seeking to explore new areas of experimental research. The chair of each PhD-granting astronomy, chemistry, or physics department in the United States and Canada may nominate up to two tenured faculty members annually who are without major research funding ($30,000 per year). Following internal review, proposals will be invited from selected nominees. Institutional matching funds for direct costs on a one-for-two dollar basis are required. Nominations are accepted in the spring or fall of each year with deadlines on the first working day of May or October. Guidelines are available online.
Restrictions: Overhead, indirect costs, faculty salary, tuition, common supplies, library costs, page charges, and secretarial services will not be provided.
Samples: To be divided among five scientists at U.S. graduate institutions—to support experimental faculty research in astronomy, chemistry, and physics, $249,147 (2005); To be divided among three U.S. scientists—to support experimental research in astronomy, chemistry, and physics, $150,000 (2004).
Amount of Grant: $50,000 maximum plus $25,000 maximum matching grant
Contact: Patricia Maguire, Executive Assistant to the Vice President, (520) 571-1111; fax: (520) 571-1119; email: pmaguire@rescorp.org
Internet: http://www.rescorp.org/roa/roa.htm
Sponsor: Research Corporation
101 N Wilmot Road, Suite 250
Tucson, AZ 85711-3332

Research Corporation Special Opportunities in Science Awards **5079**
Awards offer support for projects that fall outside regular programs, but that enhance science research or that bear on of the infrastructure of science. Proposals may occasionally be encouraged for research projects that are exceedingly novel, that run counter to scientific paradigms and are unlikely to achieve support from more traditional sources. Only the most challenging and innovative proposals are likely to succeed. After internal evaluation, formal applications will be invited from those proposals that are of further interest.
Requirements: Science projects from nonprofit institutions of higher learning and scientific associations are eligible for support.
Samples: Arlene Russell, U of California (Los Angeles, CA)—to support Partners in Science, $112,000.; Carl Akerlof, U of Michigan (MI)—for research on optical counterparts of gamma-ray bursts, $50,000.
Contact: Dr. Silvia Ronco, Program Officer, (520) 571-1111; fax: (520) 571-1119; email sronco@rescorp.org
Internet: http://www.rescorp.org/sos/sos.htm
Sponsor: Research Corporation
101 N Wilmot Road, Suite 250
Tucson, AZ 85711-3332

Research Program at Earthwatch Grants **5080**
The program considers proposals for field research in any discipline that can gainfully employ nonspecialists in the implementation of a carefully constructed pure or applied research project. The volunteers, who are recruited and screened to meet scientists' needs, are highly educated citizens dedicated to improving environmental understanding. The center encourages proposals that are interdisciplinary and/or transnational. Specific requests for proposals include the life sciences, earth sciences, social sciences, and interdisciplinary. A typical project would employ five to 10 volunteers each on three to six sequential teams. Teams normally spend one to three weeks in the field. Shorter and longer teams are encouraged where appropriate, as are larger or smaller teams. Preliminary proposals should be submitted 12 to 14 months in advance of anticipated field dates. Full proposals are invited upon review of preliminary materials and will be peer reviewed. Proposals are accepted and reviewed year-round.
Requirements: Earthwatch primarily supports postdoctoral or equivalent scholarship, including principal investigators with commensurate life experience. Proposals are welcomed from advanced scholars of any nationality, covering any geographic region. Applicants intending to conduct research in foreign countries must include host-country nationals in their research staffs.
Contact: Center for Field Research, (800) 776-0188 or (978) 461-0081; fax: (978) 461-2332; email: research@earthwatch.org
Internet: http://www.earthwatch.org/research/grantslists.html
Sponsor: Earthwatch Institute
3 Clocktower Pl, Suite 100, Box 75
Maynard, MA 01754

Rett Syndrome Research Foundation Grants **5081**
The foundation (RSRF) awards research grants and postdoctoral fellowships for international biomedical research for the treatment and cure of Rett Syndrome. Relevant disciplines include but are not limited to developmental biology, molecular biology, genetics, biochemistry, cellular physiology, neurology, immunology, pharmacology, gastroenterology, orthopedics, and experimental neurotherapeutics. Treatment-related research is especially encouraged. RSRF seeks to encourage research that encompasses innovative approaches and techniques that will lead to NIH and other funding sources. The foundation is committed to funding senior investigators as well as junior, provided they have a supportive environment. Teams working together from different disciplines, institutions as well as industry are encouraged to apply. Grants are made for technical support, supplies, equipment, and relevant travel. No salary support is provided for the Principal Investigator. Indirect costs of up to 10 percent of direct costs may be included. Letters of intent must be received by the listed application deadline; full proposals are by invitation. Guidelines are available online.
Samples: Dr. Bruria Ben Zeev and Dr. Eva Gak, Sheba Medical Ctr (Israel)— for research, Molecular Diagnosis of Rett Syndrome: An Alternative Scheme, $50,000 one-year award (2003); Dr. Reiko Maki Fitzsimonds, Yale U School of Medicine (New Haven, CT)—for research, Relationship Between Synaptic Structure and Plasticity in Dendritic Neuropathologies Associated with Rett Syndrome, $100,000 two-year award (2003); Dr. Michael Greenberg, Children's Hospital Boston (MA)—for research Neuronal Activity-Dependent Regulation of MeCP2, $100,000 two=year award.
Amount of Grant: $50,000 maximum per year ($40,000 salary plus $10,000 of expendables)
Date(s) Application is Due: Jan 31
Contact: Monica Coenraads, Director of Research, (203) 445-9233; fax: (203) 445-9234; email: monica@rsrf.org
Internet: http://www.rsrf.org/researchers/4.7.1.html
Sponsor: Rett Syndrome Research Foundation
4600 Devitt Drive
Cincinnati, OH 45246

Reuter Foundation Grants **5082**
The Reuter Foundation's areas of interest lie in Social Services to help the poor, needy, homeless and disadvantaged in the Cleveland OH and Collin County TX areas.
Requirements: There is no special form for grant applications. However, each proposal must include a one-page cover letter describing the project, the amount of funds requested, a statement that you are not a 509(a)(3) agency, the contact person, and email address. The preparation of this cover letter is important because if it is not worded convincingly, the remainder of the proposal may not be read. The cover letter and proposal should be sent to: Proposal@ReuterFdn.org and signed by the organizations Executive Director.
Restrictions: Except in unusual circumstances grants do not support endowment programs; construction or renovation of buildings; equipment purchases; artistic or civic projects; debt reductions; benefit functions; conferences, travel, or media projects; graduate studies, scholarships; or to influence legislation or participate in any political campaigns.
Geographic Focus: Ohio; Texas
Samples: The Wilkerson Center, Dallas, TX $5,000 - for general operating expenses related to their new Pleasant Grove facility in support of poor and needy area families,; New Life Community, Cleveland, OH $15,000 - to help support homeless, jobless single mothers and their children with shelter, job training, job retention, money management, parenting, budgeting, and preparation for GED.; Suicide Prevention

Alliance, Beachwood, OH $5,000 - to support instructor salaries in presenting teenage suicide prevention and depression identification in local high schools.;
Contact: Grants Administrator; (216) 961-1141; fax: (216) 651-1777; email: mail@ReuterFdn.org
Internet: http://www.reuterfdn.org/apply.htm
Sponsor: Reuter Foundation
14430 Indian Creek Drive
Cleveland OH 44130

RGK Foundation Grants **5083**
RGK Foundation awards grants in the broad areas of Education, Community, and Medicine/Health. The foundation's primary interests within Education include programs that focus on formal K-12 education (particularly mathematics, science and reading), teacher development, literacy, and higher education. Within Community, the foundation supports a broad range of human services, community improvement, abuse prevention, and youth development programs. Human service programs of particular interest to the foundation include children and family services, early childhood development, and parenting education. The foundation supports a variety of Community Improvement programs including those that enhance non-profit management and promote philanthropy and voluntarism. Youth development programs supported by the foundation typically include after-school educational enrichment programs that supplement and enhance formal education systems to increase the chances for successful outcomes in school and life. The foundation is also interested in programs that attract female and minority students into the fields of mathematics, science, and technology. The foundation's current interests in the area of Medicine/Health include programs that promote the health and well-being of children, programs that promote access to health services, and Foundation-initiated programs focusing on ALS.
Requirements: Although there are no geographic restrictions to the foundation's grantmaking program, the foundation no longer accepts unsolicited requests for international agencies or programs. All applicants must complete an electronic Letter of Inquiry from the Web site as the first step. RGK Foundation will entertain one electronic Letter of Inquiry (LOI) per organization in a twelve-month period. While the foundation occasionally awards grants for operating expenses, capital campaigns, endowments, and international projects, such grants are infrequent and usually initiated by the foundation. Grants are made only to nonprofit organizations certified as tax exempt under Sections 501(c)3 or 170(c) of the IRS code. Hospitals, educational institutions, and governmental institutions meeting these requirements are eligible to apply.
Restrictions: The foundation refrains from funding annual funds, galas or other special-event fundraising activities; debt reduction; emergency or disaster relief efforts; dissertations or student research projects; indirect/administrative costs; sectarian religious activities, political lobbying, or legislative activities; institutions that discriminate on the basis of race, creed, gender, or sexual orientation in policy or in practice; or loans, scholarships, fellowships, or grants to individuals; unsolicited requests for international organizations or programs or for ALS research projects.
Samples: 2008: COMMUNITY - Ark House Foundation (Richardson, TX) to provide low cost, temporary housing for patients and their families while undergoing long-term medical treatment in the Dallas area.; Rider University (Lawrenceville, NJ) to expand the Minding Our Business after-school and summer entrepreneurship program serving 6th-8th grade students in Trenton public schools.; YWCA of Greater Atlanta (Atlanta, GA) expansion of Teen Girls in Technology Program to middle school girls at Title One schools. EDUCATION - BookSpring (Austin, TX) to build regional alliance's capacity following merger with Reading is Fundamental of Austin.; Generations Incorporated (Boston, MA) to expand the Classroom Literacy program using older adults as literacy aids in Boston Public elementary schools.; The Field Museum (Chicago, IL) to improve curriculum in Chicago Public Schools as part of the High School Transformation Project.; MEDICINE/HEALTH - Ambulatory Surgery Access Coalition (San Francisco, CA) to provide essential outpatient surgery to uninsured through a volunteer network.; Marathon Kids (Austin, TX) to enhance national program capacity through technology upgrades for the website.; Travis County Medical Society Foundation (Austin, TX) support program and ongoing enhancements for Project Access that will add dental services and Bilingual Patient Service Coordinator.
Amount of Grant: $1000-$25,000 average
Contact: Suzanne Haffey, Grants Associate; (512) 474-9298; fax (512) 474-7281; email: shaffey@rgkfoundation.org
Internet: http://www.rgkfoundation.org/public/guidelines
Sponsor: RGK Foundation
1301 West 25th Street, Suite 300
Austin, TX 78705-4236

Rhode Island Foundation Grants **5084**
The foundation seeks to promote philanthropic activities that will improve the living conditions and well-being of the inhabitants of Rhode Island. Grants for capital and operating purposes principally to agencies working in the fields of education, health care, the arts and cultural affairs, youth, the aged, social services, urban affairs, historic preservation, and the environment.
Requirements: Rhode Island nonprofits organizations may apply. See Foundations website for additional guidelines.
Restrictions: The foundation does not make grants for endowments, research, religious groups for religious purposes, hospital equipment, capital needs of health organizations, or to educational institutions for general operating expenses.
Geographic Focus: Rhode Island
Samples: Ocean State Action Fund, Cranston, RI, $50,000 - for continued funding for grassroots organizing, coalition work, and public communications in support of affordable, quality health care for all Rhode Islanders (2008); Community Housing Land Trust of Rhode Island, Providence, RI, $50,000 - for new program targeting foreclosed properties or those close to being foreclosed upon (2008); Sakonnet Preservation Association, Little Compton, RI, $25,000 - for three-community partnership to preserve drinking water quality and open space in Newport County (2008);
Contact: Owen Heleen, Vice President for Grant Programs; (401) 427-4009 or (401) 274-4564; fax: (401) 331-8085; email: oheleen@rifoundation.org
Internet: http://www.rifoundation.org/Nonprofits/GrantOpportunities/tabid/175/Default.aspx
Sponsor: Rhode Island Foundation
One Union Station
Providence, RI 02903

Rich Foundation Grants **5085**
The foundation awards grants to nonprofit organizations in Atlanta, GA, to support community funds, performing arts, cultural programs, higher education, social services, homeless, youth, hospitals, heart disease, theater, and AIDS. Types of support include annual campaigns, building funds, continuing support, equipment acquisition, operating budgets, technical assistance, and research. Contact the office for application forms. The board meets quarterly to review requests.
Requirements: Nonprofit organizations in the Atlanta, GA, area are eligible for grant support.
Geographic Focus: Georgia
Amount of Grant: $5000-$100,000 average
Date(s) Application is Due: Mar 15; Jun 15; Sep 15; Dec 15
Contact: Anne Poland Berg, Grant Consultant, (404) 262-2266
Sponsor: Rich Foundation
11 Piedmont Ctr, Suite 204
Atlanta, GA 30305

Richard and Rhoda Goldman Fund Grants **5086**
The fund is interested in supporting programs that will have a significant positive impact in an array of fields, including culture, the environment, population, Jewish affairs, violence prevention, children and youth, the elderly, and social and human services. Types of support include general operating support, continuing support, capital campaigns, program development, and seed money. Funds will be allocated primarily in the San Francisco Bay area of California and to organizations benefiting this area. There are no application deadlines.
Requirements: Nonprofit organizations serving primarily the San Francisco Bay area of California are eligible.
Restrictions: The fund does not accept applications for research or award grants or scholarships for individuals, conferences, documentary films, or fund-raisers. Unsolicited proposals for support of arts organizations or institutions of primary, secondary, or higher education will not be accepted.
Geographic Focus: California
Samples: San Francisco Beautification (CA)—for efforts to increase green schoolyards, $10,000 (2005); American Jewish Joint Distribution Committee (New York, NY)—for relief efforts following the earthquake and tsunamis that struck Southeast Asia in 2004, $30,000 (2005); United States Public Interest Research Group Education Fund (Washington, DC)—for efforts to increase attention to the role of hard money in federal political campaigns, $20,000 (2005); Community Awareness and Treatment Services (San Francisco, CA)—for support of this 24-hour shelter and transitional-housing program for women, $25,000 (2005).
Amount of Grant: $1000-$500,000
Contact: Sam Salkin, Executive Director, (415) 345-6300; fax: (415) 345-9686; email: info@goldmanfund.org
Internet: http://www.goldmanfund.org
Sponsor: Richard and Rhoda Goldman Fund
P.O. Box 29924, 211 Lincoln Boulevard
San Francisco, CA 94129

Richard and Susan Smith Family Foundation Grants **5087**
The foundation awards grants to Massachusetts nonprofit organizations in its areas of interest, including arts and culture, children and youth, disadvantaged (economically), education (early childhood, elementary, and higher), minorities, biomedical research, health care and social services delivery. Types of support include annual campaigns, building construction/renovation, capital campaigns, curriculum development, general operating grants, program/project support, seed grants, and service delivery programs. There are no application forms.
Requirements: Massachusetts nonprofit organizations serving the greater Boston area are eligible.
Restrictions: Grants do not support political activities; religious activities; individuals; or requests for deficit financing or endowment funds.
Geographic Focus: Massachusetts
Samples: Compassionate Care ALS, West Falmouth, MA $25,000 - for the purchase of a cargo van to transport adaptive equipment to and from patient homes (2009); United Teen Equality Center, Lowell, MA $944,780 - funding provides violence prevention, youth development, education and community organizing programs for the youth center, the grant is payable over 5 years (2009);
Amount of Grant: $5000-$238,000 range
Date(s) Application is Due: Mar 15; Aug 15
Contact: David Ford, Executive Director; (617) 278-5200; fax: (617) 278-5250; email: dford@smithfamilyfoundation.net

Internet: http://www.smithfamilyfoundation.net
Sponsor: Richard and Susan Smith Family Foundation
1280 Boylston Street, Suite 100
Chestnut Hill, MA 02467

Richard Carley Hunt Memorial Postdoctoral Fellowships 5088
A limited number of nonrenewable postdoctoral fellowships are awarded each year to scholars within five years after receipt of the doctorate in anthropology to aid in the write-up of research results for publication. Funds may be used for any items necessary to the completion and/or write-up of research. Applicants who require funds during the first half of the following year should meet the May 1 deadline. Projects scheduled to begin in July or later may be submitted by the November 1 deadline. Decisions for each application cycle will be announced six to eight months after the deadline date. Only one application may be submitted during any 12-month period. Eligible individuals may request application forms.
Requirements: Doctoral-level students and researchers and postdoctoral scholars in anthropology and closely related disciplines are eligible to apply. Qualified scholars are eligible without regard to nationality or institutional affiliation. Applicants must hold the PhD at the time of application.
Restrictions: Funds do not provide for salary and/or fringe benefits of applicant; nonproject personnel; travel to meetings; or institutional overhead. Expenses incurred prior to the effective date of an award also will not be covered.
Amount of Grant: $40,000 maximum
Date(s) Application is Due: May 1; Nov 1
Contact: Program Contact, (212) 683-5000; email: info@wennergren.org
Internet: http://www.wennergren.org/programsirg.html
Sponsor: Wenner-Gren Foundation for Anthropological Research
470 Park Avenue S, 8th Fl
New York, NY 10016-6819

Richard Lounsbery Foundation Grants 5089
The foundation aims to enhance national strengths in science and technology through support of programs in the following areas: science and technology components of key U.S. policy issues; elementary and secondary science and math education; historical studies and contemporary assessments of key trends in the physical and biomedical sciences; and start-up assistance for establishing the infrastructure of research projects. Among international initiatives, the foundation has a long-standing priority in Franco-American scientific cooperation. The foundation generally provides seed money or partial support, rarely renews grants for continuing activities, does not normally fund endowments or laboratory research, and aims to achieve high impact by funding novel projects and forward-looking leaders.
Restrictions: Requests are denied for support of capital or building funds, loans, or individuals.
Amount of Grant: $10,000-$50,000 average
Contact: Grants Administrator, (202) 872-8080; fax: (202) 872-9292; email: proposals@rlounsbery.org
Internet: http://www.rlounsbery.org
Sponsor: Richard Lounsbery Foundation
1020 19th Street NW, Suite LL60
Washington, DC 20036

Richard P. Covert Scholarship 5090
Two scholarships are awarded annually, one each at the undergraduate and graduate levels, in the field of health care information and management systems. Funding is provided by HIMSS. Completed application form, available from the society, must be accompanied by relevant transcripts, a technical paper and/or an essay, and a faculty recommendation. Annual deadline dates may vary; contact the program office for exact dates.
Requirements: Applicant must be at least a junior when the scholarship is awarded and must be a student member of HIMSS. Previous winners are not eligible, although an undergraduate winner is eligible if pursuing a graduate degree.
Amount of Grant: $5000
Date(s) Application is Due: Oct 31
Contact: John Page, Executive Director, HIMSS Scholarship Committee, (312) 644-4467; fax: (312) 664-6143; email: jpage@himss.org
Internet: http://www.himss.org/ASP/scholarship_ms.asp
Sponsor: Healthcare Information and Management Systems Society
230 E Ohio Street, Suite 500
Chicago, IL 60611-3269

Richard P. McCormick Prize for Scholarly Publication 5091
This prize is given to the author of an outstanding scholarly work on New Jersey history published during the preceding two years. The prize is awarded in odd-numbered years. Guidelines and applications are available on the Web site.
Amount of Grant: $1000
Date(s) Application is Due: Jan 2
Contact: Mary Murrin, Director Grants Program, (609) 984-0954; fax: (609) 633-8168; email: mary.murrin@sos.state.nj.us
Internet: http://www.state.nj.us/state/history/grants.html
Sponsor: New Jersey Historical Commission
225 W State Street, 4th Fl
Trenton, NJ 08625

Righteous Persons Foundation Grants 5092
The foundation, established by Steven Spielberg, makes grants nationally for projects to revitalize Jewish life, help young people learn about Judaism, and promote tolerance among people of all faiths and ethnicities. Many of the awards are for projects that explore new ideas and opportunities with a special emphasis on young people and that include many projects that may have otherwise had a difficult time obtaining funding. Projects that are national in scope or can serve as models for other communities will receive preference.
Requirements: U.S. 501(c)3 nonprofits are eligible.
Restrictions: Grants are not awarded to support endowments, capital campaigns, building funds, university faculty chairs, individual synagogues or day schools, research, or the publication of books or magazines.
Samples: Raoul Wallenberg Committee of the United States (New York, NY)—to help bring its A Study of Heroes curriculum to additional U.S. schools and communities, $20,000 matching grant over two years.; Anti-Defamation League (Los Angeles, CA)—for its Stop the Hate Program, $50,000 matching grant.; Clark U (Worcester, MA)—to support a doctoral fellow in its program on the history of the Holocaust, $60,000.; Jewish Television Network (Los Angeles, CA)—to produce a television program on Jewish art and culture, $50,000.
Amount of Grant: $5000-$400,000 range
Contact: Rachel Levin, Program Officer, (310) 314-8393; fax: (310) 314-8396
Sponsor: Righteous Persons Foundation
2800 28th Street, Suite 105
Santa Monica, CA 90405

Robert A. Levy Fellowship in Law and Liberty 5093
The School of Law at the university invites applications from PhD and ABD economists and political scientists who wish to pursue a JD at George Mason University. Qualified applicants will be evaluated for their promise as law and economics scholars capable of making significant contributions to understanding the institutions of a free society.
Requirements: To be eligible for a Levy Fellowship, an applicant must either have earned a PhD in economics, philosophy, political science, finance, or a related field from an accredited university or have successfully completed all course work in one of the above fields and have passed a general exam for a PhD in one of those fields from an accredited university. Candidates must take the LSAT no later than February 2006, register with the Law School Data Assembly Service (LSDAS), and apply for admission to the George Mason University School of Law.
Amount of Grant: $22,000 per year plus tuition waiver
Date(s) Application is Due: Apr 1
Contact: Director, (703) 993-8066; fax: (703) 993-8088; email: djohnsen@gmu.edu
Internet: http://www.gmu.edu/departments/law/admission/levy.html
Sponsor: George Mason University
301 N Fairfax Drive
Arlington, VA 22201-4498

Robert and Patricia Switzer Foundation Fellowships 5094
The foundation awards 20 environmental fellowships to graduate students in New England and California each year. The foundation seeks to identify and support individuals with the potential to contribute significantly to improving the natural environment. All applicants for fellowships must be nominated in writing by a faculty member, advisor, or an environmental professional before application materials are made available to the applicant. The foundation has a partnership with The San Francisco Foundation to manage the nomination, application and selection process for the program in California (225 Bush Street, Suite 500; San Francisco, CA 94104-4224). Nominations are due on the listed application deadline.
Requirements: To be eligible for a fellowship in California or New England, a candidate must be a U.S. citizen; have completed either one semester of course work at the master's level or one year of doctoral course work; be enrolled in an accredited institution in California or New England (Connecticut, Maine, Massachusetts, New Hampshire, Rhode Island, and Vermont); and have superior academic qualifications.
Amount of Grant: $13,000
Date(s) Application is Due: Jan 6
Contact: Erin Lloyd, (207) 338-5654; email: erin@switzernetwork.org
Internet: http://www.switzernetwork.org/fellowship-overview.html
Sponsor: Robert and Patricia Switzer Foundation
P.O. Box 293
Belfast, ME 04915-0293

Robert B McMillen Foundation Grants 5095
The Robert B McMillen Foundation is a non-profit charitable foundation. The Foundation offers funding in the states of Washington and Alaska. The two areas of intent are: medical and social enhancement. Fifty percent of the annual funding is earmarked for medical research. The Foundation will consider making grants to non-profit organizations involved in researching cardiology, lipid and organ transplants. Twenty-five percent of annual giving is earmarked for social areas including, but not limited to, Goodwill, Salvation Army & United Way. Preference is given to organizations and/or programs that use art as the vehicle to impact communities and change individual lives.
Requirements: The Foundation does not accept uninvited medical or social proposals. Contact the Foundations office for information on, how to be invited to make a medical or social proposal to the McMillen Foundation.
Restrictions: The Foundation do not make grants to: religious, fraternal or other organizations primarily benefiting their own members; other private foundations nor

political organizations; for-profit entities; other than scholarships, the foundation will not make multi-year grants.
Geographic Focus: Alaska; Washington
Samples: Seattle Children's Hospital - Pediatric Heart Failure, Organ Transplantation and Neovascularization (2008); ArtsEd Washington - PAL Program and Arts Partners Project (2008); Seattle Goodwill - General Support (2008).
Contact: Cassandra Town, President; (425) 313-5711; fax: (425) 313-8955; email: cassandra@mcmillenfoundation.org
Internet: http://www.mcmillenfoundation.org/mission.htm
Sponsor: Robert B McMillen Foundation
55 1st Place NW, Suite 2, P.O. Box 1523
Issaquah, WA 98027

Robert Bosch Foundation Fellowships **5096**
The foundation annually sponsors an intensive work fellowship program in Germany to provide young American professionals with a comprehensive overview of the political, economic, and cultural environment of Europe, especially Germany. The program, running from September through May, will consist of an intensive German language course with a seminar on German affairs; practical internships in ministries of the German federal government and later in local government or private industry; and special seminars in Berlin, Paris, Brussels, and Stuttgart. Internships are arranged by the Bosch Foundation in cooperation with fellows in accordance with their professional interests and backgrounds. The Robert Bosch Foundation is in Stuttgart, Germany, and is represented in the United States by CDS International Inc (formerly Carl Duisberg Society). Program-related travel, limited insurance, and monthly supplement for spouses are provided.
Requirements: Eligible to apply are U.S. citizens between the ages of 23 and 34 holding a graduate or professional degree or equivalent professional experience in business administration, economics, public administration, political science, law, journalism, or mass communications. Prerequisites include an outstanding academic background, public commitment, and character traits indicative of leadership qualities. Some professional experience is desirable. German fluency is required at the start of the program. If needed, language training will be provided by the foundation in the U.S. and Germany as needed.
Restrictions: Non-U.S. citizens, including permanent residents, are not eligible.
Date(s) Application is Due: Oct 14
Contact: U.S. Representative, (212) 497-3518; fax: (212) 497-3535; email: bosch@cdsintl.org
Internet: http://www.cdsintl.org/fromusa/bosch.htm
Sponsor: CDS International Inc
871 United Nations Plaza
New York, NY 10017

Robert Bowne Foundation Grants **5097**
The foundation awards grants to support out-of-school literacy programs for disadvantaged children and youth. Funding within this grantmaking focus includes program support, research, special projects, staff development, technical assistance, evaluation, and advocacy. Priority is given to programs that provide participants with a variety of reading and writing experiences. The foundation also supports adult basic education and literacy programs, family and youth services, and certain educational institutions. Types of support include general operating support, continuing support, program development, conferences and seminars, seed money, curriculum development, research, technical assistance, and matching funds. Application forms are not required, and there are no application deadline dates.
Requirements: Nonprofit organizations in the greater New York City area, with emphasis on the boroughs outside Manhattan, may submit letters of application.
Restrictions: Grants are not awarded to religious groups, primary or secondary schools, colleges or universities, individuals, or to support capital campaigns or endowments.
Geographic Focus: New York
Samples: Coalition for Hispanic Family Services (NY)—to support the Arts and Literacy after-school program, $25,000.; Forest Hills Community House (NY)—to support its after-school program, $25,000.
Amount of Grant: $20,000-$30,000 average
Contact: Anne Lawrence, Program Officer, (212) 229-7227; fax: (212) 886-0400; email: alawrence@robertbownefoundation.org
Internet: http://www.robertbownefoundation.org/index.php
Sponsor: Robert Bowne Foundation
345 Hudson Street
New York, NY 10014

Robert E. and Evelyn McKee Foundation Grants **5098**
The foundation awards grants in the following categories: civic, cultural, and religion; education and scholarships; hospitals; medical and medical research; welfare, rehabilitation, and mental health; United Way; and youth activities. Grants awarded for scholarships are made to a few selected high schools and universities, who select scholarship recipients. Religious contributions are basically limited to the local Episcopal churches. The majority of funding is made to local charitable organizations or through local affiliates of national or statewide organizations.
Requirements: 501(c)3 tax-exempt organizations are eligible.
Restrictions: Funding is not available to organizations outside the United States or those limited by race or ethnic origin. The foundation refrains from awarding grants for endowments or deficit financing. Grants are not made to other foundations, except for a local community foundation.
Geographic Focus: Texas
Samples: Saint Francis on the Hill Episcopal Church (El Paso, TX)—$1000 (2003); U of Texas at El Paso (TX)—for Department of Engineering scholarships, $8000 (2003); Texas Tech U Health Sciences Ctr (El Paso, TX)—for the University Breast Care Center, $3000.; Special Olympics, Area 19 (El Paso, TX)— $1000 (2003).
Amount of Grant: $1000-$10,000
Date(s) Application is Due: Dec 15
Contact: Grants Administrator, (915) 581-4025; fax: (915) 833-3714; email: mckeefoundation@directway.com
Internet: http://www.mckeefoundation.org
Sponsor: Robert E. and Evelyn McKee Foundation
5835 Cromo Drive, Suite 1
El Paso, TX 79912

Robert H. and Clarice Smith Predoctoral Fellowship **5099**
The fellowship funds productive scholarly work in Dutch or Flemish art history, intended for the advancement or completion either of a doctoral dissertation or of a resulting publication. The fellow may use the grant to study either in the United States or abroad; there are no residence requirements at the National Gallery. Application must be made through the chair of the graduate department of art history or other appropriate department. Departments should limit nominations to one candidate. The fellowship grant begins on September 1 and is not renewable.
Requirements: Applicants must have completed their residence requirements and coursework for the PhD before the date of application and know two languages related to the topic of the dissertation. Applicants must be either U.S. citizens or enrolled in universities in the United States.
Samples: Catherine Lusheck, U of California, Berkeley—for dissertation, Studies in Rubens' Drawings: Mode, Manner, and the Problems of Stylistic Variety.
Amount of Grant: $24,000 for one year
Date(s) Application is Due: Nov 15
Contact: Fellowships Program, Center for Advanced Study in the Visual Arts, (202) 842-6482; fax: (202) 789-3026; email: advstudy@nga.gov
Internet: http://www.nga.gov/resources/casvapre.htm
Sponsor: National Gallery of Art - The Fellowship Program
Fourth Street and Constitution Avenue NW
Washington, DC 20565

Robert J. Kleberg Jr. and Helen C. Kleberg Foundation Grants **5100**
The foundation awards grants on a national basis for medical research, community services, education, health services, arts/humanities, wildlife, veterinary, and animal sciences. Local community organizations also are supported. Types of support include building construction/renovation, equipment acquisition, conferences and seminars, research, and matching funds. There are no application deadlines or forms. Initial approach should be by letter on organization?s letterhead. Send one copy of proposal. The board usually meets in June and December, with final notification in six months.
Requirements: 501(c)3 tax-exempt organizations are eligible. Requests from universities must be transmitted by the appropriate offices, such as the chancellor, president, or director of development.
Restrictions: Grants do not support community organizations outside Texas; organizations limited by race or religion; individuals; general purposes; endowments; deficit financing; ongoing operating expenses; overhead or indirect costs; scholarships or fellowships; or loans.
Samples: U of Texas MD Anderson Cancer Ctr (Houston, TX)—to establish a center that will focus on investigating molecular markers, the genes and proteins that may help identify cancer in its early stages or help predict how cancer will respond to treatment, $5 million (2005).
Contact: Robert Washington, Grants Coordinator, (210) 271-3691
Internet: http://www.texascancercouncil.org/cancerfunding/state/04.html
Sponsor: Robert J. Kleberg Jr. and Helen C. Kleberg Foundation
700 N Saint Mary?s Street, Suite 1200
San Antonio, TX 78205

Robert L. Baker Summer Research Fellowship for Second Language Acquisition in an Environment of Immersion **5101**
The Middlebury College Language Schools invite applications for eight-week summer fellowships to conduct research in second-language acquisition or applied linguistics. Fellows will live on the Vermont campus during the summer session and conduct research on the Middlebury immersion program in one or more of the nine language schools: Arabic, Chinese, French, German, Italian, Japanese, Portuguese, Russian or Spanish. Research results will be made available to the language teaching community at Middlebury and beyond. Contact the office for guidelines. Applications should be submitted to: Michael R. Katz, Dean of Language Schools and Schools Abroad, Middlebury College, Middlebury, Vermont 05753.
Requirements: Fellows will be named on the basis of a description of the research project, a plan of action, a resume, and at least two letters of recommendation.
Amount of Grant: $3000-$4000 plus room and board
Date(s) Application is Due: Mar 15
Contact: Grants Administrator, (802) 443-5510; email: languages@middlebury.edu
Internet: http://www.middlebury.edu/academics/ls/resources/fellowships_scholarships/baker.htm
Sponsor: Middlebury College Language Schools
Middlebury, VT 05753

Robert M. Hearin Foundation Grants 5102
The foundation awards grants to four-year colleges and universities in Mississippi. Areas of interest include program development and improvement, research, scholarships, and faculty development. There are no application deadlines or forms.
Geographic Focus: Mississippi
Samples: U of Mississippi, School of Business Administration (MS)—for faculty development, professional staff support, and undergraduate services, $8.5 million over five years (2002); Mississippi State U, College of Engineering (MS)—to conduct research in communications, energy, and human systems, and to expand undergraduate programs and outreach, $3.6 million over four years (2001); Mississippi State U, School of Architecture (MS)—for graduate fellowships, student recruitment, new technology and educational materials, and equipment maintenance and replacement, $400,000 over four years $400,000 (2001).
Amount of Grant: $200,000-$650,000
Contact: Daisy Blackwell, (601) 366-8363; fax: (601) 961-6876
Sponsor: Robert M. Hearin Foundation
P.O. Box 16505
Jackson, MS 39207

Robert Mapplethorpe Foundation Grants 5103
The foundation provides grants in two unrelated major areas: medical research to advance the cure and treatment of AIDS and HIV infection and the creation or expansion of photography departments in museums and other public institutions. Types of support include equipment acquisition, program development, publication, and research. A formal application is not required. The foundation board meets regularly to consider requests.
Restrictions: Individual artists will not receive support. The foundation does not award grants for social services.
Contact: Grants Administrator, (212) 755-3025; fax: (212) 941-4764
Internet: http://www.mapplethorpe.org/foundation-guidelines.html
Sponsor: Robert Mapplethorpe Foundation
120 Wooster Street, 4th Fl
New York, NY 10012-0000

Robert Penn Warren Center for the Humanities Visiting Fellowship 5104
The Warren Center invites applications from established scholars interested in participating in a faculty seminar as a visiting fellow during one or both semesters. Seminars, which focus on medicine, health, and society, meet once a week and allow the visiting fellow time to pursue a major project. Application forms are available on the Web site.
Requirements: Applicants should have an advanced degree and significant publications in a relevant discipline.
Amount of Grant: $40,000 maximum
Date(s) Application is Due: Jan 18
Contact: Mona Frederick, Executive Director, (615) 343-6060; email: mona.frederick@vanderbilt.edu or rpw.center@vanderbilt.edu
Internet: http://www.vanderbilt.edu/rpw_center/fellows.htm
Sponsor: Vanderbilt University
VU Station B, No 351534
Nashville, TN 37235

Robert R. McCormick Tribune Foundation Grants Programs 5105
The foundation funds programs in the areas of community development, citizenship, journalism, and education. The communities program is designed to combine the efforts of the foundation and various corporate entities to increase local philanthropy and improve the social and economic conditions of communities. The citizenship program awards grants to promote patriotism, service, and volunteerism; encourage effective philanthropy; and honor the U.S. armed forces. The journalism program supports research, development, and education to assist and inform journalism's leaders, encourage informed coverage of national security issues, and support freedom of expression. The education program is designed to focus on the quality of early childhood education, increase the number of accredited preschool programs, strengthen the skills of preschool program directors, improve the professional development system for preschool teachers, and heighten public understanding of the importance of early childhood education. Types of support include building/renovation funds, equipment, capital campaigns, operating budgets, technical assistance, land acquisition, matching funds, program-related investments, and special projects. Preproposal discussion with staff is encouraged. Applications are accepted at any time.
Requirements: The foundation funds education programs in the Chicago metropolitan area only. Community programs are funded in Anaheim; Atlanta; Chicago; Cleveland; Denver; El Paso; Escondido; Fort Lauderdale; Los Angeles; Manhattan, KS; New York; Newport News, VA; Orlando; and Phoenix.
Geographic Focus: Arizona; California; Colorado; Connecticut; Florida; Georgia; Illinois; Kansas; Louisiana; Massachusetts; New York; Ohio; Texas; Virginia; Washington
Samples: Gettysburg National Battlefield Museum Foundation (York, PA)—for a gallery that will showcase the contributions of Illinois troops during the Civil War's Battle of Gettysburg, $1 million (2005); National-Louis U, Ctr for Early Childhood Leadership (Chicago, IL)—for new programs for early-childhood administrators, a national public-awareness campaign on early-childhood issues, and an endowment that will support research, scholarships, and special projects, $2.1 million (2005). Rush U Medical Ctr (Chicago, IL)—for a state-of-the-art emergency-services facility, $7.5 million (2005).
Date(s) Application is Due: Jun 1 (education); Sep 1 (citizenship); Oct 1 (journalism)
Contact: Mark Hallett, Senior Program Officer, (312) 222-3512; fax: (312) 222-3523; email: mhallett@tribune.com
Internet: http://www.rrmtf.org/mtf/mtfinfo.htm
Sponsor: Robert R. McCormick Tribune Foundation
435 N Michigan Avenue, Suite 770
Chicago, IL 60611

Robert Schalkenbach Foundation Grants 5106
The foundation seeks to support projects that advocate ideas and programs compatible with Henry George's philosophy of economic justice. George's central thesis was that policies should avoid the perpetuation of unearned privilege, particularly those forms that allow economic rent, or socially-produced surplus, to be privatized. Applicants who have an understanding of these issues and who recognize that Georgist philosophy represents a 'third way' that is distinct from capitalism and socialism or from progressive and libertarian ideals. The foundation will consider two types of projects: 1) Research - projects that deepen understanding of how Georgist principles are connected with issues and debates in the social science or humanities disciplines. 2) Policy - policy-oriented projects that demonstrate how Georgist philosophy yields practical results in a variety of policy settings. Traditionally, the Foundation has primarily supported projects which advocate a gradual tax shift from buildings to land within the context of local property tax reform. At present, the foundation wishes to expand the areas in which it will consider policy-oriented applications. Proposals of this type should demonstrate the capacity not only to evaluate the likely consequences of a policy but also to identify the businesses, advocacy groups, and other organizations that might be mobilized on its behalf, and the arguments that might be used to persuade them to join a coalition.
Requirements: Applicants who are are not prior recipients of a Schalkenbach funding should first submit a one-page letter of inquiry by email. The letter of inquiry should indicate the general nature of the proposed project and the amount of funding that will be requested. Above all, it should indicate specifically how the project will advance the aim of the foundation in promoting the ideas of Henry George either in policy-oriented activities or intellectual research. The foundation also encourages applicants to pursue a variety of topics and issues rather than focusing on the two-rate property tax as the sole practical consequence of Georgist philosophy. Proposals that promote that idea will continue to be considered, but they will be given lower priority than other policy-oriented proposals.
Restrictions: The foundation is not primarily interested in projects that deal specifically with Henry George as a person. Also, it not generally receptive to applications for projects that are primarily libertarian in orientation.
Amount of Grant: $2,000-$6,000
Contact: Mark A. Sullivan, Administrative Director; (212) 683-6424; fax: (212) 683-6454; email: msullivan@schalkenbach.org
Internet: http://www.schalkenbach.org/funding-program.html
Sponsor: Robert Schalkenbach Foundation
90 John Street, Suite 501
New York, NY 10038

Robert Sterling Clark Foundation Grants 5107
The foundation supports projects directed toward strengthening the management of cultural institutions in New York City (one-time projects designed to improve internal management, reduce operating costs through resource sharing, and/or increase earned income or individual contributions); improving the performance of public institutions (projects designed to improve the effectiveness or accountability of New York City and State agencies, and promote public/private partnerships in developing New York's economic base); and protecting reproductive freedom (projects designed to protect or expand access to family planning services nationally or in New York). Types of support include continuing support, program development, research, technical assistance, and consulting services. Applications are accepted at any time.
Requirements: U.S. nonprofits are eligible. Arts grants support New York City programs only. Government accountability programs should pertain to New York City or State.
Restrictions: Grants are not awarded to support individuals or for annual campaigns, seed money, emergency funds, deficit financing, capital or endowment funds, matching gifts, scholarships, fellowships, conferences, or films.
Samples: Natural Resources Defense Council (New York, NY)—to provide research, advocacy, public education, and assistance to policy makers in order to strengthen New York City's recycling program, $50,000 (2005); Gotham Chamber Opera (New York, NY)—to expand and strengthen the company's Board of Directors and to develop a professional individual-giving campaign, $20,000 (2005); Friends of Hudson River Park (New York, NY)—for efforts to complete the construction of the Hudson River Park, $25,000 (2005); Advocates for Youth (Washington, DC)—to train and mobilize young people to promote comprehensive sexuality education at the federal, state, and local levels, $65,000 (2005).
Amount of Grant: $1000-$200,000
Contact: Margaret Ayers, Executive Director, (212) 288-8900; fax: (212) 288-1033
Internet: http://fdncenter.org/grantmaker/rsclark
Sponsor: Robert Sterling Clark Foundation
135 E 64th Street
New York, NY 10021

Robert W. Deutsch Foundation Grants 5108
The foundation provides support to technology programs in institutions of higher education in Maryland. Types of support include: continuing support, curriculum development, fellowships, general operating support, internship funds, organizational

development, program development, research, and seed grants. There are no application deadlines or forms; submit initial inquiries by email or postal mail.
Restrictions: No support for religious or political organizations.
Geographic Focus: Maryland
Samples: University of Maryland-College Park, College Park, MD. $250,000 - For nanotechnology research, payable over 1 year (2007).
Amount of Grant: $5,000 - $210,000 range
Contact: Jane Brown, Executive Director; (410) 252-9244; fax: (410) 252-0997; email: rdeutsch@rwd.com
Sponsor: Robert W. Deutsch Foundation
5521 Research Park Drive
Baltimore, MD 21228-4664

Roberto Longhi Foundation Fellowships **5109**
Fellowships are available for nine months at the institute for those who are seriously dedicated to research in the history of art, particularly Italian painting from the 13th to the 18th centuries. Eligible are Italian citizens who possess degrees from Italian universities with theses in the history of art, and non-Italian citizens who have fulfilled the preliminary requirements for doctoral degrees in the history of art at an accredited university or institution of equal standing. Fellows begin their research in October; they must collaborate on a specific group research project selected by the scientific committee while doing research on their own projects. Fellows are also required to attend seminars, lectures, and other activities arranged by the institute.
Requirements: Application must contain the candidate's biographical data, transcript of undergraduate and graduate records, copy of degree thesis (if any) and other original works, a curriculum vita indicating knowledge of foreign languages spoken and written, two letters of reference from persons acquainted with the candidate's work, the subject of the research that the candidate is interested in pursuing, and two passport photographs. Fellows must live in Florence for the duration of the fellowship.
Restrictions: Applicants must be under thirty-two before the deadline date. Fellows may not make arrangements with any other institution during the period of the fellowship.
Date(s) Application is Due: May 15
Contact: Phillip Grosshans, Assistant Director: (919) 681-5132; fax: (919) 684-2418; email: phillipg@duke.edu
Internet: http://researchfunding.duke.edu/detail.asp?OppID=4988
Sponsor: Duke University
P.O. Box 104010
Durham, NC 27708

Rockefeller Archive Center Ehrlich Grants **5110**
Through a generous gift, the Rockefeller Archive Center has available funds to support short-term research in the Paul Ehrlich Collection at the Archive Center. The Ehrlich Collection is the largest and most important body of original and facsimile materials documenting the life and scientific-medical research program of Nobel Laureate Paul Ehrlich (1854-1915). Application to the Ehrlich grant program may be made at any time and, if judged worthy, awards will be made within a month of the receipt of the applicant's two letters of recommendation. Reimbursement of up to $2,500 will be made upon receipt of original expense receipts after the completion of the grant recipient's research visit.
Requirements: Application to this program is open to U.S. citizens and citizens of foreign countries. Prospective applicants are required to contact the Archive Center with an initial description of the applicant's research.
Restrictions: Grant recipients must commence their research at the Center within one year of being notified of the award. The grant recipients are also, requested to submit a short report on their research at the Center. Edited versions of or excerpts from these reports may be used in the Center's publications. Recipients also are asked to provide the Center with a copy of any publication resulting from research conducted as a result of the grant.
Amount of Grant: $2,500
Contact: Camilla Harris, Grants Administrator; (914) 366-6311 or (914) 631-4505; fax: (914) 631-6017; email: charris@rockarch.org or archive@rockarch.org
Internet: http://www.rockarch.org/grants/ehrlichgrant.php
Sponsor: Rockefeller Archive Center,
15 Dayton Avenue
Sleepy Hollow, NY 10591-1598

Rockefeller Archive Center Grants in Aid Program **5111**
Rockefeller Archive Center's competitive grant-in-aid program provides awards for reimbursement of up to $5000 for certain receipted expenses related to research conducted at the RAC. The research grants are designed to foster, promote, and support research by serious scholars in the collections located at the Rockefeller Archive Center, which include the records of the Rockefeller Family, The Rockefeller University, The Rockefeller Foundation, and other philanthropies and associated individuals. The deadline for applications is November 15th each year, and awards are announced at the end of the following February. Grant recipients must commence their research at the Center within one year of being notified of the award. Grant recipients are requested to submit a short report on their research at the Center. Edited versions of or excerpts from these reports may be used in the Center's publications. Recipients also are asked to provide the Center with a copy of any publication resulting from research conducted as a result of the grant.
Requirements: Application to the grant-in-aid program is open to U.S. citizens and citizens of foreign countries. Certain U.S. Government requirements will apply to non-U.S. citizens.
Restrictions: Awards are made to individuals. Institutions are not eligible to apply. The grant does not support research at other institutions, and it does not provide general tuition support.
Amount of Grant: $5,000
Date(s) Application is Due: Nov 15
Contact: Camilla Harris, Grants Administrator; (914) 366-6311 or (914) 631-4505; fax: (914) 631-6017; email: charris@rockarch.org or archive@rockarch.org
Internet: http://www.rockarch.org/grants/generalgia.php
Sponsor: Rockefeller Archive Center, Rockefeller University
15 Dayton Avenue
Sleepy Hollow, NY 10591-1598

Rockefeller Archive Center Research Grants **5112**
The foundation awards grants for travel and research at the Rockefeller Archive Center, which contains records of the Rockefeller Foundation, Rockefeller University, and other affiliated philanthropic organizations and individuals. The center also will award grants to support research on the history of basic medical research, the life of the Nobel laureate Paul Ehrlich, and on postcolonial studies.
Requirements: Scholarly reseearchers, particularly students preparing dissertations, are eligible.
Restrictions: The Archive Center's programs do not support research at other institutions, and they do not provide general tuition support.
Contact: Camilla Harris, Grants Administrator; (914) 366-6311 or (914) 631-4505; fax: (914) 631-6017; email: charris@rockarch.org or archive@rockarch.org
Internet: http://www.rockarch.org/grants/
Sponsor: Rockefeller Foundation
15 Dayton Avenue
Sleepy Hollow, NY 10591-1598

Rockefeller Archive Center Visiting Archivist Fellowship **5113**
The Rockefeller Archive Center invites applications for a Visiting Archivist Fellowship for a professional archivist from: South and Central America, including Mexico; the Caribbean; Africa; south and central Asia; China; Korea; Taiwan; the Philippines; Indonesia; and the Pacific Ocean states. The Rockefeller Archive Center will host a visiting archivist for at least two weeks, but no longer than one month for the purposes of enhancing professional development, and extending knowledge of the Center's collections. Applications must be postmarked by January 31st each year, and awards will be announced at the end of April. Upon completion of the fellowship, the visiting archivist can receive reimbursement for up to $8000 in receipted expenses for round-trip Economy Class air fare; housing, food and local transportation.
Requirements: Applicants must be, or recently have been, employed as archivists, and demonstrably able (1) to improve their archival skills through a residency at the Center, and (2) to contribute to a better understanding of the Center's documentation of the applicant's nation or region. The successful applicant will have a B.A. or equivalent, professional archival training, and 2-5 years experience as an archivist. He/she should be seeking to expand his/her experience in order to develop new services or technologies in his/her home institution. The applicant must have a strong command of spoken and written English, and must provide evidence to that effect. Potential applicants are required to contact an archivist at the Archive Center by email at archive@rockarch.org or by fax at (914) 631-6017 well in advance of completing their applications with a description of their fellowship project.
Amount of Grant: $8,000
Date(s) Application is Due: Jan 31
Contact: Camilla Harris, Grants Administrator; (914) 366-6311 or (914) 631-4505; fax: (914) 631-6017; email: charris@rockarch.org or archive@rockarch.org
Internet: http://www.rockarch.org/grants/visitingarchivist.php
Sponsor: Rockefeller Archive Center
15 Dayton Avenue
Sleepy Hollow, NY 10591-1598

Rockefeller Brothers Fund Fellowship in Nonprofit Law **5114**
The fellowship permits one fellow each year to spend one year in residence at the Vera Institute of Justice in New York City, working closely with Vera's general counsel and special counsel on legal issues. The fellowship will provide a law school graduate with exposure to a wide variety of legal and organizational issues encountered by nonprofit organizations. It also will provide the opportunity to identify an emerging or changing area of law with particular significance for nonprofits. Fellows will be selected without regard to race, color, religion, gender, political beliefs, national origin, disability, age, or sexual orientation.
Requirements: Graduates of accredited U.S. law schools are eligible.
Amount of Grant: $47,000 plus benefits
Date(s) Application is Due: Dec 5
Contact: Erin Bainbridge Ortiz, Program Coordinator; (212) 998-6168; fax: (212) 995-3149; email: ncpl.info@nyu.edu
Internet: http://www.rbf.org/grantsdatabase/grantsdatabase_show.htm?doc_id=616213
Sponsor: National Center on Philanthropy and the Law, New York University
110 W. Third Street, Room 205
New York, NY 10012

Rockefeller Foundation Humanities Fellowships **5115**
Fellowships in the humanities and the study of culture will be offered at 21 host institutions that were selected for their potential to promote new work in the humanities.

The fellowships are meant to serve scholars who are testing disciplinary boundaries or moving into newer fields of inquiry within the humanities. Although the majority of the fellow's time will be spent pursuing his or her own research toward publication, the residency may involve participation in seminars, conferences, or other collaborative activities within the host program. Complete information about eligibility, stipends, and procedures for application is available directly from current host institutions (listed in application). Individuals apply for resident fellowships at the host institutions.
Requirements: Residential fellowships are meant to serve scholars who are testing disciplinary boundaries or moving into newer fields of inquiry within the humanities. Applications are invited from programs in colleges and universities and other institutions, such as libraries, museums, and community cultural spaces, with engaged communities of scholars and other resources of value to emerging fields of the humanities.
Restrictions: Awards may not be used for the completion of graduate studies, advanced training, curriculum development projects, or for writing poetry or fiction.
Contact: Grants Administrator; (212) 869-8500; fax: (212) 764-3468
Internet: http://www.rockfound.org/bellagio/bellagio.shtml
Sponsor: Rockefeller Foundation
420 Fifth Avenue
New York, NY 10018-2729

Rockefeller Foundation MAP Fund Grants 5116
MAP works to build a thriving, risk-welcoming contemporary performance field by providing project-specific funding to playwrights, choreographers, directors, composers and performers experimenting in any performance tradition or discipline. MAP believes that exploration drives human progress, in art as in science and medicine. Sometimes the path of progress is lighted by an act of the imagination on the part of the artist. MAP exists to facilitate these acts and help bring them to their fullest realization. Up to 40 grants per annual cycle, ranging from $10,000 to $45,000. The average award amount is $25,000.
Requirements: Applications must come from organizations based in the United States that have current nonprofit federal tax status (501c3). Unincorporated artists or ensembles may apply to MAP through a fiscal sponsor. Organizations and artists must demonstrate at least 2 years professional experience. MAP supports only projects that contain a live performance. Eligible projects must not have premiered anywhere in the world before September 1.
Restrictions: Artists will not receive support for more than one project per funding cycle. Artists supported through the fund for the past two consecutive years will not be eligible for support in the current year but may reapply the following year.
Amount of Grant: $50,000 maximum; $10,000-$45,000 average
Date(s) Application is Due: Jan 11
Contact: Moira Brennan, Program Director; (212) 226-1677; fax: (212) 226-7665; email: mapinfo@mapfund.org
Internet: http://www.mapfund.org/apply.html
Sponsor: Rockefeller Foundation
73 Spring Street, Suite 401
New York, NY 10012

Rockefeller Foundation Scholarly Residencies 5117
A Bellagio residency provides time for critical thinking, disciplined work, individual reflection, and collegial engagement, uninterrupted by the usual professional and personal demands. The Center typically offers one-month residencies for no more than 12 scholars and scientists at a time. Individuals in any discipline - and from any part of the world - are welcome to apply. Space is reserved for both academic projects, as well as projects that align with the Foundation's mission to expand opportunities for poor or vulnerable people and to help ensure that globalization, benefits are shared more widely. Residency cohorts include artists, writers, non-governmental organization practitioners, policymakers, scholars and scientists from around the world. Meals and informal presentations of work provide an opportunity for residents to engage with each other. During special dinners, residents also have the chance to interact with the participants of international conferences that are hosted in various buildings on the Center, property. The combination of private time for work and reflection and informal gatherings is an important part of life at the Center.
Restrictions: Accommodations are not available for children and family members other than spouses/life partners, nor for friends or pets.
Amount of Grant: $35,000-$40,000 average, plus $2000-$5000 toward travel, benefits, and relocation costs
Contact: Residencies Administrator; (212) 869-8500; fax: (212) 764-3468
Internet: http://www.rockfound.org/bellagio/bel_scholar.shtml
Sponsor: Rockefeller Foundation
420 Fifth Avenue
New York, NY 10018-2702

Rockwell International Corporate Trust Grants Program 5118
Grants are awarded in the areas of education and youth development, with emphasis in math, science, and engineering; and culture and the arts, with emphasis on youth educational programs. Rockwell also contributes to health, human services, and civic organizations and also gives special consideration to qualifying organizations in which employees are involved as volunteers. Types of support include capital grants, challenge/matching grants, conferences and seminars, development grants, endowments, research grants, scholarship funds, and general operating grants. Awards include single-year, multiple-year, and provisional continuing support of specific programs. Applicants should submit full, detailed proposals. Proposals are accepted at any time. At the Web site, click on About Us, then Corporate Citizenship, and then Rockwell Automation Corporation Trust.
Restrictions: Organizations are not eligible if they have not received a permanent, tax-exempt ruling determination from the federal government; if they cannot provide current full, certified, audited financial statements; or if they are private foundations. Funding will not be considered for the following purposes: general endowments, deficit reduction, grants to individuals, federated campaigns, organizations or projects outside the United States, religious organizations for religious purposes, or fraternal or social organizations.
Amount of Grant: $1000-$50,000 average
Contact: Rockwell International Corporation General Information; (562) 797-3311
Internet: http://www.rockwell.com
Sponsor: Rockwell International Corporate Trust
1201 S Second Street
Milwaukee, WI 53204

Rocky Mountain Mineral Law Foundation Grants 5119
Grants are awarded to persons or organizations for meritorious projects in natural resources law such as the preparation of teaching materials; specialized faculty or student research; law school seminars, lectures, short courses, and symposia; and publication of substantive articles. Preference will be given to Member Law Schools of the Rocky Mountain Mineral Law Foundation (see the website for a full listing).
Requirements: No special application form is required; submit a letter and supporting data including: your qualifications to undertake the project; a detailed explanation of your proposal; a budget of your total anticipated expenses, exclusive of project salary/remuneration for application; and, the amount of support you are requesting from the foundation. Applications should be submitted electronically to csherwood@rmmlf.org.
Restrictions: Personal remuneration or honoraria are not within the scope of these grants.
Geographic Focus: Arizona; California; Colorado; Idaho; Kansas; Louisiana; Montana; Nebraska; Nevada; New Mexico; North Dakota; Oklahoma; Oregon; South Dakota; Texas; Utah; Washington; WyomingCanada
Contact: Carla Sherwood; (303) 321-8100 ext 107; fax: (303) 321-7657; email: csherwood@rmmlf.org
Internet: http://www.rmmlf.org/geninfo/schgrant.htm
Sponsor: Rocky Mountain Mineral Law Foundation
9191 Sheridan Boulevard, Suite 203
Westminster, CO 80031

Roeher Institute Research Grants 5120
The institute, through the support of the Scottish Rite Charitable Foundation of Canada, offers research grants to associates, associations, and agencies in a broad range of fields relating to human services and intellectual disability. The renewable grants are offered for one year. Research projects must be under the direction of the Roeher Institute academic associates or other university faculty or under researchers approved by the Roeher Institute. Areas constituting research priorities for the Roeher Institute are issues affecting people with intellectual disabilities; integration of people who have an intellectual disability into society; prevention, early identification, and minimization of disabling conditions; and strategies for social change that improve the quality of life of persons with intellectual disabilities. Applicants are asked to submit a summary of their research project, budget, and financial requirements.
Requirements: Applications are invited from university faculty, department heads, supervisors, Roeher Institute associates, and/or consultants.
Amount of Grant: $C10,000 maximum
Contact: Cameron Crawford, President; (416) 661-9611; fax: (416) 661-5701; email: cameronc@roeher.ca or cameronc@worldchat.com
Internet: http://www.roeher.ca/english/about/about.htm
Sponsor: Roeher Institute
4700 Keele Street, York University, Kinsmen Bldg.
North York, ON M3J 1P3 Canada

Roger Adams Award in Organic Chemistry 5121
This award, sponsored by Organic Reactions, Inc. and Organic Syntheses, Inc., and administered by the American Chemical Society, is given biennially in odd-numbered years to recognize and encourage outstanding contributions to research in organic chemistry defined in its broadest sense. The award will be presented biennially in odd-numbered years at the ACS National Meeting. The recipient will deliver a lecture at the Biennial National Organic Chemistry Symposium of the American Chemical Society National Meeting and Symposium. The award is granted to an individual without regard to nationality. Nominations will be accepted by the indicated deadline date in even-numbered years.
Requirements: Any individual, except a member of the award committee, may submit one nomination or seconding letter for the award in any given year. The nominating documents consist of a letter of not more than 1000 words containing an evaluation of the nominee's accomplishments and a specific identification of the work to be recognized, a biographical sketch including date of birth, and a list of publications and patents authored by the nominee.
Restrictions: Self-nominations are not accepted.
Samples: Samuel J. Danishefsky, $25.000 (2007); Jerrold Meinwald, $25,000 (2005); Albert Eschenmoser, $25,000 (2003).
Amount of Grant: $25,000 and travel allowance to award symposium
Date(s) Application is Due: Nov 1

Contact: Felicia Dixon, Awards Administrator; (800) 227-5558 or (202) 872-4408; fax: (202) 776-8008; email: f_dixon@acs.org or awards@acs.org
Internet: http://portal.acs.org/portal/acs/corg/content?_nfpb=true&_pageLabel=PP_ARTICLEMAIN&node_id=1319&content_id=CTP_004550&use_sec=true&sec_url_var=region1
Sponsor: American Chemical Society
1155 Sixteenth Street NW
Washington, DC 20036

Roger K. Summit Scholarship **5122**
Three awards are made in North America; Europe, the Middle East, and Africa; and Asia-Pacific to graduate students in library and information science who have demonstrated outstanding interest or performance in electronic information services. Scholarship recipients are selected based on academic achievement; demonstrated interest in electronic information services, based on course work, research and experience; proficiency on Dialog's online services and products; and faculty recommendations. Application and guidelines are available online.
Requirements: Graduate students in library and information science are eligible.
Amount of Grant: $5000
Contact: Betty Jo Hibberd, Administrator, email: Betty.Jo.Hibberd@dialog.com
Internet: http://gep.dialog.com/scholarship
Sponsor: Dialog Corporation
11000 Regency Pkwy, Suite 10
Cary, NC 27511

Rohm and Haas Company Grants **5123**
The company awards grants in five key philanthropic categories: education, environment, civic and community, health and human services, and arts and culture. Education grants support programs linking education in science, technology, and math to workplace and career opportunities; and educational enrichment for students, particularly in the areas of math, science, technology, and the environment either during or after school hours (including weekends, summer, and other times when school is not in session). Environment grants support after-school programs that build environmental awareness and understanding; sustainability programs that educate and promote development that meets the needs of the present society without compromising the ability of future generations to meet their own needs; pollution prevention and waSuite management, environmental conservation and biodiversity, and energy and water conservation; and community beautification. Civic and community grants support regional efforts that build local competitiveness and strengthen the economic and social base in Rohm and Haas host communities; programs that cut across multiple giving categories and/or that impact key stakeholders in Rohm and Haas communities; and programs that focus on volunteerism. Health and human services grants support public safety awareness; disaster preparedness; structured youth development such as leadership training; and structured youth activities and recreation. Arts and culture grants support organizational capacity building in the form of plan implementation—technology implementation and/or enhancement, communications and marketing, outreach and audience development, financial management and fundraising, board development, strategic development, professional development, and educational program development. Application and guidelines are available online.
Requirements: Eligible national and international organizations include performing and visual arts, humanities and historical societies, and museums. Preference is given to requests from Chicago, IL, and the greater Philadelphia area.
Restrictions: Grants are not awarded to individuals or for fundraising, advertising, or testimonials.
Amount of Grant: $250-$50,000
Contact: Alexandra Samuels, (215) 592-3644; fax: (215) 592-6808; email: alexandra_samuels@rohmhaas.com
Internet: http://www.rohmhaas.com/community/giving/guidelines/guidelines.html
Sponsor: Rohm and Haas Company
100 Independence Mall W
Philadelphia, PA 19106-2399

Rolex Awards for Enterprise Program **5124**
The program rewards individuals undertaking groundbreaking work in any of five fields: science and medicine, technology and innovation, exploration and discovery, the environment, and cultural heritage. The awards support individuals in carrying out their work rather than acknowledging success after the project's completion. Projects must expand knowledge of the world, improve the quality of life, or contribute in some way to the betterment of humankind. Deadline dates are May 31, for projects from Asia, the Pacific, and the Middle East; July 31, for projects from Africa and North, Central, and South America; and September 30, for projects from Western and Eastern Europe.
Requirements: The program is open to anyone of any age or nationality.
Amount of Grant: $100,000
Date(s) Application is Due: May 31; Jul 31; Sep 30
Contact: Corporate Contributions, (212) 758-7700; fax: (212) 759-6521; email: secretariat@rolexawards.com
Internet: http://www.rolexawards.com/awards/how-to-apply/index.html
Sponsor: Rolex Watch U.S.A
665 Fifth Avenue
New York, NY 10022-5385

Romanell-Phi Beta Kappa Professorship **5125**
The Professorship is awarded annually to scholars in the field of philosophy, without restriction to any one school of philosophical thought. The professorship is intended to recognize not only distinguished achievement but also the recipient's contribution or potential contribution to public understanding of philosophy. The professorship carries with it a stipend of $7,500, which will be awarded in addition to the recipient's regular salary.
Requirements: Although the awardee need not be a member of Phi Beta Kappa, he or she must be on the faculty of an institution sheltering a chapter of Phi Beta Kappa.
Samples: Marilyn Frye, Michigan State University, $7,500 (2007-08); Alexander Rosenberg, Duke University, $7,500 (2006-07); Linda Zagzebski, University of Oklahoma, $7,500 (2005-06).
Amount of Grant: $7,500
Contact: Sandra Beasley, Awards Coordinator, (202) 265-3808; fax: (202) 986-1601; email: SBeasley@pbk.org or info@pbk.org
Internet: http://staging.pbk.org/AM/Template.cfm?Section=Romanell_Professorship&Template=/CM/HTMLDisplay.cfm&ContentID=2295
Sponsor: Phi Beta Kappa Society
1606 New Hampshire Avenue, NW
Washington, DC 20009

Ronald Breslow Award for Achievement in Biomimetic Chemistry **5126**
The purpose of the Award is to recognize outstanding contributions to the field of biomimetic chemistry. Special consideration will be given to individuals who are scientifically active. The award consists of $5,000 and a certificate. Reasonable travel expenses to the meeting at which the award will be presented will be reimbursed.
Requirements: Any individual, except a member of the award committee, may submit one nomination or seconding letter for the award in any given year. The nominating documents consist of a letter of not more than 1,000 words containing an evaluation of the nominee's accomplishments and a specific identification of the work to be recognized, a biographical sketch including date of birth, and a list of publications and patents authored by the nominee.
Restrictions: Self-nominations are not accepted.
Samples: James P. Collman—award recipient, $5,000.; Joanna Aizenberg—award recipient, $5,000 (2008); Francois Diederich—award recipient, $5,000 (2007); Barbara Imperiali—award recipient, $5,000 (2006).
Amount of Grant: $5,000, plus traveling expenses
Date(s) Application is Due: Nov 1
Contact: Felicia Dixon, Awards Administrator; (800) 227-5558 or (202) 872-4408; fax: (202) 776-8008; email: f_dixon@acs.org or awards@acs.org
Internet: http://portal.acs.org/portal/acs/corg/content?_nfpb=true&_pageLabel=PP_ARTICLEMAIN&node_id=1319&content_id=CTP_004551&use_sec=true&sec_url_var=region1
Sponsor: American Chemical Society
1155 Sixteenth Street, NW
Washington, DC 20036

Ronald McDonald Children's Charities Grants **5127**
Ronald McDonald House Charities Board of Trustees is most interested in national and/or international organizations that have a specific program related to children's health and well being. Organizations seeking funding should have a specific program that directly improves the health and well being of children (ages 0-18 years); addresses a significant funding gap or critical opportunity; has long-term impact in terms of replication or reach; and produces measurable results. Types of support include capital grants, challenge/matching grants, formula grants, professorships, program grants, research grants, scholarships (to students from the New York Tri-State McDonald's Restaurants and from the charities), seed money grants, and visiting scholars. The board meets quarterly to review, select, and award grants to organizations that have demonstrated an ability to respond to the needs of specific groups of children in a definitive, hands-on manner that yields measurable results.
Requirements: Nonprofit 501(c)3 tax-exempt organizations are eligible.
Restrictions: The charity does not fund advertising or fund-raising drives; partisan, political, or denominational programs; general and administrative costs; intermediary funding agencies; endowment campaigns; medical research; programs that are strictly local in focus and not replicable; ongoing salaries or travel expenses; or requests that are not in writing.
Samples: Elizabeth Glaser Pediatric AIDS Foundation (Santa Monica, CA)—for its program to help prevent mother-to-child HIV transmission in Cameroon, $300,000 (2005).
Amount of Grant: $100-$10 million; $250,000 average
Date(s) Application is Due: May 3; Aug 2; Dec 15
Contact: Grants Manager, (630) 623-7048; fax: (630) 623-7488
Internet: http://www.rmhc.org/grant/index.html
Sponsor: Ronald McDonald Children's Charities
1 Kroc Drive
Oak Brook, IL 60521

Roothbert Fund Scholarships **5128**
The fund awards grants for undergraduate or graduate study at U.S. accredited colleges or universities. The fund does not emphasize any particular form of religious practice or worship, but seeks to provide support to persons motivated by spiritual values. Scholarships have been awarded to persons entering a wide range of careers; preference is given to applicants who can satisfy high scholastic requirements and are considering

careers in education. Applicant interviews are scheduled on extremely short notice in March and are held in New York, New Haven, Philadelphia and Washington, DC. The fund typically awards 20 new scholarships each year. Scholarships may be renewed. Applications are available each year between November 1 and January 25.
Requirements: As a rule, the fund awards scholarships to applicants whose current or permanent address is located in one of the following states and applicants planning to move to one of such states prior to March 1 of the year in which the application is made: Maine, New Hampshire, Vermont, Rhode Island, Massachusetts, Connecticut, New York, New Jersey, Pennsylvania, Ohio, Delaware, Maryland, District of Columbia, Virginia, West Virginia, or North Carolina.
Geographic Focus: Connecticut; Delaware; District of Columbia; Maine; Maryland; Massachusetts; New Hampshire; New Jersey; New York; North Carolina; Ohio; Pennsylvania; Rhode Island; Vermont; Virginia; West Virginia
Amount of Grant: $2000-$3000
Date(s) Application is Due: Jan 31
Contact: Scholarship Administrator, (212) 870-3116; email: mailroothbertfund.org
Internet: http://www.roothbertfund.org/scholarships.php
Sponsor: Roothbert Fund
475 Riverside Drive, Room 252
New York, NY 10115

Rosalinde and Arthur Gilbert Foundation/AFAR New Investigator Awards in Alzheimer's Disease 5129

The major goal of this program is to support important research in areas in which more scientific investigation is needed to improve the prevention, diagnosis, and treatment of Alzheimer's Disease. The program will also serve to encourage junior investigators in the United States and Israel to pursue research and academic careers in the neurosciences, and Alzheimer's disease in particular. Projects in basic and translational research related to Alzheimer's disease (AD) that are clinically relevant, will be considered. Examples of promising areas of research include, but are not limited to: basic mechanisms of aging in the central nervous system; genetics of AD; neuroimaging and precursors of AD; cellular and Molecular pathways of AD; biological Markers of AD; exercise and dietary factors; neurogenesis and AD; impact of environmental agents in CNS aging and early AD; inflammation; cardiovascular and cerebrovascular factors; drug discovery. It is anticipated that 5 grants of up to $75,000 each will be awarded. Applicants may propose to use the award over the course of one or two years as justified by the proposed research. Funds may not be requested for overhead or indirect costs. Funding will begin July 1.
Requirements: The applicant must be an independent investigator with assigned independent space and must have received a junior faculty appointment (instructor, assistant professor or equivalent) no sooner then July 1st. The proposed research must be conducted at any type of not-for-profit setting in the United States or Israel. Five criteria are used to determine the merit of an application: qualifications of the applicant; quality of the proposed research; relevance of the proposal to how mechanisms of aging may lead to AD; excellence of the research environment; likelihood that the project will advance the applicant's career in basic research on the mechanisms of aging and AD. All candidates must submit applications endorsed by their institution. The deadline for receipt of applications and supporting materials from candidates based in the U.S. is December 16th, at 5:00 p.m. EST. The deadline for receipt of applications and supporting materials from candidates based in Israel is December 30th. See website for instruction sheet and application for complete application procedures. Incomplete applications cannot be considered.
Restrictions: The Gilbert/AFAR Research Grant Program does not provide support for: Postdoctoral fellows in the laboratory of a senior investigator; investigators who have already received major extramural funding for research on Alzheimer's disease and/or aging (such as an R01 grant); Senior faculty, i.e. at the rank of Associate Professor level or higher; NIH Intramural program employees. Applicants cannot apply for the AFAR Research Grant Program or Glenn/AFAR BIG Program.
Geographic Focus: Alabama; Alaska; Arizona; Arkansas; California; Colorado; Connecticut; Delaware; District of Columbia; Florida; Georgia; Guam; Hawaii; Idaho; Illinois; Indiana; Iowa; Kansas; Kentucky; Louisiana; Maine; Maryland; Massachusetts; Michigan; Minnesota; Mississippi; Missouri; Montana; Nebraska; Nevada; New England; New Hampshire; New Jersey; New Mexico; New York; North Carolina; North Dakota; Northeast; Northwest; Ohio; Oklahoma; Oregon; Pennsylvania; Puerto Rico; Rhode Island; South Carolina; South Dakota; Tennessee; Texas; U.S. Virgin Islands; Utah; Vermont; Virginia; Washington; West Virginia; Wisconsin; WyomingIsrael
Amount of Grant: $75,000
Date(s) Application is Due: Dec 16 U.S.A; Dec 30 Israel
Contact: Grants Manager; (212) 703-9977 or (888) 582-2327; fax: (212) 997-0330; email: grants@afar.org or info@afar.org
Internet: http://afar.org/Gilbert.html
Sponsor: American Federation for Aging Research
55 West 39th Street, 16th Floor
New York, NY 10018

Rosalynn Carter Institute John and Betty Pope Fellowships 5130

The John and Betty Pope Fellowship in Caregiving is a unique program of the Rosalynn Carter Institute for Caregiving at Georgia Southwestern State University. The fellowship provides financial support for outstanding individuals pursuing training and careers in fields related to caregiving, and provides a structure and course of study intended to prepare them for leadership in the field. The certificate program of study includes courses in psychology, nursing, sociology and special education taken over up to four years of study at GSW. The amount of the award is determined annually and is currently set at $3750 per academic year.
Requirements: Award of Pope Fellowships is currently limited to students pursuing a degree program at Georgia Southwestern State University and who are or will be enrolled in the Caregiver Specialist Certificate at GSW. Applicants must have a grade point average of 3.0 as well as an established interest in academic study related to caregiving or a record of outstanding professional or informal caregiving. In addition, Pope Fellows must: enroll for minimum of 12 credit hours each semester (undergraduate) or 9 credit hours (graduate) for a full award; attend Pope Fellowship orientation and other required meetings; and make steady progress in fulfilling requirements of the Caregiver Specialist Certificate.
Amount of Grant: $6000 per year graduate Pope fellows; $3000 per year undergraduate Pope fellows; $3000 per year graduate scholarships; $1500 per year undergraduate scholarships
Date(s) Application is Due: Feb 15
Contact: Leisa Easom, Pope Fellowship Director; (229) 928-1234; fax: (229) 931-2663; email: leasom@gsw.edu or rci@canes.gsw.edu
Internet: http://www.rosalynncarter.org/PopeFellowship/
Sponsor: Georgia Southwestern State University
800 GSW Drive
Americus, GA 31709-4379

Rosenthal Writer-in-Residence Program 5131

The Rosenthal Writer-in-Residence Program brings professional nonfiction writers to the Shorenstein Center to provide an opportunity for them to conduct research and work on a specific project, as well as interact with a community of scholars and students. In addition to a writing project, the Writer will teach student workshops and participate in Shorenstein Center events. Candidates for the program should be writers with an established project and a solid history of published books or essays in the area of press, policy or public affairs. Writers who wish to be considered for the Rosenthal Writer-in-Residence Program should send a one-page letter describing their writing project, along with their curriculum vitae.
Contact: Alex S. Jones; (617) 495-8269; fax: (617) 495-8696; email: alex_jones@harvard.edu
Internet: http://www.hks.harvard.edu/presspol/fellowships/fellowship_rosenthal.html
Sponsor: John F. Kennedy School of Government, Harvard University
79 JFK Street
Cambridge, MA 02138

Ross Foundation Grants 5132

The Ross Foundation, established in 1966, supports organizations involved with education, arts and cultural enrichment, community beautification and improvement, historical preservation, mental health and people who are developmentally disabled, and forestry research and conservation management. Giving is limited to the Arkadelphia and Clark County, Arkansas region. Contact the Foundation for an application form to begin your proposal.
Requirements: Non-profits operating in the Arkadelphia and Clark County, Arkansas region are eligible to apply for funding. Applicants submit the following: copy of IRS Determination Letter; brief history of organization and description of its mission; copy of most recent annual report/audited financial statement/990; listing of board of directors, trustees, officers and other key people and their affiliations; detailed description of project; amount of funding requested, and 6 copies of the proposal.
Geographic Focus: Arkansas
Amount of Grant: $3,000 - $20,000 range
Date(s) Application is Due: Jan1; Mar 31; May 1; July 31; Sep 1; Nov 30;
Contact: Mary Elizabeth Eldridge, Director of Programs; (870) 246-9881; fax: (870) 246-9674; email: meeldridge@rossfoundation.us
Sponsor: Ross Foundation
P.O, Box 335
Arkadelphia, AR 71923

Rotary Foundation Ambassadorial Scholarships 5133

Scholarships further international understanding and friendly relations among peoples of different countries. Academic-year scholarships support one academic year of study or training in another country where Rotary clubs are located. During the time of study abroad, Rotary scholars are expected to be outstanding ambassadors of goodwill to the people of the host country. Scholarships include graduate, undergraduate, or vocational study. Applicant must be nominated by a Rotary club and district. Upon completion of the scholarship, scholars are expected to share the experiences of understanding acquired during the study year with the people of their home countries. Multiyear ambassadorial scholarships may be awarded for two to three years of degree-oriented study abroad and carry flat awards of $12,000 or its equivalent per year. Cultural ambassadorial scholarships cover the cost of three to six months of foreign language study and cultural immersion. Round-trip transportation between home and study city, all required academic fees, some necessary educational supplies, room and board, limited educational travel during award year, and limited contingency expenses up to $19,000 are provided. Applications are considered for candidates interested in studying Arabic, English, French, German, Hebrew, Italian, Japanese, Korean, Mandarin Chinese, Polish, Portuguese, Russian, Spanish, Swahili, and Swedish. Some Rotary districts may only offer one type of scholarship; applicants must check with the local club regarding availability. Application deadline dates vary by local Rotary club. Applications are available on the Web site.
Requirements: Applicants must have completed two years of university work or appropriate professional experience before starting scholarship studies. Applicants must be citizens of countries in which there are Rotary clubs.

Restrictions: The following are ineligible to apply: a Rotarian or honorary Rotarian; an employee of a club, district, or other Rotary entity or of Rotary International; a spouse, a lineal descendent, or an ancestor of any living person in the foregoing categories; or a spouse of a lineal descendent.
Amount of Grant: $12,000-$25,000
Contact: Scholarships Administrator, (847) 866-3000; fax: (847) 328-8554; email: scholarshipinquiries@rotaryintl.org
Internet: http://www.rotary.org/foundation/educational/amb_scho/index.html
Sponsor: Rotary Foundation of Rotary International
One Rotary Ctr, 1560 Sherman Avenue
Evanston, IL 60201

Rotch Traveling Scholarship **5134**
The scholarship, one of the oldest and most prestigious in the United States, provides a stipend to the winner of a two-stage design competition for eight to 10 months of travel throughout the world. An additional payment is awarded upon completion of a report to be added to the scholarship's archives. A detailed statement of eligibility requirements, history of the scholarship, and the year's calendar will be sent with the application form. Requests for application forms must be made in writing and received no later than January 3.
Requirements: An applicant must be a U.S. citizen under 35 years of age on January 1 of the year of the competition. In addition, the applicant must have a degree from an accredited school of architecture plus one full year of professional experience in an architectural office in Massachusetts or have a degree from an accredited Massachusetts school of architecture plus one full year of professional experience in any U.S. architectural firm.
Amount of Grant: $35,000 first prize; $15,000 second prize
Date(s) Application is Due: Jan 3
Contact: Hugh Shepley, Vice President, (617) 951-1433; fax: (617) 951-0845
Internet: http://www.rotchscholarship.org/competition/index.htm
Sponsor: Rotch Traveling Scholarship Committee
52 Broad Street, 4th Fl
Boston, MA 02109

Roy J. Carver Charitable Trust Higher Education Grants **5135**
The trust supports Iowa and western Illinois colleges and universities. The Trust, interests in this area extend to those projects that strengthen institutions and offer innovative learning opportunities to students enrolled in baccalaureate and graduate programs throughout the grant-making region of Iowa and parts of Illinois. Typically, funding is awarded to capital or programmatic initiatives that seek to substantively improve instruction or increase access to educational resources, with the goal of better preparing students for future academic and professional challenges. Projects incorporating interdisciplinary approaches and demonstrating the potential for lasting impact among a diverse range of students are especially encouraged. A majority of awards have centered on teaching and research in the natural sciences.
Requirements: Grants are made to 501(c)3 tax-exempt colleges and universities located in Iowa and western Illinois.
Restrictions: The trust does not support annual campaigns or ongoing operations, direct grants to individuals, religious activities, organizations without 501(c)3 status, fund-raising benefits or program advertising, or political parties or candidates.
Geographic Focus: Illinois; Iowa
Samples: Coe College, Cedar Rapids, IA, Summer research program, $71,250 (2006); Cornell College, Mount Vernon, IA, Academic counseling and library services, $88,000 (2006); Iowa State University, Ames, IA, Research study on secondary-level mathematics instruction, $111,458 (2006).
Amount of Grant: $10,000-$150,000 average
Date(s) Application is Due: Feb 15; May 15; Aug 15; Nov 15
Contact: Dr. Troy Ross, Executive Administrator, (563) 263-4010; fax: (563) 263-1547; email: info@carvertrust.org
Internet: https://www.carvertrust.org/index.php?page=35
Sponsor: Roy J. Carver Charitable Trust
202 Iowa Avenue
Muscatine, IA 52761-3733

Roy J. Carver Charitable Trust Medical and Science Research Grants **5136**
The trust supports Iowa and western Illinois nonprofit organizations and projects in the area of medical and scientific research. The goal of the program in medical and scientific research is to provide support for innovative investigation that may hold great promise for advancing scientific knowledge and improving human health. In this context, the Trust has recognized a variety of university-based scientific endeavors, most notably in the fields of medicine, engineering and the natural sciences. Although grants are occasionally awarded to address institutional capital needs, the focus of this program area is typically on the work of specific laboratories, often emphasizing multidisciplinary approaches and the development of untried, inherently risky research directions. Applicants are encouraged to contact staff prior to the development of a proposal. The board considers requests at quarterly meetings in January, April, July, and October.
Requirements: Grants are made to 501(c)3 tax-exempt scientific organizations in Iowa and western Illinois.
Restrictions: The trust does not support annual campaigns or ongoing operations, direct grants to individuals, religious activities, organizations without 501(c)3 status, fund-raising benefits or program advertising, or political parties or candidates.
Geographic Focus: Illinois; Iowa
Samples: Iowa State University, Ames IA—A study of high-intensity ultrasound technology for clinical applications, $166,458 (2006); University of Illinois, Urbana, IL—Enhance bioinformatics & mass spectrometry resources, $1,804,293 (2006); University of Iowa Foundation, Iowa City, IA—Continuing support for program, endowment and capital needs at the Carver College of Medicine, $6,104,365 (2006).
Amount of Grant: $10,000-$150,000 average
Date(s) Application is Due: Feb 15; May 15; Aug 15; Nov 15
Contact: Dr. Troy Ross, Executive Administrator; (563) 263-4010; fax: (563) 263-1547; email: info@carvertrust.org
Internet: https://www.carvertrust.org/index.php?page=56
Sponsor: Roy J. Carver Charitable Trust
202 Iowa Avenue
Muscatine, IA 52761-3733

Royal Belgian Numismatic Society Research Grants **5137**
Each year the society awards a number of small research grants, available to fellows and nonfellows, for any field of numismatic research. There is no standard application form, but applicants should set out the purpose of their project; what they plan to achieve; detailed costings; and what other sources of finance they have available, have applied for, or intend to apply for. Applications in writing, with the names of two referees, should be submitted; the awards should be announced within two months of the closing date.
Requirements: Contenders must be no more than 35 years old as of the deadline date; participation is not restricted to holders of diplomas, titles, nor to nationality.
Amount of Grant: Bf500
Date(s) Application is Due: Dec 31
Contact: c/o Department of Coins and Medals, email: RNS@dircon.co.uk
Internet: http://www.rns.dircon.co.uk/grants.htm
Sponsor: Royal Belgian Numismatic Society
Great Russell Street, British Museum
London WC1B 3DG United Kingdom

Royal Caribbean Cruises Ocean Fund **5138**
The mission of the Ocean Fund is to support efforts to restore and maintain a healthy marine environment, minimize the impact of human activity on this environment, and promote awareness of ocean and coastal issues and respect for marine life. Ocean Fund grants are made to a variety of nonprofit groups and institutions whose activities are directly related to marine conservation. This includes initiatives in research, education, and innovative technologies. A committee of marine experts and cruise line executives meets twice annually to award grants.
Samples: Alaska SeaLife Ctr (AK)—to develop a behind-the-scenes, interactive video program, teaching visitors and satellite-linked groups about ongoing research and marine wildlife rehabilitation, $50,000.; Florida Aquarium (FL)—to upgrade the Beaches and Bays exhibit, $25,000.; U of Miami (FL)—for a graduate fellowship, $40,000.; National Marine Sanctuary Foundation (Key West, FL)—for construction of a multipurpose theater at the new Dr. Nancy Foster Florida Keys Eco-Discovery Ctr, $200,000.
Amount of Grant: $30,000-$50,000 average
Date(s) Application is Due: Sep 30
Contact: Ocean Fund, (305) 539-6573
Internet: http://www.oceanfund.org/about.html
Sponsor: Royal Caribbean Cruises
1050 Caribbean Wy
Miami, FL 33132-2096

Royal Scottish Geographical Society Expedition Grants **5139**
The RSGS is an educational charity the aims of which are to advance the science of Geography and create a greater understanding of the wider world, in the belief that geographical knowledge makes a significant difference to the way in which we view and shape the world we live in. The Society wishes to encourage expeditions having a geographical research and exploration component, where the research projects are integrated in an overall objective. The expedition grants committee will give preference to projects with a substantial geographical content. If the expedition is not in the fields of Human or Physical Geography, there should be a geographical aspect of a related discipline with the emphasis on spatial distributions, ecology, environmental science, ethnology, surveying or exploration.
Requirements: The closing date is January 31 for expeditions planned to take place in the following 15 months. Essential criteria include: Expedition composed of at least three persons; University expeditions to spend a minimum of three weeks in the field, and school expeditions a minimum of two weeks; Expedition should originate in Scotland or have at least one Scottish member; Geographical / scientific research program planned; Relevant training / expertise among party; Realistic logistical and financial planning underway. The person nominated as your contact person should also be readily accessible, and should respond promptly to any correspondence from the Society. Download the required application form from the sponsor's website.
Restrictions: Grants are not available for purely sporting or adventure expeditions. Expeditions composed of teams whose members are each pursuing research on individual unrelated topics are not favoured, and no support is available for post-graduate field-work for a higher degree, nor for the support of post-doctoral research.
Amount of Grant: 900 GBP
Date(s) Application is Due: Jan 31
Contact: Bruce Maclachlan, Marketing and Publicity Officer; 0[0 44]1361 882 444; email: bruce.maclachlan@virgin.net
Internet: http://www.geo.ed.ac.uk/rsgs/expedits/
Sponsor: Royal Scottish Geographical Society
40 George Street
Glasgow, Scotland G1 1QE United Kingdom

Royal Scottish Geographical Society Small Research Grants 5140

The program is designed to provide individual members of the Society in higher education institutions with a grant not exceeding 1000 GBP in support of geographical research. Typical grants can be in the region of 250 - 500 GBP. This award is given specifically to help with the cost of data collection and fieldwork. Applications are particularly welcomed from postgraduates and those early in their research careers.

Requirements: Research grants are available to individuals only, and applicants must be members of the Royal Scottish Geographical Society. (Membership is open to all.) The required application form can be downloaded from the RSGS website.

Restrictions: No more than one application can be accepted from an individual per year, and re-applications for the same project are not acceptable. Awards cannot be made retrospectively.

Samples: (2006) Leonel Arturo Lopez Toledo, Aberdeen University - Distribution, abundance, and conservation status of the endangered tree species Guaicum sanctum in Mexico. (2005) Dr. Kate T. Smith, University of Edinburgh - A study of catastrophic volcanic floods in Iceland, to assist in future hazard management. Ruth Mottram, University of Street Andrews - Calving glacier dynamics in south-east Iceland. Innes Keighren, University of Edinburgh - A study of the work of the American geographer and author Ellen Churchill Semple. (2004) Iain Cameron, University of Edinburgh - Ecological survey of neotropical savannah to investigate vegetation topography relationships. Andrew Morrison, University of Nottingham - Canada's home children : cultural geographies of displacement, loss, and postcolonial recovery. Kerry Anne Mairs, University of Edinburgh - Geomorphology and settlement in the Faroe Islands : an analysis of human-environment interactions. Edward Wallington, University of Edinburgh - Synthetic aperture radar for neotropical vegetation height retrieval. (2003) Alistair Geddes, Pennsylvania State University - Historical geography of Scottish agricultural statistics, with a view to assessing how statistical projects shaped the development of farming into a modern economic sector (involving extensive use of the Society's Coppock Collection).

Amount of Grant: 1000 GBP maximum

Date(s) Application is Due: Jan 31

Contact: Small Research Grants Administrative Officer; 0141 552 3330; fax: 0141 552 3331; email: rsgs@strath.ac.uk

Internet: http://www.geo.ed.ac.uk/rsgs/grants1.htm

Sponsor: Royal Scottish Geographical Society
40 George Street
Glasgow, Scotland G1 1QE United Kingdom

Royal Scottish Geographical Society Small Travel Grants 5141

The program is designed to provide individual members of the Society in higher education institutions with grants up to a maximum of 500 GBP in support of geographical research. Awards are given specifically to help with the cost of travel and subsistence associated with attendance at conferences and symposia where the applicant intends to present a research paper. Applications for travel grants are welcomed particularly from postgraduates and those early in their research careers. Typical grants are in the region of 250 - 500 GBP.

Requirements: Grants are available to individuals only, and applicants must be members of the Royal Scottish Geographical Society. (Membership is open to all.) The required application form can be downloaded from the RSGS website.

Restrictions: No more than one application can be accepted from an individual per year, and re-applications for the same project are not acceptable. Awards cannot be made retrospectively.

Samples: (2006) Philippa Ascough, University of St. Andrews - International Radiocarbon Conference - Oxford; Dr. Rhian Thomas, University of Glasgow - International Conference on Riverine Hydroecology : Advances in Research and Applications - Stirling; Kerry-Anne Mairs, University of Edinburgh - The View from Here: History and Sociology of the North Atlantic Region Conference - Laval University, Quebec, Canada

Amount of Grant: 500 GBP maximum

Date(s) Application is Due: Jan 31

Contact: Travel Grants Administrative Officer

Internet: http://www.geo.ed.ac.uk/rsgs/grants2.htm

Sponsor: Royal Scottish Geographical Society
40 George Street
Glasgow, Scotland G1 1QE United Kingdom

RRF General Program Grants 5142

The program funds service, education, research, and advocacy projects. The foundation is particularly interested in innovative projects that have the potential to change practice, policy, or delivery systems. The foundation's programs seek to improve the availability and quality of community-based and institutional long-term care programs; expand opportunities for older persons to play meaningful roles in society; support selected basic, applied, and policy research into the causes and solutions of significant problems of the aged; and increase the number of professionals and paraprofessionals adequately prepared to serve the elderly. Proposals fall into three general categories: research, model projects and service, and education and training. Guidelines are available online.

Requirements: Direct service projects that seek to improve the availability and quality of community-based and institutional long-term care programs; expand opportunities for older persons to play meaningful roles in society; support selected applied and policy research into the causes and solutions of significant problems of the aged; increase the number of professionals and paraprofessionals adequately prepared to serve the elderly.

Restrictions: The foundation will only consider proposals from applicants in Illinois, Indiana, Iowa, Kentucky, Missouri, Wisconsin, and Florida. Priority is given to nonprofit organizations serving the Chicago metropolitan area.

Geographic Focus: Florida; Illinois; Indiana; Iowa; Kentucky; Missouri; Wisconsin

Amount of Grant: $15 million total

Date(s) Application is Due: Feb 1; May 1; Aug 1

Contact: Marilyn Hennessy, President, (773) 714-8080; fax: (773) 714-8089; email: info@rrf.org

Internet: http://www.rrf.org/overview/fundinginterests/genprogram.html

Sponsor: Retirement Research Foundation
8765 W. Higgins Road, Suite 430
Chicago, IL 60631-4170

RSC Abbyann D. Lynch Medal in Bioethics 5143

The award was created to honour bioethicist and Past-President of AMS, Professor Abbyann D. Lynch, C.M., Ph.D., for her substantial achievements and support of biomedical ethics in Canada. Its purpose is to raise the profile of ethics in medicine and science by honouring leading Canadian exponents in these fields and bring recognition to their work. The bronze medal and a cash amount of $2,000 are offered every year if there is a suitable candidate.

Requirements: The medal is awarded for a major contribution in bioethics by a Canadian. The contribution may be a book, a report, a scholarly article, a monograph or series of articles, which will normally be published in the two years preceding the nomination. Candidates shall be Canadian citizens or persons having permanent resident status, and who have lived in Canada during the three years preceding the date of nomination. Nominations shall be made by three persons with the initiator being a Fellow of the Society. The form for nomination can be downloaded from the sponsor's website.

Restrictions: Self-nomination for an award is not permitted.

Samples: 2007 - Eike-Henner W. Kluge; 2006 - Christine Overall; 2005 - Jocelyn Downie; 2003 - Hubert Doucet.

Amount of Grant: $2,000 CAN

Date(s) Application is Due: Dec 1

Contact: Erin Hearty, Officer, Fellowship & Awards; (613) 991-6990; fax: (613) 991-6996; email: fellowship@rsc-src.ca

Internet: http://www.rsc.ca/index.php?page_id=61&lang_id=1&award_id=16

Sponsor: Royal Society of Canada
170 Waller Street
Ottawa, ON K1N 9B9 Canada

RSC Alice Wilson Award 5144

At the behest of the Committee for Advancement of Women in Scholarship, the award is given yearly to a woman of outstanding academic qualifications who is entering a career in scholarship or research at the postdoctoral level. The recipient is chosen from the current year's female winners of postdoctoral fellowships from the Medical Research Council, the Social Sciences and Humanities Research Council, and the Natural Sciences and Engineering Research Council, each council in turn by year. The award consists of a diploma and cash award.

Requirements: Society fellows may nominate Canadian citizens and persons who have been Canadian residents for the five years preceding the date of the attribution of the award.

Samples: 2007 - Melanie Ann Huntley (NSERC); 2006 - Teresa Liu-Ambrose (CIHR); 2005 - Candice L. Odgers (SSHRC); 2004 - Miaki Ishii (NSERC); 2003 - M. Joan Saary (CIHR)

Amount of Grant: $1,000 CAN

Date(s) Application is Due: Dec 1

Contact: Jeanne Salo, Officer, Fellowship and Awards; (613) 9007; email: jsalo@rsc.ca

Internet: http://www.rsc.ca/index.php?page_id=61&lang_id=1&award_id=11

Sponsor: Royal Society of Canada
170 Waller Street
Ottawa, ON K1N 9B9 Canada

RSC Bancroft Award 5145

Awarded every two years if there is a suitable candidate, this award is given for publication, instruction, and research in the geological and geophysical sciences that have noticeably contributed to public understanding and appreciation of the subject. Nominations must be received by the listed application deadline. A diploma is included with the award.

Requirements: Candidates are to be to Canadian citizens or persons who have been Canadian residents for the five years prior to award date. Nominations must be made by three persons with the initiator being a fellow of the society and consist of citation of not more than 250 words stating concisely why the nominee should be considered, curriculum vita, list of publications, and detailed appraisal of nominee's work. The nomination dossier will remain valid for five years from the year of first nomination.

Restrictions: Self-nomination for an award is not permitted.

Samples: 2006 - David J. Dunlop; 2004 - William Richard Peltier; 2002 - John J. Clague; 2000 - Jan Veizer.

Amount of Grant: $2,500 CAN

Date(s) Application is Due: Dec 1

Contact: Jeanne Salo, Officer, Fellowship and Awards; (613) 9007; email: jsalo@rsc.ca

Internet: http://www.rsc.ca/index.php?page_id=61&lang_id=1&award_id=11

Sponsor: Royal Society of Canada
170 Waller Street
Ottawa, ON K1N 9B9 Canada

RSC Jason A. Hannah Medal 5146

The Jason A. Hannah Medal was established in 1976 by RSC with the assistance of Associated Medical Services Incorporated, through its Hannah Institute for the History of Medicine. The medal was established to honor the late President and Managing Director of AMS Inc., Dr. Jason A. Hannah, and to bring recognition to the work of Canadian research in the history of medicine. The award is normally made for a book published in the two years preceding its nomination. The bronze medal and a cash amount are offered every year if there is a suitable nomination.
Requirements: The medal is awarded for an important Canadian publication in the history of medicine. In this context, the word 'Canadian' refers to the citizenship or residence of the author, or to a content clearly relevant to Canadian medicine and health care. Nominations shall be made by three persons with the initiator being a Fellow of the Society.
Samples: (2007) Michael Bliss, 'Harvey Cushing: A Life in Surgery'. (2006) James Opp, 'Home, Work, and Play: Situating Canadian Social History, 1840-1980'. (2005) David and Rosemary R. Gagan, 'or Patients of Moderate Means: A Social History of the Voluntary Public General Hospital in Canada, 1890-1950'. (2004) F. J. Paul Hackett, 'A Very Remarkable Sickness: Disease and Epidemics in the Petit Nord, 1670-1846'.
Amount of Grant: $1,500 CAN
Date(s) Application is Due: Mar 1
Contact: Erin Hearty, Officer, Fellowship & Awards; (613) 991-6990; fax: (613) 991-6996; email: fellowship@rsc-src.ca
Internet: http://www.rsc.ca/index.php?page_id=61&lang_id=1&award_id=14
Sponsor: Royal Society of Canada
170 Waller Street
Ottawa, ON K1N 9B9 Canada

RSC John L. Synge Award 5147

The award, established to honor John Lighton Synge, one of the first mathematicians working in Canada to obtain international recognition by research in mathematics, is given for outstanding research in any of the branches of the mathematical sciences. The award consists of a diploma and a cash amount, and is awarded at irregular intervals if there is a suitable candidate.
Requirements: The award is given for outstanding research in any of the branches of the mathematical sciences. Some preference is given to candidates whose age is not over 40 in the year of the award. Society fellows may nominate Canadian citizens or persons having permanent resident status, and who have lived in Canada during the three years preceding the date of nomination.
Restrictions: Self-nomination for an award is not permitted.
Samples: (2006) Stephen A. Cook, Department of Computer Science, University of Toronto; (1999) George A. Elliott, Department of Mathematics, University of Toronto; (1996) Joel S. Feldman.
Amount of Grant: $2,500 CAN
Date(s) Application is Due: Dec 1
Contact: Jeanne Salo, Officer, Fellowship and Awards; (613) 9007; email: jsalo@rsc.ca
Internet: http://www.rsc.ca/english/awards_synge.html
Sponsor: Royal Society of Canada
170 Waller Street
Ottawa, ON K1N 9B9 Canada

RSC McLaughlin Medal 5148

This medal and cash award are given annually for important research of sustained excellence in any branch of medical sciences. Nominations must be received by the listed application deadline.
Requirements: Candidates are to be Canadian citizens or persons who have been Canadian residents for the five years prior to award date. Nominations must be made by three persons, with the initiator being a fellow of the society, and consist of citation of not more than 250 words stating concisely why the nominee should be considered, curriculum vita, list of publications, and detailed appraisal of nominee's work. The nomination dossier will remain valid for three years from the year of first nomination.
Restrictions: Self-nomination for an award is not permitted.
Samples: 2007 - John R. G. Challis. 2006 - Michael D. Tyers. 2005 - Robert E. W. Hancock. 2004 - John J.M. Bergeron. 2003 - Robert G. Korneluk.
Amount of Grant: $2,500 CAN
Date(s) Application is Due: Dec 1
Contact: Erin Hearty, Officer, Fellowship & Awards; (613) 991-6990; fax: (613) 991-6996; email: fellowship@rsc-src.ca
Internet: http://www.rsc.ca/index.php?page_id=61&lang_id=1&award_id=17
Sponsor: Royal Society of Canada
170 Waller Street
Ottawa, ON K1N 9B9 Canada

RSC Miroslaw Romanowski Medal 5149

The medal is awarded for significant contributions to the resolution of scientific aspects of environmental problems or for important improvements to the quality of an ecosystem in all aspects, terrestrial, atmospheric, and aqueous, brought about by scientific means. The award, which consists of a medal and cash award, is made every year if there is a suitable candidate. Nominations must be received by the listed application deadline. An annual lecture series for the award recipient is also associated to the medal.
Requirements: Society fellows may nominate Canadian citizens and persons who have been Canadian residents for the five years preceding the date of the attribution of the award.
Restrictions: Self-nomination for an award is not permitted.
Samples: 2007 - Christopher M. Wood. 2006 - Richard Peltier. 2005 - Robie Macdonald. 2004 - Derek C.G. Muir. 2003 - Peter J. Dillon.
Amount of Grant: $3000 CAN
Date(s) Application is Due: Dec 1
Contact: Jeanne Salo, Officer, Fellowship and Awards; (613) 9007; email: jsalo@rsc.ca
Internet: http://www.rsc.ca/index.php?page_id=61&lang_id=1&award_id=6
Sponsor: Royal Society of Canada
170 Waller Street
Ottawa, ON K1N 9B9 Canada

RSC Pierre Chauveau Medal 5150

The Pierre Chauveau Medal was established in 1951 to honour the memory of Pierre J.O. Chauveau (1820-1890), FRSC, writer, orator, educator, Canadian statesman and the second President of RSC (1883-1884). He was the first Premier of Quebec (1867-1872) and Speaker of the Senate (1874). The medal is awarded for a distinguished contribution to knowledge in the humanities other than Canadian literature and Canadian history. The silver medal is offered every two years if there is a suitable candidate.
Requirements: Candidates shall be Canadian citizens or persons having permanent resident status, and who have lived in Canada during the three years preceding the date of nomination. Nominations shall be made by three persons with the initiator being a Fellow of the Society. The form for nomination can be downloaded from the sponsor's website.
Restrictions: Self-nomination for an award is not permitted.
Samples: 2005 - Shana Poplack; 2003 - Gilles Bibeau; 2001 - Paul-Hubert Poirier; 1999 - Michael Millgate.
Date(s) Application is Due: Dec 1
Contact: Jeanne Salo, Officer, Fellowship and Awards; (613) 9007; email: jsalo@rsc.ca
Internet: http://www.rsc.ca/index.php?page_id=61&lang_id=1&award_id=3
Sponsor: Royal Society of Canada
170 Waller Street
Ottawa, ON K1N 9B9 Canada

RSC Rutherford Memorial Medals 5151

Two awards are given annually, one in physics and one in chemistry, if there are suitable candidates for outstanding research in any branch of physics and chemistry. In recognition of Lord Rutherford's own research carried out in Canada at a relatively young age, some preference will be given to candidates whose age is not over 40 in the year of the award. Whenever possible, the society subsidizes the travel expenses of the award winners to attend the annual meeting at which they are to receive the awards. Nominations must be received by the listed application deadline. Each gold plated silver medal is accompanied by a cash amount.
Requirements: Candidates are to be Canadian citizens or persons having permanent resident status, and who have lived in Canada during the three years preceding the date of nomination. Nominations must be made by three persons, with the initiator being a fellow of the society, and consist of a citation of not more than 250 words stating concisely why the nominee should be considered, curriculum vita, list of publications, and detailed appraisal of nominee's work. The nomination dossier will remain valid for three years from the year of first nomination.
Restrictions: Self-nomination for an award is not permitted.
Samples: 2007 - Gregory D. Scholes (Chem), Victoria Michelle Kaspi (Phys); 2006 - Molly Shoichet (Chem), Aephraim M. Steinberg (Phys); 2005 - Jillian M. Buriak (Chem), Peter Grutter (Phys); 2004 - Andrew Woolley (Chem), Sajeev John (Phys).
Amount of Grant: $2,500 CAN
Date(s) Application is Due: Dec 1
Contact: Jeanne Salo, Officer, Fellowship and Awards; (613) 9007; email: jsalo@rsc.ca
Internet: http://www.rsc.ca/index.php?page_id=61&lang_id=1&award_id=7
Sponsor: Royal Society of Canada
170 Waller Street
Ottawa, ON K1N 9B9 Canada

RSC Sir John William Dawson Medal 5152

The award honors Sir John William Dawson, the first president of the society and the foremost Canadian scientist and educator of his day: geologist, naturalist, teacher, author, man of religion, educational administrator, and promoter of learning. The award is made every two years, if there is a suitable candidate, and honors important and sustained contributions by one individual in at least two different domains in the general areas of interest of the society or in interdisciplinary research. Nominations must be received by the listed application deadline. The award includes a silver medal and a cash amount
Requirements: Candidates shall be Canadian citizens or persons having permanent resident status, and who have lived in Canada during the three years preceding the date of nomination. Nominations shall be made by three persons with the initiator being a Fellow of the Society. The form for nomination can be downloaded from the sponsor's website.
Restrictions: Self-nomination for an award is not permitted.
Samples: 2007 - Roderick Alexander Macdonald; 2005 - Jean-Charles Chebat; 2003 - Martin L. Friedland; 2001 - Maurice Ptito.
Amount of Grant: $2500 CAN
Date(s) Application is Due: Dec 1
Contact: Jeanne Salo, Officer, Fellowship and Awards; (613) 9007; email: jsalo@rsc.ca
Internet: http://www.rsc.ca/index.php?page_id=61&lang_id=1&award_id=4
Sponsor: Royal Society of Canada
170 Waller Street
Ottawa, ON K1N 9B9 Canada

RSC Thomas W. Eadie Medal 5153

The medal is awarded annually in recognition of major contributions to any field through engineering or applied science, with preference given to those having an impact on communications, in particular the development of the Internet. Nominations must be received by the listed application deadline. The award includes a bronze medal and a cash amount.
Requirements: Candidates are to be Canadian citizens or persons who have been Canadian residents for the five years prior to award date. Nominations must be made by three persons, with the initiator being a fellow of the society, and consist of citation of not more than 250 words stating concisely why the nominee should be considered, curriculum vita, list of publications, and detailed appraisal of nominee's work. The nomination dossier will remain valid for five years from the year of first nomination.
Restrictions: Self-nomination for an award is not permitted.
Samples: 2007 - Hussein T. Mouftah; 2006 - Alberto Leon-Garcia; 2005 - Norman C. Beaulieu; 2004 - Vijay K. Bhargava; 2003 - Morrel P. Bachynski.
Amount of Grant: $3000 CAN
Date(s) Application is Due: Dec 1
Contact: Erin Hearty, Officer, Fellowship & Awards; (613) 991-6990; fax: (613) 991-6996; email: fellowship@rsc-src.ca
Internet: http://www.rsc.ca/index.php?page_id=61&lang_id=1&award_id=5
Sponsor: Royal Society of Canada
170 Waller Street
Ottawa, ON K1N 9B9 Canada

RTKL Traveling Fellowship 5154

The RTKL Traveling Fellowship is intended to encourage and support foreign travel undertaken to further education toward a professional degree. The fellowship was established by a $25,000 contribution to the American Architectural Foundation by the architecture firm RTKL in honor of one of its founders, Francis T. Taliaferro. Each year, one $2,500 fellowship is awarded to a student submitting the winning proposal outlining a foreign itinerary which is directly relevant to his or her educational goals.
Requirements: Applicants must complete travel prior to graduation. See website for application form.
Geographic Focus: Washington
Amount of Grant: $2,500
Date(s) Application is Due: Feb 16
Contact: Mary Felber, Program Director, (202) 626-7511; email: mfelber@aia.org; fax: (202) 626-7420
Internet: http://www.archfoundation.org/aaf/aaf/Programs.Fellowships.htm
Sponsor: American Architectural Foundation
1799 New York Avenue NW
Washington, DC 20006

RTNDF Capitol Hill News Internships 5155

The foundation sponsors internships in Washington, DC, for recent college graduates whose career objective is electronic journalism and are interested in political reporting. Preference is given to minority students. The first deadline is for spring internships; second deadline is for summer internships.
Contact: Karen Jackson Bullitt, Manager, Education Programs, (202) 467-5218; fax: (202) 223-4007; email: karenb@rtndf.org
Internet: http://www.rtnda.org/asfi/internships/internships.shtml#capitolhill
Sponsor: Radio and Television News Directors Foundation
1600 K Street NW, Suite 700
Washington, DC 20006-2838

Rural Student Scholarships 5156

Two one-year scholarships are awarded annually to offer practicing rural teachers an opportunity to pursue education and training that would not otherwise be affordable within their districts. The scholarships will be awarded to rural educators who work with students with disabilities or who are currently retooling from a regular education career to work with students with disabilities. Scholarships are to be used during the academic year following notification of awards.
Requirements: Applicants must be currently employed by a rural school district as a certified teacher in regular or special education; working with students with disabilities, or with regular education students and retooling for work in special education; and pursuing a goal of increasing skills in special education, or retooling from a regular education to a special education career.
Amount of Grant: $1000 maximum
Date(s) Application is Due: Feb 25
Contact: Dr. David Forbush, email: davidf@cc.usu.edu
Internet: http://extension.usu.edu/acres/scholarships.html
Sponsor: American Council on Rural Special Education
Utah State University, 2865 Old Main Hill
Logan, Utah 84322-2865

Russell Berrie Foundation Grants 5157

The foundation supports nonprofit organizations primarily in Israel and the New York and New Jersey metropolitan areas. Grants are awarded for education (colleges and universities, Jewish education, divinity schools, parochial elementary and secondary schools, and religious higher education), health care (hospitals and respite homes), international programs (development programs, ministries, and missions), religious organizations (religious welfare, synagogues, and temples), social issues, and youth welfare. Types of support include capital support and general operating support. Applicants should submit a brief letter of inquiry and include a description of the program and the organization, including a financial statement. There are no application deadlines.
Requirements: Christian, Jewish, and Roman Catholic organizations in New York, New Jersey, and Israel are eligible.
Geographic Focus: New Jersey; New YorkIsrael
Samples: Pave the Way Foundation (New York, NY)—for unrestricted support of this organization that promotes good will and reconciliation between different religions, $100,000; American Technion Society (New York, NY)—to establish a nanotechnology institute at the Technion-Israel Institute of Technology, in Haifa, $26 million.
Amount of Grant: $5000-$20,000 typically
Contact: Susan Strunk, Administrative Director, (201) 928-1880
Internet: http://www.russberrie.com/foundation.html
Sponsor: Russell Berrie Foundation
300 Frank Burr Boulevard, Building E, 7th Fl
Teaneck, NJ 07666-6704

Russell Family Foundation Environmental Sustainability Grants 5158

The foundation is committed to improving protection of the environment in western Washington, with an emphasis on the waters of Puget Sound. The foundation focuses its grantmaking in this program on Puget Sound, Environmental Education and Green Business and supports systemic and community-centered strategies that encourage collaboration across sectors and, where possible, leverage other support, build partnerships and reach new audiences. Though not a requirement for funding, they welcome projects that effectively address and connect several of the above-mentioned areas of interest.
Requirements: A Letter of Inquiry is required prior to any formal application. If the foundation is interested in considering a grant request, they will ask for a more detailed proposal. Full proposals are accepted by invitation only.
Restrictions: Requests are accepted from 501(c)3 organizations. Nonprofit entities such as public schools and school districts also may apply. Organizations must be located in and/or provide services within the Puget Sound region. The region includes 12 counties: Clallam, Jefferson, Kitsap, Mason, Thurston, Pierce, King, Snohomish, Skagit, Whatcom, Island, and San Juan.
Geographic Focus: Washington
Amount of Grant: $1,000 - $35,000
Date(s) Application is Due: Jan 8; Jul 16
Contact: Grant Administrator, email: info@trff.org
Internet: http://www.trff.org/grant_programs/env_education.asp
Sponsor: Russell Family Foundation
P.O. Box 2567
Gig Harbor, WA 98335

Russell Sage Foundation Grants 5159

The foundation pursues four programs: a program of research on the future of work concerned with the causes and consequences of the decline in demand for low-skill workers in advanced economies; a program of research on current U.S. immigration that focuses on the adaptation of the second generation to American society; the social psychology of cultural contact that focuses on improving relations between racial and ethnic groups in schools, workplaces, and neighborhood settings; and the examination of social inequality on a number of dimensions, including family well-being, educational quality and opportunity, health care and coverage, legal services and criminal justice, political participation and representation, banking and credit, neighborhoods and housing, pension provision, environmental quality, and even access to computers and the Internet. The board meets in June and November. Unless otherwise specified in the request for proposals in a particular research competition, the deadline for submissions to the November funding round is mid-August, and to the June round is mid-March.
Requirements: Higher education institutions, research groups, and researchers with PhD degrees may submit grant applications. All funded research is U.S.-focused.
Restrictions: Support is not granted for predoctoral research.
Amount of Grant: $35,000 - $500,000
Contact: Grants Administrator; (212) 750-6000; fax: (212) 371-4761; email: info@rsage.org
Internet: http://www.russellsage.org/programs/index.shtml
Sponsor: Russell Sage Foundation
112 East 64th Street
New York, NY 10021

Ruth and Vernon Taylor Foundation Grants 5160

The foundation awards grants to nonprofit organizations in the areas of arts and humanities, civic and public affairs, education, environment, health, and social services. Types of support include general operating support, building construction/renovation, endowment funds, and research. The foundation suggests that initial contact be made in writing.
Requirements: Organizations located in Colorado, Illinois, Montana, New Jersey, New York, Pennsylvania, Texas, or Wyoming are eligible.
Restrictions: Grants are not awarded to individuals.
Geographic Focus: Colorado; Illinois; Montana; New Jersey; New York; Pennsylvania; Texas; Wyoming
Amount of Grant: $1000-$20,000 average
Contact: Friday Green, Trustee, (303) 893-5284
Sponsor: Ruth and Vernon Taylor Foundation
518 17th Street, Suite 1670
Denver, CO 80202

RWJF Active Living Research Grants 5161

Active Living Research supports research examining how environments and policies impact physical activity, especially among ethnic minorities and children living in low-income communities. Findings are expected to inform environmental and policy changes that will promote active living among children and families. the Foundation encourages experts in fields such as public health, public administration, law, economics, transportation, medicine, recreation, urban planning, architecture, the behavioral sciences, child care and education to work together to identify promising approaches for increasing physical activity and preventing obesity among children and families.
Requirements: Requirements vary from one Call for Proposals to the next.
Restrictions: The Program does not provide support for international activities or programs, or institutions in other countries.
Contact: James F. Sallis, Ph.D., Program Director; (619) 260-5534; fax: (619) 260-1510; email: ALR@projects.sdsu.edu
Internet: http://www.activelivingresearch.org/grantsearch
Sponsor: Robert Wood Johnson Foundation
3900 Fifth Avenue, Suite 310
San Diego, CA 92103

RWJF Bridging the Gap: Research Informing Practice and Policy for Healthy Youth Behavior 5162

RWJF has developed three integrated strategies to reverse the childhood obesity epidemic: evidence, action and advocacy. The Bridging the Gap: Research Informing Practice and Policy for Healthy Youth Behavior program is intended to improve understanding of school, community, state and national policies and environmental factors affecting youth diet, physical activity, obesity, and tobacco, alcohol and drug use, and to evaluate the effectiveness of interventions to prevent youth obesity and tobacco use.
Requirements: Eligible applicants for research grants are scholars associated with educational institutions, research organizations, health care providers or other public or nonprofit organizations. Preference will be given to public entities or nonprofit organizations tax exempt under Section 501(c)3 of the Internal Revenue Code. Applicants may be independent of or associated with organizations currently receiving grants. All proposals must be submitted only through the RWJF Grantmaking Online system.
Contact: C. Tracy Orleans; (609) 627-5962 or (877) 843-7953; email: cto@rwjf.org
Internet: http://www.rwjf.org/grants/npo.jsp?FUND_ID=55100
Sponsor: Robert Wood Johnson Foundation
College Road East and Route 1, P.O. Box 2316
Princeton, NJ 08543-2316

RWJF Changes in Health Care Financing and Organization Grants 5163

The program supports policy analysis, research, evaluation and demonstration projects that provide policy leaders timely information on health care policy and financing issues. Supported projects include: examining significant issues and interventions related to health care financing and organization and their effects on health care costs, quality and access; and exploring or testing major new ways to finance and organize health care that have the potential to improve access to more affordable and higher quality health services. The grant period is 12 or fewer months for small grants and 12 or more months for large grants. There are no application deadlines. Program information and the online application process are available online.
Requirements: Researchers, as well as practitioners and public and private policy-makers working with researchers, are eligible to apply. Projects may be initiated from within many disciplines, including health services research, economics, sociology, political science, public policy, public health, public administration, law and business administration. Multidisciplinary teams and researchers who are just beginning their careers,Äîperhaps teaming with a more senior researcher to develop the analytic approach,Äîare especially encouraged to apply. Preference will be given to applicants that are public agencies or are tax-exempt under Section 501(c)3 of the Internal Revenue Code and are not private foundations as defined under Section 509(a). Proposals for this program must be submitted through the RWJF Grantmaking Online system.
Restrictions: Grant funds may not be used to subsidize individuals for the cost of health care, to construct or renovate facilities, for lobbying, or as a substitute for funds currently being used to support similar activities.
Samples: U of Southern Maine (ME)—to support the research of Dr. J. William Thomas, Using Physician Profiling Software to Evaluate the Practice Efficiency of Physician Specialists, $102,829.; Georgetown U (WA)—to support the research of Mila Kofman, JD, Private Insurance Markets: The Missing Link-Association Health Plans and Other Pooled Purchasing Arrangements, $230,547 (2003); U of Kansas Medical Center Research Institute (KS)—to support the research of Dr. Mary Zimmerman, Local Funding for Health Services in Rural Counties, $100,000.
Amount of Grant: $100,000 maximum small grants; $100,000 minimum large grants
Contact: Program Officer; (202) 292-6700; fax: (202) 292-6800; email: hcfo@academyhealth.org
Internet: http://www.rwjf.org/applications/solicited/cfp.jsp?ID=19274
Sponsor: Robert Wood Johnson Foundation
College Road East and Route 1, P.O. Box 2316
Princeton, NJ 08543-2316

RWJF Clinical Scholars Program Grants 5164

The program fosters the development of leaders who will transform U.S. health and health care. These future leaders will be equipped to work with communities and to conduct innovative research as a result of the didactic and experiential learning they receive at each of four chosen training sites (University of California, Los Angeles; University of Michigan; University of Pennsylvania; and Yale University). The program's focus emphasizes community-based research and leadership training. The goal of the program is the integration of scholars' clinical training with skills in program development and research methodology to help them find solutions for issues in public policy, community health, and health services research. The program offers graduate-level study and research as part of a university-based postresidency training program. Up to 29 scholars will be selected for appointments beginning the following year. Under the program, the universities provide Scholars with their stipends and health insurance. Stipends currently range from $49,000 to $51,500 per year, depending on the training site. Limited financial support is also available at each university for expenses associated with Scholars' research, including travel.
Requirements: To be eligible, physicians must be U.S. citizens; be committed to a career consonant with the program's purposes and priorities; be highly regarded by those responsible for their clinical training; and plan to complete the clinical requirements of their residency/fellowship training by the date of entry into the program (except for surgeons). Scholars may not hold appointments as sub-specialty fellows during their tenure in the program.
Amount of Grant: $49,000 - $51,500 per year
Date(s) Application is Due: Feb 15
Contact: Desmond Kimo Runyan, Program Director; (919) 843-8261; fax: (650) 566-2340; email: drunyan@med.unc.edu or rwjcsp_admin@med.unc.edu
Internet: http://rwjcsp.unc.edu/
Sponsor: Robert Wood Johnson Foundation
333 MacNider Hall
Chapel Hill, NC 27599-7105

RWJF Evaluating Innovations in Nursing Education Grants 5165

Evaluating Innovations in Nursing Education (EIN) will support evaluations of interventions that expand teaching capacity or promote faculty recruitment and retention in nursing schools. The program aims to increase the number of nursing school graduates by evaluating strategies that address the nurse faculty shortage. An important activity of the program will be disseminating successful strategies so they can be replicated in other nursing education settings.
Requirements: Nursing schools or health care organizations who will act as the applicant institution and subcontract to the evaluation team are eligible. Preference will be given to applicant organizations that are tax-exempt under Section 501(c)3 of the Internal Revenue Code and are not private foundations as defined under Section 509(a). The Robert Wood Johnson Foundation makes grants to organizations in the United States and its territories only.
Amount of Grant: $300,000 maximum
Date(s) Application is Due: Nov 24
Contact: Program Director; (732) 932-4670; email: info@evaluatinginnovationsinnursing.org
Internet: http://www.evaluatinginnovationsinnursing.org/
Sponsor: Robert Wood Johnson Foundation
College Road East and Route 1, P.O. Box 2316
Princeton, NJ 08543-2316

RWJF Finding Answers: Disparities Research for Change Grants 5166

The program was created to support rigorous evaluations of practical and replicable solutions designed to reduce and eliminate disease specific racial and ethnic health care disparities. It seeks to improve the quality of health care provided to patients affected by racial and ethnic disparities by: Granting funds to discover and evaluate practical and replicable solutions designed to reduce and eliminate disease specific racial and ethnic health care disparities; Focusing on interventions aimed at health care delivery for one or more of the following health concerns: cardiovascular disease, depression, or diabetes; Conducting systematic reviews of the literature regarding racial and ethnic health care disparities interventions; and Disseminating results from these research efforts and systematic reviews to encourage health care systems to address racial and ethnic gaps in care.
Requirements: For-profit organizations are eligible to apply, however preference will be given to applicants that are public agencies or are tax-exempt under Section 501(c) (3) of the Internal Revenue Code and are not private foundations as defined under Section 509(a). Proposals must focus on innovative interventions that reduce racial and ethnic disparities in the treatment of cardiovascular disease, depression, and diabetes. There are no age restrictions for the target populations of the interventions proposed for evaluation using program funds. Although there are no restrictions on the age of target populations, proposals targeting pediatric and geriatric populations are encouraged. Applications are accepted in two phases. The first phase is a brief proposal submission open to all interested research teams. The second phase is a full proposal submission and will be accepted by invitation only. All applications, both the brief and full proposals, must be submitted through the Robert Wood Johnson Foundation, Grantmaking Online system.
Restrictions: Individuals are not eligible to apply. Stand-alone prevention interventions such as tobacco cessation or exercise programs are not eligible unless they are one part of a larger, more comprehensive, intervention. The program will only directly fund health care delivery institutions. However, eligible institutions may partner/sub-contract with academic or research groups.
Amount of Grant: $100,000 - $275,000
Contact: Marshall H. Chin, Director; (866) 344-9800 or (773) 702-4769; email: mchin@medicine.bsd.uchicago.edu
Internet: http://www.solvingdisparities.org/
Sponsor: Robert Wood Johnson Foundation
College Road East and Route 1, P.O. Box 2316
Princeton, NJ 08543-2316

RWJF Generalist Physician Faculty Scholars 5167

This national career development program is designed to develop leaders in primary care academics by providing funding for junior faculty research. Each year the program provides $300,000 for 15 junior faculty members in family medicine, internal medicine and general pediatrics, to conduct four-year research projects nationwide. Under the program, the scholars, sponsored by their medical schools, receive support to enhance their research productivity, develop innovative mentoring programs for medical students, and participate in the design and implementation of primary education initiatives. The program coordinates several activities including an annual meeting for the scholars to present their research findings and conference calls for the scholars to discuss professional issues.

Requirements: Accredited colleges and universities within the U S. are eligible to apply.
Samples: Dartmouth Medical School Center for the Evaluative Clinical Sciences, $240,906.; University of California, San Francisco, School of Medicine, $300,000.; University of Colorado at Denver and Health Sciences Center, $299,980.; Mount Sinai School of Medicine of New York University, $300,000.; University of Pennsylvania School of Medicine, $299,993.
Amount of Grant: $300,000 maximum
Contact: Cynthia D. Mulrow; (210) 877-2275; email: mulrowc@uthscsa.edu
Internet: http://www.rwjf.org/applications/solicited/npo.jsp?FUND_ID=55121
Sponsor: Robert Wood Johnson Foundation
College Road East and Route 1, P.O. Box 2316
Princeton, NJ 08543-2316

RWJF Harold Amos Medical Faculty Development Scholars 5168

The Harold Amos Medical Faculty Development Program supports basic/biomedical, clinical, and health services/epidemiology research. The purpose of the AMFDP award is to facilitate the transition of the newly-trained clinician who wishes to develop into an independent investigator. A completed application consists of two parts: the completion of the online proposal process, and submission of other (paper) supporting documents.
Requirements: To be eligible, applicants must be physicians from historically disadvantaged backgrounds (ethnic, financial, or educational) who are U.S. citizens or permanent residents at the time of application deadline; have excelled in their education; are now completing or have completed their formal clinical training; are prepared to devote four consecutive years to research; are committed to pursuing academic careers; and are committed to improving the health status of the underserved, decreasing health disparities, or serving as role models for students and faculty from historically disadvantaged backgrounds Preference will be given to physicians who have recently completed their formal clinical training. In order to pursue the advanced research training required by this program, applicants must identify faculty mentors with whom a research plan will be developed. The mentor's experience in the supervision of trainees and the adequacy of the mentor's research environment and support will be strongly considered. Deadline for receipt of completed online applications - March 23. Deadline for postmark of all supporting documents and hard copies - March 24. Complete guidelines and application instructions are available at the website.
Amount of Grant: $104,140
Date(s) Application is Due: Mar 23; Mar 24
Contact: Nina Ardery, Program Deputy Director; (301) 565-4080 or (317) 278-0500; fax: (301) 565-4088; email: nardery@erols.com or amfdp@starpower.net
Internet: http://www.amfdp.org/applicat.htm
Sponsor: Robert Wood Johnson Foundation
College Road East and Route 1, P.O. Box 2316
Princeton, NJ 08543-2316

RWJF Health & Society Scholars 5169

The program is a two-year fellowship designed to build the nation's capacity for research, leadership, and action to address the broad range of factors affecting health. The goal of the program is to improve health by training scholars to rigorously investigate the connections between biological, behavioral, environmental, economic, and social determinants of health; and develop, evaluate, and disseminate knowledge and interventions based on integration of these determinants. Each year the program will enable up to 18 outstanding individuals who have completed doctoral training to engage in an intensive two-year program at one of six nationally prominent universities. Applications and guidelines are available online.
Requirements: Individuals must apply only through the foundation's online system (http://www.hsscholars.org). To be eligible, scholars must have completed doctoral training by the time of entry into the program (August or September) in one of a variety of fields including, but not limited, to the behavioral and social sciences, the biological and natural sciences, health professions, public policy, public health, history and ethics; have significant research experience; clearly connect their research interests to substantive population health concerns; and be citizens or permanent residents of the United States or its territories. Applications from candidates with diverse backgrounds are encouraged.
Restrictions: Individuals placed in the program must be willing to relocate to the institution designated during the match process for the duration of the two-year program.
Amount of Grant: $89,000 plus health insurance
Date(s) Application is Due: Oct 2
Contact: Pamela G. Russo, Program Officer; (212) 419-3566; fax: (212) 419-3569; email: hss@nyam.org
Internet: http://www.healthandsocietyscholars.org
Sponsor: Robert Wood Johnson Foundation
College Road East and Route 1, P.O. Box 2316
Princeton, NJ 08543-2316

RWJF Health Care Costs: Research to Inform Policy Grants 5170

The Robert Wood Johnson Foundation, through its Changes in Health Care Financing and Organization (HCFO) initiative, is issuing a special topic solicitation for proposals to conduct research and analyses on the complex causes and potential solutions to rising health care costs. The solicitation seeks research contributing to an evidence base that defines and addresses the problem of rising health care costs, and how it relates to health insurance coverage. Research should focus on identifying sources of continued cost growth, examining promising cost control efforts, and exploring the relationship between health care costs and health insurance. The research questions are broken down into four major categories (Cost Controls, Affordability, Pricing, and Cost Estimates). The questions are not an exhaustive list, but have been identified among the priorities for the foundation. However, these should not be viewed as prescriptive.
Requirements: Researchers should have a demonstrated track record in health services research, economics, financing, or public policy. Junior researchers may apply as the principal investigator. However, to be competitive, the proposal must provide evidence that there is sufficient senior research oversight and support. All applicants are required to submit a brief proposal.
Restrictions: Grants are not available for ongoing general operating expenses or existing deficits; endowment or capital costs, including construction, renovation, or equipment purchases; basic biomedical research; conferences, symposia, publications, or media projects, unless they are clearly related to the foundation's goals; research on unapproved drug therapies or devices; international programs and institutions; or direct support to individuals.
Contact: Bonnie J. Austin, Deputy Director; (202) 292-6700; fax: (202) 292-6800; email: Bonnie.Austin@academyhealth.org or hcfo@academyhealth.org
Internet: http://www.hcfo.net/grants.htm
Sponsor: Robert Wood Johnson Foundation
College Road East and Route 1, P.O. Box 2316
Princeton, NJ 08543-2316

RWJF Health Impact Project Grants 5171

The Health Impact Project, a collaboration of the Robert Wood Johnson Foundation and The Pew Charitable Trusts, encourages the use of health impact assessment (HIA) to help decision-makers better assess proposed policies, projects and programs with respect to their impact on health, in order that they may avoid adverse health consequences and costs and improve health. The foundations are issuing this call for proposals to demonstrate the effectiveness of HIAs and promote their incorporation into local, state, tribal and federal decision-making. Grants will range from $25,000 to $150,000 and must be completed within 24 months. Proposals for more than $150,000 may be considered under rare and exceptional circumstances.
Requirements: Eligible applicant organizations include: state, tribal or local agencies; tax-exempt educational institutions; or publicly supported charitable organizations that are exempt from federal income tax as an organization described by Section 501(c)3 of the Code. Applicant organizations must be located in the United States or its territories at the time of application.
Amount of Grant: $25,000-$150,000 range
Contact: Program Director; (202) 540-6012; email: healthimpactproject@pewtrusts.org
Internet: http://www.rwjf.org/applications/solicited/cfp.jsp?ID=20921
Sponsor: Robert Wood Johnson Foundation
College Road East and Route 1, P.O. Box 2316
Princeton, NJ 08543-2316

RWJF Health Policy Fellowships 5172

The purpose of this fellowship is to give outstanding mid-career health professionals in academic and community-based settings an opportunity to learn more about the legislative process through direct participation. Fellows actively contribute to the formulation of national health policies and accelerate their careers as leaders in health policy. Up to six grants of up to $165,000 each will be made. Each fellow will receive up to $94,000 for the Washington stay in salary plus fringe benefits or fellowship stipend. Fellows will receive an additional allowance for relocation subject to limitations provided in detail on the program's Web site. November 13 is the deadline for receipt of applications submitted online, with January 29 the deadline for receipt of full proposals.
Requirements: Applicants may have backgrounds in the following disciplines: allied health professions; biomedical sciences; dentistry; economics or other social sciences; health services organization and administration; medicine; nursing; public health; or social and behavioral health. Applicants must be U.S. citizens or have permanent residency status at the time of application. Potential applicants must obtain documented sponsorship from an eligible institution to support their application. Nonprofit health care organizations and academic centers are eligible sponsors. Eligible institutions may sponsor more than one candidate, but no more than one individual will be selected from any one sponsoring organization.
Restrictions: No additional direct costs (such as laptops or PDAs) or indirect costs are paid for during the first year of this program.
Amount of Grant: $165,000 maximum over three years
Date(s) Application is Due: Nov 13; Jan 29
Contact: Marie Michnich, Director; (202) 334-1506; fax: (202) 334-3862; email: mmichnich@nas.edu
Internet: http://www.rwjf.org/applications/solicited/cfp.jsp?ID=20904
Sponsor: Robert Wood Johnson Foundation
College Road East and Route 1, P.O. Box 2316
Princeton, NJ 08543-2316

RWJF Healthy Eating Research Grants 5173

Healthy Eating Research is a national program that will support research to identify, analyze, and evaluate environmental and policy strategies that can promote healthy eating among children and prevent childhood obesity. Special emphasis will be given to research projects that benefit children in the low-income and racial/ethnic populations at highest risk for obesity. The program will fund two types of research grants: studies to identify and/or evaluate promising school food environment and policy changes (12- to 18-month awards up to $100,000 each, and 18- to 36-month awards up to $400,000 each); and analyses of macro-level policy or system determinants of school food environments and policies (12- to 18-month awards up to $75,000 each). Applications must be submitted through the online process. Calls for Proposals and guidelines are available online.

Requirements: Preference will be given to those applicants that may be either public entities or nonprofit organizations that are tax-exempt under Section 501(c)3 of the Internal Revenue Code. Applicant organizations must be based in the United States or U.S. Territories. The focus of this program is the United States; studies of policies in other countries will be considered only to the extent that they may directly inform U.S. policy. The experience and qualifications of the research team is one of the primary criteria for proposal review. A doctorate or terminal degree is preferred for the principal investigator. Deadline for receipt of brief proposals - February 13. Select applicants will be invited to submit full proposals. Deadline for receipt of full proposals - May 23.

Restrictions: This program does not fund demonstration projects. The foundation does not award grants to private individuals or applicants from outside the U.S. or U.S. Territories. No faxed, emailed or mailed proposals will be accepted.

Amount of Grant: $400,000 maximum

Contact: Kathy Kosiak, Research Coordinator; (800) 578-8636; fax: (612) 624-9328; email: healthyeating@umn.edu

Internet: http://www.healthyeatingresearch.org/

Sponsor: Robert Wood Johnson Foundation
College Road East and Route 1, P.O. Box 2316
Princeton, NJ 08543-2316

RWJF Interdisciplinary Nursing Quality Research Initiative 5174

The initiative awards two-year grants to support the generation, dissemination, and translation of research to improve the contributions of nursing to the quality of hospital care. It will support interdisciplinary teams of nurse scholars and scholars from other disciplines to address gaps in knowledge about the relationship between nursing and health care quality. Topic areas include: identifying, developing, and testing new measures to advance quality nursing research; and testing the effects of implementing the nursing-sensitive performance measures endorsed by the National Quality Forum. Investigator-initiated ideas that are consistent with the broader set of program goals also will be considered. The two-phase application process begins with online submittal of a brief proposal dur by December 4; full proposals are by invitation, and are due by April 9. Guidelines are available online.

Requirements: Teams of researchers who are U.S. citizens or permanent residents are eligible. Each team must have at least one nursing researcher and at least one researcher from another discipline. Preference will be given to 501(c)3 applicants that are not 509(a) private foundations.

Amount of Grant: $300,000 maximum over two years

Date(s) Application is Due: Dec 4; Apr 9

Contact: Lori A. Melichar, Grants Administrator; (301) 571-8161 or (800) 978-8309; email: INQRIhelpdesk@constellagroup.com

Internet: http://www.inqri.org

Sponsor: Robert Wood Johnson Foundation
College Road East and Route 1, P.O. Box 2316
Princeton, NJ 08543-2316

RWJF Investigator Awards in Health Policy Research 5175

The program supports researchers whose cross-cutting and innovative ideas promise to contribute meaningfully to improving health and health policy in America. Each year, the program awards 24- to 36-month grants of up to $335,000 to highly qualified individuals to fund approximately 10 research projects that have national policy relevance. The program funds projects with a broader scope than traditional NIH or NSF grants. Data-driven health services research focusing on a narrowly defined research question is not appropriate for this program, nor are pilot projects that are designed to evaluate clinical care in a specific geographic area or to demonstrate service delivery programs or financing strategies.

Requirements: Individuals in the United States who are affiliated either with a U.S.-based educational institution or a 501(c)3 tax-exempt organization are eligible. Applications are welcomed from investigators in fields such as anthropology, business, demography, economics, engineering, ethics, genetics, health and social policy, history, journalism, law, medicine, nursing, political science, psychology, public health, science policy, social work, sociology and others. The program seeks a diverse group of applicants including minorities, researchers early in their careers, and individuals in nonacademic settings. Key Dates: Deadline for receipt of letters of intent - March 28; Applicants will be notified by mail if they have been selected to submit a full proposal. Deadline for receipt of full proposals - July 27.

Restrictions: Past awardees and teams of more than two researchers are not eligible to apply.

Samples: Michael D. Cohen, Ph.D. - Handoffs in Hospitals: Research for the Design of Better Practices; Julie Fairman, Ph.D., FAAN, R.N. - Practice Politics: History of Nursing 1975 to the Present; Joseph J. Fins, M.D. - Minds Apart: Severe Brain Injury and Health Policy; Dominick L. Frosch, Ph.D., Jose A. Pagan, Ph.D. - Direct-to-Consumer Advertising: Do Television Pharmaceutical Ads Prompt More Than Just Prescription Requests?

Amount of Grant: up to $335,000

Date(s) Application is Due: Mar 28; Jul 27

Contact: Cynthia Church, Program Administrator; (732) 932-3817; fax: (732) 932-3819; email: depdir@ifh.rutgers.edu

Internet: http://www.investigatorawards.org

Sponsor: Robert Wood Johnson Foundation
College Road East and Route 1, P.O. Box 2316
Princeton, NJ 08543-2316

RWJF Jobs to Careers: Promoting Work-Based Learning for Quality Care - Research Grants 5176

The program will support partnerships to advance and reward the skill and career development of incumbent (currently employed) workers providing care and services on the front lines of our health and health care systems. The program seeks to develop and redesign systems that support and institutionalize learning and career advancement and test new models of work-based learning. The research grants will support applied research to inform these efforts. Research funded through this program will complement and supplement learnings from the Jobs to Careers' implementation grants, which will fund partnerships of health and health care employers and educational institutions to promote the program, objectives. Grant funds may be used for project staff salaries, consultant fees, data collection and analysis, meetings, supplies, project-related travel, and other direct project expenses, including a limited amount of equipment essential to the project.

Requirements: Eligible applicants for research grants through Jobs to Careers are scholars associated with educational institutions, research organizations, health care providers or other public or nonprofit organizations. Preference will be given to public entities or nonprofit organizations tax exempt under Section 501(c)3 of the Internal Revenue Code. Applicants may be independent of or associated with organizations currently receiving Jobs to Careers implementation grants. All proposals must be submitted only through the RWJF Grantmaking Online system.

Restrictions: Grant funds may not be used to subsidize individuals for the costs of their health care, to support clinical trials of unapproved drugs or devices, to construct or renovate facilities, for lobbying, or as a substitute for funds currently being used to support similar activities.

Amount of Grant: up to $200,000

Date(s) Application is Due: Jun 19

Contact: Maria Kniesler Flynn, Program Director; (617) 728-4446; email: mflynn@jff.org

Internet: http://www.jobs2careers.org

Sponsor: Robert Wood Johnson Foundation
College Road East and Route 1, P.O. Box 2316
Princeton, NJ 08543-2316

RWJF New Connections Initiative Grants 5177

This program is designed to expand the diversity of perspectives that inform RWJF programming and introduce new researchers and scholars to the Foundation, while simultaneously helping to meet staff needs for data that can measure progress towards programming objectives. The initiative will provide funding for up to 12 awards per year. Senior Consultants are eligible to receive up to $50,000 in either contracts or grants. Junior Investigators are eligible for grants only of up to $55,000, including funding for a consultant and methodological training. All funding is expected to last 12 months.

Requirements: This program seeks scholars who have historically been underrepresented in research activities supported by RWJF. This includes people from ethnic or racial minority groups, first-generation college graduates and people from low-income communities. The initiative will support research activities that address specific questions identified by RWJF programming teams and portfolios. Contact the sponsor for the programming focus. There are two stages in the application process. The first is the submission of a brief proposal that describes the project (deadline - December 28). Brief proposals will be reviewed and selected applicants will be invited to submit full proposals in a second stage (deadline - March 7). All proposals must be submitted online through the RWJF Grantmaking Online system. Guidelines and application instructions are available at the website.

Restrictions: Full proposal submission is by invitation only.

Samples: Dr. Bettina M. Beech, Vanderbilt University School of Medicine - An Examination of Policies to Support Healthy Eating Among Rural High School Students in the Delta; Dr. Rhonda BeLue, The Pennsylvania State University - Investigating Racial and Ethnic Disparities in the Quality and Location of Care; Dr. Ronica Rooks, Kent State University - Explaining Racial Disparities in Quality of Care and Health Care Utilization among Black and White Adults: the Chronic Illness and Caregiving Survey

Amount of Grant: $50,000 - $55,000

Contact: Debra Joy P?'rez, Ph.D., Senior Program Officer; (617) 495-9003; fax: (617) 495-8696; email: djperez@fas.harvard.edu

Internet: http://www.rwjf-newconnections.org

Sponsor: Robert Wood Johnson Foundation
College Road East and Route 1, P.O. Box 2316
Princeton, NJ 08543-2316

RWJF Partners Investing in Nursing, Future Grants 5178

The program is a collaborative initiative of the Robert Wood Johnson Foundation (RWJF) and the Northwest Health Foundation (NWHF), with the intention of

addressing nursing issues at the community level through funding partnerships with local and regional foundations. Up to 10 awards of up to $250,000 each will be made for projects lasting up to 24 months. Local foundations, with other partners (such as state workforce investment boards, hospital and long-term-care associations, foundations and other funding sources) will match these awards with at least $1 for every $2 provided by the program. Funding will be commensurate with the size and scope of the proposed activity and the foundation, experience with nursing issues. Funding is not intended to support the continuation of ongoing projects, but it may support new programs led by foundations that are already working in the field of nursing. October 29 is the deadline for receipt of brief proposals, and January 5 is the deadline for receipt of invited full proposals.
Requirements: Eligible institutions include local or regional private, family or community foundations and public charities. Eligible foundations are those classified as tax-exempt under Section 501(c)3 as a public charity or private foundation, a nonexempt charitable trust treated as a private foundation under Section 4947(a)(1), or organizations that claim status as private operating foundations under Section 4942(j)(3) or (5) of the Internal Revenue Code. Government entities, corporate grantmakers and others may participate as part of funding collaboratives; however, they may not serve as the applicant institution. All proposals to this program must address one or more of these topics: (1) Diversity; (2) Educational Infrastructure and Faculty Development; (3) Public Health; (4) Long-Term Care; and/or (5) Mid-Level Management. Proposals for planning as well as implementation projects will be considered. Deadline for receipt of brief proposals - November 9. Deadline for receipt of invited full proposals - February 1.
Restrictions: This program accepts only proposals submitted through the RWJF Grantmaking Online system at http://grantmaking.rwjf.org/pin2
Amount of Grant: up to $250,000
Date(s) Application is Due: Oct 29; Jan 5
Contact: Judith L. Woodruff, J.D., Program Director; (971) 230-0093; email: judith@nwhf.org
Internet: http://www.rwjf.org/applications/solicited/cfp.jsp?ID=20903
Sponsor: Robert Wood Johnson Foundation
College Road East and Route 1, P.O. Box 2316
Princeton, NJ 08543-2316

RWJF Pioneer Portfolio Grants 5179

The program aims to support novel, high-return ideas that may have far-reaching impact on people's health, the quality of care they receive and the systems through which that care is provided. It seeks ideas not only from the mainstream of health and health care but also looks to sources outside of these fields for innovations that might have transformative impact. Several projects apply approaches from diverse sectors such as finance, design and entertainment to forge new solutions in health and health care arenas.
Requirements: Preference is given to public agencies, public charities or are tax-exempt under section 501(c)3 of the Internal Revenue Code. Organizations must be located in the United States or one of its territories.
Restrictions: The organization does not fund any of the following: Ongoing general operating expenses or existing deficits; Endowment or capital costs; Basic biomedical research; Research on drug therapies or devices; International programs; Direct support of individuals; Lobbying of any kind. Additionally, RWJF no longer funds projects in the following areas: End-of-life care; Long-term care; Specific chronic conditions (that are not part of the strategies of other Program Areas); Physical activity for adults age 50 or older.
Samples: Ramanan Laxminaryan, Extending the Cure: Laxminaryan is reframing how we fight drug-resistant diseases by suggesting we manage antibiotics as a shared societal resource. Working with regulatory, health, pharmaceutical and economic experts, he is building evidence and consensus around incentive-based policy alternatives to ensure a vital pipeline of effective antibiotics and combat 'superbugs' that cannot be eliminated with existing drugs.; Project HealthDesign: A national program that supports interdisciplinary teams to look beyond stand-alone personal health records (PHR) to design next-generation PHR systems that empower patients to better manage their health and health care.; Influenza Prediction Markets: University of Iowa business school professors teamed up with an infectious disease expert to test whether the Iowa Electronic Markets,Äìprediction markets that aggregate information and are best known for predicting election and film box office results,Äìcan target where influenza may strike in advance.; Myelin Repair Foundation (MRF): MRF's innovative research model provides the infrastructure for five scientific teams,Äìspread across different university labs in the U.S. and Canada,Äìto work virtually toward treatment breakthroughs for patients suffering from multiple sclerosis.
Contact: Chinwe Onyekere, Program Officer; (609) 627-7504 or (888) 631-9989; email: conyeke@rwjf.org
Internet: http://www.rwjf.org/pioneer/submission.jsp
Sponsor: Robert Wood Johnson Foundation
College Road East and Route 1, P.O. Box 2316
Princeton, NJ 08543-2316

RWJF Scholars in Health Policy Research Fellowships 5180

The program is a two-year fellowship targeted to outstanding new PhDs in economics, political science, and sociology to advance their involvement in health policy research. Each year, the program enables up to 12 individuals to spend two years at one of three nationally prominent universities—Harvard University, the University of California at Berkeley (in collaboration with the University of California at San Francisco), and the University of Michigan. Those selected as scholars will have access to the full range of university resources, including libraries, computers, databases and research support, and will receive stipends from their university of $80,000 for the first year and $83,000 for the second year. Application and guidelines are available online.
Requirements: Applicants must be U.S. citizens or permanent residents at the time of application and have received a doctoral degree after January 2002 but no later than July 15, 2007. Applicants must have received a doctoral degree in economics, political science, or sociology. Preference will be given to applicants who have not previously worked in the areas of health or health policy research.
Samples: Elizabeth Armstrong, Fetal Personhood: the Raw Edge of Obstetrical Practice and Ethics; Michael Greenstone,a project on the association between particulates in air pollution and infant mortality rates; Cathy Cohen, Race, Politics, and Adolescent Health: Understanding the Health Attitudes and Behaviors of African American Youth; Daniel Carpenter, Reputation and Regulation: A Study of Pharmaceutical Policy at the FDA
Amount of Grant: $89,000
Date(s) Application is Due: Oct 21
Contact: Program Coordinator; (617) 353-9220; fax: (617) 353-9227; email: rwjf@bu.edu
Internet: http://www.rwjf.org/applications/solicited/cfp.jsp?ID=20841
Sponsor: Robert Wood Johnson Foundation
College Road East and Route 1, P.O. Box 2316
Princeton, NJ 08543-2316

RWJF Young Epidemiology Scholars (YES) Competition 5181

The competition for original student research is designed to inspire talented students to investigate the many behavioral, biological, environmental and social factors that affect health and, based upon this knowledge, to identify ways to improve the health of the public. The competition awards up to 120 college scholarships each year to high school juniors and seniors who conduct outstanding research projects that apply epidemiological methods of analysis to a health-related issue.
Requirements: The student competition is open to high-school juniors and seniors who are citizens or permanent residents of the United States and enrolled in a high school in the United States, Puerto Rico, Guam, U.S. Virgin Islands American Samoa, Wake and, Midway Islands or the Marianas. Home-schooled students are also eligible. To enter the competition, you must register online and upload a written report of your research project. Reports sent via email, fax or mail will not be accepted.
Restrictions: Only one project per student will be accepted. Team projects are not eligible. If you are a previous national finalist or scholarship winner, you may not submit the same topic as a previous year unless you have significantly advanced your research (e.g., adding substantive new data, analysis or other material).
Amount of Grant: up to $50,000
Date(s) Application is Due: Feb 1
Contact: YES Program Coordinator; The College Board; 11911 Freedom Drive, Suite 300; Reston VA 20190; (800) 626-9795, ext. 5849; fax: (703) 707-5599; email: yes@collegeboard.org
Internet: http://www.collegeboard.com/yes
Sponsor: Robert Wood Johnson Foundation
College Road East and Route 1, P.O. Box 2316
Princeton, NJ 08543-2316

Ryan Gibson Foundation Grants 5182

The Foundation welcomes any and all research grant requests from around the country. The main focus is on blood related cancers. Applicants should submit a one to two page proposal with the following information: description of the procedure and or drug that is to be tested/developed, and what the desired/expected result will be; some background on the researcher and the facility that will be used (past projects success and or failures should be included); and a budget of anticipated costs.
Samples: Memorial Sloan-Kettering Cancer Center, $50,000 (2006); University of Arkansas Medical Sciences, $40,000 (2005); Cell and Gene Therapy Division of the Baylor College of Medicine, $50,000 (2005).
Amount of Grant: $50,000 average
Contact: Sarah V. Wilson, National Director, (214) 206-3812; fax: (214) 206-3813; email: mail@ryangibsonfoundation.org
Internet: http://ryangibsonfoundation.org/research/grants.asp
Sponsor: Ryan Gibson Foundation
2920 Taylor Street
Dallas TX 75226

S.E.VEN Fund Annual Grants 5183

The sponsor is a virtual non-profit entity run by entrepreneurs whose strategy is to markedly increase the rate of diffusion of enterprise-based solutions to poverty. This is accomplished by targeted investment that fosters thought leadership through books, films and websites; supporting role models - whether they are entrepreneurs or innovative firms - in developing nations; and shaping a new discourse in government, the press and the academy around private-sector innovation, prosperity and progressive human values. This program is intended to catalyze, support and disseminate research on questions of economic development, prosperity and entrepreneurship. Funding is particularly targeted toward new frontiers in enterprise-based solutions to poverty and innovative ideas that are unlikely to be supported by conventional funding sources.
Requirements: S.E.VEN welcomes the participation of entrepreneurs, business people, investors, academics, laypeople, and philanthropists. To begin the process, add your name to the database online. Further information will be emailed to you.

Samples: Entrepreneurs In Kenya and India, Michael Kremer, Ph.D.; Private Enterprise Solutions for Poverty Alleviation: Approaches of the U.S. Military, Mark B. Milstein, Ph.D.; The Moral Imperative of Free Enterprise: The Only Known Solution to Poverty, Jay W. Richards, Ph.D.; Aspirations and Well-Being Outcomes in Ethiopia, Tanguy Bernard and Alemayehu Seyoum Taffesse, Ph.D.
Amount of Grant: $100,000 maximum
Contact: Elizabeth Hooper, Executive Director; email: info@sevenfund.org
Internet: http://www.sevenfund.org/research-funding-areas.html
Sponsor: Social Equity Venture Fund (S.E.VEN)
1770 Massachusetts Avenue, 247
Cambridge, MA 02140

S.E.VEN Fund Mini-Grants **5184**
The sponsor is a virtual non-profit entity run by entrepreneurs whose strategy is to markedly increase the rate of diffusion of enterprise-based solutions to poverty. This is accomplished by targeted investment that fosters thought leadership through books, films and websites; supporting role models - whether they are entrepreneurs or innovative firms - in developing nations; and shaping a new discourse in government, the press and the academy around private-sector innovation, prosperity and progressive human values. Mini-grant requests should align with the S.E.VEN Fund's funding priorities and do not exceed $10,000. The sponsor considers funding requests for mini-grants on an ongoing basis; the competition runs three times annually.
Requirements: This program's funding is intended to catalyze, support and disseminate research on questions of economic development, prosperity and entrepreneurship. Funding is particularly targeted toward new frontiers in enterprise-based solutions to poverty and innovative ideas that are unlikely to be supported by conventional funding sources. Funded projects are typically research-based or involve the creative use of the media - books, film, the Web, organizing conferences, etc. You can apply for funding by filling out the online mini-grant request form.
Restrictions: The sponsor is unable to provide funding for direct services.
Samples: Maternova - a new web portal, that will bridge the fields of social entrepreneurship and international maternal & neonatal health.; Empowerment of Students Through the Entrepreneurial Approach - This project focuses on empowering students in Las Vegas, New Mexico to become future entrepreneurs through structured curricula and applied projects.; Top30Under30 Innovative Entrepreneurs Exhibition (InEx) - to support InEx Fellowships enabling young African entrepreneurs to meet with leading finance and investment institutions.; The Role of Universities in Building A Culture of Civic Responsibility, Interdependence, and Prosperity - Roundtable event held in Kigali, Rwanda focusing on laying the groundwork for higher education to play a central role in developing citizens to sustain a strong civic culture and develop the entrepreneurial skills necessary for creating a successful and equitable economy.
Amount of Grant: up to $10,000
Contact: Elizabeth Hooper, Executive Director; email: info@sevenfund.org
Internet: http://www.sevenfund.org/submit-mini-grants.php
Sponsor: Social Equity Venture Fund (S.E.VEN)
1770 Massachusetts Avenue, 247
Cambridge, MA 02140

S.E.VEN Fund Open Enterprise Solutions to Poverty RFP Grants **5185**
The RFP competition focuses on academic and scientific research in the field of Enterprise-based Solutions to Poverty. The competition will award up to two (2) research grants of no more than $100,000 each for a duration of no more than twelve months. There is no geographic or other limitation on who may compete for these funds - anyone may compete for these funds.
Requirements: Anyone may submit an Initial Proposal in order to compete for a SEVEN RFP grant as long as the proposal reflects the scope and guidelines stated in the call for RFPs. The winning parties will be required to enter into a contract with SEVEN prior to final award. Since this is a program run by both practitioners and academics, SEVEN may be a bit more flexible than traditional funding agencies in the definition of a Principal Investigator. Thus, they may accept applications from non-academics as the PI or a co-investigator if his/her experience and background warrant it. Acceptable use of grant funds include: student or postdoctoral salary and benefits for part of the academic year; summer salary and teaching buyout for academics; support for specific projects during sabbaticals; assistance in writing or publishing books; modest allowance for justifiable equipment, computers, publication charges, and other supplies; modest travel allowance; development of workshops, conferences, or lecture series for professionals; or, justified overhead of at most 20%.
Samples: Microrate Inc. - Improving Flows of Foreign Capital to the Microfinance Industry
Amount of Grant: $100,000
Date(s) Application is Due: Oct 15
Contact: Elizabeth Hooper, Executive Director; email: info@sevenfund.org
Internet: http://www.sevenfund.org/enterprise-solutions-poverty/index.php
Sponsor: Social Equity Venture Fund (S.E.VEN)
1770 Massachusetts Avenue, 247
Cambridge, MA 02140

S.E.VEN Fund WHY Prize **5186**
The sponsor's objective with The WHY Prize is to stimulate innovation in the specific domain of Enterprise Solutions to Poverty by encouraging collaboration; specifically, they will support collaboration between macroeconomists and anthropologists. The WHY Prize intends to catalyze cross-fertilization across disciplines and spur progress in finding sustainable solutions to poverty. The prize is a one-time award of $50,000.
Requirements: Anyone may submit a nomination in order to compete for The WHY Prize as long as the proposal reflects the scope and guidelines stated in the call for nominations. Individuals may nominate themselves or others for consideration. Nominated teams must include one individual who holds a masters or doctoral degree in economics and anthropology respectively. Other disciplines may join the collaboration to broaden the perspective beyond these two disciplines. The WHY prize will reward a team that has published in either a major venue like the Atlantic Monthly, New York Times Magazine, Financial Times, Businessweek, Forbes, Foreign Policy, Foreign Affairs etc. or in a peer reviewed journal, with a view toward producing further work such as a book or additional articles in a peer reviewed format. The published piece must add to the dialogue around enterprise solutions to poverty by proposing solutions that represent the best integration of actionable insights and thought leadership. Nominees may submit an article, paper, or book for consideration by applying on the sponsor's website.
Restrictions: Only applications submitted through the form on the sponsor's website are accepted.
Amount of Grant: $50,000
Date(s) Application is Due: Oct 7
Contact: Elizabeth Hooper, Executive Director; email: info@sevenfund.org
Internet: http://www.sevenfund.org/press-release-mar-2008-WHY-prize.html#qanda
Sponsor: Social Equity Venture Fund (S.E.VEN)
1770 Massachusetts Avenue, 247
Cambridge, MA 02140

Sachs Foundation Scholarships **5187**
The foundation uses its funds to improve the well-being of African American residents of Colorado. Funds are provided for undergraduate studies. Graduate grants are available to individuals who received undergraduate grants. The foundation will finance each awardee for four years. Applications are released and accepted from January 1 until March 1 each calendar year; applicants are informed in May regarding their eligibility for an interview in June.
Requirements: All applicants must be African American U.S. citizens and residents of Colorado for at least five years. Funds are only available for young students graduating from accredited high schools with at least a 3.4 grade point average.
Geographic Focus: Colorado
Amount of Grant: $4,000 approximately per year up to four years
Date(s) Application is Due: Mar 1
Contact: Grants Administrator; (719) 633-2353; email: sachs@frii.com
Internet: http://www.sachsfoundation.org/scholarships.htm
Sponsor: Sachs Foundation
90 South Cascade Avenue, Suite 1410
Colorado Springs, CO 80903

Sage Foundation Grants **5188**
The foundation awards grants nationwide, with emphasis on the Midwest, to nonprofit organizations in support of higher education, legal education, and secondary education; seniors; hospitals; disabled; Catholic giving and welfare; youth; child welfare; social services; and cultural programs. Types of support include annual campaigns, building construction/renovation, capital campaigns, challenge grants, endowments, equipment acquisition, general operating grants, multi-year support, project development, research, and scholarship funds. There are no application deadlines or forms; submit a letter of inquiry. The board meets quarterly.
Amount of Grant: $1,000-$250,000
Contact: Melissa Sage Fadim, Grants Administrator; (810) 227-7660 or (212) 737-7311
Sponsor: Sage Foundation
P.O. Box 1919
Brighton, MI 48116

Saint Mary's College Research Fellowships **5189**
The Center for Women's InterCultural Leadership seeks to increase awareness of the importance of intercultural dialog in today's world, to highlight the leadership roles women have adopted in such dialog, and to foster the next generation of women leaders. In pursuit of these goals, CWIL offers a number of fellowships, which are joint appointments with academic departments, including fellows, senior fellows, postdoctoral fellows, dissertation fellows, and annual fellowships. All appointments are for one year with the possibility of a renewal for a second year. CWIL fellows have three general types of responsibilities: research; teaching courses relating to cultural or women's leadership issues; and assisting the CWIL and/or the department in its programmatic and community activities. The relative weight of these activities will vary depending on the type of fellowship and the needs of the sponsoring department.
Date(s) Application is Due: Feb 1
Contact: Center for Women's InterCultural Leadership, (574) 284-4182; fax: (574) 284-4141; email: cwil@saintmarys.edu
Internet: http://www.saintmarys.edu/~cwil/php/research.scholarship/fellows/index.php
Sponsor: Saint Mary's College
9 Havican Hall
Notre Dame, IN 46556-5001

Saison Foundation Grants **5190**
This program provides support to projects that will contribute to increasing the awareness of contemporary Japanese theater and dance in the international community through conferences, symposia, translation (from Japanese to other languages), and publication.

Particular attention will be given to translation projects for contemporary Japanese plays that will be staged for audiences outside of Japan. Principal individuals/organizations of the projects are eligible to apply. Applicants planning to hold conferences or symposia in Tokyo as part of their projects may also apply for the use of Morishita Studio up to a month between June of this year and May of next year.
Amount of Grant: 500,000 yen-1.5 million yen
Contact: Masao Katayama, Secretary-General, 81-3-3535-5566; fax: 81-3-3535-5565; email: foundation@saison.or.jp
Internet: http://www.saison.or.jp/english/grant/index.htm
Sponsor: Saison Foundation
1-16-1 Ginza
Chuo-ku, Tokyo 104104-0061 Japan

Salem Foundation Grants 5191
The primary purpose of the Foundation is support of higher education, cultural activities, health services, hospital building funds, and Christian giving. Its fields of interest include: arts; Christian agencies and churches; the disabled education (secondary and higher); the environment; family services; federated giving programs; health care (particularly heart and circulatory disease research); hospitals; human services; museums; and Protestant federated giving programs. Types of support include: annual campaigns, building/renovation projects, capital campaigns, general operating support, and program development. There is no formal application process and no deadlines. Please contact via letter.
Restrictions: Giving primarily in Florida, Maine and Minnesota. No grants to individuals.
Geographic Focus: Florida; Maine; Minnesota
Amount of Grant: $10,000 maximum
Contact: Robert S. Parish, President, (952) 476-6292
Sponsor: Salem Foundation, Inc.
2181 Springwood Road
Wayzata, MN 55391-2254

Sally Kress Tompkins Architectural History Fellowship 5192
Jointly awarded each year by the Society and the Historic American Buildings Survey (HABS) of the National Park Service, the fellowship allows an architectural history student to work as a summer intern on an HABS project. A certificate and a stipend are presented at the annual meeting.
Requirements: Applicants should be pursuing graduate studies in architectural history or other related fields.
Amount of Grant: $10,000
Date(s) Application is Due: Feb 1
Contact: Lisa Davidson, Coordinator, (202) 354-2179; fax: (202) 371-6473; email: lisa_davidson@nps.gov
Internet: http://www.cr.nps.gov/habshaer/joco/sktf.htm
Sponsor: Society of Architectural Historians
1849 C Street NW, 2270
Chicago, IL 60610-2144

Samuel Eliot Morison Naval History Scholarship 5193
This scholarship program was established to promote development of a broad understanding of naval history within the American national experience. The scholarship provides a cash award to an active duty commissioned officer of the U.S. Navy or Marine Corps with demonstrated leadership potential and high academic qualifications who is already pursuing graduate study in history, international relations, or a related field. The award will help defray expenses related to research, travel, and books or other educational materials. Annual deadline dates may vary; contact program staff for exact dates.
Requirements: Commissioned officers should submit a letter of application via their commanding officer. Selection will be based on relevance of the chosen area of study to U.S. naval history, demonstrated professional performance with particular emphasis on the officer's specialty, academic ability including baccalaureate record, the career needs of the officer, and potential for professional growth.
Amount of Grant: $5000
Date(s) Application is Due: Jul 31
Contact: Senior Historian, Washington Navy Yard, (202) 433-7230
Internet: http://www.history.navy.mil/prizes/prize3.htm
Sponsor: Naval Historical Center
805 Kidder Bresse SE
Washington, DC 20374-5060

Samuel Freeman Charitable Trust Grants 5194
The trust has broad funding interests in supporting U.S. charitable, scientific, literary, and education programs. The trust awards grants to support cancer research and treatment; the preservation, exhibition, and operation of historical railway equipment; and secondary schools and universities. Types of support include general operating support, continuing support, annual campaigns, and seed money. There are no application deadlines; proposals are reviewed throughout the year.
Requirements: U.S. nonprofit organizations may apply.
Amount of Grant: $1000-$150,000 average
Contact: Linda Franciscovich, c/o U.S. Trust Company, (212) 852-3683; fax: (212) 852-3377
Sponsor: Samuel Freeman Charitable Trust
114 W 47th Street
New York, NY 10036-1532

Samuel H. Kress Foundation Grants 5195
The foundation concentrates grants on projects and programs that promote the understanding and enjoyment of European art and architecture, especially in relation to its historical context. Grants support the documentation, care, and display of European art and architecture. This area includes two special programs: Old Masters in Context, and the European Preservation Program. Grants also support programs that develop and utilize the skills and experience of trained professionals. Emphasis is placed on two program areas: Essential Resources, and Sharing Expertise. The board meets in January, May, and October; applications are reviewed throughout the year. Guidelines are available online.
Restrictions: The foundation does not make grants to support living artists, films, art history programs below the predoctoral level, or to purchase works of art.
Samples: Yale U (New Haven, CT)—for salary support of a curatorial fellow at its Lewis Walpole Library, and for a meeting to plan an exhibition about Strawberry Hill, Walpole's house near London, $37,500 (2005); Chancellerie d'Orleans (Paris, France)—to construct a scale model of this 18th-century building, with the goal of persuading the French government to finance its reconstruction, $42,000 (2004); Church of the Holy Virgin, Monastery at Slivnica (Macedonia)—to conserve frescoes, $19,000 (2004); Royal Garden of Prague Castle, Summer Palace (Prague, Czech Republic)—for efforts to restore reliefs in this 16th-century Italian Renaissance palace, $50,000 (2004).
Contact: Lisa Ackerman, Vice President, (212) 861-4993; fax: (212) 628-3146
Internet: http://www.kressfoundation.org/kressorg/grant.html
Sponsor: Samuel H. Kress Foundation
174 E 80th Street
New York, NY 10021

Samuel Roberts Noble Foundation Grants Program 5196
The foundation is engaged in basic plant biology research as well as agricultural research, consultation, and development. Grants are made in a number of areas, with primary focus on capital funding for higher education and on health research and delivery systems. Requests are initiated with a letter describing the organization and the project for which funds are requested. Applications are provided if the request is in a current area of interest. Public affairs requests are initiated with a two-page summary of the organization and the project. Applications must be submitted a month preceding a Trustee meeting. Requests may not be submitted by fax.
Requirements: Generally, grants are geographically limited to the southwestern United States, with primary emphasis in Oklahoma. Matching funds are required.
Samples: Heritage Foundation (Washington, DC)—for its research and educational programs and to finance the purchase of its building in Washington's Capitol Hill neighborhood, $5 million.
Amount of Grant: $5 million-$12 million total approximately
Date(s) Application is Due: Mar 1; Jun 1; Sep 1; Dec 1
Contact: Donna Windel, (580) 223-5810 ext 6245; fax: (580) 221-7362; email: dkwindel@noble.org
Internet: http://www.noble.org/Admin/Granting/grGuides.html
Sponsor: Samuel Roberts Noble Foundation
P.O. Box 2180
Ardmore, OK 73402

Samueli Foundation Health Grants 5197
The Foundation considers grants to agencies, primarily in Orange County, California, that serve the community, and whose programs meet the guidelines listed. Grants are usually approved for a defined period of time, but may be paid over a multi-year period. In the area of Health, its goals are to: support programs promoting alternative and integrative medicine; and utilize creative and innovative approaches to education, prevention, treatment and research of disease and illness.
Requirements: 501(c)3 tax-exempt organizations are eligible.
Restrictions: The foundation does not fund umbrella fund raising organizations, political campaigns, or grants to individuals.
Amount of Grant: $1,000-$50,000 average
Date(s) Application is Due: Nov 24
Contact: Maggie Sherer, Grants Administrator; (949) 760-4400; fax: (949) 759-5707; email: Info@samueli.org
Internet: http://www.samueli.org/health.aspx
Sponsor: Samueli Foundation
2101 East Coast Highway, 3rd Floor
Corona del Mar, CA 92625

Samueli Institute Scientific Research Grants 5198
The Institute operates in conjunction with the Uniformed Services University of Health Sciences and Universitatsklinikum Freiburg Institute. Its mission is is to transform health care through the scientific exploration of healing. The Institute is one of an elite group of organizations in the nation with a track record in complementary and integrative medicine, healing relationships and military medical research. The major purpose of funding is to investigate healing processes and their application in promoting health and wellness, preventing illness, and treating disease. Researchers should respond to Requests for Applications or Program Announcements.
Amount of Grant: $2,000-$600,000 range
Contact: Matt Fritts; (703) 299-4800; email: mfritts@SamueliInstitute.org
Internet: http://www.siib.org/funding/investigator-home/218-SIIB.html
Sponsor: Samueli Institute
1737 King Street, Suite 600
Alexandria, VA 22314

Samuels Family Scholarships 5199

The program awards scholarships to doctoral students interested in pursuing careers related to or in elementary and secondary schools. Scholarships are solicited annually and renewable for up to three years. An application form is available online.
Date(s) Application is Due: Jan 15
Contact: Dr. Robert Ronau, (502) 852-0593; email bob@louisville.edu
Internet: http://www.louisville.edu/edu/fainfo.htm
Sponsor: University of Louisville
Louisville, KY 40292

San Antonio Area Foundation Grants 5200

Grants are usually awarded in the areas of health care and biomedical research, community and social services, arts and culture, education (early childhood education, higher education, medical schools, nursing schools, adult basic education and literacy), and animal services. Types of support include operating budgets, continuing support, annual campaigns, seed grants, emergency funds, equipment acquisition, matching funds, scholarship funds, research, lectureships, professorships, and building renovations. Applicants must submit a letter of intent to apply; if the letter is approved, the foundation will send a request for a proposal to the applicant. The foundation only reviews full proposals from applicants whose letters of intent have been approved. If asked to submit a proposal, the applicant will be sent an application package with a letter of notification in early February.
Requirements: Grants are made to organizations in the San Antonio, Texas, area.
Geographic Focus: Texas
Samples: San Antonio Symphony (TX)—for the Young People's Concert series, $25,000.; Southwest Foundation for Biomedical Research (San Antonio, TX)—to purchase equipment, $15,000.
Amount of Grant: $250-$250,000 average
Date(s) Application is Due: Nov 15
Contact: Lydia R. Saldana, Program Officer; (210) 228-3753 or (210) 225-2243; fax: (210) 225-1980; email: lsaldana@saafdn.org or info@saafdn.org
Internet: http://www.saafdn.org/NetCommunity/Page.aspx?pid=254
Sponsor: San Antonio Area Foundation
110 Broadway Street, Suite 230
San Antonio, TX 78205-1948

San Diego Women's Foundation Grants 5201

The San Diego Women, Foundation (SDWF) strengthens and improves women's capacities to engage in significant philanthropy in the San Diego region. The foundation chooses one focus area of grant making each year. Its primary focus areas include: education, arts and culture, environment, employment and economic development, civil society, and health and human services. Check the web site each September for that year's focus guidelines.
Requirements: Only 501(c)3 organization within San Diego County should apply.
Geographic Focus: California
Samples: AjA Project in Community Heights, $40,000; Eveoke Dance Theater: Young Artists Program, $28,000; La Jolla Playhouse: Enriching Children's Lives Through Theater, $17,000.
Amount of Grant: $25,000 minimum
Date(s) Application is Due: Dec 6
Contact: Tracy Johnson, Director; (619) 814-1374 or (619) 235-2300; fax: (619) 239-1710; email: tracy@sdfoundation.org
Internet: http://www.sdwomensfoundation.org/
Sponsor: San Diego Women's Foundation
2508 Historic Decatur Road, Suite. 200
San Diego, CA 92106

Sandler Program for Asthma Research Grants 5202

The mission of the foundation is to develop important new pathways of investigation in basic research regarding asthma. The program, which is open to investigators from all fields, particularly encourages applications from investigators not currently studying asthma. Innovation and risk are strongly encouraged. The program especially (but not exclusively) seeks applications in the following areas: development of new animal models for asthma; studies in humans or of human materials (excluding clinical trials); genetics of asthma; and technology development relevant to asthma. Three-year senior investigator awards and three-year junior investigator awards will be made. Awards to junior investigators are intended to support investigators who have demonstrated exceptional early accomplishment as independent investigators, permitting the expansion of their work into the field of asthma; they are not start-up funds. The proposed work should be directed toward uncovering basic mechanisms in the pathogenesis of asthma.
Requirements: Senior investigators will have well-established research programs and an international reputation for their research. They will usually hold a full-time academic appointment as professor, tenured associate professor, or the equivalent. Junior investigators will have already established an independent research program. They will usually hold a full-time academic appointment as an assistant professor, early associate professor, or the equivalent. Investigators may apply from any U.S. or Canadian nonprofit organization that can and will provide the necessary facilities and infrastructure for the research, and that will assure compliance with guidelines for animal and human studies established by the U.S. Department of Health and Human Services. Applicants from Canadian institutions must also provide evidence that the institution has been designated as a public charity by the U.S. Internal Revenue Service.
Samples: Robinna G. Lorenz, M.D., Ph.D., University of Alabama at Birmingham (2006); Boris Reizis, Ph.D., Columbia University (2006); John B. Wallingford, Ph.D., University of Texas at Austin (2006).
Amount of Grant: $250,000 per year senior investigator award; $150,000 per year junior investigator award
Date(s) Application is Due: Feb 8
Contact: William E. Seaman, Program Director, (415) 514-0730; fax: (415) 514-0734; email: seamanb@sandlerresearch.org
Internet: http://www.sandlerresearch.org
Sponsor: Sandler Foundation
Box 0509, UCSF
San Francisco, CA 94143-0509

Sarah Rebecca Reed Scholarship 5203

This scholarship is awarded to a graduate student who has been accepted into an ALA-accredited library science program. An application form must be obtained from the society.
Requirements: Applicant must include transcripts, five references, and completed application form.
Amount of Grant: $2000
Date(s) Application is Due: Mar 15
Contact: Dr. Jane Robbins, Executive Director, School of Information Studies, (850) 644-3907; fax: (850) 644-9763; email: Beta_Phi_Mu@lis.fsu.edu
Internet: http://www.beta-phi-mu.org/scholarships.html#reed
Sponsor: Beta Phi Mu International Library and Information Science Honor Society
101 Louis Shores Building, Florida State University
Tallahassee, FL 32306-2048

Sarah Scaife Foundation Grants Program 5204

Approximately 80 percent of the annual grants are made to public policy programs that address major domestic and international issues. There are no geographic restrictions. Other grants are made in the fields of education and culture, health and medicine, scientific research, public affairs, and recreation and equipment in all geographic areas, but primarily in western Pennsylvania. Types of support include general operating support, continuing support, program development, conferences and seminars, publication, seed grants, fellowships, research, and matching funds. Applications are accepted at any time. Foundation staff considers requests in February, May, September, and November.
Restrictions: Grants are not made to individuals or to national organizations for general fund-raising purposes.
Samples: U of Pittsburgh (Pittsburgh, PA)—to construct facilities for the planned Pittsburgh Institute for Neurodegenerative Diseases, $5.4 million.
Amount of Grant: $25,000-$200,000 average
Contact: Michael Gleba, Vice President of Programs, (412) 392-2900
Internet: http://www.scaife.com/sarah.html
Sponsor: Sarah Scaife Foundation
1 Oxford Ctr, 301 Grant Street, Suite 3900
Pittsburgh, PA 15219

Sarasota Memorial Healthcare Foundation Grants 5205

The Foundation is an organization with the mission to improve the delivery of health care in the Sarasota area through the acquisition and utilization of philanthropic funds. Although the Foundation primarily supports Sarasota Memorial Health Care System, it also awards grants to other area nonprofit organizations to meet the community's changing health care needs. Grants totaling more than $32,000,000 have been provided since the Foundation, establishment in 1976. Current fields of interest include: health care; medical care; and community health systems.
Requirements: Health care programs and organization in the Sarasota, Florida, region are eligible to apply.
Geographic Focus: Florida
Contact: Alexandra Quarles, President: (941) 917-1286; email: alex-quarles@smh.com
Internet: http://www.smhf.org
Sponsor: Sarasota Memorial Healthcare Foundation
1515 S. Osprey Avenue, Suite B4
Sarasota, FL 34239

Sarkeys Foundation Grants 5206

Major areas of foundation support include education and higher education, health care and medical research, children and family services, and cultural and humanitarian programs of regional significance. Types of support include building construction/renovation, capital campaigns, endowments, equipment, matching/challenge support, professorships, program development, and research grants. Grant proposals are considered at the October and April meetings of the board of trustees. The trustees may consider proposals at other regular or special meetings as circumstances warrant. The board may determine to condition awarding of a grant upon the fulfillment of an institution's additional fund-raising obligation for the project within a specified time period and may withdraw the grant if the institution fails to meet the required time allotted.
Requirements: 501(c)3 institutions located in Oklahoma are eligible.
Restrictions: Needs normally outside the range of the foundation's support include direct mail solicitations, annual campaigns, hospitals, operating expenses, vehicle purchases, permanent financing of a program, start-up funds for new organizations, or feasibility studies. Sarkeys Foundation will not accept proposals for new construction and/or major renovation projects this year and until further notice. In addition, Sarkeys will not accept any requests over $50,000

Geographic Focus: Oklahoma
Samples: Rogers State U (Claremore, OK)—to endow a chair in business information technology, $250,000 matching grant.
Amount of Grant: $50,000 maximum
Date(s) Application is Due: Feb 1; Aug 1
Contact: Susan Frantz, Program Officer, (405) 364-3703; fax: (405) 364-8191; email: Sarkeys@sarkeys.org
Internet: http://www.sarkeys.org/grants.htm
Sponsor: Sarkeys Foundation
530 E Main Street
Norman, OK 73071

Sasakawa Fellowships 5207
The Fellowships allow faculty to attend the National Faculty Development Institute on Incorporating Japanese Studies into the Undergraduate Curriculum, held on the San Diego State University campus. Up to 20 awards will be granted to full-time faculty members who teach undergraduate students, academic administrators, and librarians at member institutions of the American Association of State Colleges and Universities (AASCU). The Institute provides a $ 5,500 Sasakawa Fellowship for each participant selected. That Fellowship includes the administrative and academic costs of the Institute and campus housing. Participants receive up to $600 for transportation and an $800 stipend for meals and other expenses. Institutions whose faculty are accepted for the fellowships pay a partial tuition of $500.
Requirements: Participants must be full-time faculty or administrators at one of the more than 400 member institutions of the association and have no prior training in Japanese studies.
Amount of Grant: $5500
Date(s) Application is Due: Dec 13
Contact: Patricia Fesci, Sasakawa Fellowships, (202) 478-4668; fax: (202) 296-5819; email: fescip@aascu.org
Internet: http://www.aascu.org/programs/jsi
Sponsor: American Association of State Colleges and Universities
1307 New York Avenue NW, 5th Fl
Washington, DC 20005-4701

SASS Travel Grants 5208
The Norwegian Ministry of Foreign Affairs and the Norwegian Information Service in the United States offer these travel grants. Grants are meant as financial assistance for teachers and students visiting Norway for study and research purposes; within this general limitation, the awards may be used in accordance with personal preferences. Applications must contain a description of the study project(s) in addition to personal data, curricula, and possible references.
Requirements: Travel grants are offered to members of the Society for the Advancement of Scandinavian Study who are citizens and residents of the United States and are university or college teachers of Norwegian or other courses in Norwegian culture and society and to graduate students who have passed their preliminary examinations in these fields.
Amount of Grant: $750-$1500
Date(s) Application is Due: Feb 1
Contact: Grants and Scholarships, (212) 421-7333; fax: (212) 754-0583; email: cg.newyork@mfa.no
Internet: http://www.norway.org/Education
Sponsor: Norwegian Information Service in the 825 Third Avenue, 38th Fl
New York, NY 10022-7584

Savoy Foundation Post-Doctoral and Clinical Research Fellowships 5209
Fellowships of the Savoy Foundation are available to Canadian researchers and to foreign nationals or for projects conducted in Canada. Post-doctoral and clinical research fellowships of $30,000 will be awarded to scientists or medical specialists (Ph.D. or M.D.) wishing to carry out a full time research project in the field of epilepsy. The grant will be for one (1) year, but will be renewable, exceptionally, once upon request. As a rule, the modalities of attribution and amounts awarded will be similar to those of governmental funding agencies (MRC or FRSQ). The fellow with the highest mark will received the prestigious Steriade-Savoy award of an additional $1,500.
Restrictions: Fellows funded by the Savoy Foundation must not cumulate two major awards.
Amount of Grant: $30,000
Date(s) Application is Due: Jan 15
Contact: Fellowship Coordinator; (450) 358-9779; fax: (450) 346-1045; email: epilepsy@savoy-foundation.ca
Internet: http://www.savoy-foundation.ca/eng/mission/defaut.htm
Sponsor: Savoy Foundation
230 Foch Street
Street Jean Sur Richelieu, Quebec J3B 6B2

Savoy Foundation Research Grants 5210
A small number of research grants are available each year to clinicians and/or established scientists working on epilepsy or related subjects. These grants will be attributed for the following purposes: launching of a new research project, preliminary studies in preparation of a more substantial request to be addressed to another agency, continuation or completion of a project, sharing of expenses, or a limited project not requiring funding in excess of the limit. As a rule, these grants are not renewable. Contact the foundation for application forms.
Requirements: Research grants are available to Canadian citizens or for projects conducted in Canada.
Amount of Grant: $25,000
Date(s) Application is Due: Jan 15
Contact: Grants Administrator; (450) 358-9779; fax: (450) 346-1045; email: epilepsy@savoy-foundation.ca
Internet: http://www.savoy-foundation.ca/eng/mission/defaut.htm
Sponsor: Savoy Foundation
230 Foch Street
Street Jean Sur Richelieu, Quebec J3B 6B2 Canada

Savoy Foundation Studentships 5211
Studentships of the Savoy Foundation are available to Canadian researchers and to foreign nationals or for projects conducted in Canada. Studentships will be awarded to meritorious applicants wishing to acquire training and pursue research in a biomedical discipline, the health sciences or social sciences related to epilepsy. These studentships last one year, but will be renewable three times, upon request, for a maximum duration of four years. The stipend will be $15,000 per year. The student with the highest mark will receive the prestigious Van Gelder-Savoy award of an additional $1,500. An annual sum of $1,000 will be allocated to the laboratory or institution as additional support for the research project.
Requirements: To be eligible, the candidate must have a good university record (B. Sc., M.D. or equivalent diploma) and have ensured that a qualified researcher affiliated to a university and/or hospital, will supervise his/her work. Concomitant registration in a graduate program (M. Sc. or Ph. D.) is encouraged.
Restrictions: Students funded by the Savoy Foundation must not cumulate two studentship awards.
Amount of Grant: $15,000
Date(s) Application is Due: Jan 15
Contact: Caroline Savoy, (450) 358-9779; fax: (450) 346-1045; email: epileps@savoy-foundation.ca
Internet: http://www.savoy-foundation.ca/eng/mission/defaut.htm
Sponsor: Savoy Foundation
230 Foch
Street Jean Sur Richelieu, Quebec J3B 2B2 Canada

SC Johnson Fund Grants 5212
The fund awards grants every year locally and globally to improve health, education, social progress, and environmental protection. Types of support include building construction/renovation, capital campaigns, employee matching gifts, equipment acquisition, and seed grants.
Geographic Focus: Wisconsin
Amount of Grant: $9.5 million total
Contact: Program Administrator, (800) 494-4855 or (262) 260-4855
Internet: http://www.scjohnson.com/community/global.asp
Sponsor: SC Johnson Fund
1525 Howe Street, MS 066
Racine, WI 53403

Scaife Family Foundation Grants 5213
The foundation's grant program is primarily directed toward public policy programs that address major domestic and international issues. There are no geographical restrictions. Initial inquiries should be in letter form signed by the organization's president, or authorized representative, and have the approval of the organization's board of directors. The letter should include a concise description of the specific program for which funds are requested, and a program budget and timetable. Accompanying information should include a copy of the latest audited financial statements and annual report; current budget; list of officers and directors and their major affiliations; and a copy of the current determination letter from the Internal Revenue Service. Additional information may be requested, if needed, for further evaluation. The foundation meets in February, May, September, and November to consider grants. However, requests may be submitted at any time and will be acted upon as expeditiously as possible.
Requirements: 501(c)3 tax-exempt organizations are eligible.
Restrictions: The foundation does not make grants to individuals for any purpose or to nationally organized fundraising groups. Proposals for the following are usually declined: event sponsorships, endowments, capital campaigns, renovations, or government agencies.
Samples: U of Pittsburgh (PA)—for the Pittsburgh Institute for Neurodegenerative Diseases, $5.4 million.
Contact: Michael Gleba, Executive Vice President, (412) 392-2900
Internet: http://www.scaife.com/sarah.html
Sponsor: Scaife Family Foundation
301 Grant Street, Suite 3900
Pittsburgh, PA 15219-6401

Schlesinger Library Dissertation Grants 5214
The program supports doctoral dissertation research at the Arthur and Elizabeth Schlesinger Library on the History of Women in America. Only projects that use Schlesinger Library resources will be considered for funding; priority will be given to projects that require access to materials that can be used only at the library. Each proposal should include a cover page giving the applicant's name; address; affiliation; title of the proposed project; and a brief project description (of no more than seven pages) that explains the research to be undertaken, the Schlesinger Library holdings to be consulted,

and the significance of these holdings to the dissertation. A curriculum vita of no more than two pages, a work plan, and an outline budget should be attached to the proposal. Include the names and addresses of two references who have been asked to send letters of recommendation directly to the Schlesinger Library, one of whom should be the director of the dissertation.
Requirements: Applicants must have completed all course work toward the doctoral degree and have an approved dissertation topic.
Amount of Grant: $1500
Date(s) Application is Due: Apr 1
Contact: Schlesinger Library, (617) 495-8647
Internet: http://www.radcliffe.edu/schles/grants/dissertation.php
Sponsor: Radcliffe Institute for Advanced Study
10 Garden Street
Cambridge, MA 02138

Schlesinger Library Research Support Grants 5215
The institute awards research support grants for use at the Arthur and Elizabeth Schlesinger Library on the History of Women in America. These grants are intended for faculty and unaffiliated scholars who are actively pursuing research that requires or will benefit from access to the holdings of the Schlesinger Library. Each proposal should include a cover page giving the applicant's name; address; affiliation if any; the working title of the proposed project; and a brief project description (of no more than seven pages) that explains the research to be undertaken, the Schlesinger Library holdings to be consulted, the significance of these holdings to the proposed research. A curriculum vita of no more than two pages, the name of one reference who has been asked to send a supporting letter directly to the Library, and a budget should be attached. These grants cover only research-related expenses and do not cover primary living costs, such as rent or food, for local applicants.
Requirements: Awards will be made to scholars who have completed the PhD or other doctoral degree at least a year before the time of application, or who have equivalent research and writing experience. Priority will be given to scholars who have demonstrated research productivity and who are at work on projects that require access to resources available only at the Schlesinger Library.
Amount of Grant: $2000 maximum
Date(s) Application is Due: Apr 1
Contact: Grants Administrator, Schlesinger Library, (617) 495-8262
Internet: http://www.radcliffe.edu/schles/grants/index.php
Sponsor: Radcliffe Institute for Advanced Study
10 Garden Street
Cambridge, MA 02138

Schlumberger Excellence in Educational Development Connectivity Grants 5216
Schlumberger is a global technology services company with corporate offices in New York, Paris, and The Hague. SEED is a nonprofit commnity development program of the corporation that offers to disadvantaged schools located in developing countries the financial and technical assistance needed to connect them to the Internet. Schools meeting the stated eligibility requirements are encouraged to contact SEED for information on how to apply.
Requirements: In order for a school to receive a grant it must: be located in a country where Schlumberger has ongoing operations; be unable to access other sources of funding at the present time; have a defined plan to assume the monthly connectivity costs when the SEED grant expires; be IT-aware (have motivated staff members and experience with how to use the IT resources for educational purposes); have a defined plan as to how the IT resources are to be used for educational purposes; be equipped with a basic infrastructure (a secure room, stable electricity, and/or generators); and be willing to share their resources within their community and participate in collaborative projects (i.e., ReSEED plan).
Samples: Stanford U (Palo Alto, CA)—to conduct research on alternative energy supplies that would reduce greenhouse-gas emissions, through the Global Climate and Energy project, approximately $25 million over 10 years (2002).
Contact: SEED Adminstrator, (212) 350-9400
Internet: http://www.oilfield.slb.com/content/about/community.asp?
Sponsor: Schlumberger Corporation
153 E 53rd Street, 57th Fl
New York, NY 10022

Schlumberger Foundation Grants 5217
Almost all the grants are made in the area of higher education, especially in the fields of engineering, electronics, and various sciences that are related to the company's business interests. Disbursements are made to a small number of technical institutions for research programs. Smaller grants are made to support outstanding engineering students through a program of undergraduate scholarships and graduate school fellowships in engineering and the earth sciences at selected universities, the recipients being chosen by the appropriate authority at each university. The foundation also supports educational associations and special education programs. Remaining contributions are made to programs in fine arts, music, and theater; to the environment; and to museums, with primary emphasis on organizations in the New York City metropolitan area. The foundation considers requests for general operating support, program development, fellowships, scholarship funds, and research. Applications are accepted at any time; however, funds often are committed as much as a year in advance.
Requirements: Applying organizations must be 501(c)3 tax-exempt. Grants to educational institutions are made to colleges and universities located in the United States and Canada.
Restrictions: Grants are not awarded directly to individuals or to organizations outside the United States and Canada.
Amount of Grant: $1000-$15,000 average
Contact: Arthur Alexander, Executive Secretary, (212) 350-9400; fax: (212) 350-9457
Sponsor: Schlumberger Foundation
153 E 53rd Street, Fl 57
New York, NY 10022

Schnurmacher Foundations Grants 5218
The foundations award grants nationally, with emphasis on California, Connecticut, and New York. Areas of interest include arts, education, Jewish agencies and temples, human services, medical research, mental health, and religion. Types of support include capital campaigns, conferences and seminars, continuing support, curriculum development, endowments, equipment acquisition, general operating support, matching/challenge grants, research grants, and scholarship funds. There are no application forms or deadlines.
Requirements: Nonprofit organizations are eligible.
Samples: Neurosciences Institute (La Jolla, CA)—for its campaign for endowment and research operations, $1 million.; Beth Israel Medical Ctr (New York, NY)—for general operating support, $52,500.
Amount of Grant: $1000-$200,000 average
Contact: Ira Weinstein, President, (212) 838-7766
Sponsor: Schnurmacher Foundations
155 E 55th Street, Suite 6J
New York, NY 10002

Scholarships for Advanced Finnish Studies and Research 5219
Scholarships are granted to nationals of all countries for studies or research in Finnish language, literature, linguistics, ethnology, and folkloristics in a Finnish university. Scholarships are available to postgraduates and young researchers travelling to and from Finland. CIMO runs scholarship programs based on bilateral cultural agreements with over 30 countries. On the programs for visitors from abroad, the grant applications are submitted by Finnish universities. Tuition is provided. Applications for these scholarships are made preferably in Finnish in free-form or application forms available on the Web site.
Requirements: The scholarship recipient should not be older than 35 and should have a master's degree in relevant subjects or have studied at an advanced university level.
Restrictions: Scholars may not accept full-time paid work during the scholarship period.
Contact: Scholarships Administrator, 358 0 774 770 33; fax: 358 0 7747 7064; email: cimoinfo@cimo.fi
Internet: http://home.cimo.fi/oppaat/Study_in_Finland.pdf
Sponsor: Center for International Mobility (CIMO)
P.O. Box 343
Helsinki 00531
Finland

Schomburg Center for Research in Black Culture Scholars in Residence Grants 5220
The program assists scholars and professionals whose research in the black experience can benefit from extended access to the center's resources. Fellowships allow recipients to spend six months or a year in residence with access to resources at both the Schomburg Center and the New York Public Library. The program encourages research and writing in black history and culture, facilitates interaction among participating scholars, and provides widespread dissemination of findings through lectures, publications, and colloquia and seminars. It encompasses projects in African, Afro-American, and Afro-Caribbean history and culture.
Requirements: Candidates for advanced degrees must have received the degree or completed all requirements for it by the application deadline. Foreign nationals who have resided in the U.S. for three years immediately preceding the award date are eligible.
Restrictions: Persons seeking support for research leading to degrees are not eligible.
Amount of Grant: $25,000 for six months and $50,000 maximum for 12 months
Date(s) Application is Due: Dec 1
Contact: Scholars in Residence Program, (212) 491-2228
Internet: http://www.nypl.org/research/sc/scholars/aboutscholar.html
Sponsor: Schomburg Center for Research in Black Culture
515 Malcolm X Boulevard
New York, NY 10037-1801

School of American Research Advanced Seminars 5221
Advanced seminars promote in-depth communication among scholars who are at a critical stage of research on a common topic or on interrelated questions and whose interaction has the potential to move the discipline of anthropology forward with new insights into human culture and society, behavior, evolution, or critical contemporary issues. Participants appraise ongoing research, assess recent innovations in theory and methods, and share data and knowledge relevant to broad anthropological problems. Proposals are accepted throughout the year but those received after the deadline will be considered during the next review cycle.
Date(s) Application is Due: May 1
Contact: Director of Academic Programs, (505) 954-7200; email: seminars@sarsf.org
Internet: http://www.sarweb.org/seminars/seminargiude.htm
Sponsor: School of American Research
P.O. Box 2188
Santa Fe, NM 87504-2188

School of American Research Resident Scholar Fellowships 5222

The resident scholar program, supported by NEH, the Weatherhead Foundation, and the Katrin H. Lamon Endowment for Native American Art and Education, is designed to foster the intellectual growth of scholars pursuing significant research and writing projects in anthropology or from allied perspectives, such as those of history, sociology, art, law, and philosophy. Both humanistically and scientifically oriented scholars are encouraged to apply. In an atmosphere that combines freedom from outside academic responsibilities with the lively exchange of ideas found in an active community of scholars, the program offers the time and facilities essential to creativity and intellectual productivity. Through NEH, three positions are available for scholars who hold the PhD at the time of application and whose work falls within the humanities, broadly defined. Two positions supported by the Weatherhead Foundation are available for either pre- or postdoctoral scholars whose work may be either humanistic or scientific. The Katrin H. Lamon fellowship is open to a Native American pre- or postdoctoral scholar whose research meets the same criteria as those of the other resident scholar categories. The length of tenure is nine months, September 1 through May 31. Applications must be received by December 1 for tenure beginning the following September. A furnished apartment, separate office on campus, and library assistance are provided. Contact the office for application guidelines.

Requirements: Applications are welcome from scholars working in any world area or on any topic within anthropology and related disciplines in the humanities and social sciences. Fellowships are awarded to PhD candidates, postdoctoral scholars, and retired scholars. Predoctoral applicants must be nominated by their degree-granting departments, and only one nominee will be considered from each department.

Amount of Grant: $30,000 stipend maximum

Date(s) Application is Due: Nov 15

Contact: Director of Academic Programs, (505) 982-2919; email: scholar@sarsf.org

Internet: http://www.sarweb.org/scholars/description.htm

Sponsor: School of American Research

P.O. Box 2188

Santa Fe, NM 87504-2188

School of Visual Arts National Scholarship 5223

SVA has set aside scholarship funding for applicants to participate in four-year degree programs in advertising, graphic design, illustration, cartooning, animation, fine arts, photography, film and video, interior design, art education, and art therapy. Additional scholarship aid is awarded through the admissions process, based upon a review of merit and need. Partial merit scholarships are awarded to graduate students through the admissions process. Need- and merit-based grants are also available to continuing students through the offices of the chairs of the university's various programs. Contact the program office for application deadline dates.

Requirements: To be considered, an applicant must submit a complete application and an application fee ($35 for domestic students and $50 for international students); an official copy of high school transcripts; SAT or ACT scores; a letter of recommendation from high school or college art chairperson; and an interview and portfolio review of original work, if the student resides within 250 miles of New York City. Graduate scholarships are awarded by the graduate program chairs.

Contact: Office of Student Financial Services, (212) 592-2030; fax: (212) 592-2029; email: sfs@adm.schoolofvisualarts.edu

Internet: http://www.schoolofvisualarts.edu/adm/index.jsp?sid0=4&sid1=58&sid2=97

Sponsor: School of Visual Arts

209 E 23rd Street

New York, NY 10010

Schramm Foundation Grants 5224

The foundation awards grants to Colorado nonprofit organizations in its areas of interest, including arts and culture, civic affairs, community development, education (elementary, secondary, and higher), healthcare, housing, humanities, medical research, science, social services delivery, technology, women's issues, and youth. Types of support include building construction/renovation, continuing support, equipment acquisition, general operating support, matching/challenge grants, program development, and scholarship funds. Applications are accepted from July 1 through the listed application deadline.

Requirements: Colorado 501(c)3 nonprofit organizations are eligible. Preference is given to requests from the Denver area.

Restrictions: Grants do not support advertising, advocacy organizations, individuals, international organizations, political organizations, religious organizations, school districts, special events, or veterans organizations.

Geographic Focus: Colorado

Samples: U of Denver (CO)—to renovate Spruce Hall of the Graduate School of Social Work, $100,000.

Amount of Grant: $500-$26,500 range

Date(s) Application is Due: Aug 31

Contact: Gary Kring, President, (303) 797-6718

Sponsor: Schramm Foundation

800 Grant Street, Suite 404

Denver, CO 80203

Scientific Communities Initiative Grants 5225

The coalition is awarding grants to digital ventures offering information in the areas of science, technology, and medicine (STM). The purpose of the grants is to stimulate and accelerate the creation of new university-based information communities serving users in key STM fields. Proposed projects should offer strategic alternatives to inefficiencies in the current, traditional scientific communication process. Projects should provide increased responsiveness to the needs of scientists—as authors and information users—and to libraries by offering an effective new electronic venue for publication of research; aggregating content/links needed by targeted users in sufficient breadth and mass; addressing important objectives of universities and colleges; delivering convenient, efficient access to this information in a Web environment optimized to the needs of scientists in the target discipline; and fostering a more open and competitive marketplace in which academe exercises an increased role in assuring broad dissemination of research. Partnerships and collaborations among universities and professional societies are encouraged. Letters of intent must meet the listed application deadline; full applications will be requested.

Amount of Grant: $100,000 minimum; $500,000 total

Contact: Heather Joseph, (202) 296-2296; email: sparc@arl.org

Internet: http://www.arl.org/sparc/core/index.asp?page=f17

Sponsor: Scholarly Publishing and Academic Resources Coalition

21 Dupont Circle, Suite 800

Washington, DC 20036

Scleroderma Foundation New Investigator Grants 5226

The foundation invites applications from promising new investigators who hold faculty or equivalent positions and who wish to pursue a career in research related to systemic sclerosis (SSc, scleroderma). This grant will support promising research that is likely to lead to individual research project grants. Before completion of the grant, investigators are encouraged to seek continuing support for research through other grants mechanisms. Applicants may request support for up to three years.

Requirements: Applications may be submitted by domestic nonprofit organizations, public and private, such as universities, colleges, hospitals, and laboratories. Applicants must have a doctorate degree in medicine, osteopathy, veterinary medicine, or one of the sciences, and must have completed a postdoctoral fellowship by the grant award date.

Restrictions: Foreign organizations and institutions are not eligible. New investigator grants may not be used to support thesis or dissertation research or fellowship training. Applicants who have been a principal investigator on foundation grants or other national, private, or government agencies other than fellowship grants are not eligible. Applicants may not simultaneously apply for this grant and the Physician Scientist Development Award, which is cofunded by the Arthritis Foundation.

Samples: Dr. Constantin Bona, Mount Sinai School of Medicine (NY)—for the project Molecular Mechanisms IL-4 Induced Fibrosis in Scleroderma, $150,000,; Dr. Joseph Korn, Boston U School of Medicine (MA)—for the project Role of Integrin Alpha 11 in Fibroblast Function, $149,998.

Amount of Grant: $50,000 maximum per year for up to three years

Date(s) Application is Due: Sep 15

Contact: Carolyn Weller, (800) 722-4673 or (978) 750-4499; fax: (978) 750-9902; email: sfinfo@scleroderma.org

Internet: http://www.scleroderma.org/research/researchhome.htm

Sponsor: Scleroderma Foundation

12 Kent Way, Suite 101

Byfield, MA 01922

Searle Scholars Program Grants 5227

The program's aim is to identify junior faculty members who have demonstrated a potential for doing innovative research in the biomedical sciences. Approximately 15 scholars are selected each year to receive the award. Scholars will receive a grant to support their research over the next three years. Application forms are distributed through invited institutions.

Requirements: Applicants will be expected to be pursuing independent research careers in biochemistry, cell biology, genetics, immunology, neuroscience, pharmacology, related areas in chemistry, medicine, and the biological sciences. Candidates should be in their first or second year at the assistant professor level. This appointment must be a tenure-track position and must be in an academic department of an invited, degree-granting institution. Potential applicants whose institutions do not have tenure-track appointments should consult with the scientific director of the program prior to preparing an application.

Restrictions: Applications from any institution not invited to participate will not be considered. No more than two applications will be accepted from any one institution.

Amount of Grant: $80,000 per year for three years

Date(s) Application is Due: Sep 30

Contact: Mary Tarpey, Program Contact, (847) 714-1702; fax: (847) 714-1716; email: mary.tarpey@kinshipcorp.com

Internet: http://www.searlescholars.net/apply

Sponsor: Searle Scholars Program

400 Skokie Boulevard, Suite 300

Northbrook, IL 60062

Seaver Institute Grants 5228

The institute supports a variety of areas including the arts, private education, youth, health concerns, public affairs, the disabled, literacy, and science. Seed grants are awarded to nonprofit organizations for projects that offer the potential for a significant advancement in their fields. The institute seeks proposals that incorporate state-of-the-art scientific research, exceptional educational development, unique approaches to the creative arts, and the ramifications of shifting societal realities. There are no application deadlines. Interested applicants should submit an inquiry letter of two or three pages and include the goals of the project, a preliminary budget, and the qualifications of the organization and principal investigator. If interested, the institute will invite a full proposal.

Requirements: U.S. 501(c)3 nonprofits are eligible.

Restrictions: Grants are not made for construction, endowments, ongoing projects, operating budgets, deficits, or individual scholarships.
Amount of Grant: $1000-$35,000 average
Date(s) Application is Due: Apr 1; Nov 1
Contact: Victoria Seaver Dean, President, (310) 979-0298; email: vsd@theseaverinstitute.org
Sponsor: Seaver Institute
11611 San Vicente Boulevard, Suite 545
Los Angeles, CA 90049

SeaWorld and Busch Gardens Conservation Fund Grants 5229
The fund's grants support wildlife conservation projects conducted by recognized charitable organizations and noted scientists throughout the world. Grants support projects in the areas of species research, terrestrial; species research, aquatic; habitat protection; animal rescue and rehabilitation; and conservation education. Grants are awarded on a semiannual basis. Both guidelines and application are available online.
Requirements: Charitable organizations and scientists worldwide are eligible.
Amount of Grant: $5,000-$25,000
Date(s) Application is Due: Jun 1; Dec 1
Contact: Fred Jacobs, Senior Director; (314) 613-6077; fax: (314) 613-6079; email: fred.jacobs@anheuser-busch.com or mailbox@swbgfund.org
Internet: http://www.swbg-conservationfund.org/get_involved.html
Sponsor: SeaWorld and Busch Gardens Conservation Fund
231 S. Bemiston Avenue, Suite 600
Clayton, MO 63105

SEG Grants for Small Geoscience Departments 5230
SEG awards grants to faculty in small geoscience departments (no more than five full-time faculty) that emphasize undergraduate education to attend SEG-sponsored meetings, short courses, and/or field trips. Types of support include publication, seed grants, and grants to individuals.
Amount of Grant: $1000
Contact: Executive Director, (720) 981-7882; fax: (720) 981-7874; email: seg@segweb.org
Internet: http://segweb.org
Sponsor: Society of Economic Geologists Foundation
7811 Shaffer Pkwy
Littleton, CO 80127

Self Foundation Grants 5231
The Self Family Foundation primarily serves the Greenwood, South Carolina and surrounding counties. The Self Family Foundation will consider requests from other regions of South Carolina, if they have statewide impact. The Foundation supports non-profit organizations in the following areas of interest: Education - enhancing children's school readiness and ability to achieve educational goals with success; working together with schools, families, youth-serving organizations and other community organizations to support the intellectual and social development of young people; Arts, Culture & History - increasing awareness of and access to activities that broaden the horizons and knowledge of all residents, within Greenwood, South Carolina; Civic & Community - improving neighborhood resources and community attitudes and services, emphasizing the utilization of existing facilities, skills and talents to create a stronger, more unified community through collaboration with organizations, businesses, public institutions and residents; Health and Human Services - promoting individual and community wellness, prevention and literacy.
Requirements: IRS 501(c)3 tax-exempt or 509(a) nonprofit organizations in South Carolina are eligible. Letters of Inquiry or requests for information may be emailed to: application@selffoundation.org or mailed to Foundation. Contact Foundation Program Officer by email or phone prior to submitting proposal. All applicants must have the financial potential to sustain the project on a continuing basis after any funding is approved.
Restrictions: No grants to individuals and no loans available.
Geographic Focus: South Carolina
Amount of Grant: $1500-$50,000
Date(s) Application is Due: Feb 15; May 15; Aug 15; Nov 15
Contact: (864) 953-2441 or (864) 941-4036; fax: (864) 941- 4091; email: info@selffoundation.org
Internet: http://www.selffoundation.org/grantmaking4.htm
Sponsor: Self Family Foundation
P.O. Box 1017
Greenwood, SC 29648-1017

Senate Legislative Fellows Program 5232
Positions are available for eligible graduate students and recent graduates at the master's- and doctoral-degree levels. The program offers two types of awards: one fellowship is for applicants demonstrably in pursuit of careers in journalism/public relations; another is for persons exceptionally well-suited for placement in high-level leadership offices. The application form does not distinguish between the programs, but applicants indicate and rank their preferences. In addition to the award stipend, paid in biweekly installments over the term of the fellowship year, fellows are eligible for some benefits and options, such as health, dental, eye, and life insurance and retirement provisions. Applicants are advised to work with experienced campus liaison officers from whom applications and advice are available. Deadline is the first friday in May. Telephone in advance to indicate your intent to apply.
Requirements: Applicant must be a resident of New York State or attend a New York institution on a full-time basis and be matriculated in a recognized graduate program at an accredited college or university.
Restrictions: Former enrollees in Senate-sponsored programs administered by the Student Programs Office are ineligible.
Geographic Focus: New York
Amount of Grant: $29,500 stipend
Date(s) Application is Due: May 6
Contact: Dr. Russell Williams, Director, Senate Student Programs Office, (518) 455-2611; fax: (518) 426-6827; email: students@senate.state.ny.us
Internet: http://www.senate.state.ny.us/graduate.nsf/b80d007f27ecd52285256eca0046e344/59f81cd3056e769505256b8700700873?OpenDocument
Sponsor: New York State Senate
90 S Swan Street, Suite 401
Albany, NY 12247

Seneca Nation Higher Education Grants 5233
This program seeks research information that can be used to revitalize and stabilize the Seneca Language. This research will examine the following influences on second-language learning: curriculum, instruction, standards, assessment, school and classroom settings, and educational leadership and teacher professional development, as well as, family and community collaboration with schools.
Requirements: Enrolled tribal members of the Seneca Nation of Indians may apply.
Amount of Grant: $5000 maximum
Contact: Carrie Haynes, Director, (716) 532-3341; fax: (716) 532-3269
Internet: http://www.sni.org/eae.html
Sponsor: Seneca Nation of Indians
12861 Route 438
Irving, NY 14081

Sensient Technologies Foundation Grants 5234
The foundation awards grants to eligible nonprofit organizations in its areas of interest, including arts, community development, education, family services, federated giving programs, food services, general charitable giving, higher education, homelessness, human services, hospitals (general), medical research, mental health/crisis services, minorities, nutrition, performing arts, public policy, research, residential/custodial care, hospices, urban/community development, and voluntarism promotion. Types of support include annual campaigns, building construction/renovation, capital campaigns, emergency funds, employee matching gifts, endowments, general operating support, matching/challenge support, program development, research, and scholarship funds. There are no application forms or deadlines.
Requirements: Indiana, Missouri, and Wisconsin nonprofits, including school districts, are eligible.
Geographic Focus: Indiana; Missouri; Wisconsin
Amount of Grant: $25-$100,000 range
Contact: Doug Arnold, (414) 347-3727; fax: (414) 347-4783
Sponsor: Sensient Technologies Foundation, Inc.
777 E Wisconsin Avenue
Milwaukee, WI 53202-5304

Sequoyah Fellowship Program for Native American and Alaska Native Graduate Students 5235
These fellowships provide one-year, one-time unrestricted stipends to Native Americans and Alaska Natives. In past years, fellowships have been awarded to eligible recipients pursuing graduate degrees in such diverse fields as anthropology, microbiology, education, law, mechanical engineering, psychology, and business administration at universities throughout the United States. Approximately 10 fellowships are awarded each year around October 1. Student should provide the association with a biography describing background; tribal affiliation and a certificate showing tribal membership; university, year, and subject of study; career goals; interests; financial situation; two letters of recommendation; social security number; and the name, address, and phone number of the financial aid officer at the school/university he or she attends. Applications will be accepted from July 1 through October 1 each year.
Requirements: To be eligible, a graduate student must be an enrolled member of a Native American tribe or Alaskan Native corporation. Applicant must be minimally one-fourth degree Indian blood. A certificate of tribe enrollment and blood quantum from the tribe or BIA should be included.
Amount of Grant: $1500
Date(s) Application is Due: Aug 1
Contact: Lisa Wyzlic, Scholarship Coordinator, (240) 314-7155; fax: (240) 314-7159; email: lw.aaia@verizon.net
Internet: http://www.indian-affairs.org
Sponsor: Association on American Indian Affairs
966 Hungerford Drive, Suite 12-B
Rockville, MD 20850

Sexually Transmitted Diseases Postdoctoral Fellowships 5236
The ASHA Research Fund fellowships program is intended to expand the number of trained scientists working in the field of sexually transmissible diseases research by providing postdoctoral candidates with the opportunity to spend two years in a challenging, stimulating, and instructive STD research environment under the sponsorship of an established researcher. Applications for research training relative to any aspect of the sexually transmissible diseases are welcome. ASHA has supported

projects in chancroid, human papilloma virus (HPV), group B streptococcal infection, herpes, gonorrhea, chlamydia, HIV, pelvic inflammatory disease (PID), and syphilis. In funding HIV-related research, ASHA gives priority to researchers exploring the relationship between HIV and other STDs. Although applications covering fellowship training in basic investigations are not discouraged, it is the intent of this program to focus on research of short- or medium-term duration. Requests for applications are usually issued in May of odd-numbered years, with an application deadline of October 15, and the award being given the following spring for a fellowship period commencing on July 1.
Requirements: Applications are requested from MDs, PhDs, DSc who will be conducting two-year research projects under the sponsorship of established STD investigators. Applicants are required to study at established U.S. institutions.
Restrictions: ASHA fellowships may not be used in support of faculty positions unless approved by ASHA's management. Current ASHA fellows are not eligible to apply for other ASHA fellowships.
Amount of Grant: $35,500 first year; $36,000 second year
Date(s) Application is Due: Nov 15
Contact: Grant Administrator; (919) 361-8400; fax: (919) 361-8425
Internet: http://www.ashastd.org/programs/arf.html
Sponsor: American Social Health Association
P.O. Box 13827
Research Triangle Park, NC 27709

SFI Principal Investigator Program Grant 5237
Science Foundation Ireland advances national scientific progress by awarding grants for research on a competitive basis in those fields of science, engineering, software engineering, computer science, and mathematics that underpin biotechnology and information and communications technology. Grants support the research program costs of the principal investigator and research group. Eligible costs include relevant research expenses, including personnel, equipment, consumables, collaboration, and travel; contributions to salaries for staff hired specifically to carry out the research program; and access to necessary special facilities, such as synchrotron beam lines or high-speed computing. Grants may be held for five years. Application and guidelines are available online.
Requirements: All applicants will be expected to have a proven record of internationally recognized independent research accomplishments and to have at least three years of independent research experience beyond the PhD or equivalent.
Amount of Grant: Euro 50,000 to Euro 200,000 typically
Contact: Program Contact, fax: 353 (0)1 607 3285; fax: 353 (0)1 607 3201; email: info-@-sfi.ie
Internet: http://www.sfi.ie/content/content.asp?section_id=510&language_id=1
Sponsor: Science Foundation Ireland
Wilton Park House, Wilton Place
Dublin 2
Ireland

Shalem Graduate Fellowships 5238
The fellowships support the work of a select group of students pursuing independent research in diverse areas of Jewish national life at the Shalem Center Jerusalem. Fellows research theoretical and practical problems facing Israel and develop innovative approaches to solving them. In consultation with senior staff of the center, each fellow researches a particular problem or set of problems in Jewish public life, culminating in significant papers for publication and presentation in public forums. Possible topics of research include constitutional reform, economic deregulation, education, government reform, Zionist theory and history, Israeli foreign policy, Jewish social thought, Israeli culture and cultural institutions, and other topics relevant to the public life of the Jewish nation. The Shalem Center will not be accepting candidates for the 2003-2004 academic year of the Shalem Graduate Fellowships Program.
Requirements: Candidates holding the BA or higher degree in political science, economics, history, philosophy, religion, law, literature, and other relevant disciplines may apply.
Amount of Grant: $15,000
Date(s) Application is Due: Feb 1
Contact: Grants Administrator, (887) 298-7300; email: shalem@shalem.org.il
Internet: http://www.shalem.org.il/felo.htm
Sponsor: Shalem Center
5505 Connecticut Avenue NW, #1140
Washington, DC 20015

Shastri Indo-Canadian Institute Faculty Research Fellowships 5239
Faculty fellowships support research and training in India by Canadian scholars with full- or part-time appointments in a Canadian institution of higher education. Retired faculty who maintain affiliation with a Canadian institution of higher education are eligible to apply if, upon return to Canada, they will continue to contribute to India studies through either teaching or research. Fellowships are awarded on the basis of detailed proposals, which are peer reviewed. Faculty research fellowships are for scholars who have demonstrated excellence in a specific field of India studies. Faculty training fellowships are for scholars who have had little or no previous involvement with India studies and wish to undertake an intensive program of study at an Indian institution; wish to change their disciplinary focus within India studies; wish to develop or enhance their knowledge of an Indian language; or wish to carry out comparative research in India.
Requirements: Eligible are Canadian citizens and landed immigrants with full- or part-time appointments in Canadian institutions who plan to return to Canada following the tenure of the fellowship.
Restrictions: Eligibility is restricted to citizens and permanent residents of Canada who are a minimum of 18 years of age.
Amount of Grant: Rs19,200 living allowance per month, Rs4,800 maximum per month for research expenses, and return airfare to India
Date(s) Application is Due: Jun 30
Contact: Anita Dennis, Program Officer; (403) 220-7467; fax: (403) 289-0100; email: sici@ucalgary.ca
Internet: http://www.sici.org/programmes/details/faculty-research-fellowship/
Sponsor: Shastri Indo-Canadian Institute
1402 Education Tower, 2500 University Drive NW
Calgary, AB T2N 1N4 Canada

Shastri Indo-Canadian Institute Librarian Fellowships 5240
Librarian Fellowships are awarded for Canadian university librarians to undertake: an intensive program of study at an Indian Institution; enhancement of knowledge of and Indian Language; or enhancement of research support for south Asian studies. Academic Fellowships are for a minimum of three to a maximum of twelve months. Arts Fellowships are for a minimum of four months and a maximum of twelve months. The awards are made for the fiscal year (April 1 - March 31) following the annual competition and must begin before April 1.
Requirements: Faculty Fellowships support research and training in India by Canadian scholars with full- or part-time appointments in a Canadian institution of higher education. Retired faculty who maintain affiliation with a Canadian institution of higher education are eligible to apply if upon return to Canada they will continue to contribute to India studies through either teaching or research. They are awarded on the basis of detailed proposals, which are peer reviewed.
Restrictions: Eligibility is restricted to citizens and permanent residents of Canada who are a minimum of 18 years of age.
Amount of Grant: Rs19,200 living allowance per month, Rs4,800 maximum per month for research expenses, and return airfare to India
Date(s) Application is Due: Jun 30
Contact: Anita Dennis, Program Officer; (403) 220-7467; fax: (403) 289-0100; email: sici@ucalgary.ca
Internet: http://www.sici.org/programmes/details/librarian-fellowship/
Sponsor: Shastri Indo-Canadian Institute
1402 Education Tower, 2500 University Drive NW
Calgary, AB T2N 1N4 Canada

Shastri Indo-Canadian Institute Millennium Development Research Grants 5241
Funds will be provided to a Canadian or Indian member university on behalf of a binational research teams to undertake public policy relevant research toward one of the eight United Nations Millennium Development Goals. Grant recipients will be required to disseminate their findings, both to the academic and to the broader community. Acceptable expenses include: travel and accommodation for participants, meals, communications costs (audio, video conferencing, promotional materials), organizational support, translation, copying and printing, dissemination of research, research assistant allowance, and reasonable out of pocket expenses. Travel expenses may either be for Canadians going to India or for Indians coming from India or domestic travel for research purposes. Total awards are $80,000 maximum.
Requirements: The lead applicant may be an individual researcher/faculty member or research team affiliated with a Canadian or Indian member institution of the Shastri Indo-Canadian Institute. Applicant must have a partner in the other country. Grant recipients will be required to integrate consideration of gender equality and environmental sustainability in their research model. (Preferences will be given to: applicants who partner with business, and or government; and applicants who also partner with business, civil society organizations and/or government.
Restrictions: The Shastri Institute will not pay institutional overhead.
Amount of Grant: $80,000 maximum
Date(s) Application is Due: Dec 1
Contact: Anita Dennis, Program Officer; (403) 220-7467; fax: (403) 289-0100; email: sici@ucalgary.ca
Internet: http://www.sici.org/programmes/details/shastri-millennium-development-research-grants/
Sponsor: Shastri Indo-Canadian Institute
1402 Education Tower, 2500 University Drive NW
Calgary, AB T2N 1N4

Shastri Indo-Canadian Institute Partnership Development Seed Grants 5242
The purpose of the grant is to be a catalyst for scholarly activities which will foster the creation of new knowledge and contribute to formal institutional research agreements or other long-term partnerships. Funds may be used to support a binational seminar, conference, meeting, or other activities that: build awareness of potential collaboration opportunities with Indian institutions; promote sharing of knowledge, information, resources and expertise; showcase existing Canada-India research collaborations and their impact on society; and assist researchers to take advantage of Canada-India collaborative opportunities. The Shastri Institute will provide five grants to Canadian or Indian institutions of higher learning to support catalyst activities such as workshops, seminars, conferences, scholarly discussions and planning meetings that are expected to lead to significant research collaborations with universities or research institutions from the other country. Acceptable expenses include: travel and accommodation for speakers and participants, meals, communication costs (audio, video conferencing, promotional materials), organizational support, translation, copying and printing. Travel expenses may either be for Canadians going to India or for Indians coming to Canada.

Applications which promote partnerships with business, civil society organizations and/or government are especially welcome.
Requirements: Individual researcher/faculty member or research teams affiliated with a Canadian or Indian university are eligible.
Restrictions: The Shastri Institute will not pay institutional overhead.
Amount of Grant: $20,000 maximum
Date(s) Application is Due: Oct 31
Contact: Anita Dennis, Program Officer; (403) 220-7467; fax: (403) 289-0100; email: sici@ucalgary.ca
Internet: http://www.sici.org/programmes/details/partnership-development-seed-grant/
Sponsor: Shastri Indo-Canadian Institute
1402 Education Tower, 2500 University Drive NW
Calgary, AB T2N 1N4 Canada

Shastri Indo-Canadian Institute Postdoctoral Research Fellowships 5243
Post-doctoral Research Fellowships are open to scholars who have completed a Ph.D. and wish to do research in India. Applicants need not have an academic position. These awards will not normally be made if it has been greater than five years since completion of a doctoral degree. Academic Fellowships are for a minimum of three to a maximum of twelve months. The awards are made for the fiscal year (April 1 - March 31) following the annual competition and must begin before April 1.
Requirements: Faculty Fellowships support research and training in India by Canadian scholars with full- or part-time appointments in a Canadian institution of higher education. Retired faculty who maintain affiliation with a Canadian institution of higher education are eligible to apply if upon return to Canada they will continue to contribute to India studies through either teaching or research. They are awarded on the basis of detailed proposals, which are peer reviewed.
Restrictions: Eligibility is restricted to citizens and permanent residents of Canada who are a minimum of 18 years of age. The government of India does not permit research in strategic areas or on sensitive regional, political, and social themes.
Amount of Grant: Rs19,200 living allowance per month, Rs4,800 maximum per month for research expenses, and return airfare to India
Date(s) Application is Due: Jun 30
Contact: Anita Dennis, Program Officer; (403) 220-7467; fax: (403) 289-0100; email: sici@ucalgary.ca
Internet: http://www.sici.org/programmes/details/post-doctoral-fellowship/
Sponsor: Shastri Indo-Canadian Institute
1402 Education Tower, 2500 University Drive NW
Calgary, AB T2N 1N4 Canada

Shastri Indo-Canadian Institute Student Language Training Fellowships 5244
Student Language Training Fellowships are awarded to students enrolled in graduate degree programmes outside of India who have taken training in an Indian language in Canada and wish to pursue advanced studies in that language in India. The award is normally for a period of twelve months and tenable only at recognized educational institutions that award appropriate diplomas or certificates. The awards are made for the fiscal year (April 1 - March 31) following the annual competition and must begin before April 1. Academic Fellowships are for a minimum of three to a maximum of twelve months.
Requirements: Student fellowships are awarded to students specializing in India who are at the thesis level in a graduate degree program.
Restrictions: Eligibility is restricted to citizens and permanent residents of Canada who are a minimum of 18 years of age.
Amount of Grant: Rs19,200 living allowance per month, Rs4,800 maximum per month for research expenses, and return airfare to India
Date(s) Application is Due: Jun 30
Contact: Anita Dennis, Program Officer; (403) 220-7467; fax: (403) 289-0100; email: sici@ucalgary.ca
Internet: http://www.sici.org/programmes/details/student-language-training-fellowship/
Sponsor: Shastri Indo-Canadian Institute
1402 Education Tower, 2500 University Drive NW
Calgary, AB T2N 1N4 Canada

Shastri Indo-Canadian Institute Student Research Fellowships 5245
Student Research Fellowships enable Canadian citizens enrolled in master's or doctoral programs at institutions outside India to conduct research in India towards their degree. The Institute offers support only for the period spent in India and only for a maximum of one year. Students must affiliate with an Indian institution. Academic Fellowships are for a minimum of three to a maximum of twelve months. The awards are made for the fiscal year (April 1 - March 31) following the annual competition and must begin before April 1.
Requirements: Student Fellowships are awarded to students specializing in India entering or enrolled in a graduate degree program.
Restrictions: Eligibility is restricted to citizens and permanent residents of Canada who are a minimum of 18 years of age.
Amount of Grant: Rs19,200 living allowance per month, Rs4,800 maximum per month for research expenses, and return airfare to India
Date(s) Application is Due: Jun 30
Contact: Anita Dennis, Program Officer; (403) 220-7467; fax: (403) 289-0100; email: sici@ucalgary.ca
Internet: http://www.sici.org/programmes/details/student-research-fellowship/
Sponsor: Shastri Indo-Canadian Institute
1402 Education Tower, 2500 University Drive NW
Calgary, AB T2N 1N4 Canada

Shell Oil Products U.S. Corporate Contributions 5246
The company supports organizations that work to improve the quality of life and the general welfare of its employees and the citizens of their home communities. Contributions are made in the areas of social betterment and welfare, education, health, culture, civics, and certain unclassified causes. In the area of social betterment and welfare, in addition to local community funds, support may be given to institutions for the handicapped, associations for the blind, or other well-established organizations that enjoy wide community support and receive no support from local United Funds. Financial aid to education is directed primarily to accredited colleges and universities where the company has significant operations, where it recruits, or where research of interest or applicability to its business is being conducted. Approximately one-half of Pennzoil's corporate giving is directed toward educational support. Corporate contributions to hospitals, clinics, rehabilitation centers, and other health-related organizations are restricted to building programs, equipment additions, and unusual research programs. Contributions to campaigns conducted for the more prevalent diseases or disabilities may also be considered. Contributions are made to organizations engaged in the performing or visual arts, museums, zoos, libraries, and other cultural endeavors. Support will be given to selected civic organizations working to improve the quality of life in local Pennzoil communities. There may be causes outside the preceding classifications that are deserving of support and to which consideration will be given on an individual basis.
Requirements: Contributions are made to tax-exempt educational, charitable, health, and social betterment organizations. Those interested in applying for funding from Shell should be advised that Shell's contributions budget is completed in the fall of each year. Organizations making proposals following the completion of the budgeting process are not likely to be considered for funding during the following year.
Restrictions: Shell does not generally support strictly sectarian or denominational religious activities, secondary schools, veterans' organizations, fraternal organizations, charitable advertising, donations of products, or donations that are not tax deductible.
Samples: Volunteer Florida Foundation (Tallahassee, FL)—for the Hurricane Charley Disaster Relief Fund, given in conjuction with Motiva Enterprises LLC, $50,000 (2004).
Contact: Manager, Constituency & Community Relations, (713) 546-8590
Internet: http://www.shelloilproductsus.com
Sponsor: Shell Oil Products U.S.
1100 Louisiana, Suite 2346
Houston, TX 77002

Shelley and Donald Rubin Foundation Grants 5247
The foundation's mission is to support innovative efforts to transform society's institutions and to empower individuals to develop their full potential. The foundation is primarily interested in supporting the inclusion of art from non-Western European cultures into the mainstream of scholarship and display. In addition, the foundation supports research, action, and other projects designed to reveal and understand barriers to the full access of all people to American society and the larger international community. Areas of particular interest include, but are not limited to: access to health care; AIDS and its effects on society's institutions; the celebration of ethnic and cultural diversity that simultaneously encourages intergroup understanding; and cultural and arts programs that encourage individual and community identity. Projects in other areas that are most often funded are those that in themselves serve as catalysts for social change, addressing emerging problems as well as better known, long-standing problems in new ways. On a limited basis, programs eligible for funding may also include evaluation, technical assistance, and demonstration projects leading to the development of innovative models. Some examples of technical assistance of interest are the establishment of Web sites, the development of Web/TV productions, and related outreach activities.
Requirements: Art-related projects eligible for funding most often fall within the areas of enhancing the foundation's web site, tibetart.org; preserving Himalayan art; supporting educational activities reaching a broad constituency; supporting Himalayan art and architecture restoration projects as well as traveling exhibits bringing Himalayan art to new audiences.
Restrictions: Grants do not support individuals, capital funds, operating funds, or funding for the delivery of direct services. The foundation does not accept letters of intent/proposals via email.
Samples: Tibetan Children's Education Foundation (Helena, MT)—for a series of exhibitions on Tibetan tangka paintings in the larger context of raising awareness of Tibetan art and culture in the Midwestern United States, $6000 (2004); National Coalition Against Censorship (New York, NY)—to support the development of a national youth Free Speech and Democracy Film Contest, $3000 (2004); Kidsave International (Los Angeles, CA)—to implement Weekend Miracles, a program designed to match African-American foster parents with difficult-to-place African American children, $10,000 (2004); Commission on the Public's Health (New York, NY)—to develop and maintain an information network about the health care safety net in local neighborhoods in New York City, $2000 (2004).
Amount of Grant: $5000-$60,000; $5000-$15,000 average
Contact: Evelyn Jones Rich, Executive Director, (212) 780-2035; email: erich@sdrubin.org
Internet: http://www.sdrubin.org/guidelines.htm
Sponsor: Shelley and Donald Rubin Foundation
115 Fifth Avenue
New York, NY 10003

SHGAPE Vincent P. DeSantis Prize 5248

The prize will be awarded in odd-numbered years for the best book treating any aspect of United States history in the period 1865-1920 published during the two years preceding the year of the award. It must be the author, first book. Books written in English and published between January 1, 2007 and December 31, 2008 are eligible for the 2009 prize. Authors or their publishers should send a copy of the book to each committee member by October 1, 2008. The prize will be awarded at the SHGAPE luncheon held at the Organization of American Historians annual meeting.

Amount of Grant: $500

Date(s) Application is Due: Dec 31

Contact: Thomas R. Pegram, (410) 617-2856; email: tpegram@loyola.edu

Internet: http://www.h-net.msu.edu/announce/show.cgi?ID=156867

Sponsor: Society for Historians of the Gilded Age and Progressive Era (SHGAPE)
310 Auditorium Building, Michigan State University
East Lansing, MI 48824-1120

Shorenstein Center Fellowships 5249

The Joan Shorenstein Center offers a one-semester fellowship at Harvard's Kennedy School of Government. Fellows are domestic or international journalists, scholars, and/or policymakers who are interested in the influence of the press on public policy and politics. The purpose of the Fellowship Program is to advance existing research in press/politics and to provide an opportunity for distinguished experts to reflect on their discipline. The goal is to foster a collegial and intellectual environment that will enrich and complement one's knowledge of the field. The primary focus for a fellow in residence is a paper (approximately 15 pages in length) on a topic examining and analyzing the influence of the press on politics and public policy in the domestic or international arena. A stipend of $30,000 disbursed in four installments over the semester. Office space, computer, printer and telephone are provided.

Requirements: Domestic and international applicants are eligible.

Restrictions: Travel and living expenses are not covered by the Center.

Amount of Grant: $30,000 average

Date(s) Application is Due: Feb 1

Contact: Edith Holway, Fellowship Program Director; (617) 495-8269; fax: (617) 495-8696; email: edith_holway@harvard.edu

Internet: http://www.hks.harvard.edu/presspol/fellowships/index.html

Sponsor: John F. Kennedy School of Government, Harvard University
79 JFK Street
Cambridge, MA 02138

Short-Term Travel to Japan Grants 5250

Grants of a maximum of $3,000 are available to cover expenses while in Japan conducting a specific project explicitly related to Japan which can be accomplished in the period of time requested. These grants are intended for short-term research trips by scholars who are already familiar with Japan and with their topic, but who need time in Japan in order to complete their work.

Requirements: Grants are made only to people with a Ph.D. or comparable professional qualification. Ph.D. candidates are not eligible for this program.

Restrictions: Grantees must use American carriers for any international transportation to be reimbursed under this program. Applicants must not have received funds in this category within the past five years.

Amount of Grant: $3,000

Date(s) Application is Due: Feb 1, Oct 1

Contact: Teresa Spence, Director, (734) 665-2490; fax (734) 665-3801; email: tspence@asian-studies.org

Internet: http://www.asian-studies.org/grants/main.htm#NEAC-JAPAN

Sponsor: Northeast Asia Council/Association for Asian Studies
1021 E Huron Street
Ann Arbor, MI 48104

SHRM Dissertation Awards 5251

These awards are given to four promising researchers in the field of human resource management at the dissertation stage of their doctoral education. This annual competition is funded by the SHRM Foundation, in partnership with the HR Division of the Academy of Management and is intended to support the dissertation research of doctoral candidates. The winners of these dissertation grants each receive a plaque, $5,000 from the SHRM Foundation, and are honored at the Academy of Management, annual meeting.

Requirements: The applicant, dissertation must address a phenomenon that is of importance to the field of human resources and have potential applicability to HR practice. Applicants must be in good standing in a doctoral program in human resource management or a closely related field. Both the applicant and their dissertation advisor (or faculty sponsor in cases where there is not a formal advisor) must be members of the HR Division of the Academy of Management. Applicants must have had their dissertation proposals approved by their dissertation committees prior to application (in schools where there is not a formal proposal defense, documentation (an email from a faculty advisor/sponsor) is needed showing that the student, proposed research design has been deemed acceptable.

Restrictions: Applicants must be in the dissertation stage of their doctoral degree. Applicants who will have defended their dissertation by the submission deadline or who will do so within 6 months after the deadline are ineligible.

Amount of Grant: $5,000

Date(s) Application is Due: Feb 1

Contact: Lynn M. Shore, San Diego State University, (619) 594-4309; email: lshore@mail sdsu.edu

Internet: http://www.shrm.org/foundation/EducationGrants.asp

Sponsor: Society for Human Resource Management Foundation
5500 Campanile Drive
San Diego, CA 92128

SHRM Foundation Research Grants 5252

The SHRM Foundation seeks to fund high impact research in the human resource management field with a focus on addressing current challenges or understanding emerging trends. Research topics should enhance the effectiveness of (a) HR professionals, (b) specific HR practices, functions, or systems, or (c) organizations through HR practices. The SHRM Foundation looks to fund high impact HR research, whether the focus is on addressing current challenges or understanding emerging trends. The grant must be for original rigorous empirical academic research projects that advance the HR profession. While aimed at an academic audience, funded research should also have clear applicability for HR practice. As such, projects submitted for funding should have a high likelihood of both adding value to the HR academic literature (i.e., be suitable for leading academic journals) and yielding practical implications for HR managers (i.e., applied outlets should be interested in the research results). The SHRM Foundation is open to funding research using any type of research methodology as long as the proposed methodology is sound and appropriate for the proposed research question(s). Those research questions typically (but not solely) take the form of theoretically derived hypotheses.

Amount of Grant: $200,000 maximum

Date(s) Application is Due: Jan 24; Apr 20; Sep 21

Contact: Beth McFarland, (703) 535-6371; email: bmcfarland@shrm.org

Internet: http://www.shrm.org/foundation/fundingguidelines.asp

Sponsor: Society for Human Resource Management Foundation
1800 Duke Street
Alexandria, VA 22314

SHRM Michael R. Losey Human Resource Research Award 5253

The Award recognizes outstanding contributions by a human resources researcher or professional to the human resource field. The award acknowledges significant past research accomplishments and facilitates continuing contributions by an individual to the human resource management field.

Restrictions: The Award is open to everyone, excluding current SHRM, SHRM Foundation and Human Resource Certification Institute Board memebers and members of SHRM volunteer panels.

Samples: Gary P. Latham, Ph.D. (2006); Frank Schmidt (2005); Dr. Lee Dyer (2004).

Amount of Grant: $50,000

Date(s) Application is Due: Jan 22

Contact: Nancy Woolever, (703) 535-6457; email: loseyaward@shrm.org

Internet: http://www.shrm.org/loseyaward

Sponsor: Society for Human Resource Management
1800 Duke Street
Alexandria, VA 22314

Sid W. Richardson Foundation Grants 5254

Grants are provided to tax-exempt organizations in Texas in the areas of education (museums, learning centers, day schools, K-12 schools, higher education institutions, and business and economic education), health (medical schools, organ donor registries, hospitals, disease prevention, health science centers, and nursing associations), arts (arts councils, visual and performing arts festivals, museums, ballet, symphony orchestra, and arts education programs), and human services (boys' and girls' clubs, united funds, the elderly, crime prevention, the disabled, housing opportunities, food programs, and drug and alcohol abuse prevention); specific goals within these larger areas change periodically. Types of support include operating budgets, seed grants, building construction funds, equipment acquisition, endowment funds, research, publications, conferences and seminars, matching funds, continuing support, and projects/programs. Award amounts vary depending on proposed projects.

Requirements: Proposals must cover programs and projects within Texas.

Restrictions: Grants are not made to individuals.

Geographic Focus: Texas

Samples: All Saints Episcopal School (Fort Worth, TX)—for its capital campaign, $1 million (2005).

Amount of Grant: $10,000-$100,000 typically

Date(s) Application is Due: Jan 15

Contact: Valleau Wilkie Jr., Executive Director, (817) 336-0494; fax: (817) 332-2176

Internet: http://www.sidrichardson.org/GrantInfo.html

Sponsor: Sid W. Richardson Foundation
309 Main Street
Fort Worth, TX 76102

Sidgmore Family Foundation Grants 5255

The foundation supports programs that ensure the health, education, and general well-being of children, youth, and young adults. The foundation is interested in funding organizations with programs, projects, and research that utilize sports to promote the physical, mental, and spiritual development of young people; recognize that an impoverished environment limits the possibilities for children to develop and thrive; improve the quality of education available to youth, with special emphasis on programs designed to improve teacher training and teaching methods; and further advance

knowledge and research in the medical fields of vision, hearing, and allergies. There are no application deadlines. Guidelines are available online.
Requirements: Nonprofit organizations are eligible. Preference is given to organizations that serve residents in the Washington metropolitan area, Maryland, and Virginia.
Restrictions: Grants do not support government agencies, national health organizations, political activities, or public policy advocacy groups.
Contact: Grants Administrator, (516) 541-2713; fax: (516) 541-2803; email: SidgmoreFound@aol.com
Internet: http://fdncenter.org/grantmaker/sidgmore/prior.html
Sponsor: Sidgmore Family Foundation
71 Leewater Avenue
Massapequa, NY 11758

Sidney and Hadassah Musher Publication Prize **5256**
The grant facilitates publication of a doctoral dissertation relating to Jewish life in Israel or America, 1880 to the present. The award is offered in even-numbered years.
Requirements: Applicant must have completed his/her dissertation and have a commitment for publication from an academic or university press.
Samples: Arieh Bruce Saposnik—for From Babel and Uganda to the Promised Land: The Creation of a National Hebrew Culture in the Jewish Yishuv in Palestine, 1903-1914, $4000 (2004).
Amount of Grant: $4000
Date(s) Application is Due: Apr 1
Contact: Grants Administrator, (212) 629-0500 ext 219; fax: (212) 629-0508; email: Grants@JewishCulture.org
Internet: http://www.jewishculture.org/jewish_scholarship/jewish_scholarship_musher_prize.html
Sponsor: National Foundation for Jewish Culture
330 Seventh Avenue, 21st Fl
New York, NY 10021

Sidney Stern Memorial Trust Grants **5257**
The foundation awards grants to nonprofit organizations in its areas of interest, including civil rights, children and youth, community development, disabled, education, health and science, social services, and Native Americans. Types of support include annual campaigns, building construction/renovation, deficit financing, emergency funds, endowment funds, equipment acquisition, land acquisition, scholarship funds, seed grants, special projects, and research. Requests from California receive preference. There are no application deadlines. The board meets monthly, except August.
Amount of Grant: $2500-$8000 average
Contact: Program Contact, (310) 459-2117
Sponsor: Sidney Stern Memorial Trust
P.O. Box 893
Pacific Palisades, CA 90272

Sigma Xi Grants in Aid of Research **5258**
Research awards are made to support specific scientific projects in any field of science or engineering. Awards are limited to the graduate and undergraduate levels of study. Awards are made payable to the individual recipients. No part of the award may be used for payment of indirect costs to recipient's institution. Applications are accepted online only.
Requirements: Eligible to apply are nationals of any country who are students in undergraduate- or graduate-degree programs with a clearly outlined project of scientific investigation. Applicants need not be members of the society. See website for complete listing of requirements and restrictions.
Samples: Seamus Alois Decker, Emory U (GA)—for research on determinants of human stress hormone levels, $400.; Kathleen Noel Kannass, U of Iowa (IA)—for research on infants' understanding of the physical world, $500.; Carol Jeanne Ormand, U of Minnesota (MN)—for research on the kinematics of fault-bend folding, $500.
Amount of Grant: $2500 maximum in astronomy, vision related research; $1000 maximum in all other fields
Date(s) Application is Due: Mar 15; Oct 15
Contact: Programs Coordinator, (800) 243-6534 or (919) 549-4691; fax: (919) 549-0090; email: giar@sigmaxi.org
Internet: http://www.sigmaxi.org/programs/giar/index.shtml
Sponsor: Sigma Xi, The Scientific Research Society
P.O. Box 13975
Research Triangle Park, NC 27709

Sigrid Juselius Foundation Grants **5259**
The Sigrid Juselius Foundation, offers funding for domestic medical research. To enable established foreign medical researchers to work in a Finnish institution for 1-12 months, the Foundation also offers grants to nationals of other countries. Resources are allocated to senior and post doc researchers. The grants cover all disciplines of medical science. The grants can cover personal expenses (living costs) and travel.
Requirements: The Foundation does not award direct grants to foreign medical research. A foreign scientist interested in coming over to Finland has to get in touch with a Finnish scientist of his particular field, and the application should be made on the part of the Finnish host together with the proposed visiting researcher. A electronic application is available on the Sigrid Juselius Foundation website.
Restrictions: The funds are not available for studies or doctoral theses.
Date(s) Application is Due: Apr 15; Sep 15
Contact: Funding Officer; 358 20 710 9082 or 358 20 710 9083; fax: 358 20 710 9089
Internet: http://www.sigridjuselius.fi/index.php?page_id=48
Sponsor: Sigrid Juselius Foundation
Aleksanterinkatu 48 B
Helsinki FI-00100 Finland

Simone and Cino del Duca Grand Prix Awards **5260**
The Simone and Cino del Duca Foundation of the Institut de France awards each year a scientific Grand Prix based on the recommendations of a jury of members of the Institut de France. The amount of the Grand Prix will be 364,000 euros: 250,000 euros to finance the work of the award-winning group, 50,000 euros awarded to the director of the research group, and an additional 64,000 euros towards hiring one or more French or foreign postdoctoral researchers to be selected by the group director.
Requirements: The Grand Prix will reward an internationally recognized French or European mathematician proposing an ambitious research project based on the collective work of a group under his/her direction. The proposal should describe the scientific objectives and the means that will be used to implement the project (duration of the project, invitations, meetings, seminars, etc.).
Amount of Grant: 364,000 EU
Date(s) Application is Due: Jan 25
Contact: Grants Administrator; 01 47 66 01 21; fax: 01 46 22 45 02; email: fondation-del-duca@institut-de-france.fr
Internet: http://www.enseignementsup-recherche.gouv.fr/cid23192/call-for-applications-for-the-grand-prix-scientifique-simone-and-cino-del-duca-2009.html
Sponsor: Institut de France
23, quai Conti
Paris 75270 France

Simons Postdoctoral Fellowship in Disarmament and Non-Proliferation of Weapons of Mass Destruction and Their Delivery Systems **5261**
The program invites applications for this postdoctoral fellowship in Disarmament and Non-Proliferation of Weapons of Mass Destruction and Their Delivery Systems. Current issues of highest concern are: nuclear disarmament as the driving motivation of all nuclear nonproliferation and arms control; missile proliferation, especially nuclear-capable delivery systems; missile defense development and the weaponization of space, especially as pertaining to Canadian foreign policy choices; biological, chemical, and radiological weapons proliferation, both among states and to non-state organizations; the Nuclear Nonproliferation Treaty (NPT) Review Conference; impact of U.S. nuclear policy on global disarmament efforts; nuclear proliferation challenges from East Asia through South Asia to the Middle East; multilateral institutions and regimes sustaining nuclear disarmament efforts; and international cooperative security mechanisms and domestic regime reform as pertaining to promoting human security and improving global governance. The fellowship, which is awarded for one year and renewable for a second year on an evaluated basis, carries an annual stipend with an additional research and travel expense allowance. Applicants holding other awards tenable at the University of British Columbia will receive an adjusted stipend. Guidelines are available online.
Requirements: Eligibility includes PhD awarded no more than three calendar years prior to the anticipated commencement date of the fellowship. PhD awarded at the time of application is preferred; all requirements for the doctoral degree must be completed before commencement of the fellowship.
Amount of Grant: $40,000 stipend plus $5000 research and travel expense allowance
Contact: Fellowship Administrator, University of British Columbia, email: simons.centre@ubc.ca
Internet: http://www.ligi.ubc.ca/collateral/common/index.cfm?fuseaction=view&pageName=peaceAndDisarmament&contentID=395&section=Centres&subSection=Disarmament%20%26amp%3B%20Non%2DProliferation
Sponsor: Liu Institute for Global Issues
6476 NW Marine Drive
Vancouver, BC V6T 1Z2 Canada

Simpson Fund Grants **5262**
The fund serves a broad range of organizations in its operating communities that address key community needs such as education, health services, economic development, the arts, and more. Simpson also provides support to environmental and conservation projects, such as environmental education, habitat restoration, research relevant to the timber industry, as well as other projects that impact the timber business. The fund makes an extra effort to support organizations of interest to, or recommended by, Simpson employees. Support is generally committed for one year at a time. Capital projects are given priority over requests for contributions to operating funds. The fund supports the United Way in all of its operating communities. Guidelines are available online.
Requirements: Contributions are limited to Simpson communities: Washington—Pierce, Mason, Thurston and Grays Harbor Counties; Oregon—Lincoln and Tillamook Counties; and California—Del Norte and Humboldt Counties.
Geographic Focus: California; Oregon; Washington
Amount of Grant: $1000-$10,000
Contact: Grants Administrator, (206) 224-5198; email: cmusgra@simpson.com
Internet: http://www.simpson.com/communitycontribute.cfm
Sponsor: Simpson Fund
1301 Fifth Avenue, Suite 2800
Seattle, WA 98101-2613

Sinfonia Foundation Research Assistance Grants **5263**
The grants give American musicians or music educators the opportunity to do postgraduate research in American music or music education.

Requirements: Grants are open to applicants of all nationalities who show evidence of previous successful writing and scholarly research or to those who show evidence of unusual knowledge or competence in the field to be researched.
Amount of Grant: $1000 maximum
Date(s) Application is Due: Mar 1
Contact: Cheri Faith Spicer, Administrative Coordinator, (800) 473-2649 or (812) 867-2433; email: lyrecrest@sinfonia.org
Internet: http://www.sinfoniafoundation.org/research.html
Sponsor: Sinfonia Foundation
10600 Old State Road
Evansville, IN 47711

Sioux Falls Area Community Foundation Grants **5264**
The purpose of the foundation's grantmaking program is to provide support primarily in the Sioux Falls area to a broad range of activities, such as charitable and educational programs, scientific and social research, care of the needy, improvement of social services and educational facilities, prevention and alleviation of social ills, encouragement of the arts, and protection of the environment. The board will consider grants to programs that address the underlying causes of specific problems of the community; programs that work to combat racism, sexism, and other forms of discrimination; new, innovative programs that require start-up funds; community studies, programmatic research, and original artistic works; established organizations that continue to be relevant to the needs of the community; and educational programs. Grant applications are due in February, April, June, August, October, and December; contact the office for specific grant deadlines. Initial inquiries may be made by phone or letter.
Requirements: Grants are made to 501(c)3 nonprofit organizations within the Sioux Falls, SD, area.
Restrictions: The board will not consider requests for capital or construction drives, ongoing operational support, elimination of an organization's deficit, computer hardware or software, payment of staff salaries, sectarian religious purposes, endowments, political advocacy projects, telephone solicitations, national fund-raising efforts, or travel for groups or individuals.
Geographic Focus: South Dakota
Samples: Hope Haven International Ministries (SD)—to produce 500 CDs to recruit and train volunteers and offsets transportation costs for elderly volunteers who are unable to drive themselves to the distribution center, $4100 (2003); Lennox Area Historical Society (SD)—to purchase supplies and display cases for an exhibit commemorating the community's 125 years of main street business, $2000 (2003); Augustana College/Ctr for Western Studies (SD)—to support production of an updated history of the state of South Dakota and distribute it to libraries and schools throughout the state, $3500 (2003); Dakota Academy of Performing Arts (SD)—for middle school students to attend a performance of Live Wire, a play focused on AIDS education, $2600 (2003).
Amount of Grant: $3 million annually
Contact: Sue Brown, President, (605) 336-7055; fax: (605) 336-0038; email: sbrown@sfacf.org
Internet: http://www.sfacf.org/sfacf/aboutgrants.aspx
Sponsor: Sioux Falls Area Community Foundation
300 N Philips Avenue, Suite 102
Sioux Falls, SD 57104-6006

Sir Dorabji Tata Trust Individual Education and Travel Grants **5265**
The trust offers several scholarships/education grants for higher education: General Education Grants, Sir Dorab Tata Merit Scholarships, Dr SR Moolgaokar Scholarship, M. Phil scholarships, and a scholarship of Rs 6,000 per annum is also given to a trainee at the Training Ship Rajendra Chanakya. The Trust also offers Travel grants for short-term studies abroad as well as for attending conferences, presentation of papers for research and sports related activities. The grants range from Rs 20,000 to Rs 50,000. Applications are received throughout the year and the grants are processed based on an internal appraisal process. A panel of educationists interview candidates recommended by their respective colleges. These scholarships are provided till the student completes his or her first degree.
Requirements: The application may be made by way of a letter, by email, or in person. The following documents as may be applicable should accompany the form: Marksheets from SSC onwards (college studies) / school report (school studies); Proof of family income - salary certificate or tax returns or similar proof; Proof of residence (ration card); Student's Identity Card; Copy of the fee receipts; Proof of the college being approved by Association of Indian Universities and the course being approved by Accreditation Board of AICTE. Each scholarship/award also has additional specific guidelines - see the website for more details.
Restrictions: Forms received late will not be accepted. Limited to India.
Date(s) Application is Due: Sep 30
Contact: R.M. Lala, Director; +91 (22) 6665-8282; fax: +91 (22) 2204-5427 / + 91 (22) 2282 6092; email: sdtt@sdtatatrust.com
Internet: http://www.dorabjitatatrust.org/ind_grants/ind_ietg.asp
Sponsor: Sir Dorabji Tata Trust
Bombay House, 24 Homi Mody Street
Mumbai 400 001 India

Sir Halley Stewart Trust Grants **5266**
The trust awards grants worldwide to organizations and individuals in the areas of medical research, social and education, and religion. Medical research grants focus on problems of the elderly, preventing disability in children, tropical diseases, and Crohn's disease. In the education and social category, grants support research and innovative development projects that attempt to prevent and resolve conflicts, promote reconciliation, and increase understanding within families and across racial, cultural, class, religious, and professional divides. Religion grants advance Christianity and include theological training, teaching in the UK about Christianity, encouragement of specific groups of people (e.g., the elderly, people with disability, students in higher education, etc.) to explore their spiritual needs and strengths, communicating the Christian message through the media, projects that attempt to help disadvantaged people, etc. Applications will not be accepted by fax or email.
Requirements: Nonprofit organizations worldwide are eligible.
Restrictions: Grants do not support general appeals of any kind; personal education fees or fees for taught courses—unless connected with research that falls within current priority areas; educational or 'GAP' year travel projects for young people; purchase, erection, or conversion of buildings; capital costs; running costs of established organizations; or university overhead charges.
Amount of Grant: @L875,000 total
Contact: Sue West, Administrator, 44-1954-260707
Internet: http://www.sirhalleystewart.org
Sponsor: Sir Halley Stewart Trust
22 Earith Road
Willingham, Cambridge CB4 5LS United Kingdom

Sir John Soane's Museum Foundation Traveling Fellowship **5267**
The purpose of the program is to enable scholars to pursue research projects related to the work of Sir John Soane, 19th-century British architect and collector, or Sir John Soane's Museum and its collections. The foundation was formed for the purpose of promoting art and architecture in the United States and abroad. Application and guidelines are available online.
Requirements: Application is open to individuals enrolled in a graduate-degree program in the history of art, architecture, decorative arts, interior design, or a field appropriate to the foundation's purpose.
Amount of Grant: $5000
Date(s) Application is Due: Mar 1
Contact: Executive Director, (212) 223-2012; fax: (212) 223-9366; email: info@soanefoundation.com
Internet: http://www.soane.org/foundation/f_fellowrules.htm
Sponsor: Sir John Soane's Museum Foundation
1040 First Avenue, No 311
New York, NY 10022

Siragusa Foundation Health Services & Medical Research Grants **5268**
The Siragusa Foundation, established in 1950 by Ross D. Siragusa, is a private family foundation that is committed to honoring its founder by sustaining and developing Chicago's extraordinary nonprofit resources. The Foundation's Health Services and Medical Research Program is balanced between medical research organizations, which are concerned with scientific inquiry and the exploration of treatment methods, and health service organizations, which provide direct social and clinical care. The Foundation primarily funds fellowships, single-disease research programs and clinical treatment services. The Foundation believes it is critical to fund research efforts, health support services, patient care services and treatment, and advocacy programs for both adults and youth, all of which are highly complementary to each other.
Requirements: Nonprofit 501(c)3 organization in Chicago are eligible to apply. Before submitting a formal proposal to the Foundation, prospective applicants should submit a two-page letter of inquiry outlining the background of the organization and the proposed project. The letter of inquiry should describe succinctly the project for which funding is requested, including how it relates to the Foundation's interests, the target audience, the estimated budget and request amount. After reviewing the letter of inquiry, the Foundation may or may not request a formal proposal.
Restrictions: The Foundation limits potential grantees from applying for funds to once per calendar year. The Foundation does not accept unsolicited proposals from outside the Chicago area and does not support individuals or political advocacy. The foundation typically does not support capital expenditures or endowments.
Geographic Focus: Illinois
Samples: Arthritis Foundation, Greater Chicago Chapter, Chicago, Illinois, $5,000 (2008); Children's Memorial Foundation, Chicago, Illinois, $215,000 (2008); Rehabilitation Institute of Chicago, Chicago, Illinois, $16,666 (2008).
Contact: Kristen M. Buerster, Adminstrative Assistant/ Grants Administrator; (312) 755-0064; fax: (312) 755-0069
Internet: http://www.siragusa.org/pages/program_areas/13.php
Sponsor: Siragusa Foundation
One East Wacker Drive, Suite 2910
Chicago, IL 60601

Skaggs Foundation Grants **5269**
The Skaggs Foundation was established in 1962 in California, and currently provides funding in Alaska. The foundations area of interest include: animals/wildlife; preservation/protection; children services; special education; environmental education; environment; natural resources; marine science. Types of support available with these funds include: continuing support; endowments; equipment; general/operating support; internship funds; land acquisition; matching/challenge support; program evaluation; research; technical assistance.
Requirements: Nonprofit organizations are eligible, preference is given to requests from Alaska. The initial approach should be a one or two page letter of inquiry. There is no formal application form. Applicants should submit the following: copy of IRS

Determination Letter; copy of current year's organizational budget and/or project budget; one copy of the proposal.
Restrictions: No grants to individuals.
Geographic Focus: Alaska
Date(s) Application is Due: May 1; Oct 1
Contact: Samuel D. Skaggs Jr., President; (907) 463-4843
Sponsor: Skaggs Foundation
P.O. Box 20510
Juneau, AK 99802-0510

Skin Cancer Foundation Research Grants 5270
Funding is provided annually for basic research, clinical studies, and educational programs related to skin cancer. The foundation's grants program is intended for pilot projects, which, if successful, can be further developed in order to be eligible for larger grants from other sources. Grants include: The Dr. Patricia Wexler Research Grant Awards (one-year, $10,000 grant for a research project); The Theodore S. Tromovitch Memorial Award (one-year, $5,000 grant for a research project); The Melissa K. Bambino Memorial Award (one-year, $8,000 grant for a research project); The Robert N. Cooke Memorial Award (one-year, $3,000 grant for a research project); Pevonia Botantica Research Award (one-year, $10,000 grant for a research project)
Requirements: Candidates must comply with all directions on the application form (available at the website). All projects are selected on the basis of their applicability to the foundation's mission of reducing the incidence, morbidity and mortality of skin cancer. Research must be conducted at institutions within the U.S. Preference is given to projects that address, at the basic science and clinical level, improved methods of prevention, detection, and treatment of skin cancers.
Restrictions: Overhead or indirect costs are not eligible.
Samples: Galina V. Shurin, PhD, University of Pittsburgh - Phosphatidylserine-based ointment for the topical therapy of skin cancer; Paul Nghiem, MD, PhD, University of Washington - Merkel cell carcinoma: Building a tumor and cell line toolbox; David Polsky, MD, PhD, NYU Medical Center - The use of chloroquine to reduce ultraviolet light-induced DNA damage in the skin
Amount of Grant: $3,000 - $10,000
Date(s) Application is Due: Oct 1
Contact: Grants Administrator; (800) 754-6490; email: info@skincancer.org
Internet: http://www.skincancer.org/research-grants-2010.html
Sponsor: Skin Cancer Foundation
149 Madison Avenue, Suite 901
New York, NY 10016

SLA Affirmative Action Scholarship 5271
One scholarship per academic year is given for graduate study leading to a master's degree at a recognized school of library or information science in the United States or Canada. Preference will be given to those applicants interested in pursuing careers in special librarianship.
Requirements: Applicant must be a member of a minority group, defined by the U.S. government as African American, Hispanic, Asian or Pacific Islander, Native American, or Alaskan Native.
Amount of Grant: $6000 maximum
Date(s) Application is Due: Oct 31
Contact: Scholarship Committee, (703) 647-4900; fax: (703) 647-4901; email: sla@sla.org
Internet: http://www.sla.org/content/learn/scholarship/sch-index/index.cfm#aascholar
Sponsor: Special Libraries Association
331 S Patrick Street
Alexandria, VA 22314-3501

SLA ISI Scholarships 5272
This scholarship is for beginning doctoral candidates who have been accepted into the program of accredited schools of library science, information science, or related fields of study in the United States or Canada. Preference is given to applicants interested in special librarianship. Awards are made without regard to race, sex, age, religion, or ethnic background. The scholarship winner will be notified in May. Official announcement of the award will be made at the association's annual conference in June.
Requirements: Applicant must be a beginning PhD candidate enrolled in a course of study by the time the award is given, have an interest in special librarianship, be a U.S. citizen, have work experience in a special library at the master's degree level, have a good academic record, and show financial need.
Restrictions: Because ISI awards medical librarianship scholarships through the Medical Library Association, persons planning careers in medical librarianship cannot be considered.
Amount of Grant: $1000 maximum
Date(s) Application is Due: Oct 31
Contact: Scholarship Committee, (703) 647-4900; fax: (703) 647-4901; email: sla@sla.org
Internet: http://www.sla.org/content/learn/scholarship/sch-index/index.cfm#isi
Sponsor: Special Libraries Association
331 S Patrick Street
Alexandria, VA 22314-3501

SLA Mary Adeline Connor Professional Development Scholarships 5273
Each year, one or more scholarships will be granted for post-MLS-certificate or degree programs in any subject area, technological skill, or managerial expertise relevant to applicant's career needs and goals in special librarianship. Preference will be given to individuals who display aptitude for special library work, who show promise of accomplishment, and who are judged capable of making specific contributions to the library profession. Study program proposals to be completed within two years of date of the award will receive special consideration. Scholarships may include travel stipend. The scholarship winner is announced during the association's annual conference in June.
Requirements: The applicant should be a member of SLA who has the MLS degree and who has worked in special libraries for at least five years. Applicant must be accepted into a certificate or degree program offered by an accredited college or university or by a professional association or society. Evidence of high academic achievement and financial need must be submitted.
Amount of Grant: $6000
Date(s) Application is Due: Oct 31
Contact: Scholarship Committee, (703) 647-4900; fax: (703) 647-4901; email: sla@sla.org
Internet: http://www.sla.org/content/learn/scholarship/sch-index/index.cfm#macscholar
Sponsor: Special Libraries Association
331 S Patrick Street
Alexandria, VA 22314-3501

SLA Plenum Scholarships 5274
This scholarship, made possible by funds from the Plenum Publishing Corporation, is awarded for graduate study leading to a doctoral degree at a recognized school of library or information science. Preference will be given to those applicants who display an aptitude for special library work, who show promise of accomplishment, and who are judged capable of making specific contributions to the library profession. Awards are made without regard to race, sex, age, religion, or ethnic background. The scholarship winner is notified in May. Official announcement of the award is made at the association's annual conference in June.
Requirements: Applicants must be doctoral candidates enrolled in a course of study by the time the award is given. They must submit evidence of high academic achievement, financial need, and dissertation topic approval. Additionally, applicants should be SLA members who have worked in special libraries.
Amount of Grant: $1000
Date(s) Application is Due: Oct 31
Contact: Scholarship Committee, (703) 647-4900; fax: (703) 647-4901; email: sla@sla.org
Internet: http://www.sla.org/content/learn/scholarship/sch-index/index.cfm#plenum
Sponsor: Special Libraries Association
331 S Patrick Street
Alexandria, VA 22314-3501

SLA Scholarships 5275
At least five SLA scholarships per academic year are awarded for graduate study leading to the master's degree at a recognized school of library or information science in the United States or Canada. Extra consideration will be given to members of SLA and to persons who have worked in and for special libraries, but the scholarships are not restricted to these persons. Awards are made without regard to race, sex, age, religion, or ethnic background. Scholarship winners are notified in May. Official announcement and presentation of the awards are made at the association's annual conference in June.
Requirements: Applicants must be college graduates or college seniors of any nationality with an interest in special librarianship, have definite interest in and aptitude for special library work, have good academic records, and demonstrate financial need.
Amount of Grant: $6000
Date(s) Application is Due: Oct 31
Contact: Scholarship Committee, (703) 647-4900; fax: (703) 647-4901; email: sla@sla.org
Internet: http://www.sla.org/content/learn/scholarship/index.cfm
Sponsor: Special Libraries Association
331 S Patrick Street
Alexandria, VA 22314-3501

SLA Steven I. Goldspiel Memorial Research Grants 5276
The purpose of the research fund is to support projects that promote research on and advancement of library sciences, focusing on projects that address the following goals: future and current user issues, measures of productivity and value, client/user satisfaction measures, and staffing. Grants applications are reviewed by SLA's research committee and are evaluated based on the purpose and objectives of the proposed project, the significance of the topic to the profession, the project's methodology, qualifications of the staff, and appropriateness of the project's budget and timetable. Recipients must file quarterly status reports, and the SLA reserves the right of first publication of any written reports about or resulting from the project. The award includes travel connected with the project, personnel, computer time, printing, and dissemination. Application materials are available on the Web site. Annual deadline dates may vary; contact program staff for exact dates.
Requirements: Applicants must be members of the Special Libraries Association.
Samples: Andrew Dillon, School of Library and Information Science, Indiana U—for the project Understanding Users in Digital Environments: A Longitudinal Study of Genre Influences in Information Work.; Claire McInerney, NY Ctr for Technology in Government (NY)—for the project Using Information in the Virtual Office: How Special Libraries are Serving Telecommuters.
Amount of Grant: $16,000-$20,000

Date(s) Application is Due: Feb 2
Contact: John Latham, Director, Knowledge Exchange, (703) 637-4915; email: jlatham@sla.org
Internet: http://www.sla.org/content/learn/scholarship/goldspiel/index.cfm
Sponsor: Special Libraries Association
1700 18th Street NW
Washington, DC 20009-2514

SME Manufacturing Education Plan Grants 5277
The program provides grants to institutions that can respond to the direct needs of the manufacturing industry. Proposals should address one or more of the competency gaps and integrate two or more of the following areas: capital equipment, student outreach and development, faculty development, curriculum development, and research initiation. Past grants have helped create new Associates and Bachelors degrees in manufacturing, new manufacturing discovery laboratories, and learning centers for interdisciplinary education.
Requirements: Community colleges, four-year institutions, and universities in North America are eligible to receive grants. Proposal must be submitted in electronic format - PDF or MS Word.
Samples: University of South Florida (Tampa, FL) - Virtual MedLab: A Virtual Medical Device Laboratory for Manufacturing Education — addresses the problem of maintaining the supply of highly trained workforce for advanced medical manufacturing and product development. University of Central Missouri (Warrensburg, MO) - Advance Manufacturing Research and Development Initiative — Through focused study into the advanced manufacturing automation technologies of robotics, programmable logic controllers, computer integrated machining equipment, quality assurance measurement components, automatic storage and retrieval systems, barcode readers, and assembly equipment, students will be trained and prepared to work in companies seeking to optimize product throughput in tandem with becoming more globally competitive.
Amount of Grant: $500,000 maximum
Date(s) Application is Due: Dec 5
Contact: Grants Administrator, (313) 425-3300; fax: (313) 425-3411; email: foundation@sme.org
Internet: http://www.sme.org/cgi-bin/smeefhtml.pl?/foundation/grants/grants.htm@&SEF&
Sponsor: Society of Manufacturing Engineers Education Foundation
One SME Drive, Box 930
Dearborn, MI 48121-0930

SME Research Initiation Grants 5278
The program funds projects that will expand the research activities of manufacturing educators. These research activities have value both in their contributions to the knowledge of manufacturing and in their enrichment of the education experience. The purpose of Research Initiation grants is to give faculty the resources to conduct preliminary research on an idea that has the potential to be submitted to a major funding agency.
Requirements: Subjects for consideration include, but are not limited to, the following: Manufacturing Processes, Manufacturing Systems, and Human Factors. Indirect or burdened rate costs may be included in the funding request. Faculty salary benefits, student assistant salaries, supplies, and materials required for prototype development may also be included. Proposal must be submitted in electronic format - PDF or MS Word.
Restrictions: Grants do not support student assistant salaries, capital equipment, overhead, or indirect costs. fax or hard copy proposals will not be accepted.
Samples: University of Southern California, $15,000 to fund a research project titled A DLP-Projector based Testbed for Manufacturing Polymer, Ceramic and Metal Parts; Washington State University, $15,000 to fund a research project titled The Investigation of Mg-Based Metallic Glass Wool for Manufacturing and Commercialization Potential; Illinois Institute of Technology, $15,000 to fund a research project titled Computing tip-specimen shape interaction in scanning probe microscopy: Toward nanoscale dimensional metrology of general three-dimensional structures.
Date(s) Application is Due: Feb 1
Contact: Grants Coordinator, (313) 271-1500; fax: (313) 425-3400; email: foundation@sme.org
Internet: http://www.sme.org/cgi-bin/smeefhtml.pl?/foundation/grants/frinit.htm@&SEF&
Sponsor: Society of Manufacturing Engineers Education Foundation
P.O. Box 930, One SME Drive
Dearborn, MI 48121-0930

SMH Heart Failure Research Grants 5279
The program is intended to foster clinical research that seeks to understand methods for improving the quality of care for patients hospitalized with a primary (or secondary) diagnosis of heart failure. A long-term goal is to foster the development of collaborations of hospitalists and other personnel within and across sites, particularly research collaborations with a high likelihood of leading to subsequent research applications to governmental agencies (e.g., NIH). Funds may only be used to support work directed at evaluation of the program, and cannot be applied to program implementation, support for clinical personnel, rental of space, or to support administrative personnel. Nonbinding letters of intent are requested by June 15; applications are due August 31 for applications. Guidelines are available online.
Requirements: Eligible institutions are for-profit or nonprofit acute-care hospitals or acute-care healthcare systems; or public or private institutions, such as universities, colleges, and physician practice groups. Individuals eligible to become principal investors are SHM members in good standing, as of July 1; and have the skills, knowledge, and resources necessary to carry out the proposed research, with an invitation to work with his or her institution to develop an application for support.
Amount of Grant: $25,000 per year, plus 9 percent maximum indirect costs
Date(s) Application is Due: Jun 15; Aug 31
Contact: Charisse Braithwaite, (800) 843-3360 or (215) 351-2473; fax: (215) 843-2536; email: Cbraithwaite@hospitalmedicine.org
Internet: http://www.hospitalmedicine.org//AM/Template.cfm?Section=Home
Sponsor: Society of Hospital Medicine
190 N Independence Mall W
Philadelphia, PA 19106-1572

Smith Family New Investigator Awards 5280
These awards are made to promote basic research in the areas of AIDS, cancer, cardiovascular disease, diabetes, neuroscience, autism, and related disorders. The program favors innovative research, clearly presented proposals, realistic objectives, qualified personnel, and financial viability. The program will make six to eight two-year awards.
Requirements: Massachusetts medical groups may submit proposals.
Geographic Focus: Massachusetts
Amount of Grant: $100,000 per year
Contact: Grants Advisory Services, (617) 451-0049 ext 702; fax: (617) 423-4619; email: research@tmfnet.org
Internet: http://www.tmfnet.org/grantmake.html
Sponsor: The Medical Foundation
95 Berkeley Street, Suite 201
Boston, MA 02116

Smith Richardson Foundation Domestic Public Research Fellowship 5281
The foundation's Domestic Public Policy Program seeks to support the work of the next generation of public policy researchers and analysts. The foundation will award at least three research grants to individuals who are interested in conducting research and writing on domestic public policy issues. Grantees are expected to produce a book or an article suitable for publication in a peer-reviewed journal. The grant can be used to cover the salary costs of the researcher and to underwrite research costs, such as travel, research assistance, and data acquisition. Each grant will be paid directly to, and should be administered by, the institution at which the researcher works. Applicants should submit to the Foundation a one-page executive summary and a 10-page research proposal. Guidelines are available online.
Requirements: Fellowships will only be awarded to individual researchers and not to research teams. An applicant must have received a PhD after January 1, 1998. He or she must hold a position as a full-time faculty member at a college or university or as a full-time fellow at a public policy think tank or research organization in the United States. An applicant should explain how he or she meets these requirements in the cover letter to his or her research proposal.
Amount of Grant: $60,000
Date(s) Application is Due: Jul 1
Contact: Public Policy Research Program, (203) 222-6222
Internet: http://www.srf.org
Sponsor: Smith Richardson Foundation
60 Jesup Road
Westport, CT 06880

Smith Richardson Foundation International Security and Foreign Policy Grants 5282
The objective of the International Security and Foreign Policy Program is to assist the U.S. policy community in developing effective national security strategies and foreign policies. The Foundation is committed to supporting projects that help the policy community face the fundamental challenge of ensuring the security of the United States, protecting and advancing American interests and values abroad, and enhancing international order. The International Security and Foreign Policy Program has a number of priority areas of interest. In recent years, the Foundation has sought to assist the policy community, efforts to combat global terrorism by supporting projects on critical issues, such as improving intelligence gathering. Because the battle against terrorism will also be fought on the ideological front, the Foundation has supported projects on improving U.S. public diplomacy in order to promote democracy and to give foreign publics a better understanding of U.S. policies.
Requirements: The Smith Richardson Foundation has a rigorous proposal review process. The first step in the process is the submission of a concept paper. Concept papers should not exceed five pages, follow concept paper template (see website) when preparing a submission. Upon receipt of a grant application, the Foundation will either mail or email a confirmation of receipt to the grant applicant. Requests for grants of $50,000 or less are reviewed on an ongoing basis and are handled as promptly as possible. Requests for grants greater than $50,000 and for multi-year grant support are made at one of the regular board meetings. If the staff determines that a project warrants further consideration under the Foundation, guidelines, an applicant will be asked to submit a full proposal that conforms to a proposal template provided by the Foundation.
Contact: Dr. Nadia Schadlow, Senior Program Officer; (203) 222-6222; fax (203) 222-6282; email: Schadlow@srf.org
Internet: http://www.srf.org/grants/international.php
Sponsor: Smith Richardson Foundation
60 Jesup Road
Westport, CT 06880

Smith Richardson Foundation Junior Faculty Research Grants **5283**
The foundation's International Security and Foreign Policy Program sponsors an annual grant competition to support junior faculty research on American foreign policy, international relations, international security, military policy, and diplomatic and military history. The foundation will award at least three research grants to support tenure-track junior faculty engaged in the research and writing of a scholarly book on an issue or topic of interest to the policy community. These grants are intended to buy-out up to one year of teaching time and to underwrite research costs (including research assistance and travel). Each grant will be paid directly to, and should be administered by, the academic institution at which the junior faculty member works. Projects in military and diplomatic history are especially encouraged.
Requirements: An applicant must have a PhD, preferably in political science, public policy, policy analysis, international political economy, or history. He or she also must hold a position as a full-time tenure-track faculty member of a U.S. college or university. An applicant should explain how he or she meets all of these requirements in a cover letter to the proposal.
Restrictions: Group or collaborative projects will not be considered.
Amount of Grant: $60,000
Date(s) Application is Due: Jul 1
Contact: Grants Administrator, (203) 222-6222
Internet: http://www.srf.org
Sponsor: Smith Richardson Foundation
60 Jesup Road
Westport, CT 06880

Smithsonian Graduate Student Research Fellowships **5284**
Students actively engaged in graduate study at any level may apply to study and conduct research for 10-week periods under the guidance of Smithsonian staff members. Applicants must propose to conduct research in disciplines that are pursued at the Smithsonian (American history, art, and culture; history of science and technology; history of art, design, crafts, and decorative arts; anthropology, archaeology, linguistics, and ethnic studies; evolutionary, systematic, behavioral, and environmental biology and conservation; earth, mineral, and planetary sciences; and materials characterization and conservation) and must plan to spend their entire tenure in residence. Applications are posted to the Web site in September.
Requirements: Students must be formally enrolled in graduate programs, have completed at least one semester, and have not yet been advanced to candidacy if in PhD programs.
Amount of Grant: $4500
Date(s) Application is Due: Jan 15
Contact: Program Contact, Office of Fellowships and Grants, (202) 275-0655; fax: (202) 275-0489; email: siofg@si.edu
Internet: http://www.si.edu/ofg/fell.htm
Sponsor: Smithsonian Institution
750 Ninth Street NW, Suite 9300
Washington, DC 20560

Smithsonian Institution Libraries Resident Scholar Program Grants **5285**
The Smithsonian Institution Libraries currently offers two Resident Scholar Programs. The Dibner Library of the History of Science and Technology Resident Scholar Program, supported by The Dibner Fund. The program supports individuals working on a topic relating to the history of science and technology who can make substantial use of collections in the Dibner Library. Scholars must be in residence at the Dibner Library during the award period. The Baird Society Resident Scholar Program awards stipends to individuals working on a topic relating to the holdings of the libraries' special collections. Scholars must be in residence at the Smithsonian during the award period.
Requirements: The program is open to historians, librarians, doctoral students, and postdoctoral scholars.
Amount of Grant: $2500 per month for up to six months
Date(s) Application is Due: Mar 1
Contact: Resident Scholar Programs, (202) 633-3872; email: SILResidentScholars@si.edu
Internet: http://www.sil.si.edu/researchintern/index.htm
Sponsor: Smithsonian Institution Libraries
P.O. Box 37012, NMAH 1041, MRC 672
Washington, DC 20013-7012

Smithsonian James Renwick Fellowships in American Crafts **5286**
This program awards fellowships for scholarly research in the modern American craft movement. The fellowships are intended for beginning and advanced scholars knowledgeable in the history of 20th-century American art, craft, or design. Proposals on post-1930 craft developments are especially encouraged. Preference will be given to proposals dependent upon the Smithsonian resources. Fellowships are available for three to 12 months of study and research. Award includes allowances.
Requirements: Fellowships are for graduate students in doctoral programs and postdoctoral scholars or candidates with substantial records of professional accomplishment.
Amount of Grant: $35,000 per year
Date(s) Application is Due: Jan 15
Contact: SAAM Fellowship Office, (202) 275-1557; email: fellowships@saam.si.edu
Internet: http://www.si.edu/ofg/fell.htm#fsaam
Sponsor: Smithsonian Institution
National Museum of American Art
Washington, DC 20560-0510

Smithsonian Latino Studies Fellowships **5287**
Fellowships provide opportunities to U.S. Hispanic predoctoral students and postdoctoral and senior scholars to pursue research topics that relate to Latino art, culture, and history. Interdisciplinary subjects are encouraged and can be undertaken at more than one of the Smithsonian museums and/or research units, and advised by one or more of the Smithsonian research staff members. A research and travel allowance is made available to cover additional costs of spending up to one-third of the fellowship tenure away from the Smithsonian, if appropriate and necessary, but not at the fellow's home institution. Applications are posted to the Web site in September.
Requirements: Postdoctoral fellowships are offered to scholars who have held a PhD or equivalent for less than seven years. Senior fellowships are offered to scholars who have held a PhD or equivalent for seven years or more. Predoctoral fellowships are offered to doctoral candidates who have completed preliminary course work and examinations, and have been advanced to candidacy.
Amount of Grant: $20,000 plus allowances each year for predoctoral fellowships; $35,000 plus allowances for postdoctoral and senior fellowships
Date(s) Application is Due: Jan 15
Contact: Program Contact, (202) 275-0655; email: siofg@si.edu
Internet: http://www.si.edu/ofg/fell.htm
Sponsor: Smithsonian Institution
P.O. Box 37012
Washington, DC 20013-7012

Smithsonian Minority Students Internships **5288**
Internships are available for U.S. minority students to participate in research or museum-related activities for periods of 10 weeks. Interns work on projects ranging from researching and preparing traveling exhibits to assisting collections management staff, to studying archaeological material and plant relationships in tropical rain forests. All activities are under the supervision of professional staff members at Smithsonian museums, research institutes and offices, and the projects are designed to match educational background, interests, skills, and career goals of students with the needs of staff members. The appointment carries a weekly stipend, and travel allowances may be provided. Application and guidelines are available online.
Requirements: U.S. minority undergraduate and graduate students are invited to apply.
Amount of Grant: $400 per week for graduates
Date(s) Application is Due: Feb 1; Oct 15
Contact: Program Contact, Office of Fellowships and Grants, (202) 275-0655; fax: (202) 275-0489; email: siofg@si.edu
Internet: http://www.si.edu/ofg/ofgapp.htm
Sponsor: Smithsonian Institution
P.O. Box 37012, Victor Building, Suite 9300, MRC 902
Washington, DC 20013-7012

Smithsonian National Museum of American Art Fellowships **5289**
Fellowships support independent research, dissertation research, or a combination of dissertation and curatorial research at the National Museum of American Art. Only one application is necessary; applicants for fellowships in American art will automatically be considered in all relevant award categories. Applicants are evaluated on their academic standing, scholarly qualifications, experience, and the quality of the research project or study proposed and its suitability to the museum's collections, facilities, and programs. The Patricia and Phillip Frost Fellowship and the Sheila W. and Richard J. Schwartz Fellowship are offered to pursue research in American art and visual culture. Topics that match the interests of the museum and its research staff are supported. The Sara Roby Fellowship in Twentieth Century American Realism is awarded to a scholar whose research topic matches the Sara Roby Foundation's interest in American realism. Stipends include research and travel allowances. Fellowships begin on or after June 1.
Amount of Grant: $35,000 senior/postdoctoral fellowship; $20,000 predoctoral fellowship; plus research and travel allowances for both
Date(s) Application is Due: Jan 15
Contact: Office of Fellowships, (202) 275-0655; email: siofg@si.edu
Internet: http://www.si.edu/ofg/fell.htm
Sponsor: Smithsonian Institution National Museum of American Art
National Museum of American Art
Washington, DC 20560-0210

Smithsonian Native American Internships **5290**
Internships in residence at the Smithsonian are awarded to undergraduate and graduate Native American students to participate in research or museum activities related to Native American studies for periods of 10 weeks. The appointments carry a weekly stipend, and a travel allowance may be provided. This award is granted more than once in a year.
Requirements: Native American undergraduate and graduate students are eligible.
Amount of Grant: $400 per week up to 10 weeks
Date(s) Application is Due: Feb 1; Oct 1
Contact: Program Contact, Office of Fellowships and Grants, (202) 275-0655; fax: (202) 275-0489; email: siofg@si.edu
Internet: http://www.si.edu/ofg/intern.htm#iofg
Sponsor: Smithsonian Institution
750 Ninth Street NW, Suite 9300
Washington, DC 20560

Smithsonian Native American Visiting Student Awards 5291
Appointments are available for currently enrolled advanced Native American graduate students. Visiting students pursue independent research in association with Smithsonian staff. Stipends are per day or per week and include a travel allowance and small research allowance. This award is granted three times a year, depending upon when the funds are requested.
Requirements: Native American graduate students are eligible.
Amount of Grant: $100 per day for up to 21 days; $400 per week for three to 10 weeks
Date(s) Application is Due: Feb 1; Oct 1
Contact: Program Contact, Office of Fellowships and Grants, (202) 275-0655; email: siofg@.si.edu
Internet: http://www.si.edu/ofg/fell.htm
Sponsor: Smithsonian Institution
P.O. Box 37012, Victor Building, Suite 9300, MRC 902
Washington, DC 20013-7012

Smithsonian Postdoctoral Research Fellowships 5292
Fellowships support independently conceived and designed research study by persons at advanced academic levels working with Smithsonian staff members and making use of Smithsonian collections, facilities, and other special resources. Proposals for research may be offered in fields of study conducted at the institute (American history, art, and culture; history of science and technology; history of art, design, crafts, and decorative arts; anthropology, archaeology, linguistics, and ethnic studies; evolutionary, systematic, behavioral, and environmental biology and conservation; earth, mineral, and planetary sciences; and materials characterization and conservation). The fellowship term is three to 12 months. Applications are posted on the Web site each September.
Requirements: Applicants must have a PhD or equivalent degree for less than seven years.
Amount of Grant: $35,000 stipend per year
Date(s) Application is Due: Jan 15
Contact: Program Contact, Office of Fellowships and Grants, (202) 275-0655; fax: (202) 275-0489; email: siofg@si.edu
Internet: http://www.si.edu/ofg/fell.htm
Sponsor: Smithsonian Institution
750 Ninth Street NW, Suite 9300
Washington, DC 20560

Smithsonian Predoctoral Research Fellowships 5293
Fellowships support independently conceived and designed research study by predoctoral students to work with Smithsonian staff members and use Smithsonian collections, facilities, and other special resources. Proposals are accepted for research in the areas of study conducted at the institute (American history, art, and culture; history of science and technology; history of art, design, crafts, and decorative arts; anthropology, archaeology, linguistics, and ethnic studies; evolutionary, systematic, behavioral, and environmental biology and conservation; earth, mineral, and planetary sciences; and materials characterization and conservation). The fellowship term is three to 12 months.
Requirements: Candidates must have completed all coursework and examinations and be ready to engage in dissertation research.
Amount of Grant: $20,000 stipend per year
Date(s) Application is Due: Jan 15
Contact: Program Contact, Office of Fellowships and Grants, (202) 275-0655; fax: (202) 275-0489; email: siofg@si.edu
Internet: http://www.si.edu/ofg/fell.htm
Sponsor: Smithsonian Institution
750 Ninth Street NW, Suite 9300
Washington, DC 20560

Smithsonian Rockefeller Foundation Humanities Fellowships 5294
The Smithsonian is hosting up to six fellows for each of three years to work on the theoretical development of the concept of cultural heritage. The fellowships aim to expand and refine the theoretical framework for cultural heritage to include grassroots voices. Reflecting the perspectives of academic specialists, civil society groups, and public cultural organizations, fellows' work will inform dialogs and practice across social, political, and disciplinary boundaries, as well as indicate future directions for policy. Applicants will focus their critical inquiry on such issues as cultural capital, intellectual property, community-based rights, cultural tourism, employment, and poverty reduction, and will demonstrate originality of approach and significant potential for making a contribution to the formulation of cultural heritage policy and practice. Fellowships include a stipend and an allowance for travel to and from Washington, DC, as necessary. The fellowship tenure is six weeks to five months, between September 1 and July 31. The listed application deadline is for letters of intent. Annual deadline dates may vary; contact program staff for exact dates.
Requirements: Applicants need not be U.S. citizens to be eligible, and approximately half of the fellows are drawn from outside the United States. Applicants whose native language is not English are expected to have a sufficiently good command of spoken English so that they can fully discuss, debate, and exchange ideas and practices about cultural heritage.
Restrictions: These fellowships are not intended to support undergraduate or graduate studies, or research. Projects will not be considered for the rewriting of dissertations editing of texts, papers, or the preparation of textbooks or anthologies.
Date(s) Application is Due: Jan 14
Contact: Center for Folklife and Cultural Heritage, (202) 275-1150; fax: (202) 275-1119, email: culturalheritagefellows.si.edu or folklife-info@si.edu
Internet: http://www.folklife.si.edu/opportunities/fellowships_RF.html
Sponsor: Smithsonian Institution
P.O. Box 37012, Victor Building, Suite 4100
Washington, DC 20013-7012

Smithsonian Senior Fellowships 5295
Each year, an occasional appointment is made to distinguished senior scientists for research in the fields of anthropology, natural history, environmental science and ecological processes, tropical biology, African art and culture, oriental art, astrophysics and related topics, history of science and technology, American cultural and social history, or art history. The fellowship term is three to 12 months. Prior to submitting applications, interested senior scientists should consult with Smithsonian staff in their chosen fields. Applications are posted to the Web site in September.
Requirements: Applicants must have held a PhD or equivalent for at least seven years.
Amount of Grant: $35,000 per year
Date(s) Application is Due: Jan 15
Contact: Office of Fellowships, (202) 275-0655; email: siofg@si.edu
Internet: http://www.si.edu/ofg/fell.htm
Sponsor: Smithsonian Institution
P.O. Box 37012
Washington, DC 20013-7012

Smithsonian Tropical Research Institute Short Term Fellowships 5296
The Smithsonian Tropical Research Institute (STRI), is a division of the Smithsonian Institution in Washington DC and maintains research facilities for marine and terrestrial research at various locations on the Isthmus of Panama. Disciplines include ecology, anthropology, paleontology, paleoecology, evolutionary biology, molecular phylogenetics, biogeography, animal behavior, neurobiology, soils sciences, and physiology of tropical plants and animals. Support is provided to carry out short-term research projects in the tropics. Fellowships may provide a modest stipend to cover living expenses while at STRI (currently $750/month); a modest research allowance that rarely exceeds $2000 and is usually less; and roundtrip coach airfare. Applicants are encouraged to seek additional sources of funding, and partial awards are often given.
Requirements: Before submitting a formal application all applicants should consult with STRI research staff, who will serve as potential advisors, to confirm that they are willing to supervise and support the proposed project. Proposals without the support of named advisors will not be funded. When corresponding with advisors applicants are strongly encouraged to ask for advice on the scientific merit of the proposed research, methodology and logistics, and for written comments on preliminary drafts of the proposals. Proposals may be submitted in English or Spanish. Download the complete guidelines from the sponsor's website.
Restrictions: Projects usually are complete in themselves and their maximum duration is 3 months; extensions are awarded only in exceptional circumstances.
Amount of Grant: up to $750/month
Date(s) Application is Due: Feb 15; May 15; Aug 15; Nov 15
Contact: Adriana Bilgray, STRI Fellowships & Internships; (507) 212-8031; fax: (507) 212-8150; email: fellows@si.edu
Internet: http://www.stri.org/english/education_fellowships/fellowships/index.php
Sponsor: Smithsonian Tropical Research Institute
Office of Academic Programs, Unit 0948
AP.O , AA 34002-0948
Washington, DC 20560

SNM Molecular Imaging Research Grant For Junior Medical Faculty 5297
The objective of this program is to provide salary support for one junior faculty member in an academic/research setting to enable them to engage in Molecular Imaging research related to diagnostic or therapeutic applications. The grant will support new, innovative pilot projects in any area of molecular imaging research. The grant may supplement other funds but not duplicate other support.
Requirements: Applicant Eligibility: applicants must have MD degree (or equivalent) and have completed a residency and be certified in their specialty; applicants must be practicing in an academic/research setting as a faculty member in the United States; applicants must be within 5 years of their initial faculty appointment with an academic rank of instructor or assistant professor (or equivalent); applicants must provide a detailed proposal of the proposed research activity; applicants must not have received grant/contract amounts totaling $50,000.00 or more in a single calendar year as the principle investigator; preference will be given to individuals who have demonstrated great potential for a research career in the field of molecular imaging; multidisciplinary projects are encouraged; only one application per faculty member is accepted per grant; no grantee may receive more than one SNM research grant in any one year. Likewise no other research grant request or continuation will be considered until a satisfactory summary of an earlier grant is received; applicants are required to be a SNM member at the time of application.
Restrictions: The grant will not pay institutional overhead costs or indirect costs
Amount of Grant: $100,000 over two years
Date(s) Application is Due: Jan 20
Contact: Nicole Kern, SNM Program Manager; (703) 652-6795 ext. 1255; fax: (703) 708-9015; email: nkern@snm.org
Internet: http://interactive.snm.org/index.cfm?PageID=7252
Sponsor: Society of Nuclear Medicine
1850 Samuel Morse Drive
Reston, VA 20190

SNM Pilot Research Grants **5298**

The pilot research grants are designed to help basic or clinical scientists in the early stages of their career conduct research that may lead to further funding. The Mitzi and William Blahd, MD Pilot Research Grant is awarded to the applicant with the highest-ranked proposal. Preference will be given to individuals who have demonstrated great potential for a research career in the field of nuclear medicine/molecular imaging.
Requirements: Applicant Eligibility: basic or clinical scientists employed by academic and research-oriented organizations in North America are eligible; applicants must not be no more than five years post training.
Amount of Grant: $25,000
Date(s) Application is Due: Feb 20
Contact: Nicole Mitchell, Program Manager; (703) 652-6795; email: nkern@snm.org
Internet: http://interactive.snm.org/index.cfm?PageID=2255
Sponsor: Society of Nuclear Medicine
1850 Samuel Morse Drive
Reston, VA 20190

SNM Pilot Research Grants in Nuclear Medicine/Molecular Imaging **5299**

The pilot research grants are designed to help basic or clinical scientists in the early stages of their career conduct research that may lead to further funding.
Requirements: Applicant Eligibility: basic or clinical scientists employed by academic and research-oriented organizations in North America are eligible; applicants must be no more than five years post training; applicant must not have served as the principal investigator of a peer-reviewed grant for more than $50,000 in a single calendar year; preference will be given to individuals who have demonstrated great potential for a research career in the field of nuclear medicine/molecular imaging.
Restrictions: The grant will not pay salaries, institutional overhead costs or indirect costs. Funds may supplement other funds, but not duplicate other support. Grants will be limited to a time span of one year. No grantee may receive more than one grant in any one year.
Amount of Grant: $25,000
Date(s) Application is Due: Feb 20
Contact: Nicole Kern, SNM Program Manager; (703) 652-6795 ext. 1255; fax: (703) 708-9015; email: nkern@snm.org
Internet: http://interactive.snm.org/index.cfm?PageID=2255
Sponsor: Society of Nuclear Medicine
1850 Samuel Morse Drive
Reston, VA 20190

SNM Postdoctoral Molecular Imaging Scholar Grants **5300**

The postdoctoral program will support a two-year research endeavor that promotes integration of molecular imaging into the career of the trainee. The funds maybe used for new, innovative pilot projects in any area of molecular imaging research. The grant may supplement other funds, but not duplicate other support. A summary report and an accounting of funds will be requested at the halfway point and upon completion of the research project. Grantees are expected to submit results of the research as an abstract for presentation at the SNM Annual Meeting and/or as a scientific manuscript to the Journal of Nuclear Medicine. A final report is due within 60 days after completion of the research project.
Requirements: Applicant Eligibility: MD, MD/PhD, and PhD applicants are eligible; applicants from any country may apply; applicants must hold a full-time position in an educational institution when the award starts; applicants must have completed all basic training and be no more than 2 years post training; applicants must not have served as a principal investigator on peer-reviewed grants totaling $50,000 or more in a single calendar year; preference will be given to individuals who have demonstrated great potential for a research career in the field of molecular imaging; only one application per trainee is accepted; no grantee may receive more than one SNM research grant in any one year. Likewise no other research grant request or continuation will be considered until a satisfactory summary of an earlier grant is received.
Restrictions: Grant extensions are not often allowed, but will be reviewed and decided upon by the SNM Grants and Awards Committee if requested. The grant will not pay institutional overhead costs or indirect costs.
Amount of Grant: $60,000 over two years
Date(s) Application is Due: Jan 20
Contact: Nicole Kern, SNM Program Manager; (703) 652-6795 ext. 1255; fax: (703) 708-9015; email: nkern@snm.org
Internet: http://interactive.snm.org/index.cfm?PageID=7252
Sponsor: Society of Nuclear Medicine
1850 Samuel Morse Drive
Reston, VA 20190

SNM Student Fellowship Awards **5301**

The SNM Student Fellowships are designed to stimulate interest in molecular imaging/nuclear medicine by supporting their full-time participation in clinical and basic research activities for three months (or less). These Fellowships are made possible by support of the Education and Research Foundation for SNM.
Requirements: Candidates from any country may apply. Students must be enrolled in a medical, pharmacy or graduate school, or be an undergraduate student demonstrating outstanding competence in nuclear medicine and/or molecular imaging research. Fellowship must be carried out during a three-month that starts no earlier then March and completed no later then December 31st.
Amount of Grant: $3000
Date(s) Application is Due: Jan 23
Contact: Nicole Kern, SNM Program Manager; (703) 652-6795, ext. 1255; fax: (703) 708-9015; email: nkern@snm.org
Internet: http://interactive.snm.org/index.cfm?PageID=2254
Sponsor: Society of Nuclear Medicine
1850 Samuel Morse Drive
Reston, VA 20190

SNM Tetalman Memorial Young Investigator Awards **5302**

Established by the family and friends of Dr. Tetalman, this award honors the work of a young investigator who is pursuing a career in nuclear medicine. It is named in memory of a highly respected and productive clinician and researcher, Mark Tetalman, M.D., whose career was cut tragically short. The award is based on a submitted paper supporting current research efforts and accomplishments, teaching, clinical service and administration. Service to the Society and a commitment to academic nuclear medicine will also be considered.
Requirements: Applicants must be 36 years old or younger as of July 1 of the application year and must have obtained certification in Nuclear Medicine or Nuclear Radiology or have completed a PhD program within the last seven years. Complete applications must be must be received, not simply post-marked by November 14th. One original unstapled copy of the completed application must be sent to: SNM Development Office, Attention: SNM Committee on Awards
Amount of Grant: $2500
Date(s) Application is Due: Nov 14
Contact: Nicole Kern, SNM Program Manager; (703) 652-6795 ext. 1255; fax: (703) 708-9015; email: nkern@snm.org
Internet: http://interactive.snm.org/index.cfm?PageID=2253
Sponsor: Society of Nuclear Medicine
1850 Samuel Morse Drive
Reston, VA 20190

SNM/ Covidien Seed Grant in Molecular Imaging/Nuclear Medicine Research **5303**

The Society of Nuclear Medicine is pleased to offer the SNM/Covidien Seed Grant in Nuclear Medicine Research. This grant is designed to assist researchers in conducting new and innovative pilot projects that have potential for future support from foundations, corporations or government agencies.
Requirements: Applicant Eligibility: basic or clinical scientists employed by academic and research-oriented organizations; applicants from any country may apply; applicants must hold a full-time faculty position in an educational institution when the award period begins; applicants must have completed all advanced training and be no more than five years post training; applicants must not have served as a principal investigator on peer-reviewed grants totaling $50,000 or more in a single calendar year; preference will be given to individuals who have demonstrated great potential for a research career in the field of nuclear medicine/molecular imaging.
Restrictions: No grantee may receive more than one SNM research grant in any one year. The grant will not pay institutional overhead costs or indirect costs. No salary support for principal investigator or co-principal investigators. Grants will be limited to a time span of one year; no extensions are allowed
Amount of Grant: $25,000
Date(s) Application is Due: Feb 20
Contact: Nicole Kern, SNM Program Manager; (703) 652-6795 ext. 1255; fax: (703) 708-9015; email: nkern@snm.org
Internet: http://interactive.snm.org/index.cfm?PageID=2255
Sponsor: Society of Nuclear Medicine
1850 Samuel Morse Drive
Reston, VA 20910

SOBP A.E. Bennett Research Award **5304**

The awards, supported by the A.E. Bennett Neuropsychiatric Research Foundation, are in recognition of basic science and clinical science research in biological psychiatry by young investigators. Candidates must be actively engaged in the research for which the award is sought. Applicants need only write a brief description of their research, submit 2-3 published or submitted papers and arrange for two letters of support (at least one of the letters needs to be from a member of the Society).
Requirements: Submissions are welcomed from laboratory researchers in any country who have not passed their 45th birthday or have not been engaged in research for greater than 10 years following award of their terminal degree or the end of formal clinical/fellowship training, whichever is later by deadline date; candidates need not be current members of the society.
Amount of Grant: $2,000
Date(s) Application is Due: Jan 15
Contact: Maggie Peterson; (904) 953-2842; fax: (904) 953-7117; email: maggie@mayo.edu or sobp@sobp.org
Internet: http://www.sobp.org/i4a/pages/index.cfm?pageID=3379
Sponsor: Society of Biological Psychiatry
4500 San Pablo Road
Jacksonville, FL 32224

Social Security Research and Demonstration Grants **5305**

Grants support social, economic, and demographic research on topics of importance to the Social Security and Supplemental Security Income (SSI) programs and the current and future well-being of their beneficiaries; to demonstrate effective, ongoing, and transferable approaches for identifying needy elderly, blind, and disabled individuals potentially eligible for SSI and assisting them through the SSI application process; and

to demonstrate innovative methods of referral, treatment (with a heavy emphasis on vocational rehabilitation), and monitoring SSI beneficiaries who are mentally impaired chemical abusers. Projects awarded include innovative research and demonstrations that add to existing knowledge and improve methods and techniques for the management, administration, and effectiveness of social security administration programs. Closing dates for applications are announced in the Federal Register.
Requirements: Applicants applying for grant funds may include state and local governments, educational institutions, and other nonprofit and for-profit organizations.
Restrictions: For-profit organizations may apply with the understanding that grant funds may not be paid as profit to any grant recipient. Individuals are not eligible to apply. Grant funds may not be used for construction or major renovation of buildings.
Amount of Grant: $20,000-$4 million range
Contact: Phyllis Smith, Grants Management Officer, Office of Operations Contracts and Grants, (410) 965-9518; email: phyllis.y.smith@ssa.gov
Internet: http://www.ssa.gov/oag/grants/ssagrant_info.htm
Sponsor: Social Security Administration
1710 Gwynn Oak Avenue, 1-E-4 Gwynn Oak Building
Baltimore, MD 21207

Society for Biomedical Screening Small Grants **5306**
The society's small grants program will provide start-up funding for research projects with potential to advance the disciplines, technologies, and skills involved in screening and molecular discovery, including the support of graduate or undergraduate training in these areas. The program is intended to support projects that may have an applied focus or those that may involve radically new concepts. Applications are considered without regard to race, religion, national origin, physical handicap, or age of the candidates.
Requirements: Applicants must be employed full time in a college university, or nonprofit research institute in order to serve as principal investigators.
Date(s) Application is Due: Feb 18
Contact: David Roman, (203) 743-1336; fax: (203) 748-7557; email: droman@sbsonline.org
Internet: http://www.sbsonline.org/awards_grants
Sponsor: Society for Biomedical Screening
36 Tamarack Avenue, #348
Danbury, CT 06811

Society for Imaging Informatics in Medicine (SIIM) Emeritus Mentor Grants **5307**
The mission of SIIM, The Society for Imaging Informatics in Medicine, is to advance computer applications and information technology in medical imaging through education and research. In support of this mission, the SIIM Research and Development Committee offers a $20,000 grant for a retired, or semi-retired, expert in imaging informatics to act as a mentor for a trainee (student, resident, fellow) or junior faculty member for hypothesis-driven research. Any area of research involving medical imaging informatics is eligible for support, as well as development of new hardware, software, or processes to support electronic imaging practice, education, or research.
Requirements: A senior, experienced investigator jointly applies with a resident, fellow, graduate student, or junior faculty for a SIIM emeritus mentor grant. Preference will be given to projects at SIIM institutions and to projects that contribute to the development of careers of new or prospective researchers. This grant may not be applied for in conjunction with another SIIM grant covering the same subject material.
Amount of Grant: $20,000
Date(s) Application is Due: Aug 3
Contact: Grant Manager; (703) 723-0432; fax: (703) 723-0415; email: grants@siimweb.org or info@siimweb.org
Internet: http://www.scarnet.org/index.cfm?id=63
Sponsor: Society for Imaging Informatics in Medicine (SIIM)
19440 Golf Vista Plaza, Suite 330
Leesburg, VA 20176-8264

Society for Imaging Informatics in Medicine (SIIM) Micro Grants **5308**
The mission of SIIM, The Society for Imaging Informatics in Medicine, is to advance computer applications and information technology in medical imaging through education and research. In support of this mission, the SIIM Research and Development Committee offers a $5,000 grant for hypothesis-driven experiments of short duration and high risk. Prior preliminary data is not required. An imaging informatics project that needs resources falls between what a researcher is willing to fund personally or is more than his/her department provides in seed money and what is typically requested in grant applications. Examples include trying unconventional hardware or software in an imaging informatics experiment, trying a particular piece of hardware for a user-interface design, paying for short duration consulting assistance, or paying for an educational course that the researcher needs to enhance a particular study.
Requirements: There are no restrictions on who may apply for the grant. Trainees (residents, fellows, and graduate students) are required to include a letter from a faculty member, acting as an advisor, to show that the applicant's department supports the project. Preference will be given to projects at SIIM institutions and to projects that may yield high value results or provide enough preliminary data to justify additional work and application for traditional grant funding.
Amount of Grant: $5,000
Date(s) Application is Due: Aug 3
Contact: Grant Manager; (703) 723-0432; fax: (703) 723-0415; email: grants@siimweb.org or info@siimweb.org
Internet: http://www.scarnet.org/index.cfm?id=63
Sponsor: Society for Imaging Informatics in Medicine (SIIM)
19440 Golf Vista Plaza, Suite 330
Leesburg, VA 20176-8264

Society for Imaging Informatics in Medicine (SIIM) Research Grants **5309**
The mission of SIIM, The Society for Imaging Informatics in Medicine, is to advance computer applications and information technology in medical imaging through education and research. In support of this mission, the SIIM Research and Development Committee offers two $50,000 grants for original research. Any area of research involving medical imaging informatics is eligible for support, as well as development of new hardware, software, or processes to support electronic imaging practice, education, or research.
Requirements: Residents, fellows, graduate students, or faculty may apply for SIIM research grants. Preference will be given to projects at SIIM institutions and to projects that contribute to the development of careers of new or prospective researchers. Residents, fellows, and students must conduct their projects under the guidance of experienced investigators. Grant recipients will not be eligible for concurrent support through other SIIM grants.
Amount of Grant: $50,000
Date(s) Application is Due: Aug 3
Contact: Grant Manager; (703) 723-0432; fax: (703) 723-0415; email: grants@siimweb.org or info@siimweb.org
Internet: http://www.scarnet.org/index.cfm?id=63
Sponsor: Society for Imaging Informatics in Medicine (SIIM)
19440 Golf Vista Plaza, Suite 330
Leesburg, VA 20176-8264

Society for Imaging Informatics in Medicine (SIIM) Small Grant for Product Support Development **5310**
The mission of SIIM, The Society for Imaging Informatics in Medicine, is to advance computer applications and information technology in medical imaging through education and research. In support of this mission, the SIIM Research and Development Committee offers a $20,000 grant to support the development of a product or tool for use in the field of imaging informatics. Any area of product or tool development for use in the field of medical imaging informatics is eligible for support (e.g., open source application development, work flow measurement tools, replacement PACS change management tools).
Requirements: Residents, fellows, graduate students, or faculty may apply for product support development grants. Preference will be given to projects at SIIM institutions and to projects that contribute to the development of careers of new or prospective researchers. Residents, fellows, and students must conduct their projects under the guidance of experienced investigators.
Amount of Grant: $20,000
Date(s) Application is Due: Aug 3
Contact: Grant Manager; (703) 723-0432; fax: (703) 723-0415; email: grants@siimweb.org or info@siimweb.org
Internet: http://www.scarnet.org/index.cfm?id=63
Sponsor: Society for Imaging Informatics in Medicine (SIIM)
19440 Golf Vista Plaza, Suite 330
Leesburg, VA 20176-8264

Society for Imaging Informatics in Medicine (SIIM) Small Training Grants **5311**
The mission of SIIM, The Society for Imaging Informatics in Medicine, is to advance computer applications and information technology in medical imaging through education and research. In support of this mission, the SIIM Research and Development Committee offers a $20,000 grant targeted at providing training opportunities in the field of imaging informatics for students and imaging informatics professionals. Any area of training within medical imaging informatics is eligible for support. The grant is designed to support hands-on training or experience with a designated mentor or sponsor in the imaging informatics arena. It is not intended to support tuition or other traditional education-related expenses. The period of training must last a minimum of three months and may extend no longer than twelve months.
Requirements: Residents, fellows, graduate students, or faculty members at recognized institutions may apply for SIIM training grants. Preference will be given to training experiences at SIIM institutions and to those that contribute to the careers of new or prospective imaging informatics personnel.
Amount of Grant: $20,000
Date(s) Application is Due: Aug 3
Contact: Grant Manager; (703) 723-0432; fax: (703) 723-0415; email: grants@siimweb.org or info@siimweb.org
Internet: http://www.scarnet.org/index.cfm?id=63
Sponsor: Society for Imaging Informatics in Medicine (SIIM)
19440 Golf Vista Plaza, Suite 330
Leesburg, VA 20176-8264

Society for Investigative Dermatology Career Development Award **5312**
The society makes one-year career development awards to encourage trainees in skin-related fields to become skin scientists. Applications will be considered in the following areas: skin biology, skin disease, skin research publishing, and presentation of skin research at a forum. Application forms may be obtained from the Dermatology Foundation, 1563 Maple Avenue, Evanston, IL 60201. Contact program staff to determine availability of awards.
Requirements: Eligible applicants include MDs and PhDs engaged in projects related to the skin and skin diseases.
Amount of Grant: $40,000 per year

Contact: Program Contact, Department of Dermatology, Columbia University, (216) 579-9300; fax: (216) 579-9333; email: sid@sidnet.org
Internet: http://www.sidnet.org
Sponsor: Society for Investigative Dermatology Inc
820 W Superior Avenue, 7th Fl
Cleveland, OH 44113-1800

Society for Research in Child Development Executive Branch Policy Fellowships 5313
The fellowships are designed to provide greater interaction between the developmental research community and federal research programs and policies. Fellows spend a year in Washington working as resident scholars in a federal agency that sponsors or uses developmental research, providing advice and guidance to programs, and learning about federal science policy processes.
Requirements: Applicants must have doctoral-level degrees and exceptional competence in an area of child development research. PhDs, MDs, EdDs, and DSWs are encouraged to apply.
Amount of Grant: $45,000-$65,000 stipends
Date(s) Application is Due: Dec 15
Contact: Director, SRCD Office for Policy and Communications, (202) 336-5926; fax: (202) 336-5953; email: SRCD@apa.org
Internet: http://www.srcd.org/policyfellowships.html
Sponsor: Society for Research in Child Development
750 First Street NE,
Washington, DC 20002-4242

Society for the Arts in Healthcare Environmental Arts Research Grant 5314
The American Art Resources and Society for the Arts in Healthcare Environmental Arts Research Grant program offers a cash award sponsored by American Art Resources in Houston, Texas. The program is provided to a Principal Investigator to conduct or supplement Arts in Healthcare research. Selection is made on the merit, innovation and feasibility of the research proposal.
Requirements: Applicant organizations must be based in the United States.
Contact: Anita Boles, Executive Director; (202) 299-9770; fax: (202) 299-9887; email: anita@thesah.org
Internet: http://www.thesah.org/template/page.cfm?page_id=570
Sponsor: Society for the Arts in Healthcare (SAH)
2437 15th Street, NW
Washington, DC 20009

Society for the History of Technology Abbott Payson Usher Prize 5315
This prize was established to encourage the publication of original research of the highest standard. It is awarded annually to the author of the best scholarly work published during the preceding three years under the auspices of the Society for the History of Technology. A certificate is included with the $400 prize.
Amount of Grant: $400
Contact: Amy Bix, Secretary; (515) 294-8469; fax: (515) 294-6390; email: shot@iastate.edu
Internet: http://www.historyoftechnology.org/awards/usher.html
Sponsor: Society for the History of Technology
Iowa State University, 618 Ross Hall
Ames, IA 50011

Society for the History of Technology Brooke Hindle Postdoctoral Fellowships 5316
THIS Fellowship in the History of Technology honors the contribution of Brooke Hindle to the work of the Society for the History of Technology. The fellowship is for $10,000 and may be used for any purpose connected with research or writing in the history of technology for a period of not less than four months during the year following the award. Applicants should be intending either to prepare a dissertation for publication as articles or as a monograph, whether or not this involves fresh primary research, or to develop a new project based on primary research.
Requirements: Applicants must hold a doctorate in the history of technology or a related field, normally awarded within the preceding four years, or expect to have graduated by the time of the award.
Amount of Grant: $10,000
Date(s) Application is Due: Apr 15
Contact: Amy Bix, Secretary; (515) 294-8469; fax: (515) 294-6390; email: shot@iastate.edu
Internet: http://www.historyoftechnology.org/awards/hindle.html
Sponsor: Society for the History of Technology
Iowa State University, 618 Ross Hall
Ames, IA 50011

Society for the History of Technology Dibner Award for Excellence in Museum Exhibits 5317
The Dibner Award for Excellence in Museum Exhibits was established in 1985 to recognize excellence in museums and museum exhibits that interpret the history of technology, industry, and engineering to the general public. Winning exhibits, in addition to being well designed and produced, should raise pertinent historical issues. Artifacts and images should be used in a manner that interests, teaches, and stimulates both the general public and historians. The award consists of a plaque and up to $1,000 to cover expenses for a member of the design team to accept the award at the SHOT awards banquet. The Society especially encourages nominations from local and regional historical societies. Virtual exhibits are not eligible for the award.
Requirements: Anyone, including the institution or individual responsible for its creation, may nominate an exhibit for the Dibner Award.
Restrictions: Exhibits are eligible for this award if they have been open to the public for no more than 24 months before the deadline for nominations.
Amount of Grant: $1,000
Date(s) Application is Due: Apr 1
Contact: Amy Bix, Secretary; (515) 294-8469; fax: (515) 294-6390; email: shot@iastate.edu
Internet: http://www.historyoftechnology.org/awards/dibner.html
Sponsor: Society for the History of Technology
Iowa State University, 618 Ross Hall
Ames, IA 50011

Society for the History of Technology IEEE Life Members Prize in Electrical History 5318
This prize is awarded annually to the best paper in electrical history published during the previous year. The prize is sponsored by the life members of the IEEE. The competition is open to any historical paper published in a learned journal or magazine if it treats the art or engineering aspects of electrotechnology and its practitioners. Electrotechnology encompasses power, electronics, telecommunications, and computer science. The cash portion of the prize is shared among joint authors; individual certificates are presented to each author. The prize consists of a cash award of $500 and a certificate.
Requirements: The article must be written in English.
Samples: Eden Medina, Journal of Latin American Studies—Designing Freedom, Regulating a Nation: Socialist Cybernetics in Allende's Chile (2007); Martin Collins, History and Technology—One World...One Telephone: Iridium, One Look at the Making of a Global Age (2006).
Amount of Grant: $500
Date(s) Application is Due: Apr 15
Contact: Amy Bix, Secretary; (515) 294-8469; fax: (515) 294-6390; email: shot@iastate.edu
Internet: http://www.historyoftechnology.org/awards/ieeeprize.html
Sponsor: Society for the History of Technology
Iowa State University, 618 Ross Hall
Ames, IA 50011

Society for the History of Technology International Scholars Program 5319
The program is intended to foster a stronger international community for the study of the history of technology by identifying and helping non-U.S. historians of technology to participate in the meetings and governance of the society and receive recognition from their respective governments and national academic communities. Appointments will be for two-year terms beginning each January. International scholars will be made full members of the society and receive complimentary subscriptions to the society's journal. They will be strongly encouraged to attend the annual meetings during their terms. In addition, they will be asked to prepare at least one report or essay review on the history of technology scene in their country for possible presentation to, or publication by, the society. Up to seven scholars are appointed each year.
Requirements: Scholars must reside and work outside the United States; they may be both junior and more advanced scholars. An individual may be nominated, or may nominate her/himself, by sending a letter containing a brief statement about the individual (current status, employment, field of interest and research, etc.), a brief curriculum vita, a list of publications, and a fax number or email address.
Date(s) Application is Due: Apr 15
Contact: Amy Bix, Secretary; (515) 294-8469; fax: (515) 294-6390; email: shot@iastate.edu
Internet: http://www.historyoftechnology.org/awards/scholars.html
Sponsor: Society for the History of Technology
Iowa State University, 618 Ross Hall
Ames, IA 50011

Society for the History of Technology Joan Cahalin Robinson Prize 5320
This prize is awarded annually for the best paper presented at the SHOT annual meeting by a historian under 30 years of age or by a person past 30 who is an accredited graduate student or candidate for a higher degree and who is presenting his or her first paper at a SHOT annual meeting. Candidates for the award are judged not only on the quality of the historical research and scholarship demonstrated in the written papers submitted to the Robinson Prize Committee before the meeting, but also on the effectiveness of their oral presentations. A certificate is included with the prize.
Amount of Grant: $350
Date(s) Application is Due: Jul 1
Contact: Amy Bix; (515) 294-8469; fax: (515) 294-6390; email: shot@iastate.edu
Internet: http://www.historyoftechnology.org/awards/robinson.html
Sponsor: Society for the History of Technology
Iowa State University, 618 Ross Hall
Ames, IA 50011

Society for the History of Technology Melvin Kranzberg Dissertation Fellowship 5321
The Fellowship is presented annually to a doctoral student engaged in the preparation of a dissertation on the history of technology, broadly defined. This award is in memory of the co-founder of the Society and honors Melvin Kranzberg's many contributions

to developing the history of technology as a field of scholarly endeavor and SHOT as a professional organization. The $4,000 award is unrestricted and may be used in any way that the winner chooses to advance the research and writing of his or her dissertation. Possible uses include underwriting the costs of travel to archival collections; photocopying or microfilming; translation of documents; and so on.
Requirements: Students from institutions of higher learning anywhere in the world who are working on projects in the history of technology are eligible to apply; doctoral candidates from outside the United States are especially encouraged to submit application materials. Applicants must have completed all requirements for their doctorate except for the dissertation by September 1 of the year the award is made.
Restrictions: The award may not be used for university tuition or fees.
Amount of Grant: $4,000
Date(s) Application is Due: Apr 15
Contact: Amy Bix, Secretary; (515) 294-8469; fax: (515) 294-6390, email: shot@iastate.edu
Internet: http://www.historyoftechnology.org/awards/kranzberg.html
Sponsor: Society for the History of Technology
Iowa State University, 618 Ross Hall
Ames, IA 50011

Society for the History of Technology NASA Fellowship 5322
The NASA Fellowship in the History of Space Technology, offered by SHOT and supported by the National Aeronautics and Space Administration (NASA) History Division, will fund one predoctoral or postdoctoral fellow for up to one academic year to undertake a research project related to the history of space technology. The fellowship may support advanced research related to all aspects of space history, leading to publications on the history of space technology broadly considered, including cultural and intellectual history, institutional history, economic history, history of law and public policy, and history of engineering and management. The fellowship carries a stipend of $17,000, paid quarterly. Fellows will carry out their research projects using personal office space, equipment, and supplies.
Requirements: Applicants must possess a doctorate in history of technology or in a closely related field, or be enrolled as a student in a doctoral degree program and have completed all requirements for the Ph.D., except the dissertation, in history of technology or a related field. Eligibility is not limited to U.S. citizens or residents
Restrictions: Funds may not be used to support tuition or fees.
Amount of Grant: $17,000
Date(s) Application is Due: Apr 1
Contact: Amy Bix, Secretary; (515) 294-8469; fax: (515) 294-6390; email: shot@iastate.edu
Internet: http://www.historyoftechnology.org/awards/nasa.html
Sponsor: Society for the History of Technology
Iowa State University, 618 Ross Hall
Ames, IA 50011

Society for the History of Technology Samuel Eleazar and Rose Tartakow Levinson Prize 5323
This prize is awarded annually by the Society for the History of Technology for an original essay in the history of technology that is the author's first work intended for publication. The essay must explicitly examine in some detail a technology or technological device within the framework of social or intellectual history. Write for details on how to present the essay for consideration.
Requirements: Manuscripts already published or accepted for publication are not eligible. In order to be considered, manuscripts must be in English and of a length suitable for publication as an article in Technology and Culture,Äîapproximately 7,500 words.
Amount of Grant: $450
Date(s) Application is Due: Apr 15
Contact: Amy Bix, Secretary; (515) 294-8469; fax: (515) 294-6390; email: shot@iastate.edu
Internet: http://www.historyoftechnology.org/awards/levinson.html
Sponsor: Society for the History of Technology
Iowa State University, 618 Ross Hall
Ames, IA 50011

Society for the Humanities Postdoctoral Fellowships 5324
The society supports and encourages creative research in the humanities, especially investigations that deal with essential humanistic concepts, stressing the methods common to the several branches of the humanities. Fellows must spend the fellowship year in residence at Cornell. Applicant's research should be related to the year's theme, which is Historicizing the Global Postmodern. Fellows spend most of their time doing research and writing but are encouraged to offer informal seminars related to their research. Application materials, including letters of recommendation, must be postmarked on or before the listed application deadline.
Requirements: Applicants must have received the PhD degree before January 1, 2005.
Amount of Grant: $40,000 for academic year
Date(s) Application is Due: Oct 1
Contact: Program Administrator, (607) 255-9274; email: humctr-mailbox@cornell.edu
Internet: http://www.arts.cornell.edu/sochum/html/index.html
Sponsor: Cornell University
27 East Avenue
Ithaca, NY 14853-1101

Society of Architectural Historians Rosann S. Berry Fellowship 5325
This fellowship is awarded each year by the society to enable a student engaged in advanced graduate study to attend the annual meeting of the society. A citation and an award are presented at the annual meeting.
Requirements: Applicant must have been a member of SAH for at least one year before applying for this award and must be engaged in advanced graduate study (normally beyond the master's level) that involves some aspect of the history of architecture or of one of the fields closely allied to it (e.g., city planning, landscape architecture, decorative arts, or historic preservation).
Amount of Grant: $500 stipend to offset costs of travel, lodging, and meals directly related to the meeting
Date(s) Application is Due: Nov 15
Contact: Pauline Saliga, Executive Director, (312) 573-1365; fax: (312) 573-1141; email: psaliga@sah.org
Internet: http://www.sah.org/index.php?module=ContentExpress&func=display&ceid=171&meid=-1
Sponsor: Society of Architectural Historians
1365 N Astor Street
Chicago, IL 60610-2144

Society of Cosmetic Chemists Allan B. Black Award Sponsored by Presperse 5326
Dedicated to the advancement of cosmetic science, the Society strives to increase and disseminate scientific information through meetings and publications. By promoting research in cosmetic science and industry, and by setting high ethical, professional and educational standards, it strives to improve the qualifications of cosmetic scientists. Its mission is to further the interests and recognition of cosmetic scientists while maintaining the confidence of the public in the cosmetic and toiletries industry. With the Allan B. Black Award, a scroll and honorarium are awarded annually for the Best Paper on make-up technology either presented at an annual meeting, a seminar, or published in the Journal of Cosmetic Science, the Official Journal of the Society of Cosmetic Chemists.
Contact: Theresa Cesario, Executive Director; (212) 668-1500; fax: (212) 668-1504; email: TCesario@SCConline.org
Internet: http://www.scconline.org/website/about_scc/awards_fellowships.shtml
Sponsor: Society of Cosmetic Chemists
120 Wall Street, Suite 2400
New York, NY 10005-4088

Society of Cosmetic Chemists Award Sponsored by McIntyre Group, Ltd. 5327
Dedicated to the advancement of cosmetic science, the Society strives to increase and disseminate scientific information through meetings and publications. By promoting research in cosmetic science and industry, and by setting high ethical, professional and educational standards, it strives to improve the qualifications of cosmetic scientists. Its mission is to further the interests and recognition of cosmetic scientists while maintaining the confidence of the public in the cosmetic and toiletries industry. This award, given in the form of a scroll and honorarium, are awarded for the best paper presented at the Society's Annual Scientific Meeting.
Contact: Theresa Cesario, Executive Director; (212) 668-1500; fax: (212) 668-1504; email: TCesario@SCConline.org
Internet: http://www.scconline.org/website/about_scc/awards_fellowships.shtml
Sponsor: Society of Cosmetic Chemists
120 Wall Street, Suite 2400
New York, NY 10005-4088

Society of Cosmetic Chemists Award Sponsored by The HallStar Company 5328
Dedicated to the advancement of cosmetic science, the Society strives to increase and disseminate scientific information through meetings and publications. By promoting research in cosmetic science and industry, and by setting high ethical, professional and educational standards, it strives to improve the qualifications of cosmetic scientists. Its mission is to further the interests and recognition of cosmetic scientists while maintaining the confidence of the public in the cosmetic and toiletries industry. Offered in the form of a scroll and honorarium, this honor is awarded for the Best Paper that makes the greatest scientific contribution to knowledge in the field of protecting against or ameliorating damage to human skin caused by exposure to UV radiation presented at the Society's Annual Scientific Seminar.
Contact: Theresa Cesario, Executive Director; (212) 668-1500; fax: (212) 668-1504; email TCesario@SCConline.org
Internet: http://www.scconline.org/website/about_scc/awards_fellowships.shtml
Sponsor: Society of Cosmetic Chemists
120 Wall Street, Suite 2400
New York, NY 10005-4088

Society of Cosmetic Chemists Chapter Best Speaker Award 5329
Dedicated to the advancement of cosmetic science, the Society strives to increase and disseminate scientific information through meetings and publications. By promoting research in cosmetic science and industry, and by setting high ethical, professional and educational standards, it strives to improve the qualifications of cosmetic scientists. Its mission is to further the interests and recognition of cosmetic scientists while maintaining the confidence of the public in the cosmetic and toiletries industry. The Chapter Best Speaker Award is given for the best scientific paper presented before each Chapter during a given calendar year. Recommendations for this award originates with the Chapters and are submitted to the National Office. A scroll and honorarium ($250 for a manuscript submission and $150 for an abstract submission) is provided by National for presentation by the Chapter.

Contact: Theresa Cesario, Executive Director; (212) 668-1500; fax: (212) 668-1504; email: TCesario@SCConline.org
Internet: http://www.scconline.org/website/about_scc/awards_fellowships.shtml
Sponsor: Society of Cosmetic Chemists
120 Wall Street, Suite 2400
New York, NY 10005-4088

Society of Cosmetic Chemists Chapter Merit Award 5330
Dedicated to the advancement of cosmetic science, the Society strives to increase and disseminate scientific information through meetings and publications. By promoting research in cosmetic science and industry, and by setting high ethical, professional and educational standards, it strives to improve the qualifications of cosmetic scientists. Its mission is to further the interests and recognition of cosmetic scientists while maintaining the confidence of the public in the cosmetic and toiletries industry. The Chapter Merit Award is awarded to a Chapter Member for outstanding service, dedication and voluntary services over a period of time. A scroll is provided by National for presentation by the Chapter. Recommendations for this award originates with each Chapter and are submitted to the National Office.
Contact: Theresa Cesario, Executive Director; (212) 668-1500; fax: (212) 668-1504; email: TCesario@SCConline.org
Internet: http://www.scconline.org/website/about_scc/awards_fellowships.shtml
Sponsor: Society of Cosmetic Chemists
120 Wall Street, Suite 2400
New York, NY 10005-4088

Society of Cosmetic Chemists Frontier of Science Award 5331
Dedicated to the advancement of cosmetic science, the Society strives to increase and disseminate scientific information through meetings and publications. By promoting research in cosmetic science and industry, and by setting high ethical, professional and educational standards, it strives to improve the qualifications of cosmetic scientists. Its mission is to further the interests and recognition of cosmetic scientists while maintaining the confidence of the public in the cosmetic and toiletries industry. The Frontier of Science Award is an honorarium given to a speaker who has achieved exceptional national or international stature in the scientific community for delivering a lecture at the Annual Scientific Meeting.
Contact: Theresa Cesario, Executive Director; (212) 668-1500; fax: (212) 668-1504; email: TCesario@SCConline.org
Internet: http://www.scconline.org/website/about_scc/awards_fellowships.shtml
Sponsor: Society of Cosmetic Chemists
120 Wall Street, Suite 2400
New York, NY 10005-4088

Society of Cosmetic Chemists Hans Schaeffer Award 5332
Dedicated to the advancement of cosmetic science, the Society strives to increase and disseminate scientific information through meetings and publications. By promoting research in cosmetic science and industry, and by setting high ethical, professional and educational standards, it strives to improve the qualifications of cosmetic scientists. Its mission is to further the interests and recognition of cosmetic scientists while maintaining the confidence of the public in the cosmetic and toiletries industry. The Hans Schaeffer Award, offered in the form of a scroll and honorarium, are awarded annually for the Most Innovative Paper presented at either an annual meeting or a seminar.
Contact: Theresa Cesario, Executive Director; (212) 668-1500; fax: (212) 668-1504; email: TCesario@SCConline.org
Internet: http://www.scconline.org/website/about_scc/awards_fellowships.shtml
Sponsor: Society of Cosmetic Chemists
120 Wall Street, Suite 2400
New York, NY 10005-4088

Society of Cosmetic Chemists Joseph P. Ciaudelli Award 5333
Dedicated to the advancement of cosmetic science, the Society strives to increase and disseminate scientific information through meetings and publications. By promoting research in cosmetic science and industry, and by setting high ethical, professional and educational standards, it strives to improve the qualifications of cosmetic scientists. Its mission is to further the interests and recognition of cosmetic scientists while maintaining the confidence of the public in the cosmetic and toiletries industry. The Joseph P. Ciaudelli Award, offered in the form of a scroll and honorarium, are bestowed annually for the best article appearing in the Journal of Cosmetic Science on the subject of hair care technology.
Contact: Theresa Cesario, Executive Director; (212) 668-1500; fax: (212) 668-1504; email: TCesario@SCConline.org
Internet: http://www.scconline.org/website/about_scc/awards_fellowships.shtml
Sponsor: Society of Cosmetic Chemists
120 Wall Street, Suite 2400
New York, NY 10005-4088

Society of Cosmetic Chemists Keynote Award Lecture Sponsored by Ruger Chemical Corporation 5334
Dedicated to the advancement of cosmetic science, the Society strives to increase and disseminate scientific information through meetings and publications. By promoting research in cosmetic science and industry, and by setting high ethical, professional and educational standards, it strives to improve the qualifications of cosmetic scientists. Its mission is to further the interests and recognition of cosmetic scientists while maintaining the confidence of the public in the cosmetic and toiletries industry. The Keynote Award is an honorarium given to a speaker who has achieved exceptional national or international stature in the scientific community for delivering a lecture at the Annual Scientific Meeting.
Contact: Theresa Cesario, Executive Director; (212) 668-1500; fax: (212) 668-1504; email: TCesario@SCConline.org
Internet: http://www.scconline.org/website/about_scc/awards_fellowships.shtml
Sponsor: Society of Cosmetic Chemists
120 Wall Street, Suite 2400
New York, NY 10005-4088

Society of Cosmetic Chemists Literature Award 5335
Dedicated to the advancement of cosmetic science, the Society strives to increase and disseminate scientific information through meetings and publications. By promoting research in cosmetic science and industry, and by setting high ethical, professional and educational standards, it strives to improve the qualifications of cosmetic scientists. Its mission is to further the interests and recognition of cosmetic scientists while maintaining the confidence of the public in the cosmetic and toiletries industry. Through the Literature Award, a scroll and honorarium are presented annually to the author or authors of scientific papers in basic research judged to be an outstanding contribution to cosmetic science and technology.
Contact: Theresa Cesario, Executive Director; (212) 668-1500; fax: (212) 668-1504; email: TCesario@SCConline.org
Internet: http://www.scconline.org/website/about_scc/awards_fellowships.shtml
Sponsor: Society of Cosmetic Chemists
120 Wall Street, Suite 2400
New York, NY 10005-4088

Society of Cosmetic Chemists Maison G. de Navarre Medal 5336
Dedicated to the advancement of cosmetic science, the Society strives to increase and disseminate scientific information through meetings and publications. By promoting research in cosmetic science and industry, and by setting high ethical, professional and educational standards, it strives to improve the qualifications of cosmetic scientists. Its mission is to further the interests and recognition of cosmetic scientists while maintaining the confidence of the public in the cosmetic and toiletries industry. The Society's highest honor, the Maison G. de Navarre Medal, recognizes individuals for their technical contributions to cosmetic science. The award is presented annually.
Contact: Theresa Cesario, Executive Director; (212) 668-1500; fax: (212) 668-1504; email: TCesario@SCConline.org
Internet: http://www.scconline.org/website/about_scc/awards_fellowships.shtml
Sponsor: Society of Cosmetic Chemists
120 Wall Street, Suite 2400
New York, NY 10005-4088

Society of Cosmetic Chemists Merit Award 5337
Dedicated to the advancement of cosmetic science, the Society strives to increase and disseminate scientific information through meetings and publications. By promoting research in cosmetic science and industry, and by setting high ethical, professional and educational standards, it strives to improve the qualifications of cosmetic scientists. Its mission is to further the interests and recognition of cosmetic scientists while maintaining the confidence of the public in the cosmetic and toiletries industry. The Merit Award is presented annually for outstanding service and distinguished leadership in Society activities.
Contact: Theresa Cesario, Executive Director; (212) 668-1500; fax: (212) 668-1504; email: TCesario@SCConline.org
Internet: http://www.scconline.org/website/about_scc/awards_fellowships.shtml
Sponsor: Society of Cosmetic Chemists
120 Wall Street, Suite 2400
New York, NY 10005-4088

Society of Cosmetic Chemists Robert A. Kramer Lifetime Service Award 5338
Dedicated to the advancement of cosmetic science, the Society strives to increase and disseminate scientific information through meetings and publications. By promoting research in cosmetic science and industry, and by setting high ethical, professional and educational standards, it strives to improve the qualifications of cosmetic scientists. Its mission is to further the interests and recognition of cosmetic scientists while maintaining the confidence of the public in the cosmetic and toiletries industry. The Robert A. Kramer Lifetime Service Award, given at the discretion of the Board of Directors, goes to an individual who has given extraordinary service over the course of his or her membership.
Contact: Theresa Cesario, Executive Director; (212) 668-1500; fax: (212) 668-1504; email: TCesario@SCConline.org
Internet: http://www.scconline.org/website/about_scc/awards_fellowships.shtml
Sponsor: Society of Cosmetic Chemists
120 Wall Street, Suite 2400
New York, NY 10005-4088

Society of Cosmetic Chemists Shaw Mudge Award 5339
Dedicated to the advancement of cosmetic science, the Society strives to increase and disseminate scientific information through meetings and publications. By promoting research in cosmetic science and industry, and by setting high ethical, professional and educational standards, it strives to improve the qualifications of cosmetic scientists. Its mission is to further the interests and recognition of cosmetic scientists while maintaining the confidence of the public in the cosmetic and toiletries industry. The Shaw Mudge

Award, given in the form of a scroll and honorarium, are awarded for the best paper presented at the Society's Annual Scientific Seminar.
Contact: Theresa Cesario, Executive Director; (212) 668-1500; fax: (212) 668-1504; email: TCesario@SCConline.org
Internet: http://www.scconline.org/website/about_scc/awards_fellowships.shtml
Sponsor: Society of Cosmetic Chemists
120 Wall Street, Suite 2400
New York, NY 10005-4088

Society of Cosmetic Chemists Student Poster Awards **5340**
Dedicated to the advancement of cosmetic science, the Society strives to increase and disseminate scientific information through meetings and publications. By promoting research in cosmetic science and industry, and by setting high ethical, professional and educational standards, it strives to improve the qualifications of cosmetic scientists. Its mission is to further the interests and recognition of cosmetic scientists while maintaining the confidence of the public in the cosmetic and toiletries industry. The Student Poster Award, sponsored by DD-Chemco, Inc., is given annually in conjunction with the Annual Scientific Seminar. Honorariums are given for first, second, third and fourth place Best Student Poster.
Contact: Theresa Cesario, Executive Director; (212) 668-1500; fax: (212) 668-1504; email: TCesario@SCConline.org
Internet: http://www.scconline.org/website/about_scc/awards_fellowships.shtml
Sponsor: Society of Cosmetic Chemists
120 Wall Street, Suite 2400
New York, NY 10005-4088

Society of Cosmetic Chemists Young Scientist Awards **5341**
The Young Scientist Awards are presented each year to recognize the outstanding achievements of new cosmetic scientists throughout the industry. Each year every manufacturer, supplier or other organization in the personal care industry may select an individual in their employ as a recipient of this award. The criteria for selection is made by the individual company. Information is sent in March of each year and is posted on the SCC Website. The submission deadline is July 1st each year. The awards may be presented at an SCC Chapter Meeting or through an internal corporate presentation. Recipient names are listed in the SCC Winter Newsletter.
Requirements: Recipients be an active member of the Society for at least one year and working in the industry less than 10 years.
Date(s) Application is Due: Jul 1
Contact: Theresa Cesario, Executive Director; (212) 668-1500; fax: (212) 668-1504; email: TCesario@SCConline.org
Internet: http://www.scconline.org/website/about_scc/awards_fellowships.shtml
Sponsor: Society of Cosmetic Chemists
120 Wall Street, Suite 2400
New York, NY 10005-4088

Society of Medical Friends of Wine Research Award **5342**
One award is given biennially for published original research identifying substances in wine; ascertaining the effects of wine components on living cells, tissues, or organs; or indicating appropriate clinical applications of wine in treatment or prevention of disease. Applications are accepted at any time.
Amount of Grant: $2000
Contact: Susan Guerguy, Executive Secretary, (925) 933-9691; fax: (925) 939-5224; email: marq@inreach.com
Internet: http://www.medicalfriendsofwine.org
Sponsor: Society of Medical Friends of Wine
511 Jones Pl
Walnut Creek, CA 94597

Society of Naval Architects and Marine Engineers Scholarships **5343**
The society annually awards both graduate and undergraduate scholarships to encourage study in naval architecture, marine engineering, ocean engineering or marine industry related fields. All applicants must plan to study in the United States or Canada. The listed application deadline is for graduate scholarships; undergraduate scholarships are available at and administered through select universities. Guidelines are available online.
Requirements: Applicants are required to be U.S. or Canadian citizens with character, leadership, and general promise for a future in the marine industry, as well as scholastic ability and desire to pursue advanced study.
Amount of Grant: $10,000
Date(s) Application is Due: Feb 1
Contact: Philip Kimball, Executive Director, (201) 798-4800 ext 3024; fax: (201) 798-4975; email: efaustino@sname.org
Internet: http://www.sname.org/membership/Scholar.html
Sponsor: Society of Naval Architects and Marine Engineers
601 Pavonia Avenue
Jersey City, NJ 07306

Society of Neuroscience Minority Fellowships **5344**
The goal of the program is to increase the diversity of the pool of individuals participating in mental-health related neuroscience research and teaching programs. The program recruits and trains outstanding individuals of traditionally underrepresented racial and ethnic minorities. Training stipends are available for predoctoral (up to three years) and postdoctoral (up to two years) fellowships. The program includes training stipends in accordance with NRSA standard guidelines; travel assistance and registration to attend the annual meeting, in addition to SFN member resources; enrichment programs including funds to participate in activities outside of the fellow's home laboratory; and mentoring opportunities, with a mentor from among SFN memberships. Details and application are available online.
Date(s) Application is Due: Jan 1; Apr 1; Sep 1
Contact: Jessica Pearce, Program Coordinator,(202) 462-6688; fax: (202) 462-9740; email: jessica@sfn.org
Internet: http://www.sfn.org/content/Programs/DiversityinNeuroscience/mnfp/index.html
Sponsor: Society for Neuroscience
11 Dupont Circle NW, Suite 500
Washington, DC 20036

Sol Goldman Charitable Trust Grants **5345**
The trust supports the arts, education, the environment, health organizations, human services, and Jewish and Protestant agencies. Preference is given to requests from the New York metropolitan area. There are no application deadlines or forms.
Samples: Johns Hopkins Medicine (Baltimore, MD)—to support a collaborative research initiative focused on developing novel means of earlier diagnosis and treatment of pancreatic cancer, $10 million (2005); Yale U (New Haven, CT)—to endow the deanship of its law school, $2.5 million (2005).
Contact: Grants Administrator, (212) 265-2280; fax: (212) 582-0186
Sponsor: Sol Goldman Charitable Trust
640 Fifth Avenue
New York, NY 10019

South Florida Student Shark Program (SFSSP) **5346**
The South Florida Student Shark Program (SFSSP) is a collaborative, multi-disciplinary research and education program supporting the career development for graduate, undergraduate as well as high school students. Founded in 2006, the SFSSP is a partnership among the University of Miami Rosenstiel School, The Explorers Club, the NOAA LMRCSC & the Herbert W. Hoover Foundation. Focusing on the study and conservation of Floridian shark species, mangrove fish habitat and the Florida watershed, the program encourages students to take an active role in modern scientific education & research. The SFSSP, full-immersion approach allows students to actively grow as future scientists, while supporting the important ongoing research crucial to shark conservation. Primary objectives are to: provide practical, hands-on marine science education and self initiated research project opportunities for high school, undergraduate and graduate college students in the marine science field. Educational and training opportunities accruing from the proposed work will also especially be made available to underrepresented (gender, ethnicity, disability) students; conduct numerous studies that will aid in the monitoring of biological and environmental conditions of Floridian marine ecosystems; eliminate general misconceptions about sharks and the Florida watershed by disseminating the results of this study and creating public awareness of the perils faced by sharks and the current threats to the Florida watershed; and help students (high school, undergraduate and graduate college) gain and develop useful skills in marine science and play an active role in creating, conducting and disseminating the program through service learning interactions. Contact the office for further details.
Contact: David Die, Co-Director; (305) 421-4607; fax: (305) 361-4457; email: ddie@rsmas.miami.edu; Neil Hammerschlag, Co-Director; (305) 421-4356 or (954) 815-0920; email: nhammerschlag@rsmas.miami.edu
Internet: http://cufer.rsmas.miami.edu/wordpress/?page_id=105
Sponsor: Rosenstiel School, University of Miami
4600 Rickenbacker Causeway
Miami, FL 33149

Southern Baptist Oral History Grants **5347**
The project of the Southern Baptist Historical Library and Archives encourages the development and pursuit of historical research projects that involve interviews with Baptists. Qualified projects would relate to any aspect of Baptist studies but must include interviews with Baptists that would be of interest to the Southern Baptist Historical Library and Archives for inclusion in its collection. Interested applicants must complete a signed application form by the stated deadline. Grants support travel costs, transcription fees, and supplies. There are no application deadlines. All requests received by April 1 of each year will be compiled and evaluated during April and applicants notified of decision in May.
Requirements: The program is open to anyone doing research projects in the area of Baptist studies.
Amount of Grant: $2000 maximum
Date(s) Application is Due: Apr 1
Contact: Bill Sumners, Director, Library and Archives, (615) 244-0344; email: bill@sbhla.org
Internet: http://www.sbhla.org/oralhistorygrant.htm
Sponsor: Southern Baptist Historical Library and Archives
901 Commerce Street, Suite 400
Nashville, TN 37203-3630

Southern Baptist Study Grants **5348**
These grants are used to encourage research in the unique resources of the Southern Baptist Historical Library and Archives operated by the Historical Commission of the Southern Baptist Convention. The grants, which reimburse travel expenses to Nashville, TN, are available to qualified researchers making extensive use of the resources of

the library and archives. Priority is usually given to Southern Baptist research topics. Applications are available on the Web site.
Requirements: This program is intended to aid graduate students, professors, historians, and other researchers who wish to make use of the resources of the Southern Baptist Historical Library and Archives.
Samples: Derek Chang, Duke U (Durham, NC)—for Shackles of Hierarchy: Race, Religion, and Citizenship in American Baptist Home Missions, 1865-1900, $500.; Dr. Ed Crowther, Adams State College (Alamosa, CO)—for Baptist Chaplains during the Civil War, $500.
Amount of Grant: $750 maximum
Date(s) Application is Due: Apr 1
Contact: Bill Sumners, Director, Library and Archives, (615) 244-0344; email: bill@sbhla.org
Internet: http://www.sbhla.org/sg_info.htm
Sponsor: Southern Baptist Historical Library and Archives
901 Commerce Street, Suite 400
Nashville, TN 37203-3630

Southern Jewish Historical Society Grants **5349**
The grant is awarded annually for completion of a project in Southern Jewish history and culture. The grant focuses on completion/publication of a project in its final stage.
Restrictions: The program does not fund research, travel, or writing.
Samples: Birmingham Jewish Community Ctr (Birmingham, AL)—for a project to preserve, protect, and share local history, $1900.; Susan Gross, Archives Committee of Congregation Agudath Achim (Shreveport, LA)—for publication of a book detailing the synagogue history, $1100.
Amount of Grant: $4500
Date(s) Application is Due: Aug 1
Contact: Dr. Scott M. Langston, Grants Committee, (417) 328-1742; fax: (417) 328-1771; email: SLangsto@sbuniv.edu or info@jewishsouth.org
Internet: http://www.jewishsouth.org/grants_new.htm
Sponsor: Southern Jewish Historical Society
P.O. Box 5024
Atlanta, GA 30302-5024

Southern Jewish Historical Society Student Essay Prizes **5350**
The society annually makes one award for the best paper dealing with Southern Jewry by a current graduate student and one award for the best paper by an undergraduate college student. Submission should focus on a subject related to Southern Jewish history and use primary sources and appropriate documentation. The submitter's name, address, and academic affiliation (graduate or undergraduate) must appear only on the cover letter and not in the body of the paper. Award presentations are made at the society's annual meeting in November. Contact Dr. Mark Greenberg for additional submission information.
Amount of Grant: $500 for graduate essay; $250 for undergraduate essay
Contact: Dr. Mark Greenberg, Director, Resource Center for Florida History & Politics, University of South Florida Libraries, (813) 974-4141; fax: (813) 974-3417; email: mgreenbe@lib.usf.edu
Internet: http://www.jewishsouth.org/essay.htm
Sponsor: Southern Jewish Historical Society
4202 E Fowler Avenue, LIB 122
Tampa, FL 33620-5400

Soy Health Research Program Incentive Awards **5351**
This program supports the development of complete research proposals on the topic of soy and bone health, soy and prostate health, and the dose-response relationship between isoflavone consumption and health and efficacy.
Amount of Grant: $10,000
Date(s) Application is Due: Mar 15
Contact: Steve Veile, Project Manager, (573) 635-3265 or (888) 772-8449; email: stevev@communiqueinc.com or soy@communiqueinc.com
Internet: http://www.talksoy.com/Shrp/default.htm
Sponsor: United SoyBean Board
P.O. Box 237, 512 E Capitol Avenue
Jefferson City, MO 65102

Special Olympics Health Professions Student Grants **5352**
The purpose of this program is to engage health professions students to work with persons with intellectual disabilities as a way of filling in a gap that exists in most health program curricula. The program promotes short-term projects exploring issues that impact the health and well-being of all persons with intellectual disabilities, including, but not limited to, Special Olympics athletes. Projects may include: data collection and analysis on issues impacting persons with intellectual disabilities; measurement of attitudes, opinions and behaviors of health professionals, coaches, family/caregivers and athletes; follow-up assessments of existing programs; or health promotion projects. Projects that involve collaborations with Special Olympics Programs or other CDC grant recipients (e.g., state and local health departments) are encouraged. Special Olympics is not currently prioritizing specific research areas for funding, as the program???s primary goal is to engage health professions students to work with the population of persons with intellectual disabilities. All awards will be made to the institution on behalf of the student grantee. Funding must include all expenses and no indirect costs may be covered by this grant. Applications are accepted year-round. Application and guidelines are available online.
Requirements: Health professions students at the bachelor, master, doctorate and postdoctorate levels from accredited institutions or programs in the United States or other countries are eligible. Fields or disciplines may include, audiology, dentistry, pre-med, medicine, nursing, optometry, physical therapy, podiatry, public health, social work, law, public administration, or other related health areas. All students??? projects will require a faculty advisor.
Amount of Grant: $5,000 maximum
Contact: Darcie Mersereau, Grants Administrator; (202) 628-3600 or (800) 700-8585; fax: (202) 824-0200; email: 2006hastudentgrant@specialolympics.org
Internet: http://info.specialolympics.org/Special%20Olympics%20Public%20Website/English/Initiatives/Research/Health%20Student%20Grants%20Available.htm
Sponsor: Special Olympics
1133 19th Street, NW
Washington, DC 20036-3604

Spencer Foundation Education Dissertation Fellowships **5353**
The foundation supports research that gives promise of yielding new knowledge leading to the improvement of education. The program will award approximately 30 nonrenewable fellowships to support completion of the dissertation. The foundation is interested in a wide variety of disciplinary and interdisciplinary approaches, although by direction of its charter it gives emphasis to the behavioral sciences. Education is defined broadly to include all the situations and institutions in which it proceeds across the entire lifespan, in the United States, and around the world. Specific areas include studies of education as a social institution, studies of human development that affect the nature and outcome of the educative process, studies in the processes of teaching and learning, studies in the history of the behavioral sciences as they have affected educational theory and practice, and efforts to assist in the continuing renewal of the educational community. Candidates should be interested in pursuing further research in education once the doctorate is attained. Fellowship applicants may request or download current application forms and instructions at any time between April and October.
Requirements: Grants are made only to institutions, not directly to individuals. Applicants nominated by their institutions must be candidates for the doctoral degree in any field of study at a graduate school in the United States. Dissertation topics must concern education. All predissertation requirements must be completed by June 1 of the upcoming year.
Restrictions: Awards are not designed for individuals in the fields of guidance or curriculum development, nor for those planning to be school administrators.
Amount of Grant: $20,000
Date(s) Application is Due: Nov 1
Contact: Spencer Dissertation Fellowships, (312) 274-6526; fax: (312) 337-0282; email: fellows@spencer.org
Internet: http://www.spencer.org/programs/index.htm
Sponsor: Spencer Foundation
875 N Michigan Avenue, Suite 3930
Chicago, IL 60611-1803

Spencer Foundation Major Research Grants **5354**
Funded research projects vary widely, ranging from medium-sized studies that can be completed within a year by an individual researcher to more extensive collaborative studies that last several years. There are no funding priorities for subjects of research; projects originate from research ideas initiated in the field by scholars and other researchers. Preliminary proposals are welcome at any time and will be responded to with in one month of receipt. Guidelines are available online.
Requirements: Ordinarily, principal investigators applying for a Major Research Grant must be affiliated with a school district, a college or university, a research facility, or a cultural institution. Researchers must also have an earned doctorate in an academic discipline or professional field or appropriate experience in the teaching profession.
Restrictions: Grantees may not receive two research grants simultaneously from the foundation. The foundation does not fund direct interventions or evaluations of programs. The foundation does not pay government-approved overhead rates on research grants; overhead requests of more than $75,000 may not exceed 15 percent of the requested direct costs. The foundation does not pay indirect costs on research grants of $75,000 or less.
Samples: Research for Action (Philadelphia, PA)—to study how teachers and administrators at Philadelphia schools use data to inform instructional practices, $437,000 over three years (2005).
Amount of Grant: $75,000 maximum
Contact: Susan Dauber, Senior Program Officer, (312) 274-6511; fax: (312) 337-0282; email: information@spencer.org or majgrant@spencer.org
Internet: http://www.spencer.org/programs/index.htm
Sponsor: Spencer Foundation
875 N Michigan Avenue, Suite 3930
Chicago, IL 60611

Spencer Foundation Research Grants **5355**
The foundation provides funding for investigations that promise to yield new knowledge about education in the United States or abroad. Funding ranges in size from smaller grants that can be completed within a year, to larger, multi-year endeavors. Guidelines are available online. Proposed programs should align with the one or more of the foundation's Areas of Inquiry: The Relationship Between Education and Social Opportunity; Organizational Learning in Schools, School Systems, and Higher Education Institutions; Teaching, Learning, and Instructional Resources; and, Purposes and Values of Education.

Requirements: Initial proposals are accepted at any time, regardless of the level of funding requested. Applicants must possess an earned doctorate and be affiliated with a school, college, university, or cultural institution. Educators currently employed in K-12 settings who are not pursuing a graduate degree are also eligible.
Restrictions: Principal investigators are not permitted to receive two research grants simultaneously. Projects may not extend beyond a year in duration, and no extensions will be granted. Grants do not support indirect costs.
Amount of Grant: $1000-$40,000 range
Contact: Annie Brinkman, Program Administrator, (312) 274-6511; email: abrinkman@spencer.org
Internet: http://www.spencer.org/programs/grants/research_grants.htm
Sponsor: Spencer Foundation
625 N. Michigan Avenue, Suite 1600
Chicago, IL 60611

SPIE A.E. Conrady Award 5356
The A. E. Conrady Award is and sponsored Optical Research Associates. The awards given annually in recognition of exceptional contributions in design, construction, and testing of optical systems and instrumentation. The recognition of this award is based on developments of new equipment, techniques, and applications for designing, testing, analyzing, and/or evaluating optical systems, components, and theories.
Requirements: Nominations must be accompanied by the following: completed award nomination form; brief citation stating the accomplishments that make the nominee deserving of this particular award; a one-page narrative description of the significant aspects of the nominee's career. Curriculum vitae listing educational background, positions held, publications, awards, honors, activities and offices held within the Society; a minimum of two (2) letters of reference from other than the nominator. Diversity of support is preferred; not all letters of recommendation should come from nominee's organization.
Samples: Roland Winston, University of California, California, $2,000 Honorarium. .;2009.;
Amount of Grant: $2,000 Honorarium
Date(s) Application is Due: Oct 1
Contact: Awards Coordinator, (360) 676-3290 or (888) 504-8171; fax: (360) 647-1445; email: awards@spie.org
Internet: http://spie.org/app/awards/nominate.aspx?redir=app/awards/nominate.aspx&ID=x2894
Sponsor: International Society Advancing Light-Based Research.
1000 20th Street (P.O. Box 10)
Bellingham, WA 98225

SPIE Chandra S. Vikram Award in Optical Metrology 5357
The Chandra S. Vikram Award in Optical Metrology is given annually in recognition of exceptional contributions to the field of optical metrology. The award may be presented for a specific achievement, development, or invention of significant importance to optical metrology, or may be given for lifetime achievement.
Requirements: Nominations must be accompanied by the following: completed award nomination form; brief citation stating the accomplishments that make the nominee deserving of this particular award; a one-page narrative description of the significant aspects of the nominee's career. Curriculum vitae listing educational background, positions held, publications, awards, honors, activities and offices held within the Society; a minimum of two (2) letters of reference from other than the nominator. Diversity of support is preferred; not all letters of recommendation should come from nominee's organization.
Amount of Grant: $2,000 Honorarium
Date(s) Application is Due: Oct 1
Contact: Awards Coordinator, (360) 676-3290 or (888) 504-8171; fax: (360) 647-1445; email: awards@spie.org
Internet: http://spie.org/app/awards/nominate.aspx?redir=app/awards/nominate.aspx&ID=x2894
Sponsor: International Society Advancing Light-Based Research.
1000 20th Street (P.O. Box 10)
Bellingham, WA 98225

SPIE Dennis Gabor Award 5358
The Dennis Gabor Award is presented annually in recognition of outstanding inventive accomplishments in electro-optical systems, especially those which further the development of holographic imaging, and metrology applications.
Requirements: Nominations must be accompanied by the following: completed award nomination form; brief citation stating the accomplishments that make the nominee deserving of this particular award; a one-page narrative description of the significant aspects of the nominee's career. Curriculum vitae listing educational background, positions held, publications, awards, honors, activities and offices held within the Society; a minimum of two (2) letters of reference from other than the nominator. Diversity of support is preferred; not all letters of recommendation should come from nominee's organization.
Samples: Leonid Glebov, College of Optics & Photonics, University of Central Florida, Florida, $2,000 Honorarium .;2008.;
Amount of Grant: $2,000 Honorarium
Date(s) Application is Due: Oct 1
Contact: Awards Coordinator, (360) 676-3290 or (888) 504-8171; fax: (360) 647-1445; email: awards@spie.org
Internet: http://spie.org/app/awards/nominate.aspx?redir=app/awards/nominate.aspx&ID=x2894
Sponsor: International Society Advancing Light-Based Research.
1000 20th Street (P.O. Box 10)
Bellingham, WA 98225

SPIE Frits Zernike Award 5359
The Frits Zernike Award for Microlithography is given annually for outstanding accomplishments in microlithographic technology, especially those furthering the development of semiconductor lithographic imaging solutions.
Requirements: Nominations must be accompanied by the following: completed award nomination form; brief citation stating the accomplishments that make the nominee deserving of this particular award; a one-page narrative description of the significant aspects of the nominee's career. Curriculum vitae listing educational background, positions held, publications, awards, honors, activities and offices held within the Society; a minimum of two (2) letters of reference from other than the nominator. Diversity of support is preferred; not all letters of recommendation should come from nominee's organization.
Amount of Grant: $2,000 Honorarium
Date(s) Application is Due: Oct 1
Contact: Awards Coordinator, (360) 676-3290 or (888) 504-8171; fax: (360) 647-1445; email: awards@spie.org
Internet: http://spie.org/app/awards/nominate.aspx?redir=app/awards/nominate.aspx&ID=x2894
Sponsor: International Society Advancing Light-Based Research.
1000 20th Street (P.O. Box 10)
Bellingham, WA 98225

SPIE George W. Goddard Award 5360
The George W. Goddard Award is given annually in recognition of exceptional achievement in optical or photonic instrumentation for aerospace applications.
Requirements: Nominations must be accompanied by the following: completed award nomination form; brief citation stating the accomplishments that make the nominee deserving of this particular award; a one-page narrative description of the significant aspects of the nominee's career. Curriculum vitae listing educational background, positions held, publications, awards, honors, activities and offices held within the Society; a minimum of two (2) letters of reference from other than the nominator. Diversity of support is preferred; not all letters of recommendation should come from nominee's organization.
Samples: Neil Gehrels, Astropartical Physics Laboratory, NASA, U.S.A, $2,000 Honorarium .;2009.;
Amount of Grant: $2,000 Honorarium
Date(s) Application is Due: Oct 1
Contact: Awards Coordinator, (360) 676-3290 or (888) 504-8171; fax: (360) 647-1445; email: awards@spie.org
Internet: http://spie.org/app/awards/nominate.aspx?redir=app/awards/nominate.aspx&ID=x2894
Sponsor: International Society Advancing Light-Based Research.
1000 20th Street (P.O. Box 10)
Bellingham, WA 98225

SPIE Gold Medal of the Society Award 5361
The Gold Medal of the Society is the highest honor the Society bestows. Beginning in 1977, it has been awarded annually in recognition of outstanding engineering or scientific accomplishments in optics, electro-optics, or photographic technologies or applications. The recipient(s) shall have made an exceptional contribution to the advancement of relevant technology.
Requirements: Nominations must be accompanied by the following: completed award nomination form; brief citation stating the accomplishments that make the nominee deserving of this particular award; a one-page narrative description of the significant aspects of the nominee's career. Curriculum vitae listing educational background, positions held, publications, awards, honors, activities and offices held within the Society; a minimum of two (2) letters of reference from other than the nominator. Diversity of support is preferred; not all letters of recommendation should come from nominee's organization.
Samples: M. J. Soileau, College of Optics & Photonics, University of Central Florida, Florida, $10,000 Honorarium .;2008.;
Amount of Grant: $10,000 Honorarium
Date(s) Application is Due: Oct 1
Contact: Awards Coordinator, (360) 676-3290 or (888) 504-8171; fax: (360) 647-1445; email: awards@spie.org
Internet: http://spie.org/app/awards/nominate.aspx?redir=app/awards/nominate.aspx&ID=x2894
Sponsor: International Society Advancing Light-Based Research.
1000 20th Street (P.O. Box 10)
Bellingham, WA 98225

SPIE Technology Achievement Award 5362
The SPIE Technology Achievement Award is given in recognition of exceptional contributions in design, construction, and testing of optical systems and instrumentation.
Requirements: Nominations must be accompanied by the following: completed award nomination form; brief citation stating the accomplishments that make the nominee

deserving of this particular award; a one-page narrative description of the significant aspects of the nominee's career. Curriculum vitae listing educational background, positions held, publications, awards, honors, activities and offices held within the Society; a minimum of two (2) letters of reference from other than the nominator. Diversity of support is preferred; not all letters of recommendation should come from nominee's organization.
Samples: Jame. G. Grote, Air Force Research Laboratory, Wright-Patterson Air Force Base, U.S.A, $2,000 Honorarium .;2009.;
Amount of Grant: $2,000 Honorarium
Date(s) Application is Due: Oct 1
Contact: Awards Coordinator, (360) 676-3290 or (888) 504-8171; fax: (360) 647-1445; email: awards@spie.org
Internet: http://spie.org/app/awards/nominate.aspx?redir=app/awards/nominate.aspx&ID=x2894
Sponsor: International Society Advancing Light-Based Research.
1000 20th Street (P.O. Box 10)
Bellingham, WA 98225

Spina Bifida Young Investigator Awards 5363
The association (SBA) and the Spina Bifida Foundation award fellowships to support spina bifida research by promising young investigators. The purpose of the award is to provide initial or continuing financial support to a fellow in an established research program. The institution at which the established research program is located will be the awardee and will be expected to administer the award on behalf of the fellow, who will be the principal investigator. The grant will begin at the awardee institution's discretion between August 1 and December 1, and will be for a one-year period, unless prior written approval for a no-cost extension is given by SBA. Application and guidelines are available online.
Requirements: The fellow must possess an MD, PhD, DSW, DScN, or equivalent terminal degree prior to beginning the research under the award, but may not have held a faculty title of assistant professor or higher for more than 24 months at the time the fellowship is to begin.
Amount of Grant: $25,000 fellowships
Date(s) Application is Due: May 9
Contact: Young Investigator Awards, (800) 621-3141 or (202) 944-3285; email: jgiffels@sbaa.org
Internet: http://www.sbaa.org/site/PageServer?pagename=research_young_investigator_2005_RFP
Sponsor: Spina Bifida Association
4590 MacArthur Blvd NW, Suite 250
Washington, DC 20007-4226

SPSSI Applied Social Issues Internships 5364
Proposals are invited for the application of social science principles to social issues in cooperation with a community, city, or state government organization; public interest group; or other nonprofit entity. Awards cover research costs, community organizing, summer stipends, and related expenses. Complete application includes a two- to five-page proposal including a budget, short resume, letter from a faculty sponsor/supervisor of the project, and letter from an organizational sponsor. Applications should be submitted to SPSSI central office at the address listed. Annual deadline dates may vary; contact the office for exact dates.
Requirements: College seniors, graduate students, and first-year postdoctorates in psychology, applied social science, and related disciplines are eligible.
Amount of Grant: $300-$2500
Date(s) Application is Due: May 1
Contact: Applied Social Issues Internship Program, (202) 675-6956; fax: (202) 675-6902; email: awards@spssi.org
Internet: http://www.spssi.org/ASIflyer.html
Sponsor: Society for the Psychological Study of Social Issues
208 I Street NE
Washington, DC 20002

SPSSI Grants in Aid Program 5365
Grants in aid fund scientific research related to the basic interests and goals of SPSSI and particularly those that are not likely to receive support from traditional sources. The committee encourages proposals involving unique and timely research opportunities; underrepresented institutions and new investigators; volunteer research teams; and actual, not pilot, projects. Funding up to $1000 is available for graduate student research, with strong preference given to applications from students at the dissertation stage of the graduate career. Proposals must be accompanied by an appropriate official university agreement matching the amount requested.
Requirements: An interested person should submit five copies of a statement that includes a cover sheet stating title of proposal, name of investigator, address, phone, and fax number; an abstract of 100 words or less summarizing the proposed research; project purposes, rationale, and research methodology and analytical procedures to be employed; relevance to SPSSI goals and grant criteria; status of human subjects review process; resume of investigator; and specific amount requested including a budget. A recommended length for the first four categories above is five to seven double-spaced, typed pages.
Restrictions: Funds are not normally provided for travel to conventions, travel or living expenses while conducting research, stipends of principal investigators, costs associated with manuscript preparation, or the indirect costs of institutions.
Amount of Grant: $1000 predoctoral average; $2000 postdoctoral average
Date(s) Application is Due: May 1; Nov 1
Contact: Grants in Aid Program, (734) 662-9130; fax: (734) 662-5607; email: awards@spssi.org
Internet: http://www.spssi.org/GIAflyer.html
Sponsor: Society for the Psychological Study of Social Issues
P.O. Box 1248
Ann Arbor, MI 48106-1248

SPSSI Louise Kidder Early Career Award 5366
The award recognizes social issues researchers who have made substantial contributions to the field early in their careers. Nominees should be investigators who have made substantial contributions to social issues research within five years of receiving a graduate degree and who have demonstrated the potential to continue such contributions. Nominees need not be current SPSSI members.
Requirements: Person's nominating candidates should send a cover letter outlining the nominee's accomplishments to date and anticipated future contributions, the nominee's current curriculum vita, and three letters of support.
Amount of Grant: $500 maximum
Date(s) Application is Due: May 1
Contact: Kidder Award Program, (202)-675-6956; fax: (202)-675-6902; email: awards@spssi.org
Internet: http://www.spssi.org/kidder_flyer.html
Sponsor: Society for the Psychological Study of Social Issues
208 I Street NE
Washington, DC 20002

SPSSI Social Issues Dissertation Award 5367
The award is for the best doctoral dissertation in psychology (or in a social science with psychological subject matter). Criteria for acceptance are scientific excellence and potential application to social problems. Applicant is required to send applications via electronic submission to awards@spssi.org. As the subject of the email, write Dissertation Award. Attach a 500-word summary of the dissertation labeled Summary by [your last name], [your first name], preferably in Word format. The summary should include title, rationale, methods, and results of dissertation, as well as its implications for social problems. In the text of the email state: title of the dissertation, your name, postal and email addresses, phone number and university granting the degree.
Requirements: Any doctoral dissertation in psychology or in a social science with psychological subject matter, accepted after March 1 and before March 1 of the following year, is eligible.
Amount of Grant: $750 first prize; $550 second prize
Date(s) Application is Due: May 1
Contact: Social Issues Dissertation Award, (202) 675-6956; fax: (202)-675-6902; email: awards@spssi.org
Internet: http://www.spssi.org/Dissertationflyer.html
Sponsor: Society for the Psychological Study of Social Issues
208 I Street NE
Washington DC 20002-4340

SRC Education Alliance Undergraduate Research Assistants Program 5368
The Undergraduate Research Assistants (URA) program seeks to attract academically qualified students early, i.e., at the freshman/sophomore level, and build their interest and enthusiasm for disciplines of interest to the semiconductor industry by providing valid research experiences. The program was piloted from 2001 to 2003 with funding from the Semiconductor Industry Association and SIA Board companies. Although the overall numbers in the pilot were relatively small, data indicate that students involved early-on in valid research projects are much more likely to continue to graduate school in disciplines of interest to the semiconductor industry. The Program provides continuous support through undergraduate school for students who maintain a high level of interest and participation in the program. Mentoring by faculty, graduate students, and industry is a key element of the program as are internships provided by sponsoring companies.
Contact: Program Director; (919) 941-9400; fax: (9191) 941-9450; email: srcea@src.org
Internet: http://srcea.src.org/programs/default.asp
Sponsor: Semiconductor Research Corporation
P.O. Box 12053
Research Triangle Park, NC 27709-2053

SRC Education Science Grants for Postdoctoral Positions in Sweden 5369
The grant will enable education science researchers with Swedish or non-Swedish doctorates (PhDs or equivalent) to work at Swedish higher education institutions or research establishments. The maximum fellowship period of 24 months is reduced by the length of time for which the applicant, since obtaining a doctorate, has held a postdoctoral position, a similar grant or a fellowship, at a higher education institution or faculty or company other than the one where the doctorate, or equivalent, was awarded. The host university/institution or equivalent is the employer and determines which form of employment the position will have.
Requirements: Individual education science researchers may apply. Swedish doctorate or foreign degree equivalent to a doctorate is required. The doctorate must have been awarded within three years before and latest at the same day as the deadline for applications. Sick leave, service in the Armed Forces, doctor's employment as a pre-registration house officer, work for trade unions and student organizations, parental leave and other circumstances of this kind are not included in the three years. There is no age limit for applicants.
Date(s) Application is Due: Feb 20

Contact: Helene Sundewall, Research Administrator, Educational Sciences; +46-8-546-44-159; fax: +46-8-546-44-180; email: helene.sundewall@vr.se
Internet: http://www.vr.se/mainmenu/applyforgrants/callforproposals/grantsview.4.64a4ae3e1092b3475b6800063.html?resourceId=-1392&languageId=2
Sponsor: Swedish Research Council
Klarabersgviadukten 82
Stockholm, Sweden SE-103 78

SRC Educational Science Conference Grants **5370**
The intent of the Program is to foster links between the research field of Educational Science and international research, and to encourage mobility in Swedish higher education. The grant caters for Swedish arrangers of international conferences relating to the research field of Educational Science. Grants for invitations to foreign lecturers of international repute are (with respect to travel and accommodation costs) given priority. To receive a grant, it is important that the congress, conference or symposium is considered to be of manifest interest for the research area of education science.
Requirements: The applicant with scientific responsibility must have a doctorate or corresponding skills.
Restrictions: Support for conferences is awarded only to cover costs (travel and lodging) for a limited number of invited foreign lecturers. Grants are not provided for other participants, for administration or for social-security contributions.
Amount of Grant: 100,000 SEK (Swedish kronor) maximum
Date(s) Application is Due: Feb 15
Contact: Helene Sundewall, Research Administrator, Educational Sciences; +46-8-546-44-159; fax: +46-8-546-44-180; email: helene.sundewall@vr.se
Internet: http://www.vr.se/mainmenu/applyforgrants/callforproposals/grantsview.4.64a4ae3e1092b3475b6800063.html?resourceId=-1418&languageId=2
Sponsor: Swedish Research Council
Klarabersgviadukten 82
Stockholm, Sweden SE-103 78

SRC Educational Science Doctoral Program Grants **5371**
The Program's function is to promote research of high scientific quality, relevant to teacher training and professional teaching activity. A number of national doctoral programs have already been established (in mathematics, specializing in subject teaching; adult learning; aesthetic learning processes; and Swedish language specializing in subject teaching). The Research Council's Committee for Educational Science now seeks, in particular, to encourage applications for admission to national, multidisciplinary doctoral programs with the emphasis on Educational Science. Funds are granted for a maximum of four years.
Requirements: The program must: include several cooperating HEIs, which should include at least one with teacher training; have a formalized organization with a program director or coordinator in charge of implementation; offer a coherent study program for doctoral students on a well-developed theme, or a well-defined research program; be responsible for its own admission, curricula and monitoring; and must provide a relevant range of courses for doctoral students. The HEIs taking part must jointly offer a good research environment for the specific theme or program concerned. The number of doctoral students funded with grants from the Research Council's CES must not exceed five, and the total amount applied for should not exceed SEK 2 million.
Contact: Helene Sundewall, Research Administrator, Educational Sciences; +46-8-546-44-159; fax: +46-8-546-44-180; email: helene.sundewall@vr.se
Internet: http://www.vr.se/responsibilities/researchfunding/educationalscience.4.69f66a93108e85f68d48000693.html
Sponsor: Swedish Research Council
Klarabersgviadukten 82
Stockholm, Sweden SE-103 78

SRC Educational Science Grants **5372**
The Swedish Government set up the Committee for Educational Science (CES) as a special expert body in the Swedish Research Council. The Committee's function is to promote research of high scientific quality, relevant to teacher training and professional teaching activity. Accordingly, this means research on learning, knowledge formation, education and teaching. During the CES's existence, the pressure of applications has increased by almost 300 percent. For the current year, CES had a budget of approximately 128 million SEK. It also supports the development of research closely associated with practice, i.e. closely connected with teaching activity. Research support in Educational Science is subject to a requirement of co-funding from, and cooperation between, participating higher education institutions (HEIs).
Contact: Helene Sundewall, Research Administrator, Educational Sciences; +46-8-546-44-159; fax: +46-8-546-44-180; email: helene.sundewall@vr.se
Internet: http://www.vr.se/responsibilities/researchfunding/educationalscience.4.69f66a93108e85f68d48000693.html
Sponsor: Swedish Research Council
Klarabersgviadukten 82
Stockholm, Sweden SE-103 78

SRC Educational Science Grants for Invitation of Visiting Scientists **5373**
The grant is intended to promote the internationalization of Swedish research in educational sciences. Costs for guest researcher's travel expenses, board and lodging, and in exceptional cases a stipend payment are allowable. Grants are normally administered by a Swedish university/institution or another Swedish public organization with research as its purpose. Applications should be submitted electronically.
Requirements: Individual researchers at Swedish university departments may apply. The applicant with scientific responsibility must have obtained a doctorate or have an equivalent qualification. The application should be written in Swedish or English.
Date(s) Application is Due: Mar 3
Contact: Helene Sundewall, Research Administrator, Educational Sciences; +46-8-546-44-159; fax: +46-8-546-44-180; email: helene.sundewall@vr.se
Internet: http://www.vr.se/mainmenu/applyforgrants/callforproposals/grantsview.4.64a4ae3e1092b3475b6800063.html?resourceId=-1382&languageId=2
Sponsor: Swedish Research Council
Klarabersgviadukten 82
Stockholm, Sweden SE-103 78

SRC Educational Science Grants for Travels to Foreign Institutions as a Guest Scientist **5374**
The intent is to foster links between the research field of Educational Science and international research, and to encourage mobility in Swedish higher education. Costs for guest researcher's travel expenses, board and lodging, and in exceptional cases a stipend payment are allowable. The grant is intended for longer periods of time (up to six months). The abstract in the online form should include a description of: the foreign seat of learning which the researcher wishes to visit; the period envisaged; brief reasons for the visit, and the program planned. Applications should be submitted electronically.
Requirements: Individual researchers at Swedish university departments may apply. The application should be written in Swedish or English.
Date(s) Application is Due: Mar 3
Contact: Helene Sundewall, Research Administrator, Educational Sciences; +46-8-546-44-159; fax: +46-8-546-44-180; email: helene.sundewall@vr.se
Internet: http://www.vr.se/mainmenu/applyforgrants/callforproposals/grantsview.4.64a4ae3e1092b3475b6800063.html?resourceId=-1385&languageId=2#kontakt
Sponsor: Swedish Research Council
Klarabersgviadukten 82
Stockholm, Sweden SE-103 78

SRC Educational Science Junior Research Position Grants **5375**
The Program's function is to promote research of high scientific quality, relevant to teacher training and professional teaching activity. This grant relates mainly to salary costs and usually covers four years. An applicant for a junior research position must, in the application, propose the subject area for and location of the position sought. If an application is successful, the applicant is appointed at the HEI or research institute concerned.
Contact: Helene Sundewall, Research Administrator, Educational Sciences; +46-8-546-44-159; fax: +46-8-546-44-180; email: helene.sundewall@vr.se
Internet: http://www.vr.se/responsibilities/researchfunding/educationalscience.4.69f66a93108e85f68d48000693.html
Sponsor: Swedish Research Council
Klarabersgviadukten 82
Stockholm, Sweden SE-103 78

SRC Educational Science Language Checking Grants **5376**
The Council supports basic research in educational science (research into training, education, teaching and learning). Basic research/curiosity-driven research refers to research that provides new knowledge and contributes explanations and understanding. Scientific quality is the deciding factor in the assessment of applications. Educational science research may deal with, for example, different kinds of learning, the link between the nature of the content, teaching and learning or about learning in different social, cultural and institutional contexts including working life. It may concern diversity aspects of learning and memory such as gender, ethnicity, social background, talent, disabilities, etc. or learning at different stages of life, such as lifelong learning. Other examples are research into the effects of educational innovations, educational reforms or different educational arrangements, and the way the education system and informal learning function under the influence of different social, economic and political conditions. Important aspects are the possibilities for research to shed light on topical and fundamental issues in educational practice and the teaching profession. Applications must be made entirely electronically.
Requirements: The applicant with scientific responsibility must have obtained a doctorate or have an equivalent qualification. The application must be written in Swedish or English. Grants are normally administered by a university/institution (host university/institution). A full application comprises the filled-in online form and appendices. Note that Appendix S, signed, must be submitted to the Swedish Research Council in paper form.
Restrictions: Grants are not awarded to studies of a more developmental nature, development of an activity or follow-up.
Amount of Grant: 15,000 SEK (Swedish kronor)
Date(s) Application is Due: Oct 1
Contact: Helene Sundewall, Research Administrator, Educational Sciences; +46-8-546-44-159; fax: +46-8-546-44-180; email: helene.sundewall@vr.se
Internet: http://www.vr.se/mainmenu/applyforgrants/callforproposals/grantsview.4.64a4ae3e1092b3475b6800063.html?resourceId=-1480&languageId=2#kontakt
Sponsor: Swedish Research Council
Klarabersgviadukten 82
Stockholm, Sweden SE-103 78

SRC Educational Science Medium-Cost Scientific Equipment Grants 5377
The Program's function is to promote research of high scientific quality, relevant to teacher training and professional teaching activity. This grant is for purchase of equipment costing between 170,000 and 2,000,000 SEK, and is disbursed as a lump sum. One of these applicants receives project funds from the Research Council for the activity to which the equipment relates.
Amount of Grant: 170,000 - 2,000,000 SEK (Swedish kronor)
Contact: Helene Sundewall, Research Administrator, Educational Sciences; +46-8-546-44-159; fax: +46-8-546-44-180; email: helene.sundewall@vr.se
Internet: http://www.vr.se/responsibilities/researchfunding/educationalscience.4.69f66a93108e85f68d48000693.html
Sponsor: Swedish Research Council
Klarabersgviadukten 82
Stockholm, Sweden SE-103 78

SRC Educational Science Network Grants 5378
The Program's function is to promote research of high scientific quality, relevant to teacher training and professional teaching activity. This grant is for a network of researchers in specific subject areas within Educational Science.
Contact: Helene Sundewall, Research Administrator, Educational Sciences; +46-8-546-44-159; fax: +46-8-546-44-180; email: helene.sundewall@vr.se
Internet: http://www.vr.se/responsibilities/researchfunding/educationalscience.4.69f66a93108e85f68d48000693.html
Sponsor: Swedish Research Council
Klarabersgviadukten 82
Stockholm, Sweden SE-103 78

SRC Educational Science Postdoctoral Fellowships 5379
These fellowships are designed to enable educational science researchers with Swedish doctorates or exams from EUI (European University Institute) to stay at foreign universities or research establishments. The Committee wants to contribute to strengthening current research as well as supporting research within less researched fields, preferably interdisciplinary or multidisciplinary. The Committee especially welcomes applications from young researchers who wish to qualify in the field of educational science. The maximum fellowship period of 24 months is reduced by the length of time for which the applicant, since obtaining a doctorate, has held a postdoctoral position, a similar grant or a fellowship, at a higher education institution or faculty or company other than the one where the doctorate, or equivalent, was awarded. The proposed research project must be essentially different from the applicant's previous research. Essentially different means a new direction, new ways of thinking and/or new orientation of the research. The amount of the fellowship will be the standard grant rate, calculated according to costs in the host country. Applications to the Swedish Research Council must be made entirely electronically, and the application may be written in either Swedish or English.
Requirements: A Swedish doctorate or exam from EUI (European University Institute) is required. The doctorate must have been awarded within three years before and latest at the same day as the deadline for applications. Sick leave, service in the Armed Forces, doctor's employment as a preregistration house officer, work for trade unions and student organizations, parental leave and other circumstances of this kind are not counted in the three years. There is no age limit for applicants.
Date(s) Application is Due: Feb 20
Contact: Helene Sundewall, Research Administrator, Educational Sciences; +46-8-546-44-159; fax: +46-8-546-44-180; email: helene.sundewall@vr.se
Internet: http://www.vr.se/mainmenu/applyforgrants/callforproposals/grantsview.4.64a4ae3e1092b3475b6800063.html?resourceId=-1445&languageId=2
Sponsor: Swedish Research Council
Klarabersgviadukten 82
Stockholm, Sweden SE-103 78

SRC Educational Science Project Research Grants 5380
The Program's function is to promote research of high scientific quality, relevant to teacher training and professional teaching activity. This grant is for research during a specific period of time, to carry out a defined research task. Normally, the grant is awarded for up to three years. A doctoral student¬¥s position may be included. Up to SEK 170,000 in equipment costs may be applied for.
Restrictions: Salary costs for a person employed as a junior researcher or assistant lecturer is not funded.
Contact: Helene Sundewall, Research Administrator, Educational Sciences; +46-8-546-44-159; fax: +46-8-546-44-180; email: helene.sundewall@vr.se
Internet: http://www.vr.se/responsibilities/researchfunding/educationalscience.4.69f66a93108e85f68d48000693.html
Sponsor: Swedish Research Council
Klarabersgviadukten 82
Stockholm, Sweden SE-103 78

SRC Educational Science Travel Grants 5381
The intent is to foster links between the research field of Educational Science and international research, and to encourage mobility in Swedish higher education. To accomplish this, the Committee for Educational Science (CES) has allocated funds for conference and travel grants. Travel grants are intended first and foremost to enable post-doc researchers at Swedish seats of learning who have graduated not more than five years before the proposed travel to establish personal contacts with foreign researchers, and enjoy a short stay within a foreign research environment. CES also aims to promote opportunities for researchers, especially those with newly awarded PhDs, to stay briefly in other research environments in Sweden. Grants are provided primarily for researchers with recent doctorates at Swedish higher education institutions (HEIs), for the following purposes: attending congresses, conferences or symposia - provided that the researcher has received a personal invitation to offer an academic contribution; and spending brief periods (not exceeding one month) at other Swedish HEIs or universities abroad. Types of cost include travel costs, board and lodgings and specified conference expenses.
Requirements: The applicant with scientific responsibility must have a doctorate or corresponding skills.
Contact: Helene Sundewall, Research Administrator, Educational Sciences; +46-8-546-44-159; fax: +46-8-546-44-180; email: helene.sundewall@vr.se
Internet: http://www.vr.se/mainmenu/applyforgrants/callforproposals/grantsview.4.64a4ae3e1092b3475b6800063.html?resourceId=-1461&languageId=2#kontakt
Sponsor: Swedish Research Council
Klarabersgviadukten 82
Stockholm, Sweden SE-103 78

SRC Guest Professorship Program 5382
The Swedish Research Council provides funding for three Guest Professorships, as well as a professorship in the Swedish language at the University of Tartu, Estonia. The Olof Palme Guest Professorship is intended for research of importance for efforts to bring about peace. The Kerstin Hesselgren Guest Professorship, instituted in memory of Sweden's first female member of Parliament (the Riksdag), is awarded to an eminent woman researcher from outside of Sweden. The Guest Professorship for Artistic Development is intended to support arts and humanities programs. Deadlines vary, and applicants should contact the SRC for further details.
Contact: Berit Angusti, Senior Research Administrator; +46 (0)8 546 44 315; fax: +46-8-546-44-180; email: berit.angusti@vr.se
Internet: http://www.vr.se/mainmenu/organisation/aboutus/scientificcouncilforhumanitiesandsocialsciences.4.69f66a93108e85f68d4800011579.html;jsessionid=C3A13BBA053853B83ABC978C5F432BA5?contactPage=4.69f66a93108e85f68d4800011579&state=showCreateMessage&sv.url=12.20a
Sponsor: Swedish Research Council
Klarabersgviadukten 82
Stockholm, Sweden SE-103 78

SRC Humanities and Social Science Postdoctoral Fellowships 5383
These fellowships are designed to enable humanities and social science researchers with Swedish doctorates or exams from EUI (European University Institute) to stay at foreign universities or research establishments. The maximum fellowship period of 24 months is reduced by the length of time for which the applicant, since obtaining a doctorate, has held a postdoctoral position, a similar grant or a fellowship, at a higher education institution or faculty or company other than the one where the doctorate, or equivalent, was awarded. The proposed research project must be essentially different from the applicant's previous research. Essentially different means a new direction, new ways of thinking and/or new orientation of the research. The amount of the fellowship will be the standard grant rate, calculated according to costs in the host country. Applications to the Swedish Research Council must be made entirely electronically, and the application may be written in either Swedish or English.
Requirements: A Swedish doctorate or exam from EUI (European University Institute) is required. The doctorate must have been awarded within three years before and latest at the same day as the deadline for applications. Sick leave, service in the Armed Forces, doctor's employment as a preregistration house officer, work for trade unions and student organizations, parental leave and other circumstances of this kind are not counted in the three years. There is no age limit for applicants.
Date(s) Application is Due: Feb 20
Contact: Berit Angusti, Senior Research Administrator; +46 (0)8 546 44 315; fax: +46-8-546-44-180; email: berit.angusti@vr.se
Internet: http://www.vr.se/mainmenu/applyforgrants/callforproposals/grantsview.4.64a4ae3e1092b3475b6800063.html?resourceId=-1439&languageId=2
Sponsor: Swedish Research Council
Klarabersgviadukten 82
Stockholm, Sweden SE-103 78

SRC Humanities and Social Sciences Grants 5384
Researchers in this Program study the workings of society and the creation and dissemination of cultural values. A researcher can obtain a project research grant to perform a research task, as a rule for three years, to cover salary, materials, travel, publication and other expenses. Long-term support ‚Äì for up to six years ‚Äì is also given to leading researchers who are expected to increase, develop and convey knowledge in their research fields. Professors can apply for funds for sabbatical terms, i.e. remuneration for six months full-time research. Grants may also be provided for temporary research, junior research and doctoral student positions in fields assigned priority by the Scientific Council. Examples of the Research Council's international involvement are its collaboration within the EU Sixth Framework Programme, in the thematic area of Citizens and governance in a knowledge-based society.
Contact: Berit Angusti, Senior Research Administrator; +46 (0)8 546 44 315; fax: +46-8-546-44-180; email: berit.angusti@vr.se
Internet: http://www.vr.se/responsibilities/researchfunding/humanitiesandsocialsciences.4.69f66a93108e85f68d480001122.html
Sponsor: Swedish Research Council
Klarabersgviadukten 82
Stockholm, Sweden SE-103 78

SRC Humanities and Social Sciences Grants for Postdoctoral Positions In Sweden 5385

The grant will enable researchers with Swedish or non-Swedish doctorates (PhDs or equivalent) to work at Swedish higher education institutions or research establishments. The maximum fellowship period of 24 months is reduced by the length of time for which the applicant, since obtaining a doctorate, has held a postdoctoral position, a similar grant or a fellowship, at a higher education institution or faculty or company other than the one where the doctorate, or equivalent, was awarded. The host university/institution or equivalent is the employer and determines which form of employment the position will have.
Requirements: Individual researchers may apply. Swedish doctorate or foreign degree equivalent to a doctorate is required. The doctorate must have been awarded within three years before and latest at the same day as the deadline for applications. Sick leave, service in the Armed Forces, doctor's employment as a pre-registration house officer, work for trade unions and student organizations, parental leave and other circumstances of this kind are not included in the three years. There is no age limit for applicants to the Swedish Research Council.
Date(s) Application is Due: Feb 20
Contact: Berit Angusti, Senior Research Administrator; +46 (0)8 546 44 315; fax: +46-8-546-44-180; email: berit.angusti@vr.se
Internet: http://www.vr.se/mainmenu/applyforgrants/callforproposals/grantsview.4.64a4ae3e1092b3475b6800063.html?resourceId=-1387&languageId=2
Sponsor: Swedish Research Council
Klarabersgviadukten 82
Stockholm, Sweden SE-103 78

SRC IC Design Challenge 5386

The key objective of the Design Challenge is for university teams to create novel, high performance circuit designs that make end products more competitive. These products can be digital, analog, mixed-signal or wireless. Using the provided technology and design kit, successful contest teams will design circuits that clearly demonstrate potential for high performance for a target application offering advantages over existing designs. Specific circuitries of particular interest to the sponsors are LNAs, mixers, high speed I/Os, low power, high resolution graphics and thermal management; however the contest is not limited to these areas. Sponsors are looking for creativity, innovation, and the best usage of the given technology to implement an important new or improved circuit or circuit subsystem. A secondary objective is to assist faculty in stimulating greater interest in IC design careers among students, both graduate and undergraduate and from diverse populations. The contest will consist of two phases: Phase One, a design phase, and Phase Two, a fabrication and evaluation phase. The top three Phase One entries will win cash prizes for their university departments of $10,000, $5,000, and $3,000 for first, second, and third places. Five of the Phase One teams will be invited to prove their designs in the Jazz 180nm SiGe process. Fabricated chips will be packaged for evaluation by Quick Pak. The value of this Phase Two award is approximately $100K per team (includes library/design kits, mask and wafer fab, packaging and support). Judges will select three top winners from Phase Two to be awarded cash prizes of $25,000, $15,000, and $10,000 for first, second and third places.
Requirements: This Design Challenge is open to North American university teams composed of faculty and students. Multiple teams may enter the contest from each university. Teams may consist of graduate and undergraduate students and must have a faculty team leader; an assistant faculty team leader is optional. Teams are encouraged to have one or more undergraduates from diverse populations participate on the team with objective of creating interest in careers in microelectronics. Industry personnel may participate as consultants to the teams on a limited basis with approval from contest leaders.
Date(s) Application is Due: Sep 1
Contact: Contest Administrator; (919) 941-9400; fax: (919) 941-9450
Internet: http://www.src.org/ICcontest/default.asp
Sponsor: Semiconductor Research Corporation
P.O. Box 12053
Research Triangle Park, NC 27709-2053

SRC Medical Research Grants 5387

The Scientific Council for Medicine supports research throughout the discipline of Medical Science, i.e. Medicine, Odontology, Pharmaceutics and Health Sciences. Medical research supported by the Scientific Council ranges from studies at molecular and cell level, and research involving laboratory animal science and patients, to epidemiological studies on groups of human subjects. The Program offers grants for research and junior research positions; grants for research periods for clinical researchers; grants for research stays (sabbaticals) abroad; project research grants (distributed, as a rule, for three years for salary, materials, travel, publication, etc); international postdoctoral fellowships and grants for postdoctoral positions in Sweden; networks; symposium; research equipment collaboration grant and grant for inviting a visiting scientist.
Contact: Paola Norlin, Research Administrator; +46 (0)8 546 44 311; fax: +46-8-546-44-180; email: paola.norlin@vr.se
Internet: http://www.vr.se/responsibilities/researchfunding/medicine.4.69f66a93108e85f68d48000837.html
Sponsor: Swedish Research Council
Klarabersgviadukten 82
Stockholm, Sweden SE-103 78

SRC Medical Science Grants for Large International Conferences in Sweden 5388

The Council has decided to call for grant applications to arrange occasional, large international conferences addressing topics in medical science in Sweden. Applications must be received no later than May 30. Large international conferences play an important role in the research process since they provide a forum for researchers from different countries to disseminate research results and exchange research findings. For applicants to be granted economic support, the scientific goals and conference program must be judged to have high scientific priority. Furthermore, the scientific content of the conference should contribute to the advancement of an area supported and prioritized by the Council. The grants will be framework grants that may be used to cover expenses, e.g. for travel and per diem costs for participants, venue rental, meeting publications, and other costs relevant to the scientific program. The number of conference participants is unrestricted.
Requirements: The applicant should be the individual responsible for the scientific content of the conference and shall have a doctorate or corresponding qualifications. Other members of the organizing committee or initiators of the conference may be listed as associates. The application shall be written in English and submitted electronically via VR-Direct.
Amount of Grant: 1,000,000 SEK (Swedish kronor) maximum
Date(s) Application is Due: May 30
Contact: Agneta Axelsson, Executive Secretary, Research Administrator; +46 (0)8 546 44 275; fax: +46-8-546-44-180; email: agneta.axelsson@vr.se
Internet: http://www.vr.se/mainmenu/applyforgrants/callforproposals/grantsview.4.64a4ae3e1092b3475b6800063.html?resourceId=-1477&languageId=2
Sponsor: Swedish Research Council
Klarabersgviadukten 82
Stockholm, Sweden SE-103 78

SRC Medical Science Grants for Postdoctoral Positions in Sweden 5389

The grant will enable Medical Science researchers with Swedish or non-Swedish doctorates (PhDs or equivalent) to work at Swedish higher education institutions or research establishments. The maximum fellowship period of 24 months is reduced by the length of time for which the applicant, since obtaining a doctorate, has held a postdoctoral position, a similar grant or a fellowship, at a higher education institution or faculty or company other than the one where the doctorate, or equivalent, was awarded. The host university/institution or equivalent is the employer and determines which form of employment the position will have.
Requirements: Individual researchers may apply. Swedish doctorate or foreign degree equivalent to a doctorate is required. The doctorate must have been awarded within three years before and latest at the same day as the deadline for applications. Sick leave, service in the Armed Forces, doctor's employment as a pre-registration house officer, work for trade unions and student organizations, parental leave and other circumstances of this kind are not included in the three years. There is no age limit for applicants.
Date(s) Application is Due: Feb 20
Contact: Paola Norlin, Research Administrator; +46 (0)8 546 44 311; fax: +46-8-546-44-180; email: paola.norlin@vr.se
Internet: http://www.vr.se/mainmenu/applyforgrants/callforproposals/grantsview.4.64a4ae3e1092b3475b6800063.html?resourceId=-1389&languageId=2
Sponsor: Swedish Research Council
Klarabersgviadukten 82
Stockholm, Sweden SE-103 78

SRC Medical Science Postdoctoral Fellowships 5390

These fellowships are designed to enable medical science researchers with Swedish doctorates or exams from EUI (European University Institute) to stay at foreign universities or research establishments. The maximum fellowship period of 24 months is reduced by the length of time for which the applicant, since obtaining a doctorate, has held a postdoctoral position, a similar grant or a fellowship, at a higher education institution or faculty or company other than the one where the doctorate, or equivalent, was awarded. The proposed research project must be essentially different from the applicant's previous research. Essentially different means a new direction, new ways of thinking and/or new orientation of the research. The amount of the fellowship will be the standard grant rate, calculated according to costs in the host country. Applications to the Swedish Research Council must be made entirely electronically, and the application may be written in either Swedish or English.
Requirements: A Swedish doctorate or exam from EUI (European University Institute) is required. The doctorate must have been awarded within three years before and latest at the same day as the deadline for applications. Sick leave, service in the Armed Forces, doctor's employment as a preregistration house officer, work for trade unions and student organizations, parental leave and other circumstances of this kind are not counted in the three years. There is no age limit for applicants.
Date(s) Application is Due: Feb 20
Contact: Paola Norlin, Research Administrator; +46 (0)8 546 44 311; fax: +46-8-546-44-180; email: paola.norlin@vr.se
Internet: http://www.vr.se/mainmenu/applyforgrants/callforproposals/grantsview.4.64a4ae3e1092b3475b6800063.html?resourceId=-1442&languageId=2
Sponsor: Swedish Research Council
Klarabersgviadukten 82
Stockholm, Sweden SE-103 78

SRC Natural and Engineering Science Conference Grants 5391

The main purpose of the program is to enable eminent foreign lecturers to be invited to and attend international conferences. The conference program must be of high quality,

and the lecturer invited must be a researcher of international repute. An application for a conference grant may be submitted at any time of the year, though applications are not dealt with between 15 May and 15 August. There are no special forms for applications for conference grants. The maximum conference grant is SEK 100,000.
Requirements: Applicants must have obtained their PhDs by the date of the application. The conference must be international and to be organized in Sweden, another Nordic country or one of the Baltic countries by Swedish arrangers, and it must cover a research field that is supported, or could be supported, by the Swedish Research Council. An application for a conference grant must be received not later than two months before the conference is to take place.
Restrictions: Support for conferences is awarded only to cover costs (travel and lodging) for a limited number of invited foreign lecturers. Grants are not provided for other participants, for administration or for social-security contributions.
Amount of Grant: 100,000 SEK (Swedish kronor)
Contact: Mona Berggren, Research Administrator, Natural and Engineering Sciences; +46-8 546-44-246; fax: +46-8-546-44-180; email: mona.berggren@vr.se
Internet: http://www.vr.se/mainmenu/applyforgrants/callforproposals/grantsview.4.64a4ae3e1092b3475b6800063.html?resourceId=-357&languageId=2#kontakt
Sponsor: Swedish Research Council
Klarabersgviadukten 82
Stockholm, Sweden SE-103 78

SRC Natural and Engineering Science Grants for Postdoctoral Positions in Sweden **5392**

The grant will enable natural and engineering science researchers with Swedish or non-Swedish doctorates (PhDs or equivalent) to work at Swedish higher education institutions or research establishments. The maximum fellowship period of 24 months is reduced by the length of time for which the applicant, since obtaining a doctorate, has held a postdoctoral position, a similar grant or a fellowship, at a higher education institution or faculty or company other than the one where the doctorate, or equivalent, was awarded. The host university/institution or equivalent is the employer and determines which form of employment the position will have.
Requirements: Individual researchers may apply. Swedish doctorate or foreign degree equivalent to a doctorate is required. The doctorate must have been awarded within three years before and latest at the same day as the deadline for applications. Sick leave, service in the Armed Forces, doctor's employment as a pre-registration house officer, work for trade unions and student organizations, parental leave and other circumstances of this kind are not included in the three years. There is no age limit for applicants.
Date(s) Application is Due: Feb 20
Contact: Annika Johansen, Research Administrator; +46 (0)8 546 44 251; fax: +46-8-546-44-180; email: annika.johansen@vr.se
Internet: http://www.vr.se/mainmenu/applyforgrants/callforproposals/grantsview.4.64a4ae3e1092b3475b6800063.html?resourceId=-1390&languageId=2#kontakt
Sponsor: Swedish Research Council
Klarabersgviadukten 82
Stockholm, Sweden SE-103 78

SRC Natural and Engineering Science Postdoctoral Fellowships **5393**

These fellowships are designed to enable natural and engineering science researchers with Swedish doctorates or exams from EUI (European University Institute) to stay at foreign universities or research establishments. The maximum fellowship period of 24 months is reduced by the length of time for which the applicant, since obtaining a doctorate, has held a postdoctoral position, a similar grant or a fellowship, at a higher education institution or faculty or company other than the one where the doctorate, or equivalent, was awarded. The proposed research project must be essentially different from the applicant's previous research. Essentially different means a new direction, new ways of thinking and/or new orientation of the research. The amount of the fellowship will be the standard grant rate, calculated according to costs in the host country. Applications to the Swedish Research Council must be made entirely electronically.
Requirements: A Swedish doctorate or exam from EUI (European University Institute) is required. The doctorate must have been awarded within three years before and latest at the same day as the deadline for applications. Sick leave, service in the Armed Forces, doctor's employment as a preregistration house officer, work for trade unions and student organizations, parental leave and other circumstances of this kind are not counted in the three years. There is no age limit for applicants.
Date(s) Application is Due: Feb 20
Contact: Annika Johansen, Research Administrator; +46 (0)8 546 44 251; fax: +46-8-546-44-180; email: annika.johansen@vr.se
Internet: http://www.vr.se/mainmenu/applyforgrants/callforproposals/grantsview.4.64a4ae3e1092b3475b6800063.html?resourceId=-1443&languageId=2
Sponsor: Swedish Research Council
Klarabersgviadukten 82
Stockholm, Sweden SE-103 78

SRC Natural and Engineering Science Senior Research Positions in Sweden **5394**

With its Senior Research Positions in Sweden, the Council seeks to recruit senior researchers and renew Swedish research. This kind of position is intended as a step on the career ladder for an independent, established researcher at senior lecturer (Associate Professor/Docent) level, who is in an active research phase. In the Swedish system this phase is, as a rule, the period between a junior research position/forskarassistentanstallning and permanent employment as a senior university lecturer or professor. The primary obligation of appointees is to conduct research. In addition, these researchers must supervise doctoral students, assess the doctoral students' performance and teach, mainly in postgraduate studies. It is also possible to carry out research outside Sweden for brief spells during the grant period, provided this is done in accordance with the host higher education institution (HEI) and that the research is of benefit to Swedish research in the Natural and Engineering Sciences. Research periods abroad lasting longer than six months in any given three-year period must be approved by the Swedish Research Council. One researcher will be appointed for each position listed.
Requirements: A person who has obtained a doctorate (PhD) or has a non-Swedish degree deemed equivalent to a doctorate, for example a Candidate of Sciences in Russia, is eligible for a Senior Research Position. The doctorate must have been awarded within the ten years prior to the end of the application period. A person whose doctorate was awarded more than ten years previously may also be considered if special reasons apply. Special reasons are: leave of absence due to ill-health; service in the Armed Forces; and functions in trade unions and student organizations; parental leave.
Date(s) Application is Due: Feb 20
Contact: Annika Johansen, Research Administrator; +46 (0)8 546 44 251; fax: +46-8-546-44-180; email: annika.johansen@vr.se
Internet: http://www.vr.se/mainmenu/applyforgrants/callforproposals/grantsview.4.64a4ae3e1092b3475b6800063.html?resourceId=-1378&languageId=2
Sponsor: Swedish Research Council
Klarabersgviadukten 82
Stockholm, Sweden SE-103 78

SRC Natural and Engineering Science Travel Grants **5395**

Travel grants are available for natural science and engineering science researchers both domestic and abroad. Grants for researchers' attendance at congresses, conferences or symposia are allowable, provided that the researcher concerned has received a personal invitation by the arrangers to hold a lecturer, and if the meeting is deemed to be of obvious interest to Swedish Natural and/or Engineering Sciences. An application for a travel grant must be received not later than two months before the date of departure. Applications for grants below 6,000 SEK will not be considered, and no applications are dealt with between May 15 and August 15.
Requirements: Doctoral students can obtain grants only in exceptional cases. This does not apply to travel relating to observations or experiments at facilities with which the Swedish Research Council has special agreements.
Restrictions: Poster presentations are not approved. Grants are not provided for accompanying family members.
Amount of Grant: 6,000 SEK minimum (Swedish kronors)
Contact: Mona Berggren, Research Administrator, Natural and Engineering Sciences; +46-8 546-44-246; fax: +46-8-546-44-180; email: mona.berggren@vr.se
Internet: http://www.vr.se/mainmenu/applyforgrants/callforproposals/grantsview.4.64a4ae3e1092b3475b6800063.html?resourceId=-857&languageId=2#kontakt
Sponsor: Swedish Research Council
Klarabersgviadukten 82
Stockholm, Sweden SE-103 78

SRC Natural and Engineering Sciences Grants **5396**

The Scientific Council for Natural and Engineering Sciences is an expert body that takes decisions on financial support for Swedish basic research. Members of the Council are appointed by the research community and the Government. Their priority ranking of applications is based on assessment of the quality of the research project and the applicant's skill. Research for which support is provided may be conducted throughout the spectrum of basic research in the natural and engineering sciences. Examples are: the Earth and our climate; biology and biotechnology; biodiversity; chemistry and chemical engineering; physics and applied physics; electronics; system engineering; astronomy; materials science; computer science; and mechanical engineering.
Amount of Grant: 600,000 SEK (Swedish kronor) average
Contact: Mona Berggren, Research Administrator; +46 (0)8 546 44 246; fax: +46-8-546-44-180; email: mona.berggren@vr.se
Internet: http://www.vr.se/responsibilities/researchfunding/naturalandengineeringsciences.4.69f66a93108e85f68d48000652.html
Sponsor: Swedish Research Council
Klarabersgviadukten 82
Stockholm, Sweden SE-103 78

SRC Research Infrastructures Grants **5397**

The Program supports the establishment and use of infrastructure for Swedish research of the highest scientific quality. Research infrastructures include, for example, research installations and facilities for register research or extensive data networks, used nationally or internationally by several research groups, that are required for high-quality research in several research fields. The Program processes grant applications for: expensive scientific equipment; large databases; planning; and operations. Four review groups deal with applications within the Program's sphere of responsibility. These are the groups for: infrastructure for astronomy and subatomic research; infrastructure for molecular, cell and material research; infrastructure for research on the Earth and its near surroundings; and infrastructure for e-Science.
Amount of Grant: 650,000,000 SEK (Swedish kronor) total
Contact: Finn Karlsson, Head of Division; +46 (0)8 546 44 247; fax: +46-8-546-44-180; email: finn.karlsson@vr.se
Internet: http://www.vr.se/responsibilities/researchfunding/researchinfrastructures.4.69f66a93108e85f68d480001169.html
Sponsor: Swedish Research Council
Klarabersgviadukten 82
Stockholm, Sweden SE-103 78

SRC Simon Karecki Fellowship 5398

The Fellowship will be awarded to a doctoral student who is performing or is planning to perform research through the Center. The Fellowship will be awarded competitively on the basis of quality, relevance, and timeliness of the proposed research; demonstrated academic and leadership ability, and the appropriateness of the institutional environment for the research. The Fellowship will be awarded annually; however, the current recipient may re-compete for follow-on funding. Selection of the Fellowship recipient will be determined by consensus of the Governing Board. The Awards are based on academic excellence, demonstrated concern for the Earth, environment including significant contribution to research in environmentally benign semiconductor manufacturing, demonstrated ability for technical communication, and demonstrated ability to work with others and foster cooperation.

Contact: Fellowship Administrator, (919) 941-9400; fax: (919) 941-9450; email: srcea@src.org

Internet: http://srcea.src.org/programs/karecki/default.asp

Sponsor: Semiconductor Research Corporation
P.O. Box 12053
Research Triangle Park, NC 27709-2053

SSHRC Aboriginal Research: Development Grants 5399

The primary purpose of the program is to help teams of Aboriginal community organizations and university-based researchers develop research partnerships and proposals to investigate issues of concern to Aboriginal peoples. Up to $25,000 is available per applicant for up to 2 years.

Requirements: Eligible institutions must have active degree-granting status in the SSHRC fields at the undergraduate level or beyond, hold membership in the Association of Universities and Colleges of Canada, have fewer than 250 full-time faculty in the social sciences and humanities, and be independent of the federal government for the purpose of faculty employment status.

Amount of Grant: $25,000 maximum for up to 2 years

Contact: Garry Pinard, Program Officer; (613) 992-5129; fax: (613) 947-0223; email: garry.pinard@sshrc.ca

Internet: http://www.sshrc.ca/web/apply/program_index_e.asp

Sponsor: Social Sciences and Humanities Research Council of Canada
350 Albert Street, P.O. Box 1610
Ottawa, ON K1P 6G4 Canada

SSHRC Aboriginal Research: Research Grants 5400

The primary intent of the program is to support university-based researchers and Aboriginal community organizations to conduct research on issues of concern to Aboriginal peoples. Up to $100,000 is available annually to each successful applicant, with a maximum of $250,000 to be granted over a 3 year period.

Requirements: Eligible institutions must have active degree-granting status in the SSHRC fields at the undergraduate level or beyond, hold membership in the Association of Universities and Colleges of Canada, have fewer than 250 full-time faculty in the social sciences and humanities, and be independent of the federal government for the purpose of faculty employment status.

Amount of Grant: $100,000 annually; $250,000 over 3 years

Contact: Garry Pinard, Program Officer; (613) 992-5129; fax: (613) 947-0223; email: garry.pinard@sshrc.ca

Internet: http://www.sshrc.ca/web/apply/program_index_e.asp

Sponsor: Social Sciences and Humanities Research Council of Canada
350 Albert Street, P.O. Box 1610
Ottawa, ON K1P 6G4 Canada

SSHRC Aid to Research Workshops and Conferences in Canada 5401

Grants are available to help defray the costs of organizing workshops and conferences. Eligible expenses include administrative costs, travel and subsistence fees (but not honoraria), and the publication of workshop and conference proceedings. The value and period of tenure of a workshop grant is up to $25,000 for a workshop held within 12 months following the announcement of competition results. The values and periods of tenure of conference grants are: up to $20,000 for a conference of up to 200 participants that is held within 12 months following the announcement of competition results; up to $35,000 for a conference of 201 to 400 participants that is held within 12 months following the announcement of competition results; and up to $50,000 for a conference of more than 400 participants that is held within the 12 months following the announcement of competition results. There are two competitions each year.

Amount of Grant: $50,000 maximum

Date(s) Application is Due: May 1; Nov 1

Contact: Bryde Kelly, Program Officer; (613) 947-2060; fax: (613) 992-7635; email: bryde.kelly@sshrc.ca

Internet: http://www.sshrc.ca/web/apply/program_descriptions/conferences_e.asp

Sponsor: Social Sciences and Humanities Research Council of Canada
350 Albert Street, P.O. Box 1610
Ottawa, ON K1P 6G4 Canada

SSHRC Aid to Scholarly Journals 5402

The objective of this program is to increase access to, and the readership of, original research results in the social sciences and humanities through Canadian scholarly journals. Through the Aid to Scholarly Journals program, SSHRC will contribute to the broad dissemination and impact of original research results in social science and humanities scholarship via Canadian scholarly journals. The program will also assist journals as they seek to take advantage of advances in communication technologies. Funds will be awarded to help defray the costs of publishing scholarly articles, to assist with distribution costs, and to support journal organizations in their transitions to digital media. Eligible expenses include costs related to: management of the peer-review process (including honoraria provided to support staff); editing (including staff salaries, release time and travel expenses); computer hardware and software; copy preparation (including typesetting, copy-editing and translation); document layout (including image presentation and converting images to digital formats); marketing and other promotional activities; fees paid to electronic-publication service providers; printing and mailing; technical assistance; and technological training for journal staff. Grants are tenable for three years. The maximum value of a grant is $30,000 per annum.

Requirements: To be eligible, a journal must have as its main objective the publishing of the results of advanced research and scholarly work by specialists for a readership of specialists or practitioners, focusing primarily on a discipline or disciplines in the social sciences and humanities.

Amount of Grant: $30,000 maximum annually

Date(s) Application is Due: Jun 30

Contact: Michele Dupuis, Program Officer; (613) 947-2045; fax: (613) 992-7635; email: michele.dupuis@sshrc.ca

Internet: http://www.sshrc.ca/web/apply/program_descriptions/scholarly_journals_e.asp

Sponsor: Social Sciences and Humanities Research Council of Canada
350 Albert Street, P.O. Box 1610
Ottawa, ON K1P 6G4 Canada

SSHRC Aid to Scholarly Publications (ASPP) 5403

The council's objective of sharing research results in the social sciences and humanities as widely as possible is supported through the ASPP, which subsidizes the publication of individual works that make important contributions to the advancement of knowledge in these fields. This program is administered by the Humanities and Social Sciences Federation of Canada (HSSFC). More information may be obtained by writing to the HSSFC at 151 Slater Street, 4th Floor, Ottawa, ON K1P 5H3; or by contacting the Aid to Scholarly Publications Programme at (613) 238-6112 ext 350; email: secaspp@aspp.hssfc.ca; or Web site: http://www.fedcan.ca/english/aspp/aspp.html.

Amount of Grant: $C10,000 for Special Initiative grant; $C30,000 maximum per year for a general grant

Contact: Michele Dupuis, Program Officer; (613) 947-2045; fax: (613) 992-7635; email: michele.dupuis@sshrc.ca

Internet: http://www.sshrc.ca/web/apply/program_index_e.asp

Sponsor: Social Sciences and Humanities Research Council of Canada
350 Albert Street, P.O. Box 1610
Ottawa, ON K1P 6G4 Canada

SSHRC Aid to Small Universities 5404

The objective of this program is to enable small universities to develop and strengthen focused research capacity in the social sciences and humanities. Examples of activities that the ASU program supports: start-up costs/partial funding of research centers; stipends to doctoral students, provided the program of studies is related to the ASU grant; agenda-setting seminars; visiting scholars (travel and stipend); organization of colloquia or symposia; and seed funding for collaborative research or the development of partnerships.

Requirements: Eligible institutions must have active degree-granting status in the SSHRC fields at the undergraduate level or beyond, hold membership in the Association of Universities and Colleges of Canada, have fewer than 250 full-time faculty in the social sciences and humanities, and be independent of the federal government for the purpose of faculty employment status.

Amount of Grant: $C30,000 maximum per year for up to 3 years

Date(s) Application is Due: Dec 1

Contact: Fellowships and Institutional Grants Division; (613) 943-7777; fax: (613) 943-1329; email: fellowships@sshrc.ca

Internet: http://www.sshrc.ca/web/apply/program_descriptions/aid_small_universities_e.asp

Sponsor: Social Sciences and Humanities Research Council of Canada
350 Albert Street, P.O. Box 1610
Ottawa, ON K1P 6G4 Canada

SSHRC Aileen D. Ross Fellowship 5405

The Aileen D. Ross Fellowship is awarded to an outstanding SSHRC Doctoral award or Postdoctoral Fellowship holder conducting research in sociology, specifically for research on poverty. No specific application form is required; SSHRC will inform successful Postdoctoral or Doctoral Fellowship recipients. For doctoral level applicants registered at a Canadian university, the deadline is set by the university; for all others, it is November 13. For the postdoctoral level, the deadline is October 24.

Requirements: To be eligible, fellowship recipients must: be a citizen or permanent resident of Canada; have earned a first doctorate from a recognized university no earlier than October 2, 2005 (date all degree requirements were completed, not the date of graduation) or have completed a first doctorate no earlier than October 2, 2002, but have had their career interrupted or delayed for the purpose of child rearing; not hold a tenure or tenure-track faculty position; have finalized arrangements for affiliation with a recognized university or research institution; have applied not more than twice before for a SSHRC Postdoctoral Fellowship.

Amount of Grant: $10,000 in addition to the value of the fellowship

Date(s) Application is Due: Oct 24; Nov 13

Contact: Fellowships and Institutional Grants Division; (613) 943-7777; fax: (613) 943-1329; email: fellowships@sshrc.ca
Internet: http://www.sshrc.ca/web/apply/program_index_e.asp
Sponsor: Social Sciences and Humanities Research Council of Canada
350 Albert Street, P.O. Box 1610
Ottawa, ON K1P 6G4 Canada

SSHRC Bora Laskin National Fellowship in Human Rights Research 5406
The purpose of this fellowship is to support research, preferably of a multidisciplinary or interdisciplinary nature, and to develop Canadian expertise in the field of human rights, with emphasis on themes and issues relevant to the Canadian human rights scene. The Fellowship supports research in all areas of humanities and social sciences, including journalism and law. Preference is given to multidisciplinary or interdisciplinary research projects. One non-renewable fellowship, tenable for 12 months in Canada, is awarded every year. The award consists of a $45,000 stipend plus $10,000 for research and research-related travel expenses.
Requirements: Applicants must be affiliated with a Canadian postsecondary institution. Preference will be given to applicants with at least five years of proven research experience in their field, as this Fellowship is not intended for scholars just beginning their research careers. Accordingly, this award may not be used to complete the requirements for a degree program. Please note that applicants, and anyone else participating in any capacity in the proposed project, must not be under SSHRC sanction for financial or research misconduct.
Amount of Grant: $C45,000 plus $C10,000 research and travel costs
Date(s) Application is Due: Oct 1
Contact: Mylene Seguin, Program Officer; (613) 992-5358; fax: (613) 943-1329; email: Mylene.Seguin@sshrc.ca
Internet: http://www.sshrc.ca/web/apply/program_descriptions/fellowships/bora_laskin_e.asp
Sponsor: Social Sciences and Humanities Research Council of Canada
350 Albert Street, P.O. Box 1610
Ottawa, ON K1P 6G4 Canada

SSHRC Canadian Environmental Issues: Community University Research Alliances (CURA) Grants 5407
The purpose of the program is to support the creation of alliances between community organizations and postsecondary institutions which, through a process of ongoing collaboration and mutual learning, will foster innovative research, training and the creation of new knowledge on topics relating to SSHRC's Canadian Environmental Issues special call. Specific objectives of this call are to: promote sharing of knowledge, resources and expertise between postsecondary institutions and organizations in the community; enrich research, teaching methods and curricula in postsecondary institutions; reinforce community decision-making and problem-solving capacity; and enhance students' education and employability by means of diverse opportunities to build their knowledge, expertise and work skills through hands-on research and related experience. Eligible applicants must submit, jointly with their partners, a Letter of Intent application form by the November 20 due date. Applicants whose Letters of Intent are approved by the selection committee will be invited to submit, again jointly with their partners, a Formal Application.
Requirements: Applications must be submitted jointly by one or more postsecondary institutions and one or more organizations from the community. A community may refer to either a geographic focus or an issue/interest focus.
Amount of Grant: $200,000 maximum
Date(s) Application is Due: Nov 20
Contact: Gail Zboch, Program Officer; (613) 943-1148; fax: (613) 947-0223; email: gail.zboch@sshrc.ca
Internet: http://www.sshrc.ca/web/apply/program_descriptions/environment_cura_e.asp
Sponsor: Social Sciences and Humanities Research Council of Canada
350 Albert Street, P.O. Box 1610
Ottawa, ON K1P 6G4 Canada

SSHRC Canadian Environmental Issues: Public Outreach Grants 5408
Public Outreach Grants for the Canadian Environmental Issues special call are designed to mobilize existing and ongoing research results on topics relating to the Canadian Environmental Issues special call. In this area, both Dissemination and Workshop/Conference grants are available. Public Outreach grants will be tenable for one year only. Grants have no maximum value. Visit the website for further details.
Requirements: Applications may be submitted by individual researchers or teams of researchers affiliated with Canadian postsecondary institutions. SSHRC encourages those researchers not affiliated with a postsecondary institution to participate as collaborators.
Date(s) Application is Due: Nov 17
Contact: Michele Dupuis, Program Officer; (613) 947-2045; fax: (613) 992-7635; email: michele.dupuis@sshrc.ca
Internet: http://www.sshrc.ca/web/apply/program_descriptions/environment_po_e.asp
Sponsor: Social Sciences and Humanities Research Council of Canada
350 Albert Street, P.O. Box 1610
Ottawa, ON K1P 6G4 Canada

SSHRC Canadian Initiative on Social Statistics Data Training Schools Grants 5409
The Data Training Schools (DTS) serves postsecondary institution researchers, graduate and undergraduate students, policy analysts and data librarians. DTS courses cover topics in quantitative methods and the use of Canadian micro-data. The DTS initiative contributes to increasing the number of researchers engaged in quantitative research. Data Training Schools offer short courses in social data analysis, which are adapted to Canadian data and research concerns. These include a series of intermediate level courses in data analysis as well as more advanced courses in the analysis of longitudinal and/or multilevel data. The specific objectives of the DTS courses are to: increase participants' awareness of, and familiarity with, currently available Statistics Canada data and surveys; teach participants how to use and apply specific techniques to real data, which means they can manage data, scale variables, create composite variables, handle missing data, use design weights and use more complex statistical methods appropriate to the analysis of longitudinal or multilevel data sets; build expertise that focuses on Canadian issues and Canadian data; and aid in the identification of policy questions that can be illuminated through quantitative research. Eligible expenses include: travel and accommodation for resource personnel and course attendees (excluding Statistics Canada personnel); administrative expenses related to the organization of the DTS; and fees paid to main speakers and course instructors, but not to attendees or Statistics Canada personnel. The program offers three awards. Each is valued at up to $50,000 annually for three years.
Requirements: A DTS must be hosted in a Canadian postsecondary institution for no less than seven days. Sessions should be intensive and focus on formats that encourage the active contribution of all participants.
Amount of Grant: $50,00 annually for three years
Date(s) Application is Due: Dec 7
Contact: Garry Pinard, Program Officer; (613) 992-5129; fax: (613) 947-0223; email: garry.pinard@sshrc.ca
Internet: http://www.sshrc.ca/web/apply/program_descriptions/ciss_data_training_e.asp
Sponsor: Social Sciences and Humanities Research Council of Canada
350 Albert Street, P.O. Box 1610
Ottawa, ON K1P 6G4 Canada

SSHRC Doctoral Fellowships 5410
Doctoral fellowships are offered to assist in developing research and scholarly resources in the arts, social sciences, and humanities. Fellowships are for two years, renewable, tenable in Canada or abroad. The deadline for applicants registered at Canadian universities is in the fall on the date set by the head of the department (contact faculties of graduate studies). For other applicants, the deadline is November 13. For information on college and university campuses, contact the council or the dean of graduate studies, the officer of research administration, the registrar, or the student awards officer.
Requirements: Candidates must be Canadian citizens or have permanent resident (landed immigrant) status at the time of application and must have completed one year of doctoral study or a master's degree by the date of taking the award.
Amount of Grant: $20,000 per annum
Date(s) Application is Due: Nov 13
Contact: Fellowships and Institutional Grants Division; (613) 943-7777; fax: (613) 943-1329; email: fellowships@sshrc.ca
Internet: http://www.sshrc.ca/web/apply/program_descriptions/fellowships/doctoral_e.asp
Sponsor: Social Sciences and Humanities Research Council of Canada
350 Albert Street, P.O. Box 1610
Ottawa, ON K1P 6G4 Canada

SSHRC Image, Text, Sound and Technology (ITST) Grants 5411
The program will provide funding for researchers and their graduate students to develop expertise in applying advanced technology through summer institutes, workshops and conferences. The ITST program will help researchers refine their ability to apply digital multimedia technologies to their research on image, text and sound. The overall objectives of the ITST program are to: reflect on, interpret, and analyze new digital media, multimedia, and text-based computing technologies and integrate these into humanities and social sciences research; bring together theorists, experimentalists, and technologists from different disciplines, to share and nurture ideas and methods that challenge research to advance through the use of audio-visual and text-based technologies; and facilitate the creation of national and international networks of, and partnerships among, researchers, industries, governments, and individuals that will promote and sustain social sciences and humanities research and resources worldwide. Up to $50,000 is available for twelve months.
Requirements: Canadian postsecondary institutions are eligible to apply. The summer institute, workshop or conference must be held at the applying institution. Detailed instructions accompany the application form.
Amount of Grant: $50,000 maximum
Date(s) Application is Due: Sep 15
Contact: Garry Pinard, Program Officer; (613) 992-5129; fax: (613) 947-0223; email: garry.pinard@sshrc.ca
Internet: http://www.sshrc.ca/web/apply/program_descriptions/itst/workshops_e.asp
Sponsor: Social Sciences and Humanities Research Council of Canada
350 Albert Street, P.O. Box 1610
Ottawa, ON K1P 6G4 Canada

SSHRC Institutional Grants 5412
Under this program, SSHRC helps eligible Canadian postsecondary institutions fund small-scale research activities by their faculty in the social sciences and humanities.

The purpose of this program is to assist Canadian universities to: develop, increase or strengthen research excellence in the social sciences and humanities; assist, in particular, researchers embarking on their research career to become competitive in grant competitions at the national level; assist established researchers with modest funding requirements or those wishing to reorient and strengthen their research programs; and support national and international dissemination and collaboration.
Requirements: To be eligible for this program, an institution must: be an institutional member of the AUCC (including all private universities that are institutional members of the AUCC); have active degree-granting status in at least one of the social sciences and humanities disciplines or be an institutional member of the AUCC and be affiliated or federated with an eligible institution, but grant their own degrees, receive their own operating budget directly from the provincial government and have their own board of directors; and offer programs in at least one of the social sciences and humanities disciplines.
Amount of Grant: $C5,000 minimum per year
Date(s) Application is Due: Dec 1
Contact: Fellowships and Institutional Grants Division; (613) 943-7777; fax: (613) 943-1329; email: fellowships@sshrc.ca
Internet: http://www.sshrc.ca/web/apply/program_descriptions/institutional_grants_e.asp
Sponsor: Social Sciences and Humanities Research Council of Canada
350 Albert Street, P.O. Box 1610
Ottawa, ON K1P 6G4 Canada

SSHRC International Opportunities Fund Grants **5413**
The International Opportunities Fund (IOF) program will help researchers develop, participate in, and lead diverse international collaborative research activities. The objectives of the program are to: help Canadian researchers to initiate and develop international research collaborations; and facilitate Canadian participation or leadership in current or planned international research initiatives that offer outstanding opportunities to advance Canadian research. Development Grants are worth up to $25,000 for one year. Project Grants are worth up to $75,000 for one year. For each grant type, SSHRC will consider extensions for up to one additional year on a case-by-case basis.
Requirements: SSHRC invites applications from individual researchers or research teams.
Amount of Grant: $25,000 development grants; $75,000 project grants
Date(s) Application is Due: May 8; Nov 7
Contact: Mika Oehling, Program Officer; (613) 992-4227; fax: (613) 947-0223; email: mika.oehling@sshrc.ca
Internet: http://www.sshrc.ca/web/apply/program_descriptions/iof_e.asp
Sponsor: Social Sciences and Humanities Research Council of Canada
350 Albert Street, P.O. Box 1610
Ottawa, ON K1P 6G4 Canada

SSHRC Joseph-Armand Bombardier CGS Doctoral Scholarships **5414**
The J.-Armand Bombardier CGS Doctoral Scholarships programs aim to develop research skills and assist in the training of highly-qualified personnel by supporting students who demonstrate a high standard of scholarly achievement in undergraduate and graduate studies in the social sciences and humanities. Scholarships are valued at $35,000 per annum for 36 months.
Requirements: To qualify to apply for the J.-Armand Bombardier CGS Doctoral Awards programs, you must: be a citizen or permanent resident of Canada; be applying for support to pursue your first PhD (or equivalent); be pursuing doctoral studies (or a combined MA/PhD) in the social sciences or humanities; not have already received a scholarship or fellowship from SSHRC, NSERC or CIHR to undertake or complete a doctoral degree or combined MA/PhD; not be applying in the upcoming academic year to NSERC or CIHR; or not be pursuing a degree program that combines undergraduate and graduate degrees.
Amount of Grant: $35,000 per annum
Date(s) Application is Due: Nov 13
Contact: Fellowships and Institutional Grants Division; (613) 943-7777; fax: (613) 943-1329; email: fellowships@sshrc.ca
Internet: http://www.sshrc.ca/web/apply/program_descriptions/fellowships/doctoral_e.asp
Sponsor: Social Sciences and Humanities Research Council of Canada
350 Albert Street, P.O. Box 1610
Ottawa, ON K1P 6G4 Canada

SSHRC Jules and Gabrielle Leger Fellowships **5415**
This fellowship is offered in odd-numbered years to assist in developing research and scholarly resources in the social sciences and humanities. The fellowship is awarded for research and writing on the historical and contemporary contribution of the Crown and its representatives, federal and provincial; and the political, constitutional, cultural intellectual, and social life of the country. For further information on university campuses, contact the dean of graduate studies, the officer of research administration, the registrar, or the student awards officer; or contact the council.
Requirements: Applicants must be Canadian citizens or permanent residents of demonstrated scholarly competence who will be able to devote their full time for at least eight months. The fellowship is tenable at a recognized university or research institute.
Amount of Grant: $C40,000 plus $C10,000 for research and travel costs
Date(s) Application is Due: Oct 1
Contact: Mylene Seguin, Program Officer; (613) 992-5358; fax: (613) 943-1329; email: Mylene.Seguin@sshrc.ca
Internet: http://www.sshrc.ca/web/apply/program_descriptions/fellowships/leger_e.asp
Sponsor: Social Sciences and Humanities Research Council of Canada
350 Albert Street, P.O. Box 1610
Ottawa, ON K1P 6G4 Canada

SSHRC Knowledge Impact in Society Grants **5416**
The broad objective of the KIS pilot program is to explore the potential of an institutional funding model to support effective knowledge exchange and mobilization. The KIS pilot program will fund post-secondary institution-based strategic knowledge mobilization initiatives that systematically enable non-postsecondary institution stakeholder communities to benefit from existing academic research knowledge in the social sciences and humanities. SSHRC therefore invites proposals from postsecondary institutions to undertake strategic knowledge mobilization initiatives,Âisystematic and sustained activities or sets of complementary activities that: draw on accumulated social sciences and humanities expertise in areas where an institution has demonstrated research strengths; create or expand relationships and knowledge exchange between social sciences and humanities researchers and non-academic stakeholders; and facilitate knowledge impact for the broader common good.
Requirements: Eligible institutions must have active degree-granting status in the SSHRC fields at the undergraduate level or beyond, hold membership in the Association of Universities and Colleges of Canada, have fewer than 250 full-time faculty in the social sciences and humanities, and be independent of the federal government for the purpose of faculty employment status. This program offers institutional grants, which are available only to postsecondary institutions. Individual postsecondary institutions may apply, or two or more may submit a joint application. Letters of Interest are due on December 16, and full proposals should arrive by January 27.
Amount of Grant: $100,000 maximum
Date(s) Application is Due: Dec 16; Jan 27
Contact: Michele Dupuis, Program Officer; (613) 947-2045; fax: (613) 992-7635; email: michele.dupuis@sshrc.ca
Internet: http://www.sshrc.ca/web/apply/program_descriptions/knowledge_impact_e.asp
Sponsor: Social Sciences and Humanities Research Council of Canada
350 Albert Street, P.O. Box 1610
Ottawa, ON K1P 6G4 Canada

SSHRC Major Collaborative Research Initiatives **5417**
Grants are awarded for up to seven years to carry out large-scale collaborative research. Collaboration is defined as an ongoing, active, and substantial relationship between two or more researchers, in the same or different disciplines and/or institutions, who bring different perspectives to bear on major research projects requiring long-term funding. Partnerships between universities and private and public sector groups are encouraged. Requests for renewal of support for an additional period of up to five years may be considered. Eligible proposals must be of such a scope that they require a minimum budget of $100,000 for each of the seven years. The maximum budget is $2.5 million over seven years. Letters of Intent are due January 31, with full proposals due by September 1.
Requirements: Research partners are expected to contribute in cash or in kind.
Amount of Grant: $700,000-$2,500,000 over seven years
Date(s) Application is Due: Jan 31; Sep 1
Contact: Brenda Werrell, MCRI Program Coordinator; (613) 947-9659; fax: 613-992-7635; email: brenda.werrell@sshrc.ca
Internet: http://www.sshrc.ca/web/apply/program_descriptions/mcri_e.asp
Sponsor: Social Sciences and Humanities Research Council of Canada
350 Albert Street, P.O. Box 1610
Ottawa, ON K1P 6G4 Canada

SSHRC Management, Business and Finance Grants **5418**
The overall purpose of the National Forum on Management, Business and Finance is to stimulate an interdisciplinary and multi-sectoral national dialogue around innovative research and training in management, business and finance. This forum will communicate the results and impacts of research in these areas in Canada over the past decade and will identify future research directions and opportunities. Grants are available under three categories in this program: National Forum; Public Outreach Dissemination; and Public Outreach Workshops and Conferences. See the website for specific details.
Requirements: Applications may be submitted by an eligible institution. Applications may also be submitted by a consortium of institutions, as long as the organization administering the funds is eligible.
Amount of Grant: $400,000 maximum
Date(s) Application is Due: Oct 1; Oct 6
Contact: Eric Bastien, Senior Program Officer; (613) 992-3407; fax: (613) 947-0223; email: eric.bastien@sshrc.ca
Internet: http://www.sshrc.ca/web/apply/program_index_e.asp
Sponsor: Social Sciences and Humanities Research Council of Canada
350 Albert Street, P.O. Box 1610
Ottawa, ON K1P 6G4 Canada

SSHRC Postdoctoral Fellowships **5419**
The Postdoctoral Fellowships program supports the most promising new scholars in the social sciences and humanities and assists them in establishing a research base at an important time in their research careers. Postdoctoral Fellowships provide stipendiary support to recent PhD graduates who are: undertaking original research; publishing

research findings; developing and expanding personal research networks; broadening teaching experience; and preparing to become competitive in national research grants competitions. Posdoctoral Fellowships are valued at approximately $38,000 per year, plus an accountable research allowance of up to $5,000 per award. These are non-renewable fellowships, tenable for a minimum of 12 months and a maximum of 24 months. Fellowships will normally be awarded to candidates affiliated with a university other than that which awarded the PhD.
Requirements: To apply to this program, applicants must: be a citizen or permanent resident of Canada; have earned a first doctorate from a recognized university no earlier than October 2, 2005 (date all degree requirements were completed, not the date of graduation) or have completed a first doctorate no earlier than October 2, 2002, but have had their career interrupted or delayed for the purpose of child rearing; not hold a tenure or tenure-track faculty position; have finalized arrangements for affiliation with a recognized university or research institution; have applied not more than twice before for a SSHRC Postdoctoral Fellowship.
Amount of Grant: $38,000 per year, plus an accountable research allowance of up to $5,000 per award
Date(s) Application is Due: Oct 24
Contact: Fellowships and Institutional Grants Division; (613) 943-7777; fax: (613) 943-1329; email: fellowships@sshrc.ca
Internet: http://www.sshrc.ca/web/apply/program_descriptions/fellowships/postdoctoral_e.asp
Sponsor: Social Sciences and Humanities Research Council of Canada
350 Albert Street, P.O. Box 1610
Ottawa, ON K1P 6G4 Canada

SSHRC Postdoctoral Prize **5420**
The SSHRC Postdoctoral Prize is an annual award for the most outstanding SSHRC Postdoctoral Fellowship recipient. $10,000 is given in addition to the value of the postdoctoral fellowship. There is no particular application form, and SSHRC will inform the successful postdoctoral fellowship recipient.
Requirements: To apply to this program, applicants must: be a citizen or permanent resident of Canada; have earned a first doctorate from a recognized university no earlier than October 2, 2005 (date all degree requirements were completed, not the date of graduation) or have completed a first doctorate no earlier than October 2, 2002, but have had their career interrupted or delayed for the purpose of child rearing; not hold a tenure or tenure-track faculty position; have finalized arrangements for affiliation with a recognized university or research institution; have applied not more than twice before for a SSHRC Postdoctoral Fellowship.
Amount of Grant: $10,000 in addition to the value of the postdoctoral fellowship
Contact: Fellowships and Institutional Grants Division; (613) 943-7777; fax: (613) 943-1329; email: fellowships@sshrc.ca
Internet: http://www.sshrc.ca/web/apply/program_index_e.asp
Sponsor: Social Sciences and Humanities Research Council of Canada
350 Albert Street, P.O. Box 1610
Ottawa, ON K1P 6G4 Canada

SSHRC Queen's Fellowships **5421**
Awards assist in developing research and scholarly resources in the social sciences and humanities. Recipients of the fellowships are selected from successful doctoral candidates. Tuition fees and travel allowances in addition to the value of the doctoral fellowship will be provided. For further information on university campuses, contact the dean of graduate studies, the officer of research administration, the registrar, or the student awards officer; or contact the council. Tuition fees and allowances for travel to and from the main place of tenure and for research purposes under certain conditions are provided.
Requirements: Fellowships are open to students who are Canadian citizens or permanent residents by the deadline date.
Amount of Grant: $C15,000
Contact: Fellowships and Institutional Grants Division; (613) 992-0691 or (613) 943-7777; fax: (613) 992-1787; email: fellowships@sshrc.ca
Internet: http://www.sshrc.ca/web/apply/program_index_e.asp
Sponsor: Social Sciences and Humanities Research Council of Canada
350 Albert Street, P.O. Box 1610
Ottawa, ON K1P 6G4 Canada

SSHRC Research Development Initiatives **5422**
The overall objective of the RDI program is to support research in its initial stages by supporting the development of new ways of analyzing, structuring, integrating and transferring knowledge in the humanities and the social sciences. The specific objectives of the program are to help researchers to: develop new research questions; explore conceptual and methodological perspectives and directions; and critically analyze and assess research, including its achievements, impacts, strengths and state of development. The maximum value of an RDI grant is $40,000 over two years.
Requirements: Applicants may be individual researchers or groups of researchers affiliated with Canadian universities or other postsecondary institutions, as well as scholarly associations in the humanities and social sciences. Foreign researchers not affiliated with a Canadian postsecondary institution are eligible to apply as co-applicants.
Restrictions: Federal, provincial, territorial and municipal government departments and for-profit organizations are not eligible to administer SSHRC funds.
Amount of Grant: $40,000 maximum
Date(s) Application is Due: Apr 7; Oct 7
Contact: Garry Pinard, Program Officer; (613) 992-5129; fax: (613) 947-0223; email: garry.pinard@sshrc.ca
Internet: http://www.sshrc.ca/web/apply/program_descriptions/rdi_e.asp
Sponsor: Social Sciences and Humanities Research Council of Canada
350 Albert Street, P.O. Box 1610
Ottawa, ON K1P 6G4 Canada

SSHRC Research/Creation Grants in Fine Arts **5423**
The broad purpose of the pilot program is to support and develop excellence in research in artistic disciplines. The program invites artist-researchers affiliated with Canadian postsecondary institutions, within a broad range of disciplines, to apply for research grants of up to three years' duration to support research/creation. Grants are available for programs of research/creation of up to three years. The value of these grants is up to a maximum of $100,000 per year, but totalling not more than $250,000 in a three-year period. SSHRC welcomes both small- and large-scale projects.
Requirements: Only those who fulfill the definition of artist-researcher are eligible to apply to this program. Humanities and social science scholars affiliated with Canadian postsecondary institutions who engage in research/creation are also eligible to apply.
Amount of Grant: $100,000 annually; $250,000 over 3 years
Date(s) Application is Due: Oct 20
Contact: Mathieu Ravignat, Senior Program Officer; (613) 947-3724; fax: (613) 947-0223; email: mathieu.ravignat@sshrc.ca
Internet: http://www.sshrc.ca/web/apply/program_descriptions/fine_arts_e.asp
Sponsor: Social Sciences and Humanities Research Council of Canada
350 Albert Street, P.O. Box 1610
Ottawa, ON K1P 6G4 Canada

SSHRC Sport Participation Research Doctoral Award Supplements **5424**
Applicants for a SSHRC doctoral award who indicate their interest on the application form and who are working in one or more of the target areas identified below will be considered for a Sport Canada Doctoral Award Supplement. Those who already hold a SSHRC doctoral award and are working in one or more of the target areas are also eligible to apply for this supplement. For applicants registered at a Canadian postsecondary institution, the postsecondary institution sets the deadline; for all others, November 13 is the deadline.
Requirements: Applicants for, and holders of, a SSHRC doctoral award who are working in one or more of the target areas of interest, and who will be entering the third or fourth year of a doctoral program at a Canadian university, are eligible to apply.
Amount of Grant: $10,000
Date(s) Application is Due: Nov 13
Contact: Jacques Critchley, Senior Program Officer; (613) 992-5145; fax: (613) 947-0223; email: jacques.critchley@sshrc.ca
Internet: http://www.sshrc.ca/web/apply/program_descriptions/sport_can_e.asp
Sponsor: Social Sciences and Humanities Research Council of Canada
350 Albert Street, P.O. Box 1610
Ottawa, ON K1P 6G4 Canada

SSHRC Sport Participation Research Grants **5425**
The partners seek to stimulate research in sport participation in Canada by increasing the funding available through the Standard Research Grants (SRG) program. Sport Canada has funding available for those standard research grant applications that propose programs of research relevant to its policy priorities and that the SRG adjudication committee has recommended for funding, but which, due to budgetary constraints, did not receive a regular Standard Research Grant. This program has two broad objectives. The first is to promote Canadian research that will develop better understanding, based on empirical evidence, of Canadians' participation in sport, and inform those programs and policies,Äîincluding the Canadian Sport Policy,Äîthat are intended to promote and enhance Canadians' participation in sport. The second is to build Canada's capacity to conduct research on and related to participation in sport, specifically in the target areas described in the following section. Up to $100,000 per annum is available, but not totaling more than $250,000 in a three-year period.
Requirements: A minimum budget of $7,000, not including any research time stipend request, is required in at least one of the three years (except when the applicant is a researcher at a Canadian postsecondary institution that does not receive a SSHRC Institutional Grant). The partners invite proposals from individual researchers and research teams affiliated with Canadian postsecondary institutions who are working in one or more of the target areas of interest. Teams may consist of an applicant (principal investigator), and one or more co-applicants, research collaborators, non-governmental and community partners, student assistants, other assistants and support staff.
Amount of Grant: $100,000 maximum per annum
Date(s) Application is Due: Oct 15
Contact: Jacques Critchley, Senior Program Officer; (613) 992-5145; fax: (613) 947-0223; email: jacques.critchley@sshrc.ca
Internet: http://www.sshrc.ca/web/apply/program_descriptions/sport_can_e.asp
Sponsor: Social Sciences and Humanities Research Council of Canada
350 Albert Street, P.O. Box 1610
Ottawa, ON K1P 6G4 Canada

SSHRC Standard Research Grants **5426**
The broad purpose of the Standard Research Grants program is to support research and develop excellence in research activities in the social sciences and humanities. The specific objectives of the program are to: support high quality independent programs of research, as proposed by scholars and judged by their peers; provide opportunities for the training

of future researchers; contribute to the development or elaboration of new theoretical or methodological approaches to research; maintain and develop vigorous disciplinary research activities; foster and develop vigorous collaborative, multidisciplinary research activities among researchers in the social sciences and humanities; and assist the communication of research results both within and beyond the academic community. The value of these grants is up to a maximum of $100,000 per year, but not totaling more than $250,000 in a three-year period. A minimum budget of $7,000, not including any research time stipend.
Requirements: These grants are open to individual scholars or groups of scholars who are formally affiliated with or occupy positions at Canadian universities or other Canadian postsecondary institutions.
Amount of Grant: $C100,000 maximum per year, $C250,000 maximum over three years
Date(s) Application is Due: Oct 15
Contact: Program Administrator; (613) 992-3145; fax: (613) 992-1787; email: z-info@sshrc.ca
Internet: http://www.sshrc.ca/web/apply/program_descriptions/standard_e.asp
Sponsor: Social Sciences and Humanities Research Council of Canada
350 Albert Street, P.O. Box 1610
Ottawa, ON K1P 6G4 Canada

SSHRC Strategic Knowledge Clusters Grants **5427**
The overall objective of the Strategic Knowledge Clusters program is to build upon and add value to research supported through SSHRC, other programs by supporting Canadian researchers in their efforts to develop and sustain creative, innovative knowledge mobilization networks that lead to increasing the impact of research on policy and program development. Through such support, the program seeks to foster and enhance: synthesis and application of humanities and social sciences research knowledge in areas in which Canadian researchers demonstrate strength and that have importance for Canadian society within a global context; collaborative knowledge partnerships able to leverage external funding that will support new research in the humanities and social sciences; and enhancement of the international impact of Canadian social sciences and humanities research. Strategic Knowledge Clusters Grants are worth up to $300,000 annually for up to seven years (i.e., a maximum of $2.1 million over seven years). Up to seven grants will be awarded in this competition.
Requirements: Applications must be submitted by the cluster director on behalf of all the cluster participants. The cluster director, whom the application must identify as the administrative lead, must satisfy all the regular eligibility requirements for an applicant (principal investigator). In addition, the director must be affiliated with a Canadian postsecondary institution.
Amount of Grant: $300,000 maximum annually
Date(s) Application is Due: Nov 20
Contact: Michele Dupuis, Program Officer; (613) 947-2045; fax: (613) 992-7635; email: michele.dupuis@sshrc.ca
Internet: http://www.sshrc.ca/web/apply/program_descriptions/knowledge_cluster_e.asp
Sponsor: Social Sciences and Humanities Research Council of Canada
350 Albert Street, P.O. Box 1610
Ottawa, ON K1P 6G4 Canada

SSHRC Therese F. Casgrain Fellowship **5428**
This program is administered by SSHRC on behalf of the Therese F. Casgrain Foundation, which was established in 1982 to perpetuate the memory of Therese F.-Casgrain by encouraging others to pursue her thought and action in the field of social justice. This fellowship supports research in the field of social justice, particularly in defense of individual rights and the promotion of the economic and social interests of Canadian women. The Foundation is particularly interested in research that includes a discussion of public policy options and makes concrete recommendations for change.
Requirements: Applicants must be Canadian citizens or permanent residents of Canada, hold a doctorate degree or equivalent advanced professional degree at the time of the award, have proven research experience, and not be under SSHRC sanction for financial or research misconduct.
Amount of Grant: $C40,000, of which up to $C10,000 may be used for travel and research expenses
Date(s) Application is Due: Oct 1
Contact: Mylene Seguin, Program Officer; (613) 992-5358; fax: (613) 943-1329; email: Mylene.Seguin@sshrc.ca
Internet: http://www.sshrc.ca/web/apply/program_descriptions/fellowships/casgrain_e.asp
Sponsor: Social Sciences and Humanities Research Council of Canada
350 Albert Street, P.O. Box 1610
Ottawa, ON K1P 6G4 Canada

SSHRC William E. Taylor Fellowship **5429**
Established in memory of William E. Taylor, SSHRC past president, this award is given every year to the most outstanding SSHRC doctoral fellowship candidate. In addition to the regular SSHRC doctoral award, the fellow receives a $C5000 stipend and allowances to cover tuition fees, travel to the main place of tenure, and travel for research purposes. There is no application form.
Requirements: Fellowships are open to students who are Canadian citizens or permanent residents by deadline date.
Amount of Grant: $C5000
Contact: Fellowships and Institutional Grants Division; (613) 943-7777; fax: (613) 943-1329; email: fellowships@sshrc.ca
Internet: http://www.sshrc.ca/web/apply/program_index_e.asp
Sponsor: Social Sciences and Humanities Research Council of Canada
350 Albert Street, P.O. Box 1610
Ottawa, ON K1P 6G4 Canada

SSMRF Julia's Hope Pilot Grants **5430**
The foundation supports research toward a cure for Sanfilippo Syndrome Type A, an inherited neurodegenerative disorder for which there is no treatment or cure, and related neurodegenerative disorders. Investigators are encouraged to utilize MPS-IIIA animal models. The grants are designed to bring sponsored research to a level at which additional, ongoing support may be secured through larger grants from NIH or other sources. Applications are available online.
Amount of Grant: $50,000
Date(s) Application is Due: Mar 1
Contact: Program Contact, (402) 330-8636; email: mail@juliashope.org
Internet: http://www.juliashope.org/grants.htm
Sponsor: Sanfilippo Syndrome Medical Research Foundation
P.O. Box 571
Boys Town, NE 68010

SSRC Abe Fellowship Program **5431**
The Fellowship is designed to encourage international multidisciplinary research on topics of pressing global concern. The program seeks to foster the development of a new generation of researchers who are interested in policy-relevant topics of long-range importance and who are willing to become key members of a bilateral and global research network built around such topics. Rather than seeking to promote greater understanding of a single country, the Abe Fellowship Program encourages research on the experiences and future challenges of the U.S. and Japan in a comparative or global perspective. The Program committee seeks applications for research focusing explicitly on policy-relevant and contemporary issues that have a comparative or transnational perspective and that draw the study of the U.S. and Japan into wider disciplinary or theoretical debates. Applicants are invited to submit proposals for research in the social sciences or the humanities relevant to any one or combination of the following three themes: global issues; problems common to industrial and industrializing societies; or issues that pertain to U.S.-Japan relations. Applications are welcome from scholars and non-academic research professionals. Funding for the program is provided by the Japan Foundation Center for Global Partnership.
Requirements: Applications are welcome from scholars and non-academic research professionals. Competition is open to citizens of the United States and Japan as well as to nationals of other countries who can demonstrate strong and serious long-term affiliations with research communities in Japan or the United States. Applicants must hold a Ph.D. or the terminal degree in their field, or have attained an equivalent level of professional experience. Applications from researchers in professions other than academia are encouraged
Samples: Dave Arase, associate professor of politics, Pomona College—for the project The Japan Sea Initiative—The Role of Local Governments in Subregional Cooperation.;Tsutomu Kono, visiting scholar, United Nations U—for the project Epistemic Communities and UN Reform.
Date(s) Application is Due: Sep 1
Contact: Mary Byrne McDonnell, Program Officer; (212) 377-2700, ext. 420; fax: (212) 377-2727; email: abe@ssrc.org or exec@ssrc.org
Internet: http://www.ssrc.org/fellowships/abe
Sponsor: Social Science Research Council
810 Seventh Avenue
New York, NY 10019

SSRC Collaborative Grants in Media and Communications **5432**
This program is premised on the belief that advancing public-interest agendas requires not just the scaling up of political activity, but also a more robust and better-integrated process of knowledge production in and around issues of media, communications policy, and the public sphere. The SSRC offers two types of Collaborative Grants for academic-advocacy partnerships in media and communications. Small Grants provide up to $7,500 for short-term academic research in support of advocacy and activism in media and communications. The deadline is October 15, and opportunities for subsequent competitions will be available at roughly 4-5 month intervals. Large Grants provide up to $30,000 in support for academic-advocacy research collaborations designed to change media/telecommunications infrastructure, practices, or policies. Letter of Inquiry (LOI) for Large Grants are due February 8th, with full proposals due April 30th. General areas of interest for the program include: measuring the success or failure of mainstream media in advancing different public interest goals or values; measuring the impact of existing alternative or community media systems on communities, public discourse, or democratic processes; developing better, actionable accounts of the role of new media in people's lives; analyzing policymaking and/or regulatory systems; analyzing emerging systems, frameworks, or models of media and communications that transcend the current regulatory framework; analyzing economic models, industry structure, markets, or audiences for different kinds of media; creating analytical tools or research resources for use by advocates, communities, or the public; and documenting or evaluating advocacy or organizing strategies around communications and media issues. Both large and small grants are awarded through competitive application processes, with recipients selected by an independent committee of researchers and advocates.

Requirements: Projects must involve substantive collaboration between: a researcher based at a university, college, or other academically-oriented research institution (advanced graduate students are eligible); or a U.S.-based non-profit advocacy, organizing or community group working on media and/or telecommunications issues.
Amount of Grant: $7,500 maximum for small grants; $30,000 maximum for large grants
Date(s) Application is Due: Feb 8; Apr 30; May 1; Oct 15
Contact: Joe Karaganis, Program Director; (212) 377-2700, ext. 469; fax: (212) 377-2727; email: karaganis@ssrc.org
Internet: http://programs.ssrc.org/media/
Sponsor: Social Science Research Council
810 Seventh Avenue
New York, NY 10019

SSRC Dissertation Proposal Development Fellowships (DPDF) **5433**
The Fellowship is a strategic fellowship program designed to help graduate students in the humanities and social sciences formulate doctoral dissertation proposals that are intellectually pointed, amenable to completion in a reasonable time frame, and competitive in fellowship competitions. The program is organized around distinct research fields, sub-disciplinary and interdisciplinary domains with common intellectual questions and styles of research. Each year, an SSRC Faculty Advisory Committee selects five fields proposed by pairs of research directors who are tenured professors at different doctoral degree-granting programs at U.S. universities. Research directors receive a stipend of $7500. Graduate students in the early phase of their research, generally 2nd and 3rd years, apply to one of five research fields led by the two directors; each group is made up of ten to twelve graduate students. Fellows participate in two workshops: one in the late spring that helps prepare them to undertake pre-dissertation research on their topics; and one in the early fall, designed to help them synthesize their summer research and to draft proposals for dissertation funding. Fellows are eligible to apply for up to $5000 from SSRC to support pre-dissertation research during the summer.
Requirements: Students in the humanities and social sciences undertaking doctoral dissertation research may apply for one of the five annual research fields named. The program is designed for second and third year students who have not yet had their dissertation proposals approved by their thesis directors and their home institutions. Only in exceptional circumstances will the DPDF entertain fellowship applications from first-year students.
Date(s) Application is Due: Jan 30
Contact: Peter Sahlins, Director of Academic Programs; (212) 377-2700, ext. 476; fax: (212) 377-2727; email: sahlins@ssrc.org
Internet: http://programs.ssrc.org/dpdf/
Sponsor: Social Science Research Council
810 Seventh Avenue
New York, NY 10019

SSRC Eurasia Program Dissertation Write-Up Fellowships **5434**
These awards of $22,000 for one academic year provide support to graduate students near the completion of their doctoral programs in the social sciences and related humanities. These fellowships are intended for applicants who have completed all formal components of their dissertation field research or data collection, and who have made significant progress in outlining emergent, innovative contributions to scholarship. Regions and countries currently supported by the program include Armenia, Azerbaijan, Belarus, Georgia, Kazakhstan, Kyrgyzstan, Moldova, the Russian Federation, Tajikistan, Turkmenistan, Ukraine and Uzbekistan.
Requirements: Applicants must be citizens of the United States and have completed all requirements for the PhD except for the dissertation at the time the fellowship commences. Applicants must have obtained ABD status (must have completed all of the requirements for the PhD degree except for the dissertation) by the application submission deadline.
Restrictions: Funding is presently not available for research solely on the Baltic States (Estonia, Latvia, and Lithuania).
Amount of Grant: $22,000
Date(s) Application is Due: Nov 13
Contact: Seteney Shami, Program Director; (212) 377-2700, ext. 512; fax: (212) 377-2727; email: shami@ssrc.org or eurasia@ssrc.org
Internet: http://programs.ssrc.org/eurasia/fellowships/dissertation_writeup_fellowships/
Sponsor: Social Science Research Council
810 Seventh Avenue
New York, NY 10019

SSRC Eurasia Program Predissertation Training Fellowships **5435**
These awards invite graduate students in the early stages of their graduate careers (preference is given to those in their first or second years) to enhance their research skills for projects on the New States of Eurasia, the Soviet Union and/or the Russian Empire. Awards of up to $7,000 are intended to support the following types of training: language learning at a recognized program in the U.S. or abroad (when justifiable, independent language study may be acceptable); formal training away from one, home institution to acquire analytical or methodological skills normally unavailable to the candidate; and well-defined exploratory research expressly leading to the formulation of a dissertation proposal. Applicants will be asked to select one of two options when applying for this fellowship: Predissertation Training Fellowship with Language Component (if their project includes a language component of any kind, even if it is only a portion of their full project); or Pre-dissertation Training Fellowship without Language Component (if they are not proposing to undertake any form of language training).
Requirements: All applicants must be citizens or permanent residents of the United States as of the application submission deadline. Applicants must be enrolled in a doctoral program or equivalent degree in the social sciences or humanities, not yet having sat for their qualifying exams or advanced to the Ph.D. candidacy. It is expected that the applicant will complete all fellowship activities before advancing to candidacy.
Restrictions: ABD's are not eligible for these fellowships.
Amount of Grant: $7,000 maximum
Date(s) Application is Due: Nov 13
Contact: Seteney Shami, Program Director; (212) 377-2700, ext. 512; fax: (212) 377-2727; email: shami@ssrc.org or eurasia@ssrc.org
Internet: http://programs.ssrc.org/eurasia/fellowships/predissertation_training_fellowships/
Sponsor: Social Science Research Council
810 Seventh Avenue
New York, NY 10019

SSRC Eurasia Program Teaching Fellowships **5436**
The Eurasia Program Teaching Fellowship is designed to encourage and support faculty members at all career levels in their efforts to further impart their knowledge and expertise to their students. These awards of up to $7,000 support the creation of original and innovative course curricula. Funds support the rethinking and reframing of courses in the humanities and social sciences that directly relate to the whole or part of Eurasia. The supported courses are wholly new, or substantial revisions of a course previously taught.
Requirements: All applicants must be U.S. citizens or permanent residents teaching at U.S. institutions. A strong candidate will have a proven track record of research and teaching in his/her field of Eurasian studies. Fellowships are particularly appropriate for faculty with heavy teaching loads and with proven desires to push the teaching of Eurasian studies in innovative directions and to incorporate contemporary research and thinking on Eurasian studies into new teaching curricula for use in classrooms.
Restrictions: Funding is presently not available for research solely on the Baltic States (Estonia, Latvia, and Lithuania).
Amount of Grant: $7,000
Date(s) Application is Due: Nov 13
Contact: Seteney Shami, Program Director; (212) 377-2700, ext. 512; fax: (212) 377-2727; email: shami@ssrc.org or eurasia@ssrc.org
Internet: http://programs.ssrc.org/eurasia/fellowships/Teaching_Fellowships/
Sponsor: Social Science Research Council
810 Seventh Avenue
New York, NY 10019

SSRC Eurasian Program Postdoctoral Research Fellowships **5437**
The program sponsors fellowships for research on the Russian Empire, the Soviet Union, and the New States of Eurasia. Regions and countries currently supported by the program include Armenia, Azerbaijan, Belarus, Georgia, Kazakhstan, Kyrgyzstan, Moldova, Russian Federation, Tajikistan, Turkmenistan, Ukraine, and Uzbekistan. In most cases, scholars would use approximately $15,000 to release themselves from teaching and administrative obligations in their departments (or as a living stipend for independent scholars) and $5,000 for research expenses. All funds must be expended within no more than 18-24 months of receiving the award.
Requirements: Eligible applicants must be from a relevant field in a social science or humanities discipline, either at an accredited university or working as an independent scholar.
Restrictions: Funding is presently not available for research solely on the Baltic States (Estonia, Latvia, and Lithuania).
Amount of Grant: $20,000 average
Date(s) Application is Due: Nov 13
Contact: Seteney Shami, Program Director; (212) 377-2700, ext. 512; fax: (212) 377-2727; email: shami@ssrc.org or eurasia@ssrc.org
Internet: http://programs.ssrc.org/eurasia/fellowships/postdoctoral_fellowships/
Sponsor: Social Science Research Council
810 Seventh Avenue
New York, NY 10019

SSRC HIV/AIDS and Public Health Policy Research in Africa Fellowships **5438**
The program is open to African researchers, policy analysts, program planners and practitioners to support research on health and social policy in Eastern and Southern Africa relating to: political economy of care-giving and HIV/AIDS with particular emphasis on gender analysis; or sexual violence and HIV/AIDS. Two senior fellowships of up to $25,000 and six associate fellowships of up to $10,000 will be awarded. Primary goals are to: increase knowledge and understanding about the implications of HIV/AIDS in Eastern and Southern Africa as it relates to care-giving and sexual violence; improve public health policy responses to the pandemic and increase understanding about the interaction between local and global health systems; and foster collaboration among African institutions, researchers, policy analysts, program planners and practitioners both within and outside of the continent.
Requirements: Fellowships will be awarded to African researchers (master, degree and above) and mid-career policy analysts, program planners and practitioners with at least eight years of field experience. Preference will be given to applicants based in Eastern and Southern Africa. Applications will be reviewed on the basis of their academic quality and

rigor; policy relevance; feasibility; potential contribution to the field; and the applicant, professional and research background and achievements.
Amount of Grant: $10,000 for associate fellowships; $25,000 for senior fellowships
Date(s) Application is Due: Jul 15
Contact: Alex de Waal, Program Director; (212) 377-2700, ext. 619; fax: (212) 377-2727; email: dewaal@ssrc.org
Internet: http://www.ssrc.org/programs/hiv-aids-program/
Sponsor: Social Science Research Council
One Pierrepont Plaza, 15th Floor
Brooklyn, NY 11201

SSRC International Dissertation Research Fellowships 5439
The program supports distinguished graduate students in the humanities and social sciences conducting dissertation research outside the United States. Seventy-five fellowships will be awarded with funds provided by the Andrew W Mellon Foundation. The program is committed to scholarship that advances knowledge about non-U.S. cultures and societies grounded in empirical and site-specific research (involving fieldwork, research in archival or manuscript collections, or quantitative data collection). The program promotes research that is situated in a specific discipline and geographical region and is engaged with interdisciplinary and cross-regional perspectives. Fellowships will provide support for nine to twelve months of dissertation research. Individual awards will be approximately $20,000.
Requirements: The program is open to full-time graduate students in the humanities and social sciences — regardless of citizenship — enrolled in doctoral programs in the United States. Applicants must complete all Ph.D. requirements except on-site research by the time the fellowship begins or by December, whichever comes first.
Restrictions: Proposals that identify the United States as a case for comparative inquiry are welcome; however, proposals that require no field research outside the United States are not eligible.
Amount of Grant: $20,000 maximum
Date(s) Application is Due: Nov 6
Contact: Nicole Stahlmann, Program Director; (212) 377-2700, ext. 447; fax: (212) 377-2727; email: stahlmann@ssrc.org
Internet: http://www.ssrc.org/programs/idrf
Sponsor: Social Science Research Council
810 Seventh Avenue
New York, NY 10019

SSRC Japan Society for the Promotion of Science (JSPS) Fellowships 5440
The Program for U.S. researchers provides promising and highly qualified recent PhDs with opportunities to conduct research in Japan. JSPS guidelines target the applicant who wishes to conduct cooperative research under the leadership of a host researcher, thereby advancing the Fellow's own research and at the same time stimulating Japanese academic circles through close collaboration with young Japanese researchers. The SSRC Japan Advisory Board recommends to JSPS up to 20 candidates annually from the social sciences and humanities. Award offers are made by JSPS in early-summer for fellowships that must commence between April 1 and November 30 (long-term) or April 1 and March 31 (short-term). Applications are welcome from all social science and humanities disciplines and need not be explicitly related to the study of Japan. Projects must include work with colleagues and resources in Japan and propose a single, continuous stay in Japan from 1 to 12 months (short-term) or 1 to 2 years (long-term).
Requirements: Applicants must be U.S. citizens or permanent residents at the time of application and submit proof of affiliation with an eligible host research institution in Japan as part of the application packet. Permanent residents must provide a copy of a permanent resident card (green card). Citizens of other countries are eligible for short-term fellowships if they have completed a master, or PhD course at an institution of higher education in the U.S. and, upon completing the course, have for at least three continuous years conducted high-level research at a university in the U.S. Priority will be given to researchers who are not Japanese citizens.
Restrictions: JSPS does not extend fellowships to scholars employed in the institutions under jurisdiction of the U.S. Department of Defense. Scholars who have previously received funding from JSPS for 12 months or longer are not eligible to apply for JSPS fellowships.
Date(s) Application is Due: Dec 1
Contact: Program Officer; (212) 377-2700; fax: (212) 377-2727; email: japan@ssrc.org
Internet: http://fellowships.ssrc.org/japan/postdoc/
Sponsor: Social Science Research Council
810 Seventh Avenue
New York, NY 10019

SSRC Japan Studies Dissertation Workshop Grants 5441
The Workshop seeks to create a sustained network of advanced graduate students and faculty by providing the opportunity to give and receive critical feedback on dissertations in progress. The annual workshop draws participants from all fields in the social sciences and humanities and from institutions throughout the U.S. and Canada. Individual students lead discussions of their projects with mentor faculty and student peers from various disciplines to receive creative and critical input on improving their fieldwork plans or writing strategies. Based on narrative project descriptions submitted as part of the application packet, selected students will prepare a synthetic essay incorporating all projects from which discussions of broader methodological and thematic issues will be developed and incorporated into the four-day agenda. In most cases, the SSRC will fully cover participants' travel, lodging and meals for the duration of the workshop. Limited funds are available for participants traveling from outside the United States.
Requirements: Eligible to apply are full-time advanced graduate students, regardless of citizenship, who are enrolled at U.S. institutions. Applicants must have an approved dissertation prospectus, but cannot have completed writing for final submission. A narrative description of the dissertation topic and a letter of reference from the student's advisor are required as part of the application.
Date(s) Application is Due: Oct 1
Contact: Program Officer; (212) 377-2700; fax: (212) 377-2727; email: japan@ssrc.org
Internet: http://fellowships.ssrc.org/japan/dissertation_workshop/
Sponsor: Social Science Research Council
810 Seventh Avenue
New York, NY 10019

SSRC Mellon Mays Predoctoral Research Grants 5442
The SSRC-Mellon Mays Graduate Initiatives Program offers three grants (Graduate Studies Enhancement, Predoctoral Research Development and Dissertation Completion) designed to meet specific needs during the doctoral degree process.
Requirements: Applicants must have been selected as Mellon Mays Fellows as undergraduates. Fellows may apply for one grant per year and must be enrolled in a doctoral program in a Mellon-designated field of study or have filed a petition for inclusion of another field. See website for a complete Field of Study List and, additional guidelines.
Amount of Grant: $5,000 maximum
Date(s) Application is Due: Nov 1
Contact: Cally L. Waite, Program Director; (212) 377-2700, ext. 602 or (718) 517-3676; fax: (212) 377-2727; email: waite@ssrc.org
Internet: http://www.ssrc.org/fellowships/mellon-mays-predoctoral-research-grants/
Sponsor: Social Science Research Council
One Pierrepont Plaza, 15th Floor
Brooklyn, NY 11201

SSRC-ESRC Collaborative Visiting Fellowships 5443
The ESRC and the SSRC have a common mission of promoting, funding, and disseminating important and socially useful knowledge in the social sciences. This is the fourth round of a scheme designed to encourage communication and cooperation between social scientists in Great Britain and the Americas, and to explore and develop possibilities for future exchanges to be organized by the ESRC and the SSRC. In this round the geographical coverage is extended to all the Americas.
Requirements: Applicants from the Americas should have received a Ph.D in one of the social sciences (including history) by the time the proposed visiting fellowship would start. They should have been based in the country from which they are applying for at least two years before the application deadline of April 16th. The fellowship scheme is open to scholars from universities, colleges, independent research organizations, and public agencies in the Americas and to scholars at ESRC-supported centers, programs, groups, networks, as well as holders of Large Grants awards or Professorial Fellowships in Britain.
Amount of Grant: $9,500 maximum
Date(s) Application is Due: Apr 16
Contact: Fellowships Director; (212) 377-2700, ext. 500; fax: (212) 377-2727; email: migration@ssrc.org or fellowships@ssrc.org
Internet: http://www.ssrc.org/fellowships/esrc
Sponsor: Social Science Research Council
810 Seventh Avenue
New York, NY 10019

SSSR Distinguished Article Award 5444
The annual award was established to recognize the most outstanding article published by a SSSR member or members within the past year (Scholars and graduate students from all fields are invited to join). The award will be presented at the annual meeting. It includes a plaque and $1,000.
Requirements: An article may be nominated by the editor of the journal in which it appeared, by its author, or by any other SSSR member deeming it to be of outstanding quality. Articles must have been published in refereed journals. Nominated articles must be sent by May 15 to the committee chair. The committee will judge each article's quality of scholarship, importance to the field, and relevance to the larger society. Any article involving some aspect of the scientific study of religion is eligible.
Restrictions: Submissions published solely in edited volumes will not be considered.
Amount of Grant: $1,000
Date(s) Application is Due: May 15
Contact: Peter Beyer; Chair of Distinguished Article Award Committee; email: pbeyer@uottawa.ca
Internet: http://www.sssrweb.org/Awards.cfm#8
Sponsor: Society for the Scientific Study of Religion (SSSR)
IUPUI Center for the Study of Religion and American Culture, 425 University Boulevard
Indianapolis, IN 46202-5140

SSSR Student Paper Award 5445
The award recognizes the best paper scheduled for presentation by a student at the annual meeting. (Scholars and graduate students from all fields are invited to join.) SSSR wishes to (1) encourage scholarly work and participation by students in SSSR meetings, and (2) give students the opportunity to receive constructive feedback and criticism on their work by professional colleagues and mentors. The award will be presented at the annual meeting. It includes a plaque and $1,000.

Requirements: Authors of papers submitted for this competition must be graduate or undergraduate students at institutions of higher learning anywhere in the world, but papers must be submitted in English. Any co‚Äìauthors must also be students. To be eligible, a paper must not have been published or accepted for publication. At least one author for the paper must be registered for the SSSR conference. Submission of the paper implies a commitment for an author to appear and present the paper at the conference. Student authors who intend to enter this competition may wish to notify the Student Paper Award Committee Chair via email at the time of their abstract submission, but they are not required to do so.
Restrictions: Neither faculty nor post‚Äìdoctoral co‚Äìauthors are permitted.
Amount of Grant: $1,000
Date(s) Application is Due: Apr 15; Aug 1
Contact: Carolyn Warner; Chair of Student Paper Award Committee; email: cwarner@asu.edu
Internet: http://www.sssrweb.org/Awards.cfm#9
Sponsor: Society for the Scientific Study of Religion (SSSR)
IUPUI Center for the Study of Religion and American Culture, 425 University Boulevard
Indianapolis, IN 46202-5140

SSSS Student Research Grants **5446**
The society awards three grants annually for graduate research in areas of human sexuality. The purpose of the research is to advance the knowledge about sexuality. An applicant must send one copy each of a completed application fact sheet (available on the Web site) and a letter from the department chairperson stating student status within the institution and the educational purpose of the research. Also, an applicant must send seven copies each of a 10-page description of the proposed research, a proposed budget for the project, and a 150-word abstract.
Requirements: Students enrolled in a graduate degree-granting program working toward their master's degree or doctorate are eligible to apply, although the research need not be for the thesis.
Amount of Grant: $1000
Date(s) Application is Due: Feb 1; Sep 1
Contact: Program Contact, (610) 530-2483; fax: (610) 530-2485; email: thesociety@inetmail.att.net
Internet: http://www.sexscience.org/awards/index.php?category_id=427
Sponsor: Society for the Scientific Study of Sexuality
P.O. Box 416
Allentown, PA 18105

Stackner Family Foundation Grants **5447**
The foundation awards grants concentrated in the Milwaukee metropolitan area for building programs and equipment acquisition, medical research, programs at all levels of education, social welfare, child welfare, and programs that assist the physically disabled. Additional categories of support include general operating support, continuing support, annual campaigns, capital campaigns, and program development. The board meets each year in January, April, July, and October. Application forms are not required.
Requirements: Nonprofit organizations in the greater Milwaukee, WI, area may request grant support.
Geographic Focus: Wisconsin
Samples: Horizon House (Milwaukee, WI)—for an outpatient program, $4000.;Penfield Childrens Ctr (Milwaukee, WI)—for the capital campaign, $10,000.
Amount of Grant: $25-$25,000 range
Date(s) Application is Due: Mar 15; Jun 15; Sep 15; Dec 15
Contact: Paul Tillman, fax: (414) 646-5409; email: Stackner@MSH.com
Sponsor: Stackner Family Foundation
411 E Wisconsin Avenue
Milwaukee, WI 53202-4497

Stackpole-Hall Foundation Grants **5448**
The foundation awards grants in Pennsylvania in support of higher/secondary education, and literacy and vocational education and projects; social services, including youth and child welfare agencies; the arts and cultural programs; health services, including mental health and drug abuse; community development, including civic affairs and leadership development, and volunteerism; and environmental conservation and protection. Grants are awarded in support of annual and capital campaigns, building construction/renovation, equipment acquisition, program and project development, matching funds, and seed money. Application forms are not required.
Requirements: Nonprofit organizations in Elk County, PA, may submit grant proposals.
Restrictions: Grants are not awarded to individuals or for scholarships, fellowships, operating budgets, loans, or endowment funds.
Geographic Focus: Pennsylvania
Amount of Grant: $200-$50,000 range
Contact: William Conrad, Executive Secretary, (814) 834-1845; fax: (814) 834-1869; email: stackpolehall@alltel.net
Sponsor: Stackpole-Hall Foundation
44 S Saint Mary's Street
Saint Mary's, PA 15857

Staglin Family Music Festival NARSAD Schizophrenia Research Award **5449**
The award supports an investigator with the MD, PhD, or doctoral degree and postdoctoral training applicable to schizophrenia. The candidate should be at the assistant or associate professor level, and be age 45 or younger. Applications should be sent electronically (as one document) via email. Guidelines are available online.
Requirements: The candidate should be nominated by the dean, department chair, or head of his/her scientific program of the university in which he or she is employed.
Amount of Grant: $250,000
Date(s) Application is Due: Dec 2
Contact: email: grants@narsad.org
Internet: http://www.narsad.org/research/apply
Sponsor: National Alliance for Research on Schizophrenia
60 Cutter Mill Road, Suite 404
Great Neck, NY 11021

Staney Foundation Treatment Trials for Schizophrenia and Bipolar Disorder **5450**
The foundation is interested in supporting studies of compounds not supported by the pharmaceutical industry in the treatment of schizophrenia and bipolar disorder, including novel agents and studies of off-label use of drugs approved for other conditions. Grants are awarded for a maximum of three years, depending on design. Applicant may apply online.
Amount of Grant: $300,000 maximum per year for a period of up to three years
Date(s) Application is Due: Mar 1; Oct 1
Contact: Xiomara Herrera, Grants Administrator, (301) 571-0760 ext 124; fax: (301) 571-0769; email: herrerax@stanleyresearch.org
Internet: http://www.stanleyresearch.org/application
Sponsor: Stanley Foundation
5430 Grosvenor Ln, Suite 200
Bethesda, MD 20814

Stanford Center on Adolescence Youth Purpose Research Grants **5451**
Small grants for dissertation and postdoctoral research will be awarded in March for research to be completed by September. The Stanford Center on Adolescence aims to spread the recognition of youth purpose as an important area of scientific research by encouraging all inquiries that shed light on topics related to purpose among young people, such as adolescent motivation, personal meaning, or positive youth development, in an array of settings found in contemporary society. The Center operationally defines purpose as a stable and generalized intention to accomplish something that is meaningful to the self and that leads to productive engagement with some aspect of the world beyond the self. Through its own research, the Center seeks to address questions about how purpose develops, the antecedents of purpose, the difference that purpose makes in the lives of young people, and how to best cultivate purpose in youth. The Center encourages researchers from all disciplines to submit proposals for projects that will contribute to the scientific study of this research topic. Applicants may request up to $10,000 in funding for research-related expenses.
Requirements: Applicants must be affiliated with an accredited college or university and should be U.S citizens or permanent residents. Postdoctoral applicants must have received their Ph.D. within the past five years.
Amount of Grant: $10,000 maximum
Date(s) Application is Due: Jan 17
Contact: Lisa Staton, Administrative Associate; (650) 725-8205; fax: (650) 725-8207; email: stanfordcoa@stanford.edu
Internet: http://www.stanford.edu/group/adolescent.ctr/researchawards
Sponsor: Stanford Center on Adolescence
505 Lasuen Mall
Stanford, CA 94305-3083

Stanford Humanities Center External Faculty Fellowships **5452**
External fellowships fall into two categories: senior fellowships for well-established scholars, and junior fellowships for scholars who during the fellowship year will be at least three but no more than 10 years beyond receipt of the PhD. The fellowship program is primarily designed to offer research opportunities both to members of humanities departments as traditionally defined and to other scholars seriously interested in humanistic issues. Fellows are expected to make contributions to the intellectual life of the Stanford community and to remain in residence, living in the immediate area of the university, during the regular academic year. Annual deadline dates may vary; contact program staff for exact dates.
Requirements: External fellowships are intended primarily for persons currently teaching or affiliated with academic institutions, but others may apply. Applications will be judged on the promise of the specific research project being proposed, the originality and intellectual distinction of the candidate's previous work, the research project's potential interest to scholars in different fields of the humanities, and the applicant's ability to engage in collegial interaction.
Restrictions: Scholars who have received major research grants within the last three years generally will not be as competitive as other applicants.
Samples: Susan Cahn, Assistant Professor of History, State U of New York (Buffalo, NY)—to continue researching and writing her book Sexual Reckonings: Adolescent Girlhood in the Modern South, 1920-1960.; Angela Harris, Professor Law, Boalt Hall School of Law, U of California (Berkeley, CA)—to complete a first draft of her manuscript, What We Talk About When We Talk About Race.
Amount of Grant: $35,000 maximum annual stipend (junior fellows); $50,000 maximum annual stipend (senior fellows); $12,500 maximum stipend for housing and moving allowance
Date(s) Application is Due: Oct 15
Contact: Chi Elliott, Fellowship Administrator, (650) 723-3054; fax: (650) 723-3918; email: chiyuma@stanford.edu

Internet: http://shc.stanford.edu/shc/external.html
Sponsor: Stanford University
424 Santa Teresa Street, Stanford Humanities Ctr
Stanford, CA 94305-4015

Stanley Foundation Schizophrenia and Bipolar Disorder Research Grants 5453
The program supports researchers at all levels of development who are already in the field of schizophrenia and bipolar disorder, as well as researchers in other fields of medicine and biology (e.g., neurology, immunology, biochemistry, and epidemiology) who wish to initiate new research projects in schizophrenia and bipolar disorder. Up to two years of research will be supported.
Amount of Grant: $75,000 per year
Date(s) Application is Due: Mar 1
Contact: Rhoda Marte, (301) 571-0760 ext 118; fax: (301) 571-0769; email: marter@stanleyresearch.org
Internet: http://www.stanleyresearch.org/application
Sponsor: Stanley Foundation
5430 Grosvenor Ln, Suite 200
Bethesda, MD 20814

Starr Foundation Grants 5454
The foundation awards grants to nonprofit organizations in the areas of education, medicine and health care, human needs, public policy, and culture. Education grants support scholarships for deserving students, groups that offer need-based financial aid, and international-exchange programs. Human needs grants support food programs for the poor, job training, literacy, adequate housing, and programs for the disabled. Medicine and health care grants include capital grants to hospitals, significant research grants, and grants to assist in the provision of health care to underserved communities. Public policy grants support international relations and democratic institutions around the world. Culture grants go to exchange programs and provide services to the elderly and/or disabled. The foundation rarely funds local charities outside of New York City but may fund national organizations that serve communities outside of New York. Types of support include multiyear grants, general operating grants, and capacity-building grants. There are no application deadlines. The board meets throughout the year.
Samples: World Trade Ctr Memorial Foundation (New York, NY)—to build the World Trade Center Memorial and a museum to remember the September 11, 2001, and February 26, 1993, terrorist attacks, $25 million (2005); Johns Hopkins U (Baltimore, MD)—to endow a professorship and support other needs at the South Asia Studies program of the university's Nitze School of Advanced International Studies, $3 million (2005); Dickinson College (Carlisle, PA)—to augment a scholarship endowment, $1 million (2005); For relief efforts in South Asia and Africa—$2.5 million distributed (2005).
Amount of Grant: $70 million total
Contact: Florence Davis, President, (212) 770-6882; email: florence.davis@starrfdn.org or grants@starrfoundation.org
Internet: http://fdncenter.org/grantmaker/starr
Sponsor: Starr Foundation
70 Pine Street, 14th Fl
New York, NY 10270

State Farm Companies Doctoral Dissertation Award 5455
Doctoral dissertation awards are designed to stimulate research and development of knowledge of use to the insurance industry and to increase the number of qualified teachers of insurance in the colleges and universities of the country. Applications are sent to deans and directors of doctoral programs at select colleges and universities in December. Completed applications, nomination forms, and required attachments must be received in the Foundation office by March 31. A grant is given to any college that submits eligible nominations.
Requirements: Awards are available to doctoral candidates who have completed a major portion of their doctoral programs and who are at the dissertation stage and are majoring in insurance or a related field of study. Applicants must be U.S. citizens.
Restrictions: Only applicants nominated by the director of the school's doctoral program qualify as candidates. Direct applications will not be considered.
Samples: Illinois Wesleyan U—for academic enrichment and scholarship programs for minority students, $918,340.
Amount of Grant: $10,000; $3000 for winner's graduate institution
Date(s) Application is Due: Mar 31
Contact: Assistant Secretary, (309) 766-2039 or 766-2161; fax: (309) 766-3700; email: info@statefarm.com
Internet: http://www.statefarm.com/about/part_spos/grants/doctoral.asp
Sponsor: State Farm Companies Foundation
One State Farm Plaza, SC3
Bloomington, IL 61710-0001

State Scholarships Foundation Grants 5456
The foundation awards scholarships for postgraduate/postdoctoral studies in Greece to nationals from Western Europe, United States, Canada, Australia, and Japan. The scholarships support postgraduate/doctoral studies (MA/Msc, up to three years); PhD studies (up to four years); postdoctoral research studies (six months to one year); further education in the Greek language, literature, philosophy, history, and art (for professors of Greek in foreign universities, for six months up to one year); and specialization in fine arts (for up to two years).
Requirements: Application is open to nationals of the Balkan countries, countries of Western Europe, United States, Canada, Australia, and Japan. In most cases, the applicant must hold a graduate degree from a foreign university (outside of Greece) and a first postgraduate degree or equivalent. Medical doctors should have either completed their specialization or hold a postgraduate degree. For postdoctoral research studies, applicants must hold a PhD degree. Their first university degree must be from a foreign university. For further education in the Greek language, literature, philosophy, history, or art, applicants must be professors in a department of Greek studies of a foreign university; and a validated certificate of the relevant university department abroad attesting the profession and employment of the applicant must also be submitted. Applicants must have an excellent knowledge of English or French.
Amount of Grant: 600,00 euros plus taxes for initial expenses
Date(s) Application is Due: Feb 28
Contact: M. Mikedaki-Drakaki, Chief of Section, 3010-3726300; fax: 3010-3312759
Internet: http://www.iky.gr
Sponsor: State Scholarships Foundation
14 Lyssicratous Street
Athens 105 58
Greece

STC Research Grants 5457
For many years, the Society for Technical Communication has offered research grants to talented individuals doing research on topics that are of concern and interest to the Society's mainstream membership. To date nearly $300,000 has been awarded for research grants. Along with other STC initiatives to increase member satisfaction and educational opportunities, the STC Research Grants program is currently being re-envisioned and reformulated. STC remains committed to funding quality research of value to academic and industry practitioners and teachers.
Requirements: Proposals may be submitted by student members, regular members, and nonmembers of the STC.
Amount of Grant: $10,000 maximum
Date(s) Application is Due: Jun 30; Dec 1
Contact: Rachel Spilka, Manager, STC Research Grants Committee; (414) 962-7254; email: spilka@uwm.edu
Internet: http://www.stc.org/edu/grantsResearch01.asp
Sponsor: Society for Technical Communication
4520 N. Sheffield Avenue
Shorewood, WI 53211

STC Scholarships in Technical Communication 5458
Each year the society awards two scholarships to undergraduates and two scholarships to graduates to assist students who are pursuing established degree programs for careers in some area of technical communication. The completed application form must be accompanied by a certified copy of the latest transcript from the school currently attended and two letters of recommendation from faculty members in the technical communication field.
Requirements: Full-time undergraduate students having upper-class standing in a two- or four-year degree program in technical communication and full-time graduate students working toward a master's or doctoral degree in technical communication may apply.
Amount of Grant: $1,500
Date(s) Application is Due: Feb 15
Contact: Elreatha Matthews, Education Manager; (571) 366-1903 or (703) 522-4114; fax: (703) 522-2075; email: elreatha.matthews@stc.org or stc@stc.org
Internet: http://www.stc.org/edu/scholarshipInfo01.asp
Sponsor: Society for Technical Communication
901 North Stuart Street, Suite 904
Arlington, VA 22203-1822

Stefan and Wanda Wilk Prizes for Research in Polish Music 5459
These prizes are sponsored by the Polish Music Reference Center and School of Music at U.S.C and are intended to stimulate research on Polish music in academic circles outside of Poland. The prizes are awarded in two independent competitions, each held biennially (in different years). The prizes are awarded to authors of the best papers reflecting original research on some aspect of the music of Poland. The competition is comprised of an essay prize and a book prize. The essay prize is divided into two categories, student and professional. The winning essay by an author who is a student will receive a prize of $500. Professionals (who have completed music studies and hold a doctorate or its equivalent) will compete for a prize of $1000.
Requirements: Both essay competitions are open to all authors outside of Poland. All submissions should be in English. For the book prize, the competition is open to works published outside of Poland in English, French, or German during the past 5 years.
Restrictions: Essays previously published in any form will not be considered. Polish scholars or students in Poland are ineligible for the prizes.
Amount of Grant: $500-$1000 for the essay prize; $2000 for the book prize
Date(s) Application is Due: Jun 30
Contact: Dr. Maria Anna Harley, Wilk Prizes in Polish Music, (213) 740-9369; fax: (213) 740-3217; email: polmusic@usc.edu
Internet: http://www.usc.edu/go/polish_music/wilkprizes/wprizes.html
Sponsor: Polish Music Reference Center
University of Southern California, School of Music, 840 W 34th Street
Los Angeles, CA 90089-0851

Stella and Charles Guttman Foundation Grants 5460
The majority of grants are made to organizations providing services to people in the New York City metropolitan area. Contributions also are made to organizations performing scientific research or research relating to the delivery of social services without geographical limitations. The foundation supports programs that address the educational and social service needs of at-risk children and youth, including arts and cultural programs, drug prevention programs, family planning services, health and mental health services, intergenerational programs, programs for the physically disabled, and youth employment programs. The foundation also supports programs that promote the general welfare of the elderly and enhance their quality of life. Charities that conduct activities in Israel, including programs that foster improved relations between Jews and Arabs, also receive support. The foundation maintains a special grant category through which it can respond to special projects and address critical needs outside its principal program areas. Types of support include general operating support, continuing support, program development, and matching funds.
Requirements: Charitable 501(c)3 or 170(b)1 organizations are eligible with a strong emphasis on New York City and Israel.
Restrictions: The foundation does not make grants directly to individuals or to organizations not qualified as charitable, for foreign travel or study, to initiate or defend public interest litigation, to support anti-vivisectionist causes, or to religious organizations for religious observances.
Geographic Focus: New York
Samples: Educational Alliance (New York, NY)—for general support, $25,000.; Creative Ctr for Women with Cancer (New York, NY)—for general support, $25,000.; Givat Haviva Educational Foundation (New York, NY)—for the Children Teaching Children program and Ibn Rouchd School in Jat, Israel, $16,000.
Amount of Grant: $5000-$30,000 average
Contact: Elizabeth Olofson, Executive Director, (212) 371-7082; fax: (212) 371-8936; email: info@guttmanfoundation.org
Internet: http://www.guttmanfoundation.org
Sponsor: Stella and Charles Guttman Foundation
122 E 42nd Street, Suite 2010
New York, NY 10168

Stella Blum Research Grant 5461
This research grant will be given to an undergraduate or graduate student to support a single research project focusing on any aspect of North American costume. This grant supports student scholarships and increases visibility for the recipients and their schools.
Requirements: Candidates must be Costume Society members in good standing; students matriculating, at the time of their funded research, in a degree program at an accredited institution; doing research projects in the field of North American costume.
Amount of Grant: $3000 maximum plus travel component of up to $500 to attend National Symposium to present completed research
Date(s) Application is Due: May 1
Contact: Stella Blum Grant Administrator, (800) 272-9447 or (410) 275-1619; fax: (410) 275-8936; email: national.office@costumesocietyamerica.com
Internet: http://costumesocietyamerica.com/GrantsAwards/stellablum.html
Sponsor: Costume Society of America
P.O. Box 73, 55 Edgewater Drive
Earleville, MD 21919

Stephen S. Chang Award 5462
This award is given annually to recognize a scientist, technologist, or engineer who has made distinguished accomplishments in basic research, and the knowledge the person thus produced must have been utilized by industries for the improvement or development of products related to lipids. A prospective recipient must agree to be present for acceptance of the award and to deliver an award address at an AOCS annual meeting.
Requirements: Eligibility for nomination shall be demonstrated through high-quality scientific papers, patents, or the marketing of commercial products developed through the nominee's scientific or technological efforts. The award shall be made without regard for national origin, place of residence, race, color, creed, or gender.
Amount of Grant: $6000 approximately
Date(s) Application is Due: Oct 15
Contact: Program Contact, (217) 359-2344; fax: (217) 351-8091; email: membership@aocs.org
Internet: http://www.aocs.org/member/awards/awd-chang.asp
Sponsor: American Oil Chemists' Society
P.O. Box 3489
Champaign, IL 61826-3489

Sterling-Turner Charitable Foundation Grants 5463
The charitable foundation awards grants to Texas organizations, primarily for higher and secondary education, adult basic education and literacy, social services, youth, the elderly, fine and performing arts groups and other cultural programs, church support and religious programs (Catholic, Jewish, and Protestant), hospitals and health services, hospices, research, conservation, and civic and urban affairs. Grants are awarded for general operating support, annual campaigns, capital campaigns, continuing support, building construction/renovation, equipment acquisition, endowment funds, program and project development, conferences and seminars, curriculum development, fellowships, scholarship funds, research, and matching funds. The board meets in April.
Requirements: Nonprofit Texas organizations in Tom Green, Kerr, Fort Bend, Harris and Travis County, may submit written requests for grants.
Restrictions: Individuals are ineligible.
Geographic Focus: Texas
Samples: Kinkaid School (Houston, TX)—for program support, $10,000.; Houston Baptist U (Houston, TX)—for program support, $25,000.; Museum of Fine Arts of Houston (Houston, TX)—for operating support, $166,667.
Amount of Grant: $5000-$25,000 average
Date(s) Application is Due: Mar 1
Contact: Patricia Moser, Executive Assistant, (713) 237-1117; fax: (713) 223-4638; email: patricia@sterlingturnerfoundation.org
Internet: http://sterlingturnerfoundation.org
Sponsor: Sterling-Turner Charitable Foundation
815 Walker Street, Suite 1543
Houston, TX 77002-5724

Stoody-West Fellowship for Graduate Study in Religious Journalism 5464
The grant is given to assist a Christian student engaged in religious journalism, or one planning to enter this field, in taking graduate study at an accredited school or department of journalism of his/her choice. The purpose of the fellowship is to enhance the recipient's professional competence and thereby help to perpetuate the standards exemplified by Ralph Stoody and Arthur West. Religious journalism is broadly interpreted to include audiovisual, electronic, and print journalism. Half of the grant is paid in August or September after the recipient is enrolled in the graduate program; the second half is paid in December. Applications are available on the Web site.
Requirements: Applicant should submit three letters of recommendation, an official academic transcript mailed directly by the school, a statement about interest in religious journalism and future plans, four examples of journalistic work, a list of published articles, and a recent photograph.
Restrictions: Funds are not paid for summer sessions.
Amount of Grant: $6000
Date(s) Application is Due: Mar 15
Contact: Fellowship Committee, (888) 278-4862; email: scholarships@umcom.org
Internet: http://umcom.org/pages/news.asp?class=6&ID=427&type=2&product_id=0
Sponsor: United Methodist Communications
P.O. Box 320
Nashville, TN 37202-0320

Strake Foundation Grants 5465
The purpose of the foundation's grant program is to fund primarily Roman Catholic-affiliated associations, including hospitals and higher and secondary educational institutions. Additional areas of interest include adult basic education and literacy, museums, and arts and culture. Support is provided for operating budgets, continuing support, annual campaigns, special projects, matching funds, scholarship funds, building construction/renovation, and research.
Requirements: Awards are made to organizations located only in the United States, primarily in Texas.
Restrictions: Awards are not made to support elementary schools or individuals, nor for deficit financing, consulting services, technical assistance, publications, or loans.
Geographic Focus: Texas
Samples: College of Saint Thomas More (Fort Worth, TX)—to support courses and materials related to the college's new bachelor of arts degree, $10,000.
Amount of Grant: $1000-$20,000 average
Date(s) Application is Due: Apr 1; Oct 1
Contact: George Strake Jr., President, (713) 216-2400; fax: (713) 216-2401; email: foundation@strake.org
Sponsor: Strake Foundation
712 Main Street, Suite 3300
Houston, TX 77002

Streisand Foundation Grants 5466
The foundation awards grants for national programs affecting women and at-risk youth. Areas of support include programs for children and youth, with a focus on the economically disadvantaged; civil rights, with emphasis on race relations between African Americans and Jews; AIDS research, advocacy, service, and litigation; nuclear disarmament; and environmental preservation. Types of support include general operating grants, technical assistance, and project-specific support. Applicants are asked to submit one- to two-page letters of inquiry along with a copy of the 501(c)3 determination letter. Applications are accepted between September 1 and the listed deadline. Guidelines are available online.
Requirements: 501(c)3 nonprofits and California projects in the Los Angeles area are eligible.
Restrictions: Grants are not made to individuals nor do they support start-up organizations, endowments, or capital campaigns.
Samples: Crossroads Community Foundation (Santa Monica, CA)—to support free quality arts and enrichment programming for children at Walgrove Elementary School in Venice, CA, $5000.; Women Against Gun Violence (Culver City, CA)—to galvanize support for the reduction and prevention of gun violence, $5000.; League of Conservation Voters Education Fund (Washington, DC)—to help grassroots environmental organizations motivate their members to be involved in the decision-making process, $20,000.
Amount of Grant: $10,000-$20,000 average
Date(s) Application is Due: Dec 1

Contact: Program Contact, (310) 535-3767; fax: (310) 314-8396; email: stfnd@aol.com
Internet: http://www.barbrastreisand.com/bio_streisand_foundation.html
Sponsor: Streisand Foundation
2800 28th Street, Suite 105
Santa Monica, CA 90405

Stroke Association Allied Health Professional Research Bursaries 5467
The Research Bursary Awards are primarily intended for the purpose of training nurses and therapists, but consideration will be given to other health professionals. These bursaries will be awarded to UK departments that can demonstrate a track record and current participation in stroke research. The host department will be required to provide appropriate supervision and a research training program suitable for equipping the candidate for a career in stroke research, and to enable them to obtain a post graduate qualification. Bursaries may be taken up on a part-time, pro-rata basis.
Requirements: Research must take place in the United Kingdom.
Samples: Linda Hammett, St. George's, University of London (2007);Andrea Atzori, University of Southampton (2007); Emma Jones, University of Sheffield (2007).
Amount of Grant: 30,000 Britsh pounds per annum for up to two years (up to a maximum of 60,000 British pounds)
Date(s) Application is Due: Jan 5
Contact: Research Department, (020) 7566-0345; fax: (020) 7490-2686; textphone: (020) 7251-9096; email: research@stroke.org.uk
Internet: http://www.stroke.org.uk/research/apply_for_funding/allied_health.html
Sponsor: Stroke Association
Stroke House, 240 City Road
London, England EC1V 2PR United Kingdom

Stroke Association Clinical Fellowships 5468
Fellowships are awarded to individuals for training in stroke medicine, and are intended to enable a specialist registrar to gain the appropriate clinical experience and accreditation required for a career in the prevention, treatment or rehabilitation of stroke.
Requirements: Applicants are required to nominate a supervisor and training program in a host institution that has been approved for stroke medicine training by the Stroke Medicine Sub-specialty Advisory Committee (SSAC). The funding is not to be taken as part of a one year taught course or for those who have completed the SpR training program and have certification.
Amount of Grant: 40,000 British pounds
Date(s) Application is Due: Jan 5
Contact: Research Department, (020) 7566-0345; fax: (020) 7490-2686; textphone: (020) 7251-9096; email: research@stroke.org.uk
Internet: http://www.stroke.org.uk/research/research_we_fund/fellowships/index.html
Sponsor: Stroke Association
Stroke House, 240 City Road
London, England EC1V 2PR United Kingdom

Stroke Association Research Project Grants 5469
The Association is particularly interested in receiving project applications related to stroke prevention, acute treatment, assessment, ethnic factors, epidemiology, rehabilitation, psychology and stroke in people of working age. Applications in the fields of epidemiology, natural history, prevention, diagnosis, imaging, therapy both medical and surgical, assessment and rehabilitation are particularly welcome.
Requirements: Research must take place in the United Kingdom.
Restrictions: Project grants do not cover: senior lecturer (or equivalent) or consultant salaries; routine laboratory equipment (unless the volume of work created by the project would be problematic); computers (unless justified, and only up to 1000 British pounds); maintenance contracts; non-contract breakdown and maintenance costs; travel to conferences and meetings (unless project is a large trial); routine secretarial support; NHS extra treatment costs; redundancy costs (unless included and agreed at the time of application); general institutional overheads; or costs for open access publishing over a total of 2000 British pounds for the entire award.
Amount of Grant: 60,000 British pounds per annum for up to three years (up to a maximum of 180,000 British pounds)
Date(s) Application is Due: Jul 20
Contact: Research Department, (020) 7566-0345; fax: (020) 7490-2686; textphone: (020) 7251-9096; email: research@stroke.org.uk
Internet: http://www.stroke.org.uk/research/apply_for_funding/research_project.html
Sponsor: Stroke Association
Stroke House, 240 City Road
London, England EC1V 2PR United Kingdom

STTI Alpha Eta Collaborative Research Grants 5470
The purpose of the STTI Alpha Eta Collaborative Research Grant is to provide up to U.S.$2000 research grant to Tau Lambda-at-large chapter member(s) for a new research project in Africa. Funds for this grant are provided by Alpha Eta Chapter.
Requirements: Eligibility Criteria: member of Tau Lambda-at-Large residing in Africa, submit online a completed research application package and signed research agreement via the online submission system; ready to implement research project when funding is received; submit a final report; submit completed abstract to Virginia Henderson International Nursing Research Library and credit research grant partners in all publications and presentations of the research.
Amount of Grant: $2,000
Date(s) Application is Due: Dec 1
Contact: Research Services; (888) 634-7575 or (317) 634-8171; fax: (317) 634-8188; email: research@stti.iupui.edu
Internet: http://www.nursingsociety.org/Research/SmallGrants/Pages/AlphaEta.aspx
Sponsor: Sigma Theta Tau International
550 West North Street
Indianapolis, IN 46202

STTI Doris Bloch Research Awards 5471
The purpose of the STTI Doris Bloch Research Awards is to encourage nurses to contribute to the advancement of nursing through research. Allocation of funds is based on the quality of the proposed research, the future promise of the applicant and the applicant's research budget. Applications from novice researchers who have received no other national research funds are encouraged and will receive preference for funding, other aspects being equal. Preference will be given to Sigma Theta Tau International members, other qualifications being equal. Funding is provided by a gift to the honor society's Research Endowment from the Doris Bloch estate.
Requirements: Eligibility Criteria: registered nurse with current license; hold a master's or doctoral degree or be enrolled in a doctoral program; submission of a completed research application package and signed research agreement via on-line submission system; ready to implement research project when funding is received.
Amount of Grant: $5,000 max.
Date(s) Application is Due: Dec 1
Contact: Research Services; (888) 634-7575 or (317) 634-8171; fax: (317) 634-8188; email: research@stti.iupui.edu
Internet: http://www.nursingsociety.org/Research/SmallGrants/Pages/grant_bloch.aspx
Sponsor: Sigma Theta Tau International
550 West North Street
Indianapolis, IN 46202

STTI Emergency Nurses Association Foundation Grant 5472
The ENA Foundation and Sigma Theta Tau International, Inc. (STTI) have combined resources to offer this annual research grant. The ENA Foundation/STTI research grant provides funding for research that will advance the specialized practice of emergency nursing. All relevant research topics will be considered. Priority will be given to research studies relating to the ENA/ENA Foundation research initiatives.
Requirements: Eligibility Criteria: principal investigator must be a registered nurse; team members may be from other disciplines; principal investigator must have a master's degree; submit a completed research application package and signed research agreement; be ready to start, or have already started, the research project. An electronic Research Grant Application is available on the ENA Foundation website. All applications MU.S.T be returned to ENA Foundation.
Restrictions: Application packets submitted to Sigma Theta Tau International will not be reviewed.
Samples: Sarah Lynn Anderson, MSN, PhD., University of Virginia - Project Title: "Changes in Genital Injuries in Women Based on Skin color after Consensual Intercourse" (2008);
Amount of Grant: $6000
Date(s) Application is Due: Mar 1
Contact: Department of Research; (847) 460-4119; fax: (847) 460-4005; email: res@ena.org; Suling Li, Director of Research; (847) 460-4110; email: sli@ena.org
Internet: http://www.nursingsociety.org/Research/SmallGrants/Pages/grant_ena.aspx
Sponsor: Emergency Nurses Association Foundation
915 Lee Street
Des Plaines, IL 60016 -6569

STTI Environment of Elder Care Nursing Research Grants 5473
The Environment of Elder Care Nursing Research Grant co-sponsored by Hill-Rom Company will advance the science of nursing through research focused on critical aspects of elder care including clear lungs, no falls, safe skin, patient comfort and ease-of-use. Applications from novice researchers who have received no other national research funds are encouraged. Preference will be given to Sigma Theta Tau International members, other qualifications being equal. Allocation of funds is based on the quality of the proposed research, the future promise of the applicant, and the applicant's research
Requirements: Eligibility Criteria: registered nurse with a current license; hold at minimum a Master's degree or its equivalent, and/or be enrolled in a Doctoral program; submission of completed research application packet and a signed research agreement via our on-line submission system; ready to implement research project when funding is received; submit to STTI a final report; submit completed abstract to STTI's Virginia Henderson International Nursing Research Library; and submit an abstract for publication to Hill-Rom regarding the research done as a product of the grant, and credit research grant partners in all publications and presentations of the research.
Amount of Grant: $9,000 max.
Date(s) Application is Due: Jul 1
Contact: Research Services; (888) 634-7575 or (317) 634-8171; fax: (317) 634-8188; email: research@stti.iupui.edu
Internet: http://www.nursingsociety.org/Research/SmallGrants/Pages/EnvironmentofCare.aspx
Sponsor: Sigma Theta Tau International
550 West North Street
Indianapolis, IN 46202

STTI Joan K. Stout RN Research Grants 5474

The purpose of the Sigma Theta Tau International/Joan K. Stout, RN, Research Grant is to advance ongoing evidence-based study by nurse researchers on the impact of the practice of simulation education in schools of nursing and clinical care settings. The allocation of funds is based upon a research project that is ready for implementation. The proposed research project should be designed to ensure the ongoing practice of nurse-led simulation in improving quality of care in clinical and/or academic settings with the potential for further funding and ongoing research. Funds provided by interest from a research endowment established at Sigma Theta Tau International Foundation for Nursing, with a donation by the Hugoton Foundation.
Requirements: Eligibility Criteria: registered nurse with current license; have a master's degree, and/or be enrolled in a Doctoral program; submit a completed research application via online submission system; ready to implement research project when funding is received; complete project within one (1) year of funding; submit to STTI a final report; submit completed abstract to STTI's Virginia Henderson International Nursing Research Library.
Amount of Grant: $5,000
Date(s) Application is Due: Jul 1
Contact: Research Services; (888) 634-7575 or (317) 634-8171; fax: (317) 634-8188; email: research@stti.iupui.edu
Internet: http://www.nursingsociety.org/Research/SmallGrants/Pages/JoanKStout.aspx
Sponsor: Sigma Theta Tau International
550 West North Street
Indianapolis, IN 46202

STTI National League for Nursing Grants 5475

Funds for this grant are made possible from a gift from Dr. Diane Billings to Sigma Theta Tau International Foundation for Nursing's research endowment. The purpose of this grant is to support research that advances the science of nursing education and learning through the use of technology in dissemination of knowledge.
Requirements: Eligibility Criteria: preference given to Sigma Theta Tau International and/or National League for Nursing members; grant proposals must advance the science of nursing education through the use of technology in the dissemination of knowledge; registered nurse with current license; hold a master's or doctoral degree or be enrolled in a doctoral program; submit a completed research application package and signed research agreement; ready to implement research project when funding is received. All applications must be submitted via the online submission system (see STTI website)
Amount of Grant: $5,000
Date(s) Application is Due: Jun 1
Contact: Research Services; (888) 634-7575 or (317) 634-8171; fax: (317) 634-8188; email: research@stti.iupui.edu
Internet: http://www.nursingsociety.org/Research/SmallGrants/Pages/grant_nln.aspx
Sponsor: Sigma Theta Tau International
550 West North Street
Indianapolis, IN 46202

STTI Rosemary Berkel Crisp Research Awards 5476

The purpose of the STTI Rosemary Berkel Crisp Research Awards is to support nursing research in the critical areas of women's health, oncology, and pediatrics. The allocation of funds is based upon a research project that is ready for implementation in the area of women's health, oncology, or pediatrics; the quality of the proposed research, future potential of the application, appropriateness of the research budget, and feasibility of time frame. Funds provided by interest from a gift from The Harry L. Crisp, II and Rosemary Berkel Crisp Foundation to the honor society's Research Endowment.
Requirements: Eligibility Criteria: registered nurse with current license; hold a master's or doctoral degree or be enrolled in a doctoral program; submit a completed research application package and signed research agreement via the online submission system; ready to implement research project when funding is received; member of Sigma Theta Tau International; Note: Some preference will be given to applicants residing in Illinois, Missouri, Arkansas, Kentucky, and Tennessee.
Amount of Grant: $5,000 max.
Date(s) Application is Due: Dec 1
Contact: Research Services; (888) 634-7575 or (317) 634-8171; fax: (317) 634-8188; email: research@stti.iupui.edu
Internet: http://www.nursingsociety.org/Research/SmallGrants/Pages/grant_crisp.aspx
Sponsor: Sigma Theta Tau International
550 West North Street
Indianapolis, IN 46202

STTI Small Grants 5477

The purpose of the Sigma Theta Tau International Small Grants is to encourage nurses to contribute to the advancement of nursing through research. Allocation of funds is based on the quality of the proposed research, the future promise of the applicant, and the applicant's research budget. Applications from novice researchers who have received no other national research funds are encouraged and will receive preference for funding, other aspects being equal. Preference will be given to Sigma Theta Tau members, other qualifications being equal.
Requirements: Eligibility Criteria: registered nurse with current license; have a master's or doctoral degree or be enrolled in a doctoral program; submit a completed research application package and signed research agreement via the online submission system; ready to implement research project when funding is received.
Amount of Grant: $5000 maximum
Date(s) Application is Due: Dec 1
Contact: Research Services; (888) 634-7575 or (317) 634-8171; fax: (317) 634-8188; email: research@stti.iupui.edu
Internet: http://www.nursingsociety.org/Research/SmallGrants/Pages/small_grants.aspx
Sponsor: Sigma Theta Tau International
550 West North Street
Indianapolis, IN 46202-2156

STTI Virginia Henderson Clinical Research Grants 5478

The purpose of the Sigma Theta Tau International Virginia Henderson Clinical Research Grant is to encourage the research career development of clinically based nurses through support of clinically oriented research. Allocation of funds is based on a research project ready for implementation, the quality of the proposed research, the future potential of the applicant, appropriateness of the research budget and feasibility of the time frame. Funds for this biennial grant are provided by interest from the Virginia Henderson Clinical Research Endowment Fund.
Requirements: Eligibility Criteria: registered nurse actively involved in some aspect of health care delivery, education or research in a clinical setting; hold a master's or doctoral degree or be enrolled in a doctoral program; submit a completed research application package and signed research agreement via the online submission system; ready to implement research project when funding is received; member of Sigma Theta Tau International.
Amount of Grant: $5,000 max.
Date(s) Application is Due: Dec 1
Contact: Research Services; (888) 634-7575 or (317) 634-8171; fax: (317) 634-8188; email: research@stti.iupui.edu
Internet: http://www.nursingsociety.org/Research/SmallGrants/Pages/grant_VHL.aspx
Sponsor: Sigma Theta Tau International
550 West North Street
Indianapolis, IN 46202

STTI/ATI Educational Assessment Nursing Research Grants 5479

The purpose of this grant, co-sponsored by Assessment Technologies Institute (ATI), with the Honor Society of Nursing, Sigma Theta Tau International will support research that encourages the appropriate use of a standardized assessment program in a nursing education curriculum, including, but not limited to: use of formative assessment to drive remediation; use of summative assessment to guide curricular evaluation and revision; timing of assessment administration within a program of study; identification of key markers of at-risk student status.
Requirements: Eligibility Criteria: registered nurse with current license; have a master's or doctoral degree or be enrolled in a doctoral program; submit a completed research application package and signed research agreement via the online submission system; ready to implement research project when funding is received; submit a final report; submit completed abstract to the Virginia Henderson Research Library and credit research grant partners in all publications and presentations of the research.
Amount of Grant: $4,500
Date(s) Application is Due: Jul 1
Contact: Research Services; (888) 634-7575 or (317) 634-8171; fax: (317) 634-8188; email: research@stti.iupui.edu
Internet: http://www.nursingsociety.org/Research/SmallGrants/Pages/STTIATI.aspx
Sponsor: Sigma Theta Tau International
550 West North Street
Indianapolis, IN 46202

STTI/Council for the Advancement of Nursing Science Grants 5480

The purpose of the Sigma Theta Tau International/The Council for Advancement of Nursing Science research grant is to encourage qualified nurses to improve the health of the world, people through research. Proposals for clinical, educational or historical research, including plans for broadly disseminating the research findings, may be submitted for the grant.
Requirements: Eligibility Criteria: preference given to Sigma Theta Tau International member; hold at minimum a Master's degree or equivalent; submission of completed research application via on-line submission system; ready to implement research project when funding is received; submit a final report; submit completed abstract to Virginia Henderson International Nursing Research Library; submit completed project, abstract for presentation at Sigma Theta Tau or Council for Advancement of Nursing Science research meeting; and credit research grant partners in all publications and presentations of the research.
Amount of Grant: $5,000
Date(s) Application is Due: Jul 1
Contact: Research Services; (888) 634-7575 or (317) 634-8171; fax: (317) 634-8188; email: research@stti.iupui.edu
Internet: http://www.nursingsociety.org/Research/SmallGrants/Pages/CANS.aspx
Sponsor: Sigma Theta Tau International
550 West North Street
Indianapolis, IN 46202

Stuart Foundation Grants 5481

The foundation's mission is to help children and youth in California and the state of Washington become capable, responsible citizens. Areas of interest include strengthening communities that serve families (community building, service integration, institutional and community partnerships, professional development, and technical assistance), strengthening the child welfare system (permanency, supportive services for youth, comprehensive assessment and services, prevention, or volunteering), strengthening public schools (teacher development, public policy, early learning, systemic school reform, and other opportunities), and policy analysis and advocacy. The foundation is interested in funding programs that promote collaboration and integrated services to improve children's outcomes. Grants also include assistance with community engagement, conflict resolution, research, technology, and training. Types of support include seed money, operating support, and special projects. Programs receive funding each year; renewal is subject to board and staff decisions. Obtain a guidelines brochure from the office.

Requirements: The foundation supports programs serving California and Washington.

Restrictions: The foundation does not make grants to support political activities, endowments, building campaigns, fundraising events, material acquisition, or operating funds.

Geographic Focus: California; Washington

Samples: Washington Education Foundation (Issaquah, WA)—for the Foster Care to College Partnership, which seeks to expand access to college for children in Washington State who are in foster care, $234,751 (2005); Partnership for Learning (Seattle, WA)—for its efforts to increase public understanding of, and support for, standards-based education reforms in Washington State, $100,000 (2005); Parent Services Project (San Rafael, CA)—for an institute that works to build the leadership skills of parents in Marin County, CA, $25,000 (2005); Foundation Consortium/San Francisco Foundation Community Initiative Funds (San Francisco, CA)—for general support as this organization ceases operations, documents its work, and transitions its projects to other entities, $160,000 (2005).

Amount of Grant: $5000-$100,000 range

Date(s) Application is Due: Mar 1; Jun 1; Sep 1; Dec 1

Contact: Stephanie Titus, Grants Manager, (415) 393-1551; fax: (415) 393-1552

Internet: http://www.stuartfoundation.org/how.html

Sponsor: Stuart Foundation
50 California Street, Suite 3350
San Francisco, CA 94111-4735

Stuart L. Bernath Book Prize 5482

Prizes are given annually to an author for a first book dealing with any aspect of the history of American foreign relations. The purpose of the award is to recognize and encourage distinguished research and writing by scholars of American foreign relations. Books may be nominated by the author, the publisher, or by any member of the society. Winning books should have interpretive and analytical qualities of high levels, demonstrate mastery of primary material and relevant secondary works, and be examples of careful organization and distinguished writing.

Requirements: Book must be a history of international relations; biographies of statespeople and diplomats are included. A nominating letter by the author, the publisher, or any member of the Society for Historians of American Foreign Relations explaining why the book deserves consideration must accompany each entry in the competition. Five copies of each book must be submitted with a letter of nomination.

Restrictions: General surveys, autobiographies, editions of essays and documents, and works that are primarily representative of social science disciplines other than history are not eligible.

Samples: Christopher Endy—prize recipient for Cold War Holidays: American Tourism in France, $2500 (2005); David Engerman—prize recipient for Modernization from the Other Shore: American Intellectuals and the Romance of Russian Development, $2500 (2004).

Amount of Grant: $2500

Date(s) Application is Due: Dec 1

Contact: Prize Administrator, (614) 292-1951 or (614) 292-7200; fax: (614) 292-2282; email: shafr@osu.edu

Internet: http://www.shafr.org/prizes.htm

Sponsor: Society for Historians of American Foreign Relations
Ohio State University, Dept of History, 106 Dulles Hall, 230 W 17th Avenue
Columbus, OH 43210

Stuart L. Bernath Lecture Prize 5483

The lecture, to be delivered at the annual meeting of the Organization of American Historians, will be given by a younger scholar with an excellent reputation for research and teaching. Each lecturer will not address specifically his/her own research interests but broad issues of concern to students of American foreign policy. Inquiries may also be directed to William O. Walker III, Department of History, Florida International University, University Park, Miami, FL 33199.

Requirements: The prize is open to any person under 41 years of age or within 10 years of the receipt of the PhD whose scholarly achievements represent excellence in teaching and research. Nominations may be made by any member of SHAFR or of any other established history, political science, or journalism department or organization.

Samples: Kristin Hoganson—prize recipient, $500 (2005); Kurk Dorsey—prize recipient, $500 (2004).

Amount of Grant: $500

Date(s) Application is Due: Feb 28

Contact: Prize Administrator, (614) 292-1951 or (614) 272-7200; fax: (614) 292-2282; email: shafr@osu.edu

Internet: http://www.shafr.org/prizes.htm

Sponsor: Society for Historians of American Foreign Relations
Ohio State University, Dept of History, 106 Dulles Hall, 230 W 17th Avenue
Columbus, OH 43210-1367

Stuart L. Bernath Scholarly Article Prize 5484

The purpose of the prize is to recognize and encourage distinguished research and writing by young scholars in the field of diplomatic relations. The competition is open to any article or essay appearing in a scholarly journal or edited book on any topic in U.S. foreign relations published during the year preceding the award. All articles appearing in the journal Diplomatic History will be automatically considered without nomination. Other nominations should be submitted by the author or by any member of the society.

Requirements: The author must not be over 40 years of age or if more than 40, must be within 10 years of receiving the PhD at the time of acceptance for publication. The article or essay must be among the first six publications by the author.

Restrictions: Previous winners of this prize are not eligible.

Samples: Lawrence A. Peskin—prize recipient, $1000 (2005); Max Paul Friedman—prize recipient, $1000 (2004).

Amount of Grant: $1000

Date(s) Application is Due: Feb 1

Contact: Prize Administrator, (614) 272-7200; fax: (614) 292-2282; email: shafr@osu.edu

Internet: http://www.shafr.org/prizes.htm

Sponsor: Society for Historians of American Foreign Relations
Ohio State University, Dept of History, 106 Dulles Hall, 230 W 17th Avenue
Columbus, OH 43210

Summerfield G. Roberts Award 5485

A cash award is presented at the annual membership meeting in the spring to the author of a work of creative writing on the Republic of Texas. The purpose of this award is to encourage literary effort and research about historical events and personalities during the days of the Republic of Texas, 1836-1846, and to stimulate interest in this period. The judges will determine which manuscript best portrays the spirit, character, strength, and deeds of men and women who lived in the Republic of Texas. The manuscripts must be written or published during the calendar year for which the award is given. There is no word limit on the material submitted for the award. The manuscripts may be either fiction or nonfiction, poems, essays, short stories, novels, or biographies. Judges for a given year are the winners of the award in the last three years. Five copies of each entry must be mailed to the office shown for the use of the judges and to add to the archives; no copies are returned.

Requirements: The competition is open to all writers everywhere; they need not reside in Texas, nor must the publishers be in Texas.

Samples: New Orleans and the Texas Revolution, By Edward L. Millier, $2500 (2004); Frontier Blood, the Saga of the Parker Family by Jo Ella Powell Exley.

Amount of Grant: $2500

Date(s) Application is Due: Jan 15

Contact: Secretary, (979) 245-6644; fax: (979) 244-3819; email: srttexas@srttexas.org

Internet: http://www.srttexas.org/sumfield.html

Sponsor: Sons of the Republic of Texas
1717 Eighth Street
Bay City, TX 77414

Sun Microsystems Academic Equipment Grants 5486

The program grants equipment to academic and research institutions for creative projects that address Sun Microsystems's priorities and create partnerships for success. The primary investment priorities are higher education—teaching of Sun technologies, Web-based learning, scientific and engineering computing, and business collaborations; and K-12— primary and secondary education, and university outreach to K-12. Application form and guidelines are available online.

Requirements: U.S. 3304(f) or 501(c)3 tax-exempt organizations that are degree-granting institutions or involved in basic research are eligible. Non-U.S. institutions must either be degree-granting or involved in basic research and be recognized as noncommercial enterprises by their national governments. 501(c)3 nonprofits that are neither degree-granting nor directly involved with basic research may be eligible if they directly support the academic and research communities in their mission to create and disseminate knowledge, and if they demonstrate financial need.

Restrictions: This program is for hardware donations only. Grants do not include maintenance, service, or technical support.

Date(s) Application is Due: Mar 2; Jun 2; Sep 2; Dec 2

Contact: Sabina Chitkara, AEG Program Office, (650) 786-8775; fax: (650) 786-3451; email: sabina.chitkara@eng.sun.com

Internet: http://www.sun.com/edu/grants

Sponsor: Sun Microsystems Foundation
901 San Antonio Road, MS MPK15-204
Palo Alto, CA 94303

Sunny and Abe Rosenberg Foundation Grants 5487

The foundation awards grants to New York City area nonprofit organizations in its areas of interest, including health organizations—mental health, cancer research, and programs for the young, elderly, or physically impaired; education—excellence in public schools, libraries, and groups that fight discrimination; artistic and cultural—

museums, performing arts, and arts education in public schools; recreation—city parks and youth camps; social services—food distribution to the needy and elderly, AIDS prevention, family planning, youth employment, and alcohol and drug rehabilitation; and international—orphanages and libraries. Grants generally are awarded for one year, but multiyear funding will be considered for exceptional projects.
Requirements: New York City-area nonprofit organizations are eligible.
Geographic Focus: New York
Amount of Grant: $10,000-$50,000 range
Contact: Karen McQuade, Executive Assistant, (212) 660-1214; fax: (212) 660-2180; email: info@rosenbergfoundation.org
Internet: http://www.rosenbergfoundation.org
Sponsor: Sunny and Abe Rosenberg Foundation
888 7th Avenue, 24th Fl
New York, NY 10019

Sunshine Natural Wellbeing Foundation Grants 5488
The foundation awards grants to nonprofit organizations for existing or upstart programs relating to alternative treatments and research for asthma, immune disorders, and other similar illnesses, as well as children's educational programs relative to alternative health. Submit a proposal with specific goals, project budget, organization budget, audited financial reports, 501(c)3 letter, IRS 990 for previous year, list of current contributors, and list of board members.
Requirements: 501(c)3 nonprofit organizations are eligible. Giving is primarily made in Florida.
Restrictions: Individuals are not eligible.
Geographic Focus: Florida
Date(s) Application is Due: Sep 1
Contact: Delores Wooldridge, (941) 346-5297
Sponsor: Sunshine Natural Wellbeing Foundation
6547 Midnight Pass Road, Suite 19
Sarasota, FL 34242

Supelco AOCS Research Award 5489
This award is given annually to a qualified scientist for outstanding original research on fats, oils, lipid chemistry, and/or biochemistry, the results of which have been published in technical papers of high quality. A travel and expense allowance and a plaque are provided. Application materials are available upon request.
Amount of Grant: $8000
Date(s) Application is Due: Nov 1
Contact: Awards Officer, (217) 359-2344; fax: (217) 351-8091; email: membership@aocs.org
Internet: http://www.aocs.org/member/awards/awd-research.asp
Sponsor: American Oil Chemists' Society
P.O. Box 3489
Champaign, IL 61826-3489

Supreme Court Fellows Commission Fellowships 5490
The Supreme Court Fellows Commission seeks outstanding individuals who are interested in working in Washington, DC, within the federal judiciary. Fellows spend one calendar year (late August through early September) at the Supreme Court of the United States, the Federal Judicial Center, the Administrative Office of the United States Courts, or the U.S. Sentencing Commission working on various projects concerning the federal court system and the administration of justice. Up to four fellowships are awarded each year. Candidates must submit a resume, 700-word essay explaining their interest in the program, copies of two publications or other writing samples, and three reference letters. Additional information, an online application, and complete instructions are available online.
Requirements: Candidates must be familiar with the judicial system and have at least one postgraduate degree and two or more years of professional experience with high achievement. Multidisciplinary training and experience is desirable.
Amount of Grant: $$110,878
Date(s) Application is Due: Nov 10
Contact: Administrative Director, Judicial Fellows Program, (202) 479-3415
Internet: http://www.fellows.supremecourtus.gov/index.html
Sponsor: Supreme Court of the 1 First Street NE
Washington, DC 20543

Surdna Foundation Arts Teachers Fellowships 5491
The foundation invites arts teachers from specialized, public arts high schools to apply for funding for artistic development through its arts teachers fellowship program. The program offers teachers the opportunity to immerse themselves in their own creative work, interact with other professional artists, and stay current with new practices. Download complete program information, guidelines, and application material, which are available online.
Requirements: Permanently assigned full- and part-time arts teachers in specialized, public arts high schools who have been teaching for at least five years are eligible. Dance, music, theater-arts, and visual-arts teachers are eligible.
Amount of Grant: $5000; $1500 to each fellow's school
Date(s) Application is Due: Nov 18
Contact: Edward Skloot, Executive Director, (212) 557-0010; fax: (212) 557-0003; email: artsfellowship@surdna.org
Internet: http://www.surdna.org/programs/artsteachersfellowships.html
Sponsor: Surdna Foundation
330 Madison Avenue, 30th Fl
New York, NY 10017

Surdna Foundation Grants 5492
The foundation makes grants in five program areas: the environment, community revitalization, effective citizenry, the arts, and nonprofit-sector support. The environmental program emphasizes projects in four areas: biological diversity and the human communities that depend on it, human systems, transportation and urban/suburban land use, and energy. The citizenry program focuses on dispute resolution and mediation programs, youth development, encouraging community participation, and helping people work together to solve serious social and civic problems. Grants in the community revitalization field focus on access to low-cost housing, helping low-income people become more self-sufficient, and strengthening families and early-childhood education. Funding in the arts focuses on education grants to support general curriculum and professional development tailored to assist students ages 11 through 18. Nonprofit-sector grants concentrate on strengthening the sector's role in American public life and in public-policy creation; promoting the study of overlapping issues such as tax exemption and public funding; galvanizing philanthropic activity; and enhancing nonprofits' management, finance, leadership, and use of technology. The foundation makes both project and general support grants. The foundation recommends potential applicants to submit a letter of inquiry before submitting proposals. Grants are approved in February, May, and September.
Requirements: 501(c)3 organizations may apply. The foundation urges applicants to send two-to three-page letters of intent before sending proposals. IRS nonprofit status certification, recent audited financial statements, and project budget should also be included.
Restrictions: Grants are not made to individuals or given to organizations outside the United States. The foundation generally does not fund capital campaigns or building construction. Unsolicited proposals are not accepted in the arts program.
Samples: National 4-H Council (Chevy Chase, MD)—to start up the Youth in Governance Initiative, which will prepare and place young people in decision-making bodies within the state and local structures of the initiative's three pilot states, $110,000 (2005); Maryland Assoc of Nonprofit Organizations (Baltimore, MA)—for its Standards for Excellence project, which details ethical practices and policies that nonprofit groups should adhere to, $100,000 over two years (2004).
Contact: Edward Skloot, Executive Director, (212) 557-0010; fax: (212) 557-0003; email: request@surdna.org
Internet: http://www.surdna.org
Sponsor: Surdna Foundation
330 Madison Avenue, 30th Fl
New York, NY 10017-5001

Susan G. Komen Breast Cancer Foundation Basic, Clinical, and Translational Research Grants 5493
The foundation awards funding for basic, clinical, and translational breast cancer research. Grants offer two years of support for a variety of research projects in such areas as complementary and alternative medicine; detection and diagnosis, prognosis; risk, prevention, and epidemiology; local treatment; psychosocial treatment; systemic treatment; and tumor cell biology. The foundation gives preference to projects that have potential for high impact in their field, may not be considered by other funders, and benefit the medically underserved in communities. Application deadlines are August 1 for electronic submittal of abstracts; August 8 for electronic submittal of applications; and August 15 for hard-copy applications. Information and application forms are posted on the foundation's Web site.
Requirements: A wide range of nonprofits, including research groups and health and human service agencies, are eligible to apply.
Samples: Dr. Eliot Rosen, Long Island Jewish Medical Ctr (NY)—for the study BRCA1 as a Target for Chemoprevention Agents, $249,939 (2003); Dr. Margot Ip, Roswell Park Cancer Institute—for the study Stromal-Vascular Precursors as a Target for Breast Cancer Prevention, $249,987 (2003); Dr. Georgia Chenevix-Trench, Queensland Institute of Medical Research—for the study The role of the ATM Gene in Familial Breast Cancer, $249,300 (2003); Dr. Amanda Spurdle, Queensland Institute of Medical Research—for the study Evaluation of BRCA1 and BRCA2 Unclassified Sequence Variants, $248,250 (2003).
Amount of Grant: $250,000 maximum over two years
Date(s) Application is Due: Aug 1; Aug 8; Aug 15
Contact: Grants Manager, (888) 300-5582; fax: (972) 855-1640; email: grants@komen.org
Internet: http://www.komen.org/intradoc-cgi/idc_cgi_isapi.dll?IdcService=SS_GET_PAGE&nodeId=390
Sponsor: Susan G. Komen Breast Cancer Foundation
5005 LBJ Fwy, Suite 370
Dallas, TX 75244

Susan G. Komen Breast Cancer Foundation Brinker Award for Scientific Distinction 5494
The award recognizes outstanding work that has advanced basic research concepts, or affected clinical or social-behavioral applications, in the field of breast cancer research, screening, treatment, or prevention. The intent is to recognize scholars for a specific contribution, a consistent pattern of contributions, or leadership in the field that has had a substantial impact on breast cancer. Each award will include an honorarium, citation, and inscribed crystal statuette designed by Tiffany & Co. In the event that more than one person is selected for an award, the honorarium will be divided between the

recipients and each will receive the citation and a statuette. It is strongly suggested that an individual scholar be nominated; however, researchers who have worked as a team or who have each made similarly outstanding contributions in the field may be nominated together in an award category. Awardees are expected to present their important findings in a lectureship at the annual San Antonio Breast Cancer Symposium. Nomination form and guidelines are available online.
Requirements: Breast cancer researchers who are affiliated with nonprofit institutions (academic, industry, government, or other) in the United States and abroad are eligible.
Amount of Grant: $20,000 honorarium
Date(s) Application is Due: Jun 30
Contact: Award Administrator (972) 855-1600; fax: (972) 855-1605
Internet: http://www.komen.org/intradoc-cgi/idc_cgi_isapi.dll?IdcService=SS_GET_PAGE&nodeId=560
Sponsor: Susan G. Komen Breast Cancer Foundation
5005 LBJ Freeway, Suite 250
Dallas, TX 75244

Susan G. Komen Breast Cancer Foundation Interdisciplinary Fellowships **5495**
The foundation serves as a catalyst to develop fellowship training for interdisciplinary programs in breast care that will prepare physicians for careers devoted to serving the multispecialty needs of the breast cancer patient; offer an interdisciplinary curriculum integrated into a comprehensive program; and provide special emphasis on enhancing the physician's understanding of the patient with benign and malignant breast disease while developing a better treatment environment for future patients. The program will fund two consecutive one-year fellowships. Guidelines are available online.
Requirements: The host institution must be an accredited medical institution in the United States or similarly recognized non-U.S. institution.
Restrictions: For-profit organizations are not eligible.
Amount of Grant: $35,000 per year
Date(s) Application is Due: Feb 7
Contact: Dianne McDonald, Research Grants Manager, (888) 300-5582; email: grants@komen.org
Internet: http://www.komen.org/intradoc-cgi/idc_cgi_isapi.dll?IdcService=SS_GET_PAGE&nodeId=395
Sponsor: Susan G. Komen Breast Cancer Foundation
5005 LBJ Freeway, Suite 250
Dallas, TX 75244

Susan G. Komen Breast Cancer Foundation Population Specific Research Grants **5496**
The foundation supports research project grants in support of innovative efforts focusing on the epidemiology of breast cancer within specific populations at risk for the disease, such as African American, Asian/Pacific Islander, Hispanic, Native American, lesbian, and low-literacy populations. Grants provide two years of support. Annual deadline dates may vary; contact program staff for exact dates.
Requirements: U.S. nonprofit organizations and international institutions are eligible.
Samples: Dr. Gail Tomlinson, U of Texas Southwestern Medical Ctr (TX)—for Decision Making and BRCA Testing in African Americans, $250,000 (2003); Dr. Deborah Erwin, U of Arkansas for Medical Sciences (AR)—for Development Of A Latina Witness Project Model, $248,480 (2003); Anna Wu, Ph.D. University of Southern California (CA)—for Soy, Tamoxifen, and Breast Cancer in Asian-Americans, $250,000 (2003).
Amount of Grant: $250,000 over two or three years
Contact: Director, Grants and Sponsored Programs, (972) 855-1600; fax: (972) 855-1605; email: grants@komen.org
Internet: http://www.komen.org/grants/available.asp
Sponsor: Susan G. Komen Breast Cancer Foundation
5005 LBJ Fwy, Suite 250
Dallas, TX 75244

Susan G. Komen Breast Cancer Foundation Postdoctoral Fellowships **5497**
Postdoctoral fellowships offer three years of support for basic, clinical, or translational cancer research and public health or epidemiology studies. Detailed program-specific eligibility requirements and application information are available directly from the foundation or its Web site. Abstracts submitted electronically are due August 1; applications submitted electronically are due August 8; hard-copy applications are due August 15.
Requirements: The fellowship candidate must be no more than three years post completion of PhD or equivalent or five years post completion of MD or comparable doctoral degree in the clinical health profession. U.S. citizenship or residency is not required.
Samples: Dr. Carolyn Bertozzi, U of California at Berkeley (CA)—fellowship recipient for Disruption of 4-Epimerase: A Novel Anti-Metastasis Strategy for Breast Cancer $135,000 (2003); Dr. Christopher Contag, Stanford U—fellowship recipient, for In Vivo Recruitment and Activation of NK-T Cells into Tumors, $135,000 (2003); Dr Carol Lange, U of Minnesota Cancer Ctr (MN)—fellowship recipient, for Molecular Regulation of Breast Tumor Kinase, $135,000 (2003); Dr. Wayne Marasco, Dana-Farber Cancer Institute—fellowship recipient, for Isolation and Characterization of Human Anti-Human CXCR4 Antibodies for Immunotherapy of Breast Cancer Metastasis, $135,000 (2003).
Amount of Grant: $45,000 per year
Date(s) Application is Due: Aug 1; Aug 8; Aug 15
Contact: Director, Grants and Sponsored Programs, (888) 300-5582; fax: (972) 855-1640; email: grants@komen.org
Internet: http://www.komen.org/intradoc-cgi/idc_cgi_isapi.dll?IdcService=SS_GET_PAGE&nodeId=391
Sponsor: Susan G. Komen Breast Cancer Foundation
5005 LBJ Fwy, Suite 250
Dallas, TX 75244

Swann Foundation Fellowships **5498**
The fellowship is awarded annually to assist ongoing scholarly research and writing projects in the field of caricature and cartoon. Although research must be in the field of caricature and cartoon, there is no limitation regarding the place or time period covered. There is also no restriction on which university department might oversee a project proposed for the fellowships, provided the subject pertains to caricature or cartoon art. This program is administered by the Library of Congress Prints and Photographs Division. Application forms are available on the Web site.
Requirements: To be eligible, an applicant must be a candidate for an MA or PhD degree in a university of the United States, Canada, or Mexico working toward the completion of a dissertation or thesis for that degree or engaged in postgraduate research within three years of receiving his or her MA or PhD.
Amount of Grant: $15,000
Date(s) Application is Due: Feb 15
Contact: Martha Kennedy, (202) 707-9115; fax: (202) 707-6647 or email: swann@loc.gov
Internet: http://lcweb.loc.gov/rr/print/swann/swannhome.html
Sponsor: Swann Foundation Fund
9140 E Hampton Drive
Capitol Heights, MD 20743-3809

SWE Scholarships **5499**
SWE administers approximately 70 scholarships annually. All grants are announced in the spring or summer for use in the following academic year. Grants are made at the beginning of the fall term upon proof of registration. Scholarship information and applications are sent to the deans of accredited engineering schools in October and March of each year; they can also be obtained by sending a self-addressed, stamped envelope to the society's headquarters. Applications for freshman and reentry scholarships have a May 15 postmark deadline date; applications for sophomore, junior, senior, and graduate scholarships have a February 1 postmark deadline date.
Requirements: All SWE-administered scholarships are open only to women majoring in engineering in a school, college, or university with an accredited engineering program, and who will be in a specified year of study during the academic year following presentation of the grant. Reentry scholarships are limited to women who have been out of the engineering job market for a minimum of two years and who will return to school for an engineering program, undergraduate or graduate.
Amount of Grant: $1000-$10,000
Date(s) Application is Due: Feb 1; May 15
Contact: Scholarship Program, (312) 596-5223; fax: (312) 644-8557; email: hq@swe.org
Internet: http://www.societyofwomenengineers.org/scholarships/brochure.aspx
Sponsor: Society of Women Engineers
230 E Ohio Street, Suite 400
New York, NY 10005

Swedish Natural Science Research Council Grants **5500**
The Swedish Natural Science Research Council offers grants for senior research positions, postgraduate fellowships, junior researchers, lectureships, adjunct professorships, and council professorships. Applications for grants from NFR for financing employment at the universities are due in mid-May each year. For more information, contact the senior scientific secretaries of the committees: scientific council for humanities and social sciences, scientific council for medicine, scientific council for natural and engineering sciences, and committee for educational science.
Contact: Grants Administrator; +46-8-546 44 000; fax: +46-8-546 44 180; email: vetenskapsradet@vr.se
Internet: http://www.vr.se/mainmenu/applyforgrants/callforproposals.4.64a4ae3e1092b3475b6800055.html
Sponsor: Swedish Natural Science Research Council
Regeringsgatan 56
Stockholm SE-103 78 Sweden

Swedish Women's Educational Association International Scholarships **5501**
SWEA is a global non-profit organization for Swedish speaking women who live or have lived abroad. Its primary objective objective is to protect the Swedish language, support and inform about culture and traditions. Each year a scholarship is awarded for studies at the graduate level in Swedish literature, language, or linguistics. The award is directed specifically toward candidates who are well into the process of writing their dissertation.
Requirements: Students of any nationality may apply. A good knowledge of the Swedish language, written or spoken, is required. Applicant must reside permanently outside Sweden and be studying at a non-Swedish university.
Amount of Grant: $7,000
Date(s) Application is Due: Mar 1
Contact: Scholarship Committee; (702) 873-7399; fax: (702) 873-0492; email: office@swea.org

Internet: http://www.swea.org/inenglish___3175.aspx
Sponsor: Swedish Women's Educational Association International
P.O. Box 30190
Las Vegas, NV 89173-0190

Swiss National Science Foundation DORE (DO REsearch) Conference Grants 5502
DORE (DO REsearch) is the Foundation's promotional tool for practical research at Universities of Applied Sciences and Universities of Teacher Education. DORE supports practical research in the fields of social work, health, music and theater, art, education, applied psychology and applied linguistics. Awards are granted for research projects, publications, courses for junior scientists, and scientific conferences. A maximum amount of 10,000 Swiss francs is awarded for one scientific conference to primarily pay for invited speakers' travel and accommodation fees. Applications are to be submitted no later than three months prior to the planned event.
Amount of Grant: 10,000 Swiss francs maximum per conference
Contact: Secretariat of International Relations; +41 (0) 31 308 22 22; fax: +41 (0) 31 301 30 09; email: dore@snf.ch
Internet: http://www.snf.ch/e/funding/conferences/doreconferences/seiten/default.aspx
Sponsor: Swiss National Science Foundation (SNSF)
Wildhainweg 3, P.O. Box 8232
Bern CH-3001 Switzerland

Swiss National Science Foundation Fellowships for Advanced Researchers 5503
Fellowships for advanced researchers enable young scientists planning to follow an academic career to benefit from a stay abroad in order to increase their knowledge and scientific reputation. These fellowships for overseas visits in any discipline are supported by the Foundation, except experimental and clinical medicine. Funding includes personal maintenance, a fixed sum for travel expenses and may include a contribution towards research and conference expenses. The amount of the fellowship is based on personal details, family obligations and cost of living in the host country and runs for a period of 12 to 36 months.
Requirements: Applicant requirements include; a doctorate degree; independent research activity, with at least one year of research experience at post-doc level at the time the application is submitted; Swiss nationality, residence or settlement permit, cross-border commuter permit; and connections with Swiss research.
Date(s) Application is Due: Feb 1; Aug 1
Contact: Secretariat of International Relations; +41 (0) 31 308 22 22; fax: +41 (0) 31 301 30 09; email: pri@snf.ch
Internet: http://www.snf.ch/e/funding/individuals/advancedresearchers/seiten/default.aspx
Sponsor: Swiss National Science Foundation (SNSF)
Wildhainweg 3, P.O. Box 8232
Bern CH-3001 Switzerland

Swiss National Science Foundation Fellowships for Prospective Researchers 5504
Fellowships for prospective researchers offer young scientists, who are at the point of starting their research career, the possibility to stay at a research institution abroad. These fellowships for a stay oversea are available in all disciplines supported by the Swiss National Science Foundation (SNSF). Funding includes personal maintenance, a fixed sum for travel expenses and may include a contribution towards research and conference expenses. The amount of the fellowship is based on personal details, family obligations and cost of living in the host country. For informations about the fellowships duration, an applicant should contact the local Research Commission.
Requirements: Applicants must have: a doctorate (lic., diploma or master's degree may be acceptable in some cases at the end of a doctoral thesis); for a postdoctoral fellowship, at the latest 3 years after obtaining the doctorate; for a doctoral fellowship at the end of the thesis, at the earliest 2 years after obtaining the licentiate, diploma or master's degree; Swiss nationality, residence or settlement permit, or cross-border commuter permit; and connections with Swiss research.
Contact: Secretariat of International Relations; +41 (0) 31 308 22 22; fax: +41 (0) 31 301 30 09; email: pri@snf.ch
Internet: http://www.snf.ch/e/funding/individuals/prospective/seiten/default.aspx
Sponsor: Swiss National Science Foundation (SNSF)
Wildhainweg 3, P.O. Box 8232
Bern CH-3001 Switzerland

Swiss National Science Foundation Marie Heim-Vogtlin (MHV) Grants 5505
To increase the number of women in research, the Foundation annually awards a limited number of Marie Heim-Vogtlin (MHV)-subsidies. These subsidies are awarded after a two-phase selection procedure. The subsidies aim to facilitate the recipients research and integration in a host institution at a Swiss University. The program provides financial support where no other means of such integration exists or where female researchers cannot qualify for other funding programs. The subsidies are aimed for doctoral and postdoctoral candidates within all disciplines funded by the SNSF. The objective is to support well-qualified female scientists who are or were forced to accept an interruption or reduction of their research activities due to family obligations or a change of residence as a result of their partner, career development. Applications for these subsidies may only be submitted in response to a specific call.
Requirements: A candidate's occupation degree for the project must be at least 50%. In order to achieve sustained integration, the host institution must furnish not only a written declaration of acceptance and job confirmation but also a continuing commitment following expiry of the MHV-subsidy.
Contact: Secretariat of International Relations; +41 (0) 31 308 22 22; fax: +41 (0) 31 301 30 09; email: pri@snf.ch
Internet: http://www.snf.ch/e/funding/individuals/mhv/seiten/default.aspx
Sponsor: Swiss National Science Foundation
Wildhainweg 3, P.O. Box 8232
Bern CH-3001 Switzerland

Swiss National Science Foundation Scientific Conferences Grants 5506
The Swiss National Science Foundation (SNSF) provides funding for scientific conferences (especially colloquiums and symposiums) if these are of obvious interest to the Swiss research community. Funding is mainly intended to cover travel and accommodation expenses for guest speakers from abroad. Applications for conference grants are not subject to a deadline. However, organizers must submit their applications at least six months before the date of the planned event.
Requirements: The conference venues must be in Switzerland.
Amount of Grant: 10,000 Swiss francs maximum per conference
Contact: Secretariat of International Relations; +41 (0) 31 308 22 22; fax: +41 (0) 31 301 30 09; email: pri@snf.ch
Internet: http://www.snf.ch/e/funding/conferences/scientificconferences/seiten/default.aspx
Sponsor: Swiss National Science Foundation (SNSF)
Wildhainweg 3, P.O. Box 8232
Bern CH-3001 Switzerland

Sybil G. Jacobs Award for Outstanding Use of Tobacco Industry Documents 5507
The award recognizes a person who has made a significant and well-recognized contribution to the health of the public in the recent past through use of tobacco documents. The award also honors innovation in the use and application of tobacco industry documents to improve the public's health and, where applicable, to further the goals of tobacco prevention and control in order to help build a world where young people reject tobacco and anyone can quit. Those nominated should be individuals who have made a notable impact through innovative use of tobacco industry documents as applied to research, policy, or advocacy.
Requirements: At least one of the following two criteria must be met by the nominated individual: Nominees must have made a remarkable research, policy, or advocacy contribution with the use of tobacco industry documents, and/or; Nominees must have employed innovative, creative approaches to the employment of tobacco industry documents that result in an improvement in the health or public awareness of a community or nation. Nominations may be made by individuals such as colleagues, peers, coworkers, instructors/professors, governments, or CBOs that are qualified to make such a nomination because of their familiarity with the nominee's work and contribution. The preferred method of submission is through the Legacy Awards website at www.americanlegacy.org/awards/nominate.asp.
Restrictions: Nominees must not have any affiliation with the tobacco industry. Members of the awards committee may not nominate potential recipients, although they may provide written support as part of a nomination package. Employees and directors of the American Legacy Foundation and their relatives are not eligible.
Samples: 2006: Ray Goldstein, freelance paralegal; Ruth E. Malone, R.N., Ph.D., professor of nursing and health policy in the Department of Social & Behavioral Sciences, School of Nursing, University of California, San Francisco.
Amount of Grant: $7,500
Date(s) Application is Due: Jul 11
Contact: Jennifer Bramble, (202) 454-5555; fax: (202) 454-5599; email: awards@americanlegacy.org
Internet: http://www.americanlegacy.org/49.htm
Sponsor: American Legacy Foundation
Legacy Awards, 2030 M Street, NW, 6th Floor
Washington, DC 20036

Syracuse University Graduate Fellowships 5508
Syracuse University makes available 100 graduate fellowships annually, created to support graduate training for students with outstanding qualifications and the potential to make significant contributions in their future professions. The competition is open to applicants in all fields of graduate study at Syracuse University. Full tuition for an academic year and a stipend are provided.
Requirements: Full-time students must be nominated by an academic department.
Amount of Grant: $11,350 for master's students, $19,000 for doctoral students
Contact: Fellowships Administrator, Graduate Enrollment Management Center, (315) 443-4492; email: grad@syr.edu
Internet: http://gradsch.syr.edu/student/FinResor.htm
Sponsor: Syracuse University
207 Bowne Hall
Syracuse, NY 13244

T. James Kavanagh Foundation Grants 5509
A large percentage of funding supports the Catholic church and religious associations, along with support for U.S. Roman Catholic schools, with emphasis on Pennsylvania and southern New Jersey. The foundation also makes music awards. Types of support include annual campaigns, continuing support, equipment, general operating support, program development, research, and scholarship funds. Proposals should be submitted preferably by the end of February, July, or October. The board meets in March, August, and November to consider requests.

Requirements: Giving is strictly limited to the U.S., with emphasis on southern NJ and PA. Any Roman Catholic affiliate, church, school, college, or hospital will be considered.
Restrictions: Grants will not be awarded outside of the United States, not even to U.S. missions or to help their agencies abroad. No funding will be given for individuals; endowment funds; seed money; deficit financing; land acquisition; publications; conferences; scholarships or fellowships; matching gifts; or loans.
Samples: Saint Joseph's R.C. School (Sharon, PA)—for the learning assistance program, $10,000.; Opera Company of Philadelphia (Philadelphia, PA)—$10,000.; Merion Mercy Academy (Merion Station, PA)—$5000.
Amount of Grant: $2000-$10,000 average
Contact: Thomas Kavanagh, Trustee, fax: (610) 356-4606
Sponsor: T. James Kavanagh Foundation
234 E State Street
Sharon, PA 16146

T.R. Higgins Lectureship Award 5510
The award recognizes an outstanding lecturer and author whose technical research paper or papers may be identified as an outstanding contribution to engineering literature on fabricated structural steel published during the eligibility period. The award will be made to a nominated individual on the basis of his/her reputation as a lecturer combined with an evaluation of the originality, the contribution to engineering knowledge, the projected future significance, the clarity of presentation, and the value to the fabricated structural steel industry, as found by the jury in the paper or papers named in the nomination. Nominations must be received by the listed deadline. Contact the institute to verify deadline date.
Requirements: The author must be a permanent resident of the United States. The paper or papers must have been published in a professional journal that provides for a discussion of papers by the engineering profession within the recent five-year period.
Amount of Grant: $10,000
Date(s) Application is Due: Aug 1
Contact: Award Administrator, (312) 670-2400; fax: (312) 670-5403
Internet: http://www.aisc.org/Content/NavigationMenu/About_AISC/Competition_and_Awards/T_R__Higgins_Lectureship/T_R__Higgins_Lectureship.htm
Sponsor: American Institute of Steel Construction
One E Wacker Drive, Suite 3100
Chicago, IL 60601-2001

Takeda Techno-Entrepreneurship Award 5511
The award is presented to research proposals that show the greatest promise of contribution to human well-being, through the creation and application of new engineering intellect and knowledge. The practical application or commercialization of this engineering knowledge (i.e., techno-entrepreneurship), which will generate substantial value for people in the near term, is the foundation's mission. Included within the scope of the award are research projects that are already in development, as well as research proposals for future projects. Three application fields are covered by the awards: information and electronics, life sciences, and world environment. Applicants take part in competitive cyber workshops held on the Internet. Online discussions during the workshops are considered to be part of the selection process. Three session subjects are selected each year (in each application field) for the award. A period of time is allocated for the proposal of subjects in order for the Takeda Foundation to decide which themes are most important each year. Once the session subjects are determined, there is a call for research proposals related to these three subjects. One award winner and four finalist commendations are awarded per session. The February 28 deadline is for submission of session topics; the May 30 deadline is for submission of research proposals.
Requirements: The award program is open to individuals of all nationalities and all organizational affiliations, including corporations, universities, and other organizations.
Amount of Grant: $56,000
Date(s) Application is Due: Feb 28; May 30
Contact: Mariko Tsusaka or Kazuyoshi Shimada, +81-3-3549-2781; fax: +81-3-3549-2787; email: ttea2003@takeda-foundation.jp
Internet: http://www.takeda-foundation.jp/en/award/tech/info/index.html
Sponsor: Takeda Foundation
Saint Luke 's Tower 32F, MS33, 8-1 Akashi-cho
Chuo-ku, Tokyo 104-6591 Japan

Takemi Program in International Health Research and Advanced Training Fellowships 5512
This midcareer program offers a limited number of fellowship positions for research and advanced training on critical issues of international health, especially those relating to developing countries. This interdisciplinary program has four major objectives: to create better methods for mobilizing and using health resources in both rich and poor countries; to promote cooperative research and comparative analysis of health policies and programs in different countries; to study transnational causes of health conditions such as population migration, disease transmission, and air and water pollution; and to bring together leading health professionals and scholars from many nations for research and training. Fellowship positions cover estimated costs for the program, which includes 10 months in residence at the Harvard School of Public Health. Contact program staff for deadlines.
Requirements: Applicants should have completed graduate degrees and demonstrated potential leadership capacity in their home countries. They are expected to show strong promise and appropriate preparation to enable them to benefit from a period at Harvard. Applicants may come from the disciplines of medicine, law, public health, economics, management, and social sciences. Applicants should submit a curriculum vita and publications list, a proposal of research and writing to be undertaken during the fellowship, and three letters of reference. The proposal should not exceed five typed pages and should include a clear statement of objectives, significance or relevance, data sources, and methodology.
Restrictions: Clinical research will not be funded. Individuals enrolled in degree-granting programs are ineligible.
Contact: Program Coordinator, Takemi Program in International Health, Harvard School of Public Health, (617) 432-0686; fax: (617) 432-1251; email: takemi@hsph.harvard.edu
Internet: http://www.hsph.harvard.edu/takemi
Sponsor: Harvard University
665 Huntington Avenue, Building 1-1207
Boston, MA 02115-6021

Talbert and Leota Abrams Foundation 5513
The foundation awards grants to organizations in Michigan in its fields of interest, including higher education, adult basic education and literacy programs, libraries and reading, and community funds. The grants support research, scholarship funds, and program development. The foundation does not fund the maintenance of existing programs or services, but seeks innovative opportunities. The board meets in June to consider requests. Application may be made by letter.
Requirements: Nonprofit organizations in Michigan, primarily central Michigan, are eligible. Churches may apply for nonsectarian, nonreligious grants.
Restrictions: Grants do not support athletic activities, churches for sectarian use, individuals, operating expense, loans, or traveling expenses.
Geographic Focus: Michigan
Samples: Michigan Dyslexia Institute (Lansing, MI)—program support, $64,000.; Lansing Community College Foundation (Lansing, MI)—program support, $2500.; Junior Achievement, Mid-Michigan (Lansing, MI)—program support, $1000.
Amount of Grant: $3000-$174,000
Date(s) Application is Due: May 31
Contact: Joe Foster Jr., Secretary, (517) 706-5791
Sponsor: Talbert and Leota Abrams Foundation
P.O Box 27337
Lansing, MI 48909-7337

Tanner Humanities Center Research Fellowships 5514
The Tanner Humanities Center awards fellowships to encourage, support, and disseminate important humanistic research. Each fellow will spend one year or one semester conducting research full-time, in residence at the center. In addition, fellows will present work-in-progress talks throughout the year. The deadline for faculty applicants is October 1 to their department chair and October 15 to the center. Visiting faculty must apply by December 1. The deadline for graduate students to apply for doctoral fellowships is March 1. Guidelines and application forms for each fellowship are available online.
Requirements: Applicants must have received the PhD by August two years prior to the year of application. Eligible fields include anthropology, communication, history (including art history and criticism, film history, musicology, and theater history), comparative religion, ethnic studies, jurisprudence, languages and linguistics, literature, philosophy, women's studies, and historical or philosophical applications of the social and natural sciences.
Restrictions: The center does not consider applications for projects leading to completion of advanced degrees.
Date(s) Application is Due: Mar 1; Oct 1; Oct 15; Dec 1
Contact: Obert C. and Grace A. Tanner Humanities Ctr, (801) 581-7989; fax: (801) 585-3510; email: Tanner.Humanities@hum.utah.edu
Internet: http://vegeta.hum.utah.edu/humcntr/applications.html
Sponsor: University of Utah
380 S 1400 E, Room 201
Salt Lake City, UT 84112-0312

TD4HIM Foundation Grants 5515
The TD4HIM Foundation primarily contributes and supports the growth of religious organizations, local Boys & Girls Clubs, community service organizations, and national organizations that promote cancer research. The foundation, named in honor of Trevin Dilfer, the 5-year-old son of football quarterback Trent Dilfer, also supports programs that conduct heart disease research.
Samples: Pajaro Valley Unified School District (2006).
Contact: TD4HIM Program Coordinator, (408) 829-3381
Sponsor: TD4HIM Foundation
60 Shoreline Circle
Incline Village, NV 89451-9529

Teachers College Minority Postdoctoral Fellowships 5516
The program aims to increase the number of minority faculty members at graduate schools of education. It provides recent doctorate recipients the opportunity to develop a program of research and participate in a mentored program at a graduate research university for one year. Fellows conduct research as outlined in the application proposals, provide written reports of the research programs, and participate in an academic department with the option of teaching a course or providing a series of special lectures, to be decided. Fellowships may provide campus housing and limited relocation costs.

Requirements: Candidates must have earned the doctorate degree in any area within the past three years and be committed to research and teaching in an education-related area.
Amount of Grant: $30,000 plus $3000 for research
Date(s) Application is Due: Jan 6
Contact: Young Jin Jo, Office of the Dean, (212) 678-3998; email: jo@exchange.tc.columbia.edu
Internet: http://www.tc.edu/dean/mpost.htm
Sponsor: Columbia University Teachers College
525 W 120th Street, Box 54
New York, NY 10027

Technology Excellence Fellowship Program 5517
The program offers competitive grants to cover a share of the costs of hosting mid- to senior level managers and executives for six weeks of hands-on professional training in U.S. business practices. The program encourages joint ventures and investment in the high-tech sectors of the United States and Israel by facilitating partnerships between U.S. and Israeli business communities. Participating U.S. organizations have the choice of either nominating a candidate from their existing Israeli business contacts or working with TEF to identify a qualified candidate. Grants cover round-trip airfare from trainee's home country to the training site; per diem for trainee; and housing allowance (structured as a reimbursable award). The program is bidirectional and also accepts applications from Israeli organizations to host American interns. Interested organizations apply after the opening announcement is published. The opening announcement is broadcast on the Internet and via email.
Requirements: Any U.S. profit or nonprofit organization or institution may apply to host interns.
Amount of Grant: $4000 per eight-week period for housing costs; $50 per diem for trainee
Contact: Valerie Herold, (202) 419-0430; fax: (202) 419-0435; email: valerie@usistf.org
Internet: http://www.usistf.org/03_A_02.html
Sponsor: U.S.-Israel Science and Technology Foundation
1130 17th Street NW, Suite 312
Washington, DC 20036

Temple University George D. McDowell Fellowship 5518
The fellowship provides a stipend and access to photographs and newspaper clippings in the archives of the Philadelphia Evening Bulletin newspaper, a resource for research on the Philadelphia metropolitan area. The program provides support for a six-week fellowship to do research and study that results in a scholarly work. Application and guidelines are available online.
Requirements: Undergraduate and graduate students, independent scholars, educators, writers, filmmakers, and anyone interested in the history of Philadelphia are invited to apply.
Restrictions: Current employees of Temple University are ineligible.
Amount of Grant: $500 per week
Date(s) Application is Due: Feb 15
Contact: Margaret Jerrido, Head of Urban Archives, (215) 204-5750; fax: (212) 204-3681; email: urban@library.temple.edu
Internet: http://library.temple.edu/collections/urbana
Sponsor: Temple University
1210 W. Berks Street
Philadelphia, PA 19122

Temple-Inland Foundation Grants 5519
The foundation makes grants to IRS 501(c)3 tax-exempt organizations located in geographic operating areas of Temple-Inland Inc's subsidiary companies. Areas of interest include services for children and youth, Christian churches and organizations, education at all levels, health care and organizations, human services, and arts and culture. Types of support include general operating support, research, employee matching gifts, and employee-related scholarships. Requests should be received two weeks prior to quarterly board meetings. Submit a written request for guidelines and specific application deadlines. The listed deadline date is for scholarships.
Restrictions: Ineligible applicants include fraternal, veterans, political, local social, and service organizations.
Amount of Grant: $1000-$30,000 average
Date(s) Application is Due: Mar 15
Contact: Richard Warner, (936) 829-1721; fax: (936) 829-7727
Internet: http://www.templeinland.com
Sponsor: Temple-Inland Foundation
1300 S Mopac
Austin, TX 78749

Templeton Advanced Research Grants 5520
The institute invites proposals to develop research projects in one of three topical areas: religion, spirituality, healing, and health outcomes—research that advances scientific understanding of the link, if any, between religion, spirituality, healing, and health outcomes; religion, spirituality, and human flourishing—research that scientifically explores how religion and spirituality contribute to the virtues and human strengths that reflect humanity's highest aspirations and noble qualities; and competitive dynamics and cultural evolution of religions and God concepts—inquiries that approach religion and spirituality as an important creative or causal factor in the formation of society. Successful proposals will demonstrate, among other qualities, the promise of taking these fields in new directions, the possibility of addressing a question or shortcoming proposed in the field analyses, methodological innovation, and the potential for significant impact in the research community. The listed application deadline is for letters of intent. Guidelines are available online.
Requirements: Scholars and researchers from fields such as anthropology, economics, history, political science, psychology, religious studies, and sociology are encouraged to apply.
Amount of Grant: $1 million health outcomes; $1 million human flourishing; $150,000 God concepts
Date(s) Application is Due: Aug 1
Contact: Christopher Stawski, (215) 789-2200; fax: (215) 789-2222; email: tarp@metanexus.net
Internet: http://www.metanexus.net/tarp
Sponsor: Metanexus Institute on Religion and Science
3624 Market Street, Suite 301
Philadelphia, PA 19104

Tennessee Arts Commission Arts Projects Grants 5521
This grant category provides funds for a wide variety of quality arts projects and programs to arts organizations. Funds must be matched dollar-for-dollar. Projects typically supported under this category include projects that involve and promote Tennessee artists; visiting artists conducting master classes; workshops, festivals, and conferences; presentation of touring programs, including Southern Arts Federation touring presentations; public performances, productions, and exhibitions produced by the applicant; promotion, publicity, and newsletters; administrative and artistic staff support; research and documentation as part of a project or program development; consultancies and residencies for administrative and artistic activities; improved program accessibility for special constituencies; ticket endowment; extensions of literary projects, journals with continuing publication, or juried anthologies; technical assistance projects; and apprenticeship programs. Funds awarded to a single organization in this category range from $500 to $6000 for statewide organizations and $500 to $8000 for other organizations. Annual deadline dates are usually in late January or early February; contact program staff for exact dates.
Requirements: Applicants must be nonprofit organizations chartered in Tennessee that have proof of federal tax-exempt status. In addition, organizations must be service or single-discipline arts organizations that have a statewide mission. An organization may receive funds for no more than two projects in the same fiscal year. Multiple events, such as a performance series, should be submitted as one application. Prior to submitting an application, qualifying organizations must discuss their project proposal with staff members.
Geographic Focus: Tennessee
Amount of Grant: $500-$8000
Date(s) Application is Due: Jan 24
Contact: Kim Leavitt, (615) 532-5934; fax: (615) 741-8559; TDD: (615) 741-1701; email: kim.leavitt@state.tn.us
Internet: http://www.arts.state.tn.us/grantprograms.htm
Sponsor: Tennessee Arts Commission
401 Charlotte Avenue
Nashville, TN 37243-0780

TERC Research Fellowships 5522
The program invites applications for a two-year residential research fellowship focused on issues related to improving learning and teaching in the sciences and mathematics for children from historically underserved communities. Applications are sought from recent PhDs or EdDs who are members of groups historically underrepresented in the sciences and mathematics. Individuals whose research connects theory and practice, with a focus on issues related to learning and teaching, whether in school or out, in the domains of mathematics, the sciences, and technology are particularly encouraged to apply. Individuals whose research focuses on academic literacies more generally are also encouraged to apply. Fellows receive an annual stipend and are eligible for additional funds to cover travel, participation in professional conferences, and research related expenses. One fellow will be chosen for residence. The fellow will work collaboratively with a mentor from ERC senior staff. The fellow will also have the opportunity to collaborate with ongoing ERC projects and to share his or her emerging research and writing with the broader ERC community.
Requirements: Applicants must be within three years of receiving their PhD or EdD degree. The degree must be completed by July.
Amount of Grant: $55,000 stipend
Date(s) Application is Due: Mar 14
Contact: Hollee Freeman, (617) 547-0430; fax: (617) 349-3535; email: Hollee_Freeman@terc.edu
Internet: http://www.terc.edu/template/jobs/item.cfm?jobid=129
Sponsor: TERC Inc
2067 Massachusetts Avenue
Cambridge, MA 02140

Terwilliger Fellowship 5523
This fellowship is named for Enterprise Trustee J. Ronald Terwilliger, chairman and CEO of Trammel Crow Residential. The one-year Terwilliger Fellowship plays a leadership role in launching the new Enterprise Innovation Unit, which has primary responsibility for innovation: research, development and planning of potential new Enterprise programs, products and partnerships. The fellow works with the Enterprise

executive management team and key partners to develop new solutions to advance Enterprise, mission.
Contact: Kathy Holmes, Program Associate; (410) 772-2411; fax: (410) 964-1918; email: kholmes@enterprisecommunity.org
Internet: http://www.enterprisecommunity.org/programs/awards_and_fellowships/
Sponsor: Enterprise Community Partners / Enterprise Foundation
10227 Wincopin Circle, Suite 500
Columbia, MD 21044

Tessie Agan Award Competition 5524
This is an annual award competition open to undergraduate and graduate students to promote and encourage careers in the field of housing education. Students submit a research or position paper on some aspect or current issue in housing. The awards committee of the AAHE selects both graduate and undergraduate winners. Award winners must present their papers at the AAHE national conference. The call for papers is in March; deadlines for submission are posted in the In House newsletter. Specific guidelines are available through the AAHE headquarters. Contact program staff for availability.
Amount of Grant: $250 undergraduate award; $750 graduate award
Date(s) Application is Due: Apr 30
Contact: JoAnn Emmell, Director, (540) 231-9259; fax: (540) 231-1697; email: jemmel@vt.edu.
Internet: http://www.housingeducators.org/MEMAREA/HeraAwards.shtml
Sponsor: American Association of Housing Educators
5060 FCS Dept, Illinois State University
Normal, IL 61790-5060

Texas Instruments Foundation Grants 5525
The foundation supports civic, research, educational, health, welfare, charitable, and cultural movements. Consideration is based on the merit of each individual request. Grants are awarded in five categories: operating funds and building funds; hospital building funds and funds for medical research; operating funds for charitable and educational community service organizations; miscellaneous charitable, education-related, scientific, cultural, and arts organizations; and matching gifts to colleges, universities, and arts and cultural organizations. Generally, education (pre-K through graduate) receives about 75 percent of the annual grants. A majority of support goes to programs in plant cities, including Hunt Valley, MD; Attleboro, MA; Dallas, Houston, McKinney, Sherman, and Versailles, TX; and Santa Cruz and Tustin, CA. Applications are accepted at any time; individual requests are considered by the Board of Directors during scheduled meetings throughout the year.
Requirements: Applicants must be 501(c)3 tax-exempt organizations. National organizations are eligible to apply.
Restrictions: The foundation does not support political activities, sectarian or denominational religious groups, veterans organizations, fraternal or labor organizations, courtesy advertising, benefit entertainment/sponsorships, or tax-supported institutions.
Samples: American National Red Cross (Washington, DC)—to assist tsunami-relief efforts, $500,000 (2005); U of Texas Southwestern Medical Ctr at Dallas (TX)—to recruit new assistant professors who are conducting research on advanced imaging technologies, $1.5 million (2004); Collin Community College District (Plano, TX)—to endow a scholarship fund for students pursuing four-year degrees in engineering and related fields, $1 million (2004).
Contact: Ann Pomykal, Director of Corporate and Foundation Giving, (214) 480-3221; fax: (214) 480-6820
Internet: http://www.ti.com/corp/docs/company/citizen/foundation/index.shtml
Sponsor: Texas Instruments Foundation
P.O. Box 660199, MS 8656
Dallas, TX 75266-0199

Texas Library Association Scholarship 5526
Scholarships are awarded to Texas residents for graduate library study in ALA-accredited degree programs. The Ray C. Janeway scholarship is available to a student who has been accepted as a graduate student to a Texas ALA-accredited library education program. The TLA Summer School Scholarship is available for a graduate student who is pursuing a master's degree during summer sessions only. The Garrett Scholarship is available to a student who is concentrating his or her studies in children's, young adult, or school librarianship. This scholarship is awarded in odd-numbered years. The Van Dusen-Brady-Tobin Scholarship is available to a student who is pursuing graduate studies leading to a career as an elementary school or children's librarian. This scholarship is awarded in even-numbered years. Application forms are available on the Web site.
Requirements: All applicants must be Texas residents who attained at least a B average during the last two years of a baccalaureate degree program and have been accepted as graduate students at a Texas ALA-accredited library education program.
Geographic Focus: Texas
Amount of Grant: $2000 Ray C. Janeway scholarship; $1000 Garrett scholarship and Van Dusen-Trady-Tobin scholarship; $500 summer school scholarship
Date(s) Application is Due: Jan 31
Contact: Catherine Lee, TLA Scholarships, (512) 328-1518 ext 20 or (800) 580-2852; fax: (512) 328-8852; email: catherinel@txla.org or tla@txla.org
Internet: http://www.txla.org/html/awards/scholar.html
Sponsor: Texas Library Association
3355 Bee Cave Road, No 401
Austin, TX 78746-6763

Texas Tech Fine Arts PhD Fellowships 5527
Interdisciplinary fine arts PhD fellowships are available in art, music, and theater combined with studies in the philosophy of art. Fellowships include stipend plus benefits and additional opportunities for add-on fellowships.
Amount of Grant: $15,000 maximum
Date(s) Application is Due: Feb 1
Contact: Carol Anders, Academic Program Assistant, (806)-742-3825; fax: (806) 742-1971; email: carol.m.anders@ttu.edu
Internet: http://www.art.ttu.edu/home.html
Sponsor: Texas Tech University
P.O. Box 42081
Lubbock, TX 79409-2081

The World Food Prize 5528
The international prize recognizes — without regard to race, religion, nationality, or political beliefs — the achievements of individuals who have advanced human development by improving the quality, quantity, or availability of food in the world. The award recognizes contributions in any field involved in the world food supply—food and agriculture science and technology, manufacturing, marketing, nutrition, economics, poverty alleviation, political leadership, and the social sciences. The prize emphasizes the importance of a nutritious and sustainable food supply for all people. Nomination form and complete details are available online. Annual deadline dates may vary; contact program staff for exact dates.
Samples: Dr. Per Pinstrup-Andersen, Director General of the International Food Policy Research Institute—for research that enabled several governments to reform their food subsidy programs, dramatically increasing food availability to the most poor in each country, and for the establishment of programs in which families receive food subsidies when children stay in school, $250,000.
Amount of Grant: $250,000
Date(s) Application is Due: Mar 1
Contact: Judith Pim, Director of Secretariat Operations, (515) 245-3796; fax: (515) 245-3785; email: jpim@worldfoodprize.org
Internet: http://worldfoodprize.org/Home/tonominate.htm
Sponsor: World Food Prize Foundation
666 Grand Avenue, Suite 1700
Des Moines, IA 50309

The World is My Parish United Methodist History Research Grant 5529
This grant program was created to enable the collection of written and oral history and encourage research regarding the development of The United Methodist Church and its antecedents in any of the conferences not in the United States. Grants totaling $3,000 will be available with each grant being at least $1,000 awarded at the discretion of the committee of judges. A reference copy of the final product will be donated to the General Commission on Archives and History of The United Methodist Church. The Commission will send copies to other United Methodist general agencies as appropriate.
Requirements: The applicant must submit an application which includes the following: (if at all possible the application should be in English but the applicant, language will be accepted) Biographical information; description of the project, including its significance, what resources will be used, and how it will be made available to the public; expected date of completion; and, an itemized budget. The applicant must submit three letters of recommendation from persons who can provide evidence of his/her capability.
Date(s) Application is Due: Dec 31
Contact: General Secretary, General Commission on Archives and History; (973) 408-3189; fax: (973) 408-3909; email: research@gcah.org
Internet: http://www.gcah.org/site/c.ghKJI0PHIoE/b.4619067/k.B158/World_is_My_Parish_Research.htm
Sponsor: United Methodist Church
P.O. Box 127
New York, NY 10115

Theodore Lentz Postdoctoral Fellowship in Peace and Conflict Resolution 5530
The fellowship is awarded to candidates who have completed their PhD requirements and wish to focus their research on peace and conflict resolution in the international or intercultural arenas. In addition to research, the fellow will teach one class each semester.
Requirements: Applicants should provide a curriculum vita, evidence of completion of the PhD, three letters of recommendation, and a research proposal of approximately 750 words.
Amount of Grant: $23,400 stipend plus $1000 for research travel and expenses
Date(s) Application is Due: Apr 15
Contact: Dr. Joel Glassman, Director, Center for International Studies, (314) 516-5753; fax: (314) 516-6757; email: jglassman@umsl.edu
Internet: http://www.umsl.edu/services/cis/research/lentz_fellow.html
Sponsor: Center for International Studies
University of Missouri, 366 Social Sciences and Business Building, One University Boulevard
Saint Louis, MO 63121-4499

ThinkQuest Scholarship Competition 5531
The program provides an opportunity for students and educators to work collaboratively in teams to create high-quality, innovative, and content-rich Web sites that are made freely available to others via the Internet. Applicants must be members of a team consisting of two to six student members and one to two coaches. Teams may submit their entries

in any of five categories: arts and literature; science and mathematics; social sciences; sports and health; and interdisciplinary—a combination of subjects from two of the other categories, or it may not fit into any category. All student members are expected to make significant contributions to the success of the team. Coaches help with team formation, provide guidance and encouragement, and help students establish relationships with subject experts, local industry, and/or government in support of their efforts. Judges will select the winning entries in each of three divisions: grades 4, 5, and 6; grades 7 and 8; and grades 9, 10, 11, and 12. Awards include first, second, and third place; and a collaborative award. The March deadline is for registration; the May deadline is for entries. Annual deadline dates may vary; contact program staff for exact dates.
Requirements: A student is eligible if he or she is a student in grades 4 through 12 (or equivalent) on the registration deadline in any public, private, or parochial school in the five boroughs of New York City or any home school in New York City that meets all requirements of the jurisdiction(s) having authority over the education of its student(s). Additionally, a team may consist of students from the same and/or different grades and/or schools. A coach is eligible if he or she is a teacher, librarian, or other employee on the professional staff of a school attended by at least one of the student team members; a teacher at a home school attended by at least one of the student team members who is approved by the parent or guardian of each student on the team; a professional staff member of a nonprofit organization, such as a Boys or Girls club, community technology center, or similar organization, one of whose missions is to assist in the teaching of and learning by students in grades 4 through 12 and who is approved by the parent or guardian of each student on the team; able to give regular help and coaching assistance to the student(s) he or she has agreed to coach; and is over the age of 18 on the application deadline.
Amount of Grant: $100-$2000
Date(s) Application is Due: Mar 31; May 5
Contact: Lisa Ernst, Executive Director, (646) 773-1519; email: ernst@tqnyc.org
Internet: http://www.tqnyc.org/rules
Sponsor: ThinkQuest Inc
1350 Avenue of the Americas, Suite 1120
New York, NY 10019

Third Millennium Foundation Fellowships 5532
Through the Third Millennium Foundation's Fellowship programs, college students, graduate students and activists receive financial and professional support that enables them to build their capacity across a full range of human rights issues. In doing so, the Foundation aims to develop a cadre of young activists from different countries whose understanding of, commitment to, and effectiveness in human rights work has been enhanced. These Fellowships provide a unique opportunity for in-depth exposure to and reflection on a broad range of perspectives, situations and experiences related to human rights work.
Contact: Fellowships Manager; (212) 421-5244; fax: (212) 421-5243; email: tmf.usa@verizon.net
Internet: http://www.seedsoftolerance.org/initiative-fellowships.html
Sponsor: Third Millennium Foundation
340 West 12th Street
New York, NY 10014

Third Millennium Foundation Grants 5533
The Third Millennium Foundation's principal goal is to support programs and initiatives designed to promote tolerance, particularly among the young. Its work is focused on developing young children's understanding of and respect for the differences that exist among themselves, especially those related to culture, ethnicity, gender, race, and socioeconomic status. The Foundation emphasizes art in its grant making and continues to explore creative new avenues for unlearning intolerance. Foundation programs build on research findings demonstrating that gaining self-control and the ability to relate positively to others is developed in early childhood and that family, educators, and child caregivers play significant roles in shaping lifelong attitudes. The Foundation makes grants in the United States and internationally, and supports arts and human rights programs. Its programmatic emphasis is on education, children's rights and capacity building for grassroots organizations.
Contact: Grants Manager; (212) 421-5244; fax: (212) 421-5243; email: tmf.usa@verizon.net
Internet: http://www.seedsoftolerance.org/initiative-grantmaking.html
Sponsor: Third Millennium Foundation
340 West 12th Street
New York, NY 10014

Thomas and Agnes Carvel Foundation Grants 5534
This foundation awards support for general charitable purposes as well as for nutrition research and youth programs to nonprofit organizations in the tri-state area comprising New York, New Jersey, and Connecticut. Types of support include general operations grants, capital grants, grants in aid, program/project grants, matching gifts, and research grants. Application should be by letter and should outline the project, budget, and staffing of the project for which funding is being requested.
Requirements: IRS 501(c)3 organizations in New York, New Jersey, and Connecticut are eligible.
Restrictions: Grants are not made to individuals.
Geographic Focus: Connecticut; New Jersey; New York
Samples: Seton Hall (South Orange, NJ)—to provide full scholarships to 10 high-school students from Newark, NJ, who commit to a year of volunteer service in Newark upon graduation, $50,000.
Amount of Grant: $2000-$200,000 range
Date(s) Application is Due: Oct 1
Contact: Ann McHugh, Vice President, (914) 793-7300
Sponsor: Thomas and Agnes Carvel Foundation
35 E Grassy Sprain Road
Yonkers, NY 10710

Thomas and Dorothy Leavey Foundation Grants 5535
The foundation awards grants to nonprofit organizations in its areas of interest, including hospitals, medical research, higher and secondary education, and Catholic church groups. Types of support include general operating support and employee-related scholarships.
Requirements: The foundation gives primarily in southern California.
Geographic Focus: California
Samples: Loyola High School of Los Angeles (CA)—to construct a science building, $2.5 million challenge grant (2004).
Amount of Grant: $10,000-$250,000 average
Contact: Kathleen Leavey McCarthy, Chair, (310) 551-9936
Sponsor: Thomas and Dorothy Leavey Foundation
10100 Santa Monica Boulevard, Suite 610
Los Angeles, CA 90067

Thomas C. Ackerman Foundation Grants 5536
The foundation awards grants to California nonprofits in its areas of interest, including arts and culture, community development, education, and health and human services. Types of support include general operating support, emergency grants, seed grants, scholarship funds, and matching grants. Contact the office for application forms. Deadlines listed are for letters of intent. Annual deadline dates may vary; contact program staff for exact dates.
Requirements: California nonprofits, primarily in San Diego County, are eligible.
Restrictions: The foundation does not award grants to religious organizations for religious purposes; individuals; or for brick-and-mortar projects, conferences, symposia, or human medical or biological research.
Geographic Focus: California
Samples: San Diego Senior Olympics (San Diego, CA)—for Website Online Registration Development, $3500.; San ElderHelp of San Diego (San Diego, CA)—for Shared Housing Program, $10,000.; Museum of Contemporary Art (San Diego, CA)—for the program Off Broadway: Emerging Artists in Downtown San Diego, $7500.
Amount of Grant: $300-$25,000 typically
Date(s) Application is Due: Jan 31; Apr 29; Jul 29; Oct 31
Contact: Program Contact, (619) 699-5411; fax: (619) 232-8311; email: Info@AckermanFoundation.org or pgarcia@luce.com
Internet: http://www.ackermanfoundation.org
Sponsor: Thomas C. Ackerman Foundation
600 W Broadway, Suite 2600
San Diego, CA 92101-3391

Thomas Foundation Grants 5537
The foundation awards grants to eligible nonprofit organizations in its areas of interest, including medical research and charitable giving. Grants are awarded for one year and may be renewed. Submit a written proposal and include tax-exemption documentation.
Requirements: 501(c)3 tax-exempt organizations serving southeastern Michigan and New York are eligible.
Geographic Focus: Michigan; New York
Amount of Grant: $1000-$115,000 range
Contact: Jay Brody, Director, (248) 528-1111
Sponsor: Thomas Foundation
201 W Big Beaver Road, Suite 600
Troy, MI 48084-4161

Thomas J. Watson Foundation Fellowships 5538
The program was inaugurated to give exceptional college graduates the freedom to engage in a year of independent study and travel abroad. Administered in cooperation with 51 U.S. private colleges and universities, up to 60 fellows may be selected. The fellowship experience enables fellows to immerse themselves in other cultures for an entire year. In selection of fellows, the foundation seeks to identify individuals who demonstrate integrity, strong ethical character, intelligence, the capacity for vision and leadership, and potential for humane and effective participation in the world community. The proposed project should involve investigation into an area of demonstrated concern and personal interest. A candidate's academic record and extracurricular activities will be considered. Fellows may travel alone or with a spouse or child. Institutions' nominations and their nominees' completed applications must reach the program office by November 2.
Requirements: Graduating seniors at participating institutions are eligible for nomination by their institution. The foundation welcomes applicants from a diverse range of backgrounds and academic disciplines.
Samples: Annemarie Ackerman, Macalester College—for travel to Brazil, Cape Verde, Mozambique, India, and Portugal for her study Documenting the Portuguese Empire: Photography in the Lusophone World. Leisy Abrego, Ponoma College—for travel to Mexico, Panama, Peru, and Uruguay to study oral histories of minority women in Latin America.; Jared Johnson, Oberlin College—for travel to United Kingdom to study choral music in English society.
Amount of Grant: $22,000 for single fellow; $31,000 if traveling with spouse or child
Date(s) Application is Due: Nov 1

Contact: Beverly Larson, Executive Director, (401) 274-1952; fax: (401) 274-1954; email: tjw@watsonfellowship.org
Internet: http://www.watsonfellowship.org/site/what/what.html
Sponsor: Thomas J. Watson Foundation
293 S Main Street
Providence, RI 02903

Thomas Jefferson Agricultural Institute Food and Society Policy Fellowships 5539
This is a national program of part-time, one-year professional fellowships offered to improve communications about food and agriculture issues in the United States. Fellows use media, scholarship, public education, and outreach to promote change in the policies and structures that shape the U.S. food system. The program seeks fellows who can develop communication pieces that address the connections between youth, diet, health, and how our food is produced. Within this fellowship theme, examples of relevant topics include farm-to-school programs, community food systems, and the role of food choices in youth obesity. Past fellowship recipients have included university professors, nutrition educators, public health professionals, and staff at nonprofit organizations working on food and agriculture issues. Farmers or ranchers with a strong interest and experience in communicating about how to foster healthy youth through the production of healthy food are also encouraged to apply. Individuals involved in education, research, or program development in food and agriculture will receive preference over full-time communications professionals. Early career professionals (defined as 35 years of age or younger) will receive preference for half of the fellowships. Guidelines are available online.
Requirements: Applicants should be U.S. residents actively involved in a professional career that involves two or more of the following: agriculture, health promotion, youth development, food production, and/or policy analysis.
Amount of Grant: $30,000 stipend
Contact: Sue Roberts, (515) 965-3859; email: susan@susan-roberts.net
Internet: http://foodandsocietyfellows.org/pages/guidelines/guidelines.html
Sponsor: Thomas Jefferson Agricultural Institute
601 W Nifong Boulevard, Suite 1D
Columbia, MO 65203

Thompson Charitable Foundation Grants 5540
Established in 1987, the Foundation supports organizations involved with education, Christian organizations, Autism research, health, human services, youth, including funding for capital and building improvements for human service organizations and educational institutions. Giving is limited to Bell, Clay, Laurel, and Leslie counties, Kentucky; Anderson, Blount, Knox, and Scott counties, Tennessee; and Buchanan and Tazewell counties, Virginia.
Requirements: Nonprofits operating in Bell, Clay, Laurel, and Leslie counties, Kentucky; Anderson, Blount, Knox, and Scott counties, Tennessee; and Buchanan and Tazewell counties, Virginia are eligible to apply. Initial contact should be a letter, no more then 2 pages long. The letter should include: a statement of the problem, the project will address; a detailed description of project, and amount of funding requested.
Restrictions: No support for religious or political organizations, budget deficits or endowments.
Geographic Focus: Kentucky; Tennessee; Virginia
Amount of Grant: $2,000-$500,000
Date(s) Application is Due: Mar 31; Sep 1
Contact: Debbie Black, Exec. Director; (865) 588-0491; fax: (865) 588-4496; email: debbie@cf.org
Sponsor: The Thompson Charitable Foundation
800 South Gay Street, Suite 2021
Knoxville, TN 37929-9710

Thornton Foundation Grants 5541
The foundation awards grants to Calilfornia nonprofit organizations in its areas of interest, including arts/ and culture and higher/secondary education. Types of support include annual campaigns, building construction/renovation, continuing support, endowment funds, and research. There are no application deadlines.
Requirements: California nonprofit organizations are eligible.
Geographic Focus: California
Amount of Grant: $1000-$200,000 range
Contact: Charles Thornton Jr., President, (213) 629-3867
Sponsor: Thornton Foundation
523 W Sixth Street, Suite 636
Los Angeles, CA 90014

Thrasher Research Fund Grants 5542
The fund awards grants in support of research that seeks to improve children's health where insufficient research and investigation have been performed. In particular, the fund supports medical research that seeks to prevent or cure children's critical illnesses, injuries, and disabilities. Projects should be scientifically sound and culturally appropriate; document methods and practices that have sustainable benefits; have specific aims and a well-designed methodology; and evaluate the significance of impact on children's health. The fund normally supports projects no longer than three years. Financial support may be provided for supplies, minor equipment (items less than U.S.$1000), and technical personnel assistance related to a specific fund-sponsored project. The fund requires the principal investigator to submit semiannual project progress and financial reports. The initial application procedure is to submit a five-page prospectus, which should include a brief background statement justifying the proposed project; hypothesis to be tested; specific aims of the project; design and methodology of the proposed project, including evaluation component; time schedule and detailed budget; and investigator(s) curriculum vita (in addition to four-page prospectus). Applicants are encouraged to consult program staff prior to submitting a prospectus.
Restrictions: The fund does not award grants for general operations, construction or renovation of buildings or facilities, loans, scholarships, or the purchase of major equipment. Grants generally are not awarded for conferences, workshops, or symposia. Proposals in the areas of sexually transmitted diseases, abortion or reproductive physiology, and contraceptive technology will not be considered. In addition, proposals involving the use of therapeutically aborted human fetal tissue will not be considered.
Samples: Dr. Rebecca Stoltzfus, Johns Hopkins School of Hygiene and Public Health (Baltimore, MD)—for a study to improve iron status and growth of preschool children in East Africa, $141,529 total.; Paul Calvert, Intermediate Technology (Rugby, England)—for study on environmental health and sanitation for fishing communities, $117,250 total.
Amount of Grant: $100,000-$350,000 average
Contact: Julie Busse, (801) 240-4720; fax: (801) 240-1625; email: BusseJ@thrasherresearch.org
Internet: http://www.thrasherresearch.org
Sponsor: Thrasher Research Fund
15 W South Temple Street, Suite 1650
Salt Lake City, UT 84101

Thurgood Marshall Dissertation Fellowships for African American Scholars 5543
The goal of these fellowships is to increase the number of African American faculty in American higher education by supporting them in completing the dissertation; and by bringing to the college more role models for potential African American scholars among Dartmouth undergraduates. Fellows are allowed access to outstanding library and computing facilities and will have the opportunity to participate in classroom activities with research faculty dedicated to undergraduate teaching. Each fellow is expected to complete the dissertation during the one-year residency period and to participate in teaching one 10-week course during the year, either as primary instructor or as part of a team. The award includes stipend, office space, library privileges, and a research assistance fund. Contact the office for application materials.
Requirements: African American scholars who have completed all other PhD requirements except the dissertation are eligible. Fellows may be taking the PhD degree in any discipline taught in the Dartmouth undergraduate curriculum.
Amount of Grant: $25,000 stipend and $2500 research assistance
Date(s) Application is Due: Feb 1
Contact: Sandy Spiegel, Director of Graduate Recruiting and Diversity, (603) 646-6578; email: Sandra.J.Spiegel@Dartmouth.edu
Internet: http://www.dartmouth.edu/~gradstdy/funding/fellowships/cem.html
Sponsor: Dartmouth College
6062 Wentworth, Room 304
Hanover, NH 03755-3526

TIAA-CREF Institute Research Grant Program 5544
The program was established to encourage scholars to pursue research in areas of interest to TIAA-CREF, such as retirement income security issues and strategic issues in higher education. esearch findings from many grant projects have been published in journals such as the American Economic Review, Journal of Finance, the Journal of Economic Theory, and the Journal of Financial Services Research.
Requirements: Researchers who are interested in applying for a grant are strongly advised to submit a brief letter of inquiry to determine whether the Institute's present interests and budget allow consideration of the request. The letter should be submitted via email. The letter should include the following four items : 1) a short description of the project, 2) estimated overall budget, 3) period of time for which funds are requested, and 4) names of researchers who will be conducting the research. The Institute will respond to the inquiry and may ask the grant seeker to submit a formal proposal.
Samples: 'Critical Mass on Corporate Boards: Why Three or More Women Enhance Governance', Vicki W. Kramer, V. Kramer & Associates Alison M. Konrad, Richard Ivey School of Business, University of Western Ontario, Sumru Erkut, Wellesley Centers for Women
Contact: Paul Yakoboski, Principal Research Fellow, 800 842-2733 x3663; email: pyakoboski@tiaa-cref.org
Internet: http://www.tiaa-crefinstitute.org/research/grants/index.html
Sponsor: TIAA-CREF Institute
730 Third Avenue
New York, NY 10017

TIAA-CREF Paul A. Samuelson Award 5545
The award recognizes outstanding scholarly writing on issues related to lifelong financial security. It is given each year in recognition of an outstanding research publication containing ideas that the public and private sectors can use to maintain and improve Americans financial well being. The work must be theoretical or empirical in nature and must cover a subject directly relevant to lifelong financial security.
Requirements: Individuals who have published research, in either book or article form, between January 1 of last year and June 20 of the current year are eligible.
Samples: Robert Dammon, Chester Spatt, and Harold Zhang—for their article titled Optimal Asset Location and Allocation with Taxable and Tax-Deferred Investing (The Journal of Finance, June 2004), $10,000 (2004).
Amount of Grant: $10,000 to author of winning publication
Date(s) Application is Due: Sep 13

Contact: Douglas Fore, Grants Administrator, (800) 842-2733, ext. 7768; email: dfore@tiaa-cref.org
Internet: http://www.tiaa-crefinstitute.org/awards/samuelson/index.html
Sponsor: TIAA-CREF Institute
730 Third Avenue
New York, NY 10017-3206

Tibor T. Polgar Fellowships **5546**

This research program is conducted in cooperation with the New York State Department of Environmental Conservation and provides a summertime grant and limited research funds for eight graduate or undergraduate students to conduct research on the Hudson River. The objectives of the program are to gather important information on all aspects of the river and to train students in conducting estuarine studies and public policy research. The fellowships may be awarded for studies anywhere within the tidal Hudson estuary from the Federal Dam at Troy, NY, to New York Harbor. Because of the training and educational aspects of the program, each potential fellow must be sponsored by a primary advisor. The advisor must be willing to commit sufficient time for supervision of the research and to attend at least one meeting to review the progress of the research. Advisors receive a $500 stipend.
Requirements: Graduate and undergraduate students are eligible.
Amount of Grant: $3800 per fellowship
Date(s) Application is Due: Feb 28
Contact: Helena Andreyko, Program Contact, (212) 483-7667; fax: (212) 924-8325; email: info@hudsonriver.org
Internet: http://www.hudsonriver.org/polgar.htm
Sponsor: Hudson River Foundation
17 Battery Pl, Suite 915
New York, NY 10004

Tides Foundation California Wildlands Grassroots Fund **5547**

The program supports conservationists advocating for the permanent protection of intact wildlands on both public and private lands in order to help preserve California's wilderness and native biological diversity. Priority is given to small, nonprofit organizations with budgets of $100,000 or less, and individual activists with a sponsoring organization; and geographic areas and advocacy efforts that have not received significant foundation support. The fund targets project-specific requests and does not typically provide general support (such as overhead costs including rent, staff salaries). Grant-supported activities include, but are not limited to, mailings; paid advertising; advocacy-related travel; research; hiring technical, legal, or scientific experts; consultants; equipment purchase or rental; training; retreats; monitoring; mapping; and advocacy-related personal expenses (i.e., childcare). Guidelines and application are available online.
Requirements: California 501(c)3 and 501(c)4 tax-exempt organizations with budgets of $100,000 or less are eligible.
Geographic Focus: California
Amount of Grant: $5000 maximum; $1000-$3500 typically
Contact: Malisa Tsang , (415) 561-6400; fax: (415) 561-6401; email: mtsang@tides.org
Internet: http://www.calwildlandsfund.org/index.cfm
Sponsor: Tides Foundation
P.O. Box 29903
San Francisco, CA 94129-0903

Tides Foundation Death Penalty Mobilization Fund Grants **5548**

The fund supports efforts to abolish the death penalty in individual states and as a federal policy. Projects supported may focus on intermediate steps aimed at mitigating aspects of the death penalty, including working toward a moratorium on executions. The fund supports short-term, concrete projects rather than providing general operating or organizational development support. Priority is given to projects that entail collaboration between groups and aim to achieve maximum impact in a short time. Examples include media campaigns, research such as polling leading to a larger community education and organizing strategy, and short-term projects within larger organizational goals. Preference is given to projects that commence within six months of funding rather than those that have already begun. The proposal deadline is the first day of each month at 5 pm Pacific Time. Grant decisions will be made by the first day of the following month.
Requirements: Strategic collaborations of local, regional, and national nonprofit 501(c)3 and 501(c)4 organizations working on the death penalty are eligible.
Amount of Grant: $5000 maximum
Contact: Kathleen Lee or Michelle Coffey, (212) 509-1049; fax: (212) 509-1059; email: mcoffey@tides.org; Jane Lin, Program Assistant, (415) 561-7810; email: jlin@tides.org
Internet: http://www.tidesfoundation.org/dpmf_rapidresp_rfp.cfm
Sponsor: Tides Foundation
40 Exchange Pl, Suite 1111
New York, NY 10005

Times Mirror Foundation Grants Program **5549**

The foundation makes grants to nonprofit organizations in operating communities in its areas of interest, including journalism, education, literacy, community enrichment, and art and culture. The foundation focuses its support on programs that improve the quality of journalism, education and literacy, strengthen the fabric of the community, and enhance cultural appreciation and understanding. In addition, the foundation approves grants for unrestricted operating support, capital campaigns, seed grants, and special projects. Qualified organizations seeking grant support should send a two- to three-page letter briefly describing the organization, its purpose, programs, any special project to be considered, and its governing boards. Proof of tax-exempt status, list of current supporting organizations and amount of support, an organizational budget for current and upcoming fiscal years, and a copy of the most recent financial statements should be included.
Requirements: The foundation's geographical area is restricted to Southern California (Los Angeles, Orange, Ventura, San Bernardino and Riverside counties); Stamford, Greenwich and Hartford, Connecticut; and Allentown, Pennsylvania, and Queens and Long Island, New York. An applicant organization must include evidence of 501(c)3 status when requesting funds.
Restrictions: The foundation will not consider grants for production costs of films, videos, or television programs; religious or fraternal purposes; publications; conferences; or to individuals.
Geographic Focus: California; Connecticut; New York; Pennsylvania
Samples: Maryland African American Museum (Baltimore, MD)—for its capital campaign, $80,000 (2005); Archdiocese of Baltimore Chancery (Baltimore, MD)—for its Reading Scholars program, $25,000 (2005); Ctr for Nonprofit Management (Los Angeles, CA)—for its resource library of nonprofit-related materials, $15,000 (2005).
Amount of Grant: $1000-$25,000 average
Date(s) Application is Due: Apr 15; Oct 15
Contact: Michelle Williams, Executive Director, (213) 237-3945; fax: (213) 237-2116
Internet: http://www.timesmirrorfoundation.org
Sponsor: Times Mirror Foundation
202 W First Street
Los Angeles, CA 90053

Timoshenko Medal **5550**

This medal is bestowed in recognition of distinguished contributions to applied mechanics without restriction to nationality or profession. A certificate is included with the award. Listed deadline is for nomination.
Requirements: Any individual member, group of members, or committee may nominate candidates for this award.
Samples: Grigory Isaakovich Barenblatt—award winner, $1000 (2005); Morton E. Gurtin—award winner, $1000 (2004); Lambert B. Freund—award winner, $1000 (2003).
Amount of Grant: $1000
Date(s) Application is Due: Oct 1
Contact: Program Contact, (212) 591-7736; fax: (212) 705-7739; email: ditullio@asme.org or infocentral@asme.org
Internet: http://www.asme.org/honors/ms71/saa/timoshenko.html
Sponsor: American Society of Mechanical Engineers
3 Park Avenue
New York, NY 10016

Tinker Foundation Field Research Grants **5551**

Grants must be used to support individual research by outstanding graduate students and junior faculty in Iberia or Latin America. The grants fund the travel costs (international and in-country) of recipients to provide them with the opportunity to acquire a comprehensive knowledge of language and culture, to gather research data, and to develop contacts with scholars and institutions in their fields. Each university center or institute that receives an award must conduct a competition to select the recipients. The awards are to be used for brief periods of field research and should reflect primarily the major interests of the foundation—economic policy and governance and those targeted social science disciplines having strong public policy implications, international relations, and environmental policy studies. Up to three grants will be awarded annually.
Requirements: The program is open to all recognized centers or institutes of Ibero-American or Latin American studies with graduate doctoral programs at accredited U.S. universities.
Samples: To enable graduate students at nine U.S. universities to conduct research in Latin America, Spain, and Portugal, $130,000 total.
Amount of Grant: $10,000-$15,000 per annum
Date(s) Application is Due: Oct 1
Contact: Margaret Cushing, Program Officer, (212) 421-6858; fax: (212) 223-3326; email: mcushing@tinker.org or tinker@tinker.org
Internet: http://fdncenter.org/grantmaker/tinker/field.html
Sponsor: Tinker Foundation
55 E 59th Street
New York, NY 10022

Tinker Foundation Institutional Grants **5552**

Grants are made for topics and activities related to Ibero-America, Spain, Portugal, and Antarctica. Priority is given to work in economic policy, governance, and environmental policy. Collaboration of the nongovernmental environmental community with the private sector is preferred. Emphasis is placed on those activities that offer innovative solutions to many of the problems facing these regions and incorporate new mechanisms for addressing environmental, economic, political, and social issues. Activities may include research projects and conferences/workshops.
Requirements: Community organizations in the United States and abroad are eligible.
Restrictions: Grants are not awarded to support annual appeals, building construction, major equipment purchases, health or medical issues, production costs for films or radio, arts and humanities projects, endowments, or general operations.
Samples: Centro de Divulgacion del Conocimiento Economico (Caracas, Venezuela)—for a project designed to identify effective ways to manage informal urban street vendors in Caracas, $50,000; Johns Hopkins U (Baltimore, MD)—to support a policy research project on China's economic, political, and strategic interests in South America, to

be conducted by the Western Hemisphere Program at the Paul H. Nitze School of Advanced International Studies, $100,000 two-year grant; Marine Stewardship Council Limited (Seattle, WA)—for targeted fishery outreach and promotion of the MSC eco-labeling program in Spain and Portugal, $115,000 two-year grant.
Amount of Grant: $25,000-$100,000
Contact: Renate Rennie, President, (212) 421-6858; fax: (212) 223-3326; email: rrennie@tinker.org or tinker@tinker.org
Internet: http://fdncenter.org/grantmaker/tinker/index.html
Sponsor: Tinker Foundation
55 E 59th Street
New York, NY 10022

Tobias Landau Fellowships in Marine Biology 5553
The foundation awards field study support for marine science research in the United States and Israel. Americans will work at the Heinz Steinitz Marine Biology Laboratory in Eilat, Israel, and Israelis at marine stations in the United States. Cooperative support is welcomed. Research proposals are required; application forms giving details are available from the foundation. All applications are reviewed by an international panel on marine biology. Awards are made on the basis of scientific merit, novelty of the research ideas, and their suitability to the resources and location of the Eilat Station or the appropriate American station if the applicant is an Israeli.
Amount of Grant: $5000 maximum per year
Contact: Dr. Helen Hayes, Director, Tobias Landau Fellowships, (828) 692-5170
Sponsor: Tobias Landau Foundation of the United Jewish Endowment Fund
105 Echo Drive
Hendersonville, NC 28739

Tofias PC Charitable Foundation Grants 5554
The foundation currently focuses on strengthening the organization skills of health-focused nonprofit organizations. Grants will be awarded to support organizational skills, including performance management, financial planning, fundraising and revenue generation, and building and management; and organizations working to improve people's health through education, research, awareness, advocacy, and/or direct services. The foundation is particularly interested in organizations with annual budgets up to $25 million; strong, financially stable organizations where a small grant could address a specific need; proposals with clearly defined infrastructure goals and achieveable implementation and evaluation plans; and organizations with strong leadership and an active volunteer network. One-year grants will be awarded. Applicants should submit a letter of intent (two-page maximum); full proposals are by invitation. Letters of intent are accepted on an ongoing basis.
Requirements: Organizations in the following Massachusetts counties are eligible: Bristol, Essex, Middlesex, Norfolk, Plymouth, Suffolk, and Worcester. Organizations in the following Rhode Island counties are eligible: Bristol, Kent, Providence, Newport, Washington.
Restrictions: The foundation will not support endowments, scholarships, or political offices.
Geographic Focus: Massachusetts; Rhode Island
Amount of Grant: $2000-$5000 average
Contact: Grants Administrator, (617) 761-0600 or (888) 761-3834 ext 637; email: foundation@tofias.com
Internet: http://www.tofias.com/foundation
Sponsor: Tofias PC Charitable Foundation
350 Massachusetts Avenue
Cambridge, MA 02139

Tourette Syndrome Association Post-Doc Fellowships 5555
The Fellowship program is interested in supporting any research that has the potential to contribute to the understanding of Tourette Syndrome, including genetics, pathogenesis, pathophysiology, clinical treatments and animal model development. Post-doctoral research fellowships of up to $40,000 for one year are awarded. For preliminary screening, a Letter of Intent briefly describing the scientific basis and relevance of the proposed project should be sent to tsadevel@earthlink.net by October 1. The cover page to accompany each letter can be downloaded from the website. Applicants will be notified electronically as to whether or not to proceed with a full grant proposal. If invited to apply for a research grant, detailed instructions will be provided, and full proposals are due by December 9th.
Requirements: Candidates must have a M.D., Ph.D. or equivalent. Previous experience in the field of movement disorders is desirable, but not essential. Applicants may reside outside the United States.
Amount of Grant: $40,000 maximum
Date(s) Application is Due: Oct 1; Dec 9
Contact: Fellowship Administrator; (718) 224-2999, ext. 222; fax: (718) 279-9596; email grantadministrator@tsa-usa.org
Internet: http://tsa-usa.org/research.html
Sponsor: Tourette Syndrome Association
42-40 Bell Boulevard
Bayside, NY 11361-2820

Tourette Syndrome Association Research Grants 5556
The Program is interested in receiving applications from any research study that has the potential to contribute to the understanding of Tourette Syndrome, including genetics, pathogenesis, pathophysiology, clinical treatments and animal model development. Research grants of up to $75,000 for one year are awarded. Applicants may apply for a second year of funding, which is reviewed competitively. For preliminary screening, a Letter of Intent briefly describing the scientific basis and relevance of the proposed project should be sent as an email attachment to tsadevel@earthlink.net by October 1st. The cover page to accompany each letter can be downloaded from the website. Applicants will be notified electronically as to whether or not to proceed with a full grant proposal. If invited to apply for a research grant, detailed instructions will be provided, and full proposals are due by December 9th.
Requirements: Candidates must have a M.D., Ph.D. or equivalent. Previous experience in the field of movement disorders is desirable, but not essential. Applicants may reside outside the United States.
Amount of Grant: $75,000
Date(s) Application is Due: Oct 1; Dec 9
Contact: Heather Cowley, Manager; (718) 224-2999, ext. 247; fax: (718) 279-9596; email: Heather.Cowley@tsa-usa.org or grantadministrator@tsa-usa.org
Internet: http://tsa-usa.org/research.html
Sponsor: Tourette Syndrome Association
42-40 Bell Boulevard
Bayside, NY 11361

TRDRP California Research Awards 5557
This program is focused on research projects that address questions specific to tobacco-related disease or tobacco control issues in California. The awards will provide funding for average annual direct costs of $140,000; $170,000 per year for research involving human subjects. Full indirect costs are paid to eligible institutions. Maximum duration is 3 years. (This is a new TRDRP award mechanism as of 2009.)
Requirements: Principal Investigators from California not-for-profit organizations are eligible for funding. Research undertaken with TRDRP funds must be conducted primarily in California. The Principal Investigator must conduct and supervise the research project directly and in person. Underrepresented racial and ethnic investigators are encouraged to submit applications. To be competitive, proposals must present a compelling justification, including evidence if available, supporting the California-specific criterion. The proposal should be fully developed, scientifically rigorous, and include sound background information, hypothesis, and promising preliminary studies or preliminary data.
Restrictions: All submitted applications will be screened for their direct relevance to tobacco use or tobacco-related disease. Proposed research that is not relevant will not undergo peer review for scientific merit. It is incumbent on the applicant to make a compelling case that the proposed work is directly relevant to the mission of TRDRP.
Geographic Focus: California
Samples: Examples of previously-funded TRDRP grants that addressed California-specific research questions (but not limited to): Diabetes & tobacco exposure in So CA American Indians - Study tobacco use patterns with California Indian tribal members and tribal tobacco control policies.; Environmental contexts of smoking for Southeast Asians - Investigate the relationship between the social environment and smoking among Southeast Asians in California urban areas.; The geography of tobacco related disease in California - Describe the spatial distribution of hospital admitted tobacco-related disease prevalence rates and smoking prevalence in California.
Amount of Grant: $140,000 - $170,000 per year
Date(s) Application is Due: Dec 5
Contact: Kamlesh Asotra, Ph.D., (510) 287-3366, kamlesh.asotra@ucop.edu; M.F. Bowen, Ph.D., (510) 987-9811, mf.bowen@ucop.edu; Phillip Gardiner, Dr.P.H., (510) 987-9353, phillip.gardiner@ucop.edu
Internet: http://www.trdrp.org/fundingopps/callawardmechs.asp#calrchawd
Sponsor: Tobacco-Related Disease Research Program (TRDRP)
University of California,ÄìOffice of the President, 300 Lakeside Drive, 6th Floor
Oakland, CA 94612-3550

TRDRP Dissertation Research Awards 5558
This award is intended to support the dissertation research of a doctoral candidate pursuing tobacco-related research. The award is designed for students advanced to candidacy no later than July 1, and initiating their dissertation research. Maximum of $20,000 annual direct costs averaged over the duration of the award (maximum 2 years) for stipend, supplies, and domestic travel. This award also provides an additional maximum of $10,000 per year for tuition/enrollment fee remission, fringe benefits, and health insurance.
Requirements: Applications in all relevant research areas are welcomed, but applications in the social/behavioral sciences and in public policy are encouraged. The applicant and principal mentor must be affiliated with an academic research institution. U.S. citizenship is not a requirement. The candidate must commit a minimum of 80 percent time to the research project.
Restrictions: No equipment purchases are allowed. No indirect costs are allowed.
Geographic Focus: California
Amount of Grant: $30,000 per year
Date(s) Application is Due: Dec 5
Contact: Kamlesh Asotra, Ph.D., (510) 287-3366, kamlesh.asotra@ucop.edu; M.F. Bowen, Ph.D., (510) 987-9811, mf.bowen@ucop.edu; Phillip Gardiner, Dr.P.H., (510) 987-9853, phillip.gardiner@ucop.edu
Internet: http://www.trdrp.org/fundingopps/callawardmechs.asp
Sponsor: Tobacco-Related Disease Research Program (TRDRP)
University of California,ÄìOffice of the President, 300 Lakeside Drive, 6th Floor
Oakland, CA 94612-3550

TRDRP Exploratory/Developmental Research Awards 5559

The purpose of these grants is to gather preliminary data or demonstrate proof-of-principle (i.e., pilot projects), or to conduct a research project within the specified limits of money and time. The ultimate goal of these awards is to provide the foundation for proposals for fully-developed research project awards from other funding programs. Total direct costs of $250,000 for the entire project duration (maximum of 2 years). Indirect costs are paid to eligible institutions.

Requirements: Principal Investigators from California not-for-profit organizations are eligible for funding. Research undertaken with TRDRP funds must be conducted primarily in California. The Principal Investigator must conduct and supervise the research project directly and in person. Underrepresented racial and ethnic investigators are encouraged to submit applications. Although Principal Investigators may submit as many proposals for these awards as they wish for separate projects, no more than one Exploratory/Developmental Research Award will be made to any individual PI in the same year. Only proposals that address one of TRDRP, Primary research area will be eligible and will undergo peer review. Exploratory/Developmental Research Awards are for new projects only; that is, competing renewal applications will not be accepted.

Restrictions: All submitted applications will be screened for their direct relevance to tobacco use or tobacco-related disease. Proposed research that is not relevant will not undergo peer review for scientific merit. It is incumbent on the applicant to make a compelling case that the proposed work is directly relevant to the mission of TRDRP.

Amount of Grant: $250,000

Date(s) Application is Due: Dec 5

Contact: Kamlesh Asotra, Ph.D., (510) 287-3366, kamlesh.asotra@ucop.edu; M.F. Bowen, Ph.D., (510) 987-9811, mf.bowen@ucop.edu; Phillip Gardiner, Dr.P.H., (510) 987-9853, phillip.gardiner@ucop.edu

Internet: http://www.trdrp.org/fundingopps/callawardmechs.asp#explordev

Sponsor: Tobacco-Related Disease Research Program (TRDRP)
University of California,ÄîOffice of the President, 300 Lakeside Drive, 6th Floor
Oakland, CA 94612-3550

TRDRP Grants 5560

TRDRP will consider unsolicited requests to support research-related activities that build research infrastructure, disseminate research findings, or stimulate new research directions. Support can be requested for scientific conferences to assess tobacco's impact on California's population; or to allow tobacco investigators to evaluate, in a timely manner, new and breaking trends in tobacco control or tobacco-related disease research. The TRDRP Scientific Advisory Committee will make recommendations regarding funding. These opportunities will be limited in number, scope, cost, and duration. Please contact a TRDRP research administrator regarding the appropriateness of your proposal prior to submission.

Requirements: In order to qualify for funding, the planned activities must be directly related to one or more of TRDRP's Priority research areas. The activity must primarily take place in California, involve California investigators, and include, where applicable, discussants and speakers funded by TRDRP. Unsolicited requests may be submitted at any time. Requests will be evaluated expeditiously by ad-hoc peer review when appropriate.

Restrictions: All submitted applications will be screened for their direct relevance to tobacco use or tobacco-related disease. Proposed research that is not relevant will not undergo peer review for scientific merit. It is incumbent on the applicant to make a compelling case that the proposed work is directly relevant to the mission of TRDRP.

Geographic Focus: California

Contact: Kamlesh Asotra, Ph.D., (510) 287-3366, kamlesh.asotra@ucop.edu; M.F. Bowen, Ph.D., (510) 987-9811, mf.bowen@ucop.edu; Phillip Gardiner, Dr.P.H., (510) 987-9853, phillip.gardiner@ucop.edu

Internet: http://www.trdrp.org/fundingopps/callawardmechs.asp

Sponsor: Tobacco-Related Disease Research Program (TRDRP)
University of California,ÄîOffice of the President, 300 Lakeside Drive, 6th Floor
Oakland, CA 94612-3550

TRDRP Participatory Research Awards: CARA/SARA 5561

The purpose of the Community Academic Research Awards (CARA) and the School Academic Research Awards (SARA) is to stimulate and support collaborations between community-based organizations/ schools with academic investigators. These awards are for a collaborative partnership to perform scientifically rigorous research into tobacco control issues that are identified as important and meaningful to specific communities/ schools in the state.

Requirements: The applicant partners must demonstrate the use of methods that are relevant, culturally sensitive, and appropriate in terms defined and accepted by the participating communities/schools. Applicants are encouraged to consider TRDRP's Primary research areas and and the TEROC recommended research strategies that lend themselves to a participatory research approach. (See http://www.trdrp.org/fundingopps/call.asp#teroc.)

Restrictions: All submitted applications will be screened for their direct relevance to tobacco use or tobacco-related disease. Proposed research that is not relevant will not undergo peer review for scientific merit. It is incumbent on the applicant to make a compelling case that the proposed work is directly relevant to the mission of TRDRP.

Geographic Focus: California

Date(s) Application is Due: Dec 5

Contact: Kamlesh Asotra, Ph.D., (510) 287-3366, kamlesh.asotra@ucop.edu; M.F. Bowen, Ph.D., (510) 987-9811, mf.bowen@ucop.edu; Phillip Gardiner, Dr.P.H., (510) 987-9853, phillip.gardiner@ucop.edu

Internet: http://www.trdrp.org/fundingopps/callawardmechs.asp

Sponsor: Tobacco-Related Disease Research Program (TRDRP)
University of California,ÄîOffice of the President, 300 Lakeside Drive, 6th Floor
Oakland, CA 94612-3550

TRDRP Postdoctoral Fellowship Awards 5562

These are awards for individuals to obtain postdoctoral research training under a designated mentor. The application must be prepared and submitted exclusively by the fellow and must outline an original research project (separate from the project of a mentor). Letters of support addressing the candidate, training, potential, and the commitment of the mentor and the department to the candidate, career development are essential.

Requirements: Principal Investigators from California not-for-profit organizations are eligible for funding. Research undertaken with TRDRP funds must be conducted primarily in California. The Principal Investigator must conduct and supervise the research project directly and in person. Underrepresented racial and ethnic investigators are encouraged to submit applications. To be eligible, the candidate must be recognized by the applicant institution as a postdoctoral fellow no later than July 1 of the year intended for the fellowship. U.S. citizenship is not a requirement. The fellow must commit a minimum of 75 percent time to the research project. Direct costs of $45,000 per year averaged over the duration of the award. No indirect costs are allowed. Maximum duration is 3 years.

Restrictions: All submitted applications will be screened for their direct relevance to tobacco use or tobacco-related disease. Proposed research that is not relevant will not undergo peer review for scientific merit. It is incumbent on the applicant to make a compelling case that the proposed work is directly relevant to the mission of TRDRP.

Geographic Focus: California

Amount of Grant: $45,000 per year

Date(s) Application is Due: Dec 5

Contact: Kamlesh Asotra, Ph.D., (510) 287-3366, kamlesh.asotra@ucop.edu; M.F. Bowen, Ph.D., (510) 987-9811, mf.bowen@ucop.edu; Phillip Gardiner, Dr.P.H., (510) 987-9853, phillip.gardiner@ucop.edu

Internet: http://www.trdrp.org/fundingopps/callawardmechs.asp#postdoc

Sponsor: Tobacco-Related Disease Research Program (TRDRP)
University of California,ÄîOffice of the President, 300 Lakeside Drive, 6th Floor
Oakland, CA 94612-3550

Trevithick Book Prize Award 5563

This award is given each year for the encouragement of innovative excellence in research in the Committee on Resources and Environment program. The application should consist of a concise 1000-word description of the candidate's research. Clearly drafted diagrams (maximum of four) may be used. The supervisor should append his/her comments on the innovation and quality of the work and its importance to the field. A rigid format is not defined, but the following topics should be covered: aim, methods, results, interpretation of results, and implications for and relations to the general field of study.

Requirements: The competition is open to all graduate students currently registered in the RESR program who are completing their theses prior to April 1 of the current academic year. Those completing their theses after this date are eligible for the competition the following year. A student may enter the competition only once.

Amount of Grant: $C200 maximum

Date(s) Application is Due: Jun 30

Contact: Connie Baines, Faculty of Graduate Studies, (403) 220-5690; email: cbaines@ucalgary.ca

Internet: http://www.grad.ucalgary.ca/funding/internal_scholarships/lev_4/trevithick.htm

Sponsor: University of Calgary
2500 University Dr NW, Earth Sciences Building, Room 720
Calgary, AB T2N 1N4 Canada

Trinity College Ann Plato Fellowship 5564

The fellowship, named for a 19th-century African-American poet, essayist, and teacher, supports a doctoral student who is an American citizen and is engaged in writing his or her dissertation. Applicants from historically underrepresented groups, including African Americans, Asian Americans, Native Americans, Pacific Islanders, and U.S. Latinos, are especially encouraged to apply. Preference will be given to candidates working in fields currently supported by the college. The fellow enjoys faculty status, delivers a formal, public lecture in the fall semester, and teaches one course in the spring semester. The fellow is expected to be in residence during the fellowship tenure and to become engaged in the Trinity College community. The fellowship provides a stipend; a campus apartment; an office; use of a computer; library privileges at Trinity, including the Watkinson Library, and its consortial colleges; and ready access to Hartford-area archives, including the Connecticut Historical Society, the Wadsworth Atheneum, the state library, the Cities Data Center, and the Harriet Beecher Stowe Center. Appointment is for one academic year. Applicants should send a letter of interest, curriculum vita, a copy of their dissertation proposal, a 10 to 20 page writing sample, and three letters of recommendation to the Ann Plato Search Committee.

Amount of Grant: $32,000 stipend

Date(s) Application is Due: Dec 15

Contact: Janet Marotto, email: gradgrnt@indiana.edu

Internet: http://www.indiana.edu/~gradgrnt/v14n2/plato.html

Sponsor: Trinity College
Williams 232
Hartford, CT 06106

Trinity College Senior Research Fellowships 5565

The college is accepting applications for senior research fellowships in the arts or humanities. The fellowships are of academic standing comparable to that of a distinguished research professorship at a major university, and applicants are expected to have a correspondingly distinguished record of achievement in research. The stipend, for a fellow not receiving regular remuneration from other sources, will fall within the scale of stipends paid by Cambridge University to holders of university professorships. Guidelines are available online.
Date(s) Application is Due: Oct 17
Contact: Fellowship Administrator, 01223-338-400; fax: 01223-338-564
Internet: http://www.trin.cam.ac.uk/index.php?pageid=46&vacid=52
Sponsor: Trinity College
Cambridge CB2 1TQ United Kingdom

Triological Society Research Career Development Awards 5566

The purpose of this award is to provide support for the research career development of otolaryngologists (head and neck surgeons) who have made a commitment to focus their research endeavors on patient-oriented research, e.g., clinical trials, translational research. Research training supported by this award may be related to any research questions relevant to the specialty of otolaryngology, as long as it is demonstrated that the training will have a direct impact on the applicant, ability to pursue his/her long-range research objectives. Grants can be for one to two years, are non-renewable, and have a $40,000 maximum award limitation. There are approximately six grants awarded annually. Letters of Intent are due by December 15, with full proposals received by January 15 annually.
Requirements: Otolaryngologists-head and neck surgeons who hold full-time, part-time or contributed service medical school faculty appointments and who have made a commitment to focus their research endeavors on patient-oriented research may apply. Applicants must be sponsored by the Chair of his/her division or department and by an official representative of the institution which would administer the award and in whose name the application is formally submitted.
Amount of Grant: $40,000 maximum
Date(s) Application is Due: Dec 15; Jan 15
Contact: Gail Binderup; (402) 346-5500; fax: (402) 346-5300; email: info@triological.org; Stephanie L. Jones, Assistant Director; (703) 519-1586 or (703) 836-4444; email: sljones@entnet.org
Internet: http://www.triological.org/researchgrants.htm
Sponsor: Triological Society
555 North 30th Street
Omaha, NE 68131

TSA Research Grant Awards 5567

The association accepts applications for grants in the following categories: proposals in basic neuroscience specifically relevant to Tourette syndrome (TS); clinical studies related to the etiology, pathophysiology, and treatment of TS; and one postdoctoral fellowship. Relevant scientific fields include animal models, basal ganglia physiology, biochemistry, epidemiology, genetics, molecular biology, neuroanatomy, neuroimaging, neurology, neuropathology, neuropsychology, neurophysiology, pharmacology, psychiatry, and psychology. As a means of preliminary screening to assess the relevance and quality of intended proposals, a letter of intent briefly describing the scientific basis of the proposed project is requested. An approximate level of project funding should be included. Letters of intent must meet the application deadline. All applicants will be promptly informed as to whether to proceed with a final proposal. Deadline listed is for letter of intent.
Requirements: Investigators already studying TS and those currently working in areas of science that can be applied to the problems of TS are invited to apply.
Amount of Grant: $75,000 maximum PhD and MD applicants; $40,000 maximum postdoctoral fellowships
Date(s) Application is Due: Oct 17
Contact: TSA Scientific Advisory Board, (718) 224-2999 ext 222; fax: (718) 279-9596; email: grantadministrator@tsa-usa.org
Internet: http://www.tsa-usa.org
Sponsor: Tourette Syndrome Association
42-40 Bell Boulevard
Bayside, NY 11361-2874

Tull Charitable Foundation Grants Program 5568

Priority interest areas for awarding grants, in Georgia, are education, health and human services, youth development, and the arts. Grants support capital projects (e.g., buildings and endowments) in these areas. The amount of award varies depending on the project for which funding is requested.
Requirements: Grants are provided to organizations in Georgia.
Restrictions: Grants are not available for operating support, research, conferences and seminars, special events, or individuals.
Geographic Focus: Georgia
Samples: Clark Atlanta U (GA)—for a classroom building, $100,000.
Amount of Grant: $25,000-$50,000 average
Date(s) Application is Due: Jan 1; Apr 1; Jul 1; Oct 1
Contact: Program Contact, (404) 659-7079; email: jan@tullfoundation.org
Internet: http://www.tullfoundation.org/app_procedures.asp
Sponsor: Tull Charitable Foundation
50 Hurt Plz, Suite 1245
Atlanta, GA 30303

Twentieth-Century Japan Research Awards 5569

Two $1,500 grants are available to support research in the library's Prange Collection and East Asia Collection on topics related to the period of the Allied Occupation of Japan and its aftermath, 1945-1960. Holders of the Ph.D. or an equivalent degree are eligible to apply, as are graduate students who have completed all requirements for the doctorate except the dissertation. The competition is open to scholars in all parts of the world and from any discipline, but historical topics are preferred. Applicants should send a curriculum vitae and a two- to three-page description of their chosen research project.
Restrictions: University of Maryland faculty, staff, and students may not apply.
Samples: Kyoko Omori—The Culture of Japanese Vernacular Modernism, 1920-1950 (2006-07); Maki Umemura—The Pharmaceutical Industry in Japan since 1945 (2006-07); Takashi Nishiyama—Former Military Scientists and Engineers, 1919-1964 & Labor Activism in the Japan National Railways, 1945-1955 (2005-06.; Alicia Volk—Democratizing Japanese Art, 1945-1960 (2005-06).
Amount of Grant: $1,500
Date(s) Application is Due: Nov 9
Contact: Program Director, (301) 405-8739; fax: (301) 314-9399; email: historycenter@umd.edu
Internet: http://www.history.umd.edu/HistoryCenter/awards.htm
Sponsor: Nathan and Jeanette Miller Center for Historical Studies
Department of History, 2115 Francis Scott Key Hall, University of Maryland
College Park, MD 20742-7315

TWS Foundation Grants 5570

The foundation awards grants to eligible nonprofit organizations in its areas of interest, including community development, arts, health care, public policy, research, and human services. Types of support include general operating support and scholarship funds. There are no application forms or deadlines.
Requirements: Nonprofit organizations, primarily in Connecticut, New York, Rhode Island, and Texas, are eligible.
Geographic Focus: Connecticut; New York; Rhode Island; Texas
Samples: National Center for Policy Analysis (Dallas, TX)—$200,000.; Westerly Hospital Foundation (Westerly, RI)—$100,000.; Philharmonic-Symphony Society of New York (New York, NY)—$10,000.
Amount of Grant: $100-$230,000 range
Contact: Thomas Smith, Trustee
Sponsor: TWS Foundation
323 Railroad Avenue
Greenwich, CT 06830-6306

U.S. Department of Education American Overseas Research Centers 5571

This program provides grants to establish or operate overseas research centers that promote postgraduate research, exchanges, and area studies. Grants may be used to pay for all or a portion of the cost of establishing or operating a center or program. Costs may include faculty and staff stipends and salaries; faculty, staff, and student travel; operation and maintenance of overseas facilities; teaching and research materials; the acquisition, maintenance, and preservation of library collections; travel for visiting scholars and faculty members who are teaching or conducting research; preparation for and management of conferences; and the publication and dissemination of material for the scholars and general public.
Requirements: Eligible Applicants are Consortia of institutions of higher education that: receive more than 50 percent of their funding from public or private U.S. sources; have a permanent presence in the country in which the center is located; and, are tax-exempt nonprofit organizations described in Section 501(c)3 of the Internal Revenue Code. Note: Competitions for funding are held every four years. The most recent competition: FY 2007.
Samples: FY 2007 - FY 2010: American Institute for Sri Lankan Studies, $60,511.; American Research Center in Egypt, $72,582.; American School of Classical Studies at Athens, $91,598.; Center for Khmer Studies, $82,909.; American Institute for Maghrib Studies, $109,406.; American Institute of Indian Studies, $109,650.; American Research Center in Turkey, $109,406.; West Africa Research Association, $106,195.; Palestinian American Research Center, $88,181.; American Institute of Pakistan Studies, $63,893.; American Center for Mongolian Studies, $105,669.
Amount of Grant: $80,000 - $100,000
Date(s) Application is Due: Mar 15
Contact: Cheryl Gibbs; (202) 502-7634; fax: (202) 502-7859; email: cheryl.gibbs@ed.gov
Internet: http://www.ed.gov/programs/iegpsaorc/index.html
Sponsor: U.S. Department of Education
1990 K Street, NW, 6th Floor
Washington, DC 20006-8521

U.S. Department of Education Disability and Rehabilitation Research and Related Projects 5572

The purpose of the program is to plan and conduct research, demonstration projects, training, and related activities to improve the lives of individuals with disabilities. The projects are quite varied, though all are aimed at fulfilling NIDRR's (National Institute on Disability and Rehabilitation Research) overarching goals of inclusion, integration, employment, and self-sufficiency for people with disabilities.
Requirements: Institutions of Higher Education, States, Nonprofit Organizations are eligible to apply. Public or private agencies—including for-profit agencies, Indian tribes, and tribal organizations may also apply.

Contact: Donna Nangle; (202) 245-7462; fax: (202) 245-7323; email: donna.nangle@ed.gov
Internet: http://www.ed.gov/programs/drrp/index.html
Sponsor: U.S. Department of Education
400 Maryland Ave. SW, Room 6030, PCP
Washington, DC 20202-2700

U.S. Department of Education Foreign Languages and Area Studies Fellowships 5573
The mission of the International Education Programs Service (IEPS) is to meet the national needs for expertise and competence in foreign languages and area or international studies. This program provides grants to a select number of public and private institutions of higher education to enable them to offer competitive academic year and summer fellowships to graduate students annually. The disciplinary structure of a FLAS program may be interdisciplinary or multi-disciplinary, including the humanities, social sciences, or other professional studies, and must include the study of the languages of the geographical area of specialization. Institutions compete for fellowship allocations every four years.
Requirements: Eligible applicants include Institutions of higher education that offer performance-based language instruction. A student is eligible to receive a fellowship if he or she: is a graduate student; is a citizen, national, or permanent resident of the United States; is accepted for enrollment or is enrolled in a program that combines modern foreign language training with area or international studies; and shows potential for high academic achievement based on grade point average, class ranking, or similar measures determined by the institution. Institutions should email questions to ope_nrc-flas@ed.gov or visit the IEPS Web site at www.ed.gov/programs/iegpsflasf for more information, including: application deadline, application package, technical assistance, workshop and meeting schedules, and program statute and regulations. Students should visit the IEPS Web site at http://www.ed.gov/programs/iegpsflasf/awards.html for a list of currently funded FLAS institutions and contact the schools directly regarding selection procedures and competition schedules.
Contact: Peter Baker (202) 219-7060, peter.baker@ed.gov; Cheryl E. Gibbs (202) 502-7634, cheryl.gibbs@ed.gov; Rod Schwartz (202) 502-7805, rod.schwartz@ed.gov; Amy E. Wilson (202) 502-7689, amy.wilson@ed.gov
Internet: http://www.ed.gov/programs/iegpsflasf/index.html
Sponsor: U.S. Department of Education
International Education Programs Service; 1990 K Street, NW, 6th Floor
Washington, DC 20006-8521

U.S. Department of Education Graduate Assistance in Areas of National Need 5574
This program provides fellowships, through academic departments and programs of IHEs, to assist graduate students with excellent records who demonstrate financial need and plan to pursue the highest degree available in their course study at the institution in a field designated as an area of national need. Grants are awarded to programs and institutions to sustain and enhance the capacity for teaching and research in areas of national need.
Requirements: Institutions of Higher Education (IHEs) may apply. Academic departments and programs of IHEs that provide courses of study leading to a graduate degree may apply. Nondegree granting institutions may submit joint proposals with degree-granting IHEs.
Restrictions: Students are ineligible to apply for funding.
Amount of Grant: $130,656 - $261,312
Date(s) Application is Due: Mar 16
Contact: Gary Thomas or Rebecca Green; (202) 502-7767 or (202) 502-7779; fax: (202) 502-7859; email: ope_gaann_program@ed.gov
Internet: http://www.ed.gov/programs/gaann/index.html
Sponsor: U.S. Department of Education
1990 K Street, NW, 6th Floor
Washington, DC 20006-8524

U.S. Department of Education Jacob K. Javits Fellowships 5575
This program provides fellowships to students of superior academic ability,Äìselected on the basis of demonstrated achievement, financial need, and exceptional promise,Äìto undertake study at the doctoral and Master of Fine Arts level in selected fields of arts, humanities, and social sciences. Panels of experts appointed by the Javits Fellowship Board (Board) select fellows according to criteria established by the Board. Students must demonstrate financial need by filing the Free Application for Federal Student Aid (FAFSA). The selected fields of study supported by the Javits Fellowship Program are available at http://www.ed.gov/programs/jacobjavits/javits-fieldsofstudy.html. For fiscal year 2008, the maximum stipend was $30,000, and the institutional payment was $12,891.
Requirements: Eligibility is limited to U.S. citizens or nationals, permanent residents of the U.S., or citizens of any one of the Freely Associated States. Applicants must be eligible to be accepted to or currently attending a graduate program leading to a doctorate or a terminal master's degree in an eligible field of study at an institution of higher education approved by an accrediting agency recognized by the Secretary of the Department of Education. Eligibility is limited to individuals who, as Javits fellows, will be eligible to receive any grant, loan, or work assistance pursuant to Section 484 of the Higher Education Act of 1965, as amended.
Restrictions: Applicants attending a foreign institution are ineligible. Students who have already received a Javits fellowship in previous years are not eligible.
Date(s) Application is Due: Oct 3
Contact: Carmen Gordon and Sara Starke; Jacob K. Javits Fellowships Program; (202) 502-7542; fax: (202) 502-7859; email: OPE_Javits_Program@ed.gov
Internet: http://www.ed.gov/programs/jacobjavits/index.html
Sponsor: U.S. Department of Education
1990 K Street, NW, 6th Floor
Washington, DC 20006-8524

U.S. Department of Education National Resource Centers (NRC) 5576
The program provides grants to establish, strengthen, and operate language and area or international studies centers that will be national resources for teaching any modern foreign language. Grants support: instruction in fields needed to provide full understanding of areas, regions or countries; research and training in international studies; work in the language aspects of professional and other fields of study; and instruction and research on issues in world affairs.
Requirements: Institutions of higher education or consortia of institutions of higher education may apply. A comprehensive or undergraduate National Resource Center (NRC): 1) teaches at least one modern foreign language; 2) provides (a) instruction in fields necessary to provide a full understanding of the areas, regions, or countries in which the languages are commonly used; (b) resources for training and research in international and foreign language aspects of professional and other fields of study; or(c) opportunities for training and research on issues in world affairs that concern one or more countries; 3) provides outreach and consultative services on a national, regional and local basis; 4) maintains linkages with overseas institutions of higher education and other organizations that may contribute to the teaching and research of the Center. Funds for undergraduate travel are allowable only in conjunction with a formal program of supervised study in the subject area on which the center focuses.
Restrictions: Equipment costs exceeding ten percent of the grant are not allowable. Grant funds may not be used to supplant.
Amount of Grant: $90,000 - $500,000
Date(s) Application is Due: Nov 14
Contact: Peter Baker (202) 219-7060, peter.baker@ed.gov; Cheryl E. Gibbs (202) 502-7634, cheryl.gibbs@ed.gov; Rod Schwartz (202) 502-7805, rod.schwartz@ed.gov; Amy E. Wilson (202) 502-7689, amy.wilson@ed.gov
Internet: http://www.ed.gov/programs/iegpsnrc/index.html
Sponsor: U.S. Department of Education
1990 K Street, NW, 6th Floor
Washington, DC 20006-8521

U.S. Department of Education Native Hawaiian Education Grants 5577
The purpose of this program is to develop innovative education programs to assist native Hawaiians and to supplement and expand programs and authorities in the area of education. Authorized activities include, among others: early education and care programs; family-based education centers; beginning reading and literacy programs; activities to address the needs of gifted and talented native Hawaiian students; special education programs; professional development for educators; and activities to enable native Hawaiian students to enter and complete postsecondary education programs.
Requirements: Nonprofit Organizations, Native Hawaiian education organizations; Native Hawaiian community-based organizations; public and private nonprofit organizations, agencies, and institutions (including State Education Agencies [SEAs], Local Education Agencies [LEA]s, and Institutions of Higher Education [IHEs]) with experience in developing or operating Native Hawaiian programs or programs of instruction in the Native Hawaiian language; and consortia thereof may apply.
Date(s) Application is Due: Mar 24
Contact: Joanne Osborne, (202) 401-1265, email: Joanne.osborne@ed.gov; Stephen Balkcom, (202) 260-2737, email: Stephen.Balkcom@ed.gov
Internet: http://www.ed.gov/programs/nathawaiian/index.html
Sponsor: U.S. Department of Education
400 Maryland Ave. S.W., Room 3E214
Washington, DC 20202-6200

U.S. Department of Education Rehabilitation Engineering Research Centers Grants 5578
RERCs support activities that: (1) lead to the development of methods, procedures, and devices that will benefit individuals with disabilities, especially those with the most severe disabilities; or (2) involve technology for the purposes of enhancing opportunities for meeting the needs of and addressing the barriers confronted by individuals with disabilities in all aspects of their lives. Types of activities supported include: the development of technological systems for persons with disabilities; stimulation of the production and distribution of equipment in the private sector; and clinical evaluations of equipment. Awards are for five years, except that grants to new recipients or to support new or innovative research may be made for fewer than five years.
Requirements: States, Institutions of Higher Education (IHEs), Nonprofit Organizations, public or private agencies - including for-profit agencies, Indian tribes, and tribal organizations may apply. RERCs must be operated by or in collaboration with one or more institutions of higher education or nonprofit organizations.
Amount of Grant: $850,000 - $1,000,000
Date(s) Application is Due: Apr 1; Aug 1
Contact: Donna Nangle; (202) 245-7462; fax: (202) 245-7323; email: donna.nangle@ed.gov
Internet: http://www.ed.gov/programs/rerc/index.html
Sponsor: U.S. Department of Education
OSERS, 400 Maryland Avenue, S.W., Room 6030, PCP
Washington, DC 20202-2700

U.S. Department of Education Rehabilitation Research Training Centers 5579
The RRTCs conduct coordinated and advanced programs of research, training, and information dissemination. Each RRTC has a major program of research in a particular area, such as mental illness, vocational rehabilitation (VR), or independent living, which is specified by the National Institute on Disability and Rehabilitation Research (NIDRR). The RRTCs must serve as centers of national excellence and national or regional resources for providers and individuals with disabilities and their representatives. RRTC awards are for five years, except that grants to new recipients or to support new or innovative research may be made for fewer than five years.
Requirements: Note: Each year, competitions are held in specific areas that determine the types of projects. Eligible applicants include Institutions of Higher Education and Nonprofit Organizations. States, public or private agencies - including for-profit agencies, Indian tribes, and tribal organizations may also apply. Rehabilitation research and training centers must be operated by or in collaboration with: (1) one or more institutions of higher education or (2) one or more providers of rehabilitation or other appropriate services.
Amount of Grant: up to $675,000
Date(s) Application is Due: Aug 21; Aug 22
Contact: Donna Nangle; (202) 245-7462; fax: (202) 245-7323; email: donna.nangle@ed.gov
Internet: http://www.ed.gov/programs/rrtc/index.html
Sponsor: U.S. Department of Education
OSERS, 400 Maryland Avenue, SW, Room 6030, PCP
Washington, DC 20202-2700

U.S. Department of Education Ronald E. McNair Postbaccalaureate Achievement Grants 5580
This program prepares participants for doctoral studies through involvement in research and other scholarly activities. Participants are from disadvantaged backgrounds and have demonstrated strong academic potential. Institutions work closely with participants as they complete their undergraduate requirements. Institutions encourage participants to enroll in graduate programs and then track their progress through to the successful completion of advanced degrees. The goal is to increase the attainment of Ph.D. degrees by students from underrepresented segments of society. Projects provide: academic counseling, financial aid assistance, mentoring, research opportunities, seminars, summer internships, and tutoring. Guidance for students seeking admission and financial aid for graduate programs also is supported.
Requirements: Institutions of higher education and combinations of institutions of higher education may apply. Students who qualify for McNair must be enrolled in a degree-granting program at an eligible institution. In all projects, at least two-thirds of the participants must be low-income, potential first-generation college students. The remaining participants may be from groups that are underrepresented in graduate education.
Amount of Grant: $200,000 - $300,000 per year
Contact: Eileen Bland; Team Leader, College and University Support Team (CU.S.T); (202) 502-7730; email: eileen.bland@ed.gov
Internet: http://www.ed.gov/programs/triomcnair/index.html
Sponsor: U.S. Department of Education
1990 K Street, NW, Suite 7000
Washington, DC 20006-8510

U.S. Department of Education United States-Russia Program: Improving Research and Educational Activities in Higher Education 5581
The program provides grants that demonstrate partnerships between Russian and American institutions of higher education that contribute to the development and promotion of educational opportunities between the two nations, particularly in the areas of mutual foreign language learning and the cooperative study of mathematics and science. Collaborative projects comprised of partnerships of Russian and American universities and colleges will be supported.
Requirements: Institutions of Higher Education (IHEs) or combinations of IHEs and other public and private nonprofit institutions and agencies are eligible to apply. Applicants must select one of three academic discipline as the subject area for their grant proposal: (A) Environmental Science Studies; (B) Biotechnology; or, (C) Any discipline, other than (A) and (B). Applications are invited from institutions of higher education with the capacity to contribute to a collaborative project in the areas listed with a Russian institution. The consortium partners, through promoting the study of and communication in foreign languages, are expected to increase awareness and understanding of the two cultures, and to strengthen the professional and scholarly ties between the two countries. Russian institutions will apply to The Russian Ministry of Education and Science for funding.
Restrictions: Any application that proposes a budget exceeding $150,000 for a single budget period of 12 months will be rejected. The Assistant Secretary for Postsecondary Education may change the maximum amount through a notice published in the Federal Register.
Geographic Focus: Alabama, Alaska; Arizona; Arkansas; California; Colorado Connecticut; Delaware; District of Columbia; Florida; Georgia; Guam; Hawaii; Idaho, Illinois; Indiana; Iowa; Kansas; Kentucky; Louisiana; Maine; Maryland; Massachusetts, Michigan; Minnesota; Mississippi; Missouri; Montana; Nebraska; Nevada; New Hampshire; New Jersey; New Mexico; New York; North Carolina; North Dakota; Ohio, Oklahoma; Oregon; Pennsylvania; Puerto Rico; Rhode Island; South Carolina; South Dakota; Tennessee; Texas; U.S. Virgin Islands; Utah; Vermont; Virginia; Washington; West Virginia; Wisconsin; WyomingRussia
Samples: 2008: Stanford University, CA - $188,718
Amount of Grant: $189,000 average
Date(s) Application is Due: Jul 7
Contact: Krish Mathur, Coordinator, U.S.-Russia Program; (202) 502-7512; fax: (202) 502-7877; email: Krish.Mathur@ed.gov
Internet: http://www.ed.gov/programs/fipserussia/index.html
Sponsor: U.S. Department of Education
1990 K Street, N.W., Room 6155
Washington, DC 20006-8544

UCLA / RAND Corporation Post-Doctoral Fellowships 5582
The post-doctoral training program is jointly housed in the RAND Health Sciences program and the UCLA School of Public Health. Post-doctoral fellows have the option of pursuing a Masters degree in the UCLA Department of Health Services. The grant covers all tuition and fees, as well as a stipend, which currently ranges from $35,568 to $51,036, depending on the number of years of post-doctoral experience. Travel money is also available for fellows to attend professional meetings.
Requirements: To be eligible, candidates must have obtained a Ph.D., medical degree, or other professional doctoral degree. Awards will last two years. Fellows will be required to attend at least 8 courses at UCLA or RAND, including three courses in health services research methods, two in statistics, and three in health services issue areas. Those pursuing a Masters degree must complete a total of 12 courses. Upon arrival, fellows will align with an ongoing research project at RAND or UCLA. RAND and UCLA have numerous ongoing projects in the health services research area. These fall into several areas, including: access/utilization among the underserved; assessing quality, effectiveness, outcomes; health behaviors; alternative delivery systems; hospital and physician payment; health care cost containment; technology assessment; private health insurance; international comparisons. By the end of the first year, the fellow will identify a research idea and write a proposal. The proposal may be addressed to the principal investigator of an ongoing research project, or to an outside funding agency. By the end of the program, the fellow will prepare an article on research conducted during the fellowship, and submit it to a peer reviewed journal. This can either be done alone, or in collaboration with the research director or other member of the fellow's research team.
Amount of Grant: $35,568 - $51,036
Contact: RAND Health Communications; (310) 393-0411, ext. 7775; email: RAND_Health@rand.org
Internet: http://www.ph.ucla.edu/hs/post_doctoral.html
Sponsor: RAND Corporation
1776 Main Street, P.O. Box 2138
Santa Monica, CA 90407-2138

UCLA International Institute Global Fellowships 5583
The institute awards postdoctoral global fellowships, tenable for one to two years, beginning September 1. Fellows will be drawn from the arts and humanities (including literary studies and philosophy), the natural sciences (including engineering, medicine, and public health), and the social sciences (including law, public policy, and management). Global fellows will pursue their own research and teach one undergraduate seminar. Fellows are expected to produce working papers, journal articles, and books, or works in other media, during their tenure at UCLA. Successful applicants will be of the highest intellectual caliber and will be doing pathbreaking work, preferably of multidisciplinary interest, on (or with obvious relevance to) the cultural, economic, social, and political dynamics of the contemporary world. Among the topics of particular interest to the institute are: new forms of governance and coordination, whether in the governmental, nongovernmental, or private spheres; security concerns that transcend traditional interstate conflict; the causes and consequences of globalization; transnational cultures and multiple identities (or resistance to these trends); and public health, the environment, and economic development. Applications and nominations should be submitted, on-line, between 1 October and 1 November. There are no citizenship requirements.
Requirements: Candidates will only be considered if they received their PhDs (or equivalent) between 1 January 1998 and 1 April 2005.
Amount of Grant: $50,000 stipend
Date(s) Application is Due: Jan 14
Contact: Global Fellows Program, (310) 825-4921; email: globalfellows@international.ucla.edu
Internet: http://www.international.ucla.edu/globalfellows
Sponsor: UCLA International Institute
11349 Bunche Hall, UCLA
Los Angeles, CA 90095-1487

UCLA Postdoctoral/Visiting Scholar Fellowship in Ethnic Studies 5584
The university's Institute of American Cultures, in cooperation with the four Ethnic Studies Research Centers, offers fellowships in support of research on African Americans, American Indians, Asian Americans, and Chicanas/os. The IAC will offer one or two postdoctoral fellowships that focus on intergroup or comparative research on two or more of these communities in a local, national, or global context. The fellowship includes a stipend (which can be used as a sabbatical supplement), research support, and health benefits. Appointments generally are for a nine-month period beginning on September 1 or October 1.
Requirements: Individuals must be U.S. citizens or permanent residents and hold a PhD from an accredited college or university at the time of the appointment. Those appointed as visiting scholars must have completed their doctorate more than five years ago and hold tenure-track positions in academe.
Restrictions: UCLA faculty, staff, and currently enrolled students are ineligible.

Amount of Grant: $30,000-$35,000 plus $4000 maximum research support and health benefits
Date(s) Application is Due: Jan 14
Contact: Program Contact, (310) 825-1233; email: IACcoordinator@gdnet.ucla.edu
Internet: http://www.gdnet.ucla.edu/iacweb/iachome.htm
Sponsor: University of California, Los Angeles
1237 Murphy Hall, Box 951419
Los Angeles, CA 90095-1419

UCLA Research Grant Program in Ethnic Studies 5585
The institute, in conjunction with the the American Indian Studies Center, the Asian American Studies Center, Bunche Center for African American Studies, and the Chicano Studies Research Center, invites applications for support of research on African Americans, American Indians, Asian Americans, and Chicanas/os. The IAC also is working with the University of California Committee on Latino Research and the Chicano Studies Research Center to offer research grants in Latino Policy Studies. Prior to submission of the application, applicants should discuss their proposal with the assistant director or director of the appropriate center, or in the case of interethnic proposals, with the assistant directors or directors of the centers involved. The program is on a reimbursement basis only. Guidelines and application are available online.
Requirements: UCLA faculty, staff, graduate students, and IAC postdoctoral fellows are eligible.
Restrictions: Funds for the purchase of permanent equipment will be provided only under exceptional circumstances. Conference travel, whether the applicant is presenting or attending, is ineligible.
Amount of Grant: $7000 maximum faculty grants; $3000 maximum graduate student grants
Date(s) Application is Due: Apr 28
Contact: Ethnic Studies, (310) 825-1233; email: IACcoordinator@gdnet.ucla.edu
Internet: http://www.gdnet.ucla.edu/iacweb/rgweber.htm
Sponsor: Institute of American Cultures
1237 Murphy Hall
Los Angeles, CA 90095

UCT Mentally Handicapped Citizens Teachers Scholarships 5586
The purpose of these scholarships is to help teachers of the mentally handicapped in the continental United States and Canada obtain a degree and/or additional certification. In addition to requesting an application, the student must provide a resume of past work experience in the mentally handicapped field, indicate current college rank or standing, and state future educational and/or career goals. Application forms are available upon request. Applications are due prior to class registration and must be renewed annually.
Requirements: To be eligible, applicants must meet one or more of the following criteria: be a teacher of the mentally handicapped who needs additional coursework to be certified or to retain certification; be an experienced teacher who wishes to become certified to teach the mentally handicapped; have a bachelor's or master's degree and plan to pursue graduate work in the field of mental retardation; be a college junior or senior undergraduate whose course of study is mental retardation and who is planning to teach the mentally handicapped or have plans to become an instructor under a structured trade, vocational, or recreational program at a facility for the mentally retarded; and have a letter of recommendation from the instructor at the facility where the person intends to become a teacher in the field of mental retardation.
Restrictions: Applicants teaching in areas of special education other than mental retardation are ineligible.
Amount of Grant: $750 maximum annually; $3000 maximum total
Contact: Ann Marshall, Fraternal Department, (800) 848-0123 ext 126 or (614) 341-6994; email: amarshall@uct.org
Internet: http://www.uct.org/scholarships.html
Sponsor: Order of United Commercial Travelers of America
632 N Park Street
Columbus, OH 43215

UE Explore Undergraduate Research Grants 5587
This an undergraduate research program designed to provide students and faculty the opportunity to work together on research projects. The ideas for the research projects may be generated by the students or the faculty. There are both summer projects and semester projects, which are performed during the academic year. For the on-campus summer projects, students receive a stipend for 10 weeks, free residence housing during that period, and grant money to pay for literature searches, books, and equipment. Students receive grant money for the semester projects. A wide range of projects have been funded that cover the humanities, arts, sciences, and professional schools. Learn more about who is eligible and how to obtain funds through the web site.
Requirements: Students who satisfy and one of the following criteria are welcome to apply: students in good standing in the Honors Program; students who have a 3.5 GPA or above (based on at least 60 semester hours of which 30 must be earned at UE) (these hour requirements must be satisfied before the beginning of the project); academic scholarship winners (trustee, president or faculty); students recommended by consensus of as academic department or school (the consensus must be verified in writing from the department chair or dean).
Amount of Grant: $3500 stipend and $2000 maximum grant; $1000 maximum grant semester projects
Contact: Program Contact, (800) 423-8633; email: admission@evansville.edu
Internet: http://www2.evansville.edu/undergraduateresearch
Sponsor: University of Evansville
1800 Lincoln Avenue
Evansville, IN 47722

UEF Advancement of Engineering Grants 5588
The purpose of these grants is to provide either start-up funding for promising programs, or incremental funding for significant opportunities in established programs. Major impact must be possible within the level of UEF's, possibly leveraged, funding. National programs are preferable to regional ones. Interengineering programs are preferable to single discipline programs. Proposals in the following areas are encouraged but all proposals are welcome: precollegiate technical programs; midcareer assistance for engineers; opportunities for retired engineers to volunteer their skills; public awareness of engineers and engineering. Listed deadline is for concept letters; full proposals are by invitation.
Amount of Grant: $20,000-$50,000
Date(s) Application is Due: Jun 1
Contact: Dr. David Belden, Executive Director, (973) 244-2328; fax: (973) 882-5155; email: engfnd@aol.com
Internet: http://www.uefoundation.org
Sponsor: United Engineering Foundation
P.O. Box 70
Mount Vernon, VA 22121-0070

UI-UC Center for Latin American Studies Summer Visiting Scholars Grants 5589
This is an annual summer visiting scholars competition for faculty from nonresearch U.S. universities and colleges. Recipients research and write on a Latin American topic and enjoy access to university libraries and resources for one month during the summer at either or both the University of Illinois at Urbana/Champaign or University of Chicago. Awards include a stipend for residence and for travel to and from the recipient's home institution to either school. Residence must be completed by August 15. Applicants should send a letter of interest, a curriculum vita, one letter of reference, and a brief project proposal (approximately 500 words). There is no additional application form.
Amount of Grant: $2500 (for residence at Urbana), $3000 (for residence at University of Chicago), plus $500 travel
Date(s) Application is Due: Mar 15
Contact: Dain Borges, Center for Latin American Studies, (773) 702-1234; fax: (773) 702-1755; email: clas@uchicago.edu
Internet: http://clas.uchicago.edu/outreach/summervisiting.html
Sponsor: University of Chicago
5848 S University Avenue, Kelly Hall 310
Chicago, IL 60637

UIC Rockefeller Foundation Resident Fellowships in the Humanities 5590
The Latin American and Latino Studies Program at the university will host a Rockefeller Humanities Residency site. The program seeks one junior and one senior fellow whose research interests explore the diversification and internal hybridity of Latino communities in Chicago. Themes explored may include interlatino relations, interracial dynamics, and interactions and transculturations between native-born Latinas/os and more recent Latin American immigrants. The fellowships include stipend, fringe benefits, and travel and research expenses. Guidelines are available online.
Amount of Grant: $37,100 junior fellowship; $47,700 senior fellowship
Date(s) Application is Due: Feb 1
Contact: Lorena Garc'a, (312) 996-8749; fax: (312) 996-1796; email: lorena@uic.edu
Internet: http://www.uic.edu/las/latamst/rockefeller1.htm
Sponsor: University of Illinois at Chicago
1527 University Hall, 601 S Morgan Street
Chicago, IL 60607-7115

UICC American Cancer Society International Fellowships for Beginning Investigators 5591
The international fellowships fund 12-month basic or clinical research projects in the areas of epidemiology, prevention, detection, diagnosis, treatment, and psycho-oncology that foster a bidirectional flow of knowledge, experience, expertise, and innovation to and from the United States. Candidates should be in the early stages of their careers. Applications that are geared to the development of specific cancer control measures in developing and central and east European countries are particularly encouraged. Six to eight fellowships, with tenures beginning in April, will be awarded each year.
Requirements: Eligible candidates should hold assistant professorships or similar positions at their home institutes and have, ordinarily, a minimum of two and a maximum of ten years of postdoctoral experience after obtaining their MD or PhD degrees or equivalents.
Restrictions: Candidates who are physically present at the proposed host institute while their applications are under consideration are not eligible for UICC fellowships.
Samples: Paula Ximena Fernandez-Calotti, from Laboratory of Oncohematology Institute of Hematology Research, M.R. Castex, Argentina, to the Department of Biochemistry and Molecular Biology, University of Barcelona, Spain (2007); Nikita A. Makretsov, from the Department of Pathology & Laboratory Medicine, Memorial University of Newfoundland, Canada, to the Department of Oncology Hutchinson / MRC Research Centre, United Kingdom (2007).
Amount of Grant: $45,000
Date(s) Application is Due: Dec 1
Contact: Beate Vought, Fellowships Coordinator, +41 22 809 1843; fax: +41 22 809 1810; email: fellows@uicc.ch
Internet: http://www.uicc.org/index.php?id=928&L=0%2Fgyn.html

Sponsor: International Union Against Cancer / American Cancer Society
62, route de Frontenex
Geneva 1207 Switzerland

UICC International Cancer Technology Transfer (ICRETT) Fellowships 5592
The program helps to facilitate rapid international transfer of cancer research techiques and technology and clinical management skills. It is hoped that funding will support the exchange of knowledge and enhance skills in basic, clinical, behavioral, and epidemiological areas of cancer research, in cancer control and prevention (including tobacco control). Participants should acquire appropriate clinical management, diagnostic and therapeutic skills for effective application and use in the home organization upon their return. There are 120 to 150 awards available annually, with applications accepted at any time and notification generally within 60 days of registration.
Requirements: Eligible applicants include appropriately qualified investigators, pathologists, epidemiologists, cancer registrars, behavioral scientists, tobacco control activists, laboratory technicians, clinicians and general practitioners, generally in the early stages of their careers.
Amount of Grant: $3,000
Contact: Beate Vought, Fellowships Coordinator, +41 22 809 1843; fax: +41 22 809 1810; email: fellows@uicc.ch
Internet: http://www.uicc.org/index.php?id=925&L=0%2Fgyn.html
Sponsor: International Union Against Cancer
62, route de Frontenex
Geneva 1207 Switzerland

UICC Yamagiwa-Yoshida Memorial International Cancer Study Grants 5593
These study grants enable cancer investigators from any country to establish bilateral research projects abroad that exploit complementary materials or skills, including advanced training in experimental methods or special techniques. There are 14 to 16 three-to six-month grants awarded each year. Selection takes place twice annually, with recipients notified in mid-April and mid-October. Grants should be taken up within six months following notification. Travel awards contribute toward the least expensive international two-way airfare or other appropriate form of transport. Financial support is not provided for dependents. Application materials are available on the Web site.
Requirements: Candidates should have appropriate scientific or medical qualifications and be actively engaged in cancer research.
Samples: Dr. Hakan Akbulut, from Department of Medical Oncology, Cebeci Hospital, Turkey, to Sydney Kimmel Cancer Center, U.S.A (2007); Dr. Gil Ast, from Department of Human Molecular Genetics & Biochemistry, Sackler School of Medicine, Israel, to Department of Biochemistry, School of Medicine, University of Washington, U.S.A (2007).
Amount of Grant: $10,000 average for three months
Date(s) Application is Due: Jan 1; Jul 1
Contact: Beate Vought, Fellowships Coordinator, +41 22 809 1843; fax: +41 22 809 1810; email: fellows@uicc.ch
Internet: http://www.uicc.org/index.php?id=930&L=0%2Fgyn.html
Sponsor: International Union Against Cancer
62, route de Frontenex
Geneva 1207 Switzerland

UK Foundation for the Study of Infant Deaths Grants 5594
The foundation awards grants in England, Wales, and Northern Ireland for research into sudden infant death syndrome. Specific areas of interest include epidemiology, immunology, pathology, statistics, and infection. Graduate researchers of all nationalities are considered.
Amount of Grant: @L100,000 maximum
Date(s) Application is Due: Mar 1
Contact: Secretary-General, 0870 787 0885; fax: 0870 787 0725; email: research@sids.org.uk
Internet: http://www.sids.org.uk/fsid/grants.htm
Sponsor: UK Foundation for the Study of Infant Deaths
Artillery House, 11-19 Artillery Row
London SW1P 1RT United Kingdom

UNC Minority Presence Doctoral Study Grants 5595
The Minority Presence Grant Program for Doctoral Study, Law, and Veterinary Medicine provides black residents of North Carolina with stipends for the academic year, with an option of additional support for study in the summer session. Recipients must be full-time students pursuing doctoral degrees, law degrees or degrees in veterinary medicine at East Carolina University, North Carolina State University, the University of North Carolina at Chapel Hill, the University of North Carolina at Greensboro, or the University of North Carolina at Charlotte.
Amount of Grant: $4000 maximum; $500 summer support
Contact: Grants Administrator, (919) 962-7813 or (919) 962-5106
Internet: http://www.law.unc.edu/SearchDetails.aspx?ID=30
Sponsor: University of North Carolina
100 Ridge Road, CB3380
Chapel Hill, NC 27599-3380

UNC Minority Presence Grant Program 5596
This program provides funds for grants to white students at predominantly African American institutions and to African American students at predominately white institutions. Students must be residents of North Carolina who are enrolled for at least three hours of degree credit coursework and demonstrate financial need. Part I includes funds for minority presence grants for students attending the North Carolina Central University School of Law. General Program II consists of grant funds for Native Americans, Hispanics, and Asians.
Requirements: Grant recipients must be residents of North Carolina who are enrolled for at least three hours of degree credit coursework and demonstrate financial need.
Geographic Focus: North Carolina
Contact: Grants Administrator, (919) 962-1590 or (919) 962-5106
Internet: http://www.law.unc.edu/SearchDetails.aspx?ID=30
Sponsor: University of North Carolina School of Law
100 Ridge Road, CB3380
Chapel Hill, NC 27599-3380

UNC School of Information and Library Science Doctoral Fellowships 5597
The School of Information and Library Science at the university supports doctoral candidates with an interest in teaching and research in academic librarianship. For the first two years of study, fellows receive full tuition and health coverage; a research assignment in one of the Triangle Research Library Network Libraries; and an annual stipend. Application and guidelines are available online.
Requirements: To apply, follow the school's regular doctoral admissions procedures (http://sils.unc.edu/programs/phd/admissions.html).
Amount of Grant: $25,000 stipend
Date(s) Application is Due: Feb 15
Contact: Heidi Madden, (919) 966-5903; email: heidi_madden@unc.edu
Internet: http://www.trln.org
Sponsor: University of North Carolina-Chapel Hill
Campus Box 3360, Manning Hall
Chapel Hill, NC 27599-3360

UNCC TIAA-CREF Doctoral Fellowships 5598
The university awards doctoral fellowships in three areas of study: optical science and engineering, applied mathematics, and information technology. Fellowships provide one year of support, stipend plus tuition support for the academic year, and an opportunity for a paid internship with TIAA-CREF during the course of their academic programs at the university. Persons who have not previously earned a doctoral degree are eligible to apply. Applicants from minority groups historically underrepresented in the disciplines of science, mathematics, and technology are especially encouraged to apply. Write for further information and application materials.
Requirements: Applicants must be admitted to full standing to one of the following degree programs: PhD in optical science and engineering, PhD in applied mathematics, or PhD in information technology. Applicants must enroll as full-time graduate students for the fall and spring terms.
Amount of Grant: $25,000 stipend
Date(s) Application is Due: Feb 15
Contact: Maryanne Maree-Sams, (704) 687-3368; email: MMareesa@email.uncc.edu
Internet: http://www.uncc.edu/gradmiss/index.asp
Sponsor: University of North Carolina at Charlotte
9201 University City Boulevard, 220 Cato Hall
Charlotte, NC 28223

UNCFSP NAI-MIRS Summer Sabbatical Awards 5599
The NASA Astrobiology Institute Minority Institution Research Support (NAI-MIRS) Summer Sabbatical Awards provide opportunities for researchers working in minority institutions to collaborate with NASA Astrobiology Institute (NAI) Research Teams. In 2005, NAI partnered with Tennessee State University (TSU) to broaden the program by organizing summer sabbaticals for faculty from all minority serving institutions at selected NAI Team sites. The goals of NAI-MIRS are to give faculty at minority serving institutions an opportunity to expand their research, to increase diversity at all levels of the NAI, and to strengthen the participation of minority institutions in the field of astrobiology through research opportunities and ongoing collaborations. See website for application information.
Requirements: Applicants must be: faculty members at a U.S. college or university designated by the U.S. Department of Education as a Minority Serving Institution; able to provide a commitment from their institution to include astrobiology in university programs; qualified to conduct astrobiology competitive research in the host and their home institution; U.S. Citizen or U.S. Permanent Resident.
Amount of Grant: $26,500
Date(s) Application is Due: Mar 16
Contact: Jacqueline Howard-Matthews, Director; (703) 205-8145 or (703) 677-3400; fax: (703) 205-7645; email: jhowardmatthews@uncfsp.org
Internet: http://www.uncfsp.org/spknowledge/default.aspx?page=program.view&areaid=5&contentid=570&type=ANMT&nview=y
Sponsor: United Negro College Fund Special Programs Corporation
2750 Prosperity Avenue, Suite 600
Fairfax, VA 22031

UNCFSP NASA Science and Technology Institute Summer Scholars 5600
The NASA Science and Technology Institute (NSTI) Summer Scholars Program, a component of the Institute, is a ten-week summer research experience for undergraduates in science, technology, engineering, and mathematics (STEM) disciplines at a NASA Center. NSTI Scholars will be matched with a mentor to conduct NASA-relevant research. Scholars will be placed at one of three NASA centers including: Ames Research

Center in Moffett Field, CA, Johnson Space Center in Houston, TX or Glenn Research Center in Cleveland, OH.
Requirements: Applicants must meet the following: must be a member of an underrepresented group, which includes women, ethnic minorities and persons with disabilities majoring in a Science, Technology, Engineering or Math (STEM) discipline; be a current college freshmen, sophomore or junior; maintain a minimum 3.0 on a 4.0 grading scale;
Date(s) Application is Due: feb 20
Contact: Melissa C. Green, Director; (703) 205-7636 or (703) 677-3400; fax: (703) 205-7645; email: Melissa.Green@uncfsp.org
Internet: NASA Science and Technology Institute (NSTI) Summer Scholars Program
Sponsor: United Negro College Fund Special Programs Corporation
2750 Prosperity Avenue, Suite 600
Fairfax, VA 22031

UNCFSP NASA Administrator's Fellowship Program **5601**
As one of UNCFSP, premier fellowships, the NASA Administrator, Fellowship Program (NAFP) enhances the professional development for NASA employees as well as STEM faculty from Minority Institutions (MIs) while building and strengthening the STEM infrastructure at MIs. NASA employee fellows teach and/or conduct research and STEM faculty fellows conduct research at NASA Centers. All NAFP fellows engage in a comprehensive, integrated professional development curriculum to equip them with the skills to become progressive change agents in their organizations and institutions.
Requirements: Open to U.S. citizens, including residents of Puerto Rico, the Virgin Islands, and other U.S. territories. Applicants must hold a Ph.D., Sc.D, or equivalent (less than 20 years) in a STEM discipline OR have expertise in a NASA-related discipline.
Date(s) Application is Due: Jan 21
Contact: Melissa C. Green, Director; (703) 205-7636 or (703) 677-3400; fax: (703) 205-7645; email: Melissa.Green@uncfsp.org
Internet: http://www.uncfsp.org/nasa/nafp
Sponsor: United Negro College Fund Special Programs Corporation
2750 Prosperity Avenue, Suite 600
Fairfax, VA 22031

UNCFSP NASA Harriett G. Jenkins Predoctoral Fellowships **5602**
The mission of the program is to increase the numbers of women, minorities, and persons with disabilities participating in the math, science, engineering, and technology (MSET) disciplines. Up to 20 fellows will be selected annually to receive three-year fellowships to support graduate education leading to doctoral degrees in NASA-related disciplines, including aeronautics/aerospace, environnemental sciences, astronomy, life sciences, bioengineering, materials sciences, biology, mathematics, chemistry, meteorology, computers/computer science, physical sciences, earth sciences, physics/health physics, engineering, and science education. Application and guidelines are available online.
Requirements: Applicants must meet the following: must be a member of an underrepresented group, which includes women, ethnic minorities and persons with disabilities majoring in a Science, Technology, Engineering or Math (STEM) discipline; must either be within the first 3 years of a graduate program at an accredited, U.S. university OR be in the process of applying to an accredited STEM graduate program; cannot participate in any other federally-funded fellowship or scholarship program; fellows are required to be enrolled as a full-time student. Although the fellowship is not renewable during the three-year period, the availability of support for subsequent years is 1) subject to availability of funds and 2) contingent upon certification to NASA by the fellow's institution that the fellow is making satisfactory academic progress toward an advanced degree
Amount of Grant: $16,000 - $22,000 Annual stipends, $8,500 Annual tuition
Date(s) Application is Due: Feb 15
Contact: Melissa C. Green, Director; (703) 205-7636 or (703) 677-3400; fax: (703) 205-7645; email: Melissa.Green@uncfsp.org
Internet: http://www.uncfsp.org/spknowledge/default.aspx?page=program.view&areaid=1&contentid=177&typeid=jpfp
Sponsor: United Negro College Fund Special Programs Corporation
2750 Prosperity Avenue, Suite 600
Fairfax. VA 22031

UNESCO Fellowships Program **5603**
UNESCO fellowships and study grants are awarded to candidates especially from developing countries nominated by their governments and are not, in principle, for the pursuit of academic studies leading to degrees or diplomas. UNESCO fellowships generally are provided for training at the postgraduate level and rarely exceed one year. Study grants for short observation visits abroad, lasting not more than two months, may be provided to candidates occupying positions of high professional responsibility in their country. Fields of study are those relating to education, natural sciences, social and human sciences, culture, communication, and information and informatics. For more ample information, please contact the National Commission for UNESCO in your country.
Requirements: Applicants must have exceptional merit, and have demonstrated that they possess outstanding potential to contribute to the social, cultural and/or economic development of their country to be considered for a UNESCO fellowships.
Restrictions: The UNESCO Secretariat cannot entertain requests from individuals. Undergraduate students who wish to pursue studies in their own country cannot be beneficiaries of UNESCO fellowships.
Contact: Fellowship Program Officer, 33 (0) 1 45 68 1 00; fax: 33 (0) 1 45 66 94 29; email: fellowships@unesco.org
Internet: http://www.unesco.org/general/eng/about/fellowship.shtml
Sponsor: United Nations Educational, Scientific, and Cultural Organization
7 place de Fontenoy
Paris 07 SP 75352
France

Union Carbide Foundation Grants **5604**
The foundation supports nonprofits in areas of company operations, including Danbury, CT; Kanawha Valley, WV; central New Jersey; Taft, LA; and Texas City and Seadrift, TX. Grants also are awarded outside of the United States. The majority of grant supports goes for restructuring elementary and secondary education: developing math and science programs, advancing technology through research, preparing women and minorities for technical careers, and environmental protection. Additional areas of consideration include health and social services, civic affairs, the environment, and higher education. Grants are awarded for general operations, emergency funds, in-kind gifts, and programs/projects. The board meets in the spring and fall to consider grant proposals.
Requirements: Nonprofits in company-operating areas in Connecticut, West Virginia, New Jersey, Louisiana, and Texas may apply.
Geographic Focus: Connecticut; Louisiana; New Jersey; Texas; West Virginia
Samples: PhD Project (Montvale, NJ)—to attract successful African American, Hispanic American, and American Indian businesspeople to teaching careers at business schools where they can serve as role models for underrepresented students, $50,000.
Amount of Grant: $5000-$45,000
Contact: Grants Administrator, (203) 794-2000
Internet: http://www.dow.com/facilities/namerica/texops/community/grants.htm
Sponsor: Union Carbide Foundation
39 Old Ridgebury Road, L-4
Danbury, CT 06817-0001

United Airlines Foundation Education Grants **5605**
The foundation's primary mission is to help elementary, middle, and high school students prepare for challenging and rewarding careers. The foundation invites proposals year-round for projects that enhance education opportunities beyond the classroom, including programs involving museums and other cultural entities. Other major education focus areas include mentoring and tutoring; and adopt-a-school and school-to-work programs, which support students via job shadowing, mentoring, internships, and apprenticeships. Cultural and civic institutions receive consideration for funding of programs with a school outreach component. Grants support specific projects, annual campaigns, general operating budgets, research grants, employee-related scholarships, and employee matching gifts. The board meets in March, June, September, and December. Applicants should call for availability and deadlines.
Requirements: Nonprofits and school districts in Boston, MA; Chicago, IL; Denver, CO; Los Angeles and San Francisco, CA; New York, NY; Miami, FL; and Seattle, WA are eligible.
Restrictions: The foundation does not support capital grants, in-kind donations, or development campaigns.
Geographic Focus: California; Colorado; Florida; Illinois; Massachusetts; New York; Washington
Amount of Grant: $1000-$50,000 average
Contact: Grants Administrator, fax: (708) 700-7345; email: uafoundation@ual.com
Internet: http://www.united.com/page/article/0,,1367,00.html
Sponsor: United Airlines Foundation
P.O. Box 66100
Chicago, IL 60666

United Cerebral Palsy Research Grants **5606**
The foundation is interested in stimulating and providing funding for pilot studies on research important to the prevention of cerebral palsy and to improvement in the quality of life of persons with disabilities due to cerebral palsy and related developmental disorders. Research grants generally are awarded in support of basic and clinical science pilot studies or for time-limited clinical studies or trials. High priority is given to prevention of prematurity, decrease of perinatal neurologic morbidity, prevention of central nervous system infections of the newborn and to medical, surgical, and bioengineering methods that may improve the functioning of persons with cerebral palsy addressing issues such as impaired mobility, impaired communication skills, swallowing, drooling, and effects of aging on further diminution of function. Funds are awarded initially for a period not to exceed two years and may be renewed for a one- or two-year period, to a total of four years.
Requirements: Grants are awarded to institutions or organizations on behalf of a principal investigator in support of biomedical and bioengineering research in areas that have a significant relationship to cerebral palsy.
Amount of Grant: $50,000 maximum annually
Contact: Medical Director, (202) 973-7140; fax: (202) 776-0414; email: lsmithslade@ucp.org
Internet: http://www.ucp.org/ucp_generaldoc.cfm/1/4/23/23-23/113
Sponsor: United Cerebral Palsy Research and Educational Foundation
1660 L Street NW, Suite 700
Washington, DC 20036-5602

United Hospital Fund of New York Grants Program **5607**
The fund supports innovative programs in improving and expanding access to healthcare services, including those for children and youth; health care planning and management; health economics and hospital finance; management training in the health industries;

and AIDS-related service programs. Most grants are awarded to organizations in New York City.
Requirements: Nonprofits in New York City are eligible.
Geographic Focus: New York
Samples: Coalition of Voluntary Mental Health Agencies (New York, NY)—to help ensure the viability of nonclinical mental-health options for adults in New York State as community-based mental-health groups shift to a new state licensing and payment model, $50,000 (2005); eHealth Initiative (New York, NY)—to develop a policy agenda in New York State that promotes adoption and the use of health-information technology, $75,000 (2005); Medicaid Simplification and Information Technology Project (New York, NY)—to help develop an electronic Medicaid application for widespread public use in New York City, $50,000 (2005); New York City Health and Hospitals Corp (NY)—for a project to promote quality-improvement efforts at adult critical-care units, $45,000 (2005).
Amount of Grant: $4 million total; $100,000-$175,000 typically
Date(s) Application is Due: Feb 15; Jun 15; Oct 15
Contact: Anya Nawrocky, Grants Manager, (212) 494-0767; fax: (212) 494-0800; email: anawrocky@uhfnyc.org
Internet: http://www.uhfnyc.org/info-url3156/info-url.htm
Sponsor: United Hospital Fund of New York
350 Fifth Avenue, 23rd Fl
New York, NY 10118

United Methodist John Harrison Ness Memorial Award 5608
The General Commission on Archives and History of The United Methodist Church offers an annual award of $500 (first prize) and $300 (second prize) to the students enrolled in the M. Div. program (or its equivalent) in United Methodist or other seminaries accredited by the Association of Theological Schools who submit the best papers on some aspect of United Methodist history broadly conceived. Guidelines are available online.
Requirements: The paper must have been judged by its author's professor as one of quality. Once the paper has passed his or her critical judgment, the professor will submit it to the General Commission on Archives and History. The paper must be properly footnoted and bear evidence of thorough and reliable handling of sources using the annotation standards accepted at the seminary. A bibliography also must be included. The paper should normally range in length from 4,000 to 8,000 words. Three copies of the paper are to be submitted for judging. The winning paper(s) may be published in 'Methodist History', at the discretion of the editor.
Restrictions: The decision of the judges will be final. No prize will be awarded if the judges decide that no paper submitted has the required quality of scholarship for an award.
Date(s) Application is Due: Feb 1
Contact: General Secretary, General Commission on Archives and History; (973) 408-3189; fax: (973) 408-3909; email: research@gcah.org
Internet: http://www.gcah.org/site/c.ghKJI0PHIoE/b.3526379/k.8DA6/John_Ness_Award.htm
Sponsor: United Methodist Church
P.O. Box 127
Madison, NJ 07940

United States-Latin American-Canadian Relations Internships 5609
This organization is concerned with the full range of inter-American economic, trade, political, and diplomatic relations. Its activities include monitoring human rights violations, evaluating U.S. policy initiatives and activities in the hemisphere and formulating viable alternatives, seeking a diplomatic rather than a military solution to disputes, and surveying legislation that affects Latin America and Canada. The council offers internships to highly qualified undergraduate and graduate students as well as those who already have received degrees. Approximately 60 interns per year are accepted. Internships are voluntary and are awarded on a competitive basis with summer positions being the most sought after and most difficult to achieve. Full-time interns are usually given preference in the selection process. Interns conduct research and write articles for many publications; maintain liaison with national and international media; make research inquiries to legislative and executive agency offices; attend meetings, hearings, seminars, and discussions on topics relevant to COHA's work; perform telephone and personal interviews with Washington personalities concerned with hemispheric affairs; assist in the production of COHA's publications, press releases, monographs, and promotional mailings; and assist in office administration. Interns have the opportunity to pursue research topics of their own interest and to have their work published in the national media. An intern can expect to have two-thirds of her/his time devoted to research/writing activities and one-third to administrative function. Applications are accepted at any time.
Requirements: Candidates should have some background and/or interest in U.S.-Latin American affairs, international relations, and should display strong research and writing skills. Language ability, particularly in Spanish, Portuguese, or French, although not essential, is desired. A strong interest/major in journalism may be substituted for the normal background requirements. Candidates who do not have these qualifications are often selected if they exhibit proven research and writing skills. To apply, send resume transcript or description of grades and relevant courses taken, and a writing sample.
Contact: Program Contact, (888) 317-7654 or (202) 223-4975; fax: (202) 223-4979; email: coha@coha.org
Internet: http://www.coha.org/about_internships.php
Sponsor: Council on Hemispheric Affairs
1250 Connecticut Avenue NW, Suite 1C
Washington, DC 20036

United Theological Seminary Merit-Based Scholarships 5610
Merit-based scholarships are awarded to students who demonstrate exceptional academic ability. This aid covers up to 60 percent of tuition and can be awarded together with need-based aid to provide up to 100 percent of tuition. Scholarships include: the George D. Dayton Scholarship—granted up to four years to Masters of Divinity and Master of Arts in Religious Leadership students preparing for positions of professional church leadership; and the Presidential Scholarship—granted for up to three years to students in Master of Arts in Religion and Theology, Master of Arts in Theology and the Arts, or Master of Arts in Women Studies.
Requirements: Successful candidates must hold a baccalaureate degree from a regionally accredited college or university; be admitted in good standing to United Theological Seminary as a degree candidate; take the Graduate Record Examination (GRE) test; and demonstrate general excellence in academic and intellectual achievement, personal characteristics and qualities, leadership capacity in both the interpersonal and group contexts, service orientation, and faith commitment.
Geographic Focus: Minnesota
Date(s) Application is Due: Jan 31
Contact: Michelle Turnau, Director of Financial Aid and Housing; (651) 255-6117 or (651) 633-4311; fax: (651) 633-4315; email: mturnau@unitedseminary.edu
Internet: http://www.unitedseminary.edu/admissions/financialaid.asp
Sponsor: United Theological Seminary of the Twin Cities
300 Fifth Street NW
New Brighton, MN 55112-2598

University Center for Human Values Laurance S. Rockefeller Visiting Fellowships 5611
Princeton University, University Center for Human Values, invites applications for these fellowships. The fellowships are awarded for one academic year to outstanding teachers and scholars to devote a year in residence at Princeton writing about ethics and human values. Fellows participate in a fellows' seminar, a seminar in ethics and public affairs, and other programs of the university center. They enjoy access to Firestone Library and a wide range of activities throughout the university. A major part of time during the fellowship period, September through May, is devoted to research on ethics and human values.
Requirements: Applicants are expected to have a doctorate or a professional postgraduate degree.
Date(s) Application is Due: Nov 1
Contact: Stephen Macedo, Director, (609) 258-4763; fax: (609) 258-4772; email: macedo@princeton.edu
Internet: http://www.princeton.edu/values
Sponsor: Princeton University
5 Ivy Ln, University Ctr for Human Values
Princeton, NJ 08544

University of British Columbia Graduate Student Fellowships 5612
The Faculty of Graduate Studies offers merit-based fellowship awards to full-time graduate students registered at the University of British Columbia. Fellowship awards are open to all disciplines. Application forms are available on the Web site. All UGF application materials must be received by the deadline set by the department.
Requirements: Applicants must be graduate students with a first class standing in each of the last two years of full-time study. Students from any country and from all fields are eligible.
Contact: Faculty of Graduate Studies, (604) 822-8501 or 822-0976; fax: (604) 822-5802; email: grad-award@mercury.ubc.ca
Internet: http://www.grad.ubc.ca/awards/index.asp?menu=004,000,000,000
Sponsor: University of British Columbia
180-6371 Crescent Road
Vancouver, BC V6T 1Z2 Canada

University of Calgary Nursing Alumni Scholarship 5613
One scholarship per year is given to a Canadian citizen or landed immigrant of high academic standing who is registered in or admitted to a full-time master's program, who intends to make a career in nursing, and who is a registered member of the University of Calgary Faculty of Nursing Alumni Association. The scholarship may be used for any area or discipline at the master's level related to nursing. Annual deadline dates may vary; contact program staff for exact dates.
Requirements: To be eligible, applicants must be Canadian citizens or landed immigrants; graduate students in a master's program who intend to make a career in nursing; and registered members of the University of Calgary Faculty of Nursing Alumni Association.
Amount of Grant: $C1000
Date(s) Application is Due: Dec 2
Contact: Maria Lavoie, Graduate Scholarship Officer, (403) 220-8483; email: lavoiem@ucalgary.ca or ucawards@ucalgary.ca
Internet: http://www.grad.ucalgary.ca/funding/internal_scholarships/lev_4/nurs_alum_burs.htm
Sponsor: University of Calgary
2500 University Dr NW, Earth Sciences Building, Room 720
Calgary, AB T2N 1N4 Canada

University of California President's Postdoctoral Fellowships 5614
The fellowship, originally designed to encourage applications from minority and women scholars in areas where they were severely underrepresented, is now open to all qualified

candidates. The program provides postdoctoral fellowships to qualified persons who intend to pursue academic careers and to enhance their prospects for appointment to faculty positions at the University of California and other institutions of higher education. In addition to financial support, fellowships include mentorship and guidance in areas critical to success in an academic career. Application materials are available upon request.
Requirements: Applicants must be U.S. citizens or permanent residents and must hold a PhD from an accredited university. Applicants are required to have earned their PhD degree by June 30 of the year for which they are applying.
Amount of Grant: $40,000-$50,000 (depending on experience) annual stipend and up to $4000 for research-related expenses
Date(s) Application is Due: Nov 15
Contact: Kim Adkinson, President's Fellowships, (510) 987-9503; fax: (510) 587-6077; email: kim.adkinson@ucop.edu
Internet: http://www.ucop.edu/acadadv/ppfp
Sponsor: University of California Office of the President
1111 Franklin Street, 11th Fl
Oakland, CA 94607-5200

University of Cincinnati Visiting Professorship in Women's Studies 5615
The Center for Women's Studies provides one visiting professorship for one academic quarter per year to provide a new academic environment for the recipient and to avail the faculty at the center with her/his expertise. The scholar teaches one course, holds a faculty seminar, and participates in community activities. Stipend, accommodations, and travel expenses are provided.
Requirements: Applicants must hold the doctoral degree or equivalent and have considerable teaching and research experience in women's studies.
Samples: 2002-2003 LeAnne Howe, Native American novelist and author of Shell Shaker. American Indian Women: Saviors, Guides, Law Makers.
Contact: Anne Sisson Runyan, Director, (513) 556-6776; fax: (513) 556-6771; email: womens.studies@uc.edu or runyanas@email.uc.edu
Internet: http://asweb.artsci.uc.edu/womens_studies/Faculty_Section/index.html
Sponsor: University of Cincinnati
P.O. Box 210164, Ctr for Women's Studies
Cincinnati, OH 45221-0164

University of Colorado Doctoral Dissertation Fellowship in Media, Religion, and Culture 5616
Grants are designed to coincide with the development of dissertation research in the early stages of candidacy for students engaged in research projects involving media, religion, and culture. Fellows will be selected from various universities and disciplines on the basis of their fellowship proposals and will be expected to reside in their current location (the fellowship does not require relocation). Fellows will be brought together for master seminars at which leading scholars in media, religion, and culture will interact with fellows, their advisors, and the research team at the University of Colorado's School of Journalism and Mass Communication. The fellowships stipend may be used to cover research and other expenses as desired by the recipient.
Amount of Grant: $12,000; $32,000 postdoctoral fellowship
Date(s) Application is Due: Mar 31
Contact: Monica Emerich, Fellowship Coordinator, email: monica.emerich@colorado.edu
Internet: http://www.colorado.edu/journalism/mcm/mrc/mrc-fellowships.htm
Sponsor: University of Colorado
1171 University Avenue, Box 478
Boulder, CO 80309-0478

University of Colorado Ted Scripps Fellowship in Environmental Journalism 5617
The fellowships in environmental journalism help dedicated professional journalists acquire knowledge to cover the environment more effectively and enrich the public's understanding of this crucial subject. This flexible, nondegree program allows fellows to deepen their knowledge of the environment through classes, weekly seminars, and field trips. They also engage in independent study expected to lead to a significant piece of journalistic work. Fellows select at least two courses each semester from a broad array of environmental science, policy, and law courses offered by individual departments and as part of the Interdisciplinary Graduate Certificate Program in Environmental Policy. Prior experience in covering the environment is not required. Five fellowships are awarded each year.
Requirements: U.S. citizens with a minimum of five years full-time professional journalism experience and who have completed an undergrduate degree are eligible. Applicants may include general assignment reporters, editors, producers, environmental reporters and full-time freelancers.
Amount of Grant: $44,000 stipend for nine-month acdemic year
Date(s) Application is Due: Mar 1
Contact: Ted Scripps Fellowship Coordinator, (303) 492-4114; email: cej@colorado.edu
Internet: http://www.colorado.edu/journalism/cej
Sponsor: University of Colorado, School of Journalism and Mass Communication
1511 University Avenue, 478 UCB
Boulder, CO 80309-0478

University of Delaware—Hagley Program Fellowships 5618
The focus of the University of Delaware-Hagley Program is the history of industrialization, broadly defined to include such topics as gender and technology; work and society; industrial architecture and design; culture and political economy; and the process of de-industrialization. Students develop flexible programs of study. Most students focus on American topics, but the program's strong faculty in European and non-Western societies also supports the study of industrialization in a comparative global context. There are eight fellows in the program, and approximately three new fellowships are awarded each year. In addition to all tuition for courses at the University of Delaware, master's and doctoral candidates receive a yearly stipend. Fellowships may be renewed once for those seeking a terminal master's degree and three times beyond the initial year for those seeking the doctorate. In addition, students in the master's program are required to serve as teaching fellows for two semesters and those in the PhD program for four semesters. Fellows also receive support for travel to conferences, archives, and museum.
Requirements: Students must have a BA, MA, or equivalent degree. The aptitude portion of the GRE is required and must be taken no later than December to assure full consideration of the application.
Amount of Grant: $12,000 stipend
Date(s) Application is Due: Jan 15
Contact: Coordinator, (302) 831-8226; fax: (302) 831-1538
Internet: http://www.udel.edu/History/hist/text/gradframe.html
Sponsor: University of Delaware—Hagley Museum and Library
236 John Munroe Hall, Dept of History
Newark, DE 19716-2547

University of Georgia Counseling and Psychological Services Postdoctoral Fellowship 5619
The program is designed to develop advanced skills in the practice of clinical and counseling psychology. The postdoctoral program provides fellows an opportunity to refine and expand experience in the following areas: clinical practice with individuals, couples and groups, assessment, crisis intervention, workshops, and educational programming. Postdoctoral fellows will work in close collaboration with senior staff and will function as professionals with many of the same responsibilities. Fellows attend all case conferences, professional development workshops, and staff meetings. In addition to the stipend, fellows are also eligible for staff benefits, including health and dental insurance, retirement, vacation, sick leave, professional development, exercise facility, library privileges, and licensing preparation time. Guidelines are available online. Annual deadline dates may vary; contact program staff for exact dates.
Requirements: Admission requirements include completion of all professional degree requirements from a regionally accredited institution of higher education or an APA/CPA'accredited program and completion of an APA approved predoctoral internship. Fellow must have received diploma before first day of fellowship or have a letter from Director of Graduate Studies verifying that all degree requirements have been met. Previous counseling center experience is preferred.
Amount of Grant: $25,000 stipend
Date(s) Application is Due: Mar 11
Contact: Dr. Al Clarke, University Health Center, (706) 542-2273; email: aclarke@uhs.uga.edu
Internet: http://www.uhs.uga.edu/CAPS/postdoc/index.html
Sponsor: University of Georgia
University of Georgia Health Ctr
Athens, GA 30602-1755

University of Kentucky Rockefeller Foundation Humanities Fellowships 5620
The university's Appalachian Center and the Committee on Social Theory invite applications for a three-year program of humanities fellowships on the theme Civic Professionalism and Global Regionalism: Justice, Sustainability, and the Scaling Up of Community Participationon. The program addresses questions of globalization, democracy, and environmental sustainability. The goal of the fellowships is to bring together scholars and citizen activists and/or artists who are at the leading edge of global innovation in new models for partnership between communities, academics, and government. There are two types of fellowships: one for people working with citizen, grassroots, and nonprofit organizations in the U.S. Appalachian region; and the other for academic scholars working on questions related to fellowship program themes. Academic fellows are eligible for fellowships of one or two semesters. For the fall semester, fellowships must fall within the time period of mid-August to mid-December. For the spring semester, fellowships must fall between the second week of January and the first week of May. In addition to a stipend, the fellowship provides benefits equal to approximately 20 percent of the stipend. Fellows will be provided with office space, a computer, Internet access, data and research services, as well as assistance in networking with university faculty and others as needed for fellows' projects, travel and relocation expenses to Lexington for the fellowship residency and a final conference, housing in the Lexington area, and supplemental lodging and travel costs for family members. Guidelines and applications are available online.
Requirements: Applicants should apply in the activist track or academic track. Scholars from the global South—particularly, Asia, Latin American, Africa, and the Pacific rim—are especially encouraged.
Restrictions: Awards may not be used for graduate studies, advanced training, or writing poetry or fiction. Current employees from the University of Kentucky (UK) or with advanced degrees from UK are not eligible.
Amount of Grant: $35,000 stipend for nine months, or $3889 for each month of fellowship
Date(s) Application is Due: Feb 3
Contact: Pam Webb, Budget Officer, Appalachian Center, (859) 257-4851 or (859) 257-4852; fax: (859) 257-3903; email: rock@uky.edu

Internet: http://www.uky.edu/RGS/AppalCenter
Sponsor: University of Kentucky
624 Maxwelton Ct
Lexington, KY 40506-0347

University of London Visiting Fellowships 5621

Visiting fellowships support work in any one or more of the areas covered by the university's research institutes and programs, including advanced legal studies, classical studies, Commonwealth studies, English studies, Germanic and romance studies, historical research, study of the Americas, the Warburg Institute, and the philosophy program. The fellowships, tenable for up to six consecutive months between September and June, offer travel, accommodation, and research expenses. Application forms are available from the office.
Requirements: Fellowships are available to applicants of professorial or equivalent status for the purpose of supporting research in London.
Amount of Grant: @L16,000
Contact: Program Contact, School of Advanced Study, 020-7862-8659; fax: 020-7862-8657; email: Deans.Office@sas.ac.uk
Internet: http://www.sas.ac.uk/School_Fellowships
Sponsor: University of London
Senate House, Malet Street, Room 310
London WC1E 7HU United Kingdom

University of Massachusetts Rockefeller Foundation Humanities Fellowships 5622

The four-year residential fellowship program focuses on (re)constructions of identity and place in the Vietnamese diaspora after 1975, and asks scholars to explore how diverse constructions of Vietnamese identity and community as well as Vietnamese history, literature, and culture are being shaped and reshaped in the contemporary postwar and postrefugee era across generations. The program supports individual scholars and seeks to catalyze a network of scholars with collective commitment to innovative, interdisciplinary research relevant to the Vietnamese diaspora. During the residency period, fellows working in either English or Vietnamese will conduct research and interact with faculty, staff, visiting scholars and writers, students, and local Vietnamese community members affiliated with five collaborative institutions.
Requirements: Scholars, writers, and researchers of any country are eligible. Those with interdisciplinary interests are especially invited to apply.
Amount of Grant: $35,000 maximum
Date(s) Application is Due: Jan 31
Contact: Nguyen Ba Chung, Project Director, William Joiner Center, (617) 287-5850; fax: (617) 287-5855; email: chung.nguyen@umb.edu
Internet: http://rockfound.org/display.asp?context=1&Collection=1&DocID=622&Preview=0&ARCurrent=1
Sponsor: University of Massachusetts
100 Morrissey Boulevard
Boston, MA 02125-3393

University of Oslo International Summer School (ISS) Scholarships 5623

A limited number of scholarships are available, mainly to applicants from developing countries, Eastern Europe, and the former Soviet Union. Applications must be made on a special scholarship application form that may be obtained from the ISS Head Office in Oslo. Students from Europe may apply for scholarships offered under the bilateral Cultural Agreements between Norway and their home country. Scholarship application procedures vary from country to country; inquire about them and request an application form at the Norwegian Embassy or Consulate General, Ministry of Education, or Ministry of Foreign Affairs in the home country. There are also some scholarships available to U.S. and Canadian students through several different private organizations, including Sons of Norway. Contact the ISS North American Branch Office for further details.
Requirements: Under the cultural agreements, scholarships are open to students with good academic records from Austria, Belgium, Bulgaria, Czech Republic, Egypt, Germany, Great Britain, Greece, Hungary, Ireland, Italy, Mexico, Netherlands, Poland, Portugal, Romania, Russia, Slovakia, Spain, and Turkey, typically. For scholarships to applicants from developing countries, preference is given to graduate level candidates from NORAD's main partner countries (i.e., Afghanistan, Angola, Bangladesh, Bhutan, China, Cuba, East Timor, Eritrea, Ethiopia, Guatemala, India, Indonesia, Iran, Laos, Malawi, Maldives, Mali, Mozambique, Nepal, Nicaragua, Nigeria, Pakistan, Palestinian Territories, South Africa, Sri Lanka, Tanzania, Uganda, Vietnam, Zambia and Zimbabwe).
Restrictions: Funding for transatlantic travel is not provided. Those ineligible to apply for scholarships include applicants for Norwegian language courses level I and II; medical doctors or nurses applying for Norwegian language courses; applicants to Norwegian language courses at any level residing in Norway; candidates from developing countries applying for undergraduate courses.
Date(s) Application is Due: Feb 1
Contact: Charman Theresa Garibaldi, Administrative Assistant, 47-22-85-63-85; fax: 47-22-85-41-99
Internet: http://www.uio.no/iss/iss.html
Sponsor: University of Oslo International Summer School
P.O. Box 1082 Blindern
Oslo NO-0317 Norway

University of Queensland Travel Awards for International Collaborative Research 5624

The University of Queensland provides funds annually to facilitate visits to the university by scholars from other countries. Travel assistance is available to scholars actively engaged in research who will contribute substantially to research activity in the nominated school or center at the university. The assistance consists of: an air fare to Brisbane, inclusive of visa fees levied by the Australian government for the visitor only; and, subject to the visitor spending at least four weeks during semester at the university, a return air fare for the visitor only. In exceptional circumstances, and at the discretion of the selection committee, there is provision for a limited contribution to subsistence on a basis identified as genuine merit of proposal and genuine need, to be proportionate to the length of stay. Grants are normally reserved for suitably qualified staff of universities and internationally recognized research establishments. Other persons, however, may be considered where special circumstances exist and comparable standards are satisfied.
Requirements: Applicant must spend at least four weeks at the university during the semester. Applicants must be qualified members of a recognized university who are engaged in academic work that will contribute substantially to the department to which they are attached at the University of Queensland.
Restrictions: Applications will not be accepted from persons already in Australia.
Date(s) Application is Due: Aug 1
Contact: Office of Research, Postgraduate Studies, 61-7-336-53560 or 61-7-336-54584; fax: 61-7-336-54455; email: oper@research.uq.edu.au
Internet: http://www.uq.edu.au/research/orps/?id=4793&pid=4547
Sponsor: University of Queensland
University of Queensland, Street Lucia
Brisbane, Queensland 4072 Australia

Upjohn Institute Dissertation Award 5625

The Upjohn Institute supports and conducts policy-relevant research on employment and unemployment-related issues. This award, for the best PhD dissertation on employment-related issues, may come from any discipline. Dissertations are evaluated by a panel of economists using the following criteria: policy relevance, technical quality of the research, potential impact on real-world problems, and presentation. Applicants are advised that they will need to supply a copy of the entire dissertation if they are selected as a finalist. A first prize and up to two honorable mention awards will be made.
Requirements: Any individual whose dissertation has been accepted during the 24-month period of July 1 to June 30, 2006, is eligible.
Amount of Grant: $2000 first prize, $750 honorable mention
Date(s) Application is Due: Jul 7
Contact: Institute Grant Committee, (616) 343-5541; fax: (616) 343-3308; email: webmaster@upjohninstitute.org
Internet: http://www.upjohninst.org/dissert.html
Sponsor: W.E. Upjohn Institute for Employment Research
300 S Westnedge Avenue
Kalamazoo, MI 49007-4686

Upjohn Institute for Employment Policy Research Grants 5626

The institute supports policy-relevant research on employment and unemployment at the national, state, and local levels. Grants made under this program are expected to result in research of a rigorous nature and publication of a monograph by the institute. Although the institute encourages applicants to submit proposals that consider any policy-relevant issue related to employment or unemployment, the following topics are of particular interest: employment relationships—between workers and firms, how they are changing, and the policy implications of these changes; low wages and public policy—research on policy responses to the problems of low earnings of the poor in the United States and to the sluggish wage growth of the American middle class; and studies of social insurance systems. The institute also offers mini-grants to conduct research and write papers on innovative research topics. Mini grants provide a small amount of money to meet special funding needs that, without support, would prevent researchers from pursuing the project. Prospective applicants are encouraged to send brief descriptions of ideas by the listed application deadline dates to determine institute interest. Application deadlines and instructions are posted on the Web site.
Requirements: Applicants must submit eight copies of both a three-page summary outlining their proposed research and a vita that describes their professional qualifications. The selection committee will evaluate the preliminary proposal and invite a subset of applicants to submit a full 15-page proposal for evaluation and selection. The invitation to submit a full proposal may include suggestions and comments by reviewers on ways to strengthen the proposal.
Restrictions: Dissertation research, multiyear projects, or conferences not initiated by the institute will not be funded. Proposals will not be accepted by fax or email.
Samples: Richard Butler, U of Minnesota; Yong-Seung Park, Kyung Hee University; and Kate Kimpan, Minnesota Department of Labor and History—Human Resource Management (HRM) Policy, Safety Practices and Workers' Compensation Costs.; Christopher Flinn, New York University—for The Minimum Wage and Labor Market Outcomes: Theory and Evidence.
Amount of Grant: $75,000 maximum; $5000 maximum for mini-grants
Date(s) Application is Due: Feb 1; Aug 2
Contact: Institute Grant Committee, (616) 343-5541; fax: (616) 343-3308; email: webmaster@upjohninstitute.org
Internet: http://www.upjohninst.org/grantann.html#researchg
Sponsor: W.E. Upjohn Institute for Employment Research
300 S Westnedge Avenue
Kalamazoo, MI 49007-4686

UPS Foundation Grants 5627

The foundation focuses its support on national programs in the areas of human welfare, including programs for families and children in crisis, the economically or culturally disadvantaged, the physically or mentally challenged, and community development programs; K-12 educational programs, including academic research, programs that raise the level of educational effectiveness, innovative programs to enhance the quality of instruction, family learning opportunities, and school involvement projects; and major initiatives, adult literacy, and the distribution of prepared and perishable food. The foundation matches employee gifts to educational and cultural organizations. Qualified organizations are encouraged to submit a proposal in concise letter form (two-page maximum) containing a description and mission of the organization, description of program/project for which funding is requested, statement of program/project need and goals, description of how program/project goals will be attained and evaluated, total costs of the program/project, list of committed alternate funding sources, and total amount requested from the foundation.

Requirements: IRS 501(c)3 tax-exempt organizations are eligible.

Restrictions: The foundation does not award grants to individuals, religious organizations or theological functions, or church-sponsored programs limited to church members. Grants supporting capital campaigns, endowments, or operating expenses are seldom approved.

Samples: Children's Hunger Alliance (Columbus, OH)—to expand access to school breakfasts for needy children, $250,000 (2005); National Conference of State Legislatures (Washington, DC)—to educate state legislators on various ways to improve nutrition and fight obesity, $250,000 (2005); Student Coalition for Action in Literacy (Chapel Hill, NC)—to provide leadership training and development support for student literacy projects, $137,977 (2005); Teachers of English in Speakers of Other Languages (Arlington, VA)—to provide training resources to teachers of English as a second language, $237,000 (2005).

Amount of Grant: $38.4 million total; $1000-$200,000 average

Date(s) Application is Due: Sep 1

Contact: Evern Cooper, President, (404) 828-6374; fax: (404) 828-7435

Internet: http://www.pressroom.ups.com/mediakits/socialresponsibility/foundation/0,1387,,00.html

Sponsor: UPS Foundation
55 Glenlake Pkwy NE
Atlanta, GA 30328

Ursula Thrush Peace Seed Grants 5628

Ursula Thrush founded the Maria Montessori School of the Golden Gate and Teacher Training Center in San Francisco and helped to establish several other Montessori programs, including The Science of Peace Task Force and the Montessori Peace Academy. Her dedication to fulfilling Maria Montessori's vision for peace through children opened doors to many Montessori educators, inspiring them to include peace education in their classrooms and communities. Since 2004, the AMS Peace Committee has awarded annual Peace Seed Grants to encourage educators to promote peace through their teaching. Lesley Nan Haberman, headmistress of The Family Schools in New York City, initiated the grant in memory of her dear friend Ursula Thrush.

Requirements: Montessorians who have formulated a project fostering education for peace are invited to apply. Applicants must have a Montessori background. The project must further education for peace and must reach a significant number of children and/or educators. The recipient of the grant must share the results of his/her project with the AMS community. The project goal must be accomplished in the year following receipt of the grant. The recipient must show how he/she will be accountable for the use of the funds.

Samples: 2009: Alyson Peterson (founder, Montessori Island School, Livingston, MT) Peace Garden for Sustainable Agriculture - to help expand the Peace Garden at Montessori Island School, which is part of her initiative to create, with her students, sustainable, local agriculture. Michele Melvin (founding directress of Montessori Habitat School in Champaign, IL) Spirit Stirring: Fostering Appreciation for Humanity and Exploration of Cosmological Role in Montessori Middle School Students - Michele developed 16 experimental methods targeted at nurturing the spirits of adolescents, including their appreciation for humanity and understanding of their role in the universe, and has been measuring their efficacy in her own middle school classroom. With the winning grant funds, she will broaden her research to include several more middle schools across the nation in her research.

Amount of Grant: $1,100

Date(s) Application is Due: Feb 1

Contact: Judi Bauerlein; (212) 358-1250; fax: (212) 358-1256; email: Judi4trees@sbcglobal.net

Internet: http://www.amshq.org/peaceseedconnection.htm#thrushGrant

Sponsor: American Montessori Society
281 Park Avenue South
New York, NY 10010-6102

U.S. Constitution Study Fellowships for Teachers and Prospective Teachers 5629

The fellowships support in-service secondary school teachers (grades 7-12) of American history, American government, social studies, and graduating or graduated collegians who wish to become secondary school teachers of the same subjects. The awards cover tuition, fees, books, room, and board associated with study leading to master's degrees in American history, political science, or education with concentrations in the framing, principles, and history of the U.S. Constitution. Stipends cover five years of part-time study by teachers or two years of full-time study by recent baccalaureates. Applications should be sent to P.O. Box 4030, 2201 N Dodge Street, Iowa City, IA 52243-4030; (800) 525-6928.

Requirements: College seniors and college graduates without teaching experience (junior fellows) and experienced secondary school teachers in grades 7-12 (senior fellows) who will normally enroll in full-time and part-time graduate study leading to a master's degree in history or political science; a master of arts in teaching degree in history or social studies; or a related master's degree in education that permits a concentration in history, government, social studies, or political science are eligible fellowship candidates.

Restrictions: Fellowships do not support study toward doctoral degrees or teacher certificates.

Amount of Grant: $24,000 maximum

Date(s) Application is Due: Mar 1

Contact: Fellowship Program, (800) 525-6928; fax: (202) 653-6045; email: madison@act.org

Internet: http://www.jamesmadison.com

Sponsor: James Madison Memorial Fellowship Foundation
2000 K Street NW, Suite 303
Washington, DC 20006

U.S. CRDF CGP ETF 2009 Energy Research Competition For Joint U.S.-Estonian Research Teams 5630

The U.S. Civilian Research and Development Foundation (CRDF) is a nonprofit organization authorized by the U.S. Congress and established in 1995 by the National Science Foundation. This unique organization promotes international scientific and technical collaboration through grants, technical resources, and training. In cooperation with the Estonia Science Foundation (ETF), CRDF is announcing a joint 2009 Energy Research Competition which provides awards of approximately $50,000 for up to two years of support to joint research teams of U.S. and Estonian scientists and engineers. Applications must fall under one of the following two (2) areas of energy research and development: Clean Fossil Fuels Energy and Renewable Energy.

Requirements: All proposals must be related to renewable energy or clean fossil fuel energy. The program will consider proposals from all natural sciences, specifically from any of the following disciplines: Biology, Chemistry, Geology, Mathematics, Physics, or Electronic, Material, Mechanical, and Manufacturing Engineering. CRDF strongly encourages applicants to contact a CGP Program Manager at cgp@crdf.org if they are unsure whether their research topic is eligible for this competition. The total maximum grant size is $50,000 for up to two years of research and development. The Estonian team is eligible to receive 78% of the funds for their expenses, including institutional support. The U.S. team is eligible for 22% of the award total. These funds will be allocated separately and administered directly to the individuals and institutions involved. Proposals submitted to the competition must have one Estonian Principal Investigator and one U.S. Principal Investigator, who share the overall responsibility for the project in their respective countries and coordinate all project participants and institutions. Each Principal Investigator must: possess the degree of Ph.D., M.D., or the equivalent research experience; have at least five (5) scientific publications in peer-reviewed scientific literature; and, work full-time in a civilian research environment. The Estonian Principal Investigator and participants on the Estonian team must be citizens or permanent residents of Estonia. All Estonian participants must reside in Estonia during the project period. The U.S. Principal Investigator and participants of the U.S. team must be U.S. citizens or permanent residents of the U.S. and must reside in the U.S. during the project period. U.S. graduate students may be foreign nationals, but they must be enrolled in an accredited degree program at a U.S. institution. All projects must be oriented toward non-military objectives and must be carried out in a civilian research environment. U.S. project participants are encouraged to travel to the partner labs to ensure an effective exchange of ideas and research findings. For U.S. For-Profit companies only: Letter of Commitment to contribute 10% of Estonian expenses and coverage of all U.S. expenses is required. For Estonian For-Profit companies only: Letter of Commitment to contribute up to 5-10% of total Estonian expenses is required. For Proposals with in-kind contributions: Letter of Commitment from contributing U.S. and/ or Estonian institution is required.

Restrictions: CRDF reserves the right to restrict the participation of any individual or institution from its programs. CRDF complies with all U.S. laws and regulations pertaining to export control and the participation of foreign nationals or institutions in its activities. It is the policy of CRDF not to conduct any transactions with U.S. restricted entities without appropriate authorization from the U.S. Government.

Geographic Focus: Alabama; Alaska; Arizona; Arkansas; California; Colorado; Connecticut; Delaware; District of Columbia; Florida; Georgia; Guam; Hawaii; Idaho; Illinois; Indiana; Iowa; Kansas; Kentucky; Louisiana; Maine; Maryland; Massachusetts; Michigan; Minnesota; Mississippi; Missouri; Montana; Nebraska; Nevada; New Hampshire; New Jersey; New Mexico; New York; North Carolina; North Dakota; Ohio; Oklahoma; Oregon; Pennsylvania; Puerto Rico; Rhode Island; South Carolina; South Dakota; Tennessee; Texas; U.S. Virgin Islands; Utah; Vermont; Virginia; Washington; West Virginia; Wisconsin; WyomingEstonia

Amount of Grant: $50,000 approximately

Date(s) Application is Due: Dec 1

Contact: Mr. Stuart Politi, Senior Program Manager; (703) 526-9720; fax: (703) 526-9721; email: cgp@crdf.org

Internet: http://www.crdf.org/funding/funding_show.htm?doc_id=1014817

Sponsor: U.S. Civilian Research and Development Foundation
1530 Wilson Boulevard, 3rd Floor
Arlington, VA 22209

U.S. CRDF CGP-Climate Change and Energy Competition For Joint U.S.-Ukrainian Research Teams 5631

The U.S. Civilian Research and Development Foundation (CRDF) is a nonprofit organization authorized by the U.S. Congress and established in 1995 by the National Science Foundation. This unique organization promotes international scientific and technical collaboration through grants, technical resources, and training. CRDF is currently accepting proposals for the 2009 Cooperative Grants Program (CGP) Climate Change and Energy Competition. This competition is open only to joint teams of scientists from the United States and Ukraine. This program provides awards of up to $90,000 U.S. Dollars (U.S.D) for up to two years of support to joint research teams of U.S. and Ukrainian scientists.

Requirements: CRDF will accept applications from all scientific disciplines as long as the research proposed is basic in nature. The program will consider proposals from any of the following disciplines: Biology, Chemistry, Geology, Mathematics, Physics, or Electronic, Material, Mechanical, and Manufacturing Engineering. CRDF strongly encourages applicants to contact a CGP Program Manager at cgp@crdf.org if they are unsure whether their research topic is eligible for this competition. Eligible research projects should be submitted to one of the two Focus Areas of this competition: Climate Change Impacts and Climate Change Solutions. Proposals eligible for Focus Area I: Climate Change Impacts should i) address the impacts of climate change on human and biological systems ii) measure, monitor, and model the processes that will provide accurate future projections of climatic and environmental changes or iii) study regionally specific feedbacks associated with the climate change. Proposals eligible for Focus Area II: Climate Change Solutions should address solutions to climate change such as developing technologies in energy, land and water management, and materials science that may mitigate the impact of climate change. Each proposal submitted to CGP must have one Ukrainian Principal Investigator and one U.S. Principal Investigator, who share overall responsibility for the project in their respective countries, coordinating all project participants and institutions. Each Principal Investigator must: possess the degree of Ph.D., M.D., or the equivalent research experience; have at least five (5) scientific publications in peer-reviewed scientific literature (if the applicant has received his or her Ph.D., M.D., or the equivalent in the past six (6) years, three (3) publications are required), and, work full-time in a civilian research environment. The Ukrainian Principal Investigator must be a citizen and permanent resident of Ukraine. All Ukrainian participants must reside in Ukraine for the majority of the grant period. The U.S. Principal Investigator and participants on the U.S. team must be U.S. citizens or permanent residents of the U.S. and must reside in the U.S. during the entire course of the project. Graduate students on the U.S. team may be foreign nationals, but they must be enrolled in an accredited degree program at a U.S. institution. All projects must be oriented toward non-military objectives and must be carried out in a civilian research environment. (U.S. Government Laboratory employees working on civilian-oriented research and development projects are eligible to apply to this program.) For Ukrainian applicants only: Ukrainian-Language Text of Project Narrative, Expert Conclusion, and Author's Statement submitted to CRDF counterpart agency. For U.S. For-Profit companies only: Letter of Commitment to contribute 10% of Ukrainian expenses and coverage of all U.S. expenses. For Ukrainian For-Profit companies only: Letter of Commitment to contribute up to 5-10% of total Ukrainian expenses. For proposals with in-kind contributions: Letter of Commitment from contributing U.S. and/ or Ukrainian institution is required. For proposals involving Human Subjects research: Human Subjects Protocol Summary Form and supporting information, such as Institutional Review Board/Internal Ethical Committee Approval, Human Subjects Governmental Approval, and Confirmation of Federal-Wide Assurance. No more than 84% ($76,000) of the funds awarded to each project must be used for the Ukrainian team, expenses, including institutional support. No more than 16% ($14,000) may be used for the U.S. team, expenses.

Geographic Focus: Alabama; Alaska; Arizona; Arkansas; California; Colorado; Connecticut; Delaware; District of Columbia; Florida; Georgia; Guam; Hawaii; Idaho; Illinois; Indiana; Iowa; Kansas; Kentucky; Louisiana; Maine; Maryland; Massachusetts; Michigan; Minnesota; Mississippi; Missouri; Montana; Nebraska; Nevada; New Hampshire; New Jersey; New Mexico; New York; North Carolina; North Dakota; Ohio; Oklahoma; Oregon; Pennsylvania; Puerto Rico; Rhode Island; South Carolina; South Dakota; Tennessee; Texas; U.S. Virgin Islands; Utah; Vermont; Virginia; Washington; West Virginia; Wisconsin; WyomingUkraine

Amount of Grant: $90,000

Date(s) Application is Due: Dec 1

Contact: Mr. Stuart Politi, Senior Program Manager; (703) 526-9720; fax: (703) 526-9721; email: cgp@crdf.org

Internet: http://www.crdf.org/funding/funding_show.htm?doc_id=1014809

Sponsor: U.S. Civilian Research and Development Foundation
1530 Wilson Boulevard, 3rd Floor
Arlington, VA 22209

U.S. CRDF CGP-RFBR Climate Change and Energy Competition For Joint U.S.-Russian Research Teams 5632

The U.S. Civilian Research and Development Foundation (CRDF) is a nonprofit organization authorized by the U.S. Congress and established in 1995 by the National Science Foundation. This unique organization promotes international scientific and technical collaboration through grants, technical resources, and training. CRDF and the Russian Foundation for Basic Research (RFBR) are currently accepting proposals for the 2009 Cooperative Grants Program (CGP) - RFBR Climate Change and Energy Competition. This competition is open only to joint teams of scientists from the United States and the Russian Federation. The total maximum grant amount is $55,000 U.S. Dollars (U.S.D) from CRDF and 1,200,000 Russian Rubles (RUB) from RFBR (a combined total of approximately $93,000 U.S.D) for up to two years of research and development.

Requirements: CRDF and RFBR will accept applications from all scientific disciplines as long as the research proposed is basic in nature. All proposals must be submitted to both CRDF and RFBR. The program will consider proposals from any of the following disciplines: Biology, Chemistry, Geology, Mathematics, Physics, or Engineering. CRDF strongly encourages applicants to contact a CGP Program Manager at cgp@crdf.org if they are unsure whether their research topic is eligible for this competition. Eligible research projects should be submitted to one of the two Focus Areas of this competition: Climate Change Impacts and Climate Change Solutions. Proposals eligible for Focus Area I: Climate Change Impacts should i) address the impacts of climate change on human and biological systems ii) measure, monitor, and model the processes that will provide accurate future projections of climatic and environmental changes or iii) study regionally specific feedbacks associated with the climate change. Proposals eligible for Focus Area II: Climate Change Solutions should address solutions to climate change such as developing technologies in energy, land and water management, and materials science that may mitigate the impact of climate change. Each proposal submitted to the CGP-RFBR Climate Change and Energy competition must have one Russian Principal Investigator and one U.S. Principal Investigator, who share overall responsibility for the project in their respective countries, coordinating all project participants and institutions. Each Principal Investigator must: possess the degree of Ph.D., M.D., or the equivalent research experience; have at least five (5) scientific publications in peer-reviewed scientific literature (if the applicant has received his or her Ph.D., M.D., or the equivalent in the past six (6) years, three (3) publications are required); and, work full-time in a civilian research environment. (U.S. Government Laboratory employees working on civilian-oriented research and development projects are eligible to apply to this program.) The Russian Principal Investigator and participants of the Russian team must be citizens or permanent residents of the Russian Federation. All Russian participants must reside in the Russian Federation for the majority of the grant period. The U.S. Principal Investigator and participants on the U.S. team must be U.S. citizens or permanent residents and must reside in the U.S. during the course of the CGP-RFBR project. Graduate students on the U.S. team may be foreign nationals, but they must be enrolled full-time in an accredited degree program at a U.S. institution. All projects must be oriented toward non-military objectives and must be carried out in a civilian research environment. Special *Requirements*: Clearance for Former Weapons Scientists; Proposal Overview Form submitted to CRDF Moscow office; U.S. For-Profit companies only: Letter of Commitment to contribute 10% of Russian expenses and coverage of all U.S. expenses; Russian For-Profit companies only: Letter of Commitment to contribute up to 5-10% of total Russian expenses; Proposals with In-Kind contributions: Letter of Commitment from contributing U.S. and/ or Russian institution. For proposals involving Human Subjects research: Protocol Summary Form and supporting information, such as Institutional Review Board/Internal Ethical Committee Approval, Human Subjects Governmental Approval, and Confirmation of Federal-Wide Assurance.

Restrictions: Proposals oriented towards applied research will not be accepted.

Geographic Focus: Alabama; Alaska; Arizona; Arkansas; California; Colorado; Connecticut; Delaware; District of Columbia; Florida; Georgia; Guam; Hawaii; Idaho; Illinois; Indiana; Iowa; Kansas; Kentucky; Louisiana; Maine; Maryland; Massachusetts; Michigan; Minnesota; Mississippi; Missouri; Montana; Nebraska; Nevada; New Hampshire; New Jersey; New Mexico; New York; North Carolina; North Dakota; Ohio; Oklahoma; Oregon; Pennsylvania; Puerto Rico; Rhode Island; South Carolina; South Dakota; Tennessee; Texas; U.S. Virgin Islands; Utah; Vermont; Virginia; Washington; West Virginia; Wisconsin; WyomingRussia

Amount of Grant: $93,000

Date(s) Application is Due: Dec 1

Contact: Mr. Stuart Politi, Senior Program Manager; (703) 526-9720; fax: (703) 526-9721; email: cgp@crdf.org

Internet: http://www.crdf.org/funding/funding_show.htm?doc_id=1014806

Sponsor: U.S. Civilian Research and Development Foundation
1530 Wilson Boulevard, 3rd Floor
Arlington, VA 22209

U.S. CRDF Leishmaniasis: Collaborative Research Opportunities in North Africa and the Middle East 5633

The U.S. Civilian Research and Development Foundation (CRDF) is a nonprofit organization authorized by the U.S. Congress and established in 1995 by the National Science Foundation. This unique organization promotes international scientific and technical collaboration through grants, technical resources, and training. CRDF, with funding from the National Institute for Allergy and Infectious Diseases (NIAID), announces a targeted, collaborative grant competition on Leishmaniasis in the North Africa and the Middle East (MENA) Region. This grant competition is a follow-on activity to the June 2009 interdisciplinary research conference in Tunisia sponsored by NIAID that brought together regional and international scientists to discuss research opportunities on sand fly biology, leishmania parasites, immunology, leishmaniasis epidemiology, animal reservoirs, clinical manifestations, genetics, diagnostics, treatment, and prevention of leishmaniasis in North Africa and the Middle East. These follow-on grants will provide up to one- year, support to joint research teams of U.S. and MENA scientists, with the possibility of up to a one year no-cost extension, if justified. CRDF and NIAID anticipate that the research implemented through these awards will lay the foundation for future grant proposals to NIAID and other funding agencies. The average anticipated grant will provide up to $30,000 U.S. dollars for a period of one year. At least 80% of the funds awarded to each project will be used for project-related expenses of the MENA component of the research team, including institutional support. No more than 20% may be used for U.S. team expenses. As part of the grant, the CRDF

strongly encourages a member of the MENA team to visit the U.S. team's laboratory, and members of the U.S. team are encouraged to visit the MENA laboratory.
Requirements: Each proposal submitted to the CRDF must have at least one MENA Principal Investigator and at least one U.S. Principal Investigator, who share overall responsibility for the project in their respective countries, coordinating all project participants and institutions. Each proposal must meet each of the following eligibility criteria: 1) Only workshop attendees and their institutions are eligible to apply. 2) Each Principal Investigator must work full-time in a civilian research environment. (U.S. Government Laboratories working on civilian-oriented R&D projects are eligible to apply to this program.) 3) The U.S. Principal Investigator and participants of the U.S. team may be foreign nationals, but must reside in the U.S during the course of the grant award. Graduate students on the U.S. team may be foreign nationals, but they must be enrolled in an accredited degree program at a U.S. institution. A PI from Navy Medical Research Unit 3 (NAMRU3) may serve as either a U.S. or MENA researcher. 4) All projects must be in one of the following subject areas: Sand fly biology; Leishmania Parasites; Immunology; Leishmaniasis Epidemiology; Animal reservoirs; Clinical manifestations, treatment, and prevention of leishmaniasis in North Africa and the Middle East (but not clinical trials involving interventions); Genetics; Diagnostics. 5) Only one proposal will be accepted from a U.S. or MENA PI. 6) All projects must be oriented toward non-military objectives and must be carried out in a civilian research environment. Use of human subjects requires appropriate approvals from all institutions involved including Local Ethics Committee (LEC) and U.S. IBR. Middle East countries: Afghanistan, Behrain, Iran, Iraq, Israel, Jordan, Kyrgyzstan, Kuwait, Lebanon, Oman, Pakistan, Palestine, Qatar, Saudi Arabia, Syria, Tajikistan, Turkey, Turkmenistan, United Arab Emirates, Uzbekistan, Yemen. North African countries: Algeria, Egypt, Libya, Morocco, Sudan, Tunisia, Western Sahara.
Restrictions: No clinical trials involving interventions will be accepted for this competition.
Amount of Grant: up to $30,000
Date(s) Application is Due: Sep 30
Contact: Mr. Stuart Politi, Senior Program Manager; (703) 526-9720; fax: (703) 526-9721; email: cgp@crdf.org
Internet: http://www.crdf.org/funding/funding_show.htm?doc_id=949719
Sponsor: U.S. Civilian Research and Development Foundation
1530 Wilson Boulevard, 3rd Floor
Arlington, VA 22209

U.S. CRDF Science and Technology Entrepreneurship Program (STEP) Business Partnership Grant Competition Armenia **5634**

The U.S. Civilian Research and Development Foundation (CRDF) is a nonprofit organization authorized by the U.S. Congress and established in 1995 by the National Science Foundation. This unique organization promotes international scientific and technical collaboration through grants, technical resources, and training. The STEP Business Partnership Grants Competition, conducted by CRDF, the Ministry of Economy of RA and Enterprise Incubator Foundation (EIF), promotes research and development (R&D) partnerships between companies (Company) and teams of researchers/scientists (Technology Team) to develop new commercial opportunities of economic benefit to both parties.
Requirements: Projects must have R&D as their core task and must include preliminary market assessment, customer needs analysis and business development components. Projects that demonstrate their ability to enable revenue creation and/or attract non-STEP Partner organization sources of funding will be given priority consideration in the selection process. All goals and outcomes of the projects must be defined and measurable. STEP BPG projects are limited to a maximum of 12 months. Funds for projects are provided by the CRDF and the Ministry of Economy of RA. Funds are limited to $15,000. At least 10% of company cash contribution is mandatory. New Science Team/Company Partnership is preferable. All funds must be used for Science Team expenses. At first, written proposals should be submitted to EIF for review on eligibility and completeness of information provided. Complete proposals must be submitted to CRDF via electronic proposal submission (EPS) and emailed to the EIF by (18:00) local time on the following submission deadlines to be eligible. Proposals for EIF should be both in English and Armenian. Each proposal submitted to the STEP BPG is required to have a single Principal Investigator (PI) from a Company and a single Principal Investigator (PI) from the Science Team who will be responsible for coordinating all participating personnel, organizations, and work. Scientists and engineers may participate concurrently on more than one EIF/CRDF grant, but cannot be the PI on more than one active EIF/CRDF project or exceed full-time work status on the combined EIF/CRDF projects. Projects are required to be in research and development fields that are oriented toward commercial, non-military objectives and must be carried out in a civilian research environment. The participants of the Science Team are required to be scientists or engineers at institutes, universities, companies, or other organizations involved in research and development. The PIs and all participants of the Science Team are required to be citizens of and must currently reside and work in Armenia. The primary partner of the Company Team is required to be a for-profit company based in Armenia. In order to be considered Armenia-based, a company must have registered offices in Armenia and its facilities and personnel associated with the project must be located in this country. Joint U.S. - Armenia for-profit ventures are eligible to participate. The PI of the Company Team is required to be a citizen of Armenia and must currently reside and work in Armenia.
Restrictions: The Company and the organization employing the team of scientists may not be subsidiaries of or owned, either wholly or partially, by one another. Organizations and individuals of the Science Team working for organizations that are barred by the U.S. Government from receiving U.S. financial support, materials, or services are not eligible to participate in BPG program.
Geographic Focus: Alabama; Alaska; Arizona; Arkansas; California; Colorado; Connecticut; Delaware; District of Columbia; Florida; Georgia; Guam; Hawaii; Idaho; Illinois; Indiana; Iowa; Kansas; Kentucky; Louisiana; Maine; Maryland; Massachusetts; Michigan; Minnesota; Mississippi; Missouri; Montana; Nebraska; Nevada; New Hampshire; New Jersey; New Mexico; New York; North Carolina; North Dakota; Ohio; Oklahoma; Oregon; Pennsylvania; Puerto Rico; Rhode Island; South Carolina; South Dakota; Tennessee; Texas; U.S. Virgin Islands; Utah; Vermont; Virginia; Washington; West Virginia; Wisconsin; WyomingUkraine
Amount of Grant: $15,000
Date(s) Application is Due: Oct 5
Contact: Natalia Pipia, Senior Program Manager; (703) 526-9720; fax: (703) 526-9721; email: step@crdf.org
Internet: http://www.crdf.org/funding/funding_show.htm?doc_id=990740
Sponsor: U.S. Civilian Research and Development Foundation
1530 Wilson Boulevard, 3rd Floor
Arlington, VA 22209

U.S. CRDF STEP Business Partnership Grant Competition Ukraine **5635**

The U.S. Civilian Research and Development Foundation (CRDF) is a nonprofit organization authorized by the U.S. Congress and established in 1995 by the National Science Foundation. This unique organization promotes international scientific and technical collaboration through grants, technical resources, and training. The Science and Technology Entrepreneurship Program Business Partnership Grants competition (STEP BPG), conducted by CRDF and the Ministry of Education and Science (MES) promotes research and development (R&D) partnerships between companies (Company) and teams of scientists (Science Team) to develop new commercial opportunities of economic benefit to both parties. At the first stage of the STEP Program, two workshops will be carried out in Lviv and Sevastopol, Ukraine which will provide basic training on technology commercialization, and intellectual property protection. Workshops will also further clarify STEP program goals and objectives and will introduce the BPG competition terms and conditions.
Requirements: STEP BPG projects must have research and development (R&D) as their core task and must include a preliminary market assessment; a customer needs analysis, and business development components. Projects that demonstrate their ability to enable revenue creation and/or attract non-STEP Partner organization sources of funding will be given priority consideration in the selection process. All goals and outcomes of the projects must be well defined and measurable. STEP BPG projects are limited to a maximum of 12 months. Funds for projects are provided by the CRDF and MES. Funds are limited to $24,000. All funds must be used for Science Team expenses. In support of the goals of the Business Partnership Grant competition to increase the number of linkages between the science and business sectors, preference will be given to newly-formed science- business partnerships. Preference will also be given to projects with at least 10% company cash contribution. Each proposal submitted to the STEP BPG competition is required to have a single Principal Investigator (PI) from a Company and a single Principal Investigator (PI) from the Science Team who will be responsible for coordinating all participating personnel, organizations, and work. Scientists and engineers may participate concurrently on more than one CRDF grant, but cannot be the PI on more than one active CRDF project or exceed full-time work status on the combined CRDF projects. The STEP BPG projects are required to be in research and development fields that are oriented toward commercial, non-military objectives and must be carried out in a civilian research environment. All proposals are required to be submitted to CRDF in English. Two copies of all proposals should be submitted in printed form to the MES in Ukrainian. The participants of the Science Team are required to be scientists or engineers at institutes, universities, companies, or other organizations involved in research and development. The PIs and all participants are required to be citizens of and must currently reside and work in Ukraine. The primary partner is required to be a for-profit company based in Ukraine or in the U.S; thus, a company must have registered offices in Ukraine or in the U.S. and its facilities and personnel associated with the project must be located in these countries. Joint U.S.-Ukrainian for-profit ventures are eligible to participate. The PI of the Company Team is required to be a citizen of and must currently reside and work in Ukraine. Note: The MES and CRDF discourage the inclusion of an 'administrative' superior in a proposal in any role for which individual financial support is requested. An 'Administrative Superior' is defined as an individual who has direct administrative authority over the Principal Investigator, and who works within the same division, laboratory, or unit as the Principal Investigator. Institutional leadership including university rectors and vice rectors and institute directors and deputy directors are also defined as 'administrative superiors.' This definition does not apply to other individuals of higher 'rank' within the institution but who work outside the division, laboratory or unit of the Principal Investigator.
Restrictions: The Company and the organization employing the team of scientists may not be subsidiaries of or owned, either wholly or partially, by one another. Organizations and individuals working for organizations that are barred by the U.S. Government from receiving U.S. financial support, materials, or services are not eligible to participate in CRDF programs. No grant funds can be used for company expenses.
Geographic Focus: Alabama; Alaska; Arizona; Arkansas; California; Colorado; Connecticut; Delaware; District of Columbia; Florida; Georgia; Guam; Hawaii; Idaho; Illinois; Indiana; Iowa; Kansas; Kentucky; Louisiana; Maine; Maryland; Massachusetts; Michigan; Minnesota; Mississippi; Missouri; Montana; Nebraska; Nevada; New Hampshire; New Jersey; New Mexico; New York; North Carolina; North Dakota; Ohio; Oklahoma; Oregon; Pennsylvania; Puerto Rico; Rhode Island; South Carolina; South

Dakota; Tennessee; Texas; U.S. Virgin Islands; Utah; Vermont; Virginia; Washington; West Virginia; Wisconsin; WyomingUkraine
Amount of Grant: $24,000
Date(s) Application is Due: Dec 20
Contact: Natalia Pipia, Senior Program Manager; (703) 526-9720; fax: (703) 526-9721; email: step@crdf.org
Internet: http://www.crdf.org/funding/funding_show.htm?doc_id=1034127
Sponsor: U.S. Civilian Research and Development Foundation
1530 Wilson Boulevard, 3rd Floor
Arlington, VA 22209

U.S.-Mexico Fund for Culture Grants **5636**
The fund's aim is to strengthen the cultural exchange and collaboration between Mexico and the United States by encouraging a creative, fertile dialog between their artistic and cultural communities. The fund provides economic support for projects of excellence that reflect the artistic and cultural diversity of Mexico and the United States and that are capable of leading to close, lasting collaboration and exchange among artists, researchers and scholars, independent groups, and related institutions of both countries. Applications are accepted for institutional projects—from nonprofit cultural institutions (museums, libraries, universities, historical archives, etc.) and planning and/or research projects—from joint initiatives of individuals, groups, and nonprofit cultural institutions for the development of the planning and/or research stages of projects aiming to promote or improve collaboration programs between the artistic communities of Mexico and the United States. Grants are awarded in the disciplines of dance, theater, music, visual arts, literature, media arts, cultural studies, and libraries. Application forms and procedures are available on the Web site.
Requirements: Application may be submitted for individual or group projects presented by artists, curators, producers, independent groups, nonprofit cultural organizations, universities, and presenters.
Restrictions: Grants may not be used for general operating support or capital improvements, such as expansion of permanent facilities, purchase of equipment or regular salaries, or to cover honoraria of the project manager or of any other member of the institution. In the case of universities and other institutions in the United States, the resources granted by the fund can not be applied to overhead expenses.
Amount of Grant: $2000-$30,000
Date(s) Application is Due: Jul 18
Contact: Grants Administrator, 5-592-53-86 or 5-535-67-3; fax: 5-566-80-71
Internet: http://www.culturalcommons.org/directorydetail.cfm?ID=1361
Sponsor: U.S.-Mexico Fund for Culture
Londres 16 PB
Col Juarez 06600 DF Mexico

U.S.A Funds Access to Education Scholarships **5637**
The scholarships offer financial support to students with economic need. The program invites residents of the state of Hawaii to apply for U.S.A Funds Hawaii Silver Anniversary Scholarships. Residents of all other states, the District of Columbia, Puerto Rico, Guam, the U.S. Virgin Islands, and all U.S. territories and commonwealths may apply for U.S.A Funds Access to Education Scholarships. Up to 50 percent of the awards are targeted to members of ethnic-minority groups or individuals with physical disabilities. Scholarship awards are based on the applicant's past academic performance and future potential; leadership and participation in school and community activities; work experience; career and educational aspirations and goals; and unusual personal or family circumstances. Guidelines and application are available online.
Requirements: Eligibility includes high-school seniors, currently enrolled college students, or incoming college students who plan to enroll or are enrolled in full- or half-time undergraduate or full-time graduate coursework at an accredited two- or four-year college, university, or vocational/technical school; have an annual adjusted gross family income of $35,000 or less, as documented on a federal income-tax return on which the applicant is claimed as a federal tax exemption; are U.S. citizens or eligible noncitizens (permanent U.S. residents with a Green Card, or refugees with a Departure Record (Form I-94) from the U.S. Immigration and Naturalization Service; and are not currently in default on a federal education loan.
Amount of Grant: $6000 maximum; $1500 full-time students and $750 to half-time undergraduates
Date(s) Application is Due: Mar 15
Contact: Scholarship Management Services, Scholarship America, (866) 497-8723; email: contact@usafunds.org
Internet: http://www.usafunds.org/about_usa_funds/index.html
Sponsor: U.S.A Funds
P.O. Box 297, 1 Scholarship Wy
Saint Peter, MN 56082

U.S.AID Advanced Training for Leadership and Skills (ATLAS) Fellowships **5638**
The program's purpose is to strengthen leadership and technical abilities and enhance the professional performance of individuals serving in African public and private sector entities, including universities, research centers, and other key development institutions. The program is administered by U.S.AID and the African American Institute and focuses on sub-Saharan Africa. The grants program provides support to Africa-based voluntary organizations, including alumni associations, professional organizations, and women's groups whose members include U.S.AID-trained participants. Grants are given for a variety of activities, including workshops, newsletters, and the development of professional activities.
Contact: Ted Clarke, (202) 712-1847; fax: (202) 216-3229; email: clarke@usaid.gov; Program Contact, African-American Institute, (212) 949-5666; fax: (212) 818-9505
Internet: http://www.info.usaid.gov/educ_training/atlas.htm
Sponsor: United States Agency for International Development
1300 Pennsylvania Avenue NW, Ronald Reagan Building
Washington, DC 20523

U.S.C Annenberg/Getty Arts Journalism Fellowships **5639**
Funded by a grant from the J. Paul Getty Trust, the U.S.C Annenberg/Getty Arts Fellowship seeks to establish a new standard of excellence in arts and culture coverage. The program's philosophy is guided by a core belief in first-hand encounters with artists, arts administrators, and journalism colleagues. The three-week residencies in Los Angeles familiarize the fellows with the city's vibrant cultural scene as a means to strengthen their intellectual and analytical journalism skills, fill them with new story ideas, and make them consider their hometowns in new lights. The program period is April 22 through May 13. Application and guidelines are available online.
Requirements: Requirements for Eligibility: candidates must be mid-career critics, reporters, writers, columnists, editors, broadcasters or producers who specialize in arts, culture or entertainment for or a newspaper, magazine, broadcast or online media organization; must have at least three years of steady professional arts journalism in your background; the program is NOT intended for entry-level journalists, students or journalism educators. # Freelancers and/or self-employed individuals who regularly publish online, in print or over the airwaves ARE eligible, as long as their application is accompanied by a letter of recommendation by a newsroom editor/manager at an organization that regularly publishes or broadcasts their work OR by an experienced journalist who can speak to their work in convincing detail; work examples for the application must have been published or produced between May - June of previous year; this is a national and international program, applicants from anywhere MAY apply; advanced degrees or specialized training in the arts are NOT required; we do not discriminate on the basis of race, religion, national or international origin, color, sex, age, veteran status, or disability. It is our intention that all qualified applications be given equal opportunity and that our selection is based upon merit alone.
Samples: NATE CHINEN - The New York Times, freelance music critic and JazzTimes columnist. As a regular contributor to the Times, he produces original material for its website, podcast and blog to supplement his print pieces (2008); DAVID SILLITO - British Broadcasting Corporation News, London, arts and culture correspondent. A general reporter from 1989 to 1998, Sillito has reported for last ten years on all aspects of arts, entertainment, media and social and cultural change for main national and international outlets on BBC Television. Although television is his principal medium, he also regularly provides arts coverage for BBC Radio 4 and BBC Online (2008);
Date(s) Application is Due: Jan 27
Contact: Sasha Anawalt, (213) 740-3914; emal: uscgetty@usc.edu; fax: (213) 740-8624
Internet: http://annenberg.usc.edu/Home/CentersandPrograms/ProfessionalEducation/GettyArtsJourn.aspx
Sponsor: University of Southern California
300 S Grand Avenue, Suite 3950
Los Angeles, CA 90071

U.S.C Art History Fellowships **5640**
The program is designed for people pursuing the master's of fine arts degree and pursuing experience in teaching art on a college level. The fellowships include stipend, full tuition, monthly living expenses, and a private studio on campus.
Requirements: Applicants must have a BA, preferably in art history, for the MA program and an MA in art history for the PhD program; must have taken the GRE within five years of applying to the programs; and must have a working knowledge of one language for the MA, two languages for the PhD, usually French or German.
Date(s) Application is Due: Feb 1
Contact: Director of Admissions, (213) 740-7296; fax: (213) 740-7577
Internet: http://afaweb.esd.usc.edu/dcpage2.cfm?PageBaseID=50231
Sponsor: University of Southern California
Watt Hall 104, School of Fine Arts
Los Angeles, CA 90089-0292

U.S.C Center for Religion and Civic Culture Undergraduate Awards **5641**
The center, at the University of Southern California, awards small grants for independent research on religion conducted by U.S.C undergraduates under the guidance of a faculty member. These awards are designed to encourage faculty-student collaboration, in which the student has an important intellectual investment. Twelve awards for undergraduates and for faculty mentors are available.
Requirements: These awards are only available to U.S.C undergraduates and faculty members.
Amount of Grant: $1500 for undergraduates and $500 for faculty mentors
Date(s) Application is Due: Sep 22
Contact: Grants Administrator, (213) 740-8562; fax: (213) 740-5810; email: crcc@usc.edu
Internet: http://www.usc.edu/schools/college/crcc/academic
Sponsor: University of Southern California
835 W 34th Street, Suite 106
Los Angeles, CA 90089-0751

U.S.C Pentecostal and Charismatic Research Initiative Individual Grants 5642

The Pentecostal and Charismatic Research Initiative (PCRI) request for proposals (RFP) for Individual Grants is designed to provide a scholarly framework based in the social sciences and financial support for research into Pentecostalism and charismatic Christianity in four specific geographic regions: Africa, Asia, Latin America, and the former Soviet Union. PCRI will award up to 15 grants in this area. $1,500,000 in grant funding is available for the individual grants RFP.

Requirements: Doctoral students and scholars affiliated with universities, colleges, and research centers may apply. Applicants may request up to $100,000 over two years (24 months) to support scholarly research, although not all projects can be supported at this level. Research projects should engage topics of interest not only to Pentecostal specialists, but to a broader group of scholars interested in the dynamics of social change and the potentially dynamic role of religion in this process. Proposals must be focused on one of four regions (Africa, Asia, Latin America, or the former Soviet Union). Comparative research projects are also eligible, including projects that explore the connections between Western countries and one or more of the four regions mentioned above. PCRI is particularly interested in the economic implications of this fast growing movement. However, scholars from all social science disciplines are encouraged to apply (e.g., anthropologists, sociologists, political scientists, etc.). Grants must be administered through a scholarly institution such as a university, seminary, or research center.

Restrictions: Projects involving only theological questions are not eligible under this RFP, but collaborative research involving theologians may be appropriate. No more than 10% of a grant may be devoted to a combination of indirect costs, overhead or administration.

Amount of Grant: up to $100,000

Date(s) Application is Due: Aug 1

Contact: Ms. Brie Loskota, Managing Director; (213) 740-8562; fax: (213) 740-5810; email: pcri@usc.edu

Internet: http://www.usc.edu/schools/college/crcc/pcri/individuals/

Sponsor: Center for Religion and Civic Culture, University of Southern California (U.S.C)
825 Bloom Walk, ACB 439
Los Angeles, CA 90089-1481

U.S.C Pentecostal and Charismatic Research Initiative Regional Center Grants 5643

The Pentecostal and Charismatic Research Initiative (PCRI) request for proposals (RFP) for regional centers is designed to provide a scholarly framework based in the social sciences and financial support for research into Pentecostal and charismatic Christianity in four specific geographic regions: Africa, Asia, Latin America, and the former Soviet Union. PCRI will award up to seven center grants. $2,000,000 in grant funding is available for the regional center RFP. Grants will be awarded to a single institution, but a project may include scholars from surrounding cities ‚Äî or even countries ‚Äî in the region.

Requirements: Grants are intended to strengthen or establish research hubs in a region and create new clusters of scholars working together. These centers should develop consultations and conferences, and provide logistical support for the associated scholars, graduate students, and research staff. Research projects should engage topics of interest not only to Pentecostal specialists, but to a broader group of scholars interested in the dynamics of social change and the potentially dynamic role of religion in this process. Funding will be limited to research that has the potential to significantly impact the 'big picture' understanding of Pentecostal and charismatic Christianity and that addresses the role of religion in human society. Proposals must be focused on one of four regions (Africa, Asia, Latin America, or the former U.S.SR). Comparative research projects are also eligible, including projects that explore the connections between Western countries and one or more of the four regions mentioned above. While proposals may include scholars from North America and Europe, the core component of research staff must be located in the region of focus. Also, at least one of the Principal Investigators must reside in Africa, Asia, Latin America, or the former Soviet Union. Grants will be awarded based on academic merit and not on the affiliation (secular, denomination, etc.) of the research center.

Restrictions: No more than 10% of a grant may be devoted to a combination of indirect costs, overhead or administration.

Amount of Grant: up to $500,000

Date(s) Application is Due: Aug 1

Contact: Ms. Brie Loskota, Managing Director; (213) 740-8562; fax: (213) 740-5810; email: pcri@usc.edu

Internet: http://www.usc.edu/schools/college/crcc/pcri/centers/

Sponsor: Center for Religion and Civic Culture, University of Southern California (U.S.C)
825 Bloom Walk, ACB 439
Los Angeles, CA 90089-1481

U.S.DA National Research Initiative Competitive Grants 5644

The Program supports research, education, and extension projects that address key problems of National, regional, and multi-state importance in sustaining all components of agriculture (farming, ranching, forestry, aquaculture, rural communities, human nutrition, processing, etc.). Emerging agricultural issues of national concern, including food safety, honey bee colony collapse disorder (CCD), and sustainable alternative energy resources, drives NRI programs to advance scientific discoveries. Providing this support requires that NRI advance fundamental sciences in support of agriculture and coordinate opportunities to build on these discoveries. Building on these discoveries will necessitate new efforts in education and extension that deliver science-based knowledge to people, allowing them to make informed practical decisions. The NRI will accept applications for fundamental research, mission-linked research, and integrated research, education, and extension projects. Application deadlines vary by program. For additional information, review each program description in the Request for Applications (RFA) for program priorities, funding levels, and application deadlines.

Amount of Grant: $5,000-1,500,000

Contact: Grants Program Management; (202) 401-5022; fax: (202) 401-6488; email: nricgp@csrees.usda.gov

Internet: http://www07.grants.gov/search/search.do;jsessionid=HVhSLncmJhTRFdrMMJjdFLwCR9LQrmpF7M427YBZTfd45y7yP7cV!1602867008?oppId=15487&flag2006=true&mode=VIEW

Sponsor: Department of Agriculture; Cooperative State Research, Education, and Extension Service (CSREES)
1400 Independence Avenue SW., Stop 2201
Washington, DC 20250-2201

U.S.F Irvine Minority Scholar Dissertation Fellowships 5645

The university invites applications for dissertation-year fellowships from minority students in one of the following programs: College of Arts and Sciences—dance or visual and performing arts, history (ancient Mediterranean, Asian, California, Middle Easter), philosophy, psychology, sociology, and theology and religious studies; College of Professional Studies—organizational development/behavior, information systems management; and School of Education—teacher education, counseling psychology; and School of Nursing—medical/surgincal nursing. U.S.F senior faculty serve as mentors and consultants while the scholar completes the dissertation and initiates an ongoing program of scholarly or creative work, develops a repertoire of potential course offerings, and familiarizes him-/herself with the service responsibilities of a university faculty member. Scholars teach one undergraduate course in the discipline each semester, serve as resources to students and faculty, and contribute to U.S.F's multicultural living/learning environment through activities that celebrate ethnic diversity and cultural pluralism. The program provides a stipend and limited support for relocation and research-related expenses plus housing support, office space and computer, and library privileges.

Requirements: Scholars must be U.S. citizens or permanent residents and members of the following ethnic/racial groups: African Americans, Asian Americans, Hispanics/Latinos, or American Indians.

Amount of Grant: $30,000

Date(s) Application is Due: Jan 15

Contact: Dr. Gerardo Marin, Associate Dean, College of Arts and Sciences

Internet: http://www.usfca.edu/hr/employment/faculty11.html

Sponsor: University of San Francisco
2130 Fulton Street
San Francisco, CA 94117-1080

U.S.F Richard F. Pride Research Fellowship 5646

The fellowship is designed to address the underrepresentation of minorities in college and university faculties. It provides the opportunity for individuals of demonstrated ability to pursue a PhD degree at the University of South Florida in liberal arts disciplines, including applied anthropology, marine sciences, biology, mathematics, chemistry, psychology, philosophy, English, and communication. Tuition for up to four academic years is provided. Application materials may be obtained from the office.

Requirements: Applicants must be U.S. citizens. Preference will be given to African American candidates.

Restrictions: Fellowship can only be used at the University of South Florida.

Amount of Grant: $20,000 per year for up to four years

Date(s) Application is Due: Jan 15

Contact: Institute on Black Life, (813) 974-4727; fax: (813) 974-5042; email: ibl@ibl.usf.edu

Internet: http://usfweb2.usf.edu/ibl/pride.htm

Sponsor: Institute on Black Life
University of South Florida, 4202 E Fowler Avenue, SVC 1087
Tampa, FL 33620-6911

U.S.G Foundation Grants 5647

The foundation supports nonprofits in company-operating areas in the categories of arts/culture, higher education, hospitals and health care, social welfare, youth agencies, federated giving, public policy research, and government/public administration. Types of support include general operating support, continuing support, annual campaigns, capital campaigns, building construction/renovation, equipment acquisition, scholarship funds, research, technical assistance, employee matching gifts and employee-related scholarships, and matching grants. There are no required application forms or deadlines. The board meets quarterly to consider requests. Giving primarily in areas of company operations.

Restrictions: Grants are not made to individuals.

Samples: Child Welfare League of America (Washington, DC)—to conduct the Positive Parenting for Homeless Families of Young Children program, $100,000.

Amount of Grant: $1000-$40,000 average range

Contact: Harold Pendexter, Program Contact, (312) 606-4297; fax: (312) 606-5316

Sponsor: U.S.G Foundation
125 S Franklin Street
Chicago, IL 60606

U.S.HMM Summer Research Workshops for Scholars 5648

The Center for Advanced Holocaust Studies invites teams of scholars to conduct two-week workshops at the museum during June through August. Support defrays the

travel, lodging, meals, and incidental costs of teams of six to 10 scholars interested in conducting research, discussing research methodology, and sharing research results on similar or closely related Holocaust studies topics. The center welcomes a variety of approaches by scholars in history, political science, economics, philosophy, religion, sociology, literature, psychology, and other disciplines. Fields of inquiry include, but are not limited to, historiography and documentation of the Holocaust, ethics, law, comparative genocide studies, and the impact of the Holocaust on contemporary society and culture. Other areas of interest include resistance, refuge and rescue, and propaganda and mass media as they relate to genocide.
Requirements: Workshop participants may include, but are not limited to, faculty members or dissertation-stage graduate students at universities and colleges; scholars and writers working independently or in institutions with research or educational missions, such as museums, memorials, libraries, and historical associations; and government and nongovernmental organizations.
Amount of Grant: $20,000 maximum
Date(s) Application is Due: Jan 31
Contact: University Programs Division; fax: (202) 479-9726; email: university_programs@ushmm.org
Internet: http://www.ushmm.org/research/center/seminars
Sponsor: United States Holocaust Memorial Museum Research Institute
100 Raoul Wallenberg Pl SW
Washington, DC 20024-2150

U.S.HMM Visiting Scholars Fellowships **5649**
The center awards fellowships to support research and writing about the Holocaust. Awards are granted on a competitive basis. Fields of inquiry include, but are not limited to, historiography and documentation of the Holocaust, comparative genocide studies, and the impact of the Holocaust on contemporary society and culture as well as other underresearched subject areas that focus on the experiences of Holocaust victims and bystanders. The center welcomes a variety of approaches by scholars in history, political science, literature, philosophy, religion, sociology, psychology, and other disciplines. The center encourages applications that utilize the Museum's recent archival acquisitions from regional and state archives of Belgium, Croatia, the Czech Republic, France, Poland, Romania, Slovakia, and Ukraine; as well as records of Jewish provenance from the Ringelblum Archive, the World Jewish Congress, and other Jewish rescue and relief organizations (international, state, and local). They can also utilize the Museum's library; oral history collection; music, film and photo archives, and other resources available in the Washington, DC, area. Biweekly fellows' meetings and senior seminar programs provide unique settings for debate and discussion.
Requirements: Fellowships are awarded to candidates working on their dissertations (ABD), postdoctoral researchers, senior scholars, and professionals holding degrees from accredited academic and research institutions worldwide.
Amount of Grant: $3000 maximum monthly stipend
Date(s) Application is Due: Nov 26
Contact: Visiting Fellows Program, Center for Advanced Holocaust Studies, (202) 314.7829; fax: (202) 479-9726; email: visiting_scholars@ushmm.org
Internet: http://www.ushmm.org/research/center/fellowship
Sponsor: United States Holocaust Memorial Museum Research Institute
100 Raoul Wallenberg Pl SW
Washington, DC 20024-2126

U.S.IA Bosnia and Herzegovina Undergraduate Development Program Grants **5650**
This program is designed to allow Bosnian students an opportunity to obtain knowledge, insight, and cultural enrichment through academic studies at American colleges and universities. Four-year colleges and universities may host between two and five Bosnian students in a one-year, nondegree undergraduate program. The U.S.IA strongly encourages institutions to guide students to courses in American studies or other courses that emphasize democracy, market economy, and civic society. Funding supports one academic year of study in the fields of agriculture, business administration, civic education, criminal justice, economics, education, environmental resource management, journalism/mass communications, political science, and public administration. The academic-year program is followed by a four- to 12-week internship in the students' field of specialization. The program also includes culturally enriching activities including, but not limited to, community outreach and service projects, a welcome orientation program at the host institution, and U.S.IA-sponsored midyear and end-of-year workshops in Washington, DC.
Requirements: Four-year universities and colleges are eligible to apply. Preference is given to institutions with demonstrated experience in hosting Bosnian students, partnerships with Bosnian higher education institutions, or expertise and interest in the region; strong international student advising offices with experience dealing with cultural, educational and adjustment issues for foreign students; accessibility to and opportunities for cultural and social activities; and diverse, multiethnic student populations. Preference will also be given to institutions that can provide cost sharing toward tuition, fees, room and board expenses, and/or other direct participant expenses. Cost sharing may also be in the form of direct administrative and program costs.
Date(s) Application is Due: Mar 1
Contact: Office of Academic Programs, European Branch, (202) 205-0525 or (202) 647-3525; fax: (202) 619-6565; email: sgovatsk@usia.gov
Internet: http://www.americancouncils.org/program.asp?PageID=121&ProgramID=44
Sponsor: United States Information Agency
301 Fourth Street SW
Washington, DC 20547

U.S.IA NIS Educational Advising Centers **5651**
The goal of this program is to operate educational advising centers in NIS, including Erevan, Armenia; Baku, Azerbaijan; Minsk, Belarus; Tbilisi, Georgia; Almaty, Kazakstan; Bishkek, Kyrgyzstan; Chisinau, Moldova; Moscow, Novosibirsk, Saint Petersburg, and Vladivostok, Russia; Dushanbe, Tajikistan; Ashgabat, Turkmenistan; Kyiv, Ukraine; and Tashkent, Uzbekistan. These centers provide comprehensive and unbiased information to interested students, scholars, and other individuals about study opportunities in the United States.
Requirements: Public and private nonprofit organizations meeting the provisions described in IRS regulation 26 CFR 1.501(c) may submit proposals for administering advising centers in NIS.
Date(s) Application is Due: Oct 31
Contact: Educational Information and Resources Branch (ECA/A/S/A), (202) 619-5434; fax (202) 401-1433; email: advise@pd.state.gov
Internet: http://exchanges.state.gov/education/educationusa
Sponsor: United States Information Agency
301 Fourth Street SW
Washington, DC 20547

U.S.IP International Peace Projects Solicited Grants **5652**
Each year the institute supports research in the fields of international peace and conflict management. Solicited grants are awarded for special priority topics, which are identified by the institute. Topics for the current competition include Promoting Private Sector Economic Recovery in Countries Emerging from Violent Conflict; and Electoral Politics and Islamist Political Parties and Groups in Muslim Majority Countries. Application materials are available online.
Requirements: U.S. and international nonprofit organizations and official public institutions are eligible. Individuals also are eligible; affiliation with an institution is preferred.
Amount of Grant: $25,000-$45,000
Date(s) Application is Due: Mar 1; Oct 1
Contact: Grants Administrator, (202) 429-3842; fax: (202) 429-6063; TTY: (202) 457-1719; email: grants@usip.org
Internet: http://www.usip.org/grants/solicited.html
Sponsor: United States Institute of Peace
1200 17th Street NW, Suite 200
Washington, DC 20036-3011

U.S.IP International Peace Projects Unsolicited Grants **5653**
Unsolicited grants are offered across a broad range of relevant disciplines, skills, and approaches. Topic areas of interest to the institute include, but are not restricted to international conflict resolution; diplomacy; negotiation theory; functionalism and track two diplomacy; methods of third-party dispute settlement; international law; international organizations and collective security; deterrence and balance of power; arms control; psychological theories about international conflict; the role of nonviolence and nonviolent sanctions; moral and ethical thought about conflict and conflict resolution; and theories about relationships among political institutions, human rights, and conflict. The institute welcomes proposals of an interdisciplinary or multidisciplinary nature. The required application form and guidelines are available online.
Requirements: Nonprofit organizations, official public agencies, and individuals are eligible to apply.
Restrictions: Funds are not provided for projects that intervene in ongoing disputes or involve lobbying.
Amount of Grant: $25,000-$45,000 average
Date(s) Application is Due: Mar 1; Oct 1
Contact: Unsolicited Grants Program, (202) 429-3842; fax: (202) 429-6063; email: grants@usip.org
Internet: http://www.usip.org/grants/unsolicited.html
Sponsor: United States Institute of Peace
1200 17th Street NW, Suite 200
Washington, DC 20036-3011

U.S.IP Jennings Randolph Peace Scholar Dissertation Fellowships **5654**
These fellowships are intended to support the research and writing of dissertations addressing the sources and nature of international conflict and the full range of ways to prevent or end conflict and to sustain peace. Dissertation projects from all disciplines are welcome. Priority is given to projects that contribute knowledge relevant to the formulation of policy on international peace and conflict issues. Peace Scholars carry out their work at their universities or other appropriate research sites.
Requirements: Doctoral students at U.S. universities who will have completed all requirements for doctoral degree except the dissertation are eligible to apply. Applicants may be citizens of any country.
Restrictions: Awards may not be made for projects that constitute policymaking for a government agency or private organization, focus to any substantial degree on conflicts within U.S. domestic society; or adopt a partisan, advocacy, or activist stance.
Amount of Grant: $17,000 for 12 months
Date(s) Application is Due: Jan 9
Contact: Jennings Randolph Program Fellowships, (202) 429-3886; fax: (202) 439-6063; TTY (202) 457-1719; email: jrprogram@usip.org
Internet: http://www.usip.org/fellows/index.html
Sponsor: United States Institute of Peace
1200 17th Street NW, Suite 200
Washington, DC 20036-3011

U.S.IP Jennings Randolph Peace Scholar Senior Fellowships 5655

The institute awards fellowships to enable outstanding scholars, policymakers, journalists, and other professionals from around the world to conduct research at the U.S. Institute of Peace in Washington on important issues concerning international conflict and peace. The program integrates the work of senior fellows into the institute's education, research, and training activities. It also works closely with the institute's offices of publications, and congressional and public affairs to disseminate knowledge from these projects among policymakers, researchers, educators, and the general public through a combination of policy briefings, public events, media appearances, and published materials—including books and reports. The 10-month fellowship includes a stipend based on the fellow's previous year's salary and professional standing. Fellows also are provided partial reimbursement of health insurance premiums, an office with a computer and voice mail, and the services of a part-time research assistant. Application and guidelines are available online.
Requirements: Candidates must have an undergraduate degree from an accredited university. Postgraduate education is not required. Fellowships are open to citizens of any country.
Restrictions: Awards may not be made for projects that constitute policymaking for a government agency or private organization; focus to any substantial degree on conflicts within U.S. domestic society; or adopt a partisan, advocacy, or activist stance.
Amount of Grant: $80,000 maximum for a 10-month fellowship
Date(s) Application is Due: Sep 15
Contact: Jennings Randolph Program Fellowships, (202) 429-3886; fax: (202) 429-6063; TTY: (202) 457-1719; email: jrprogram@usip.org
Internet: http://www.usip.org/fellows
Sponsor: United States Institute of Peace
1200 17th Street NW, Suite 200
Washington, DC 20036-3011

Utah Arts Council Technical Assistance Grants 5656

Funding support is available on a cash-matching basis to advance artistic development, technical capabilities, and management skills. Examples of projects that qualify are support for consultant services to further management skills, artistic or technical production, and artistic development; local arts needs-assessment, survey design, implementation, and interpretation; conferences and workshops and travel to them; design and implementation of audience development and promotional campaigns or fund-raising drives to augment existing programming; and research and design of cultural facilities. Annual deadline dates may vary; contact program staff for exact dates. First time applicants, call the AE office at (801)236-7542 to schedule a consultation.
Requirements: Utah arts organizations are eligible. Matching money must be in-hand at the time of application.
Geographic Focus: Utah
Amount of Grant: $1000 maximum
Date(s) Application is Due: Sep 1
Contact: Sherry Waddingham, Grants Officer, (801) 236-7550; fax: (801) 236-7556; TDD: (800) 346-4128; email: swaddingham@utah.gov
Internet: http://arts.utah.gov/grants_program
Sponsor: Utah Arts Council
617 E South Temple
Salt Lake City, UT 84102-1177

V. Kann Rasmussen Foundation Grants 5657

The foundation has three principal areas of funding interest: environment—to strengthen environmental research and education, and to further the involvement of an informed public in environmental decision making; Greenwood and South Carolina—especially in the environmental, social, and educational areas; and medical research—medical research projects, with particular emphasis on eye research. Preference will be given to projects involving collaboration among people, among institutions, and across disciplines. The foundation considers funding proposals twice each year, at meetings held in March-April and October-November. The foundation requires a letter of inquiry by email; full proposals are by invitation. The deadline listed is for the letter of intent. The schedule for each round of grantmaking will be updated regularly. Guidelines are available online.
Samples: Massachusetts Institute of Technology (MA)—$6.2 million (2004); Teaching Kids About the Environment (SC)—$238,000 (2004); Eye Research Collaborative-Retinopathy—$3.62 million (2004); National Council of Churches—$50,000 (2004).
Date(s) Application is Due: May 25
Contact: Martin Kaplan, Trustee, c/o Wilmer Cutler Pickering Hale and Dorr LLP, (617) 526-6610; fax: (617) 526-5000; email: mkaplantrustee@vkrf.org
Internet: http://www.vkrf.org
Sponsor: V. Kann Rasmussen Foundation
60 State Street
Boston, MA 02109

Vada and Barney Oldfield Fellowship for National Security Reporting 5658

Applicants for this fellowship must be employed in electronic news and have 10 years or less experience in the field. The award recognizes excellence in national defense and security issues by a reporter or producer. Fellowship award includes stipend plus an all expense-paid trip to the annual RTNDF international conference. Application and guidelines are available online.
Requirements: Applicant should write a letter, endorsed by his/her news director, setting forth the reasons for seeking the fellowship and how he/she intends to use the grant money. This letter should be accompanied by supporting material exemplifying the applicant's best work as a member of the news staff. Format and length are not prescribed. Script, tape, or preferably both are acceptable. There is no application fee. Entry materials, including tapes, become the property of RTNDF and cannot be returned.
Amount of Grant: $1000 fellowship plus expenses-paid trip to annual conference
Date(s) Application is Due: May 9
Contact: Carol Knopes, (202) 467-5215; email: carolk@rtndf.org
Internet: http://www.rtndf.org/asfi/index.asp
Sponsor: Radio and Television News Directors Foundation
1600 K Street NW, Suite 700
Washington, DC 20006

Van Alen Institute New York Prize Fellowship 5659

The fellowship program supports emerging research and experimental practice in public architecture. Fellows are given an opportunity to pursue advanced independent study and to generate projects that are accessible to the general public. Projects may be an ongoing component or an outcome of fellows' research. They can range across media, temporalities, dimensions and scales in the form of exhibitions, installations, performances, symposia and other formats. The institute supports projects that provoke new ways of thinking about and making public space, that critically interrogate how and for whom spaces operate as 'public', and that engage public audiences in imaginative and productive ways.
Requirements: he fellowship program is open to emerging practitioners and scholars from the design and planning disciplines, and other fields in the arts, humanities and sciences that foster critical and creative inquiry into contemporary forms of public space. The fellowship is international in scope - the institute welcomes applications from around the world, across the country, and around the region. Candidates may be early to mid-career individuals in the design and planning disciplines, arts, humanities or sciences. Design professionals, artists and other practitioners must hold at least a Bachelors' degree. Scholars should hold a terminal degree as defined by their field of specialty. Candidates must be able to spend a minimum of three months in continuous residence in New York City; must be able to develop and present projects within the fellowship period; and, must possess English proficiency. Joint applications are welcome, provided all parties meet eligibility requirements.
Restrictions: Students are not eligible. Work for the completion of a degree is not eligible.
Amount of Grant: $5,000 - $15,000
Date(s) Application is Due: May 9
Contact: Jessica Blaustein, Fellowship Program Coordinator; (212) 924-7000, ext. 5; email: jblaustein@vanalen.org
Internet: http://www.vanalen.org/nyprize/fellowship_overview.html
Sponsor: Van Alen Institute
30 W. 22nd Street, 6th Floor
New York, NY 10010

Van Ameringen Foundation Grants 5660

The foundation supports mental health treatment and crisis intervention services in the metropolitan New York and Philadelphia areas. Grants are awarded in the following categories: bereavement; education and media; emotionally disturbed children and adolescents; frail and isolated elderly; housing; medically ill/dually diagnosed; prisons and jails; recent immigrants; rehabilitation centers; research and training; structural reform and advocacy; victims of domestic abuse; and vocational services. Contact the office for an annual report prior to applying.
Requirements: Nonprofits in the New York and Philadelphia areas are eligible.
Restrictions: Grants do not support proposals for mental retardation, physical disabilities, drugs/alcohol, endowments, capital projects, annual fundraising drives, or international activities.
Geographic Focus: New York; Pennsylvania
Samples: Disability Advocates (New York, NY)—for advocacy and litigation services for mentally ill inmates in New York State, $75,000 (2005); Jewish Board of Family and Children's Services (New York, NY)—for early childhood group-therapy programs at two Head Start centers, $100,000 (2005); VIP Community Services (New York, NY)—for salary support of a mental-health social worker who assists addicted and recovering clients, $50,000 (2005); Baruch College Fund (New York, NY)—to expand services at the Counseling Center, which treats immigrant, low-income students who are emotionally disturbed, $50,000 (2005).
Amount of Grant: $10,000-$50,000
Date(s) Application is Due: Apr 1; Jul 1; Dec 1
Contact: Henry van Ameringen, President, (212) 758-6221
Internet: http://www.vanamfound.org
Sponsor: Van Ameringen Foundation
509 Madison Avenue, Suite 2010
New York, NY 10022

Van Courtlandt Elliott Prize 5661

This prize is awarded annually in North America for a first-published article in the field of medieval studies. The applicant must submit three copies of offprints of the article, which must be at least five pages in length and have a publication date of two years before the prize award date (the April following the deadline date).
Requirements: Authors submitting articles for consideration must have their primary residence in North America. Author must also submit a statement that the submission is his or her first published article in the medieval field.
Samples: Carol Symes—for her article, The Appearance of Early Vernacular Plays: Forms, Functions, and the Future of Medieval Theater, Speculum 77 (2002), 778-831, $500.
Amount of Grant: $500

Date(s) Application is Due: Oct 15
Contact: Program Contact, (617) 491-1622; fax: (617) 492-3303; email: speculum@medievalacademy.org
Internet: http://www.medievalacademy.org/grants/gradstudent_elliottprize.htm
Sponsor: Medieval Academy of America
104 Mt Auburn Street, 5th Fl
Cambridge, MA 02138

Vancouver Foundation Grants and Community Initiatives Program 5662
Vancouver Foundation's Grants and Community Initiatives program provides Community Impact Grants in nine fields of interest: Animal Welfare; Arts and Culture; Children, Youth and Families; Education; Environment; Health and Social Development; Health and Medical Research; Youth Homelessness; Youth Philanthropy. The Vancouver Foundation has a two-stage application process. The first stage is a letter of intent. This consists of a brief proposal to determine basic suitability. The second stage is submission of a full grant application by a specific date. The letter of intent may be submitted online. You will be notified, if you are invited to submit a full grant application. The following is a list of deadline dates for the letters of intent and applications. Animal Welfare, Arts and Culture, Children, Youth and Families, Education, Environment, and Health and Social Development grants have a deadline date of: Letter of Intent, June 18, Application, August 6. Youth Homelessness grants: Letter of Intent, August 6, Application, September 24. Youth Philanthropy grants: Letter of Intent, September 17, Application, October 1. Health and Medical Research grants deadline dates are undetermined, at this time.
Requirements: Under our Grants and Community Initiatives program, Vancouver Foundation funds eligible applicants, which include: registered charities and qualified donees under the Income Tax Act; First Nations that may be considered a public body performing a function of government.
Restrictions: The Foundation does not fund the following: 100% of a proposal's costs; an organization's ongoing operational or core expenses; retroactive funding, or for any expenses to be incurred prior to the Foundation's decision date; debt retirement or reserves; mortgage pay-downs; conferences, competitions, symposia, annual events, or travel to/attendance at such events; office equipment and furniture; activities of religious organizations that serve primarily their membership and/or their direct religious purposes, unless the community at large will benefit significantly; sabbatical leaves, student exchanges; medical facilities or equipment; league-based sports and recreation programs; library acquisitions and construction; school construction; publication of books; academic or dissertation research; research with human subjects; endowment matching grants; capital requests, with the exceptions noted above; activities previously supported through government funding; bursaries, scholarships and awards; school trips, annual events or equipment; grants to individuals or businesses; funding requests if reports on any previous grants are overdue.
Contact: Mark Gifford, Director, Grants and Community Initiatives; (604) 629-5362; email: markg@vancouverfoundation.ca; Catharine Hume, Grants and Community Initiatives; (604) 629-5361; email: catharineh@vancouverfoundation.ca
Internet: http://www.vancouverfoundation.ca/grants/communityinitiatives.htm
Sponsor: Vancouver Foundation
555 West Hastings Street, Suite 1200
Vancouver, BC V6B 4N6 Canada

Vanguard Public Foundation Grant Funds 5663
Vanguard provides funding to groups in northern California under four categories; the Social Justice Fund, the Community Institution Building Program, the Technical Assistance and Capacity Building Program, and the Social Justice Sabbatical Fund. Types of support include seed money, technical assistance, and operating expenses. Priority is given to projects without access to traditional funding sources because they may be thought to be risky, controversial, or of low priority. Vanguard strongly encourages coalitions that emphasize joint strategies and projects. Visit the web site for application guidelines. The next deadline to submit a proposal for the Social Justice Fund is October 14 and for a Social Justice Sabbatical is December 15. Annual deadline dates may vary; contact program staff for exact dates.
Requirements: Vanguard funds new or existing organizations involved in direct organizing or advocacy that are based in northern California (i.e., all counties north of Monterey). At the time a grant is made, an organization must be tax-exempt or have a fiscal sponsor.
Restrictions: Vanguard does not fund capital campaigns or equipment purchase; organizations involved in direct services, research, or education, unless the proposed project clearly has an organizing component and cannot be supported within the general program budget; production costs for film; out-of-state travel one-time conferences or events not integrally related to ongoing organizing; organizations with access to traditional funding sources and budgets of over $200,000 (unless the project for which funds are requested is unlikely to attract support because of its risky or controversial character); and costs already incurred.
Geographic Focus: California
Samples: Global Fund for Women (CA)—for general operating support, $11,000.; Cannery Workers' Assoc (CA)—for community development, $8000.; National Coalition of Education Activists (CA)—for general operating support, $8075.
Date(s) Application is Due: Oct 14; Dec 15
Contact: Grants Director, (415) 487-2111; fax: (415) 487-2124; email: danielle@vanguardsf.org or vpf@vanguardsf.org
Internet: http://www.vanguardsf.org
Sponsor: Vanguard Public Foundation
383 Rhode Island Street, Suite 301
San Francisco, CA 94103

VanLobenSels/RembeRock Foundation Grants 5664
The foundation provides grants to organizations that serve vulnerable and underserved populations in Northern California. Proposals will be considered for project and program start-up funding for new ideas that demonstrate the potential for improvement and enhancement of life in our communities; projects and programs designed to provide fair treatment and equal access to the law for disadvantaged residents and newcomers; projects and programs targeted toward underserved and at-risk populations; and projects and programs that enhance access to services for underserved populations. Types of support include project development, conferences and seminars, publication, seed money, research, and matching funds. Proposals are considered every six to eight weeks, and grants are awarded throughout the year.
Requirements: Organizations must be located in California.
Restrictions: The foundation does not entertain applications that require medical, scientific, or other technical knowledge; request funding for an individual or family; or request funding for scholarships or stipends.
Geographic Focus: California
Samples: Black Women Organized for Educational Development (Oakland, CA)—for the Mentor Leadership program, $17,000.; Ctr on Juvenile and Criminal Justice (San Francisco, CA)—for expansion of the entrepreneurship center, $10,000.; Coalition for Alternatives in Mental Health (Berkeley, CA)—for trainee programs, $5000.
Contact: Toni Rembe, President, (415) 512-0500; fax: (415) 371-0227; email: info@vlsrr.org
Internet: http://www.vlsrr.org/frameset_guidelines.html
Sponsor: VanLobenSels/RembeRock Foundation
131 Steuart Street, Suite 301
San Francisco, CA 94105

Vernon K. Krieble Foundation Grants 5665
The foundation awards grants worldwide in its areas of interest, including arts and performing arts, Canada, civil rights, education and higher education, government/public administration and public affairs, hospitals, leadership development, and youth programs. Limited funds are available to applying organizations; most grants are awarded to preselected organizations.
Requirements: Nonprofit organizations are eligible.
Samples: American Foreign Policy Council, (Washington, DC)—for general support, $10,000 (2003); Connecticut Childrens Medical Ctr (Hartford, CT)—for general support, $5000 (2003); U of Hartford, Hartford Art School (West Hartford, CT)—for general support, $2000 (2003); American Enterprise Institute for Public Policy Research (Washington, DC)—for general support, $1000 (2003).
Amount of Grant: $1000-$5000 average
Date(s) Application is Due: Jun 1; Dec 1
Contact: Helen Krieble, President, (303) 758-3956; fax: (303) 488-0068
Sponsor: Vernon K. Krieble Foundation
1777 S Harrison Street, Suite 807
Denver, CO 80210

VFH African American History Mini Grants 5666
This grant program seeks to increase public understanding of African American history and culture; to support research and documentation of Virginia's African American historic sites; to strengthen the institutions that interpret African American history in the state; and to provide a solid intellectual foundation for African American heritage tourism in every region of the Commonwealth. Grants may be used to undertake new research; develop interpretive plans; create new promotional or educational materials; train interpretive staff and volunteers; plan and carry out new education programs such as lectures, community forums, or radio programs; and plan and carry out new educational programs that will continue beyond the grant period such as permanent exhibits. Especially encouraged are proposals on the history of African American education in Virginia.
Geographic Focus: Virginia
Amount of Grant: $3000 maximum
Date(s) Application is Due: Feb 1
Contact: David Bearinger, (434) 924-3296; fax: (434) 296-4714; email: aahv@virginia.edu
Internet: http://www.web.virginia.edu/vfhdb/aahv/grants.html
Sponsor: Virginia Foundation for the Humanities
145 Ednam Drive
Charlottesville, VA 22903-4629

VFH and Institute on Violence and Survival Resident Fellowships 5667
The Institute on Violence and Survival conducts research and outreach on violence and its survival. In addition to more academic applications, the institute funds projects to develop workshops, curriculum units, and other outreach materials to disseminate new research findings—and a humanities perspective—to those who work with survivors and perpetrators of violence. Fellowships are awarded for one semester, a full year, or for the summer session only. Fellows have university faculty status while in residence. Former VFH fellows must wait three years before applying for another fellowship. Former fellows are welcome to apply for space (without stipend) at any time. Application guidelines are available online.
Requirements: Fellowship opportunities are open to independent and affiliated scholars, professionals, and others working on projects in the humanities. Postdoctoral applicants are strongly encouraged to apply for projects other than dissertation revisions.
Amount of Grant: $15,000 per semester
Date(s) Application is Due: Oct 15

Contact: Virginia Foundation for the Humanities, (434) 924-3296; fax: (434) 296-4714; email: aspencer@virginia.edu
Internet: http://www.virginiafoundation.org/institutes/vands/featuredfellow/fellowships.htm
Sponsor: Virginia Foundation for the Humanities and Public Policy
145 Ednam Drive
Charlottesville, VA 22903-4629

VFH Humanities in the Public Interest Fellowships 5668
The VFH supports scholarly work on the humanities in the public interest. The fellowship program offers time, space, and resources to scholars bringing the humanities to visibility—applying the tools of history, philosophy, ethics, cultural studies, and literary criticism to matters of public concern. Applications are available on the Web site.
Requirements: Independent and university-affiliated scholars and other professionals are encouraged to apply.
Restrictions: The foundation does not support work toward a degree.
Amount of Grant: $15,000 maximum stipend per semester
Date(s) Application is Due: Oct 15
Contact: Virginia Foundation for the Humanities, (434) 924-3296; fax: (434) 296-4714; email: aspencer@virginia.edu
Internet: http://www.virginiafoundation.org/fellowships/essay.html
Sponsor: Virginia Foundation for the Humanities and Public Policy
145 Ednam Drive
Charlottesville, VA 22903-4629

VFW Ladies Auxiliary Cancer Research Postdoctoral Fellowships 5669
The VFW Ladies Auxiliary offers three postdoctoral research fellowships annually so that recipient researchers can devote their full time for one year to the challenge of cancer research. The fellowship is not renewable. The grant is not to be used for lab equipment or supplies; it is a stipend for personal living expenses of the accredited researcher. The application must be filed on appropriate application form and must describe the proposed research project that the nominee will be conducting, including abstract and brief outline of methods, procedures, and significance. The proposed research must be the principal professional responsibility of the recipient during the stated period of the award. The application must be accompanied by a letter of recommendation.
Requirements: Nominees must be U.S. citizens who are qualified in a specific cancer-related specialty and must have earned a doctoral level degree by the time the fellowship is activated (July 1). Colleges and research institutions are invited to nominate researchers; institutions may nominate more than one person.
Amount of Grant: $50,000
Date(s) Application is Due: Mar 1
Contact: Judy Millick, Program Contact, (816) 561-8655; fax: (816) 931-4753; email: jmillick@vfw.org or info@ladiesauxvfw.com
Internet: http://www.ladiesauxvfw.com/cancer%20fellowships.html
Sponsor: Veterans of Foreign Wars Ladies Auxiliary
406 W 34th Street
Kansas City, MO 64111

Vice Admiral Edwin B. Hooper Research Grants 5670
The center offers grants to assist scholars in the research or writing of books or articles by helping to defray the costs of travel, living expenses, and document duplication related to the research process. In accepting the awards, scholars must commit themselves to work on a study of U.S. naval history intended for publication. The number of awards depends on availability of funds. Applications are available on the Web site.
Requirements: Applicants must be citizens of the United States and hold a PhD degree from an accredited university, awarded no later than January 31 of the award year, or equivalent attainment as a published author.
Samples: Dr. Roger Dingman, U of Southern California—for an analysis of the U.S. Navy's presence in Japan from 1945 to the present.; Dr. David Skaggs, Professor Emeritus, Bowling Green U—to complete his book, Freshwater Commodores.
Amount of Grant: $2500
Date(s) Application is Due: Feb 28
Contact: Senior Historian, Washington Navy Yard, (202) 433-7230
Internet: http://www.history.navy.mil/prizes/prize2.htm
Sponsor: Naval Historical Center
805 Kidder Breese SE
Washington, DC 20374-5060

Victor E. Speas Foundation Grants 5671
The foundation awards grants to nonprofit organizations in the Kansas City, MO, area to support health care, particularly medically related higher education. Funding also supports nonprofits serving the health care needs of the disabled, the elderly, and youth, such as substance abuse, AIDS, and services for children and women. Types of support include seed grants, capital campaigns, building construction/renovation, program development, general operating support, matching grants, and emergency funds.
Requirements: Nonprofit organizations in Missouri counties, including Jackson, Clay, Platte, and Cass, are eligible.
Geographic Focus: Missouri
Amount of Grant: $150-$225,000 range
Contact: David Ross, Senior Vice President, Bank of America, (800) 357-7094
Sponsor: Victor E. Speas Foundation
P.O. Box 831041
Dallas, TX 75283-1041

Viets Medical Student Myasthenia Gravis Research Fellowships 5672
This fellowship is for the purpose of initiating and furthering the interests of potential investigators from colleges and medical schools in the area of causative or treatment research in myasthenia gravis (MG), including clinical investigations as well as laboratory studies concerned with research on MG, underlying mechanisms of neuromuscular transmission, and the pathogenesis of MG. A qualified candidate desiring training in MG-related research should first find a scientist with an established research program who will agree to supervise her/his training and the research to be carried out. The research can be basic or clinical and conducted in the student's own medical school or in one close to the student's residence. The number of fellowships awarded each year is limited by the amount of funds available. Application guidelines are available online.
Requirements: Submit eight copies of the following: cover letter, summary of the research proposal and its association to Myasthenia Gravis or related neuromuscular conditions, curriculum vita of applicant and sponsoring preceptor, and letter of recommendation from preceptor that indicates acceptance of the candidate and outlines the proposed work plan for the research study.
Samples: Dan Chao, Stanford U, under the direction of Kang Shen, MD, PhD (CA)—for Elucidating Molecular Mechanisms of Synapse Specificity, $3000 (2004); Sidharth Ghosh, U of Texas Medical Branch at Galveston, under the direction of Premkumar Christadoss, MD (TX)—for Role of NK Cells in Secondary Immune Response to Acetylcholine Receptor in Experimental Autoimmune Myasthenia Gravis, $3000 (2004); Joshua Mozes, U of California at Davis, under the direction of David Richman, MD (CA)—for Efficient Production of Recombinant Anti-Acetylcholine Receptor Monoclonal Antibodies, $3000 (2004).
Amount of Grant: $3000
Date(s) Application is Due: Mar 15
Contact: Fellowships Administrator, (800) 541-5454 or (651) 917-6256; fax: (651) 917-1835; email: mgfa@myasthenia.org
Internet: http://www.myasthenia.org/research
Sponsor: Myasthenia Gravis Foundation
1821 University Avenue W, Suite S256
Saint Paul, MN 55104

Viking Children's Fund Grants 5673
The fund supports the needs of children in Minnesota and the upper Midwest. Grants support research at the University of Minnesota Department of Pediatrics into the major diseases and disorders of childhood; and nonprofits that focus efforts in the areas of health education and family services supporting the well-being of children in the upper Midwest.
Requirements: Nonprofit organizations in Minnesota and the upper Midwest, including Iowa, North Dakota, South Dakota, and Wisconsin, are eligible.
Geographic Focus: Iowa; Minnesota; North Dakota; South Dakota; Wisconsin
Amount of Grant: $500-$5000
Date(s) Application is Due: Apr 1
Contact: Patrick Leopold, Viking Children's Fund Coordinator, (952) 828-6500; email: info@vikings.com
Internet: http://www.vikings.com/commchildfund.html
Sponsor: Minnesota Vikings Football Club
9520 Viking Drive
Eden Prairie, MN 55344

Viola W. Bernard Foundation Grants 5674
The foundation supports innovative programs that address the interplay between social conditions and the psychological health of children and families. Grants are generally limited to projects or programs for children and young people with a mental health component. Recognizing that new social situations create new psychological needs, the foundation anticipates grant requests for mental health programs that evolve out of emerging issues and changing needs, requiring innovative approaches (i.e., the transition from welfare to work, from dependency to independence, from delinquency and rebellion to cooperation). Opportunities for grant support continuously emerge around foci of social conflict that often involve the police, school support, and social agencies. Only one proposal from an organizational entity will be funded at a given time. There is a preference for projects and programs in the New York City tri-state area, where site visits are practicable if deemed desirable. Applications may be submitted throughout the year and are reviewed biannually, in the fall and spring.
Restrictions: Applications from individuals will not generally be considered, although in rare cases an exception may be granted. faxed and email applications will not be accepted.
Amount of Grant: $200,000 total annually
Contact: Dr. Perry Ottenberg
Internet: http://www.violawbernardfoundation.org
Sponsor: Viola W. Bernard Foundation
210 W Washington Sq, Suite 750
Philadelphia, PA 19106

Virginia A. Hodgkinson Research Prize 5675
In partnership with the Association for Research on Nonprofit Organizations and Voluntary Action, the research prize is awarded annually in recognition of a book that informs nonprofit policy and practice. Recipients are researchers and practitioners who work in or with the voluntary sector, and through their work, have achieved local, regional, national, or international impact and recognition. Submissions from researchers and practitioners representing any discipline, nonprofit field, or organization are welcome. Multidisciplinary approaches, submissions from young scholars and practitioners

working in collaboration with others, and from researchers and practitioners at the start of their careers are encouraged. Two research prizes will be awarded.
Requirements: Each submission will be judged on the contribution the research makes to policy and practice. The research article, report, or book must have been published within three calendar years prior to the year of the award. Books can be considered more than once. Entries must be published by a university press or commercial publishing house. Submissions from scholars representing any discipline are welcome. In addition, submissions are encouraged from young scholars, practitioners working in collaboration with university-based researchers, and scholars working on multi-disciplinary approaches. Self-nominations will be accepted.
Restrictions: Prior prize recipients are ineligible for a five-year period after their award is received.
Samples: 2006 - Francie Ostrower, Attitudes and Practices Concerning Effective Philanthropy; 2005 - (First Prize) Pippa Norris and Ronald Inglehart, Sacred and Secular: Religion and Politics Worldwide. (Honorable Mentions: Richard Chait, Bill Ryan, and Barbara Taylor, Governance as Leadership: Reframing the Work of Nonprofit Boards, and Steven Dubb and Gar Alperovitz of The Democracy Collaborative, Building Wealth: The New Asset-Based Approach to Solving Social and Economic Problems.
Amount of Grant: $2000 first prize; $1000 second prize
Date(s) Application is Due: Jun 30
Contact: Thomas Jeavons, Executive Director, Association for Research on Nonprofit Organizations and Voluntary Action (ARNOVA); (317) 684-2120; email: tjeavons@arnova.org
Internet: http://www.independentsector.org/programs/research/vah_research_prize.html
Sponsor: Independent Sector
1200 Eighteenth Street, NW, Suite 200
Washington, DC 20036

Virginia and Derrick Sherman Emerging Scholar Lecture Grant 5676
This lecture series, an annual event during the week marking United Nations Day, is designed to provide a forum for promising new scholars (untenured assistant professors or young researchers) to present their perspectives on current issues in the fields of modern history, politics, and international relations to the university community and to the public. Topic for the current year is Transition to Democracy in Eastern Europe: Lessons from Recent History. The university will subsequently publish the lecture. The Sherman scholar also will meet with graduate and undergraduate classes, share his or her expertise with faculty members in history and political science, and be available to local media. The grant includes an honorarium to the scholar.
Amount of Grant: $5000 honorarium
Date(s) Application is Due: May 30
Contact: Dr. Susan McCaffray, Department of History, (910) 962-3308
Internet: http://www.uncw.edu
Sponsor: University of North Carolina Wilmington
601 S College Road
Wilmington, NC 28403

Virginia Environmental Endowment Kanawha and Ohio River Valleys Program 5677
Currently limited to the Kanawha and Ohio River Valleys of Kentucky and West Virginia, this program supports research, education, and community action on water quality and the effects of water pollution on public health and the environment. Emphasis is given to collaboration by citizens, government, and industry to develop effective public policies and to increasing public participation in water quality monitoring and watershed management.
Requirements: Constructive, result-oriented projects that are conducted by existing nonprofit organizations are funded. Activities that unite business, government, and civic interests for environmental improvement are encouraged, as are projects that serve as models for other communities. Organizations must be in Kentucky or West Virginia.
Geographic Focus: Kentucky; West Virginia
Samples: Kentucky Waterways Alliance (Munfordville, KY)—for a statewide citizens' group that works to monitor and protect water quality and to raise public awareness about related issues, $40,000 over two years (2004); West Virginia Rivers Coalition (Elkins, WV)—to develop scientifically sound criteria for determining the level of nutrients in West Virginia rivers, $17,200 (2004).
Date(s) Application is Due: Jun 15
Contact: Gerald McCarthy, Executive Director, (804) 644-5000; email: info@vee.org
Internet: http://www.vee.org/programs.cfm
Sponsor: Virginia Environmental Endowment
P.O. Box 790, 1051 E Cary Street, 3 James Ctr, Suite 1400
Richmond, VA 23218-0790

Virginia Environmental Endowment Mini Grant Program 5678
Mini grants are awarded in the areas of environmental education, and water protection quality. The focus of environmental education grants is to increase training in environmental education for educators and school administrators and to expand the use of existing educational resources (curricula, multi-media programs, publications, field studies) in schools and community programs. Projects must focus on water quality, land use, and/or sustainable communities; and must include training for teachers/project leaders and service learning activities that engage students and community members in research to solve environmental problems. Water quality protection grants are awarded to increase public involvement in the restoration and protection of Virginia's water resources through water quality testing and monitoring, pollution prevention, and public education. Matching funds from other sources are usually required. Mini grants may include support for training, research, data collection and analysis, monitoring equipment, supplies, curriculum materials, publications, printing, and promotion. Requests for computer and multi-media equipment and instructional software must be accompanied by specific plans for related teacher training, curriculum integration, and long-term maintenance. Guidelines are available online.
Requirements: Primary, intermediate, and secondary schools (public and private), as well as nongovernmental, nonprofit community organizations in Virginia, are eligible to apply.
Restrictions: Local, state, and federal government agencies and programs are not eligible. Grants will not be provided for salaries or benefits of school personnel or for field trips outside Virginia. Grants will not be awarded for political action programs, capital drives and building funds, boats, scholarships, endowments, lawsuits and litigation, or overhead costs.
Geographic Focus: Virginia
Samples: Fishburn Park Elementary School (Roanoke, VA)—for curriculum programs, teacher training, and habitat design in wetlands and water protection, $4425 matched by $3955.; Turner Ashby High School (Bridgewater, VA)—for water quality education programs, $3483 matched by $4347.
Amount of Grant: $1000-$5000
Date(s) Application is Due: Jun 15; Dec 1
Contact: Gerald McCarthy, Executive Director, (804) 644-5000; fax: (804) 644-0603
Internet: http://www.vee.org/programs.cfm
Sponsor: Virginia Environmental Endowment
P.O. Box 790, 1051 E Cary Street, Suite 1400
Richmond, VA 23206-0790

Virginia Environmental Endowment Virginia Program Grants 5679
The program focuses specifically on improving the quality of the environment for the direct benefit of the citizens of the Commonwealth of Virginia. Emphasis is given to the following programs: pollution prevention, natural resources conservation, sustainable communities, and environmental education; to enable citizens to identify local environmental problems, develop creative solutions, and share results with other communities; and to support legal research and education on other policies and laws that affect Virginia's environment. Types of support include operating budgets, continuing support, seed grants, matching funds, scholarship funds, project development, research, conferences and seminars, and publications. Applicants must submit four copies of full proposals for consideration. A copy of the annual report may be requested, wherein explicit application instructions are listed.
Requirements: Eligible for funding are constructive, result-oriented projects that are conducted by existing Virginia nonprofit organizations. Activities that unite business, government, and civic interests for environmental improvement are encouraged, as are projects that serve as models for other communities.
Geographic Focus: Virginia
Samples: Nature Conservancy-Virginia Program Office (Charlottesville, VA)—for its campaign to increase financing for Virginia's natural resources, $90,000 (2005); Shenandoah Valley Pure Water Forum (Harrisonburg, VA)—to help implement a user-friendly resource on technical and scientific data related to water quality and quantity in the Shenandoah Valley, $24,344 (2005); Rivanna Conservation Society (Palmyra, VA)—to protect Virginia's Rivanna River and its tributaries and to increase the society's membership and member participation, $15,000 over two years (2005).
Date(s) Application is Due: Jun 15; Dec 1
Contact: Gerald McCarthy, Executive Director, (804) 644-5000; email: info@vee.org
Internet: http://www.vee.org/programs.cfm
Sponsor: Virginia Environmental Endowment
P.O. Box 790, 1051 E Cary Street
Richmond, VA 23206-0790

Virginia G. Piper Charitable Trust Grants 5680
The trust awards grants to nonprofit organizations, primarily serving residents of Maricopa County, AZ, that have actively operated as a Section 501(c)3 organization for at least three years, in its areas of interest, including early childhood, youth, and arts and culture. The trust seeks proposals that link health, medical research and/or educational programs specifically to the needs of children, youth, and the elderly, as well as faith-based organizations that can use their considerable capacity and presence to address needs in these areas. Types of funding support include program grants, capital grants, endowment grants, challenge grants, and single- and multi-year grants. There are no application forms. Letters of inquiry can be submitted at any time.
Requirements: Arizona nonprofit organizations, primarily in Maricopa County, are eligible.
Restrictions: Individuals are not eligible.
Geographic Focus: Arizona
Samples: East Valley Children's Theatre (Mesa, AZ)—for marketing activities related to the theater's move to the new Mesa Arts Center, $12,000 (2005); Saint Mary's Food Bank (Phoenix, AZ)—to enable three local food banks to gather and distribute food donated by a chain of grocery stores, $225,000 (2005); Mount Claret Retreat Ctr (Phoenix, AZ)—to renovate and construct facilities at this small retreat and conference center owned by the Diocese of Phoenix, $250,000 (2005); Mesa Community College (AZ) and Scottsdale Community College (AZ)—to establish centers at which older adults can learn about health and physical fitness, as well as educational, paid, and volunteer opportunities, $533,946 over three years, and $347,010 over two years, respectively (2005).
Amount of Grant: $5000-$2 million; $65,000 median grant

Contact: Karin Bishop, Grants Manager, (480) 948-5853; fax: (408) 348-1316; email: info@pipertrust.org
Internet: http://www.pipertrust.org
Sponsor: Virginia G. Piper Charitable Trust
6720 N Scottsdale Road, Suite 350
Scottsdale, AZ 85253

Virginia Historical Society Research Fellowships 5681
The society offers fellowships of up to four weeks a year to help defray travel expenses for commuting researchers who wish to conduct research using collections of the society. Awards include the Andrew W. Mellon Research Fellowships, the Betty Sams Christian Fellowships in business history, the Frances Lewis Fellowships in women's studies, and the Reese Fellowships in American Bibliography and the History of the Book in the Americas. Recipients are expected to work on a regular basis in the society's reading room. Applicants whose research promises to result in significant publication, including the Virginia Magazine of History and Biography, will receive primary consideration. Applicants should send an original and three copies of the following: cover letter, resume, two letters of recommendation (which may be sent separately), and a description of their research project not longer than two double-spaced pages that also states the length of the award requested.
Requirements: Applications are welcome from doctoral candidates who reside outside of the Richmond metropolitan area.
Restrictions: Undergraduates, master's students, and graduate students not yet admitted to PhD candidacy are not eligible.
Geographic Focus: Virginia
Amount of Grant: $150-$500 per week
Date(s) Application is Due: Feb 1
Contact: Nelson Lankford, (804) 342-9672; fax: (804) 355-2399; email: nlankford@vahistorical.org
Internet: http://www.vahistorical.org/research/fellowships.htm
Sponsor: Virginia Historical Society
428 N Boulevard
Richmond, VA 23220

Virginia Tech Dissertation Year/Postdoctoral Fellowship for Academic Diversity 5682
The program seeks to recruit and mentor new minority faculty members. The program provides young professionals with an opportunity to be mentored by experienced faculty, to make progress on the dissertation, or to advance research past the dissertation. ABD fellows will teach a course during the term of the fellowship, usually in the second semester, and will deliver a minimum of one seminar on their research. Applications are due the first week of January. Applications received after that date will be considered only if openings become available. Applicants will be notified by the first week of March.
Requirements: Minority doctoral students are eligible. This fellowship is open to doctoral students who have completed all degree requirements except the dissertation (ABD) or who have recently (within the last year) graduated. The university seeks individuals who have been or who are currently under-represented in their disciplines. Applicants should be academically strong and seeking a career in academe. Applicants must be able to demonstrate the potential for success as a faculty member and be a U.S. citizen.
Amount of Grant: $25,000
Contact: Virginia Tech Graduate School, (540) 231-4558; fax: (540) 231-3714; email: grads@vt.edu
Internet: http://www.grads.vt.edu/financial/fellowship_scholarship/index.html
Sponsor: Virginia Tech University
213 Sandy Hall (0325)
Blacksburg, VA 24061

Virginia Tuition Assistance Grant Program 5683
Grants are awarded to any full-time undergraduate or graduate (including medical or law) student, regardless of financial need, who is a domiciliary resident of Virginia enrolled at an eligible private college or university in Virginia. Completed applications must be submitted to the financial aid office of the institution on, or before, July 31 prior to the fall semester of enrollment. Applications submitted after July 31, but no later than December 1, will be considered for an award only if funds are available.
Requirements: Applicant must be a domiciliary resident of Virginia; enrolled full-time as an undergraduate, graduate, or professional student; and enrolled at an eligible institution pursuing an eligible degree that is not in religious training or theological education.
Geographic Focus: Virginia
Amount of Grant: $2500 undergraduate; $1900 graduate
Date(s) Application is Due: Jul 31; Dec 1
Contact: Elizabeth Waddy, Student Aid Program Specialist, (804) 225-2632 or (804) 225-2600; fax: (804) 225-2604; email: ElizabethWaddy@schev.edu
Internet: http://www.schev.edu/students/factsheetvtag.asp?from=schevs
Sponsor: State Council of Higher Education for Virginia
101 N 14th Street
Richmond, VA 23219

Viscount Bennett Fellowship 5684
One fellowship per year is granted for study in the field of legal education, training, and ethics tenable at an institution in Canada, the United States, England, or France. Fellowships are awarded for one year and are not renewable. Application forms may be obtained from the association.
Requirements: Canadian citizens who have graduated from an approved law school in Canada or who are in the final year of a law school program are eligible.
Samples: Talha Syed (Camrose, AB)—award recipient.; Pascale Fournier (Sainte-Foy, QC)—award recipient.
Amount of Grant: $C25,000
Date(s) Application is Due: Nov 15
Contact: Stephen Hanson, Senior Director of Communications, (613) 237-2925; fax: (613) 237-0185; email: stephenh@cba.org
Internet: http://www.cba.org/CBA/Awards/viscount_bennett
Sponsor: Canadian Bar Association
865 Carling Avenue, Suite 500
Ottawa, ON K1S 5S8 Canada

Visiting Fellowship in Irish Studies at National University Galway 5685
A fellowship of four months or more at National University at Galway, Ireland, is awarded to a scholar normally resident in the United States whose work deals with Irish studies. The fellow must give one faculty seminar on the subject of his or her research at the completion of the fellowship. The fellowship includes transatlantic travel, office accommodations, and other privileges appropriate to the status of a visiting faculty member. Contact the office for appropriate forms.
Requirements: Fellowships are open to scholars of all disciplines relating to Irish studies.
Amount of Grant: $13,000 stipend
Date(s) Application is Due: Dec 31
Contact: Grants Administrator, (973) 605-1991; email: info@iaci-usa.org
Internet: http://www.iaci-usa.org/fellowship.html
Sponsor: Irish-American Cultural Institute
1 Lackawanna Pl
Morristown, NJ 07960

Visiting Research Fellowships at the University of Edinburgh 5686
Approximately 15 postdoctoral fellowships are offered annually to promote advanced research within the field of the humanities, especially literature, language, linguistics, history, art, archaeology, and philosophy. Fellows study and research at the Institute for Advanced Studies in the Humanities for periods from two to six months, nonrenewable. Use of study and access to libraries are provided. Forms are available and should be completed and returned no later than the end of November each year. Fellows undertake individual research programmes approved by the Fellowship Election Committee; fellows make public presentations of their work, and are expected to publish the results in due course.
Requirements: Candidates must be scholars of established reputation or younger scholars holding a doctorate. All nationalities may apply.
Restrictions: A candidate for a doctorate should not apply.
Date(s) Application is Due: Dec 1
Contact: Secretary, Institute for Advanced Studies in the Humanities, 44-0131-650 4671; fax: 44-0131-668 2252; email: iash@ed.ac.uk
Internet: http://www.ed.ac.uk/iash/visiting.fellowships.html
Sponsor: University of Edinburgh
Hope Park Sq
Edinburgh, Scotland EH8 9NW United Kingdom

Visiting Scholar in Archaeology Program 5687
This research program for postdoctoral archaeologists at the Center for Archaeological Investigations at SIU supports resident scholars pursuing problem-oriented research in some aspect of prehistory. The visiting scholar organizes and conducts a conference. The 11-month appointment begins in August. Scholar receives office and laboratory space in the center and the use of the research facilities, clerical staff, and the university library. Fringe benefits are included. The visiting scholar is selected on the basis of a five-page proposal outlining the nature and structure of the conference, as well as on the strength of her/his vita and references.
Requirements: Applicants must hold a PhD in anthropology or related discipline with a specialization in archaeology. Degree must be completed by beginning of appointment.
Amount of Grant: $2500 per month
Date(s) Application is Due: Feb 1
Contact: Dr. Heather Lapham, Center for Archaeological Investigations, (618) 453-5031; fax: (618) 453-8467; email: hlapham@siu.edu
Internet: http://www.siu.edu/~cai/vsposition.htm
Sponsor: Southern Illinois University at Carbondale
Center for Archaeological Investigations, 3479 Faner Hall, MC 4527
Carbondale, IL 62901-4527

Visiting Scholars in African-American Studies Research Grants 5688
The W.E.B. DuBois Institute, a center for research on all aspects of African American life, history, and culture at Harvard University, invites applications for visiting scholar research positions. Scholars pursue their own research while interacting on a regular basis with 10 to 15 other visiting scholars working in the field. Weekly colloquia offer fellows the opportunity to share their work with institute colleagues and with interested faculty and students. Fellows have access to computers, the Internet, research assistance, and the extensive research and library resources of Harvard University. The Institute also has the Franklin D. and Wendy F. Raines Library in African American Studies, a small working reference library in African American Studies. Appointments are made for not less than one semester. Applications and guidelines are available online.
Date(s) Application is Due: Jan 20

Contact: Lisa Gregory, (617) 495-8508; fax: (617) 496-2871; email: lgregory@fas.harvard.edu
Internet: http://www.fas.harvard.edu/~du_bois
Sponsor: Harvard University
12 Quincy Street
Cambridge, MA 02138

Volkswagen of America Corporate Contributions Grants 5689
The corporation accepts proposals broadly from programs servicing needs in healthcare, wellness, children, medical research, education and diversity.
Requirements: There are no specific deadlines, however the review committee does meet quarterly to go over proposals received. The committee considers the type of organization, its mission and the intended use of the funds.
Restrictions: Grants are available to 501(c)3 organizations only.
Amount of Grant: $100,000 total per year; $1000-$2500 average
Contact: Tara Jones, email: tara.jones@vw.com; (248) 754-4693
Sponsor: Volkswagen of America, Inc.
3800 Hamlin Road
Auburn Hills, MI 48326

Vulcan Materials Company Foundation Grants 5690
The company foundation supports nonprofit organizations in company-operating areas. Grants support a broad array of programs and projects in the areas of health and welfare, community and economic development, education (including an employee matching grants program), the environment, law and justice, women and children, arts and humanities, minorities, and literacy. Requests will be considered for support of general operating expenses, capital and annual campaigns, scholarship funds, research, continuing support, and seed money. There are no application deadlines.
Requirements: The foundation awards grants only to public charities and units of government, such as public schools and parks. A public charity is any Section 501(c)3 charitable organization that is not a private foundation because it meets one of the three internal Revenue Code definitions: 509(a)1, 509(a)2 or 509(a)3.
Restrictions: The foundation does not fund individuals; organizations outside the United States; telephone or mass-mail appeals, political organizations; testimonial dinners; sectarian religious activities; organizations which have discriminatory practices; or athletic, labor, fraternal and veterans associations. The Foundation generally will not consider requests from organizations located in communities where Vulcan has no operations, offices or employees.
Geographic Focus: Alabama; Arizona; Arkansas; California; Florida; Georgia; Illinois; Indiana; Iowa; Kansas; Kentucky; Louisiana; Maryland; Mississippi; Missouri; New Mexico; North Carolina; Pennsylvania; South Carolina; Tennessee; Texas; Virginia; Wisconsin
Samples: Auburn U Foundation (Auburn U, AL)—for scholarships and fellowships, $45,000.; Family and Child Services (Birmingham, AL)—for youth programs, $20,000.; Metropolitan Arts Council (Birmingham, AL)—for general support, $79,233.
Amount of Grant: $2.8 million total
Contact: Program Contact, (205) 298-3222; email: giving@vmcmail.com
Internet: http://www.vulcanmaterials.com/social.asp?content=guidelines
Sponsor: Vulcan Materials Company Foundation
P.O. Box 385014
Birmingham, AL 35238-5014

W. C. Griffith Foundation Grants 5691
The W. C. Griffith Foundation was established in 1959 by a donation of William C. Griffith, and Ruth Perry Griffith. The Foundation gives primarily in Indianapolis-Carmel, Indiana area. The Foundation provides support primarily for hospitals, health associations, medical and cancer research, the arts, including music and museums, community funds and development, higher, secondary, and other education, family planning services, child welfare, the homeless, the environment, libraries, and Christian religious organizations.
Requirements: Applicants should submit the following: detailed description of project and amount of funding requested; 1 Copy of the proposal.
Geographic Focus: Indiana
Samples: Eiteljorg Museum of American Indians and Western Art, Indianapolis, IN, $25,000 (2007); Providence Cristo Rey High School, Indianapolis, IN, $10,000 (2007);
Amount of Grant: $1,000 - $45,000 range
Contact: Trust Coordinator; (317) 693-2504
Sponsor: W. C. Griffith Foundation
101 West Ohio Street, Suite 1450
Indianapolis, IN 46204-1998

W. David Smith, Jr. Graduate Student Paper Award 5692
This award, sponsored by the DuPont Chemical Company Division of AIChE, is given annually to an individual for published work in the application of computing and systems technology to chemical engineering. The work must have been done by the individual while pursuing graduate or undergraduate studies in chemical engineering. A plaque is included with the award. Submit nominations to Dr. Karen High, School of Chemical Engineering - 423 EN, Oklahoma State University, Stillwater, OK 74078; high@okstate.edu; phone: (405) 744-9112; fax: (405)744-6338
Requirements: All members as well as other interested persons are urged to nominate deserving candidates. All information and supporting documentation are to be included with completed forms; nominating forms and instructions may be obtained from the institute.
Amount of Grant: $1500
Date(s) Application is Due: Apr 15
Contact: AIChE Awards Programs, (212) 591-7096; fax: (212) 591-8898; email: awards@aiche.org
Internet: http://students.aiche.org/honors/awarddetail.asp?AwardID=30
Sponsor: American Institute of Chemical Engineers
3 Park Avenue
New York, NY 10016-5910

W. Frank Blair Award 5693
This award, to encourage excellence in written data presentation, is given annually to a student for the best paper, published or unpublished, pertaining to the natural sciences in the Chihuahuan Desert Region. Papers must be based on original research performed by the author and must be in English, 7500 words or less.
Requirements: Undergraduate and graduate students are eligible for the award. Entries must be accompanied by a letter from a department chairperson that certifies the author's status as a student. Entries must be accompanied by a letter from a department chairperson stating that the entrant is enrolled as a full-time student at the time the entry is submitted.
Amount of Grant: $500
Date(s) Application is Due: Nov 1
Contact: Blair Award Competition, (432) 364-2499; fax: (432) 364-2686; email: manager@cdri.org
Internet: http://www.cdri.org/Blair%20Award/Blair.html
Sponsor: Chihuahuan Desert Research Institute
P.O Box 905
Fort Davis, TX 79734

W.C. Griffith Foundation Grants 5694
The foundation awards grants, primarily in Indiana, to nonprofit organizations in its areas of interest, including arts and culture, cancer and medical research, children and youth services, Christian churches and organizations, community development, education, environmental programs, health organizations and hospitals, higher education, homeless services, human reproduction and fertility, library science and libraries, minorities, museums, music and performing arts, secondary education, and social services delivery. Types of support include building construction/renovation, capital campaigns, and continuing support. There are no application forms or deadlines. The board meets in June and November; proposals should be received in May and October.
Requirements: Nonprofit organizations are eligible. Preference is given to requests from Indianapolis, Indiana.
Amount of Grant: $500-$50,000 range
Contact: Curt Farran, c/o National City Bank of Indiana, (317) 267-7262
Sponsor: W.C. Griffith Foundation
101 W Washington Street, Suite 600E
Indianapolis, IN 46255

W.K. Kellogg Nonprofit Research Fellowship 5695
The Michigan Program of the Nonprofit Sector Research Fund offers a fellowship for doctoral students undertaking research on the nonprofit sector in Michigan. The fellowship recipient also is involved in coordinating the Michigan Program. As part of their applications, prospective fellows should develop a proposal for a specific project that has the endorsement of a host organization which will house them during their fellowship. The project may take a variety of different forms; original research, synthesis of existing research, development of a resource manual for use by practitioners and/or policy makers, etc. The fund does not require that the fellowship project be the same as the applicant's doctoral dissertation. The fellow should spend an average of 10 to 20 hours per week on the fellowship project. The fellow will also spend up to seven hours a week assisting the national Nonprofit Sector Research Fund staff in coordinating the activities of the Michigan Program. Contact the program office or see Web site for program deadlines.
Samples: Jessica T. Kovan, Department of Fisheries and Wildlife, Michigan State U—for Maintaining Motivation and Effectiveness: A Focus on Environmental Staff of Nonprofit Organizations, $20,000.
Amount of Grant: $20,000 maximum, with an additional $5000 for the institution hosting the fellow
Contact: Nonprofit Sector Research Fund, (202) 736-5800; fax: (202) 293-0525; email: nsrf@aspeninst.org
Internet: http://www.aspeninstitute.org/site/c.huLWJeMRKpH/b.612415/k.42C8/Nonprofit_Sector_Research_Fund.htm
Sponsor: Aspen Institute
1 Dupont Cir NW, Suite 700
Washington, DC 20036

W.M. Keck Foundation Fellowships for Young Scholars 5696
The Huntington Library welcomes applications from scholars for awards to help them carry on significant research in the collections. The intent of the program is to encourage outstanding young scholars to pursue their own lines of inquiry (completing the research necessary to revise a dissertation or begin a new project); discuss their work with other scholars in residence and with the permanent research staff of the library; and take advantage of the unique intellectual opportunities in southern California. In selecting persons to receive awards, attention is paid to the value of the project, the ability of

the scholar, and the degree to which special strengths of the library will be utilized. Holders of awards are expected to be in residence throughout the tenure. One month is the minimum period for which awards are made. Applicants for the Keck fellowships are automatically considered for the Huntington fellowships. There is no need to apply for both. Applications are received from October 1 to December 15 of each year for awards within the 12-month period beginning on the following June 1; selection announcements will be made no later than April 1.
Requirements: The fellowships are restricted to young scholars working in the fields of British and American history, literature, and art. Candidates must be nontenured faculty or doctoral candidates at the dissertation stage.
Amount of Grant: $2300 per month
Date(s) Application is Due: Dec 15
Contact: Director of Research, (626) 405-2194; fax: (626) 449-5703; email: cpowell@hungtington.org
Internet: http://www.huntington.org/ResearchDiv/Fellowships.html
Sponsor: Huntington Library and Art Gallery
1151 Oxford Road
San Marino, CA 91108

W.M. Keck Foundation Grants Program **5697**
The foundation concentrates on strengthening studies and programs in accredited universities and colleges and major, independent medical research institutions in the areas of science, engineering, and medicine. Consideration also is given to liberal arts education and elementary and secondary education. Additionally, some consideration, focused in southern California, is given to organizations in the categories of arts and culture, civic and community, health care, early education (0 to five years old) and precollegiate education. Grants have been awarded to support medical research, purchase needed equipment, and renovate laboratory space at premier institutions of higher education and research across the country. The foundation considers applications that reflect institutional priorities as identified by the applicant's CEO. Grants are awarded twice each year, in June and December. Initial application must be by letter of inquiry, not to exceed three pages, accompanied by evidence of tax-exemption; and most recent, full, certified, audited financial statements. Letters of inquiry are accepted year-round. Submission of full proposal is upon invitation of the foundation only after review of the inquiry letter; unsolicited proposals are not accepted at any time.
Requirements: U.S. four-year colleges and universities, medical schools, and major independent research institutions are eligible. Applicant organizations must be classified 501(c)3 and not as a private foundation under 509(a) of the IRS code.
Restrictions: Grants are not considered for individuals, routine expenses, deficit reduction, fund-raising events, dinners or mass mailings, conferences or seminars, publications or films, or public policy research. Support is not given to conduit organizations or organizations lacking permanent tax-exempt ruling determination from the federal government and state of California (for California-based applicants). Electronic applications are not accepted.
Samples: U of Wisconsin at Madison (WI)—for a three-year project involving stem-cell research, $1.25 million (2005); Carnegie Mellon U (Pittsburgh, PA)—for research into the ways in which the brain deciphers language, which may yield advances in the treatment of such neurological disorders as autism and dyslexia, $750,000 (2005); Keck Graduate Institute of Applied Life Sciences (Claremont, CA)—to increase its endowment and augment its general operating fund, $20,000 over seven years (2005).
Amount of Grant: $81.7 million total
Date(s) Application is Due: May 15; Nov 15
Contact: Dr. Sandra Glass or Joan DuBois, Program Vice Presidents, (213) 680-3833; fax: (213) 614-0934; email: info@wmkeck.org
Internet: http://www.wmkeck.org
Sponsor: W.M. Keck Foundation
550 S Hope Street, Suite 2500
Los Angeles, CA 90071

W.W. Smith Charitable Trust Grants **5698**
The trust awards grants regionally in its areas of interest, including basic medical research dealing with cancer, AIDS, and heart disease; scholarship programs for undergraduate students at four-year universities and colleges; and proposals that address specifically food, clothing, and shelter for children and the aged. Resources are concentrated in areas where the most need can be discerned and where government or private assistance has not been available. Types of support include capital projects, challenge/matching grants, general operating grants, program grants, and research grants. Requests for grants for food, clothing, and shelter for children and the aged are considered throughout the year. Application deadlines are June 15 for cancer and AIDS research; and September 15 for heart research. Grants may be for more than one year but not for more than three years.
Requirements: Grant recipients are limited to organizations within the five-county area of Pennsylvania including Buck, Chester, Delaware, Montgomery, and Philadelphia Counties. Organizations must be tax-exempt and not classified as private foundations or private operating foundations.
Restrictions: After three consecutive years of funding, at least two years must elapse before further applications from the organization may be considered.
Geographic Focus: Pennsylvania
Samples: La Salle U (Philadelphia, PA)—for financial aid for students, $98,000.; Philadelphia College of Bible (Langhorne, PA)—for financial aid for students, $58,000.
Amount of Grant: $15,000-$84,000 average
Date(s) Application is Due: Jun 15; Sep 15
Contact: Frances Pemberton, Trust Administrator, (610) 397-1844; fax: (610) 397-1680
Internet: http://www.wwsmithcharitabletrust.org
Sponsor: W.W. Smith Charitable Trust
200 Four Falls Corporate Ctr, Suite 300
West Conshohocken, PA 19428

W.W. Smith Charitable Trust Grants for Medical Research in Cancer, AIDS, and Heart Disease **5699**
Concentration of the grant program in this area is toward basic scientific research that can help uncover and solve some of the mysteries of cancer, heart disease, and AIDS. Types of support include challenge/matching grants, general operating grants, program development grants, research grants, and scholarships. Funding is directed to projects that are new or headed by little-known investigators who are not yet qualified for large government grants. The trust also supports promising research outside the conventional pathways. Grant applications should be submitted no later than June 15 for cancer and AIDS research and no later than September 15 for cardiovascular research. Each application should be accompanied by an audited financial statement for the institution of the researcher. Projects requested should be one to three years in length and should not request funds for equipment alone. A special application form is required.
Requirements: Institutions in the general geographic area of New Jersey, eastern Pennsylvania, and Maryland may submit proposals. Specialized application form and specific guidelines should be obtained from the trust prior to submitting a proposal.
Restrictions: Only 10 percent of indirect costs are allowed on all grant proposals. Funds to support students, fellows, and postdoctoral candidates should be supported by curriculum vita. Applications should be submitted for the work of individual investigators and not for general work of research centers.
Geographic Focus: New Jersey
Samples: Thomas Jefferson U, Jefferson Medical College (Philadelphia, PA)—to endow a chair in cardiac research, $1.5 million.; Coriell Institute for Medical Research—to support cancer research, $165,000.; Fox Chase Cancer Ctr—to support AIDS research, $166,645.
Date(s) Application is Due: Jun 15; Sep 15
Contact: Frances Pemberton, Trust Administrator, (610) 397-1844; fax: (610) 397-1680; email: info@wwsmithcharitabletrust.org
Internet: http://www.wwsmithcharitabletrust.org
Sponsor: W.W. Smith Charitable Trust
200 Four Falls Corporate Center, Suite 300
West Conshohocken, PA 19428

Wabash Center Project Grants **5700**
The Center provides grants for activities that enhance teaching and learning in the fields of religion and theology. Proposals need to demonstrate that the project improves practical application of teaching and learning methods, creates a supportive environment for teachers, promotes a sustained conversation about pedagogy, or supports research and study of pedagogical issues. Projects might involve meetings of scholars across schools, regions, denominations, or subject areas to think abut syllabi, teaching strategies, student learning styles, or teaching as a vocation. Projects might involve a single department experimenting with creative ways to enhance teaching or a doctoral institution helping their students prepare for teaching careers. Project Grants require a full application process and are awarded at three different times during the year for up to $20,000.
Requirements: Full-time faculty members in religion departments of colleges and universities in North America (United States and Canada) and in theological schools accredited by the Association of Theological Schools are eligible to apply for these grants.
Restrictions: Activities not funded include routine curriculum revision or course development; international travel for research, attendance at disciplinary conferences, or the publication of conference papers; or the preparation of textbooks.
Samples: Jon Levisohn, Brandeis University, Waltham, Massachusetts—for Teaching Rabbinic Literature: A Conference on Bridging Scholarship and Pedagogy (2007); Butler Community College, El Dorado, Kansas—Assessing Religion Assessment Tools in Kansas Community Colleges: Colloquium III (2007).
Amount of Grant: $20,000 maximum
Date(s) Application is Due: Jan 1; May 1; Sep 1
Contact: Rev. Dr. Paul O. Myhre, Associate Director; (765) 361-6436 or (800) 655-7117; fax: (765) 361-6051; email: myhrep@wabash.edu
Internet: http://www.wabashcenter.wabash.edu/grants/types.aspx
Sponsor: Wabash Center for Teaching and Learning in Theology and Religion
Wabash College, 301 W. Wabash Avenue, P.O. Box 352
Crawfordsville, IN 47933

Wabash Center Small Project Grants **5701**
The Center provides grants for activities that enhance teaching and learning in the fields of religion and theology. Proposals need to demonstrate that the project improves practical application of teaching and learning methods, creates a supportive environment for teachers, promotes a sustained conversation about pedagogy, or supports research and study of pedagogical issues. Projects might involve meetings of scholars across schools, regions, denominations, or subject areas to think abut syllabi, teaching strategies, student learning styles, or teaching as a vocation. Projects might involve a single department experimenting with creative ways to enhance teaching or a doctoral institution helping their students prepare for teaching careers. Small Project Grants have a short application procedure and can be approved anytime throughout the year for: up to $2,500; and group or individual projects.

Requirements: Full-time faculty members in religion departments of colleges and universities in North America (United States and Canada) and in theological schools accredited by the Association of Theological Schools are eligible to apply for these grants.
Restrictions: Activities not funded include routine curriculum revision or course development; international travel for research, attendance at disciplinary conferences, or the publication of conference papers; or the preparation of textbooks.
Amount of Grant: $2,500 maximum
Contact: Rev. Dr. Paul O. Myhre, Associate Director; (765) 361-6436 or (800) 655-7117; fax: (765) 361-6051; email: myhrep@wabash.edu
Internet: http://www.wabashcenter.wabash.edu/grants/index.html
Sponsor: Wabash Center for Teaching and Learning in Theology and Religion
Wabash College, 301 W. Wabash Avenue, P.O. Box 352
Crawfordsville, IN 47933

Wabash Center Waorkshop and Colloquy Participant Grants 5702
The Center provides funds for projects that enhance teaching and learning in the fields of religion or theology. It seeks to fund: projects that improve the practical application of teaching and learning methods; projects that create supportive environments for teachers; and projects that promote a sustained conversation about pedagogy. In this category, the Center offers particular grant programs to workshop and colloquy participants, such as: the Summer Workshop Fellowship and Grant Program; Colloquy Follow-up Grants; and Workshop on Teaching College Introductory Courses Departmental Grants. Only participants in these workshop or colloquy programs are eligible for these grants.
Requirements: Full-time faculty members in religion departments of colleges and universities in North America (United States and Canada) and in theological schools accredited by the Association of Theological Schools are eligible to apply for these grants.
Restrictions: Activities not funded include routine curriculum revision or course development; international travel for research, attendance at disciplinary conferences, or the publication of conference papers; or the preparation of textbooks.
Contact: Rev. Dr. Paul O. Myhre, Associate Director; (765) 361-6436 or (800) 655-7117; fax: (765) 361-6051; email: myhrep@wabash.edu
Internet: http://www.wabashcenter.wabash.edu/grants/index.html
Sponsor: Wabash Center for Teaching and Learning in Theology and Religion
Wabash College, 301 W. Wabash Avenue, P.O. Box 352
Crawfordsville, IN 47933

Waksman Foundation for Microbiology Grants 5703
The purpose of the foundation is to promote, encourage, and aid scientific investigations, publication, teaching, and research in microbiology. A wide variety of programs will be supported, including awards/prizes, exchange programs, professorships, seed money grants, and grants to support lectureships, courses, unusual publications, etc. in the field of microbiology. Innovative educational programs dealing with microbiologic topics and making use of contemporary communication techniques are of special interest, as are programs concerned with enhancing public awareness of science, including grades K-12 teaching programs that make use of microorganisms. Application is made by a simple letter (not to exceed five pages, plus attachments) to Dr. Waksman, setting forth details of the proposal and a budget. There is no fixed application deadline; proposals are considered approximately four times annually. First proposals should request no more than two years support. Discussion by telephone or email of applicant's plans, before submission of proposal, is encouraged.
Requirements: Grants are awarded to qualified organizations with appropriate interests. To qualify for support, activities must be expected to address a national or an international audience. Innovative educational programs dealing with microbiological topics and making use of contemporary communication techniques are a special focus of interest. Also, programs concerned with enhancing public awareness of science, including K-12 teaching programs that make use of microorganisms, are of special interest.
Restrictions: No grants are made to individuals, or for fellowships, conventional research projects, travel, general or operating support, or capital or endowment funds. Funding is not provided for indirect costs.
Samples: American Society for Microbiology—for the development of microbiology Web site, $20,000.; Dr. Carl Woese, U of Illinois at Urbana-Champaign (IL)—winner of the The Selman A. Waksman Award for excellence in microbiology, $5000.; Yale U (New Haven, CT)—to support the Committee on International Health, $5000.
Amount of Grant: $500-$20,000 typically
Contact: Dr. Frederick Neidhardt, President, (734) 995-2951; fax: (734) 995-9071; email: fcneid@umich.edu
Internet: http://www.waksmanfoundation.org
Sponsor: Waksman Foundation for Microbiology
1150 W Medical Ctr Drive
Ann Arbor, MI 48109-0620

Walker Cowen Eighteenth-Century Studies Manuscript Prize 5704
The university sponsors this biennial prize and awards it to the author of an unpublished book-length manuscript (60,000-150,000 words) dealing with scholarship in 18th-century studies in history, literature, philosophy, or the arts. The award includes publication of the manuscript by the University Press of Virginia. To be eligible, manuscripts should be submitted in triplicate. The winning author will be offered an advance publishing contract and a payment of $2,000. The other half of the prize money will be disbursed upon acceptance of a final revised manuscript for publication.
Requirements: The manuscript must be in English; however, the author may be of any nationality. Only manuscripts that are not under review elsewhere will be considered.
Amount of Grant: $5000
Date(s) Application is Due: Nov 1
Contact: Cowen Award Judges, c/o University Press of Virginia, (434) 924-3361; fax: (434) 982-2655, email: mmm5w@virginia.edu or upressva@virginia.edu
Internet: http://www.upress.virginia.edu/news.html
Sponsor: University of Virginia
P.O. Box 400318
Charlottesville, VA 22904-4318

Wallace E. Stegner Fellowships in Creative Writing 5705
Each year, Stanford University offers 10 Stegner fellowships; five are in fiction and five are in poetry. They are designed for persons of exceptional talent who are determined to be professional writers. Fellows devote their entire time to writing, and thus their fellowship obliges them to register for and to attend only the graduate workshop in fiction or poetry. Fellowships begin in late September at the beginning of the autumn quarter. Application forms must be obtained from the Secretary of Creative Writing, English Department. Applications are available after September 1. Workshop tuition is included in the fellowship. Deadline dates vary; contact program staff for exact date.
Requirements: The fellowship is awarded on the basis of a portfolio of writing; previous publication is not necessary.
Amount of Grant: $56,000 for two years
Contact: Virginia Hess, Program Coordinator, Creative Writing Program, (650) 725-1208; email: vfhess@stanford.edu
Internet: http://www.stanford.edu/dept/english/cw/fellowship.html
Sponsor: Stanford University
Department of English, Stanford University
Stanford, CA 94305-2087

Wallace Foundation Grants 5706
The foundation has three current objectives: strengthen education leadership to improve student achievement; improve after-school learning opportunities; and expand participation in arts and culture. Specifically, the fund seeks to develop effective school leaders to improve student learning, create new standards for community participation in the arts, and provide high-quality informal learning opportunities in communities for children and families. Most grants are awarded as part of funds-initiated programs, and unsolicited proposals are rarely funded. Nevertheless, organizations may submit a one- to two-page letter of inquiry describing the project, the organization, the estimated total for the project, and the portion requiring funding.
Requirements: U.S. nonprofit organizations are eligible.
Restrictions: Grants do not support religious or fraternal organizations; international programs; conferences; historical restoration; health, medical, or social service programs; environmental/conservation programs; capital campaigns; emergency funds or deficit financing; private foundations; or individuals.
Samples: National Governors Assoc (Washington, DC)—for efforts to create and implement ways to improve high school-graduation and college-preparedness rates in 13 states, $1 million (2005); Mayor's Fund to Advance New York City (NY)—to help strengthen out-of-school programs for school children in New York, $12 million over five years (2005); Education Partnership, Providence After School Alliance (RI)—for efforts to create a system of high-quality after-school learning opportunities, $5 million over five years (2004); Stanford U (CA)—for research on how training and education programs can strengthen principals' effectiveness, led by Linda Darling-Hammond and Debra Meyerson of Stanford U and the Finance Project in Washington, $1.245 million over three years (2004).
Contact: Program Contact, (212) 251-9700; fax: (212) 679-6990
Internet: http://www.wallacefoundation.org/about/fundGuides.cfm
Sponsor: Wallace Foundation
2 Park Avenue, 23rd Fl
New York, NY 10016

Wallace H. Coulter Foundation Grants 5707
The foundation awards grants to support research and education in the fields of engineering, medicine, and science. Past grants also have supported humanitarian interests. There are no application deadlines or forms.
Restrictions: Grants do not support religious organizations or individuals.
Samples: Boston U, Biomedical Engineering Dept (MA)—to help advance promising biomedical technologies to the clinical-application phase, $2.9 million (2005).
Amount of Grant: $35,000-$200,000 average
Contact: Wayne Barlin, Vice President, (305) 559-2991; fax: (305) 559-5490; email: wb@whcf.org
Internet: http://www.whcf.org
Sponsor: Wallace H. Coulter Foundation
790 NW 107th Avenue, Suite 215
Miami, FL 33172

Walter H. Capps Center for the Study of Religion and Public Life Research Fellowships 5708
Each year the Capps Center awards several research fellowships worth to advanced PhD students at UCSB with approved dissertation topics related to religion and public life. Topics of particular interest include religion's influence on the public order; moral and ethical issues of broad concern; and public policy in a religiously pluralist world. Students from any division of the university may apply for the fellowships. The annual deadline is May 1. Those receiving the fellowships are designated Junior Fellows of the Center and are expected to participate in the Capps Seminar. To apply, students must

provide a prospectus, a budget detailing how funds would be used, and the names of three references.
Samples: J. Shawn Landres, for Intimacy, Memory, and Ideology at Generation X Seeker Services (2003); Mary Ingram, for Claiming Controversial Science: Competing Religious Discourses on Human Embryonic Stem Cell Research and Human Therapeutic Cloning (2003); William Robert, for The Public Dimension of Religion in the Life and Work of Simone Weil (2003).
Amount of Grant: $2500-$5000
Date(s) Application is Due: May 1
Contact: Wade Clark Roof, (805) 893-2562; email: info@cappscenter.ucsb.edu
Internet: http://www.cappscenter.ucsb.edu/fellow.html
Sponsor: University of California at Santa Barbara
Santa Barbara, CA 93106

Walter J. Jensen Fellowship for French Language, Literature, and Culture **5709**
The Fellowship is awarded for at least six months of study in France. The purpose of the award is to help educators and researchers improve education in standard French language, literature and culture and in the study of standard French in the United States. The fellowship is awarded annually and has a stipend of at least $10,000. Phi Beta Kappa will also cover a single round-trip, economy-class ticket for the recipient to travel to France; some additional support may be available to those with dependents.
Requirements: Candidates must be U.S. citizens under the age of 40 who can demonstrate their career does or will involve active use of the French language. They must have earned a bachelor's degree from an accredited four-year institution with a 3.0 minimum GPA in French language and literature as a major. They must demonstrate superior competence in French, according to the standards established by the American Association of Teachers of French.
Samples: Rachel Corkle, New York University, $10,000 (2008); Thomas Tilden Daniels, University of Pennsylvania, $10,000 (2007); Laura Spear, University of Illinois at Urbana-Champaign, $10,000 (2006).
Amount of Grant: $10,000 plus air fare to France
Contact: Sandra Beasley, Awards Coordinator, (202) 265-3808; fax: (202) 986-1601; email: SBeasley@pbk.org or info@pbk.org
Internet: http://staging.pbk.org/AM/Template.cfm?Section=Jensen_Fellowship&Template=/CM/HTMLDisplay.cfm&ContentID=2648
Sponsor: Phi Beta Kappa Society
1606 New Hampshire Avenue, NW
Washington, DC 20009

Walter Nicolai Prize in Biomedical Gerontology **5710**
This award is given annually for meritorious research by a graduate or medical student in the area of biomedical gerontology. For purposes of the prize, biomedical gerontology is defined as basic biological or clinical research involving normal aging or age-associated diseases. The research should be oriented toward improving the quality of later life. There is one prize plus awards to the first two runners-up to help enable them to attend the annual meeting of the American Aging Association (AGE). Nomination should include four copies each of the following: six- to 10-page (double-spaced) report of the research with name and departmental affiliations in the format used by the AGE journal, sponsor's letter commenting on the significance of the work, and candidate's curriculum vita with current professional address. The sponsor's letter will serve as the nominating document. The recipient of the prize, as well as the two runners-up, will be notified and the prizes will be presented at the annual meeting.
Requirements: A nomination may be made by any member of a society associated with aging research. The sponsor may reside in the United States, Canada, or Mexico.
Amount of Grant: $500 first prize
Contact: Program Contact, (610) 627-2626; fax: (610) 565-9747; email: ameraging@aol.com
Internet: http://www.americanaging.org
Sponsor: American Aging Association (AGE)
110 Chesley Drive
Media, PA 19063

Walter S. Johnson Foundation Grants **5711**
The foundation invests in programs to help children and youth throughout northern California, with particular interest in the San Francisco Bay area and in Washoe County, Nevada. The foundation's grants reflect one or more of three goals: ensure the well-being of children and youth, strengthen public education, and assist young people in the transition to adulthood. To strengthen public education, grants are made for the professional development of educators—especially recruitment, preparation, and support of new teachers. It also helps reform how schools prepare students for college. In supporting youth in their transition to adulthood, grants are made for school-to-career training, teaching, and support activities. The foundation helps nonprofits focusing on positive youth attributes and explores opportunities to work with adjudicated youth. Additional funding interests include improving support services in low-income neighborhoods and helping families in crises, such as homelessness, divorce, or domestic violence situations. From time to time, the foundation makes grants that are outside the regular guidelines. Some reflect special board interests, others may respond to major natural disasters, and still others provide an opportunity to explore potential new areas for future grant making. Types of support include general operating grants, research grants, and special projects. Potential applicants are encouraged to call before submitting an application.
Requirements: Proposals are accepted from non-profit organizations throughout Northern California and Washoe County, Nevada.
Restrictions: The foundation does not consider proposals for grants to individuals or to religious organizations for sectarian purposes. Funds are not provided for capital projects, organizational deficits, or general operating support.
Geographic Focus: California; Nevada
Samples: Forum for Youth Investment Impact Strategies (Washington, DC)—to provide technical assistance to 11 district-community partnerships of a high school reform initiative, $100,000 (2004); Markkula Center/Character Based Literacy Project, Santa Clara University (Santa Clara, CA)—to continue implementing the Character Based Literacy curriculum, document student gains, and develop and implement a business plan, $200,000 (2004); Berkeley Biotechnology Education (CA)— for continuing support of its school-to-career program, $60,000 (2003); Catholic Charities of the East Bay (Oakland, CA)—to create a coordinated continuum of services for youth who are leaving the foster care system, $75,000 (2003); YouthBuild U.S.A (Somerville, MA)—to strengthen its youth-development programs in California, $200,000 (2003).
Amount of Grant: $3000-$50,000 average
Contact: Donna Takasuka, Grants Administrator, (650) 326-0485; fax: (650) 326-4320; email: dtakasuka@wsjf.org
Internet: http://www.wsjf.org
Sponsor: Walter S. Johnson Foundation
525 Middlefield Road, Suite 160
Menlo Park, CA 94025

Warren Alpert Foundation Prize **5712**
The annual prize recognizes scientists and researchers from around the world for their contributions to humanity and their breakthroughs in the understanding and curing of major diseases. Many of the world's foremost scientific and medical professionals, working in a variety of disciplines, are nominated by their colleagues for this prestigious honor. Winners are presented with a monetary award at the foundation's annual luncheon. Recipients are selected by the foundation's scientific advisory board, which comprises internationally renowned biomedical scientists.
Samples: David Goeddel, PhD; Sidney Pestka, MD; and Charles Weissmann, MD, PhD—award recipients, for their pioneering work on the purification, characterization, and cloning of human interferon-alpha, $150,000 each (2003).
Amount of Grant: $100,000 award
Contact: Patrick Gilligan, (617) 432-2094; email: patrick_gilligan@hms.harvard.edu
Internet: http://www.hms.harvard.edu/fa/alpertprize
Sponsor: Warren Alpert Foundation
25 Shattuck Street, Room 206
Boston, MA 02115

Washington Women's Foundation Grants **5713**
The grant committee of the Pooled Fund researches community needs in five predetermined issue areas: cultural arts, education, environment, health, and social services. The foundation funds projects that address at least one of the following criteria: responses to urgent and critical need; bold new ventures; and/or new approaches to time-worn problems. The foundation funds broadly; it does not limit funding to women's causes, and there are no geographic restrictions to eligibility. Beginning in January of each year, work groups identify giving priorities, compile data, and analyze information on organizations of interest. Agencies are selected to receive a formal request for proposal. Unsolicited proposals are not accepted. Organizations may submit an online letter of inquiry form to be considered.
Requirements: 501(c)3 nonprofit organizations are eligible.
Samples: Academy of Children's Theatre of Richland (WA)—for a newly acquired building to house theater-art classes, outreach programs, and performance space for children in the Tri-Cities/Mid-Columbia Valley area, $50,000 (2005); Community for Youth (Seattle, WA)—to strengthen and expand its mentoring and after-school programs to help underachieving yet promising high-school students stay in school, $100,000 (2005); Jubilee Women's Ctr (Seattle, WA)—to provide safe, affordable, long-term housing in Seattle for women who are homeless or face an economic crisis, $85,000 (2005); Okanogan Family Planning (WA)—for medical equipment, furnishings, and other supplies at a new family-planning clinic, $85,000 (2005).
Amount of Grant: $50,000-$100,000 over two to three years
Contact: Grants Administrator, (206) 340-1710; fax: (206) 407-2177; email: info@wawomensfoundation.or
Internet: http://www.wawomensfoundation.org/grants/guide.html
Sponsor: Washington Women's Foundation
2100 24th Avenue S, Suite 330
Seattle, WA 98144

Wasie Foundation Grants **5714**
The foundation limits grantmaking activities to organizations in Minnesota and currently makes grants in the areas of higher education, specifically endowed scholarship programs, and the care and treatment of individuals afflicted with schizophrenia. Types of support include equipment acquisition, endowment funds, program development, conferences and seminars, scholarship funds, research, and matching funds. Call or write for application procedures and current deadlines; the foundation encourages phone contact.
Requirements: Organizations in Florida and Minnesota are eligible.
Restrictions: Grant making is limited to the Saint Paul/Minneapolis metropolitan area.
Geographic Focus: Florida; Minnesota
Amount of Grant: $15,000-$200,000 average
Contact: Jan Preble, Program Associate, (612) 455-6880; fax: (612) 455-6888

Sponsor: Wasie Foundation
4999 France Avenue S, Suite 250
Minneapolis, MN 55410

Watson Davis Prize **5715**
The prize honors books in the history of science directed to a wide public (including undergraduate instruction). They should be introductory in assuming no previous knowledge of the subject and in being directed to audiences of beginning students and general readers. They should introduce an entire field, a chronological period, a national tradition, or the work of a noteworthy individual. Multi-authored or edited books are eligible, whereas unrevised reprints of previously published works are not. The prize may not be split between two books. Books published during the three years prior to the award date are eligible for nomination. The committee appointed to select the winner will determine the books to be considered by reviewing lists of recently published books and soliciting nominations from major scholarly publishers. Members and others are also urged to nominate books for consideration by sending their suggestions to the HSS executive secretary. Annual deadline dates may vary; contact program staff for exact dates.
Restrictions: Unrevised reprints of previously published works are not eligible.
Amount of Grant: $1000
Date(s) Application is Due: Apr 1
Contact: Program Contact, (352) 392-1677; fax: (352) 392-2795; email: info@hssonline.org
Internet: http://www.hssonline.org/society/awards/mf_awards.html
Sponsor: University of Washington
Box 351330
Seattle, WA 98195

Watson-Brown Foundation Grants **5716**
The foundation focuses its giving within the U.S. Southeast and awards grants to nonprofit organizations and institutions that advance education. The majority of grants support higher education by funding programs (usually associated with the humanities) within colleges and universities and publications by scholarly presses. Areas of interest include history, literature, law, agricultural education, and historic preservation. Types of support include operating support, capital projects, endowments, professorships, publication underwriting, research grants, program grants, and challenge grants. Grant application forms and support materials should be requested by phoning the office.
Requirements: 501(c)3, 509(a)1, 509(a)2, and 509(a)3 nonprofit organizations are eligible. Generally, the foundation supports only organizations, institutions, or programs located in the southeastern region of the United States.
Restrictions: Grants do not support individuals (except for scholarships awarded through the foundation's internal scholarship program); religious programs, or political/lobbying activities; debt reduction; advertising or marketing efforts; social events such as sports tournaments or galas; or individual projects for more than one year.
Geographic Focus: South
Samples: Georgia Trust for Historic Preservation (Atlanta, GA)—to develop a network of, and provide training to, staff members working at historic houses and sites in Georgia, $30,000 (2003).
Date(s) Application is Due: Apr 15
Contact: Grants Administrator, (706) 595-8886; fax: (706) 595-3948; email: mcoussons@bellsouth.net
Internet: http://www.watson-brown.org/grants/index.html
Sponsor: Watson-Brown Foundation
310 Tom Watson Wy
Thomson, GA 30824

WAWH Barbara Penny Kanner Award **5717**
The Western Association of Women Historians will award $500 to the best scholarly bibliographical and historical guide to research focused on women or gender history. The Kanner Award is intended to promote the practice of bibliomethodology or autobiography in historical context. The bibliomethodology award should reflect the critical tools of the historian's craft as they have been developed to provide research guides rather than library catalogs. The autobiography in historical context award should reflect the craft of history as developed and interpreted in individual lives. Book-length submissions are preferred but substantial guides in other forms (articles or book chapters) may also be considered. The bibliographical and the autobiography awards will be given in alternate years, with submissions for each award considered for two-year intervals from the date of publication.
Requirements: Recipient must be a member of the association.
Samples: Candace Falk—for Guide to Papers of Emma Goldman, $500.
Amount of Grant: $500
Date(s) Application is Due: Jan 15
Contact: Regina Lark, UCLA Center for the Study of Women, (310) 206-5898; email: rlark@women.ucla.edu
Internet: http://www.wawh.org/2005awardinfo.html#Kanner
Sponsor: Western Association of Women Historians
P.O. Box 758
Knights Landing, CA 95645

WAWH Founders Dissertation Fellowship **5718**
The association promotes the professional achievement of female historians working in any field of history. Usually one fellowship is offered per year to members of the association.
Requirements: Applicants must be members of the WAWH, advanced to candidacy, writing the dissertation at the time of the application, and expecting to receive the PhD.
Samples: Jennifer Green—for Charity in the Province of Girona in the 12th Century, $1000.
Amount of Grant: $1000
Date(s) Application is Due: Jan 15
Contact: Alexandra Nickliss, Dept of Social Science, City College of San Francisco, (415) 239-3283; email: anicklis@ccsf.org
Internet: http://www.wawh.org/awardsandprizes.html#FDF
Sponsor: Western Association of Women Historians
City College of San Francisco, 50 Phelan Avenue
San Francisco, CA 94112

WAWH Sierra Award **5719**
The award is given each year for the best monograph published by a Western Association of Women Historians member during the previous year. The monograph must be based on original research (not anthology or edited work) in any field of history.
Requirements: Individuals submitting works for the award should either be members residing in the western states (Alaska, Arizona, California, Colorado, Hawaii, Idaho, Kansas, Montana, Nebraska, Nevada, New Mexico, North Dakota, Oklahoma, Oregon, South Dakota, Texas, Utah, Washington, Wyoming, and Canadian western provinces [Alberta, British Columbia]); or, if resident elsewhere, individuals should have been paid WAWH members for three years (though not necessarily consecutive years) prior to submission.
Samples: Patricia Schechter—for Ida B. Wells-Barnett and American Reform, $250.; Jessica Weiss—for To Have and To Hold: Marriage, the Baby Boom, and Social Change, $250.
Amount of Grant: $250
Date(s) Application is Due: Jan 15
Contact: Jessica Weiss, California State University, (510) 885-3239, email: jessica.weiss@csueastbay.edu
Internet: http://www.wawh.org/awardsandprizes.html#Sierra
Sponsor: Western Association of Women Historians
25800 Carlos Bee Boulevard
Hayward, CA 94542-3014

Wayne and Gladys Valley Foundation Grants **5720**
The foundation awards grants to California nonprofits in its primary areas of interest, including education at all levels, health care and medical research, youth and families, human services, sciences and engineering, and religion. Types of support include general operating support, building construction/renovation, program development, professorships, research, and matching funds. There are no application forms or deadlines. The board meets in February, May, September, and November to consider requests.
Requirements: California nonprofits in Alameda, Contra Costa, and Santa Clara Counties are eligible.
Geographic Focus: California
Samples: U of the Pacific (Stockton, CA)—for its new biological-sciences center, $2.5 million (2005); Santa Clara U (CA)—to construct a new library, $15 million (2004); Dominican U of California (San Rafael, CA)—for a new science and technology center, $2.25 million (2004).
Contact: Michael Desler, Executive Director, (510) 466-6060; fax (510) 466-6067; email: info@wgvalley.org
Sponsor: Wayne and Gladys Valley Foundation
1939 Harrison Street, Suite 510
Oakland, CA 94612-3532

Weizmann Institute of Science Postdoctoral Fellowships **5721**
The institute operates in Israel and internationally in the fields of education and science (biology, biophysics-biochemistry, chemistry, mathematical sciences, and physics). Postdoctoral fellowships and visiting professorships are available. Applications and information are posted on the Web site.
Requirements: The postdoctoral fellowships are open to citizens of all countries. Candidates must have received a PhD degree or equivalent within three years of the submission deadline.
Date(s) Application is Due: Jan 1; May 15
Contact: Fellowships Program, 972-8-9344553; fax: 972-8-9344114; email: postdoc@weizmann.ac.il
Internet: http://www.weizmann.ac.il/feinberg/postdoc_fell_weiz.shtml
Sponsor: Weizmann Institute of Science
P.O. Box 26
Rehovot 76100
Israel

Welch Award in Chemistry **5722**
The purpose of the award is to foster and encourage basic chemical research and to recognize, in a substantial manner, the value of chemical research contributions for the benefit of mankind. Any person can be considered for the award. The award is intended to recognize contributions that have not previously been rewarded in a similar manner. All nominations must include a brief biographical sketch of the proposed nominee, a list of 25 of the nominee's most important scientific publications, a statement of the research areas in chemistry on which the nomination is based, and not more than three

seconding nomination letters. Nominations with supporting documents should be sent to the office.
Requirements: Nominations will be solicited from selected scientific organizations and individuals. Nominations received from other organizations and individuals will also be considered, though no person may nominate oneself.
Restrictions: Directors and officers of the foundation and members of the scientific advisory board of the foundation are not eligible.
Samples: George Whitesides—award winner, for his research at Harvard U, $300,000 (2005); Dr. Allen J. Bard, The University of Texas at Austin (Austin, TX)—award winner, $300,000 (2004).
Amount of Grant: $300,000
Date(s) Application is Due: Feb 1
Contact: Carla Atmar, Grant Programs Coordinator, (713) 961-9963, ext. 13 or (713) 961-9884; fax: (713) 961-5168
Internet: http://www.welch1.org/Awards/WelchAwardinChemist0943/index.asp
Sponsor: Welch Foundation
5555 San Felipe, Suite 1900
Houston, TX 77056-2732

Welch Foundation Departmental Grants **5723**
The purpose of the departmental grant is to support chemical research by members of the chemistry department faculty, and to provide an opportunity for students to study chemistry in a less structured way. The grant is also designed to provide students with the tools they need to gain and develop an interest in pursuing a career in chemistry. This funding also ensures that students will have scholarships, lab equipment and chemicals and can travel to industry meetings and conferences. These grants are awarded to smaller and medium-sized educational institutions in Texas.
Restrictions: Applications are restricted to smaller and medium-sized educational institutions in Texas.
Geographic Focus: Texas
Amount of Grant: $150,000 minimum
Date(s) Application is Due: Feb 1
Contact: Carla Atmar, Grant Programs Coordinator, (713) 961-9963, ext. 13 or (713) 961-9884; fax: (713) 961-5168
Internet: http://www.welch1.org/GrantsPrograms/DepartmentalGrants/index.asp
Sponsor: Welch Foundation
5555 San Felipe, Suite 1900
Houston, TX 77056

Welch Foundation Endowed Chairs **5724**
The purpose of an endowed chair is to provide a faculty position for an eminent scientist, thus further increasing the basic scientific research in chemistry and other sciences in Texas. Historically, the institution must be willing to match the endowment funds dollar for dollar. The Welch Foundation accepts grant requests throughout the grant year, and the Foundation has no required application form for an endowed chair proposal.
Requirements: To be considered for an endowed chair, a Texas educational institution must have a Ph.D. program in chemistry.
Geographic Focus: Texas
Contact: Carla Atmar, Grant Programs Coordinator, (713) 961-9963, ext. 13 or (713) 961-9884; fax: (713) 961-5168
Internet: http://www.welch1.org/GrantsPrograms/EndowedChairs/index.asp
Sponsor: Welch Foundation
5555 San Felipe, Suite 1900
Houston, TX 77056

Welch Foundation Norman Hackeman Award in Chemical Research **5725**
The purpose of the award is to encourage and recognize young chemical scientists in Texas for their past research endeavors. The award is meant not only to recognize the work of a scientist, but to encourage those who are embarking on careers dedicated to increasing fundamental understanding of chemistry. The award is not intended to finance any further research or activities of the recipient, nor is the recipient required to render any future services as a condition of receiving the award.
Requirements: A candidate for the award must be 40 years of age or younger on June 30 for the year in which nominated. Any Texas college or university may nominate two candidates for any given year. The nomination should normally be by letter from the institution in which the individual is a faculty member. To be eligible, a nominee should have spent the most recent three years ending June 30 doing chemical research in Texas.
Restrictions: Directors and officers of the foundation and members of the scientific advisory board of the foundation are not eligible. Self-nominations are not accepted.
Geographic Focus: Texas
Samples: Dr. Patrick G. Harran, The University of Texas Southwestern Medical Center at Dallas, Dallas, TX (2007); Paul S Cremer, Ph. D., Department of Chemistry, Texas A&M University (2006); Zhijian James Chen, University of Texas Southwestern Medical Center at Dallas (2005).
Amount of Grant: $100,000
Date(s) Application is Due: Sep 15
Contact: Carla Atmar, Grant Programs Coordinator, (713) 961-9963, ext. 13 or (713) 961-9884; fax: (713) 961-5168
Internet: http://www.welch1.org/Awards/NormanHackermanAwar11BD/index.asp
Sponsor: Welch Foundation
5555 San Felipe, Suite 1900
Houston, TX 77056-2732

Welch Foundation Research Grants **5726**
The foundation supports fundamental chemical research at educational institutions within the state of Texas. Each research grant provides a minimum of $150,000 in funding over a three year period to support research in chemistry by a full-time tenured or tenure-track faculty member who serves as principal investigator. Each research grant may utilize its funds on stipends, fellowships, scholarships, permanent scientific equipment, expendable scientific items, services, equipment maintenance, publication expenses, and travel.
Restrictions: Applications are restricted to universities, colleges, or other educational institutions located in the state of Texas. Only one proposal may be submitted naming a particular principal investigator.
Geographic Focus: Texas
Samples: Southern Methodist U (Dallas, TX)—to support chemistry research, $600,000 (2003); U of Texas, Houston Health Science Ctr (Houston, TX)—to support chemistry research, $300,000 (2003); Texas A&M U (College Station, TX)—to support chemistry research, $3.51 million (2003).
Amount of Grant: $150,000 minimum
Date(s) Application is Due: Feb 1
Contact: Carla Atmar, Grant Programs Coordinator, (713) 961-9963, ext. 13 or (713) 961-9884; fax: (713) 961-5168
Internet: http://www.welch1.org/GrantsPrograms/ResearchGrants/index.asp
Sponsor: Welch Foundation
5555 San Felipe, Suite 1900
Houston, TX 77056-2732

Wellesley Centers for Women Postdoctoral Fellowships **5727**
The program provides systematic research training for candidates who want to develop expertise in research on childhood and adolescence and investigate variations in race and ethnicity, gender, and social class and how these interact with risk and resiliency factors in human development. The training program consists of three components: training and career development; collaborating with the preceptor on the preceptor's externally funded research projects; and independent research of the fellow's choosing conducted under the guidance of his/her preceptor. One of the program's goals is to identify and develop scholars from underrepresented groups who will launch independent research careers through external funding. Guidelines are available online.
Requirements: Candidates must be U.S. citizens or permanent residents. Researchers with PhDs, MDs, or comparable doctoral degrees from accredited foreign institutions also are eligible.
Amount of Grant: $40,000 minimum
Date(s) Application is Due: Jan 1
Contact: Sumru Erkut, (781) 283-2533; email: serkut@wellesley.edu
Internet: http://www.wcwonline.org/postdoc/index.html
Sponsor: Wellesley College
106 Central Street
Wellesley, MA 02481

Wenner-Gren Foundation African Fellowships **5728**
The Fellowships are intended to provide support for African students undertaking study leading to a Ph.D. at a South African university that can provide them with international-level training in anthropology (including biological anthropology and archaeology). The annual fellowship is $17,500 and can be used towards travel, living expenses, tuition, student fees, insurance, books, research expenses, and any other relevant categories of expenditure while studying at the Host institution. The fellowship is renewable for up to two additional years upon successful completion of each preceding year's study. A separate application can be made for an additional year of funding to support dissertation write-up. The Program has a two-stage application process. The first stage is a preliminary inquiry. The deadline for the preliminary inquiry is December 1. If a candidate is considered eligible based on the review of the preliminary inquiry he/she will be invited to submit a full application. The deadline for the full application is December 15. Final decisions are made by January 15.
Requirements: The Fellowship is intended for young African scholars who may not otherwise be able to pursue a doctoral degree in anthropology. Normally, applicants will be under 35 years of age at the time they begin their fellowship. They must be citizens and residents of an African country at the time of application. They must also be members of an underrepresented group in academic anthropology/archaeology.
Amount of Grant: $17,500
Date(s) Application is Due: Dec 1; Dec 15
Contact: Leslie C. Aiello, President; (212) 683-5000; fax: (212) 683-9151; email: laiello@wennergren.org
Internet: http://www.wennergren.org/programs/programs_show.htm?doc_id=377114&attrib_id=13232
Sponsor: Wenner-Gren Foundation
470 Park Avenue South, 8th Floor
New York, NY 10016-6819

Wenner-Gren Foundation Conference and Workshop Grants **5729**
In accordance with the mission of the Foundation, priority is given to events that foster the creation of an international community of research scholars in anthropology and advance significant and innovative anthropological research. Conferences are defined as public events that are comprised primarily of oral and poster presentations to a larger audience of anthropologists. Priority is given to major conferences sponsored by large international anthropological organizations. Workshops are defined as working meetings that focus on developing and debating topical issues in theoretical anthropology. Priority

is given to those workshops that devote the majority of time to discussion and debate rather than to the presentation of papers. Application deadlines for Conference and Workshop Grants are June 1 and December 1.
Requirements: Applications must be made by a professional anthropologist who is the primary organizer of the proposed conference or workshop. A co-applicant can be specified on the application.
Restrictions: Applications are not accepted from students without a co-applicant who has a doctorate and holds an established academic position. Individual requests for travel support to attend meetings are not accepted. Symposia within larger professional meetings are generally not supported. Limited funds are available to aid the participation of Third World scholars in international conferences, which are provided in the form of grants to the organizer(s).
Samples: David William Cohen, U of Michigan, and Luise White, Emory U—conference grant recipients for Words and Voices: Critical Practices of Orality in Africa and in African Studies at the Bellagio Conference Ctr in Bellagio, Italy $11,000.
Amount of Grant: $15,000 maximum
Date(s) Application is Due: Jun 1; Dec 1
Contact: Leslie C. Aiello, President; (212) 683-5000; fax: (212) 683-9151; email: laiello@wennergren.org
Internet: http://www.wennergren.org/programs/programs_show.htm?doc_id=370402&attrib_id=13235
Sponsor: Wenner-Gren Foundation
470 Park Avenue South, 8th Floor
New York, NY 10016-6819

Wenner-Gren Foundation Dissertation Fieldwork Grants **5730**
These grants are awarded to individuals to aid doctoral dissertation or thesis research. The program contributes to the Foundation's overall mission to support basic research in anthropology and to ensure that the discipline continues to be a source of vibrant and significant work that furthers understanding of humanity's cultural and biological origins, development, and variation. The Foundation supports research that demonstrates a clear link to anthropological theory and debates, and promises to make a solid contribution to advancing these ideas. There is no preference for any methodology, research location, or subfield. The Foundation particularly welcomes proposals that employ a comparative perspective, can generate innovative approaches or ideas, and/or integrate two or more subfields. Grants are non-renewable and provide a maximum of U.S $25,000. Applicants should submit application materials using the Foundation's online application submission procedure as well as send printed copies to the Foundation by regular mail.
Requirements: Students must be enrolled in a doctoral program (or equivalent, if applying from outside the United States) at the time of application. Students of all nationalities are eligible to apply. There is no time limit on the duration of the grant, and funding may be requested to cover distinct research phases.
Restrictions: Funding is not provided for salary and/or fringe benefits of applicant, tuition, non-project personnel, travel to meetings, institutional overhead, or other support for writing, publication assistance, and film- or video making (unless inherent to the research). Expenses incurred prior to the effective date of an award will not be covered.
Amount of Grant: $25,000 maximum
Date(s) Application is Due: May 1; Nov 1
Contact: Leslie C. Aiello, President; (212) 683-5000; fax: (212) 683-9151; email: laiello@wennergren.org
Internet: http://www.wennergren.org/programs/programs_show.htm?doc_id=367834&attrib_id=13232
Sponsor: Wenner-Gren Foundation
470 Park Avenue South, 8th Floor
New York, NY 10016-6819

Wenner-Gren Foundation Historical Archives Grants **5731**
The objective of this program is to encourage the preservation of unpublished records and other materials of value for research on the history of anthropology. Grants are offered for two purposes: to assist senior scholars (or their heirs) holding unpublished research materials and personal papers of significant value for the study of the history of Anthropology, with the expenses of preparing and transferring them for archival deposit; and to aid oral-history interviews with senior anthropologists. Applicants must first submit a Preliminary Inquiry to determine their eligibility. If the project is deemed eligible for consideration, an application form will be sent. There is no fixed deadline. Inquiries and applications are reviewed as they are received.
Restrictions: Funds are not provided to establish archival facilities or develop collections already held in archival deposit.
Samples: Vida Chenoweth (Enid, OK)—to aid preparation of documentary materials for audio tapes to be deposited with the American Folklife Ctr at the Library of Congress, $5000.; Dr. Diane Gifford-Gonzalez, U of California (Santa Cruz, CA)—to aid oral history interviews with Dr. Jia Lanpo of the Institute of Paleontology and Paleoanthropology in Beijing, China, $5000.
Amount of Grant: $15,000 maximum
Contact: Leslie C. Aiello, President; (212) 683-5000; fax: (212) 683-9151; email: laiello@wennergren.org
Internet: http://www.wennergren.org/programs/programs_show.htm?doc_id=369096&attrib_id=13236
Sponsor: Wenner-Gren Foundation for Anthropological Research
470 Park Avenue South, 8th Floor
New York, NY 10016-6819

Wenner-Gren Foundation Hunt Postdoctoral Fellowships **5732**
The fellowships support the writing-up of already completed research. The fellowship is awarded to scholars in the earlier stages of their careers, when they frequently lack the time and resources to develop their research for publication. Scholars with a Ph.D. in hand for no more than ten years (from the application deadline) are eligible to apply. A maximum of eight Hunt Postdoctoral Fellowships are awarded annually.
Requirements: Applicants must have a Ph.D. or equivalent at the time of application. Applicants must have received a Ph.D. or equivalent within ten years of the application deadline.
Restrictions: Aid is not provided for salary and/or fringe benefits of applicant, family expenses, or institutional overhead. Support is not normally provided for master's studies.
Samples: Adolfo Neves de Oliveira Jr., U of Brasilia (Brasilia, Brazil)—to aid training in social anthropology at St.Andrews U, St. Andrews, Scotland, under supervision of Dr. Joanna Overing, $12,500.; Kaberi Kar Gupta, Wildlife Institute of India (Dehradun, India)—to aid study in biological anthropology at Arizona State U, Tempe, AZ, supervised by Dr. Leanne Nash—$12,500, renewal.
Amount of Grant: $40,000
Date(s) Application is Due: May 1; Nov 1
Contact: Leslie C. Aiello, President; (212) 683-5000; fax: (212) 683-9151; email: laiello@wennergren.org
Internet: http://www.wennergren.org/programs/programs_show.htm?doc_id=368531&attrib_id=13233
Sponsor: Wenner-Gren Foundation
470 Park Avenue South, 8th Floor
New York, NY 10016-6819

Wenner-Gren Foundation International Collaborative Research Grants **5733**
Grants are available to assist anthropological research projects undertaken jointly by two (or more) investigators from different countries. These grants are renewable for a second period of research. The purpose of the program is to encourage collaborations in which the principal investigators bring different and complementary perspectives, knowledge, and/or skills. Priority is given to new collaborations, but an independent project within a larger, long-term collaboration can be considered.
Requirements: Projects must involve at least one principal investigator from outside the United States, Canada, Western Europe, Australia, or New Zealand. Both investigators must meet the qualification for Post-PhD Grants (i.e., holding the doctorate or equivalent in anthropology or a related discipline).
Restrictions: Projects primarily for other purposes, such as training, education, or writing, are not eligible under this program.
Samples: Dr. Francis Bulbeck, U of Western Australia, and Dr. H. Darmawan, Balai Arkeologi, (Ujung Pandang, Indonesia)—to aid collaborative archaeological and historical research on Luwuk kingdom and the origin of Bugis culture, in South Sulawesi, Indonesia, $18,000.; Dr. Mark Hall, U of California (Berkeley, CA) and Dr. Damdisurenguiin Tseveendorj and Dr. Zagd Batsaikhan, Mongolian Academy of Science (Ulaanbataar, Mongolia)—to aid collaborative research on the rise of pastoral complexity in Northeast Eurasia, $23,400.
Amount of Grant: $30,000 maximum
Date(s) Application is Due: Jun1; Dec 1
Contact: Leslie C. Aiello, President; (212) 683-5000; fax: (212) 683-9151; email: laiello@wennergren.org
Internet: http://www.wennergren.org/programs/programs_show.htm?doc_id=368683&attrib_id=13233
Sponsor: Wenner-Gren Foundation
470 Park Avenue South, 8th Floor
New York, NY 10016-6819

Wenner-Gren Foundation Post-PhD Research Grants **5734**
Grants are awarded to individuals holding a Ph.D. or equivalent degree to support individual research projects. The program contributes to the Foundation's overall mission to support basic research in anthropology and to ensure that the discipline continues to be a source of vibrant and significant work that furthers our understanding of humanity's cultural and biological origins, development, and variation. The Foundation supports research that demonstrates a clear link to anthropological theory and debates, and promises to make a solid contribution to advancing these ideas. There is no preference for any methodology, research location, or subfield. The Foundation particularly welcomes proposals that employ a comparative perspective, can generate innovative approaches or ideas, and/or integrate two or more subfields. Grants are for a maximum of U.S. $25,000 towards research expenses.
Requirements: Applications are accepted from scholars who are close to completing their doctorate or equivalent degree; however, the grant cannot be awarded until the degree is completed.
Amount of Grant: $25,000 maximum
Date(s) Application is Due: May 1; Nov 1
Contact: Leslie C. Aiello, President; (212) 683-5000; fax: (212) 683-9151; email: laiello@wennergren.org
Internet: http://www.wennergren.org/programs/programs_show.htm?doc_id=368389&attrib_id=13233
Sponsor: Wenner-Gren Foundation
470 Park Avenue South, 8th Floor
New York, NY 10016-6819

Wenner-Gren Foundation Short-Term Fellowships 5735

The Fellowships are intended to provide support for doctoral students or post-doctoral scholars who are within ten years of receiving their doctorate to enhance their research capacity in the field of anthropology. The fellowship may be used to: pursue research in a library outside the applicant's home country; or take advantage of a training opportunity outside the applicant's home country. The fellowships provide funding of up to $5000 and are not renewable. The length of the library or training visit should normally be no more than 3 months in duration. The award can be used towards relevant expenses for travel, subsistence, fees, insurance, books, etc. There is no deadline for the Wadsworth Short-Term Fellowships; however, there is a two-stage application process. The first stage is the preliminary inquiry. If an applicant is considered eligible based on the review of the preliminary inquiry, he/she will be invited to submit a full application.
Requirements: Applicants must be from countries where anthropology is underrepresented and where there are limited resources to send scholars overseas for training or library research. Student applicants must be registered for a doctoral degree in anthropology or a related sub-discipline at a university in a qualifying country. Post-doctoral applicants must be within ten years of completing their doctoral degree and hold an academic position in a university in a qualifying country.
Amount of Grant: $5,000
Contact: Leslie C. Aiello, President; (212) 683-5000; fax: (212) 683-9151; email: laiello@wennergren.org
Internet: http://www.wennergren.org/programs/programs_show.htm?doc_id=376103&attrib_id=13232
Sponsor: Wenner-Gren Foundation for Anthropological Research
470 Park Avenue South, 8th Floor
New York, NY 10016-6819

Wenner-Gren Foundation Wadsworth International Fellowships 5736

The Fellowships are intended to provide support for students undertaking study leading to a Ph.D. or equivalent doctoral degree at universities where they can receive international-level training in anthropology. These fellowships are available only to students from countries where anthropology is underrepresented and where there are limited resources to send students overseas for training. Priority is given to applicants who have not already begun graduate training outside of their home countries. Applications from individuals who are more advanced in their doctoral studies will, however, be considered. The annual fellowship is $17,500 and can be used towards travel, living expenses, tuition, income taxes, student fees, insurance, books, research expenses and any other relevant categories of expenditure while studying at the Host Institution. The fellowship is renewable for up to two additional years upon successful completion of each preceding year's study or research. A separate application can be made for an additional year of funding to support dissertation write-up. The Program has a two-stage application process. The first stage is a preliminary inquiry. The deadline for the preliminary inquiry is February 1. If a candidate is considered eligible based on the review of the preliminary inquiry he/she will be invited to submit a full application. The deadline for the full application is March 1. Final decisions are made by May 1.
Requirements: Applicants must be from countries where anthropology is underrepresented and where there are limited resources to send students overseas for training. Normally, applicants will be under 35 years of age at the time they begin their fellowship.
Restrictions: Applicants wishing to study anthropology at universities in their home country are not eligible to apply.
Amount of Grant: $17,500
Date(s) Application is Due: Feb 1; Mar 1
Contact: Leslie C. Aiello, President; (212) 683-5000; fax: (212) 683-9151; email: laiello@wennergren.org
Internet: http://www.wennergren.org/programs/programs_show.htm?doc_id=368103&attrib_id=13232
Sponsor: Wenner-Gren Foundation
470 Park Avenue South, 8th Floor
New York, NY 10016-6819

West Virginia Artists Fellowships 5737

This program is designed to reward excellence in artistic production and expression. The awards are based upon reviews of examples of recent work in the artists' chosen discipline and media. Each year awards will be made in the areas of the visual, performing, and literary arts; specific genres and media will alternate yearly. Applicants creating work under the supervision of a master artist or instructor will be reviewed on a case-by-case basis. Only one application in a single medium will be accepted from any artist. Fellowship grantees may not reapply in any medium for three years.
Requirements: Individual artists, who are at least eighteen years old and legal residents of West Virginia for at least one year before the application deadline are eligible.
Geographic Focus: West Virginia
Amount of Grant: $3500
Date(s) Application is Due: Sep 1
Contact: Grants Coordinator, (304) 558-0240; fax: (304) 558-2779; TDD: (304) 558-0220; email: barbie.anderson@wvculture.org
Internet: http://www.wvculture.org/arts/grantbook/direct.htm
Sponsor: West Virginia Commission on the Arts
1900 Kanawha Blvd E
Charleston, WV 25305-0300

West Virginia Wesleyan College Teaching Fellowship 5738

The Department of History and Political Science invites applications for a teaching fellowship. Candidates should have expertise in at least one of the following areas in history or political science: Latin America, Asia, Africa, women's history, African-American history, parties and elections, or public administration. Teaching responsibilities of two courses per semester may include U.S. history surveys, state and local government, U.S. federal government, and one history/political science course in the candidate's area of expertise. Applicants should send letter of interest, curriculum vita, representative syllabi and student evaluations, unofficial transcripts, and three letters of reference that must include a letter from the dissertation advisor with evidence of ABD status.
Requirements: Preference is given to ABDs with teaching experience who are nearing completion of the PhD.
Date(s) Application is Due: Mar 11
Contact: Robert Rupp, (304) 473 - 8561; email: rupp_r@wvwc.edu
Internet: http://www.wvwc.edu/aca/hst/hstfront.htm
Sponsor: West Virginia Wesleyan College
59 College Avenue
Buckhannon, WV 26201-2995

Westar Energy Foundation Grants 5739

The foundation awards grants to nonprofit organizations in company operating areas in Kansas. Areas of interest include children and youth, particularly educational programs related to enhancing employment opportunities; health programs related to the general well-being of the community; seniors, particularly with energy assistance programs; and environmental programs, particularly programs cooperating with federal, state, and local government units in nonpartisan efforts to improve the environment or educate the public about environmental issues. Types of support include building funds, capital campaigns, continuing support, emergency funds, equipment, fellowships, in-kind gifts, general operating budgets, renovation projects, research, scholarship funds, and seed money. There are no application deadlines; the board meets quarterly to consider requests.
Requirements: Nonprofit organizations in company-operating areas of Kansas are eligible to apply.
Geographic Focus: Kansas
Samples: Exploration Place (Wichita, KS)—for its capital campaign, and for the Season of Creativity exhibit, $100,000 and $40,000, respectively.
Amount of Grant: $300-$10,000 average
Contact: Cynthia McCarvel, President, (785) 575-1544; fax: (785) 785-6399
Internet: http://www.wstnres.com/corp_com/contentmgt.nsf/publishedpages/foundation?opendocument&menu=5
Sponsor: Westar Energy Foundation
P.O. Box 889
Topeka, KS 66601-0089

Western European Architectural Foundation Gabriel Prize 5740

The award is conferred annually in honor of the French architect Ange Jacques Gabriel (1698-1782) for the study of classical architecture and landscape in France. While abroad, Gabriel laureates focus on some particular aspect of French architecture. Candidates may apply online through the Gabriel Prize Web site.
Requirements: The candidate must be a U.S. citizen.
Amount of Grant: $17,500
Date(s) Application is Due: Dec 31
Contact: Gabriel Prize Coordinator, (210) 829-4040; fax: (210) 829-4049; email: contact@gabrielprize.org
Internet: http://www.gabrielprize.org
Sponsor: Western European Architectural Foundation
306 West Sunset, Suite 119
San Antonio, TX 78209

Western Regional Graduate Program 5741

Residents of 14 western states are eligible to attend distinguished graduate programs at reduced tuition rates through the Western Regional Graduate Program, a program coordinated by WICHE. Qualified residents of Alaska, Arizona, Colorado, Hawaii, Idaho, Montana, Nevada, New Mexico, North Dakota, Oregon, South Dakota, Utah, Washington, and Wyoming may enroll in any of the qualified graduate programs at institutions in those states (list available on the Web site). Students pay resident tuition at public institutions or substantially reduced tuition at private institutions. Application deadlines are set by the institution and department. Students should apply directly to the programs, not WICHE; however, WICHE will provide an information brochure upon request.
Geographic Focus: Alaska; Arizona; Colorado; Hawaii; Idaho; Montana; Nevada; New Mexico; North Dakota; Oregon; South Dakota; Utah; Washington; Wyoming
Date(s) Application is Due: Nov 18
Contact: Program Coordinator, (303) 541-0200; email: info-sep@wiche.edu
Internet: http://wiche.edu/SEP/WRGP
Sponsor: Western Interstate Commission for Higher Education
P.O. Box 9752
Boulder, CO 80301-9752

Westinghouse Charitable Giving Grants 5742

The foundation makes charitable contributions to community priorities primarily where Westinghouse has a presence. Areas of emphasis include education—elementary, secondary, and high school programs that emphasize math and science (some non-fine arts programs); health and welfare—to improve quality of life in the community; and civic and social affairs—to support economic development, environmental quality, and the preservation of public safety. Within each area, the foundation actively encourages programs that meet the needs of special populations, such as the disadvantaged, the

young, the elderly, the gifted, people with disabilities, minorities, and women. Because of the foundation's approach to funding in Westinghouse plant locations, all organizations, except colleges and universities, should submit their requests for support to the nearest Westinghouse location. Applications are accepted on an ongoing basis.
Requirements: 501(c)3 tax-exempt organizations are eligible.
Restrictions: The following organizations and activities are ineligible: individuals; political contributions; religious organizations; highly specialized health, medical, or welfare programs; fine arts organizations/programs; organizations that discriminate by race, color, creed, gender, or national origin; hospitals; colleges, universities, and two-year institutions; capital improvement or building projects; medical education; chairs or professorships; liberal arts, fine arts, or similar educational programs; educational research programs, regardless of academic discipline; graduate education; general endowment; equipment purchases at universities; United Way-affiliated agencies; tickets; or memberships.
Samples: Regional Math/Science Collaborative (Pittsburgh, PA)—to support education reform and international economic education, $20,000.; NAACP Benton-Franklin County Chapter (Pasco, WA)—to encourage academic excellence in the sciences, humanities, and visual and performing arts for African American ninth-12th grade students, $10,000.
Amount of Grant: $5000 maximum
Contact: Program Officer, Contributions and Community Affairs, (412) 374-6824; fax: (412) 642-4874
Internet: http://www.westinghousenuclear.com/E2.asp
Sponsor: Westinghouse Electric Corporation
P.O. Box 355, ECE 575C
Pittsburgh, PA 15230-0355

Weyerhaeuser Company Foundation Contributions Program 5743
The foundation's mission is to improve the quality of life in communities where Weyerhaeuser has a major presence and to provide leadership that increases public understanding of issues where society's needs intersect with the interests of the forest products industry. Community service grants include awards in the fields of education and youth, health and welfare, civic and community improvement, and culture and the arts. In addition to community grants, industry-related awards are made to educational institutions, environmental groups, and professional organizations that promote further understanding of how the forest products industry responds to a changing society. Types of support include seed money, building construction/renovation, equipment acquisition, employee-related scholarships, publication, conferences and seminars, fellowships, lectureships, operating budgets, research, program development, employee matching gifts, and technical assistance. Applicants should send a short letter that introduces the project and sponsoring organization and provide tax-exempt status evidence. If further consideration is warranted, the foundation may ask for additional information or a formal proposal. Proposals may be submitted at any time. Applications and guidelines may be requested from the Web site.
Requirements: Applying organizations must have nonprofit, tax-exempt status. The foundation concentrates on direct services in the communities where Weyerhaeuser has significant land holdings and operating facilities, including Alabama, Arkansas, Mississippi, North Carolina, Oklahoma, Oregon, and Washington. A limited number of smaller awards are also made to other locales where fewer employees are based.
Restrictions: Grants are not awarded to individuals or for political campaigns, activities that influence legislation, religious organizations seeking funds for theological purposes, or funds to purchase tickets or tables at fundraising benefits.
Geographic Focus: Alabama; Arkansas; Mississippi; North Carolina; Oklahoma; Oregon; Washington
Samples: World Forestry Ctr (Portland, OR)—for its capital campaign, including exhibit renovation and a marketing study, $1 million (2002).
Contact: Elizabeth Crossman, President, (253) 924-3159; fax: (253) 924-3658; email: foundation@weyerhaeuser.com
Internet: http://www.weyerhaeuser.com/citizenship/philanthropy/default.asp
Sponsor: Weyerhaeuser Company Foundation
P.O. Box 9777
Federal Way, WA 98063-9777

Whitaker International Fellows and Scholars Program 5744
The Program consists of two branches: Whitaker Fellows and Whitaker Scholars. The goal of both Programs is to assist in the development of professional leaders who are not only superb engineers and scientists, but who also will lead and serve the profession with an international outlook. The term of the Fellow awards will be one academic year (as defined by the academic calendar of the host country). The term of the Scholar awards can be for as little as one academic semester or as long as a full academic year. Awards will typically begin between July 1st and October 1st, but may begin as late as March 31st.
Requirements: U.S. citizens and non-citizen permanent residents may apply. Fellows must have a BS or MS degree in biomedical engineering (or bioengineering, which is considered here synonymous if the program emphasizes biomedical engineering), and they will not have a doctorate at the time they start the grant. The Fellows applicant's most recent degree may not have been obtained more than two years prior to the start of the grant. Scholars must have a either degree in biomedical engineering and a doctorate which may not have been obtained more than two years prior to the start of the grant, or must be advanced PhD candidates who have been admitted to candidacy and who will receive the PhD prior to departure.
Amount of Grant: $15,000 maximum
Date(s) Application is Due: Jan 25
Contact: Vijay Renganathan, Program Manager; (212) 984-5442; email: whitaker@iie.org
Internet: http://whitakerawards.org/program-overview
Sponsor: Institute of International Education
809 United Nations Plaza
New York, NY 10017

White House Fellowships 5745
The purpose of the program is to provide gifted and highly motivated young Americans with some firsthand experience in the process of governing the nation. Fellows spend a year as full-time, paid assistants to senior White House staff and top-ranking government officials; write speeches, help draft and review proposed legislation, answer congressional inquiries, chair meetings, and conduct briefings; and meet with U.S. officials and military leaders, foreign heads of state, and world leaders from the worlds of business, the arts, science and technology, and the media. Fellows also travel to cities and regions outside of Washington and to other countries to see U.S. policy in action, learn about other cultures, and see the U.S. and federal government from differing points of view. Application and guidelines are available online.
Requirements: Candidates are judged on their professional, academic, and other accomplishments that indicate remarkable achievement in the early stages of their career; their current leadership skills and evidence of growth potential in those skills; and their strong commitment to public service, exceptional writing ability, and strong organizational and management skills.
Amount of Grant: $83,000
Date(s) Application is Due: Feb 1
Contact: Sheila Coates, Program Contact, (202) 606-1818; fax: (202) 395-6179
Internet: http://www.whitehouse.gov/fellows/about/purpose.html
Sponsor: President's Commission on White House Fellowships
1900 E Street NW, Room B431
Washington, DC 20415

Whitehall Foundation Neurobiology Research Grants 5746
Through its program of grants and grants in aid, the foundation assists scholarly research in the life sciences. The foundation is currently interested in basic research in invertebrate and vertebrate (excluding clinical) neurobiology, specifically investigations of neural mechanisms involved in sensory, motor, and other complex functions of the whole organism as these relate to behavior. Research grants are available to established scientists of all ages working at accredited institutions and will be awarded for up to three years, with possible renewal on a competitive basis for a maximum of two years. Grants in aid are designed for researchers at the assistant professor level who experience difficulty in competing for research funds because they have not yet become firmly established. Grants in aid will also be made to senior scientists. The foundation's policy is to assist those dynamic areas of basic biological research that are not heavily supported by federal agencies or other foundations with specialized missions. Types of support include personnel (salary support for students and technicians); permanent equipment, materials, and supplies needed to complete the project; travel for research; publication costs; and overhead (no more than 25 percent or supplies). Applicant should initiate the proposal process by submitting a one-page letter written in technical language identifying the nature of the research proposal and indicating whether the request is for a grant or grant in aid. Grant application forms will be sent to those applicants whose proposed project matches the foundation's current funding interests. Deadlines listed are for letters of intent.
Requirements: Accredited U.S. institutions are eligible. The foundation prefers to support scientists at the beginning of their career and those senior scientists who have maintained productivity. The principal investigator must hold no less than the position of assistant professor or the equivalent.
Samples: Daniel Pollen, U of Massachusetts (MA)—for research on the cerebral integration in the primate visual system, $40,000 of three-year grant.; Richard Ivry, U of California (CA)—for research on the role of the cerebellum in temporal processing, $24,000 final payment on three-year grant.; Paul Black, Boston U (Boston, MA)—to support research on brain-immune system interactions, $130,020 final payment on three-year grant.
Amount of Grant: $30,000-$75,000 typically; $30,000 maximum grants in aid
Date(s) Application is Due: Jan 15; Apr 15; Oct 1
Contact: Program Director, (561) 655-4474; fax (561) 655-4978; email: email@whitehall.org
Internet: http://www.whitehall.org
Sponsor: Whitehall Foundation
P.O. Box 3423
Palm Beach, FL 33480

Whiting Foundation Grants 5747
The grant program awards funds for broad purposes: in Michigan, to further secular and religious education and research of all kinds; to aid and improve the physical, financial, mental, and moral condition of the poor, the sick, the young, the aged, and the disabled among all classes; and nationally for cancer research. Types of support include general operating support and program development. Although funding is on a national scale the majority of funding is primarily made in the greater Flint, MI, area. Application forms are not required.
Samples: Cancer Research Institute (New York, NY)—$10,000 (2003); Insight Inc (Flint, MI)—$10,000 (2003); Christ Episcopal Ctr (Flint, MI)—$7500 (2003); Whaley Childrens Ctr (Flint, MI)—$30,000 (2003).
Amount of Grant: $500-$500,000 range

Date(s) Application is Due: Apr 30
Contact: Donald Johnson, Jr., President, (810) 767-3600
Sponsor: Whiting Foundation
901 Citizens Bank Building
Flint, MI 48502

Whitley Awards **5748**
The scheme supports pragmatic, long-lasting field-based nature conservation projects. A major award from the scheme should form the bulk of a project's required funding. The program is interested in wide nature conservation issues, not simply individual wildlife species preservation projects.
Requirements: Applications from leaders of teams undertaking appropriate projects anywhere in the world are encouraged. Close involvement of host country researchers and local institutions is essential. Conservationists from any country working anywhere in the world are welcome to apply. Projects involving a single visit to a country are unlikely to be funded.
Restrictions: PhD project and undergraduate expeditions are ineligible.
Amount of Grant: @L30,000
Date(s) Application is Due: Nov 18
Contact: Whitley Award Scheme, +44 (0)20 7602 3443; fax: +44 (0)20 7603 3935; email: info@whitleyaward.org
Internet: http://www.whitley-award.org
Sponsor: The Whitley Laing Foundation
50 Queensdale Road
London W11 4SA United Kingdom

WHO Sexual Violence Research Initiative **5749**
The purpose of the program is to facilitate the strengthening of research partnerships and proposal development in the field of sexual violence. The research should fall under the broad heading of improving sexual violence services. Examples of possible research issues include understanding the magnitude and nature of sexual violence; exploring alternative community-based responses (including justice) to sexual violence; examining the medico-legal response to sexual violence; examining the medical response to sexual violence; and looking at prevention of sexual violence against women. Seed funding is awarded to support activities related to the development of a joint proposal such as a meeting among partners to finalize a proposal and, in special cases, limited field testing/piloting that is needed to prepare a compelling proposal for funding. Each partnership should electronically submit a brief cover letter and a concept paper. Guidelines are available online.
Requirements: Each concept note must be submitted by at least three entities (i.e., academic/research institution, service delivery institution, and nongovernmental organization). At least one partner must be a nongovernmental organization with a gender and woman-centered perspective. At least one of the partners must demonstrate a track record of proven research capacity, research expertise, and institutional support. There should be an individual with demonstrated previous research experience involved in the research at each study site. Preference will be given to groups that are based in lower-and middle-income countries.
Restrictions: The program does not fund interventions or the research itself.
Amount of Grant: $10,000-$40,000 typically
Date(s) Application is Due: Jul 8
Contact: Program Administrator, c/o FCH/GWH Batiment X, fax: (41) 22 791 15; email: svrimail@who.int
Internet: http://www.who.int/svri/en
Sponsor: World Health Organization
Avenue Appia 20
Geneva 1211
Switzerland

WHOI Arnold B. Arons Award for Excellence in Teaching, Advising, and Mentoring **5750**
The Award will be conferred periodically every two to five years as circumstances warrant to a WHOI Scientific Staff or Senior Technical Staff member who has demonstrated any or all of the following qualities during his/her participation in the education programs of the Institution: sustained excellence in teaching; sustained excellence in advising and/or mentoring; and/or exceptional promise in teaching, advising, or mentoring during early career involvement in the Institution's education programs. The Award will consist of a plaque engraved with the recipient's name and a cash award.
Requirements: All members of the WHOI Scientific Staff and Senior Technical Staff (Research Specialists, Information System Specialists, Senior Engineers, Senior Research Specialists, Senior Information Systems Specialists and Principal Engineers), with the exception of the members of the Award Committee, are eligible for the award.
Samples: Dr. Joseph Pedlosky, Senior Scientist, Physical Oceanography (2005); Dr. Lauren Mullineaux, Senior Scientist, Biology (2004).
Contact: Shelley M. Dawicki, (508) 289-2270; fax: (508) 457-2180; email: sdawicki@whoi.edu
Internet: http://www.whoi.edu/page.do?pid=9839
Sponsor: Woods Hole Oceanographic Institution
Fenno House MS#40
Woods Hole, MA 02543

WHOI Excellence in Ocean Science Journalism Award **5751**
The award recognizes work that has made a significant and lasting contribution to public awareness and understanding of issues and advances in ocean sciences, and their impact on society as a whole. The award will be given in two categories: print and broadcast. The cash prize and award will be presented in Woods Hole in the fall, with the recipients presenting a lecture or seminar on an appropriate science journalism topic at that time.
Requirements: Entries must be books, articles or reports, or programs that were published or broadcast in any medium within the past five years, intended for and available to the general public, and written in English (or include an English translation). A nominee must be the author of the submitted work, and must be an employee of a print or electronic publication or a broadcast or cable station or network or be a freelance writer or producer.
Samples: Gareth Cook, The Boston Globe (2005); Beth Daley, The Boston Globe (2005); Paul Kennedy, CBC Radio 1 (2005).
Amount of Grant: $5,000
Date(s) Application is Due: May 13
Contact: Shelley Dawicki, 289-3340; fax: (508) 457-2180; email: media@whoi.edu
Internet: http://www.whoi.edu/page.do?pid=8897
Sponsor: Woods Hole Oceanographic Institution
Fenno House MS#40
Woods Hole, MA 02543

WHOI George P. Panteleyev Award **5752**
The George 'Gera' Pavlovich Panteleyev Award is conferred annually on the MIT/WHOI or WHOI Graduate Student in the graduating class for that academic year who best exemplifies the commitment to improving the graduate education experience and graduate student life at Woods Hole Oceanographic Institution. The Award consists of a plaque with the recipient's name and the date engraved, and a cash award, awarded at the annual WHOI Graduate Reception.
Requirements: Nominations are welcome from current and graduating Joint Program students and WHOI faculty and staff.
Samples: Kristina Dahl, Geology and Geophysics (2006); Margaret Boettcher, Geology and Geophysics (2005); Astri Kvassnes, Geology and Geophysics (2004).
Date(s) Application is Due: May 15
Contact: Marsha Gomes, (508) 289-2219; fax: (508) 457-2188; email: education@whoi.edu
Internet: http://www.whoi.edu/page.do?pid=9840
Sponsor: Woods Hole Oceanographic Institution
Academic Programs Office, MS #31
Woods Hole, MA 02543

WHOI Ocean Science Journalism Fellowship **5753**
The program introduces participants to the interdisciplinary nature of ocean sciences and engineering. Through seminars, lab visits and fieldwork, Ocean Science Journalism Fellows will be introduced to a broad range of current and future research programs in ocean engineering, marine biology, marine chemistry and geochemistry, marine geology and geophysics, and physical oceanography. Topics of discussion will range from harmful algal blooms to deep-sea hydrothermal vents, the ocean's role in climate, ocean observatories both permanent and drifting, new underwater vehicles being developed to explore the seas, and marine policy issues related to society's growing use of the oceans. Travel, room, and board are covered for fellows. The program will take place in September each year.
Requirements: The fellowship is designed for professional science writers and editors with at least two years experience. While freelancers will be considered, preference will be given to journalists in staff reporting, editing/producing, and news or editorial management positions.
Date(s) Application is Due: Jun 1
Contact: Shelley M. Dawicki, (508) 289-2270; fax: (508) 457-2180; email: sdawicki@whoi.edu
Internet: http://www.whoi.edu/home/news/media_jfellowship.html
Sponsor: Woods Hole Oceanographic Institution
Fenno House MS#40
Woods Hole, MA 02543

WICHE Professional Student Exchange Program (PSEP) **5754**
This program is open to students who are residents of Arizona, Colorado, Hawaii, Idaho, Montana, Nevada, New Mexico, North Dakota, Oregon, Utah, Washington, and Wyoming. The program has been developed to facilitate interstate resource sharing by providing means for states without specific professional schools to send residents to institutions in other WICHE states. Professional schools are those of medicine, dentistry, veterinary medicine, physical therapy, occupational therapy, optometry, podiatry, pharmacy, architecture, law, graduate library studies, graduate nursing education, public health, maritime technology, osteopathic medicine, and physician assistant. The student will pay (approximately) in-state or partial private tuition, and the sending state pays the receiving school an additional support fee. More information on the program, including eligibility requirements and restrictions, can be found on the Web site, or by contacting the program coordinator. In most states, the deadline for receiving completed applications for certification is October 15 of the year preceding admission. However, a few states have earlier deadlines or accept late applications. Contact program staff for exact dates.
Requirements: The program is limited to professional fields requested by member states (Alaska, Arizona, Colorado, Hawaii, Idaho, Montana, Nevada, New Mexico, North Dakota, Oregon, Utah, Washington, and Wyoming) and approved by WICHE. Students must meet state requirements. Support is limited to states that appropriate funds to assist specific numbers of students in specific fields.

Geographic Focus: Alaska; Arizona; Colorado; Hawaii; Idaho; Montana; Nevada; New Mexico; North Dakota; Oregon; Utah; Washington; Wyoming
Date(s) Application is Due: Oct 15
Contact: Program Coordinator, (303) 541-0214; email: sjackson@wiche.edu or info-sep@wiche.edu
Internet: http://www.wiche.edu/SEP/index.asp
Sponsor: Western Interstate Commission for Higher Education
P.O. Box 9752
Boulder, CO 80301-9752

Widener Memorial Foundation in Aid of Handicapped Children Grants 5755
The foundation exists to support research into the causes, treatment, and prevention of diseases and conditions that handicap children orthopedically and to aid and assist public and private charitable organizations in the care, education, and rehabilitation of children with disabilities. Types of support include building/renovations, equipment, seed money, research, and program/project development. There are no application forms. The board meets each year in May and November to consider requests.
Requirements: Nonprofit organizations in Delaware Valley, PA, may apply for grant support.
Restrictions: Grants are not awarded to individuals for scholarships or fellowships or for loans, endowment funds, or matching gifts.
Geographic Focus: Pennsylvania
Samples: West Chester U (PA)—to renovate an elevator in Mitchell Hall to improve access for disabled people, $100,000.
Amount of Grant: $15,000-$100,000 average
Date(s) Application is Due: Apr 15; Oct 15
Contact: F. Eugene Dixon Jr., President, (215) 836-7500
Sponsor: Widener Memorial Foundation in Aid of Handicapped Children
P.O. Box 178, 665 Thomas Road
Lafayette Hill, PA 19444-0178

WILD Foundation Grants 5756
Founded in 1974, WILD is the only international organization dedicated entirely and explicitly to wilderness protection around the world. WILD works to protect the planet's last wild places and the wildlife and people who depend upon them, because wilderness areas provide essential social, spiritual, biological and economic benefits. The WILD Foundation focuses on four main program areas: World Wilderness Congress, wilderness policy and research, communications and field projects. Refer to the Foundations website for more detailed information on each program of interest. There are no specific deadlines with which to adhere. Contact the Foundation for further application information and guidelines.
Amount of Grant: Over $900,000 total
Contact: Charlotte Baron, Chairperson; (303) 442-8811; fax. (303) 442-8877; email: info@wild.org
Internet: http://www.wild.org/main/about/
Sponsor: The WILD Foundation
717 Poplar Avenue
Boulder, CO 80304

Wildlife Conservation Research Fellowships 5757
This program is designed to support individual research projects that lead directly to conservation of threatened wildlife and wildlife habitat. Projects must have a scientific foundation and should incorporate clearly identifiable and specific conservation goals. The RFP supports field research in Africa, Asia, and Latin America. Traditionally the RFP has not supported research in North America, Australia, or Europe, or their territories. However, the RFP has just begun to accept applications from Native Americans (U.S.) and First Nation Peoples (Canada) who intend to conduct work on native lands on issues of direct relevance to wildlife. Application materials are available on the Web site. The principal researcher must write the proposal. Applications written on behalf of another individual are not allowed. Translation by a third party is permitted, however, applications are accepted in English, French, and Spanish.
Requirements: Individuals, not organizations, are eligible. Nationals of the country of research are strongly encouraged to apply.
Restrictions: The program does not support conferences, expeditions, travel to scientific meetings, legal actions, construction of permanent field station, tuition, salaries overhead costs, costly laboratory analyses, gene storage, vehicle purchase, computer purchase, or captive breeding. Organizations are not eligible for funding. Previous research fellows are not eligible for funding for the same project. Faculty and/or research advisors should not be listed as principal investigators unless they plan to conduct and achieve the majority of the field work.
Amount of Grant: $25,000 maximum; $11,000 average
Date(s) Application is Due: Mar 1; Sep 1
Contact: Program Coordinator, Research Fellowship Program, International Conservation Program, (718) 220-5155; fax: (718) 364-4275; email: fellowship@wcs.org
Internet: http://wcs.org/sw-around_the_globe/rfp/fellowship
Sponsor: Wildlife Conservation Society
185th Street and Southern Boulevard
Bronx, NY 10801

Wiley Prize in Biomedical Sciences 5758
The prize is intended to recognize contributions that have opened new fields of research or advanced novel concepts or their applications in a particular biomedical discipline. The award may recognize a specific contribution or a series of contributions that demonstrate the nominee's significant leadership in the development of research concepts or their clinical application. The award will consist of a grant, and each year's recipient will deliver a lecture at Rockefeller University.
Requirements: Wiley invites and encourages the nomination of exceptional PhD and MD scientists whose research has set the standard for excellence. More than one nomination can be made from the same organization. International nominations are eligible, and the number of years in a position is not relevant to the nomination.
Amount of Grant: $25,000
Date(s) Application is Due: Jul 31
Contact: Deborah Wiley, (212) 748-6783; fax: (212) 748-6940; email: dwiley@wiley.com
Internet: http://www.wiley.com/legacy/wileyfoundation/nominations.html
Sponsor: Wiley Foundation
111 River Street
Hoboken, NJ 07030

Wilhelm Sander Foundation Grants 5759
The purpose of the Foundations grants are to support research in the field of human medicine with a focus on cancer research.
Restrictions: The grant is limited to Germany and Switzerland.
Contact: Dr. J. Koppenhofer: (089) 544-1870; fax: (089) 544-18720; email: info@sanst.de
Internet: http://www.wilhelm-sander-stiftung.de
Sponsor: Wilhelm Sander Foundation
Goethe route 74
Munich 80336 Germany

Willamette University Minority Graduate Fellowship 5760
The fellowship supports a year-long residence at the university for a minority scholar to complete his or her dissertation. The PhD may be in any discipline in Williamette's undergraduate liberal arts curriculum. The program is intended to prepare minority scholars for careers in American higher education and also exposes such students to doctoral studies. The fellowship extends from approximately September 1 through August 31, during which time the dissertation is to be completed. It also may include teaching, either individually or as part of a team. Applications are available on the Web site and will be accepted until the position is filled.
Requirements: Applicants must be U.S. citizens and have completed all PhD requirements.
Amount of Grant: $25,000 stipend and $2500 research fund
Contact: Dr. Lawrence Cress, Dean, College of Liberal Arts, (503) 370-6285 or (503) 370-6300
Internet: http://www.willamette.edu/mgf
Sponsor: Willamette University
900 State Street
Salem, OR 97301

Willard and Pat Walker Charitable Foundation Grants 5761
The foundation awards grants to Arkansas nonprofit organizations in its areas of interest, including arts, health organizations, higher education, and social services. Types of support include building construction/renovation, general operating grants, and program development. There are no application forms or deadlines.
Requirements: Arkansas nonprofit organizations are eligible.
Restrictions: Individuals are not eligible.
Geographic Focus: Arkansas
Samples: Ouachita Baptist (Arkadelphia, AR)—for its capital campaign, $2.5 million (2004); U of Arkansas for Medical Sciences (Little Rock, AR)—to expand research and patient care in the areas of Alzheimer's disease, eye diseases, and mental illness, $21.5 million (2004); U of Arkansas (Fayetteville, AR)—to construct a building as part of the College of Business, $8 million (2003).
Contact: Tommy Karr, (501) 756-7031
Sponsor: Willard and Pat Walker Charitable Foundation
P.O. Box 5000
Springdale, AR 72765-5000

Willary Foundation Grants 5762
The foundation supports projects that are interesting, creative, and imaginative and that benefit communities in northeastern Pennsylvania. The foundation is particularly interested in projects that support leadership and the development of leadership in business, the economy, education, human services, government, the arts, media, and research. The foundation gives preference to efforts that have a ripple effect in the community and those that are conducted in conjunction with other sources of funding. Willary seeks to foster both individuals and groups with unique, innovative, or unusual ideas and efforts. Applications available on Web site.
Requirements: Individuals and nonprofits in Lackawanna and Luzerne counties, PA, are eligible to apply.
Restrictions: Generally, Willary will not consider applications for capital campaigns and annual drives.
Geographic Focus: Pennsylvania
Samples: Tri-County Human Services Center (Carbondale, PA)—for a childhood development program, $10,032. Scranton Cultural Ctr (PA)—for a program in arts and culture, $100,000.
Amount of Grant: $1000-$25,000 range

Contact: Program Contact, (570) 961-6952; fax: (570) 961-7269; email: info@willary.org
Internet: http://www.willary.org
Sponsor: Willary Foundation
P.O. Box 283
Scranton, PA 18501

William A. Klinger Research Award **5763**

The AGC Education and Research Foundation is proud to support construction industry-related research through its William A. Klinger Research Award. Research topics are developed by the Foundation Board of Directors in cooperation with the Associated General Contractors of America's Education Committee, and RFPs are sent to universities across the country for consideration.
Requirements: RFPs for upcoming Klinger Awards are made available in August and sent to universities across the country for consideration. To be placed on the mailing list for future RFPs, please send an email to Floretta Slade, Director of Programs (see contact info for email address).
Samples: 2006: Michigan State University, project geared toward enabling construction constructors to successfully navigate the LEED compliance process and to become active team members with owners and designers from the early stages of the project. 2005: M.E. Rinker, Sr. School of Building Construction at the University of Florida on a research project entitled, 'Moisture Control Construction Checklist (MC3)' provides the construction industry with a simplified, PC-based moisture control construction checklist that will assist building contractors in minimizing the liabilities associated with mold-related hazards.
Amount of Grant: $25,000
Date(s) Application is Due: Nov 1
Contact: Floretta Slade, Director of Programs, (703) 837-5342; fax: (703) 837-5402; email: sladef@agc.org
Internet: http://www.agc.org/page.ww?section=AGC+Foundation&name=Industry-Related+Research
Sponsor: AGC Education and Research Foundation
2300 Wilson Boulevard, Suite 400
Arlington, VA 22201

William Adlington Cadbury Charitable Trust Grants **5764**

The trust was established to support general charitable purposes in the West Midlands, the rest of the United Kingdom including projects in Northern Ireland, and UK charities working overseas. Areas of interest include social welfare, medical and health-care projects including medical research, education and training, The Religious Society of Friends (Quakers), places of religious worship and associated social projects, conservation of the environment including the preservation of listed buildings and monuments, arts, and penal affairs. Applications are considered on a continuing basis throughout the year. Application and guidelines are available online.
Restrictions: The trust does not fund individuals (whether for research, expeditions or educational purposes); projects concerned with travel, adventure, sports, or recreation; organizations that do not have UK charity registration (except those legally exempt from registration); or overseas charities not registered in the United Kingdom.
Contact: Carolyn Bettis, Trust Administrator, 44-121-472-1464
Internet: http://www.wa-cadbury.org.uk
Sponsor: William Adlington Cadbury Charitable Trust
Bristol Road, Selly Oak
Birmingham B29 6QF United Kingdom

William and Charlotte Parks Foundation for Animal Welfare Grants **5765**

The foundation awards grants to support animal protection to both animal protection organizations and to individual scholars who are pursuing a course of study that might advance the goals of the trust founders. Organizations may receive funds for capital projects, for general operating expenses, or for specific projects. Doctoral or post-doctoral candidates may receive funds to support a specific research project. These funds must be distributed through an organization with 501(c)3 status. Deadlines are May 1 for project, operating, and capital grants; and December 1 for PhD/veterinary fellowships. Application and guidelines are available online.
Requirements: 501(c)3 tax-exempt organizations are eligible. A non-U.S. applicant must be registered as a charitable organization in the home country.
Restrictions: Grants will not normally be awarded to improve animal health, for local spay/neuter assistance, to save endangered species, to rehabilitate wildlife, or to support political candidates. Grants focused on the conservation or protection of wild animal populations will not be funded. Grants will not ordinarily be made to organizations with an annual income of more than $1 million or a large asset base, with the exception of institutions of higher learning.
Amount of Grant: $10,000 project grants; $30,000 maximum for doctoral research
Date(s) Application is Due: May 1; Dec 1
Contact: Donna Pease, fax: (301) 548-7726; email: info@parksfoundation.org
Internet: http://www.parksfoundation.org
Sponsor: William and Charlotte Parks Foundation
700 Professional Drive
Gaithersburg, MD 20879

William and Flora Hewlett Foundation Education Grants **5766**

Grants in the education program are made to promote long-term institutional development, reform, or knowledge creation in five priority areas: improving instruction, California reform, technology, universal education, and opportunity. Grants to improve instruction should increase student achievement by developing, demonstrating, evaluating, and disseminating approaches to systematically improve classroom instruction in urban school systems; initiatives in California reform incorporate research, policy analyses, public information strategies, and other means to help improve the quality of the state, schools and community colleges; open educational resources improve access to exemplary postsecondary and K-12 educational content by making high-quality open educational resources freely available on the Web; and opportunity grants are made that support the overall goals of the Foundation yet are not directly aligned with one of the other program areas. Types of support include policy studies, research, development, demonstrations, evaluations, dissemination, and public engagement to accomplish its objectives. An individual grant should develop knowledge usable beyond its boundaries and should add clear and substantial value to accomplishing the goals of a Program priority. Grants must be approved by the board of directors, which meets quarterly. Meeting dates are available upon request. The foundation does not accept proposals via fax or email.
Requirements: Initial contact should be by letter of inquiry addressed to the president providing a straightforward statement of needs and aspirations for support. Applicants who receive favorable responses will be invited to submit formal proposals.
Restrictions: Requests will not be considered to fund student aid, construction, equipment and computer purchases, education research, basic scientific research, health research, or health education programs.
Geographic Focus: California
Samples: U of Maryland at College Park (MA)—for the Center for International Development and Conflict Management, $100,000 (2005); U of California at Berkeley (CA)—for a series of education-policy seminars for California state legislators and staff members, $125,000 (2005); Stanford U (CA)—to create and test Web-based resources designed to support the teaching and learning of American history, $500,000 (2005); Carnegie Mellon U (Pittsburgh, PA)—to expand the number of online courses, $1.5 million (2005).
Contact: Program Contact, (650) 234-4500; fax: (650) 234-4501; email: info@hewlett.org
Internet: http://www.hewlett.org/Grantseekers
Sponsor: William and Flora Hewlett Foundation
2121 Sand Hill Road
Menlo Park, CA 94025

William and Flora Hewlett Foundation Environmental Grants **5767**

The program has three goals: to save the great ecosystems of the North American West; to reduce the environmental impacts of fossil-fuel energy systems by promoting energy efficiency and renewable energy sources; and to build an infrastructure for environmental concerns among California's growing but still underrepresented populations in the LA basin and the Central Valley. Types of support include general operating funds, seed money, continuing support, emergency funds, land acquisition, special projects, matching funds, employee matching gifts, and endowment funds. Grants must be approved by the board of directors, which meets quarterly. Meeting dates are available upon request. The foundation does not accept proposals via fax or email.
Requirements: Initial contact should be by letter of inquiry addressed to the president providing a straightforward statement of needs and aspirations for support. Applicants who receive favorable responses will be invited to submit formal proposals.
Restrictions: The foundation usually does not support proposals in the areas of basic research; capital construction; conferences, symposia, or workshops; environmental education (K-12 or adult); museum facilities, exhibits, or programs; or specific media projects. The foundation does not make awards to individuals, organizations outside the United States, or local land trusts, nor does it support community organizing, advocacy, or litigation.
Samples: Centro Interdisciplinario de Biodiversidad y Ambiente (Mexico)—to design and implement a rapid-transit bus corridor in Mexico City, $400,000 (2005); David Suzuki Foundation (Vancouver, Canada)—for the Turning Point Initiative, which works on land-use issues on the North Pacific Coast, $750,000 (2005); Nature Conservancy (Arlington, VA)—to raise money for the protection of Great Bear Rainforest, in western Canada, $345,000 (2005); Relational Culture Institute (Fresno, CA)—for an air-quality project in California's San Joaquin Valley, $90,000 (2005).
Contact: Hal Harvey, Program Director, (650) 234-4500 ext 5647; fax: (650) 234-4501; email: info@hewlett.org
Internet: http://www.hewlett.org/Programs/Environment/envHome.htm
Sponsor: William and Flora Hewlett Foundation
2121 Sand Hill Road
Menlo Park, CA 94025

William Andrews Clark Predoctoral Fellowships **5768**

Three-month residential fellowships are awarded to advanced doctoral candidates of the University of California whose dissertations concern any area represented in the Clark's collections or are linked to Center/Clark programs supported by these collections or by others at UCLA. The Center for 17th- and 18th-Century Studies, an organized research unit of UCLA, provides a forum for the discussion of central issues in the study of the 17th and 18th centuries, facilitates research and publication, supports scholarship, and encourages the creation of interdisciplinary, cross-cultural programs that advance the understanding of this period. The application deadline is for fellowships to be carried out between July 1 of one year and June 30 of the next.
Requirements: University of California graduate students are eligible.
Amount of Grant: $6000 for three months
Date(s) Application is Due: Feb 1
Contact: Fellowship Coordinator, fax: (310) 206-8577; email: clarklib@humnet.ucla.edu

Internet: http://www.humnet.ucla.edu/humnet/c1718cs/applic6.htm
Sponsor: William Andrews Clark Memorial Library
310 Royce Hall, 405 Hilgard Avenue
Los Angeles, CA 90095-1404

William B. Dietrich Foundation Grants 5769

The foundation awards grants to nonprofits in Pennsylvania in its areas of interest, including elderly services; AIDS and AIDS research; children and youth services; education and higher education; historic preservation; human services; museums; and secondary education. Types of support include general operating support, capital campaigns, building construction/renovation, program development, research, and matching funds. There are no application forms or deadlines.
Requirements: Pennsylvania nonprofits are eligible to apply.
Restrictions: No grants are provided for individuals.
Geographic Focus: Pennsylvania
Samples: Woodmere Art Museum (Philadelphia, PA)—to expand its administrative offices, $200,000.; Wellness Community of Philadelphia (PA)—to restore and renovate an 18th-century barn that will serve as a new facility for this group that offers programs and services for people with cancer and their families, $350,000.
Contact: William Dietrich, President, (215) 979-1919
Sponsor: William B. Dietrich Foundation
P.O. Box 58177
Philadelphia, PA 19102-8177

William Caspar Graustein Memorial Fund Grants 5770

The fund supports nonprofits focusing on children's education in Connecticut. It focuses on projects that expand the supply of high quality early care and education in communities; that improve the quality of existing early care and education; that build strong connections between early care and elementary education; and those that improve students' social, emotional and academic performance. A brief letter of inquiry concisely stating the project's goals should be the route of initial approach.
Requirements: Connecticut nonprofits focusing on children's education are eligible for grant support.
Restrictions: In general, the fund does not support: non 501(c)3 organizations, individuals, religious organizations for sectarian purposes, capital campaigns, political causes and activities, or scholarships.
Geographic Focus: Connecticut
Samples: Ctr for Youth as Resources (Washington, DC)—for a model youth-development program in Bridgeport, CT, that will involve students in community service activities designed to prevent violence and improve overall community life, $50,000.
Contact: Program Contact, (203) 230-3330; email: gmfmail@wcgmf.org
Internet: http://www.wcgmf.org/code/grants/grant.html
Sponsor: William Caspar Graustein Memorial Fund
2319 Whitney Avenue, Suite 2B, 1 Hamden Ctr
Hamden, CT 06518

William G. McGowan Charitable Fund Grants 5771

The fund awards grants to eligible nonprofit organizations that are devoted to three specific philanthropic goals: developing the talents and gifts of the very young, especially those who have been disenfranchised by virtue of low income status, inner-city conditions or family situations; funding for selected areas of medical research directed toward finding cures for, or relieving the pain and suffering of, those afflicted with debilitating conditions or diseases; and the McGowan Scholars program, which provides financial assistance to selected college and university students interested in pursuing an undergraduate or graduate degree at a college or university business school accredited by the International Association for Management Education or Association of Collegiate Business Schools and Programs.
Requirements: Nonprofit organizations in the following states and cities are eligible: Chicago, IL; District of Columbia and its Virginia suburbs; metropolitan Kansas City, western New York; and northeast Pennsylvania.
Restrictions: The fund does not support multiyear grants; accept requests online; or fund art or theatrical activities, building funds, church renovation campaigns, or endowments.
Geographic Focus: California; District of Columbia; Illinois; Missouri; New York; Pennsylvania; Virginia
Samples: Rochester Institute of Technology, College of Applied Science (NY)—to construct the Center for Telecommunications, $2 million (2004); U of the Pacific (Stockton, CA)—for a scholarship for a business student, $18,000 (2004); Foundation for the National Archives (Washington, DC)—for the National Archives Experience, which will make important American documents accessible to the public and teach visitors about the value of archives, $5 million (2003).
Amount of Grant: $8.1 million total; $10,000-$100,000 average
Date(s) Application is Due: Jan 2; May 1; Sep 1
Contact: Bernard Goodrich, Executive Director, (301) 320-8570; fax: (301) 320-8627; email: goodric@aol.com or info@mcgowanfund.org
Internet: http://www.mcgowanfund.org/guidelines.html
Sponsor: William G. McGowan Charitable Fund
P.O. Box 40515
Washington, DC 20016-0515

William K. Warren Foundation Grants 5772

The foundation awards grants to eligible nonprofit organizations in its areas of interest, including Catholic healthcare facilities, Catholic social services, education, and medical research. Types of support include building construction/renovation, endowment campaigns, general operating support, program development, and research grants. Grants are awarded primarily in Oklahoma. There are no application forms or deadlines. The board meets semiannually.
Geographic Focus: Oklahoma
Samples: Salvation Army of Tulsa (OK)—for general operating support, $135,000.; Saint Anthony Hospital Foundation (Oklahoma City, OK)—for general support, $100,000.
Amount of Grant: $1000-$50,000 average
Contact: Grants Administrator, (918) 492-8100
Sponsor: William K. Warren Foundation
P.O. Box 470372
Tulsa, OK 74147-0372

William N. and Myriam Pennington Foundation Grants 5773

The foundation supports education, health, and medical research in Nevada. Funding priority is given to organizations serving children and youth, the elderly, and the economically disadvantaged. Types of support include general operating support, building construction/renovation, equipment acquisition, and scholarships to individuals.
Requirements: Nevada nonprofits may apply for grant support.
Geographic Focus: Nevada
Samples: Assistance League of Reno-Sparks (Reno, NV)—for general support, $15,000.; U of Nevada, Department of Speech Pathology (Reno, NV)—for equipment, $17,500.
Amount of Grant: $2500-$200,000 range
Contact: Kent Green, Foundation Manager, (775) 333-9100
Sponsor: William N. and Myriam Pennington Foundation
441 W Plumb Ln
Reno, NV 89509

William Schallek Memorial Graduate Study Fellowships 5774

The fellowship supports an advanced graduate student who is writing a PhD dissertation in any relevant discipline dealing with late-medieval Britain (ca. 1350-1500). The fellowship helps defray research and living expenses for the equivalent of an academic year of study. The fellowship recipient must devote full time to the dissertation project and may not hold any job or teaching position or work on another project during the term of the fellowship. Forms may be obtained from the society or at its Web site.
Requirements: Applicants must be graduate students whose dissertation proposals have been approved by their dissertation committees. They must be members of the Medieval Academy as of 15 September of the year in which they apply. In accordance with the terms of the gift to the Richard III Society, applicants must be citizens or permanent residents of the United States or Canada.
Amount of Grant: $30,000
Date(s) Application is Due: Oct 15
Contact: Sheryl Mullane-Corvi, (617) 491-1622; fax: (617) 492-3303; email: speculum@medievalacademy.org
Internet: http://www.medievalacademy.org/grants/gradstudent_grants_schallek.htm
Sponsor: American Branch of the Richard III Society
Medieval Academy of America , 104 Mt Auburn Street, 5th Fl
Cambridge, MA 02138

William T. Grant Foundation Distinguished Fellows Awards 5775

The goal of the program is to increase the supply of, demand for, and use of high-quality research in the service of improved youth outcomes and well-being. To accomplish this goal, the program gives researchers the opportunity to immerse themselves in practice or policy settings and allows influential mid-career practitioners and policymakers to work and be mentored in research settings. Each Fellow will receive up to $175,000 for the total duration of the fellowship, which may range between six months and two years, depending on the proposed design. The time structure of the fellowship is deliberately flexible. Fellowships are permitted to extend up to two years provided that the fellowship activity amounts to a minimum of half of a person-year spent at the fellowship site(s) over the duration of the project. (Thus, the minimum duration is full-time for six months at the fellowship site.) The foundation will also provide small grants up to $25,000 to the fellowship site(s) to defray the costs associated with hosting a Distinguished Fellow.
Requirements: Influential mid-career practitioners, policymakers, and researchers are encouraged to apply. Support is restricted to researchers seeking to work in policy or practice settings, or policymakers and practitioners seeking to work in research settings. The deadline for receipt of Letters of Inquiry is January 10. All fellowships must support activities that help the Distinguished Fellow advance the foundation, Current Research Priorities (download the most current listing at the website). Grant awards will be made to the applicant's employer, which must be a non-profit private or governmental organization.
Restrictions: The foundation reserves the right to terminate the award if the Fellow's project changes substantially from that originally proposed by the Fellow. Additionally, approval will be required from the foundation if the Fellow wishes to transfer the award to another institution.
Samples: Tamera Coyne-Beasley, M.D., M.P.H., Associate Professor of Pediatrics and Internal Medicine, University of North Carolina at Chapel Hill — Host Organizations: National Students Against Violence Everywhere (SAVE) and Center for the Prevention of School Violence (CPSV), part of the North Carolina Department of Juvenile Justice and Delinquency Prevention; David DuBois, Ph.D., Professor, Community Health Sciences, School of Public Health, University of Illinois at Chicago — Host Organizations: Big Brothers Big Sisters of Metropolitan Chicago (BBBSMC) and Big Brothers Big Sisters of America (BBBSA); Abram Rosenblatt, Ph.D., Professor, Department of Psychiatry,

University of California, San Francisco — Host Organizations: Santa Cruz County, Probation and Substance Abuse Departments and the New California Network.
Amount of Grant: $175,000 maximum; $25,000 maximum to fellowship site
Date(s) Application is Due: Jan 8
Contact: Sharon Brewster, ODF Grants Coordinator; (212) 752-0071; fax: (212) 752-1398; email: sbrewster@wtgrantfdn.org
Internet: http://www.wtgrantfoundation.org/funding_opportunities/fellowships/william_t__grant_distinguished_fellows
Sponsor: William T. Grant Foundation
570 Lexington Avenue, 18th Floor
New York, NY 10022-6837

William T. Grant Foundation Research Grants **5776**
The foundation supports research to improve the lives of youth ages 8 to 25 in the United States. This is done primarily by investing in high quality empirical studies. The foundation's Current Research Interests are understanding and improving social settings (i.e., families, schools, peer groups, organizations, programs, etc.), their effects on youth, and the use and influence of scientific evidence. Within these interests, the foundation is interested in descriptive and intervention studies. There are no geographical boundaries for its support of research projects.
Requirements: The grants are limited, without exception, to established tax-exempt, private organizations and institutions. To apply for a major grant, the Principal Investigator must submit a Letter of Inquiry to the grants coordinator briefly describing the project or program and the financial needs. Letters of Inquiry are due on the deadline dates listed. If it is determined that the project falls within the current program interests and priorities, a full proposal will be requested. Full proposal deadlines are four to six months prior to the quarterly board meetings in March and October.
Restrictions: The foundation does not support or make contributions to building funds, fundraising drives, endowment funds, general operating budgets, or scholarships. Grants are made to organizations or institutions, not individuals.
Amount of Grant: $200,000 - $500,000
Date(s) Application is Due: Nov 1; Jan 3; Apr 1
Contact: Nancy Rivera-Torres, Research Grants Coordinator; (212) 752-0071; fax: (212) 752-1398; email: nrivera@wtgrantfdn.org
Internet: http://www.wtgrantfoundation.org/funding_opportunities/research_grants
Sponsor: William T. Grant Foundation
570 Lexington Avenue, 18th Floor
New York, NY 10022-6837

William T. Grant Foundation Scholars Program **5777**
The program supports promising postdoctoral researchers from diverse disciplines. Studies from these researchers deepen and broaden the knowledge base in areas that contribute to creating a society that values young people as a resource and helps them live up to their potential. Primary research areas are youth development; programs, policies, and institutions affecting young people; and adults' attitudes about and perceptions of young people, along with the consequences of those attitudes and perceptions. The foundation focuses on young people ages 8-25, and is particularly interested in research that is interdisciplinary, examines young people in social, institutional and cultural contexts, and addresses issues that are relevant to youth-related programs and policies. Applicants do not need to be in a tenure-track position or affiliated with a university to apply for the program. On the Web site, click on funding opportunities and then William T. Grant Scholars.
Requirements: Applicants at all non-profit institutions, both in the United States and abroad, are eligible. International applicants should be pursuing research that has clear, compelling policy or practice implications for the settings of young people ages 8 to 25 in the United States. Any discipline is eligible. The award is designed for early career researchers. Applicants must have received their terminal degree within seven years of submitting their application. In many scholarly disciplines this translates to a maximum of seven years following the award of the doctoral degree and includes time spent as a post-doctoral fellow. In medicine, the seven-year maximum is dated from the completion of the first residency.
Restrictions: The award may not be used as a post-doctoral fellowship.
Amount of Grant: $350,000 over five years
Date(s) Application is Due: Jul 8
Contact: Sharon Brewster, ODF Grants Coordinator; (212) 752-0071; fax: (212) 752-1398; email: sbrewster@wtgrantfdn.org
Internet: http://www.wtgrantfoundation.org/funding_opportunities/fellowships/william_t__grant_scholars/william_t_grant_scholars
Sponsor: William T. Grant Foundation
570 Lexington Avenue, 18th Floor
New York, NY 10022-6837

William T. Grant Foundation Understanding the Acquisition, Interpretation, and Use of Research Evidence in Policy and Practice Grants **5778**
The William T. Grant Foundation has a longstanding interest in supporting research that can inform policy and practice. The Foundation's particular focus is on policies and practices that affect youth ages 8 to 25 in the United States. In this area, there are significant gaps between research and policy, and between research and practice. Researchers express frustration that policymakers and practitioners do not use or misuse research findings. Policymakers and practitioners suggest that research is often not relevant to their work or is not easily accessible or understood. Many researchers, research funders, and intermediary organizations have sought to address these gaps by encouraging the production of more rigorous research evidence, better research syntheses, and improved approaches to disseminating research evidence. Policymakers have also tried to improve the connection between research and practice by mandating the use of research findings through law or regulation. Relatively little research attention has been devoted to understanding the user side,Äìthat is, studying what affects policymakers' and practitioners' acquisition, interpretation, and use of research evidence. At the Foundation, we believe studies of this topic will increase our understanding of how to improve the production and subsequent use of research for and in policy and practice. For the next several years, the Foundation anticipates supporting a group of research projects, with award amounts ranging from $100,000 to $600,000, covering direct and indirect costs for two to three years of work. Applicants should submit letters of inquiry by May 12. Selected applicants will be nvited to submit full proposals, which will be due by October 6. Funding decisions will be made at the Board of Trustees meeting in June and awards will be made available shortly thereafter.
Requirements: To be eligible for consideration, applicants must: be employed at a nonprofit institution, either in the United States or abroad. In most instances, these organizations will have 501(c)3 tax exempt status, set up to receive and process grant awards. However, in rare cases, colleges or universities (such as Arizona State University) will not have 501(c)3 tax exemption. If your institution fits into this category, the Foundation can instead accept an IRS determination letter which states that it is classified as a school under sections 509a1 and 170b1Aii of the IRS Code. Submit a project that is consistent with the stated goals of the RFP; and reflect high standards of evidence and rigorous methods, commensurate with the RFP's goals.
Amount of Grant: $100,000 - $600,000 two - three years
Date(s) Application is Due: May 12
Contact: Nancy Rivera-Torres, Research Grants Coordinator; (212) 752-0071; fax: (212) 752-1398; email: nrivera@wtgrantfdn.org
Internet: http://www.wtgrantfoundation.org/funding_opportunities/research_grants
Sponsor: William T. Grant Foundation
570 Lexington Avenue, 18th Floor
New York, NY 10022-6837

Williams Companies Foundation Grants **5779**
The foundation provides opportunities that improve the quality of life in company-operating locations. Areas of interest include education—local and national educational institutions, libraries, research projects, and more; health and human services—programs that contribute to the social, economic, and cultural well-being of communities; arts and humanities—national and local arts organizations, museums, cultural centers, public broadcasting, and more; and civic and community—national and local organizations that work to combat social problems, elevate the family, and break down barriers to economic and social growth. Applicants should submit proof of nonprofit status and a letter with an outline describing the organization, the amount and purpose of the request, and whether the request is for general operating support or for a specific program. There are no application deadlines.
Requirements: Nonprofits in Williams operating communities are eligible.
Geographic Focus: Alabama; Louisiana; New Jersey; North Carolina; Pennsylvania; South Carolina; Texas; Virginia
Contact: Community Relations Coordinator, (918) 573-2912; email: communityrelationstulsa@williams.com
Internet: http://www.williams.com/community/foundation.asp
Sponsor: Williams Companies Foundation
1 Williams Ctr
Tulsa, OK 74172

Winn Feline Foundation Grants **5780**
The foundation supports research studies into medical problems affecting cats. Grants have supported research on feline asthma, anemia, inflammatory bowel disease, hyperthyroidism, FIP, megacolon, polycystic kidney disease, transdermal drug delivery, liver disease, and vaccine-associated fibrosarcoma. Studies general in scope are primarily encouraged; however, the foundation is also interested in projects that address problems unique to individual breeds. Continuation of grants awarded will be considered. Application guidelines are available online.
Requirements: Applicants may be faculty veterinarians, postdoctoral fellows, practicing veterinarians, or veterinary students.
Restrictions: The foundation does not fund salaries of principal investigators, major equipment expenditures, travel or indirect costs.
Samples: Thomas Schermerhorn, VMD, DACVIM, Kansas State U (MO)—for the study Development of a Lab Technique to Help Understand Feline Diabetes: Development of a Real-Time RT-PCR Technique to Study Feline Glucose Metabolism, $9638 (2003); Stephen Boyle and Beverly Purswell, Virginia-Maryland Regional College of Veterinary Medicine, Virginia Tech (VA)—for the study Feline Herpes Virus—Oral Delivery System for an Immunocontraceptive Vaccine Specific for Cats, $14,957 (2003); Diane Addie, Dept of Veterinary Pathology, U Glasgow (Scotland)—for the study Investigation of Coronavirus Carrier Cats: Viral Replication in Lifelong Feline Coronavirus Carriers, $15,000 (2003).
Amount of Grant: $15,000 maximum
Date(s) Application is Due: Dec 1
Contact: Grants Administrator, (732) 528-9797; email: winn@winnfelinehealth.org
Internet: http://www.winnfelinehealth.org
Sponsor: Winn Feline Foundation
P.O. Box 1005
Manasquan, NJ 08736-0805

Winn-Dixie Stores Foundation Grants 5781
Winn-Dixie makes all donations through the foundation, focusing on the stores' trade area in the southern states. Winn-Dixie's giving is accomplished primarily through its matching gifts program. The majority of these funds have generally gone to United Way, with direct grants to United Way campaigns in many of its market communities. Grants for education are given to colleges, associated college funds, and elementary and secondary schools. Besides United Way, other health and human services donations include support for local branches of national disease organizations, hospices, mental health programs, and hospitals. Youth-related grantees include 4-H clubs, Scouts, YMCA/YWCA, and Junior Achievement. Additional types of support include continuing support, annual campaigns, building construction/renovation, equipment acquisition, program development, conferences and seminars, scholarship funds, research grants, and matching funds. There are no formal guidelines or application forms.
Requirements: IRS 501(c)3 tax-exempt organizations in Alabama, Florida, Georgia, southern Indiana, Kentucky, Louisianna, Mississippi, southern Ohio, Oklahoma, Texas, and Virginia are eligible, including national organizations.
Restrictions: Grants will not be made to individuals or to religious or political organizations. Matching contributions to fraternal, professional, social, or recreation groups will not be made. No grants except for matching contributions are made outside the trade territories served by Winn-Dixie: Alabama, Florida, Georgia, Kentucky, Louisiana, North Carolina, South Carolina, Texas, and Virginia.
Samples: YMCA, Florida's First Coast Metropolitan (Jacksonville, FL)—$30,000.; Urban League of Greater New Orleans (LA)—$25,000.
Amount of Grant: $100-$10,000 average
Contact: D.F. Henry, President, (904) 783-5000; P.O. Box B, Jacksonville, FL 32203-0297
Sponsor: Winn-Dixie Stores Foundation
5050 Edgewood Ct
Jacksonville, FL 32254-3699

Winston Churchill Foundation Scholarships 5782
The program offers American citizens of exceptional ability and outstanding achievement the opportunity to pursue graduate studies in engineering, mathematics, or the sciences at the University of Cambridge. The foundation awards at least thirteen scholarships and hopes to increase that number to fifteen soon. The one-year awards lead to the Masters of Philosophy (MPhil), the Certificate of Post-Graduate Study (CPGS in different fields), the Certificate of Advanced Study (CAS in theoretical or applied mathematics), and a Diploma (in Computer Science). On occasion the foundation offers the award to an individual who has made outstanding contributions exemplifying Churchill's attributes and ideals. is worth between $44,000 and $50,000. It covers all University and College tuition and fees (currently about $25,000). In addition, students receive a living allowance of 10,000 GBP if enrolled in a nine-month academic program and 12,000 GBP if enrolled in a full-year academic program. They also receive an allowance of up to $1,000 for travel to and from the United Kingdom. The foundation also offers the possibility of a Special Research Grant of up to $2,000; this grant may cover travel for presentations at international conferences, short stays at another university or institute for special research, and other activities. Married students should consult with the Foundation about the possibility of additional support.
Requirements: Applicants for a Churchill Scholarship must be citizens of the United States and must be a senior who is enrolled in one of the institutions participating in the program competition or a student who has recently graduated from one of those institutions. (List available at www.winstonchurchillfoundation.org/Institutions.html.) Upon taking up a scholarship, a Churchill Scholar must be between the ages of 19 and 26, hold a bachelor's degree or its equivalent, and may not have attained a doctorate. Instructions, forms and applications are available at the sponsor's website.
Samples: (2008-2009) Matthew Biancalana, University of Chicago, Biochemistry.; Lisa Bond, University of North Carolina, Clinical Biochemistry.; Byron Drury, Haverford College, Applied Mathematics and Theoretical Physics.; Jeremy Harris, Johns Hopkins University, Chemistry.; Stephanie Jones, University of North Carolina, Chemistry.; Katherine Leitch, Michigan State University, Physiology.; Alison Miller, Harvard University, Mathematics.; Vivek Narsimhan, California Institute of Technology, Applied Mathematics and Theoretical Physics.; Andrew Niles, University of Rochester, Mathematics.; Courtney Ondeck, Carnegie Mellon University, Chemistry.; Anand Oza, Princeton University, Applied Mathematics and Theoretical Physics.; Aaron Pixton, Princeton University, Mathematics.; Benjamin Safdi, University of Colorado, Applied Mathematics and Theoretical Physics
Amount of Grant: $44,000 - $50,000
Date(s) Application is Due: Nov 12
Contact: Peter C. Patrikis, Executive Director; (212) 752-3200; fax: (212) 246-8330; email: info@winstonchurchillfoundation.org
Internet: http://www.winstonchurchillfoundation.org/Scholarships.html
Sponsor: Winston Churchill Foundation of the 600 Madison Avenue, 16th floor
New York, NY 10022-1615

Winterthur Museum Lois F. McNeil Dissertation Fellowships 5783
The fellowships are for doctoral candidates engaged in research on American decorative arts and material culture. The award is for either one or two semesters fellowships. Applications are available on the Web site.
Requirements: Applicants must have completed coursework, passed qualifying exams, fulfilled language requirements, and have an approved prospectus. The fellowship is open to students in departments of history, anthropology, folklore, art history, American studies, African American history, historic preservation, and related fields.
Amount of Grant: $7000 per semester
Date(s) Application is Due: Jan 15
Contact: Gretchen Townsend Buggeln, Director, Winterthur Research Fellowship Program, (302) 888-4649; fax: (302) 888-4870; email: academicprograms@winterthur.org
Internet: http://www.winterthur.org/research/fellowship.asp?sub=fellowships_avail
Sponsor: Winterthur Museum, Garden, and Library
Winterthur, DE 19735

Winterthur Research Fellowships 5784
Fellowships are awarded from one to six months to encourage research in America's artistic, cultural, intellectual, and social history. Applicants need not apply for a specific named fellowship, but designated awards include: Faith Andrews Fellowships—for the study of Shaker life and material culture; Robert Lee Gill Fellowships—for research on American decorative arts, painting, architecture, or historic preservation; and Dwight P. Lanmon Fellowships—for the study of glass and ceramics. Grants are available to academic, museum, and independent scholars, and to support dissertation research. Resources include a library of approximately one-half million imprints, manuscripts, visual materials, and printed ephemera supporting interdisciplinary study of American life into the early 20th century. Furnished rental housing is available on the grounds. Fellows are provided with a library carrel, 24-hour access to the library's circulating collections, and circulation privileges at the nearby Morris Library of the University of Delaware. Applications are available on the Web site.
Requirements: Applications are welcome from professors, museum and public history professionals, doctoral students, and independent scholars.
Amount of Grant: $1500 per month
Date(s) Application is Due: Jan 15
Contact: Gretchen Townsend Buggeln, Director, Research Fellowship Program, (302) 888-4640; email: gbuggeln@winterthur.org
Internet: http://www.winterthur.org/research/fellowship.asp
Sponsor: Winterthur Museum, Garden, and Library
Winterthur Museum
Winterthur, DE 19735

Wisconsin American Legion Auxiliary Scholarships 5785
The American Legion Auxiliary in Wisconsin offers scholarships each year to high school students, college students, and graduate students to provide financial support for education to students with outstanding scholastic ability who are in financial need. Various named scholarships are available as well as scholarships for special areas such as nursing, health careers, and special education. When requesting an application form, the applicant should specify if he or she is a high school student, college student, or graduate student; there are three different forms. Some of the special area scholarships have additional requirements; ask for guidelines when requesting the forms.
Requirements: Students must be residents of Wisconsin; be the daughter, son, wife, or widow of a veteran; have a minimum 3.2 GPA on a 4.0 scale; and be in need of financial aid to continue education. The granddaughter or great-granddaughter may apply if she is an American Legion Auxiliary member.
Geographic Focus: Wisconsin
Amount of Grant: $18,000 total; $750-$1000 scholarships for individuals
Date(s) Application is Due: Mar 15
Contact: Grants Administrator, (608) 745-0124; fax (608) 745-1947; email: alawi@amlegionauxwi.org
Internet: http://www.amlegionauxwi.org
Sponsor: Wisconsin American Legion Auxiliary
P.O. Box 140
Portage, WI 53901-0140

Wisconsin Humanities Council Mini-Grants for Scholarly Research 5786
The Wisconsin Humanities Council (WHC) awards mini-grants to support scholarly research that will lead to public humanities programs. The public humanities program may take the form of an exhibit, media production, reading/discussion series, public lecture, or any other means otherwise eligible for funding by the WHC. Research grant funds may be used for expenses related to research, including sabbatical support, travel, student workers, and secretarial assistance. Guidelines are available on the Web site.
Requirements: Proposals should be submitted jointly by a scholar, as the project director, and by a nonprofit organization (other than a college or university) eligible to receive funds from the WHC. Although the scholar will ordinarily write the grant proposal, the nonprofit organization will work with the scholar to develop the research project and to ensure that both the research and the consequent public humanities program address the questions and concerns of the public audience that the project is designed to reach. The nonprofit organization will also provide the fiscal agent. Scholars must have an advanced degree in an appropriate humanities discipline, but may be K-12 educators or independent scholars not employed by a college or university.
Restrictions: Grants cannot be used for purchase of permanent equipment or for travel to academic conferences.
Geographic Focus: Wisconsin
Amount of Grant: $2000 maximum
Date(s) Application is Due: Jan 1; Mar 1; May 1; Jul 1; Sep 1; Nov 1
Contact: Dena Wortzel, Director of Programs, (608) 262-0706; fax: (608) 263-7970; email: dwortzel@facstaff.wisc.edu or jebecker2@facstaff.wisc.edu
Internet: http://www.wisconsinhumanities.org
Sponsor: Wisconsin Humanities Council
222 S Bedford Street, Suite F
Madison, WI 53703-3688

Women in United Methodist History Research Grant 5787

The grant provides seed money (travel, secretarial services, etc.) for research projects relating specifically to the history of women in the United Methodist Church or its antecedents. Grants totaling $2,500 will be available with each grant being at least $1,000 awarded at the discretion of the committee of judges. A copy of the final product will be deposited with the General Commission on Archives and History of The United Methodist Church.

Requirements: Selection is made by a committee consisting of three persons who are historians of women in United Methodism. Proposals on women of color and on history at the grass roots level are especially encouraged. Application guidelines are available online. The candidate must submit an application which includes the following: vita or resume; description of the project, including the significance of the research; description of the final product - applicants are encouraged to consider a variety of formats (written, audiovisual, oral history, bibliographies, archival guides, etc.); timetable for the project, including the expected date of completion; budget; indication of how the research results will be disseminated. The candidate must submit three letters of recommendation from persons who can provide evidence of his/her scholarly capability.

Restrictions: The grant money is not to be used for equipment, publications costs, or researcher's salary.

Date(s) Application is Due: Dec 31

Contact: General Secretary, General Commission on Archives and History; (973) 408-3189; fax: (973) 408-3909; email: research@gcah.org

Internet: http://www.gcah.org/site/c.ghKJI0PHIoE/b.3526385/k.D947/Women_in_UM_History_Research.htm

Sponsor: United Methodist Church
P.O. Box 127
Madison, NJ 07940

Women in United Methodist History Writing Award 5788

One award is given each year for an original manuscript on the history of women in the United Methodist Church or its antecedents. Consideration will be given to completed, original manuscripts of article length. The winning manuscript is considered for publication in Methodist History. A copy of the winning entry will be deposited with the General Commission on Archives and History of the United Methodist Church. Guidelines are available online.

Requirements: Manuscripts should not be longer than twenty double-spaced, typewritten pages. Footnotes and bibliography should follow the Kate Turabian Manual of Style. Selection is made by a committee consisting of three persons who are historians of women in United Methodism.

Amount of Grant: $500

Date(s) Application is Due: Dec 31

Contact: General Secretary, General Commission on Archives and History; (973) 408-3189; fax: (973) 408-3909; email: research@gcah.org

Internet: http://www.gcah.org/site/c.ghKJI0PHIoE/b.3526381/k.7DAA/Women_in_UM_History_Award.htm

Sponsor: United Methodist Church
P.O. Box 127
Madison, NJ 07940

Women's Foundation of California Grants 5789

The Women, Foundation of California envisions a future where communities in California are healthy, safe and economically prosperous. The Foundation achieve's this vision by focusing on women and girls as change agents through their central role in families and communities. The Foundation supports organizations that meet the following objectives: organizations mobilizing communities to change policies and systems; to improve the lives of women and girls; developing the leadership of women and girls working to change policies and systems and create stronger movements for social justice; promoting the human rights of women and girls in all sectors of society, including economic, political, social, environmental, sexual, cultural and civil rights.

Requirements: The foundation awards grants to: California 501(c)3 nonprofit corporations or California nonprofit groups with a 501(c)3 fiscal sponsor; organization/program prioritize low-income women and girls in California; health and safety: this includes reproductive health and justice, environmental health and justice and gender-based violence; economic prosperity: this includes efforts towards economic security and economic development. Organizations or programs that utilize one or more of the following strategies to bring about measurable and sustainable change: policy advocacy; community organizing; leadership development; media and/or public education; civic participation; participatory action research; youth-led programs and initiatives. Contact the Foundation for further application information and guidelines.

Restrictions: The Women, Foundation of California does not make grants to or for the following: individuals; for-profit organizations or small businesses; start-up loans; capital projects; direct services; fundraising events; shelters; scholarships; capital improvements; endowments; loans; debt reduction; expenses that accrued prior to the date of the grant award. Funding for publications and one-time events such as retreats and conferences will only be considered if they are integrated within a larger strategy for social change.

Geographic Focus: California

Amount of Grant: $10,000 - $40,000 range

Contact: Maya Thornell-Sandifor, Senior Program Officer; (415) 837-1113; fax: (415) 837-1144; email: mayat@womensfoundca.org

Internet: http://www.womensfoundca.org/grantmaking.html

Sponsor: Women's Foundation of California
340 Pine Street, Suite 302
San Francisco, CA 94104

Women's International League for Peace and Freedom Internships 5790

WILPF works to achieve through peaceful means world disarmament, full rights for women, racial and economic justice, an end to all forms of violence, and to establish those political, social, and psychological conditions which can assure peace, freedom, and justice for all. Internships, which occasionally provide stipends, are available to women throughout the year, including the Jane Addams Internship, in Philadelphia, PA; United Nations Internship—in New York, NY) and WILPF International Internship, in Geneva, Switzerland. Internships are available from January 1 to March 30; June 12 to August 30; and September 15 to December 15. To apply, send a resume writing sample, internship goals, and date available with desired internship placement.

Requirements: Internships are for women, preferably between the ages of 25 and 30. Preference is given to WILPF members.

Restrictions: The internship does not provide for health or other insurance, or health-related expenses.

Date(s) Application is Due: Apr 1; Jul 15; Nov 15

Contact: Jody Dodd, (215) 563-7110; fax: (215) 563-5527; email: jdodd@wilpf.org

Internet: http://www.wilpf.org/get-involved/internjob.htm

Sponsor: Women's International League for Peace and Freedom
1213 Race Street
Philadelphia, PA 19107

Women's Studies Dissertation Fellowships 5791

Dissertation fellowships in women's studies are available to applicants from any of the humanities or social sciences who demonstrate strong research and teaching interests in race, class, gender, and cultural difference. Application for these nine-month residential awards is strongly encouraged from racial, ethnic, and other women candidates. Scholars are required to be in residence during the entire fellowship period. To apply, send curriculum vita, a brief description of dissertation project, a writing sample (approximately 25 pages), and three letters of reference.

Requirements: Candidates must be advanced to candidacy and expect completion of dissertation during term of residence.

Amount of Grant: $20,000 approximately for nine-month residency

Date(s) Application is Due: Jan 28

Contact: Chair, Women's Studies Program, (805) 893-4330; fax: (805) 893-8676; email: wmst@womst.ucsb.edu

Internet: http://www.womst.ucsb.edu

Sponsor: University of California, Santa Barbara
4631 S Hall
Santa Barbara, CA 93106

Women's Studies in Religion Program Visiting Lecturers and Research Associates 5792

The program is designed to explore women's distinctive religious experience and perspectives in their cultural and racial diversity, and to examine male-centeredness and sexism in religious traditions of the world. The program encourages and guides the development of women's studies in religion to help transform the study of religion so it more adequately reflects the study of human experience. Each year, the school appoints as visiting lecturers and research associates in the program five scholars whose work provides leadership in women's studies in the various fields of religion. These scholars conduct individual research projects on topics related to the history and function of gender in the symbolization of religious traditions; the institutionalization of roles in religious communities; and the dynamics of the interaction between religious systems of belief and the personal, social, and cultural conditions of women. Each is also a member of one of the school's departments and offers in one of the three curricular areas a semester course related to this research project. Full-time residence at Harvard Divinity School is required while carrying out proposed research projects. In the spring semester each associate gives a public lecture on her/his research. Application guidelines and deadlines are available from the Harvard Divinity School.

Requirements: Candidates with doctorates in the fields of religion and those with primary competence in other humanities, social science, and public policy fields who have serious investment in religion are eligible.

Amount of Grant: $40,000 and benefits for a full-time appointment

Date(s) Application is Due: Nov 15

Contact: Women's Studies in Religion Program, Divinity School, (617) 495-5705; fax: (617) 496-8564; email: admin@wsrp.harvard.edu

Internet: http://www.hds.harvard.edu/wsrp/appform/application.htm

Sponsor: Harvard University
45 Francis Avenue, Andover Hall
Cambridge, MA 02138

Woodrow Wilson Center Fellowships 5793

The center awards 20-25 four- to 10-month residential fellowships annually in an international competition. Proposals may be submitted in a broad range of the social sciences and humanities on national and/or international issues—topics that intersect with questions of public policy or provide the historical framework to illumine policy issues of contemporary importance. Fellows should be prepared to interact with policymakers in Washington and with Wilson Center staff who are working on similar topics.

Requirements: For academic participants, eligibility is limited to the postdoctoral level. Normally it is expected that academic candidates will have demonstrated their scholarly development with publications beyond the PhD dissertation. For other applicants, an equivalent level of professional achievement is expected. Applicants should have a good command of spoken English.

Amount of Grant: $26,200-$35,000 stipend
Date(s) Application is Due: Oct 1
Contact: Woodrow Wilson International Center for Scholars, (202) 691-4170; email: fellowships@wwic.si.edu
Internet: http://www.wilsoncenter.org/index.cfm?fuseaction=Fellowships.welcome
Sponsor: Woodrow Wilson International Center for Scholars
1300 Pennsylvania Avenue NW, 1 Woodrow Wilson Plaza
Washington, DC 20004-3027

Woodrow Wilson Doctoral Dissertation Fellowships in Women's Studies 5794
The Woodrow Wilson Dissertation Fellowship in Women, Studies encourages original and significant research about women that crosses disciplinary, regional, or cultural boundaries. The Women, Studies Fellowships are provided to Ph.D. candidates at institutions in the United States who will complete their dissertations during the fellowship year. Fellows receive $3,000 to be used for expenses connected with the dissertation. These may include, but are not limited to, travel, books, microfilming, taping, and computer services.
Requirements: Applications will be judged on originality and significance to women, studies, scholarly validity, the applicant, academic preparation and ability to accomplish the work, and whether the dissertation will be completed within a reasonable time period. Evidence of commitment to women, studies is also considered. Collaborative work is not eligible. Applications must be filed using the online form provided on the Woodrow Wilson National Fellowship Foundation website. A completed application comprises the online application form, two letters of recommendation, and a current transcript.
Amount of Grant: $3,000
Date(s) Application is Due: Oct 13
Contact: Dana DiLullo, Fellowship Director; (609) 452-7007; fax: (609) 452-0066; email: WWTeachingFellowships@woodrow.org
Internet: http://www.woodrow.org/fellowships/women_gender/index.php
Sponsor: Woodrow Wilson National Fellowship Foundation
5 Vaughn Drive, Suite 300
Princeton, NJ 08540-6313

Woodrow Wilson MMUF Dissertation Grants 5795
The MMUF Dissertation Grants are available to graduate students who participated in the Mellon Mays Undergraduate Fellowship Program. The grants provide graduate students at the critical juncture of completing their graduate degrees with support to spend a year finishing the writing of the dissertation. The dissertation grant offers up to $20,000 for a 12-month period. Fellows are encouraged to apply for dissertation grants from their home institutions. Winners who do not receive dissertation support from their own institutions are eligible to receive the full amount. The grant will be reduced by funding from other sources to ensure that each awardee receives a maximum award from all sources of $20,000. Woodrow Wilson will ask graduate schools to waive tuition for winners of the MMUF Dissertation Grant.
Requirements: Applicants must be Mellon Mays Undergraduate Fellows and candidates for the Ph.D. degree in fields recognized under the terms of the Mellon Mays Undergraduate Fellowship Program. Candidates must have completed all pre-dissertation requirements preceding the application deadline. Specifically, each applicant must have passed all comprehensive examination, completed all coursework for the degree, received approval of the dissertation topic, and completed most or all of the fieldwork and/or research necessary to begin writing. Applications must include the following: a three-page prospectus; a draft dissertation chapter; a three-page personal statement; an official graduate transcript; two letters of recommendation,Äìone from the dissertation director and one from another academic knowledgeable about the Fellow academic performance and/or familiar with the Fellow, dissertation project and its contribution to the field of scholarship; the official budget form; and a declaration of previous fellowship awards. Applicants who have held a similar (non-institutional) dissertation award for the final year of dissertation writing are not eligible. At present, applications for the MMUF Dissertation Grants may be submitted in hard copy only; there is no online application as yet. If you are interested in applying to the MMUF Dissertation Grants Program or need more information, please email Sylvia Sheridan, Assistant Director.
Amount of Grant: $20,000
Date(s) Application is Due: Dec 19
Contact: Sylvia Sheridan, Assistant Director; (609) 452-7007; fax: (609) 452-0066; email: mmuftr@woodrow.org
Internet: http://www.woodrow.org/fellowships/opportunity/dissertation/index.php
Sponsor: Woodrow Wilson National Fellowship Foundation
5 Vaughn Drive, Suite 300
Princeton, NJ 08540-6313

Woodrow Wilson MMUF Travel & Research Grants 5796
The MMUF Travel & Research Grants are available to graduate students who participated in the Mellon Mays Undergraduate Fellowship Program. The grants provide eligible graduate students with the financial means to complete their research prior to the start of dissertation writing. Funding is for one summer or one semester of travel and/or research support. The MMUF Travel & Research grants provide up to $5,000. Grants may be used to cover dissertation research, travel to/from research sites, and photocopying of documents and/or the purchase of access to databases for research purposes.
Requirements: Eligibility: applicants must be Mellon Mays Undergraduate Fellows and candidates for the Ph.D. degree in fields recognized under the terms of the Mellon Mays Undergraduate Fellowship Program; must have passed all comprehensive examinations, completed all course work for the degree, and selected a dissertation topic that has been approved by the dissertation advisor. Applications must include the following: the official application form; a resume; a personal statement as outlined in the application; an official budget form. Funding may be applied to travel to research sites, cost of meals and lodging at research sites, photocopying or microfilming of documents, purchase of access to research databases, fees for use of research facilities, and other research-related expenses; an official graduate transcript; and a brief letter of recommendation (two-page maximum) from the dissertation advisor endorsing the request for funding.
Restrictions: The award may not be used to purchase computer software or hardware, and may not be used for general living expenses.
Amount of Grant: $5,000
Date(s) Application is Due: Mar 20
Contact: Sylvia Sheridan, Assistant Director; (609) 452-7007; fax: (609) 452-0066; email: pickeringfaf@woodrow.org
Internet: http://www.woodrow.org/fellowships/opportunity/research/index.php
Sponsor: Woodrow Wilson National Fellowship Foundation
5 Vaughn Drive, Suite 300
Princeton, NJ 08540-6313

Woodrow Wilson National Fellowship Foundation Travel and Research Grants 5797
The grants provide eligible graduate students with the financial means to complete their research prior to the start of dissertation writing. Funding is for one summer or one semester of travel and/or research support. Awards are based on the scholarly merit of individual applications received. If you are interested in the MMUF Travel & Research Grants Program or need more information, email Sylvia Sheridan, Assistant Director.
Requirements: Applicants must be Mellon Mays Undergraduate Fellows and candidates for the Ph.D. degree in fields recognized under the terms of the Mellon Mays Undergraduate Fellowship Program. All candidates must have passed all comprehensive examinations, completed all course work for the degree, and selected a dissertation topic that has been approved by the dissertation advisor. Applications must include the following: the official application form; a resume; a personal statement as outlined in the application; an official budget form. Funding may be applied to travel to research sites, cost of meals and lodging at research sites, photocopying or microfilming of documents, purchase of access to research databases, fees for use of research facilities, and other research-related expenses; an official graduate transcript; and a brief letter of recommendation (two-page maximum) from the dissertation advisor endorsing the request for funding.
Restrictions: The award may not be used to purchase computer software or hardware, and may not be used for general living expenses.
Amount of Grant: $5,000
Date(s) Application is Due: Mar 20
Contact: Sylvia Sheridan, Assistant Director; (609) 452-7007; fax: (609) 452-0066; email: mmuftr@woodrow.org
Internet: http://www.woodrow.org/fellowships/opportunity/research/index.php
Sponsor: Woodrow Wilson National Fellowship Foundation
5 Vaughn Drive, Suite 300
Princeton, NJ 08540-6313

Woodrow Wilson Practicum Grants 5798
The program supports Humanities PhD students who have created public scholarship internships, engaging their scholarship in a context outside of college teaching and research. As part of the Humanities at Work initiative, Practicum Grants address three challenges: to expand the career horizons of doctoral students in the humanities; to bring the insight of the humanities to all aspects of American life; and to bring the life of the larger community into the academy. The range of internship possibilities is unlimited. A successful applicant must demonstrate a strong relationship between the proposed internship and his/her field of study. Internships should be completed by the end of the summer. Guidelines and application are available online.
Requirements: Applicants must be currently enrolled PhD students who are making timely progress toward completion of their degree in the humanities or humanistic social sciences. Applicants must be U.S. citizens or permanent residents. The Internship site must be located in the United States.
Amount of Grant: $2000 maximum
Date(s) Application is Due: Apr 8
Contact: Woodrow Wilson Practicum Grants, (609) 452-7007; fax: (609) 452-0066; email: practicum@woodrow.org
Internet: http://www.woodrow.org/phd/Practicum/practicum_grants_faq.html
Sponsor: Woodrow Wilson National Fellowship Foundation
P.O. Box 5281
Princeton, NJ 08543-5281

Woodward Fund Grants 5799
Grants support nonprofit institutions, corporations, and associations that are located in Georgia or one of its neighboring states and are organized and operated exclusively for religious, educational, and charitable and scientific purposes. Grantmaking focuses on capital projects. The distribution committee makes its decisions based on materials provided in the grant request. The trust does not require a formal application form, but requests a proposal outlining the project or program and containing the following information: project/program description; complete itemized project budget, including project schedule; need for the project; other funding sources, including the amount received from each source; names and qualifications of those conducting the project; objectives and how they will be achieved; a brief description of the applicant organization; method and criteria for project evaluation; list of officers, board of directors, or trustees; copy of the 501(c)3 designation letter; and plans for recognition of the trust in the project. Applicants should not contact committee members personally concerning a proposal or anticipated approach to the fund.

Requirements: 501(c)3 nonprofit organizations in Georgia and its neighboring states are eligible. Government agencies to which contributions by individuals are made deductible from income by IRS laws also are eligible.
Restrictions: Grants are not awarded to individuals or for scholarships or student loans.
Geographic Focus: Alabama; Florida; Georgia; North Carolina; South Carolina; Tennessee
Samples: Egleston Children Hospital at Emory U (Atlanta, GA)—for program support, $100,000.; Paidei School (Atlanta, GA)—for program support, $50,000.
Amount of Grant: $5000-$300,000
Date(s) Application is Due: May 1; Nov 1
Contact: Alice Sheets, c/o Wachovia Bank, NA Trustee, email: grantinquiriesga@wachovia.com
Internet: http://www.wachovia.com/corp_inst/charitable_services/0,,3302_3312,00.html
Sponsor: David, Helen, and Marian Woodward Fund
3414 Peachtree Road, MC: GA8023
Atlanta, GA 30326

Worcester Reed Warner Medal Award **5800**
The medal is awarded annually for outstanding contributions to the permanent literature of engineering. Contributions may be single papers, treatises or books, or a series of papers. They are to deal with progressive ideas relative to engineering, scientific, and industrial research associated with mechanical engineering; the design and operation of mechanical and associated equipment; industrial engineering or management, organization, operation, and the concomitants of each; or other subjects closely associated with the foregoing. A certificate is included with the award.
Requirements: To qualify as having permanent value, any paper or treatise must be at least five years old. Recipients may be nonmembers of ASME.
Samples: Ephraim Suhir—award recipient, $2000 (2004); Tsu-Wei Chou—award recipient, $2000 (2002).
Amount of Grant: $2000
Date(s) Application is Due: Mar 1
Contact: Gilda DiTullio, Manager, (212) 705-7736; fax: (212) 705-7739; email: ditulliog@asme.org or infocentral@asme.org
Internet: http://www.asme.org/honors/ms71/list.html
Sponsor: American Society of Mechanical Engineers
345 E 47th Street
New York, NY 10017-2392

World Bank Graduate Scholarship Program **5801**
With grant funding from the government of Japan, the World Bank has established scholarships for graduate studies leading to a higher degree in a development-related social science. The program is aimed at increasing the number of high-quality professionals in economic and social development. Scholarships are available to applicants from developing countries to pursue a career in a development field. It is anticipated that each recipient will return to his/her home country at the conclusion of studies and apply the knowledge and skills acquired to help accelerate the pace of economic and social development. The scholarship provides full tuition, the cost of medical and accident insurance obtained through the university, a monthly subsistence allowance for one person, and paid economy-class travel. Contact the organization for application packet. Application forms for the Regular Program in English, French, Spanish, and Portuguese are available from the JJ/WBGSP Secretariat from September through January. Forms are available at World Bank offices in many countries, as well as online.
Requirements: Qualified applicants must be nationals of a World Bank member country currently eligible to borrow, younger than age 45 (and preferably under age 35), of good character, and in good health. They must have attained the bachelor's degree or equivalent in law, business or public administration, or any other development-related technical discipline. Candidates must have a superior academic record with at least three, preferably four to five, years of recent experience in a public agency or public service organization engaged in development-related activities. Admittance to two universities renowned for development research and teaching where the recipient will pursue a master's or doctorate degree is also required.
Restrictions: The program does not sponsor undergraduate studies, distance learning programs, short-term training, conferences, seminars, thesis writing, research projects, and fields of studies not related to development. Executive directors, their alternates, and staff of the World Bank Group (the World Bank, the International Finance Corporation, and the International Development Association), including consultants, as well as close relatives, are ineligible. MBA applicants are ineligible. Those with refugee or asylum status are ineligible.
Date(s) Application is Due: Mar 31
Contact: JJ/WBGSP Secretariat, Economic Development Institute, (202) 473-1000; fax: (202) 477-6391; email: opportunities@worldbank.org
Internet: http://go.worldbank.org/NV4RFT1BP0
Sponsor: World Bank
1818 H Street NW
Washington, DC 20433

World Press Institute Fellowships **5802**
The institute offers four-month journalism fellowships, which provide immersion into the governance, politics, business, media, and culture of the United States for experienced international journalists through a demanding schedule of study and travel throughout the country. Emphasis is placed on the role and responsibilities of a free press in a democracy.
Requirements: Fellowships are offered to highly recommended foreign journalists between 25 and 35 years of age who are full-time professionals. Requirements for applicants, who must have demonstrated a potential for leadership, include: at least five years of full-time news experience; fluency in all aspects of English; several written essays; three letters of recommendation; at least three work samples.
Date(s) Application is Due: Dec 31
Contact: A. Kristine Mortensen, Program Director, (612) 696-6360; fax: (612) 696-6306; email: wpi@macalester.edu
Internet: http://www.worldpressinstitute.org/fellowship.htm
Sponsor: World Press Institute
1576 Summit Avenue
Saint Paul, MN 55105

WOTRO Postdoctoral Fellowships **5803**
The Netherlands Foundation for the Advancement of Tropical Research (WOTRO) funds research proposals on development-related research in developing countries. The following criteria apply to the substantive appraisal of an application: scientific relevance; originality of the topic; clarity of the problem definition; effectiveness of the proposed research methods and techniques; research design and feasibility (including financial feasibility); and institutional support and supervision. The following criteria are used in order to determine development-relevance: institutional support in the relevant country/countries; quality of the development cooperation relationship between the institute(s) in question and the relevant institute in the Netherlands; promotion of development-related research capacity in the relevant country/ countries; social importance of the research (i.e. usability of the results of the research by NGOs, grass-roots movements or government agencies in the relevant country/ countries, at local or macro level); effectiveness of the proposed dissemination of the research results in the relevant country/countries, at both academic and administrative/political level; and usability of the results for combating poverty, for sustainable development, environmental control, and improving the position of women in the relevant country/countries. The listed application deadline is for submission of preliminary proposals.
Requirements: Applicant must be a citizen of one of the developing countries, must be an employee of a local institute within a developing country, and must have an MA or MSc or equivalent degree.
Date(s) Application is Due: Jun 15
Contact: Dr. Yilma Jobre Makonnen, Training Programmes Officer, (251) 1-463215 ext 172; fax: (251) 1-461252 or 464645; email: y.jobre@cgiar.org or ILRI-Kenya@cgiar.org
Internet: http://www.ilri.cgiar.org
Sponsor: International Livestock Research Institute
P.O. Box 5689
Addis Ababa 00100
Ethiopia

WREI Congressional Fellowships **5804**
The institute administers this congressional fellowship program designed to train women as potential leaders in public policy formation and to examine issues from the perspective of the experiences and needs of women. A fellow works 30 hours per week in a congressional office as a legislative aide on policy issues affecting women. Applications are available after November 1 and are due in mid-February. Selection is based on academic performance, experience with community groups, and interests in the analysis of gender differences as they affect federal laws and legislating. Annual deadline dates may vary; contact program staff for exact dates.
Requirements: All students enrolled in graduate or professional programs anywhere in the United States are eligible. It is recommended that students have completed nine hours of graduate coursework or equivalent and have demonstrated an interest in research or political activity relating to women's social and political status. Only currently enrolled graduate and professional students are eligible. The applicant must be enrolled in an accredited U.S. institution.
Amount of Grant: $1300 monthly stipend for academic year, plus $500 for health insurance and $1500 maximum for tuition
Date(s) Application is Due: May 20
Contact: Congressional Fellowship Administrator, (202) 628-0444; fax: (202) 628-0458; email: wrei@wrei.org
Internet: http://www.wrei.org/fellowships/index.htm
Sponsor: Women's Research and Education Institute
1750 New York Avenue NW, Suite 350
Washington, DC 20006

Wright Center Fellowships **5805**
The center makes awards to teachers in grades six-12 who have made significant contributions to their school's science curriculum and whose innovations have improved student understanding of science. Fellows may take free courses at Tufts and study at the school's Center for Science and Mathematics Teaching, the Teacher Resource Library, and the Science Visualization Laboratory.
Requirements: Applicants must be full-time teachers, at the elementary, middle-school, or high-school level. They should have had a minimum of five years' science teaching experience.
Amount of Grant: $45,000 stipend plus $500 for books and travel
Date(s) Application is Due: Feb 1
Contact: Zach Smith, Program Contact, (617) 627-5394; email: wright_center@tufts.edu
Internet: http://www.tufts.edu/as/wright_center/fellowships/fellopp.html

Sponsor: Tufts University
4 Colby Street, Science and Technology Ctr, Suite 267
Medford, MA 02155

Wright Center Mini-Fellows 5806
K-12 teachers come to the center individually or in teams of two. Teacher teams need not be based at the same school. When teachers apply as a team they may both have teaching assignments in science or one teacher is a science specialist, while the other member of the team is from another discipline. In the past the second teacher has often been a nonscience instructor, for example, an art, English, or history teacher.
Amount of Grant: $750 honorarium plus $250 for books
Contact: Zach Smith, Program Contact, (617) 627-5394; fax (617) 627-3995; email: zach.smith@tufts.edu
Internet: http://www.tufts.edu/as/wright_center/fellowships/fellopp.html
Sponsor: Tufts University
4 Colby Street, Science and Technology Ctr, Suite 267
Medford, MA 02155

WWF Russell E. Train EFN Alumni Grants 5807
EFN Alumni Grants support ongoing education and training for former Russell E. Train Fellows or Scholars who have earned a degree through EFN. There is no deadline for Alumni grants; they are awarded on a first-come, first-served basis. Funds may be used to attend conferences, workshops, and short-term training courses offered by universities and other institutions or to conduct research within a WWF priority place.
Requirements: To be eligible for an EFN Alumni Grant, applicants must: be a former Russell E. Train Fellow or Scholar; have submitted all reports required by their grant contracts; have completed their two-year work requirement in their home country; have a bank account in their own name that can receive U.S. dollars; not have received an EFN Alumni Grant from EFN in the past three years. To apply go to: http://www.wwfapply.com. EFN alumni from any country are eligible to apply.
Restrictions: Grant requests may not exceed U.S. $3,500 and alumni must observe a three year waiting period before requesting another grant.
Amount of Grant: $3,500 max.
Contact: Mercedes Gonzalez, Program Officer; (202) 778-9612; email: mercedes.gonzalez@wwfus.org
Internet: http://www.worldwildlife.org/science/fellowships/train/item1829.html
Sponsor: World Wildlife Fund
1250 24th Street, North West, P.O. Box 97180
Washington, DC 20090-7180

WWF Russell E. Train EFN Professional Development Grants 5808
EFN's Professional Development Grants provide support for mid-career conservationists to pursue short-term, non-degree training to upgrade their knowledge and skills through short courses, workshops, conferences, and study tours, or through practical training such as internships and professional attachments. Professional Development Grants are awarded throughout the year, on a first-come, first-served basis, given the availability of funds. EFN Professional Development Grants support for all training-related costs including registration fees and tuition, meals and accommodations, books and materials, international travel, and local transportation. Training may take place anywhere in the world. Typical grants range in amount from $1,500 to $5,000.
Requirements: To be eligible for an EFN Professional Development Grant, you must meet the following criteria: you must be a citizen or legal resident of: Belize; Bhutan; Bolivia; Brazil; Cambodia; Cameroon; Central African Republic; Colombia; Dem. Republic of Congo; Ecuador; Gabon; Guatemala; Guyana; Honduras; Indonesia; Kenya; Laos; Malaysia; Mexico; Mozambique; Namibia; Nepal; Peru; Philippines; Republic of Congo; Suriname; Tanzania; Vietnam; you must have at least three years work experience (paid or unpaid) in conservation or a related field; your current work and proposed training must help advance conservation in the previously listed countries; your proposed training must be completed within 6 months and must not lead to an academic degree; you must currently be working in a conservation-related field and have promise of continued employment at the completion of your proposed training from your employer; you must have approval from your employer to participate in the proposed training; you must have a bank account in your own name that can receive U.S. dollars; you must not have received a Professional Development Grant from EFN in the past three years; you must not have received a Train Fellowship or an EFN scholarship in the past. You may complete the EFN Professional Development Grant application online at: http://www.wwfapply.com or you may also request an application form by email at: efn@wwfus.org. In your request for an application form, specify your name, the country where you currently reside, and the name and dates of the training course or conference you wish to attend.
Restrictions: Individuals currently enrolled in an academic degree program are not eligible. Applicants who require funds in less than 90 days of submitting their applications will not be considered.
Amount of Grant: $1,500-$5,000 range
Contact: WWF Member Services; (800) 960-0993;
Internet: http://www.worldwildlife.org/science/fellowships/train/item1828.html#pdgapply
Sponsor: World Wildlife Fund
1250 24th Street, North West, P.O. Box 97180
Washington, DC 20090-7180

WWIC/National Service Agency Civil Society Nonprofit Scholars Grants 5809
The Corporation for National and Community Service and the Woodrow Wilson International Center for Scholars created the Civil Society Nonprofit Scholars program. This annual international competition names up to three scholars who wish to spend nine months conducting applied research that will illuminate contemporary public policy issues and increase the body of research on the intersection of civil society, the nonprofit sector, volunteerism, and public policy. Scholars will receive a stipend based on their current salary, be located at the corporation in Washington, and develop relationships with other Washington-area organizations. Each scholar will also have access to the extensive research resources afforded by the corporation, the center, and the Washington area. In addition, scholars will be encouraged to participate in corporation-sponsored activities as well as seminars and other events offered by the center for the larger community of fellows and scholars. After the program period, scholars will be expected to disseminate their research findings through various public media, including presentations, open forum discussions, publications, and World Wide Web postings. Applications should be sent to Civil Society Nonprofit Scholars, 1201 New York Avenue NW, Room 8100-H, Washington, DC, 20525.
Requirements: Scholars from a wide variety of backgrounds (including government, the corporate world, the professions, and academia) will be chosen for their outstanding capabilities and relevant experience. Applications are welcome from any country, but research proposals must explore preferred themes within an American context. For academic applicants, eligibility is limited to the postdoctoral level and these candidates should demonstrate scholarly development—as evidenced through publications—beyond their doctoral dissertation. For other applicants, an equivalent level of professional achievement is expected.
Restrictions: Applicants working on degrees at the time of the application (even if the degree is to be awarded prior to the proposed scholarship year) are not eligible unless there is an equivalent level of professional achievement evidenced in their work experience. In addition, applicants for Woodrow Wilson fellowships will not be considered for the civil society nonprofit scholars grants.
Amount of Grant: $85,000 maximum
Contact: Grants Administrator, (202) 691-4170; email: civilsocietyscholars@wwic.si.edu
Internet: http://wwics.si.edu/index.cfm?fuseaction=sf.civil
Sponsor: Woodrow Wilson International Center for Scholars
1300 Pennsylvania Avenue NW
Washington, DC 20004-3027

WWICS Humanities and Social Sciences Fellowships 5810
The center awards approximately 20 residential fellowships each year for advanced research in the humanities and social sciences. Fellows are provided offices, access to the Library of Congress, manuscript typing services, computers, and research assistants. Selected works will be published through the Woodrow Wilson Center Press. Where appropriate, fellows are associated with the center's programs on Asia; Eastern and Western Europe; the historical, cultural, and literary studies program; the international studies program; the Kenan Institute for Advanced Russian Studies; the Latin American program; or the United States program. Fellowships are normally for an academic year, although a few fellowships are available for shorter or longer periods, with a minimum of four months and a maximum of 12. The award includes stipend, travel expenses, and 75 percent of health insurance premiums for fellows, their spouses, and dependent children.
Requirements: Applications from any country are welcome. Individuals with outstanding capabilities and experience from a wide variety of backgrounds (including government, the corporate world, and the professions, as well as academia) are eligible for appointment. For academic participants, eligibility is limited to the postdoctoral candidates who have demonstrated their scholarly development by publication beyond the PhD dissertation. An equivalent degree of professional achievement is expected for nonacademics.
Restrictions: Applicants working on a degree at the time of application (even if it is to be completed at the time of the proposed fellowship year) are not eligible.
Amount of Grant: $26,200-$85,000 range
Date(s) Application is Due: Oct 1
Contact: Program Coordinator, (202) 691-4170; fax: (202) 691-4001; email: fellowships@wwic.si.edu
Internet: http://www.wilsoncenter.org
Sponsor: Woodrow Wilson International Center for Scholars
1 Woodrow Wilson Plaza, 1300 Pennsylvania Avenue NW
Washington, DC 20004-3027

WWNFF Dissertation Grants in Women's Studies 5811
The foundation offers dissertation grants to encourage original and significant research about women on such topics as the evolution of women's role in society, women in history, the psychology of women, and women as seen in literature and art. Grants cover research expenses, which may include, but are not limited to, travel, books, microfilming, taping, and computer services. Applications are available from the Web site.
Requirements: Students in doctoral programs who have completed all predissertation requirements in any field of study at U.S. graduate schools are eligible.
Restrictions: Candidates who are within a few months of completing their work should not apply.
Amount of Grant: $3000
Date(s) Application is Due: Oct 10
Contact: Women's Studies Dissertation Fellowships, (609) 452-7007; fax: (609) 452-7828; email: wswhrec@woodrow.org
Internet: http://www.woodrow.org/womens-studies/purpose.html#awards
Sponsor: Woodrow Wilson National Fellowship Foundation
P.O. Box 5281
Princeton, NJ 08543-5281

WWNFF Dissertation Grants in Women's Health 5812
Dissertation grants encourage original and significant research on issues related to women's health. This program emphasizes the implications of research for the understanding of women's lives and significance for public policy or treatment. Previous grant recipients have explored such topics as postpartum depression; contraceptive law; sex hormones and ovarian cancer; dietary determinants of morbidity; sex and violence in everyday life; and sexuality on the Internet. Grant funds may be used for expenses connected with the dissertation, including but are not limited to, travel, books, microfilming, taping, and computer services. Ten awards will be made. Guidelines are available online. Annual deadline dates may vary; contact program staff for exact dates.
Requirements: Students in doctoral programs such as nursing, public health, anthropology, history, sociology, psychology, and social work, at graduate schools in the United States are eligible to apply. Candidates must have completed all pre-dissertation requirements by October 24 and expect to complete their dissertations by the summer of next year.
Restrictions: Candidates who are within a few months of completing their work should not apply.
Amount of Grant: $6000
Date(s) Application is Due: Oct 11
Contact: Women's Health Dissertation Fellowships, (609) 452-7007; fax: (609) 452-0066; email: contact@woodrow.org
Internet: http://www.woodrow.org/womens-studies/health/purpose.html
Sponsor: Woodrow Wilson National Fellowship Foundation
P.O. Box 5281
Princeton, NJ 08543-5281

Wyeth Predoctoral Fellowship 5813
One fellowship is awarded annually for 24 months. The fellowship is intended for the advancement and completion of a doctoral dissertation that concerns the traditions of American art. The fellow is expected to spend one year of the fellowship period on dissertation research in the United States or abroad, and one year at the center to complete the dissertation.
Requirements: Applicants must have completed their residence requirements and coursework for the PhD and general or preliminary examinations before the date of application, and know two foreign languages related to the topic of the dissertation. Applicants must be either U.S. citizens or enrolled in a university in the United States.
Samples: Branden Joseph, Harvard U—for dissertation, I'll Be Your Mirror: Robert Rauschenberg and Andy Warhol, 1952-1968. Pamela Lee, Harvard U—for dissertation, Object to Be Destroyed: The Work of Gordon Matta-Clark.
Contact: Fellowships Program, Center for Advanced Study in the Visual Arts, (202) 842-6482; fax: (202) 789-3026; email: advstudy@nga.gov
Internet: http://www.nga.gov/resources/casvapre.htm
Sponsor: National Gallery of Art
2000B South Club Drive
Landover, MD 20785

Xerox Foundation Grants 5814
The foundation's goals are to preserve and invigorate its communities, to help educate and prepare tomorrow's workers and leaders (including enhancing learning opportunities for minorities and disadvantaged), to promote volunteerism among Xerox people, and to give the company a dimension of difference. Giving is broad in scope, encompassing education/work force preparedness, science/technology, employee/community affairs, culture, and national affairs. Xerox grants tend to complement the corporation's overall business goals, i.e., to support research and education in fields of science and technology directly applicable to the business; focus on geographical areas where large numbers of Xerox people live or where there are major facilities; involve a partnership among private, government, and nonprofit organizations; address current and relevant issues; offer creative approaches to long-standing problems; and support well-defined goals. Applications are reviewed on a monthly basis.
Requirements: Grants are made only to 501(c)3 and 509(a) organizations.
Restrictions: The foundation declines requests to support individuals; capital grants (new construction or renovation); endowments or endowed chairs; organizations supported by United Way, unless permission has been granted by United Way to a member agency to conduct a capital fund drive or a special benefit; political organizations or candidates; religious or sectarian groups; or municipal, county, state, federal, or quasi-government agencies.
Samples: National Underground Railroad Freedom Ctr (Cincinnati, OH)—to establish the John Parker Library, which will house a variety of research material, including a digital archive of material related to the Underground Railroad and a genealogy-research area, $1 million (2004).
Amount of Grant: $500-$50,000 range
Contact: Joseph Cahalan, Vice President, (203) 968-3445; fax: (203) 968-3330
Internet: http://www.xerox.com/foundation
Sponsor: Xerox Foundation
P.O. Box 1600, 800 Long Ridge Road
Stamford, CT 06904

Xoran Technologies Resident Research Grant 5815
The purpose of this grant is to stimulate original resident research in otolaryngology projects that are well-conceived and scientifically valid, with potential to clarify the role of head and neck computed tomography in diagnosis or treatment. Awards are for one year, are non renewable, and have a $10,000 maximum limit. Generally, a single award is made available annually. Letters of Intent are due on December 15, with full proposals due by January 15.
Requirements: Application is open to residents of an accredited otolaryngolgoy-head and neck surgery training program in the U.S. and Canada for projects that are well-conceived and scientifically valid, with potential to clarify the role of the head and neck computed tomography in diagnosis or treatment.
Amount of Grant: $10,000 maximum total direct costs
Date(s) Application is Due: Dec 15; Jan 15
Contact: Stephanie L. Jones, Assistant Director; (703) 519-1586 or (703) 836-4444; email: sljones@entnet.org
Internet: http://www.entnet.org/EducationAndResearch/coreGrants.cfm
Sponsor: Xoran Technologies, Inc.
One Prince Street
Alexandria, VA 22314-3357

xyz.net Corporate Giving Grants 5816
The corporate giving program makes small cash donations and gifts of products and services to Alaskan nonprofit organizations for the betterment of the Alaskan community. Areas of interest include youth/youth sports, technology/medical research, women's issues, education, arts and culture, and volunteerism as it relates to any of the areas already mentioned. The board meets monthly to review requests. An application is available online.
Requirements: Alaskan 501(c)3 tax-exempt organizations serving Anchorage, Kenai, Soldotna, and Homer are eligible. Applicants must be subscribers to xyz for at least three months, and an employee from xyz must be involved in the organization or willing to bring the request forward to the company.
Restrictions: Individuals, organizations limited by race or religion, political campaigns, and organizations that are not tax-exempt are ineligible. Grants do not support facilities (building funds) or indirect costs.
Geographic Focus: Alaska
Contact: Teresa Rivas, Corporate Giving Coordinator, (907) 565-2200; fax: (907) 562-6047; email: trivas@alaska.net
Internet: http://www.xyz.net/corpgiving.shtml
Sponsor: Internet Alaska Inc
4050 Lake Otis Pkwy, Suite 108
Anchorage, AK 99508

Yale Center for British Art Visiting Fellowships 5817
The Yale Center for British Art offers a limited number of resident fellowships for one to four months to scholars in postdoctoral or equivalent research related to British art and to museum professionals whose research interests include British art. The fellowships allow scholars of either literature, history, the history of art, or related fields to study the center's holding of paintings, drawings, prints, and rare books and to make use of its research facilities. Recipients are required to be in residence in New Haven during the fellowship period. Applications should be submitted along with curriculum vita, a brief outline of the proposed research, the preferred month of tenure, and two confidential letters of recommendation. Fellowship awards include the cost of travel to and from New Haven and also provide accommodation and a living allowance. One fellowship per annum is reserved for a member of the American Society for Eighteenth-Century Studies.
Date(s) Application is Due: Jan 13
Contact: Director, Yale Center for British Art, (203) 432-7192; fax: (203) 432-9628; email: bacinfo@yale.edu
Internet: http://www.yale.edu/ycba/education/index.htm
Sponsor: Yale Center for British Art
Box 208280
New Haven, CT 06520-8280

Yale Special Collections Humanities Fellowship 5818
The Yale University Library invites applications for its postdoctoral humanities fellowship program. The two-year fellowship offers recipients an opportunity for in-depth research using the library's special and archival collections, as well as for limited undergraduate teaching responsibilities. Preference will be given to applicants who demonstrate an interest in multidisciplinary or multicollection endeavors. Guidelines are available online.
Requirements: The fellowships are open to recent PhD recipients (within five years of earning their degrees).
Amount of Grant: $44,300 annual stipend; $1500 annual travel/research allowance
Date(s) Application is Due: Apr 2
Contact: Alice Prochaska, Librarian
Internet: http://www.library.yale.edu/special_collections/spcfellowships.html
Sponsor: Yale University Library
P.O. Box 208240
New Haven, CT 06520-8240

Yampa Valley Community Foundation Grants 5819
The foundation will consider proposals from organizations engaged in nonprofit activities in Routt County that benefit Routt County citizens. The foundation will make grants for general operations. An online request form and an application form are available on the Web site. Applications are generally accepted in the spring, and recipients are notified in the early summer. Call for specific dates. Application forms are available on the Web site.
Requirements: Nonprofit organizations in Steamboat Springs, CO, may apply.
Restrictions: The foundation does not make grants for debt reduction, endowments, political, or religious purposes.

Geographic Focus: Colorado
Amount of Grant: $500-$2500 typically
Date(s) Application is Due: Mar 1; Jun 1; Aug 1; Nov 1
Contact: Brooke Sanchez, Program Director, (970) 879-8632; fax: (970) 871-0431; email: brooke@yvcf.org
Internet: http://www.yvcf.org
Sponsor: Yampa Valley Community Foundation
P.O. Box 881869, 465 Anglers Drive
Steamboat Springs, CO 80438

YIVO Abraham and Rachela Melezin Fellowship **5820**
This fellowship helps support research on Jewish educational networks in pre-war Vilna and the Vilna region. The fellowship, which carries a stipend, is for a period of one to three months to conduct research at YIVO's library and archives. The recipient is expected to deliver a public lecture based on his/her research and submit a paper for possible publication in the YIVO Annual or YIVO-bleter. Applicants should send or fax a resume, two letters of support, and a research proposal of no more than three pages. Annual deadline dates vary; contact program staff for exact dates.
Amount of Grant: $1500
Date(s) Application is Due: Dec 31
Contact: Program Contact, (212) 246-6080; fax: (212) 292-1892
Internet: http://www.yivoinstitute.org/jstudies/jstudies_fr.htm
Sponsor: YIVO Institute for Jewish Research
15 W 16th Street
New York, NY 10011

YIVO Aleksander and Alicja Hertz Memorial Fellowship **5821**
This fellowship helps support research on Polish-Jewish history in the modern period, particularly Jewish-Polish relations and Jewish contributions to Polish literature and culture. The fellowship, which carries a stipend, is for a period of one to three months to conduct research at YIVO's library and archives. The recipient is expected to deliver a public lecture based on his/her research and submit a paper for possible publication in the YIVO Annual or YIVO-bleter. Applicants should send or fax a resume, two letters of support, and a research proposal of no more than three pages. Annual deadline dates may vary; contact program staff for exact dates.
Amount of Grant: $1500
Date(s) Application is Due: Dec 31
Contact: Director of Research, (212) 246-6080; fax: (212) 292-1892
Internet: http://www.yivoinstitute.org/jstudies/jstudies_fr.htm
Sponsor: YIVO Institute for Jewish Research
15 W 16th Street
New York, NY 10011

YIVO Institute for Jewish Research Scholarships **5822**
Scholarships are offered to facilitate participation in the Uriel Weinreich Program in Yiddish Language, Literature, and Culture, a six-week intensive language course offered during the summer session at Columbia University. The program is designed to produce substantial skills in speaking, reading, and writing Yiddish. Three college credits are available to participants; the intensive nature of the program, however, precludes employment or enrollment in other courses.
Date(s) Application is Due: Mar 15
Contact: Dr. Paul Glasser, (212) 246-6080; fax: (212) 292-1892; email: pglasser@yivo.cjh.org
Internet: http://www.yivoinstitute.org/courses/urielw_fr.htm
Sponsor: YIVO Institute for Jewish Research
15 W 16th Street
New York, NY 10011

YIVO Maria Salit-Gitelson Tell Memorial Fellowship **5823**
This fellowship helps support original research in the field of Lithuanian Jewish history, the city of Vilnius in particular. The fellowship, which carries a stipend, is for a period of one to three months during the spring semester to conduct research at YIVO's library and archives. The recipient is expected to deliver a public lecture based on his/her research and submit a paper for possible publication in the YIVO Annual or YIVO-bleter. Applicants should send or fax a resume, two letters of support, and a research proposal of no more than three pages. Annual deadline dates may vary; contact program staff for exact dates.
Amount of Grant: $1500
Date(s) Application is Due: Dec 31
Contact: Director of Research, (212) 246-6080; fax: (212) 292-1892
Internet: http://www.yivoinstitute.org/jstudies/jstudies_fr.htm
Sponsor: YIVO Institute for Jewish Research
15 W 16th Street
New York, NY 10011

YIVO Natalie and Mendel Racolin Memorial Fellowship **5824**
This fellowship helps support original research in the field of East European Jewish history. The fellowship, which carries a stipend, is for a period of one to three months to conduct research at YIVO's library and archives. The recipient is expected to deliver a public lecture based on his/her research and submit a paper for possible publication in the YIVO Annual or YIVO-bleter. Applicants should send or fax a resume, two letters of support, and a research proposal of no more than three pages. Annual deadline dates may vary; contact program staff for exact dates.
Amount of Grant: $1500
Date(s) Application is Due: Dec 31
Contact: Director of Research, (212) 246-6080; fax: (212) 292-1892
Internet: http://www.yivoinstitute.org/jstudies/jstudies_fr.htm
Sponsor: YIVO Institute for Jewish Research
15 W 16th Street
New York, NY 10011

YIVO Professor Bernard Choseed Memorial Fellowship **5825**
This fellowship helps support original research in the field of East European studies. The fellowship, which carries a stipend, is for a period of one to three months during the spring semester to conduct research at YIVO's Library and Archives. The recipient is expected to lead a minimum of two YIVO faculty seminars; deliver a public lecture based on the research; and submit a paper for possible publication in the YIVO Annual or YIVO-bleter. Applicants should send or fax a resume, two letters of support, and a research proposal of no more than than three pages. Annual deadline dates may vary; contact program staff for exact dates.
Amount of Grant: $7500
Date(s) Application is Due: Dec 31
Contact: Director of Research, (212) 246-6080; fax: (212) 292-1892
Internet: http://www.yivoinstitute.org/jstudies/jstudies_fr.htm
Sponsor: YIVO Institute for Jewish Research
15 W 16th Street
New York, NY 10011

YIVO Rose and Isidore Drench Fellowship in American Jewish History **5826**
The fellowship will support a postdoctoral scholar conducting research in American Jewish history at the YIVO library and archives. Special consideration will be given to scholars working on some aspect of the Jewish labor movement. The fellowship is for a period of one to three months. The recipient is expected to present a public paper at YIVO based on the research and submit a transcript of the paper for possible publication in the YIVO Annual. Annual deadline dates may vary; contact program staff for exact dates.
Requirements: Applicants should send or fax a resume, two letters of support, and a research proposal of no more than three pages to the director of research.
Amount of Grant: $2500
Date(s) Application is Due: Dec 31
Contact: Program Contact, (212) 246-6080; fax: (212) 292-1892
Internet: http://www.yivoinstitute.org/jstudies/jstudies_fr.htm
Sponsor: YIVO Institute for Jewish Research
15 W 16th Street
New York, NY 10011

YIVO Vivian Lefsky Hort Fellowship in Yiddish Literature **5827**
The fellowship supports original research in the field of Yiddish literature for a period of one to three months during spring of the upcoming year at YIVO's library and archives. The recipient is expected to deliver a lecture based on the research and submit a paper for possible publication in the YIVO Annual or YIVO-bleter. Annual deadline dates may vary; contact program staff for exact dates.
Requirements: Applicants should send or fax a resume, two letters of support, and a research proposal of no more than three pages to the director of research.
Amount of Grant: $2000
Date(s) Application is Due: Dec 31
Contact: Program Contact, (212) 246-6080; fax: (212) 292-1892
Internet: http://www.yivoinstitute.org/jstudies/jstudies_fr.htm
Sponsor: YIVO Institute for Jewish Research
15 W 16th Street
New York, NY 10011

Young Scientists Summer Program **5828**
IIASA, annual 3-month Young Scientists Summer Program (YSSP) offers research opportunities to talented young researchers whose interests correspond with IIASA, ongoing research on issues of global environmental, economic and social change. Many of IIASA's National Member Organizations (NMOs) provide special grants to enable young scientists from their countries, or foreign students studying in their country, to participate in the program. From June through August accepted participants work within the Institute, Research Programs under the guidance of IIASA scientific staff. The YSSP provides a unique opportunity for participants to: advance their research under the direct supervision of an experienced IIASA scientist, and at the same time contribute to IIASA, ongoing scientific agenda; broaden their research interests by working in IIASA, interdisciplinary and international research environment; build contacts with IIASA, worldwide network of collaborators and with other YSSP fellows. The summer program will run from 1 June until 31 August.
Requirements: Interested candidates should have research experience corresponding to a level typical of a researcher about two years prior to receiving a PhD or equivalent degree.
Date(s) Application is Due: Jan 15
Contact: Barbara Hauser, YSSP Coordinator, 43 2236 807 0; fax: 43 2236 71 313; email: hauser@iiasa.ac.at
Internet: http://www.iiasa.ac.at/Admin/YSP/IIASA_YSSP.html
Sponsor: International Institute for Applied Systems Analysis
Norton's Woods, 136 Irving Street
Cambridge, MA 02138-1996

Yves Rocher Foundation Terra Ficaria Prize for Scientific Research **5829**

Since its inception, the Foundation has upheld and promoted Yves Rocher's strongest conviction that the future of humanity depends upon its ability to respect and live in harmony with nature. To that end, the Foundation spearheads international programs to protect endangered species. In 2004, the Foundation further demonstrated its commitment to the environment by creating the Terra Ficaria Scientific Award to recognize research in plant studies for the betterment of humanity. Since then, scientists have been honored for their contribution with a monetary prize of 20,000 Euros to continue the development of their research.

Samples: Doyle McKey—for researching ways of maintaining the genetic diversity of Yucca plants in order to preserve their adaptive potential (2006); Pierre Cabalion—for ethnobotanical study of the sclerophyllous forests or dry woodland of New Caledonia with a view to stopping their extinction (2005); Evelyne Leterme—for the study of the fruit tree heritage in France from a cultural, agronomic and genetic perspective (2004).

Amount of Grant: 20,000 Euro

Contact: Christina Hane; (001) 450-442-9555, ext. 2158; email: rpr@yrnet.com

Internet: http://www.yves-rocher-fondation.org/ca/en/preserving_nature/terra_ficaria/

Sponsor: Yves Rocher Foundation
2199 Fernand Lafontaine
Longueuil, Quebec J4G 2V7 Canada

Zonta International Foundation Amelia Earhart Fellowships **5830**

This program awards 35-40 Fellowships annually to assist women in advanced study and research in a suitable area of aerospace science or engineering. These Fellowships are open to women of any nationality and may be used at any institution of higher learning offering accredited graduate courses and degrees in aerospace studies. The Fellowship may be used for tuition, books and fees, or living expenses. Requests for a second award by current fellows undergo the same evaluation procedures as the initial application; a third award may be made in exceptional cases. The successful candidate identifies the relevance of her field of study to aerospace-related science or aerospace-related engineering, demonstrates outstanding academic credentials, and specifically outlines her professional or career goals as well as detailing an original focus in her research project.

Requirements: To qualify, a woman must have at the time of application a bachelor's degree in a qualifying area of science or engineering, a superior academic record and evidence of potential at a recognized institution of higher learning, and completion of one year of graduate school.

Amount of Grant: $6,000 maximum

Date(s) Application is Due: Nov 15

Contact: Foundation Assistant; (630) 928-1400; fax: (630) 928-1559; email: zontafdtn@zonta.org

Internet: http://www.zonta.org/site/PageServer?pagename=prog_ae_fellowships

Sponsor: Zonta International Foundation
1211 W 22nd Street, Suite 900
Oak Brook, IL 60523

Subject Index

NOTE: Numbers refer to entry numbers

Biology

Biology Education

Biology, Behavioral

Biology, Cellular

Biomedical Research Resources

Biomedical Research Training

Biomedicine

Biometry

Cancer Prevention

Cancer/Carcinogenesis

Carbohydrates

Carbon Dioxide

Cardiology

Chemistry Education

Chemistry, Analytical

Classical Studies

Classroom Instruction

Climatology

Clinical Medicine, General

Clinical Psychology

Clinical Trials

Clinics

Cloning of Cells

Clouds

Coal Processing

Coastal and Ocean Engineering

Coastal Processes

Communicative Disorders, Hearing

Communicative Disorders, Speech

Community and School Relations

Community College

Community Development

Community Outreach Programs

Community Service Programs

Computer Science Education

Computer Simulation/Modeling

Computer Software

Cultural Diversity

Cultural Heritage Preservation

Economics Education

Ecosystems

Ecuador

Editing

Education

Fishman Family Foundation Grants, 2358

SC Johnson Fund Grants, 5212
USF Irvine Minority Scholar Dissertation Fellowships, 5645

Environmental Law

Environmental Planning/Policy

Environmental Programs

Environmental Research

Environmental Studies

Enzymology

Epidemiology

Epigraphy

Epilepsy

Episcopal Church

Genetics, Molecular

Genocide

Genomics

Geochemistry

Geodesy

Geographic Quantitative Methods

Health Care Access

Health Records

Health Research

Health Sciences

Health Services Delivery

Health, Allied Fields

Hearing

Hearing Impairments

Heart Disease

Heat Transfer

Heating, Ventilation, and Air Conditioning

Hematologic Diseases

Hematologic System

Hematology

Hemophilia

Hepatitis

Higher Education Administration

Higher Education Studies

Higher Education, Private

Higher Education, Public

Hispanic Education

Hispanic Studies

Hispanics

Historic Preservation

History

Humanities Education

International Relations

Journalism Education

Judicial/Law Administration

Junior and Community Colleges

Junior High School Education

Jurisprudence

Juvenile Delinquency

Juvenile Law

Kazakhstan

Kidney Diseases and Disorders

Mathematics and Statistics

Mathematics Education

Mathematics, Applied

Matrix Theory

Mechanical Engineering

Medical Ethics

Medical Informatics

Medical Physics

Medical Programs

Minority Education

Music Appreciation

Music Composition

Music Education

Music History

Music Therapy

Music, Instrumental

Music, Vocal

Musicians in Residence

Musicology/Music Theory

Mutagenics

Myasthenia Gravis

Mycology

Myeloma

Nanoscale Science

Nanoscience

Nanostructures

Nanosystems

National Disease Organizations

National Planning/Policy

Nursing Education

Nursing Homes

Nutrition

Nutrition Education

Nutrition/Dietetics

Oceanography, Physical

Oceans and Seas

Odontology

Oils and Fats

Olfaction

Oncology

Opera/Musical Theater

Operations Research

Ophthalmology

Optical Communications

Optics

Optometry

Oral Diseases

Oral Health and Hygiene

Oral History

Oral-Motor Disorders

Orchestras

Pharmacy

Pharmacy Education

Philadelphia

Philanthropy

Philippines

Philosophy

Physical Sciences Education

Physics

Physics Education

Physics, Applied

Physics, Atomic

Physics, Computational

Physics, Fluid

Physics, High Energy

Physics, Molecular

Physics, Nuclear

Physics, Optical

Physics, Particle

Physics, Physics, High Energy, Physics, Particle

Physics, Plasma

Psychology Education

Psychology of Aging

Psychometrics

Psychopathology

Psychosomatic Disorders

Psychotherapy

Public Administration

Public Affairs

Public Broadcasting

Public Education

Public Health

Religion

Religious Studies

Religious Welfare Programs

Remedial Education

Remote Sensing

Renaissance History

Renewable Energy Sources

Repetitive Motion Disorders

Reproduction

Reproductive Disorders

Research Resources (Health/Safety/Medical)

Respiratory Diseases

Respiratory System

Restoration and Preservation

Restoration and Preservation, Art Works/Artifacts

Restoration and Preservation, Manuscripts/Books/ Music Scores

School Dental Programs

School Food Programs

School Health Programs

School-to-Work Transition

Science

Security

Sedimentology

Seismology

Semantics

Semiconductors

Seminars

Senegal

Senile Dementia

Senior Citizen Programs and Services

Sensing

Sensing Devices and Transducers

Sensors

Sensory System

Social Sciences

Water Resources, Environmental Impacts

Water Resources, Management/Planning

Water Supply

Water Treatment

Waterways and Harbors

Weapons

Welding

Welfare Reform

Welfare-to-Work Programs

Wellness

Wetlands

Wildlife

Women's Education

Women's Employment

Women's Health

Women's Rights

Women's Studies

End Index by subjects

Program Type Index

NOTE: Numbers refer to entry numbers

Basic Research

Building Construction and/or Renovation

Centers: Research/Demonstration/Service

Citizenship Instruction

Community Development

Consulting/Visiting Personnel

Cultural Outreach

Dissertation/Thesis Research Support

Emergency Programs

Exhibitions, Collections, Performances, Video/ Film Production

Faculty/Professional Development

Fellowships

International Grants

Job Training/Adult Vocational Programs

Matching/Challenge Funds

Materials/Equipment Acquisition (Computers, Books, Videos, etc.)

Preservation/Restoration

Publishing/Editing/Translating

Religious Programs

Scholarships

Seed Grants

Service Delivery Programs

Symposiums, Conferences, Workshops, Seminars

Technical Assistance

Training Program/Internships

Travel Grants

Vocational Education

Geographic Index

Note: This index lists grants for which applicants must be residents of or located in a specific geographic area. Numbers refer to entry numbers.

United States

Alabama

Alaska

Delaware

Georgia

Guam

Iowa

Maryland

Minnesota

Mississippi

New York

North Carolina

North Dakota

Ohio

South Dakota

Tennessee

US Virgin Islands

Utah

Wisconsin

Outside United States and Canada

Africa

Argentina

Asia

Australia

Belgium

Bolivia

Brazil

LaVergne, TN USA
15 January 2010
170175LV00001B/2/P

9 780984 172535